Structural Engineering Handbook

The photos on the front cover are courtesy of Thornton Tomasetti (Chicago office) and Jacobs (New York office). Mustafa Mahamid is grateful to the two firms for their contributions.

Structural Engineering Handbook

MUSTAFA MAHAMID

EDWIN H. GAYLORD, JR.

CHARLES N. GAYLORD

Fifth Edition

Mc Graw Hill

New York Chicago San Francisco Athens London
Madrid Mexico City Milan New Delhi
Singapore Sydney Toronto

Library of Congress Control Number: 2020934154

Structural Engineering Handbook, Fifth Edition

1 2 3 4 5 6 7 8 9 LWI 25 24 23 22 21 20

ISBN 978-1-260-11598-7
MHID 1-260-11598-4

This book is printed on acid-free paper.

Sponsoring Editor
Ania Levinson

Editing Supervisor
Stephen M. Smith

Production Supervisor
Pamela A. Pelton

Acquisitions Coordinator
Elizabeth M. Houde

Project Manager
Sarika Gupta,
Cenveo® Publisher Services

Copy Editors
Syed Haider and Girish Sharma,
Cenveo Publisher Services

Proofreader
Bruce Owens

Indexer
Arc Indexing, Inc.

Art Director, Cover
Jeff Weeks

Composition
Cenveo Publisher Services

About the Editors

Mustafa Mahamid, Ph.D., S.E., P.E., P.Eng., F.SEI, F.ASCE, F.ACI, is a clinical associate professor in the Department of Civil and Materials Engineering at the University of Illinois—Chicago and is a member of the AccessEngineering Faculty Advisory Board. He previously worked as a practicing structural engineer at GRAEF and Skidmore, Owings & Merrill and currently is a consultant for various structural engineering firms in the United States and abroad. He is a Fellow of ASCE, a Fellow of ACI, and a Fellow of ASCE's Structural Engineering Institute (SEI) and is active on various ACI and ASCE/SEI technical committees.

The late **Edwin H. Gaylord, Jr.** was a professor of civil engineering at the University of Illinois—Urbana-Champaign.

The late **Charles N. Gaylord** was a professor of civil engineering at the University of Virginia.

Contents

Contributors *xv*
Preface *xix*

Chapter 1. Structural Loads *Mustafa Mahamid, David A. Fanella* 1
1.1 Introduction 1
1.2 Dead Loads 1
1.3 Live Loads 1
1.4 Rain Loads 7
1.5 Soil Lateral Loads 8
1.6 Snow and Ice Loads 9
1.7 Wind Loads 20
1.8 Earthquake Loads 49
1.9 Flood and Tsunami Loads 49
1.10 Load Combinations 57
References 60

Chapter 2. Structural Analysis *Aly M. Said, Anas S. Issa, M. Shahria Alam* 61
PART A FUNDAMENTALS AND APPLICATIONS TO STRUCTURAL FRAMEWORKS 61
2.1 Introduction 61
2.2 Fundamentals of Analysis 61
2.3 Energy Principles 63
2.4 Analysis of Statically Determinate Structures 67
2.5 Analysis of Statically Indeterminate Structures 72
2.6 Force Method 74
2.7 Displacement Method 77
PART B ARCHES AND RIGID FRAMES 91
2.8 Introduction 91
2.9 Analysis of Arches 93
2.10 Design of Arches 94
2.11 Design of Frames 101
2.12 Special Topics 111
2.13 Construction and Details 112
References 116

Chapter 3. The Finite-Element Method *Craig D. Foster, Sheng-Wei Chi* 119
3.1 Overview of Theory of Finite-Element Analysis 119
3.2 Outline of Steps in a Finite-Element Analysis 122
3.3 Types of Structural and Mechanical Analyses 122
3.4 Elements and Nodes 124
3.5 Meshing 125
3.6 Applying Loads and Displacement Boundary Conditions 126
3.7 Material Models and Other Properties 126
3.8 Solving the Quasi-Static Problem 126

3.9 Postprocessing—Solving for Strain, Stress, and Other Quantities 126
3.10 Dynamic Finite-Element Analysis 126
3.11 Nonlinear Finite-Element Analysis 128
3.12 Verification and Validation 128
3.13 Issues and Pitfalls in Finite-Element Analysis 128
3.14 Introduction to Finite Elements for Thermal, Thermomechanical, and Other Problems 130
References 131

Chapter 4. Computer Applications in Structural Engineering
Raoul Karp, Bulent N. Alemdar, Sam Rubenzer **133**
4.1 Introduction 133
4.2 Computer Structural Analysis Simulation 133
4.3 Structural Finite Elements 135
4.4 Foundations 139
4.5 Verifying Analysis Results 142
4.6 Building Information Modeling and Interoperability 142
4.7 Summary 143
References 143

Chapter 5. Earthquake-Resistant Design *S. K. Ghosh* **145**
5.1 Overview 145
5.2 Nature of Earthquake Motion 145
5.3 Design Philosophy 146
5.4 Seismic Design Requirements of the 2018 IBC/ASCE 7-16 147
References 162

Chapter 6. Fracture and Fatigue *Kedar S. Kirane, Zdeněk P. Bažant, J. Ernesto Indacochea, Vineeth Kumar Gattu* **165**
PART A CONCRETE AND COMPOSITES 165
6.1 Introduction to Quasibrittle Fracture 165
6.2 Concrete 166
6.3 Fiber-Reinforced Composites 169
References 173
PART B STRUCTURAL STEELS 175
6.4 Fracture of Structural Steels 175
6.5 Fatigue of Structural Steels 182
References 193

Chapter 7. Soil Mechanics and Foundations *Joseph W. Schulenberg, Krishna R. Reddy* **195**
7.1 Soil Behavior 195
7.2 Shallow Foundation Analyses 199
7.3 Deep Foundations 204
7.4 Retaining Structures 210
7.5 Investigations 218
7.6 Soil Improvement 219
7.7 Monitoring 221
References 221

Chapter 8. Design of Structural Steel Members *Jay Shen, Bulent Akbas, Onur Seker, Charlies J. Carter* **225**
8.1 Design of Steel and Composite Members 225
8.2 Seismic Design of Steel Members in Moment and Braced Frames 264
8.3 Concluding Remarks 303
References 303

Chapter 9. Design of Cold-Formed Steel Structural Members
Nabil A. Rahman, Helen Chen, Cheng Yu **305**
9.1 Shapes and Applications 305
9.2 Materials 306

9.3 Corrosion Protection 308
9.4 Member Design Methodology 309
9.5 Structural Stability in Determining Member Forces 311
9.6 Member Design 311
9.7 Assemblies and Systems 315
9.8 Connections 317
References 320

Chapter 10. Aluminum Structural Design *J. Randolph Kissell* **321**
10.1 Introduction 321
10.2 Design Overview 322
10.3 Determining Required Forces 322
10.4 Axial Tension 323
10.5 Axial Compression 324
10.6 Flexure 325
10.7 Shear and Torsion 326
10.8 Combined Forces 328
10.9 Connections 328
10.10 Serviceability 331
10.11 Fabrication and Erection 331
10.12 Fatigue 331
References 331

Chapter 11. Design of Reinforced-Concrete Structural Members
Mustafa Mahamid, David A. Fanella **333**
11.1 Concrete 333
11.2 Reinforcement 333
11.3 Specifications, Codes, and Standards 334
11.4 General Requirements for Strength and Serviceability 334
11.5 General Principles of the Strength Design Method 338
11.6 General Principles and Requirements 339
11.7 One-Way Slabs, Two-Way Slabs, and Beams 340
11.8 Columns 352
11.9 Tension Members 354
11.10 Members Subjected to Flexure and Axial Load 354
11.11 Walls 356
11.12 Footings 360
11.13 Two-Way Slab Systems 365
11.14 Seismic 367
References 368

Chapter 12. Design of Prestressed-Concrete Structural Members
Sri Sritharan, Maryam Nazari **369**
Notation 369
12.1 Introduction 370
12.2 Prestress Losses 371
12.3 Analysis Concepts 374
12.4 Design for Flexure 380
12.5 Design for Shear, Bond, and Bearing 386
12.6 Design of Continuous Beam 389
12.7 Design for Torsion 393
12.8 Design for Seismic Loads 396
References 397

Chapter 13. Masonry Construction *Richard Bennett,*
Sam Rubenzer **399**
Notation 399
13.1 Masonry Materials 400
13.2 Masonry Analysis 401
13.3 Unreinforced Masonry 402
13.4 Reinforced Masonry 403

13.5 Seismic Design 407
13.6 Specifying Masonry and Testing Masonry 409
13.7 Masonry Fire Walls 409
References 410

Chapter 14. Timber Structures *Thomas Williamson, Mustafa Mahamid* **411**
14.1 Wood as an Engineering Material 411
14.2 Allowable Stress Design (ASD) vs. Load and Resistance Factor Design (LRFD) 413
14.3 Structural Wood Products 413
14.4 Member Design 422
14.5 Connection Design 425
14.6 Lateral Load Resistance 430
References 435

Chapter 15. Bridge Engineering *Soliman Khudeira, Tony Shkurti, Eric Stone, James Carter III, Jamshid Mohammadi, Bora Jang* **437**
PART A STEEL AND CONCRETE BRIDGES 437
15.1 General 437
15.2 Loads 437
15.3 Steel Bridges 453
15.4 Bearing and Expansion Details 459
15.5 Beam and Plate-Girder Bridges 462
15.6 Truss Bridges 469
15.7 Concrete Bridges 470
15.8 Slab Bridges 471
15.9 T-Beam Bridges 473
15.10 Box-Girder Bridges 483
15.11 Prestressed-Concrete Bridges 493
15.12 Bridge Railings 505
15.13 Administration of a Bridge System 506
References 507
PART B STEEL BOX-GIRDER BRIDGES 508
15.14 Benefits 508
15.15 Internal Elements 509
15.16 Limit States and Combinations 509
15.17 Analysis of Box Girders 511
15.18 Design of Box-Section Members 512
15.19 Internal Element Design 512
15.20 Concluding Remarks 512
References 513
PART C CURVED STEEL I-GIRDER BRIDGES 514
15.21 General 514
15.22 Design Considerations 514
15.23 Curved Girder Mechanics 517
References 524
PART D CURVED STEEL BOX-GIRDER BRIDGES 525
15.24 General 525
15.25 Design Considerations 526
15.26 Box-Girder Mechanics 529
References 531
PART E CURVED CONCRETE BOX-GIRDER BRIDGES 532
15.27 General 532
15.28 Concrete Box-Girder Bridge Types 532
15.29 Significance of Structural Configuration and Details 533
15.30 Review of Theory 535
15.31 Design Considerations 539
References 547

Chapter 16. Railroad Bridge Design *Mohsen Issa, Ahmad Hammad* **549**
16.1 Introduction 549
16.2 General Requirements 549
16.3 Loading 549
16.4 Timber 550
16.5 Steel 551
16.6 Concrete 552
References 557

Chapter 17. Industrial Buildings *Jules Van de Pas, John Rolfes* **559**
17.1 Planning Industrial Buildings 559
17.2 Code Requirements and Industrial Loads 563
17.3 Framing Systems 566
References 573

Chapter 18. Tall Buildings *Charles Besjak, Brian McElhatten, Preetam Biswas* **575**
18.1 Definition of Tall Building 575
18.2 General Design Considerations 575
18.3 Structural Design Considerations 576
18.4 Serviceability Considerations 577
18.5 Structural Systems for Tall Buildings 577
18.6 System Conceptualization 579
18.7 System Parameters and Choices 581
18.8 Analysis Parameters and Techniques 591
18.9 Performance-Based Design for Wind 595
18.10 Performance-Based Design for Seismic 599
18.11 Creep and Shrinkage 600
18.12 High-Rise Building Glossary 601
18.13 Additional Credits 601
Acknowledgments 601
References 601

Chapter 19. Thin-Shell Concrete Structures *Edmond Saliklis, Mustafa Mahamid, David P. Billington, Julian A. Dumitrescu* **603**
19.1 Introduction and Historical Overview 603
19.2 Shells of Rotation 609
19.3 Domes 613
19.4 Cylindrical Barrel Shells 626
19.5 Folded Plates 632
19.6 Translation Shells of Double Curvature 638
19.7 Dimensioning 647
19.8 Stability and Safety 647
19.9 Construction 649
19.10 Appearance 649
References 649

Chapter 20. Cable-Supported Roofs *Paul A. Gossen, Keith M. MacBain* **651**
20.1 Introduction 651
20.2 Cable Materials and Construction 651
20.3 Structural Systems 656
20.4 Dynamics 658
20.5 Analysis and Design 658
20.6 Erection 661
References 661

Chapter 21. Reinforced-Concrete Silos *Ramez B. Gayed, Mustafa Mahamid, Amin Ghali* **663**
21.1 General 663
21.2 Design of Walls 676
21.3 Design of Bottoms 681
References 695

Chapter 22. Design of Steel Tanks *Leslie D. Scott* **697**
22.1 Introduction 697
22.2 Design Considerations 701
22.3 Design of Reservoirs and Standpipes 707
22.4 Elevated Tanks 717
22.5 Accessories 723
22.6 Materials 726
22.7 Commercially Available Computer Programs 726
22.8 The Engineer's Role 727
References 727

Chapter 23. Electrical Transmission and Substation Structures *Michael D. Miller, Robert E. Nickerson* **729**
23.1 Introduction 729
23.2 Referenced Standards and Manuals of Practice 729
23.3 Transmission Line Structures 729
23.4 Substation Structures 737
References 740

Chapter 24. Chimneys *Shu-Jin Fang, Victor Bochicchio* **741**
24.1 Introduction 741
24.2 Design Loads 741
24.3 Steel Stacks 745
24.4 Reinforced-Concrete Chimneys 750
24.5 Liners and Linings 752
24.6 Foundations 754
References 754

Chapter 25. Health Monitoring and Nondestructive Testing *Didem Ozevin, Farhad Ansari* **757**
25.1 Introduction 757
25.2 Components of NDE Methods 757
25.3 Waveform-Based NDE Methods and Applications 760
25.4 The Applications of Waveform-Based NDE Methods to Infrastructure Systems 765
References 767

Chapter 26. Building Information Modeling (BIM) for Structural Engineering *Nawari O. Nawari* **771**
26.1 Introduction 771
26.2 BIM Fundamentals 771
26.3 BIM Processes in Practice 776
26.4 Structure and Architecture Synergy (SAS) Framework 778
26.5 Conclusions 786
References 786

Chapter 27. Structural Fire Engineering *V. K. R. Kodur, M. Z. Naser* **789**
27.1 General 789
27.2 Designing Structures for Fire Safety 789
27.3 Fire Growth and Development 789
27.4 Properties of Construction Materials at Elevated Temperatures 791
27.5 Fire Resistance Evaluation 796
27.6 Prescriptive-Based Approaches 796

27.7 Rational Design Approaches 798
27.8 Summary 800
References 800

Chapter 28. Disproportionate Collapse and Blast-Resistant Design *Shalva Marjanishvili, Robert Smilowitz* **801**
PART A STRUCTURAL ROBUSTNESS AND DISPROPORTIONATE COLLAPSE 801
28.1 Introduction 801
28.2 Disproportionate Collapse Mitigation Strategies 802
28.3 Progressive Collapse Modeling 802
28.4 Progressive Collapse Analysis Example 806
References 809
PART B BLAST-RESISTANT DESIGN 809
28.5 Introduction 809
28.6 Blast Phenomena 810
28.7 Dynamic Design Approach 812
28.8 Structural Behavior 813
28.9 Primary Structure: Material Selection and Design for Blast Resistance 814
28.10 Secondary Structure: Material Selection and Design for Blast Resistance 815
28.11 Conclusion 815
References 815

Chapter 29. FRP Strengthening of Reinforced-Concrete Members *Hayder A. Rasheed* **817**
29.1 FRP Properties for Strengthening Applications 817
29.2 Flexural Strengthening Design for Beams and Slabs 818
29.3 Shear Strengthening Design for Beams 822
29.4 Confinement Strengthening Design for Circular Columns 827
References 831

Chapter 30. Structural Glass and Glazing *Rui de S. Camposinhos* **833**
30.1 Introduction 833
30.2 Glass: Production and Properties 834
30.3 Glass as a Structural Material 835
30.4 Actions 839
30.5 Codes and Standards 842
30.6 Plate Buckling 850
30.7 Lateral-Torsional Buckling 854
30.8 Glass Columns 856
References 858

Chapter 31. Machine Foundations *O. Salem Ali* **861**
31.1 Background 861
31.2 Classification of Machine Based on Machine Type 861
31.3 Classification of Machine Based on Type of Excitation Force 861
31.4 Classification of Machine Based on Foundation Type 861
31.5 Classification of Machine Load Transfer Mechanism 861
31.6 Design Limits for Machine Foundations 862
31.7 Effect of the Supporting Soil 862
31.8 Energy Transfer Mechanism 863
31.9 Effect of Embedment of Foundation 863
31.10 Reduction in Permissible Soil Stress 864
31.11 Damping in Soil 864
31.12 Modeling Techniques for Machine Foundations 865
31.13 Block-Type Foundation 865
31.14 Mat Foundations 866
31.15 Elevated Machine Foundation 867

31.16 Three-Dimensional Finite-Element Modeling 868
31.17 Soil Modeling.. 868
31.18 Current Approach of Soil Modeling... 869
31.19 Methods to Compute Dynamic Impedance Functions 870
31.20 Foundations Supported on Piles... 870
31.21 Piles Subjected to Lateral Vibrations.. 871
31.22 Elastic Continuum 872
31.23 Piles Subjected to Vertical Vibrations.. 875
31.24 Pile Group Effect.. 881
References 882

Chapter 32. Value Methodology ***Muthiah Kasi, Charles A. Bartlett*** **885**
32.1 Introduction.. 885
32.2 VM Job Plan 885
32.3 The Key Features of VM Job Plan: Function Analysis 889
32.4 Example 1: Parapet Joint Detail 890
32.5 Evaluation Phase 893
32.6 Implementation 894
32.7 Summary... 894
References 894

Chapter 33. Stone Cladding ***Rui de S. Camposinhos*** **895**
33.1 Introduction: Natural Stone Cladding... 895
33.2 Natural Stone Depiction 896
33.3 Mechanical Properties 898
33.4 Cladding Systems and Methods 899
33.5 Limit States Design 901
33.6 Dowel Anchorage... 905
33.7 Kerf Anchorage... 909
33.8 Undercut Anchorage.. 911
References 915

Index 919

Contributors

Bulent Akbas, Ph.D. *Department of Civil Engineering, Gebze Technical University, Gebze, Turkey* (Chap. 8)

M. Shahria Alam *Professor, School of Engineering, University of British Columbia, Kelowna, British Columbia, Canada* (Chap. 2)

Bulent N. Alemdar, Ph.D., PE *Principal Research Engineer, Bentley Systems, Inc., Carlsbad, California* (Chap. 4)

O. Salem Ali, Ph.D., PE *Engineering Manager, Structural Technologies* (Chap. 31)

Farhad Ansari *Professor, Department of Civil and Materials Engineering, University of Illinois at Chicago* (Chap. 25)

Charles A. Bartlett, PE, CVS (Chap. 32)

Zdeněk P. Bažant *McCormick Institute Professor and Walter P. Murphy Professor of Civil and Environmental Engineering, Mechanical Engineering and Material Science and Engineering, Northwestern University, Evanston, Illinois* (Chap. 6)

Richard Bennett *Professor and Director of Engineering Fundamentals, University of Tennessee, Knoxville, Tennessee* (Chap. 13)

Charles Besjak, PE, SE *Skidmore, Owings & Merrill* (Chap. 18)

David P. Billington *Professor Emeritus of Civil Engineering, Princeton University (Deceased)* (Chap. 19)

Preetam Biswas, PE *Skidmore, Owings & Merrill* (Chap. 18)

Victor Bochicchio *Executive Vice President, Hamon Custodis, Inc., Somerville, New Jersey* (Chap. 24)

Rui de S. Camposinhos *Coordinator Professor with Aggregation, Polytechnic of Porto, Portugal* (Chaps. 30, 33)

Charlies J. Carter, Ph.D., PE, SE *American Institute of Steel Construction, Chicago, Illinois* (Chap. 8)

James Carter III *Consulting Engineer, Chicago, Illinois* (Chap. 15)

Helen Chen, Ph.D., PE *American Iron and Steel Institute, Washington, D.C.* (Chap. 9)

Sheng-Wei Chi, Ph.D. *Department of Civil Engineering, University of Illinois at Chicago* (Chap. 3)

Julian A. Dumitrescu *Consulting Structural Engineer, Raytheon Engineers and Constructors (Retired)* (Chap. 19)

David A. Fanella, Ph.D., SE, PE *Concrete Reinforcing Steel Institute* (Chaps. 1, 11)

Shu-Jin Fang *Consultant, and Former Associate and Senior Manager, Sargent & Lundy, Chicago, Illinois* (Chap. 24)

Craig D. Foster, Ph.D. *Department of Civil Engineering, University of Illinois at Chicago* (Chap. 3)

Vineeth Kumar Gattu *Chemical and Fuel Cycle Technologies, Argonne National Laboratory* (Chap. 6)

Ramez B. Gayed *Adjunct Professor, University of Calgary, and Senior Structural Engineer, thyssenkrupp Industrial Solutions (Canada) Inc.* (Chap. 21)

Amin Ghali *Professor Emeritus, University of Calgary* (Chap. 21)

S. K. Ghosh *S. K. Ghosh Associates LLC, Palatine, Illinois* (Chap. 5)

Paul A. Gossen, PE *Geiger Engineers, Suffern, New York* (Chap. 20)

Ahmad Hammad, Ph.D., PE, SE *Assistant Vice President and Senior Engineering Manager, WSP USA* (Chap. 16)

J. Ernesto Indacochea *Professor Emeritus, Civil and Materials Engineering, University of Illinois at Chicago* (Chap. 6)

Anas S. Issa *Postdoctoral Fellow, Applied Laboratory for Advanced Materials & Structures (ALAMS), University of British Columbia, Kelowna, British Columbia, Canada* (Chap. 2)

Mohsen Issa, Ph.D., PE, SE *Professor, Department of Civil and Materials Engineering, University of Illinois at Chicago* (Chap. 16)

Bora Jang *Sharma and Associates, Inc.* (Chap. 15)

Raoul Karp, PE, SE *Vice President of Design Engineering Analysis, Bentley Systems, Inc., Carlsbad, California* (Chap. 4)

Muthiah Kasi, PE, SE, CVS-Life (Chap. 32)

Soliman Khudeira *Illinois Institute of Technology and Chicago DOT* (Chap. 15)

Kedar S. Kirane *Assistant Professor of Mechanical Engineering, Stony Brook University, Stony Brook, New York* (Chap. 6)

J. Randolph Kissell, PE (Chap. 10)

V. K. R. Kodur, Ph.D., P.Eng. *University Distinguished Professor and Chairperson, Department of Civil and Environmental Engineering, Michigan State University, East Lansing, Michigan* (Chap. 27)

Keith M. MacBain, Ph.D., PE *Geiger Engineers, Suffern, New York* (Chap. 20)

Mustafa Mahamid, Ph.D., SE, PE, P.Eng. *Clinical Associate Professor, Department of Civil and Materials Engineering, University of Illinois at Chicago* (Chaps. 1, 11, 14, 19, 21)

Shalva Marjanishvili, D.Sc., PE, SE *Technical Director, Hinman Consulting Engineers, Inc., San Francisco, California* (Chap. 28)

Brian McElhatten, PE, SE *Arup* (Chap. 18)

Michael D. Miller, PE, P.Eng. *VP Engineering, SAE Towers* (Chap. 23)

Jamshid Mohammadi *Illinois Institute of Technology* (Chap. 15)

M. Z. Naser, Ph.D., PE *Assistant Professor, Glenn Department of Civil Engineering, Clemson University, Clemson, South Carolina* (Chap. 27)

Nawari O. Nawari, Ph.D., PE *School of Architecture, University of Florida, Gainesville, Florida* (Chap. 26)

Maryam Nazari *Assistant Professor, Department of Civil and Geomatics Engineering, California State University, Fresno, California* (Chap. 12)

Robert E. Nickerson, PE *Consultant* (Chap. 23)

Didem Ozevin *Associate Professor, Department of Civil and Materials Engineering, University of Illinois at Chicago* (Chap. 25)

Nabil A. Rahman, Ph.D., PE *FDR Engineers, Raleigh, North Carolina* (Chap. 9)

Hayder A. Rasheed, Ph.D., PE *Professor, Kansas State University, Manhattan, Kansas* (Chap. 29)

Krishna R. Reddy *Professor of Civil Engineering, University of Illinois at Chicago* (Chap. 7)

John Rolfes *Vice President, CSD Structural Engineers, Milwaukee, Wisconsin* (Chap. 17)

Sam Rubenzer, PE, SE *Founder and Structural Engineer, FORSE Consulting, Eau Claire, Wisconsin* (Chaps. 4, 13)

Aly M. Said *Associate Professor, Architectural Engineering, Pennsylvania State University, University Park, Pennsylvania* (Chap. 2)

Edmond Saliklis *California Polytechnic State University (Cal Poly), San Luis Obispo* (Chap. 19)

Joseph W. Schulenberg *Assistant Clinical Professor of Civil Engineering, University of Illinois at Chicago* (Chap. 7)

Leslie D. Scott *Chief Engineer, Tank Industry Consultants, Inc., Indianapolis, Indiana* (Chap. 22)

Onur Seker, Ph.D. *Department of Civil Engineering, Gebze Technical University, Gebze, Turkey* (Chap. 8)

Jay Shen, Ph.D., PE, SE *Department of Civil, Construction and Environmental Engineering, Iowa State University, Ames, Iowa* (Chap. 8)

Tony Shkurti *Consulting Engineer, Chicago, Illinois* (Chap. 15)

Robert Smilowitz, Ph.D., PE *Senior Principal, Thornton Tomasetti* (Chap. 28)

Sri Sritharan *Wilkinson Chair of Interdisciplinary Engineering and Professor of Structural Engineering, Department of Civil, Construction and Environmental Engineering, Iowa State University, Ames, Iowa* (Chap. 12)

Eric Stone *Consulting Engineer, Chicago, Illinois* (Chap. 15)

Jules Van de Pas *Vice President, CSD Structural Engineers, Greenwood Village, Colorado* (Chap. 17)

Thomas Williamson, PE *Retired Vice President of Quality and Technical Services, APA–The Engineered Wood Association* (Chap. 14)

Cheng Yu, Ph.D., PE *University of North Texas, Denton, Texas* (Chap. 9)

Preface

As a practicing structural engineer and as an educator, I have always believed that structural engineers and architects should have knowledge of the design of the various types of structures and of their components, various analysis and design methods, the technologies used in this analysis, and the design and production of engineering drawings. The *Structural Engineering Handbook* provides established engineers, young engineers preparing for license exams, architects, and civil engineering students a comprehensive reference on the planning and design of a variety of engineered structures. It also gives the designer the information likely needed for all design phases.

The handbook covers various types of structures, such as tall buildings, industrial buildings, bridges including railroad bridges, thin-shell structures, arches, cable-supported roofs, steel tanks for liquids, retaining structures, blast-resistant structures, bins and silos for granular material, steel transmission towers and poles, and chimneys. Structural loads for the various types of structures are also covered, and there is comprehensive coverage of classical structural analysis methods, finite-element analysis, and computer applications in structural engineering. Additionally, earthquake-resistant design has been covered based on the most recent codes and standards. Design of reinforced concrete, prestressed concrete, structural steel, cold-formed steel, masonry, wood, and aluminium are covered. A chapter on soil mechanics, soil exploration, and foundation design is also provided. Design against fatigue and fracture is covered for concrete, composites, and steel.

In this fifth edition, all chapters have been rewritten, some chapters in previous versions of the handbook have been removed due to recent developments in design or construction practices, and 12 new chapters have been added. The new chapters cover structural loads, fracture mechanics of concrete and composites, railroad bridges, health monitoring of structures, building information modeling (BIM), structural fire engineering, progressive collapse and blast-resistant design, strengthening of concrete using fiber-reinforced polymer (FRP), structural glass, design of foundations for machines, value engineering, and stone cladding.

The 33 chapters of the handbook have been written by 66 contributors. They have presented their material in a ready-to-use form with flowcharts to show step-by-step procedures wherever possible. Therefore, derivations of formulas are omitted in all but a few instances, and many worked-out examples are given. Background information, descriptive matter, and explanatory material have been condensed or omitted. Because each chapter treats a subject that is broad enough to fill a book by itself, the contributors have had to select the material that, in their judgment, is likely to be the most useful to the greatest number of users. References and sources of additional material are noted for most of the topics that could not be treated in sufficient detail.

I am very grateful to the contributors for their tremendous efforts in writing, reviewing, and editing their work, and for their patience during the time it has taken to complete the fifth edition.

Mustafa Mahamid, Ph.D., S.E., P.E., P.Eng.
University of Illinois at Chicago

Structural Engineering Handbook

Chapter 1
Structural Loads

BY
MUSTAFA MAHAMID, Ph.D., SE, PE, P.Eng. *University of Illinois at Chicago*
DAVID A. FANELLA, Ph.D., SE, PE *Concrete Reinforcing Steel Institute*

1.1 INTRODUCTION

Applicable nominal loads on a structure are determined from the general building code under which the project is to be designed and constructed. Chapter 16 of the IBC (Ref. 1) contains the minimum magnitudes of some nominal loads and references ASCE/SEI 7 (Ref. 2) for others. For a specific project, the governing local building code should be consulted for any variances from the IBC or ASCE/SEI 7.

It is common for nominal loads to be referred to as service loads. These loads are multiplied by load factors in the strength design method. Exceptions are the wind load effect W and the earthquake load effect E: Both are defined to be strength-level loads where the load factor is equal to 1.0.

Table 1.1 contains a list of loads from the IBC and ASCE/SEI 7. Comprehensive information on the determination of structural loads can be found in Ref. 3.

1.2 DEAD LOADS

Nominal dead loads D are the actual weights of construction materials and fixed service equipment that are attached to or supported by the building or structure. Specific examples of such loads are listed under the definition of "dead load" in IBC 202.

Dead loads are considered to be permanent loads because their magnitude remains essentially constant over time.

Superimposed dead loads are permanent loads other than the weights of the structural members and include the following: floor finishes and/or topping; walls; ceilings; heating, ventilating, and air-conditioning (HVAC) and other service equipment; fixed partitions; and cladding.

Minimum design dead loads for various types of common construction components are provided in ASCE/SEI Table C3.1-1a, and minimum densities for common construction materials are given in ASCE/SEI Table C3.1-2. In cases where information on dead load is unavailable, values of dead loads used in design must be approved by the building official (IBC 1606.2).

1.3 LIVE LOADS

1.3.1 General

Live loads are transient in nature and vary in magnitude over the life of a structure. These loads are produced by the use and occupancy of a building or structure and do not include construction loads, environmental loads (such as wind loads, snow loads, rain loads, earthquake loads, and flood loads), or dead loads (IBC 202).

IBC Table 1607.1 contains nominal design values of uniformly distributed and concentrated live loads L_o as a function of occupancy or use. The occupancy description listed in the table is not necessarily group-specific (occupancy groups are defined in IBC Chapter 3). For example, an office building with a Business Group B classification may

Table 1.1 Summary of Loads Addressed in the IBC and ASCE/SEI 7

Notation	Load	Code section
D	Dead load	IBC 1606
D_i	Weight of ice	Chap. 10 of ASCE/SEI 7
E	Combined effect of horizontal and vertical earthquake-induced forces as defined in ASCE/SEI 12.4.2	IBC 1613 and ASCE/SEI 12.4.2
E_m	Maximum seismic load effect of horizontal and vertical forces as set forth in ASCE/SEI 12.4.3	IBC 1613 and ASCE/SEI 12.4.3
F	Load due to fluids with well-defined pressures and maximum heights	-------
F_a	Flood load	IBC 1612
H	Load due to lateral earth pressures, ground water pressure, or pressure of bulk materials	IBC 1610 (soil lateral loads)
L	Live load, except roof live load, including any permitted live load reduction	IBC 1607
L_r	Roof live load including any permitted live load reduction	IBC 1607
R	Rain load	IBC 1611
S	Snow load	IBC 1608 and Chapter 7 of ASCE/SEI 7
T	Cumulative effects of self-straining forces and effects	See ASCE/SEI 2.3.4 and 2.4.4
W	Load due to wind pressure	IBC 1609 and Chapters 26 to 31 of ASCE/SEI 7
W_i	Wind-on-ice load	IBC 1614 and Chapter 10 of ASCE/SEI 7

also have storage areas that may warrant live loads of 125 or 250 psf (6.0 or 12.0 kN/m^2) depending on the type of storage, which are greater than the prescribed office live loads. Structural members are designed on the basis of the maximum effects due to application of either a uniform load or a concentrated load and need not be designed for the effects of both loads applied at the same time. The building official must approve live loads that are not specifically listed in the table.

Partitions that can be relocated (i.e., those types that are not permanently attached to the structure) are considered to be live loads in office and other buildings. A live load equal to at least 15 psf (0.72 kN/m^2) must be included for movable partitions if the nominal uniform floor live load is less than 80 psf (3.8 kN/m^2).

IBC Table 1607.1 prescribes a minimum roof live load of 20 psf (0.96 kN/m^2) for typical roof structures; larger live loads are required for roofs used as gardens or places of assembly.

ASCE Table 4.3-1 also contains minimum uniform and concentrated live loads, and some of these values differ from those in IBC Table 1607.1. ASCE Tables C4.3-1 and C4.3-2 can be used as a guide in establishing live loads for some commonly encountered occupancies.

1.3.2 Reduction in Live Loads

Because live loads are transient in nature, the probability that a structural member will be subjected to the full effects from nominal live loads decreases as the area supported by the member increases. Except for uniform live loads on roofs, the minimum uniformly distributed live loads L_o in IBC Table 1607.1 are permitted to be reduced in accordance with the methods in IBC 1607.11.1 or 1607.11.2. The general method of live load reduction in IBC 1607.11.1 is also given in ASCE/SEI 4.7. Reduction of roof live loads must conform to IBC 1607.13.2.

1.3.3 General Method of Live Load Reduction

IBC Equation (16-23) can be used to obtain a reduced live load L for members that support an area $K_{LL}A_T \geq 400$ ft^2 (37.2 m^2):

$$L = L_o\left(0.25 + \frac{15}{\sqrt{K_{LL}A_T}}\right)$$

$$\geq 0.50L_o \text{ for members supporting one floor}$$

$$\geq 0.40L_o \text{ for members supporting two or more floors}$$

In SI Units

$$L = L_o\left(0.25 + \frac{4.75}{\sqrt{K_{LL}A_T}}\right)$$

In this equation, K_{LL} is the live load element factor given in IBC Table 1607.11.1, and A_T is the tributary area supported by the member in square feet (square meters).

The live load element factor K_{LL} converts the tributary area A_T into an influence area, which is considered to be the adjacent floor area from which the member derives its load. In other words,

$$K_{LL} = \text{influence area/tributary area}$$

Figure 1.1 illustrates how the reduction multiplier $0.25 + 15/(\sqrt{K_{LL}A_T})$ varies with respect to the influence area $K_{LL}A_T$. Included in the figure are the minimum influence area of 400 square feet and the limits of 0.5 and 0.4, which are the maximum permitted reductions for members supporting one floor and two or more floors, respectively.

One-Way Slabs

Live load reduction on one-way slabs is permitted provided that the tributary area, A_T, does not exceed an area equal to the slab span times a width normal to the span of 1.5 times the slab span (i.e., an area with an aspect ratio of 1.5). The live load will be somewhat higher for a one-way slab with an aspect ratio of 1.5 than for a two-way slab with the same aspect ratio. This recognizes the benefits of higher redundancy that results from two-way action.

ASCE/SEI 4.7.6 has the same requirements for live load reduction on one-way slabs as that in IBC 1607.11.1.1.

Heavy Live Loads

According to IBC 1607.11.1.2, live loads that are greater than 100 psf must not be reduced except for the following:

1. Live loads for members supporting two or more floors are permitted to be reduced by a maximum of 20 percent, but L must not be less than that calculated by IBC 1607.11.1.
2. In occupancies other than storage, additional live load reduction is permitted if it can be shown by a registered design professional that such a reduction is warranted.

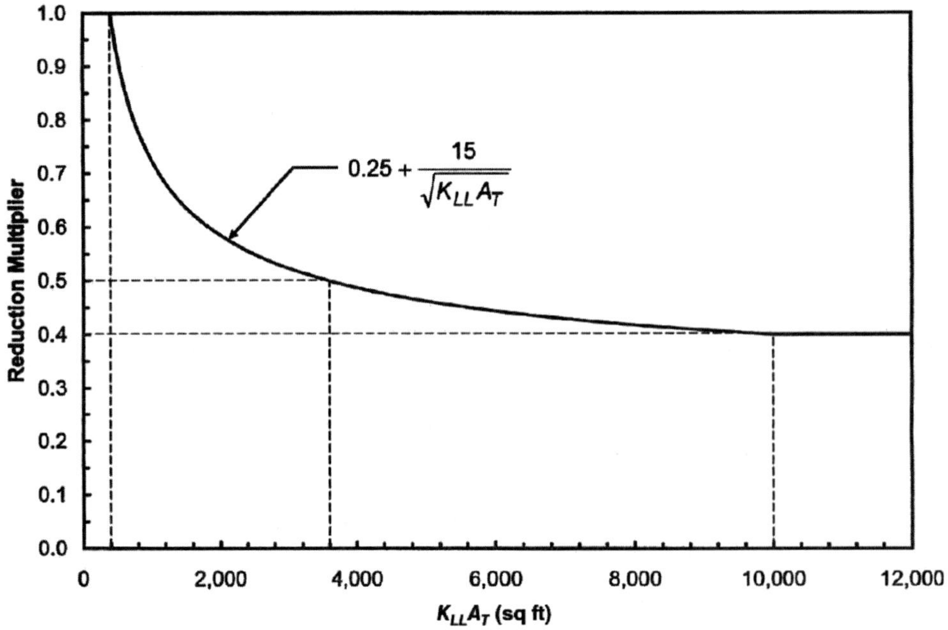

Figure 1.1 Reduction multiplier for live load in accordance with IBC 1607.11.1.

In buildings that support relatively large live loads, such as storage buildings, several adjacent bays may be fully loaded; as such, live loads should not be reduced in those situations. Data in actual buildings indicate that the floor in any story is seldom loaded with more than 80 percent of the nominal live load. Thus, a maximum live load reduction of 20 percent is permitted for members that support two or more floors, such as columns and walls.

Passenger Vehicle Garages

The live load in passenger vehicle garages is not permitted to be reduced, except for members supporting two or more floors; in such cases, the maximum reduction is 20 percent, but L must not be less than that calculated by IBC 1607.11.1 (IBC 1607.11.1.3). Thus, live load reduction is not permitted except for members that support two or more floors.

Group A (Assembly) Occupancies

Due to the nature of assembly occupancies, there is a high probability that the entire floor is subjected to full uniform live load. According to Footnote m in IBC Table 1607.1, live load reduction is not permitted in assembly areas, except for follow spot, projection, and control rooms, unless specific exceptions of IBC 1607.11 apply.

Flowchart 1 shown in Fig. 1.2 can be used to determine basic uniform live load reduction in accordance with IBC 1607.11.1.

1.3.4 Alternative Uniform Live Load Reduction

An alternative method of uniform live load reduction, which is based on provisions in the 1997 Uniform Building Code (Ref. 4), is given in IBC 1607.11.2. IBC Equation 16-24 can be used to obtain a reduction factor

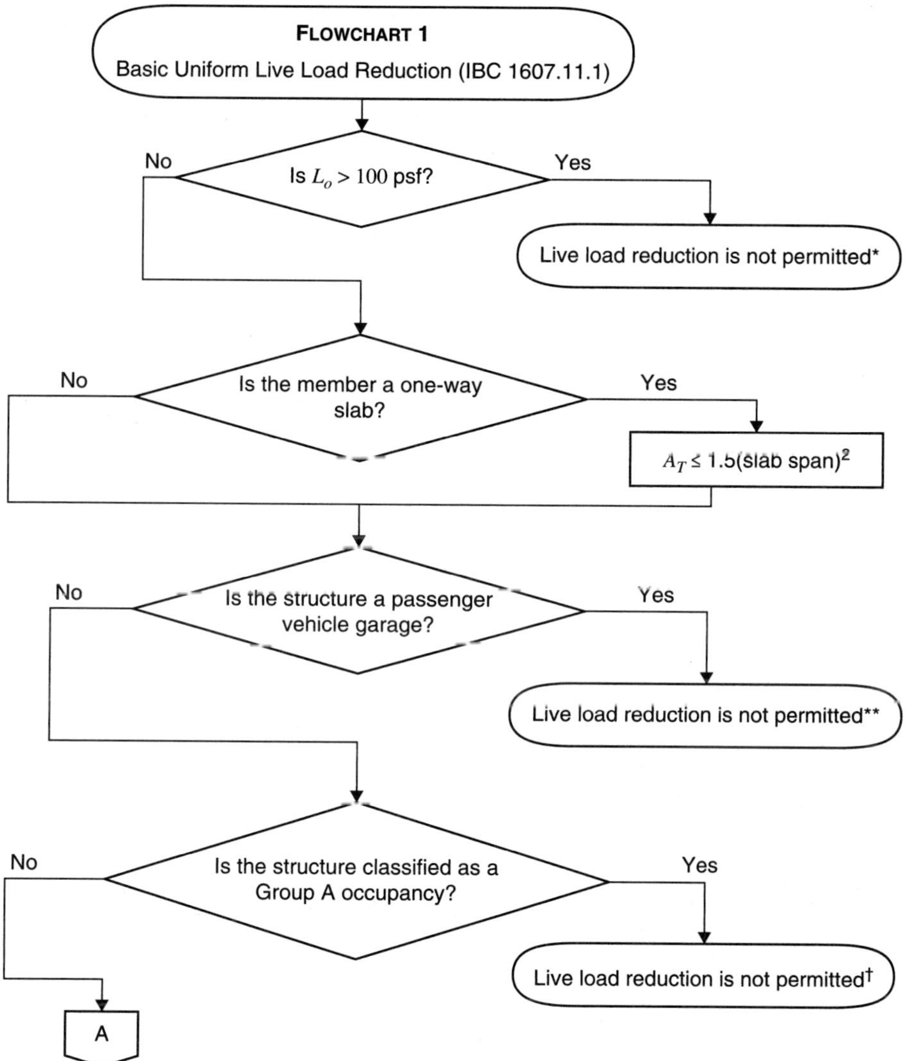

* See IBC 1607.11.1.2 for two exceptions to this requirement.

** Live loads for members supporting two or more floors are permitted to be reduced by a maximum of 20 percent (IBC 1607.11.1.3).

† Live loads for members supporting follow spot, projections, and control rooms are permitted to be reduced (see Footnote m in IBC Table 1607.1).

Figure 1.2 Basic uniform live load reduction in accordance with IBC 1607.11.1 (Flowchart 1).

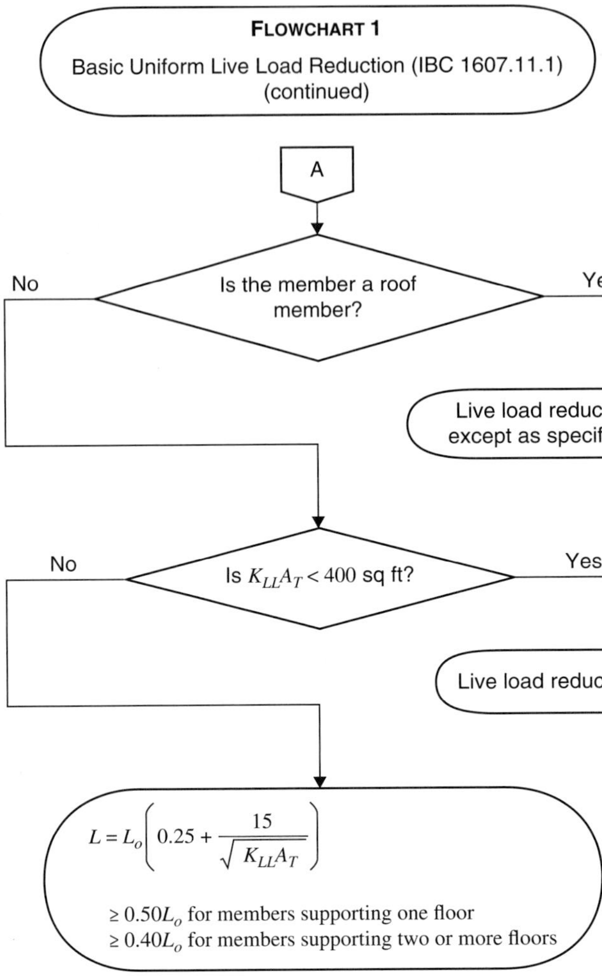

Figure 1.2 *(Continued)*

R for members that support an area greater than or equal to 150 square feet where the live load is less than or equal to 100 psf.

Flowchart 2 shown in Fig. 1.3 can be used to determine alternative uniform live load reduction in accordance with IBC 1607.11.2.

1.3.5 Roof Loads

In general, roofs are to be designed to resist dead, live, wind, and, where applicable, rain, snow, and earthquake loads. A minimum roof live load of 20 psf is prescribed in IBC Table 1607.1 for typical roof structures, while larger live loads are required for roofs used as gardens or places of assembly.

IBC 1607.13.2 permits nominal roof live loads on flat, pitched, and curved roofs and awnings and canopies other than fabric construction supported by a skeleton frame to be reduced in accordance with IBC Equation 16-26:

$$L_r = L_o R_1 R_2; \quad 12 \leq L_r \leq 20$$

where L_o = unreduced roof live load per square foot of horizontal roof projection supported by the member

L_r = reduced roof live load per square foot of horizontal roof projection supported by the member

$$R_1 = \begin{cases} 1 & \text{for } A_t \leq 200 \text{ square feet} \\ 1.2 - 0.0001A_t & \text{for } 200 \text{ square feet} < A_t < 600 \text{ square feet} \\ 0.6 & \text{for } A_t \geq 600 \text{ square feet} \end{cases}$$

$$R_2 = \begin{cases} 1 & \text{for } F \leq 4 \\ 1.2 - 0.05F & \text{for } 4 < F < 12 \\ 0.6 & \text{for } F \geq 12 \end{cases}$$

A_t = tributary area (span length multiplied by effective width) in square feet supported by a member

F = the number of inches of rise per foot for a sloped roof = the rise-to-span ratio multiplied by 32 for an arch or dome

No live load reduction is permitted for members supporting less than or equal to 200 square feet as well as for roof slopes less than or equal to 4:12. In no case is the reduced roof live load to be taken less than 12 psf. The minimum load determined by this equation accounts for occasional loading due to the presence of workers and materials during repair operations.

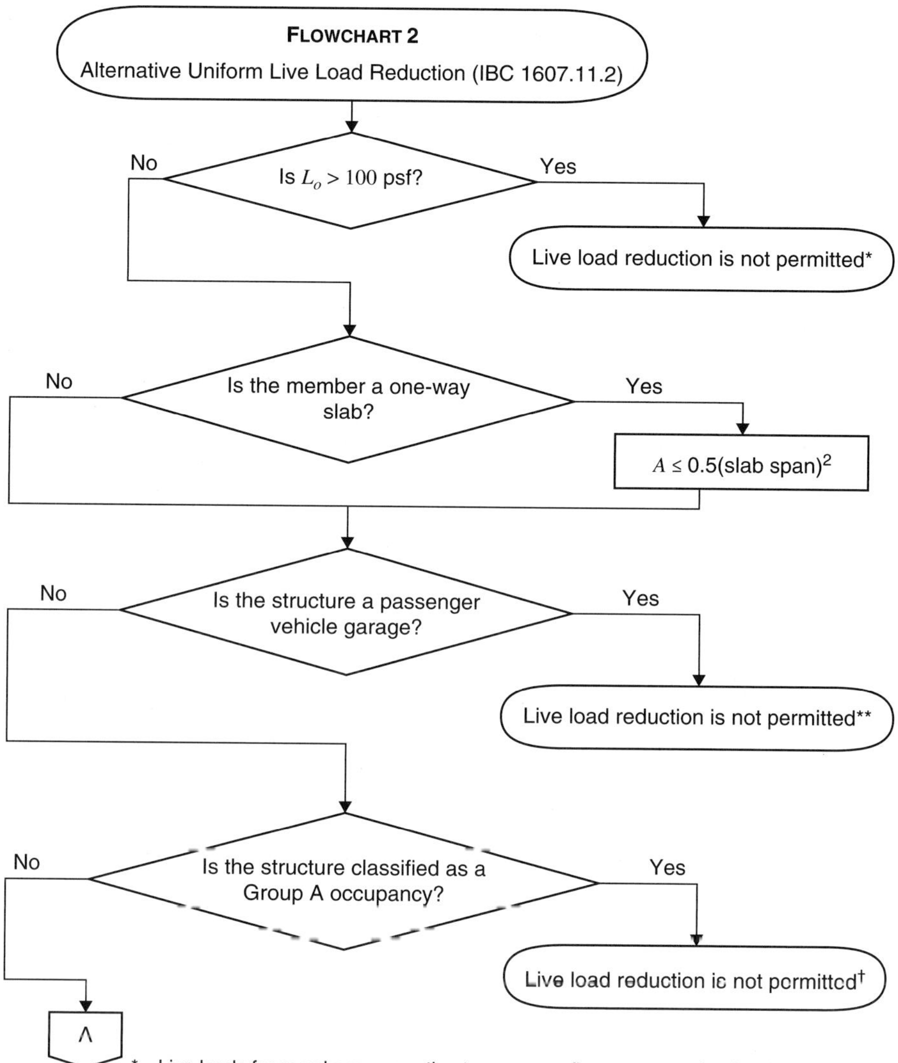

* Live loads for members supporting two or more floors are permitted to be reduced by a maximum of 20 percent [IBC 1607.11.2(1); also see exception in that section].

** Live loads for members supporting two or more floors are permitted to be reduced by a maximum of 20 percent [IBC 1607.11.2(2)].

† Live loads for members supporting follow spot, projections, and control rooms are permitted to be reduced (see Footnote m in IBC Table 1607.1).

Figure 1.3 Alternative uniform live load reduction in accordance with IBC 1607.11.2 (Flowchart 2).

Live loads are permitted to be reduced on areas of occupiable roofs using the provisions of IBC 1607.11 for floor live loads (IBC 1607.13.3). Live loads that are greater than or equal to 100 psf at areas of roofs that are classified as Group A (assembly) occupancies are not permitted to be reduced unless specific exceptions of IBC 1607.11 apply (see Footnote m in IBC Table 1607.1).

A minimum roof live load of 20 psf is required in unoccupied landscaped areas on roofs (IBC 1607.13.3.1). The weight of landscaping material is considered a dead load and must be determined based on the saturation level of the soil.

A minimum roof live load of 5 psf is required for awnings and canopies in accordance with IBC Table 1607.1 (IBC 1607.13.4). Such elements must also be designed for the combined effects of snow and wind loads in accordance with IBC 1608 and 1609.

1.3.6 Crane Loads

Design provisions for runway beams that support moving bridge cranes and monorail cranes are given in IBC 1607.14. In general, the support structure of the crane must be designed for the maximum wheel load, vertical impact, and horizontal impact as a simultaneous load combination.

A typical top-running bridge crane is depicted in Fig. 1.4. The trolley and hoist move along the crane bridge, which is supported by the

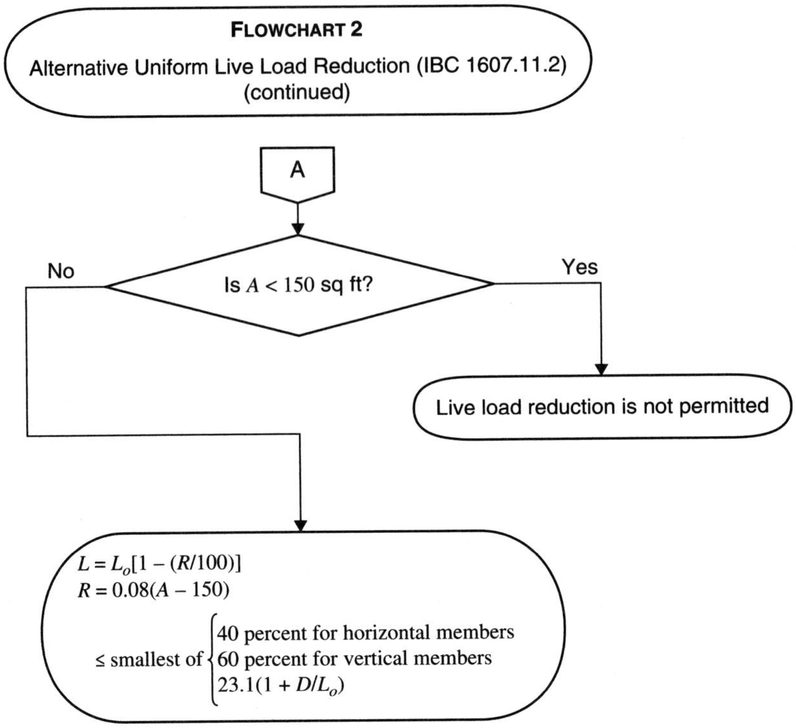

Figure 1.3 *(Continued)*

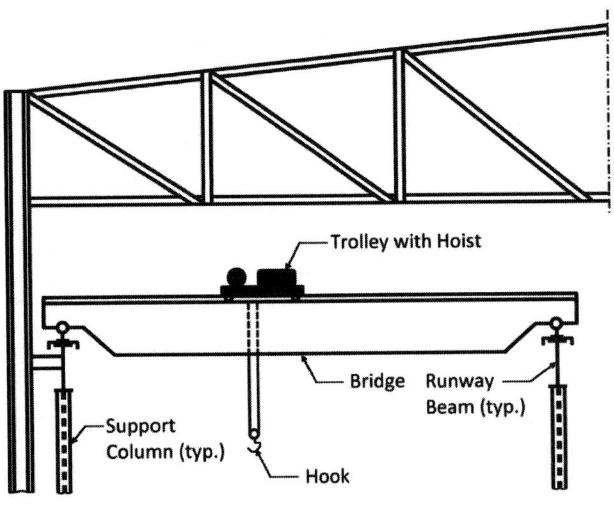

Figure 1.4 Top-running bridge crane.

runway beams and support columns. The entire crane assemblage can also move along the length of the runway beams.

The maximum wheel loads that are to be used in the design of the supporting members are due to the weight of the bridge plus the sum of the rated capacity and the weight of the trolley. The trolley is to be positioned on its runway at the location where the resulting load effect is maximum; generally, this occurs when the trolley is moved as close to the supporting members as possible.

The maximum wheel loads must be increased by the percentages given in IBC 1607.14.2 to account for the vertical impact force that is caused by the starting and stopping movement of the suspended weight from the crane and by the movement of the crane along the rails.

A lateral force that acts perpendicular to the crane runway beams is generated by the transverse movement of the crane, that is, by movement that occurs perpendicular to the runway beam (see Fig. 1.5). According to IBC 1607.14.3, the magnitude of this load on crane

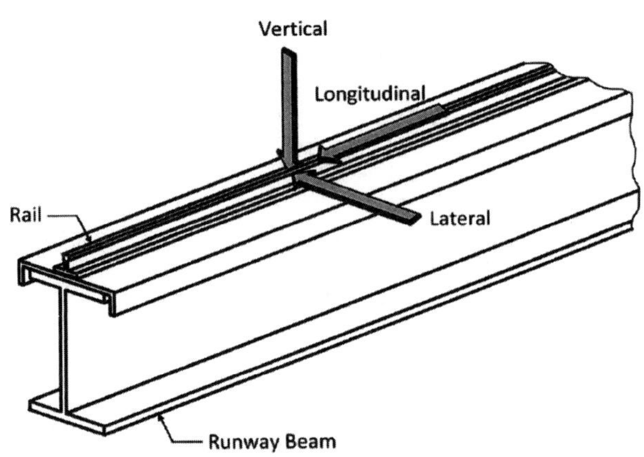

Figure 1.5 Crane loads on a runway beam.

runway beams with electronically powered trolleys is to be taken equal to 20 percent of the sum of the rated capacity of the crane and the weight of the hoist and trolley. It is assumed that this load acts horizontally at the traction surface of the runway beam and that it is distributed to the runway beam and supporting structure (such as columns) based on the lateral stiffness of the members.

A longitudinal force is generated on a crane runway beam by acceleration, deceleration, and braking of the crane bridge beam (see Fig. 1.5). This load is taken as 10 percent of the maximum wheel loads of the crane and is assumed to act horizontally at the traction surface of the runway beam in either direction parallel to the beam (IBC 1607.14.4). Bridge cranes with hand-geared bridges are exempt from this provision.

Reference 5 contains additional information on the determination of crane loads for industrial buildings.

1.3.7 Interior Walls and Partitions

Interior walls and partitions (including their finishing materials) that are greater than 6 feet in height are required to be designed for a horizontal load of 5 psf (IBC 1607.15). This requirement is intended to provide sufficient strength and durability of the wall framing and its finished construction when subjected to nominal impact loads, such as those from moving furniture or equipment and from HVAC pressurization.

Requirements for fabric partitions that exceed 6 feet in height are given in IBC 1607.15.1. A horizontal load equal to 5 psf is to be applied to the partition framing. Additionally, a 40-pound load must be applied over an 8-in.-diameter area of the fabric face at a height of 54 in. above the floor. This condition is meant to simulate the load caused by a person leaning against the fabric using their hand as the point of contact.

1.4 RAIN LOADS

1.4.1 General

IBC 1611 and ASCE/SEI Chapter 8 contain requirements for design rain loads. The nominal rain load R is determined based on the amount of water that can accumulate on a roof assuming that the primary roof drainage system is blocked. When this occurs, water will rise above the primary roof drain until it reaches the elevation of the roof edge or the secondary drainage system. The depth of water above the primary drain at the design rainfall intensity is based on the flow rate of the secondary system, which varies widely depending on the type of secondary system that is used.

The type and location of the secondary drains and the amount of rainwater above their inlets under design conditions must be known in order to determine R. Coordination among the design team (architectural, structural, and plumbing) is very important when establishing rain loads.

Chapter 11 of the International Plumbing Code® (IPC®) contains requirements on the design of roof drainage systems, including the required size and number of drains based on the area that is serviced (Ref. 6). Rainfall rates are given for various cities in the United States in Appendix B of the IPC. These rates are based on the maps in IPC Figure 1106.1, which have the same origin as the maps in the IBC (see the following discussion).

1.4.2 Design Rain Load

IBC Equation 16-35 or ASCE/SEI Equation 8.3-1 is used to determine the rain load R:

$$R = 5.2(d_s + d_h)$$

The constant in this equation is equal to the unit load per inch depth of rainwater, that is, the density of water divided by 12 in. per foot: 62.4/12 = 5.2 psf per inch.

The total depth of rainwater on a roof that is to be used in determining R consists of two parts:

1. The depth of water on the undeflected roof up to the inlet of the secondary drainage system when the primary drainage system is blocked (d_s).
2. The additional depth of water on the undeflected roof above the inlet of the secondary drainage system at its design flow (d_h).

The static head d_s is the distance from the primary drain to the secondary drain and is determined in the design of the combined drainage system.

The rainwater depth d_h (which is also referred to as the hydraulic head) is a function of the rainfall intensity i at the site, the area of roof A that is serviced by that drainage system, and the size of the drainage system.

IBC Figure 1611.1 provides the rainfall rates for a storm of 1-hour duration that has a 100-year return period. These rates have been determined by a statistical analysis of weather records. Both the primary and the secondary drainage systems must be designed for the prescribed rainfall rate (see Section 1108 of the IPC on how to size secondary drainage systems). It is always good practice to check with local building authorities to ensure that the proper rainfall rate is used in design.

In general, d_h depends on the type and size of the secondary drainage system and the flow rate Q it must handle. The following equation from C8.5 can be used to determine the flow rate through a single secondary drainage system:

$$Q = 0.0104Ai$$

In this equation, Q is in gallons per minute, A is in square feet, and i is in inches per hour. The constant 0.0104 is obtained based on the units associated with the variables in the equation:

$$\text{Constant} = \text{ft}^2 \times \frac{\text{in.}}{\text{hr}} \times \frac{1\ \text{ft}}{12\ \text{in.}} \times \frac{7.48\ \text{gal}}{\text{ft}^3} \times \frac{1\ \text{hr}}{60\ \text{min}} = 0.0104$$

The following equations relate the flow rate Q in gallons per minute to the hydraulic head d_h in inches for channel- and closed-type scuppers (see Fig. 1.6 and Ref. 7):

$$\text{Channel-type scuppers}: Q = 2.9\, b d_h^{1.5} \text{ when } h \geq d_h$$

$$\text{Closed-type scuppers}: Q = 2.9\, b\left[d_h^{1.5} - (d_h - h)^{1.5}\right] \text{ when } h < d_h$$

where b and h are the width and depth of the scupper, respectively, in inches. Note that the channel-type scuppers equation is also applicable to closed-type scuppers when $h \geq d_h$.

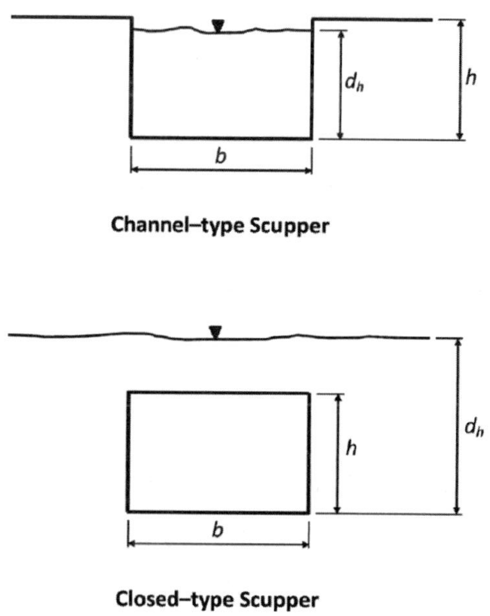

Figure 1.6 Roof scuppers.

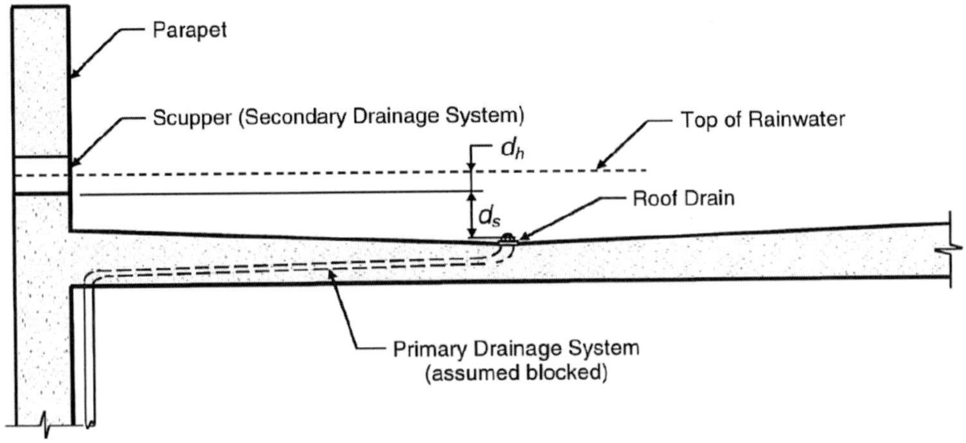

Figure 1.7 Water depths, d_s and d_h, in accordance with IBC 1611 for typical perimeter scuppers.

ASCE/SEI Table C8-1 gives flow rates in gallons per minute and corresponding hydraulic heads for circular drain, channel-type scupper, and closed-type scupper drainage systems.

Figure 1.7 illustrates the rainwater depths d_s and d_h that are to be used in determining R for the case of a scupper secondary drainage system, while Fig. 1.8 illustrates these water depths for a typical interior circular secondary drainage system.

Where buildings are configured such that rainwater will not collect on the roof, no rain load is required in the design of the roof, and a secondary drainage system is not needed. What is important to note is that the provisions of IBC 1611 and ASCE/SEI Chapter 8 must be considered wherever the potential exists that water may accumulate on the roof.

1.4.3 Ponding Instability

In situations where roofs do not have adequate slope or have insufficient and/or blocked drains to remove water due to rain or melting snow, water will tend to pond in low areas, which will cause the roof structure to deflect. These low areas will subsequently attract even more water, leading to additional deflection. Sufficient stiffness must be provided so that deflections will not continually increase until instability occurs, resulting in localized failure.

The provisions in ASCE/SEI 8.4 are provided to help ensure that susceptible bays have adequate stiffness to preclude progressive deflection and adequate strength to resist the ponding load plus the applicable design rain load (IBC 1611.2). Susceptible bays are defined in ASCE/SEI 8.4 and illustrated in ASCE/SEI Figure C8.4-1.

ASCE/SEI Figure C8.4-2 shows a roof with interior primary drains and perimeter overflow (secondary) drains. All bays are susceptible in this situation regardless of the roof slope because water can be impounded on the roof when the primary drainage system is blocked.

1.4.4 Controlled Drainage

In some jurisdictions, the rate of rainwater flow from roofs into storm drains is limited. Controlled-flow drains are typically utilized in such cases and are designed in accordance with IPC 1110 (Ref. 6). The drainage flow rate provided by the primary drainage system (controlled-flow drains) is less than the rainfall rate and water intentionally accumulate on the roof.

According to IBC 1611.3 and ASCE/SEI 8.5, a secondary drainage system at a higher elevation must be provided when controlled-flow drains are used as the primary drainage system; this is meant to limit the accumulation of water on the roof above that elevation. Note that controlled drainage hardware must not be used on the secondary drainage system because the secondary system is the backup in case the primary system is blocked.

Roofs must be designed in such cases to support the stormwater temporarily supported on them, similar to traditional roof drainage systems.

Ponding instability in accordance with IBC 1611.2 or ASCE/SEI 8.4 must also be checked.

1.5 SOIL LATERAL LOADS

1.5.1 General

Foundation walls of a building or structure and retaining walls must be designed to resist the lateral loads caused by adjacent soil. A geotechnical investigation is usually undertaken to determine the magnitude of the soil pressure. In cases where the results of such an investigation are not available, the lateral soil loads in IBC Table 1610.1 are to be used (similar design lateral loads are provided in ASCE/SEI Table 3.2-1).

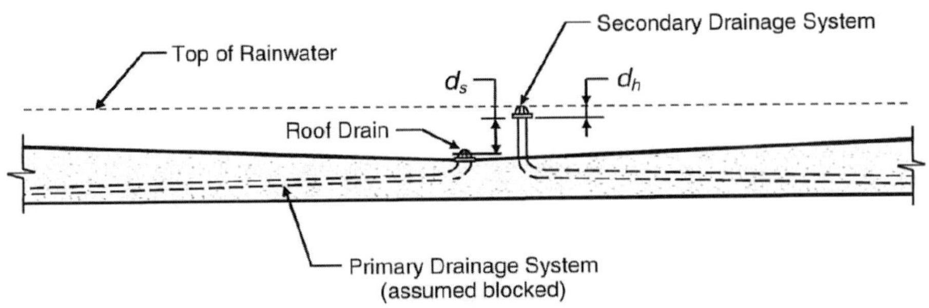

Figure 1.8 Water depths, d_s and d_h, in accordance with IBC 1611 for typical interior drains.

The design lateral soil load, H, depends on the type of soil and the boundary conditions at the top of the wall. Walls that are restricted to move at the top are to be designed for the at-rest pressures tabulated in IBC Table 1610.1, while walls that are free to deflect and rotate at the top are to be designed for the active pressures in that table.

Foundation walls that do not extend more than 8 feet below grade and are laterally supported at the top by flexible diaphragms are permitted to be designed for the active pressure values given in the table.

Lateral soil pressures are not provided for the expansive soils identified by Note b in IBC Table 1610.1 because these soils have unpredictable characteristics. These soils absorb water and tend to shrink and swell to a higher degree than other soils. As these soils swell, they are capable of exerting relatively large forces on the soil-retaining structure. As such, expansive soils are not suitable as backfill (see Note b).

In addition to lateral pressures from soil, walls must be designed to resist the effects of hydrostatic pressure due to undrained backfill (unless a drainage system is installed) and to any surcharge loads that can result from sloping backfills or from driveways or parking spaces that are close to a wall. Submerged or saturated soil pressures include the weight of the buoyant soil plus the hydrostatic pressure.

ASCE/SEI 3.2.2 contains requirements for the design of any horizontal element supported directly on soil, such as slabs on grade and basement slabs. Full hydrostatic pressure must be applied over the entire area of such elements where applicable. Elements that are supported by expansive soils must be designed to accommodate the upward loads caused by the expansive soil, or the expansive soil is to be removed or stabilized around and beneath the structure.

1.6 SNOW AND ICE LOADS

1.6.1 Introduction

Structural members that are a part of roofs, balconies, canopies, and similar structures that are exposed to the environment must be designed for the effects of snow loads in those geographic areas where snowfall can occur.

Loads on buildings and other structures due to snow are determined based on the anticipated ground snow load, the occupancy of the building, the exposure, the thermal resistance of the roof structure, and the shape and slope of the roof. Partial loading, unbalanced snow loads due to roof configuration, drift loads on lower or adjacent roofs and on projections such as parapets and mechanical equipment, sliding snow loads, and rain-on-snow loads must also be considered when designing for the effects from snow.

IBC 1608.1 requires that design snow loads be determined by the provisions of Chapter 7 of ASCE/SEI 7. In-depth information is given in the following sections on these provisions along with pertinent background information on the methodologies that are utilized.

In certain parts of the United States, atmospheric ice loads must be considered in the design of all structures and structural elements that are exposed to the elements. Chapter 10 of ASCE/SEI 7-16 contains provisions on how to determine ice loads due to freezing rain on a variety of structural shapes, objects, and configurations as a function of the design ice thickness.

1.6.2 Flat Roof Snow Loads

Once a ground snow load has been established, a flat roof snow load, p_f, is determined by Equation 7.3-1 of ASCE/SEI 7:

$$p_f = 0.7C_eC_tI_sp_g$$

The 0.7 factor in the equation above is a conservative ground-to-roof snow load conversion factor. C_e is the exposure factor, accounts for the wind at the site, and is related to the type of terrain and the exposure of the roof. Values of C_e are given in ASCE/SEI 7 Table 7.3-1 as a function of the surface roughness category and the type of roof exposure. Surface roughness categories B, C, and D are defined in 26.7 for wind design (see Table 1.2).

Table 1.2 Surface Roughness Categories

Surface roughness category	Description
B	Urban and suburban areas, wooded areas, or other terrain with numerous closely spaced obstructions having the size of single-family dwellings or larger
C	Open terrain with scattered obstructions having heights generally less than 30 feet; this category includes flat open country and grasslands
D	Flat, unobstructed areas and water surfaces; this category includes smooth mudflats, salt flats, and unbroken ice

Roof exposures are defined as fully exposed, partially exposed, and sheltered (see Footnote a in Table 7.3-1). A fully exposed condition exists where a roof is exposed on all sides with no shelter provided by adjoining terrain, higher structures, or trees. Roofs with large mechanical equipment, parapets that extend above the height of the balanced snow load, or other similar obstructions are not considered to be fully exposed because such conditions can provide some shelter to the wind.

Obstructions are defined as providing shelter when they are located within a distance of $10h_o$ from the roof, where h_o is the height of the obstruction above the roof (see Footnote b in Table 7.3-1). The conifers depicted in Fig. 1.9 provide shelter to the building roof if the distance, x, between the centerline of the building and the tree line is less than or equal to $10h_o$. In cases where deciduous trees that are leafless in winter surround the site, the fully exposed category is applicable.

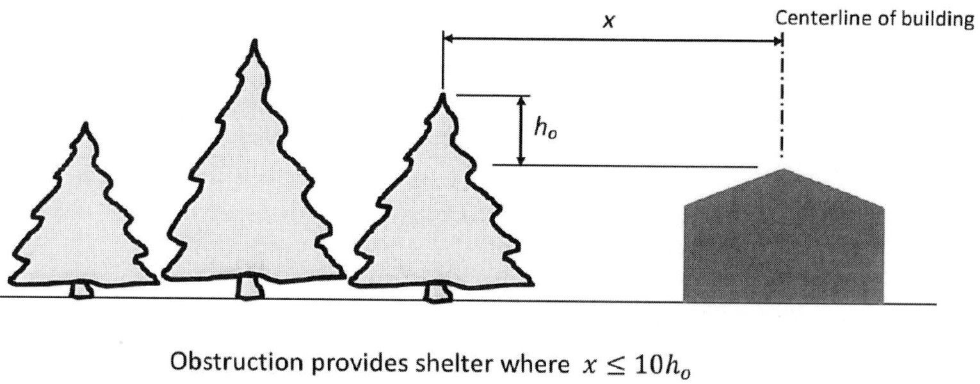

Figure 1.9 Sheltered roof exposures for snow loads.

Partially exposed roof exposures are to be used where fully exposed and sheltered conditions do not apply. This is generally the most common roof exposure.

Similar to surface roughness categories, a roof exposure condition must represent the conditions that are expected during the life of the building.

C_t is the thermal factor; it accounts for the amount of heat loss through the roof. Values of C_t are given in ASCE/SEI 7 Table 7.3-2 as a function of the thermal condition. Like terrain categories and exposure conditions, the thermal condition that is chosen must represent the anticipated conditions during winters for the life of the building.

I_s is the importance factor; it adjusts the snow load based on the occupancy of the structure. Risk categories for buildings and other structures are defined in IBC Table 1604.5, and values of I_s are given in ASCE/SEI Table 1.5-2.

1.6.3 Minimum Snow Loads for Low-Slope Roofs

ASCE/SEI 7.3.4 contains provisions for minimum snow loads, p_m, for low-slope roofs, which are defined as follows:

- Monoslope, hip, and gable roofs with slopes less than 15 degrees.
- Curved roofs where the vertical angle from the eaves to the crown is less than 10 degrees.

Flowchart 3 shown in Fig. 1.10 shows the procedure for finding the flat roof snow load.

1.6.4 Sloped Roof Snow Loads

Design snow loads for all structures are based on the sloped roof snow load, p_s, which is determined by the equation

$$p_s = C_s p_f$$

The slope factor, C_s, depends on the slope and temperature of the roof, the presence or absence of obstructions, and the degree of slipperiness of the roof surface.

Figure 7.4-1 of ASCE/SEI 7 contains graphs of C_s for warm and cold roof conditions, and C7.4 contains equations for C_s. Flowchart 4 in Fig. 1.11 shows the procedure for finding the roof slope factor snow load, and Flowchart 5 in Fig. 1.12 shows the procedure for finding the sloped roof snow load.

1.6.5 Partial Loading

The partial loading provisions of 7.5 of ASCE/SEI 7 must be satisfied for continuous roof framing systems and all other roof systems where removal of snow load on one span (by wind or thermal effects) causes an increase in stress or deflection in an adjacent span. For example, an increase in bending moment and deflection will occur in the span of a cantilevered roof member that is adjacent to the cantilever span where half the snow load is removed.

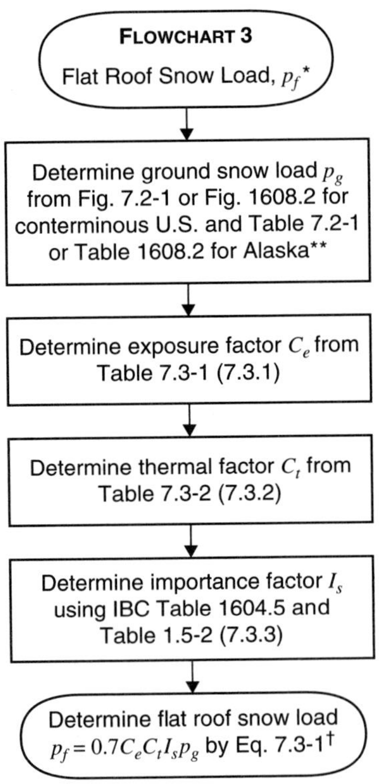

* A flat roof is defined as a roof with a slope that is less than or equal to 5 degrees.

** "CS" in the maps signifies areas where a site-specific study must be conducted to determine p_g. Numbers in parentheses represent the upper elevation limit in feet for the ground snow load values given below. Site-specific studies are required at elevations not covered in the maps.

† Minimum snow loads for low-slope roofs p_m are specified in 7.3.4. Low-slope roofs are defined in 7.3.4.

Figure 1.10 Flat roof snow load (Flowchart 3).

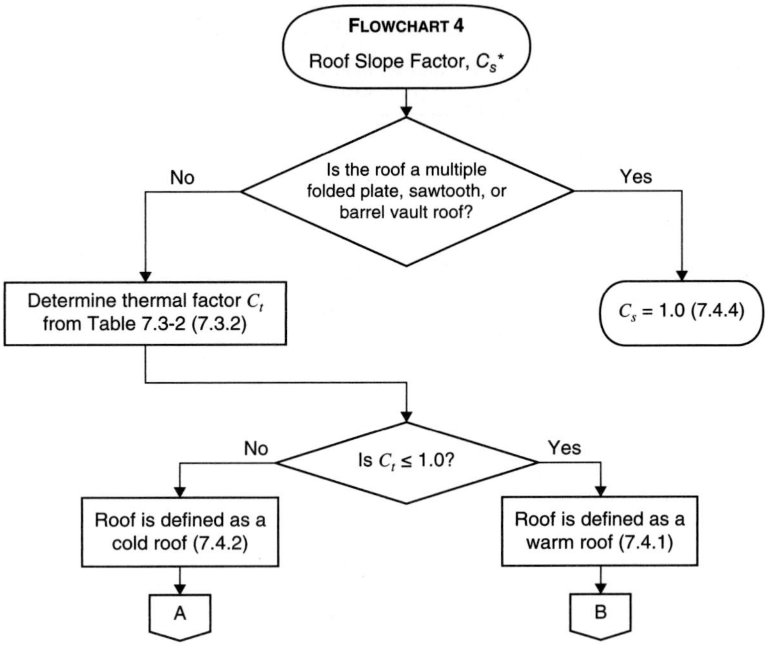

* Portions of curved roofs having a slope exceeding 70 degrees shall be considered free of snow load, that is, $C_s = 0$ (7.4.3).

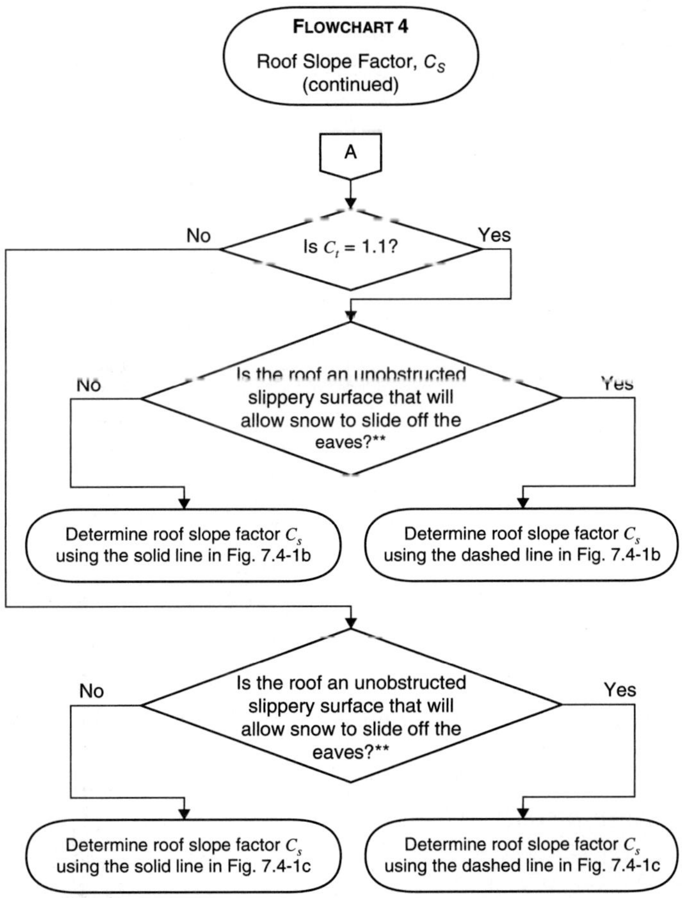

Figure 1.11 Roof slope factor (Flowchart 4).

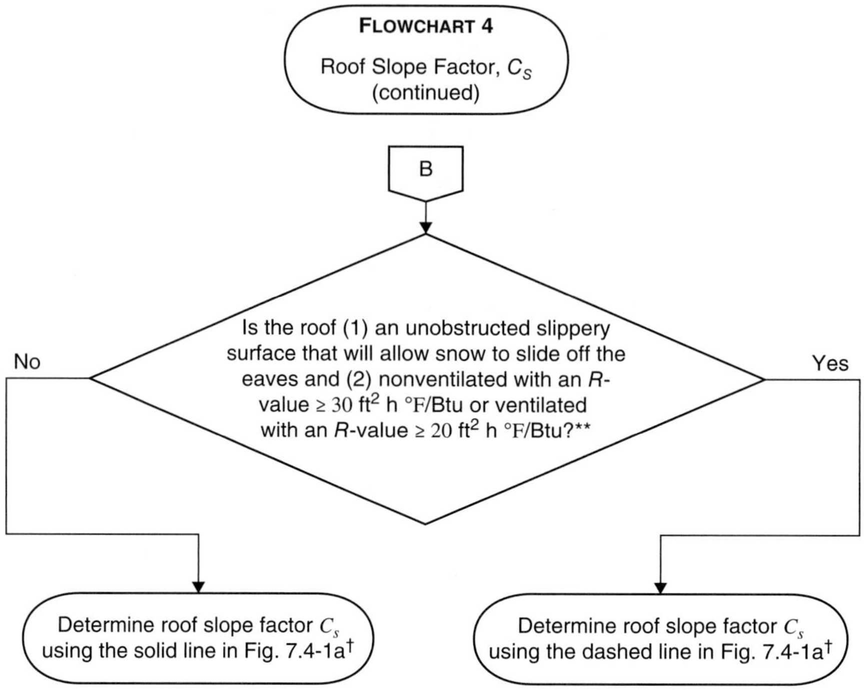

** See 7.4 for definitions of unobstructed and slippery surfaces. An *R*-value for a roof is defined as its thermal resistance.

† See 7.4.5 for an additional uniformly distributed load that is to be applied on overhanging portions of warm roofs due to formation of ice dams and icicles along eaves.

Figure 1.11 *(Continued)*

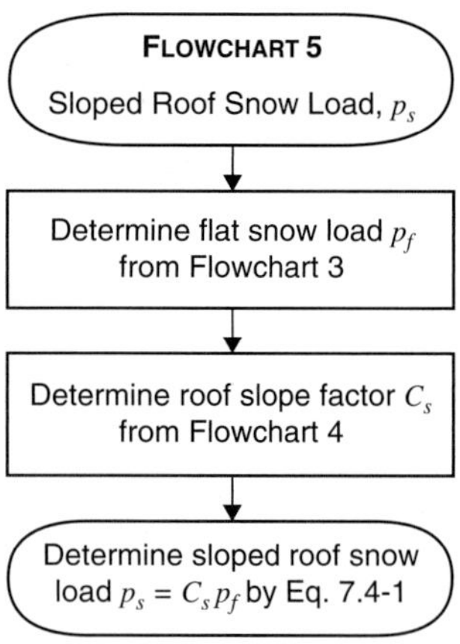

Figure 1.12 Sloped roof snow load (Flowchart 5).

For simplicity, only the three load cases given in Figure 7.5-1 of ASCE/SEI 7 need to be investigated; comprehensive alternate span (or checkerboard) loading analyses are not required:

- Full balanced snow load on either of the exterior spans and half of the balanced snow load on all other spans.
- Half of the balanced snow load on either exterior span and full balanced snow load on all other spans.
- All possible combinations of full balanced snow load on any two adjacent spans and half of the balanced snow load on all other spans. There will be $(n - 1)$ possible combinations for this case where n is equal to the number of spans in the continuous beam system.

Case 1 can occur when two different snow events that correspond to half of the balanced snow load are separated by an event that prevents the lower half of the roof snow from drifting. The intervening event can be sleet or freezing rain, for example. After the second snowfall occurs, a strong wind blows across the roof, removing snow from all of the spans and depositing it on all of the downwind spans. Case 1 in Figure 7.5-1 depicts the case for wind blowing from right to left.

Wind blowing from left to right must also be considered, and this is covered in Case 2 for a wind of shorter duration than that in Case 1. In particular, the wind event in Case 2 is long enough to remove the snow from the first span and deposit it on the adjoining spans.

The load pattern depicted for Case 3 in Figure 7.5-1 could correspond to a case where there is an obstruction on the roof (such as a parapet) that blocks the wind from blowing the snow behind the obstruction.

Partial loading provisions need not be considered for structural members that span perpendicular to the ridgeline of gable roofs with slopes between 2.38 degrees (½ on 12) and 30.3 degrees (7 on 12) because this loading case is addressed in the unbalanced snow load provisions of 7.6.1, which are covered in the next section.

1.6.6 Unbalanced Roof Snow Loads

Overview

Unbalanced snow load occurs on sloped roofs from wind and sunlight and in most cases can be considered a drift load. Wind tends to reduce the snow load on the windward portion of a roof and increase the snow load on the leeward portion. Provisions for unbalanced snow loads are given in 7.6.1 for hip and gable roofs; in 7.6.2 for curved roofs; in 7.6.3

for multiple folded plate, sawtooth, and barrel vault roofs; and in 7.6.4 for dome roofs. When determining unbalanced snow loads on a roof, it is important to consider wind from all directions.

Hip and Gable Roofs

Depicted in Fig. 1.13 is unbalanced snow load on a hip and gable roof of a building due to wind. The shape of the roof drift can typically be characterized as a triangle located close to the ridgeline: the surcharge is essentially zero at the ridge, and the top surface of the surcharge is basically horizontal. Loading conditions meant to represent this behavior are given in 7.6.1.

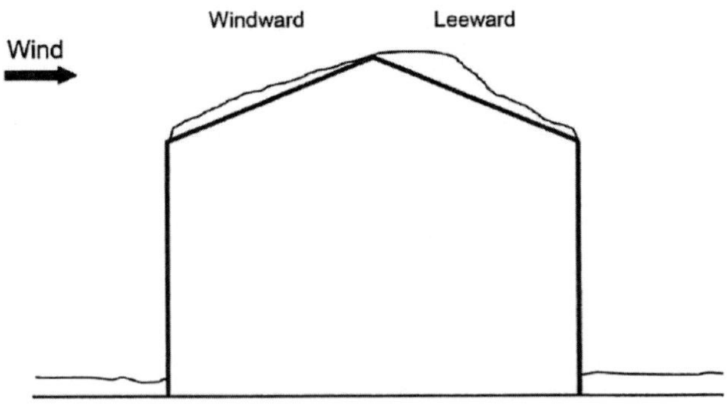

Figure 1.13 Unbalanced snow load on a hip or gable roof.

Unbalanced snow loads must be considered for roofs with slopes of ½ on 12 (2.38 degrees) through 7 on 12 (30.3 degrees). Drifts typically do not form on roofs with slopes less than and greater than these limiting values. Balanced and unbalanced load conditions that must be addressed for hip and gable roofs are illustrated in Figure 7.6-2 (see Fig. 1.14).

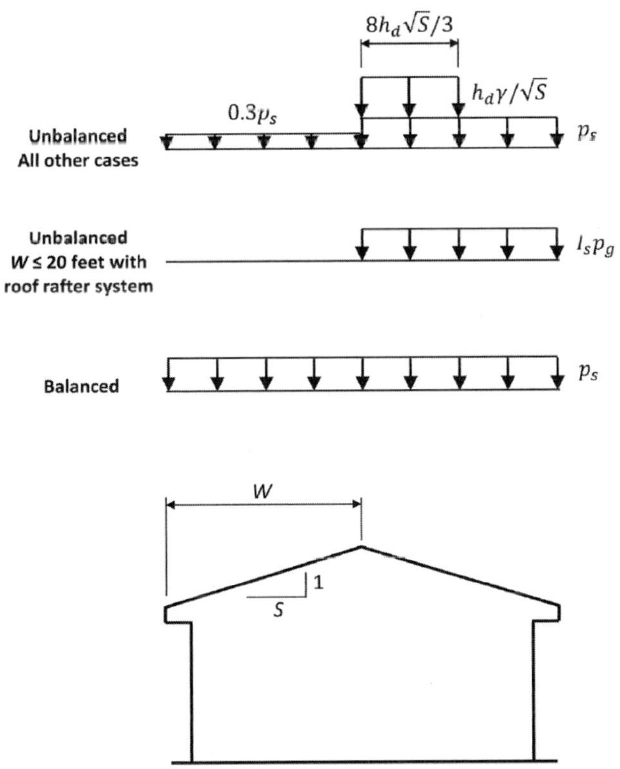

Figure 1.14 Balanced and unbalanced snow loads for hip and gable roofs.

In cases where W is less than 20 feet, a value of W equal to 20 feet is used in Equation 4.3. In such cases, the snow drift height, h_d, determined by Equation 4.3 need not be taken greater than

$$\sqrt{I_s p_g W / 4\gamma}$$

where W is the actual fetch distance, not the minimum fetch distance of 20 feet.

The density of snow, γ, is determined by Equation 7.7-1:

$$\gamma = 0.13\, p_g + 14 \le 30 \text{ pcf}$$

Flowchart 6 in Fig. 1.15 can be used to determine the unbalanced roof snow loads for hip and gable roofs.

Curved Roof

Provisions for balanced and unbalanced snow loads on curved roofs are given in 7.6.2. Any portion of a curved roof that has a slope that exceeds 70 degrees can be considered free of snow loads. Note that the roof slope is measured from the horizontal to the tangent of the curved roof at that point. In cases where the roof slope exceeds 70 degrees, the point on the roof at a slope of 70 degrees is considered to be the eave.

Balanced and unbalanced load cases that are applicable to curved roofs are given in Figure 7.4-2 as a function of the slope of the roof at the eave.

It is important to note that these provisions are not applicable to curved roofs that are concave upward. In such cases, as well as in cases for other roof geometries and complicated site conditions, wind tunnel model studies should be used to establish design snow loads (see C7.14).

The unbalanced load is very similar to Case 2: the windward portion of the roof between the windward eave and the crown is assumed to be free of snow, and there are two cases to consider for the leeward portion of the roof depending on whether or not the ground or another roof abuts the curved roof within 3 feet of its eaves (see Figure 4.8 in Ref. 3).

Flowchart 7 in Fig. 1.16 can be used to determine the unbalanced roof snow loads for curved roofs.

Multiple Folded Plate, Sawtooth, and Barrel Vault Roofs

Provisions for unbalanced snow loads on folded plate, sawtooth, and barrel vault roofs are given in 7.6.3. Unbalanced loads need to be considered when the slope exceeds ⅜ in./ft (1.79 degrees).

Similar to curved roofs, unbalanced loads are equal to $0.5p_f$ at the crown or ridge of the roof and are equal to $2p_f/C_e$ at the valleys. Illustrated in Figure 7.6-3 are the balanced and unbalanced loads for a sawtooth roof.

Flowchart 8 in Fig. 1.17 can be used to determine the unbalanced roof snow loads for folded plate, sawtooth, and barrel vault roofs.

Dome Roofs

According to 7.6.4, unbalanced snow loads for dome roofs are to be determined in the same manner as for curved roofs. Unbalanced loads are to be applied to the downwind 90-degree sector of the dome in plan (see Fig. 1.18). The load decreases linearly to zero over 22.5-degree sectors on either side of this sector. No snow load is taken on the remaining 225-degree upwind sector.

The balanced and unbalanced load distributions depend on the slope of the roof at the eaves. In the unbalanced load case, the snow load at the eave or where the slope is equal to 70 degrees decreases linearly to zero over the 22.5-degree sector on each side of the 90-degree downwind sector of the roof.

1.6.7 Drifts on Lower Roofs

Overview

Section 7.7 contains provisions for snow drifts that can occur on lower roofs of a building due to the following:

1. Wind depositing snow from higher portions of the same building or an adjacent building or terrain feature (such as a hill) to a lower roof
2. Wind depositing snow from the windward portion of a lower roof to the portion of the lower roof adjacent to a taller part of the building

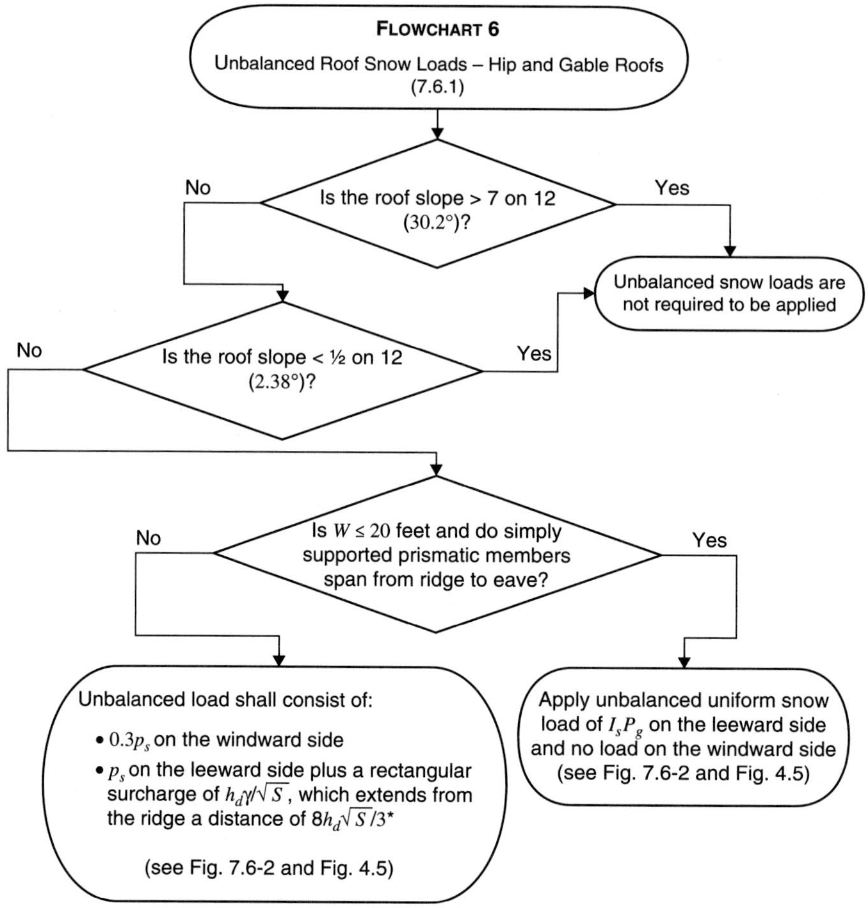

* h_d is the drift height from Fig. 7.6-1 with W substituted for ℓ_u, γ = snow density determined by Eq. 7.7-1, and S = roof slope run for a rise of one.

Figure 1.15 Unbalanced roof snow loads—hip and gable roofs (Flowchart 6).

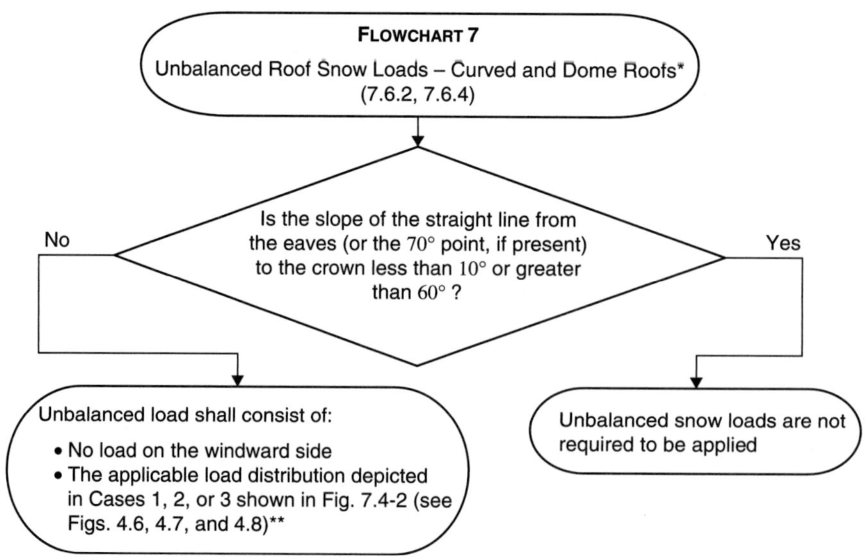

* Portions of curved roofs having a slope > 70° shall be considered free of snow.

** See Fig. 4.10 for application of unbalanced snow loads on a dome roof.

Figure 1.16 Unbalanced roof snow loads—curved and dome roofs (Flowchart 7).

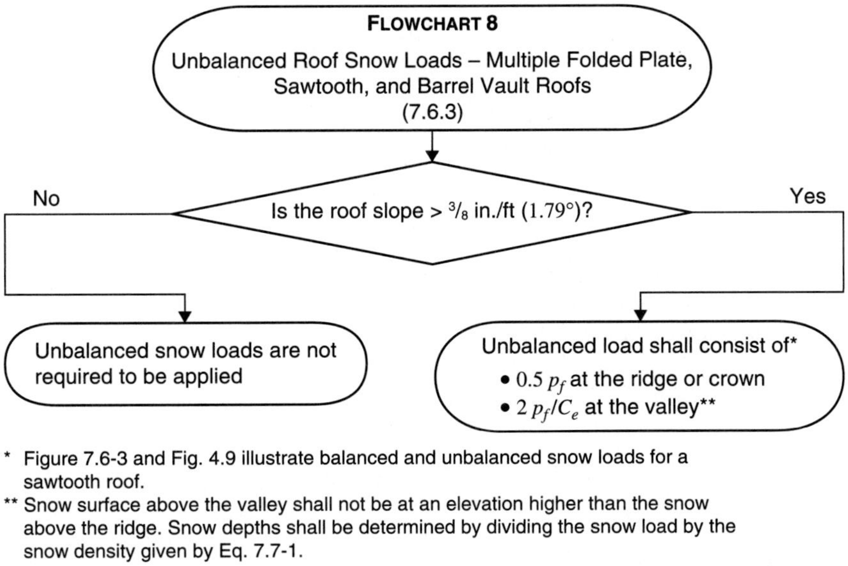

Figure 1.17 Unbalanced roof snow loads—multiple folded plate, sawtooth, and barrel vault roofs (Flowchart 8).

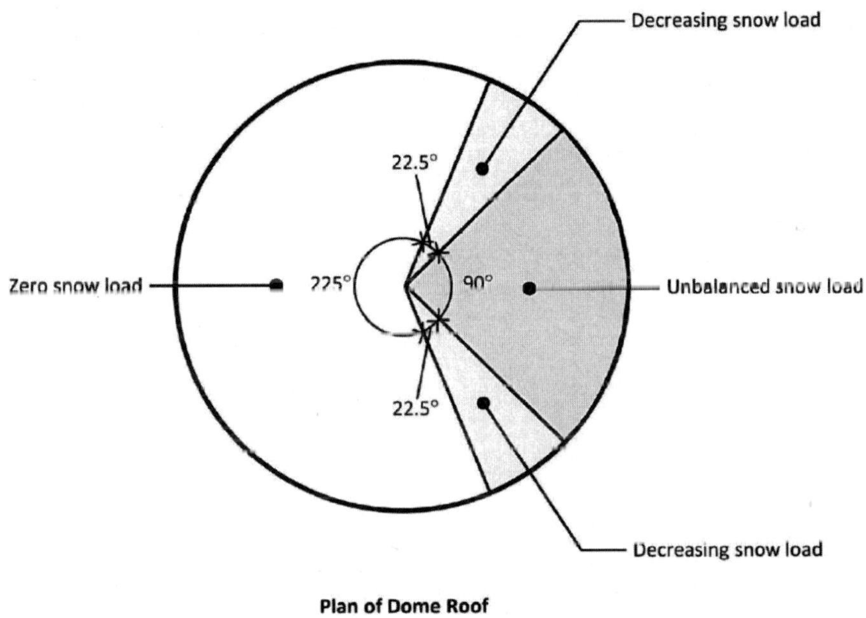

Figure 1.18 Unbalanced loads for a dome roof.

The first type of drift is called a leeward drift, and the second is a windward drift. Both types of drifts are illustrated in Figure 7.7-1 (see Fig. 1.19).

In cases where there is a change in elevation between roofs, snow will have a tendency to accumulate at this location (see Fig. 1.19). Depending on wind direction, either a windward or a leeward drift will form. A triangular shape is used to characterize windward drifts as well.

The configuration that is to be taken for snow drifts on lower roofs is depicted in Figure 7.7-2 (see Fig. 1.20).

In the case of leeward drifts, the height of the drift, h_d, at the roof step is determined from Figure 7.6-1 by substituting the length of the upper roof, l_{upper}, for l_u and using the snow importance factor, I_s:

$$\frac{h_{dleeward}}{\sqrt{I_s}} = \left[0.43(l_{upper})^{\frac{1}{3}}(p_g + 10)^{\frac{1}{4}}\right] - 1.5$$

The clear height, h_c, shown in Fig. 1.20 is equal to the height of the roof step minus the height of the balanced snow, h_b. It is determined

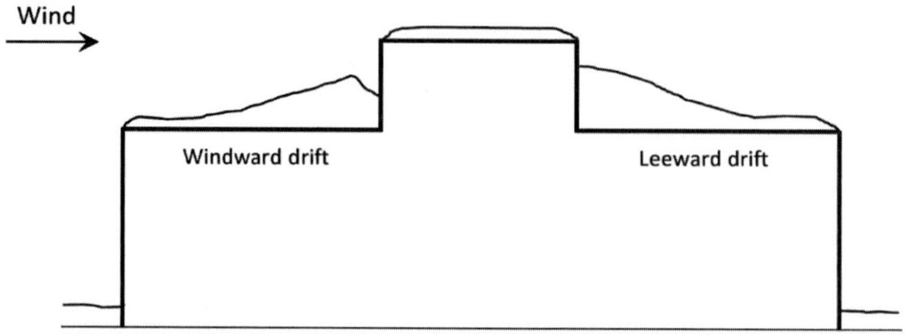

Figure 1.19 Windward and leeward snow drifts.

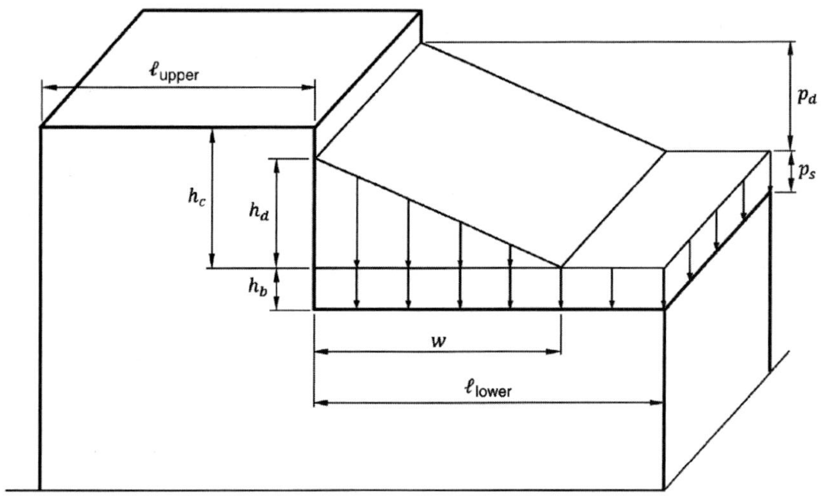

Figure 1.20 Configuration of snow drifts.

in this way assuming that the upper roof is blown clear of snow in the vicinity of the drift. While this assumption is generally valid for windward drifting, it is not necessarily accurate for leeward drifting because there may still be snow on the upper level roof when the drift has stopped growing. For simplicity, the same assumption is used for the clear height regardless of the type of drift that is formed.

The height of the balanced snow, h_b, is determined using the magnitude of the balanced snow load, p_s, and the density of the snow, γ:

$$h_b = \frac{p_s}{\gamma} = \frac{p_s}{0.13p_g + 14}$$

where γ need not be taken greater than 30 pcf (see Equation 7.7-1).

The width of the drift, w, depends on the clear height, h_c, as follows:

$$w = \begin{Bmatrix} 4h_d & \text{where } h_d \le h_c \\ \dfrac{4h_d^2}{h_c} \le 8h_c & \text{where } h_d > h_c \end{Bmatrix}$$

If it is found that the governing drift height, h_d, exceeds the clear height, h_c, then the drift height is set equal to h_c. The drift width in this case has been established by equating the cross-sectional area of the triangular drift that is limited to h_c (i.e., $h_c \times w/2$) with the cross-sectional area of the triangular drift that is not limited to h_c [i.e., $(h_d \times 4h_d/2)$]. The upper limit of drift width 8_{hc} is based on studies that showed that additional snow accumulation is not expected within a rise-to-run ratio range of 1:6.5 to 1:10.

There may be cases where the drift width, w, exceeds the length of the lower roof, ℓ_{lower}. In these situations, the drift load is to taper linearly to zero at the far end of the lower level roof. This provision may be applicable to canopies over an entranceway to a building. Figure 1.21 shows the proper load distribution to use where $w > \ell_{\text{lower}}$.

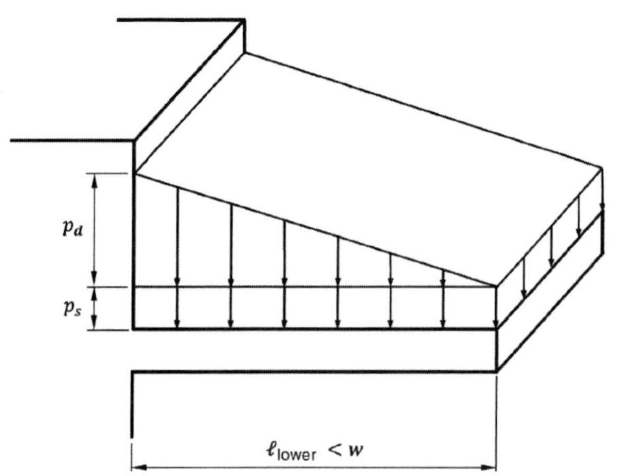

Figure 1.21 Load configuration where the drift width is greater than the length of the lower roof.

According to 7.7.1, drift loads need not be considered where the ratio h_c/h_b is less than 0.2. In such cases, the roof step is relatively small so that the drift formation is negligible.

Adjacent Structures

Leeward drifts can form on the roof of a structure that is close enough to an adjacent one that has a higher roof. According to 7.7.2, leeward drifts form on lower roofs of adjacent buildings when the horizontal separation distance, s, between the two is less than 20 feet and is less than 6 times the vertical separation distance, h. The drift surcharge is determined using the provisions of 7.7.1. See Ref. 3, Section 4.2, for details on drift calculations at adjacent structures.

Intersecting Drifts at Low Roofs

ASCE/SEI 7.7.3 contains provisions for snowdrifts that can occur at reentrant corners and parapet wall corners. Three-dimensional drifts form at such locations due to wind acting in multiple directions. The provisions in 7.7.1 are to be used to determine the individual two-dimensional snowdrift geometries for both leeward and windward intersecting snow drifts. Reference 3 Section 4.2 contains details on the determination of intersection drifts at low roofs.

Roof Projections and Parapets

Drift loads on sides of roof projections (including rooftop equipment) and at parapet walls are determined by the provisions of 7.8, which are based on the drift requirements of 7.7.1.

The drift that is formed adjacent to a parapet wall is a windward drift, and the height of the drift is equal to three-quarters of h_d from Figure 7.6-1.

Sliding Snow

The load caused by snow sliding off of a sloped roof onto a lower roof is determined by the provisions of 7.9. Such loads are superimposed on the balanced snow load of the lower roof and need not be used in combination with drift, unbalanced, partial, or rain-on-snow loads.

Sliding snow loads are applicable for upper roofs that are slippery with slopes greater than ¼ on 12 and for upper roofs that are not slippery with slopes greater than 2 on 12.

Rain-on-Snow Surcharge Load

The snow load provisions in the discussions above consider load effects due to light rain on snow; effects due to heavy rain are not directly taken into account in the 50-year ground snow loads. At locations where the ground snow load is greater than 20 psf, it is assumed that because of the relatively deep snow pack, heavy rains have a less likely chance of permeating through the snow pack and draining away; as such, rain-on-snow load effects have been captured in the ground snow load measurements, and an additional surcharge to account for this is not required.

A rain-on-snow surcharge load of 5 psf is to be added on all roofs that meet the conditions of 7.10. This surcharge load applies only to the balanced load case and need not be used in combination with drift, sliding, unbalanced, minimum, or partial loads.

Ponding Instability

Provisions for ponding instability and progressive deflection of roofs are given in 7.11 and 8.4 of ASCE 7. Susceptible bays, that is, bays with a roof slope less than ¼ in. per foot or those where the primary drain system is blocked and the secondary drain system is functional, must be analyzed for the effects from the larger of the snow load or the rain load assuming the primary drainage system is blocked. The roof structure in these bays must be designed with adequate stiffness to preclude ponding instability.

Roof surfaces with a slope greater than or equal to ¼ in. per foot toward points of free drainage need not be considered susceptible to ponding instability.

Existing Roofs

Requirements for increased snow loads on existing roofs due to additions and alterations are covered in 7.12. New structures built adjacent to existing ones within a horizontal distance of 20 feet have the potential to increase the snow loads on the existing roof when the new roof is higher than the existing one. In particular, both drift and sliding snow loads must be considered on the existing roof due to the presence of the new roof. Additionally, the existing roof will most likely be partially exposed or sheltered by the new building or alteration; if it were fully exposed previously, an increase in snow load will be realized due to this change in exposure.

ASCE/SEI 7.12 requires that owners or agents for owners of an existing building with a lower roof be notified for the potential of increased snow loads when a new building or alteration with a higher roof is to be located within 20 feet of the existing building.

General Procedure to Determine Snow Loads

The following general procedure, which is based on that given in C7, can be used to determine design snow loads in accordance with Chapter 7 of ASCE/SEI 7:

1. Determine ground snow load, p_g, (7.2).
2. Determine flat roof snow load, p_f, by Equation 7.3-1 (7.3). The procedure is shown in Fig. 1.10 (Flowchart 3).
3. Determine sloped roof snow load, p_s, by Equation 7.4-1 (7.4). The procedure is shown in Figs. 1.11 and 1.12 (Flowchart 4 and Flowchart 5).
4. Consider partial loading (7.5).
5. Consider unbalanced snow loads (7.6). The procedure is shown in Fig. 1.15 (Flowchart 6) for hip and gable roofs, Fig. 1.16 (Flowchart 7) for curved and dome roofs, and Fig. 1.17 (Flowchart 8) for multiple folded plate, sawtooth, and barrel vault roofs.
6. Consider snow drifts on lower roofs (7.7) and roof projections (7.8). The procedure to determine snow drifts is shown in Fig. 1.22 (Flowchart 9).
7. Consider sliding snow (7.9).
8. Consider rain-on-snow loads (7.10).
9. Consider ponding instability (7.11).
10. Consider existing roofs (7.12).

1.6.8 Ice Loads

Overview

An *ice-sensitive structure* is defined as one in which the effects due to atmospheric ice loading governs the design of part or all of the structure (IBC 202 and ASCE/SEI 10.2). Examples include the following: (1) lattice structures, (2) guyed masts, (3) overhead lines, (4) light suspension and cable-stayed bridges, (5) aerial cable systems (e.g., ski lifts), (6) amusement rides, (7) open catwalks and platforms, (8) flagpoles, and (9) signs.

Freezing rain is rain or drizzle that falls into a layer of subfreezing air in the earth's surface and freezes on contact with the ground or any other exposed surface to form glaze (clear, high-density) ice. Compared to *in-cloud icing* (which occurs when a supercooled cloud or fog droplets that are carried by the wind freeze on impact with objects) and snow, freezing rain is considered the cause of the most severe ice loads in most of the contiguous United States. Because values of ice thickness for in-cloud icing and snow are not currently available in a form that is suitable for inclusion in ASCE/SEI 7, only data for freezing rain are given in Chapter 10.

Ice that is formed on exposed surfaces increases the size of the surface, which increases the projected area that is exposed to wind. Chapter 10 contains requirements that address the proper wind loads that must be used on ice-covered structures.

The provisions of Chapter 10 do not apply to structures that are covered by national standards (e.g., electric transmission systems and communication towers and masts). In such cases, the standards and guidelines in 10.1.3 are to be used where applicable.

Design for dynamic load effects resulting from galloping, ice shedding, and aeolian vibrations, to name a few, are not covered in Chapter 10. Such effects must be considered in certain types of ice-sensitive structures.

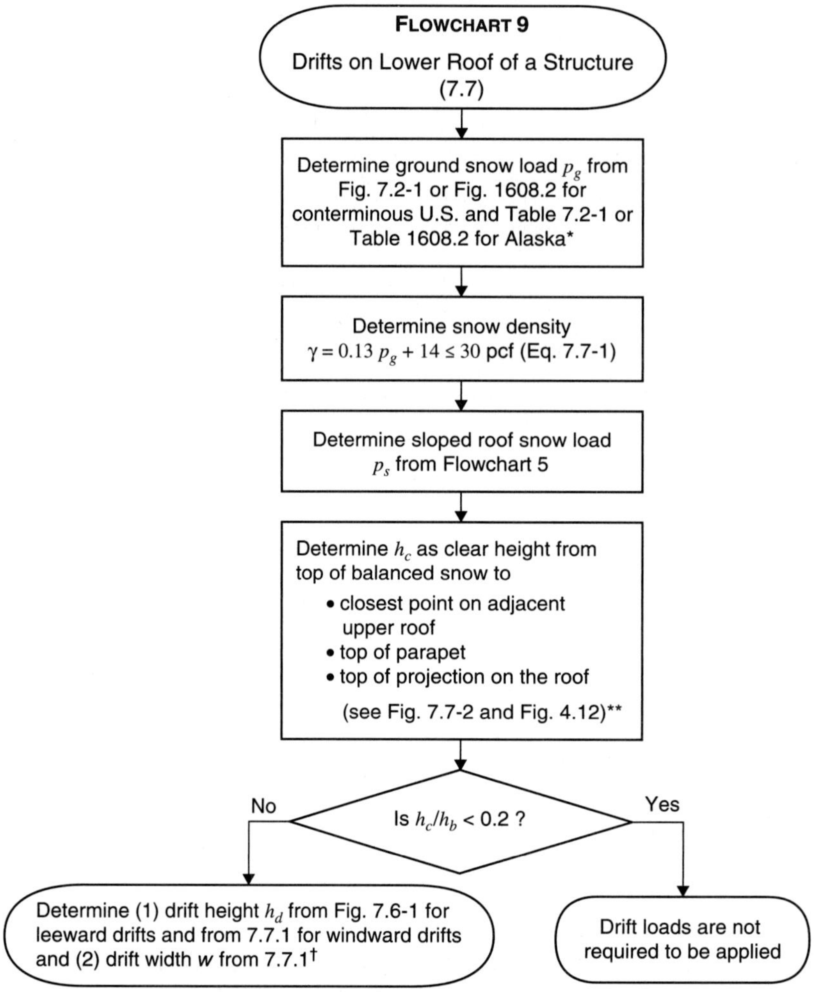

* "CS" in the maps signifies areas where a site-specific study must be conducted to determine p_g. Numbers in parentheses represent the upper elevation limit in feet for the ground snow load values given below. Site-specific studies are required at elevations not covered in the maps.

** Height of balanced snow $h_b = p_s/\gamma$ or p_f/γ (7.7.1).

† See 7.7.2 and Figs. 4.14 and 4.15 for drift loads caused by adjacent structures and terrain features. See 7.8 for drift loads on roof projections and parapet walls.

Figure 1.22 Drift on lower roof of a structure (Flowchart 9).

Ice Loads Due to Freezing Rain

Nominal Ice Thickness Figures 10.4-2 through 10.4-6 contain an equivalent uniform radial thickness, t, of ice due to freezing rain at a height of 33 feet above the ground for the contiguous 48 states and Alaska based on a 500-year mean recurrence interval. The concurrent 3-second gust wind speeds are also given in the figures; these wind speeds correspond to the winds that occur during the freezing rain storm and those that occur between the time the freezing rain stops and the temperature rises to above freezing. Special icing regions are also identified on the maps (gray shaded areas) and occur in the western mountainous regions and in the Appalachian Mountains.

The following factors are considered in the calculations of the ice load:

- **Height Factor**

The height factor, f_z, is determined by Equation 10.4-4, and it is used to increase the radial thickness of ice for any height above ground, z:

$$f_z = \left(\frac{z}{33}\right)^{0.10} \quad \text{for } 0 \text{ feet} < z \le 900 \text{ ft}$$

$$= 1.4 \quad \text{for } z > 900 \text{ ft}$$

- **Importance Factor**

Importance factors, I_i, adjust the nominal ice thickness and concurrent wind pressure based on the occupancy of the structure. Risk categories for buildings and other structures are defined in IBC Table 1604.5, and values of I_i are given in ASCE/SEI Table 1.5-2.

- **Topographic Factor**

Because of wind speed-up effects, the ice thickness and concurrent wind speed are larger for buildings and structures that are situated on hills, ridges, and escarpments compared to those located on level terrain. To account for these effects, the nominal ice thickness is modified by $(K_{zt})^{0.35}$, where K_{zt} is the topographic factor determined by Equation 26.8-1.

Design Ice Thickness for Freezing Rain The design ice thickness, t_d, that is to be used in calculating ice weight is the nominal ice thickness, t, multiplied by the modification factors noted above (see Equation 10.4-5):

$$t_d = tI_i f_z (K_{zt})^{0.35}$$

Ice Weight Ice load is determined using the ice weight, D_i, that is formed on all exposed surfaces of structural members, guys, components, appurtenances, and cable systems. The cross-sectional area of ice, A_i, to be used in the determination of D_i in such cases is determined by Equation 10.4-1:

$$A_i = \pi t_d (D_c + t_d)$$

The volume of ice on flat plates and large three-dimensional objects, such as domes and spheres, is determined by Equation 10.4-2:

$$V_i = \pi t_d A_s$$

For flat plates, A_s is the area of one side of the plate. For domes or spheres, A_s is the projected area of the dome or sphere and is determined by Equation 10.4-3:

$$A_s = \pi r^2$$

In this equation, r is the radius of the maximum cross section of a dome or a radius of a sphere.

The ice volume given by Equation 4.16 is for a flat plate or projected surface that is oriented perpendicular to the path of the raindrops. ASCE/SEI 10.4.1 permits Equation 4.16 to be multiplied by 0.8 for vertical plates and 0.6 for horizontal plates.

Once A_i or V_i have been computed, the corresponding ice weight, D_i, is determined by multiplying A_i or V_i by the density of ice. According to 10.4.1, the density of ice must not be taken less than 56 pcf.

1.6.9 Wind on Ice-Covered Structures

Overview

Ice that has formed on structural members, components, and appurtenances increases the projected area that is exposed to wind and changes the structure's wind drag coefficients. Ice accretions tend to round sharp edges, thereby reducing drag coefficients for members like angles and rectangular bars.

Ice-sensitive structures must be designed for the wind loads determined by the provisions in ASCE/SEI Chapters 26 through 31 using increased projected area and the modifications set forth in 10.5.1 through 10.5.5. The loads determined in this fashion are defined as the wind-on-ice loads, W_i.

Wind on Ice-Covered Chimneys, Tanks, and Similar Structures

For chimneys, tanks, and other similar types of structures, design wind loads are determined by Equation 29.4-1:

$$F = q_z G C_f A_f$$

The velocity pressure, q_z, is determined by Equation 26.10-1 using the topographic factor, K_{zt}, determined in accordance with 10.4.5 and the concurrent wind speed, V_c:

$$q_z = 0.00256 K_z K_{zt} K_d K_e V_c^2$$

The velocity pressure exposure coefficient, K_z, is determined from Table 26.10-1, and the wind directionality factor, K_d, is defined in 26.6. The ground elevation factor K_e is defined in 26.9.

The gust factor, G, is determined in accordance with 26.11. For structures with square, hexagonal, and octagonal cross sections, the force coefficients, C_f, are given in Figure 29.4-1.

Similarly, wind on ice-covered chimneys, tanks, and similar structures is determined by Equation 29.4-1:

$$F = q_z G C_f A_f$$

For wind on ice-covered solid freestanding walls and solid signs, wind loads are determined for solid freestanding walls and solid signs using Equation 29.3-1:

$$F = q_h G C_f A_s$$

where q_h is determined by Equation 29.3-1 at height h, which is defined in Figure 29.3-1, and G is determined in accordance with 26.11.

The force coefficients, C_f, given in Figure 29.3-1 and the gross area, A_s, of the solid freestanding wall or solid sign must be based on the dimensions including ice.

Wind on Ice-Covered Open Signs and Lattice Frameworks

Wind loads are determined for these types of structures using Equation 29.4-1. C_f, given in Figure 29.4-2, are to be used in this equation. In all cases, the solidity ratio e (which is equal to the ratio of solid area to gross area) that is defined in Figure 29.4-2 is to be based on the projected area, including ice.

Wind on Ice-Covered Trussed Towers

Wind loads are determined using Equation 29.4-1 using the force coefficients, C_f, from Figure 29.4-3 and the gross area, A_f, based on the dimensions, including ice.

Wind on Ice-Covered Guys and Cables

Because there is very little published experimental data for force coefficients for ice-covered guys and cables, a single value of 1.2 is permitted to be used in all cases (10.5.5). The rationale behind using this value is given in C10.5.5.

Design Temperatures for Freezing Rain

Figures 10.6-1 and 10.6-2 give temperatures concurrent with ice thickness due to freezing rain for the contiguous 48 states and Alaska, respectively.

The design temperature for ice and wind-on-ice that is to be used is the temperature from Figures 10.6-1 and 10.6-2 or 32°F, whichever gives the maximum load effects. For Hawaii, the temperature must be 32°F.

1.6.10 Partial Loading

It has been found that variations in ice thickness due to freezing rain at a given elevation are usually small over distances of about 1000 feet. Thus, partial loading from freezing rain does not usually produce maximum load effects. However, in certain types of structures, partial ice loads can produce maximum effects, and this must be considered in design. Additional information on this topic can be found in C10.7.

General Procedure to Determine Atmospheric Ice Loads

The following general procedure, which is based on that given in 10.8, can be used to determine design ice loads in accordance with Chapter 10 of ASCE/SEI 7:

1. Determine nominal ice thickness, t; the concurrent wind speed, V_c; and the concurrent temperature from Figures 10.4-2 through 10.4-6 or from a site-specific study in accordance with 10.1.1 (10.4.2, 10.6).
2. Determine height factor, f_z, for each design segment of the structure (10.4.3).
3. Determine importance factor, I_i, (10.4.4).
4. Determine topographic factor, K_{zt}, (10.4.5).
5. Determine design ice thickness, t_d, by Equation 10.4-5 (10.4.6).
6. Determine the weight of ice, D_i, using the applicable equations in 10.4.1 (10.4.1).
7. Determine the velocity pressure, q_z, for wind speed, V_c, in accordance with 26.10.
8. Determine the applicable wind force coefficients, C_f, in accordance with 10.5.
9. Determine the gust effect factor, G, in accordance with 26.11.
10. Determine the design wind force, W_i, in accordance with the applicable provisions of Chapter 29 (10.5).

1.7 WIND LOADS

1.7.1 Introduction

NATURE OF WIND LOADS

In general, wind loading is the effect of the atmosphere passing by a stationary structure attached to the earth's surface. An in-depth discussion on the mechanics of atmospheric circulations can be found in Ref. 8.

Loads on buildings and other structures due to the effects from wind are determined by considering both atmospheric and aerodynamic effects. These effects form the basis of the methodologies given in ASCE/SEI 7 for the determination of wind loads.

Atmospheric Effects The atmospheric factors that have a direct impact on the magnitude of wind loading on a building or other structure are obtained from meteorological and boundary layer effects.

Meteorology Meteorology is the study of the atmosphere, and wind climatology is a branch within meteorology that focuses on the prediction of storm conditions. In particular, extreme wind speeds associated with different types of storms and the probability of occurrence of such extreme values are analyzed at specific geographical locations. This information is used in developing design wind speed maps in the IBC and ASCE/SEI 7. Wind velocity is used in calculating the maximum design wind loads that can be expected on a building or structure during its life span.

Wind speed measurements that are utilized in creating the design wind speed maps in the IBC and ASCE/SEI 7 are obtained from essentially two different sources. In non-hurricane regions, wind speed data is collected from standard three-cup anemometers that are located 33 feet above the ground in open terrain (such as at airports) at 485 National Weather Service weather stations throughout the United States that have at least 5 years of available data. Design wind speeds in hurricane regions are obtained from statistical simulations based on historical data since actual weather data on hurricanes at any particular location are generally limited.

The wind speed that was recorded at the aforementioned weather stations for many years was the *fastest-mile wind speed*, which is the maximum wind speed averaged over 1 mile of wind passing through an anemometer. From this definition, it follows that the corresponding averaging time is equal to 3600 divided by the velocity of wind in miles per hour. As the National Weather Service phased out the older equipment, the fastest-mile wind speed was replaced with a *3-second gust speed*. The 3-second averaging time is based on the response characteristics of the newer instruments that are utilized at the weather stations.

Thunderstorms, hurricanes, tornadoes, and special regional effects are the climatological events that are of primary interest when designing buildings and structures in the United States. Figure 1.23 illustrates the controlling climatological events that produce extreme wind speeds in regions of the conterminous United States. It is evident from the figure that the prevailing wind speeds are generated by thunderstorms in most of the country.

Wind speeds generated by tornadoes are not included in the wind speed maps in the IBC or ASCE/SEI 7. The primary reason for this has to do with the significantly lower probability of occurrence of tornadoes compared to that for basic wind speeds. Additional information on how to design for tornadoes is covered in Ref. 3.

Special wind regions can have wind speeds significantly greater than those in surrounding areas. Such regional effects include wind blowing over mountain ranges or through gorges or valleys. These regions are identified as special wind regions on the basic wind speed maps.

Boundary Layer Fluid Dynamics The layer of the earth's atmosphere that is located from the surface of the earth to approximately 3300 feet above the surface is known as the boundary layer. The fluid dynamic effects that occur within this layer have an important impact on the magnitude of wind loads on buildings and other structures.

In general, the surface of the earth exerts a horizontal drag force on wind, which impedes its flow. The more frictional resistance is experienced, the closer the wind flow is to the surface; thus, wind velocity is smaller at or near the ground level compared to levels above the surface. Similarly, at a given height above the surface, wind velocity is smaller over rougher surfaces compared to smoother ones because of friction. Figure 1.24 depicts the variation of wind speed with respect to height and surface roughness. Both phenomena are captured in the current wind load provisions using a modified version of the power-law methodology that was first introduced in Ref. 9. Note that the wind velocities become constant above certain heights for the different roughness categories; these heights are defined as gradient heights in ASCE/SEI 7.

Aerodynamic Effects

Overview Wind flow is disturbed due to the presence of a building or structure in its path. The resulting responses due to this disturbance are governed by the laws of aerodynamics. In the case of a building or other structure that has an essentially block-like shape (which is referred to as a bluff body), bluff body aerodynamics is used to predict the effects caused by placing the bluff body in the flow of wind.

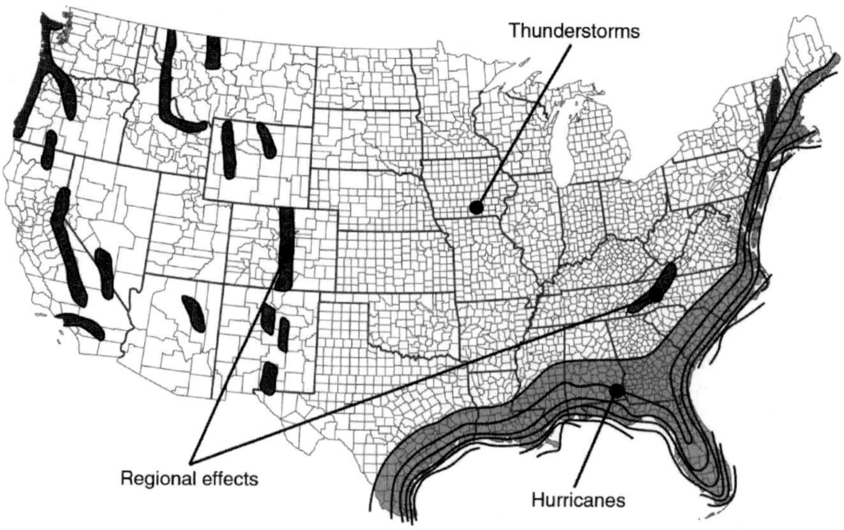

Figure 1.23 The controlling climatological events that produce extreme wind speeds in regions of the conterminous United States.

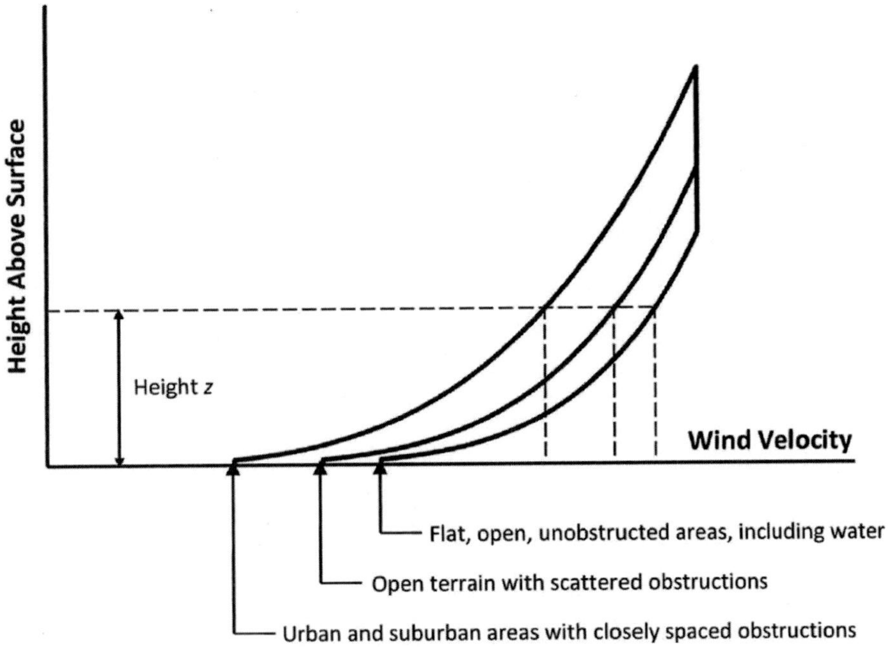

Figure 1.24 The variation of wind speed with respect to height and surface roughness.

In general, when wind comes into contact with a building or other structure, the following pressures are created:

1. *External pressures*, which act on all exterior surfaces; these pressures are caused by the effects that are generated when the wind strikes the building or other structure.
2. *Internal pressures*, which act on all interior surfaces; these pressures are due to leakage of air through the exterior surface to the interior space.

It is assumed that the external and internal pressures act perpendicular to the exterior and interior surfaces of the building or other structure, respectively. A pressure is defined as positive when it acts toward the surface and negative when it acts away from the surface. Positive pressure is commonly referred to as just pressure, and negative pressure is also identified as suction.

External Pressure Figure 1.25 depicts idealized wind flow around a gable roof building. Wind velocity increases at distances above the surface, as shown in the following equation below. The pressure p_s at this point can be theoretically determined using Bernoulli's equation:

$$p_s = \frac{1}{2}\rho V^2$$

In this equation, ρ is the atmospheric air density, and V is the wind velocity at the elevation of the stagnation point.

The leeward wall and the side walls all have negative pressure (the negative pressure on the leeward wall is referred to as the wake).

A generic representation of the external wind pressures on the walls and roof of the building depicted in Fig. 1.25 is given in Fig. 1.26.

The dimensionless pressure C_p, which is commonly referred to as a *pressure coefficient*, is defined as the pressure p at any point on the building or structure divided by the stagnation pressure:

$$C_p = \frac{p}{\rho V^2/2}$$

Internal Pressure and Enclosure Classification Internal pressures act on all interior surfaces of a building or structure and are due to leakage of air (pressure) through the exterior surface to the interior space. The number and size of openings in the envelope of the structure play key roles when determining the effects caused by internal pressures. ASCE/SEI 7 defines an *opening* as an aperture or hole in the building envelope that allows air to flow through the building envelope and that is designed as open during design winds. Doors, operable windows, exhausts for ventilation systems, and gaps (deliberate or otherwise) around cladding are just a few examples of openings.

A number of different cases need to be examined in order to properly understand the effects that internal pressure can have on a building or structure. In the case of a building that has a relatively large opening on the windward wall, the wind flow will try to inflate the building, resulting in internal pressure that is positive (see Fig. 1.27). Conversely, an opening on the leeward wall, side walls, or roof will try to deflate the building and will result in negative internal pressure.

An *open building* is defined by ASCE/SEI 7 as structures having at least 80 percent open at each wall.

A *partially enclosed* building has openings that are large enough to affect internal pressure and have background porosity low enough to allow internal pressure to build up.

An *enclosed* building is one that is neither open nor partially enclosed.

The effects of internal pressure are accounted for in ASCE/SEI 7 by internal pressure coefficients, the magnitudes of which are based on the enclosure classification of the building or other structure. It is evident from Fig. 1.27 that internal pressure does not contribute to the overall horizontal wind pressures that act on a structure because such pressures cancel out. However, internal pressures must be considered in the design of individual components (such as walls and roof framing) and cladding; the type of internal pressure (positive or negative) that results in the critical load combination must be used.

Gust Effects Wind velocity typically changes dramatically with time; numerous peaks and valleys normally occur over relatively short time spans. In general, the average wind speed obtained from a wind event is larger when a shorter averaging time is used. The peaks in wind velocity are called *gusts*, and these effects must be considered in design.

Gust-effect factors are used in ASCE/SEI 7 to account for this phenomenon. In short, a gust-effect factor relates the peak to mean response in terms of an equivalent static design load or load effect. The method that is used to determine such factors is based on the following

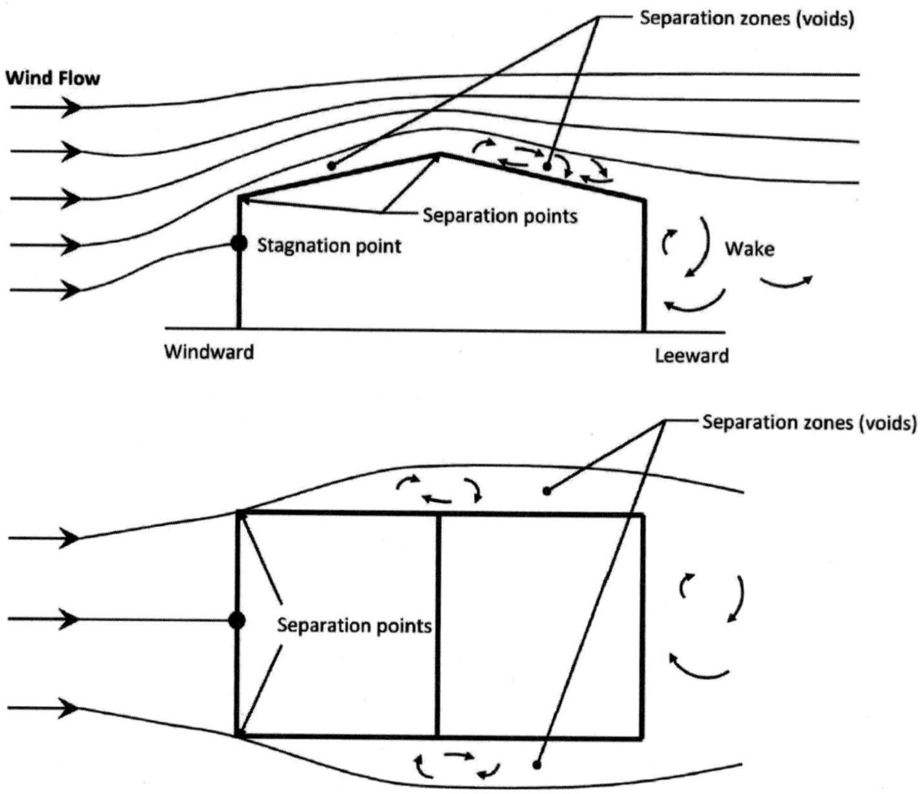

Figure 1.25 Idealized wind flow around a gable roof building.

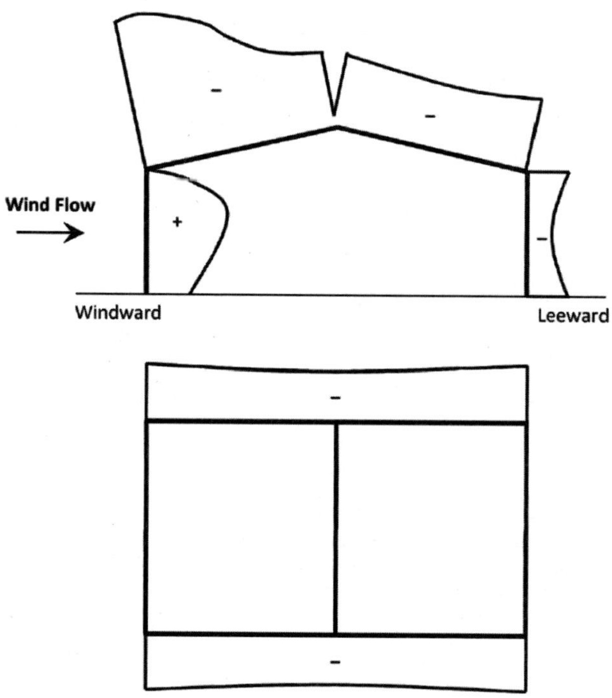

Figure 1.26 Wind pressures on a gable roof building.

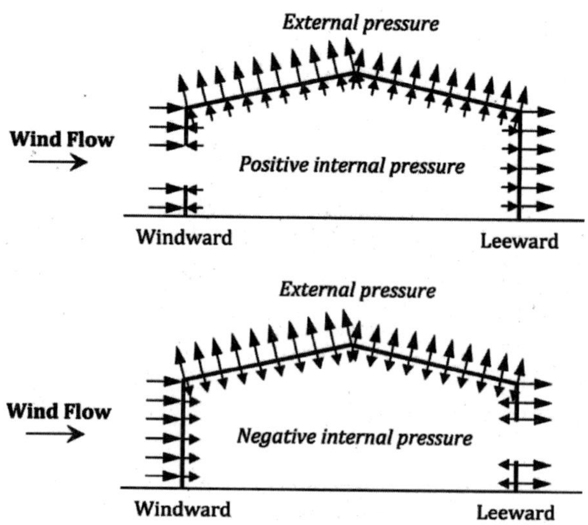

Figure 1.27 Effects of openings on internal pressure distribution.

assumption: the fundamental mode of vibration of a building or structure has an approximately linear mode shape. Additional information on the pioneering work in gust-effect factors can be found in Ref. 10.

It is important to note that the gust-effect factors in ASCE/SEI 7 for both rigid and flexible buildings (i.e., buildings with a fundamental frequency greater than or equal to 1 Hz and less than 1 Hz, respectively) account for the effects in the along-wind direction only. In the case of flexible buildings or structures, along-wind effects due to dynamic amplification are also accounted for in the gust-effect factor.

Certain types of buildings and structures—especially those that are relatively tall and slender—are susceptible to one or more of the following: (1) across-wind load effects, (2) vortex shedding, (3) instability due to galloping or flutter, or (4) dynamic torsional effects. The gust-effect factors in ASCE/SEI 7 do not account for the loading effects caused by these phenomena; wind tunnel tests should be performed in such cases to properly capture these effects.

Overview of Code Requirements

According to IBC 1609.1.1, wind loads on buildings and structures are to be determined by the provisions of Chapters 26 to 30 of ASCE/SEI 7-16 or by the alternate all-heights method of IBC 1609.6. Five exceptions are given in IBC 1609.1.1 that permit wind loads to be determined on certain types of structures using industry standards other than ASCE/SEI 7, and one exception is given that permits the use of wind tunnel tests that conform to the provisions of Chapter 31 of ASCE/SEI 7-16.

Wind is assumed to come from any horizontal direction, and its effects are applied in the form of pressures that act normal to the surfaces of a building or other structure (IBC 1609.1.1). Positive wind pressure acts toward the surface and is commonly referred to as just pressure. Negative wind pressure, which is also called suction, acts away from the surface.

Positive pressure acts on the windward wall of a building, and negative pressure acts on the leeward wall, the side walls, and the leeward portion of the roof (see Fig. 1.28).

In general, pressures must be considered on the *main wind-force resisting system* (MWFRS) and *components and cladding* (C&C) of a building or other structure. The MWFRS consists of structural elements that have been assigned to resist the effects from the wind loads for the overall structure. Shear walls, moment frames, and braced frames are a few examples of different types of MWFRSs.

Table 1.3 contains the procedures that are available in ASCE/SEI 7-16 to determine wind loads on MWFRSs and C&C. In the *directional procedure*, wind loads for specific wind directions are determined using external pressure coefficients that are based on wind tunnel tests of prototypical building models for the corresponding direction of wind.

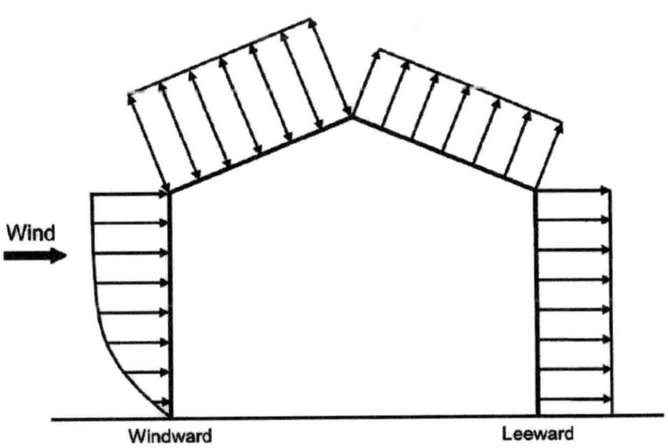

Figure 1.28 Application of wind pressures on a building with a gable or hip roof in accordance with IBC and ASCE/SEI 7.

Table 1.3 Summary of Wind Load Procedures in ASCE/SEI 7-16

System	ASCE/SEI 7-16 chapter	Description
MWFRS	27	Directional procedure for buildings of all heights
	28	Envelope procedure for low-rise buildings[1]
	29	Directional procedure for building appurtenances and other structures
	31	Wind tunnel procedure for any building or other structure
C&C	30	• Envelope procedure in Parts 1 or 2, or • Directional procedure in Parts 3, 4, and 5 • Building appurtenances in Part 6
	31	• Wind tunnel procedure for any building or other structure

[1]Low-rise buildings are defined in ASCE/SEI 26.2 as enclosed or partially enclosed buildings that have (1) a mean roof height that is less than or equal to 60 feet and (2) a mean roof height that is less than or equal to the least horizontal dimension of the building.

The provisions of this procedure cover a wide range of buildings and other structures.

In the *envelope procedure*, pseudo-external pressure coefficients derived from wind tunnel tests on prototypical building models that are rotated 360 degrees in a wind tunnel are used to determine wind loads; the pressure coefficients envelope the maximum values that are obtained from all possible wind directions.

The alternate all-heights method of IBC 1609.6, which is a simplification of the directional procedure in ASCE/SEI 7, is another method that is available to determine wind loads on buildings and other structures that meet the conditions of that section. Contrary to its name, this method is applicable to a certain class of buildings that have a height less than or equal to 75 feet.

With the exception of the wind tunnel procedure, all of these procedures and methods in Table 1.3 and the alternate all-heights method are static methods for estimating wind pressures. Static methods generally yield very accurate results for low-rise buildings.

Figure 26.1-1 provides an outline of the process that is required for determining wind loads. In-depth information on all of these procedures and methods, including the general requirements in ASCE/SEI Chapter 26 and the alternate all-heights method of IBC 1609.6, is given in the following sections.

1.7.2 General Requirements

Overview

Chapter 26 of ASCE/SEI 7-16 contains the following general requirements for determining wind loads on MWFRS and C&C:

- Basic wind speed (Figures 26.5-1 and 26.5-2)
- Wind directionality (26.6)
- Exposure (26.7)
- Topographic effects (26.8)
- Ground elevation effects (26.9)
- Velocity pressure (26.10)
- Gust effects (26.11)
- Enclosure classification (26.12)
- Internal pressure coefficients (26.13)

These requirements are used in conjunction with the methods and procedures contained in Chapters 27 through 31.

The following sections discuss these requirements and provide additional background information on fundamental concepts.

Wind Hazard Map

Regardless of the wind load procedure that is employed to determine wind pressures, the basic wind speed V must be determined at the location of the building or other structure (note that in the IBC, the basic wind speed is designated V_{ult} and is defined as the ultimate design wind speed).

Figures 1609.3(1) through 1609.3(8) in the IBC and Figures 26.5-1A through 26.5-1D and Figures 26.5-2A through 26.5-2D in ASCE/SEI 7 are identical and provide basic wind speeds based on 3-second gusts at 33 feet above ground for Exposure C for different risk categories, which are defined in IBC Table 1604.5.

Since the wind speeds are ultimate design wind speeds, the wind load factors in the design load combinations are equal to 1.0 (see load combinations in Sec. 1.10 of this chapter). Table 1.4 provides a summary of the information associated with these maps.

Table 1.4 Summary of Basic Wind Speed Maps in the 2018 IBC and ASCE/SEI 7-16

Location	Figure no.		Risk category*	Return period (years)
	IBC	ASCE/SEI 7		
Conterminous United States Alaska Puerto Rico Guam Virgin Islands American Samoa	1609.3(4) 1609.3(1) 1609.3(2) 1609.3(3)	26.5-1A 26.5-1B 26.5-1C 26.5-1D	I II III IV	300 700 1700 3000
Hawaii	1609.3(8)	26.5-2A	I	300
	1609.3(5)	26.5-2B	II	700
	1609.3(6)	26.5-2C	III	1700
	1609.3(7)	26.5-2D	IV	3000

*See IBC Table 1604.5 for definitions of risk categories.

The shaded areas on the wind speed maps are designated as special wind regions. These are areas where unusual wind conditions exist. The local authority having jurisdiction over the project should be consulted to obtain the local design wind speed (ASCE/SEI 26.5.2).

The ultimate design wind speeds on the wind hazard maps do not include effects of tornadoes (ASCE/SEI 26.5.4). However, ASCE/SEI C26.5.4 contains references and a tornadic gust wind speed map of the United States that corresponds to a return period of 100,000 years. The information presented in this section can be used as a guide in developing and designing buildings and other structures for the effects of tornadoes. Reference 11 also contains design guidance for tornadoes.

Where required, the basic design wind speeds in Figures 1609.3(1) through 1609.3(8) can be converted to allowable stress design wind speeds, V_{asd}, using Equation 16-33 in IBC 1609.3.1:

$$V_{asd} = V_{ult}\sqrt{0.6}$$

Values of V_{asd} are tabulated in IBC Table 1609.3.1 for various V_{ult}.

Wind Directionality

The wind directionality factor K_d that is given in 26.6 accounts for the statistical nature of wind flow and the probability of the maximum effects occurring at any particular time for any given wind direction.

Table 26.6-1 contains values of K_d as function of structure type. This factor is equal to 0.85 for the MWFRS and C&C of buildings.

Exposure

Wind Direction and Sectors According to IBC 1609.4 and ASCE/SEI 26.7, an exposure category must be determined upwind of a building or other structure for each wind direction that is considered in design. Wind must be assumed to come from any horizontal direction when determining wind loads (IBC 1609.1.1 and ASCE/SEI 26.5.1). One rational way of satisfying this requirement is to assume that there are eight wind directions: four that are perpendicular to the main axes of the building or other structure and four that are at 45-degree angles to the main axes. Figure C26.7-8, which is reproduced here as Fig. 1.29, shows the sectors that are to be used to determine the exposure for a selected wind direction (IBC 1609.4.1 and ASCE/SEI 26.7.1).

Surface Roughness Categories Surface roughness categories are defined in IBC 1609.4.2 and ASCE/SEI 26.7.2 and are summarized in Table 1.5.

Table 1.5 Surface Roughness Categories

Surface roughness category	Description
B	Urban and suburban areas, wooded areas or other terrain with numerous closely spaced obstructions having the size of single-family dwellings or larger
C	Open terrain with scattered obstructions having heights generally less than 30 feet; this category includes flat open country and grasslands
D	Flat, unobstructed areas, and water surfaces; this category includes smooth mudflats, salt flats, and unbroken ice

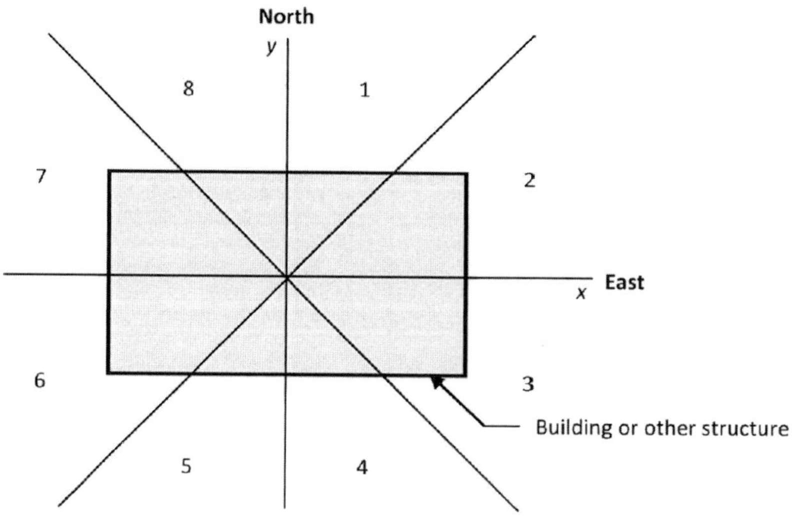

Figure 1.29 Sections for determining exposures.

Exposure Categories

Exposure categories are based on the surface roughness categories defined above and essentially account for the boundary layer concept of surface roughness. Table 1.6 contains definitions of the three exposure categories given in IBC 1609.4.3 and ASCE/SEI 26.7.3. Definitions of Exposure B and Exposure D are illustrated in Figures C26.7-1 and C26.7-2, respectively.

Table 1.6 Exposure Requirements

Exposure category	Definition
B	• Mean roof height $h \leq 30$ feet Surface roughness category B prevails in the upwind direction for a distance > 1500 feet • Mean roof height $h > 30$ feet Surface roughness category B prevails in the upwind direction for a distance > 2600 feet or 20 times the height of the building, whichever is greater
C	Applies for all cases where Exposure B or D do not apply
D	• Surface roughness category D prevails in the upwind direction for a distance > 5000 feet or 20 times the height of the building, whichever is greater • Surface roughness immediately upwind of the site is B or C, and the site is within a distance of 600 feet or 20 times the building height, whichever is greater, from an Exposure D condition as defined above

According to ASCE/SEI 26.7.3, the exposure category that results in the largest wind loads must be used for sites that are located in transition zones between exposure categories. However, the exception in 26.7.3 permits an intermediate exposure category to be used provided that it is determined by a rational analysis method defined in recognized literature. An example of such an analysis is given in C26.7.

Exposure Requirements

ASCE/SEI 26.7.4 contains exposure requirements that must be satisfied for all of the wind load procedures that are available in ASCE/SEI 7-16. A summary of these requirements is given in Table 1.7.

Table 1.7 Exposure Requirements

Wind load procedure	Chapter	Requirements
Directional	27	• MWFRS of enclosed and partially enclosed buildings Use an exposure category determined in accordance with 26.7.3 in each wind direction • Open buildings with monoslope, pitched, or troughed free roofs Use the exposure category determined in accordance with 26.7.3 from the eight sectors that results in the highest wind loads
Envelope	28	• MWFRS of all low-rise buildings designed using this procedure Use the exposure category determined in accordance with 26.7.3 from the eight sectors that results in the highest wind loads
Directional	29	• Building appurtenances and other structures Use an exposure category determined in accordance with 26.7.3 in each wind direction
C&C	30	• C&C Use the exposure category determined in accordance with 26.7.3 from the eight sectors that results in the highest wind loads

Topographic Effects Buildings or other structures that are sited on the upper half of an isolated hill, ridge, or escarpment can experience significantly higher wind velocities than those sited on relatively level ground. The *topographic factor* K_{zt} in 26.8 accounts for this increase in wind speed, which is commonly referred to as wind speed-up.

Wind speed-up must be considered only when all of the five conditions in 26.8.1 are satisfied. When all of these conditions are met, K_{zt} is determined by Equation 26.8-1:

$$K_{zt} = (1 + K_1K_2K_3)^2$$

Values of K_1, K_2, and K_3 are obtained from Figure 26.8-1. K_1 accounts for the shape of the topographic feature and the maximum speed-up effect, K_2 accounts for the reduction in wind speed-up with respect to horizontal distance, and K_3 accounts for the reduction of wind speed-up with respect to height above the local terrain.

Ground Elevation Effects The ground elevation factor, K_e, adjusts the velocity pressure, q_z, determined in accordance with 26.10 based on the reduced mass density of air at elevations above sea level. Table 26.9-1 contains values of K_e, which can be calculated using the equation in Note 2 of the table where z_g is the ground elevation above sea level. A more complete version of Table 26.9-1 that includes air density values is provided in Table C26.9-1.

The constant 0.00256 in Equation 26.10-1 for q_z is used to convert a wind speed pressure based on the mass density of air for the standard atmosphere, which is defined as a temperature of 59°F and a sea level pressure of 29.92 in. of mercury. Values of air density other than the standard atmosphere values are adjusted using K_e. It is permitted to take $K_e = 1.0$ for all elevations, which is conservative except for elevations below sea level; however, using $K_e = 1.0$ to calculate q_z for all areas below sea level in the United States is not unconservative.

Velocity Pressure

The velocity pressure q_z at height z above the ground surface is determined by Equation 26.10-1; this is essentially Bernoulli's equation, and it converts the basic wind speed V to a velocity pressure:

$$q_z = 0.00256K_zK_{zt}K_dK_eV^2$$

The terms in the equation above are discussed below. Note that at the mean roof height of the building, the velocity pressure is denoted q_h, and the velocity pressure coefficient is denoted K_h; that is, the subscript changes from z to h.

Air Density The constant 0.00256 in the equation is related to the mass density of air for the standard atmosphere (59°F and sea level pressure of 29.92 in. of mercury) and is obtained as follows (constant = one-half times the density of air times the velocity squared, where the velocity is in miles per hour and the pressure is in pounds per square foot):

$$\text{Constant} = 0.5\left[\frac{0.0765\dfrac{\text{lb}}{\text{ft}^3}}{32.2\dfrac{\text{ft}}{\text{sec}^2}}\right]\times\left[\left(1\frac{\text{mi}}{\text{hr}}\right)\times 5280\frac{\text{ft}}{\text{mi}}\times\frac{1\text{ hr}}{3600\text{ sec}}\right]^2 = 0.00256$$

The numerical constant of 0.00256 should be used except where sufficient weather data are available to justify a different value.

Velocity Pressure Exposure Coefficient, K_z This coefficient modifies wind velocity (or pressure) with respect to exposure and height above ground. Values of K_z for Exposures B, C, and D at various heights above ground level are given in Table 26.10-1. In lieu of linear interpolation and for heights greater than 500 feet above the surface, K_z may be calculated at any height z using the equations at the bottom of that table:

$$K_z = \begin{cases} 2.01\left(\dfrac{15}{z_g}\right)^{\frac{2}{\alpha}} & \text{for } z \leq 15\text{ ft} \\ 2.01\left(\dfrac{z}{z_g}\right)^{\frac{2}{\alpha}} & \text{for } 15\text{ ft} \leq z \leq z_g \end{cases}$$

The constant α is the 3-second gust speed power law exponent, which defines the approximately parabolic shape of the wind speed profile for each exposure (see Fig. 1.24). The nominal height of the atmospheric boundary layer, which is also referred to as the gradient height, is denoted as z_g. Values of α and z_g are given in Table 26.11-1 as a function of exposure.

The above discussion on the determination of K_z is valid for the case of a single roughness category (i.e., uniform terrain). Procedures on how to determine K_z for a single roughness change or multiple roughness changes are given in C27.3.1.

Topographic Factor K_{zt} This factor modifies the velocity pressure exposure coefficients for buildings located on the upper half of an isolated hill or escarpment. See the "Topographic Effects" section.

Wind Directionality Factor K_d This factor accounts for the statistical nature of wind flow and the probability of the maximum effects occurring at any particular time for any given wind direction. See Figure 26.8-1 and wind directionality section above for more information on how to determine K_d.

Ground Elevation Factor K_e This factor adjusts the velocity pressure, q_z, determined in accordance with 26.10 based on the reduced mass density of air at elevations above sea level.

Basic Wind Speed V The basic wind speed, V, is the 3-second gust speed at 33 feet above the ground in Exposure C (see IBC 1609.3 or ASCE/SEI 26.5.1).

Flowchart 10 (Fig. 1.30) can be used to determine the velocity pressures, q_z and q_h.

Gust-Effect Factor

The effects of wind gusts must be included in the design of any building or other structure. The *gust-effect factor* defined in 26.11 accounts for both atmospheric and aerodynamic effects in the along-wind direction.

The gust-effect factor depends on the natural frequency n_1 of the structure. In particular, the method in which the gust-effect is determined is contingent on whether the structure is rigid or flexible. By definition, a *rigid building or other structure* is one where $n_1 \geq 1$ Hz, and a *flexible building or other structure* is one where $n_1 < 1$ Hz (26.2).

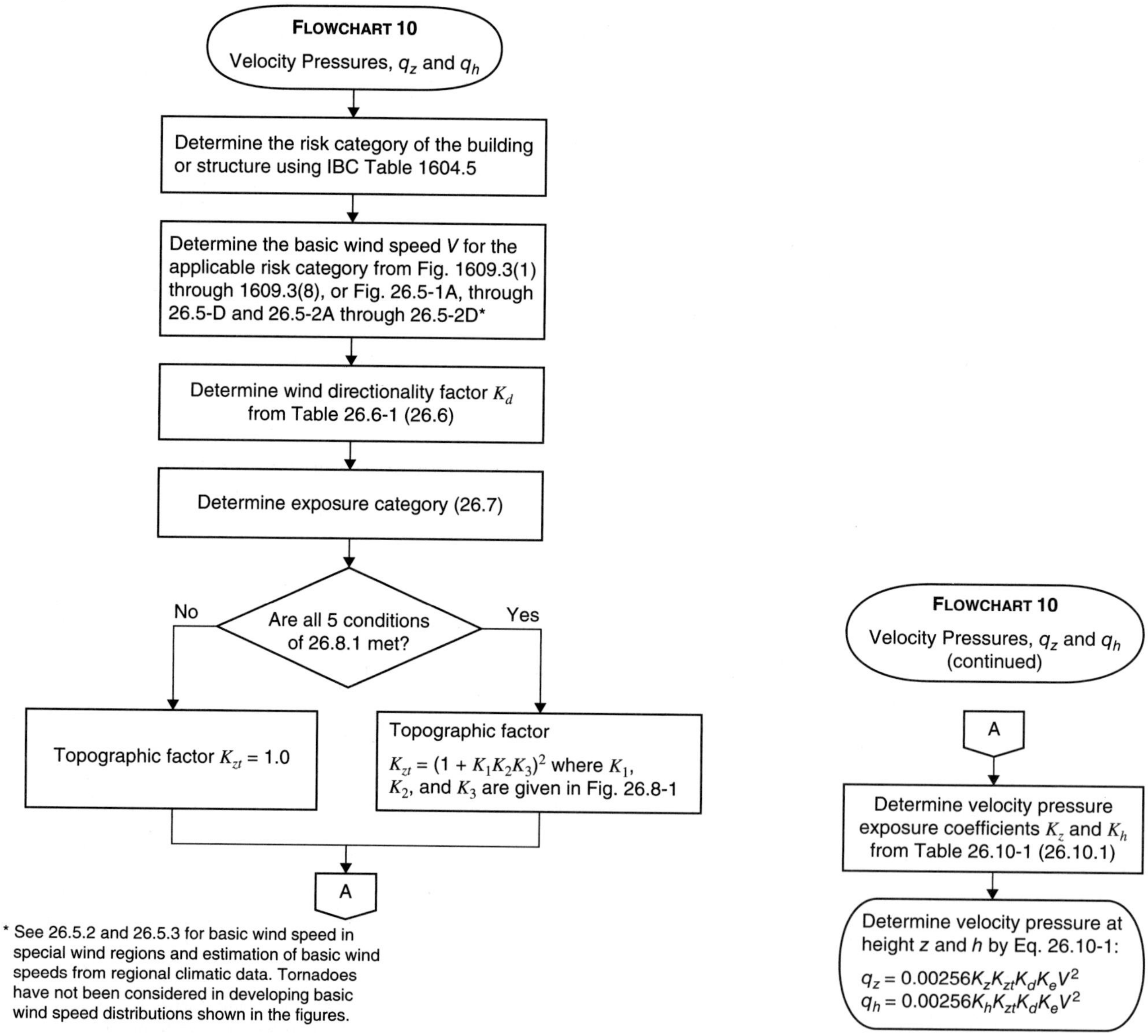

Figure 1.30 Determinations of velocity pressures, q_z and q_h (Flowchart 10).

Note that low-rise buildings that satisfy the definition in 26.2 (i.e., buildings with a mean roof height $h \le 60$ feet and $h \le$ least horizontal dimension of building) are permitted to be considered rigid (26.11.2).

Figure 1.31 (Flowchart 11) contains step-by-step procedures on how to determine the gust-effect factor for both rigid and flexible structures.

Approximate Natural Frequency

Many tools are available to determine the fundamental frequency n_1 of a structure. Most computer programs that are used to analyze structures can provide an estimate of n_1 based on member sizes and material properties that are used in the model. In the preliminary design stages, this information may not be known. Thus, 26.11.3 provides equations to determine an approximate natural frequency n_a for concrete and steel buildings that meet the height and slenderness conditions of 26.11.2.1.

1. Building height must be less than or equal to 300 feet.
2. Building height must be less than four times its effective length L_{eff}, which is determined by Equation 26.11-1.

Enclosure Classification

The following discussion covers definitions for each type of classification and the requirements for protecting glazed openings in wind-borne debris regions.

Any building or other structure must be classified as enclosed, partially enclosed or open based on the definitions in 26.2. A summary of these definitions is given in Table 1.8.

The quantities in Table 1.8 are as follows (see Fig. 1.32):

A_o = total area of openings in a wall that receives positive external pressure

A_g = gross area of wall in which A_o is identified

A_{oi} = sum of the areas of openings in the building envelope (walls and roof) not including A_o

A_{gi} = sum of the gross surface areas of the building envelope (walls and roof) not including A_g

Requirements for buildings that comply with more than one classification are given in 26.12.4. For a building that meets both open and partially enclosed definitions, the building is to be classified as open.

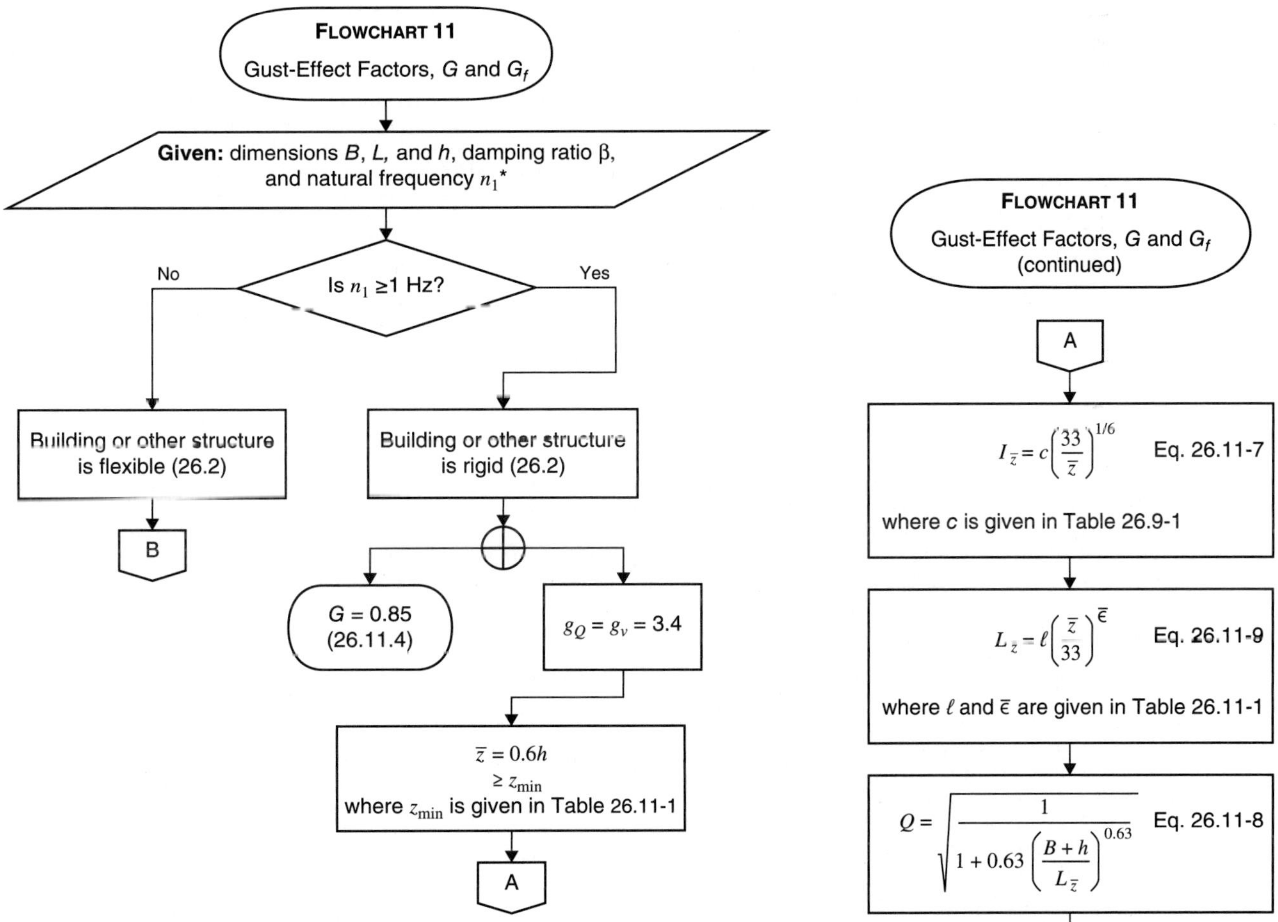

* Notes:

1. Information on structural damping can be found in C26.11.
2. n_1 can be determined from a rational analysis or estimated from approximate equations given in 26.11.3.

Figure 1.31 Gust-effect factors, G and G_f (Flowchart 11).

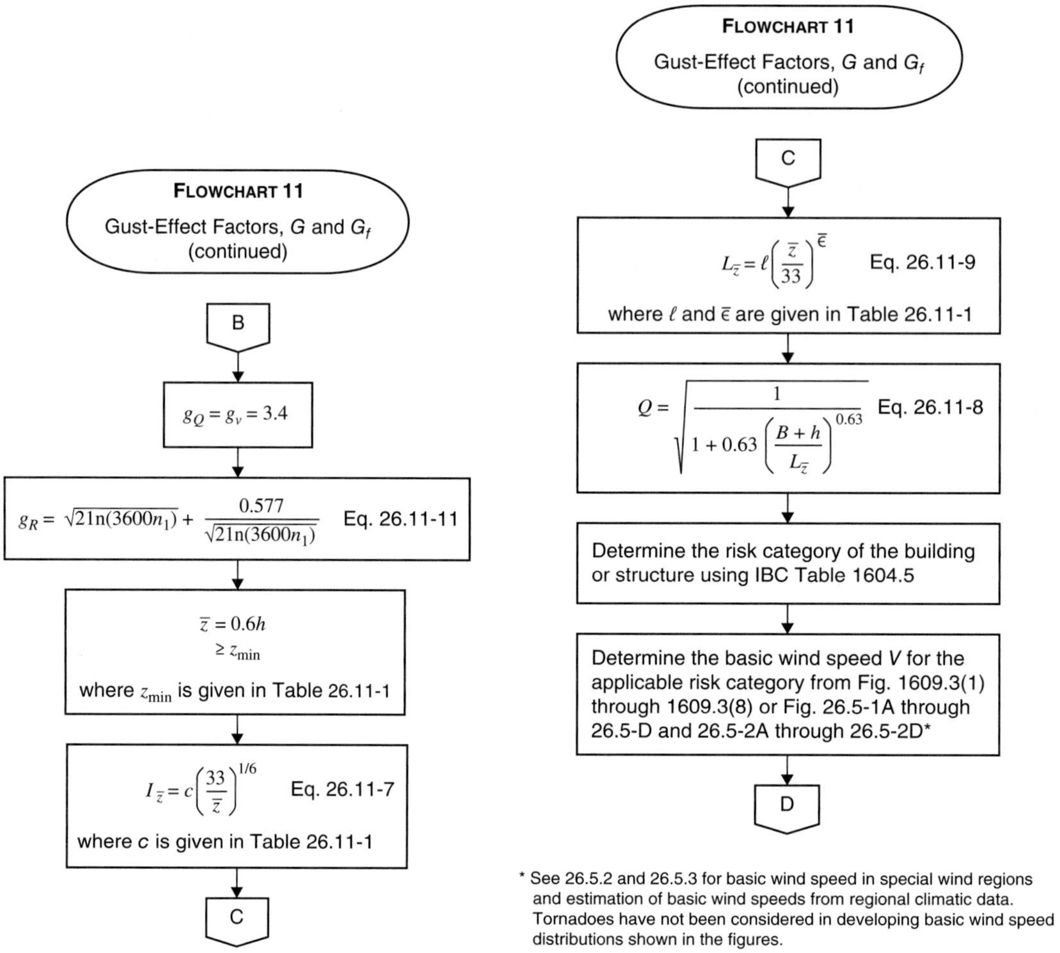

Figure 1.31 *(Continued)*

The situation is significantly different in certain regions where hurricanes can occur. *Hurricane-prone regions* are located along the Atlantic Ocean and Gulf of Mexico coasts where the basic wind speed for Risk Category II buildings is greater than 115 mph. Hawaii, Puerto Rico, Guam, Virgin Islands, and American Samoa are also classified as hurricane-prone regions. *Wind-borne debris regions* are in hurricane-prone regions and are located as follows (IBC 202 and ASCE/SEI 26.12.3.1):

1. within 1 mile of the coastal mean high water line where the basic wind speed (ultimate design wind speed—IBC) is greater than or equal to 130 mph or
2. in areas where the basic wind speed (ultimate design wind speed—IBC) is greater than or equal to 140 mph.

Actual locations of wind-borne debris regions are to be based on the wind speeds that are in the IBC and ASCE/SEI 7 figures, which are summarized in Table 1.9 (IBC 202 and ASCE/SEI 26.12.3.1).

Special requirements are given in IBC 1609.1.2 and ASCE/SEI 26.10.3.2 for the protection of glazed openings in wind-borne debris regions.

Internal Pressure Coefficient

Internal pressure coefficients (GC_{pi}) are given in Table 26.13-1 and are based on the enclosure classifications defined in 26.12. These coefficients have been obtained from wind tunnel tests and full-scale data and are assumed to be valid for a building of any height even though the wind tunnel tests were conducted primarily for low-rise buildings. Gust and aerodynamic effects are combined into one factor (GC_{pi}); in accordance with 26.11.7, the gust-effect factor shall not be determined separately in the analysis.

For partially enclosed buildings that contain a single, relatively large volume without any partitions, the reduction factor R_i calculated by Equation 26.13-1 may be used to reduce the applicable internal pressure coefficient. This reduction factor is based on research that has shown that the response time of internal pressure increases as the volume of a building without partitions increases; as such, the gust factor associated with the internal pressure is reduced, resulting in lower internal pressure.

1.7.3 Main Wind-Force Resisting Systems

Overview

Chapters 27, 28, 29, and 31 in ASCE/SEI 7-16 contain design requirements for determining wind pressures and loads on MWFRSs of buildings and other structures (see Table 1.3). The provisions in Chapters 27 through 29 are discussed in the following sections. Chapter 31, which contains the requirements for wind tunnel procedures, is covered in this chapter.

Directional Procedure for Buildings (Chapter 27)

Scope The Directional Procedure of Chapter 27 applies to the determination of wind loads on the MWFRS of enclosed, partially enclosed and open buildings of all heights that meet the conditions and limitations given in 27.1.2 and 27.1.3, respectively.

FLOWCHART 11

Gust-Effect Factors, G and G_f (continued)

D

$$\bar{V}_{\bar{z}} = \bar{b}\left(\frac{\bar{z}}{33}\right)^{\bar{\alpha}}\left(\frac{88}{60}\right)V \quad \text{Eq. 26.11-16}$$

where $\bar{b}$ and $\bar{\alpha}$ are given in Table 26.11-1

$$N_1 = \frac{n_1 L_{\bar{z}}}{\bar{V}_{\bar{z}}} \quad \text{Eq. 26.11-14}$$

$$R_n = \frac{7.47N_1}{(1 + 10.3N_1)^{5/3}} \quad \text{Eq. 26.11-13}$$

$$R_h = \frac{1}{\eta} - \frac{1}{2\eta^2}\left(1 - e^{-2\eta}\right) \text{ for } \eta > 0$$

$R_h = 1$ for $\eta = 0$

where $\eta = 4.6 n_1 h/\bar{V}_{\bar{z}}$

E

Figure 1.31 *(Continued)*

FLOWCHART 11

Gust-Effect Factors, G and G_f (continued)

E

$$R_B = \frac{1}{\eta} - \frac{1}{2\eta^2}\left(1 - e^{-2\eta}\right) \text{ for } \eta > 0$$

$R_B = 1$ for $\eta = 0$

where $\eta = 4.6 n_1 B/\bar{V}_{\bar{z}}$

$$R_L = \frac{1}{\eta} - \frac{1}{2\eta^2}\left(1 - e^{-2\eta}\right) \text{ for } \eta > 0$$

$R_L = 1$ for $\eta = 0$

where $\eta = 15.4 n_1 L/\bar{V}_{\bar{z}}$

$$R = \sqrt{\frac{1}{\beta} R_n R_h R_B (0.53 + 0.47R_L)} \quad \text{Eq. 26.11-12}$$

$$G_f = 0.925\left(\frac{1 + 1.7I_{\bar{z}}\sqrt{g_Q^2 Q^2 + g_R^2 R^2}}{1 + 1.7 g_v I_{\bar{z}}}\right) \quad \text{Eq. 26.11-10}$$

Table 1.8 Enclosure Classifications

Classification	Definition
Open building	For each wall in the building, $A_o \geq 0.8A_g$
Partially enclosed building	A building that complies with all of the following conditions: • $A_o \geq 1.1A_{oi}$ • $A_o >$ lesser of 4 square feet or $0.01A_g$ • $A_{oi}/A_{gi} \leq 0.2$
Enclosed building	A building that complies with the following conditions: $A_o \leq$ smaller of {$0.01A_g$; 4 square feet}
Partially open building	A building that does not comply with the requirements for open, partially enclosed or enclosed buildings

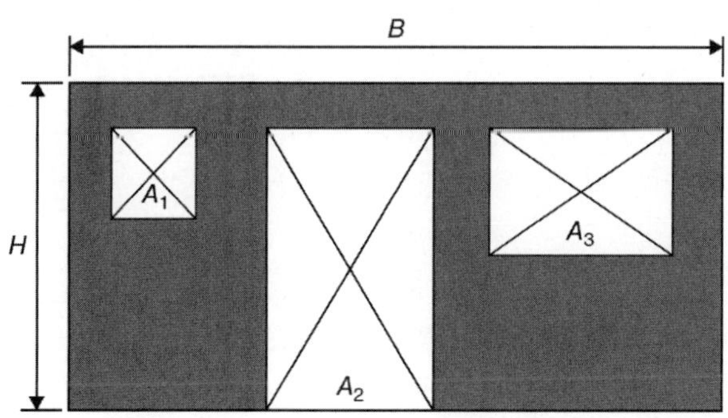

$A_g = B \times H$

$A_o = A_1 + A_2 + A_3$

Figure 1.32 Definition of wall openings for determination of enclosure classification.

Table 1.9 Wind-Borne Debris Region Wind Speed Figures

Classification	Figure no.
• Risk Category II buildings and other structures • Risk Category III buildings and other structures, except health care facilities	Figures 1609.3(1) and 1609.3(5) Figures 26.5-1B and 26.5-2B
• Risk Category III health facilities	Figures 1609.3(2) and 1609.3(6) Figures 26.5-1C and 26.5-2C
• Risk Category IV buildings and other structures	Figures 1609.3(3) and 1609.3(7) Figures 26.5-1D and 26.5-2D

A summary of the wind load procedures and their applicability for MWFRSs in accordance with Chapter 27 are given in Table 1.10.

Part 1 in this chapter is applicable to enclosed, partially enclosed, and open buildings of all heights; a wide range of buildings is covered by the provisions in this part. In general, wind pressures are determined as a function of wind direction using equations that are appropriate for each surface of the building. A simplified method for a special class of buildings up to 160 feet in height is provided in Part 2, which is based on the provisions in Part 1.

In order to apply these provisions, buildings must be regularly shaped (i.e., must have no unusual geometrical irregularities in spatial form) and must exhibit essentially along-wind response characteristics. Buildings of unusual shape that do not meet these conditions must be designed by either recognized literature that documents such wind load effects or by the wind tunnel procedure in Chapter 31 (27.1.3).

Reduction in wind pressures due to apparent shielding of surrounding buildings, other structures, or terrain is not permitted (27.1.4). Removal of such features around a building at a later date could result in wind pressures that are much higher than originally accounted for; as such, wind pressures must be calculated assuming that all shielding effects are not present.

Minimum design wind pressures and loads are given in 27.1.5, which are applicable to buildings designed using Part 1 or 2. In the case of enclosed or partially enclosed buildings, wind pressures of 16 psf and 8 psf must be applied simultaneously to the vertical plane normal to the assumed wind direction over the wall and roof area of the building, respectively. Application of these minimum wind pressures are illustrated in Fig. 1.33 for wind along the two primary axes of the building. For open buildings, the minimum wind force is equal to 16 psf multiplied by the area of the open building either normal to the

Table 1.10 Summary of Wind Load Procedures in Chapter 27 of ASCE/SEI 7-16 for MWFRSs

ASCE/SEI chapter	Part	Applicability: Building type	Applicability: Height limit	Conditions
27	1	Enclosed Partially enclosed Open	None	• Regular-shaped building • Building does not have response characteristics making it subject to across-wind loading, vortex shedding, instability due to galloping or flutter • Building is not located at a site where channeling effects or buffeting in the wake of upwind obstructions warrant special consideration
	2	Enclosed, simple diaphragm	$h \leq 160$ feet	• Same conditions as in Part 1 • Building must meet the conditions for either a Class 1 or Class 2 building: Class 1 Building: 1. $h \leq 60$ feet 2. $0.2 \leq L/B \leq 5.0$ Class 2 Building: 1. 60 feet $< h \leq 160$ feet 2. $0.5 \leq L/B \leq 2.0$ 3. $n_1 \geq 75/h$ • Building is an enclosed simple diaphragm building as defined in 26.2

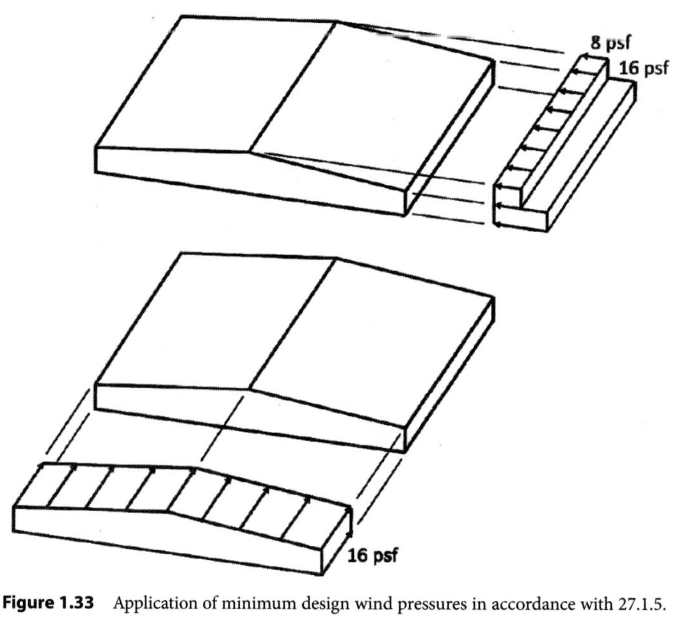

Figure 1.33 Application of minimum design wind pressures in accordance with 27.1.5.

wind direction or projected on a plane normal to the wind direction. It is important to note that minimum design pressures or loads are load cases that must be considered separate from any other load cases that are specified in Part 1 or 2.

Part 1—Enclosed, Partially Enclosed, and Open Buildings of All Heights

Overview Part 1 of Chapter 27 is applicable to buildings with any general plan shape, height, or roof geometry that matches the figures provided in this chapter. This procedure entails the determination of velocity pressures, gust-effect factors, external pressure coefficients, and internal pressure coefficients for each surface of a rigid, flexible, or open building. Table 27.2-1 contains the overall steps that can be used to determine wind pressures on such buildings.

Design Wind Pressures

Enclosed and Partially Enclosed Rigid and Flexible Buildings Design wind pressures p are calculated by Equation 27.3-1 for the MWFRS of enclosed and partially enclosed rigid buildings of all heights:

$$p = qGC_p - q_i(GC_{pi})$$

This equation is used to calculate the wind pressures on each surface of the building: windward wall, leeward wall, side walls, and roof. The pressures are applied simultaneously on the walls and roof, as depicted in Figure 27.3-1 (see also Fig. 1.28). The first part of the equation is the external pressure contribution, and the second part is the internal pressure contribution. External pressure varies with height above ground on the windward wall and is a constant on all of the other surfaces based on the mean roof height.

The gust-effect factor G for rigid buildings may be taken equal to 0.85 or may be calculated by Equation 26.11-6. For flexible buildings, G_f determined in accordance with 26.11.5 is to be used in Equation 27.3-1 instead of G.

External pressure coefficients C_p capture the aerodynamic effects that are discussed in Sec. 1.7.1 of this chapter and have been determined experimentally through wind tunnel tests on buildings of various shapes and sizes. These coefficients reflect the actual wind loading on each surface of a building as a function of wind direction.

Figure 27.3-1 contains C_p values for windward walls, leeward walls, side walls, and roofs for buildings with gable and hip roofs, monoslope roofs, and mansard roofs. Wall pressure coefficients are constant on windward and side walls and vary with the plan dimensions of the building (i.e., vary with the aspect ratio of the building L/B) on the leeward wall. The table in the upper part of this figure also designates which velocity pressure to use—q_z or q_h—on a particular wall surface. Roof pressure coefficients vary with the ratio of the mean roof height to the plan dimension of the building (h/L) and with the roof angle (θ) for a given wind direction (normal to ridge or parallel to ridge). All of these pressure coefficients are intended to be used with q_h, and the parallel to ridge wind direction is applicable for flat roofs. It is evident from the figure that negative roof pressures increase as the ratio h/L increases. Also, as θ increases, negative pressure decreases until a roof angle is reached where the pressure becomes positive; this is consistent with the aerodynamic effect of the separation zone (see Fig. 1.25). Where two values of C_p are listed in the figure, the windward roof is subjected to either positive or negative pressure, and the structure must be designed for both. Other important information on the use of this figure is given in the notes below the tabulated pressure coefficients.

The external pressure coefficients in Figure 27.3-2 for dome roofs are adapted from the 1995 edition of the Eurocode (Ref. 12) and are based on data obtained from a modeled atmospheric boundary layer flow that does not fully comply with the wind tunnel testing requirements given in Chapter 31. Two load cases must be considered. In Case A, pressure coefficients are determined between various locations on the dome by linear interpolation along arcs of the dome parallel to the direction of wind; this defines maximum uplift on the dome in many cases. In Case B, the pressure coefficient is assumed to be a constant value at a specific point on the dome for angles less than or equal to 25 degrees and is determined by linear interpolation from 25 degrees to other points on the dome; this properly defines positive pressures for some cases, which results in maximum base shear. Wind tunnel tests are recommended for domes that are larger than 200 feet in diameter and in cases where resonant response can be an issue (C27.3.1).

The pressure and force coefficients in Figure 27.3-3 for arched roofs are the same as those that were first introduced in 1972 (Ref. 13). These coefficients were obtained from wind tunnel tests conducted under uniform flow and low turbulence. References 9 and 14 can be consulted for pressure coefficients that are not specified in this figure.

The velocity pressure for internal pressure determination q_i is used in capturing the effects caused by internal pressure. On all of the surfaces of enclosed buildings and for negative internal pressure evaluation in partially enclosed buildings, q_i is to be taken as the velocity pressure evaluated at the mean roof height q_h. For positive internal pressure evaluation, 27.3.1 permits q_i to be set equal to q_z in partially enclosed buildings, where q_z is the velocity pressure evaluated at the location of the highest opening in the building that could affect positive internal pressure. Note that it is conservative to set q_i equal to q_h in all cases where positive internal pressure is evaluated. In the case of low-rise buildings, the distance between the uppermost opening and the mean roof height is usually relatively small, and this approximation yields reasonable results. However, this approximation can be overly conservative in certain cases, especially for taller buildings where the distance between the uppermost opening and the mean roof height is relatively large. For buildings located in wind-borne debris regions with glazing that does not meet the protection requirements of 26.12.3.2, q_i is to be determined assuming that the glazing will be breached.

The velocity pressure q_i is multiplied by the internal pressure coefficient (GC_{pi}). Both positive and negative values of (GC_{pi}) must be considered in order to establish the critical load effects.

Open Buildings with Monoslope, Pitched, or Troughed Free Roofs Design wind pressures p are calculated by Equation 27.3-2 for the MWFRS of open buildings with monoslope, pitched, or troughed roofs:

$$p = q_h GC_N$$

In this equation, q_h is the velocity pressure at the mean roof height determined by the equation for velocity pressure in Sec. 1.7.2, and G is the gust-effect factor determined in accordance with 26.11. Net pressure coefficients C_N are given in Figures 27.3-4 through 27.3-7, which are based on the results from wind tunnel studies. Two load cases are identified in the figures: Load Case A and Load Case B. Both load cases must be considered in order to obtain the maximum load effects for a particular roof slope and blockage configuration.

For structures with free roofs that contain fascia panels where the angle of the plane of the roof from the horizontal is less than or equal to 5 degrees, the fascia panels are to be considered as an inverted parapet. The contribution of the wind loads on the fascia panels to the wind loads on the MWFRS is to be determined using the provisions of ASCE/SEI 27.3.4 with q_p in Equation 27.3-3 taken as q_h.

Roof Overhangs In the case of roof overhangs, the positive external pressure on the bottom surface of a windward roof overhang is determined using the external pressure coefficient for the windward wall (C_p = 0.8). This pressure is combined with the top surface pressures determined in accordance with Figure 27.3-1 (see Fig. 1.34).

Parapets Design wind pressures p_p for the effects of parapets on the MWFRS of rigid or flexible buildings with flat, gable, or hip roofs are calculated by Equation 27.3-3:

$$p_p = q_p(GC_{pn})$$

In this equation, q_p is the velocity pressure evaluated at the top of the parapet, and (GC_{pn}) is the combined net pressure coefficient, which is equal to +1.5 for a windward parapet and −1.0 for a leeward parapet. It is important to note that p_p is the combined net pressure due to the combination of the net pressures from the front and back surfaces of the parapet.

The pressures on the front and back of the parapet have been combined into one pressure, which is captured by the combined net pressure

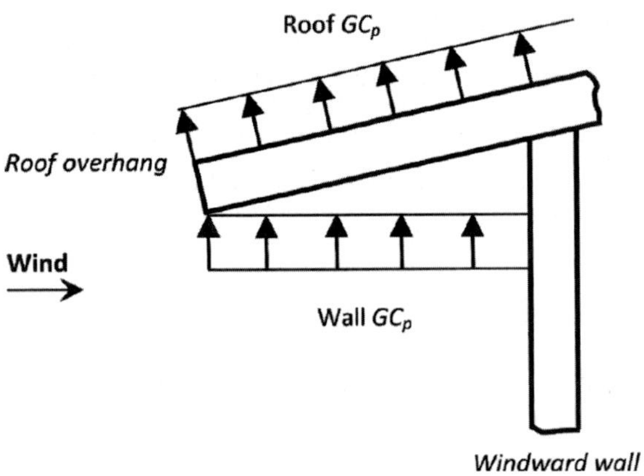

Figure 1.34 Application of wind pressures on a roof overhang, Part 1 of Chapter 27.

coefficients (GC_{pn}) for windward and leeward parapets. Since the wind can occur in any direction, a parapet must be designed for both sets of pressures. Note that the internal pressures inside the parapet cancel out in the determination of the combined pressure coefficient.

The pressures determined on the parapets are combined with the external pressures on the building to obtain the total wind pressures on the MWFRS.

Design Wind Load Cases Buildings subjected to the wind pressures determined by Chapter 27 must be designed for the load cases depicted in Figure 27.3-8, which are reproduced here in Fig. 1.35 (in this figure, the subscripts x and y refer to the principal axes of the building, and w and l refer to the windward and leeward faces, respectively). In Load Case 1, design wind pressures are applied along the principal axes of a building separately.

Load Cases 2 accounts for the effects of nonuniform pressure on different faces of the building due to wind flow; these pressure distributions have been documented in wind tunnel tests. Nonuniform pressures introduce torsion on the building, and this is accounted for in design by subjecting the building to 75 percent of the design wind

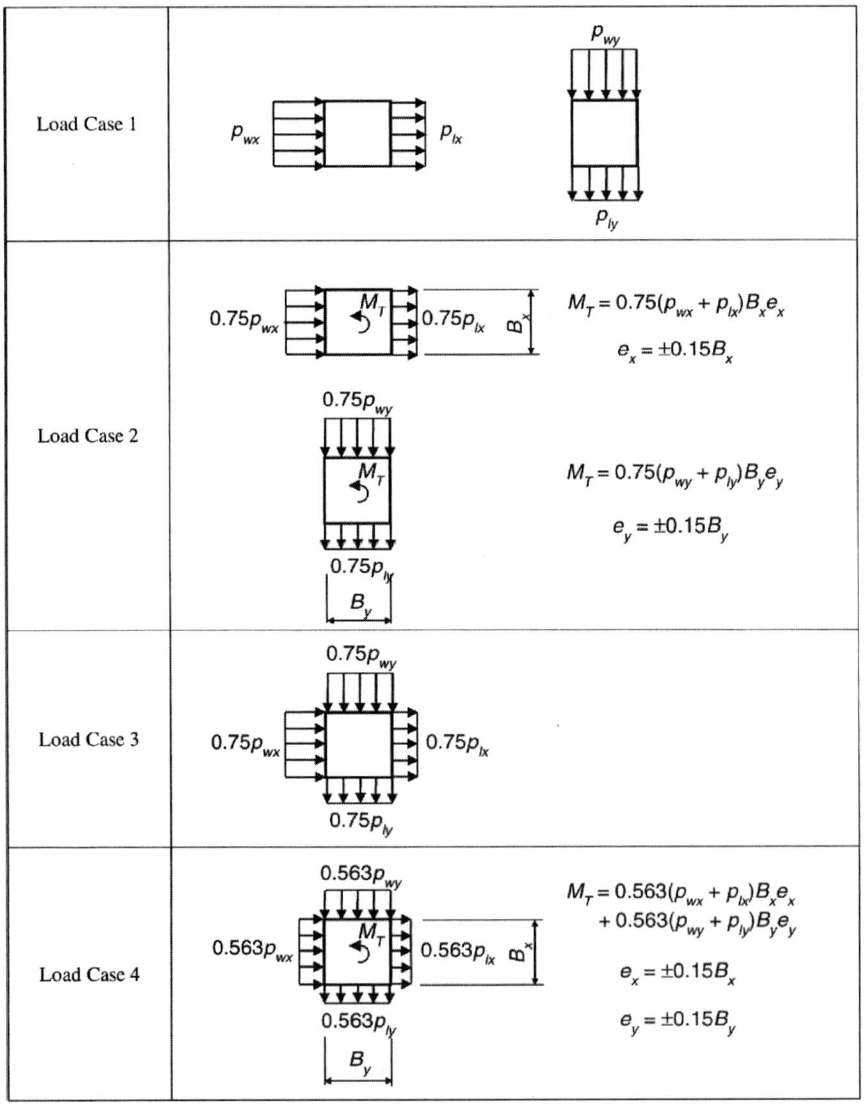

Figure 1.35 Design wind load cases in Part 1 of Chapter 27.

pressures applied along the principal axis of the building plus a torsional moment M_T that is determined using an eccentricity equal to 15 percent of the appropriate plan dimension of the building. Torsional effects are determined in each principal direction separately.

A critical load case can occur when the design wind load acts diagonally to a building. This is accounted for in Load Case 3, where 75 percent of the maximum design wind pressures are applied along the principal axes of a building simultaneously.

Load Case 4 considers the effects due to diagonal wind loads and torsion. Seventy-five percent of the wind pressures in Load Case 2 are applied along the principal axes of a building simultaneously, and a torsional moment is applied, which is determined using 15 percent of the plan dimensions of the building.

In the case of flexible buildings, dynamic effects can increase the effects from torsion. Equation 27.3-4 accounts for these effects. The eccentricity e determined by this equation is to be used in the appropriate load cases in Fig. 1.35 in lieu of the eccentricities e_x and e_y that are given in that figure for rigid structures. An eccentricity must be considered for each principal axis of the building, and the sign of the eccentricity must be plus or minus, whichever causes the more severe load case.

Flowchart 12 (Fig. 1.36) can be used to determine design wind pressures on the MWFRS of buildings in accordance with Part 1 of Chapter 27.

Part 2—Enclosed, Simple Diaphragm Buildings with $h \le 160$ Feet

Overview Part 2 of Chapter 27 is applicable to enclosed, simple diaphragm buildings with mean roof heights less than or equal to 160 feet that meet the additional conditions in 27.4.2 for either a Class 1 or Class 2 building. Based on the procedures in Part 1, it is meant to be a simplified method for determining wind pressures on such buildings.

According to 26.2, a *simple diaphragm building* is one in which both windward and leeward wind loads are transmitted by roof and vertically spanning wall assemblies through continuous floor and roof assemblies to the MWFRS. In other words, the wind loads are delivered to the elements of the MWFRS via roof and floor diaphragms. As such, internal pressures cancel out in the determination of the total wind load in the direction of analysis. Thus, in order for this approach to be valid, no structural expansion joints are permitted in the system: expansion joints interrupt the continuity of the diaphragm, resulting in internal pressures that do not cancel out. Also, no girts or other horizontal members should be present that transfer significant wind loads directly to the vertical members of the MWFRS.

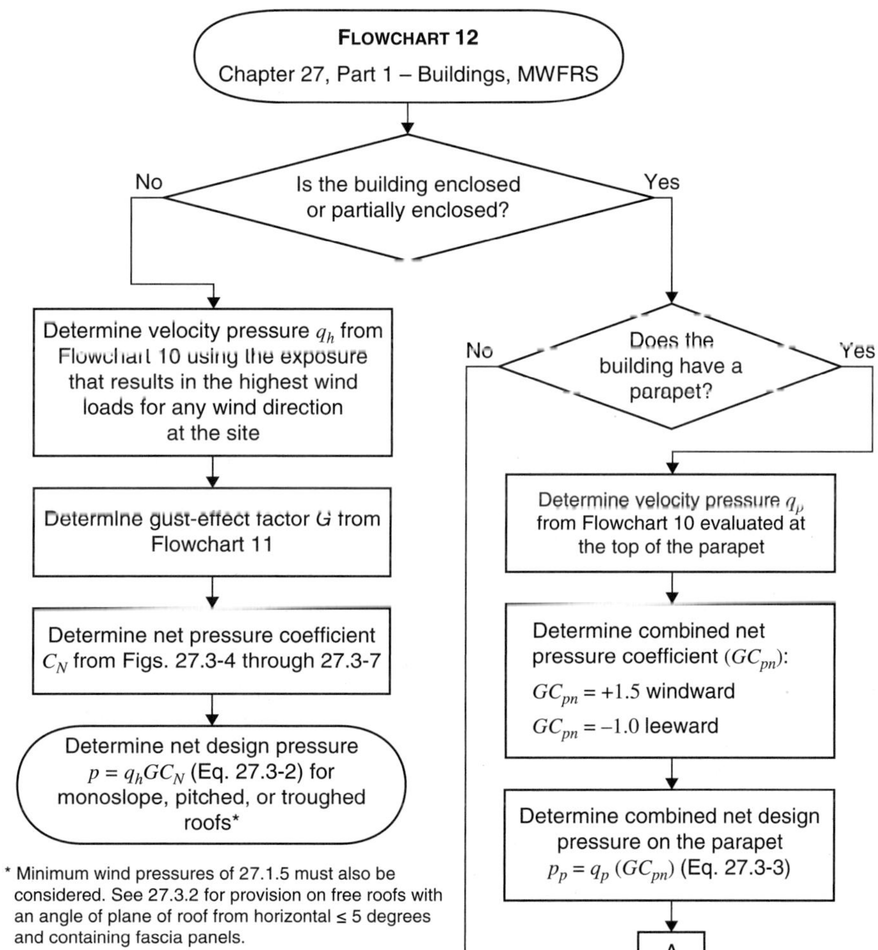

Figure 1.36 Design wind pressures on the MWFRS of buildings in accordance with Part 1 of Chapter 27 (Flowchart 12).

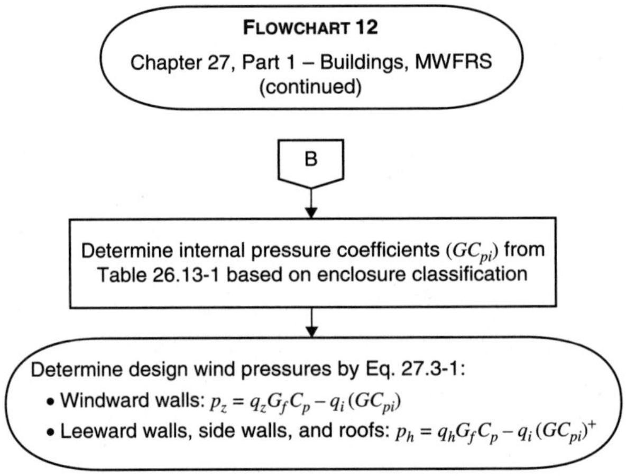

+ Notes:

1. See 27.3.5 and Fig. 27.3-8 for the load cases that must be considered.
2. Minimum wind pressures of 27.1.5 must also be considered.

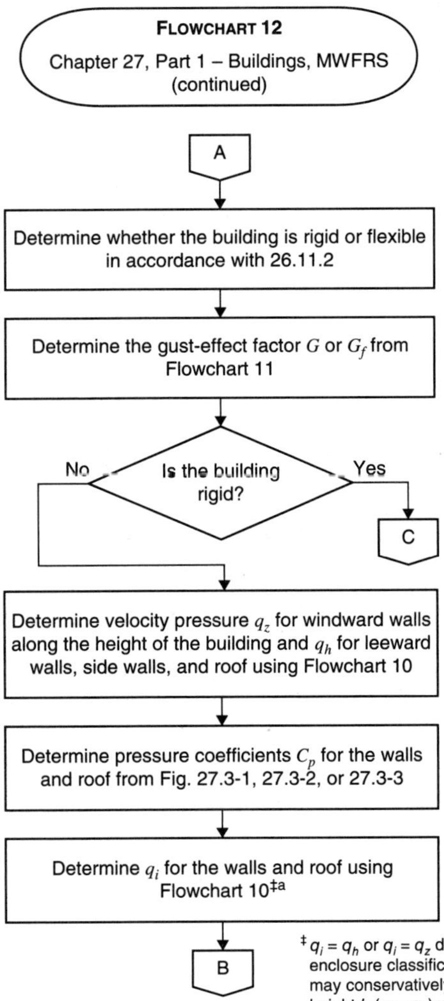

‡ $q_i = q_h$ or $q_i = q_z$ depending on enclosure classification (see 27.3.1). q_i may conservatively be evaluated at height h ($q_i = q_h$) where applicable.

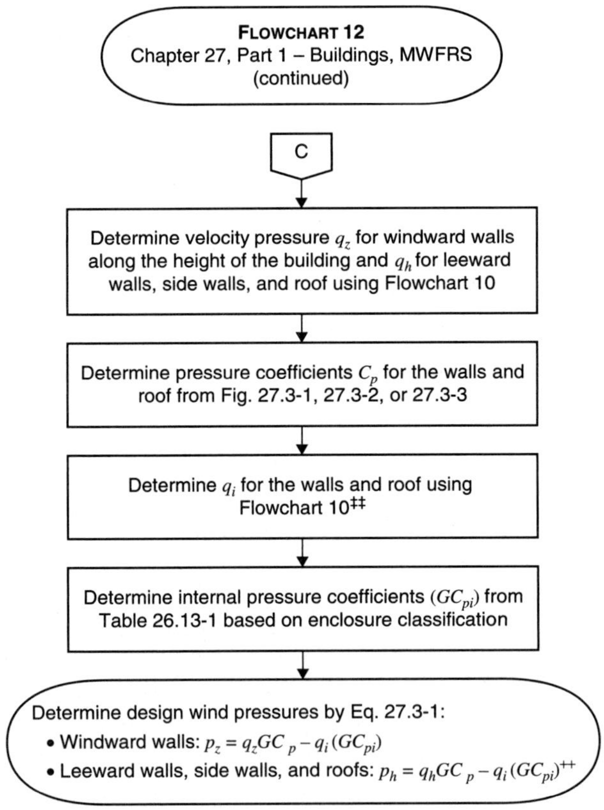

‡‡ $q_i = q_h$ or $q_i = q_z$ depending on enclosure classification (see 27.3.1). q_i may conservatively be evaluated at height h ($q_i = q_h$) where applicable.

++ Notes:

1. See 27.3.5 and Fig. 27.3-8 for the load cases that must be considered.
2. Minimum wind pressures of 27.1.5 must also be considered.
3. See 27.3.3 for wind pressure on roof overhangs.

Figure 1.36 *(Continued)*

The conditions that define Class 1 and Class 2 buildings are summarized in Table 1.10 and are illustrated in Figure 27.4-1. Since Class 1 buildings are limited to a mean roof height that is less than or equal to 60 feet, it assumed that these buildings are rigid and a gust-effect factor of 0.85 has been used to calculate the tabulated wind pressures in this part (see 26.10 and 26.11.1). For Class 2 buildings, the requirement that the natural frequency must be greater than or equal to $75/h$ (where h is the mean roof height in feet) is needed to ensure that the gust-effect factor that has been calculated and built into the design procedure is consistent with the tabulated wind pressures. This frequency ($75/h$) is meant to represent a reasonable lower bound to values of frequencies found in practice.

The design procedures in this part apply to buildings with either rigid or flexible diaphragms. For consideration of wind loading, diaphragms constructed of untopped metal deck, concrete-filled metal deck, and concrete slabs can be idealized as rigid (27.4.5). Diaphragms constructed of wood panels can be considered flexible.

Table 27.4-1 contains the overall steps that can be used to determine wind pressures on Class 1 and Class 2 buildings.

Design Wind Pressures

Wall and Roof Surfaces Net design wind pressures for the walls and roof surfaces of Class 1 and 2 buildings can be determined directly from Tables 27.5-1 and 27.5-2, respectively.

Net wall pressures are tabulated for Exposures B, C, and D as a function of wind velocity V, mean roof height h, and building aspect ratio L/B. The top pressure in the table is defined as p_h, and the bottom pressure is defined as p_o. Interpolation between these values is permitted (see Note 5 in Table 27.5-1).

Along-wind net wind pressures are distributed over the height of the building as shown in Table 27.5-1 and Figure 27.5-1. The net pressures are applied normal to the projected area of the building walls in the direction of wind.

The pressure on the side walls is determined using Note 2 in Table 27.5-1. In particular, the side wall pressures are calculated as a percentage of the along-wind pressure p_h at the top of the building based on the aspect ratio L/B.

A method of determining the distribution of the tabulated net wall pressures between the windward and leeward wall surfaces of a building is provided in Note 4 of Table 27.5-1. Having such a distribution can be useful when designing floor and roof diaphragm elements such as collectors as well as shear walls that are a part of the MWFRS.

Tabulated roof pressures are given in Table 27.5-2 for Exposure C as a function of V, h, and roof slope. Exposure adjustment factors are provided in the table for Exposures B and D; the tabulated roof pressures are to be multiplied by the appropriate adjustment factors.

Roof pressures are applied perpendicular to the roof surfaces as shown in Figure 27.5-1. The different zones over which these pressures are to be distributed are identified in Table 27.5-2 for flat, gable, hip, monoslope, and mansard roofs.

Roof pressure is given for two load cases, and both must be investigated where applicable. Load Case 2 is required when investigating maximum overturning effects on the building due to the wind pressures.

According to 27.5.1, pressures on the walls and the roof must be applied simultaneously to the building as shown in Figure 27.5-1. Also, the MWFRS must be designed for the load cases defined in Figure 27.3-8. Load Cases 2 and 3 (torsional load cases) need not be considered in the following cases for buildings that meet the requirements of Appendix D.

Parapets According to 27.5.2, the additional pressure on the MWFRS due to roof parapets is equal to 2.25 times the wall pressure tabulated in Table 27.6-1 using an aspect ratio of $L/B = 1.0$ and a height h_p equal to the distance from the ground to the top of the parapet. This net horizontal pressure accounts for both the windward and leeward parapet loading on both the windward and the leeward building surfaces and is applied to the projected area of the parapet surface simultaneously with the net wall and roof pressures (see Figure 27.5-2).

Roof Overhangs ASCE/SEI 27.6.3 contains provisions to account for the effects of vertical wind pressures on roof overhangs. According to this section, a positive wind pressure equal to 75 percent of the roof edge pressure from Table 27.5-2 for Zone 1 or Zone 3, whichever is applicable, must be applied to the underside of the windward overhang (see Figure 27.5-3).

Flowchart 13 (Fig. 1.37) can be used to determine design wind pressures on the MWFRS of buildings in accordance with Part 2 of Chapter 27.

Envelope Procedure for Buildings (Chapter 28)

Scope The Envelope Procedure of Chapter 28 applies to the determination of wind loads on the MWFRS of enclosed or partially enclosed low-rise buildings that meet the conditions and limitations given in 28.1.2 and 28.1.3, respectively. This procedure is the former "low-rise buildings" provision in Method 2 of ASCE/SEI 7-05 for MWFRSs. The simplified method in this chapter is based on the provisions of Method 1 of ASCE/SEI 7-05 for simple diaphragm buildings up to 60 feet in height.

A *low-rise building* is defined in 26.2 as an enclosed or partially enclosed building with a mean roof height less than or equal to 60 feet and a mean roof height that does not exceed the least horizontal dimension of the building.

A summary of the wind load procedures and their applicability for MWFRSs in accordance with Chapter 28 is given in Table 1.11.

Part 1 in this chapter is applicable to enclosed and partially enclosed low-rise buildings that have a flat, gable, or hip roof. Like the method provided in Part 1 of Chapter 27, wind pressures are determined as a function of wind direction using equations that are appropriate for each surface of the building.

In order to apply these provisions, buildings must be regularly shaped (i.e., must have no unusual geometrical irregularities in spatial form) and must exhibit essentially along-wind response characteristics. Buildings of unusual shape that do not meet these conditions must be designed by either recognized literature that documents such wind load effects or the wind tunnel procedure in Chapter 31 (28.1.3).

Reduction in wind pressures due to apparent shielding of surrounding buildings, other structures, or terrain is not permitted (28.1.4). Removal of such features around a building at a later date could result in wind pressures that are much higher than originally accounted for; as such, wind pressures must be calculated assuming that all shielding effects are not present.

Part 1—Enclosed and Partially Enclosed Low-Rise Buildings

Overview As noted above, Part 1 of this chapter is applicable to enclosed, partially enclosed, and open low-rise buildings that have a flat, gable, or hip roof. This procedure entails the determination of the velocity pressure at the mean roof height (which is determined as a function of exposure, height, topographic effects, wind directionality, wind velocity, ground elevation, and building occupancy), combined gust-effect factors and external pressure coefficients and internal pressure coefficients. Table 28.2-1 contains the overall steps that can be used to determine wind pressures on such buildings.

Velocity Pressure Velocity pressure, q_z, is determined by Equation 26.10-1, where all the quantities in that equation are determined in accordance with the appropriate sections in Chapter 26 (see Sec. 1.7.3 of this chapter).

Design Wind Pressures

MWFRS Design wind pressures p are determined by Equation 28.3-1 for the MWFRS of low-rise buildings that satisfy the conditions and limitations of Part 1 of this chapter:

$$p = q_h[(GC_{pf}) - (GC_{pi})]$$

In this equation, q_h is the velocity pressure determined by Equation 26.10-1 evaluated at the mean roof height h, (GC_{pf}) are external pressure coefficients, and (GC_{pi}) are internal pressure coefficients, which are determined by Table 26.13-1. Wind pressures are determined on each of the building surfaces identified in Figure 28.3-1.

The external pressure coefficients (GC_{pf}) combine both a gust-effect factor and external pressure coefficients for low-rise buildings and are not allowed to be separated (28.3.1.1). Unlike the external pressure

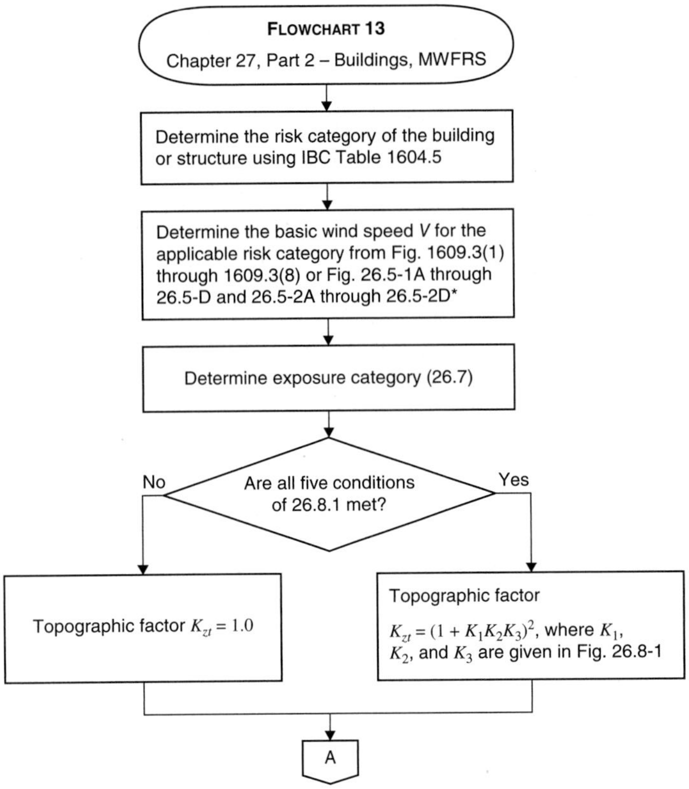

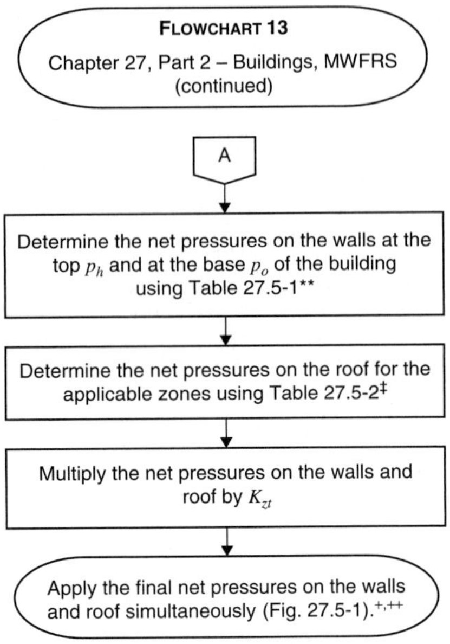

* See 26.5.2 and 26.5.3 for basic wind speed in special wind regions and estimation of basic wind speeds from regional climatic data. Tornadoes have not been considered in developing basic wind speed distributions shown in the figures.

** Notes:
1. See Table 27.5-1 for distribution of net pressure over the height of the building.
2. See the notes in Table 27.5-1 for pressures on side walls and for other important information.

‡ The net pressures from the table must be multiplied by the exposure adjustment factors in this table for Exposures B and D.

\+ Minimum wind pressures of 27.1.5 must also be considered.

++ See 27.5.2 and Fig. 27.5-2 for additional load on MWFRS from parapets and 27.5.3 and Fig. 27.5-3 for wind on roof overhangs.

Figure 1.37 Design wind pressures on the MWFRS of buildings in accordance with Part 2 of Chapter 27 (Flowchart 13).

Table 1.11 Summary of Wind Load Procedures in Chapter 28 of ASCE/SEI 7-16 for MWFRSs

ASCE/SEI chapter	Part	Applicability		Conditions
		Building type	Height limit	
28	1	Enclosed, low-rise Partially enclosed, low-rise	$h \le 60$ feet $h \le$ least horizontal dimension	• Regular-shaped building • Building does not have response characteristics making it subject to across-wind loading, vortex shedding, instability due to galloping or flutter • Building is not located at a site where channeling effects or buffeting in the wake of upwind obstructions warrant special consideration
	2	Enclosed, simple diaphragm, low-rise		• Same conditions as in Part 1 • $n_1 \ge 1$ Hz • Building has an approximately symmetrical cross section in each direction with either a flat roof or a gable or hip roof with $\theta \le 45$ degrees • Building is exempted from torsional load cases as indicated in Note 5 of Figure 28.4-1, or the torsional load cases defined in Note 5 do not control the design of any of the MWFRSs of the building

coefficients given in Figure 27.3-1 that reflect the actual pressure on each surface of a building as a function of wind direction, the coefficients in Figure 28.3-1 are essentially "pseudo" pressure conditions that, when applied to a building, envelop the desired structural actions independent of wind direction.

The "pseudo" values of (GC_{pf}) were determined from the output of wind tunnel tests, which measured bending moments, total horizontal forces, and total uplift as a building was rotated 360 degrees in the wind tunnel (see Ref. 15 and Figure C28.3-1). Thus, values of (GC_{pf}) produce maximum measured structural actions and are not the actual surface pressures.

In order to capture all appropriate structural actions, a building must be designed for all wind directions by considering in turn each corner of the building as the windward (or reference) corner; these conditions are illustrated in Figure 28.3-1. At each corner, two load cases must be considered (Load Case A and Load Case B), one for each range of wind direction. In general, a total of 16 separate load cases must be evaluated since both positive and negative internal pressure must be considered. For symmetrical buildings, some of these load cases will be repetitive and can be eliminated. Figure 1.38 illustrates Load Case A and Load Case B for the same windward corner of a low-rise building.

The torsional load cases given in Figure 28.3-1 must be considered in the design of all low-rise buildings except for the following (see Note 5 in this figure):

- One-story buildings with a mean roof height of less than or equal to 30 feet
- Buildings two stories or less framed with light-frame construction (i.e., structural systems made up of repetitive wood or cold-formed steel framing members or subassemblies)
- Buildings two stories or less with flexible diaphragms

It is evident from Equation 28.3-1 that the effects from internal pressure cancel out when evaluating the total horizontal wind pressure on the MWFRS of a low-rise building. It can be critical, however, in one-story buildings with moment-resisting frames and in the top story of buildings with moment-resisting frames.

Parapets The design wind pressure for the effect of parapets on the MWFRS of low-rise buildings with flat, gable, or hips roofs is determined by Equation 28.3-2, which is identical to Equation 27.3-3 in Part 1 of Chapter 27.

Roof Overhangs Positive external pressures on the bottom surface of windward roof overhangs are to be determined using the pressure coefficient $C_p = 0.7$ in combination with the top surface pressures determined by Figure 28.3-1. Application of this pressure is similar to that shown in Fig. 1.34.

Minimum Design Wind Loads ASCE/SEI 28.3.4 prescribes the minimum wind pressures in the design of an MWFRS for enclosed or partially enclosed low-rise buildings. The pressures of 16 psf on the projected area of the walls and 8 psf on the projected area of the roof are considered a separate load case from any of the other load cases specified in this part (see Fig. 1.33).

Flowchart 14 (Fig. 1.39) can be used to determine design wind pressures on the MWFRS of buildings in accordance with Part 1 of Chapter 28.

Part 2—Enclosed, Simple Diaphragm, Low-Rise Buildings

Overview Part 2 is applicable to enclosed, simple diaphragm, low-rise buildings with flat, gable, or hip roofs that meet the conditions in 28.5.2. Although there are eight conditions that need to be satisfied (see Table 1.10), a large number of typical low-rise buildings meet these criteria.

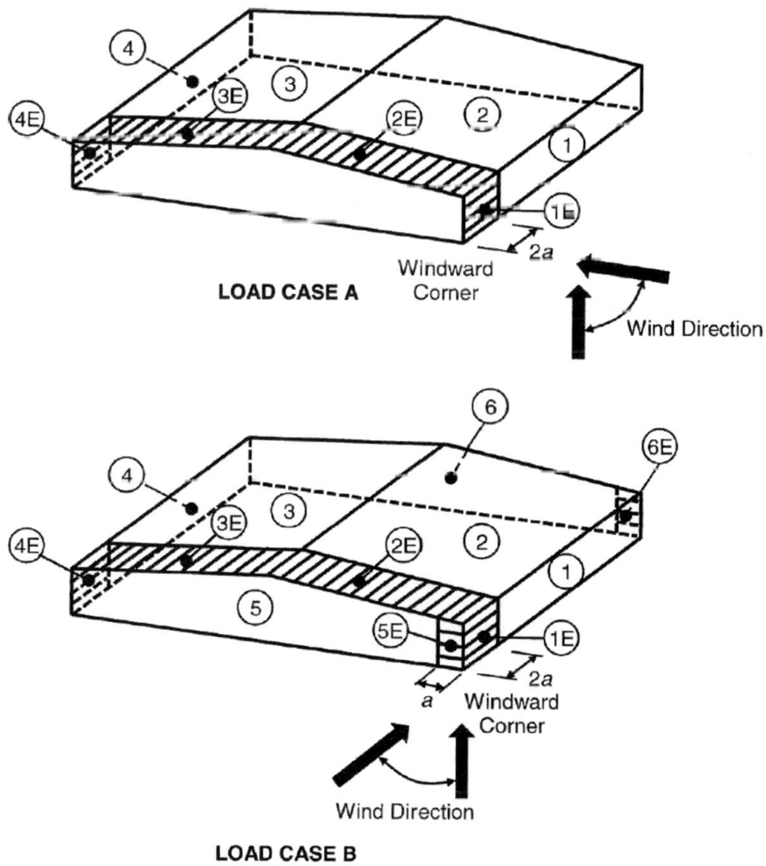

Figure 1.38 Basic load cases for low-rise buildings, Part 1 of Chapter 28.

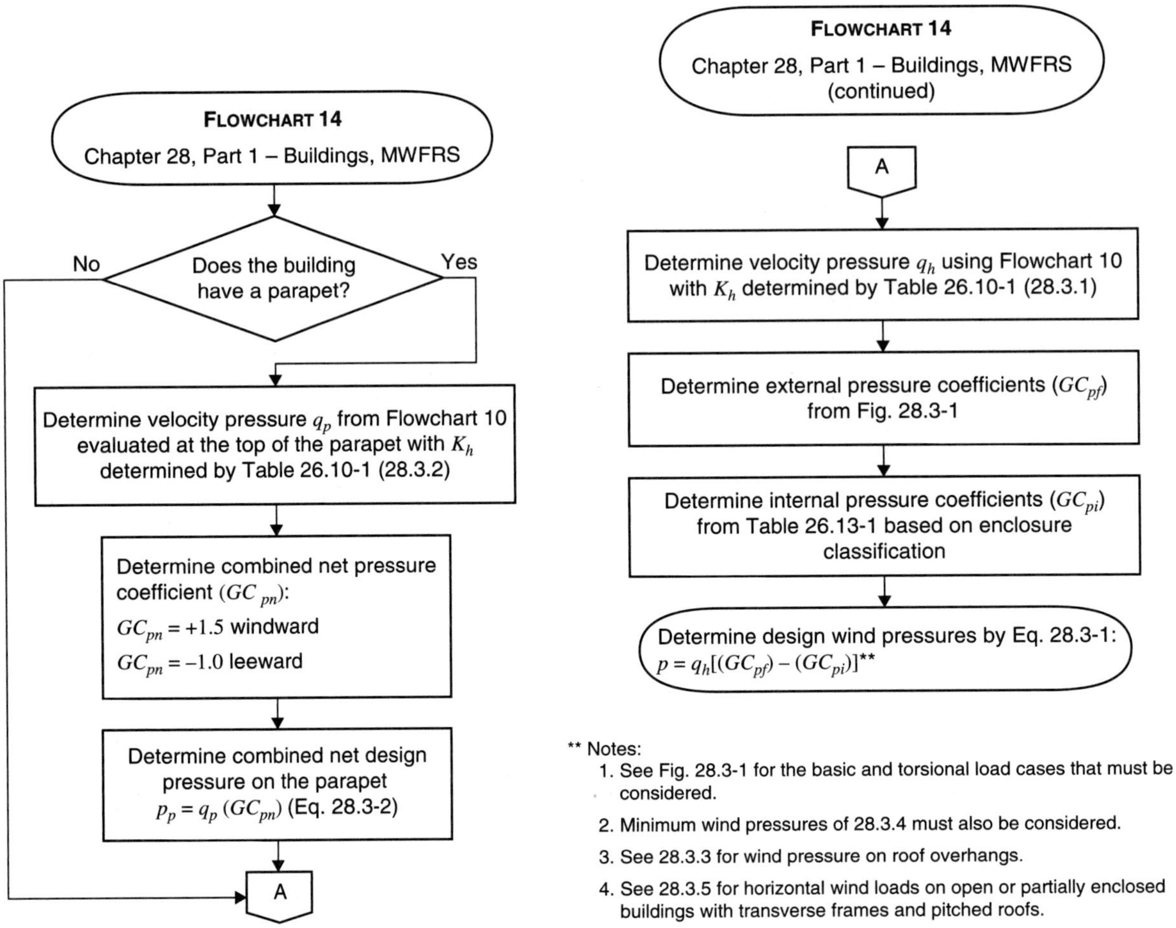

Figure 1.39 Design wind pressures on the MWFRS of buildings in accordance with Part 1 of Chapter 28 (Flowchart 14).

This method is based on Part 1 of Chapter 28 for simple diaphragm buildings. Simple diaphragm buildings are structures where wind loads are delivered to the elements of the MWFRS via roof and floor diaphragms. As such, internal pressures cancel out in the determination of the total wind load in the direction of analysis. In order for this approach to be valid, no structural expansion joints are permitted in the system (expansion joints interrupt the continuity of the diaphragm resulting in internal pressures that do not cancel out), and no girts or other horizontal members should be present that transfer significant wind loads directly to the vertical members of the MWFRS.

According to condition number 8 in 28.5.2, only buildings that are exempted from torsional load cases as indicated in Note 5 of Figure 28.3-1 or those where the torsional load cases defined in this note do not control the design of any of the MWFRSs of the building are permitted to be designed by this simplified procedure. The torsional loading in Figure 28.3-1 was considered to be too complicated to include in a simplified method of determining wind pressures.

Design Wind Pressures

MWFRS Simplified design wind pressures p_{s30} for walls and roofs located at various zones on a building are tabulated in Figure 28.5-1 as a function of the basic wind speed V and roof angle for buildings with a mean roof height of 30 feet that are located on primarily flat ground in Exposure B. Modifications are made to these tabulated pressures based on actual building height and exposure using the adjustment factor λ that is given in the figure. Such pressures must also be modified by the topographic factor K_{zt} where applicable.

Horizontal pressure p_s is determined by Equation 28.5-1:

$$p_s = \lambda K_{zt} p_{s30}$$

Horizontal wall pressures on Zones A and C are the net sum of the windward and leeward pressures on vertical projections of the wall (see Figure 28.5-1, which is reproduced here as Fig. 1.40). Horizontal roof pressures on Zones B and D are the net sum of the windward and leeward pressures on the vertical projection of the roof, and the vertical roof pressures on Zones E, F, G, and H are the net sum of the external and internal pressures (using an internal pressure coefficient of ±0.18 for enclosed buildings) on the horizontal projection of the roof. The pressure coefficients that were used to generate these pressures are from Figure 28.3-1.

Due to the enveloped nature of the loads obtained by this method for roof members, such members that are part of the MWFRS that span at least from the eave to the ridge or that support members spanning at least from the eave to the ridge need not be designed for the higher end zone loads depicted in Fig. 1.40. The interior zone loads are applicable in such cases.

Minimum Design Wind Loads ASCE/SEI 28.5.4 prescribes the minimum wind pressures in the design of an MWFRS for buildings designed by this method. The minimum load cases is defined by assuming that the pressures p_s for Zones A and C are equal to +16 psf, for Zones B and D are equal to +8 psf, and for Zones E, F, G, and H are equal to zero.

Flowchart 15 (Fig. 1.41) can be used to determine design wind pressures on the MWFRS of buildings in accordance with Part 2 of Chapter 28.

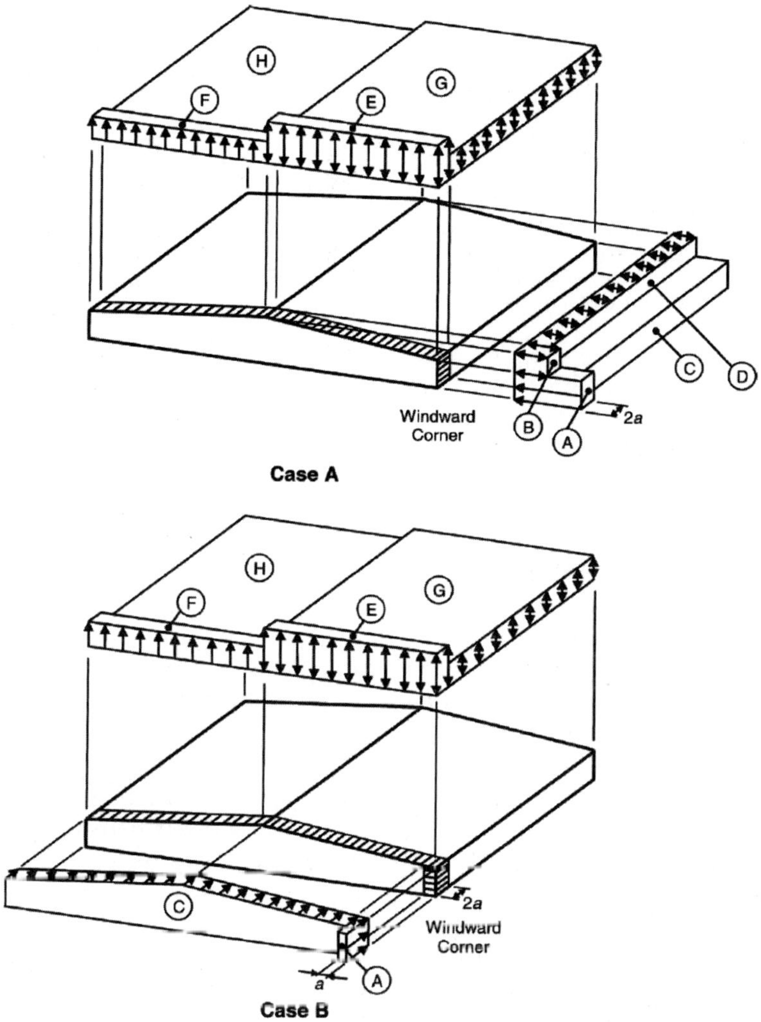

Figure 1.40 Application of wind pressures for low-rise buildings, Part 2 of Chapter 28.

Directional Procedure for Building Appurtenances and Other Structures (Chapter 29)

Scope Chapter 29 is applicable to the determination of wind loads on the MWFRSs of building appurtenances (including rooftop structure and rooftop equipment) and other structures (including solid freestanding walls, freestanding solid signs, chimneys, tanks, open signs, lattice framework, and trussed towers). The conditions of 29.1.2 and the limitations of 29.1.3 must be satisfied in order for these provisions to be applied.

Wind pressures are calculated using specific equations that are part of the Directional Procedure in Chapter 27.

Solid Freestanding Walls and Solid Signs

Solid Freestanding Walls and Solid Freestanding Signs The design wind for F for solid freestanding walls and solid freestanding signs is determined by Equation 29.3-1:

$$F = q_h G C_f A_s$$

In this equation, q_h is the velocity pressure determined by Equation 29.10-1 at height h defined in Figure 29.3-1.

The gust-effect factor G is determined from 26.11. The term A_s is the gross area of the solid freestanding wall or sign.

Net force coefficients C_f are given in Figure 29.3-1 as a function of the geometrical properties of the wall or sign. These coefficients are based on the results of boundary layer wind tunnel studies (see C29.3.1 for more information on these studies as well as for an equation to that can be used to determine C_f).

Three load cases must be investigated: Cases A, B, and C. Case A and Case B consider the resultant wind load acting normal through the geometric center of the wall or sign and normal through a point at an eccentricity from the geometric center, respectively. Case C accounts for the higher pressures that occur near the windward edge of a freestanding wall or sign when it is subjected to an oblique wind direction (i.e., at a direction that is not normal to the face of the wall or sign). Wind tunnel tests and full-scale test data have shown that the net pressures at the windward edges are significantly reduced where return corners are present. Reduction factors that can be used to account for this reduced pressure are given under Case C in the figure as a function of the return length.

Signs and walls are considered to be solid in cases where any openings comprise less than 30 percent of the gross area. Force coefficients

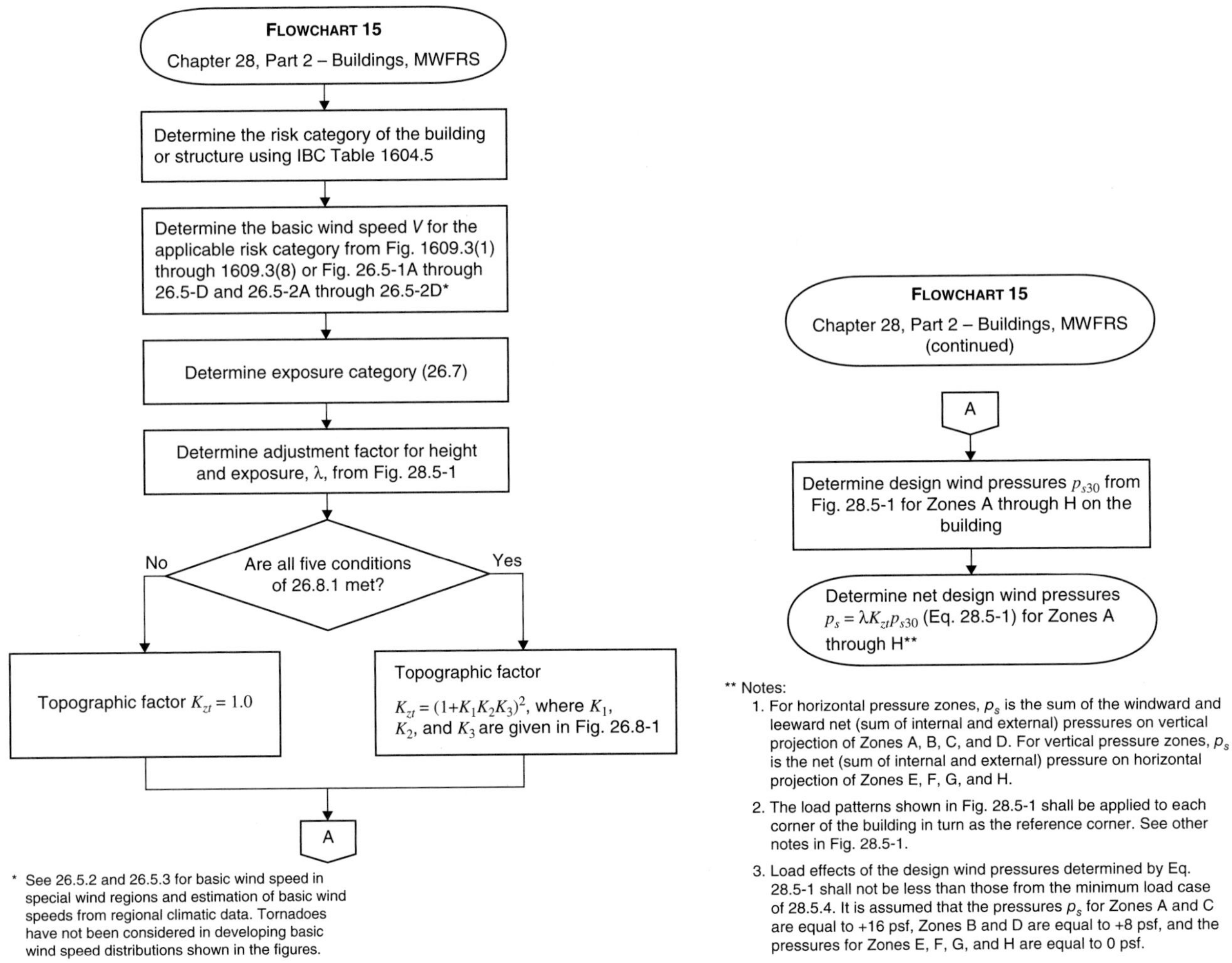

Figure 1.41 Design wind pressures on the MWFRS of buildings in accordance with Part 2 of Chapter 28 (Flowchart 15).

are permitted to be reduced by the factor in Note 2 of Figure 29.3-1 where such openings occur.

Solid Attached Signs The provisions of Chapter 30 (C&C) must be used to determine wind pressures on solid signs that are attached to the wall of a building that meet the following conditions:

- Plane of the sign is parallel to and in contact with the plane of the wall.
- The sign does not extend beyond the side or top edges of the wall.

This procedure is also applicable to signs that are not in direct contact with the wall provided that the gap between the sign and the wall is less than or equal to 3 feet and the edge of the sign is at least 3 feet from the edges of the wall (i.e., side and top edges and bottom edges of elevated walls). In any case, the internal pressure coefficient (GC_{pi}) is to be set equal to zero when calculating the pressures in accordance with Chapter 30. In essence, the attached sign should experience approximately the same external pressure as the wall to which it is attached.

Other Structures

The design wind force F for other structures—including chimneys, tanks, rooftop equipment for buildings with $h > 60$ feet, open signs, lattice frameworks, and trussed towers—is determined by Equation 29.4-1:

$$F = q_z G C_f A_f$$

In this equation, q_z is the velocity pressure determined by Equation 29.10-1 at height z of the centroid of the area A_f, and G is the gust-effect factor determined in accordance with 26.11. The area A_f is either the projected area normal to the wind or the actual surface area depending on how the force coefficient C_f is specified.

Force coefficients C_f are given in the following figures:

- Figure 29.4-1: Chimneys, tanks, rooftop equipment, and similar structures
- Figure 29.4-2: Open signs and lattice frameworks
- Figure 29.4-3: Trussed towers

The coefficients in Figures 29.4-1 and 29.4-2 are the same as those originally reported in Ref. 13.

Values of C_f in Figure 29.4-1 are given for square, hexagonal, and round cross sections as a function of the height to cross-sectional dimension of the section. It is important to note that this figure is valid for rooftop equipment on buildings that have a mean roof height greater than 60 feet; the requirements of 29.4-2 must be used in cases where $h \le 60$ feet.

The force coefficients in Figure 29.4-2 are applicable to open signs, that is, signs with openings comprising more than 30 percent of the gross area. Signs that do not meet this criterion are classified as solid, and the force coefficients in Figure 29.3-1 must be used.

Force coefficients for trussed towers with square and triangular cross sections are given in Figure 29.4-3. These simplified coefficients are consistent with those in Ref. 16.

Rooftop Structures and Equipment for Buildings with $h \le 60$ Feet

Both lateral and vertical wind loads must be determined for structures and equipment located on the rooftop of buildings with mean roof heights less than or equal to 60 feet. The lateral force F_h is determined by Equation 29.4-2:

$$F_h = q_h(GC_r)A_f$$

In this equation, q_h is the velocity pressure determined by Equation 29.10-1 evaluated at the mean roof height of the building, and A_f is the vertical projected area of the rooftop structure or equipment on a plane normal to the direction of wind (see Fig. 1.42, where $A_f = h_1B_1$).

The combined gust-effect factor and pressure coefficient (GC_r) accounts for higher wind pressures due to higher correlation of pressures across the structure surface, higher turbulence on the roof of the building, and accelerated wind speed on the roof.

Equation 29.4-3 is used to calculate the uplift force F_v on rooftop structures and equipment:

$$F_v = q_h(GC_r)A_r$$

This force is distributed over the horizontal projected area of the rooftop structure or equipment A_r, which is equal to L_1B_1 in Fig. 1.42.

Parapets

ASCE/SEI 29.5 refers to 27.3.5 and 28.3.2 for wind pressures on parapets for buildings of all heights and for low-rise buildings, respectively.

Roof Overhangs

ASCE/SEI 29.6 refers to 27.3.4 and 28.3.3 for wind pressures on parapets for buildings of all heights and for low-rise buildings, respectively.

Minimum Design Wind Loading

The minimum design wind load for other structures must be greater than or equal to 16 psf multiplied by the projected area normal to the wind A_f (29.7). This load case is to be applied as a separate load case in addition to the other load cases specified in Chapter 29.

Flowchart 16 (Fig. 1.43) can be used to determine design wind pressures on the MWFRS of other structures and building appurtenances in accordance with Chapter 29.

1.7.4 Components and Cladding

Overview

Chapter 30 contains design wind load provisions for C&C. These requirements may be used in the design of such elements provided that the conditions and limitations of 30.1.2 and 30.1.3 are satisfied. The following sections discuss the six parts that are contained in this chapter.

A summary of the wind load procedures and their applicability for C&C in accordance with Chapter 30 are given in Table 1.12.

Similar to MWFRSs, reduction in wind pressures due to apparent shielding of surrounding buildings, other structures, or terrain is not permitted in the design of C&C (30.1.4). Removal of such features around a building at a later date could result in wind pressures that are much higher than originally accounted for; as such, wind pressures must be calculated assuming that all shielding effects are not present.

Design wind pressures determined by Chapter 30 are permitted to be used in the design of air permeable roof or wall cladding. Examples of this type of cladding include siding, pressure-equalized rain screen walls, shingles, tiles, concrete roof pavers, and aggregate roof surfacing. In general, this type of cladding allows partial air pressure equalization between their exterior and interior surfaces. Additional information can be found in C30.1.5.

General Requirements

Minimum Design Wind Pressures According to 30.2.2, the design wind pressure for C&C shall not be less than a net pressure of 16 psf acting

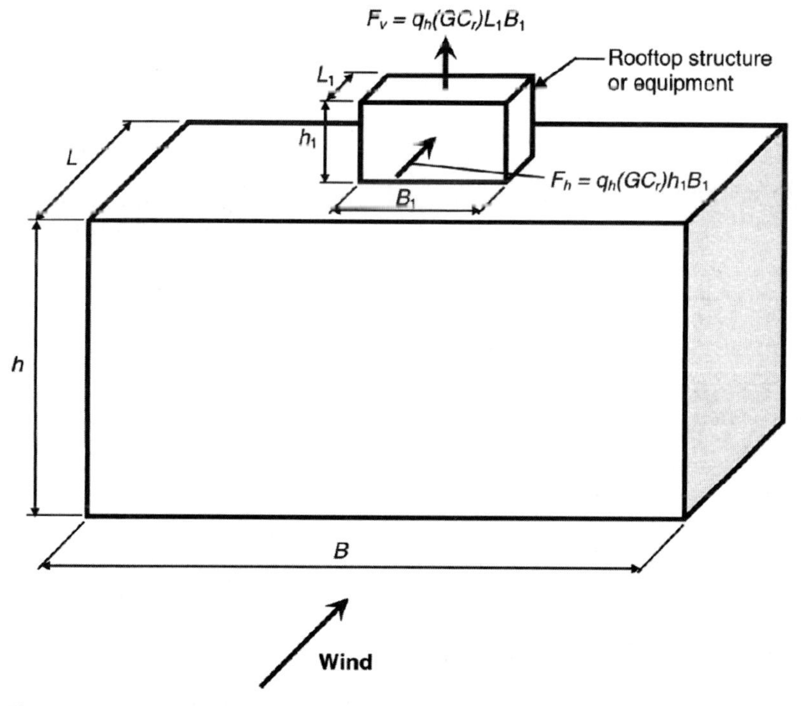

Figure 1.42 Lateral and vertical wind loads on rooftop structures and equipment.

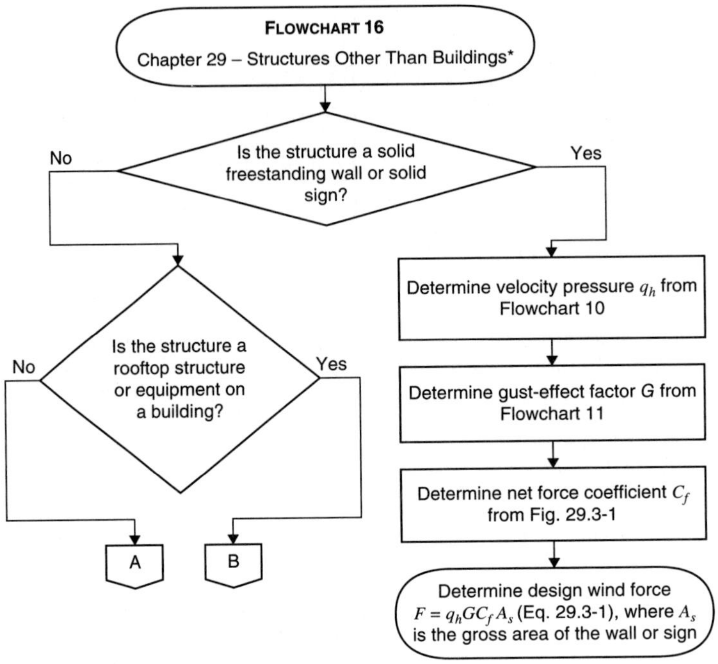

*See 29.4.2 for circular bins, silos, and tanks; 29.4.3 and 29.4.4 for rooftop solar panels; 29.5 for parapets; and 29.6 for roof overhangs.

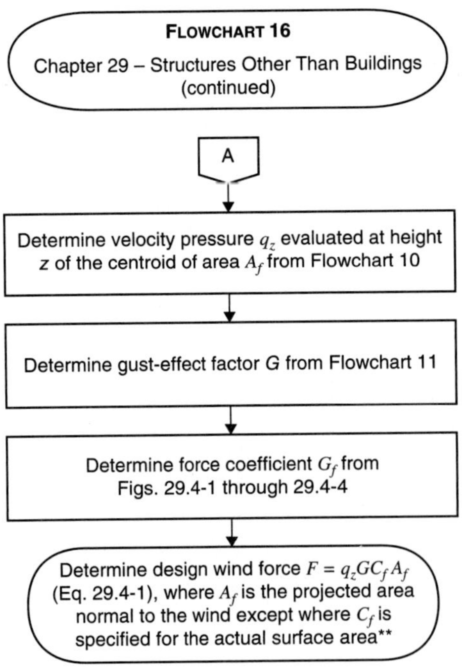

** The design wind force for other structures shall be not less than 16 psf multiplied by A_f (29.7).

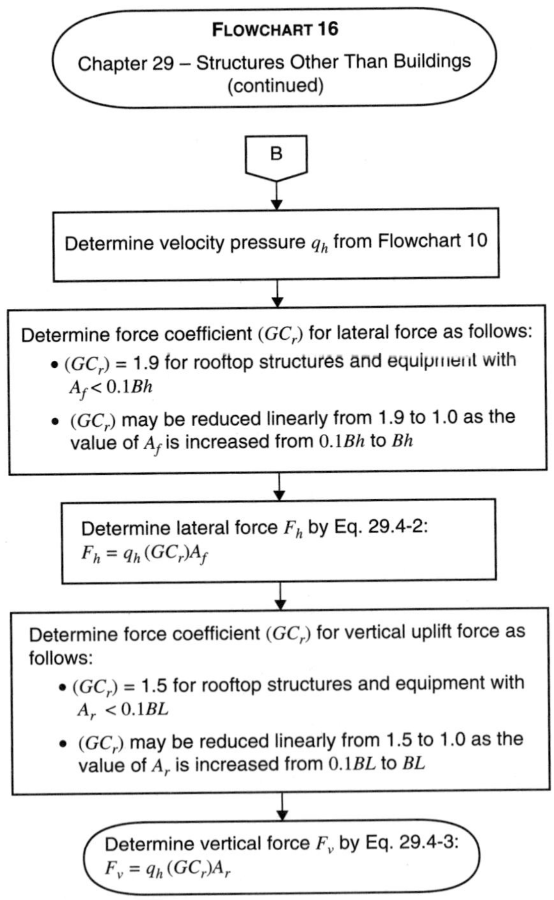

Figure 1.43 Design wind pressures on the MWFRS of other structures and building appurtenances in accordance with Chapter 29 (Flowchart 16).

Table 1.12 Summary of Wind Load Procedures in ASCE/SEI 7-16 for C&C

ASCE/SEI chapter	Part	Applicability: Building/element type	Applicability: Height limit	Conditions
30	1	Enclosed, low-rise Partially enclosed, low-rise Enclosed with $h \le 60$ feet Partially enclosed with $h \le 60$ feet	$h \le 60$ feet and $h \le$ least horizontal dimension $h \le 60$ feet	• Regular-shaped building • Building does not have response characteristics making it subject to across-wind loading, vortex shedding, instability due to galloping or flutter • Building is not located at a site where channeling effects or buffeting in the wake of upwind obstructions warrant special consideration • See additional conditions on selected figure(s) referenced in this part
	2	Enclosed, low-rise Enclosed with $h \le 60$ feet	$h \le 60$ feet and $h \le$ least horizontal dimension $h \le 60$ feet	• Same first three conditions as in Part 1 • Building has with either a flat roof, a gable roof with $\theta \le 45$ degrees, or a hip roof with $\theta \le 27$ degrees
	3	Enclosed Partially enclosed	$h > 60$ feet	• Same first three conditions as in Part 1 • See additional conditions on selected figure(s) referenced in this part
	4	Enclosed	$h \le 160$ feet	• Same first three conditions as in Part 1
	5	Open	None	• Same first three conditions as in Part 1 • See additional conditions on selected figure(s) referenced in this part
	6	Parapets	---	• Same first three conditions as in Part 1 • All building types except enclosed buildings with $h \le 160$ feet for which the provisions of Part 4 are used
		Roof overhangs	---	• Same first three conditions as in Part 1 • All building types except enclosed buildings with $h \le 160$ feet for which the provisions of Part 4 are used
		Roof structures and equipment	$h \le 60$ feet	• Same first three conditions as in Part 1

in either direction (positive or negative) normal to the surface. Like in the case of MWFRSs, this is a load case that needs to be considered in addition to the other required load cases in this chapter.

Tributary Areas Greater Than 700 Square Feet C&C elements that support a tributary area greater than 700 square feet are permitted to be designed for wind pressures using the provisions for MWFRSs. The 700-square-foot tributary area is deemed sufficiently large enough so that the localized wind effects are not pronounced as is the case of C&C; as such, the wind pressures on these elements are essentially equal to those determined by the method for MWFRSs.

External Pressure Coefficients Numerous figures are provided in this chapter that give values for the combined gust-effect factor and pressure coefficient (GC_p) for C&C. The gust-effect factor and pressure coefficients are not permitted to be separated (30.2.4). Additional information on these coefficients is given in the following sections.

Velocity Pressure The velocity pressure q_z evaluated at height z is determined by Equation 26.10-1. Additional information on how to determine q_z is given in Sec. 1.7.3 of this chapter.

Envelope Procedures

Scope Parts 1 and 2 in 30.3 and 30.4, respectively, contain methods to determine wind loads on C&C of low-rise buildings that meet the conditions of 30.3.1 and 30.4.1, respectively. Both of these parts are based on the Envelope Procedure in Chapter 28. Different methods were utilized in the development of these procedures than those that were developed for MWFRSs. Additional information is provided in the following sections.

Part 1—Low-Rise Buildings

Overview Part 1 is applicable to enclosed and partially enclosed low-rise buildings and buildings with a mean roof height less than or equal to 60 feet.

This procedure entails the determination of the velocity pressure at the mean roof height of the building (which is determined as a function of exposure, topographic effects, wind directionality, wind velocity, and building occupancy), combined gust-effect factors, and external pressure coefficients and internal pressure coefficients. Design wind pressures are obtained for various designated zones on the walls and roof of buildings.

Table 30.3-1 contains the overall steps that can be used to determine wind pressures on C&C of such buildings.

Design Wind Pressures Design wind pressures p are determined by Equation 30.3-1:

$$p = q_h[(GC_p) - (GC_{pi})]$$

In this equation, q_h is the velocity pressure evaluated at the mean roof height of the building as determined by Equation 26.10-1, and (GC_{pi}) are the internal pressure coefficients given in Table 26.13-1. Combined gust-effect factor and pressure coefficients (GC_p) are given in the following figures for walls and roofs:

- Walls: Figure 30.3-1
- Roofs
  - Flat, gable, and hip roofs: Figures 30.3-2A through 30.3-2C
  - Stepped roofs: Figure 30.3-3
  - Multispan gable roofs: Figure 30.3-4
  - Monoslope roofs: Figures 30.3-5A and 30.3-5B
  - Sawtooth roofs: Figure 30.3-6
  - Domed roofs: Figure 30.3-7
  - Arched roofs: Figure 27.3-3, Note 4

Except for domed roofs, external pressure coefficients are provided in these figures for various surfaces of the walls and roofs as a function of the *effective wind area*, which is defined in 26.2. This area does not necessarily correspond to the area of the building surface that contributes to the load under consideration. Two cases can arise when determining the effective wind area of C&C. In the first case, the effective wind area is equal to the total area that is tributary to the element. The second case occurs where components are spaced relatively closely together (such as wall studs or roof trusses). Due to the close spacing, the load is distributed and shared amongst adjoining components. To account for this load distribution, an effective width equal to one-third the span length is used in determining the effective wind area, which in this case is equal to the span length multiplied by the effective width. This area is usually greater than the tributary area of the component, which is generally long and narrow. The larger of the effective areas from two these cases is used in determining the pressure coefficients.

Consider the basic representation of a glazing system depicted in Fig. 1.44. Glazing panels that have a height h_2 span between mullions that are spaced a distance s on center. The mullions span the full story

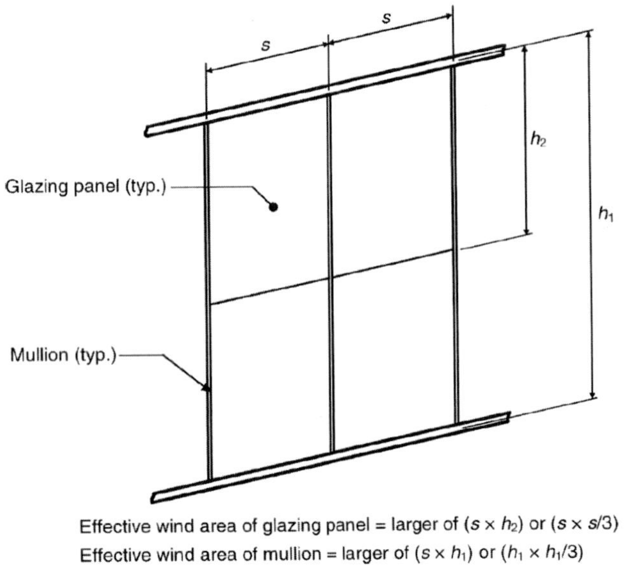

Figure 1.44 Effective wind areas for C&C.

height h_1 between floor slabs. The following are the effective wind areas for these two components:

- Glazing panel: Larger of ($s \times h_2$) or ($s \times s/3$)
- Mullion: Larger of ($s \times h_1$) or ($h_1 \times h_1/3$)

For both components, the first area is the area tributary to the component, and the second is that associated with the effective width. The effective wind area, which is defined as the larger of these two areas, depends on the relative magnitudes of the spans and heights for both components.

Flowchart 17 (Fig. 1.45) can be used to determine design wind pressures on C&C of buildings in accordance with Part 1 of Chapter 30.

Part 2—Low-Rise Buildings, Simplified

Overview The simplified method in Part 2 is applicable to enclosed low-rise buildings and buildings with a mean roof height less than or equal to 60 feet with flat, gable or hip roofs. Wind pressures on C&C located at various building surfaces are determined from the tabulated values in Figure 30.4-1 for a building with a mean roof height of 30 feet located at a site classified as Exposure B. These pressures are adjusted for different building heights and exposures using the adjustment factor in the figure.

Table 30.4-1 contains the overall steps that can be used to determine wind pressures on C&C of such buildings.

Design Wind Pressures Simplified design wind pressures p_{net30} for C&C located at various zones on a building are tabulated in Figure 30.4-1 as a function of the basic wind speed V and roof angle for buildings with a mean roof height of 30 feet that are located on primarily flat ground in Exposure B. Modifications are made to these tabulated pressures based on actual building height and exposure using the adjustment factor λ that is given in the figure. Such pressures must also be modified by the topographic factor K_{zt} evaluated at $0.33h$ where applicable.

Net design wind pressure p_{net} is determined by Equation 30.4-1:

$$p_{\text{net}} = \lambda K_{zt} p_{\text{net30}}$$

The methodology of determining the wind pressures on C&C is essentially the same as that for MWFRS. The tabulated wind pressures in the shaded areas of Figure 30.4-1 are less than the minimum wind pressure of 16 psf prescribed in 30.2.2. The note at the bottom of this figure points out that the final net pressure determined by the equation above must be greater than or equal to 16 psf.

FLOWCHART 17
Chapter 30, Part 1 – Buildings, C&C*

↓

Determine velocity pressure q_h from Flowchart 10

↓

Determine external pressure coefficient (GC_p) for zones on the walls and roof from Figs. 30.3-1 through 30.3-7 and 27.3-3 based on the effective wind area**

↓

Determine internal pressure coefficients (GC_{pi}) from Table 26.13-1 based on enclosure classification

↓

Determine design wind pressures by Eq. 30.3-1:
$p = q_h[(GC_p) - (GC_{pi})]^‡$

* C&C elements with tributary areas greater than 700 square feet may be designed using the provisions for MWFRSs (30.2.3).

** See 26.2 for definition of effective wind area.

‡ Minimum wind pressures of 30.2.2 must also be considered.

Figure 1.45 Design wind pressures on C&C of buildings in accordance with Part 1 of Chapter 30 (Flowchart 17).

Flowchart 18 in Fig. 1.46 can be used to determine design wind pressures on C&C of buildings in accordance with Part 2 of Chapter 30.

Directional Procedures

Scope Parts 3 through 7 contain methods to determine wind loads on C&C of buildings, parapets, roof overhangs, building appurtenances, rooftop structures and equipment, attached canopies, and nonbuilding structures that meet the conditions in Table 1.12. These parts are based on the Directional Procedure in Chapter 27.

Part 3—Buildings with $h > 60$ Feet

Overview Part 3 is applicable to enclosed or partially enclosed buildings with a mean roof height greater than 60 feet with various types of roof shapes. Design wind pressures are calculated using the appropriate external pressure coefficients for walls and roofs given in Figures 30.5-1, 27.3-3, and 30.3-7.

This procedure entails the determination of the velocity pressure (which is determined as a function of exposure, height, topographic effects, wind directionality, wind velocity, and building occupancy), combined gust-effect factors, and external and internal pressure coefficients. Design wind pressures are obtained for designated zones on the surfaces of buildings.

Table 30.5-1 contains the overall steps that can be used to determine wind pressures on C&C of buildings designed by this method.

Design Wind Pressures Design wind pressures p on C&C are determined by Equation 30.5-1:

$$p = q(GC_p) - q_i(GC_{pi})$$

All of the terms in this equation, except for the external pressure coefficients (GC_p), are the same as those for MWFRSs. External pressure coefficients can be found in Figure 30.5-1 for walls and flat roofs, in Note 4 of Figure 27.3-3 for arched roofs, in Figure 30.3-7 for domed roofs, and in Note 6 of Figure 30.5-1 for other roof angles and geometries.

The exception in 30.5.2 permits design wind pressures on the C&C of buildings with mean roof heights greater than 60 feet and less than

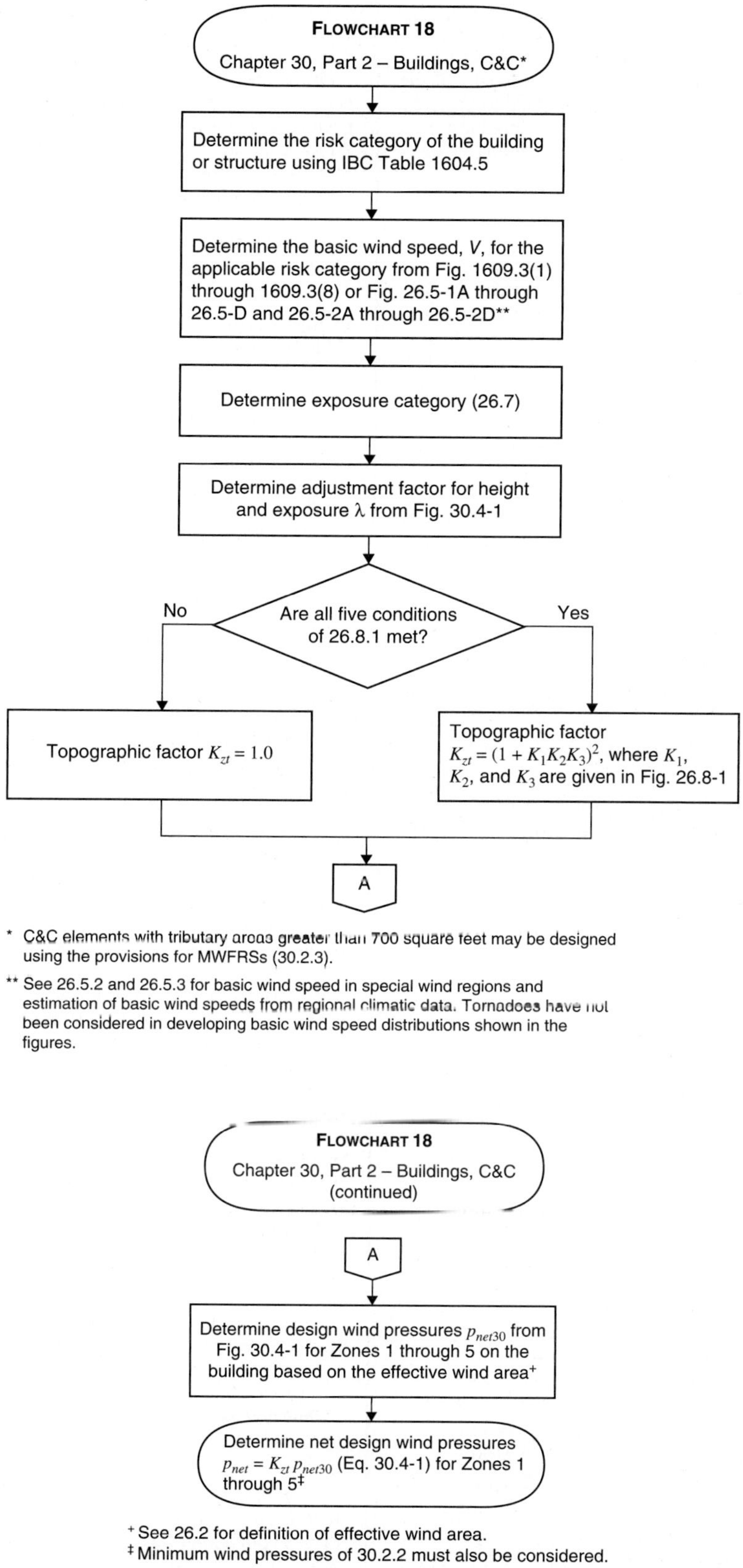

Figure 1.46 Design wind pressures on C&C of buildings in accordance with Part 2 of Chapter 30 (Flowchart 18).

90 feet to be determined using the external pressures coefficients (GC_p) from Figures 30.3-1 through 30.3-6 in Part 1 provided that the height-to-width ratio of the building is less than or equal to one. The external pressure coefficients from the envelope procedure have been deemed adequate to predict design wind pressures for buildings of these proportions.

Flowchart 19 (Fig. 1.47) can be used to determine design wind pressures on C&C of buildings in accordance with Part 3 of Chapter 30.

Part 4—Buildings with 60 Feet < h ≤ 160 Feet

Overview Part 4 provides a simplified method of determining wind loads on C&C of enclosed buildings with a mean roof height less than or equal to 160 feet. Wind pressures on C&C located on various surfaces can be read directly from Table 30.7-2 for a building site classified as Exposure C and an effective wind area of 10 square feet. These pressures are modified by an effective area reduction factor, exposure adjustment factor, and the topographic factor where applicable (see Table 30.7-2 and Equation 30.7-1).

Table 30.7-1 contains the overall steps that can be used to determine wind pressures on C&C of buildings designed by this method.

Design Wind Pressures

Wall and Roof Surfaces Design wind pressures on designated zones of wall and roof surfaces are determined from Table 30.7-2 as a function of the basic wind speed V, mean roof height, and roof angle for buildings that are located on primarily flat ground in Exposure C. These tabulated pressures are valid for an effective wind area of 10 square feet and have been determined using the applicable external pressure coefficients from Part 3 (namely, Figure 30.6-1 for flat roofs; Figure 30.4-2A, 2B, and 2C for gable and hip roofs; and Figure 30.4-5A and 5B for monoslope roofs) and an internal pressure coefficient of ±0.18 for enclosed buildings. Modifications are made to these tabulated pressures based on the actual exposure and effective wind area. Design wind pressures p are determined by Equation 30.7-1:

$$p = p_{\text{table}}(EAF)(RF)K_{ZT}$$

In this equation, (EAF) is the exposure adjustment factor given in Table 30.7-2, which modifies the tabulated pressures in cases where the exposure at the site is different than Exposure C.

The effective area reduction factor (RF) is also given in Table 30.7-2 and modifies the tabulated pressures for effective wind areas greater than 10 square feet. Reduction factors, which are based on the graphs of the external pressure coefficients in the figures in Part 3, are provided for designated zones on walls and roofs for five different roof shapes and for roof overhangs.

Flowchart 20 (Fig. 1.48) can be used to determine design wind pressures on C&C of wall and roof surfaces in accordance with Part 4 of Chapter 30.

Parapets Equation 30.7-1 in conjunction with the applicable edge and corner pressures p_{table} in Table 30.7-2 are used to determine parapet C&C wind pressures. Pressures are applied to the parapet in accordance with Figure 30.7-1. Two load cases must be considered, namely, Load Case A, which is applicable to windward parapets, and Load Case B, which is applicable to leeward parapets:

- Load Case A: Positive wall pressure from Zone 4 or 5 is applied to the windward (front) surface of the parapet, and negative roof pressure from Zone 2 or 3 is applied to the leeward (back) surface of the parapet.
- Load Case B: Positive wall pressure from Zone 4 or 5 is applied to the windward (back) surface of the parapet, and negative wall pressure from Zone 4 or 5 is applied to the leeward (front) surface of the parapet.

The height that is to be used in determining p_{table} from Table 30.7-2 is the height to the top of the parapet. The pressures obtained by this method are slightly conservative compared to those determined by Part 3.

Roof Overhangs Equation 30.7-1 in conjunction with the applicable pressures p_{table} in Table 30.7-2 are used to determine roof overhang C&C wind pressures. Pressures are applied to the overhang in accordance with Figure 30.7-2.

In Zones 1 and 2, the pressures on the top surface of the roof overhang are set equal to the tabulated roof pressures in those zones, and in Zone 3, the pressures are set equal to 1.15 times the tabulated roof pressures in this zone. On the underside of the overhang, the pressure is set equal to the adjacent wall pressure. In all cases, these pressures are slightly conservative and are based on the external pressure coefficients in Figures 30.4-2A to 30.4-2C of Part 3.

Part 5—Open Buildings

Overview Part 5 is applicable to open buildings with pitched, monoslope, or troughed roofs. Net design wind pressures on C&C are determined by Equation 30.8-1 using the appropriate net pressure coefficients in Figures 30.8-1, 30.8-2, and 30.8-3.

Table 30.8-1 contains the overall steps that can be used to determine wind pressures on C&C of buildings designed by this method.

Design Wind Pressures Net design wind pressures on the C&C of open buildings are determined by Equation 30.8-1:

$$p = q_h GC_N$$

In this equation, q_h is velocity pressure evaluated at the mean roof height of the building, and G is the gust-effect factor determined in accordance with 26.9.

The net pressure coefficients C_N are given in Figure 30.8-1 for monosloped roofs, Figure 30.8-2 for pitched roofs, and Figure 30.8-3

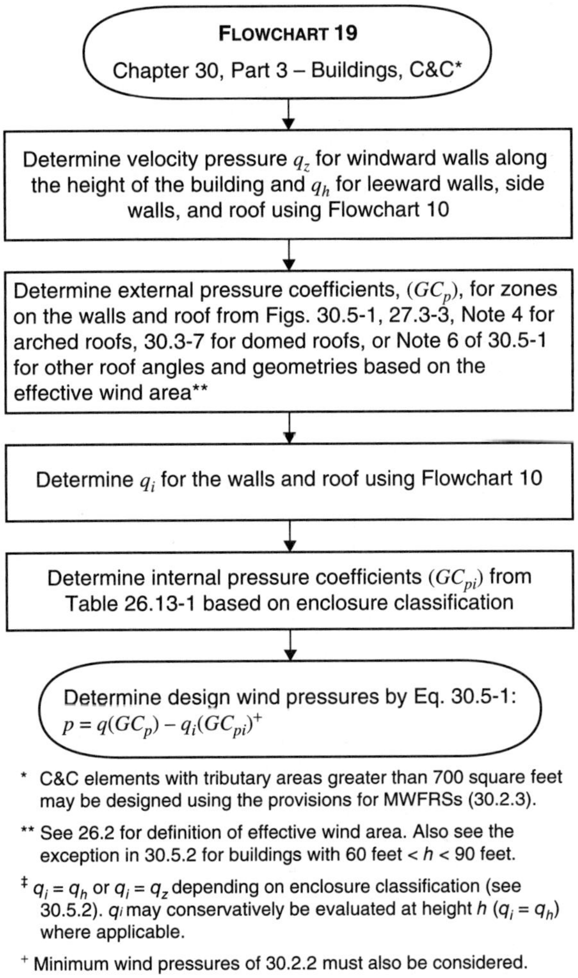

Figure 1.47 Design wind pressures on C&C of buildings in accordance with Part 3 of Chapter 30 (Flowchart 19).

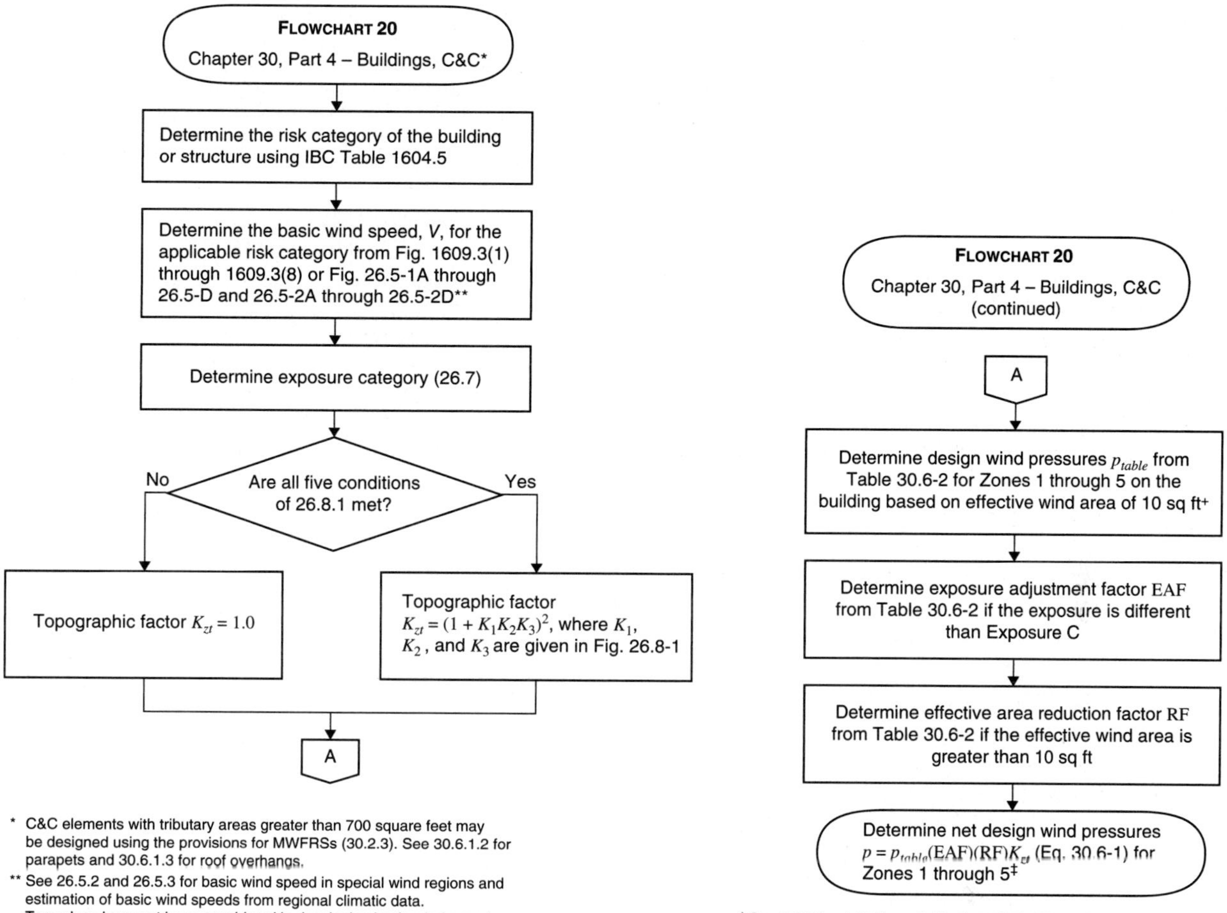

* C&C elements with tributary areas greater than 700 square feet may be designed using the provisions for MWFRSs (30.2.3). See 30.6.1.2 for parapets and 30.6.1.3 for roof overhangs.

** See 26.5.2 and 26.5.3 for basic wind speed in special wind regions and estimation of basic wind speeds from regional climatic data. Tornadoes have not been considered in developing basic wind speed distributions shown in the figures.

† See 26.2 for definition of effective wind area.

‡ Minimum wind pressures of 30.2.2 must also be considered.

Figure 1.48 Design wind pressures on C&C of wall and roof surfaces in accordance with Part 4 of Chapter 30 (Flowchart 20).

for troughed roofs based on the effective wind area. These coefficients include contributions from both the top and bottom surfaces of the roof, which implies that the element receives pressure from both surfaces.

Flowchart 21 in Fig. 1.49 can be used to determine design wind pressures on C&C of wall and roof surfaces in accordance with Part 5 of Chapter 30.

Part 6—Building Appurtenances and Rooftop Structures and Equipment

Overview Part 6 contains methods to determine wind pressures on C&C of parapets (30.9), roof overhangs (30.10), and rooftop structures and equipment for buildings with $h \leq 60$ feet (30.11).

This procedure entails the determination of velocity pressure (which is determined as a function of exposure, height, topographic effects, wind directionality, wind velocity, and building occupancy), combined gust-effect factors, and external pressure coefficients and internal pressure coefficients. Design wind pressures are obtained for designated zones on the surfaces of buildings.

Design Wind Pressures

Parapets Design wind pressures p for C&C elements of parapets for all building types and heights, except for enclosed buildings with a mean roof height less than or equal to 160 feet (see Part 4) are calculated by Equation 30.9-1:

$$p = q_p[(GC_p) - (GC_{pi})]$$

In this equation, q_p is the velocity pressure evaluated at the top of the parapet, and (GC_{pi}) are the internal pressure coefficients from Table 26.11-1 based on the porosity of the envelope of the parapet. The external pressure coefficients (GC_p) are determined from the same figures as those in Parts 1 and 3 for walls and various roof configurations (see 30.9 for a comprehensive list of applicable figures).

Similar to the requirements of Part 4, Load Case A and Load Case B must be considered for the windward and leeward parapets, respectively. Figure 30.9-1, which illustrates these load case, is essentially the same as Figure 30.7-1 in Part 4 with the exception that the pressures in Part 6 must be determined from the applicable figures noted in 30.9.

Table 30.9-1 contains the overall steps that can be used to determine wind pressures on C&C of parapets designed by this method.

Flowchart 22 (Fig. 1.50) can be used to determine design wind pressures on C&C of parapets in accordance with Part 6 of Chapter 30.

Roof overhangs. Design wind pressures p for C&C elements of roof overhangs are determined by Equation 30.10-1:

$$p = q_h[(GC_p) - (GC_{pi})]$$

In this equation, q_h is the velocity pressure measured at the mean roof height, and (GC_{pi}) are the internal pressure coefficients from Table 26.11-1.

The external pressure coefficients (GC_p) for overhangs are given in Figures 30.4-2A to 30.4-2C (flat roofs, gable roofs, and hip roofs,

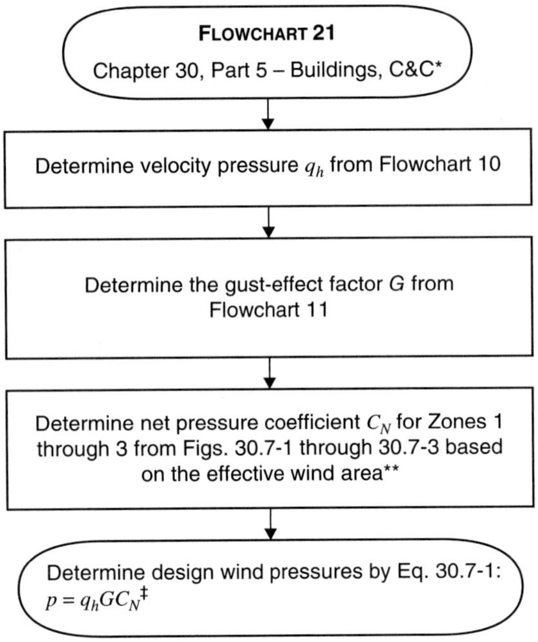

* C&C elements with tributary areas greater than 700 square feet may be designed using the provisions for MWFRSs (30.2.3).
** See 26.2 for definition of effective wind area.
‡ Minimum wind pressures of 30.2.2 must also be considered.

Figure 1.49 Design wind pressures on C&C of wall and roof surfaces in accordance with Part 5 of Chapter 30 (Flowchart 21).

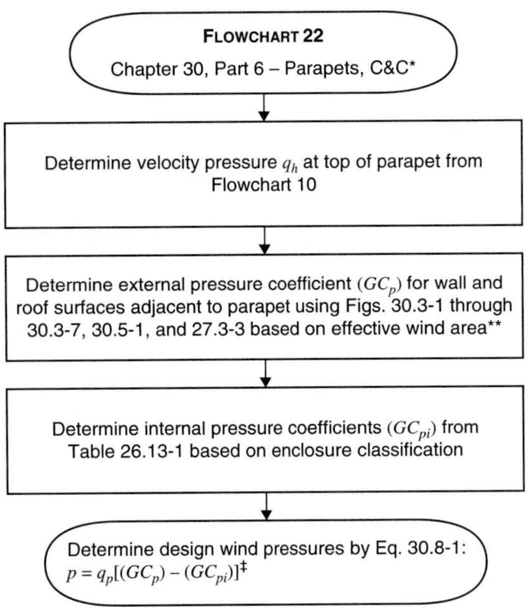

* C&C elements with tributary areas greater than 700 square feet may be designed using the provisions for MWFRSs (30.2.3).
** See 26.2 for definition of effective wind area.
‡ Two load cases must be considered (see Fig. 30.8-1). Also, minimum wind pressures of 30.2.2 must also be considered.

Figure 1.50 Design wind pressures on C&C of parapets in accordance with Part 6 of Chapter 30. (Flowchart 22).

respectively) and include contributions from top and bottom surfaces of overhang.

Table 30.10-1 contains the overall steps that can be used to determine wind pressures on C&C of roof overhangs designed by this method.

Flowchart 23 (Fig. 1.51) can be used to determine design wind pressures on C&C of roof overhangs in accordance with Part 6 of Chapter 30. Flowchart 24 (Fig. 1.52) can be used to determine the design wind pressures on C&C of attached canopies.

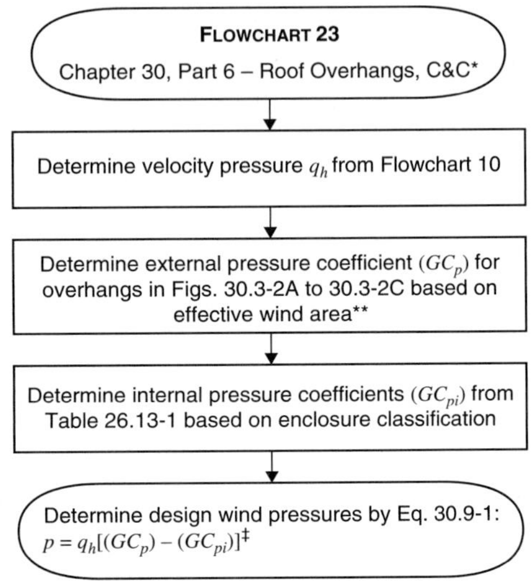

* C&C elements with tributary areas greater than 700 square feet may be designed using the provisions for MWFRSs (30.2.3).
** See 26.2 for definition of effective wind area.
‡ Minimum wind pressures of 30.2.2 must also be considered.

Figure 1.51 Design wind pressures on C&C of roof overhangs in accordance with Part 6 of Chapter 30 (Flowchart 23).

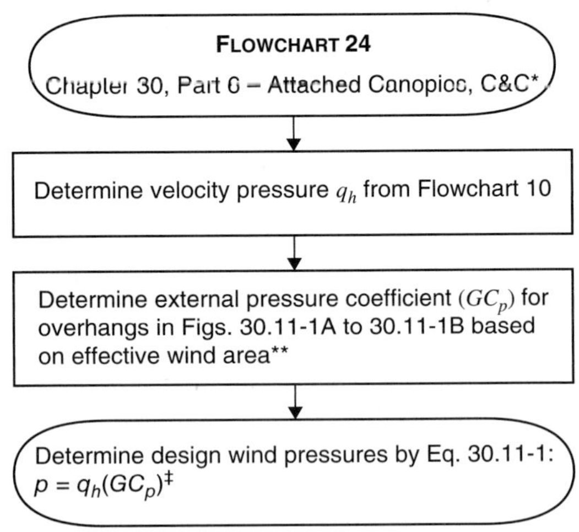

* C&C elements with tributary areas greater than 700 square feet may be designed using the provisions for MWFRSs (30.2.3).
** See 26.2 for definition of effective wind area.
‡ Minimum wind pressures of 30.2.2 must also be considered.

Figure 1.52 Design wind pressures on C&C of attached canopies (Flowchart 24).

Rooftop Structures and Equipment for Buildings with h ≤ 60 Feet The provisions of 29.5.1 that are applicable in the design of MWFRSs are to be used in determining the design wind pressures on the C&C of rooftop structures and equipment for buildings with a mean roof height less than or equal to 60 feet.

The design wind pressures on the walls and roof are determined by dividing the horizontal and vertical wind loads obtained from Equations 28.5-1 and 29.3-1, respectively, by the corresponding projected areas of the structure or equipment. The wall pressures can act inward or outward while the roof pressure acts outward.

1.7.5 Wind Tunnel Procedure

Overview

The Wind Tunnel Procedure in Chapter 31 can be utilized to determine wind loads on MWFRSs and C&C of any building or other structure in lieu of any of the procedures in the Chapters 27 through 30 and must be used where the conditions of these procedures are not satisfied (in particular, where a structure contains any of the characteristics defined in 27.1.3, 28.1.3, 29.1.3, or 30.1.3).

Wind tunnel tests should be seriously considered where buildings or other structures are not regularly shaped, are flexible and/or slender, have the potential to be buffeted by upwind buildings or other structures, or have the potential to be subjected to accelerated wind flow from channeling by buildings or topographic features. In the case of tall, slender buildings, only a wind tunnel test can properly capture any possible effects due to vortex shedding, galloping, or flutter. For buildings in the heart of a city, a wind tunnel test is mandatory since Exposures B through D cannot properly capture the conditions in such cases. Every project has its own unique characteristics, and engineering judgment also plays a role in the decision-making process. When determining whether a wind tunnel test is required, it is always very important to keep in mind the limitations in Chapters 27 through 30, especially the general one related to along-wind response.

Of all of the methods that are contained in ASCE/SEI 7-16, the wind tunnel procedure is generally considered to produce the most accurate results. For certain types of buildings, the results from a wind tunnel test will be significantly smaller than those from any of the other methods. On the other hand, wind tunnel tests will yield results that are greater than those obtained from the other methods under certain conditions; as such, it is important to understand when such tests are required in order to adequately design the building or other structure for the effects of wind.

Information on the three basic types of wind tunnel test models that are commonly used is given in C31. Wind tunnel tests can also provide valuable information on snow loads, the effects of wind on pedestrians, and concentrations of air-pollutant emissions, to name a few. References are provided in C31 that provide detailed information and guidance for the determination of wind loads and other types of design data by wind tunnel tests.

Test Conditions

Basic requirements for test conditions of wind tunnel tests or any other tests that employ a fluid other than air are given in 31.2. These seven conditions must be followed when any such test is conducted.

Additional information on the basic procedures of conducting a wind tunnel test can be found in the references in C31.

Dynamic Response

The test conditions of 31.2 are to be used when determining the dynamic response of a building or other structure. Mass distribution, stiffness, and damping must be properly accounted for in the model and in the subsequent analysis.

Load Effects

ASCE/SEI 31.4.2 prescribes limitations on the wind speed used in the tests, and 31.4.3 gives lower limits on the magnitude of the principal loads that are to be applied to a building or structure for both the MWFRS and C&C. Two conditions are given that permit the limiting values to be reduced. C31 provides a comprehensive discussion on the reasons behind these limitations.

1.8 EARTHQUAKE LOADS

See Chap. 5 of this handbook.

1.9 FLOOD AND TSUNAMI LOADS

1.9.1 Overview of Flood and Tsunami Loads

In general, flooding is the overflow of excess water from a body of water (river, stream, lake, ocean, etc.) onto adjoining land. Depending on local topography, one or more bodies of water can contribute to flooding at a particular site. Storms usually generate the most significant flood hazards in areas other than those where tsunamis need to be considered.

IBC 202 defines flooding as a general or temporary condition of partial or complete inundation of normally dry land from

1. the overflow of inland or tidal waters or
2. the unusual or rapid accumulation or runoff of surface waters from any source.

Along coastlines and the shorelines of large lakes, flooding is caused by wind-driven surges and waves that push water onshore, while along streams and rivers, flooding results from the accumulation of rainfall runoff that drains from upland watersheds.

In order to correctly characterize the potential hazards caused by floodwaters, the following parameters must be investigated at any site regardless of the source:

- Origin of flooding
- Frequency of flooding
- Depth of floodwaters
- Velocity of floodwaters
- Direction of floodwaters
- Duration of flooding
- Effects due to waves
- Effects due to erosion and scour
- Effects due to flood-borne debris

It is evident from the aforementioned parameters that a variety of load types can be exerted on buildings due to flooding. The effects of these loads can be intensified by erosion and scour, which can lower the ground surface around foundations, causing loss of bearing capacity and resistance to uplift due to lateral loads. Because it is not always possible to quantify each of the above parameters for a particular site, conservative estimates need to be made in order to define the corresponding flood loads.

A tsunami is basically a series of waves with variable long periods, which typically result from tsunamigenic earthquakes. These types of earthquakes occur along major subduction zone plate boundaries, such as those bordering the Pacific Ocean, where a tectonic plate is subducting under an adjoining plate (see Figure C6.7-2). Tsunamis can also be generated by landslides near a coast or by underwater and undersea volcanic eruptions.

The effects on structures due to this coastal flooding hazard can be catastrophic, especially for one- and two-family dwellings and low-rise buildings of light-frame construction. According to C6.1, it is not practical to include tsunami design requirements for these types of low-rise structures because, based on forensic engineering surveys of major tsunami events of the past decade, such structures do not survive significant tsunami loading.

Tsunami loads and effects on structures are based on inundation depth and flow velocities of water during inflow and outflow stages at the site. The types of loads that need to be considered are like those required for flooding, as noted above. Because the characteristics of future tsunamis cannot be adequately determined using only historical records, a probabilistic design approach is used instead of a deterministic one.

1.9.2 Overview of Code Requirements

All structures and portions of structures located in flood hazard areas must be designed and constructed to resist the effects of flood hazards and flood loads (IBC 1612.1). IBC 1612.2 requires that the design and construction of buildings and structures located in flood hazard areas, including Coastal High Hazard Areas and Coastal A Zones, must be in accordance with Chapter 5 of ASCE/SEI 7-16 and Ref. 11. Where a building or structure is located in more than one flood zone or is partially located in a flood zone, the entire building or structure must be designed and constructed according to the requirements of the more restrictive flood zone.

In the case of tsunamis, IBC 1615.1 requires that the design and construction of Risk Category III and IV buildings and structures located in the Tsunami Design Zones defined in the Tsunami Design Geodatabase must be in accordance with Chapter 6 of ASCE/SEI 7-16. Appendix M in the IBC provides tsunami vertical evacuation planning criteria for those coastal communities where a tsunami hazard, as shown in a Tsunami Design Zone Map, is evident.

Hazards and loads that need to be considered for structures located in flood hazard areas and in tsunami-prone regions are given in the following sections.

Flood Loads

Flood Hazard Areas The first step in the design for flood loads is to determine if a building or structure is located in a flood hazard area or not. By definition, a *flood hazard area* is the greater of the following two areas (IBC 202):

1. The area within a floodplain subject to a 1 percent or greater chance of flooding in any year
2. The area designated as a flood hazard area on a community's flood hazard map or otherwise legally designated

The first of these two areas is typically acquired from Flood Insurance Rate Maps (FIRMs), which are prepared by the Federal Emergency Management Agency (FEMA) through the National Flood Insurance Program (NFIP). Conducting flood hazard studies and providing FIRMs and Flood Insurance Studies (FISs) for participating communities are major activities undertaken by the NFIP. Included in FISs are the FIRM, the Flood Boundary and Floodway Map (FBFM), the base flood elevation (BFE), and supporting technical data.

A FIRM is the official map of a community on which FEMA has delineated both the special flood hazard areas and the risk premium zones that are applicable to the community. In general, FIRMs show flood hazard areas along bodies of water where there is a risk of flooding by a base flood, that is, a flood having a 1 percent chance of being equaled or exceeded in any given year. Sample FIRMs are shown in Figure C1-3 of Ref. 11 and Fig. 1.53.

A FIRM that is produced in a digital format is designated a Digital Flood Insurance Rate Map (DFIRM). A sample DFIRM is given in Fig. 1.53. FIRMs and DFIRMs for specific areas can be obtained from the FEMA Flood Map Service Center website (https://msc.fema.gov/portal).

In regards to the return period for floods, the term "100-year flood" is a misnomer. Contrary to popular belief, it is not the flood that will occur once every 100 years but rather the flood elevation that has a 1 percent chance of being equaled or exceeded in any given year. The "100-year flood" could occur more than once in a relatively short period of time. The flood elevation that has a 1 percent chance of being equaled or exceeded in any given year is the standard used by most government agencies and the NFIP for floodplain management and to determine the need for flood insurance.

In addition to showing the extent of flood hazards, the FIRMs and DFIRMs also show BFEs and floodways. The BFE is the height to which floodwaters will rise during passage or occurrence of the base flood relative to the datum that is used on the flood hazard map. Statistical methods and computer methods that take into account the shape and nature of the floodplain (ground contours and the presence of any buildings, bridges, and culverts) were used by FEMA to obtain the BFEs along rivers and streams. In such cases, the BFEs are provided next to river cross sections on the flood hazard maps (see Fig. 1.54). Along coastal areas, BFEs include wave heights and are established considering historical storm and wind patterns. The BFEs for coastal and lake areas are shown in parentheses immediately below the flood zone on the flood hazard maps (see Figs. 1.53 and 1.54).

Floodways are channels of a river, creek or other watercourse and adjacent land areas that must be reserved in order to discharge the base floodwaters without cumulatively increasing the water surface elevation by more than a designated height. As such, floodways must be kept clear of encroachments (such as fill or buildings) so that the base flood can be discharged without increasing the water surface elevations by more than the designated height given in floodway data tables in FISs. A floodway schematic is given in Figure C1-2 of Ref. 11.

Some local jurisdictions develop and subsequently adopt flood hazard maps that are more extensive than FEMA maps. In such cases, flood design and construction requirements must be satisfied in the

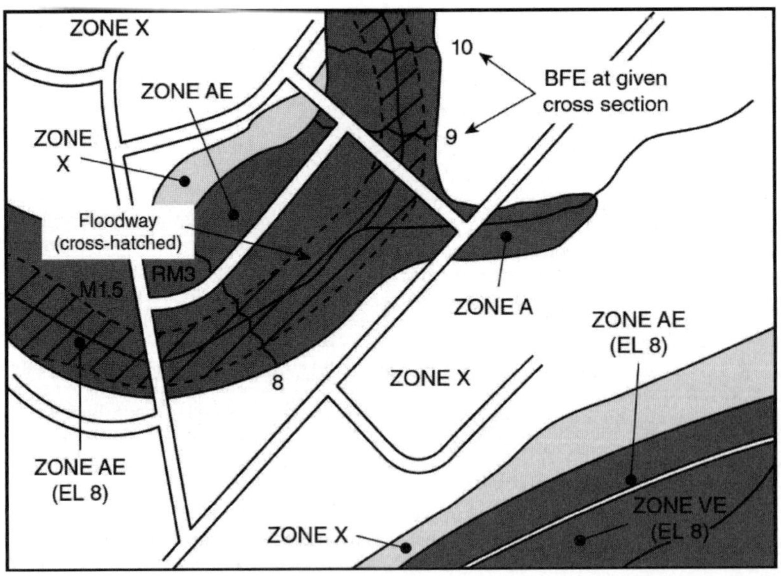

Figure 1.53 Sample flood insurance rate map (Source: FEMA).

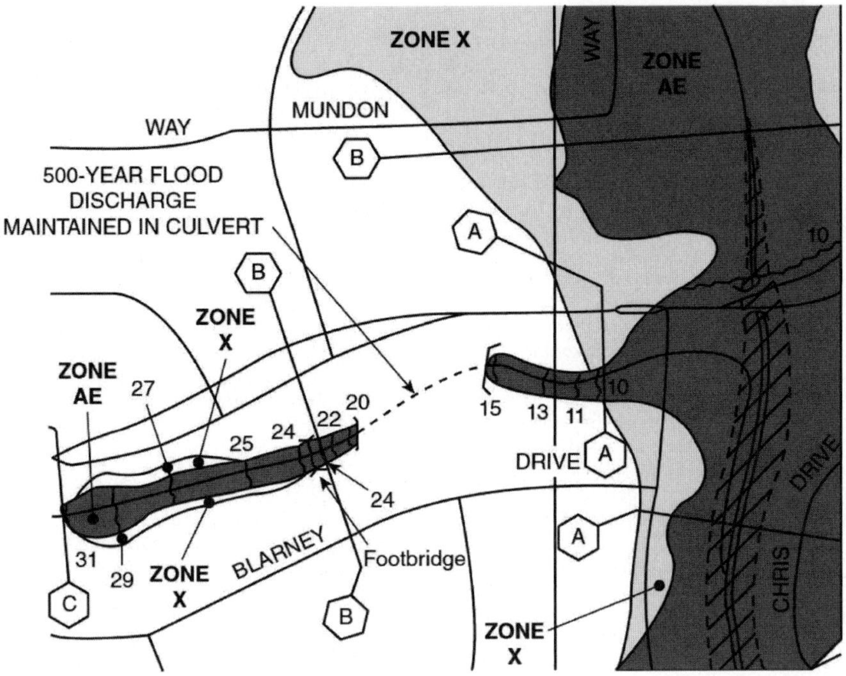

Figure 1.54 Sample digital flood insurance rate map (Source: FEMA).

areas delineated by the more extensive maps. Thus, a design flood is a flood associated with the greater of the area of a base flood based on a 1 percent or greater chance of flooding in any year or the area legally designated as a flood hazard area by a community.

Flood Hazard Zones Flood hazard zones are areas designated on the flood hazard maps that indicate the magnitude and severity of the flood hazards. Such zones are identified in FIRMs.

Table 1.13 contains general descriptions of flood hazard zones and their designations. Comprehensive definitions for all of these zones can be found on the FEMA flood zone website (https://www.fema.gov/flood-zones).

A designation of "X" is given to those areas with low to moderate risk of flooding. Areas of minimal flood hazard (areas above the 500-year flood level) are designated by the unshaded Zone X.

A shaded Zone X is an area of moderate flood hazard (areas between the limits of the 100-year and 500-year floods). On older FIRMs, Zone B corresponds to shaded Zone X and Zone C corresponds to unshaded Zone X.

Special Flood Hazard Areas (high-risk areas) begin with the letter "A" or "V." A Zones are those areas within inland or coastal floodplains where high-velocity wave action is not expected during the base flood. In contrast, V Zones, which are also designated as Coastal High Hazard Areas, are those areas within a coastal floodplain where high-velocity wave action from storms or seismic sources can occur during the base flood event.

A High-Risk Flood Hazard Area is an area where one or more of the following hazards may occur: alluvial fan flooding, flash floods, mudslides, ice jams, high-velocity flow, coastal wave heights greater than or equal to 1.5 feet, or erosion.

The concept of a Coastal A Zone is introduced in Chapter 5 of ASCE/SEI 7 and in Ref. 11 to facilitate application of load combinations in Chapter 2 of ASCE/SEI 7. IBC 1612.2 references Chapter 5 of ASCE/SEI 7 and Ref. 11 for the design and construction of buildings and structures located in flood hazard areas. The requirements in these documents are covered below. The NFIP regulations do not differentiate between Coastal and Noncoastal A Zones.

Table 1.13 FEMA Flood Hazard Zones (Flood Insurance Zones)

Zone	Description
Moderate to Low-Risk Areas	
X*	These zones identify areas outside of the flood hazard area. • Shaded Zone X identifies areas that have a 0.2 percent probability of being equaled or exceeded during any given year (500-year flood). • Unshaded Zone X identifies areas where the annual exceedance probability of flooding is less than 0.2 percent.
High-Risk Areas	
A, AE, A1-30, A99, AR, AO, and AH	These zones identify areas of flood hazard that are not within the Coastal High Hazard Area; i.e., these areas are subject to flooding from riverine (noncoastal) sources.
High-Risk Coastal Areas	
V, VE, and V1-V30	These zones identify the Coastal High Hazard Area, which extends from offshore to the inland limit of a primary frontal dune along an open coast and any other portion of the flood hazard zone that is subject to high-velocity wave action from storms or seismic sources and to the effects of severe erosion and scour. Such zones are generally based on wave heights (3 feet or greater) or wave run-up depths (3 feet or greater).

*Zone B on older FIRMs corresponds to shaded Zone X on more recent FIRMs. Zone C on older FIRMs corresponds to unshaded Zone X on more recent FIRMs.

According to ASCE/SEI 5.2, a Coastal A Zone is an area located within a flood hazard area that is landward of a V Zone or landward of an open coast without mapped V Zones (such as the shorelines of the Great Lakes). Wave forces and erosion potential should be taken into consideration when designing a structure for the effects of flood loads in such zones.

To be classified as a Coastal A Zone, the principal source of flooding must be from astronomical tides, storm surges, seiches, or tsunamis and not from riverine flooding. Additionally, stillwater flood depths must be greater than or equal to 2.0 feet, and breaking wave heights must be greater than or equal to 1.5 feet during the base flood conditions (see ASCE/SEI C5.2). Stillwater depth is the vertical distance between the ground and the stillwater elevation, which is the elevation that the surface of water would assume in the absence of waves. The stillwater elevation is referenced to the North American Vertical Datum (NAVD), the National Geodetic Vertical Datum (NGVD), or other datum, and it is documented in FIRMs.

The principal sources of flooding in Noncoastal A Zones are runoff from rainfall, snowmelt, or a combination of both.

According to Section 4.1.1 of Ref. 11, Coastal A Zones are those locations (1) landward of a V Zone or shoreline and seaward of the Limit of Moderate Wave Action (LiMWA) if the LiMWA is delineated on a FIRM or (2) designated by the authority having jurisdiction. The LiMWA is defined as the line shown on a FIRM that indicates the inland limit of the 1.5-foot breaking wave height during the base flood. FEMA indicates LiMWA lines on revised FIRMs prepared after December 2009. A schematic of a Coastal A Zone is given in Figure C1-1 of Ref. 11. The definition of a Coastal A Zone in Ref. 11 is the same as that in ASCE/SEI 7.

It is recommended to check with the local building official for the most current information on flood hazard areas prior to designing a structure in a flood-prone area.

Design and Construction

According to IBC 1612.2, the design and construction of buildings and structures located in flood hazard areas must be in accordance with Chapter 5 of ASCE/SEI 7 and ASCE/SEI 24-14 (Ref. 11). Section 1.6 of ASCE/SEI 24 requires that design flood loads and their combination with other loads be determined by ASCE/SEI 7 (ASCE/SEI 24 references ASCE/SEI 7-10, which was the current edition of the standard at that time).

The provisions of ASCE/SEI 24 are intended to meet or exceed the requirements of the NFIP. Figures 1-1 and 1-2 in ASCE/SEI 24 illustrate the application of the standard and the application of the chapters in the standard, respectively.

The provisions contained in IBC Appendix G, Flood-Resistant Construction, are intended to fulfill the floodplain management and administrative requirements of NFIP that are not included in the IBC. IBC appendix chapters are not mandatory unless they are specifically referenced in the adopting ordinance of the jurisdiction.

Other provisions related to construction in flood hazard areas worth noting are found in IBC Chapter 18. IBC 1805.5 contains requirements for grading and fill in flood hazard areas. IBC 1805.1.2.1 requires raised floor buildings in flood hazard areas to have the finished grade elevation under the floor (such as at a crawlspace) to be equal to or higher than outside finished grade on at least one side. The exception permits under-floor spaces in Group R-3 residential buildings to comply with Ref. 17. This bulletin provides guidance on crawlspace construction and gives the minimum NFIP requirements for crawlspaces constructed in Special Flood Hazard Areas.

Design Flood Elevation

The design flood elevation (DFE) is used in the determination of flood loads and is defined as the elevation of the design flood including wave height (see Fig. 1.55). For communities that have adopted minimum NFIP requirements, the DFE is identical to the BFE. The DFE exceeds the BFE in communities that have adopted requirements that exceed minimum NFIP requirements. Freeboard is defined as the additional depth of water above the BFE that has been approved by a local jurisdiction. This is essentially a factor of safety to account for uncertainties in the determination of flood elevations. As such, it provides an increased level of flood protection, which could reduce flood insurance premiums for structures located in flood hazard areas.

The DFE is also used in determining the minimum elevation of the top of the lowest floor of a building or structure. In general, the IBC and ASCE/SEI 24 require that the lowest floor be elevated to or above the DFE. Table 2-1 in ASCE/SEI 24 contains the minimum elevation of the top of the lowest floor as a function of the Flood Design Class of Buildings and Structures, which are defined in Table 1-1, for flood hazard areas

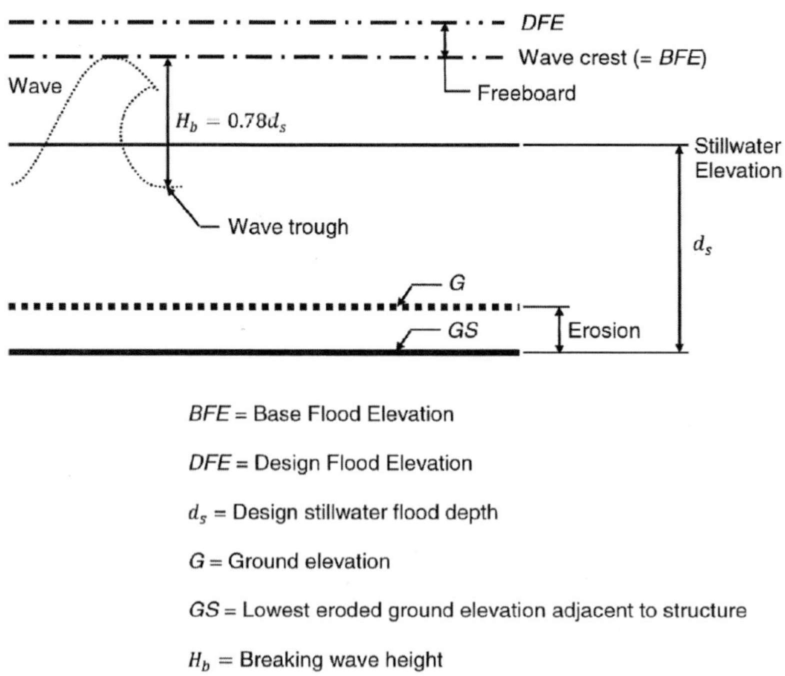

Figure 1.55 Flood parameters.

Table 1.14 Lowest Floor Reference

Zone	Reference	Code section
A	Top of the lowest floor (walking surface)	IBC 1612.4 ASCE/SEI 24 Section 2.3
V Coastal A	Bottom of the lowest horizontal structural member supporting the lowest floor	IBC 1612.4 ASCE/SEI 24 Section 4.4

other than Coastal High Hazard Areas, Coastal A Zones, and High-Risk Flood Hazard Areas. Similar information is provided in Table 1.14 of ASCE/SEI 24 for structures located in Coastal High-Hazard Areas and Coastal A Zones. Table 1.14 contains a summary of the lowest floor reference based on flood hazard zones.

Stillwater Flood Depth

The stillwater flood depth, d_s, is also used in calculating flood loads. This is defined as the vertical distance between the eroded ground elevation and the stillwater elevation. Unless local jurisdictions have adopted a more severe design flood, the stillwater elevation in riverine and lake areas is equal to the BFE published in the FIS and shown on the FIRM.

In coastal areas, the stillwater elevation is the average water level including waves and is published in the FIS (such values are not shown on FEMA FIRMs). Note that the stillwater elevation must be referenced to the same datum that was used in establishing the BFE and the DFE (see Fig. 1.55). Guidance on estimating erosion in coastal areas can be found in Ref. 18. Also, any freeboard that has been added to the 100-year flood should not be used to increase d_s because the load factors in ASCE/SEI 7 are developed for the 100-year flood load.

Breaking Wave Height

Design breaking wave height, H_b, is an important design parameter in the calculation of flood loads at coastal sites. Wave heights at such sites are to be calculated as the heights of depth-limited breaking waves (i.e., wave heights that are limited by the depth of the water), which are equivalent to 78 percent of the stillwater depth (see Fig. 1.55). Seventy percent of the breaking wave height lies above the stillwater elevation.

Design Flood Velocity

Design flood velocity, V, is used in determining hydrodynamic flood loads.

The following equations can be used to estimate design flood velocities in coastal areas (Ref. 18 and ASCE/SEI C5.4.3):

$$V = \frac{d_s}{t}$$

$$V = (gd_s)^{0.5}$$

These two equations provide lower and upper bound estimates of the design flood velocity, respectively. In these equations, $t = 1$ second and g is the acceleration due to gravity (32.2 ft/sec^2). The lower bound velocity should be used if the site is

- distant from the flood source,
- located in Zone A,
- on flat or gently sloping terrain, or
- unaffected by other buildings or obstructions.

The upper bound velocity is appropriate if the site is

- near the flood source,
- in Zone V,
- in Zone AO adjacent to Zone V,
- in Zone A subject to velocity flow and wave action,
- on steeply sloping terrain, or
- adjacent to other large buildings or obstructions that will confine or redirect floodwaters and increase local flood velocities.

Shown in Fig. 1.56 are the lower and upper bound velocities, V, as a function of the stillwater flood depth, d_s. The preceding discussion on the determination of flood velocity is applicable to non-tsunami situations.

Types of Flood Loads

Floodwaters can create loads or pressures on surfaces of buildings and structures that are of two basic types: hydrostatic and hydrodynamic. Wave loads are generally considered a special type of hydrodynamic load. Objects that are transported by moving floodwaters can strike a structure causing an impact load.

The following loads are referenced in ASCE/SEI 5.4:

- Hydrostatic loads (ASCE/SEI 5.4.2): Hydrostatic loads occur when stagnant or slowly moving water (velocity less than 5 feet per second; see ASCE/SEI C5.4.2) comes into contact with a building or building component. The water can be above or below the ground surface.
- Hydrodynamic loads (ASCE/SEI 5.4.3): Hydrodynamic loads are caused by water moving at a moderate to high velocity above the ground level. Similar to wind loads, the loads produced by moving water include an impact load on the upstream face of a building, drag forces along the sides, and a negative force (suction) on the downstream face.
- Wave loads (ASCE/SEI 5.4.4): Wave loads result from water waves propagating over the surface of the water and striking a building or other object. Such loads can be separated into the following four categories:
    - Nonbreaking waves: The effects of nonbreaking waves can usually be determined using the procedures in ASCE/SEI 5.4.2 for hydrostatic loads on walls and in ASCE/SEI 5.4.3 for hydrodynamic loads on piles.
    - Breaking waves: These loads are caused by waves breaking on any portion of a building or structure. Although these loads are of short duration, they generally produce the largest magnitude of all of the different types of wave loads.

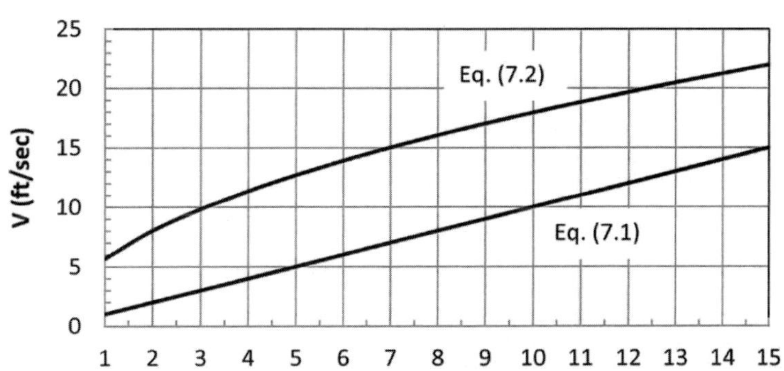

Figure 1.56 Velocity versus design stillwater flood depth.

- Broken waves: The loads caused by broken waves are similar to hydrodynamic loads caused by flowing or surging water.
- Uplift: Uplift effects are caused by wave run-up striking any portion of a building or structure, deflection or peaking against the underside of surfaces.

- Impact loads (ASCE/SEI 5.4.5): Impact loads occur where objects carried by moving water strike a building or structure. The magnitude of these loads is very difficult to predict; however, reasonable allowances can be made for them considering a number of different inherent uncertainties.

Reference 19 divides impact loads into three categories: (1) normal impact loads, (2) special impact loads, and (3) extreme impact loads. Normal impact loads result from the isolated impacts of commonly encountered objects. The size, shape, and weight of waterborne debris may vary according to region; thus, that which is common in one area may not be common in another. Special impact loads result from large objects such as broken up ice floats and accumulations of waterborne debris. The loads are caused by these objects either striking or resting against the building or structure.

Extreme impact loads result from very large objects (such as boats, barges, or parts of collapsed buildings) striking a building or structure. Design for such extreme loads is usually not practical in most cases unless the probability that such an impact load during the design flood is high.

Because most impact loads generated by waterborne debris are of short duration, a dynamic analysis of the building or structure may be appropriate. In cases where the natural period of the structure is much greater than 0.03 second, the impact load can be treated as a static load applied to the structure.

Determination of these loads is based on the design flood, which is defined as the greater of the base flood (a flood having a 1 percent chance of being equaled or exceeded in any given year) and the flood adopted by the local jurisdiction.

Loads on walls that are required by ASCE/SEI 24 to break away (i.e., breakaway walls) are given in ASCE/SEI 5.3.3. The minimum design load must be the largest of the following loads: (1) wind load in accordance with ASCE/SEI Chapter 26, (2) seismic load in accordance with ASCE/SEI Chapter 12, or (3) 10 psf. The maximum permitted collapse load is 20 psf, unless the design meets the conditions of ASCE/SEI 5.3.3.

Reference 3 covers the flood load types in detail.

Flood Load Combinations

While Chapter 5 of ASCE/SEI 7 provides the methods to determine the various types of loads due to floodwaters, it does not specify how these individual flood loads get combined into the flood load, F_a, that is to be used in the appropriate load combinations specified in IBC 1605.2.1 and ASCE/SEI 2.3.2. The following discussion provides insight on how F_a should be determined for the building or structure as a whole and for individual structural members.

It is evident from the information presented in the previous sections that the presence of an individual flood load depends on the flood zone in which the building or structure is sited as well as the building or structure type. Thus, the proper determination of F_a depends on these attributes as well.

Flood Load Combinations in A Zones In an A Zone where the source of the floodwater is from a river or lake, it is reasonable to assume for the analysis of the entire (global) foundation with n piles or columns that one of the piles or columns will be subjected to the impact load, F, and the remaining piles or columns will be subjected to the hydrodynamic load, F_{dyn}; thus, the total flood load is $F_a = F + (n - 1)F_{dyn}$.

The flood load, F_a, on an individual pile or column that supports the structure is equal to the sum of the hydrodynamic load, F_{dyn}, and the impact load, $F(F_a = F_{dyn} + F)$. These same individual flood loads are used to determine F_a on a solid foundation wall with flood openings that conform to the requirements of ASCE/SEI 24 Section 2.7.1.

In the case of a solid foundation wall without flood openings that meets the dry-floodproofed requirements of ASCE/SEI 24 Section 6.2, F_a is the sum of the hydrostatic load, F_{sta}; the hydrodynamic load, F_{dyn}; and the impact load, $F(F_a = F_{sta} + F_{dyn} + F)$.

Breaking wave loads need not be considered in A Zones except at sites that are subject to coastal waves loads. Determination of F_a for such sites is the same as that for Coastal A Zones and V Zones where the breaking wave load is equal to F_D for piles and columns or F_t, F_{nv}, or F_{oi}, as applicable, for walls.

Flood Load Combinations in Coastal A Zones and V Zones In a Coastal A Zone or a V Zone, hydrodynamic loads, breaking wave loads, and impact loads must be considered.

For the analysis of the entire (global) foundation with n piles or columns, it is reasonable to assume that one of the piles or columns will be subjected to the impact load, F, and that the remaining piles or columns will be subjected to the hydrodynamic load, F_{dyn}; thus, the total flood load is $F_a = F + (n - 1)F_{dyn}$.

It is usually unrealistic to assume that impact loads occur on all piles or columns at the same time as breaking wave loads. As such, for the case of an individual pile or column in the front row (i.e., in the row that experiences the initial effect of the floodwaters), F_a is equal to the larger of F_{dyn} and F_D plus $F[F_a = \text{larger of } (F_{dyn}, F_D) + F]$.

As noted previously, solid foundation walls are not permitted by the IBC or ASCE/SEI 24 in Coastal A Zones or V Zones.

Tsunami Loads and Effects

Overview ASCE/SEI 6.2 defines a tsunami as a series of waves with variable long periods that typically result from earthquake-induced uplift or subsidence of the seafloor. Landslides near a coast or underwater volcanic eruptions can also generate tsunamis.

When a tsunamigenic earthquakes occurs, the seafloor is uplifted and down-dropped, which pushes the column of water above the fault up and down. The potential energy that results from pushing the water above mean sea level is transferred into kinetic energy in the form of horizontal propagation of two oppositely traveling tsunami waves.

As the waves travel outward from the source, amplification of the waves occurs. The first part of the wave to reach the shoreline is a trough, which is followed by very strong and fast-moving tides. Run-up occurs when the wave travels past the shoreline onto land and is a measure of the height of water onshore above a reference datum. The maximum horizontal inland extent of flooding due to a tsunami is defined as the inundation limit (see Fig. 1.57, which is based on ASCE/SEI Figure 6.2-1).

Run-up elevation, R, is the vertical distance above ground elevation at the maximum tsunami inundation limit. This must include any relative change in sea level, which could be caused by ocean rise and/or ground subsidence due to the tsunami. According to ASCE/SEI 6.2, run-up elevation is measured with respect to the North American Vertical Datum 1988 (NAVD 88) reference datum. Effects of relative sea level change and seismic subsidence on tsunami inundation are summarized in Figure C6.1-2.

The *grade plane* is the horizontal reference plane at the site of the structure; it represents the average elevation of finished ground level adjoining the structure at all exterior walls. In cases where the finished ground level slopes away from the exterior walls, the grade plane is determined using the lowest points that are within the area between the structure and the property line. If the property line is more than 6 feet from the structure, the area to be used in determining the grade plane is that between the structure and 6 feet from the structure. The depth of water at the grade plane of a structure, including any relative sea level change, is defined as the *inundation depth.*

General Requirements The tsunami requirements in Chapter 6 of ASCE/SEI 7-16 are applicable to the following states: Alaska, California, Hawaii, Oregon, and Washington. These states have designated Tsunami-Prone Regions (TPR), which, by definition, are coastal regions where tsunamigenic earthquakes determined in accordance with the Probabilistic Tsunami Hazard Analysis method in Chapter 6 can cause a tsunami inundation hazard with a run-up greater than 3 feet.

Certain structures in TPR must be analyzed and designed for inundation based on the Maximum Considered Tsunami (MCT), which

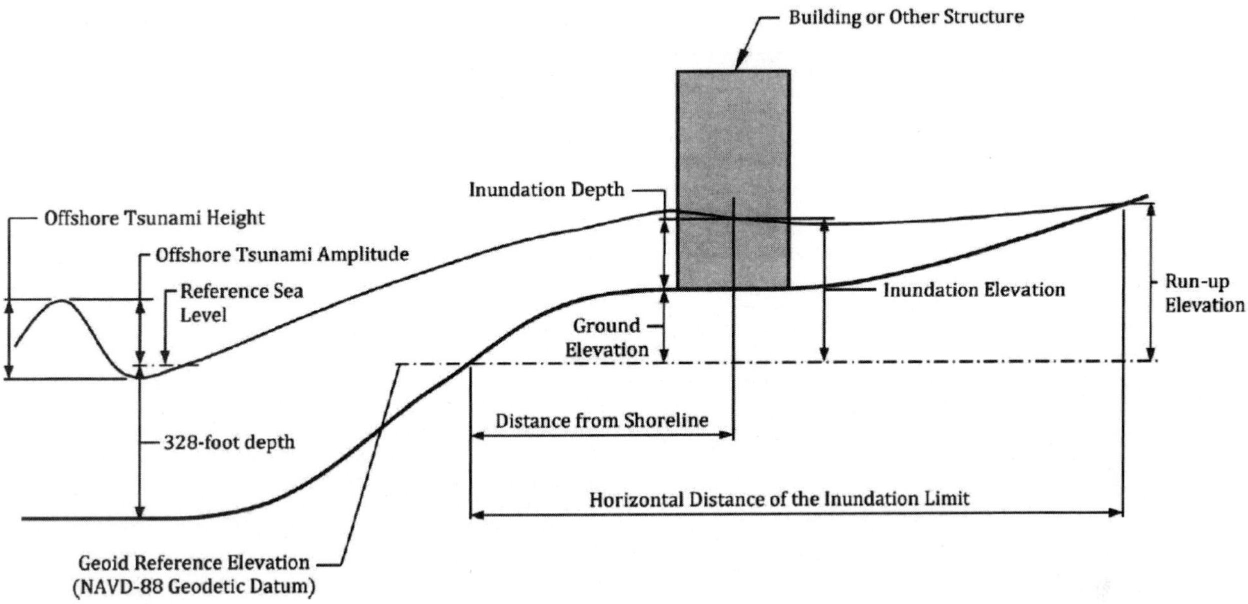

Figure 1.57 Tsunami parameters for a typical topographic flow transect.

is defined as a probabilistic tsunami having a 2 percent probability of being exceed in a 50-year period (i.e., a 2,475-year mean recurrence interval). More information on the MCT can be found in C6.1.

The Tsunami Design Zone Map (TDZM) in Figure 6.1-1 designates the potential horizontal inundation limits of the MCT for the five states. The Tsunami Design Zone (TDZ) is an area identified on the TDZM between the shoreline and the inundation limit.

The ASCE Tsunami Design Geodatabase contains maps that graphically depict the extent of inundation (i.e., the TDZ) due to the MCT (https://asce7tsunami.online). A site can be located by address, by latitude and longitude, or by clicking on the map. The distance from the shoreline to the inundation limit can be obtained by utilizing the "Transect" feature of the program. In addition to the TDZ, run-up elevations and inundations depths are also provided at numerous points (which are designated by red triangles and red circles, respectively). Note that a state or local jurisdiction may have adopted a TDZ and other data that are different from those in the Geodatabase, so it is good practice to check with the local authorities prior to performing the analysis.

ASCE 7-16 Tsunami Design Zone Maps for Selected Locations (https://ascelibrary.org/doi/book/10.1061/9780784480748) contain TDZM in high-resolution PDF format for the 62 locations indicated by circles in Figures C6.1-1(a) to (i). For the purpose of identifying the TDZ, these downloadable PDF maps are considered equivalent to the results obtained from the ASCE Tsunami Design Geodatabase.

According to 6.1.1, the following buildings and other structures located within a TDZ must be designed for the MCT:

- Tsunami Risk Category IV buildings and structures,
- Tsunami Risk Category III buildings and structures with an inundation depth greater than 3 feet at any location within the footprint of the structure, and
- where required by a state or locally adopted building code statue to include design for tsunamis effects, Tsunami Risk Category II buildings with a mean height above the grade plane greater than the height designated in the statue and with an inundation depth greater than 3 feet at any location within the footprint of the building. Tsunami Risk Categories are defined in 6.4.

Two exceptions to these requirements are given in 6.1.1. In the first exception, tsunami loads and effects determined in accordance with Chapter 6 need not be applied to Tsunami Risk Category II single-story buildings of any height without (1) mezzanines or any occupiable roof level and (2) any critical equipment or systems. The second exception permits the TDZ, inundation limits, and run-up elevations to be determined by (1) the site-specific procedures of 6.7 for coastal regions that are subject to tsunami inundation that are not covered in Figure 6.1-1 or (2) the procedures in 6.5.1.1 for Tsunami Risk Category II or III structures.

Tsunami Risk Categories

According to 6.4, Tsunami Risk Categories are the same as the Risk Categories defined in 1.5 with the following modifications:

1. Federal, state, local or tribal governments are permitted to designate Critical Facilities (such as power-generating stations, water-treatment facilities for potable water, wastewater treatment facilities, and other public utility facilities not included in Risk Category IV) to Tsunami Risk Category III. Additional information on this modification can be found in C6.4.
2. The following structures need not be included in Tsunami Risk Category IV, and state, local, or tribal governments are permitted to designate them to Tsunami Risk Category II or III:
   a. Fire stations, ambulance facilities, and emergency vehicle garages
   b. Earthquake or hurricane shelters
   c. Emergency aircraft hangars
   d. Police stations that do not have holding cells and that are not uniquely required for postdisaster emergency response as a Critical Facility
3. Tsunami Vertical Evacuation Refuge Structures (TVERS) must be included in Tsunami Risk Category IV.

The Essential Facilities listed in Item 2 above do not need to be included in Tsunami Risk Category IV because these facilities should be evacuated prior to the arrival of the tsunami. Additional information on these modifications can be found in C6.4.

A TVERS is a structure designated and designed to serve as a point of refuge where people can evacuate above a tsunami (6.14). More information is given in C6.14.

Analysis of Design Inundation Depth and Flow Velocity

Analysis Procedures Two procedures are given in Chapter 6 to determine the inundation depth and flow velocities due to the MCT at a site:

Table 1.15 Inundation Depth and Flow Velocity Analysis Procedures Where Run-Up Is Given in Figure 6.1-1*

Analysis procedure	Tsunamic risk category			
			IV	
	II	III	Excluding TVERS	TVERS
Energy Grade Line Analysis (6.6)	Required	Required	Required	Required
Site-Specific Probabilistic Tsunami Hazard Analysis (6.7)	Permitted	Permitted	Required**	Required

*Chapter 6 does not apply to Tsunami Risk Category II and III buildings and other structures where the MCT inundation depth ≤ 3 feet.

**A site-specific PTHA need not be performed where the inundation depth resulting from EGLA is determined to be less than 12 feet at any point within the location of the structure.

(1) Energy Grade Line Analysis (EGLA) in 6.6 and (2) site-specific Probabilistic Tsunami Hazard Analysis (PTHA) in 6.7. Table 1.15 indicates the required analysis procedures in accordance with the provisions of 6.5 based on Tsunami Risk Category for locations in Figure 6.1-1, which include run-up.

For TPR that are not covered in Figure 6.1-1, the exception in 6.5.1 requires that the procedures of 6.5.1.1 must be used to determine the ratio of tsunami run-up elevation to Offshore Tsunami Amplitude, *R*/*HT*, for Tsunami Risk Category II and III buildings and other structures. The Offshore Tsunami Amplitude is the MCT amplitude relative to the reference sea level, which is measured where the undisturbed water depth is equal to 328 feet (see Fig. 1.57). Values of *HT* are obtained from the Geodatabase.

Table 1.16 indicates the required inundation depth and flow velocity procedures in accordance with the provisions of 6.5 based on Tsunami Risk Category for locations where (1) the inundation depth is not shown in the TDZM of Figure 6.1-1 and (2) the Offshore Tsunami Amplitude is given in Figure 6.7-1.

See Ref. 3 for details on the analysis procedures in Table 1.16.

Table 1.16 Inundation Depth and Flow Velocity Analysis Procedures Where Run-Up Is Calculated from Figure 6.7-1*

Analysis procedure	Tsunamic risk category			
			IV	
	II	III	Excluding TVERS	TVERS
R/H Analysis (6.5.1.1.)	Required	Required	Not permitted	Not permitted
Energy Grade Line Analysis (6.6)	Required	Required	Required	Required
Site-Specific Probabilistic Tsunami Hazard Analysis (6.7)	Permitted	Permitted	Required	Required

*Chapter 6 does not apply to Tsunami Risk Category II and III buildings and other structures where the MCT inundation depth is ≤ 3 feet.

Structural Design Procedures for Tsunami Effects

Overview The requirements of 6.8 must be satisfied for all structures, components, and foundations that are subjected to the loads and effects of the MCT.

Buildings and other structures assigned to Tsunami Risk Category II and III must be designed as a minimum to meet Collapse Prevention Structural Performance criteria; that is, the damaged structural members and components of a building or other structure must be able to support gravity loads after a design tsunami event. In such cases, the structure retains little or no margin against collapse.

Tsunami Risk Category III Critical Facilities and Tsunami Risk Category IV buildings and other structures located within the TDZ must be designed to satisfy the following three requirements:

1. Operational nonstructural components, equipment required for essential functions, and the bottom of any horizontal structural members supporting such components and equipment must be located above the inundation elevation of the MCT.
2. Structural components and connections in occupiable levels and foundations must be designed in accordance with Immediate Occupancy Structural Performance criteria; that is, the structure is safe to be occupied after a design basis tsunami event. Also, occupiable levels are permitted where the elevation of such levels equals or exceeds the MCT inundation elevation.
3. Buildings and other structures designated as TVERS must also comply with 6.14.

Structural Performance Evaluation

Structures must be designed for strength and stability considering the three tsunami load cases in 6.8.3.1. These load cases are based on inundation depths and associated flow velocities.

Load Case 1 (Buoyancy): At an exterior inundation depth that is less than or equal to the maximum inundation depth or the lesser of one story or the height of the top first-story windows, combined hydrodynamic force with buoyant force must be evaluated with respect to the depth of water in the interior. The interior water depth is evaluated in accordance with 6.9.1. According to the exception to these requirements, open structures and structures where the soil properties or foundation and structural design prevent detrimental hydrostatic pressurization on the underside of the foundation and the lowest structural slab need not be designed for Load Case 1.

The primary purposes of Load Case 1 are to check (1) overall stability of the structure and (2) foundation anchorage against net uplift. The maximum buoyant force on the structure is based on the associated hydrodynamic force. The uplift force, which is calculated in accordance with 6.9.1, depends on the differential inundation depth exterior to the structure and the flooded depth within the structure.

Load Case 2 (Maximum Velocity): Two-thirds of maximum inundation depth when flow velocity and specific momentum flux (hu^2) are maximum. Both incoming and receding directions of water flow must be considered.

The maximum hydrodynamic forces on the structure are calculated in Load Case 2.

Load Case 3 (Maximum Inundation Depth): Maximum inundation depth when the flow velocity is assumed to be at one-third of its maximum. Both incoming and receding directions of water flow must be considered.

Hydrodynamic forces associated with maximum inundation depth are calculated in Load Case 3.

Tsunami Importance Factors

Tsunami importance factors, I_{tsu}, are given in Table 6.8-1 (see Table 1.17). These importance factors are to be applied to the tsunami hydrodynamic and impact loads in 6.10 and 6.11, respectively. Information on the derivation of the importance factors is given in C6.8.3.2.

Load Combinations

Tsunamis are considered rare events, so structural systems and members need to be designed for a set of load effects that would be expected

Table 1.17 Tsunami Importance Factors, I_{tsu}

Tsunami risk category	I_{tsu}
II	1.0
III	1.25
• Tsunami Risk Category IV • Vertical Evacuation Refuges • Tsunami Risk Category III Critical Facilities	1.25

during such an event. Principal tsunami forces and effects are to be combined using Equations 6.8-1a and 6.8-1b:

$$0.9D + F_{TSU} + H_{TSU}$$

$$1.2D + F_{TSU} + 0.5L + 0.2S + H_{TSU}$$

where F_{TSU} = tsunami load effect for incoming and receding directions of flow

H_{TSU} = load caused by tsunami-induced lateral foundation (earth) pressures developed under submerged conditions (i.e., caused by water seepage through the soil). In cases where the net effect of H_{TSU} counteracts the principal load effect, the load factor for H_{TSU} must be taken as 0.9.

The tsunami load combinations are consistent with the Extraordinary Load Combinations of ASCE/SEI 7 Section 2.5 (see Sec. 1.10 of this chapter). The inundation depths and velocities for determining F_{TSU} and H_{TSU} must be consistent with the hazard level of the MCT.

Reference 3 covers lateral systems acceptance criteria, Structural Component Acceptance Criteria, Flow Velocity Amplification, Directionality of Flow, and other requirements.

Structural Loads and Effects

Like buildings subjected to flood effects, the following must be considered in the design of buildings and other structures subjected to tsunami effects:

1. Hydrostatic loads
2. Hydrodynamic loads
3. Debris impact loads
4. Scour effects

Reference 3 covers detailed calculations of the tsunami load types mentioned above, foundation design, and requirements.

1.10 LOAD COMBINATIONS

1.10.1 Introduction

In accordance with IBC 1605.1, structural members of buildings and other structures must be designed to resist the load combinations of IBC 1605.2, 1605.3.1, or 1605.3.2. Load combinations that are specified in Chapters 18 through 23 of the IBC, which contain provisions for soils and foundations, concrete, aluminum, masonry, steel, and wood, must also be considered. The structural elements identified in ASCE/SEI Chapters 12, 13, and 15 must be designed for the load combinations with overstrength of ASCE/SEI 2.3.6 or 2.4.5.

IBC 1605.2 contains the load combinations that are to be used when strength design or load and resistance factor design is utilized. Load combinations using allowable stress design are given in IBC 1605.3. Both sets of combinations are covered in this section of this chapter. The combinations of IBC 1605.2 or 1605.3 can also be used to check overall structural stability, including stability against overturning, sliding, and buoyancy (IBC 1605.1.1).

In ASCE/SEI 7-16, the load combinations with seismic load effects have been removed from ASCE/SEI Chapter 12 and placed in ASCE/SEI Chapter 2 in sections separate from the basic load combinations.

1.10.2 Load Effects

The load effects that are included in the IBC and ASCE/SEI 7 load combinations are summarized in Table 1.18. More details on these load effects can be found in those documents as well as in subsequent sections of this chapter.

1.10.3 Load Combinations Using Strength Design or Load and Resistance Factor Design

The basic load combinations where strength design or, equivalently, load and resistance factor design is used are given in IBC 1605.2 and summarized in Table 1.19. These equations establish the minimum required strength that needs to be provided in the members of a building or structure.

Table 1.18 Load Effects

Notation	Load effect	Notes
D	Dead load	See IBC 1606
D_i	Weight of ice	See IBC 1614 and Chapter 10 of ASCE/SEI 7
E	Seismic load effect defined in ASCE/SEI 12.4.2	See IBC 1613, ASCE/SEI 12.4.2
E_m	Seismic load effect including overstrength defined in ASCE/SEI 12.4.3	See IBC 1613, ASCE/SEI 12.4.3
F	Load due to fluids with well-defined pressures and maximum heights	—
F_a	Flood load	See IBC 1612, Chapter 5 of ASCE/SEI 7
H	Load due to lateral earth pressures, groundwater pressure or pressure of bulk materials	See IBC 1610
L	Roof live load greater than 20 psf or less	See IBC 1607
L_r	Roof live load of 20 psf or less	See IBC 1607
R	Rain load	See IBC 1611
S	Snow load	See IBC 1608, Chapter 7 of ASCE/SEI 7
T	Cumulative effects of self-straining forces and effects	See ASCE/SEI 2.3.4 and 2.4.4
W	Load due to wind pressure	See IBC 1609, Chapters 26 to 31 of ASCE/SEI 7
W_i	Wind-on-ice load	See IBC 1614, Chapters 10 of ASCE/SEI 7

These load combinations apply only to strength limit states; serviceability limit states for deflection, vibration, drift, camber, expansion, and contraction and durability are given in Appendix C of ASCE/SEI 7.

The seismic load effect, E, that is to be used in IBC Equation 16-5 (ASCE/SEI load combination 6) is equal to the following (see ASCE/SEI 12.4.2):

$$E = E_h + E_v$$

where E_h = horizontal seismic load effect defined in ASCE/SEI 12.4.2.1 = Q_E

E_v = vertical seismic load effect defined in ASCE/SEI 12.4.2.2 = $0.2S_{DS}$

ρ = redundancy factor defined in ASCE/SEI 12.3.4

Q_E = effects of horizontal seismic forces applied to the structure

S_{DS} = design spectral response acceleration parameter at short periods

Table 1.19 Summary of Load Combinations Using Strength Design or Load and Resistance Factor Design (IBC 1605.2)

IBC equation no.	Load combination
16-1	$1.4(D + F)$
16-2	$1.2(D + F) + 1.6(L + H) + 0.5(L_r \text{ or } S \text{ or } R)$
16-3	$1.2(D + F) + 1.6(L_r \text{ or } S \text{ or } R) + 1.6H + (f_1L \text{ or } 0.5W)$
16-4	$1.2(D + F) + 1.0W + f_1L + 1.6H + 0.5(L_r \text{ or } S \text{ or } R)$
16-5	$1.2(D + F) + 1.0E + f_1L + 1.6H + f_2S$
16-6	$0.9D + 1.0W + 1.6H$
16-7	$0.9(D + F) + 1.0E + 1.6H$

f_1 = 1 for places of public assembly live loads in excess of 100 psf and for parking garages = 0.5 for other live loads.
f_2 = 0.7 for roof configurations (such as sawtooth) that do not shed snow off the structure = 0.2 for other roof configurations.

Thus, IBC Equation 16-5 (ASCE/SEI load combination 6) can be written as follows:

$$(1.2 + 0.2\ S_{DS})D + 1.2F + \rho Q_E + f_1L + 1.6H + f_2S$$

In IBC Equation 16-7 (ASCE/SEI load combination 7), the seismic load effect that is to be used is $E = E_h - E_v$ (see ASCE/SEI 12.4.2). Therefore, this equation can be written as follows:

$$(0.9 - 0.2\ S_{DS})D + 0.9F + \rho Q_E + 1.6H$$

Refer to ASCE/SEI 7 12.4.2.2 for exceptions related to vertical seismic load effects.

Fluid load effects, F, occur in tanks and other storage containers due to stored liquid products. The stored liquid is generally considered to have characteristics of both a dead load and a live load. It is not a purely permanent load because the tank or storage container can go through cycles of being emptied and refilled. The fluid load effect is included in IBC Equations 16-1 through 16-5, where it adds to the effects from the other loads. It is also included in IBC.

Equation 16-7 counteracts the effects from uplift due to seismic load effects, E. Because the wind load effects, W, can be present when the tank is either full or empty, F is not incorporated into IBC Equation 16-6; that is, the maximum effects occur when F is set equal to zero.

The load combinations given in IBC 1605.2 are the same as those in ASCE/SEI 2.3.1 with some indicated in these sections.

According to IBC 1605.2.1, the load combinations of ASCE/SEI 2.3.2 are to be used where flood loads, F_a, must be considered in design (flood loads are determined by Chapter 5 of ASCE/SEI 7). In particular, the following modifications are to be made:

- V Zones or Coastal A Zones
  $1.0W$ in IBC Equations 16-4 and 16-6 must be replaced by $1.0W + 2.0F_a$.
- Noncoastal A Zones
  $1.0W$ in IBC Equations 16-4 and 16-6 must be replaced by $0.5W + 1.0F_a$.

Definitions of Coastal High Hazard Areas (V Zones) and Coastal A Zones are given in ASCE/SEI 5.2.

The load factors on F_a are based on a statistical analysis of flood loads associated with hydrostatic pressures, pressures due to steady overland flow, and hydrodynamic pressures due to waves, all of which are specified in ASCE/SEI 5.4.

In cases where self-straining loads, T, must be considered, their effects in combination with other loads are to be determined by ASCE/SEI 2.3.4 (IBC 1605.2.1). Instead of calculating self-straining effects based on upper bound values of this variable like other load effects, the most probable effect expected at any arbitrary point in time is used. More information, including load combinations that should be considered in design, is given in ASCE/SEI C2.3.4.

IBC 1605.2.1 requires that the load combinations of ASCE/SEI 2.3.3 be used where atmospheric ice loads must be considered in design. The following modifications to the load combinations must be made when a structure is subjected to atmospheric ice and wind-on-ice loads (atmospheric and wind-on-ice loads are determined by Chapter 10 of ASCE/SEI 7; see IBC 1614):

- $0.5(L_r$ or S or $R)$ in ASCE/SEI combination 2 (IBC Equation 16-2) must be replaced by $0.2D_i + 0.5S$.
- $1.0W + 0.5(L_r$ or S or $R)$ in ASCE/SEI combination 4 (IBC Equation 16-4) must be replaced by $D_i + W_i + 0.5S$.
- $1.0W$ in ASCE/SEI combination 5 (IBC Equation 16-6) must be replaced by $D_i + W_i$.
- $1.0W + L + 0.5(L_r$ or S or $R)$ in ASCE/SEI combination 4 (IBC Equation 16-4) must be replaced by D_i.

See ASCE/SEI C2.3.3 for more information on the load factors used in these equations. ASCE/SEI 2.3.5 provides information on how to develop strength design load criteria where no information on loads or load combinations is given in ASCE/SEI 7 or where performance-based design in accordance with ASCE/SEI 1.3.1.3 is being utilized. Detailed information on how to develop such load criteria that is consistent with the methodology used in ASCE/SEI 7 can be found in ASCE/SEI C2.3.5.

1.10.4 Load Combinations Using Allowable Stress Design

Overview

The basic load combinations where allowable stress design (working stress design) is used are given in IBC 1605.3. A set of basic load combinations is given in IBC 1605.3.1, and a set of alternative basic load combinations is given in IBC 1605.3.2. Both sets are examined below.

Basic Load Combinations

The basic load combinations of IBC 1605.3.1 are summarized in Table 1.20.

Table 1.20 Summary of Basic Load Combinations Using Allowable Stress Design (IBC 1605.3.1)

Equation no.	Load combination
16-8	$D + F$
16-9	$D + H + F + L$
16-10	$D + H + F + (L_r$ or S or $R)$
16-11	$D + H + F + 0.75L + 0.75(L_r$ or S or $R)$
16-12	$D + H + F + (0.6W$ or $0.7E)$
16-13	$D + H + F + 0.75(0.6W) + 0.75L + 0.75(L_r$ or S or $R)$
16-14	$D + H + F + 0.75(0.7E) + 0.75L + 0.75S$
16-15	$0.6D + 0.6W + H$
16-16	$0.6(D + F) + 0.7E + H$

These load combinations apply to the design of all members in a structure and also provide for overall stability of a structure.

The seismic load effect, E, is a strength-level load. A factor of 0.7, which is approximately equal to 1/1.4, is applied to E in IBC Equations 16-12, 16-14, and 16-16 to convert the strength-level effects to service-level effects. Similarly, a factor of 0.6 is applied to W in IBC Equations 16-12, 16-13, and 16-15.

The seismic load effect, E, that is to be used in IBC Equations 16-12 and 16-14 (ASCE/SEI load combinations 8 and 9) is equal to $E = E_h + E_v$.

Thus, IBC Equations 16-12 and 16-14 (ASCE/SEI load combinations 8 and 9) can be written as follows:

IBC Equation 16-12:

$$(1 + 0.14\ S_{DS})D + H + F + 0.7\rho Q_E$$

IBC Equation 16-14:

$$(1 + 0.105\ S_{DS})D + H + F + 0.525\rho Q_E + 0.75L + 0.75S$$

In IBC Equation 16-16 (ASCE/SEI load combination 10), the seismic load effect that is to be used is $E = E_h - E_v$. Therefore, this equation can be written as follows:

$$(0.6 - 0.14\ S_{DS})D + 0.6F + 0.7\rho Q_E + H$$

The exceptions applicable to these load combinations are given in ASCE/SEI 12.4.2.2 and in IBC 1605.3.1. See Ref. 3 details of these exceptions.

Increases in allowable stresses that are given in the materials chapters of the IBC or in referenced standards are not permitted when the load combinations of IBC 1605.3.1 are used (IBC 1605.3.1.1). According to IBC 1605.3.1.2, the load combinations of ASCE/SEI 2.4.2 are to be used where flood loads, F_a, must be considered in design. In particular, the following modifications are to be made:

- V Zones or Coastal A Zones
  $1.5F_a$ must be added to the other loads in IBC Equations 16-12, 16-13, 16-14, and 16-15, and E is set equal to zero in IBC Equations 16-12 and 16-14.
- Noncoastal A Zones
  $0.75F$ must be added to the other loads in IBC Equations 16-12, 16-13, 16-14, and 16-15 and is equal to zero in IBC Equations 16-12 and 16-14.

Where self-straining loads, T, must be considered in design, the provisions of ASCE/SEI 2.4.4 are to be used to determine the proper combination of T with other loads (IBC 1605.3.1.2). ASCE/SEI C2.4.4 provides load combinations for typical situations.

IBC 1605.3.1.2 requires that the load combinations of ASCE/SEI 2.4.3 be used where atmospheric ice loads must be considered in design. The following modifications to the load combinations must be made when a structure is subjected to atmospheric ice and wind-on-ice loads:

- $0.7D_i$ must be added to ASCE/SEI combination 2 (IBC Equation 16-9).
- (L_r or S or R) in ASCE/SEI combination 3 (IBC Equation 16-10) is to be replaced by $0.7D_i + 0.7W_i + S$.
- $0.6W$ in ASCE/SEI combination 7 (IBC Equation 16-15) is to be replaced by $0.7D_i + 0.7W_i$.
- $0.7D_i$ must be added to ASCE/SEI combination 1 (IBC Equation 16-8).

Alternative Basic Load Combinations

The alternative basic load combinations can be found in IBC 1605.3.2 and are summarized in Table 1.21.

Table 1.21 Summary of Alternative Basic Load Combinations Using Allowable Stress Design (IBC 1605.3.2)

Equation no.	Load combination
16-17	$D + L + (L_r$ or S or $R)$
16-18	$D + L + 0.6\omega W$
16-19	$D + L + 0.6\omega W + S/2$
16-20	$D + L + S + 0.6\omega W/2$
16-21	$D + L + S + E/1.4$
16-22	$0.9D + E/1.4$

These load combinations are based on the allowable stress load combinations that appeared in Ref. 20 for many years.

Unlike the basic load combinations of IBC 1605.3.1, allowable stresses are permitted to be increased or load combinations are permitted to be reduced where permitted by the material chapters of the IBC (Chapters 18 through 23) or by referenced standards when the alternative basic load combinations of IBC 1605.3.2 are used. This applies to those load combinations that include wind or earthquake loads.

The alternative allowable stress design load combinations do not include a load combination comparable to IBC Equation 16-15 for dead load counteracting wind load effects. Instead of a specific load combination, IBC 1605.3.2 states that for load combinations that include counteracting effects of dead and wind loads, only two-thirds of the minimum dead load that is likely to be in place during a design wind event is to be used in the load combination.

As noted in the preceding discussion, the seismic load effect, E, is a strength-level load. This strength-level load is divided by 1.4 in IBC Equations 16-21 and 16-22 to convert it to a service-level load. Similarly, W is multiplied by 0.6 in IBC Equations 16-18 through 16-20.

The coefficient ω in IBC Equations 16-18 through 16-20 is equal to 1.3, where wind loads are calculated in accordance with ASCE/SEI Chapters 26 through 31 and where allowable stresses have been increased or load combinations have been reduced as permitted by the material chapters of the IBC or the referenced standards in IBC Chapter 35. In all other cases, the coefficient ω is to be taken as 1.0. It is shown in Sec. 1.7 of this chapter that the wind directionality factor, which is equal to 0.85 for building structures, is explicitly included in the velocity pressure equation for wind. In earlier editions of ASCE/SEI 7 and in the legacy codes, the directionality factor was part of the load factor, which was equal to 1.3 for wind. Thus, for allowable stress design, $\omega = 1.3 \times 0.85 \approx 1.0$, and for strength design, $\omega = 1.6 \times 0.85 \approx 1.3$.

ASCE/SEI 12.13.4 permits a reduction of foundation overturning due to earthquake forces, provided that the criteria of that section are satisfied. Such a reduction is not permitted when the alternative basic load combinations are used to evaluate sliding, overturning, and soil bearing at the soil–structure interface. Also, the vertical seismic load effect, E_v, in ASCE/SEI Equation 12.44 may be taken as zero when proportioning foundations using these load combinations.

The two exceptions in IBC 1605.3.2 for crane hook loads and for combinations of snow loads, roof live loads, and earthquake loads are the same as those in IBC 1605.3.1, which were discussed previously.

IBC 1605.3.2.1 requires that where F, H, or T must be considered in design, each applicable load is to be added to the load combinations in IBC Equations 16-17 through 16-22. As noted previously, the effects of T in combination with other loads must be determined in accordance with ASCE/SEI 2.4.4.

ASCE/SEI 7 does not contain provisions for the alternative basic load combinations of IBC 1605.3.2.

Load Combinations with Overstrength

The following load combinations, which are given in ASCE/SEI 2.3.6 and 2.4.5, must be used where required by ASCE/SEI Chapters 12, 13, and 15 instead of the corresponding load combinations in IBC 1605.2 and 1605.3 (see IBC 1605.1 and ASCE/SEI 12.4.3):

Basic Combinations for Strength Design with Overstrength

IBC Equation 16-5:

$$(1.2 + 0.2\,S_{DS})D + \Omega_O Q_E + f_1 L + f_2 S + 1.2F + 1.6H$$

IBC Equation 16-7:

$$(0.9 - 0.2\,S_{DS})D + \Omega_O Q_E + 0.9F + 1.6H$$

Basic Combinations for Allowable Stress Design with Overstrength

IBC Equation 16-12:

$$(1 + 0.14\,S_{DS})D + 0.7\Omega_O Q_E + H + F$$

IBC Equation 16-14:

$$(1 + 0.105\,S_{DS})D + 0.525\Omega_O Q_E + 0.75L + 0.75S + H + F$$

IBC Equation 16-16:

$$(0.6 - 0.14\,S_{DS})D + 0.7\Omega_O Q_E + 0.6F + H$$

Alternative Basic Combinations for Allowable Stress Design with Overstrength

IBC Equation 16-21:

$$\left(1 + \frac{0.2\,S_{DS}}{1.4}\right)D + \frac{\Omega_O Q_E}{1.4} + L + S$$

IBC Equation 16-22:

$$\left(0.9 - \frac{0.2\,S_{DS}}{1.4}\right)D + \frac{\Omega_O Q_E}{1.4}$$

where $E_m = E_{mh} + E_v = \Omega_0 Q_E + 0.2S_{DS}D$ for use in IBC Equations 16-5, 16-12, 16-14, and 16-21 (see ASCE/SEI 12.4.3)

$= E_{mh} - E_v = \Omega_0 Q_E - 0.2S_{DS}D$ for use in IBC Equations 16-7, 16-16, and 16-22 (see ASCE/SEI 12.4.3)

Ω_0 = system overstrength factor obtained from ASCE/SEI Table 12.21 for a particular seismic force resisting system

When the simplified procedure of ASCE/SEI 12.14 is utilized, the seismic load effects defined in ASCE/SEI 12.14.3 are to be used in the referenced load combinations.

ASCE/SEI 2.4.5 permits allowable stresses to be increased by a factor of 1.2 where allowable stress design is used with seismic load effect including overstrength. This increase is not to be combined with increases in allowable stresses or reductions in load combinations that are otherwise permitted in ASCE/SEI 7 or in other referenced materials standards.

The provisions of ASCE/SEI 12.3.3.3 apply to structural members that support discontinuous frames or shear wall systems where the discontinuity is severe enough to be deemed a structural irregularity.

General Structural Integrity

Provisions for structural integrity are given in IBC 1616 and are applicable to buildings classified as high-rise buildings in accordance with IBC 403 and assigned to Risk Category III or IV with frame structures or bearing wall structures. A high-rise building is defined in IBC 202 as a building with an occupied floor located more than 75 feet above the lowest level of fire department vehicle access.

ASCE/SEI 1.4 also contains general structural integrity requirements that are applicable to all structures.

These minimum strength criteria help ensure that structural integrity is maintained for anticipated and minor unanticipated loading events that have a reasonable chance of occurring during the life of the structure. Guidelines for providing general structural integrity are given in ASCE/SEI C1.4.

The load combinations for general structural integrity loads are given in ASCE/SEI 2.6 for strength design and allowable stress design:

Strength Design Notional Load Combinations

a. $1.2D + 1.0N + L + 0.2S$
b. $0.9D + 1.0N$

Allowable Stress Design Notional Load Combinations

a. $D + 0.7N$
b. $D + 0.75(0.7N) + 0.75L + 0.75(L_r \text{ or } S \text{ or } R)$
c. $0.6D + 0.7N$

The effects from the loads specified in ASCE/SEI 1.4.1 through 1.4.4 are defined as notional load effects, N. Note that wind load effects, W, and seismic load effects, E, are not included in these load combinations.

ASCE/SEI C1.4 also contains information on general collapse and limited local collapse and provides case studies of the former.

Extraordinary Loads and Events

Requirements for extraordinary loads and events are given in ASCE/SEI 1.4.5, which references ASCE/SEI 2.5. That section provides minimum requirements for strength and stability of a structure where it has been required by the owner or code having jurisdiction that the structure be able to withstand the effects from extraordinary events.

In general, extraordinary events arise from service or environmental conditions that are not traditionally considered in the design of ordinary buildings because their probability of occurrence is low and their duration is short. Fires, explosions, vehicular impact, and tornadoes are all examples of such events. The purpose of these requirements is to help ensure that buildings and structures have sufficient strength and ductility and are adequately tied together so that damage caused by the extraordinary event is relatively small.

A load combination for checking the capacity of a structure or structural element to withstand the effects of an extraordinary event is given in ASCE/SEI Equation 2.5-1:

$$(0.9 \text{ or } 1.2)D + A_k + 0.5L + 0.2S$$

In this equation, A_k is the load or load effect that results from the extraordinary event A. Similar to what is done in the strength design load combinations, the load factor on A_k is set equal to 1.0 because it can be considered a strength-level load.

A factor of 0.9 is to be used on the dead load effect, D, if the dead load has a stabilizing effect; otherwise, a load factor of 1.2 should be used. The load factors on L and S correspond approximately to the mean of the yearly maximum live and snow loads. Roof live loads, L_r, and rain loads, R, are not included in this load combination because they have short durations in comparison to S, and thus the probability of their occurring with A_k is negligible.

ASCE/SEI Equation 2.5-2 is to be used to check the residual load-carrying capacity of a structure or structural element following the occurrence of a damaging event. In particular, selected load-bearing members are to be removed from the structure assuming that they have been critically damaged (i.e., they have essentially no load-carrying capacity), and the capacity of the damaged structure is to be evaluated by the following load combination:

$$(0.9 \text{ or } 1.2)D + 0.5L + 0.2(L_r \text{ or } S \text{ or } R)$$

REFERENCES

1. International Code Council (ICC). 2018. International Building Code. IBC, Washington, DC.
2. Engineering Institute (SEI) of the American Society of Civil Engineers (ASCE). 2016. *Minimum Design Loads for Buildings and Other Structures.* ASCE/SEI 7-16. ASCE/SEI, Reston, VA.
3. Fanella, D. 2018. *Structural Loads 2012 IBC and ASCE/SEI 7-10,* International Code Council (ICC). ICC, Washington, DC.
4. International Conference of Building Officials (ICBO). 1997. *Uniform Building Code.* ICBO, Whittier, CA.
5. American Institute of Steel Construction (AISC). 2004. *Industrial Buildings—Roof to Anchor Rods, Steel Design Guide No. 7,* 2nd ed. AISC, Chicago.
6. International Code Council (ICC). 2012. *International Plumbing Code.* ICC, Washington, DC.
7. Factory Mutual Insurance Company (FMIC). 2011. *Roof Loads for New Construction.* FM Global Property Loss Prevention Data Sheets 1-54. FMIC, Johnston, RI.
8. Simiu, E., and Scanlan, R.H. 1996. *Wind Effects on Structures: Fundamentals and Applications to Design.* 3rd ed. John Wiley & Sons, New York.
9. Davenport, A.G. 1960. "Rationale for Determining Design Wind Velocities." *Journal of the Structural Division* 88:39–68.
10. Smith, B.S., and Coull, A. 1991. *Tall Building Structures: Analysis and Design.* John Wiley & Sons, New York.
11. Federal Emergency Management Agency (FEMA). 2011. *Taking Shelter from the Storm: Building a Safe Room for Your Home or Small Business.* FEMA 320, 3rd ed. FEMA, Washington, DC.
12. European Committee for Standardization. 1995. *Eurocode 1: Basis of Design and Actions on Structures, Part 2–4: Actions on Structures—Wind Actions,* Brussels, Belgium.
13. American National Standards Institute (ANSI). 1982. *Minimum Design Loads for Buildings and Other Structures.* ANSI A58.1-1982. ANSI, New York.
14. Swiss Society of Engineers and Architects (SIA). 1956. *Normen fur die Belastungsannahmen, die Inbetriebnahme und die Uberwachhung der Bauten.* SIA Technische Normen No. 160. SIA, Zurich, Switzerland.
15. American Society of Civil Engineers. 1961. *Wind Forces on Structures. Transactions of the ASCE,* 126(2): 1124–1198.
16. Davenport, A.G., Surry, D., and Stathopoulos, T. 1978. *Wind Loads on Low-Rise Buildings.* Final Report on Phase III, BLWT-SS4. University of Western Ontario, London, Ontario, Canada.
17. International Association for Shell and Spatial Structures. 1991. *Structural Standards for Steel Antenna Towers and Antenna Supporting Structures.* ANSI/EIA/TIA-222-E-1991.
18. American Society of Civil Engineers (ASCE). 2014. *Flood Resistant Design and Construction.* ASCE 24-14. ASCE, Reston, VA.
19. Federal Emergency Management Agency (FEMA). 2001. *Crawlspace Construction for Buildings Located in Special Flood Hazard Areas.* FEMA/FIA-TB-11. FEMA, Washington, DC.
20. Federal Emergency Management Agency (FEMA). 2011. *Coastal Construction Manual.* FEMA P-55. FEMA, Washington, DC.
21. Walton, T.L., Jr., Ahrens, J.P., Truitt, C.L., and Dean, R.G. 1989. *Criteria for Evaluating Coastal Flood Protection Structures.* Technical Report CERC 89-15. U.S. Army Corps of Engineers, Waterways Experiment Station.
22. U.S. Army Corps of Engineers. 1995. *Flood Proofing Regulations.* EP 1165-2-314. Office of the Chief of Engineers.
23. Ellingwood, B. 1981. "Wind and Snow Load Statistics for Probabilistic Design." *Journal of the Structural Division* 107(7): 1345–1350.
24. Galambos, T.V., Ellingwood, B., MacGregor, J.G., and Cornell, C.A. 1982. "Probability-Based Load Criteria: Assessment of Current Design Practice." *Journal of the Structural Division* 108(5): 959–977.
25. International Conference of Building Officials. (ICBO) 1997. *Uniform Building Code.* ICBO, Whittier, CA.
26. Applied Technology Council (ATC). ATC Windspeed by Location. https://www.atcouncil.org/windspeed.

Chapter 2
Structural Analysis

BY

ALY M. SAID *Associate Professor, Architectural Engineering, Pennsylvania State University, University Park, Pennsylvania*

ANAS S. ISSA *Postdoctoral Fellow, Applied Laboratory for Advanced Materials & Structures (ALAMS), University of British Columbia, Kelowna, British Columbia, Canada*

M. SHAHRIA ALAM *Professor, School of Engineering, University of British Columbia, Kelowna, British Columbia, Canada*

PART A
FUNDAMENTALS AND APPLICATIONS TO STRUCTURAL FRAMEWORKS

2.1 INTRODUCTION

The initial step in structural analysis is to idealize, or *model*, the actual structure. This step depends on the complexity of the structure, the purpose for which the analysis is being made, and the available computational tools. It also invariably depends on the time spent on modeling versus the degree of accuracy achieved.

With the continuous progress of personal computers and structural analysis software, it is practicable to model structures for analysis with significant accuracy even into the nonlinear range, while accounting inelastic material behavior and geometry changes under load. Dynamic response of a structure to loads such as earthquake, wind, and waves can be routinely evaluated. Such analyses can provide essential information on postelastic stress redistribution and identify potential failure mechanisms, thus providing better estimates of safety margins. Nonetheless, analyses are made invariably under the assumption that the structure behaves in a linear elastic manner, that the deflections are small, and that the loads are applied slowly. Despite the fact that these assumptions may not all be reasonable approximations in a given case, a linear elastic static analysis is often made as a first step since such an analysis often provides useful information and also provides a context within which the results of more advanced analyses can be evaluated and interpreted.

Structural modeling for analysis must be done with a clear understanding of the structural load-carrying mechanisms that are involved. Experience-based engineering judgment should be exercised to identify essential aspects of structural behavior versus unimportant detail.

2.1.1 Classification of Structures

Structures can be classified for analytical purposes in several ways. They can be considered as a number of structural components that may carry a system of forces. This breakdown into simpler components can be achieved at several levels; thus, a structure may be regarded as consisting of a number of subassemblages or *substructures* until finally arriving at the basic components which are *structural members* or *elements*. This hierarchical breakdown is useful for computational purposes as well as interpretation of behavior.

Structural members or elements can be classified based on the essential force systems they carry and based on their shapes and dimensions. A classification scheme can divide elements into the following three categories:

1. *Line elements:* in which the cross-sectional dimensions are small compared to the length of the member. For the case of the force system being *axial*, either tension or compression, the elements are *bars*, *rods*, *columns*, or *cables*. Alternatively, if the force system has bending or a combination of bending, shear and axial load, the elements can be *beams*, *arches*, or *beam-columns*.
2. *Surface elements* (two-dimensional equivalent of line elements): in which the thickness is small compared to the surface dimensions. For the case of the force system being two-dimensional and lying in the element's own plane, the elements are panels, deep beams, and tension and compression membranes. For the case of flexural surface, the elements are referred to as plates or shell.
3. *Solid elements:* in which all three dimensions are of comparative magnitude. These elements are typically reserved for special structures such as raft foundations and dams, where three-dimensional stress is of importance in the analysis.

2.2 FUNDAMENTALS OF ANALYSIS

2.2.1 Description of Structural Configuration

The state of the structure may be expressed in terms of its *degrees of freedom*. These are described as a set of independent *coordinates*, *minimum in number*, required to uniquely describe the shape and orientation of the structure. As a general case, a rigid body in space has six degrees of freedom, which are three Cartesian components of translations and three rotations with a given system of coordinates. A planar truss has $2j$ degrees of freedom: the two Cartesian components of displacement of each joint, where j is the number of joints (a special case based on the absence of bending moments in trusses). A planar frame has $3j$ degrees of freedom: two displacements and a rotation for each joint. In each of these cases, the degree of freedom is a physical, natural choice used to define the deformed shape of the structure. However, the degrees of freedom can be chosen in a rather more abstract manner, which is very beneficial in some cases. For instance, a degree of freedom may simply be a multiplier or amplitude of a particular deflected shape or configuration of the structure. A good example in structural dynamics is the set of mode shapes of vibration of the structure. A general deflected shape of the structure can be specified as a linear combination of these mode shapes. The amplitude of each of these mode shapes in that linear combination is a degree of freedom. Another example is a simply supported beam. The beam rotations at either end supports can be defined as the (two) degrees of freedom for the beam. Otherwise, the deflected shape can be defined by the superposition of two special shapes: a symmetric deflection shape in which the rotations at the ends of the beam are of equal magnitude but in opposite directions and an antisymmetric deflection shape in which the rotations are of equal magnitude and in the same direction. The contribution of each of these shapes that is existing in the general shape can be designated as a degree of freedom. In these latter cases, the independent shapes (the mode shapes, or the symmetric/antisymmetric beam deflection shapes) serve the same function that the Cartesian unit vectors do in the case of the truss joint displacements.

Other Structural Variables

For example, in the case of a planar truss under load, it is evident that if the two Cartesian coordinates of every joint are known, then the position of each joint in the loaded configuration is known. Accordingly, member lengths in the deformed configuration can be calculated, which then can provide member elongations and/or strains. Based on these strains and knowledge of the material behavior, the stresses or, alternatively, member axial forces can be calculated. Hence, the specification of the degrees of freedom directly determines other structural variables of interest. In the procedure just outlined, the degrees of freedom are viewed as the fundamental variables and the strains and stresses as derived variables. Nonetheless, since unique relationships between these variables exist for elastic structures, it is possible that, for example, member forces may be considered as the fundamental variables and the joint displacements and strains as derived variables. The selection of different fundamental variables leads to different approaches of structural analysis. The displacement method and the force method are the two best known such general methods. Various hybrid methods have also been developed, particularly for the finite-element analysis of plates and shells.

2.2.2 Basic Principles

When solving any structural analysis problem, the structural variables must satisfy three basic physical laws, which are described here in the context of a truss:

Equilibrium: The member forces (stresses) and the externally applied loads must satisfy the equations of static (or dynamic) equilibrium.

Compatibility: The member elongations (strains) and the joint displacements must be geometrically compatible to ensure the preservation of structural continuity.

Constitutive laws: The axial force (stress) and elongation (strain) of each member must satisfy the behavior laws for the particular material of which the member or structure is constructed.

For the simple planar truss shown in Fig. 2.1, these relationships are

$$\begin{aligned} f_1 + f_2 \cos a &= p_{x4} \\ f_2 \sin a + f_3 &= p_{y4} \end{aligned} \tag{2.1}$$

$$\begin{aligned} e_1 &= u_{x4} \\ e_2 &= u_{x4} \cos a + u_{y4} \sin a \\ e_3 &= u_{y4} \end{aligned} \tag{2.2}$$

$$\begin{aligned} f_1 &= \frac{A_1 E_1}{L_1} e_1 \\ f_2 &= \frac{A_2 E_2}{L_2} e_2 \\ f_3 &= \frac{A_3 E_3}{L_3} e_3 \end{aligned} \tag{2.3}$$

where f_k = axial tension
e_k = elongation of member k
u_{xi}, u_{yi} = Cartesian components of displacement of joint i
P_{xi}, P_{yi} = external loads acting on joint i

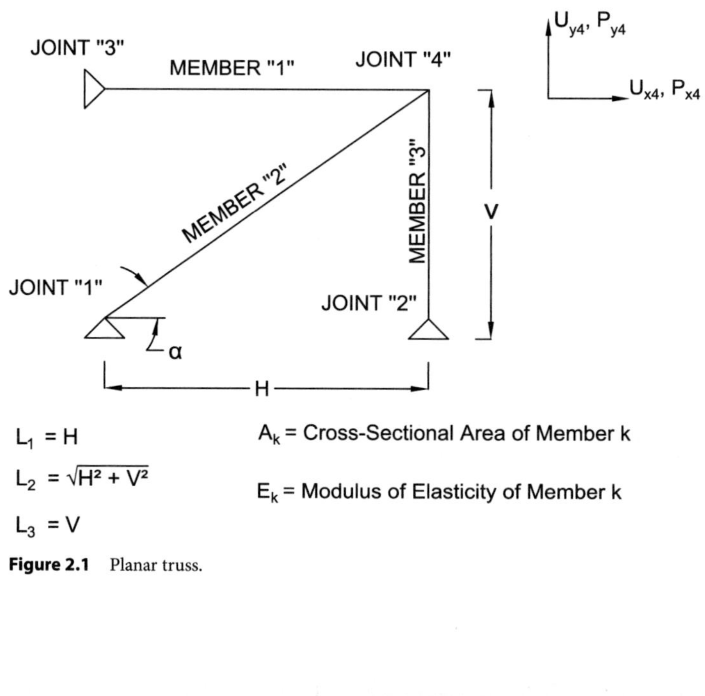

Figure 2.1 Planar truss.

In the general case of a plane or space truss, Eqs. (2.1) through (2.3) can be written as follows:

$$\mathbf{Hf} = \mathbf{p} \tag{2.4}$$

$$\mathbf{e} = \mathbf{Cu} \tag{2.5}$$

$$\mathbf{f} = \mathbf{Ee} \tag{2.6}$$

in which the degrees of freedom are encompassed in the displacement vector **u**, the member forces (stresses) in **f**, and the member elongations (strains) in **e**. If the plane truss has j joints, b members, and $r \geq 3$ restrained degrees of freedom, the *equilibrium matrix* **H** is $(2j - r \times b)$, the *compatibility matrix* **C** is $(b \times 2j - r)$, and the matrix **E,** which contains member stiffness information, is $(b \times b)$. Equations (2.4) through (2.6) display the general approach which appears throughout structural analysis, from simple trusses to finite-element models of very complex structures. In the latter case = (usually numerical), integrations over element volumes are necessary. The force vector **Hf** is often called the *internal resisting force vector* **i**. It contains simply of the member internal forces resolved into the directions of the global degrees of freedom.

2.2.3 Static Determinacy and Stability

Several important special cases can be identified from Eqs. (2.4) through (2.6). If the equilibrium matrix **H** of a plane truss is *square* $(2j - r = b)$ and *nonsingular*, the truss is *statically determinate* and *statically stable* (with appropriate geometric configuration); the member forces **f** can be calculated from the equations of joint equilibrium, Eq. (2.4), and the member elongations and joint displacements can be consequently determined. If $2j - r > b$, the truss is unstable; there are insufficient member forces to satisfy statics for all possible loading conditions. If $2j - r < b$, the truss is *statically indeterminate.* It is possible that a truss is unstable even if $2j - r \geq b$. This condition corresponds to a geometric arrangement of members and/or supports that produces instability.[1] Instability, as examined here, refers to the possibility of (possibly small) rigid-body motions of the entire structure or of substructures within the structure taking place without requiring member deformations. Such rigid-body modes are called *kinematic modes*. This type of instability must be distinguished from elastic instability or *buckling*.

If the truss is statically indeterminate, the compatibility and stress-strain relations must be used in the solution. In this case, the joint displacements can be expressed as

$$\mathbf{u} = \mathbf{K}^{-1}\mathbf{p} \tag{2.7}$$

in which the matrix **K** = **HEC** is the *structure stiffness matrix*. The relation $\mathbf{H}^T = \mathbf{C}$ can be verified for the example shown in Fig. 2.1 and Eqs. (2.1) through (2.3). It is true in general, as will be shown later. The matrix **E** is symmetric and positive definite. Consequently, the structure stiffness matrix **K** has the same properties.

Consequences of Statical Indeterminacy

In statically indeterminate structures, particular systems of internal forces can exist in the absence of external loads. These systems of internal forces are sometimes referred to as *prestress, residual stress, autostress,* or *self-stress*. The classification of a structure as statically determinate or indeterminate is not simply for the purposes of analysis. Both types of structures behave in a fundamentally different manner. Some of these differences in behavior (and their consequences) can be concluded from Eqs. (2.4) through (2.7). For instance, in a statically determinate structure undergoing small displacements, the internal forces are determined directly and uniquely from the external loads. If the loads are zero, so are the internal forces. No states of residual stress are possible. Thus, in a statically determinate structure, no internal stresses are generated by, for example, differential temperature changes, differential support movements, or a lack of fit of the structural members. Alternatively, in an indeterminate structure, these effects can produce internal stresses, which can be significant. On the other hand, it is clear from Eqs. (2.4) through (2.7) that relative member stiffnesses influence internal force distributions in indeterminate structures. This has implications on member dimensions for design, which must usually be performed in an iterative design/analysis cycle. A second major implication is that, when inelastic behavior occurs, causing changes in localized stiffness of the structure, a redistribution of internal stresses occurs if the structure can stand the associated deformations. Alternative load paths can accordingly develop in indeterminate structures. Determinate structures, in contrast, have no such ability for stress redistribution and therefore do not possess any additional safety margin.

2.2.4 Superposition

The main characteristic of linear systems is that solutions can be superposed. Accordingly, if loads $\mathbf{p}_1$ and $\mathbf{p}_2$ produce displacements $\mathbf{u}_1$ and $\mathbf{u}_2$, respectively, then a load $\mathbf{p}_1 + \mathbf{p}_2$ produces a displacement $\mathbf{u}_1 + \mathbf{u}_2$. Consequently, the principle of superposition applies for small displacements of elastic systems. This principle is used in developing many of the methods of linear structural analysis and also as a mean of simplifying calculations.

2.2.5 Symmetry

Symmetry properties can be used as an organizing concept for understanding the behavior of particular classes of structures. Furthermore, these properties of structures are routinely used in analysis in order to reduce the amount of necessary computation, especially in the case of large structures. Symmetry is often utilized by the engineer when the external load also possesses some degree of symmetry. However, although the structure must possess some particular geometric symmetry properties in order for these methods to be utilized, the external load need satisfy no particular symmetry conditions.

The simple portal frame shown in Fig. 2.2*a* possesses symmetry about a vertical plane x–x'. The applied loading can be split into two loading conditions: one symmetric with respect to the symmetry plane and one antisymmetric with respect to this plane.

The *reduced structure* (one-half of the original structure) can then be analyzed under two different sets of boundary conditions at the plane of symmetry, as shown in Fig. 2.2*b*, and the results superposed. While reflection symmetry is most common, many other kinds of symmetry occur in structures. Rotational symmetry, for example, describes the synthesis of a structure by rotation of a substructure. Symmetry in the calculated results can provide an additional check, even when it is not used to reduce the amount of computation.

2.3 ENERGY PRINCIPLES

Energy principles provide alternative means of viewing some of the basic results and methods of structural analysis.[2] They also lead to extremely capable techniques for obtaining *approximate solutions*, for example, in the finite-element analysis of plates and shells and general two- and three-dimensional solids.

2.3.1 Principle of Virtual Work

The principle of virtual work states that if a body (or mechanical system) in equilibrium under the action of a system of forces is given a small (or virtual) displacement, the "virtual" work done by the system of forces during this displacement is equal to zero. This principle can be regarded as an alternative, parallel statement of the conditions of equilibrium. The principle is not restricted to any particular material behavior assumption, which makes it equally applicable to inelastic and elastic systems.

The concepts involved in this principle can be shown by studying the equilibrium of a rigid particle in a plane (Fig. 2.3). The particle is considered to have zero dimensions; that is, it is a point in two-dimensional space. A system of forces $\bar{F}_1, \bar{F}_2, \ldots, \bar{F}_n$ acts on the particle. The resultant of this system of forces is $\bar{R} = \sum_{i=1}^{n} \bar{F}_i$. The *virtual displacement* $\delta\bar{u}$ is defined as an arbitrary reversible displacement consistent with the constraints. The virtual displacement is arbitrary in both magnitude and direction. Therefore, since the rigid particle in the plane has two degrees of freedom, there are two independent values needed to specify the

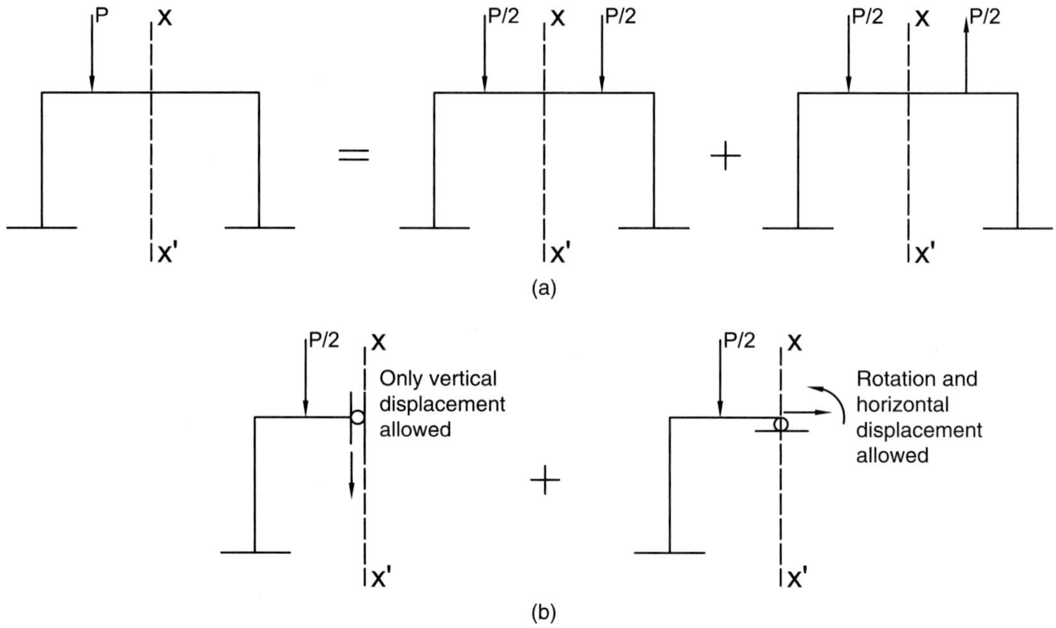

Figure 2.2 Use of symmetry: (*a*) decomposition of load into symmetric and antisymmetric components and (*b*) symmetric and antisymmetric structures of reduced size.

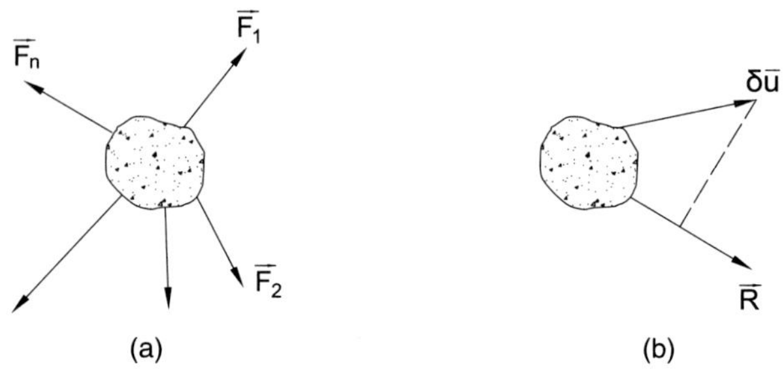

Figure 2.3 Principle of virtual work for a particle: (*a*) rigid particle in two dimensions under a system of forces and (*b*) resultant force and virtual displacement.

virtual displacement. Rather than magnitude and direction, these could instead be taken as the components $\delta\vec{u}_x$ and $\delta\vec{u}_y$ of $\delta\vec{u}$ with respect to Cartesian axes (x, y). The work done by any force $\vec{F}_i$ were it to move through the displacement $\delta\vec{u}$ would be $\vec{F}_i \cdot \delta\vec{u}$. Therefore, the *virtual work* δW done by the system of forces is

$$\delta W = \sum\nolimits_{i=1}^{n} (\vec{F}_i \cdot \delta\vec{u}) = \vec{R} \cdot \delta\vec{u}$$

The principle of virtual work is that the particle is in equilibrium if and only if $\delta W = 0$ for an *arbitrary* $\delta\vec{u}$. The latter condition is essential, as it excludes the special case in which $\delta\vec{u}$ is orthogonal to $\vec{R}$. Therefore, $\delta W = 0$ requires that $\vec{R} = \vec{0}$ for equilibrium. The virtual work can also be expressed as

$$\delta W = \left(\sum\nolimits_{i=1}^{n} F_{xi}\right)\delta u_x = \left(\sum\nolimits_{i=1}^{n} F_{yi}\right)\delta u_y$$

Therefore, $\delta W = 0$ is completely equivalent to $\sum\nolimits_{i=1}^{n} F_{xi} = 0$ and $\sum\nolimits_{i=1}^{n} F_{yi} = 0$, the two equations of static equilibrium. Each equation of equilibrium corresponds to a degree of freedom of the system. The coefficient of each virtual displacement in the expression for δW is called a *generalized force*.

A rigid body with finite dimensions, shown in Fig. 2.4, has three degrees of freedom in two dimensions, two translations δu_x and δu_y in the plane, and a rotation $\delta\theta_z$ about an axis perpendicular to the plane through point O. Therefore, the components of the virtual displacement $\delta\vec{u}_i$ at the point of application (x_i, y_i) of one of the forces $\vec{F}_i$ can be written in terms of the three independent quantities δu_x, δu_y, and $\delta\theta_z$ as

$$\begin{aligned} \delta u_{xi} &= \delta u_x - y_i \cdot \delta\theta_z \\ \delta u_{yi} &= \delta u_x + x_i \cdot \delta\theta_z \end{aligned} \tag{2.8}$$

The virtual work is now given by the expression $\delta W = \sum\nolimits_{i=1}^{n} (\vec{F}_i \cdot \delta\vec{u}_i)$, or

$$\delta W = \left(\sum\nolimits_{i=1}^{n} F_{xi}\right) \cdot \left(\sum\nolimits_{i=1}^{n} F_{yi}\right) \cdot \delta u_y + \left[\sum\nolimits_{i=1}^{n} (-y_i F_{xi} + x_i F_{yi})\right] \cdot \delta\theta_z$$

which gives the three equations of static equilibrium in the plane: $\sum\nolimits_{i=1}^{n} F_{xi} = 0$, $\sum\nolimits_{i=1}^{n} F_{yi} = 0$, and $\sum\nolimits_{i=1}^{n} M_{zi} = 0$.

The body under consideration need not be rigid for the foregoing relations to apply. The virtual displacement was simply chosen as if the body were rigid, but this virtual displacement can equally be imposed on a deformable body. Thus, imposition of rigid-body virtual displacements on a deformable body generates the overall equations of static equilibrium.

A deformable body may also be subjected to virtual displacement patterns, which, if they were actual displacements, would induce deformations in it. Such virtual displacement patterns offer information about the internal forces. For instance, the simple plane truss shown in Fig. 2.1 has two degrees of freedom: u_{x4} and u_{y4}. Let virtual displacements δu_{x4} and δu_{y4} be applied to the structure. These *nodal* virtual displacements define a deflected shape of the entire structure that is a perturbation from the equilibrium position, as illustrated in Fig. 2.5. Such a virtual displacement pattern would, if it were actually applied to the structure, produce deformations in some or all of the structural members. Accordingly, the internal forces, which would resist such changes, must contribute to the virtual-work expression in this case.

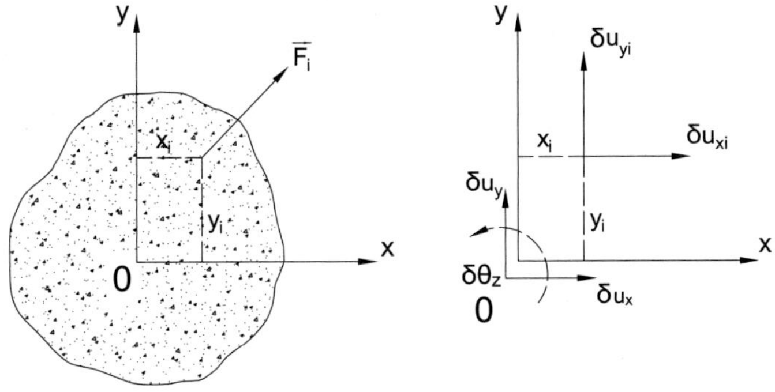

Figure 2.4 Rigid-body virtual displacements.

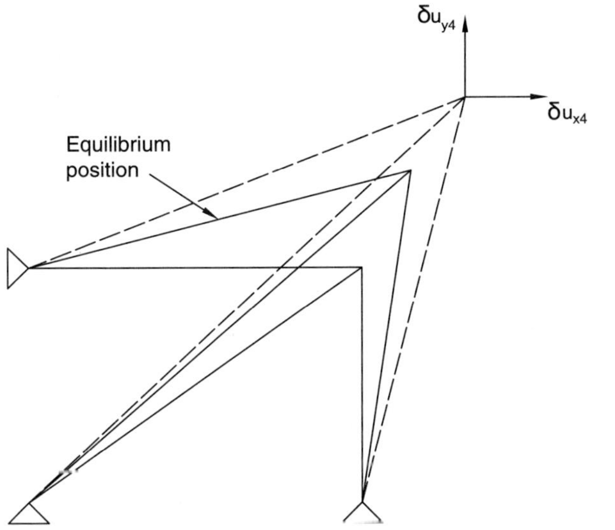

Figure 2.5 Virtual displacement from equilibrium position.

It is important to note that the *forces do not change during the virtual deformation* of the structure.

Since the joints of the truss can be regarded as rigid particles, the preceding relations can be directly applied to them. Equilibrium at joint 4 can be expressed using Eq. (2.1). Thus, the virtual work can be written as

$$\delta W = [p_{x4} - (f_1 + f_2 \cos a)] \cdot \delta u_{x4} + [p_{y4} - (f_2 \sin a + f_3)] \cdot \delta u_{y4} \qquad (2.9)$$

This expression can be separated into two components: the first accounting for external loads, called the *external virtual work* δW_{ext}, and the second accounting for internal forces, called the *internal virtual work* δW_{int}:

$$\delta W_{\text{ext}} = p_{x4} \cdot \delta u_{x4} + p_{y4} \cdot \delta u_{y4} \qquad (2.10a)$$

$$\begin{aligned} \delta W_{\text{int}} &= -[f_1 \cdot \delta u_{x4} + f_2 \cdot (\delta u_{x4} \cos a + \delta u_{y4} \sin a) + f_3 \cdot \delta u_{y4}] \\ &= -[f_1 \cdot \delta e_1 + f_2 \cdot \delta e_2 + f_3 \cdot \delta e_3] \end{aligned} \qquad (2.10b)$$

In the general case, Eqs. (2.4) through (2.6), the virtual work can be expressed as

$$\delta W = \delta \mathbf{u}^T \cdot (\mathbf{p} - \mathbf{Hf}) \qquad (2.11)$$

and the external and internal virtual work can be written as

$$\delta W_{\text{ext}} = \delta \mathbf{u}^T \cdot \mathbf{p} \qquad (2.12a)$$

$$\delta W_{\text{int}} = -\delta \mathbf{u}^T \cdot \mathbf{Hf} = -\delta(\mathbf{H}^T \mathbf{u})^T \cdot \mathbf{f} = -\delta \mathbf{e}^T \cdot \mathbf{f} \qquad (2.12b)$$

Comparison of the latter two expressions in Eq. (2.12*a*) illustrates that $e = \mathbf{H}^T\mathbf{u}$. Since $e = \mathbf{Cu}$ from Eq. (2.5), it follows that $\mathbf{C} = \mathbf{H}^T$. The principle of virtual work can accordingly be expressed as

Compatibility

$$\delta \mathbf{u}^T \cdot \mathbf{p} = \delta \mathbf{e}^T \cdot \mathbf{f} \qquad (2.13)$$

Equilibrium

In Eq. (2.13), the internal force **f** and the external load **p** satisfy equilibrium, Eq. (2.4). The (virtual) joint displacements $\delta\mathbf{u}$ and the corresponding (virtual) member elongations $\delta\mathbf{e}$ satisfy compatibility, Eq. (2.5). Material behavior descriptions are not involved in this relation, and it thus applies to inelastic materials equally. Equation (2.13) may be considered as a *general reciprocal relation* that relates *any compatible set of kinematic variables* ($\delta\mathbf{u}$, $\delta\mathbf{e}$) and *any equilibrated set of static variables* (**p**, **f**). For some purposes, it is convenient to consider these compatible kinematic variables ($\delta\mathbf{u}$, $\delta\mathbf{e}$) as the actual displacements and elongations and the equilibrated static variables (**p**, **f**) as a conveniently chosen auxiliary load system. This is the foundation for the *unit dummy load method* (Article 2.13).

2.3.2 Principle of Minimum Potential Energy

The principle of minimum potential energy handles the displacements or degrees of freedom of the structure as primary variables. It is similar to the displacement or stiffness methods of structural analysis. The (total) *potential energy* Π is equal to the sum of the (internal) *strain energy U* and the load potential, or *external potential energy*, Ω. These quantities are defined as follows.

Strain Energy

The strain energy is the stored energy of deformation. For a one-dimensional structural member with an internal force vs. elongation relation as shown in Fig. 2.6, the strain energy is the work done by the force in generating the current elongation. Thus,

$$U(e) = \int_0^e f(\bar{e})d\bar{e}$$

which is the area under the curve up to the current elongation *e*. The gradual change of force with elongation is accounted for in this calculation.

This quantity involves not only the material response but also the cross-sectional area and length of the member. Accordingly, it is better instead to define a strain energy per unit volume (density), U_0, which

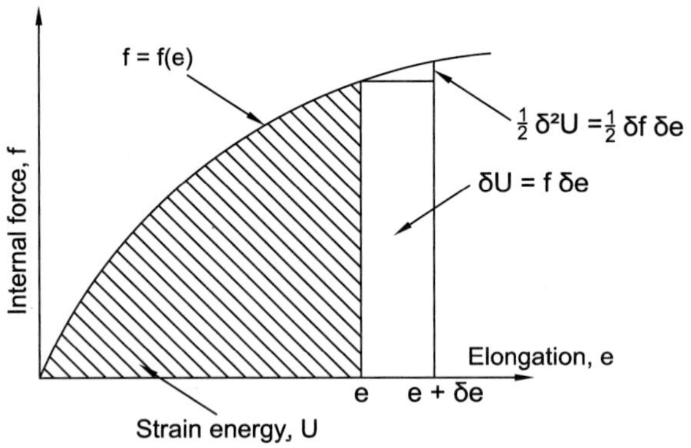

Figure 2.6 Strain energy.

involves only the material response regardless of member dimensions. Thus, in terms of stress σ and strain ε, the strain-energy density

$$U_0(\varepsilon) = \int_0^e \sigma(\bar{\varepsilon})d\bar{\varepsilon}$$

For a linear elastic material under uniaxial stress, $U_0(\varepsilon) = \frac{1}{2}\sigma(\varepsilon)\varepsilon = \frac{1}{2}E\varepsilon^2$, where E is the modulus of elasticity. The first-order change in strain energy due to a change in elongation or strain is expressed as δU and is given by the expression $\delta U = f\delta e$ as shown in Fig. 2.6; therefore, the relationship between the first-order change in strain energy and the internal virtual work is expressed as $\delta U = \delta W_{\text{int}}$.

These relations can be generalized to multiaxial states of stress as

$$U = \frac{1}{2}\int_v [\sigma_x\varepsilon_x + \sigma_y\varepsilon_y + \tau_{xy}\gamma_{xy} + \cdots]dV = \frac{1}{2}\int_v \varepsilon^T \cdot \sigma\, dV \tag{2.14a}$$

$$\delta U = \int_v [\sigma_x\,\delta\varepsilon_x + \sigma_y\,\delta\varepsilon_y + \tau_{xy}\,\delta\gamma_{xy} + \cdots]dV = \int_v \delta\varepsilon^T \cdot \sigma\, dV \tag{2.14b}$$

The expression for U given in Eq. (2.14*a*) is valid for three-dimensional linear elastic systems. The expression for δU in Eq. (2.14*b*) is general.

Expressions for U and δU for different types of structural elements are obtained by introducing in Eq. (2.14) the appropriate assumptions regarding strain distributions in each particular case. For example, in traditional beam theory, it is assumed that plane sections remain plane. The strains, ε_x, are accordingly assumed to vary linearly through the beam depth, so $\varepsilon_x = \varepsilon_{x0} - yk_y - zk_z$, in which the x-axis is the longitudinal axis of the beam and the y- and z-axes are the principal axes of the cross section, ε_x is the axial strain at the beam centroidal axis, and $k_y = 1/R_y$ and $k_z = 1/R_z$ are the beam curvatures for bending in the x-z and the x-y planes, respectively. This strain distribution describes a beam under axial force and biaxial bending about its principal axes. Additionally, transverse shear deformations are neglected. Consequently, introducing these constraints in Eq. (2.14) and separating the volume integration into an integration over the cross section followed by integration along the length of the beam, the following expressions emerge:

$$U_{\text{beam}} = \frac{1}{2}\int_{x=0}^{L} [N_x\varepsilon_{x0} + M_y k_y + M_z k_z]dx \tag{2.15a}$$

$$\delta U_{\text{beam}} = \int_{x=0}^{L} N_x\,\delta\varepsilon_{x0} + M_y\,\delta k_y + M_z\,\delta k_z]dx \tag{2.15b}$$

in which

$$N_{x0} = \int_A \sigma_x\, dA \qquad M_y = -\int_A y\sigma_x\, dA \qquad M_Z = -\int_A z\sigma_x\, dA \tag{2.16}$$

For the truss-type structural system discussed in Eqs. (2.4) through (2.7), the strain energy is described by the following equation:

$$U = \frac{1}{2}\mathbf{e}^T \cdot \mathbf{f} = \frac{1}{2}\mathbf{u}^T(\mathbf{C}^T\mathbf{E}\mathbf{C})\mathbf{u} = \frac{1}{2}\mathbf{u}^T\mathbf{K}\mathbf{u} \tag{2.17}$$

This equation for strain energy in terms of the stiffness matrix **K** is general.

External Potential Energy

For conservative external loadings, a potential energy Ω of the loads can be defined such that $\Omega = -\mathbf{u}^T \cdot \mathbf{p}$.

Potential Energy

The total potential energy Π of the structural system is $\Pi = U + \Omega$. It is a function of the displacements and, consequently, of the degrees of freedom of the structure. The principle of minimum potential energy supplies the criterion for choosing the actual displacements of the structure (i.e., those that also satisfy equilibrium) from the set of all possible displacements that satisfy the constraints (boundary conditions). It may be expressed in the following way: Of all displacements that satisfy the constraints (boundary conditions on displacements), those that also satisfy equilibrium make the potential energy an absolute minimum.

The potential energy is expressed as

$$\Pi = \frac{1}{2}\mathbf{u}^T\mathbf{K}\mathbf{u} - \mathbf{u}^T\mathbf{p} \tag{2.18}$$

If a small displacement increment $\delta\mathbf{u}$ is applied from the equilibrium configuration **u**, the potential energy changes from Π to $\Pi + \Delta\Pi$. The change in potential energy is equal to

$$\begin{aligned}\Delta\Pi &= \delta\mathbf{u}^T(\mathbf{K}\mathbf{u} - \mathbf{p}) + \frac{1}{2}\delta\mathbf{u}^T\mathbf{K}\mathbf{u} \\ &= \delta\Pi + \frac{1}{2}\delta^2\Pi\end{aligned} \tag{2.19}$$

According to the principle, $\Delta\Pi$ must be positive since any displacement away from the equilibrium position is associated with an increase in potential energy. Consequently, for an equilibrium position, $\delta\Pi = 0$, which leads to the equilibrium equations of the stiffness method $\mathbf{K}\mathbf{u} = \mathbf{p}$. The additional requirement for *stable equilibrium* is that $\delta^2\Pi > 0$, which provides analytical conditions for the study of structural stability.

Application of the minimum potential energy method yields the same equations as does the virtual-work method. Both methods lead to effective approximation methods.

2.3.3 Principle of Minimum Complementary Potential Energy

The principle of minimum complementary energy handles the stresses (internal forces) as primary variables. It is akin to the force or flexibility approaches of structural analysis. It provides the criterion for selecting the actual stresses (i.e., those corresponding to compatible strains) from the set of all possible stresses that satisfy equilibrium and stress boundary conditions. It may be expressed in the following way: Of all stresses (internal forces) that satisfy equilibrium and stress boundary conditions, those that also satisfy the associated strain compatibility conditions make the complementary potential energy an absolute minimum.

2.3.4 The Reciprocal Theorem

The reciprocal theorem relates two separate loaded configurations of a linearly elastic structure.[1] Figure 2.7 shows two separate loading conditions acting on the same linearly elastic structure. The displacements are shown in the figure with subscripts that have the following meaning: $\mathbf{d}_{ij}$ is the displacement at point i due to the action of load $\mathbf{P}_j$. The displacements are the components of the resultant structural displacements in the directions of the loads at the same points; that is, the respective force and displacement variables are *work-conjugate*.

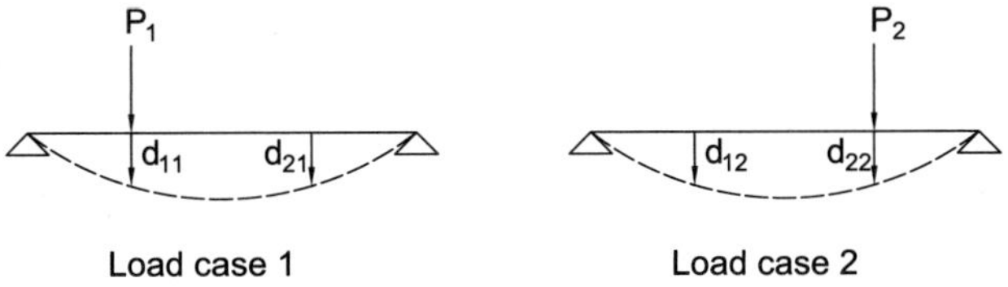

Figure 2.7 The reciprocal theorem.

The reciprocal theorem indicates that $\mathbf{P}_1\mathbf{d}_{12} = \mathbf{P}_2\mathbf{d}_{21}$. It is sometimes stated in terms of unit loads, that is, for the special case $\mathbf{P}_1 = \mathbf{P}_2 = 1$, where it is known as *Maxwell's law*. Moments and rotations can be considered in this relationship as well as forces and displacements. The terms that appear must be work-conjugate pairs.

The two configurations need not be restricted to single loads. The generalization to multiple loads is usually called *Betti's law*. It can be stated in the following form. The work done by the first loading system acting through the displacements produced by the second loading is equal to the work done by the second loading system acting through the displacements produced by the first loading. The main use of these relationships is in demonstrating symmetry of stiffness and flexibility coefficients and in developing a method (the Müller-Breslau principle[2]) for constructing influence lines.

2.4 ANALYSIS OF STATICALLY DETERMINATE STRUCTURES

Statically determinate *models* of structures are often useful. For example, a truss model in which each member carries only axial force can capture the important aspects of structural behavior in many cases, even though the assumption of frictionless pin joints connecting the members is far from accurate. Furthermore, the detailed analysis of members is often carried out as a final (postprocessing) step after the analysis of the full (statically indeterminate) structure. In this case, the indeterminate structural analysis provides sufficient information so that the detailed member analysis can proceed as if the member or substructure were statically determinate.

2.4.1 Plane and Space Trusses

Idealized trusses have frictionless pin joints at which the centroidal axes of connecting members meet (no eccentricities) and are subjected to loads only at these joints. As a result, only axial force, either tension or compression, is transmitted by each member. One main requirement is that loads are applied only at the joints. Truss analysis with both a rigid-jointed frame model and an ideal truss model of the same structure generally yield axial forces that agree closely with each other when loads are applied only at the joints. Additional internal forces (shears and bending moments) exist in the truss, but these are due only to compatibility effects and are not required for equilibrium. The additional stresses due to these bending effects are called *secondary stresses* and are often in the range of 15 to 20 percent or less of the stresses due to the axial forces.

Plane and space trusses are described in general by Eqs. (2.4) through (2.6). For plane trusses, the criterion for static determinacy is that $b + r = 2j$, in which b = number of members, r = number of independent reaction components, and j = number of joints. For space trusses, the criterion for static determinacy is that $b + r = 3j$. The degree of static indeterminacy is $(b + r) - 2j$ and $(b + r) - 3j$, respectively, when these indices are positive. These criteria govern whether or not the complete set of reactions and internal forces can be determined from the equations of equilibrium (statical determinacy). In some cases, external reactions can be calculated, or some member forces can be calculated, even if the structure is statically indeterminate overall, as determined by these criteria.

Certain special truss configurations, called *simple* trusses and *compound* trusses, which are statically determinate overall, do not require a simultaneous solution of the equations of joint equilibrium, Eq. (2.4). A simple truss can be defined as one that can be assembled by starting with a simple triangular-shaped panel and then individually connecting new joints, using two new members for each joint. A compound truss consists of simple-truss substructures. For these two particular configurations, the reactions and member forces can be calculated using the *method of joints* and the *method of sections*. These two methods are also useful for verification of results of analyses of indeterminate structures.

2.4.2 Beams and Frames

Rigid frames, or moment frames, consist of members connected together by joints that are assumed to transfer moments between connecting members. In some cases, more complicated structural models may be used in which some of the joints are assumed to be frictionless pins or in which there are other types of force or moment *releases*. Each such release limits or constrains the internal force system at a given location in the structure and accordingly provides additional information that augments the equations of equilibrium. These additional equations are sometimes referred to as *equations of condition*.

For plane frames, the criterion for static determinacy is that $3b + r = 3j + c$, in which c = number of releases and the remaining symbols are as defined in Article 2.4.1. For space frames, the criterion for static determinacy is that $6b + r = 6j + c$. The degree of static indeterminacy is $(3b + r) - (3j + c)$ and $(6b + r) - (6j + c)$, respectively, when these indices are positive. These criteria govern whether or not the complete set of reactions and internal forces can be calculated from the equations of statics (statical determinacy). As with trusses, it is possible that external reactions can be calculated (if $r = 3 + c$ for plane frames or $r = 6 + c$ for space frames) or that some member forces can be determined, even if the structure is statically indeterminate overall, as determined by these criteria.

Reaction Calculations Using Virtual Work

Throughout structural analysis, there are two alternative approaches: (1) direct application of the basic principles (equilibrium, compatibility, constitutive laws) and (2) a work-energy approach, using virtual work as the fundamental tool. In the case of statically determinate structures, reactions can be determined by directly using the equations of static equilibrium. Alternatively, virtual work can be used to determine external reactions by applying suitably chosen rigid-body virtual displacements that violate the constraint corresponding to the reaction under consideration, that is, to allow the appropriate reaction to do some virtual work. For complex but statically determinate structures with force releases, the virtual-work approach is likely to be more convenient than the direct equilibrium approach.

Figure 2.8 shows a simple portal frame that is pinned at the base and that has a moment release at one of the beam-to-column connections. In order to calculate the vertical reaction R, a virtual-displacement pattern is chosen in which the right-hand support is moved vertically. The resulting pattern is such that the angle between column and girder at the moment of release can change; however, the angle between column and girder at the rigid joint must remain a right angle. The virtual work of the reaction is $R \cdot \delta u$, while the virtual work of the applied distributed load is $-wH \cdot ½\, H/L \cdot \delta u$. These are both contributions to the external virtual work δW_{ext}. The internal virtual work is zero in this case since the virtual-displacement pattern is that of a rigid body. Thus, by setting $\delta W = \delta W_{ext} = 0$, the reaction is found to be $R = wH^2/2L$. A main advantage of the virtual-work approach is that it is often possible to avoid calculating auxiliary information that may not be of interest.

The full analysis of a frame can be considered to cover the determination of external reactions and internal forces in each member at the joints of the frame. The subsequent determination of internal forces in a given member at a given member cross section can be handled as beam analysis. The relations between load, shear, and bending moment in the

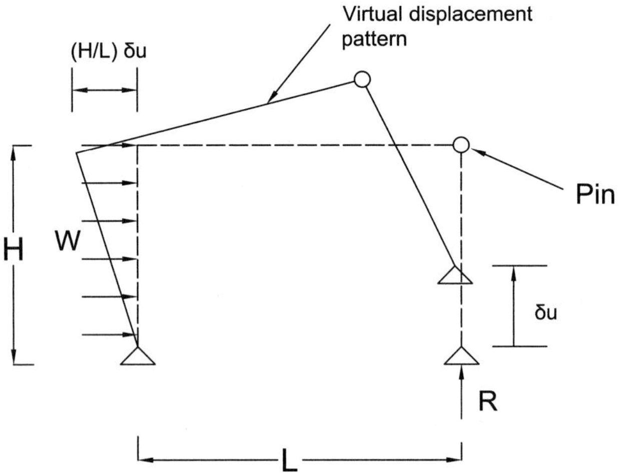

Figure 2.8 Frame reaction using virtual work.

beam are determined by examining a free body of a differential element of the beam (Fig. 2.9), which gives

$$\frac{dV}{dx} = p(x)$$
$$\frac{dM}{dx} = V(x) + m(x) \tag{2.20}$$

where $p(x)$ = distributed lateral load
$m(x)$ = distributed applied moment
The sign convention is shown in Fig. 2.9.

Equation (2.20) provide expressions for the slopes of the shear and moment diagrams at any axial coordinate in the beam. They can be integrated between two arbitrary cross sections of the beam $x = a$ and $x = b$ to provide the following relations:

$$V(b) - V(a) = \int_a^b p(x)dx + \sum_n P_{Ci}$$
$$M(b) - M(a) = \int_a^b V(x)dx + \int_a^b m(x)dx + \sum_n M_{Ci} \tag{2.21}$$

which show that the change in shear between two cross sections is equal to the resultant external load acting on the beam between those two cross sections, and the change in bending moment between two cross sections is equal to the area under the shear diagram plus the resultant external moment acting on the beam between those two cross sections.

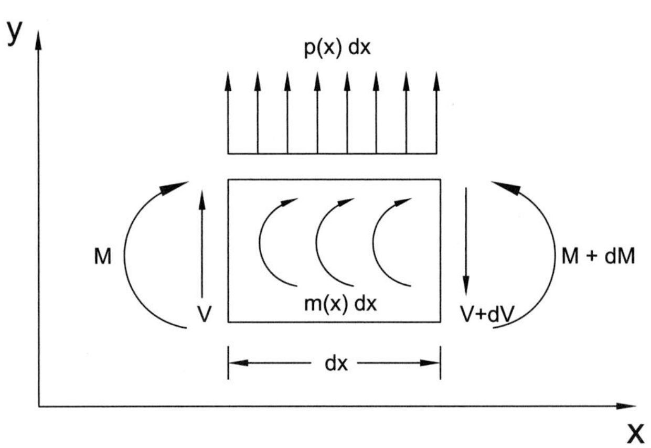

Figure 2.9 Load, shear, and bending moment sign convention.

2.4.3 Beam Deflections

Methods for the calculation of beam deflections are of two general types: they are based either on (1) the integration of the differential equation for beam deflections (using double integration or one of its many equivalents, such as the moment-area method or conjugate beam) on (2) work-energy methods (e.g., dummy unit-load method).

Differential Equation for Beam Deflections

Equations (2.20) relate load to shear and shear to bending moment. Two additional differential relations complete the definition of beam response as follows:

$$\frac{d\theta}{ds} = \frac{M}{EI}$$

$$\frac{dy}{dx} = \tan\theta$$

where EI = flexural stiffness of beam
θ = angle between x-axis and tangent to beam axis
If rotations of the beam axis are small, $\theta^2 << 1$ (in radians), the curvature $d\theta/ds$ can be approximated as $d\theta/dx$ and the slope $dy/dx = \tan\theta$ as θ so that these equations can be written as

$$\frac{d\theta}{dx} = \frac{M(x)}{EI}$$
$$\frac{dy}{dx} = \theta(x) \tag{2.22}$$

Equations (2.22) are often combined into the single relation:

$$\frac{d^2y}{dx^2} = \frac{M(x)}{EI}$$

For a statically determinate beam, the equation for bending moment can be determined by statics, and then the deflected shape from this equation can be determined by double integration or a number of alternative equivalent methods.

Table 2.1 gives some common beam formulas, showing in most cases maximum values of shear, moment, slope, and deflection.

EXAMPLE

For the first case in Table 2.1, the bending moment $M(x) = M_0(1 - x/L)$. Two successive integrations of $d^2y/dx^2 = m(x)/El$ give

$$y(x) = \frac{M_0}{EI}\left(\frac{x^2}{2} - \frac{x^3}{6L}\right) + C_1x + C_2$$

Constants C_1 and C_2 are found from the boundary conditions $y(0) = 0$ and $y(L) = 0$. Then

$$y(x) = \frac{M_0L^2}{6\,EI}\left[-2\left(\frac{x}{L}\right) + 3\left(\frac{x}{L}\right)^2 - \left(\frac{x}{L}\right)^3\right]$$

When the boundary conditions are such that the beam is not statically determinate, Eqs. (2.20) and (2.22) must be solved together rather than sequentially.

Classical Analogs for Double Integration

Equations (2.20) and (2.22) have a similar structure that produces some of the approaches used for calculating beam deflections. For instance, slope, θ, and deflection, y, can be analogous with shear, V, and bending moment, M, respectively. Consequently, analyzing a fictitious beam with M/EI as a distributed load can yield information on slope and deflection. This is the basis for the *conjugate beam* method.[3] Other well-known methods, such as the *moment-area method*,[4] also use

Table 2.1 Formulas for Beams

Structure	Shear	Moment	Slope	Deflection
		Simply supported beam		
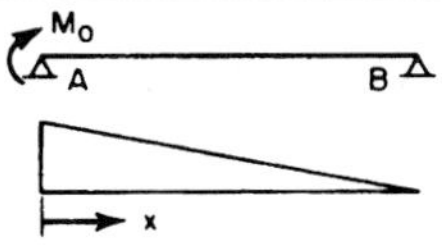	$S_A = -\frac{M_o}{L}$	M_o	$\theta_A = \frac{M_o L}{3EI}$ $\theta_B = -\frac{M_o L}{6EI}$	$y_{max} = 0.062\frac{M_o L^2}{EI}$ at $x = 0.422L$
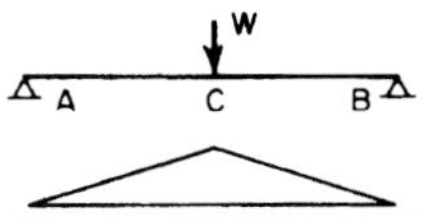	$S_A = \frac{W}{2}$	$M_C = \frac{WL}{4}$	$\theta_A = -\theta_B = \frac{WL^2}{16EI}$	$y_C = \frac{WL^3}{48EI}$
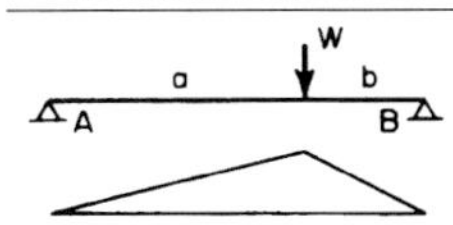	$S_A = \frac{Wb}{L}$ $S_B = -\frac{Wa}{L}$	$M_a = \frac{Wab}{L}$	$\theta_A = \frac{Wab}{6EIL}(L+b)$ $\theta_B = -\frac{Wab}{6EIL}(L+a)$	$y_a = \frac{Wa^2b^2}{3EIL}$
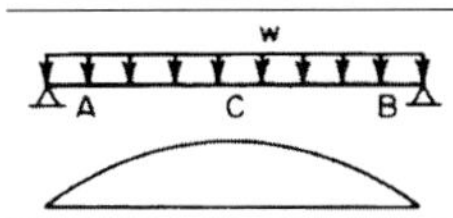	$S_A = \frac{wL}{2}$	$M_C = \frac{wL^2}{8}$	$\theta_A = -\theta_B = \frac{wL^3}{24EI}$	$y_C = \frac{5wL^4}{384EI}$
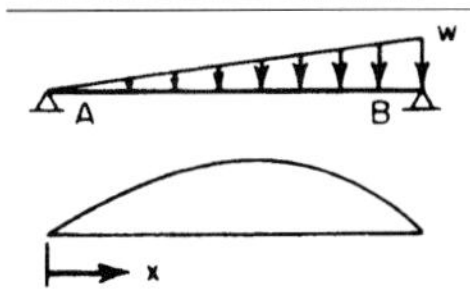	$S_A = \frac{wL}{6}$ $S_B = -\frac{wL}{3}$	$M_{max} = 0.064\ wL^2$ at $x = 0.577L$	$\theta_A = \frac{7wL^3}{360EI}$ $\theta_B = -\frac{8wL^3}{360EI}$	$y_{max} = 0.00652\frac{wL^4}{EI}$ at $x = 0.519L$
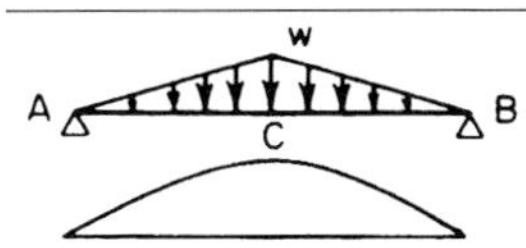	$S_A = \frac{wL}{4}$	$M_C = \frac{wL^2}{12}$	$\theta_A = -\theta_B = \frac{5wL^3}{192EI}$	$y_C = \frac{wL^4}{120EI}$
		Fixed beam		
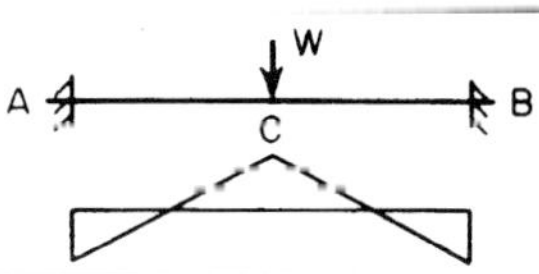	$S_A = \frac{W}{2}$	$M_C = \frac{WL}{8}$	$\theta_A = \theta_B = 0$	$y_C = \frac{WL^3}{192EI}$
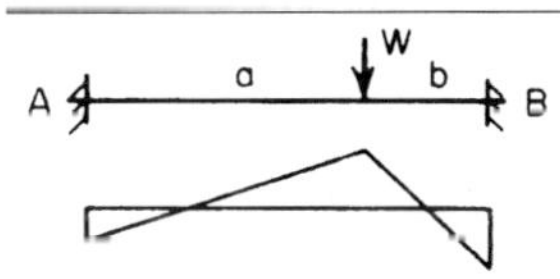	$S_A = \frac{Wb^2}{L^3}(3a+b)$ $S_B = -\frac{Wa^2}{L^3}(3b+a)$	$M_A = -\frac{Wab^2}{L^2}$ $M_B = -\frac{Wba^2}{L^2}$	$\theta_A = \theta_B = 0$	$y_a = \frac{Wa^3b^3}{3EIL^3}$
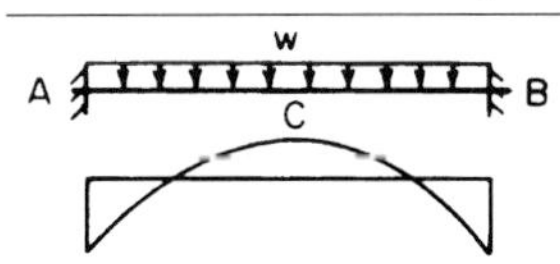	$S_A = \frac{wL}{2}$	$M_A = M_B = -\frac{wL^2}{12}$	$\theta_A = \theta_B = 0$	$y_C = \frac{wL^4}{384EI}$
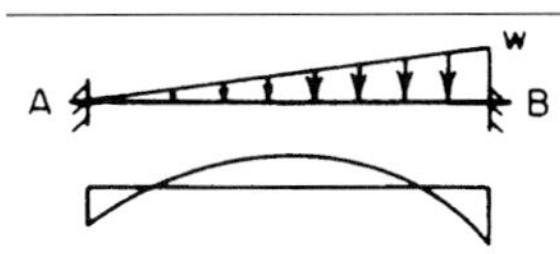	$S_A = \frac{3wL}{20}$ $S_B = -\frac{7wL}{20}$	$M_A = -\frac{wL^2}{30}$ $M_B = -\frac{wL^2}{20}$	$\theta_A = \theta_B = 0$	$y_{max} = 0.00131\frac{wL^4}{EI}$ at $x = 0.525L$
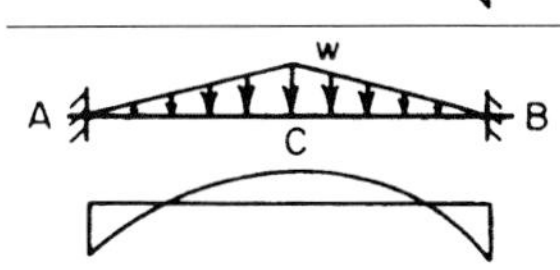	$S_A = \frac{wL}{4}$	$M_A = M_B = -\frac{5wL^2}{96}$	$\theta_A = \theta_B = 0$	$y_C = \frac{0.7wL^4}{384EI}$

(Continued)

Table 2.1 Formulas for Beams (*Continued*)

Structure	Shear	Moment	Slope	Deflection
Cantilever beam				
	o	M_o	$\theta_A = \frac{M_O L}{EI}$	$y_A = -\frac{M_O L^2}{2EI}$
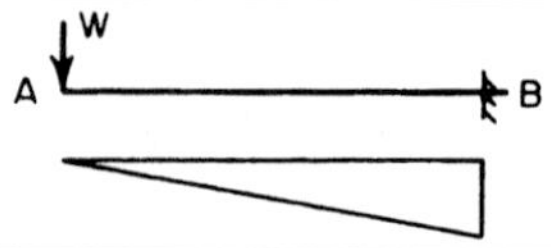	w	$M_B = -WL$	$\theta_A = -\frac{WL^2}{2EI}$	$y_A = \frac{WL^3}{3EI}$
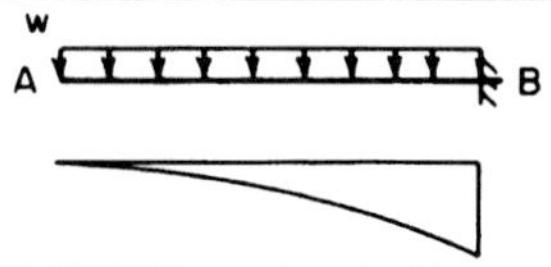	$S_B = -wL$	$M_B = -\frac{wL^2}{2}$	$\theta_A = -\frac{wL^3}{6EI}$	$y_A = \frac{wL^4}{8EI}$
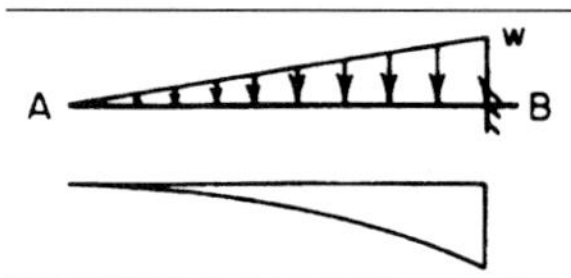	$S_B = -\frac{wL}{2}$	$M_B = -\frac{wL^2}{6}$	$\theta_A = -\frac{wL^3}{24EI}$	$y_A = \frac{wL^4}{30EI}$
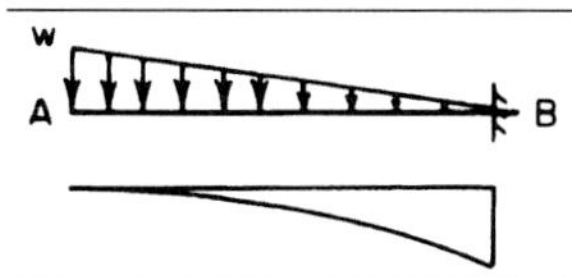	$S_B = -\frac{wL}{2}$	$M_B = -\frac{wL^2}{2}$	$\theta_A = -\frac{wL^3}{8EI}$	$y_A = \frac{11wL^4}{120EI}$
Propped cantilever				
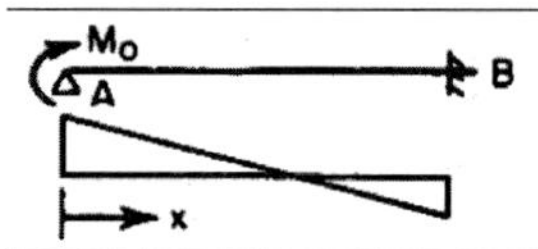	$S_A = -\frac{3M_O}{2L}$	$M_B = -\frac{M_O}{2}$	$\theta_A = \frac{M_O L}{4EI}$	$y_{max} = \frac{M_O L^2}{27EI}$ at $x = \frac{L}{3}$
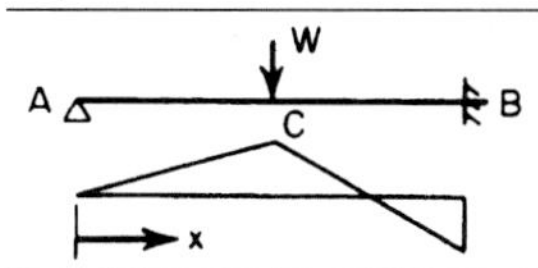	$S_A = \frac{5W}{16}$	$M_B = -\frac{3WL}{16}$ $M_C = \frac{5WL}{32}$	$\theta_A = \frac{WL^2}{32EI}$	$y_{max} = 0.00932\frac{WL^3}{EI}$ at $x = 0.447L$
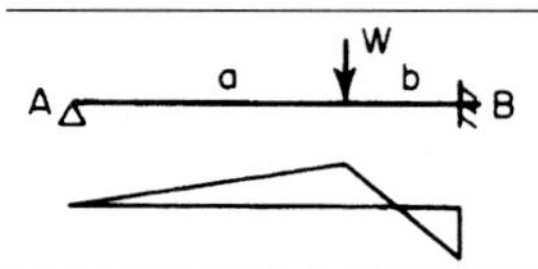	$S_A = \frac{Wb^2}{2L^3}(a+2L)$ $S_B = -\frac{Wa}{2L^3}(3L^2-a^2)$	$M_B = -\frac{Wab}{L^2}\left(a+\frac{b}{2}\right)$	$\theta_A = \frac{Wab^2}{4EIL}$	$y_a = \frac{Wa^2b^3}{12EIL}(3L+a)$
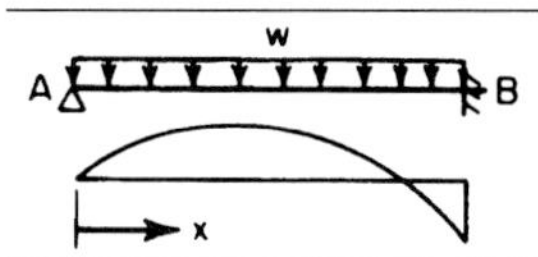	$S_A = \frac{3wL}{8}$	$M_B = -\frac{wL^2}{8}$	$\theta_A = -\frac{wL^3}{48EI}$	$y_{max} = 0.0054\frac{wL^4}{EI}$ at $x = 0.422L$
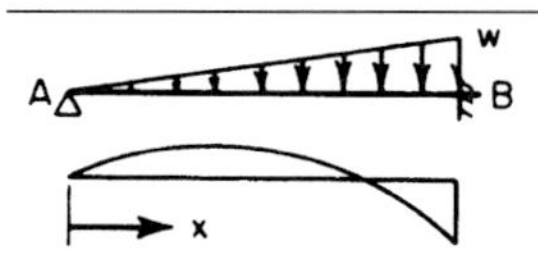	$S_A = \frac{wL}{10}$	$M_{max} = 0.03wL^2$ at $x = 0.447L$ $M_B = -\frac{wL^2}{15}$	$\theta_A = \frac{wL^3}{120EI}$	$y_{max} = 0.00239\frac{wL^4}{EI}$ at $x = 0.447L$
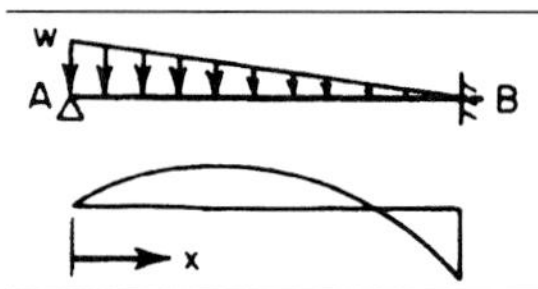	$S_A = \frac{11wL}{40}$	$M_{max} = 0.0423wL^2$ at $x = 0.329L$ $M_B = -\frac{7wL^2}{120}$	$\theta_A = \frac{wL^3}{80EI}$	$y_{max} = 0.00305\frac{wL^4}{EI}$ at $x = 0.402L$

this analogy. The moment-area method is presented in a geometrical approach that emphasizes visualization of the beam's deflected shape. It is usually stated in the form of two moment-area principles that specify procedures for computing (1) changes $\Delta\theta$ in the beam tangent angle between two beam cross sections and (2) deflection of the beam at a given section with respect to the tangent drawn from another section. The methods used for calculating these quantities in the moment-area method are equivalent to Eq. (2.21) for calculating changes in shear and bending moment in the beam, using the curvature M/EI as the load. These methods can be used to compute deflections due to general inelastic strains by substituting the actual curvature, however it is caused, for the term M/EI.

The Dummy Unit-Load Method

Deflection can be also calculated using work-energy methods. The dummy unit-load method is one of the most versatile of this class of methods. Alternatives such as Castigliano's second theorem[1] are similar to the dummy unit-load method. The dummy unit-load method uses the principle of virtual work, displayed in Eq. (2.13) for trusses. The actual strains and displacements of the structure are chosen as the compatible set of kinematic variables. The equilibrated set of static variables is selected as an external load of unit magnitude, the dummy unit-load (placed on the structure so that it is work-conjugate to the desired displacement), and its associated internal forces. Thus, the expression for the external virtual work reduces to $\delta\mathbf{u}^T \cdot \mathbf{p} = 1 \cdot \Delta$, where Δ is the desired displacement. The dummy unit-load method can then be expressed in the general form

$$\Delta = \int_V \varepsilon^T \cdot \sigma dV \tag{2.23}$$

in which the dummy unit load and its associated stresses σ satisfy equilibrium and the displacement Δ results from the actual strains ε.

The displacement that is calculated is actually a *generalized* displacement—any type of deformation response (rotations, relative deflections, etc.) can be computed by applying the appropriate work-conjugate dummy unit load. A rotation can be computed using the dummy unit load as a concentrated moment. Relative deflections can be computed using pairs of dummy unit loads.

A characteristic of the dummy unit-load method is that unnecessary intermediate results need not usually be computed. On the other hand, a single calculation gives the generalized displacement at one location only in the structure, and if a more complete description of the deflected shape is needed, other approaches may be more practical.

In order to use the dummy unit-load method, the internal virtual-work expression given in general form in Eq. (2.23) must be made more specific.

Internal-Virtual-Work Expressions

The internal-virtual-work expressions for specific types of structural members are derived from Eq. (2.23) by imposing the appropriate kinematic assumptions. For beams subjected to axial force and biaxial bending, the result is Eq. (2.15). For the case when torsional rotations and transverse shear deformations are also taken into account, the internal virtual work in a single member can be expressed as

$$\delta W_{\text{int}} = -\int_{x=0}^{L} (N_x\, \delta\varepsilon_{x0} + M_y\, \delta k_y + M_z\, \delta k_z + V_y\, \delta\gamma_y + V_z\, \delta\gamma_z + T\, \delta\phi) dx \tag{2.24}$$

The internal virtual work for the structure is calculated by summing member contributions.

For members of a truss, the axial force N_x is constant along the member length, and

$$\int_{x=0}^{L} \delta\varepsilon_{x0}\, dx = \delta e$$

Accordingly, the internal virtual work for the truss whole structure is $-\sum_k N_k \delta_{ek}$, as indicated in Eq. (2.13).

The dummy unit-load formula given by Eq. (2.23) can, therefore, be written as

$$\Delta = \sum_k \left[n_k e_k + \int_{x=0}^{L_k} (m\kappa + \upsilon\gamma + t\phi) dx \right] \tag{2.25}$$

for the case of planar bending, in which the (constant) axial force n, bending moment m, shear force υ, and torsional moment t are the internal force system associated with the dummy unit load and e, κ, θ, and ϕ are the corresponding actual deformations causing the generalized deflection Δ. The summation is over all members in the structure.

The deflection formula, Eq. (2.25), separates the total deflection into distinct terms, axial length changes, and flexural, shear, and torsional deformations. Additionally, the actual deformations need not be caused by loads—they may be due to, for example, temperature changes in the structure. In the case of load-produced deformations, the deformations are related to the internal forces N, M, V, and T by

$$e = \frac{NL}{Ae} \quad \kappa = \frac{M}{EI} \quad \gamma = \frac{V}{A_s G} \quad \phi = \frac{T}{GJ} \tag{2.26}$$

The effective shear area A_s that appears in Eq. (2.26) accounts for the nonuniform distribution of shear stress over the beam cross section. For rectangular cross sections, it is typically taken as $A_s = A/1.2$. For WF sections, a good approximation is $A_s = A_{\text{web}}$. The calculation of torsional deformations does not account for warping torsion effects.

EXAMPLE

To calculate the vertical deflection at midspan of the simple, statically determinate truss shown in Fig. 2.10, a unit load is placed at point C as shown. For trusses, Eq. (2.25) reduces to

$$\Delta = \sum_k n_k e_k \tag{2.27}$$

If the member elongations are caused only by its internal forces, then Eq. (2.27) specializes further to

$$\Delta = \sum_k n_k \frac{N_k L_k}{A_k E_k} \tag{2.28}$$

The calculations are summarized in Table 2.2. The member length changes due to the actual load are shown in the column headed "e." The member forces n_k due to the dummy unit load are shown in the column headed "n." The quantity n_k is the vertical deflection of joint C due to a unit length change of member k with all other members remaining unchanged in length. Accordingly, if member AF were to be elongated by, say, 0.1 in. (due to any cause), then joint C would deflect vertically by $(-0.707)(0.1) = -0.0707$ in. (the negative sign indicates that the deflection is in the opposite direction to the unit load, which is upward in this case). The vertical component of the deflection of joint C is calculated as 0.59 in.

EXAMPLE

The calculation of flexural deflections shown by the simple portal frame in Fig. 2.11*a*. The horizontal deflection of joint C is sought. Accordingly, a dummy unit load is placed at C in the horizontal direction (Fig. 2.11*c*). Ignoring axial and shear deformations, Eq. (2.25) becomes

$$\Delta = \sum_{\text{k}} \left(\int_{x=0}^{L_k} m \frac{M}{EI} dx \right) \tag{2.29}$$

in which M/EI are the member curvatures due to internal bending moments and m are the bending moments corresponding to the dummy unit load. The flexural stiffness EI is taken as constant and is the same for all members. The curvatures are shown in Fig. 2.11*b* and the bending moments due to the dummy unit load in Fig. 2.11*c*.

Values of typical integrals that appear in Eq. (2.29) are given in Table 2.3. For the frame of Fig. 2.11, they have the values $5/24wH^4/EI$ for AB, $wLH^3/6EI$ for BC, and 0 for CD. Thus, the horizontal component of deflection of joint C based solely on flexure of the frame is $\Delta_C = (5/24wH^4/EI)\,(1 + 0.8L/H)$.

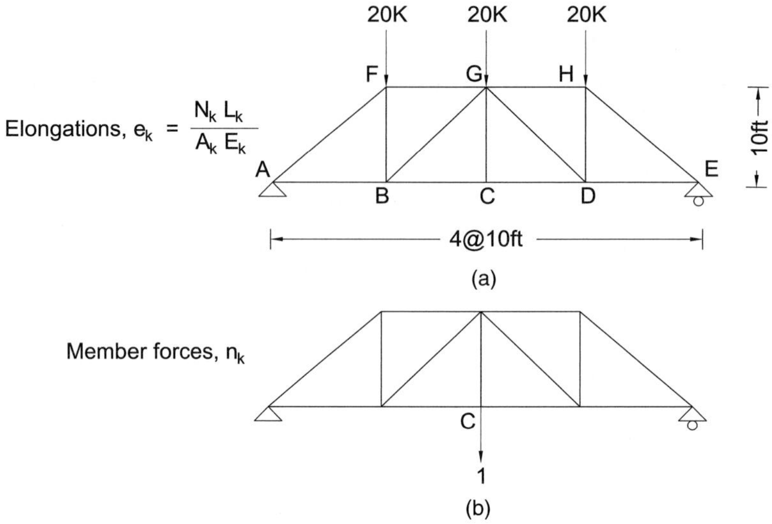

Figure 2.10 Truss deflection by dummy unit-load method. (*a*) Actual load. (*b*) Dummy unit load.

Table 2.2 Deflection of Truss of Fig. 2.10*

Member	*N*, kips	*L*, in	*A*, in^2	*e*, in	*n*	*ne*, in
AB, DE	+30.0	120	1.5	0.080	+0.5	+0.040
BC, CD	+40.0	120	2.0	0.080	+1.0	+0.080
FG, GH	−30.0	120	2.0	−0.060	−0.5	+0.030
AF, EH	−42.4	169.7	2.5	−0.096	−0.707	+0.068
BC, DG	−14.1	169.7	1.0	−0.080	−0.707	+0.057
BF, DH	+10.0	120	1.0	0.040	+0.5	+0.020
CG	+0.0	120	1.0	0.000	+1.0	+0.000
					$\Sigma ne =$	+0.59

*E = 30,000 ksi for all members.

2.5 ANALYSIS OF STATICALLY INDETERMINATE STRUCTURES

The two fundamental approaches of structural analysis of statically indeterminate structures are the *force method* and the *displacement method.* The force method is also referred to as the *compatibility* method or the *flexibility* method. The displacement method is also referred to as the *equilibrium* method or the *stiffness* method.

2.5.1 Comparison of Force and Displacement Methods

The force method and the displacement method are analogous methods with corresponding steps. In the force method, equilibrium is satisfied at all stages, and the governing equations of the method result from imposing compatibility conditions. In the displacement method, compatibility is satisfied at all stages, and the governing equations of the method result from imposing equilibrium equations. The force method is closely related to the principle of minimum complementary potential energy, and the displacement method is closely related to the principle of minimum potential energy.

The displacement method is easily adapted in simple computer codes for structural analysis because of a number of practical advantages. Nonetheless, the concepts embodied in the force method are fundamental to a full understanding of structural theory.

The three-span continuous beam shown in Fig. 2.12 illustrates the fundamentals of these methods.

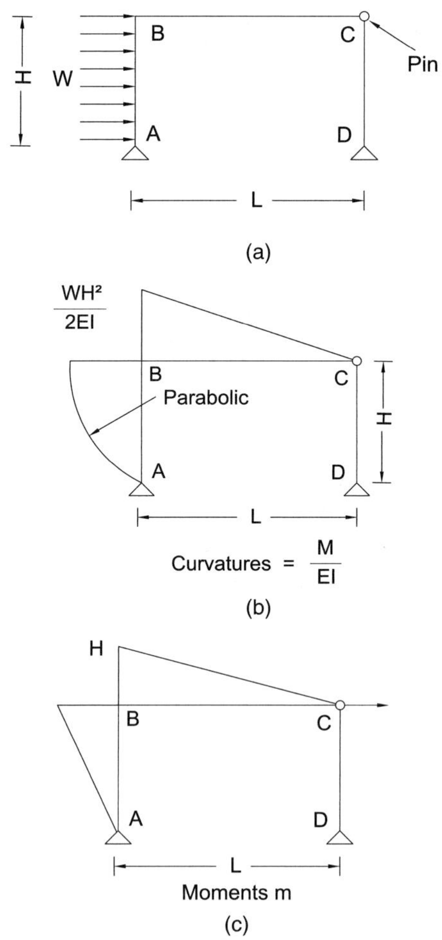

Figure 2.11 Frame deflection by dummy unit-load method.

Table 2.3 Values of $\int_0^L L\,m\,M\,dx$

	Linear M diagrams				Parabolic M diagrams		
	m, L	m_0, L	m_1, L	m_0, m_1, L	Origin, M_1, L/2, L/2	Origin, M_1, L	Origin, M_0, M_1, L
M, L	mML	$\frac{1}{2}mM_0L$	$\frac{1}{2}mM_1L$	$\frac{1}{2}mL(M_0+M_1)$	$\frac{2}{3}mM_1L$	$\frac{1}{3}mM_1L$	$\frac{1}{3}mL\,(2M_0-M_1)$
M_0, L	$\frac{1}{2}m_0ML$	$\frac{1}{3}m_0M_0L$	$\frac{1}{6}m_0M_1L$	$\frac{1}{6}m_0L(2M_0+M_1)$	$\frac{1}{3}m_0M_1L$	$\frac{1}{12}m_0M_1L$	$\frac{1}{12}m_0L(5M_0-M_1)$
M_1, L	$\frac{1}{2}m_1ML$	$\frac{1}{6}m_1M_0L$	$\frac{1}{3}m_1M_1L$	$\frac{1}{6}m_1L(2M_1+M_0)$	$\frac{1}{3}m_1M_1L$	$\frac{1}{4}m_1M_1L$	$\frac{1}{4}m_1L(M_0-M_1)$
M_0, M_1, L	$\frac{1}{2}ML(m_0+m_1)$	$\frac{1}{6}M_0L(2m_0+m_1)$	$\frac{1}{6}M_1L(m_0+2m_1)$	$\frac{L}{6}[m_0(2M_0+M_1)$ $+m_1(2M_1+M_0)]$	$\frac{1}{3}M_1L(m_0+m_1)$	$\frac{1}{12}M_1L(m_0+3m_1)$	$\frac{L}{12}\begin{bmatrix} m_0(5M_0-M_1) \\ +3m_1(M_0-M_1) \end{bmatrix}$

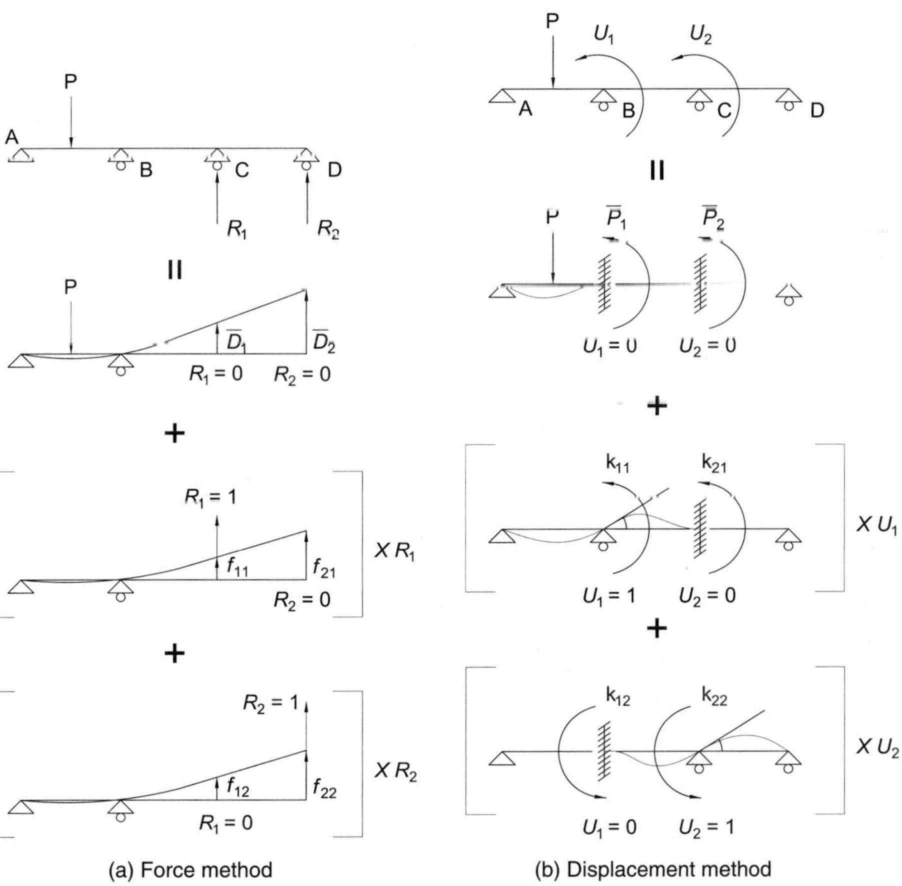

Figure 2.12 Comparison of (*a*) force and (*b*) displacement methods.

FORCE METHOD

Consider a structure that is statically indeterminate to the nth degree. For the beam shown in Fig. 2.12a, $n = 2$. A statically determinate stable structure, the *primary structure*, is produced by introducing sufficient releases that render the structure statically determinate. The generalized forces corresponding (in a work-conjugate sense) to these releases are the *redundants*. The primary structure is now subjected to $n + 1$ loading cases: the first of these is the external loading condition, and each of the remaining n loading cases corresponds to a unit value of one redundant with all other redundants set equal to zero. The generalized displacements at the releases in each of these loading cases are defined as the flexibility coefficients. The flexibility coefficient f_{ij} is the generalized displacement work-conjugate to redundant R_j due to a unit value of redundant R_j with all other redundants equal to zero.

A superposition of the $n + 1$ loading cases is constructed to generate expressions for the total generalized displacement at each release, as indicated in Fig. 2.12a. For this example, these two generalized displacements are the vertical displacements at supports C and D, which must be zero in the indeterminate beam. The equations for this example are

$$\begin{aligned} f_{11}R_1 + f_{12}R_2 + \bar{D}_1 = D_1^0 = 0 \\ f_{21}R_1 + f_{22}R_2 + \bar{D}_2 = D_2^0 = 0 \end{aligned} \tag{2.30}$$

Thus, the equations of the force method define compatibility conditions that are satisfied by choosing the appropriate combination (i.e., the values of the redundants R_i) of the n loading cases to superpose on the external load case.

In general, the equations of the force method have the form

$$\begin{aligned} \mathbf{F}\cdot\mathbf{R} + \bar{\mathbf{D}} = \mathbf{D^0} \\ \mathbf{F}\cdot\mathbf{R} = \mathbf{D} \end{aligned} \tag{2.31}$$

where $\mathbf{F}$ = symmetric, positive definite structure flexibility matrix

$$\mathbf{D} = \mathbf{D}^0 - \bar{\mathbf{D}}\cdot\bar{\mathbf{D}}$$

= generalized displacement of primary structure due to external load

$\mathbf{D}^0$ = specified generalized displacement at releases (often equal to zero)

DISPLACEMENT METHOD

Consider a structure that is kinematically indeterminate to the nth degree. This means that there are n unknown degrees of freedom U_i. Kinematic indeterminacy in the displacement method corresponds to static indeterminacy in the force method. In the beam shown in Fig. 2.12b, the degree of kinematic indeterminacy (the number of degrees of freedom) is $n = 2$. There is, in general, no particular correlation between the degree of indeterminacy of a structure and the number of degrees of freedom it has.

The first step in the displacement method is to produce a kinematically *determinate* structure by restraining all degrees of freedom. This fixed-end structure is the entity in the displacement method that corresponds to the primary structure in the force method. It is usually highly statically indeterminate.

For simplification, external loads are separated into two types of loads: element loads that are applied to the members and joint loads. The fixed-end structure is now subjected to $n + 1$ load cases. The first consists of the external member loads acting on the fully restrained structure. Each of the remaining n loading cases corresponds to a unit value of one of the degrees of freedom with all other degrees of freedom equal to zero. The generalized forces at the degrees of freedom in each of these loading cases are known as the *stiffness coefficients*. The stiffness coefficient k_{ij} is the generalized force work-conjugate to the degree of freedom U_i which is due to a unit value of degree of freedom U_j with all other degrees of freedom equal to zero.

A superposition of the $n + 1$ cases is constructed to generate expressions for the total generalized force at each degree of freedom. The element loads give rise to a set of generalized forces $\bar{P}_i$ often called the *fixed-end forces* or *fixed-end moments*. The generalized force at each degree of freedom must be equal to the specified (externally applied) generalized force—the joint load. In the example shown in Fig. 2.12b, the two degrees of freedom are joint rotations at B and C. The equations are

$$\begin{aligned} k_{11}U_1 + k_{12}U_2 + \bar{P}_1 = P_1^0 = 0 \\ k_{21}U_1 + k_{22}U_2 + \bar{P}_2 = P_2^0 = 0 \end{aligned} \tag{2.32}$$

Thus, the equations of the displacement method express equilibrium conditions that are satisfied by choosing the appropriate combination (i.e., the values of the degrees of freedom U_i) of the n loading cases to superpose on the external load case.

In general, the equations of the displacement method have the form

$$\begin{aligned} \mathbf{K}\cdot\mathbf{U} + \bar{\mathbf{P}} = \mathbf{P^0} \\ \mathbf{K}\cdot\mathbf{U} = \mathbf{P} \end{aligned} \tag{2.33}$$

where $\mathbf{K}$ = symmetric, positive definite *structure stiffness matrix*

$$\mathbf{P} = \mathbf{P}^0 - \bar{\mathbf{P}}$$

$\mathbf{P}$ = generalized force in fixed-end structure due to element loads

$\mathbf{P}^0$ = specified (externally applied) generalized force at degrees of freedom, that is, joint loads

2.6 FORCE METHOD

The steps involved in the force method are (1) selection of redundants, (2) calculation of generalized displacements in the primary structure for the external loading $\bar{D}_i$, (3) calculation of the generalized displacements (the flexibility coefficients f_{ij}) for the n unit redundant loading cases, (4) imposition of compatibility conditions work-conjugate to the redundants, (5) solution of the simultaneous equations for the values of the n redundants R_i, and (6) back substitution for member forces and deformations.

Difficulties arise in setting up a procedure for making a standardized, systematic choice of redundants. This resulted in a strong preference for displacement methods in computer codes for structural analysis. Nonetheless, for specific classes of structures, the force method can be very efficient.

2.6.1 Structure Flexibility Coefficients

The generalized displacements $\bar{D}_i$ and f_{ij} of the primary structure can be calculated using any method for calculating displacements in statically determinate structures. Both direct methods and work-energy methods are usable. The dummy unit-load method is illustrated in the following article.

2.6.2 Dummy Unit-Load Method Examples

STATICALLY INDETERMINATE TRUSS

The planar truss shown in Fig. 2.13 is two degrees statically indeterminate ($r = 3$, $j = 8$, $b = 15$; $b + r - 2j = 2$). The redundants R_1 and R_2 are selected as the internal forces (tension positive) in members CF and CH. Accordingly, releases are inserted in members CF and CH, consisting of infinitesimal cuts. The generalized displacements that are work-conjugate to the redundants are therefore the relative displacements (i.e., opening or closing) of the cuts. Whenever internal generalized forces (as opposed to external reactions) are chosen as redundants, the corresponding generalized displacements are relative quantities of this type. The relative displacements due to the external loads are $\bar{D}_1$ and $\bar{D}_2$.

To calculate the relative displacement $\bar{D}_1$, the required dummy unit-load system is a pair of unit forces that induce bar forces n_i^1. Thus, $\bar{D}_i = \sum n_i^1 e_i$, where e_i are the member elongations in the primary structure due to the external actions. If these external actions consist only of mechanical loads, then $e_i = N_i\, L_i/A_i\, E_i$, where N_i are the bar forces due to the external loads. The same dummy unit-load system is used to calculate the relative displacements f_{11} and f_{12} due to the loadings $R_1 = 1$,

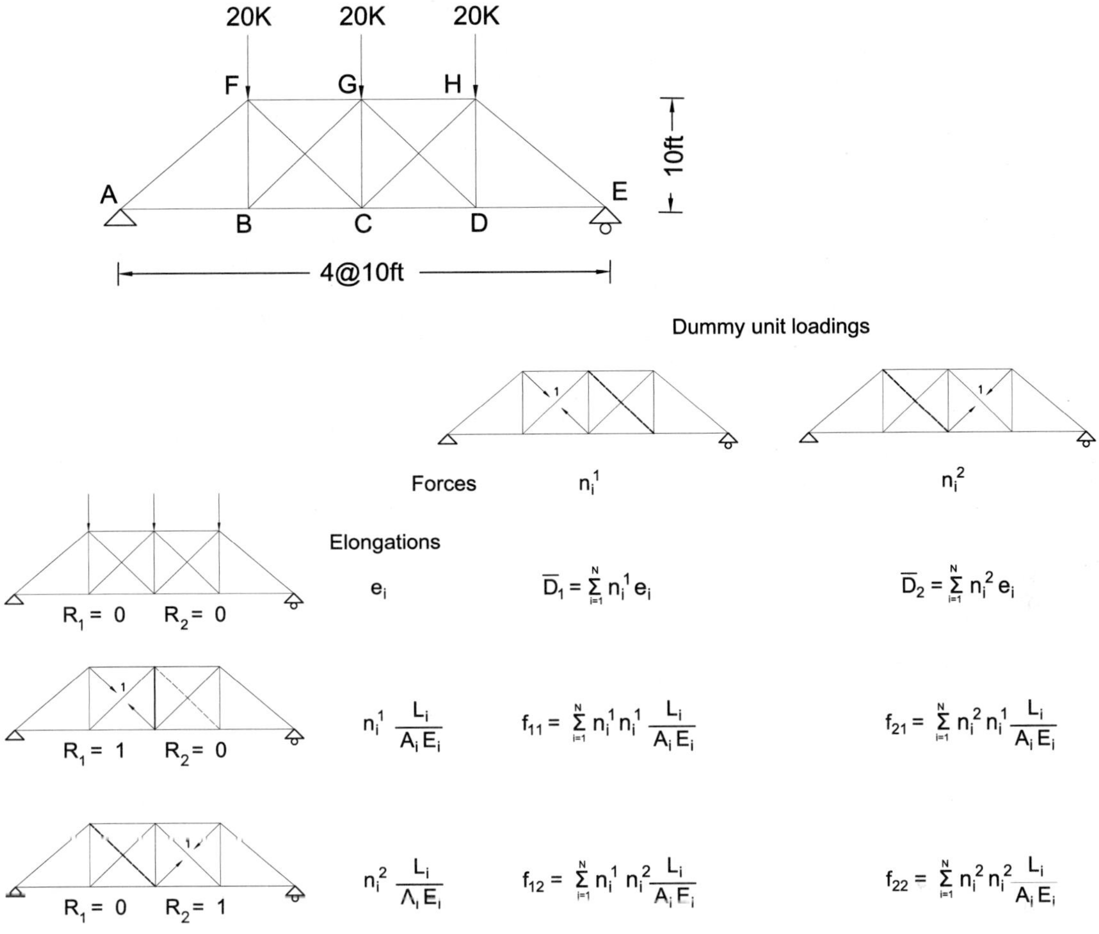

Figure 2.13 Statically indeterminate truss by force method.

$R_2 = 0$ and $R_1 = 0$, $R_2 = 1$, respectively. This dummy unit load system is the same as the load system for $R_1 = 1$, $R_2 = 0$. The relative displacements $\bar{D}_2$, f_{21}, and f_{22} are calculated in a similar way using a dummy unit-load system consisting of a pair of unit forces that induce bar forces n_i^2. This second dummy unit-load system is the same as the load system for $R_1 = 0$, $R_2 = 1$. The formulas for the flexibility coefficients shown in Fig. 2.13 verify that $f_{11} > 0$, $f_{22} > 0$, and $f_{12} = f_{21}$.

Table 2.4 summarizes the calculations. The redundants R_1 and R_2 are found from

$$\begin{bmatrix} 0.01732 & 0.00200 \\ 0.00200 & 0.01732 \end{bmatrix} \begin{bmatrix} R_1 \\ R_2 \end{bmatrix} = \begin{bmatrix} 0.1224 \\ 0.1224 \end{bmatrix} s$$

$$\begin{bmatrix} R_1 \\ R_2 \end{bmatrix} = \begin{bmatrix} 6.34 \\ 6.34 \end{bmatrix}$$

The total member forces in the indeterminate truss are found as a final step from $\bar{N}_i = N_i + n_i^1 R_1 + n_i^2 R_2$.

Statically Indeterminate Frame

The portal frame shown in Fig. 2.14 is one degree statically indeterminate. The horizontal reaction at the base of the columns is chosen as the redundant R_1. The dummy unit-load system that is required in order to calculate the horizontal displacements $\bar{D}_1$ and f_{11} at the support D in the primary structure consists of a unit horizontal force at D. Accounting solely for flexural deformations, the horizontal displacement at the support D in the primary structure can be calculated as indicated in Fig. 2.14. Table 2.3 gives formulas for integrals $\int M_1 M_2 \, dx$. Accordingly, $\bar{D}_1 = wL^3H/12E_bI_b$ and $f_{11} = 2H^3/3E_cI_c + H^2L/E_bI_b$. Hence, the redundant force R_1 is

$$R_1 = \frac{1}{H}\frac{wL^2}{H}\frac{1}{1+2a/3}$$

in which $a = (E_bI_b/L)/(E_cI_c/H)$ represents the beam-to-column relative stiffness. The maximum positive and maximum negative bending moments in the frame are shown in Fig. 2.15 as a function of the beam-to-column relative stiffness. Shear and axial deformations can be accounted for by adding the appropriate terms shown in Eq. (2.25).

2.6.3 Three-Moment Equation for Continuous Beams

The three-moment equation is a special force-method solution for continuous beams in which the bending moments over the interior supports are selected as redundants. The general form of the equation relates bending moments in the continuous beam at three adjacent interior supports. This results in a standard solution for continuous beams that can be simply applied, yielding well-conditioned simultaneous equations with a narrow bandwidth (the bandwidth is 3).

The primary structure, created by releasing internal bending moments over all interior supports, consists of a number of adjacent simply supported spans (Fig. 2.16*b*). Accordingly, the basic element

Table 2.4 Statically Indeterminate Truss (Fig. 2.13) by Force Method*

Mem.	L_i	A_i	N_i	n_i^1	n_i^2	e_i	$n_i^1 e_i$	$n_i^2 e_i$	f_{11}	f_{12}	f_{22}	$\bar{N}_i$
AB	120	1.5	+30.0			0.080						+30.0
DE	120	1.5	+30.0			0.080						+30.0
BC	120	2.0	+40.0	−0.707		0.080	−0.0566		0.001			+35.5
CD	120	2.0	+40.0		−0.707	0.080		−0.0566			0.001	+35.5
FG	120	2.0	−30.0	−0.707		−0.060	0.0424		0.001			−34.5
GH	120	2.0	−30.0		−0.707	−0.060		0.0424			0.001	−34.5
AF	169.7	2.5	−42.4			−0.096						−42.4
EH	169.7	2.5	−42.4			−0.096						−42.4
BG	169.7	1.0	−14.1	+1.0		−0.080	−0.0800		0.00566			−7.8
DG	169.7	1.0	−14.1		+1.0	−0.080		−0.0800			0.00566	−7.8
BF	120	1.0	+10.0	−0.707		0.040	−0.0283		0.002			+5.5
DH	120	1.0	+10.0		−0.707	0.040		−0.0283			0.002	+5.5
CG	120	1.0	+ 0.0	−0.707	−0.707	0.000			0.002	0.002	0.002	−9.0
CF	169.7	1.0	+ 0.0	+ 1.0		0.000			0.00566			+6.3
CH	169.7	1.0	+ 0.0		+ 1.0	0.000					0.00566	+ 6.3
						Σ =	−0.1224	−0.1224	0.01732	0.002	0.01732	
							$= \bar{D}_1$	$= \bar{D}_2$				

*Force units are kips; length units are inches; E = 30,000 ksi.

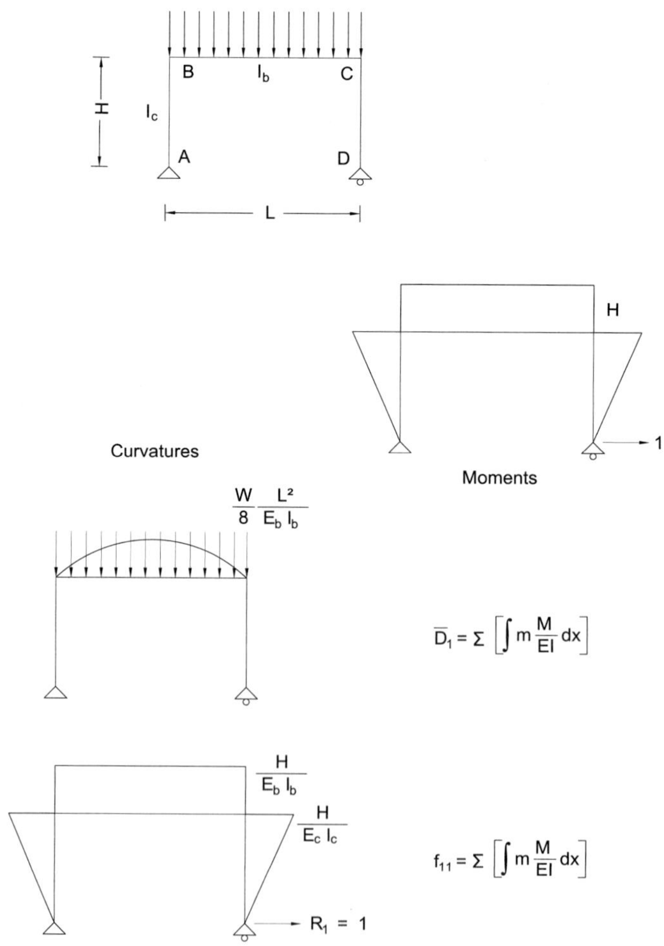

Figure 2.14 Statically indeterminate portal frame by force method.

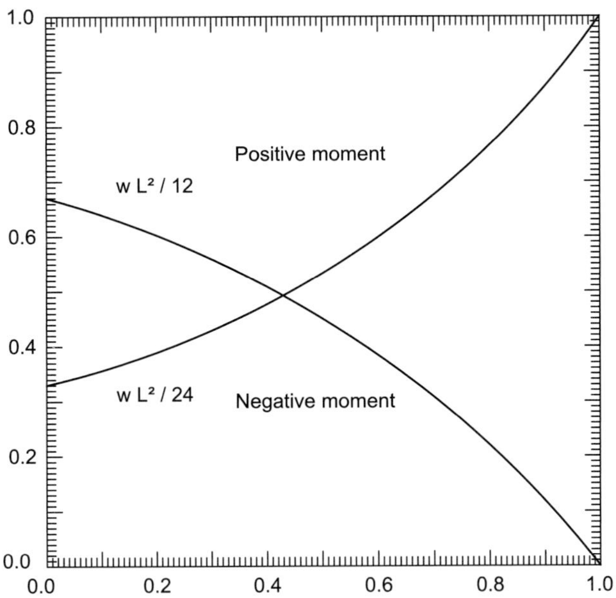

Figure 2.15 Effects of proportions on maximum bending moments in hinged-base portal frame.

in this approach is a simply supported beam, and the generalized displacements D_1 and f_{ij} of the force method are constructed from calculated rotations at end supports. A small core of standard results suits many problems of practical interest (Fig. 2.17). Incidentally, the three-moment equation enjoys the advantageous characteristics typically associated with displacement methods.

The example shown in Fig. 2.16 is a four-span continuous beam with equal span lengths L and constant stiffness EI. The external load P is applied at the center of the span AB. The flexibility coefficients f_{11}, f_{21}, f_{31} are the relative rotations at the releases for the unit-load system $R_1 = 1$, $R_2 = 0$, $R_3 = 0$ (Fig. 2.16*c*). These are found from Fig. 2.17 as $f_{11} = L/3EI + L/3EI$, $f_{21} = L/6EI$, $f_{31} = 0$. The two contributions to f_{11} are the end rotations at B of spans AB and BC, respectively, under the pair of unit moments $R_1 = 1$. If the spans AB and BC have different lengths and elastic properties, then $f_{11} = (L/3EI)_{AB} + (L/3EI)_{BC}$, $f_{21} = (L/6EI)_{BC}$, and $f_{31} = 0$. The generalized displacement D_1 of the primary structure is the rotation at B of span AB due to the external load P. This is given in Fig. 2.17 as $PL^2/16EI$. Additional loads are treated similarly.

Once the redundants R_i are calculated, the bending moment diagram for the continuous beam is constructed using superposition as shown in Fig. 2.16.

2.7 DISPLACEMENT METHOD

The displacement method can be defined through the following steps: (1) identification of the degrees of freedom, (2) calculation of generalized forces P_i in the fully restrained structure due to element loads (the fixed forces), (3) calculation of generalized forces (the stiffness coefficients k_{ij}) due to the n unit degree of freedom displacement cases, (4) imposition of equilibrium conditions work-conjugate to the n degrees of freedom (assembly of stiffness equations), (5) solution of the simultaneous equations for the values of the n degrees of freedom U_i, and (6) back substitution for the member forces and deformations.

The displacement method is generally preferred over the force method, as it is easily adoptable into computer code. This is mainly due to the fact that its operations can be systematically performed so that all structural elements, whether they represent trusses, frames, plates, shells, or solids, can be handled in the same way using the same procedures and algorithms. Details that are specific to a particular type of structural member can be kept to a small separate part of the computer code.

2.7.1 Structure Stiffness Coefficients

The stiffness coefficients k_{ij} of a structure are determined by summation of stiffness coefficients of individual members. This is illustrated by the planar truss shown in Fig. 2.1. The two degrees of freedom, U_1 and U_2, are defined as the horizontal and vertical displacements, respectively, of joint 4. The stiffness coefficients are determined from the basic formulas for the truss shown in Eqs. (2.1) through (2.3). Compatibility relations shown in Eq. (2.2) determine member elongations given the imposed joint displacements, constitutive relations (elastic behavior) shown in Eq. (2.3) determine member forces from elongations, and equilibrium relations shown in Eq. (2.1) determine the stiffness coefficients from the member forces.

The stiffness coefficients are displayed in Fig. 2.18 in matrix form, in which the separate contributions of the three members appear as *member stiffness matrices*. Member 2 is oriented at an angle a with respect to the Cartesian coordinate system (x, y). Member stiffness matrices are usually expressed in a local or *element coordinate system*, which in this case would have a local x-axis pointing along the member axis. Subsequently, coordinate transformation is used to define member stiffness information in the *global coordinate system*.

2.7.2 Member Stiffness Matrices

The member stiffness matrix is the fundamental building block of the displacement method. Member stiffness information[5] is essential for every category of structural member (model) from which a given structural model is to be formulated.

For planar trusses, the stiffness matrix shown in Fig. 2.19 is referenced to four degrees of freedom to comprise the most general case. The four degrees of freedom u_1', u_2', u_3', u_4' are defined relative to the member local coordinate system. The truss member has no stiffness associated with displacements perpendicular to its centerline provided that the displacements are small. The transformation matrix $\mathbf{T}$ relates the displacements in the two coordinate systems as $\mathbf{u}' = \mathbf{T} \cdot \mathbf{u}$.

Since $\mathbf{u}'^T \cdot \mathbf{p}' = \mathbf{u}^T\mathbf{p}$, the relation between forces in the two coordinate systems follows. The transformation

$$\mathbf{u}' = \mathbf{T} \cdot \mathbf{u} \quad \mathbf{p} = \mathbf{T}^T \cdot \mathbf{p}' \tag{2.34}$$

is known as a *contragredient transformation*. The stiffness matrix of the member in the global coordinate system is consequently expressed as

$$\mathbf{k} = \mathbf{T}^T \cdot \mathbf{k}' \cdot \mathbf{T} \tag{2.35}$$

For a planar frame, the degrees of freedom at a joint comprise a rotation in addition to two displacements. The transformation matrix $\mathbf{T}$ is accordingly a 6 × 6 rather than a 4 × 4 matrix, which is derived from the matrix shown in Fig. 2.19 by adding rows and columns corresponding to degrees of freedom 3 and 6, with 1 on the diagonal and 0 elsewhere.

Member response in the displacement method is separated into two components: (1) the response of the fixed-end member subjected to member loads and (2) the response of the unloaded member to the degrees of freedom. The stiffness coefficients are correlated to this latter part of the member deformation. Both the direct approach and the work-energy approach can be used to develop the stiffness coefficients for one-dimensional members (line elements). For plates, shells, and solids, however, the work-energy approach is necessary.

The stiffness coefficients for a plane beam can be synthesized using the approach shown in Fig. 2.20. Axial displacements are not considered. The calculation of the stiffness coefficients entails the solution of a statically indeterminate beam, which can be performed by the force method. To calculate the stiffness coefficients k_{33} and k_{43}, the force method gives

$$\begin{bmatrix} f_{33} & f_{43} \\ f_{43} & f_{44} \end{bmatrix} \begin{bmatrix} k_{33} \\ k_{43} \end{bmatrix} = \begin{bmatrix} 1 \\ 0 \end{bmatrix}$$

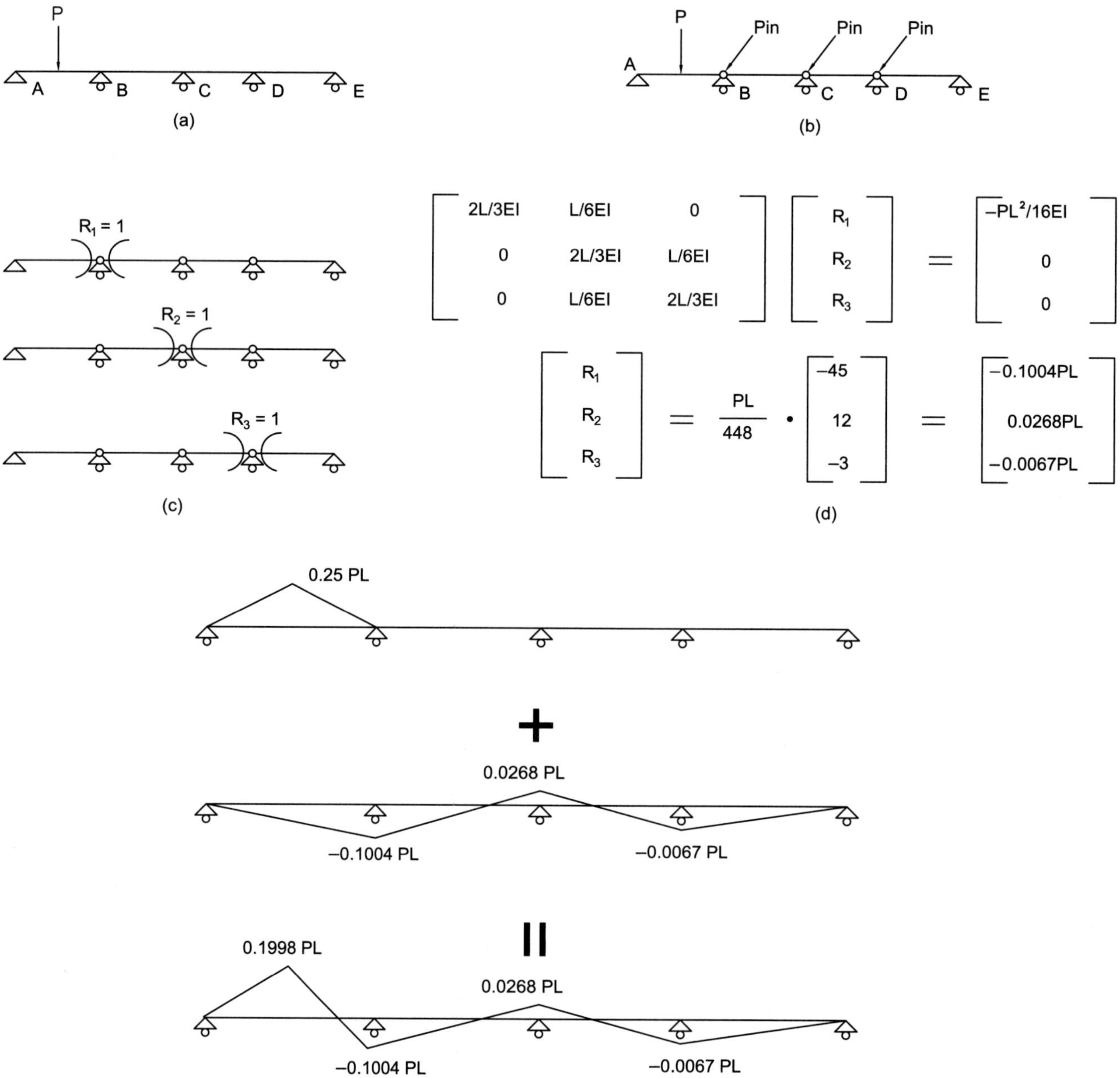

Figure 2.16 Three-moment equation for continuous beam: (*a*) four-span continuous beam, (*b*) primary structure, (*c*) unit-load systems, (*d*) compatibility equations of force method, and (*e*) contruction of bending-moment diagram.

and for the calculation of the stiffness coefficients k_{34} and k_{44},

$$\begin{bmatrix} f_{33} \; f_{43} \\ f_{43} \; f_{44} \end{bmatrix} \begin{bmatrix} k_{34} \\ k_{44} \end{bmatrix} = \begin{bmatrix} 1 \\ 0 \end{bmatrix}$$

In these two sets of simultaneous equations, the flexibility coefficients are assumed to be known, and the stiffness coefficients are sought. The flexibility coefficient matrix is the same in both cases, so they can be combined into

$$\begin{bmatrix} f_{33} \; f_{43} \\ f_{43} \; f_{44} \end{bmatrix} \begin{bmatrix} k_{33} \; k_{34} \\ k_{43} \; k_{44} \end{bmatrix} = \begin{bmatrix} 10 \\ 01 \end{bmatrix} \text{or } \mathbf{f}_{bb} \cdot \mathbf{k}_{bb} = \mathbf{1}$$

from which it is found that $\mathbf{k}_{ab} = \mathbf{f}_{bb}^{-1}$. The flexibility coefficients can be found by using the dummy unit-load method, in which, for example, shear deformations can be included, as can flexural deformations.

The full 4×4 stiffness matrix is found by using equilibrium and the fact that the stiffness matrix must be symmetric. Equilibrium entails that $\mathbf{k}_{ab} = \mathbf{T}_{ab} \cdot \mathbf{k}_{bb}$ and $\mathbf{k}_{aa} = \mathbf{T}_{ab} \cdot \mathbf{k}_{ba}$, where

$$\mathbf{T}_{ab} = \begin{bmatrix} -1 & 0 \\ -L & -1 \end{bmatrix}$$

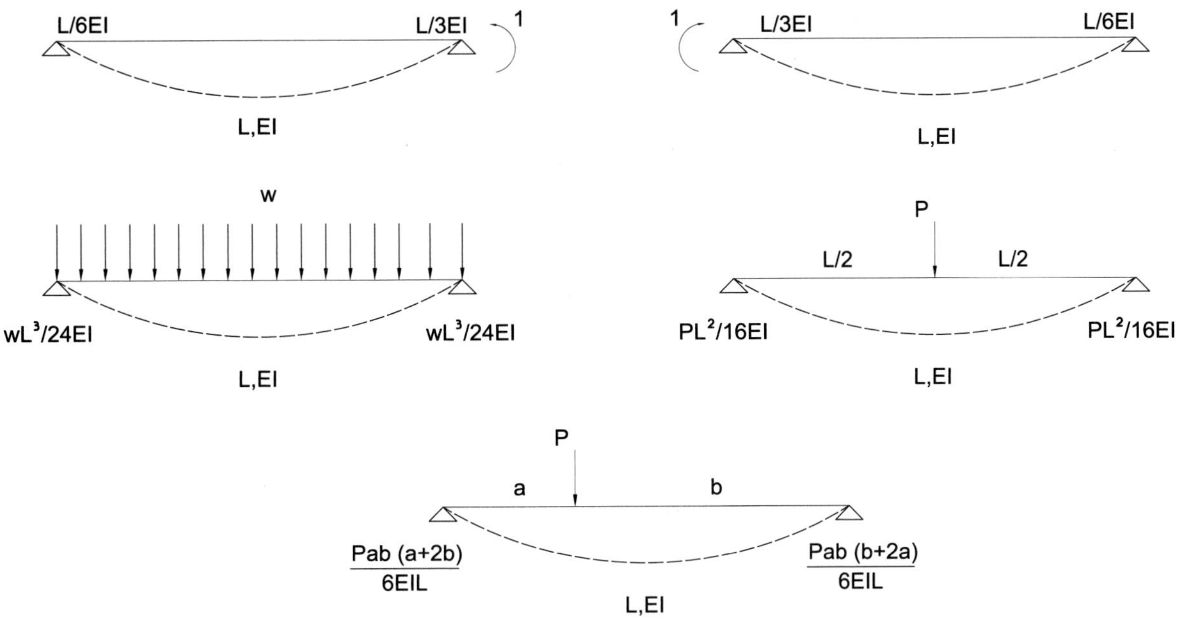

Figure 2.17 Support rotations in a simply supported beam.

and symmetry entails that $\mathbf{k}_{ab} = \mathbf{k}_{ba}^T$. Accordingly, the 4 × 4 stiffness matrix for the planar beam is

$$\mathbf{k} = \begin{bmatrix} \mathbf{T}_{ab}\mathbf{k}_{bb}\mathbf{T}_{ab} & \mathbf{T}_{ab}\mathbf{k}_{bb} \\ \mathbf{k}_{bb}\mathbf{T}_{ab}^T & \mathbf{k}_{bb} \end{bmatrix} \tag{2.36}$$

Figure 2.21 shows the explicit form of this stiffness matrix expanded to a 6 × 6 matrix to encompass the axial degrees of freedom.

Direct Method for Determining Member Response

The response of one-dimensional (line) elements can also be determined directly from the differential relations based on equilibrium, compatibility, and the constitutive law. For a planar beam in which only flexural deformations are considered, Eqs. (2.20) and (2.22) are the initial point. In case shear deformation effects are to be considered, the second differential relation in Eq. (2.22) is modified to include the transverse shear strain γ, which is assumed to be constant through the beam cross section. Additionally, the constitutive relation between shear force and shear strain is used. The result is

$$\frac{d\theta}{dx} = \frac{M(x)}{EI} \qquad \frac{dy}{dx} = \theta(x) - \gamma \qquad \gamma(x) = \frac{V(x)}{A_s G} \tag{2.37}$$

In this theory, also called *Timoshenko beam* theory, the rotation θ of the line that is originally perpendicular to the centerline differs from the rotation *dy/dx* of the centerline by the shear angle γ.

The solutions of Eqs. (2.20) and (2.37) present the complete response of the beam. The part of the response involving the applied load $p(x)$ determines the fixed-end forces. The part of the response for $p(x) = 0$ determines the stiffness coefficients.

Figure 2.22 shows the solutions of Eqs. (2.20) and (2.37) for $p(x) = 0$ and with nodal degrees of freedom $u_1 = y(0)$, $u_2 = \theta(0)$, $u_3 = y(L)$, $u_4 = \theta(L)$. The internal generalized beam forces are determined from Eq. (2.37) as $V(x) = A_s G_\gamma(x)$, $M(x) = EI\, d\theta(x)/dx$, and the nodal generalized forces are determined from $p_1 = V(0)$, $p_2 = -M(0)$, $p_3 = -V(L)$, $p_4 = M(L)$. The coefficients of the generalized coordinates u_1, u_2, u_3, u_4 in these expressions for p_1, p_2, p_3, p_4 are the stiffness coefficients shown in Fig. 2.21. Axial displacements are included, inserting the axial stiffness coefficients in the stiffness matrix as in Fig. 2.21. The analytical expressions for the beam deformations $y(x)$, $\theta\ (x)$, and $\gamma(x)$ in terms of the degrees of freedom involve *shape functions*, shown in Eq. (2.38) for the beam deflection $y(x)$:

$$y(x) = [N_1^y, N_2^y, N_3^y, N_4^y] \begin{bmatrix} u_1 \\ u_2 \\ u_3 \\ u_4 \end{bmatrix} \tag{2.38}$$

Shape functions N_i^θ for $\theta(x)$ and N_i^γ for $\gamma(x)$ are defined in a similar manner as Eq. (2.38), using the information given in Fig. 2.22.

The shape functions N_i^y each contain two distinct terms, the first proportional to $1/(1 + \phi)$, representing the *flexural deflection*, and the second proportional to $\phi(1 + \phi)$, representing the *shear deflection*. The parameter $\phi = 12EI/A_s GL^2$ determines the importance of shear deformation, $\phi \to 0$ represents the flexural beam theory, and $\phi \to \infty$ represents the case of a purely shear-deformable beam.

The flexural shape functions for beam deflection, expressed simply as N_i are shown in Fig. 2.23. These shape functions are part of a family of interpolation polynomials called *Hermitian shape functions*.[6,7] Hermitian interpolation uses nodal values of displacement derivatives (u_2, u_4) in addition to nodal values of the displacements (u_1, u_3) themselves. These shape functions, along with the *Lagrangian shape functions*,[6,7] play a significant role in finite-element analysis. (Lagrangian interpolation uses only nodal function values.)

2.7.3 Member Fixed-End Forces

The fixed-end member response to element loads, that is, $p(x) \neq 0$ with $u_i = 0$, is superposed on the unloaded member response to the degrees of freedom, that is, $p(x) = 0$ with $u_i \neq 0$. Accordingly, the member generalized forces are expressed as shown in Eq. (2.39) in terms of the stiffness coefficients and the fixed-end forces:

$$\begin{bmatrix} p_1 \\ p_2 \\ p_3 \\ p_4 \end{bmatrix} = \begin{bmatrix} k_{11} & k_{12} & k_{13} & k_{14} \\ k_{21} & k_{22} & k_{23} & k_{24} \\ k_{31} & k_{32} & k_{33} & k_{34} \\ k_{41} & k_{42} & k_{43} & k_{44} \end{bmatrix} \begin{bmatrix} u_1 \\ u_2 \\ u_3 \\ u_4 \end{bmatrix} + \begin{bmatrix} \bar{p}_1 \\ \bar{p}_2 \\ \bar{p}_3 \\ \bar{p}_4 \end{bmatrix} \tag{2.39}$$

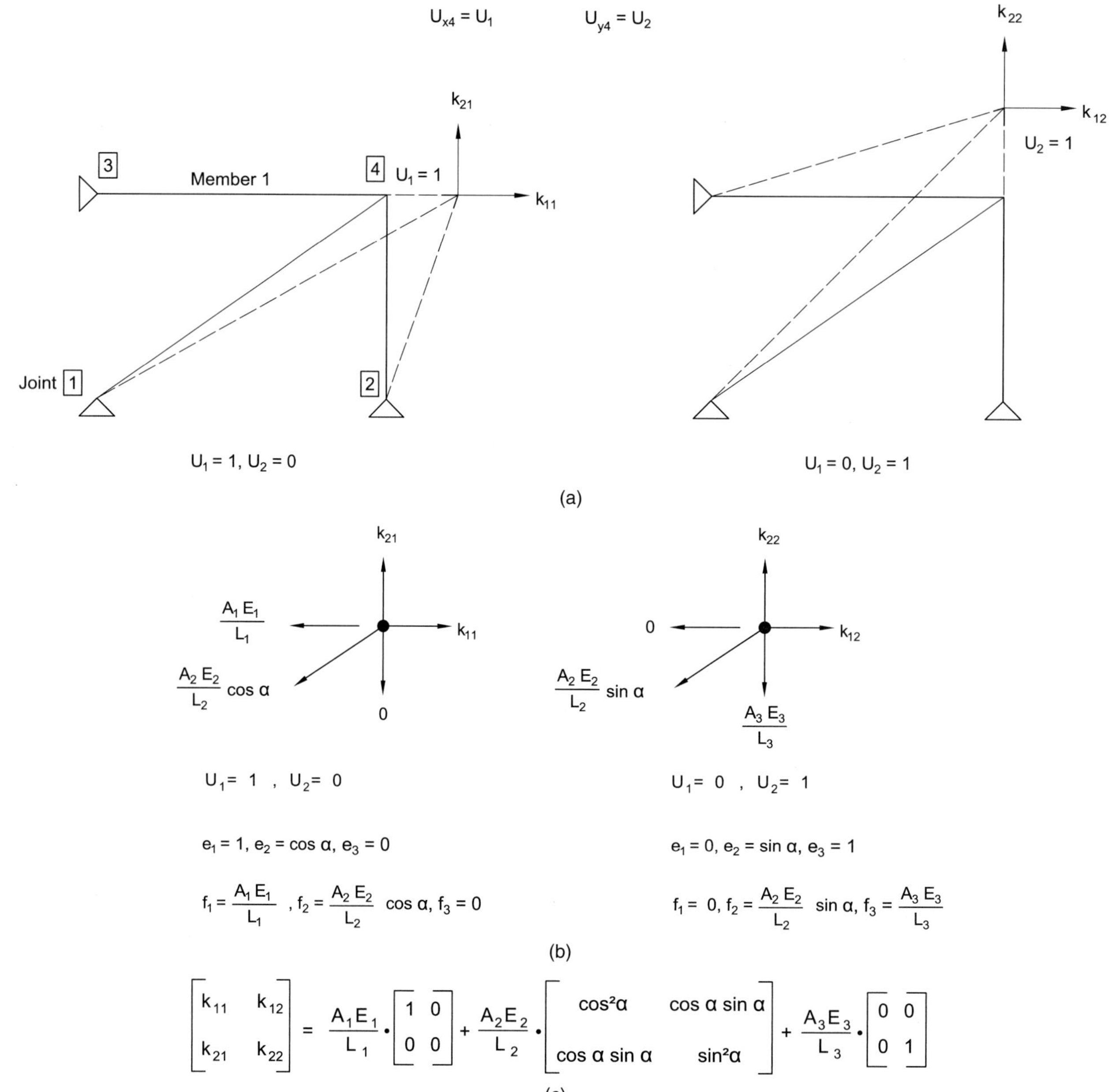

Figure 2.18 Structure stiffness coefficients: (*a*) unit degrees of freedom and definition of structure stiffness coefficients, (*b*) joint equilibrium, and (*c*) structure stiffness assembled from member contributions.

The member fixed-end forces are found from the solution of a statically indeterminate structure, the fixed-ends beam, subjected to the specified loading. The specified element loading may encompass mechanical loads (distributed or concentrated forces and distributed or concentrated moments), distortions (relative displacements or rotations, useful for generating influence lines), or inelastic deformations, such as those due to temperature changes.

The solution for member fixed-end forces can be performed using any of the methods used to determine the stiffness coefficients. A convenient formulation is obtained using the reciprocal theorem (Article 2.3.4) considering, in turn, each of the unit displacement cases shown in Fig. 2.23, paired with the fixed-end loaded member. This leads to an expression for member fixed-end forces due to a distributed force $p(x)$:

$$\begin{bmatrix} p_1 \\ p_2 \\ p_3 \\ p_4 \end{bmatrix} = -\int_0^L \begin{bmatrix} N_1^y \\ N_2^y \\ N_3^y \\ N_4^y \end{bmatrix} p(\bar{x})d\bar{x} \tag{2.40}$$

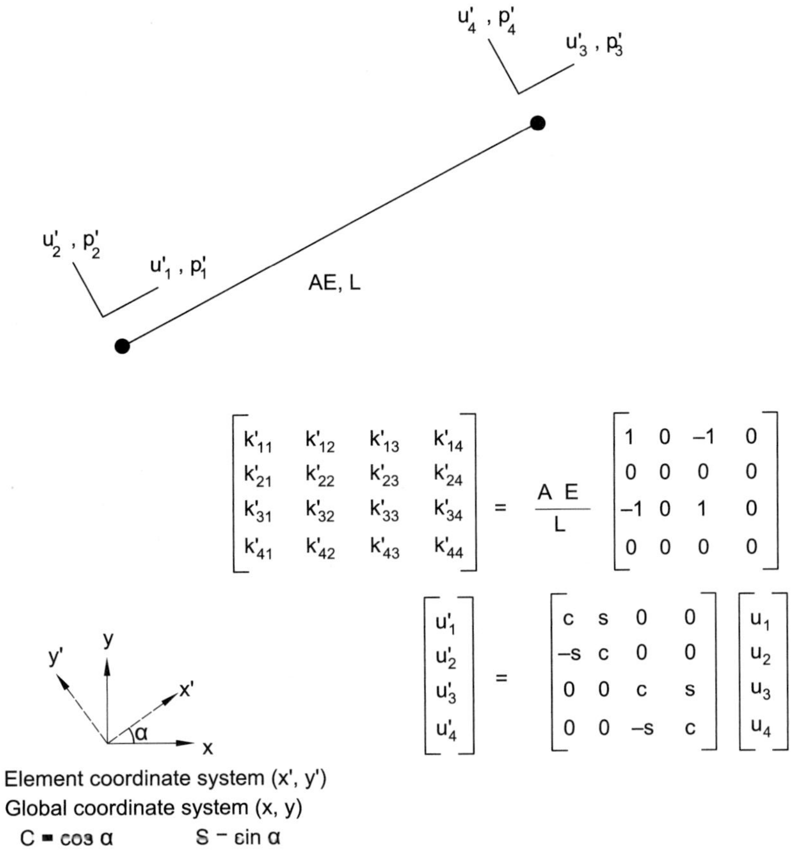

Figure 2.19 Stiffness matrix for planar truss member.

Figure 2.24 illustrates member fixed-end forces in a planar flexural beam (ignoring shear deformations) for common loading, including two cases of unit distortions. Figure 2.25 summarizes member fixed-end forces in a planar shear beam (ignoring flexural deformations) for the same loading conditions. In the general case in which there are both flexural and shear deformations, the fixed-end forces are

$$\bar{p} = \frac{1}{1+\phi} \cdot \bar{p}_{\text{flex}} + \frac{\phi}{1+\phi} \cdot \bar{p}_{\text{shear}} \tag{2.41}$$

in which $\phi = A_s G/12EIL^2$ is zero when there is no shear deformation and infinite when there is no flexural deformation.

2.7.4 Portal Frame by Displacement Method

The portal frame shown in Fig. 2.26 has fixed supports and carries a horizontal force P at girder level. The unsupported frame (i.e., before applying boundary conditions at A and D) has 12 degrees of freedom (Fig. 2.27*b*). The boundary conditions constrain six of these to zero, and there remain six degrees of freedom (two displacements and a rotation at B and at C) as shown in Fig. 2.27*c*. Generally, in analyses of such structures, it is assumed that axial length changes in the members are negligible; this assumption introduces an additional three constraints, one of which is a relative constraint; that is, the horizontal displacements at B and at C are equal. For an analysis in which axial length changes are assumed to be zero, there are three degrees of freedom as shown in Fig. 2.27*d*.

The structure stiffness coefficients shown in Fig. 2.28 are determined as the sums of the appropriate member contributions for each of the imposed unit degrees of freedom. Axial deformations of the members are taken into account, but shear deformations are disregarded.

The joint equilibrium equations for the portal frame are illustrated in Fig. 2.29. If axial deformations are assumed to be zero, then the joint equilibrium equations are as shown in Fig. 2.30.

2.7.5 Slope-Deflection Equations

The slope-deflection equations are a set of stiffness relations for a planar flexural beam. The expression for the end moment p_2, Eq. (2.39), is $p_2 = (4EI/L)u_2 + (2EI/L)u_4 + (6EI/L^2)(u_3 - u_1) + \bar{p}_2$. In the usual notation associated with the slope-deflection method, $p_2 = M_{ab}$, $u_2 = \theta_a$, $u_4 = \theta_b$, and the *chord rotation* ψ_{ab} is defined as $\psi_{ab} = (u_3 - u_1)/L$. The fixed-end moment $\bar{p}_2$ is usually written as M^F_{ab}. With these changes in notation, the generic slope-deflection equation is expressed as

$$M_{ab} = \frac{2EI}{L}(2\theta_a + \theta_b - 3\psi_{ab}) + M^F_{ab}$$

Expressions for the end shears are usually not written explicitly in terms of the degrees of freedom; they are subsequently found from equilibrium. Axial deformations are neglected, and joint moment equilibrium equations are formulated. When joint translations take place (sway), the corresponding equilibrium equations are derived by first expressing joint forces in terms of member end moments.

2.7.6 Moment Distribution

The moment-distribution method[8] is a quick numerical solution technique used for certain classes of framed structures. It is well suited as a

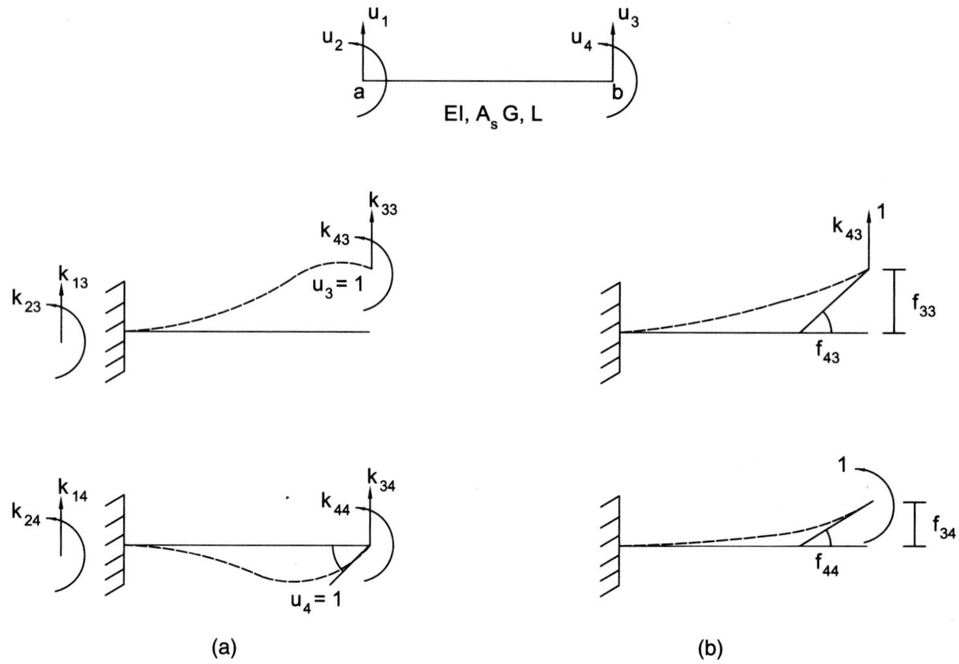

$$f_{bb} = \begin{bmatrix} f_{33} & f_{34} \\ f_{43} & f_{44} \end{bmatrix} = \begin{bmatrix} \frac{L^3}{3EI}(1+\frac{\phi}{4}) & \frac{L^2}{2EI} \\ \frac{L^2}{2EI} & \frac{L}{EI} \end{bmatrix}$$

$$k_{bb} = \begin{bmatrix} k_{33} & k_{34} \\ k_{43} & k_{44} \end{bmatrix} = \frac{1}{1+\phi} \cdot \frac{12EI}{L_3} \begin{bmatrix} 1 & \frac{-L}{2} \\ \frac{-L}{2} & \frac{L^2}{3}(1+\frac{\phi}{4}) \end{bmatrix}$$

$$\phi = \frac{12EI}{A_s G L^2}$$

Figure 2.20 Coefficients for plane beam: (*a*) stiffness and (*b*) flexibility.

hand-calculation technique for small structures, especially continuous beams, in which joint translations (sway) do not occur. The method starts by first calculating the fixed-end moments in the fully restrained structure. Then one joint at a time allowed to rotate to an equilibrium position (unlocked), with all other joints restrained. This unlocking of a single joint initiates a redistribution of bending moments and shears in members connected to that joint. This is called *balancing* the joint. The balanced joint is then relocked in its new equilibrium position, and the process is replicated for a new joint. Each cycle of joint balancing further satisfies equilibrium in the structure until the process is terminated. The procedure is approximate only and is usually concluded when there is small equilibrium imbalance. This is one of the appealing features of the method; approximate results can be achieved very quickly, and additional cycles of joint balancing can be performed if higher accuracy is needed.

The balancing of a single joint is achieved using *distribution factors* and *carryover factors*, which are determined as shown in Fig. 2.31. The equilibrium imbalance, say $-\overline{M}_i$, at a given joint i is eliminated by applying a concentrated moment $\overline{M}_i$ to the joint. This creates end moments in each member connected to joint i. The sum of these member end moments is $\overline{M}_i$, and the distribution factor determines the fraction of the imbalance to be transferred to each member; the distribution factor for a member at a joint is the ratio of its rotational stiffness to the total rotational stiffness of the joint. For prismatic members, the member rotational stiffness is $4EI/L$. Since the far end of each member connected to joint i is fixed against rotation while the balancing is done, a moment is generated there as well. This moment is calculated as a carryover factor times the member end moment at the balanced joint. For prismatic members, the carryover factor is 0.5. An example of the use of moment distribution for a continuous beam is shown in Fig. 2.32.

Various refinements of the basic moment-distribution procedure have been developed: an approach of stiffness condensation can be utilized when the bending moment in a member at a joint is known; member rotational stiffnesses can be altered to account for symmetry or antisymmetry.[8]

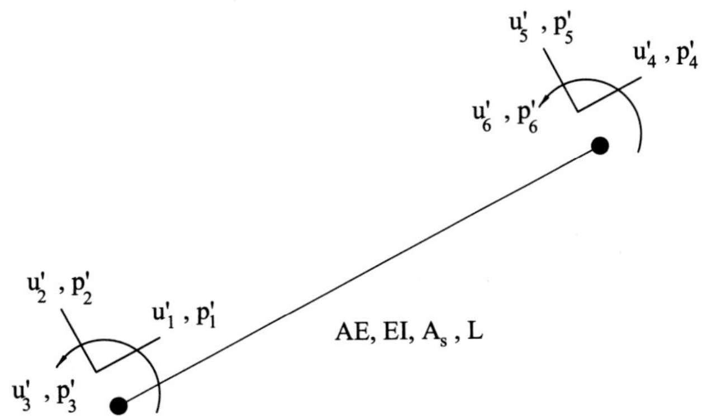

	1	2	3	4	5	6
1	$\frac{AE}{L}$			$\frac{-AE}{L}$		
2		$\frac{12EI}{L^3(1+\phi)}$	$\frac{6EI}{L^2(1+\phi)}$		$\frac{-12EI}{L^3(1+\phi)}$	$\frac{6EI}{L^2(1+\phi)}$
3		$\frac{6EI}{L^2(1+\phi)}$	$\frac{(4+\phi)EI}{(1+\phi)L}$		$\frac{-6EI}{L^2(1+\phi)}$	$\frac{(2-\phi)EI}{(1+\phi)L}$
4	$\frac{-AE}{L}$			$\frac{AE}{L}$		
5		$\frac{-12EI}{L^3(1+\phi)}$	$\frac{-6EI}{L^2(1+\phi)}$		$\frac{12EI}{L^3(1+\phi)}$	$\frac{-6EI}{L^2(1+\phi)}$
6		$\frac{6EI}{L^2(1+\phi)}$	$\frac{(2-\phi)EI}{(1+\phi)L}$		$\frac{6EI}{L^2(1+\phi)}$	$\frac{(4+\phi)EI}{(1+\phi)L}$

Note: $\phi = \frac{12EI}{A_s G L^2}$

Figure 2.21 Stiffness matrix for plane-frame element.

2.7.7 Matrix Formulation of the Displacement Method

The displacement method can be formulated into a systematic procedure that applies to all types of structures.[9,10] The details of element formulation are confined to the most basic level of the procedure in which element stiffnesses and fixed-end forces (or element *equivalent nodal loads*) are calculated and element stresses are back-calculated. *Transformations* of element stiffnesses and loads from the element coordinate system to the global coordinate system, *assembly* of the structure stiffness matrix and structure load vector, and *solution* of the system of joint equilibrium equations are fundamental algorithms used for all structures.

The element nodal displacements $\mathbf{u}'_{(e)}$ in the local element coordinate system are transformed to the global structure coordinate system by the transformation matrix $\mathbf{T}_{(e)}$ using

$$\mathbf{u}'_{(e)} = \mathbf{T}_{(e)} \cdot \mathbf{u}_{(e)} \tag{2.42a}$$

For a plane truss member, the transformation matrix is shown in Fig. 2.19.

The relation between the $n \times 1$ nodal displacement vector of an element and the $N \times 1$ structure nodal displacement vector is

$$\mathbf{u}_{(e)} = \mathbf{L}_{(e)} \cdot \mathbf{U} \tag{2.42b}$$

in which the $n \times N$ *localizing matrix* $\mathbf{L}_{(e)}$ encompasses only 1s and 0s and is very sparse. This set of matrices, one for each element, incorporates the element connectivity information needed to assemble the structure stiffness matrix and load vector. In practice, this information is stored for each element in a compact form, the $n \times 1$ *destination vector*, rather than in the sparse localizing matrix. The destination vector is a list encompassing the structure degree of freedom corresponding to each element degree of freedom.

A virtual-work approach is preferred since it can be applicable to all types of structural elements. The virtual work of an assembly of

$$
\begin{aligned}
y(x) = {} & u_1\left[\frac{1}{1+\phi}\left(1-3\xi^2+2\xi^3\right)+\frac{\phi}{1+\phi}(1-\xi)\right] \\
& +Lu_2\left[\frac{1}{1+\phi}\left(\xi-2\xi^2+\xi^3\right)+\frac{\phi}{1+\phi}\frac{1}{2}\left(\xi-\xi^2\right)\right] \\
& +u_3\left[\frac{1}{1+\phi}\left(3\xi^2-2\xi^3\right)+\frac{\phi}{1+\phi}(\xi)\right] \\
& +Lu_4\left[\frac{1}{1+\phi}\left(-\xi^2+\xi^3\right)+\frac{\phi}{1+\phi}\frac{1}{2}\left(-\xi+\xi^2\right)\right] \\
L\theta(x) = {} & u_1\left[\frac{1}{1+\phi}\left(-6\xi+6\xi^2\right)\right] \\
& +Lu_2\left[\frac{1}{1+\phi}\left(1-4\xi+3\xi^2\right)+\frac{\phi}{1+\phi}(1-\xi)\right] \\
& +u_3\left[\frac{1}{1+\phi}\left(6\xi-6\xi^2\right)\right] \\
& +Lu_4\left[\frac{1}{1+\phi}\left(-2\xi+3\xi^2\right)+\frac{\phi}{1+\phi}(\xi)\right] \\
L\gamma(x) = {} & u_1\left[\frac{1}{1+\phi}(1)\right] \\
& +Lu_2\left[\frac{1}{1+\phi}\left(\frac{1}{2}\right)\right] \\
& +u_3\left[\frac{1}{1+\phi}(-1)\right] \\
& +Lu_4\left[\frac{1}{1+\phi}\left(\frac{1}{2}\right)\right]
\end{aligned}
$$

$$\text{Note}: \phi = \frac{12EI}{A_c GL^2}$$

Figure 2.22 Beam shape functions.

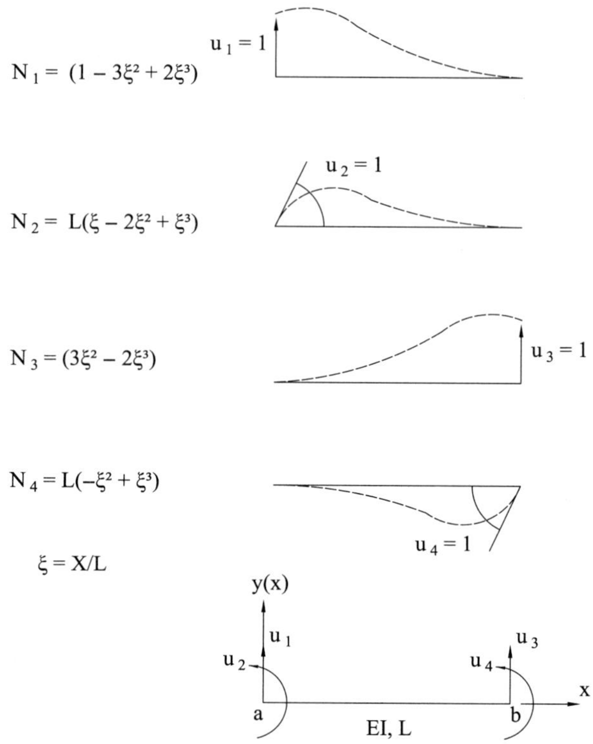

Figure 2.23 Beam (Hermitian) shape functions.

elements (a structure or substructure) is the sum of the virtual-work contributions of the individual elements. Thus,

$$
\begin{aligned}
\delta W &= \sum_e \delta W_{(e)} \\
\delta W_{\text{int}} &= \sum_e \delta W_{(e)\text{int}} \\
\delta W_{\text{ext}} &= \sum_e \delta W_{(e)\text{ext}}
\end{aligned}
\tag{2.43}
$$

The internal virtual work of an element is expressed as

$$\delta W_{(e)\text{int}} = -\int_{V_{(e)}} \delta\varepsilon_{(e)}^{T}\cdot\sigma_{(e)}dV_{(e)} \tag{2.44}$$

in which $\varepsilon_{(e)}$ is the generalized strain vector for the element and $\sigma_{(e)}$ is the generalized stress vector, Eq. (2.14).

The element displacements are interpolated from the element nodal displacements using shape functions (Article 2.7.2), which are typically polynomials, in the form

$$u(x) = \mathbf{N}\cdot\mathbf{u}'_{(e)} \tag{2.45}$$

in which $u(x)$ stands for any relevant displacement component in the interior of the element. In some cases, there may be more than a single displacement quantity that is interpolated. For example, in a two-dimensional continuum, displacements $u(x)$ and $v(x)$ are needed to describe deformation. The interpolation functions for a planar beam are shown in Fig. 2.23, in which the beam displacement $y(x)$ perpendicular to its centerline is interpolated in terms of the four generalized nodal displacements u_1, u_2, u_3, and u_4. The interpolation functions for one-dimensional line elements are often chosen as the exact solutions for the unloaded member subjected to imposed nodal degrees of freedom. Cubic polynomials of this type, shown in Fig. 2.23, are exact solutions for an unloaded beam. Likewise, for a truss element, the shape functions are taken as linear polynomials $N_1 = 1 - x/L$, $N_2 = x/L$. For this class of elements, the direct approach yields the same result as virtual work. However, for plates, shells, and two- or three-dimensional solids, the direct approach is not applicable; exact solutions are not obtainable for element deformed shapes under imposed unit nodal degrees of freedom. It is for this latter class of structural elements that the virtual-work approach is necessary. Now, Eq. (2.45) represents an assumed *approximate element displacement field*. The strain-displacement relations for the particular structural element are then used to obtain the element generalized strains as

$$\varepsilon_{(e)} = \mathbf{B}_{(e)}\cdot\mathbf{u}'_{(e)} \tag{2.46}$$

Equations (2.45) and (2.46) represent the main step in the finite-element method. The element shape functions in Eq. (2.45) cannot be chosen arbitrarily; they must fulfill certain requirements to provide reasonable results.

The internal virtual work for the element can now be

$$\delta W_{(e)\text{int}} = \delta\mathbf{u}'^{T}_{(e)}\cdot\int_{V(e)}\mathbf{B}^{T}_{(e)}\cdot\sigma_{(e)}dV_{(e)}$$

or

$$\delta W_{(e)\text{int}} = -\delta\mathbf{u}'^{T}_{(e)}\cdot\mathbf{I}'^{T}_{(e)} = -\delta\mathbf{u}^{T}_{(e)}\cdot\mathbf{I}_{(e)} \tag{2.47}$$

in which the *element internal resisting force vector* $\mathbf{I}_{(e)}$ is the nodal force vector, which is equivalent, in a work-energy sense, to the element stresses $\sigma_{(e)}$. The element internal resisting force vector is defined as

$$\mathbf{I}_{(e)} = \mathbf{T}^{T}_{(e)}\cdot\int_{V(e)}\mathbf{B}^{T}_{(e)}\cdot\sigma_{(e)}dV_{(e)} \tag{2.48}$$

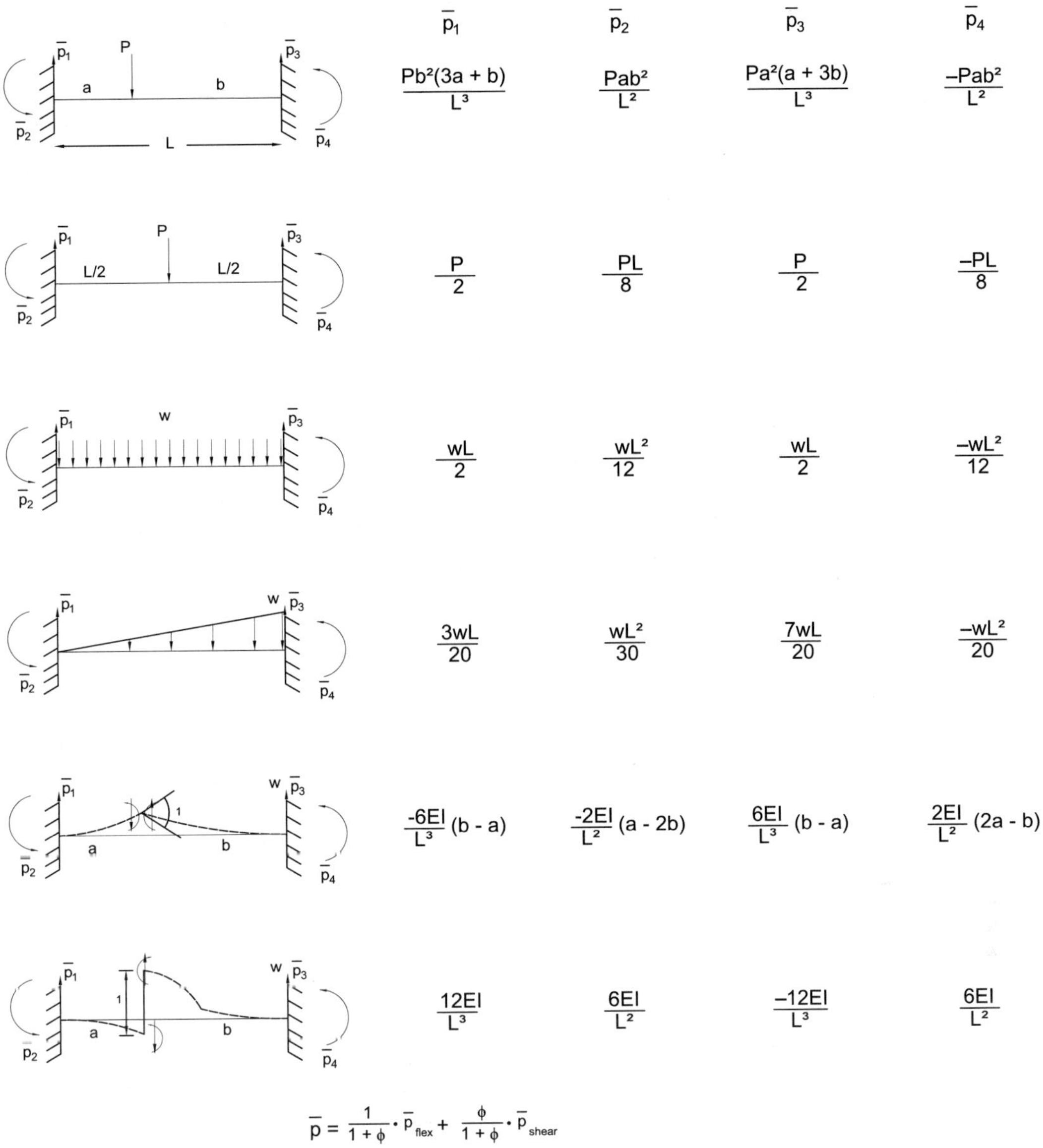

Figure 2.24 Fixed-end member forces for planar flexural beam.

The element external virtual work can be

$$\delta W_{(e)\text{ext}} = \delta u'^{T}_{(e)} \cdot (-\mathbf{p}'_{(e)}) = \delta \mathbf{u}^{T}_{(e)} \cdot (-\bar{\mathbf{p}}_{(e)}) \tag{2.49}$$

in which $\bar{p}_{(e)}$ are the element fixed-end forces and $-\bar{p}'_{(e)}$ are the element *equivalent nodal loads*. The equivalent nodal loads are the fixed-end forces with reversed sign since the equivalent nodal loads are simply the work equivalents of the applied element loads and the fixed-end forces are the reactions to these applied element loads.

The virtual work for the entire structure can be then assembled from element contributions as

$$\begin{aligned} \delta W_{\text{int}} &= \sum_{e} \delta W_{(e)\text{int}} = -\delta \mathbf{U}^{T} \cdot \sum_{e} \mathbf{L}^{T}_{(e)} \cdot \mathbf{I}_{(e)} = -\delta \mathbf{U}^{T} \cdot \mathbf{I} \\ \delta W_{\text{ext}} &= \sum_{e} \delta W_{(e)\text{ext}} = \delta \mathbf{U}^{T} \cdot \left[\mathbf{P}^{0} + \sum_{e} \mathbf{L}^{T}_{(e)} \cdot (-\bar{\mathbf{P}}_{(e)}) \right] = -\delta \mathbf{U}^{T} \cdot \mathbf{P} \end{aligned} \tag{2.50}$$

in which

$$\begin{aligned} \mathbf{I} &= \sum_{(e)} \mathbf{L}^{T}_{(e)} \cdot \mathbf{I}_{(e)} \\ &= \sum_{(e)} \mathbf{L}^{T}_{(e)} \cdot \mathbf{T}^{T}_{(e)} \cdot \int_{V(e)} \mathbf{B}^{T}_{(e)} \cdot \sigma_{(e)} dV_{(e)} \\ \mathbf{P} &= \mathbf{P}^{0} + \mathbf{L}^{T}_{(e)} \cdot \mathbf{T}^{T}_{(e)} \cdot (-\bar{\mathbf{P}}'_{(e)}) \end{aligned} \tag{2.51}$$

and the applied joint load vector is $\mathbf{P}^0$.

The principle of virtual work now gives

$$\delta W = -\delta \mathbf{U}^{T} \cdot (\mathbf{I} - \mathbf{P}) = 0 \tag{2.52}$$

from which the basic set of equilibrium equations of the displacement method is achieved as

$$\mathbf{I} = \mathbf{P} \tag{2.53}$$

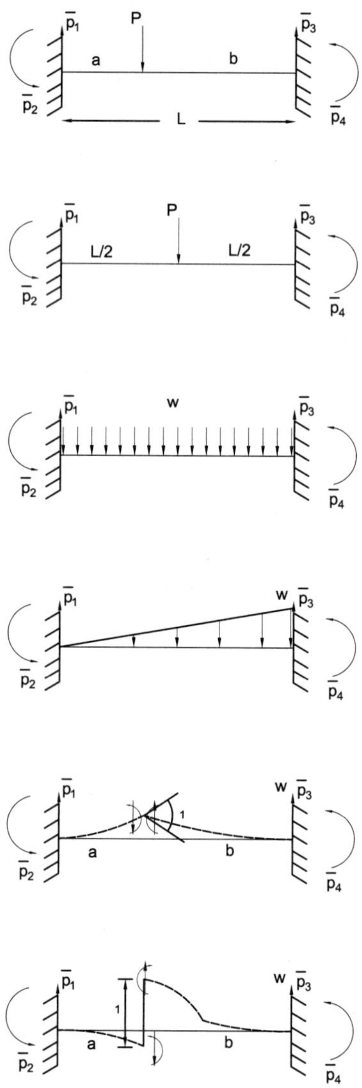

$\bar{p}_1$	$\bar{p}_2$	$\bar{p}_3$	$\bar{p}_4$
$\frac{Pb}{L}$	$\frac{1}{2}\frac{Pab}{L}$	$\frac{Pa}{L}$	$-\frac{1}{2}\frac{Pab}{L}$
$\frac{P}{2}$	$\frac{PL}{8}$	$\frac{P}{2}$	$\frac{-PL}{8}$
$\frac{wL}{2}$	$\frac{wL^2}{12}$	$\frac{wL}{2}$	$\frac{-wL^2}{12}$
$\frac{wL}{6}$	$\frac{wL^2}{24}$	$\frac{wL}{3}$	$\frac{-wL^2}{24}$
0	0	0	0
0	0	0	0

$$\bar{p} = \frac{1}{1+\phi}\cdot \bar{p}_{flex} + \frac{\phi}{1+\phi}\cdot \bar{p}_{shear}$$

Figure 2.25 Fixed-end member forces for planar shear beam.

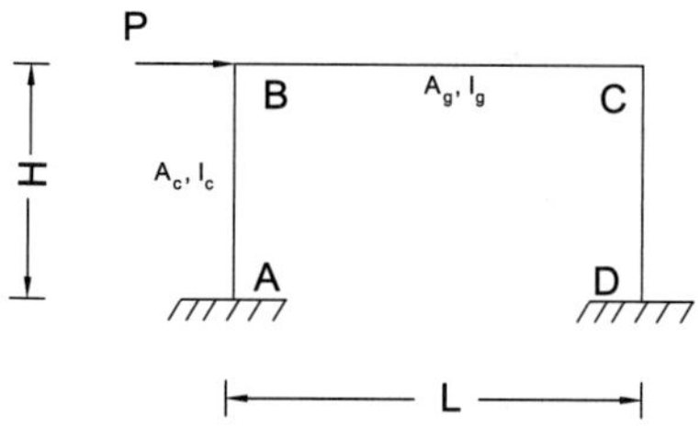

Figure 2.26 Fixed-base portal frame.

Equation (2.53), in which the internal resisting force vector **I** and the load vector **P** are defined in terms of element contributions by Eq. (2.51), depends only on equilibrium and compatibility relations. It is valid for any type of material constitutive law.

If the element material behavior is elastic, element stresses and strains can be related by a constitutive relation, and the element internal virtual work can be expressed as

$$\sigma_{(e)} = \mathbf{E}_{(e)} \cdot \varepsilon_{(e)} \tag{2.54}$$

and the element internal virtual work can be expressed as

$$\begin{aligned} \delta W_{(e)\text{int}} &= \delta \mathbf{u}_{(e)}^{\prime T} \cdot \left(\int_{V_{(e)}} (\mathbf{B}_e^T \cdot \mathbf{E}_{(e)} \cdot \mathbf{B}_{(e)}) dV_{(e)} \right) \cdot \mathbf{u}_{(e)}^{\prime} \\ &= -\delta \mathbf{u}_{(e)}^{\prime T} \cdot \mathbf{k}_{(e)}^{\prime} \cdot \mathbf{u}_{(e)}^{\prime} \end{aligned} \tag{2.55}$$

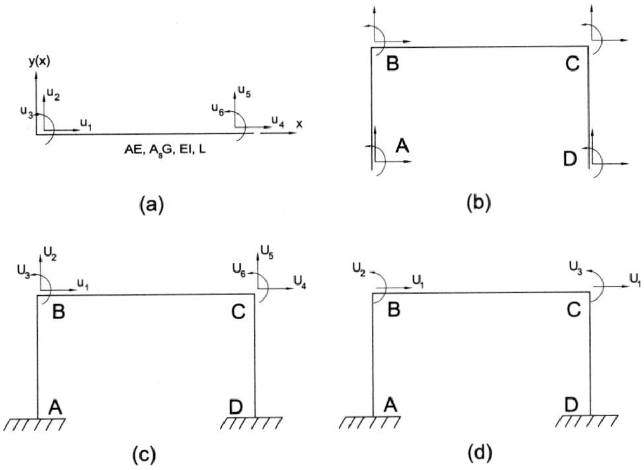

Figure 2.27 Degress of freedom in portal frame: (*a*) member, (*b*) unsupported frame, (*c*) supported frame, and (*d*) supported frame with axial length changes constrained.

where $\mathbf{k}'_{(e)}$ is the element stiffness matrix in element local coordinates expressed as

$$\mathbf{k}'_{(e)} = \int_{V_{(e)}} (\mathbf{B}_e^T \cdot \mathbf{E}_{(e)} \cdot \mathbf{B}_{(e)}) dV_{(e)} \tag{2.56}$$

The internal resisting force vector becomes

$$\mathbf{I} = \mathbf{K} \cdot \mathbf{U} \tag{2.57}$$

where the structure stiffness matrix is

$$\mathbf{K} = \sum_{(e)} \mathbf{L}_{(e)}^T \cdot \mathbf{k}_{(e)} \cdot \mathbf{L}_{(e)} \tag{2.58}$$

and the element stiffness matrix in global coordinates is

$$\mathbf{k}_{(e)} = \mathbf{T}_{(e)}^T \cdot \mathbf{k}'_{(e)} \cdot \mathbf{T}_{(e)} \tag{2.59}$$

Equation (2.58) shows the structural *stiffness assembly* process. It is performed very effectively in practice using destination vectors instead of the sparse matrix multiplications shown in Eq. (2.58). In nonlinear analysis, a similar technique is used for load vector assembly or internal resisting force vector assembly instead of the matrix multiplications shown in Eq. (2.50).

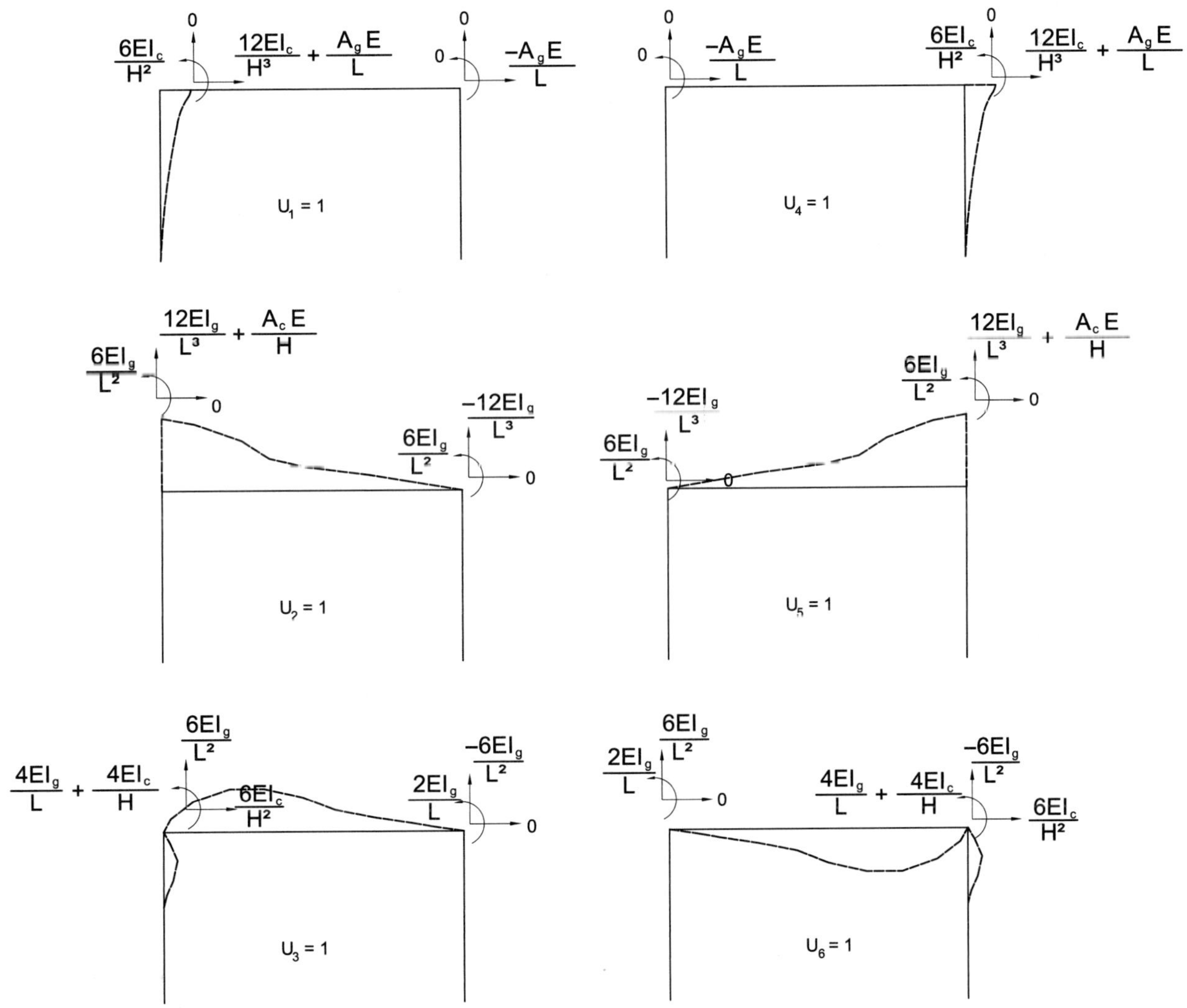

Figure 2.28 Structural stiffness coefficients for portal frame.

$$
\begin{array}{c}
\\ 1 \\ 2 \\ 3 \\ 4 \\ 5 \\ 6
\end{array}
\begin{bmatrix}
1 & 2 & 3 & 4 & 5 & 6 \\
\frac{12EI_c}{H^3}+\frac{A_gE}{L} & & \frac{6EI_c}{H^2} & \frac{-A_gE}{L} & & \\
 & \frac{12EI_g}{L^3}+\frac{A_cE}{H} & \frac{6EI_g}{L^2} & & \frac{-12EI_g}{L^3} & \frac{6EI_g}{L^2} \\
 & \frac{6EI_g}{L^2} & \frac{4EI_g}{L}+\frac{4EI_c}{H} & & \frac{-6EI_g}{L^2} & \frac{2EI_g}{L} \\
\frac{-A_gE}{L} & & & \frac{12EI_c}{H^3}+\frac{A_gE}{L} & & \frac{6EI_c}{H^2} \\
 & \frac{-12EI_g}{L^3} & \frac{-6EI_g}{L^2} & & \frac{12EI_g}{L^3}+\frac{A_cE}{H} & \frac{-6EI_g}{L^2} \\
 & \frac{6EI_g}{L^2} & \frac{2EI_g}{L} & & \frac{-6EI_g}{L^2} & \frac{4EI_g}{L}+\frac{4EI_c}{H}
\end{bmatrix}
\begin{bmatrix} U_1 \\ U_2 \\ U_3 \\ U_4 \\ U_5 \\ U_6 \end{bmatrix}
=
\begin{bmatrix} P \\ 0 \\ 0 \\ 0 \\ 0 \\ 0 \end{bmatrix}
$$

Figure 2.29 Joint-equilibrium equations for portal frame.

$$
\begin{bmatrix}
1 & 2 & 3 \\
\frac{24EI_c}{H^3} & \frac{6EI_c}{H^2} & \frac{6EI_c}{H^2} \\
\frac{6EI_c}{H^2} & \frac{4EI_g}{L}+\frac{4EI_c}{H} & \frac{2EI_g}{L} \\
\frac{6EI_c}{H^2} & \frac{2EI_g}{L} & \frac{4EI_g}{L}+\frac{4EI_c}{H}
\end{bmatrix}
\begin{bmatrix} U_1 \\ U_2 \\ U_3 \end{bmatrix}
=
\begin{bmatrix} P \\ 0 \\ 0 \end{bmatrix}
$$

$$U_1 = \frac{PH^3}{6EI_c} \cdot \frac{1+1.5a}{1+6a} \qquad U_2 = U_3 = \frac{-PH^2}{4EI_c} \cdot \frac{1}{1+6a}$$

$$M_{base} = \frac{PH}{2} \cdot \frac{1+3a}{1+6a} \qquad M_{top} = \frac{PH}{2} \cdot \frac{3a}{1+6a}$$

$$\text{Note } a = \frac{EI_g/L}{EI_c/H}$$

Figure 2.30 Portal frame with zero axial deformations.

Substitution of Eq. (2.57) into the equilibrium equations Eq. (2.53) yields the set of simultaneous linear equations

$$K \cdot U = P \tag{2.60}$$

Following the solution of Eq. (2.60) for the nodal displacements of the structure, element displacements, strains, and stresses are calculated using back-substitution.

2.7.8 Frame Element Stiffness Properties and Equivalent Loads Using Virtual Work

Derivation of element properties by the virtual-work approach provides a unified basis for the displacement method, even when alternative procedures are available (i.e., line elements). The stiffness formulation using virtual work described in Article 2.7.7 is applied to a plane flexural beam element in this article.

The beam displacement $v(x)$ is interpolated using the cubic Hermitian polynomials shown in Fig. 2.23:

$$v(x) = [N_1, N_2, N_3, N_4] \begin{bmatrix} u_1 \\ u_2 \\ u_3 \\ u_4 \end{bmatrix} \tag{2.61}$$

Based on Eq. (2.24) the expression for the internal virtual work of the planar flexural beam becomes

$$\delta W_{\text{int}} = -\int_{x=0}^{L} M \, \delta\kappa \, dx \tag{2.62}$$

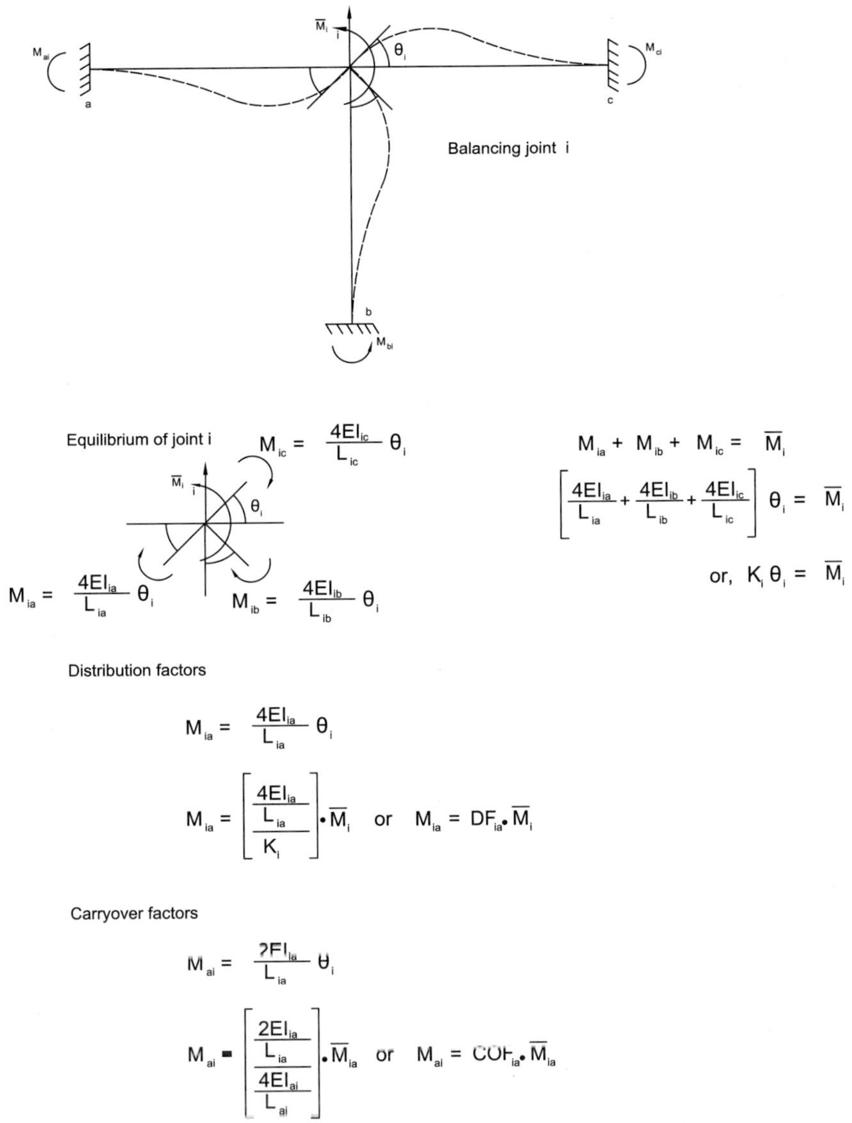

Figure 2.31 Distribution and carryover factors for moment distribution.

The beam curvature is the generalized strain and is determined from

$$\kappa = +v(x)'' = +[N_1'', N_2'', N_3'', N_4''] \begin{bmatrix} u_1 \\ u_2 \\ u_3 \\ u_4 \end{bmatrix} \tag{2.63}$$

in which ( )″ denotes $d^2(\ )/dx^2$. The nodal displacement-to-element strain transformation $\mathbf{B}_{(e)}$ is

$$\mathbf{B}_{(e)} = +\frac{1}{L^2}[(-6+12\xi)L(-4+6\xi)(6-12\xi)L(-2+6\xi)] \tag{2.64}$$

The elastic constitutive matrix $\mathbf{E}_{(e)} = EI$ is a scalar. Therefore, the element stiffness matrix is expressed as

$$\mathbf{k}'_{(e)} = EIL\int_{\xi=0}^{1} \mathbf{B}_{(e)}^T \cdot \mathbf{B}_e \, d\xi \tag{2.65}$$

The stiffness matrix given by Eq. (2.65) is the same as that shown in Fig. 2.21 for the four flexural degrees of freedom, with the shear parameter ϕ set equal to zero.

The element equivalent load vector corresponding to a distributed load $p(x)$ is calculated as follows:

$$-\overline{p'}_{(e)} = L\int_{\xi=0}^{1} \begin{bmatrix} N_1 \\ N_2 \\ N_3 \\ N_4 \end{bmatrix} p(\xi)d\xi \tag{2.66}$$

The load vector $-\overline{p'}_{(e)}$ given by Eq. (2.65) is shown for multiple cases in Fig. 2.24.

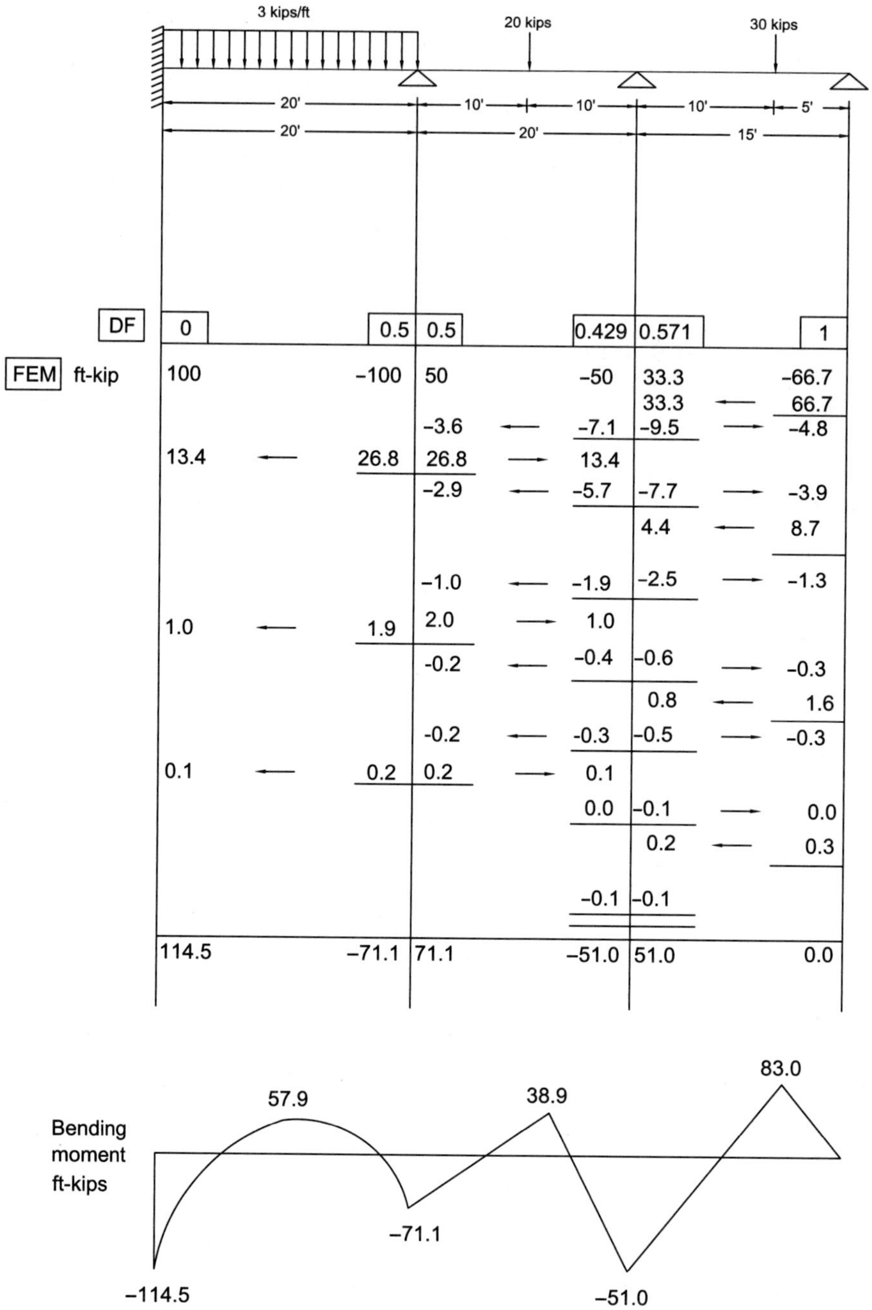

Figure 2.32 Moment distribution for continuous beam.

PART B
ARCHES AND RIGID FRAMES

2.8 INTRODUCTION

Arches and rigid frames are commonly categorized by the development under vertical loads of inclined rather than vertical responses. Mixed construction is selected by the main characteristic; therefore, Fig. 2.33 is referred to as a rigid-frame bridge. Rings are completely bounded arches or rigid frames.

Cross sections are designed for moment, shear, and thrust, with magnitudes determined by the location of the pressure line as shown in Fig. 2.34. In a uniformly loaded parabolic arch, if the pressure line overlaps with the axis of the structure, all cross sections are subjected to compression force, with no moment or shear. If the pressure line falls within the kern of the section, there will exist bending moment, shear, and thrust but no tension force on the cross section. If, finally, as in Fig. 2.34*c* and 2.34*d*, the shape of the structure differs from the pressure line, moment may become dominant. A survey of proximate approaches based on sketching deflected shapes is given in Ref. 11.

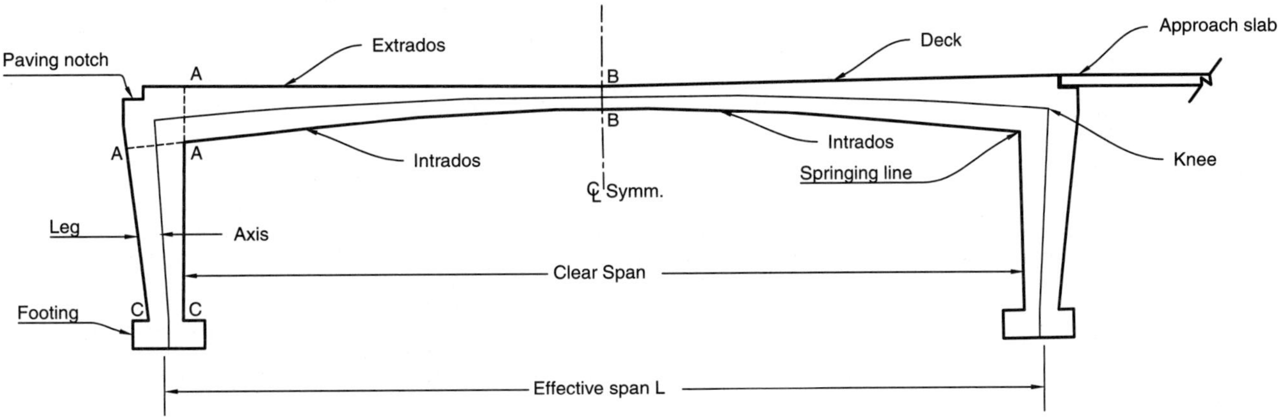

Figure 2.33 Rigid-frame bridge.

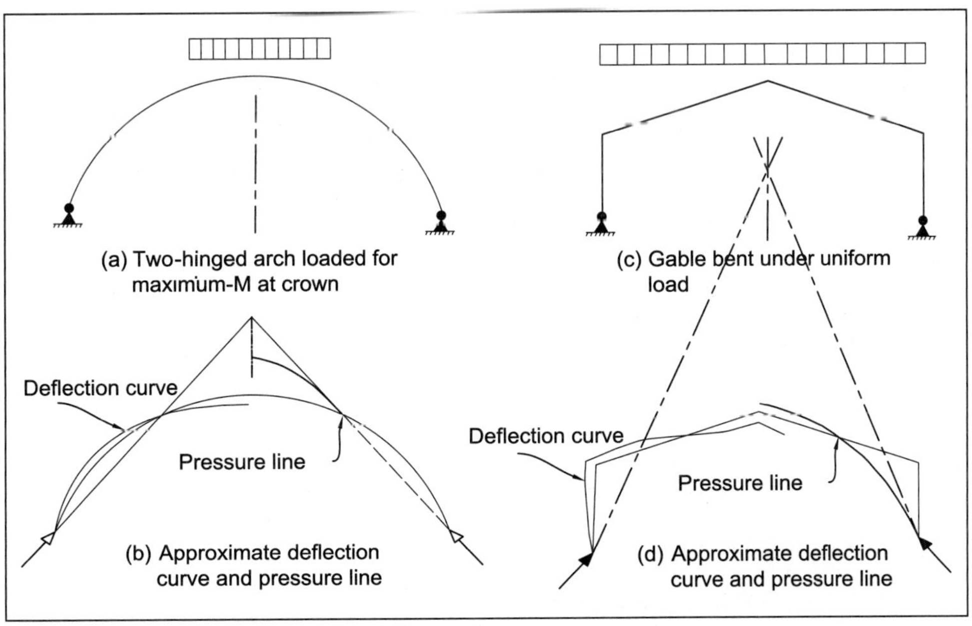

Figure 2.34 Rigid-frame bridge.

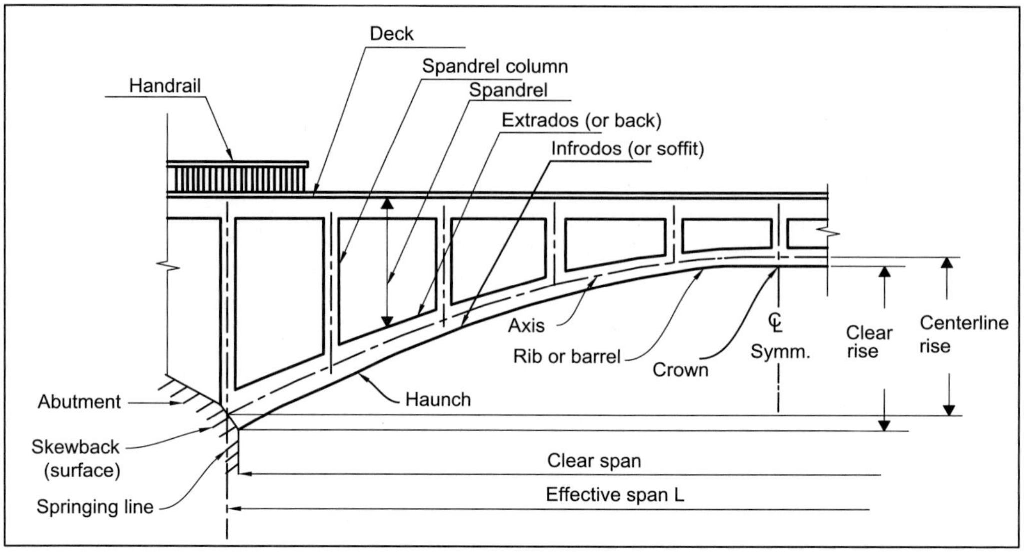

Figure 2.35 Open-spandrel arch.

2.8.1 Classification

Arches may consist of multiple parallel *ribs* or of a single curved sheet or *barrel* (Fig. 2.35). Components-wise, the croton is the highest point of the rib or barrel; the soffit and back are the under surface and the top surface of the arch, respectively; and the springing line is the intersection of the abutment or pier and the soffit. *Intrados* and *extrados* are lines of intersection of soffit and back with a vertical plane through the crown and springing lines. The *haunch* is the midsection of the arch rib or barrel between the crown and springing line. The *clear rise* is the height of intrados at the crown above the springing line. The *skewback* is the inclined surface on which the arch rests. The *centerline rise* is the height of the axis of the arch at the crown above the axis at the skewback.

In bridges, the *spandrel* is the space between the back of arch and roadway. Depending on whether this space is left relatively open or filled with earth, one has either an *open-spandrel arch* (Fig. 2.35) or a *filled-spandrel arch*. Arches in which a structural tie is built between reaction points to take the horizontal thrust are called *tied arches*. Types of arches are shown in Table 2.5.

Table 2.5 Arch Classifications

Fixed	Three degrees indeterminate. Requires good foundations. Relatively rigid and theoretically slightly more economical. High-temperature stresses with flat arches. Most common usage in reinforced concrete.
One-hinged	Two degrees indeterminate. Rarely used.
Two-hinged	One degree indeterminate. Common usage in steel, timber, aluminum. Less rigid than fixed arch, but thrust line is definitely located at abutments. Relatively insensitive to rotations and to moderate differential deflections of supports.
Three-hinged	Statically determinate. Common usage in steel and timber. Free from temperature stresses and insensitive to foundation settlements.
Two- or three-hinged, tied	Does not require massive abutments. Not affected by settlements of supports.
Through and half-through	Overhead lateral bracing may be objectionable aesthetically. Overcome by using slender ribs with relatively heavier tie girder deck (Langer arch-stiffened girder).
Deck	Rib usually carries all the load, though occasionally deck girder made to participate (Langer arch-stiffened girder).
Solid rib	Loads brought to individual ribs by beams, shells, or other transverse elements.
Barrel	In bridges, the spandrel fill gives excessive dead weight. Path of loads to barrel uncertain.
Braced rib (lattice)	
Open spandrel	Clearer path of load to ribs.
Spandrel-braced	Analysis as a trussed framework.
Semicircular	
Segmental (arc of circle)	
Multicentered	Use of segments of circles facilitates field layout.
Parabolic	Funicular line for uniformly distributed load.
Elliptical	
Catenary	Shape of inverted free-hanging string.
Gothic	
Flat	Large horizontal thrust. Large effect of rib shortening and temperature.
High-rise	Reduced horizontal thrust. Reactions may approach vertical.

Rings. Similar to arches and rigid frames, the nomenclature and classification of rings follow the same pattern. Cross sections may be constant or variable, curvatures may be constant or variable, and members may be straight or curved. The ring may rest on point supports or, as in the case of ducts, on a continuous earth support having either rigid, elastic, semielastic, or plastic characteristics.

Bents are rigid frames composed of columns supporting transverse members. Several types are presented in Fig. 2.36. Similar to rings, frames may have members of constant or variable cross section (Fig. 2.33). Rigidity of joints, in terms of continuity of members of the frame, is essential for structural stability. In case elasticity or inelasticity of joints has to be taken into account, special methods of frame analysis for *semirigid* joints are used.

Rigid frames are shaped to optimize the utilization of the interior space better than arches, as would be the case with a bridge crossing a major road with restricted vertical clearances. The approximately straight deck, rigidly connected to piers and abutments, highlights practical features for artistic treatment.

Although moments are amplified at the laps of rigid frames, the moments at the center of span are small compared to a simple span of the same length. The narrower depth at midspan gives better clearance and appearance, lowers the approach lengths of the overpassing road, and increases the underpassing road, and ridge excavation or fill is reduced, land procurement is lowered, and the total cost is reduced. Additionally, maintenance and replacement costs are low, and widening is easily executed in the future.

2.9 ANALYSIS OF ARCHES

2.9.1 Assumptions

Analysis of a final design is required to validate that the shape and cross sections are to produce stresses/deformations within allowable values under the given system of loads. While analysis is thus a terminal step, it also provides the basis for many estimates that are useful in preliminary design.

Stress analysis on curved members is based on the assumption that the cross-sectional dimensions of the member are small relative to the radius of curvature; therefore, stresses may be calculated by formulas that are valid for straight members. For very sharp curvatures, curved-beam formulas for angle changes and stresses have to be used. It is also presumed, except where otherwise noted, that the arch can be analyzed by first-order, or so-called elastic, theory. This theory states that the effect of deformations on the bending moments may be neglected and that the principle of superposition will apply. For buckling investigations in long spans, deformation theory (second-order theory) has to be applied.

2.9.2 Kern Relationships

In designing/checking cross sections for combined axial loads and bending moments, it is usually helpful to use the kern relationships. The limiting points of the kern are the "kern points," which have eccentricities e_k from the center of gravity given by

$$e_k = \frac{r^2}{c} \tag{2.67}$$

where r = radius of gyration
c = distance to extreme fiber

The extreme fiber stresses f for an unsymmetrical section (Fig. 2.37) are given by the following equations:

$$\text{Upper flange:}\quad f_u = \frac{N}{A} + \frac{Mc_u}{I} = N\left(e + \frac{r^2}{c_u}\right)\frac{c_u}{I} = M_{ku}\frac{c_u}{I} \tag{2.68}$$

$$\text{Lower flange:}\quad f_l = \frac{N}{A} + \frac{Mc_l}{I} = N\left(e + \frac{r^2}{c_l}\right)\frac{c_l}{I} = M_{kl}\frac{c_l}{I} \tag{2.69}$$

Each stress f may be maximized by creating the influence line for the suitable M_k.

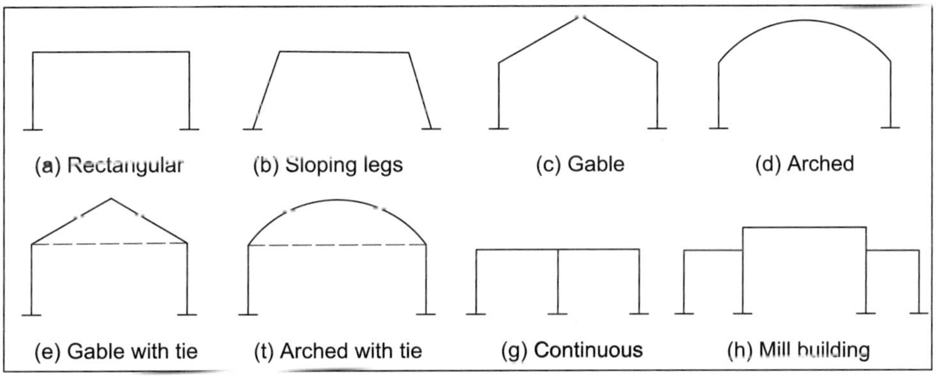

Figure 2.36 Types of bents.

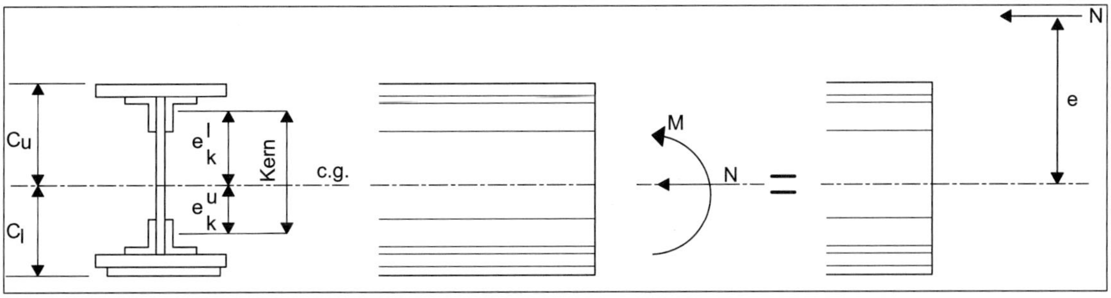

Figure 2.37 Kern points.

2.9.3 Finite-Element Analysis

In finite element analysis, a structure is divided into segments linked at the joints, and equations of continuity, compatibility, and boundary conditions which are set up in matrix form, based on the elastic properties of the structure and the external loading conditions. The solution of the equations generates the deflections and forces at all joints, which can be used for the final design or verification of existing sections. This analysis technique is now mostly adopted in computer application programs. Input commonly consists of the following data:

- Support conditions
- Loading conditions
- Description and type of structure
- Member orientation and properties
- Geometry of structure as defined by the joint coordinates

It should be highlighted that although these programs are easy to use, the engineer has to understand the behavior of the structure and practice engineering judgment in setting up the structural model in order to appropriately assess the results obtained from the output.

2.9.4 Energy Methods

With a typical fixed arch (Fig. 2.38), indeterminate to the third degree,

$$\begin{aligned} 1\cdot\delta_a &= \delta_{a0} + X_a\delta_{aa} + X_b\delta_{ab} + X_c\delta_{ac} + \delta_{at} \\ 1\cdot\delta_b &= \delta_{b0} + X_a\delta_{ba} + X_b\delta_{bb} + X_c\delta_{bc} + \delta_{bt} \\ 1\cdot\delta_c &= \delta_{c0} + X_a\delta_{ca} + X_b\delta_{cb} + X_c\delta_{cc} + \delta_{ct} \end{aligned} \tag{2.70}$$

where X_a, X_b, X_c = three redundants
$\delta_a, \delta_b, \delta_c$ = displacements in the directions of X_a, X_b, and X_c, respectively
$\delta_{a0}, \delta_{b0}, \delta_{c0}$ = displacements in the directions of X_a, X_b, and X_c, respectively, due to the applied external loads acting on the determinate base system
δ_{mn} = displacement in the direction of X_m due to $X_n = 1$
$\delta_{at}, \delta_{bt}, \delta_{ct}$ = displacements of the base system in the directions of X_a, X_b, and X_c, respectively, due to uniform temperature change $t°$ and/or a differential temperature $\Delta t°$ between top and bottom fibers

The displacements δ are given by

$$\delta_{mn} = \delta_{nm} = \int \frac{M_m M_n ds}{EI} + \int \frac{N_m N_n ds}{EA} + \int \frac{T_m T_n ds}{GJ} + \int \frac{kV_m V_n ds}{AG} \tag{2.71}$$

For temperature effects $t°$ and $\Delta t°$,

$$\delta_{mt} = \int \frac{M_m \varepsilon_t \Delta t^{o} ds}{h} + \int N_m \varepsilon_t t^{o}\, ds \tag{2.72}$$

In the above expressions, M_m and M_n are the bending moments in the base system due to $X_m = 1$ and $X_n = 1$, respectively. Likewise, N, T, and V are the normal forces, torques, and shears in the base system subjected to the unit redundant designated by the respective subscripts. The temperature coefficient of expansion is ε_t. κ the shape coefficient (in the shear-energy expression). For the case of common arches and loads,

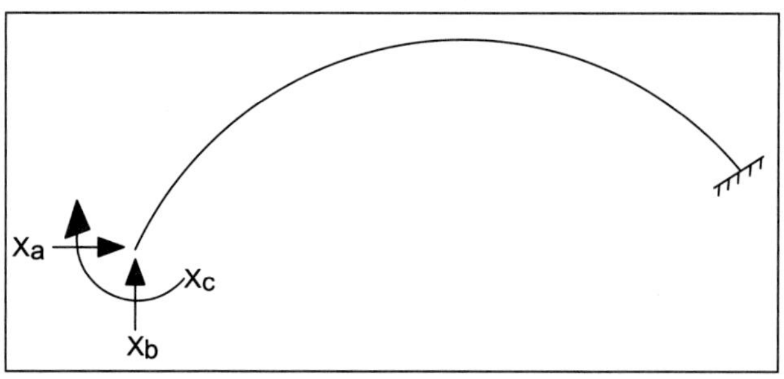

Figure 2.38 Fixed-ended-arch redundant.

the shear-energy effects are ignored, the torsional term does not apply, and the equations reduce to a consideration of normal forces, flexural effects, and temperature.

Equation (2.70) may be written in matrix form:

$$\begin{bmatrix} \delta_{aa} & \delta_{ab} & \delta_{ac} \\ \delta_{ba} & \delta_{bb} & \delta_{bc} \\ \delta_{ca} & \delta_{cb} & \delta_{cc} \end{bmatrix} \begin{bmatrix} X_a \\ X_b \\ X_c \end{bmatrix} = \begin{bmatrix} \delta_a - \delta_{a0} - \delta_{at} \\ \delta_b - \delta_{b0} - \delta_{bt} \\ \delta_c - \delta_{c0} - \delta_{ct} \end{bmatrix}$$

or

$$AX = K$$

which may be solved by inversion:

$$X = A^{-1}K$$

2.10 DESIGN OF ARCHES

2.10.1 General Procedure

Design of a ring, arch, or rigid frame includes a series of phases each introducing a refinement of the previous one. This is mainly true for indeterminate structures, in which preliminary but controlled calculations of design shears, moments, and thrusts are required, whereas in determinate systems, the statical values may be obtained accurately at the beginning.

The first phase includes planning with respect to the character of the spanned opening, rise, span, loading, clearance requirements, architecture, and aesthetics.

The second phase proceeds with finding the general shape and estimated design shears, moments, and thrusts from which tentative cross sections are obtained. Such values are typically derived from estimated influence-line data. On the other hand, the cross sections can be assumed from data on similar designs that have demonstrated satisfactory performance or from empirical relationships based on experience. Recurrently, as in the case of concrete arches and steel rigid frames, an intermediate phase of design refinement is possible at this point, using available design charts or tables, some of which are comprehensive enough to yield a final design needing very slight future modification.

The final phase of design for indeterminate structures is the formal analysis of the structure, to validate deformations and stresses, and to make slight adjustments if needed.

2.10.2 Preliminary Selection of Shape

To decrease bending moments, the arch axis should conform as closely as possible to the pressure line or equilibrium polygon of the loads. For uniformly distributed loads, such curve is usually parabolic. However, dead load generally increases in the direction of the abutments, resulting in a rising of the pressure line between crown and springing and an increase in the inclination at the springing line. The pressure line also differs as live load is superimposed on the dead load. This difference will take place with live load over a part of the span or even with full or partial live load over the entire span. Many designers overlook the fact that uniform live load over the full span has the effect of producing a more closely parabolic pressure line than does dead load alone.

The AASHTO specifications state that the lever rule may be used for the distribution of gravity loads in trusses and arches when analyzed as planar structures. If a space analysis is used, either the lever rule or direct loading through the deck or deck system may be used.

Arch ribs are most generally of variable thickness, increasing in depth from the crown to the springing. A rough estimate is to relate this variation to the secant of the angle of inclination of the arch axis to the horizontal. The assumption that moments of inertia differ directly as the secant of the angle is suitable mainly for developing simple approximate formulas. It has been found that results are comparatively insensitive to moderate deviations from this assumption.

2.10.3 Approximations for Special Shapes

The following estimates are valid to all constructional (elastic) materials.

Three-Hinged Arch

The influence line for the horizontal reaction $H_a = H_b$ is shown in Fig. 2.39.

Symmetrical Two-Hinged Arch

For a parabolic or flat circular shape ($f/L \le 1/8$) with the difference in moment of inertia given by $I = I_c \sec \phi$, where I_c is the moment of inertia at the crown and ϕ the inclination of the tangent with the horizontal (Fig. 2.40), the influence line for the horizontal reaction is given by

$$H = \frac{5L}{8f}(k - 2k^3 + k^4)v \tag{2.73}$$

where

$$v = \frac{1}{1 + \frac{15}{8} I_c / A_c f^2}$$

The parameter v represents the effect of rib shortening as a result of axial forces and can be ignored, excluding very flat arches (for arches with $f/L = 1/7$ to $1/9$, $v = 0.98$). At midspan, $k = 0.5$, so $H_{max} = 0.195vL/f$.

The above formula may be further approached by a parabolic influence line

$$H = \frac{3Lk(1-k)v}{4f} \tag{2.74a}$$

with $H_{max} = 3Lv/16f$. For full-span uniform load w,

$$H = \frac{wL^2 v}{8f} \tag{2.74b}$$

The formula can also be estimated by a semicubic parabola with central ordinate $H_{max} = 0.20vL/f$, the equation of which is

$$\begin{aligned} H &= \frac{L}{5f}[1 - (1-2k)^{3/2}]v \qquad & 0 \le k \le 0.5 \\ H &= \frac{L}{5f}[1 - (2k-1)^{3/2}]v \qquad & 0.5 \le k \le 1.0 \end{aligned} \tag{2.75}$$

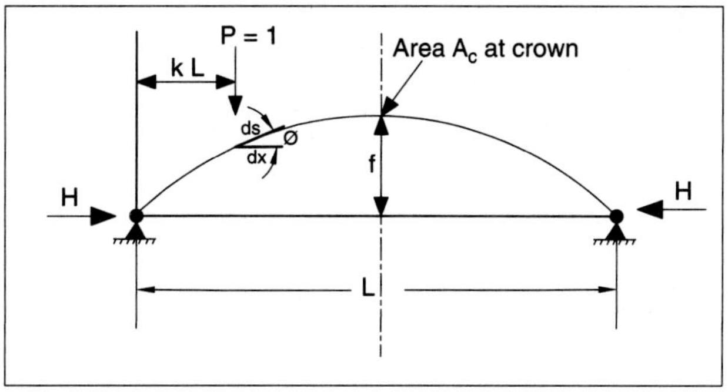

Figure 2.40 Symmetrical two-hinged arch.

The semicubic parabola is also a practical estimate for H for the two-hinged spandrel-braced arch.

For a uniform temperature change $t°$,

$$H = \frac{\varepsilon_t t^{o} E I_c}{\frac{8}{15} f^2 + I_c / A_c} \tag{2.76}$$

or, if the normal forces, represented by the term I_c/A_c, are neglected,

$$H = \frac{15}{8} \varepsilon_t t^{o} E \frac{I_c}{f^2} \tag{2.77}$$

Symmetrical Tied Arch

For a parabolic or flat circular shape ($f/L \le 1{:}8$) with $I = I_c \sec \phi$, the influence line for H is given by Eq. (2.7), except that the parameter v is substituted by

$$v' = \frac{1}{1 + \left(\frac{15}{8} f^2\right)(I_c / A_t + I_c / A_c)} \approx 1 \tag{2.78}$$

where A_T = area of tie.

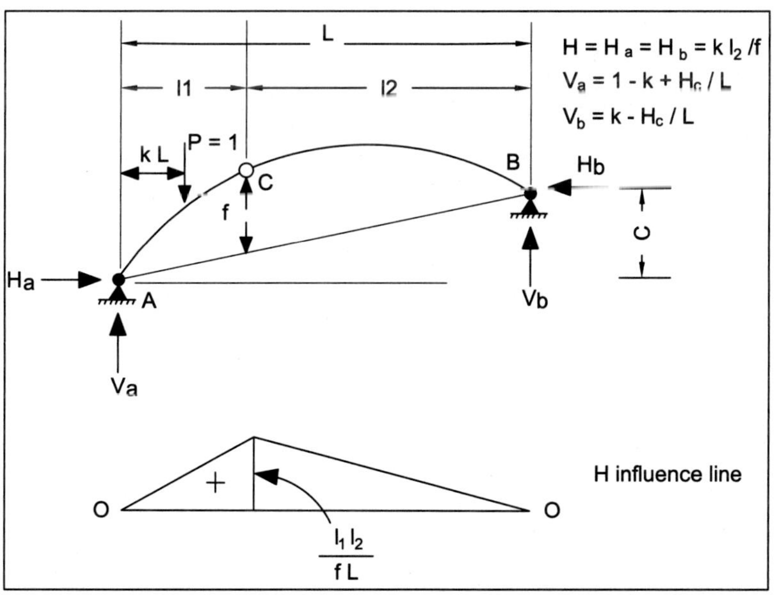

Figure 2.39 Unsymmetrical three-hinged arch.

For a uniform temperature rise t^o of both arch and tie, $H = 0$. If, however, the tie is warmed to t^o and the arch to $t^o + \Delta t^o$, then

$$H = \frac{\varepsilon_t \Delta t^o E I_c}{\frac{8}{15} f^2 + I_c(1/A_c + 1/A_T)} \tag{2.79}$$

SYMMETRICAL FIXED ARCH

For a parabolic or flat circular shape ($f/L \le 1{:}8$), with $I = I_c \sec \phi$, the influence lines for the reactions (Fig. 2.41) are given by

$$V_A = (1+2k)(1-k)^2$$
$$H_A = \frac{15}{4}\frac{L}{f}(k-k^2)^2 v$$
$$M_A = \frac{L}{2}(1-k)^2(5k^2 v - 2k) \tag{2.80}$$

where

$$v = \frac{1}{1 + 45/4\, I_c/A_c f^2}$$

Since $v \approx 1$, it typically can be neglected.

For full-span uniform vertical load w per unit length of span,

$$H = \frac{wL^2 v}{8f}$$
$$V_A = \frac{wL}{2}$$
$$M_A = M_B = \frac{1}{12} wL^2 (1-v)$$
$$M_C = \frac{1}{24} wL^2 (1-v) \tag{2.81}$$

For uniform full-rise lateral load w per unit vertical rise, acting from left to right,

$$H_A = \frac{11wf}{14} \qquad H_B = \frac{3wf}{14}$$
$$V_A = \frac{wf^2}{4L} \qquad V_B = \frac{wf^2}{4L} \tag{2.82}$$
$$M_A = \frac{51wf^2}{280} \qquad M_B = \frac{19wf^2}{280}$$

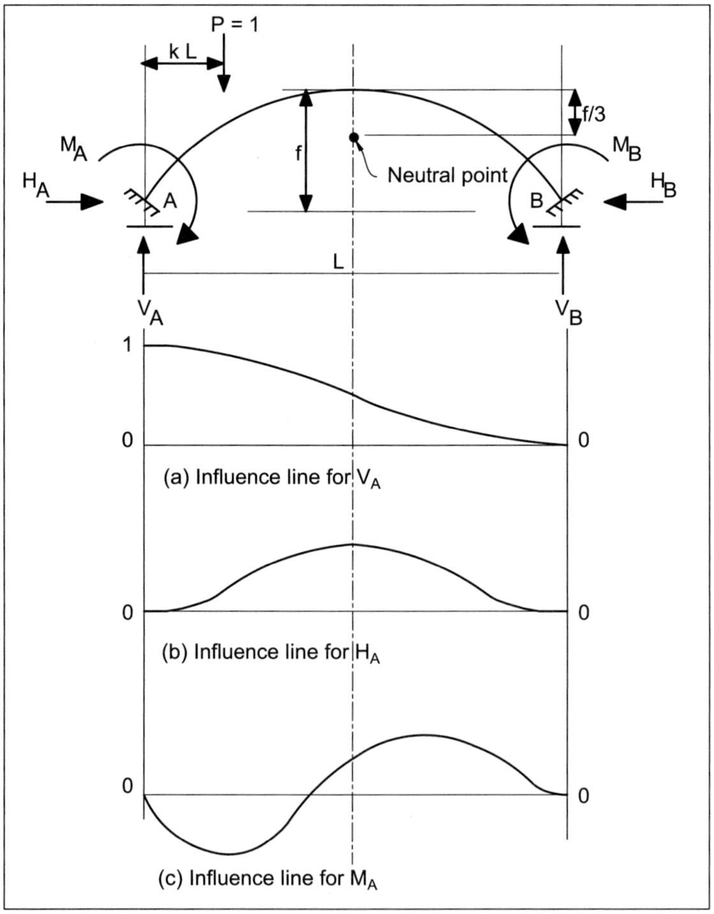

Figure 2.41 Symmetrical fixed circular arch.

For a temperature change $t°$,

$$H = \frac{45}{4}\frac{\varepsilon_t t^{\circ} E I_c v}{f^2}$$
$$V = 0 \qquad (2.83)$$
$$M_A = \frac{2}{3} f H$$

Fixed Circular Arch of Constant Section

Exact values of reactions and moments for a fixed circular arch of constant cross section are presented in Table 2.6.

Fixed Concrete Highway Arch Bridges

Certain empirical expressions in common use, which are helpful in choosing first trial sections for concrete highway arches, are noted below.

1. Whitney variation of rib thickness (Fig. 2.42)

$$d_k = d_c c \sqrt[6]{1 + \tan^2 \phi} \qquad (2.84a)$$

where d_c = depth at crown
d_k = depth at kL from crown

$$c = [1 - 2k\,(1 - I_c/I_s \cos\phi_x)]^{-1/3} \qquad (2.84b)$$

2. F. F. Wald formula (barrel arches)

$$d_c = \sqrt{L} + \frac{L}{10} + \frac{w_L}{200} + \frac{w_c}{400} \qquad (2.85)$$

where d_c = depth at crown, in.
L = clear span, ft
W_L = live load, psf
W_c = dead load, psf, above arch at crown

Wald recommends that the depth of springing be taken at from 1.5 to 3 times d_c, increasing with the flatness of the arch.

Table 2.6 Coefficients for Circular Fixed Arch Constant I*

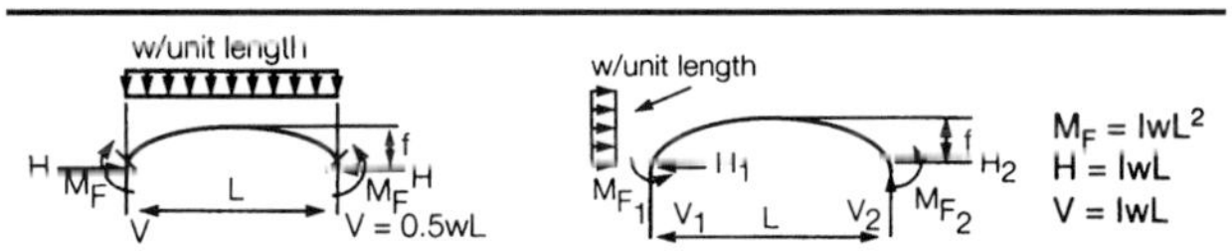

f/L	M_F	H	M_{F1}	M_{F2}	H_1	H_2	V_1, V_2
0.10	0.001230	1.2584	0.001831	0.000689	0.07857	0.02143	0.002480
0.12	0.001680	1.0507	0.002647	0.000994	0.09433	0.02567	0.003559
0.14	0.00224	0.9027	0.003598	0.001377	0.10990	0.03010	0.004825
0.16	0.002898	0.7924	0.004709	0.001817	0.12556	0.03444	0.006274
0.18	0.003659	0.7069	0.005981	0.002320	0.14125	0.03875	0.007899
0.20	0.00450	0.6388	0.007404	0.002900	0.15689	0.04311	0.009696
0.22	0.005439	0.5833	0.008989	0.003551	0.17254	0.04746	0.011660
0.24	0.006452	0.5372	0.010734	0.004282	0.18817	0.05183	0.013784
0.26	0.007549	0.4983	0.012643	0.005094	0.20378	0.05622	0.016062
0.28	0.008725	0.4652	0.014719	0.005994	0.21938	0.06062	0.018488
0.30	0.009981	0.4366	0.016963	0.006983	0.23496	0.06504	0.021054
0.32	0.011315	0.4177	0.019380	0.008065	0.25053	0.06947	0.023755
0.34	0.012726	0.3898	0.021972	0.009245	0.26609	0.07391	0.026583
0.36	0.014213	0.3705	0.024740	0.010527	0.28163	0.07837	0.029533

*From Ref. 11.

3. W. J. Douglas formula (barrel arches)

$$\begin{array}{ll} L = 20 \text{ ft} & d_c = 0.03(6 + L) \text{ ft} \\ L = 20 \text{ to } 50 \text{ ft} & d_c = 0.015(30 + L) \text{ ft} \\ L = 50 \text{ to } 150 \text{ ft} & d_c = 0.0001(11{,}000 + L^2) \text{ ft} \\ L = 150 \text{ ft} & d_c = 0.016(75 + L) \text{ ft} \end{array} \qquad (2.86)$$

4. V. A. Cochrane formulas for bridges (Fig. 2.8)

Open-spandrel arch:

$$y = \frac{8f}{6 + 5f/L}\left(3k^2 + \frac{10k^4 f}{L}\right) \qquad (2.87)$$

$$\tan\phi_s = \frac{8f/L}{6 + 5f/L}\left(3 + \frac{5f}{L}\right) \qquad (2.88)$$

Filled-spandrel arch:

$$y = \frac{4f}{1 + 3f/L}\left(k^2 + \frac{24k^5 f}{L}\right) \qquad (2.89)$$

$$\tan\phi_s = \frac{4f/L}{1 + 3f/L}\left(\frac{1 + 7.5f}{L}\right) \qquad (2.90)$$

Table 2.7 presents the dimensions of a typical highway arch bridge of reinforced concrete, which may be of value in initial proportioning of arch bridges.

Circular and Parabolic Arches of Constant Section

In roofs, bridges, and so on, arch ribs of prismatic section usually are more desired for the ease of fabrication and their appearance.

2.10.4 Intermediate Design

Tables for circular and elliptical arches of varying thickness are available.[12] Results are presented in tables and diagrams for reactions due to dead loads, live loads, and temperature and rib-shortening effects. Results also present tables and diagrams together with complete influence lines for springing line, moments at crown, and quarter points. Also, they include diagrams giving minimum and maximum moments with corresponding thrusts for crown, springing line, and quarter points. The data are shown in terms of two parameters: the *shape coefficient N*, which is the ratio of the drop of the arch axis at the quarter point to the rise *f*, and the *form coefficient m*, which characterizes the variation of the cross section of the rib (Fig. 2.42). These parameters, together with the span rise and thickness or moment of inertia of the rib at the crown, totally create the rib and its reactions under live load and temperature change.

In design, the value of $g = w_s/w_c$ is calculated first. In considering a design that is already completed, the value of N is calculated straight from the ordinates of the rib axis, after which g is computed from

$$g = \frac{1}{2}\left(\frac{1}{N} - 2\right)^2 - 1 \qquad (2.91)$$

Shrinkage effects can generally be calculated as equivalent to a 15°C temperature drop.

The horizontal thrust due to rib shortening under dead load is typically found distinctly in terms of the horizontal thrust due to dead load:

$$H_{RS} = -H_d \frac{I_c}{A_c C'_m C f^2} \qquad (2.92)$$

where H_{RS} = horizontal thrust due to rib shortening under dead load
H_d = horizontal thrust under dead load
A_c = area of cross section at crown
C = coefficient from Table 2.8
C'_m = coefficient from Fig. 2.43

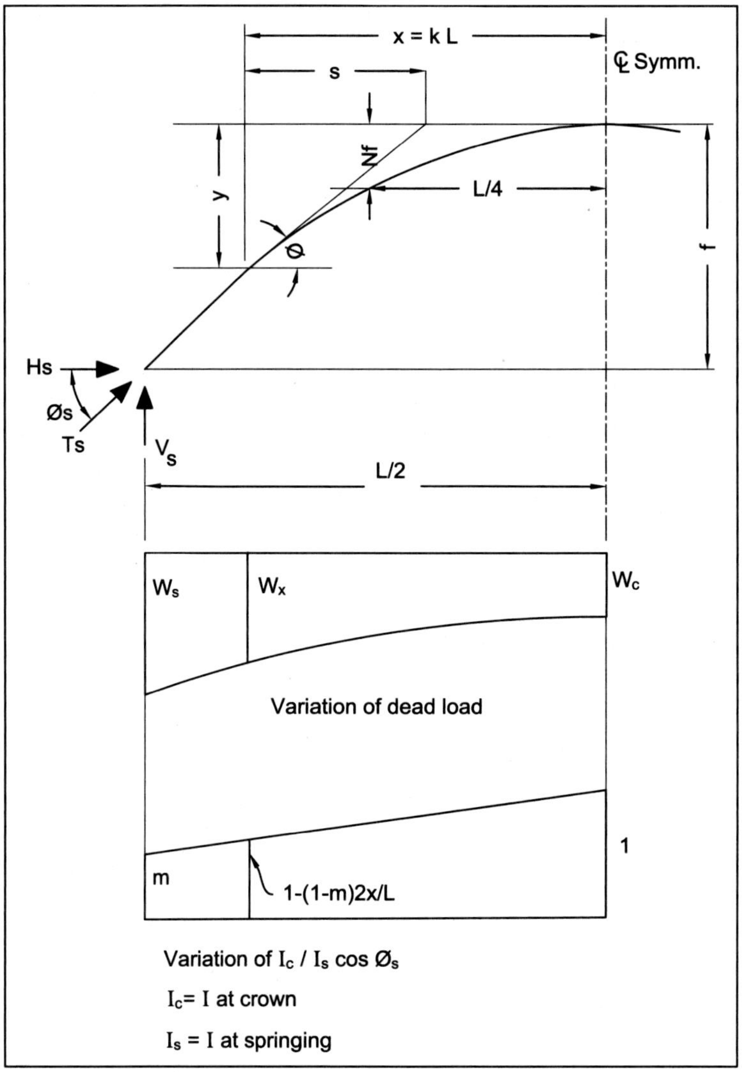

Figure 2.42 Whitney-arch notation.

Table 2.7 Typical Concrete Highway Arch Bridges

Span, ft	Rise, ft	Roadway, ft	Sidewalks	Crown depth and width	Springing depth and width	Type
50	9	23	2 at 5 ft 0 in.	10 in. × 36 ft 6 in.	2 ft 6 in. × 36 ft 6 in.	Filled spandrel
60	12	20	None	12 in. × 23 ft 0 in.	2 ft 6 in. × 23 ft 0 in.	Filled spandrel
70	14	19	None	1 ft 3 in. × 22 ft 0 in.	3 ft 7½ in. × 22 ft 0 in.	Filled spandrel
80	14	24		2 ft 0 in. × 4 ft 0 in.	3 ft 6 in. × 4 ft 0 in.	Open spandrel, 2 ribs
90	21	27	None	1 ft 6 in. × 5 ft 0 in.	3 ft 5 in. × 5 ft 0 in.	Open spandrel, 3 ribs
100	22	20	2 at 5 ft 0 in.	3 ft 0 in. × 3 ft 0 in.	6 ft 0 in. × 3 ft 0 in.	Open spandrel, 2 ribs
120	16	36	2 at 8 ft 0 in.	2 ft 1 in. × 6 ft 0 in.	4 ft 11 in. × 6 ft 0 in.	Open spandrel, 4 ribs
145	21	18	None	3 ft 0 in. × 4 ft 3 in.	5 ft 3 in. × 4 ft 3 in.	Open spandrel, 2 ribs
160	46	18	None	3 ft 2 in. × 4 ft 0 in.	6 ft 4 in. × 4 ft 0 in.	Open spandrel, 2 ribs
190	36	20	1 at 5 ft 0 in.	2 ft 3 in. × 10 ft 0 in.	5 ft 0 in. × 10 ft 0 in.	Open spandrel, 2 ribs
205	31	21	None	4 ft 0 in. × 4 ft 0 in.	8 ft 0 in. × 4 ft 0 in.	Open spandrel, 2 ribs

Table 2.8 Parameter C for Rib Shortening*

N	g	$m = 0.15$	$m = 0.20$	$m = 0.25$	$m = 0.30$	$m = 0.40$	$m = 0.50$
0.25	1.000	0.0329	0.0370	0.0410	0.0448	0.0520	0.0588
0.24	1.347	0.0320	0.0361	0.0400	0.0437	0.0509	0.0577
0.23	1.756	0.0312	0.0352	0.0390	0.0428	0.0498	0.0566
0.22	2.240	0.0302	0.0342	0.0380	0.0417	0.0486	0.0553
0.21	2.814	0.0294	0.0332	0.0370	0.0407	0.0475	0.0541
0.20	3.500	0.0285	0.0323	0.0361	0.0396	0.0466	0.0530
0.19	4.324	0.0276	0.0314	0.0351	0.0386	0.0455	0.0519
0.18	5.321	0.0267	0.0305	0.0341	0.0376	0.0443	0.0507
0.17	6.536	0.0258	0.0295	0.0331	0.0366	0.0432	0.0496
0.16	8.031	0.0250	0.0286	0.0322	0.0356	0.0422	0.0484
0.15	9.889	0.0241	0.0277	0.0312	0.0346	0.0411	0.0472

*From Ref. 3.

Moments due to the above rib shortening are found from

$$M_{RS} = -H_{RS}y' \tag{2.93}$$

where y' is the vertical distance between the elastic center and the point about which moments are considered.

2.10.5 Approximations of Whitney Data

Hardy Cross[13] derived average values of maximum moments at crown and springing, which may be applied as a shorter, different, intermediate fine-tuning for the initial design, as follows:

$$\begin{aligned} \int \frac{ds}{EI} &= A = \frac{L}{EI_c}\left(\frac{1+m}{2}\right) \\ \int x^2 \frac{ds}{EI} &= \frac{1}{12}AL^2\left(\frac{1+3m}{2+2m}\right) \\ \int y^2 \frac{ds}{EI} &= \frac{4}{45}Af^2\left(\frac{1+3m}{2+2m}\right) \end{aligned} \tag{2.94}$$

If ds/I = constant, the Whitney form coefficient m is unity (i.e., 1), and the parenthetical multipliers in the above formulas cancel out, giving exact values for a parabolic arch with this particular variation of cross section. For uniform live load, the kern moments are

$$\begin{aligned} \text{Max } M_s^k &= \pm\frac{1}{45}w_l L^2 \\ \text{Max } M_c^k &= \pm\frac{1}{225}w_l L^2 \end{aligned} \tag{2.95}$$

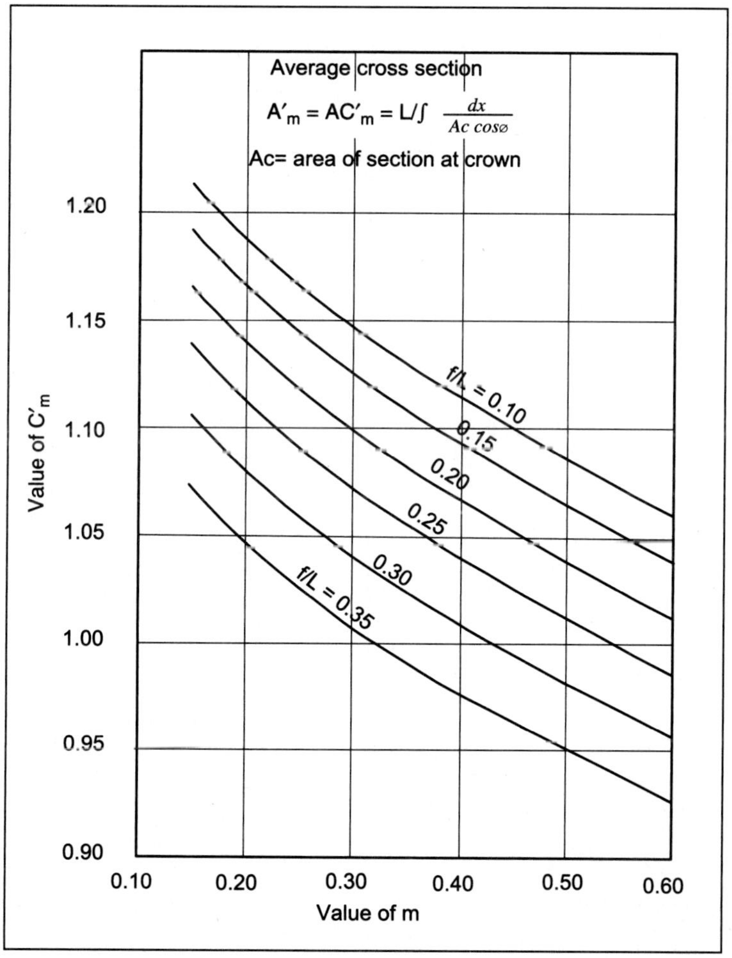

Figure 2.43 Parameter C'_m for rib shortening. (*From Ref. 3.*)

For max $$H = 0.10 w_L \frac{L^2}{f}$$

For max $$H_c = \frac{1}{8}\frac{WL}{f}$$

For dead load, as an approximate value,

$$H_c = \frac{1}{8}\frac{WL}{f} \tag{2.96}$$

$$T_s = H_c \sec\phi_s \tag{2.97}$$

where W = total dead load.

For temperature,

$$H_t = \frac{\varepsilon_t E t^{\circ} L}{I_y} = \frac{45}{4}\varepsilon_t E t^{\circ}\frac{I_c}{f^2} \approx 200 t^{\circ}\frac{I_c}{f^2}\frac{4}{1+3m} \tag{2.98}$$

$$M_c = H_t y_c = \frac{15}{4}\varepsilon_t E t^{\circ}\frac{I_c}{f^2} \approx 67.5 t^{\circ}\frac{I_c}{f^2}\frac{2+2m}{1+3m} \tag{2.99}$$

$$f_t = \frac{M_c y}{I} = \frac{15}{8}\varepsilon_t E t^{\circ}\frac{d}{f} \approx 33 t^{\circ}\frac{d_c}{f}\frac{2+2m}{1+3m} \tag{2.100}$$

Where live loads are uniform, the estimated locations for maximum moments are indicated in Fig. 2.44.

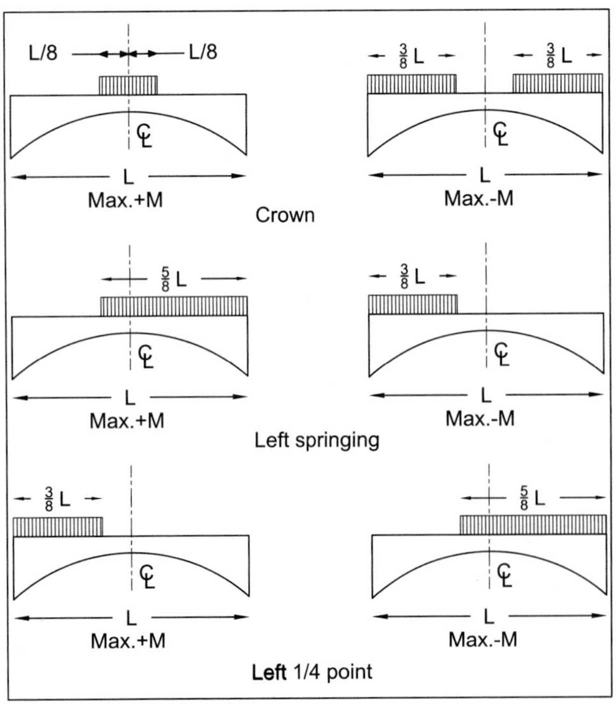

Figure 2.44 Approximate uniform live-load positioning for maximum moments.

As a guide in selection of sections for stress, Cross recommended proportioning permissible concrete stress are given as follows:

Stress due to dead load	50% (range 40–60%)
Stress due to live-load compression	5%
Stress due to live-load flexure	30% (range 20–50%)
Stress due to flexure from rib shortening, shrinkage, and temperature change	15% (range 5–20%)
Total stress	100%

2.10.6 Final Design

The final design of an arch typically includes a proper analysis to confirm the initial or transitional design under several combinations of load, shrinkage, temperature change, rib shortening, and so on, with such final corrections to reinforcement, cross section, and shape as are dictated thereby. It should be noted that allowable stresses for combined compression and bending are variable and a function of the relative value of moment to thrust. Also, reference has to be made to the specific interaction or combined-stress formulas of the specifications involved. It is also to be noted that for loading combinations including shrinkage, temperature change, rib shortening, and so on, an increase in allowable stresses is permitted by the different codes.

2.10.7 Unsymmetrical Arches

The design of unsymmetrical arches is hindered by the lack of simple estimates, which are fairly difficult to formulate. Whitney[14] has proposed and presented an extension to unsymmetrical arches of the methods described in Article 9. Several designers use a trial-and-error approach, starting from a preliminary shaping to fit the pressure line for dead load or dead load plus one-half live load.

2.10.8 Ultimate Design of Concrete Arches

Reports of ACI Committee 312, now separated, had pursued developing a general method of design of concrete arches that would consider all important factors, such as dead and live loading, abutment movements, effects of volume changes (shrinkage, temperature effects, and plastic flow), and arch deflection.[15] Because of the nonlinear relation between loads and deflection, the committee preferred a fundamentally ultimate-strength method of design. This has been implemented in the AASHTO specifications.

The first step of the proposed design procedure involves determining moments and thrusts in the rib for each of the stated loading conditions, assuming the rib to perform as an elastic member and ignoring the influence of deflections on moments. This agrees with the standard methods of elastic analysis.

The second step includes the calculation of the deformation of the arch rib by standard elastic methods, giving the angular rotations and deflections at all points of the rib. Secondary moments are obtained and deflections and rotations calculated again. The procedure is repeated to convergence. The effect of horizontal deflections may also be considered but can be ignored for arches that have a rise-to-span ratio that is less than 1:5.

Values of modulus of elasticity of 2,000,000 psi are proposed for determination of deflection moments and 4,000,000 psi for stability volumetric changes. It is also suggested that standard ultimate-strength design procedures for combined axial and bending thrust be used in the proportioning and design of concrete-arch ribs.

The plasticity of concrete subjected to high stresses or stresses of long duration similarly results in redistribution of moment between sections. Plastic flow of concrete is fast in the initial stages but slows with time. It causes a reduction in the stresses in the concrete and an increase in the stress in the reinforcement. It is preferable to decenter arches as soon as possible if sufficient reinforcing is provided to avoid overstress in the steel.

The 1932 and 1940 reports of ACI Committee 312 constitute significant data from which possible effects of shrinkage, temperature change, and plastic flow can be assessed. While shrinkage, temperature, and abutment-movement stresses are relieved by cracking of the ribs and plastic flow, deformation stresses are amplified. Records confirm that cracking and plastic flow have caused failure months after decentering. To avoid this, ribs must be balanced to control initial volumetric and displacement stresses, and adequate steel must be provided to avoid extreme flow under long time periods. A minimum of 1 percent reinforcement is required in the AASHTO specifications.

2.11 DESIGN OF FRAMES

2.11.1 Steel Frames

For buildings constructed of single-span rigid frames, the following will typically deliver good economy for average roof loads[16]:

Span, ft	Frame spacing, ft
30–40	16
40–60	18
60–100	20
> 100	1/5 to 1/6 of span

Structures that have bases completely fixed against rotation are not that common. Hence, rigid frames are often designed as if hinged at their column bases. Expensive pin-connected base details are seldom required; in general, a regular flat base plate with a single line of anchor bolts set perpendicular to the span at the column centerline will be sufficient.

Frames should be appropriately braced for lateral stability. Korn[16] proposes the use of portals, laced struts, or sway frames as follows:

- Spans up to 40 ft: no lateral bracing is necessary where purlins or a deep metal deck is used. (Rod bracing would be required for one or more bays.)
- Spans from 41 to 60 ft: bracings are mandatory on a diagonal centerline of a haunch or knee and at a ridge. (Intermediate bracing members would be desirable for the 60-ft span.)
- Spans over 60 ft: bracings are mandatory at haunches and along ridges, with added intermediate sway frames spread out at a distance equal to 90 times the least radius of gyration and not to exceed 100 times that dimension.

2.11.2 Concrete Rigid-Frame Bridges

The following proportions are recommended[17] for frames of the type shown in Fig. 2.33:

1. Thickness at center of span (*B-B*) equal to about *L*/35. However, this value can be reduced to *L*/40 when the frame is founded on a practically unyielding foundation; it must be increased where the footings rest on extremely compressible soils.
2. Thickness at *A-A* equal to about *L*/15.
3. Thickness at base (*C-C*) equal to around 1 ft 6 in. for 30-ft spans, 2 ft 6 in. for 60-ft spans, and 3 ft 4 in. for 90-ft spans.

2.11.3 Design

By drawing the deflected shape of the structure under certain combinations of load, one can roughly locate the inflection points, which make the structure statically determinate. The section near the center of span would be governed by

- dead plus live plus impact,
- the same combination plus shrinkage plus temperature change, and
- the total of all these plus spread of the footings (often specified or otherwise arbitrarily assumed as 1/2 in.). The knee section would be governed by dead plus live plus impact plus active lateral earth pressure on the legs or by this combination plus rise in temperature.

EXAMPLE

A frame subjected to a concentrated load is depicted in Fig. 2.45. The deflected location is sketched, and the inflection points *E, F, C*, and *H* are placed. By considering the partial structure *EBFGCH* in (*c*), the reaction $RD = Pd1/d2 = 3 \times 4/7.8 = 0.51P$. Equally, $R_A = 0.74P$. Then

$$V_D = \frac{5}{\sqrt{5^2+5^2}} \times 0.51P = 0.361P$$

$$H_D = \frac{5}{\sqrt{5^2+5^2}} \times 0.51P = 0.361P$$

$$M_D = 0.5 \times 10 \times H_D = 1.80P$$

Inconsistencies in these reactions are due to estimates in locating the points of inflection.

Estimated positions of the inflection points for rectangular bents are shown in Fig. 2.46. These positions are based on members with constant moment of inertia and the same stiffnesses for columns and beams. If these two conditions are not met, the point of inflection has to be relocated toward the stiffer joint (by about 5 to 15 percent of the designated distance to that joint) to attain a better first estimate.

Charts for horizontal reactions in two-legged rigid frames with hinged bases are shown in Figs. 2.47 to 2.50. Also, moment and reaction tables

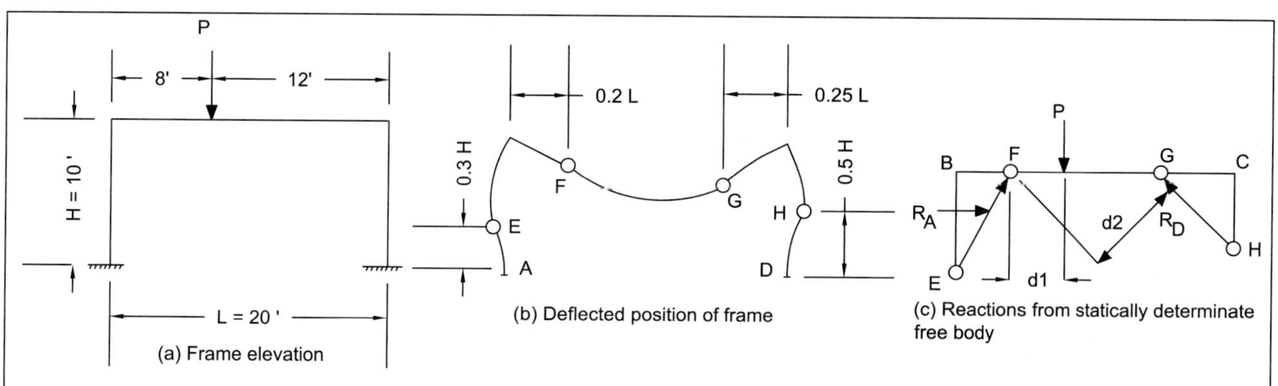

Figure 2.45 Frame analysis by locating points of inflection.

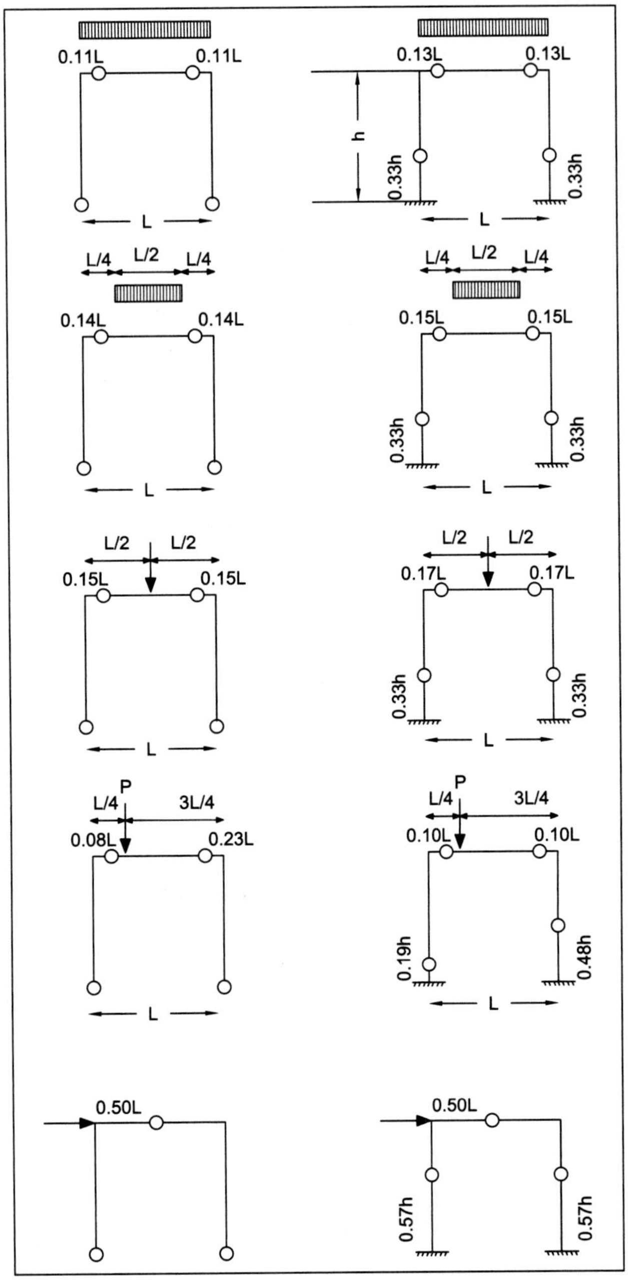

Figure 2.46 Approximate location of points of inflection.

in Ref. 18 list the corresponding lightest W section in A36 steel. Comparable aids for the first estimate of forces and moments are presented in tabular form.[19] Formulas for a large variety of framed structures are also available.[20]

Member sizes and dimensions of specific frames, like pier bents for grade-separation structures, are proportioned for standardization of design or for appearance. In such cases, a final analysis may be conducted without recourse to an initial design.

EXAMPLE 1—DESIGN OF STEEL-BUILDING RIGID FRAME

A single-story building needs a clear span of about 100 ft and a minimum height of 18 ft. Welded A36 steel gabled frames, 20 ft on centers, will satisfy the requirements. The AISC specification will be followed. The dimensions of the frame are presented in Fig. 2.52. The bases are assumed to be hinged.

Loads and load combinations:

1. Dead load *D*:

Roofing	= 4 psf
Insulation	= 3
Metal deck	= 2
Miscellaneous	= 4
Purlins	= 4
	17 psf of sloping surface

$$0.017 \times \frac{52.7}{50} \times 20 = 0.36 \text{ kip/ft of frame}$$

$$\text{Frame} = 0.10$$

$$D = 0.46 \text{ kip/ft}$$

2. Snow load *S*: at 30 psf, $S = 0.03 \times 20 = 0.6$ kip/ft
3. Drift snow load *Dr*: 0.6 kip/ft on half span only
4. Wind load *W*: at 20 psf, $W = 0.02 \times 20 = 0.4$ kip/ft

Various combinations of these loads are studied (Fig. 2.52). Usually, combination 1 governs the design of the frame. The other combinations may control the design of details. A 33 percent rise in allowable stresses is allowed for combinations 3 to 5.

Initial Analysis Figures 2.47 and 2.48 are used to find out the horizontal reactions. It is assumed that $I_2/I_1 = 0.6$. This ratio will be checked later:

$$K = \frac{I_2}{I_1}\frac{h}{m} = 0.6\frac{20}{52.7} = 0.228$$

$$Q = \frac{f}{h} = \frac{16.67}{20} = 0.833$$

For uniform vertical load *w*, Fig. 2.47 results in

$$C_1 = 0.059 \quad H_A = A_E = \frac{C_1 w L^2}{h} = 28w$$

For horizontal loads, Fig. 2.48 results in

$$C_6 = 0.56 \quad H_E = C_6 h w = 11.2w$$

Loading

D	$H_A = H_E = 28 \times 0.46 = 12.9$ kips
S	$H_A = H_E = 16.8$ kips
DR	$H_A = H_E = 8.4$ kips
	$H_E = 4.5$ kips
W	$H_A = 36.67 \times 0.4 - 4.5 = 10.2$ kips

The moments and forces at different points of the frame can now be calculated and are summarized in Table 2.9.

Design of Members

Rafter: $M = +338$ ft-kips
$N = +31.1$ kips (not corrected for moment redistribution)
$V = 1.4$ kips (not corrected for moment redistribution)
Unbraced lengths: $L_x = 44.27$ ft (assume $K_x = 1.5$) $L_y = 6.32$ ft

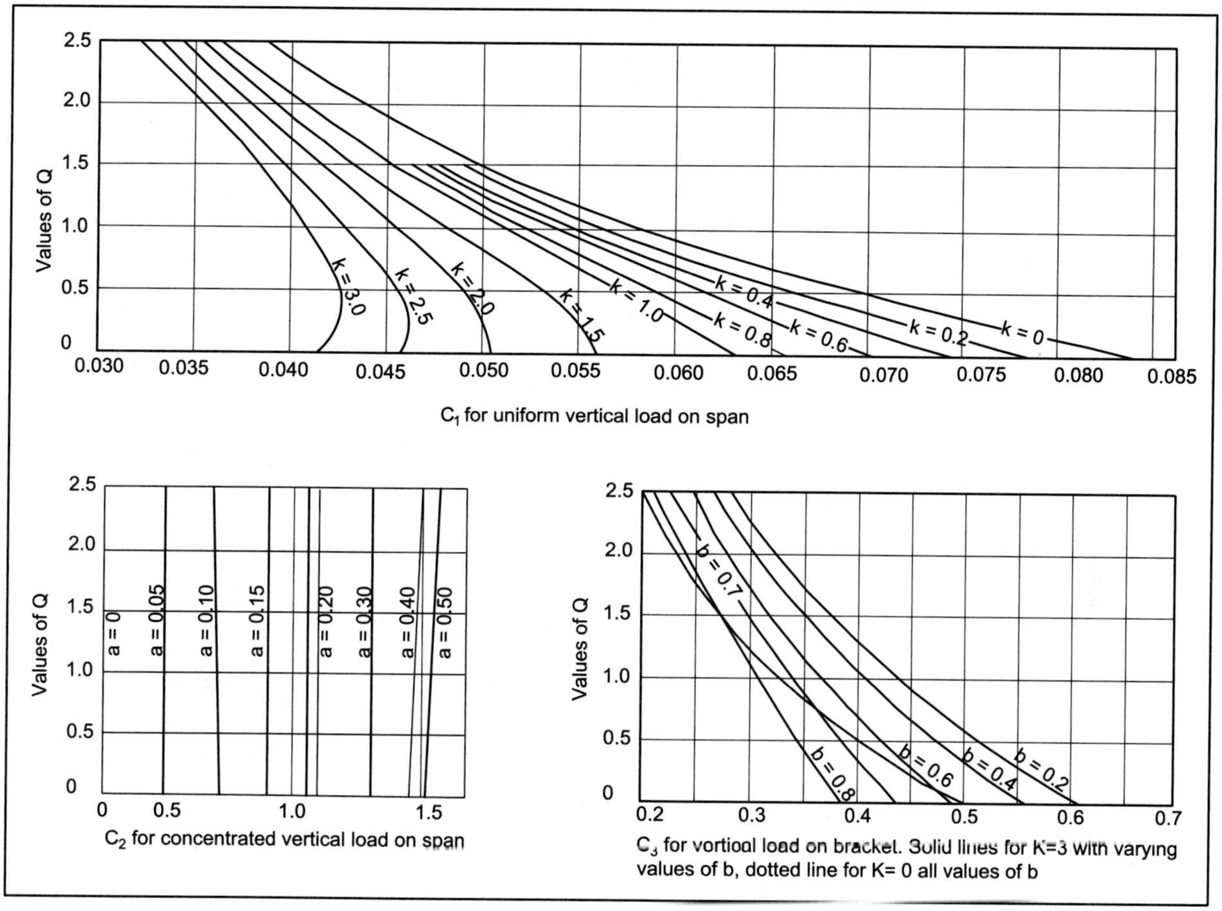

Figure 2.47 Horizontal reactions for ridge and rectangular frames, hinged bases. (*Courtesy of American Institute of Steel Construction.*)

Try $W24 \times 84$:

$$r_x = 9.79 \text{ in.} \quad \left(\frac{KL}{r}\right)_x = 81.5 \quad F_a = 15.2$$

$$\frac{L_y d}{A_f} = 263 < 500 \quad \frac{76 b_f}{\sqrt{F_y}} = 114 < L_y \quad F_b = 24 \text{ ksi}$$

$$f_a = \frac{31.1}{24.7} = 126 \quad f_b = \frac{338 \times 12}{196} = 20.6 \text{ ksi}$$

$$\frac{f_a}{F_a} = \frac{1.26}{15.2} = 0.083 < 0.15$$

$$\frac{f_b}{F_b} = \frac{20.6}{24.0} = 0.859$$

$$0.083 + 0.859 = 0.942 < 1.0$$

Column

Use $M = -417$ ft-kips (no moment redistribution)

$N = +53.0$ kips

$V = 29.7$

Unbraced lengths: $L_x = 14$ ft $\quad L_y = 14$ ft $\quad K_y = 1.0$

For effective length coefficient K_x:

At top of column: $G_T = \dfrac{I_1/h}{I_2/(2m)} = \dfrac{1}{0.6}\,\dfrac{2 \times 52.7}{20} = 8.8$

At bottom of column: $G_B = 10$

$K_x = 2.9$

Try $W30 \times 99$:

$$r_x = 11.7 \quad \left(\frac{KL}{r}\right)_x = 42$$

$$r_y = 2.1 \quad \left(\frac{KL}{r}\right)_y = 84 \quad F_a = 14.9 \text{ ksi}$$

$$\frac{76}{\sqrt{F_y}} b = 132 \text{ in.} < 14 \text{ ft} \quad F_b = \frac{417 \times 12}{269} 18.6 \text{ ksi}$$

$$f_a = \frac{53}{29.1} = 1.82 \text{ ksi} \quad f_b = \frac{417 \times 12}{269}$$

$$\frac{f_a}{F_a} = \frac{1.82}{14.9} = 0.12 < 0.15$$

$$\frac{f_b}{F_b} = \frac{18.6}{22.0} = 0.84$$

$$0.12 + 0.84 = 0.96 < 1.0$$

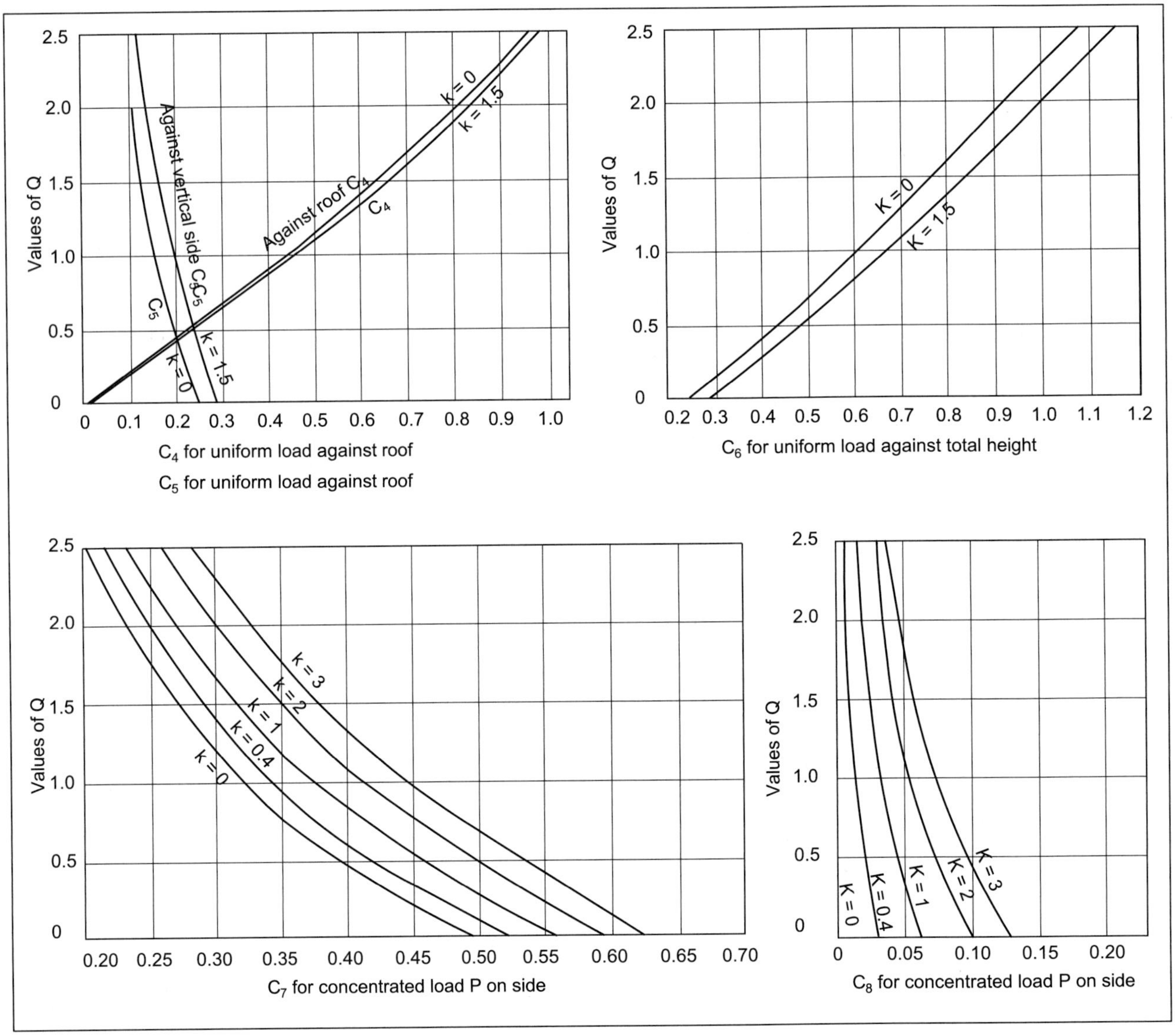

Figure 2.48 Horizontal reactions for ridge and rectangular frames, hinged bases. (*Courtesy of American Institute of Steel Construction.*)

Final Analysis Check assumed sections:

$$\frac{I_2}{I_1} = \frac{2364.3}{3988.6} = 0.594 \approx 0.6$$

Haunch design (Fig. 2.53): Try haunch makeup:

2 flange plates $10\frac{1}{2} \times \frac{11}{16}$	14.45	5550
1 web plate $38\frac{1}{2} \times \frac{1}{2}$	19.25	2380
	33.70	7930

Use loads at point *B* (Table 2.9).
At section *eb*: $M = -542$ kips (Fig. 2.54)

$$N = +53 \text{ kips}$$
$$V = 29.7 \text{ kips}$$

Unbraced length, $L = 6.0 - 1.77 = 4.23$ ft = 51 in.

AISC formula (F1–6): $r_y^2 = \dfrac{b^2}{12} \dfrac{A_f}{A_f + A_w/6} = 7.52$

$$r_y = 2.74 \quad \frac{L}{r_y} = 18.6 < \sqrt{\frac{102{,}000 C_b}{F_y}}$$

AISC formula (F1–8): $\dfrac{Ld}{A_f} = 282 < 545 \qquad F_b\ 22$ ksi

$$f_b = \frac{N}{A} + \frac{Mc}{I} = \frac{53}{33.7} + \frac{542 \times 12 \times 20}{7930} = 18.0 \text{ ksi}$$

At section *bf*, the haunch is checked correspondingly and found to be satisfactory ($f_b = 17.7$ ksi).

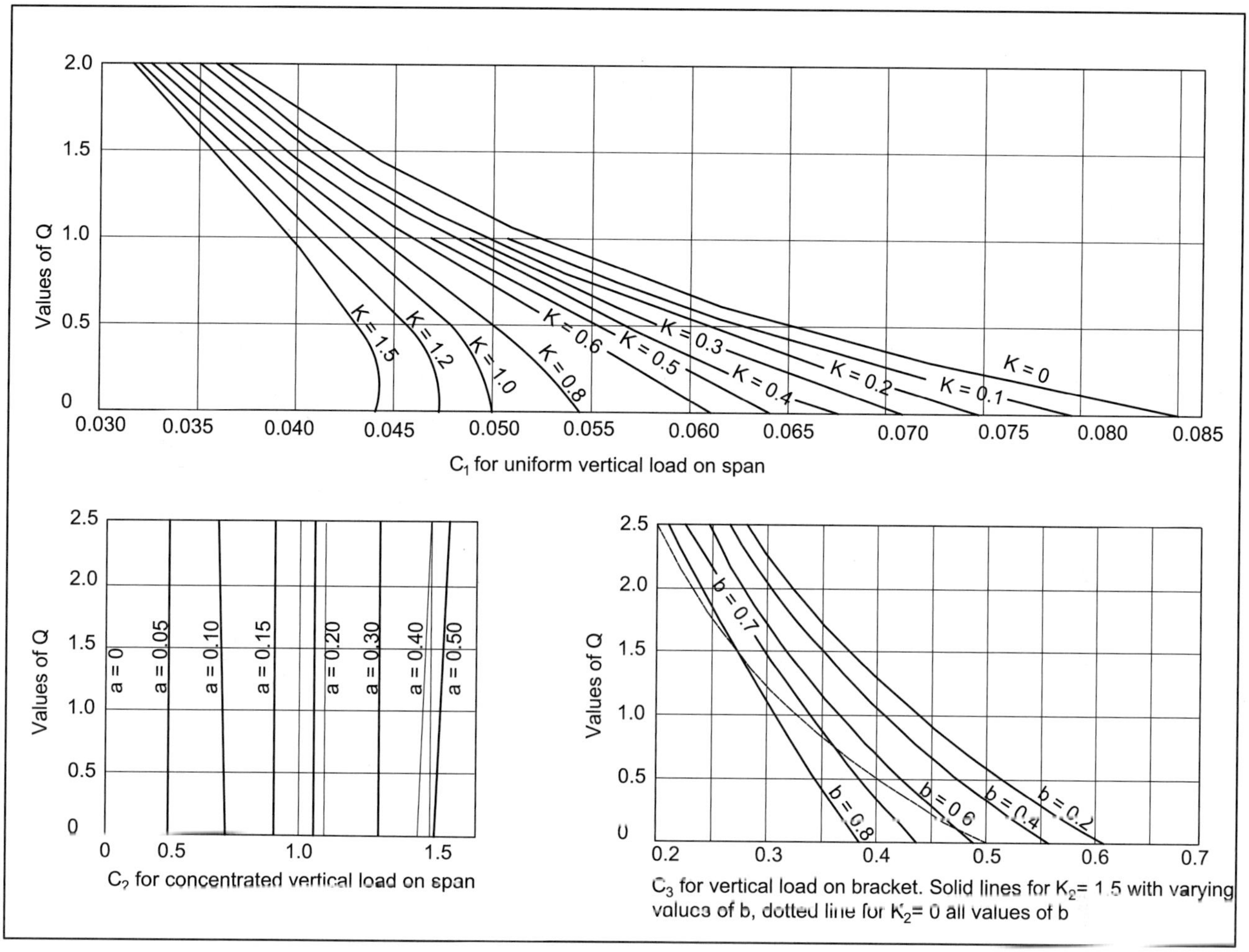

Figure 2.49 Horizontal reactions for parabolic and rectangular frames, hinged bases. (*Courtesy of American Institute of Steel Construction.*)

For the diagonal stiffener required at *b*, the resultant of the flange loads in rafter and column is 117 kips:

$$A_s = 117/22 = 5.3 \text{ in.}^2\text{; use two plates } 5 \times 9/16.$$

2.11.4 Arched Bents; Continuous Arches on Elastic Piers

Arches integral with and resting on relatively flexible columns are used in both bridge and building work. They are commonly found in buildings with thin-barrel concrete-shell roofs, where the arches perform as single or continuous end or intermediate frames to take the loads of the roof shell. Building arches are usually made with constant-section members of regular geometric shape, while bridge arches may be of variable section and shape. Tabular solutions for circular arched bents and gable bents are available.[21,22]

Arched frames can be analyzed using the general method of virtual work. Single-span frames are most simply done using the column analogy. However, in continuous arched frames, numerical procedures, such as those including slope deflection or moment distribution, are more common. The moment-distribution procedure is described mainly because it is similar to that used for multistory rectangular bents, in which fixed-end moments are distributed while joints translation is prevented and corrections are done for the actual displacement of the joints. The corrections are done by applying unit horizontal displacements consecutively at each joint and writing simultaneous equations equal in number to the total of such joints having horizontal freedom of movement. This is done in order to balance the horizontal thrusts. The simultaneous-equation approach is particularly adaptable to cases with several different loading conditions, the benefit of which is not common for other differences of this procedure and which may even offer difficulties in convergence.

The main difference in working with curved members is that flexural stiffness, fixed-end moments, and moment carryover factors are changed. However, there exist fixed-end thrusts due to loads, a thrust produced by an applied moment at one end when the far end is free to rotate, and a moment produced by horizontal displacement when there is no end rotation. The following summarizes the use of the column analogy for obtaining these basic constants.

From Fig. 2.55 and the formula $f = P/A + M_x y_c/I_x + M_y x_c/I_y$

$$K = \text{moment stiffness} = f_a = \frac{1}{A} + \frac{y_c^2}{I_x} + \frac{x_c^2}{I_y} \tag{2.101}$$

$$CK = \text{carryover moment} = f_b = \frac{1}{A} + \frac{y_c^2}{I_x} - \frac{x_c^2}{I_y} \tag{2.102}$$

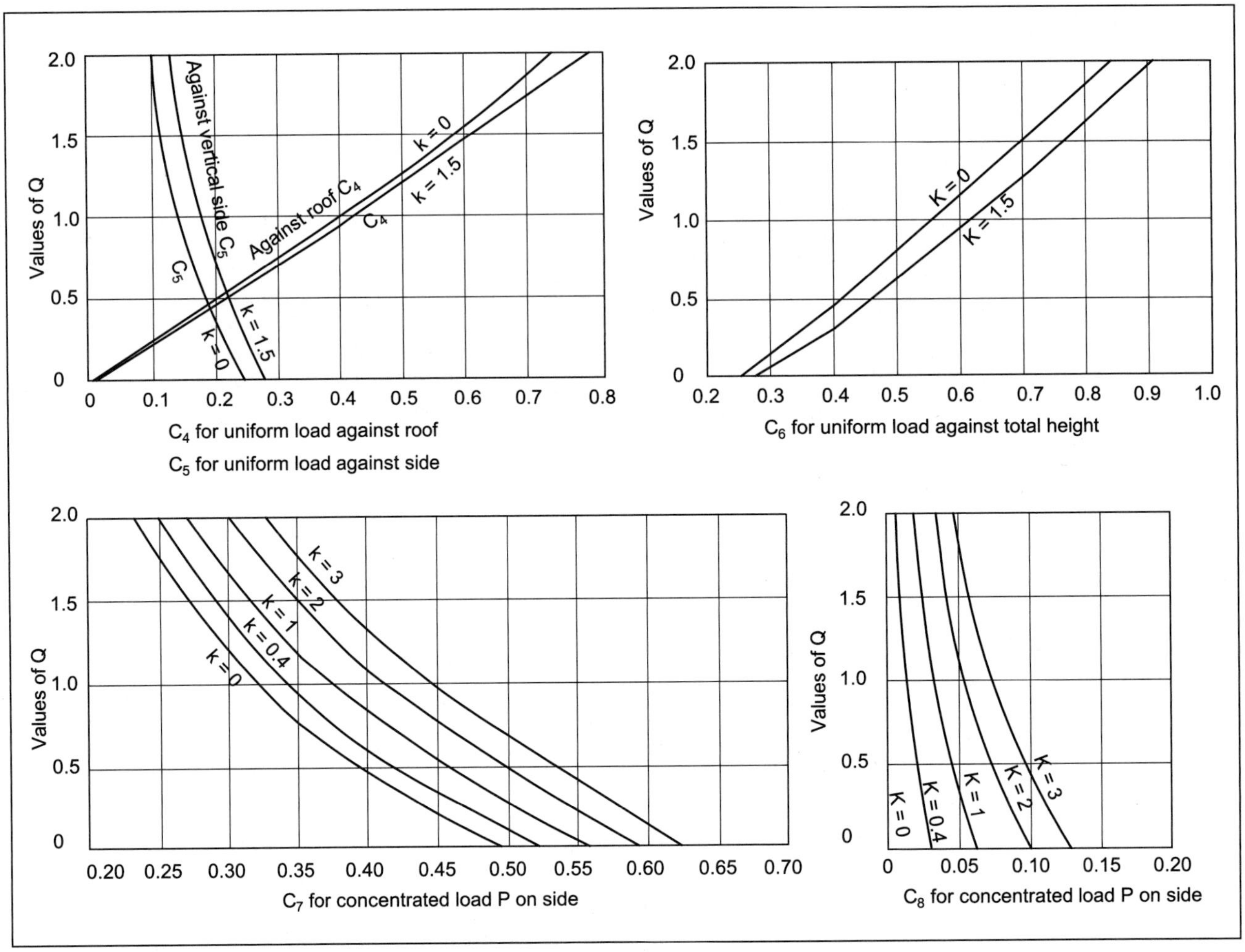

Figure 2.50 Horizontal reactions for parabolic and rectangular frames, hinged bases. (*Courtesy of American Institute of Steel Construction.*)

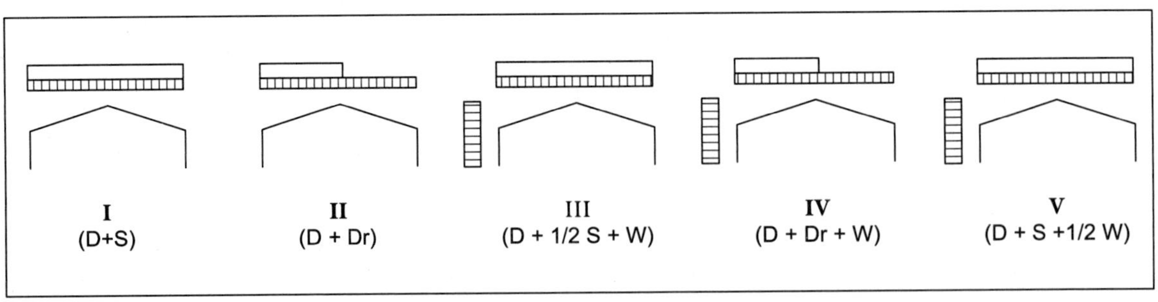

Figure 2.51 Loading combinations, Example 1.

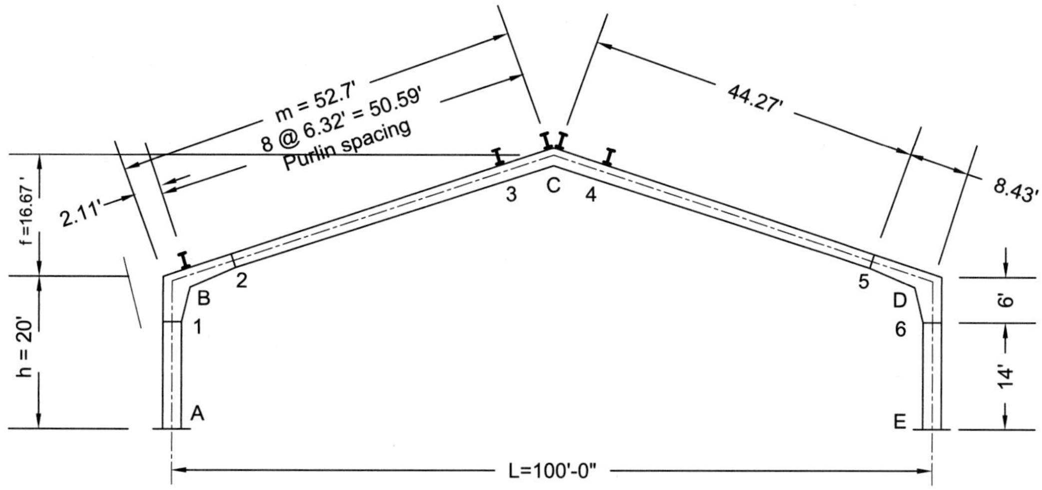

Figure 2.52 Frame elevation, Example 1.

Table 2.9 Moments, Axial Forces, and Shears for Frame of Fig. 2.52

Points / Loading	A		1			B			2			3			C		
	R	*H*	*M**	*N*†	*V*	*M*	*N*	*V*	*M*	*N*	*V*	*M*	*N*	*V*	*M*	*N*	*V*
D	23.0	12.9	−181	23.0	12.9	−258	23.0	12.9	−123	18.4	14.2	+121	13.5	0.6	+ 102	12.3	4.1
S	30.0	16.8	−236	30.0	16.8	−336	30.0	16.8	−160	24.0	18.5	+158	17.6	0.8	+133	16.0	5.3
Dr (on right half)	7.5	8.4	−118	7.5	8.4	−168	7.5	8.4	−130	10.4	4.4	+29	10.4	4.4	+67	10.4	4.4
Dr (on left half)	22.5	8.4	−118	22.5	8.4	−168	22.5	8.4	−29	13.1	12.6	+130	7.1	−5.3	+67	5.6	−9.8
W (on right side)	2.7	4.5	−63	2.7	4.5	−90	2.7	4.5	−80	5.2	1.2	−40	5.2	1.2	−30	5.2	1.2
W (on left side)	−2.7	−10.2	+104	−2.7	−4.6	+124	−2.7	−2.2	+106	−1.9	−2.3	+3	2.3	−3.7	30	3.1	4.0
D + *S*		−417	53.0	29.7	594	53.0	29.7	−283	42.4	32.7	+279	31.1	1.4	+235	28.3	9.4	

*Moments in ft-kips.
†N in kips; negative sign denotes tension.

$$C = \text{carryover factor } \frac{f_b}{f_a} \tag{2.103}$$

M_v = moment due to vertical displacement (no end rotation)

$$= f_a \text{ for } P = 0, M_x = 0, M_y = 1$$

$$= \frac{x_c}{I_y} \tag{2.104}$$

M_u = moment due to unit horizontal displacement

$$= f_a \text{ for } P = 0, M_x = 1, M_y = 0$$

$$= \frac{y_c}{I_x} \tag{2.105}$$

H_m = thrust due to unit moment (far end fixed)

$$= \frac{y_c/I_x}{K} \tag{2.106}$$

H'_m = thrust due to unit moment (far end pinned)

$$= \frac{H_m}{1+C} \tag{2.107}$$

For arches of regular shape and constant cross section, the above values, in addition to fixed-end moments and thrusts, can be obtained from Tables 2.7 and 2.10. For arches of variable section, Whitney's tables can be used or the constants computed using the above shown formulas.

Example 2 demonstrates the application of moment distribution to arched frames. A single-span circular arch frame under uniform vertical load is used only for demonstration; it is simpler to resolve this case by column analogy or other methods. Two degrees of freedom (points *A* and *B*) with respect to horizontal movement are available. Nevertheless, because of symmetry, only one equation is needed. Where asymmetry of loading or geometry exists, two sidesway corrections have to be made in a single-bay frame, needing two simultaneous equations.

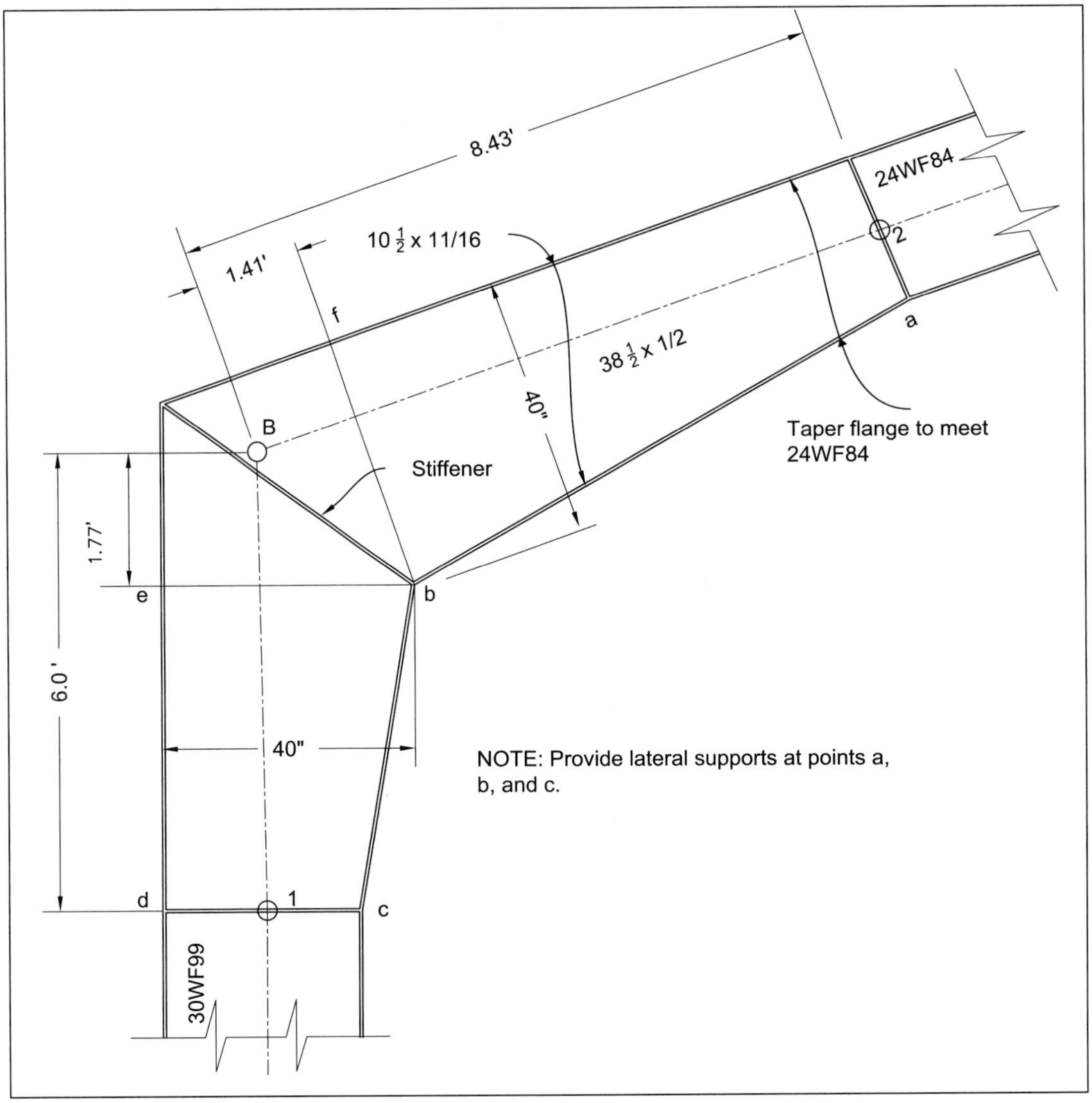

Figure 2.53 Haunch detail, Example 1.

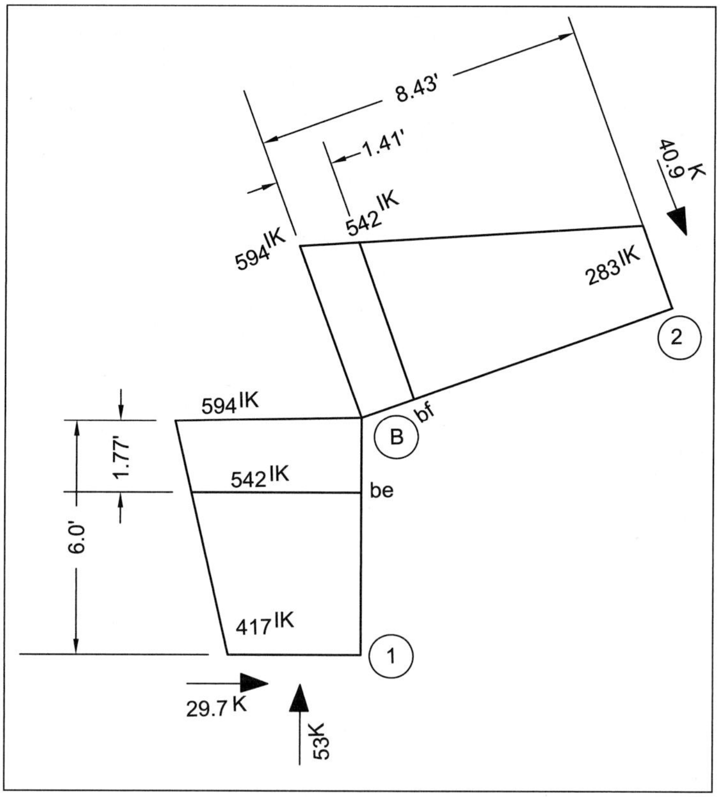

Figure 2.54 Moments and forces at haunch.

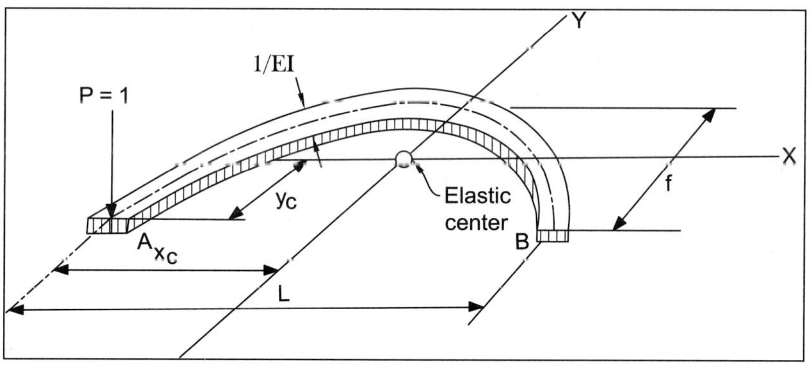

Figure 2.55 Analogous column loaded to calculate moment stiffness and carryover.

Table 2.10 Properties of Arches of Constant Cross Section

Arch shape	f/L	y_c	Moment stiffness K	Moment carryover factor C	Thrust due to unit moment (other end free to rotate) H'_m	Moment due to unit horizontal displacement (no end rotation) M_u	Moment due to unit vertical displacement (no end rotation) M_v
Parabolic	0.1	0.006L	8.55 EI/L	0.330	6.28L	71.42EI/L^2	5.73EI/L^2
	0.2	0.129	7.54	0.322	3.17	31.64	5.11
	0.3	0.187	6.46	0.316	2.15	18.27	4.42
	0.4	0.242	5.53	0.312	1.64	11.89	3.80
	0.5	0.295	4.79	0.311	1.33	8.34	3.30
Circular	0.1	0.067	8.71	0.343	6.27	73.23	5.72
	0.2	0.132	7.80	0.357	3.14	33.17	5.02
	0.3	0.196	6.69	0.383	2.10	19.42	4.13
	0.4	0.258	5.59	0.415	1.58	12.52	3.27
	0.5	0.318	4.63	0.450	1.27	8.55	2.55

*From J. Michalos, *Theory of Structural Analysis and Design*, The Ronald Press Company, New York, 1958.

EXAMPLE 2—ARCH BENT ANALYSIS

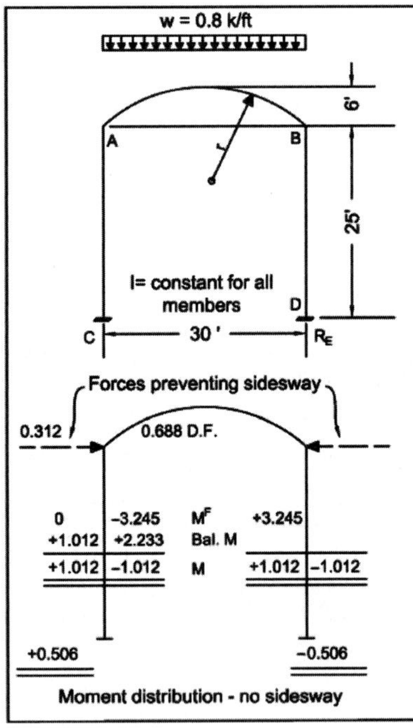

All units in feet and kips. Positive moments rotate the joint clockwise. Thrusts to right are positive.

Columns

$K = 4E\,I/L = 0.160\,EI$
$C = +0.5$

Arch (Tables 3 and 6)

$f/L = 0.20$
$M^F = 0.004507 \times 0.8 \times 30^2 = 3.245$
$H^F = 0.6388 \times 0.8 \times 30 = 15.33$
$K = 7.80\,EI/30 = 0.260\,EI$
$C = -0.357$
$K_{symm} = 0.260\,EI\,(1+0.357) = 0.353\,EI$
$H_m = 3.14/30 = 0.105$
$M_u = 33.17\,EI/30^2 = 0.0369\,EI$

Thrusts at A	at B
$+15.33$	$-15.33 = H^F$
-0.47	$+0.47$ = in arches due to rotation $= 3.245 - 1.012 = 2.33$ reduction
	$\times$ 2 ends
	4.466
	$\times 0.105 = H'_m$
	0.47
$+0.06$	-0.06 = in columns due to rotation $(1.012 + 0.506)/25 = 0.06$
$+14.92$	-14.92 = total thrust preventing sidesway

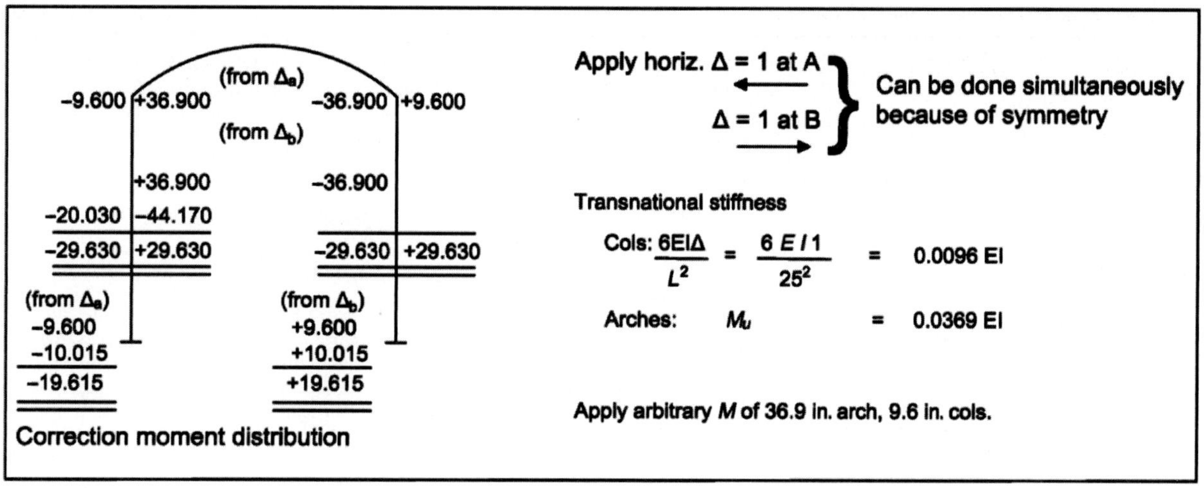

Thrusts at A   at B $y_C = 3.96$

-18.64 $+18.64$ = thrust due to ratation, $73.800/3.96 = 18.64$

$+9.28$ -9.28 = in arches due to rotation $29.630 - 73.800 = -44.170$ reduction

2 ends

88.340

$0.105 = H'_m$

9.28

-1.97 $+1.97$ = in columns due to rotation $(29.630 + 19.615)/25 = 1.97$

-11.33 $+11.33$ = total thrust at A and B

To eliminate unbalanced thrusts at A and B
$+14.92 - 11.33\,\times = 0, \times = 1.317$
Final $M_A = -1.012 + 1.317(+29.630) = +38.0$ft-kips $= -M_D$
$M_C - 0.506 + 1.317(-19.615) = -25.3$ft-kips $= -M_D$
$H_c = -H_D = -0.06 + 1.317(+1.97) = +2.53$ kips

2.12 SPECIAL TOPICS

2.12.1 Second-Order Theory

The analysis methods deliberated in Articles 5 and 9 are first order analyses. Such analyses discard the secondary moments as a result of deflection of the arch or frame. For example, to compute the deflections Δ_1 related to a first-order analysis of the two-hinged arch, moment increments $dM_1 = H\Delta_1$ could be determined. These, sequentially, generate additional deflections and the second increment of moment dM_2 and so on. Calling $dM_1/M_1 = \alpha_1$ and $dM_2/dM_1 = \alpha_2$ and assuming that subsequent ratios dM_3/dM_2 and so on are equal to α_2, the final moment can be estimated by

$$M = M_1(1+\alpha_1+\alpha_1\alpha_2+\alpha_1\alpha_2^2+\cdots) = M_1\left(1+\frac{\alpha_1}{1-\alpha_2}\right)$$

If M_1 is considered as an incremental moment, the formula becomes

$$M = M_1\left(\frac{1}{1-\alpha}\right)$$

Charts for estimating the amplification of stress resulting from secondary moments in flexible steel arches are useful for the initial design of long-span arches. Moreover, it helps in avoiding proportions, which might be dangerous as result of instability.[23] Discussions on second-order theory are given in Refs. 24 and 25.

2.12.2 Interaction of Arch and Deck

In tied arches (tie is a continuous deck girder) and in spandrel arches (deck is continuous), the moment at any section will be divided between the arch rib and the deck girder. The division can be estimated as if the two elements were a flitched beam, with common deflections and angular variations. This problem is discussed in Refs. 26, 27, and 28.

2.12.3 Buckling of Arches

Discussion of arch buckling can be found in the paper by Wastlund.[29] Common cases of critical loads are summarized below:

1. Circular ring, of radius r and constant section, under radial pressure p per unit length (antisymmetric buckling):

$$\text{Critical axial force } S_{cr} = \frac{3EI}{r^2}$$

2. Two-hinged circular arch with rise f, radius r, central angle θ, and of constant section, under radial pressure p per unit length (antisymmetric buckling):

$$\text{Critical axial force } S_{cr} = \left(\frac{\pi^2}{\theta^2}-1\right)\frac{EI}{r^2}$$

3. Parabolic arch of rise f and span L:

$$H_{cr} = \frac{\beta EI}{L^2}$$

where β is given in Table 2.11.

Table 2.11 Values of β for Buckling of Parabolic Arches*

	f/L	0	0.1	0.2	0.3	0.4	0.5	Source
3-hinged	I = const.	29.8	28.5	24.9	20.2	15.4	...	Stuessi
					(19.8)	(13.6)		
	$I = I_c \sec \phi$	29.7	29.4	27.7	25.3	22.6	19.8	Dischinger
					(25.1)	(19.4)	(15.0)	
2-hinged	I = const.	39.4	35.6	28.4	19.4	13.7	9.6	Lockschin
	$I = I_c \sec \phi$	39.4	37.2	31.6	25.1	19.4	15.0	Dischinger
Fixed	I = const.	80.8	75.8	63.1	47.9	34.8	...	Stuessi
	$I = I_c \sec \phi$	80.8	78.4	70.8	61.1	51.1	41.8	Dischinger

*Values are for symmetrical buckling. Figures in parentheses are for unsymmetrical buckling.

2.12.4 Laterally Loaded Arches and Frames

Laterally loaded arches and frames (i.e., subjected to wind and/or earthquake) can be analyzed by the general method discussed in Article 2.9.4 [Eq. (2.70)], considering the effect of torsion on the sections. Self-supporting ribs without bracing are fairly uncommon. This is because it is expected in bridge work to tie the ribs by lateral struts. Such struts can vary in flexural and torsional stiffness from very flexible elements that significantly divide the load alike between ribs to very stiff struts that alter the rib system into a Vierendeel truss fixed at the abutments.

The analysis of such laterally loaded systems is very complicated. A numerical approach proposed by Baron and Michalos[30] offers an arranged and systematic calculation. Michalos[31] also provides influence-line data for laterally loaded parabolic arches as a function of the ratios of rise to span and of bending to torsional stiffness. The availability of such data is satisfactory for designing normal arches estimating a parabolic shape and for the initial design of long-span arches.

2.12.5 Skewed Barrel Arches and Rigid-Frame Slabs

A fixed, right barrel arch or rigid-frame slab, analyzed as a plane structure of unit width, is an example of a three degrees statically indeterminate structure. On the other hand, the corresponding skewed structure comprises a space structure of six degrees redundancy, having three unknown internal moments and three unknown internal thrusts at any cut section in the arch (Fig. 2.56). A skewed two-hinged barrel arch has four redundants. For extreme skews, multiple spans, or long spans, shearing stresses due to torsion become significant. While solution using the virtual-work method is totally feasible, substantial difficulty is experienced because of the torsional moments.

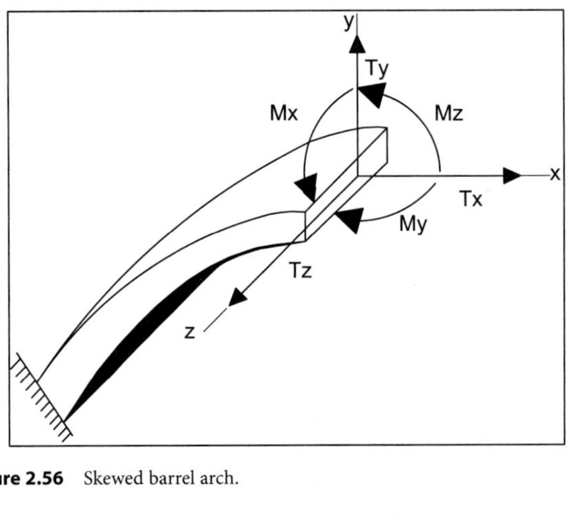

Figure 2.56 Skewed barrel arch.

Contemporary methods to skew-barrel analysis assess the effects of skew separately as a kind of secondary stress. In this analysis, the ordinary stresses, T_x, T_y, M_z, for the rectangular structure of the same span and the same elastic and geometric properties are used as principal stresses. The skew stresses (T_z, M_x, M_y) are then calculated by simple estimates based on several theoretical and experimental investigations. Such methods of analysis already have been developed.[32,33]

2.13 CONSTRUCTION AND DETAILS

2.13.1 Concrete Arches and Frames

REINFORCEMENT

The reinforcement of arch rib and rigid frame is typically provided in the form of parallel intradosal and extradosal bars, with some transverse distribution and temperature reinforcement in the case of barrel arches, stirrups as spacing hoops, and for shear reinforcement whenever needed. Fabricated structural-steel sections, such as plates, channels, or single or paired angles, can also be used as top and bottom longitudinal reinforcement. Such reinforcement may be tied together with latticing or battens to form a truss. For stability purposes, such trusses can be made as three-hinged arches between abutments with transverse bracing. Rarely, the trusses are built suitably stiff and strong to support the arch forms and the wet concrete without the need for centering.

FALSEWORK OR CENTERING

Two common methods of supporting the arch rib or barrel during placing of concrete are pipe scaffolding and timber or steel-trussed centering. The reuse of centering for various sections or ribs, bridges, or building roofs is enabled by dismantling or movement on rails or skids or by crane. Centering has to be designed carefully to allow for camber, shrinkage, and settlement of the centering. Release or striking of the centering has to take place only after the concrete has reached its design strength. Release is achieved using wedges, jacks, or other devices allowing a controlled sequence of operations. Usually in such an operation, the crown is released first, then the haunches concurrently on each side.

SEQUENCE OF PLACING CONCRETE

To reduce shrinkage stresses, concrete arches are poured in different lateral sections, preferably with small keyways between the sections where the keyways being placed last. Also, the arrangement should consider deflections of the arch centering, with the abutment sections preferably placed last. Other sections have to be placed symmetrically with respect to the center of the bridge span. In the case of wide barrel arches that require a longitudinal joint, to avoid relative settlements, the sections on each side of such joint should be poured on independent centering.

Concreting of rigid frames is typically done in the order: footings, legs, and deck. Proper shear keys are placed between consecutive pours of the mentioned elements. In low frames with heavy legs or in continuous frames where shrinkage and temperature stresses are important, the deck can be poured with a shrinkage keyway in the spans. This keyway is filled with concrete after the main shrinkage in the deck has taken place.

DRAINAGE AND WATERPROOFING

To eliminate frost damage and the surcharge of accumulated water, drainage of filled-spandrel arches by suitable tile or pipes to an outlet beyond the abutments is required. Spandrel-filled barrel arch bridges must be treated using an asphaltic membrane or surface-coat waterproofing on the extrados and backs of spandrel walls.

EXPANSION JOINTS

The handrails, deck, spandrel walls, and other elements that may contribute to the longitudinal action of the arch have to be provided with expansion joints. Joints at crown and springing are acceptable for spans to about 70 R, but for longer spans, five or more joints have to be used.

Deck Participation

In the case of spandrel walls and the deck near the crown, an interaction of the rib with the superstructure occurs, which stiffens the rib of a spandrel-filled arch. Alternatively, the interaction tends to convert the arch into a Vierendeel truss in the case of an open-spandrel system having stiff columns near the crown. In practice, there is a tendency to deal with participation stresses as secondary effects and not to account for them to contribute to the carrying capacity of the rib. Consequently, measures such as jointing, mentioned previously, have to be taken to avoid interaction; otherwise, objectionable cracking may happen. Also, it is possible to form the superstructure after striking the arch centering; nonetheless, this will eliminate the participation only with respect to dead load. In longer spans, with columns of lesser stiffness, the interaction can be reduced. Commonly, control joints are located at ends of spandrel columns to produce articulation without spoiling the appearance, but some measures have to be taken in order to carry traction forces of the deck.

Temporary and Permanent Hinges

Different devices can be used to recompense for the stresses produced by the displacement of the pressure line as a result of rib shrinkage, drop in temperature, shortening, and foundation settlements. The use of temporary hinges at crown and abutments is most popular on long-span concrete arches, mostly with low rise-to-span ratios. This is done to form a three-hinged arch under dead load so that on removal of centering, there is no displacement of the pressure line due to rib shortening. Simultaneously, by placing the hinges with calculated eccentricities, the effects of temperature drop and shrinkage can be countered by stresses of opposite sign induced in the rib. All these produce considerable economies.

All hinges are sequentially concreted solid so that under live load, the structure acts as a hingeless arch. Hinges can be made out of concrete, using the Mesnager or Considére designs (Fig. 2.57), or they can be made of cast steel similar to standard hinges for steel bridges. However, using steel will be at a considerable increase in cost over concrete hinges. The Mesnager hinge is typically designed with hinge opening h equal to or somewhat greater than the vertical thickness t. Hinge bars are limited to an L/r of about 20 to 40 based on the inclined length between faces of concrete. The two inclined hinge bars are designed to take, as a truss, the normal and shear forces from the rib acting at their intersection, with areas selected so that it does not exceed 30 percent of the steel yield stress. Also, lateral stirrups and ties have to be provided to resist the tendency of the concrete to burst. Additionally, they have to be fully developed at a distance of 8 bar diameters from the face. Mesnager hinges should not be used for high shear-to-thrust ratios, and limits of V/H of 0.024, 0.195, and 1.0 have been suggested for angles $\theta = 15°$, $30°$, and $45°$, respectively.

The Considére hinge is a short section of spiral column where its dimension results in a resistance to rotation that is small relative to that of the main connecting members. Considére hinges are designed first to find the percentage of longitudinal/spiral steel from spiral-column formulas and then to limit the deformation in the extreme fibers at maximum rotation to an allowable percentage of the ultimate value.

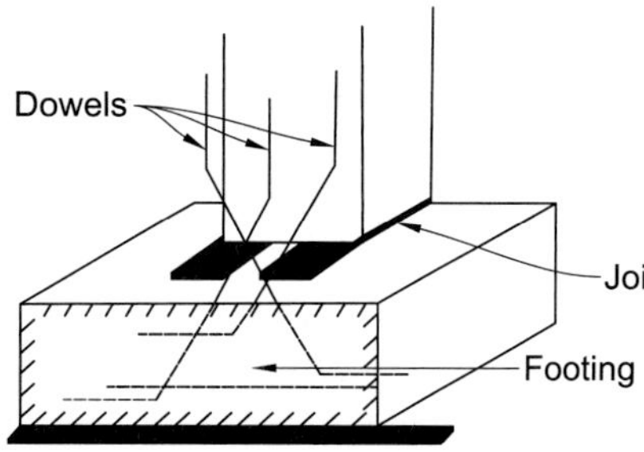

Figure 2.58 Typical detail of hinged base for concrete rigid frame. (*Courtesy of Portland Cement Association.*)

Details and charts for design of both types of hinges are presented in Ref. 34.

As desired, stress recompense can be affected by a system of inserting hydraulic jacks at the crown in a temporary gap to adjust the line of pressure before concreting the gap (Ref. 35). Other means of recompense used by Freyssinet[35] induce residual stresses by casting the arch rib in layers.

The details of a hinged base for a concrete rigid frame are shown in Fig. 2.58.

Foundations

The importance of stable, rigid foundations for arches is very significant. Factors to be considered include protection against erosion and scour, proper anchorage of reinforcement, vertical and lateral stability of bearing foundations and pile foundations, and proper design of hinges.

2.13.2 Steel Arches and Rigid Frames

Knees

Elementary beam-flexure theory cannot be used to analyze the knee of a rigid frame. Extensive experimental studies have been conducted both on full-scale steel knees and by models photo elastic analyses. Figures 2.59 and 2.60 illustrate stresses in tests of an actual structure and specify the shift of the neutral axis toward the inside corner.[36]

Analysis of rigid-frame steel knees can be carried out by estimate approaches given by Bleich[37] and by Olander.[38] Olander's formulas, based on the *developed* cross sections shown by the circular arcs, estimated the stresses fairly closely (Fig. 2.61).

Wind Bracing

To achieve lateral stability in bridges, wind bracing has to be placed between the ribs. In order to form with longer spans a curved Vierendeel truss capable of resisting the lateral forces, bracing typically consists of struts rigidly connected to the ribs. Another alternative is to use a truss of diagonals and struts between the ribs or of crossed diagonals alone.

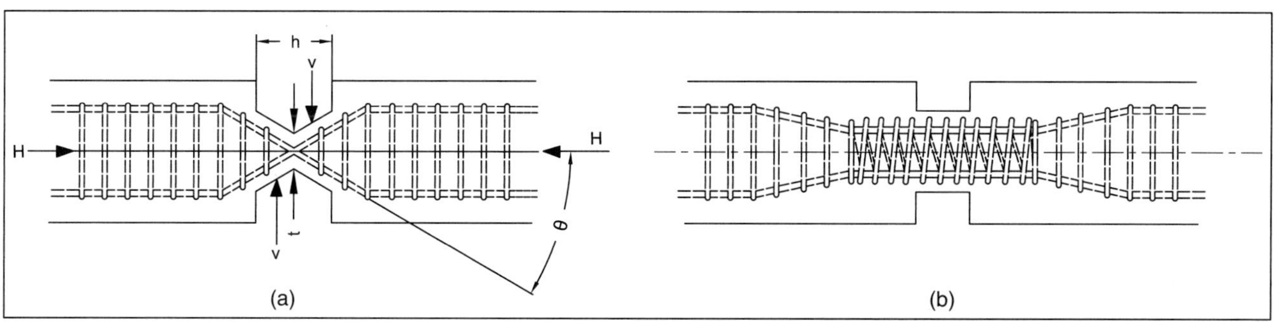

Figure 2.57 Concrete hinges: (*a*) Mesnager and (*b*) Considére.

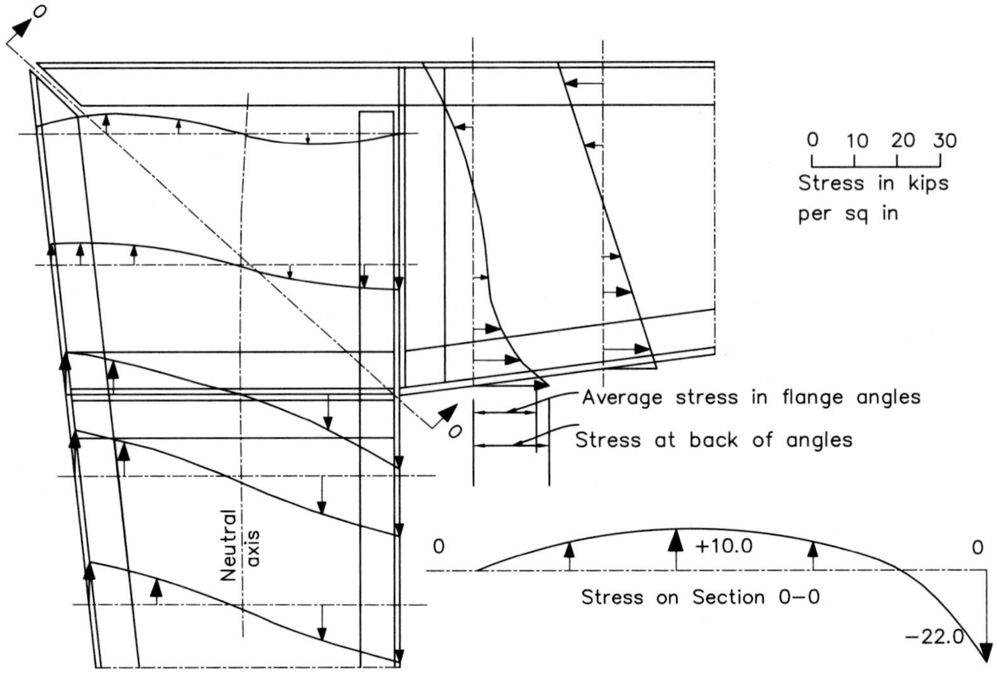

Figure 2.59 Normal stresses in square-knee frame. (*From Ref. 24.*)

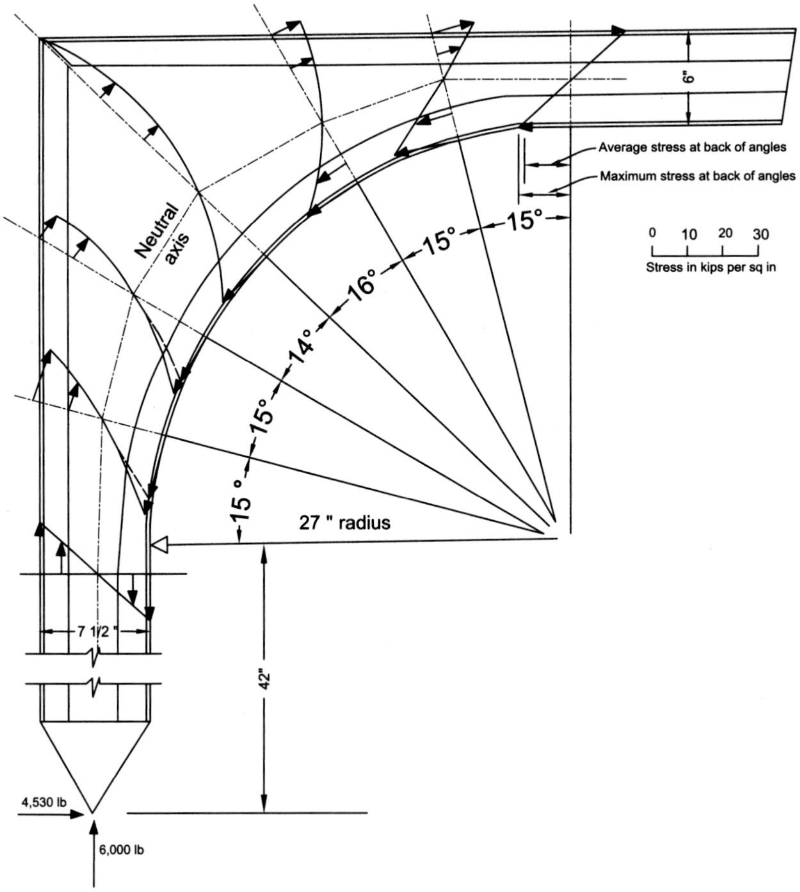

Figure 2.60 Normal stresses in curved-knee frame. (*From Ref. 24.*)

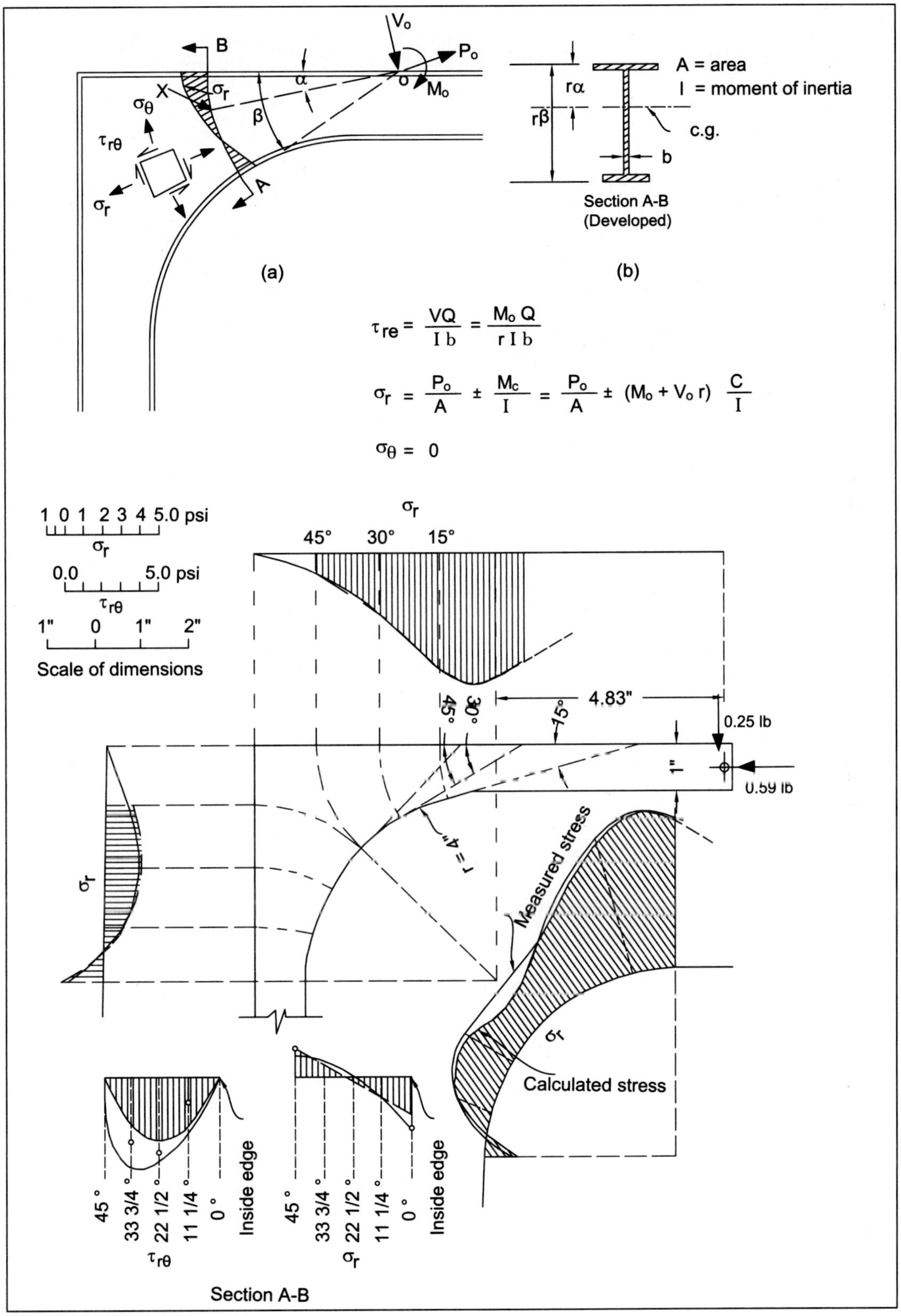

Figure 2.61 Normal stresses in curved knees. (*From Ref. 26.*)

Figure 2.62 Basket-handle arch bridge, Rio Blanco, Mexico. (*Kavanagh and Piccone.*)

In arch bridges of the Langer-girder type, the Vierendeel type or truss with heavy overhead lateral bracing is unappealing. The effective use of the arches to brace themselves, as employed in the 250-ft-span Rio Blanco Bridge (Figs. 2.62 and 2.63) and in the 815-ft span of the Fehrmarnsund Bridge on the International Highway Route between Germany and Denmark, demonstrates the lightness that such lateral bracing can take. The Rio Blanco bridge is comprised basically of two tie girders and two box-type steel arch ribs, open on the underside, with 30 wire-rope hangers. A reinforced-concrete slab on a gridwork of diagonal floor beams is used for the floor. The framework was manufactured in sections and field-welded on timber falsework. The bridge has three traffic lanes and two walks.

2.13.3 Economics

Costs of arches and rigid frames are regularly governed by practical and functional factors rather than those of structural efficiency. For instance, costs of long-span arch bridges may be dictated by the economy and feasibility of the large falsework requirements.

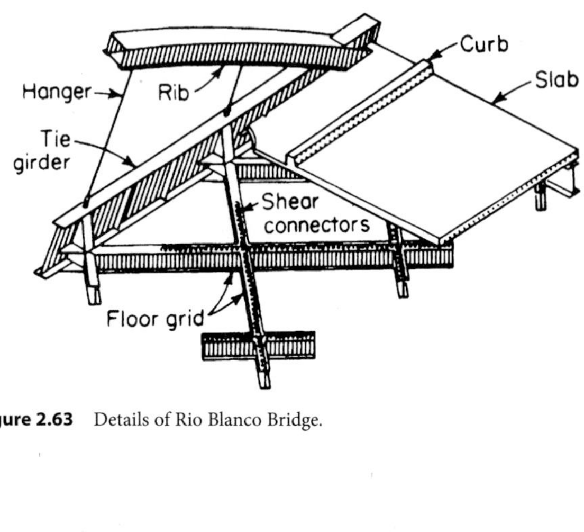

Figure 2.63 Details of Rio Blanco Bridge.

REFERENCES

1. Timoshenko, S. P., and D. H. Young. *Theory of Structures*, 2d ed. New York: McGraw-Hill, 1965.
2. Langhaar, H. L. *Energy Methods in Applied Mechanics*. New York: John Wiley & Sons, 1962.
3. Norris, C. H., J. B. Wilbur, and S. Utku. *Elementary Structural Analysis*, 3d ed. New York: McGraw-Hill, 1976.
4. West, H. H. *Analysis of Structures*, 2d ed. New York: John Wiley & Sons, 1989.
5. Przemieniecki, J. S. *Theory of Matrix Structural Analysis*. New York: McGraw-Hill, 1968.
6. Cook, R. D. *Finite Element Modeling for Stress Analysis*. New York: John Wiley & Sons, 1995
7. Zienkiewicz, O. C., and R. L. Taylor. *The Finite Element Method, Vol. 1, Basic Formulation and Linear Problems*. New York: McGraw-Hill, 1989.
8. Lightfoot, E. *Moment Distribution*. London: E. & F. M. Spon Ltd., 1961.
9. McGuire, W., and R. H. Gallagher. *Matrix Structural Analysis*. New York: John Wiley & Sons, 1979.
10. Holzer, S. M. *Computer Analysis of Structures*. New York: Elsevier, 1985.
11. Benjamin, J. R. *Statically Indeterminate Structures*. New York: McGraw-Hill, 1959.
12. Design constants for circular and elliptical arches of variable thickness, *Adv. Eng. Bull.* (8), Portland Cement Assoc., Chicago, 1963.
13. Cross, H., and N. D. Morgan. *Continuous Frames of Reinforced Concrete*. New York: John Wiley & Sons, 1932.
14. Whitney, C. S. Analysis of unsymmetrical concrete arches, *Trans. ASCE* (99) 1268, 1934.
15. Whitney, C. S. Plain and reinforced concrete arches, Report of Committee 312, *J. ACI* (47), 681, May 1951; (37), 1, September 1940; (28), 479, March 1932.
16. Korn, M. P. *Steel Rigid Frames Manual—Design and Construction*. Ann Arbor, Mich.: J. W. Edwards, 1953.

17. *Analysis of Rigid Frame Concrete Bridges*. Chicago: Portland Cement Association.
18. *Steel Gables and Arches*. Chicago: American Institute of Steel Construction, 1963.
19. Griffiths, J. D. *Single Span Rigid Frames in Steel*. Chicago: American Institute of Steel Construction, 1948.
20. Kleinlogel, A. *Rahmenformeln*. Berlin: Wilhelm Ernst & Sohn, KG, 1943.
21. Design constants for circular arch bents, *Adv. Eng. Bull.* (7), Portland Cement Assoc., Chicago, 1963.
22. Design constants for continuous gable frames, *Adv. Eng. Bull.* (15), Portland Cement Assoc., Chicago, 1965.
23. Rowe, R. S. Amplification of stress in flexible steel arches, *Trans. ASCE* (119), 910, 1954.
24. Asplund, S. O. Deflection theory of arches, *Trans. ASCE* (128), 307, 1963.
25. Freudenthal, A. Deflection theory for arches, *Int. Assoc. Bridge Struct. Eng. Mem.* (3), 100, 1935.
26. Garrelts, J. M. Design of St. Georges tied arch span, *Trans. ASCE* (108), 543, 1943.
27. Hardesty, S., and J. M. Garrelts. Rainbow Arch Bridge over Niagara Gorge: design, *Trans. ASCE* (110), 6, 1945.
28. Freyssinet, E. Three monumental bridges built in Venezuela, *Civil Eng.*, March 1953, 157.
29. Wastlund, G. Stability problems of compressed steel members and arch bridges, *Proc. ASCE* (86), ST6, 47, June 1960.
30. Baron, F., and J. P. Michalos. Laterally loaded plane structures and structures curved in space, *Trans. ASCE* (117) 279, 1952.
31. Michalos, J. Effects of lateral loads on arches, *J. ACI* (47) 377, 1951.
32. Barron, M. Reinforced concrete skewed rigid-frame and arch bridges, *Trans. ASCE* (116) 999, 1951.
33. Michalos, J. Analysis of skewed rigid frames and arches, *J. ACI* (48) 437, 1952.
34. Ernst, G. C. Design of hinges and articulations in reinforced concrete structures, *Trans. ASCE* (106) 862, 1941.
35. Freyssinet, E. Three monumental bridges built in Venezuela, *Civil Eng.*, March 1953, 157.
36. Lyse, I., and W. E. Black. An investigation of steel rigid frames, *Trans. ASCE* (107) 127, 1942.
37. Bleich, F. *Design of Rigid Frame Knees*. Chicago: American Institute of Steel Construction, 1959.
38. Olander, H. C. Stresses in the corners of rigid frames, *Trans. ASCE* (119) 797, 1954.

Chapter 3

The Finite-Element Method

BY

CRAIG D. FOSTER, Ph.D. *Department of Civil Engineering, University of Illinois at Chicago*
SHENG-WEI CHI, Ph.D. *Department of Civil Engineering, University of Illinois at Chicago*

3.1 OVERVIEW OF THEORY OF FINITE-ELEMENT ANALYSIS

The finite-element method is, at its most essential, a numerical technique for solving differential equations. The method works particularly well for even-ordered differential equations, including those for solid mechanics analysis (second order) and beam, plates, and shells (fourth order). In this chapter, we will focus on the mechanical analysis of continuum bodies, plates, and shells, with some comments on beams and frames. The analysis of beams and frames, however, is remarkably similar to matrix structural analysis, and we will leave detailed discussion of frames to that chapter.

This section gives an outline of the theory of finite-element analysis so that the reader has a basic understanding of the underpinnings of the method. Many details are omitted. There are many excellent textbooks on the subject covering the theory in greater depth, including Bathe, 2014; Fish and Belytschko, 2007; Hughes, 1987; Kim and Sankar, 2008; Logan, 2011; Reddy, 2005; Zienckiewicz et al., 2013.

The method works by breaking up the domain of interest into smaller pieces, called elements, as shown in Fig. 3.1. On each element, the primary variable of interest, the displacement here, is approximated using simple functions. These functions are created by tracking the displacement at a finite number of points, called nodes, and using interpolation

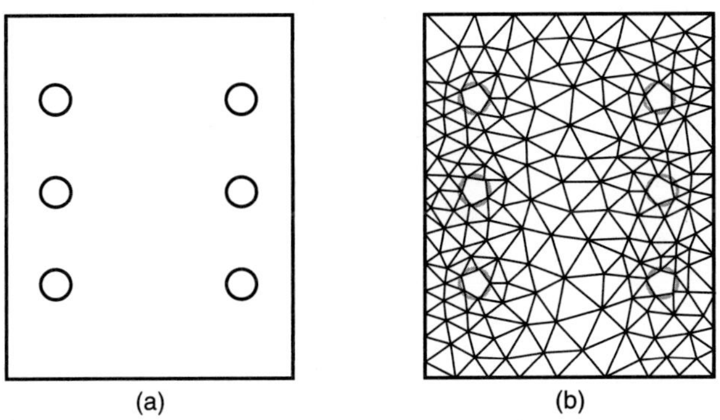

Figure 3.1 A plate meshed with an unstructured mesh of straight-edged triangles.

functions, known as shape functions, to determine the entire displacement field. Typically, the element geometry is mapped to a standard shape in a so-called parent domain, or set of natural coordinates, as shown in Tables 3.1 and 3.2. In this parent domain, the shape functions are easy to define, usually with simple low-order polynomials. The coordinate mapping, which often uses the same shape functions as the displacement, allows for a variety of shapes and orientations in physical space. The ability to break an object into triangles, quadrilaterals, tetrahedra, or hexahedra of arbitrary shape (within some limits, as discussed later) is one of the features that make finite elements so successful and popular as compared to many other numerical methods.

The most general method of manipulating the finite-element equations is the method of weighted residuals. However, in mechanics, the problem can be more easily approached by the principle of virtual displacements. In the quasi-static case, the resulting equation is

$$\int_{\Omega}\{\delta\varepsilon\}\cdot\{\sigma\}d\Omega = \int_{\Omega}\delta\mathbf{u}\cdot\mathbf{b}\,d\Omega + \int_{\Gamma}\delta\mathbf{u}\cdot\mathbf{t}\,d\Gamma \tag{3.1}$$

where $\delta\mathbf{u}$ is the virtual displacement, $\mathbf{b}$ is a body force per unit volume (such as gravitational or magnetic force), $\mathbf{t}$ is the surface traction vector, Ω is the domain of interest, and Γ is its boundary. The vectors $\{\sigma\}$ and $\{\delta\varepsilon\}$ are the stress and virtual strain, respectively, in vector (Voigt) form. The left-hand side of the equation is the internal virtual strain energy. The right-hand side is the external virtual work, with the first term arising from body forces, that is, forces that act at a distance such as gravity, and the second term arising from forces applied to the surface.

Because the displacement field is characterized by a finite number of points, we can factor out the (initially unknown) nodal displacement values. The finite-element approximation displacement $\mathbf{u}^h$ can be written

$$\mathbf{u}^h = \mathbf{N}\mathbf{d} \tag{3.2}$$

where $\mathbf{N}$ is a matrix of the shape function values and $\mathbf{d}$ is the vector of unknown nodal displacements. Similarly, the virtual displacement may be written

$$\delta\mathbf{u}^h = \mathbf{N}\mathbf{c} \tag{3.3}$$

where $\mathbf{c}$ are the unknown nodal values of virtual displacements. The strain vector in the small strain case just contains spatial derivatives of the displacement field. Since the vectors $\mathbf{d}$ and $\mathbf{c}$ are not functions of

Table 3.1 Common Types of Two-Dimensional Elements with Some Properties

Example physical shape	Shape in parent domain	Number of nodes	Name	Order of accuracy
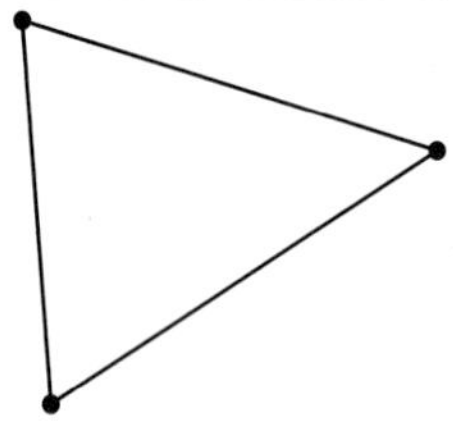	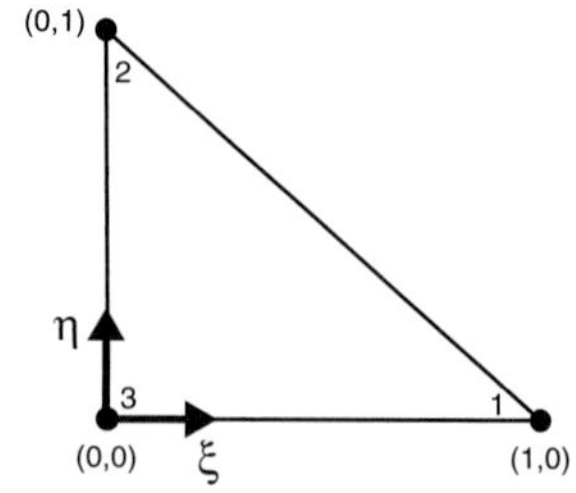	3	Linear triangle, constant strain triangle	1
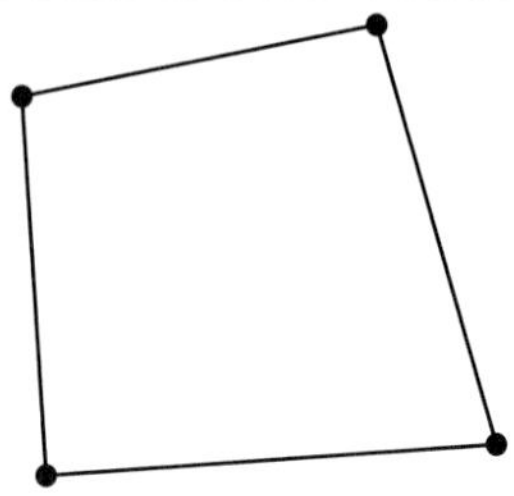	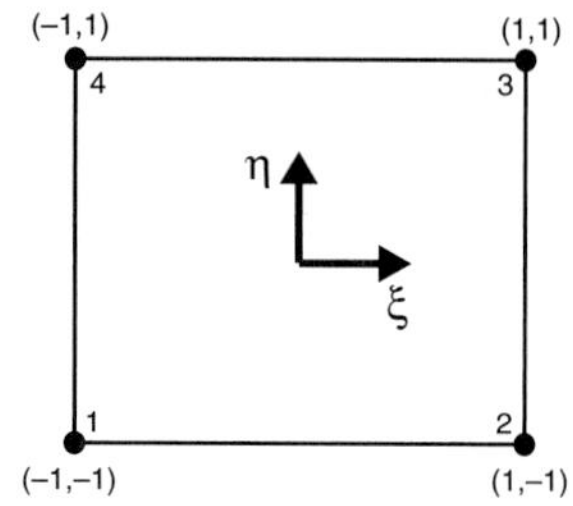	4	Bilinear quadrilateral	1
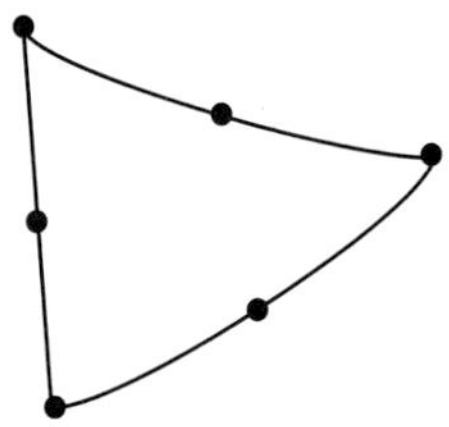	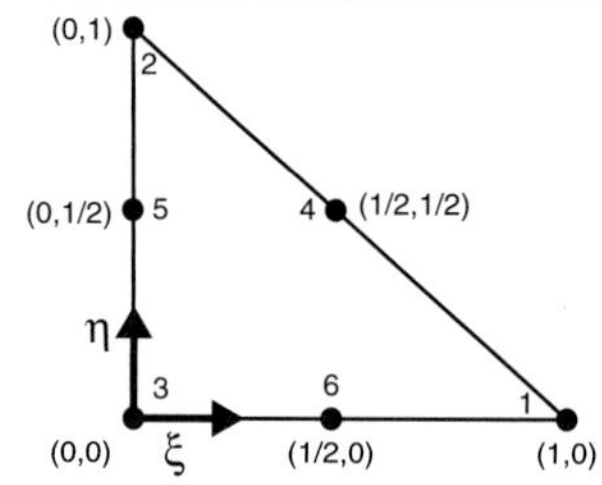	6	6-node triangle	2
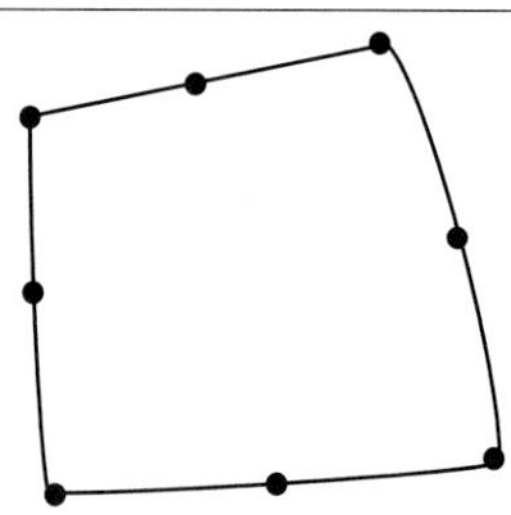	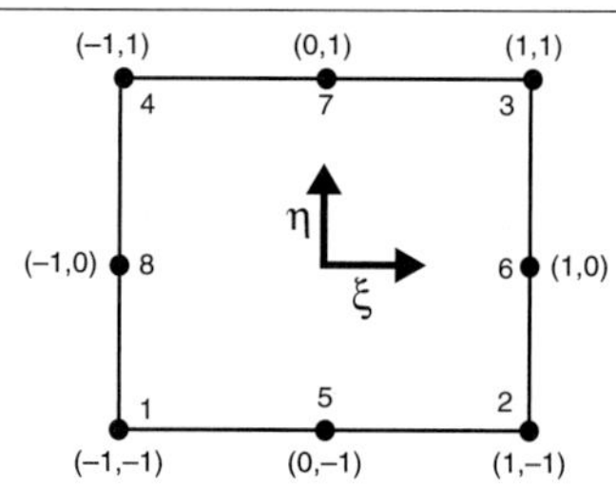	8	Serendipity quadrilateral	2
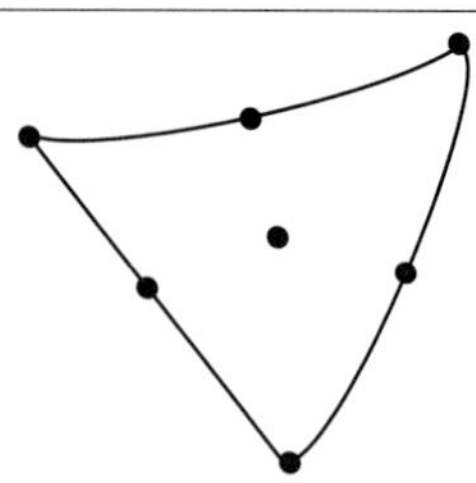	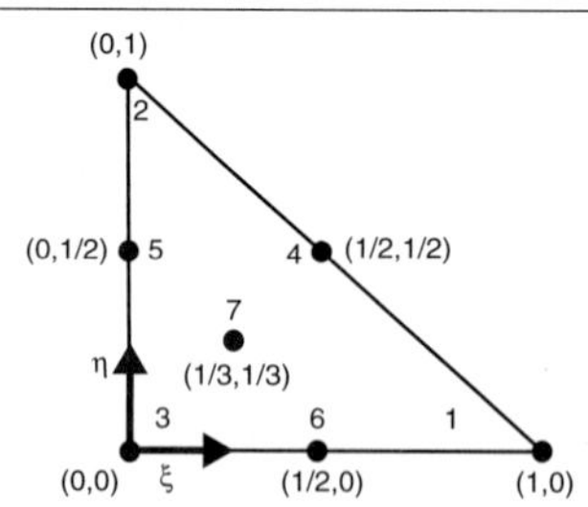	7	7-node triangle	2
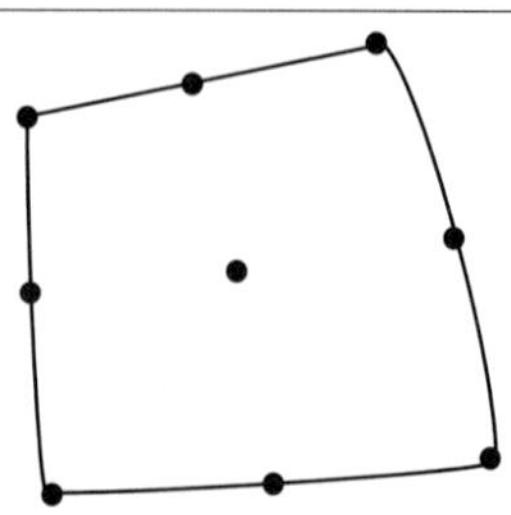	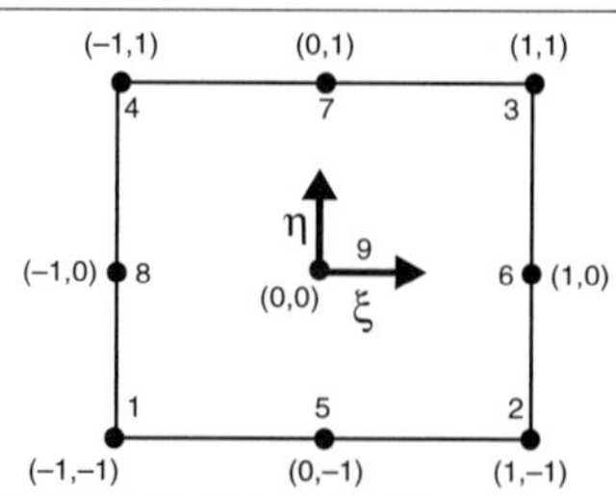	9	Biquadratic quadrilateral	2

Table 3.2 Common Types of Three-Dimensional Elements with Some Properties; Coordinate Limits in the Parent Domain for the 10-Node Tet and Serendipity Brick Are the Same as for the Linear Tet and Trilinear Brick, Respectively; Coordinates Have Been Omitted for Figure Clarity

Example physical shape	Shape in parent domain	Number of nodes	Name	Order of accuracy
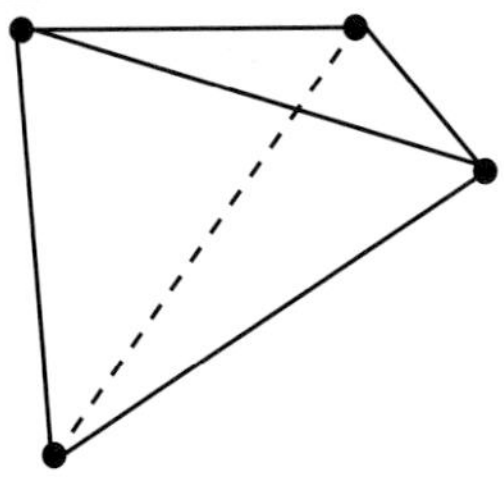	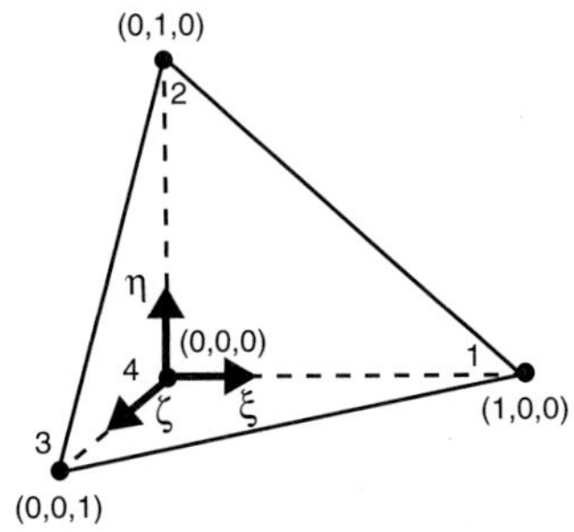	4	Linear tetrahedron, constant strain tetrahedron	1
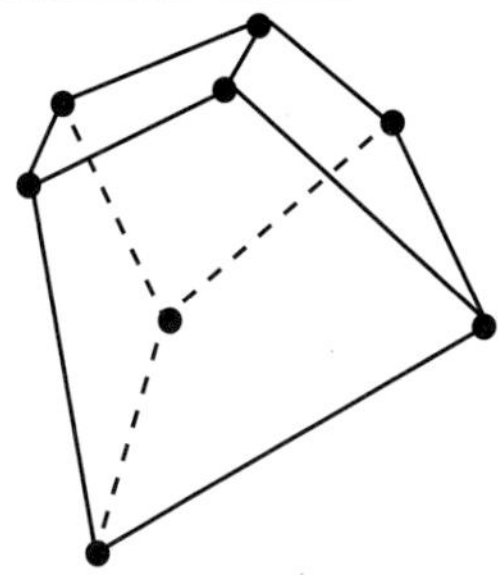	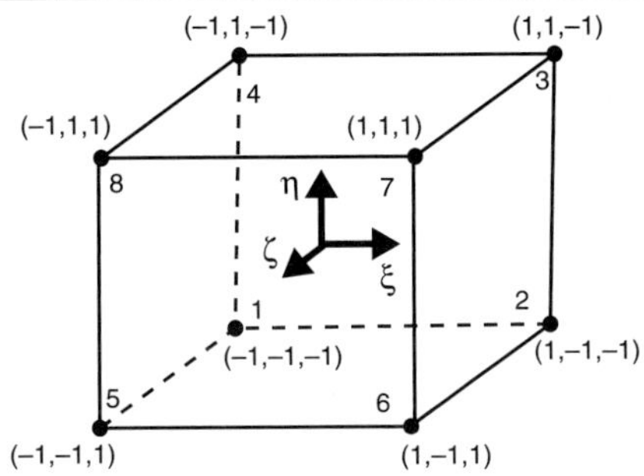	8	Trilinear hexahedron, trilinear brick	1
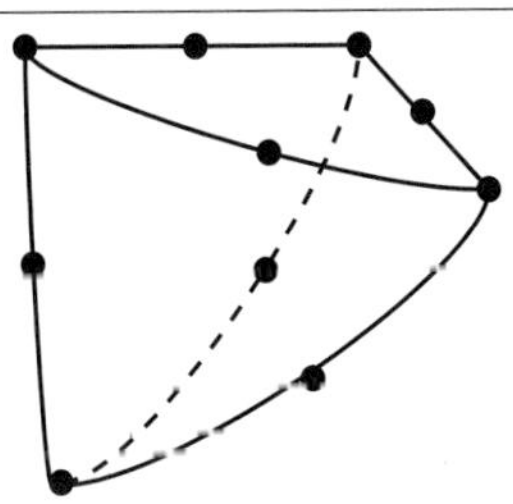	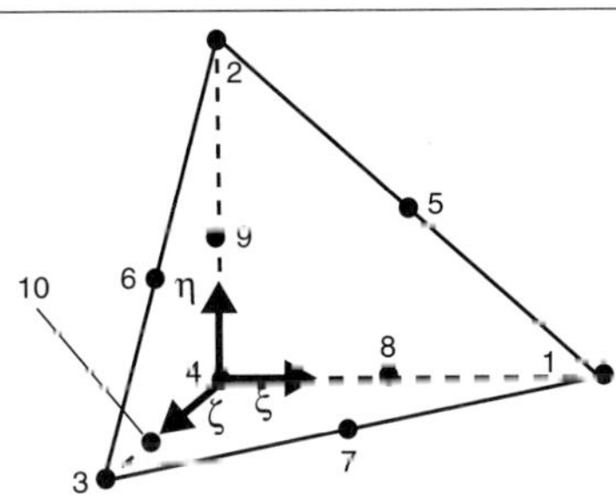	10	10-node tetrahedron	2
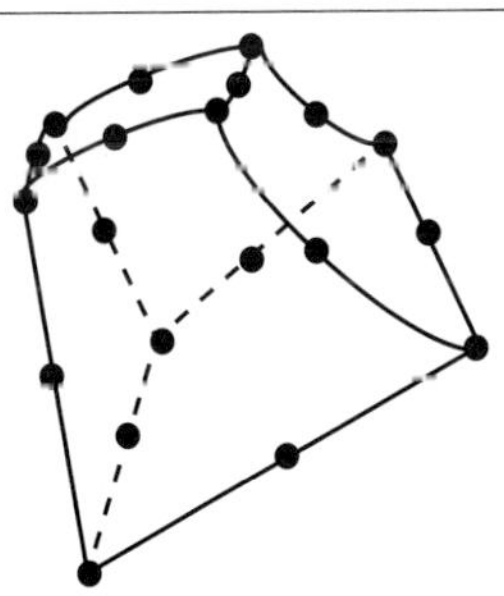	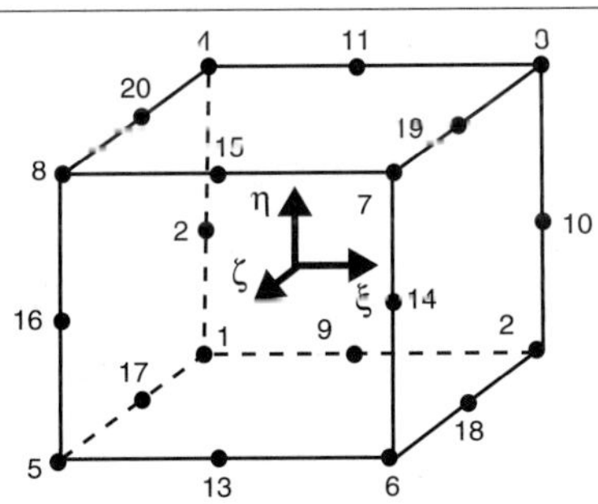	20	Serendipity hexahedron, serendipity brick	2

the spatial variables (i.e., x, y, and z), the strain and virtual strain may be written

$$\{\varepsilon^h\} = \mathbf{Bd} \tag{3.4}$$

$$\{\delta\varepsilon^h\} = \mathbf{Bc} \tag{3.5}$$

where **B** is a matrix that contains the derivatives of the shape functions. Finally, we note that the stress and strain are related through the elastic modulus matrix, $\{\sigma\} = \mathbf{C}^e\{\varepsilon\}$. Substituting these values into Eq. (3.1), we find

$$\mathbf{c}^T \int_\Omega \mathbf{B}^T\mathbf{C}^e\mathbf{B}\, d\Omega\mathbf{d} = \mathbf{c}^T \int_\Omega \mathbf{N}^T\mathbf{b}\, d\Omega + \mathbf{c}^T \int_\Gamma \mathbf{N}^T\mathbf{t}\, d\Gamma \tag{3.6}$$

Letting $\int_\Omega \mathbf{B}^T\mathbf{C}^e\mathbf{B}\, d\Omega = \mathbf{K}$ and $\int_\Omega \mathbf{N}^T\mathbf{b}\, d\Omega + \int_\Gamma \mathbf{N}^T\mathbf{t}\, d\Gamma = \mathbf{f}$, this equation becomes

$$\mathbf{c}^T\mathbf{Kd} = \mathbf{c}^T\mathbf{f} \tag{3.7}$$

Since the nodal values of the virtual displacement are arbitrary (and, in fact, we never solve for them), the only way to ensure this equation holds is if $\mathbf{Kd} = \mathbf{f}$. This matrix equation is solved for the unknown nodal displacements, and once those are known, other quantities like stress and strain can be solved. The integrals for the stiffness matrix $\mathbf{K}$ and the force vector $\mathbf{f}$ are typically carried out by numerical integration, with a small number of integration points per element. Gaussian quadrature is typically used for quadrilateral and hexahedral (distorted cube) elements, and other schemes have been developed for triangles and tetrahedra. While the number of points used is usually determined by matches order of the polynomial to be integrated, sometimes using fewer integration points than theoretically optimal for part or all of the integration can improve the results. This approach is discussed in more detail later.

A number of details have been omitted above, and the equations can be extended to include nonzero specified displacements, material and geometric nonlinearity, and other complications. Many commercial software packages now exist that handle the mathematics internally. However, some knowledge of the working of finite elements is necessary in order to ensure accurate results. A novice can easily obtain results that appear convincing but are far from the true solution. Hence, some care should be taken to create the proper model of a structure.

3.2 OUTLINE OF STEPS IN A FINITE-ELEMENT ANALYSIS

A user of a finite-element software package must make numerous decisions in building and running finite-element models. The first is what type of analysis should be performed. For mechanical or structural analysis, is a dynamic analysis necessary, or will a quasi-static one suffice? Should the analysis be linear or nonlinear?

Once the type of analysis is decided, the user has to decide how the geometry will be handled. Three-dimensional analysis may be necessary, or a two-dimensional approximation may be accurate enough. In some cases, it may be more efficient to model structures with "reduced continua," using plate, shell, beam, or truss elements.

The user defines the geometry of the model. Sometimes a computer-aided design (CAD) model may be imported into the commercial code, but CAD models can contain information and hidden features that can confuse a finite-element code. The geometry can also be constructed within the finite-element code or in another software specifically for preprocessing finite-element geometries.

The geometry must then be meshed. Choosing an appropriate element type or types, the model is broken into small pieces. Most commercial codes have semiautomatic procedures for meshing. Meshing can be complicated, though, and generally most of the user's time is spent developing the geometrical model and meshing it. Specialized software that focuses only on meshing exists and may be helpful for complex geometries. The mesh information can then be imported directly into the finite-element analysis software.

The loads, boundary conditions, and material properties must be specified. These may be applied either to the solid model before meshing or to the elements. Generally, the former is preferred, as the mesh may be altered over the course of several simulations.

For trusses, frames, plates, shells, and other reduced continua, geometric properties must also be specified. These are quantities such as cross-sectional area in bars, moment of inertia in beams, and thickness in plates and shells that are not explicitly modeled in the geometry but are necessary to determine the behavior of the structure under loading.

When the entire model has been built and all the loads and boundary conditions have been applied, the program may be run to solve for the displacements. The program usually also calculates strains, stresses, and other quantities. The user may then list results, plot them, or export them to other software for further analysis and visualization.

3.3 TYPES OF STRUCTURAL AND MECHANICAL ANALYSES

In structural and mechanical analysis, either a quasi-static or a dynamic analysis may be performed. We focus first on quasi-static analyses, those where the loading is slow enough that we can neglect inertial effects.

3.3.1 Three-Dimensional Analysis

The most general way to model the geometry is to explicitly model the whole structure in three dimensions. It is often necessary to model the whole structure in three dimensions to obtain accurate results. However, meshing can also be challenging in three dimensions for some geometries. It is also usually the most costly in terms of computation time. That said, with current technology, many three-dimensional quasi-static problems can be run even on a single processor.

3.3.2 Two-Dimensional Analysis

All problems in the real world are three-dimensional. For some problems, though, a two-dimensional approximation provides sufficiently accurate results. Clearly, creating both the geometric model and the mesh is easier. With fewer nodes and degrees of freedom per node, the solution time is also much faster. However, the geometry, material properties, and loading must all act in a way where such an approximation is reasonably accurate.

Plane Strain

For a plane strain model, every section of the body parallel to the plane considered should have the same geometry and material properties. The forces and displacement boundary conditions should all be in this plane. Furthermore, there should be no strain in the out-of-plane direction. From Poisson effects, stress may develop in the out-of-plane direction, however. This happens when the body is constrained out of plane, such as a concrete dam across a canyon, or approximately when the object is comparatively long in the out-of-plane direction. Examples of this case include strip footings and retaining walls. There may be minor edge effects in this case, but for the most part, there is no displacement and hence no strain out of plane.

In plane strain finite-element analyses, the analysis is typically performed per unit thickness. Hence, all the forces must be input per unit thickness. Some commercial codes also allow a "plane strain with thickness" option, where thickness is included. In this case, the absolute forces should be used.

Plane Stress

Plane stress is similar to plane strain, but instead of zero strain in the out-of-plane direction, there is zero stress. An example of this is a thin plate, where the stress on the faces is zero. While there is no stress in the out-of-plane direction, strain may develop from Poisson effects. Because the out-of-plane strain may vary depending on the in-plane stresses, a small amount of shear stress may develop out of plane. Hence, this simplification is approximate except in special cases, and in objects that are too thick, the plane stress assumption generally loses its validity.

As with plane strain, in plane stress, the analysis is typically performed per unit thickness. In this case, forces should be applied per unit thickness. It may also be applied with a given thickness. Total force should be applied if a plane stress with thickness option is used.

In many cases, plane stress and plane strain solutions can be significantly different, especially for values of Poisson's ratio far from zero. Hence, it is important for the engineer to understand which case applies to their work, if either does. Plane stress is more flexible than plane strain, resulting in higher displacements in load-controlled problems and lower stresses in displacement-controlled problems. Table 3.3 shows in more detail how the plane stress and plane strain functions.

Table 3.3 Comparison of Plane Stress and Plane Strain; Assumes Out of Plane Direction Is the *z*-Direction

	Plane stress	Plane strain
Out-of-plane condition	$\sigma_{33} = \sigma_{13} = \sigma_{23} = 0$	$\varepsilon_{33} = \varepsilon_{13} = \varepsilon_{23} = 0$
Generally	$\varepsilon_{33} \neq 0$	$\sigma_{33} \neq 0$
If isotropic	$\varepsilon_{13} = \varepsilon_{23} = 0$	$\sigma_{13} = \sigma_{23} = 0$
Used when	Material is relatively thin and unconstrained out of plane	Material is relatively thick or constrained out of plane
Examples	Shear wall, connection plate, beam in bending	Retaining wall, strip footing

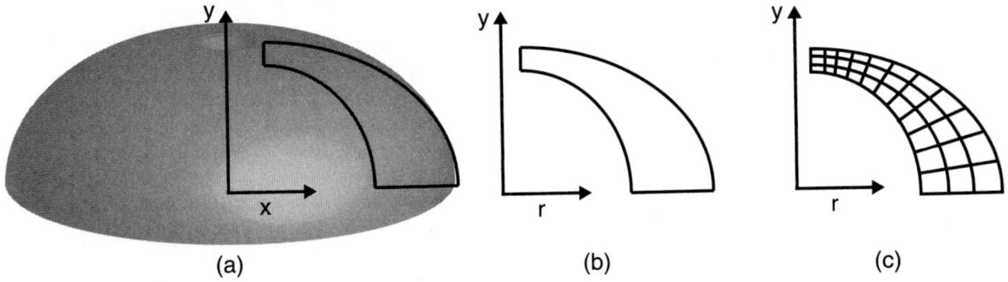

Figure 3.2 (*a*) A dome with an oculus to be modeled under gravity loading, (*b*) cross-sectional geometry to be modeled in the finite-element code, and (*c*) meshed geometry, far more efficient than a three-dimensional model.

AXISYMMETRY

A third type of two-dimensional approximation is the axisymmetric model. Axisymmetry may be applied when a model is rotationally symmetric about some axis. The geometry, material properties, loads, and boundary conditions must be symmetric about this axis. If all these conditions hold, each section behaves the same. An example is a dome under gravity loading, as shown in Fig. 3.2. A dome under wind loading would not be applicable, as the load is not the same around the axis. Since every section is under the same loads, it behaves the same, and only one side of the section need be modeled, as shown in the figure. The displacement of any point on the axis of rotation is zero orthogonal to the axis. There are both strain and stress perpendicular to the plane modeled; the normal strain in this direction may be determined as

$$\varepsilon_{\theta\theta} = u_r / r$$

where r is the distance from the axis of rotation and u_r is the displacement in that direction. This component of the strain is known as the "hoop strain," and the corresponding stress is the "hoop stress."

3.3.3 Reduced Continua

Bars, trusses, beams, frames, plates, and shells are sometimes referred to as reduced continua. Instead of explicitly modeling the full three-dimensional geometry, only a part of the geometry is modeled with geometric properties used to approximate the behavior out of plane. These geometric properties include cross-sectional area for bars, moment of inertia and cross-sectional area for beams, and thickness in plates and shells. These properties are implicit in formulation but do not need to be explicitly modeled in the geometry or meshed. For example, a bar can be treated as a line element with cross-sectional area as a property rather than modeling the bar geometry with three-dimensional elements. With some assumptions, analytical models can be developed as to how the elements will behave.

BARS, TRUSSES, BEAMS, AND FRAMES

Bars and beams can be treated as line elements, with the behavior influenced by geometry of the cross section. For one-dimensional geometries, with proper choice of shape functions, the exact deformation may be found at the endpoints of the element (not internally, however). Hence, the elements for many analyses do not need to be refined.

For bars, the cross-sectional area is generally the only geometric parameter necessary. The formulation assumes that internal force is distributed equally over a given cross section, though the cross section is allowed to vary along the length of the bar. The formula for the element stiffness then becomes

$$\mathbf{k}^e = \int_{\Omega^e} \mathbf{B}^T \mathbf{C}^e \mathbf{B}\, d\Omega = \int_0^L \mathbf{B}^T EA\mathbf{B}\, dx \qquad (3.8)$$

where E is Young's modulus, A is the cross-sectional area, and L is the length of the bar. For a bar element with one degree of freedom at each end and constant cross-sectional area, this reduces to the familiar stiffness terms of

$$\mathbf{k}^e = \frac{EA}{L}\begin{bmatrix} 1 & -1 \\ -1 & 1 \end{bmatrix} \qquad (3.9)$$

More sophisticated bar elements exist that account for changing cross-sectional area with deformation and nonlinear material models. It is possible to create elements with interior nodes as well, but they are not commonly used in practice.

Beam elements are also treated as line elements but with deformation out of plane. Beams can be two-dimensional or three-dimensional, depending on whether the loading is in one plane or not. The moment of inertia is required for these elements, and shear coefficients are required for more advanced formulations. Because the differential equation for the beam is fourth order, not only displacements but also slopes or rotations must remain continuous. Physically, a discontinuity in the slope, that is, a kink in the beam, would result in infinite strain energy. In finite elements, this is solved by having two degrees of freedom at each node, a displacement, and a rotation or slope. The simplest shape functions used for a beam are the Hermitian cubics, and these capture the behavior of an Euler–Bernoulli beam exactly at the nodes. The stiffness matrix for a two-dimensional Euler–Bernoulli beam with constant moment of inertia is

$$\mathbf{k}^e = \frac{EI}{L^3}\begin{bmatrix} 12 & 6L & 12 & 6L \\ 6L & 4L^2 & -6L & 2L^2 \\ -12 & -6L & 12 & -6L \\ 6L & 2L^2 & -6L & 4L^2 \end{bmatrix} \qquad (3.10)$$

where I is the moment of inertia of the beam element. These stiffness terms are identical to the ones commonly derived in structural analysis. The beam element often also has stretch in the axial direction, using the bar formulation above. In the simplest approximation, these are uncoupled, which is reasonable for small deformations. More sophisticated formulations account for the interaction of the two types of deformation.

Beam element formulations can also be extended to account for other factors. For relatively thick beams, shear deformation becomes important and can be added, as done in Timoshenko beams. Nonlinear material models are also possible; however, in the most general case, this requires fairly expensive numerical integration across the cross section. Most beam elements have nodes only at the ends. Some elements with center nodes have been developed, but they are not commonly used.

By using proper orthogonal matrices or rotation matrices, bar and beam elements can be joined at different angles. For bars, the result is a truss. For beams, the result is a frame. These can be three-dimensional or plane structures. The resulting formulations are the same as or very similar to structural analysis formulations.

PLATES AND SHELLS

Plates and shells are thin solid structures that are bounded by two parallel surfaces with an offset, known as thickness, much smaller than lateral dimensions of the surfaces. They are called "plates" when the two surfaces are flat and "shells" otherwise. Plates and shells can be modeled as two-dimensional continua with deformation of mid-surface. In the finite-element procedure, the domain of plate or shell will be subdivided into small elements. Flat element patches make a good approximation for shell structures when the element patches are refined; therefore, they can be used for both plate and shell structures. These elements are generally called "shell elements" in the finite-element method terminology.

Geometrically, the plate is similar to the two-dimensional elasticity problem describe in Sec. 3.3.2; however, the plate is subjected to out-of-plane loads, and bending moments need to be considered. The moments of inertia about in-plane directions are needed in the formulation.

Based on classical plate theory, under the assumptions of infinitesimal strains and orthotropic materials, the in-plane displacements are uncoupled from the transverse deflections, w. The strain vector for a plate, therefore, can be expressed as

$$\bar{\varepsilon} = \varepsilon - z\kappa \tag{3.11}$$

where $\kappa = \left[\frac{\partial^2 w}{\partial x^2}, \frac{\partial^2 w}{\partial y^2}, \frac{\partial^2 w}{2\partial x \partial y} \right]^T$ and ε is the same as the one in two-dimensional plane stress analysis; x and y denote the in-plane coordinates, and z denotes the out-of-plane one, and w is the transverse displacement field. The equation of motion of the plate, considering only the static case, can be derived using the principle of virtual displacement (Reddy, 2018):

$$\int_\Omega \{\delta\bar{\varepsilon}\} \cdot \{\bar{\sigma}\}\, d\Omega = -\int_\Omega \delta w f \, d\Omega + \int_\Gamma \frac{\partial \delta w}{\partial n} M_{nn}\, d\Gamma - \int_\Gamma \delta w \left(V_{nz} + \frac{\partial M_{nm}}{\partial m} \right) d\Gamma \tag{3.12}$$

Considering isotropic homogeneous elastic constitutive law, $\bar{\sigma} = \mathbf{C}^e\varepsilon - z\mathbf{C}^e\kappa$ with $\mathbf{C}^e$ the elastic modulus matrix for plane stress condition. If the in-plane deformation is negligible, the stiffness matrix can be derived from the left-hand side of Eq. (3.12) after introducing finite-element approximation for transverse displacement, $w^h \equiv \mathbf{N}_w \mathbf{q}$,

$$\mathbf{k}^e = -\frac{t^3}{12}\int_{\Omega^e} \mathbf{B}_w^T \mathbf{C}^e \mathbf{B}_w \, d\Omega^e$$

where $\mathbf{B}_w$ is the derivative matrix such that $\kappa^h = \mathbf{B}_w \mathbf{q}$. Since second derivatives of w^h appear in the weak form, not only displacement w but also its derivatives must remain continuous. Hermitian cubic two-dimensional shape functions are usually used in $\mathbf{N}_w$.

The right-hand side of Eq. (3.12) represents external forces. The first term is the work done by f, the transverse load per unit area. The last two terms describe the external force due to bending moment and shear on the boundary, where $\mathbf{n}$ and $\mathbf{m}$ denote the normal and tangential unit vectors on the boundary, respectively; $\frac{\partial(\cdot)}{\partial n} \equiv \frac{\partial(\cdot)}{\partial x} n_x + \frac{\partial(\cdot)}{\partial y} n_y$; $M_{nm} \equiv \mathbf{n}^T \mathbf{A} \mathbf{m}$ and $\mathbf{A} = \begin{bmatrix} M_{xx} & M_{xy} \\ M_{xy} & M_{yy} \end{bmatrix}$; and $V_{nz} = \left(\frac{\partial M_{xx}}{\partial x} + \frac{\partial M_{xy}}{\partial y} \right) n_x + \left(\frac{\partial M_{xy}}{\partial x} + \frac{\partial M_{yy}}{\partial y} \right) n_y$

The above classical plate formulation, known as Kirchhoff plate, is the analog of the Euler–Bernoulli beam in two dimensions. The plate formulation can be extended to account for shear deformation by introducing rotational degrees of freedom, which is known as the Mindlin plate and is an analog of the Timoshenko beam in two dimensions. Nonlinear material models subjected to large deformation are also possible. This requires sophisticated formulations. Interested readers are referred to Zienkiewicz et al. (2013) and Bathe (2014). Most plate elements are quadrilateral and have nodes only at the ends. Some elements with center nodes have been developed, but they are not commonly used.

3.4 ELEMENTS AND NODES

After deciding on a type of analysis, the user must select an appropriate element or elements for the mesh. Elements may have only corner nodes, or they may contain additional nodes. The choice depends on the type of analysis being conducted and the specific geometry. For example, two-dimensional models are typically meshed with either triangles or quadrilaterals. It is generally easier to create a mesh of triangular elements, except if the geometry can be easily subdivided into nearly rectangular substructures. Mid-side nodes generally improve accuracy for quasi-static simulations but may produce worse results for dynamic simulations.

Elements with only corner nodes are often referred to as low-order elements, as the interpolation functions are less than quadratic. Elements with mid-side and perhaps additional nodes that can reproduce quadratic or better interpolation in the displacement are referred to as higher-order elements.

The remainder of this section is a list of commonly used elements and their uses by analysis type.

3.4.1 Two-Dimensional Elements—Plane Strain, Plane Stress, and Axisymmetry

Linear triangles, also known as three-node triangles and constant strain triangles, are the simplest two-dimensional elements. They have only three nodes per element, one at each corner. The resulting interpolation of the displacement field is linear, meaning that the derivative and hence strain are constant. The low order of strain can cause overly stiff behavior compared to the true solution in the case of nearly incompressible materials in plane strain and axisymmetry, bending, and some other situations (see Secs. 3.13.1 and 3.13.2 for more detail). Note that in plane stress, locking for nearly incompressible materials is not an issue because the body is free to contract out of plane. This poor performance is termed locking, and these elements are usually not used. They can be successfully employed in simple situations by analysts experienced enough to watch for these issues. Solutions to relieve volumetric locking exist, including the F-bar patch (de Souza Neto et al., 2005), but they are not widely available in commercial codes.

Bilinear or four-node *quadrilaterals* also have only corner nodes and are low-order elements. Some commercial software packages refer to these as linear quadrilaterals, though the interpolation actually has a little more freedom than the purely linear interpolation. They are, however, subquadratic. Using strictly the interpolation functions derived from their placement, these elements also lock for nearly incompressible materials in plane strain and axisymmetry as well as in bending. However, because of slight extra freedom, one of several formulations can be applied to relieve locking. In most commercial codes, one of these formulations is the default to relieve volumetric locking. See Sec. 3.13.1 for details. The relief of locking in bending is more computationally intensive and is generally not the default choice; see Sec. 3.13.2. The downside of quadrilaterals is that it is more challenging to mesh complicated geometries with quadrilaterals than triangles. In two dimensions, though, reasonable meshing usually can still be accomplished without great difficulty.

Six-node triangles have quadratic interpolation. These elements have nodes in the center of each edge in addition to each vertex. The interpolation provided by the mid-side nodes allows the elements to have curved edges. They are second-order accurate. Hence, if the element linear dimensions are cut in half, the error in the displacement will be roughly a fourth of the original mesh. Three-node triangles and four-node quads, by contrast, are only first-order accurate. The same mesh refinement will result in only half the error. Therefore, these elements are often the element of choice for quasi-static analysis. However, obtaining an effective mass matrix is difficult for higher-order elements, and they are therefore often avoided in dynamic analysis.

Eight-node quadrilaterals, often called *serendipity quadrilaterals*, are also quadratic elements. Like six-node triangles, these elements have nodes at the vertices and middle of the edges. They may have curved edges, have second-order accuracy, and also have issues with the mass matrix in dynamic analysis. These elements are the most popular quadrilaterals for quasi-static analysis, and a good choice for bodies that are easily meshed with rectangular elements.

Other elements have been formulated but are not often used except for special cases. Such elements include *seven-node triangles* and *nine-node* (biquadratic) *quads*, which have an interior node. Generally, these have the same order of accuracy as six-node triangles and eight-node quads, with more degrees of freedom. They are therefore less efficient elements.

3.4.2 Three-Dimensional Elements

For the most part, the common three-dimensional elements are used in the same ways as their two-dimensional counterparts. Tetrahedra, sometimes called tets, are analogous to triangles. Hexahedra, distorted cubes, are often referred to as bricks. It should be noted that three-dimensional meshing is significantly more complicated than two-dimensional mesh, and brick meshing can be quite challenging unless the geometry is easily subdivided into nearly rectangular solid subdomains. Because of this difficulty, bricks may be degenerated into wedge elements or pyramid elements with quadrilateral base.

Four-node, or linear, *tetrahedra* are the simplest three-dimensional elements. Like three-node triangles, they exhibit locking for nearly incompressible materials and bending-dominated deformation. While they can produce acceptable results in other cases, they are generally not favored and often warned against in commercial codes. Methods for alleviating volumetric locking exist (de Souza Neto et al., 2005) but are not widely available in commercial codes.

Eight-node, or trilinear, *bricks* are analogous to four-node quads. Without modification, they exhibit locking in bending and for nearly incompressible materials. However, modifications discussed in Secs. 3.13.1 and 3.13.2 alleviate these issues. One of the solutions for volumetric locking is usually the default in most commercial codes, but the complexity of solutions for locking in bending usually means that this option is not a default. These elements are the most popular for three-dimensional dynamic analysis, though for some bodies, meshing can be challenging with bricks.

Ten-node tetrahedra are quadratic elements with four nodes at the vertices and six more at the center of each edge. Like six-node triangles, these elements are second-order accurate, and edges may be curved. They are very popular for quasi-static analysis.

Twenty-node bricks, also called serendipity bricks, are popular brick elements for quasi-static analysis where brick meshing makes sense. They have eight-corner nodes and 12 at the centers of the edges. The elements are second-order accurate. Edges may be curved.

Other three-dimensional elements have been developed, including *27-node bricks* with three-by-three-by-three nodal placement, and *15-node tets*. They are generally less popular than the elements above. The most common three-dimensional elements are shown in Table 3.2.

3.4.3 Truss and Frame Elements

In both two and three dimensions, most *truss elements* have two nodes, one at each end, with linear interpolation between. Elements with a center node have been developed but are not usually used in practice. Similarly, most *beam* and *frame elements* have two nodes. For transverse displacements, each node has both a translation and a rotational degree of freedom; therefore, cubic interpolation is used for the transverse displacement. Linear interpolation is typically used for axial displacement as well as torsion for three-dimensional elements. Higher-order beam and frame elements have been created by adding either additional nodes or more degrees of freedom per node. These elements are, however, uncommon in practice.

3.4.4 Plate and Shell Elements

Geometrically, plate and shell bending elements are same as two-dimensional elements described in Sec. 3.4.1. They are mainly triangular or rectangular elements. However, since the interelement continuity condition should be enforced, the shape functions and required degrees of freedom for each node are different from those two-dimensional elements.

Four-node nonconforming bending elements have four nodes located at corners. Each node has three degrees of freedom in control of w, $\frac{\partial w}{\partial x}$, and $\frac{\partial w}{\partial y}$. The shape functions for this element are Hermitian cubics. However, the normal slope condition between element boundaries cannot be satisfied, and thus this element is called nonconforming. Nonetheless, this element performs well in practice and is the most popularly used element for plate and shell analysis.

Four-node conforming bending elements have four nodes located at corners as in the nonconforming counterpart mentioned above. In contrast, each node has four degrees of freedom in control of w, $\frac{\partial w}{\partial x}$, $\frac{\partial w}{\partial y}$, and $\frac{\partial^2 w}{\partial x \partial y}$. The shape functions are the tensor product of one-dimensional Hermitian cubics. The interelement continuity conditions are exactly satisfied; thus, it is called conforming.

Three-node nonconforming bending elements are triangular elements for shell and plate. Similar to quadrilateral elements, there are three degrees of freedom per node. Although they are beneficial for modeling complicated geometry, numerical stability issues must be properly handled.

Higher-order elements, eight-node/nine-node quadrilateral, and six-node triangular elements, can also be used for plate and shell modeling. With high-order shape functions, the interelement continuity conditions can be enforced but with high computational costs.

Degenerated shell elements are another family of plate and shell elements used in commercial software. In modeling point of view, they are three-dimensional brick elements. These brick elements discussed in Sec. 3.4.2 could be used as degenerated shell elements. However, their kinematic and constitutive behaviors are similar to conventional shell elements.

3.5 MESHING

For the finite-element analyst, meshing can be one of the most time-consuming parts of a given job. Very coarse meshes are inaccurate, while overly fine meshes may take the finite-element solver too long to run. Highly distorted elements are also inaccurate, and creating a mesh of elements relatively free of distortion can be a challenge if the geometry is complex.

Fortunately, many software programs for meshing exist, some built into the finite-element codes and some dedicated to just creating a mesh. The process is not always automatic, but several algorithms are available to the user for different mesh types.

3.5.1 Element Choice

The first task of meshing is to select a proper element or elements for the analysis. Mixing elements of equal order is possible, but mixing elements of unequal order may result in increased error unless properly designed transition elements are used. These elements are not always available in commercial codes. As mentioned previously, it is generally easier to create a mesh of triangular than quadrilateral elements, and it is easier to mesh tetrahedral elements than hexahedral elements unless the geometry is easily broken into pieces that are nearly rectangular or rectangular solids. However, for dynamic analysis, low-order elements are preferred, and low-order triangles and tetrahedra often do not perform well. For quasi-static analysis, second-order elements are generally preferred.

3.5.2 Selecting an Initial Mesh

To create an initial mesh, it is generally good practice to start with a coarse mesh that runs quickly to check for errors in the analysis. A finer mesh may be required in regions where you the geometry is complicated, but try to avoid a mesh that is too fine, i.e. one with too many elements.

3.5.3 Refinement and Convergence Checks

Once an initial mesh is run and checked for errors, one can refine the solution to ensure the results are accurate. Refinement is the process of splitting elements into smaller elements, often by bisecting the edges.

Further adjustments are made to fit the geometry and produce elements with less distortion. As the mesh is refined, the solution should approach the true solution. If the initial mesh is coarse, it may be wise to refine everywhere once. Where the stress and strain do not change much, this refinement will be sufficient. Further refinement may be necessary in regions where the stress and strain are changing greatly. Novice users often make the mistake of refining where stress is high. In fact, if the stress is high but relatively constant, no extreme refinement is necessary. The regions where refinement is necessary are those where the gradient of stress and strain is high, that is, where the stress and strain change dramatically over a short distance.

It may be necessary as well to refine regions where a very accurate solution is needed, and this may include regions of high stress.

Ideally, refinement is performed until the difference in the solution between two refinements is within acceptable limits of accuracy.

3.6 APPLYING LOADS AND DISPLACEMENT BOUNDARY CONDITIONS

Two types of forces are applied to the body. Forces may be applied to the surface. These include concentrated loads, pressure, and other distributed loads. Distributed forces are integrated over each element surface to create equivalent nodal forces. In commercial codes, the loads may be applied either as surface loads to the geometry or directly to the elements or nodes. The former approach is generally preferable, as the nodal forces change with refinement. Most finite-element codes determine the nodal forces from the applied loads and mesh automatically. Otherwise, the user has to calculate the nodal forces for each refinement, and this can be tricky if the geometry is complicated.

The second type are body forces, such as gravity or magnetic force, that act at a distance. These again are integrated over each element and replaced with equivalent nodal forces. In most codes, body forces may be applied to volumes (or areas in two dimensions) or directly to elements. Again, it is usually advisable to apply body forces to volumes so that the forces are automatically applied to the elements in the proper manner.

Proper displacement boundary conditions are essential to an accurate finite-element analysis. In quasi-static analysis, enough displacement boundary conditions are required to ensure structural stability, following the principles of statics. In two dimensions, the displacement of at least one point must be specified in each direction, plus an additional restriction to prevent rigid body rotation. In three dimensions, displacement must be specified in three directions for at least one point, and three more conditions must be arranged in a way to prevent rigid body rotation about the three coordinate axes. Failure to apply stable boundary conditions will result in a singular stiffness matrix. Most commercial codes will produce errors or warnings that the stiffness matrix is nearly singular, not invertible, or that a small pivot has been detected. Of course, boundary conditions may be applied to lines and areas as well as nodes. For beams, plates, and shells, rotational boundary restrictions may also be applied.

One should take care to apply realistic boundary conditions as well. Boundary conditions applied only to create stability artificially can induce artificial stresses that are not present in the real structure.

3.7 MATERIAL MODELS AND OTHER PROPERTIES

The equations developed in Sec. 3.1 assume linear elasticity. Anisotropic materials can be modeled simply by using a different matrix $\mathbf{C}^e$. Most commercial codes simply require the input of the material constants, such as Young's modulus and Poisson's ratio for isotropic materials.

For many applications, a linear material model is sufficient, for example, to determine the maximum stress and see whether yielding may occur and where stress is most critical. In other cases, such as pushover analysis, a nonlinear model may be necessary. Nonlinear material models create a more complicated and time-consuming solution scheme, and hence the complexity of the material model must be balanced with the solution time.

To implement a material model in a finite-element code, two formulae are usually required. The first is the formula that determines the stress from the strain or strain increment as well as potentially other quantities. These other quantities may include strain rate and history variables, such as those used in plasticity or damage models. The variety of material models that have been implemented in finite-element codes is enormous, and the user should consult relevant sources on the types of models of interest.

The other quantity often needed is the tangent modulus, the matrix of derivatives of the components of the stress with respect to the components of the strain. This matrix replaces $\mathbf{C}^e$ in the formulation of the stiffness matrix. For some material models, these derivatives can be quite cumbersome to calculate, and therefore some algorithms have been developed to work around the use of the tangent modulus.

3.8 SOLVING THE QUASI-STATIC PROBLEM

Once the geometry, mesh, loads, and material properties have been specified, the problem can be run. This involves solving the matrix equation $\mathbf{Kd} = \mathbf{f}$ for the unknown nodal displacements $\mathbf{d}$. If the matrix $\mathbf{K}$ is ill conditioned or singular, then a solution may not be able to be obtained. Usually, this is the result of the structure being unstable, but occasionally other issues may cause ill conditioning, as discussed in Sec. 3.13.4.

A number of algorithms may be chosen to solve the system of equations, either direct solvers, often sparse solvers, or iterative methods, such as preconditioned conjugate gradient. For small systems, direct solvers are usually the safer choice. For models with a large number of degrees of freedom, iterative solvers are usually faster when they converge, but they may not converge to a solution.

3.9 POSTPROCESSING—SOLVING FOR STRAIN, STRESS, AND OTHER QUANTITIES

Once the nodal displacement vector is solved, those displacements can be transferred to the elements, and strain can be solved from the equation $\{\varepsilon^h\} = \mathbf{Bd}$. For linear problems, the stress is $\{\sigma\} = \mathbf{C}^e\{\varepsilon\}$, while for nonlinear problems, the stress-strain relationship is specified by some other equation.

Two important notes should be made about stress and strain. First, strain and strain are calculated at integration points, interior points on the element, not at the nodes. In fact, the stress and strain are discontinuous at the element boundaries for standard elements. When plotting stress and strain, many commercial codes allow such element quantities to be plotted either in the elements or at the nodes. The nodal values are generally obtained by some averaging process, and plots tend to look smoother. It is tempting to prefer such plots. However, the element plots give a more accurate picture of the finite-element solution and show where the mesh may need refinement. Any areas with large jumps in stress and strain across element boundaries are places that should be refined.

Second, stress and strain approximations are usually less accurate than the displacement solution. Since these quantities are related to the derivatives of the displacement, convergence is slower. Hence, if stress or strain is the quantity of interest for the simulation, the analyst must make sure that this is not changing significantly during refinement rather than just the displacement.

3.10 DYNAMIC FINITE-ELEMENT ANALYSIS

In many problems, such as earthquake and sometimes wind loading, inertia effects are important. In this case, the equations are time dependent and usually take the form

$$\mathbf{M}\ddot{\mathbf{d}} + \mathbf{C}\dot{\mathbf{d}} + \mathbf{Kd} = \mathbf{f} \tag{3.13}$$

Here, $\mathbf{M}$ is the mass matrix, $\mathbf{C}$ is the damping matrix, and $\dot{\mathbf{d}}$ and $\ddot{\mathbf{d}}$ are the nodal velocities and accelerations, respectively. Unlike the quasi-static case, in dynamic analysis resulting from the finite-element discretization are a set of ordinary differential equations that are typically solved using a finite difference scheme, covered in Sec. 3.10.3.

The resulting nodal displacements, velocities, and acceleration are solved at discrete points in time.

It is not necessary to restrict rigid body motion to solve a dynamic problem (the body can accelerate in time), but the initial displacement and velocity must be specified at every node.

3.10.1 Mass Matrices

The mass matrix takes the mass distributed over the structure and discretizes it to the nodes. The mass matrix is symmetric and generally positive definite. Both consistent and lumped mass matrices have been used in finite-element analysis, with the latter usually preferred.

Consistent Mass

The consistent mass is derived from (and therefore consistent with) the variational equation. It takes the form

$$\mathbf{M} = \int_{\Omega} \mathbf{N}^T \rho \mathbf{N}\, d\Omega \tag{3.14}$$

where ρ is the material density. The sum of the terms in the mass matrix for each direction (x, y, and z) is equal to the total mass of the system. The matrix is not diagonal, however, with some of the mass "between" two of the nodes. In very coarse meshes, this can lead to unintuitive behavior of the material, but this disappears on sufficient refinement.

Lumped Mass Approximations

Despite the mathematical basis of the consistent mass, lumped, or diagonal, mass matrices are often a simpler and more effective option. Lumped mass matrices replace the distributed mass of the system with point masses at the nodes. There are several different schemes to determine how much of the mass goes to each node, but the total mass again summed for all the degrees of freedom in a given direction should equal the mass of the body, as with consistent mass matrices. Lumped mass matrices are, however, diagonal, which gives them far more efficient performance in some time-stepping schemes, as discussed in Sec. 3.10.3.

Most of the common schemes for lumping mass to the nodes give similar results for low-order elements. Despite the less rigorous mathematical basis, they tend to give better results than consistent mass matrices. They also tend to have slightly better stability properties (see Sec. 3.10.3). Because of the improved performance, lumped mass is usually preferred over consistent mass.

For higher-order elements, some lumping schemes give highly incorrect results, with negative masses at some nodes (these are not usually implemented in commercial codes). Even with consistent mass matrices or so-called special lumping schemes, which have limited mathematical justification, these elements usually produce worse results than lower-order elements. For this reason, bilinear quadrilaterals and trilinear hexahedra are often preferred for dynamic analysis.

3.10.2 Damping

The physical processes behind damping are still not very well understood. Therefore, the models in common usage are still phenomenological but convenient algorithmically. The most common model is viscous damping, where the damping is strain-rate dependent. In the linear case, the damping can be written as a constant damping matrix **C** times the velocity $\dot{\mathbf{d}}$.

A common assumption is Rayleigh damping, which assumes that the damping matrix is additively decomposed into a part proportional to the stiffness and a part proportional to the mass:

$$\mathbf{C} = a\mathbf{M} + b\mathbf{K} \tag{3.15}$$

The constants a and b are often determined from the two most important modes. Rayleigh damping allows the problem to be decomposed modally by generalized eigenvector analysis. If some other form of C is used, the problem cannot be solved modally.

Displacement-based damping, or Coulomb damping, is also sometimes used. More sophisticated damping models are also sometimes used in nonlinear analysis, and there are many variations on such models.

3.10.3 Time Discretization

The ordinary differential equations in time are generally solved using a finite difference scheme, which approximates the time integration at chosen points in time. An alternative approach is to use space-time finite elements, which treat time as another dimension to be meshed. While this technique has several advantages, including fewer time points where the solution is not changing rapidly in time, the method is more complex and not generally available in commercial codes.

Discretization Techniques

The most common time discretization technique used in dynamic finite-element analysis is the Newmark method. This method approximates the time integration as a weighted average of the acceleration at the two endpoints of the time interval

$$\mathbf{v}_{n+1} = \mathbf{v}_n + \Delta t_{n+1}[(1-\gamma)\mathbf{a}_n + \gamma \mathbf{a}_{n+1}] \tag{3.16}$$

$$\mathbf{d}_{n+1} = \mathbf{d}_n + \Delta t_{n+1}\mathbf{v}_n + \frac{(\Delta t_{n+1})^2}{2}[(1-2\beta)\mathbf{a}_n + 2\beta \mathbf{a}_{n+1}] \tag{3.17}$$

Here, $\mathbf{a}_n$, $\mathbf{v}_n$, and $\mathbf{d}_n$ are the finite difference approximations for the acceleration, velocity, and displacement, respectively, at time t_n and $\Delta t_{n+1} = t_{n+1} - t_n$ is the amount of time between the two steps. It is not necessary to keep Δt constant, and in fact many finite-element programs automatically increase the time step size if the progression appears to be stable and accurate. Along with the step size, the two parameters α and β control accuracy and stability of the algorithm. These equations can be thought of approximating the integration in time. It is clearer in the top equation, which, if exact, would be

$$\mathbf{v}_{n+1} = \mathbf{v}_n + \int_{t_n}^{t_{n+1}} \mathbf{a}(\mathbf{t})\, dt \tag{3.18}$$

In the finite difference approximation, the integral is replaced with a weighted average of the acceleration at the two endpoints of the time interval multiplied by the length of time. While not exact, the weighted average gives a good approximation to the velocity when Δt_{n+1} is small. While intuitively $\gamma = ½$ is most accurate, there are reasons to choose other values, as discussed below. A similar though more complicated argument can be made for the second equation, with $2\beta = ½$, or $\beta = ¼$, being most accurate.

A common extension to the Newmark method is the Hilber–Hughes–Taylor (HHT) α-method. This technique modifies the balance of momentum equation and can be used to numerically damp high-frequency modes. High-frequency modes usually do not contribute much to the solution, can cause stability issues, and are not as well approximated by the finite-element discretization. Therefore, it is often desirable to minimize their effects. The modified balance of momentum equation is

$$\begin{aligned}\mathbf{Ma}_{n+1} + (1+\alpha)\mathbf{Cv}_{n+1} - \alpha\mathbf{Cv}_n + (1+\alpha)\mathbf{Kd}_{n+1} - \alpha\mathbf{Kd}_n \\ = \mathbf{f}((1+\alpha)t_{n+1} - \alpha t_n)\end{aligned} \tag{3.19}$$

Typically, $-\frac{1}{3} \le \alpha \le 0$. For $\alpha < 0$, the acceleration at time t_{n+1} becomes partly a function of the force, displacement, and velocity of the previous time, causing a slight drag on the acceleration that damps the higher modes. By setting $\alpha = 0$, the Newmark method is recovered.

Convergence, Stability, and Efficiency

Convergence in time is a function on the time step size as well as the choice of α, β, and γ. The Newmark method is second-order accurate if $\gamma = ½$; that is, the error is proportional to the square of the time step. If the time step size is cut in half, then the error will be roughly one-fourth of what it was previously. If $\gamma \neq ½$, then the method is first-order accurate, meaning that the error is proportional to the step size.

In addition, we need ensure that the solution is stable. A solution is numerically unstable if it tends to infinity over time when the true solution does not. This phenomenon is a function of the finite difference scheme and not related to resonance or other physical phenomena when the solution does tend to grow uncontrolled in time. The Newmark scheme is unconditionally stable if $2\beta \geq \gamma \geq ½$. If $\gamma \leq ½$, the solution is never stable, and hence this choice is never made. If $\gamma \geq ½$ and $2\beta < \gamma$, the solution is stable if the time step is small enough. For no damping, we must have

$$\Delta t \leq \frac{\gamma/2-\beta}{\omega} \tag{3.20}$$

for stability. Here, ω is the largest natural frequency of the finite-element model, which is always at least that of the smallest element. Damping does not hurt the critical time step and in some cases helps. It should be noted that as the element size decreases, ω increases. Hence, as the mesh is refined, the time step must also be refined for conditionally stable methods, and very small time steps may be necessary for fine meshes, especially for stiffer materials.

The HHT α-method is unconditionally stable and second-order accurate if $-\frac{1}{3} \leq \alpha \leq 0$, $\beta = (1-\alpha^2)/4$, and $\gamma = (1 - 2\alpha)/2$.

If the mass and damping matrices are diagonal, as they are in the case of lumped mass and mass-proportional damping, then the displacement, velocities, and accelerations can be solved one equation at a time for each time step, without solving a set of matrix equations. This approach is known as explicit dynamics and is popular due to its efficiency at each time step. To use explicit dynamics with the Newmark method, we must have $\beta = 0$. From the stability discussion earlier, we can see that a small time step is necessary, and this is a drawback of this technique. However, the fast per step calculation and simplicity of the approach continues to make it popular.

Some of the popular choices for Newmark methods are listed in Table 3.4. The most useful are central difference for explicit dynamics and average acceleration for unconditional stability.

Table 3.4 Characteristics of Common Choices for the Newmark Family of Methods

β	γ	Common names	Order of accuracy	Stability	Implicit/explicit
0	½	Central difference	Second	If $\omega\Delta t \leq 2$	Explicit (if **M** and **C** diagonal)
1/12	½	Fox-Goodwin	Second	If $\omega\Delta t \leq 2\sqrt{6}$	Implicit
1/6	½	Linear acceleration	Second	If $\omega\Delta t \leq 2\sqrt{3}$	Implicit
¼	½	Average acceleration, trapezoid rule	Second	Unconditional	Implicit

Modal Analysis

Rather than perform a full dynamic analysis, one can perform a modal analysis. This analysis returns the fundamental frequencies, which are generalized eigenvalues of the mass and stiffness, and the mode shapes, which are the generalized eigenvectors. Knowing the fundamental frequencies can be very useful in a variety of applications, including earthquake analysis, offshore structures, and other vibration applications.

Though more common in structural analysis than finite-element analysis, it is possible to select a small number of important modes and solve the dynamic problem individually on those modes, then use them to reconstruct the motion of the entire structure.

3.11 NONLINEAR FINITE-ELEMENT ANALYSIS

Nonlinearity stems from two sources: large deformation and material behavior. Material nonlinearity occurs when a material does not have a linear stress-strain curve. Examples include plasticity and damage in steel and concrete and nonlinear elasticity in soils and concrete under some conditions. Geometric nonlinearity occurs because the strain cannot be written as a linear function of the gradient of the displacement, which is assumed in linear elasticity. For strains under roughly 4 percent and no large rotations, the effects of geometric nonlinearity can usually be ignored.

In finite-element analysis, any nonlinearity means that the equation $\mathbf{Kd} = \mathbf{f}$ is replaced $\mathbf{f}_{int}(\mathbf{d}) = \mathbf{f}_{ext}$. The equations can be extended to the dynamic case, but regardless, a nonlinear solver, such as Newton–Raphson, sometimes with line search, or the arc length method, is necessary. Most general-purpose finite-element codes have such solvers. The methods are iterative, however, and can be considerably more time consuming than linear solvers.

3.12 VERIFICATION AND VALIDATION

Mistakes in finite-element analysis can have severe consequences and can be easy to make. In 1991, the Sleipner A platform off the coast of Norway sunk in large part due to an error in the finite-element analysis, which underpredicted the stress. In addition, in some situations, the finite-element method itself performs poorly, as discussed in Sec. 3.13. Hence, it is important to check the results, never blindly accepting the computer output.

There are two methods to check a finite-element analysis: verification and validation. Verification is comparison to another analytical solution or a different numerical method. Clearly, if the analytical solution were very good, there would be no need to perform a finite-element analysis. Such solutions are often a little simplified, but the more accurate solution, the better. At a minimum, such solutions should be able to catch errors in values or units of loads or material properties. The goal is to make sure the solution is behaving more or less as expected.

Validation is comparison to an experimental or observed result. This type of check is always a good idea but is essential in the case where the analyst is a little unsure of the physics of the problem, for example, if they are not sure that the problem will behave in a linear elastic fashion. Again, experiments cannot reveal all the detail that the finite-element model can, but the more ways to check the two solutions, the better.

3.13 ISSUES AND PITFALLS IN FINITE-ELEMENT ANALYSIS

While extremely effective for many problems, the finite-element method does not always provide accurate answers, which is why verification and validation are important. First, the differential equations and material models used are just models of real behavior, not exact re-creations. Material properties and loads are never exactly known, and often some simplifications in geometry are necessary to obtain a tractable model. That said, a solution can usually be obtained within a reasonable error.

Analysts should be aware of certain numerical issues, however, that can cause serious inaccuracies in the solution. The most common are briefly discussed below. It is the responsibility of the analyst to check their work to ensure that the results make sense.

3.13.1 Incompressible and Nearly Incompressible Materials

As mentioned previously, low-order elements exhibit overly stiff behavior for materials that are incompressible or nearly so. There are several ways to view this issue. One is that the low-order elements do not have enough degrees of freedom to allow deformation without volume change in many cases. More details of the underlying issue can be found in Hughes (1987), among other places.

There are three popular solutions: mixed methods, strain projection, and reduced integration. Mixed methods break out the mean stress as a separate variable and use a different interpolation scheme for this variable. The interpolation need not be continuous, and a lower-order interpolation works better. There is significant mathematical analysis of the choices, but for four-node quads and eight-node bricks, constant pressure is often used with some modification. While very flexible, the method has extra degrees of freedom, although they can often be solved at the element level.

Strain projection, or B-bar methods, reduce the order of the volumetric part of the strain-displacement matrix **B**. Again, for four-node quads and eight-node bricks, a spatially constant volumetric strain across each element is often assumed, leading to a constant volumetric part of **B**.

Reduced integration uses fewer Gauss points than necessary to integrate part or all of the stiffness. Selective reduced integration uses fewer integration points for only part of the stiffness, usually the volumetric part. A single integration point is used for the volumetric stiffness in the case of four-node quads and eight-node bricks. All three of the above methods produce very similar results and exactly the same results in the case of linear elasticity and constant Jacobian. One of these approaches is usually the default in commercial codes, and users are warned against standard Gaussian quadrature.

An alternative approach is uniform reduced integration, using fewer integration points for the stiffness. The underintegration results in spurious zero-energy modes, known as hourglass modes, which create local oscillations in the solution. These generally do not affect the overall solution greatly, but the oscillations look odd locally. Methods known as hourglass control detect and try to minimize these modes with reasonable success. Fully reduced integration is also often available in commercial codes, and while hourglass control is not perfect, this method has the advantage of reducing the number of integration points within the element, which can save computation time.

3.13.2 Locking in Bending

Low-order elements also perform poorly in bending-dominated problems. Modified elements have been developed, though they are fairly sophisticated and are more time consuming than the modified elements described above. Therefore, they are usually not used by default in commercial codes, and the user needs to specify this option. In quasi-static problems, higher-order elements are usually preferred.

3.13.3 Reflex Angles

The theory of elasticity predicts that bodies with sharp reflex angles, angles greater than 180 degrees, will experience infinite stress under some loading conditions. In the finite-element solution, the stress at these corners will become higher as the mesh is refined, with no limit. There is no issue with the finite-element performance here; rather, the issue is with the assumption of elasticity.

There are a few solutions to this issue. The easiest is to recognize that the problem is local and that likely the solution away from this area is still valid. The second is to introduce plasticity or some other material model that prevents the stress from increasing indefinitely. This approach will also give the user some idea of the plastic zone induced by such a corner. The third approach, which is also a good idea from a manufacturing perspective, is to round the corner or introduce a fillet to alleviate the stress singularity.

3.13.4 Element Node Numbering

There are conventions for element node numbering in finite elements. Meshes that are generated by meshing software or commercial codes always follow the correct order. Occasionally, though, it still makes sense to generate meshes by hand. If a mesh is created by hand, it must follow the node numbering convention, or, a negative element Jacobian can arise.

The convention for two-dimensional elements is as follows: First, the corner nodes are labeled in the counterclockwise direction (which corner is first is unimportant). Second, mid-side nodes are labeled counterclockwise, starting with the node between nodes 1 and 2. Finally, the interior node is labeled if it exists.

For three-dimensional elements, again corner nodes are labeled first. For bricks, the "bottom" face is labeled counterclockwise, then the top face. Node 5 is directly "above" node 1, and counterclockwise should be considered looking down on the element from above. Mid-side nodes are labeled on the bottom face, then the top face, using the same convention as in two dimensions. Then the mid-side nodes between the two faces are numbered.

The node numbering for common elements is shown in Tables 3.1 and 3.2. Most commercial finite-element codes and good development codes will issue warnings if the element Jacobian is negative, giving the user a chance to locate and fix labeling errors.

3.13.5 Conditioning

Conditioning is a general issue in solving large matrix equations numerically. Because of machine round-off in computations, inaccurate results occur for solving some systems of equations, particularly if there is a great difference in the magnitude of the eigenvalues. Finite-element software may detect this issue and report that the stiffness matrix is nearly singular or has a small pivot. The analyst should first check that boundary conditions are properly applied and the components are properly connected, that is, that the structure is stable. Structural instability is by far the most common source of these errors.

In some cases, however, poorly conditioned stiffness matrices do arise. There are some ways in which conditioning can be improved for certain problems. The condition number, the ratio of the magnitudes of the maximum and minimum eigenvalues for direct methods, measures this. As the condition number approaches roughly 10^{16} for double-precision arithmetic, the solution loses accuracy.

The analyst should avoid modeling materials of stiffnesses differing by several orders of magnitude. Roughly, each order-of-magnitude difference leads to an order-of-magnitude difference in the condition number. In the case where one material is vastly stiffer than another, it may be possible to consider the material as rigid or reduce the stiffness difference to within an acceptable tolerance, at least in quasi-static problems.

Elements of vastly different sizes should be avoided as well. Each order-of-magnitude difference in each linear dimension increases the condition number by roughly an order of magnitude. While selective refinement, that is, refinement in local regions, is a good strategy for maintaining reasonable simulation times, overrefinement can lead to poor conditioning. More than one or two orders of magnitude may lead to poorly conditioned matrices, and if there are issues, this difference should be reduced further. Coarse meshes overall tend to have fewer issues with conditioning, so starting with a coarse mesh may help troubleshoot this issue.

If using the penalty method in contact and constraints, each order of magnitude of the penalty parameter over the stiffness roughly adds an order of magnitude to the condition number. If there appear to be conditioning issues, making the penalty parameter as small as possible while still enforcing the constraint is desirable. Depending on the accuracy needed, this value may be as low as two to three orders of magnitude above the stiffness matrix. The effect of the penalty parameter may be tested on a coarse mesh initially to find the lowest value that is not greatly affecting the solution.

3.13.6 Distorted Elements

The reader will recall that the physical element geometry is mapped to a more regular local coordinate system. There is no need to match the local geometry exactly, but if the elements are very distorted, the mapping becomes inaccurate. Even if there are only a few, these elements can significantly affect the problem solution. Most commercial codes and meshing software warn if some elements are highly distorted. Remeshing with finer elements in the region of the distorted elements can help. Obtaining a mesh free of distorted elements is, however, one of the biggest challenges for many finite-element problems.

In the extreme case, quadrilateral or brick elements may have reflex angles, angles greater than 180 degrees, in physical space. These angles create a negative Jacobian in part of the element in the coordinate map, which can lead to nonphysical behavior. These shapes should be avoided altogether.

3.13.7 Compatibility

In order to ensure convergence, elements must be both complete and compatible. Completeness for mechanics means that the element must be able to reproduce a linear displacement field within each element. All commonly used finite elements are complete.

Compatibility is a continuity condition that must be enforced between elements as well as within. For continuum models, the displacement must be continuous. This means that elements must have the same interpolation across element boundaries. Hence, elements with linear displacement along the edges, such as four-node quads, should not be mixed with elements with quadratic displacement along the edges, such as eight-node quads, without special treatment. Transition elements have been developed that allow for transition between regions, though they are not often available in commercial codes.

Beam, plate, and shell elements should also maintain continuous slopes across their bending degrees of freedom, but this is satisfied for most elements.

There are cases in which compatibility can be violated with success, but it is not recommended without mathematical justification.

3.13.8 Mixed Structural and Continuum Elements

In some cases, it makes sense to model part of a structure using continuum elements and part using structural elements. In this case, the connections between the continuum and structural elements must be treated with care. Truss elements, which rotate freely with respect to the continuum structure, are not an issue. But because continuum elements lack rotational degrees of freedom, where a beam element connects to a continuum structure, the default behavior is that of a ball-and-socket joint. Constraint formulations can couple the beam end rotation to the solid deformation but are not always available in commercial codes.

An alternate approach is to extend the beam into the solid and attach it to the solid nodes so that it rotates with the solid deformation. It should be noted that unless this reproduces the physical problem, the stresses in the solid may not be physical around this area. Another approach is to attach the beam to a plate that is attached to the solid structure. This approach is excellent if there is a physical plate in the attachment. See Fig. 3.3 for an example.

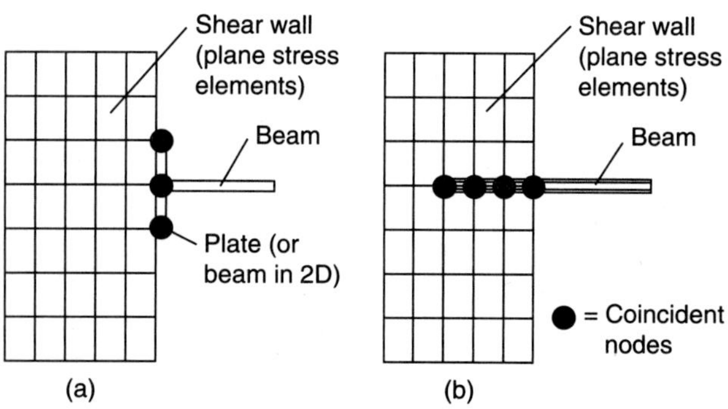

Figure 3.3 Connecting a beam to a shear wall (*a*) via a plate element or (*b*) by embedding the beam in the wall. The latter approach may give inaccurate stresses near the connection if it is not physically connected in this way.

3.14 INTRODUCTION TO FINITE ELEMENTS FOR THERMAL, THERMOMECHANICAL, AND OTHER PROBLEMS

In addition to mechanical and structural problems, finite-element analysis has shown great success in modeling heat condition, electrical conduction, fluid flow in porous media, and other problems. The method is far more easily applied for even-order differential equations.

The formulation for heat conduction is similar to that of the mechanical problem but is simpler since it is a scalar problem. In the steady-state case, the problem still reduces to $\mathbf{KT} = \mathbf{f}$, with $\mathbf{T}$ as the vector of nodal temperature values:

$$\mathbf{K} = \int_{\Omega} \mathbf{B}^T \boldsymbol{\kappa} \mathbf{B}\, d\Omega \tag{3.21}$$

and

$$\mathbf{f} = \int_{\Omega} \mathbf{N}^T q\, d\Omega + \int_{\Gamma_h} \mathbf{N}^T \phi\, d\Gamma \tag{3.22}$$

Here, $\boldsymbol{\kappa}$ is the thermal conductivity matrix (which is equal to the scalar conductivity times the identity matrix for isotropically conducting materials), q is heat production per unit volume (from chemical reactions, mechanical dissipation, etc.), and ϕ is heat flux across the boundary on the places where that is specified. Once the temperatures are solved, secondary quantities, such heat flux and heat flow, can be found during postprocessing.

In the transient case, the equation becomes

$$\mathbf{M}\dot{\mathbf{T}} + \mathbf{K}\mathbf{T} = \mathbf{f} \tag{3.23}$$

where $\mathbf{M} = \int_{\Omega} \mathbf{N}^T \rho c \mathbf{N}\, d\Omega$ is the thermal capacity matrix. Here, ρ is the density and c is the specific heat of the material. Like dynamic analysis, a lumped mass approximation is often used. The set of ordinary differential equations in time is also solved using a finite difference scheme. However, since this is a first-order set of ordinary differential equations in time, instead of second-order, like dynamics, a simpler scheme like the generalized trapezoid rule, is usually used:

$$\mathbf{T}_{n+1} = \mathbf{T}_n + \Delta t[\alpha \dot{\mathbf{T}}_n + (1-\alpha)\dot{\mathbf{T}}_{n+1}] \tag{3.24}$$

For $\alpha = 0$ and lumped mass, the equations become explicit but first-order accurate and conditionally stable. With $\alpha \geq \frac{1}{2}$, the equations become unconditionally stable but may exhibit nonphysical oscillations for large time steps unless $\alpha = 1$. The common choices of α and their properties are listed in Table 3.5.

Table 3.5 Characteristics of Common Choices of α for the Generalized Trapezoid Rule

α	Common names	Order of accuracy	Stability	Numerical oscillations
0	Explicit Euler	First	Conditional	If Δt too large
½	Crank–Nicolsen, trapezoid rule	Second	Unconditional	If Δt too large
1	Implicit Euler	First	Unconditional	No

Explicit methods are less popular in thermal analysis than mechanical, largely because there are fewer of degrees of freedom, only one per node, and the problems are solved more quickly. That said, for large problems, explicit analysis can be used.

Nonlinear extensions include temperature-dependent thermal conductivity. Film coefficients are often used to approximate convection boundaries with air or other fluids.

With a simple change in the meaning of the variables, the same equations can be used to solve problems in fluid flow through porous media like soils, electrical conduction, and chemical reactions. The nonlinear extensions are sometimes different, but sometimes commercial code developers are aware that thermal models are used for other problems and include models suitable for porous flow and other problems.

Finite-element models have also been widely applied to coupled physics problems, such as thermomechanical or poromechanical problems. In the former case, heat could cause thermal expansion, inducing stress or strain in the body. Inelastic deformation, on the other hand, can cause temperature increases in the body. These types of coupled problems can be solved simultaneously in many commercial finite-element codes.

It is worth mentioning that sometimes one physics affects the other but not vice versa. In thermomechanics, if the deformation is assumed to be purely elastic (no damping, plasticity, etc.), then one can solve the thermal problem first and use the results as input to the mechanical problem. Otherwise, the two problems must be solved simultaneously. This approach, while it can be solved, is more challenging, creating larger stiffness matrix, longer solution times, and possible conditioning issues.

In thermomechanical problems, the same elements may be used for heat conduction as for the mechanical problem. However, in poromechanics, the coupling causes locking issues for equal-order elements, very similar to that found in incompressible materials. There are many solutions to the locking problem, including using lower-order interpolation for the porous flow problem and various stabilization schemes, but the analyst should take care to employ an effective method for such problems. Only a brief overview is given here, so performing some background investigation in the literature before beginning is recommended.

REFERENCES

KJ Bathe. *Finite Element Procedures*. 2nd Ed. Prentice Hall, 2014.

J Fish and T Belytschko. *A First Course in Finite Elements*, Wiley, 2007.

EA de Souza Neto, FM Andrade Pires, and DRJ Owen. "F-bar-based linear triangles and tetrahedra for finite strain analysis of nearly incompressible solids. Part I: formulation and benchmarking." *International Journal for Numerical Methods in Engineering* 62:3, 353–383, 2005.

TJR Hughes. *The Finite-Element Method: Linear Static and Dynamic Analysis*. Dover, 1987.

NH Kim and BV Sankar. *Introduction to Finite-Element Analysis and Design*. Wiley, 2008.

DL Logan. *A First Course in the Finite Element Method*. 5th Ed. CL Engineering, 2011.

JN Reddy. *Introduction to the Finite Element Method*. 3rd Ed. McGraw-Hill Education, 2018.

OC Zienkiewicz, RL Taylor, and JZ Zhu. *The Finite-Element Method: Its Basis and Fundamentals*. 7th Ed. Butterworth-Heinemann, 2013.

Chapter 4

Computer Applications in Structural Engineering

BY

RAOUL KARP, PE, SE *Vice President of Design Engineering Analysis, Bentley Systems, Inc., Carlsbad, California*
BULENT N. ALEMDAR, Ph.D., PE *Principal Research Engineer, Bentley Systems, Inc., Carlsbad, California*
SAM RUBENZER, PE, SE *Founder and Structural Engineer, FORSE Consulting, Eau Claire, Wisconsin*

4.1 INTRODUCTION

The end of the 20th century brought with it major advances in computation that have opened the door to the age of advanced simulation and automation in the structural engineering profession. This ability allows engineers to design more resilient, innovative, and economical structures that are increasingly more necessary to meet global growth and sustainability demands. In this chapter, the authors explore standard industry practices in computational engineering simulation to enable these critical societal outcomes.

4.2 COMPUTER STRUCTURAL ANALYSIS SIMULATION

The most common mathematical solution used to perform computational simulation is known as the finite-element analysis method (Clough 1960, Bathe 1996, Cook et al. 1989, McGuire et al. 2000). The finite-element analysis method involves idealization of a physical structure (Fig. 4.1), its support conditions, and its applied loading, in a mathematical framework, to determine the structural behavior of the structure.

4.2.1 Finite-Element Analysis: A Primer

In the finite-element analysis method, individual components of a structure are mathematically idealized as discrete (finite) line, surface, and volumetric elements. Finite elements are connected to each other at nodes. Loads and boundary constraints are applied at locations of nodes (loads can be also applied to elements directly, but eventually, they are later resolved to nodes). The primary unknowns are the displacements (translations and rotations), called degrees of freedom (DOFs), at each node. For a two-dimensional problem, there are three DOFs defined at each node (two translational and one rotational), and for a three-dimensional model, six DOFs are defined at each node (Fig. 4.2). The governing static and dynamic equations are expressed in the form of algebraic equations at nodes.

Each finite element is mathematically represented by a local stiffness matrix that defines a relationship between element displacements and forces. All element local stiffnesses are assembled together in a global stiffness matrix (**K**) by combining stiffness term from each element contributing to a specific DOF. Similarly, loads applied at nodes and on elements are combined and assembled into global load vector (**F**) for each nodal DOF. The final algebraic equation **Ku** = **F** is obtained. The numerical solution of this equation for each load vector yields the displacement vector **u**, which holds translation and rotations magnitudes at each node (DOF). These nodal displacements, along with element local stiffness matrix and applied loads on elements, are used to calculate stresses, forces, and deformations within elements.

Practical Insight: As finite-element models solve for displacement, it is often the case that modeling errors can be easily identified by reviewing the displaced shape from both gravity and lateral load cases, load combinations, or modal periods of the model to look for unexpected deformations or mode shapes.

4.2.2 Types of Finite-Element Analysis

The purpose of a structural analysis is to obtain sufficient and accurate information about behavior of structure under prescribed loads. This information is used to design structural members for stiffness and strength requirements. Based on magnitude and type of loading and expected level of structural response, an engineer decide to run a linear (elastic) or nonlinear analysis.

A structural analysis is either static or dynamic. Static analyses consider the structural response in the absence of any dynamic actions; that is, loads are applied so slowly that they do not induce any inertia forces on structure, such as dead, live, or thermal loads. A dynamic analysis simulates the effects of dynamic forces from earthquakes, blast, wind, and so on structures. Under dynamic loading, inertia forces develop that, are varied with time. This condition necessitates a solution for each time step. It is also possible to obtain approximate results without pursuing a complete history of response: the Response History Analysis method provides peak values of forces and deformations, which are good estimates for structural responses and sufficient for a preliminary design.

Linear (Elastic) Analysis

This common type of analysis assumes that the structure undergoes small displacements and that all members remain elastic throughout the applied loading. A linear analysis determines the elastic stiffness

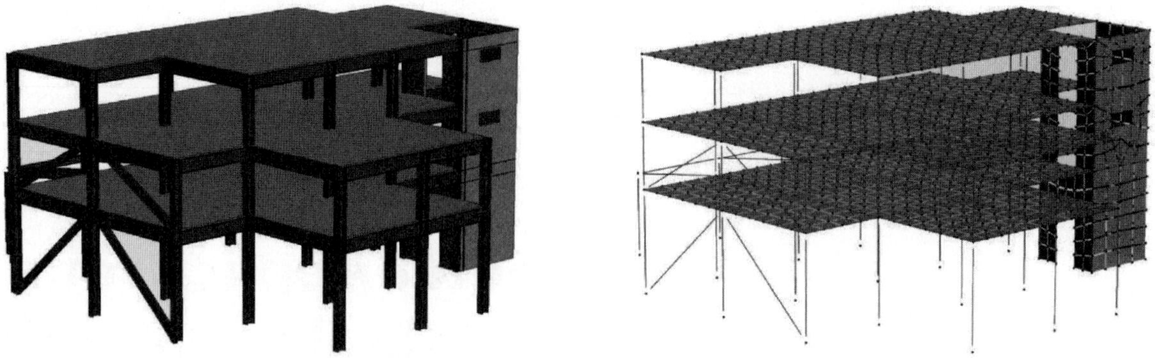

Figure 4.1 Physical structure idealized in a three-dimensional finite-element analytical model. (*Images from STAAD.Pro, Courtesy Bentley Systems.*)

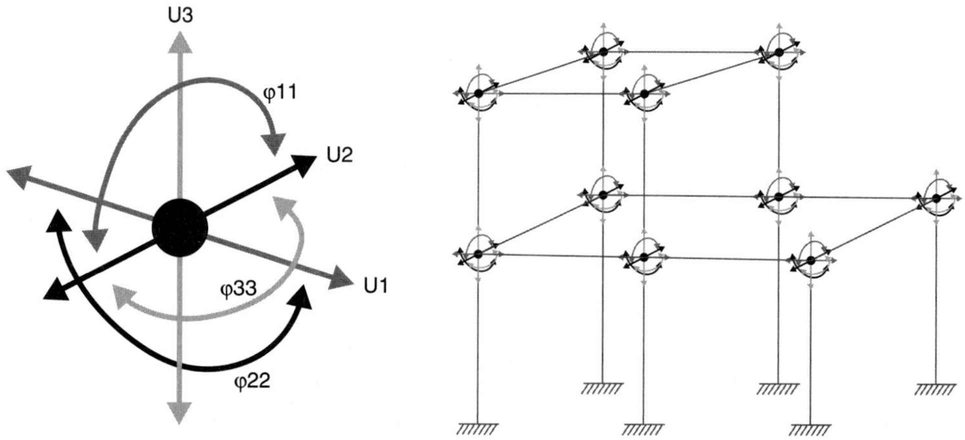

Figure 4.2 Six DOFs at each node in a three-dimensional finite-element model.

of each element, which is a function of the member material, cross section, and initial geometry. The relationship between loads and displacements is proportional and linear. A linear analysis is appropriate if the displacements are small relative to the initial geometry of the structure. An advantage of linear analysis is that the displacement and forces from every load condition can be determined in isolation. The resulting displacements and forces can then be factored and combined with other load condition results as prescribed by the building code load combinations.

This type of analysis is usually used to proportion member sizes and to gain a general understanding of the force-transfer mechanism of structures under applied loads. As member forces are a function of member stiffness (cross section) and the design of a member is in turn a function of the forces of the member, it stands to reason that the analysis and design process is iterative.

For some cases, materials properties are modified to account for some level of material nonlinear effects within a linear analysis. For instance, ACI 318 typically prescribes crack factors that act to reduce the bending and axial stiffness to account for expected cracking under elastic loading conditions. Another example is for steel members in which the direct analysis method of AISC 360 allows reduction of modulus of elasticity to account for residual stresses developed during the manufacturing process of steel components.

Practical Insight: A linear analysis is often appropriate even for load conditions such as seismic loading, which will result in nonlinear behavior in the structure. Where the building codes do not mandate a nonlinear analysis (ASCE 7), they determine an equivalent static or dynamic loading that takes into consideration the ductility of the lateral system to reduce the loading by reducing the base shear with a response modification factor, which is a function of the structure type and ductility. The structural analysis is performed with this reduced base shear, but the engineer is required to detail the structure according to code requirements to ensure that the ductility assumed in the loading can be achieved in the actual loading event.

Nonlinear Analysis

A nonlinear analysis should be utilized when a structure undergoes large displacements (nonlinear geometric effects) or when materials go beyond their elastic limit (yielding or cracking in material) and a more accurate simulation of structural behavior is required. The nonlinear interaction between loads and structural response prohibits the use of superposition of individual load condition results, and analysis must be performed for each combination of gravity and lateral loads acting together.

This type of analysis involves an incremental-iterative numerical solution in which structural response is traced for incremental load increments. For each step, the structural response is based on current elements stiffness as well as the deformation history of the structure. This approach solves for equilibrium at each load increment. Once an equilibrium state is reached, displacements in nodes and forces in members are updated, and the solution is advanced to the next load increment.

Nonlinear material effects are handled according to where they are assumed to be occurring within the elements (Fig. 4.3). One approach for the beam-type element is the *plastic hinge* or *concentrated plasticity* approach, in which material nonlinear effects are assumed to be

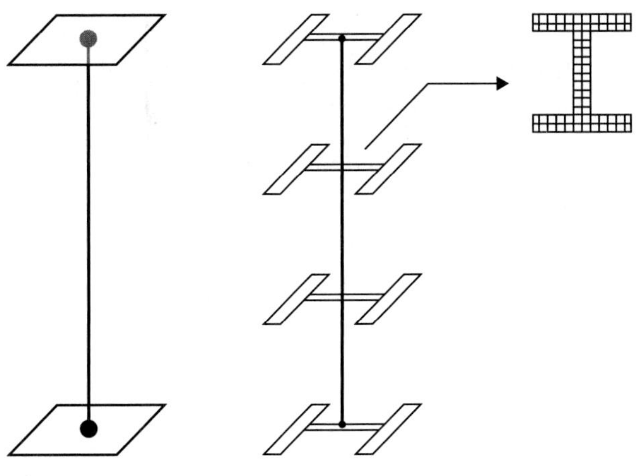

Figure 4.3 Plastic hinge and distributed plasticity model idealization for a beam-column.

concentrated at the ends of elements while the rest of the element is assumed to be elastic (White et al. 1991, Liew et al. 1993, Ziemian and McGuire 2002). Another approach which considers gradual spread of inelastic behavior within the element is called *distributed plasticity* or *plastic zone* model (Alemdar 2001, Teh and Clarke 1998). The former method accommodates a zero-length plastic hinge element inserted at the ends when the element reaches its plastic capacity. The latter method calculates the spread of plasticity by the numerical integration across the discretized cross section (fiber mesh) located at the selected locations along the element. Although the conventional plastic hinge method requires less computational effort, it may overestimate the capacity of a structure when significant distributed yielding takes places in members subjected to large moments but small compressive loads.

Specially formulated nonlinear material models are developed to capture material behavior beyond their initial elastic ranges. These models are employed either at the location of the plastic hinge element or at each fiber of the discretized cross section for the distributed plasticity method. A nonlinear force-deformation relationship is defined by a series of rules described for monotonic or cyclic (hysteretic) behavior (Fig. 4.4).

Practical Insight: It is possible that a nonlinear numerical solution may converge slowly or may even diverge. To address these conditions, smaller finite elements, smaller loading increments, or adjusting converge tolerance may result in improved performance. When the analysis is not converging, it may be an indication of a true structure instability at the level of applied loads.

The structural geometry or member properties will likely need to be adjusted to address this situation.

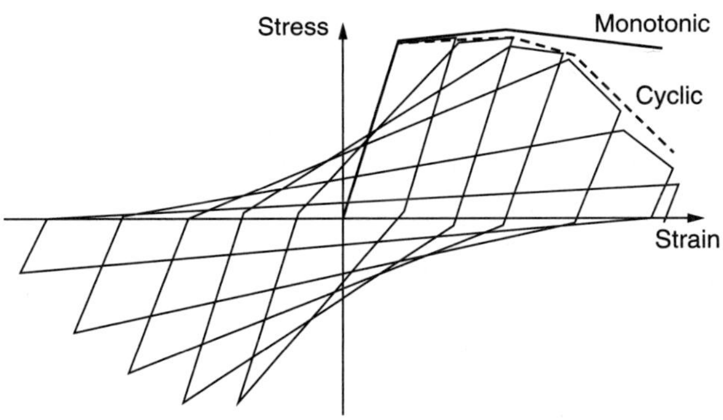

Figure 4.4 Material nonlinear stress-strain (backbone) envelope.

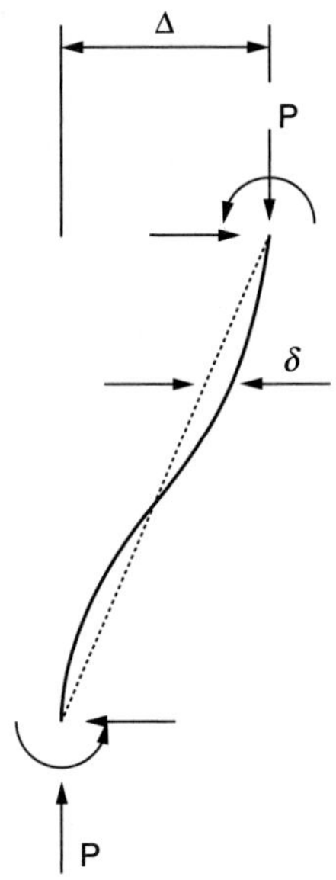

Figure 4.5 P-Δ and P-δ effects shown for a beam-column.

Geometric nonlinear effects stem from elements subjected to large displacements and rotations. In these situations, the structure displacements are considered large, and equilibrium equations need to be solved in the deformed position of the structure. In the analysis of building structures, P-Δ (P-large delta) and P-δ (P-small delta) are the most common sources of such effects (Fig. 4.5). They introduce additional moments due to member axial forces. It should be noted that geometric nonlinear effects, in general, have a destabilizing effect. This means that a structure may reach an instability state at the limit of having significant geometric nonlinear effects.

4.3 STRUCTURAL FINITE ELEMENTS

Physical members such as beams, columns, walls, and foundations carry loads and are mathematically idealized in the finite-element analysis. In the idealization, each member is defined based on its characteristic response. A beam and column element carries loads through axial and bending action, a brace is typically considered to transfer loads through its axial stiffness, and walls and diaphragms carry loads through in-plane and out-of-plane actions.

Beam-column members are typically idealized with linear analytical elements with six DOFs at each end. Specially formulated line elements can capture complex behaviors, such as material inelasticity, residual stresses, and effects of geometric imperfections. Internal forces are calculated in terms of axial, shear, bending, and torsional forces, which are meaningful for engineers.

Wall and diaphragms are idealized with a mesh of discrete triangular or quadrilateral plate, membrane or shell finite elements. A membrane element provides in-plane stiffness, which is suitable to model a wall, primarily providing stiffness through in-plane shear action. A floor resisting out-of-plane forces is modeled with plate-type elements. A shell element is considered a combination of both in-plane (membrane) and out-of-plane (bending) behavior, and they can be used to model walls and diaphragms as well.

4.3.1 Beams, Columns, Braces, and Joists

Beams, columns, and braces carry loads through their bending and axial stiffness. They are connected to each other at joints, and force-transfer behavior between adjacent beams and columns is determined by how these elements are connected. A beam is considered *free* in bending for a particular axis at its end if it does not transfer moment in that axis

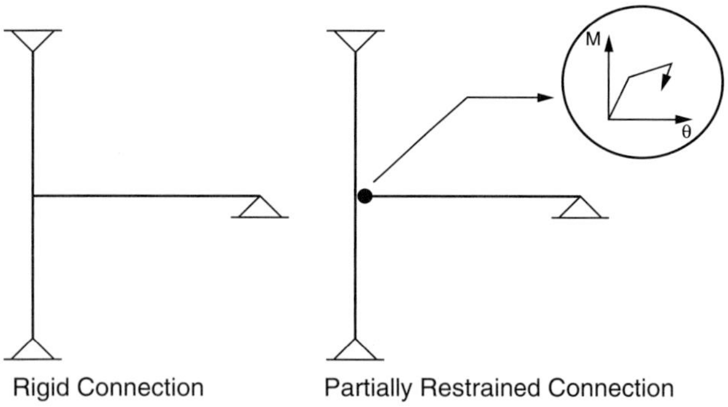

Figure 4.6 Beam and column are connected rigidly or partially.

(DOF) to adjacent column. This modeling approach is usually exploited for gravity beams that are assumed to provide no lateral stiffness. If a full moment transfer is to be modeled for a specific DOF, a *fixed* end condition for that DOF element is considered. These modeling simplifications are regarded as ideal conditions. In reality, a connection between a beam and a column is neither perfectly *fixed* nor *free* but commonly is something in between.

If necessary, a more realistic approach to model force transfer between a beam and a column is to insert a connection element at element ends (Fig. 4.6). A connection element is classified as rigid (fully restrained), semirigid (partially restrained), and flexible (unrestrained). For beams and columns, a connection element is usually represented with a plastic-hinge element in which load-capacity and load-deformation properties are expressed in terms of moment-rotation (or curvature) or P-M-M curves (axial-moments coupling). It should be noted that a connection element can be modeled to exhibit linear or nonlinear behavior.

A centerline approach is usually adapted in modeling beams and columns. This means that an analytical model is created at the centerline geometry of members. This approach is sufficient to define beam and column geometry situations. There are several special cases in which a more elaborated solution is needed. A few examples for such special cases are demonstrated in Fig. 4.7. For example, where a diagonal brace

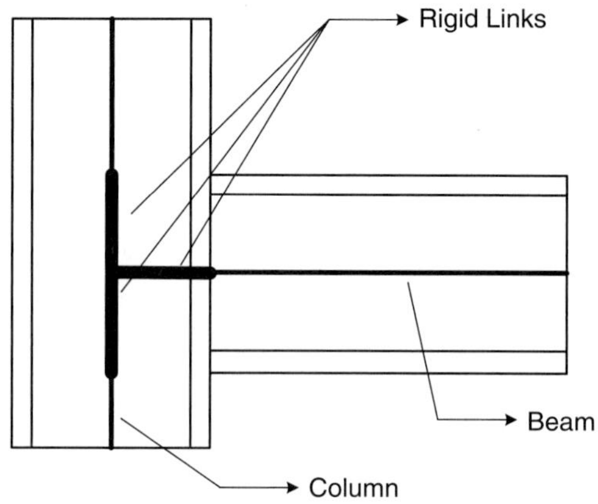

Figure 4.8 Modeling beam-column intersection with rigid link elements.

member is eccentrically connected to a beam or a beam framing off-center into a column. It is also common to consider a cardinal point on the cross section to align with the modeled analytical member. Such special cases are modeled with rigid links inserted at offset locations, between the analytical node and geometric center of the member, or they are directly handled within element formulation via translating forces at element ends to the face of joints.

While analytically members are modeled centerline to centerline, physically they are connected at the face of members. To idealize the behavior of the physical member, it is common practice to provide a rigid end link or to modify the element formulation to account for this physical shortening of members at the joint.

In a beam-column joint, the panel zone is the region in the column web limited by the extension of beam flanges into the column (Fig. 4.8). This intersection zone may be subjected to high shear forces under lateral loads. Deformations in panel zone impact the elastic or inelastic behaviors of structures.

A full rigid link at the ends of the beams and columns would assume that no deformation occurs in the panel zone region. To idealize true

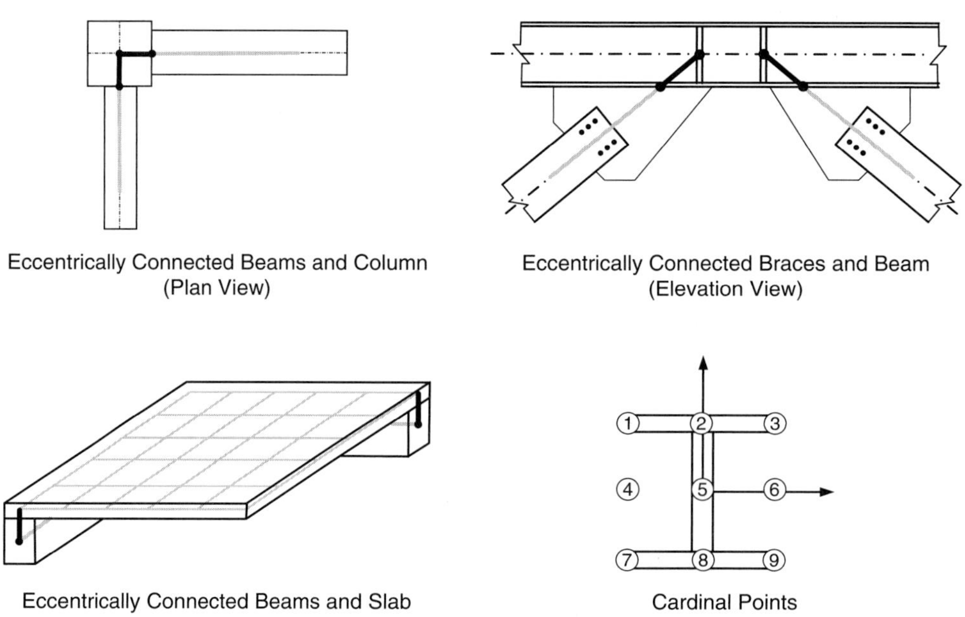

Figure 4.7 Examples of elements connected eccentrically (element centerlines are indicated with dashed lines).

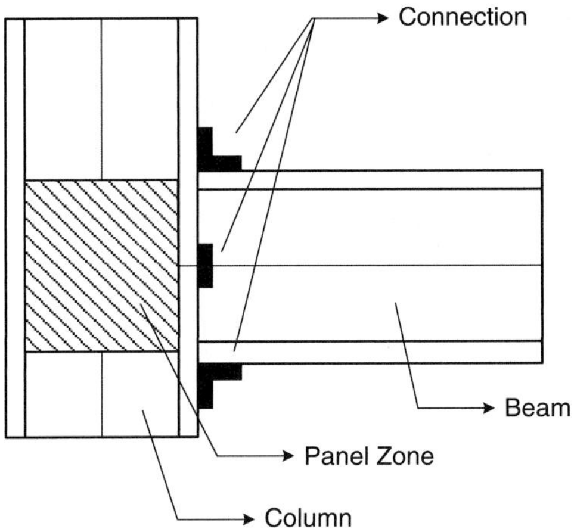

Figure 4.9 Panel zone.

structure flexibility, it is common practice to use a 50 percent reduction in the rigid link length to account more accurately for the deformation of the panel zone. A more advanced solution that considers stiffness and strength of the panel zone is to use Krawinkler or scissors panel zone models (Charney and Downs 2004, Krawinkler 1978). These models provide a more realistic approach in terms of accessing panel zone deformations directly in analysis.

4.3.2 Modeling Coupling Beams in Shear Walls

Coupling beams are the members used to couple two shear wall systems. Force transfer between shear walls is through moments and shear in the coupling beam. Depending on the depth of a coupling beam, it is classified as a slender, moderately deep, or deep coupling beam.

For a shallow deep beam, the behavior is controlled by bending. In this case, a slender coupling beam zone can be idealized by a beam element connected to shear walls at a single node, as shown in Fig. 4.10. If a wall element is modeled with shells including drilling DOFs, then moment and shear are transferred from beam to shear walls at a single node. If this is not the case, the coupling beam is pinned connected to the wall. If moment continuity is required, one practical insight is to model a rigid beam across the width of the wall (a rigid beam is stiff in bending but has a small axial stiffness).

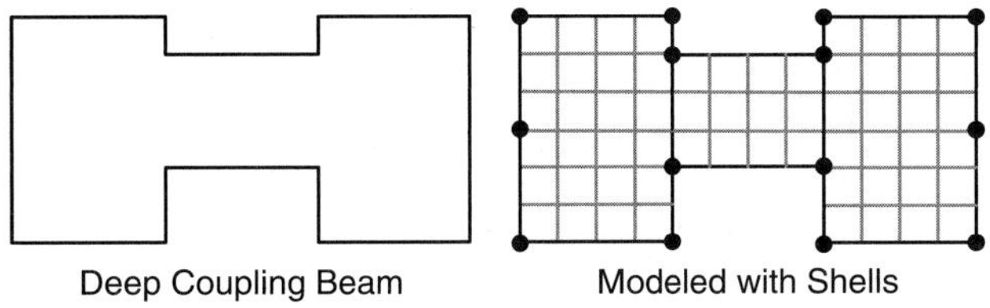

Figure 4.11 Modeling deep coupling beams.

In deep coupling beams, the behavior is controlled through shearing action. In this case, a deep coupling beam segment should be modeled using shell elements (Fig. 4.11).

For moderately deep beams, the behavior is controlled by bending or by bending and shear together. To model the moderately deep beam zone, it is recommended to provide a beam accompanied by a special connection element inserted horizontally or vertically at the end of the beam. The special connection element is rigid in bending but has small stiffness in axial (Fig. 4.12).

4.3.3 Modeling Diaphragms

Structural diaphragms are critical elements in a structure's response to lateral loads. Inertial forces from seismic loads and external wind or other lateral forces must be transferred through the diaphragm to the lateral force-resisting system (LFRS). For seismic loads, the diaphragm is typically designed to remain elastic, ensuring that energy dissipation occurs in appropriately designed LFRS.

The lateral distribution of forces through a diaphragm is a function of the diaphragm's stiffness and deformation relative to that of the LFRS. There are generally three different ways a diaphragm is idealized (Fig. 4.13), namely *flexible*, where the diaphragm is assumed to have no stiffness, loads are directly distributed to the LFRS based on area tributary to each frame, and all lateral frames behave independently of each other; *rigid*, where the diaphragm is assumed infinitely stiff in-plane (with no axial, shear, or in-plane flexural deformation), lateral loads are applied at a single point on each diaphragm, loads are distributed to the LRFS based on their relative stiffness, but displacements are coupled, as all frames are rigidly connected to each other; and *semirigid*, where the actual bending, shear, and axial stiffness of the diaphragm is idealized, lateral loads are applied directly to multiple nodes on the diaphragm, and forces are distributed according to the relative stiffness of the diaphragm and the lateral frames.

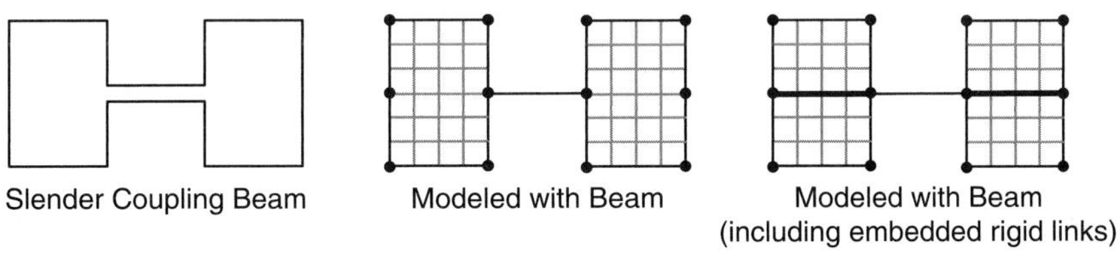

Figure 4.10 Modeling slender coupling beams.

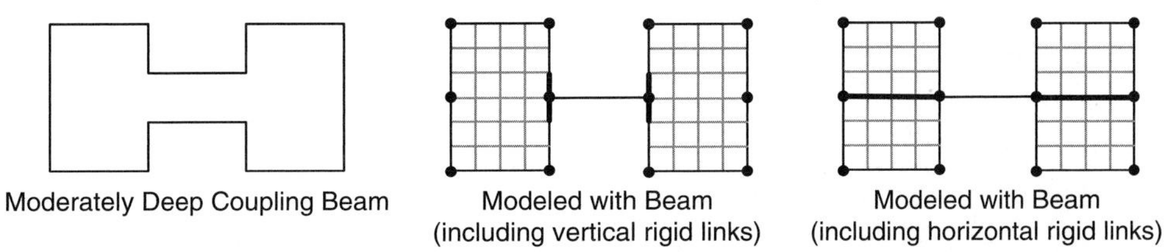

Figure 4.12 Modeling moderately deep coupling beams.

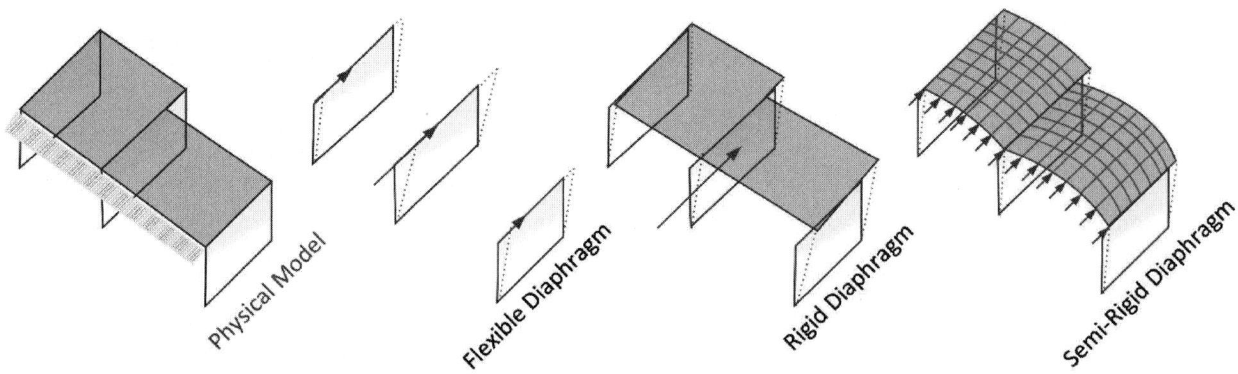

Figure 4.13 Impact of diaphragm type on loading, deformation, and lateral frame connection.

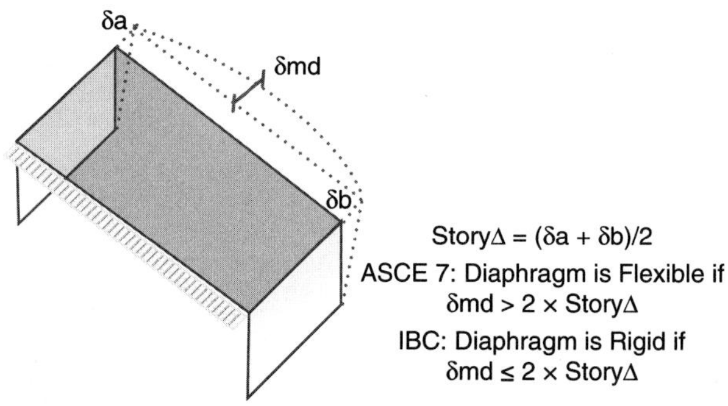

Figure 4.14 Analytical guidelines to determine diaphragm flexibility.

Building codes such as ASCE 7, IBC, and EN-1 EC 8 provide prescriptive and analytical guidelines to determine if a diaphragm should be considered flexible, rigid, or semirigid in a simulation (see Fig. 4.14).

Flexible Diaphragms

Bare steel deck and wood structural panel diaphragms can be assumed flexible where the LFRS is relatively stiff, such as in the case of brace frames, shear walls, or low-rise residential wood framing (ASCE 7). Flexible diaphragms are not explicitly part of the computer simulation. Lateral loads are applied directly to each disconnected lateral frame. This system is efficient from a computational speed perspective but does not capture any interaction between the lateral frame.

Rigid Diaphragms

Regular-shaped concrete slabs and concrete on metal deck with aspect ratios less than 3 (ASCE 7) can automatically be considered rigid. Rigid diaphragms are analyzed by condensing all DOFs to a single "node" with two translational and one rotational DOF perpendicular to the plane of the diaphragm (the master node). The displacements calculated at this single master node are extrapolated to calculate the displacement of all nodes (slaves) in the diaphragm. Each frame is rigidly coupled to each other and will resist lateral force proportional to its stiffness and the location where the loads are applied.

Any linear element connected to the diaphragm (i.e., beam, collector) does not incur any axial forces, as all nodes are rigidly connected to each other. If axial forces are of interest (collectors), nodes must be disconnected, at least at one end of the member, to allow axial deformation to occur.

Because all DOFs are condensed to a single node, rigid diaphragms can exist in only a single plane. Where more than one rigid diaphragm intersect one another, the common edge nodes should be alternately disconnected from each diaphragm, respectively. In some applications, a horizontal constraint may be acceptable even for sloped diaphragms if the slope is shallow and spreading components on diaphragm supporting members are not significant.

Semirigid Diaphragms

Semirigid diaphragms are explicitly considered in the analysis of the structure. The diaphragm material properties, including thickness, Poisson's ratio (typically 0.2–0.3), and modulus of elasticity, are required to determine the diaphragm in-plane stiffness. In addition, most analysis computer applications allow the engineer to specify modifiers or crack factors that reduce the axial, shear, and bending stiffness of the diaphragm. These factors allow the engineer to specify different properties in orthogonal diaphragm directions, such as for concrete planks or where concrete in the ribs of metal deck may be considered in one axis but not in the perpendicular.

Most analysis products allow the engineer to modify the size of the diaphragm mesh. An example of a coarse and a fine meshed diaphragm is given in Fig. 4.15.

Practical Insight: There is a trade-off between increased accuracy and slower computational speed as the mesh size decreases. A coarse mesh may be in the order of span/6 with a fine mesh around span/12. However, there is evidence (Adams 2017) that a relatively coarse mesh size of 4 to 8 feet (1.0–2.5 m) produces acceptable results for lateral force distribution and cross-section force determination. If cross-section forces are required through narrow areas of the diaphragm, then a finer mesh to ensure that at least four or more elements exist between the edges of the diaphragm is recommended as illustrated in Fig. 4.15. Also, where stresses in adjacent elements differ significantly (or stresses within element change substantially), a finer mesh may be desired.

Lateral loading is applied directly at the nodes on the diaphragm (Fig. 4.16). For wind loads, these are applied along the edges of the diaphragm nodes; for seismic loading, they are applied throughout the diaphragm where diaphragm mass exists.

Accidental eccentricity of lateral loading is a common requirement to account for uncertainty in the location of the floor mass or wind loads. Lateral loads at each node need to be modified to produce the desired resultant eccentricity.

Load transfer between the diaphragm and intersecting walls and other vertical elements needs to be considered carefully. The diaphragm must be "connected" to all vertical elements at coincident nodes, or continuity and load transfer is not ensured (see Fig. 4.17). Some analysis software provide a line or point constraint between intersecting elements to ensure continuity between intersecting elements.

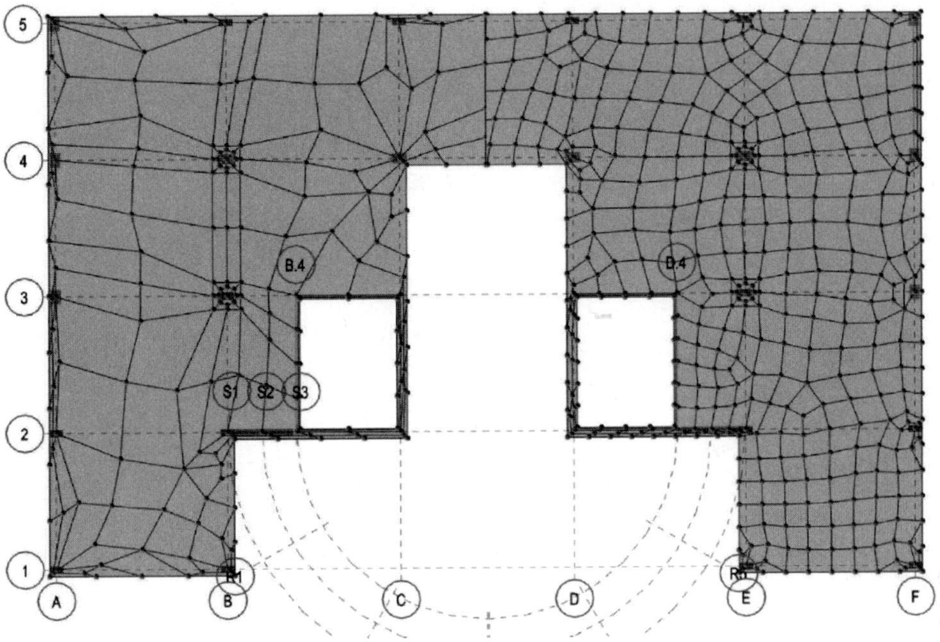

Figure 4.15 Impact of mesh density coarse on left, fine on right. (*RAM Structural System, Courtesy Bentley Systems.*)

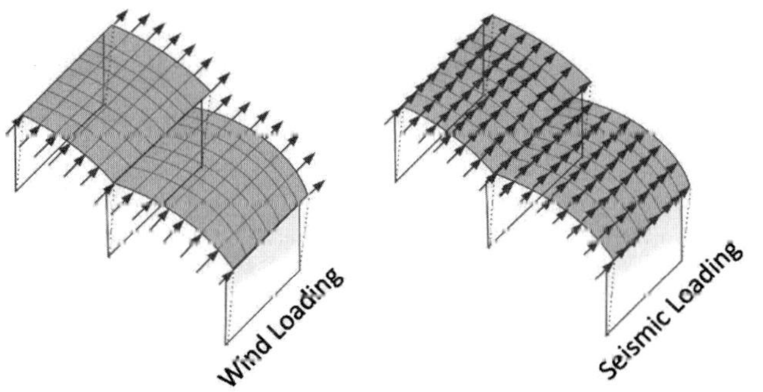

Figure 4.16 Lateral wind and seismic loading conditions.

4.4 FOUNDATIONS

The analysis of foundations should consider the interaction of superstructure, foundation, and soil. The introduction of soil springs at foundation boundary conditions is commonly used to simulate the impact of soil-structure interaction on the superstructure behavior. Figure 4.18 shows different types of structure foundations.

4.4.1 Isolated or Spread Footings

Isolated footings spread the supported loads of individual members through direct bearing on supporting soil. The foundation is typically assumed to be rigid to produce a uniform soil pressure profile on the soil. The soil-structure behavior can be idealized in the finite-element analysis with vertical, horizontal, and rotational springs (Fig.4.19).

The soil spring stiffness is calculated from the modulus of subgrade reaction K_s, which is a function of the foundation geometry, the soil properties, the load distribution, and the stiffness of the foundation (Terzaghi 1955). This elastic spring approach is valid under conditions of small displacements within the modulus elastic range.

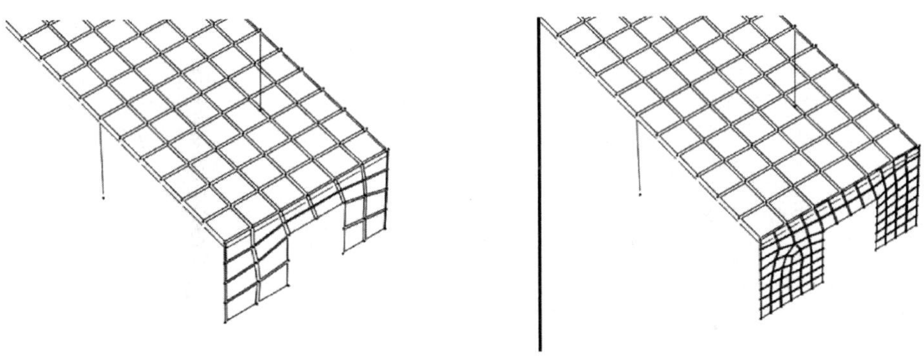

Figure 4.17 Compatible and incompatible intersecting surface meshes.

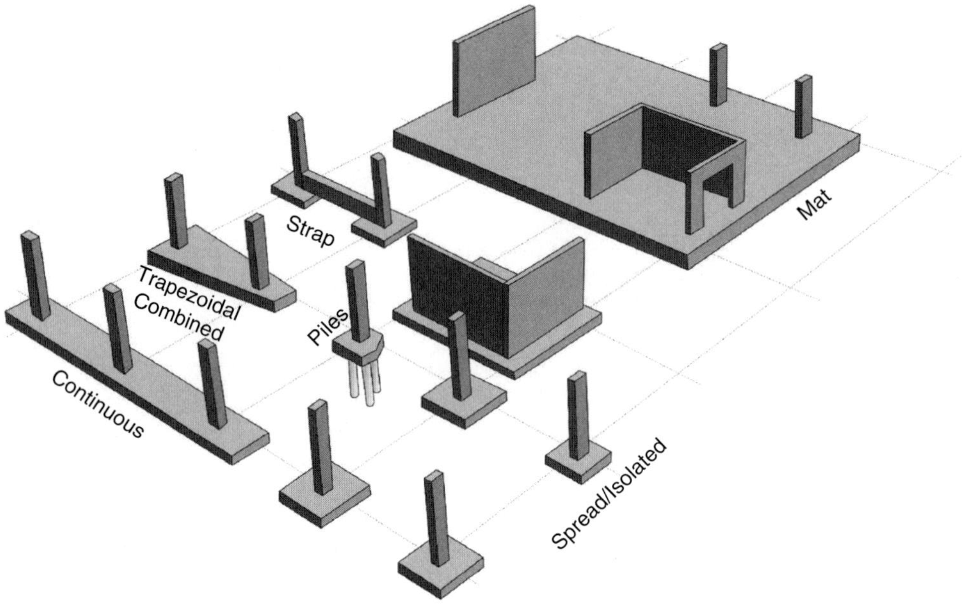

Figure 4.18 Typical foundation types.

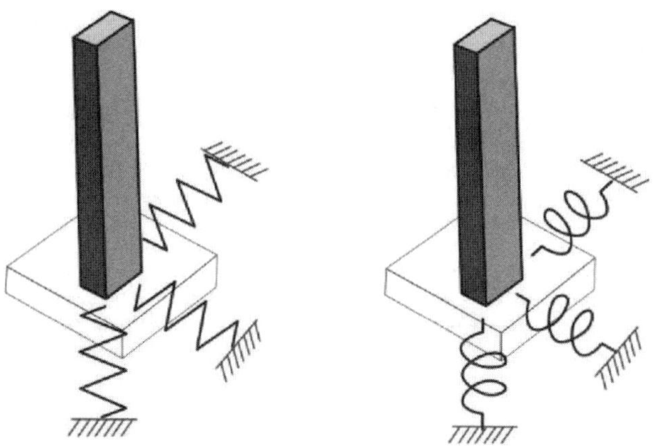

Figure 4.19 Translational and rotational springs at isolated footings.

The Taylor Method (Taylor 1967) is one method to calculate the rotational spring stiffness assuming a rigid footing. Bowles (1996), however, recommends that a foundation stiffness less than rigid be considered, for which Tettinek and Matl (1953) provide some guidance on calculating the foundation stiffness.

4.4.2 Combined and Continuous Footings

Combined footings are used where large nonuniform bearing pressure is likely to occur on an isolated footing, such as where a column is supported on the edge of a footing, perhaps where it is constrained to a property boundary. By engaging adjacent columns through a strap beam or thick foundation, the eccentric footing moment is distributed to produce a more uniform bearing pressure. Continuous footings typically carry one or more lateral load–carrying elements imposing large vertical and moment forces on the foundation (Fig. 4.20). The foundation continuity dissipates these forces to improve structural deformation and reduce soil-bearing pressure. The simulation of the behavior of combined and continuous footings is commonly achieved through the inclusion of the foundation as a series of beam finite elements, supported on an elastic foundation idealized as a series of vertical spring supports at each node (Winkler 1871, Hetenyi 1946).

The soil boundary spring stiffness values are calculated utilizing the modulus of subgrade reaction acting over the area tributary to each node.

Practical Insight: To more accurately capture the soil continuity and empirical evidence, it is common practice to double the spring stiffness at the ends of the footing (Newmark 1943, Bowles 1996). In the event that the foundation experiences uplift, an iterative analysis approach removing tension springs should be performed. At points of bearing stress concentration, such as at wall ends, it is advisable to use shorter finite elements and additional soil springs to avoid unrealistic bearing stress concentrations. However, keep adjacent finite-element lengths

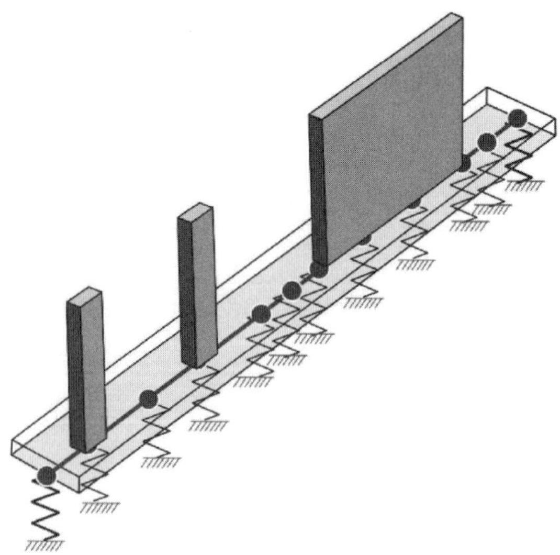

Figure 4.20 Beam on elastic foundation idealization of continuous footing.

within a ratio of no more than 3:1 in length (Bowles 1996). Where bearing stresses exceed capacity, the foundation thickness can be increased to distribute stresses more evenly along the foundation, or the width can be increased to reduce the bearing pressure by increasing the surface area. One limitation of the beam on elastic foundation approach is that it does not consider flexibility in the direction of the footing width that may reduce minor axis design moments.

4.4.3 Pile/Pier Foundations

A pile foundation is used where soil in lower strata needs to be engaged to support the structure. Piles resist vertical loads through skin friction and/or end bearing. Lateral loads are resisted through passive soil resistance on the piles and pile caps and through bending resistance of piles rigidly attached to the pile cap. The lateral resistance of soil on the piles is highly nonlinear and requires calculation of p-y soil resistance values to be used in the computer simulation (Fig. 4.21). Piles are typically analyzed in isolation for each load condition considering the given pile size and soil properties along its length. The number, size, and location of a pile in a pile group and the direction of loading will reduce the p-y resistance of each pile (Rollins and Olson 2006). It is common practice to average the reduction factor for all piles in a group for each load condition, to be applied to p-y values for the computer simulation (Valley M, FEMA P-751, 2012).

Forces at the top of each pile are calculated by assuming that the superstructure shear and moment reaction are applied at the top of the pile cap and distributed to each pile through the rigid cap.

Practical Insight: It is common to assume a fixed base boundary condition at the location of pile caps in a structural analysis. This assumption can be unconservative in seismic condition where resonance between the dynamic response of the foundation and structure would amplify the structural response or where differential foundation displacements may result. Some code standards (ASCE 41) now prohibit the fixed base assumption in certain structures. In these situations, an equivalent translational and rotational stiffness from the pile group can be applied to the superstructure analysis. The nonlinear foundation stiffness suggests bounding the superstructure analysis with the extreme foundation spring stiffnesses as appropriate.

4.4.4 Mat/Raft Foundations

A mat foundation is commonly used in industrial applications where equipment loads are large or the location of loads may change, in buildings where loads are large or have deep basements (particularly if

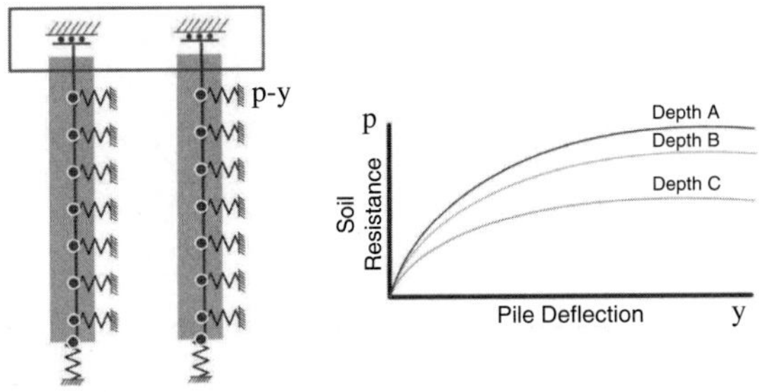

Figure 4.21 Pile cap and pile analysis model and p-y curves.

submerged below the water table), or where soil conditions are inconsistent or volatile (as in clay heave) and likely to result in differential settlements in uncoupled foundations. A mat foundation may be supported on piles where settlement or soil capacity warrants.

To simulate the behavior of mat foundations, a slab on elastic foundation finite-element approach is utilized. A two-dimensional plate finite element is used with behavior based on the slab thickness, material properties, and geometry. The soil is idealized with soil springs at each finite element node representative of the contributory surface area to the node (Fig. 4.22).

Practical Insight: Modeling a constant modulus of subgrade reaction over the area of the foundation is inaccurate and potentially unconservative. Various authors (Aristorenas and Gomez 2014, Bowles 1996) suggest approaches to simulate the larger stiffness at the edge of the foundation. It is recommended that the engineer consult with a geotechnical engineer to determine the appropriate subgrade modulus to use at different zones in the foundation.

If the mat experiences undesired uplift (or bearing stress concentrations), it is recommended to first decrease the mesh size [fine mesh can be considered to approximately span 12 in. (30 cm) in size], solve the analysis interactively by removing springs in tension, or increase the slab thickness to engage additional weight. Structural design is performed along middle and column strips to more appropriately address stress concentrations at any one point. By first setting line constraints at the edges of these design strips, the finite-element mesh will produce a higher-fidelity mesh to facilitate force integration as illustrated in Fig. 4.23.

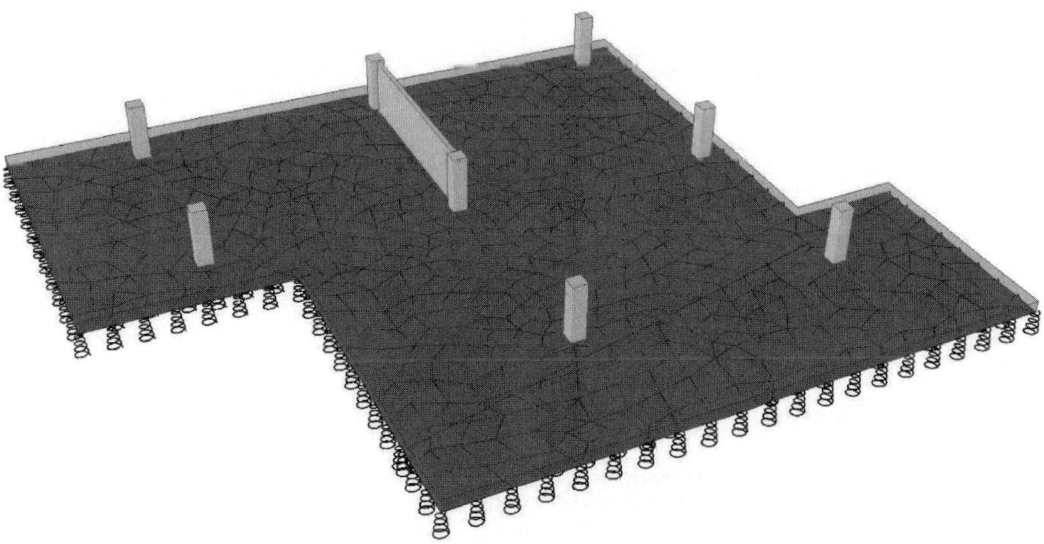

Figure 4.22 Mat foundation with soil springs. (*Image RAM Concept courtesy Bentley Systems.*)

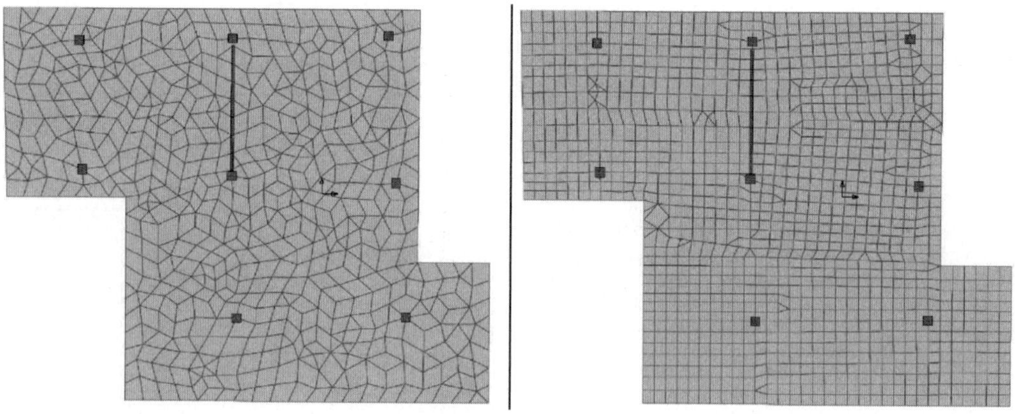

Figure 4.23 Mesh integrity before and after defining design strips. (*RAM Concept, Courtesy Bentley Systems.*)

ADVANCED SOIL-STRUCTURE ANALYSIS

With advances in computation capacity, the behavior of soil can now explicitly be included in the simulation of the structural behavior. This can be achieved through the idealization of the soil as solid (three-dimensional) finite elements or through emulation with a series of nonlinear springs and dampers (NIST GCR 12-917-21 2012, NEHRP 2009). These analyses approaches are not yet widely used in the industry as of this writing.

4.5 VERIFYING ANALYSIS RESULTS

A critical step in utilizing computer applications is verifying the results prior to applying design checks to the software results.

Often, computer applications are used to generate loads for a structure based on the general project location and the geometry of the building project. Determining the overall load outside the software and comparing the nodal reaction for critical load cases is an important and simple way to verify that the structure has the proper magnitude of load.

As stated previously, finite-element analysis is based on nodal displacements; therefore, a key to verifying the results is determining the anticipated deflected shape and then comparing the software results to that shape. Based on anticipated lateral strength, the displaced shape in translation and rotation should be able to be anticipated. This is a qualitative check more than a quantitative check but is no less important.

In terms of dynamic analysis, a good check on the validity of the computer results can start with code approximations based on the size of the building and structural system.

Finally, engineers must determine whether the software application checks the capacity of all the results. Some of the results that are stored within results data are checked against code-determined capacity equations, and other results are unchecked. This is generally not an oversight in the computer application but simply a limitation of the program. Much care must be used to determine whether the unchecked residual results are significant and need to be checked outside the computer application. A common example of unchecked residual results is out-of-plane horizontal bending in finite-element plates used to represent wall elements. Often (but not always), in-plane shear forces or vertical out-of-plane bending forces are checked against code-determined capacities.

4.6 BUILDING INFORMATION MODELING AND INTEROPERABILITY

Project success and productivity dictates close collaboration and information sharing among many organizations and domain specialists. Building information modeling (BIM) has rapidly become one of the most widely used approaches to achieve this collaboration. BIM is founded on the notion of a common data environment (CDE) through which disparate parties can share information on a single project. In the past decade, over 70 percent of design firms in North America have reported some level of BIM adoption (SmartMarket 2012), and globally, the World Economic Forum (2018) reports one-third of infrastructure and urban development organizations using a CDE on projects (World Economic Forum, 2018).

The key drivers of BIM and interoperability (SmartMarket 2012) in the engineering field are the following:

Productivity—Automate drawing production, visual communication, and collaboration through interoperability. Schedules have been shown to be improved by up to 30 percent (Parvan, 2012).

Risk Reduction—Reduce data reentry risk through interoperability, improved version control, and geometric accuracy.

Value Engineering—Cost estimation, clash detection, scheduling, and cost reduction.

INTEROPERABILITY THE CATALYST FOR BIM SUCCESS

The glue that enables collaboration on BIM and a CDE is interoperability. It used to be that a common drawing format was the lingua franca of collaboration; today there are many BIM environments with both proprietary and open standard forms of interoperability. Most software vendors offer an application programmers interface (API) to allow information sharing directly with the solution (Bentley Systems—OpenBuildings Designer™, Autodesk–Revit™, GRAPHISOFT–ARCHICAD™, Nemetschek–Allplan™, and others). Digital twins, allowing for immersive cloud visualization, visual change tracking and status, and data insights, are furthering the promise of BIM from design operations (Bentley iTwin, Nemetshek Bimplus). Maintaining interoperability with each unique API is costly for BIM ecosystem software providers. Open standards are a more cost-effective solution where multiple BIM providers support the same standard. Some of the most common standards include buildingSMART–IFC (Industry Foundation Classes), NIBS–COBIE (Construction Operations Building Information Exchange), and CIS/2 (CIMSteel Integration Standards for steel detailing). Engineers should be aware of which interoperability format they will be interoperating with when developing their analytical models. With open standards, engineers need to determine the IFC model view definition, which defines a specific set of exchange capabilities, and the level of development (LOD) that defines the level of detailed information to be shared (typically at LOD 300 for analytical work flows—BIMFORUM 2017) to be used on the project. These requirements of interoperability may require that a higher-fidelity physical-analytical modeling (Karp and Taylor 2017) be developed by the engineer.

4.7 SUMMARY

Advances in software and computational power over the past 50 years have enabled engineers to meet the increasing global demand for more resilient, innovative, and economical structures. Advances in the finite-element method have allowed for better understanding of complex structural behavior through mathematically idealized linear and nonlinear behavior.

Technology continues to evolve at a rapid pace. New advancements in hardware, software, and technology, including cloud computing and machine learning, will continue to improve the engineer's ability to simulate more complex structures, loadings, and design alternatives. It is important to remember, though, that analytical models created within computer applications are approximations of real structures. Engineers must strive to improve these models and validate the results to understand the structures and true behavior and to provide safe and efficient designs.

REFERENCES

Adams, Leonard Allen. 2017. "Modeling and Analysis of Diaphragms." Bentley Systems, Inc.

AISC 360-16. "Specification for Structural Steel Buildings," American Institute of Steel Construction.

Alemdar, Bulent N. 2001. "Distributed Plasticity Analysis of Steel Building Structural Systems." PhD thesis, Georgia Institute of Technology, Atlanta.

Aristorenas, G., and Gomez, J. 2014. "Subgrade Modulus Revisited." *Structure*.

ASCE 7-16. "Minimum Design Loads for Buildings and Other Structures." American Society of Civil Engineers.

ACI Committee 318-14. "Building Code Requirement for Structural Concrete." American Concrete Institute.

ASCE 41-17. "Seismic Evaluation and Retrofit of Existing Buildings," American Society of Civil Engineers.

Bathe, Klaus-Jurgen. 1996. *Finite Element Procedures*. Prentice Hall.

Bowles, J. 1996. *Foundation Analysis and Design*. McGraw-Hill.

Charney, Finley A., and Downs, W. D. 2004. "Modeling Procedures for Panel Zone Deformations in Moment Resisting Frames." Connections in Steel Structures V—Amsterdam, June 3–4.

Clough, R. W. 1960. "The Finite Element Method in Plane Stress Analysis." Proc. 2nd ASCE Conf. on Electronic Computation, Pittsburgh, PA.

Cook, R. D., Malkus, D. S., and Plesha, M. E. 1989. *Concepts and Applications of Finite Element Analysis*. 3rd ed. John Wiley & Sons.

Parvan, K. 2012. "Estimating the Impact of BIM Utilization on Building Project Performance." Dissertation, Department Civil and Environmental Engineering University of Maryland.

EN 1998-1:2013 EC 8. "Design of Structures for Earthquake Resistance—Part 1: General Rules." Seismic Actions and Rules for Buildings.

IBC-2015. "International Building Code," International Code Council.

Hetenyi, M. 1946. *Beams on Elastic Foundation*. University of Michigan Press.

Krawinkler, H. 1978. "Shear in Beam-Column Joints in Seismic Design of Steel Frames." *Engineering Journal*, Third Quarter, 82–91.

Karp R., and Taylor J. 2017. "Analytical-Physical Modeling and Interoperability Propels BIM Advancement." Bentley Systems, Inc.

Liew, J. Y., White, D. W., and Chen, W. F. 1993. "Second-Order Refined Plastic Hinge Analysis for Frame Design: Part I." *Journal of Structural Engineering*, Vol. 119, No. 11, 3196–3216.

McGuire, W. Gallagher, R. H., and Ziemian, R. D. 2000. *Matrix Structural Analysis*. 2nd ed. Wiley.

Newmark, N. M. 1943. "Numerical Procedure for Computing Deflections, Moments and Buckling Loads." *Transactions of the American Society of Civil Engineers*, Vol. 108, 1161–1234.

NIST GCR 12-917-21. 2012. "Soil-Structure Interaction for Building Structures." Grant/Contract Reports (NISTGCR)-12-917-21.

NEHRP. 2009. "Recommended Provisions for Seismic Design of New and Other Structures," Part 2. FEMA P-750.

Parvan, Kiavash. 2012. "Estimating the Impact of BIM Utilization on Building Project Performance," dissertation, University of Maryland.

Rollins, Kyle, and Olsen, G. 2006. "Pile Spacing Effects on Lateral Pile Group Behavior," *Journal of Geotechnical and Geoenvironmental Engineering*, Vol. 132, Issue 10.

SmartMarket. 2012. *The Business Value of BIM in North America*. McGraw-Hill.

Taylor, P. W. 1967. "Design of Spread Footings for Earthquake Loading." Proc. 5th Australia-New Zealand Conference.

Teh, L., and Clarke, J. M. 1998. "Plastic-Zone Analysis of 3D Steel Frames Using Beam Elements." *Journal of Structural Engineering*, Vol. 125, No. 11.

Terzaghi, K. 1955. "Evaluation of Coefficients of Subgrade Reaction." *Geotechnique*, Vol. 5, No. 4.

Tettinek, W., and Matl, F. 1953. "A Contribution to Calculating the Inclination of Eccentrically Loaded Foundations." 3rd ICSFME, Vol. 1.

Valley, Michael. 2012. "2009 NEHRP Recommended Seismic Provisions", FEMA P-751, 2012.

White, D. W., Liew, J. Y. R., and Chen, W. F. 1991. "Second-Order Inelastic Analysis for Frame Design." Rep. No., CE-STR-91-12, School of Civil Engineering, Purdue University, West Lafayette, IN.

Winkler E. 1871. "Der Eisnbahn-Oberbau." Dominicus Prag.

World Economic Forum. 2018. "An Action Plan to Accelerate Building Information Modeling (BIM) Adoption." Boston Consulting Group.

Ziemian, R. D., and McGuire, W. 2002. "Modified Tangent Modulus Approach: A Contribution to Plastic Hinge Analysis." *Journal of Structural Engineering*, Vol. 128, No. 10, 1301–1307.

Chapter 5
Earthquake-Resistant Design

BY
S. K. GHOSH *S. K. Ghosh Associates LLC, Palatine, Illinois*

5.1 OVERVIEW

The purpose of this chapter is to assist in the seismic design of structures using the provisions of the 2018 edition of the *International Building Code* (IBC) [5.1]. The legal building codes of most local jurisdictions (cities, counties, states) in the United States are currently based on one edition or another of the IBC.

The IBC is issued by the International Code Council (ICC). There were three former model code bodies: Building Officials and Code Administrators International Inc. (BOCA), publishers of *The BOCA National Building Code* (BOCA/NBC); International Conference of Building Officials (ICBO), publishers of the *Uniform Building Code* (UBC); and the Southern Building Code Congress International Inc. (SBCCI), publishers of the *Standard Building Code* (SBC), which merged to form the ICC. The BOCA/NBC, mostly used in the northeastern quarter of the country; the UBC, mostly adopted in the western half of the country; and the SBC, mostly used in the southeastern quarter of the country, are now referred to as the *legacy* codes. The first edition of the IBC, dated 2000, was the culmination of years of cooperative effort by committees representing the three model code bodies and other organizations. Following the 2000 IBC, a new IBC edition has been published at regular 3-year intervals, the 2018 IBC being the last one as of this writing.

Chapter 16 of the IBC addresses the design requirements for various types of loads and design load combinations. The ASCE Standard *Minimum Design Loads and Associated Criteria for Buildings and Other Structures* [5.2] is the basis of all provisions related to nonseismic forces. The seismic design provisions of the 2018 IBC are also those of ASCE 7-16, which in turn used the 2015 *NEHRP Provisions* [5.3] as a resource document.

Throughout this chapter, section numbers from the 2018 IBC are referenced; Section 1613 of the 2018 IBC is cited as IBC 1613. Similarly, section references from ASCE 7-16 appear, for example, as ASCE 6.5.

5.2 NATURE OF EARTHQUAKE MOTION

Ground motion resulting from earthquakes presents unique challenges to the design of structures. The forces that a structure must resist in an earthquake result directly from the distortions caused by the motion of the ground that supports it. The response—magnitude and distribution of forces and displacements—of a structure resulting from such ground motion is influenced by the properties of the structure and its foundation as well as the character of the exciting motion.

Earthquakes produce large-magnitude forces of short duration that must be resisted by a structure without causing collapse and preferably without significant damage to the structural members. Lessons from past earthquakes and research have provided technical solutions that will minimize loss of life and property damage associated with earthquakes. For materials such as concrete that lack inherent inelastic deformability or ductility, a critical part of the solution is to provide special detailing of the reinforcement to ensure a ductile response to lateral forces. Inelastic deformability is the ability of a structure to sustain gravity loads as it deforms laterally beyond the stage where the deformations are recoverable, that is, beyond the stage where no residual deformations remain in a structure once the earthquake motion subsides. Irrecoverable deformations, also called residual deformations, are associated with damage, while recoverable deformations cause no damage.

Figure 5.1 illustrates a simplified representation of a building during an earthquake. As the ground on which the building rests is displaced, the base of the building moves with it. However, the inertia of the building mass resists this motion and causes the building to suffer a distortion (greatly exaggerated in the figure). This distortion wave travels along the height of the structure. The continued shaking of the base causes the building to undergo a complex series of oscillations.

It is important to draw a distinction between forces due to wind and those produced by earthquakes. These forces are often thought of as being similar just because codes specify design wind as well as earthquake forces in terms of equivalent static forces. Although both wind and earthquake forces are dynamic in character (meaning they vary with time), a basic difference exists in the manner in which they are induced in a structure. Whereas wind loads are externally applied loads and hence are proportional to the exposed surface of a structure, earthquake forces are essentially inertia forces. The latter result from the distortion produced by both the earthquake motion and the inertial resistance of the structure. Their magnitude is a function of the mass of the structure rather than its exposed surface. Also, in contrast to structural response to essentially static gravity loading or even to wind loads, which can often be validly treated as static loads, the dynamic character of the response to earthquake excitation can seldom be ignored.

The earthquake ground motion quantity most commonly used in analytical studies is the timewise variation of the ground acceleration in the immediate vicinity of a structure. At any point, the ground acceleration may be described by horizontal components along two perpendicular directions and a vertical component. In addition, rocking and twisting (rotational) components may be present; however, these

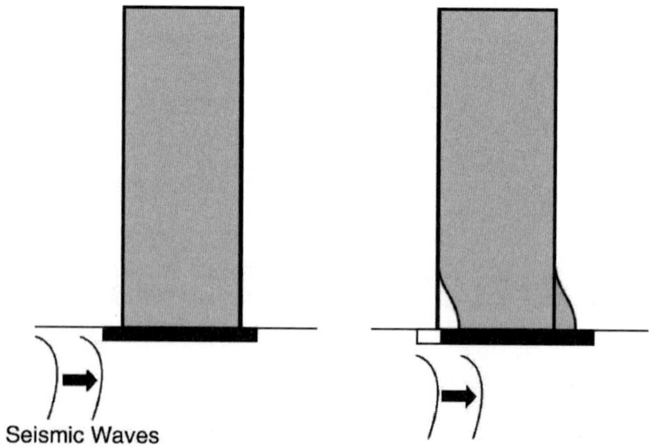

Figure 5.1 Simplified representation of building behavior during an earthquake.

are usually negligible. Because buildings and most other structures are most sensitive to horizontal or lateral distortions, it has been the practice in most instances to consider structural response to the horizontal components of ground motion only. The effects of the vertical components of ground motion were until relatively recently not considered significant enough to merit special attention except as they influence gravity loads. In many instances, a further simplification of the actual three-dimensional response of structures is made by assuming the horizontal acceleration components to act nonconcurrently in the direction of each principal plan axis of a building. (ASCE 7-16 indicates quite specifically when this is not appropriate anymore.) It is implicitly assumed that a building designed by this approach will have adequate resistance against the resultant acceleration acting in any direction.

The complete system of inertia forces in a structure can be determined only by evaluating the acceleration of every mass particle. The analysis can be greatly simplified if the deflections of the structure can be specified adequately by a limited number of displacement components or ordinates. The number of displacement components required to specify the positions of all significant mass particles in a structure is called the number of degrees of freedom of the structure. In the so-called lumped-mass idealization, the mass of the structure is assumed concentrated at a number of discrete locations. Because the floor and roof elements (diaphragms) in a building are relatively heavy, a large proportion of the building mass is concentrated in these elements. For structural analysis purposes, the mass of other building components, such as walls and columns, and that associated with the superimposed dead loads are normally assumed concentrated at the floor and roof levels.

A multi–degree-of-freedom system possesses as many natural modes of vibration as there are degrees of freedom. The distinguishing feature of a mode of vibration is that a dynamic system can, under certain circumstances, vibrate in that mode alone. During such vibration, the ratio of the displacements of any two masses remains constant with time. These ratios define the characteristic shape of the mode; the absolute amplitude of motion is arbitrary. Each natural mode shape m has a natural period of vibration T_m associated with it. The period T_m is the time required for one cycle of motion in the deflected shape characteristic of the mth natural mode of vibration. The term *natural* is used to qualify each vibration quantity to emphasize the fact that these are natural properties of the structure, depending on its stiffness and mass, when it is allowed to vibrate freely without any external excitation. An idealized n-story building possesses n natural periods of vibration $(T_1, T_2, \ldots, T_n)$ arranged from the largest to the smallest, corresponding to natural modes 1 through n. The longest period is that of the first, or the fundamental, mode and is designated T_1. Figure 5.2 shows that any arbitrary displaced shape of a structure may be expressed in terms of the amplitudes of the mode shapes. The equation of motion of any mode m of a multidegree system is equivalent to the equation of motion for a single-degree of freedom system.

Modal periods are functions of the dynamic properties of mass, stiffness, and, beyond the stage of elastic response, also strength. The seismic response of short, stiff buildings is dominated by the first, or fundamental, mode of response; most of the building mass vibrates in that mode. The higher modes of vibration contribute significantly to seismic response only when a building is relatively tall and flexible or contains geometric horizontal or vertical irregularities.

5.3 DESIGN PHILOSOPHY

When a structure responds elastically to ground motion during a severe earthquake, the maximum response accelerations may be several times the maximum ground acceleration. It is generally uneconomical and unnecessary to design a structure to respond in the elastic range to the maximum earthquake-induced inertia forces. Thus, the design seismic horizontal forces prescribed in ASCE 7-16 [5.2], as well as forces prescribed in prior model building codes, standards, and resource documents, are generally less than the elastic response inertia forces induced by the design earthquake.

It is expected that structures would undergo fairly large deformations when subjected to a major earthquake. The intent is that these large deformations will be accompanied by yielding in some of the members of the structure. It should be evident that the use of code-prescribed

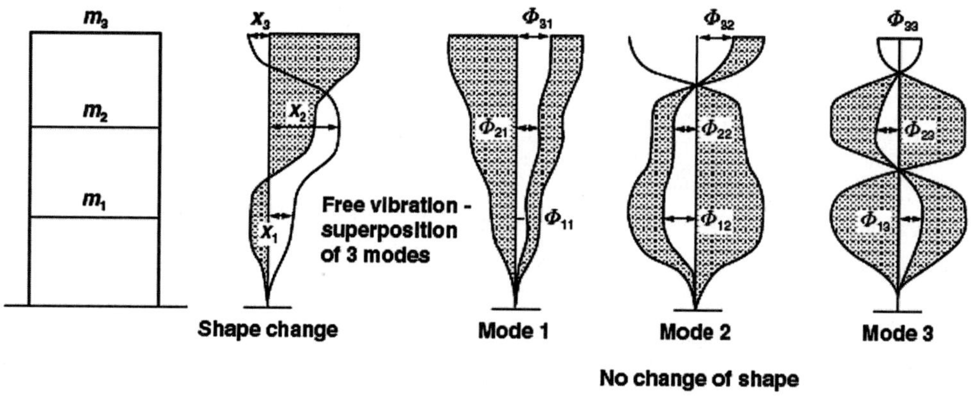

Figure 5.2 Mode superposition analysis of earthquake response.

seismic design forces implies that critical regions of certain members should have sufficient inelastic deformability to enable the structure to survive without collapse when subjected to several cycles of deformations well into the inelastic range. This means avoiding all forms of brittle failure and achieving adequate inelastic deformation via the yielding of localized regions of certain members or connections in the structure. These regions are often called plastic hinges. This explains why the material chapters of the IBC contain detailing and other requirements that go hand in hand with the code-prescribed seismic forces. For concrete structures, satisfying the design and detailing requirements in Chapter 19 of the 2018 IBC, which are essentially the requirements in Chapter 18 of ACI 318-14 [5.4] for moderate to high seismic design categories, provides the levels of inelastic deformability that are required by the design force levels of the IBC/ASCE 7.

Experience from recent earthquakes has shown that structures designed to the level of seismic forces prescribed by codes can survive major earthquake shaking. This is mainly due to the ability of well-designed structures to dissipate seismic energy by inelastic deformations in particular regions of certain members in the structure. Decrease in structural stiffness caused by accumulating damage and soil-structure interaction also helps at times.

Figure 5.3 shows the idealized force-displacement relationship of a structure subjected to the design earthquake of the 2018 IBC and ASCE 7-16, as defined in IBC 202 and ASCE 11.2. On the horizontal axis are the earthquake-induced displacements. The quantity V along the vertical axis is the code notation for design base shear, a global force quantity. The curve in the figure may be thought of as the envelope or the backbone curve of hysteretic force-displacement loops that describe the response of a structure subjected to reversed cyclic displacement histories of the type imposed by earthquake ground motion.

The base shear V is to be distributed along the height of the structure as required by ASCE 12.8.3 (discussed later). The distribution results in a series of lateral forces concentrated at the various floor levels. Next, a mathematical model of the structure is to be elastically analyzed under these lateral forces. The quantity δ_{xe} represents the lateral displacement at floor level x obtained from this analysis, and Q_E represents the member forces (bending moments, shear forces, axial forces, etc.). This procedure is called the equivalent lateral force procedure.

As the structure responds inelastically to the design earthquake defined in IBC 202 and ASCE 11.2, the lateral displacement at floor level x increases from δ_{xe} to $C_d\delta_{xe}$, and the member forces increase from Q_E to $\Omega_o Q_E$. Both the deflection amplification factor C_d and the overstrength factor Ω_o depend on the structural system used for earthquake resistance and are given in ASCE Table 12.2-1. Quantities V_E and δ_{xe} are the base shear and the lateral displacement at floor level x, respectively, corresponding to the hypothetical elastic response of the structure to the design earthquake defined in IBC 202 and ASCE 11.2. Figure 5.3 suggests that a response modification factor R of 2 used in design would result in an essentially elastic response of a structure to the design earthquake. The basis for this is explained in [5.5].

5.4 SEISMIC DESIGN REQUIREMENTS OF THE 2018 IBC/ASCE 7-16

The seismic design provisions in the 2018 IBC are contained in Section 1613. This section simply references ASCE 7-16 and does not contain much text. Section 1613 still retains as much information as is needed for the establishment of the seismic design category (SDC) of a structure. (Note: A discussion on how to determine the SDC of a structure is given later in this chapter.) These provisions are reproduced from ASCE 7-16. Provisions beyond that are to be found only in ASCE 7-16. Adopted specifically by reference by the 2018 IBC are ASCE 7-16 Chapters 11, Seismic Design Criteria; 12, Seismic Design Requirements for Building Structures; 13, Seismic Design Requirements for Nonstructural Components; 15, Seismic Design Requirements for Nonbuilding Structures; 17, Seismic Design Requirements for Seismically Isolated Structures; and 18, Seismic Design Requirements for Structures with Damping Systems. Chapter 14 of ASCE 7-16, Material-Specific Seismic Design and Detailing Requirements, adopts materials standards such as ACI 318 and makes certain modifications to them. By not adopting Chapter 14 of ASCE 7-16, the 2018 IBC requires that concrete design be done by Chapter 19 of the 2018 IBC and not by Chapter 14 of ASCE 7-16. By the same token, steel design must be done by Chapter 22 of the 2018 IBC and not by Chapter 14 of ASCE 7-16. Note, however, that Chapter 19 does specifically require precast concrete diaphragms in buildings assigned to SDC C, D, E, or F to be designed by ASCE 7 16 Section 14.2.4.

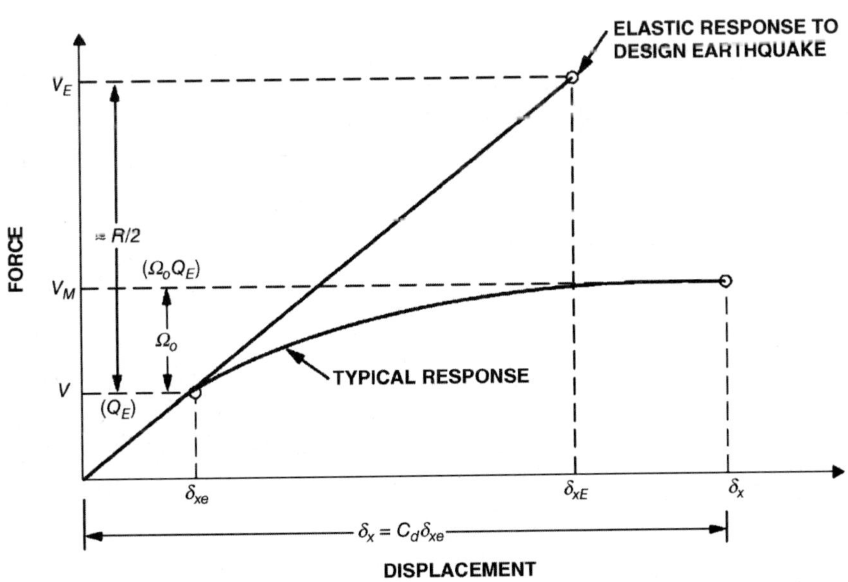

Figure 5.3 Idealized force-displacement relationship of a structure subjected to the design earthquake of the 2018 IBC or ASCE 7-16.

The ASCE 7-16 seismic chapters that are not *specifically* adopted by the 2018 IBC are 16, Nonlinear Response History Analysis; 19, Soil-Structure Interaction for Seismic Design; 20, Site Classification Procedures for Seismic Design; 21, Site-Specific Ground Motion Procedures for Seismic Design; 22, Seismic Ground Motion, Long Period Transition, and Risk Coefficient Maps; and 23, Seismic Design Reference Documents. Chapter 23 is not adopted because the IBC has its own Chapter 35 where all referenced documents are listed alphabetically by the promulgating organization. Some of the chapters, such as 16 and 21, are referenced from the other chapters, such as 12, that are specifically adopted.

5.4.1 General Requirements

IBC 1613.1 and ASCE 11.1.2 require that every structure and its components be designed for the effects of earthquake motions. However, these sections do permit a number of important exceptions. The IBC exceptions are the following:

1. Detached one- and two-family dwellings in Seismic Design Categories (SDC) A, B, and C or located where the mapped short-period spectral response acceleration S_S (discussed later) is less than 0.4*g* are totally exempt from all seismic design requirements.
2. The seismic force–resisting system of wood frame buildings that conform to the provisions of IBC 2308 (conventional light frame construction) are not required to be analyzed as specified in IBC 1613. It is felt that the requirements of IBC 2308 are adequate to provide safety based on the history of such light-frame construction—invariably low-rise structures—in earthquakes. There is a similar exemption in ASCE 7-16, which reads, Detached one- and two-family wood-frame dwellings not included in Exemption 1 with not more than two stories above grade plane, satisfying the limitations of and constructed in accordance with the IRC (the *International Residential Code*) [5.6].
3. Agricultural storage structures intended only for incidental human occupancy are exempt from all seismic design requirements because of the exceptionally low risk to life involved.
4. Structures that require special consideration of their response characteristics and environment that are not addressed by the IBC or ASCE 7 and for which other regulations provide seismic criteria, such as vehicular bridges, electrical transmission towers, hydraulic structures, buried utility lines and their appurtenances, and nuclear reactors.

ASCE 7-16 has an additional exemption for piers and wharves that are not accessible to the general public.

ASCE 11.4.2 provides that structures located where the mapped spectral response acceleration at short periods (discussed later), $S_S \le 0.15g$, and the mapped spectral response acceleration at 1-second period (also discussed later), $S_1 \le 0.04g$, need only comply with ASCE 11.7. Areas with $S_S \le 0.15g$ and $S_1 \le 0.04g$ are darkened on the map of the United States in Fig. 5.4. The darkened areas are, in a sense, analogous to Zone 0 of the 1997 UBC [5.7].

IBC 101.4.7 contains the following requirement for existing buildings: The provisions of the *International Existing Building Code* (IEBC) [5.8] shall apply to matters governing the repair, alteration, change of occupancy, addition to and relocation of existing buildings.

IBC 1604.9 is also important: This section requires that appropriate seismic detailing requirements and limitations are to be satisfied even where wind forces govern the design.

Structures in SDC A need only comply with ASCE 11.7. For structures exposed to such a low seismic risk, that section refers to ASCE 1.4, which requires only that a complete lateral force–resisting system be provided and that all elements of the structure be tied together. The lateral force–resisting system must be proportioned to resist a lateral force at every floor level equal to 1 percent of the weight at that floor level as depicted in Fig. 5.5 (ASCE 1.4.2).

5.4.2 Site Ground Motion

The IBC, before its 2012 edition, and ASCE 7, before its 2010 edition, used to define a maximum considered earthquake (MCE) ground motion, which, for most of the United States, was an earthquake with an approximately 2500-year return period (2 percent probability of exceedance in 50 years). However, in coastal California, the MCE was the largest earthquake that can be generated by the known seismic sources (deterministic earthquake). In ASCE 7-10, this uniform-hazard ground motion (outside of deterministic areas) was replaced by risk-targeted ground motion. The MCE ground motion, corresponding to a 2 percent in 50-year hazard level, was replaced by MCE_R ground motion, corresponding to a 1 percent in 50-year collapse risk target. Two other significant changes were made in mapped ground motion in ASCE 7-10. A switch was made from "geo-mean" ground motion (square root of the product of ground motions recorded in any two orthogonal directions) to maximum direction ground motion or ground motion in a direction that causes maximum structural response. Also, deterministic ground motion was changed from mean plus one standard deviation or 150 percent of median

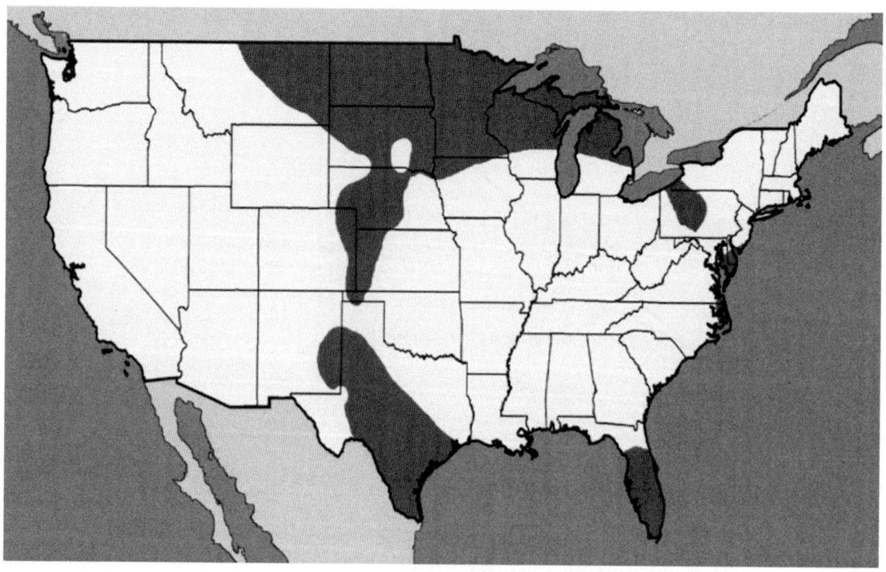

Figure 5.4 Areas of the United States with $S_S \le 0.15g$ and $S_1 \le 0.04g$.

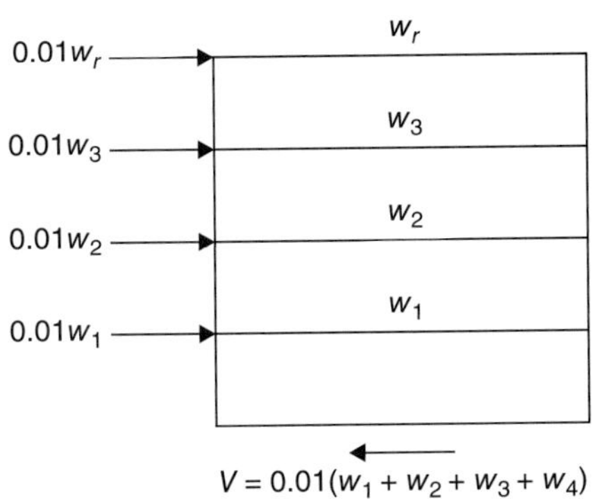

Figure 5.5 Design seismic force distribution for structures assigned to SDC A.

ground motion to mean plus two standard deviations or 180 percent of median ground motion.

IBC Figures 1613.2.1 (1) and 1613.2.1 (2) and ASCE Figures 22-1 and 22-2 contain contour maps of the conterminous United States giving S_S and S_1, respectively. The quantities S_S and S_1 are the mapped MCE_R spectral response accelerations at periods of 0.2 seconds and 1.0 seconds, respectively, at Site Class B-C interface. Prior to ASCE 7-16, the mapped quantities used to be for Site Class B. For the concept of the spectral acceleration, refer to [5.9]. Spectral acceleration is directly related to base shear (base shear equals spectral acceleration times mass or spectral acceleration divided by g times weight, where g is the acceleration due to gravity). IBC Figures 1613.2.1(3) through 1613.2.1(8) and ASCE Figures 22.3 through 22.8 contain similar contour maps for Hawaii, Alaska, Puerto Rico, Guam, and American Samoa.

Site Classes B and C are two of six site classes defined by ASCE Table 20.3-1. Site classification is based on one of three measured properties of the soils in the top 100 feet of the site: shear wave velocity $\bar{v}_s$, standard penetration resistance (or blow count) $\bar{N}$ (ASTM D 1586-99 [5.10]), or undrained shear strength $\bar{s}_u$ (ASTM D 2166-00 [5.11] or D 2850-039 [5.12]). Where site-specific data are not available to a depth of 100 feet, appropriate soil properties are permitted to be estimated by the registered design professional preparing the soils report based on known geologic conditions.

When the soil properties are not known in sufficient detail to determine the site class, ASCE 7 editions through 2010 required Site Class D to be used unless the building official had determined that Site Class E or F soil was likely to be present at the site. ASCE 7-16 retains this provision but additionally requires that, where Site Class D is selected as the default site class per ASCE 11.4.3, the acceleration-related site coefficient F_a (see below) be not less than 1.2. This, in effect, makes C the default site class in areas of strong shaking.

At the beginning of seismic design, S_S and S_1 must be determined for the site. This can be done by interpolating between contours on the maps included in the IBC or ASCE 7; however, this is neither simple nor accurate. A better and more accurate approach would be to use the Web-based hazard tool available on the ASCE website (https://asce7hazardtool.online) that provides the S_S and S_1 values based on the latitude/longitude information or the street address of the building location. This, however, requires a yearly subscription. A similar but free Web-based tool is also available on the website of the Applied Technology Council (https://hazards.atcouncil.org).

Once S_S, S_1, and the site class at a site are determined, a short-period or acceleration-dependent site coefficient, F_a, and a long-period or velocity-dependent site coefficient, F_v, must be determined from IBC Tables 1613.2.3(1) and 1613.2.3(2) or ASCE Tables 11.4-1 and 11.4-2, respectively. The coefficient F_a is a function of the site class and S_S, while F_v is a function of the site class and S_1. This dependence of the site coefficients on the seismicity at the site creates an elaborate scheme: There is a row of values of each site coefficient for a particular site class. It should be noted that the values of F_a and F_v are higher in areas of low seismicity and lower in areas of high seismicity. This is in line with observations that weak subsurface rock motion is amplified to a much larger extent by overlying soft soil deposits than is strong subsurface rock motion. The tables for F_a and F_v were stable and essentially unchanged from 1994 UBC [5.7] through ASCE 7-10. In ASCE 7-16, the tables have undergone significant revisions. Both F_a and F_v used to be equal to 1.0 for all values of S_S and S_1, respectively, for site class B, which used to be the benchmark site class. Now the values of F_a and F_v are unity at Site Class B-C interface; the values are smaller than one for Site Class B. Site-specific ground motion procedures are required wherever the site class is F and also for Site Class D and E in situations spelled out in ASCE 11.4.8.

Having determined F_a and F_v, S_S must be modified by F_a, resulting in $S_{MS} = F_aS_S$, the soil-modified MCE_R spectral response acceleration at short periods (IBC Equation 16-36, ASCE Equation 11.4-1). Likewise, S_1 must be modified by F_v, resulting in $S_{M1} = F_vS_1$, the soil-modified MCE spectral response acceleration at 1-second period (IBC Equation 16-37, ASCE Equation 11.4-2).

For design purposes, two-thirds of these soil-modified MCE values are used. The 5 percent damped design spectral response accelerations at short periods, S_{DS}, and at 1-second period, S_{D1}, are determined by IBC Equations 16-38 and 16-39 or ASCE Equations 11.4-3 and 11.4-4, respectively:

$$S_{DS} = \frac{2}{3}S_{MS}$$

$$S_{D1} = \frac{2}{3}S_{M1}$$

Note that ⅔ is the reciprocal of 1.5, the lower-bound margin of safety that is acknowledged to have been built into seismic design by model codes of the recent past. In other words, a code-designed structure is thought highly unlikely to collapse under ground motion that is one and a half times as strong as the design earthquake ground motion. The design earthquake of all codes of the recent past was the 500-year-return-period earthquake (10 percent probability of nonexceedance in 50 years). In coastal California, where MCE ground motion was no more than 150 percent as strong as ground motion in the 500-year-return period earthquake, collapse in the MCE was prevented by code-based design. In the Midwest and the East, where MCE ground motion could be four to five times as strong as ground motion in the 500-year-return-period earthquake safety against collapse in the MCE was not ensured by code-based design. The design approach of the 2018 IBC and ASCE 7-16 ensures a uniform collapse risk of 1 percent in 50 years for structures across the United States in the risk-targeted maximum considered earthquake because all structures must be designed for two-thirds of the MCE_R ground motion and because a margin of safety of at least one and a half is still built in.

A procedure for determining design response spectrum is contained in ASCE 11.4.6. This section provides a method for obtaining a 5 percent damped response spectrum from S_{DS} and S_{D1}. The response spectrum consists of two curves representing the short-period range (a region of constant spectral response acceleration) and the long-period range (a region of constant spectral response velocity). A very long period range, which is a region of constant spectral response displacement, is also included; it starts at the period T_L. ASCE Figure 11.4-1 depicts the design response spectrum based on the provisions in ASCE 11.4.6.

The long-period transition period T_L is given on maps in Figures 22-14 through 22-17, which are similar to zone maps, for all 50 states and U.S. territories. One must locate one's site on this map to determine T_L, which ranges between 4 and 16 seconds, depending on the location. The constant displacement branch of the design response spectrum is significant in determining the slosh height in tanks and the design seismic forces of long-period buildings.

The ramp building up to the flat top of the design spectrum is defined by specifying that the spectral-response acceleration at zero period is equal to 40 percent of the spectral-response acceleration corresponding to the flat top, S_{DS}, and that the period T_0 at which the ramp ends is 20 percent of the period, T_S, at which the constant acceleration and the constant velocity portions of the spectra meet. That period,

$$T_S = S_{D1}/S_{DS}$$

is solely a function of the seismicity and the soil characteristics at the site of the structure. It also serves as the dividing line between short- and long-period responses.

A procedure for classifying a site is contained in ASCE 20.3 and Table 20.3-1. The procedure depends on average soil properties, which are defined in ASCE 20.4.

While the site-specific ground motion procedures, now set forth in ASCE Chapter 21, can be used to determine ground motions for any structure, ASCE 11.4.8 requires "site-specific" analysis to be carried out in a number of different situations: (1) for structures located on Site Class F soils, (2) for seismically isolated structures and structures with damping systems on sites that are "near-source" ($S_1 \geq 0.6g$), (3) for structures on Site Class E sites with S_S greater than or equal to 1.0, and (4) for structures on Site Class D and E sites with S_1 greater than or equal to 0.2, subject to important exceptions listed in ASCE 11.4.8. In order to clarify what is required for these different situations, the site-specific procedure section in ASCE 7-16 is divided into (1) Section 21.1 Site Response Analysis (applicable to Site Class F soils) and (2) Section 21.2 Ground Motion Hazard Analysis (applicable to the other situations listed above).

5.4.3 Seismic Design Category

Every structure shall be assigned to a seismic design category (SDC) in accordance with IBC 1613.2.5 or ASCE 11.6. The SDC is used in the IBC and ASCE 7 to determine permissible structural systems, limitations on structural height and irregularity, the components of the structure that must be designed for seismic resistance, and the types of lateral force analysis that must be performed. Very importantly, the SDC is also used to determine the level of detailing that is required to be done for a structure: ordinary, intermediate, or special. Note that an SDC need not be determined for those structures for which earthquake effects need not be considered (see discussion above and the exceptions in IBC 1613.1, ASCE 11.1.2).

The seismic design category is a function of occupancy or use (use category) and of soil-modified seismic risk at the site of the structure in the form of the design spectral response accelerations at short periods, S_{DS}, and at 1-second period, S_{D1}. A structure located where $S_1 \geq 0.6g$ is assigned to SDC E if its risk category is I, II, or III and to SDC F if its risk category is IV. For structures not assigned to SDC E or F, the SDC needs to be determined twice—first as a function of S_{DS} by Table 1613.2.5(1) (same as ASCE 7-16 Table 11.6-1) and a second time as a function of S_{D1} by Table 1613.2.5(2) (same as ASCE 7-16 Table 11.6-2); the more severe category governs.

The point has been made that it is unnecessary and wasteful to require that the seismic design category of a short-period structure be determined by long-period ground motion. Therefore, where S_1 is less than 0.75, the SDC is permitted to be determined from IBC Table 1613.2.5(1) or ASCE Table 11.6-1 alone where all of the following apply:

1. In each of the two orthogonal directions, the approximate fundamental period of the structure, T_a, is less than $0.8T_s$, where $T_s = S_{D1}/S_{DS}$.
2. In each of two orthogonal directions, the fundamental period of the structure used to calculate the story drift is less than T_s.
3. The upper-bound design base shear is used in design (i.e., $V = S_{DS} W/(R/I_e)$).
4. The diaphragms are rigid or are permitted to be idealized as rigid by ASCE 12.3.1, or, for diaphragms permitted to be idealized as flexible by ASCE 12.3.1, the distances between vertical elements of the seismic force–resisting system do not exceed 40 feet.

The period $T_s = S_{D1}/S_{DS}$ is the period at which the short-period or constant-acceleration part of the design spectrum transitions into the long-period or velocity-governed part of the spectrum. ASCE 7 does not impose an upper limit on the period used to calculate story drift, and as a result, it probably is a more realistic representation of the actual period of the structure than T_a or C_uT_a (the limit that cannot be exceeded by rationally computed period except for period used to calculate story drift). Thus, condition 2 above ensures that the relaxation in SDC determination applies only to short-period structures. It probably renders condition 1, which is intended for the same purpose, superfluous. Condition 3 above is intended to impose a design force penalty on a structure that may stray into the long-period range despite all the other safeguards. The final condition is included because when the diaphragm span is no more than 40 feet, the dynamics of the diaphragms are unlikely to drastically influence the period of the structure, which is typically the concern with flexible-diaphragm structures.

Whether there is any advantage to be gained from the relaxation in question depends strictly on the relationship between S_{DS} and S_{D1} at the site of a structure. For many locations across the United States, there is no advantage because the short- and long-period ground motion parameters yield the same seismic design category. However, in other situations, the relaxation in question may yield a one- or even two-category reduction in the seismic design category. In other words, while the SDC may be D on the basis of S_{D1}, it may be only C on the basis of S_{DS}. In that case, only intermediate rather than special detailing would be required for a structure.

Relevant information on the six SDCs of the IBC and ASCE 7 is given in Table 5.1. Reference 5.13 contains a lucid step-by-step procedure to determine the SDC for any structure according to the provisions of the 2018 IBC.

Table 5.1 Seismic Design Categories of the 2018 IBC and ASCE 7-16

Seismic design category (SDC)	Use category	Description
A	All	Structures of any occupancy where anticipated ground motions are minor, even for very long-return periods ($S_{DS} < 0.167$, $S_{D1} < 0.067$).
B	I, II, III	Structures in regions where moderately destructive ground shaking is anticipated ($0.167 \leq S_{DS} < 0.33$, $0.067 \leq S_{D1} < 0.133$).
C	IV	Structures in regions where moderately destructive ground shaking may occur ($0.167 \leq S_{DS} < 0.33$, $0.067 \leq S_{D1} < 0.133$).
	I, II, III	Structures in regions with somewhat more severe than moderately destructive ground shaking potential ($0.33 \leq S_{DS} < 0.50$, $0.133 \leq S_{D1} < 0.20$).
D	IV	Structures in regions with somewhat more severe than moderately destructive ground shaking potential ($0.33 \leq S_{DS} < 0.50$, $0.133 \leq S_{D1} < 0.20$).
	All	Structures in regions expected to experience destructive ground shaking ($0.50 \leq S_{DS}$) but not located close to major active faults ($S_1 < 0.75$).
E	I, II, III	Structures in regions located close to major active faults ($S_1 \geq 0.75$).
F	IV	Structures in regions located close to major active faults ($S_1 \geq 0.75$).

Structures assigned to SDC E or F are not allowed to be sited "where a known potential exists for an active fault to cause rupture of the ground surface at the structure" (ASCE 11.8.1).

5.4.4 Building Configuration

Past earthquakes have repeatedly shown that buildings having irregular configurations in plan and/or elevation suffer greater damage than those having regular configurations. In irregular structures, inelastic behavior can concentrate in certain localized regions, resulting in rapid deterioration of structural elements in these areas. In contrast, inelastic demands tend to be better distributed throughout a regular structure. Elastic analysis methods typically used to analyze a structure are not capable of accurately predicting the distribution of earthquake demands in an irregular structure. The requirements in ASCE 12.3.2 encourage the use of buildings with regular configurations and prohibit the use of highly irregular buildings located on sites close to active faults.

Horizontal Structural Irregularities

Five different plan structural irregularities are defined in ASCE Table 12.3-1 (see Table 5.2):

- Torsional irregularity
- Reentrant corner irregularity
- Diaphragm discontinuity irregularity
- Out-of-plane offset irregularity
- Nonparallel system irregularity

Structures having horizontal structural irregularities in ASCE Table 12.3-1 must comply with the requirements of the applicable sections referenced in that table.

Vertical Irregularities

Five different types of vertical structural irregularities are contained in ASCE Table 12.3-2 (see Table 5.3):

- Stiffness—soft story irregularity
- Weight (mass) irregularity
- Vertical geometric irregularity
- In-plane discontinuity in vertical lateral force—resisting element irregularity
- Discontinuity in lateral strength—weak story irregularity

Vertical configuration irregularities affect seismic response at various floor levels and induce forces at these levels that depart significantly from the distribution assumed in the equivalent lateral force procedure in ASCE 12.8, which is discussed below. Structures having vertical irregularities as defined in ASCE Table 12.3-2 must comply with the requirements of the applicable sections referenced in that table.

5.4.5 Analysis Procedures

According to ASCE 12.1.1, "The design seismic forces, and their distribution over the height of the building structure, shall be established in accordance with one of the applicable procedures indicated in Section 12.6 and the corresponding internal forces and deformations in the members of the structure shall be determined. An approved alternative procedure shall not be used to establish the seismic forces and their distribution unless the corresponding internal forces and deformations in the members are determined using a model consistent with the procedure adopted." ASCE 12.1.1 does make an exception to this requirement: "As an alternative, the simplified design procedures of Section 12.14 is permitted to be used in lieu of the requirements of Sections 12.1 through 12.12, subject to all the limitations contained in Section 12.14." The following is a summary of the simplified design procedures, followed by a discussion of the equivalent lateral force procedure.

Simplified Analysis

The applicability of the simplified procedure in the stand-alone Section 12.14 is clearly defined through a list of limitations. In effect, the procedure is limited in its applicability to simple and redundant structures falling under Risk Categories I and II, not exceeding three stories in height, where the seismic force–resisting elements are arranged in a torsion-resistant, regular layout. Furthermore, only bearing wall and building frame systems are permitted to be designed by this procedure. Because the simplified procedure does not require a drift check as part of the design, moment frame systems are excluded since drift is a major concern in their design. The simplified design procedure allows drift to be taken as 1 percent when an estimate of drift is required for determining structural separations or to meet specific design requirements. For easier and faster navigation, parts of ASCE 7-16 Table 12.2-1 are reproduced in ASCE 12.14 as Table 12.14-1 to provide the R-values for only the structural systems that are permitted to be designed by the simplified method. However, Table 12.14-1 does not include values for the overstrength factor, Ω_o, or the deflection amplification factor, C_d, because the simplified analysis procedure uses a blanket 2.5 value for Ω_o for all permitted structures while not requiring any drift calculation using C_d. Furthermore, the load combination equations are reproduced in this section, with the modification that the overstrength factor Ω_o in the equations is taken as 2.5, while ρ is taken as 1.0. The seismic base shear V is determined in accordance with ASCE Equation 12.14-12:

$$V = \frac{FS_{DS}}{R}W$$

where R is the response modification factor contained in ASCE Table 12.14-1 for seismic force–resisting systems and W is the effective seismic weight of the structure, which is defined in ASCE 12.7.2, with the definition repeated in ASCE 12.14.8.1 (a more complete discussion on these two quantities is given in the "Equivalent Lateral Force Procedure" section):

$$S_{DS} = \frac{2}{3}F_a S_s$$

where F_a is permitted to be taken as 1.0 for rock sites or 1.4 for soil sites or determined in accordance with ASCE 11.4.3. For the purpose of simplified design, sites are permitted to be considered as rock if there is no more than 10 feet of soil between the rock surface and the bottom of spread footing or mat foundation. For the purposes of simplified design, S_S need not be taken any larger than 1.5.

The seismic base shear of simplified analysis varies (ASCE 7-16 Equation 12.14-12) with the building height and is 20 percent higher than the upper-bound design base shear of the equivalent lateral force procedure for a three-story building. The importance factor, I_e, is equal to 1.0 since only Risk Category I and II structures may be designed by this procedure. Also, because only short-period structures may be designed by this procedure, the upper-bound design base shear of the equivalent lateral force procedure (next section) governs.

The factor F accounts for the difference between the triangular distribution of the equivalent lateral force procedure for low-rise buildings and the distribution of the design base shear given by ASCE Equation 12.14-13, which is in proportion to the weights at the various floor levels (for the same weight at every level, it is a uniform distribution). For a one-story building, there is no difference between the two distributions; therefore, $F = 1.0$. For a three-story building, the difference is the largest; therefore, $F = 1.2$. For a two-story building, the difference is in between and $F = 1.1$.

The vertical distribution of the base shear is determined by ASCE Equation 12.14-13:

$$F_x = \frac{FS_{DS}}{R}w_x$$

where F_x is the seismic force applied at level x and w_x is the portion of the effective seismic weight of the structure W at level x. The vertical distribution of the base shear V is depicted in Fig. 5.6.

Equivalent Lateral Force Procedure

The equivalent lateral force (ELF) procedure is contained in ASCE 12.8. This analysis procedure can be used for all structures assigned to SDC B and C (ASCE 12.6) as well as for many structures assigned to SDC D, E, and F (see ASCE Table 12.6-1 for the analysis procedures that are to be utilized for SDC D, E, or F structures).

Table 5.2 Horizontal Structural Irregularities

Irregularity type			Description
1a	Torsional irregularity	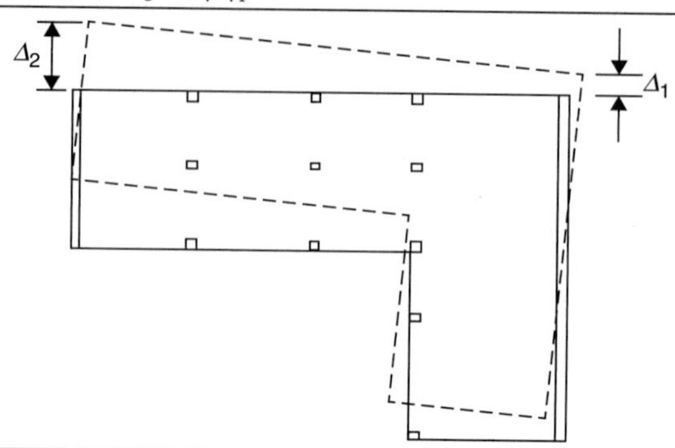	$\Delta_2 > 1.2\left(\frac{\Delta_1+\Delta_2}{2}\right)$
1b	Extreme torsional irregularity		$\Delta_2 > 1.4\left(\frac{\Delta_1+\Delta_2}{2}\right)$
2	Reentrant corners	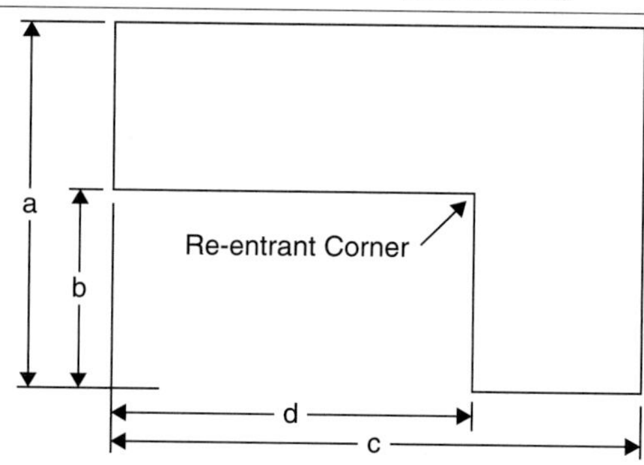	• Projection b > 0.15a and • Projection d > 0.15c
3	Diaphragm discontinuity	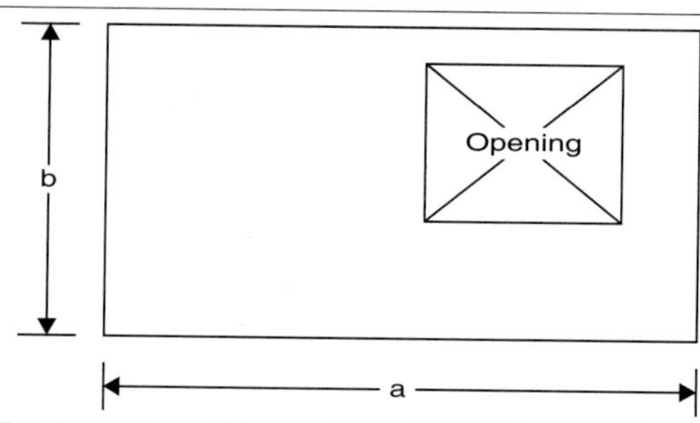	• Area of opening > 0.5ab or • Changes in effective diaphragm stiffness > 50% from one story to the next
4	Out-of-plane offsets	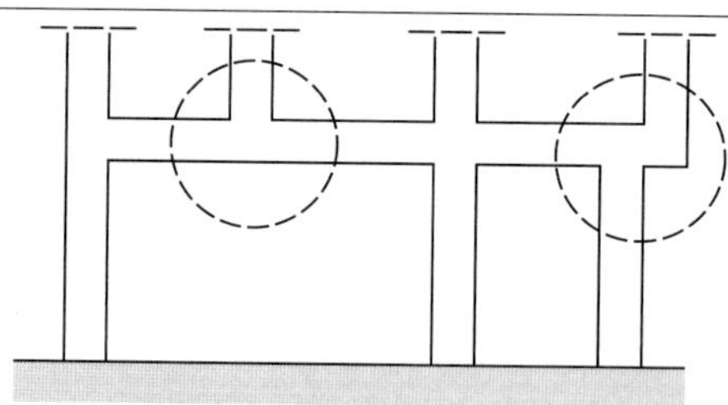	Discontinuities in lateral force–resisting path, such as out-of-plane offsets of vertical elements
5	Nonparallel systems	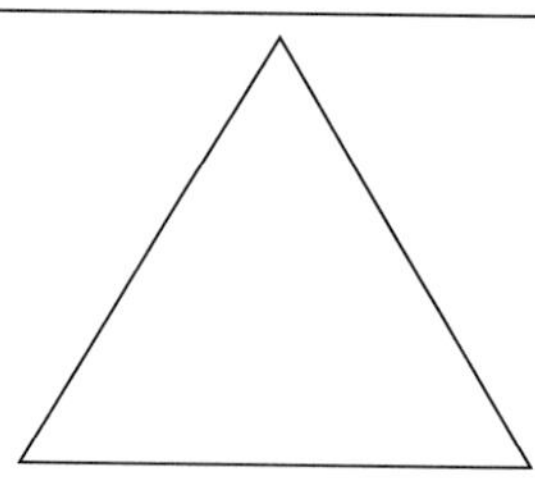	Vertical lateral force–resisting elements are not parallel to the major orthogonal axes of the lateral force–resisting system

Table 5.3 Vertical Structural Irregularities

		Irregularity type	Description
1a	Stiffness irregularity—soft story	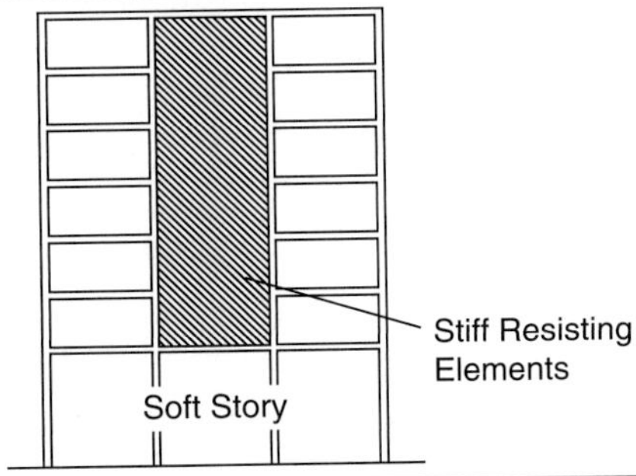	Soft story stiffness < 70% (story stiffness above) Or < 80% (average stiffness of 3 stories above)
1b	Stiffness irregularity—extreme soft story		Soft story stiffness < 60% (story stiffness above) Or < 70%(average stiffness of 3 stories above)
2	Weight (mass) irregularity	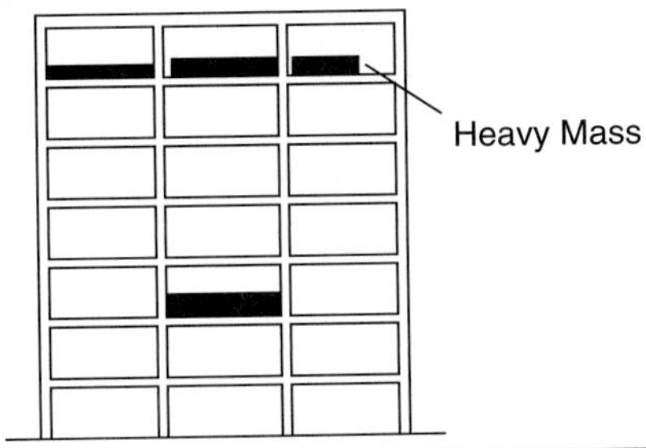	Story mass > 150% (adjacent story mass) (a roof that is lighter than the floor below need not be considered)
3	Vertical geometric irregularity		Horizontal dimension of lateral force–resisting system in story > 130% of that in adjacent story
4	In-plane discontinuity in vertical lateral force–resisting elements	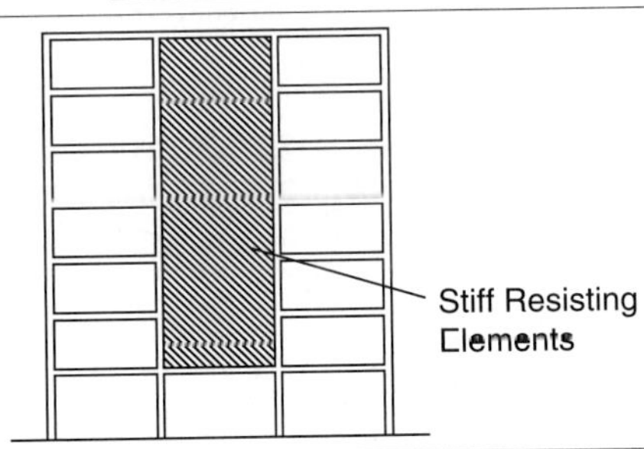	There is an in-plane offset of a vertical seismic force–resisting element resulting in overturning demands on supporting structural members.
5a	Discontinuity in lateral strength—weak story	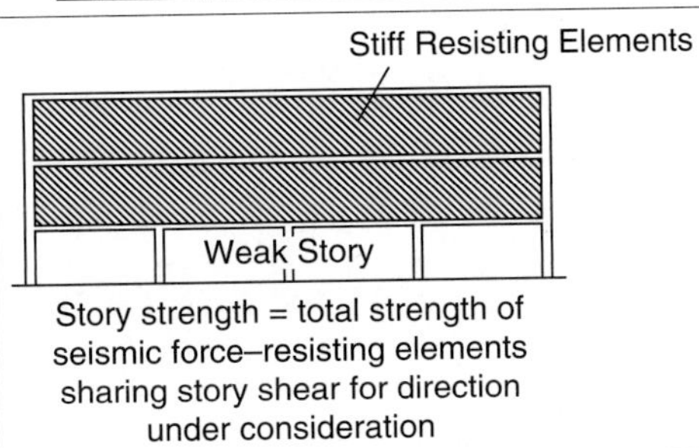	Weak story strength < 80% (story strength above)
5b	Discontinuity in lateral strength—extreme weak strong		Weak story strength < 65% (story strength above)

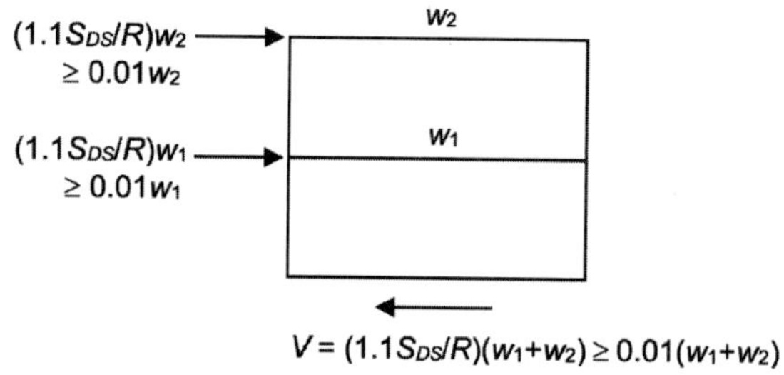

Figure 5.6 Design seismic force distribution according to the simplified method in ASCE 12.14.

Design Base Shear, V In a given direction, the design base shear V is determined by ASCE Equation 12.8-1:

$$V = C_s W$$

where C_s is the seismic response coefficient determined in accordance with ASCE 12.8.1.1 and W is the effective weight of the structure, which includes the total dead load and the following other loads (ASCE 12.7.2):

- In areas used for storage, a minimum of 25 percent of the reduced floor live load (floor live load in public garages and open parking garages need not be included)
- When an allowance for partition load is included in the floor load design, the actual partition weight or a minimum weight of 10 psf of floor area, whichever is greater
- Total operating weight of permanent equipment
- 20 percent of the uniform design snow load, regardless of actual roof slope, where the flat roof snow load exceeds 30 psf.

Seismic Response Coefficient, C_s The seismic response coefficient, C_s, is determined by ASCE Equation 12.8-3:

$$C_s = \frac{S_{D1}}{T(R/I_e)} \text{ for } T \le T_L$$

and by ASCE Equation 12.8-4:

$$C_s = \frac{S_{D1}T_L}{T^2(R/I_e)} \text{ for } T > T_L$$

where S_{D1} = design spectral response acceleration at 1-second period
R = response modification factor determined from ASCE Table 12.2-1
I_e = seismic importance factor determined from ASCE Table 1.5-2
T = elastic fundamental period of the structure determined in accordance with ASCE 12.8.2
T_L = long-period transition period determined from ASCE Figures 22-14 through 22-17

The value of C_s need not exceed that from ASCE Equation 12.8-2:

$$C_s = \frac{S_{DS}}{(R/I_e)}$$

where S_{DS} = design spectral response acceleration at short periods (0.2 second)

Also, C_s must not be less than that determined by ASCE Equation 12.8-5:

$$C_s = 0.044 S_{DS} I_e \ge 0.01$$

For structures located where S_1 is equal to or greater than 0.6g, C_s shall not be taken less than that determined by ASCE Equation 12.8-6:

$$C_s = \frac{0.5 S_1}{(R/I_e)}$$

The design spectrum defined by ASCE Equations 12.8-1 and 12.8-2 through 12.8-6 is depicted in Fig. 5.7. Equation 12.8-2 represents the constant-acceleration portion of the spectrum, while Equation 12.8-3 represents the constant-velocity portion and Equation 12.8-4 the constant-displacement portion. The design force level defined by Equations 12.8-1 and 12.8-2 through 12.8-4 and 12.8-6 is based on the assumption that a structure will undergo several cycles of inelastic deformation during major earthquake ground motion; therefore, the force level is related to the type of structural system and the structure's estimated ability to

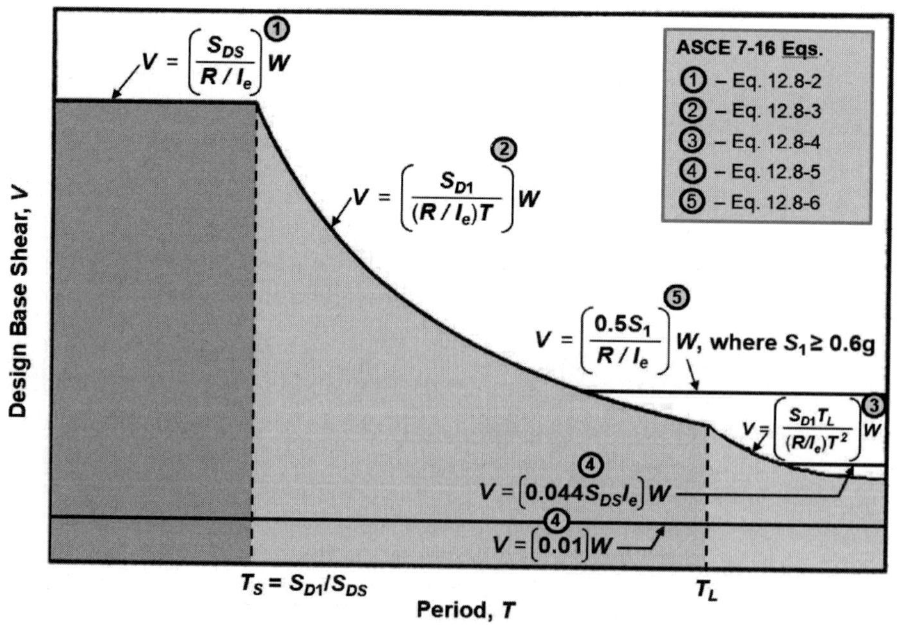

Figure 5.7 Design response spectrum according to the equivalent lateral force procedure in ASCE 12.8.

sustain these deformations and dissipate energy without collapse. It is important to note that a force level equal to that defined by Equations 12.8-1 and 12.8-2 through 12.8-6 is also used as the lower bound for the modal response spectrum analysis procedure of ASCE Section 12.9.

The minimum design base shear of $0.044S_{DS}I_eW$, applicable in all seismic design categories, is included in view of the uncertainty and the lack of knowledge of actual structural response of long-period structures subject to earthquake ground motion. The absolute minimum design base shear of 1 percent of the seismic weight is simply an SDC A minimum based on structural integrity considerations.

Following the Northridge earthquake of 1994, a second lower-bound on the design base shear, applicable in Seismic Zone 4 only, was added to the 1997 UBC [5.7]. This second minimum was specifically intended to account for the large displacement and velocity pulses observed in near-fault ground motions in the Northridge earthquake. A corresponding minimum, applicable to all structures located where the mapped MCE spectral response acceleration at 1-second period, S_1, equals or exceeds 0.6*g*, which is roughly equivalent to Zone 4 of the 1997 UBC, is required by ASCE 7-16.

Elastic Fundamental Period, *T* The design base shear is dependent on the elastic fundamental period, *T*, for buildings in the intermediate height range. However, *T*, which is a function of the mass and the stiffness of a structure, cannot be determined until a structure has been designed since, until then, the stiffness and the mass cannot be evaluated. In view of this, building codes include approximate period formulas, the purpose of which is to get the design process started. The approximate period formulas deliberately produce estimated periods that are shorter than the "real" periods of actual structures, the idea being that if the initial period estimate is later not refined, the design should still be safe. Since design base shear is inversely proportional to period (ASCE Equations 12.8-1 and 12.8-3) or to square of the period (ASCE Equations 12.8-1 and 12.8-4), a shorter period means a higher base shear used in design.

ASCE 12.8.2.1 provides Equation 12.8-7 to determine an approximate fundamental period T_a:

$$T_a = C_t h_n^x$$

where C_t = parameter given in ASCE Table 12.8-2
h_n = height above the base to the highest level of the building

Coefficients C_t and x are listed in Table 12.8-2 for concrete and steel moment–resisting frames where the frames resist 100 percent of the design lateral force and are not enclosed or adjoined by more rigid elements that may prevent the frames from deflecting freely when subjected to seismic forces. Coefficients are also listed for eccentrically braced frames and buckling-restrained braced frames of steel. For all other structural systems,

$$T_a = 0.020h_n^{0.75}$$

The base is the level at which the horizontal seismic ground motions are considered to be imparted to a structure. Having started design based on the approximate period and having gone part of the way through, it is possible to refine the initial period estimate if desired. The code permits this to be done.

Period may be estimated by any rational procedure as long as it is in conformance with the principles of mechanics (ASCE 12.8-2). There is, however, scope for potential abuse here. The rationally computed period of a concrete or a masonry building is very much dependent on what stiffness assumption is made in the period computation. Gross section stiffness versus cracked section stiffness makes a big difference; how low the cracked section stiffness is taken to be obviously has a major impact. In order to ensure that an unreasonably low design base shear is not used, based on an unduly long period calculated using unrealistic stiffness assumptions, the code imposes a limit on rationally computed period. According to ASCE 12.8-2, the rationally computed *T* may not be taken any longer than a multiplier C_u, which is obtained from ASCE Table 12.8-1, times the approximate period T_a (this restriction typically does not apply to drift computations). In ASCE 7-16, C_u depends on the long-period design spectral response acceleration, S_{D1}, and varies from 1.4 for $S_{D1} \geq 0.4$ to 1.7 for $S_{D1} \leq 0.1$.

The period $T_S = S_{D1}/S_{DS}$ in Fig. 5.7 is the dividing line between "short-period" and "long-period" response. If the period of a structure *T* is less than or equal to the transition period, T_s, its response is governed by the "flat top" or period-independent part of the spectrum, making it a short-period structure. If, on the other hand, *T* is larger than or equal to T_s, its response is governed by the period-dependent part of the spectrum, making it a long-period structure.

ASCE Equation 12.8-8 may also be used to determine the approximate fundamental period:

$$T_a = 0.1N$$

where *N* is the number of stories in a building. This equation is applicable to concrete or steel moment–resisting frame buildings not exceeding 12 stories in height and having a minimum story height of 10 feet. This approximate equation has long been in use for low- to moderate-height frames.

The approximate fundamental period for masonry or concrete shear wall structures can be determined from ASCE Equation 12.8-9 as follows:

$$T_a = \frac{0.0019}{\sqrt{C_w}} h_n$$

where C_w is calculated by ASCE Equation 12.8-10 as follows:

$$C_w = \frac{100}{A_B} \sum_{i=1}^{x} \frac{A_i}{\left[1 + 0.83\left(\frac{h_n}{D_i}\right)^2\right]}$$

where A_B = base area of the structure
A_i = area of shear wall i
D_i = length of shear wall i
h_n = height above base to highest level of structure
x = the number of shear walls in the building effective in resisting lateral forces in the direction under consideration

Seismic Importance Factor, I_e ASCE Table 1.5-2 contains the seismic importance factor, I_e, for each risk category. As indicated above, design seismic forces are directly proportional to I_e. The forces increased by I_e-factors larger than 1 are intended to increase the likelihood that structures belonging to high-occupancy and essential categories would continue to function during and following an earthquake.

Response Modification Factor, *R* The response modification factor, *R*, is intended to account for differences in the inelastic deformability or energy dissipation capacity of various structural systems. It reflects the reduction in structural response caused by damping, overstrength, and inelasticity.

It has been suggested in Section 5.3 above that an *R*-value of 2 used in design would result in essentially elastic response of a structure to the design earthquake of the IBC. By contrast, the *R*-values assigned by ASCE 7-16 to seismic force–resisting structural systems range from 1 to 8, as can be seen from ASCE Table 12.2-1. An *R*-value of 8 (for special reinforced-concrete or steel moment frames) represents one-quarter ($^2/_8$) of the strength level that would have been needed for elastic response to the design earthquake of the IBC. An *R*-value of 1 (for a cantilever column system detailed to conform to the requirements for ordinary concrete moment frames) represents elastic response to the design earthquake of the IBC, with a margin of safety built in.

The *R*-values in ASCE Table 12.2-1 are based largely on engineering judgment of the performance of various materials and systems in past earthquakes. For ATC 3-06 [5.14], the predecessor document to the NEHRP *Provisions* [5.3] in which *R* was first introduced, certain agreed-on reference structures were selected. Two systems having high and low expected levels of performance were chosen to be "a steel ductile frame and a box type masonry or concrete building, respectively." In today's terminology, these would be the special moment frame of steel and a bearing wall system consisting of masonry or concrete shear walls, respectively. The R-values for these two systems were chosen considering the seismic design forces assigned to them by older editions of the UBC. No compelling arguments were offered to change the design basis loads for these systems or to change their interrelationship. The expected performances of other systems were then evaluated relative to these reference systems in order to determine the other R-values. Considerations focused on the following issues:

1. The degree to which the system can be allowed to go beyond the elastic range, its degree of energy dissipation in so doing, and the stability of the vertical load-carrying system during inelastic response due to maximum expected ground motion.
2. The effect of failure or partial failure of vertical elements of the seismic force–resisting system on the vertical load-carrying capacity and stability of the total building system.
3. The inherent redundancy of the system that would allow some progressive inelastic excursions without overall failure. Localized failure of a part must not lead to failure of the system.
4. Where dual systems are employed, important performance characteristics include the ability of the secondary (backup) system to maintain vertical support when the primary system suffers significant damage at the maximum deformation response. The backup system can serve to redistribute lateral loads when the primary system undergoes degradation and should stabilize the building in the event that the primary system is badly damaged.

Vertical Distribution of Seismic Forces Once the design base shear V has been determined, the lateral force F_x to be applied at level x of the structure is determined by ASCE Equations 12.8-11 and 12.8-12:

$$F_x = C_{vx} V$$

$$C_{vx} = \frac{w_x h_x^k}{\sum_{i=1}^{n} w_i h_i^k}$$

where C_{vx} = vertical distribution factor
k = distribution exponent related to the building period
= 1 for $T \le 0.5$ second
= 2 for $T \ge 2.5$ second
= 2, or is to be determined by linear interpolation between 1 and 2 for 0.5 second < T < 2.5 second
h_i, h_x = height from the base to level i or x, respectively
w_i, w_x = portion of W located or assigned to level i or x, respectively

For structures with $T \le 0.5$ second, V is distributed linearly over the height, varying from zero at the base to a maximum value at the top. For 0.5 second < T < 2.5 second, a linear interpolation between a linear and a parabolic distribution is permitted, or a parabolic distribution is also allowed (see Fig. 5.8). When $T \ge 2.5$ second, a parabolic distribution is to be used. The larger the value of k, the higher the proportion of V distributed to the upper portions of a structure. This produces more overturning moment for the same base shear, which is characteristic of flexible building response.

Horizontal Distribution of Seismic Forces The seismic design story shear V_x in any story x is the sum of the lateral forces acting at the floor or roof level supported by that story and all the floor levels above, including the roof (ASCE Equation 12.8-13):

$$V_x = \sum_{i=x}^{n} F_i$$

The distribution of V_x to the vertical elements of the seismic force–resisting system (shear walls and frames) in story x is determined by the flexibility of the supported diaphragm.

Per ASCE 12.3, certain diaphragms are "prescriptively" flexible (meaning flexible without any deflection calculation—ASCE 12.3.1.1). Certain other diaphragms are "prescriptively" rigid (ASCE 12.3.1.2). ASCE 12.3 also defines diaphragms that are flexible by calculation.

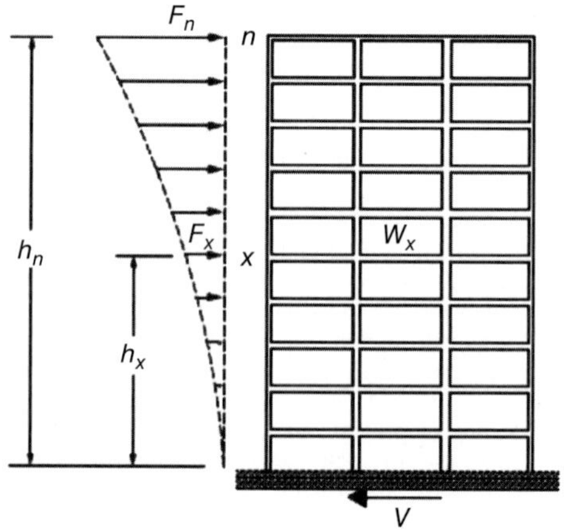

Figure 5.8 Vertical distribution of seismic forces according to ASCE 12.8.3.

Per ASCE 12.3.1.3, diaphragms not satisfying the conditions of Section 12.3.1.1 or 12.3.1.2 are permitted to be idealized as flexible provided that

$$\frac{\delta_{MDD}}{\Delta_{ADVE}} > 2 \qquad \text{(ASCE Equation 12.3-1)}$$

where δ_{MDD} and Δ_{ADVE} are as shown in Fig. 5.9. The loading used in this calculation is required to be that prescribed in Section 12.8.

According to ASCE 12.3.1.1, "Unless a diaphragm can be idealized as either flexible or rigid in accordance with Section 12.3.1.1, 12.3.1.2, or 12.3.1.3, the structural analysis shall explicitly include consideration of the stiffness of the diaphragm (i.e., semirigid diaphragm assumption). This is not a requirement of the 2018 IBC. According to 2018 IBC Section 1604, "A diaphragm is rigid for the purpose of distribution of story shear and torsional moment when the lateral deformation of the diaphragm is less than or equal to two times the average story drift." In other words, a diaphragm that is not flexible by calculation can be treated as a rigid diaphragm in analyzing for the distribution of the total lateral force to various vertical elements of the seismic force–resisting system.

For flexible diaphragms, the seismic design story shear V_x is distributed to various vertical elements based on the area of the diaphragm tributary to each line of resistance.

For rigid diaphragms, V_x is distributed to the various vertical elements of the seismic force–resisting system in the story under consideration based on the relative lateral rigidities of the vertical resisting elements and the diaphragm (ASCE 12.8.4). Virtually all computer programs utilized for structural analysis assume diaphragms to be rigid (by default), unless otherwise specified.

Torsion, Including Accidental Torsion Where diaphragms are not flexible, provisions must be made for the increased horizontal forces induced in vertical elements of the seismic force–resisting system resulting from inherent torsion, as it is called in ASCE 7-16, due to eccentricity between the center of application of the lateral forces (center of mass) and the center of rigidity of the seismic force–resisting system (through which the resultant of the resistances to the lateral forces acts). Forces are not to be decreased due to torsional effects.

The torsional design moment at a given story is required to be the moment resulting from eccentricities between applied design lateral forces at levels above that story and the center of rigidity of the vertical resisting elements in that story (ASCE 12.8.4.1) *plus* an accidental torsion (ASCE 12.8.4.2). To compute the accidental torsion, the mass at each level must be assumed to be displaced from the calculated center of mass in each direction a distance equal to 5 percent of the building

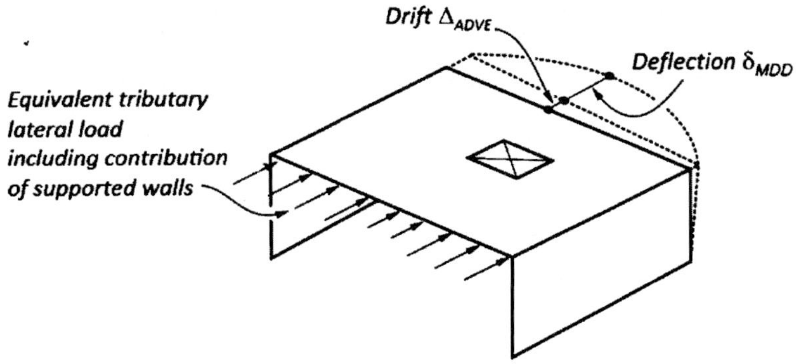

Figure 5.9 Flexible diaphragm.

plan dimension at that level perpendicular to the direction of the force under consideration. For many editions of ASCE 7 through ASCE 7-10, the consideration of accidental torsion was always required. In an important change, ASCE 7-16 states the following:

> Accidental torsion shall be applied to all structures for determination if a horizontal irregularity exists as specified in Table 12.3-1. Accidental torsion moments (M_{ta}) need not be included when determining the seismic force effects E in the design of the structure and in the determination of the design story drift ..., except for the following structures:
>
> 1. Structures assigned to Seismic Category B with Type 1b horizontal structural irregularity.
> 2. Structures assigned to Seismic Category C, D, E, and F with Type 1a or Type 1b horizontal structural irregularity.

ASCE 12.8.4.3 further requires that where a torsional irregularity or an extreme torsional irregularity exists, as defined in ASCE Table 12.3-1, the effects must be accounted for by multiplying the accidental torsion at each level by an amplification factor A_x (≤ 3.0), given by ASCE Equation 12.8-14. This last requirement applies to buildings in Seismic Design Categories C, D, E, and F only.

Overturning According to ASCE 12.8.5, structures are to be designed for the effects of overturning caused by the seismic forces determined from ASCE 12.8.3. The overturning moment M_x at level x is determined by

$$M_x = \sum_{i=x}^{n} F_i(h_i - h_x)$$

where F_i = portion of V induced at level i
h_i, h_x = height from the base to level i or x, respectively

ASCE 12.13.4 allows for certain reductions in overturning effects at the soil-foundation interface.

Drift Limitation Drift computation starts with δ_{xe}, the elastically computed lateral deflection at floor level x under code-prescribed seismic forces (the design base shear V, distributed along the height of the structure in the manner prescribed by the code). Recognizing that the deflections δ_{xe} are much lower than the actual lateral deflections the various floor levels would undergo if the structure were to be subjected to the design earthquake of the 2018 IBC or ASCE 7-16 (two-thirds of the risk-targeted maximum considered earthquake), the deflections δ_{xe} are multiplied by the deflection amplification factor C_d, producing estimated design earthquake displacements at the various floor levels. At the same time, the δ_{xe} values are divided by the seismic importance factor I_e by which the code-prescribed seismic forces (under which the deflections δ_{xe} were calculated) were increased for structures belonging to the higher risk categories. This is necessary because the limiting values of interstory drift in ASCE 7-16 are more stringent for structures in the higher risk categories. Without the division by I_e, structures with I_e larger than 1.0 would be doubly penalized. Thus, the deflection δ_x at level x is determined by ASCE Equation 12.8-15:

$$\delta_x = \frac{C_d \delta_{xe}}{I_e}$$

The design story drift Δ is computed as the difference of the deflections δ_x at the centers of mass of the diaphragms at the top and the bottom of the story under consideration (see Fig. 5.10). Where centers of mass do not align vertically, it is permitted to compute the deflection at the bottom of the story based on the vertical projection of the center of mass at the top of the story. For structures assigned to SDC C or higher, with torsional or extreme torsional irregularity (ASCE Table 12.3-1), Δ is computed as the largest difference of the deflections along any of the edges of the diaphragms at the top and the bottom of the story under consideration.

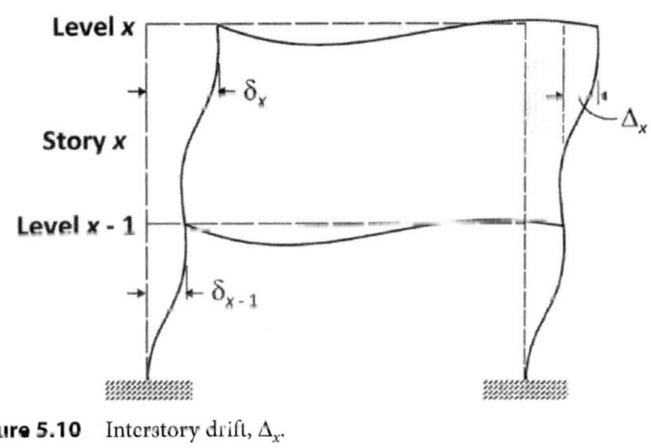

Figure 5.10 Interstory drift, Δ_x.

Deflections δ_{xe} must be determined based on the elastic properties of all elements of the seismic force–resisting system, considering the spatial distribution of the mass and the stiffness of the structure. For concrete elements, stiffness properties must include the effects of cracked sections.

For the purposes of drift computation, the value of the base shear V need not be limited by the upper-bound limitation on the computed fundamental period T ($T \leq C_u T_a$) (ASCE 12.8.6.2).

The design story drift Δ_x must be increased by $1.0/(1 - \theta)$ when P-delta effects are determined to be significant (see the following section on that topic). When calculating drift, the redundancy factor ρ (see later discussion) is to be taken as 1.0.

Once the design story drifts are computed, they are to be compared with the allowable story drift Δ_a in ASCE Table 12.12-1 (ASCE 12.12-1). Buildings subjected to earthquakes need drift control to restrict damage to partitions, shaft and stair enclosures, and glass and other fragile nonstructural elements and, more important, to minimize differential movement demand on the seismic safety elements. The ASCE 7-16 limitations on story drift depend on the risk category, and generally become more restrictive for the higher risk categories, to provide a higher level of performance. The limits also depend on the type of structure. The design story drifts must not exceed the allowable values.

***P*-Delta Effects** Member forces and story drifts induced by *P*-delta effects must be considered in the evaluation of overall structural stability if such effects are significant. *P*-delta effects need not be considered when the ratio of secondary moment to primary moment does not exceed 0.10. The ratio may be evaluated for any story as the product of the dead, floor live, and snow load above the story (P_x) times the unamplified drift (Δ_x/C_d) in the story amplified by the seismic importance factor (I_e) divided by the product of the seismic shear in the story (V_x) and the story height (h_{sx}) (see ASCE Equation 12.8-16 and Fig. 5.11):

$$\theta = \frac{P_x \Delta_x I_e}{V_x h_{sx} C_d}$$

If θ is greater than θ_{max}, the structure is potentially unstable and must be redesigned. θ_{max} is given by ASCE Equation 12.8-17 as follows:

$$\theta_{max} = \frac{0.5}{\beta C_d} \leq 0.25$$

where β is the ratio of shear demand to shear capacity for the story between levels x and x–1. If β is not calculated, β is to be taken equal to 1.0.

For $0.10 < \theta \leq \theta_{max}$, interstory drift and element forces must be computed including *P*-delta effects. This can be accomplished by multiplying displacements and member forces by $[1.0/(1-\theta)]$.

5.4.6 Seismic Force–Resisting Systems

The basic structural systems that may be used to resist earthquake forces are listed in ASCE Table 12.2-1. Included in the table are the response modification factors R to be used in determining the base shear V, the overstrength factor Ω_o to be used in determining maximum element design forces, and the deflection amplification factor C_d to be used in determining design story drift. Also given in the table are height limits for the various structural systems, which depend on the SDC. It is the height limits portion of the table that indicates that when it comes to concrete moment frames, ordinary detailing is permitted only for SDC B (and, of course, A); intermediate or special detailing is required for SDC C; and special detailing is the minimum requirement for SDC D, E, or F. Also included in Table 12.2-1 is a column that lists ASCE 7 sections where detailing requirements for the various structural systems are specified. This column is not particularly useful because most of the references are to ASCE 7 Chapter 14 section numbers. For all reinforced-concrete systems, reference is made to ASCE 14.2. That section references ACI 318-14, which is where the actual detailing requirements are found. A general description of each major category of seismic force–resisting systems is given below.

For structural members within a building assigned to SDC D, E, or F that are not proportioned to resist forces induced by earthquake motions, the deformation compatibility requirements of ASCE 12.12.5 must be satisfied. In short, every structural component not included in the seismic force–resisting system in the direction under consideration must be designed to be adequate for vertical load–carrying capacity and the induced bending moments and shear forces resulting from the design story drift Δ_x. Note that reinforced-concrete frame members not designed as part of the seismic force–resisting system are deemed to satisfy deformation compatibility requirements if they comply with Section 18.14 of ACI 318-14 [5.4].

Moment-Resisting Frame Systems

Figure 5.12*a* depicts a moment-resisting frame system. This is a structural system with moment frames providing support for gravity loads. Lateral forces are resisted primarily by flexural action of the frame members. All the moment frames or some of them or selected portions thereof may be designated as the seismic force–resisting system; the members of the seismic force–resisting system must be designed and detailed accordingly based on the SDC.

See ASCE Table 12.2-1 Row C for moment-resisting frame systems specifically recognized by ASCE 7-16. Detailing requirements for ordinary, intermediate, and special moment frames of concrete and steel are given in ACI 318 [5.4] and AISC 341[5.15], respectively. Detailing requirements for steel-concrete composite moment frames are also given in AISC 341 [5.15].

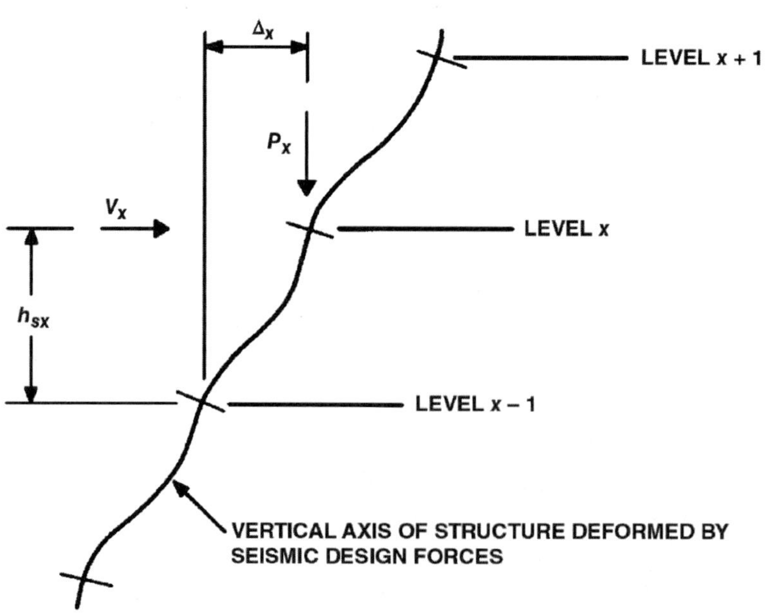

Figure 5.11 *P*-delta effect.

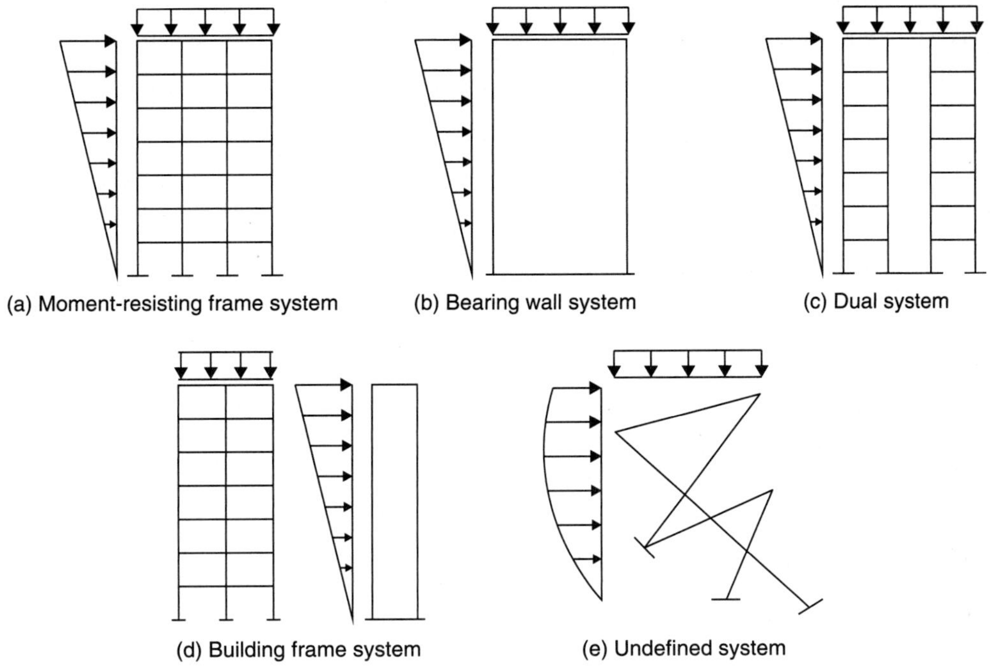

Figure 5.12 Seismic force–resisting structural systems.

Bearing Wall Systems

The bearing wall system depicted in Fig. 5.12*b* is a structural system without an essentially complete space frame that provides support for the gravity loads. Bearing walls provide support for all or most of the gravity loads. Resistance to lateral forces is provided by the same bearing walls acting as shear walls. See ASCE Table 12.2-1 Row A for bearing wall systems of various materials specifically recognized by ASCE 7-16.

Dual Systems

A dual system, depicted in Fig. 5.12(*c*), is a structural system with the following essential features:

1. An essentially complete space frame provides support for gravity loads.
2. Resistance to lateral forces is provided by moment-resisting frames capable of resisting at least 25 percent of the design base shear and by shear walls or braced frames (ASCE 12.2.5.1).
3. The two subsystems (moment-resisting frames and shear walls) are designed to resist the design base shear in proportion to their relative rigidities (ASCE 12.2.5.1).

ASCE 7-16 separately recognizes dual systems in which the moment-resisting frame consists of special moment frames (ASCE Table 12.2-1 Row D) and dual systems in which the moment-resisting frame consists of intermediate moment frames (ASCE Table 12.2-1 Row E).

The concept of the dual system loses its validity in buildings assigned to SDC A and B since it is questionable whether the moment frames, which are required to have only ordinary detailing, can act as a backup to the ordinary reinforced-concrete shear walls (the inelastic deformability of both systems are comparable). In areas of low seismicity, utilizing a shear wall-frame interactive system (ASCE Table 12.2-1 Row F) is more logical. In this system, defined in ASCE 12.2.5.8, the shear walls and frames resist the lateral forces in proportion to their rigidities, considering interaction between the two subsystems at all levels. In addition, the shear strength of the shear walls must be at least 75 percent of the design story shear at each story. The frames of the shear wall-frame interactive system must be capable of resisting at least 25 percent of the design story shear in every story.

Building Frame Systems

A building frame system is depicted in Fig. 5.12(*d*). This is a structural system with an essentially complete space frame that supports the gravity loads. Resistance to lateral forces is provided by shear walls or braced frames. No interaction between the shear walls or braced frames and the moment frames is considered in the lateral load analysis; all of the lateral forces are allocated to the walls. Deformation compatibility between the gravity frame and the seismic force–resisting shear walls or braced frames must be ensured through the design earthquake displacements.

Similar to dual systems, the concept of the building frame system loses its appeal for structures assigned to SDC A or B since there is little to be gained from assigning the entire lateral resistance to the shear walls in the absence of any special detailing requirements for the frames. As noted above, a shear wall-frame interactive system may be more practical and economical in such cases.

Cantilevered Column Systems

Cantilevered column systems are defined in ASCE 11.2 as seismic force–resisting systems in which lateral forces are resisted entirely by columns acting as cantilevers from the base. These structures have little redundancy and overstrength, and inelastic behavior is concentrated at their bases. As a result, they have substantially less energy dissipation capacity than other systems.

Undefined Structural Systems

Undefined structural systems are any systems not listed in ASCE Table 12.2-1. The coefficients R, Ω_o, and C_d are to be substantiated based on approved cyclic test data and analysis (ASCE 12.2.1).

5.4.7 Seismic Force Effects

Once the seismic force–resisting system has been chosen and the design base shear V has been determined and distributed over the height of the building, an analysis of a mathematical model (ASCE 12.7.3) of the structure is performed (yielding the effects of horizontal seismic forces, Q_E), and the structural members are designed for the combined effects of gravity and seismic forces. The effects of wind are to be considered as well.

Load Combinations

Basic load combinations for strength design are given in IBC 1605.2 and ASCE 2.3.1 and 2.3.6. The following load combinations of IBC 1605.2, which are somewhat different from those of ASCE 2.3.1 and 2.3.6 and take precedence over them, are utilized in the design of structural members:

$1.4(D + F)$ (IBC Equation 16-1)

$1.2(D + F) + 1.6(L + H) + 0.5(L_r \text{ or } S \text{ or } R)$ (IBC Equation 16-2)

$1.2(D + F) + 1.6(L_r \text{ or } S \text{ or } R) + (f_1 L \text{ or } 0.8W)$ (IBC Equation 16-3)

$1.2(D + F) + 1.6W + f_1 L + 1.6H + 0.5(L_r \text{ or } S \text{ or } R)$ (IBC Equation 16-4)

$1.2(D + F) + 1.0E + f_1 L + 1.6H + f_2 S$ (IBC Equation 16-5)

$0.9D + 1.0W + 1.6H$ (IBC Equation 16-6)

$0.9(D + F) + 1.0E + 1.6H$ (IBC Equation 16-7)

where D, F, L, H, L_r, S, R, W, and E are the effects of dead loads, fluid pressures, live loads, lateral earth pressures, roof live loads, snow loads, rain loads, wind loads, and seismic forces, respectively. The quantity f_1 is equal to either 1.0 for places of public assembly, for live loads in excess of 100 psf, and for parking garage live loads or to 0.5 for other live loads, and f_2 is equal to 0.7 for roof configurations (such as sawtooth) that do not shed snow off the structure or to 0.2 for other roof configurations.

Seismic Force Effect, E

According to IBC 1602, Notations, E is the combined effect of horizontal and vertical earthquake forces, $E_h \pm E_v$, as defined in ASCE 2.3.6. According to ASCE 2.3.6, for use in ASCE load combination 6, IBC Equation 16-5, $E = E_h + E_v$. With E_h and E_v as defined by ASCE Equations 12.4-3 and 12.4-4a, respectively, when the effects of gravity and seismic ground motion are additive,

$$E = \rho Q_E + 0.2S_{DS}D$$

where Q_E = effect of the F_x forces obtained by distributing the design base shear V along the height of the structure in the manner prescribed by ASCE Equations 12.8-11 and 12.8-12

ρ = redundancy factor determined in accordance with ASCE 12.3.4 for structures assigned to SDC D, E, or F

= 1.0 for structures assigned to SDC A, B, or C

Also according to ASCE 2.3.6, for use in ASCE load combination 7, IBC Equation 16-7, $E = E_h - E_v$. With E_h and E_v as defined by ASCE Equations 12.4-3 and 12.4-4a, respectively, where the effects of gravity and seismic ground motion are counteractive,

$$E = \rho Q_E - 0.2S_{DS}D$$

Redundancy

The basic premise of the redundancy provisions ASCE 12.3.4 is that the most logical way to determine lack of redundancy is to check whether a component's failure results in an unacceptable amount of story strength loss or in the introduction of extreme torsional irregularity. In ASCE 7-16, the redundancy factor, ρ, is equal to either 1.0 or 1.3, depending on whether an individual element can be removed (deemed to have failed or lost its moment-resisting capabilities) from the seismic force–resisting system without causing the remaining structure to suffer a reduction in story strength of more than 33 percent or creating an extreme torsional irregularity (Horizontal Structural Irregularity Type 1b in ASCE 7 Table 12.3-1).

Braced frame, moment frame, shear wall, and cantilever column systems have to conform to redundancy requirements. Dual systems are included also but in most cases are inherently redundant. Shear walls with a height-to-length ratio greater than 1.0 are included in redundancy considerations to help ensure that an adequate number of wall elements is included or that the proper redundancy factor is applied. Stories resisting more than 35 percent of the base shear only are considered.

The above way of determining ρ obviously involves a number—possibly a significant number—of structural analyses. There is an alternative way of determining ρ essentially by inspection. If a structure is regular in plan and there are at least two bays of seismic force–resisting perimeter framing on each side of the structure in each orthogonal direction (Fig. 5.13) at each story resisting more than 35 percent of the base shear, then ρ = 1. Significantly, the number of bays for a shear wall is to be calculated as the length of the shear wall divided by the story height or two times the length of shear wall divided by the story height for light-framed construction. Thus, the shear wall length plays a major role, and light-framed construction is given an advantage.

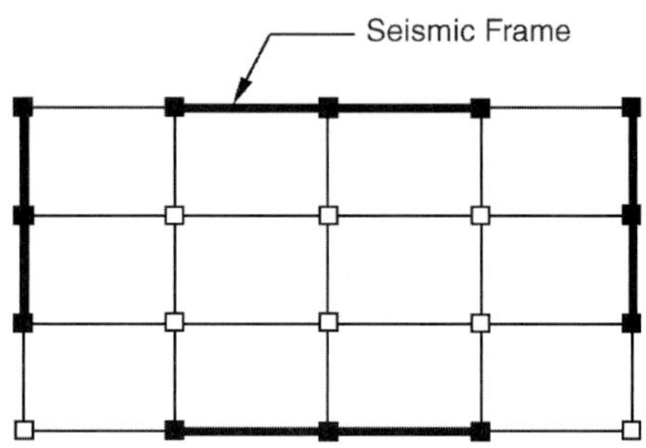

Figure 5.13 An example plan view of building with ρ = 1.0.

The 1.3 value of ρ is based on the results of reliability and redundancy studies as well as engineering judgment and is considered to provide adequate safeguard for a nonredundant building without causing serious overdesign. The threshold story shear of 35 percent of the base shear was arrived at through parametric studies and roughly includes all stories in buildings up to five or six stories tall and about 87 percent of the stories in taller buildings. The redundancy provision is mostly critical for the design of moment frames and braced frames, although shear walls with height-to-length ratios greater than 1.0 and dual systems are also included in the provision. It must be noted, however, that the requirement that the collector elements and their connections be designed for a force that incorporates overstrength generally results in the use of multiple shear walls to reduce the force in the collector elements, making a building with shear walls inherently redundant. Also, dual systems are generally redundant because of the requirements governing their design that are already in place.

ASCE 7-16 adds a user-friendly feature of conveniently listing situations where the redundancy factor, ρ, may be taken as 1.0. The value of ρ is permitted to equal 1.0 for the following:

1. SDC B or C buildings
2. Buildings assigned to SDC D through F that are regular in plan and have at least two bays of seismic force–resisting framing on each side of the building perimeter at each story that resists at least 35 percent of the base shear
3. Drift calculation and P-delta effects
4. Design of nonstructural components
5. Design of nonbuilding structures that are not similar to buildings
6. Design of collector elements, splices, and their connections for which the load combinations with overstrength factor, as provided in ASCE 12.4.3, are required to be used
7. Design of members or connections for which the load combinations with overstrength factor, as provided in ASCE 12.4.3, are required to be used
8. Diaphragm loads determined from ASCE Equation 12.10-1
9. Structures with damping systems designed in accordance with ASCE Chapter 18
10. Design of structural walls for out-of-plane forces, including their anchorage

MAXIMUM SEISMIC FORCE EFFECT, E_M

IBC 1605.1 requires that buildings and other structures and portions thereof be designed to resist the seismic load effects including overstrength factor in accordance with ASCE 2.3.6 and 2.4.5 where required by ASCE Chapters 12, 13, and 15. There are three sections in Chapter 12 that require design to resist seismic load effects including overstrength factor: 12.2.5.2—foundation and other elements used to provide overturning resistance at the base of cantilever column elements (SDC B or higher); 12.3.3.3—elements supporting discontinuous walls or frames of structures having certain horizontal or vertical irregularities (SDC B or higher); and 12.10.2.1—diaphragm collector elements, splices, and their connections to resisting elements (SDC C and above). The load combinations including overstrength factor in ASCE 2.3.6 are

$$1.2D + E_v + E_{mh} + L + 0.2S \qquad \text{(ASCE 2.3.6 Combination 6)}$$

$$0.9D - E_v + E_{mh} \qquad \text{(ASCE 2.3.6 Combination 7)}$$

where $E_{mh} \pm E_v = E_m$ is the maximum effect of horizontal and vertical forces determined from

$E_m = \Omega_o Q_E + 0.2S_{DS}D$ (when effects of gravity and seismic forces are additive; ASCE Equations 12.4-5, 12.4-7, and 12.4-4a)

$E_m = \Omega_o Q_E - 0.2S_{DS}D$ (when effects of gravity and seismic forces counteract; ASCE Equations 12.4-6, 12.4-7, and 12.4-4a)

The overstrength factor Ω_o, which is given in ASCE Table 12.2-1 for the various seismic force–resisting systems, increases the effects of code-prescribed seismic forces to represent the actual forces that may be experienced in a structural member as a result of the design earthquake ground motion (see Fig. 5.3, which also shows the corresponding displacement that is expected when Q_E are amplified by Ω_o). The term $\Omega_o Q_E$ need not exceed the maximum force that can be transferred to the element by the other elements of the seismic force–resisting system.

5.4.8 Diaphragms, Chords, and Collectors

DIAPHRAGMS

For structure assigned to SDC B and above, floor and roof diaphragms are required to be designed to resist the design seismic forces from structural analysis (F_i at floor level i) but not less than F_{px} forces given by ASCE Equation 12.10-1:

$$0.2S_{DS}I_e w_{px} \le F_{px} = \frac{\sum_{i=x}^{n} F_i}{\sum_{i=x}^{n} w_i} w_{px} \le 0.4S_{DS}I_e w_{px}$$

where F_i is the design level force applied to level i (force at level i obtained by distributing V along the height of the structure in accordance with ASCE 12.8.3), w_i is the weight tributary to level i, and w_{px} is the weight tributary to the diaphragm at level x. The above formula relates the design force for the diaphragm to the forces from the vertical distribution of the base shear and the floor weights. Any forces due to offsets in the vertical seismic force–resisting system or changes in lateral stiffness of the vertical elements must be added to the force determined from ASCE Equation 12.10-1.

In older codes and standards, V_{px} used to denote the portion of the seismic shear force at the level of the diaphragm that is required to be transferred through the diaphragm due to offsets or changes in the stiffness of the vertical elements of the seismic force–resisting system above and below the diaphragm. An example of an offset in the vertical elements of the seismic force–resisting system is illustrated in Fig. 5.14. The diaphragm must be designed for the additional force due to this offset.

The redundancy factor, ρ, theoretically applies to the design of diaphragms in structures assigned to SDC D, E, or F. However, for inertial forces calculated in accordance with Equation 12.10-1, the redundancy factor is equal to 1.0. For transfer forces, the redundancy factor ρ for the structure used to be applied. The multiplier was changed to Ω_o, the overstrength factor for the seismic force–resisting system utilized for the structure in the direction of the transfer forces.

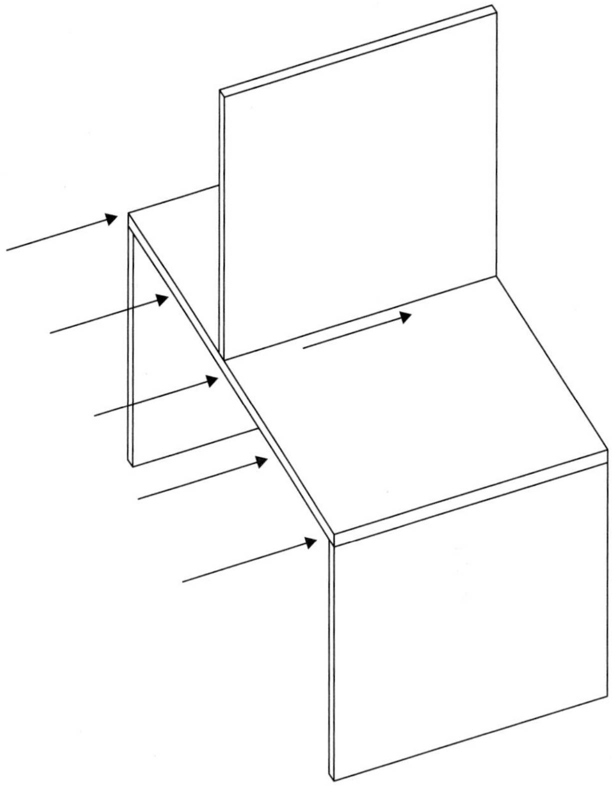

Figure 5.14 Example of offset of vertical components of seismic force–resisting system.

COLLECTOR ELEMENTS

In general, collector elements transfer seismic forces from the diaphragm to the vertical elements of the seismic force–resisting system. Collectors (or drag struts) are required, for example, when shear walls do not extend the full length of the diaphragm in the direction of loading. It is essential that the seismic forces are transferred to the shear walls in order to guarantee a continuous load path. Therefore, IBC 1605.1 and ASCE 12.10.2.1 require that for structures assigned to SDC C or higher, collector elements, splices, and their connections be designed to resist the load combinations with overstrength factor in ASCE 2.3.6.

ALTERNATIVE DIAPHRAGM DESIGN FORCE LEVEL

ASCE 7-16 includes a new Section 12.10.3, Alternative Design Provisions for Diaphragms Including Chords and Collectors, within Section 12.10, Diaphragm Chords and Collectors. The new section provides for an alternative determination of diaphragm design force level, which is mandatory for precast concrete diaphragms in buildings assigned to SDC C, D, E, or F. The alternative is permitted to be used for other precast concrete diaphragms, cast-in-place concrete diaphragms, and wood-sheathed diaphragms on wood framing. Section 12.10.3 does not apply to steel deck diaphragms. ASCE 7-16 also includes newly added Section 14.2.4, containing detailed seismic design provisions for precast concrete diaphragms and including a connector qualification protocol. 2018 IBC Chapter 19 requires precast concrete diaphragms assigned to SDC C or above to be designed by ASCE 14.2.4

ASCE 7-16 12.10.3 presents a near-elastic diaphragm design force as the statistical sum of first mode effect and higher mode effects [5.16]. The first–mode effect is reduced by the R-factor of the seismic force–resisting system but then amplified by the overstrength factor, Ω_o, because vertical element overstrength will generate higher first-mode forces in the diaphragm. The effect caused by higher-mode response is not reduced. In recognition of the deformation capacity and overstrength of the diaphragm, the elastic diaphragm force from the first and higher modes of response is then reduced by a diaphragm force reduction factor, R_s.

In structures assigned to SDC C, D, E, or F; collectors; and their connections, including connections to vertical elements, are required to be designed to resist 1.5 times the diaphragm inertial forces from ASCE 12.10.3.2 plus 1.5 times the design transfer forces. The 1.5 is a smaller multiplier than the Ω_o used with the design force level of ASCE 12.10.1 because the diaphragm forces given in ASCE 12.10.3 are felt to be more accurately determined.

5.4.9 Bearing Walls, Shear Walls, and Their Anchorage

In addition to seismic forces in the plane of the wall, structural walls (bearing walls or shear walls, as defined in ASCE 11.2) in structures assigned to SDC B and higher are to be designed to resist an out-of-plane force F_p equal to $0.4S_{DS}I_e$ times the weight of the wall with a minimum force of 10 percent of the weight of the wall (ASCE 12.11.1).

Additionally, structural walls are to be anchored to the roof and floor members that provide lateral support for the wall. The anchorage of structural walls to supporting construction shall provide a direct connection capable of resisting the following:

$$F_p = 0.4S_{DS}\, k_a\, I_e\, W_p \qquad \text{(ASCE Equation 12.11-1)}$$

F_p shall not be taken less than $0.2k_a\, I_e W_p$

$$k_a = 1.0 + L_f/100 \qquad \text{(ASCE Equation 12.11-2)}$$

k_a need not be taken larger than 2.0

where k_a = amplification factor for diaphragm flexibility

L_f = the span, in feet, of a flexible diaphragm that provides the lateral support for the wall; the span is measured between vertical elements that provide lateral support to the diaphragm in the direction considered; use zero for rigid diaphragms (Fig. 5.15)

W_p = the weight of the wall tributary to the anchor

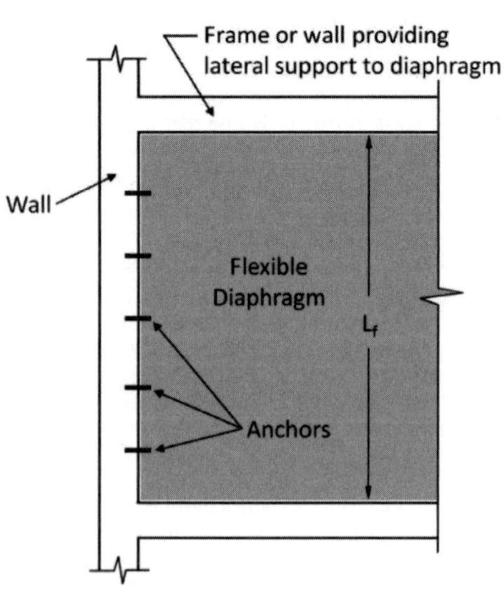

Figure 5.15 Anchorage of walls to flexible diaphragm.

Where the anchorage is not located at the roof and all diaphragms are not flexible, the value from ASCE Equation 12.11-1 is permitted to be multiplied by the factor $(1 + 2z/h)/3$, where z is the height of the anchor above the base of the structure and h is the height of the roof above the base.

Structural walls shall be designed to resist bending between anchors where the anchor spacing exceeds 4 feet.

5.4.10 Direction of Seismic Load

ASCE 12.5.1 requires for all structures that the seismic forces be applied in a direction that will produce the most critical load effects in the members. For structures assigned to SDC B, this requirement is satisfied if the design seismic forces are applied separately and independently in each of two orthogonal directions (ASCE 12.5.2).

SDC C structures that have horizontal irregularity Type 5 in ASCE Table 12.3-1 (nonparallel systems irregularity) are required to satisfy the provisions in ASCE 12.5.3. This section requires that such a structure be analyzed for seismic forces using a three-dimensional representation in accordance with item 1 or 2 below:

1. The structure must be analyzed using the equivalent lateral force analysis procedure of ASCE 12.8, the modal response spectrum analysis procedure of ASCE 12.9.1, or the linear response history analysis procedure of ASCE 12.9.2, with the loading applied independently in any two orthogonal directions. The critical direction requirement of ASCE 12.5.1 is deemed to be satisfied if the members are designed for the following orthogonal combination of the applied loads: 100 percent of the forces for one direction plus 30 percent of the forces for the perpendicular direction (each member is designed for the combination producing the critical load effects).
2. The structure must be analyzed using the linear response history analysis procedure of ASCE 12.9.2 or the nonlinear response history analysis procedure of ASCE Chapter 16, with orthogonal pairs of ground motion acceleration histories applied simultaneously.

For structures assigned to SDC D or higher, the orthogonal combination procedure is additionally required by ASCE 12.5.4 for columns or walls that form part of two or more intersecting seismic force–resisting systems and are subjected to axial load due to seismic forces greater than or equal to 20 percent of the axial load design strength. The orthogonal combination procedure is the 100 + 30 percent procedure of ASCE 12.5.3 described above.

5.4.11 Detailing Requirements

As discussed previously, the design seismic forces determined by the analysis procedures described in Section 5.4.5 are only a fraction of the actual forces that an elastic structure may experience during the design earthquake of ASCE 7. For design purposes, the seismic forces are reduced since it is impractical as well as uneconomical to design a structure to remain elastic in such situations. In order to prevent collapse, structures must be properly detailed so that they are able to dissipate the earthquake energy through inelastic deformations.

The 2018 IBC references the requirements of ACI 318-14 for the design of reinforced-concrete structures, with a few modifications. Chapter 18 of ACI 318-14 contains the design and detailing requirements for structures assigned to SDC B or higher. A summary of the sections of Chapter 18 to be satisfied as a function of the SDC for various types of components resisting earthquake effects is given below in Table 5.4.

REFERENCES

5.1. International Code Council, *International Building Code,* 2018 IBC, Publications, Country Club Hills, IL, 2018.

5.2. American Society of Civil Engineers, *ASCE Standard Minimum Design Loads and Associated Criteria for Buildings and Other Structures,* ASCE 7-16, Reston, VA, 2016.

5.3. Building Seismic Safety Council, *NEHRP (National Earthquake Hazards Reduction Program) Recommended Provisions for Seismic Regulations for New Buildings and Other Structures,* FEMA P-1050, 2015 Edition, Washington, DC, 2015.

Table 5.4 Sections of ACI 318 Chapter 18 to Be Satisfied*

Component resisting earthquake effect, unless otherwise noted	Seismic design category			
	A (None)	B (18.2.1.3)	C (18.2.1.4)	D,E,F (18.2.1.5)
Analysis and design requirements		18.2.2	18.2.2	18.2.2, 18.2.4
Materials		None	None	18.2.5–18.2.8
Frame members		18.3	18.4	18.6, 7, 8, 9
Structural walls and coupling beams		None	None	18.10
Precast structural walls		None	18.5	18.5,[†] 18.11
Structural diaphragms and trusses	None	None	None	18.12
Foundations		None	None	18.13
Frame members not proportioned to resist forces induced by earthquake motions		None	None	18.14
Anchors		None	18.2.3	18.2.3

*In addition to requirements of Chapters 1 through 17 and 19 through 26 and ACI 318.2, except as modified by Chapter 18. Section 14.1.4 also applies in SDC D, E, and F.

[†]As permitted by the legally adopted general building code of which this Code [ACI 318] forms a part.

5.4. American Concrete Institute, *Building Code Requirements for Structural Concrete (ACI 318-14) and Commentary (ACI 318R-14)*, American Concrete Institute, Farmington Hills, MI, 2014.

5.5. Ghosh, S. K., and Henry, J., *IBC Handbook—Structural Provisions*, International Code Council, Country Club Hills, IL, 2009.

5.6. International Code Council, *International Residential Code*, International Code Council, Country Club Hills, IL, 2018.

5.7. International Conference of Building Officials, *Uniform Building Code*, International Conference of Building Officials, Whittier, CA, 1994, 1997.

5.8. International Code Council, *International Existing Buildings Code*, International Code Council, Country Club Hills, IL, 2018.

5.9. Ghosh, S. K., *Seismic Design Using Structural Dynamics (2015 IBC/ASCE 7-10/ACI 318-14)*, International Conference of Building Officials, Country Club Hills, IL, 2016.

5.10. American Society for Testing and Materials, *Specification for Penetration Test and Split-Barrel Sampling of Soils*, ASTM D 1586-04, ASTM, West Conshohocken, PA, 2004.

5.11. American Society for Testing and Materials, *Test Method for Unconfined Compressive Strength of Cohesive Soil*, ASTM D2166-00, ASTM, West Conshohocken, PA, 2000.

5.12. American Society for Testing and Materials, *Test Method for Unconsolidated, Undrained Triaxial Compression Test on Cohesive Soils*, ASTM D2850-03A, ASTM, West Conshohocken, PA, 2003.

5.13. S. K. Ghosh Associates LLC, *CodeMaster—Seismic Design Category*, S. K. Ghosh Associates LLC, Palatine, IL, 2018.

5.14. Applied Technology Council, *Tentative Provisions for the Development of Seismic Regulations for Buildings*, ATC Publication ATC 3-06, NBS Special Publication 510, NSF Publication 78-8, U.S. Government Printing Office, Washington DC, 1978.

5.15. American Institute of Steel Construction, *Seismic Provisions for Structural Steel Buildings*, AISC 341-16, Chicago, IL, 2016.

5.16. Rodriguez, M., Restrepo, J. I., and Carr, A. J., "Earthquake Induced Floor Horizontal Accelerations in Buildings," *Earthquake Engineering—Structural Dynamics*, Vol. 31, 2002, pp. 693–718.

Chapter 6
Fracture and Fatigue

PART A
CONCRETE AND COMPOSITES

BY

KEDAR S. KIRANE *Assistant Professor of Mechanical Engineering, Stony Brook University, Stony Brook, New York*
ZDENĚK P. BAŽANT *McCormick Institute Professor and Walter P. Murphy Professor of Civil and Environmental Engineering, Mechanical Engineering and Material Science and Engineering, Northwestern University, Evanston, Illinois*

6.1 INTRODUCTION TO QUASIBRITTLE FRACTURE

Concrete and fiber-reinforced composites are two of the most widely used structural materials. Both are composite materials consisting of multiple constituents and have a highly heterogeneous microstructure. Concrete typically consists of aggregate bonded by hardened cement paste while fiber composites typically consist of a weaker matrix material reinforced by strong fibers, which may be discontinuous or continuous, consisting of parallel fibers of a fabric. The fracturing behavior of these materials differs from metals. They exhibit almost no plasticity. They fail by propagation of a macrocrack having at its front a fracture process zone (FPZ), which contains microcracks and microslips and normally has a size not negligible compared to structural dimensions.

The fracture propagation starts either from a preexisting macrocrack or by formation of an FPZ in a region with a stress concentration, often at the boundary. The large size of the FPZ, which is dictated by material heterogeneity and is approximately a constant material property, endows the material with a significant capability to dissipate energy, which can provide certain measure of structural ductility. So, even though these materials are colloquially referred to as brittle, in scientific discourse they are properly termed "quasibrittle." But note that this term is relative. When the structure becomes so large that the FPZ size is negligible, a quasibrittle structure becomes "brittle." Vice versa, when a brittle structure, made, for example, of fine grained mortar, becomes sufficiently small, its behavior becomes quasibrittle. There are many quasibrittle materials—aside from concrete and fiber composites, such as most rocks, coal, wood, sea ice, coarse-grained ceramics, rigid foams, paper, carton, bone, and bio- and bio-inspired materials.[1]

To assess the strength of structures with cracks, fracture mechanics must be used.[2] Today there are three types of fracture mechanics: (1) linear elastic fracture mechanics (LEFM), which was originated by in 1921 by Griffith[3] and characterizes brittle failure; it deals with sharp cracks whose FPZ is negligible compared to structure dimension (e.g., the FPZ in fatigued steel has micrometer dimensions). (2) The ductile fracture mechanics,[2] was developed fully during the 1960s, in which there is a long and wide plastic (or plastic-hardening) zone in front of the crack but the FPZ is still negligible; and (3) quasibrittle fracture mechanics,[1] there is a large FPZ in front of the crack, with almost no plastic zone surrounding it. The ductile, as well as quasibrittle, fracture mechanics is approximately treated by various adaptations of LEFM.

In LEFM, the crack propagation criterion can be stated either in terms of the fracture energy, G_f, which represents the energy required to extend the crack by a unit area, or in terms of the critical stress intensity factor, called the fracture toughness, K_c. Their relation, due to Irwin, is

$$K_c = \sqrt{G_f E'} \tag{6.1}$$

where, for plane stress, $E' = E$ = Young's modulus of elasticity and, for plane strain, $E' = E/(1 - \nu^2)$ where ν is the Poisson's ratio. The crack can grow and possibly cause structural failure when K becomes equal to K_c.

For quasibrittle materials, however, due to the FPZ, a significant part of the structure volume acts nonlinearly, making LEFM inapplicable. A simple but effective treatment of this nonlinearity is the equivalent LEFM approach.[1,2] In this approach, the nonlinear FPZ is accounted for by assuming that its effect on the compliance and load capacity of the structure is essentially the same as if an LEFM sharp crack were extended, by distance c_f, roughly into the middle of the FPZ.[1] The longer crack, of length $a_e = a_0 + c_f$, where c_f is approximately constant, is called the effective or equivalent crack (see Fig. 6.1). The remaining treatment is similar to LEFM, as used for example, for fatigue-embrittled steel. The equivalent LEFM can be applied for crack initiation even when only the FPZ, but no actual crack, is as yet present, which is not the case with LEFM.

Some rules have to be introduced to express how the equivalent crack extends under increased loads. For instance, the fracture energy G_f for equivalent LEFM (alternately interpreted as the internal material resistance R to crack growth) is not a constant but varies with the crack length, becoming a constant for larger crack sizes. The plot of this internal resistance R versus the crack extension Δa from the notch is referred to as the R-curve[1,2] (or resistance curve). This curve can be empirically determined from lab tests.

An important aspect that must be considered is the structure size effect on its nominal strength, defined as $\sigma_N = P_{max}/A$, where P_{max} is the load capacity (maximum load) and A the structural cross-section area (which can be taken arbitrarily but must be homologous for various sizes). A host of past studies[1,4] have established the close interplay between failure load, FPZ size, and the resulting scaling in the structural strength.

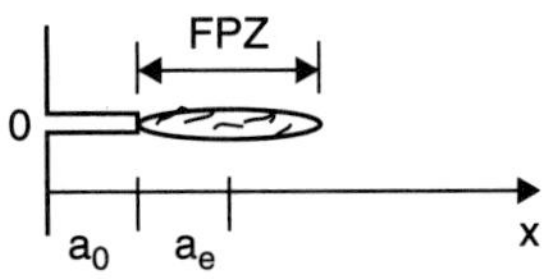

Figure 6.1 Equivalent LEFM crack.

Size effects of two basic simple types can be distinguished. Consider first the *type I* size effect,[4] which occurs in unnotched or un-precracked structures that fail when a macrocrack initiates from a smooth surface in the presence of a stress gradient. It is explained by the stress redistribution due to FPZ formation, the equivalent LEFM crack becoming nonzero.[5,6] In the absence of the statistical part, which is important for very large structures, the size effect on the nominal structure strength is

$$\sigma_N = \sigma_0\left(1+\frac{rl_0}{D}\right)^{1/r} \tag{6.2}$$

Here D is the structure size, σ_0 is the asymptotic strength for a structure of infinite size, l_o is a length constant proportional to the material characteristic length scale, and r is a dimensionless empirical constant with typical value 1.45.

For very large sizes ($D \rightarrow \infty$) this equation predicts a vanishing size effect. However, in reality, in structures of sizes much larger than the FPZ, the strength in type I failures follows the Weibull statistical size effect.[7] A smooth transition to the statistical size effect can be incorporated in Eq. 6.2 by modifying it to a generalized energetic-statistical size effect[5] law as follows:

$$\sigma_N = \sigma_0\left(\left(\frac{l_0}{D}\right)^{rn/m}+\frac{rl_0}{D}\right)^{1/r} \tag{6.3}$$

Here n is the number of dimensions of the failure mode (1, 2, or 3) and m is the Weibull modulus, typically 24 for concrete and 8 to 10 for fiber composites.[8–10] A best fit of Eq. 6.3 is shown in Fig. 6.2 to a variety of test data on concrete and composites.

The degree of brittleness (opposite of ductility) is important for the probabilistic distribution function (pdf) of structural strength and of lifetime under cyclic and static fatigue. For brittle materials, the pdf of strength and fatigue lifetimes is of Weibull type,[6] while for ductile materials it is Gaussian (or normal). In the central range both pdfs are hard to distinguish, but the very low failure probability point (P_f~10^{-6}) in the Weibull pdf is almost twice as far from the mean as for the Gaussian [for the same mean and the typical coefficient of variation of errors (CoV)]. For quasibrittle structures, the pdf is a hybrid of both.[11–13] It consists of a Weibull tail grafted on the left onto a Gaussian core. For structures including one or a few FPZs, the grafting point is at $P_f \sim 0.001$. As the structure size increases, the grafting point moves to the right and when the structure volume becomes >10^5 times the FPZ, the entire pdf becomes Weibullian. This is important for setting the safety factors for design, which should ensure $P_f < 10^{-6}$ per lifetime (which is about 10^4 times lower than one's probability of dying in a car accident and about the same as the probability of being killed by a lightning).

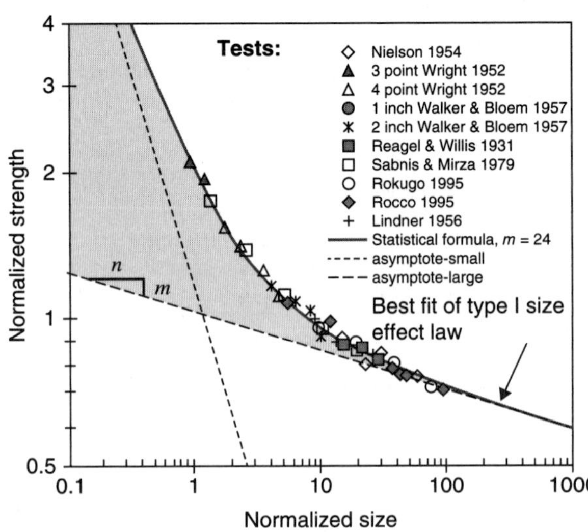

Figure 6.2 Type I size effect (measured for concrete and composites)[5] [reproduced with permission from American Concrete Institute (ACI)].

In structures with a large traction-free crack at peak load (or a sharp notch) the *type II* size effect is observed.[1,4,11,14–22] This is a purely deterministic size effect on the mean strength (while material randomness affects only the CoV). It becomes the strongest for large sizes and corresponds to LEFM (i.e., $\sigma_N \propto D^{-1/2}$). As long as the failure modes for different sizes of geometrically scaled structures are also geometrically similar (which is often the case), the strength follows the Bažant size effect law[1,4,11]

$$\sigma_N = Bf_t'\left(1+\frac{D}{D_0}\right)^{-1/2} \tag{6.4}$$

Here B is an empirical, dimensionless material constant f_t' is the local tensile strength of material, and D_0 is a transitional size proportional to the material characteristic length scale, which in turn is proportional to the FPZ size. The size effect factor in Eq. 6.4, proposed to ACI (American Concrete Institute) in 1984, was in 2019 incorporated into the ACI design code (ASI Standard 318-2019) for beam shear, punching and strut-and-tie model.

The size effects of types I and II have been observed in concrete and in fiber-reinforced composites, aside from other quasibrittle materials[1,4,11] (see Figs. 6.2 and 6.3). Both size effect laws consist of a length scale determined by the FPZ size, which is a material property. This length is essential for realistic mathematical prediction of structure strength. The rest of this chapter describes the fracturing behavior of concrete and fiber composites under various loading conditions which leads to the formation of the FPZ. Most of these failure modes are quasibrittle, exhibit the R curve, transitional size effects on strength, and hybrid Gauss–Weibull probability distributions of strength.

6.2 CONCRETE

Concrete in its cured form may be treated as isotropic and homogeneous in the macroscopic sense. On the microstructural level it is highly heterogeneous, full of cracks of sizes at all scales from the nanoscale up.[23] Its mechanical response depends highly on the confinement. The uniaxial strength in tension, typically 4–5 MPa, is normally 8–10 times weaker than in compression. This is because the aggregate pieces can withstand large compressive stresses, but under tension the aggregate-cement bond is much weaker.[24,25]

6.2.1 Stress-Strain Curve

The typical stress-strain curve of concrete[24] under uniaxial compression as well as tension is shown in Fig. 6.4. It consists of two stages—prepeak and postpeak. The initial part is linear elastic. As the load increases, formation of microcracks and frictional microslips induces a prepeak nonlinearity in the stress-strain curve, more pronounced in compression than tension.

Beyond the peak, the uniaxial stress decreases gradually at increasing strain, which is called postpeak softening and can be observed in a stable manner only in a testing machine with a sufficiently stiff frame. The stability of postpeak testing can be greatly enhanced by controlling some displacement that is monotonically increasing (e.g., the crack opening displacement). Complete failure corresponds to stress reduction to zero. The area under the stress-strain curve represents the energy dissipated per unit volume.

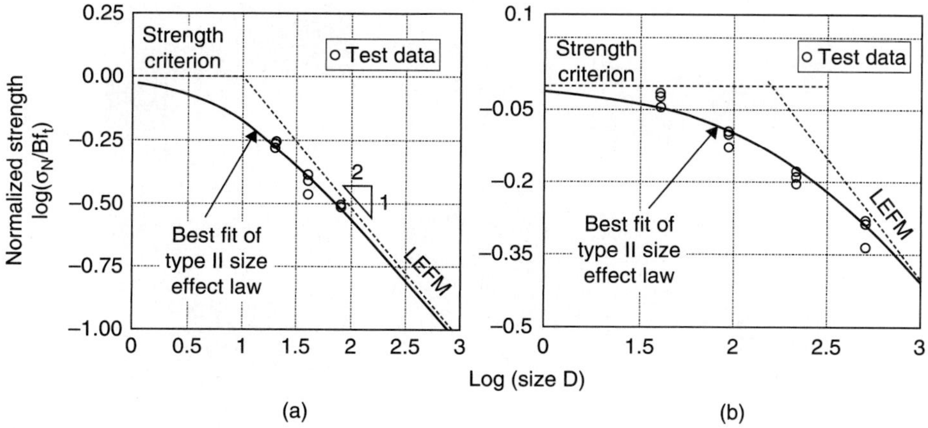

Figure 6.3 Type II size effect. (*a*) carbon fiber composite[22] test data of 2016 and (*b*) concrete[18] test data of 2014 [reproduced with permission from Elsevier].

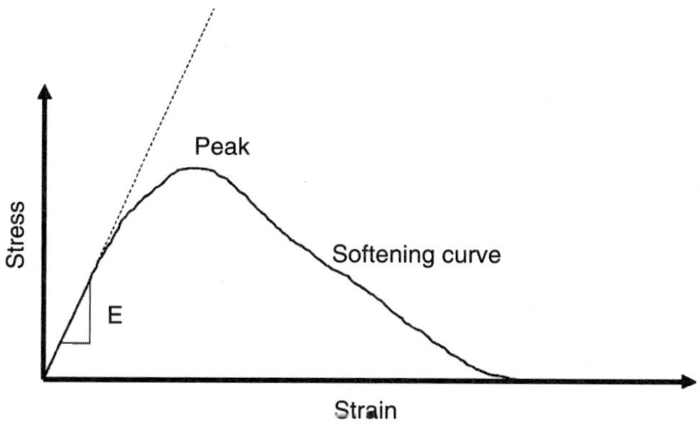

Figure 6.4 Sketch of a typical uniaxial stress-strain curve of concrete.

In specimens or structures larger than a few FPZs, the postpeak softening is inevitably accompanied by localization of fracturing damage into a band of a finite width.[1,26] Under uniaxial tension, this band forms along a plane normal to the loading direction, as shown in Fig. 6.5(*a*); see also Fig. 6.6(*a*) for tensile fracture under bending. Under uniaxial compression, various failure mechanisms are possible depending on the boundary conditions. If the ends are sliding, a single dominant axial splitting crack forms where damage localizes along a normal plane due to tensile strains generated in the lateral direction (perpendicular to the loading direction). The splitting is caused by elastic Poisson effect but mainly inelastic mechanisms such as inclined slips and wedging of one aggregate between two,[21] as shown in Fig. 6.5(*b*) and Fig. 6.6(*b*). The axial splitting fracture causes no size effect because no strain energy is released from outside the band.[1,2] But if one or both ends are fixed, a compression-shear failure band forms where the damage localizes along one or two inclined planes as shown in Fig. 6.5(*c*) and (*d*). This failure actually consists of parallel small axial splitting cracks arranged into an inclined band and does cause a size effect.[1]

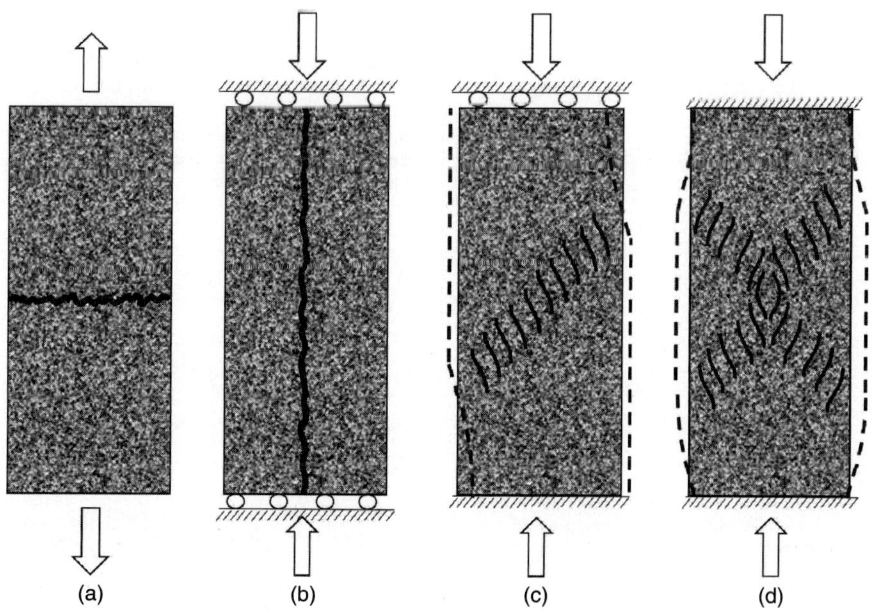

Figure 6.5 (*a*) Damage localization under uniaxial tension. (*b*) Axial splitting failure under uniaxial compression with sliding ends. (*c*) Compression-shear failure under uniaxial compression with one end fixed. (*d*) Compression-shear failure under uniaxial compression with both ends fixed.[1]

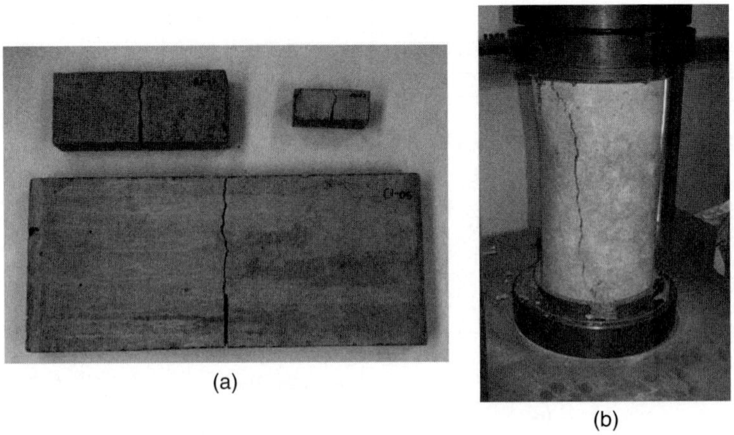

Figure 6.6 (*a*) Tensile failure in concrete under three-point bending. (*b*) Axial splitting compressive failure of a concrete cylinder [reproduced with permission from Elsevier].

6.2.2 Multiaxial Fracture

In majority of structures, concrete is subjected to multiaxial stress. The biaxial compression strength of concrete is higher than uniaxial, especially in the presence of end restraints.[24,25] Biaxial tension makes little difference for strength from uniaxial. Under combined tension and compression, concrete is the weakest. The failure under biaxial loads also occurs by localization along preferred orientations, typically perpendicular to the maximum principal tensile stress,[27] and also exhibits size effects.

Triaxial compressive loading of concrete can cause a major increase in its strength. Under hydrostatic compression as well as uniaxial compressive strain with virtually rigid confinement, concrete never fails and only densifies by pore collapse. The postpeak behavior varies dramatically depending on the degree of confinement. Increasing confinement from none, the response becomes increasingly ductile and eventually switches from softening to hardening[21,24,25,28] as shown in Fig. 6.7. At the microscale, frictional slip becomes dominant over microcracking.

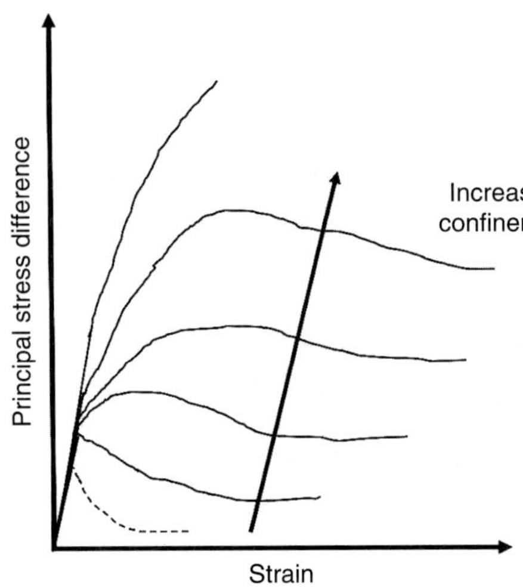

Figure 6.7 Schematic behavior of concrete under confined compression (dashed line indicates uniaxial compression).

6.2.3 Fatigue

Ideally, the strength of reinforced concrete structures depends on steel bars, and even though large hairline cracks, traversing 50 to 80 percent of the structure width, typically develop, they have little effect on structural strength under monotonic as well as cyclic loading. Thus, reinforced concrete is generally much less sensitive to fatigue than metals. Nevertheless, like in all materials, cracks grow under cyclic as well as sustained loads, caused, for example, by traffic, wind, rotating machinery, or earthquakes.[28] This can be quite important for some situations (plain concrete, zones of low reinforcement, development of excessive crack width allowing ingress of corrosive agents, etc.).

In fatigue, concrete is markedly different from metals.[1,26–30] Fatigue occurs by activation of preexisting microcracks and microslips within the cyclic FPZ. Since similar microcracks also form during monotonic loading, the fatigue crack surfaces in concrete have no peculiar topography unlike metals, and are harder to distinguish visually. The only difference is that the cyclic FPZ is smaller in size compared to the monotonic FPZ.[31,32]

The fatigue behavior of concrete depends on the strength of the material, the loading (magnitude, waveform, frequency, etc.) and the environmental conditions. Concrete undergoes fatigue under both tension and compression and in fact compression fatigue bands can grow transversely to compressive stress.[29] At the microstructure level, fatigue consists of irreversible microcracking damage. The load displacement curve exhibits for each unload-reload cycle a hysteresis loop (Fig. 6.8) whose area equals the energy dissipated in that cycle. The elastic modulus degrades gradually with increasing inelastic deformation in each cycle. The cyclic load-displacement curve is enveloped by the monotonic one.[33–35]

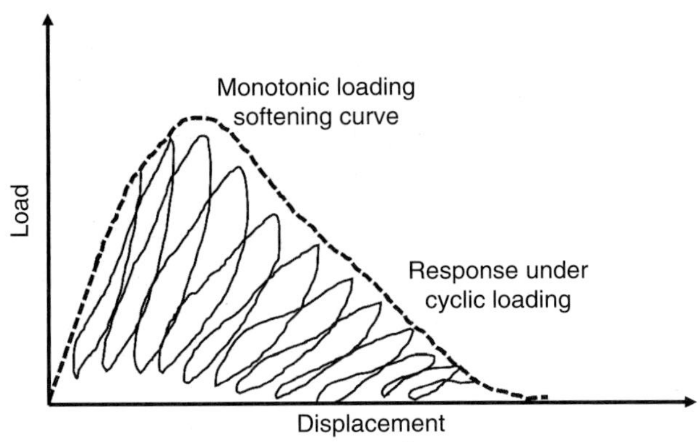

Figure 6.8 Schematic of unload-reload hysteresis loops in concrete fatigue.[10]

6.2.4 Design for Fatigue

Widely accepted for the fatigue assessment of concrete (under uniaxial compression, without stress reversal) is the Aas-Jakobsen equation[28,36] for the so-called S-N curve (stress S vs. number N of cycles)

$$\frac{\sigma_{c,\max}}{f_c} = 1 - \beta(1-R)\log N \tag{6.5}$$

which represents a simplified, total life approach and does not consider the propagation of damage; $\sigma_{c,\max}$ is the maximum stress level of the cyclic loading, f_c is the compressive strength of concrete, R is the load ratio ($\sigma_{\min}/\sigma_{\max}$), N is the number of cycles to failure and β is an empirical material constant whose typical value is 0.0685. See Fig. 6.9 for a typical S-N curve for plain concrete.

There are alternate expressions for fatigue lifetimes in various design codes:[38]

(i) For pure compression

$$\log N = \frac{8}{Y-1}(S_{c,\max}-1) \text{ if } \log N < 8$$

$$\log N = 8 + \frac{8\ln(10)}{Y-1}(Y - S_{c,\min})\log\left(\frac{S_{c,\max}-S_{c,\min}}{Y-S_{c,\min}}\right) \text{ if } \log N > 8 \tag{6.6}$$

$$\text{with } Y = \frac{0.45+1.8S_{c,\min}}{1+1.8S_{c,\min}-0.3S_{c,\min}^2} \text{ and } S_{c,\min} = \frac{|\sigma_{c,\min}|}{f_{c,\text{fat}}};\ S_{c,\max} = \frac{|\sigma_{c,\max}|}{f_{c,\text{fat}}} \tag{6.7}$$

(ii) For compression-tension with $\sigma_{ct,\max} < 0.026|\sigma_{c,\max}|$,

$$\log N = 9(1 - S_{c,\max}) \tag{6.8}$$

The fatigue reference compressive strength $f_{c,\text{fat}}$ has been introduced to take into account the increase of fatigue sensitivity of concrete with increasing compressive strength. It is given by

$$f_{c,\text{fat}} = 0.85\beta_{cc}(t)\beta_{c,\text{sus}}f_c\left(1-\frac{f_c}{400}\right) \tag{6.9}$$

Here $\beta_{cc}(t)$ is an empirical coefficient which depends on the age t of concrete (in days) when fatigue loading starts, and $\beta_{c,\text{sus}}$ is another empirical coefficient which takes into account the effect of high mean stresses during loading and its typical value is 0.85 when all the stresses are expressed in MPa.

(iii) For pure tension and tension-compression with $\sigma_{ct,\max} \geq 0.026|\sigma_{c,\max}|$

$$\log N = 12(1 - S_{ct,\max}) \text{ where } S_{ct,\max} = \frac{|\sigma_{ct,\max}|}{f_t} \tag{6.10}$$

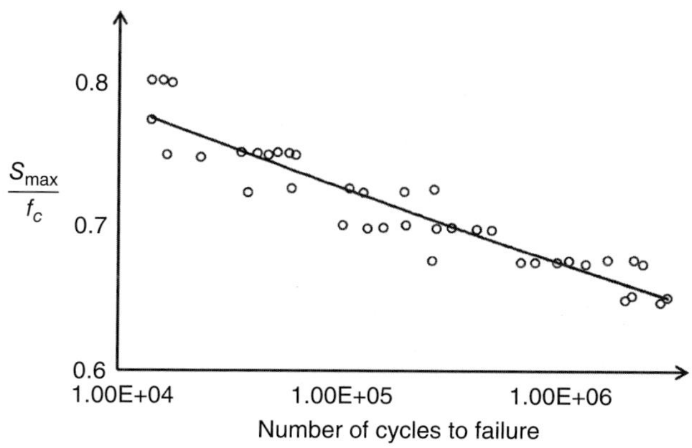

Figure 6.9 S-N curve for typical plain concrete (compression)[37] [reproduced with permission from American Concrete Institute (ACI)].

Here $\sigma_{ct,\max}$ is the maximum tensile stress and f_t is the minimum characteristic tensile strength. Thus, Eq. 6.10 prescribes the fatigue life under tension to depend only on the maximum stress level. Under variable amplitude loading, the Miner rule[29] may be applied.

The total life approach is much simpler but conservative and highly empirical. The less conservative, damage-tolerant approach is usually not pursued for concrete, although it is possible. This is because, similar to metals, fatigue cracks in concrete also grow according to the Paris law.[39] This has long been considered as an empirical law relating the rate of crack growth per cycle, da/dN, to the amplitude ΔK of the stress intensity factor K through a power law (although recently it was derived from the activation energy–controlled breakage rate of interatomic bonds[40]). The Paris law reads

$$\frac{da}{dN} = C(\Delta K)^m \tag{6.11}$$

The typical value of exponent m for concrete is 8 to 12,[41] which is much higher than what is seen for metals (2 to 4). The coefficient C has been found[29] to be sensitive to environmental conditions (such as temperature), loading frequency as well as the load ratio R. Under tension-tension fatigue, a distinct size effect has also been reported in the Paris law for concrete which appears to affect the coefficient C but not the exponent m.[31,41] This makes the Paris law plots of concrete specimens of different sizes appear as parallel lines spaced apart on log-log scale. It is expected that this size effect would vanish for very large sizes but providing direct experimental evidence is difficult since this is beyond most lab-scale specimen sizes (as the cyclic FPZ is about 1 foot long). This experimental observation also implies a significant size effect in the fatigue lifetimes of concrete structures. The size effect in fatigue of concrete is a topic of ongoing research.[11,31]

6.3 FIBER-REINFORCED COMPOSITES

Due to their stiffness, strength, corrosion resistance, fatigue resistance, and, most important light weight, fiber-reinforced composites are widely used in aerospace, naval, and automotive engineering. They are also becoming increasingly prevalent in infrastructure applications—mainly for structural upgrades such as rapid repair, rehabilitation, and seismic retrofitting of existing structures.

Their basic constituents are glass or carbon fibers and a polymer matrix. The fibers carry most of the load and provide most of the stiffness and tensile strength of the material. The matrix and the fiber-matrix interface bond contribute strongly to the composite's shear strength, transverse strength, and, also, fatigue strength.[42] The matrix transfers the load between the individual fibers and acts as the agent holding the fiber structure in place. Since the fibers have specific directionality, a composite is always anisotropic (or orthotropic) at the macroscale, unlike concrete, which is essentially isotropic. In structural applications, the anisotropy is often mitigated by using quasi-isotropic composite lay-ups consisting of individual layers (or pre-pregs) oriented in various directions.

Composites undergo fracture through a variety of mechanisms depending on the applied load and its angle with the fiber direction. The fracture is progressive in nature and occurs at a number of length scales, from the barely observable phenomenon of fiber-matrix debonding on the microscale to the matrix cracking, fiber breaking and delamination[43] on the macroscale. Often one failure mechanism triggers some others, depending on the specific loading scenario. Numerical modeling and prediction of failure of composites under multiaxial loads is not a straightforward task.

6.3.1 Longitudinal Fracture

When the applied load is tensile, in-plane and aligned with the fibers, the dominant mode of failure is fiber breaking. The longitudinal tensile strength of the composite lamina is governed by the tensile strength of fibers. Ideally, if all the fibers had equal strength, they would break simultaneously when the applied load became equal to their strength. However, in reality, the fibers have a strength distribution. So, typically, the weakest fibers fail first. The stress then gets redistributed to adjacent fibers, which makes them more susceptible to failure. This usually also triggers other failure modes such as localized matrix shear failure around the fiber breaks and fiber-matrix debonding. Eventually the adjacent fibers break, and the smaller cracks merge to produce one large crack.[42] This crack is essentially the localized zone of damage that propagates to cause the final failure, as shown in Fig. 6.10(*a*). At the end, additional mechanisms, such as fiber pullout, also appear [see Fig. 6.11(*a*)].

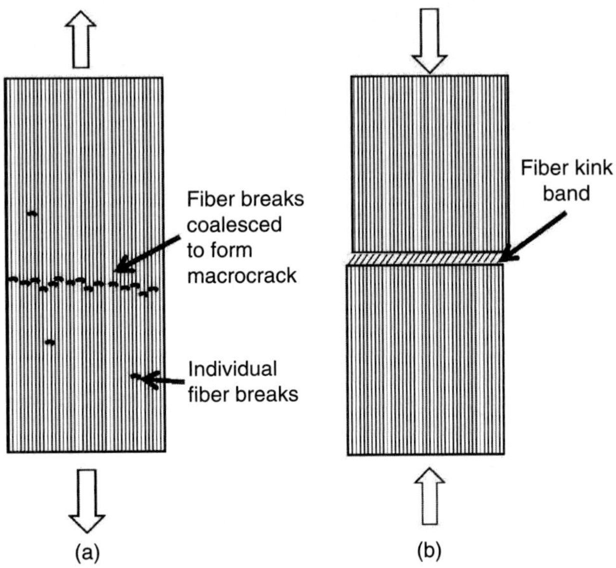

Figure 6.10 Failure of unidirectional composite lamina under uniaxial load (*a*) tension and (*b*) compression.

Under compression, the response differs vastly from tension, just like in concrete. However, unlike concrete, the strength of composites in compression is usually lower than in tension. This is due to the fact that under compression, composites typically fail by formation of a kink band[43–49] whose onset is triggered by a buckling instability in the fibers, as shown in Fig. 6.10(*b*) and Fig. 6.11(*b*). Thus, the ultimate compressive load capacity of composites depends less on the compressive strength of fibers than the shear strength of the matrix, whose main role is to resist the microbuckling of fibers under compression.

A compression kink band fails not simultaneously but by propagation of the fiber buckling front, causing a size effect, similar to tensile fracture.[1,50] This size effect is similar to that caused by a localized damage band propagation under tension and is important to consider especially for design of larger structures.

Unlike concrete and rock, measuring the postpeak softening curve of fiber composites in a stable fashion has never succeeded until recently.[51] In the 1960s, stable measurement of postpeak in concrete and rock was made possible by the discovery that the testing frame had to be an order of magnitude stiffer than previously used, and stability of test was further helped by developing fast servo-control and by controlling the test by a crack mouth or crack tip gauge.[51] However, for composites it has not work until 2016 when it was found that the grips (or fixtures) for tensile loading of fracture tests of composites supplied by manufactures were far too soft and light (note that no such grips are needed for concrete since the specimen is loaded by contact with the loading frame head or glued in the case of tension). Upon designing loading grips that are two orders of magnitude stiffer and one order of magnitude heavier than those traditionally supplied by the manufacturer, a stable postpeak in textile composite fracture is now found to be observed easily[52] (even under load point control; U.S. Patent US 10.416.053 B2, 2019).

6.3.2 Transverse and Shear Fracture

When the load is transverse to the fibers, the failure becomes matrix dominated. Stress concentrations arise at the fiber-matrix interface leading to debonding and matrix microcracking and microslips. This can occur at a variety of locations. Eventually the microcracks propagate and coalesce becoming one large macrocrack, which ultimately leads to failure, as shown in Fig. 6.12. Interface debonding and fiber splitting may also develop. Under the in-plane shear loading, too, the failure is governed by the properties of the matrix but in that case, the cracks often form parallel to the fibers.

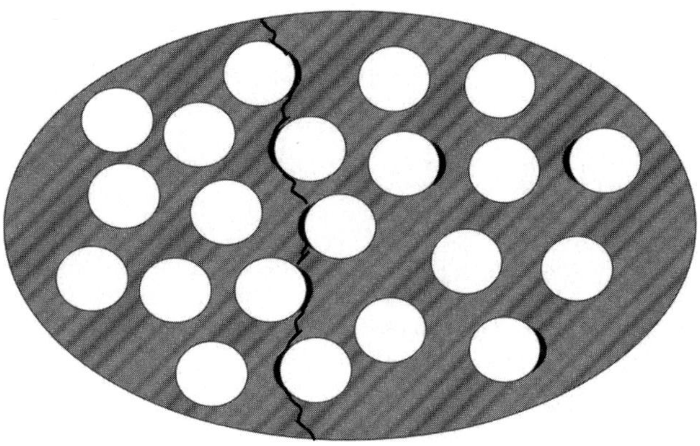

Figure 6.12 Debonding-induced microcracks coalesced into a larger crack under transverse load.

6.3.3 Multiaxial Failure Theories and Challenges

Composites are often subjected to multiaxial loads. Besides, due to their anisotropy, off-axis uniaxial loads can lead to a multiaxial state of stress in the material. Much research effort has been devoted to developing failure theories and models to predict the multiaxial strength envelope for composites. There exist a large number of theories

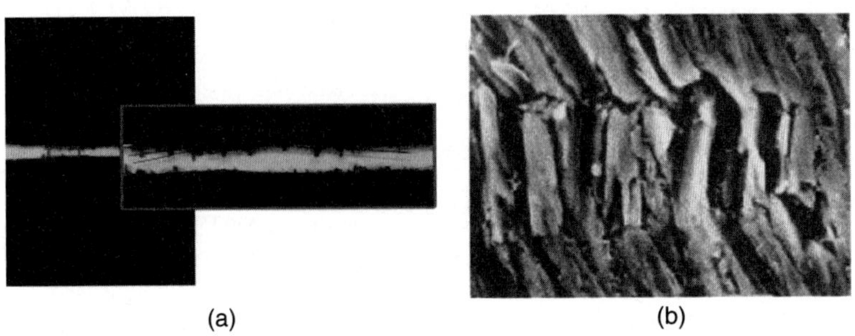

Figure 6.11 (*a*) Tensile fracture showing broken fibers indicative of fiber pullout. (*b*) Fiber kink band [reproduced with permission from Elsevier].

with varying degrees of empiricism. Recently, a worldwide exercise was conducted to systematically evaluate the prominent ones.[53–57] While varying degrees of success were obtained for these theories (see Fig. 6.13), overall the predictions were found to differ by more than +200 percent and −50 percent from the mean. Also none of the theories were found to predict the complete envelope of the failure behavior for various multidirectional laminates. From a structure designer's perspective choosing a failure theory is thus a rather difficult task. Another weakness of this exercise was that the size effect was not tested.

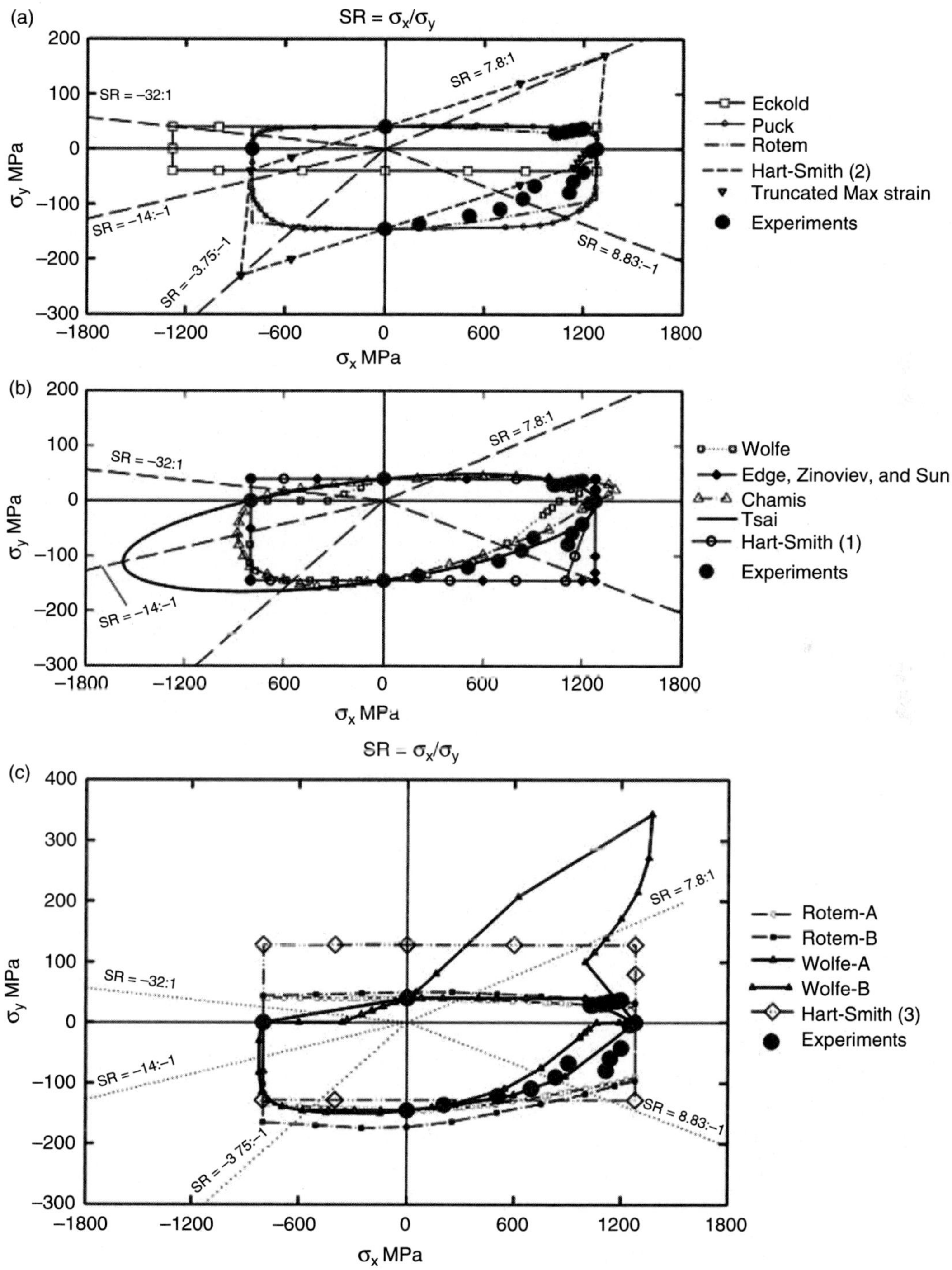

Figure 6.13 Multiaxial strength envelopes for an E-glass/epoxy composite—measured and predicted via a variety of failure theories[53] [reproduced with permission from Elsevier].

The simplest lamina failure theories are based on the maximum stress or maximum strain. They essentially involve determining the principal stresses in the in-plane lamina and comparing each stress component to the corresponding strength. These theories do not account for interactions between stress components and are really plastic limit analysis theories. Tensor polynomial theories such as the Tsai–Hill and the Tsai–Wu criterion[58] aim to provide one equation for the entire failure envelope and try to account for the interactions. These are in general slightly more effective in predicting the strength envelope—especially the Tsai–Wu criterion. It is expressed as

$$f_i\sigma_i + f_{ij}\sigma_i\sigma_j = 1 \tag{6.12}$$

where f_i and f_{ij} are tensors consisting of constants related to the material strength and $i, j = 1, 2,\ldots, 6$ (following Voigt notation for the stress tensor σ, e.g., $\sigma_1 = \sigma_{11}, \sigma_6 = \sigma_{12}$).

It should be noted that the application of these failure theories to structures must be accompanied by structural analysis using analytical tools such as the classical laminate theory, or numerical methods such as finite-element analysis. This would yield the state of stress at each point within the structure, on which the suitably chosen failure theory would predict whether failure would occur at that point. However, the approaches based on the multiaxial strength envelope can work only for predicting the initiation of failure. Applying them to predict progressive or propagating failure leads to a loss of objectivity. In other words, changing the mesh size leads to different results.[59] This is because these theories do not consider the energy required for fracture propagation. Consequently, they miss the quasi brittleness of the composite and, most important the size effect. These problems can be overcome by adapting more advanced damage models with a localization limiter, which simulate progressive degradation of the moduli and are calibrated such that the predicted energy dissipation would be consistent with the fracture energy of the material.[60,61]

6.3.4 Delamination Fracture

Out-of-plane stresses are generated in a composite laminate, especially when adjacent layers have differing fiber orientations. These stresses are often the highest at the free edges and tend to separate the layers, leading to another failure mode called delamination.[42] This can occur under three basic modes, viz. modes I, II, and III shown in Fig. 6.14.

Unlike intralaminar fracture, the fracture energy for delamination under each mode can be measured in the lab in a stable fashion without very stiff grips. This is because the out of plane stiffness and strength of the composite usually not very high. There exists a standard test of delamination fracture energy, although it needs to be improved to obtain information on the size effect, which is significant.[62,63,75]

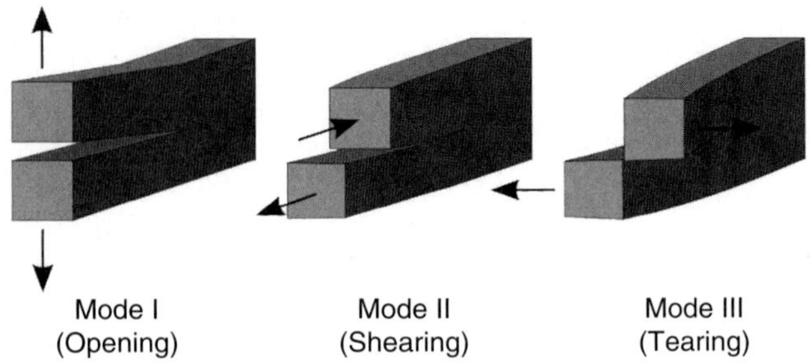

Figure 6.14 Modes of delamination.

6.3.5 Fatigue

Composites in general have a better fatigue resistance than metals, but the fatigue process is a lot more complicated. Unlike metals where one well-defined fatigue crack grows in a self-similar manner, in composites multiple damage mechanisms can occur and progress simultaneously. The factors that influence the fatigue of composites are the fiber type, matrix type, reinforcement architecture (unidirectional, fabric, braided, etc.), composite lay-up sequence, environmental conditions (temperature, humidity), stress ratio R of cyclic load, and the frequency.

While in metals the fatigue crack growth is almost undetectable throughout most of the fatigue lifetime, until sudden acceleration toward the end, in composites it is detectable even in the early stages as shown in Fig. 6.15.[64–66] The crack growth leads to progressively reduced stiffness and strength. Thus, the tests of the residual stiffness and the residual strength provide a simple measure of the remaining useful lifetime of the structure (though only the former is nondestructive). Many phenomenological models for fatigue life prediction are based on this idea.[64,67–69] Most design practices rely on phenomenological approaches, such as the S-N curve for the total life, because the complexity of crack growth hinders developing a mechanistic predictive model. S-N curves for various carbon fiber–reinforced polymer composites are shown in Fig. 6.16.[70] Whereas many models exist,[71–74] they lack general applicability in terms of loading and material system. Further complications arise due to random scatter of strength and of fatigue lifetimes, as the scatter is much higher than that observed in metals. A comprehensive probabilistic approach to fatigue of composites still awaits development.

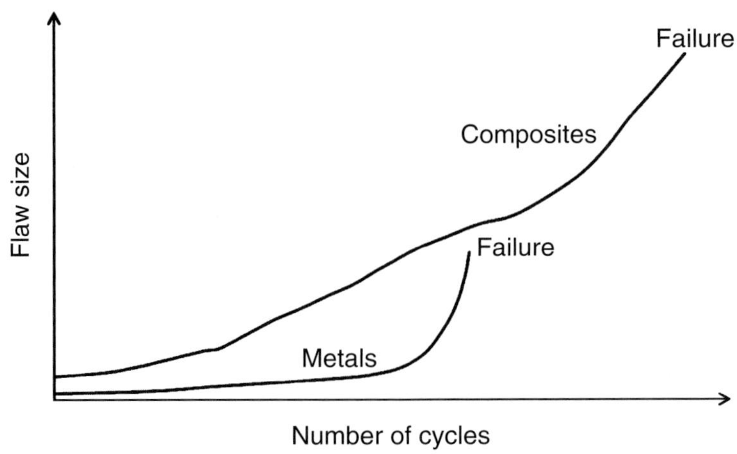

Figure 6.15 Comparison of fatigue damage evolution in composites and metals.

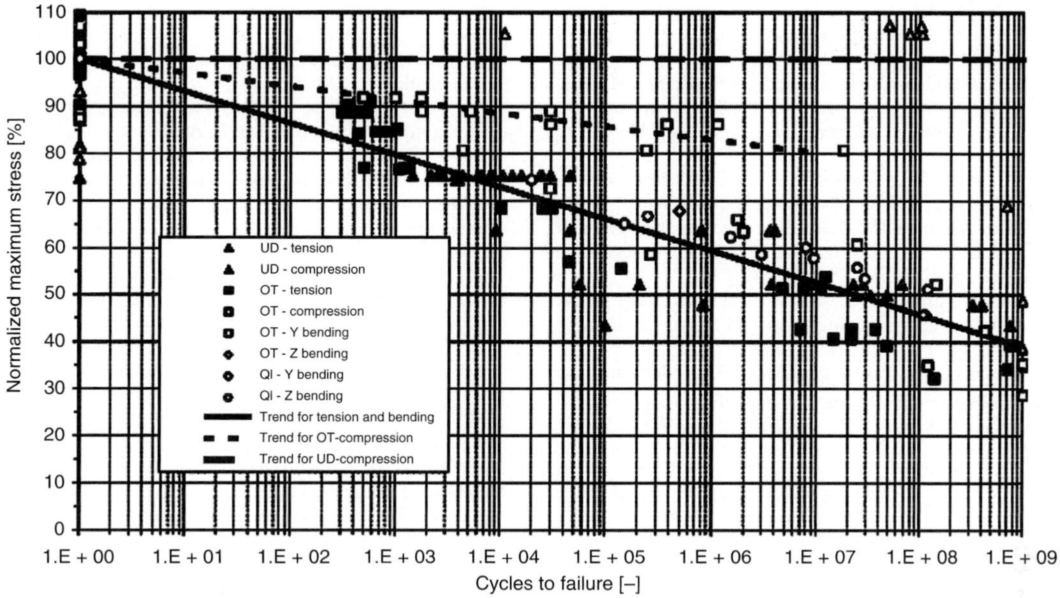

Figure 6.16 S-N curves for various carbon fiber–reinforced polymer composites[70] [reproduced with permission from Elsevier].

REFERENCES

1. Bažant, Z.P., and Planas, J. (1998). *Fracture and Size Effect in Concrete and Other Quasibrittle Materials*. CRC Press, Boca Raton, FL.
2. Anderson, T.L. (1995). *Fracture Mechanics: Fundamentals and Applications*. CRC Press, Boca Raton, FL.
3. Griffith, A.A. (1920). "The phenomena of rupture and flow in solids." Philos. Trans. Roy. Soc. A, 221, 163–198.
4. Bažant, Z.P. (2002). *Scaling of Structural Strength*. Hermes Penton Science Kogan Page Science, London.
5. Bažant, Z.P., and Novák, D. (2000). "Energetic-statistical size effect in quasibrittle failure at crack initiation." *ACI Mater. J.*, 97(3), 381–392.
6. Bažant, Z.P., and Li, Z. (1995). "Modulus of rupture: Size effect due to fracture initiation in boundary layer." J. Struct. Engrg., 121(4), 739–746.
7. Weibull, W. (1939). "The phenomenon of rupture in solids." Proc. Roy. Swed. Inst. Engg. Res. Stockholm, 153, 1–55.
8. Harb, N., Bezzazi, B., Mehrez, S., Hamitouche, K., and Dilmi, H. (2017). "Probabilistic analysis of the behavior of polymer matrix composite materials reinforced by different types of fibers." IOP Conf. Series: Materials Science and Engineering, 264(2017), 012021, doi:10.1088/1757-899X/264/1/012021.
9. Harlow, D.G., and Phoenix, S.L. (1979). "Bounds on the probability of failure of composite materials." Int. J. Fract., 15(4), 312–336.
10. Mahesh, S., and Phoenix, S.L. (2004). "Lifetime distributions for unidirectional fibrous composites under creep-rupture loading." Int. J. Fract., 127, 303–360.
11. Bažant, Z.P., and Le, Jia-Liang. (2017). *Probabilistic Mechanics of Quasibrittle Structures: Strength, Lifetime, and Size Effect*. Cambridge University Press, Cambridge.
12. Bažant, Z.P., and Pang, S.-D. (2007). "Activation energy based extreme value statistics and size effect in brittle and quasibrittle fracture." J. Mech. Phys. Solids, 55, 91–134.
13. Le, J.-L., Bažant, Z.P., and Bažant, M.Z. (2011). "Unified nanomechanics based probabilistic theory of quasibrittle and brittle structures: I. Strength, static crack growth, lifetime and scaling." J. Mech. Phys. Solids, 59, 1291–1321.
14. Bažant, Z.P., Gettu, R., and Kazemi, M.T. (1991). "Identification of nonlinear fracture properties from size-effect tests and structural analysis based on geometry-dependent R-curves." Int. J. Rock Mech. Min. Sci., 28(1), 43–51.
15. Yu, Qiang, Le, Jia-Liang, Hubler, H.H., Wendner, R., Cusatis, G., and Bažant, Z.P. (2016). "Comparison of main models for size effect on shear strength of reinforced and prestressed concrete beams." Structural Concrete (fib), 17(5), 778-789, doi: 10.1002/suco.201500126.
16. Dönmez, A., and Bažant, Z.P. (2017). "Size effect on punching strength of reinforced concrete slabs with and without shear reinforcement." ACI Struct J., 114(4), 876–886.
17. Bažant, Z.P. (1984). "Size effect in blunt fracture: Concrete, rock, metal." J. Engrg. Mech. 110(4), 518–535.
18. Hoover, C.G., Bažant, Z.P., Vorel, J., Wendner, R., and Hubler, M.H. (2013). "Comprehensive concrete fracture tests: Description and results." Eng. Fract. Mech., 114, 92–103.
19. Hoover, C.G., and Bažant, Z.P. (2014). "Cohesive crack, size effect, crack band and work-of-fracture models compared to comprehensive concrete fracture tests." Int. J. Fracture, 187(1), 133–143.
20. Bažant, Z.P., and Xiang, Yuyin. (1997). "Size effect in compression fracture: Splitting crack band propagation." J. Engrg. Mech. 123(2), 162–172.
21. Caner, F.C., and Bažant, Z.P. (2002). "Lateral confinement needed to suppress softening of concrete in compression." J. Engrg. Mech. 128(12), 1304–1313.
22. Salviato, M., Kirane, K., Ashari, SE, Bažant, Z.P., and Cusatis, G. (2016). "Experimental and numerical investigation of intra-laminar energy dissipation and size effect in two-dimensional textile composites." Compos. Sci. Technol., 135, 67–75.
23. Mehta, P.K., and Monteiro, P.J.M. (2014). *Concrete: Microstructure, Properties, and Materials*. 4th ed. McGraw-Hill Education, New York.
24. Van Mier, J.G.M. (1998). "Failure of concrete under uniaxial compression: An overview." Fracture Mechanics of Concrete Structures, Proceedings FRAMCOS-3, AEDIFICATIO Publishers, Freiburg, Germany.
25. Geel, van, H.J.G.M. (1998). "Concrete behaviour in multiaxial compression: Experimental research." Eindhoven Technische Universiteit Eindhoven, doi: 10.6100/IR515170.
26. Pijaudier-Cabot, G., and Dufour, F., eds. (2012). *Damage Mechanics of Cementitious Materials and Structures*. Wiley, New York, doi: 10.1002/9781118562086.

27. Kirane, K., Bažant, Z.P., and Zi, G. (2014). "Fracture and size effect on strength of plain concrete disks under biaxial flexure analyzed by microplane model M7." J. Engrg. Mech., 140(3), 604–613.

28. Torrenti, J.M., Pijaudier-Cabot, G., and Reynouard, J.M., eds. (2010). *Mechanical Behavior of Concrete*. Wiley-ISTE, New York.

29. Suresh, S. (1991). *Fatigue of Materials*. Cambridge University Press, Cambridge.

30. Rice J. "Mechanics of crack tip deformation and extension by fatigue." Am. Soc. Test. Mater., 415, 247–311.

31. Kirane, K., and Bažant, Z.P. (2016). "Size effect in Paris law and fatigue lifetimes for quasibrittle materials: Modified theory, experiments and micro-modeling." Int. J. Fatigue, 83, 209–220.

32. Kirane, K., and Bažant, Z.P. (2014). "Microplane damage model for fatigue of quasibrittle materials: Subcritical crack growth, lifetime and residual strength." Int. J. Fatigue, 70, 93–105.

33. Horii H., Shin H.C., and Pallewatta T.M. (1992). "Mechanism of fatigue crack growth in concrete." Cem. Concr. Compos., 14, 83–89.

34. Bahn, B.Y., and Hsu, C.T. (1998). "Stress-strain behavior of concrete under cyclic loading." ACI Mater. J., 95(2), 178–193.

35. Hordijk, D., and Reinhardt, H. (1991). "Growth of discrete cracks under fatigue loading." In *Toughening Mechanisms in Quasi-Brittle Materials*, edited by S. Shah. Springer, New York, 541–554.

36. Aas Jakobsen K., and Lenshow R. (1973). "Behavior of reinforced columns subjected to fatigue loading." ACI J., 70, 199–206.

37. ACI Committee 215. (1992). *Considerations for Design of Concrete Structures Subjected to Fatigue Loading (ACI 215R-92) (Reapproved 1997)*. American Concrete Institute, Farmington Hills, MI.

38. *The FIB Model Code for Concrete Structures 2010*, Ernst & Sohn GmbH & Co. KG, Berlin.

39. Paris, P., and Erdogan, F. (1963). "A critical analysis of crack propagation laws." J. Basic Engrg., 85, 528–534.

40. Le, Jia-Liang, and Bažant, Z.P. (2011). "Unified nano-mechanics based probabilistic theory of quasibrittle and brittle structures: II. Fatigue crack growth, lifetime and scaling." J. Mech. Phys. Solids, 59, 1322–1337.

41. Bažant, Z.P., and Xu, K. (1991). "Size effect in fatigue fracture of concrete." ACI Mater. J., 88(4), 390–399.

42. Daniel, I.M., and Ishai, O. (1994). *Engineering Mechanics of Composite Materials*. Oxford University Press, New York.

43. Talreja, R., and Varna, J. (2016). *Modeling Damage, Fatigue and Failure of Composite Materials*. Woodhead Publishing, London.

44. Budiansky, B., and Fleck, N.A. (1993). "Compressive failure of fibre composites." J. Mech. Phys. Solids, 41(1), 183–211.

45. Fleck, N.A., Jelf, P.M., and Curtis, P.T. (1995). "Compressive failure of laminated and woven composites." J. Compos. Tech. Res., 17–13, 212–220.

46. Cox, B.N., Dadkhah, M.S., Inman, R.V., Morris, W.L., and Zupon, J. (1992). "Mechanisms of compressive failure in 3D composites." Acta Metall. Mater., 40(12), 3285–3298.

47. Budiansky, B., and Fleck, N.A. (1993). "Compressive failure of fibre composites." J. Mech. Phys. Solids, 41(1), 183–211.

48. Gutkin, R., and Pinho, S.T. (2014). "Combining damage and friction to model compressive damage growth in fibre-reinforced composites." J. Comp. Mater., doi: 10.1177/0021998314549614.

49. Svensson, D., Alfredsson, K.S., Stigh, U., and Jansson, N.E. (2016). "Measurement of cohesive law for kink-band formation in unidirectional composite." Engrg. Fract. Mech., 151, 1–10.

50. Bažant, Z.P., Kim, J.-J.H., Daniel, I.M., Becq-Giraudon, E., and Zi, G. (1999). "Size effect on compression strength of fiber composites failing by kink band propagation." Int. J. Fract 95, 103–141 (special issue).

51. Bažant, Z.P., and Cedolin, L. (2010). "*Stability of Structures: Elastic, Inelastic, Fracture and Damage Theories*. #rd ed. third edition, World Scientific, London.

52. Salviato, M., Chau, V.T., Li, W., Bažant, Z.P., and Cusatis, G. (2016). "Direct testing of gradual postpeak softening of fracture specimens of fiber composites stabilized by enhanced grip stiffness and mass." J. Appl. Mech., 83–11, 111003:1-11.

53. Hintona, M.J., Kaddourb, A.S., and Soden, P.D. (2002). "A comparison of the predictive capabilities of current failure theories for composite laminates, judged against experimental evidence." Compos. Sci. Technol., 62, 1725–1797.

54. Soden, P.D., Hinton, M.J., and Kaddour, A.S. (1998). "Lamina properties, lay-up configurations and loading conditions for a range of fibre-reinforced composite laminates." Compos. Sci. Technol., 58, 1011–1022.

55. Sun, C.T., Tao, J.X., and Kaddour, A.S. (1998). "The prediction of failure envelopes and stress/strain behavior of composite laminates: Comparison with experimental results." Compos. Sci. Technol., 62, 1673–1682.

56. Cuntze, R.G., and Freund, A. (2004). "The predictive capability of failure mode concept-based strength criteria for multidirectional laminates." Compos. Sci. Technol., 64, 343–377.

57. Talreja, R. (2014). "Assessment of the fundamentals of failure theories for composite materials." Compos. Sci. Technol., 105, 190–201.

58. Tsai, S.W., and Wu, E.M. (1970). "A general theory of strength for anisotropic materials." J. Comp. Mater., 5, 58–80.

59. Bažant, Z.P., Belytschko, T.B., Chang, T. (1984). "Continuum theory for strain softening." J. Eng. Mech., 110(12), 1666–1692.

60. Kirane, K., Salviato, M., and Bažant, Z.P. (2016). "Microplane-triad model for elastic and fracturing behavior of woven composites." J. Appl. Mech., 83(April), 041006-1–041006-14.

61. Talreja, R. (2016). "Physical modelling of failure in composites." Philos. Trans. Roy. Soc. A, 374, 20150280.

62. ASTM D5045 Standard Test Methods for Plane-Strain Fracture Toughness and Strain Energy Release Rate of Plastic Materials. American Society for Testing and Materials, West Conshohocken, PA, 1999.

63. Salviato, M., Kirane, K., Bazant, Z.P., and Cusatis, G. (2019). "Mode I and II interlaminar fracture in laminated composites: A size effect study." Compos. Struct., 86(September), 091008-1-8.

64. Salkind, M.J. (1972). "Fatigue of Composites," Composite Materials: Testing and Design (Second Conference). ASTM STP 497. American Society for Testing and Materials, West Conshohocken, PA, 143–169.

65. U.S. Department of Defense. (2002). *Composite Materials Handbook. Volume 3, Polymer Matrix Composites: Materials, Usage, Design, and Analysis*. U.S. Department of Defense, Washington, DC.

66. *Marine Composites*. 2nd ed. Eric Greene Associates, Annapolis, MD, 1999.

67. Post, N., Bausano, J., Case, S., and Lesko, J. (2006). "Modeling the remaining strength of structural composite materials subjected to fatigue." Int. J. Fatigue, 28(10), 1100–1108.

68. Khan, Z., Al-Sulaiman, F.A., Farooqi, J.K., and Younas, M. (2001). "Fatigue life predictions in woven carbon fabric/polyester composites based on modulus degradation." J. Reinf. Plast. Compos., 20(5), 377–398.

69. Van Paepegem, W., and Degrieck, J. (2002). "A new coupled approach of residual stiffness and strength for fatigue of fibre-reinforced composites." Int. J. Fatigue, 24(7), 747–762.

70. Michel, A.S., Kieselbach, R., and Jörg Martens, H. (2006). "Fatigue strength of carbon fibre composites up to the gigacycle regime (gigacycle-composites)." Int. J. Fatigue, 28(3), 261–270.

71. Sevenois, R.D.B., and Van Paepegem, W. (2015). "Fatigue damage modeling techniques for textile composites: Review and comparison with unidirectional composite modeling techniques." Appl. Mech. Rev., 67(2), 020802, doi: 10.1115/1.4029691.

72. Degrieck, J., and Van Paepegem, W. (2001). "Fatigue damage modeling of fibre-reinforced composite materials: Review." Appl. Mech. Rev., 54(4), 279–300.

73. Quaresimin, M., Susmel, L., and Talreja, R. (2010). "Fatigue behaviour and life assessment of composite laminates under multiaxial loadings." Int. J. Fatigue, 32(1), 2–16.

74. Chen, H., Shivakumar, K., and Abali, F. (2006). "A comparison of total fatigue life models for composite laminates." Fatigue Fract. Eng. Mater. Struct., 29(1), 31–39.

75. Salviato, M., Kirane, K., Bažant, Z. P., & Cusatis, G. (2019). Mode I and II Interlaminar Fracture in Laminated Composites: A Size Effect Study. Journal of Applied Mechanics, 1-36.

PART B
STRUCTURAL STEELS

BY

J. ERNESTO INDACOCHEA *Professor Emeritus, Civil and Materials Engineering, University of Illinois at Chicago*
VINEETH KUMAR GATTU *Chemical and Fuel Cycle Technologies, Argonne National Laboratory*

6.4 FRACTURE OF STRUCTURAL STEELS

6.4.1 Introduction

Most structural steel members exhibit considerable ductility at normal tensile service loadings and temperatures. If failures should occur, it is only after considerable elongation and reduction in area, with a large amount of energy absorption. Brittle fractures, on the other hand, occur with little or no elongation or reduction in area and with very small energy absorption. Brittle fracture is a type of catastrophic failure that usually occurs without prior plastic deformation and at extremely high speeds, as fast as 2000 m/s (7000 ft/s) in steels.[1]

Schematic examples of the stress-strain relations for these two types of failure are presented in Fig. 6.17. When speaking of a material strength it is considered to be its resistance to plastic flow. Strength generally increases with plastic strain which is defined as *work or strain hardening*, reaching a maximum at the ultimate engineering tensile strength, σ_{UTS}. The area under the whole stress-strain curve up to fracture represents the work to cause rupture and the toughness of the material defined as the resistance to the propagation of a crack. Most structural materials exhibit considerable strain (deformation) before reaching the tensile or ultimate strength σ_{UTS}. In contrast, brittle materials exhibit almost no deformation before failure. However, under conditions of low temperature, rapid loading, and high constraint, even ductile materials may not exhibit deformation before fracture. Obviously, ductile behavior is much more desirable than brittle behavior because the high absorption energy and large deformation that occur before ductile failure is important to prevent any catastrophic brittle fracture.

Ductile failures are characterized by a slow crack extension and larger applied stresses needed for the crack to propagate. Such cracks are considered stable and are characterized by substantial plastic deformation and high energy absorption as observed in Fig. 6.17. Figure 6.18*a* shows the typical features of a ductile fracture in a Charpy V-notch (CVN) specimen after testing, characterized by large shear lips and substantial local deformation, and Fig. 6.18*b* presents the SEM micrograph that shows equiaxed dimples. Brittle fractures extend rapidly with very little plastic deformation and without increase in applied load, which makes them unstable cracks. These cracks are usually characterized by a flat fracture surface with little or no ductility as seen in Fig. 6.18*c*, they

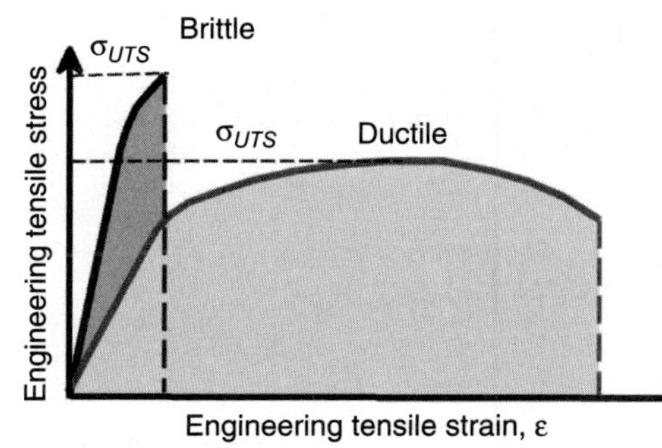

Figure 6.17 Schematic representations of tensile stress-strain behavior for ductile and brittle metals loaded to fracture.

propagate in specific crystallographic planes (Fig. 6.18*d*) and they are known as cleavage fracture. Brittle fractures are not as common as yielding, buckling, or fatigue failures but, when they do occur, they may be more costly in terms of human life and property damage.

Figure 6.19 shows a brittle failure of an alloy steel vessel that failed during hydrotesting operations. Factors that increase the susceptibility to brittle fracture are (*a*) metallurgical degradation that may occur in some steels during heat treatments that may lead to temper embrittlement due to carbide precipitation, sigma phase embrittlement, etc.; (*b*) grain coarsening and impurities in the steel (e.g., for example, large grain sizes and steel contaminants such as sulfur and phosphorous typically reduce steel toughness, which is a measure of resistance to brittle fracture; and (*c*) material thickness, that is, the thicker the component the higher the susceptibility to brittle fracture due to triaxial stress conditions. Thicker materials produce a state of higher constraint and therefore are less likely to deform under stress as opposed to crack initiation and propagation.

In the 1994 Northridge earthquake, several fractures occurred in moment connections of highly constrained joints in buildings, which is further evidence that brittle fractures can occur in ductile materials when high-constraint conditions exist.

6.4.2 Introduction to Fracture Mechanics

Many large and complex structures such as bridges, ships, aircraft, and pressure vessels can have cracks, sharp notches, or discontinuities of different types. Using fracture mechanics, an engineer can quantitatively establish allowable stress levels and inspection requirements to design against the occurrence of brittle fractures in such structures. In addition, fracture mechanics may be used to analyze the growth of small cracks to critical size by fatigue loading or stress corrosion crack growth. Therefore, fracture-mechanics testing and analysis techniques have several advantages over traditional notch-toughness test methods and offer the designer a method of quantitative design to prevent brittle fracture in structures.

Using fracture mechanics principles, an expression has been developed that relates this critical stress for crack propagation (σ_c) and crack length (a) as

$$K_c = Y\sigma_c\sqrt{\pi a}$$

In this expression K_c is the fracture toughness, a property that is a measure of a material's resistance to brittle fracture when a crack is present. K_c has units of MPa$\sqrt{m}$ or psi$\sqrt{in}$. Here, Y is a dimensionless parameter or function that depends on both crack and specimen sizes and geometries as well as the manner of load application.

The critical value of the stress-intensity factor at failure, K_c, is a material property analogous to the resistance force in structural design. The K_c value for thick specimens is known as the plane strain fracture toughness, K_{Ic} it is also defined by the fracture toughness cited for most situations. The "*I*" subscript in K_{Ic} denotes that the plane strain fracture toughness is for mode I (tensile) crack displacement.

Equations that describe the elastic-stress field in the vicinity of a crack tip in a body subjected to tensile stresses can be used to establish the relation between K_I, σ, and crack size, a, for different structural configurations, as shown in Fig. 6.20. K_I values for other crack geometries and different structural configurations are available in the literature.[1]

Figure 6.18 Charpy V-notch samples showing examples of (*a* and *b*) ductile failure and (*c* and *d*) brittle failure in low-carbon steel specimens.

From testing, the critical value of K_I at failure (K_c, K_{Ic}, $K_{Ic\ int}$, or K_{Id}) can be determined for a given metal at a particular thickness and at a specific temperature and loading rate. Using this critical material property, the designer can determine theoretically the flaw size that can be tolerated in structural members for a given design stress level, temperature, and loading rate. Conversely, the engineer can determine the design stress level that can be safely used for a flaw size that may already be present in an existing structure.

The several critical stress-intensity factors for each structural material are highly dependent on service conditions such as temperature, loading rate, and constraint. They are:

K_c = critical stress-intensity factor for slow loading rates and thin plates. Generally varies with specimen thickness, as shown in Fig. 6.21.

K_{Ic} = critical stress-intensity factor for slow loading rates and thick plates. Applies only to thick plates, and it is shown in Fig. 6.21.

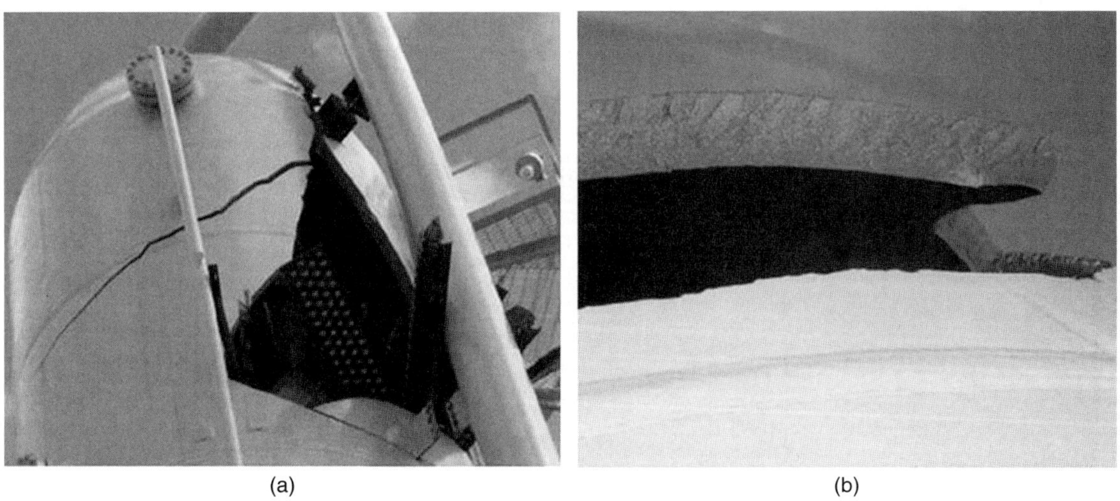

Figure 6.19 (*a*) Brittle fracture of a steel pressure vessel after hydrostatic testing. (*b*) Close-up of the fracture showing no deformation at the wall thickness.[2]

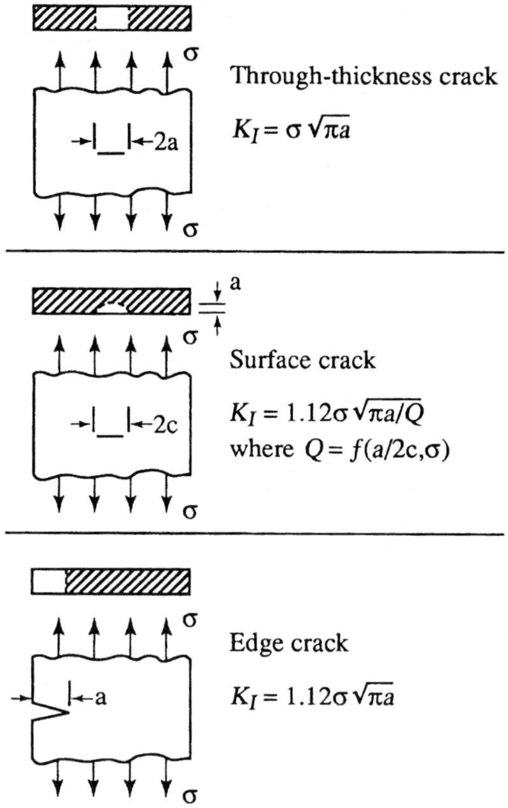

Figure 6.20 K_I values for different crack geometries.

K_{Id} = critical stress-intensity factor for impact loading rates and thick plates. Values are generally less than K_{Ic} at the same temperature, as shown in Fig. 6.22.

$K_{Ic\ int}$ = critical stress-intensity factor for loading rates that are intermediate between slow and impact, Fig. 6.22, such as the loading rate for bridges.

These critical values are obtained by testing actual structural materials to failure at various temperatures and loading rates.

Examples of K_c, K_{Ic}, $K_{Ic\ int}$, and K_{Id} values for a typical structural steel having room-temperature yield strength of 50 ksi (345 kPa) are presented in Fig. 6.22. These results, obtained at three different loading rates, show the large effect that temperature and loading rate can have on the critical stress-intensity factors for a particular structural material.

6.4.3 Fracture Toughness

Fracture toughness tests have been used in design to prevent brittle failure by using principles of fracture mechanics. Notch toughness, on the other hand, is usually measured as the amount of energy, expressed in units of joules or foot-pounds, required to fracture a notched specimen at a particular temperature and loading rate. Notch toughness is based on a stress analysis in the vicinity of a notch or crack. It does not depend on the use of extensive service experience to translate laboratory results into practical design information as long as the engineer can obtain or determine:

1. The fracture toughness of a material using fracture-mechanics-types of tests or correlations with notch-toughness tests such as the CVN impact test.
2. The nominal stress on the structural material being analyzed.
3. Flaw size and geometry of the structural material being analyzed.

One of the most widely used tests to measure notch toughness is the CVN impact test. It consists of a test machine that includes a pendulum used to impact and measure the absorbed energy at various temperatures. The absorbed energy is plotted as a function of temperature. Typical CVN results for common structural materials are given in Fig. 6.23 that shows the transition from brittle to ductile behavior under conditions of

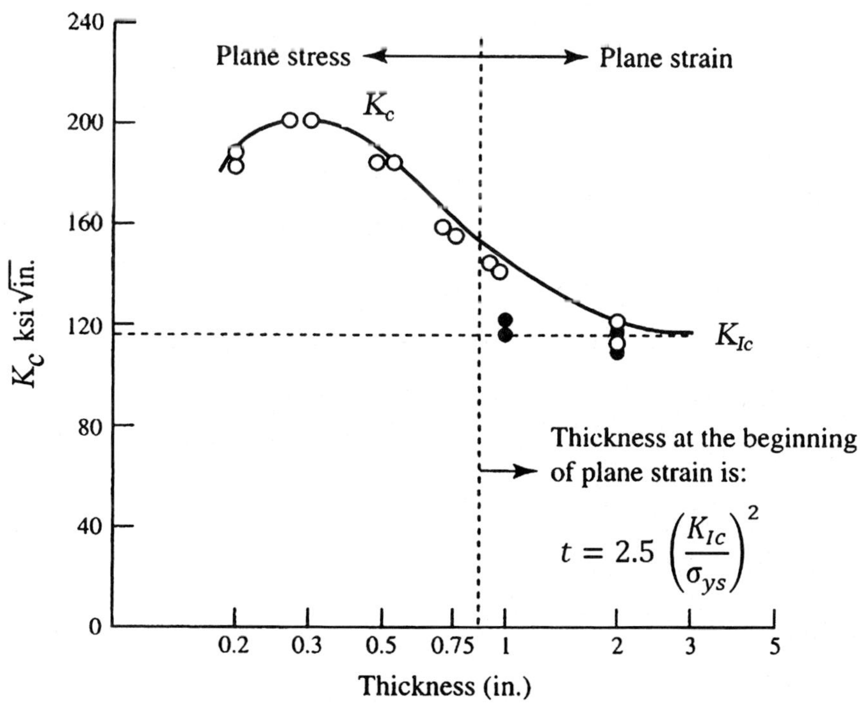

Figure 6.21 Effect of thickness on K_c behavior.

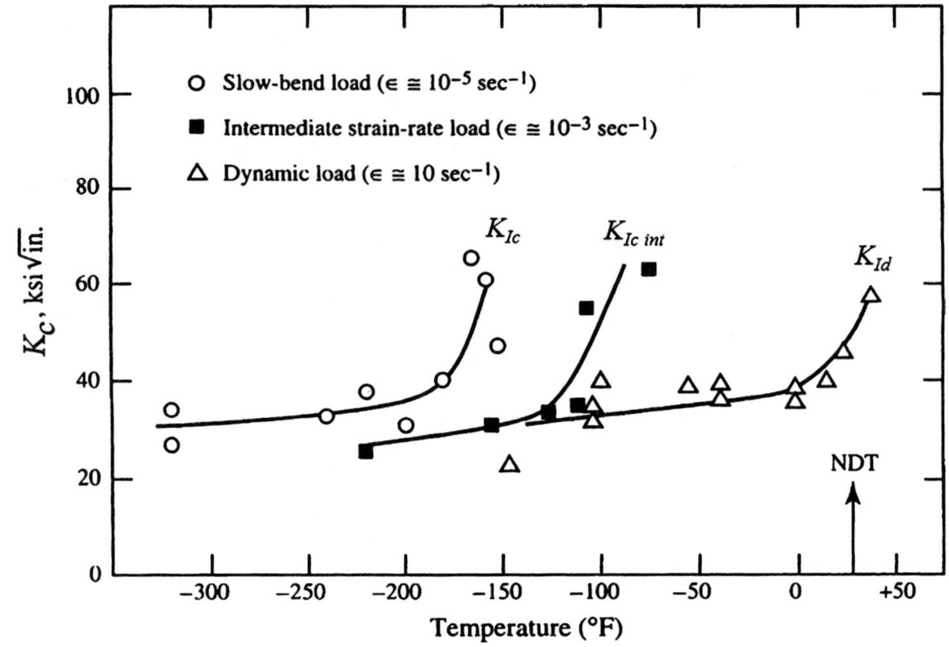

Figure 6.22 Effect of temperature and strain rate on fracture toughness of 50-ksi- (345-MPa-) yield strength structural steel.

temperature. The CVN impact values shown at the lower left of Fig. 6.23 are representative of low levels of notch toughness or brittle behavior, while the values at higher temperatures (upper right) are representative of ductile-type behavior. It should be noted that some materials (such as aluminum and very high-strength steel) do not exhibit a distinct transition behavior. In addition, some materials have low notch toughness at all temperatures (e.g., 75 ksi- or 517.0 MPa-yield aluminum or high-strength steels); whereas some materials have a high level of notch toughness at all temperatures (e.g., 304L stainless steel).

The change in absorbed energy, ductility (lateral expansion or contraction at the root of the notch), and fracture appearance (as measured by percent shear on the surface) for a typical structural steel are shown in Fig. 6.24. At 60°C (140°F), full ductile behavior is observed. At −128°C (−200°F), complete brittle behavior is observed. The region between these two extremes is called the *transition region*. Note that the transition region is different for the two different loading rates. Loading rate has a very significant effect on the fracture behavior of structures, as observed in Fig. 6.24.

Various *transition temperatures* are often established as an indication of the notch toughness of a structural material. For example, the 20.3 Joule

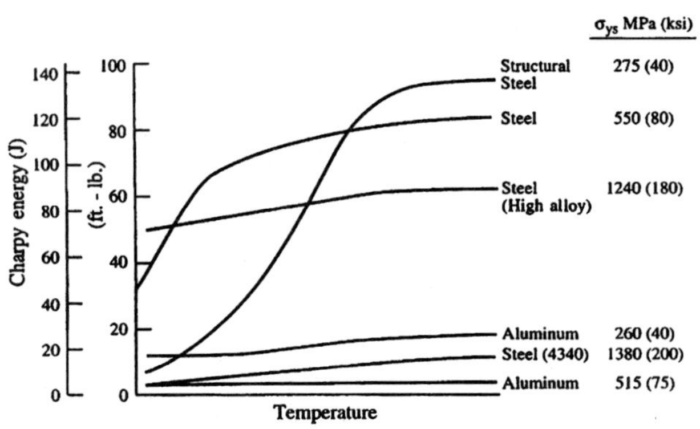

Figure 6.23 Charpy V-notch impact energy versus temperature for selected structural metal alloys.

(15-ft-lb) impact transition temperature for the steel shown in Fig. 24*a* is about −1°C (30°F). Also, the 50 percent impact fracture-appearance transition temperature for this steel is about −1°C (30°F), as shown in Fig. 6.24*c*. Obviously, these transition temperatures will vary from material to material, depending on the particular notch-toughness characteristics of each material. One traditional method of design to prevent brittle fracture in a member is to specify that it can be used only above some particular transition temperature. That is, for the material shown in Fig. 6.24, one fracture-control design method has been to restrict the use of this steel to temperatures above the 20.3 Joule (15-ft-lb) impact transition temperature for those structures subject to impact loading.

The nil-ductility temperature (NDT) test is another American Society for Testing and Materials (ASTM)[3] method to predict behavior of structural steels. Below the NDT temperature, the steel is considered to be brittle under conditions of impact loading. At slow or intermediate loading rates, the steel can still exhibit satisfactory toughness at lower temperatures, as shown in Fig. 6.24.

6.4.4 Importance of Loading Rate

In addition to the effects of notches and temperature on the fracture behavior of structural materials, the rate of load application also can be significant. That is, if notched members are loaded slowly, the transition from brittle to ductile behavior may occur at much lower temperatures than for impact loading. Figure 6.24 shows that a slow loading rate caused the transition behavior to be shifted to lower temperatures. Many structural materials, primarily low-strength structural steels, exhibit loading-rate effects.

For members or components subjected to impact or dynamic rates of loading, the impact-loading-rate transition temperature curve should be used to measure notch toughness. A similar statement can be made for a member loaded at an intermediate loading rate (between slow and impact). Loads can be applied to structural members either very slowly [over 10 to 100 seconds (see Fig. 6.22) as in the case of a slow-moving tractor crossing a bridge], very rapidly (such as an impact load in 0.01 to 0.001 second from a falling weight), or, as is generally the case, at some intermediate rate (such as a truck crossing a short-span bridge at a speed of 50 to 60 mph in about 1 second). Earthquake loadings generally are slow to intermediate. Typical load-time plots for loads applied to structural members are shown in Fig. 6.25.

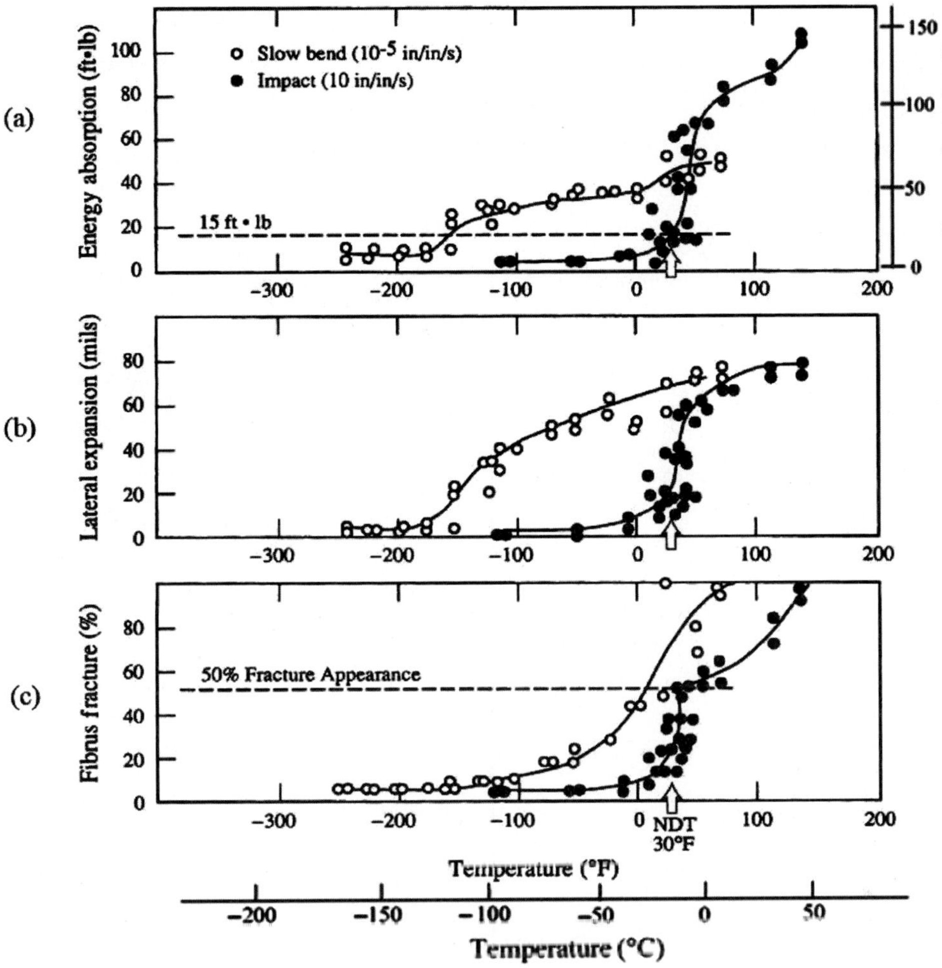

Figure 6.24 Charpy V-notch energy absorption, lateral expansion, and fibrous fracture for impact and slow bend test of standard CVN samples for a typical low-strength structural steel.

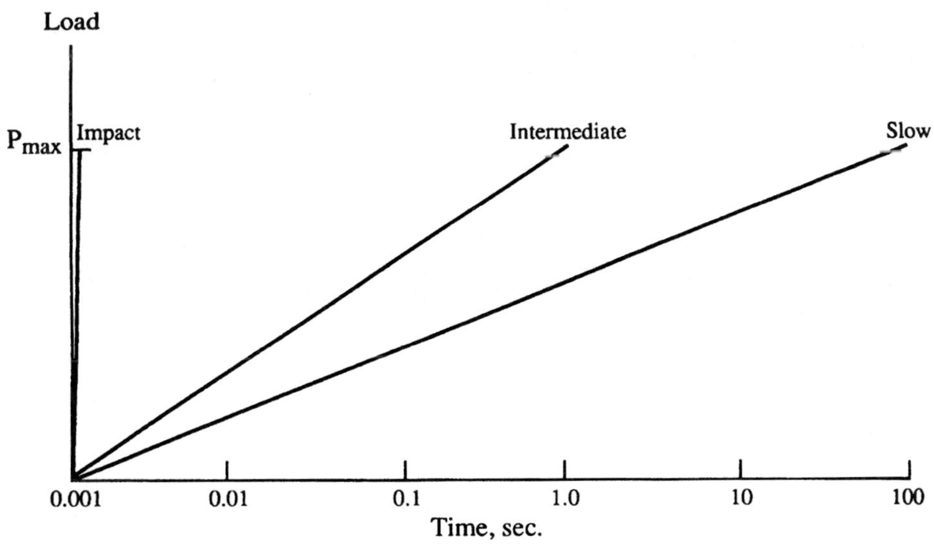

Figure 6.25 Schematic of load-time plots for loads applied to structural members.

6.4.5 Correlations between Various Fracture Test Results

The dynamic critical stress-intensity factor, K_{Id}, can be related to CVN impact results in the lower half of the transition region (below the ductile-to-brittle transition temperature) as follows:

$$K_{Id} = \sqrt{5 \cdot \text{CVN} \cdot E}$$

With CVN given in ft-lb or Joules and E (the Modulus of Elasticity) in psi or MPa, the units of K_{Id} are psi$\cdot\sqrt{in}$ (or MPa$\sqrt{m}$). Each of these K_{Id} values (at a specific test temperature) can then be "shifted" to account for the difference in loading rate between K_{Id} and K_{Ic} using the following relation:

$$T_s = 215 - 1.5\sigma_{ys}$$

where σ_{ys} = yield strength, ksi or MPa.

The temperature shift between K_{Id} and K_{Ic} int is shown in Fig. 6.26 as $0.75Ts$. Thus, K_{Ic}, $K_{Ic\ int}$, or K_{Id} values can be approximated from CVN impact test results as shown in Fig. 6.26. Other relations exist between elastic-plastic fracture test results (such as the J-integral or CTOD test results) and K_{Id} values, and these are described in the literature.[1]

6.4.6 Fracture Mechanics Design

The fundamental design approach to preventing brittle fracture in structural materials is to keep the calculated stress-intensity factor K_I (the driving force) below the critical stress-intensity factor, K_c (the resistance force). This is analogous to keeping $\sigma < \sigma_{ys}$ to prevent yielding.

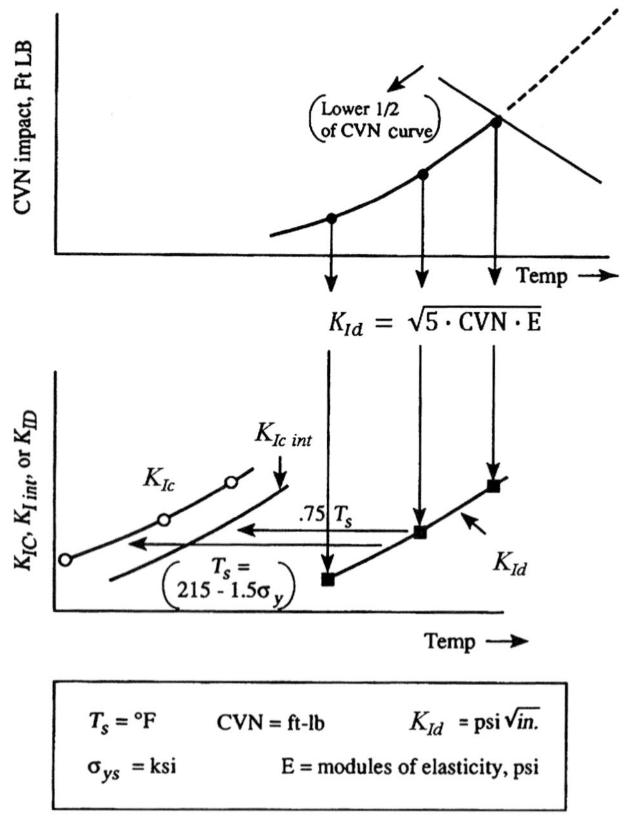

Figure 6.26 Relationship between CVN-K_{Id}-$K_{Ic\ int}$ and K_{Ic} test results.

A general design procedure to prevent brittle fracture in structural members is as follows:

1. Calculate the maximum nominal stress σ for the member being analyzed.
2. Estimate the most likely flaw geometry and initial crack size a_o. To design against brittle fracture during the expected lifetime of a structure, estimate the maximum probable crack size during the expected lifetime.
3. Calculate K_I for these conditions of stress σ and flaw size a using the appropriate K_I relation. (Examples of various K_I relations are given in Fig. 6.22.)
4. Determine or estimate the critical stress-intensity factor K_c, K_{Ic}, $K_{Ic\ int}$, or K_{Id} for the material from which the member is to be built. These critical stress-intensity values are a function of the expected service temperature and loading rate and can be obtained from fracture mechanics tests. Approximate critical stress-intensity values can be estimated from CVN impact test results as shown in Fig. 6.26. Frequently, CVN impact test results are specified for specific structural cases such as the AASHTO Fracture Control Plan. Estimates of K_c can be made for specific cases using the procedure shown in Fig. 6.26.
5. Compare K_I with K_{Ic}, $K_{Ic\ int}$, or K_{Id}. To design against brittle fracture, ensure that K_I will be less than the critical stress-intensity factor throughout the entire life of the structure. This may require the selection of a different material or reduction of the maximum nominal service stress. Also, it may require better quality control during fabrication, or periodic inspection for cracks through the life of the structure.

The general relationship among material toughness K_c, K_{Ic}, $K_{Ic\ int}$, or K_{Id}, nominal stress, σ, and crack size a is shown schematically in Fig. 6.27. If, for a particular combination of stress and crack size in a structure, K_I reaches the critical K_c level, brittle fracture can occur. Thus, there are many combinations of stress and flaw size which may cause fracture in a structure that is fabricated from a material having a particular value of K_c at a particular service temperature and loading rate. Conversely, there are many combinations of stress and flaw size that will not cause fracture of a particular structural material.

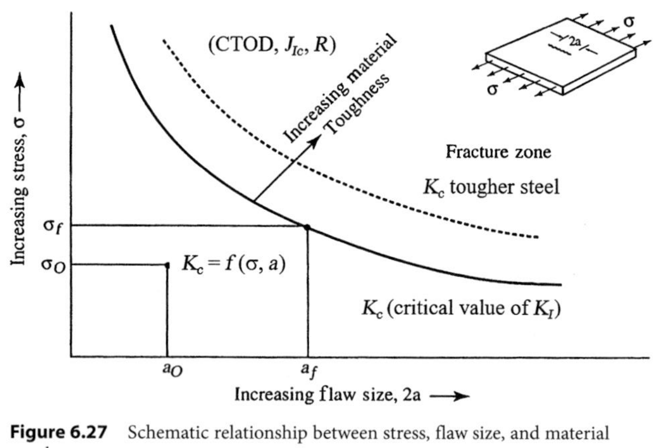

Figure 6.27 Schematic relationship between stress, flaw size, and material toughness.

As an example of the design application of fracture mechanics, consider the equation relating K_I to the applied stress and flaw size for a through-thickness crack in a wide plate (Fig. 6.28). Assume that laboratory test results show that for a particular structural steel with a yield strength of 80 ksi (552.0 MPa) the K_c is 60 ksi $\sqrt{in}$ (66.0 MPa $\sqrt{m}$) at the service temperature, loading rate, and plate thickness. Also assume that the design stress is 20 ksi (138 MPa). Substituting $K_I = K_c$ = 60 ksi $\sqrt{in}$(66.0 MPa $\sqrt{m}$) and σ = 20 ksi (138 MPa) results in $2a_{CR}$ = 5.7 in. (0.145 m). Thus, for these conditions, the maximum tolerable flaw size would be about 5.7 in. (0.145 m). For a design stress of 45 ksi (310 MPa), the same material could tolerate a flaw size of only about 1.1 in. (0.028 m). If residual stresses, such as might be caused by welding, are present so that the local stress in the vicinity of a crack is 80 ksi (552.0 MPa), the tolerable flaw size is reduced considerably. Note that if

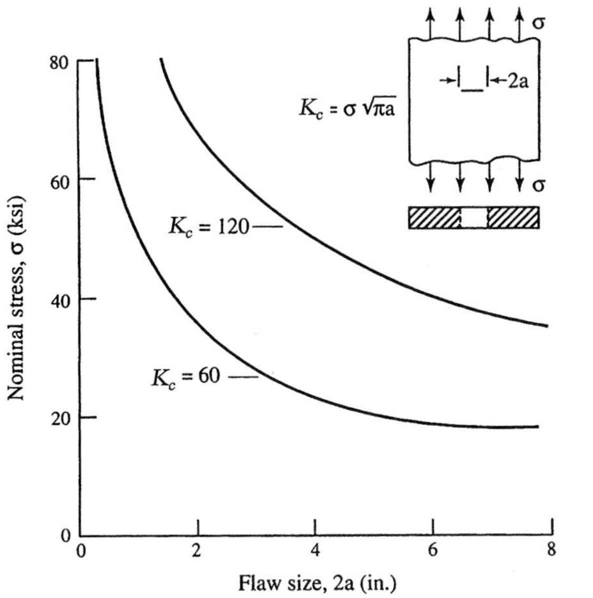

Figure 6.28 Stress–flaw size relationship for through-thickness crack for materials with $K_C = 60$ ksi $\sqrt{in}$ (66 MPa $\sqrt{m}$) and $K_C = 120$ ksi $\sqrt{in}$ (132 MPa $\sqrt{m}$).

a tougher material is used, say with $K_C = 120$ ksi $\sqrt{in}$ (132.0 MPa $\sqrt{m}$), the tolerable flaw sizes at all stress levels are significantly increased. If the notch toughness of a material is sufficiently high, brittle fractures will not occur, and failures under tensile loading occur by general yielding, unless the conditions of constraint are similar to those of the Ingram Barge, Fig. 6.29, or possibly the Northridge moment connections. Fortunately, most structural materials have a reasonably high level of notch toughness at service temperatures and loading rates and the constraint is not too rigid.

An analogy that may be useful in understanding the fundamental aspects of fracture-mechanics design is the comparison with Euler

Figure 6.29 Photograph of the I.O.S.3301 Martha Ingram Barge brittle fracture. Charpy V-notch toughness of 55 ft-lb (74.7 Joules) at service temperature.

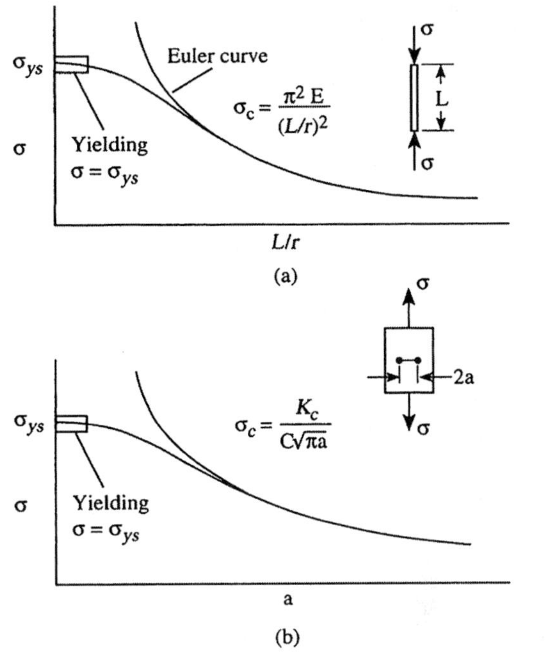

Figure 6.30 Analogy between (*a*) column instability and (*b*) crack instability.

column instability (Fig. 6.30). The stress level required to cause instability in a column (buckling) decreases as the *L*/*r* ratio increases (Fig. 6.30*a*). Similarly, the stress level to cause instability (fracture) in a flawed or cracked tension member decreases as the flaw size increases (Fig. 6.30*b*). As the stress level in either case approaches the yield strength, both the Euler analysis and the K_c analysis are invalid because of general yielding. To prevent buckling, the actual stress and *L*/*r* values must be below the Euler curve. To prevent fracture, the actual stress and flaw size must be below the K_c level shown in Fig. 6.30.

6.4.7 Prevention of Brittle Fracture

A. General To prevent brittle fractures in steel structures, it is usually sufficient to make sure that the steels properties meet the requirements of the recognized standard specifications, to proportion the details of the structure with reasonable care, and to insist on good practice in fabrication and manufacture. However, it is important to note that the lower the expected minimum service temperature, the more urgent these precautions become. A designer who adopts a standard specification as part of the design must remember that whatever assurance the specification gives against brittle fracture is based, in part, on experience with the particular type of structure and service for which the specification was written. Extrapolation to different situations must be made with great care and thorough analysis.

B. Material Specifications Toughness provisions of ASTM A709 for bridge steel and the AASHTO specifications ensure suitable materials for the intended service. Materials are available for three temperature regions, identified as Zones 1, 2, and 3. The corresponding CVN requirements are specified in the ASTM A709 specification. AASHTO prescribes the following lowest service temperatures:

Zone 1: 0°F and above (≥ −17°C)
Zone 2: −1° to −30°F (−18°C to −34°C)
Zone 3: −31° to −60°F (−35°C to −51°C)

For each of these zones, AASHTO specifies a minimum CVN impact toughness that is a function of plate thickness (see AASHTO Standard Specifications for Transportation Materials and Methods of Sampling and Testing). These levels of toughness requirements have been established with due consideration on the intermediate strain rates to which bridges are loaded as shown in Figs. 6.22, 6.24, and 6.25.

AISC specifications do not include explicit toughness requirements, since few brittle fractures have occurred in buildings, and loading rates are slow. Some failures, which occurred during construction, could have been avoided with tougher steels or, in some cases, with better design and/or fabrication details. For example, cracking in beam-to-column connections during the Northridge earthquake is an example of the need for a better understanding of how constraint affects the behavior of welded joints.

Steels for low-temperature applications, such as power-transmission poles, oil or gas storage tanks, water storage tanks, and pipelines, must be selected with an appropriate consideration of toughness.

If brittle fracture may be a problem in the case of a structure for which there is no adequate specification or experience, the general fracture mechanics design procedure described previously should be followed.

C. Design Considerations Because brittle fractures usually start at points of stress concentration, details in the design are of great importance and should include the following factors:

1. *Improved geometry.* Stress concentrations should be minimized, large radii should be used at changes in sections, and notches must be eliminated.
2. *Elimination of cracks and discontinuities.* These include laps, seams, cracks, splits, laminations, pits, inclusions, undercut, weld cracks, and incomplete weld fusion.
3. *Provision of flexibility in the structure to reduce stress concentrations.* Highly constrained joints can lead to premature cracking.
4. *Reduction of residual stresses that increase the possibility of fracture initiations.*
5. *Use of crack arresters in critical applications.* Crack arresters are inserts of plates or shapes that have very high toughness levels so that any brittle cracks are "arrested" before they propagate through the entire structure. They may be bolted or welded.
6. *Design for multiple load paths or redundancy.*
7. *Cracks may initiate at attachments of secondary or nonstructural members.* Crack may propagate across a joint and damage the main structural member. Thus, such attachments also should be designed carefully.

D. Design Stresses Design stresses based on the elastic properties of structural steels are usually about 60 percent of the yield strength. To protect against crack propagation by controlling the stress level alone, research has shown that the maximum design stress should be no greater than 20 percent of the yield strength of the steel [5 to 10 ksi (34.5 MPa to 69.0 MPa) for mild structural steels]. However, attempting to control brittle fracture using low-stress levels may not be feasible; thus for proper design, consideration must be given to all factors discussed above.

E. Fabrication For the fabrication of steel structures susceptible to brittle fractures it is extremely important to cut, form, and assemble the material with great care.

Cutting:

1. Shearing or punching holes in plates work hardens the material drastically at the shear edge or edge of the hole, reducing ductility, and notch toughness. The risk is eliminated if the edges are reamed; this requires removing about $^1/_{16}$ or $^1/_8$ in. (1.5 to 3.0 mm) of the diameter edge, depending on the thickness of the material and the diameter of the holes.
2. Flame-cut edges of plates, if they are carefully machine-cut, generally provide a satisfactory smooth surface, particularly if the edge is to be welded. However, blowout cuts can produce harmful irregularities.
3. Notches produced by access openings, change in section sizes, should be modified by introducing a curvature at the notch with as large a radius as it is practical.

Forming:

1. Local forming or cold working can be extremely detrimental. It can reduce ductility significantly and decrease notch toughness.
2. Heating after cold working will often produce recovery and recrystallization causing the material to become soft, thus reducing its strength. Steels that have been previously cold-worked by rolling, bending, shearing, punching, or straightening, may be seriously affected if they are heated in the range about 400 to 850°F (204 to 454°C) as by galvanizing, adjacent welding, heat straightening, and so forth.

Welding:

1. Sudden steep temperature changes in small spots on relative heavy members caused by arc strikes to initiate the welding arc are susceptible to cracking. The arc strike should be done in a separate plate or in a place within the weld joint that will be remelted as the weld is made.
2. Preheating can be used effectively to reduce residual stresses and lessen weld cracking and thus increase notch toughness of large weldments.
3. Post-weld heat treatments are also effective in improving notch toughness of large weldments, particularly in the heat-affected zone (HAZ).
4. Peening should generally be avoided and it is not permitted by most specifications. It can reduce toughness in the weld metal and it is not very effective in reducing residual stresses.

F. Inspection Because of the importance of workmanship on the low-temperature performance of structural steels, adequate inspection is essential. Nondestructive techniques that can be used include visual, radiographic, ultrasonic, magnetic-particle, and dye-penetrant procedures.

6.5 FATIGUE OF STRUCTURAL STEELS

6.5.1 Fatigue

It is a form of failure that occurs in structures subjected to dynamic and fluctuating stresses (e.g., bridges, aircraft, and machine components). Under these circumstances, it is possible for failure to occur at a stress level considerably lower than the tensile or yield strength for a static load. The term *fatigue* is used because this type of failure normally occurs after a lengthy period of repeated stress or strain cycling. Fatigue is important inasmuch as it is the single largest cause of failure in metals, estimated to be involved in approximately 90 percent of all metallic failures; polymers and ceramics (except for glasses) are also susceptible to this type of failure. Furthermore, fatigue is catastrophic and dangerous, occurring very suddenly and without warning. A fatigue failure usually originates at a point of stress concentration such as a sharp corner or notch, or at a metallurgical inclusion or flaw. Once nucleated, the crack propagates across the part under the cyclic or repeated stresses. During this stage of the fatigue process, clamshell or "beach" marks are created, as shown in Fig. 6.31. Finally, the remaining section becomes so small that it can no longer support the load, and complete fracture occurs. Thus there are usually two distinct types of surface areas that can be recognized: (1) a smooth surface region due to the rubbing action between the open surface region as the crack propagates across the

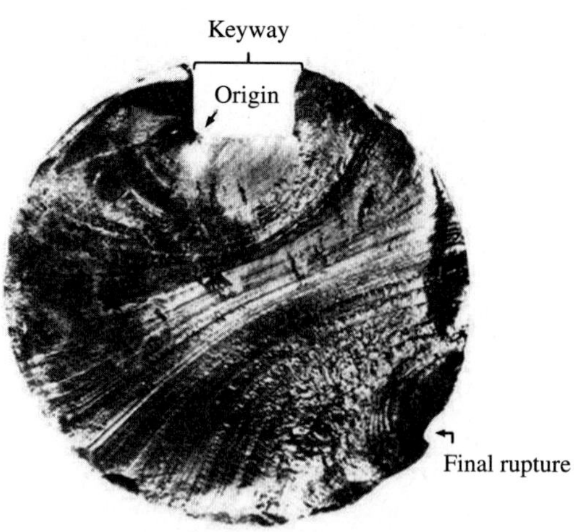

Figure 6.31 Light fractograph of the fatigue-fracture surface of a keyed shaft of 1040 steel (hardness - Rockwell C 30). The fatigue crack originated at the left bottom corner of the keyway and extended almost through the entire cross section before final rupture occurred.[4]

section and (2) a rough surface area formed by the fracture when the load becomes too high for the remaining cross section.

6.5.2 Fatigue in Structural Steels

Many evaluations have been made of the several factors affecting the fatigue behavior of a metal alloy, member, or structure.[1,5] Based on these studies, requirements have been provided in design specifications to prevent such failures. The principal parameters are the magnitude of the alternating stresses and the number of cycles necessary to produce failure. Data from this test are plotted in the form of S-N curves in which the stress S to cause failure is plotted against the number of cycles N at which failure occurs. Figure 6.32 shows typical S-N curves for high-carbon steel and a high-strength aluminum alloy. For the aluminum alloy, the stress to cause failure decreases as the number of cycles is increased. For the carbon steel, there is first a decrease in fatigue strength as the number of cycles is increased and then there is leveling off in the curve, with no decrease in fatigue strength as the number of cycles is increased. This horizontal part of the S-N plot is called the fatigue or endurance limit and lies between 10^6 and 10^{10} cycles. Many ferrous alloys exhibit an endurance limit that is about one-half their tensile strength. Nonferrous alloys such as aluminum alloys do not have an endurance limit and may have fatigue strengths as low as one-third their tensile strength.

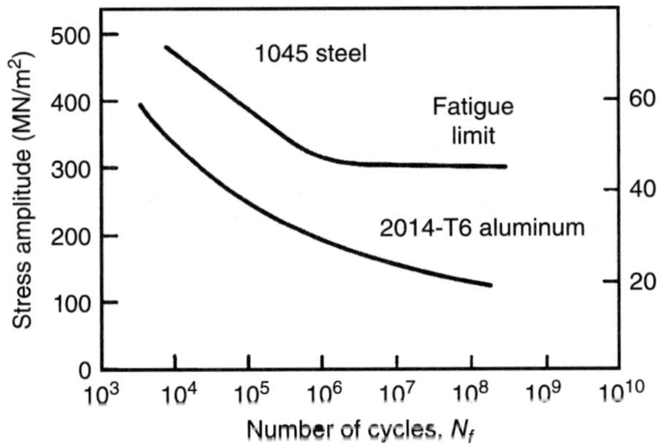

Figure 6.32 Stress amplitude versus number of cycles (S-N) curves for medium-carbon steel 1045 and aluminum alloy 2014-T6.[6]

The relationship between the fatigue strength to failure for a given number of cycles may be described on an S-N, or Wohler, curve (Fig. 6.33). Such a diagram, presented on a log-log basis, portrays the behavior of a specific detail for a given loading condition and can be expressed by

$$F_n = S\left(\frac{N}{n}\right)^K \tag{6.13}$$

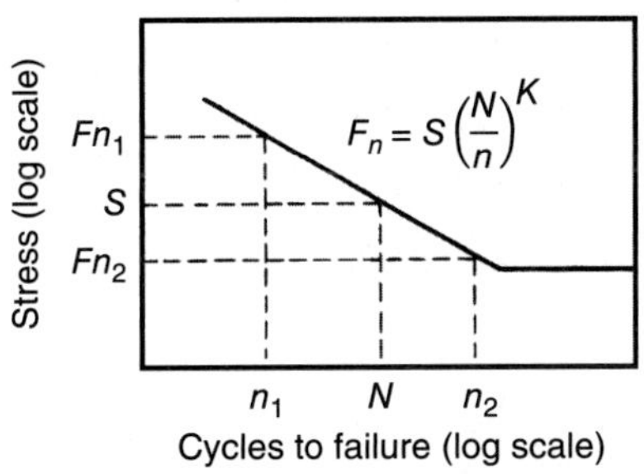

Figure 6.33 Schematic of the S-N diagram.

where F_n = fatigue strength computed for failure at n cycles
S = stress which produced failure in N cycles
K = slope of the straight-line S-N curve

The curve becomes essentially horizontal at about 10^6 cycles, generally referred to as the *fatigue limit.* Such relationships can be used in design, but they are of greater value in the form of a modified Goodman fatigue diagram (Fig. 6.34), which provides a relatively complete picture of the fatigue behavior of a given connection, member, or structure. The ordinate is the maximum stress, and the abscissa is the minimum stress (either tension or compression as shown). The radial lines indicate the stress ratio. The curves n_1, n_2, etc., represent failure at various lives. Design stresses can then be presented in the form of Eq. (6.14).

$$\sigma_{\max} = \frac{\sigma}{1-K_1R} \le \sigma_a \tag{6.14}$$

where $s_{\max}$ = maximum allowable repeated stress
σ = stress parameter which is a function of the ultimate strength of the steel, the life, and the type of member
K_1 = coefficient which is a function of life and the stress ratio
R = stress ratio (ratio of minimum to maximum stress, taken algebraically)
σ_a = basic allowable static stress for the member under consideration

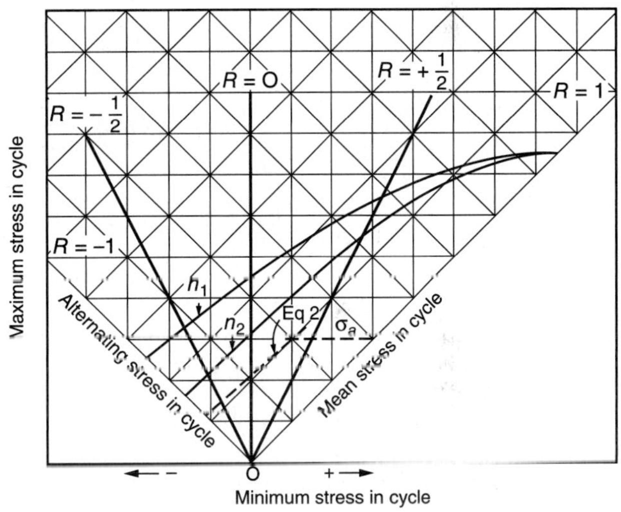

Figure 6.34 Modified Goodman diagram for various lives and stress ranges.

To further simplify design, K_1 can be taken equal to 1. This corresponds to a constant range of stress for a given life and is the basis for many design specifications.

6.5.3 Major Factors Affecting the Fatigue Strength of a Metal or Alloy

The fatigue strength is affected by factors other than the metal or alloy composition. Some of the most important are:

(a) Stress concentration: the fatigue strength is significantly reduced by the presence of stress raisers such as notches, holes, keyways, or sharp changes in cross sections. Fatigue failures can be minimized by careful design to avoid stress raisers whenever possible.

(b) Surface roughness: In general the smoother the surfaces finish on the metal sample, the higher the fatigue strength. Rough surfaces create stress raisers that facilitate fatigue crack formation.

(c) Surface condition: Since most fatigue failures originate at the metal surface, any major change in the surface condition will affect the fatigue strength of the metal. For example, surface-hardening treatments for steels, such as carburizing and nitriding, which harden the surface, increase fatigue life. Decarburizing, on the other hand, which softens a heat-treated steel surface, lowers fatigue life. Introduction of a favorable compressive residual stress pattern on the metal surface also increases fatigue life.

(d) Environment: If a corrosive environment is present during the cyclic stress of a metal, the chemical attack greatly accelerates the rate at which fatigue cracks propagate. The combination of corrosion attack and cyclic stresses on a metal is known as *corrosion fatigue.*

In general, the fatigue resistance of a structural steel is proportional to its ultimate strength. However, this proportionality can vary considerably, or even be nonexistent, depending upon the stress concentrations in the member being considered and the number of stress cycles to which it is subjected. Small, rotating members of steel (rotating-beam specimens) with polished surfaces generally exhibit a fatigue limit in reversal equal to approximately 50 percent of the ultimate tensile strength of the material. Such a value may be significant in the design of some machine parts. However, when such members contain severe stress concentrations, the percentage will be much smaller. Furthermore, corroding environments can be expected to produce further reductions. The effects of geometry and corrosive environment on the relationship between fatigue strength and ultimate strength of steels are shown in Fig. 6.35.

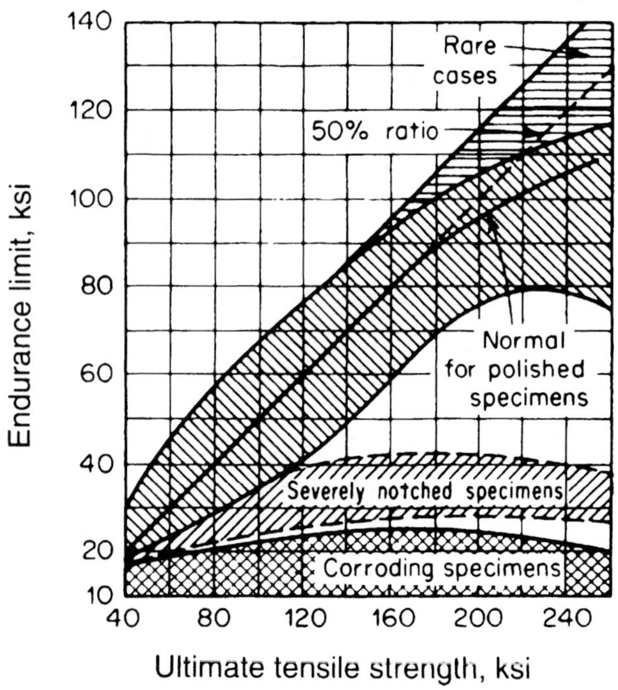

Figure 6.35 Relationship between the fatigue limit and ultimate tensile strength of various steels.[7]

As-rolled plates and plain rolled structural steel members also have fatigue strengths proportional to the ultimate strength of the material. This relationship, shown in Fig. 6.36 for a zero-to-tension loading, may be represented approximately by

$$F_{2,000,000} = 20 + \frac{\text{ultimate tensile strength, ksi}}{4} \tag{6.15}$$

For other conditions of loading, fatigue resistance may be determined from the following approximate empirical equation:

$$F = \frac{3.0F_r}{1.7 - R} \tag{6.16}$$

where F_r = fatigue strength for complete reversal of stress
F = fatigue strength for particular value of R

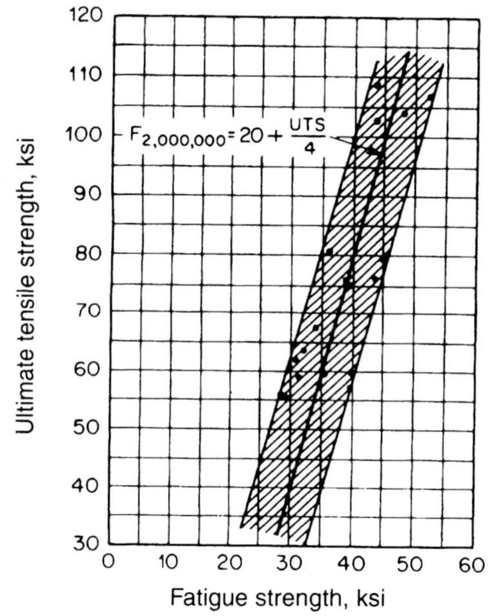

Figure 6.36 Relationship between the fatigue strength of plain plates will mill scale on surfaces, zero-to-tension stressing.[7]

These equations give an estimate of the basic fatigue strength of the plain material. To relate them to a design or working stress of the form shown in Eq. (6.14), an appropriate factor of safety must be incorporated.

The rate or frequency of cyclic loading has no significant effect on fatigue strength when the applied stresses are relatively low (long life) and the frequency is less than 3000 cycles per minute (cpm). But, if the stresses are high enough to produce plastic deformation with each cycle of loading (low life), an increase in speed of loading will produce an increase in apparent fatigue strength. The magnitude of this effect can be determined only by tests.

Laboratory tests are generally conducted at selected magnitudes of minimum and maximum cyclic stress, but many structures are subjected to maximum and minimum stresses which vary in magnitude from cycle to cycle. Extensive research suggests that many factors affect the manner in which fatigue damage accumulates under this type of loading. Nevertheless, the concept suggested by Palmgren and applied by Miner[8] appears to provide a simple and reasonably reliable prediction of fatigue behavior under random loadings. For failure,

$$\sum_i \frac{n_i}{N_i} = \frac{n_1}{N_1} + \frac{n_2}{N_2} + \frac{n_3}{N_3} + \Lambda = 1 \tag{6.17}$$

where n_i = number of stress cycles at stress level σ_i
N_i = number of stress cycles to produce failure at σ_i

A major limitation of the Palmgren–Miner rule is that it does not consider sequence effects, that is, the order of the loading makes no difference in this rule. Sequence effects are definitely observed in many cases. A second limitation is that the Palmgren–Miner rule says that the damage accumulation is independent of stress level. This can be seen from the modified S-N diagram (Fig. 6.37) where the entire curve is shifted the same amount, regardless of stress amplitude.

A fourth factor to be considered is the effect of residual stresses. Views differ markedly concerning this factor, and the effect may in fact vary from one instance to another, depending upon the alloy composition, fabrication condition (e.g., whether it is welded) and geometry of the member, the states and magnitudes of residual and applied stress, and, perhaps, other factors. In general, the introduction of residual compressive stresses at a location that is critical in fatigue can be expected to increase the fatigue resistance at this location for

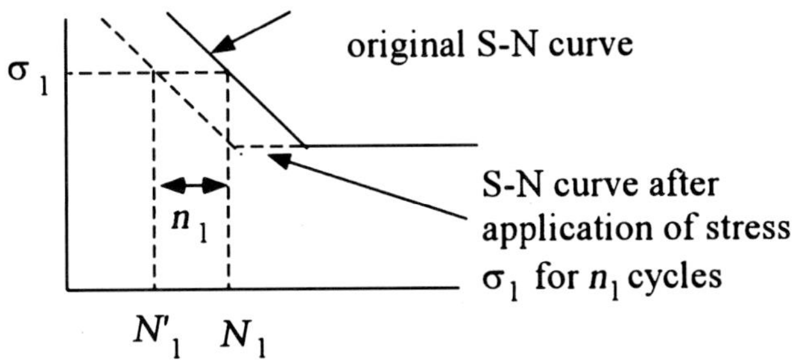

Figure 6.37 Modified S-N diagram showing the entire curve shifting the same amount regardless of stress amplitude.

low levels of stress (long life). However, for levels of stress sufficient to cause yielding (short life) the effect of residual stresses will be small or nonexistent. Residual tensile stresses, in general, are found to have little or no effect, unless they can be introduced in such a manner as to reduce the resulting range of stress under service loadings, which can be beneficial. Thus, caution must be exercised in taking advantage of any effect of residual stresses.

Laboratory studies show that the fatigue resistance in flexure of relatively small members increases as the diameter or depth decreases; this is a result of the effect of size and the accompanying strain gradient. Under axial loadings the size effect is generally found to be small or nonexistent. In relatively large flexural members (depths greater than 6 in.) the effect of strain gradient is relatively small. The factor which is generally of greatest importance is the geometry, either internal or external, and the accompanying stress concentration. The magnitude of this effect varies with the sharpness of the change in geometry (the notch radius), the material composition and fabrication process, and the life or number of cycles to failure. Sharp notches produce very large effective strain concentrations and result in extremely low-fatigue resistance. The magnitude of the effective strain concentration for a given detail increases with the ultimate strength of the material (Fig. 6.35) but decreases as the life decreases, that is, as the magnitude of the applied stress is increased. Consequently, great care must be used in designing structures that are subjected to repeated loadings, and in the selection of the structural details.

Other variables, such as temperature and rest periods, may also affect the fatigue resistance of a structure. However, these are generally considered to be of secondary importance for most steel structures and are seldom considered in design.

6.5.4 Structural Members

Fatigue behavior in axially loaded members is markedly different under compressive and tensile loadings. Repeated compressive loadings alone will generally not produce fatigue failures. Fatigue resistance of tensile members is generally controlled by the behavior of their connections. Only if no special precautions are taken to protect the connections, or if unusually severe stress concentrations exist in the member itself, will the tension member be critical in fatigue. In this case, fatigue behavior is determined by the factors discussed in Fig. 6.34.

Properly fabricated flexural members, beams, and girders have a high-fatigue resistance. However, details such as cover plates, splices, and stiffeners may affect their behavior markedly and must be considered in the design.

6.5.5 Riveted Connections

Laboratory studies demonstrate that the fatigue resistance of riveted connections can be relatively low as a result of stress concentrations produced by rivet holes, by the eccentricity of loading in the connections, and the bearing of the rivets in the holes. The type of steel generally has little effect, so there is little or no advantage to high-strength steels under severe fatigue conditions. They may be advantageous only where the number of repeated loads is relatively small and/or the stress range is relatively low. Variations in joint details and rivet patterns are also found to have little effect.

The clamping force of the fasteners can be quite important. Tight, well-driven rivets are essential to maximum fatigue resistance. Loose rivets may decrease fatigue resistance by as much as 25 percent over that of joints with tight rivets. Riveted structures are seldom used now. However, because of the age of most of the existing riveted structures, it occasionally may be necessary to use the existing fatigue information and loading history data to evaluate the extent of fatigue damage to which these structures have been subjected.[9]

6.5.6 Bolted Connections

The fatigue strength of the ASTM A325 or A490 bolt subjected to direct axial tension is related directly to the clamping force in the fastener. When properly tightened (to the proof load of the fastener), the fatigue limit can be expected to approximate 80 percent of the proof load. However, if the fastener is loose, or loses its clamping force, fatigue resistance may be as low as 20 percent of the proof load.

In many connections in which the fasteners are loaded in tension, there is a prying action, which must be taken into account. The resulting increase in fastener tension may reduce the fatigue limit to as little as 40 percent of the proof load. Thus, eccentricities in the loading of bolted tension connections should be reduced to a minimum, insofar as is practicable.

The clamping force is also important in shear-type connections. Adequate clamping (70 percent of specified minimum tensile strength in A32S or A490 bolts) will provide a fatigue strength in a bolted connection of A36 steel 25 percent greater than that of a similar, properly riveted connection. In structures where slip in the connections would be objectionable and must be avoided, frictional resistance rather than fatigue resistance may control the design.

Occasionally it is necessary to use single lap joints rather than butt-type joints. The fatigue resistance of the lap joint is comparable with that of the butt-type joint under tensile loading. However, under compressive loadings the eccentricity in the connection produces a bending which will reduce the fatigue resistance significantly and should be considered in the design.

6.5.7 Welded Connections

Welded structures can have relatively low-fatigue resistance if they are improperly designed and/or fabricated. Extensive data on the fatigue resistance of welded structural connections have been collected.[8,10] The geometry of the structure or connection and the associated stress concentration are of greatest importance among the many factors affecting the fatigue strength of welded structural members and connections.

The fatigue properties of welded structural connections depend on the tensile strength of the steel used for their fabrication. However, the magnitude of this variation depends on the severity of the stress concentrations produced from welding and the severity of the fatigue loading conditions. Under conditions where a connection is subjected to high stresses and relatively few applications of loading, the higher strength materials will prove advantageous. However, under relatively low levels of stress applied a large number of times, and with severe stress concentrations, the benefits of high-strength materials may be relatively small or nonexistent. Thus, the more severe the fatigue conditions, the less advantage there is to the use high-strength materials.

An understanding of the stress flow through the structure or connection is important in designing for resistance to fatigue. Butt-welded joints, because of their relatively smooth flow of stress, provide excellent resistance. However, fillet-welded joints or lap joints, wherein a marked change in direction of stress or load transfer occurs, have relatively low resistance. Thus, care must be exercised in providing the best geometry. Whenever possible, butt-welded connections should be employed and strap plates or doubler plates avoided. Such added material, although it may increase the static strength, will reduce markedly the connection's fatigue resistance. Fatigue resistance of a butt-welded connection may generally be increased by grinding the weld reinforcement flush with the surface of the connected members, provided the weld is sound.

6.5.8 Design for Fatigue

The principal design specifications[11–14] for bridges and buildings include provisions for repeated loadings based on the results of numerous laboratory tests of riveted, bolted, and welded members and connections.

For the general purposes of fatigue design, members and connections can be grouped into the 42 structural details shown in Fig. 6.38 and described in Table 6.1. The mean fatigue resistances (stress ranges) of the members shown in Fig. 6.38 are given in Table 6.2. These values are for constant stress-range loading and can be applied only for very limited conditions, since in most instances the load will vary randomly.

Four significantly different theoretical loading histograms are shown in Fig. 6.39 in terms of a frequency distribution of ratios of range in cyclic stress to the maximum range in cyclic stress.[15] These four conditions are defined in Table 6.3, the first three of which represent beta-distribution probability density functions that have shape factors q and r as shown. The bridge-loading histogram of Fig. 6.40 is given as an example for an actual structure. (Note that the value of the loading coefficient, CL, is between that of load types I and II of Fig. 6.40.) Other histograms of highway- and railway-bridge loadings are shown in references 16 and 17. Additional loading frequency distributions are presented in reference 18.

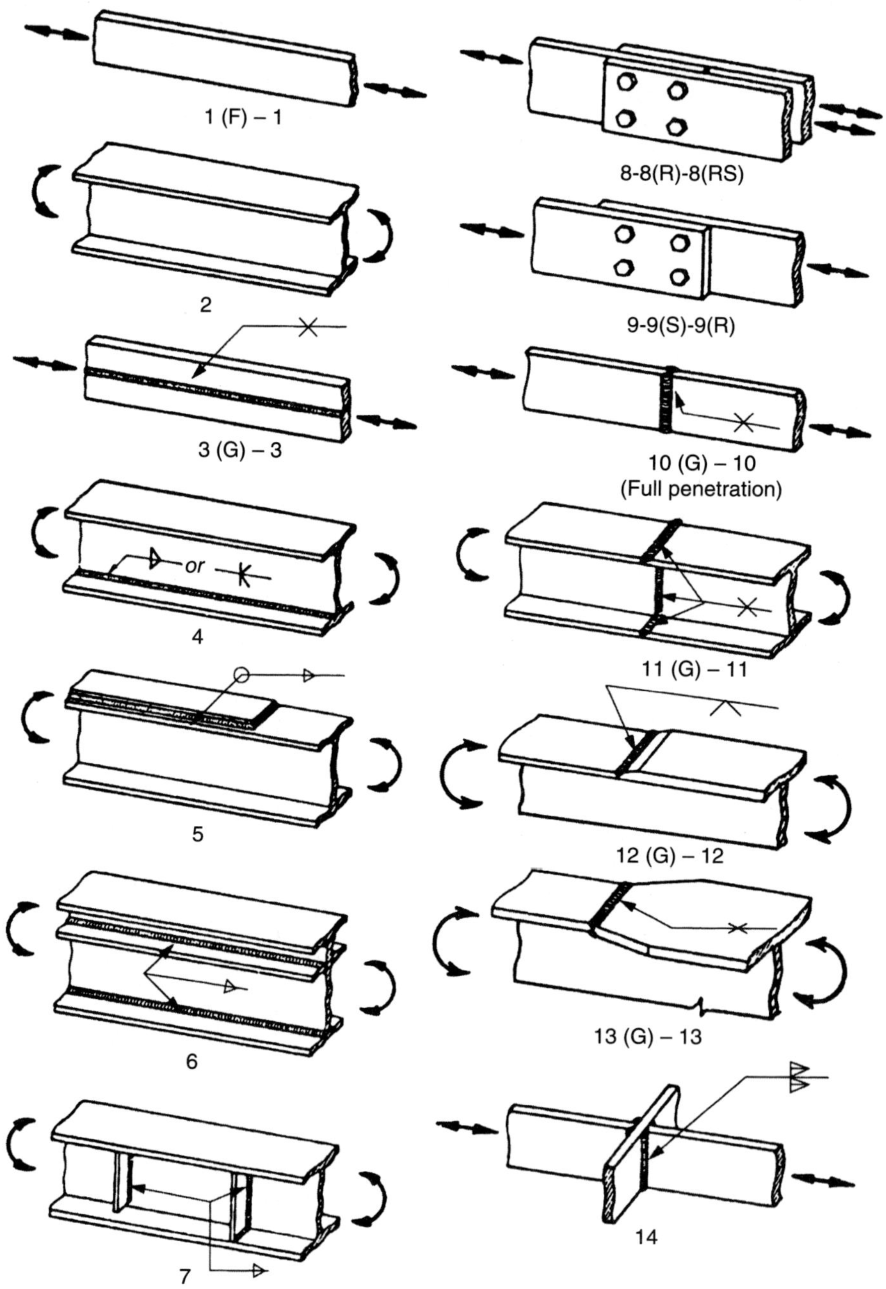

Figure 6.38 Structural details for fatigue design requirements of Tables 6.1 and 6.2.[9,10]

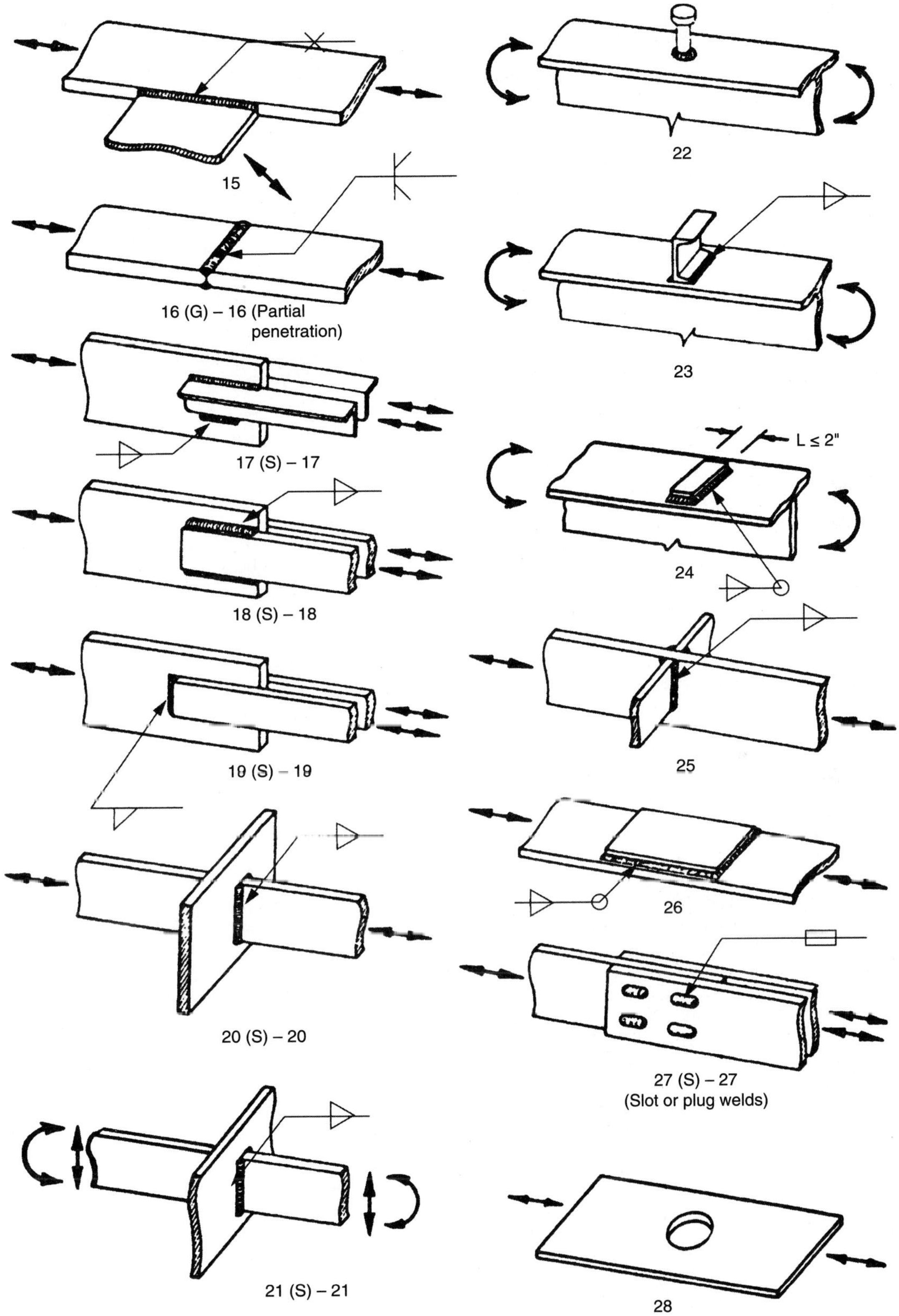

Figure 6.38 (*Continued*)

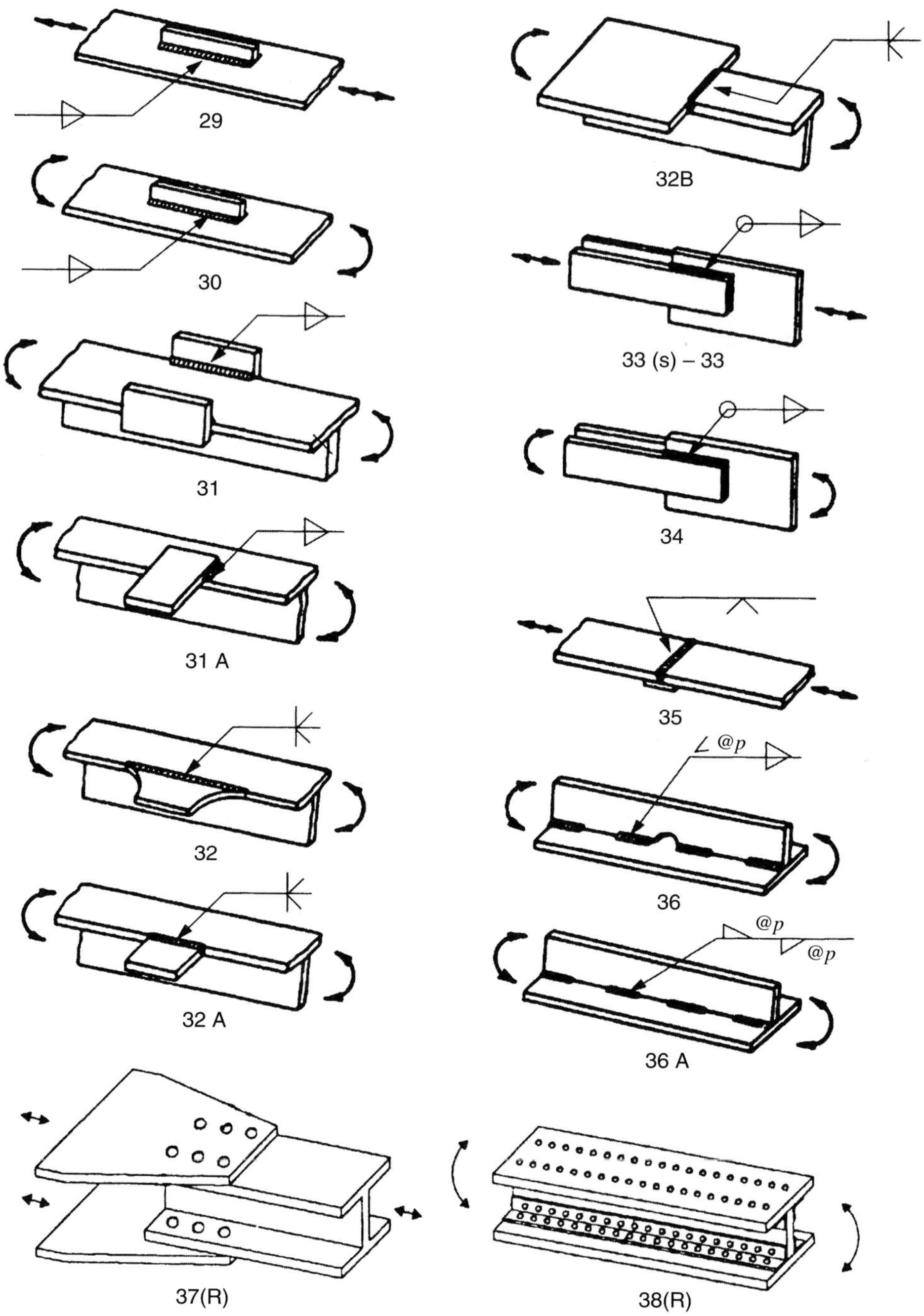

Figure 6.38 (*Continued*)

The allowable fatigue design stress range S_R may be determined from

$$S_R = S_r C_L R_F \tag{6.18}$$

where, S_r = mean constant-cycle fatigue stress range for desired life (cycles of loading), Table 6.2
C_L = loading coefficient to be selected for load type, Table 6.3
R_F = reliability factor based on a statistical fatigue analysis for probability of survival, Table 6.4

Values of R_F are based on the total uncertainty in fatigue life (scatter or uncertainty in fatigue data in fatigue damage model, in stress analysis, and in fabrication) and on the assumption that the distribution of fatigue life can be represented by a Weibull distribution.[10] Factors for reliabilities of 90, 95, and 99 percent are given in Table 6.4. Values of C_L based on a linear log S-log N fatigue relationship, the linear-damage rule of Eq. (6.17), and the loading frequency-distribution functions of 22 are given in Table 6.3.[15] Thus, using the loading history in Fig. 6.40 and a 95 percent level of

Table 6.1 Description of Structural Details for Fatigue Design

Type of detail	Example[a] (see Fig. 6.38)
Plain Material	
• Base metal with rolled or cleaned surface	1, 2, 28
Build-Up Weldments	
• Base metal and weld metal in members, without attachments, built up of plates or shapes connected by continuous fillet welds or full-penetration groove welds parallel to the direction of applied stress	3, 4, 6
• Base metal at end of partial-length welded cover plates having square or tapered ends, with or without welds across the ends	5
• Calculated flexural stress at toe of welds on girder webs or flanges adjacent to welded transverse stiffeners	7
Bolted Connections	
• Base metal at net section of high-strength bolted connections, except bearing-type connections an axially loaded joints which induce out-of-plane bending in connection material	8
• Base metal at net section of other mechanically fastened joints[b]	9[b]
• Shear on fasteners of mechanically fastened joints (little or no clamping)[c]	9s[d]
Riveted Connections	
• Base metal at net section of riveted double-shear connections, axially loaded, no out-of-plane bending	8(R)[b]
• Base metal at net section of riveted single-shear connections in plates, truss-type members, and gusset plates	9(R)[b], 37(R)[b]
• Base metal in net section of built-up beams and beams with cover plates, flexural loading	38(R)[b]
• Shear on fasteners of riveted double-shear connections, axially loaded, no out-of-plane bending	8(RS)[b]
Groove Welds	
• Base metal and weld metal, when reinforcement is not removed, in or adjacent to fullpenetration groove-welded splices with or without transitions. Transitions must have slopes not greater than 1 to 2½	10, 11, 12, 13
• Base metal or weld metal in or adjacent to full penetration groove welds in tee or cruciform joints	14
• Base metal at details attached by groove welds subject to transverse and/or longitudinal loading	15, 32, 32A, 32B
• Weld metal of partial-penetration transverse groove welds, based on effective throat area of the weld or welds	16
• Base metal and weld metal at weld on backing strip	35
Fillet-Welded Connections	
• Base metal at ends of intermittent fillet welds	36, 36A
• Base metal at junction of axially loaded members with fillet-welded end connections. Welds must be disposed about the axis of the member so as to balance weld stresses	17, 18, 19, 20, 21, 33, 34
• Continuous longitudinal and/or transverse fillet welds subject to shear parallel to the member axis or in combination with shear due to flexure	17s, 18s, 33s
• Transverse fillet welds in tee joints, or end member[c]	19s, 20s, 21s
Miscellaneous Details	
• Base metal adjacent to short (2-in. maximum length in direction of stress) fillet-welded attachments	22, 23, 24, 25
• Base metal adjacent to longer fillet-welded attachments	26, 29, 30, 31, 31A
• Base metal at plug or slot welds	27[h]
• Shear on plug or slot welds	27s[e]

[a]These examples are given as guidelines and are not intended to exclude other reasonably similar situations
[b]This detail provides for stress in the base metal.
[c]Where stress reversal is involved, A307 bolts are not recommended.
[d]This detail provides for shear on the fasteners.
[e]This detail provides for stress on the throat of the weld.

reliability, the allowable design stress for a beam with partial length cover plates should be 95.8 MPa (13.9 ksi) for an expected life of 2,000,000 cycles and 57.9 MPa (8.4 ksi) for an expected life of 10^7 cycles.

The allowable fatigue design stress range is obtained as follows:

1. From Fig. 6.38 and Table 6.1, identify the type of detail at which fatigue cracking is most likely to occur.
2. Identify the loading frequency distribution which best represents the loading history to which the detail will be subjected during its life (Fig. 6.39). If no information is available, the type IV distribution can be used and will be conservative.
3. Determine the number of cycles of loading expected during the life of the structure.
4. Obtain the basic mean stress range S_r from Table 6.2 for the member or detail.
5. Determine the loading coefficient C_L, based on load type from Table 6.3 (or from a loading distribution function as discussed inreference.[15]).
6. Select a reliability factor R_F, based on importance of the detail to the safety of the structure, from Table 6.4.
7. Determine the maximum allowable fatigue design stress range from Eq. (6.18). The maximum stress based on this stress range should not exceed the maximum allowable static design stress for the detail in question.

6.5.9 Protecting against Fatigue Failure

The designer should:[17]

1. Avoid details of design that produce severe stress concentrations or poor stress distribution.
2. Provide gradual changes in the section and avoid reentrant notch-like corners.
3. Avoid abrupt changes of section or stiffness in members or components.
4. Align parts so as to eliminate eccentricities or reduce them to a minimum.
5. Avoid making attachments on parts subjected to severe fatigue loadings.
6. Use continuous welds rather than intermittent welds.
7. Avoid details that introduce high-localized constraint.
8. Provide suitable inspection to guarantee proper riveting, adequate clamping in high-strength bolts, and the deposition of sound welds.
9. Provide for suitable inspection during the fabrication and erection of structures.
10. When fatigue cracks are discovered, take immediate steps to prevent their propagation into the structure.

Table 6.2 Mean Fatigue Strength Stress Range, S_n ksi[a]

	Cycles of loading			
Detail no.[b] (Fig. 6.38)	100,000	500,000	2,000,000	10,000,000
1[c]	46.2	40.5	36.2	31.7
1(F)	42.5	32.1	25.2	19.0
2[c]	54.0	42.3	34.3	26.9
3	44.6	34.0	27.0	20.5
3(G)	44.9	34.9	28.0	21.8
4	42.5	32.0	25.0	18.8
5	26.3	16.1	10.5	6.4
6	42.5	32.0	25.0	18.8
7	44.8	29.2	20.2	13.2
8	55.8	43.6	35.3	27.6
8(RS)	37.1	31.9	27.9	24.0
8(R)	33.9	28.0	23.7	19.6
9	32.6	27.6	23.9	20.2
9(S)	27.1	21.8	18.1	14.6
9(R)	23.9	20.4	17.7	15.1
10[c]	34.1	27.5	23.0	18.6
10(G)	38.6	31.6	26.6	21.8
11	33.2	25.1	19.7	14.9
11(G)	29.4	24.4	20.8	17.3
12	33.2	23.0	16.7	11.6
12(G)	40.8	30.7	24.0	18.1
13	33.2	23.0	16.7	11.6
13(G)	48.3	33.0	23.8	16.3
14	40.0	29.0	22.0	16.0
15	24.4	16.6	12.0	8.2
16	32.8	23.2	17.1	12.1
16(G)	32.8	26.1	21.3	16.9
17	27.8	18.1	12.2	8.1
17(S)	28.2	22.9	19.2	15.6
18	20.3	13.6	9.7	6.5
18(S)	25.7	21.6	18.6	15.6
19	23.1	18.6	15.5	12.5
19(S)	27.5	22.2	18.5	14.9
20	26.5	18.7	13.9	9.8
20(S)	27.5	21.7	17.7	13.9
21	21.0	18.9	17.3	15.6
21(S)	42.4	34.1	28.2	22.7
22	39.8	23.9	15.4	9.2
23	35.4	21.4	13.8	8.3
24	35.4	21.4	13.8	8.3
25	33.2	26.5	21.8	17.4
26	34.0	21.0	13.9	8.6
27	25.0	15.0	9.6	5.8
27(S)	21.8	16.1	12.4	9.1
28	40.1	32.5	27.2	22.1
29	34.7	20.8	13.4	8.1
30	45.6	28.3	18.7	11.6
31	20.2	13.9	10.1	7.0
31A	30.6	19.2	12.9	8.1
32	27.0	18.5	13.4	9.2
32A	24.5	14.8	9.6	5.8
32B	21.5	13.6	9.2	5.8
33	21.3	13.8	9.4	6.1
33(S)	25.5	21.9	19.1	16.4
34	41.2	33.2	27.5	22.1
35	32.4	21.2	14.8	9.7
36	28.6	22.7	18.6	14.8
36A	33.6	24.6	18.8	13.8
37(R)	23.9	20.4	17.7	15.1
38(R)	28.9	21.6	16.8	12.6

[a] Constant stress range, 0.50 reliability.
[b] (F) = flame cut edges, (G) = weld surface ground flush, (S) = shear stress on fasteners or welds, (R) = riveted, (RS) = rivets in shear.
[c] For these details, high-strength low-alloy and quenched and tempered steels may provide a higher fatigue strength.

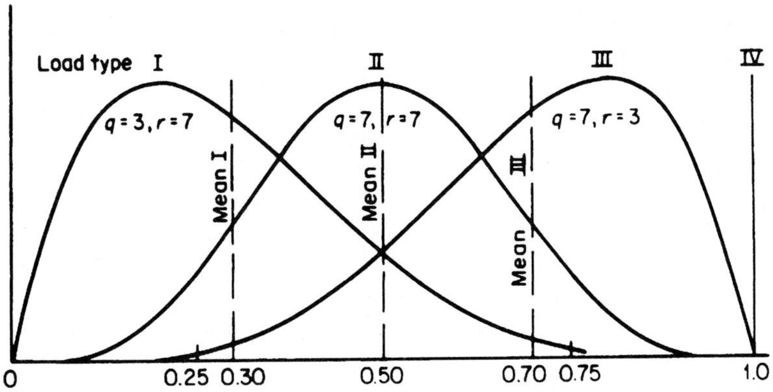

Figure 6.39 Loading frequency distributions.[15]

Table 6.3 Random Loading Coefficients C_L

Type	Load description (see Fig. 6.23)	Coefficient C_L
I	Primarily light loading cycles: mean range of stress 30% of maximum ($q = 3$, $r = 7$)	2.75
II	Medium loading cycles: mean range of stress 50% of maximum ($q = 7$, $r = 7$)	1.85
III	Primarily heavy loading cycles: mean range of stress 70% of maximum ($q = 7$, $r = 3$)	1.35
IV	Constant loading cycles: stress range constant and equal to 100% of maximum	1.00

Table 6.4 Reliability Factors R_F

Level of reliability	Structural importance of detail	Reliability factor R_F
90%	Secondary details for which fatigue cracking is of little structural significance	0.67
95%	Major structural details for which fatigue cracking is important: members in redundant structures	0.60
99%	Major structural details in fracture-critical members where fatigue cracking is critical	0.45

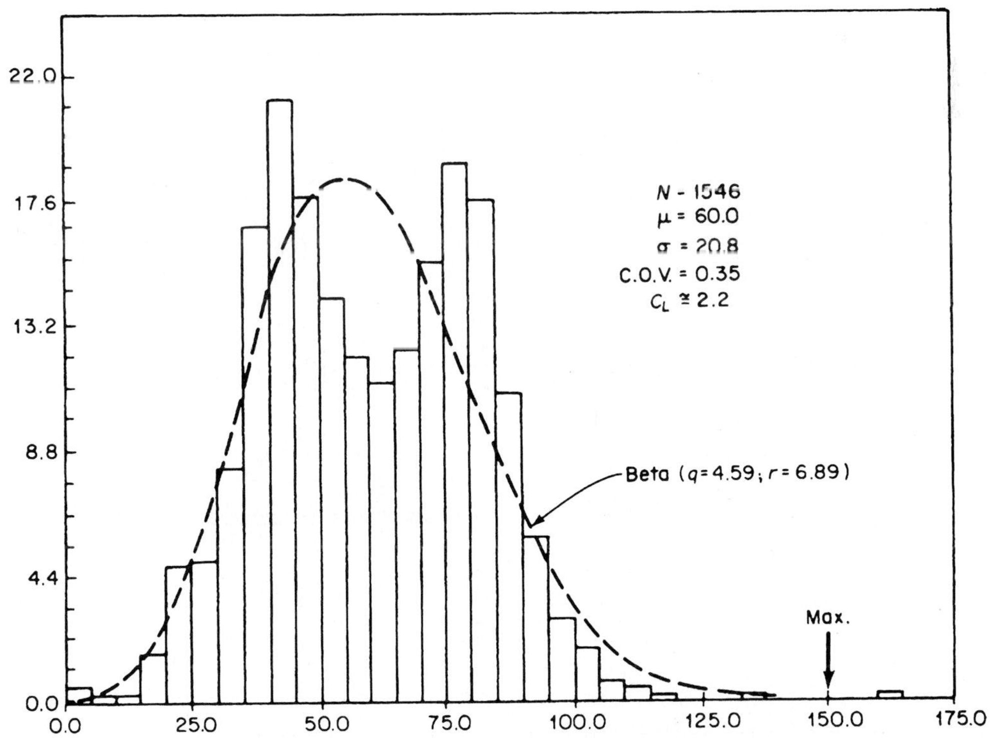

Figure 6.40 Histogram for bottom-flange strain range, beam bridge, heavy truck traffic.[16]

6.5.10 Lamellar Tearing

Another type of failure of great importance, where cracking occurs in the base metal beneath the weld (Fig. 6.41). It is caused by a combination of high, localized stresses from weld contraction and poor through-thickness ductility in the steel. The tearing is initiated by the separation of the interface between inclusions and metal (delamination) or by fracture of an inclusion itself. The cracks expand by linkage of the delaminations in the same plane and by shear steps that join cracks on different planes. This type of material fracture is the separation of the material resulting from through-thickness strains induced by weld metal shrinkage or by applied loads (Fig. 6.41).[19] The fatigue resistance, fracture strength, and toughness of a welded joint all can be significantly affected by lamellar tearing.[20–25]

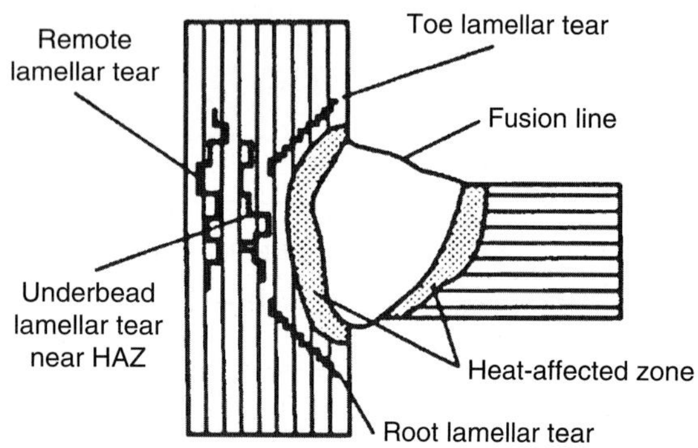

Figure 6.41 Orientation of lamellar tears in a base metal below the weld.

The controlling factors in lamellar tearing are: (a) quantity and morphology of low melting nonmetallic inclusions (usually sulfides or silicates), and (b) magnitude of the induced stresses normal to the plate surface. The magnitude of the induced stresses depends on the joint design, the degree of restraint imposed on the joint, the plate thickness, the orientation of the weld, and the size of the weld. Restrained corner or T-joints are the most susceptible to lamellar tearing because the fusion boundary is roughly parallel to the plate surface (Fig. 6.42) and through-thickness contraction stresses are high. Lamellar tearing in butt welds is rare. Thick-plate, high-imposed restraint (e.g., rigid clamping), and large weld beads all increase residual stresses and the likelihood of tearing.

Hydrogen has also been found to increase susceptibility to lamellar tearing. For this reason, a moderate preheat and the use of low-hydrogen consumables are beneficial in reducing the tendency to lamellar tearing. Use of welding electrodes that provide welds with the lowest acceptable yield strength but of sufficient strength to carry the design loads is another alternative to consider to reduce lamellar tearing, as well as using weld sequences that eliminate or minimize restraint and through-thickness weld shrinkage strains. Extreme caution in making weld repairs of highly restrained connections is important since weld repairs are usually made under greater restraint than the original welds. Ultimately, use of steels in which nonmetallic inclusions have been minimized—steels with reduced sulfur and oxygen content and steels with a reduction of area in the through thickness direction of at least 20 percent as determined in the specified ASTM-A770 test (see ASTM A770/A770M-86).[1]

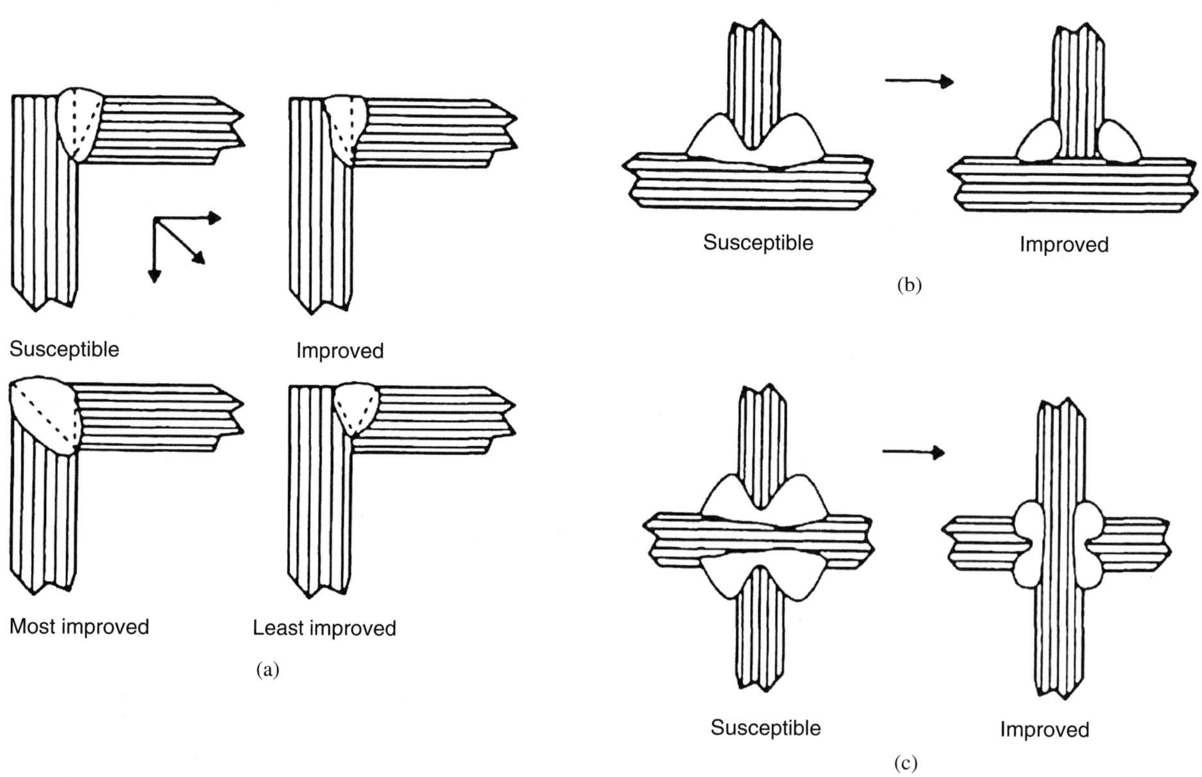

Figure 6.42 Methods used to reduce lamellar tearing due to welding.

REFERENCES

1. Rolfe, Stanley T., and Barsom, John M. (1999). Barsom. *Fracture and Fatigue Control in Structures, Applications of Fracture Mechanics.* 3rd ed. Butterworth-Heinemann, Oxford.

2. Reynolds, J. (2007). "99 Diseases of pressure equipment: Brittle fracture." Vol 13. Issue 1 *Inspectioneering J.*

3. American Society for Testing and Materials. (1995). *ASTM Book of Standards.* American Society for Testing and Materials, West Conshohocken, PA.

4. American Society for Metals. (1974). *Metals Handbook.* American Society for Metals, Vol. 9, 8th ed. American Society for Metals, Cleveland, OH.

5. Livieri, P., and Lazzarin, P. (2005). "Fatigue strength of steel and aluminum welded joints based on generalized stress intensity factors and local strain energy values. *Int. J Fract.,* 133(3), 247–276.

6. Hayden, H.W., Moffatt, W.G., and Wulff, J. (1965). *The Structure and Properties of Materials.* Vol. 3. Wiley, New York.

7. Battelle Memorial Institute (1946). *Prevention of Failure of Metals under Repeated Stress.* John Wiley, New York.

8. Miner, M.A. (1945). "Cumulative damage in fatigue." *J. Appl. Mech,* 12, A159–A164.

9. Munse, W.H. (1992). *Estimating the Remaining Life of Riveted Railway Bridges.* Report of study conducted for the Association of American Railroads. December.

10. Munse, W.H., Wilbur, T.W. Tellalian, M.L. Nicoll, K., and Wilson, K. (1983). *Fatigue Characterization of Fabricated Ship Details for Design,* Ship Structure Committee, SSC-318.

11. American Institute of Steel Construction. *Specification for the Design, Fabrication and Erection of Structural Steel for Buildings,* 8th ed., 1989, and *Load and Resistance Factor Design Specification for Structural Steel Buildings,* 2nd ed., 1993. American Institute of Steel Construction, Chicago.

12. American Railway Engineering Association. (1994). *Specifications for Steel Railway Bridges.* American Railway Engineering Association, Chicago.

13. American Association of State Highway and Transportation Officials. *Standard Specification for Highway Bridges,* 15th ed., 1992, and *LFRD Bridge Design Specifications,* 1st ed. American Association of State Highway and Transportation Officials, Washington, DC.

14. American Welding Society. Structural Welding Code, D1.1-94. American Welding Society, Miami, FL,1994.

15. Ang, A. H-S., and Munse, W.H. (1975). *Practical Reliability Basis for Structural Fatigue.* ASCE National Structural Engineering Convention, April 14–18.

16. Ruhl, J.A., and Walker, W.H. (1975). *Stress Histories for Highway Bridges Subjected to Traffic Loading.* University of Illinois Urbana, Civil Engineering Studies, Structural Research Series 416. March.

17. Fatigue of Welded Joints Committee—Welding Research Council. (1954). "Designing and making welded structural steel members for cyclic loading." *Welding J.,* 38, Res. Suppl., August.

18. Munse, W.H. (1964). *Fatigue of Welded Steel Structures.* Welding Research Council, New York.

19. American Institute of Steel Construction. (1973). "Commentary on highly restrained welded connections." *AISC Eng. J.,* 10, 61–73.

20. Kaufmann, E.T., and Stout, R.D. (1983). "The toughness and fatigue strength of welded joints with buried lamellar tears." *Welding J.,* 62(11), 301s–306s.

21. Ganesh, S., and Stout, R.D. (1977). "Effect of welding variables on lamellar tearing susceptibility in the Lehigh test." *Welding J.,* 56(3), 78s–87s.

22. Kaufmann, E.J., Pense, A.W., and Stout, R.D. (1981). "An evaluation of factors significant to lamellar tearing." *Welding J.,* 60(3), 43s–49s.

23. Ganesh, S., and Stout, R.D. (1977). "Material variables affecting lamellar tearing susceptibility in steels." *Welding J.,* 55(11), 341s–355s.

24. Jubb, J.E.M. (1971). *Lamellar Tearing.* Welding Research Council Bulletin No. 168. December. Welding Research Council, Shaker Heights. OH.

25. Farrar, J.C.M., and Dolby, R.E. (1972). *Lamellar Tearing in Welded Steel Fabrication.* The Welding Institute, Cambridge.

Chapter 7

Soil Mechanics and Foundations

BY

JOSEPH W. SCHULENBERG *Assistant Clinical Professor of Civil Engineering, University of Illinois at Chicago*

KRISHNA R. REDDY *Professor of Civil Engineering, University of Illinois at Chicago*

7.1 SOIL BEHAVIOR

Soil behavior is based on the soil type and the rate of loading. There are two main soil types: frictional and cohesive. There are four main types of loads: long term, short term, dynamic, and transient. Additionally, there can be creep deformation of clays under constant load over long periods of time.

7.1.1 Identification

Though identification is often based on grain size, one of the primary ways that grain size impacts soils is permeability. Clays, which have very low permeability, on the order of 1×10^{-7} cm/s, do not allow water to readily dissipate. Granular soils have permeability that may be one or more orders of magnitude higher than clays. As such, they more readily dissipate pore-water pressure and therefore feel the effects of loading more quickly than clays. Therefore, discussion of soil behavior will begin with permeability.

Permeability

Permeability refers to the relative ease or difficulty with which a liquid flows through a medium such as soil. The coefficient of permeability k is a velocity of flow under a unit hydraulic gradient and is expressed in terms of distance/time, frequently feet per minute. It is dependent on the flowing fluid and its temperature. Typical values of k for various soils are shown in Table 7.1.

The permeability of soil is often of importance in the selection of suitable materials for filter drains, free-draining base courses, and relatively impervious earth cores for dams. It also affects the quantity of groundwater flow into excavations, wells, and underground drains. The size and number of pumps and the layout of a well-point system for drainage will depend on the permeability of the soils at a given site. Moreover, on vertical sand-drain installations, the coefficients of permeability of natural deposits in both the vertical and the horizontal directions are required to estimate the relative time rate of consolidation when both radial drainage toward the sand drains and vertical drainage to more pervious horizontal soil layers are involved.

Table 7.1 Typical Values of Coefficient of Permeability *k* for Soils

Soil type	Typical coefficient of permeability, ft/day
Well-graded clean gravels, gravel-sand mixtures	75
Poorly graded clean gravels, gravel-sand-silt	180
Silty gravels, poorly graded gravel-sand-silt	1.5×10^{-3}
Clayey gravels, poorly graded gravel-sand-clay	1.5×10^{-4}
Well-graded clean sands, gravelly sands	4×10^{-2}
Poorly graded clean sands, sand-gravel mix	4×10^{-2}
Silty sands, poorly graded sand-silt mix	2×10^{-2}
Sand-silt clay mix with slightly plastic fines	3.0×10^{-3}
Clayey sands, poorly graded sand-clay mix	7.5×10^{-4}
Inorganic silts and clayey silts	1.5×10^{-3}
Mixture of inorganic silt and clay	3.0×10^{-4}
Inorganic silts of low to medium plasticity	1.5×10^{-4}
Organic silts and silt-clays, low plasticity	Quite variable
Inorganic clayey silts, elastic silts	1.5×10^{-4}
Inorganic clays of high plasticity	1.5×10^{-4}
Organic clays and silty clays	Quite variable
1 cm/s ≈ 2840 ft/day ≈ 2 ft/min	
1 ft/year ≈ 1×10^{-6} cm/s	

Soil permeability may be determined either in the field by, for example, well pumping tests or in the laboratory on small and, it is hoped, representative samples. Field-well pumping tests are relatively expensive and time consuming and require specially trained personnel to plan, conduct, and analyze the results. They are seldom justified except where large projects are involved or where underground water supplies are to be developed by a system of wells. However, the field-well pumping test does give more representative and reliable estimates of permeability since a much larger mass of the actual soil in an undisturbed condition is involved.

For most jobs, engineers resort to laboratory tests of small samples to determine k. Two tests, the falling-head test for the more impervious fine sands, silts, and clays and the constant-head test for more pervious gravels and clean sands, are more commonly employed. These tests attempt to determine k within an order of magnitude, and for most practical situations this is sufficiently accurate.

FRICTIONAL (GRAVELS, SANDS, SILTS)

Most coarse-grained soils, such as sands and gravels, can be determined by the percentage passing the number 200 sieve (0.075 mm). If 50 percent or more is retained on the number 200 sieve, the soil is considered coarse and may be a sand or a gravel.

However, silts, which present construction difficulties, are determined as materials both with more than 50 percent passing the number 200 sieve and having a combination of liquid limit and plastic limit such that they behave as silts. The liquid limit is the water content dividing line where wetter soils behave as a liquid and drier soils behave as a solid. The plastic limit is the water content dividing line where wetter soils behave plastically and drier soils behave as semisolid. The procedure for determining the liquid limit and plastic limit is defined by ASTM D4318-17 "Standard Test Methods for Liquid Limit, Plastic Limit, and Plasticity Index of Soils."

Coarse-grained soils are typically considered pervious. However, only a relatively small percentage of fines can make a coarse-grained soil behave as a fine-grained soil with respect to permeability. As a reference, soils with 30 percent fines are expected to behave as fine-grained soils even though they may have less than 50 percent passing the number 200 sieve.

The short- and long-term strengths of coarse-grained soils are taken to be the same.

COHESIVE (CLAYS)

In addition to having 50 percent of the material passing the number 200 sieve, cohesive materials have plasticity. The engineering description of this plasticity is described by the Atterberg limit chart shown in Fig. 7.1.

Though the permeability of fine-grained soils is often taken to be 1×10^{-7} cm/s, fissures and natural discontinuities can make clays more permeable than this in practice. Therefore, naturally placed clays should not be considered impermeable barriers without additional study to confirm.

Clays have short-term strength governed by their stress history and clay minerology. Long-term strength of clays is governed by their effective friction angle. It should be noted that stiff fissured clays exhibit considerably less frictional strength than would be predicted by laboratory studies (Stark and Eid, 1994). Stiff fissured soil strength can be predicted by shear strength measured in the lab for remolded samples taken to large strains.

7.1.2 Rate of Loading

LONG TERM

For long-term loading, excess pore pressure has dissipated, and strength is determined by the frictional strength of the soil.

Granular Soils In the case of granular soils, many silts, and normally consolidated claylike soils that are loaded slowly enough to permit the drainage of water from the soil mass, the relationship between shearing strength and consolidation pressure is essentially linear (AB in Fig. 7.2*a*). If a representative sample of soil in a large mass is effectively confined by a pressure due to the surrounding soils, an increment of vertical normal stress (which might be the result of a foundation load acting some distance above the sample) can be applied before failure is imminent (Fig. 7.2*b*). At failure, the magnitude of the resultant stress on the theoretical surface of failure OC is represented by the length of line AC. This resultant stress may be resolved into a normal component [σ_n and a shearing component s (AD and CD, respectively)]. The shearing component can be expressed in terms of the normal component by

$$s = \sigma_n \tan\phi \tag{7.1}$$

where ϕ = effective angle of internal friction

The effective angle of internal friction is not constant for a given soil or for soils in general but is dependent on a number of factors, such as void ratio, moisture content, confining pressure, grain shape, grain size, and gradation. In practice, it must be estimated or determined experimentally. For small jobs, the angle of internal friction can be estimated for sands by using Table 7.2 or the following relationship:

$$\phi = 28.5° + \frac{N}{4} \tag{7.2}$$

where N = number of blows to advance standard penetration sampler (SPT) 1 foot

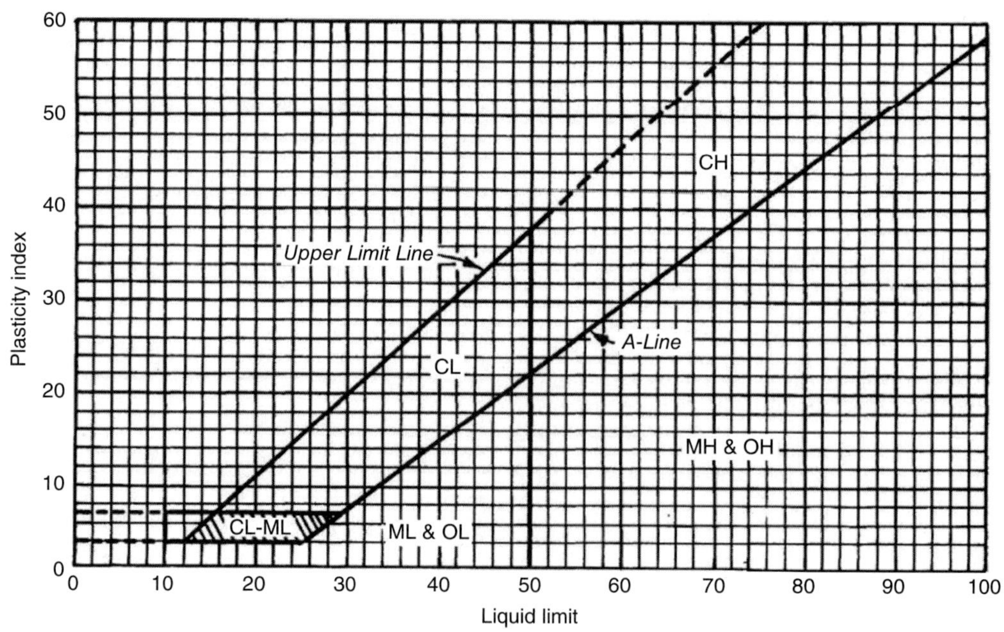

Figure 7.1 Plasticity chart showing classification group symbols: source: U.S. Army Corps of Engineers, EM 1110-2-1906 (20 August 1986).

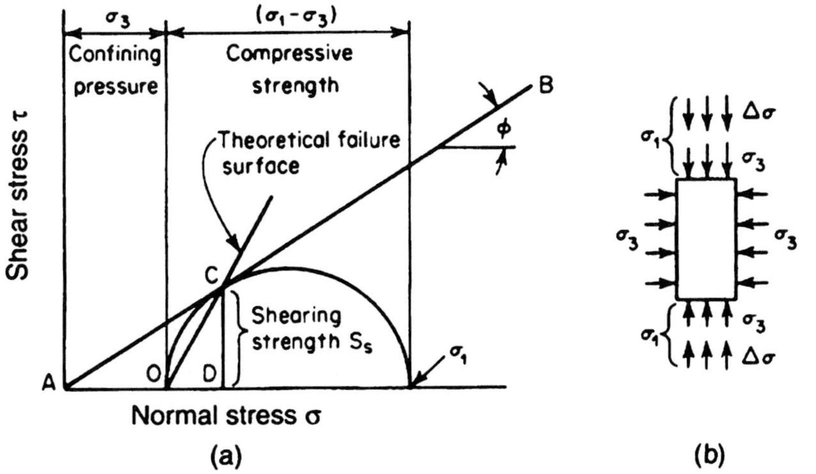

Figure 7.2 Strength relationship for sands, silts, and drained normally consolidated clays. (*a*) Mohr-coulomb failure criteria and (*b*) location of principal stresses and stress increments.

Table 7.2 Typical Values of ϕ for Dry Sand Composed Primarily of Quartz

Very loose	Loose	Medium	Dense	Very dense
$N < 4$	$5 < N < 10$	$10 < N < 30$	$30 < N < 50$	$50 < N$
$\phi < 28.5°$	$28.5° < \phi < 32°$	$32° < \phi < 36°$	$36° < \phi < 41°$	$41° < \phi < 46°$

For wet sands, these values should be decreased by 1 to 2 degrees. For gravels and crushed rock with similar states of density, they should be increased from 2 to 6 degrees.

For dry silts and very silty sands, values of ϕ' are usually 2 to 6 degrees. less than those shown in Table 7.2. This difference is more pronounced as the deposit becomes more dense. For silts and very silty sands below the groundwater table, values of ϕ' are, for the great majority of cases, considerably less (one-third to one-half) than the values for dry material. This difference between the dry and saturated silt and silty sands is caused by their low permeability and the buildup of pressure in the pore water during the application of sudden loads to deposits of these soils.

Cohesive Soils For cohesive (claylike) soils, the role played by low permeability and the excess pore pressure caused by superimposed loads is even more pronounced and important. For saturated clays that have not been preloaded by overburden or undergone a period of drying, the strength relationship is similar to that for sands and silts (Fig. 7.2*a*). However, the angle of internal friction for these clays is usually less than that for sand. Where the rate of loading is slow enough to permit full drainage of the pore water and thereby relieve the excess pore-water pressure, ϕ for clays will range from 20 to 30 degrees. Soils under the stacks of an expanding library are believed to be subjected to this type of loading. In most cases, however, the rate of loading of the soil will be more rapid, and the excess pore-water pressure will not be fully relieved by internal drainage. In cases of this type, the angle of internal friction will range from 10 to 20 degrees.

For clays that have been subjected to loads in excess of their present overburden loads or for clays that have undergone a period of drying, the strength characteristics may be represented approximately by line AB in Fig. 7.3. This line is neither exactly straight nor fixed in its location even for a given soil. For most practical problems, however, it is sufficiently correct to assume the linear relationship. The position and slope of the line will be influenced by the nature and rate of the loading to be placed on the soil. Line AB, for example, might represent quite well the strength characteristics of a preloaded or precompressed clay that subsequently is to carry additional increments of building or overburden load that will be applied slowly enough, over a period of years, to permit internal drainage and to prevent the buildup of excess pore-water pressures. The gradual buildup of a fill or slag dump on a precompressed clay deposit could represent such a loading. If, however, load increments are applied to the same soil more quickly over a period of weeks or a few months and drainage is essentially thwarted, line DE would better represent the strength characteristics. The loads applied by most buildings and fills during and immediately after construction are examples of loadings for which DE would best represent the strength characteristics of the underlying precompressed soil.

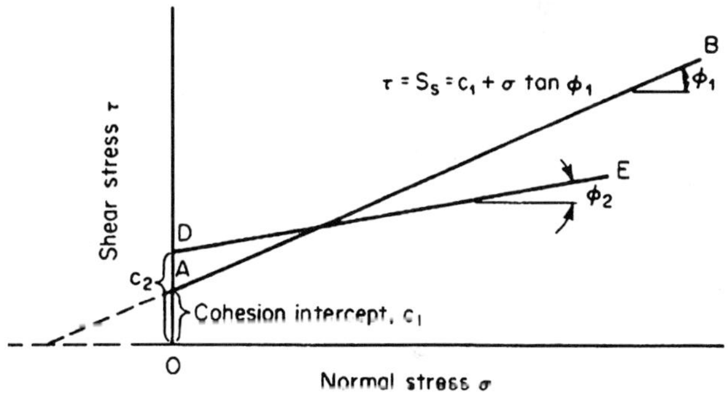

Figure 7.3 Strength relationship for preloaded (precompressed) clays.

The two parameters *c* (cohesion intercept) and ϕ (Fig. 7.3) are of great importance in determining the strength characteristic of a claylike soil. They are used to estimate the bearing capacity of a soil, to determine the stability of a slope or an earth mass underlying a proposed building site, and so on. They are not constant, even for a given soil, and should be evaluated by laboratory tests that duplicate as nearly as possible the most unfavorable field conditions that can be anticipated.

Short Term

Short-term loading is the period of loading before excess pore-water pressure dissipates. For sand, the shear strength is the same for short- and long-term loading because excess pore pressure dissipates immediately on loading.

For clays, short-term loading is dominated by the soil's cohesion and is typically assumed to be a constant value.

In many instances, engineers need to know the peak strengths, ultimate strengths, and deformation moduli of the soils with which they are working. These strength characteristics should be obtained under the conditions of confinement and loading anticipated during the following stages: (1) preconstruction (excavation or filling), (2) construction, and (3) postconstruction. Reproducing these field conditions exactly in the laboratory is impossible in many instances. This has led to the use of results from simpler tests that are thought to give conservative results. For example, in many practical situations, the results of a simple compression test, similar to the test on a concrete cylinder but employing a smaller specimen, may be used to determine the peak strength and deformation modulus of any soil that will cohere enough to retain a given shape (usually cylindrical). The results of such a test may be plotted as shown in Fig. 7.4 and the peak strength q_u determined. The subscript u indicates that the sides of the sample are not subjected to a confining pressure; that is, the sample is unconfined. The relative stiffness of the soil may be obtained from the slope of the initial tangent E_i or from the slope of a secant E_s drawn to an arbitrary point on the curve (for example, $q_u/3$).

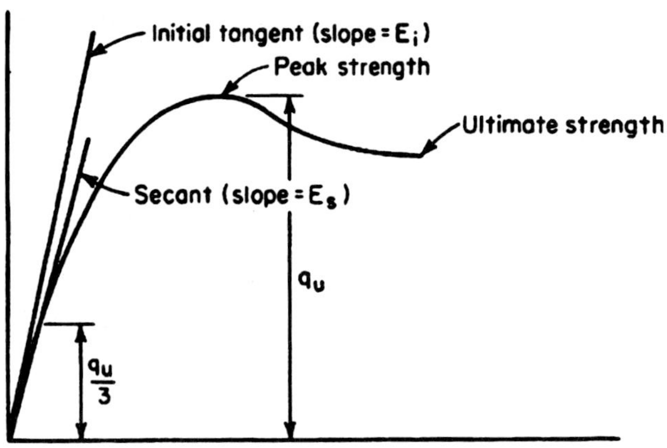

Figure 7.4 Stress-strain relationship for soil.

For claylike soils, Table 7.3 gives representative values of unconfined compressive strength (q_u) for six states of consistency. Unconfined values of Young's modulus (E_s) will usually range from 5 tons/ft^2 for soft clay to 100 tons/ft^2 for stiff clays. Values of E_s are commonly 10 to 50 percent less. Normally, the period of loading to failure in these tests ranges from 5 to 20 minutes. For more rapid or transient rates of loading, both strength and moduli normally show an increase.

Table 7.3 Consistency of Cohesive Soils

Consistency	Unconfined compressive strength q_u, tons/ft^2	Field characteristics	Blows/ft N in standard penetration test
Very soft	Less than 0.25	Easily penetrated several inches by fist	Less than 2
Soft	0.25–0.50	Easily penetrated several inches by thumb	2–4
Medium	0.50–1.0	Can be penetrated several inches by thumb with moderate effort	5–8
Stiff	1.0–2.0	Readily indented by thumb but penetrated only with great effort	9–15
Very stiff	2.0–4.0	Readily indented by thumbnail	16–30
Hard	Greater than 4.0	Indented by thumbnail with difficulty	More than 30

Also shown in Table 7.3 are the corresponding number of blows N for each of the six states of consistency as obtained from the standard penetration test. On many small to medium-sized jobs, where the cost of laboratory soil testing may be difficult to justify, engineers frequently use the penetration test as a guide to foundation types and bearing capacities and, at times, the need for additional drilling and laboratory testing.

The unconfined compressive strength of a claylike soil is of practical value in many instances. For example, it can be shown that, for a factor of safety of approximately 3, the allowable bearing capacity q_a of many claylike soils under shallow foundations is essentially equal to q_u, where q_u is an average of several tests run on samples taken within a depth below the footing equal to the width of the footing. This relationship is, for claylike soils, essentially independent of the shape and depth of the footing. However, pressures of this magnitude may cause excessive footing settlements if the clay is normally consolidated, that is, has never been subjected to a pressure in excess of its present overburden pressure. These normally consolidated clays can usually be identified quite readily since their natural water contents are closer to their liquid limit than to their plastic limit. The liquid limit is a moisture content for which a remolded soil begins to change from a plastic state to a semiliquid state as its moisture content is increased. The plastic limit is a moisture content for which a remolded soil begins to change from a semisolid to a plastic state as the moisture content is increased. These limits are readily obtained by means of the Atterberg limits test. The relationship $q_a = q_u$ is not applicable in the case of claylike soils with unusual secondary structures, that is, hairline cracks and slickensides (slick internal surfaces). The bearing capacity of these soils is best determined by plate bearing tests.

Dynamic Loading

Dynamic loads, through either impact, vibration loads, or earthquake loads, are considered more complex and require special study. For dynamic loading, coarse-grained soils may behave as undrained because the pore-water pressure does not have time to dissipate. For clays, the strength is typically assumed to be the same as for undrained. For both clays and sands, there may be strength reduction due to cyclic loads.

The structural engineer is often interested in the strength and deformation characteristics of soil under loads that are transient, vibratory, or repeated. Several studies of the effects of each of these loadings have been made in recent years, but no general relationship to cover all possibilities has as yet been developed. However, enough has been learned to indicate the general nature of such a relationship.

Shown in Fig. 7.5 are typical stress-strain curves for cohesive and cohesionless soils under the action of various types of loading. Gradually applied load curves are generally used in most engineering applications. However, studies have shown that both the stiffness and the ultimate strength of soils under the action of repeated loads, where impact is not a factor, are usually less than they are under the action of gradually applied loads. For claylike soils, the ultimate strength for repeated loads is usually only 80 to 90 percent of that for gradually applied loads, while for granular materials, the range is from 0 to 80 percent. The stiffness of cohesive soils in a repeated-load test, as measured by a slope modulus, ranges from 50 to 95 percent of that obtained in gradually applied load tests. For cohesionless soils, the repeated-load modulus is often only about 50 percent of that for gradually applied loads.

Transient Loads

Where the load duration to failure is measured in a matter of seconds or fractions of seconds, the loads are called transient. Studies thus far have shown that both the stiffness and the strength of saturated clays and shales, which are not at the time subject to heavy static loading, will increase as the time of loading to failure decreases. Increases ranging from 0 to 100 percent have been measured for claylike soils. However, the strength and deformation moduli for dry granular materials obtained

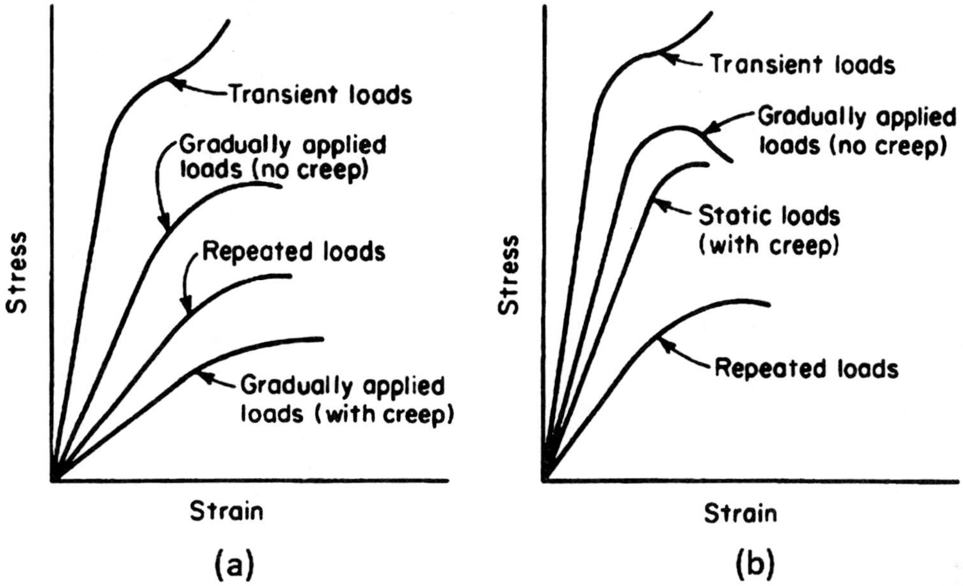

Figure 7.5 Probable stress-strain curves for (*a*) cohesive and (*b*) cohesionless soils under various loadings.

in transient tests have not been greatly different from those obtained in conventional, gradually applied load tests.

CREEP

When the applied loads are sustained over a long period of time so that creep becomes a factor, there is evidence to indicate that the ultimate strength of saturated claylike soils may be only 30 to 70 percent of that for a gradually applied load that is not sustained. For granular materials, these figures are believed to be somewhat greater. The stiffness of saturated clays as obtained from creep tests may range from 20 to 40 percent of that obtained from conventional strength tests in which creep is not a factor.

7.2 SHALLOW FOUNDATION ANALYSES

7.2.1 Deformation Types

The principal phenomena that cause footing or foundation movements may be classified as elastic-plastic deformations or shear displacement.

7.2.2 Elastic-Plastic Deformation

DESCRIPTION

The construction of fills and structures, lowering of the groundwater table, and so forth normally cause an increase in the load carried by an underlying soil deposit. Conversely, excavations and the removal of structures result in a decrease in the intensity of the load. These or any other changes in loading will cause such a deposit to deform.

When a soil deposit that is, in part, confined by surrounding soil is subjected to an increasing load from a footing or plate, the soil and footing may settle in a manner similar to one of the curves in Fig. 7.6. In each of the cases represented, the soil deformation may be characterized by an elastic portion, a plastic portion, and finally a shearing or flow portion. Normally, the elastic portion of soil deformation or settlement is rather small and occurs during construction, so it can often be neglected. It results from the cumulative elastic deformation of the soil's primary framelike structure and the deformation of individual soil grains constituting this structure. This portion of the total deformation usually can be recovered if the applied load is released. It is of small

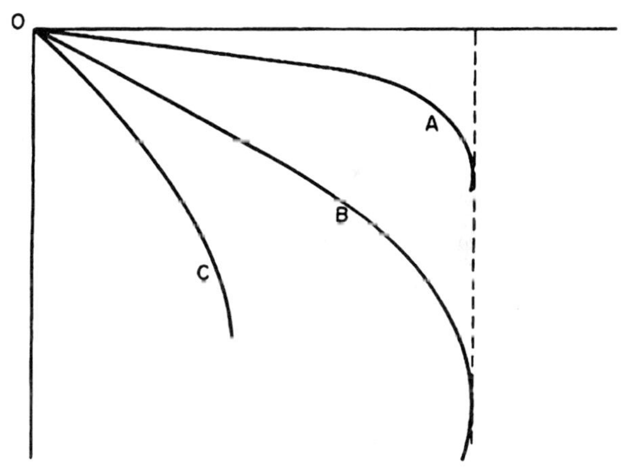

Figure 7.6 Load-settlement curves for soils.

consequence unless the applied loads are repeated, as is the case in highway pavements, for example.

As the load is increased beyond the elastic range, a nonrecoverable deformation develops. This portion may be called plastic and can, for many soils, become quite large and significant. Normally, it is larger in the case of soft clays, silts, and loose sands (curve C of Fig. 7.6). For hard or stiff clays and dense sands, it is normally much smaller (curve A).

If the load continues to increase, the deformation and breakup of the soil's natural structure may lead to complete rupture of the soil mass. Present theories of bearing capacity are sufficiently reliable so that rupture failures need not occur under building foundations. However, examples of these failures are not uncommon under fills and heavily loaded storage elevators founded on soft clays. Even though two soils (curves A and B of Fig. 7.6) may not be loaded heavily enough to cause a rupture failure of either, one (curve B) may deform elastically and plastically much more than the other (curve A) even though their ultimate load-carrying capacity is essentially the same.

Allowable Settlement

The amount of differential settlement that can be permitted for various buildings has never been clearly established. Terzaghi and Peck (1967) suggest 3/4 in. as the amount of permissible differential settlement between adjacent columns spaced about 20 feet apart in ordinary buildings. Thornley (1959) divides structures into five classes and suggests permissible gross, differential, and net settlements for each (Table 7.4). Sowers (1979) has published a table of limiting settlements based on structural considerations. Table 7.5 shows values from that portion of his table that is devoted to differential settlements.

Settlement of Granular Soils

Sands On the basis of certain theoretical relationships between settlement, footing width, contact pressures, and the observed settlement of loaded plates and footings placed at the surface of granular deposits, Peck, Hanson, and Thornburn (1974) have prepared a chart that enables one to choose allowable soil pressures that, for sands, will keep the elastic-plastic settlement to a tolerable magnitude. This chart

Table 7. 4 Structures Classified According to Permissible Foundation Settlements*

Requirements	Structures
Class I	
1. Differential settlements under working load must be held within a maximum limit of ⅛ in. 2. Gross settlement under working load, maximum ½ in. As an indication that these requirements will be met— 3. Gross settlement under 200% working load, maximum 1½ in. 4. Net settlement after application and removal of 200% working load, maximum limit ½ in. Foundation must develop full capacity by direct bearing without dependence upon transference of load from structure to soil by friction	*a.* Monumental structures, with interiors or exteriors of marble or other material in which cracking is readily observable *b.* Cathedrals and large power plants, steam and hydraulic *c.* Foundations for heavy machinery, without severe vibration or impact *d.* Structures designed to remain in serviceable condition for exceptionally long periods, such as monuments *e.* Grain elevators, storage bins, and other structures subject to wide changes in loading lasting over considerable periods *f.* Large concrete tanks *g.* Office buildings, hotels, and stores, all of 10 or more stories in height and all of reinforced-concrete construction or structural steel and concrete construction *h.* Warehouses of multiple-story, heavy-load type of reinforced concrete *i.* Retaining walls
Class II	
1. Differential settlements under working load, maximum ⅛ in. 2. Gross settlement under working load, maximum ¼ in. As an indication that these requirements will be met— 3. Gross settlement under 200% working load, maximum ¾ in. 4. Net settlement after 200% working load released, maximum ½ in. Foundation should be capable of developing load capacity by direct bearing without dependence upon friction for transference of load	*a.* Foundation for machinery causing heavy vibration or impact *b.* Buildings housing delicately balanced instruments, such as automatic telephone exchanges, telescopes, testing equipment *c.* Concrete arches for bridges, hangars, and the like
Class III	
1. Differential settlements under working load, maximum ¼ in. 2. Gross settlement under working load, maximum ¾ in. As an indication that these requirements will be met— 3. Gross settlement under 200% working load, maximum 1½ in. 4. Net settlement after 200% working load has been applied and released, maximum 1 in. These structures may be of equal money value with those of Class I and Class II but will be less affected by settlement. They should develop their bearing at least principally by direct loading in or on a noncompressible stratum not underlain by materials of lesser bearing value	*a.* Bridges, structural steel or suspension types *b.* Steel-frame buildings *c.* Steel tanks *d.* Piers and docks
Class IV	
1. Differential settlements under working load, maximum ½ in. 2. Gross settlement under working load, maximum 1½ in. As an indication that these requirements will be met— 3. Gross settlement under 200% working load, maximum 2 in. 4. Net settlement after 200% working load released, maximum 1½ in. Loads in these foundation units may be transferred to bearing strata either directly or by friction, or by a combination of both	*a.* Factories *b.* Stores *c.* Apartment buildings *d.* Hotels of less than, say, 15 stories, steel-frame type *e.* Churches *f.* Schools *g.* Warehouses of medium load capacity *h.* Machine shops not housing massive machinery or extra-heavy cranes or gaging devices of high degrees of delicacy *i.* Recreational buildings *j.* Highway structures, grade eliminations
Class V	
Permissible settlements vary too widely to tabulate	Temporary structures of all types such as military bridges, falsework for concrete arches, wood-frame buildings, etc., would be included in this class

*From Thornley (1959).

Table 7.5 Maximum Allowable Differential Movements for Various Structures*

Type of structure	Max. differential movement
High continuous brick walls	0.0005–0.001*L*
One-story brick mill building, wall cracking	0.001–0.002*L*
Plaster cracking (gypsum)	0.001*L*
Reinforced-concrete building frame	0.0025–0.004*L*
Reinforced-concrete building curtain walls	0.003*L*
Steel frame, continuous	0.002*L*
Simple steel frame	0.005*L*

Note: L is the distance between adjacent columns or between any two points that settle differentially.

*Sowers (1978).

(Fig. 7.7) employs the results of the standard penetration-resistance tests. As a rough guide, 10 blows per foot in the standard penetration test produce an allowable pressure of 1 ton/ft^2, 20 blows produce 2 tons/ft^2, and so on.

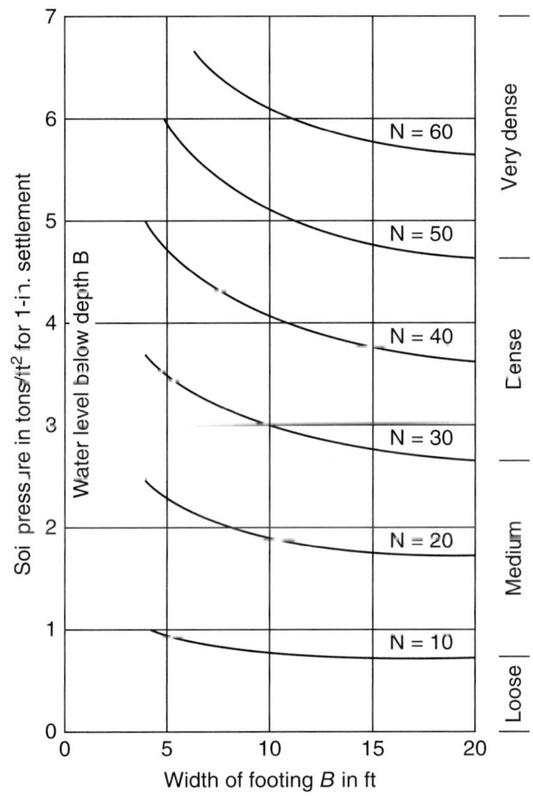

Figure 7.7 Settlement for foundations on sand (Thornburn et al., 1974).

Silts and Loess For silty soils, the allowable bearing capacity, as determined by permissible elastic-plastic deformation, will depend on the nature of the silt deposit. For medium to dense nonplastic silts or rock flours, the allowable bearing capacity can be determined by assuming that they act much like very fine or silty sands. For medium and stiff plastic silts, the allowable bearing capacity can be obtained as for clays.

Footings on silt may be proportioned by assigning the silt to the category of either a sand or a clay and proceeding with the corresponding method for determining the foundation size. For important structures, a more elaborate procedure may be indicated wherein the shear strength of the silt is determined by either field vane tests or appropriate laboratory triaxial tests.

The behavior of loess deposits may be quite different from that of either sand or clay, and the final foundation design requires a program of standard load tests. The allowable soil pressure should not exceed one-third the failure load, as represented by the poorest load-settlement curve, or one-half the load at which ½ in. of settlement was obtained in the load test, whichever is smaller.

Inasmuch as loess deposits usually have a relatively loose structure that is likely to collapse under moderate loads if the natural water content increases, unusual precautions must be taken to ensure adequate surface drainage to prevent, insofar as possible, changes in the water content. A procedure reported by Kezdi (1964) as developed in Hungary is based on an index of collapsibility:

$$i = \frac{\Delta e}{1+e_1} \tag{7.3}$$

where Δe is the change in void ratio in a consolidation test as a consequence of adding water when the consolidation pressure is 3 tons/ft^2 and e_1 is the void ratio just prior to adding water (Fig. 7.7). Experience suggests that for values of i less than 0.02, the structure of the loess is not dangerous for the support of footing foundations, but if it exceeds 0.02, precautions are necessary. This may require the use of deep foundations.

Settlement of Clays: Consolidation

Clays too behave either plastically or elastically. If the maximum applied stress is below the maximum stress the soil has experienced, the soil is considered "overconsolidated," has a relatively low displacement under load, and behaves plastically over time. If the soil begins load at its maximum past pressure, that is, if the stress the soil experiences is the same as the maximum stress it has ever experienced, then the soil is "normally consolidated," has a relatively high displacement under load, and behaves plastically over time. However, if the soil is below the maximum stress the soil has experienced and is loaded beyond the maximum stress it has experienced, the displacement will initially be relatively low until the maximum past pressure is reached, at which point the displacement will be relatively high. Finally, the deformation will be plastic.

Overconsolidated Clays For most claylike soils that have been precompressed, the elastic-plastic deformations that develop for properly chosen footing loads are likely to be tolerable for most ordinary buildings. The average unconfined compressive strength q_u of samples obtained within a zone equal to the minimum width of the largest footing or foundation is often used as the allowable bearing pressure q_a for these clays. When soil samples are not available for testing, the results of standard penetration tests run in the same zone, though often erratic, are often used to estimate the allowable bearing capacity of precompressed clays. The average of several penetration tests run at intervals of from 2 to 5 feet in depth in the zone of influence below the footing or foundation is used. As a rule, each 6 blows per foot for the penetration-resistance test is equivalent to 1 ton/ft^2 of allowable bearing capacity in precompressed clays. For example, a soil with an average penetration resistance of 12 blows per foot would normally be expected to have an allowable bearing capacity of 2 tons/ft^2.

Underconsolidated and Normally Consolidated Clays Settlements of footings placed on normally consolidated or underconsolidated clays will be much larger and usually excessive for allowable pressures chosen by the rules for preconsolidated clays. The amount and rate of settlement can frequently be estimated by means of a theory of consolidation.

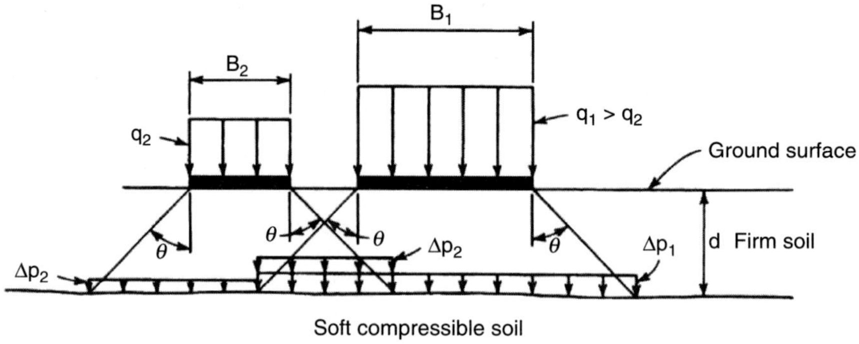

Figure 7.8 Pressures on subsurface stratum.

Fortunately, normally consolidated and underconsolidated clays are not often encountered in bridge and building foundation work. They are most likely to be found in coastal areas, former and existing lakes, and alluvial deposits. Where encountered, they may be preloaded with fill to reduce future settlement, and vertical sand drains may be employed to speed up the settlement process. In some circumstances, an amount of soil that is essentially equal to the weight of the planned structure is removed from the basement area to reduce future settlements to a tolerable amount. In other cases, piles or piers are driven or drilled through the normally consolidated clay layer to reach a more firm and unyielding material below, thereby effectively bypassing the more compressible material.

Total Settlement The amount of settlement or consolidation in a confined layer of normally loaded claylike soil of low to medium sensitivity[1] may be determined from the following equation:

$$S = \frac{DC_c}{1+e_o}\log\frac{p_o+\Delta p}{p_o} \tag{7.4}$$

where S = settlement of the compressible laver
D = thickness of the compressible layer, ft
C_c = dimensionless compression index of the soil in the compressible layer
p_o = present effective overburden pressure, psf, acting on the compressible layer, usually calculated at mid-depth of the layer
e_o = original or field void ratio of the compressible layer where void ratio is the ratio of volume of voids to volume of solids (V_v/V_s)
Δp = increase in pressure, psf, at the center of the soft compressible layer, caused by footing or building loads, fills, lowering the groundwater table, and so on

Values of D, C_c, and e_o are obtained by drilling into and through the deposit (noting the thicknesses of various layers) to secure undisturbed samples and subsequently testing selected and typical samples in the laboratory. The tests usually include the Atterberg limits, natural water content, undisturbed and remolded strength, and consolidation.

Stress Distribution

Determining Δp. The term Δp in Eq. (7.3) represents the increase in pressure at the mid-depth of the compressible laver caused by lowering of the groundwater table or by proposed building loads and so on. In any normal situation where the groundwater table above the compressible layer is to be lowered, either permanently or for an extended period, Δp will be equal to $h\gamma_w$, where h is the depth in feet that the groundwater table will be lowered and γ_w is the unit weight of water.

The most common type of construction or building load is the uniform load. Where this type of loading can be assumed, it is often sufficiently accurate to estimate Δp by assuming that the surface load is distributed uniformly at each level below the surface on an area that becomes larger with depth as determined by the angle θ (Fig. 7.8). This angle is arbitrarily chosen; the value of 30 degrees is often used (Boston Code method) and is believed to be conservative in most cases.

Taylor (1948) gives the following equations for Δp:

For square footings,

$$\frac{\Delta p}{q} = \left(\frac{B/d}{B/d+2\tan\theta}\right)^2 \tag{7.5}$$

For long narrow footings,

$$\frac{\Delta p}{q} = \frac{B/d}{B/d+2\tan\theta} \tag{7.6}$$

Taylor gives curves for these pressures when θ = 30 degrees (Figs. 7.9 and 7.10). Either the figure or the equations may be used to estimate the value of Δp on or within a compressible layer beneath either a single isolated footing or a group of footings if the pressure beneath

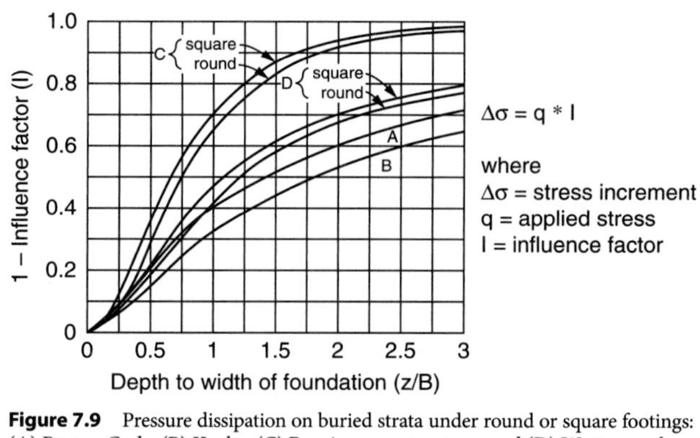

Figure 7.9 Pressure dissipation on buried strata under round or square footings: (A) Boston Code, (B) Kogler, (C) Bousinesq center stress, and (D) Westergaard center stress (after D. W. Taylor).

[1]Sensitive soils are those whose undisturbed compressive strength is greater than four times the remolded compressive strength. If this ratio exceeds 8, the soil is called extrasensitive. Remolding involves the complete breakup and mixing of the soil at unaltered water content.

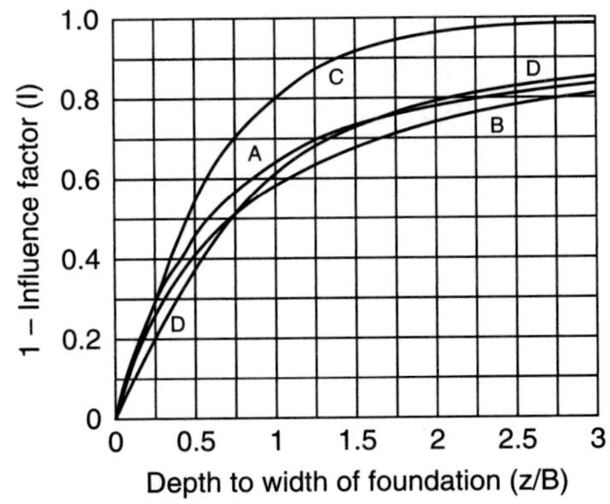

Figure 7.10 Pressure dissipation on buried strata under long wall footings: (A) Boston Code, (B) Kogler, (C) Bousinesq center stress, and (D) Westergaard center stress (after D. W. Taylor).

each footing is uniformly distributed. The combined effect of a series of isolated footings $\Delta p_1 + \Delta p_2$ (Fig. 7.8) may be obtained by assuming that the pressure from each footing is dispersed and distributed as shown and by combining their separate effects for a selected location in the compressible layer.

More sophisticated methods for estimating Δp, based on elastic relationships, have been developed by Boussinesq, Westergaard, and others and are widely used in the case of soil deposits. Curves for Δp under the center of the loaded area, for both the Boussinesq and the Westergaard solutions, are shown in Figs. 7.9 and 7.10.

In the preceding discussion of Eq. (7.4), only the normally consolidated clay, for which the present overburden and preconsolidation pressures are equal, that is, $p_o = p_c$, was considered since it is for this case that the equation is most reliable. For precompressed or overconsolidated soils, where the preconsolidation pressure p_c exceeds the present overburden pressure p_o, settlements predicted by Eq. (7.4) may be several times greater than the actual settlements depending on the ratio of $\Delta p/(p_c - p_o)$. Terzaghi and Peck (1967) indicate that the actual settlement will be only one-fourth to one-tenth of the predicted value [Eq. (7.4)] if $\Delta p/(p_c - p_o)$ is less than 50 percent. For the underconsolidated case ($p_c > p_o$), predicted settlements may be considerably smaller than the actual settlements. For these two cases, as well as for the extrasensitive clays, a quantitative estimate of settlement is difficult to make.

Time-Dependent Settlement

In the case of soils that are fully saturated (soils located below the groundwater table), there will usually be a time delay between the application of a foundation load and the resulting total elastic-plastic deformation. This delay is caused by the fact that the voids in the soil structure are filled with water, which is relatively incompressible. The water, which undergoes an increase in hydrostatic pressure on application of the load, cannot quickly escape from a relatively impervious soil. This thwarting of the impending elastic-plastic soil deformation results in a time-dependent deformation that has been called consolidation when the lateral deformation is small in comparison with the vertical movement.

Terzaghi's theory of consolidation may be used to predict the time rate of consolidation. This theory assumes (1) one-dimensional consolidation wherein the soft compressible layer is so restrained by the confining layers above and below that it cannot move laterally (that is, the net deformation within the compressible layer is vertical) and vertical movement of water from the compressible layer; (2) a completely saturated and homogeneous layer of compressible material; (3) constant values for certain soil parameters that actually vary with pressure; (4) that Darcy's law is valid for the flow of water through soil; (5) an ideal linear relationship between pressure and void ratio; (6) that secondary consolidation that occurs to varying degrees in most soils is negligible; (7) that small samples of soil tested in the laboratory will be representative of large masses of soil in the field; and (8) certain other assumptions that are normally believed to be of lesser importance.

The differential equation governing the time rate of consolidation is

$$\frac{\partial u}{\partial t} = c_v \frac{\partial^2 u}{\partial z^2} \tag{7.7}$$

In this equation, u is a function that represents the excess pore-water pressure, z is a function that represents the distance to a point in the layer, C_v is the coefficient of consolidation as obtained from a laboratory consolidation test, and t represents time. One solution of primary interest is

$$U = 1 - \sum_{m=0}^{m=\infty} \frac{2}{M^2} e^{-M^2 T} \tag{7.8}$$

In this expression, U is the average percent consolidation within the layer, T a time factor, and M a dimensionless number that is a function of m, an integer. This has been solved, and values of U and T are listed in Table 7.6. Although the T versus U relationship is a function of the shape of the initial excess-pore-pressure-distribution curve throughout the depth of the compressible layer and the conditions of drainage, the values shown in Table 7.5 are adequate for most typical cases. With the theoretical relationship between the time factor T and the average percent of consolidation U known and evaluated, the time required for a given percent of consolidation to occur in a compressible layer may be calculated by

$$t = \frac{TH^2}{C_v} \tag{7.9}$$

Table 7.6 Values of Average Consolidation *U* and Corresponding Time Factor *T*

U	0.1	0.2	0.3	0.4	0.5	0.6	0.7	0.8	0.9
T	0.008	0.031	0.071	0.126	0.197	0.287	0.403	0.567	0.848

In this expression, t is the time required to reach a certain average percent consolidation U. The term H is the half thickness of the compressible layer when there are drainage surfaces at both the top and the bottom of the layer or the full thickness where a drainage layer is located at only one of the surfaces, top or bottom, of the compressible deposit. The term C_v is the coefficient of consolidation obtained from a laboratory consolidation test. It is not a constant for a given soil, but varies with each increment of pressure. T is the dimensionless time factor corresponding to the percent consolidation and may be obtained from Table 7.6. If time t is expressed in years, H in feet, and C_v in cm²/s, Eq. (7.9) becomes

$$t = 29.5 \times 10^{-6} \frac{TH^2}{C_v} \tag{7.10}$$

Predictions of settlement of typical and critical points under a structure can be made by constructing curves similar to that in Fig. 7.11. For example, assume that the ultimate settlement predicted by Eq. (7.8) is 4.3 in.; then, for a settlement of 2.15 in., $U = 0.50$. The corresponding value of T is 0.197 (Table 7.6). Using this value and the appropriate value of C_v in Eq. (7.10) gives $t = 4.8$ years, which is one point on the time-settlement curve. Additional points may be obtained in a similar manner. Actual measured column footing settlements may be subsequently plotted on the time-settlement curve, and although the predicted and actual settlement rates may not be the same, the theoretical curve can often be fitted to the actual curve to establish a more reliable projection of the future rate of settlement.

For more permeable materials (clean sands and gravels), the time delay between application of a load and the resulting elastic-plastic deformation is normally insignificant; the deformation occurs during construction. For the more impervious soils (silts and clays), the time

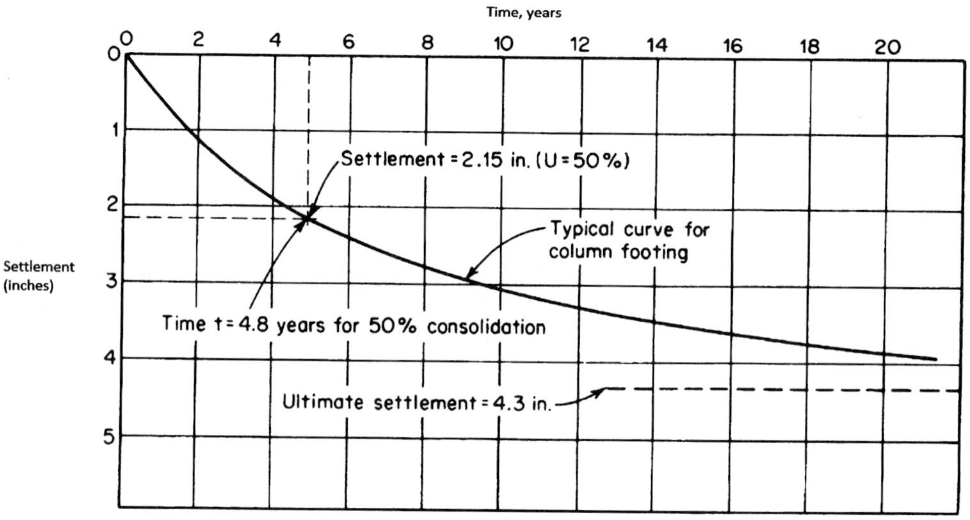

Figure 7.11 Time-settlement curve.

required for a consummation of the total elastic-plastic deformation (consolidation) can involve years and even decades. Both the magnitude and the time rate of this consolidation can often be estimated by employing principles of soil mechanics.

7.2.3 Bearing Capacity

The second criterion that must be considered when choosing an allowable bearing pressure involves what has been termed a shear or rupture displacement of a mass of soil beneath the footing. The pressure applied to the soil by the footing must not be great enough to cause this displacement. A footing near the surface of a soil deposit and the nature of the soil displacement that occurs when the footing load Q_D becomes excessive are shown in Fig. 7.12. Both theoretical and experimental studies have shown that a small wedge of soil usually forms below a footing and tends to act as a part of the footing. Masses of soil below and on either side of this wedge undergo both lateral and vertical displacement, and this soil displacement is accompanied by a downward movement of the footing when the footing pressures become excessive.

Reasonably reliable solutions have been developed for long, narrow, shallow footings. For $D_f < 1.5B$, Terzaghi and Peck (1967) give the following equations for bearing capacity. For general shear failures, which usually occur in stiff or very stiff clays and dense to very dense sands, the ultimate load Q_D in pounds per foot of footing is

$$Q_D = B\left(cN_c + \gamma_1 D_f N_q + \frac{1}{2}\gamma_2 B N_\gamma\right) \tag{7.11}$$

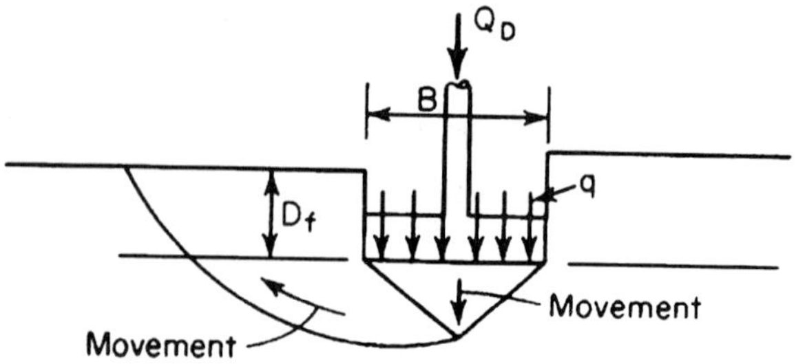

Figure 7.12 Shallow continuous footing showing probable mode of shear or rupture displacement.

and, for local shear failures, which usually occur in very loose sands or soft clays, the ultimate load Q'_D in pounds per foot of footing is

$$Q'_D = B\left(\frac{2}{3}cN'_c + \gamma_1 D_f N'_q + \frac{1}{2}\gamma_2 B N'_\gamma\right) \tag{7.12}$$

where
- B = minimum width of footing, ft
- c = cohesive strength of soil, psf
- γ_1, γ_2 = unit weights of soil, pcf, above and below the base of footing, respectively
- D_f = minimum depth of footing below the adjacent surface, ft
- N_c, N_q, N_γ = dimensionless factors for general shear failures (solid curves in Fig. 7.13)
- N'_c, N'_q, N'_γ = dimensionless factors for local shear failures (dashed curves in Fig. 7.13)

For square footings of width B,

$$Q_{ds} = B^2(1.3cN_c + \gamma_1 D_f N_q + 0.4\gamma_2 B N_\gamma) \tag{7.13}$$

For circular footings of radius r,

$$Q_{dr} = \pi r^2(1.3cN_c + \gamma_1 D_f N_q + 0.6\gamma_2 r N_\gamma) \tag{7.14}$$

In these equations, Q_{ds} and Q_{dr} are total loads for the general shear failure. For a local shear condition, these equations must be modified by using N'_c, N'_γ, and N'_q values from the dashed curves of Fig. 7.13 and the factor 2/3 with cohesion values c, as in Eq. (7.12).

A factor of safety of 3 is normally employed when these equations are used to obtain an estimate of the allowable bearing capacity. Although other solutions of this problem have been developed, some of which are slightly more refined, any difference usually becomes academic when one considers the variation in soil strength that can exist in a soil mass, our present methods of evaluating soil strength, and the factor of safety of 3. Moreover, a shear failure is seldom critical when a factor of safety of 3 is used. Only in those cases where a narrow footing ($B < 5$ feet) is at or near the surface of a deposit of loose sand and for which the groundwater table is at or quite near the surface is the shear or rupture criterion likely to govern.

7.3 DEEP FOUNDATIONS

When shallow soils do not have sufficient bearing capacity to support a building foundation, deep foundations are required. Types of deep foundations include caissons, piers, and piles.

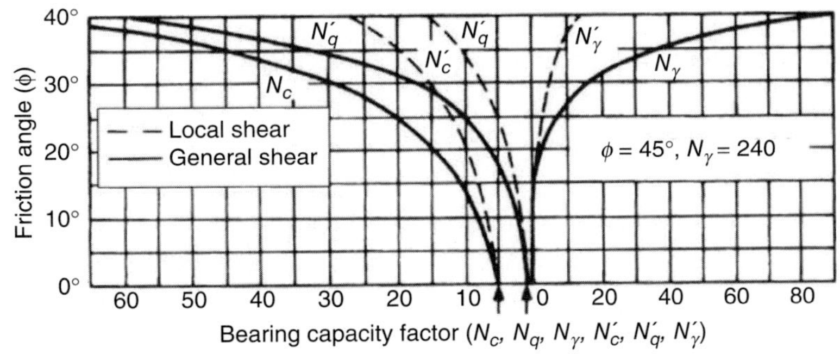

Figure 7.13 Bearing capacity for general and local shear.

7.3.1 Caissons

A caisson is a special type of pier consisting of a hollow shell that is sunk into position to form a major part of the completed foundation. There are three principal types: box, open, and pneumatic.

Box Caissons

Box caissons are open at the top and closed at the bottom. They are usually constructed on land, floated to the site, and sunk on a previously prepared bearing surface. Sinking is generally accomplished by filling the box with concrete, stone masonry, or, exceptionally, sand and gravel.

Open Caissons

Open caissons are open at both top and bottom and are sunk by dredging out the enclosed material. They are usually the size of the completed foundation and must be provided with a number of wells, extending from top to bottom, large enough to provide easy passage for the excavating buckets. Construction is started directly over the area where the caisson is to be permanently located. If the ground surface is below water, the caisson may be raised by forming a sand island and starting it on the fill. As the caisson is sunk, the upper portion is built up with concrete.

Caissons must be fairly massive in order to overcome the side friction, and occasionally it is necessary to reduce side friction by jetting. When the founding elevation has been reached, the bottom is carefully cleaned, sometimes with the assistance of a diver, and a concrete seal of thickness sufficient to resist uplift of the external water is cast prior to pumping out the caisson. After dewatering, the surface of the seal is inspected before concreting the remainder of the caisson. Open caissons have been sunk to depths in excess of 200 feet.

Pneumatic Caissons

Pneumatic caissons are closed at the top and open at the bottom and are filled with compressed air. They are generally used where the depth below water is between 40 and 110 feet, which is about the maximum depth people can work advantageously under compressed air. Air locks must be provided for passage of workers and materials, and the bottom must have clearance sufficient to provide adequate headroom in the working chamber for the muckers. Transition decompression chambers must be provided. The pneumatic caisson can be controlled more precisely than the open caisson because selective excavation is possible and the bearing surface can be more suitably prepared and inspected before concrete is placed.

7.3.2 Pier Foundations

Pier foundations are often used where it is necessary to carry the foundation to a considerable depth in order to reach a suitable bearing stratum. The principal difference between a footing and a pier foundation is that the depth of the latter is generally greater than $4B$. To reach the bearing stratum, a variety of construction procedures have been developed. These are important because the behavior of a pier foundation may depend primarily on the details of construction. Perhaps the most common method of pier construction is in an open excavation. This is often accomplished by making a braced excavation to the bearing stratum and forming a pier therein.

Care must be taken in the design of the bracing to provide adequate support for the earth pressures and added hydrostatic pressures that may exist. In the case of large piers that are surrounded by water, excavation is frequently accomplished within the protection of a cellular cofferdam.

A common method of pier construction by hand excavation was developed in Chicago, where it is inappropriately called a caisson. A circular shaft corresponding to the final dimensions of the pier is excavated and lined with vertical boards, or lagging, that are held in place by circular steel rings that are placed as the excavation advances. At the bearing stratum, the bottom may be enlarged or belled out to increase the bearing area. On completion, the entire excavation is filled with concrete; the rings and lagging are normally left in place. This method is suitable only for clays that do not contain water-bearing seams or pockets. Where such materials are encountered, the sheeting is sometimes driven ahead of the bottom of the excavation to seal off the water. Although timber sheet piling is often used for this purpose, the Gow method may be more suitable. Circular steel shells about 6 feet long are driven into the ground, and the soil within the shell is excavated. Succeeding sections about 2 in. smaller in diameter are then driven, and the process is repeated until the water-bearing layers are passed. The excavations can then be advanced as in the Chicago method, including the formation of a bell. If it is necessary to drive sheeting or a casing to advance through a water-bearing stratum, it is essential that this protection remain in place as the concrete shaft is cast. The pier must extend far enough into the bearing stratum to ensure that the soil in which the bell is to be formed can stand without support until the concrete can be placed.

Drilled Piers

Several different types of machines have been developed that can drill holes varying in diameter from about 12 in. to as much as 12 feet. When the bearing stratum has been reached, a special belling attachment can be used to enlarge the bottom of the hole and form a bell that may be up to three times the diameter of the pier shaft. Under ideal conditions, a shaft 50 feet deep can be drilled in about 30 minutes, and a 50- to 75-foot pier is easily completed in one 8-hour shift. When it is necessary to provide lateral support to prevent the intrusion of soft clays or silts, a steel lining may be used in conjunction with the drilling process.

Special techniques have been developed for advancing drilled excavations through running sands located below the water table. Although it is common practice to salvage the lining, this may be somewhat hazardous to the integrity of the pier shaft because of the possibility that the running sand and silt or soft clay may squeeze the pier shaft or actually intrude completely across it should there be momentary arching of the concrete within the casing as the latter is removed. The presence of an intrusion is usually not known until the structure begins to show signs of distress. Therefore, if a lining is required to prevent the intrusion of cohesionless or very soft soils, it generally should be left in place to become a part of the completed pier. Care should be exercised that the bottom of the bell is free of loose or compressible materials that may otherwise cause settlement of the completed pier.

Drilled Piers on Clay The allowable bearing pressure q_a for a pier founded on clay is

$$q_a = 1.25 q_u \left(1 + \frac{0.2B}{L}\right) \tag{7.15}$$

where q_u = average unconfined compressive strength within a depth below pier equal to its width
B = width of base
L = length of base

This provides a factor of safety of 3 against a bearing-capacity failure. No additional support should be assigned because of the shearing resistance that might develop between the pier shaft and the adjacent soil.

The settlement of a pier foundation in clays may be predicted by the procedure for footing foundations. However, in general, a bearing stratum that is adequate for the support of pier foundations is likely to be overconsolidated and not cause differential settlement between adjacent footings in excess of ¾ in.

Drilled Piers on Sand The bearing pressure for a pier on sand is usually governed by considerations of settlement because the effect of depth is to increase the factor of safety against a bearing-capacity failure. Experience has shown that the settlement of a pier is about half that of a footing of the same dimensions at the same unit pressures, provided the relative densities of the subsoil are equal. Therefore, the allowable soil pressure can be determined from Fig. 7.7 appropriately corrected for the position of the water table. As a consequence of the smaller settlement of a pier compared with a footing, these values may be doubled unless there is a possibility that scour may remove most of the materials above the base of the pier.

Foundation Requirements The allowable bearing pressures for caisson foundations are essentially the same as those for piers. However, because of their size and importance, caisson foundations are generally founded on very hard soils or rock where the strength of the subsoil may be secondary in importance to the allowable stresses permitted in the concrete of the caisson itself.

7.3.3 Pile Foundations

Bearing pile foundations are columns that transmit load to some depth in soil. Piles are classified as shown in Table 7.7. They may be used singly or in groups of several, although it should be recognized that the capacity of a group is not necessarily that of a single pile times the number of piles in the group. Because the final location of any pile may be 3 in. or more from its desired location, it is not good practice to use fewer than three piles for the support of a column unless lateral structural framing capable of withstanding the bending moments due to the possible eccentricity is provided. Piles supporting walls are customarily driven in pairs or staggered. The minimum center-to-center spacing of piles is usually 30 in., although they are customarily driven at a spacing of about three times the butt diameter.

Pile-Driving Equipment

Table 7.8 shows a number of hammer types and their principal characteristics. Drop hammers are frequently used for relatively small jobs where their slowness is not important. However, single-acting hammers are more common on larger projects. These are operated by either steam or compressed air and differ from double-acting hammers in that their ram falls by gravity whereas that of the double-acting hammer is forced down by the operating gas under pressure. The double-acting hammer is somewhat faster. However, the energy of its blow depends on the gas pressure and is therefore somewhat less certain than that of a single-acting hammer. However, there is considerable uncertainty as to the energy delivered to the pile in either type.

The energy delivered to the cushion block by a diesel hammer is difficult to evaluate in the field because the stroke of the hammer varies with the driving resistance. Therefore, care is required in its selection and use

Table 7.7 Classification of Bearing Piles

<table>
<tr><td colspan="7">Load-transfer action: friction, point bearing</td></tr>
<tr><td colspan="7">Attitude</td></tr>
<tr><td colspan="3">Battered</td><td colspan="4">Vertical</td></tr>
<tr><td colspan="3">Between 1 horizontal to 6 vertical and about 5 horizontal to 12 vertical</td><td colspan="4"></td></tr>
<tr><td colspan="7">Placement</td></tr>
<tr><td colspan="3">Driven</td><td colspan="2">Vibrated</td><td colspan="2">Bored</td></tr>
<tr><td colspan="3">Soil compacted by vibration and displacement
Jetted
Spudded
Preexcavated</td><td colspan="2">Minimal compaction or vibration of adjacent soil</td><td colspan="2">No compaction or vibration</td></tr>
<tr><td colspan="7">Material of construction</td></tr>
<tr><td colspan="2">Timber,* sawn† or uncut</td><td colspan="2">Concrete</td><td colspan="2">Steel</td><td>Composite</td></tr>
<tr><td>Treated‡</td><td>Untreated§</td><td>Precast</td><td>Cast-in-place</td><td>Pipe</td><td>H sections</td><td></td></tr>
<tr><td>Douglas fir
Red oak
Southern pine</td><td>Cedar
Cypress
Douglas fir
Oak Pine
Spruce</td><td>Tapered
Parallel sides
Cylinder (require reinforcing)
Prestressed
Post-tensioned</td><td>With shells
Driven
Dropped in
Without shells
Drilled
Rammed
Pedestal
Mixed-in-place</td><td>Open end
Closed end</td><td></td><td>Timber with cast-in-place concrete
Timber with precast concrete
Pipe with cast-in-place concrete</td></tr>
</table>

*Subject to attack by marine borers.
†Not common in the United States.
‡Coal-tar creosote (16 pcf or refusal).
§Must be permanently below water table.

Table 7.8 Characteristics of Pile Hammers

Hammer characteristics	Drop	Hammer types					
		Steam or air			Diesel		
		Single-acting	Double-acting	Differential	Single-acting (open end)	Double-acting (closed end)	Vibratory
Rated energy (ft-lb)	7000 to 60,000	7000 to 1,800,000	4000 to 165,000	15,000 to 50,000	4000 to 280,000	8000 to 65,000	na
Impact velocity (ft/s)	23 to 31	8 to 16	15 to 19	12.5 to 14	Depends on stroke	Depends on chamber pressure	na
Blows/min	4 to 8	35 to 60	95 to 300	98 to 300	40 to 60	80 to 105	750 to 2000 pulses/min
Hammer suitability for types of piles	All types except concrete piles	Versatile for pile	Timber, steel H, and pipe piles		All types of piles		Steel H and pipe end bearing piles

if it is necessary to evaluate the driving energy for use in a pile-driving formula without supplementary load tests.

In general, the hammer should be as large as can be safely used without damaging the pile; it is better to select a hammer on the heavy side. Hammers can be compared by computing ultimate pile capacities (Table 7.9) for various assumed final driving resistances. The computed capacity may be used to evaluate the driving stresses in the pile, which should not exceed 40 percent of its yield-point strength. However, present practice is to use the wave equation solution provided by commercial software, such as GRL WEAP (Pile Dynamics).

There are two types of vibratory pile-driving systems: one vibrates the pile at 10 to 30 cps; the other, which is called a resonant driver, vibrates it at the natural frequency of the pile (average about 100 cps). The vibrators utilize counterrotating shafts to obtain a positive vertical force and to balance the lateral forces. The vertical force generated by the vibratory driver is transmitted to the pile, which will accept more power if the force is increased or if the frequency of the vibrator approaches the natural frequency of the pile in longitudinal vibration. To transmit this force, the vibratory unit must be firmly attached to the pile. Vibratory methods can drive piles at a rate many times faster than is possible with conventional equipment. They are much quieter, as the only noise is from the motor and the hum of the vibrator, and ground vibrations generally extend only a few feet away. However, mechanical problems frequently require considerable time for maintenance, which may materially offset the high driving rate when total production is considered.

Conventional pile driving is often assisted by supplemental operations. Jetting is used to loosen the soil around the pile and facilitate penetration into or through medium to dense sands. It requires a discharge pipe with a 1½- to 3-in. diameter, a 1- to 1¼-in. nozzle, and a pumping capacity of about 500 gal/min at a pressure of 100 to 300 psi. Jetting should always be followed by driving to recompact the loosened granular materials.

Spudding is the driving of a heavy mandrel into the soil and withdrawing it to minimize the driving required on the pile placed in the spudded opening. Although it is not often used, spudding may facilitate the penetration of resistant strata because of its more efficient transfer of driving energy.

Preexcavation, which is accomplished with augers or specially developed equipment to remove soil representing the bulk of the volume of the pile to be driven, may facilitate driving, but it is used primarily to control the displacements and heave associated with displacement piles.

Field Methods to Determine Pile Capacity

Pile-Driving Formulas Pile-driving formulas (Table 7.9) are intended to relate the driving resistance to the supporting capacity under static loading. Every formula (and there are many) includes one or more factors that must be determined experimentally. Experience has demonstrated that pile-driving formulas are entirely misleading. As a result, wave equation analyses should be used instead. The wave equation is solved by several commercially available software packages.

Wave Equation The wave equation can be used with reasonable confidence to predict impact stresses in a pile during driving and to estimate its static capacity (Hirsch et al., 1970). The theoretical model requires the evaluation of all factors that may influence the transmission of a stress wave initiated by the pile hammer ram. Thus, the properties of the hammer ram, cushion blocks, cap block, pile, and soil must be known. The solution requires an electronic computer and is best presented as curves relating ultimate resistance and maximum pile stress to the driving resistance. The ultimate capacity should be adjusted for freeze or relaxation on the basis of a pile load test to failure.

The energy delivered to the pile is an important parameter for determining capacity using dynamic formulas or wave equation analysis. However, the actual energy delivered to a pile can vary significantly from its rated energy. For example, tests in Michigan (1965) showed that the actual energy delivered to a steel pile ranged from 0.26 to 0.65 times the rated energy of the pile hammer. The average value was 0.45. Hammer energy can be measured in the field using a pile-driving monitor (PDM). A PDM can determine the energy delivered to a pile by measuring the variation of axial strain and acceleration with time during pile driving. Strain gages and accelerometers are mounted on the pile near the pile head. PDM results can be used directly to estimate bearing capacity or can be used to back-calculate static and dynamic soil pile properties to be used as input for wave equation analyses.

Load Tests It is essential to distinguish between test piles and pile load tests. Test piles are prototypes driven to determine the lengths of piles required at a particular site. The driving resistance is used as the basis for ordering piles of appropriate length. Test piles should be located near a soil-boring representative of the poorest conditions known at the site, and a record of the penetration resistance (blows per foot) should be made for the entire length of the test pile. The driving record should also indicate the characteristics of the driving equipment, including the cushion block.

Pile load tests are used to determine their design capacity. The test load is seldom carried to a sufficient intensity to determine the ultimate capacity. It is usually required that the test load be carried to at least twice the design load and that the resulting load-settlement curve be corrected for the elastic shortening of the pile before an analysis of the data is attempted. The elastic shortening is of particular importance for relatively long point-bearing piles.

Many procedures are used to interpret the results of a pile load test. If the pile has been loaded to failure, the design load should be based on a factor of safety of at least 2.5. A failure load, Q_u, may be determined as the load at which the load-settlement curve approaches a vertical tangent or as the intersection of tangents to the load-settlement curve near the point where the settlement per unit load increases markedly

Table 7.9 Selected Pile-Driving Formulas

Engineering News (Wellington, 1893) (poor, should be discontinued)

$$Q_u = \frac{WH}{s+c}$$

$c = 1.0$ for gravity hammer
$c = 0.1$ for other hammers
Recommended factor of safety = 6

Hiley (1922) (good)

$$Q_u = \frac{e_f\ WH}{s + \frac{1}{2}(C_1 + C_2 + C_3)} \times \frac{W + n^2Wp}{W + Wp}$$

n = coefficient of restitution
e_f = efficiency of hammer
C_1, C_2, C_3 = temporary compression of pile cap and head, pile, and soil, respectively, in inches
$C_1 = 0.02$ to 0.50, $C_2 = 0.05$ to 0.5, $C_3 = 0$ to 0.10
(for guide to selection, see Chellis, Ref. 6)
Recommended foctor of safety = 2.75

Janbu (1953) (very good)

$$Q_u = \frac{WH}{k_u s}$$

$$k_u = C_d\left[1 + \sqrt{1 + \frac{\lambda}{C_d}}\right]$$

$$C_d = 0.75 + 0.15\frac{Wp}{W}, \quad \lambda = \frac{WHL}{AEs^2}$$

k_u = (approximate) from chart

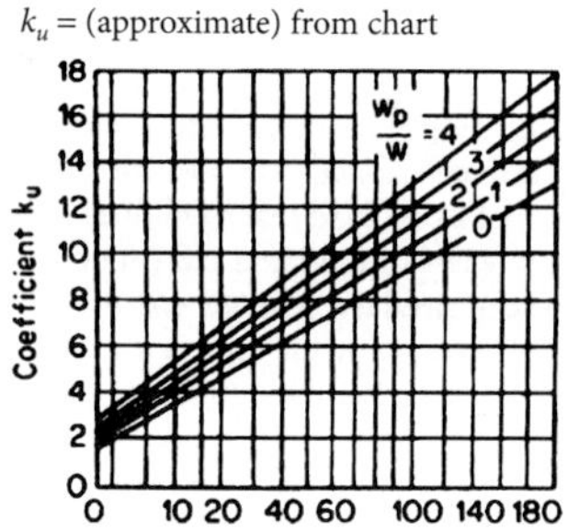

$\frac{WHL}{AEs^2}$ (after Janbu, 1953)

Recommended factor of safety = 3

Pacific Coast (Uniform Building Code, Los Angeles, 1955) (fair)

$$Q_u = \frac{E_n \frac{W + kWp}{W + Wp}}{s + \frac{Q_u L}{AE}}$$

$k = 0.25$ for steel piles, $k = 0.1$ for others
Recommended factor of safety = 4

Vibrated piles* (M. T. Davison, 1963) (tentative)

$$Q_u = \frac{550\ H.P.}{r_p + fs_t}$$

$H.P.$ = horsepower, ft-lb/s
r_p = rate of penetration, in./s
f = frequency, cycles/s
s_t = temporary compression loss factor
= 0.02 to 0.07 in. (determine from load test to failure)

Q_u = ultimate capacity, lb
W = weight of pile hammer, lb
H = drop of hammer, in.
s = final pile penetration, in.
E_n = energy of driving (WH), in.-lb
L = length of pile, in.
W_p = weight of pile, lb
A = net cross-sectional area of driven pile, sq in.
E = modulus of elasticity of pile, psi
timber = 1.2×10^6
concrete = 2.5×10^6
steel = 30×10^6

*From M. T. Davisson (1963).

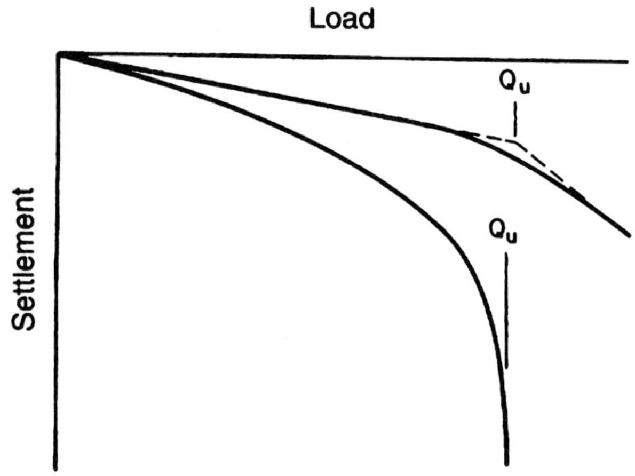

Figure 7.14 Failure loads—pile-load tests.

(Fig. 7.14). Perhaps the most common criterion is that the design load shall not exceed one-half the test load that causes a settlement of 0.01 in./ton of test load. Where a failure load has been reached, the design load should be the smaller obtained from these two criteria.

Although pile load tests are often made at the start of construction to verify the design assumption, it is desirable that they be made early enough for the data to serve as design criteria. It is always desirable to conduct at least two tests in any load-testing program and to base the analysis on the most unsatisfactory result. Averaging should be avoided.

If piles are to be driven through a soft layer and into a firm stratum, the support provided by the soft layer may disappear as the layer consolidates so that all the load is transmitted to the firm stratum. The capacity of the latter may be determined by conducting a pair of load tests, one on a pile driven into the firm stratum and the second, which must be carried to failure, on a pile stopping 2 to 3 feet above the firm stratum. By subtracting the load carried in the soft layer from the load for the pile penetrating the firm stratum, the support of the latter is obtained. A careful analysis of these conditions is required. If the soft layer is likely to settle as a consequence of the new construction, it may ultimately tend to drag down the piles with a load equivalent to the temporary support it originally provided. This is known as negative skin friction and further reduces the usable capacity of the pile support provided by the firm stratum.

Lateral Capacity

Present practice utilizes commercial programs to compute lateral capacity, such as LPILE (Ensoft, Inc.). The following discussion describes other methods for determining lateral capacity that can be used to check computer results.

Where a pile-supported structure is subjected to lateral loads, the vertical piles may provide more lateral resistance than is commonly realized. Prevailing rules of thumb commonly permit an arbitrary lateral load per pile—often 1000 pounds—without any consideration as to the type of pile or the soil in which it is driven. Since a pile-supported structure does not transmit load directly to the soil beneath the pile cap, frictional resistance should not be assumed between the base of the structure and the underlying soil. Therefore, the piles must be adequate to resist all lateral loads. The ultimate lateral bearing pressure per unit length of pile at a given depth in clay is

$$Q_d = 9cB = 4.5q_u B \tag{7.16}$$

and in sands is

$$Q_d = 3\gamma' z \frac{1 + \sin\phi}{1 - \sin\phi} \tag{7.17}$$

where Q_d = ultimate load per unit length of pile, lb/ft
c = cohesion, psf
q_u = unconfined compressive strength, psf
B = width of pile, ft
γ' = effective unit weight of soil, pcf
z = depth, ft
ϕ = angle of shearing resistance

The working load should not exceed $Q_d/2$ beyond a depth of $4B$ under any circumstance. For lateral loads smaller than $Q_d/2$, the soil reaction at any depth is given by

$$w = \frac{C_w Q_{hg}}{T} \tag{7.18}$$

where w = soil reaction, lb/in.
C_w = soil reaction coefficient from Fig. 7.15
Q_{hg} = shear at ground surface, lb
T = relative stiffness of pile ($\sqrt[5]{EI/n_h}$), in.
EI = flexural stiffness of pile, lb-in.^{2}
n_h = constant of horizontal subgrade reaction, lb/in.^{3}

The modulus of subgrade reaction $k = w/y$ is the ratio of the total reaction per unit length of pile to the corresponding deflection. For granular soils, k is directly proportional to the depth x, and it has been shown that k is also proportional to depth for normally consolidated clays and silts. However, for overconsolidated clays, k is usually assumed to be a constant, and the corresponding relative stiffness is ($\sqrt[4]{EI/k}$). Since k is proportional to depth for most soils of interest, the case of constant k is not considered here.

The constant of horizontal subgrade reaction is $n_h = k/x$. Typical values of n_h are presented in Table 7.10. Actual values can be determined experimentally by driving two instrumented piles relatively close together and jacking them apart. By measuring the loads and deflections at various depths (with a tiltmeter), k can be determined.

Table 7.10 Typical Values of n_h*

Soil type	N	n_h, lb/in.^{3} Dry	Submerged
Sand			
Loose	<10	9.4	5.3
Medium	10–30	28	19
Dense	>30	75	45
Very loose under repeated loading	<5	....	1.5
Silt, very soft, organic			
Clay	<3	....	0.4–1.0
Very soft	<3		
Static loads	....	....	2
Repeated loads	....	....	1

*M. T. Davisson (1963).

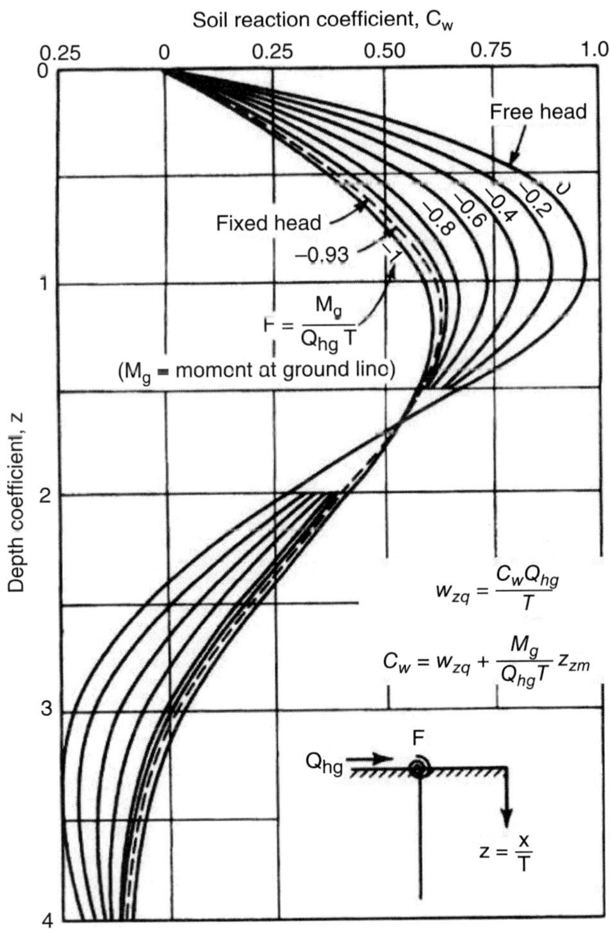

Figure 7.15 C_w versus z for different fixity of head (Prakash, 1962).

The lateral deflection y, in inches, given by

$$y = C_y \frac{Q_{hg} T^3}{EI} \tag{7.19}$$

where C_y = deflection coefficient from Fig. 7.16 and may also be used to evaluate k from a jacking test as well as to determine anticipated deflections.

The moment in a laterally loaded pile (in.-lb) is

$$M = C_m Q_{hg} T \tag{7.20}$$

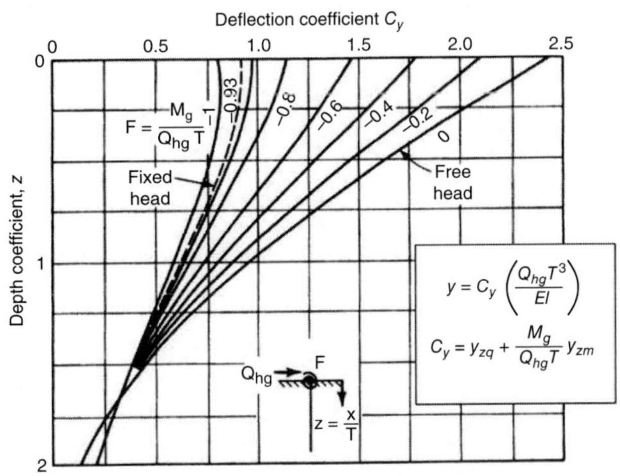

Figure 7.16 Nondimensional coefficients for lateral deflection of a pile, assuming soil modules proportional to depth, or $E_s = kx$ (Matlock and Reese, 1961).

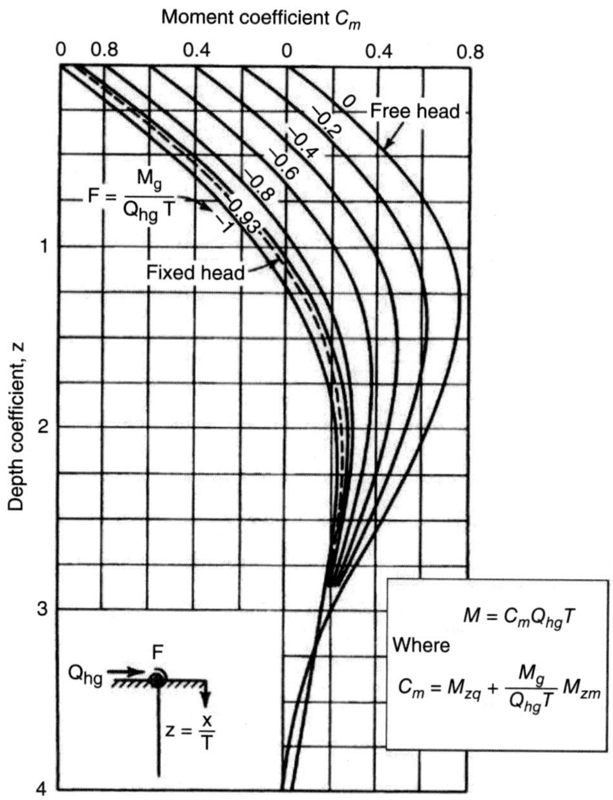

Figure 7.17 Moment coefficients for laterally loaded piles (Prakash, 1962).

where C_m = moment coefficient from Fig. 7.17. It is noted in Figs. 7.15, 7.16, and 7.17 that a fixed-head pile is generally more favorable than one with a free head. The fixity depends on the structural characteristics of the pile cap and the connection between the pile and the pile cap. The latter is attained by embedding the pile at least 24 in. into the pile cap.

The preceding analyses are for single piles, which corresponds to a minimum spacing of eight pile diameters in the direction of the lateral load and three pile diameters normal to the direction of load. Closer spacings cause a reduction in the modulus of subgrade reaction and an increase in the relative stiffness. From the limited information available, it is recommended that a minimum spacing of $3B$ be maintained normal to the load and that T be increased linearly to a limiting value of $1.3T$ as the pile spacing in the direction of the load decreases from $8B$ to $2.5B$.

7.4 RETAINING STRUCTURES

7.4.1 Types of Retaining Structures

Types of retaining walls include gravity, semigravity, cantilever, counterfort, crib, and mechanically stabilized earth.

Gravity Wall

The gravity wall (Fig. 7.18*a*) depends entirely on the weight of masonry and any soil resting thereon for stability. It must be of sufficient thickness to resist the forces acting on it without developing tensile stresses. Concrete gravity walls usually contain a nominal amount of reinforcement near the exposed surfaces to control temperature cracking. Because of the small amount of reinforcement and the relatively simple formwork, gravity walls may prove to be economical for heights up to about 10 feet.

Semigravity Wall

The semigravity wall (Fig. 7.18*b*) has largely supplanted the gravity wall. It is somewhat more slender and more conservative of materials. However, it requires vertical reinforcement along the inner face and into the footing to resist the rather small tensile forces that develop in these locations. A nominal amount of temperature steel must be provided to control surface cracking. It may be economical for heights up to about 20 feet.

Cantilever Wall

The cantilever wall (Fig. 7.18*c*) is the most common type. It consists of a base slab and a stem that are fully reinforced to resist the moments and shears to which they are subjected. The cantilever wall is relatively thin and economical of materials.

Counterfort Wall

The counterfort wall (Fig. 7.18*d*) consists of a relatively thin concrete slab that is supported at intervals on the back side by vertical counterforts connected to the base. The loads on the face slab are carried into the counterforts and in turn to the base slab. The space between the counterforts and above the base slab is part of the backfill. All the elements of a counterfort wall are fully reinforced.

The primary reinforcement is longitudinal in the face slab, both longitudinal and transverse in the base, and extends in several directions in the counterforts. The counterfort wall requires considerable formwork but has an economic advantage when relatively high walls of some length are required.

Crib Wall

Retaining walls are sometimes formed of stacked rectangular elements to form cells that are filled with soil (Fig. 7.18*e*). Their stability depends on the weight of the crib units and their filling and on the strength of the filling material. Crib walls are relatively inexpensive. They are usually made of reinforced concrete, although both timber and fabricated metal crib walls are not uncommon.

Gabion Wall

A gabion is a type of crib wall. It is made of compartmented rectangular wire-mesh baskets filled with large gravel, boulders, or rock. They are typically 3 feet wide and 6, 9, or 12 feet long, with heights of 1, 1½, and 3 feet, and may be stacked.

Mechanically Stabilized Earth Walls

A mechanically stabilized earth (MSE) wall is an internally reinforced, gravity-type retaining structure (Fig. 7.18*f*). The wall consists of horizontal layers of reinforcing elements embedded in the backfill and a wall facing. Reinforcing elements are typically metallic (strips, wire meshes, or rebar grids) or polymeric (strips, sheets, or grids). The primary function of the wall facing is to prevent loss of backfill from between the reinforcing elements. The facing is typically made of precast concrete architectural elements connected to the reinforcing elements. For MSE walls reinforced with sheets, grids, and wire mesh, the reinforcement can also be used as the wall facing by extending it beyond the face and wrapping the extra length back into the wall. Design guidance for MSE walls can be found in FHWA-NHI-10-024 (FHWA, 2009).

7.4.2 Gravity Structures

Earth Pressures

The states of stress in an earth mass may be conveniently represented on a Mohr rupture diagram (Fig. 7.19), wherein the normal stresses on any plane in a mass of soil are plotted as abscissas and the corresponding shear stresses as ordinates. In this diagram, the principal stresses plot on the horizontal axis, which also represents the planes on which the principal stresses act since, by definition, a principal plane is one on which there are no shear stresses. In a semi-infinite mass of soil with a horizontal surface, the normal stress on any horizontal plane is a principal stress equal to the

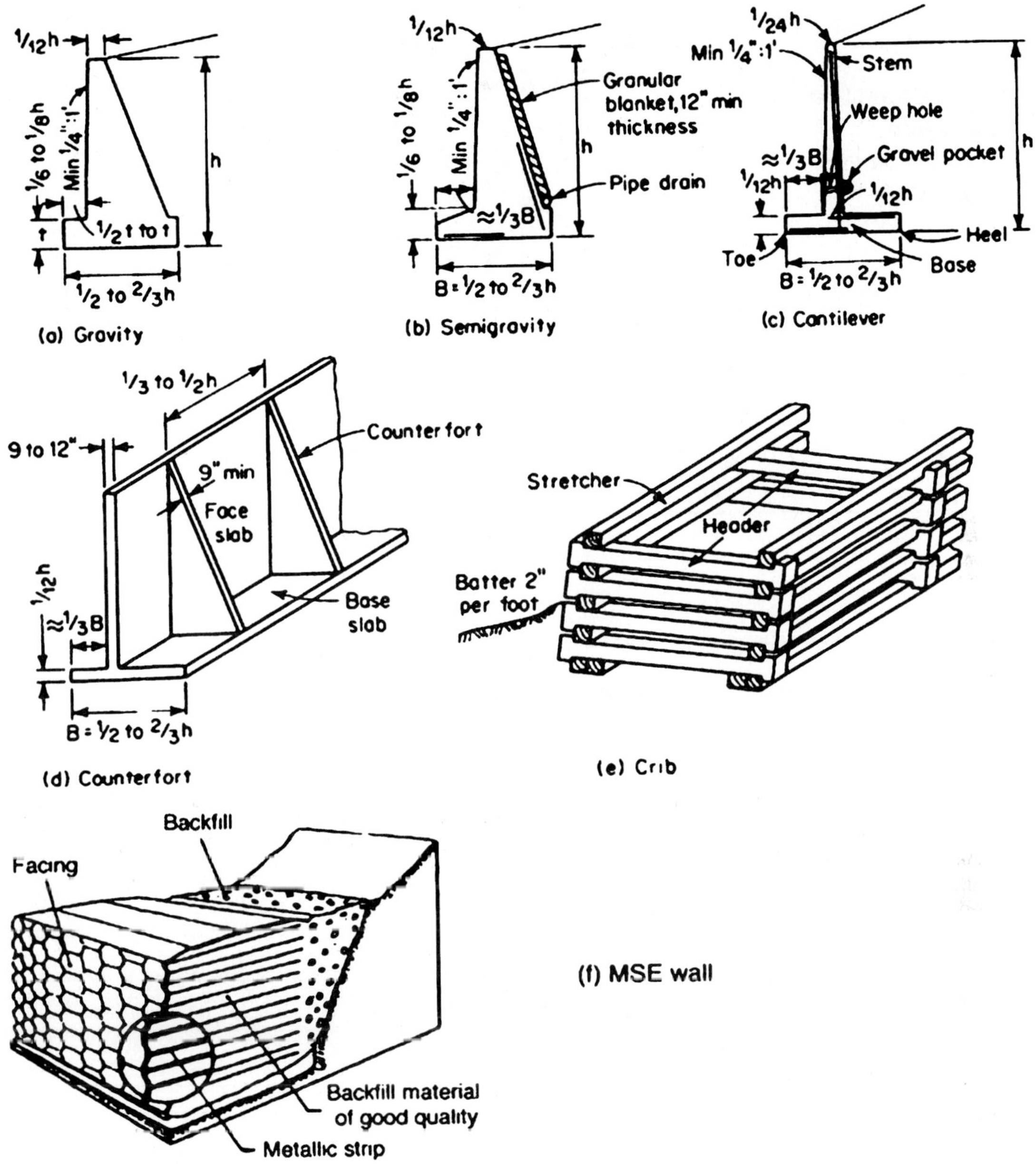

Figure 7.18 Retaining-wall types (with trial dimensions).

weight of the overlying material. A second principal plane is oriented at 90 degrees to the former, which corresponds to a vertical plane in this case.

A Mohr circle gives the stresses on every plane through a point in the soil mass. The major and minor principal stresses are denoted by σ_v and σ_h, respectively. The major principal stress is given by

$$\sigma_v = \gamma h \tag{7.21a}$$

where γ = unit weight of soil and z = depth. The minor principal stress is given by

$$\sigma_h = \sigma_v k_o \tag{7.21b}$$

where k_o is an empirical constant known as the coefficient of earth pressure at rest. Values of k_o range from about 0.7 for dense sand to 0.55 for loose sand and may be increased to about 0.8 by tamping in layers. Overconsolidated clays may have a value k_o greater than unity, in which case σ_v becomes the minor principal stress.

Any change in the value of k for a particular mass of soil must result from either expansion or compression of the entire mass in a horizontal direction. The vertical pressure σ_v does not change because the weight of soil above any horizontal section is unaffected by these deformations. However, the horizontal pressure decreases if the mass of soil expands and increases if it is compressed. The lower limiting value $\sigma_h = \sigma_v k_a$ (Fig. 7.19) is controlled by the properties of the material as they influence the location of the rupture line, which is the locus of failure stresses. The rupture line makes the angle ϕ with the horizontal, where ϕ is the angle of internal friction. The corresponding stress circle C represents a failure condition, and any further expansion is associated

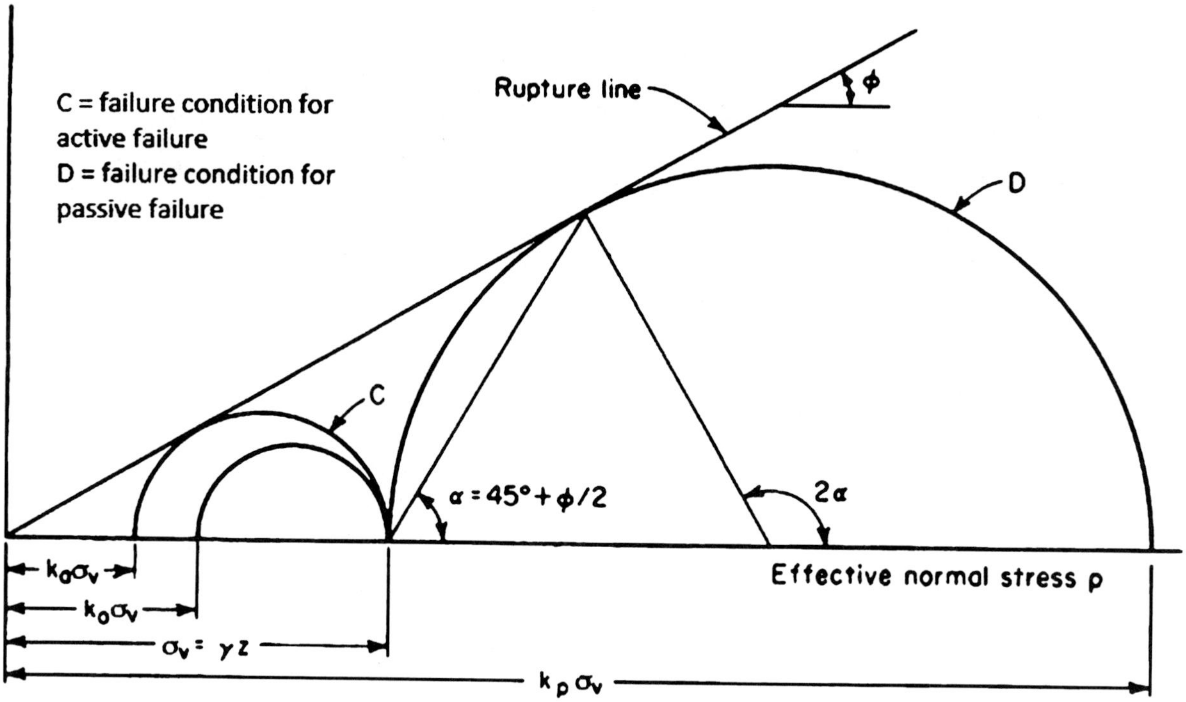

Figure 7.19 Shear-stress relationship for earth pressures.

with sliding along two plane surfaces within the soil mass. These failure surfaces intersect the plane on which the major principal stress acts at an angle of 45 degrees + $\phi/2$.

The maximum value $\sigma_h = \sigma_v k_p$ is also controlled by the location of the rupture line. The corresponding stress conditions are represented by circle D. Any further compression results in sliding along two sets of plane surfaces that make an angle of 45 degrees + $\phi/2$ with the plane on which the major principal stress acts. In this case, the horizontal pressure is the major principal stress, so the plane on which it acts is vertical.

The parameters k_a and k_p are called the coefficients of active and passive earth pressure, respectively. Since the stresses corresponding to the active and passive states represent a failure condition, these conditions are said to result in a state of plastic equilibrium. Every intermediate state, including the state of rest, is referred to as a state of elastic equilibrium. It can be demonstrated from the geometry in Fig. 7.19 that

$$k_a = \frac{1-\sin\phi}{1+\sin\phi} = \tan^2\left(45 - \frac{\phi}{2}\right) \tag{7.22}$$

$$k_a = \frac{1+\sin\phi}{1-\sin\phi} = \tan^2\left(45 + \frac{\phi}{2}\right) \tag{7.23}$$

For a soil where the rupture line intercepts a shear strength at zero normal pressure called cohesion c (Fig. 7.3), these relationships must be modified as shown below:

$$k_a = \tan^2\left(45 - \frac{\phi}{2}\right) - \frac{2c}{\gamma z}\tan\left(45 - \frac{\phi}{2}\right) \tag{7.24}$$

$$k_a = \tan^2\left(45 + \frac{\phi}{2}\right) - \frac{2c}{\gamma z}\tan\left(45 + \frac{\phi}{2}\right) \tag{7.25}$$

Rankine's Theory Rankine's earth-pressure theory corresponds to the stress and deformation conditions for the states of plastic equilibrium described by the American Railway Engineering Association (AREA, 1954) (Fig. 7.20).

The resultant active pressure on a vertical plane of height h through a semi-infinite mass of soil whose surface is inclined at an angle β to the horizontal is

$$P_a = \frac{1}{2}\gamma h^2\left(\cos\beta\frac{\cos\beta - \sqrt{\cos^2\beta - \cos^2\phi}}{\cos\beta + \sqrt{\cos^2\beta - \cos^2\phi}}\right) \tag{7.26}$$

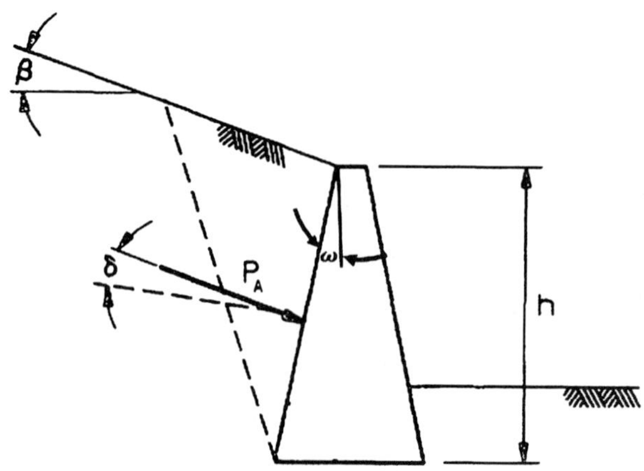

Figure 7.20 Angles utilized for Coulomb and Rankine pressure calculations (EM 1110-2-2502, 29 September 1989).

The resultant force P_a is parallel to the ground surface and acts through the lower third point of the vertical plane. For the special case when the ground surface is horizontal, this equation simplifies to

$$P_a = \frac{1}{2}\gamma h^2 \frac{1-\sin\phi}{1+\sin\phi} = \frac{1}{2}\gamma h^2 k_a \tag{7.27}$$

If a vertical plane rotates about a point at depth z in a soil mass, a wedge-shaped zone develops the state of plastic equilibrium corresponding to Rankine's theory. This wedge of soil is bounded by plane surfaces. Rankine's theory deals strictly with the equilibrium conditions of such a wedge when the shear resistance of the soil is fully mobilized. Theoretically, it is applicable to retaining walls only if the wall does not interfere with the formation of any part of the wedge on either side of the vertical surface that passes through the point of rotation. Nevertheless, Rankine's theory is commonly used to compute the earth pressure against a vertical plane through the heel of a wall.

Coulomb's Theory The Coulomb earth-pressure theory for the active state of stress yields the resultant pressure against the back of a retaining wall. It assumes that the soil slides on the back of the wall and mobilizes the shearing resistance between the wall and the soil. The resultant is inclined at an angle δ, the angle of friction between the soil and the wall, with respect to the normal. The other side of the wedge is a failure surface in the soil, which implies that the wall yields sufficiently to develop an active state of stress. It is easily demonstrated that the surface of rupture must be curved in order for the forces involved to be in static equilibrium.

Coulomb's formula for the resultant active earth pressure against the back of a retaining wall is

$$P_a = \frac{1}{2}\gamma' h^2 \left(\frac{\cos^2(\phi-\omega)}{\cos^2\omega\cos(\delta+\omega)\left[1+\sqrt{\frac{\sin(\delta+\phi)\sin(\phi-\beta)}{\cos(\delta+\phi)\cos(\omega-\beta)}}\right]^2} \right) \tag{7.28}$$

where γ' = effective unit weight of backfill
h = vertical projection of pressure surface
ϕ' = effective friction angle of backfill
ω = angle between vertical plane and back of wall
δ = angle of friction between the soil and wall
β = slope of backfill with respect to horizontal

Equivalent Fluid Method The equivalent fluid method is a modification of the Rankine active earth-pressure theory for the case of the backfill with a horizontal ground surface [Eq. (7.20b)]. The equivalent fluid pressure is assigned a value $\gamma_f = \gamma_s * (1-\sin\phi)/(1+\sin\phi)$. Typical values for γ_f range from 30 to 45 pcf. A specific value within this range is often used as a standard in many offices. Although the equivalent-fluid method is widely used, it has little to commend it. There is indeed a danger that it may be misused for the case where the ground surface is not horizontal. Furthermore, it leads to misunderstanding and a tendency to obscure the real nature of the earth-pressure problem. Finally, water pressure should also be added to this equivalent fluid for cases where water acts on the structure, which are many.

Trial-Wedge Method The practical problem often involves boundary conditions that cannot be readily incorporated into formulas based on the Rankine or the Coulomb theories. For these cases, as well as for the usual case, a graphical solution known as the trial-wedge method can be used. The method can be adapted to give earth pressures in accordance with either the Rankine or Coulomb theory, depending on the assumptions introduced at the outset. One of its principal advantages is that every force involved is represented in both direction and magnitude. Thus, the designer is forced to contemplate each of them independently, and the influence of the shear strength of the backfill material can hardly pass unnoticed. The trial-wedge method is applicable for backfills of soils possessing cohesion, internal friction, or both; for backfills having any configuration of ground surface; and for surcharges located in any position on the backfill.

In the trial-wedge method, plane surfaces of rupture may be assumed in all cases. The error introduced by assuming a plane surface when the actual surface of rupture is curved is not significant in active earth-pressure computations.

Rankine's theory is used in the trial-wedge example of Fig. 7.21. A surface of rupture, such as $BF4$ in a, is assumed. This surface is vertical through the depth of cracking h_o, which is given by

$$h_o = \frac{2c}{\gamma}\tan\left(45° + \frac{\phi}{2}\right) \tag{7.29}$$

Cracking occurs only in the case of a cohesive soil. The weight of the wedge of soil between the vertical plane AB and the trial rupture surface is computed for a unit length of wall. This weight is held in equilibrium by the forces acting on the plane AB and on the surface of rupture. The active earth pressure P_a must be parallel to the ground surface to satisfy the conditions of static equilibrium. If the ground surface is broken within the active zone, as shown in Fig. 7.21a, the AREA (1954) has adopted the following procedure for approximating the direction of P_a: Point b is located on the ground surface at a distance $2h$ measured horizontally from the back edge of the top of the wall. This point is connected to the back of the wall by a straight line, which is assumed to represent the direction of the active earth pressure. The block of soil between the wall and the vertical plane on which the Rankine pressure acts is usually considered in the analysis as a part of the wall.

On the surface of the rupture through the backfill, the resultant R of the normal reaction and the friction is inclined at an angle ϕ (Fig. 7.21a). To this must be added the shearing resistance C due to cohesion along the same surface. Since the directions of all the forces and the magnitudes of all but the active earth pressure and the resultant force on the surface of rupture are known, it is possible to draw a force polygon from which the active earth pressure can be obtained. Thus, for the wedge $ABF4$, the force polygon is $B4T'4B$, where $B4$ is the weight of the wedge including surcharge, $4T'$ the cohesion C, $T'4$ the active pressure P_a, and $4B$ the resultant R on the surface of rupture. The procedure is repeated for other trial surfaces, usually by superimposing the force polygons, and the locus of the computed earth pressures is plotted. The maximum value obtained from this locus is the active earth pressure against the vertical surface AB. In this method, surcharge loads can be readily added to the appropriate weight vectors as they are encountered by different trial surfaces of rupture.

Location of Resultant Pressure. Because of the influence of surcharge loads and an irregular ground surface, the point of application of the resultant pressure cannot be arbitrarily taken at the lower third point. Its location can be determined as follows:

1. Divide the pressure surface $6'B$ below the zone of tension cracks into four equal parts of height h_1 (Fig. 7.21a).
2. Determine the active pressures P, P_1, P_2, and P_3 as if the base of the wall were, in turn, at B, E', D', and C'. This requires three additional trial-wedge polygons corresponding to Fig. 7.21b.
3. Compute the increments of pressure $P - P_1$, $P - P_2$, and so on and divide these differences by the height h_1 of the parts to determine the average unit pressure in each part.
4. Determine the elevation $\bar{h}$ of the centroid of the resulting approximate-pressure diagram. This is the approximate point of application of the resultant earth pressure.

The foregoing procedure results in the following formula:

$$\bar{h} = \frac{h_1}{3P}(4P_3 + 2P_2 + 4P_1 + P) \tag{7.30}$$

Thus, for practical applications, Eq. (7.30) eliminates the calculations in steps 3 and 4 above. Although the formula is applicable only when $6'B$ is divided into four equal parts, use of a larger number of divisions does not appear to be justified.

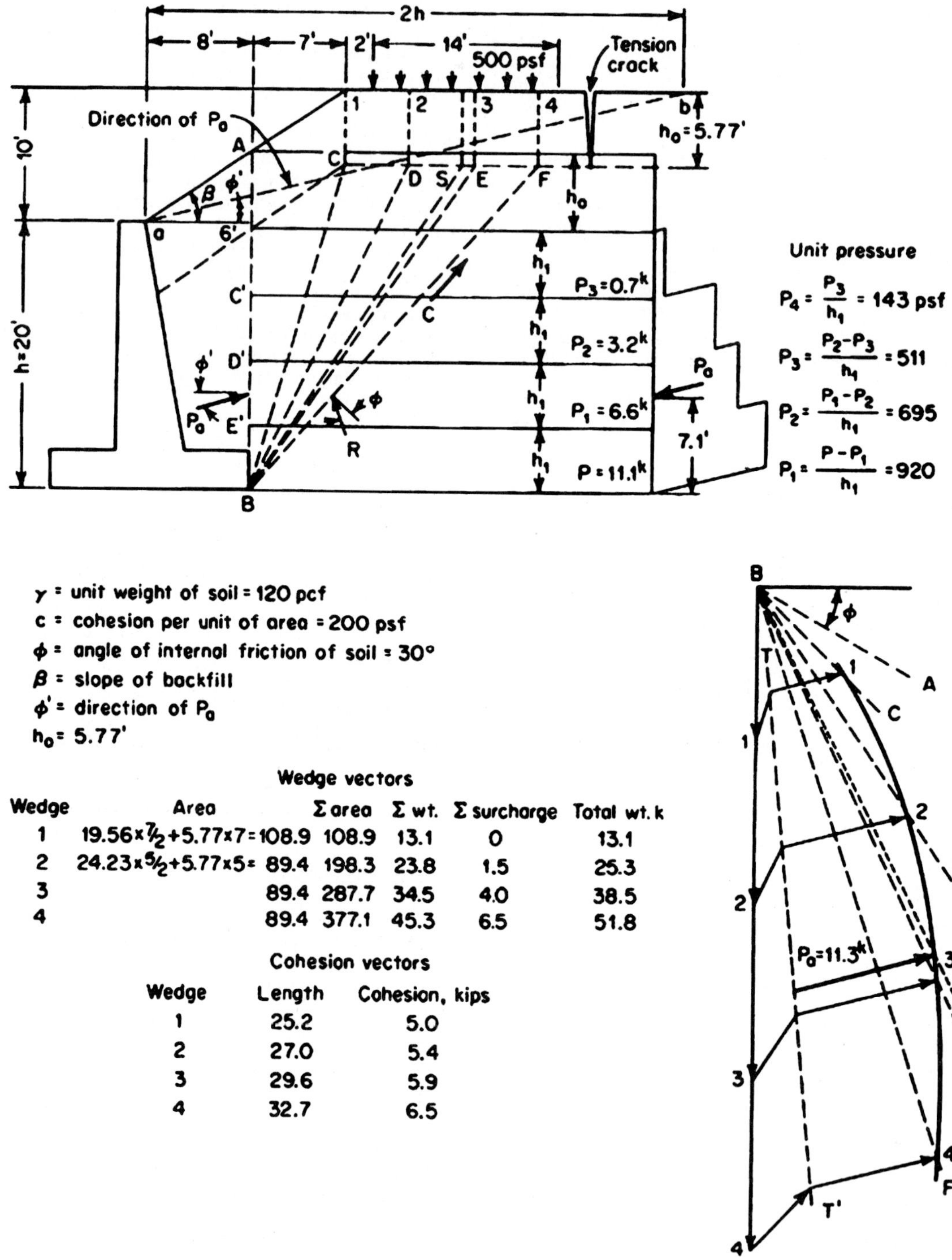

Wedge vectors

Wedge	Area	Σ area	Σ wt.	Σ surcharge	Total wt. k
1	19.56×7/2+5.77×7=108.9	108.9	13.1	0	13.1
2	24.23×5/2+5.77×5= 89.4	198.3	23.8	1.5	25.3
3	89.4	287.7	34.5	4.0	38.5
4	89.4	377.1	45.3	6.5	51.8

Cohesion vectors

Wedge	Length	Cohesion, kips
1	25.2	5.0
2	27.0	5.4
3	29.6	5.9
4	32.7	6.5

Figure 7.21 Trial-wedge method for earth-pressure computations (AREA, 1954).

Determining Earth Pressure

The properties of the retained soil must be known in order to determine the magnitude of the loads that the retaining structure must be designed to resist. It is usually possible to learn in advance the general type of backfill material and to classify it into one of the five categories shown in Table 7.11. For these classifications, charts for estimating the pressure of the backfill against retaining walls less than about 20 feet in height were prepared by Terzaghi and Peck (1967) (Figs. 7.21 and 7.22).

Table 7.11 Type of Backfill for Retaining Walls

Type	Description
1	Coarse-grained soil without admixture of fine soil particles, very free-draining (clean sand, gravel, or broken stone)
2	Coarse-grained soil of low permeability due to admixture of particles of silt size
3	Fine silty sand; granular materials with conspicuous clay content; or residual soil with stones
4	Soft or very soft clay; organic silt; or soft silty clay
5	Medium or stiff clay that may be placed in such a way that a negligible amount of water will enter the spaces between the chunks during floods or heavy rains

These charts, partly based on theory and partly empirical, have been adopted by the AREA (1954). For the cases covered, reliable estimates of earth pressures can be made without recourse to earth-pressure theories.

For walls higher than about 20 feet, for cases where Figs. 7.22 and 7.23 are not applicable (for example, surcharged backfill), and for cases where more refined analyses are desired, properties of the backfill should be determined by laboratory tests and earth pressures computed by methods described previously.

Base Pressures

Figure 7.24 shows a cross section through a cantilever retaining wall and the forces acting thereon. The earth pressure is denoted by P_a. The forces W_s and W_c are the weight of soil and the weight of concrete, respectively. It is customary to analyze a 1-foot length of retaining wall except for the case of the counterfort wall, where a panel length equal to the spacing of the counterforts is commonly analyzed. Since a retaining wall is subjected to lateral forces, the foundation is usually eccentrically loaded. The eccentricity e of the resultant force on the base, measured from the center of the base, is determined by taking moments at some convenient point in the plane of the base, such as a, of all the forces acting above the plane of the base.

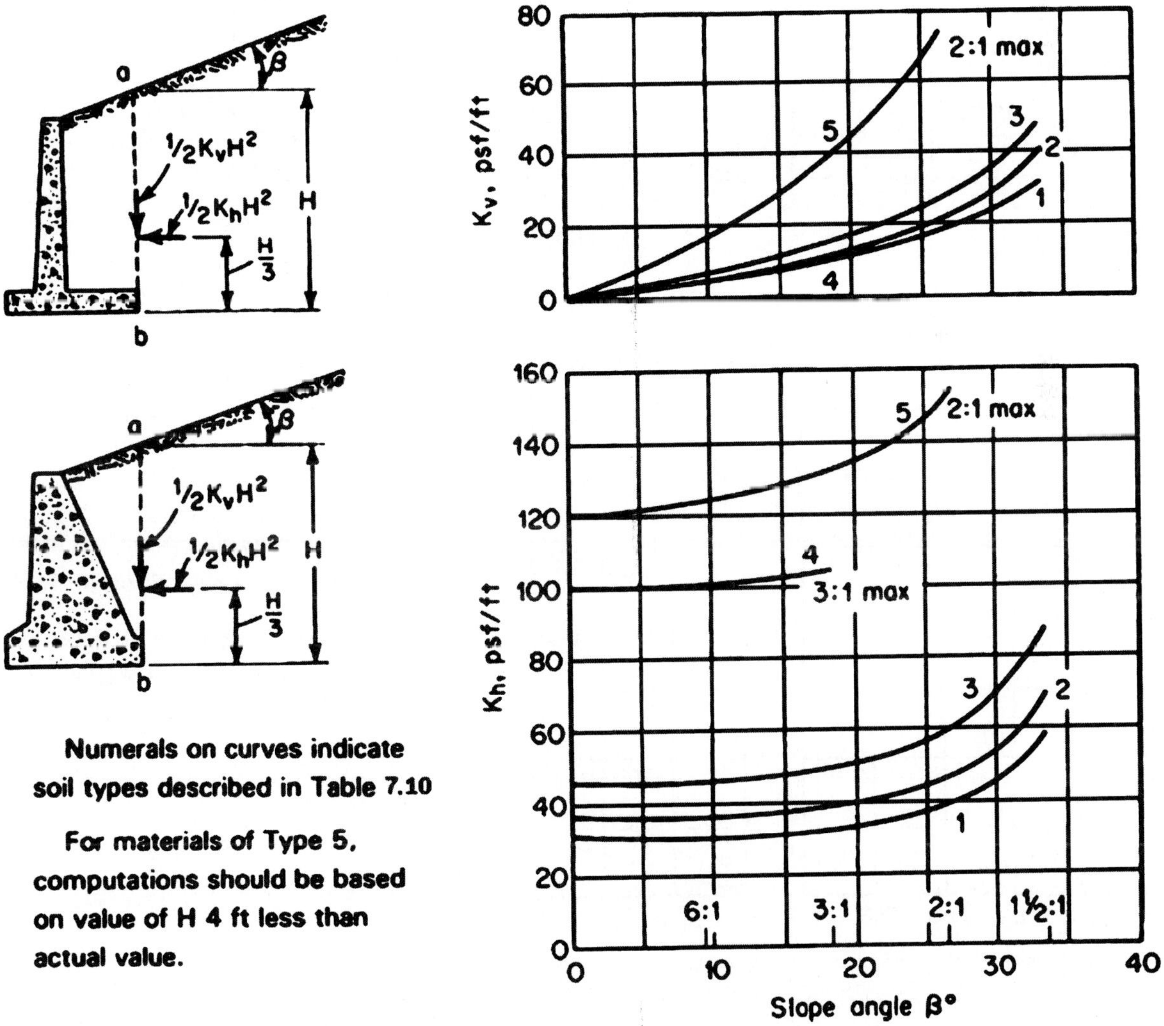

Figure 7.22 Earth-pressure charts for retaining walls less than 20 feet high. See Table 7.10 for soil types (AREA, 1954).

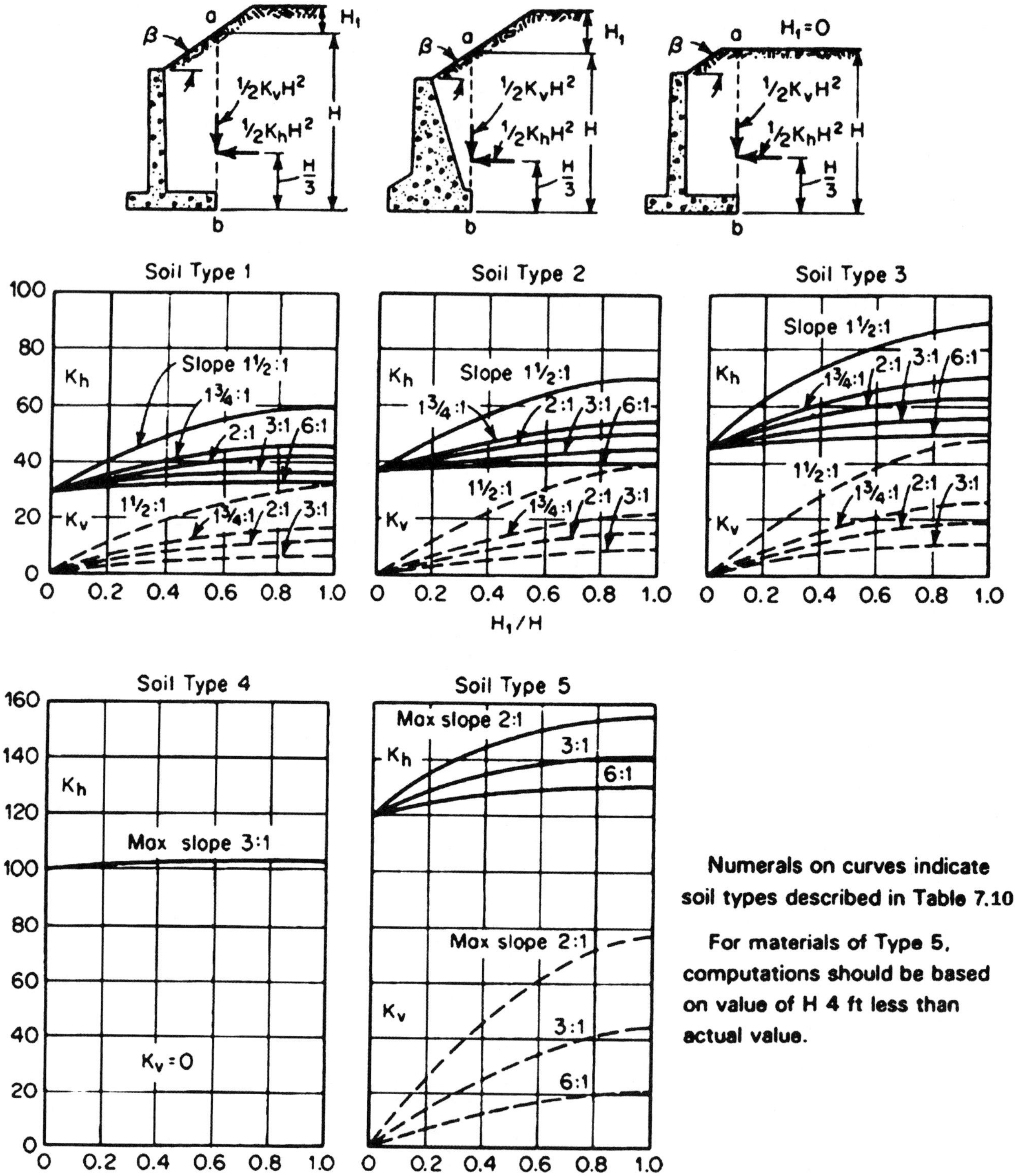

Figure 7.23 Earth-pressure charts for retaining walls less than 20 feet high. See Table 7.10 for soil types (AREA, 1954).

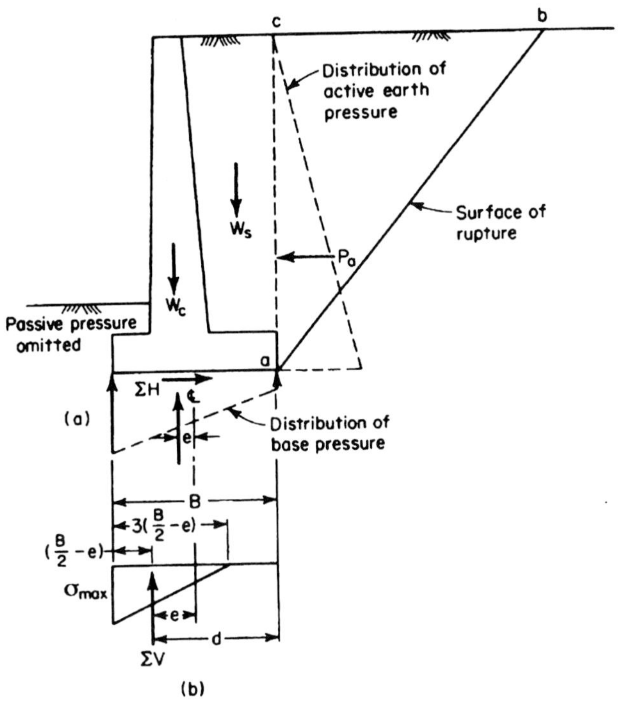

Figure 7.24 Resultant force on base of retaining wall.

If the resultant lies within the middle third of the base, the pressures on the base are

$$\sigma_{\max} = \frac{\sum V}{BL}\left(1+\frac{6e}{B}\right) \tag{7.31}$$

$$\sigma_{\min} = \frac{\sum V}{BL}\left(1-\frac{6e}{B}\right) \tag{7.32}$$

where L = length of wall considered.

If the resultant is outside the middle third, the soil-pressure diagram is triangular across a length $3(B/2 - e)$ of the base (Fig. 7.23*b*), and the maximum soil pressure is

$$\sigma_{\max} = \frac{2\sum V}{3L\left(\frac{B}{2}-3\right)} \tag{7.33}$$

Settlement of Retaining Walls

The behavior of a retaining wall or abutment is a problem in soil-structure interaction. The lateral loads depend on the nature of the soil in contact with the structure and the deformations that occur. The backfill is an essential part of a retaining structure, and the properties and characteristics of the backfill are just as important as the properties of the concrete in the structure. A retaining structure must support its backfill without detrimental lateral movement, and the surface of the backfill must not settle unduly. These deformations depend to a large extent on the properties of the backfill and the subsoil. If the subsoil contains compressible layers, these are likely to consolidate under the weight of the backfill and lead to differential settlements. Since the pattern of settlement tends to be bowl-shaped, it is not uncommon for settlement to result in the retaining wall tilting toward the backfill. More commonly the wall rotates forward or away from the backfill as a result of the lateral forces involved. The settlement at the surface of the backfill includes not only the consolidation that takes place in the subsoil but any reduction in thickness of the backfill as well. Settlement within the backfill may be minimized by careful compaction control during construction.

7.4.3 Slopes

Presently software such as SlopeW (GeoSlope) can be used to determine slope stability. Slope stability is covered my many references and texts (for example, Duncan et al., 2014; Abramson, 2001; EM 1110-2-1902, 31 Oct 2003). The following provides simplified analyses useful provided the simplifying assumptions hold.

Cohesionless materials, such as uncemented clean sands and gravels, will stand at their angle of repose, which is approximately the angle that develops when a pile of the loose material is formed by pouring from a container held a few inches above the pile. Angles of repose of 28 to 35 degrees are commonly used for clean sands and gravels. The angle of repose is essentially independent of time and the depth of excavation or embankment.

Very loose and saturated deposits of granular materials are often quite unstable under the action of transient, blast, and shock loads. Slope angles one-fifth to one-third of those indicated above, may be required for stability of these materials, unless the deposit can be made denser by vibration prior to construction. Vibrations caused by pile driving are often used.

For cohesive materials such as clays and sandy or silty clays, the concept of angle of repose cannot be employed since the angle or slope to which a cohesive material will stand is a function of the depth of excavation or embankment. The depth to which a cut above the groundwater table can be made in a cohesive soil depends on the shearing strength and unit weight of the soil, the slope angle, and time. The role played by time in the slope-stability problem has not been quantitatively evaluated except for a few isolated deposits. However, it is known that the shearing strength of a soil is affected by time. In some soils (for example, precompressed and certain marine-deposited clays), the strength may decrease with time, while in other cases (for example, certain normally consolidated clays and recent fills), it may increase with time. From the standpoint of stability of the soil, it is normally wise to complete construction in a temporary excavation and backfill as quickly as possible. Slopes of permanent excavations, such as highway cuts, are made flatter than those of temporary excavations. Both these practices reflect an appreciation for the influence of time in slope-stability problems.

The stability of slopes of embankments and cuts in homogeneous deposits of cohesive soils bounded by a plane horizontal top surface and a plane inclined slope and underlain by a horizontal surface of a firm stratum (Fig. 7.25) can be determined by using Fig. 7.26 or 7.27. The effects of seepage are not directly accounted for in the solutions on which these figures are based. Given the proposed depth of cut (or height of embankment) H, the overall depth of the deposit to the firm horizontal layer $n_d H$, the unit weight of the soil γ, the slope angle β, and the strength parameters c and ϕ (Terzaghi and Peck, 1967), the stability factor N_s is determined from Fig. 7.26 or 7.27.

The cohesive strength required to provide a factor of safety of 1 can be computed from the following equation:

$$C_{\text{reqd}} = \frac{\gamma H}{N_s} \tag{7.34}$$

This value can then be compared with the available cohesive strength. The factor of safety will depend on the length of time that the cut must stand open, the threat of losses in life and to adjacent property that might result from failures, the reliability of the soil-strength data, and the probability that the top surface may need to carry live loads. For temporary cuts, a factor of safety of 1.5 to 2.0 is often used in ordinary situations. A larger factor of safety is normally required for permanent cuts since the effects of climate, time, and so on play a greater role.

Three positions of the failure surface were considered in preparing Figs. 7.26 and 7.27: (1) toe surfaces (circles) that pass through the base of the slope or toe, (2) surfaces that break out on the slope above the toe,

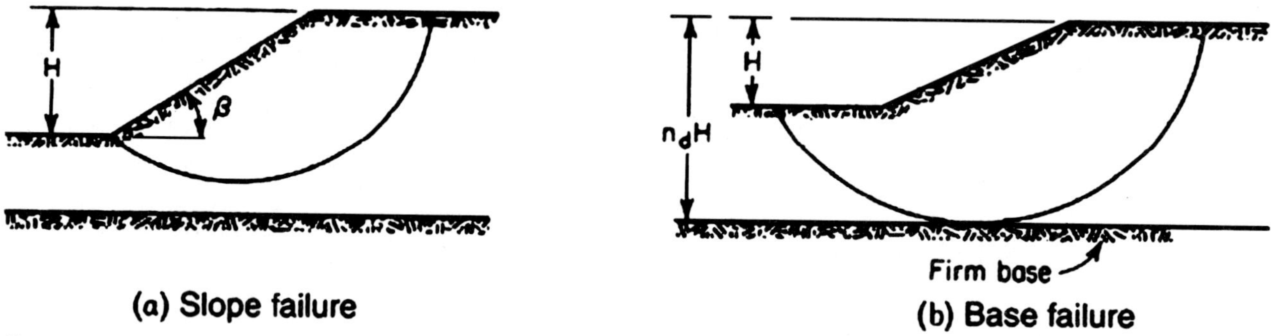

Figure 7.25 Types and positions of critical slide circles (Terzaghi and Peck, 1967).

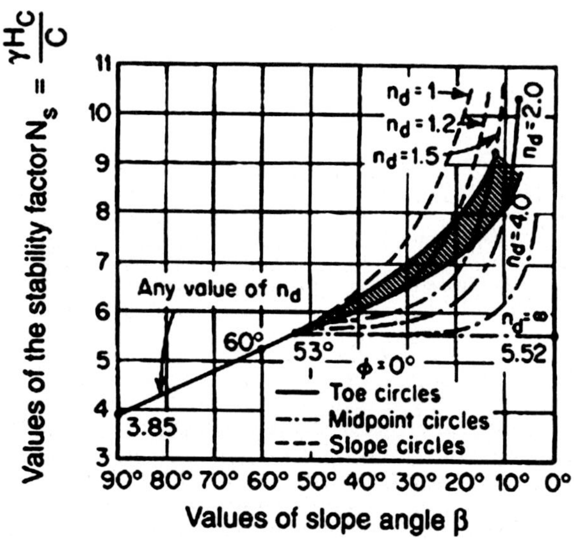

Figure 7.26 Relationship between stability factor N_s, slope angle β, and depth factor n_d for a purely cohesive soil (ϕ = 0) soil (Terzaghi and Peck, 1967).

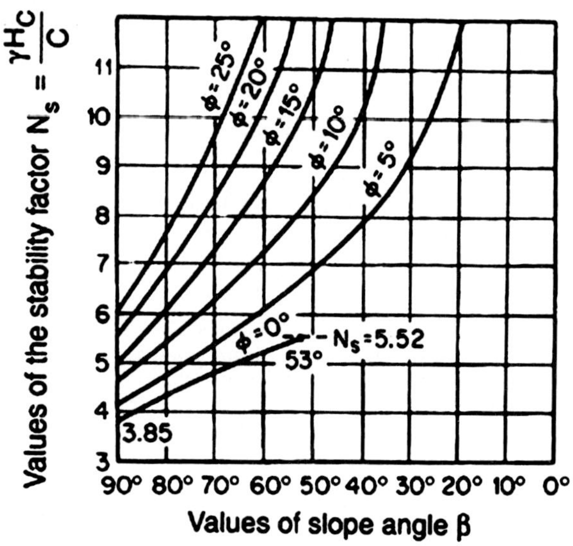

Figure 7.27 Relationship between stability factor N_s, slope angle β, and ϕ for a soil with friction and cohesion (ϕ ≥ 0, $c \neq 0$) (Terzaghi and Peck, 1967).

and (3) midpoint or deep surfaces that break the ground beyond the base or toe of the cut or embankment. When ϕ = 0, the failure surface will always pass through the toe if β ≥ 53 degrees. When ϕ = 0 and β ≤ 53 degrees, any of the three types of failure may occur, depending on the value of β and the depth factor n_d. Figure 7.15 shows that the failure surface will be (1) a slope circle for combinations of β and n_d that fall above the shaded area, (2) a toe circle for combinations that fall within the shaded area, and (3) a midpoint or deep circle for combinations that fall below the shaded area. For ϕ > 0 (Fig. 7.27), the chance of failures being other than toe failures is quite small; hence, the factor n_d loses much of its importance.

The limitations on Figs. 7.26 and 7.27 often seriously restrict their use in practice. When unusual site conditions exist that are not covered by these figures, a more involved stability analysis is required.

VERTICAL CUTS

The unconfined compressive strength (q_u) is also used to estimate the depth to which temporary vertical cuts can be made in claylike soils. For example, the ultimate or critical depth of a vertical cut H_c in such a soil is given approximately by the formula

$$H_c = \frac{q_u}{\gamma} \tag{7.35}$$

where γ is the unit weight of the clay, which will usually range from 115 to 125 pcf. When a factor of safety of 2 is used, the permissible depth of cut is

$$H_c = \frac{q_u}{2\gamma} \tag{7.36}$$

The unconfined compressive strength of claylike soils is often used in analyzing landslides and the stability of slopes with nonvertical sides.

7.5 INVESTIGATIONS

7.5.1 Geology

An investigation of site geology is imperative prior to any field investigation and prior to design. There is a wealth of past investigations that can be drawn on. These include quaternary maps, maps of bedrock, aquifers, karst deposits, and loess. Simple maps, such as U.S. Geological Survey quadrangles, can be useful in identifying potentially objectionable foundation materials, such as those found in wetland areas.

7.5.2 Field Methods and Testing

SAMPLING

Sampling is most often performed with standard penetration test samplers. These are useful for providing soil samples for index testing but disturb samples, making them less useful for soil strength. Thin-walled samplers can be used to take soils with less disturbance. Thin-walled samplers are typically 3 in. in diameter. Larger samplers up to 5 in. in diameter are used for very soft material, then trimmed in the laboratory for testing.

Soil samples provide the advantage of a physical sample and are considered superior to in situ tests. In situ tests, such as the cone penetrometer test (CPT), vane shear test (VST), pressuremeter test (PMT), and dilatometer test (DMT), are useful. However, they should be correlated with physical samples and laboratory tests.

Laboratory Testing

A few laboratory tests can go a long way to help classify soils. Because most soils, particularly cohesive soils, are saturated or near saturation below the ground surface, water content tests can be used to make reliable estimations of unit weights for cohesive soils. Atterberg limit tests, in addition to being used for classification, can be used to estimate stress history. Soils with water contents near the liquid limit are likely normally consolidated. Soils with water contents near the plastic limit are likely overconsolidated to heavily overconsolidated.

Water contents of granular soils should not be trusted because water drains or evaporates after sampling and during transporation. Unit weights should be estimated for these soils using blow counts or other indicator tests.

Triaxial testing is highly recommended for significant structures. The key to effectively specifying these tests is to specify confining stresses near the stresses the soil is likely to experience during load. The limitation of triaxial tests are due mainly to their sparcity during investigations. Tests in only a few locations will tell the engineer only what the soil strengths are in those locations. Another limitation is that only some soils may be sampled. Very hard material will not likely be able to be sampled because the thin-walled sampler will be crushed by hard sample and the associated bolders and cobbles. Very soft material (below q_u of 0.25 tsf) are considered too soft to sample. Lack of triaxial tests showing very soft material does not mean that these materials are not present. In these cases, the engineer should pay particular attention to the geotechnical report, which will often fill in the gaps left by lack of testing. Another area to gain information on materials that are too soft to test include blow counts, which will show drill roads advancing with 2 or less blows per foot, or with weight of hammer.

Indicator Tests

Indicator tests, such as the CPT, VST, PMT, and DMT, have several distinct advantages over standard penetration testing and laboratory sampling. However, each must be treated with caution and include confirmatory testing.

Cone Penetrometer The CPT (ASTM D5778-12) permits continuous testing. Additionally with the piezocone pore-water pressure measurement can be taken. The piezocone permits performing pore-water pressure dissipation tests, which can be used to estimate permeability. Additionally, after pore-water pressure has dissipated, water pressure in a formation can be determined.

Vane Shear VSTs (ASTM D2573/D2573M-15) are valuable in soft to medium stiff clays. They permit development of an in situ shear strength profile. Care should be taken in interpreting vane test data by comparing to normalized shear strength values and expected consolidation history.

Dilatometer DMTs (ASTM D6635-15) can be used for classification as well as to estimate a host of geotechnical parameters. Of particular use is the coefficient of lateral earth pressure, which the DMT simulates most readily.

Pressuremeter The PMT (ASTM D4719-07) has gained wider use lately. It operates similarly to the dilotometer except that it uses an inflatable membrane. It can be used to determine soil properties similar to the dilatometer, particularly elastic properties of soils.

7.6 SOIL IMPROVEMENT

7.6.1 Shallow Soils

Soils used as backfill around structures and buried utilities; support for pavements and structures; embankments for highways, railroads, parking areas, levees, dams, and so on; and linings for canals and reservoirs are compacted so that their strength and stability will be increased, their permeability decreased, their resistance to frost action and erosion enhanced, and their compressibility decreased.

On projects where compaction is anticipated, representative samples of typical soils at the site should be obtained at the time subsurface explorations are being conducted and subjected to one or more of the following laboratory tests: Atterberg limits, laboratory compaction, natural water content, swell, and permeability.

The laboratory compaction test is performed to determine an optimum moisture content and dry density for each soil. The standard Proctor tests (AASHTO T99, ASTM D698-12, or British Standard 1377-4) are generally specified where light to medium rollers are employed. The modified Proctor test (AASHTO T180, ASTM D1557) is usually used where greater compactive effort is required.

Shown in Table 7.12 are typical values of standard Proctor optimum moisture content and density for various soils. Optimum moisture content for the standard test is usually 2 to 5 percentage points less than

Table 7.12 Typical Values of Optimum Moisture Content and Density for Various Soils

Unified classification		Range of max dry unit weight, pcf	Range of optimum moisture content, %
Group symbol	Description		
GW	Well-graded clean gravels, gravel-sand mixtures	125–135	11–8
GP	Poorly graded clean gravels, gravel-sand mix	115–125	14–11
GM	Silty gravels, poorly graded gravel-sand-silt	120–135	12–8
GC	Clayey gravels, poorly graded gravel-sand-clay	115–130	14–9
SW	Well-graded clean sands, gravelly sands	110–130	16–9
SP	Poorly graded clean sands, sand-gravel mix	100–120	21–12
SM	Silty sands, poorly graded sand-silt mix	110–125	16–11
SM-SC	Sand-silt clay mix with slightly plastic fines	110–130	15–11
SC	Clayey sands, poorly graded sand-clay mix	105–125	19–11
ML	Inorganic silts and clayey silts	95–120	24–12
ML-CL	Mixture of inorganic silt and clay	100–120	22–12
CL	Inorganic clays of low to medium plasticity	95–120	24–12
OL	Organic silts and silt-clays, low plasticity	80–100	33–21
MH	Inorganic clayey silts, elastic silts	70–95	40–24
CH	Inorganic clays of high plasticity	75–105	36–19
OH	Organic clays and silty clays	65–100	45–21

a soil's plastic limit, while for the modified Proctor, it is usually 5 to 8 percentage points less. Modified Proctor optimum densities are approximately 7 to 10 pcf greater than standard Proctor values.

Shown in Table 7.13 are typical compaction requirements for various structures based on the standard Proctor laboratory test. Where highly expansive soils are encountered or when large percentages (exceeding about 20 to 30 percent) of gravel are present in the soil, special compaction requirements exist.

With the start of (and during) construction, field density tests (AASHTO T147) should be made on the following bases:

1. one test for every 200 to 500 yd^3 of backfill around structures or in trenches, depending on the total quantity of material involved
2. one test for every 2000 yd^3 of material placed for mass earthwork
3. one test for every 500 yd^3 of highway or airfield pavement subgrade
4. at least one test for every 1000 yd^3 of material placed in relatively thin sections, such as canal or reservoir linings
5. one test whenever there is definite suspicion about the quality of moisture control and compaction

Close moisture-content control is practiced if two-thirds of all field values fall within a range ±1 percent of the median moisture content specified. Erratic yet often satisfactory control is indicated if only two-thirds of all field values fall in a range ±3 percent.

Satisfactory compaction procedures are generally being employed if two-thirds of all densities fall within 2 or 3 percent above or below the percent maximum density required. Unsatisfactory or erratic compaction is evidenced where only two-thirds of all field values fall in a range ±5 percent.

Unless test sections have been built or the performance of a given piece of equipment on a particular soil is already known, it is usually better to specify the required percent of density, moisture-content limits, and lift thickness and allow the contractor some leeway in the selection of compaction equipment and methods.

7.6.2 Compressible Clayey Soils

If the subsoil is unusually soft and compressible or consists of organic soils to a relative shallow depth, it is frequently economical to remove these undesirable soils and backfill with select granular materials compacted in layers. However, if the cost of excavation and replacement is prohibitive, the soft compressible materials can frequently be improved by consolidating them with a surcharge fill. Where permanent fill is required, this portion of the fill should be compacted to a height adequate to include the predicted settlements that will occur. Additional fill, equivalent in magnitude to the proposed structural loads, may then be placed to force consolidation of the underlying soft materials. When the desired degree of consolidation has been accomplished, the excess fill may be removed and the structure built with the assurance that the majority of the settlement has already taken place under the surcharge load. However, this approach takes time.

Vertical sand drains are often installed to accelerate the consolidation process. After removing the vegetation, a pervious gravel or sand blanket of adequate thickness to support construction equipment is placed on the surface of the compressible soil. Since this pervious blanket must function as a drain and also serve as a working platform, it should be at least 2 feet thick.

Two types of vertical drains are common: prefabricated drains and sand drains. Prefabricated drains are about 4 in. wide, 0.5 in. thick, and as long as necessary for penetration through the compressible soil. A prefabricated drain consists of a polymer core wrapped by a geotextile. The core has a geometric pattern of ribs, ridges, or nubs to provide an open channel for water to flow along its long axis. The geotextile wrap is typically a non-woven one that prevents the soil from intruding the core while allowing water to flow into the core. The nonwoven geotextile may also keep flow channels open in the core by spanning between ridges, ribs, and nubs.

A prefabricated drain is installed using the displacement method. The drain is threaded through a long, hollow steel mandrel, typically rectangular in cross section, and then wrapped around a disposable base plate at the tip of the mandrel. The mandrel and attached base plate are then advanced into the ground to the desired depth. On extraction of the mandrel, the base plate and prefabricated drain remain in place and form a vertical drain. Speed and economy are advantages of using prefabricated drains.

The vertical sand drains are installed by driving a mandrel of 18- to 24-in. diameter through the compressible soil, filling it with relatively clean sand, and withdrawing the casing while the sand is forced out by the application of gas pressure inside the mandrel. Alternatively, an augered hole is backfilled with sand. The piles are driven in a pattern at a spacing of 8 to 15 feet. As fill is placed over the drainage blanket, the added load forces water from the subsoil into the sand drains, from which it escapes through the drainage blanket.

7.6.3 Loose Granular Soils

Sometimes, granular soils can be improved by injecting them with grout. However, grouting is often attempted with materials that have little likelihood of penetrating the material to be improved. In general, the penetration of grouts into natural soils depends on the effective size D_{10} of the material to be grouted. Table 7.14 shows the relationship between various types of grout and the limiting size of the material to be grouted.

Since loose soils are far more readily grouted than dense ones, the grouting of soils approaching the finest sizes indicated in Table 7.14 should be approached with some skepticism. Inasmuch as they penetrate the void spaces of the soil, grouts are of principal value in decreasing the soil permeability and are often used only for this purpose. Some of the more desirable chemical grouts are relatively expensive, and therefore they are used only in special or unusual circumstances. In addition, many chemical grouts have been declared carcinogenic.

Table 7.13 Typical Compaction Requirements Based on the Standard Proctor Laboratory Compaction Test

Purpose of fill	Density required as % of optimum laboratory density		Suggested range in moisture content dry (−) or wet (+) of optimum	Suggested lift thickness after compaction, in. (for hand tamping, use one-half of these values)	
	Coarse-grained cohesionless soils	Fine-grained cohesive soils		Coarse-grained cohesionless soils	Fine-grained cohesive soils
Support of structure	100	98	−1 to +2	10	6
Backfill around structure or in pipe or utility trenches	98	95	−2 to +2	8	6
Subgrade of excavation for structure and for earth dams greater than 50 feet high	98	98	−1 to +2	12	6
Earth dams less than 50 feet high and for support of highway or airfield pavement	95	95	−1 to +3	12	6
Canal or small reservoir lining	…	95	−2 to +2	…	6
Drainage blanket or filter	98	…	+2 to +4	10	
Other fills requiring nominal amount of strength and incompressibility	93	90	−3 to +3	10	6

Table 7.14 Criteria for Grout Injection

Grout	Material to be grouted	Finest size that can be grouted, mm
Cement	Fissures	0.10
	Loose sand	$D_{10} = 0.5$
	Dense sand	$D_{10} = 1.4$
	Soil	$D_{10} = 0.65$
Silicates	Soil	$D_{10} = 0.1$
Organic resins	Soil	$D_{10} = 0.07$

Loose free-draining sand may be improved by making it denser. Under suitable circumstances, this may be accomplished with explosives, and satisfactory results can also be obtained with compaction piles. More commonly, vibroflotation is used. This is a patented process in which a large vibrator unit, known as a vibroflot, is lowered into the sand with the assistance of water jetting. Under the combined action of the jetting and vibration, the sand is compacted to form a crater at the ground surface that is filled with sand. The vibroflot is usually inserted at a spacing of 6 to 8 feet in the area to be treated. Supplemental tests are required after treatment to determine its effectiveness. However, it is difficult to verify compliance with a specification requiring that a certain percentage of relative density must be achieved because indirect methods are usually required. It is probably more realistic to determine, in the field, the results that can be achieved for a particular vibroflot spacing and proceed with foundation design on the basis of the actually achieved results.

Dynamic compaction accomplished by dropping a 10- to 40-ton pounder from a height of 40 feet or more has been used to improve existing soils (FHWA-SA-95-07, Geotechnical Engineering Circular No. 1, Dynamic Compaction).

7.7 MONITORING

7.7.1 Groundwater

Critical for groundwater is choosing the appropriate monitoring tool. Open wells work for sandy soils but not for clayey materials. This is because of the time it takes for water to fill the well space. A guide for selecting the appropriate piezometer for the formation was proposed by Terzaghi and Peck (1967) and can be found in EM 1110-2-1908 (USACE, 30 June 1995) as shown in Fig. 7.28.

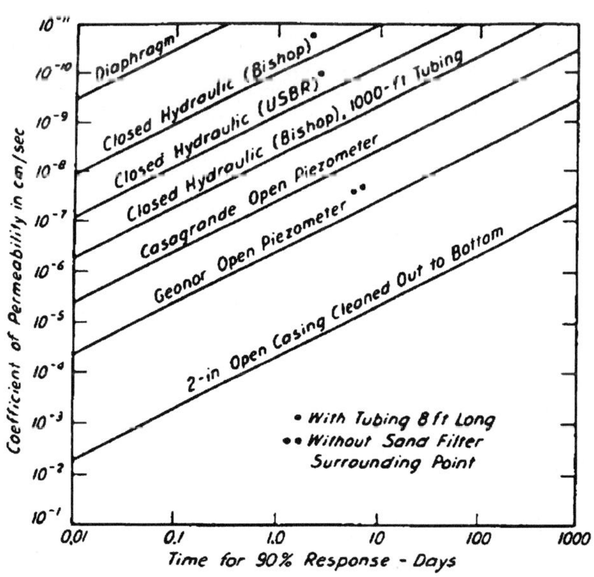

Figure 7.28 Approximate response time for various types of piezometers (Terzaghi and Peck, 1967).

7.7.2 Movement

Some methods for monitoring movement can include inclinometers, time domain reflectometry, settlement plates, surveying, photogrammetry, and bathymetry. A standard reference for instrumentation is Geotechnical Instrumentation for Monitoring Field Performance (J. Dunnicliff, 1993). Additionally, "Instrumentation of Embankment Dams and Levees" (EM 1110-2-1908, 30 June 1995) should be consulted for monitoring for soil movement.

REFERENCES

1. Abramson, L. W., T. S. Lee, S. Sharma, and G. M. Boyce. *Slope Stability and Stabilization Methods.* 2nd ed. New York: John Wiley & Sons, 2001.

2. Acker, W. L., III. *Basic Procedures for Soil Sampling and Core Drilling.* Scranton, PA: Acker Drill Co., 1974.

3. American Railway Engineering Association. *Manual of Recommended Practice, Pt. 5, Retaining Walls and Abutments.* Washington, DC: American Railway Engineering Association, 1954.

4. American Association of State Highway Transportation Officials. T99 Standard Method of Test for Moisture–Density Relations of Soils Using a 2.5-kg (5.5-lb) Rammer and a 305-mm (12-in.) Drop. 1 January 2017.

5. American Association of State Highway Transportation Officials. T180 Standard Method of Test for Moisture–Density Relations of Soils Using a 4.54-kg (10-lb) Rammer and a 457-mm (18-in.) Drop. 1 January 2017.

6. American Society for Testing and Materials. Annual Book of ASTM Standards, Vols. 04.08 and 04.09, Soil and Rock; Geosynthetics. Philadelphia: ASTM, 1995.

7. American Society for Testing and Materials. D4719-07 Standard Test Methods for Prebored Pressuremeter Testing in Soils.

8. American Society for Testing and Materials. Standard Test Method for Electronic Friction Cone and Piezocone Penetration Testing of Soils. D5778-12 Standard Test Method for Electronic Friction Cone and Piezocone Penetration Testing of Soils. 1 January 2012.

9. American Society for Testing and Materials. ASTM D698-12 Standard Test Methods for Laboratory Compaction Characteristics of Soil Using Standard Effort (12,400 ft-lbf/ft³ (600 kN-m/m³)). 1 May 2012.

10. American Society for Testing and Materials. ASTM D1557 Standard Test Methods for Laboratory Compaction Characteristics of Soil Using Modified Effort (56,000 ft-lbf/ft³ (2,700 kN-m/m³)). 1 May 2012.

11. American Society for Testing and Materials. D2573/D2573M-15 Standard Test Method for Field Vane Shear Test in Saturated Fine-Grained Soils. 1 July 2015.

12. American Society for Testing and Materials. D6635-15 Standard Test Method for Performing the Flat Plate Dilatometer. 1 November 2015.

13. American Society for Testing and Materials. D4318-17 Standard Test Methods for Liquid Limit, Plastic Limit, and Plasticity Index of Soils. July 2017.

14. Baguelin, F., J. F. Jezequel, and D. H. Shields. *The Pressuremeter and Foundation Engineering.* Clausthal, Germany: Trans Tech Publications, 1978.

15. Bally, A. W., and A. R. Palmer. *Geology of North America: An Overview.* New York: Geological Society of America, 1989.

16. Benoit, J., and G. W. Clough. Self-boring pressuremeter tests in soft clay. Journal of the Geotechnical Engineering Division (January 1986), 60–78.

17. Bishop, A. W. A new sampling tool for use in cohesionless sands below groundwater level. Geotechnique (1)2, December 1948, 125–131.

18. British Standard. 1377-4 Methods of test for soils for civil engineering purposes. Compaction-related tests. 1990.

19. Chellis, R. D. *Pile Foundations.* 2nd ed. New York: McGraw-Hill, 1961.

20. Christopher, B. R., S. A. Gill, J. P. Giroud, I. Juran, J. K. Mitchell, F. Schlosser, and J. Dunnicliff. *Reinforced Soil Structures: Vol. 1.*

Design and Construction Guidelines. Federal Highway Administration Publication No. FHWA-RD-89-043. Washington, DC: U.S. Department of Transportation, November 1990.

21. Davisson, M. T. Estimating buckling loads for piles. Proc. 2d Panamerican Conf. Soil Mech Found. Eng., Brazil (1) (1963).

22. Davisson, M. T., and P. Shamsher. A review of soil-pole behavior. Highway Research Record (39) (1963).

23. Davisson, M. T. Foundations in difficult soils—State of the practice deep foundations-driven piles. Seminar on Foundations in Different Soils. New York: Metropolitan Section, ASCE, April 1989.

24. Density changes of sand caused by sampling and testing, Waterways Expt. Sta. Potomology Investigations Rept. 12–1. Vicksburg, MS: Waterways Experiment Station, 1952.

25. Design Manual 7.1, Soil Mechanics, Section 7 Penetration Resistance Tests. Alexandria, VA: Naval Facilities Engineering Command, 1982.

26. Duncan, M. J., S. G. Wright, and T. L. Brandon. *Soil Strength and Slope Stability*. 2nd ed. New York: John Wiley & Sons, 2014.

27. Dunnicliff, J. *Geotechnical Instrumentation for Monitoring Field Performance*. New York: John Wiley & Sons, 1988.

28. Federal Highway Administration. *Static Testing of Deep Foundations*. Publication No. FHWA-SA-91-042. Washington, DC: U.S. Department of Transportation, 1992.

29. Federal Highway Administration. *Geotechnical Engineering Circular No. 1, Dynamic Compaction*. Publication No. FHWA-SA-95-07. Washington, DC: U.S. Department of Transportation, 1995.

30. Federal Highway Administration. *Design and Construction of Mechanically Stabilized Earth Walls and Reinforced Soil Slopes, Volume I*. Publication No. FHWA-NHI-10-024. Washington, DC: U.S. Department of Transportation, 2009.

31. Geologic Map of the United States, Scale 1:2,500,000. Washington, DC: U.S. Geological Survey, 1932.

32. Glacial Map of North America, Scale 1:4,555,000. New York: Geological Society of America, 1945.

33. Glacial Map of the United States East of the Rocky Mountains, Scalel: 1,750,000. New York: Geological Society of America, 1959.

34. Goble, G. G., G. Likins, and F. Rausche. *Bearing Capacity of Piles from Dynamic Measurements*. Columbus, OH: DOT Report 05–75 (1975), 76.

35. *Ground Water Manual*. Washington, DC: U.S. Bureau of Reclamation, 1977.

36. *Groundwater and Wells*. St. Paul, MN: Johnson Division, 1986.

37. Hiley, A. The efficiency of the hammer blow, and its effect with reference to piling. London: Engineering (2 June 1922) 673.

38. Hirsch, T. J., L. L. Lowery, J. M. Coyle, and C. H. Samson Jr. Pile-driving analysis by one-dimensional wave theory: State of the art. Highway Research Record (333) (1970), 33–54.

39. Hunt, C. B. *Surficial Deposits of the United States*. New York: Van Nostrand Reinhold, 1986.

40. Huntington, W. C. *Earth Pressures and Retaining Walls*. New York: John Wiley & Sons, 1957.

41. Hvorslev, M. J. *Subsurface Exploration and Sampling of Soils for Civil Engineering Purposes*. Vicksburg, MS: Waterways Experiment Station, 1949.

42. In situ measurement of soil properties, vols. 1 and 2. Proc. ASCE Geotechnical Specialty Conf., New York, 1976.

43. Janbu, N. *Use Analyse Energetique du Battage des Pieux a l'aide de Parametres sand Dimension*. Publication No. 3. Oslo: Norwegian Geotechnical Institute, 1953.

44. Kezdi, A. *Bodenmechanik*. Budapest: Akademiai Kiado, 1964.

45. King, Ruth R., et al. Bibliography of North American Geology, 1959. U.S. Geological Survey Bulletin (1145). Washington, D.C., 1961.

46. Leonards, G. A., ed. *Foundation Engineering*. New York: McGraw-Hill, 1962.

47. Liao, S. S. C., and R. V. Whitman. Overburden correction factors for SPT in sand. Journal of the Geotechnical Engineering Division (March 1986), 373–377.

48. Lutenegger, A. J. Suggested method for performing the borehole shear test. Geotechnical Testing Journal (10)1 (March 1987), 19–25.

49. American Association of State Highway and Transportation Officials. *Manual on Subsurface Investigations*. Washington, DC: American Association of State Highway and Transportation Officials, 1988.

50. Matlock, H., and L. C. Reese. Foundation analysis of off-shore pile-supported structures. Proc. 5th Int. Conf: Soil Mech. Found. Eng. (2) (1961).

51. Mayne, P. W., and J. K. Mitchell. Profiling of overconsolidation ratio in clays by field vane. Canadian Geotechnical Journal (25)1 (March 1988), 150–157.

52. Mesri, G. Coefficient of secondary compression. Journal of the Soil Mechanics and Foundations Division (99)SM1, January 1973, 123–137.

53. Michigan State Highway Commission. A performance investigation of pile driving hammers and piles. March 1965.

54. Milligan, V. Field measurement of permeability in soil and rock. In Situ Measurement of Soil Properties (2) (1975), 3–36.

55. Mitchell, J. K. In-situ techniques for site characterization. Site Characterization & Exploration (1979), 107–129.

56. Mitchell, J. K., and J. K. Zornberg. Reinforced soil structures with poorly draining backfills, pt. II: Case histories and applications. Geosynthetics International (2)1 (1995), 265–307.

57. Mitchell, J. K., and W. C. B. Villet. *Reinforcement of Earth Slopes and Embankments*. NCHRP Report No. 290. Washington, DC: Transportation Research Board, 1987.

58. Newmark, N. M. Influence charts for computation of vertical displacements in elastic foundations. University of Illinois Bulletin (44)45 (23 March 1947), 14.

59. Peck, R. B., and H. O. Ireland. Backfill guide. Journal of the Structural Division (83)4 (July 1957), 1–10.

60. Peck, R. B., W. E. Hanson, and T. H. Thornburn. *Foundation Engineering*. New York: John Wiley & Sons, 1974.

61. Pihlainen, J. A., and G. H. Johnston. Guide to a field description of permafrost. National Research Council of Canada, Associate Committee on Soil and Snow Mechanics, Tech. Mem. 79, 1963.

62. Prakash, S. *Behavior of Pile Groups Subjected to Lateral Loads*. PhD thesis, University of Illinois, 1962.

63. Prakash, S., and H. D. Sharma. *Pile Foundations in Engineering Practice*. New York: John Wiley & Sons, 1990.

64. Rausche, F., G. G. Goble, and G. Likins. Dynamic determination of pile capacity. Journal of Geotechnical Engineering (3)3 (1985), 367–383.

65. Reeves, R. G., A. Anson, and D. Landen, eds. *Manual of Remote Sensing*. Vols. 1 and 2. Falls Church, VA: American Society of Photogrammetry, 1975.

66. Robertson, P. K. In situ testing and its application on foundation engineering. Canadian Geotechnical Journal (23)2 (June 1986), 573–594.

67. Robertson, P. K., and R. G. Campanella. Interpretation of cone penetration tests, part I: Sand. Canadian Geotechnical Journal (20)4 (December 1983), 718–733.

68. Sanglerat, G. *The Penetrometer and Soil Exploration*. Amsterdam: Elsevier, 1972.

69. Seed, H. B., and P. DeAlba. Use of SPT and CPT tests for evaluating the liquefaction resistance of sands, use of in situ tests in geotechnical engineering. ASCE Geotechnical Special Publication No. 6, 1986, 281–302.

70. Skempton, A. W. Standard penetration test procedures and the effects in sands of overburden pressure, relative density, particle size, aging and overconsolidation. Geotechnique (36)3 (September 1986), 425–447.

71. Skempton, A. W., and V. A. Sowa. The behavior of saturated clays during sampling and testing, Geotechnique (13)4 (December 1963), 269–290.

72. Sowers, G. B. *Introductory Soil Mechanics and Foundations*. 4th ed. New York: Macmillan, 1979.

73. Sowers, G. B., and G. F. Sowers. *Introductory Soil Mechanics and Foundations*. 3rd ed. New York: Macmillan, 1970.

74. American Association of State Highway and Transportation Officials. *Standard Specification for Highway Bridges*. 15th ed. Washington, DC: American Association of State Highway and Transportation Officials, 1992.

75. American Society for Testing and Materials. ASTM D4595, Annual Book of ASTM Standards 04(08), Standard Test Method for Tensile Properties of Geosynthetics by the Wide Width Strip Method. Philadelphia: ASTM, 1992, 809–819.

76. Stark, H. T., and H. T. Eid. Drained residual strength of cohesive soils. Journal of Geotechnical Engineering (120)5 (May 1994), 856–871.

77. Survey and treatment of marsh deposits. Highway Research Board Bibliography 15, 1954.

78. Sutherland. Granular materials, review paper, session 1. Proceedings of Conference on Settlement of Structures. Cambridge: British Geotechnical Society; London: Pentech Press, 1975, 473–499.

79. Taylor, D. W. *Fundamentals of Soil Mechanics.* New York: John Wiley & Sons, 1960.

80. Terzaghi, K. Anchored bulkheads, Transactions of the ASCE (119) (1954), 1243–1280.

81. Terzaghi, K., and R. B. Peck. *Soil Mechanics in Engineering Practice.* 2nd ed. New York: John Wiley & Sons, 1967.

82. Thornburn, T. H. Geology and pedology in highway soil engineering. Reviews in Engineering Geology (2), 1968, 17–58.

83. Thornley, J. H. *Foundation Design and Practice, and Economic View.* New York: Columbia University Press, 1959.

84. Timber piles and construction timbers. ASCE Manuals of Engineering Practice, no. 17. New York: ASCE.

85. Tomlinson, M. J. *Foundation Design and Construction.* 5th ed. New York: Longman Scientific & Technical/John Wiley & Sons, 1987, 28–32, 526–530.

86. Tomlinson, M. J. *Foundation Design and Construction.* New York: John Wiley & Sons, 1963.

87. Tomlinson, M. J. *Pile Design and Construction Practice.* London: Palladian Publications, 1987, 344–370.

88. Tschebotarioff, G. P. *Foundations, Retaining and Earth Structures.* New York: McGraw-Hill, 1973.

89. *Uniform Building Code.* Los Angeles: Pacific Coast Building Officials Conference, Vol. 1 (1955), 207–208.

90. United States Army Corps of Engineers. *Geophysical Exploration.* EM 1110-1-1802. Washington, DC: U.S. Government Printing Office, 1979.

91. United States Army Corps of Engineers. *Laboratory Soil Testing.* EM 1110-2-1906. Washington, DC: U.S. Government Printing Office, 20 August 1986.

92. United States Army Corps of Engineers. *Retaining and Floodwalls.* EM 1110-2-2502, Washington, DC: U.S. Government Printing Office, 29 September 1989.

93. United States Army Corps of Engineers. *Instrumentation of Embankment Dams and Levees.* EM 1110-2-1908. Washington, DC: U.S. Government Printing Office, 1995.

94. United States Army Corps of Engineers. *Slope Stability.* EM 1110-2-1902. Washington, DC: U.S. Government Printing Office, 31 October 2003.

95. Waterways Experiment Station. Undisturbed sand sampling below the water table. Waterways Experiment Station Bulletin (35). Vicksburg, MS: Waterways Experiment Station, 1950.

96. Way, D. S. *Terrain Analysis.* Stroudsburg, PA: Dowden, Hutchinson and Ross, 1973.

97. Wellington, A. M. Discussion of the Iron Wharf at Fort Monroe, Va., by J. R. Duncklee. Transactions of the ASCE (27), paper no. 543 (August 1892), 129–137.

98. Woods, R. D., ed. *Geophysical Characterization of Sites.* New York: International Science Publishers, 1994.

Chapter 8

Design of Structural Steel Members

BY

JAY SHEN, Ph.D., PE, SE *Department of Civil, Construction and Environmental Engineering, Iowa State University, Ames, Iowa*

BULENT AKBAS, Ph.D. *Department of Civil Engineering, Gebze Technical University, Gebze, Turkey*

ONUR SEKER, Ph.D. *Department of Civil Engineering, Gebze Technical University, Gebze, Turkey*

CHARLIES J. CARTER, Ph.D., PE, SE *American Institute of Steel Construction, Chicago, Illinois*

8.1 DESIGN OF STEEL AND COMPOSITE MEMBERS

8.1.1 Introduction

There are three main structural steel materials available in steel building and bridge construction shown in Table 8.1. Typical stress-strain curve for structural steel for tension is shown in Fig. 8.1.

Structural steel shapes are either hot-rolled or built-up shapes. The standard hot rolled shapes are wide-flange W-shape, M-shape, S-shape, C-shapes (or channels), angles, WT-shapes (or structural tees), hollow sections, and pipes.

LRFD Design Philosophy

The current AISC specification AISC 360 provides two design methods for the design of structural members, namely Allowable Strength Design (ASD) and Load and Resistance Factor Design (LRFD). Both methods intend to provide a certain level of probabilistic safety against failure that might occur in a member, connection, or structural system. The specification uses the term "failure" instead of "limit state." Limit states are defined for strength and serviceability as:

- *Strength Limit States*

Plastic strength is used and ductility, buckling, fatigue, fracture, etc. is considered for design. Either LRFD or ASD method should be used for design.

- *Serviceability Limit States*

The element is not allowed to go beyond F_y. The corresponding limit states are related to the daily use of structures such as displacement, floor vibration, permanent deformation, cracking, etc. Neither LRFD nor ASD is needed for the design of these limit states.

The LRFD approach follows the general engineering design philosophy and stated as follows:

$$\underbrace{R_u = \sum Y_i Q_i}_{\text{demand}} \leq \underbrace{\phi R_n}_{\text{capacity}} \tag{8.1}$$

Table 8.1 Structural Steel Types and Their Designation

Type	ASTM designation	F_y Minimum yield stress ksi	F_u Tensile strength ksi	Notes
Carbon steels	A36	36	58–80	primarily used in buildings
	A53 Grade B	35	60	for pipes
	A500 (Grade A, B, or C)	33, 42, 46	45, 58, 62	for structural tubing
High-strength low-alloy steels	A242	42–50	63–70	in bridge construction where corrosion resistance is desired
	A572 Gr. 50	50	65	primarily used in buildings
	A588	50	70	corrosion resistance is 4 times that of A36
	A992	50	65	having a maximum yield-to-tensile
	A913	50	60	strength ratio of 0.85 for all structural shapes
Alloy steels	A514 Gr. 100	100	110–130	Not available in hot-rolled shapes
	A709 Gr. 100	90	100–130	

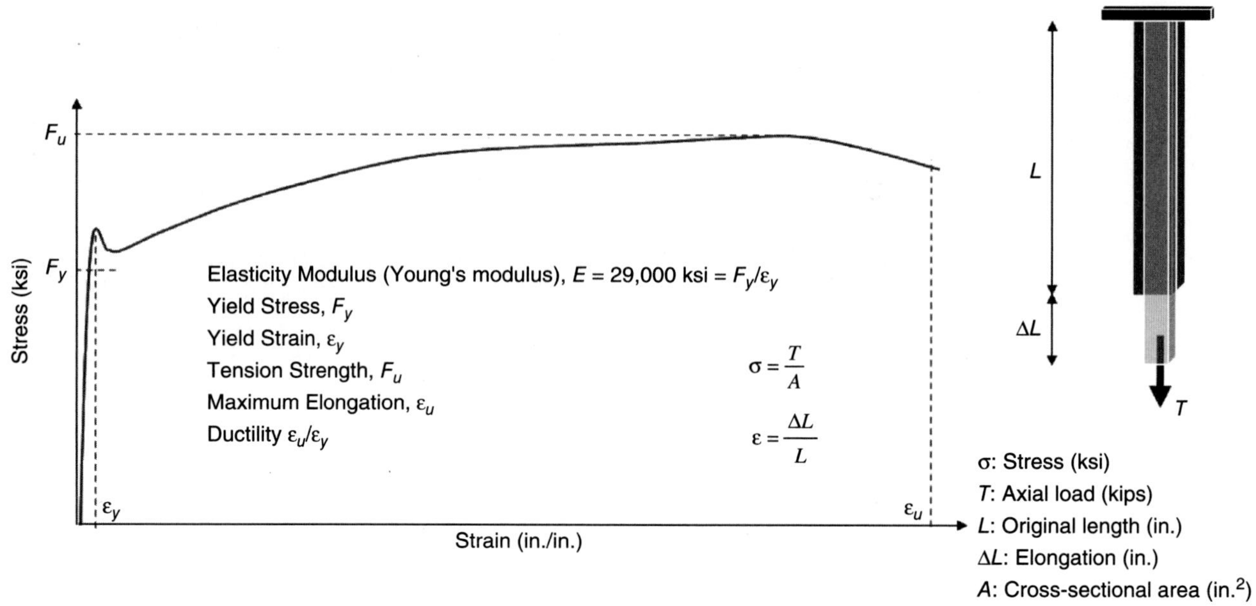

Figure 8.1 Typical stress-strain curve for structural steel.

where R_u = required strength
R_n = nominal strength
Q_i = load effects
ϕ = resistance factor
γ_i = load factor
ϕR_n = design strength

The left side of Eq. (8.1) represents the demand side of the design equation which should come from structural analyses, whereas the right side of the equation is the capacity and defined in the specification. The main difference between LRFD and ASD design philosophies is that ASD uses a single load factor, where LRFD defines load factors for each load effect probabilistically. The nominal strength in both LRFD and ASD design methodologies assume that the entire cross section of the structural member reaches yield stress (see Fig. 8.2).

8.1.2 Tension Members

Tension members are those structural components that primarily carry tension load. Tension members can be found in many structures, such as bottom chords of trusses, bracings and uplifting columns, cables, eyebars, and so on. Tension members can be divided into the following categories:

- Rolled shapes and built-up sections
- Pin-connected members and eyebars
- Tie bars and rods
- Cables

Typical cross sections of tension member are shown in Fig. 8.3.

Design of a tension member is to identify the controlling failure mode (the limit state with the smallest capacity), and to select the section or connection to ensure the safety. When a structural component

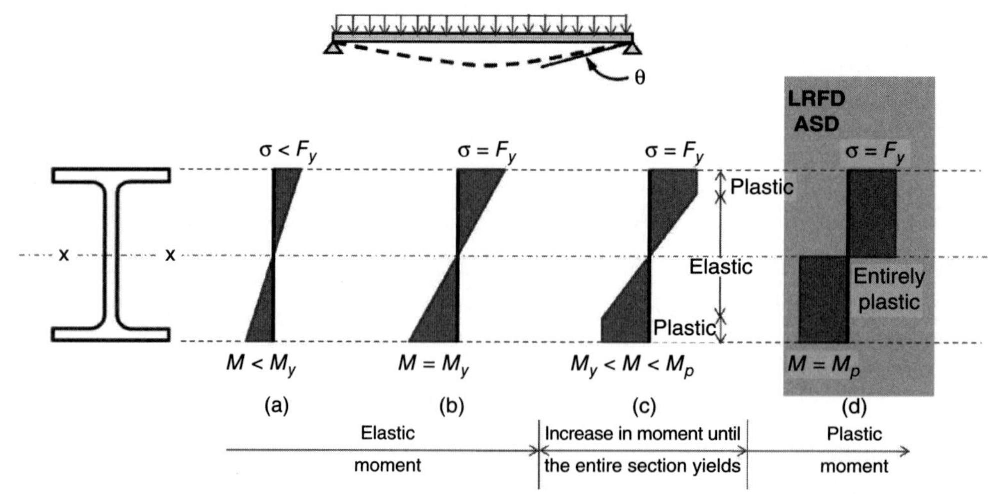

Figure 8.2 Normal stress distribution for elastic, partially inelastic, and fully plastic loading of an I-shaped section.

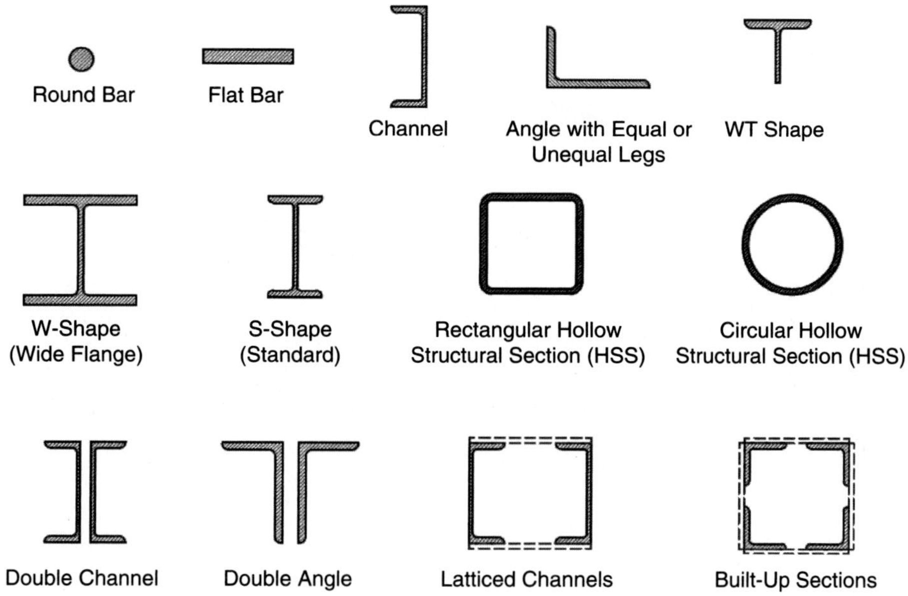

Figure 8.3 Cross section of typical tension members.

is subject to a tension force, it behaves elastically until reaching one of the following limit states:

- Tensile yielding in the gross section
- Tensile rupture in the net section
- Block shear rupture
- Connection failure

Even though there is no slenderness (length over minimum radius of gyration) limit for design of tension members, too slender tension members are not desirable. Therefore, a slenderness ratio of less than 300 is preferred for tension members other than rods to avoid sagging and vibration as well as problems during construction.

Tensile yielding in the gross section:

This limit state defines the yielding of the gross section of a tension member without holes and excessive deformation of the member is prevented.

$$P_n = F_y A_g \qquad \textit{(Eq. D2-1, AISC 360)}$$

where A_g is the gross area of member, and F_y is the specified minimum yield stress.

Tensile rupture in the net section:

This limit state defines fracture in the net section due to the reduction in the gross section (Fig. 8.4). Thus,

$$P_n = F_u A_e \qquad \textit{(Eq. D2-2, AISC 360)}$$

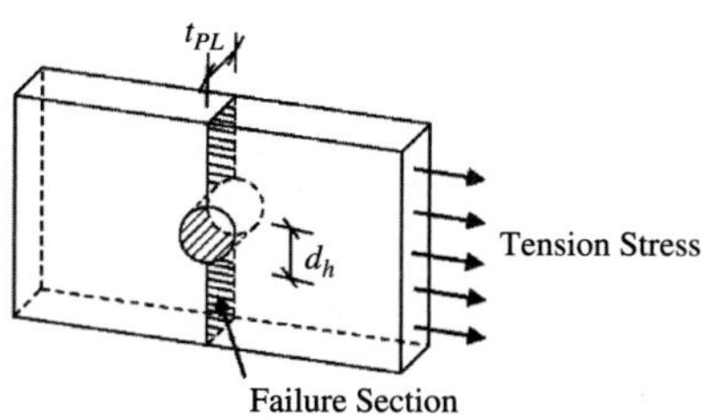

Figure 8.4 Limit state of fracture in the effective area.

$$A_e = UA_n \qquad \textit{(Eq. D3-1, AISC 360)}$$

$$U = 1 - \frac{\bar{x}}{l} \qquad \textit{(Table D3-1, AISC 360)}$$

where A_e = effective net area, in.²
F_u = specified minimum tensile strength, ksi
U = the shear lag factor
A_n = net area, in.² ($A_n = A_g - d_h t_{pl}$)

The shear lag factor, U, accounts for the nonuniform stress distribution when some of the elements of a tension member are not directly connected as in the angle bar connection in Fig. 8.5. In such case, the tensile force is not uniformly distributed over the net area of the member. Therefore, a reduction coefficient is introduced to account for such case.

Block shear rupture:

Failure might occur by shear along a plane through the fasteners plus tension along a perpendicular plane on the area effective in resisting tearing failure. This type of failure is called as "block shear failure." Nominal block shear strength is determined by the summation of the shear and tension terms.

$$P_n = 0.6F_u A_{nv} + U_{bs} F_u A_{nt} \le 0.6F_y A_{gv} + U_{bs} F_u A_{nt} \qquad \textit{(Eq. J4-5, AISC 360)}$$

where A_{nt} = net area in tension, in.²
A_{gv} = gross area subject to shear, in.²
A_{nv} = net area subject to shear, in.²
U_{bs} = reduction coefficient (1.0 if tension stress is uniform, 0.5 if not)

Failure of connection elements:

Although it is undesirable, connection failure might be possible for tension members due to limit states of fracture, bearing and tearing out of connection elements (e.g., bolts or welds).

EXAMPLE 1

An angle L5 × 3 × ⅜ is connected to a gusset plate with five ¾" A325-N bolts, as shown in Fig. 8.6. Compute LRFD design strength of L5 × 3 × ⅜ considering all possible limit states for the angle.

L5 × 3 × ⅜: A_g = 2.86 in.²; F_y = 36 ksi; F_u = 58 ksi

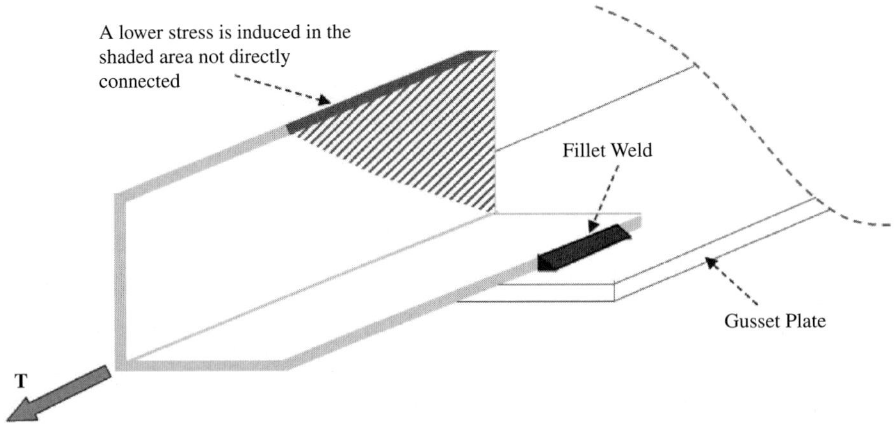

Figure 8.5 Shear lag.

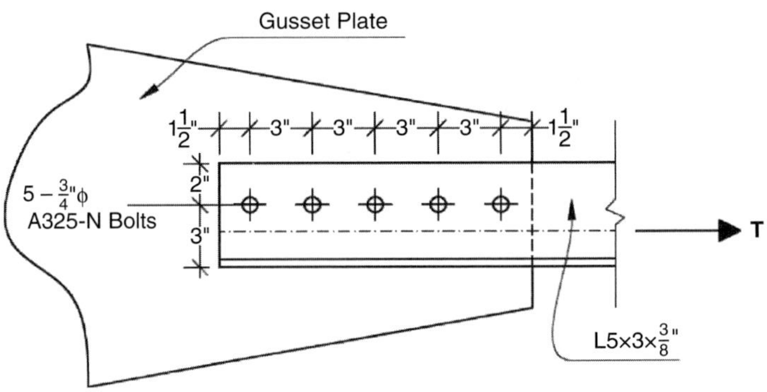

Figure 8.6 Example 1.

Yielding in the gross section:

$$P_n = 36^{\text{ksi}} \times 2.86^{\text{in.}^2} = 103^{\text{kips}} \qquad \phi_t = 0.9$$

$$\phi_t P_n = 0.9 \times 103 = 92.7^{\text{kips}}$$

Tensile rupture in the net section:

$$A_n = A_g - \Delta A = 2.86 - \frac{3}{8} \times \left(\frac{3}{4} + \frac{1}{8}\right) = 2.53 \text{ in.}^2$$

$$U = 1 - \frac{0.698}{12} = 0.942 > 0.80 \rightarrow U = 0.942 \text{ (see Fig. 8.7)}$$

$$A_e = 0.942 \times 2.53 = 2.38 \text{ in.}^2$$

$$P_n = 58^{\text{ksi}} \times 2.38^{\text{in.}^2} = 138.2^{\text{kips}} \qquad \phi_t = 0.75$$

$$\phi_t P_n = 0.75 \times 138.2 = 103.5^{\text{kips}}$$

Block shear rupture:

$$A_{gv} = 0.375 \times 13.5 = 5.063 \text{ in.}^2$$

$$A_{nv} = 0.375 \times (13.5 - 4.5 \times 0.875) = 3.586 \text{ in.}^2 \text{ (see Fig. 8.8)}$$

$$A_{nt} = 0.375 \times (2 - 0.5 \times 0.875) = 0.586 \text{ in.}^2 \text{ (see Fig. 8.8)}$$

$$P_n = \{0.6F_u A_{nv};\ 0.6F_y A_{gv}\}_{\min} + U_{bs} F_u A_{nt}$$

$$P_n = \{0.6 \times 58 \times 3.586;\ 0.6 \times 36 \times 5.063\}_{\min} + 1.0 \times 58 \times 0.586$$

$$P_n = \{124.8;\ 109.4\}_{\min} + 34 = 143.4^{\text{kips}} \qquad \phi_t = 0.75$$

$$\phi_t P_n = 0.75 \times 143.4 = 107.6^{\text{kips}}$$

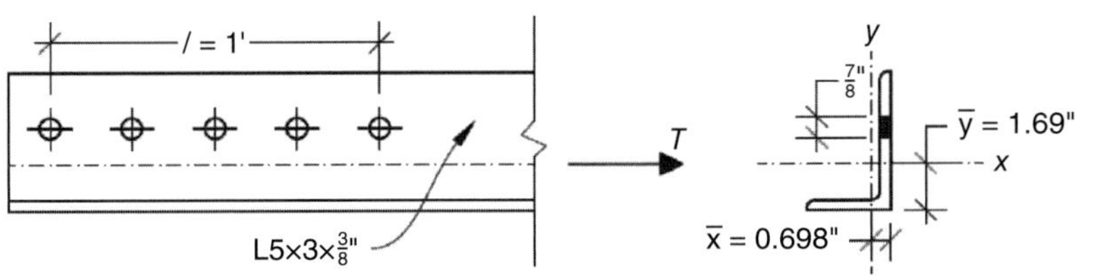

Figure 8.7 Net area.

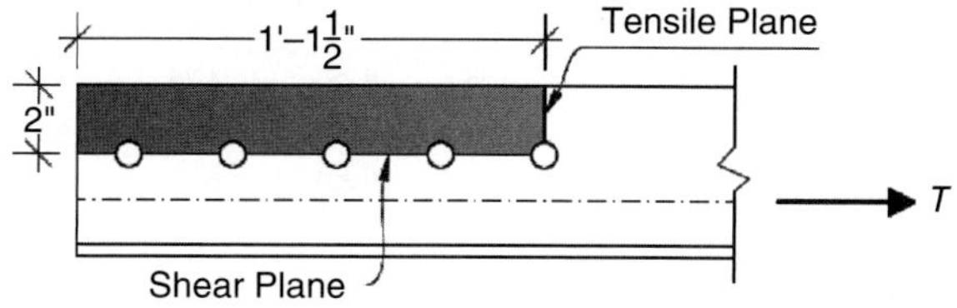

Figure 8.8 Block shear rupture.

Limit state	$\phi_t P_n$ (kips)
Yielding in the gross section	92.7
Tensile rupture in the net section	103.5
Block shear rupture	107.6
Design strength	**92.7**

EXAMPLE 2

A truss member made of a double-channel 2C5 × 6.7 connected to a ½-in. gusset plate with eight ⅝-in. diameter A325-N bolts in standard holes, as shown in Fig. 8.9. The service loads are 50 kips dead load and 25 kips live load. Determine the adequacy of this member using LRFD.

2C5 × 6.7: $A_g = 3.34$ in.^{2}; $t_w = 0.190$ in.; $F_y = 36$ ksi; $F_u = 58$ ksi

Determine the Design Strength

Yielding in the gross section:

$$P_n = 36^{\text{ksi}} \times (2 \times 3.34^{\text{in.}^2}) = 240.5^{\text{kips}} \qquad \phi_t = 0.9$$

$$\phi_t P_n = 0.9 \times 240.5 = 216.4^{\text{kips}}$$

Tensile rupture in the net section:

$$A_n = A_g - \Delta A = (2 \times 3.34) - 4 \times 0.190 \times \left(\frac{5}{8} + \frac{1}{8}\right) = 6.11 \text{ in.}^2$$

$$U = 1 - \frac{\bar{x}}{l} = 1 - \frac{0.484}{6} = 0.92$$

$$A_e = UA_n = 0.92 \times 6.11 = 5.62 \text{ in.}^2$$

$$P_n = F_u A_e = 58^{\text{ksi}} \times 5.62^{\text{in.}^2} = 325.7^{\text{kips}} \qquad \phi_t = 0.75$$

$$\phi_t P_n = 0.75 \times 325.7 = 244.3^{\text{kips}}$$

Block shear rupture:

$$A_{gv} = 4 \times (0.19 \times 7) = 5.32 \text{ in.}^2$$

$$A_{nv} = 5.32 - 14 \times \left[0.19 \times \left(\frac{5}{8} + \frac{1}{8}\right)\right] = 3.325 \text{ in.}^2 \text{ (Fig. 8.10)}$$

$$A_{nt} = 2 \times [0.19 \times (2 - 0.75)] = 0.475 \text{ in.}^2 \text{ (Fig. 8.10)}$$

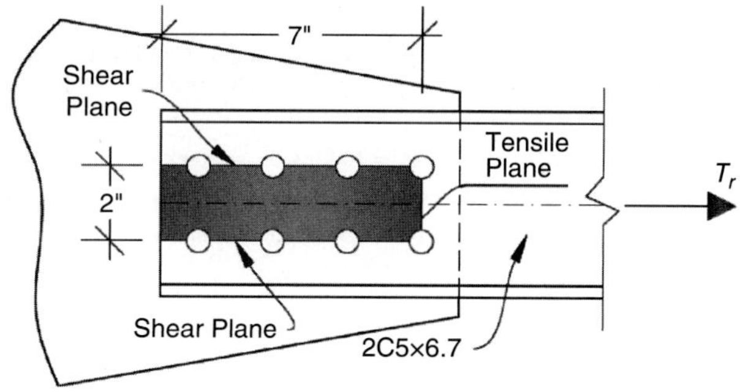

Figure 8.10 Block shear failure mode of the double channel.

$$P_n = \{0.6F_u A_{nv};\ 0.6F_y A_{gv}\}_{\min} + U_{bs} F_u A_{nt}$$

$$P_n = \{0.6 \times 58 \times 3.325;\ 0.6 \times 36 \times 5.32\}_{\min} + 1.0 \times 58 \times 0.475$$

$$P_n = \{115.7;\ 114.9\}_{\min} + 27.6 = 142.5^{\text{kips}} \qquad \phi_t = 0.75$$

$$\phi_t P_n = 0.75 \times 142.5 = 106.9^{\text{kips}}$$

Limit state	$\phi_t P_n$ (kips)
Yielding in the gross section	216.4
Tensile rupture in the net section	244.3
Block shear rupture	106.9
Design strength	**106.9**

Determine the Required Strength According to *ASCE 7, Section 2.3.1*, the basic load combinations that include dead and live loads are as follows:

(1) 1.4D
(2) 1.2D + 1.6L

where $P_D = 50$ kips and $P_L = 25$ kips.

$$Combination\#1 \rightarrow 1.4P_D = 1.4 \times 50 = 70^{\text{kips}}$$

$$Combination\#2 \rightarrow 1.2P_D + 1.6P_L = 1.2 \times 50 + 1.6 \times 25 = 100^{\text{kips}}$$

$$P_r = \{70^{\text{kips}};\ 100^{\text{kips}}\}_{\max} = 100^{\text{kips}}$$

Determine If the Member Is Adequate

$$\phi_t P_n = 106.9^{\text{kips}} > P_r = 100^{\text{kips}}$$

Thus, the tension member is adequate for LRFD.

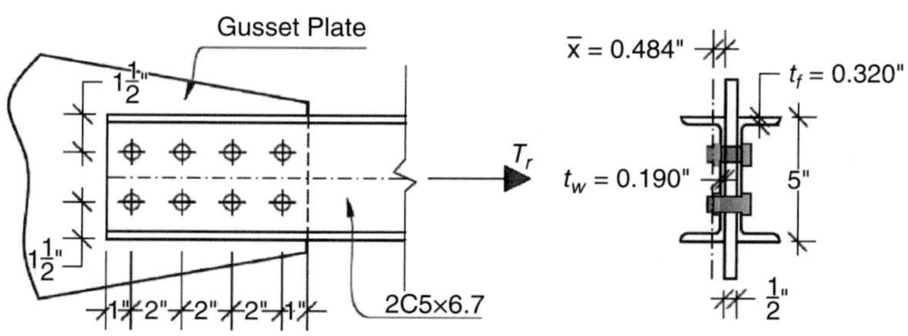

Figure 8.9 Example 2.

EXAMPLE 3

A fillet welded bracing member is subject to dead, live, and wind loads, as shown in Fig. 8.11. Select a square hollow structural section (HSS) using LRFD.

Given: $P_D = 20$ kips; $P_L = 10$ kips; $P_W = 200$ kips.

Determine the Required LRFD Strength According to *ASCE 7*, the basic load combinations that include dead, live, and wind loads are as follows:

$$(1)\ 1.4P_D = 1.4\times 20 = 28^{\text{kips}}$$
$$(2)\ 1.2P_D + 1.6P_L = 1.2\times 20 + 1.6\times 10 = 40^{\text{kips}}$$
$$(3)\ 1.2P_D + (P_L \text{ or } 0.5P_W) = 1.2\times 20 + 0.5\times 200 = 124^{\text{kips}}$$
$$(4)\ 1.2P_D + 1.0P_W + P_L = 1.2\times 20 + 1.0\times 200 + 10 = 234^{\text{kips}}$$
$$(5)\ 0.9P_D + 1.0P_W = 0.9\times 20 + 1.0\times 200 = 218^{\text{kips}}$$

$$P_r = 234^{\text{kips}}$$

Select a Trial Section

Based on yielding in the gross section $\phi_t P_n = \phi_t F_y A_g \ge P_r$

$$A_g \ge \frac{P_r}{\phi_t F_y} = \frac{234}{0.9\times 50} = 5.2 \rightarrow A_{g,\min} = 5.2 \text{ in.}^2$$

Based on fracture in the net section $\phi_t P_n = \phi_t F_u A_e \ge P_r$

$$A_e \ge \frac{P_r}{\phi_t F_u} = \frac{234}{0.75\times 62} = 5.03 \rightarrow A_{e,\min} = 5.03 \text{ in.}^2$$

Possible shapes are given in the following table.

Shape	Weight (lb/ft)	A_g (in.^{2})	$t_{d,\text{wall}}$ (in.)	A_n (in.^{2})*	A_e (in.^{2})**
HSS5 × 5 × 5/16	19.08	5.26	0.291	4.86	4.13
HSS5 × 5 × 3/8	22.37	6.18	0.349	5.70	4.85
HSS6 × 6 × 5/16	23.34	6.43	0.291	6.03	5.12
HSS6 × 6 × 3/8	27.48	7.58	0.349	7.10	6.04

*Deduction for the slotted portion is shown in Fig. 8.12.
**Effective area is assumed to be 85% of the net area.

Try HSS6 × 6 × 5/16 (ASTM A500 Gr. C)
HSS6 × 6 × 5/16: $A_g = 6.43$ in.^{2}; $t_{d,\text{wall}} = 0.291$ in.; $F_y = 50$ ksi; $F_u = 62$ ksi

Determine LRFD Strength

Yielding in the gross section:

$$P_n = 50^{\text{ksi}} \times 6.43^{\text{in.}^2} = 321.5^{\text{kips}} \qquad \phi_t = 0.9$$
$$\phi_t P_n = 0.9\times 321.5 = 289.4^{\text{kips}}$$

Tensile rupture in the net section:

$$A_n = A_g - \Delta A = 6.43 - 2\times\left(\frac{5}{8}+\frac{1}{16}\right)\times 0.291 = 6.03 \text{ in.}^2$$

(see Fig. 8.12, t_{gusset} = 5/8")

$$\bar{x} = \frac{B^2 + 2BH}{4(B+H)} = \frac{6^2 + 2\times(6\times 6)}{4\times(6+6)} = 2.25 \text{ in.}$$

$$l = 10 \text{ in.} > H = 6 \text{ in.} \rightarrow U = 1 - \frac{\bar{x}}{l} = 1 - \frac{2.25}{10} = 0.775$$

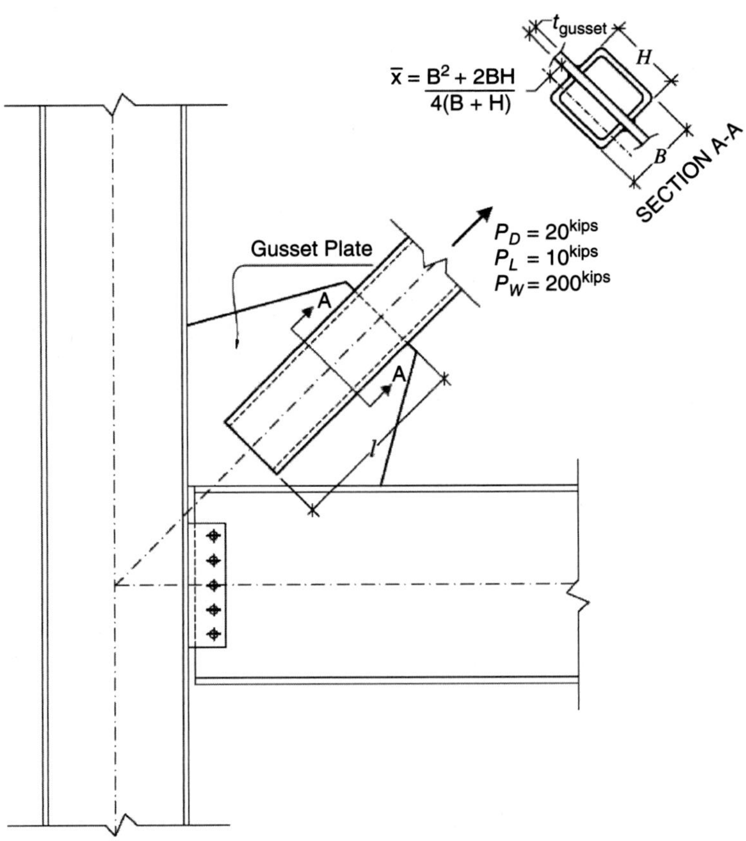

Figure 8.11 Example 3.

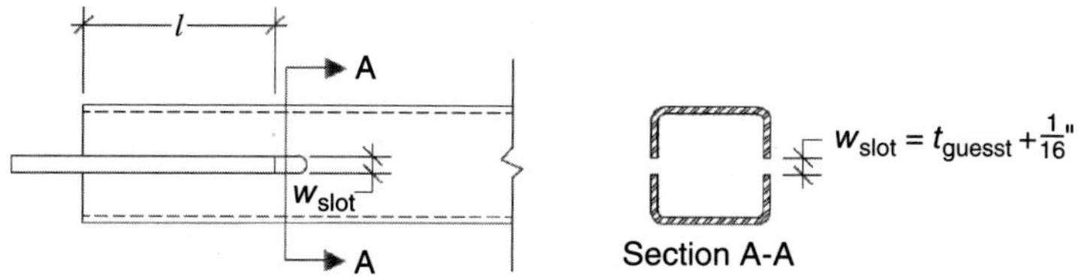

Figure 8.12 Net section at the slotted portion.

Note that the length of the longitudinal fillet welds is calculated based on t_{weld} = 3/16 in. ($F_{EXX} = 70^{ksi}$).

$$A_e = UA_n = 0.775 \times 6.03 = 4.67 \text{ in.}^2$$

$$P_n = F_u A_e = 62^{ksi} \times 4.67^{in.^2} = 289.7^{kips} \qquad \phi_t = 0.75$$

$$\phi_t P_n = 0.75 \times 289.7 = 217.3^{kips}$$

Block shear rupture:

$$A_{gv} = A_{nv} = 4 \times (l \times t_{d,wall}) = 4 \times (10 \times 0.291) = 11.64 \text{ in.}^2;$$

$$A_{nt} = 0 \text{ (see Fig. 8.13)}$$

$$P_n = \{0.6F_u A_{nv};\ 0.6F_y A_{gv}\}_{min} + U_{bs} F_u A_{nt} = 0.6F_y A_{gv}$$

$$P_n = 0.6 \times 50 \times 11.64 = 349.2^{kips} \qquad \phi_t = 0.75$$

$$\phi_t P_n = 0.75 \times 349.2 = 261.9^{kips}$$

Determine If the Member Is Adequate

$$\phi_t P_n = 217.3^{kips} < P_r = 234^{kips}$$

Thus, HSS6 × 6 × 5/16 is *not* adequate for LRFD.

EXAMPLE 4

A building canopy is supported by a pin-connected A36 steel rod, as shown in Fig. 8.14. Check the dimensional requirements and the adequacy of the pin connection for the given loading conditions.

Given: w_{Dead} = 1.2 kips/ft; w_{Snow} = 0.6 kips/ft; t_{plate} = ½" (A36)

Select a Tension Rod Diameter

(1) $1.4w_D = 1.4 \times 1.2 = 1.68^{kips/ft}$

(2) $1.2w_D + 1.6w_S = 1.2 \times 1.2 + 1.6 \times 0.6 = 2.4^{kips/ft}$

$$P_r = \frac{w \times L_{beam}}{\sin\theta} = \frac{2.4^{kips/ft} \times 6.25^{ft}}{\sin 33°} = 27.5^{kips}$$

$$\phi_t P_n = \phi_t F_y A_g \geq P_r$$

$$A_{g,req} \geq \frac{P_r}{\phi_t F_y} = \frac{27.5}{0.9 \times 36} = 0.85 \text{ in.}^2$$

Use ϕ7/8" A36 steel rod.

Check Dimensional Requirements According to *AISC 360, Section D5.2*, dimensional requirements are as follows:

1. $a \geq 1.33b_e$
2. $w \geq 2b_e + d$
3. $b_e = 2t + ⅝" \leq b$

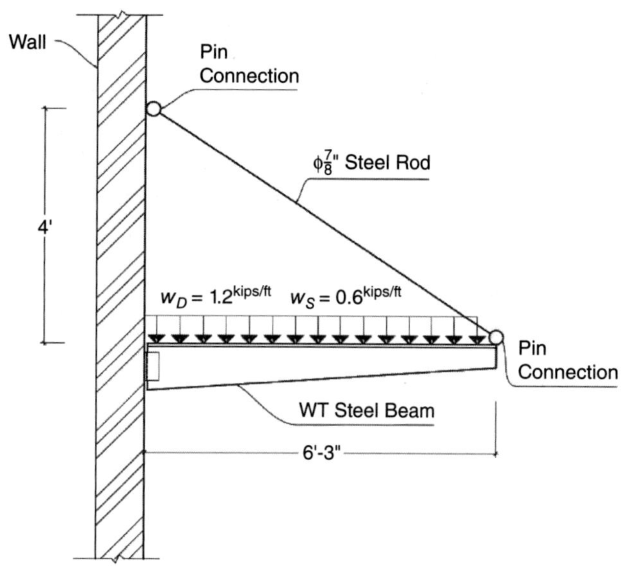

Figure 8.14 A building canopy.

4. The pinhole shall be located midway between the edges of the member in the direction normal to the applied force.

where a = shortest distance from edge of the pin hole to the edge of the member measured parallel to the direction of the force.

w = plate width perpendicular to rod centerline.

b_e = 2t + 0.63 in. but not more than the actual distance from the edge of the hole to the edge of the part measured in the direction normal to the applied force.

d = diameter of pin.

t = thickness of plate.

Dimensions of the pin connection shown in Fig. 8.15 satisfy the requirements summarized above.

$$b_e = 2t + 5/8" = 2 \times \frac{1}{2} + \frac{5}{8} = 1.625 \text{ in.}$$

$$a = 2.25" > 1.33b_e = 1.33 \times 1.625 = 2.16 \text{ in.}$$

$$b = 2.0" > b_e = 1.625 \text{ in.}$$

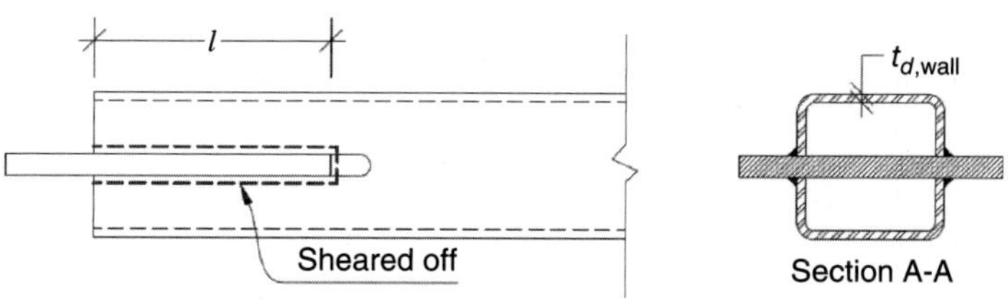

Figure 8.13 Block shear rupture.

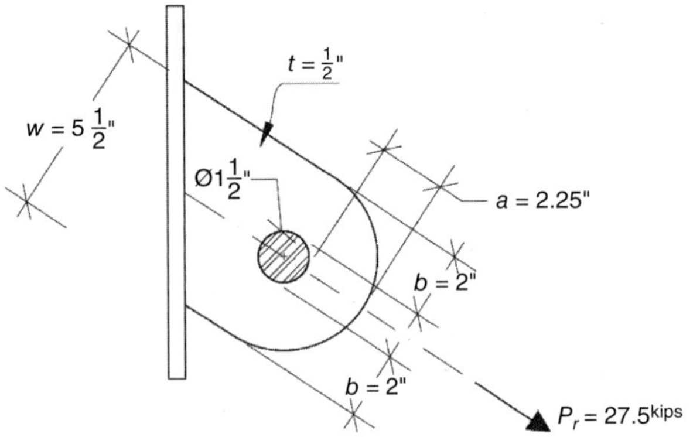

Figure 8.15 Dimensions of the pin connection.

Tensile rupture on the net effective area

$$P_n = F_u(2tb_e) \qquad \text{(Eq. D5-1, AISC 360)}$$

$$P_n = 58\times\left(2\times\frac{1}{2}\times 1.625\right) = 94.3^{\text{kips}} \qquad \phi_t = 0.75$$

$$\phi_t P_n = 0.75\times 94.3 = 70.7^{\text{kips}}$$

Shear rupture on the effective area

$$P_n = 0.6F_uA_{sf} = 0.6F_u 2t(a+d/2) \qquad \text{(Eq. D5-2, AISC 360)}$$

$$P_n = 0.6\times 58\times\left[2\times\frac{1}{2}\times(2.25+1.5/2)\right] = 104.4^{\text{kips}} \qquad \phi_t = 0.75$$

$$\phi_t P_n = 0.75\times 104.4 = 78.3^{\text{kips}}$$

Bearing strength

$$R_n = 1.8F_yA_{pb} \qquad \text{(Eq. J7-1, AISC 360)}$$

where A_{pb} = projected area in bearing.

$$R_n = 1.8\times 36\times\left(1.5\times\frac{1}{2}\right) = 48.6^{\text{kips}} \qquad \phi_t = 0.75$$

$$\phi_t R_n = 0.75\times 48.6 = 36.5^{\text{kips}}$$

Thus, bearing strength controls the design.

$$\phi_t P_n = 36.5^{\text{kips}} > P_r = 27.5^{\text{kips}}$$

8.1.3 Compression Members

The primary failure mode in a column member is global buckling. Compressive strength of a column is essentially dependent on overall slenderness ratio (KL/r) and width-to-thickness ratio of the elements of a column member along with the uncertainties such as, initial imperfection and residual stresses. In any steel column, an initial curvature and residual stresses would be introduced during the process of making the column. Due to the initial curvature and/or residual stresses before axial load is applied, the maximum load that can be applied is reduced from the buckling load. This is mainly because that the stress distribution is not uniform from the very beginning, and some areas might have reached yield stress ($\sigma = F_y$) earlier than others.

Composite Columns

Composite columns may be constructed with rolled or built-up steel shapes, encased in concrete or with concrete placed inside steel pipes or tubes (Fig. 8.16). A composite columns can carry significantly higher loads than that of the R/C columns of the same size. For a few decades, the encasing concrete was not considered in the calculations, but it is mainly used for fire and corrosion protection.

For low-rise buildings, the steel columns are often encased in concrete for having a good appearance or fire protection, corrosion protection, and from vehicles when used in garages. Considering the effect of concrete, smaller steel shapes might be used. For high-rise buildings, it is possible to have more floor space, since the sizes of composite columns are smaller than is required for R/C columns for the same loads.

Limitations and detailing requirements for composite columns:

The following limitation must be satisfied in order to qualify as an encased composite column subject to axial force:

a. The cross-sectional area of the steel core, A_s, should make up at least 1 percent of the total composite cross-sectional area, A_g.

b. For concrete encasement:

- Longitudinal bars has to be continuous at story levels,
- Lateral ties or spirals have also to be used. In cases where lateral ties are used, a minimum of No. 3 bar spaced at a maximum of 12 in. on center, or a No. 4 bar spaced at a maximum of 16 in. on center should be used.

c. Maximum spacing of lateral ties may not exceed 0.5 times the least lateral column dimension.

d. The minimum reinforcement ratio for continuous longitudinal reinforcement, $\rho_{sr} = A_{sr}/A_g$ is 0.004, where A_{sr} is area of continuous longitudinal reinforcement.

e. Clear spacing of at least 1.5 reinforcing bar diameter or 1.5 in. is required between the steel core and longitudinal reinforcement.

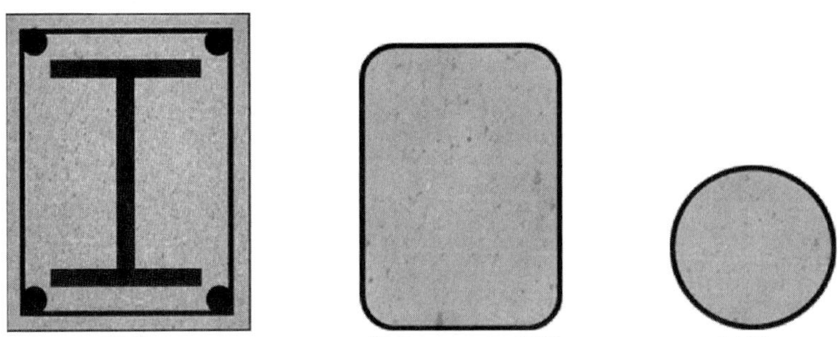

(a) W-Shape (b) Rectangular HSS (c) Pipe

Figure 8.16 Composite columns.

f. Minimum width-to-wall thickness ratio for pipes and rectangular HSS to be classified as a compact section:

- For pipes $D/t \le \dfrac{0.15E}{F_y}$
- For rectangular HSS $b/t \le 2.26\sqrt{\dfrac{E}{F_y}}$

Design strength of composite columns subject to axial force:

The design compressive strength, $\phi_c P_n$, of an encased composite column subject to axial force is the sum of the axial strengths of the steel section, the reinforcing bars, and the concrete and is defined as

a. When

$$\frac{P_{no}}{P_e} \le 2.25$$

$$P_n = P_{no}\left(0.658^{\frac{P_{no}}{P_e}}\right) \tag{8.2}$$

b. When $\dfrac{P_{no}}{P_e} > 2.25$

$$P_n = 0.877P_e \tag{8.3}$$

where A_c = area of concrete, in.^{2}
A_c = cross-sectional area of steel section, in.^{2}
E_c = elastic modulus of concrete, $= w_c^{1.5}\sqrt{f'_c}$, ksi
EI_{eff} = effective stiffness of composite section, kip-in.^{2}
$= E_sI_s + E_sI_{sr} + C_1E_cI_c$
E_s = elastic modulus of steel, 29,000 ksi
F_{ysr} = specified minimum yield stress of reinforcing bars, ksi
I_c = moment of inertia of the concrete section about the elastic neutral axis of the composite section, in.^{4}
I_s = moment of inertia of the steel shape about the elastic neutral axis of the composite section, in.^{4}
I_{sr} = moment of inertia of reinforcing bars about the elastic neutral axis of the composite section, in.^{4}
$L_c = KL$
P_n = axial strength of a composite compression member, kips
P_e = elastic critical buckling load, kips
$= \pi^2(EI_{\text{eff}})/L_c^2$
$P_{no} = F_yA_s + F_{ysr}A_{sr} + 0.85f'_cA_c$
$C_1 = 0.25 + 3\left(\dfrac{A_s + A_{sr}}{A_g}\right) \le 0.7$
f'_c = specified compressive strength of concrete, ksi
w_c = unit weight of concrete, lb/ft^3 ($90 < w_c \le 155$ lb/ft^3)
$\phi_c = 0.75$

EXAMPLE 5

Determine the effective length factor, K, for the first and second story columns in the moment frame shown in Fig. 8.17. Moment frame is braced at each story level in the perpendicular direction.

W12 × 58: $I_x = 475$ in.^{4}; W12 × 30: $I_x = 238$ in.^{4}; W16 × 40: $I_x = 518$ in.^{4}

Determine G_A, G_B, and G_C

$$G = \frac{\sum(E_cI_c/L_c)}{\sum(E_gI_g/L_g)} = \frac{\sum(I_c/L_c)}{\sum(I_g/L_g)}$$

$G_A = 1.0$ (Fixed support)

$$G_B = \frac{2\times(475/144)}{518/240} = 3.06$$

$$G_C = \frac{475/144}{238/240} = 3.33$$

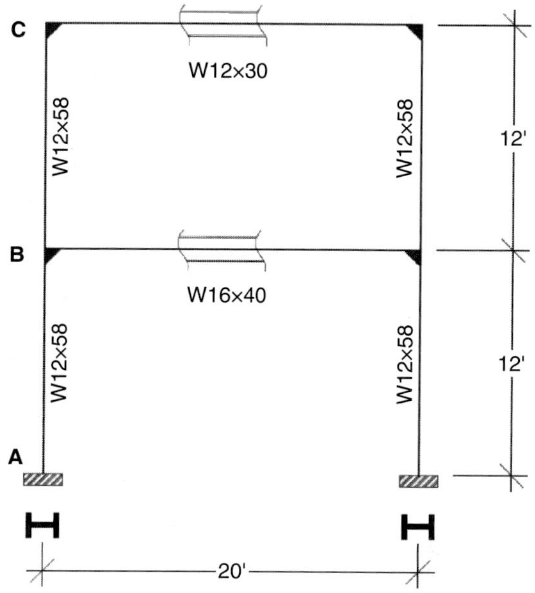

Figure 8.17 A two-story one-bay moment frame.

Enter the Alignment Chart for Sway Frames with G_A, G_B, and G_C Values

$$\text{Member } AB \rightarrow \left|\begin{matrix} G_A = 1.0 \\ G_B = 3.06 \end{matrix}\right| \rightarrow K_{x,AB} \approx 1.55 \text{ (Fig. 8.18)}$$

$$\text{Member } BC \rightarrow \left|\begin{matrix} G_B = 3.06 \\ G_C = 3.33 \end{matrix}\right| \rightarrow K_{x,BC} \approx 1.88 \text{ (Fig. 8.19)}$$

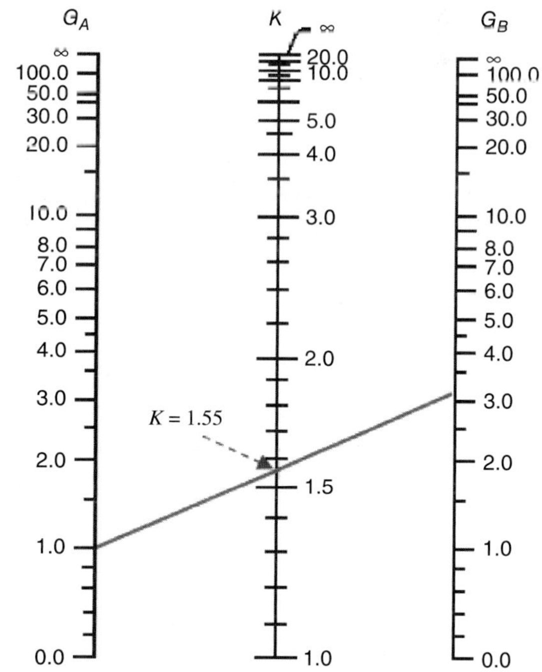

Figure 8.18 Effective length factor for member *AB*.

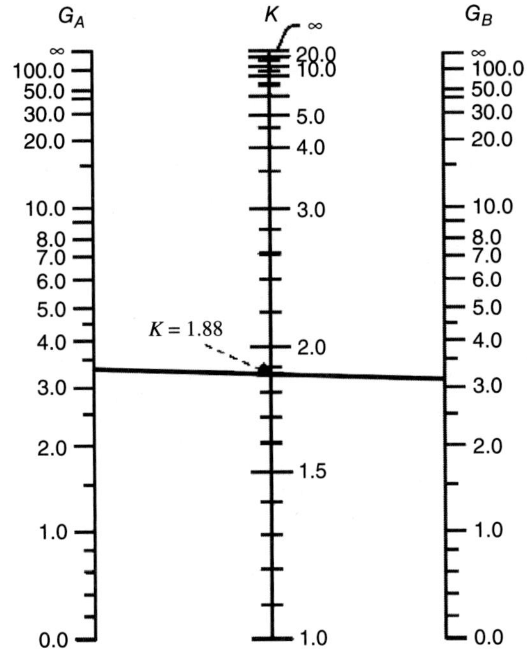

Figure 8.19 Effective length factor for member *BC*.

EXAMPLE 6

Determine the compressive design strength of column AB in Example 5.

W12 × 58: A_g = 17 in.²; r_x = 5.28 in.; r_y = 2.51 in.; F_y = 50 ksi

Check Width-to-Thickness Ratio according to Table B4.1a, AISC 360

$$\lambda_{r,f} = 0.56\sqrt{\frac{E}{F_y}} = 0.56\sqrt{\frac{29{,}000}{50}} = 13.5 > b/2t_f = 7.82$$

$$\lambda_{r,w} = 1.49\sqrt{\frac{E}{F_y}} = 1.49\sqrt{\frac{29{,}000}{50}} = 35.9 > h/t_w = 28.1$$

W12 × 58 is a non-slender compression member and the unbraced torsional and flexural buckling lengths are the same. Thus, the limit state of flexural buckling applies (*Section E3, AISC 360*).

$$P_n = F_{cr}A_g \qquad \textit{(Eq. E3-1, AISC 360)}$$

where,

$$F_{cr} = \left(0.658^{F_y/F_e}\right)F_y \text{ when } \frac{L_c}{r} \le 4.71\sqrt{\frac{E}{F_y}} \qquad \textit{(Eq. E3-2, AISC 360)}$$

$$F_{cr} = 0.877F_e \text{ when } \frac{L_c}{r} > 4.71\sqrt{\frac{E}{F_y}} \qquad \textit{(Eq. E3-3, AISC 360)}$$

Determine the Compressive Strength

$$K_x = 1.55;\ K_y = 1.0$$

$$\left(\frac{L_c}{r}\right)_x = \frac{1.55\times(12\times12)}{5.28} = 42.3 < \left(\frac{L_c}{r}\right)_y = \frac{1.0\times(12\times12)}{2.51} = 57.4$$

$$\frac{L_c}{r} = 57.4 < 4.71\sqrt{\frac{29{,}000}{50}} = 113.4$$

Thus, inelastic flexural buckling occurs about *y-y* axis.

$$F_e = \frac{\pi^2E}{(L_c/r)^2} = \frac{\pi^2\times29{,}000}{(57.4)^2} = 86.9^{\text{ksi}}$$

$$F_{cr} = (0.658^{F_y/F_e})F_y = (0.658^{50/86.9})\times50 = 39.3^{\text{ksi}}$$

$$P_n = F_{cr}A_g = 39.3\times17 = 668.1^{\text{kips}} \qquad \phi_c = 0.90$$

$$\phi_cP_n = 0.9\times668.1 = 601.3^{\text{kips}}$$

EXAMPLE 7

Select a double-angle for the top chord of the truss given in Fig. 8.20 using LRFD. The service load, $P_{\text{service}} = P = 40$ kips consists of 60 percent dead load and 40 percent snow load.

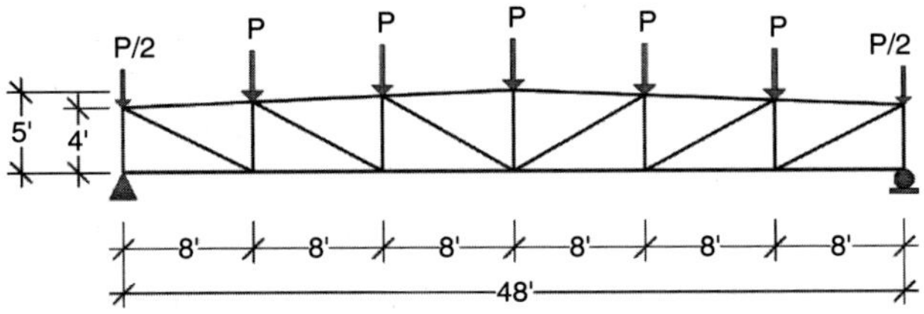

Figure 8.20 A roof truss of an industrial building.

Determine the Required Strength

$$P_{\text{LRFD}} = 1.2P_D + 1.6P_S = 1.2\times(0.6\times P_{\text{service}}) + 1.6\times(0.4\times P_{\text{service}})$$
$$= 1.2\times(0.6\times40) + 1.6\times(0.4\times40) = 54.4^{\text{kips}}$$

The maximum axial force obtained from the analysis of the truss is $P_{\max} = 7.21P$ in compression. Thus, the required strength, P_r can be computed as follows:

$$P_r = 7.21P_{\text{LRFD}} = 7.21\times54.4 = 392^{\text{kips}}$$

Select a Trial Section (Assume $F_{cr} = 0.75F_y$)

$$A_g \ge \frac{P_r}{\phi_cF_{cr}} = \frac{392}{0.9\times(0.75\times36)} = 16.1 \text{ in.}^2$$

Try 2L8 × 4 × ¾ Separated by ¾ in. Connectors (Two Angles Are Oriented with the Long Legs Back-to-Back, LLBB, see Fig. 8.21)

A_g = 17 in.²; *s* = ¾ in.; r_x = 2.55 in.; r_y = 1.69 in.; r_z = 0.85 in.; $\bar{r}_0$ = 3.99 in.; *H* = 0.587 in.; *J* = 3.22 in.⁴; F_y = 36 ksi

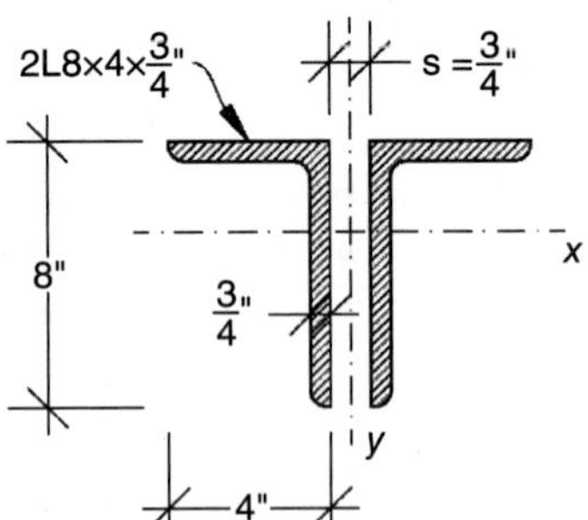

Figure 8.21 2L8 × 4 × ¾ separated by ¾ in.

Check Limiting Width-to-Thickness Ratio according to Table B4.1*a*, AISC 360

$$\lambda_r = 0.45\sqrt{\frac{E}{F_y}} = 0.45\sqrt{\frac{29{,}000}{36}} = 12.8 > b/t = \frac{8}{3/4} = 10.7$$

2L8 × 4 × ¾ is a non-slender compression member. Thus, the limit state of flexural buckling about x-axis or flexural-torsional buckling about y- and z-axes applies (*Sections E3 and E4, AISC 360*).

Compute Elastic Buckling Stress about x-x Axis (K_xL_x = 8 ft)

$$\left(\frac{L_c}{r}\right)_x = \frac{(8\times12)}{2.55} = 37.6 \rightarrow F_{ex} = \frac{\pi^2 E}{(L_c/r)_x^2} = \frac{\pi^2\times 29{,}000}{(37.6)^2} = 202.5^{\text{ksi}}$$

Compute Elastic Buckling Stress about y-y Axis (K_yL_y = 8 ft) The two angles are connected with two welded intermediate connector stitches (a = 32 in.). In order to prevent one of the individual angles from buckling between the connectors, *Section E6, AISC 360* requires the use of the modified slenderness ratio, $(L_c/r)_m$. In conformance with *Section E6.1-b, AISC 360*, the modified slenderness ratio is defined as the slenderness ratio of a single angle between the connectors.

$$\left(\frac{L_c}{r}\right)_m = \left(\frac{L_c}{r}\right)_0 \text{ when } \frac{a}{r_i} \le 40 \qquad (\textit{Eq. E6-2a, AISC 360})$$

$$\left(\frac{L_c}{r}\right)_m = \sqrt{\left(\frac{L_c}{r}\right)_0^2 + \left(\frac{K_i a}{r_i}\right)^2} \text{ when } \frac{a}{r_i} > 40 \qquad (\textit{Eq. E6-2b, AISC 360})$$

where K_i = 0.50 for angles back-to-back
= 0.75 for channels back-to-back
= 0.86 for all other cases
a = distance between connectors
r_i = minimum radius of gyration of individual component
$(L_c/r)_0$ = slenderness ratio of built-up member acting as a unit in the buckling direction being addressed

$$\frac{a}{r_i} = \frac{32}{0.85} = 37.6 < 40 \rightarrow \left(\frac{L_c}{r}\right)_m = \left(\frac{L_c}{r}\right)_0$$

$$\left(\frac{L_c}{r}\right)_0 = \left(\frac{L_c}{r}\right)_y = \frac{(8\times12)}{1.69} = 56.8 \rightarrow F_{ey} = \frac{\pi^2 E}{(L_c/r)_y^2} = \frac{\pi^2\times 29{,}000}{(56.8)^2}$$
$$= 88.7^{\text{ksi}}$$

To make sure that the compression member acts as a unit, AISC requires that the slenderness ratio of an individual component does not exceed ¾ times the governing slenderness ratio of the built-up member (*Section E6-2, AISC 360*).

$$\frac{a}{r_i} = \frac{32}{0.85} = 37.6 < \frac{3}{4}\times 56 = 42$$

Compute the Elastic Torsional and Flexural-Torsional Buckling Stress

$$F_{ez} = \left(\frac{\pi^2 EC_w}{L_{cz}^2} + GJ\right)\frac{1}{A_g\bar{r}_0^2} \qquad (\textit{Eq. E4-7, AISC 360})$$

where G = shear modulus of elasticity of steel
J = torsional constant
r_0 = polar radius of gyration about the shear center

Note that the term with C_w can be excluded from *Eq. E4-7, AISC 360* when double angles are used. Thus, the expression becomes

$$F_{ez} = \frac{GJ}{A_g\bar{r}_0^2} = \frac{11{,}200\times3.22}{17\times3.99^2} = 133.3^{\text{ksi}}$$

Per *Section E4, AISC 360*, for singly symmetric members twisting about the shear center where y is the axis of symmetry, the torsional or flexural-torsional elastic buckling stress, F_e, is determined as follows:

$$F_e = \left(\frac{F_{ey}+F_{ez}}{2H}\right)\left[1-\sqrt{1-\frac{4F_{ey}F_{ez}H}{(F_{ey}+F_{ez})^2}}\right] \qquad (\textit{Eq. E4-3, AISC 360})$$

$$F_e = \left(\frac{F_{ey}+F_{ez}}{2H}\right)\left[1-\sqrt{1-\frac{4F_{ey}F_{ez}H}{(F_{ey}+F_{ez})^2}}\right]$$
$$= \left(\frac{88.7+133.3}{2\times0.587}\right)\left[1-\sqrt{1-\frac{4\times88.7\times133.3\times0.587}{(88.7+133.3)^2}}\right] = 64.1^{\text{ksi}}$$

Since the elastic flexural and torsional buckling stresses, F_{ex}, F_{ey}, and F_{ez} are greater than their flexural-torsional buckling counterpart, the flexural-torsional buckling stress, F_e, controls. Therefore, the critical stress, F_{cr}, is determined as follows:

$$\frac{F_y}{F_e} = \frac{36}{64.1} = 0.56 < 2.25$$
$$F_{cr} = (0.658^{F_y/F_e})F_y = (0.658^{36/64.1})\times 36 = 28.5^{\text{ksi}}$$
$$P_n = F_{cr}A_g = 28.5\times17 = 483.8^{\text{kips}} \qquad \phi_c = 0.90$$
$$\phi_c P_n = 0.9\times483.8 = 435.4^{\text{kips}} > P_r = 392^{\text{kips}}$$

Use 2L8 × 4 × ¾ – ¾ in. LLBB with two welded intermediate connectors within the 8-foot length.

EXAMPLE 8

Select a pair of 15-in. channels made of A36 steel for the column shown in Fig. 8.22. The required LRFD strength, P_r, is 500 kips.

Select a Trial Section (Assume $F_{cr} = 0.80F_y$)

$$A_{g,\text{required}} = \frac{P_r}{\phi_c F_{cr}} = \frac{500}{0.9\times(0.8\times36)} = 19.3 \text{ in.}^2$$

Try Two Channels C15 × 33.9 (A_g = 10 in.²; I_x = 315 in.⁴; I_y = 8.07 in.⁴; r_x = 5.61 in.; r_y = 0.901 in.; $\bar{x}$ = 0.788 in.; g = 2 in.)

$$I_x = 2\times315 = 630 \text{ in.}^4;\ r_x = 5.61 \text{ in.}$$
$$I_y = 2\times[8.07+10\times(7.5-0.788)^2]$$
$$= 917.2 \text{ in.}^4 \rightarrow r_y = \sqrt{\frac{I_y}{A}} = \sqrt{\frac{917.2}{2\times10}} = 6.77 \text{ in.}$$

Check Limiting Width-to-Thickness Ratio according to Table B4.1a, AISC 360

$$\lambda_{r,f} = 0.56\sqrt{\frac{E}{F_y}} = 0.56\sqrt{\frac{29{,}000}{36}} = 15.9 > b/t_f = 5.23$$

$$\lambda_{r,w} = 1.49\sqrt{\frac{E}{F_y}} = 1.49\sqrt{\frac{29{,}000}{36}} = 42.3 > h/t_w = 31.8$$

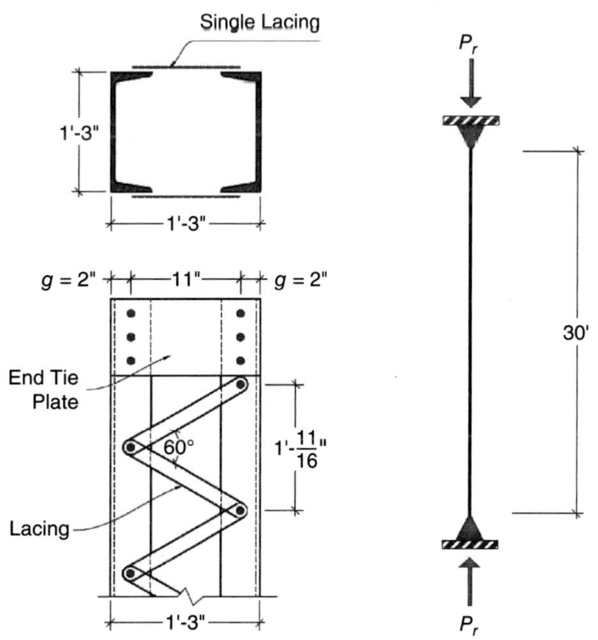

Figure 8.22 Built-up column with end ties and single lacing.

Determine LRFD Design Strength The torsional unbraced length does not exceed the flexural buckling lengths about *x-x* and *y-y* axes. Thus, flexural buckling stress controls the compressive strength of the built-up section.

$$K_x = K_y = 1.0$$

$$\frac{L_c}{r} = \left(\frac{L_c}{r}\right)_x = \frac{1.0\times(30\times12)}{5.61} = 64.2$$

$$F_e = \frac{\pi^2 E}{(L_c/r)^2} = \frac{\pi^2\times29{,}000}{(64.2)^2} = 69.5^{\text{ksi}}$$

$$F_{cr} = (0.658^{F_y/F_e})F_y = (0.658^{36/69.5})\times36 = 29^{\text{ksi}}$$

$$\phi_c P_n = 0.9\times29\times(2\times10) = 522^{\text{kips}} > P_r = 500^{\text{kips}}$$

Use 2C15 × 33.9.

Dimensional Requirements for Lacing and End Tie Plates *Section E6-2, AISC 360* states that laced built-up compression members shall meet the following requirements and recommendations:

- AISC suggests that single lacing bars are preferably inclined at a 60-degree angle or more with respect to the axis of the member (Fig. 8.22).
- Lacing can be selected among flat bars or any hot-rolled shapes (e.g., angles or channels). However, AISC recommends that it is ideal that the lacing consists of double flat bars or angles when the distance between the fasteners is greater than 15 in. Since the distance (1'-11/16") is less than 15 in., single lacing made of flat bars can be used.
- Per *Section E6-2, AISC 360*, it is required that the slenderness ratio of each channel between the fasteners should be less than 75 percent of the governing slenderness ratio of the built-up member so that each part will not buckle before the compression member individually.

$$\frac{a}{r_i} = \frac{a}{r_y} = \frac{12.688}{0.901} = 14.1 < \frac{3}{4}\times64.2 = 48.2$$

- Force on lacing bar should be determined based on a shearing strength normal to the axis of the member equal to 2 percent of the available compressive strength of the member.

$$P_c = \phi_c P_n = 522^{\text{kips}} \rightarrow V_{u,\text{lacing}} = 0.02\times P_c = 0.02\times522 = 10.44^{\text{kips}}$$

Note that $V_{u,\text{lacing}}$ is resisted by the two plates on both sides. Therefore, the axial force acting on each plate can be computed as:

$$F_{\text{plate}} = \pm\frac{V_{u,\text{lacing}}}{2\cos30°} = \pm\frac{10.44^{\text{kips}}}{2\times\cos30°} = \pm6.03^{\text{kips}}$$

Positive and negative signs represent tension and compression, respectively. It should be noted that the size of the lacing plate should be determined based on the 6.03 kips compressive force acting on the bar while the fasteners are designed using the tension force of 6.03 kips.

- The slenderness ratio of single lacing bars should be less than 140. Thus, the minimum plate thickness can be

$$\frac{L}{r_{\text{plate}}} \le 140 \rightarrow r_{\text{plate}} \ge \frac{12.688}{140} = 0.091 \text{ in.}$$

$$r_{\text{plate}} = \sqrt{\frac{bt^3/12}{bt}} = 0.289t \ge 0.091 \text{ in.}$$

$$t_{\min} = 0.315 \text{ in.}$$

Use ⅜-in. flat bar.

To determine the design strength of the lacing plate,

$$\frac{L}{r_{\text{plate}}} = \frac{12.688}{0.289\times3/8} = 117$$

$$F_e = \frac{\pi^2 E}{(L_c/r)^2} = \frac{\pi^2\times29{,}000}{(117)^2} = 20.9^{\text{ksi}}$$

$$F_{cr} = 0.877F_e = 0.877\times20.9 = 18.3^{\text{ksi}}$$

$$A_{g,\text{required}} = \frac{F_{\text{plate}}}{\phi_c F_{cr}} = \frac{6.03}{0.9\times18.3} = 0.37 \text{ in.}^2 \rightarrow b_{\text{plate}} = 1 \text{ in.}$$

Use Pl. 1" × ⅜ in. for lacing.

- ss of the end tie plates should not be less than 0.02 times the distance between the fasteners.

Min. tie plate thickness $= 0.02\times(15-2\times2) = 0.22$ in.

Use ¼-in. thick plate for end tie plates.

The other dimensions of the lacing and end tie plates can be determined by using the required spacing between the fasteners.

EXAMPLE 9

A composite column shown in Fig. 8.23 subject to axial load consists of a W12 × 87 encased in a 20 in. × 20 in. concrete column. Eight #7 bars are used for longitudinal reinforcement and #3 lateral ties spaced at 12 in. centerline to centerline are used for lateral reinforcement. F_y = 50 ksi for steel shape and Grade 60 reinforcing bars are used. Normal weight concrete with f_c' = 6 ksi is used. Compute the design strength for an effective length of 18 ft and 9 ft about *x*- and *y*-axes, respectively.

$$A_s = 25.6 \text{ in.}^2 \ge 0.01A_g = 0.01\times20\times20 = 4 \text{ in.}^2 \text{ OK.}$$

$$F_y = 50 \text{ ksi} < F_y = 75 \text{ ksi}$$

$$F_{ysr} = 55 \text{ ksi} < F_y = 80 \text{ ksi}$$

$$E_s = 29{,}000 \text{ ksi}$$

$$E_c = w_c^{1.5}\sqrt{f_c'} = 145^{1.5}\sqrt{6} = 4277 \text{ ksi}$$

$$3 \text{ ksi} \le f_c' = 6 \text{ ksi} \le 10 \text{ ksi}$$

For lateral ties #3 spaced at 12 in. OK.

$$\rho_{sr} = A_{sr}/A_g = \left(8\times\frac{\pi0.875^2}{4}\right)/(20\times20) = 0.012 \ge 0.004 \text{ OK.}$$

$$A_c = 20\times20 - A_s - A_{sr}$$

$$= 400 - 25.6 - \left(8\times\frac{\pi0.875^2}{4}\right)$$

$$= 400 - 25.6 - 4.81$$

$$= 369.59 \text{ in.}^2$$

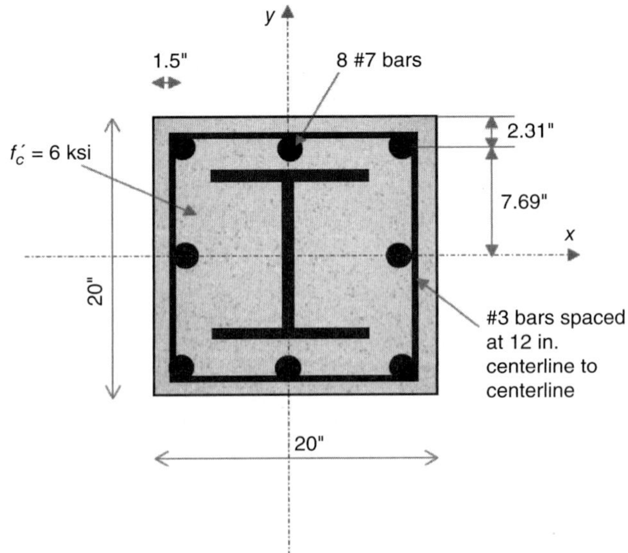

Figure 8.23 Example 9.

$$P_{no} = F_y A_s + F_{ysr} A_{sr} + 0.85 f_c' A_c$$

$$= 50 \times 25.6 + 55 \times \left(8 \times \frac{\pi 0.875^2}{4}\right) + 0.85 \times 6 \times 369.59$$

$$= 3429.5 \text{ kips}$$

$$C_1 = 0.25 + 3\left(\frac{A_s + A_{sr}}{A_g}\right) \le 0.7$$

$$C_1 = 0.25 + 3\left(\frac{25.6 + 4.81}{20 \times 20}\right) = 0.478 \le 0.7$$

Effective Stiffness about x-Axes

$$EI_{\text{eff},x} = E_s I_{s,x} + E_s I_{sr,x} + C_1 E_c I_{c,x}$$

$$I_{s,x} = 740 \text{ in.}^4$$

$$I_{sr,x} = \left(8 \times \frac{\pi 0.875^4}{64}\right) + 6 \times \left(\frac{\pi 0.875^2}{4}\right) \times 7.69^2 = 213.6 \text{ in.}^4$$

$$I_{c,x} = \left(\frac{20 \times 20^3}{12}\right) - 740 - 213.6$$

$$= 12{,}379.7 \text{ in.}^4$$

$$EI_{\text{eff},x} = 29{,}000 \times 740 + 29{,}000 \times 213.6 + 0.478 \times 4277 \times 12{,}379.7$$

$$= 52{,}963{,}533 \text{ kip-in.}^2$$

Effective Stiffness about y-Axes

$$EI_{\text{eff},y} = E_s I_{s,y} + E_s I_{sr,y} + C_1 E_c I_{c,y}$$

$$I_{s,y} = 241 \text{ in.}^4$$

$$I_{sr,y} = \left(8 \times \frac{\pi 0.875^4}{64}\right) + 6 \times \left(\frac{\pi 0.875^2}{4}\right) \times 7.69^2 = 213.6 \text{ in.}^4$$

$$I_{c,y} = \left(\frac{20 \times 20^3}{12}\right) - 241 - 213.6$$

$$= 12{,}878.7 \text{ in.}^4$$

$$EI_{\text{eff},y} = 29{,}000 \times 241 + 29{,}000 \times 213.6 + 0.478 \times 4277 \times 12{,}878.7$$

$$= 39{,}512{,}692 \text{ kip-in.}^2$$

$$P_{e,x} = \pi^2 (EI_{\text{eff},x})/L_{c,x}^2 = \pi^2(52{,}963{,}533)/(1.0 \times 18 \times 12)^2$$

$$= 11{,}204 \text{ kips} \rightarrow \text{buckling about x-axis controls}$$

$$P_{e,y} = \pi^2 (EI_{\text{eff},y})/L_{c,y}^2 = \pi^2(39{,}512{,}692)/(1.0 \times 9 \times 12)^2 = 33{,}434 \text{ kips}$$

$$\frac{P_{no}}{P_e} = \frac{3429.5}{11{,}204} = 0.306 \le 2.25$$

$$P_n = P_{no}\left(0.658^{\frac{P_{no}}{P_e}}\right) = 3{,}429.5\left(0.658^{\frac{3{,}429.5}{11{,}204}}\right) = 3{,}017.1 \text{ kips}$$

$$\phi_c P_n = 0.75 \times 3427.5 = 2262.8 \text{ kips}$$

8.1.4 Beams

Non-Composite Beams

Beam is a member that is primarily subjected to transverse loading, often gravity (dead and live) load effects. Thus primary internal actions at its cross section are bending moment, M, kip-ft, and shear force, V, kip. Most commonly used sections for beams: I-shaped wide flange sections, often called W-shapes. I-shaped members are more efficient for bending resistance because its large percentage of areas of steel is located far away from strong axis subjected to bending moment. Such efficiency might be compromised if flanges and/or web are made too slender, or the beam is over a long span without lateral supports, due to the fact that local and/or lateral torsional buckling limit states might occur before available plastic moment strength is reached. It is also noteworthy that steel sections for beams are usually made of open shapes, consisting of plates (such as web and flanges) in the rolled shapes, or welded together in built-up sections. This type of shape is particularly efficient for bending about its strong axis. However, it is extremely vulnerable to torsion.

Table 8.2 Local Buckling Limit State-Flexural Strength Relationship

Classification of compression element	Limiting width-to-thickness ratio (Table B4.1b)	Failure mode	Nominal flexural capacity
Compact	$\lambda \le \lambda_p$	Plastic hinge formation	$M_n = M_p$
Non-compact	$\lambda_p < \lambda \le \lambda_r$	Inelastic local buckling	$M_y < M_n < M_p$
Slender	$\lambda_r < \lambda$	Elastic local buckling	$M_n < M_y$

λ_p = Limiting width-thickness ratio for "Compact/Non-compact section" λ_r = Limiting width-thickness ratio for "Non-compact/Slender section."

Possible limit states in pure bending are plastic hinge formation, local buckling (LB) of flange and/or web under compressive stresses induced by bending deformation, and lateral torsional buckling (LTB). Plastic hinge formation, which occurs when all other buckling modes are prevented takes place when the entire cross-section yields, either in tensile or compressive stresses. Therefore, flexural capacity of a beam is controlled by the buckling modes. The interaction between the limit state of local buckling and flexural capacity of a beam is summarized in Table 8.2.

LTB is affected by lateral bracing condition of the beam, as well as beam's section properties. The variation of the flexural strength with the increasing unbraced length is as follows:

$$L_b \le L_p \Rightarrow M_n = M_p \text{ (Plastic Hinge Formation. The limit state of LTB does not apply.)}$$

$$L_p < L_b \le L_r \Rightarrow M_n = C_b\left[M_p - (M_p - 0.7F_y S_x)\left(\frac{L_b - L_p}{L_r - L_p}\right)\right] \le M_p = Z_x F_y \text{ (Inelastic LTB)} \rightarrow \text{AISC 360-16, F2-2}$$

$$L_r < L_b \Rightarrow M_n = F_{cr} S_x \le M_p \text{ (Elastic LTB)} \rightarrow \text{AISC 360-16, F2-3}$$

where L_b = laterally unbraced length
L_p = limiting laterally unbraced length to prevent LTB
L_r = limiting laterally unbraced length to prevent elastic LTB

It's apparent that the smaller laterally unbraced length, L_b, the less likely LTB occurs. Table 8.3 summarizes the relation between laterally unbraced length, L_b, and failure mode, together with moment capacity.

Table 8.3 Lateral-Torsional Buckling Limit State-Flexural Strength Relationship

Laterally unbraced length	Failure mode	Nominal flexural capacity
$L_b \le L_p$	Plastic hinge formation	$M_n = M_p$
$L_p < L_b \le L_r$	Inelastic LTB	$M_y < M_n < M_p$
$L_r < L_b$	Elastic LTB	$M_n < M_y$

Composite Beams

A composite beam is composed of steel beam and reinforced concrete. Composite beams provide greater economy and increase functionality. Ribs of the metal deck can either be perpendicular or parallel to beams (Fig. 8.24). In a typical floor framing system, the ribs would be perpendicular to floor beams and parallel to the supporting girders. This way,

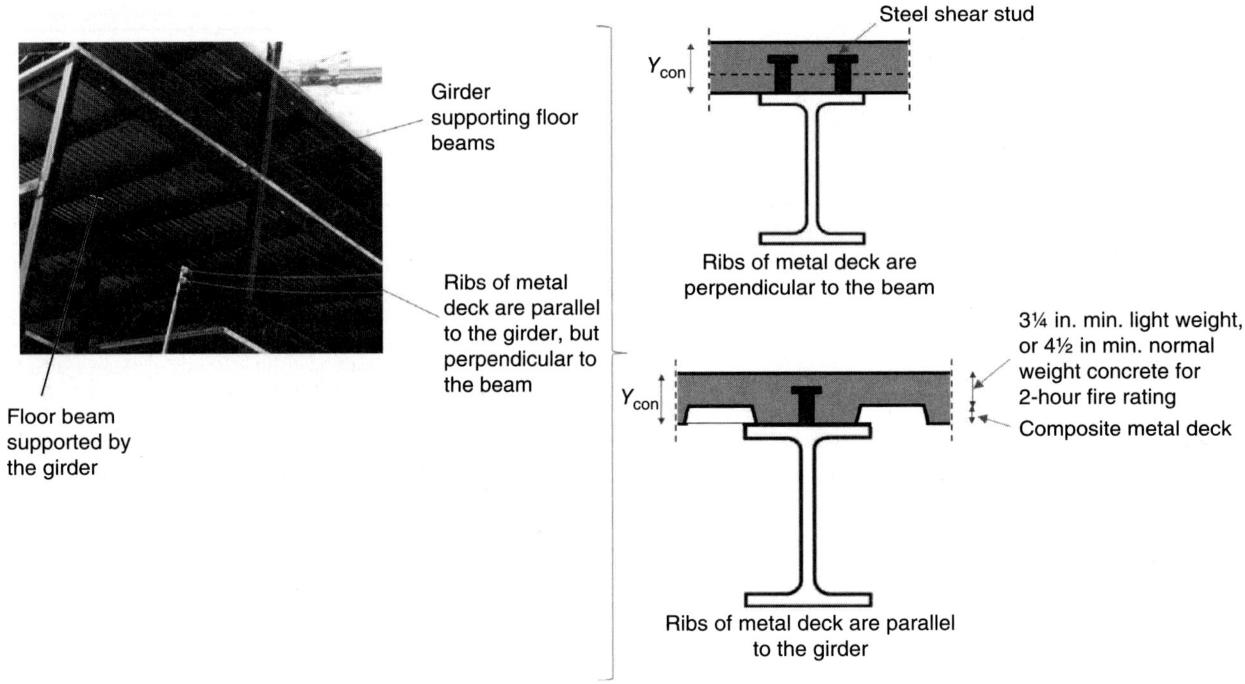

Figure 8.24 A typical steel building with composite floor.

the beam is considered to be laterally braced by the floor deck system continuously, and the girder is assumed to be laterally braced by the beams it supports at certain intervals. In order to have concrete slab and steel beam work together to resist bending under gravity loads, the slippage between them should be prevented by shear connectors (i.e., shear studs) (Fig. 8.24).

Plastic Strength of Fully and Partially Composite Beams for the Limit State of Yielding The nominal bending (flexural) strength of a composite beam, M_n, will be reached at a limit state (failure mode) when the entire cross-section yields in either tension or compression and concrete crushes in compression. The corresponding stress distribution on the composite beam section at this limit state is called plastic stress distribution (Fig. 8.25).

where A_s = cross-section area of the steel beam
C = compression in the concrete, kips
F_y = minimum yield stress of the steel, ksi
L = span length
M_n = nominal bending (flexural) strength, kip-ft
ϕM_n = design bending strength, kip-ft
PNA = plastic neutral axis
T = tension in the steel, kips
Y_2 = the distance between the center of compression area and the top of beam flange
a = the depth of compressive stress distribution in the concrete slab
$= \dfrac{F_y A_s}{0.85 f_c' b_{\text{eff}}}$ for PNA in concrete slab
b_1, b_2 = half span lengths on both sides of the beam
b_f = flange width of the steel shape
b_{eff} = effective width of the concrete slab ($\min\{L/4; (b_1 + b_2)/2\}$)
e = distance between the centers of compression and tension areas (or moment arm of internal resisting couple C and T)
$= d/2 + Y_2$
f_c' = 28-day compressive strength of the concrete, ksi
$3 \text{ ksi} \le f_c' \le 10$ ksi for normal weight concrete
$3 \text{ ksi} \le f_c' \le 6$ ksi for light weight concrete
t_c = slab depth (for deck ribs perpendicular to beam or girder, concrete below top of the deck is ignored; for deck ribs parallel to beam or girder, average thickness, of concrete and deck can be used)
$\phi = 0.9$

Note that PNA is determined by the fact the total compressive force acting in the compression area (above PNA) should be equal to the total tension force acting in the tension area (below PNA), that is, $C = T$.

In many cases, the flexural strength of a fully composite beam is significantly higher than the required strength, and a certain size of steel beam has to be used for construction strength and stiffness during an unshored construction. For economic purpose, a partially composite beam, which has fewer number of shear connectors than needed to prevent slip, might be a good alternative. Reduction in the number of the shear connectors would reduce flexural strength of the composite beam to a certain degree, but has potential to lower the cost due to installation of fewer shear connectors. There is significant slip between concrete and steel beam in a partially composite beam (Fig. 8.26).

Limit state for a fully composite beam is that the concrete crushes under compression and the steel beam yields under tension with sufficient number of shear connectors preventing the slip between the concrete and steel beam. Composite section can reach its maximum flexural strength with the concrete crushed in compression and the steel beam yielded in tension. For a partially composite beam, limit state is that the shear connectors fail before the concrete crushes under compression, the stress distribution over the concrete and steel beam might vary depending on the number of shear connectors, and the steel beam yielding, with no slip between the concrete and steel beam. In a partially composite beam, the number of shear connectors is less than sufficient, the shear connectors would fail before the full flexural strength is reached, and the slip between the concrete slab and steel beam would occur.

Elastic Behavior and Moment of Inertia of Composite Beam for Deflections Calculation Deflection is calculated based on elastic behavior of the composite beam. For a composite beam section with the steel beam and concrete slab being connected by connectors to

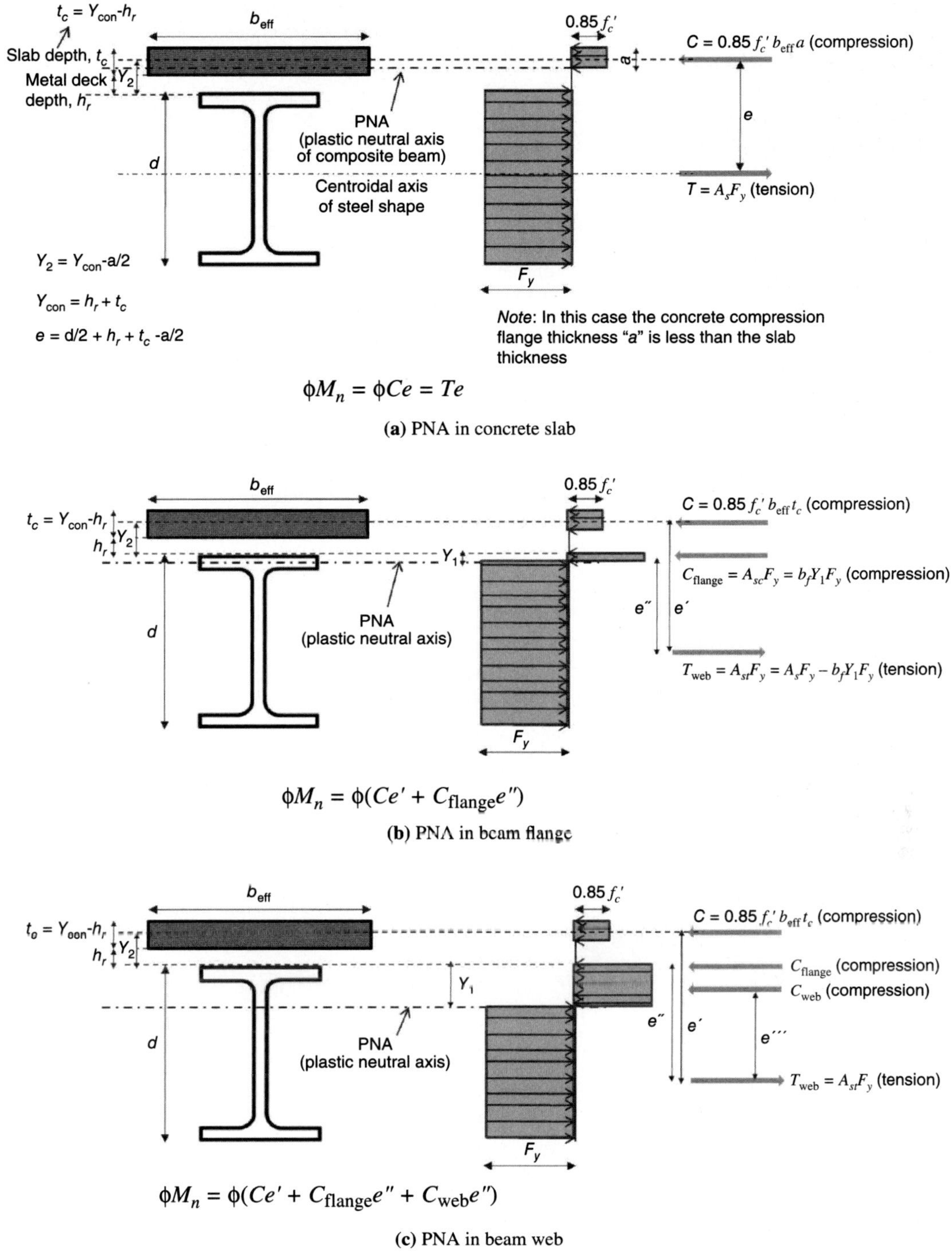

Figure 8.25 Stress distribution in a composite beam.

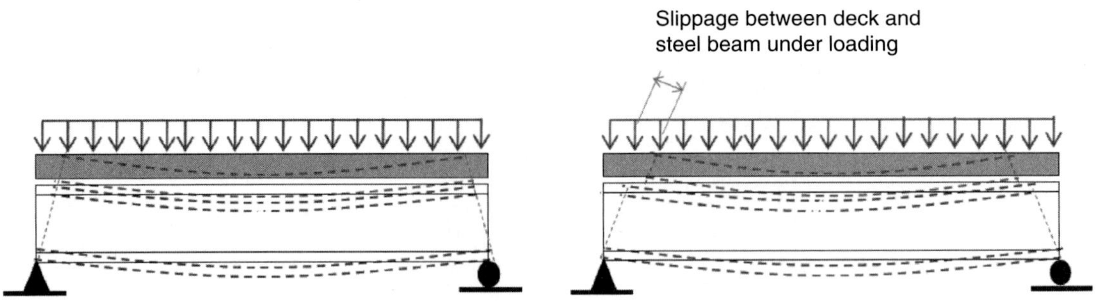

(a) Fully composite beam

(b) Partially composite beam

Figure 8.26 Fully and partial composite beams.

prevent slippage between them, the section should remain plane after bending with a linear strain distribution (Fig. 8.27). The elastic stress in concrete (ε_c) and steel (ε_s) at a location y above and below the neutral axis are equal to:

$$\varepsilon_s = \varepsilon_c \tag{8.4}$$

$$\sigma_s = E_s\varepsilon_s \text{ and } \sigma_c = E_c\varepsilon_c \tag{8.5}$$

where σ_s = elastic stress in steel
E_s = elastic modulus of steel, 29,000 ksi
σ_c = elastic stress in concrete
E_c = elastic modulus of concrete = $w_c^{1.5}\sqrt{f_c'}$, ksi
w_c = unit weight of concrete (lb/ft³)
f_c' = 28-day compressive strength of concrete, ksi

Since E_s is far larger than E_c, $\sigma_s >> \sigma_c$ (Fig. 8.27*d*), even though the strains are the same. In other words, the elastic stress distribution is no longer linear, that is, cannot be represented with a single straight line), but nonlinear, that is, can be represented by two lines. This is because there are two different material properties, E_S and E_C. To have an equivalent linear stress distribution in order to calculate beam deflection, we can replace concrete by "steel," temporarily. If we make the strain in the concrete at any point be equal to the strain in any replaced steel at that point:

$$\varepsilon_s = \varepsilon_c \quad \text{or} \quad \frac{f_c}{E_c} = \frac{f_s}{E_s} \tag{8.6}$$

$$f_s = \frac{E_s}{E_c} f_c = nf_c \tag{8.7}$$

where f_c = stress in concrete, ksi
f_s = stress in steel, ksi
$n = \frac{E_s}{E_c}$ = modular ratio (is about 8 to 10)

Equation 8.5 indicates "n" in.² area of concrete are needed to resist the same force as 1.0 in.² area of steel. So, if we divide the concrete area by "n," and the resulting area is called transformed area from concrete to equivalent "steel" area (Figs. 8.27*a* and *b*). The most convenient way to reduce the concrete area by "n times" is to reduce effective width of the concrete slab to b_{eff}/n. The stress distribution becomes linear in the transformed section (Fig. 8.27*e*). In other words, the transformed section (Fig. 8.27*b*) is a homogeneous steel section for elastic behavior.

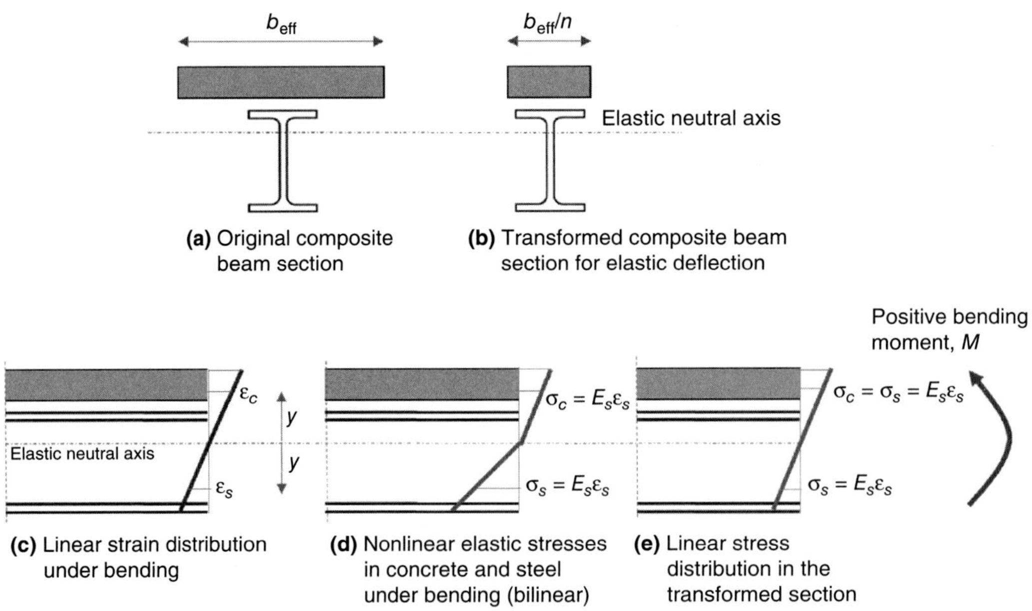

Figure 8.27 Elastic stress distributions in original composite and transformed sections.

Elastic stresses in steel and concrete can be determined based on the transformed section as follows:

a. Locate the elastic neutral axis (the same as centroidal axis for a homogeneous section) and moment of inertia of the transformed section (I_{tr}).

b. Find the elastic stresses at the top and bottom of the steel beam.

$$f_{st} = \frac{My_t}{I_{tr}} \quad (8.8)$$

$$f_{sb} = \frac{My_b}{I_{tr}} \quad (8.9)$$

c. Find the elastic stresses in the concrete at the top of slab.

$$f_c = \frac{My_c}{nI_{tr}} \quad (8.10)$$

where M = applied moment
I_{tr} = moment of inertia of the transformed section
y_t = distance from the neutral axis to the top of the steel section
y_b = distance from the neutral axis to the bottom of the steel section
y_c = distance from the neutral axis to the top of the concrete slab

Shear Connectors

Shear connectors are needed to prevent slippage between the concrete slab and steel beam (Fig. 8.28). The horizontal shear, created by compressive force C (Fig. 8.28) acting on the concrete at any cross section at the interface between the steel beam and the concrete slab, needs to be transferred by shear connectors. Due to the fact that the bending moment changes from section to section, and shear force transferred by shear connectors varies through entire beam. For a simple design, the shear connector design has been based on the design models in Figs. 8.28*b* and *c*.

We will consider a segment of composite beam between the point of maximum positive moment and the point of zero moments, segment AB (Fig. 8.28*a*). It is assumed that the total horizontal shear force, V', between the concrete slab and steel beam will be uniformly distributed among equally spaced shear connectors between A and B. This total horizontal shear force V' is determined based on three limit states: concrete crushing, tensile yielding of steel section, or strength of shear connectors. In other words,

$$V' = \text{The smallest of} \begin{cases} C \\ T \\ \Sigma Q_n \end{cases} \quad (8.11)$$

where $C = 0.85 f'_c A_c$ crushing of the concrete
$T = F_y A_s$ · tensile yielding of the steel beam
A_c = area of concrete slab within effective width times the smaller of "a" and thickness of concrete slab, in.^{2}
A_s = area of steel cross section, in.^{2}
ΣQ_n = sum of nominal strengths of shear connectors between the point of the maximum positive moment and the point of zero moment, kips (given by I3-3 and I3-4 for stud and channel shear connectors, respectively in AISC 360-16, based on experimental data)

In a partially composite beam, $\Sigma Q_n < A_s F_y$, and stress distribution in the composite section is unknown. However, equivalently, for a given ΣQ_n defined by the total number of shear connectors over a length of beam between $M = 0$ and $M = M_{max}$, AISC Table 3.19 in the 14th AISC Manual can also be used in this case ($Y_1 > 0$):

$$\sum F_x = 0, \text{ i.e., } \sum Q_n + A_{\bar{s}} F_y = (A_s - A_{\bar{s}}) F_y \quad (8.12)$$

where A_s = area of the steel beam
$A_{\bar{s}}$ = the portion of steel area that is needed to satisfy equilibrium in beam length along x direction

Detail limitations for shear studs are given in Fig. 8.29.

EXAMPLE 10

Figure 8.30 shows the typical plan of a four-story steel office building. Lateral-force resisting frames are arranged on the perimeter. On floors, dead load is estimated as 85 psf, and live load is 50 psf for floors. Determine the uniformly distributed factored load on beam B1. Select a W-shape for the floor secondary beam, B1 using LRFD. A992 steel is used.

Determine the Factored Load on the Beam

$$w_D = 85^{psf} \times 10^{ft} / 1000^{lb/kips} = 0.85 \text{ kip/ft}$$

According to *Section 4.7, ASCE 7*, floor live load, L_0, can be reduced if the product of live load element factor and the tributary area of the member, $K_{LL}A_T$ is equal or greater than 400 ft^2.

$$A_T^{B1} = 10 \times 25 = 250 \text{ ft}^2 < 400 \text{ ft}^2$$

Thus, it is not permitted to reduce the live load on *B*1.

$$w_L = 50^{psf} \times 10^{ft} / 1000^{lb/kips} = 0.50 \text{ kip/ft}$$
$$w_u = 1.2w_D + 1.6w_L = 1.2 \times 0.85 + 1.6 \times 0.50 = 1.82 \text{ kip/ft}$$

Select a Trial Section (Assume $M_n = M_p$)

$$M_r = \frac{w_u L^2}{8} = \frac{1.82 \times 25^2}{8} = 142.2 \text{ kip-ft}$$
$$\phi_b M_p = \phi_b F_y Z_x \geq M_r$$
$$Z_{x,\text{required}} = \frac{M_r}{\phi_b F_y} = \frac{142.2 \times 12^{in./ft}}{0.9 \times 50} = 37.9 \text{ in.}^3$$

Try *W*8 × 40 (Z_x = 39.8 in.^{3}; I_x = 146 in.^{4}; $b/2t_f$ = 7.21; h/t_w = 17.6)
Including the beam weight,

$$w_u = 1.2w_D + 1.6w_L = 1.2 \times (0.85 + 0.04) + 1.6 \times 0.50 = 1.87 \text{ kip/ft}$$
$$M_r = \frac{w_u L^2}{8} = \frac{1.87 \times 25^2}{8} = 146.1 \text{ kip-ft}$$

Check Limiting Width-to-Thickness Ratio according to Table B4.1b, AISC 360

$$\lambda_{p,f} = 0.38\sqrt{\frac{E}{F_y}} = 0.38\sqrt{\frac{29{,}000}{50}} = 9.15 > \lambda_f$$
$$= b/2t_f = 6.08 \text{ Compact flange}$$

$$\lambda_{p,w} = 3.76\sqrt{\frac{E}{F_y}} = 3.76\sqrt{\frac{29{,}000}{50}} = 90.6 > \lambda_w = h/t_w = 36.9 \text{ Compact web.}$$

Thus, the possible limit states are either yielding (plastic hinge formation) or LTB.

Check the Limit State of LTB As shown in Fig. 8.30, steel deck ribs are perpendicular to the compression flange of the beam. Therefore, the deck is assumed to provide continuous lateral support.

$$L_p > L_b = 0 \rightarrow \text{The limit state of LTB does not apply.}$$

where L_b = laterally unbraced length
L_p = the limiting laterally unbraced length for the limit state of yielding

Thus, the limit state of plastic hinge formation controls ($M_n = M_p$).

Determine the Design Flexural Strength

$$M_n = M_p = F_y Z_x = 50^{ksi} \times 39.8^{in.^3} / 12^{in./ft} = 165.8 \text{ kip-ft} \qquad \phi_b = 0.9$$
$$\phi_b M_p = 0.9 \times 165.8 = 149.2 \text{ kip-ft} > M_r = 146.1 \text{ kip-ft}$$

Determine the Design Shear Strength Per *Section G2, AISC 360*, the nominal shear strength is,

$$V_n = 0.6 F_y A_w C_{v1} \qquad \text{(Eq. G2-1, AISC 360)}$$

where A_w = area of web, the overall depth times the web thickness
C_{v1} = web shear strength coefficient

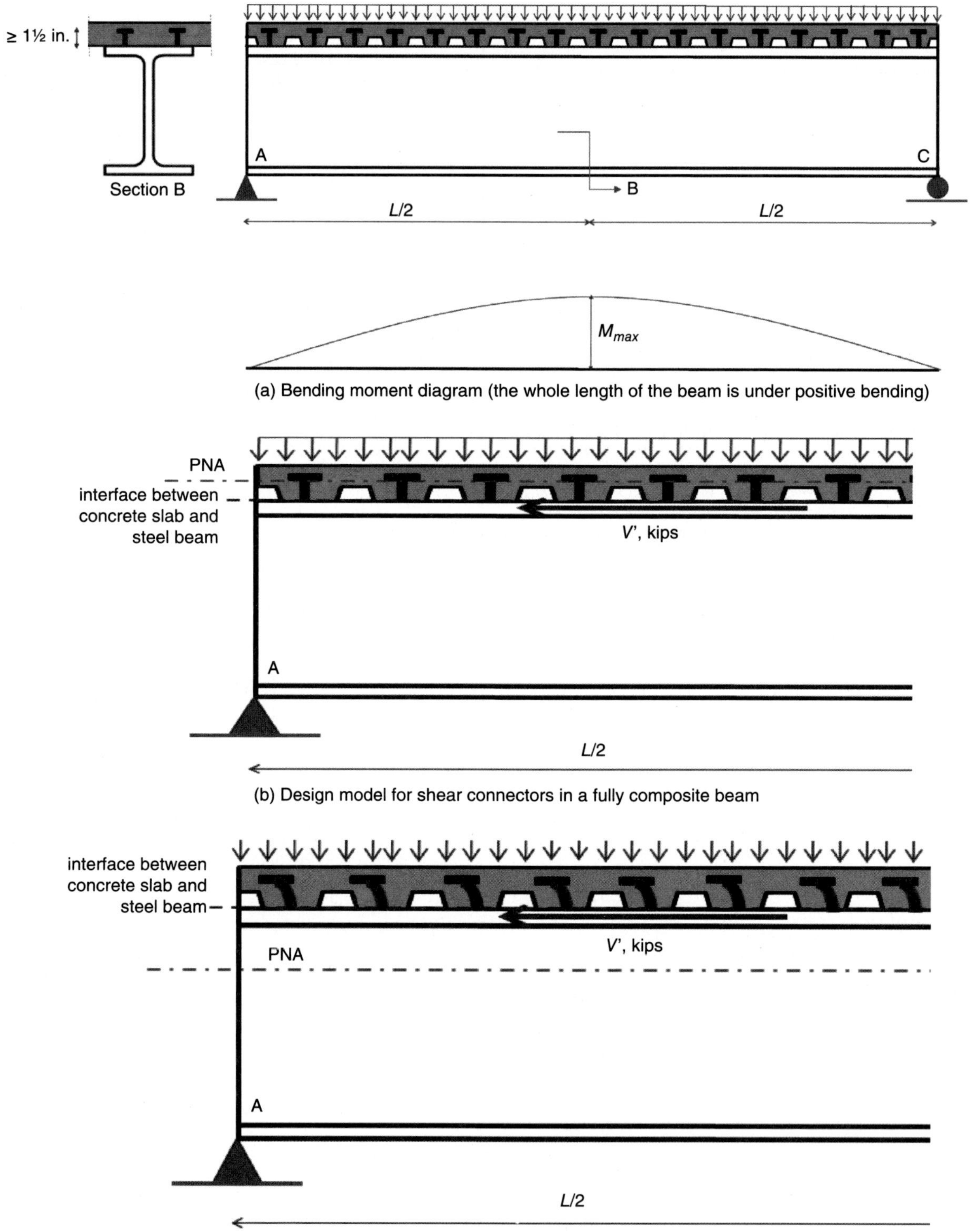

(a) Bending moment diagram (the whole length of the beam is under positive bending)

(b) Design model for shear connectors in a fully composite beam

(c) Design model for shear connectors in a partially composite beam (PNA moves farther down into the steel beam due to the slip between the concrete and steel beam when stear studs are less sufficient)

Figure 8.28 A simply supported composite beam, its bending moment diagram, and design models for shear connectors.

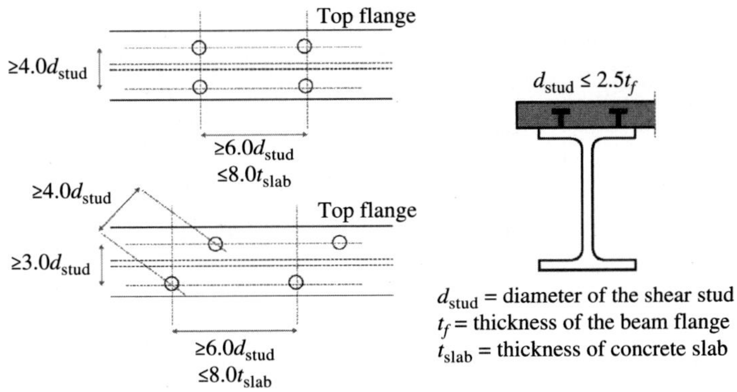

Figure 8.29 Detailed limitations for shear studs.

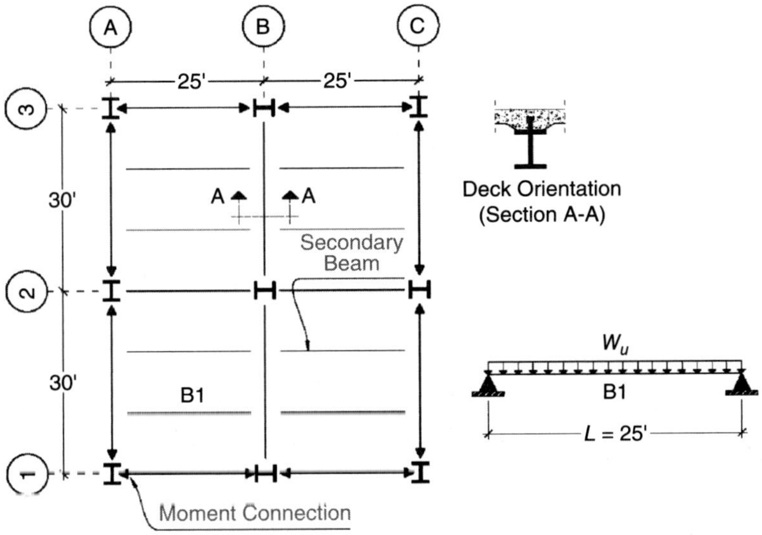

Figure 8.30 Typical floor plan of an office building.

For webs of hot-rolled I-shapes, C_{v1} and ϕ_v vary depending on the width-to-thickness ratio of the web.

$$\phi_v = 1.0 \text{ and } C_{v1} = 1.0 \text{ when } \frac{h}{t_w} < 2.24\sqrt{E/F_y} \qquad \textit{(Eq. G2-2, AISC 360)}$$

$$\frac{h}{t_w} = 36.9 < 2.24\sqrt{E/F_y} = 2.24\sqrt{29{,}000/50} = 53.9 \Rightarrow \phi_v = 1.0;\ C_{v1} = 1.0$$

$$A_w = d \times t_w = 8.25 \times 0.36 = 2.97^{\text{in.}^2}$$

$$V_n = 0.6F_yA_wC_{v1} = 0.6 \times 50 \times 2.97 \times 1.0 = 89.1^{\text{kips}}$$

$$\phi_v V_n = 1.0 \times 89.1 = 89.1^{\text{kips}} > V_r = \frac{w_u L}{2} = \frac{1.87 \times 25}{2} = 23.4^{\text{kips}}$$

W10 × 17 is adequate for strength requirements.

Check the Vertical Live Load Deflection

$$w_L = 0.50 \text{ kip/ft}$$

$$\Delta_{max} = \frac{5w_L L^4}{384EI} = \frac{5 \times (0.5/12^{\text{in./ft}}) \times (25 \times 12^{\text{in./ft}})^4}{384 \times 29{,}000 \times 146}$$

$$= 1.04^{\text{in.}} > \Delta_{limit} = \frac{L}{360} = \frac{25 \times 12^{\text{in./ft}}}{360} = 0.83^{\text{in.}}$$

The stiffness of W8 × 40 is not sufficient. Select a deeper shape.

$$\Delta_{max} = \frac{5w_L L^4}{384EI} \geq \Delta_{limit} = \frac{L}{360}$$

$$I_{x,\text{required}} = \frac{1800w_L L^3}{384E} = \frac{1800 \times (0.5/12) \times (25 \times 12)^3}{384 \times 29{,}000} = 182 \text{ in.}^4$$

Try W12 × 30 ($Z_x = 43.1$ in.^{3}; $I_x = 238$ in.^{4}; $b/2t_f = 7.41$; $h/t_w = 41.8$)

$$M_r = \frac{w_u L^2}{8} = \frac{1.856 \times 25^2}{8} = 145 \text{ kip-ft}$$

$$\phi_b M_p = \phi_b F_y Z_x = 0.9 \times 50^{\text{ksi}} \times 43.1^{\text{in.}^3}/12^{\text{in./ft}}$$

$$= 161.6 \text{ kip-ft} > M_r = 145 \text{ kip-ft}$$

$$\phi_v V_n = \phi_v(0.6F_yA_wC_{v1}) = 1.0 \times (0.6 \times 50 \times 3.19 \times 1.0)$$

$$= 95.9^{\text{kips}} > V_r = \frac{w_u L}{2} = \frac{1.856 \times 25}{2} = 23.2^{\text{kips}}$$

$$\Delta_{max} = \frac{5w_L L^4}{384EI} = \frac{5 \times (0.5/12^{\text{in./ft}}) \times (25 \times 12^{\text{in./ft}})^4}{384 \times 29{,}000 \times 238}$$

$$= 0.64^{\text{in.}} < \Delta_{limit} = \frac{L}{360} = 0.83^{\text{in.}}$$

Thus, W12 × 30 is adequate.

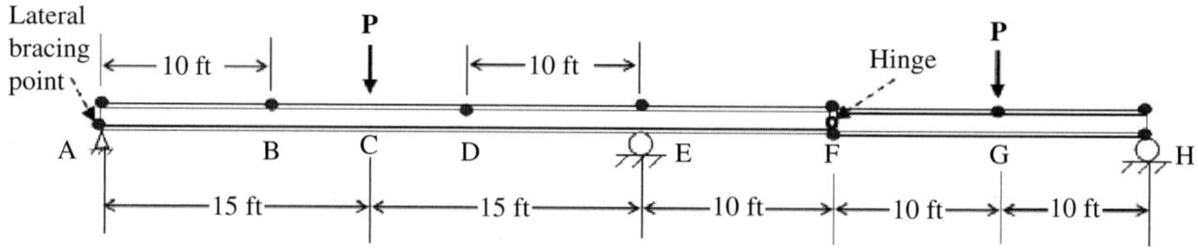

Figure 8.31 A continuous beam made of W27 × 102 (A992).

EXAMPLE 11

As seen in Fig. 8.31, a W27 × 102, $F_y = 50$ ksi, beam is loaded with the service load, P as shown. Both top and bottom flanges at A, F, and H have lateral bracing. However, only the top flange at B, D, E, and G has lateral bracing. The locations (top or bottom flanges) of bracing are shown in the figure by a solid dot. Find the maximum service load P (consists of 60 percent dead load and 40 percent live load) that can be applied to the beam based on LRFD. Neglect the beam weight.

Structural Analysis Under Service Loads The bending moment diagram of the continuous beam is drawn on compression side, as given in Fig. 8.32.

Determine the Flexural Strength of W27 × 102 Considering LTB and Local Buckling Limit States

(1) Check the limit state of local buckling

$\lambda < \lambda_p$, compact section. The limit state of local buckling does not apply.

(2) Check the limit state of LTB

W27 × 102: $L_p = 7.59$ *ft,* $L_r = 22.3$ *ft,* $Z_x = 305$ *in.*^{3}*,* $S_x = 267$ *in.*^{3}*,* $F_y = 50$ *ksi.*

Section E is not laterally braced due to the fact that the flange in compression at section E is its bottom flange, but the only bracing point is at the top flange. As shown in Table 8.4, segment DF has the largest L_b and segment BD has the smallest C_b. Thus, segments BD and DF are the critical segments.

- Flexural capacity of Segment DF (the segment with the largest L_b, Fig. 8.33)

Table 8.4 LTB Limit States for Each Segment

Segment	C_b	L_b (ft)	$M_{service}$	Limit state
AB	1.67	10	10P/3	Inelastic LTB might be possible ($L_p < L_b < L_r$)
BD	<1.67	10	5P	Inelastic LTB might be possible ($L_p < L_b < L_r$)
DF*	<1.67	20	5P	Inelastic LTB might be possible ($L_p < L_b < L_r$)
FG	1.67	10	5P	Inelastic LTB might be possible ($L_p < L_b < L_r$)
GH	1.67	10	5P	Inelastic LTB might be possible ($L_p < L_b < L_r$)

*The most critical segment.

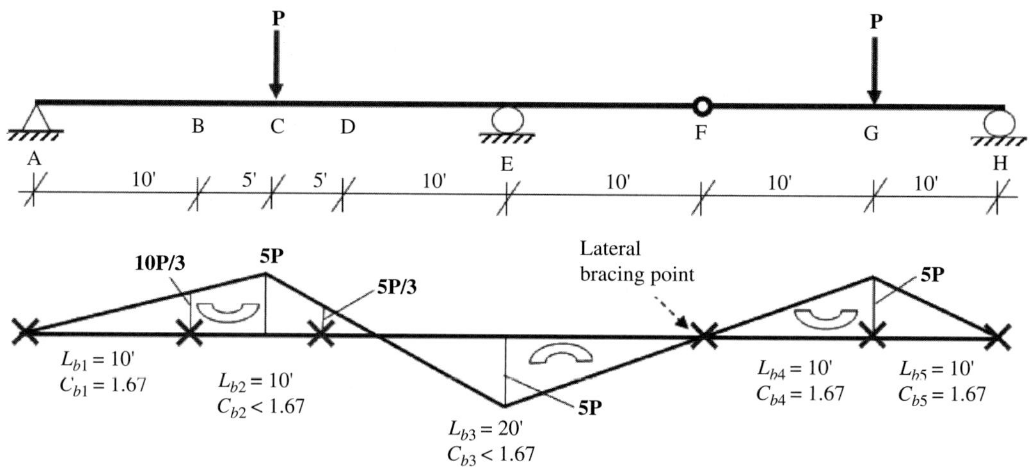

Figure 8.32 Bending moment diagram due to service loads.

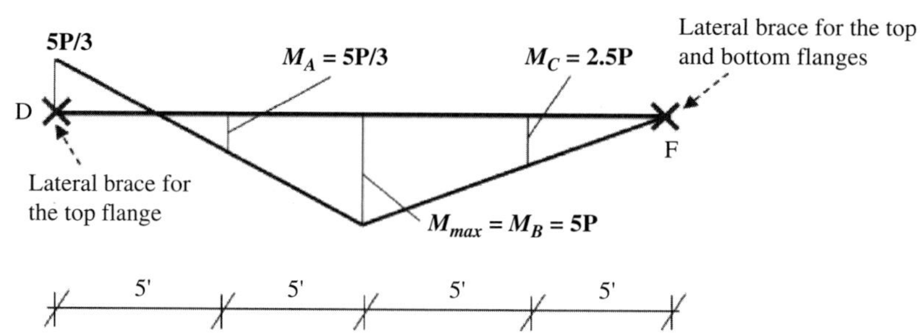

Figure 8.33 Moment diagram of segment DF for C_b calculation.

$L_p = 7.59$ ft $< L_b = 20$ ft $\le L_r = 22.3$ ft Inelastic LTB would occur.

$$M_n = C_b\left[M_p - (M_p - 0.7F_yS_x)\left(\frac{L_b - L_p}{L_r - L_p}\right)\right] \le M_p \quad (Eq.\ F2\text{-}2,\ AISC\ 360)$$

$$C_b = \frac{12.5M_{max}}{2.5M_{max} + 3M_A + 4M_B + 3M_C} \quad (Eq.\ F1\text{-}1,\ AISC\ 360)$$

where M_{max} = maximum moment in the unbraced segment
M_A = moment at ¼ point of the unbraced segment
M_B = moment at ½ point of the unbraced segment
M_C = moment at ¾ point of the unbraced segment

$$C_{b,DF} = \frac{12.5\times 5P}{2.5\times 5P + 3\times(5P/3) + 4\times 5P + 3\times 2.5P} = 1.39$$

$$M_n = 1.39\times\left[1270 - \left(1270 - \frac{0.7\times 50\times 267}{12^{\text{in./ft}}}\right)\left(\frac{20-7.59}{22.3-7.59}\right)\right]$$

$$= 1189 \text{ kip-ft} < M_p = 1270 \text{ kip-ft}$$

$$M_n = 1189 \text{ kip-ft}$$

- Flexural capacity of Segment BD (the segment with the smallest C_b, Fig. 8.34)

$L_p = 7.59$ ft $< L_b = 10$ ft $\le L_r = 22.3$ ft Inelastic LTB would occur.

$$C_{b,BD} = \frac{12.5\times 5P}{2.5\times 5P + 3\times(25P/6) + 4\times 5P + 3\times(10P/3)} = 1.14$$

$$M_n = 1.14\times\left[1270 - \left(1270 - \frac{0.7\times 50\times 267}{12^{\text{in./ft}}}\right)\left(\frac{10-7.59}{22.3-7.59}\right)\right]$$

$$= 1356 \text{ kip-ft} > M_p = 1270 \text{ kip-ft}$$

$$M_n = M_p = 1270 \text{ kip-ft}$$

Find P_{max} for LRFD The smallest nominal flexural capacity is obtained from segment DF. Therefore, the flexural design strength of the beam is calculated as follows:

$$M_c = \phi_b M_n = 0.9\times 1189 = 1070 \text{ k-ft}$$

The required LRFD strength, M_r

$$M_r = M_u = 1.2M_D + 1.6M_L = 1.2\times(0.6\times 5P) + 1.6\times(0.4\times 5P) = 6.8P$$

The maximum service load, P is determined using $M_c \ge M_r$

$$1070 \ge 6.8P$$

$$157.3^{\text{kips}} \ge P \Rightarrow P_{max} = 157.3 \text{ kips}$$

EXAMPLE 12

A 60-foot span simply supported plate girder made of A36 steel shown in Fig. 8.35 is subject to a uniformly distributed of w_u and concentrated load of P_u. Uniformly distributed dead and live loads are estimated as 1.25 kip/ft. and 0.75 kip/ft., respectively. Concentrated dead and live loads are 50 kips and 30 kips. Determine the adequacy of the girder using LRFD.

Analyze the Beam under the Factored Loads (LRFD)

$$w_u = 1.2w_D + 1.6w_L = 1.2\times 1.25 + 1.6\times 0.75 = 2.70 \text{ kip/ft}$$

(Girder weight is included)

$$P_u = 1.2P_D + 1.6P_L = 1.2\times 50 + 1.6\times 30 = 108 \text{ kips}$$

Shear force and bending moment diagrams of the girder due to the factored loads are given in Fig. 8.36.

The plate girder, shown in Fig. 8.37, is subjected to bending moment and shear force. The plate girder will be examined under the effect of shear and concentrated force actions first. A special attention to local limit states of the web will be paid, mainly due to its large depth over web thickness ratio.

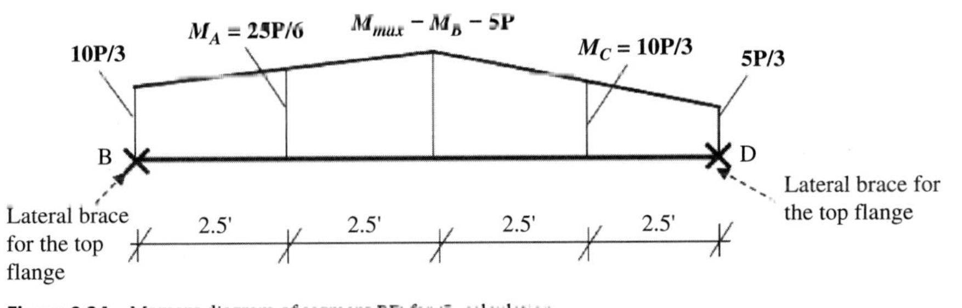

Figure 8.34 Moment diagram of segment BD for C_b calculation.

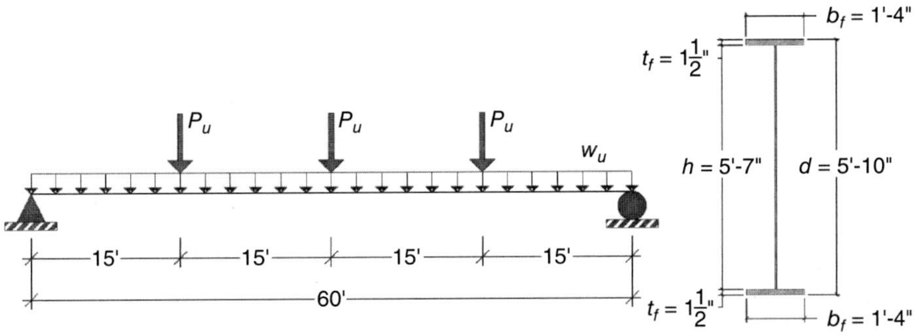

Figure 8.35 Simply supported plate girder.

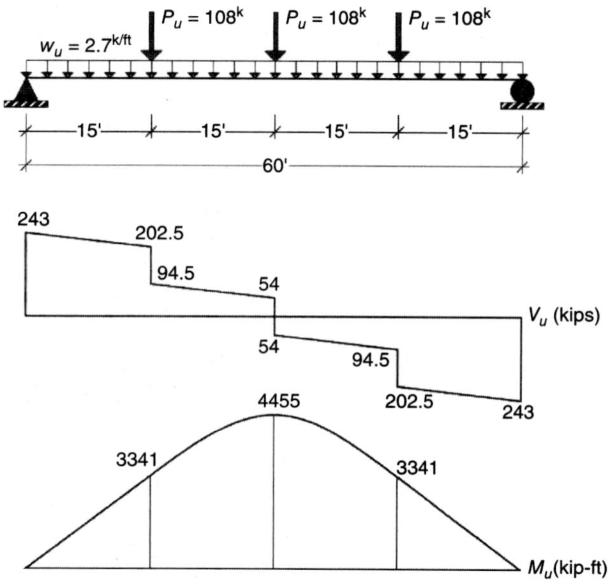

Figure 8.36 Shear force and bending moment diagram due to the factored loads.

Check Dimensional Requirements for the Web

$$\frac{h}{t_w} = \frac{d-2t_f}{t_w} = \frac{70-2\times1.5}{3/8} = 178 > 5.70\sqrt{E/F_y}$$

$$= 5.70\times\sqrt{29{,}000/36} = 161.8$$

The web of the girder is slender as per *Table B4.1b, AISC 360.* In accordance with *Section F13-2, AISC 360,* the web should satisfy the following slenderness limitation:

$$\left(\frac{h}{t_w}\right)_{\max} = 12\sqrt{E/F_y} \text{ when } \frac{a}{h} \le 1.5 \qquad \textit{(Eq. F13-3, AISC 360)}$$

$$\left(\frac{h}{t_w}\right)_{\max} = \frac{0.40E}{F_y} \text{ when } \frac{a}{h} > 1.5 \qquad \textit{(Eq. F13-3, AISC 360)}$$

where a is the clear distance between transverse stiffeners.

Since design of the stiffeners is not complete, the clear distance between the stiffeners is assumed to be 15 ft.

$$\frac{a}{h} = \frac{15\times12}{67} = 2.68 > 15 \rightarrow \left(\frac{h}{t_w}\right)_{\max} = \frac{0.40E}{F_y} = \frac{0.40\times29{,}000}{36} = 322.2$$

$$\left(\frac{h}{t_w}\right)_{\max} = 322.2 > \frac{h}{t_w} = 178.6 \text{ The web slenderness satisfies the limit.}$$

Check the Web under Concentrated Force according to Section J10, AISC 360 The applicable limit states are the web local yielding, web local crippling, and/or web compression buckling, since single concentrated compressive force is applied to the girder.

- Web local yielding *(Section J10-2, AISC 360)*

$$R_n = F_{yw}t_w(5k + l_b) \text{ when } e > d \qquad \textit{(Eq. J10-2, AISC 360)}$$

$$R_n = F_{yw}t_w(2.5k + l_b) \text{ when } e \le d \qquad \textit{(Eq. J10-3, AISC 360)}$$

where F_{yw} = specified minimum yield stress of the web material
k = distance from outer face of the flange to the web toe of the fillet
l_b = length of bearing
t_w = thickness of web
e = distance from member end to applied force

Use ½-in. thick intermittent fillet welds to connect the web and flanges of the girder.

$$k = t_f + a_w = 1.5\text{"} + 1/2\text{"} = 2\text{"}; \text{ Bearing length: } l_b = 15\text{"}$$

For the interior sections subject to the concentrated force of P_u = 108 kips:

$$\phi R_n = F_{yw}t_w(5k + l_b) = 1.0\times(36\times3/8\times[5\times(1.5+1/2)+15])$$

$$= 337.5^{\text{kips}} > P_u = 108^{\text{kips}}$$

For the end sections subject to the support reaction of R = 243 kips:

$$\phi R_n = F_{yw}t_w(2.5k + l_b) = 1.0\times(36\times3/8\times[2.5\times(1.5+1/2)+15])$$

$$= 270^{\text{kips}} > R = 243^{\text{kips}}$$

Bearing stiffeners are not required for the limit state of web local yielding.

- Web local crippling *(Section J10-3, AISC 360)*

When $e \ge d/2$

$$R_n = 0.80t_w^2\left[1+3\left(\frac{l_b}{d}\right)\left(\frac{t_w}{t_f}\right)^{1.5}\right]\sqrt{\frac{EF_{yw}t_f}{t_w}}Q_f \qquad \textit{(Eq. J10-4, AISC 360)}$$

When $e < d/2$ and $l_b/d \le 0.2$

$$R_n = 0.40t_w^2\left[1+3\left(\frac{l_b}{d}\right)\left(\frac{t_w}{t_f}\right)^{1.5}\right]\sqrt{\frac{EF_{yw}t_f}{t_w}}Q_f \qquad \textit{(Eq. J10-5a, AISC 360)}$$

When $e < d/2$ and $l_b/d > 0.2$

$$R_n = 0.40t_w^2\left[1+\left(\frac{4l_b}{d}-0.2\right)\left(\frac{t_w}{t_f}\right)^{1.5}\right]\sqrt{\frac{EF_{yw}t_f}{t_w}}Q_f$$

(Eq. J10-5b, AISC 360)

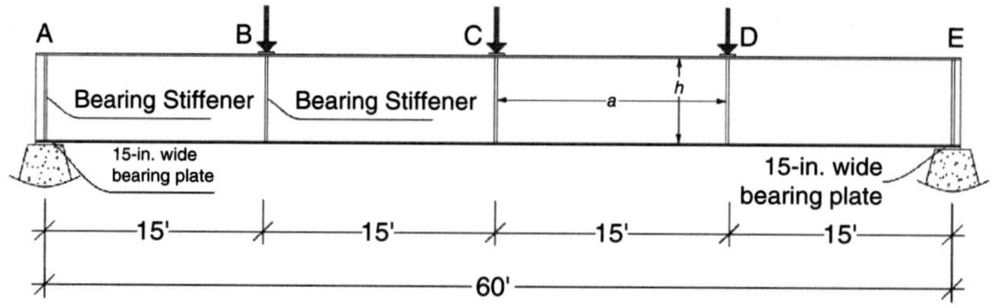

Figure 8.37 Plate girder with bearing stiffeners.

For the interior sections subject to the concentrated force of $P_u = 108$ kips ($e > d/2$):

$$\phi R_n = \phi 0.80 t_w^2 \left[1 + 3\left(\frac{l_b}{d}\right)\left(\frac{t_w}{t_f}\right)^{1.5}\right]\sqrt{\frac{EF_{yw}t_f}{t_w}}Q_f$$

$$= 0.75 \times 0.80 \times (3/8)^2 \left[1 + 3\left(\frac{15}{70}\right)\left(\frac{3/8}{1.5}\right)^{1.5}\right]\sqrt{\frac{29{,}000 \times 36 \times 1.5}{3/8}} \times 1.0$$

$$= 186.3^{\text{kips}} > P_u = 108^{\text{kips}}$$

Bearing stiffeners are not required for the interior sections.

For the end sections subject to the support reaction of $R = 243$ kips ($e < d/2$):

$$l_b/d = 15/70 = 0.21 > 0.2$$

$$\phi R_n = \phi 0.40 t_w^2 \left[1 + \left(\frac{4l_b}{d} - 0.2\right)\left(\frac{t_w}{t_f}\right)^{1.5}\right]\sqrt{\frac{EF_{yw}t_f}{t_w}}Q_f$$

$$= 0.75 \times 0.40 \times (3/8)^2 \times \left[1 + \left(\frac{4 \times 15}{70} - 0.2\right)\left(\frac{3/8}{1.5}\right)^{1.5}\right]$$

$$\times \sqrt{\frac{29{,}000 \times 36 \times 1.5}{3/8}} \times 1.0$$

$$= 93.3^{\text{kips}} < R = 243^{\text{kips}}$$

Use a pair of transverse stiffeners for the end sections.

- Web compression buckling (*Section J10-5, AISC 360*)

$$\phi R_n = \phi\left(\frac{24 t_w^3 \sqrt{EF_{yw}}}{h}\right)Q_f \qquad (\textit{Eq. J10-8, AISC 360})$$

$$= 0.9 \times \left(\frac{24 \times (3/8)^3 \times \sqrt{29{,}000 \times 36}}{67}\right) \times 1.0$$

$$= 17.4^{\text{kips}} < P_u = 108^{\text{kips}} \text{ and } R = 243^{\text{kips}}$$

Use a pair of stiffeners for the end and interior sections.

Determine the Shear Strength according to Section G, AISC 360
The web is subjected to a compressive normal stress induced by shear force, V, and the web would buckle along its diagonal in the direction of this compressive stress. After the diagonal with a certain bandwidth buckles, the web does not necessarily lose its function completely since the other diagonal portion (tension field) of the web is subjected to tension, and a post-buckling strength might be possible. With proper support from the stiffeners, the tension field, behaving like a diagonal "truss" member connected to the stiffeners and flanges, might significantly increase the shear strength of the plate girder after web buckling due to tension field action. Therefore, shear strength of webs varies based on the contribution of tension field action. Tension field action is possible when the web plate is adequately supported on all its four sides by flanges and stiffeners. In the following cases, tension field action is not permitted to be used:

(1) Tension field cannot be fully developed in an end panel due to the fact that absence of one of the adjacent panels reduces the horizontal component of the diagonal force produced by the tension field.

(2) The resistance provided by the tension field reduces as the panel aspect ratio becomes smaller. Therefore, when the panel aspect ratio, a/h, is greater than three, it is not permitted to use tension field action.

(3) Furthermore, I-shaped girders with relatively small flange-to-web area ratio or large height-to-flange width ratio are allowed to use limited tension field action shear strength.

Initially, the transverse stiffeners are spaced evenly every 15 ft, as shown in Fig. 8.37.

- Shear strength of the end panels (AB and DE)

Shear strength of webs in end panels are computed without considering tension-field action as follows:

$$\text{When } \frac{h}{t_w} \le 1.10\sqrt{k_v E/F_y} \rightarrow C_{v1} = 1.0 \qquad (\textit{Eq. G2-3, AISC 360})$$

$$\text{When } \frac{h}{t_w} > 1.10\sqrt{k_v E/F_y} \rightarrow C_{v1} = \frac{1.10\sqrt{k_v E/F_y}}{h/t_w} \qquad (\textit{Eq. G2-4, AISC 360})$$

Since transverse stiffeners are used, the web plate buckling coefficient, k_v, is determined according to *Eq. G2-5, AISC 360.* Note that the clear distance between the stiffeners, a, is conservatively taken as 15 ft.

$$a/h = 15^{\text{ft}} \times 12^{\text{in./ft}} / 67^{\text{in.}} = 2.687 < 3 \rightarrow k_v = 5 + \frac{5}{(a/h)^2}$$

$$= 5 + \frac{5}{(15 \times 12/67)^2} = 5.693$$

$$\frac{h}{t_w} = \frac{67}{3/8} = 178.6 > 1.10\sqrt{k_v E/F_y} = 1.10 \times \sqrt{\frac{5.693 \times 29{,}000}{36}} = 74.5$$

$$C_{v1} = \frac{1.10\sqrt{k_v E/F_y}}{h/t_w} = \frac{74.5}{178.6} = 0.417$$

Thus, the nominal shear strength, V_n, is computed as,

$$V_n = 0.6 F_y A_w C_{v1} = 0.6 \times 36 \times (67 \times 3/8) \times 0.417 = 226.3^{\text{kips}}$$

$$\phi_v V_n = 0.9 \times 226.3 = 203.7^{\text{kips}} < V_r = V_u = 243^{\text{kips}}$$

To meet the strength requirement, the spacing between the intermediate stiffeners is reduced by adding an additional stiffener between the lateral bracing points A and B, as shown in Fig. 8.38. Therefore, the centerline-to-centerline distance from stiffener to stiffener, a, is reduced to 5 ft for the end panels to increase the shear strength.

$$a/h = 5^{\text{ft}} \times 12^{\text{in./ft}} / 67^{\text{in.}} = 0.896 < 3 \rightarrow k_v = 5 + \frac{5}{(a/h)^2} = 5 + \frac{5}{0.896^2} = 11.228$$

$$\frac{h}{t_w} = \frac{67}{3/8} = 178.6 > 1.10\sqrt{k_v E/F_y} = 1.10 \times \sqrt{\frac{11.228 \times 29{,}000}{36}} = 104.6$$

$$C_{v1} = \frac{1.10\sqrt{k_v E/F_y}}{h/t_w} = \frac{104.6}{178.6} = 0.586$$

Thus, the nominal shear strength, V_n, is computed as,

$$V_n = 0.6 F_y A_w C_{v1} = 0.6 \times 36 \times (67 \times 3/8) \times 0.586 = 318^{\text{kips}}$$

$$\phi_v V_n = 0.9 \times 318 = 286^{\text{kips}} > V_r = V_u = 243^{\text{kips}}$$

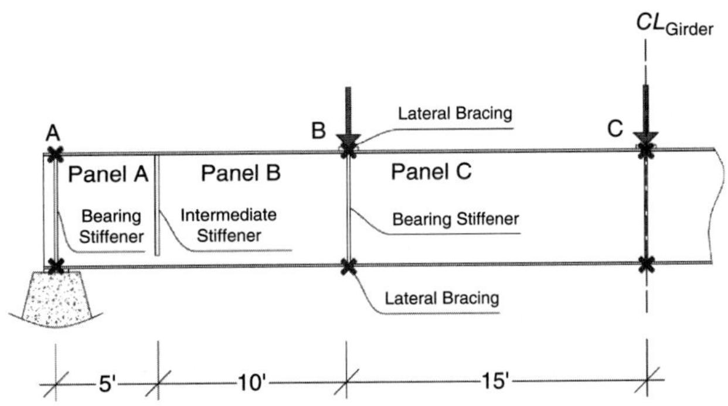

Figure 8.38 Plate girder with additional intermediate stiffeners.

- Shear strength of the interior panels with 10 ft stiffener spacing (Panel B)

Shear strength of interior panels due to tension-field action is a function of the width-to-thickness ratio and flange-to-web proportion. To determine whether or not tension-field action can be included in the shear strength of webs in interior panels, the panel aspect ratio, a/h, should be examined first.

$$a/h = 10^{\text{ft}} \times 12^{\text{in./ft}} / 67^{\text{in.}} = 1.79 < 3$$

Thus, inclusion of tension-field action is permitted. The following equations show the level of reduction in the shear strength due to web shear buckling based on the confinement provided by the flanges:

(a) When $\frac{h}{t_w} \le 1.10\sqrt{k_v E/F_y} \rightarrow V_n = 0.6F_y A_w$ *(Eq. G2-6, AISC 360)*

(b) When $\frac{h}{t_w} > 1.10\sqrt{k_v E/F_y}$

(b1) When $\frac{2A_w}{A_{fc} + A_{ft}} \le 2.25$,

$$\frac{h}{b_{fc}} \le 6.0 \text{ and } \frac{h}{b_{ft}} \le 6.0 \rightarrow V_n = 0.6F_y A_w \left[C_{v2} + \frac{1 - C_{v2}}{1.15\sqrt{1+(a/h)^2}}\right]$$

(Eq. G2-7, AISC 360)

(b2) Otherwise $\rightarrow V_n = 0.6F_y A_w \left[C_{v2} + \frac{1 - C_{v2}}{1.15\left[a/h + \sqrt{1+(a/h)^2}\right]}\right]$

(Eq. G2-8, AISC 360)

$$k_v = 5 + \frac{5}{(a/h)^2} = 5 + \frac{5}{(10 \times 12/67)^2} = 6.559$$

$$\frac{h}{t_w} = 178.6 > 1.10\sqrt{k_v E/F_y} = 1.10 \times \sqrt{\frac{6.559 \times 29{,}000}{36}} = 80$$

$$\frac{2A_w}{A_{fc} + A_{ft}} = \frac{2 \times (67 \times 3/8)}{2 \times (15 \times 1.5)} = 1.12 < 2.25$$

$$\frac{h}{b_{fc}} = \frac{h}{b_{ft}} = \frac{67}{15} = 4.47 < 6.0$$

$$\Rightarrow V_n = 0.6F_y A_w \left[C_{v2} + \frac{1 - C_{v2}}{1.15\sqrt{1+(a/h)^2}}\right]$$

$$C_{v2} = \frac{1.51 k_v E}{(h/t_w)^2 F_y} \text{ when } h/t_w > 1.37\sqrt{k_v E/F_y}$$ *(Eq. G2-11, AISC 360)*

$$h/t_w = 178.6 > 1.37\sqrt{k_v E/F_y} = 1.37 \times \sqrt{\frac{6.559 \times 29{,}000}{36}} = 99.6$$

$$C_{v2} = \frac{1.51 \times 6.559 \times 29{,}000}{178.6^2 \times 36} = 0.25$$

$$V_n = 0.6 \times 36 \times (67 \times 3/8) \times \left[0.25 + \frac{1 - 0.25}{1.15 \times \sqrt{1 + (1.79)^2}}\right] = 308.3^{\text{kips}}$$

$$\phi_v V_n = 0.9 \times 308.3 = 277^{\text{kips}} > V_r = 229.5^{\text{kips}}$$

- Shear strength of the interior panels with 15-ft stiffener spacing (Panel C)

$$a/h = 15^{\text{ft}} \times 12^{\text{in./ft}} / 67^{\text{in.}} = 2.687 < 3$$

Thus, inclusion of tension-field action is permitted.

$$\frac{h}{t_w} = 178.6 > 1.10\sqrt{k_v E/F_y} = 1.10 \times \sqrt{\frac{5.693 \times 29{,}000}{36}} = 74.5$$

$$\frac{2A_w}{A_{fc} + A_{ft}} = \frac{2 \times (67 \times 3/8)}{2 \times (15 \times 1.5)} = 1.12 < 2.25$$

$$\frac{h}{b_{fc}} = \frac{h}{b_{ft}} = \frac{67}{15} = 4.47 < 6.0$$

$$\Rightarrow V_n = 0.6F_y A_w \left[C_{v2} + \frac{1 - C_{v2}}{1.15\sqrt{1+(a/h)^2}}\right]$$

$$C_{v2} = \frac{1.51 k_v E}{(h/t_w)^2 F_y} \text{ when } h/t_w > 1.37\sqrt{k_v E/F_y}$$ *(Eq. G2-11, AISC 360)*

$$h/t_w = 178.6 > 1.37\sqrt{k_v E/F_y} = 1.37 \times \sqrt{\frac{5.693 \times 29{,}000}{36}} = 92.8$$

$$C_{v2} = \frac{1.51 \times 5.693 \times 29{,}000}{178.6^2 \times 36} = 0.217$$

$$V_n = 0.6 \times 36 \times (67 \times 3/8) \times \left[0.217 + \frac{1 - 0.217}{1.15 \times \sqrt{1 + (2.687)^2}}\right] = 246.7^{\text{kips}}$$

$$\phi_v V_n = 0.9 \times 246.7 = 222^{\text{kips}} > V_r = 94.5^{\text{kips}}$$

Design of Transverse Stiffeners according to Section G2-3 and J10-8, AISC 360 Transverse stiffeners should satisfy the dimensional, stiffness, compactness, and strength requirements stipulated in *Section G2-3 and J10-8, AISC 360.* A summary of the foregoing AISC requirements is as follows:

(1) Per *Section G2-3a*, transverse stiffeners are not required where $h/t_w \le 2.46\sqrt{E/F_y}$.

(2) It is permitted to use stiffeners shorter than web height to transmit a concentrated force. The distance from the near toe of the weld that connects girder flange to the web, to the end of the welds by which the stiffener is attached to the web should be between $4t_w$ and $6t_w$ (*Section G2-3b*). For the practical purposes, it is usually necessary to provide full-depth stiffeners on girder webs to attach torsional bracings between the girders.

(3) Stiffeners should satisfy the limiting width-to-thickness ratio requirement for flanges of I-shaped sections, $(b/t)_{st} \le 0.56\sqrt{E/F_{yst}}$, as per *Table B4.1a, AISC 360*, since the stiffeners are treated as column members (*Section G2-3d*).

(4) Stiffness of a transverse stiffener should be sufficient to develop web shear buckling and/or post-buckling resistance (*Section G2-3e*).

(5) Strength of transverse stiffeners should be sufficient in terms of global stability and bearing limit states.

- Bearing stiffeners at points B, C, and D

  Try 2Pl. 4 × ⅜.

$$(b/t)_{st} = \frac{4}{3/8} = 10.7 < 0.56\sqrt{\frac{E}{F_{yst}}} = 0.56 \times \sqrt{29{,}000/36} = 15.9$$

(Eq. G2-12, AISC 360)

Therefore, the selected stiffener is non-slender.

The stiffness requirement for transverse stiffeners is as follows:

$$I_{st} \ge I_{st2} + (I_{st1} - I_{st2})\rho_w$$ *(Eq. G2-13, AISC 360)*

$$I_{st1} = \frac{h^4 \rho_{st}^{1.3}}{40}\left(\frac{F_{yw}}{E}\right)^{1.5}$$ *(Eq. G2-14, AISC 360)*

$$I_{st2} = \left[\frac{2.5}{(a/h)^2} - 2\right] b_p t_w^3 \ge 0.5 b_p t_w^3$$ *(Eq. G2-15, AISC 360)*

where F_{yst} = specified minimum yield stress of the stiffener material
F_{yw} = specified minimum yield stress of the web material
I_{st} = moment of inertia of the transverse stiffeners about an axis in the web center for stiffener pairs, or about the face in contact with the web plate for single stiffeners

I_{st1} = minimum moment of inertia of the transverse stiffeners required for the development of the full shear post-buckling resistance of the stiffened web panels, $V_r = V_{c1}$

I_{st2} = minimum moment of inertia of the transverse stiffeners required for the development of the web shear buckling resistance, $V_r = V_{c2}$

V_{c1} = available shear strength calculated with V_n as defined in *Section G2.1* or *G2.2*, as applicable

V_{c2} = available shear strength, kips (N), calculated with $V_n = 0.6F_yA_wC_{v2}$

V_r = required shear strength in the panel being considered

b_p = smaller of the dimension a and h

$(b/t)_{st}$ = width-to-thickness ratio of the stiffener

ρ_{st} = larger of F_{yw}/F_{yst} and 1.0

ρ_w = maximum shear ratio within the web panels on each side of the transverse stiffener

Moment of inertia of the stiffeners, I_{st}, will include a strip of the web having a width of $25t_w$ (Fig. 8.39).

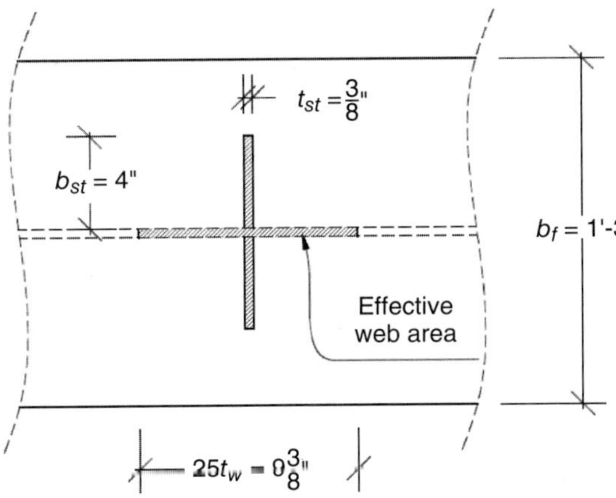

Figure 8.39 Stiffener and the effective web area.

Therefore, the moment of inertia of the shaded cruciform shape (Fig. 8.39) can be calculated as follows:

$$I_{st} = \frac{25t_w^4}{12} + 2\times\left(\frac{t_{st}b_{st}^3}{12} + t_{st}b_{st}\left(\frac{b_{st}+t_w}{2}\right)^2\right)$$

$$= \frac{25\times(3/8)^4}{12} + 2\times\left(\frac{(3/8)\times 4^3}{12} + (3/8)\times 4\times\left(\frac{(3/8)+4}{2}\right)^2\right) = 18.4$$

$$I_{st2} = \left\{\begin{array}{l}\left[\dfrac{2.5}{(a/h)^2} - 2\right]b_pt_w^3 = \left[\dfrac{2.5}{(15\times 12/67)^2} - 2\right]\times 67\times(3/8)^3 = -5.843 \\ 0.5b_pt_w^3 = 0.5\times 67\times(3/8)^3 = 1.767\end{array}\right\}$$

$$= 1.767$$

$$V_{c1} = 0.6\times 36\times(67\times 3/8)\times\left[0.217 + \frac{1-0.217}{1.15\times\sqrt{1+(2.687)^2}}\right] = 246.7^{\text{kips}}$$

$$V_{c2} = 0.6F_yA_wC_{v2} = 0.6\times 36\times(67\times 3/8)\times 0.217 = 117.8^{\text{kips}}$$

$$\rho_w = \left(\frac{V_r - V_{c2}}{V_{c1} - V_{c2}}\right) \geq 0$$

$$= \left(\frac{94.5 - 117.8}{246.7 - 117.8}\right) < 0 \rightarrow \rho_w = 0$$

$$I_{st} = 18.4 > I_{st2} + (I_{st1} - I_{st2})\rho_w = 1.767 + (I_{st1} - I_{st2})\times 0 = 1.767$$

The stiffener stiffness is adequate.

Global stability of the stiffeners is determined using an effective length of $0.75h$. Thus, the cross-sectional properties of the column member are:

$$A_{g,st} = 2(b_{st}t_{st}) + 25t_w^2 = 2\times(4\times 3/8) + 25\times(3/8)^2 = 6.52 \text{ in.}^2;$$

$$I_{st} = 18.4 \text{ in.}^4$$

$$r_{st} = \sqrt{\frac{I_{st}}{A_{g,st}}} = \sqrt{\frac{18.4}{6.52}} = 1.68 \text{ in.}$$

$$KL = 0.75h \rightarrow \frac{0.75h}{r_{st}} = \frac{0.75\times 67}{1.68} = 29.9$$

$$F_e = \frac{\pi^2 E}{(KL/r_{st})^2} = \frac{\pi^2\times 29{,}000}{(29.9)^2} = 320.2^{\text{ksi}}$$

$$F_{cr} = (0.658^{F_y/F_e})F_y = (0.658^{36/320.2})\times 36 = 34.3^{\text{ksi}}$$

$$P_n = F_{cr}A_{g,st} = 34.3\times 6.52 = 223.9^{\text{kips}} \qquad \phi_c = 0.90$$

$$\phi_cP_n = 0.9\times 223.9 = 201.5^{\text{kips}} > P_u = 108^{\text{kips}}$$

Determine Bearing Strength according to Section J7, AISC 360:

$$R_n = 1.8F_yA_{pb} \qquad \textit{(Eq. J7-1, AISC 360)}$$

where A_{pb} is the projected area in bearing.

$$R_n = 1.8\times 36\times[2\times(4-1/2)\times 3/8] = 170.1^{\text{kips}} \qquad \phi = 0.75$$

$$\phi R_n = 0.75\times 170.1 = 127.6^{\text{kips}} > P_u = 108^{\text{kips}}$$

Use *2Pl.* 4 × ⅜.

- Bearing stiffeners at point A and E

  Try *2Pl.* 6 × ½.

$$(b/t)_{st} = \frac{6}{0.5} = 12 < 0.56\times\sqrt{29{,}000/36} = 15.9 \qquad \textit{(Eq. G2-12, AISC 360)}$$

Therefore, the selected stiffener is nonslender.

Moment of inertia of the intermediate stiffeners, I_{st}, will include a strip of the web having a width of $12.5t_w$. Therefore, the moment of inertia of the cruciform shape can be calculated as follows:

$$I_{st} = \frac{12.5t_w^4}{12} + 2\times\left(\frac{t_{st}b_{st}^3}{12} + t_{st}b_{st}\left(\frac{b_{st}+t_w}{2}\right)^2\right)$$

$$= \frac{12.5\times(3/8)^4}{12} + 2\times\left(\frac{(1/2)\times 6^3}{12} + (1/2)\times 6\times\left(\frac{(3/8)+6}{2}\right)^2\right) = 79 \text{ in.}$$

$$I_{st1} = \frac{h^4\rho_{st}^{1.3}}{40}\left(\frac{F_{yw}}{E}\right)^{1.5} = \frac{67^4}{40}\times\left(\frac{36}{29{,}000}\right)^{1.5} = 22.03$$

$$I_{st2} = \left\{\begin{array}{l}\left[\dfrac{2.5}{(a/h)^2} - 2\right]b_pt_w^3 = \left[\dfrac{2.5}{(5\times 12/67)^2} - 2\right]\times 60\times(3/8)^3 = 3.535 \\ 0.5b_pt_w^3 = 0.5\times 60\times(3/8)^3 = 1.582\end{array}\right\}$$

$$= 3.535$$

$$V_{c1} = 0.6F_yA_wC_{v1} = 0.6\times 36\times(67\times 3/8)\times 0.586 = 318^{\text{kips}}$$

$$k_v = 5 + \frac{5}{(a/h)^2} = 5 + \frac{5}{(5\times 12/67)^2} = 11.235$$

$$h/t_w = 178.6 > 1.37\times\sqrt{\frac{11.235\times 29{,}000}{36}} = 130.3$$

$$C_{v2} = \frac{1.51\times 11.235\times 29{,}000}{178.6^2\times 36} = 0.428$$

$$V_{c2} = 0.6F_yA_wC_{v2} = 0.6\times 36\times(67\times 3/8)\times 0.428 = 232.3^{\text{kips}}$$

$$\rho_w = \left(\frac{243-232.2}{318-232.2}\right) = 0.126 > 0 \rightarrow \rho_w = 0.126$$

$$I_{st} = 79 > I_{st2} + (I_{st1} - I_{st2})\rho_w = 3.535 + (22.03 - 3.535)\times 0.126 = 5.87$$

The stiffener stiffness is adequate.

Global stability of the stiffeners is determined using an effective length of $0.75h$. Thus, the cross-sectional properties of the column member are:

$$A_{g,st} = 2(b_{st}t_{st}) + 12.5t_w^2 = 2\times(6\times 1/2) + 12.5\times(3/8)^2 = 7.76 \text{ in.}^2;$$

$$I_{st} = 79 \text{ in.}^4$$

$$r_{st} = \sqrt{\frac{79}{7.76}} = 3.19 \text{ in.}$$

$$KL = 0.75h \rightarrow \frac{0.75h}{r_{st}} = \frac{0.75\times 67}{3.19} = 15.75$$

$$F_e = \frac{\pi^2 E}{(KL/r_{st})^2} = \frac{\pi^2\times 29{,}000}{(15.75)^2} = 1154^{\text{ksi}}$$

$$F_{cr} = (0.658^{36/1154})\times 36 = 35.5^{\text{ksi}}$$

$$\phi_c P_n = 0.9\times 7.76\times 35.5 = 248.2^{\text{kips}} > R = 243^{\text{kips}}$$

Determine Bearing Strength according to Section J7, AISC 360:

$$\phi R_n = 0.75\times 1.8\times 36\times[2\times(6-1/2)\times 1/2] = 267.3^{\text{kips}} > R = 243^{\text{kips}}$$

Use 2*Pl.* 6 × ½.

- Intermediate stiffeners

  Try 2*Pl.* 5 × ⅜.

  The length of the stiffener:

$$l_{st} = h - 5t_w - a_w = 67 - 5\times(3/8) - 1/2 = 64\frac{5}{8} \text{ in.}$$

Strength requirement for the intermediate stiffeners is as follows:

$$A_{st} \geq P_s = \frac{F_{yw}}{F_{st}}\left[0.15DA_w(1-C_v)\frac{V_r}{\phi_v V_n} - 18t_w^2\right]$$

where P_s is the force on the intermediate stiffeners.

$$A_{st} = 2b_{st}t_{st} = 2\times(5\times 3/8) = 3.75 \text{ in.}^2$$

$$\phi_v V_n = 277^{\text{kips}}$$

$$V_r = 229.5^{\text{kips}}$$

$$D = 1.0$$

$$C_v = C_{v2} = 0.25$$

$$\Rightarrow P_s = \frac{36}{36}\times\left[0.15\times 1.0\times(67\times 3/8)\times(1-0.25)\times\frac{229.5}{277} - 18\times(3/8)^2\right]$$

$$= -0.189$$

Since P_s is negative, the intermediate stiffeners should be designed based on local buckling prevention and stiffness requirements.

Member slenderness requirement:

$$(b/t)_{st} = \frac{5}{3/8} = 13.3 < 0.56\sqrt{\frac{E}{F_{yst}}} = 0.56\times\sqrt{29{,}000/36} = 15.9$$

(Eq. G2-12, AISC 360)

Stiffness requirement:

$$I_{st} = 2\times\left(\frac{t_{st}b_{st}^3}{12} + t_{st}b_{st}\left(\frac{b_{st}+t_w}{2}\right)^2\right)$$

$$= 2\times\left(\frac{(3/8)\times 5^3}{12} + (3/8)\times 5\times\left(\frac{(3/8)+5}{2}\right)^2\right) = 34.9 \text{ in.}^4$$

$$I_{st1} = \frac{h^4\rho_{st}^{1.3}}{40}\left(\frac{F_{yw}}{E}\right)^{1.5} = \frac{67^4}{40}\times\left(\frac{36}{29{,}000}\right)^{1.5} = 22.03$$

$$I_{st2} = \begin{Bmatrix} \left[\frac{2.5}{(a/h)^2} - 2\right]b_p t_w^3 = \left[\frac{2.5}{(10\times 12/67)^2} - 2\right]\times 67\times(3/8)^3 = -4.31 \\ 0.5b_p t_w^3 = 0.5\times 67\times(3/8)^3 = 1.767 \end{Bmatrix}$$

$$= 1.767$$

$$k_v = 5 + \frac{5}{(a/h)^2} = 5 + \frac{5}{(10\times 12/67)^2} = 6.559$$

$$\frac{h}{t_w} = 178.6 > 1.10\sqrt{k_v E/F_y} = 1.10\times\sqrt{\frac{6.559\times 29{,}000}{36}} = 80$$

$$C_{v1} = \frac{1.10\sqrt{k_v E/F_y}}{h/t_w} = \frac{1.10\times\sqrt{6.559\times 29{,}000/36}}{178.6} = 0.448$$

$$V_{c1} = 0.6F_y A_w C_{v1} = 0.6\times 36\times(67\times 3/8)\times 0.448 = 243^{\text{kips}}$$

$$h/t_w = 178.6 > 1.37\sqrt{k_v E/F_y} = 1.37\times\sqrt{\frac{6.559\times 29{,}000}{36}} = 99.6$$

$$C_{v2} = \frac{1.51\times 6.559\times 29{,}000}{178.6^2\times 36} = 0.25$$

$$V_{c2} = 0.6F_y A_w C_{v2} = 0.6\times 36\times(67\times 3/8)\times 0.25 = 135.7^{\text{kips}}$$

$$\rho_w = \left(\frac{229.5-135.7}{243-135.7}\right) = 0.874$$

$$I_{st} = 34.9 > I_{st2} + (I_{st1} - I_{st2})\rho_w = 1.767 + (22.03 - 1.767)\times 0.874 = 19.48$$

The stiffener stiffness is adequate.

Therefore, 2*Pl.* 64⅝ × 5 × ⅜ intermediate stiffeners are adequate.

Determine the Flexural Strength according to Section F5, AISC 360

- Compression flange yielding (*Section F5-1, AISC 360*)

$$M_n = R_{pg}F_y S_{xc} \qquad \textit{(Eq. F5-1, AISC 360)}$$

$$R_{pg} = 1 - \frac{a_w}{1200+300a_w}\left(\frac{h_c}{t_w} - 5.7\sqrt{\frac{E}{F_y}}\right) \leq 1.0 \qquad \textit{(Eq. F5-6, AISC 360)}$$

$$a_w = \frac{h_c t_w}{b_{fc}t_{fc}} \leq 10 \qquad \textit{(Eq. F4-12, AISC 360)}$$

where R_{pg} = the bending strength reduction factor

S_{xc} = elastic section modulus referred to compression flange

h_c = twice the distance from the centroid to the inside faces of the compression flange when welds are used, for built-up sections ($h_c = h$)

b_{fc} = width of compression flange

t_{fc} = thickness of compression flange

$$I_x = \bar{I} + Ad^2 = 2\times\left[\frac{15\times 1.5^3}{12} + (15\times 1.5)\times\left(\frac{70-1.5}{2}\right)^2\right]$$

$$+ \frac{0.375\times 67^3}{12} = 62{,}195 \text{ in.}^4$$

$$S_{xc} = S_x = \frac{I_x}{c} = \frac{62{,}195}{70/2} = 1777 \text{ in.}^3$$

$$a_w = \frac{h_c t_w}{b_{fc}t_{fc}} = \frac{67\times 0.375}{15\times 1.5} = 1.117 < 10$$

$$R_{pg} = 1 - \frac{a_w}{1200+300a_w}\left(\frac{h_c}{t_w} - 5.7\sqrt{\frac{E}{F_y}}\right)$$

$$= 1 - \frac{1.117}{1200+300\times 1.117}\times\left(178 - 5.7\times\sqrt{\frac{29{,}000}{36}}\right) = 0.988 < 1.0$$

$$M_n = R_{pg}F_y S_{xc} = 0.988\times 36\times 1777/12^{\text{in./ft}} = 5268^{\text{kip-ft}}$$

- Lateral-torsional buckling (*Section F5-2, AISC 360*)

$$M_n = R_{pg}F_{cr}S_{xc} \quad \text{(Eq. F5-2, AISC 360)}$$

The LTB stress, F_{cr}, is defined as:

$$\text{When } L_p < L_b \le L_r \rightarrow F_{cr} = C_b\left[F_y - (0.3F_y)\left(\frac{L_b - L_p}{L_r - L_p}\right)\right] \le F_y \quad \text{(Eq. F5-3, AISC 360)}$$

$$\text{When } L_r < L_b \rightarrow F_{cr} = \frac{C_b\pi^2 E}{\left(\frac{L_b}{r_t}\right)^2} \le F_y \quad \text{(Eq. F5-3, AISC 360)}$$

The limiting laterally unbraced lengths, L_p and L_r, are:

$$L_p = 1.1r_t\sqrt{\frac{E}{F_y}} \quad \text{(Eq. F4-7, AISC 360)}$$

$$L_r = \pi r_t\sqrt{\frac{E}{0.7F_y}} \quad \text{(Eq. F4-7, AISC 360)}$$

$$r_t = \frac{b_{fc}}{\sqrt{12\left(1+\frac{1}{6}a_w\right)}} \quad \text{(Eq. F4-11, AISC 360)}$$

where r_t is the effective radius of gyration for the limit state of LTB.

$$r_t = \frac{b_{fc}}{\sqrt{12\left(1+\frac{1}{6}a_w\right)}} = \frac{15}{\sqrt{12\times\left(1+\frac{1}{6}\times 1.117\right)}} = 3.976 \text{ in.}$$

$$L_p = 1.1r_t\sqrt{\frac{E}{F_y}} = 1.1\times 3.976\times\sqrt{\frac{29{,}000}{36}} = 124 \text{ in.}$$

$$L_r = \pi r_t\sqrt{\frac{E}{0.7F_y}} = \pi\times 3.976\times\sqrt{\frac{29{,}000}{0.7\times 36}} = 423.7 \text{ in.}$$

Both compression and tension flanges are laterally braced every 15 ft. Therefore, $L_b = 15$ ft.

$$L_p = 10.3^{\text{ft}} < L_b = 15^{\text{ft}} \le L_r = 35.3^{\text{ft}}$$

By visual inspections $C_{b,AB} > C_{b,BC}$. Figure 8.40 shows the moment diagram where both flexural demand and C_b is maximum.

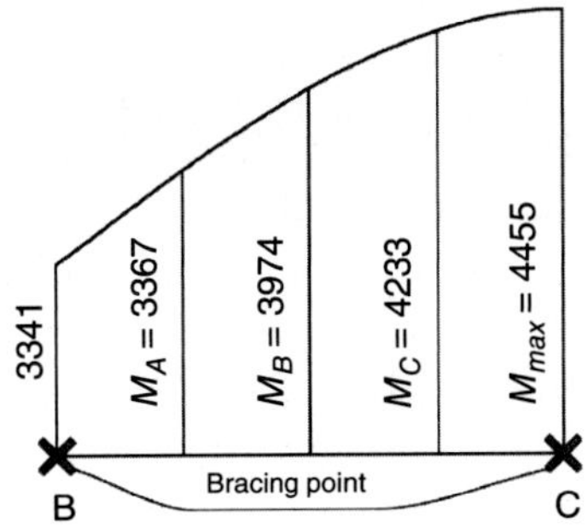

Figure 8.40 Moment diagram between B and C.

$$M_A = 3677^{\text{k-ft}}, \quad M_B = 3974^{\text{k-ft}}, \quad M_C = 4233^{\text{k-ft}}, \quad M_{\max} = 4455^{\text{k-ft}}$$

$$C_{b,BC} = \frac{12.5\times 4455}{2.5\times 4455 + 3\times 3677 + 4\times 3974 + 3\times 4233} = 1.10$$

$$F_{cr} = C_b\left[F_y - (0.3F_y)\left(\frac{L_b - L_p}{L_r - L_p}\right)\right]$$

$$= 1.10\times\left[36 - (0.3\times 36)\left(\frac{15-10.3}{35.3-10.3}\right)\right] = 37.4^{\text{ksi}} > F_y = 36^{\text{ksi}}$$

Thus, $F_{cr} = F_y = 36^{\text{ksi}}$.

$$M_n = R_{pg}F_yS_{xc} = 0.988\times 36\times 1777/12^{\text{in./ft}} = 5268^{\text{kip-ft}}$$

- Compression flange local buckling (*Section F5-3, AISC 360*)

$$\lambda_{p,f} = 0.38\sqrt{\frac{E}{F_y}} = 0.38\sqrt{\frac{29{,}000}{50}} = 9.15 > \lambda_f = b_f/2t_f = 15/(2\times 1.5) = 5$$

The limit state of flange local buckling does not apply.

- Tension flange yielding (*Section F5-4, AISC 360*)

Since $S_{xt} = S_{xc}$, the limit state of tension flange yielding does not apply.

EXAMPLE 13

A typical floor plan is shown in Fig. 8.41. The beams are ASTM A992 ($F_y = 50$ ksi, $F_u = 65$ ksi) are unshored during construction. The concrete has $f_c' = 4$ ksi. The floor has a 4½ thick normal weight concrete over

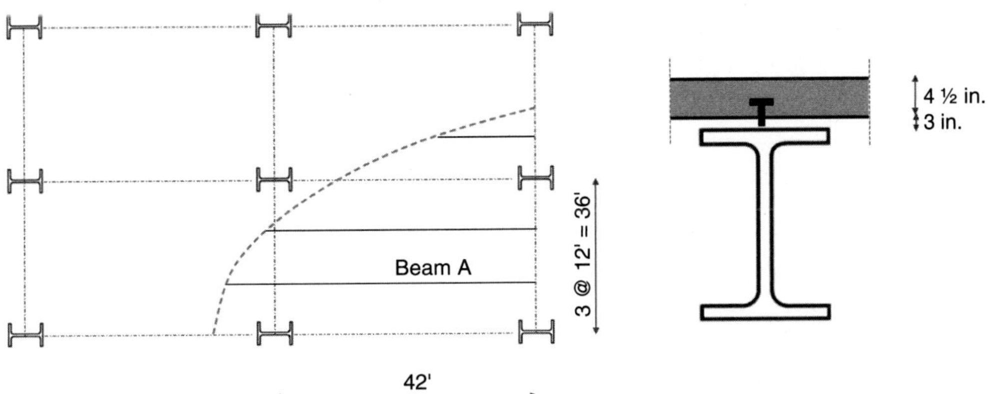

Figure 8.41 A typical floor plan and cross-section of Beam A.

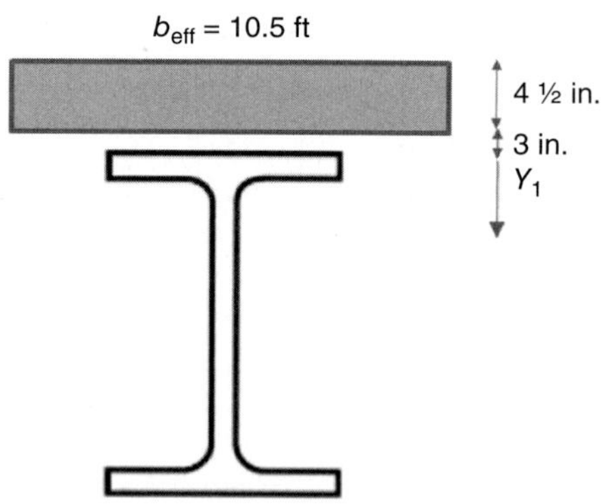

Figure 8.42 Cross-section of the composite beam.

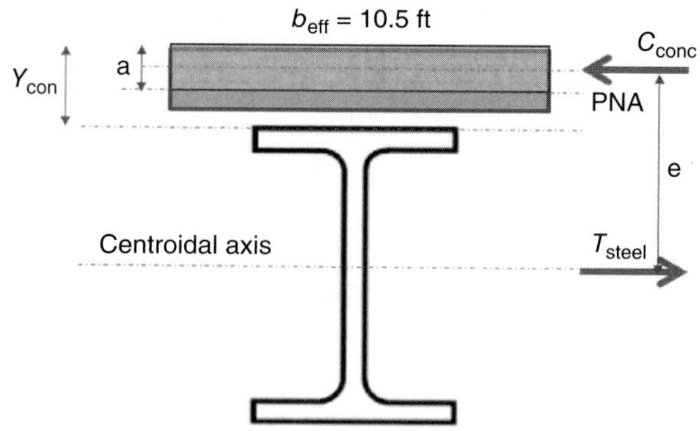

Figure 8.43 Force-couple in the cross-section.

3 in composite deck (Fig. 8.42). Assume that service live load is 100 lb/ft², and construction live load is 20 lb/ft². Select an appropriate beam size for beam A, and determine the required number of shear studs per AISC Manual. The ribs of the metal deck are perpendicular to the beam.

a. Loads and Moments:

Dead Load:

Slab = 0.075 kip/ft²
Beam weight = 0.008 kip/ft²
Miscellaneous = 0.010 kip/ft²

Live Load:

Non-reduced = 0.10 kip/ft²

Since each beam is spaced at 12 ft:

Total dead load = 0.093 × 12 = 1.12 kip/ft
Total live load = 0.10 × 12 = 1.2 kip/ft
Construction dead load (unshored)
= (0.075 + 0.008) × 12 = 1.00 kip/ft
Construction live load (unshored)
= 0.020 × 12 = 0.24 kip/ft.

The required strength:

$$w_u = 1.2 \times 1.12 + 1.6 \times 12 = 3.26 \text{ kip/ft}$$

$$M_u = \frac{1}{8} \times 3.26 \times 42^2 = 719 \text{ kip-ft}$$

$$V_u = \frac{1}{2} \times 3.26 \times 42 = 69 \text{ kip}$$

b. Select W section:

$Y_{con} = t_{slab} = 7.5$ in. (the distance from top of slab to top of steel flange, Fig. 8.42)

Assume $a = 1.0$ in. $< t_c = 4$ in., PNA is located within the concrete, $Y_1 = 0.0$

$Y_2 = Y_{con} - a/2 = 7.5 - 1.0/2 = 7.0$ in.

Enter the AISC Manual Table 3-19 with $M_u = 719$ kip-ft and $Y_2 = 7.0$ in., $Y_1 = 0.0$ (which means all steel section is subject to tension, as a fully composite beam), it seems that several W18s (46 lb and 50 lb) would work.

Try W18 × 46 ($A_s = 13.5$ in.², $d = 18.1$ in, $I_x = 712$ in.⁴)

Concrete $f_c' = 4.0$ ksi, Steel $F_y = 50$ ksi, $F_u = 65$ ksi.

c. Check flexural strength and deflection under service load:

$$b_{eff} = \min\{L/4; (b_1 + b_2)/2\} = \min\{42/4; (12 + 12)/2)\} = 10.5 \text{ ft}$$

$$a = \frac{F_y A_s}{0.85 f_c' b_{eff}} = \frac{50 \times 13.5}{0.85 \times 4 \times 10.5 \times 12} = 1.58 \text{ in.} < t_c = 4\tfrac{1}{2} \text{ in.}$$

The moment arm of internal resisting couple (Fig. 8.43) is

$$e = \frac{d}{2} + Y_{con} - \frac{a}{2} = \frac{18.1}{2} + 7.5 - \frac{1.58}{2} = 15.76 \text{ in.}$$

The nominal bending strength of a fully composite beam (with sufficient number of shear studs):

$$M_n = T \times e = F_y A_s \times e = 50 \times 13.5 \times 15.76/12 = 887 \text{ kip-ft.}$$

$$\phi_b M_n = 0.9 \times 887 = 798 \text{ kip-ft} > 719 \text{ kip-ft OK.}$$

Alternatively, from Table 3-19 in AISC Manual, W18 × 46, $Y_1 = 0.0$, $Y_2 = Y_{con} - a/2 = 7.5 - 1.58/2 = 6.71$ in.:

$$\phi_b M_n = 788 + \frac{6.71 - 6.5}{0.5}(814 - 788) = 799 \text{ kip-ft from the manual}$$

with straight line interpolation.

From the AISC Manual, Table 3-20, W18 × 46, $Y_1 = 0.0$, $Y_2 = 6.71$ in.:

$$I_{LB} = 2340 + \frac{6.71 - 6.5}{0.5}(2450 - 2340) = 2386 \text{ in.}^4 \quad \text{OK.}$$

$$\Delta = \frac{5 w_L L^4}{384 E I_{LB}} = \frac{5 \times 1.2 \times 42^4 \times 12^3}{384 \times 29{,}000 \times 2386} = 1.25 \text{ in.} < \frac{L}{360} = \frac{42 \times 12}{360} = 1.4 \text{ in. OK.}$$

d. Check flexural strength and deflection under construction load: W18 × 46, $I_x = 712$ in.⁴

The required strength:

$$w_u = 1.2 \times 1.00 + 1.6 \times 0.24 = 1.59 \text{ kip/ft}$$

$$M_u = \frac{1}{8} \times 1.59 \times 42^2 = 351 \text{ kip-ft}$$

$$\phi_b M_n = 340 \text{ kip-ft} < M_u \quad \text{Unsafe}$$

Try W18 × 55 (W18 × 50 does not work for construction load, W21 × 50 also works, let's keep W15 for its depth).

W18 × 46 ($A_s = 16.2$ in.², $d = 18.1$ in., $I_x = 890$ in.⁴, $t_w = 0.39$ in.)

Check flexural strength and deflection under service load:

$$a = \frac{F_y A_s}{0.85 f_c' b_{eff}} = \frac{50 \times 16.2}{0.85 \times 4 \times 10.5 \times 12} = 1.89 \text{ in.} < t_c = 4\tfrac{1}{2} \text{ in.}$$

The moment arm of internal resisting couple is

$$e = \frac{d}{2} + Y_{con} - \frac{a}{2} = \frac{18.1}{2} + 7.5 - \frac{1.89}{2} = 15.60 \text{ in.}$$

The nominal bending strength is:

$$M_n = T \times e = F_y A_s \times e = 50 \times 16.2 \times 15.60/12 = 1053 \text{ kip-ft.}$$

$$\phi_b M_n = 0.9 \times 1053 = 948 \text{ kip-ft} > 719 \text{ kip-ft} \quad \text{OK.}$$

Alternatively, from Table 3-19 in AISC Manual, W18 × 46, Y_1 = 0.0, $Y_2 = Y_{con} - a/2 = 7.5 - 1.72/2 = 6.64$ in.

$$\phi_b M_n = 945 + \frac{6.64-6.5}{0.5}(975-945) = 953 \text{ kip-ft from the manual}$$

with straight line interpolation.

For a fully composite beam to develop these moments, the maximum shear force between the concrete and steel beam is:

$$\sum Q_n = 0.85 \times f_c' \times b_{eff} \times a = 0.85 \times 4 \times 10.5 \times 12 \times 1.89 = 810 \text{ kips}$$

(the same as listed in Table 3-19 for W18 × 55, Y_1 = 0.0);

Also, $\Sigma Q_n = A_s F_y = 16.2 \times 50 = 810$ kips.

Notice that in this case, the construction strength controls the design (requires W18 × 55), and there is significant over-strength for a fully composite beam under service load:

$$\frac{\phi_b M_n}{M_u} = \frac{948}{719} = 1.32 \text{ (32 percent over-strength)}$$

To meet the flexural strength requirement under service load, we only need to have $\phi_b M_n \approx M_u$ = 719 kip-ft. If the same steel beam section is used, a reduced flexural strength due to a partial-composite action with reduced number of shear connectors might be an option. In other words, a partial composite beam with W18 × 55 might work well with a reduced flexural strength of 719 kip-ft as long as it satisfies the strength and deflection requirements under service loads.

Note that when we have fewer shear connectors than required to develop full flexural strength, the shear connectors might reach their collective shear strength, $\Sigma Q_n (< A_s F_y)$, and there will be a slip between concrete deck and steel beam. In such case, concrete will not reach its compressive strength f_c' before the shear connectors fail, and the equivalent PNA, at which stress = 0.0, but the concrete slab has not reached its compressive strength yet, and has to move into steel beam since total compression force in the concrete, $C' = \Sigma Q_n < A_s F_y$, is controlled by the shear connectors (Fig. 8.28c).

In a partially composite beam, characterized by the fact that $\Sigma Q_n < A_s F_y$, we can use the same equilibrium as used in a fully composite beam to determine the location (say, Y_1) of the equivalent PNA since the sum of forces along the beam length should be equal to zero, that is,

$$\sum F_z = 0, \quad \sum Q_n + A_{comp} F_y = (A_s - A_{comp}) F_y$$

$$0.85 f_c' b_{eff}\, a' = \sum Q_n$$

$$a' = \frac{\sum Q_n}{0.85 f_c' b_{eff}}$$

where A_{comp} = the area of steel above the equivalent PNA that is subjected to compression, in.^{2}

a' = equivalent depth of the concrete from the top assuming that the concrete has a uniform stress of $0.85 f_c'$

After PNA is determined (say, Y_1) is known from the above equation, we can find the flexural strength M_n by summing all forces taking moment about any point in the section. It is obvious that M_n in a partially composite section is reduced in proportion to $\Sigma Q_n / F_y A_s$ ratio. This ratio is equal to one (1.0) in a fully composite beam, and equal to zero (0.0) in a non-composite beam (with no shear connector at all). The smaller the ratio is, the lower the flexural strength. As a minimum ratio to keep any composite action, AISC suggests that $\Sigma Q_n / F_y A_s \geq 0.25$. Otherwise, the beam should be treated as non-composite bare steel beam.

In AISC Table 3-19, a fully composite beam is given as Y_1 = 0.0, which indicates $\Sigma Q_n / F_y A_s$ = 1.0 regardless where the real PNA is in the concrete. For $Y_1 > 0$, Table 3-19 might be used for partially composite beam case, where $0.25 \leq \Sigma Q_n / F_y A_s < 1.0$.

For this problem, if we make $\phi_b M_n \approx M_u$ = 719 kip-ft, we can go to the case in Table 3-19 in the specification where Y_1 is largest possible to provide $\phi_b M_n$ of about 719 kip-ft with $Y_2 \approx 6.6$ in. This occurs when Y_1 = 0.473 in. (in the top flange of the beam) and ΣQ_n = 454 kips, $\Sigma Q_n / F_y A_s = 454/810 = 0.56 > 0.25$ (OK.)

$$a' = \frac{\Sigma Q_n}{0.85 f_c' b_{eff}} = \frac{454}{085 \times 4 \times 10.5 \times 12} = 1.06 \text{ in.}$$

$$Y_2 = Y_{con} - a/2 = 7.5 - 1.06/2 = 6.97 \text{ in.}$$

$$\phi_b M_n = 765 + \frac{6.97-6.5}{0.5}(782-765) = 781 \text{ kip-ft} > 719 \text{ kip-ft} \quad \text{OK.}$$

Compute Live Load Deflection: From the AISC Manual, Table 3-20, W18 × 55, Y_1 = 0.473, Y_2 = 6.97 in.

$$I_{LB} = 2300 + \frac{6.97-6.5}{0.5}(2390-2300) = 2385 \text{ in}^4 \quad \text{OK.}$$

$$\Delta = \frac{5 w_L L^4}{384 E I_{LB}} = \frac{5 \times 1.2 \times 42^4 \times 12^3}{384 \times 29{,}000 \times 2385} = 1.2 \text{ in.} < \frac{L}{360} = \frac{42 \times 12}{360} = 1.4 \text{ in.} \quad \text{OK.}$$

Check Flexural Strength and Deflection under Construction Load:

W18 × 55, I_x = 890 in.^{4}

The required strength:

$$M_u = \frac{1}{8} \times 1.59 \times 42^2 = 351 \text{ kip-ft}$$

$$\phi_b M_n = \phi_b M_p = 420 \text{ kip-ft} > M_u \quad \text{OK.}$$

$$\Delta = \frac{5 w_L L^4}{384 E I_x} = \frac{5 \times 0.24 \times 42^4 \times 12^3}{384 \times 29{,}000 \times 890} = 0.65 \text{ in.} < 2.5 \text{ in.} \quad \text{OK.}$$

e. Design of shear studs for the partially composite beam
Use perpendicular deck with one ¾ in. diameter weak stud per rib in normal weight 4 ksi concrete.
In AISC Manual, Table 3-21, with deck perpendicular to the beam, normal weight concrete:

$$Q_n = 17.2 \text{ kips.}$$

From AISC Manual, Table 3-21, W18 × 55, Y_1 = 0.473, ΣQ_n = 454 kips

The number of shear studs for the half span of the beam = $\frac{\Sigma Q_n}{Q_n} = \frac{454}{17.2} = 26.4$ (use 27)

The total number of shear studs in the beam is 27 × 2 = 54

Spacing along the beam length: with 6 in. flute metal deck, one stud every flute (6 in. apart) in the middle 12 ft, and one stud every two flutes (12 in. apart).

Check the Spacing Requirements:

Minimum spacing = $6d_{stud}$ = 6 × ¾ = 4.5 in.

Maximum spacing = $8t_{slab}$ = 8 × 4.5 = 34 in. OK.

f. Check shear strength:
The shear strength of fully or partially composite beams with shear connectors is based on the properties of the steel section alone (neglect the contribution from the concrete for shear).

W18 × 55, d = 18. in., t_w = 0.39 in.

$$V_n = 0.6\, F_y A_w = 0.6 F_y T_w D = 0.6 \times 50 \times 0.39 \times 18.1 = 212 \text{ kips}$$

$$\phi V_n = 0.9 \times 212 = 191 \text{ kips} > V_u = 69 \text{ kips} \quad \text{OK.}$$

Final Design: W18 × 55, F_y = 50 ksi. Fifty-four ¾ in. diameter headed shear studs.

8.1.5 Beam-Column Members

When the axial load and bending moment acting on the member are both significant, the member is treated as a beam-column. The rules set forth in the previous chapters (Column Buckling, Laterally Supported Beams and Lateral Torsional Buckling of Beams) apply for beam-columns.

AISC provides interaction equations for the design of beam-column elements (*Section H, AISC 360*). Interaction between axial force, *P*, and bending moment, *M*, is the main character of a beam-column member, which affects both sides of design equation: capacity and demand. For instance, bending moment capacity reduces significantly as axial load increases or vice versa. Demand on a beam-column member is also affected by the P-M interaction due to the fact that stability for the whole structure and for each structural element is directly related to this interaction. Therefore, a second-order analysis taking into account both *P*-δ and *P*-Δ is needed. These second-order effects can be considered either by an approximate method or by direct analysis. Figure 8.44 illustrates the definition of *P*-δ and *P*-Δ and the moment amplification by the approximate second-order analysis (effective length method).

Direct analysis method (DAM) of design requires a structural engineering software and is becoming a standard method for stability analyses among structural engineers. The essentials of this method are summarized below (see also Table 8.5):

a. There is no limitation for the use of this method.

b. The axial loads acting on the gravity-only columns should be included in the structural model.

c. All stiffnesses (axial, flexural, shear) of the structural elements should be reduced 20 percent.

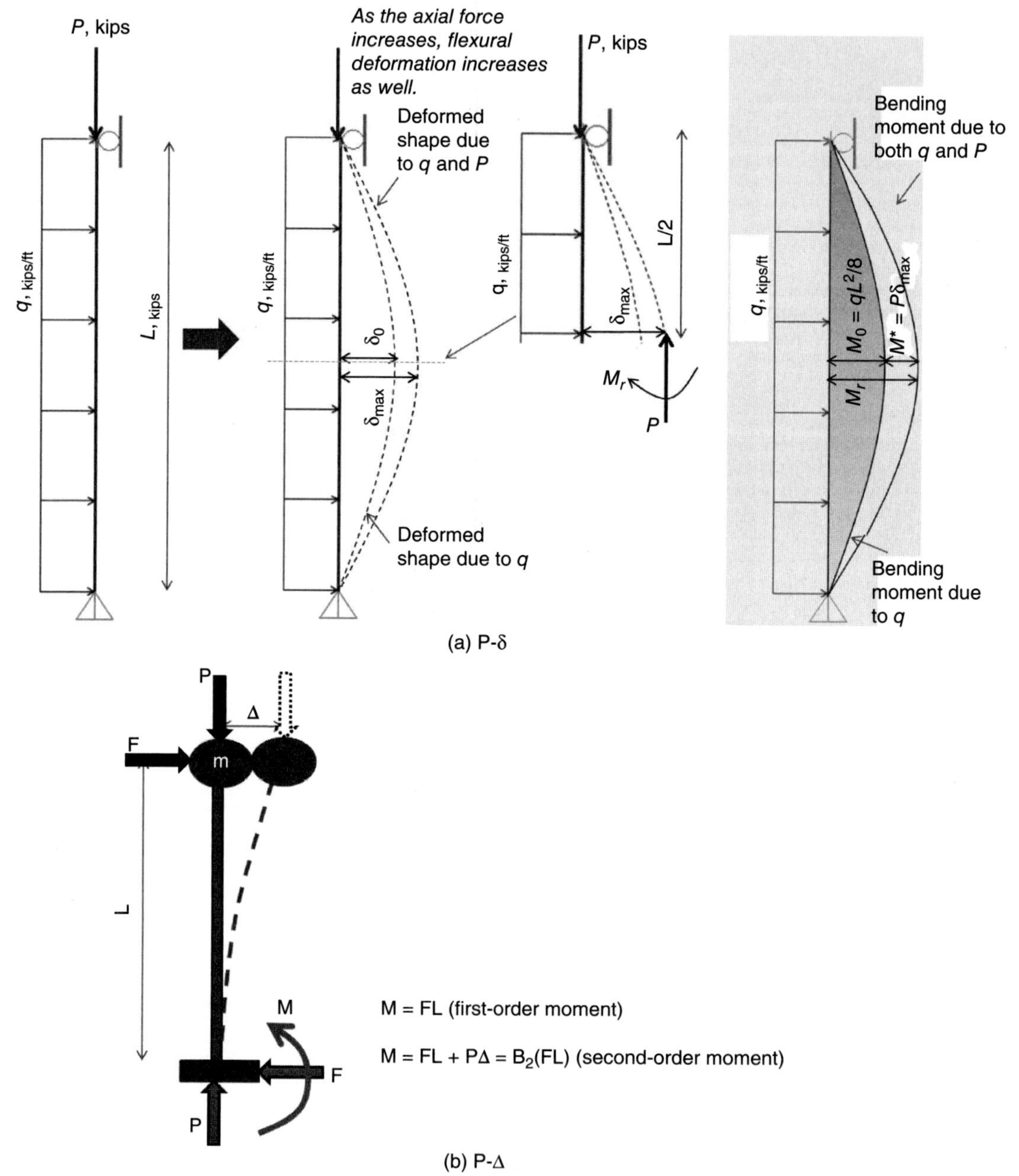

Figure 8.44 Illustration of *P*-δ and *P*-Δ effects.

Table 8.5 Notes on Direct Analysis Method of Design

Stiffness reduction parameter, τ_b	Notes
Varying	Reduced stiffnesses for second-order analysis: $EI^* = 0.8\tau_b EI$; $EA^* = 0.8EI$; $\rightarrow \tau_b = 1.0$; $\frac{P_r}{P_y} > 0.5 \rightarrow \tau_b = 4\left(\frac{P_r}{P_y}\right)\left(1-\frac{P_r}{P_y}\right)$; P_r = required axial compressive strength; $P_y = F_y A_g$; K = 1.0; $\Delta_{2nd}/\Delta_{1st} \le 1.7 \rightarrow$ notional loads in gravity-only load combinations, otherwise in all load combinations that include lateral loads; $N_i = 0.002Y_i$
Fixed	Reduced stiffnesses for second-order analysis: $EI^* = 0.8EI$; $EA^* = 0.8EI$; $\tau_b = 1.0$; K = 1.0; $\Delta_{2nd}/\Delta_{1st} \le 1.7 \rightarrow$ notional loads in gravity-only load combinations, otherwise in all load combinations that include lateral loads; $N_i = 0.003Y_i$

d. Notional loads, N_i, that consider imperfections should be added to the only gravity-only load combinations except in the case where the ratio of the maximum second-order drift to maximum first-order drift ($\Delta_{2nd}/\Delta_{1st}$) exceeds 1.7 in which case notional loads should also be added to the combinations that include other lateral loads (earthquake and wind).

$$N_i = 0.002Y_i \tag{8.13}$$

where Y_i is the gravity load applied at level i (kips).

e. Stiffness reduction parameter, τ_b, is used to adjust the flexural stiffness of all members (see Table 8.5)

EXAMPLE 14

Figure 8.45 shows the elevation of identical moment frames on lines A and C and a typical plan of a three-story building. The building is subject to dead (D), live (L), and wind (W) loads.

- Gravity loads on each floor: Dead Load = 100 psf; Live Load = 50 psf.
- Gravity loads on roof: Dead Load = 50 psf; Live Load = 20 psf.
- Wind loads acting on story levels are computed as F_1 = 67.5 kips, F_2 = 60 kips, and F_3 = 30 kips for one frame either on line A or line C.
- Governing load combination: 1.2 D + 1.0 W + 1.0 L + 0.5 L_r

Using approximate second-order analysis (effective length method), determine the adequacy of the preliminary member selected for the second-story exterior column (Column A1).

Using the tributary areas, gravity and wind load distribution on the frame is determined as given in Fig. 8.46. The frame is analyzed under the given loads and the first-order moment and axial force diagrams due to the factored gravity (1.2 D + L + 0.5 L_r) and lateral (W) loads are plotted in Figs. 8.47 to 8.50.

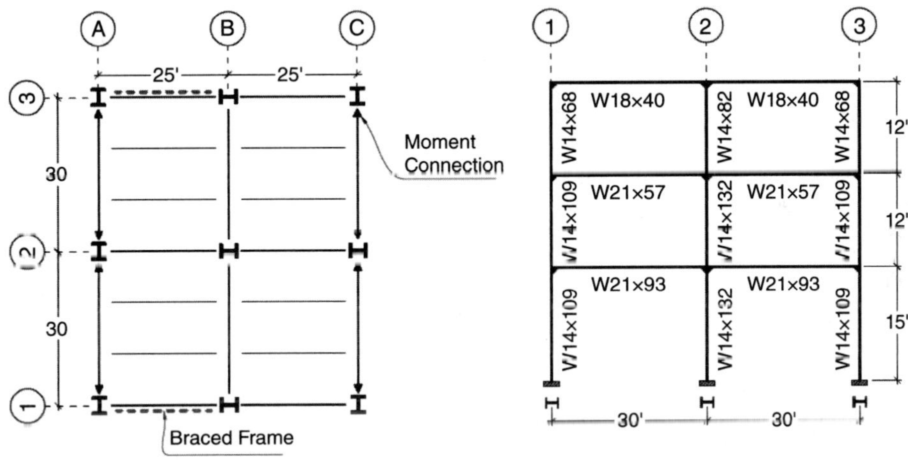

Figure 8.45 A typical plan for floors and roof (left) and elevation on line A (right).

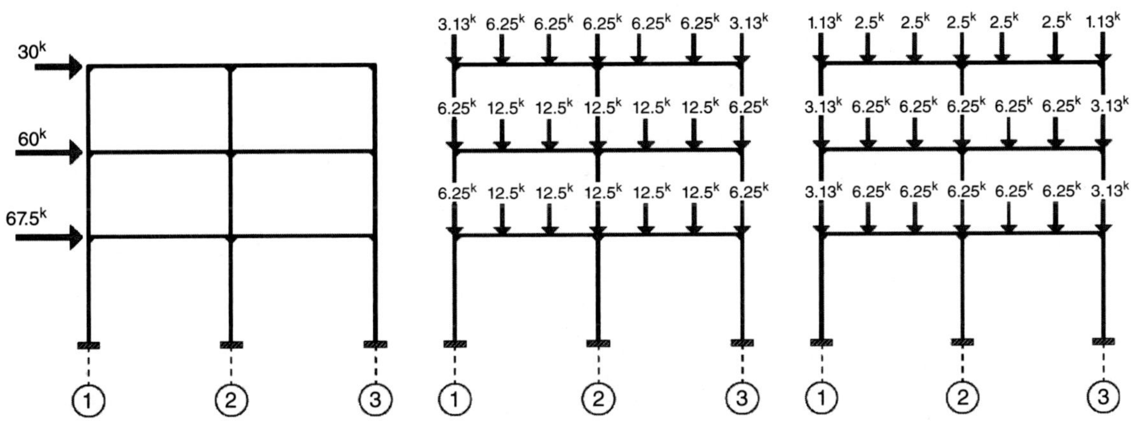

Figure 8.46 Wind (left), dead (middle), and live (right) load distribution on the frame.

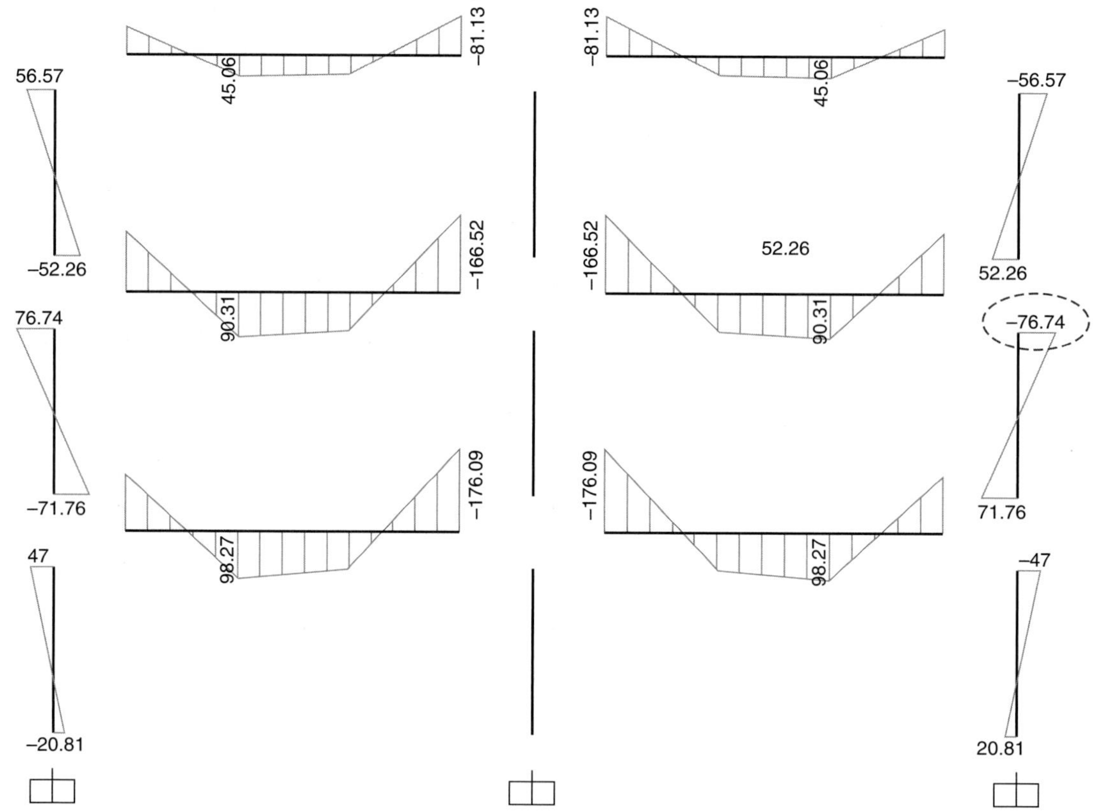

Figure 8.47 Moment diagram due to 1.2 D + 1.0 L (no translation, M_{nt}).

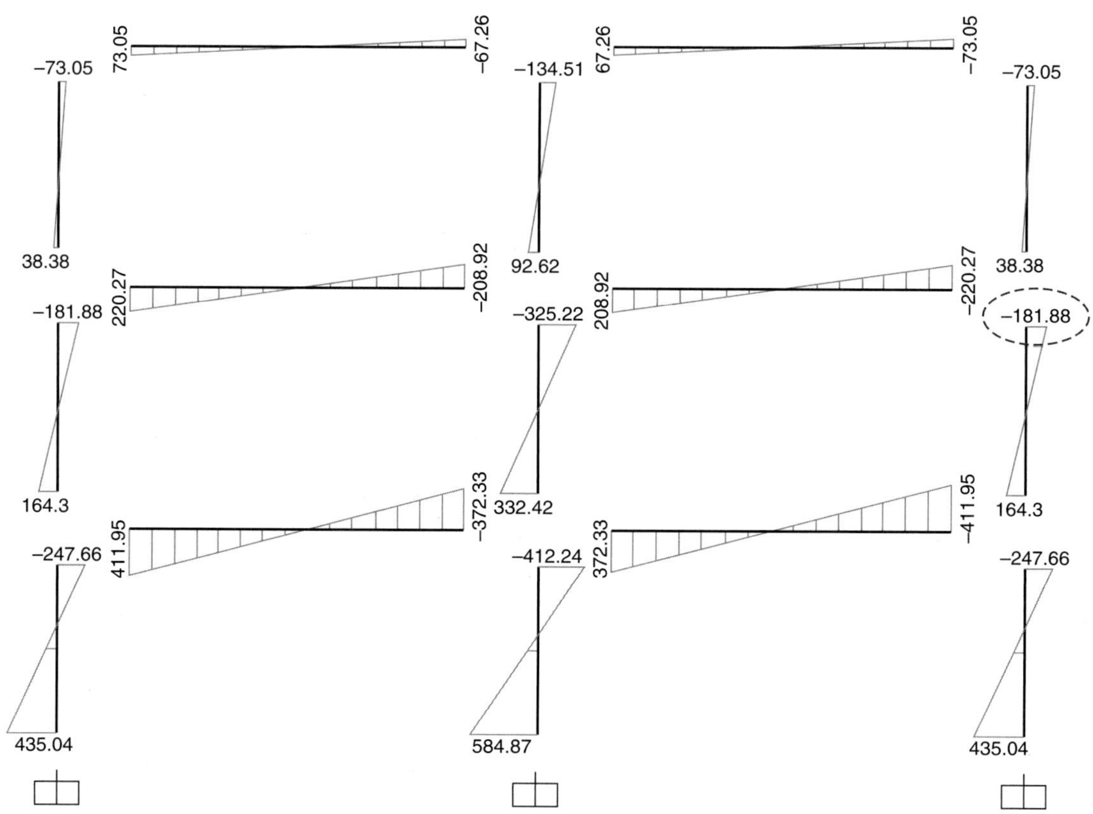

Figure 8.48 Moment diagram due to W (lateral translation, M_{lt}).

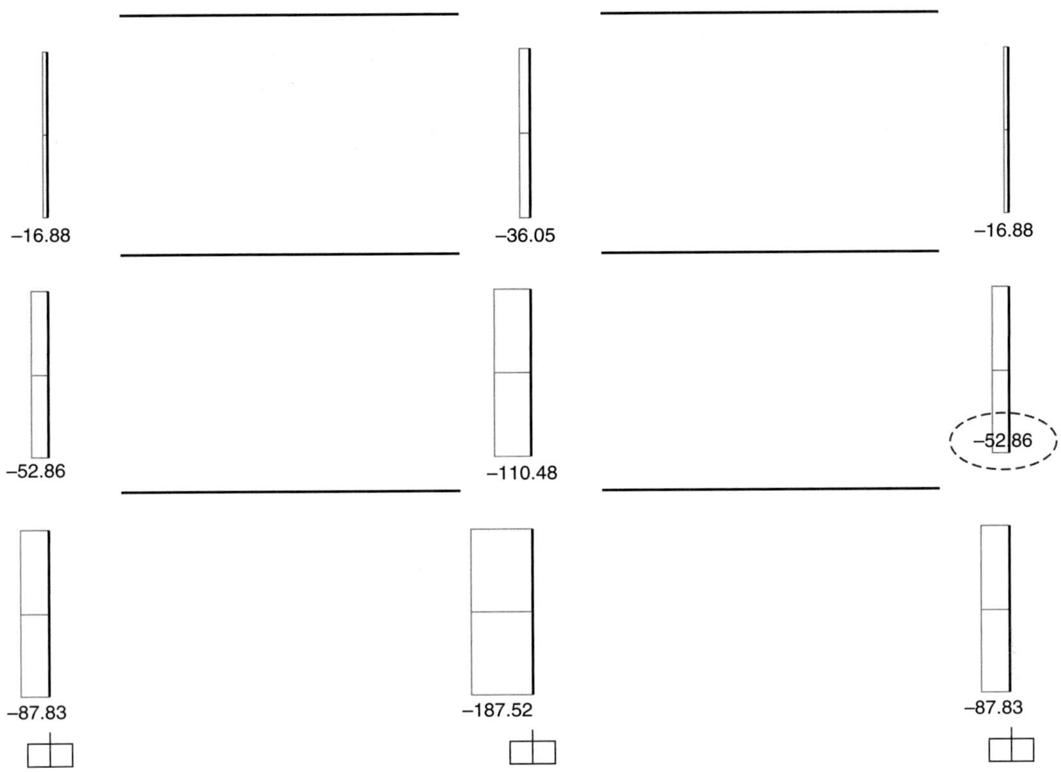

Figure 8.49 Axial force diagram due to 1.2 D + 1.0 L + 0.5 L_r (no translation, P_{nt}).

Figure 8.50 Axial force diagram due to W (lateral translation, P_{lt}).

To determine the approximate second-order bending moment and axial force demands on the column, the amplification factors that account for P-δ and P-Δ effects should be introduced first. According to *Appendix 8, AISC 360*, the required flexural strength, M_r, and axial strength, P_r, for beam-column members should be calculated using B_1 and B_2 factors, which represent the amplification due to P-δ and P-Δ effects, respectively.

$$M_r = B_1 M_{nt} + B_2 M_{lt} \qquad \textit{(Eq. A-8-1, AISC 360)}$$

$$P_r = P_{nt} + B_2 P_{lt} \qquad \textit{(Eq. A-8-2, AISC 360)}$$

where M_{lt} = first-order moment due to lateral translation of the structure only.
M_{nt} = first-order moment with the structure restrained against lateral translation.
P_{lt} = first-order axial force due to lateral translation of the structure only.
P_{nt} = first-order axial force with the structure restrained against lateral translation.

Multiplier B_1 can be calculated as,

$$B_1 = \frac{C_m}{1+\alpha P_r/P_{e1}} \geq 1.0 \qquad \textit{(Eq. A-8-3, AISC 360)}$$

$$\left.\begin{aligned} G_{top} &= \frac{\sum(E_c I_c/L_c)}{\sum(E_b I_b/L_b)} = \frac{1240/12+722/12}{1170/30} = 4.2 \\ G_{bottom} &= \frac{\sum(E_c I_c/L_c)}{\sum(E_b I_b/L_b)} = \frac{1240/12+1240/15}{2070/30} = 2.7 \end{aligned}\right\} K_x \cong 1.9$$

$$C_m = 0.6-0.4(M_1/M_2) = 0.6-0.4\times(71.76/76.64) = 0.225$$

$$P_r = P_{nt} + P_{lt} = 52.36+18.98 = 71.34^{\text{kips}}$$

$$P_{e1} = \frac{\pi^2 EI^*}{(L_{c1})^2} = \frac{\pi^2\times(29{,}000\times1240)}{(1.9\times12\times12)^2} = 4{,}741^{\text{kips}}$$

$$\alpha = 1.0 \text{ (LRFD)}$$

$$B_1 = \frac{0.225}{1+1.0\times\dfrac{71.34}{4741}} = 0.22 < 1.0$$

Thus, $B_1 = 1.0$.

Multiplier B_2 can be calculated as,

$$B_2 = \frac{1}{1-\dfrac{\alpha P_{story}}{R_M\dfrac{HL}{\Delta_H}}} \qquad \textit{(Eqs. A-8-6 and A-8-7, AISC 360)}$$

where P_{story} = total vertical load supported by the story using LRFD or ASD load combinations, as applicable, including loads in columns that are not part of the lateral force-resisting system.
$P_{e\,story}$ = elastic critical buckling strength for the story in the direction of translation being considered.
H = total story shear.
L = height of story.
ΔH = first-order interstory drift.

$$P_{story} = \text{Factored Gravity Loads} \times \frac{\text{Floor Area}}{\text{Number of moment frames}} \times \text{Number of stories above}$$

$$= \left[\left(\frac{1.2\times100^{\text{psf}}+1.0\times50^{\text{psf}}}{1000^{\text{lb/kip}}}\right)+\left(\frac{1.2\times50^{\text{psf}}+1.0\times20^{\text{psf}}}{1000^{\text{lb/kip}}}\right)\right]\times\left(\frac{50^{\text{ft}}\times60^{\text{ft}}}{2}\right)$$

$$= 375^{\text{kips}}$$

$$\Delta_H = \delta_2 - \delta_1 \text{ (see Fig. 8.51)}$$

$$= 2.2^{\text{in.}} - 1.09^{\text{in.}} = 1.12^{\text{in.}}$$

$$H = F_1 + F_2 + F_3$$

$$= 67.5+60+30 = 157.5^{\text{kips}}$$

$$R_M = 0.85$$

$$\alpha = 1.0 \text{ (LRFD)}$$

Amplification factor for P-Δ effects, B_2, is computed as follows for the second story:

$$B_2 = \frac{1}{1-\dfrac{1.0\times375^{\text{kips}}}{0.85\times\dfrac{157.5^{\text{kips}}\times(12^{\text{ft}}\times12^{\text{in./ft}})}{1.12^{\text{in.}}}}} = 1.02$$

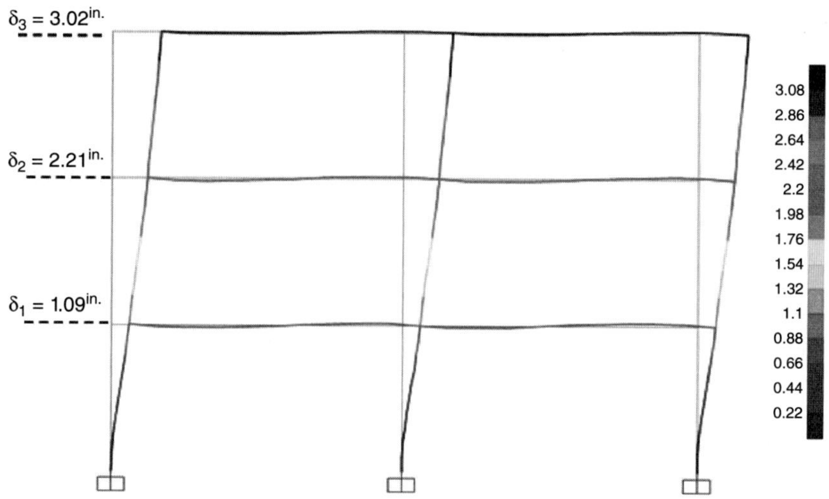

Figure 8.51 Lateral displacements due to wind load.

Using the amplification factors and the first-order analysis results, the required strengths are determined as follows:

$$\left.\begin{array}{l} M_{nt} = 76.74^{\text{kip-ft}} \\ M_{lt} = 181.88^{\text{kip-ft}} \\ B_1 = 1.00 \\ B_2 = 1.02 \end{array}\right| M_r = B_1 M_{nt} + B_2 M_{lt}$$

$$= 1.0 \times 76.74 + 1.02 \times 181.88 = 262.3^{\text{kip-ft}}$$

$$\left.\begin{array}{l} P_{nt} = 52.36^{\text{kips}} \\ P_{lt} = 18.98^{\text{kips}} \\ B_2 = 1.02 \end{array}\right| P_r = P_{nt} + B_2 P_{lt} = 52.36 + 1.02 \times 18.98 = 71.7^{\text{kips}}$$

Determine the Adequacy of W14 × 109

- Compressive strength, P_c
  W14 × 109: $A_g = 32$ in.^{2}; $r_x = 6.22$ in.; $r_y = 3.73$ in.; Compact for compression with $F_y = 50$ ksi

$$K_x = 1.90;\ K_y = 1.0$$

$$\left(\frac{L_c}{r}\right)_x = \frac{1.90 \times (12 \times 12)}{6.22} = 44 > \left(\frac{L_c}{r}\right)_y = \frac{1.0 \times (12 \times 12)}{3.73} = 38.6$$

Thus, inelastic flexural buckling occurs about *x*-*x* axis.

$$F_{cr} = (0.658^{50/147.8}) \times 50 = 43.4^{\text{ksi}}$$

$$P_c = \phi_c F_{cr} A_g = 0.9 \times 43.4 \times 32 = 1250^{\text{kips}}$$

- Flexural strength, M_c
  W14 × 109: $L_p = 13.2$ ft, $Z_x = 192$ in.^{3}; Compact for flexure with $F_y = 50$ ksi

$$L_p = 13.2^{\text{ft}} > L_b = 12^{\text{ft}} \rightarrow \text{The limit state of LTB does not apply.}$$

$$M_c = \phi_b M_p = \phi_b F_y Z_x = 0.9 \times 50 \times 192/12^{\text{in./ft}} = 720^{\text{kip-ft}}$$

- Check axial force-bending moment interaction according to *Eqs. H1-1a and H1-1b, AISC 360*

$$\left.\begin{array}{l} M_r = 262.3^{\text{k-ft}} \\ P_r = 71.7^{\text{kips}} \\ M_c = 720^{\text{k-ft}} \\ P_c = 1250^{\text{kips}} \end{array}\right\} \Rightarrow \left| \begin{array}{l} \dfrac{P_r}{P_c} = \dfrac{71.7}{1250} = 0.06 < 0.2 \rightarrow \text{Use Eq. H1-1b} \\ \dfrac{P_r}{2P_c} + \dfrac{M_r}{M_c} = \dfrac{71.7}{2 \times 1250} + \dfrac{262.3}{720} = 0.40 < 1.0 \end{array}\right|$$

Preliminary design is safe but not economical.

EXAMPLE 15

Determine the required strength of the second-story exterior column (Column A1) in the building shown in Fig. 8.45 using DAM.

Step I: Create a Model with the Leaning Columns A leaning column consisting of rigid truss elements is incorporated in the numerical model. The column is loaded with the vertical gravity loads supported by the columns that are not part of the moment frame.

$$P_{\text{leaning}}^{\text{floor}} = (12.5^{\text{ft}} \times 60^{\text{ft}}) \times 100^{\text{psf}}/1000^{\text{lb/kip}} = 75^{\text{kips}}$$

$$P_{\text{leaning}}^{\text{floor}} = (12.5^{\text{ft}} \times 60^{\text{ft}}) \times 50^{\text{psf}}/1000^{\text{lb/kip}} = 37.5^{\text{kips}}$$

Step II: Reduce All Stiffnesses That Contribute to the Stability To account for the reduction in stiffness, flexural and axial stiffnesses are reduced by 20 percent (assuming $\tau_b = 1.0$). AISC requires an additional reduction in the flexural stiffness by using a factor, τ_b, when the axial force in a member is greater than 50 percent of its axial load-carrying capacity.

$$\tau_b = 1.0 \text{ when } \frac{\alpha P_r}{P_y} \le 0.5 \quad \text{and} \quad \tau_b = 4(\alpha P_r/P_y)[1 - (\alpha P_r/P_y)]$$

when $\dfrac{\alpha P_r}{P_y} > 0.5$

Step III: Apply Notional Loads The notional load, N_i, is taken as 0.002 times the total gravity load at each story level.

$$N_i = 0.002\alpha Y_i$$

where $\alpha = 1.0$ (LRFD), $\alpha = 1.6$ (ASD), and Y_i is the total gravity load at each story level, including the loads supported by gravity frames. The load combinations adopted for the stability analysis are given in Table 8.6.

Table 8.6 Load Combinations Including Notional Load Effects

Combination	Description
Comb#1	1.4D ± 1.4ND
Comb#2	1.2D + 1.6L + 0.5L$_r$ ± (1.2ND + 1.6NL + 0.5NL$_r$)
Comb#3	1.2D + 1.0L + 1.6L$_r$ ± (1.2ND + 1.0NL + 1.6NL$_r$)
Comb#4	1.2D + 1.6L$_r$ ± 0.5W
Comb#5	1.2D ± 1.0W + L + 0.5L$_r$
Comb#6	0.9D ± 1.0W

*ND, Notional Dead Load; NL, Notional Live Load; NLr, Notional Roof Live Load.
**Notional loads are only included in gravity load combinations.

Step IV: Verify $\Delta_{2nd}/\Delta_{1st}$ *AISC 360* requires the ratio of second-order drifts to first-order drifts to be examined. If the ratio of second-order drifts to first-order drifts is less than 1.7 for all combinations, it is not required to include the notional loads in the combinations with other lateral loads (*AISC 360*, Section C2.2b). Note that each column in the frame is divided into 4-ft long element to take *P*-δ effects into account. As given in Table 8.7, second-to-first-order drift ratios are less than 1.7 for all combinations. Therefore, the notional loads are not included in combinations 4 through 6.

Table 8.7 The Ratio of Second-Order Drifts to First-Order Drifts ($\Delta_{2nd}/\Delta_{1st}$)

Combination	Story	First-order drift (in.)	Second-order drift (in.)	$\Delta_{2nd}/\Delta_{1st}$ ratio
1	1	0.009	0.009	1.03
	2	0.010	0.010	1.03
	3	0.008	0.008	1.02
2	1	0.010	0.010	1.03
	2	0.011	0.011	1.03
	3	0.007	0.008	1.03
3	1	0.009	0.009	1.03
	2	0.010	0.011	1.03
	3	0.008	0.008	1.03
4	1	0.590	0.604	1.02
	2	0.644	0.660	1.03
	3	0.479	0.491	1.02
5	1	1.179	1.211	1.03
	2	1.288	1.323	1.03
	3	0.959	0.981	1.02
6	1	1.179	1.199	1.02
	2	1.288	1.310	1.02
	3	0.959	0.974	1.02

Step V: Check if $\tau_b = 1.0$ To determine the actual τ_b values, envelope of the axial force diagrams is plotted in Fig. 8.52.

Example τ_b Calculation for the Second-Story Exterior Column (W14 × 109: $A_g = 32$ in.^{2}):

$$\left.\begin{array}{l} P_{r,\max} = 70.9^{\text{kips}} \\ P_y = A_g F_y = 32 \times 50 = 1600^{\text{kips}} \end{array}\right| \frac{\alpha P_{r,\max}}{P_y} = \frac{1.0 \times 70.9}{1600} = 0.04 < 0.50$$

Thus, $\tau_b = 1.0$

The τ_b values for the exterior and interior columns are summarized in Table 8.8.

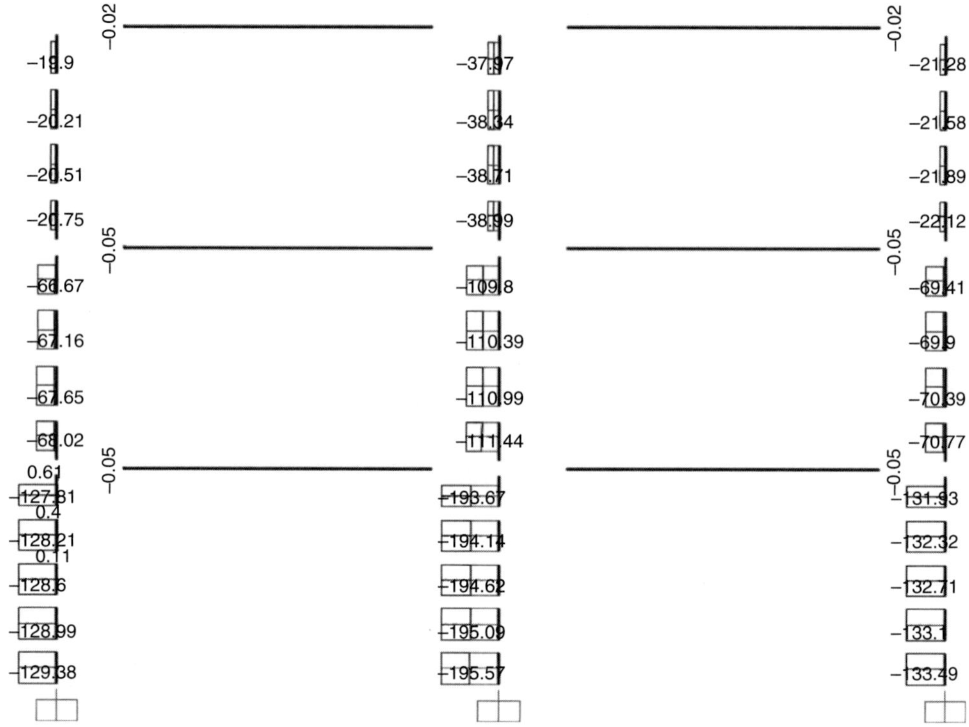

Figure 8.52 Envelope of the axial force diagrams obtained from second-order analysis.

Table 8.8 Axial Force-to-Capacity Ratio for Columns

	Exterior column				Interior column			
Story	Section	P_r (kips)	P_r/P_y (kips)	τ_b	Section	P_r (kips)	P_r/P_y (kips)	τ_b
3	W14 × 68	22.2	0.02	1.0	W14 × 82	39.1	0.03	1.0
2	W14 × 109	70.9	0.04	1.0	W14 × 132	111.6	0.06	1.0
1	W14 × 109	133.5	0.08	1.0	W14 × 132	195.6	0.10	1.0

Step VI: Determine Capacities of Members Using $K = 1.0$
Determine the Adequacy of W14 × 109

- Compressive strength, P_c

W14 × 109: A_g = 32 in.²; r_x = 6.22 in.; r_y = 3.73 in.; Compact for compression with F_y = 50 ksi

$$K_x = K_y = 1.0$$

$$\left(\frac{L_c}{r}\right) = \left(\frac{L_c}{r}\right)_y = \frac{1.0\times(12\times12)}{3.73} = 38.6$$

$$F_{cr} = (0.658^{50/192.1})\times 50 = 44.8^{\text{ksi}}$$

$$P_c = \phi_c F_{cr} A_g = 0.9\times 44.8\times 32 = 1290^{\text{kips}}$$

- Flexural strength, M_c

$$M_c = \phi_b M_p = \phi_b F_y Z_x = 0.9\times 50\times 192/12^{\text{in./ft}} = 720^{\text{kip-ft}}$$

- Check axial force-bending moment interaction (*Eqs. H1-1a and H1-1b, AISC 360*)

The required axial force and flexural strengths are obtained from the second-order analysis as follows:

$$\left.\begin{aligned} M_r &= 270.3^{\text{k-ft}} \\ P_r &= 70.9^{\text{kips}} \\ M_c &= 720^{\text{k-ft}} \\ P_c &= 1290^{\text{kips}} \end{aligned}\right\} \Rightarrow \left|\begin{aligned} &\frac{P_r}{P_c} = \frac{70.9}{1290} = 0.06 < 0.2 \rightarrow \text{Use Eq. H1-1b} \\ &\frac{P_r}{2P_c} + \frac{M_r}{M_c} = \frac{70.9}{2\times1290} + \frac{270.3}{720} = 0.40 < 1.0 \end{aligned}\right|$$

Preliminary design is safe but not economical.

EXAMPLE 16

A composite column subject to combined axial force (P_u = 600 kips) and flexural moment (M_u = 350 kip-ft) consists of a W8 × 40 encased in a 20 in. × 20-in. concrete column (Fig. 8.53). Eight #7 bars are used for longitudinal reinforcement and #3 lateral ties spaced at 12 in. centerline to centerline are used for lateral reinforcement. F_y = 50 ksi for steel shape and Grade 60 reinforcing bars are used. Normal weight concrete with f_c' = 6 ksi is used. Determine whether the composite column is adequate for the given loading. Assume an effective length of 18 ft and 9 ft about x- and y-axes, respectively.

$$A_s = 11.7 \text{ in.}^2 \geq 0.01A_g = 0.01\times 20\times 20 = 4 \text{ in.}^2 \quad \text{OK.}$$

$$F_y = 50 \text{ ksi} < F_y = 75 \text{ ksi}$$

$$F_{ysr} = 55 \text{ ksi} < F_y = 80 \text{ ksi}$$

$$E_s = 29{,}000 \text{ ksi}$$

$$Ec = w_c^{1.5}\sqrt{f_c'} = 145^{1.5}\sqrt{5} = 3904 \text{ ksi}$$

$$3 \text{ ksi} \leq f_c' = 5 \text{ ksi} \leq 10 \text{ ksi}$$

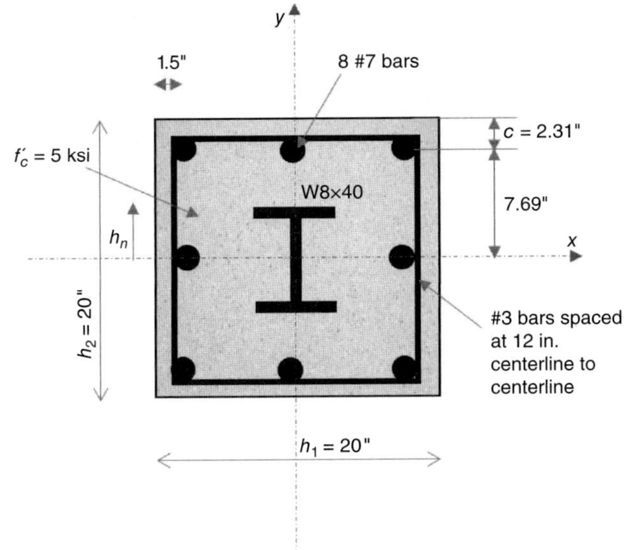

Figure 8.53 Example 16.

For lateral ties #3 spaced at 12 in. (OK.):

$$\rho_{sr} = A_{sr}/A_g = \left(8\times\frac{\pi 0.875^2}{4}\right)/(20\times 20) = 0.012 \geq 0.004 \quad \text{OK.}$$

$$A_c = 20\times 20 - A_s - A_{sr}$$

$$= 400 - 11.7 - \left(8\times\frac{\pi 0.875^2}{4}\right)$$

$$= 400 - 11.7 - 4.81$$

$$= 383.49\ \text{in.}^2$$

$$P_{no} = F_y A_s + F_{ysr} A_{sr} + 0.85 f_c' A_c$$

$$= 50\times 11.7 + 55\times\left(8\times\frac{\pi 0.875^2}{4}\right) + 0.85\times 5\times 383.49$$

$$= 2479.4\ \text{kips}$$

$$C_1 = 0.25 + 3\left(\frac{A_s + A_{sr}}{A_g}\right) \leq 0.7$$

$$C_1 = 0.25 + 3\left(\frac{11.7 + 4.81}{20\times 20}\right) = 0.374 \leq 0.7$$

Effective Stiffness about x-Axes

$$EI_{\text{eff},x} = E_s I_{s,x} + E_s I_{sr,x} + C_1 E_c I_{c,x}$$

$$I_{s,x} = 146\ \text{in.}^4$$

$$I_{sr,x} = \left(8\times\frac{\pi 0.875^4}{64}\right) + 6\times\left(\frac{\pi 0.875^4}{4}\right)\times 7.69^2 = 213.6\ \text{in.}^4$$

$$I_{c,x} = \left(\frac{20\times 20^3}{12}\right) - 146 - 213.6$$

$$= 12{,}973.7\ \text{in.}^4$$

$$EI_{\text{eff},x} = 29{,}000\times 146 + 29{,}000\times 213.6 + 0.374\times 3904\times 12{,}973.7$$

$$= 29{,}371{,}248\ \text{kip-in.}^2$$

Effective Stiffness about y-Axes

$$EI_{\text{eff},y} = E_s I_{s,y} + E_s I_{sr,y} + C_1 E_c I_{c,y}$$

$$I_{s,y} = 49.1\ \text{in.}^4$$

$$I_{sr,y} = \left(8\times\frac{\pi 0.875^4}{64}\right) + 6\times\left(\frac{\pi 0.875^4}{4}\right)\times 7.69^2 = 213.6\ \text{in.}$$

$$I_{c,y} = \left(\frac{20\times 20^3}{12}\right) - 49.1 - 213.6$$

$$= 13{,}070.6\ \text{in.}^4$$

$$EI_{\text{eff},y} = 29{,}000\times 49.1 + 29{,}000\times 213.6 + 0.374\times 3904\times 13{,}070.6$$

$$= 26{,}702{,}631\ \text{kip-in.}^2$$

$$P_{e,x} = \pi^2(EI_{\text{eff},x})/L_{c,x}^2 = \pi^2(29{,}371{,}248)/(1.0\times 18\times 12)^2 = 6213.2\ \text{kips}$$

$\rightarrow$ buckling about x-axes controls

$$P_{e,y} = \pi^2(EI_{\text{eff},y})/L_{c,y}^2 = \pi^2(26{,}702{,}631)/(1.0\times 9\times 12)^2 = 22{,}594.7\ \text{kips}$$

$$\frac{P_{no}}{P_e} = \frac{2479.4}{6213.2} = 0.40 \leq 2.25$$

$$P_n = P_{no}\left(0.658^{\frac{P_{no}}{P_e}}\right) = 2479.4\left(0.658^{\frac{2479.4}{6213.2}}\right) = 2097.2\ \text{kips}$$

$$\phi_c P_n = 0.75\times 2097.2 = 1572.9\ \text{kips}$$

Flexural Strength: Plastic section modulus of reinforcement about x-axis:

$$Z_r = (A_{sr} - A_{srs})\left(\frac{h_2}{2} - c\right)$$

$$= \left(8\times\frac{\pi 0.875^2}{4} - 2\times\frac{\pi 0.875^2}{4}\right)\left(\frac{20}{2} - 2.31\right)$$

$$= 27.74\ \text{in.}^3$$

$$A_{srs} = 2\times\frac{\pi 0.875^2}{4} = 1.203\ \text{in.}^2$$

= area of reinforcing bars at centerline of composite section

Plastic section modulus of concrete:

$$Z_c = \frac{h_1 h_2^2}{4} - Z_x - Z_r$$

Z_x = plastic section modulus of steel shape

$$Z_c = \frac{20\times 20^2}{4} - 39.8 - 27.74$$

$$= 1932.5\ \text{in.}^3$$

$$M_D = Z_x F_y + Z_r F_{ysr} + \frac{1}{2} Z_c (0.85 f_c')$$

$$= 39.8\times 50 + 27.74\times 55 + \frac{1}{2}\times 1932.5\times(0.85\times 5)$$

$$= 7622.3\ \text{kip-in.}$$

$$= 635.2\ \text{kip-ft}$$

For h_n within the flange $\left(\frac{d}{2} - t_f < h_n \leq \frac{d}{2}\right)$

$$h_n = \frac{0.85 f_c'(A_c + A_s - db_f + A_{srs}) - 2F_y(A_s - db_f) - 2F_{yr}A_{srs}}{2[0.85 f_c'(h_1 - b_f) + 2F_y b_f]}$$

$$h_n = \frac{0.85\times 5\times(383.49 + 11.7 - 8.25\times 8.07 + 1.203) - 2\times 50\times(11.7 - 8.25\times 8.07) - 2\times 55\times 1.203}{2\times[0.85\times 5\times(20 - 8.07) + 2\times 50\times 8.07]}$$

$$h_n = 3.94\ \text{in.}$$

check the assumption: $\left(\frac{8.25}{2} - 0.56 = 3.57 < h_n = 3.94\ \text{in.} \leq \frac{8.25}{2} = 4.13\ \text{in.}\right)$ OK.

$$Z_{sn} = Z_x - b_f\left(\frac{d}{2} - h_n\right)\left(\frac{d}{2} - h_n\right)$$
$$= 39.8 - 8.07\left(\frac{8.25}{2} - 3.94\right)\left(\frac{8.25}{2} - 3.94\right)$$
$$= 38.3 \text{ in.}^3$$
$$Z_{cn} = h_1 h_n^2 - Z_{sn}$$
$$= 20 \times 3.94^2 - 38.3$$
$$= 272.2 \text{ in.}^3$$
$$M_B = M_D - F_y Z_{sn} - 0.85 f_c'\left(\frac{Z_{cn}}{2}\right)$$
$$= 7622.3 - 50 \times 38.3 - 0.85 \times 5\left(\frac{272.2}{2}\right)$$
$$= 5129.9 \text{ kip-in.}$$
$$= 427.5 \text{ kip-ft}$$
$$\phi_b M_n = 0.90 \times 427.5 = 384.8 \text{ kip-ft}$$
$$\frac{P_u}{\phi_c P_n} = \frac{600}{1572.9} = 0.38 > 0.2$$

Use AISC Specification Equation H1-1a:

$$\frac{P_u}{\phi_c P_n} + \frac{8}{9}\left(\frac{M_u}{\phi_b M_n}\right) \le 1.0$$
$$0.38 + \frac{8}{9}\left(\frac{350}{384.8}\right) = 1.19 \le 1.0 \textbf{ not good}.$$

A new section is needed that satisfies the interaction equation. Alternatively, a less conservative interaction model using plastic stress distribution can be used.

a. Interaction curve for nominal strength without length effects:
Point A (M_A, P_A) (pure axial compression):

$$P_A = F_y A_s + F_{ysr} A_{sr} + 0.85 f_c' A_c$$
$$= 50 \times 11.7 + 55 \times \left(8 \times \frac{\pi 0.875^2}{4}\right) + 0.85 \times 5 \times 383.49$$
$$= 2479.4 \text{ kips}$$
$$M_A = 0$$

Point D (M_D, P_D):

$$P_D = \frac{0.85 f_c' A_c}{2}$$
$$= \frac{0.85 \times 5 \times 383.49}{2}$$
$$= 815 \text{ kips}$$
$$M_D = 635.2 \text{ kip-ft (previously calculated)}$$

Point B (M_B, P_B) (pure flexure):

$$P_B = 0$$
$$M_B = 427.5 \text{ kip-ft (previously calculated)}$$

Point C (M_C, P_C):

$$P_C = 0.85 f_c' A_c$$
$$= 0.85 \times 5 \times 383.49$$
$$= 1629.8 \text{ kips}$$
$$M_C = M_B = 427.5 \text{ kip-ft}$$

b. Interaction curve for nominal strength with length effects:

$$P_{no} = P_A$$
$$= 2479.4 \text{ kips}$$
$$\lambda = \frac{P_n}{P_{no}}$$
$$= \frac{2097.2}{2479.4}$$

= 0.846 use the same slenderness ratio for all points on the interaction curve

$$P_{A'} = \lambda P_A$$
$$= 0.846 \times 2479.4$$
$$= 2105 \text{ kips}$$
$$P_{B'} = \lambda P_B$$
$$= 0.846 \times 0$$
$$= 0$$
$$P_{C'} = \lambda P_C$$
$$= 0.846 \times 1629.8$$
$$= 1378.8 \text{ kips}$$
$$P_{D'} = \lambda P_D$$
$$= 0.846 \times 815$$
$$= 689.5 \text{ kips}$$

c. Design interaction curve ($\phi_c = 0.75$; $\phi_b = 0.90$)

$$P_{A''} = 0.75 \times 2105$$
$$= 1578.8 \text{ kips}$$
$$P_{B''} = 0.75 \times 0$$
$$= 0$$
$$P_{C''} = 0.75 \times 1378.8$$
$$= 1034.1 \text{ kips}$$
$$P_{D''} = 0.75 \times 689.5$$
$$= 517.1 \text{ kips}$$
$$M_{A''} = 0.90 \times 0$$
$$= 0$$
$$M_{B''} = 0.90 \times 427.5$$
$$= 384.8 \text{ kip-ft}$$
$$M_{C''} = 0.90 \times 427.5$$
$$= 384.8 \text{ kip-ft}$$
$$M_{D''} = 0.90 \times 635.2$$
$$= 571.7 \text{ kips}$$

It can be seen from Fig. 8.54 that the axial and flexural demands (P_u, M_u) fall within the interaction surface. Therefore, the member is adequate.

Simplified interaction curve as defined by AISC Specification Commentary Equations C-I5-1a and 1b (Fig. 8.55) can also be used as follows:

$$P_r = P_u$$
$$= 600 \text{ kips} < P_{C''} = 1034.1 \text{ kips}$$
$$\frac{M_r}{M_c} = \frac{M_u}{M_{c''}} \le 1.0$$
$$\frac{350}{384.8} = 0.91 \le 1.0 \qquad \text{OK.}$$

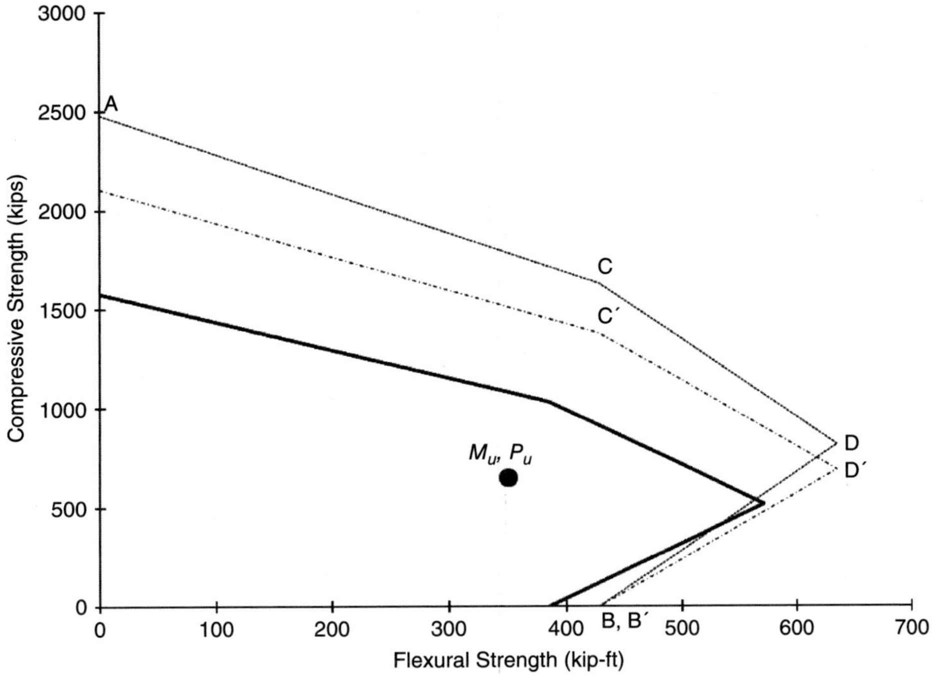

Figure 8.54 Interaction curve for nominal strength with length effects.

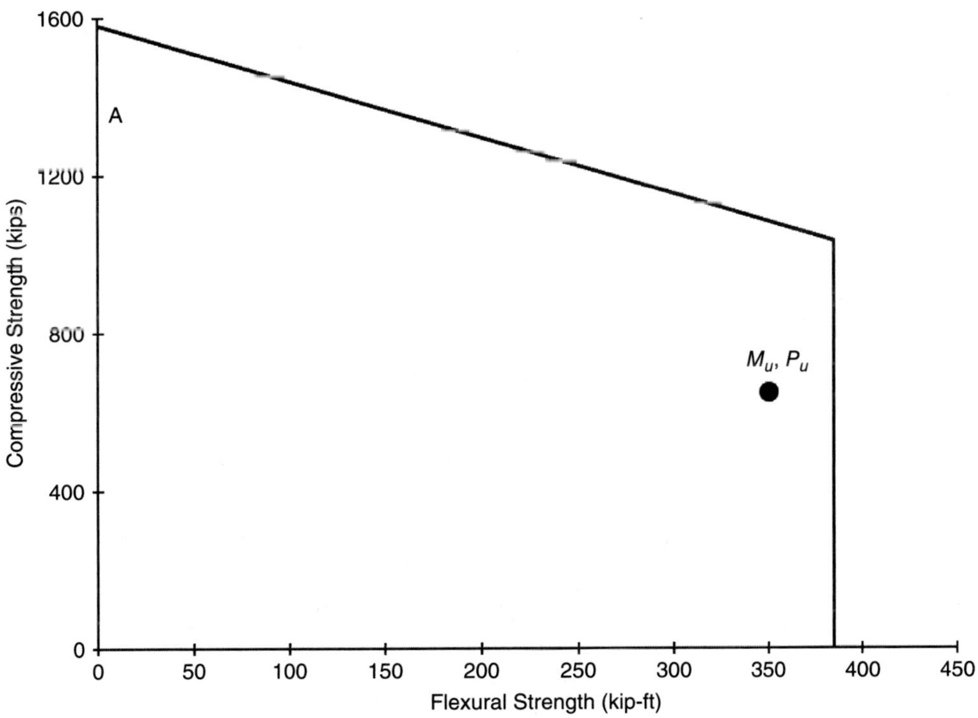

Figure 8.55 Interaction curve defined by AISC.

8.2 SEISMIC DESIGN OF STEEL MEMBERS IN MOMENT AND BRACED FRAMES

The primary concern in seismic design of steel structures is seismic (or ductile) detailing. This is simply because steel seismic force resisting systems (SFRSs) are expected to supply high ductility when subject to strong ground motions. This section discusses the basic principles of seismic detailing and how the AISC 341 (ANSI/AISC 341-16, 2016) treats them.

Steel SFRSs are classified into three main categories, namely; moment frames (MFs), braced frames, and shear walls. Braced frames can be either concentrically braced, eccentrically braced, or buckling-restrained. This section covers the design of a special moment and a concentrically braced frame in detail. A brief information about the material strength, local buckling requirement (width/thickness ratios), and basic load combinations are summarized first. Then, the design of the frames including the calculation of the seismic forces according to ASCE/SEI 7-16 (2017), sizing the structural elements using capacity-limited seismic load effects and seismic detailing (ANSI/AISC 341-16, 2016) are presented in detail. The design AISC 360s and AISC 341, ANSI/AISC 341-16 (2016), ANSI/AISC 360-16 (2016), and ASCE/SEI 7-16 (2017) will be hereafter referred to as *AISC 341*, *AISC 360*, and *ASCE 7*, respectively.

Material Strength:
Expected material strength is used when required to define the required strength for elements connecting to the structural members that are expected to supply ductility or undergo significant inelastic deformation during a seismic event (e.g., beam ends in MFs, braces in concentrically braced and buckling-restrained frames, links in eccentrically braced frames). Thus, expected material strength needs to be calculated for either adjoining member to design the member itself or connection design of the member (Fig. 8.56) and is defined as follows (*AISC 341, Section A3.2*):

$$\text{Expected yield stress} = R_y F_y \tag{8.14}$$

$$\text{Expected tensile strength} = R_t F_u \tag{8.15}$$

where F_y = minimum yield stress, ksi
F_u = minimum tensile strength, ksi
R_y = the ratio of the expected yield stress to F_y
R_t = the ratio of the expected tensile strength to F_u

The most commonly used values for R_y and R_t are given in Table 8.9.

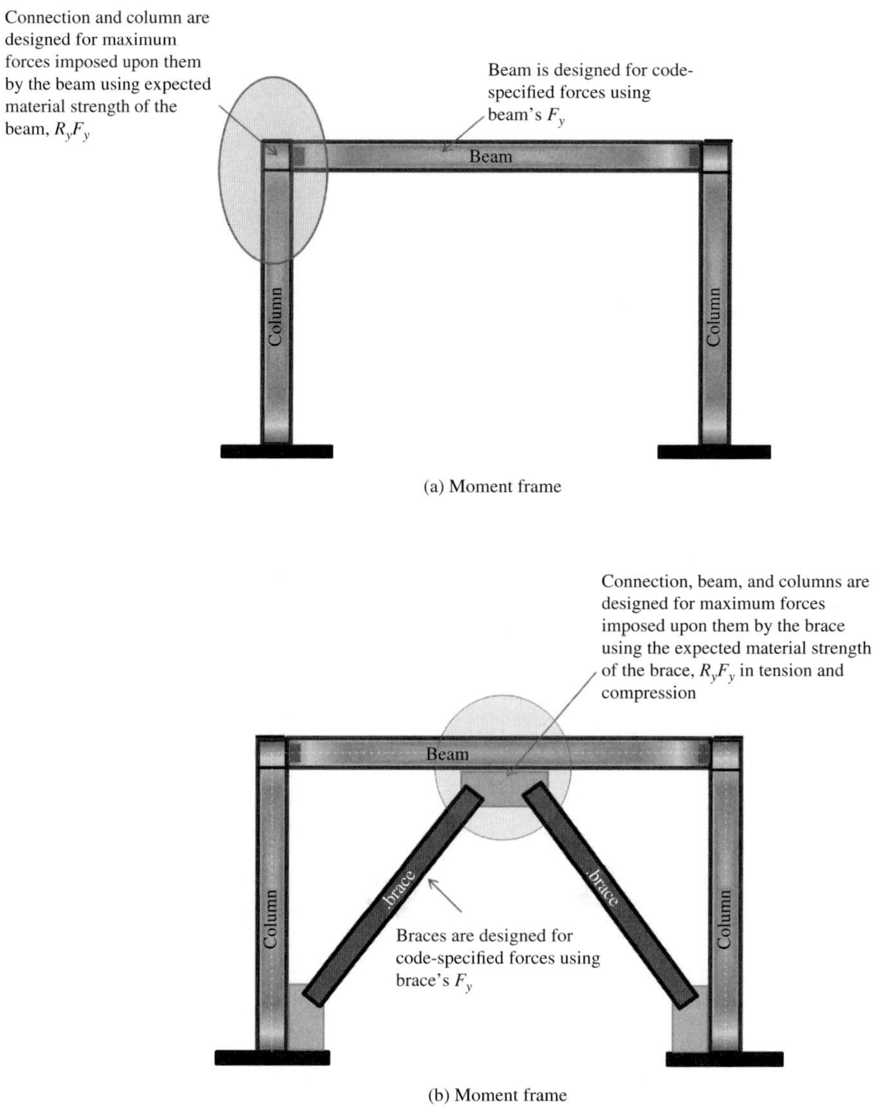

Figure 8.56 Use of expected material strength in design.

Table 8.9 R_y and R_t Values for the Most Commonly Used Steel Grades

Steel grade	R_y	R_t
ASTM A36/A36M	1.5	1.2
ASTM A572 Gr. 50	1.1	1.1
ASTM A500/A500M Gr. B	1.4	1.3
ASTM A500/A500M Gr. C	1.3	1.2
ASTM A572/A572M Gr. 42	1.3	1.0

Local Buckling (Limiting Width/Thickness Ratio) Requirement: AISC 341 also specifies member requirements in SFRS elements (beams, columns, braces, links in frames, and braced frames) to prevent local buckling until sufficient ductility is achieved. The provision classifies the members as highly ductile, λ_{hd}, or moderately ductile, λ_{md}. Table 8.10 shows the limiting width/thickness criteria for the most commonly used members in moment and braced frames. The members required to satisfy these limits in SFRS are also given in the table.

Basic Load Combinations:
The basic load combinations for strength design to be used for seismic design of moment and braced frames are as follows:

1. $1.4D$
2. $1.2D + 1.6L + 0.5L_r$
3. $1.2D + 1.6L_r + 1.0L$
4. $(1.2 + 0.2S_{DS})D + \rho Q_E + 0.5L$
5. $(0.9 - 0.2S_{DS})D + \rho Q_E$

where D is dead load; L is live load; L_r is roof live load; S_{DS} is design spectral response acceleration parameter at short periods; ρ is redundancy factor; Q_E is effects of horizontal forces resulting from seismic base shear.

Seismic loads replace the horizontal seismic loads, ρQ_E, in load combinations (4) and (5) with E_{cl} or E_{mh} $(= \Omega_0 Q_E)$ in special cases such as connection design or element design required by AISC 341, where E_{cl} is the capacity-limited horizontal seismic load effect; E_{mh} is the effect of horizontal seismic forces that includes overstrength; and Ω_0 is overstrength factor (Fig. 8.57 and Table 8.11). E_{cl} is equal to the maximum force that

Table 8.10 Limiting Width/Thickness Ratios for Members in Moment and Braced Frames

	Description of element	Width/ thickness ratio	Limiting width/thickness ratios and members in SFRS systems required to satisfy these limit values: λ_{hd} Highly ductile members	Member in SFRS systems	λ_{md} Moderately ductile members	Member in SFRS systems	Example
	Flanges of W shapes (*unstiffened element*)	b/t	$0.32\sqrt{\frac{E}{R_yF_y}}$	*Beams and columns in SMF*	$0.40\sqrt{\frac{E}{R_yF_y}}$	*Beams and columns in IMF*	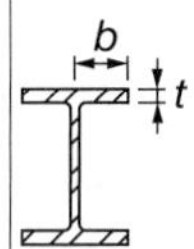
stiffened elements	Rectangular HSS used as a brace	b/t	$0.65\sqrt{\frac{E}{R_yF_y}}$	*Braces in SCBF*	$0.76\sqrt{\frac{E}{R_yF_y}}$	*Braces in OCBF and EBF*	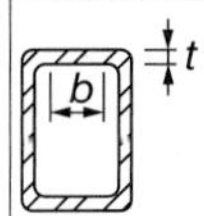
	Webs of W shapes used as a brace	h/t_w	$1.57\sqrt{\frac{E}{R_yF_y}}$	*Braces in SCBF*	$1.57\sqrt{\frac{E}{R_yF_y}}$	*Braces in OCBF and EBF*	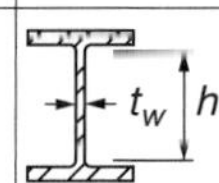
	Rectangular HSS used as a beam or column	b/t	$0.65\sqrt{\frac{E}{R_yF_y}}$	*Beams and columns in SMF, SCBF, and EBF*	$1.18\sqrt{\frac{E}{R_yF_y}}$	*Beams and columns in IMF*	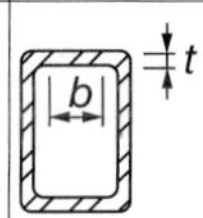
	Webs of W shapes used as a beam or column	h/t_w	For $C_a \le 0.114$ $2.57\sqrt{\frac{E}{R_yF_y}}(1-1.04\,C_a)$ For $C_a > 0.114$ $0.88\sqrt{\frac{E}{R_yF_y}}(2.68-C_a) \ge 1.57\sqrt{\frac{E}{R_yF_y}}$ $C_a = \frac{P_u}{\varnothing_c P_y}$ $Py = R_y R_y A_g$	*Beams and columns in SMF, SCBF, and EBF*	For $C_a \le 0.114$ $3.96\sqrt{\frac{E}{R_yF_y}}(1-3.04\,C_a)$ For $C_a > 0.114$ $1.29\sqrt{\frac{E}{R_yF_y}}(2.12-C_a) \ge 1.57\sqrt{\frac{E}{R_yF_y}}$	*Beams and columns in IMF*	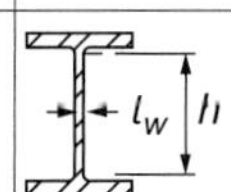
	Round HSS used as a brace	D/t	$0.053\frac{E}{R_yF_y}$	*Braces in SCBF*	$0.062\frac{E}{R_yF_y}$	*Braces in OCBF*	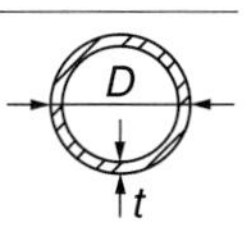

EBF, eccentrically braced frames; IMF, intermediate moment frames; OCBF, ordinary concentrically braced frames; OMF, ordinary moment frames; SCBF, special concentrically braced frames; SMF, special moment frames.

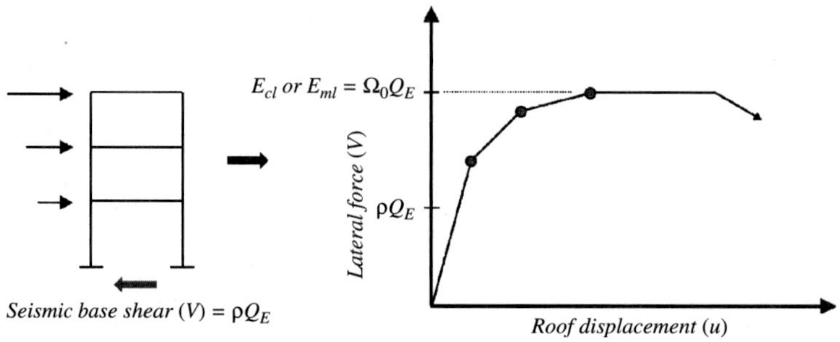

Figure 8.57 Capacity-limited horizontal seismic load effect and effect of horizontal seismic force that includes overstrength.

Table 8.11 Overstrength Factors

Seismic force resisting systems (SFRS)	Ω_0
Special, Intermediate and Ordinary Moment Frames	3
Special and Ordinary Concentrically Braced Frames	2
Eccentricall Braced Frames	2
Buckling-restrained Braced Frames	2.5

can develop in the element obtained through plastic analysis. Thus, the load combinations (4) and (5) becomes

6. $(1.2 + 0.2S_{DS})D + \Omega_0 Q_E$ (or E_{cl}) $+ 0.5L$
7. $(0.9 - 0.2S_{DS})D + \Omega_0 Q_E$ (or E_{cl})

E_{mh} in the above load combinations need not to exceed E_{cl}. It is clear that the provision allows plastic analysis to determine the maximum horizontal seismic load effect instead of using the effect of horizontal seismic forces that include overstrength and in some cases it is the only way to do so.

8.2.1 Moment Frames

Beams and columns are rigidly connected in MFs to transfer lateral forces through flexure and shear in these elements. MFs are seismically designed in such a way that yielding occurs at the beam ends through the formation of plastic hinges (Figs. 8.58 and 8.59). *AISC 341* requires that MFs sustain large inelastic deformations without failure of beam-to-column connections. These large inelastic deformations are maintained through beam flexural yielding.

Story Drift (Lateral Stiffness) Requirement

Lateral stiffness of MFs are the smallest among all steel SFRSs. That is why the design of a typical MF is governed by the story drift requirement as specified by *ASCE 7* (*Eq. 12.8-52*) as follows:

$$\delta_x = \frac{\delta_{xe} C_d}{I_e} \tag{8.16}$$

and

$$\Delta_x = \delta_x - \delta_{x-1} \tag{8.17}$$

where δ_{xe}: elastic lateral displacement of story (the lateral deflection determined by elastic analysis under seismic forces).
δ_x: amplified lateral displacement of story (approximate inelastic displacement, $\delta_x = \delta_{xe} C_d$).
C_d: deflection amplification factor.
Δ_x: inelastic story drift.
Δ_a: allowable story drift.

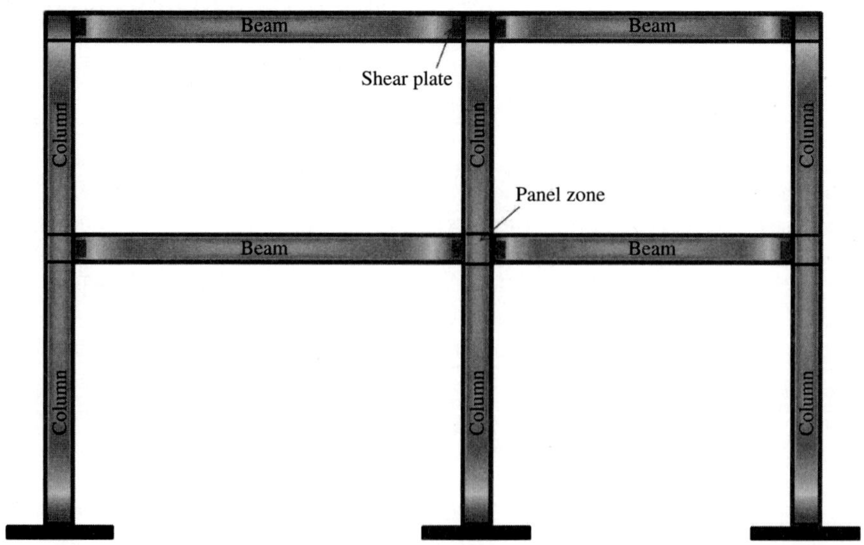

Figure 8.58 A moment frame.

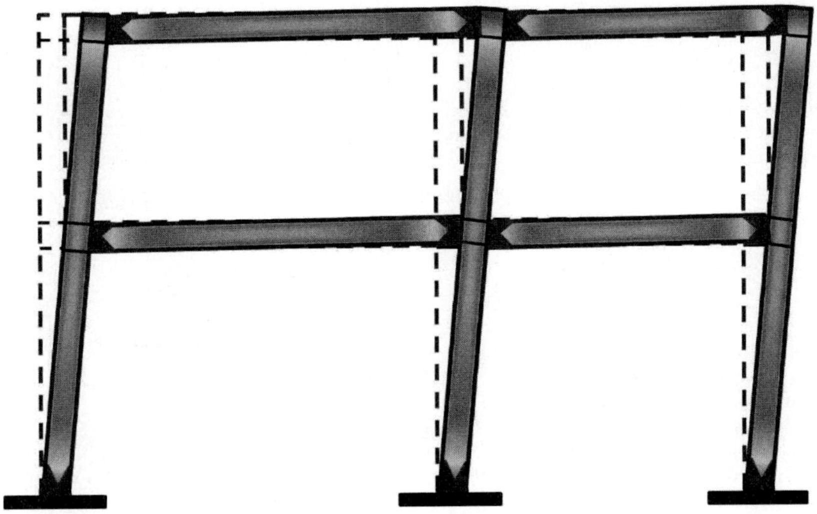

Figure 8.59 Yielding at the beams ends of a seismically detailed moment frame and formation of plastic hinges.

STRENGTH REQUIREMENT

Strength requirements for beams, girders, and columns are as follows (*AISC 341*):

Beams and Girders:
Beams and girders are required to satisfy the maximum load effect obtained from the load combinations specified in *AISC 7* (load combinations 1 through 5 in Section 8.2.1).

Columns:
The required strength of a column in a SFRS should satisfy the greater of the following requirements:

a. The load effect obtained from the load combinations specified in *ASCE 7* (load combinations 1 through 5 in Section 8.2.1).

b. The compressive axial strength and tensile strength as determined using the overstrength seismic load (load combination 6 and 7 in Section 8.2.1). Bending moments are allowed to be ignored in applying this requirement.

STRONG COLUMN—WEAK BEAM (MOMENT RATIO) REQUIREMENT

Columns at a beam-column joint must be stronger than beams. This principle is satisfied by taking the ratio of sum of plastic moment capacities of columns at the joint, $\sum M^*_{pc}$, to the sum of plastic moment capacities of beams, $\sum M^*_{pb}$ at the intersection of the beam and column centerlines (Fig. 8.60).

$$\frac{\sum M^*_{pc}}{\sum M^*_{pb}} > 1.0 \tag{8.18}$$

$\sum M^*_{pc}$ = sum of projections of the nominal flexural strengths of the columns above and below the joint to the beam centerline (reduction due to axial force in the column is required)

$= \sum Z_c(F_{yc} - P_r/A_g)$

$\sum M^*_{pb}$ = sum of projections of the expected flexural strengths of the beams at the plastic hinge locations to the column centerline.

$= \sum (M_{pr} + M_v)$

A_g = gross area of column, mm².
F_{yb} = specified minimum yield stress of beam, ksi.
F_{yc} = specified minimum yield stress of column, ksi.
M_v = additional moment due to shear amplification from the location of the plastic hinge to the column centerline, kip-in.

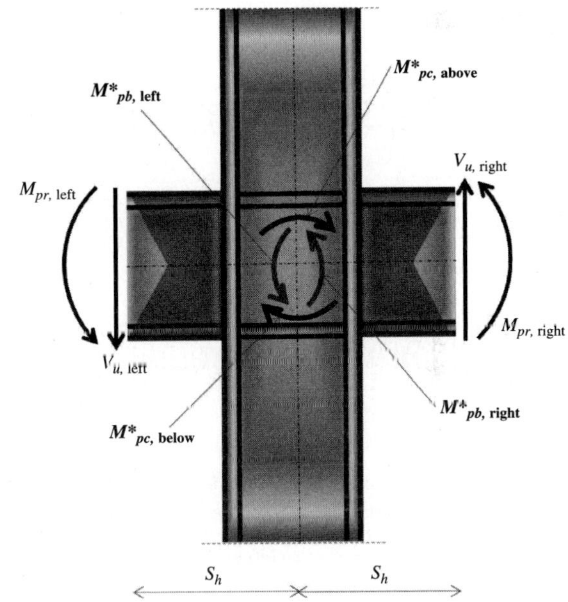

Figure 8.60 Moment ratio calculation at the joint.

P_r = required axial compressive strength of the column, kips.
Z_c = plastic section modulus of the column about the axis of bending, in³.
s_h = distance from center of plastic hinge to center of column, in.

MEMBER REQUIREMENTS

Beams and columns are required to satisfy the requirements in Table 8.10 for moderately ductile and highly ductile members for intermediate moment frame (IMF) and special moment frame (SMF), respectively. There is no width/thickness requirement for beams and columns in OMF.

BEAM-TO-COLUMN CONNECTIONS

Beam-to-column connections should sustain a story drift angle of ±0.04 rad without failure of the connection. An additional criterion for a beam-to-column connection is that it must resist a moment equal to at least $0.80M_p$ of the connected beam at a story drift angle of ±004 rad.

Shear Strength

The required shear strength of the connection, V_u, can be obtained by superposing the capacity-limited seismic load effect, E_{cl}, with the gravity load effect, $V_{gravity}$ (Fig. 8.61), as follows:

$$V_u = E_{cl} + V_{gravity} \tag{8.19}$$

where $E_{cl} = 2M_{pr}/L_h$

M_{pr} = probable maximum moment at plastic hinge $= C_{pr} R_y F_y Z_e$

$C_{pr} = \dfrac{F_y + F_u}{2F_y} \leq 1.2$

F_y = minimum yield stress, ksi

F_u = minimum tensile strength, ksi

Z_e = effective plastic section modulus of section at plastic hinge location, mm³

L_h = distance between plastic hinge locations

$V_{gravity}$ = shear force due to gravity loading (including vertical seismic load effect)$(1.2 + 0.2S_{DS})D + 0.5L$)

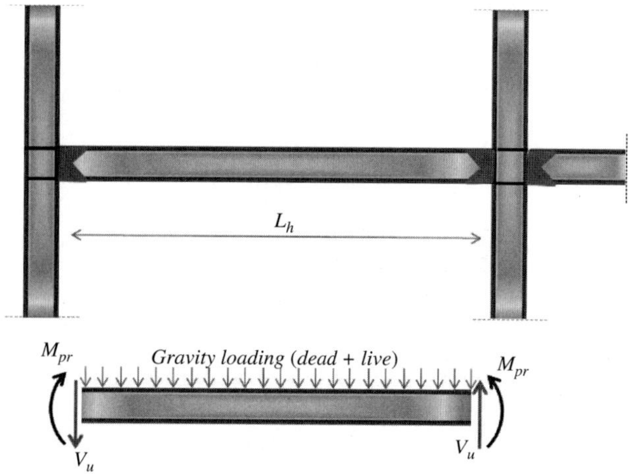

Figure 8.61 Computing the required shear strength.

Panel Zones

Shear in a typical beam-to-column joint called panel zone in an MRF is quite high due to high moment gradient in the joint region. Panel zone regions are a good source of yielding and dissipating energy due to shear yielding. Section E3.6e of the *AISC 341* specifies the minimum requirements for design of panel zones. *AISC 341* allows for limited yielding within the panel zone. The design shear strength of a panel zone is obtained considering the limit state of shear yielding as $\phi_v R_n$ ($\phi_v = 1.0$), whereas the required shear strength (R_u) can be determined as follows (Fig. 8.62):

$$R_u = \frac{\sum M_f}{d - t_f} - V_{col} \tag{8.20}$$

where $\sum M_f$ = the sum of the probable maximum moments at the plastic hinges at column faces ($M_{p,left} + M_{p,right}$)

V_{col} = shear force in the portion of the column outside the pane zone

Lateral Bracing Requirement

Lateral bracing requirement for MF beams is stated as follows (*AISC 341, Section D1.2b*):

$$L_b > 0.095 r_y E/(R_y F_y) \tag{8.21}$$

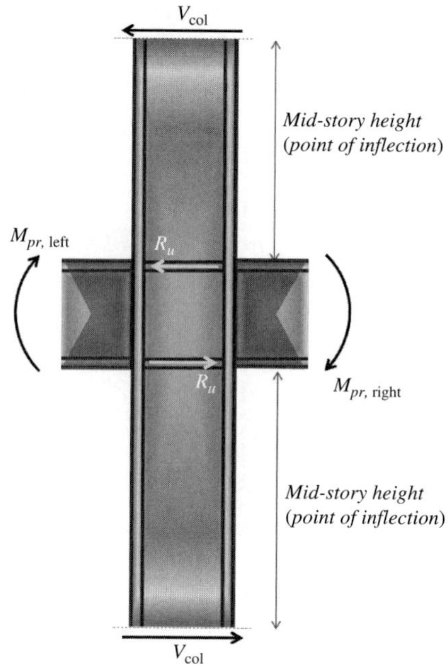

Figure 8.62 Estimating required shear strength for a panel zone, R_u.

If lateral bracing is provided as stated above, the beam can achieve ±0.04 rad drift angle without significant loss of flexural capacity (not less than $0.8M_p$).

Column Splices

Column splices in all framing types are to be arranged 4 ft or more away from the beam-to-column flange connections, where possible 4 ft will be measured from the finished floor elevation (*AISC 341, Section D2.5a*).

Prequalified Moment Connections

The basic seismic detailing requirement for a moment connection in an MRF is that the MRF should sustain large ductility without failure of the beam-to-column connection. The minimum requirements for a beam-to-column connection in SMF are specified in Chapter K of the *AISC 341*, whereas prequalified moment connections are listed in *AISC 358* standard. In the first edition of this standard, there were only two prequalified connections: Reduced beam section (RBS) and bolted (unstiffened and stiffened) extended end plate moment connections. The latest release of this standard lists 16 prequalified moment connections (*AISC 358*). RBS connection is commonly used in SMF and IMF as prequalified connections.

Continuity Plates

The use of continuity plates should be consistent with the type of prequalified moment connection. If the thickness of the column flange, t_{cf}, satisfies the following criteria, the continuity plates are not required (*AISC 341, Section E3.6f*) for W- or built-up I-shaped columns:

$$t_{cf} > \frac{b_{bf}}{6} \tag{8.22}$$

where b_{bf} is the beam flange width. Continuity plate thickness for one-sided should be at least 50 percent of the thickness of the beam flange, whereas for two-sided connections should be at least 75 percent of the thickness of the thicker beam flange on either side of the column.

8.2.2 Design Example of a Six-Story Steel Perimeter Frame Building

The building is designed in accordance with AISC 341 (2016), AISC 360 (2016), AISC 358 (2016), and ASCE 7 (2016).

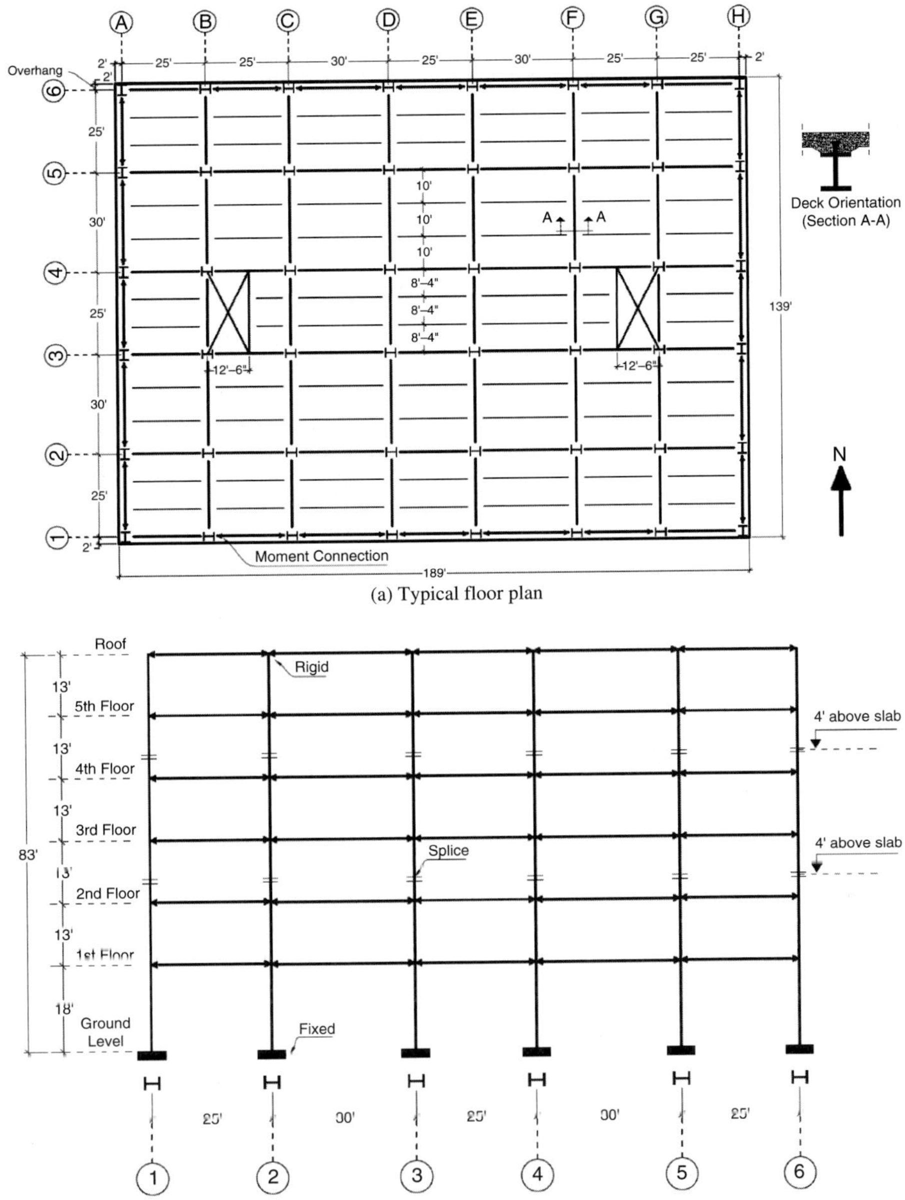

(a) Typical floor plan

(b) Elevation of frames on lines A and H

Figure 8.63 Typical floor plan and elevations of the building.

Design Steps:

(1) Check stiffness requirement (drift limitation).

(2) Check Strong Column-Weak Beam (SCWB) requirement in order to prevent possible story mechanisms.

(3) Check the strength design requirement (Capacity ≥ Demand).

In addition, all sections should be highly ductile in an SMF to prevent premature local buckling.

The example covers:

1. Determination of seismic and structural design parameters;
2. Gravity and seismic load calculations;
3. Analysis of perimeter frames;
4. Design for drift;
5. Design for strength;
6. Design of moment connections;
7. Strong-column weak-beam requirement;
8. Member requirements.

Building Description:
The office building of rectangular plan is 189 ft long and 139 ft wide in E-W and N-S directions, respectively. The building is located in Downtown, San Francisco. As shown in Fig. 8.63*a*, various span lengths of 25 ft and 30 ft are employed in both orthogonal directions. SMFs are arranged on the perimeter of the building plan. The beam-to-column connections in the interior frames in E-W and N-S directions are assumed to be shear connections. To be consistent with the conventional design procedures, the lateral resistance of the frames other than the ones located on the perimeter is neglected. With the intention of

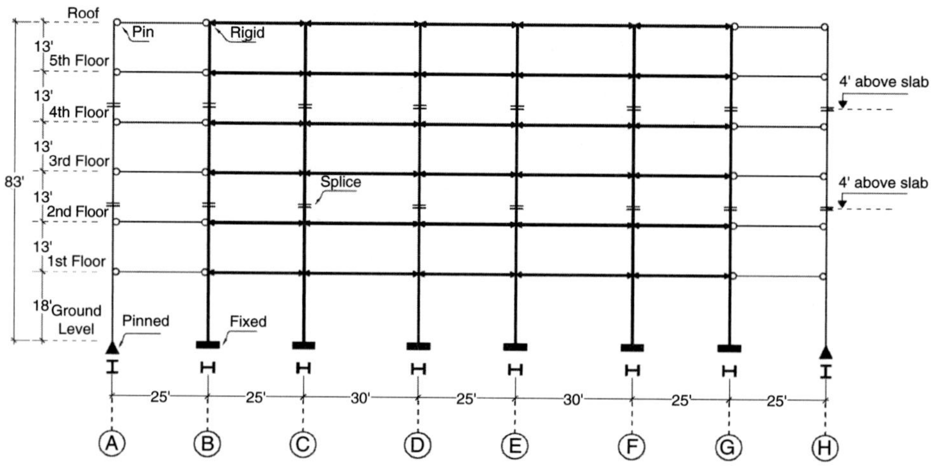

(c) Elevation of frames on lines 1 and 6

Figure 8.63 *(Continued)*

avoiding biaxial bending at the corner columns, the edge spans on lines 1 and 6 are designed as part of the gravity frames. Therefore, this example will cover the design of the identical SMFs on lines 1, 6 and A, H. The composite floor system consists of 3¼-in. thick lightweight concrete with wire mesh on 1½-in. steel deck. The orientation of the floor system is shown in Fig. 63*a*. The typical story height is 13 ft and the first story is 18 ft (Figs. 8.63*b* and *c*). The columns in the SMFs are assumed to be fixed at the base.

Determination of Gravity Load Effects

Dead Load

Dead loads consist of the self-weight of the structural and nonstructural members incorporated into the building as well as the fixed service equipment, as summarized in Table 8.12.

Table 8.12 Dead Loads

Component	Floor	Roof
3¼-in. lightweight concrete	8 psf per inch	—
1½"-in. metal deck	3 psf	3 psf
Steel Beam/Column (self-weight)	20 psf	10 psf
Roofing/Insulation/Elevator equipment	—	10 psf
Partitions	15 psf	—
Mechanical/Electrical Components	6 psf	6 psf
Flooring	10 psf	—
Suspended Ceiling	2 psf	2 psf
Fireproofing	2 psf	2 psf
Sprinkler	1 psf	1 psf
Total	**85 psf**	**34 psf**

Live Load

Uniformly distributed live loads, L_0, for floors and roof are determined in accordance with *Table 4.3-1, ASCE 7*, as follows:

Floor Live Load = 50 psf and Roof Live Load = 20 psf

Determination of Seismic Load Effects

Seismic Design Parameters The seismic design parameters for the office building located in Downtown, San Francisco, are as follows:

Risk Category = II (*Table 1.5.1, ASCE 7*)

Seismic Importance Factor, $I_e = 1.0$ (*Table 1.5.2, ASCE 7*)

Site Soil Classification: Site Class D – "Stiff Soil"

Seismic design parameters and site-specific values for the office building are summarized in Table 8.13.

Table 8.13 Site-Specific Information

Parameter	Site specific value	Reference
Mapped MCE$_R$ spectral response acceleration parameter at short periods	$S_S = 1.50$ g	USGS website
Mapped MCE$_R$ spectral response acceleration parameter at a period of 1 second	$S_1 = 0.60$ g	USGS website
Short-period site coefficient	$F_a = 1.0$	Table 11.4-1, ASCE 7
Long-period site coefficient	$F_v = 1.5$	Table 11.4-2, ASCE 7
Long-period transition period	$T_L = 12$ sec.	Fig. 22-12, ASCE 7

The design spectral response acceleration parameter at short periods, S_{DS}, and the design spectral response acceleration parameter at a period of 1 sec., S_{D1}, is determined from *Eqs. 11.4-1* through *11.4-4*, as follows:

$$S_{MS} = F_a S_S = 1.0 \times 1.50 = 1.50 \text{ g} \qquad (Eq.\ 11.4\text{-}1,\ ASCE\ 7)$$

$$S_{M1} = F_v S_1 = 1.5 \times 0.60 = 0.90 \text{ g} \qquad (Eq.\ 11.4\text{-}2,\ ASCE\ 7)$$

$$S_{DS} = \frac{2}{3} S_{MS} = 1.00 \text{ g} \qquad (Eq.\ 11.4\text{-}3,\ ASCE\ 7)$$

$$S_{D1} = \frac{2}{3} S_{M1} = 0.60 \text{ g} \qquad (Eq.\ 11.4\text{-}4,\ ASCE\ 7)$$

in which S_s and S_1 are the MCE$_R$ spectral response acceleration parameters adjusted for site class effects at short periods and at a period of 1 sec., respectively. According to *ASCE 7, Section 11.4.5*, T_0 and T_s can be obtained as 0.12 sec. and 0.60 sec., respectively, using the spectral acceleration parameters, $S_{DS} = 1.00$ g and $S_{D1} = 0.60$ g. The design response spectrum is given in Fig. 8.64.

Since $0.50 \le S_{DS}$, $0.20 < S_{D1}$ and $S_1 < 0.75$, the structure is assigned to Seismic Design Category D (*Tables 11.6-1 and 11.6-2, ASCE 7*).

Structural Parameters The structural parameters for the lateral-force-resisting system are given in Table 8.14.

Note that no limitation on building height, h_n, is specified for SMFs assigned to seismic design category D.

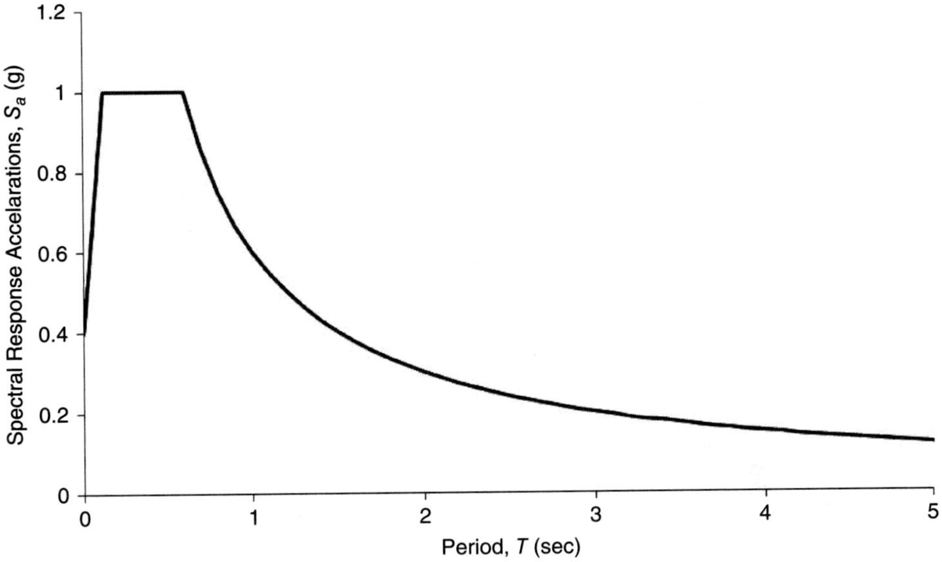

Figure 8.64 Design response spectrum.

Table 8.14 Structural Parameters for Seismic Force-Resisting Systems

Parameter	Coefficient	Reference
Seismic Force-Resisting System	Special Moment Frame (SMF)	
Building Height, h_n	83 ft	
Response Modification Coefficient, R	8	*Table 12.2-1, ASCE 7*
Overstrength Factor, Ω_0	3	*Table 12.2-1, ASCE 7*
Deflection Amplification Factor, C_d	5½	*Table 12.2-1, ASCE 7*
Approximate Period Parameter, C_t	0.028	*Table 12.8-2, ASCE 7*
Approximate Period Parameter, x	0.8	*Table 12.8-2, ASCE 7*
The Coefficient for Upper Limit on Calculated Period, C_u	1.4	*Table 12.8-1, ASCE 7*

Per *ASCE 7, Section 12.7.2*, effective seismic weight, W, of the structure consist of dead load given in Table 8.3.5. The effective seismic weight of each story is shown in Table 8.15.

Table 8.15 Seismic Effective Weight

Story	Dead load (psf)	Seismic effective weight (kips)
Roof	34	893.2
5th	85	2233.0
4th	85	2233.0
3rd	85	2233.0
2nd	85	2233.0
1st	85	2233.0
	Total	**12,058.2**

The approximate fundamental period for both orthogonal directions, as per *ASCE 7, Section 12.8.2.1*, is calculated as follows:

$$T_a = C_t h_n^x = 0.028 \times 83^{0.8} = 0.96 \text{ sec.} \qquad \textit{(Eq. 12.8-7, ASCE 7)}$$

Seismic Base Shear Seismic base shear, V, to be employed in Equivalent Lateral Force (ELF) procedure in both N-S and E-W directions is determined in compliance with *ASCE 7, Section 12.8.1*;

$$V = C_s W \qquad \textit{(Eq. 12.8-1, ASCE 7)}$$

$$C_s = \frac{S_{D1}}{T\left(\dfrac{R}{I_e}\right)} \text{ when } T < T_L \qquad \textit{(Eq. 12.8-3, ASCE 7)}$$

$S_{D1} = 0.60$ g
$T = T_a = 0.96$ sec.
$R = 8$
$I_e = 1.0$

$$C_s = \frac{0.60}{0.96 \times \left(\dfrac{8}{1.0}\right)} = 0.0781$$

C_s shall not be less than the following,

$$C_{s,\min} = 0.044 S_{DS} I_e = 0.044 \times 1.0 \times 1.0 = 0.044 \geq 0.01 \qquad \textit{(Eq. 12.8-5, ASCE 7)}$$

In addition, as specified in *ASCE 7, Section 12.8.1.1*, seismic response coefficient, C_s, shall not be less than the following for a structure located where S_1 is greater than or equal to 0.60 g.

$$C_{s,\min} = \frac{0.5 S_1}{(R/I_e)} = \frac{0.5 \times 0.60}{(8/1.0)} = 0.0375 \qquad \textit{(Eq. 12.8-6, ASCE 7)}$$

Thus,

$$C_s = \{0.0781;\ 0.044;\ 0.0375\}_{\max} = 0.0781$$

The total seismic base shear calculation for each direction is shown below.

$$V = C_s W = 0.0781 \times 12{,}058.2 = 941.8 \text{ kips}$$

The vertical force distribution along the building height is summarized in Table 8.16, using the *ASCE 7, Eqs. 12.8-11 and 12.8-12.*

$$F_x = C_{vx} V \qquad \textit{(Eq. 12.8-11, ASCE 7)}$$

$$C_{vx} = \frac{w_x h_x^k}{\sum_{i=1}^{n} w_i h_i^k} \qquad \textit{(Eq. 12.8-12, ASCE 7)}$$

Table 8.16 Vertical Force Distribution Based on T_a

Story	w_x (kips)	h_x (ft)	$W_x h_x^k$	C_{vx}	F_x (kips)
Roof	893.2	83	204,843.2	0.145489	137.0
5th	2233.0	70	415,304.0	0.294968	277.8
4th	2233.0	57	322,568.2	0.229103	215.8
3rd	2233.0	44	234,607.7	0.166629	156.9
2nd	2233.0	31	152,500.2	0.108313	102.0
1st	2233.0	18	78,141.3	0.055499	52.3
		Σ	1,407,964.7	1.00	941.8

in which,

$$k = \begin{cases} 1.0 \text{ when } T \leq 0.5 \text{ sec.} \\ 2.0 \text{ when } T \geq 2.5 \text{ sec.} \\ 0.5T + 0.75 \text{ when } 0.5 \leq T \leq 2.5 \end{cases}$$

$$\text{For } T = 0.96,\ k = 0.5 \times 0.96 + 0.75 = 1.23$$

Analysis

A 2D-numerical model of the identical SMFs on the perimeter is built in a structural analysis software, SAP2000 v.19 [5]. Columns are spliced 4 ft above the beam flange and fixed at the base. Beams and columns are modeled using elastic beam elements connected to rigid joints. Rigid end offsets are incorporated at the panel zones. The floor slabs are assumed to be rigid diaphragm in their own planes. Masses are assumed to be lumped at joints. To account for the second-order effects (*P*-Δ and *P*-δ), a leaning column consisting of rigid truss elements is introduced and the gravity loads not supported by the MFs are transferred to these columns at each story level. Columns are meshed into five equal segments to take the *P*-δ effect into account properly when conducting stability analysis. The numerical model is presented in Fig. 8.65.

Story Drift Requirement

Subsequent to determining the seismic and gravity load effects on the SMFs, the design process began by checking the stiffness requirement. As per *ASCE 7, Section 12.8.6,* amplified story drifts are to be less than the allowable limit for all stories ($\Delta_x \leq \Delta_a$).

The allowable story drift, per *ASCE 7, Table 12.12-1,* is 0.02 time the story height ($\Delta_a = 0.020h_{sx}$). Note that, according to *ASCE 7, Section 12.8.6.2,* it is permitted to determine story drifts based on the fundamental period of the structure without considering the upper limit on period. Therefore, the drift requirement is to be checked through an iterative process. In this example, first, the ELFs are to be calculated based on the C_uT_a period specified in *ASCE 7* with the purpose of reducing the number of iterations, since it is very likely that the period determined from the modal analysis of the numerical model will be greater than its approximate counterpart. After determining the member sizes that satisfy the drift requirement, a second iteration is performed using the seismic base shear computed based on the fundamental period obtained from the first iteration. Considering the strength requirement, two iterations found to be sufficient. It should also be noted that the reduction in lateral stiffness due to the flange cuts (e.g., reduced beam connections) needs to be taken into account when computing elastic drifts (*AISC 358-16, Section 5.8*). In this example, however, the seismic forces for story drift calculations are determined based on a period of 1.87 sec., while the actual fundamental period of the structure is 2.18 sec. Therefore, 14 percent difference between the base shears computed based on a period of 1.87 and 2.18 sec. is assumed to be adequate to account for the potential stiffness reduction.

1st Iteration Based on C_uT_a

1st iteration for determining the optimal section sizes that satisfy the stiffness requirement began by assuming preliminary member sizes. The preliminary members are given in Table 8.17. The seismic base shear determined based on the upper limit on period, C_uT_a, and the preliminary design of the frame are utilized as a starting point for the stiffness design.

Table 8.17 Preliminary Member Sizes

Story	Interior column	Exterior column	Girders
Roof	W 14 × 82	W 14 × 145	W 24 × 55
5th	W 14 × 82	W 14 × 145	W 24 × 84
4th	W 14 × 145	W 14 × 193	W 27 × 94
3rd	W 14 × 145	W 14 × 193	W 27 × 102
2nd	W 14 × 193	W 14 × 257	W 27 × 114
1st	W 14 × 193	W 14 × 257	W 30 × 148

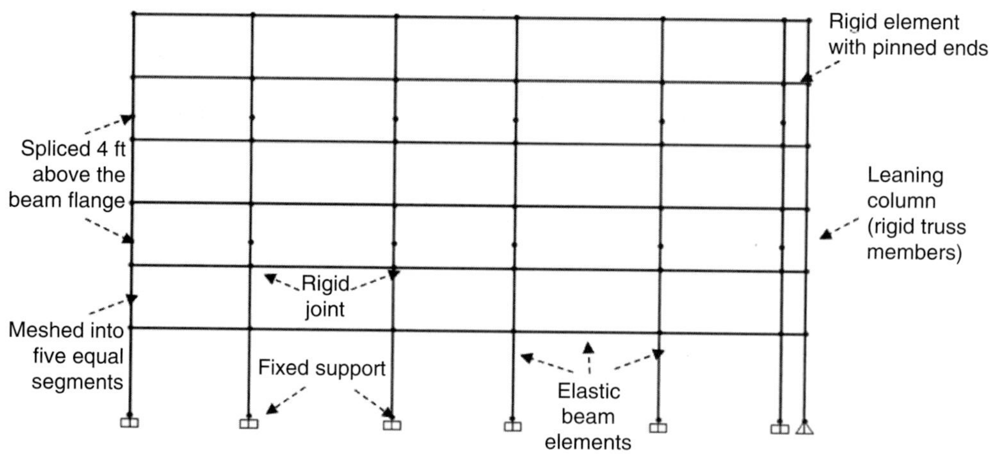

Figure 8.65 2D numerical model of the SMFs.

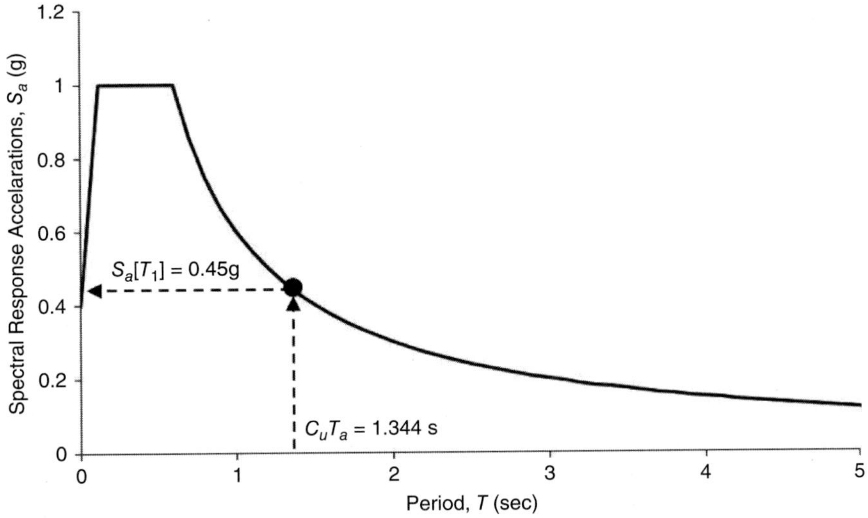

Figure 8.66 Spectral acceleration corresponding to $C_uT_a = 1.344$ (first iteration).

The spectral response acceleration corresponding to $C_uT_a = 1.344$ sec. is about 0.45 g, as demonstrated in Fig. 8.66. The total seismic base shear calculation based on C_uT_a is shown below.

$$T_u = C_uT_a = 1.4 \times 0.96 = 1.344 \text{ sec.} \qquad \textit{(Sec. 12.8.2, ASCE 7)}$$

$$C_s = \frac{0.60}{1.344 \times \left(\frac{8}{1.0}\right)} = 0.0558$$

$$V = C_sW = 0.0558 \times 12{,}058.2 = 672.9 \text{ kips}$$

Table 8.18 shows the vertical seismic forces at each diaphragm level. Due to the rigid diaphragm assumption, the seismic design story shear is distributed horizontally based on the relative stiffness of the SMFs. Therefore, the two SMFs parallel to the direction of loading (either in E-W or N-S direction) assumed to withstand the half of the total lateral force acting on the center of mass, F_{xi}. Figure 8.67 presents the lateral force distribution along the height of one SMF.

Table 8.18 Vertical Force Distribution Based on $C_uT_a = 1.344$ s

Story	w_x (kips)	h_x (ft)	k	$W_xh_x^k$	C_{vx}	F_x (kips)
Roof	893.2	83	1.422	478,502.9	0.157683	106.1
5th	2233.0	70		938,912.1	0.309403	208.2
4th	2233.0	57		701,050.9	0.23102	155.4
3rd	2233.0	44		485,160.3	0.159876	107.6
2nd	2233.0	31		294,857.3	0.097165	65.4
1st	2233.0	18		136,110.9	0.044853	30.2
			Σ	3,034,594.4	1	672.9

First, the structural model with the preliminary members (Table 8.18) is analyzed under the seismic load effects based on $C_uT_u = 1.344$ sec (Fig. 8.67). However, the computed design story drifts were not close to the allowable limit. Therefore, the member sizes were gradually reduced for a more economical design. The member sizes that satisfy the drift

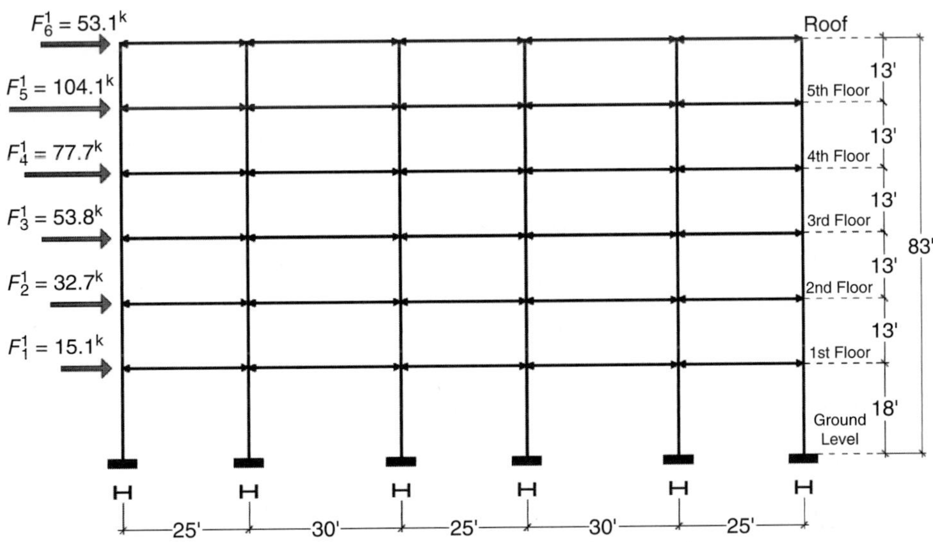

Figure 8.67 Seismic force distribution along the height of one SMF (first iteration).

requirement under the seismic forces computed based on C_uT_a = 1.344 sec. are given in Table 8.19.

Table 8.19 Member Sizes That Satisfy Drift Requirement (First Iteration Based on C_uT_a = 1.344 s)

Story	Exterior column	Interior column	Girders
Roof	W 14 × 74	W 14 × 109	W 21 × 55
5th	W 14 × 74	W 14 × 109	W 24 × 62
4th	W 14 × 176	W 14 × 193	W 24 × 84
3rd	W 14 × 176	W 14 × 193	W 27 × 84
2nd	W 14 × 193	W 14 × 283	W 27 × 94
1st	W 14 × 193	W 14 × 283	W 27 × 102

The computed elastic deflections (δ_{xe}), shown in Fig. 8.68, are amplified using a deflection amplification factor of 5.5 (C_d = 5.5) to obtain the inelastic deflections (δ_x) as well as the inelastic story drifts (Δ_x). The inelastic story drifts are compared with the allowable story drift for each story, as seen in Table 8.20. The fundamental period obtained from the frame with the structural members given in Table 8.20 is 1.87 seconds. Thus, the base shear used for determining the story drifts in the second iteration is computed based on a period of 1.87 sec.

Second Iteration Based on T_1 = 1.87 sec.
Subsequent to the first iteration, a second iteration is performed using the fundamental period obtained from the first iteration. The seismic design base shear determined using the spectral acceleration corresponding to T_1 = 1.87 sec. is 483.6 kips. The vertical distribution of the seismic base shear for the second iteration is given in Table 8.21. Likewise, the vertical distribution of the ELFs along the height of one SMF is demonstrated in Fig. 8.69.

The SMF is analyzed under the seismic loads based on T_1 = 1.87 sec. (Fig. 8.69). Following a succession of attempts to achieve the optimal design, the member sizes that meet the drift requirement under the seismic forces computed based on T_1 = 1.87 sec. are given in Table 8.22.

Table 8.23 draws a comparison between the inelastic story drifts and the allowable story drifts. Note that the fundamental period of the frame shown in Table 8.3.15 is 2.18 sec.

Table 8.20 The Story Drifts Based on C_uT_a = 1.344 s

Story	h_i (in.)	δ_{xe}	C_d	I_e	δ_x	Δ_x (in.)	Δ_a (in.)
Roof	156	3.18			17.51	1.54	3.12
5th	156	2.90			15.97	2.86	3.12
4th	156	2.38	5.50	1.00	13.11	2.85	3.12
3rd	156	1.86			10.26	3.12	3.12
2nd	156	1.30			7.13	2.96	3.12
1st	216	0.76			4.17	4.17	4.32

Table 8.21 Vertical Force Distribution Based on T_1 = 1.87 s

Story	w_x (kips)	h_x (ft)	k	$W_xh_x^k$	C_{vx}	F_x (kips)
Roof	893.2	83		1,529,685.3	0.1746	84.4
5th	2233.0	70		2,870,025.3	0.327588	158.4
4th	2233.0	57	1.685	2,030,227.3	0.231732	112.1
3rd	2233.0	44		1,312,542.2	0.149815	72.5
2nd	2233.0	31		727,511.7	0.083039	40.2
1st	2233.0	18		291,091.6	0.033226	16.1
		Σ		8,761,083.5	1	483.6

P-Δ Effects

To determine the significance of the *P*-Δ effects in the structure and to increase the stiffness of the structure accordingly, *Section 12.8.7, ASCE 7* describes an elastic stability coefficient, θ, of each story of a structure.

$$\theta = \frac{P_x \Delta I_e}{V_x h_{sx} C_d} \qquad \text{(Eq. 12.8-16, ASCE 7)}$$

$$\theta_{\max} = \frac{0.5}{\beta C_d} \leq 0.25 \qquad \text{(Eq. 12.8-76, ASCE 7)}$$

where P_x = the total vertical design load at and above Level x; where computing P_x, no individual load factor need exceeds 1.0
Δ = the design story drift occurring simultaneously with V_x
V_x = the seismic shear force acting between Levels x and x – 1
h_{sx} = the story height below Level x
β = the ratio of shear demand to shear capacity for the story between Levels x and x – 1.

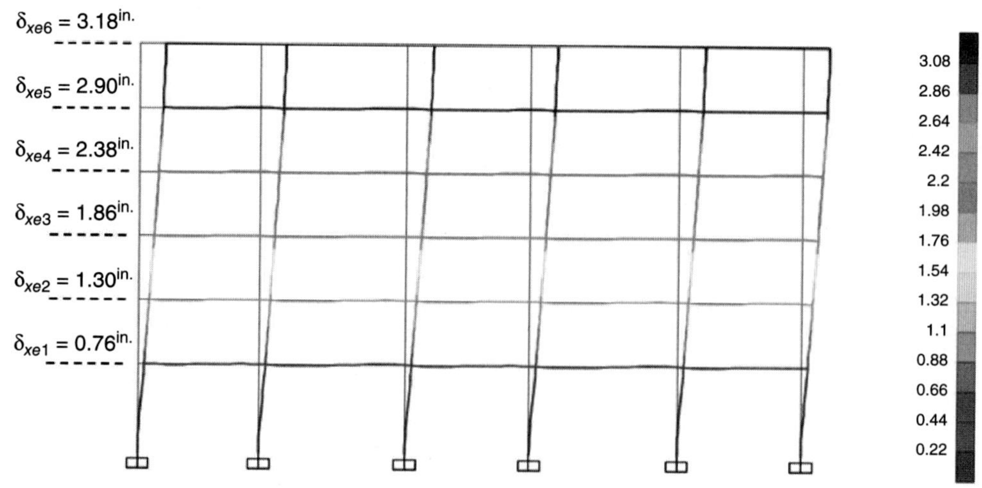

Figure 8.68 Computed elastic deflections (δ_{xe}) under seismic forces based on C_uT_a = 1.344 s.

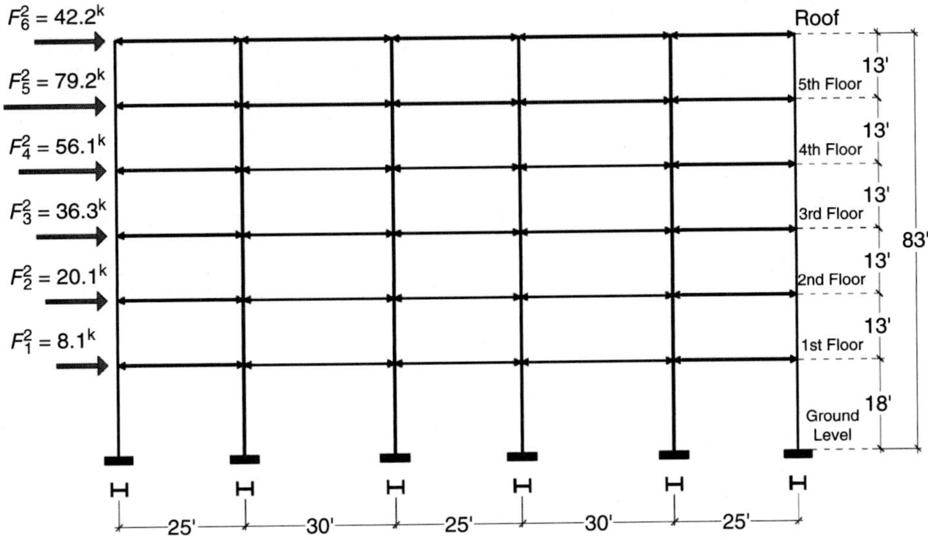

Figure 8.69 Seismic force distribution along the height of one SMF (second iteration).

Table 8.22 Final Sections Based on Drift Requirement

Story	Interior column	Exterior column	Girder
Roof	W 14 × 82	W 14 × 74	W 18 × 50
5th	W 14 × 82	W 14 × 74	W 21 × 57
4th	W 14 × 145	W 14 × 132	W 21 × 73
3rd	W 14 × 145	W 14 × 132	W 24 × 76
2nd	W 14 × 193	W 14 × 176	W 24 × 84
1st	W 14 × 193	W 14 × 176	W 24 × 94

Table 8.23 The Story Drifts Based on T_1 = 1.87 s

Story	h_i (in.)	δ_{xe}	C_d	I_e	δ_x	Δ_x (in.)	Δ_a (in.)
Roof	156	3.23			17.79	1.60	3.12
5th	156	2.94			16.20	2.74	3.12
4th	156	2.45	5.50	1.00	13.45	3.04	3.12
3rd	156	1.89			10.41	3.08	3.12
2nd	156	1.33			7.33	3.05	3.12
1st	216	0.78			4.28	4.28	4.32

Example calculation of the stability coefficient, θ, for the first story:

$$P_{x,1} = [5\times(139^{ft}\times189^{ft})\times(85^{psf}+50^{psf})+(139^{ft}\times189^{ft})\times(34^{psf}+20^{psf})]/1000^{kips/lb} = 19{,}151.6^{kips}$$

$V_{x,1} = 483.6^{kips}$ (Table 8.21 and Table 8.24)

$\Delta_1 = 4.28^{in.}$ (Table 8.23 and Table 8.24)

$$\theta_1 = \frac{P_{x,1}\Delta_1 I_e}{V_{x,1}h_{sx,1}C_d} = \frac{19{,}151.6\times4.28\times1.0}{483.6\times216\times5.5} = 0.143 < \theta_{max}$$

$$= \frac{0.5}{\beta C_d} = \frac{0.5}{0.4\times5.5} = 0.227$$

In the above equation, β is conservatively assumed to be 0.40. Note that the actual β, which is to be computed in stability analysis, is 0.15. A rigorous second-order analysis is required to determine member forces of the entire structure when one of the stability coefficients exceeds 0.10.

The summary of the computed stability coefficients for all stories is shown in Table 8.24. The stability coefficients of the first three stories exceeded 0.10, as shown in the table. Therefore, the second-order effects will be explicitly included in the strength design of the beams and columns in the following sections of this design example.

Table 8.24 Stability Coefficients, θ

Story	h_{sx} (in.)	P_x (kips)	Δ_x (in.)	I_e	V_x (kips)	C_d	θ
6th	156	1418.63	1.60	1.00	84.44	5.50	0.03
5th	156	4965.22	2.74	1.00	242.87	5.50	0.07
4th	156	8511.80	3.04	1.00	354.94	5.50	0.08
3rd	156	12,058.39	3.08	1.00	427.40	5.50	0.10
2nd	156	15,604.97	3.05	1.00	467.56	5.50	0.12
1st	216	19,151.56	4.28	1.00	483.63	5.50	0.14

Connection Design

The connections in the lateral force-resisting systems comprise RBS type moment connections (Fig. 8.70) for their superior energy-dissipation capabilities. A representative RBS design example for the first-story

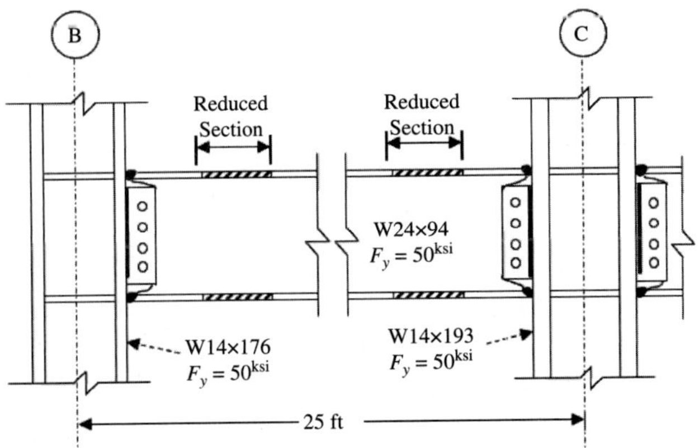

Figure 8.70 RBS connection at the exterior span of first story.

girder with 25-ft span length in accordance with *Chapter 5, AISC 358-16* is summarized below.

Cross-Sectional and Material Properties:

Beam: W24 × 94 (b_f = 9.07", t_{bw} = 0.515", t_{bf} = 0.875", d_b = 24.3", $Z_x = Z_b$ = 254 in.^{3})

Exterior Column: W14 × 176 (t_{cf} = 1.31", t_{cw} = 0.830", d_c = 15.2")

Interior Column: W14 × 193 (t_{cf} = 1.44", t_{cw} = 0.890", d_c = 15.5")

Materials: A992, Grade 50 for W shapes, and A36 or A572 for plates; Electrode – F_{exx} = 70 ksi; A325 bolts are available.

Loads on the beam: w_D = 0.354 kip/ft; w_L = 0.208 kip/ft

Check if the RBS can be used (Section 5.3.1, AISC 358-16):

Beam depth and weight are limited to W36 and 302 lb/ft, respectively (W27 × 94 < W36 × 302)

Beam flange thickness is limited to 1¾" (t_{bf} = 0.745" < 1¾")

Span length over beam depth ratio is limited to 7 (12.4 >7)

Column depth is limited to W36 (W14 < W36)

No limitation on column weight and column flange thickness

Thus, the RBS can be used.

Step-by-Step Design Process

Step 1. *Trial design values (**a, b, and c**) for the RBS (Fig. 8.71)*

$0.50\, b_{bf} \le a \le 0.75\, b_{bf}$ *AISC 358-16 (5.8-1)*

$0.65d \le b \le 0.85d$ *AISC 358-16 (5.8-2)*

$0.10\, b_{bf} \le c \le 0.25\, b_{bf}$ *AISC 358-10 (5.8-3)*

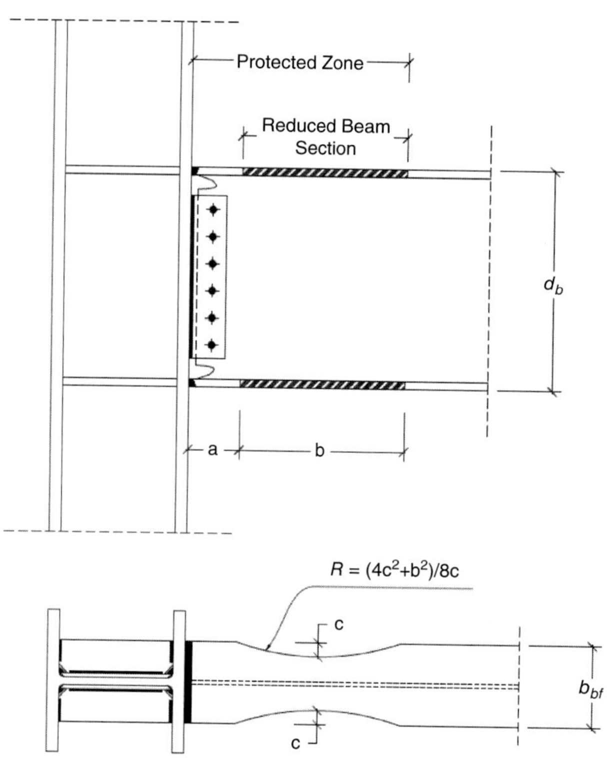

Figure 8.71 Reduced beam section connection.

First, the average values are used:

a = (0.50 + 0.75)b_{bf}/2 = (0.50 + 0.75) × 9.07/2 = 5.67 in. *Use 5-11/16"*

b = (0.65 + 0.85)d_b/2 = (0.65 + 0.85) × 24.3/2 = 18.23 in. *Use 18¼"*

c = (0.10 + 0.25)b_{bf}/2 = (0.10 + 0.25) × 9.07/2 = 1.59 in. *Use 1-9/16"*

Step 2. *Plastic section modulus of RBS, Z_{RBS}:*

$Z_{RBS} = Z_x - 2ct_{bf}(d - t_{bf})$ *AISC 358-16 (5.8-4)*

$Z_{RBS} = 254 - 2 \times 1.5625 \times 0.875 \times (24.3 - 0.875) = 190^{\text{in.}}$

(75% Z_x of W 24 × 94)

Step 3. *Compute the probable maximum moment, M_{pr}, at the center of the RBS:*

$M_{pr} = C_{pr}R_yF_yZ_{RBS}$ *AISC 358-16 (5.8-5)*

$$C_{pr} = \frac{F_y + F_u}{2F_y} = \frac{50+65}{2\times 50} = 1.15 < 1.20$$

$$M_{pr} = C_{pr}R_yF_yZ_{RBS} = 1.15\times 1.1\times 50\times 190 = 12017.5^{\text{k-in.}} = 1001.5^{\text{k-ft}}$$

Step 4. *Compute the shear force at the center of RBS (V_{RBS});*
Figure 8.72 shows the free-body diagram when plastic hinges formed at the beam ends.

$$w = 1.2w_D + 0.5w_L = 1.2\times 0.52^{\text{kip/ft}} + 0.5\times 0.31^{\text{kip/ft}} = 0.78^{\text{kip/ft}}$$

$$L_h = L - \left(2s_h + d_{c,\text{left}}/2 + d_{c,\text{right}}/2\right)$$

$$s_h = a + b/2 = 5\frac{11}{16}" + \left(18\frac{1}{4}"\right)/2 = 14\frac{13}{16}"$$

$$L_h = 25^{\text{ft}} - \left(2\times 14\frac{13}{16}" + 15.5"/2 + 15.2"/2\right)/(12^{\text{in.}}/1^{\text{ft}}) = 21.25^{\text{ft}}$$

$$V_{RBS} = \frac{2M_{pr}}{L_h} \pm \frac{wL_h}{2} = \frac{2\times 1001.5^{\text{k-ft}}}{21.25^{\text{ft}}} \pm \frac{0.78^{\text{k/ft}}\times 21.25^{\text{ft}}}{2}$$

$$= 94.3 \pm 8.3 = 102.6^{\text{kips}}/86^{\text{kips}}$$

Note that M_{pr} might change direction as ground shaking moves the structure back and forth. Therefore, a shear force of 102.6^{kips} and 86^{kips} might occur at both ends.

Step 5. *Compute the probable maximum moment at the face of the column: M_f*
Figure 8.73 shows the free-body diagram of the beam between the centerline of the RBS and the column face. Note that the gravity load on the small portion between the RBS center and the column face is neglected when determining the moment at the column face.

V_{RBS} = 102.6 kips (considering the maximum possible at each beam end)

$M_f = M_{pr} + V_{RBS}s_h$ *AISC 358-16 (5.8-6)*

$$M_f = 1001.5^{\text{k-ft}} + 102.6^{\text{kips}} \times \left(14\frac{13}{16}"\right)/(12^{\text{in.}}/1^{\text{ft}}) = 1128^{\text{k-ft}}$$

Step 6. *Compute the plastic moment capacity of the beam based on the expected yield stress, M_{pe}*

$M_{pe} = R_yF_yZ_b$ *AISC 358-16 (5.8-7)*

M_{pe} = 1.1 × 50 × 254 = 13,970$^{\text{k-in.}}$ = 1164$^{\text{k-ft}}$

Step 7. *Check the flexural strength at the face of the column flange,*

$M_f \le \phi_d M_{pe}$ *AISC 358-16 (5.8-8)*

$$M_f = 1128^{\text{k-ft}} < \phi_d M_{pe} = 1.0\times 1164^{\text{k-ft}}$$

Summary of the RBS design for each joint is summarized in Table 8.25.

Step 8. *Required shear strength of the beam and web-to-column connection:*

$V_r = V_{RBS}$ = 102.6 kips

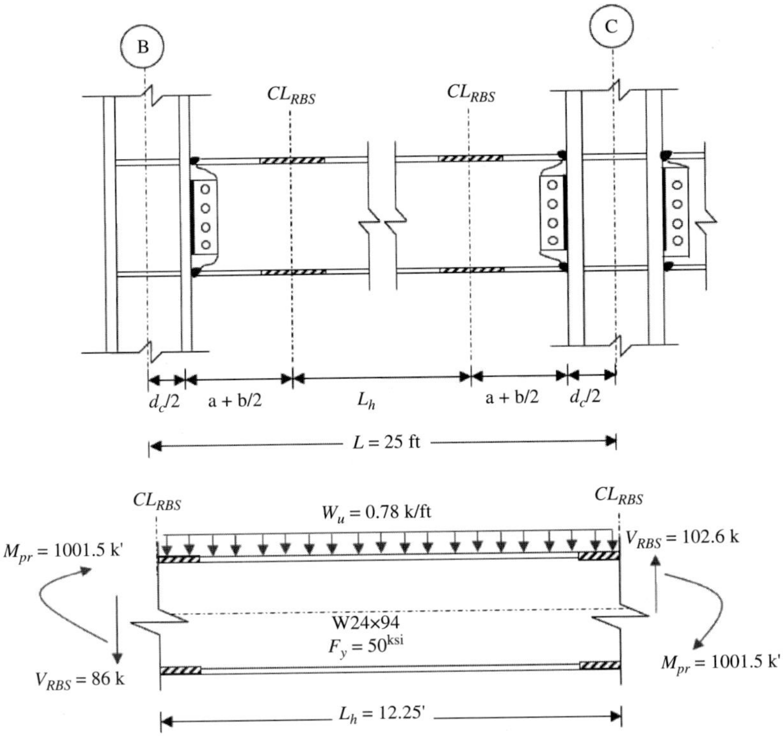

Figure 8.72 Free-body diagram of the beam with a plastic hinge at each end.

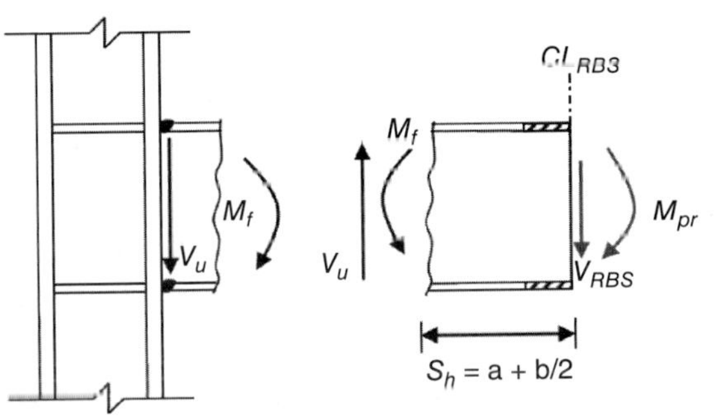

Figure 8.73 FBD of the beam between the centerline of the RBS and the column face.

Beam web strength:

$$h/t_w = 41.9 < 2.24\sqrt{E/F_y} = 53.9 \rightarrow \phi_v = 1.0; C_{v1} = 1.0$$

(*AISC 360-16, G2-2*)

$$V_c = \phi_v V_n = \phi_v 0.6 F_y A_w C_{v1}$$

(*AISC 360-16, G2-1*)

$$V_c = \phi_v V_n = 1.0 \times [(0.6 \times 50^{ksi}) \times (24.3^{in.} \times 0.515^{in.}) \times 1.0]$$

$$= 375.4^{kips} > V_r = 102.6^{kips}$$

Step 9. *Design the beam web-to-column connection*
Try shear tab plate with thickness, t_p = ½ in. > $t_{p,\min}$ = ⅜ in.
Determine the number of bolts, *n*:

A325 – N bolts, $F_{nv} = 54^{ksi}$ (*AISC 360, Table J3.2*), d_b = ¾",

Table 8.25 Summary of RBS Design

	Beam section	a (in.)	b (in.)	c (in.)	Z_{RBS}	C_{pr}	M_{pr} (k-ft)	S_h (in.)	L_h (ft)	V_{RBS} (kip)	M_f (k-ft)	ϕM_{pe} (k-ft)
Exterior Bays	W 18 × 50	4.688	13.500	1.313	74.92	1.15	394.89	11.44	21.91	39.47	432.51	462.92
	W 21 × 57	4.125	15.813	1.125	99.09	1.15	522.30	12.03	21.81	56.40	578.84	591.25
	W 21 × 73	5.188	15.875	1.438	128.47	1.15	677.15	13.13	21.58	71.15	754.98	788.33
	W 24 × 76	5.625	18.000	1.563	167.86	1.15	884.77	14.63	21.33	91.26	995.99	1026.67
	W 24 × 84	5.625	18.000	1.563	167.86	1.15	884.77	14.63	21.28	91.48	996.21	1026.67
	W 24 × 94	**5.688**	**18.250**	**1.563**	**189.95**	**1.15**	**1001.18**	**14.81**	**21.25**	**102.60**	**1127.80**	**1164.17**
Interior Bays	W 18 × 50	4.688	13.500	1.313	74.92	1.15	394.89	11.44	26.90	33.55	426.87	462.92
	W 21 × 57	4.125	15.813	1.125	99.09	1.15	522.29	12.03	26.80	49.41	571.84	591.25
	W 21 × 73	5.188	15.875	1.438	128.47	1.15	677.15	13.13	26.58	61.31	744.20	788.33
	W 24 × 76	5.625	18.000	1.563	167.86	1.15	884.77	14.63	26.33	77.46	979.18	1026.67
	W 24 × 84	5.625	18.000	1.563	167.86	1.15	884.77	14.63	26.27	77.59	979.34	1026.67
	W 24 × 94	5.688	18.250	1.563	189.95	1.15	1001.18	14.81	26.24	86.53	1107.99	1164.17

*Summary of above calculations are indicated in bold. The slight difference between the values obtained from the hand calculation and those in the table is due to round-off errors.

$$A_b = \tfrac{1}{4}\pi\, d_b^2 = \tfrac{1}{4} \times \pi \times \tfrac{3}{4}^2 = 0.442 \text{ in.}^2$$

$$\phi_n (nA_b)F_{nv} \geq V_r \rightarrow n = \frac{V_r}{\phi_n A_b F_{nv}} = \frac{102.6^{\text{kips}}}{0.75 \times 0.442^{\text{in.}^2} \times 54^{\text{ksi}}} = 5.7$$

Use six (6) ¾ in. diameter A325 – N bolts.

Check the required length for the shear tab:

Distance between two access holes, $L \approx d_b - 4t_{bf} = 24.3 - 4 \times 0.875 = 20.8$."

Note that it is permitted to extend the single shear tab connection between the weld access holes (*AISC 358, Section 5.6*).

Use L = 18 in. long shear tab plate (Fig. 8.74).

Check shear strength of 18" × 4" × ½" shear tab plate (A36 steel, $F_y = 36^{\text{ksi}}$ and $F_u = 58^{\text{ksi}}$):

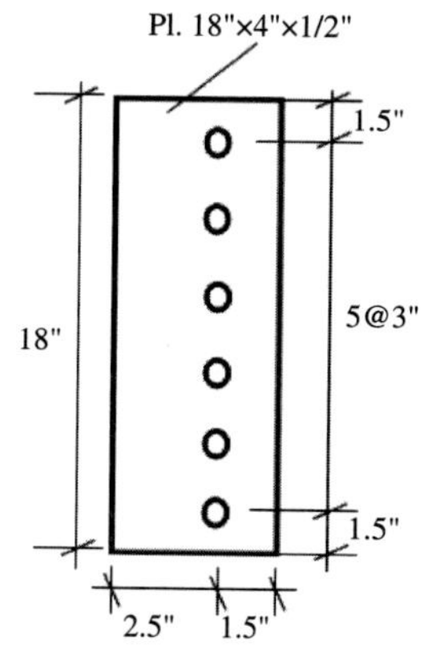

Figure 8.74 Geometry of the shear tab.

Yielding of the gross area in shear:

$$V_c = \phi_v V_n = 1.0 \times [(0.6 \times 50^{\text{ksi}}) \times (18^{\text{in.}} \times 1/2^{\text{in.}})] = 270^{\text{kips}} > V_r = 102.6^{\text{kips}}$$

Shear Rupture:

$$V_c = \phi_v V_n = 0.75 \times (0.6 \times 58^{\text{ksi}}) \times [(18^{\text{in.}} \times 1/2^{\text{in.}}) - 6 \times (3/4^{\text{in.}} + 1/8^{\text{in.}}) \times 1/2^{\text{in.}}]$$
$$= 166.4^{\text{kips}} > V_r = 102.6^{\text{kips}}$$

Bolt bearing/tear out:

Beam web strength is stronger than the shear tab for bolt bearing limit state since the web's thickness is 0.875" with A992 steel ($F_y = 50^{\text{ksi}}$ and $F_u = 65^{\text{ksi}}$) and $t_p = 0.5$" with A36 steel ($F_y = 36^{\text{ksi}}$ and $F_u = 58^{\text{ksi}}$). Thus, shear tab controls the design.

$$L_e = 1.5 - \frac{1}{2}\left(d_b + \frac{1}{16}\right) = 1.5 - \frac{1}{2} \times \left(\frac{3}{4} + \frac{1}{16}\right) = 1.09375\text{" (Edge Holes)}$$

$$L_e = 3.0 - \left(d_b + \frac{1}{16}\right) = 3.0 - \left(\frac{3}{4} + \frac{1}{16}\right) = 2.1875\text{" (Center Holes)}$$

Edge bolts:

$$\left.\begin{aligned} r_n &= 1.2L_e t_{\text{plate}} F_u < 2.4 d_b t_{\text{plate}} F_u \\ r_n &= 1.2 \times 1.09375 \times 0.5 \times 58 = 38.06 < 2.4 \times (3/4) \times 0.5 \times 58 = 52.2 \text{ kips/bolt} \end{aligned}\right\}$$

$$r_{n,\text{edge}} = 38.06 \text{ kips/bolt}$$

Center bolts:

$$\left.\begin{aligned} r_n &= 1.2L_c t_{\text{plate}} F_u \leq 2.4 d_b t_{\text{plate}} F_u \\ 1.2L_c t_{\text{plate}} F_u &= 1.2 \times 2.1875 \times 0.5 \times 58 = 76.1 \text{ kips/bolt} \\ 2.4 d_b t_{\text{plate}} F_u &= 2.4 \times (3/4) \times 0.5 \times 58 = 52.2 \text{ kips/bolt} \end{aligned}\right\}$$

$$r_{n,\text{center}} = 52.2 \text{ kips/bolt}$$

$$\phi R_n = 0.75 \times (5 \times r_{n,\text{center}} + r_{n,\text{edge}}) = 0.75 \times (5 \times 38.06^{\text{kips}} + 52.2^{\text{kips}})$$
$$= 224.7^{\text{kips}} > V_r = 102.6^{\text{kips}}$$

Block shear failure (Fig. 8.75):

$$R_n = \{0.6F_u A_{nv}; 0.6F_y A_{gv}\}_{\min} + F_u A_{nt}$$

$$A_{gv} = [(18 - 1.5) \times 1/2] = 8.25 \text{ in}^2$$

$$A_{nv} = 8.25 - [5.5 \times (3/4 + 1/16 + 1/16)] = 3.4375 \text{ in}^2$$

$$A_{nt} = [1.5 - 0.5 \times (3/4 + 1/16 + 1/16)] = 1.0625 \text{ in}^2$$

$$\phi R_n = 0.75 \times \left(0.6 \times 58^{\text{ksi}} \times 3\frac{7}{16}\text{"}; 0.6 \times 36^{\text{ksi}} \times 8\frac{1}{4}\text{"}\right)_{\min} + 0.6 \times 58^{\text{ksi}} \times 1\frac{1}{16}\text{"}$$
$$= 135.9^{\text{kips}} > V_r = 102.6^{\text{kips}}$$

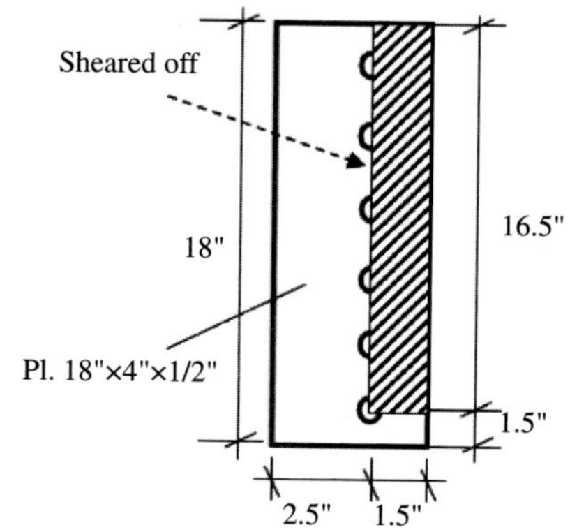

Figure 8.75 Block shear failure mode.

Two-sided fillet weld for web plate:

$$\phi R_n \geq V_r$$

$$\phi R_n = \phi \times [0.6 \times F_{EXX}(0.707 t_w) \times L_w]$$

$$t_w \geq \frac{V_r}{0.75 \times [2 \times (0.6 \times F_{EXX})(0.707) \times L_w]}$$
$$= \frac{102.6^{\text{kips}}}{0.75 \times [2 \times (0.6 \times 70^{\text{ksi}})(0.707) \times 18^{\text{in.}}]} = 0.128^{\text{in.}}$$

$t_{\min} = 3/16$" as per *Table J2.4, AISC 360*. Thus,

$$t_{\min} = \left\{0.128^{\text{in.}}; \frac{3}{16}\text{"}\right\}_{\max} \rightarrow t_{\text{weld}} = \frac{3}{16}\text{"}$$

Summary of shear connection:

Use 18 × 4 × ½" shear tab plate two-sided fillet welded to the column flange; Six (6) ¾ in. diameter A325 – N bolts (Fig. 8.76).

Step 10. *Continuity plates*

Per *AISC 358-16*, the requirements for the continuity plates are checked according to *Section 2, AISC 341*.

(a) Check continuity plate requirements:

Per *AISC 341, Section E3.6f*, all applicable limit states stipulated in *AISC 360, Section J* need to be checked. For connections with beam webs with a bolted connection to the column, P_f can be computed as follows:

$$P_f = \frac{M_f}{d^*} = \frac{M_f}{d_b - t_{fb}} = \frac{1128^{\text{k-ft}}}{(24.3^{\text{in.}} - 0.875^{\text{in.}})/(12^{\text{in.}}/1^{\text{ft}})} = 578^{\text{kips}}$$

where d^* is the distance between the flange centroids.

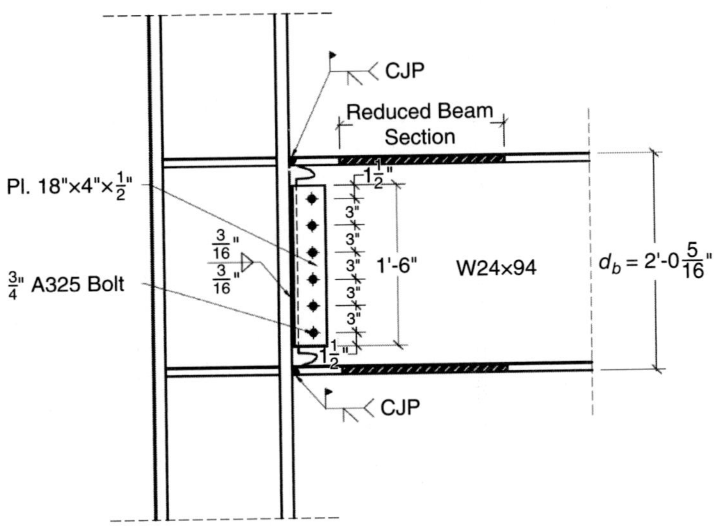

Figure 8.76 Details of the shear tab connection.

Local flange Bending:

For W14 × 176 (t_{cf} = 1.31")

$$\phi R_n = \phi(6.25 R_y F_{yf} t_f) = 0.9 \times (6.25 \times 1.1 \times 50^{\text{ksi}} \times 1.31^{\text{in.}}) = 405.3^{\text{kips}} < P_f = 578^{\text{kips}}$$

For W14 × 193 (t_{cf} = 1.44")

$$\phi R_n = \phi(6.25 R_y F_{yf} t_f) = 0.9 \times (6.25 \times 1.1 \times 50^{\text{ksi}} \times 1.44^{\text{in.}}) = 445.5^{\text{kips}} < P_f = 578^{\text{kips}}$$

Local Web Yielding:

For W14 × 176

$$\phi R_n = \phi[R_y F_{yw} t_w (5k + l_b)] = 1.0 \times [1.1 \times 50^{\text{ksi}} \times 0.83^{\text{in.}} \times (5 \times 1.91^{\text{in.}} + 0.875^{\text{in.}})] = 476^{\text{kips}} < P_f = 578^{\text{kips}}$$

For W14 × 193

$$\phi R_n = \phi[R_y F_{yw} t_w (5k + l_b)] = 1.0 \times [1.1 \times 50^{\text{ksi}} \times 0.89^{\text{in.}} \times (5 \times 2.04^{\text{in.}} + 0.875^{\text{in.}})] = 542^{\text{kips}} < P_f = 578^{\text{kips}}$$

Note that the limit state of web crippling usually does not govern the design of moment connections due to stocky webs of common W14 shapes. Furthermore, the limit state of web buckling does not apply to connections subject to lateral loads.

In addition to the limit states given in *AISC 360, Section J*, per *AISC 341, Section E3.6f.1b*, the flange thickness is limited to one-sixth of the beam flange width when the beam flange is welded to a W-shape column.

$$t_{\lim} = \frac{b_{bf}}{6} = \frac{9.07}{6} = 1.51^{\text{in.}} \qquad \textit{AISC 341 (E3-8)}$$

For W14 × 176: t_{cf} = 1.31" < $t_{\lim}$ = 1.51$^{\text{in.}}$
For W14 × 193: t_{cf} = 1.44" < $t_{\lim}$ = 1.51 $^{\text{in.}}$

Therefore, a pair of continuity plates is required for both W14 × 176 and W14 × 193.

(b) Design the continuity plates for W14 × 176 and W14 × 193

W14×193 is an interior column (two-sided connection). According to *AISC 341, E3.6f.2b*, for two-sided connections, the continuity plate thickness shall be at least equal to 75 percent of the thickness of the thicker of the two beam flanges on either side of the column. $t_s \geq t_{bf}$ = 0.75 × 0.875" = 0.656." Thus, a pair of 11/16-in. continuity plates is complete joint penetration (CJP) welded to the column flanges and web (or doubler plates) at the top and bottom flange levels of the connecting girders for the interior connection.

For one-sided (exterior) connections (W14 × 176), continuity plate thickness shall be at least one-half of the thickness of the beam flange, $t_s \geq t_{bf}/2$ = 0.875"/2 = 7/16", a pair of 7/16-in. continuity plates are CJP welded to the column flanges and web (or doubler plates) at the top and bottom flange levels of the connecting girder for the exterior connection, as presented in Fig. 8.78.

Step 11. *Check column panel zone.*

(a) Check panel Zone Strength

$$\frac{P_r}{P_y} = 0.16 < 0.75 \rightarrow \phi R_n = \phi(0.6 R_y F_y d_c t_w)\left(1 + \frac{3 b_{cf} t_{cf}^2}{d_b d_c t_w}\right)$$

(AISC 360, Eq. J10-11)

and

$$V_r = \frac{\sum M_f}{d_b - t_{fb}} - V_{u,col}$$

Example calculation for the first-story exterior joint (W14 × 176):

$$\phi R_n = 0.9 \times (0.6 \times 1.1 \times 50 \times 15.2 \times 0.83)\left(1 + \frac{3 \times 15.7 \times 1.31^2}{24.3 \times 15.2 \times 0.83}\right) = 474^{\text{kips}}$$

By assuming that the inflection points at mid-length of each story, $V_{u,col}$ can be computed as:

$$V_{u,col} = \frac{M_f}{(h_1 + h_2)/2}$$

where h_1 and h_2 are the first and second heights, respectively.

$$V_r = \frac{\sum M_f}{d_b - t_{fb}} - V_{u,col} = \frac{1128^{\text{k-ft}}}{(24.3^{\text{in.}} - 0.875^{\text{in.}})/(12^{\text{in.}}/1^{\text{ft}})} - \frac{1128^{\text{k-ft}}}{(18^{\text{ft}} + 13^{\text{ft}})/2} = 505^{\text{kips}} > \phi R_n = 474^{\text{kips}}$$

Therefore, the web requires doubler plates. A pair of fillet welded 7/16" thick doubler plates with 6-in. extensions above and below the beam flanges is used (ϕR_n = 671^{kips} after reinforcing with the doubler plates). According to *AISC 341, Section E3.6e-3*, the minimum doubler plate thickness is ¼" to avoid shear buckling. The doubler plates are placed in contact with the column web, as shown in Fig. 8.77.

(b) Check panel zone thickness

According to *AISC 341, Section E3.6e*, column web thickness is to satisfy the following:

$$t \geq (d_z + w_z)/90 \qquad \textit{AISC 341 (E3-7)}$$

where $d_z = d - 2t_f$ of the deeper beam at the connection, in.
t = thickness of column web or doubler plate, in.
w_z = width of panel zone between column flanges, in.

Interior panel zone (W14 × 193):

$$d_z = d_b - 2t_{bf} = 24.3 - 2 \times 0.875 = 22.55"$$

$$w_z = d_c - 2t_{cf} = 15.5 - 2 \times 1.44 = 12.62"$$

$$t = t_{cw} = 0.890" > (22.55 + 12.62)/90 = 0.39"$$

Web thickness is adequate.

$$t = t_p = 7/16" > (22.55 + 12.62)/90 = 0.39"$$

Doubler plate thickness is adequate.

Exterior panel zone (W14 × 176):

$$d_z = d_b - 2t_{bf} = 24.3 - 2 \times 0.875 = 22.55"$$

$$w_z = d_c - 2t_{cf} = 15.2 - 2 \times 1.31 = 12.58"$$

$$t = t_{cw} = 0.830" > (22.55 + 12.58)/90 = 0.39"$$

Web thickness is adequate.

$$t = t_p = 11/16" > (22.55 + 12.58)/90 = 0.39"$$

Doubler plate thickness is adequate.

Note that the individual doubler plate and the web thicknesses need to satisfy the thickness requirement, as specified in *AISC 341, Section E3.6e-2*, unless plug welds are used to connect doubler plates. In this example, 6-in. extended A572 doubler plates are welded to the column flanges longitudinally using ⅝" fillet welds. Details of the exterior connection are shown in Fig. 8.78.

SC/WB Requirement

Example calculations for the SC/WB requirement for the interior joint of the second story are summarized herein based on *AISC 341, Section E3.4*. Details of the interior joint are given in Fig. 8.79.

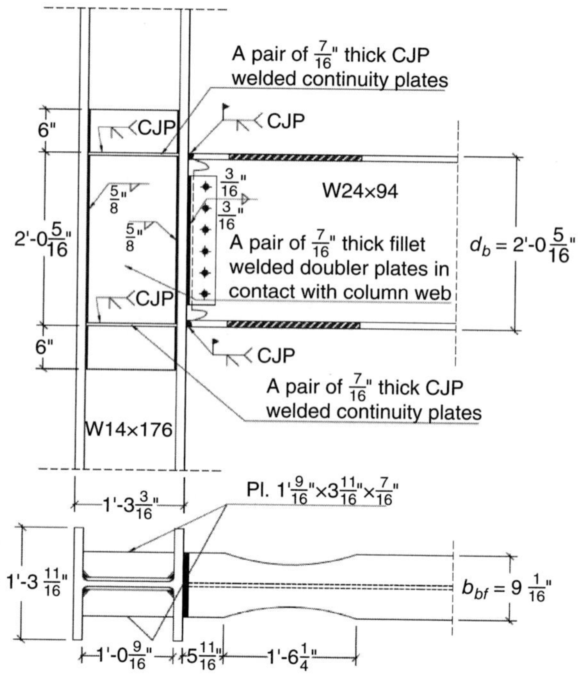

Figure 8.78 Details of the continuity and doubler plates for the exterior connection.

W14 × 193:
Z_x = 355 in.^{3}; A_g = 56.8 in.^{2}; d_c = 15.5 in.; F_y = 50 ksi; R_y = 1.1

W24 × 76:
Z_{RBS} = 167.86 in.^{3}; $V_{RBS,30\,ft}$ = 77.6 kips; $V_{RBS,25\,ft}$ = 91.5 kips; s_h = 14⅝ in.; F_y = 50 ksi; R_y = 1.1

$$\frac{\sum M^*_{pc}}{\sum M^*_{pb}} > 1.0 \qquad \textit{AISC 341 (E3-1)}$$

$$\sum M^*_{pc} = \sum Z_c(F_{yc} - P_r/A_g) \qquad \textit{AISC 341 (E3-2)}$$

$$\sum M^*_{pb} = \sum (M_{pr} + M_v) \qquad \textit{AISC 341 (E3-3)}$$

$$\sum M^*_{pc} = 355^{\text{in.}^3} \times (50^{\text{ksi}} - 284.6^{\text{kips}}/56.8^{\text{in.}^2}) + 355^{\text{in.}^3} \times (50^{\text{ksi}} - 213.4^{\text{kips}}/56.8^{\text{in.}^2}) = 32{,}388^{\text{k-in.}}$$

$$\sum M^*_{pc} = 2699^{\text{k-ft}}$$

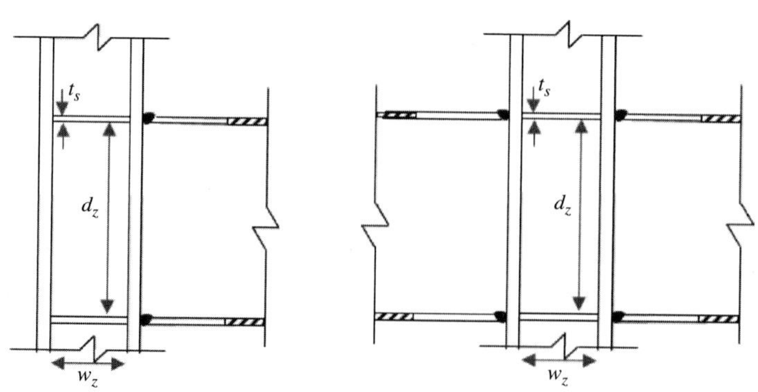

(a) One-sided (exterior) (b) One-sided (interior)

Figure 8.77 Column panel zone dimensions.

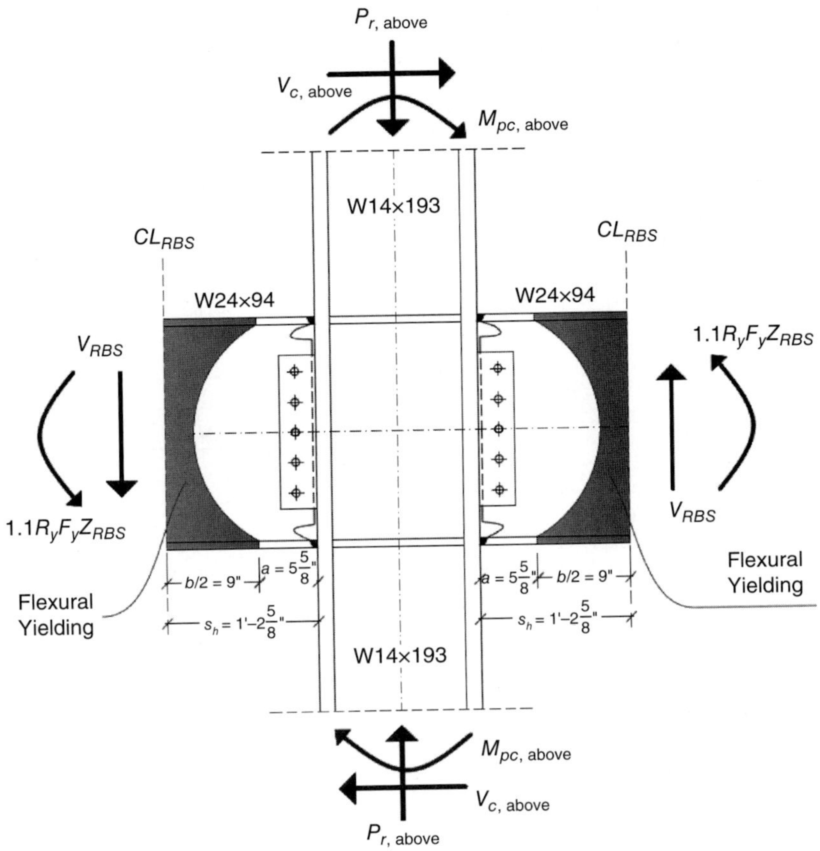

Figure 8.79 SC/WB requirement at the interior joint of the second story.

The required axial strength of the column, P_r, is determined from the greater of the following.

1. Axial force obtained from $(1.2 + 0.2S_{DS})D \pm \Omega_0 Q_E + 0.5L$
2. Axial force obtained from the second-order analysis with reduced stiffnesses, as detailed in the latter sections (DAM for stability).

$$\sum M^*_{pb} = \sum [1.1R_y F_y Z_{RBS} + V_{RBS}(S_h + d_c/2)]$$

$$\sum M^*_{pb} = [2\times(1.1\times1.1\times50^{\text{ksi}}\times167.86^{\text{in.}^3}) + (77.6^{\text{kips}} + 91.5^{\text{kips}}) \times(14.625^{\text{in.}} + 15.5^{\text{in.}}/2)] = 24{,}094^{\text{k-in.}}$$

$$\sum M^*_{pb} = 2008^{\text{k-ft}}$$

$$\frac{\sum M^*_{pc}}{\sum M^*_{pb}} = \frac{2699^{\text{k-ft}}}{2008^{\text{k-ft}}} = 1.34 > 1.0$$

Thus, SC/WB requirement is met for the joint. SC/WB check for all beam-to-column connections is summarized in Table 8.26.

To satisfy the requirement, the size of third-story interior columns is increased from W14 × 145 to W14 × 159 and the fifth-story girders are slightly reduced from W21 × 57 to W21 × 50. Note that the SC/WB

Table 8.26 Summary of SC/WB Check

	Joint	Column above	$P_{r,\text{above}}$ (kip)	$M_{pc,\text{above}}$ (k-ft)	Column below	$P_{r,\text{below}}$ (kip)	$M_{pc,\text{below}}$ (k-ft)	$\sum M^*_{pc}$	$\sum M^*_{pb}$
Exterior	6th	—	0	0.00	W 14 × 74	23.40	513.73	513.73	438.70
	5th	W 14 × 74	23.40	513.73	W 14 × 74	93.10	480.16	993.89	589.50
	4th	W 14 × 132	93.10	928.21	W 14 × 132	197.00	875.99	1804.20	769.12
	3rd	W 14 × 132	197.00	875.99	W 14 × 132	321.00	813.67	1689.66	1013.42
	2nd	W 14 × 176	321.00	1168.08	W 14 × 176	465.00	1093.95	2262.03	1015.65
	1st	W 14 × 176	465.00	1093.95	W 14 × 176	627.00	1010.55	2104.50	1149.09
Interior	6th	—	0	0.00	W 14 × 82	19.00	570.00	**570.00**	868.57
	5th	W 14 × 82	19.00	570.00	W 14 × 82	78.70	541.18	**1111.18**	**1168.32**
	4th	W 14 × 145	78.70	1043.40	W 14 × 145	148.00	1008.24	2051.64	1522.00
	3rd	W 14 × 145	148.00	1008.24	W 14 × 145	226.00	968.66	**1976.89**	**2002.31**
	2nd	W 14 × 193	226.00	1361.46	W 14 × 193	309.00	1318.23	2679.69	2007.85
	1st	W 14 × 193	309.00	1318.23	W 14 × 193	403.50	1269.01	2587.24	2270.81

*The joints that SC/WB requirement is not satisfied are indicated with bold.

Table 8.27 Summary of SC/WB Check Based on the Modified Sections

	Joint	Column above	$P_{r,above}$ (kip)	$M_{pc,above}$ (k-ft)	Column below	$P_{r,below}$ (kip)	$M_{pc,below}$ (k-ft)	$\sum M^*_{pc}$	$\sum M^*_{pb}$
Exterior	6th	—	0	0.00	W 14 × 74	24.00	513.44	513.44	437.01
	5th	W 14 × 74	24.00	513.44	W 14 × 74	92.20	480.59	994.03	509.98
	4th	W 14 × 132	92.20	928.66	W 14 × 132	196.00	876.49	1805.16	767.95
	3rd	W 14 × 132	196.00	876.49	W 14 × 132	321.00	813.67	1690.17	1012.20
	2nd	W 14 × 176	321.00	1168.08	W 14 × 176	463.00	1094.98	2263.06	1014.36
	1st	W 14 × 176	463.00	1094.98	W 14 × 176	626.00	1011.07	2106.05	1147.79
Interior	6th	—	0	0.00	W 14 × 82	19.20	569.90	569.90	868.57
	5th	W 14 × 82	19.20	569.90	W 14 × 82	79.00	541.04	1111.00	1013.43
	4th	W 14 × 159	79.00	1155.37	W 14 × 159	147.70	1120.19	2275.57	1523.22
	3rd	W 14 × 159	147.70	1120.19	W 14 × 159	224.00	1081.12	2201.31	2003.89
	2nd	W 14 × 193	224.00	1362.50	W 14 × 193	307.00	1319.27	2681.77	2007.85
	1st	W 14 × 193	307.00	1319.27	W 14 × 193	401.50	1270.05	2589.32	2270.81

requirement does not apply to the joints at the roof. The moment ratios for each joint after the changes are given in Table 8.27. The SC/WB requirement is met when the members given in Table 8.27 are employed, and thus, the stiffness (drift) requirement is examined once again (Table 8.28) to make sure that none of the stories does not exceed the allowable limit due to the slight change in the lateral stiffness.

Table 8.28 Member Sizes That Satisfy SCWB Requirement

Story	Exterior column	Interior column	Girders
Roof	W 14 × 74	W 14 × 82	W 18 × 50
5th	W 14 × 74	W 14 × 82	W 21 × 50
4th	W 14 × 132	W 14 × 159	W 21 × 73
3rd	W 14 × 132	W 14 × 159	W 24 × 76
2nd	W 14 × 176	W 14 × 193	W 24 × 84
1st	W 14 × 176	W 14 × 193	W 24 × 94

Strength Requirement

Load Combinations

1. $1.4D$
2. $1.2D + 1.6L + 0.5L_r$
3. $1.2D + 1.6L_r + 1.0L$
4. $(1.2 + 0.2S_{DS})D + \rho Q_E + 0.5L$
5. $(0.9 - 0.2S_{DS})D + \rho Q_E$

Redundancy factor, ρ, is taken as 1.0 in conformance with *ASCE 7, Section 12.3.4.*

Gravity Load Distribution To determine the vertical forces acting in the plane of the SMFs on lines 1 and 6 due to dead and live loads, first, the tributary areas for the each uniformly distributed and concentrated loads are computed. Figure 8.80 shows the typical tributary areas for each member.

The gravity loads in the plane of the SMFs are computed as follows:

Dead Loads

Uniformly Distributed Loads

$$w_D^{Roof} = (6.167^{ft}) \times 34^{psf} \times (1^{kips}/1000^{lb}) = 0.21^{kip/ft}$$
$$w_D^{Floor} = (6.167^{ft}) \times 85^{psf} \times (1^{kips}/1000^{lb}) = 0.52^{kip/ft}$$

Concentrated Loads on Interior Columns

$$P_D^{Roof} = (27.5^{ft} \times 8.33^{ft}) \times 34^{psf} \times (1^{kips}/1000^{lb}) = 7.8^{kips}$$
$$P_D^{Floor} = (27.5^{ft} \times 8.33^{ft}) \times 85^{psf} \times (1^{kips}/1000^{lb}) = 19.5^{kips}$$

Concentrated Loads on Exterior Columns

$$P_D^{Roof} = (25^{ft} \times 8.33^{ft}) \times 34^{psf} \times (1^{kips}/1000^{lb}) = 7.1^{kips}$$
$$P_D^{Floor} = (25^{ft} \times 8.33^{ft}) \times 85^{psf} \times (1^{kips}/1000^{lb}) = 17.7^{kips}$$

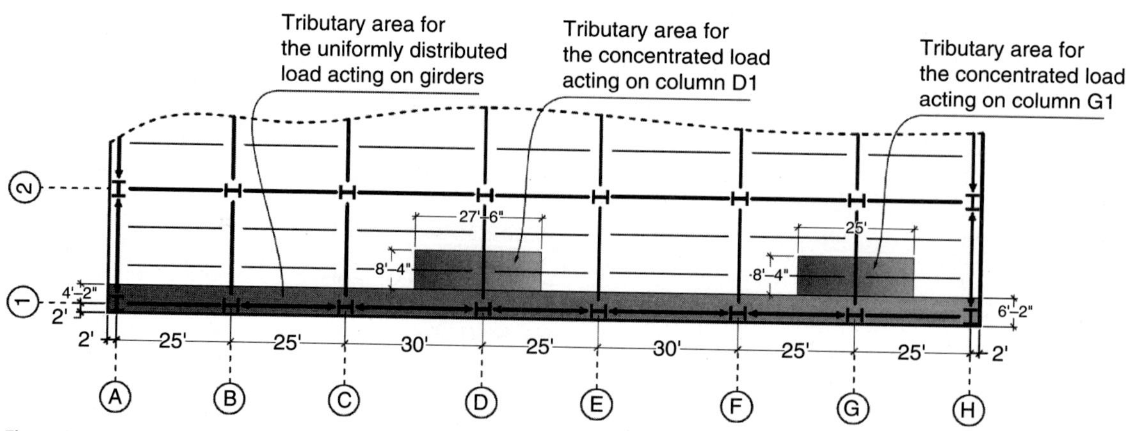

Figure 8.80 Tributary areas for the loads acting on SMF on lines 1 and 6.

Live Loads

Uniformly Distributed Loads

$$w_L^{Roof} = (6.167^{ft}) \times 20^{psf} \times (1^{kips}/1000^{lb}) = 0.12^{kip/ft}$$

$$w_D^{Floor} = (6.167^{ft}) \times 50^{psf} \times (1^{kips}/1000^{lb}) = 0.31^{kip/ft}$$

Concentrated Loads on Interior Columns

$$P_D^{Roof} = (27.5^{ft} \times 8.33^{ft}) \times 20^{psf} \times (1^{kips}/1000^{lb}) = 4.6^{kips}$$

$$P_D^{Floor} = (27.5^{ft} \times 8.33^{ft}) \times 50^{psf} \times (1^{kips}/1000^{lb}) = 11.5^{kips}$$

Concentrated Loads on Exterior Columns

$$P_D^{Roof} = (25^{ft} \times 8.33^{ft}) \times 20^{psf} \times (1^{kips}/1000^{lb}) = 4.2^{kips}$$

$$P_D^{Floor} = (25^{ft} \times 8.33^{ft}) \times 50^{psf} \times (1^{kips}/1000^{lb}) = 10.4^{kips}$$

The summary of dead and live loads acting on the frames are shown in Figs. 8.81 and 8.82, respectively.

Design of Beam-Columns

Design for Stability: Direct Analysis Method (DAM) As addressed in *Section C, AISC 360*, a stability analysis based on the DAM for calculation of the required strength of structural members is required. In order for explicitly modeling of member imperfections and inelasticity, a second-order elastic analysis that considers,

- Second-order effects (P-Δ and P-δ)
- Member deformations (flexural, shear, and axial)
- Stiffness reduction
- Initial member imperfections

is carried out. The steps for modeling and analysis are summarized below.

Step-by-step Analysis:

1. Build a realistic numerical model of the lateral force-resisting system with "leaning columns" to introduce the impact of gravity frames on second-order effects.
2. Reduce the stiffness of the lateral force-resisting system.
3. Apply notional loads in terms of the factored gravity loads or notional displacements to account for initial imperfections.
4. Conduct a second-order analysis that takes P-Δ and P-δ effects into account.
5. Determine capacities of members using K = 1.0.
6. Check the ratio of second-order drifts to first-order drifts ($\Delta_{2nd}/\Delta_{1st}$).

Step I: Create a model with the leaning columns

Step II: Reduce all stiffnesses that contribute to the stability

According to AISC *AISC 360, Section C2.3*, an additional factor, τ_b, is to be introduced for flexural stiffness adjustments based on the level of axial force-to-capacity ratio:

$$\tau_b = 1.0 \text{ when } \frac{\alpha P_r}{P_y} \leq 0.5$$

and

$$\tau_b = 4(\alpha P_r/P_y)[1-(\alpha P_r/P_y)] \text{ when } \frac{\alpha P_r}{P_y} > 0.5$$

Note that flexural stiffness is initially reduced 20 percent by assuming τ_b = 1.0. This assumption will be checked later. Axial and flexural stiffnesses are given below.

Axial Stiffness: 0.8*EA*

Flexural Stiffness: 0.8*EI*

Step III: Apply notional loads

The notional load, N_i, is taken as 0.002 times the total gravity load at each story level, which is based on an initial story out-of-straightness of 1/500 for all stories.

$$N_i = 0.002\alpha Y_i$$

where, α = 1.0 (LRFD), α = 1.6 (ASD), and Y_i is the total gravity load at each story level, including the loads supported by gravity frames. The load combinations adopted for the stability analysis are given in Table 8.29.

Step IV: Verify $\Delta_{2nd}/\Delta_{1st}$

AISC 360 requires the ratio of second-order drifts to first-order drifts to be examined when determining the load combinations in which, the

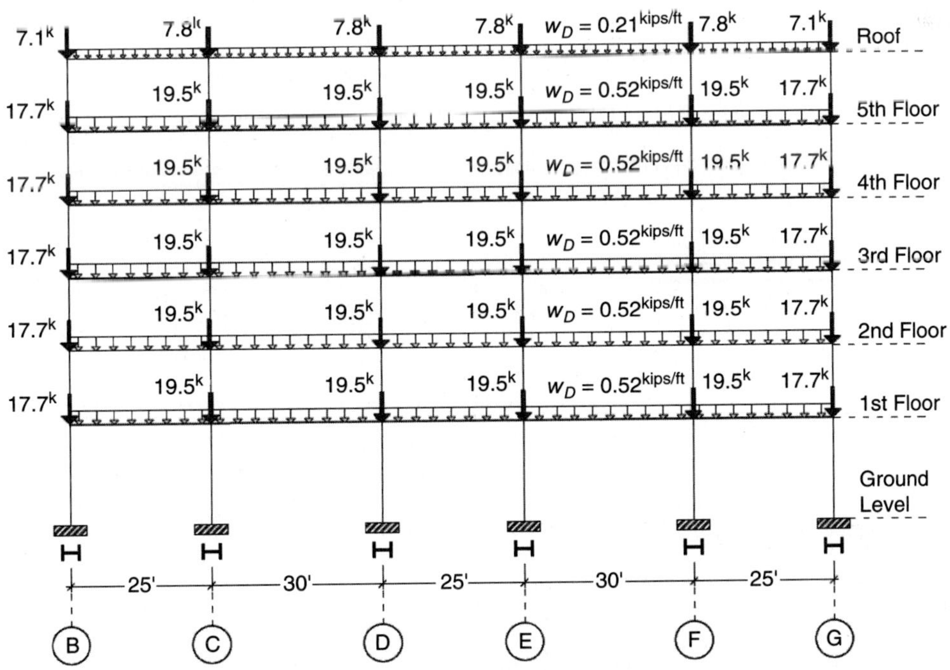

Figure 8.81 Dead load distribution to the SMFs.

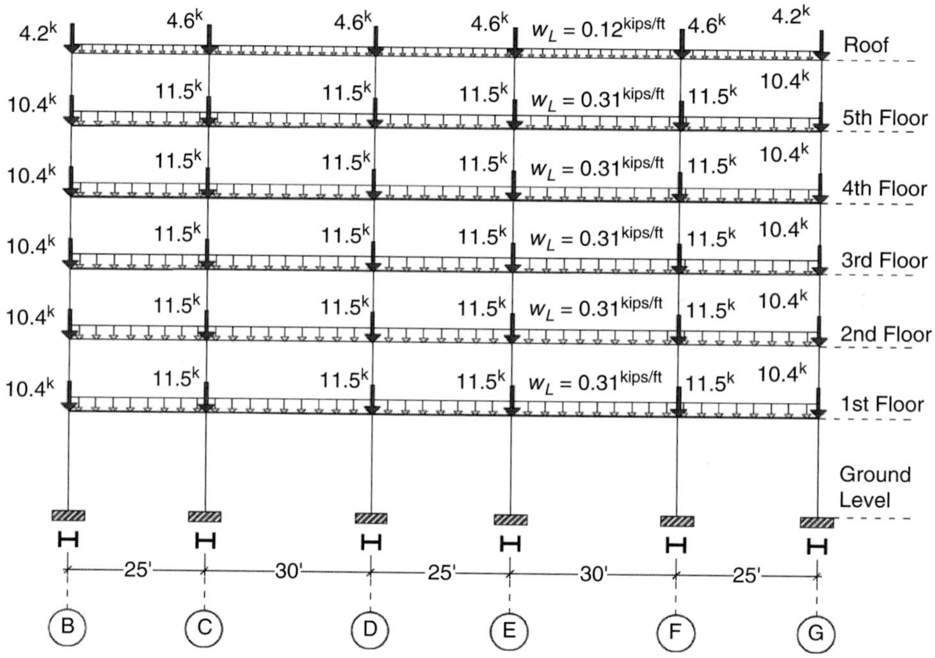

Figure 8.82 Live load distribution to the SMFs.

Table 8.29 Load Combinations Including Notional Load Effects

Combination	Description
Comb#1	$1.4D \pm 1.4ND$
Comb#2	$1.2D + 1.6L + 0.5L_r \pm (1.2ND + 1.6NL + 0.5NL_r)$
Comb#3	$1.2D + 1.0L + 1.6L_r \pm (1.2ND + 1.0NL + 1.6NL_r)$
Comb#4	$(1.2 + 0.2S_{DS})D \pm \rho Q_E + 0.5L$
Comb#5	$(0.9 - 0.2S_{DS})D \pm \rho Q_E$

*ND, Notional Dead Load; NL, Notional Live Load; NLr, Notional Roof Live Load.
**Notional loads are only included in gravity load combinations.

notional loads to be included. Table 8.30 presents the maximum drift ratios for each combination given in Table 8.29.

Table 8.30 The Ratio of Second-Order Drifts to First-Order Drifts ($\Delta_{2nd}/\Delta_{1st}$)

Combination	$\Delta_{2nd}/\Delta_{1st}$ ratio*
Comb#1	1.20
Comb#2	1.19
Comb#3	1.18
Comb#4	1.22
Comb#5	1.09

*Max. $\Delta_{2nd}/\Delta_{1st}$ ratio along the height.

Since the ratio of second-order drifts to first-order drifts is less than 1.7 for all combinations, it is not required to include the notional loads in the combinations with other lateral loads (*AISC 360, Section C2.2b*).

Step V: Check if $\tau_b = 1.0$

The actual τ_b values are to be checked to confirm that the assumption made in *Step II* is true.

Example τ_b calculation for the first-story exterior column (W 14 × 176, $A_g = 51.8$ in.^{2}):

$$\left.\begin{aligned} P_{r,\max} &= 395 \text{ kips} \\ P_y = A_g F_y &= 51.8^{\text{in.}^2} \times 50^{\text{ksi}} = 2590 \text{ kips} \end{aligned}\right\} \frac{\alpha P_{r,\max}}{P_y} = \frac{1.0 \times 395}{2590} = 0.15 < 0.50$$

Thus, $\tau_b = 1.0$

Note that $P_{r,\max} = 395$ kips is obtained from Comb#4. The τ_b values for the exterior and interior columns are summarized in Tables 8.31 and 8.32, respectively.

Table 8.31 τ_b Values for Exterior Columns

Story	Section	P_r (kips)	P_y (kips)	P_r/P_y	τ_b
6th	W 14 × 74	16.00	1090.0	0.01	1.0
5th	W 14 × 74	71.00	1090.0	0.07	1.0
4th	W 14 × 132	140.00	1940.0	0.07	1.0
3rd	W 14 × 132	216.00	1940.0	0.11	1.0
2nd	W 14 × 176	301.00	2590.0	0.12	1.0
1st	W 14 × 176	395.00	2590.0	0.15	1.0

Table 8.32 τ_b Values for Interior Columns

Story	Section	P_r (kips)	P_y (kips)	P_r/P_y	τ_b
6th	W 14 × 82	17.00	1200.0	0.01	1.0
5th	W 14 × 82	79.00	1200.0	0.07	1.0
4th	W 14 × 159	144.00	2335.0	0.06	1.0
3rd	W 14 × 159	213.00	2335.0	0.09	1.0
2nd	W 14 × 193	284.00	2840.0	0.10	1.0
1st	W 14 × 193	360.00	2840.0	0.13	1.0

Step VI: Determine capacities of members using K = 1.0
Example calculation for the available strength of the first-story exterior column is given below.

W14 × 176:
Z_x = 320 in.^{3}; S_x = 281 in.^{3}; r_x = 6.43 in.; r_y = 4.02 in.; F_y = 50 ksi; L_p = 14.2 ft; L_r = 73.2 ft

Determine column capacity, P_c:

$$\text{Since } K_x = K_y = 1.0 \text{ and } r_x = 6.43^{\text{in.}} > r_y = 4.02^{\text{in.}}$$

$$(L_c/r)_{\max} = (L_c/r)_y = \frac{1.0\times(18\times12)}{4.02}$$

$$= 53.73 < 4.71\sqrt{E/F_y} = 113 \text{ (Inelastic buckling)}$$

$$\text{Use } \textit{AISC 360-16, Eq. E3-2} \rightarrow F_{cr} = \left[0.658^{F_y/F_e}\right]F_y$$

$$F_e = \frac{\pi^2 E}{(KL/r)_y^2} = \frac{\pi^2\times 29{,}000}{53.73^2} = 99.14 \text{ ksi}$$

$$F_{cr} = (0.658^{50/99.14})\times 50 = 40.5 \text{ ksi}$$

$$\phi_c P_n = \phi_c F_{cr} A_g = 0.9\times 40.5\times 51.8 = 1887 \text{ kips}$$

Alternatively, enter AISC Manual Table 4.1;

$$\alpha = r_x/r_y = 1.60$$

$$(L_c/r)_{\max} = (L_c/r)_y = 18 \text{ ft}$$

For KL = 18 ft → $P_c = \phi_c P_n$ = 1890 kips

Note that the difference between the P_c values above caused by the round-off errors.

Determine beam capacity, M_c:
Since W14 × 176 is compact for flexure ($\lambda < \lambda_p$), the limit state of LTB need to be checked to determine the beam capacity.

Since $L_p = 14.2^{\text{ft}} < L_b = 18^{\text{ft}} < L_r = 73.2^{\text{ft}}$, inelastic LTB might occur. Thus, C_b factor is needed.

$$C_b = \frac{12.5M_{\max}}{2.5M_{\max} + 3M_A + 4M_B + 3M_C} \quad \textit{(Eqs. F1-1, AISC 360)}$$

where $M_{\max}$ = absolute value of maximum moment in the unbraced segment.
M_A = absolute value of moment at the quarter point of the unbraced segment.
M_B = absolute value of moment at the centerline of the unbraced segment.
M_C = absolute value of moment at the three-quarter point of the unbraced segment.

Using the moment diagram of the first-story exterior column given in Fig. 8.83, C_b factor can be calculated as follows:

$$C_b = \frac{12.5\times 617}{2.5\times 617 + 3\times 360 + 4\times 96 + 3\times 171} = 2.19$$

$$M_n = C_b\left[M_p - (M_p - 0.7F_yS_x)\left(\frac{L_b - L_p}{L_r - L_p}\right)\right] \le M_p$$

$$M_p = Z_xF_y = 320^{\text{in.}^3}\times 50^{\text{ksi}}/12^{\text{ft/in.}} = 1333.3^{\text{k-ft}}$$

$$M_n = 2.19$$

$$\times\left[1333.3^{\text{k-ft}} - (1333.3^{\text{k-ft}} - 0.7\times 281^{\text{in.}^3}\times 50^{\text{ksi}}/(12^{\text{in./ft}}))\left(\frac{18^{\text{ft}} - 14.2^{\text{ft}}}{73.2^{\text{ft}} - 18^{\text{ft}}}\right)\right]$$

$$= 2842^{\text{k-ft}}$$

$$M_n = 2842^{\text{k-ft}} > M_p = 1333.3^{\text{k-ft}}$$

Thus,

$$M_c = \phi_b \times M_p = 0.9 \times 1333.3 = 1200^{\text{k-ft}}$$

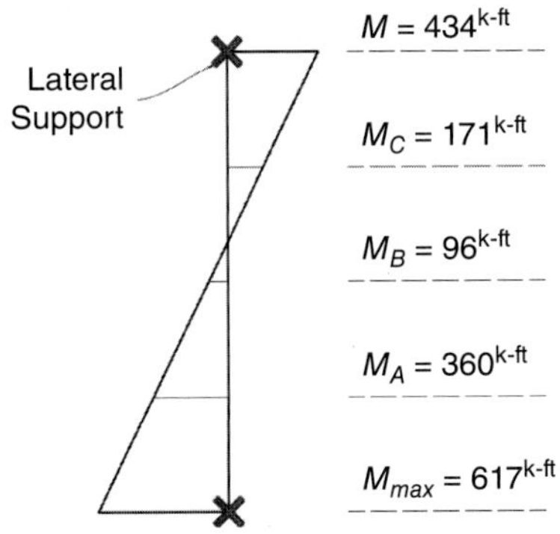

Figure 8.83 Moment diagram of the first-story exterior column.

Check P-M Interaction using Eqs. H1.1a and H1.1b, AISC 360:
The maximum demand is obtained from Comb#4:

$$P_r = 395 \text{ kips}$$

$$M_r = 617 \text{ kip-ft}$$

$$\frac{P_r}{P_c} = \frac{395}{1890} = 0.21 > 0.2 \rightarrow \textit{Use } \text{H1.1a} \frac{P_r}{P_c} + \frac{8}{9}\left(\frac{M_r}{M_c}\right) \le 1.0$$

$$\frac{395}{1890} + \frac{8}{9}\left(\frac{617}{1200}\right) = 0.67 < 1.0 \rightarrow \text{W14}\times 76 \text{ is adequate.}$$

Axial force and bending moment demand on the interior and exterior columns, as well as the capacity checks for each loading combination, are summarized in Tables 8.33 and 8.34, respectively. As indicated in the tables, the maximum demand-to-capacity ratio is 67 percent for both exterior and interior columns. In other words, the design of the columns is not governed by the strength requirement, as expected. This relatively small capacity-to-demand ratio can be attributable to the flexible nature of the MFs, which requires larger sections to limit lateral deflections than those to provide sufficient strength. It is, therefore, crucial to begin by checking the drift and SC/WB requirements in an effort to reduce the number of iterations when designing such flexible structures.

Additionally, it is important to note that even though it is not detailed herein, the overstrength seismic loads should also be considered when determining the required strength for columns in the SMFs. Per *Section D1.4a, AISC 341-16*, the required strength is the larger of the effects of the two analyses: (a) Both axial force and bending moment demands resulting from DAM and, (b) Only axial force demand from the overstrength seismic loads. When the required compressive and tensile strengths are computed using the overstrength seismic loads, it is permitted to neglect the bending moment demand. In the above calculation, only DAM is shown since Pr/Pc ratio due to the overstrength seismic loads was about 0.60, which did not govern the design of columns.

Design of Girders

Lateral Bracing of Girders To ensure the desired level of energy-dissipation, lateral bracings of highly ductile beams in SMF are spaced at a maximum spacing of $L_b = 0.095r_yE/(R_yF_y)$, per *Section D1.2, AISC 341* (Table 8.35).

The second-order analysis results of the girders under the specified loads in Comb.#4 are shown in Fig. 8.84. It should be noted that these results are only plotted for the three spans including the mid-span and two edge spans when the structure is subjected to seismic loads from one direction along with the factored gravity loads, since the structure is symmetrical. As shown in Fig. 8.84, the maximum bending moment varies from span to span. Provided that the same girder size is employed at each story level, the maximum bending moment among the girders of each story level is taken as the required strength of that story. The required flexural strength for each story level obtained from the first-order and second-order analyses are summarized and compared with the flexural capacity of the girders in Table 8.36. Similar to design of the columns, the

Table 8.33 Strength Check for Exterior Columns Based on Second-Order Analysis Results

Comb.	Story	Section	P_r (kips)	P_c (kips)	M_r (k-ft)	M_c (k-ft)	P-M Interaction
Comb.#1	6th	W 14 × 74	12.50	735.00	11.40	472.50	0.03
	5th	W 14 × 74	43.60	735.00	16.80	472.50	0.07
	4th	W 14 × 132	75.30	1540.00	19.70	877.50	0.05
	3rd	W 14 × 132	107.50	1540.00	22.90	877.50	0.06
	2nd	W 14 × 176	140.30	2090.00	27.80	1200.00	0.06
	1st	W 14 × 176	173.60	1890.00	37.40	1200.00	0.08
Comb.#2	6th	W 14 × 74	13.40	735.00	16.00	472.50	0.04
	5th	W 14 × 74	60.70	735.00	23.70	472.50	0.09
	4th	W 14 × 132	108.70	1540.00	25.50	877.50	0.06
	3rd	W 14 × 132	157.10	1540.00	28.60	877.50	0.08
	2nd	W 14 × 176	206.10	2090.00	33.90	1200.00	0.08
	1st	W 14 × 176	255.50	1890.00	36.70	1200.00	0.10
Comb.#3	6th	W 14 × 74	19.00	735.00	15.70	472.50	0.05
	5th	W 14 × 74	59.00	735.00	19.40	472.50	0.08
	4th	W 14 × 132	99.00	1540.00	22.60	877.50	0.06
	3rd	W 14 × 132	139.00	1540.00	25.20	877.50	0.07
	2nd	W 14 × 176	181.00	2090.00	30.20	1200.00	0.07
	1st	W 14 × 176	222.00	1890.00	34.80	1200.00	0.09
Comb.#4	6th	W 14 × 74	16.00	735.00	64.00	472.50	0.15
	5th	W 14 × 74	71.00	735.00	149.00	472.50	0.36
	4th	W 14 × 132	140.00	1540.00	204.00	877.50	0.28
	3rd	W 14 × 132	216.00	1540.00	256.00	877.50	0.36
	2nd	W 14 × 176	301.00	2090.00	290.10	1200.00	0.31
	1st	**W 14 × 176**	**395.00**	**1890.00**	**617.00**	**1200.00**	**0.67**
Comb.#5	6th	W 14 × 74	10.00	735.00	56.50	472.50	0.13
	5th	W 14 × 74	36.00	735.00	131.00	472.50	0.30
	4th	W 14 × 132	75.00	1540.00	180.00	877.50	0.23
	3rd	W 14 × 132	121.00	1540.00	226.00	877.50	0.30
	2nd	W 14 × 176	174.00	2090.00	252.00	1200.00	0.25
	1st	W 14 × 176	235.00	1890.00	552.00	1200.00	0.52

*Summary of the above calculations are indicated in bold.

Table 8.34 Strength Check for Interior Columns Based on Second-Order Analysis Results

Comb.	Story	Section	P_r (kips)	P_c (kips)	M_r (k-ft)	M_c (k-ft)	P-M Interaction
Comb.#1	6th	W 14 × 82	16.50	809.00	6.60	521.25	0.02
	5th	W 14 × 82	57.70	809.00	11.60	521.25	0.06
	4th	W 14 × 159	99.00	1720.00	18.00	975.00	0.05
	3rd	W 14 × 159	140.40	1720.00	23.10	975.00	0.06
	2nd	W 14 × 193	182.00	2290.00	30.80	1331.25	0.06
	1st	W 14 × 193	224.00	2080.00	42.30	1331.25	0.09
Comb.#2	6th	W 14 × 82	17.50	865.64	7.80	521.25	0.03
	5th	W 14 × 82	80.50	865.64	14.60	521.25	0.07
	4th	W 14 × 159	143.50	1770.00	19.40	975.00	0.06
	3rd	W 14 × 159	207.00	1773.14	24.20	975.00	0.08
	2nd	W 14 × 193	270.00	2365.41	31.50	1331.25	0.08
	1st	W 14 × 193	334.00	2123.52	40.40	1331.25	0.11
Comb.#3	6th	W 14 × 82	25.00	865.64	8.40	521.25	0.03
	5th	W 14 × 82	78.00	865.64	12.60	521.25	0.07
	4th	W 14 × 159	130.00	1770.00	18.00	975.00	0.06
	3rd	W 14 × 159	183.00	1773.14	22.50	975.00	0.07
	2nd	W 14 × 193	236.00	2365.41	29.50	1331.25	0.07
	1st	W 14 × 193	289.00	2123.52	38.50	1331.25	0.10
Comb.#4	6th	W 14 × 82	17.00	865.64	99.80	521.25	0.20
	5th	W 14 × 82	79.00	865.64	211.00	521.25	0.45
	4th	W 14 × 159	144.00	1770.00	344.00	975.00	0.39
	3rd	W 14 × 159	213.00	1773.14	417.50	975.00	0.49
	2nd	W 14 × 193	284.00	2365.41	506.00	1331.25	0.44
	1st	W 14 × 193	360.00	2123.52	747.00	1331.25	0.65
Comb.#5	6th	W 14 × 82	9.00	865.64	91.00	521.25	0.18
	5th	W 14 × 82	33.00	865.64	196.00	521.25	0.40
	4th	W 14 × 159	60.00	1770.00	319.00	975.00	0.34
	3rd	W 14 × 159	90.00	1773.14	380.00	975.00	0.42
	2nd	W 14 × 193	122.00	2365.41	442.00	1331.25	0.36
	1st	W 14 × 193	159.00	2123.52	664.00	1331.25	0.54

Table 8.35 Spacing of Lateral Bracings

Story	Section	r_y (in.)	$L_{b,max}$ (in.)	$L_{b,25}$*	$L_{b,30}$*
Roof	W 18 × 50	1.7	82.7	5@5"	6@5'
5th	W 21 × 50	1.3	65.1	6@4'-2"	7@4'-3"
4th	W 21 × 73	1.8	90.7	4@6'-3"	5@6'
3rd	W 24 × 76	1.9	96.2	4@6'-3"	5@6'
2nd	W 24 × 84	2.0	97.7	4@6'-3"	5@6'
1st	W 24 × 94	2.0	99.2	4@6'-3"	5@6'

*$L_{b,25}$ and $L_{b,30}$ represent the selected unbraced lengths in 25 ft and 30 ft bays, respectively.

maximum ratio of demand to capacity is about 68 percent for the girders owing to the flexibility of steel framed structures.

Member Requirements

According to *Section E3-5a, AISC 341*, beams and columns in SMF shall be highly ductile. Therefore, the requirements stipulated for highly ductile members in Section D1.1 should be met. Per *Table D1.1, AISC 341*, the limiting width-to-thickness ratio for flanges is given as follows:

$$\lambda_f = \frac{b}{2t_f} < \lambda_{hd} = 0.32\sqrt{\frac{E}{R_y F_y}}$$

Limiting width-to-thickness ratio for the web is defined as a function of the axial load to capacity ratio, C_a:

$$\lambda_{hd} = 2.57\sqrt{\frac{E}{R_y F_y}}(1.0 - 1.04C_a) \text{ when } C_a = \frac{P_u}{\phi_c P_y} \le 0.114$$

$$\lambda_{hd} = 0.88\sqrt{\frac{E}{R_y F_y}}(2.68 - C_a) \ge 1.57\sqrt{\frac{E}{R_y F_y}} \text{ when } C_a = \frac{P_u}{\phi_c P_y} > 0.114$$

Example calculation for the first-story exterior column (W14 × 176):

$$\text{W14} \times 176 \rightarrow b/2t_f = 5.97 \text{ and } h/t_w = 13.7$$

For flanges:

$$\lambda_f = \frac{b}{2t_f} = 5.97 < \lambda_{hd} = 0.32\sqrt{\frac{E}{R_y F_y}} = 0.32\sqrt{\frac{29{,}000}{1.1 \times 50}} = 7.35$$

For web:

$$C_a = \frac{P_u}{\phi_c P_y} = \frac{409.8^{\text{kips}}}{0.9 \times (1.1 \times 51.8^{\text{in.}^2} \times 50^{\text{ksi}})} = 0.16 > 0.114$$

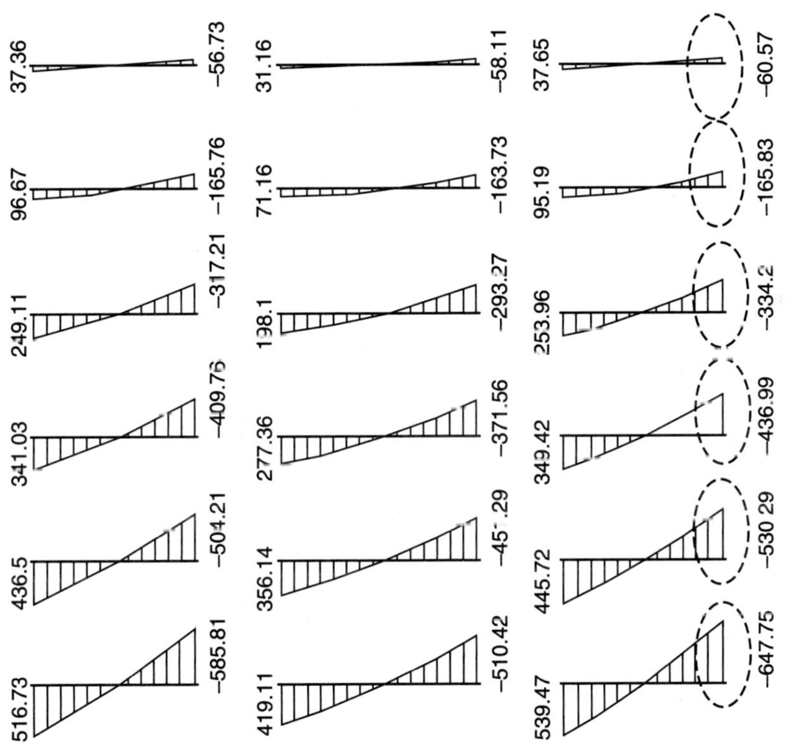

Figure 8.84 Moment diagram of girders obtained from second-order analysis under Comb.#4 (M_r^{2nd}, k-ft).

Table 8.36 Strength Checks for Girders

Story	Girder	Z_x (in.^{3})	M_p (k-ft)	M_r^{1st} (k-ft)	M_r^{2nd} (k-ft)	M_c (k-ft)	Fail/pass
Roof	W 18 × 50	101	420.8	57.6	60.6	378.8	Pass
5th	W 21 × 50	110	458.3	155.8	165.8	412.5	Pass
4th	W 21 × 73	172	716.7	300.2	334.2	645.0	Pass
3rd	W 24 × 76	200	833.3	383.2	437.0	750.0	Pass
2nd	W 24 × 84	224	933.3	455.8	530.3	840.0	Pass
1st	W 24 × 94	254	1058.3	542.1	647.8	952.5	Pass

*M_r^{1st} and M_r^{2nd} stand for the required flexural strength obtained from first-order and second-order analyses, respectively.

Table 8.37 Limiting Width-to-Thickness Ratios for Girders

Story	Girder	$b/2t_f$	λ_{hd}
Roof	W 18 × 50	6.57	7.35
5th	W 21 × 50	6.10	7.35
4th	W 21 × 73	5.60	7.35
3rd	W 24 × 76	6.61	7.35
2nd	W 24 × 84	5.86	7.35
1st	W 24 × 94	5.18	7.35

Table 8.38 Limiting Width-to-Thickness Ratios for Columns

	Column	$b/2t_f$	$\lambda_{hd,f}$	h/t_w	$\lambda_{hd,w}$
Exterior	W 14 × 74	6.41	7.35	25.4	54.8
	W 14 × 132	7.15	7.35	17.7	51.8
	W 14 × 176	5.97	7.35	13.7	54.1
Interior	W 14 × 82	5.92	7.35	22.4	54.8
	W 14 × 145	6.54	7.35	15.3	53.4
	W 14 × 193	5.45	7.35	12.8	51.5

$$h/t_w = 13.7 < \lambda_{hd} = 0.88 \times \sqrt{\frac{29{,}000}{1.1 \times 50}} \times (2.68 - 0.16)$$

$$= 54.1 > 1.57 \times \sqrt{\frac{29{,}000}{1.1 \times 50}} = 36.1$$

The summary of the member requirements for the girders and columns are presented in Tables 8.37 and 8.38, respectively.

Concluding Remarks

As listed at the beginning of this chapter, several requirements key to ensure safety, serviceability, and energy-dissipation capability of the structure are met while designing the frames. The members that satisfy strength, stiffness, and SC/WB requirements along with the compactness requirement are given in Table 8.39. In addition to the selection of the structural members, a design example of RBS connection is also presented in detail. Considering the cost of the additional plates and the rigorous detailing requirements in the panel zones, rather than introducing doubler plates to the column web, it would be cost-effective to increase the column sizes where doubler plates are required. To do so, the stiffness requirement can be checked using the seismic loads computed based on $C_u T_a$ in lieu of those based on the period obtained from the numerical model. Even though it is permitted to determine the seismic loads without considering the upper limit on the period when computing elastic drifts, the reduction in the column size would result in weak panel zones, as indicated in the example. It is very likely that the cost of the potential need for doubler plates, and their connections, in particular, would exceed the cost of increasing the column size. Therefore, it would be more economical and less time consuming to compute the drifts under the seismic loads based on the upper limit on the period for strength design.

Table 8.39 Final Design

Story	Exterior column	Interior column	Girders
Roof	W 14 × 74	W 14 × 82	W 18 × 50
5th	W 14 × 74	W 14 × 82	W 21 × 50
4th	W 14 × 132	W 14 × 159	W 21 × 73
3rd	W 14 × 132	W 14 × 159	W 24 × 76
2nd	W 14 × 176	W 14 × 193	W 24 × 84
1st	W 14 × 176	W 14 × 193	W 24 × 94

8.2.3 Concentrically Braced Frames (CBFs)

CBFs are considered to be truss systems resisting lateral forces by truss action and therefore among the most cost-effective SFRS (Fig. 8.85). Brace members in a typical CBF are expected to dissipate energy through inelastic action by yielding in tension and buckling in compression.

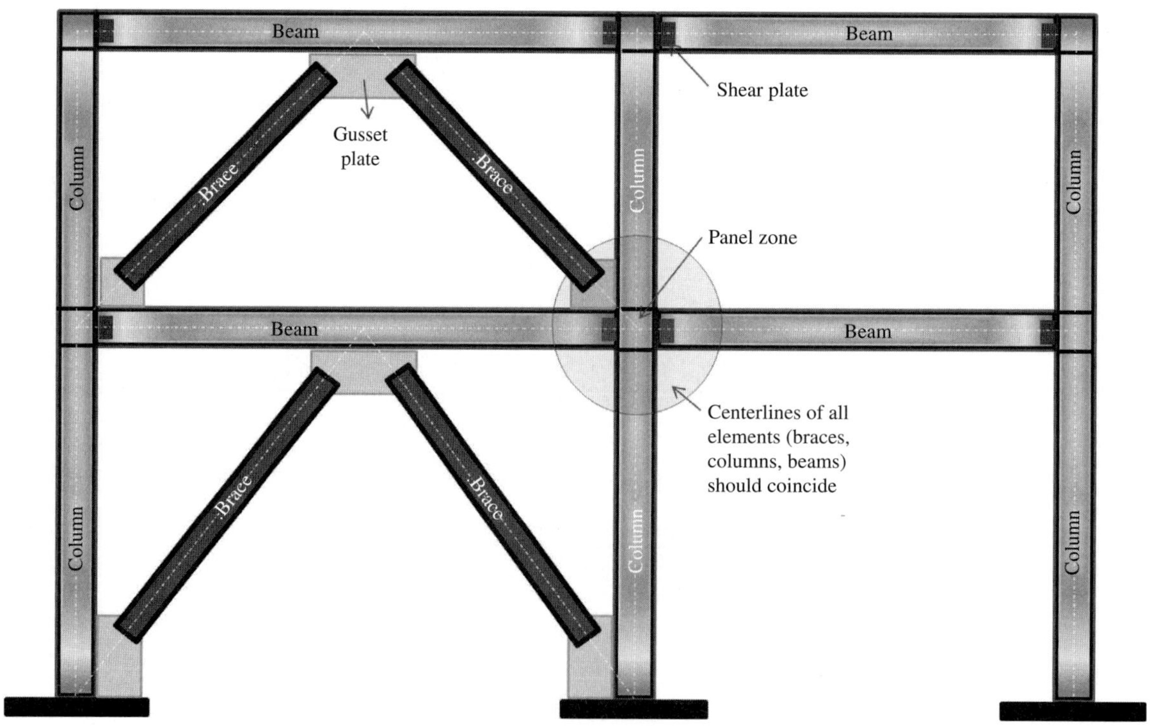

Figure 8.85 A concentrically braced frame.

Thus, seismic response of a CBF is highly dependent on the inelastic cyclic behavior of the braces. However, the braces yielding in tension are the main source of ductility, because buckling is a nonductile failure mode. CBFs are less ductile than MFs and have a lower response modification coefficient, R, values, but they meet the lateral stiffness and strength requirements easily. The design of a CBF can easily be performed by hand; no structural software is needed to calculate the internal forces in the structural elements (braces, beams, and columns).

Different bracing configurations are shown in Fig. 8.86 (V-, inverted V-, X-, diagonal, and two-story X-bracing). Even though the most common types of bracing configurations are V- and inverted V-bracing types, two-story X-bracing configuration has become quite popular recently. The main reason for this is that a large unbalanced force due to the difference in yielding and buckling strengths of braces develop in V- and inverted V-bracings which results in heavy and deep beam sections. On the other hand, in a two-story X-bracing configuration, there is no need to consider unbalanced force in the beam since the unbalanced forces below and above the beam will have the same magnitude, but act in opposite directions.

STRUCTURAL ANALYSIS

The first step in sizing the structural elements is to design the braces first in a CBF. Brace design is carried out by the lateral forces specified in *ASCE 7*. Then, beams and columns are designed for the maximum forces imposed on them by the braces. In determining the capacity-limited horizontal seismic load effects, E_{cl}, in girders and columns, the following two structural analysis cases are required to be carried out (Fig. 8.87):

a. A structural analysis in which all braces are assumed to resist forces corresponding to their expected strength in compression or tension.

b. A structural analysis in which all braces in tension are assumed to resist forces corresponding to their expected strength and all braces in compression are assumed to resist their expected postbuckling strength.

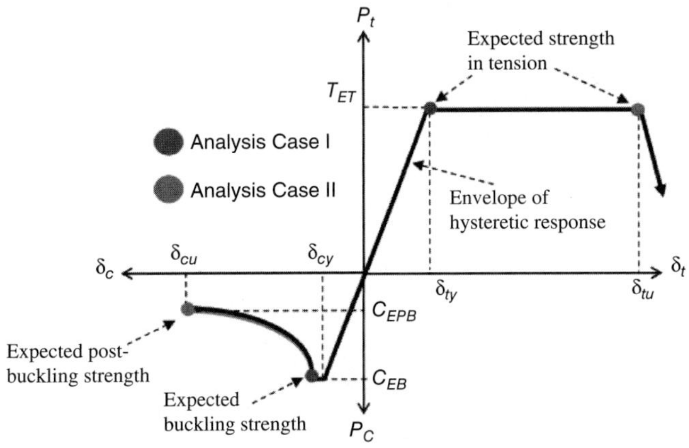

Figure 8.87 Illustration of expected brace strength in tension (T_{ET}), expected buckling (C_{EB}), and postbuckling (C_{EPB}) strengths to be used in the two structural analysis cases.

The expected brace strength in tension, T_{ET}, expected brace strength in compression (or expected buckling strength), C_{ET}, and expected post-buckling brace strength, C_{EPB}, are given as

$$T_{ET} = R_y F_y A_g$$

$$C_{ET} = \{R_y F_y A_g;\ F_{cre} A_g / 0.877\}_{\min}$$

$$C_{EPB} = 0.3\ C_{EPB}$$

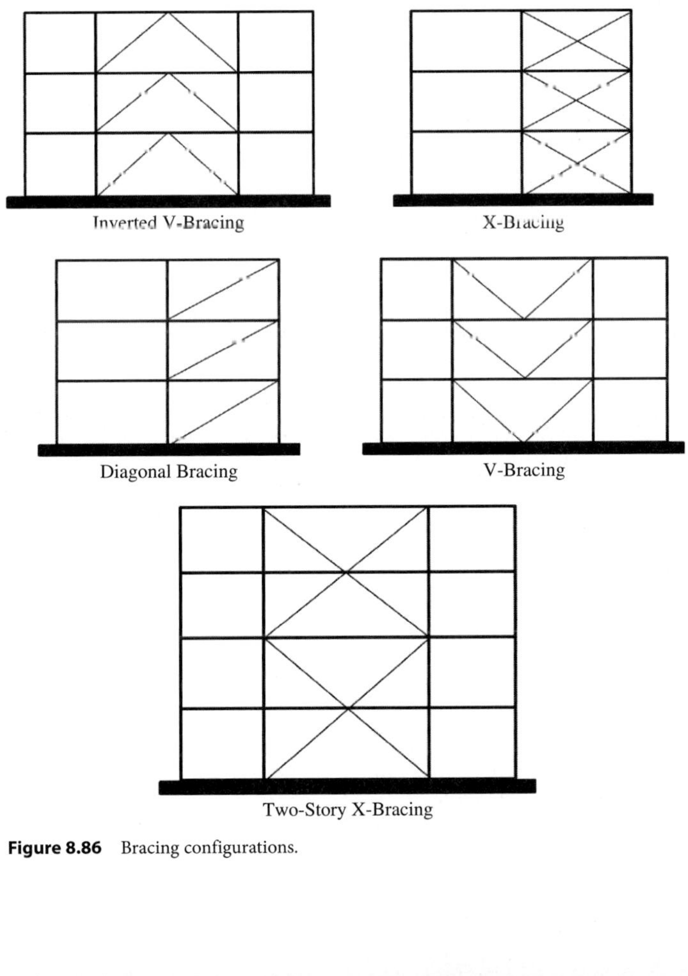

Figure 8.86 Bracing configurations.

System Requirements for V- and Inverted V-Braced Frames

Beams in V- and inverted V-braced frames should be braced to satisfy the requirements for moderately ductile members (*AISC 341, Section F2.4b*).

Member Requirement

All structural elements (beams, columns, and braces) must satisfy the requirements for highly ductile members (*AISC 341, Section F2.5a*) (Table 8.2). An additional requirement for braces is the slenderness requirement which is L_c/r should be less than 200 (L_c is effective length of brace, KL, and r is the governing radius of gyration).

8.2.4 Design Example of a Six-Story Steel Building with Special Concentrically Braced Frames

The building is designed in accordance with *AISC 341 (2016), AISC 360 (2016),* and *ASCE 7 (2016)*. The design steps for lateral force-resisting systems are as follows:

(1) Design of brace sections using seismic forces based on *ASCE 7*;
(2) Design of girders and columns in braced bays using capacity-limited seismic forces.

The example covers:

1. Determination of seismic and structural design parameters;
2. Gravity and seismic load calculations;
3. Analysis of braced frames for the design of braces;
4. Design of braces;
5. Analysis of braced frames under two different loading cases;
6. Design of girders and columns in braced bays.

Building Description The office building of rectangular plan is 189 ft long and 139 ft wide in E-W and N-S directions, respectively. The building is located in Downtown, San Francisco. As shown in Fig. 8.88*a*, various span lengths of 25 ft and 30 ft are employed in both directions. Special Concentrically Braced Frames (SCBFs) with common multi-story X bracing configuration are used as SFRSs in both orthogonal directions. Elevators located in the central bays are surrounded by the braced frames arranged on lines 3, 4 and B, D, E, and G. The self-weight of the lift and its mechanical and operating systems is supported by an additional frame not shown in Fig. 8.88. The composite floor system consists of 3¼-in. thick lightweight concrete with wire mesh on 1½-in. steel deck. The orientation of the floor system is shown in Fig. 8.88*a*. The typical story height is 13 ft and the first story is 18 ft (Figs. 8.88*b* and *c*).

Determination of Gravity and Seismic Load Effects

Gravity Loads Details of the gravity load calculations for floors and roof are presented in the MF example. Summary of the gravity loads is given in Table 8.40.

Seismic Loads

Seismic Design Parameters The seismic design parameters for the office building located in Downtown, San Francisco, are given below. Detailed calculations of the spectral acceleration parameters are summarized in Example 1.

Risk Category = II (*Table 1.5.1, ASCE 7*)

Seismic Importance Factor, $I_e = 1.0$ (*Table 1.5.2, ASCE 7*)

$S_S = 1.50$ g $\quad S_1 = 0.60$ g

$S_{DS} = 1.00$ g $\quad S_{D1} = 0.60$ g $\quad T_0 = 0.12$ sec. $\quad T_s = 0.60$ sec.

Thus, the structure is assigned to Seismic Design Category D.

The structural parameters are given in Table 8.41. As per *Table 12.2-1* and *Section 12.2.5.4* of *ASCE 7*, it is permissible for SCBFs assigned to seismic design category D to have a building height, h_n, of 240 ft except for the buildings having extreme torsional irregularity or with the frames resisting more than 60 percent of the total seismic forces in each direction. Therefore, the building height of 83 ft does not exceed the allowable limit.

The approximate fundamental period for both orthogonal directions, as per *ASCE 7, Section 12.8.2.1*, is calculated as follows:

$$T_a = C_t h_n^x = 0.02 \times 83^{0.75} = 0.55 \text{ sec.} \qquad (\textit{Eq. 12.8-7, ASCE 7})$$

Seismic Base Shear According to *Table 12.6-1, ASCE 7*, ELF procedure can be employed for the SCBFs. Seismic base shear, V, in both orthogonal directions is computed in compliance with *ASCE 7, Section 12.8.1* using $C_u T_a = 1.4 \times 0.55 = 0.77$ sec. Note that despite the elevators are supported by an additional frame vertically, the contribution of the weight of the structural and nonstructural members of the elevator shaft to the effective seismic weight is incorporated roughly by assuming that the slab is continuous along the spans where the elevators are located.

$$C_s = \frac{S_{D1}}{T(R/I_e)} = \frac{0.60}{0.77(6/1.0)} = 0.130 \qquad (\textit{Eq. 12.8-2, ASCE 7})$$

C_s shall not be less than the following:

$$C_{s,\min} = 0.044 S_{DS} I_e = 0.044 \times 1.0 \times 1.0 = 0.044 \geq 0.01$$
(*Eq. 12.8-5, ASCE 7*)

In addition, as specified in *ASCE 7, Section 12.8.1.1*, seismic response coefficient, C_s, shall not be less than the following for a structure located where S_1 is greater than or equal to 0.60 g.

$$C_{s,\min} = 0.5 S_1/(R/I_e) = 0.5 \times 0.60/(6/1.0) = 0.05 \qquad (\textit{Eq. 12.8-6, ASCE 7})$$

Thus, $C_s = \{0.130;\ 0.044;\ 0.05\}_{\max} = 0.130$

The total seismic base shear calculation for each direction is shown below.

$$V = C_s W = 0.130 \times 12{,}058 = 1567 \text{ kips} \qquad (\textit{Eq. 12.8-1, ASCE 7})$$

The vertical force distribution along the building height is summarized in Table 8.42, using the *Eqs. 12.8-11 and 12.8-12*.

$$F_x = C_{vx} V \qquad (\textit{Eq. 12.8-11, ASCE 7})$$

$$C_{vx} = \frac{w_x h_x^k}{\sum_{i=1}^{n} w_i h_i^k} \qquad (\textit{Eq. 12.8-12, ASCE 7})$$

Brace Design According to *AISC 341*, SCBFs are designed to dissipate the input energy mainly through yielding and buckling of the braces. Therefore, bracings in SCBFs are designed based on the design base shear force, which is equal to elastic base shear force divided by the response modification factor ($V_{\text{design}} = V_{\text{elastic}}/R$ where $R = 6$ for SCBFs). Seismic force distribution along the height of each braced frame is given in Fig. 8.89. It should be noted that each braced bay is assumed to withstand one-quarter of the total seismic base shear in each direction.

The braces provide the girders with a vertical support at the mid-length of the brace-intersected girders. Hence, the impact of gravity loads on brace design is determined by assuming a statically indeterminate beam to the first degree, as shown in Fig. 8.90 for the sake of simplicity. Note that the axial deformation in the four bracings located above and below the girder is neglected when calculating brace forces due to gravity loads. The total reaction force induced by gravity loads, R, represents the sum of the vertical components of the produced internal forces in braces. The magnitude of the concentrated gravity loads acting on the girders is estimated based on the tributary areas shown in Fig. 8.90. Therefore, the contribution of the gravity loads on the demand on the braces can be computed as follows:

$$P_{g,\text{floor}}^{\text{Dead}} = A_T^G \times DL = [(6'-3") \times (15')] \times 85^{\text{psf}} \times \left(\frac{1 \text{ kip}}{1000 \text{ lb}}\right) = 7.97^{\text{kips}}$$

$$P_{g,\text{floor}}^{\text{Live}} = A_T^G \times LL = [(6'-3") \times (15')] \times 50^{\text{psf}} \times \left(\frac{1 \text{ kip}}{1000 \text{ lb}}\right) = 4.69^{\text{kips}}$$

$$F_{\text{brace}}^{\text{Gravity}} = \frac{R}{4 \sin\theta} = \frac{2.37 P_g}{4 \sin\theta}$$

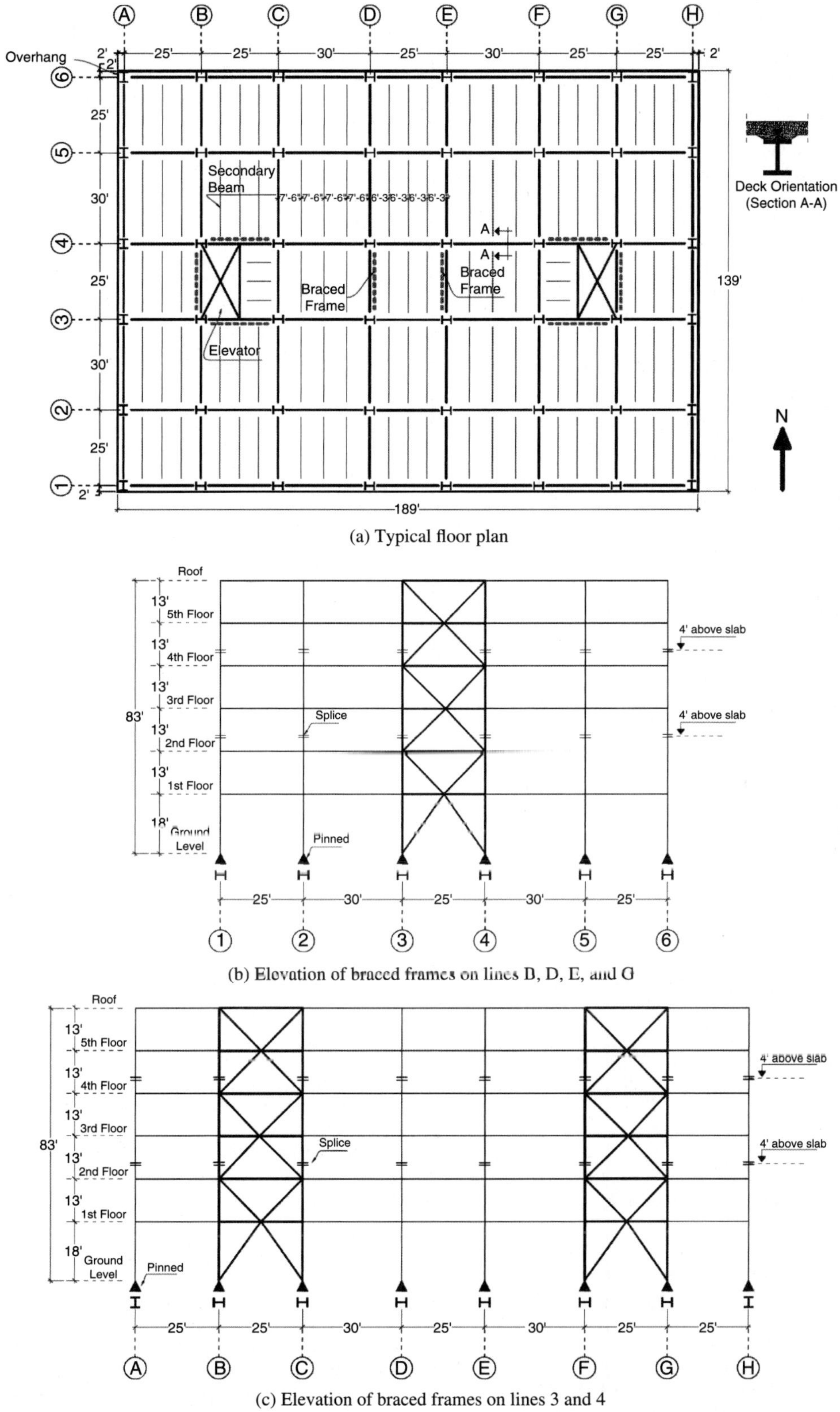

(a) Typical floor plan

(b) Elevation of braced frames on lines B, D, E, and G

(c) Elevation of braced frames on lines 3 and 4

Figure 8.88 Typical floor plan and elevations of the building.

Table 8.40 Gravity Loads for Floors and Roof

	Dead	Live
Roof	34 psf	20 psf
Floor	85 psf	50 psf

Table 8.41 Structural Parameters

Parameter	Coefficient	Reference
Seismic Force-Resisting System	Special Concentrically Braced Frame (SCBF)	
Building Height, h_n	83 ft	
Response Modification Coefficient, R	6	*Table 12.2-1, ASCE 7*
Overstrength Factor, Ω_0	2	*Table 12.2-1, ASCE 7*
Deflection Amplification Factor, C_d	5	*Table 12.2-1, ASCE 7*
Approximate Period Parameter, C_t	0.02	*Table 12.8-2, ASCE 7*
Approximate Period Parameter, x	0.75	*Table 12.8-2, ASCE 7*
The Coefficient for Upper Limit on Calculated Period, C_u	1.4	*Table 12.8-1, ASCE 7*

Table 8.42 Vertical Force Distribution Based on C_uT_a

Story	w_x (kips)	h_x (ft)	$w_x h_x^k$	C_{vx}	F_x (kips)
Roof	893.2	83	134,620	0.140	218.7
5th	2233.0	70	277,384	0.287	450.6
4th	2233.0	57	219,692	0.228	356.9
3rd	2233.0	44	163,762	0.170	266.1
2nd	2233.0	31	110,050	0.114	178.8
1st	2233.0	18	59,379	0.062	96.5
		Σ	964,887	1	1567

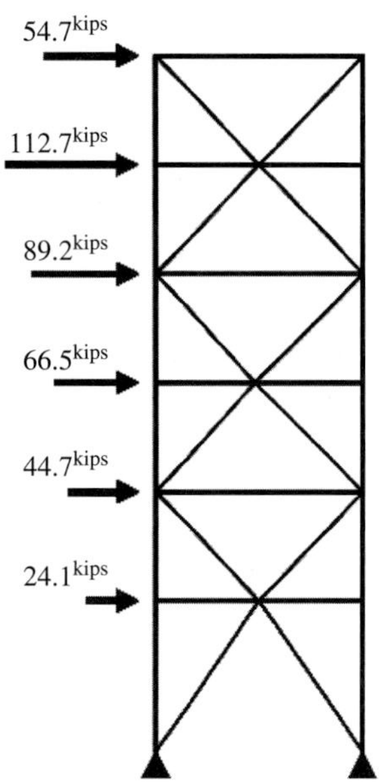

Figure 8.89 Equivalent lateral force distribution along the height of the braced frames.

Example calculation for the design of fifth-story braces:

Governing seismic load combination: $(1.2+0.2S_{DS})DL+\rho Q_E+LL$

According to *ASCE 7, Section 12.3.4.1*, it is permitted to take ρ = 1.0 when the structure is regular in plan and the SFRS on each orthogonal direction consists of two braced bays on each side of the structure. Figure 8.91 shows the seismic and gravity load effects on the braces.

$$\text{Due to dead loads} \Rightarrow P_{DL}=7.97^{\text{kips}} \rightarrow F_{b,DL}=\frac{2.37\times 7.97^{\text{kips}}}{4\times\sin 46^\circ}=6.56^k$$

$$\text{Due to live loads} \Rightarrow P_{LL}=4.69^{\text{kips}} \rightarrow F_{b,LL}=\frac{2.37\times 4.69^{\text{kips}}}{4\times\sin 46^\circ}=3.86^k$$

$$\text{Due to earthquake loads} \Rightarrow F_{b,QE}=\frac{(54.7+112.7)^{\text{kips}}}{2\times\cos 46^\circ}=\mp 120.5^k$$

$$F_{\text{brace}}^{5\text{th}}=(1.2+0.2S_{DS})F_{b,DL}+\rho F_{b,QE}+F_{b,LL}$$

$$P_r=F_{\text{brace}}^{5\text{th}}=(1.2+0.2\times 1.0)\times 6.56^k+1.0\times 120.5^k+3.86^k$$

$$=133.5^k \text{ (Compression)}$$

As seen from the above calculation, the contribution of the gravity loads in the total axial force demand on braces is relatively low compared with that of the seismic loads.

The nominal compressive strength of the fifth-story braces are calculated according to *AISC 360, Section E3*, as follows:

Try HSS6.625 × 0.312 (ASTM A500 Gr. C)

$$A_g=6.02\text{ in.}^2;\ r=2.33\text{ in.};\ F_y=46\text{ ksi};\ R_y=1.3;\ K_x=K_y=1.0;$$

$$L_{b,c}=\sqrt{12.5^2+13^2}=18.03\text{ ft}$$

$$\frac{L_c}{r}=\frac{KL_{b,c}}{r}=\frac{1.0\times 18.03^{\text{ft}}\times(12^{\text{in.}}/1^{\text{ft}})}{2.33^{\text{in.}}}$$

$$=92.86<4.71\sqrt{\frac{E}{F_y}}=4.71\times\sqrt{\frac{29{,}000}{46}}=118.3$$

Limit state of inelastic flexural buckling applies. Thus, the critical stress, F_{cr}, can be determined as per *AISC 360, Eq. E3-2, AISC 360.*

$$F_{cr}=\left(0.658^{F_y/F_e}\right)F_y \qquad \textit{(Eq. E3-2, AISC 360)}$$

$$F_e=\frac{\pi^2 E}{(L_c/r)^2}=\frac{\pi^2\times 29{,}000}{(92.86)^2}=33.19\text{ ksi} \qquad \textit{(Eq. E3-4, AISC 360)}$$

$$F_{cr}=(0.658^{46/33.19})\times 46=25.75^{\text{ksi}} \qquad \textit{(Eq. E3-1, AISC 360)}$$

$$P_n=A_gF_{cr}=6.02^{\text{in.}^2}\times 25.75^{\text{ksi}}=155^{\text{kips}}$$

$$P_c=\phi_cP_n=\phi_cA_gF_{cr}=0.9\times 155^{\text{kips}}=139.5^{\text{kips}}$$

$$P_c=139.5^{\text{kips}}>P_r=133.5^{\text{kips}}\text{ OK.}$$

AISC specifies two requirements for bracing members: (1) Section requirement; and (2) Slenderness requirement. As per *AISC 341, Section F2.5*, slenderness ratio of braces shall be less than 200 and braces shall satisfy the requirements for highly ductile members given in Table D1.1.

$$\frac{L_c}{r}=92.86<200 \qquad \textit{(F2-3, AISC 341)}$$

$$\lambda=D/t=23.6<\lambda_{hd}=0.053\frac{E}{R_yF_y}=0.053\times\frac{29{,}000}{1.3\times 46}=25.70$$

(Table D1.1, AISC 341)

The summary of brace design is given in Table 8.43. Note that to minimize the seismic demand on the brace-intersected girders, the same shape is used for the two consecutive stories above and below the girders (e.g., braces at fifth and sixth stories). Additionally, the effect of the gravity loads is neglected for the even-story braces due to their adverse impact on compressive brace forces.

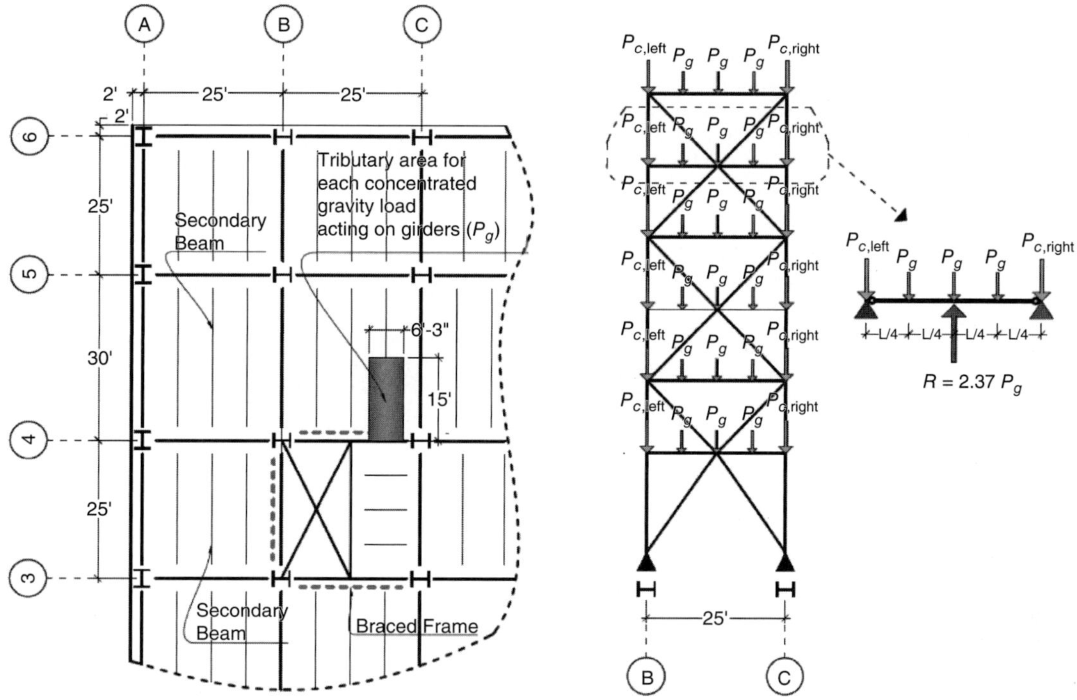

Figure 8.90 Tributary area for the gravity loads acting on the girders on line 4.

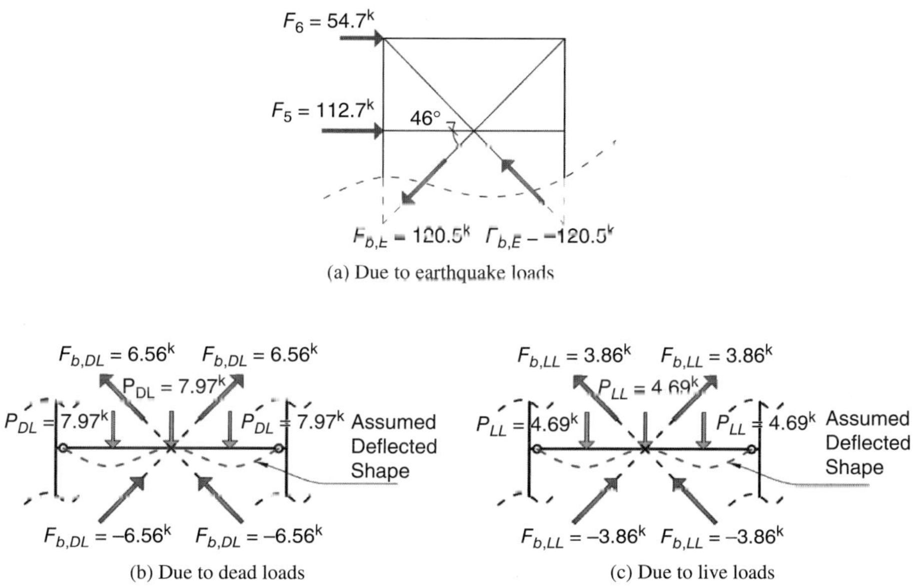

Figure 8.91 Fifth-story brace forces due to seismic and gravity loads.

Table 8.43 Summary of Brace Design

Story	Brace section*	A_g (in.^{2})	$L_{b,c}$** (ft)	L_c/r***	D/t****	Q_e (kips)	$F_{b,Qe}$ (kips)	$F_{b,DL}$ (kips)	$F_{b,LL}$ (kips)	P_r***** (kips)	P_c (kips)
6th	HSS6.875 × 0.312	6.02	18.03	92.86	23.6	54.7	39.4	0	0	39.4	139.5
5th	HSS6.875 × 0.312	6.02	18.03	92.86	23.6	112.7	120.4	6.56	3.86	**133.5**	139.5
4th	HSS7.500 × 0.500	10.3	18.03	86.89	16.1	89.2	184.7	0	0	184.7	256.6
3rd	HSS7.500 × 0.500	10.3	18.03	86.89	16.1	66.5	232.5	6.56	3.86	**245.6**	256.6
2nd	HSS10 × 0.500	13.9	18.03	64.01	21.5	44.7	264.7	0	0	264.7	436.8
1st	HSS10 × 0.500	13.9	21.91	77.79	21.5	24.1	341.6	6.56	3.86	**354.7**	383.0

*Braces are made of ASTM A500 Gr. C; **Length of braces from centerline to centerline; ***Slenderness ratios of all braces are less than 200; ****The limiting diameter-to-wall thickness ratio (D/t) for highly ductile compression members made of circular HSS is $\lambda_{hd} = 0.053E/R_yF_y = 25.70$; *****The required axial strength, P_r, is controlled by seismic load combination $1.4P_D + P_{QE} + P_{LL}$.

Girder Design Based on the capacity design approach, AISC requires that design of the members of the lateral force-resisting system other than braces (e.g., girders) shall remain elastic. Thus, the required strength of the girders is determined based on the capacity-limited seismic load effects. Per *Section F2.3, AISC 341 (2016)*, two structural analysis cases are required for the design of the girders and the columns in braced bays of SCBFs. The required strength of girders and columns shall be taken as the larger of the forces determined from the aforementioned two analyses. The capacity-limited seismic loads to be employed in each analysis case are shown in Fig. 8.92. As seen in Fig. 8.92, the two analysis cases consider the first-mode deformation when determining the sign of the internal forces.

Each analysis case represents a different stage of the hysteretic response when an SCBF is subjected to an earthquake ground motion. Brace forces to determine the capacity-limited seismic load effect on the girders and columns are stipulated in *AISC 341, F2.3* as follows:

The expected brace strength in tension: $T_{ET} = R_y F_y A_g$

The expected brace strength in compression: C_{EB} = the lesser of $R_y F_y A_g$ and (1/0.877) $F_{cre} A_g$

The expected postbuckling brace strength: $C_{EPB} = 0.30\ C_{EB}$ in which F_{cre} is the expected critical buckling stress and the factor of (1/0.877) accounts for the initial imperfections.

Design of the Brace-Intersected Girders In addition to the gravity-load-induced bending moments, the brace-intersected girders in SCBFs with V-, inverted V-, and Split-X-type bracings are subjected to seismic loads due to the unbalanced brace forces that induce both axial force and bending moment. Therefore, brace-intersected girders, in general, can be treated as beam-column members.

Analysis Case I

In this case, it is assumed that all braces reach their expected strength in tension or in compression. Therefore, the expected brace capacities can be calculated as follows:

$$\text{Brace in tension} \Rightarrow T_{ET} = R_y F_y A_g$$

$$\text{Brace in compression} \Rightarrow C_{EB} = \{R_y F_y A_g;\ (1/0.877) F_{cre} A_g\}_{\min}$$

where $F_{cre} = \left(0.658^{R_y F_y / F_e}\right) R_y F_y$

Example calculation for the design of fifth-story girder:

Governing seismic load combination: $(1.2 + 0.2S_{DS})DL + \rho Q_E + LL$

To determine the seismic load effects on the girder, first, the expected brace capacities are required.

Expected strength in tension

$$\Rightarrow T_{ET6} = T_{ET5} = R_y F_y A_g = 1.3 \times 46^{\text{ksi}} \times 6.02^{\text{in.}} = 360^{\text{kips}}$$

Expected buckling strength

$$\Rightarrow C_{EB6} = C_{EB5} = \{R_y F_y A_g;\ (1/0.877) F_{cre} A_g\}_{\min}$$

$$\frac{L_c}{r} = \frac{KL_b}{r} = \frac{0.85 \times 1.0 \times 18.03^{\text{ft}} \times (12^{\text{in.}}/1^{\text{ft}})}{2.33^{\text{in.}}} = 78.93$$

Note that when determining the expected critical stress, F_{cre}, in lieu of the distance from centerline to centerline, the distance from brace end to brace end is used as the brace length. Since the detailed geometry of the frame is not yet determined, the end-to-end distance is assumed to be 85 percent of the total brace length from centerline to centerline.

$$F_e = \frac{\pi^2 E}{(L_c/r)^2} = \frac{\pi^2 \times 29{,}000}{(78.93)^2} = 45.94^{\text{ksi}} \qquad (\textit{Eq. E3-4, AISC 360})$$

$$F_{cre} = (0.658^{1.3 \times 46/45.94}) \times 1.3 \times 46 = 34.68^{\text{ksi}}$$

$$(1/0.877) F_{cre} A_g = \frac{34.68^{\text{ksi}} \times 6.02^{\text{in.}^2}}{0.877} = 238.1^{\text{kips}}$$

$$\Rightarrow C_{EB6} = C_{EB5} = \{360^{\text{kips}};\ 238.1^{\text{kips}}\}_{\min} = 238.1^{\text{kips}}$$

Figure 8.93 shows the free-body diagram of the fifth story for determining the flexural and axial force demands on the brace-intersected girder. Then, flexural demand on the fifth-story girder due to seismic loads can be calculated as:

$$M^5_{QE1} = \frac{L}{4}[(C_{EB6} - C_{EB5}) + (T_{ET5} - T_{ET6})]\sin\alpha$$

$$M^5_{QE1} = \frac{25^{\text{ft}}}{4}[(238.1^{\text{kips}} - 238.1^{\text{kips}}) + (360^{\text{kips}} - 360^{\text{kips}})]\sin 46^\circ = 0$$

and the axial force demand can be computed as follows:

$$P^5_{QE1} = F_{5U}/2$$

$$F_{5U} = [(T_{ET5} + C_{EB5}) - (C_{EB6} + T_{ET6})]\cos\alpha$$

$$= [(360^{\text{kips}} + 238.1^{\text{kips}}) - (238.1^{\text{kips}} + 360^{\text{kips}})]\cos 46 = 0$$

$$P^5_{QE1} = 0$$

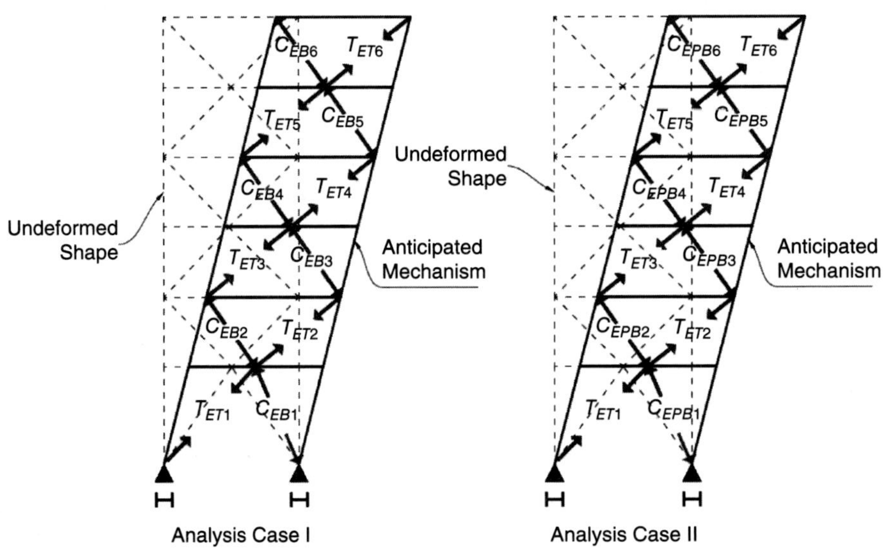

Figure 8.92 Structural analysis cases for girders and columns.

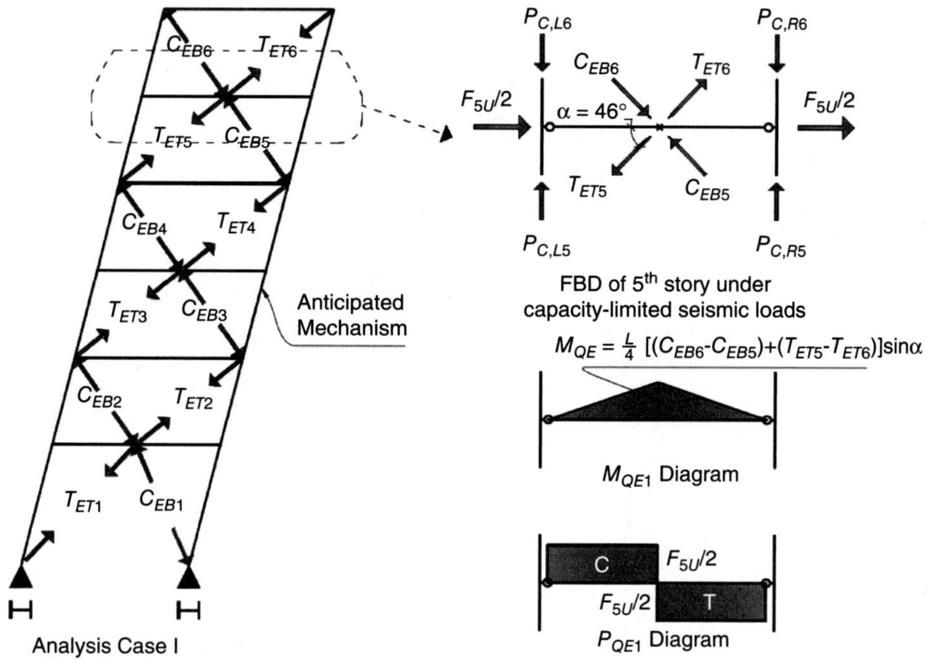

Figure 8.93 Seismic load effects on the brace-intersected girders for structural analysis case I.

As seen from the above calculation and the Free Body Diagrams (FBDs) shown in Fig. 8.93, when the same brace section is employed in the two consecutive stories, the seismic loads acting on the brace-intersected girders usually cancel each other out. In other words, the brace-intersected girders are not affected by the seismic loads and thus, can be designed to support gravity loads solely.

Thus, $P_{r1}^5 = 1.4P_{DL} + P_{Q_E} + P_{LL} = 0$

Since the axial load demand is zero, the fifth-story girder can be treated as a beam member. With the purpose of being consistent with the assumptions regarding the brace deformation in the inelastic stage, the vertical supports provided by the braces are neglected when determining the flexural demand on the girder. Assumed deflected shape of the girder under gravity loads is shown in Fig. 8.94.

$$M_{QE1}^5 = 0$$

$$M_{DL} = 0.5\times P_{DL}L_{span} = 0.5\times 7.97^{kips}\times 25^{ft} = 99.63^{k\text{-}ft}$$

$$M_{LL} = 0.5\times P_{LL}L_{span} = 0.5\times 4.69^{kips}\times 25^{ft} = 58.63^{k\text{-}ft}$$

$$M_{r1}^5 = 1.4M_{DL} + M_{QE1} + M_{LL} = 1.4\times(99.63^{k\text{-}ft}) + 0 + 58.63^{k\text{-}ft} = 198.1^{k\text{-}ft}$$

The nominal flexural strength of the fifth-story girder is computed according to *Section F2, AISC 360*, as follows:

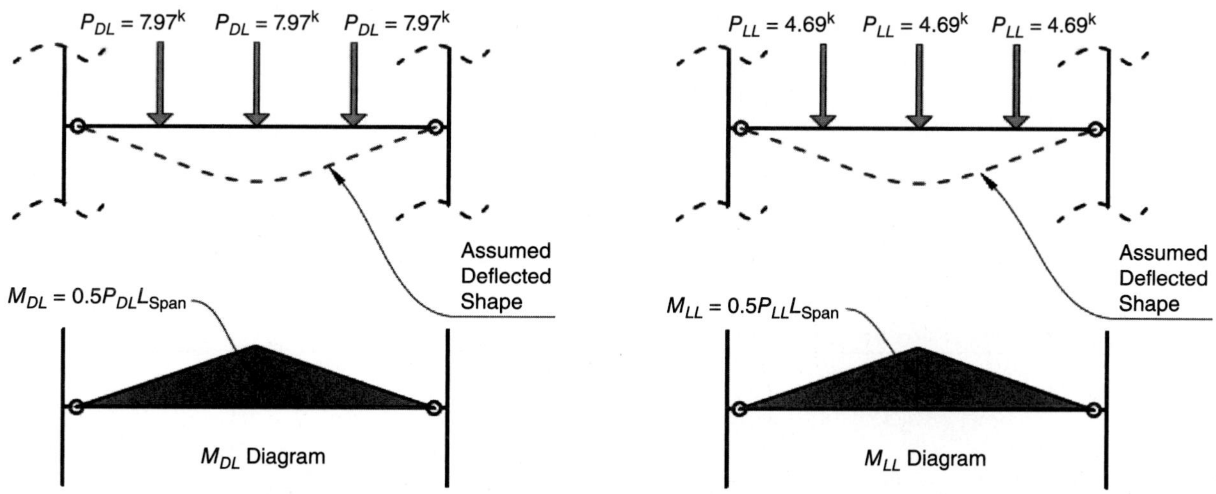

Figure 8.94 Gravity load effects on the brace-intersected girders.

To determine a trial section for the girder,

$$Z_{x,\min} = \frac{M_r}{F_y} = \frac{198.1^{\text{k-ft}} \times (12^{\text{in.}}/1^{\text{ft}})}{50^{\text{ksi}}} = 47.5^{\text{in.}^3}$$

Try W18 × 35 (ASTM A992)

$Z_x = 66.5$ in.^{3}; $S_x = 57.6$ in.^{3}; $F_y = 50$ ksi; $R_y = 1.1$; $L_p = 4.31$ ft; $L_r = 12.3$ ft

Per *AISC 341, Section F2.4b*, both flanges of the beams in SCBFs shall be laterally braced and the lateral braces shall have a maximum spacing of $0.19r_yE/(R_yF_y)$.

$$L_{b,\max} = 0.19r_yE/(R_yF_y) = \frac{0.19 \times 1.22^{\text{in.}} \times 29{,}000^{\text{ksi}}}{1.1 \times 50^{\text{ksi}}} = 122.2^{\text{in.}} > L_b = 75^{\text{in.}}$$

Selected lateral brace spacing satisfy the requirement for moderately ductile members, as per *AISC 341, Section D1.2a*.

Check the limit state of LTB:

It is assumed that the composite slab does not provide a continuous lateral support at the compression flange since the steel deck ribs are parallel to the girder (Fig. 8.88*a*). Therefore, the impact of the composite slab is conservatively neglected.

$$L_p = 4.31 \text{ ft} < L_b = 6.25 \text{ ft} < L_r = 12.3 \text{ ft} \qquad \text{and} \qquad C_b = 1.11$$

$$M_n = C_b\left[M_p - (M_p - 0.7F_yS_x)\left(\frac{L_b - L_p}{L_r - L_p}\right)\right] \le M_p$$

$$M_n = 1.11 \times \left[277.1^{\text{k-ft}} - \left(277.1^{\text{k-ft}} - \frac{0.7 \times 50^{\text{ksi}} \times 57.6^{\text{in.}^3}}{(12^{\text{in.}}/1^{\text{ft}})}\right)\left(\frac{6.25 - 4.31}{12.3 - 4.31}\right)\right]$$

$$= 278.2^{\text{k-ft}} > M_p = 277.1^{\text{k-ft}}$$

$$\phi_b M_n = 0.9 \times 271.1^{\text{k-ft}} = 249^{\text{k-ft}} > M_r = 198.1^{\text{k-ft}} \quad \text{OK.}$$

Check compactness requirement for W18 × 35:

According to *AISC 341, Section F2.5a*, beams shall meet the requirements for highly ductile members stipulated in *Table D1.1*.

$$b/2t_f = 7.06 < \lambda_{hd} = 0.32\sqrt{\frac{E}{R_yF_y}} = 0.32 \times \sqrt{\frac{29{,}000}{1.1 \times 50}} = 7.35$$

Use W18 × 35 for the fifth-story girder.

Summary of the expected brace strengths in tension and compression for structural analysis case I and the required axial force and flexural strengths for the girders are presented in Tables 8.44 and 8.45, respectively. Note that girders shall be designed based on the larger force determined from the two analysis cases. However, as indicated in Table 8.45, the capacity-limited seismic load effects on the third- and fifth-story girders are zero for both analysis cases. Hence, the required axial and flexural strengths obtained from the two analyses are identical for the brace-intersected girders other than the first-story girder, in which the angle of brace inclination above and below the girder is slightly different.

Analysis Case II

In the second case, it is assumed that braces in compression reach their expected postbuckling strength and braces in tension resist forces corresponding to their expected strength in tension. Therefore, the expected brace capacities can be calculated as follows:

$$\text{Brace in tension} \Rightarrow T_{ET} = R_yF_yA_g$$

$$\text{Brace in compression} \Rightarrow C_{EPB} = 0.3 \text{ times the } C_{EB}$$

Example calculation for the design of the first-story girder is shown below:

Expected strength in tension

$$\Rightarrow T_{ET1} = T_{ET2} = R_yF_yA_g = 1.3 \times 46^{\text{ksi}} \times 13.9^{\text{in.}} = 831.2^{\text{kips}}$$

Expected postbuckling strength

$$\Rightarrow C_{EPB} = 0.3 \times \{R_yF_yA_g;\ (1/0.877)F_{cre}A_g\}_{\min}$$

The expected postbuckling strength of first-story braces:

$$\frac{L_c}{r} = \frac{KL_b}{r} = \frac{0.85 \times 1.0 \times 21.91^{\text{ft}} \times (12^{\text{in.}}/1^{\text{ft}})}{3.38^{\text{in.}}} = 66.12$$

$$F_e = \frac{\pi^2 E}{(L_c/r)^2} = \frac{\pi^2 \times 29{,}000}{(66.12)^2} = 65.47^{\text{ksi}}$$

$$F_{cre} = (0.658^{1.3 \times 46/65.47}) \times 1.3 \times 46 = 40.8^{\text{ksi}}$$

$$(1/0.877)F_{cre}A_g = \frac{40.8^{\text{ksi}} \times 13.9^{\text{in.}^2}}{0.877} = 646.7^{\text{kips}}$$

$$\Rightarrow C_{EPB1} = 0.3 \times \{831.2^{\text{kips}};\ 646.7^{\text{kips}}\}_{\min} = 194^{\text{kips}}$$

The expected postbuckling strength of the second-story braces can be calculated in a similar manner.

$$\Rightarrow C_{EPB2} = 0.3 \times \{831.2^{\text{kips}};\ 731.6^{\text{kips}}\}_{\min} = 219.5^{\text{kips}}$$

Table 8.44 Summary of Expected Brace Strengths to Be Used in Analysis Case I

Story	Brace section	A_g (in.^{2})	L_b (ft)*	r (in.)	L_c/r	T_{ET} (kips)	C_{EB} (kips)
6th	HSS6.875 × 0.312	6.02	15.33	2.33	78.93	360.0	238.1
5th	HSS6.875 × 0.312	6.02	15.33	2.33	78.93	360.0	238.1
4th	HSS7.500 × 0.500	10.3	15.33	2.49	73.86	615.9	435.9
3rd	HSS7.500 × 0.500	10.3	15.33	2.49	73.86	615.9	435.9
2nd	HSS10 × 0.500	13.9	15.33	3.38	54.41	831.2	731.6
1st	HSS10 × 0.500	13.9	18.62	3.38	66.12	831.2	646.7

*L_b is assumed to be 85% of the brace length from centerline to centerline, $L_{b,c}$.

Table 8.45 Selected Brace-Intersected Girder Sections for Structural Analysis Case I

Story	Girder section	F_u (kips)	M_{QE1} (k-ft)	M_{DL} (k-ft)	M_{LL} (k-ft)	P_{QE1}* (kips)	M_{r1} (k-ft)	P_{r1} (kips)	$M_c = \Phi_c M_n$ (k-ft)	$P_c = \Phi_c P_n$ (kips)	Strength ratio**
5th	W18 × 35	0.0	0.0	99.6	58.6	0.0	198.1	0.0	249.0	NA	**0.80**
3rd	W18 × 35	0.0	0.0	99.6	58.6	0.0	198.1	0.0	249.0	NA	**0.80**
1st	W18 × 97	−238.0	497.0	99.6	58.6	119.0	695.1	119.0	791.3	1151	**0.93**

*Positive sign means compression; **Per *AISC 360*, Section F or Eqs. H1-1a and H1-1b.

Figure 8.95 shows the free-body diagram of the first story for determining the flexural and axial force demands on the brace-intersected girder. Then, flexural demand on the first story girder due to seismic loads can be calculated as:

$$M_{QE2}^{1} = \frac{L}{4}[(T_{ET1} - C_{EPB1})\sin\alpha_1 + (C_{EPB2} - T_{ET2})\sin\alpha_2]$$

$$M_{QE2}^{1} = \frac{25^{\text{ft}}}{4}[(831.2^{\text{kips}} - 194^{\text{kips}})\times \sin 55° + (219.5^{\text{kips}} - 831.2^{\text{kips}})\times \sin 46°]$$

$$= 512.1^{\text{k-ft}}$$

and the axial force demand can be computed as follows:

$$P_{QE2}^{1} = F_{1U}/2$$

$$F_{1U} = [(T_{ET1} + C_{EPB1})\cos\alpha_1 - (C_{EPB2} + T_{ET2})\cos\alpha_2]$$

$$F_{1U} = [(831.2^{\text{kips}} + 194^{\text{kips}})\times \cos 55° - (219.5^{\text{kips}} + 831.2^{\text{kips}})\times \cos 46°]$$

$$= -141.8^{\text{kips}}$$

$$P_{QE2}^{1} = \pm\, 141.8^{\text{kips}}/2 = \pm\, 70.9^{\text{kips}}$$

The required axial force and bending moment for the first-story girder is determined considering the seismic load combination: $(1.2+0.2S_{DS})DL + \rho Q_E + LL$ as follows:

$$\left.\begin{array}{l} P_{QE2}^{1} = 70.9^{\text{kips}} \\ P_{DL} = 0 \\ P_{LL} = 0 \end{array}\right\} P_{r2}^{1} = 1.4P_{DL} + P_{QE2}^{1} + P_{LL} = 70.9^{\text{kips}}$$

$$\left.\begin{array}{l} M_{QE2}^{1} = 512.1^{\text{k-ft}} \\ M_{DL} = 99.63^{\text{k-ft}} \\ M_{LL} = 58.63^{\text{k-ft}} \end{array}\right\} M_{r2}^{1} = 1.4M_{DL} + M_{QE2}^{1} + M_{LL} = 1.4\times(99.63^{\text{k-ft}})$$

$$+512.10^{\text{k-ft}} + 58.63^{\text{k-ft}} = 710.2^{\text{k-ft}}$$

The flexural strength of the first-story girder is computed according to *AISC 360, Section F2*, as follows:

Try W18 × 97 (ASTM A992)

Z_x = 211 in.^{3}; S_x = 188 in.^{3}; A_g = 28.5 in.^{2}; r_x = 7.82 in.; r_y = 2.65 in.; F_y = 50 ksi; R_y = 1.1; L_p = 9.36 ft

Per *AISC 341, Section F2.4b*, lateral braces are to have a maximum spacing of $0.19r_yE/(R_yF_y)$ for the brace-intersected beams in SCBFs.

$$L_{b,\max} = 0.19r_yE/(R_yF_y) = \frac{0.19\times 2.65^{\text{in.}}\times 29{,}000^{\text{ksi}}}{1.1\times 50^{\text{ksi}}} = 265.4^{\text{in.}} > L_b = 75^{\text{in.}}$$

Selected lateral brace spacing satisfy the requirement for moderately ductile members, as per *AISC 341, Section D1.2a*.

Check the limit state of LTB:
Since L_b = 6.25 ft < L_p = 9.36 ft, limit state of LTB does not apply. Thus,

$$M_c = \phi_b M_n = \phi_b M_p = 791.3^{\text{k-ft}}$$

Check compactness requirement for W18 × 97:
According to *AISC 341, Section F2.5a*, beams shall meet the requirements for highly ductile members stipulated in *Table D1.1*.

$$\text{Flange: } b/2t_f = 6.47 < \lambda_{hd} = 0.32\sqrt{\frac{E}{R_yF_y}} = 0.32\times\sqrt{\frac{29{,}000}{1.1\times 50}} = 7.35 \text{ OK.}$$

$$\text{Web: } h/t_w < \lambda_{hd} = \begin{cases} \text{For } C_a \le 0.114 \Rightarrow 2.57\sqrt{\dfrac{E}{R_yF_y}}(1-1.04C_a) \\ \text{For } C_a > 0.114 \Rightarrow 0.88\sqrt{\dfrac{E}{R_yF_y}}(2.68-C_a) \ge 1.57\sqrt{\dfrac{E}{R_yF_y}} \end{cases}$$

$$C_a = \frac{P_u}{\phi_c P_y} = \frac{70.9^{\text{kips}}}{0.9\times(1.1\times 26.5^{\text{in.}^2}\times 50^{\text{ksi}})} = 0.054 < 0.114$$

$$h/t_w = 30 < \lambda_{hd} = 2.57\times\sqrt{\frac{29{,}000}{1.1\times 50}}\times(1-1.04\times 0.054) = 55.7 \quad \text{OK.}$$

Compressive Strength: In the inelastic stage, the braces are assumed not to provide vertical support when the brace-intersected girders buckle about their major axis. Therefore, the unbraced length for major-axis flexural buckling is taken as the entire span length ($L_{c,x}$ = 25 ft). As for the minor axis flexural and torsional buckling lengths, due to the orientation of the steel deck (steel deck with ribs parallel to the beam) and the torsional bracing spacing (both flanges are braced at 6.25 ft), the girder is assumed to be unbraced between the lateral bracing points for both flexural and torsional buckling. Since the buckling lengths for flexural and torsional buckling are the same ($L_{c,y}$ = $L_{c,z}$ = 6.25 ft), the flexural

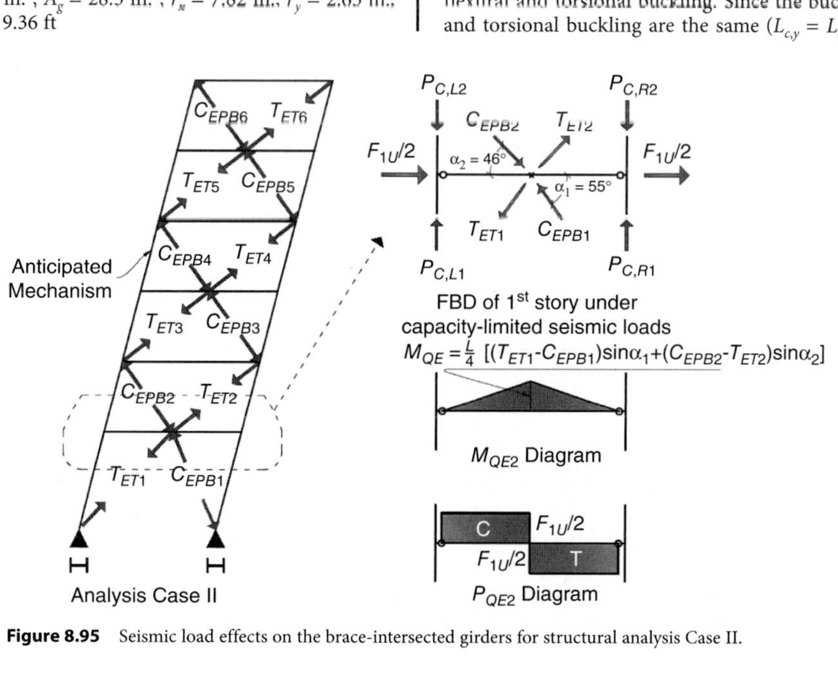

Figure 8.95 Seismic load effects on the brace-intersected girders for structural analysis Case II.

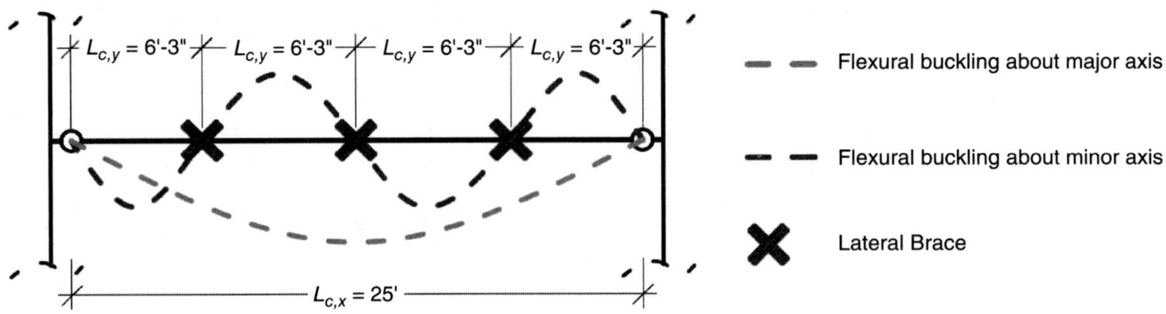

Figure 8.96 Assumed buckled shape of the brace-intersected girders about major and minor axes.

buckling strength about the minor axis will be lower than the torsional buckling strength. Thus, the compressive strength will be governed by one of the flexural buckling strengths. The buckling shapes of the girder are shown in Fig. 8.96. The buckling lengths used to determine the compressive strength of the first-story girder and the detailed calculations according to *AISC 360, Section E3*, are summarized below:

$L_{c,x}$ = 25 ft (Flexural buckling length about the x-axis)
$L_{c,y}$ = 6.25 ft (Flexural buckling length about the y-axis)
$L_{c,z}$ = 6.25 ft (Torsional buckling length about the z-axis)

Thus, the critical stress, F_{cr}, can be determined as per *AISC 360, Eq. E3-2*, based on the limit state of inelastic flexural buckling.

$$\frac{L_c}{r} = \left(\frac{L_{c,x}}{r_x}; \frac{L_{c,y}}{r_y}\right)_{\max} = \frac{25^{\text{ft}} \times (12^{\text{in.}}/1^{\text{ft}})}{7.83^{\text{in.}}} = 38.3 < 4.71 \times \sqrt{\frac{29{,}000}{50}} = 113.4$$

$$F_e = \frac{\pi^2 E}{(L_c/r)^2} = \frac{\pi^2 \times 29{,}000}{(38.3)^2} = 195.1^{\text{ksi}}$$

$$F_{cr} = (0.658^{50/195.1}) \times 50 = 44.9^{\text{ksi}}$$

$$P_c = \phi_c P_n = \phi_c A_g F_{cr} = 0.9 \times 28.5^{\text{in.}^2} \times 44.9^{\text{ksi}} = 1151^{\text{kips}}$$

Check axial force-bending moment interaction:
The interaction of axial force and bending moment is checked using *AISC 360, Eqs. H1-1a*, and *H1-1b*,

$$\left.\begin{array}{l} M_r = 710.2^{\text{k-ft}} \\ P_r = 70.9^{\text{kips}} \\ M_c = 791.3^{\text{k-ft}} \\ P_c = 1151^{\text{kips}} \end{array}\right\} \Rightarrow \left| \begin{array}{l} \dfrac{P_r}{P_c} = \dfrac{70.9}{1151} = 0.06 < 0.2 \rightarrow \text{Use Eq. H1-1b} \\ \dfrac{P_r}{2P_c} + \dfrac{M_r}{M_c} = \dfrac{70.9}{2 \times 1151} + \dfrac{710.2}{791.3} = 0.93 < 1.0 \end{array}\right|$$

Use W18 × 97 for the first story girder.
Summary of the expected brace capacities in tension and compression for structural analysis case II and the required axial force and flexural strengths for the brace-intersected girders are given in Tables 8.46 and 8.47, respectively.

Design Examples of the Girders Not Intersected by Braces The girders not intersected by bracings in the two-story X-braced frames are subjected to an axial force due to unbalanced brace forces and bending moment due to gravity loads. Therefore, the girders located in the even stories are to be designed for combined flexure and axial force under the two structural analysis cases. Similar to the design of brace-intersected girders, the heavier girder section obtained from the two analysis cases

Table 8.46 Summary of Expected Brace Strengths to Be Used in Analysis Case II

Story	Brace section	A_g (in.^{2})	L_b (ft)*	r (in.)	L_c/r	T_{ET} (kips)	C_{EPB} (kips)
6th	HSS6.875 × 0.312	6.02	15.33	2.33	78.93	360.0	71.4
5th	HSS6.875 × 0.312	6.02	15.33	2.33	78.93	360.0	71.4
4th	HSS7.500 × 0.500	10.3	15.33	2.49	73.86	615.9	130.8
3rd	HSS7.500 × 0.500	10.3	15.33	2.49	73.86	615.9	130.8
2nd	HSS10 × 0.500	13.9	15.33	3.38	54.41	831.2	219.5
1st	HSS10 × 0.500	13.9	18.62	3.38	66.12	831.2	194.0

*L_b is assumed to be 85% of the brace length from centerline to centerline, $L_{b,c}$.

Table 8.47 Selected Brace-Intersected Girder Sections for Structural Analysis Case II

Story	Girder section	F_u (kips)	M_{QE1} (k-ft)	M_{DL} (k-ft)	M_{LL} (k-ft)	P^*_{QE2} (kips)	M_{r2} (k-ft)	P_{r2} (kips)	$M_c = \Phi_c M_n$ (k-ft)	$P_c = \Phi_c P_n$ (kips)	Strength ratio**
5th	W18 × 35	0.0	0.0	99.6	58.6	0.0	198.1	0.0	249.0	NA	**0.98**
3rd	W18 × 35	0.0	0.0	99.6	58.6	0.0	198.1	0.0	249.0	NA	**0.98**
1st	W18 × 97	−141.8	512.1	99.6	58.6	70.9	710.2	70.9	791.3	1151	**0.93**

*Positive sign means compression; **Per AISC 360, Section F or Eqs. H1-1a and H1-1b.

controls the design of the girders not intersected by bracings. Example calculation for the design of the fourth-story girder is shown below using the FBDs given in Fig. 8.97.

Analysis Case I

$$\text{Expected strength in tension} \Rightarrow \begin{cases} T_{ET5} = 360^{\text{kips}} \\ T_{ET4} = 615.9^{\text{kips}} \end{cases}$$

$$\text{Expected strength in compression} \Rightarrow \begin{cases} C_{EB5} = 238.1^{\text{kips}} \\ C_{EB4} = 435.9^{\text{kips}} \end{cases}$$

Using the FBD shown in Fig. 8.97, the seismic axial force demand can be computed as follows:

$$F_{4U} = [(T_{ET4} + C_{EB4}) - (C_{EB5} + T_{ET5})]\cos\alpha$$

$$F_{4U} = [(615.9^{\text{kips}} + 435.9^{\text{kips}}) - (238.1^{\text{kips}} + 360^{\text{kips}})]\cos 46° = 315.2^{\text{kips}}$$

$$P^4_{QE1} = F_{4U}/2 + (T_{ET5} - C_{EB4})\cos\alpha = 315.2^{\text{kips}}/2 + (360^{\text{kips}} - 435.9^{\text{kips}})$$

$$\cos 46° = 104.9^{\text{kips}}$$

The required axial force and bending moment for the fourth-story girder are determined considering the seismic load combination: $(1.2 + 0.2S_{DS})DL + \rho Q_E + LL$ as follows:

$$\left.\begin{array}{l} P^4_{QE1} = 104.9^{\text{kips}} \\ P_{DL} = 0 \\ P_{LL} = 0 \end{array}\right\} P^4_{r1} = 1.4P_{DL} + P_{Q_E} + P_{LL} = 104.9^{\text{kips}}$$

$$\left.\begin{array}{l} M^4_{QE1} = 0 \\ M_{DL} = 99.63^{\text{k-ft}} \\ M_{LL} = 58.63^{\text{k-ft}} \end{array}\right\} M^4_{r1} = 1.4M_{DL} + M_{QE1} + M_{LL}$$

$$= 1.4 \times (99.63^{\text{k-ft}}) + 0 + 58.63^{\text{k-ft}} = 198.1^{\text{k-ft}}$$

Try W18 × 40 (ASTM A992)

$$M_c - \phi_b M_n = 294^{\text{k-ft}}$$

$$P_c = \phi_c P_n = 411.5^{\text{kips}}$$

Check axial force-bending moment interaction:
The interaction of axial force and bending moment is checked using *AISC 360, Eqs. H1-1a,* and *H1-1b,*

$$\left.\begin{array}{l} M_r = 198.1^{\text{k-ft}} \\ P_r = 104.9^{\text{kips}} \\ M_c = 294^{\text{k-ft}} \\ P_c = 411.5^{\text{kips}} \end{array}\right\} \Rightarrow \left| \begin{array}{l} \dfrac{P_r}{P_c} = \dfrac{104.9}{411.5} = 0.25 > 0.2 \rightarrow \text{Use Eq. H1-1a} \\ \dfrac{P_r}{P_c} + \dfrac{8}{9}\dfrac{M_r}{M_c} = \dfrac{104.9}{411.5} + \dfrac{8}{9}\dfrac{198.1}{294} = 0.85 < 1.0 \end{array}\right|$$

Check compactness requirement for W18 × 40:

$$\text{Flange: } b/2t_f = 5.73 < \lambda_{hd} = 0.32 \times \sqrt{\frac{29{,}000}{1.1 \times 50}} = 7.35 \text{ OK.}$$

$$\text{Web: } C_a = \frac{P_u}{\phi_c P_y} = \frac{104.9^{\text{kips}}}{0.9 \times (1.1 \times 11.8^{\text{in.}^2} \times 50^{\text{ksi}})} = 0.18 > 0.114$$

$$h/t_w = 50.9 > \lambda_{hd} = 0.88 \times \sqrt{\frac{29{,}000}{1.1 \times 50}} \times (2.68 - 0.18)$$

$$= 50.5 \geq 1.57 \times \sqrt{\frac{29{,}000}{1.1 \times 50}} = 36.1$$

W18 × 40 does not satisfy the requirements for highly ductile members.

Try W18 × 46 (ASTM A992)

$$\text{Flange: } b/2t_f = 5.01 < \lambda_{hd} = 0.32 \times \sqrt{\frac{29{,}000}{1.1 \times 50}} = 7.35 \text{ OK.}$$

$$\text{Web: } C_a = \frac{P_u}{\phi_c P_y} = \frac{104.9^{\text{kips}}}{0.9 \times (1.1 \times 13.5^{\text{in.}^2} \times 50^{\text{ksi}})} = 0.157 > 0.114$$

$$h/t_w = 44.6 < \lambda_{hd} = 0.88 \times \sqrt{\frac{29{,}000}{1.1 \times 50}} \times (2.68 - 0.157)$$

$$= 51 \geq 1.57 \times \sqrt{\frac{29{,}000}{1.1 \times 50}} = 36.1$$

Use W18 × 46.
Design summary and the selected girder sections for the analysis case I are shown in Table 8.48.

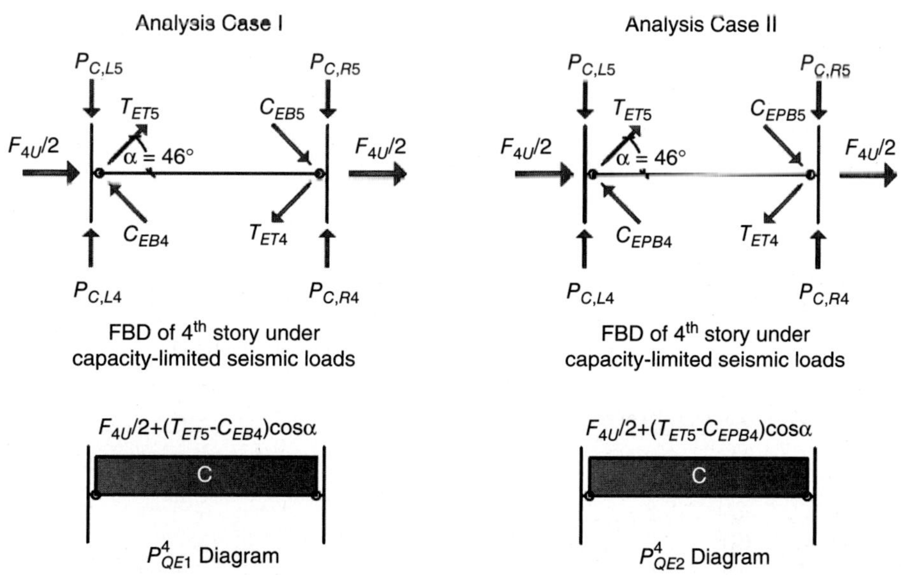

Figure 8.97 Seismic load effects on the girders not intersected by braces.

Table 8.48 Sections for the Girders Not Intersected by Braces Based on Structural Analysis Case I

Story	Girder section	F_u (kips)	M_{QE1} (k-ft)	M_{DL} (k-ft)	M_{LL} (k-ft)	P_{QE1}* (kips)	M_{r1} (k-ft)	P_{r1} (kips)	$M_c = \Phi_c M_n$ (k-ft)	$P_c = \Phi_c P_n$ (kips)	Strength ratio**
6th	W18 × 35	415.4	0.0	39.9	23.5	42.3	79.2	42.3	249.0	328.4	**0.38**
4th	W18 × 46	315.2	0.0	99.6	58.6	104.9	198.1	104.9	340.0	463.0	**0.74**
2nd	W18 × 46	355.0	0.0	99.6	58.6	97.1	198.1	97.1	340.0	463.0	**0.73**

*Positive sign means compression; **Per AISC 360, Eqs. H1-1a and H1-1b.

Analysis Case II

Expected strength in tension $\Rightarrow \begin{cases} T_{ET5} = 360^{\text{kips}} \\ T_{ET4} = 615.9^{\text{kips}} \end{cases}$

Expected postbuckling strength $\Rightarrow \begin{cases} C_{EPB5} = 71.4^{\text{kips}} \\ C_{EPB4} = 130.8^{\text{kips}} \end{cases}$

Using the FBD shown in Fig. 8.97, the seismic axial force demand can be computed as follows:

$$F_{4U} = [(T_{ET4} + C_{EPB4}) - (C_{EPB5} + T_{ET5})]\cos\alpha$$

$$F_{4U} = [(615.9^{\text{kips}} + 130.8^{\text{kips}}) - (71.4^{\text{kips}} + 360^{\text{kips}})]\cos 46° = 219.0^{\text{kips}}$$

$$P_{QE2}^4 = F_{4U}/2 + (T_{ET5} - C_{EPB4})\cos\alpha = 219.0^{\text{kips}}/2 + (360^{\text{kips}} - 130.8^{\text{kips}})$$

$$\cos 46° = 268.7^{\text{kips}}$$

The required axial force and bending moment for the fourth-story girder are determined for structural analysis case II as follows:

$$\left.\begin{matrix} P_{QE2}^4 = 268.7^{\text{kips}} \\ P_{DL} = 0 \\ P_{LL} = 0 \end{matrix}\right\} P_{r1}^4 = 1.4P_{DL} + P_{Q_E} + P_{LL} = 268.7^{\text{kips}}$$

$$\left.\begin{matrix} M_{QE2}^4 = 0 \\ M_{DL} = 99.63^{\text{k-ft}} \\ M_{LL} = 58.63^{\text{k-ft}} \end{matrix}\right\} M_{r2}^4 = 1.4M_{DL} + M_{QE1} + M_{LL} = 1.4 \times (99.63^{\text{k-ft}})$$

$$+ 0 + 58.63^{\text{k-ft}} = 198.1^{\text{k-ft}}$$

Try W18 × 55 (ASTM A992)
Check axial force-bending moment interaction:

$$\left.\begin{matrix} M_r = 198.1^{\text{k-ft}} \\ P_r = 268.7^{\text{kips}} \\ M_c = 420^{\text{k-ft}} \\ P_c = 618^{\text{kips}} \end{matrix}\right\} \Rightarrow \left| \begin{matrix} \dfrac{P_r}{P_c} = \dfrac{268.7}{618} = 0.43 > 0.2 \rightarrow \text{Use Eq. H1-1a} \\ \dfrac{P_r}{P_c} + \dfrac{8}{9}\dfrac{M_r}{M_c} = \dfrac{268.7}{618} + \dfrac{8}{9}\dfrac{198.1}{420} = 0.85 < 1.0 \end{matrix} \right|$$

Check compactness requirement for W18 × 55:

Flange: $b/2t_f = 5.98 < \lambda_{hd} = 7.35$

Web: $C_a = \dfrac{P_u}{\phi_c P_y} = \dfrac{268.7^{\text{kips}}}{0.9 \times (1.1 \times 16.2^{\text{in.}^2} \times 50^{\text{ksi}})} = 0.335 > 0.114$

$$h/t_w = 41.1 < \lambda_{hd} = 0.88 \times \sqrt{\frac{29{,}000}{1.1 \times 50}} \times (2.68 - 0.335) = 47.4 \geq 36.1$$

Use W18 × 55.

Design summary and the selected girder sections for the analysis case I is shown in Table 8.49.

Since each analysis case represents a different scenario, the heavier girder section obtained from the two analyses will be used as the final section. Table 8.50 shows the governing analysis cases and the final design for girders.

Column Design Columns in braced bays are designed based on the previously explained two structural analysis cases, in which the first-mode deformation is found to be sufficient for consideration of inelastic deformations. Due to the absence of reversed curvature along the height, the columns can be designed under the effect of axial force only. Therefore, the columns are analyzed using the capacity-limited horizontal load effects obtained from the two analyses, as shown in Fig. 8.98. Details of the horizontal load determination can be found in the girder design section.

Table 8.50 Final Sections for the Girders

Story	Girder section: Analysis Case I	Girder section: Analysis Case II	Governing analysis case	Selected girder section
6th	W18 × 35	W18 × 35	II	**W18 × 35**
5th	W18 × 35	W18 × 35	I and II	**W18 × 35**
4th	W18 × 46	W18 × 55	II	**W18 × 55**
3rd	W18 × 35	W18 × 35	I and II	**W18 × 35**
2nd	W18 × 46	W18 × 60	II	**W18 × 60**
1st	W18 × 97	W18 × 97	I and II	**W18 × 97**

FBDs for determining the axial force demand on the fourth-story columns are shown in Fig. 8.99. Even though the columns to the left are under tension, considering that the columns can be inclined in both directions, both columns are to be designed for the identical compressive forces. Example calculation for the fourth-story columns is given below.

$$\text{Analysis Case I} \Rightarrow \begin{cases} P_{C,L4}^1 = \dfrac{F_{6U} \times 26' + F_{5U} \times 13' + C_{EB4}\sin\alpha \times 25'}{25'} \\ P_{C,R4}^1 = \dfrac{F_{6U} \times 26' + F_{5U} \times 13' + T_{ET4}\sin\alpha \times 25'}{25'} \end{cases}$$

$$P_{C,L4}^1 = \frac{415.4^k \times 26' + 0^k \times 13' + 435.9 \times \sin 46 \times 25'}{25'} = 745.6^{\text{kips}} \text{ (Tension)}$$

$$P_{C,R4}^1 = \frac{415.4^k \times 26' + 0^k \times 13' + 615.9 \times \sin 46 \times 25'}{25'}$$

$$= 875.1^{\text{kips}} \text{ (Compression)}$$

Table 8.49 Sections for the Girders Not Intersected by Braces Based on Structural Analysis Case II

Story	Girder section	F_u (kips)	M_{QE2} (k-ft)	M_{DL} (k-ft)	M_{LL} (k-ft)	P_{QE2}* (kips)	M_{r2} (k-ft)	P_{r2} (kips)	$M_c = \Phi_c M_n$ (k-ft)	$P_c = \Phi_c P_n$ (kips)	Strength ratio**
6th	W18 × 35	299.7	0.0	39.9	23.5	100.2	79.2	100.2	249.0	328.4	**0.59**
4th	W18 × 55	219.0	0.0	99.6	58.6	268.7	198.1	268.7	420.0	618.0	**0.85**
2nd	W18 × 60	211.2	0.0	99.6	58.6	381.0	198.1	381.0	461.0	681.0	**0.94**

*Positive sign means compression; **Per AISC 360, Eqs. H1-1a and H1-1b.

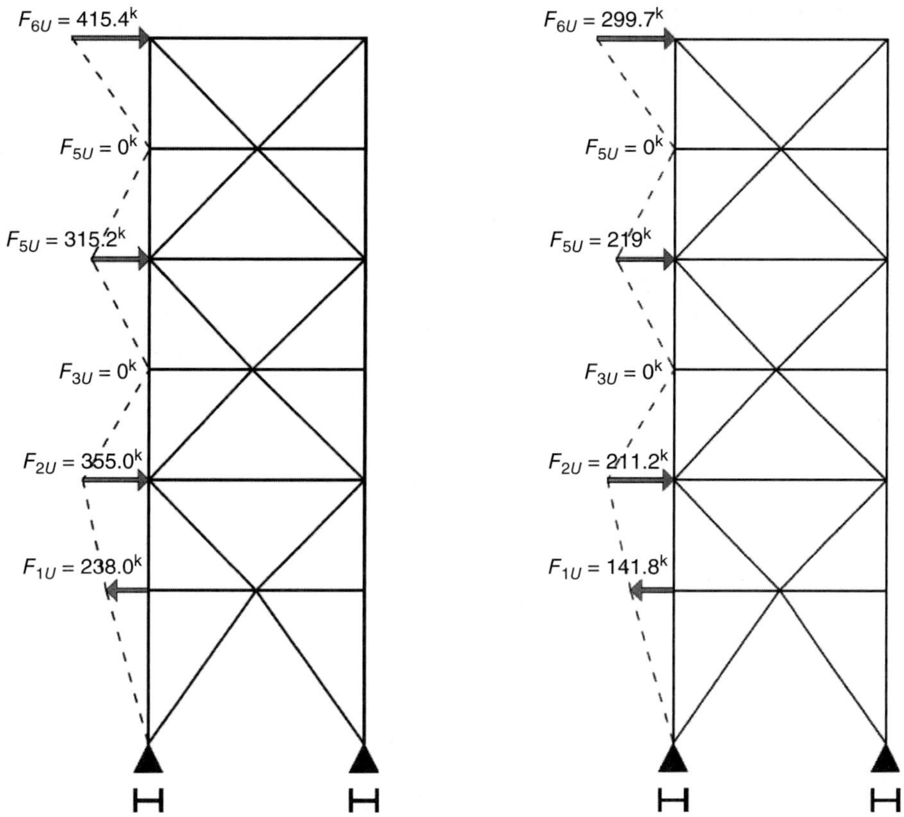

(a) Structural analysis case I (b) Structural analysis case II

Figure 8.98 Capacity-limited horizontal seismic force distribution along the height.

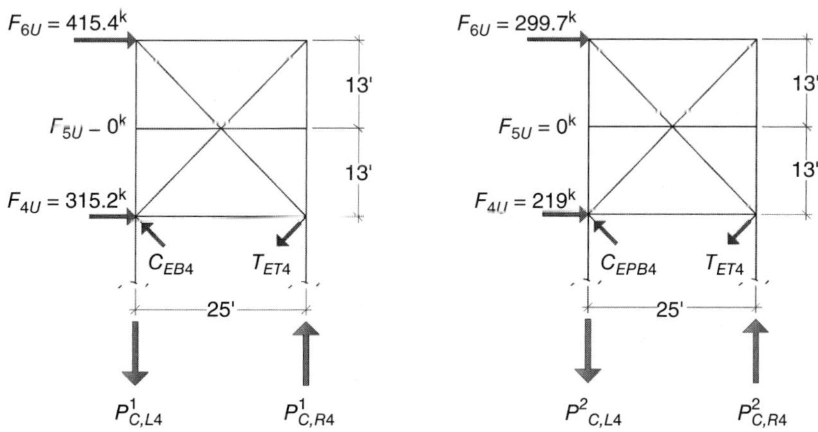

(a) Structural analysis case I (b) Structural analysis case II

Figure 8.99 FBDs of the fourth-story columns under capacity-limited seismic load effects.

Table 8.51 Summary of Column Forces for Structural Analysis Case I

Story	Brace section	T_{ET} (kips)	C_{EB} (kips)	F_u (kips)	P^1_{CL}* (kips)	P^1_{CR} (kips)	P_{QE1} (kips)
6th	HSS6.875 × 0.312	360.0	238.1	415.4	−171.3	259.0	259.0
5th	HSS6.875 × 0.312	360.0	238.1	0.0	−216.0	216.0	216.0
4th	HSS7.500 × 0.500	615.9	435.9	315.2	−745.6	875.1	875.1
3rd	HSS7.500 × 0.500	615.9	435.9	0.0	−812.0	812.0	812.0
2nd	HSS10 × 0.500	831.2	731.6	355.0	−1718.2	1789.9	1789.9
1st	HSS10 × 0.500	831.2	646.7	−238.0	−1756.5	1756.5	1756.5

*Negative sign means tension.

Summary of the rest of the column axial forces, P_{QE1}, is given in Table 8.51.

$$\text{Analysis Case II} \Rightarrow \begin{cases} P^2_{C,L4} = \dfrac{F_{6U}\times 26' + F_{5U}\times 13' + C_{EPB4}\sin\alpha\times 25'}{25'} \\ P^2_{C,R4} = \dfrac{F_{6U}\times 26' + F_{5U}\times 13' + T_{ET4}\sin\alpha\times 25'}{25'} \end{cases}$$

$$P^2_{C,L4} = \frac{299.7^k\times 26' + 0^k\times 13' + 130.8\times\sin 46\times 25'}{25'} = 405.7^{\text{kips}}\ (\text{Tension})$$

$$P^2_{C,R4} = \frac{299.7^k\times 26' + 0^k\times 13' + 615.9\times\sin 46\times 25'}{25'}$$

$$= 754.7^{\text{kips}}\ (\text{Compression})$$

The axial force demands on columns for analysis case II are summarized in Table 8.52.

Gravity loads acting on the columns in braced bays are determined based on their tributary areas.

$$P^{DL}_{C,R4} = A_T\times DL = \frac{(2\times 27.5'\times 27.5')\times 85^{\text{psf}} + (27.5'\times 27.5')\times 34^{\text{psf}}}{(1000\ \text{lb/1 kip})}$$

$$= 154.3^{\text{kips}}$$

$$P^{LL}_{C,R4} = A_T\times LL = \frac{(2\times 27.5'\times 27.5')\times 50^{\text{psf}} + (27.5'\times 27.5')\times 20^{\text{psf}}}{(1000\ \text{lb/1 kip})}$$

$$= 90.8^{\text{kips}}$$

$$P_{r4} = 1.4P_{DL} + P_{QE} + P_{LL}$$

in which P_{QE} is the larger of $\{P_{QE1};\ P_{QE2}\}$

$$P_{r4} = 1.4P^{DL}_{C,R4} + P_{QE} + P^{LL}_{C,R4} = 1.4\times 154.3^{\text{kips}} + 875.1^{\text{kips}} + 90.8^{\text{kips}} = 1182^{\text{kips}}$$

The required axial strength of columns, P_r, is given in Table 8.53. Columns are spliced at the third and fifth stories. Therefore, column sections are altered every two stories along the height. The required strengths used in column design are indicated with bold in Table 8.53.

Table 8.53 Required Axial Strength for Columns

Story	P_{QE1} (kips)	P_{QE2} (kips)	Governing analysis case	P_{QE} (kips)	P_{DL} (kips)	P_{LL} (kips)	P_r (kips)
6th	259.0	259.0	I and II	259.0	25.7	15.1	310.1
5th	216.0	155.8	I	216.0	90.0	52.9	**395.0**
4th	875.1	754.7	I	875.1	154.3	90.8	1181.9
3rd	812.0	581.4	I	812.0	218.6	128.6	**1246.5**
2nd	1789.9	1449.1	I	1789.9	282.8	166.4	2352.2
1st	1756.5	1230.7	I	1756.5	347.1	204.2	**2446.6**

Try W14 × 132 (ASTM A992) for the third- and fourth-story columns
$A_g = 38.8$ in.^{2}; $r_x = 6.28$ in.; $r_y = 3.76$ in.; $F_y = 50$ ksi

$$F_e = \frac{\pi^2 E}{(L_c/r)^2} = \frac{\pi^2\times 29{,}000}{\left(\dfrac{13\times 12}{3.76}\right)^2} = 166.3^{\text{ksi}}$$

$$F_{cr} = (0.658^{50/166.3})\times 50 = 44.1^{\text{ksi}}$$

$$P_c = \phi_c P_n = \phi_c A_g F_{cr} = 0.9\times 38.8^{\text{in.}^2}\times 44.1^{\text{ksi}} = 1540^{\text{kips}}$$

$$P_c = 1540^{\text{kips}} > P_r = 1246.5^{\text{kips}}\quad \text{OK.}$$

Check compactness requirement for W14 × 132:
According to *AISC 341, Section F2.5a*, columns shall meet the requirements for highly ductile members stipulated in Table D1.1.

$$\text{Flange: } b/2t_f = 7.15 < \lambda_{hd} = 0.32\sqrt{\frac{E}{R_y F_y}} = 7.35$$

$$\text{Web: } C_a = \frac{P_u}{\phi_c P_y} = \frac{1246.5^{\text{kips}}}{0.9\times(1.1\times 38.8^{\text{in.}^2}\times 50^{\text{ksi}})} = 0.649 > 0.114$$

$$h/t_w = 17.7 < \lambda_{hd} = 0.88\times\sqrt{\frac{29{,}000}{1.1\times 50}}\times(2.68 - 0.649) = 41.0 \geq 36.1$$

Table 8.52 Summary of Column Forces for Structural Analysis Case II

Story	Brace section	T_{ET} (kips)	C_{EPB} (kips)	F_u (kips)	P^2_{CL}* (kips)	P^2_{CR} (kips)	P_{QE2} (kips)
6th	HSS6.875 × 0.312	360.0	71.4	299.7	−51.4	259.0	259.0
5th	HSS6.875 × 0.312	360.0	71.4	0.0	−155.8	155.8	155.8
4th	HSS7.500 × 0.500	615.9	130.8	219.0	−405.7	754.7	754.7
3rd	HSS7.500 × 0.500	615.9	130.8	0.0	−581.4	581.4	581.4
2nd	HSS10 × 0.500	831.2	219.5	211.2	−1009.0	1449.1	1449.1
1st	HSS10 × 0.500	831.2	194.0	−141.8	−1230.7	1230.7	1230.7

*Negative sign means tension.

Table 8.54 Summary of Strength and Member Requirements for Columns

		Member requirement				Strength requirement		
Story	Column section	$b/2t_f$	$\lambda_{hd,f}$	h/t_w	$\lambda_{hd,w}$	P_r (kips)	P_c (kips)	Strength ratio*
6th	W14 × 53	6.11	7.35	30.9	43.8	310.1	433.0	0.72
5th	W14 × 53			30.9		**395.0**	**433.0**	**0.91**
4th	W14 × 132	7.15	7.35	17.7	41.0	1181.9	1540.0	0.77
3rd	W14 × 132			17.7		**1246.5**	**1540.0**	**0.81**
2nd	W14 × 233	4.62	7.35	10.7	39.6	2352.2	2770.0	0.85
1st	W14 × 233			10.7		**2446.6**	**2520.0**	**0.97**

*Demand to capacity ratio, P_r/P_c.

Use W14 × 132 (ASTM A992).

Summary of strength and highly ductile member requirements along with the final column sections are shown in Table 8.54.

8.3 CONCLUDING REMARKS

Design of the two-story X-braced frames (TSXBFs) is completed following the design requirements stipulated in *AISC 341, AISC 360,* and *ASCE 7.* Bracing members are designed based on the seismic base shear computed using C_uT_a = 0.77 sec., which is less than the fundamental period of the braced frame. Note that the first period of vibration obtained from the numerical model is around 1.01 sec. (T_1 = 1.01 sec.). The structural members other than braces are designed using capacity-limited seismic loads. Therefore, subsequent to the design of bracings, the rest of the structural members are analyzed under the unbalanced brace forces. However, due to the anticipated first-mode side-sway mechanism, the unbalanced braced forces imposed on the brace-intersected girders other than the first story canceled each other out. Therefore, the brace-intersected girders are designed for the gravity-load-imposed bending moment only, which resulted in very light girder sections. In the course of the design process, the load factor on live load in the governing load combination is intentionally not reduced to 0.5 with the purpose of not having such light girder sections, even though it is permitted to do so. It should be also noted that owing to the large lateral stiffness, the maximum inelastic story drift ratio along the height is found to be 0.014, which is less than the allowable limit of 0.02 radians. Thus, the design of the braced frame is controlled by strength. The final design of the structural members in the braced bays on lines 3 and 4 are given in Table 8.55 and Fig. 8.100.

Table 8.55 Final Sections for the Members in the Braced Bays on Lines 3 and 4

Story	Braces	Girders	Columns
6th	HSS6.875 × 0.312	W18 × 35	W14 × 53
5th	HSS6.875 × 0.312	W18 × 35	W14 × 53
4th	HSS7.500 × 0.500	W18 × 55	W14 × 132
3rd	HSS7.500 × 0.500	W18 × 35	W14 × 132
2nd	HSS10 × 0.500	W18 × 60	W14 × 233
1st	HSS10 × 0.500	W18 × 97	W14 × 233

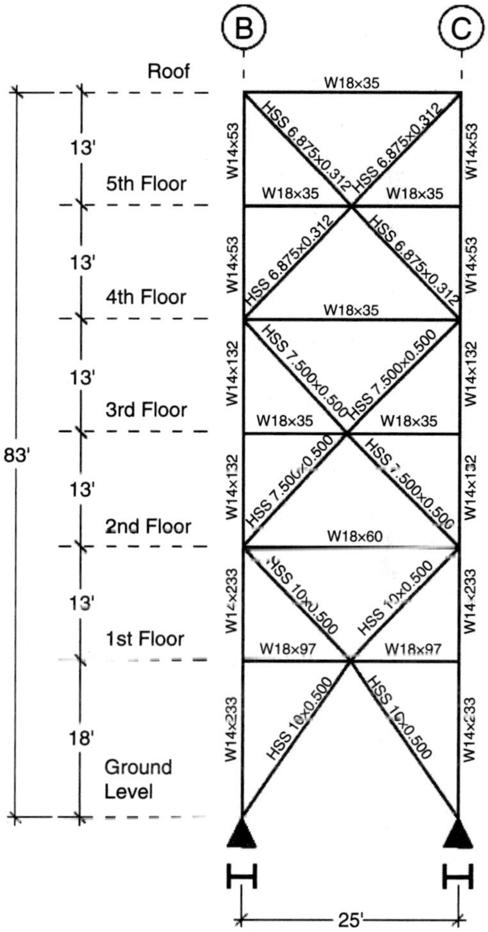

Figure 8.100 Final sections for the members in the braced bays on lines 3 and 4.

REFERENCES

1. ANSI/AISC 360-16, AISC 360 for Structural Steel Buildings, American Institute of Steel Construction, 2016.

2. ANSI/AISC 341-16, AISC 341 for Structural Steel Buildings, American Institute of Steel Construction, 2016.

3. AISC 358-16, *Prequalified Connections for Special and Intermediate Steel Moment Frames for Seismic Applications*, ANSI/AISC 358-16, American Institute of Steel Construction, IL.

4. ASCE/SEI 7-16, Minimum Design Loads and Associated Criteria for Buildings and Other Structures, American Society of Civil Engineers, 2017.

5. SAP2000, Integrated Software for Structural Analysis and Design, Computers and Structures, Inc, Berkeley, CA, 2000.

6. Engelhardt, M.D., Design *of Seismic-Resistant Steel Building Structures.* 2007. American Institute of Steel Construction, V.1.

Chapter 9

Design of Cold-Formed Steel Structural Members[a]

BY

NABIL A. RAHMAN, Ph.D., PE *FDR Engineers, Raleigh, North Carolina*
HELEN CHEN, Ph.D., PE *American Iron and Steel Institute, Washington, D.C.*
CHENG YU, Ph.D., PE *University of North Texas, Denton, Texas*

9.1 SHAPES AND APPLICATIONS

Cold-formed steel structural members are shapes commonly manufactured from steel plate, sheet, or strip material. The manufacturing process involves forming the material by either press-braking or roll forming to achieve the desired shape. Forming in press brakes is economical for moderate production volumes of a specific shape. However, the production of large quantities of identical shapes is best accomplished by roll forming. This is a fully mechanized, high-speed process that is superior for mass production. Although it is possible to produce shapes up to 1 inch thick by cold forming, cold-formed steel construction is generally restricted to the thicknesses given in Table 9.1.

Table 9.1 Thickness of Uncoated Steel Sheets

Designation thickness	Minimum base steel thickness		Design thickness	
(mils)	(in.)	(mm)	(in.)	(mm)
14	0.0141	0.355	0.0149	0.378
18	0.0179	0.455	0.0188	0.478
23	0.0226	0.574	0.0238	0.605
27	0.0269	0.683	0.0283	0.719
30	0.0296	0.752	0.0312	0.792
33	0.0329	0.836	0.0346	0.874
43	0.0428	1.087	0.0451	1.146
54	0.0538	1.367	0.0566	1.435
68	0.0677	1.720	0.0713	1.811
97	0.0966	2.454	0.1017	2.583
118	0.1180	2.997	0.1242	3.155

Sheet and strip steel are the more common materials used for cold-formed steel members and are designated by the mil thickness, which represents the minimum base steel thickness in decimal inches multiplied by 1000. The design thickness is permitted by design standards to be slightly larger than the minimum base steel thickness to reflect statistical data that show steel sheet material typically delivered with thicknesses greater than the minimums. Some trades may still use an old "gage numbers" designation for cold-formed steel members, but design standards are promoting the mil thickness designation, as it is more accurate for structural design.

Cold-formed steel members can be classified into two major categories: framing members, and panels and decks. The ease of manufacturing framing members provides a variety of cold-formed shapes. The depth of these sections generally ranges from about 0.75 to 16 in. and the thickness from 33 to 118 mils for structural applications and 18 to 30 mils for nonstructural applications. These dimensions frequently result in plane elements having large flat width-to-thickness ratios. Such slender elements are commonly stiffened with edge stiffeners or intermediate stiffeners to forestall premature local buckling. The more common shapes used in building construction for wall studs, floor joists, trusses, built-up beams, and columns are C-sections, Z-sections, hat sections, and angles. A few typical sections used as framing members are shown in Fig. 9.1. Geometric properties of several standard framing members are tabulated in the AISI D100 Design Manual (Ref. 1).

Figure 9.2 depicts several typical panel and deck sections that may be used as roof and wall panels, floor deck, roof deck, and bridge forms. These members are load-resisting shapes capable of providing a work surface and a durable enclosure against wind, rain, and snow loading. Generally, the geometry of such shapes ranges in depth from 1.5 to 7.5 in. and the thickness from 14 to 54 mils.

Although framing members are optimized to produce maximum strength and stiffness per unit weight of material, panels and decks must satisfy a variety of functional requirements of which optimum strength is only one. Other requirements are the coverage provided by a given flat width of sheet, that is, the flat width-to-thickness ratio, and the ability to function for floor electrification and telecommunication conduits. An optimum design must therefore serve multifunctional requirements.

The most typical applications for cold-formed steel structural members are stud framing for load-bearing (gravity) walls and exterior wind-bearing walls, joist framing for floors and roofs, and trusses for floors and roofs. Other structural applications include moment frames, purlins and girts for metal buildings, three-dimensional trusses for towers, corrugated sheet for decks and panels, storage racks, and storage silos. Figure 9.3 shows the application of cold-formed steel members to wall framing in a multistory building.

[a]Original material for this chapter was written by Roger A. LaBoube, Ph.D., P.E., Wei-Wen Yu Center for Cold-Formed Steel Structures.

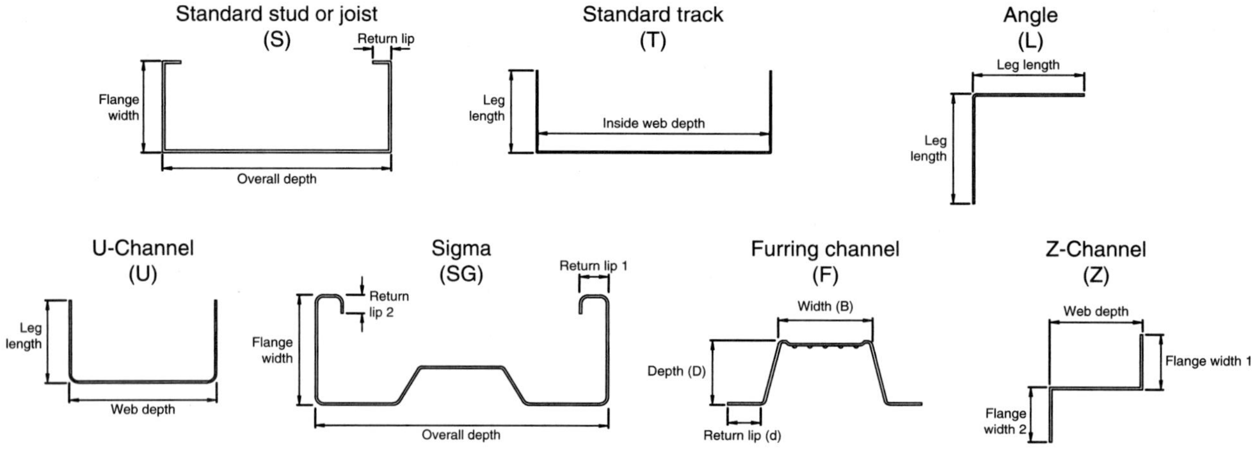

Figure 9.1 Common cold-formed steel framing members.

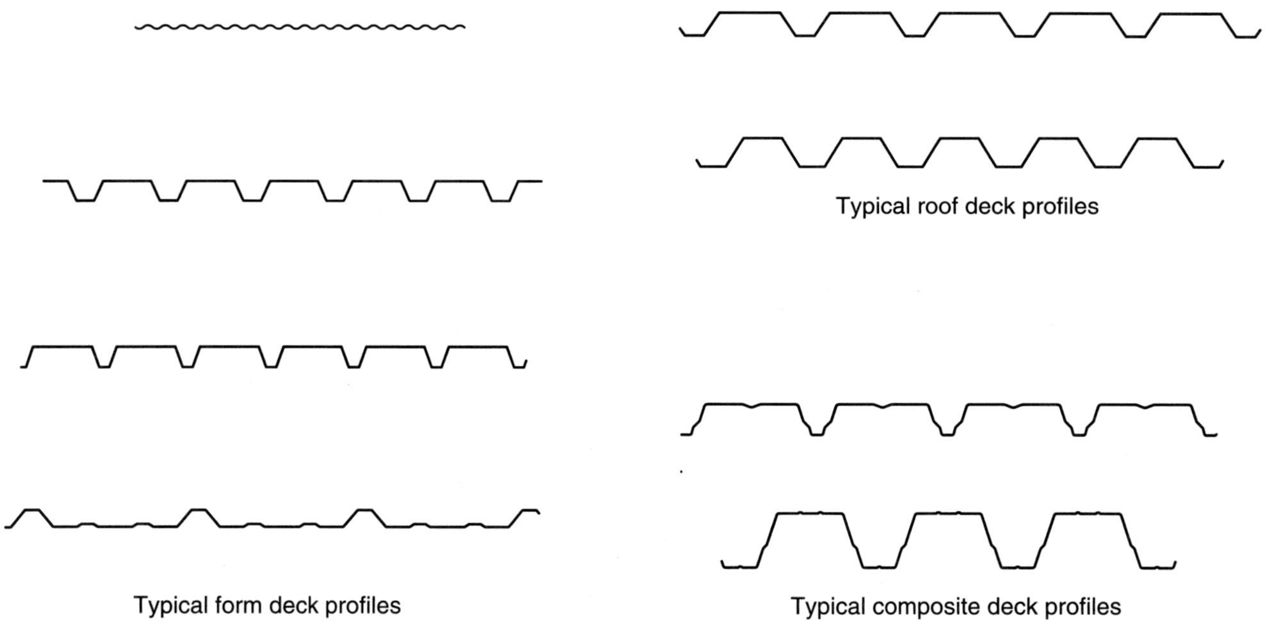

Figure 9.2 Common cold-formed steel deck and panel shapes.

Figure 9.3 Cold-formed steel wall framing in a multistory building.

Chapter 22 of the International Building Code (IBC) is the main reference code for design of cold-formed steel structural members. This code has been adopted for use as a base code standard by most jurisdictions in the United States and multiple other countries. The code refers designers to specific industry standards based on the design application: Section 2209 for steel storage racks and Sections 2210 and 2211 for framing members and decks and seismic design.

9.2 MATERIALS

A considerable variety of sheet and strip steels is available for use in cold-formed construction. To be suitable, the material must possess adequate and reliable strength and ductility. The yield point is the primary criterion for member strength under static loading, whereas the ultimate tensile strength is the primary criterion for connection strength. Tensile strength is usually less important for member strength considerations provided that there is a reasonable spread between the yield and tensile strength values of the steel. Fatigue and brittle fracture relate primarily to the tensile strength rather than the yield point and are rarely applicable considerations in cold-formed steel construction. Fatigue is more common in cyclically loaded structures like machine housings crane systems, and the slenderness of cold-formed steel members often hinders the development of the triaxial stress conditions necessary to initiate brittle fracture. Ductility is required to enable the material to be cold-formed to relatively tight radii without cracking and, in the formed member, to provide plastic stress redistribution in regions of localized stress concentrations such as bolted connections.

Elongation in a 2-in. gage length in combination with the ratio of the tensile strength to yield strength F_u/F_y is used as a measure of ductility. Generally, sufficient ductility is available if F_u/F_y is greater than or equal to 1.08 and the elongation in 2-in. gage is greater than or equal to 10 percent. Steels that do not satisfy these requirements can be used but only for shapes that require mild cold forming and that are used without highly stressed connections, for example, roof and wall panels and deck sections.

Section A3.1 of the AISI S100 Standard (Ref. 2) lists the ASTM standards that define the qualities of sheet and strip steels suitable for cold-formed construction. Certain steel grades within these ASTM standards have elongations that vary based upon the thickness of the material and therefore are excluded from the designated groups. Some of the most used ASTM steel standards are given below, while Table 9.2 shows the principal properties of these steels as given in Ref. 2:

ASTM A36/A36M, Carbon Structural Steel
ASTM A500/A500M, Cold-Formed Welded and Seamless Carbon Steel Structural Tubing in Rounds and Shapes
ASTM A529/A529M, High-Strength Carbon-Manganese Steel of Structural Quality
ASTM A572/A572M, High-Strength Low-Alloy Columbium-Vanadium Structural Steel
ASTM A653/A653M (SS Grades 33 (230), 37 (255), 40 (275), 50 (340) Class 1, Class 3 and Class 4, 55 (380) and 60 (410); HSLAS and HSLAS-F, Grades 40 (275), 50 (340), 55 (380)

Table 9.2 Principal Tensile Properties for ASTM Steels

ASTM designation	Grade	F_y ksi (min)	F_u ksi (min/max)	Percent elongation in 2 in. (min)	$\frac{F_{u(min)}}{F_{y(min)}}$
A36/A36M					
Plates and Bars	—	36	58/80	23	1.61
A500/A500M					
Round Tubing	A	33	45	25	1.36
	B	42	58	23	1.38
	C	46	62	21	1.35
	D	36	58	23	1.61
Shaped Tubing	A	39	45	25	1.15
	B	46	58	23	1.26
	C	50	62	21	1.24
	D	36	58	23	1.61
A529/A529M					
Plates and Bars	50	50	65/100	21	1.30
	55	55	70/100	20	1.27
A572/A572M					
Plates and Bars	42	42	60	24	1.43
	50	50	65	21	1.30
	55	55	70	20	1.27
	60	60	75	18	1.25
	65	65	80	17	1.23
A653/A653M					
Sheet	SS				
	33	33	45	20	1.36
	37	37	52	18	1.41
	40	40	55	16	1.38
	50 Class 1	50	65	12	1.30
	50 Class 3	50	70	12	1.40
	50 Class 4	50	60	12	1.20
	55	55	70	11	1.27
	60	60	70	10	1.17
	70	70	80	9	1.14
	80 Class 1	80	82	—	1.03
	80 Class 2	80	82	—	1.03
	80 Class 3	80	82	3	1.03
	HSLAS				
	40	40	50	22	1.25
	50	50	60	20	1.20
	55 Class 1	55	70	16	1.27
	55 Class 2	55	65	18	1.18
	60	60	70	16	1.17
	70	70	80	12	1.14
	80	80	90	10	1.13
	HSLAS-F				
	40	40	50	24	1.25
	50	50	60	22	1.20
	55 Class 1	55	70	18	1.27
	55 Class 2	55	65	20	1.18
	60	60	70	18	1.17
	70	70	80	14	1.14
	80	80	90	12	1.13
A792/A792M					
Sheet	SS				
	33	33	45	20	1.36
	37	37	52	18	1.41
	40	40	55	16	1.38
	50 Class 1	50	65	12	1.30
	50 Class 4	50	60	12	1.20
	60	60	70	10	1.17
	70	70	80	9	1.14
	80 Class 1	80	82	—	1.03
	80 Class 2	80	82	—	1.03
	80 Class 3	80	82	3	1.03
A1003/A1003M					
Sheet	ST33H	33	45	10	1.36
	ST37H	37	52	10	1.41
	ST40H	40	55	10	1.38
	ST50H	50	65	10	1.30
	ST55H	55	70	10	1.27
	ST57H	57	70	10	1.23
	ST60H	60	70	10	1.17
	ST70H	70	80	10	1.14
	ST80H	80	90	10	1.13
	ST33L	33	—	3	—
	ST37L	37	—	3	—
	ST40L	40	—	3	—
	ST50L	50	—	3	—
	ST55L	55	—	3	—
	ST60L	60	—	3	—
	ST70L	70	—	3	—
	ST80L	80	—	3	—

Class 1 and 2, 60 (410), 70 (480) and 80 (550), Steel Sheet, Zinc-Coated (Galvanized) or Zinc-Iron Alloy-Coated (Galvannealed) by the Hot-Dip Process

ASTM A792/A792M (Grades 33 (230), 37 (255), 40 (275), 50 (340) Class 1 and Class 4, and 60 (410)), Steel Sheet, 55% Aluminum-Zinc Alloy-Coated by the Hot-Dip Process

ASTM A1003/A1003M (ST Grades 50 (340) H, 40 (275) H, 37 (255) H, 33 (230) H), Steel Sheet, Carbon, Metallic- and Nonmetallic-Coated for Cold-Formed Framing Members

In addition to material ordered and produced to the ASTM standards listed in AISI S100 (Ref. 2), other sheet and strip steels are permitted to be used. To ensure the safety of such use, the minimum physical properties of such steels should be specified in a manner similar to that of ASTM steels and their structural suitability established by stipulating an adequate spread between yield and tensile strength and sufficient elongation. Adequate controls by tests and analyses must be instituted by the producer or purchaser to verify the specified properties.

Steels with yield points of 30 ksi or less are rarely used for structural applications. The bulk of the usage lies in the range of 33 to 55 ksi. Higher-strength steels are limited to specific situations.

The static strength of steel structural members depends not only on the yield point but also on the shape of the initial portion of the stress-strain diagram. Sheet and strip steel and their manufactured structural members exhibit one of the two types shown by Fig. 9.4. Figure 9.4*a* shows the behavior of a sharp-yielding steel with a practically straight line in the elastic range up to yielding, a definite horizontal inelastic yield plateau followed by a strain hardening range up to the ultimate point. Figure 9.4*b* shows the behavior of a gradually yielding steel beyond the elastic range and up to the ultimate point. For such steels, the yield strength is defined either by a 0.2 percent offset or by a stipulated amount of total elongation, usually 0.5 percent.

The strength of cold-formed steel members is generally controlled by local and overall instability (buckling). The strength of members that fail by buckling depends not only on Young's modulus E but also on the tangent modulus E_t (see Fig. 9.4*b*) determined at the magnitude of stress at which buckling occurs. Current design procedures are formulated for steels having a proportional limit (the range for which E is linear) not lower than 70 percent of the specified minimum yield point. This precludes the use of the AISI Specification for materials such as stainless steels in which the proportional limit is often considerably lower. However, ASCE has promulgated a specification for the design of stainless-steel cold-formed structural members (Ref. 3).

9.3 CORROSION PROTECTION

Cold-formed steel members are manufactured from relatively thin steel sheet, and corrosion protection is of primary interest to manufacturers and designers. Bare steel is subject to corrosion if it is not covered by an approved method of protective coating. The most common method of corrosion protective coating is metallic hot-dip zinc galvanization.

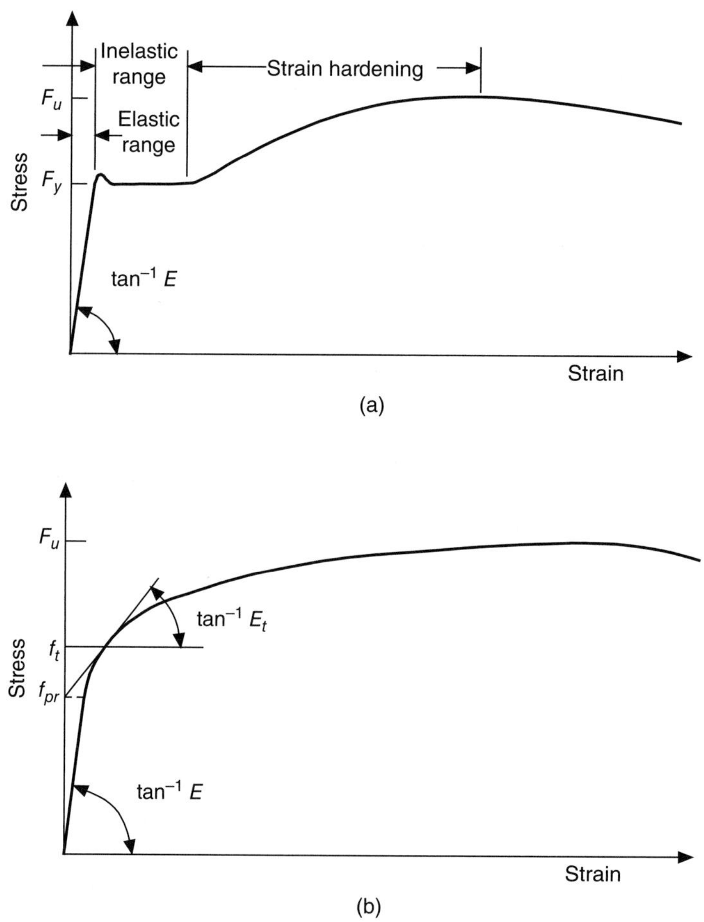

Figure 9.4 Stress-strain curves of steel sheet and strip (*a*) sharp yielding and (*b*) gradual yielding steel.

Table 9.3 Designations and Weights of Industry-Approved Protective Coatings

Coating designation	Zinc coated* oz/ft² (g/m²)	Zinc iron† oz/ft² (g/m²)	55% Al-Zinc‡ oz/ft² (g/m²)	Zinc-5%§ oz/ft² (g/m²)	Notes
Metallic coated‖	G40 [Z120]	A40 [ZF120]	AZ50 [AZM150]	GF20 [ZGF60]	Minimum coating for nonstructural framing members
	G60 [Z180]	A60 [ZF180]	AZ50 [AZM150]	GF30 [ZGF90]	Minimum coating for structural framing members
	G90 [Z275]	Not Applicable	AZ50 [AZM150]	GF45 [ZGF135]	—

*Zinc-coated steel sheet as described in ASTM A653/A653M.
†Zinc-iron alloy-coated steel sheet as described in ASTM A653/A653M.
‡55% aluminum-zinc alloy-coated steel sheet as described in ASTM A792/A792M.
§Zinc-5% aluminum alloy-coated steel sheet as described in ASTM A875/A875M.
‖1 oz/ft² = 305.15 g/m². Example G40 (0.4 oz/ft²) ≈ Z120 (122 g/m²). ASTM A1003/A1003M provides additional information for painted-metallic and nonmetallic coatings.

In this method, steel sheets are cleaned and then dipped continuously in a bath of molten zinc. The high temperature of the zinc bath enhances a reaction between steel and zinc, thereby creating a tight, metallurgical bond between the two metals. The coating of zinc will eventually oxidize, but its corrosion rate is much lower than that of carbon steel in the same air environment. The zinc coating is known to sacrificially protect any bare steel edges. Therefore, it is not required to re-treat shop or field-cut edges, punched and drilled holes, or scratches in these cold-formed steel members with zinc material. The zinc surface near a bare steel edge or a scratch will cover the area with zinc oxidation, preventing the steel from corroding until the nearby zinc is consumed.

Several other methods of protective coating for steel sheet exist in the industry and provide equivalent corrosion protection to hot-dip zinc galvanization. Table 9.3 shows industry-approved methods of protective coating and minimum coating requirements for normal exposure conditions. These normal conditions are defined as having the cold-formed steel members enclosed within a building envelope or structural assembly within a controlled environment. A heavier coating might be needed for more severe exposure conditions, such as marine or industrial atmospheres. In addition, heavier coating is recommended for connectors and fasteners at structural joints where higher levels of humidity or moisture can increase the risk of corrosion.

9.4 MEMBER DESIGN METHODOLOGY

Many design procedures that apply to hot-rolled structural members are equally applicable to cold-formed steel structural members. However, differences in behavior under load between the two types of construction necessitate differences in design methodology. The main reasons for this are the following:

(a) The design procedures for hot-rolled construction have been developed chiefly around the relatively few structural shapes and forms germane to that type of construction. In contrast, in cold-formed construction the variety of shapes that can be fabricated is almost unlimited. For this reason, the design methods must be so general that they apply to almost any possible shape.

(b) Cold-formed members are often formed by folding the flat elements to an open or closed shape. The flat width-to thickness ratio of the plane elements is frequently much larger in cold-formed than in hot-rolled construction. The relatively smaller thickness of the cross section results in much smaller torsional rigidity for cold-formed steel members as compared to hot-rolled members with similar overall cross-sectional depth and width. This is important in regard to torsional-flexural buckling of beams and columns for cold-formed steel members.

(c) The production and fabrication processes peculiar to the two types of construction affect in different ways the effective mechanical properties of the material. Thus, hot-rolling and welding cause residual cooling stresses in structural shapes, which may strongly affect the buckling strength of such members. Conversely, the cold work resulting from the cold-forming process changes the mechanical properties of sheet and strip, especially at and near corners.

9.4.1 Design Standards

Design procedures that specifically recognize the inherent complexities of cold-formed steel construction are provided in the AISI S100 Standard (Ref. 2). Design tables, charts, and examples that are developed based on this specification are provided in the AISI D100 Design Manual (Ref. 1). The AISI S240 Standard (Ref. 4) is a standard that includes provisions that are specifically applicable for cold-formed steel light-framing construction, such as structural wall stud and floor or roof joist framings. For seismic design, AISI S400 Standard (Ref. 5) should be followed. An in-depth discussion of the behavior and design of cold-formed steel members can be found in the publication *Cold-Formed Steel Design* (Ref. 6).

9.4.2 Design Basis

Both the allowable strength design (ASD) and load and resistance factor design (LRFD) methods can be used to determine member design strength. In the design standards, ASD and LRFD are founded on the same nominal strength. The allowable strength is obtained by dividing the nominal strength by a safety factor when ASD is used, and the design strength is obtained by multiplying the nominal strength by a resistance factor when LRFD is used. The required strengths, that is, member forces, determined according to ASCE 7 Standard load combinations (Ref. 7) and with consideration of structural stability (considering P-Δ and P-δ effects), should not exceed the member's allowable strength (for ASD) or design strength (for LRFD).

9.4.3 Cold-Formed Cross-Sectional Properties

The cross-sectional properties can be calculated using a so-called linear method, in which the area of the section is assumed to be concentrated along the centerline of the sheet and the elements that comprise the section, that is, flanges, webs, and corners, are replaced by straight or curved line elements. The thickness of the elements, being uniform, is accounted for after the line element properties are computed. Formulas to determine the C, Z, Hat, and Angle cross-sectional properties are provided in AISI D100 Design Manual (Ref. 1).

9.4.4 Possible Buckling Modes and Design Methodology

Due to the large flat width-to-thickness ratio of the elements comprising cold-formed steel members, these members are susceptible to local, distortional, and global buckling. Local buckling is a mode where the flat element itself buckles, while the relative orientation of the element remains unchanged as illustrated in Fig. 9.5*a*. Distortional buckling occurs when insufficient stiffness in an edge stiffener causes instability in elements that rely on the edge element for their own stability (see Fig. 9.5*b*). In addition, cold-formed members are subject to global buckling, where a member cross section translates and/or rotates around the entire section's longitudinal axis, while its cross-sectional shape remains unchanged. Depending on the member cross-sectional shape and the applied loads, the member can be subject to flexural (see Fig. 9.5*c* and *d*), torsional (see Fig. 9.5*e*), or flexural-torsional global buckling, which is a combination of both flexural and torsional buckling modes.

Two design methods can be used to determine the strengths of a cold-formed steel member.

EFFECTIVE WIDTH METHOD (EWM)

This method determines the member strength by considering the local buckling of each flat element of the cross section. The method further

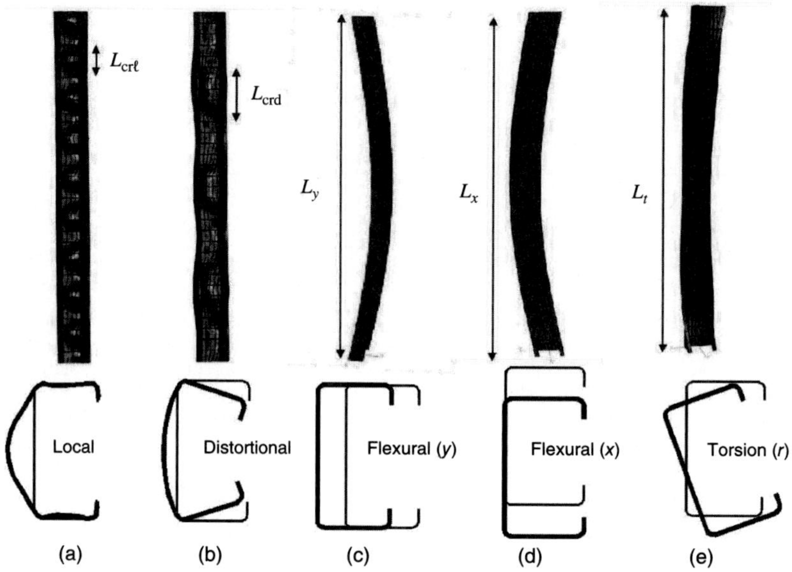

Figure 9.5 Buckling shapes.

simplifies the stress distribution by replacing the nonlinear actual stress caused by local buckling and post-buckling with a linear stress distribution on the effective widths. The effective widths are determined by equating the total force of the nonlinear stress distribution to one with a linear stress distribution on the effective widths as illustrated in Fig. 9.6.

The effective width for uniformly compressed elements can be calculated using the following equations:

$$\text{Effective width } b = \rho w \tag{9.1}$$

where w = Flat width of element

ρ = Local reduction factor

$= 1$ when $\lambda \le 0.673$

$= (1 - 0.22/\lambda)/\lambda$ when $\lambda > 0.673$ (9.2)

λ = Slenderness factor

$$= \sqrt{\frac{f}{F_{cr\ell}}} \tag{9.3}$$

f = Compressive stress in element

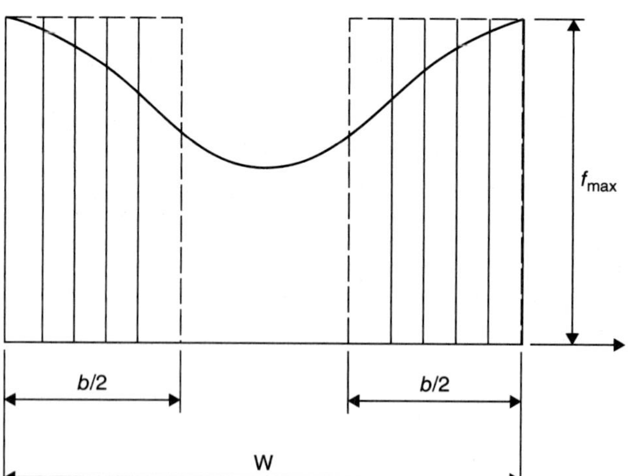

Figure 9.6 Stress distribution in stiffened compression elements.

$F_{cr\ell}$ = Element local buckling stress

$$= k\frac{\pi^2 E}{12(1-\mu^2)}\left(\frac{t}{w}\right)^2 \tag{9.4}$$

k = Plate buckling coefficient

= 4 for a stiffened element supported by a web on each longitudinal edge

= 0.43 for an element supported by a web on one longitudinal edge and free on the other end

E = Modulus of elasticity = 29,500 ksi (203,000 MPa)

μ = Poisson's ratio of steel = 0.3

t = Thickness of element

For other stress distributions and edge fixity conditions, the effective width can be determined using the provisions in Appendix 1 of the AISI S100 Standard (Ref. 2).

Direct Strength Method (DSM)

Instead of considering individual element buckling separately, this method determines the buckling loads of the whole cross section so that the interaction between the elements is considered. The method then determines member strengths due to various buckling modes. Shown in Fig. 9.7 is a signature buckling curve of a C-section member under

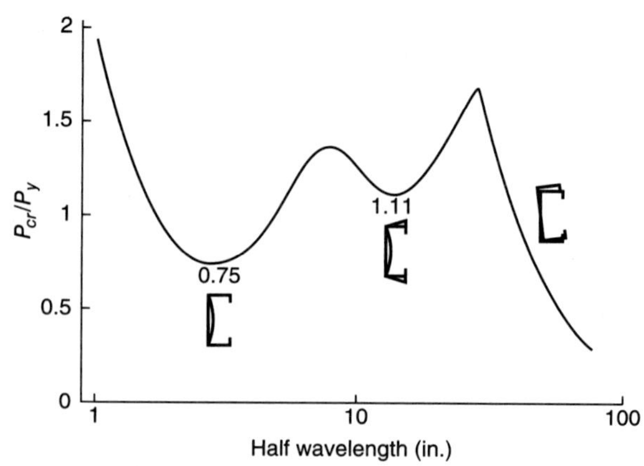

Figure 9.7 Signature buckling curve.

compression. The curve shows how the ratio of buckling load, P_{cr}, to the yield load, P_y, varies with respect to the critical unbraced lengths. The illustrations of the deformed cross section indicate that member local buckling occurs at a short buckling half-wavelength, and gradually the buckling mode is switched to distortional buckling at an intermediate buckling half-wavelength, and finally to the global buckling mode at a long buckling half-wavelength. In design, the minima of the local and distortional buckling modes are used to determine the corresponding member strength. The smallest strength due to global, local, and distortional buckling will govern the overall member strength.

9.5 STRUCTURAL STABILITY IN DETERMINING MEMBER FORCES

Structural stability can be maintained only when the structure remains balanced even in its deformed shape after being subjected to applied loads. This means that when computing the member forces, those forces caused by structural deformation, joint displacement, geometric imperfection, and stiffness reduction should be considered in the structural analysis. Three approaches are described in the AISI S100 Standard, Section C1, Design for System Stability. Readers may refer to AISI S100 for design provisions and AISI D100 Design Manual (Ref. 1) and the design examples.

9.6 MEMBER DESIGN

9.6.1 Compression Strength

Cold-formed steel compression member strength is governed by the smallest strength after considering global, local, and distortional buckling. Numerical methods (such as free-downloadable software CUFSM[b]) or analytical methods, as provided in Appendix 2 of AISI S100 Standard, can be used to determine the buckling forces. The nominal member strengths due to global, local, and distortional buckling can then be determined as follows:

(1) Nominal compression strength, P_{ne}, due to yielding and global buckling:

$$P_{ne} = A_g F_n \tag{9.5}$$

The compressive stress, F_n, is calculated as follows:

$$\text{For } \lambda_c \le 1.5,\ F_n = \left(0.658^{\lambda_c^2}\right) F_y \tag{9.6}$$

$$\text{For } \lambda_c > 1.5,\ F_n = \left(\frac{0.877}{\lambda_c^2}\right) F_y \tag{9.7}$$

$$\text{where } \lambda_c = \sqrt{\frac{F_y}{F_{cre}}} \tag{9.8}$$

where A_g = Gross cross-sectional area, F_y = Yield stress of steel, and F_{cre} = Least of the applicable global buckling stress, such as flexural, torsional, or flexural-torsional buckling. A symmetric cross section (such as I-section) member may be subject to flexural or torsional buckling, and a singly symmetric cross section (such as C-section) member may be subject to flexural or flexural-torsional buckling. The following formulas can be used to determine the corresponding buckling stress for doubly or singly symmetric sections (assume x-axis is the cross-sectional symmetric axis):

(a) Flexural buckling: $$\sigma = \frac{\pi^2 E}{(KL/r)^2} \tag{9.9}$$

(b) Torsional buckling: $$\sigma_t = \frac{1}{A_g r_o^2}\left(GJ + \frac{\pi^2 EC_w}{(K_t L_t)^2}\right) \tag{9.10}$$

[b]CUFSM is a free, open-source program using the semi-analytical finite-strip method for determination of thin-walled member stability.

(c) Flexural-torsional buckling:

$$F_{cre} = \frac{1}{2\beta}\left[(\sigma_{ex} + \sigma_t) - \sqrt{(\sigma_{ex} + \sigma_t)^2 - 4\beta\sigma_{ex}\sigma_t}\right] \tag{9.11}$$

$$\beta = 1 - (x_o/r_o)^2 \tag{9.12}$$

where K, L, and r = effective length factor for flexural buckling, unbraced length, and radius of gyration about the axis of buckling, respectively; E = modulus of elasticity = 29,500 ksi (203,000 MPa); r_o = polar radius of gyration about shear center; G = shear modulus = 11,300 ksi (78,000 MPa); J and C_w = torsional and warping constants, respectively; K_t and L_t = effective length factor and unbraced length for torsional buckling, respectively; σ_{ex} = flexural buckling stress about x-axis; and x_o = distance from the centroid to the shear center along the x-axis. For other cross-sectional types, equations are provided in Appendix 2 of the AISI S100 Standard.

(2) Nominal compression strength, $P_{n\ell}$, due to local buckling interacting with global buckling. Both EWM and DSM can be used to determine the nominal strength as follows:

(a) Effective width method: The nominal strength $P_{n\ell} = A_e F_n \le P_{ne}$, where F_n is determined in accordance with Eqs. (9.6) and (9.7) and A_e is the effective area determined according to Effective Width Method in Sec. 9.4.4.

(b) Direct strength method: The nominal strength, $P_{n\ell}$, is calculated as follows:

$$\text{For } \lambda_\ell \le 0.776;\ P_{n\ell} = P_{ne} \tag{9.13}$$

$$\text{For } \lambda_\ell > 0.776;\ P_{n\ell} = \left[1 - 0.15\left(\frac{P_{cr\ell}}{P_{ne}}\right)^{0.4}\right]\left(\frac{P_{cr\ell}}{P_{ne}}\right)^{0.4} P_{ne} \tag{9.14}$$

where $$\lambda_\ell = \sqrt{P_{ne}/P_{cr\ell}} \tag{9.15}$$

P_{ne} = Global column strength per Eq. (9.5)

$P_{cr\ell}$ = Critical elastic local column buckling load, governed by the lowest local buckling force of all the elements of the cross section. $P_{cr\ell}$ can be determined using software, such as CUFSM, or approximated by the simplified method as described in the AISI *Direct Strength Method (DSM) Design Guide* (Ref. 8) and Appendix 2 of AISI S100 (Ref. 2).

(3) Nominal strength, P_{nd}, due to distortional buckling. The nominal strength is determined as follows:

$$\text{For } \lambda_d \le 0.561;\quad P_{nd} = P_y \tag{9.16}$$

$$\text{For } \lambda_d > 0.561;\quad P_{nd} = \left[1 - 0.25\left(\frac{P_{crd}}{P_y}\right)^{0.6}\right]\left(\frac{P_{crd}}{P_y}\right)^{0.6} P_y \tag{9.17}$$

where $$\lambda_d = \sqrt{P_y/P_{crd}} \tag{9.18}$$

$$P_y = A_g F_y \tag{9.19}$$

A_g = Gross area of cross section

F_y = Yield stress

P_{crd} = Critical elastic distortional column buckling load

$$= A_g F_{crd} \tag{9.20}$$

F_{crd} = Distortional buckling stress, which can be determined per Appendix 2 of the AISI S100 Standard or using Tables III-4 to III-6 of AISI D100 Design Manual for industrial standardized studs and joists.

(4) Compression member strength: The nominal compression member strength, P_n, is governed by the smallest value of P_{ne}, $P_{n\ell}$, and P_{nd}. The available strength is P_n/Ω_c for ASD and $\phi_c P_n$ for LRFD, where Ω_c = 1.80 and ϕ_c = 0.85.

(5) Discrete bracing for compression members: When discrete bracing is used to provide lateral stability of a compression member with axial load P_{ra}, the brace should be designed to be able to take the axial

load of P_{rb} and to possess the axial stiffness of β_{rb}. Both P_{rb} and β_{rb} are calculated as follows:

$$P_{rb} = 0.01\ P_{ra} \tag{9.21}$$

For ASD,
$$\beta_{rb} = \frac{2[4-(2/n)]}{L_b}(\Omega P_{ra}) \tag{9.22}$$

where $\Omega = 2.00$ and P_{ra} is calculated per ASD load combinations:

For LRFD,
$$\beta_{rb} = \frac{2[4-(2/n)]}{L_b}\left(\frac{P_{ra}}{\phi}\right) \tag{9.23}$$

where $\phi = 0.75$; and P_{ra} is calculated per LRFD load combinations.

(6) Compression members with holes: Specific design provisions are provided in Chapter E of the AISI S100 Standard, which accounts for the effect of element holes on member strengths.

9.6.2 Flexural Strength

Similar to compression members, flexural member strength is determined by considering all of the global, local, and distortional buckling modes and the smallest strength governs the design.

(1) Nominal strength, M_{ne}, due to yielding and global buckling. The nominal member strength is calculated as follows:

$$M_{ne} = S_f F_n \le M_y \tag{9.24}$$

F_n is determined as follows:

For $F_{cre} \ge 2.78F_y$,
$$F_n = F_y \tag{9.25}$$

For $2.78F_y > F_{cre} > 0.56F_y$,
$$F_n = \frac{10}{9}F_y\left(1-\frac{10F_y}{36F_{cre}}\right) \tag{9.26}$$

For $F_{cre} \le 0.56F_y$,
$$F_n = F_{cre} \tag{9.27}$$

where S_f = Full section modulus relative to the extreme fiber in yielding about the bending axis considered, M_y = Yield moment = S_fF_y, and F_{cre} = Elastic global buckling stress. For symmetric and singly symmetric cross sections (with x-axis as the symmetric axis), lateral-torsional buckling controls global buckling, and it can be calculated as follows:

$$F_{cre} = \frac{C_b r_o A_g}{S_f}\sqrt{\sigma_{ey}\sigma_t} \quad \text{For bending about } x\text{-axis (axis of symmetry)} \tag{9.28}$$

where C_b = Bending coefficient dependent on moment gradient in the unbraced length; and σ_{ey} = Flexural buckling stress in accordance with Eq. (9.9) with K, L, and r computed about y-axis. The lateral-torsional buckling about the y-axis or buckling stresses for other type of cross sections can be obtained from Chapter F or Appendix 2 of the AISI S100 Standard.

(2) Nominal strength, $M_{n\ell}$, due to local buckling. Both EWM and DSM can be used to determine the strength $M_{n\ell}$ as follows:

(a) Effective width method: $M_{n\ell} = S_e F_n \le S_{et} F_y$ (9.29)

where S_e = Effective section modulus calculated using the effective width of the flat elements determined with the extreme fiber compressive stress at F_n computed using Eqs. (9.25) through (9.27). The effective widths of the elements can be determined in accordance with Appendix 1 of the AISI S100 Standard. S_{et} = Effective section modulus calculated with the extreme tension fiber stress at F_y.

(b) Direct strength method. The nominal strength is obtained as follows:

For $\lambda_\ell \le 0.776$,
$$M_{n\ell} = M_{ne} \tag{9.30}$$

For $\lambda_\ell > 0.776$,
$$M_{n\ell} = \left[1-0.15\left(\frac{M_{cr\ell}}{M_{ne}}\right)^{0.4}\right]\left(\frac{M_{cr\ell}}{M_{ne}}\right)^{0.4} M_{ne} \tag{9.31}$$

where
$$\lambda_\ell = \sqrt{M_{ne}/M_{cr\ell}} \tag{9.32}$$

M_{ne} = Nominal flexural strength for lateral-torsional buckling as defined in Eq. (9.24)

$$M_{cr\ell} = \text{Critical elastic local buckling moment} = F_{cr\ell}S_f \tag{9.33}$$

$F_{cr\ell}$ = Local buckling stress that is controlled by the lowest element buckling stress. $F_{cr\ell}$ can be determined numerically using software (such as CUFSM) or approximated by simplified method as described in the AISI Direct Strength Method Design Guide (Ref. 8) and Appendix 2 of AISI S100 (Ref. 2)

(3) Nominal strength, M_{nd}, due to distortional buckling. M_{nd} is determined as follows:

For $\lambda_d \le 0.673$; $M_{nd} = M_y$ (9.34)

For $\lambda_d > 0.673$;
$$M_{nd} = \left[1-0.22\left(\frac{M_{crd}}{M_y}\right)^{0.5}\right]\left(\frac{M_{crd}}{M_y}\right)^{0.5} M_y \tag{9.35}$$

where
$$\lambda_d = \sqrt{M_y/M_{crd}} \tag{9.36}$$

$$M_y = S_fF_y \tag{9.37}$$

$$M_{crd} = S_fF_{crd} \tag{9.38}$$

F_{crd} = Elastic distortional buckling stress determined per Appendix 2 of AISI S100 Standard or using Table II-5 of the AISI D100 Design Manual for standardized stud/joist sections.

(4) Member bending strength: The nominal bending member strength, M_n, is governed by the smallest value of M_{ne}, $M_{n\ell}$ and M_{nd}. The available strength is M_n/Ω_b for ASD, and ϕ_bM_n for LRFD, where $\Omega_b = 1.67$ and $\phi_b = 0.90$.

(5) Bending members with holes: Specific design provisions are provided in Chapter F of the AISI S100 Standard, which accounts for the effect of element holes on member strengths.

9.6.3 Shear Strength

Flexural member shear strength is determined based on the shear buckling and the post-buckling strengths. The nominal shear strength V_n for members without transverse web stiffeners is determined as follows:

For $\lambda_v \le 0.815$,
$$V_n = V_y \tag{9.39}$$

For $0.815 < \lambda_v \le 1.227$,
$$V_n = 0.60t^2\sqrt{Ek_vF_y} \tag{9.40}$$

For $\lambda_v > 1.227$,
$$V_n = V_{cr} = 0.904Ek_vt^3/h \tag{9.41}$$

where
$$\lambda_v = \sqrt{\frac{V_y}{V_{cr}}} \tag{9.42}$$

$$V_y = \text{Shear force of cross section at yield} = 0.6A_wF_y \tag{9.43}$$

$$A_w = \text{Area of web element} = ht \tag{9.44}$$

h = Depth of flat portion of *web* measured along plane of web
t = Web thickness
k_v = Shear buckling coefficient

$$= 4.00 + \frac{5.34}{(a/h)^2} \quad \text{when } a/h \le 1.0 \tag{9.45}$$

$$= 5.34 + \frac{4.00}{(a/h)^2} \quad \text{when } a/h > 1.0 \tag{9.46}$$

a = Shear panel length of unreinforced web element and clear distance between transverse stiffeners of reinforced web elements

The available shear strength is V_n/Ω_v for ASD and ϕ_vV_n for LRFD, where $\Omega_v = 1.60$ and $\phi_v = 0.95$.

9.6.4 Web Crippling Strength

When cold-formed steel members are subjected to transverse concentrated loads or reactions introduced by bearing on a flange, web

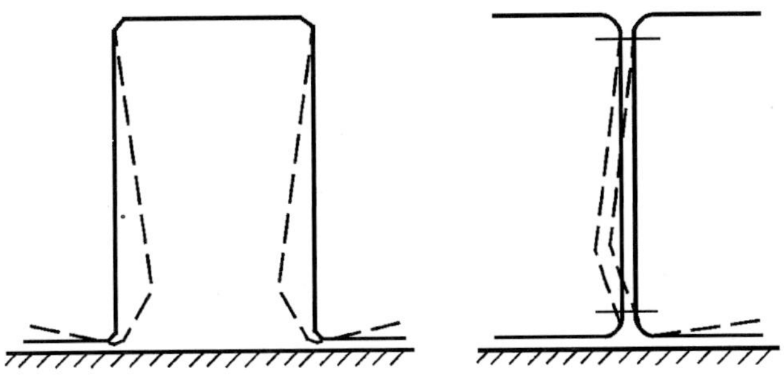

Figure 9.8 Web crippling of cold-formed steel sections.

crippling may occur due to the interaction of high local transverse web compression and longitudinal web compression from flexure as shown in Fig. 9.8.

The nominal web crippling strength, P_n, is determined by the following equation:

$$P_n = Ct^2F_y \sin\theta\left(1-C_R\sqrt{\frac{R}{t}}\right)\left(1+C_N\sqrt{\frac{N}{t}}\right)\left(1-C_h\sqrt{\frac{h}{t}}\right) \quad (9.47)$$

where C, C_R, C_N, C_h = Coefficients that are provided in AISI S100 Section G5; t = Web thickness; θ = Angle between plane of web and plane of bearing surface, $45° \le \theta \le 90°$; R = Inside bend radius; N = Bearing length [¾-in. (19-mm) minimum]; and h = Flat dimension of web measured in plane of web.

The safety and resistance factors for determining the available strengths are provided in the tables for most section types shown in Section G5 of the AISI S100 Standard. When the member does not have sufficient web crippling strength, a stiffener(s) shall be provided at the location of the applied load and/or reaction.

9.6.5 Force Combined Actions

Cold-formed steel members need to be checked for the following combined actions as applicable:

(1) Combined bending and compression

$$\frac{\overline{P}}{P_a}+\frac{\overline{M}_x}{M_{ax}}+\frac{\overline{M}_y}{M_{ay}} \le 1.0 \quad (9.48)$$

where $\overline{P}$, $\overline{M}_x$, and $\overline{M}_y$ = Required member compressive axial force and bending moments about x- and y-axis, respectively, in accordance with ASD or LRFD load combinations and with consideration of additional loading from structure deformation; P_a = Available axial strength determined in accordance with Sec. 9.6.1 above; and M_{ax}, M_{ay} = Available flexural strengths about the centroidal axes, determined in accordance with Sec. 9.6.2 above.

(2) Combined bending and shear

For beams without shear stiffeners:

$$\sqrt{\left(\frac{\overline{M}}{M_{a\ell o}}\right)^2+\left(\frac{\overline{V}}{V_a}\right)^2} \le 1.0 \quad (9.49)$$

For beams with shear stiffeners and when $\overline{M}/M_{a\ell o} > 0.5$ and $\overline{V}/V_a > 0.7$, the following interaction equation must also be satisfied:

$$0.6\left(\frac{\overline{M}}{M_{a\ell o}}\right)+\left(\frac{\overline{V}}{V_a}\right) \le 1.3 \quad (9.50)$$

where $\overline{M}$, $\overline{V}$ = Required moment and shear force, respectively, in accordance with ASD or LRFD load combinations; M_a = Available flexural strength determined in accordance with Sec. 9.6.2 above; V_a = Available shear strength determined accordance with Sec. 9.6.3 above; and $M_{a\ell o}$ = Available flexural strength for globally braced member determined per Section H2 of the AISI S100 Standard.

(3) Combined bending and web crippling

Unreinforced flat webs of shapes subjected to a combination of bending and concentrated load or reaction are required to consider the combined action of bending and web crippling. For shapes having single unreinforced webs (such as C-section members), the following equations shall be satisfied:

$$\text{For ASD: } 0.91\left(\frac{\overline{P}}{P_n}\right)+\left(\frac{\overline{M}}{M_{n\ell o}}\right) \le \frac{1.33}{\Omega};\ \Omega = 1.70 \quad (9.51)$$

$$\text{For LRFD: } 0.91\left(\frac{\overline{P}}{P_n}\right)+\left(\frac{\overline{M}}{M_{n\ell o}}\right) \le 1.33\phi,\ \phi = 0.90 \quad (9.52)$$

where $\overline{P}$, $\overline{M}$ = Concentrated load or reaction and the moment at the location determined per ASD and LRFD load combinations, respectively; P_n = Nominal web crippling strength determined per Sec. 9.6.4 above; and $M_{n\ell o}$ = Nominal flexural strength for globally braced member.

The interaction checks for an I-section formed by two C-sections or Z-section members are provided in Section H3 of the AISI S100 Standard.

Design Example 1

Check the strength of the pin-pin supported column 362S162-54 shown in Fig. 9.9 under combined compression and bending using ASCE 7 Standard load combinations (Ref. 7). The column is braced against lateral buckling about the y-axis and torsional and distortional buckling at the ends and mid-span. The gross cross-sectional properties can be calculated using the line method or obtained from the AISI D100 Design Manual. The DSM is used to determine the member strengths. The local and distortional buckling forces and moments are computed using the software CUFSM. The following buckling loads are obtained:

$$P_y = A_gF_y = (0.422)(50) = 21.1 \text{ kips} \qquad M_y = S_fF_y = (0.482)(50) = 24.1 \text{ kips}$$

$$P_{cr\ell} = 0.75P_y = 15.8 \text{ kips} \qquad M_{cr\ell} = 3.15M_y = 75.9 \text{ kip-in.}$$

$$P_{crd} = 1.11P_y = 23.4 \text{ kips} \qquad M_{crd} = 2.02M_y = 48.7 \text{ kip-in.}$$

In this example, the elastic critical global buckling limits are governed by flexural-torsional and lateral-torsional buckling, which can be calculated using Eq. (9.11) for compression and Eq. (9.28) for flexure, respectively. The following global buckling limits in compression and flexure from these equations are

$$P_{cre} = 7.8 \text{ kips} \qquad M_{cre} = 38.3 \text{ kip-in.}$$

1. Compression strength: The compression strength is governed by the smallest nominal strengths P_{ne} (global buckling), $P_{n\ell}$ (local buckling), and P_{nd} (distortional buckling). These nominal strengths can be computed using the equations provided in Sec. 9.6.1, and the results are summarized in Table 9.4.

2. Bending strength: The bending strength is governed by the smallest nominal strengths; M_{ne} (global buckling), $M_{n\ell}$ (local buckling), and M_{nd} (distortional buckling). These nominal strengths can be computed using the equations provided in Sec. 9.6.2, and the results are summarized in Table 9.5.

3. Calculate the required strength: By inspection, the ASCE 7 Standard load combination D + L (ASD) or 1.2D + 1.6L (LRFD) controls the design. The effective length method described in AISI S100 Standard, Section C1, Design for System Stability, is used in this example. Check the strength adequacy at the mid-span: since the column is pin-supported at both ends, the member forces, $P_{\ell t}$ and $M_{\ell t}$, at the mid-span due to lateral translation equal zero, and multiplier, B_2, accounting the P-Δ effect equal zero as well. The axial force and moment calculation and the compression and bending interaction check per numbered list item 1 in Sec. 9.6.5 are summarized in Table 9.6.

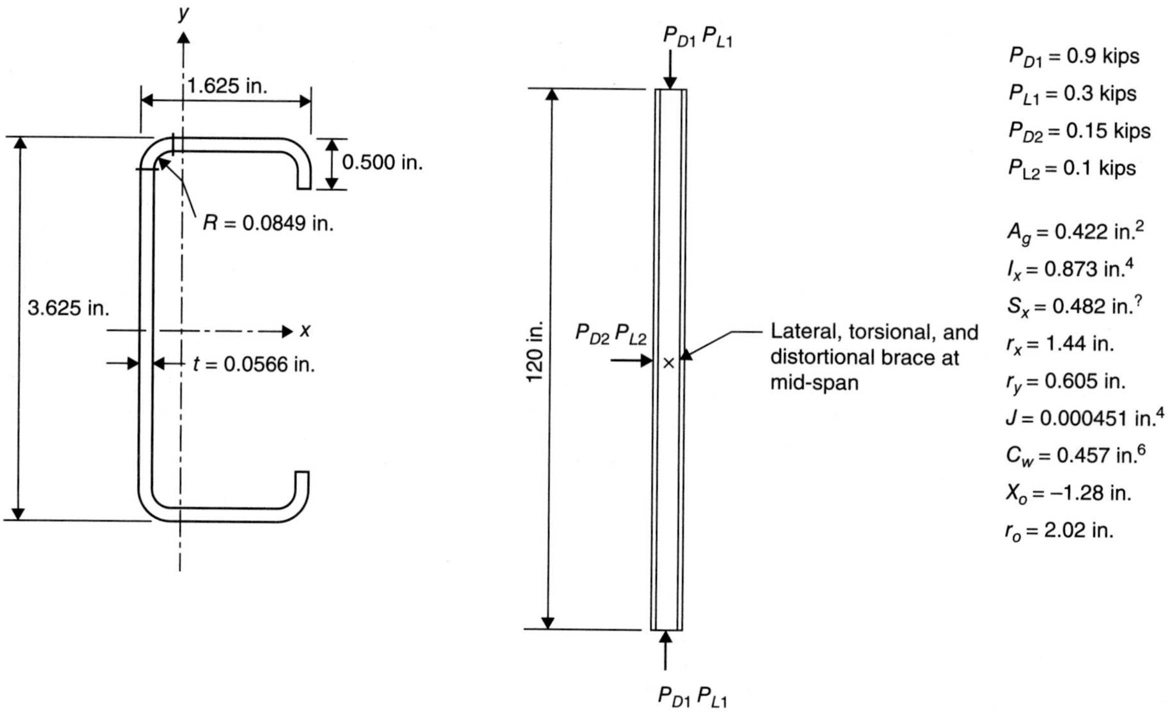

Figure 9.9 Example 1 column Section 362S162-54.

Table 9.4 Nominal Compression Strengths, P_{ne}, $P_{n\ell}$, and P_{nd} Calculation Summary

P_{ne}	$P_{n\ell}$	P_{nd}
$\lambda_c = \sqrt{\dfrac{F_y}{F_{cre}}} = 1.64 > 1.5$	$\lambda_\ell = \sqrt{\dfrac{P_{ne}}{P_{cr\ell}}} = 0.659 < 0.776$	$\lambda_d = \sqrt{\dfrac{P_y}{P_{crd}}} = 0.95 > 0.561$
$P_{ne} = A_g\left(\dfrac{0.877}{\lambda_c^2}\right)F_y = 6.86$ kips	$P_{n\ell} = P_{ne} = 6.86$ kips	$P_{nd} = \left[1 - 0.25\left(\dfrac{P_{crd}}{P_y}\right)^{0.6}\right]\left(\dfrac{P_{crd}}{P_y}\right)^{0.6} P_y = 16.5$ kips

From the values above, $P_n = 6.86$ kips, which is governed by global buckling. The available strengths are for ASD, $P_{a\_ASD} = P_n/\Omega_c =$ 3.81 kips; for LRFD, $P_{a\_LRFD} = \phi_c P_n = 5.83$ kips.

Table 9.5 Nominal Flexural Strengths, M_{ne}, $M_{n\ell}$, and M_{nd} Calculation Summary

M_{ne}	$M_{n\ell}$	M_{nd}
$2.78F_y > F_{cre} = \dfrac{M_{cre}}{S_f} = 79.4$ ksi $> 0.56F_y$	$\lambda_\ell = \sqrt{\dfrac{M_{ne}}{M_{cr\ell}}} = 0.54 < 0.776$	$\lambda_d = \sqrt{\dfrac{M_y}{M_{crd}}} = 0.703 > 0.673$
$M_{ne} = S_f \dfrac{10}{9}\left(1 - \dfrac{10F_y}{36F_{cre}}\right)F_y$ $= 22.1$ kip-in.	$M_{n\ell} = M_{ne} = 22.1$ kip-in.	$M_{nd} = \left[1 - 0.22\left(\dfrac{M_{crd}}{M_y}\right)^{0.5}\right]\left(\dfrac{M_{crd}}{M_y}\right)^{0.5} M_y$ $= 23.5$ kips

From the values above, $M_n = 22.1$ kip-in., which is governed by global buckling. The available strengths are for ASD, $M_{a\_ASD} = M_n/\Omega_b = 13.2$ kip-in.; for LRFD, $M_{a\_LRFD} = \phi_b M_n = 19.9$ kip-in.

Table 9.6 Required Strengths *P* and *M* Calculations and Interaction Check

	ASD	LRFD
Mid-span moments: $M_D = \frac{P_{D2}L}{4} = 4.5$ kip-in. $M_L = \frac{P_{L2}L}{4} = 3$ kip-in.	$M_{nt\_ASD} = M_D + M_L = 7.5$ kip-in. $P_{nt\_ASD} = P_{D1} + P_{L1} = 1.2$ kips $P_{ASD} = P_{nt\_ASD} + B_2 P_{\ell t\_ASD}$ $= 1.2$ kips	$M_{nt\_LRFD} = 1.2M_D + 1.6M_L = 10.2$ kip-in. $P_{nt\_LRFD} = 1.2P_{D1} + 1.6P_{L1}$ $= 1.56$ kips $P_{LRFD} = P_{nt\_LRFD} + B_2 P_{\ell t\_LRFD}$ $= 1.56$ kips
System Stability Calculation (AISI S100, Sec. C1): $C_m = 1$ $k_f = EI_x$; $K_1 = 1$; $L = 120$ in. $P_{e1} = \pi^2 k_f/(K_1 L)^2$ $= 17.7$ kips	$\alpha = 1.6$ $B_1 = \frac{C_m}{1-\alpha\frac{P_{ASD}}{P_{e1}}} = 1.12$ $M_{ASD} = B_1 M_{nt\_ASD} + B_2 M_{\ell t\_ASD}$ $= 8.4$ kip-in.	$\alpha = 1.0$ $B_1 = \frac{C_m}{1-\alpha\frac{P_{LRFD}}{P_{e1}}} = 1.10$ $M_{LRFD} = B_1 M_{nt\_LRFD} + B_2 M_{\ell t\_LRFD}$ $= 11.2$ kip-in.
Compression and bending check	$\frac{P_{ASD}}{P_{a\_ASD}} + \frac{M_{ASD}}{M_{a\_ASD}}$ $\frac{1.2}{3.81} + \frac{8.4}{13.2} = 0.95 < 1.0$ OK	$\frac{P_{LRFD}}{P_{a\_LRFD}} + \frac{M_{LRFD}}{M_{a\_LRFD}}$ $\frac{1.56}{5.83} + \frac{11.2}{19.9} = 0.83 < 1.0$ OK

From the interaction check results, the member strength is adequate for the applied loads.

For a complete design, the member web crippling at the pinned end supports and brace points should be checked, and the bending and web crippling interaction at the mid-span should be checked. The brace strength and stiffness should be designed per numbered list item 5 in Sec. 9.6.1.

9.7 ASSEMBLIES AND SYSTEMS

Cold-formed steel members have been used in a variety of construction applications for decades. These members are often used as part of an assembly or a system in order to attain the required strength and stiffness. It is well documented through advanced analyses and laboratory tests that the performance of cold-formed steel members in assemblies and systems surpass their performance as single elements. This section presents some of these systems and provides information on methods of analysis and design.

9.7.1 Lateral Force-Resisting Systems

Cold-formed steel lateral force-resisting systems typically fall into one of the following categories:

- Cold-formed steel light-framed shear walls with wood structural panels (plywood or oriented strand board) or steel sheet sheathing
- Cold-formed steel light-framed strap-braced wall systems
- Special bolted moment frames

A typical cold-formed steel light-framed shear wall transfers lateral loads in the plane of the wall through the sheathing and framing members and into the lower floor or foundation through the stud wall bottom track and its connection. The induced overturning forces are transferred through the vertical boundary members (end posts) to the foundation via the overturning restraint system (hold-downs) at the ends of the wall. Figure 9.10 shows a typical cold-formed steel shear wall assembly.

AISI S240 and S400 Standards (Refs. 4 and 5) recognize two basic types of cold-formed steel shear walls: Type I and Type II. A Type I shear wall is defined as a fully sheathed wall resisting in-plane forces with hold-downs at the end of each wall segment and where detailing for force transfer around the openings is provided if the wall has an opening. A Type II shear wall is defined as a wall with wood panel or steel sheet sheathing, containing multiple wall segments resisting in-plane forces, with one or more openings between wall segments and with hold-downs only at the end of the wall. There is no requirement to detail for shear transfer around openings in a Type II wall.

The tabulated shear values for cold-formed steel shear walls sheathed with wood structural panels or steel sheet sheathing are provided in the AISI S240 and AISI S400 Standards for wind load design and seismic load design, respectively. In addition to the tabulated shear values for specific assemblies, a closed-form approach, the effective strip method, can be used for calculating the nominal shear strength of Type I shear walls sheathed with steel sheet. AISI S240 and AISI S400 also provide a method for determining the deflection of Type I shear walls with wood structural panels or steel sheet sheathing. The deflection of Type I shear walls with other sheathing materials or Type II shear walls can be determined by a detailed rational analysis considering the deformation of the sheathing and its attachments.

Light-framed strap-braced walls are another commonly used lateral force-resisting system in cold-formed steel light frame construction. AISI S240 and AISI S400 Standards include design provisions for strap-braced walls for wind and seismic design, respectively. The energy dissipating mechanism of strap-braced walls is yielding of the tension straps, and the strength of the strap is determined from the expected strap yield point and the gross cross-sectional area. Strap braced walls can develop large lateral strengths by using relatively wide, thick straps. However, the response modification coefficient (R), over-strength factor (Ω_o), and deflection amplification factor (C_d) for strap-braced wall systems differ

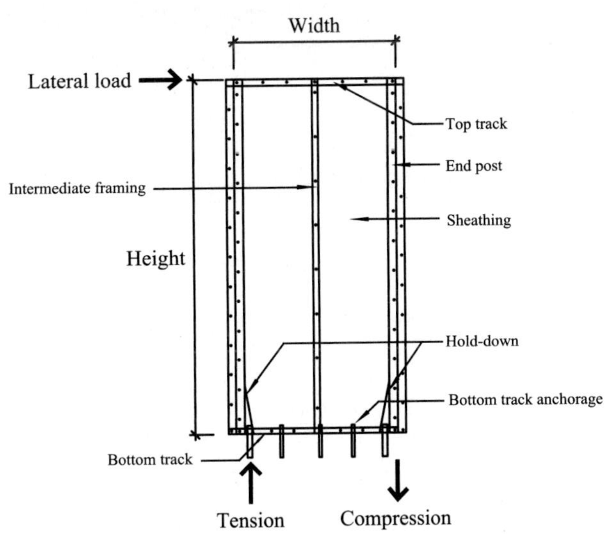

Figure 9.10 Typical Type I cold-formed steel shear wall.

significantly from those for wood structural panel or steel sheet shear walls in the ASCE 7 Standard.

For seismic design, to ensure that net section fracture of the tension straps does not occur prior to yielding of the strap gross cross section, the AISI S400 Standard places specific limitations on the strap connections. These systems are treated similarly to cold-formed steel shear walls so that the chord studs (end posts), collectors, and anchorages are required to be designed for the expected strength of the strap-braced walls. The strap connections must also be designed for the strap expected strength to ensure that the designated energy dissipating mechanism, strap yielding, can be activated prior to any other limit state being realized. The deflection of strap-braced walls can be determined from principles of mechanics with consideration of deformation of the straps, chord studs, hold-downs, and anchorages.

Cold-formed steel special–bolted moment frames (CFS–SBMF) are considered as a unique lateral force-resisting system. This type of one-story framing system uses hollow structural sections (HSS) uprights and relatively stocky C-section beams with a specially detailed bolt group at the beam-to-column connection. This specialized system is commonly used in industrial mezzanine and open-span residential structures.

CFS–SBMF systems are designed to withstand substantial inelastic deformation during significant seismic events through friction and bearing at their bolted connections. The design of CFS–SBMFs require that connections be configured so that a ductile limit state (e.g., bearing, local yielding) controls the available strength. Beams and columns are protected in order to remain elastic by capacity design principles. Drift calculations should include deflections due to the members as well as the deformation in the connections. The AISI S400 Standard includes design provisions and quality control procedures for the CFS–SBMF system.

9.7.2 Metal Building Systems

Metal building systems have evolved over the years into assemblies of structural elements that work together as a structural system. The basic elements of the metal building systems are primary rigid frames (structural steel members), secondary members composed of wall girts and roof purlins (cold-formed steel members), and metal roof and wall cladding systems (see Fig. 9.11).

Chapter I in the AISI S100 Standard includes design information on assemblies for metal building systems such as axial and flexural capacities of cold-formed C and Z shapes with cold-formed steel sheathing attached to one flange. The capacities provided are empirical and provide a single strength value that envelopes all of the pertinent flexural, flexural-torsional, lateral-torsional, distortional, and local buckling limit states. These capacities have been developed through testing performed by the Metal Building Manufacturers Association (MBMA). MBMA publishes a design manual, the Metal Building Systems Manual (Ref. 9), which includes topics for metal buildings such as load application, crane loads, drift and deflection criteria, fire ratings information, and energy conservation. The manual also includes material on common industry practices, guide specifications, and climatological data for the United States.

9.7.3 Industrial Storage Racks

Industrial steel storage racks were developed to store palletized, containerized, or large unit loads usually placed in the rack by mechanical handling equipment. The height of a rack structure is limited only by the equipment used to store or retrieve the material. The design criteria for determining the structural performance of steel rack structures are contained in the Rack Manufacturers Institute *Specification for the Design, Testing, and Utilization of Industrial Steel Storage Racks* (Ref. 10).

9.7.4 Trusses

Cold-formed steel trusses are made up of either standard C-sections or proprietary members assembled together directly or through gusset plates with screws, bolts, or welds. Trusses are typically predesigned and preengineered into customized roof shapes. Trusses are also typically prefabricated to accelerate the construction process and reduce on-site labor costs. The design of cold-formed steel truss members shall follow the AISI S100 Standard, while the AISI S240 Standard provides additional provisions on the design, quality assurance, installation, and testing of the cold-formed steel truss assemblies.

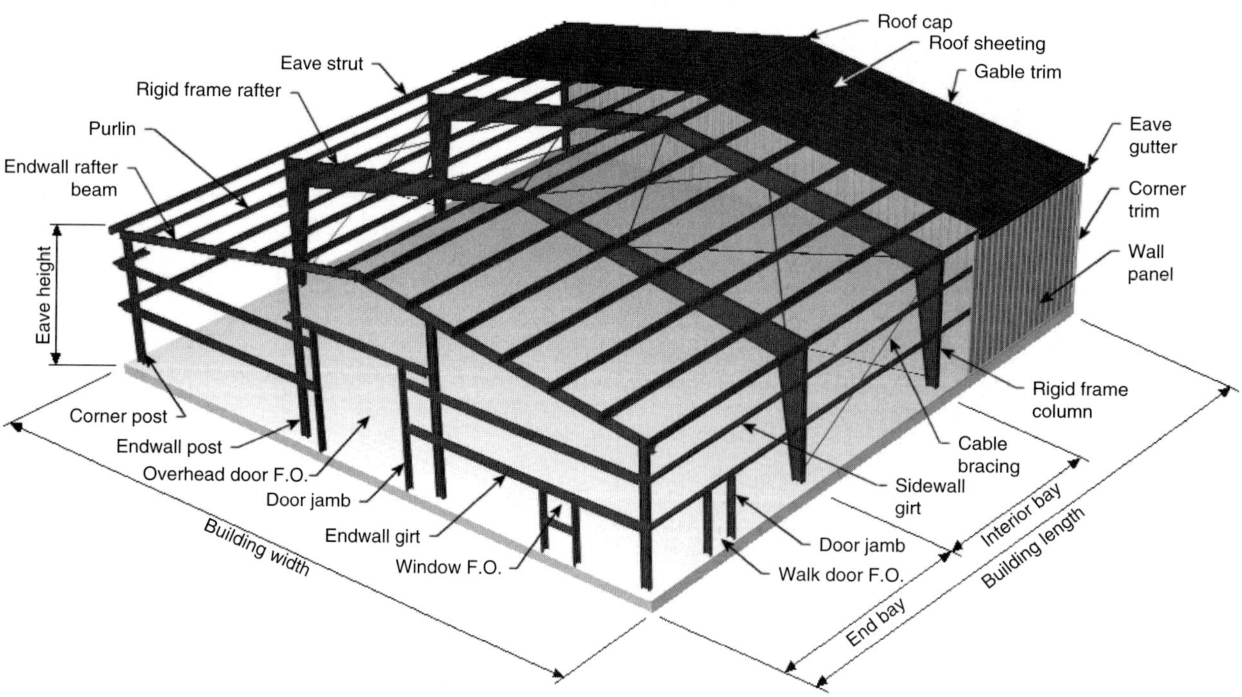

Figure 9.11 Anatomy of a metal building.

9.8 CONNECTIONS

Conventional methods of attachment, that is, fusion welds, resistance welds, and bolts, find application in cold-formed steel construction. However, because of the relatively small thickness of the members, other forms of joining, such as screws, power-actuated fasteners (PAFs), staples, adhesives, and seams formed by mechanically engaging the members, are also commonly found.

9.8.1 Welded Connections

Fusion welds are used for on-site and off-site welding to connect cold-formed steel to cold-formed steel or cold-formed steel to hot-rolled steel members. Welding of thin sheet, which is often galvanized, requires welding shapes and welding techniques that are often different from those used for welding thicker sheet, plate, and shapes. Arc welding or resistance welding may be used to connect thin sheet steel, and in the building construction industry, the arc welding process is by far the most common. Arc welding is the process of fusing material together by an electric arc, usually with the addition of weld filler metal. Resistance welds are commonly used for connecting thin sheet steels in the automotive or appliance industries.

The most common welding types to connect framing members are the fillet weld and the flare groove weld. Arc spot welds, also called puddle welds, are the standard method of attaching a floor or roof deck to structural framing. The welds are made by burning through the deck and then filling the hole with weld material. This gives the same type of connection as a plug weld in structural steel members, where the hole is predrilled or prepunched. Groove welds in butt joints are commonly used during the roll-forming process to connect the flat sheet of one coil to the next coil.

The design of welded connections for cold-formed steel construction is governed by Section J2 of the AISI S100 Standard, and the Structural Welding Code-Sheet Steel, AWS D1.3 (Ref. 11). The AISI S100 Standard includes requirements for groove welds, arc spot welds (puddle welds), arc seam welds, fillet welds, flare groove welds, and plug welds. The AWS D1.3 welding code contains detailed provisions on welding procedures, control, and inspection and should be consulted on all matters concerning fusion welding of cold-formed steel members.

9.8.2 Bolted Connections

Bolted connections in cold-formed steel construction may behave somewhat differently from bolted connections in heavier construction. The chief reason for this lies in the relative dimensions of sheets and bolts, the sheet thickness generally being a small fraction of the bolt diameter. Testing has shown that bolted cold-formed steel connections are susceptible to any of the following four failure modes:

1. Shearing of the material parallel to the direction of load.
2. Tearing of the sheet perpendicular to the direction of loading. A failure of this type is commonly referred to as a net section failure.
3. Bearing of the bolt on the sheet.
4. Shear and tension on the bolt.

Unlike connections in heavy construction, the capacity of a cold-formed steel bolted connection may be enhanced significantly by the use of washers under both the bolt head and nut. Section J3 of the AISI S100 Standard contains specific design provisions for avoiding each of the above failure modes. The AISI S100 Standard also provides specific design provisions for bolted connections when oversized or slotted holes are present. Minimum spacing between bolts is typically $3d$, where d is the nominal bolt diameter, and minimum distance from the center of the bolt to any edge or end of any connected part is $1.5d$. For oversized and slotted holes, the minimum clear distance between the edges of two adjacent holes is $2d$, and the minimum clear distance to an edge or end of any connected part is d.

9.8.3 Screw Connections

Screws are the most commonly used fasteners in cold-formed steel construction. Holes are not typically drilled in steel framing before installing the screws. Therefore, self-drilling, self-tapping screws are the most common fasteners used to frame steel members. Screws are available in diameters ranging from No. 6 to No. 14, with No. 8 to No. 12 being the most common. Lengths typically vary from ½ in. (13 mm) to as much as 3 in. (76 mm) depending on the application. Screws are generally ⅜ in. (10 mm) to ½ in. (13 mm) longer than the thickness of the connected materials so that a minimum of three threads shall extend beyond the connected material. It is important that the drill point be at least as long as the material thickness being fastened together to complete drilling of the hole prior to engagement of the threads. The correct fastener type and length for each application should be selected by consulting the screw manufacturer's technical specifications.

The screw connection requirements are fairly straightforward: capacity depends upon the thickness and tensile strength of the sheet steel, and the diameter of the screw. Design provisions of screw connections in cold-formed steel are included in Section J4 of the AISI S100 Standard. Minimum spacing between screws is typically $3d$, where d is the nominal screw diameter, and minimum distance from the center of the screw to any edge or end of any connected part is $1.5d$. The AISI S100 Standard identifies three limit states for a screw connection subjected to a shear force: bearing, tilting, and screw body shear failure. The bearing limit state generally refers to failure in the connected element/component resulting from local deformation at the loaded face of the fastener. The tilting limit state involves rotation of the screw followed by threads tearing out of the lower sheet, which reduces the connection shear capacity. The screw body shear limit state describes the mechanical strength of the screw itself in shear. For a screw connection subjected to a tension force, three limit states are also identified: pullout, pullover, and screw body tension failure. Under tension, a screw connection may fail by having the screw threads pulling out from the member not in contact with the screw head, by having the steel sheet in contact with the screw head deforming and pulling over the screw head, or by tensile facture of the screw itself. Combined shear and tension on screw connections shall also be checked if both loads are applied simultaneously. The shear and tensile strengths of the screw fastener itself are usually established and published by the screw manufacturer. When this is not the case, the AISI S100 Standard applies a 25 percent adjustment to the safety and resistance factors when the shear and tensile strengths need to be determined by testing. Figure 9.12 shows sample failure modes for screw connections.

9.8.4 Power-Actuated Fastener Connections

Power-actuated fastening is a commonly used method to attach cold-formed steel members to other substrates such as structural steel,

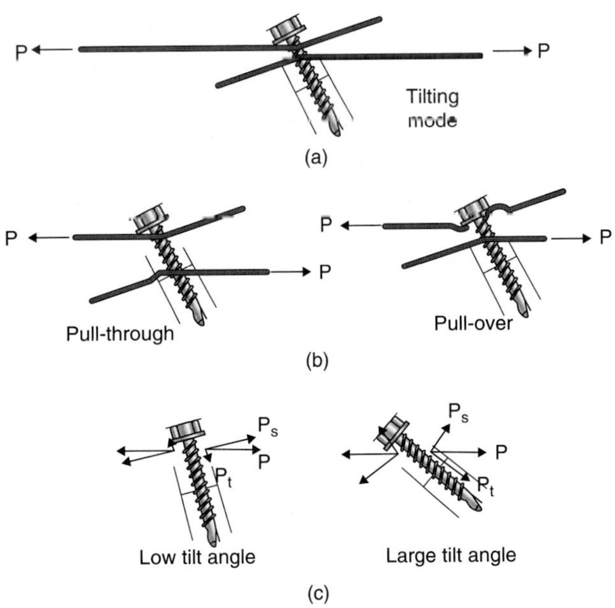

Figure 9.12 Failure modes for screw connections.

concrete, or masonry. Custom-designed PAFs are also used to fasten wood structural panels to cold-formed steel in shear wall assemblies. Although not typically employed for cold-formed steel to cold-formed steel connections, power-driven pins are also excellent for attachment of top and bottom runner tracks to slabs and beams for support of nonstructural walls and ceilings. Note that in certain seismic applications, the use of power-driven pins (and some other power-driven fasteners) in applications loaded in tension is not permitted by code. Section J5 of the AISI S100 Standard provides design provisions to address steel-to-steel connections with PAFs with specified limitations. Minimum spacing between PAFs and minimum edge distances must be met to use these design provisions. The limit states in a shear connection include sheet steel bearing, PAF tilting, PAF pullout in shear, and shear failure in the PAF itself. The design of a tension connection shall consider pullout, pullover, and tensile fracture of PAFs. The AISI S100 Standard also provides design provisions to address the combined shear and tension effect on PAF connections.

9.8.5 Other Connections

Other connection methods, such as clinching, rivets, etc., are also used in cold-formed steel construction. However, the behavior of these connections is less well understood compared to the connection types discussed above. The AISI S100 Standard does not explicitly address these special connections.

Clinching is a fastener-less connection that resists load by the interlocking of one sheet of steel into another. Crimping is another fastener-less connection but is typically only used for nonstructural drywall framing and has a relatively small connection capacity. Riveting of cold-formed steel can also be seen in roof truss connections. Rivets are pressed into both layers of sheet steel, and the legs of the rivet spread out to develop the connection capacity.

In addition to these connections, specific connectors have been developed for specific applications. Some of the most common are deflection clip connectors and deflection tracks used in non-load-bearing cold-formed steel framing to accommodate gravity and lateral deflections of the floors and roof without applying corresponding loads in the framing members. Some specialty product manufacturers have cold-formed steel-specific connectors such as hold-downs, bypass deflection connectors, joist hangers, and truss connectors. Most of these connectors use screw, bolt, or welded connections, and have connection-specific capacities listed in the technical data published by the manufacturers. A design method based on laboratory tests for clip angles used in cold-formed steel construction has been developed and presented in the AISI Research Report RP15-2 (Ref. 12).

Design Example 2

A C-shaped cold-formed steel member is connected to structural steel support using a cold-formed steel clip angle as shown in Fig. 9.13. This connection is designed to resist an in-plane vertical shear load P = 350 lb (ASD) or P = 560 lb (LRFD). In this example, screws are used at the cold-formed steel member-to-clip angle connection, while the clip angle-to-structural steel connection is checked for both screws and bolts. Spacing and edge/end distances of screws and bolts used are assumed to meet the minimum requirements of the AISI S100 Standard.

The forces applied are assumed to act concentrically with the screw group of the cold-formed member-to-clip angle connection. For the screw group (or bolt group) of the clip angle-to-steel connection, there is an eccentricity between shear force P and the screw (bolt) line. The eccentricity is assumed to be 1 in. The two connections are checked separately in this example.

Material Properties

Cold-formed steel member: 33 mil, design thickness = 0.0346 in., $F_y = 33$ ksi, $F_u = 45$ ksi

Clip angle: 33 mil, design thickness = 0.0346 in., $F_y = 33$ ksi, $F_u = 45$ ksi

Structural steel member: minimum thickness = 3/16 in., minimum, $F_u = 58$ ksi

Screws: size #10, nominal diameter, $d = 0.19$ in.

Bolts: ASTM A307, nominal diameter, $d = 0.25$ in., washer diameter $d_w = 0.625$ in.

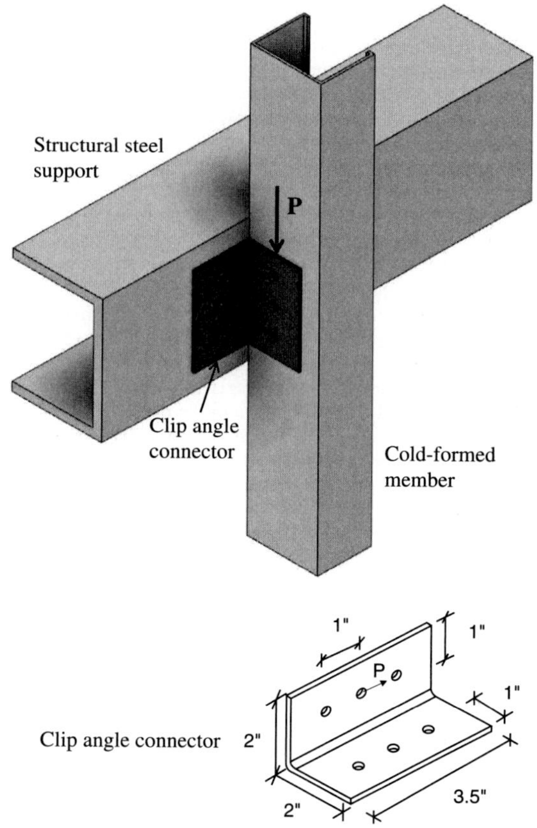

Figure 9.13 Example 2 connection design.

Cold-Formed Steel Member-to-Clip Angle Connection Assume P is applied in the plane of the vertical leg of the clip angle and is concentric with the screw group. The shear strength is governed by two limit states, tilting and bearing, and shear in the screws, per Section J4 of the AISI S100 Standard. Assume screws are driven from clip angle into the member.

Shear Strength Limited by Tilting and Bearing

Since $\frac{t_2}{t_1} \le 1$, P_{ns} shall be taken as the smallest of equations:

$$P_{ns} = 4.2(t_2^3 d)^{1/2} F_{u2} = 4.2(0.0346^3 \times 0.190)^{1/2}(45{,}000) = 530 \text{ lb}$$

$$P_{ns} = 2.7 t_1 d F_{u1} = 2.7 \times 0.0346 \times 0.190 \times 45{,}000 = 799 \text{ lb}$$

$$P_{ns} = 2.7 t_2 d F_{u2} = 2.7 \times 0.0346 \times 0.190 \times 45{,}000 = 799 \text{ lb}$$

Screw tilting governs.

$$P_{ns} = 530 \text{ lb}$$

For ASD, $\Omega = 3.00$

For LRFD, $\phi = 0.50$

Assume three screws are used at 1-in. spacing and check the shear force in each screw:

ASD

$$V = \frac{P}{3} = \frac{350}{3} = 117 \text{ lb}$$

$$V < \frac{P_{ns}}{\Omega} = \frac{530}{3} = 177 \text{ lb} \qquad \text{O.K.}$$

LRFD

$$V = \frac{P}{3} = \frac{560}{3} = 187 \text{ lb}$$

$$V < \phi P_{ns} = (0.5)(530) = 265 \text{ lb} \qquad \text{O.K.}$$

The screws have adequate shear strength in tilting.

Shear in Screws The nominal shear strength of the screws P_{ss} shall be reported by the screw manufacturer or determined by independent laboratory testing. In this example, according to the manufacturer's report, P_{ss} = 1200 lb. Check shear force in each screw:

ASD

$$V = 117 \text{ lb} < \frac{P_{ss}}{\Omega} = \frac{1200}{3} = 400 \text{ lb} \qquad \text{O.K.}$$

LRFD

$$V = 187 \text{ lb} < \phi P_{ss} = (0.5)(1200) = 600 \text{ lb} \qquad \text{O.K.}$$

Overall, the connection between the cold-formed steel member and the clip angle has adequate shear strength.

Clip Angle-to-Steel Support Connection The clip angle-to-steel support connection is in shear under in-plane shear force *P*. Both screw and bolted connections to be checked.

Screws Assume three screws are used at 1-in. spacing. The shear force in this connection is caused by in-plane shear. As discussed above, the load *P* has an eccentricity from the screw or bolt group of this connection (*e*) = 1 in. The moment caused by this eccentricity, *M* = *P e* is assumed to be resisted only by the two outer screws. These two screws will also get additional shear force perpendicular to the load *P*. The spacing of the screws is 1 in., so the distance between two outer screws (a) is 2 in. Total shear force in one of these screws is

ASD

$$V = \sqrt{\left(\frac{P}{3}\right)^2 + \left(\frac{P\,e}{a}\right)^2} = \sqrt{\left(\frac{350}{3}\right)^2 + \left(\frac{350 \times 1}{2}\right)^2} = 210 \text{ lb}$$

LRFD

$$V = \sqrt{\left(\frac{P}{3}\right)^2 + \left(\frac{P\,e}{a}\right)^2} = \sqrt{\left(\frac{560}{3}\right)^2 + \left(\frac{560 \times 1}{2}\right)^2} = 337 \text{ lb}$$

Shear Strength Limited by Bearing Since the screws are driven into hot-rolled steel, the rigidity and thickness of the hot-rolled connection prevents the screws from tilting so that the limit state of screw tilting does not apply. Only screw bearing into clip angle material is applicable:

$$P_{ns} = 2.7 t_1 d F_{u1} = 2.7 \times 0.0346 \times 0.190 \times 45{,}000 = 799 \text{ lb}$$

ASD

$$V = 210 \text{ lb} < \frac{P_{ns}}{\Omega} = \frac{799}{3} = 266 \text{ lb} \qquad \text{O.K.}$$

LRFD

$$V = 337 \text{ lb} < \phi P_{ns} = (0.5)(799) = 399 \text{ lb} \qquad \text{O.K.}$$

The screws have adequate shear strength in bearing.

Shear in Screws

ASD

$$V = 210 \text{ lb} < \frac{P_{ss}}{\Omega} = \frac{1200}{3} = 400 \text{ lb} \qquad \text{O.K.}$$

LRFD

$$V = 337 \text{ lb} < \phi P_{ss} = (0.5)(1200) = 600 \text{ lb} \qquad \text{O.K.}$$

The screws have adequate strength in shear.

Bolts For bolted connection, three 0.25-in.-diameter ASTM A307 bolts are placed at 1.0 in. on center apart. The bolt strength is checked per Section J3 of the AISI S100 Standard.

Bearing Strength with Consideration of Bolt Hole Deformation Assuming bolt hole deformation is a design consideration, the bolt hole bearing strength is

$$P_n = (4.64\alpha t + 1.53) d t F_u$$

and $\Omega = 2.22$, $\phi = 0.65$

$$P_n = (4.64 \times 0.0346 + 1.53)(0.25)(0.0346)(45{,}000) = 658 \text{ lb}$$

As discussed above for the screw connection, the eccentricity of in-plane force *P* will cause additional shear in the bolts. The total shear force in one bolt is the same as previously calculated for each screw:

ASD

$$V = 210 \text{ lb} < \frac{P_n}{\Omega} = \frac{658}{2.22} = 296 \text{ lb} \qquad \text{O.K.}$$

LRFD

$$V = 337 \text{ lb} < \phi P_n = (0.65)(658) = 428 \text{ lb} \qquad \text{O.K.}$$

The bolts have adequate shear strength in bearing.

Shear in Bolts

$$P_n = A_b F_n$$

and $\Omega = 2.00$, $\phi = 0.75$

For ASTM A307 bolts with $\frac{1}{4}$ in. $\le d < \frac{1}{2}$ in.:

$$F_{nv} = 27 \text{ ksi}$$

$$A_b = \frac{\pi d^2}{4} = \frac{\pi (0.25)^2}{4} = 0.0491 \text{ in.}^2$$

Shear strength:

$$P_{nv} = A_b F_{nv} = 0.0491 \times 27{,}000 = 1326 \text{ lb}$$

ASD

$$V = 210 \text{ lb} < \frac{P_{nv}}{\Omega} = \frac{1326}{2.00} = 663 \text{ lb} \qquad \text{O.K.}$$

LRFD

$$V = 337 \text{ lb} < \phi P_{nv} = (0.75)(1326) = 995 \text{ lb} \qquad \text{O.K.}$$

The bolts have adequate shear strength. Overall, the connection between the clip angle and the structural steel support has adequate shear strength.

Clip Angle Strength Check The clip angle is in shear under in-plane shear force *P*. AISI Research Report RP15-2 (Ref. 12) design methods are used to check the shear strength of the clip angle. In AISI RP15-2, *L* is defined as the flat width distance between the center of screw line and the bend line. In this example, *L* is conservatively set to be the distance between the center of the screw line and the outer plane of the other leg; *B* = 3.5 in., *L* = 1.0 in.

Shear Strength (AISI RP15-2)

Without Consideration of Clip Angle Deformation

$$V_n = 0.17\lambda^{-0.8} F_y B t \le 0.35 F_y B t$$

where

$$k = 2.569\left(\frac{L}{B}\right)^{-2.202} = 2.569\left(\frac{1}{3.5}\right)^{-2.202} = 40.53$$

$$F_{cr} = \frac{k\pi^2 E}{12(1-\mu^2)}\left(\frac{t}{B}\right)^2 = \frac{40.53\pi^2(29{,}500)}{12(1-0.3^2)}\left(\frac{0.0346}{3.5}\right)^2 = 105.6 \text{ ksi}$$

$$\lambda = \sqrt{\frac{F_y}{F_{cr}}} = \sqrt{\frac{33}{105.6}} = 0.559$$

So

$$V_n = 0.17(0.559)^{-0.8}(33{,}000)(3.5)(0.0346) = 1082 \text{ lb}$$

$$< 0.35F_y Bt = (0.35)(33{,}000)(3.5)(0.0346) = 1399 \text{ lb}$$

$$\Omega = 2.80,\ \phi = 0.55$$

ASD

$$V = 350 \text{ lb} < \frac{V_n}{\Omega} = \frac{1082}{2.80} = 386 \text{ lb} \qquad \text{O.K.}$$

LRFD

$$V = 560 \text{ lb} < \phi P_n = (0.55)(1082) = 595 \text{ lb} \qquad \text{O.K.}$$

With Consideration of Clip Angle Deformation
Since

$$9000\alpha\left(\frac{Bt}{L}\right) = 9000 \times 1 \times \left(\frac{3.5 \times 0.0346}{1}\right) = 1090 \text{ lb} > 1082 \text{ lb}$$

So $V_n = 1082$ lb

$$\Omega = 3.00,\ \phi = 0.55$$

ASD

$$V = 350 \text{ lb} < \frac{V_n}{\Omega} = \frac{1082}{3.00} = 361 \text{ lb} \qquad \text{O.K.}$$

LRFD

$$V = 560 \text{ lb} < \phi P_n = (0.55)(1082) = 595 \text{ lb} \qquad \text{O.K.}$$

The clip angle has adequate shear strength. Overall, the connection is adequate under in-plane shear force *P*.

REFERENCES

1. AISI (2018), AISI D100-17, *Cold-Formed Steel Design Manual*, American Iron and Steel Institute, Washington, D.C.
2. AISI (2016a), AISI S100-16, *North American Specification for the Design of Cold-Formed Steel Structural Members*, American Iron and Steel Institute, Washington, D.C.
3. ASCE (2002), ASCE 8-02, *Standard Specification for the Design of Cold-Formed Stainless Steel Structural Members*, American Society of Civil Engineers, New York.
4. AISI (2015), AISI S240-15, *North American Standard for Cold-Formed Steel Structural Framing*, American Iron and Steel Institute, Washington, D.C.
5. AISI (2016b), AISI S400-15/S1-16, *North American Standard for Seismic Design of Cold-Formed Steel Structural Systems with Supplement 1*, American Iron and Steel Institute, Washington, D.C.
6. Yu, W. W. and R. A. LaBoube and Helen H. Chen, *Cold-Formed Steel Design*, Fifth Edition, Wiley-Interscience, New York
7. ASCE (2016), ASCE 7-16, *Minimum Design Loads for Buildings and Other Structures*, American Society of Civil Engineers, Virginia.
8. AISI (2006), CF06-1, *Direct Strength Method (DSM) Design Guide*, American Iron and Steel Institute, Washington, D.C.
9. MBMA (2018), *Metal Building Systems Manual*, 2018 Edition, Metal Building Manufacturers Association, Cleveland, OH.
10. RMI (2012), *Specification for the Design, Testing, and Utilization of Industrial Steel Storage Racks*, MH16.1: 2012, Rack Manufacturers Institute, Charlotte, NC.
11. AWS D1.3 (2018), *AWD D1.3/D1.3M Structural Welding Code-Sheet Steel*, American Welding Society, Miami, FL.
12. AISI (2015), AISI RP15-2, *Load Bearing Clip Angle Design*, Research Report RP15-2, American Iron and Steel Institute, Washington, D.C.

Chapter 10
Aluminum Structural Design

BY
J. RANDOLPH KISSELL, PE

10.1 INTRODUCTION

Aluminum has several advantages for structural applications: it has a high strength-to-weight ratio, is easily formed into custom cross sections by extruding, has good corrosion resistance in most outdoor environments without coatings, and becomes more ductile at temperatures below room temperature.

10.1.1 Design Specifications

The *International Building Code* (IBC) Chapter 20 requires that aluminum structures comply with the *Specification for Aluminum Structures* (SAS). This *Specification* is Part I of the *Aluminum Design Manual* (ADM), published by the Aluminum Association (www.aluminum.org) and revised every five years, most recently in 2015. Other parts of the ADM provide a commentary to the *Specification*, section properties for common aluminum structural shapes (channels, I beams, angles, tees, zees, pipe, and round and rectangular tube), design aid tables (including buckling constants and allowable stress tables for 24 common alloy-tempers, welded and unwelded), and 31 illustrative examples demonstrating use of the *Specification*.

The AASHTO *LRFD Bridge Design Specifications* and the AASHTO *LRFD Specifications for Structural Supports for Highway Signs, Luminaires, and Traffic Signals* also address aluminum structural components. The equations for the nominal strengths of aluminum members and connections in these AASHTO specifications are very similar to those in the SAS. The strength equations in this chapter are from the SAS.

10.1.2 Material

Alloys

For structural applications, aluminum is alloyed with other elements to improve its strength. The Aluminum Association assigns a four-digit number for wrought alloys (those that are worked into shape and listed in Table 10.1) and a three-digit number, with an optional prefix letter, for cast alloys (those that are solidified from a molten state in a mold). Wrought alloys are generally more ductile and have fewer discontinuities than castings and thus are more widely used in structural applications. Aluminum alloys have greater strength but less corrosion resistance than commercially pure aluminum (the 1xxx series).

The scope of the SAS includes 27 wrought alloys.

Tempers

The strength of aluminum alloys is increased by tempering, designated by a suffix on the alloy designation. Before tempering, material is in the annealed condition, designated by the O suffix. Tempering is achieved by strain hardening (designated by an -H suffix; for example, 5052-H32) or heat treatment (designated by a -T suffix; for example, 6061-T6). All alloys can be strengthened by strain hardening, but only heat-treatable alloys (2xxx, 6xxx, and 7xxx alloys) can be strengthened by heat treatment. Heat treatment is a two-step process: first, solution heat treatment (for example, a -T4 temper) and then precipitation heat treatment (for example, a -T6 temper), also called artificial aging. Both strain hardening and heat treatment increase strength but decrease ductility.

ANSI H35.1 defines the Aluminum Association alloy and temper designation systems, which are used internationally.

Products

Aluminum products used in structural applications include wrought products such as sheet, plate, extrusions, and forgings, and castings made by the sand casting and permanent mold casting processes. Table 10.2 shows the ASTM specifications for aluminum products commonly used in structures. The SAS requires that material comply with these specifications.

Table 10.1 Wrought Alloys

Alloy	Primary alloying element	Corrosion resistance	Strength	Heat-treatable	Example
1xxx	None	Excellent	Low	No	1100
2xxx	Copper	Fair	High	Yes	2219
3xxx	Manganese	Good	Medium	No	3003
4xxx	Silicon	Good	Low		4043
5xxx	Magnesium	Good	Medium	No	5052
6xxx	Magnesium and silicon	Good	Medium	Yes	6061
7xxx	Zinc	Fair	High	Yes	7005

Table 10.2 ASTM Aluminum Product Specifications

ASTM specification	Product
B26	Sand castings
B108	Permanent mold castings
B209	Sheet and plate
B210	Drawn tube
B211	Bar, rod, and wire
B221	Extrusions
B241	Seamless pipe and seamless tube
B247	Forgings
B308	Standard structural shapes
B429	Extruded pipe and tube
B632	Rolled tread plate
B928	High-magnesium aluminum alloy sheet and plate used in corrosive environments

An ASTM aluminum product specification typically addresses multiple alloys and tempers, and all include requirements for chemical composition, dimensional tolerances, and minimum mechanical properties (usually tensile yield strength F_{ty}, tensile ultimate strength F_{tu}, and elongation). Some ASTM specifications include corrosion resistance requirements. Much of the information in these ASTM specifications for wrought products is also given in *Aluminum Standards and Data*, published by the Aluminum Association.

Properties

Material properties at room temperature are given in Table 10.3. Strengths are expressed in terms of the tensile yield strength F_{ty} and tensile ultimate strength F_{tu} since F_{ty} and F_{tu} are usually stated in product specifications.

Table 10.3 Material Properties

Property	Symbol	Value
Poisson's ratio	ν	0.33
Modulus of elasticity	E	10,100 ksi = 70,000 MPa
Shear modulus of elasticity	G	3800 ksi = 26,000 MPa
Coefficient of thermal expansion	α	13×10^{-6}/°F = 23×10^{-6}/°C
Density	γ	0.10 lb/in.^{3} = 2.7×10^3 kg/m^3
Shear yield strength	F_{sy}	0.6 F_{ty}
Shear ultimate strength	F_{su}	0.6 F_{tu}
Compressive yield strength		
Unwelded -H temper material	F_{cy}	0.9 F_{ty}
All other material	F_{cy}	F_{ty}

Density, modulus of elasticity, and coefficient of thermal expansion vary slightly by several factors, including alloy and temperature. More precise values for density and coefficient of thermal expansion are given in the ADM Part IV.

10.1.3 Effect of Temperature on Properties

Aluminum alloys that have been strain hardened or precipitation heat-treated lose strength when heated above about 200°F. The loss of strength is proportional to temperature and the cumulative time at elevated temperature. Heating to 650°F anneals most alloys, and melting occurs around 1100°F. Therefore, welding or fire exposure significantly reduces the strength of strain hardened or precipitation heat-treated aluminum.

Below room temperature, aluminum gains strength and ductility, unlike steel. For this reason, aluminum is used in cryogenic applications.

10.1.4 Welding

Welding reduces the strength of non-heat-treatable alloys to the annealed strength, and the strength of heat-treatable alloys to slightly less than the solution heat-treated strength. Yield strength (F_{tyw}) and ultimate strength (F_{tuw}) of weld-affected zones are given in the *Specification* Table A.3.3. The *Specification* designates the cross-sectional area of the weld-affected zone as A_{wz} and defines this zone as extending 1 in. from the centerline of a groove weld and the heel of a fillet weld. The actual extent of the weld-affected zone is a function of many factors, including welding process, travel speed, amperage, and material thickness.

10.2 DESIGN OVERVIEW

10.2.1 Design Methods

The SAS provides two aluminum structural design methods: allowable strength design (ASD) and load and resistance factor design (LRFD). Both require that the available strength equals or exceeds the required strength. The available and required strengths are determined differently for the two methods and depend on the type of structure being designed (for example, building-type structure or bridge-type structure). Both methods, however, use the same equations to determine the nominal strengths, which are factored to determine the available strength.

In ASD, the required strength is determined using the nominal loads, and the available strength is the nominal strength R_n divided by a safety factor Ω, which depends on the limit state. In LRFD, the required strength is determined using factored nominal loads, and the available strength is the nominal strength R_n multiplied by a resistance factor ϕ, which depends on the limit state. Safety and resistance factors for building-type structures are given in Table 10.4. Load factors are provided in codes (such as building codes) for specific structure types.

10.2.2 Buckling

The *Specification* uses the Euler buckling stress π^2E/λ^2 (where λ is the slenderness of the member or cross-sectional element) for elastic buckling, and a straight line for inelastic buckling.

For inelastic buckling, the *Specification* provides three buckling constants: B, the inelastic buckling stress for slenderness = 0; D, the slope of the inelastic buckling stress as a function of slenderness; and C, the slenderness at which the inelastic buckling stress equals the elastic buckling stress. The *Specification* provides different buckling constants for axial compression, flexural compression, and shear and different buckling constants for material that is precipitation heat-treated (in *Specification* Table B.4.2) and material that is not precipitation heat-treated (in *Specification* Table B.4.1). Buckling constants are a function of the yield strength (which varies by alloy and temper) and the modulus of elasticity (which does not vary significantly by alloy or temper) and are tabulated for various alloy-tempers in ADM Part VI Table 1-1 for unwelded material and Table 1-2 for weld-affected material.

10.3 DETERMINING REQUIRED FORCES

Because aluminum's modulus of elasticity is less than steel's, aluminum structures are more flexible than steel structures. Consequently, the stiffness of aluminum structures may change significantly as the load increases. Thus, determining the required forces and moments in aluminum structures must address potential nonlinear behavior. The *Specification* requires that the required strengths of members and connections be determined from an elastic analysis that considers the following:

1. Flexural, shear, and axial deformations, including all member and connection deformations that contribute to displacements of the structure.
2. Second-order effects, including P-Δ effects (the effect of loads acting on the displaced location of joints in a structure) and P-δ effects (the effect of loads acting on the deflected shape of a member between joints).
3. Geometric imperfections. The effect of geometric imperfections on the stability of the structure is accounted for by analyzing the structure with the joints displaced from their nominal locations by the erection tolerances.

Table 10.4 Safety and Resistance Factors for Building-Type Structures

Limit state	ϕ	Ω	SAS section
Members:			
Axial tension: rupture	0.75	1.95	D.1
Axial tension: yield	0.90	1.65	D.1
Axial compression	0.90	1.65	E.1
Flexure: rupture	0.75	1.95	F.1
Flexure: other limit states	0.90	1.65	F.1
Shear: rupture	0.75	1.95	G.1
Shear: other limit states	0.90	1.65	G.1
Torsion: rupture	0.75	1.95	H.2
Torsion: other limit states	0.90	1.65	H.2
Connections:			
Welds	0.75	1.95	J.2
Bolt tension	0.65	2.34	J.3.4
Bearing of bolts or rivets	0.75	1.95	J.3.6, J.4.6
Slip-critical bolt shear and tension	0.75	2.00	J.3.7.4
Slip-critical bolt slip (standard size and short-slotted holes perpendicular to the direction of the load)	1.00	1.50	J.3.7.5
Slip-critical bolt slip (oversized and short-slotted holes parallel to the direction of the load)	0.85	1.76	J.3.7.5
Slip-critical bolt slip (long-slotted holes)	0.70	2.14	J.3.7.5
Bolt shear; rivet shear	0.65	2.34	J.3.5, J.4.5
Screwed connection tension	0.50	3.00	J.5.4
Screwed connection shear	0.50	3.00	J.5.5
Pin bearing	0.75	1.95	J.6.5
Pin rupture from shear or flexure	0.65	2.34	J.6.4
Pin yielding from shear or flexure	0.90	1.65	J.6.4
Shear yielding of connectors	1.00	1.50	J.7.2a
Shear rupture of connectors	0.75	1.95	J.7.2b
Block shear rupture	0.75	1.95	J.7.3
Bearing on flat surfaces	0.75	1.95	J.8
Web crippling	0.75	1.95	J.9.1
Braces	0.75	2.00*	6.1

*3.0 for nodal torsional bracing of beams.

4. Member stiffness reduction due to inelasticity. This is accounted for by using a stiffness reduction factor τ_b given in the *Specification* that depends on the ratio of the axial compression force in a member to its yield strength.

5. Uncertainty in stiffness and strength, addressed by applying a factor of 0.8 to all axial, shear, and flexural stiffnesses in the structure. This can be achieved by using 80 percent of 10,100 ksi as the modulus of elasticity E for the analysis.

The analysis may be conducted for either of the following:

a) The LRFD load combinations with the results used directly to obtain the required strengths or

b) 1.6 times the ASD load combinations with the results divided by 1.6 to obtain the required strengths.

Design for seismic loads is treated differently in design codes than design for other loads, such as dead, live, wind, and snow loads, in that inelastic deformation is allowed for seismic loads. To use elastic analysis for inelastic behavior, seismic loads are divided by a seismic response modification coefficient R that is greater than 1. These coefficients are listed in ASCE 7, *Minimum Design Loads and Associated Criteria for Buildings and Other Structures*, for seismic force-resisting systems of various materials and that resist lateral force in different ways, such as braced frames or shear walls. ASCE 7 does not, however, provide seismic response modification coefficients for aluminum seismic force-resisting systems. Preliminary studies suggest that an appropriate R for aluminum braced frames would be 3. Alternatively, using an R of 1 would resist earthquakes elastically, a very conservative approach.

10.4 AXIAL TENSION

The available tensile strength of members is the lesser of the available strengths for the limit states of yielding on the gross section and rupture on the effective net section. Strengths for these limit states are provided below.

10.4.1 Yielding

For unwelded members and members with transverse welds, the axial tension strength is

$$P_{nt} = F_{ty} A_g$$

where A_g is the gross cross-sectional area. Yielding at a transverse weld or at the net section produces negligible deformation and thus is not a limit state.

For members with longitudinal welds, the axial tension strength is the sum of the yield strengths of the welded and unwelded portions of the cross section:

$$P_{nt} = F_{ty}(A_g - A_{wz}) + F_{tyw} A_{wz}$$

10.4.2 Rupture

For unwelded members, the axial tension strength is

$$P_{nt} = F_{tu} A_e / k_t$$

For welded members, the axial tension strength is

$$P_{nt} = F_{tu} (A_e - A_{ewz})/k_t + F_{tuw} A_{ewz}$$

where

A_e is the effective net area, discussed below
A_{ewz} is the weld-affected part of the effective net area
k_t is the tension coefficient, which accounts for alloy-tempers that are notch-sensitive

The rupture strength of notched specimens of such alloy-tempers is less than the yield strength of plain specimens, so their tensile strength is reduced by dividing by k_t, which is greater than 1 for those alloy-tempers. Tension coefficients are given in *Specification* Table A.3.3. For commonly used alloy-tempers, k_t is 1, except 6005-T5, for which k_t is 1.25.

10.4.3 Effective Net Area

For members with some elements of the cross section not connected at joints, shear lag causes the stress to be less in the unconnected elements than in the connected elements. For example, if an I-shaped member in axial tension is connected at its ends through its flanges and not its web, tensile stress in the web is less than in the flanges at the connection. This is addressed by using the effective net area A_e to determine the available strength of the member for the rupture limit state. The effective net area is less than the net area A_n and determined as

$$A_e = A_n\left(1 - \frac{\bar{x}}{L_C}\right)\left(1 - \frac{\bar{y}}{L_C}\right)$$

where

L_C = the length of the connection in the direction of force
x = eccentricity of the connection in the x direction (the *Specification* Commentary provides examples of determining the connection eccentricities)
y = eccentricity of the connection in the y direction

For members with only a single row of fasteners or only a transverse weld, the length of the connection L_C is zero; in this case the *Specification* requires that the effective net area be taken as the net area of the connected elements, and no part of the unconnected elements is included in the effective net area.

10.5 AXIAL COMPRESSION

The available compressive strength of members is the least of the available strengths for the limit states of member buckling, local buckling, and the interaction between member buckling and local buckling. Strengths for these limit states are provided below.

10.5.1 Member Buckling

Member compressive strength may be limited by yielding, inelastic buckling, or elastic buckling, depending on the member slenderness λ. The nominal member compressive strength P_{nc} is the product of the member stress F_c corresponding to the strength and the gross cross-sectional area A_g:

$$P_{nc} = F_c A_g$$

where

Limit state	F_c	Slenderness limits
Yielding	F_{cy}	$\lambda \le \frac{B_c - F_{cy}}{D_c}$
Inelastic buckling	$(B_c - D_c\lambda)\left(0.85 + 0.15\frac{C_c - \lambda}{C_c - \lambda_1}\right)$	$\frac{B_c - F_{cy}}{D_c} < \lambda < C_c$
Elastic buckling	$\frac{0.85\pi^2 E}{\lambda^2}$	$\lambda \ge C_c$

λ = greatest column slenderness

The 0.85 factor in the elastic buckling stress $\frac{0.85\pi^2 E}{\lambda^2}$ accounts for member out-of-straightness.

For members without welds, determine the nominal member buckling strength $P_{nc} = P_{no}$ using buckling constants B_c, D_c, and C_c (see Section 10.2.2) for unwelded material using *Specification* Table B.4.1 for material not artificially aged and B.4.2 for artificially aged tempers and F_{cy}.

For members that are fully weld-affected, determine the nominal member buckling strength $P_{nc} = P_{nw}$ using buckling constants B_c, D_c, and C_c for welded material using *Specification* Table B.4.1 and F_{cyw}.

For members with transverse welds and:

a) supported at both ends with no transverse weld farther than $0.05L$ from the member ends, $P_{nc} = P_{no}$

b) supported at both ends with a transverse weld farther than $0.05L$ from the member ends or supported at only one end with a transverse weld $P_{nc} = P_{nw}$

For members with longitudinal welds, the nominal member buckling strength P_{nc} is a weighted average of the welded and unwelded strengths:

$$P_{nc} = P_{no}(1 - A_{wz}/A_g) + P_{nw}(A_{wz}/A_g)$$

Buckling may be flexural, torsional, or flexural-torsional. Flexural buckling is buckling about a principal axis of the cross section, for which the slenderness λ is L/r, where L is the member length and r is the cross section's radius of gyration about the principal axis. Torsional buckling is buckling about the longitudinal axis of the member, manifested by twisting about the member's longitudinal axis, and may govern the compressive strength of doubly symmetric shapes, such as cruciforms. Flexural-torsional buckling is a combination of flexural and torsional buckling, experienced by unsymmetric shapes. The slenderness for torsional and flexural-torsional buckling is $\lambda = \pi\sqrt{\frac{E}{F_e}}$, where F_e is the member elastic buckling stress determined using the equations given in *Specification* Section E.2.2, which account for the member's resistance to bending, twisting, and warping.

10.5.2 Local Buckling

Local buckling is buckling of an element of a member's cross section, such as the web of an I-shaped member. The uniform compression local buckling stress for five cases (for example, the web of an I-shaped member, a flat element supported on both longitudinal edges) is given in *Specification* Section B.5.4, and the provisions are summarized in Table 10.5. When the slenderness $\lambda \le \lambda_1$, F_c is the yield strength. When

Table 10.5 Local Buckling for Axial Compression

		Support			F_c		F_c		F_c	F_c
SAS Section	λ	Element type	Edge a	Edge b	yield	λ_1	Inelastic buckling	λ_2	Elastic buckling	Postbuckling
B.5.4.1	b/t	Flat	Yes	No	F_{cy}	$\frac{B_p - F_{cy}}{5.0D_p}$	$B_p - 5.0D_p\lambda$	$\frac{C_p}{5.0}$	$\frac{\pi^2 E}{(5.0b/t)^2}$	
								$\frac{k_1 B_p}{5.0D_p}$		$\frac{k_2\sqrt{B_p E}}{5.0b/t}$
B.5.4.2	b/t	Flat	Yes	Yes	F_{cy}	$\frac{B_p - F_{cy}}{1.6D_p}$	$B_p - 1.6D_p\lambda$	$\frac{C_p}{1.6}$	$\frac{\pi^2 E}{(1.6b/t)^2}$	
								$\frac{k_1 B_p}{1.6D_p}$		$\frac{k_2\sqrt{B_p E}}{1.6b/t}$
B.5.4.3		Flat	Yes	Stiffened	See B.5.4.3: $F_c = F_{UT} + (F_{ST} - F_{UT})\rho_{ST}$					
B.5.4.4	$4.62\left(\frac{b}{t}\right)\sqrt{\frac{1 + A_s/(bt)}{1 + \sqrt{1 + \frac{10.67 I_o}{bt^3}}}}$	Flat	Yes	Yes	F_{cy}	$\frac{B_c - F_{cy}}{D_c}$	$B_c - D_c\lambda$	C_c	$\frac{\pi^2 E}{\lambda^2}$	None
			With intermediate stiffener							
B.5.4.5	$\sqrt{\frac{R_b}{t}}$	Curved	Yes	Yes	F_{cy}	$\frac{B_t - F_{cy}}{D_t}$	$B_t - D_t\lambda$	C_t	$\frac{\pi^2 E}{16\lambda^2\left(1 + \frac{\lambda}{35}\right)^2}$	None
B.5.4.6	$\pi\sqrt{\frac{E}{F_e}}$	Entire shape			F_{cy}	$\frac{B_p - F_{cy}}{D_p}$	$B_p - D_p\lambda$	$\frac{k_1 B_p}{D_p}$		$\frac{k_2\sqrt{B_p E}}{\lambda}$

the slenderness $\lambda_1 < \lambda < \lambda_2$, F_c is the inelastic buckling strength. When the slenderness $\lambda \geq \lambda_2$, F_c is the elastic buckling strength or the postbuckling strength for those elements for which postbuckling strength is recognized in the *Specification*.

The local buckling strength of the full cross section is the average of the local buckling stresses of all the elements of the cross section, weighted by their cross-sectional area.

The local buckling stress of a shape can also be determined from its elastic local buckling stress, which may be determined by an analysis such as the finite strip method, by which the member is divided into strips running the length of the member. This is useful for shapes that contain elements that are not addressed by the five cases in *Specification* Section B.5.4, which often occurs for extrusions since they may employ multiple stiffeners or other custom features. The elastic local buckling stress F_e for the entire cross section is used to determine the slenderness $\lambda = \pi\sqrt{\frac{E}{F_e}}$ of the shape for local buckling. This slenderness is then used with *Specification* Section B.5.4.6 to determine the local buckling stress of the shape. This local buckling stress may be elastic or inelastic, depending on λ. This approach not only addresses more cases than Sections B.5.4.1 through B.5.4.5 but also is more accurate.

10.5.3 Interaction between Local Buckling and Member Buckling

If the elastic local buckling stress F_e is less than the member buckling stress F_c, local buckling reduces the member buckling strength. When this interaction occurs, the member buckling strength is

$$P_{nc} = \left[\frac{0.85\pi^2 E}{\lambda^2}\right]^{1/3} F_e^{2/3} A_g$$

where λ is the member buckling slenderness. This interaction only occurs if the postbuckling strength of elements of the shape is utilized, which only occurs when the shape has slender elements.

10.6 FLEXURE

The available flexural strength of members is the least of the available strengths for the limit states of

a) yielding M_{np},
b) rupture M_{nu},
c) local buckling M_{nlb}, and
d) lateral-torsional buckling M_{nmb}

Strengths for these limit states are provided below.

10.6.1 Yielding

For the limit state of yielding, the nominal flexural strength M_{np} of wrought products is the least of $Z F_{cy}$, $1.5S_t F_{ty}$, and $1.5S_c F_{cy}$,

where

Z = plastic modulus
S_t = section modulus on the tension side of the neutral axis
S_c = section modulus on the compression side of the neutral axis

The limit on the shape factor Z/S of 1.5 precludes extreme fiber yielding at service loads. Using the plastic modulus recognizes that even when the extreme fiber yields, the material in the cross section closer to the neutral axis has not yielded and that the stress-strain curve is nonlinear once yielding occurs.

10.6.2 Rupture

For the limit state of rupture, the nominal flexural strength is

$$M_{nu} = Z F_{tu}/k_t$$

10.6.3 Local Buckling

Elements of flexural members may be in uniform compression or flexural compression. The compressive strength of elements in uniform compression in flexural members (for example, the flanges of an I-shaped member) is determined in the same manner as the compressive strength of elements in uniform compression in axial compression members using *Specification* Section B.5.4. The compressive strength of elements in flexural compression (for example, the web of an I-shaped member in major axis bending) is determined using Section B.5.5, whose provisions are shown in Table 10.6.

The local buckling strength of a shape is the sum of the local buckling stress for elements in uniform compression multiplied by the section modulus of these elements and the local buckling stress for elements in flexural compression multiplied by the section modulus of these elements.

Alternately, the local buckling stress can be determined from the flexural elastic local buckling stress of the entire cross section, which may be determined using a computer analysis such as the finite strip method. The elastic local buckling stress is then used to determine the local buckling stress using *Specification* Section B.5.5.5.

10.6.4 Lateral-Torsional Buckling

Member buckling for members in flexure is called lateral-torsional buckling because material on the compression side of the neutral axis buckles and displaces laterally, while material on the tension side of the neutral axis does not. This results in lateral displacement of the compression side and twisting of the cross section.

Table 10.6 Local Buckling for Elements in Flexure

			Support		F_c		F_c		F_c	F_c
SAS section	λ	Element type	Edge a	Edge b	Yield	λ_1	Inelastic buckling	λ_2	Elastic buckling	Post buckling
B.5.5.1	b/t	Flat	Yes	Yes	$1.5F_{cy}$	$\frac{B_{br}-1.5F_{cy}}{0.65D_{br}}$	$B_{br}-0.65D_{br}\lambda$	$\frac{C_{br}}{0.65}$	$\frac{\pi^2 E}{(0.65b/t)^2}$	
								$\frac{k_1 B_{br}}{0.65D_{br}}$		$\frac{k_2\sqrt{B_p E}}{0.65b/t}$
B.5.5.2	b/t	Flat	Yes (tension)	No (compression)	$1.5F_{cy}$	$\frac{B_{br}-1.5F_{cy}}{3.5D_{br}}$	$B_{br}-3.5D_{br}\lambda$	$\frac{C_{br}}{3.5}$	$\frac{\pi^2 E}{(3.5b/t)^2}$	None
B.5.5.4	$\sqrt{\frac{R_b}{t}}$	Curved	Yes	Yes	$B_{tb}-D_{tb}\lambda$	C_{tb}	$B_t-D_t\lambda$	C_t	$\frac{\pi^2 E}{16\lambda^2\left(1+\frac{\lambda}{35}\right)^2}$	None
B.5.5.5	$\pi\sqrt{\frac{E}{F_e}}$		Entire shape		$1.5F_{cy}$	$\frac{B_{br}-1.5F_{cy}}{D_{br}}$	$B_{br}-D_{br}\lambda$	$\frac{k_1 B_{br}}{D_{br}}$		$\frac{k_2\sqrt{B_{br} E}}{\lambda}$

LATERAL-TORSIONAL BUCKLING STRENGTH

The *Specification* provides two equations for the lateral-torsional buckling strength M_{nmb}—one for inelastic buckling, which reaches the limit state of yielding M_{np} when the slenderness is zero, and one for elastic buckling. The inelastic buckling strength equals the elastic buckling strength when the slenderness equals the buckling constant C_c.

Limit state	M_{nmb}	Slenderness limits
Inelastic buckling	$M_{np}\left(1-\frac{\lambda}{C_c}\right)+\frac{\pi^2 E\lambda S_{xc}}{C_c^3}$	$\lambda < C_c$
Elastic buckling	$\pi^2 E S_{xc}/\lambda^2$	$\lambda \geq C_c$

For lateral-torsional buckling about the *x*-axis.

The lateral-torsional buckling slenderness λ is discused below. If more than one section applies, use any applicable Section. For example, for a doubly symmetric I-shaped member, use either *Specification* Section F.4.2.1 or Section F.4.2.5 to determine the lateral-torsional buckling slenderness.

For members without welds, determine the lateral-torsional buckling strength $M_{nmb} = M_{nmbo}$ using C_c for unwelded material using F_{cy} and Table B.4.1 for material not artificially aged and Table B.4.2 for artificially aged material.

For members that are fully weld-affected, determine the lateral-torsional buckling strength $M_{nmb} = M_{nmbw}$ using C_c for welded material using Table B.4.1 and F_{cyw}.

For members with transverse welds and:

a) supported at both ends with no transverse weld farther than $0.05L$ from the member ends, $M_{nmb} = M_{nmbo}$

b) supported at both ends with a transverse weld farther than $0.05L$ from the member ends or supported at only one end with a transverse weld, $M_{nmb} = M_{nmbw}$

For members with longitudinal welds, the lateral-torsional buckling strength M_{nmb} is

$$M_{nmb} = M_{nmbo}(1 - A_{wz}/A_f) + M_{nmbw}(A_{wz}/A_f)$$

where

A_f = area of the member farther than $2c/3$ from the neutral axis, where c is the distance from the neutral axis to the extreme compression fiber

ACCOUNTING FOR MOMENT GRADIENT

The lateral-torsional buckling strength given above is for a member with uniform moment over its length. In cases where the moment varies over the member length, to account for the moment being less than the maximum moment over parts of the span, the *Specification* provides a bending coefficient C_b for singly and doubly symmetric shapes bent about the *x*-axis with $0.1 < I_{yc}/I_y < 0.9$:

$$C_b = \frac{12.5M_{max}}{2.5M_{max} + 3M_A + 4M_B + 3M_C}$$

where M_{max} = absolute value of the maximum moment in the unbraced segment

M_A = absolute value of the moment at the quarter point of the unbraced segment

M_B = absolute value of the moment at the midpoint of the unbraced segment

M_C = absolute value of the moment at the three-quarter point of the unbraced segment

I_{yc} = moment of inertia of the compression flange about the *y*-axis

I_y = moment of inertia about the *y*-axis

C_b may be conservatively taken as 1 or determined using the equation above for members between braces. C_b for other cases, such as cantilevers and shapes with various bracing and loading conditions, may be found elsewhere, such as the AISC *Specification for Structural Steel Buildings* Commentary.

Once C_b has been determined, it is used to reduce the slenderness for lateral-torsional buckling described below.

SLENDERNESS FOR LATERAL-TORSIONAL BUCKLING

The slenderness for lateral-torsional buckling depends on the member's cross-sectional properties, the location and direction of any transverse load, the loading, the unbraced length L_b, and the bending coefficient C_b. The *Specification* provides simplified equations for the slenderness of specific shapes, such as shapes symmetric about the bending axis, closed shapes, and rectangular bars, as well as the equation for slenderness of any shape. These equations are summarized in Table 10.7.

10.7 SHEAR AND TORSION

The available shear and torsional strengths of members is the least of the available strengths for the limit states of rupture, buckling, and yielding. Strengths for these limit states are provided below.

10.7.1 Shear

SHEAR RUPTURE

Table 10.8 provides shear rupture strengths for various elements.

SHEAR YIELDING AND SHEAR BUCKLING

For the limit states of shear yielding and shear buckling, the nominal shear strength V_n is

for unwelded members:

$$V_n = F_{so} A_v$$

for welded members:

$$V_n = F_{so}(A_v - A_{wz}) + F_{sw} A_{wz}$$

where

F_{so} = shear stress F_s corresponding to the shear strength for an element determined using *Specification* Section G.2, G.3, or G.4 if no part of the cross section were weld-affected. Use buckling constants for unwelded metal (Table B.4.1 or Table B.4.2) and F_{sy}.

F_{sw} = shear stress F_s corresponding to the shear strength for an element determined using *Specification* Section G.2, G.3, or G.4 if the entire cross section were weld-affected. Use buckling constants for weld-affected zones (Table B.4.1) and F_{syw}.

A_v = shear area as defined in *Specification* Section G.2, G.3, G.4, or G.5

A_{wz} = weld-affected portion of the shear area

Shear yielding and buckling limit states are tabulated in Table 10.9.

10.7.2 Torsion

TORSIONAL RUPTURE

For the limit state of torsional rupture, the shear stress F_s corresponding to the torsional strength is

for unwelded members:

$$F_s = F_{su}/k_t$$

for welded members:

$$F_s = F_{su}(1 - A_{wz}/A_g)/k_t + F_{suw}A_{wz}/A_g$$

The torsional rupture strength T_n is then calculated using F_s as shown in Table 10.10.

TORSIONAL YIELDING AND TORSIONAL BUCKLING

For the limit states of shear yielding and shear buckling, the shear stress F_s corresponding to the torsional strength is

Table 10.7 Slenderness λ for Lateral-Torsional Buckling

SAS section	Shape	λ
F.4.2.1	Symmetric about the bending axis	$\frac{L_b}{r_{ye}\sqrt{C_b}}$, where Between brace points of beams subjected to end moment only or to transverse loads applied at the beam's neutral axis, or at brace points: $r_{ye} = \sqrt{\frac{\sqrt{I_y}}{S_x}\left[\sqrt{C_w + 0.038JL_b^2}\right]}$ Between brace points of beams subjected to transverse loads applied on the extreme fiber (where the load is free to move laterally with the beam if the beam buckles): $r_{ye} = \sqrt{\frac{I_y}{S_x}\left[\pm\frac{d}{4} + \sqrt{\frac{d^2}{16} + \frac{C_w}{I_y} + \frac{0.038JL_b^2}{I_y}}\right]}$ $d/4$ is negative when the load acts toward the shear center and positive when the load acts away from the shear center where the y-axis is the principal axis in the plane of bending and I_y = moment of inertia about the y-axis S_x = section modulus about the x-axis d = depth of the beam
F.4.2.2	Singly symmetric and unsymmetric about the bending axis	If $I_{yc} \le I_{yt}$, determine the slenderness using Section F.4.2.1 where r_{ye} is calculated with I_y, S_x, and J determined as though both flanges were the same as the compression flange with the overall depth d remaining the same.
F.4.2.3	Closed shapes	$2.3\sqrt{\frac{L_b S_{xc}}{C_b\sqrt{I_y J}}}$
F.4.2.4	Rectangular bars	$\frac{2.3}{t}\sqrt{\frac{dL_b}{C_b}}$
F.4.2.5	Any shape The y-axis is the centroidal symmetry or principal axis such that the tension flange has a positive y coordinate and bending is about the x-axis. The origin of the coordinate system is the intersection of the principal axes.	$\pi\sqrt{\frac{ES_{xc}}{C_b M_e}}$, where $M_e = \frac{\pi^2 EI_y}{L_b^2}\left[U + \sqrt{U^2 + \frac{0.038JL_b^2}{I_y} + \frac{C_w}{I_y}}\right]$ where $U = C_1 g_0 + C_2\beta_x/2$ C_1 and C_2: a) If no transverse loads are applied between the ends of the unbraced segment, $C_1 = 0$ and $C_2 = 1$. b) If transverse loads are applied between the ends of the unbraced segment, $C_1 = C_2 = 0.5$. g_0 = distance from the shear center to the point of application of the load; g_0 is positive when the load acts away from the shear center and negative when the load acts toward the shear center. If there is no transverse load, $g_0 = 0$. β_x = coefficient of monosymmetry $\beta_x = \frac{1}{I_x}\left(\int_A y^3\,dA + \int_A yx^2\,dA\right) - 2y_o$ Or, for singly symmetric I shapes, $\beta_x = 0.9d_f\left(\frac{2I_{yc}}{I_y} - 1\right)\left[1 - \left(\frac{I_y}{I_x}\right)^2\right]$ where I_{yc} = moment of inertia of the compression flange about the y-axis d_f = the distance between the flange centroids y_o = distance perpendicular to the bending axis from the cross section's centroid to the shear center; positive in the direction of the load
F.5	Angles	See *Specification* Section F.5

Table 10.8 Shear Rupture Limit States

SAS section	Element	Unwelded	Welded
G.2	Flat webs supported on both edges	$V_n = F_{su}A_n/k_t$	$V_n = F_{su}(A_n - A_{wz})/k_t + F_{suw}A_{wz}$
G.3	Flat webs supported on one edge	$V_n = F_{su}A_n/k_t$	$V_n = F_{su}(A_n - A_{wz})/k_t + F_{suw}A_{wz}$
G.4	Pipes and round or oval tubes	$F_{su}A_n/(2k_t)$	$V_n = F_{su}(A_n - A_{wz})/(2k_t) + F_{suw}A_{wz}/2$
G.5	Rods	$V_n = F_{su}A_n/k_t$	$V_n = F_{su}(A_n - A_{wz})/k_t + F_{suw}A_{wz}$

Table 10.9 Shear Yielding and Buckling Limit States

SAS section	A_v	λ	λ_1	λ_2	Yield	Inelastic buckling	Elastic buckling
G.2	dt	b/t	$\frac{B_s - F_{sy}}{1.25D_s}$	$\frac{C_s}{1.25}$	F_{sy}	$B_s - 1.25D_s\, b/t$	$\frac{\pi^2 E}{(1.25b/t)^2}$
G.3	bt	b/t	$\frac{B_s - F_{sy}}{3.0D_s}$	$\frac{C_s}{3.0}$	F_{sy}	$B_s - 3.0D_s\, b/t$	$\frac{\pi^2 E}{(3.0b/t)^2}$
G.4	$\pi(D_o^2 - D_i^2)/8$	$2.9\left(\frac{R_b}{t}\right)^{5/8}\left(\frac{L_v}{R_b}\right)^{1/4}$	$\frac{1.3B_s - F_{sy}}{1.63D_s}$	$\frac{C_s}{1.25}$	F_{sy}	$1.3B_s - 1.63D_s\lambda$	$\frac{1.3\pi^2 E}{(1.25\lambda)^2}$
G.5	$\pi D^2/4$	NA	NA	NA	F_{sy}	NA	NA

Table 10.10 Torsional Rupture Limit States

SAS section	Shape	T_n
H.2.1	Pipes and round or oval tubes	$F_s J/R$
H.2.2	Rectangular tubes	$F_s C$
H.2.3	Rods	$0.262 F_s D^3$

for unwelded members:

$$F_s = F_{so}$$

for welded members:

$$F_s = F_{so}(1 - A_{wz}/A_g) + F_{sw} A_{wz}/A_g$$

where

F_{so} = shear stress corresponding to the torsional strength for an element determined using *Specification* Section H.2 if no part of the cross section were weld-affected. Use buckling constants for unwelded metal (Table B.4.1 or Table B.4.2) and F_{sy}.

F_{sw} = shear stress corresponding to the torsional strength for an element determined using *Specification* Section H.2 if the entire cross section were weld-affected. Use buckling constants for weld-affected zones (Table B.4.1) and F_{syw}.

A_{wz} = cross-sectional area of the weld-affected zone

A_g = gross cross-sectional area of the element

Torsional yielding and buckling limit states are tabulated in Table 10.11.

10.8 COMBINED FORCES

For members subject to flexure and axial force,

$$\frac{P_r}{P_c} + \frac{M_{rx}}{M_{cx}} + \frac{M_{ry}}{M_{cy}} \le 1.0$$

where all terms are positive, and

x = subscript for major principal axis bending

y = subscript for minor principal axis bending

P_r = required axial strength determined in accordance with *Specification* Chapter C

P_c = available axial strength determined in accordance with *Specification* Chapter D for axial tension and *Specification* Chapter E for axial compression

M_r = required flexural strength determined in accordance with *Specification* Chapter C

M_c = available flexural strength determined in accordance with *Specification* Chapter F

Using this equation is predicated on determining the required forces and moments in accordance with *Specification* Chapter C; otherwise, this equation may be unconservative.

10.9 CONNECTIONS

Connections may be welded or fastened. Fasteners addressed by the *Specification* are bolts, rivets, screws, and pins. Fasteners should not be designed to share loads with welds.

10.9.1 Welded Connections

The nominal strength R_n of groove, fillet, plug, slot, and stud welded joints is the lesser of the base material strength for the limit states of tensile rupture and shear rupture and the weld metal strength for the limit state of rupture as follows:

For the base metal

$$R_n = F_{nBM} A_{BM}$$

For the weld metal

$$R_n = F_{nw} A_{we}$$

where

F_{nBM} = nominal stress of the base metal corresponding to its welded ultimate strength from *Specification* Table A.3.3

F_{nw} = nominal stress of the weld metal corresponding to its ultimate strength (see below)

A_{BM} = cross-sectional area of the base metal

A_{we} = effective area of the weld

F_{nBM}, F_{nw}, A_{BM}, and A_{we} are given in Table 10.12.

Table 10.11 Torsional Yielding and Buckling Limit States

SAS section	T_n	λ	λ_1	λ_2	Yield	Inelastic buckling	Elastic buckling
H.2.1	$F_s J/R$	$2.9\left(\frac{R_b}{t}\right)^{5/8}\left(\frac{L_v}{R_b}\right)^{1/4}$	$\frac{B_s - F_{sy}}{1.25D_s}$	$\frac{C_s}{1.25}$	F_{sy}	$B_s - 1.25D_s\, b/t$	$\frac{\pi^2 E}{(1.25b/t)^2}$
H.2.2	$F_s C$	b/t	$\frac{B_s - F_{sy}}{1.25D_s}$	$\frac{C_s}{1.25}$	F_{sy}	$B_s - 1.25D_s\, b/t$	$\frac{\pi^2 E}{(1.25b/t)^2}$
H.2.3	$0.196 F_{sy} D^3$	NA	NA	NA	F_{sy}	NA	NA

Table 10.12 Nominal Strength of Welded Joints

Load type and direction relative to weld axis	Base metal: Nominal stress F_{nBM}	Base metal: Effective area A_{BM}	Weld metal: Nominal stress F_{nw}	Weld metal: Effective area A_{we}
COMPLETE-JOINT PENETRATION GROOVE WELDS				
Tension or compression normal to weld axis	F_{tuw}	$S_w L_{we}$	F_{tuw}	$S_w L_{we}$
Tension or compression parallel to weld axis	Tension or compression in parts parallel to a weld need not be considered in designing welds joining the parts			
Shear	$0.6F_{tuw}$	$S_w L_{we}$	$0.6F_{tuw}$	$S_w L_{we}$
PARTIAL-JOINT PENETRATION GROOVE WELDS				
Tension or compression normal to weld axis	F_{tuw}	$S_w L_{we}$	$0.6F_{tuw}$	$S_w L_{we}$
Tension or compression parallel to weld axis	Tension or compression in parts parallel to a weld need not be considered in designing welds joining the parts			
FILLET WELDS				
Shear	$0.6F_{tuw}$	$S_w L_{we}$	$0.6(0.85F_{tuw})$ (see note 1)	$S_{we} L_{we}$
Tension or compression parallel to weld axis	Tension or compression in parts parallel to a weld need not be considered in designing welds joining the parts			
PLUG AND SLOT WELDS				
Shear parallel to faying surface	$0.6F_{tuw}$	see J.2.3	$0.6F_{tuw}$	see J.2.3
STUD WELDS				
Shear	$0.6F_{tuw}$	$\pi D^2/4$	$0.6F_{tuw}$	$(\pi/4)(D - 1.191/n)^2$
Tension	F_{tuw}	$\pi D^2/4$	F_{tuw}	$(\pi/4)(D - 1.191/n)^2$

Notes
1. Alternately, the strength of fillet welds loaded transversely may be taken as 1.36 times the strength given above.
2. F_{tuw} for base metal is listed in *Specification* Table A.3.3.
3. F_{tuw} for filler metal is listed in Table 10.13.

With proper selection of the filler alloy to match the strength of the weld-affected base metal, the strength of a groove weld is governed by the base metal. Conversely, for almost all selections of filler and base metals, the strength of a fillet weld is governed by the strength of the filler alloy.

10.9.2 Mechanically Fastened Connections

The selection of fastener material depends on anticipated service conditions and intended service life. Because aluminum cathodically protects steel and stainless steel, fasteners of these materials are often used to connect aluminum parts. While aluminum suffers galvanic corrosion while protecting steel fasteners, because the volume of the connected parts is much greater than that of the fasteners, the galvanic current density is small, and therefore the corrosion of the connected aluminum parts in such connections is usually slight. Conversely, aluminum fasteners connecting steel parts may quickly corrode when moisture is present.

Table 10.13 Aluminum Weld Filler Strengths

Filler	Tensile ultimate strength F_{tuw} (ksi)
1100	11
2319	35
4043	24
4047	25
5183	40
5356	35
5554	31
5556	42
5654	30

Dimensions and strengths for bolted, riveted, screwed, and pinned connections are given in Table 10.14.

Bolted Connections

The only bolts addressed by the *Specification* are aluminum bolts that comply with ASTM F468, which includes the three aluminum alloys shown in Table 10.15. Although bolts of other materials are often used to connect aluminum parts, the design rules for bolts of these materials are provided not in the *Specification* but rather in the codes that address those materials such as, for steel bolts, the AISC *Specification for Structural Steel Buildings*. The *Specification* provides requirements for slip-critical connections using zinc-coated A325 bolts connecting aluminum parts with a 2.0 mil profile abrasion-blasted faying surfaces.

Riveted Connections

The only rivets addressed by the *Specification* are aluminum rivets. Riveted connections are not designed to resist tensile loads.

One-piece aluminum rivets are not commonly used in construction. Other rivets that utilize a sleeve and pin are used, but their strengths are usually determined by testing since their performance in shear is complicated by cold working and the geometry of the deformed components once they are installed.

Screwed Connections

The *Specification* addresses aluminum tapping screws with a nominal diameter D from 0.164 in. (No. 8) through 0.25 in. Screws may be thread-forming or thread-cutting, with or without a self-drilling point. Aluminum screws are often ineffective in drilling or tapping a hole, so steel screws are more often used to connect aluminum parts. Design rules for steel screws are given in other codes, such as the AISI Cold-Formed Steel Specification.

Table 10.14 Mechanically Fastened Connections

	Bolts	Rivets	Screws	Pins
Hole size D_h	$D_h \le D + 1/16''$	$D_h \le 1.04D$	Unthreaded holes: $D_h \le D + 1/16''$ Threaded holes: see Tables J.5.1, J.5.2	$D_h \le D + 1/32''$
Minimum spacing	$2.5D$	$3D$	$2.5D$	Not applicable
Minimum edge distance	$1.5D$	$1.5D$	$1.5D$	$1.5D$
Tensile strength	$(\pi(D - 1.191/n)^2/4)F_{tu}$	Not Applicable	$A_rF_{tu}/1.25$ Also see J.5.4 for pull-out and pull-over	Not applicable
Shear strength	For bolts with threads in the shear plane: $(\pi(D - 1.191/n)^2/4)F_{su}$ For bolts without threads in the shear plane: $(\pi D^2/4)F_{su}$	$\pi D_h^2 F_{su}/4$	$A_rF_{su}/1.25$ Also see J.5.5 for tilting	Shear yielding: $\pi D^2 F_{sy}/4$ Shear rupture: $\pi D^2 F_{su}/4$ Flexural yielding: $\pi D^3 F_{ty}/21.3$ Flexural rupture: $\pi D^3 F_{tu}/(21.3k_t)$
Bearing strength	For a bolt in a hole: $d_etF_{tu} \le 2DtF_{tu}$ For a bolt in a slot with the slot perpendicular to the direction of force: $1.33DtF_{tu}$	$d_etF_{tu} \le 2D_htF_{tu}$	$d_etF_{tu} \le 2DtF_{tu}$	$d_etF_{tu}/1.5 \le 1.33DtF_{tu}$

Table 10.15 Nominal Strengths of Aluminum Bolts

Alloy and temper	Shear ultimate strength F_{su} (ksi)	Tensile ultimate strength F_{tu} (ksi)
2024-T4	37	62
6061-T6	25	42
7075-T73	41	68

Aluminum screws are available in two alloys with these strengths:

Alloy-temper	F_{tu} (ksi)	F_{su} (ksi)
2024-T4	62	38
7075-T73	68	41

The nomenclature used below is that the connected part under the screw head is material 1 and that the connected part under material 1 is material 2.

The tensile strength of a screwed connection is the least of the pull-out, pull-over, and screw tensile rupture strengths.

For screws in holes, the pull-out strength varies for machine thread screws and spaced thread screws, as well as the thickness of the tapped part. Pull-out strength equations are provided in *Specification* Section J.5.4.1.1.

Screw slots are used to connect a screw to the end of an extrusion. The screw's axis is parallel to the longitudinal axis of the extrusion and the slot is C-shaped to avoid the need for a hollow in the extrusion. The nominal strength R_n for the limit state of pull-out of a screw in a screw slot with the dimensions shown in *Specification* Figure J.5.1 and Table J.5.3 is

$$R_n = 0.29DL_e F_{tu}$$

where

D = nominal diameter of the screw

F_{tu} = tensile ultimate strength of the extrusion

The screw embedment length in the screw slot L_e must be at least $2D$.

Screw chases are used to attach screws perpendicular to the longitudinal axis of an extrusion. Spaced thread screws have a greater pull-out strength than machine thread screws and flat wall chases than ribbed wall chases. The nominal strength R_n for the limit state of pull-out of a ¼-in.-diameter screw in a screw chase is

$$R_n = 0.021L_eF_{tu}(14/n)^2/w_c$$

where

F_{tu} = tensile ultimate strength of the screw chase extrusion

n = number of threads/in. of the screw

L_e = length of engagement of the screw's threads in the chase, calculated by deducting any chamfer at the top of the chase and the portion of the fastener at its point without full threads. The deduction for the point is 0.250 in. for AB screws; 0.050 in. for UNC screws; and 0.225 in. for F screws.

w_c = nominal width of the chase but no less than $0.72D$ and no greater than $0.88D$. For ribbed wall chases, the nominal width is the average of the inside and outside widths of the chase. A common chase width is $0.75D$, which, for a ¼-in.-diameter screw, is 0.188 in.

In order to minimize flexing of the chase walls (which reduces the pull-out strength), the nominal thickness of the screw chase walls should not be less than $0.6D$, and the nominal depth of the chase should not exceed $3D$.

The nominal strength R_n for the limit state of pull-over for non-countersunk screws is

$$R_n = C_{pov}t_1F_{tu1}(D_{ws} - D_h)$$

where

C_{pov} = 1.0 for valley fastening and 0.7 for crown fastening

t_1 = nominal thickness of the part in contact with the screw head or washer

F_{tu1} = tensile ultimate strength of the part in contact with the screw head or washer

D_{ws} = larger of the nominal washer diameter and the screw head diameter, but no greater than 5/8 in. Washers should be at least 0.050 in. thick.

D_h = nominal diameter of the hole in the material under the screw head

Additional provisions in the *Specification* address the pull-over strength for screws in tight-fitting holes and of countersunk screws.

The shear strength of a screwed connection is the least of the bearing, tilting, and screw shear rupture strengths. Screw tilting is only a limit state for screws connecting a thicker part under the screw head than the part below that.

Pinned Connections

Pinned connections permit the connected parts to rotate around the axis of the pin. Aluminum bars are often an economical choice for pins connecting aluminum parts. Pins are not used to resist tensile loads. Because the connected parts have a clearance between them so the parts can move independently, flexural stresses can occur in pins.

The available strength of an aluminum pin in shear is the lesser of the available strengths for the limit state of shear yielding and the limit state of shear rupture. The available strength of an aluminum pin in flexure is the lesser of the available strengths for the limit state of flexural yielding and the limit state of flexural rupture.

For pins subjected to shear and flexure,

$$\left(\frac{V_r}{V_c}\right)^3 + \left(\frac{M_r}{M_c}\right) \leq 1.0$$

where

V_r = required shear strength
V_c = available shear strength
M_r = required flexural strength
M_c = available flexural strength

10.9.3 Block Shear

The strength for the limit state of block shear rupture for mechanically fastened connections on a failure path with shear on some segments and tension on the other segments is

$$R_n = (F_{su}\,(A_{nv} + A_{gv})/2 + F_{tu}\,A_{nt})/k_t$$

where

A_{gv} = gross area in shear
A_{nv} = net area in shear
A_{nt} = net area in tension

The strength for the limit state of block shear rupture for welded connections on a failure path with shear on some segments and tension on the other segments is

$$R_n = F_{suw}\,A_{gvw} + F_{tuw}\,A_{gtw} + (F_{su}\,A_{gvo} + F_{tu}\,A_{gto})/k_t$$

where

A_{gvw} = gross weld-affected area in shear
A_{gvo} = gross area not weld-affected in shear
A_{gtw} = gross weld-affected area in tension
A_{gto} = gross area not weld-affected in tension

10.10 SERVICEABILITY

Serviceability is the preservation of a structure's function under service load combinations.

10.10.1 Vibrations

Vibrations occur when the natural frequency of a structure approximately equals the forcing frequency, such as vortex shedding due to wind. The vortex shedding frequency of tubes subjected to a constant wind velocity is given in the *Specification* Commentary Section L.5. It is usually more efficient to change the natural frequency (for example, by using dampers) than to design the structure for the number of load cycles induced by vibrations.

10.10.2 Deflections

Deflections govern the design of aluminum structures more frequently than for structures made of material with a greater modulus of elasticity. Unlike the analysis for determining required forces for the strength limit state, the analysis for determining deflections for the serviceability limit state should use the full modulus of elasticity E.

Deflection limits are not given in the *Specification*; rather, deflection limits are established elsewhere, such as in building codes or as a function of the intended service. For example, the deflection of mullions supporting windows must be limited to prevent wind from cracking the window glass.

The full width b of elements whose stress f_a exceeds the elastic local buckling stress F_e is not effective in resisting deflection. This may occur in shapes with slender elements (such as roll-formed sheet shapes) utilizing postbuckling strength. For such shapes, the moment of inertia for determining deflections should be determined using the effective width b_e of the elements of the shape, where

$$b_e = b\sqrt{F_e/f_a}$$

10.11 FABRICATION AND ERECTION

10.11.1 Tolerances

The *Specification* requires that members not deviate from straight or intended curvature by more than the member length divided by 960.

10.11.2 Welding

The *Specification* requires that welding comply with AWS D1.2 *Structural Welding Code—Aluminum.* D1.2 includes provisions for qualifying welding procedures and welders, fabricating welds, and inspecting welds but refers to the *Specification* for the structural design of aluminum welded members and welded connections. D1.2 addresses gas metal arc welding (GMAW), gas tungsten arc welding (GTAW), and friction stir welding (FSW) processes.

10.12 FATIGUE

Specification Appendix 3 addresses fatigue, the limit state of crack initiation, and growth due to the repeated application of loads. Repeated application of stress can rupture parts at stresses less than the minimum ultimate tensile strength F_{tu}.

When the number of load cycles exceeds 100,000, the *Specification* requires that the design stress range S_{rd} equal or exceed the applied stress range S_{ra}, the algebraic difference between the nominal maximum and minimum stresses applied (tension is positive and compression is negative). The design stress range is a function of the number of load cycles and the severity of detail, which ranges from category A (best; for example, extruded members) to category F (worst; for example, fillet welds). Based on the fatigue category, *Specification* Table 3.2 provides fatigue constants C_f and m that determine the design stress range $S_{rd} = C_f/N^{1/m}$ as shown in Table 10.16.

Table 10.16 Fatigue Design

Fatigue category	C_f	m	S_{rd} for N = 100,000 cycles	S_{rd} for N = 5 million cycles
A	96.5	6.85	18.0	10.2
B	130	4.84	12.0	5.4
C	278	3.64	11.8	4.0
D	157	3.73	7.2	2.5
E	160	3.45	5.7	1.8
F	174	3.42	6.0	1.9

Aluminum fatigue strengths are approximately one-third those of steel. In order to reduce the effect of stress risers that promote fatigue cracks, *Specification* Section 3.4 provides additional fabrication and erection requirements for parts subject to fatigue. This includes adding fillet welds to transverse tee and corner joints and a ¾-in. minimum radius for reentrant corners.

REFERENCES

Aluminum Association (2015), *Aluminum Design Manual*, Arlington, VA.

Kissell, J.R. and Ferry, R.L. (2002), *Aluminum Structures*, John Wiley, New York.

Chapter 11

Design of Reinforced-Concrete Structural Members

BY

MUSTAFA MAHAMID, Ph.D., SE, PE, P.Eng. *University of Illinois at Chicago*
DAVID A. FANELLA, Ph.D., SE, PE *Concrete Reinforcing Steel Institute*

11.1 CONCRETE

Concrete has a compressive strength (in standard 6 × 12-in. cylinders at 28 days) of about 3000 to 6000 psi, with a possible range from 2500 or less to 12,000 or 14,000 psi or more if special methods are used.

The tensile strength of concrete is roughly 10 percent of its compressive strength or, perhaps more precisely, $5\sqrt{f'_c}$, where f'_c is the 28-day cylinder compressive strength. For both direct tensile stress and diagonal tensile stresses of considerable magnitude that are induced by shears, steel reinforcement is ordinarily used to replace or supplement the strength of the concrete.

11.2 REINFORCEMENT

Steel reinforcement in the United States consists largely of bars, almost always deformed, in 11 sizes, standardized by the American Society for Testing and Materials (ASTM) and the U.S. Department of Commerce (Table 11.1). These bars come in a variety of grades and strengths, the

Table 11.1 Deformed Bar Designation Numbers, Nominal Weights, Nominal Dimensions, and Deformation Requirements

		Nominal dimensions*				Deformation requirements, in.		
Bar designation No.†	Nominal weight, lb/ft	Diameter, in.	Cross-sectional area, in.^{2}	Perimeter, in.	Max OD,‡ in.	Max average spacing	Min average height	Max gap (chord of 12½% of nominal perimeter)
3§	0.376	0.375	0.11	1.178	7/16	0.262	0.015	0.143
4	0.668	0.500	0.20	1.571	9/16	0.350	0.020	0.191
5	1.043	0.625	0.31	1.963	1-1/16	0.437	0.028	0.239
6	1.502	0.750	0.44	2.356	⅞	0.525	0.038	0.286
7	2.044	0.875	0.60	2.749	1	0.612	0.044	0.334
8	2.670	1.000	0.79	3.142	1⅛	0.700	0.050	0.383
9	3.400	1.128	1.00	3.544	1¼	0.790	0.056	0.431
10	4.303	1.270	1.27	3.990	1-7/16	0.889	0.064	0.487
11	5.313	1.410	1.56	4.430	1⅝	0.987	0.071	0.540
14	7.65	1.693	2.25	5.32	1-15/16	1.185	0.085	0.648
18	13.60	2.257	4.00	7.09	2½	1.58	0.102	0.864

Sizes 14 and 18 are large bars generally not carried in regular stock. They are available by arrangement with the supplier.
*The nominal dimensions of a deformed bar are equivalent to those of a plain round bar having the same weight per foot as the deformed bar.
†Bar numbers are based on the number of eighths of an inch in the nominal diameter of the bars.
‡The maximum outside diameter including deformations may be important as when punching holes in structural-steel members to accommodate bars, or in fitting couplings, or in nesting or bundling bars. Exact dimensions vary among manufacturers. Tabulated values allow for deformations, longitudinal ribs, and out of round.
§While No. 3 bars are used for stirrups, column ties, etc., it is suggested that slab bars, in general, be No. 4 or larger.

Table 11.2 Physical Requirements for ASTM Reinforcing Bars

Type of steel and ASTM specification	Size nos. inclusive	Grade	Yield strength min, psi	Tensile strength min, psi	Elongation in 8 in. min, %°	Bend test pin diameter°†
Billet steel A615	3–6	40	40,000	70,000	11–12	3.5*d*–5*d*
	3–18	60	60,000	90,000	7–9	3.5*d*–8*d*
	6–18	75	75,000	100,000	6	7*d*–9*d*
Rail steel A616	3–11	50	50,000	80,000	5–7	6*d*–8*d*
	3–11	60	60,000	90,000	4.5–6	6*d*–8*d*
Axle steel A617	3–6	40	40,000	70,000	11–12	4*d*–5*d*
	3–11	60	60,000	90,000	7–8	4*d*–8*d*
Weldable steel A706	3–18	60	60,000‡	80,000§	10–14	3*d*–8*d*

*See ASTM specification for details. Depends on bar size. Plain round bars with diameters up to 2 in. may also be furnished meeting ASTM A615, A616, and A617 requirements.
†*d* = bar nominal diameter.
‡78,000 max.
§1.25 × actual yield min., but not less than 80,000.

higher strengths usually being accompanied by a lesser ductility and vice versa. Properties of bars are covered by ASTM specifications, which are revised as need arises. Table 11.2 summarizes the main properties of the commonly available types.

Reinforcing bars are usually cut and bent in the detailed form at a fabricator's yard, tagged and delivered, ready for placement. In certain areas, the bars are delivered cut to length and are bent at the job site. Sometimes they are delivered in stock lengths of appropriate sizes and are cut and bent at the job site.

Welded-wire fabric consists of sheets or rolls of mesh made by welding each intersection of crossing layers of the same or different gage wires at equal spacings each way (square mesh) or different spacings (rectangular mesh). Sheets or rolls may have a maximum width of 12 or 13 ft (depending upon the width of welding equipment available) by whatever length can be handled.

Welded-wire fabric is designated by two numbers and two letter–number combinations, such as 6 × 8 – W8.0 × W4.0, where the first number gives the spacing in inches of the longitudinal wires and the second number gives the spacing of the transverse wires in inches. The first letter–number combination gives the type and area of the longitudinal wire and the second combination the information on the transverse wire. The Wire Reinforcing Institute (WRI) established the letter–number designation, which relates to the cross-sectional area of the wire. The letter W designates smooth wire, and the letter D describes deformed wire. The number following the letters W or D is the cross-sectional area of the wire in hundredths of a square inch. A W8.0 wire is a smooth wire with a cross-sectional area of 0.08 in.^{2}; a W4.0 wire has a cross-sectional area of 0.04 in.^{2} Commonly available styles of welded-wire fabric and their properties are given in Table 11.3.

Table 11.3 Common Stock Styles of Welded-Wire Fabric

	Style designation	Steel area, eq in./ft		Weight, approx. lb/100 ft²
		Longit.	Trans.	
Rolls:	6 × 6–W1.4 × W1.4	0.028	0.028	21
	6 × 6–W2.0 × W2.0	0.040	0.040	29
	6 × 6–W2.9 × W2.9	0.058	0.058	42
	6 × 6–W4.0 × W4.0	0.080	0.080	58
	4 × 4–W1.4 × W1.4	0.042	0.042	31
	4 × 4–W2.0 × W2.0	0.060	0.060	43
	4 × 4–W2.9 × W2.9	0.087	0.087	62
	4 × 4–W4.0 × W4.0	0.120	0.120	85
Sheets:	6 × 6–W2.9 × W2.9	0.058	0.058	42
	6 × 6–W4.0 × W4.0	0.080	0.080	58
	6 × 6–W5.5 × W5.5	0.110	0.110	80
	4 × 4–W4.0 × W4.0	1.120	0.120	85

Because fabric may be made of drawn wire and even high-strength or deformed wire and has welded cross wires, such fabric is permitted certain higher stresses in code specifications.

Some use is made of glass fibers, plastic threads, chopped wire, and similar reinforcements.

11.3 SPECIFICATIONS, CODES, AND STANDARDS

The ASTM maintains specifications for the quality of all the items incorporated into reinforced concrete and for methods of mixing and detailed methods for sampling, testing, and approving both the individual items and the finished concrete.

The American Concrete Institute publishes Building Code Requirements for Reinforced Concrete (ACI 318) and a large collection of guides and manuals, including the Detailing Manual (ACI 315) and the Manual of Inspection, SP-2. The Portland Cement Association and the Concrete Reinforcing Steel Institute both publish extensive design aids. The provisions in this chapter are per ACI 318-14 (Ref. 1).

11.4 GENERAL REQUIREMENTS FOR STRENGTH AND SERVICEABILITY

11.4.1 Introduction

This section contains the fundamental requirements for strength and serviceability that form the basis of design of all reinforced-concrete members. The basic concepts of the strength design method are presented, as are the general provisions for deflection control.

Throughout the years, there have been basically two design philosophies for reinforced-concrete members: working stress design and limit state design (referred to as the strength design method).

From the early 1900s until the early 1960s, working stress design was the primary design method for reinforced concrete. In working stress design, members are proportioned so that maximum elastic stresses due to unfactored loads (also identified as service or working loads) are equal to or less than the allowable stresses prescribed in the Code.

Limit state design involves identifying applicable limit states and determining acceptable levels of safety against occurrences of each limit state. In general, a limit state is a set of performance criteria that must be met when a structure is subjected to loads. The two fundamental limit states are as follows: (1) ultimate limit states, which correspond to the loads that cause failure (strength and stability), and (2) serviceability limit states, which correspond to the criteria that govern the service life of a structure (e.g., deflection and crack width). A structure or structural member is said to have reached a limit state (or, equivalently, is said to have "failed") when it is unable to carry out one or more of the required performance criteria. The statistical methods used to determine the level of safety required in the design process are discussed later.

The 1956 edition of the ACI Code was the first to include provisions for the "ultimate strength design method." The strength method was essentially established as the preferred design method in the 1971 Code, although an updated form of the working stress design method—referred to as the alternate design method—was still permitted to be used. It was not until 2002 that the alternate design method was officially deleted from the Code. A more comprehensive history of the evolution of the strength design method can be found in Chapter 5 of Ref. 2.

The gradual elimination of the working stress design method from the ACI Code primarily had to do with the following shortcomings of the method:

- Inability to correctly account for the variability of loads and member resistances
- Inconsistent factor of safety in member design

Chapter 4 of the Code contains the basic performance requirements for structural members and structural systems. The main limit states for reinforced-concrete structures are strength (ACI 4.6) and serviceability (ACI 4.7), and both must be considered in the design process.

The basic requirement for strength design is set forth in ACI 4.6.2 as follows:

design strength ≥ required strength

or

$$\phi S_n \geq U$$

where S_n is the nominal strength of the member.

In general, the design strength of a member, which is equal to the applicable strength-reduction factor ϕ times the nominal strength of the member S_n, must be equal to or greater than the required strength. The required strength, which is represented by the symbol U, is determined by multiplying service load effects by code-prescribed load factors. Design and required strengths for structural members are given in the Code chapters and are discussed in more detail in the following sections.

The load combinations in the 2014 edition of ACI 318 Code (Ref. 1) match those in ASCE/SEI 7 (Ref. 3) and, for the most part, the load combinations in the International Building Code (IBC) (Ref. 4). ACI load combinations are shown in Table 11.4. For more details on load combinations, see Chap. 1 of this publication.

Table 11.4 ACI Load Combinations

ACI equation number	Load combination
5.3.1a	$U = 1.4D$
5.3.1b	$U = 1.2D + 1.6L + 0.5(L_r \text{ or } S \text{ or } R)$
5.3.1c	$U = 1.2D + 1.6(L_r \text{ or } S \text{ or } R) + (1.0L \text{ or } 0.5W)$
5.3.1d	$U = 1.2D + 1.0W + 1.0L + 0.5(L_r \text{ or } S \text{ or } R)$
5.3.1e	$U = 1.2D + 1.0E + 1.0L + 0.2S$
5.3.1f	$U = 0.9D + 1.0W$
5.3.1g	$U = 0.9D + 1.0E$

11.4.2 Design Strength

Overview

The design strength of a reinforced-concrete member is equal to the nominal strength of the member S_n, which is calculated in accordance with the provisions of the Code, multiplied by a strength-reduction factor ϕ that is always less than 1. The subscript n is used to denote nominal strength. For example, the notation for nominal flexural strength of a reinforced-concrete member is M_n. Subsequent chapters of this book contain methods for determining the nominal strength. A discussion on the purpose of strength-reduction factors follows.

ACI Table 20.2.2.4a contains upper limits for the yield strength of reinforcement that can be used in design calculations. The deflection provisions of ACI 7.3.1 for one-way slabs, 8.3.1 for two-way slabs, and 9.3.1 for beams are directly related to f_y.

Strength Reduction Factors

Strength-reduction factors are commonly referred to as resistance factors or ϕ-factors and play a key role in the determination of the design strength of a reinforced-concrete member. The main purposes of these factors are as follows:

1. *To account for the understrength of a member due to variations in material strengths and dimension.* The strength of concrete can vary because concrete is a composite material made of constituent materials whose properties vary. The strength of reinforcing steel can also vary but usually to a lesser degree than concrete.

Member dimensions can differ from those specified in the construction documents because of construction and fabrication tolerances. The diameter of reinforcing bars can also fluctuate because of rolling and fabrication tolerances.

Reinforcing bars in a concrete section can be placed at locations that are different from those specified in the construction documents. Tolerances on reinforcement placement are prescribed by the ACI.

2. *To allow for inaccuracies in the design equations.* As will be shown in subsequent chapters, a number of assumptions and simplifications are made in the design equations for nominal strength. These assumptions and simplifications introduce inaccuracies that must be accounted for when determining the design strength.

3. *To reflect the degree of ductility and required reliability of a member.* Reinforced-concrete members that are more ductile, such as beams, are less sensitive to variations in concrete strength compared with members that are less ductile, such as columns.

Spiral reinforcement confines the concrete in a column better than tied reinforcement. Thus, spirally reinforced columns are more ductile and have greater toughness than tied columns.

4. *To reflect the importance of a member.* The failure of a column in a structure is usually considered to be more detrimental than failure of a beam.

ACI strength-reduction factors are given in Chapter 21. Factors for different types of actions and structural elements are contained in Table 21.2.1, and factors for members subjected to moment, axial force, or combined moment and axial force are given in Table 21.2.2. ACI 21.2.4 contains the strength-reduction factors for shear of elements in special moment frames and special structural walls.

A summary of pertinent strength-reduction factors is given in Table 11.5.

Tension-controlled and compression-controlled sections are based on the magnitude of the net tensile strain in the extreme layer of the longitudinal reinforcement at nominal strength ε_t and the value of the net tensile strain used to define a compression-controlled section ε_{ty}. In very general terms, members with tension-controlled sections are usually beams, and members with compression-controlled sections are usually columns. Lower strength-reduction factors are assigned to compression-controlled sections because they are less ductile and more sensitive to variations in concrete strength than tension-controlled sections. Also, columns tend to support areas much greater than those for beams, and the consequences of column failure in a building or structure are generally more severe than those attributed to beam failure. The advantage of using spiral reinforcement in a column is reflected in a strength-reduction factor that is greater than that for other types of lateral reinforcement.

A linear transition in the strength-reduction factor is permitted between the limits for tension-controlled and compression-controlled sections; this transition is illustrated in Fig. 11.1. Also provided in the figure are the strain diagrams corresponding to tension-controlled and compression-controlled sections for Grade 60 (Grade 420) reinforcement.

According to ACI 25.4.1.3, development lengths for reinforcement do not require a strength-reduction factor. Strength-reduction factors are also not required when determining lap splice lengths because such lengths are a function of the development length.

Information on the strength-reduction factors for shear in special moment frames and special structural walls is also given in Chapter 11 of Ref. 2.

11.4.3 Control of Deflections

Overview

Deflection control is part of the serviceability limit state that must be satisfied in the design of reinforced-concrete members. The main material parameters that affect deflections are the modulus of elasticity E_c, the modulus of rupture f_r, shrinkage, and creep.

Excessive deflections of floor or roof members can result in damage to nonstructural components of a building, such as partitions, doors,

Table 11.5 ACI Strength-Reduction Factors

Tension-controlled sections ($\varepsilon_t \geq 0.005$)		0.9
Transition ($\varepsilon_{ty} < \varepsilon_t < 0.005$)*	Members with spiral reinforcement conforming to ACI 25.7.3	$0.75+0.15\dfrac{(\varepsilon_t-\varepsilon_{ty})}{(0.005-\varepsilon_{ty})}$
	Other	$0.65+0.25\dfrac{(\varepsilon_t-\varepsilon_{ty})}{(0.005-\varepsilon_{ty})}$
Compression-controlled sections ($\varepsilon_t \leq \varepsilon_{ty}$)	Members with spiral reinforcement conforming to ACI 25.7.3	0.75
	Other	0.65
Shear	Elements not part of special moment frames or special structural walls	0.75
	Elements that are part of special moment frames or special structural walls where the nominal shear strength of the member is greater than or equal to the shear corresponding to the development of the nominal moment strength of the member†	
	Elements that are part of special moment frames or special structural walls where the nominal shear strength of the member is less than the shear corresponding to the development of the nominal moment strength of the member†	0.6
	Beam-column joints in special moment frames and diagonally reinforced coupling beams in special structural walls	0.85
Torsion		0.75
Bearing on concrete		0.65

*For sections classified as transition, it shall be permitted to use a strength-reduction factor corresponding to compression-controlled sections.
†The nominal moment strength shall be calculated considering the most critical factored axial loads that include the earthquake effects E.

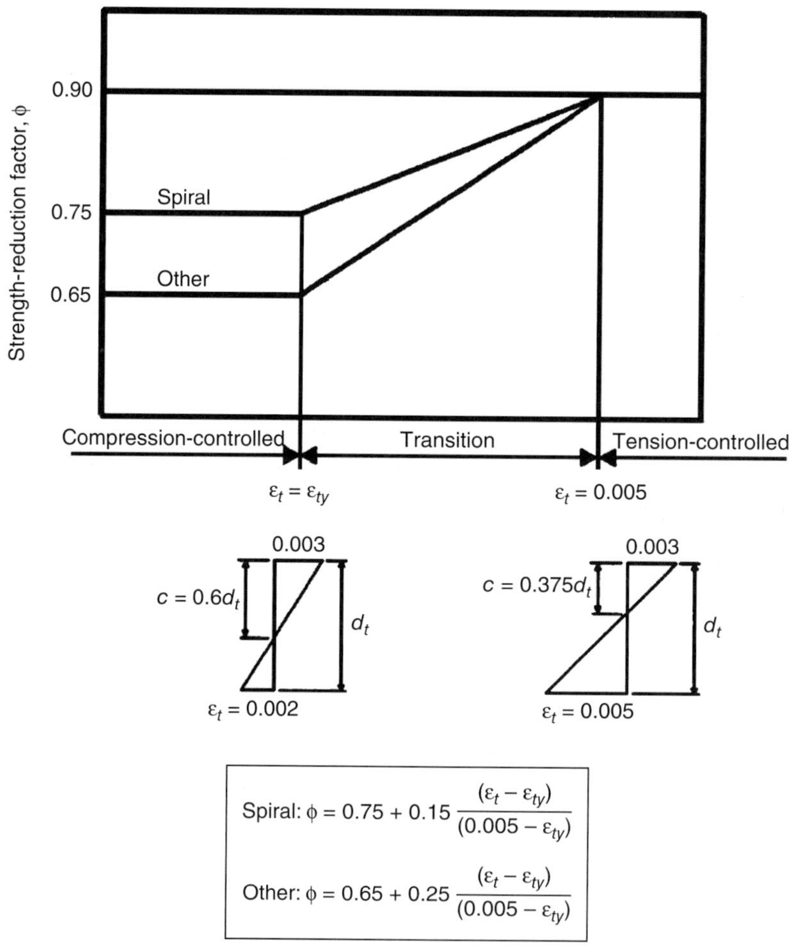

Figure 11.1 Variation of the strength-reduction factor ϕ.

or glass windows, to name a few. Such deflections can also result in the malfunction of sensitive equipment inside a building. Unsightly cracks can appear in structural members that have deflected in excess of the established limitations. Excessive deflection of roof members can cause ponding of water on the roof. Roof members will deflect under the weight of water, and if the deflection is excessive, additional water can accumulate on the roof, which can lead to even more deflection. In extreme cases, the progressive increase of deflection can result in the collapse of a roof. For these and other reasons, it is important to limit the deflection of reinforced-concrete members.

ACI 24.2 provides two methods for controlling deflections due to service loads in one-way and two-way floor and roof systems: (1) minimum thickness limitations and (2) computed deflection limitations.

One-Way Construction

The minimum thickness limitations given in ACI Tables 9.3.1.1 and 7.3.1.1, which appeared in the ACI Code at the time that the strength design method was introduced, are applicable to beams and one-way slabs, respectively, that are not attached to partitions or other construction that is likely to be damaged by relatively large deflections. Deflection criteria are essentially satisfied for members that have a thickness equal to or greater than that given in the appropriate table. A summary of the minimum thicknesses for continuous and cantilevered members is given in Fig. 11.2, where h is the overall thickness of the member.

Values of minimum thickness shown in Fig. 11.2 are applicable to members with normal-weight concrete [w_c = 145 pcf (2320 kg/m^3)] and Grade 60 (Grade 420) reinforcement. For other conditions, the values in the figure must be modified as follows:

- For structural lightweight having w_c in the range 90 to 115 pcf, multiply the values by $(1.65 - 0.005w_c) \geq 1.09$.
- For f_y other than 60,000 psi, multiply the values by $(0.4 + f_y/100{,}000)$.

In SI units, the modification factors are the following:

- For structural lightweight having w_c in the range 1440 to 1920 kg/m^3, multiply the values by $(1.65 - 0.0003w_c) \geq 1.09$.
- For f_y other than 420 MPa, multiply the values by $(0.4 + f_y/700)$.

No correction factor is provided for concrete with w_c greater than 115 pcf because research has shown that the correction factor would be close to 1, where w_c is between 115 and 145 pcf.[8] The correction factor on yield stress of reinforcement produces conservative results for values of f_y between 40,000 and 80,000 psi.

For simply supported members, minimum thickness is $\ell/20$ and $\ell/16$ for one-way slabs and beams, respectively.

The minimum thickness limitations outlined earlier are an integral part of the design procedure for typical beams and one-way slabs: The thickness of a member will usually be determined first on the basis of strength requirements and then checked against the minimum thickness requirements for deflection.

In cases where members are supporting elements that are likely to be damaged by relatively large deflections, the provisions in ACI 24.2.3 and 24.2.4 must be used to determine deflections. Immediate and long-term deflections must both be calculated, and the magnitudes of the deflections must be less than or equal to the limiting values given in ACI Table 24.2.2.

Two-Way Construction

ACI 8.3.1 contains minimum thickness requirements for two-way construction. By definition, a two-way slab system has a ratio of long-to-short spans that is less than or equal to 2.

For two-way slab systems that do not have any interior beams, the minimum slab thicknesses are given in ACI Table 8.3.1.1 as functions of f_y and the clear span length in the long direction ℓ_n. For slabs without drop panels, the thickness must be at least 5 in. (125 mm), and for slabs with drop panels, the thickness must be at least 4 in. (100 mm). Drop panels are defined in ACI 8.2.4. It has been demonstrated through the years that slabs conforming to these minimum thickness requirements have performed adequately without any problems due to short-term or long-term deflections.

For two-way slab systems with interior beams, the minimum slab thickness is determined by ACI Table 8.3.1.2. In addition to f_y and ℓ_n, h is calculated as a function of the beam stiffness along the column lines. Comprehensive coverage of these equations is provided in Chapter 7 of Ref. 2.

The deflection limits of ACI Table 24.2.2 are also applicable to immediate and long-term calculated deflections for two-way slabs. Calculation of deflections for two-way slabs is complex, and more information on the methods used to determine such deflections can be found in Chapter 7 of Ref. 2.

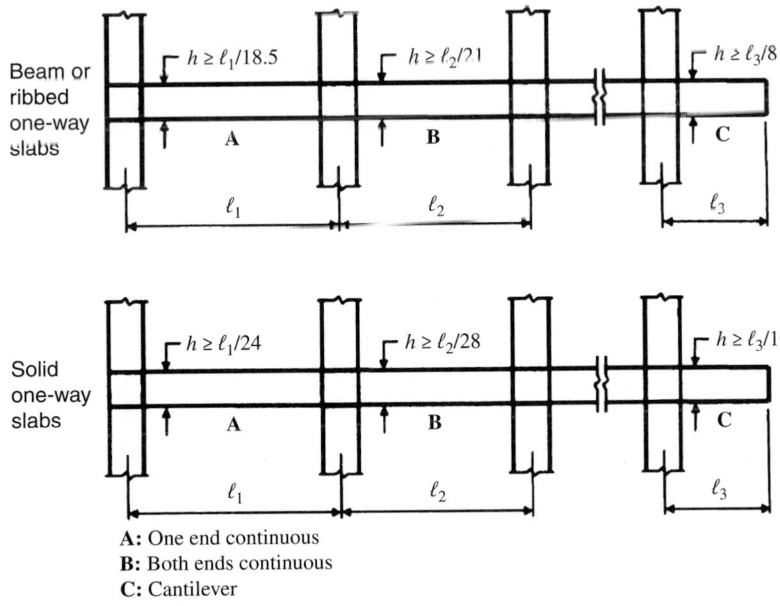

Figure 11.2 The minimum thickness of beams and one-way slabs in accordance with ACI 7.3.1 and 9.3.1.

11.5 GENERAL PRINCIPLES OF THE STRENGTH DESIGN METHOD

11.5.1 Introduction

This section covers the fundamental principles and requirements of the strength design method used in the design of reinforced-concrete structures. Presented are the design assumptions of the method and the basic techniques to determine the nominal strength of a reinforced-concrete section subjected to flexure, axial load, or a combination of both.

As a reinforced-concrete member approaches its ultimate strength, both the concrete and the reinforcing steel behave inelastically. This inelastic behavior must be captured in the design theory. It is evident from the information on material properties given in Chapter 2 of Ref. 2 that it is far easier to analytically express the inelastic behavior of reinforcing steel than that of concrete. As such, simplifying assumptions are made in the strength design method related to stress distribution in concrete.

The strength design method is based on the following two fundamental conditions (ACI 22.2.1):

1. Static equilibrium. The compressive and tensile forces acting on any cross-section of a member are in equilibrium.
2. Compatibility of strains. The strain in a reinforcing bar that is embedded in concrete is equal to the strain in concrete at that level.

The first condition must be satisfied at every cross-section of a member. It is shown later in this chapter that the basic equations of equilibrium are used in determining nominal strengths of reinforced-concrete members.

The second condition implies that there is a perfect bond between the concrete and the reinforcing steel and that both the materials act together to resist the effects from external loads. This has been proven by research, and it is due to deformed bars as well as the adhesion between the concrete and steel.

11.5.2 Design Assumptions

The design assumptions used in the strength design method are outlined in ACI 22.2.1 through 22.2.3. They are applicable in the design of members subjected to flexure, axial loads, or a combination of both. The nominal strength of a reinforced-concrete member is determined on the basis of these design assumptions.

Design Assumption No. 1

The strains in the reinforcement and the concrete shall be assumed directly proportional to the distance from the neutral axis (ACI 22.2.1.2).

For deep beams, which are defined in ACI 9.9.1.1, the strain is not linear, and a nonlinear distribution of strain must be utilized, or a strut-and-tie model as outlined in Chapter 23 of the Code may be used.

The strain distribution over the depth of a rectangular reinforced-concrete section at ultimate strength is depicted in Fig. 11.3. For illustrative purposes, it is assumed that the strains are compressive above the neutral axis and are tensile below it. The strains in the concrete and the reinforcement are directly proportional to the distance from the neutral axis, which is located at a distance c from the compression face of the section.

Because the strain distribution is linear, the strain in the concrete ε_c at the extreme compression fiber is directly proportional to the strains in the reinforcement. For example, the ratio of ε_c to the strain in the reinforcement farthest from the compression face ε_{s4} can be obtained by similar triangles:

$$\frac{\varepsilon_c}{\varepsilon_{s4}} = \frac{c}{d_4 - c} \tag{11.3}$$

Similar relationships can be established between ε_c and the other reinforcement strains and between the various reinforcement strains.

Note that the largest tensile strain occurs in the reinforcing steel farthest from the compression face.

Design Assumption No. 2

The maximum usable strain at the extreme concrete compression fiber is 0.003 (ACI 22.2.2.1).

Design Assumption No. 3

The stress in the reinforcement f_s below its specified yield strength f_y is equal to the modulus of elasticity of the steel E_s times the steel strain ε_s. The stress in the reinforcement is equal to f_y for strains ε_s greater than or equal to f_y/E_s (ACI 22.2.3.2 and 20.2.2.1). The modulus of elasticity can be taken as 29,000,000 psi for all grades of reinforcing steel (ACI 20.2.2.2).

Design Assumption No. 4

The tensile strength of concrete is neglected in the axial and flexural calculations of reinforced concrete (ACI 22.2.2.2).

The tensile resistance of concrete is used in other situations, most notably in serviceability calculations. For example, the modulus of rupture f_r, which is related to the tensile strength (see Chapter 2 of Ref. 2), is utilized in the determination of the immediate deflection of a reinforced-concrete member.

Design Assumption No. 5

The relationship between the concrete compressive-stress distribution and the concrete strain shall be assumed to be rectangular, trapezoidal, parabolic, or of any other shape that results in prediction of strength in substantial agreement with the results of comprehensive tests (ACI 22.2.2.3).

Research has shown that models using rectangular, parabolic, trapezoidal, and other-shaped compressive-stress distributions can adequately predict test results. The assumption given in ACI 22.2.2.4 permits the use of an equivalent rectangular concrete-stress distribution, which is covered under design assumption no. 6.

Design Assumption No. 6

The requirements of ACI 22.2.2.3 are satisfied by an equivalent rectangular concrete-stress distribution, which is defined in ACI 22.2.2.4.

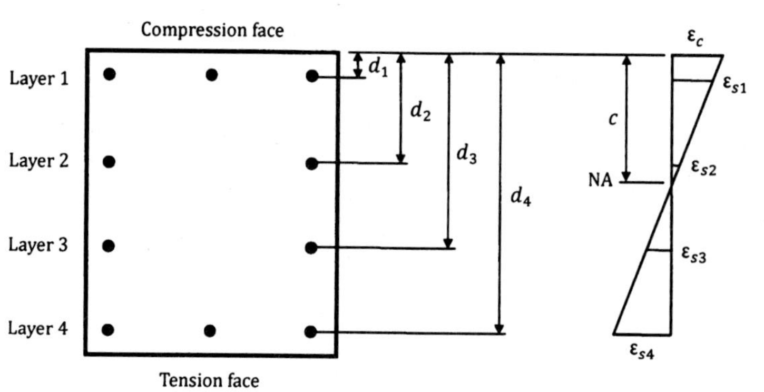

Figure 11.3 The assumed strain distribution in a reinforced-concrete section.

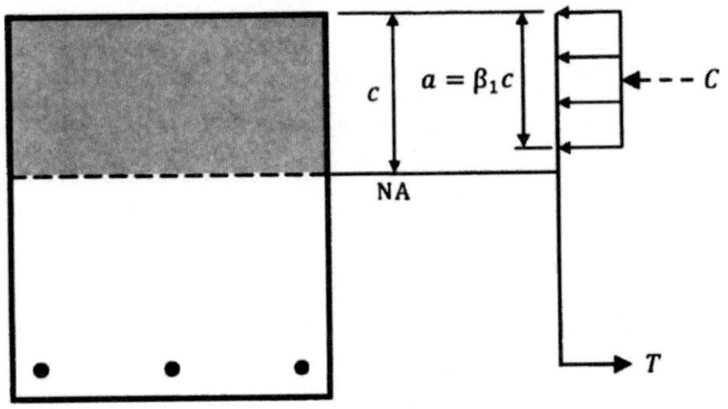

Figure 11.4 The equivalent rectangular concrete-stress distribution.

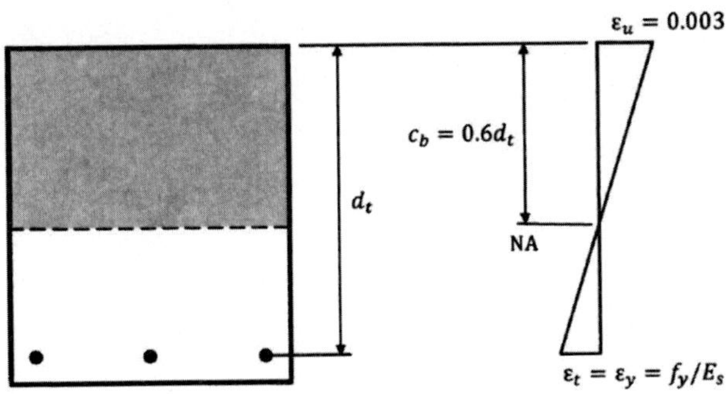

Figure 11.5 The balanced strain condition with Grade 60 (420 MPa) reinforcement.

The Code permits the use of the equivalent rectangular concrete-stress distribution defined in ACI 22.2.2.4, known as Whitney stress block, which is illustrated in Fig. 11.4.

A uniform stress equal to 85 percent of the concrete-compressive strength f_c' is distributed over the depth a, which is equal to the factor β_1 times the depth to the neutral axis c. Although this assumed stress distribution does not represent the actual compressive-stress distribution in the concrete at the ultimate state, it does provide basically the same results as those obtained from experimental investigations;[2] as noted previously, this is a requirement of the strength design method (see design assumption no. 5). The need for the factor β_1 is due to the variation in shape of the stress-strain curves for different concrete strengths. It is evident that the stress-strain curves of higher-strength concretes are more linear and exhibit less inelastic behavior than those of lower-strength concretes. Up to compressive strengths of 4000 psi, the ratio of the rectangular stress block depth a to the neutral axis depth c that best approximates the actual concrete-stress distribution is equal to 0.85, that is, $\beta_1 = 0.85$. For compressive strengths greater than 4000 psi, β_1 must be less than 0.85 in order to produce adequate results. ACI 22.2.2.4.3 requires that β_1 be reduced linearly at the rate of 0.05 for each 1000 psi in excess of 4000 psi for compressive strengths up to 8000 psi; above 8000 psi, $\beta_1 = 0.65$. The following equations define β_1.

For 2500 psi $\leq f_c' \leq 4000$ psi, $\beta_1 = 0.85$
For 4000 psi $< f_c' \leq 8000$ psi, $\beta_1 = 1.05 - 0.00005 f_c'$
For $f_c' > 8000$ psi, $\beta_1 = 0.65$

In SI:

For 17 MPa $\leq f_c' \leq 28$ MPa, $\beta_1 = 0.85$
For 28 MPa $< f_c' \leq 56$ MPa, $\beta_1 = 1.05 - 0.0071 f_c'$
For $f_c' > 56$ MPa, $\beta_1 = 0.65$

11.6 GENERAL PRINCIPLES AND REQUIREMENTS

11.6.1 Overview

The nominal strength of a reinforced-concrete member is established using the fundamental conditions of equilibrium and strain compatibility, and the assumptions presented above. Presented in this section are the basic principles utilized in the strength design method.

Design strength equations for cross-sections subjected to flexure or combined flexure and axial load were originally presented in Ref. 2 and were derived using essentially the same design assumptions as those summarized in this section. These equations form the basis of the nominal strength equations stipulated in the Code.

11.6.2 Balanced Strain Conditions

A balanced strain condition exists at a cross-section of a reinforced-concrete member when the strain in the tension reinforcement ε_s farthest from the compression face reaches the strain corresponding to yield (i.e., $\varepsilon_s = \varepsilon_y = f_y/E_s$) just as the strain in the extreme compression fiber of the concrete reaches its maximum value of 0.003. The strain distribution for the balanced condition is depicted in Fig. 11.5.

At the extreme compression fiber, the strain in the concrete is equal to the maximum value of 0.003, and the strain in the tension reinforcement, which is located at a distance d_t from the extreme compression fiber, is equal to ε_y. By definition, d_t is the distance from the extreme compression fiber to the centroid of the longitudinal tension steel that is farthest from the extreme compression fiber. With only one layer of steel, the distance d_t is the same as that from the extreme compression fiber to the centroid of the longitudinal tension steel, which is designated as d.

The ratio of the neutral axis depth c_b to the extreme depth d_t to produce a balanced strain condition in a section with tension reinforcement alone may be obtained by applying strain compatibility conditions:

$$\frac{c_b}{\varepsilon_u} = \frac{d_t - c_b}{\varepsilon_y} \tag{11.2}$$

Substituting the definitions of $\varepsilon_u = 0.003$ and $\varepsilon_y = f_y/E_s$ into Eq. (11.2) and rearranging the terms result in

$$\frac{c_b}{d_t} = \frac{\varepsilon_u}{\varepsilon_y + \varepsilon_u} = \frac{0.003}{0.003 + f_y/E_s} \tag{11.3}$$

For Grade 60 reinforcement, ACI 21.2.2.1 permits the yield strain ε_y of the reinforcement to be taken as 0.002, as opposed to 0.00207 that would be obtained by dividing the yield stress (60 ksi) by the modulus of elasticity (29,000 ksi).

11.6.3 Compression-Controlled Sections

According to ACI 21.2.2, a section is classified as compression-controlled if the net tensile strain in the extreme tension steel at nominal strength ε_t is less than or equal to the tensile strain ε_{ty} when the concrete strain in compression reaches its assumed strain limit ε_u of 0.003. For deformed reinforcement, ε_{ty} is equal to f_y/E_s, which can be taken as 0.002. as noted previously.

Note that ε_t is defined as the strain in the reinforcement after applicable strains due to creep, shrinkage, and temperature have been deducted from the total strain.

A brittle type of failure is generally expected in compression-controlled sections. This type of failure occurs suddenly with little or no warning. It is common for columns to have compression-controlled sections, that is, sections where $\varepsilon_t \leq f_y/E_s$. However, columns that are subjected to a relatively small axial-compressive force and a relatively large bending moment, for example, may have sections that are not compression-controlled.

11.6.4 Tension-Controlled Sections

A tension-controlled section is defined as a section where $\varepsilon_t \geq 0.005$ when the concrete in compression reaches its assumed strain limit ε_u of 0.003 (ACI 21.2.2). The strain distribution for the case where $\varepsilon_t = 0.005$ is shown in Fig. 11.6.

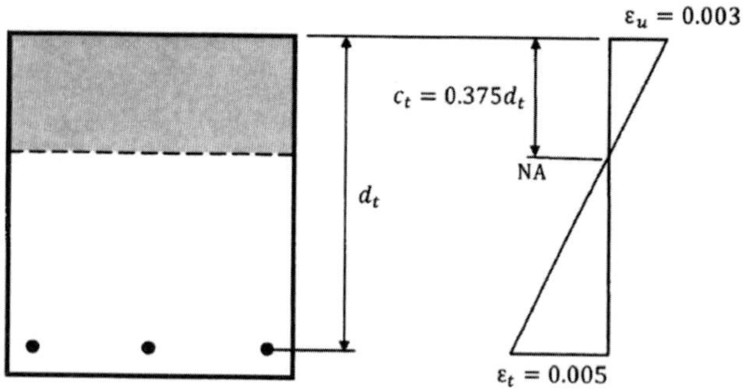

Figure 11.6 Strain distribution at the tension-controlled limit.

The ratio of the neutral axis depth c_t to the extreme depth d_t at the tension-controlled limit may be obtained by applying strain compatibility conditions. Referring to Fig. 11.4 and using similar triangles,

$$\frac{c_t}{\varepsilon_u} = \frac{d_t - c_t}{\varepsilon_t} \tag{11.4}$$

Substituting $\varepsilon_u = 0.003$ and $\varepsilon_t = 0.005$ into Eq. (11.4) and rearranging the terms result in

$$\frac{c_t}{d_t} = \frac{\varepsilon_u}{\varepsilon_y + \varepsilon_u} = \frac{0.003}{0.003 + 0.005} = 0.375 \tag{11.5}$$

Beams, one-way slabs, and two-way slabs usually have tension-controlled sections. Unlike compression-controlled sections, tension-controlled sections are ductile and generally exhibit significant deflections and cracking before failure.

Sections with ε_t between ε_{ty} and 0.005 are said to be in a transition region between compression- and tension-controlled sections. A linear transition in the strength-reduction factor occurs between the limits for compression- and tension-controlled sections. This transition is illustrated in Fig. 11.1 along with the applicable strength-reduction factors for compression- and tension-controlled sections.

Although the preceding discussion has focused on a rectangular section with one layer of tension reinforcement, the basic methods are applicable to rectangular and nonrectangular sections with more than one layer of reinforcement. The effects of cross-section shape and multiple layers of reinforcement are automatically accounted for in the strain compatibility equations that are used to determine ε_t.

11.7 ONE-WAY SLABS, TWO-WAY SLABS, AND BEAMS

11.7.1 Overview

Beams, one-way slabs, and two-way slabs are members in one-way and two-way construction that are subjected to the effects from flexure with very small or zero effects from axial forces. For purposes of discussion in this section, it is assumed that the members are primarily subjected to bending moments.

Presented next are the different types of flexural failure and the relationship between failure and the amount of reinforcement in a section. Also presented are the basic equations to determine the nominal strength of members or sections subjected primarily to flexure.

11.7.2 Types of Flexural Failure

Consider a reinforced-concrete beam with tension reinforcement only. Assume that the amount of tension reinforcement in the section is such that at failure, the reinforcing steel will yield in tension before the concrete crushes in compression. The relationship between bending moment and curvature of this beam is plotted to failure in Fig. 11.7. It is evident from the diagram that there is a long plastic region; that is, the beam exhibits a ductile response up to failure. This is commonly referred to as tension failure, and the section is tension-controlled.

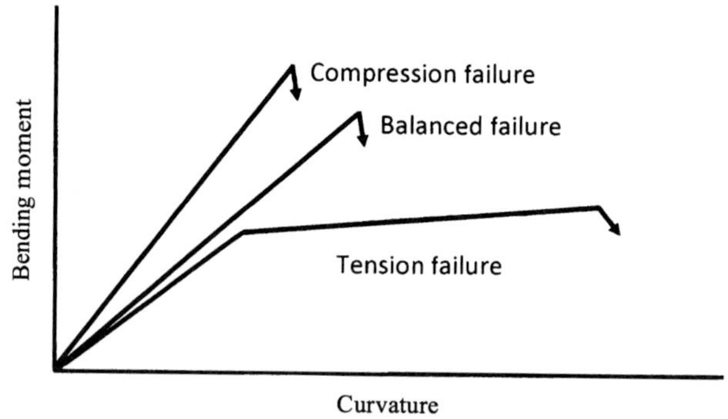

Figure 11.7 The idealized moment–curvature diagram for a reinforced-concrete beam with one layer of tension reinforcement.

In such cases, a member will typically exhibit large deformations and significant cracking prior to collapse, and it is anticipated that there will be ample warning prior to failure.

Now, assume that additional reinforcing steel is added to the section with all of the other parameters remaining the same so that at failure, the reinforcing steel yields at the same time the concrete crushes. This is commonly referred to as balanced failure. Referring to Fig. 11.5, balanced failure occurs when the strain in the extreme compression fiber of the concrete reaches the assumed crushing strain of 0.003 at the same time that the strain in the reinforcing steel reaches the yield strain. The moment-curvature relationship for balanced failure is also plotted in Fig. 11.7. It is evident that no ductility is exhibited when such a failure occurs.

Finally, assume that even more reinforcing steel is added to the section so that the total amount is greater than that corresponding to balanced failure. In such cases, the concrete in the extreme compression fiber reaches the assumed crushing strain of 0.003 prior to the reinforcing steel yielding. The moment-curvature diagram for such a member does not have the ductile post yielding response displayed by a member with an amount of reinforcement smaller than the balanced amount (see Fig. 11.7). This type of failure, which is called compression failure, occurs suddenly in a brittle manner without warning.

Tension failures are favored over compression failures in members subjected primarily to flexure. The amount of tension reinforcement is limited by the Code to ensure that it is achieved in design.

11.7.3 Maximum Reinforcement

The Code requires that one-way slabs, two-way slabs, and beams have properties that ensure that tension failure occurs.

Instead of specifying a maximum reinforcement ratio, which was stipulated in editions of the Code prior to 2002 as 75 percent of the balanced reinforcement ratio, ACI 7.3.3.1, 8.3.3.1, and 9.3.3.1 require that nonprestressed one-way slabs, two-way slabs, and beams, respectively, be designed so that $\varepsilon_t \geq 0.004$. For beams, this requirement need only be satisfied when the factored axial compressive force P_u is less than $0.10f'_cA_g$, where A_g is the gross area of the section. In essence, this requirement limits the amount of tension reinforcement that can be provided at a section: For a given cross-section and material properties, the strain in the reinforcement at nominal strength is inversely proportional to the amount of reinforcement that is provided at that section. Thus, the strain decreases as the amount of reinforcement increases. The limitation on ε_t is slightly more conservative than that required previously.

11.7.4 Minimum Reinforcement

Minimum reinforcement requirements for one-way slabs, two-way slabs, and beams are given in ACI 7.6.1, 8.6.1, and 9.6.1, respectively. These requirements are also applicable to footings (see ACI Chapter 13). Members with cross-sections that are larger than that required for strength will usually be governed by the provisions for minimum reinforcement.

To avoid this sudden type of failure, the Code prescribes a minimum amount of reinforcement $A_{s,min}$ that is to be provided at any positive or negative bending moment region of a one-way slab, two-way slab, or beam where such tensile reinforcement is required. For beams (ACI 9.6.1.2),

$$A_{s,\min} = \frac{3\sqrt{f'_c}b_w d}{f_y} \geq \frac{200 b_w d}{f_y} \quad (11.6)$$

$$\text{In SI: } A_{s,\min} = \frac{0.25\sqrt{f'_c}b_w d}{f_y} \geq \frac{1.4 b_w d}{f_y}$$

In this equation, the concrete-compressive strength f'_c has the unit of pounds per square inch (MPa), and b_w is defined as the web width of the member. The limit of 200 b_w d/f_y [1.4 $b_w d/f_y$] controls for concrete-compressive strengths less than approximately 4400 psi (30 MPa).

ACI 9.6.1.2 also contains minimum reinforcement requirements for statically determinate beams where the flange of the beam is in tension. An example of such a beam is illustrated in Fig. 11.8. The flange of the cantilever beam, which has a width of b_f, is in tension due to the uniformly distributed load that is applied along the length of the member. According to ACI 9.6.1.2, $A_{s,min}$ is determined by Eq. (11.6), where b_w in the equation is replaced by the lesser of two times the actual web width of the member ($2b_w$) or the flange width b_f. It is especially important to provide a minimum amount of reinforcement in members such as cantilever beams where bending moments cannot be redistributed to adjoining framing members.

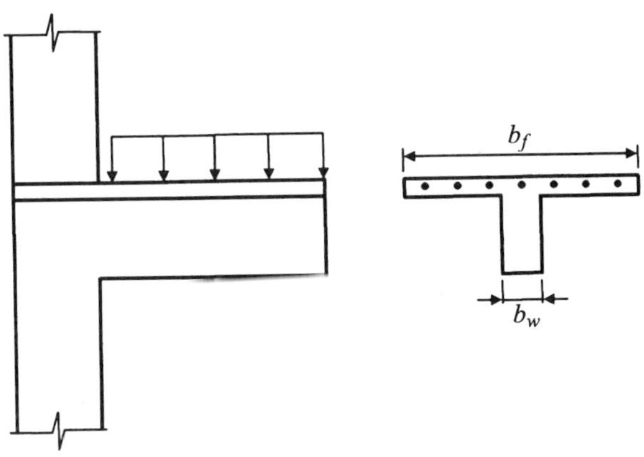

Figure 11.8 A cantilever beam with the flange in tension.

The aforementioned minimum areas of reinforcement need not be provided at any section where the provided A_s is at least one-third greater than A_s required by analysis (ACI 9.6.1.3). This exception is deemed sufficient for large members where the amount required by ACI 9.6.1.2 would be excessive.

For one-way and two-way slabs, the minimum flexural reinforcement requirements are given in ACI 7.6.1.1 and 8.6.1.1, respectively, and are the same as the minimum temperature and shrinkage reinforcement requirements of ACI 24.4.3.2:

- For deformed bars with $f_y < 60{,}000$ psi (420 MPa): $A_{s,min} = 0.0020A_g$
- For deformed bars and welded wire reinforcement with $f_y \geq 60{,}000$ psi (420 MPa): $A_{s,min}$ = greater of $0.0018 \times 60{,}000\ A_g/f_y$ or $0.0014\ A_g$
- [In SI: $A_{s,min}$ = greater of $0.0018 \times 420\ A_g/f_y$ or $0.0014\ A_g$]

A design width of 12 in. is typically used in such cases, and reinforcement is specified in square inches per foot (square millimeters per foot).

While it may be appropriate to distribute temperature and shrinkage reinforcement between two faces of a slab, the minimum flexural reinforcement should be placed as close to the face of the concrete in tension as practical.

11.7.5 Nominal Flexural Strength

OVERVIEW

The nominal flexural strength of a reinforced-concrete member is determined using the two fundamental conditions static equilibrium and compatibility of strains and the design assumptions given in this section. Methods to determine the nominal flexural strength M_n are covered in this section.

RECTANGULAR SECTIONS

Single Layer of Tension Reinforcement Consider the reinforced-concrete beam with one layer of tension reinforcement depicted in Fig. 11.9. The strain distribution and equivalent rectangular stress distribution are also shown in the figure. Because there is only one layer of reinforcement in this beam, the distance d_t from the extreme compression fiber to the centroid of the extreme layer of longitudinal tension steel is equal to the distance d from the extreme compression fiber to the centroid of the longitudinal tension reinforcement.

The ultimate strain in the concrete is 0.003, and the strain in the reinforcement is greater than the yield strain ε_y, assuming that the total area of reinforcement A_s is such that the reinforcing steel yields in tension before the concrete crushes in compression:

$$C = 0.85\ f'_c ba \quad (11.7)$$

In Eq. (11.7), b is the width of the cross-section as shown in Fig. 11.7.

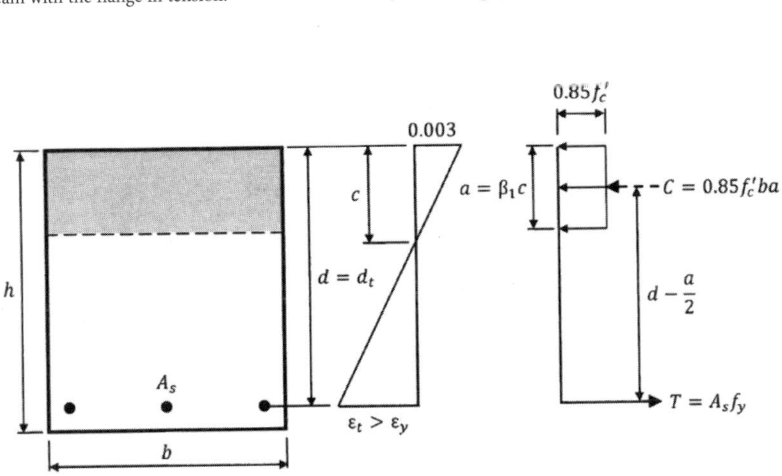

Figure 11.9 The strain and stress distributions in a rectangular beam with tension reinforcement.

The tension force T in the reinforcement is equal to the total area of reinforcement A_s times the yield strength of the reinforcement f_y:

$$T = A_s f_y \tag{11.8}$$

In order for equilibrium to be satisfied, the sum of the forces and bending moments on the section must be equal to zero. From force equilibrium, $C = T$. The depth of the equivalent stress block a can be obtained by equating Eqs. (11.7) and (11.8) and solving for a:

$$a = \frac{A_s f_y}{0.85 f_c' b} \tag{11.9}$$

The nominal flexural strength of the section M_n is obtained from moment equilibrium. Moments can be summed about any point on the section. It is usually convenient to sum moments about either C or T. Summing moments about the point of application of the resultant force C yields the following expression for M_n:

$$M_n = A_s f_y \left(d - \frac{a}{2} \right) \tag{11.10}$$

Substituting Eq. (11.9) into Eq. (11.10) results in the following:

$$M_n = A_s f_y \left(d - \frac{0.59 A_s f_y}{f_c' b} \right) \tag{11.11}$$

Define the reinforcement ratio $\rho = A_s/bd$ and the reinforcement index $\omega = \rho f_y / f_c'$.

Substituting these quantities into Eq. (11.11) results in the following nondimensional equation for M_n:

$$\frac{M_n}{bd^2 f_c'} = \omega(1 - 0.59\omega) \tag{11.12}$$

Flowchart 1 shown in Fig. 11.10 can be used to determine M_n for rectangular sections with one layer of tension reinforcement.

Multiple Layers of Tension Reinforcement Under certain conditions, the required tension reinforcement cannot adequately fit within one layer in a section (spacing requirements for reinforcing steel are covered in Chapter 6 of Ref. 2). In such cases, the bars are provided in more than one layer, as shown in Fig. 11.11.

The nominal flexural strength is determined for sections with multiple layers of tension reinforcement in the same way as that for sections with one layer. When determining the nominal flexural strength, it is important to check that all of the reinforcement yields.

Considering the section shown in Fig. 11.11, assume that the bar size and the number of bars are the same in each layer and that all of the bars are located below the neutral axis. Also assume that the yield strain ε_y occurs at a distance of d_y from the extreme compression fiber. From similar triangles, the following relationship is established between d_y and c:

$$\frac{c}{0.003} = \frac{d_y - c}{\varepsilon_y} \tag{11.13}$$

Solving for d_y results in the following:

$$d_y = c\left(1 + \frac{\varepsilon_y}{0.003}\right) \tag{11.14}$$

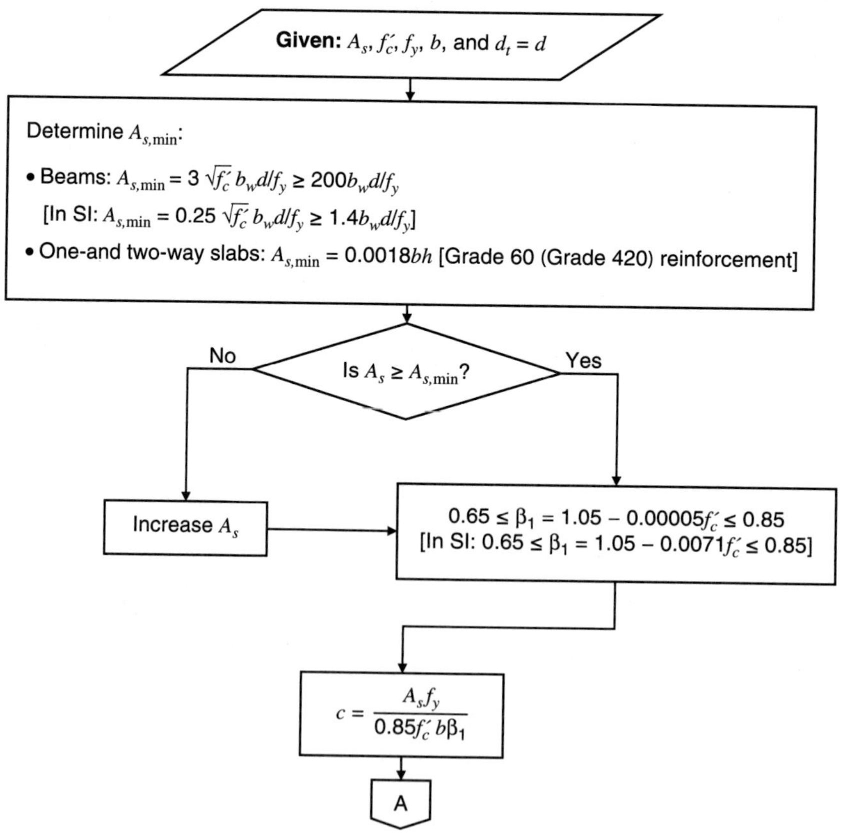

Figure 11.10 Nominal flexural strength—rectangular section with one layer of tension reinforcement.

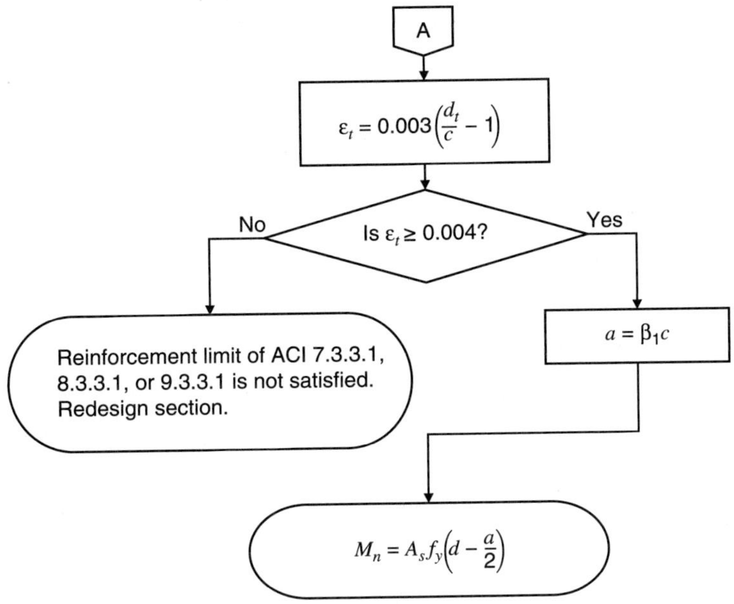

Figure 11.10 *(Continued)*

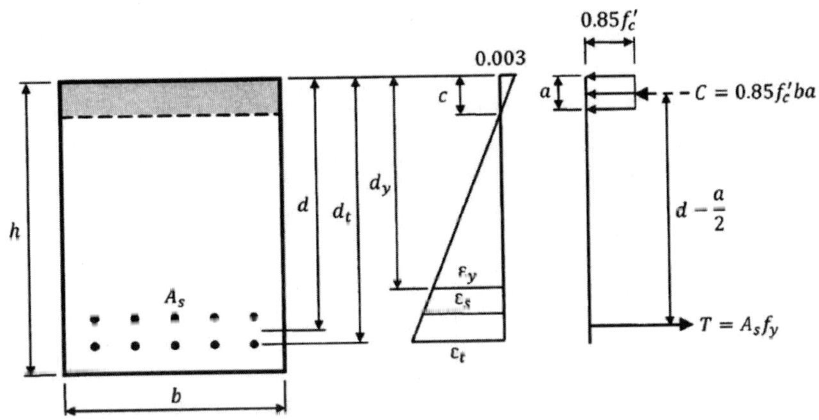

Figure 11.11 A reinforced-concrete beam with multiple layers of tension reinforcement.

For Grade 60 reinforcement ($\varepsilon_y = 60/29{,}000 = 0.00207$), Eq. (11.12) reduces to

$$d_y = 1.7c \tag{11.15}$$

Reinforcement located at a distance equal to or greater than d_y from the extreme compression fiber yields (i.e., $\varepsilon_s \geq \varepsilon_y$ and $f_s = f_y$). The nominal flexural strength is calculated by Eq. (11.10), assuming that the total area of reinforcement is concentrated at d. In situations where one or more layers of the reinforcement do not yield, the reinforcement at those levels is separated from the reinforcement in the layers that yield, and the actual stress in those bars ($f_s < f_y$) is used in the calculation of M_n.

Tension and Compression Reinforcement

Overview The beam depicted in Fig. 11.12 contains reinforcement in both the tension and compression zones of the section, where the compression reinforcement A_s' is located at a distance d' from the extreme compression fiber. Sections with both tension and compression reinforcement are commonly referred to as doubly reinforced sections.

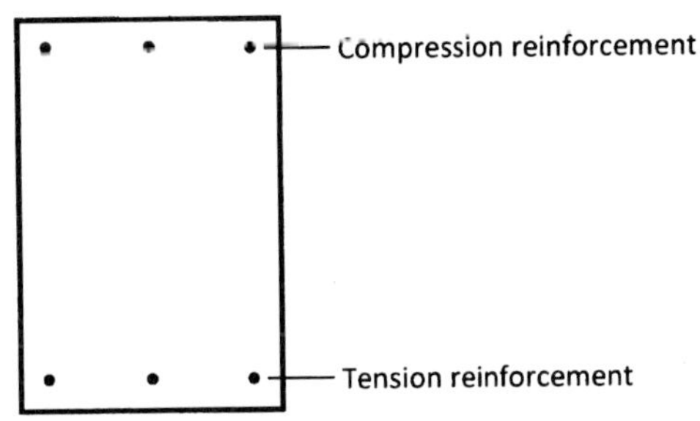

Figure 11.12 Doubly reinforced-rectangular beam.

Compression reinforcement can be added to a section to increase its design flexural strength ϕM_n (ACI 22.2.3.1). This additional strength can be achieved in situations where the dimensions of a beam are limited and the amount of tensile reinforcement that is required to resist the factored bending moments is greater than that permitted by ACI 9.3.3.1.

Reinforcement in the compression zone contributes to the total nominal flexural strength of a section, though the increase in M_n is usually relatively small. The presence of compression reinforcement in a section also results in larger values of ε_t, which essentially produces more ductile behavior. This has a direct impact on the magnitude of the strength-reduction factor ϕ because ϕ is directly proportional to ε_t (see Sec. 11.4). For example, consider a rectangular beam where the strain ε_t is equal to 0.0045. The maximum reinforcement provisions of ACI 9.3.3.1 are satisfied because $\varepsilon_t > 0.004$; however, the section is not tension-controlled because $\varepsilon_t < 0.005$. Thus, the strength-reduction factor is less than 0.90 (see Sec. 11.4). Adding a sufficient amount of reinforcement in the compression zone transforms the section from one that is in the transition region ($\phi < 0.90$) to one that is tension-controlled ($\phi = 0.90$).

Compression reinforcement is also added to help reduce long-term deflections. More information on this topic is also provided in Chapter 6 of Ref. 2.

Longitudinal reinforcement must be provided at both the top and the bottom of a reinforced-concrete beam at certain locations regardless of whether it is needed or not for flexure. It is shown in Chapter 6 of Ref. 2 that transverse reinforcement, which is usually in the form of stirrups, is required in a beam to resist the effects from shear. Stirrups must be anchored to the top and bottom longitudinal bars to properly develop them in tension (see ACI 25.7.1). Thus, longitudinal bars are needed wherever stirrups are required. Stirrups enclose the compression reinforcement and prevent it from buckling.

Nominal Flexural Strength When Compression Reinforcement Yields The strain and stress distributions in a doubly reinforced section are illustrated in Fig. 11.13. Similar to the case with multiple layers of tension steel, the nominal flexural strength of a doubly reinforced section depends on whether the compression reinforcement yields or not.

Assume that the yield strain ε_y occurs at a distance of d'_y from the extreme compression fiber. From similar triangles, the following relationship is established between d'_y and c:

$$\frac{c}{0.003} = \frac{c - d'_y}{\varepsilon_y} \tag{11.16}$$

Solving for d'_y results in the following:

$$d'_y = c\left(1 - \frac{\varepsilon_y}{0.003}\right) \tag{11.17}$$

For Grade 60 reinforcement ($\varepsilon_y = 60/29{,}000 = 0.00207$), Eq. (11.17) reduces to

$$d'_y = 0.31c \tag{11.18}$$

Compression reinforcement located a distance equal to or less than d'_y from the extreme compression fiber yields (i.e., $\varepsilon'_s \geq \varepsilon_y$ and $f'_s \geq f_y$).

When the compression steel yields, the depth of the equivalent stress block a can be obtained by satisfying force equilibrium:

$$T = C + C'_s \tag{11.19a}$$

$$A_s f_y = 0.85 f'_c ba + A'_s f_y \tag{11.19b}$$

$$a = \frac{(A_s - A'_s) f_y}{0.85 f'_c b} \tag{11.19c}$$

The total nominal flexural strength M_n is considered to be the sum of two parts. The first part M_{n1} is provided by the couple consisting of the force in the compression steel A'_s and the force in an equal area of tension steel (see Fig. 11.14):

$$M_{n1} = A'_s f_y (d - d') \tag{11.20}$$

The second part M_{n2} is provided by the couple consisting of the remaining tension steel A_s–A'_s and the compression force in the concrete C:

$$M_{n2} = (A_s - A'_s) f_y \left(d - \frac{a}{2}\right) \tag{11.21}$$

Thus, the total nominal flexural strength of a doubly reinforced section, where $f'_s = f_y$, is

$$M_n = (A_s - A'_s) f_y \left(d - \frac{a}{2}\right) + A'_s f_y (d - d') \tag{11.22}$$

Nominal Flexural Strength When Compression Reinforcement Does Not Yield When the compression reinforcement does not yield $f'_s < f_y$, the depth of the stress block a cannot be determined by Eq. (11.19c)

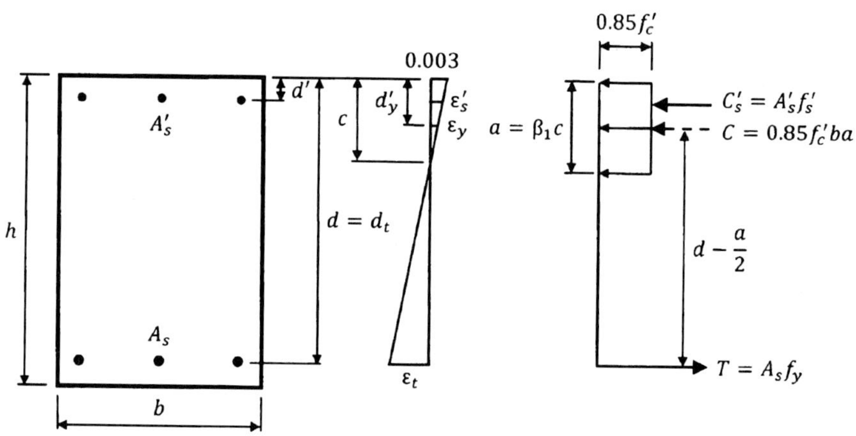

Figure 11.13 Strain and stress distributions in a doubly reinforced-concrete beam.

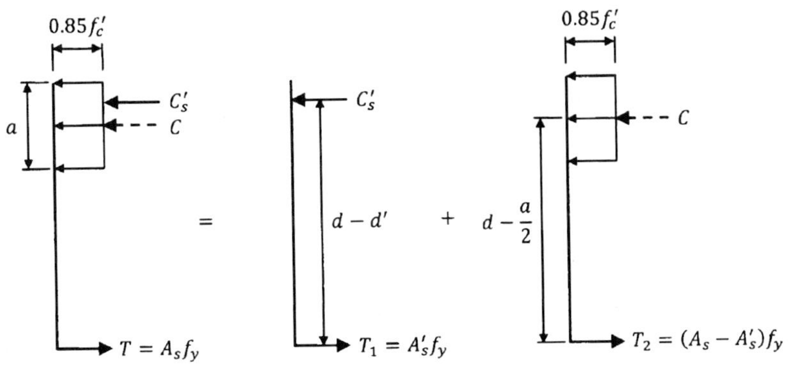

Figure 11.14 Force distribution in a doubly reinforced beam when the compression reinforcement yields.

because the magnitude of f_s' is unknown. A relationship between f_s' and the neutral axis depth c can be obtained from strain compatibility. The strain in the compression reinforcement ε_s' is related to c as follows:

$$\frac{c}{0.003}=\frac{c-d'}{\varepsilon_s'} \tag{11.23}$$

Substituting $\varepsilon_s' = f_s'/E_s$ into Eq. (11.23) and solving for f_s' results in

$$f_s'=0.003E_s\left(1-\frac{d'}{c}\right) \tag{11.24}$$

The neutral axis depth c can be obtained by satisfying force equilibrium:

$$T = C + C_s' \tag{11.25a}$$

$$A_s f_y = 0.85 f_c' ba + A_s' f_y \tag{11.25b}$$

Substituting $a = \beta_1 c$ and Eq. (11.22) into Eq. (11.23b) results in

$$A_s f_y = 0.85 f_c' b\beta_1 c + 0.003 A_s' E_s\left(1-\frac{d'}{c}\right) \tag{11.26a}$$

$$a_1c^2 + b_1c - 87A_s'd' = 0 \tag{11.26b}$$

$$c=\frac{-b_1 \mp \sqrt{b_1^2+348a_1A_s'd'}}{2a_1} \tag{11.26c}$$

where $a_1 = 0.85 f_c' b\beta_1$

$b_1 = 87A_s' - A_s f_y$

$E_s = 29{,}000$ ksi

Note that f_c' and f_y have the units of kips per square inch in the preceding equations.

In SI:

$$c=\frac{-b_1 \mp \sqrt{b_1^2+2400a_1A_s'd'}}{2a_1} \tag{11.26c}$$

where $a_1 = 0.85 f_c' b\beta_1$

$b_1 = 600A_s' - A_s f_y$

$E_s = 200{,}000$ MPa

Note that f_c' and f_y have the units of MPa in the preceding equations.

Once the neutral axis depth c is determined by Eq. (11.26c), Eq. (11.24) can be used to calculate f_s'. The nominal flexural strength is obtained by satisfying moment equilibrium. Summing moments about the centroid of the tensile reinforcement results in the following equation for M_n, where $f_s' < f_y$:

$$M_n = 0.85 f_c' ab\left(d-\frac{a}{2}\right) + A_s' f_s'(d-d') \tag{11.27}$$

Regardless of whether the compression steel yields or not, the strain ε_t must be equal to or greater than 0.004 in order to satisfy the provisions of ACI 7.3.3.1, 8.3.3.1, or 9.3.3.1.

Flowchart 2 shown in Fig. 11.15 can be used to determine M_n for rectangular sections with tension and compression reinforcement.

It is important to note that the contribution of compression reinforcement can be conservatively neglected where it is not specifically required for strength. In such cases, the nominal flexural strength can be computed using the equations developed previously for sections with only tension reinforcement.

T-Section and Inverted L-Section with Tension Reinforcement

Overview The discussions in the previous sections have focused on rectangular reinforced-concrete beams without considering the slab that is supported by the beams. In typical cast-in-place concrete construction, the beams and slabs are cast together with reinforcement that extends between the members; this forms a monolithic structure. Therefore, beams do not act alone but rather work together with the slab to resist the effects from applied loads.

A cast-in-place concrete floor/roof system is shown in Fig. 11.16. The interior beams are commonly referred to as T-beams, where the slab forms the flange of the T-beam and the concrete projection below the slab forms the web or stem of the T-beam. Similarly, the edge beam is referred to as an inverted L-beam.

Effective Flange Width The beams depicted in Fig. 11.17 are subjected to a positive moment that produces compressive stresses at the top of the section above the neutral axis. As seen in the figure, the compressive stresses are greatest over the web and decrease between the webs. In lieu of using a variable compressive stress, the maximum uniform compressive stress is assumed to act over an effective slab (flange) width. The effective width is determined such that the compressive force obtained by multiplying the maximum compressive stress by the effective width is equal to the resultant force of the actual compressive-stress distribution.

Effective flange widths depend on the geometry of the system (beam spacing, slab thickness, and span length) and are defined in ACI 6.3.2.1 for both T-beams and inverted L-beams.

The requirements of ACI 6.3.2.1 are summarized in Fig. 11.18 for the general case of varying span length, web width, and beam spacing.

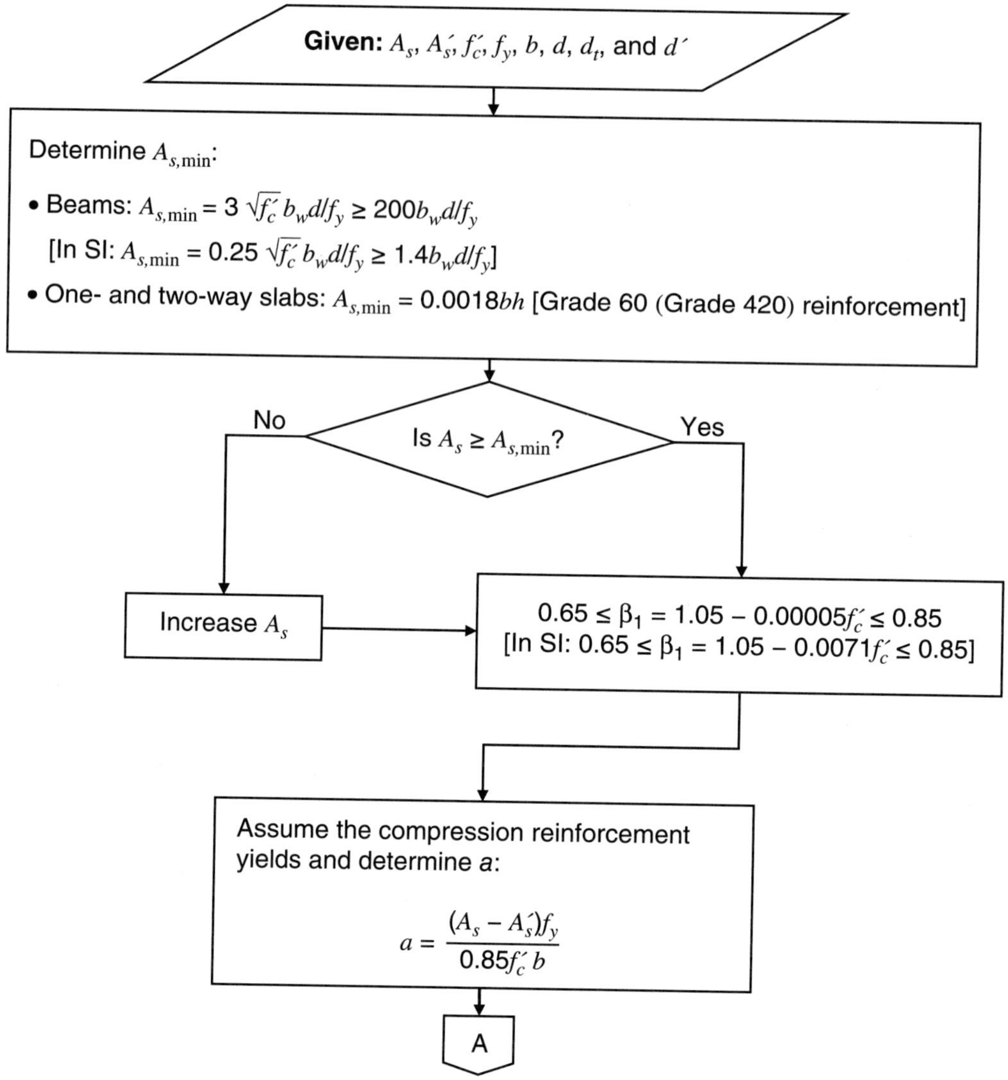

Figure 11.15 Nominal flexural strength—rectangular section with tension and compression reinforcement.

In Fig. 11.18, s_{w1} and s_{w2} are the clear distances between adjacent beam webs, and ℓ_{n1} and ℓ_{n2} are the clear span lengths of the beams. The nominal flexural strength of flanged sections is determined using this effective flange width.

Nominal Flexural Strength—Flange in Tension The flange of a T-beam or an inverted L-beam will be in tension at locations of negative moment, which in a continuous system usually occur at the faces of a support. The strain and stress distributions for a T-beam in such a case are illustrated in Fig. 11.19, where it can be seen that the compression zone falls within the web of the member. An inverted L-beam would have similar distributions.

The nominal flexural strength is determined by the equations developed previously for rectangular sections with a single layer or multiple layers of tension reinforcement (i.e., two layers of reinforcement in the slab), where $b = b_w$. Flowchart 1 shown in Fig. 11.18 can be used to determine M_n. If needed, the contribution of the compression reinforcement in the web can be included where the nominal flexural strength is determined by the equations developed previously for doubly reinforced sections (see the Flowchart 2 shown in Fig. 11.15).

Nominal Flexural Strength—Flange in Compression At locations of positive moment, which usually occur away from the faces of a support in a continuous system, a portion of the flange or the entire flange of a T-beam or an inverted L-beam will be in compression. The determination of the nominal flexural strength depends on whether the depth of the stress block a is less than or greater than the thickness of the flange. Both cases are examined next.

Depth of Stress Block Less Than or Equal to Flange Thickness ($a \leq h$) When the depth of the compression zone a falls within the flange of a T-beam or an inverted L-beam, the compressive zone is rectangular with a width equal to the effective flange width b_e (see Fig. 11.20). The nominal flexural strength of the section is determined by the equations developed earlier for rectangular sections.

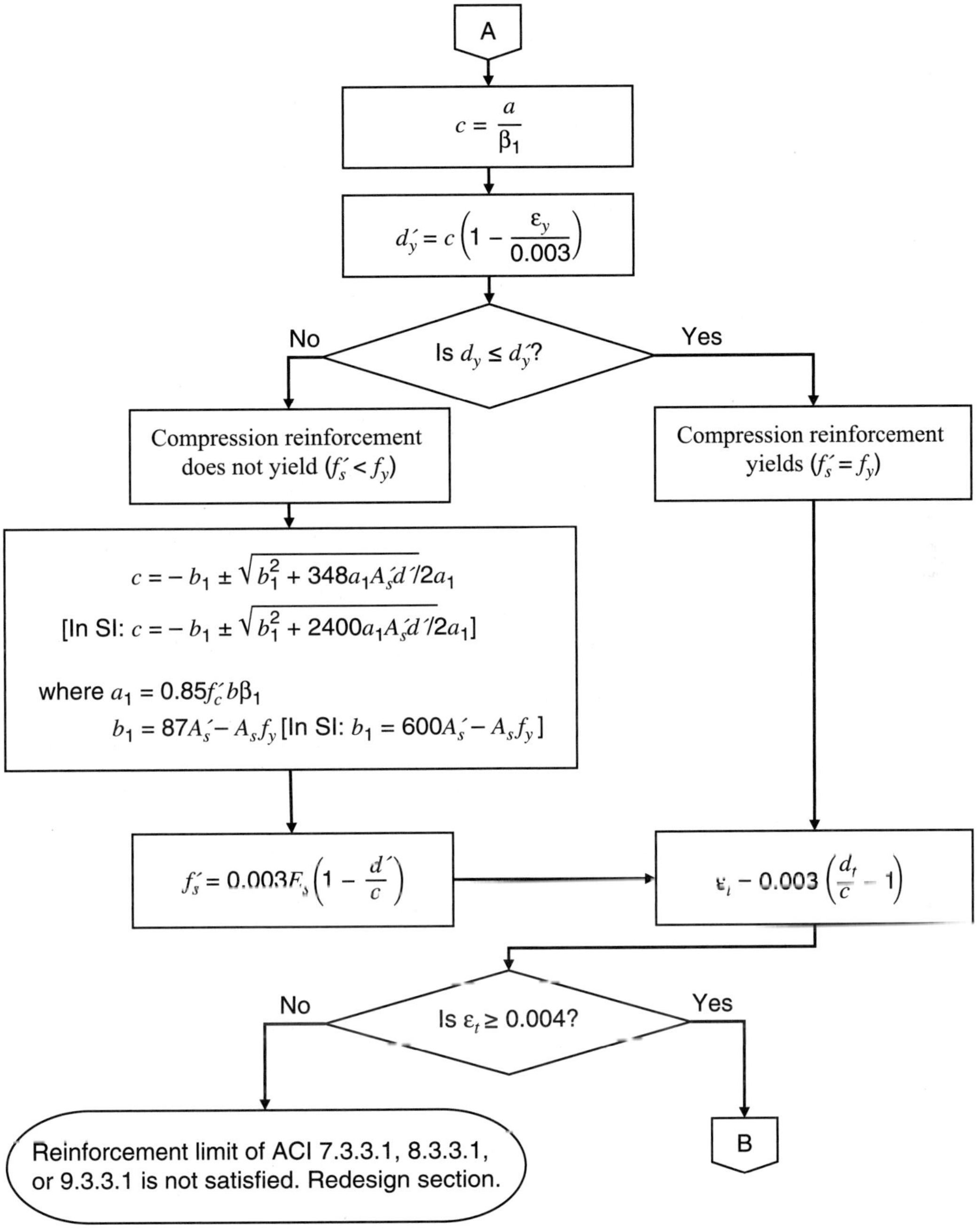

Figure 11.15 *(Continued)*

Depth of Stress Block Greater Than Flange Thickness ($a > h$) When the depth of the stress block a falls within the web of the beam, the compressive zone is T or L shaped as opposed to rectangular (see Fig. 11.21). The resultant force C is equal to $0.85f'_c$ times the area of the compressive zone and is located at its centroid. In such cases, it is convenient to divide the tensile reinforcement into two parts. The first part A_{sf} is defined as the area of steel that is required to balance the compressive force in the overhanging portions of the flange. This is depicted in Fig. 11.22. The following equation for A_{sf} is obtained from horizontal equilibrium, assuming that the tension reinforcement yields:

$$A_{sf} = \frac{0.85 f'_c(b_e - b_w)h}{f_y} \tag{11.28}$$

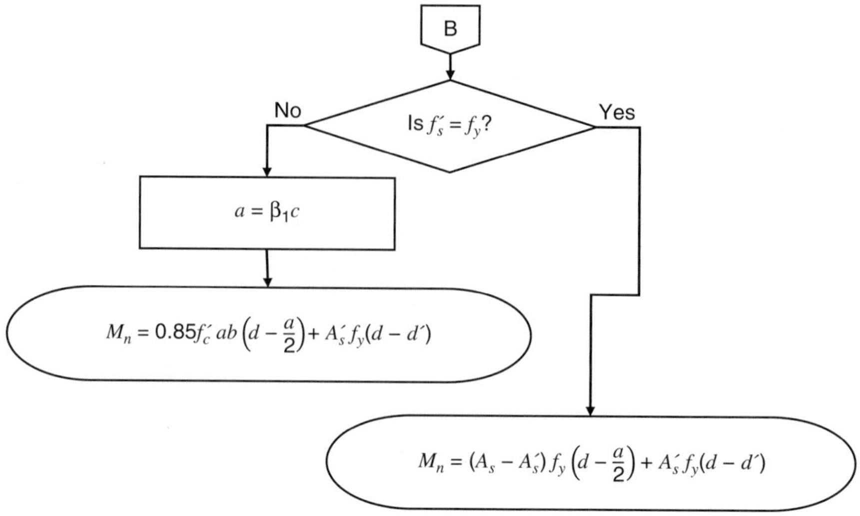

Figure 11.15 (*Continued*)

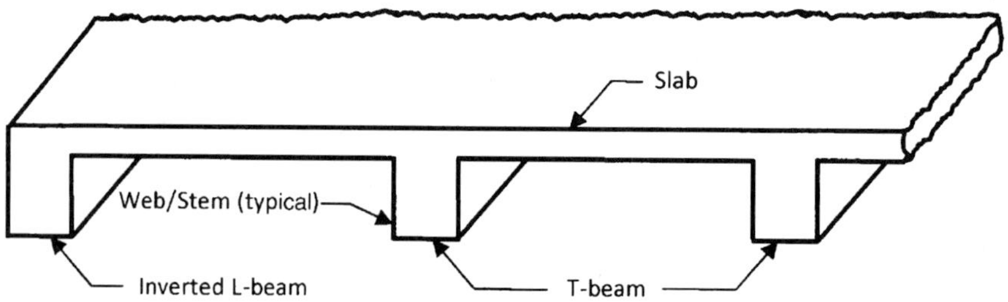

Figure 11.16 T-beam and inverted L-beam construction.

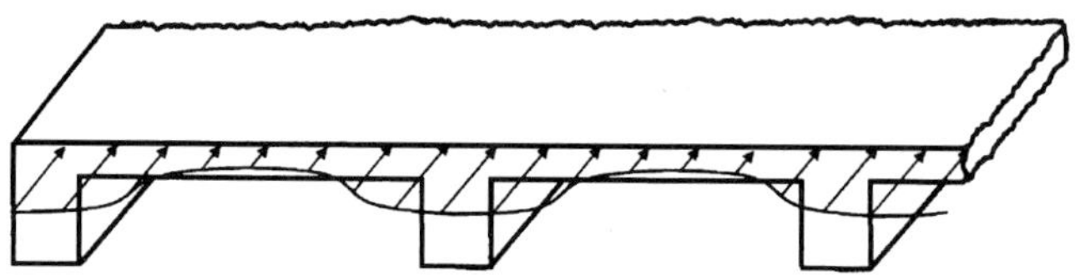

Figure 11.17 Distribution of compressive stress in a T-beam subjected to a positive moment.

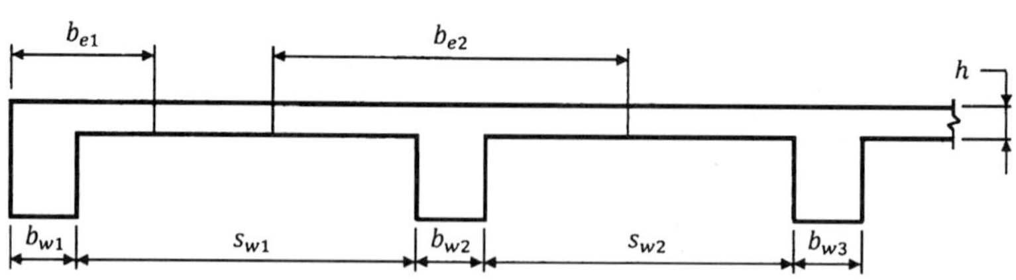

$$b_{e1} \le \begin{cases} b_{w1} + 6h \\ b_{w1} + \dfrac{s_{w1}}{2} \\ b_{w1} + \dfrac{\ell_{n1}}{12} \end{cases} \qquad b_{e2} \le \begin{cases} b_{w2} + 16h \\ b_{w2} + \dfrac{s_{w1}}{2} + \dfrac{s_{w2}}{2} \\ b_{w2} + \dfrac{\ell_{n2}}{4} \end{cases}$$

Figure 11.18 Effective flange widths for a T-beam and an inverted L-beam.

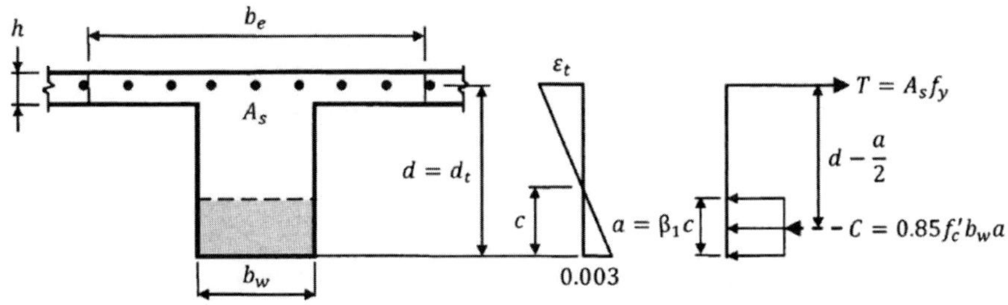

Figure 11.19 Strain and stress distributions in a T-beam with the flange in tension.

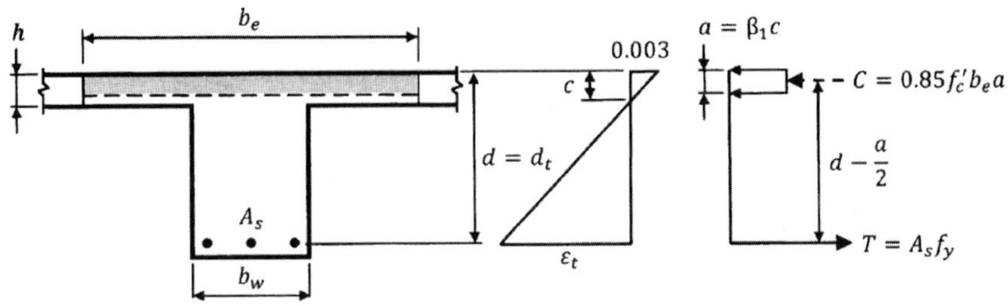

Figure 11.20 Strain and stress distributions in a T-beam with the flange in compression and $a \leq h$.

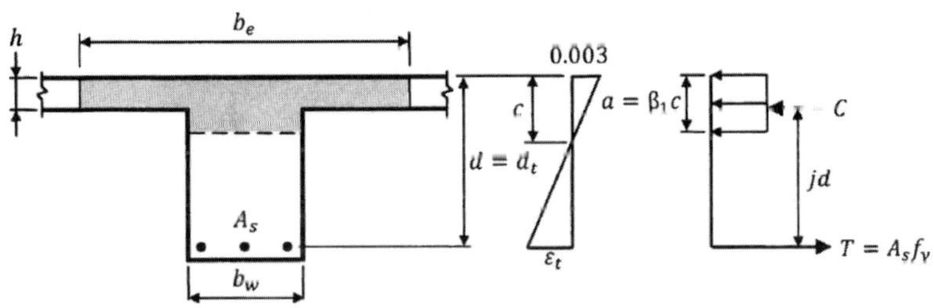

Figure 11.21 Strain and stress distributions in a T-beam with the flange in compression and $a > h$.

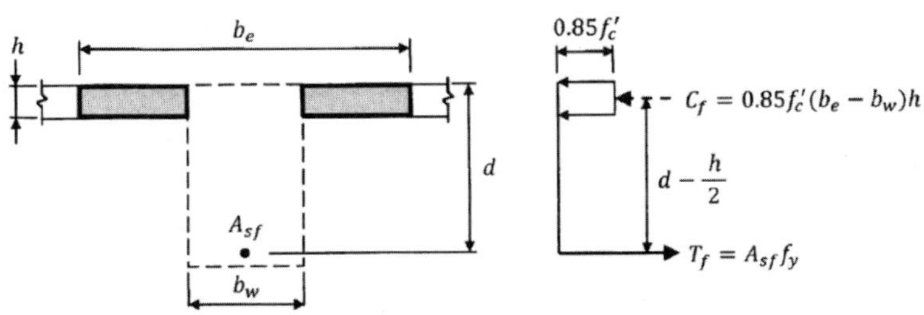

Figure 11.22 Stress distribution corresponding to overhanging flanges.

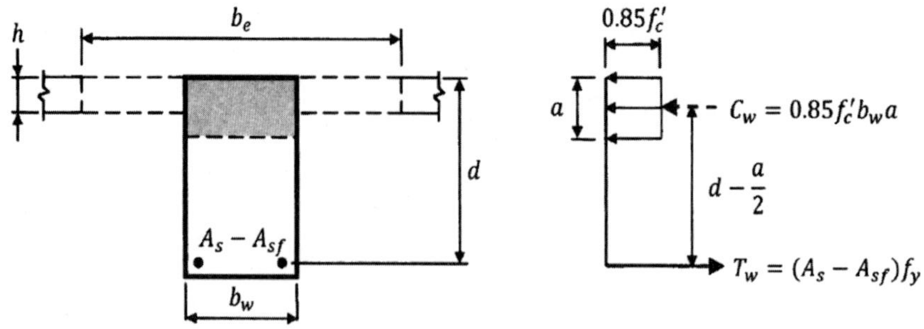

Figure 11.23 Stress distribution corresponding to the web.

The nominal flexural resistance provided by the tensile reinforcement A_{sf} is obtained from moment equilibrium:

$$M_{n1} = A_{sf} f_y \left(d - \frac{h}{2} \right) \tag{11.29}$$

The remaining part of the tensile reinforcement A_s–A_{sf} is balanced by the compression force in the rectangular part of the web (see Fig. 11.23). As usual, the depth of the stress block a is determined from horizontal equilibrium:

$$a = \frac{(A_s - A_{sf}) f_y}{0.85 f_c' b_w} \tag{11.30}$$

The nominal flexural resistance provided by the tensile reinforcement $(A_s - A_{sf})$ is obtained from moment equilibrium:

$$M_{n2} = (A_s - A_{sf}) f_y \left(d - \frac{a}{2} \right) \tag{11.31}$$

Thus, the total nominal flexural strength of the section M_n where $a > h$ is the addition of the two parts corresponding to the overhanging flanges and the web:

$$M_n = M_{n1} + M_{n2} = A_{sf} f_y \left(d - \frac{h}{2} \right) + (A_s - A_{sf}) f_y \left(d - \frac{a}{2} \right) \tag{11.32}$$

Flowchart 3 shown in Fig. 11.24 can be used to determine M_n for a T-beam or an inverted L-beam where the flange is in compression. Like in the case of rectangular sections, the minimum and maximum reinforcement requirements of ACI 9.6.1.2 and 9.3.3.1, respectively, must be satisfied.

11.7.6 Shear Strength

Introduction

In general, the design for shear consists of calculating the maximum factored shear force V_u and requiring that it is equal to or less than the design shear strength ϕV_n:

$$V_u \le \phi V_n \tag{11.33}$$

The factored shear force V_u is determined by combining the nominal shear forces determined from analysis in accordance with the load combinations given in ACI Table 5.3.1. According to ACI Table 21.2.1, the strength-reduction factor ϕ is equal to 0.75 for shear (additional requirements are given in ACI 21.2.4 for structures designed to resist earthquake effects).

The nominal shear strength V_n is determined by ACI Eq. (22.5.1.1):

$$V_n = V_c + V_s \tag{11.34}$$

In this equation, V_c is the nominal shear strength provided by concrete and V_s is the nominal shear strength provided by shear reinforcement.

Required Shear Strength

The factored shear force V_u is determined at a section using the load combinations given in ACI Table 5.3.1. An approximate shear force at the face of a support for gravity load effects can be determined using the coefficients given in ACI Table 6.5.4.

The locations of the critical sections for shear are given in ACI 7.4.3 for one-way slabs and ACI 9.4.3 for beams. It is permitted to design sections located less than a distance d from the face of a support for V_u computed at the distance d, provided the following three criteria are satisfied:

1. Support reactions in the direction of the applied shear force introduce compression into the end regions of the member.
2. Loads are applied at or near the top of the member.
3. No concentrated loads occur between the face of the support and the critical section.

Shear Strength Provided by Concrete

For members without axial force, ACI 22.5.5.1 contains two sets of equations to determine the shear strength provided by the concrete V_c:

$$V_c = 2\lambda\sqrt{f_c'}b_w d \tag{11.35}$$

$$V_c = \text{Least of} \begin{Bmatrix} \left(1.9\lambda\sqrt{f_c'} + 2500\rho_w \dfrac{V_u d}{M_u}\right) b_w d \\ (1.9\lambda\sqrt{f_c'} + 2500\rho_w) b_w d \\ 3.5\lambda\sqrt{f_c'} b_w d \end{Bmatrix} \tag{11.36}$$

where f_c' has the units of pounds per square inch (psi).

In SI:

$$V_c = 0.17\lambda\sqrt{f_c'}b_w d \tag{11.37}$$

$$V_c = \text{Least of} \begin{Bmatrix} \left(0.16\lambda\sqrt{f_c'} + 17\rho_w \dfrac{V_u d}{M_u}\right) b_w d \\ (0.16\lambda\sqrt{f_c'} + 17\rho_w) b_w d \\ 0.29\lambda\sqrt{f_c'} b_w d \end{Bmatrix} \tag{11.38}$$

where f_c' has the units of MPa.

In these equations, λ is a modification factor that reflects the reduced mechanical properties of lightweight concrete (see ACI Tables 19.2.4.2 and 19.2.4.3):

- $\lambda = 0.85$ for sand-lightweight concrete
- $\lambda = 0.75$ for all-lightweight concrete
- $\lambda = f_{ct}/6.7\sqrt{f_{cm}} \le 1.0$ [In SI: $\lambda = f_{ct}/0.56\sqrt{f_{cm}} \le 1.0$]

where the average splitting tensile strength of lightweight concrete f_{ct} has been determined by tests and f_{cm} is the measured average compressive strength of concrete.

Note that $6.7\sqrt{f_{cm}}$ is the average, measured splitting tensile strength of normal-weight concrete.

- $\lambda = 1.0$ for normal-weight concrete.

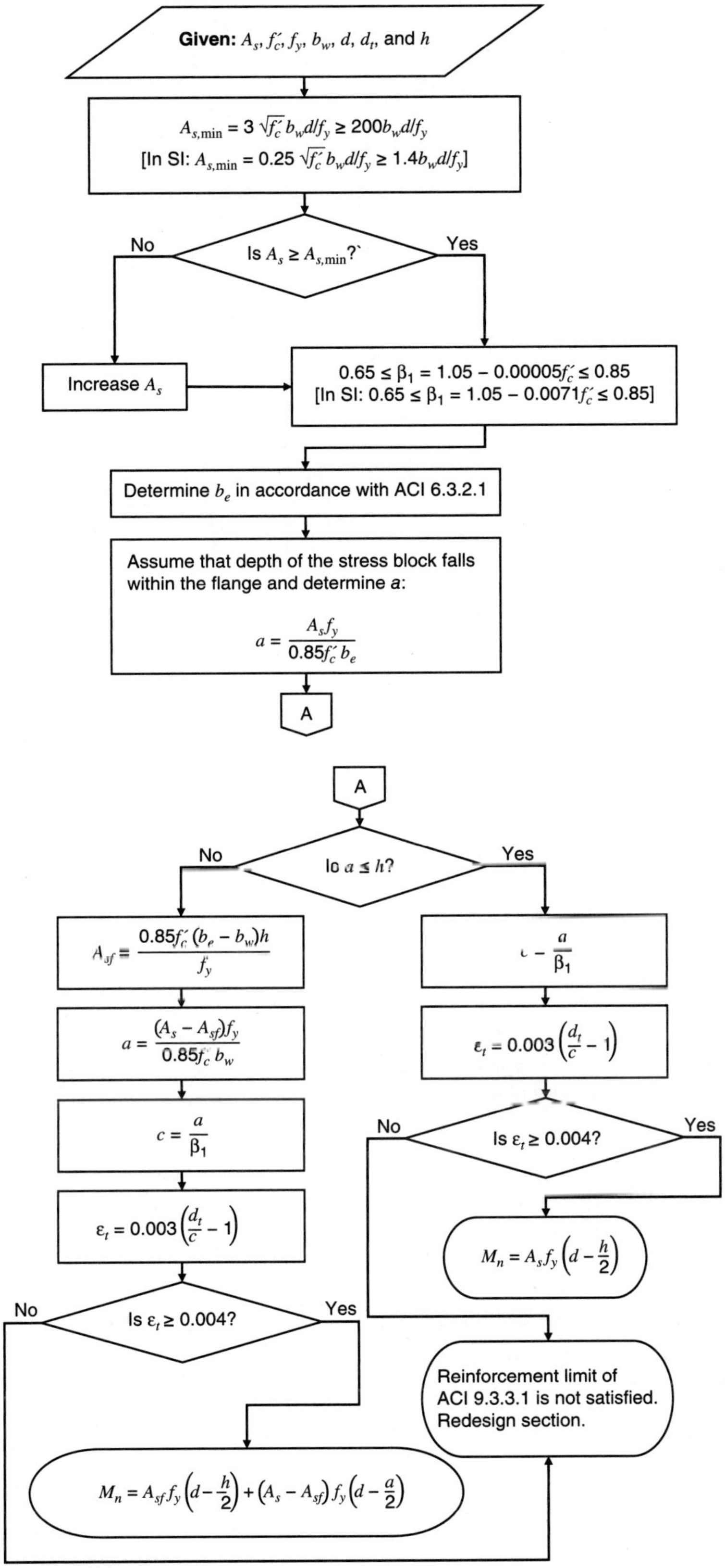

Figure 11.24 *Nominal flexural strength*—T-beam or an inverted L-beam where the flange is in compression.

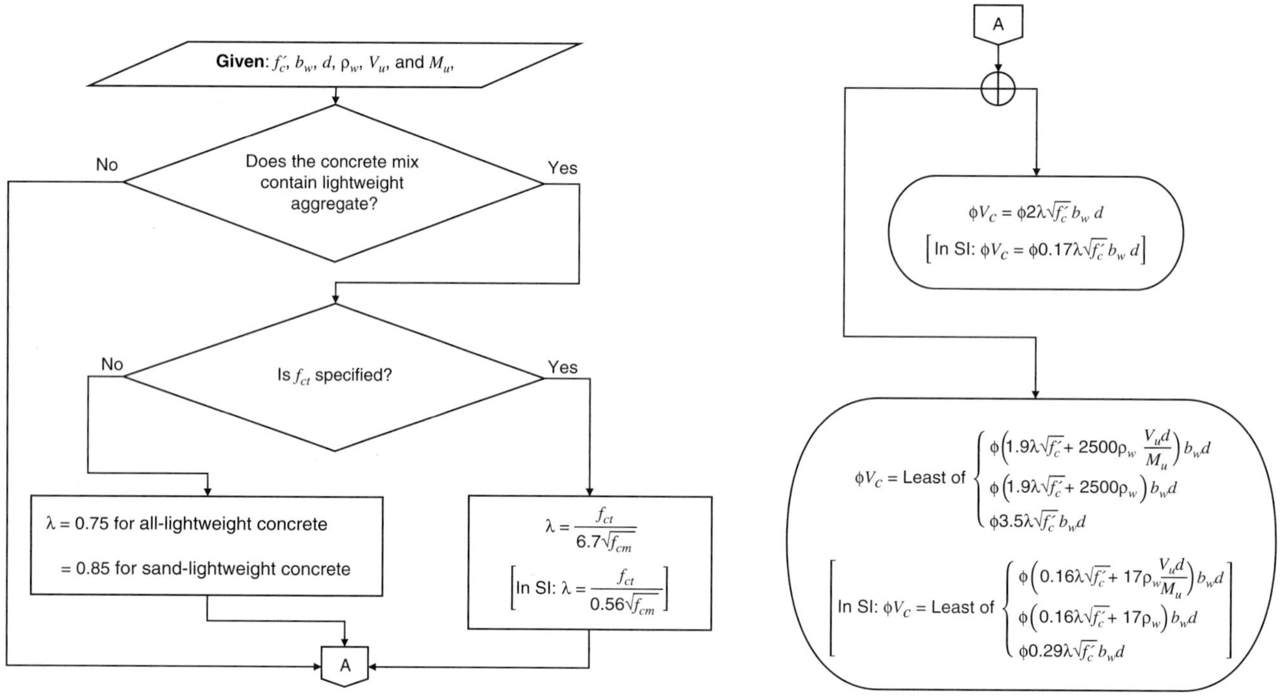

Figure 11.25 Design shear strength for concrete ϕV_c (Flowchart 4).

SHEAR STRENGTH PROVIDED BY SHEAR REINFORCEMENT

Types of Shear Reinforcement ACI 22.5.10 permits the following types of shear reinforcement:

1. Stirrups, ties, or hoops perpendicular to the longitudinal axis of the member
2. Welded wire reinforcement with wires located perpendicular to the longitudinal axis of the member
3. Spiral reinforcement
4. Inclined stirrups making an angle of at least 45 degrees with the longitudinal axis of the member and crossing the plane of the potential shear crack
5. Longitudinal reinforcement that is bent an angle of 30 degrees or more with respect to the longitudinal axis of the member

Stirrups that are oriented perpendicular to the axis of the member and are anchored to the longitudinal flexural reinforcement are the most commonly used type of shear reinforcement in beams. However, in areas of moderate to high seismic risk, spirals or hoops must be used in accordance with ACI Chapter 18. Inclined stirrups and bent longitudinal bars are rarely used in practice.

DESIGN OF SHEAR REINFORCEMENT

Shear reinforcement is required to augment the overall shear strength of a reinforced-concrete flexural member. Prior to diagonal cracking, the stress in such reinforcement is essentially zero. After diagonal cracks develop, shear reinforcement restricts crack growth and penetration into the compression zone. The width of the cracks is also controlled.

The required shear reinforcement depends on the magnitude of the factored shear force V_u and the design shear strength of the concrete ϕV_c. The Code requirements are summarized in Fig. 11.26.

Torsion is covered in Section 9.5-9.7 and Section 22.7 of ACI 318.

11.8 COLUMNS

11.8.1 Overview

This section focuses on the determination of the nominal axial strength P_n for short, reinforced-concrete columns that are subjected to essentially concentric axial loads. A *short column* is defined as one in which slenderness effects need not be considered. Slender columns are discussed in Chapter 8 of Ref. 2.

The nominal axial strength of a short column is related to the area of the column, the compressive strength of the concrete, the area and yield stress of the longitudinal reinforcement, and the type of transverse reinforcement. These quantities and their relationship to axial strength are discussed later.

11.8.2 Maximum Concentric Axial Load

Consider a reinforced-concrete column subjected to a concentric axial load P. Assume that the longitudinal reinforcement is symmetrically distributed in the section and that transverse reinforcement that meets the size and spacing requirements of ACI 10.7.6 is provided. The type of transverse reinforcement is not relevant to the discussion at this time.

When subjected to P, the length of the column L decreases by an amount equal to the longitudinal strain ε times the original length L. For a concentric axial load, ε is uniform across the section. The strains in the concrete and the steel are equal because the concrete and the longitudinal steel are bonded together. For any given ε, it is possible to compute the stresses in the concrete and longitudinal steel using the stress–strain curves of the materials (see Chapter 2 of Ref. 2). The loads in the concrete and the longitudinal steel are equal to the stresses multiplied by the corresponding areas, and the total load that a short column can carry is the sum of the maximum loads carried by the concrete and the steel.

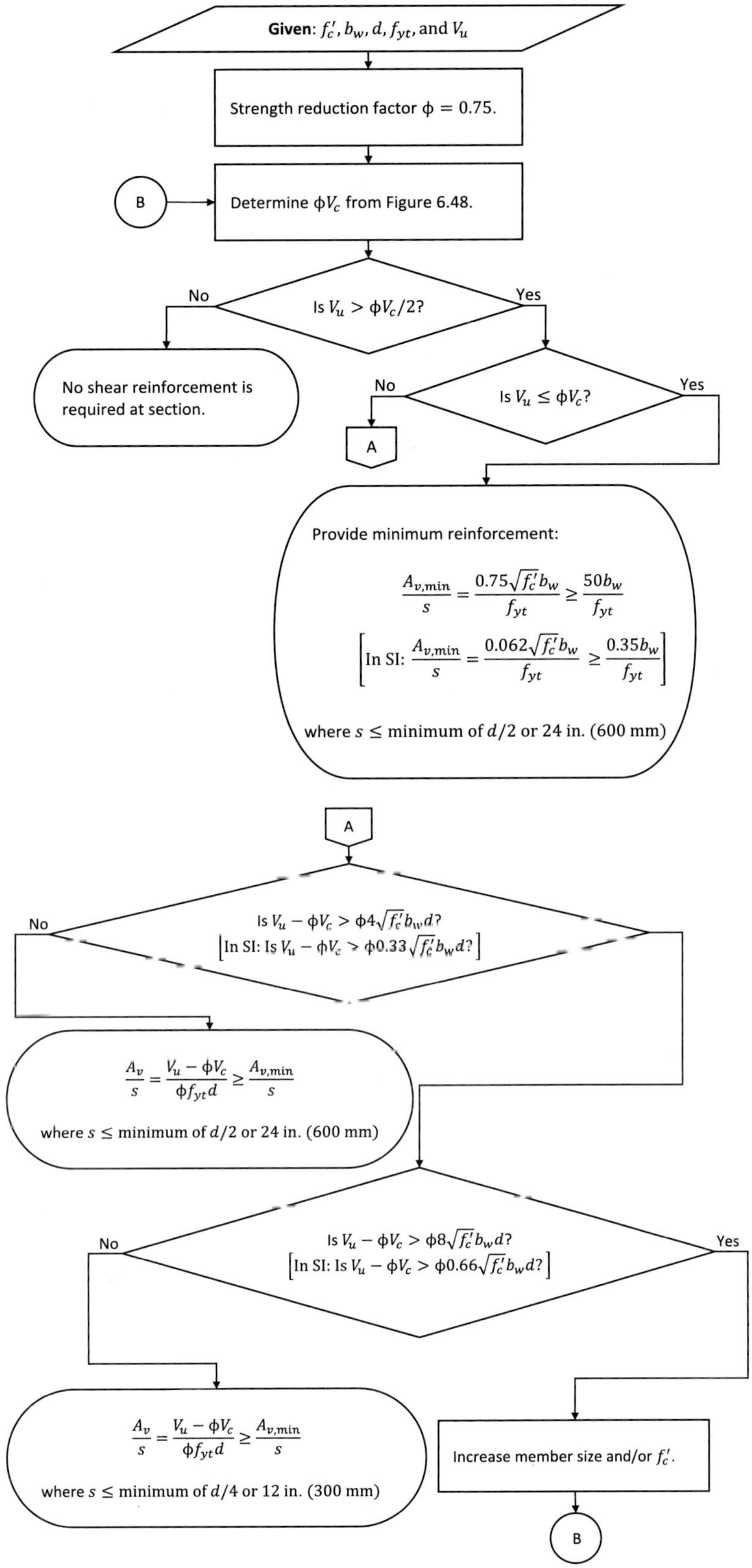

Figure 11.26 Required shear reinforcement for members subjected to shear and flexure.

The maximum compressive axial load that can be resisted by the concrete P_c is equal to the following:

$$P_c = 0.85(A_g - A_{st})f_c' \tag{11.39}$$

In this equation, A_g is the gross area of the column and A_{st} is the total area of longitudinal reinforcement in the column; thus, $(A_g - A_{st})$ is equal to the area of the concrete. The factor 0.85 is based on the results of numerous tests (Ref. 2).

The maximum axial load that can be carried by the longitudinal reinforcement P_s is equal to the area times the yield strength of the reinforcement:

$$P_s = A_{st}f_y \tag{11.40}$$

Therefore, the maximum concentric axial load P_o that can be carried by a short column is equal to the summation of the maximum loads of the concrete and the steel (see ACI 22.4.2.2):

$$P_o = 0.85(A_g - A_{st})f_c' + A_{st}f_y \tag{11.41}$$

Equation (11.33) forms the basis of the nominal axial strength, which is discussed next.

11.8.3 Nominal Axial Strength

In general, the maximum nominal axial strength $P_{n,\max}$ is equal to a constant times the concentric axial load strength P_o. The constant depends on the type of transverse reinforcement utilized in the section and accounts for any accidental eccentricities—or, equivalently, any accidental bending moments—that may exist in a column and were not considered in the analysis. These eccentricities can arise from unbalanced moments in the beams framing into the column, misalignment of columns from floor to floor, or misalignment of the longitudinal reinforcement in the column, to name a few.

ACI Table 22.4.2.1 contains a summary of the maximum nominal axial strengths based on the type of transverse reinforcement in a member. For members with spiral reinforcement conforming to ACI 22.4.2.5, the constant is equal to 0.85. Therefore,

$$P_{n,\max} = 0.85[0.85f_c'(A_g - A_{st}) + f_yA_{st}] \tag{11.42}$$

The constant 0.85 produces nominal strength approximately equal to that from earlier Codes with the axial load applied at an eccentricity equal to 5 percent of the column depth.

Similarly, for members with tie reinforcement conforming to ACI 22.4.2.4, the constant is equal to 0.80, and $P_{n,\max}$ is

$$P_{n,\max} = 0.8[0.85f_c'(A_g - A_{st}) + f_yA_{st}] \tag{11.43}$$

In this case, the axial load is applied at an eccentricity equal to approximately 10 percent of the column depth.

The design axial strength ϕP_n is obtained by multiplying the maximum nominal axial strength $P_{n,\max}$ by the corresponding strength-reduction factor ϕ. Thus,

- for members with spiral reinforcement,

$$\phi P_{n,\max} = 0.85\phi[0.85f_c'(A_g - A_{st}) + f_yA_{st}] \tag{11.44}$$

- for members with tie reinforcement,

$$\phi P_{n,\max} = 0.8\phi[0.85f_c'(A_g - A_{st}) + f_yA_{st}] \tag{11.45}$$

The strength-reduction factor ϕ is equal to 0.75 for compression-controlled sections with spiral reinforcement and 0.65 for other reinforced members including tied reinforcement (see Sec. 11.4 and ACI Table 21.2.2). It was noted in Sec. 11.4 that the larger ϕ-factor for columns with spiral reinforcement reflects the more ductile behavior of such columns compared with columns with tied reinforcement.

11.8.4 Longitudinal Reinforcement Limits

ACI 10.6.1.1 prescribes the limits on the amount of longitudinal reinforcement for columns, which are applicable to all such members regardless of the type of transverse reinforcement:

- Minimum $A_{st} = 0.01A_g$
- Maximum $A_{st} = 0.08A_g$

The lower limit is meant to provide resistance to bending, which may exist even though an analysis shows that it is not present, and to reduce the effects of creep and shrinkage of the concrete under sustained compressive stresses.

The upper limit is a practical maximum for longitudinal reinforcement in terms of economy and placement of the bars: For proper concrete placement and consolidation, the size and number of longitudinal bar sizes must be chosen to minimize reinforcement congestion, especially at beam-column joints. If column bars are lap spliced, the maximum area of longitudinal reinforcement should not exceed 4 percent of the gross column area at the location of the splice.

ACI 10.7.3.1 contains requirements on the minimum number of longitudinal bars in columns. These and other requirements are covered in Chapter 8 of Ref. 2.

Flowchart 6 shown in Fig. 11.27 can be used to determine $P_{n,\max}$.

11.9 TENSION MEMBERS

Reinforced-concrete tension members occur in certain specialty structures such as arches and trusses. As has been discussed in Sec. 11.1, the tensile strength of concrete is relatively small compared with its compressive strength. As such, the tensile strength of concrete is neglected in the design of tension members, and it is assumed that the tension load is resisted solely by the longitudinal reinforcement. Therefore, the nominal tensile strength T_n of a symmetrical reinforced-concrete tension member subjected to a concentric axial tension load is equal to the area times the yield strength of the longitudinal steel:

$$T_n = A_{st}f_y \tag{11.46}$$

11.10 MEMBERS SUBJECTED TO FLEXURE AND AXIAL LOAD

11.10.1 Overview

Columns in reinforced-concrete buildings can be subjected to the effects from more than just compressive loads. Consider columns. These members are subjected to axial compressive loads, bending moments, and shear forces from gravity loads and wind loads.

Like all reinforced-concrete members, the nominal strength of a section subjected to both flexure and axial load is determined using equilibrium, strain compatibility, and the design assumptions given in Sec. 11.5. Methods to determine the nominal strength of such members are presented later for the case of axial compression and flexure.

11.10.2 Nominal Strength

The general principles and assumptions of the strength design method are applied to the section depicted in Fig. 11.28, which is subjected to an axial compressive load and flexure. For purposes of discussion, assume that the longitudinal reinforcement in layer 1 is located closest to the extreme compression fiber of the section.

In accordance with design assumption no. 2 in Sec. 11.5, the maximum strain in the extreme compression fiber of the concrete $\varepsilon_c = 0.003$.

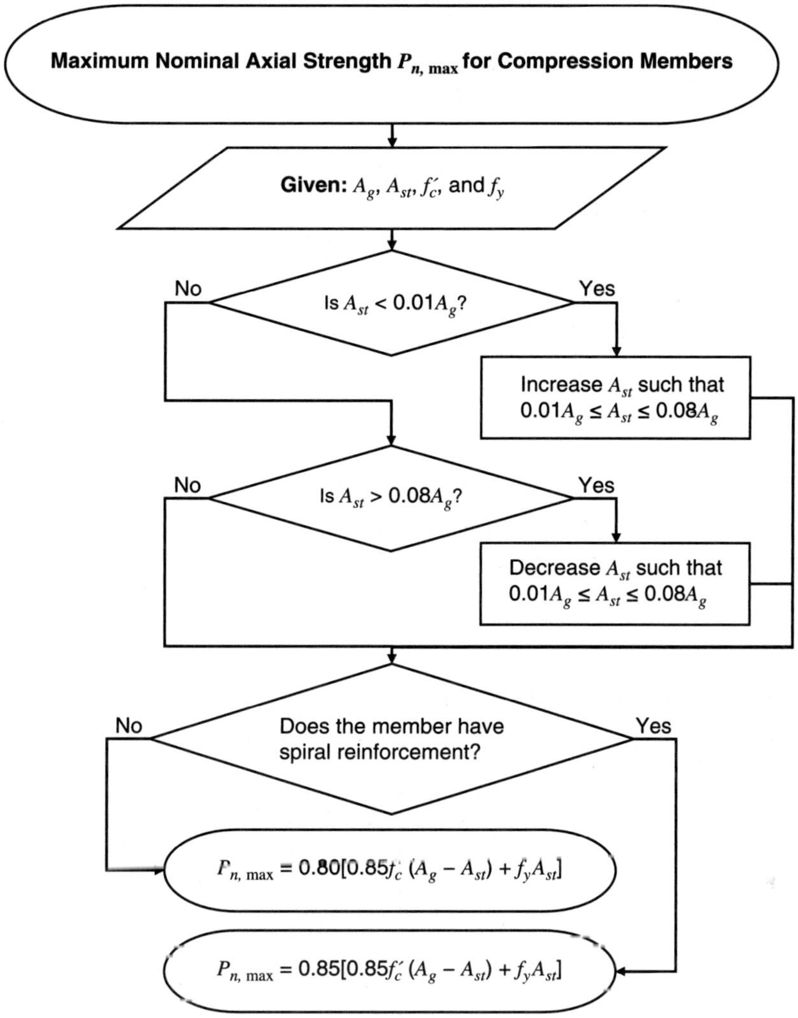

Figure 11.27 Maximum and nominal axial strength.

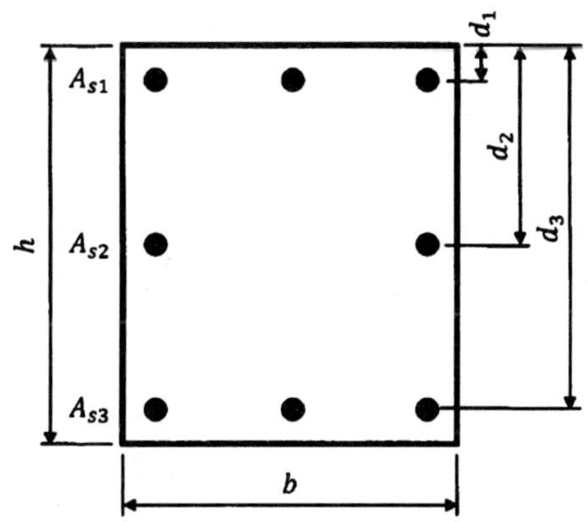

Figure 11.28 A reinforced-concrete section subjected to uniaxial compression and flexure.

By assuming a value for the strain ε_{s3} in the reinforcement in layer 3, the depth to the neutral axis c can be determined from similar triangles (see Fig. 11.29). Because the reinforcement in layer 3 is farthest from the extreme compression of the section, $\varepsilon_{s3} = \varepsilon_t$ and

$$c = \frac{0.003d_3}{\varepsilon_t + 0.003} \tag{11.47}$$

Utilizing design assumption no. 6 of Sec. 11.5, the resultant compressive force C in the concrete is obtained by multiplying the stress $0.85f_c'$ by the area under the equivalent rectangular stress block:

$$C = 0.85f_c'ab \tag{11.48}$$

In Eq. (11.40), the depth of the equivalent stress block $a = \beta_1 c$.

The strain ε_{si} in the reinforcement in layer i can also be determined from similar triangles:

$$\varepsilon_{si} = \frac{0.003(c - d_i)}{c} \tag{11.49}$$

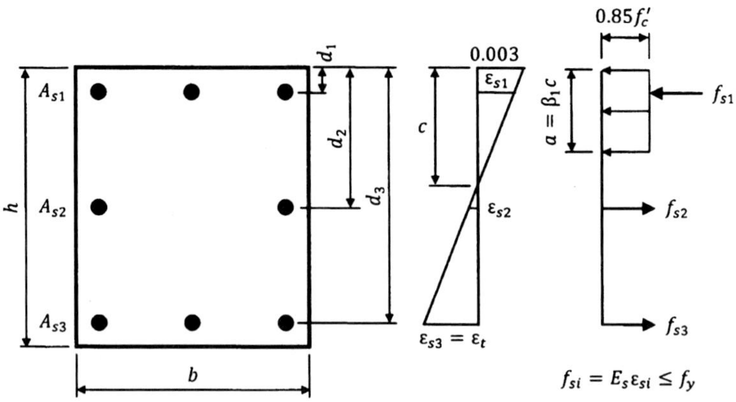

Figure 11.29 Strain and stress distributions in a section subjected to uniaxial compression and flexure.

For elastic–plastic reinforcement with the stress–strain curve defined by design assumption no. 3 given in Sec. 11.5, the stress in the reinforcement at each layer f_{si} is equal to the strain ε_{si} at that level determined by Eq. (11.49) times the modulus of elasticity of steel E_s. It is important to keep in mind that f_{si} must not exceed the yield stress f_y in tension or compression.

The magnitude of the force F_{si} in the reinforcement depends on whether the steel is in the equivalent compression zone or not:

If d_i is greater than the depth of the equivalent stress block a,

$$F_{si} = f_{si} A_{si} \tag{11.50}$$

For the section depicted in Fig. 11.25, this equation would apply to layers 2 and 3.

If d_i is less than the depth of the equivalent stress block a,

$$F_{si} = (f_{si} - 0.85 f'_c) A_{si} \tag{11.51}$$

- In this case, which is applicable to layer 1 shown in Fig. 11.29, the area of reinforcement in that layer has been included in the area ab used to compute the compressive force in the concrete C. Thus, $0.85f'_c$ must be subtracted from the steel stress f_{si} in that layer before computing the force F_{si}.

The nominal axial strength P_n for the assumed strain distribution is obtained by summing the axial forces on the section (see Fig. 11.30):

$$P_n = C + \sum F_{si} \tag{11.52}$$

Similarly, the nominal flexural strength M_n for the assumed strain distribution is determined by summing moments about the centroid of the column because this is the axis about which moments are computed in a conventional structural analysis (see Fig. 11.30):

$$M_n = 0.5C(h-a) + \sum F_{si}(0.5h - d_i) \tag{11.53}$$

Flowchart 5 shown in Fig. 11.31 can be used to determine P_n and M_n as a function of the net tensile strain in the extreme layer of longitudinal tension steel ε_t.

11.11 WALLS

11.11.1 Introduction

A wall is defined in ACI 2.3 as a member, usually vertical, that is designed to resist axial load, lateral load, or both, with a horizontal length-to-thickness ratio greater than three, and that is used to enclose or separate spaces in a building or structure. There are many different types of walls, but they are typically categorized as non–load-bearing and load-bearing. A non–load-bearing wall primarily supports its own weight. In contrast, a load-bearing wall supports dead and live loads from the floor and roof systems in addition to its own weight. IBC 202 contains more precise definitions of these wall types.

Because of their relatively large in-plane lateral stiffness, walls can attract a significant portion of the effects due to wind or earthquakes.

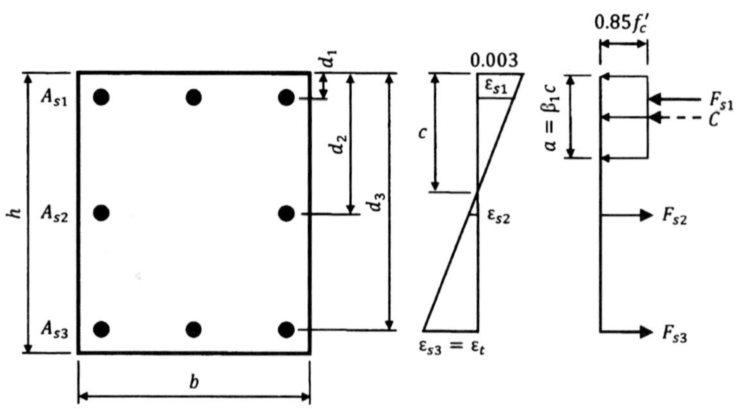

Figure 11.30 Nominal strength.

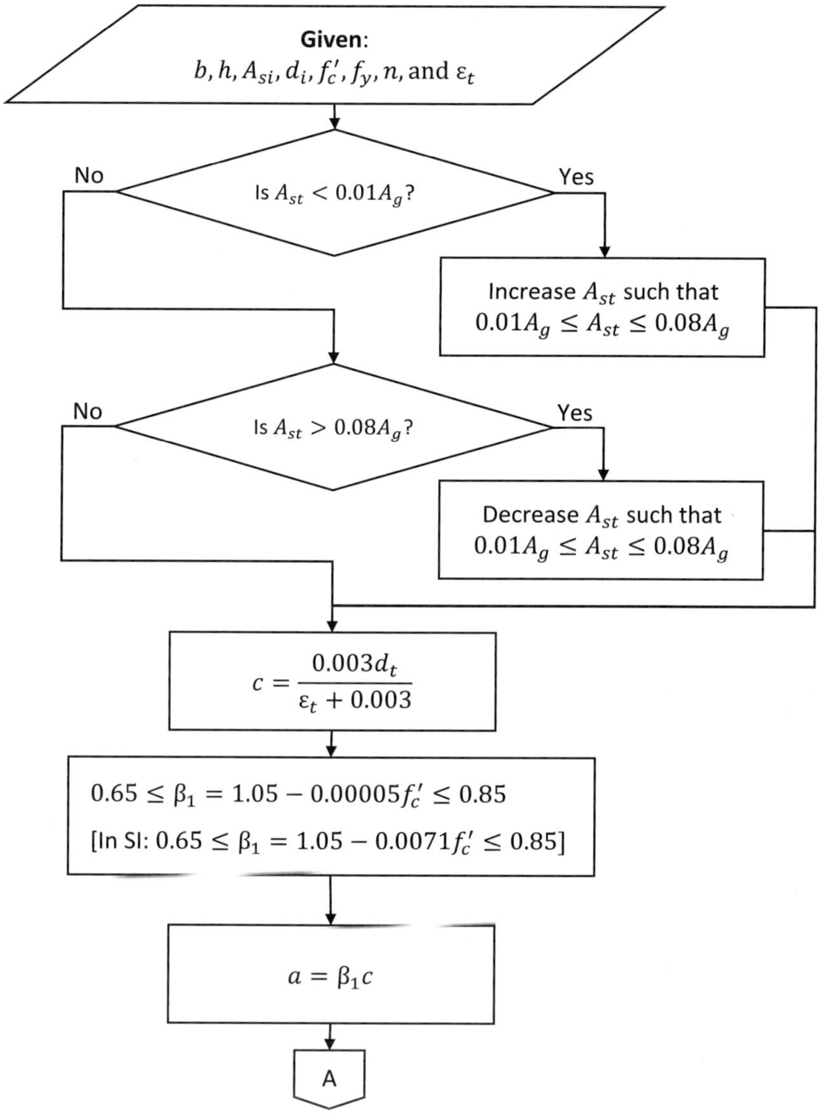

Figure 11.31 Combined flexural and axial strength.

They are used alone or in combination with moment frames to resist these load effects. Because a wall is much stiffer in the direction parallel to the plane of the wall than perpendicular to this plane, it is commonly assumed that only the walls that are oriented parallel to the direction of the lateral loads resist the lateral load effects in that direction. Such walls must be designed for combinations of axial loads, bending moments about their strong axis, and shear forces, and they are referred to as *structural walls* in ACI 318. Figure 11.32 illustrates a load-bearing reinforced-concrete wall that is subjected to axial loads and in-plane lateral loads, which cause bending about the strong axis of the wall. In other words, the in-plane loads cause an in-plane bending moment as indicated in the figure.

Walls must also be designed for any bending moments about their minor axis, caused by lateral loads applied perpendicular to the plane of the wall (e.g., a wall that is situated at the perimeter of a building is subjected to wind loads perpendicular to its face) or by axial loads acting at an eccentricity from the centroid of the wall (see Fig. 11.32*b* and *c* for these two load cases).

Basement walls and retaining walls are subjected to lateral earth pressure perpendicular to the plane of the wall. A cantilever retaining wall is designed for flexure in accordance with the strength design method presented in Sec. 11.7.5.

Properly proportioned walls can reduce lateral displacements of a building frame, which can result in the frame being designated as nonsway. This can have a significant impact on the design of the columns in the frame (especially in regard to second-order effects) and on the overall performance of the structure.

This chapter focuses on the analysis and design of walls that are subjected to axial loads, combined axial loads and bending, and shear. In general, provisions are presented for the following:

1. Sizing the cross-section
2. Determining the required amount of reinforcement
3. Detailing the reinforcement

The requirements for reinforced-concrete walls are given in ACI Chapter 11. These requirements are applicable to the design of cast-in-place, precast, and precast on-site (including tilt-up) wall systems.

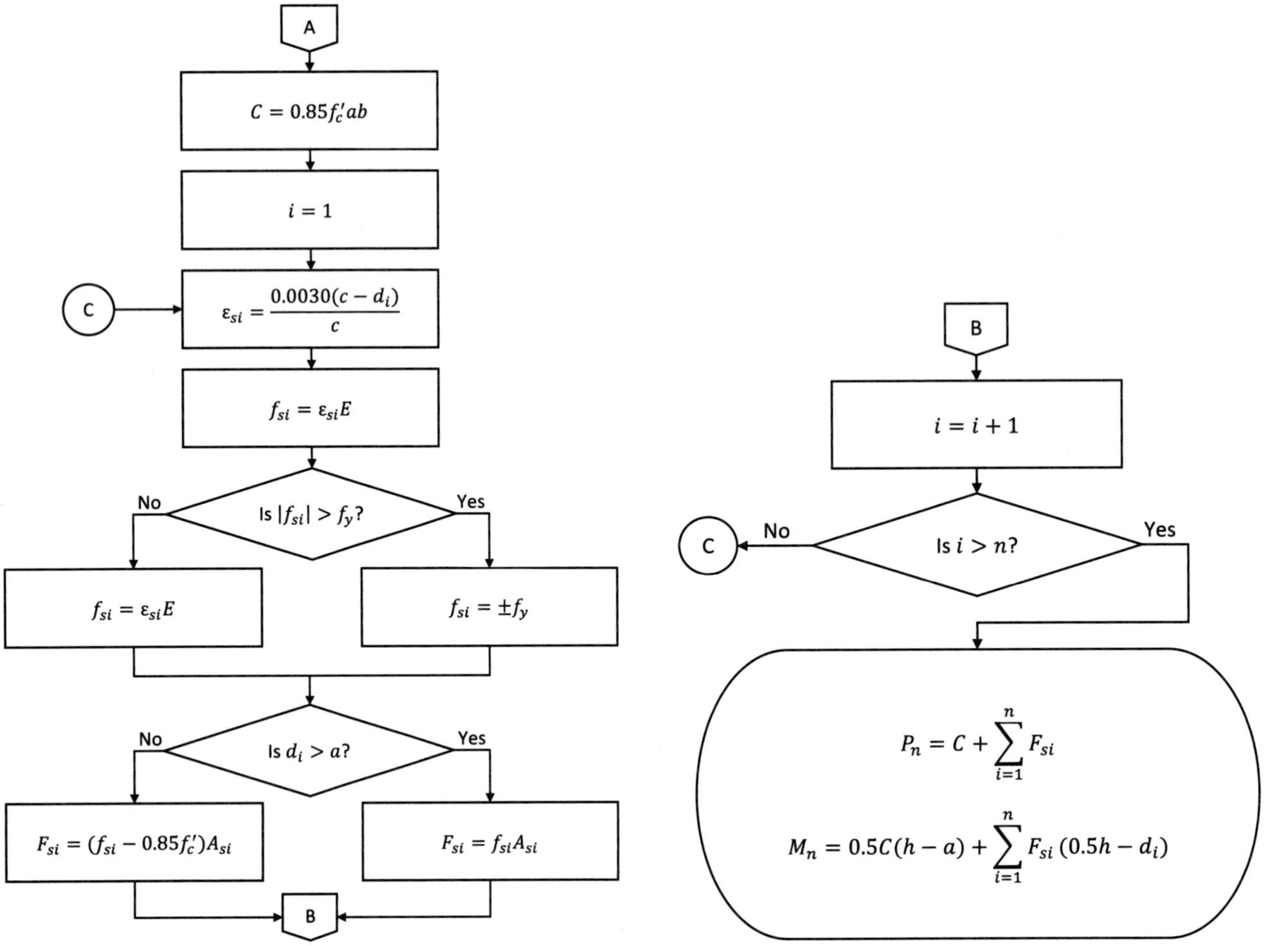

Figure 11.31 *(Continued)*

11.11.2 Design Methods for Axial Loads and Flexure

OVERVIEW

Chapter 11 of the Code gives methods that can be utilized to design reinforced-concrete walls subjected to axial loads and bending moments (both in-plane and out-of-plane). For any wall, the nominal axial strength P_n and the nominal flexural strength M_n can be determined in accordance with the provisions of ACI 22.4, which are also applicable to columns (ACI 11.5.2.1). The simplified design method of ACI 11.5.3 is permitted to be used instead of ACI 22.4 for walls subjected to axial load and out-of-plane flexure provided the limitations of ACI 11.5.3.1 are satisfied. Similar to beams, M_n is to be determined by the requirements of ACI 22.3 for non–load-bearing walls.

In general, walls are to be designed for factored axial forces P_u and factored bending moments M_u using the appropriate load combinations in ACI Table 5.3.1. Just like columns, M_u must be magnified for slenderness effects, where appropriate. The moment magnification method (ACI 6.6.4), an elastic second-order analysis (ACI 6.7), or an inelastic second-order analysis (ACI 6.8) is permitted to be used to account for slenderness effects (11.4.1.3). Note that the alternative method for out-of-plane slender wall analysis in ACI 11.8 is also permitted provided the requirements of ACI 11.8.1.1 are satisfied. This method and the moment magnification method are covered below.

11.11.3 Walls Designed Using the Provisions of ACI 22.4

Any wall may be designed using the general principles of the strength design method. The following equations must be satisfied in the design of any wall subjected to combined axial load and bending (ACI 11.5.1.1):

$$\phi P_n \geq P_u \tag{11.54}$$

$$\phi M_n \geq M_u \tag{11.55}$$

The factored axial load P_u and bending moment M_u acting on a reinforced-concrete wall must be equal to or less than the corresponding design values ϕP_n and ϕM_n in order for strength requirements to be satisfied. A design strength interaction diagram facilitates the design of a wall section.

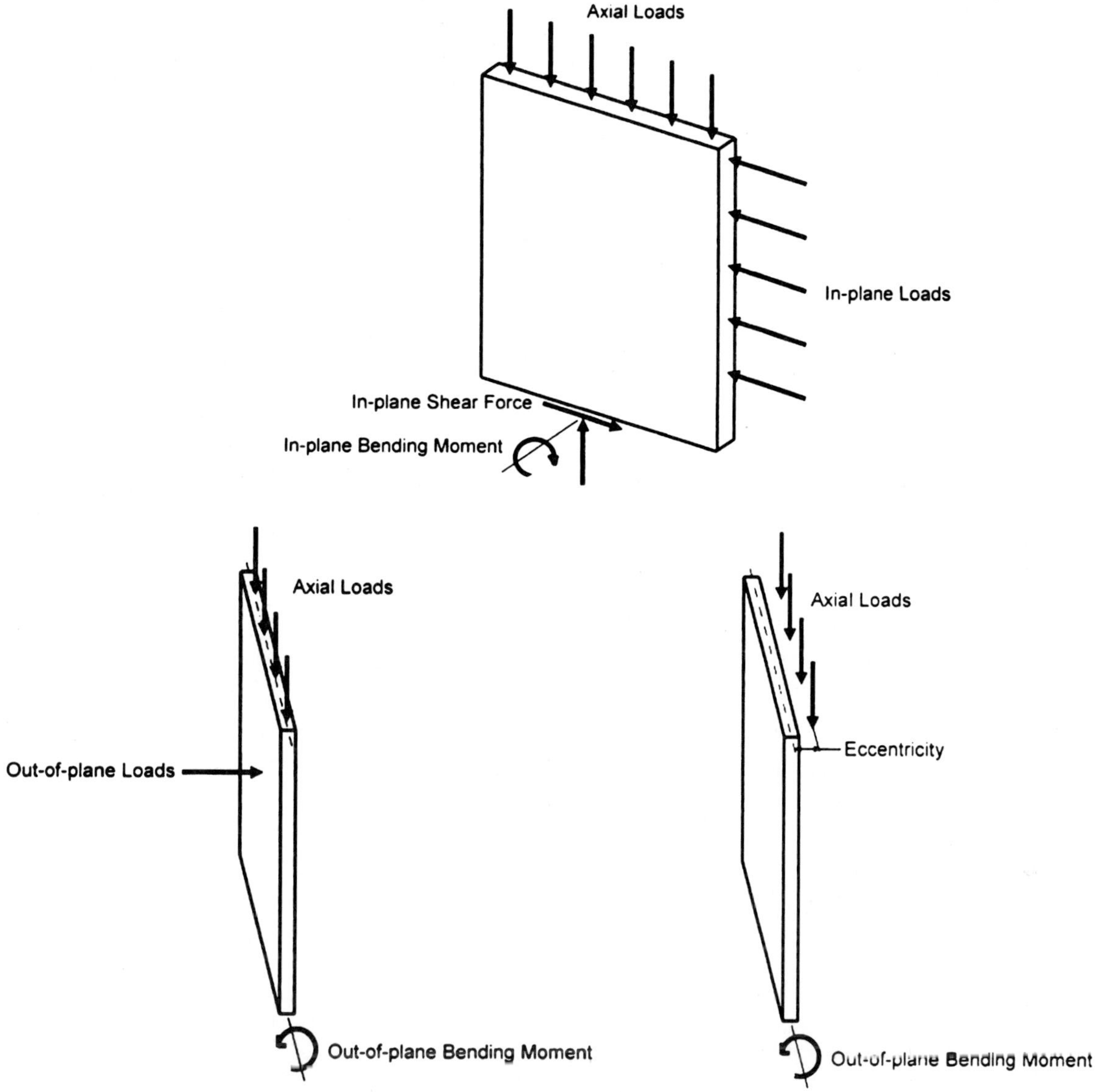

Figure 11.32 (*a*) Load-bearing wall with in-plane forces and (*b*, *c*) load-bearing wall with out-of-plane forces.

Slenderness effects must be considered in the design of walls where required. In the direction parallel to the length of the wall, slenderness effects can usually be neglected because the radius of gyration of the wall in that direction is relatively large; this results in a slenderness ratio less than the limits given in ACI 6.2.5. However, in the direction parallel to the thickness of a wall, the slenderness ratio is usually greater than the prescribed slenderness limits. As such, the factored bending moments about the minor axis of the wall must be magnified to account for slenderness effects. The three methods of analysis that are permitted for analyzing walls where slenderness effects cannot be neglected are (1) moment magnification method (ACI 6.6.4), (2) elastic second-order analysis (ACI 6.7), and (3) inelastic second-order analysis (ACI 6.8). More often than not, the frame in the direction of analysis can be considered nonsway where walls are used to resist the lateral loads.

11.11.4 Simplified Design Method

The provisions for the simplified design method are given in ACI 11.5.3. This method may be used for the design of walls where all of the following limitations are satisfied:

1. The wall has a solid, rectangular cross-section.
2. The resultant of all applicable factored loads falls within the middle third of the wall thickness.
3. The thickness of bearing walls is equal to or greater than the unsupported height or length of the wall, whichever is shorter, divided by 25, but not less than 4 in (100 mm). Furthermore, the wall thickness is equal to or greater than 7.5 in (200 mm) for exterior basement walls and foundation walls (see ACI Table 11.3.1.1).

Illustrated in Fig. 11.33 is a wall section subjected to an axial load acting at an eccentricity from the centroid of the section. In this case,

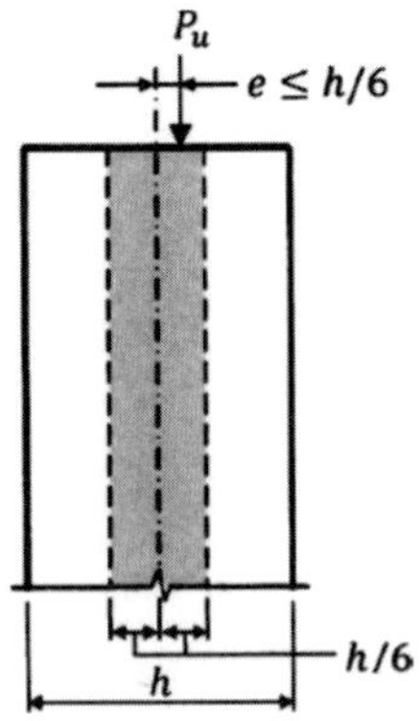

Figure 11.33 Eccentricity limitations in the simplified design method.

the second limitation of this method is satisfied where $e \le h/6$. In general, the total eccentricity caused by all applicable factored load effects, including those from lateral loads, must be compared with $h/6$.

The design axial strength ϕP_n of a wall satisfying the limitations of this method is determined by ACI Eq. (11.5.3.1):

$$\phi P_n = 0.55\phi f_c' A_g \left[1 - \left(\frac{kl_c}{32h}\right)^2\right] \qquad (11.56)$$

In this equation, A_g is the gross area of the wall and k is the effective length factor prescribed in ACI Table 11.5.3.2.

For walls that are braced against lateral translation at both ends of the wall, k is defined as follows:

- $k = 0.8$ when the wall is restrained against rotation at one or both ends.
- $k = 1.0$ when the wall is unrestrained against rotation at both ends.

A k-value of 0.8 implies that the end of the wall is attached to a member that has a flexural stiffness that is at least equal to that of the wall in the direction of analysis. Members with lesser stiffnesses cannot adequately restrain the wall against rotation.

For walls that are not braced against lateral translation, $k = 2.0$. This would be applicable, for example, to freestanding (cantilever) walls or to walls that are connected to diaphragms that undergo significant deflections when subjected to lateral loads.

Equation (11.56) takes into consideration both load eccentricity and slenderness effects. The eccentricity factor 0.55 was originally selected to give strengths comparable to those determined by ACI Chapter 22 for members with an axial load applied at an eccentricity of $h/6$. The strength-reduction factor ϕ corresponds to compression-controlled sections in accordance with ACI Table 21.2.1. Thus, ϕ is equal to 0.65 for wall sections designed by this method.

In order to satisfy strength requirements, the design strength ϕP_n determined by Eq. (11.56) must be equal to or greater than the factored axial load P_u. In essence, the design of walls by the simplified design method is carried out considering P_u is a concentric axial load.

The simplified design method is best suited for relatively short walls subjected to vertical loads. Because the total eccentricity must not exceed $h/6$, its application becomes extremely limited when lateral loads need to be considered.

Walls not meeting the limitations of this method must be designed by the other methods noted previously.

An alternative method for out-of-plane slender wall analysis is shown in ACI 318 Section 11.8 and in Chapter 9 of Ref. 2.

11.12 FOOTINGS

11.12.1 Overview

Chapter 13 of the ACI 318 Code contains the design requirements for isolated footings. Most of the provisions are also applicable to the design of combined footings and mats.

Methods are presented in the following sections on the following:

1. Sizing the base area and thickness
2. Determining the required amount of reinforcement
3. Detailing the reinforcement

11.12.2 Loads and Reactions

As in the design of any reinforced-concrete member, footings must be proportioned to resist the effects from the governing factored loads determined in accordance with ACI Table 5.3.1. This includes axial loads, bending moments, and shear forces. The base area of a footing is determined using unfactored (service) loads and the allowable soil-bearing capacity that is provided in the geotechnical report or any other referenced document (ACI 13.3.1.1). When determining base dimensions, the minimum moment requirement for slenderness considerations of ACI 6.6.4.5 need not be considered; only the computed end moment that exists at the base of a column or pedestal needs to be transferred to the footing (ACI R13.2.6.1).

The thickness of a footing and the required area of flexural reinforcement are determined using the strength design method, which utilizes factored load effects. Shear requirements must be satisfied using factored shear forces and design shear strength.

The number and arrangement of piles, drilled piers, and caissons are determined using unfactored loads (ACI 13.4.1.1). ACI 13.4.2.2 permits the moments and shears in a pile cap to be determined assuming that the reaction from any pile is concentrated at the center of the pile.

11.12.3 Sizing the Base Area

For footing depicted in Fig. 11.34 subjected to an applied service load P that acts through the centroid of the footing plan area A_f. For purposes of design, the footing is assumed to be rigid, and the resulting soil pressure q at the base of the footing is assumed to be uniform.

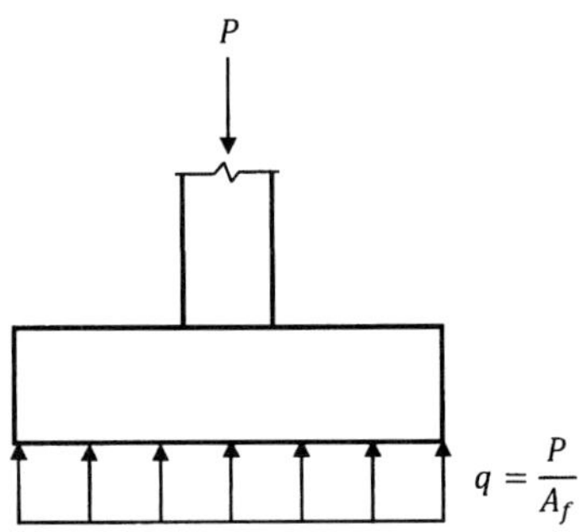

Figure 11.34 Soil pressure distribution for load acting through centroid of footing.

For footings subjected to an axial load and bending moment or, equivalently, to an axial load acting at an eccentricity e from the centroid of the footing area, the total combined stress at the base of the footing is equal to the sum of the stress due to the axial load P (axial load/footing area) and the bending moment M (bending moment/section modulus of footing). The pressure is assumed to vary linearly, as shown in Fig. 11.35. This distribution is valid where the axial load falls within the kern of the footing area, that is, where the eccentricity e is less than $L/6$. The equations in the figures for q can be used to determine the minimum and maximum pressures at the extreme edges of the footing.

11.12.4 Sizing the Thickness

Once the required area of the footing has been established on the basis of the service loads and the allowable bearing capacity of the soil, the thickness h of a footing must be determined considering both flexure and shear.

In general, a spread footing must be designed for the bending moments that are induced because of the pressure developed at the base of the footing from the factored loads. Requirements for both one- and two-way shear must also be satisfied. Methods to determine the thickness are provided in the following sections.

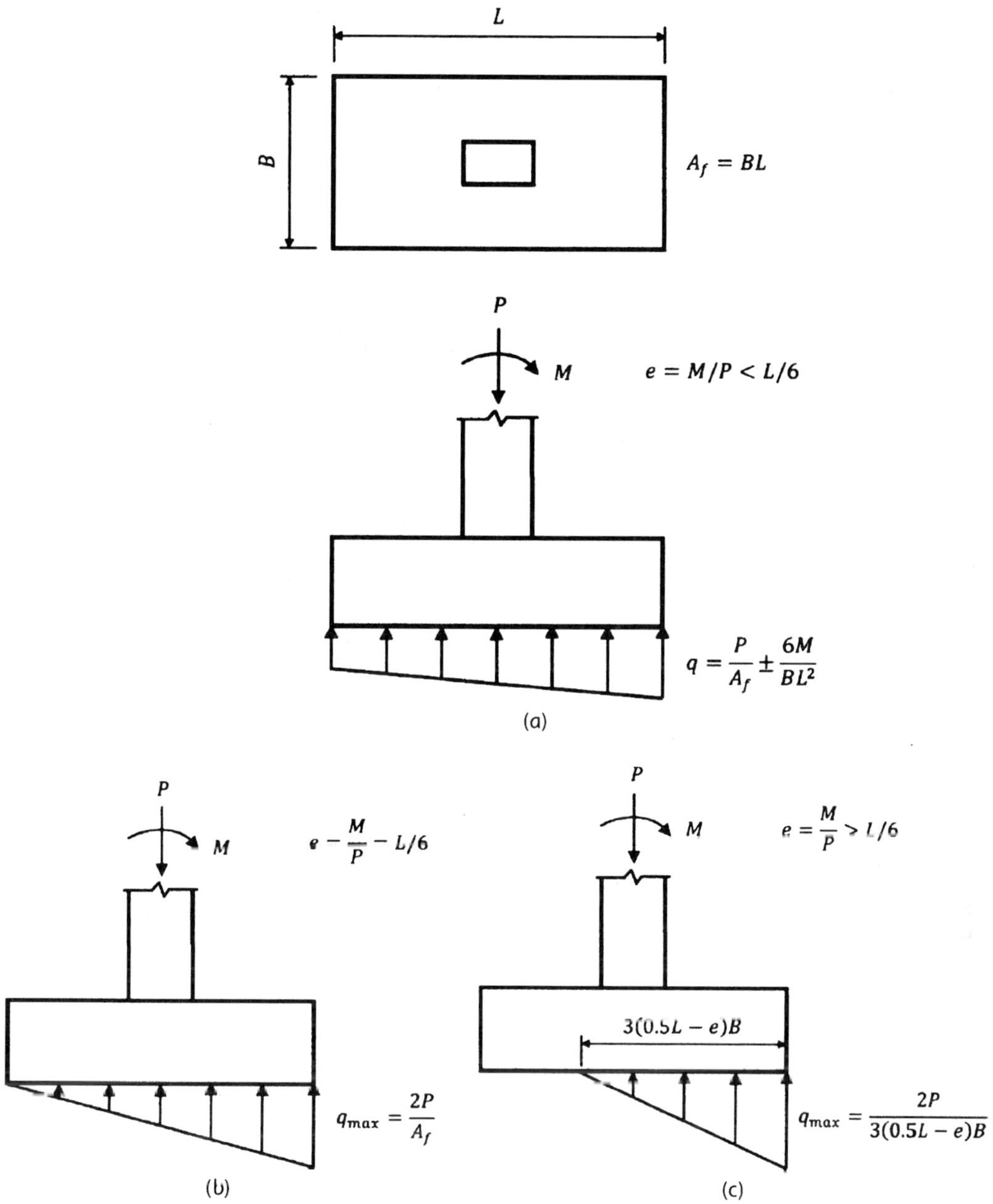

Figure 11.35 Soil pressure distribution for footing subjected to axial load and bending moment: (*a*) $e < L/6$, (*b*) $e = L/6$, and (*c*) $e > L/6$.

ACI 13.3.1.2 requires a minimum depth of 6 in. (150 mm) above the bottom reinforcement for footings. Similarly, the overall depth of a pile cap must be selected so that the effective depth of the bottom reinforcement is at least 12 in. (300 mm) (ACI 13.4.2.1). According to ACI Table 20.6.1.3.1, the minimum concrete cover to the reinforcement is equal to 3 in. (75 mm) for concrete cast against and permanently exposed to earth. Therefore, for footings the minimum overall thickness is equal to approximately 10 in. (250 mm). Similarly, the minimum overall thickness of pile caps is 16 in. (400 mm).

11.12.5 Design for Flexure

Critical Section

A spread footing must be designed for the bending moments that are induced because of the pressure developed at the base of the footing from the factored loads. Illustrated in Fig. 11.36 is an isolated spread footing subjected to a concentric factored axial load P_u. The factored pressure q_u at the base of the footing is equal to P_u divided by the area of the footing A_f.

According to ACI Table 13.2.7.1, the critical section for flexure for an isolated footing supporting a concrete column, pedestal, or wall is

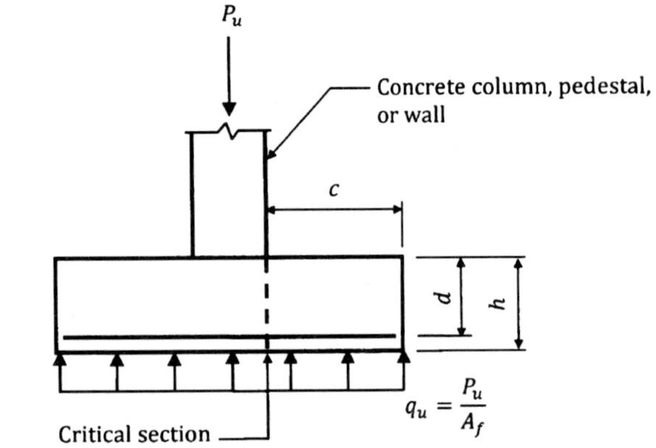

Figure 11.36 Critical section for an isolated footing supporting a concrete column, pedestal, or wall.

located at the face of the supported member. The maximum factored bending moment M_u at this critical section in this direction can be determined by the following equation, which is applicable to cantilevered members:

$$M_u = \frac{q_u c^2}{2} \tag{11.57}$$

In this equation, c is the distance from the critical section to the edge of the footing (i.e., c is the length of the cantilevered portion of the footing).

If the footing were subjected to a moment or load acting at an eccentricity, the resulting factored pressure would be nonuniform. In such cases, the bending moment at the critical section can be obtained from statics.

ACI Table 13.2.7.1 also contains critical section locations for two other cases. For footings supporting masonry walls and footings supporting columns with a steel base plate. The critical section location is shown in Fig. 11.37.

ACI 13.2.7.3 permits circular or regular polygon-shaped columns or pedestals to be replaced by an equivalent square member with the same area as the original shape for location of critical sections for moments, shear, and development of flexural reinforcement.

Determining the Required Reinforcement

Once the maximum factored moment M_u at the critical section has been determined, the required area of reinforcing steel A_s can be calculated using the strength design requirements presented in this chapter, section 5. The following equation must be satisfied for a concentrically loaded isolated footing with the nominal flexural strength M_n for a rectangular section with tension reinforcement:

$$M_u = \frac{q_u c^2}{2} \le \phi M_n = \phi A_s f_y (d - \frac{a}{2}) \tag{11.58}$$

The required strength M_u must be equal to or less than the design strength ϕM_n. An efficient design for footings would be one where the section is tension-controlled. Thus, the strength-reduction factor ϕ is equal to 0.9 in accordance with ACI Table 21.2.2. Similar equations can be derived for other pressure distributions.

According to ACI 13.3.2.1 for one-way footings and ACI 13.3.3.1 for two-way footings, the minimum area of flexural reinforcement $A_{s,\min}$ for footings of uniform cross-section must conform to that prescribed in ACI Chapter 7 for one-way slabs. For footings with Grade 60 (Grade 420) reinforcement, the minimum area of steel is equal to 0.18 percent of the gross area of the footing, which is equal to the overall thickness h times footing plan dimension B or L, depending on the direction of analysis. The maximum spacing of flexural reinforcement is the lesser of $3h$ or 18 in. (450 mm).

Detailing of Reinforcement

Requirements for the distribution of flexural reinforcement are given in ACI 13.3.2.2 for one-way footings and in ACI 13.3.3.2 and 13.3.3.3 for two-way footings. For one-way (wall) and two-way square footings, reinforcement is to be distributed uniformly across the entire width of the footing. In square footings, the reinforcement in both orthogonal layers is the same because the maximum-factored bending moments at the critical sections are the same. Uniform distribution of the reinforcement is shown in Fig. 11.38 for a one-way spread footing supporting a wall and for an isolated square spread footing supporting a column. Other reinforcement is not shown for clarity.

Flexural reinforcement in two-way rectangular footings must be distributed in accordance with ACI 13.3.3.3. Reinforcement in the long direction is uniformly distributed across the entire width of the footing. In the short direction, a portion of the total reinforcement $\gamma_s A_s$ must be uniformly distributed over a band width centered on the column or pedestal that is equal to the length of the short side of the footing. The term γ_s is determined by ACI Eq. (13.3.3.3):

$$\gamma_s = \frac{2}{\beta + 1} \tag{11.59}$$

In this equation, β is the ratio of the long side to the short side of the footing.

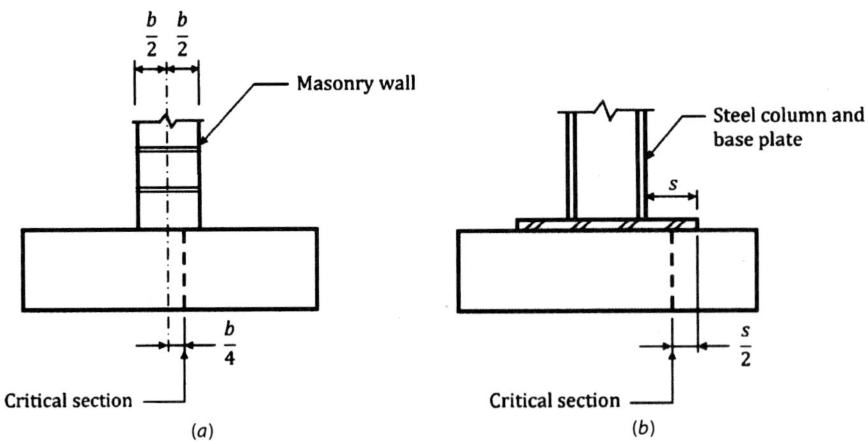

Figure 11.37 Critical section locations for (*a*) footings supporting masonry walls and (*b*) footings supporting columns with a steel base plate.

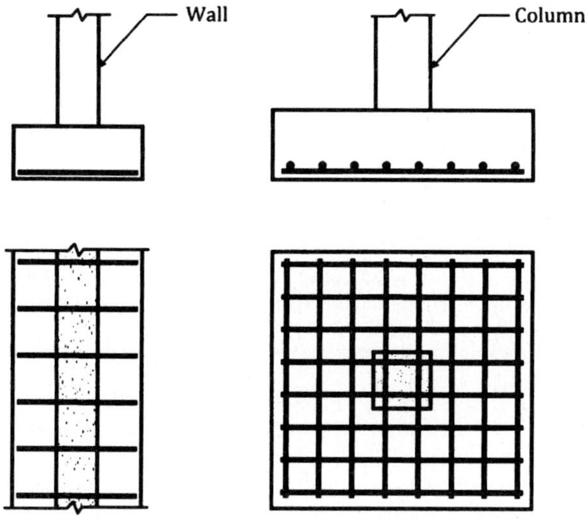

Figure 11.38 Distribution of flexural reinforcement in one- and two-way square footings.

The remainder of the reinforcement outside of the center band must be uniformly distributed. This distribution reflects the fact that the moment is largest immediately under the column and decreases with increasing distance from the column.

Figure 11.39 illustrates the provisions of ACI 13.3.3.3. In the long direction, the required area of steel A_{sB} is uniformly distributed at a spacing of s_3 over the width B. In the short direction, the portion A_{s1} of the total required area of reinforcement A_{sL} is uniformly distributed at a spacing of s_1 over a width equal to B centered on the column. Note that A_{s1} is equal to $\gamma_s A_{sL}$. The remaining reinforcement A_{s2} is uniformly distributed at a spacing of s_2 on either side of the center band. The minimum reinforcement and maximum spacing requirements of the Code must be satisfied in all portions of the footing.

In order to minimize the potential for errors while placing the bars in the short direction of a rectangular footing, a common practice is to increase the amount of reinforcement in the short direction by $2\beta/(\beta + 1)$ and space it uniformly across the long dimension of the footing instead of distributing the bars as shown in Fig. 11.39.

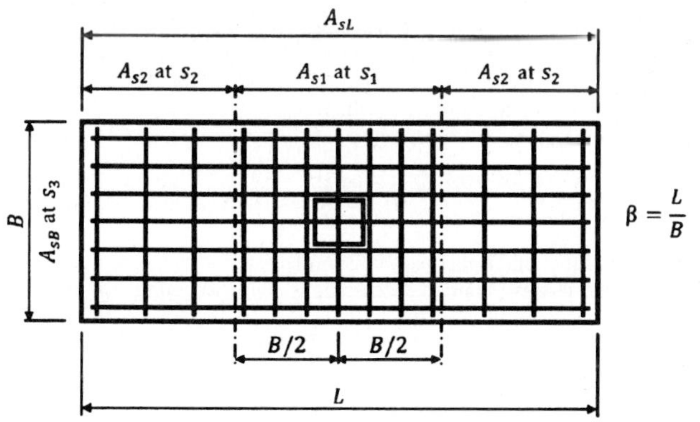

$$A_{s1} = \frac{2}{\beta + 1} A_{sL}$$

$$A_{s2} = \frac{A_{sL} - A_{s1}}{2}$$

Figure 11.39 Distribution of flexural reinforcement in a rectangular footing.

11.12.6 Design for Shear

Overview

Provisions for shear strength in footings are the same as those required for slabs. Requirements for both one- and two-way shear must be satisfied. The required footing thickness for shear is based on the severer of these two conditions.

The critical section for shear is measured from the face of a column, pedestal, or wall for footings supporting such elements (ACI 13.2.7.2). For footings supporting columns or pedestals with steel base plates, the critical section is measured from the position halfway between the face of the column or pedestal and the edge of the steel base plate.

One-Way Shear

The factored shear force V_u at the critical section, based on the factored pressure at the base of the footing within the tributary area, must be equal to or less than the design shear strength ϕV_c determined in accordance with ACI 22.5.5.1. Figure 11.40 illustrates the tributary area that is to be used in the calculation of V_u for a column supported by an isolated rectangular footing. The following equation must be satisfied at the critical section, which in this case is located a distance d from the face of the column:

$$V_u = q_u B(c - d) \le \phi V_c = \phi 2\lambda\sqrt{f_c'}Bd \qquad (11.60)$$

$$[\text{In SI}: V_u = q_u B(c - d) \le \phi V_c = \phi 0.17\lambda\sqrt{f_c'}Bd$$

According to ACI Table 22.2.1, the strength-reduction factor ϕ is equal to 0.75 for shear. One-way shear needs to be checked in the other direction as well.

Two-Way Shear

The factored shear stress v_u at the critical section, based on the factored pressure at the base of the footing within the tributary area, must be equal to or less than the design shear strength ϕv_c determined in accordance with ACI 22.6.5. The tributary area that is to be used in the calculation of v_u for a column supported by an isolated rectangular footing

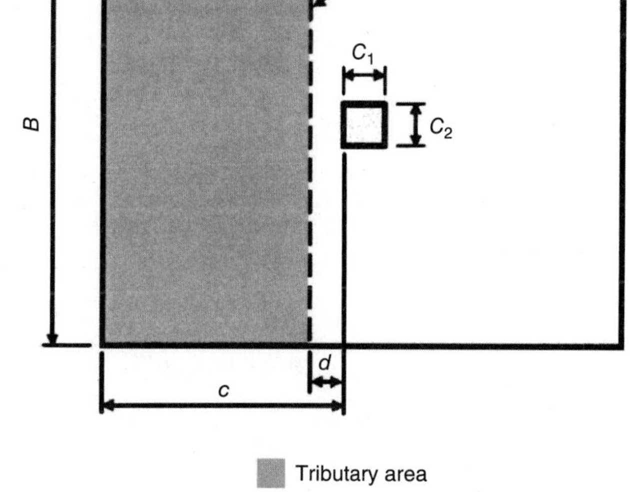

Figure 11.40 Critical section for one-way shear in a footing.

is illustrated in Fig. 11.41. The following equation must be satisfied at the critical section, which in this case is located at a distance $d/2$ from the face of the column:

$$v_u = q_u[BL-(c_1+d)(c_2+d)]/b_o d \le \phi v_c \tag{11.61}$$

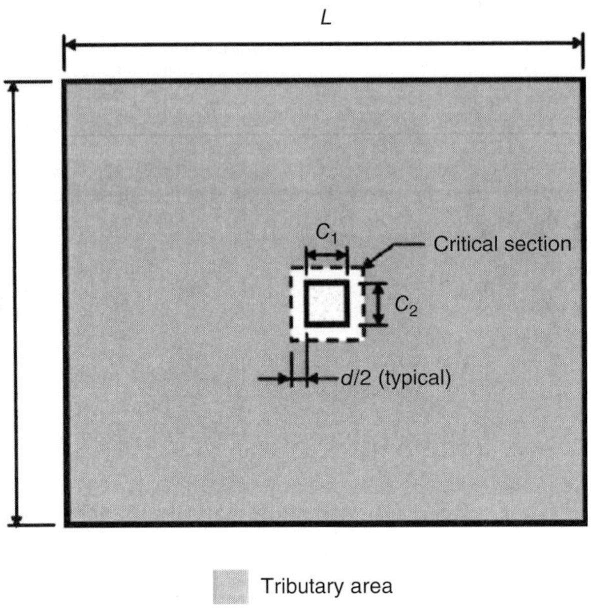

Figure 11.41 Critical section for two-way shear in a footing.

In this equation, the design shear strength of the concrete ϕv_c is the least of the values defined in ACI Table 22.6.5.2:

$$\phi v_c = \begin{Bmatrix} \phi 4\lambda\sqrt{f_c'} \\ \phi\left(2+\dfrac{4}{\beta}\right)\lambda\sqrt{f_c'} \\ \phi\left(2+\dfrac{\alpha_s d}{b_o}\right)\lambda\sqrt{f_c'} \end{Bmatrix} \tag{11.62}$$

$$\left[\text{in SI: } \phi v_c = \begin{Bmatrix} \phi 0.33\lambda\sqrt{f_c'} \\ \phi 0.17\left(1+\dfrac{2}{\beta}\right)\lambda\sqrt{f_c'} \\ \phi 0.083\left(2+\dfrac{\alpha_s d}{b_o}\right)\lambda\sqrt{f_c'} \end{Bmatrix}\right.$$

Illustrated in Fig. 11.41 is the perimeter b_o of a four-sided critical section.

For footings supporting square columns, the first of these three equations usually governs. The second equation accounts for β, which is equal to the ratio of the long side to the short side of the column, concentrated load, or reaction area. As β increases, the design shear strength decreases. The third equation accounts for the effect of b_o/d. Also included in this equation is α_s, which is equal to 40 for critical sections with four sides, 30 for critical sections with three sides, and 20 for critical sections with two sides. This equation yields the smallest design strength where $d/c_1 \le 0.25$, which rarely occurs.

11.12.7 Force Transfer at Base of Supported Members

VERTICAL TRANSFER

Bearing Stress According to ACI 16.3.3.4, the bearing strength requirements of ACI 22.8 must be satisfied for both the supported and the supporting member. For bearing on the supported member, the factored bearing load B_u must be equal to or less than the design-bearing strength ϕB_n (see ACI Table 22.8.3.2):

$$B_u \le \phi B_n = \phi 0.85 f_c' A_1 \tag{11.63}$$

In this equation, A_1 is the area of the column, wall, or pedestal that is supported by the footing and the strength-reduction factor ϕ for bearing is equal to 0.65 (ACI Table 21.2.1).

The following equation must be satisfied for bearing on a footing or pedestal:

$$B_u \le \phi B_n = \phi 0.85 f_c' A_1 \sqrt{\frac{A_2}{A_1}} \le 2\phi 0.85 f_c' A_1 \tag{11.64}$$

The term A_1 is the area of the supported member and A_2 is defined as the area of the lower base of the largest frustrum of a pyramid, cone, or tapered wedge contained wholly within the footing and having for its upper base the loaded area A_1 and having side slopes of 1 vertical to 2 horizontal.

Reinforcement across the Interface The amount of reinforcement that is required between the supported and the supporting member depends on the type of stress in the bars of the supported member under all applicable load combinations. Minimum embedment lengths into both members also depend on this stress.

Dowels are commonly used as interface reinforcement between columns or walls and footings. The dowel bars are set in the footing prior to casting the footing concrete and are subsequently spliced to the column or wall bars.

Compressive Stress in the Bars of the Supported Member Where $B_u > \phi B_n$, reinforcement must be provided to transfer the excess compressive stress from the supported member to the footing. The required area of interface reinforcement can be determined by the following equation:

$$A_s = \frac{B_u - \phi B_n}{\phi f_y} \ge A_{s,\min} \tag{11.65}$$

A minimum area of reinforcement $A_{s,\min}$ across the interface is required even where concrete-bearing strength is not exceeded. Minimum reinforcement requirements are based on the type of member that is supported.

Horizontal Transfer ACI 16.3.3.5 permits the shear-friction method of ACI 22.9 to be used for transfer of lateral loads from a supported member to a footing or pedestal. The reinforcement A_{vf} provided across the interface between the supported and the supporting member must satisfy the following equation, which is applicable where the shear-friction reinforcement is perpendicular to the interface (ACI 22.9.4.2):

$$A_{vf} \ge \frac{V_u}{\phi f_y \mu} \tag{11.66}$$

In this equation, V_u is the maximum factored shear force due to the lateral load effects obtained by the applicable load combinations given in ACI Table 5.3.1 and μ is the coefficient of friction that is determined by ACI Table 22.9.4.2. The strength-reduction factor ϕ is equal to 0.75.

11.12.8 Design Procedure

1. Determine the area of the footing
2. Determine the thickness of the footing
3. Determine the required flexural reinforcement
4. Distribute the flexural reinforcement
5. Develop the flexural reinforcement
6. Check transfer of forces at the base of the supported member

For other types of foundations, similar procedure is followed, see Chapter 10 of Ref. 2.

11.13 TWO-WAY SLAB SYSTEMS

11.13.1 Beam-Supported Slab

A solid slab supported on beams on all four sides is depicted in Fig. 11.42. This system, which was the original slab system in reinforced concrete, can accommodate a wide range of span and loading conditions. However, it is not as economical as other two-way systems with similar span and loading conditions due to formwork costs and costs associated with deeper overall floor thickness. As such, column-line beams are not used as often as they once were, except in cases where the demands for lateral force resistance are relatively large. For example, because flat-plate systems are not permitted to be the primary seismic force–resisting system (SFRSs) in areas of high seismicity, moment-resisting frames with column-line beams must be used as the SFRSs.

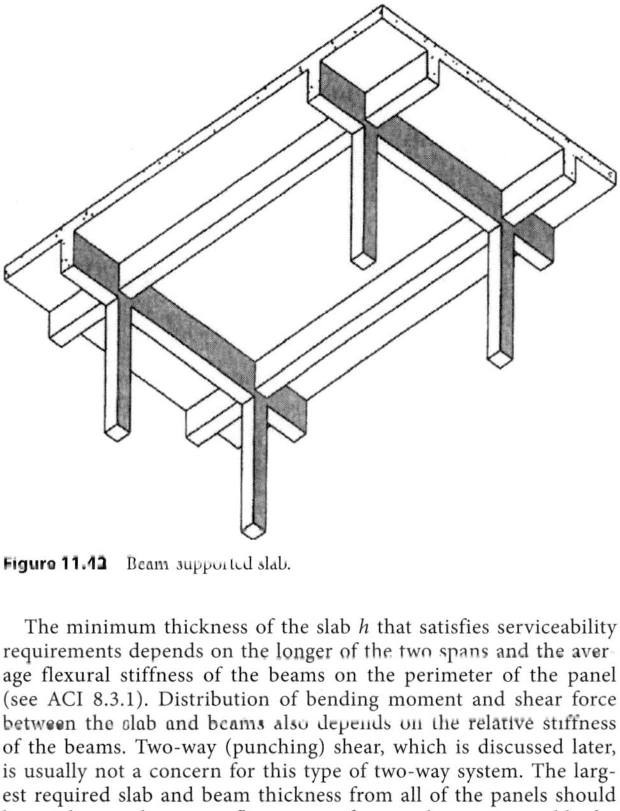

Figure 11.42 Beam-supported slab.

The minimum thickness of the slab h that satisfies serviceability requirements depends on the longer of the two spans and the average flexural stiffness of the beams on the perimeter of the panel (see ACI 8.3.1). Distribution of bending moment and shear force between the slab and beams also depends on the relative stiffness of the beams. Two-way (punching) shear, which is discussed later, is usually not a concern for this type of two-way system. The largest required slab and beam thickness from all of the panels should be used over the entire floor or roof area whenever possible for economy in formwork.

11.13.2 Flat Plate

A flat-plate floor system is a two-way concrete slab supported directly on columns with reinforcement in two orthogonal directions (Fig. 11.43). This system, which is popular in residential buildings (e.g., hotels and apartments), has the advantages of simple construction and formwork and a flat ceiling, the latter of which reduces ceiling finishing costs because the architectural finish can be applied directly to the underside of the slab. Even more significant are the cost savings associated with the low story heights made possible by the shallow floor system. Smaller vertical runs of cladding, partition walls, mechanical systems, plumbing, elevators, and a number of other items of construction translate into large cost savings, especially for medium- and high-rise buildings. Moreover, where the total height of a building is restricted, using a flat plate can result in more stories accommodated within the set height.

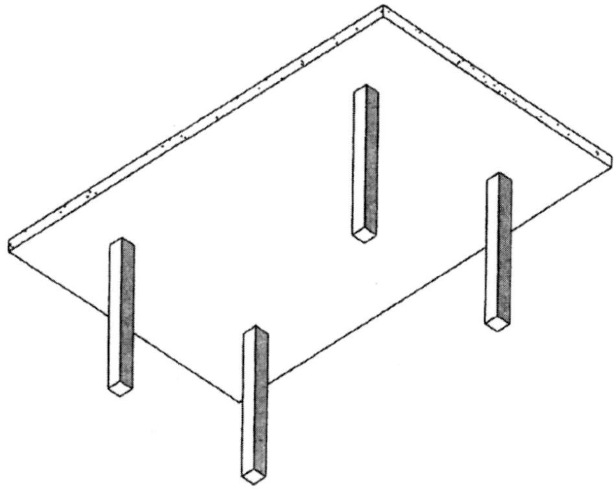

Figure 11.43 Flat plate system.

Flat plates are typically economical for span lengths between 15 and 25 ft (4.6 and 7.6 m) when subjected to moderate live loads. The thickness h of a flat plate will usually be controlled by the deflection requirements of ACI 8.3.1 for relatively short spans and live loads of 50 psf (2.4 kN/m^2) or less. Flexural reinforcement at the critical sections will be approximately the minimum amount specified in ACI 8.6.1 in such cases. Therefore, utilizing a slab thickness greater than the minimum required for serviceability is not economical because a thicker slab requires more concrete without an accompanying reduction in reinforcement. Because the minimum slab thickness requirements of ACI 8.3.1 are independent of the concrete compressive strength, a 4000-psi (28-MPa) concrete mixture is usually the most economical; using a concrete strength that is greater than 4000 psi (28 MPa) increases cost without a reduction in slab thickness.

Two-way (or punching) shear plays an important role in determining the thickness of a flat plate, especially where the spans are relatively long and/or the live load is 100 psf (4.8 kN/m^2) or greater. In order to satisfy the shear strength requirements of ACI 22.6, the required thickness is usually found to be greater than that required for serviceability. Shear stresses at edge columns and corner columns are particularly critical because relatively large unbalanced moments can occur at those locations. Providing spandrel beams significantly increases shear strength at perimeter columns, but as noted previously, there is additional material and forming costs associated with such members, and they may not fit into the architectural scheme.

Headed shear stud reinforcement provides an economical means of resisting shear stresses and helps to alleviate congestion at slab–column joints (see ACI 22.6.8).

For a live load of 50 psf (2.4 kN/m^2) or less, flat plates are economically viable for spans between 15 and 25 ft (4.6 and 7.6 m). The economical range for live loads of 100 psf (4.8 kN/m^2) is 15 to 20 ft (4.6 to 6.1 m). A flat-plate floor subjected to a 100-psf (4.8-kN/m^2) live load is only approximately 8 percent more expensive than one subjected to a 50-psf (2.4-kN/m^2) live load, primarily due to the minimum thickness requirements for deflection, which typically control for smaller live loads.

11.13.3 Flat Slab

A flat-slab floor system is similar to a flat-plate floor system, with the exception that the slab is thickened around the columns as shown in Fig. 11.44. The thickened portions of the slab are called drop panels, and they must conform to the dimensional requirements of ACI 8.2.4, which are illustrated in Fig. 11.45. In the figure, ℓ_A and ℓ_B are the

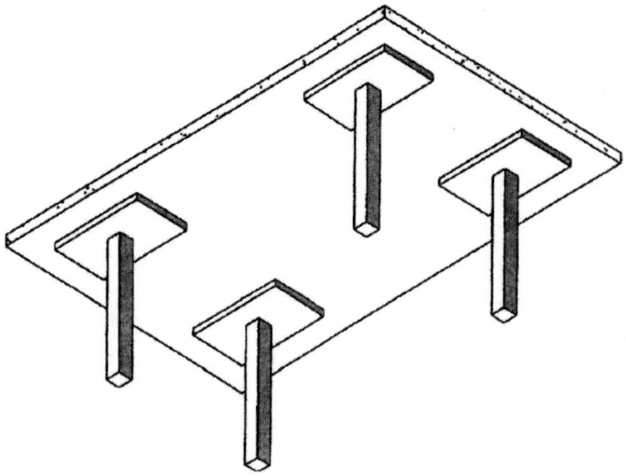

Figure 11.44 Flat-slab system.

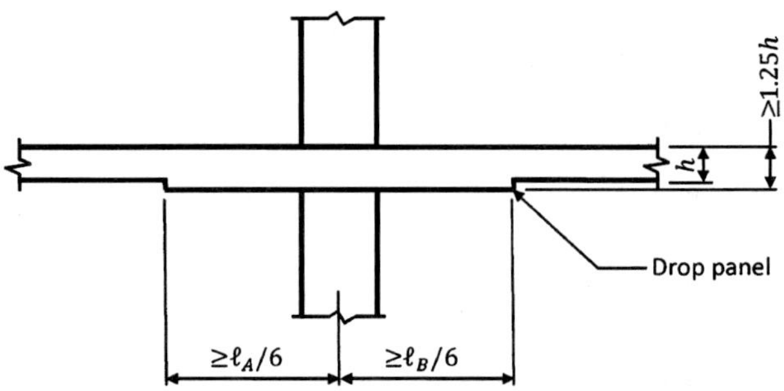

Figure 11.45 Minimum drop panel dimensions.

center-to-center span lengths in the directions shown; similar dimensional requirements must be satisfied in the orthogonal direction as well.

The main purpose of the drop panels is to increase the shear strength around the columns. Additionally, properly proportioned drop panels result in a reduction in the required amount of negative reinforcement and in the overall slab thickness (see ACI 8.5.2.2 and ACI Table 8.3.1.1).

Shear caps are thickened-concrete elements that extend horizontally below the slab at a minimum distance from the edge of the column equal to the thickness of the projection below the slab soffit (see Fig. 11.46). These elements are similar to drop panels, but they are provided exclusively to increase shear strength (see ACI 8.2.5).

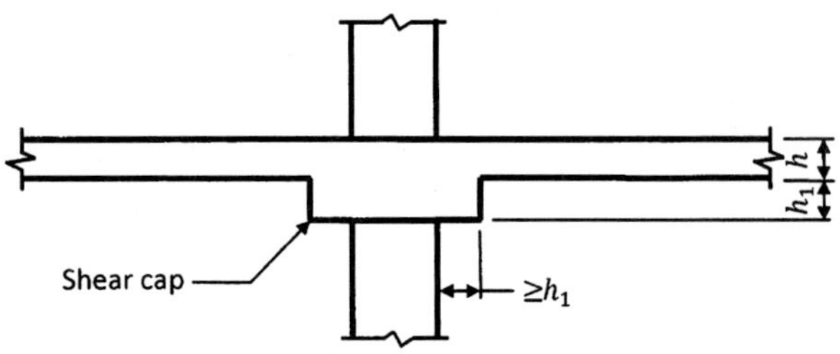

Figure 11.46 Shear cap.

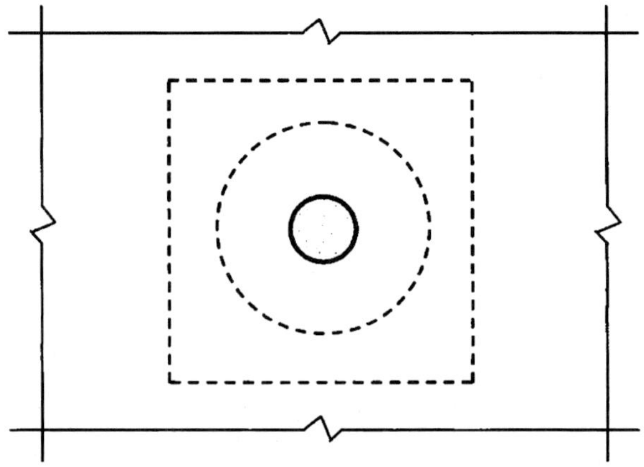

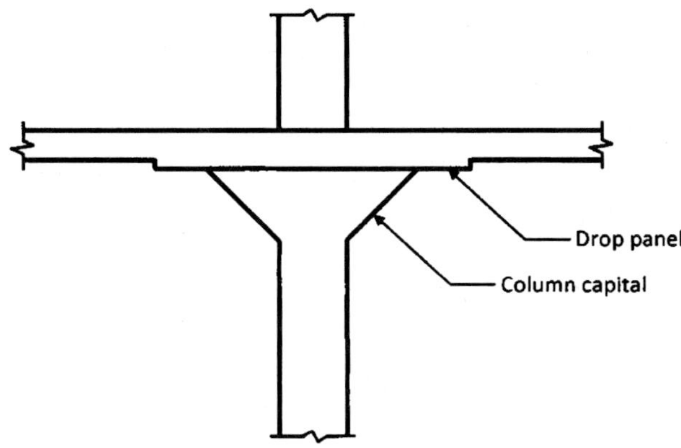

Figure 11.47 Column capitals and drop panels.

Another way to reduce shear around columns is to flare the top of a column, creating column capitals (see Fig. 11.47). For purposes of design, a column capital is part of the column, whereas a drop panel or shear cap is part of the slab. Because of relatively large formwork costs, column capitals are not commonly used anymore.

11.13.4 Two-Way Joist

Two-way joist construction, which is commonly referred to as a waffle slab system, consists of rows of concrete joists at right angles to each other with solid heads at the columns, which are needed for shear strength (see Fig. 11.48). The joists are formed by using standard square "dome" forms that are 30, 41, and 52 in. wide, resulting in 3-, 4-, and 5-ft modules, respectively. Depending on the dome width, the depth of the dome varies from 8 to 24 in. Waffle slabs are economically viable for long spans [40 to 50 ft (12.2 to 15.2 m)] with heavy loads and are used in office buildings, warehouses, libraries, museums, and industrial buildings.

For design purposes, waffle slabs are considered as flat slabs with the solid heads acting as drop panels. Thus, the minimum thickness requirements of ACI 8.3.1 must be satisfied. This is accomplished by transforming the actual cross-section of the floor system into an equivalent section of uniform thickness. In other words, a slab thickness that provides the same moment of inertia as the two-way joist section is determined.

Waffle slab construction allows a considerable reduction in dead load compared with conventional flat-slab construction because the slab thickness can be minimized owing to the short span between the joists. Thus, this system is particularly advantageous where long spans and/or

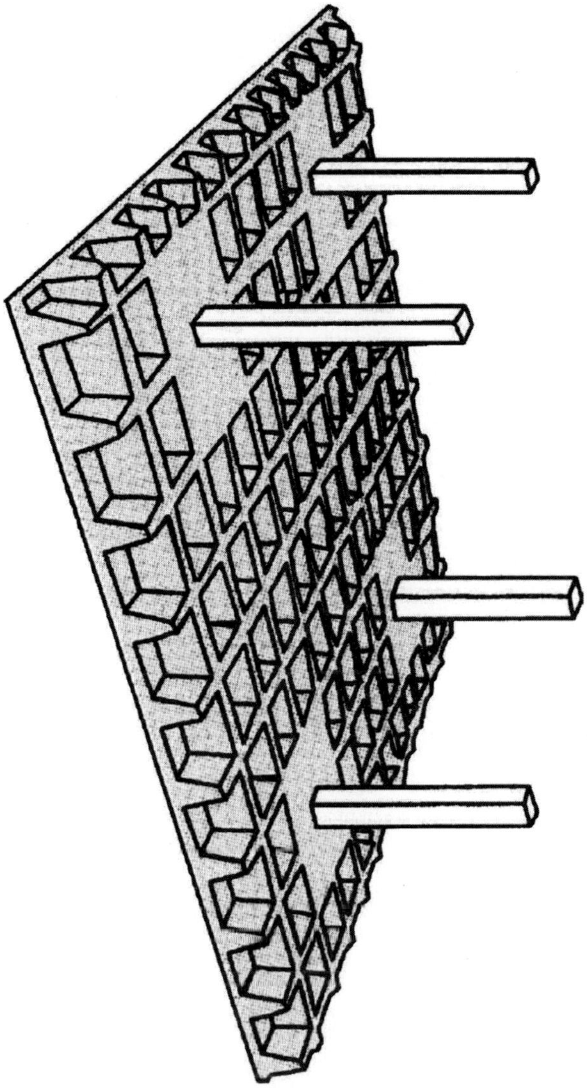

Figure 11.48 Two-way joist (waffle) system.

heavy loads are desired without the use of deepened drop panels or support beams. The geometric shape formed by the joist ribs is generally considered to be architecturally desirable and is often left exposed.

Like beam-supported slabs, waffle slabs are not specified as often as they once were. This is primarily due to the cost attributed to the formwork. Increasing the live load from 50 to 100 psf (2.4 to 4.8 kN/m^2) results in an approximately 7 percent increase in the overall cost of this system.

For details on the analysis and design methods for two-way slab systems, see Chapter 7 Ref. 2.

11.14 SEISMIC

11.14.1 Introduction

IBC 1613.1 requires that all structures, in whole and in part, be designed to resist the effects from earthquake motions. This chapter covers the design and detailing requirements for reinforced-concrete members in structures located in areas of moderate and high seismic risk. Prior to presenting the Code requirements, information is provided on seismic design criteria [including seismic design category (SDC)] and SFRSs for reinforced concrete. Chapter 5 of this handbook provides a description of how earthquake loads are calculated.

IBC 1901.2 requires that reinforced-concrete members be designed and constructed in accordance with the provisions in ACI 318-14. Chapter 18 in ACI 318-14 contains the seismic requirements that need to be satisfied for various types of members based on the SDC. These requirements, which are applicable for structures assigned to SDC B, C, D, E, and F, are covered in this chapter.

11.14.2 Seismic Force-Resisting Systems

Overview

Numerous types of reinforced-concrete systems are available to resist the effects from earthquakes. This section describes these systems and identifies their applicability and limitations with respect to the applicable SDC.

ASCE/SEI 12.2.1 requires that SFRSs conform to those indicated in ASCE/SEI Table 12.2-1. Structural system limitations and building height limits, along with other important quantities, are given in the table based on SDC.

There are no system limitations for structures assigned to SDC A. Such structures need only comply with the requirements in ASCE/SEI 11.7.

Bearing wall systems, building frame systems, moment-resisting frame systems, dual systems, shear wall-frame interactive systems, and cantilevered column systems are the general types of SFRSs given in ASCE/SEI Table 12.2-1 Each general category has systems of reinforced concrete. Descriptions of each system are given in the following sections (Note: Cantilevered column systems are not covered in this publication.) Information on the parameters contained in this table follows the system descriptions.

Bearing Wall Systems

SDC B Ordinary reinforced-concrete shear walls are permitted to be used in buildings assigned to SDC B without any limitations. Such walls must satisfy the applicable requirements of ACI Chapters 1 to 17 and 19 to 26, the provisions of Chapter 18 need not be satisfied. Detailed plain and ordinary plain concrete shear walls may also be used without limitations. Detailed plain concrete shear walls are walls complying with the requirements of ACI Chapter 14 and the additional reinforcement requirements of ASCE/SEI 14.2.2.8, and ordinary plain concrete shear walls are walls complying with the requirements of ACI Chapter 14 only.

SDC C Ordinary reinforced-concrete shear walls are permitted to be used in buildings assigned to SDC C. It is assumed that the design and detailing requirements in ACI Chapter 11 are compatible with the anticipated level of inelastic response when the walls are subjected to the effects from moderately strong ground motion.

SDC D, E, and F Special reinforced-concrete shear walls are required in buildings assigned to SDC D, E, or F. The height of a building is limited to 160 ft (49 m) for SDC D and E and is limited to 100 ft (30 m) for SDC F. The design and detailing requirements of ACI 18.2.3 through 18.2.8 and 18.10 must be satisfied for walls in buildings assigned to these SDCs.

Building Frame Systems

SDC B Ordinary plain, detailed plain, and ordinary reinforced-concrete shear walls are permitted to be used in buildings assigned to SDC B with no limitations. Building frame systems are generally not used in buildings assigned to SDC B because there is little to be gained from assigning the entire lateral resistance to the shear walls in the absence of any special detailing requirements for the frames. A shear wall-frame interactive system, which is discussed in Sec. 11.3.6 of this chapter, is usually more practical and economical in such cases.

SDC C Buildings assigned to SDC C are permitted to utilize ordinary reinforced-concrete shear walls with no limitations. Like in the case of bearing wall systems, it is assumed that the design and detailing requirements in ACI Chapter 11 are compatible with the anticipated level of inelastic response.

SDC D, E, and F Special reinforced-concrete shear walls are required to be used in a building frame system (along with the applicable building height limits) in structures assigned to SDC D and higher. It is important to note that for these SDCs, the deformational compatibility requirements of ACI 18.14 must be satisfied. The beam-column frames must be designed to resist the effects caused by the lateral deflections due to the earthquake effects because they are connected to the walls by the diaphragm at each level. The frame members, which are not designated as part of the SFRS, must be capable of supporting their gravity loads when subjected to the design displacements caused by the seismic forces.

Moment-Resisting Frame Systems

SDC B An ordinary reinforced-concrete moment can be used in buildings assigned to SDC B with no limitations. In addition to the requirements of ACI Chapters 1 to 17 and 19 to 26, the requirements of ACI 18.3 for ordinary moment frames must also be satisfied.

SDC C Buildings assigned to SDC C are permitted to utilize intermediate reinforced-concrete moment frames with no limitations. Such frames are to be designed and detailed in accordance with ACI 18.4.

SDC D, E, or F Special reinforced-concrete moment frames are required in buildings assigned to SDC D, E, or F. These frames can be used without any limitations and must be designed and detailed in accordance with ACI 18.2.3 through 18.2.8 and 18.6 through 18.8.

Dual Systems

SDC B Any of the dual systems listed in ASCE/SEI Table 12.2-1 are permitted to be used in structures assigned to SDC B. It is common for shear wall-frame interactive systems to be used in such cases (see Sec. 11.3.6).

SDC C A dual system with intermediate moment frames and ordinary reinforced-concrete shear walls is permitted as a minimum. The moment frames, which must be designed to independently resist 25 percent of the code-prescribed seismic forces, must satisfy the requirements of ACI 18.4. Only the provisions of ACI Chapters 1 to 17 and 19 to 26 need to be satisfied for the walls.

SDC D, E, or F Dual systems with special moment frames and special reinforced-concrete shear walls are required for structures assigned to SDC D, E, or F without limitations. Each component must be designed and detailed in accordance with the applicable provisions of ACI 18.2.1.6. A dual system with intermediate moment frames and special reinforced-concrete shear walls is also permitted with the following limitations: (1) For SDC D, the building height is limited to 160 ft (49 m) and (2) for SDC E and F, the building height is limited to 100 ft (30 m). Note that the use of intermediate moment frames as part of a dual system in SDC D, E, or F is not recommended (see ACI R18.2).

Shear Wall-Frame Interactive Systems

Shear wall-frame interactive systems with ordinary reinforced-concrete moment frames and ordinary reinforced-concrete shear walls are permitted to be used in buildings assigned to SDC B with no limitations. Such systems are not permitted in buildings assigned to higher SDCs; a dual system would be required in such cases. Other than ACI 18.3, no special seismic design and detailing requirements are prescribed for this system.

11.14.3 Design and Detailing Requirements

Overview

Table 11.6 contains the sections in ACI Chapter 18 that must be satisfied for different structural components in addition to those in ACI Chapters 1 through 17 and 19 through 26. These requirements are organized with respect to SDC. There are no sections in Chapter 18 that are applicable to components in structures assigned to SDC A.

For more details on the seismic provisions and detailing requirements, the reader is referred to Chapter 11 of Ref. 2.

Table 11.6 Sections of ACI Chapter 18 to Be Satisfied Based on SDC

	SDC		
Component	B (ACI 18.2.1.3)	C (ACI 18.2.1.4)	D, E, F (ACI 18.2.1.5)
Frame members	ACI 18.3	ACI 18.4	ACI 18.6–18.9
Structural walls and 6 coupling beams	None	None	ACI 18.10
Diaphragms and trusses	None	None	ACI 18.12
Foundations	None	None	ACI 18.13
Frame members not designated as part of the SFRS	None	None	ACI 18.14

REFERENCES

1. American Concrete Institute (ACI), Committee 318. 2014. Building Code Requirements for Structural Concrete and Commentary, ACI 318-14. ACI, Farmington Hills, MI.

2. David A. Fanella. Reinforced Concrete Structures: Analysis and Design, Second Edition (McGraw-Hill Education: New York, Chicago, San Francisco, Athens, London, Madrid, Mexico City, Milan, New Delhi, Singapore, Sydney, Toronto, 2016).

3. Structural Engineering Institute (SEI) of the American Society of Civil Engineers (ASCE). 2010. Minimum Design Loads for Buildings and Other Structures, ASCE/SEI 7-10. ASCE/SEI, Reston, VA.

4. International Code Council (ICC). 2015. International Building Code. ICC, Washington, DC.

5. Concrete Reinforcing Steel Institute (CRSI), Committee on Manual of Standard Practice. 2009. Manual of Standard Practice, 28th ed. CRSI, Schaumburg, IL.

Chapter 12

Design of Prestressed-Concrete Structural Members[a]

BY

SRI SRITHARAN *Wilkinson Chair of Interdisciplinary Engineering and Professor of Structural Engineering, Department of Civil, Construction and Environmental Engineering, Iowa State University, Ames, Iowa*

MARYAM NAZARI *Assistant Professor, Department of Civil and Geomatics Engineering, California State University, Fresno, California*

NOTATION

a = lever arm between C and T
A = gross cross-sectional area
A_c = net cross sectional area of concrete
A_{cp} = area enclosed by the outside perimeter of the concrete cross section
A_ℓ = total area of longitudinal reinforcement to resist torsion
A_o = gross area enclosed by torsional shear flow path
A_{oh} = area enclosed by centerline of the outermost closed transverse torsional reinforcement
A_{ps} = area of prestressing steel
A_t = transformed area
A_{to} = required total cross-sectional area of stirrups at the end of girder, to control horizontal cracking in the ends of pretensioned I girders
c_b = distance from c.g.c. to bottom fiber
c.g.c. = centroid of concrete
c.g.s. = centroid of steel
c_t = distance from c.g.c. to top fiber
C = total compressive force in concrete
C' = ultimate compressive force in concrete
C_c = coefficient of creep = δ_t/δ_i
e = eccentricity of c.g.s.
e' = eccentricity of total compressive force in concrete, C, with respect to c.g.c.
e_{sd} = change in strain as the concrete decompresses at the location of tendon
e_{sl} = strain in tendon due to additional external loading
e_{sp} = strain in tendon due to prestress
E_c = modulus of elasticity of concrete
E_{ci} = modulus of elasticity of concrete at the time of transfer
E_s = modulus of elasticity of steel
f = unit stress in general
f_c = unit stress in concrete
f_{cir} = stress in concrete at the level of the c.g.s. under the combined effect of F_0 and M_G.
f_{cds} = stress in concrete at c.g.s. of tendons due to all superimposed dead loads that are applied to the member after it is prestressed.
f_{cpa} = average compressive stress in the concrete along the member length at the c.g.s. of the tendons
f'_t, f'_b = tensile stress at top (bottom) fiber
f'_c = compressive strength of concrete, 28 days
f'_{ci} = compressive strength of concrete at transfer
f_i = initial stress in prestressing steel before transfer
f_0 = stress in prestressing steel just after transfer
f_{ps} = stress in steel at ultimate load on section
f_{pu} = ultimate stress in prestressing steel
f_{py} = yield strength of prestressing steel
f_r = modulus of rupture of concrete
f_s = steel stress in general
f_{se} = effective stress in prestressing steel after deducting all losses
f_y = yield strength of nonprestressed reinforcement
F = effective prestressing force (after losses)
F_1, F_2 = prestressing force at points 1 and 2, respectively
F_i = initial prestressing force before transfer
F_0 = prestressing force just after transfer
h = sag of cable, overall depth
I = moment of inertia of the section

[a]Much of the material for this chapter was taken from T. Y. Lin and Paul Zia, Chap. 13, *Structural Engineering Handbook*, 4th ed.

I_t = moment of inertia of transformed section
k = coefficient for depth of compression
k_b, k_t = kern distances from c.g.c. to bottom (top)
K = wobble coefficient, per ft
l_d = length of embedment
L = length of tendon
m = load factor or factor of safety
M_G = girder moment
M_T = total external moment produced by the working load
M_u = ultimate resisting moment
$n = E_s/E_c$
p_{cp} = outside perimeter of the concrete cross section
p_h = perimeter of the centerline of the outermost closed stirrup
P' = k at ultimate load
Q = statical moment of the cross-sectional area above (or below) any given level about the centroidal axis
r = radius of gyration of cross section
S_t = principal tensile stress in concrete
T = total tension in prestressing steel
T' = ultimate tension in prestressing steel
V_c = total shear carried by concrete
V_s = total shear carried by web steel
y = distance from centroidal axis
α = curvature of tendon
δ = unit elastic shortening due to transfer of prestress
δ_i = initial strain in concrete due to elastic shortening
δ_s = strain in steel due to shrinkage of concrete
δ_t = final strain in concrete including effect of creep but not of shrinkage
Δ_a = deformation of anchorage
Δf_s = loss of prestress in tendon
Δ_s = elongation of tendon
μ = coefficient of friction between tendon and surrounding material
ξ_{eq} = total equivalent damping ratio of the system expressed in percentage
θ = angle between axis of strut, compression diagonal, or compression field and the tension chord of the members

12.1 INTRODUCTION

12.1.1 Overview

Prestress concrete is conceptually similar to reinforced concrete with the exception that the steel reinforcement is tensioned, subjecting the surrounding concrete to compression in the direction of loading. In this process, desirable strains and stresses are introduced to concrete while counterbalancing undesirable strains and stresses. This approach allows a design to maximize the use of concrete and steel reinforcement while delaying the formation of tension cracks. For example, a member could be designed to be crack free at service loads. Since prestressed members can be designed to be crack free, their deflections can be easily controlled. As a result, the benefits of prestressed concrete members include increased span length, slender more aesthetically pleasing sections, and better protection against corrosion. Prestressing also promotes prefabrication of structural members, which may be accomplished on-site or more commonly at precast plants. When structures are fabricated in components, robust connections between them have to be designed and constructed so that the structures can produce dependable behavior.

In comparison to reinforced concrete, higher-strength materials make the prestressed concrete design most effective. Concrete strength in the range of 5000 to 9000 psi is commonly used, although higher strength in the 10,000 to 18,000 psi range is also possible today. In precast operation, high early strength is desired for a fast turnover. With steam curing, a transfer strength of 3500 psi is often attained overnight. A transfer strength of over 5000 psi can be attained in 3 or 4 days by using high-early-strength cement or chemical and mineral admixtures without special curing.

Two characteristics of concrete that affects the behavior and efficiency of prestressed concrete are shrinkage and creep. Shrinkage is primarily dependent on parameters such as humidity, mix design (especially water content), member thickness, member size, aggregate type, temperature, and age. The typical range of shrinkage may be expected to be in the range of 0.0002 to 0.0008. Creep, which develops under a sustained load, depends on initial stress, elastic modulus, humidity, age at loading, member thickness, aggregate type, etc. The total amount of creep strain at the end of 20 years ranges from one to five times the instantaneous elastic deformation under load (averaging about three times), the low values occurring for moist storage and for limestone aggregates. Of the total amount of creep strain, about one-fourth takes place within the first 2 weeks after application of prestress, another one-fourth within 2 to 3 months, another one-fourth within a year, and the last one-fourth in the course of many years. Upon removal of the sustained stress, roughly 80 to 90 percent of the creep will be recovered during the same length of time that it has taken to develop.

Lightweight concrete has been successfully used in prestressed-concrete construction. It has a lower modulus of rupture than normal-weight concrete and slightly less favorable shrinkage and creep characteristics. However, with the better aggregates, these properties are comparable with those of normal-weight concrete.

High-strength steel in the form of wires, strands, or bars is commonly used for prestressing with the specified material strength ranging from 150 to 300 ksi. Typical stress-strain curves for the different prestressing steel are shown in Fig. 12.1. They all have comparable modulus of elasticity in the range of 27,000 to 30,000 ksi, but none of them has a well-defined yield point. This is due to the presence of residual stresses resulting from cold drawing used in the manufacturing of these reinforcements. The yield strengths for these prestressing steel are defined using the 1 percent elongation method for wires and strands and the 0.2 percent offset method for high-strength bars. Typical elongation at rupture for these steel ranges from 5 to 6 percent.

The most commonly used prestressing steel in practice is 0.5-in.- and 0.6-in.-diameter, seven-wire grade 270 strand (i.e., with a minimum ultimate strength of 270 ksi based on the nominal area of the strand), which is produced by a combined process of low-temperature heat treatment and high tension. Because of this special process, the yield strength of the strand is raised and its relaxation is substantially reduced. This prestressing steel is thus called *low-relaxation strand*.

The prestress technology is relatively new with the first application in bridges in the United States dating back to 1949. Since then, it has been used in several different applications. They include: bridges, buildings, architectural elements, storage tanks, containments structures of nuclear power plants, towers, stadiums, earthwork, offshore production platforms, dams, piles, utility poles, pipes, and railroad ties.

12.1.2 Methods of Prestressing

Prestressing steel is typically tensioned through mechanical means regardless of it is a posttensioning or pretensioning operation. In posttensioning, hydraulic jacks are used to pull the steel against the hardened concrete; in pretensioning, the steel is tensioned against bulkheads or molds using a hydraulic jack. The capacity of these jacks varies from about 3 tons up to 2000 tons. In all cases, both the jack gage pressure and the tendon elongation are measured to ensure development of the right amount of prestress.

Pretensioning. Pretensioning in the United States is usually accomplished in the plant by the long-line process. Tendons are stretched between two bulkheads held against the ends of a stressing bed several hundred feet long. Concrete is then placed along the bed between steel, timber, or concrete forms. When the concrete has set sufficiently to carry the prestress, the tendons are detensioned either by releasing the hydraulic jack or by flame cutting, and the prestress is transferred to the members, generally through bond between steel and concrete. Since strands anchor themselves much better than wires, they are widely used in pretension operation. Devices for gripping the tendons to the bulkheads are usually made on the wedge and friction principle. Reusable quick-release grips for holding strands are routinely employed. In order to improve the behavior of prestressed beams, their tendons are often bent to desired profiles. In the long-line process, this is achieved

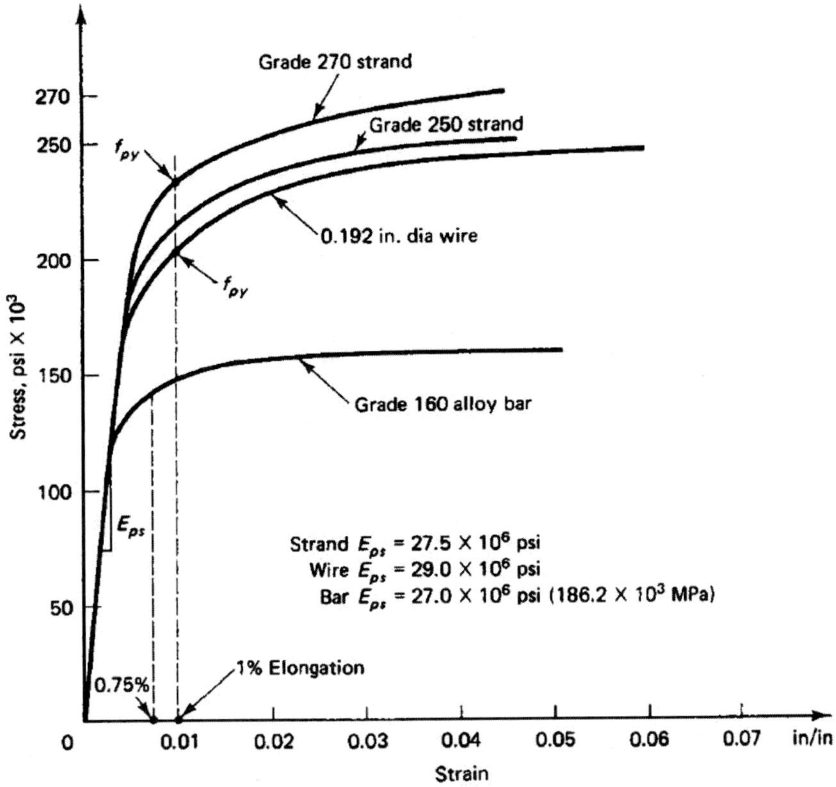

Figure 12.1 Typical stress-strain curve for prestressing steels.

by deflecting the tendons up and down along the length of the bed. This process is known as harping or draping of tendons.

Posttensioning. There are hundreds of patents and systems for posttensioning. Some are more commonly used than others. A partial list of these systems can be found in earlier editions of this handbook. More current systems are described in publications of the Post-Tensioning Institute of Farmington Hills, Michigan. Since posttensioning is typically done in the field, a qualified crew is required for the jacking operation who will provide the equipment and technical supervision. Details of various posttensioning systems can be found in the last edition of this handbook and at manufacture's websites.

Grouting. For bonding the tendons to the concrete after tensioning (in the case of posttensioning), cement grout is injected, which also serves to protect the steel against corrosion. Entry for the grout into the cableway is provided by means of access holes in the anchorage heads and cones, or pipes buried in the concrete members. The grout can be injected at one end of the member until it is forced out of the other end. For longer members, it can be applied at both ends until forced out of a center vent. Either ordinary portland cement or high-early-strength cement may be used for the grout. Coarse sand is preferred for bond and strength, but sufficient fineness is necessary considering the limited space through which the grout has to pass. To ensure good bond for small conduits, grouting under pressure is desirable; however, care should be taken to ensure that the pressure on the walls of the cable enclosure can be safely resisted. Machines for mixing and injecting the grouts are commercially available.

Where grout is applied to large void areas, some sand may be used in the grout mix. Where the space is limited, neat cement paste should be employed. Admixtures are generally used to increase workability, reduce bleeding and shrinkage, or provide expansion. Water-cement ratios typically specified are between 0.47 and 0.53. Grouting pressure generally ranges from 80 to 100 psi with a maximum pressure specified as 145 psi. After the grout has discharged from the far end, the end is plugged and the pressure is again applied at the injecting end to compact the grout. It is also good practice to wash the cables with water before grouting is started, with the excess water being removed with compressed air. When tendons are unbonded, they must be properly greased and wrapped to prevent corrosion.

A minimum grout temperature of 90°F is generally recommended. The temperature of members at the time of grouting should be maintained above 35°F until job-cured 2-in. cubes of grout reach a minimum compressive strength of 800 psi. Test cubes should be cured under temperature and moisture conditions as close as possible to those of the grout in the member. During mixing and pumping, the grout temperature should not exceed 90°F. Otherwise, difficulties may be encountered in pumping.

Plastic shielded wires and strands with diameters up to 0.6 in., prepacked with corrosion inhibitor, are commercially available for post-tensioned work. Such tendons are suggested to have a very low coefficient of friction during tensioning.

12.2 PRESTRESS LOSSES

Initial prestress that is applied to a concrete member experiences two types of losses: strain losses and friction losses. The strain losses include reduction in prestressing due to immediate elastic loss that takes place during fabrication and time-dependent losses. The friction losses occur in posttensioned tendons due to drag against the cable sheath. Due to losses, prestress force should not be treated as a constant force with time. Furthermore, concrete strength and its elastic modulus increase with time. Therefore, analysis and design of prestress concrete members should include the effective prestress force and appropriate material properties for a given stage of loading.

Descriptions of various prestress losses are given below, and guidance to estimate the amount of loss in pretensioned and posttensioned members is presented in Table 12.1.

12.2.1 Elastic Shortening of Concrete

As prestressing force is transferred to the concrete, shortening of the concrete results in loss of prestress. This mainly affects pretensioned members. In posttensioning systems, elastic shortening loss is zero if all tendons are stretched simultaneously. Some shortening loss is expected when tendons are stretched sequentially, but the loss is expected to be small compared to the value calculated in the pretensioned case.

12.2.2 Creep

Creep and shrinkage of concrete are influenced by many factors, and they are interdependent. However, treating their effects individually is accepted in engineering practice. Creep of concrete will occur over a long period of time under a sustained load. Creep is assumed to occur with the superimposed permanent dead load added to the member after

Table 12.1 Types of Prestress Losses and Evaluation

	Stage of occurrence		Prestress loss estimation	
Type of prestress loss	Pretensioned members	Posttensioned members	Pretensioned members	Posttensioned members
Elastic Shortening of Concrete (Δf_s)	At transfer	At sequential jacking	For concentrically prestressed member: Unit elastic shortening $\delta = \frac{f_c}{E_{ci}} = \frac{F_0}{A_c E_{ci}}$ $\Delta f_s = E_s \delta = \frac{E_s F_0}{A_c E_{ci}} = \frac{nF_0}{A_c}$ or $\delta_i = \frac{F_i}{A_c E_{ci} + A_{ps} E_s}$ $\Delta f_s = E_s \delta_i = \frac{E_s F_i}{A_c E_{ci} + A_{ps} E_s} = \frac{nF_i}{A_c + nA_{ps}}$ $\Delta f_s = \frac{nF_i}{A_t}$ For eccentrically prestressed member: $ES = \Delta f_s = \frac{E_s F_{cir}}{E_{ci}} = n(f_{cir})$	For n tendons or pairs of tendons sequentially tensioned with j sequential jacking steps $\Delta f_s = \frac{1}{n} \sum_{j=1}^{n} (\Delta f_s)_j$
Creep of Concrete (CR)	After transfer	After transfer	For bonded members of normal-weight concrete: $\text{CR} = K_{cr} \frac{E_s}{E_c} (f_{cir} - f_{cds})$ K_{cr} should be reduced by 20% for lightweight concrete. For unbonded members of normal-weight concrete: $\text{CR} = K_{cr} \frac{E_s}{E_c} f_{cpa}$	
Shrinkage of Concrete (SH)	After transfer	After transfer	$\text{SH} = 8.2 \times 10^{-6} K_{sh} E_s \left(1 - 0.06 \frac{V}{S}\right)(100 - \text{RH})$ K_{sh} = 1 for pretensioned members For posttensioned members, refer to **Table 12.2**	
Relaxation in Steel (RE)	Before and after transfer	After transfer	$RE = f_{si}\left[\frac{\log_{10} t}{10}\left(\frac{f_{si}}{f_{py}} - 0.55\right)\right]$ for $\frac{f_{si}}{f_{py}} > 0.55$ Replace denominator 10 with 45 for low-relaxation steel or following the ACI-ASCE method: $\text{RE} = [K_{re} - J(\text{SH} + \text{CR} + \text{ES})]C$ For K_{re}, J, and C are values refer to ACI 423.10R-16	
Anchorage Seating Loss (ANC)	—	At transfer	—	$\text{ANC} = \frac{\Delta_a E_s}{L}$
Friction (Δf_F)	—	At jacking	—	The force F_2 at any point on the tendon is given by $F_2 = F_1 e^{-\mu\theta - Kx}$ F_1 = force at the jacking end θ = angle between F_1 and F_2, radians x = length between points 1 and 2, ft μ = coefficient of friction between tendon and surrounding material (see **Table 12.3**) K = wobble coefficient, per ft (see **Table 12.3**) $\Delta f_F = -F_1(\mu\alpha + KL)$

it has been prestressed. Part of the initial compressive strain induced in the concrete immediately after transfer is reduced by the tensile strain resulting from the superimposed permanent dead load. With unbonded tendons, the losses are related to the average member strain rather than to the strain at the point of maximum moment, which should be appropriately taken into account.

12.2.3 Shrinkage

Calculations for loss resulting from shrinkage of concrete should reflect those that are most important: volume-to-surface ratio, relative humidity, and time from end of moist curing to application of prestress (see Table 12.2). It is expected that 80 percent of shrinkage will occur in the first year with the remaining 20 percent taking place over several years. As with creep, there is an upper and lower variation from the average shrinkage strain value, which can be taken as 550×10^{-6} in./in.

Table 12.2 Values of K_{sh} for Posttensioned Members

	Time after end of moist curing to application of prestress, days							
	1	3	5	7	10	20	30	60
K_{sh}	0.92	0.85	0.80	0.77	0.73	0.64	0.58	0.45

12.2.4 Relaxation in Steel

Stress relaxation in steel is the loss of stress when it is maintained at a constant strain for a period of time. Relaxation varies with steels of different compositions and treatments, but its approximate characteristics are known for most of the prestressing steels. In general, the percentage increases with increasing stress, and the relaxation is negligible when steel is subjected to a low stress. Typical curves giving the relation between relaxation and initial stress level in three types of steel wires are shown in Fig. 12.2.

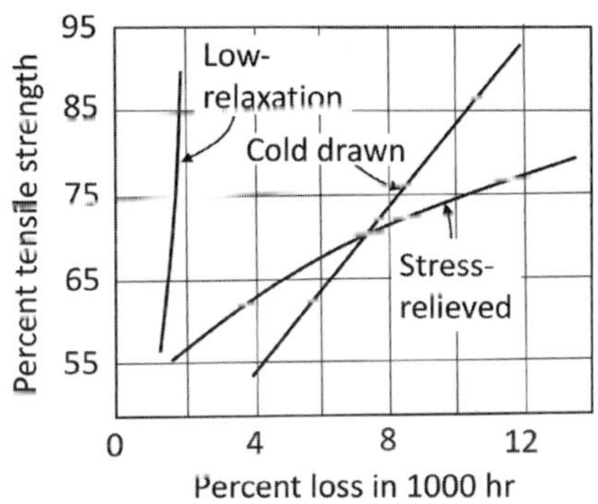

Figure 12.2 Relaxation of prestressing wires.

12.2.5 Seating Loss

When a tendon is tensioned and jack force is transferred to anchorage, some stress in prestressing steel is lost due to seating of wedges in the anchors and deformation of anchorage fixtures. The amount of settlement can be calibrated and typically provided by the manufacturer. Since this loss of prestress is caused by a fixed total amount of shortening, the percentage of loss is higher when the length of tendon is short. Hence, it is quite difficult to prestress members with a shorter length accurately, especially for those prestressing systems for which anchorage losses are relatively large. On the other hand, in the long-line process of pretensioning, this type of loss is insignificant and is not taken into consideration in design.

12.2.6 Friction

Posttensioned tendons experience friction losses due to drag against the cable sheath. Friction losses have two components: curvature effects and wobble effects. The loss of prestress due to curvature effect results from the intended curvature of the tendons. This loss is also dependent on the coefficient of friction and the pressure exerted by the tendon on the concrete. The coefficient of friction depends on the nature of the surfaces in contact, the amount and nature of lubricants, and sometimes the length of contact. The pressure between the tendon and concrete depends on the stress in the tendon and the change in angle. Since in practice the duct for a straight tendon will not be perfectly straight, some friction will exist between the tendon and its surrounding material. This is sometimes described as the wobble effect of the duct and is dependent on the length of and stress in the tendon, the coefficient of friction, and the workmanship and the method used in aligning and forming the duct. Table 12.3 may be used to determine the suitable values for curvature and wobble coefficients. There are several methods to minimize the frictional loss in the tendons. One method is to overtension them. Jacking from both ends of the member is another. Lubricants can be used to advantage for unbonded tendons. For bonded tendons, water-soluble oils can be used to reduce friction while tensioning; the lubricant is flushed off with water afterward.

Table 12.3 Coefficients for Frictional Loss

	Wobble coefficient, K	Curvature coefficient, μ
Tendons in flexible metal sheathing:		
Wire tendons	0.0010–0.0015	0.15–0.25
High-strength bars	0.0001–0.0006	0.08–0.30
7-wire strand	0.0005–0.002	0.15–0.25
Tendons in rigid metal duct:		
7-wire strand	0.0002	0.15–0.25
Mastic-coated tendons:		
Wire tendons and 7-wire strand	0.001–0.002	0.05–0.15
Pregreased tendons:		
Wire tendons and 7-wire strand	0.0003–0.002	0.05–0.15

12.2.7 Effective Prestress

The effective prestress is obtained by deducting the losses from the initial prestress. The loss of prestress varies with many factors. For a close estimate it is necessary to consider the amount of various losses in successive time intervals such as before transfer of prestress, during transfer of prestress, first year after transfer of prestress, and from first year to the end of service life of the structure. However, for the average case, the values in Table 12.4 are representative.

Table 12.4 Prestress Losses

	Pretensioning, %	Posttensioning, %
Elastic shortening of concrete	4	2
Creep of concrete	6	5
Shrinkage of concrete	7	6
Relaxation in steel	6	6
Total loss not including frictional loss	23	19

12.2.8 Elongation of Tendons

Using elastic elongation as a measure of the tension induced in the tendon is often useful during construction. If a tendon has uniform stress F along its entire length L, the total elongation is given by

$$\Delta_s = \frac{FL}{E_s A_{ps}} \tag{12.1}$$

For a tendon with uniform curvature, considering frictional loss throughout its length L, the total elongation is given by

$$\Delta_s = \frac{F_2 L}{E_s A_{ps}} \frac{e^{\mu\alpha + KL} - 1}{\mu\alpha + KL} \tag{12.2}$$

For an approximate solution,

$$\Delta_s = \frac{F_1 + F_2}{2} \frac{L}{E_s A_{ps}} \tag{12.3}$$

12.3 ANALYSIS CONCEPTS

12.3.1 Basic Concepts

Three different concepts may be used to explain and analyze the behavior of prestressed concrete under flexure. They are known as the stress concept, internal couple concept, and load balancing concept.

1. Stress Concept. In this approach, concrete is viewed as being subjected to two systems of forces, namely, internal prestress and external load. This approach is useful for analysis in the elastic range before cracking, which means that the principal of superposition is applicable. Consider a simple rectangular beam, eccentrically prestressed by a tendon (Fig. 12.3). The tendon force F produces a stress f at any cross section:

$$f = \frac{F}{A} + \frac{Fey}{I} \tag{12.4}$$

where A = area of beam cross section
e = eccentricity of tendon
y = distance from centroidal axis (c.g.c.)
I = moment of inertia of the section

If M is the moment at a section due to external load, the stress at any point on that section is

$$f = \frac{F}{A} + \frac{Fey}{I} - \frac{My}{I} \tag{12.5}$$

If prestress eccentricities and external moments exist on both principal axes,

$$f = \frac{F}{A} + \frac{Fe_y y}{I_x} + \frac{Fe_x x}{I_y} - \frac{M_x y}{I_x} - \frac{M_y x}{I_y} \tag{12.6}$$

A consistent sign convention should be adopted in the design calculations and values with appropriate signs should be used in the above equations. In this chapter, tensile stresses are assumed to be positive, so are the distances measured above the c.g.c. External moments causing concave and convex curvatures in a simply supported beam are taken as positive and negative, respectively. This implies both F and e have negative signs, but Fe is positive.

EXAMPLE 1

A prestressed-concrete rectangular beam 20 × 30 in.^{2} has a simple span of 24 ft and is loaded by a uniform load of 3 kips/ft, which includes its own weight (Fig. 12.4). The prestressing tendon is located as shown and produces an effective prestress of 360 kips. Compute fiber stresses in the concrete at the midspan section using the gross section properties.

Solution F = 360 kips, A = 20 × 30 = 600 in.^{2} (neglecting any hole due to the tendon), e = 6 in., $I = bd^3/12 = 20 \times 30^3/12$ = 45,000 in.^{4}; y = 15 in. for extreme fibers.

$$M = 3 \times \frac{24^2}{8} = 216 \text{ ft} - \text{kips}$$

$$f = \frac{-360{,}000}{600} \pm \frac{360{,}000 \times 6 \times 15}{45{,}000} \mp \frac{216 \times 12{,}000 \times 15}{45{,}000}$$

$$= -600 + 720 - 864 = -744 \text{ psi for top fiber}$$

$$= -600 - 720 + 864 = -456 \text{ psi for bottom fiber}$$

The resulting stress distribution is shown in Fig. 12.4, which suggests that an eccentric tendon is more efficient than a concentric tendon (i.e., e = 0) for counterbalancing the effects of external load. With a straight tendon and a higher value for e, the beam may experience permanent tensile stress at the ends of the span since the moments due to external loads are zero in these regions. This issue can be overcome by using a parabolic tendon with zero eccentricity at the beam ends.

2. Internal Couple Concept. This concept treats prestressed concrete as a combination of steel and concrete, with steel taking tension and concrete carrying compression. The resultant forces of these two materials create an internal couple to resist the external moment induced by external loads. The concept is useful for analysis after cracking as well as in the elastic range. The internal resisting couple C-T is shown in Fig. 12.5. When the prestressing force F and the external moment M are known and since $C = T = F$, the lever arm a, which locates the center of the compressive force C, is given by

$$a = \frac{M}{F} \tag{12.7}$$

Once C is located, the stress at any point on the section is given by

$$f = \frac{C}{A} + \frac{Ce'y}{I} \tag{12.8}$$

where e' = eccentricity of C with respect to c.g.c.

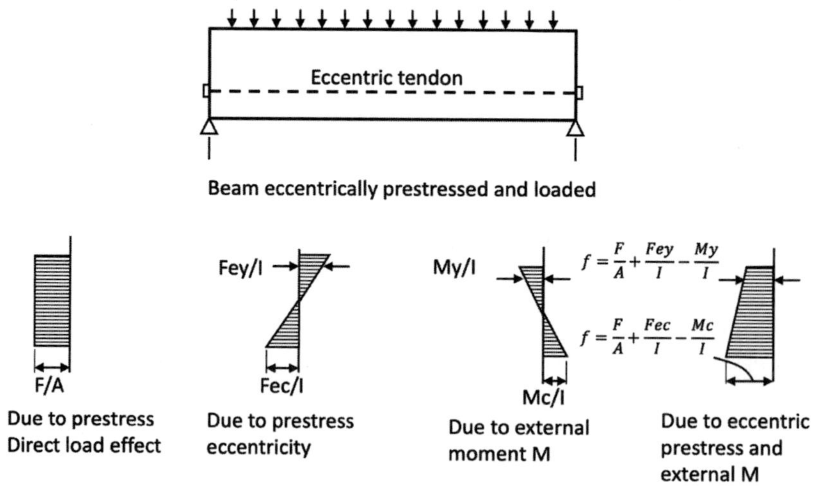

Figure 12.3 Example 1.

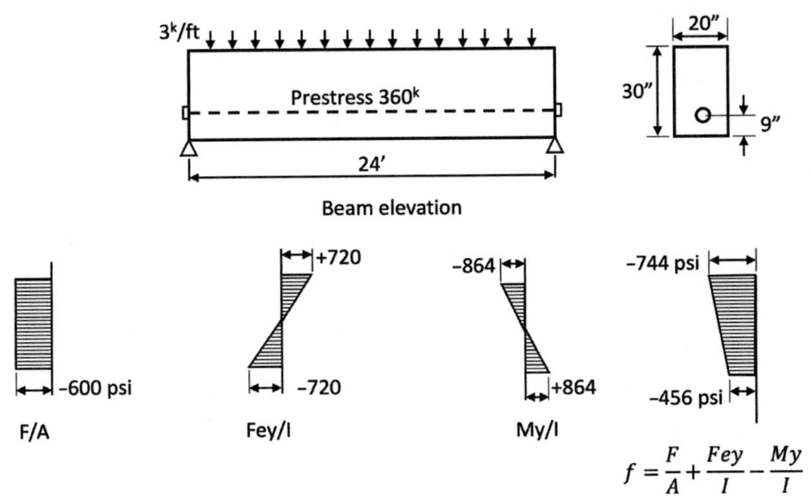

Figure 12.4 Stress distribution across an eccentrically prestressed concrete section.

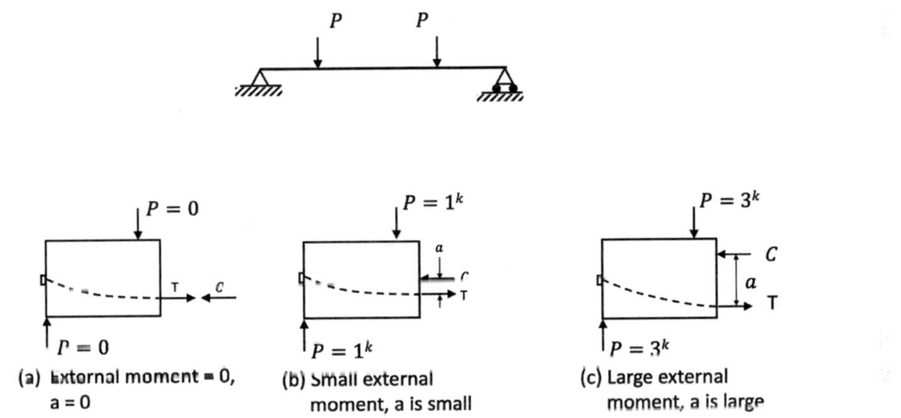

Figure 12.5 Variation of *a*.

Note that when there is no external load, *C* acts in line with *T* (i.e., $a = 0$). As the load increases, *a* will increase. The corresponding stress distribution in concrete depends on the location of *C* with respect to kern points (Fig. 12.6), which are section properties. The bottom and top kern points are defined as follows:

$$k_b = \frac{I}{Ac_t} \qquad k_t = \frac{I}{Ac_b}$$

where c_t = distance from c.g.c. to top fiber
c_b = distance from c.g.c. to bottom fiber

Following the formation of flexural cracks, there will not be any significant changes to *a*, but the internal couple will increase due to the increase in *T* and therefore *C*.

EXAMPLE 2

Same data as Example 1. Compute fiber stresses using the internal couple concept.

Solution $M = 3 \times 24^2/8 = 216$ ft-kips. The internal couple furnished by the forces $C = T = 360$ kips (Fig. 12.7) has the lever arm

$$a = \frac{216}{360} \times 12 = 7.2 \text{ in.}$$

Since *T* acts 9 in. from the bottom, *C* lies 16.2 in. from the bottom, and $e' = 16.2 - 15 = 1.2$ in. Then

$$f = \frac{-360{,}000}{600} \mp \frac{360{,}000 \times 1.2 \times 15}{45{,}000}$$
$$= -600 \mp 144$$
$$= -744 \text{ psi for top fiber}$$
$$= -456 \text{ psi for bottom fiber}$$

3. *Load Balancing Concept.* By this concept, prestressing may be viewed as an attempt to balance a portion of the external loads acting on a member. For example, prestressing in a member may be designed to sufficiently balance the gravity loads. As a result, the member will not be subjected to flexural stresses and deflections due to gravity loads, which also simplifies both the design and the analysis of the member. In its simplest form, a parabolic tendon is required in a simply supported prestressed beam to exert a uniform upward force on the beam and balance an external downward load of equal intensity, leaving the beam with a uniform compressive stress $f = F/A$ at any cross section provided that *F* acts at the beam ends with no eccentricity.

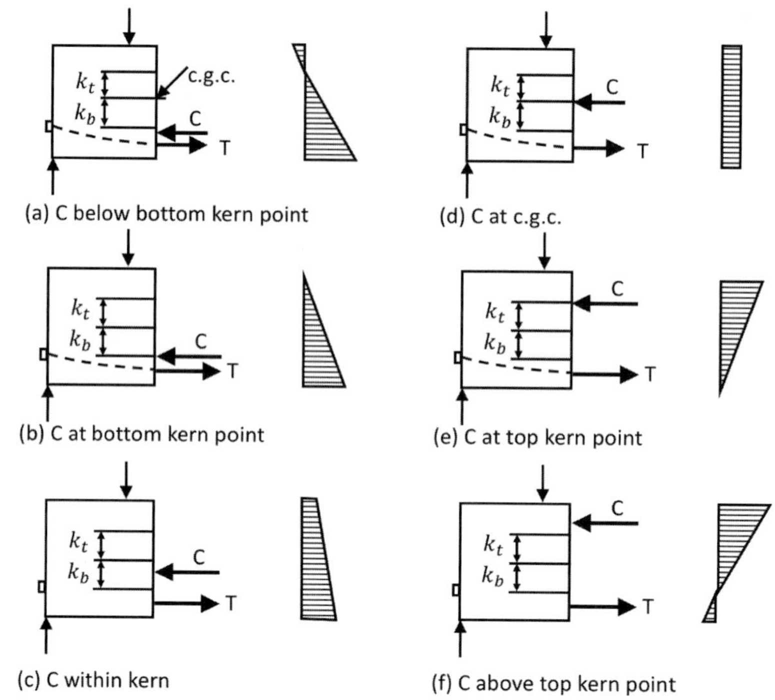

Figure 12.6 Stress distribution (elastic theory).

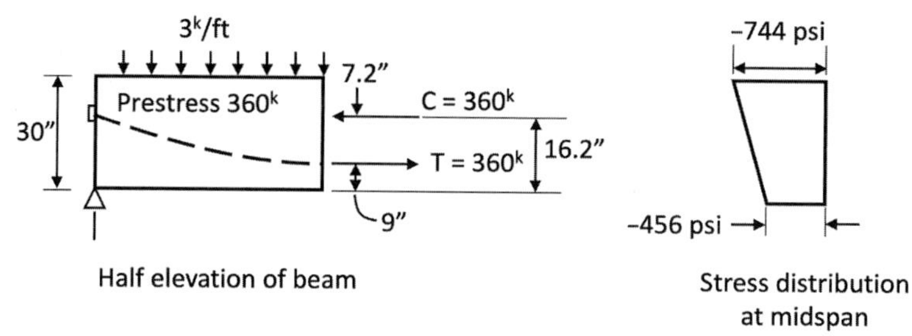

Figure 12.7 Example 2.

If the external load is not completely balanced by the upward force, the moment M of the *unbalanced* load produces an additional stress $f = My/I$.

The prestress F required to balance a uniform load w lb/ft is given by

$$F = \frac{wL^2}{8h} \tag{12.9}$$

where L = span of beam, ft
h = sag of cable, ft

EXAMPLE 3

A 20 × 30-in. concrete beam is prestressed with a parabolic cable located as shown in Fig. 12.8. Compute the prestress required to balance an external load of 2.5 kips/ft.

Solution From Fig. 12.8, $h = 6$ in.

$$F = \frac{wL^2}{8h} = \frac{2.5 \times 24^2}{8 \times 0.5} = 360 \text{ kips}$$

Under this prestress, the beam will be uniformly stressed to 360,000/(20 × 30) = 600 psi. (The horizontal component of the prestress force may be used if greater accuracy is desired.)

The beam will not experience any flexural stresses when subjected to an external load of 2.5 kips/ft. If the external load is 3 kips/ft (Examples 1 and 2), the unbalanced load is 0.5 kip/ft, and the midspan moment

$$M = \frac{wL^2}{8} = \frac{0.5 \times 24^2}{8} = 36 \text{ kips}$$

$$f = \frac{My}{I} = \mp \frac{36 \times 12{,}000 \times 15}{45{,}000} = \mp 144 \text{ psi}$$

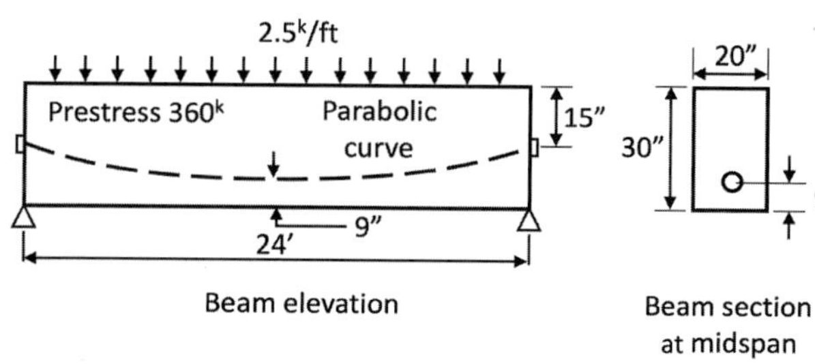

Figure 12.8 Example 3.

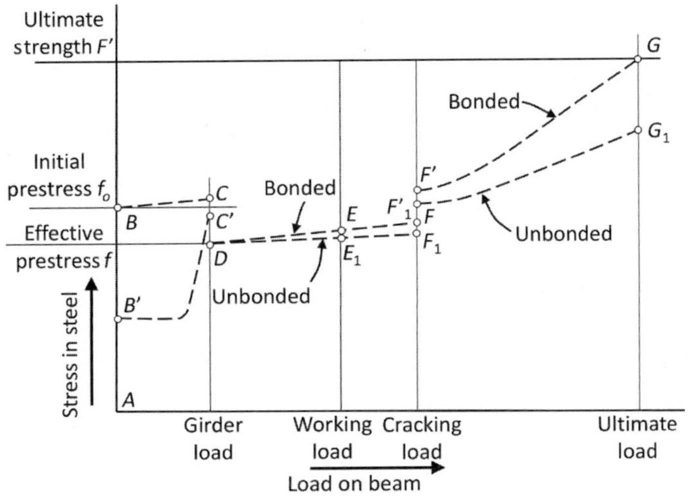

Figure 12.9 Variation of steel stress with load.

The stresses resulting from prestress and the external load of 3 kips/ft are

$$f = -600 \pm 144$$
$$= -744 \text{ psi for top fiber}$$
$$= -456 \text{ psi for bottom fiber}$$

12.3.2 Stress in Steel

Since the steel in prestressed members are tensioned prior to the member being subjected to any external load, it is important to understand the variation in steel stress, which is illustrated in Fig. 12.9. For a posttensioned beam, as the tendons are tensioned, the steel stress increases from *A* to *B*. Simultaneously, the prestress is transferred to the beam. If the beam is heavy, its full dead load will not come into play until after its falsework is removed, causing a slight increase in steel stress from *B* to *C*. If the beam is relatively light, it usually begins to camber before the steel stress reaches *B*, and its dead load comes into play immediately. Thus, the steel stress may vary from an intermediate point, say, *B′* to *C′*. Because of the camber, the tendons shorten slightly so that their stress is slightly lower, as represented by *C*. In a pretensioned beam, the steel stress generally varies from *B* to *C′* upon transfer of prestress.

Assuming the losses of prestress take place before the application of superimposed dead and live loads, the steel stress is reduced from *C* or *C′* to *D*. Only minor changes in the range *DE* are induced by superimposed dead and live loads since no flexural cracking is expected at this stage.

At cracking, stress in the steel at the crack jumps from *F* to *F′* due to concrete not being able to carry any tension force, after which it continues to increase until the ultimate load *G* is reached. Unbonded tendons would be forced to slip except for frictional resistance. This slippage would allow any strain in the unbonded tendon to distribute throughout the entire member length. Consequently, as load increases, the tendon stress will increase more slowly than that in a bonded tendon. The line $DE_1F_1F'_1G_1$ shows the stress variation in unbonded tendons, assuming the same effective prestress before addition of external load. As a result, the ultimate capacity of a beam with unbonded tendon will be lower than that of an identical beam with bonded tendon.

12.3.3 Cracking Moment

The external moment producing first hair cracks in a prestressed concrete beam is known as the cracking moment. It is a measure of the serviceability of the beam. It occurs when the tensile stress in the extreme fiber of concrete reaches its modulus of rupture f_r (Fig. 12.10):

$$\frac{F}{A} + \frac{Fec}{I} - \frac{Mc}{I} = f_r \tag{12.10}$$

Transposing, we have the value of cracking moment:

$$M = Fe + \frac{FI}{Ac} - \frac{f_r I}{c} \tag{12.11a}$$

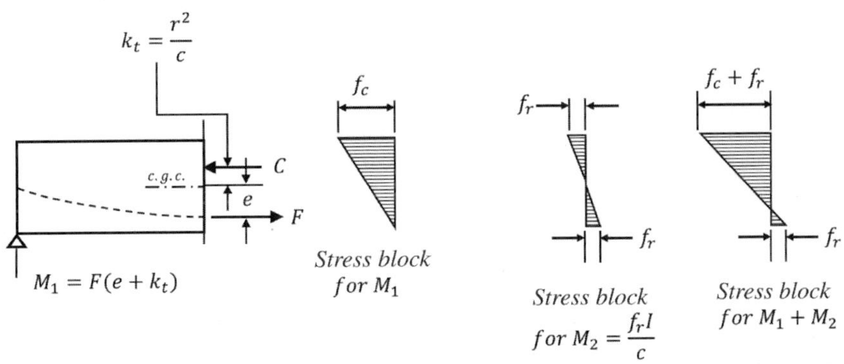

Figure 12.10 Cracking moment.

This can also be written as

$$M = F\left(e + \frac{I}{Ac}\right) - \frac{f_r I}{c} = F(e + k_t) - \frac{f_r I}{c} \qquad (12.11b)$$

which is shown as $M = M_1 + M_2$ in Fig. 12.10. The accuracy of M will depend on the accuracy of the value of f_r. Although this value has been taken as $7.5\sqrt{f'_c}$ or from modulus of rupture tests, it has been shown recently that such a value is too conservative and that the modulus of rupture depends on the member depth (Sritharan et al. 2018). A more accurate estimate of f_r can be obtained from $7.5\sqrt{f'_c}(h)^{-0.15}$ psi, where h is the member depth in feet.

12.3.4 Ultimate Moment

Underreinforced Beams. The steel in an underreinforced bonded beam is usually stressed to its ultimate strength f_{pu} under the action of the ultimate moment. Thus, the ultimate tension force is $T' = A_{ps}f_{pu}$. The compression force $C' = T'$ can be located approximately using a rectangular stress block, with an average stress of $0.85f'_c$ and a depth of $k'd$ (Fig. 12.11) such that the area of concrete A'_c within the depth $k'd$ satisfies the relation $A'_c = C'/0.85f'_c$. Equating T' and C' results in the equation

$$k' = \frac{A_{ps}f_{pu}}{0.85f'_c db} \qquad (12.12)$$

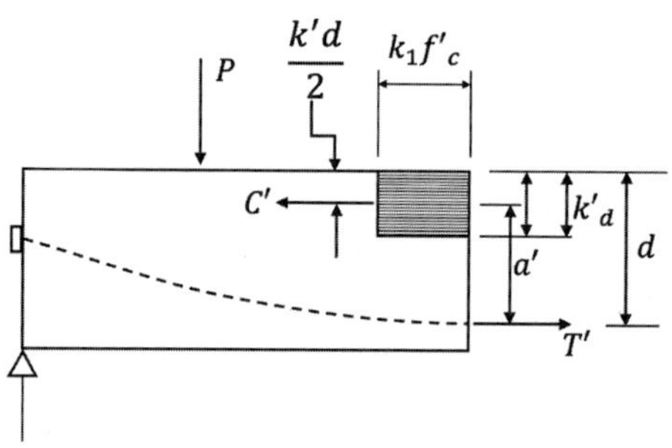

Figure 12.11 Ultimate moment.

These formulas apply even though the cross section is not rectangular, provided that the compression flange has a uniform width b for the depth $k'd$.

The ultimate resisting moment $T'a'$ (Fig. 12.11) is

$$M_u = A_{ps}f_{pu}d\left(1 - \frac{k'}{2}\right) \qquad (12.13)$$

Most building codes and bridge specifications give a slightly different and more conservative formula for rectangular sections:

$$M_u = A_{ps}f_{pu}d\left(1 - 0.6\frac{\rho_p f_{ps}}{f'_c}\right) \qquad (12.14)$$

where f_{ps} = stress in steel at failure and $\rho_p = A_{ps}/bd$, with f_{ps} being determined empirically.

For flanged sections in which the neutral axis falls outside the flange (usually where the flange thickness t is less than $1.4d\rho_p f_{ps}/f'_c$),

$$M_u = A_{sr}f_{ps}d\left(1 - \frac{k'}{2}\right) + 0.85\,f'_c(b - b')t\left(d - \frac{t}{2}\right) \qquad (12.15)$$

where b is the effective width of the flange, b' is the width of the web, and

$$A_{sr} = A_{ps} - A_{sf}$$
$$A_{sf} = 0.85f'_c(b - b')t/f_{ps}$$

Unless the value of f_{ps} is determined from detailed analysis accounting for prestrain from prestressing and the increase in strain during decompression where the tendon is located, the following is often used:

Bonded members:

$$f_{ps} = f_{pu}\left(1 - 0.5\frac{\rho_p f_{pu}}{f'_c}\right) \qquad (12.16a)$$

Unbonded members with a span-to-depth ratio of 35 or less:

$$f_{ps} = f_{se} + 10{,}000 + \frac{f'_c}{100\rho_p} \le f_{se} + 60{,}000 \le f_{py} \qquad (12.16b)$$

Unbonded members with a span-to-depth ratio greater than 35:

$$f_{ps} = f_{se} + 10{,}000 + \frac{f'_c}{300\rho_p} \le f_{se} + 30{,}000 \le f_{py} \qquad (12.16c)$$

where f_{se} is the effective stress in the prestressing steel after losses, in psi, and f_{py} is the specified yield strength of the prestressing steel.

Over the past two decades, there has been significant interest in using unbonded posttensioning in concrete walls. The motivation for this approach is to minimize the damage to walls when they are subjected to a lateral force. For this case, Henry et al. (2012) suggest that the above equation will not accurately estimate f_{ps} due to the span-to-depth ratio being very small, and therefore an alternative expression as a function of aspect ratio has been recommended.

In unbonded members, it is desirable to use a moderate amount of bonded nonprestressed reinforcement, if appropriate, as that will help distribute cracks and improve the postcracking stiffness and the ultimate strength of the beam. The minimum amount of such reinforcement required by the ACI 318-14 (2014) Code is $A_s = 0.004A$, where A is the area of that part of the cross section between the flexural tension face and the center of gravity of the cross section.

Overreinforced Beams. It is safe to assume that a rectangular section is underreinforced if $\rho_p f_{ps}/f'_c < 0.30$. If $\rho_p f_{ps}/f'_c > 0.30$, the beam is usually considered to be overreinforced, and it is often specified that the ultimate flexural strength be taken not greater than $M_u = 0.25f'_c bd^2$.

More accurate predictions for the stress in the prestressing tendon can be made with the following procedure, considering both equilibrium and strain compatibility. The maximum compressive strain in the concrete at failure is taken to be 0.003. Assuming plane sections remain plane and using a trial value of c, the strain in the steel at failure of the beam is given by $e_s = e_{sp} + e_{sd} + e_{sl}$, where e_{sp} is the strain due to prestress, e_{sd} is the change in strain as the concrete decompresses at the location of tendon, and $e_{sl} = 0.003(d - c)/c$ due to additional external loading (Fig. 12.12). The value of f_{ps} corresponding to e_s can then be obtained from the stress-strain diagram. If f_{ps} is found to be near the ultimate value f_{pu}, the section is not overreinforced. However, if f_{ps} is appreciably lower than f_{pu}, the section is overreinforced.

Knowing the value of f_{ps}, the equilibrium condition $T' = C'$ should be checked, where $T' = A_{ps}f_{ps}$ and $C' = 0.85f'_c(b)(k'd)$. The value of $k'd$ is obtained from the expression

$$k'd = \beta_1 c$$

where $\beta_1 = 0.85$ for $f'_c \le 4000$ psi. For $f'_c > 4000$ psi, the value of β_1 is reduced continuously at a rate of 0.05 for each 1000 psi in excess of 4000 psi, but β_1 shall not be taken less than 0.65.

If the equilibrium condition $T' = C'$ is not satisfied, a new trial value of c should be assumed and the procedure repeated until $T' = C'$. The moment capacity is then computed by Eq. (12.13) or (12.14).

EXAMPLE 4

RECTANGULAR SECTION

A beam of rectangular cross section is 12 in. wide by 24 in. deep. The c.g.s. of the prestressing wires is 4 in. above the bottom of the beam. Area of wires is 1.5 in.^{2}, f_{pu} = 240,000 psi, f_0 = 150,000 psi, f'_c = 5000 psi, E_c = 4,000,000 psi. Compute the ultimate resisting moment.

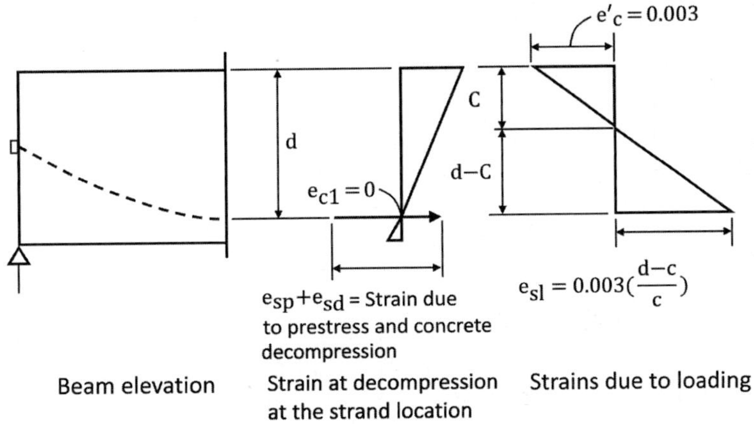

Figure 12.12 Strains at rupture.

Solution Assuming that the beam is underreinforced, the stress in the steel at the ultimate moment is $f_{ps} = f_{pu} = 240{,}000$ psi. The depth $d = 24 - 4 = 20$ in., and from Eqs. (12.12) and (12.13),

$$k' = \frac{1.5 \times 240{,}000}{0.85 \times 5000 \times 20 \times 12} = 0.353$$

$$M_u = 1.5 \times 240{,}000 \times 20\left(1 - \frac{0.353}{2}\right) = 5{,}930{,}000 \text{ in.-lb}$$

Using the more conservative formula, Eq. (12.14), the reinforcement ratio $\rho_p = 1.5/(12 \times 20) = 0.00625$, and from Eq. (12.16*a*),

$$f_{ps} = f_{pu}\left(1 - 0.5\frac{\rho_u f_{pu}}{f'_c}\right) = 240{,}000\left(1 - 0.5\frac{0.00625 \times 240}{5}\right) = 204{,}000 \text{ psi}$$

From Eq. (12.14),

$$M_u = 1.5 \times 204{,}000 \times 20\left(1 - 0.6 \times \frac{0.00625 \times 204}{5}\right) = 5{,}184{,}000 \text{ in.-lb}$$

This result is about 13 percent smaller than the previous value. Now

$$\rho_p \frac{f_{ps}}{f'_c} = 0.00625 \times \frac{204{,}000}{5000} = 0.255$$

which, since it is less than 0.3, suggests that the beam is underreinforced as assumed. If a more accurate value of M_u is desired, assume $c = 8.6$ and the maximum strain in the concrete to be 0.003 (Fig. 12.12). Then

$$e_{sl} = 0.003 \times \frac{20 - 8.6}{8.6} = 0.0040$$

With $f_0 = 150{,}000$ psi, assume the effective prestress $f_e = 125{,}000$ psi. The corresponding strain $f_e/E_s = 0.0042$. Concrete stress at the location of tendon due to prestress is

$$f_c = \frac{-187{,}500}{288} - \frac{187{,}500 \times 8 \times 8}{13{,}824}$$
$$= -651 - 868 = -1519 \text{ psi at the tendon location}$$

Hence, the decompression strain at the location of tendon is $f_c/E_c = 0.0004$. The total strain at failure is $0.0040 + 0.0042 + 0.0004 = 0.0086$. From Fig. 12.1, this corresponds to a stress of 234,000 psi. Therefore, from $T' = C'$,

$$k'd = \frac{1.5 \times 234{,}000}{0.85 \times 5000 \times 12} = 6.88$$

Since $\beta_1 = 0.80$ in this case, $c = k'd/\beta_1 = 8.60$, which is the originally assumed value. Therefore, with $f_{ps} = 234{,}000$ psi,

$$M_u = 1.5 \times 234{,}000 \times 20\left(1 - 0.6 \times \frac{0.00625 \times 234}{5}\right) = 5{,}809{,}050 \text{ in.-lb}$$

which is about 2 percent less than the first computed value. Since the value of e_{sd} is relatively small, this term may be neglected, which would have resulted in being 4 percent lower than the first computed value.

T-SECTION

Assume a T-section 24 in. deep whose flange is 20 × 3 in. and web is 5-in. thick. The c.g.s. of the prestressing wires is 4 in. above the bottom of the beam. Area of wires is 1.5 in.², $f_{pu} = 240{,}000$ psi, $f_0 = 150{,}000$ psi, and $f'_c = 5000$ psi. Compute the ultimate resisting moment.

Solution The required area in compression is (assuming $f_{ps} = f_{pu}$)

$$\frac{1.5 \times 240{,}000}{0.85 \times 5000} = 85 \text{ in.}^2$$

The flange furnishes 60 in.², leaving 25 in.² to be supplied by the web. Therefore, the compression area extends 25/5 = 5 in. below the bottom of the flange.

For an approximate solution, compression in the web can be neglected and the center of compression assumed at mid-depth of the flange. The lever arm is 24 − 4 − 1.5 = 18.5 in. and $M_u = 1.5 \times 240{,}000 \times 18.5 = 6{,}660{,}000$ in.-lb.

For a more exact solution, the centroid of the compressive area, including the web, can be determined. Thus, the distance from mid-depth of flange to the centroid is 4 × 25/(25 + 60) = 1.2 in., and the lever arm 18.5 − 1.2 = 17.3 in. The resulting moment is $M_u = 360{,}000 \times 17.3 = 6{,}220{,}000$ in.-lb, which is about 6 percent less than the approximate value.

12.3.5 Composite Sections

Figure 12.13 shows a composite section at the midspan of a simply supported beam, whose stem is precast and lifted into position with the top slab cast in place directly on the stem. If no temporary intermediate support is furnished, the weight of the stem and the liquid weight of the slab will be carried by the stem acting alone. After the slab concrete has hardened, the composite section will carry live or dead load that may be added to it. The following stress distributions are shown for various stages of working load conditions.

a. Owing to the initial prestress F_0 and the weight of the stem W_G, there will be heavy compression in the lower fibers and possibly some small tension in the top fibers. The tensile force *T* in the steel and the compressive force *C* in the concrete form a resisting couple with a small lever arm between them.

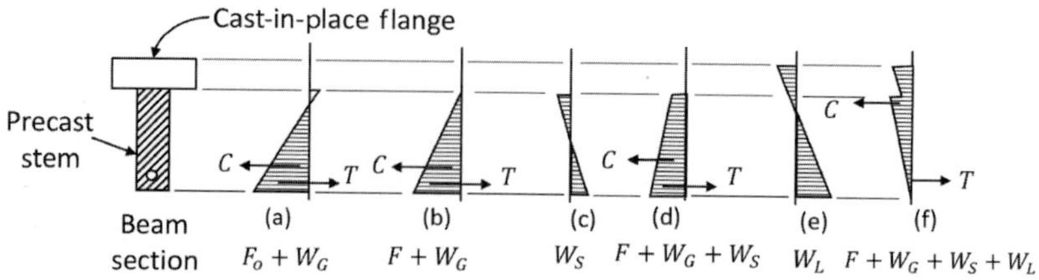

Figure 12.13 Stress distribution for a composite section.

b. After losses have taken place, the effective prestress F together with the weight of the stem will result in a slightly lower compression in the bottom fibers and some small tension or compression in the top fibers. The C-T couple will act with a slightly greater lever arm.

c. Addition of the slab weight W produces additional moment and stresses as shown. Stresses resulting from differential creep and shrinkage between the slab and the stem are neglected.

d. Adding b to c, smaller compression is found to exist at the bottom fibers and some compression at the top fibers. The lever arm for the C-T couple increases further.

e. Stresses resulting from live load W_L are shown, and the corresponding moment is resisted by the composite section.

f. Adding d to e, we have the stress block f: The C-T couple now acts with an appreciable lever arm.

The cracking moment and ultimate moment can be determined using methods similar to those previously described for noncomposite sections.

12.4 DESIGN FOR FLEXURE

12.4.1 Process

Design of prestress concrete members should ensure sufficient strength and satisfactory behavior at service conditions at all stages (i.e., jacking, transfer, and service after prestress losses). Since the consequence of cracking can increase the potential for corrosion, prestressed members are typically designed to be crack free. This goal can be more efficiently accomplished by performing the design under service loads using the elastic theory and limiting the critical stresses to specific values. Alternatively, the load balancing concept can be used to more efficiently control the deflection of the members, followed by ensuring that the stresses are within the acceptable limits. This design process, as shown in Fig. 12.14 and detailed below, can be completed with the ultimate strength (i.e., both flexure and shear) and deflection checks.

12.4.2 Beam Sections

Cross sections commonly used for prestressed-concrete beams are the rectangle, the symmetrical I, the unsymmetrical I, the T, the double T, the inverted T, and the box. The suitability of these shapes will depend on the simplicity, availability, reusability of formwork, ease in placing concrete, functional and aesthetic requirements, transportability, and theoretical considerations. The rectangular shape is the easiest to form, but its use of materials is uneconomical. Typically for a large amount of prestress with optimal eccentricities, a larger area in the compression zone is required, which makes a T-section or a wide-flange I-section more suitable. The T is suitable for high ratios of M_G/M_T, where there is little danger of overstressing at transfer and where the concrete is effectively concentrated at the compressive flange. The inverted T is good for low ratios of M_G/M_T (to avoid overstressing at transfer) but does not have a high ultimate moment. The I and box have more concrete near the extreme fibers and are efficient both at transfer and under ultimate loads but have thinner webs and require special forming. Double T-sections, with a wide top flange, are widely used in floor systems in buildings as well as in parking structures.

To increase the efficiency of precast operation, most of these products are standardized with variations in critical parameters so that they can be used cost effectively depending on the design parameters (e.g., span length). Commonly available standardized products applicable to building and foundations can be found in the PCI Design Handbook (2017). The provided details include member dimensions, section properties, and prestressing. For bridge applications, there are national standards available. However, the Department of Transportation in each state in the United States publishes their own standard girders for bridge applications.

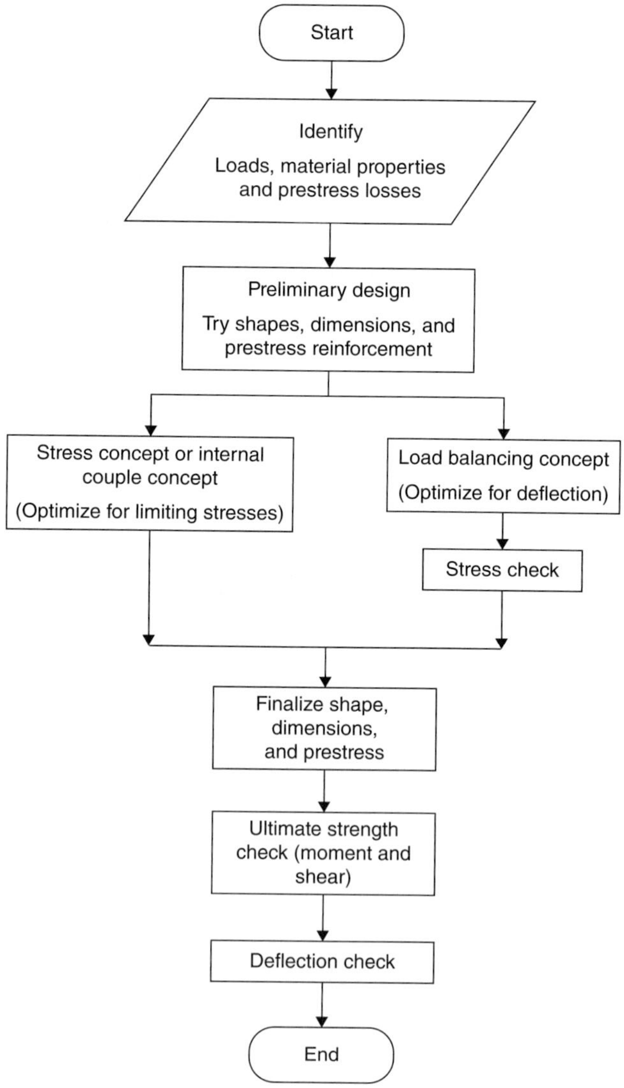

Figure 12.14 A flowchart identifying key steps in the design for flexure.

12.4.3 Span-to-Depth Ratios

For reasons of economy and aesthetics, higher span-to-depth ratios are almost always used for prestressed concrete than for reinforced-concrete members. Higher ratios are possible because prestressing facilitates better control of deflection. On the other hand, when these ratios get too high, camber (i.e., upward deflection resulting from prestressing) and deflection become quite sensitive to variations in loadings, in properties of materials, in magnitude and location of prestress, and in temperature. Furthermore, the effects of vibration become more pronounced. Care should be taken with cantilever beams since they are particularly sensitive to deflection and vibration.

Span-to-depth ratio limitations should vary with the nature and magnitude of the live load, the damping characteristics, the boundary conditions, the shape and variations of the section, the modulus of elasticity, and the span. If the structure is carefully investigated for camber, deflection, and vibration, there is no reason to adhere to any given ratio.

The limiting values in Table 12.5 may be used as a preliminary guide for building design. In general, with span-to-depth ratios of 10 percent below the tabulated values, problems of camber, deflection, and vibration are not likely to develop unless the loadings are extremely heavy and vibratory in nature. On the other hand, these ratios can be exceeded by 10 percent or more if careful study ensures acceptable behavior. The ratios are intended for both hard-rock concrete and lightweight concrete, but should be reduced by about 5 percent for lightweight concrete having E_c less than 3,000,000 psi. For long spans (say, in excess of about 70 ft) and for heavy loads (say, live loads over 100 psf), the values should be reduced by 5 to 10 percent. For in-place concrete in composite action with precast elements, the total depth may be considered in computing span-to-depth ratios.

Table 12.5 Appropriate Limits for Span-to-Depth Ratios

	Continuous spans		Simple spans	
	Roof	Floor	Roof	Floor
One-way solid slabs	52	48	48	44
Two-way solid slabs (supported on columns only)	48	44	44	40
Two-way waffle slabs (3-ft waffles)	40	36	36	32
Two-way waffle slabs (12-ft waffles)	36	32	32	28
One-way slabs withs mall cores	50	46	46	42
One-way slabs with large cores	48	44	44	40
Double tees and single tees (side by side)	44	40	40	36
Single tees (spaced 20-ft centers)	36	32	32	28

Experience with prestressed concrete railway bridges is not sufficient to establish span-to-depth ratios. Usual ratios have been in the range of 10 to 14 for box sections up to 100 ft or more. For simple-span highway bridges of the I-beam type, up to about 200 ft, a span-to-depth ratio of 20 is considered conservative, 22 to 24 is taken as norm, while 26 to 28 would be the critical limit. Box sections can have ratios about 5 to 10 percent higher than I-beams, while T-sections spaced far apart should have ratios about 5 to 10 percent lower than I-beams. Again, there is no reason to believe that a fixed span-to-depth ratio will apply to all cases.

12.4.4 Preliminary Design

Preliminary design for flexure can be based on the C-T couple. Under working load, the lever arm a varies between $0.30h$ and $0.85h$, where h = total depth of section, and averages about $0.60h$. Hence, the required effective prestressing force F can be estimated from the equation

$$F = T = \frac{M_T}{0.60h} \tag{12.17}$$

where M_T = total external moment produced by the working load.

The depth h for a prestressed section varies between 50 and 80 percent of that of an equivalent reinforced-concrete section and may be taken at 70 percent for the first trial. Having estimated the force F, the area of steel is computed by

$$A_{ps} = \frac{F}{f_s} \tag{12.18}$$

where f_s depends on the steel and may be taken as 180 ksi for strands.

The area of concrete required is estimated by

$$A_c = \frac{F}{f_{av}} \tag{12.19}$$

where f_{av}, the average precompression in the concrete, varies from 700 to 1300 psi for I- and T-beams and from 250 to 500 psi for solid slabs.

The load-balancing method, Concept 3 in Sec. 12.3.1, can also be used for preliminary design with an intension of controlling deflection in design (Sec. 12.4.12).

12.4.5 Elastic Design

Concept 1 in Sec. 12.3.1 can be used for design by the elastic theory. However, it is often more convenient to use Concept 2. Thus:

Case 1. Girder moment $M_G = 0$. Allowing no tension in the concrete, either at transfer or at maximum load, the permissible moment is (Fig. 12.15*a*)

$$M_T = F(k_t + k_b) \tag{12.20}$$

where k_t and k_b are the kern distances defined in Sec. 12.3.1. If tensile stress is allowed for the top fiber (f'_t) at transfer and bottom fiber (f'_b) during service, an additional moment can be carried:

$$M_T = F(k_t + k_b) + \frac{f'_b I}{c_b} + \frac{f'_t I}{c_t} \tag{12.21}$$

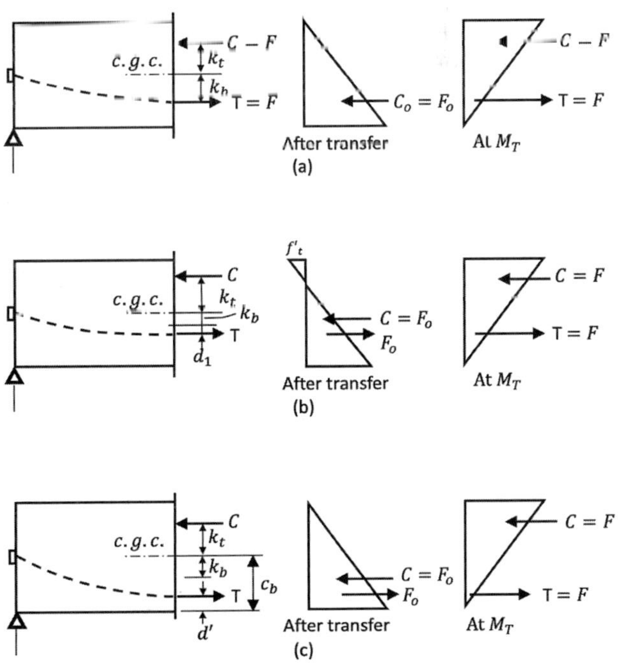

Figure 12.15 Elastic design for different ratios of M_G/M_T: (*a*) $M_G = 0$; (*b*) small M_G/M_T; (*c*) large M_G/M_T.

Case 2. When M_G is small, so that the c.g.s. cannot be located at its lowest possible position, as determined by the required concrete protection d', the distance d_1 from the bottom kern point to the c.g.s. is given by

$$d_1 = M_G + \frac{f_t' I / c_t}{F_0} \tag{12.22}$$

where f_t' is the allowable tension in the top fiber at transfer. The permissible moment is (Fig. 12.15*b*)

$$M_T = F(k_t + k_b + d_1) \tag{12.23}$$

If tensile stress f_b' is allowed for the bottom fibers,

$$M_T = F(k_t + k_b + d_1) + \frac{f_b' I}{c_b} \tag{12.24}$$

Case 3. When M_G is large so that c.g.s. is located at the lowest possible position as determined by the required concrete protection d' (Fig. 12.15*c*),

$$M_T = F(k_t + c_b - d') \tag{12.25}$$

If tensile stress f_b' is allowed for the bottom fibers,

$$M_T = F(k_t + c_b - d') + \frac{f_b' I}{c_b} \tag{12.26}$$

After the force F has been determined, the area of steel is computed by

$$A_s = \frac{F}{f_s}$$

and the extreme fiber stresses in the concrete are computed under M_G and under M_T, using the formulas in Sec. 12.3.1. If the stresses are not satisfactory, the section is revised. Direct design formulas and computer programs are also available for completing the design.

Allowable Stresses. The ACI 318–14 (2014) allowable stresses are as follows:

24.5 Permissible stresses in concrete—flexural members

24.5.3. Stresses in concrete immediately after prestress transfer (before time-dependent prestress losses) shall not exceed the following:

(a) Extreme fiber stress in compression except as permitted in (b)	$0.60 f_{ci}'$
(b) Extreme fiber stress in compression at ends of simply supported members	$0.70\, f_{ci}'$
(c) Extreme fiber stress in tension except as permitted in (d)	$3\sqrt{f_{ci}'}$
(d) Extreme fiber stress in tension at ends of simply supported members	$6\sqrt{f_{ci}'}$

Where computed tensile stresses exceed these values, bonded auxiliary reinforcement (nonprestressed or prestressed) shall be provided in the tensile zone to resist the total tensile force in concrete computed with the assumption of an uncracked section.

24.5.2 through 24.5.4. Stresses in concrete at service loads (after allowance for all prestress losses) shall not exceed the following:

(a) Extreme fiber stress in compression due to prestress plus sustained load	$0.45 f_c'$
(b) Extreme fiber stress in compression due to prestress plus total load	$0.60 f_c'$
(c) Extreme fiber stress in tension in precompressed tensile zone	$7.5\sqrt{f_c'}$
(d) Extreme fiber stress in tension in precompressed tensile zone of members (except two-way slab systems), where analysis based on transformed cracked sections and on bilinear moment-deflection relationships shows that immediate and long-time deflections comply with requirements of 24.2.3.9 and 24.2.4.2 and where cover requirements comply with 20.6.1.3	$12\sqrt{f_c'}$

24.5.1.1. Permissible stresses in 24.5.2 through 24.5.4 shall be permitted for concrete if shown by test or analysis that performance will not be impaired

20.3.2.5. Permissible stresses in prestressing tendons

20.3.2.5.1. Tensile stress in prestressing tendons shall not exceed the following:

(a) Due to tendon jacking force (but not greater than the lesser of $0.80 f_{pu}$ and the maximum value recommended by the supplier of anchorage device)	$0.94 f_{py}$
(b) Immediately after prestress transfer	$0.70 f_{pu}$

12.4.6 Ultimate Design

The ultimate moment capacity of the section must be not less than the working load moment multiplied by a load factor m, usually 1.8 for buildings and 2.0 for bridges. For underreinforced sections, the ultimate lever arm will be around $0.9d$, where d = effective depth. The area of steel required is

$$A_{ps} = \frac{mM_T}{0.9 d f_{pu}} \tag{12.27}$$

Assuming that the concrete on the compressive side is stressed to $0.85 f_c'$, the required area under compression is

$$A_c = \frac{mM_T}{0.9d \times 0.85 f_c'} \tag{12.28}$$

When designed by the above formulas, the section should be checked by the equations in Sec. 12.3.4. In addition, compressive stresses at transfer must be investigated for the tension flange, usually by the elastic theory, and checks for excessive camber, deflection, and cracking may be required.

12.4.7 Balanced-Load Design

Concept 3 in Sec. 12.3.1 can be conveniently used for design. Balancing of a uniformly distributed load by a parabolic cable was described in that section. Figure 12.16*a* illustrates the balancing of a concentrated load by bending the tendon at midspan, creating an upward component $V = 2F \sin\theta$. If this V exactly balances a concentrated load P, the fiber stress in the beam at any section (except for local stress concentrations) is given by $f = (F \cos\theta)/A_c = F/A_c$ for small values of θ. Any loading in addition to P will cause bending in an elastic homogeneous beam (up to point of cracking), and the additional stresses can be computed by $f = Mc/I$, where M is the moment produced by the additional load.

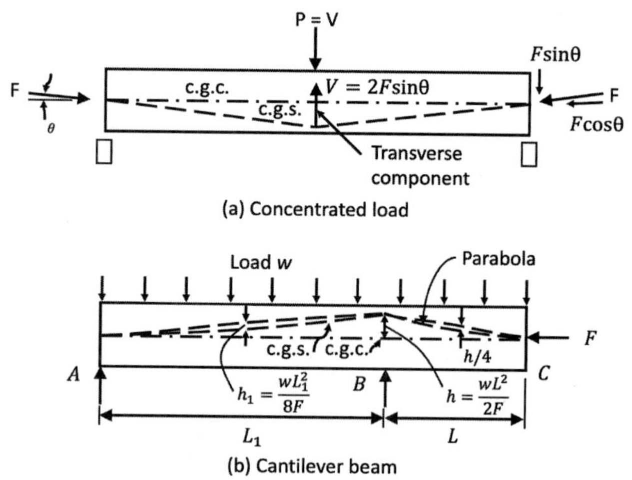

Figure 12.16 Load balancing for beams.

Now consider a cantilever beam (Fig. 12.16*b*). The conditions for load balancing become slightly more complicated because any vertical component at the cantilever end, *C*, will upset the balance unless there is an externally applied load at that end. To balance a uniformly distributed load *w*, the tangent to the c.g.s. at *C* must be horizontal. The parabola for the cantilever portion is located by computing $h = wL^2/2F$, and the parabola for the anchor arm by $h_1 = wL_1^2/8F$.

EXAMPLE 5

A double cantilever beam is to be designed so that its prestress will exactly balance a total uniform load of 1.6 kips/ft on the beam (Fig. 12.17*a*). Design the beam using the least amount of prestress, assuming that the c.g.s. must have a concrete protection of at least 3 in. If a concentrated load $P = 14$ kips is added at midspan, compute the maximum fiber stresses.

Solution To balance the load in the cantilever, the c.g.s. at the tip must be located at the c.g.c. with a horizontal tangent. To use the least amount of prestress, the eccentricity over the support should be a maximum, that is, $h = 12$ in. (or 1 ft). The prestressing force required is

$$F = \frac{wL^2}{2h} = \frac{1.6 \times 20^2}{2 \times 1} = 320 \text{ kips}$$

To balance the load on the center span, using the same prestressing force, the sag for the parabola must be

$$h_1 = \frac{wL_1^2}{8F} = \frac{1.6 \times 48^2}{8 \times 320} = 1.44 \text{ ft} = 17.3 \text{ in.}$$

Hence, the c.g.s. is located as shown in Fig. 12.17*b*, in which it is observed that the eccentricity is not maximized in the center span. However, under the combined action of the uniform load and the prestress, the beam has no deflection anywhere and is under a uniform compressive stress of

$$f = \frac{F}{A_c} = \frac{320{,}000}{360} = -889 \text{ psi}$$

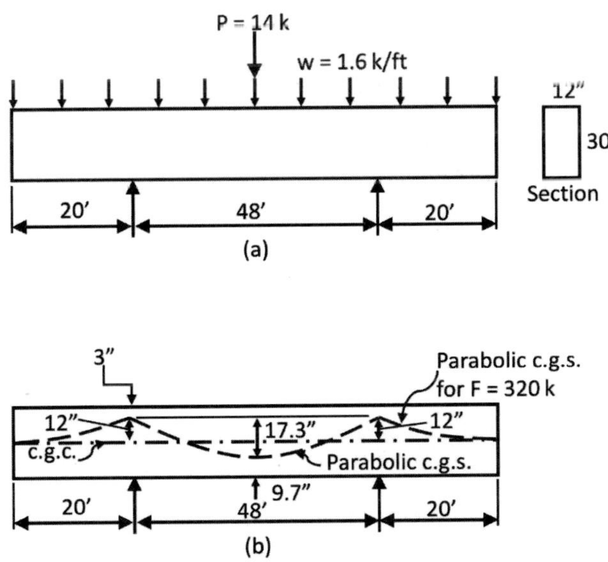

Figure 12.17 Example 5.

Owing to $P = 14$ kips, the moment *M* at midspan is

$$M = \frac{PL}{4} = \frac{14 \times 48}{4} = 168 \text{ ft-kips}$$

and the extreme fiber stresses are

$$f = \frac{Mc}{I} = \frac{6M}{bd^2} = \frac{6 \times 168 \times 12{,}000}{12 \times 30^2} = \mp 1120 \text{ psi}$$

The resulting stresses at midspan are

$$f_{\text{top}} = -889 - 1120 = -2009 \text{ psi compression}$$

$$f_{\text{bottom}} = -889 + 1120 = +231 \text{ psi tension}$$

Note that the actual cable placement may not possess the sharp bend shown over the supports, and the effect of any deviation from the theoretical position must be satisfactorily addressed in design. Also note that $F = 320$ kips is the effective prestress, so that under the initial prestress there will be a slight camber at midspan and either a camber or a deflection at the tips which can be computed.

For better stress conditions under the load *P*, it would be desirable to relocate the c.g.s. so that it would have more sag at midspan. Then a balanced condition would not exist under the uniform load *w*.

12.4.8 Cable Layouts

Typical cable layouts for pretensioned and posttensioned simple spans are shown in Figs. 12.18 and 12.19, respectively. Layouts for single and double cantilevers are shown in Figs. 12.20 and 12.21.

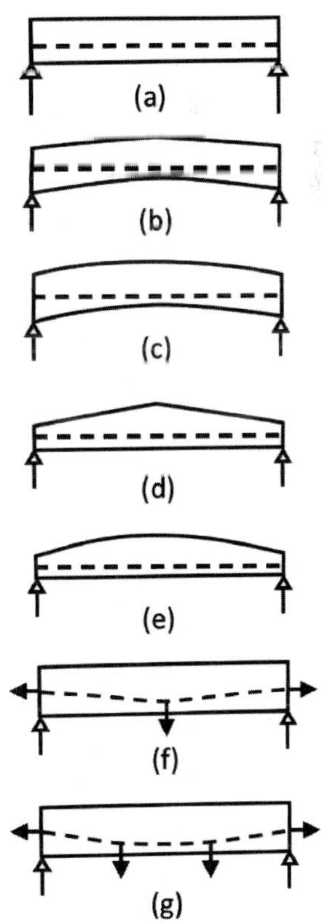

Figure 12.18 Layouts for pretensioned beams.

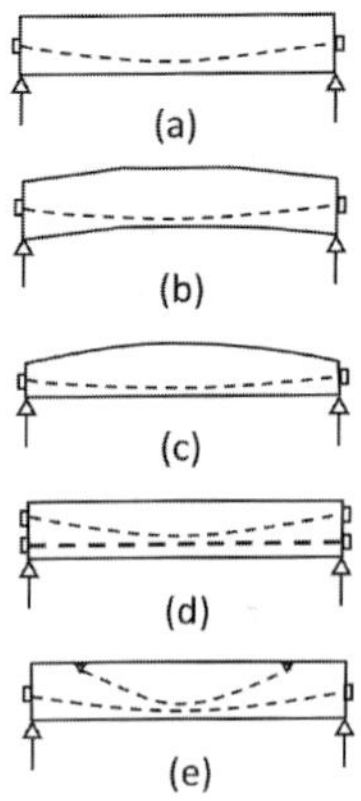

Figure 12.19 Layouts for posttensioned beams.

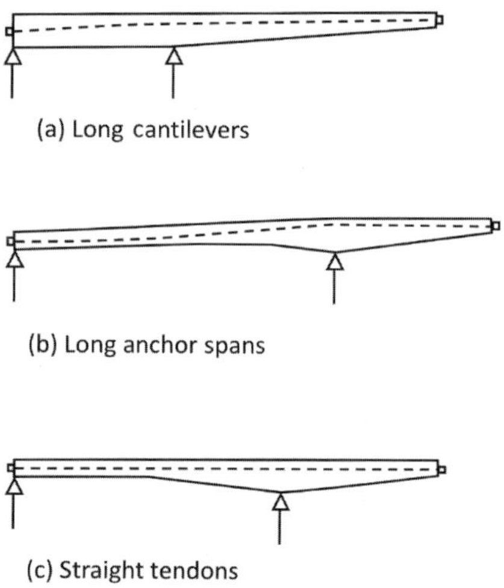

Figure 12.20 Typical layouts for single cantilevers.

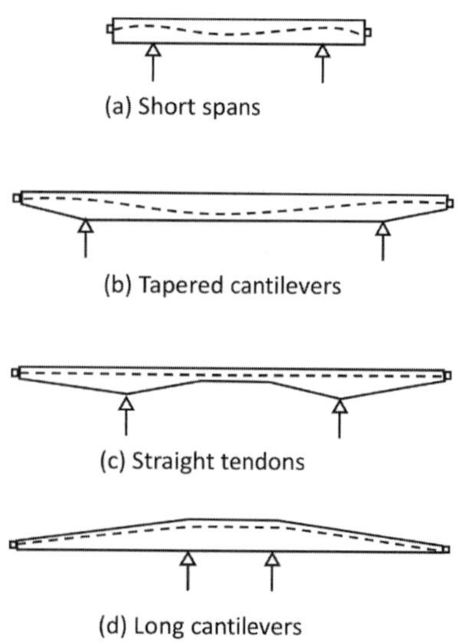

Figure 12.21 Typical Layout for double cantilevers.

12.4.9 Tendon Protection and Spacing

Minimum concrete protection for tendons is governed by requirements for fire resistance and for corrosion protection. ACI 318-14 and ACI 318.2-14 specify the following minimum thickness (inches) of concrete cover for prestressing steel, ducts, and nonprestressed steel in cast-in-place prestressed members.

	Min. cover, in.
Cast against and permanently exposed to earth	3
Exposed to earth or weather:	
Wall panels, slabs, and joists	1
Other members	1½
Not exposed to weather or in contact with the ground:	¾
Slabs, walls, joists	¾
Beams, girders, columns:	
Primary reinforcement	1½
Ties, stirrups, or spirals	1
Shells and folded-plate members:	
Reinforcement No. 5 (⅝ in.) and smaller	⅜
Reinforcement No. 6 (6/8 in.) and larger	Nominal dia.
Other reinforcement	¾

Minimum spacing of tendons is governed by several factors. First, the clear spacing between tendons, or between tendons and side forms, must be sufficient to permit easy placing of concrete. For prestressed strands at ends of member with f_c' of less than 4000 psi, ACI 318-14 calls for a minimum center-to-center spacing of the greater of $4d_b$ and $[(4/3)d_{agg} + d_b]$, where d_b and d_{agg} are nominal diameter of prestressing strand and nominal maximum size of coarse aggregate, respectively. Second, to develop the bond between steel and concrete properly, the clear distance between bars should be at least the diameter of the bars for special anchorage and 1½ times the diameter for ordinary anchorage, with a minimum of 1 in. These limitations may not be necessary for small wires and strands, which are often bundled together.

12.4.10 Partial Prestress

Partial prestress in prestressed concrete may mean one or more of the following:

1. Tensile stresses are permitted in concrete under working loads.
2. Nonprestressed reinforcement—mild steel or high tensile steel—are employed in addition to tendons.
3. Tendons are stressed to a lower level than usual.

Figure 12.22 shows load-deflection curves for a bonded beam with its concrete subjected to varying degrees of prestress. The case of full prestress permitting no tension under working load is represented by curve *b*. Partial prestressing permitting tension up to the modulus of rupture is shown by curve *c*. A nonprestressed concrete beam is shown in curve *d*. An overprestressed region is indicated between curves *a* and *b*.

The advantages of partial prestress compared with full prestress are (1) better control of camber, (2) saving in prestressing steel, (3) saving in end anchorages, (4) possible greater resilience in the structure, and (5) economical utilization of mild steel reinforcemet.

The disadvantages of partial prestress are (1) earlier appearance of cracks, (2) greater deflection under overload, (3) higher principal tensile stress under working load, and (4) slight decrease in ultimate flexural strength for the same amount of steel.

Nonprestressed reinforcements can be placed at various positions in a prestressed beam to improve its behavior and strength at different stages. Frequently, one set of reinforcement can serve to strengthen the beam in several ways:

1. To provide strength immediately after transfer of prestress:
   a. Along the compression flange, which may be under tension at transfer
   b. Along the tension flange, which may be under high compressive stress at transfer
2. To reinforce certain portions of the beam for special or unexpected loads during handling, transportation, and erection

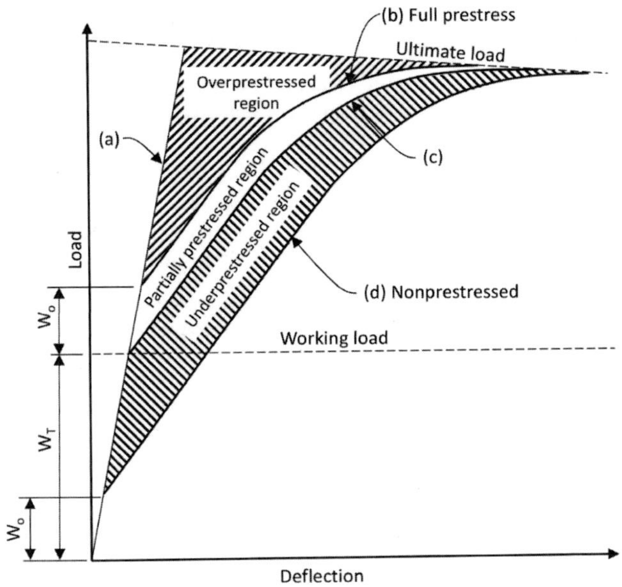

Figure 12.22 Load-deflection curves for varying degrees of prestress.

3. To distribute cracks under working loads
4. To increase ultimate capacity of the beam
5. To help carry high compression in the concrete
6. To reinforce the concrete along directions which are not prestressed: web, end block, and flange slab reinforcements

When nonprestressed reinforcements are used to carry compression, the compressive stress in the steel is generally quite high because of additional compression being induced shrinkage and creep in the concrete. When they are used to carry tension, the reinforcements cannot function effectively until the concrete has cracked. However, the design of tension reinforcements is usually made on the assumption that they will be stressed to the usual allowable values (e.g., 20,000 psi for intermediate-grade steel) and that their total tension will replace the tension in the portion of concrete that might be lost as a result of cracking.

Figure 12.23 shows the stresses and strains produced in various reinforcements under different stages of loading. It is to be noted that nonprestressed reinforcements will be stressed very little under working loads but will be effective at ultimate load, especially for underreinforced beams.

12.4.11 Combination of Prestressed and Reinforced Concrete

While a combination of prestressed and reinforced concrete is represented in the use of nonprestressed reinforcement, the flexural strength is primarily provided by the tendons, with the nonprestressed steel playing a minor role. For certain types of construction, a full combination of prestressed and reinforced concrete may be the best design, making use of the advantages of both reinforcement types. Reinforced concrete has the advantage of simplicity in construction, monolithic behavior, no camber, less creep, and reasonably high ultimate strength. Prestressed concrete utilizes high-strength steel economically, produces a favorable distribution of stress under certain conditions of loading, and controls deflection and cracking more effectively.

Certain structural elements and systems favor reinforced concrete, while prestressed concrete and partially prestressed concrete may be more appropriate for other elements and systems. Some will be best designed with a combination of reinforced and prestressed concrete having the nonprestressed steel carrying as much as 50 percent or more of the total ultimate load.

One reason for using this combination is the case of high live-load to dead-load ratio, when prestressing alone may produce excessive camber.

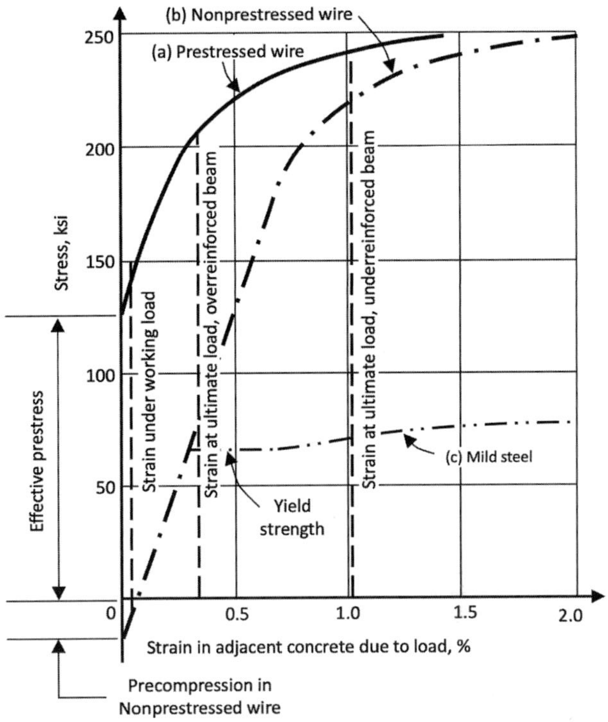

Figure 12.23 Stress-strain diagrams.

Another situation is when high amounts of added dead load requiring prestressing in stages, which may be cumbersome. A third case is the requirement of high ultimate strength or resilience to resist dynamic loadings. There is also reason to believe that a heavy amount of nonprestressed steel used in conjunction with unbonded tendons will result in economy and in developing a high ultimate stress in the tendons.

For precast members, prestressing in the form of pretensioning will help control cracking during handling, transportation, and erection and will contribute to the bending strength. Nonprestressed steel will increase both the axial load and the flexural capacity of the member. Hence, a combination may be the best solution for certain cases. The use of nonprestressed reinforcement for joints and connections is, of course, often a simple and economical solution. The use of unstressed prestressing steel for the same purpose is also being used in design.

Nonprestressed steel does not act until the concrete cracks and does not contribute toward the precracking strength. Hence, if cracking could result in a primary or a secondary failure, nonprestressed steel may be of no help. The possibility of corrosion of the prestressing steel if the member cracks too early or too often should also be investigated.

12.4.12 Deflections

While controlled deflections resulting from prestress can be advantageously used to produce desired cambers and to offset load deflection, excessive camber can cause serious trouble. Deflection owing to prestress can be computed as in Example 6.

EXAMPLE 6

A concrete beam of 32-ft simple span (Fig. 12.24) is posttensioned with 1.2 in.^{2} of high-tensile steel to an initial prestress of 140 ksi immediately after prestressing. Compute the initial deflection at midspan due to prestress and the beam's own weight, assuming $E_c = 4{,}000{,}000$ psi. Estimate the deflection after 3 months, assuming a creep coefficient $C_c = 0.8$ and an effective prestress of 120 ksi at that time.

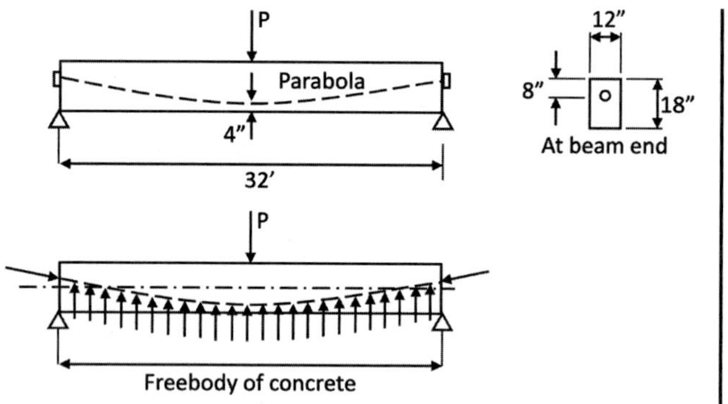

Figure 12.24 Example 6.

Solution The parabolic tendon with 6-in. midordinate (sag) is replaced by a uniform load acting along the beam with intensity

$$w = \frac{8Fh}{L^2} = \frac{8 \times 140{,}000 \times 1.2 \times 6}{32^2 \times 12} = 655 \text{ lb/ft}$$

The prestress force is eccentric 1 in. at each end of the beam, producing a moment of 140,000 × 1.2 × 1/12 = 14,000 ft-lb.

Since the weight of the beam is 225 lb/ft, the net uniform load on the concrete is 655 − 225 = 430 plf, which produces an upward deflection at midspan given by

$$\Delta = \frac{5wL^4}{384EI} = \frac{5 \times 430 \times 32^4 \times 12^3}{384 \times 4{,}000{,}000 \times (12 \times 18^3)/12} = 0.434 \text{ in.}$$

The end moments produce a downward deflection given by

$$\Delta = \frac{ML^2}{8EI} = \frac{140{,}000 \times 1.2 \times 1 \times 32^2 \times 12^2}{8 \times 4{,}000{,}000 \times (12 \times 18^3)/12} = 0.133 \text{ in.}$$

Thus, the net deflection due to prestress and beam weight at midspan is 0.434 − 0.133 = 0.301 in. upward.

To calculate the estimated deflection at 3 months, the prestress deflection is reduced proportionately for loss of prestress, and the resulting net deflection is multiplied by 1.8 to allow for creep. The camber due to initial prestress is 0.434 × 655/430 − 0.133 = 0.528 in. The initial deflection due to beam weight is 0.434 × 225/430 = 0.227 in. Thus, the expected deflection at 3 months is

$$\Delta = 1.8\left(0.528 \times \frac{120}{140} - 0.227\right) = 0.407 \text{ in. upward}$$

Deflections resulting from external load are calculated in the usual manner for homogeneous beams provided the concrete is not cracked. If the beam is bonded, the moment of inertia should be computed on the basis of the transformed section, but it can be approximated by using the gross concrete section. If the beam is unbonded, it is close enough for practical purposes to use the moment of inertia of the gross section of concrete. For computing instantaneous deflection due to live load, the following approximate values of E_c may be used:

	E_c, ksi	
Age of concrete	Hard rock	Lightweight
1 day	4000	2500
7 days	4500	3000
30 days	5000	3400
1 year	5500	3800

If live load is of long duration, the corresponding creep must be included. Also, if the load produces cracking, the elastic theory can be used only as an approximation. Accurate data concerning deflection after cracking are not available. However, the ultimate deflection can be computed accurately by establishing moment-curvature relationships at various sections along the member.

It is difficult to predict camber because it varies not only with E_c and creep of concrete but also with age, support conditions, temperature and shrinkage differential between top and bottom fibers, and variations in properties of the concrete. It is usually necessary to have experience with the product of a particular plant before accurate prediction can be made. Lacking such experience, camber computations for 1-day strength of 4000 psi may be based on E_c = 4,000,000 psi for hard rock and 2,500,000 psi for lightweight concrete. These values may then be modified for loss of prestress and creep, which can be approximated roughly by the following table:

Age of concrete	Ratio of effective to initial steel stress, %	Coefficient of flexural creep	
		Hard rock	Lightweight
1 day	94	1.0	1.0
7 days	89	1.6	1.3
30 days	86	2.0	1.5
1 year	83	2.5	1.8

It is good practice to balance the dead-load deflection by camber whenever possible. If this is done, flexural creep and variation in E_c will have little effect on camber or deflection.

Formulas for the calculation of prestress camber are given in Fig. 12.25. In these formulas, moments M_1 and M_2 are determined by multiplying the prestressing force F (more accurately, its horizontal component) by the corresponding ordinate y.

12.5 DESIGN FOR SHEAR, BOND, AND BEARING

12.5.1 Principal Tension

Under service-load conditions, a prestressed-concrete beam is generally uncracked and behaves elastically. The principal tensile stress in the web is computed as follows:

1. From the total external shear V at the section, deduct the shear V_s carried by the tendon to obtain the shear V_c carried by the concrete:

$$V_c = V - V_s \tag{12.29}$$

Occasionally, though rarely, $V_c = V + V_s$; this happens when the cable inclination is such that it adds to the shear on the concrete.

2. Compute the distribution of V_c by

$$v = \frac{V_c Q}{Ib} \tag{12.30}$$

where v = shearing unit stress at any given level
Q = statical moment of the cross-sectional area above (or below) that level about the centroidal axis
b = width of section at that level

3. Compute the fiber stress distribution for that section due to external moment M and the prestress F:

$$f_c = \frac{F}{A} + \frac{Fec}{I} - \frac{Mc}{I} \tag{12.31}$$

4. The maximum principal tensile stress S_t is given by

$$S_t = \sqrt{v^2 + \left(\frac{f_c}{2}\right)^2} - \frac{f_c}{2} \tag{12.32}$$

The greatest principal tensile stress does not necessarily occur at the centroidal axis, where the maximum vertical shearing stress exists. At some point, where f_c is diminished, a higher principal tension may exist even though v doesn't correspond to the maximum value.

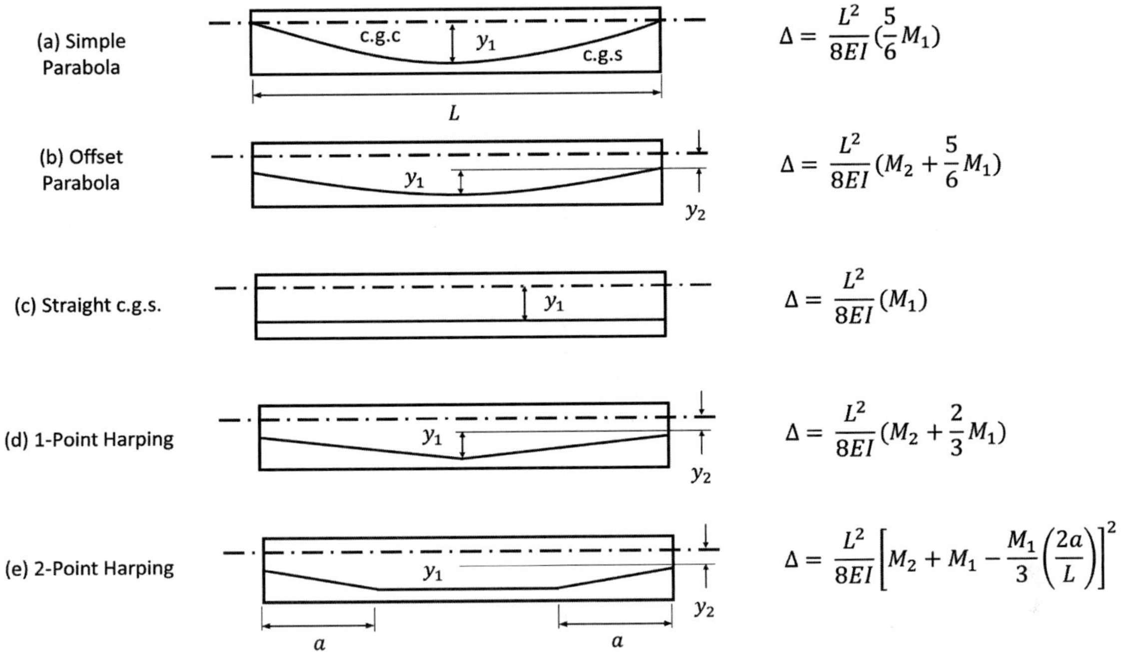

Figure 12.25 Formulas for computing midspan camber due to prestress (simple beams).

For sections without web reinforcement, S_t is often limited to $2\sqrt{f'_c}$ or less.

In composite construction, the shearing stress v between precast and in-place portions is computed from Eq. (12.30) with V = the total shear applied after the in-place portion has been cast. The allowable value of v varies from about 40 psi for smooth surfaces without ties to 160 psi for rough surfaces with adequate ties. Ties are not usually required for composite slabs or panels with large contact areas. For beams with a narrow top flange composite with in-place slabs, ties are almost always needed.

Push-off tests at the Portland Cement Association indicate an ultimate stress for composite action of about 500 psi for a rough bonded surface and 300 psi for a smooth bonded surface. About 175 psi may be added for each 1 percent stirrup reinforcement crossing the joint.

12.5.2 Web Reinforcement

Tension cracks, which may develop under combined flexure and shear, reduce moment capacity unless web reinforcement is provided. ACI 318–14 gives an empirical method based on ultimate-strength test results at the University of Illinois for design of web reinforcement. The yield strength of the stirrups extending over the length d is assumed to be effective in transmitting the shear (Fig. 12.26). Thus,

$$\frac{A_v f_y d}{s} = V_n - V_c = \frac{V_u}{\phi} - V_c \qquad (12.33)$$

where A_v = area of shear reinforcement within spacing s
f_y = yield stress of stirrup
d = effective depth of section, in.
s = longitudinal spacing of stirrups, in.
V_u = shear at ultimate load
V_c = lesser of $\{V_{ci}, V_{cw}\}$, where V_{ci} = shear at section when the vertical flexure crack starts to develop into an inclined one (i.e., flexure-shear cracking) and V_{cw} = shear at section when web cracking starts without prior flexural cracking (i.e., web-shear cracking).

The critical section is taken at $h/2$ from the theoretical point of maximum shear, where h is the overall depth of the beam.

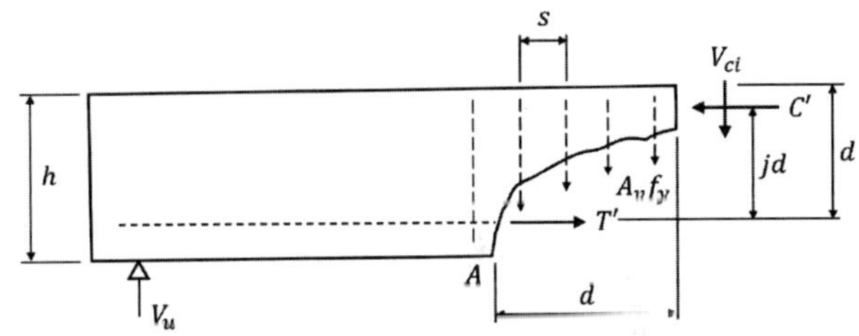

Figure 12.26 Participation of stirrups in transferring shear.

The component for a composite uniformly loaded beam may be estimated using the following equation:

$$V_{ci} = 0.6\sqrt{f'_c}\, b_w d_p + V_d + \frac{V_i M_{cre}}{M_{max}} \qquad (12.34)$$

where M_{cre} = flexural cracking moment caused by external load [Eq. (12.36)]
M_{max} = maximum moment at section due to superimposed loads ($M_{max} = M_u - M_d$)
M_u = factored moment due to the total factored load
M_d = moment due to unfactored dead load
V_i = shear coincident with M_{max} ($V_i = V_u - V_d$)
V_d = shear force at section due to unfactored dead load
b_w = width of web
d_p = distance from extreme fiber in compression to centroid of prestressing tendons or $0.8h$, whichever is larger

This equation is identical in form with ACI 318-14, Equation (22.5.8.3.1a). The last term accounts for the additional shear required to cause the vertical flexural crack to develop into an inclined one. Test results show that V_{ci} need not be taken less than $1.7\, b_w d\sqrt{f'_c}$.

The web-shear strength, V_{cw}, is given by

$$V_{cw} = \left(3.5\sqrt{f'_c} + 0.3 f_{pc}\right) b_w d_p + V_p \qquad (12.35)$$

where f_{pc} = compressive stress in concrete at centroid of section (or at junction of web and flange if centroid is in flange) due to effective prestress
V_p = vertical component of effective prestress at section

This is Equation (22.5.8.3.2) of ACI 318-14. Instead of the value given by this equation, V_{cw} may be taken as the shear at the section for the multiple of dead load plus live load that produces a principal tensile stress of $4\sqrt{f_c'}$ at the centroid or at the junction of the web and flange if the centroid is in the flange.

The cracking moment, M_{cre}, is given by

$$M_{cre} = \left(I/c_t\right)(6\sqrt{f_c'} + f_{pe} - f_d) \qquad (12.36)$$

where I = moment of inertia of beam section
c_t = distance from c.g.c. to extreme fiber in tension
f_{pe} = effective prestress at extreme fiber which is in tension due to superimposed load
f_d = stress due to unfactored dead load, at extreme fiber of section where tensile stress is caused by externally applied loads

Procedures for design of web reinforcement are the same as for nonprestressed members except that, if the effective prestress force (i.e., $A_{ps}f_{se}$) is equal to at least 40 percent of $(A_{ps}f_{pu} + A_sf_y)$, where $A_{ps}f_{pu}$ is the ultimate prestressing force of tendons and A_sf_y is the yielding force of longitudinal reinforcement, the minimum area of web reinforcement may be determined by

$$A_v = \frac{A_{ps}}{80}\frac{f_{pu}}{f_y}\frac{s}{d}\sqrt{\frac{d}{b_w}} \qquad (12.37)$$

or

$$\text{Greater of } \left\{0.75\sqrt{f_c'}\frac{b_w s}{f_y},\ 50\frac{b_w s}{f_y}\right\} \qquad (12.38)$$

whichever is smaller.

In applying Eqs. (12.34 to 12.36), the value of $\sqrt{f_c'}$ may not exceed 100 psi unless a minimum amount of web reinforcement is provided that equals $f_c'/5000$ times that given by Eq. (12.37) or (12.38). The amount required needs not to be more than three times that given by either of these two equations.

Although a minimum area of shear reinforcement shall be provided only in the regions where $V_u > 0.5\phi V_c$, use of some web reinforcement is recommended in thin-web prestressed concrete members. Providing some web reinforcement eliminates sudden brittle failure when the member is overloaded or when subjected to an unexpected axial tension resulting from local deviations from the design tendon profile.

Following the code, spacing of shear reinforcement placed perpendicular to axis of member, s, should be limited to the lesser of {3h/4 and 24 in.}. When $V_s > 4\sqrt{f_c'}\,b_wd$, the maximum spacing is reduced to ½ of that (i.e., lesser of {3h/8 and 12 in.}). It is also suggested that consideration of section enlargement is desirable when $V_s > 8\sqrt{f_c'}b_wd$.

EXAMPLE 7

Assume for the cantilever beam shown in Fig. 12.27: f_c' = 5000 psi, f_y = 40,000 psi (stirrups), effective prestress = 390 kips. Design the web reinforcement by ACI 318-14.

Solution The weight of the beam is 10 × 60 × 150/144 = 625 lb/ft. The critical section, which is shown as section A-A in Fig. 12.27, is at h/2 = 30 in. from the support, and the design shears and moments at the section are

$$V_d = 0.625 \times 6 = 3.75 \text{ kips}$$

$$M_d = 3\ V_d = 11.25 \text{ kips}$$

$$V_i = V_u - V_d = [1.2(0.625 \times 6) + 1.6(160)] - 3.75 = 256.75 \text{ kips}$$

$$M_{\max} = [1.2(0.625 \times 6 \times 3) + 1.6(160 \times 6)] - 11.25 = 1538.25 \text{ kips}$$

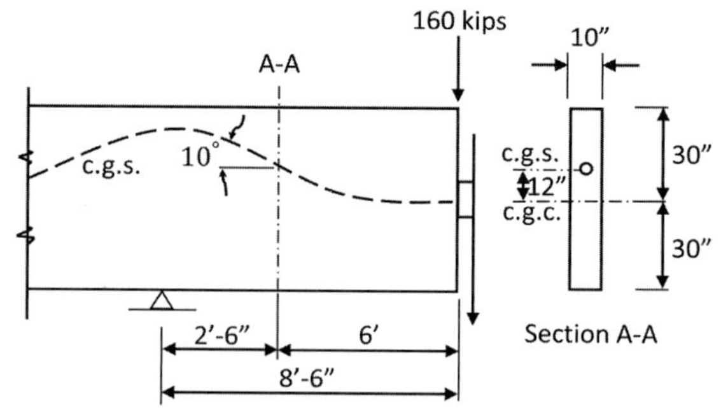

Figure 12.27 Example 7.

The section modulus I/c_t = 10 × 60²/6 = 6000 in.³, and the cross-sectional area = 60 × 10 = 600 in.² Then

$$f_d = \frac{11.25 \times 12}{6000} = 0.0225 \text{ ksi}$$

$$f_{pc} = \frac{390}{600} = 0.65 \text{ ksi}$$

The distance from c.g.c. to c.g.s. is 12 in. (Fig. 12.27). Therefore,

$$f_{pe} = f_{pc} + \frac{390 \times 12}{6000} = 1.43 \text{ ksi}$$

With $\sqrt{f_c'} = \sqrt{5000}$ = 70.7 psi = 0.0707 ksi, Eq. (12.36) gives M_{cre} = 6000 (6 × 0.0707 + 1.43 – 0.0225) = 10,990 in.-kips = 915.9 ft-kips.

The vertical component V_p of the tendon tension is 390 sin 10° = 67.7 kips. Since d_p = 42 in. and 0.8h = 48 in., use d_p = 48 in. From Eqs. (12.34) and (12.35):

$$V_{ci} = 0.6 \times 0.0707 \times 10 \times 48 + 3.75 + \frac{256.75 \times 915.9}{1538.25} = 177 \text{ kips}$$

$$V_{cw} = (3.5 \times 0.0707 + 0.3 \times 0.65) \times 10 \times 48 + 67.7 = 280 \text{ kips}$$

With the required capacity-reduction factor ϕ = 0.75, Eq. (12.33) gives

$$\frac{A_v f_y d}{s} = \frac{260.5}{0.75} - 177 = 170.3 \text{ kips}$$

Using ⅝-in. U stirrups, A_v = 0.62 in.² and

$$s = \frac{0.62 \times 40 \times 48}{170.3} = 7 \text{ in.} < \text{lesser of} \left\{3 \times \frac{60}{8} = 22.5 \text{ and } 12 \text{ in.}\right\}$$

12.5.3 Prestress Transfer Bond

Pretensioned tendons usually transfer their stress to the surrounding concrete through bond. The length of transfer varies with many factors. For plain wires, it averages about 100 diameters; for seven-wire strands, it is often taken as 50 to 75 diameters.

If the critical section for design of shear in a pretensioned member falls within the transfer length of the prestressing strands, the reduced prestress must be considered when computing V_{cw} using Eq. (12.35). The prestressing force must be assumed to vary linearly from zero to a maximum within the transfer length, assumed to be 50 times the diameter of the strand.

If a flexural crack should occur near the end of a pretensioned member, there is a tendency for the tendons to be pulled out as a result of bond slippage. For seven-wire strands without mechanical anchorage, the following formula gives the minimum length of embedment, l_d, in inches required to prevent such slippage:

$$l_d = \left(\frac{f_{se}}{3000}\right)d_b + \left(\frac{f_{ps} - f_{se}}{1000}\right)d_b$$

where f_{ps} = stress in prestressing reinforcement at nominal flexural strength, psi
f_{se} = effective prestress, psi
d_b = diameter of strand, in.

12.5.4 Anchorage

For tendons with end anchorages, where the prestress is transferred to the concrete by direct bearing, the stress may be transmitted by steel plates, steel blocks, or reinforced-concrete blocks. Stress analysis for anchorages is complicated, and as a result they are often designed by experience, tests, and usage. Since they are usually supplied by the prestressing companies, the engineer does not ordinarily have to design them.

The allowable bearing stress depends on several factors, such as the amount of reinforcement at the anchorage, the ratio of bearing area to the total area, and the assumptions made in computing the stress. Many codes and bridge design specifications prescribe the allowable bearing stress as

$$0.6f'_c\sqrt{A'/A_b} < f'_c$$

at service load and as

$$0.8f'_{ci}\sqrt{A'/A_b - 0.2} < 1.25f'_{ci}$$

at transfer load, where A_b is the bearing area of the anchor plate and A' is the area of the maximum portion of the anchorage surface that is geometrically similar to and concentric with the area A_b.

Because of strict economy in the design of end anchorages, it has not been unusual for poor concrete to fail under application of prestress. Therefore, concrete must be of high quality and must be carefully placed at the anchorages.

End Block. The portion of a prestressed member surrounding the anchorages of the tendons is often called the end block. The theoretical length of the block, sometimes called the lead length, is the distance required to transfer the prestress and distribute it throughout the entire beam cross section. End blocks are required for beams with posttensioning tendons but are not needed where all tendons are pretensioned.

In posttensioned members, a closely spaced grid of vertical and horizontal bars must be placed near the end face of the end block to resist bursting, and closely spaced vertical and horizontal reinforcement is required throughout the length of the block.

In pretensioned beams, vertical stirrups acting at a unit stress of 20,000 psi to resist at least 4 percent of the total prestressing force should be placed within the distance $d/4$ of the end of the beam, with the end stirrup as close to the end of the beam as is practicable. The Portland Cement Association Laboratories developed an empirical equation for the design of stirrups to control horizontal cracking in the ends of pretensioned I-girders:

$$A_t = 0.021\frac{T}{f_s}\frac{h}{l_t}$$

where A_t = required total cross-sectional area of stirrups at the end of girder, to be uniformly distributed over a length equal to one-fifth of the girder depth
T = total effective prestress force, lb
f_s = allowable stress for the stirrups, psi
h = depth of girder, in.
l_t = length of transfer, assumed to be 50 times the strand diameter, in.

12.6 DESIGN OF CONTINUOUS BEAM

Continuous beams may be fully cast in place with tendons continuous from one end to the other (Fig. 12.28). They may also be precast in smaller elements, which are made continuous by special posttensioning arrangements (Fig. 12.29).

When a statically determinate system is prestressed, there is no support reaction induced. However, in a continuous beam or any statically indeterminate system, support reactions are generally induced by the application of prestress because the bending of the beam due to prestress will attempt to deflect the beam at its supports. These reactions produce secondary moment in the beam.

12.6.1 Continuous-Beam C-Lines

Under the action of prestress alone, neglecting the weight of the beam and all other external loads, the *C* line (i.e., the line of pressure in the concrete) in a simple beam coincides with the c.g.s. line. In a continuous beam, the *C* line departs from the c.g.s. line in amounts required to resist the secondary moments. When the c.g.s. line is coincident with the *C* line, the condition is called "concordant." All simple beams and some continuous beams have concordant cables, while most continuous beams have nonconcordant cables.

While a nonconcordant cable usually gives a more economical solution, the concordant cable is sometimes preferred because it induces no external reactions at the supports. The concordant cable works better in a precast continuous beam. Several methods have been proposed for obtaining concordant cables. A simple rule is the following: Every real bending moment diagram for a continuous beam on nonsettling supports, produced by any combination of external loadings, whether transverse loads or moments plotted to any scale, is one location for a concordant cable in that beam. This theorem is easily proved. Since the moments in a continuous beam are computed on the basis of no deflection at the supports and since any c.g.s. line following the corresponding

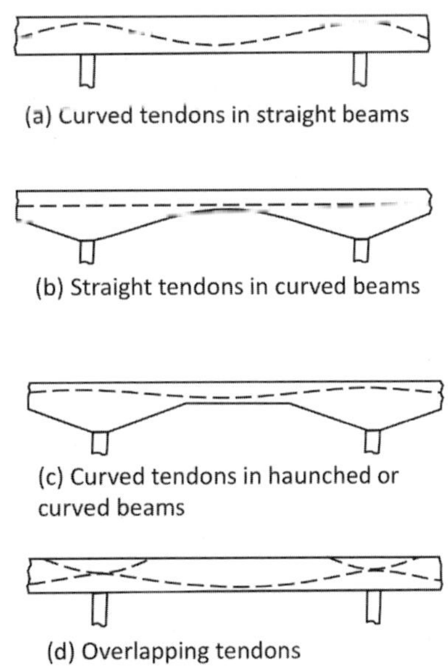

Figure 12.28 Layouts for fully continuous beams.

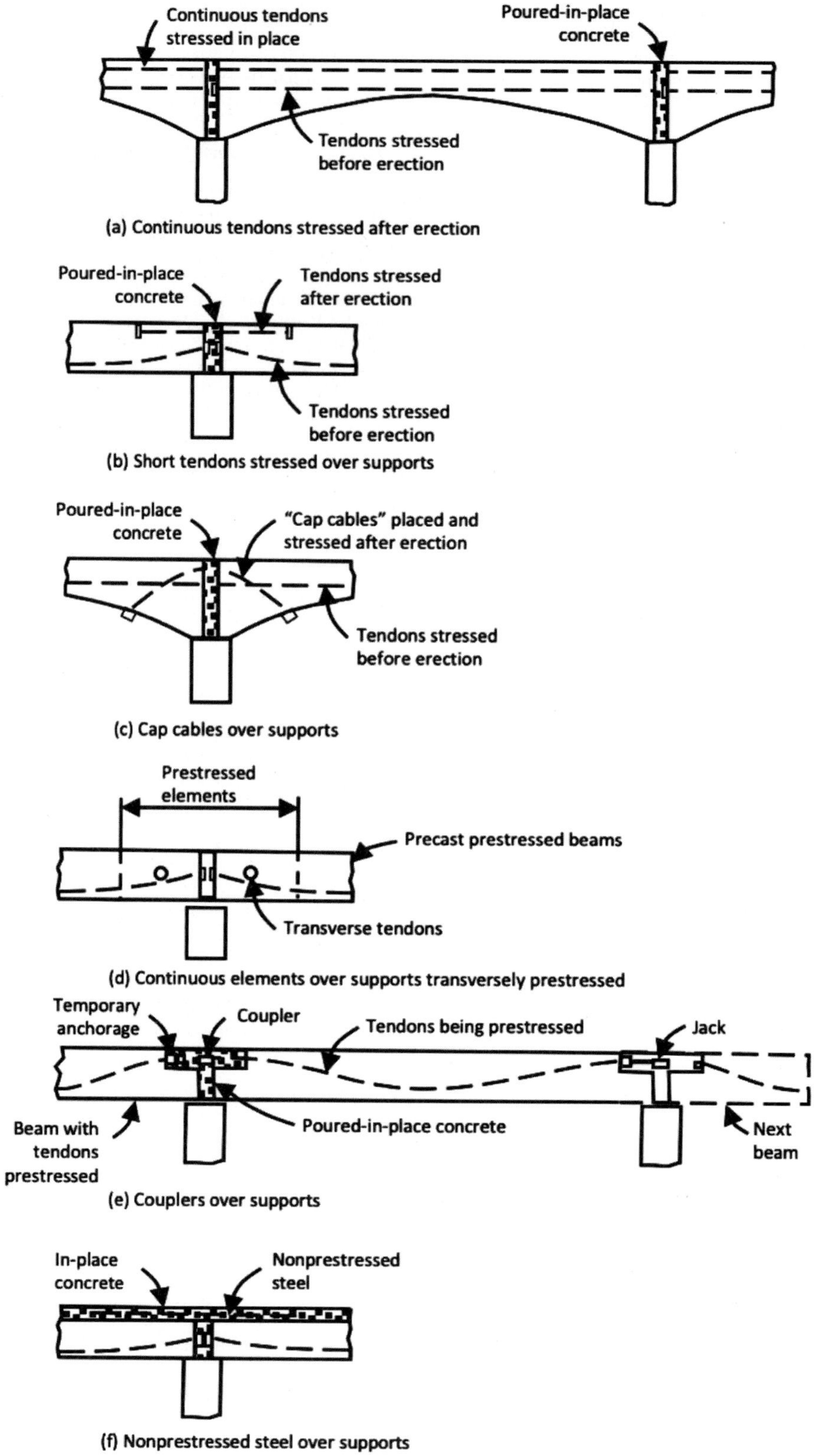

Figure 12.29 Layouts for partially continuous beams.

moment diagram will produce a similar moment diagram, that c.g.s. line will also produce no deflection over the supports; and hence, it will induce no reactions at supports, making it a concordant cable.

The *C* line can be located from the c.g.s. line by a linear transformation. Linear transformation is defined as the location of the *C* line from the c.g.s. line by displacements at the interior supports without changing its intrinsic shape within each span. Since the *C* line deviates from the c.g.s. line on account of the moments produced by the induced reactions and since such moments vary linearly within the span, it follows that the *C* line can be linearly transformed.

A simple method of determining the *C* line is shown in Examples 8 and 9.

EXAMPLE 8

A continuous prestressed-concrete beam with bonded tendons is shown in Fig. 12.30*a*. The c.g.s. is eccentric at *A*, is bent sharply at *D* and *B*, and has a parabolic curve for the span *BC*. Locate the line of pressure (i.e., *C* line) in the concrete due to prestress alone. Consider a prestress of 250 kips.

Solution The primary moment diagram due to prestress is shown in *b*. The corresponding shear diagram is shown in *c*, from which the loading diagram is drawn in *d*. The fixed-end moments are computed for this loading. At *A*, there is the additional moment $0.2 \times 250 = 50$ ft-kips resulting from the prestress eccentricity. Moment distribution is performed in *e*.

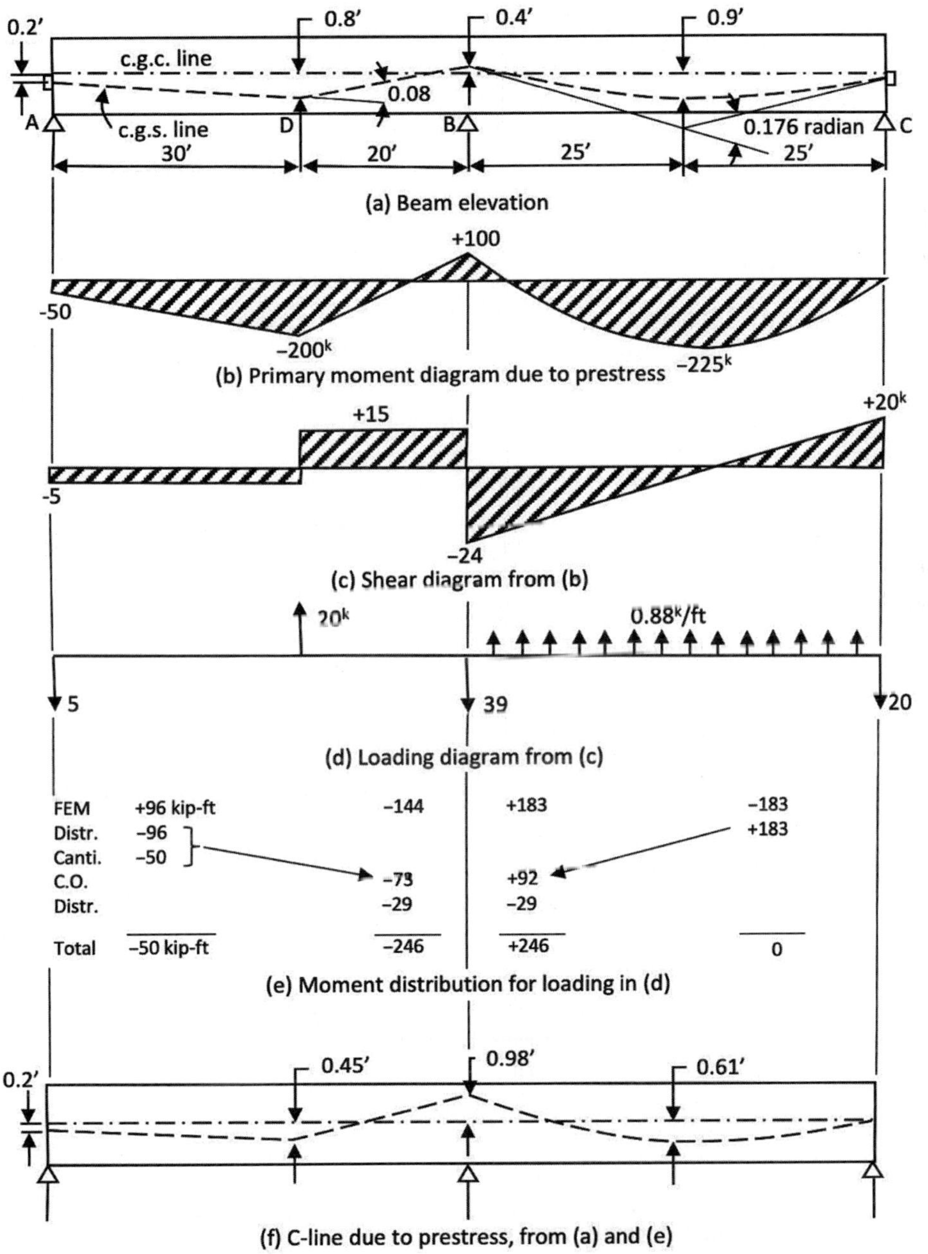

Figure 12.30 Example 8.

The eccentricity of the line of pressure at B is 246/250 = 0.98 ft. The line of pressure for the beam can be computed by plotting its moment diagram and dividing the ordinates by the value of the prestress. But this is not necessary. Since the line of pressure deviates linearly from the c.g.s. line, it is only necessary to move the c.g.s. line so that it passes through the points located over the supports, as shown in *f*. The line of pressure at D is translated upward by the amount (0.98 − 0.4) 30/50 = 0.35 ft and is now located 0.80 − 0.35 = 0.45 ft below the c.g.s. line.

At midspan of BC, the line of pressure is translated upward by the amount (0.98 − 0.4) 25/50 = 0.29 ft and is now located 0.61 ft below the c.g.c. line.

EXAMPLE 9

A uniform load of 1.2 kips/ft is applied to the beam of Example 8. Compute the stresses in the concrete at section B, where I = 39,700 in.^{4} and A_c = 288 in.^{2} (Fig. 12.31).

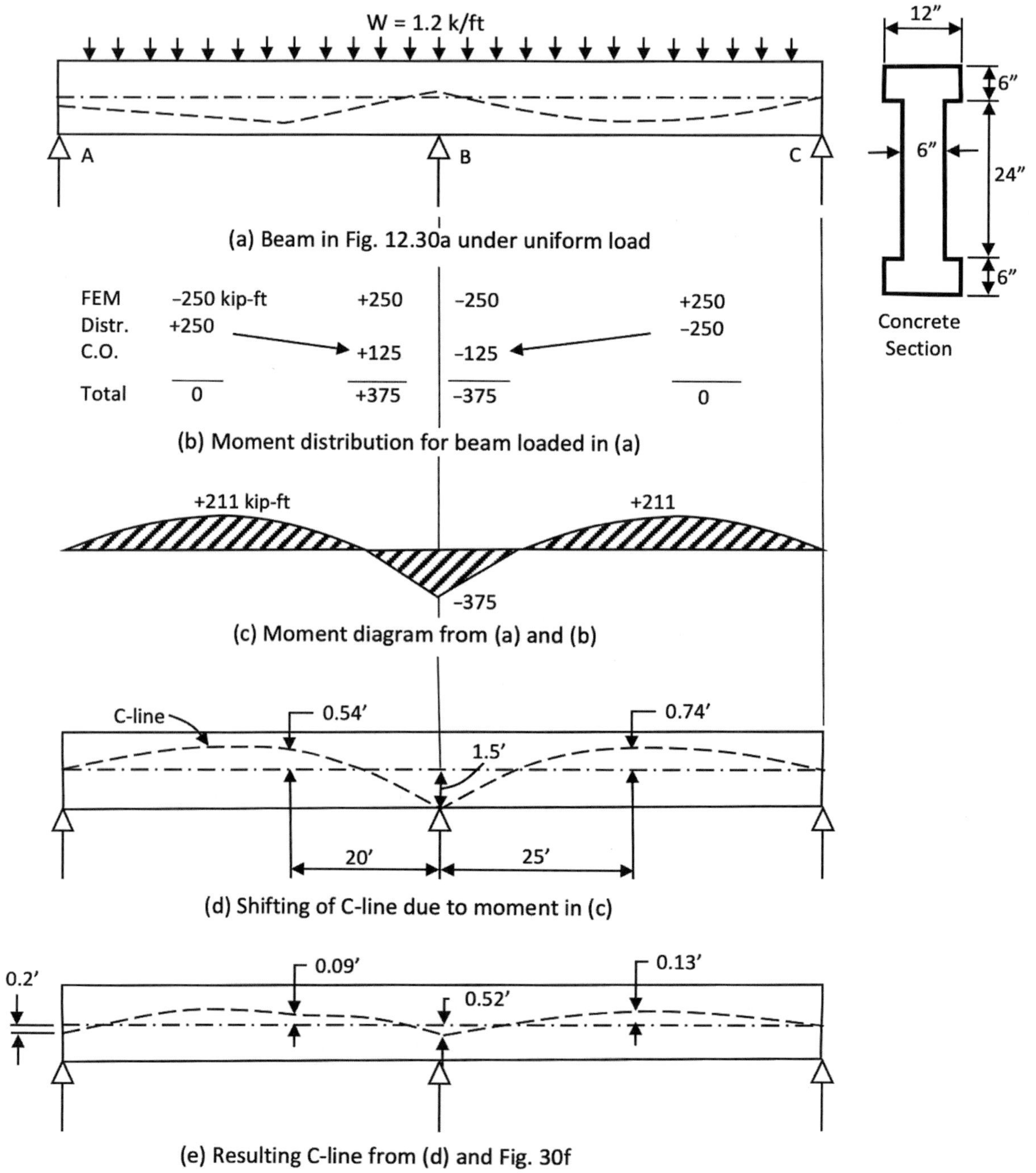

Figure 12.31 Example 9.

Solution The moment diagram for distributed load is plotted in *c*. Dividing these moments by the prestress of 250 kips gives the *C* line in *d*. Adding *d* to *f* of Example 8, the pressure line for both prestress and external load is given in *e*. The resulting moment at section *B* is $250 \times 0.52 = 130$ ft-kips, from which

$$f = \frac{-250}{288} \pm \frac{130 \times 12 \times 18}{39{,}700} = -0.867 \pm 0.707 \text{ ksi}$$
$$= -160 \text{ psi top fiber}$$
$$= -1574 \text{ psi bottom fiber}$$

12.6.2 Load-Balancing Method

This method is convenient for both the analysis and the design of prestressed continuous beams. When the external load is exactly balanced by the transverse component of the prestress, the beam is under a uniform stress $f = F/A_c$ across any section. For any change in load from that balanced condition, only the effects of change need be computed. For example, if the additional moment is *M*, the additional stresses are given by $f = My/I$. Thus, after load balancing, the analysis of prestressed continuous beams is reduced to the analysis of nonprestressed continuous beams. Since such analysis will be applied to only the unbalanced portion of the load, approximate methods may often prove sufficient.

Design by the load-balancing method gives a different visualization of the problem. It becomes relatively easy to lay out the cables in an economical manner and to compute the required prestress and the corresponding fiber stresses in concrete. This is illustrated in Example 10.

In the load-balancing method, it is often assumed that the dead load of the structure is balanced by the effective prestress. Therefore, a slight amount of camber may exist under the initial prestress. It is not always necessary to balance all the dead load since such balancing may require too much prestress in some situations, and a limited amount of deflection may not be objectionable. On the other hand, when the live load is large compared with the dead load, it may be necessary to balance a portion of the live load in addition to the dead load.

EXAMPLE 10

For the continuous beam in Fig. 12.32, determine the prestress *F* required to balance a uniform load of 1.03 kips/ft, using the most economical location of cable. Assume a concrete protection of at least 3 in. for the c.g.s. Compute the midspan section stresses and the reactions for the effect of prestress and an external load of 1.6 kips/ft.

Solution The most economical cable location is one with the maximum sag so that the least amount of prestress will be required to balance the load. A 3-in. protection is given to the c.g.s. over the center support and at midspan (a theoretical parabola based on these clearances will have slightly less than 3 in. at a point about 20 ft in from the exterior support). The c.g.s. at the beam ends should coincide with the c.g.c. and cannot be raised, not only because such raising will destroy the load balancing but also because it will not help to increase the efficiency of the cable, since unfavorable end moments will be introduced.

The cable now has a sag of 18 in., and the prestress *F* required to balance the load of 1.03 kips/ft is

$$F = \frac{wL^2}{8h} = \frac{1.03 \times 50^2}{8 \times 1.5} = 214 \text{ kips}$$

The fiber stress under this balanced-load condition is

$$f = \frac{F}{A_c} = -\frac{214{,}000}{360} = -593 \text{ psi}$$

Owing to the additional load of $1.6 - 1.03 = 0.57$ kip/ft, the negative moment at the center support is

$$M = \frac{wL^2}{8} = 0.57 \times \frac{50^2}{8} = -178 \text{ ft-kips}$$
$$f = \frac{Mc}{I} = \frac{6 \times 178 \times 12{,}000}{12 \times 30^2} = \pm 1187 \text{ psi}$$

The resulting stresses at the center support are

$$f = -593 + 1187 = +594 \text{ psi tension top}$$
$$f = -593 - 1187 = -1780 \text{ psi compression bottom}$$

The reactions due to 1.03 kips/ft can be computed from the vertical components of the cable:

$$\text{Exterior support:} \quad R_A = 1.03 \times 25 - \frac{214}{50} = 25.8 - 4.3 = 21.5 \text{ kips}$$
$$\text{Interior support:} \quad R_B = 51.6 + 2 \times 4.3 = 60.2 \text{ kips}$$

Under the action of 0.57 kip/ft load, the reactions are, by the elastic theory,

$$\text{Exterior support:} \quad R_A = 10.6 \text{ kips}$$
$$\text{Interior support:} \quad R_B = 35.6 \text{ kips}$$

Hence, the total reactions due to 1.6 kips/ft load and the effect of $F = 214$ kips are

$$\text{Exterior support:} \quad R_A = 21.5 + 10.6 = 32.1 \text{ kips}$$
$$\text{Interior support:} \quad R_B = 60.2 + 35.6 = 95.8 \text{ kips}$$

12.6.3 Ultimate Strength of Continuous Beams

The ultimate strength of prestressed continuous beams can be estimated by limit analysis. Plastic hinges form at points of maximum moment in underreinforced beams. Complete plastic hinges may not develop at cross sections where shear is large and in overreinforced beams, in which case the action will be only partly plastic.

Cracking in prestressed continuous beams can be computed by the elastic theory that cracking begins when the tensile fiber stress reaches the modulus of rupture of the concrete.

12.7 DESIGN FOR TORSION

A hollow or solid beam section subjected to torsion is idealized as a thin-walled tube both before and after cracking. The resistance to torsion is assumed to be provided by the outer skin of the cross section, the thickness of which is roughly centered on the closed stirrups and the core concrete of the solid section is neglected. Once the beam section is cracked in torsion, the torsional resistance is provided by closed stirrups and longitudinal reinforcement. It is assumed that concrete makes no contribution to the torsional resistance.

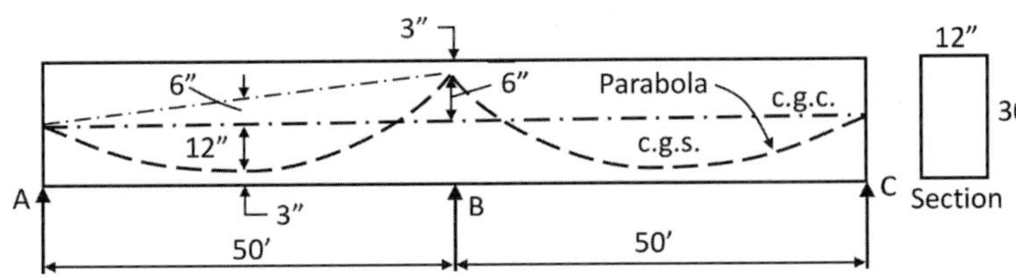

Figure 12.32 Example 10.

12.7.1 Torsional Shear Stress

In a closed thin-walled tube, torsion produces a shear flow q as shown in Fig. 12.33a. The shear flow $q = T/2A_o$ is constant at all points and acts along the path that extends around the tube at mid-thickness of the walls of the tube, where T is torque and A_o is the area enclosed by the shear flow path as shown in Fig. 12.33b, including the area of the hole if the section is hollow. The torsional shear stress at any point along the perimeter of the tube is $\tau = q/t = T/(2A_o t)$, in which t is the thickness of the wall at the point where τ is computed.

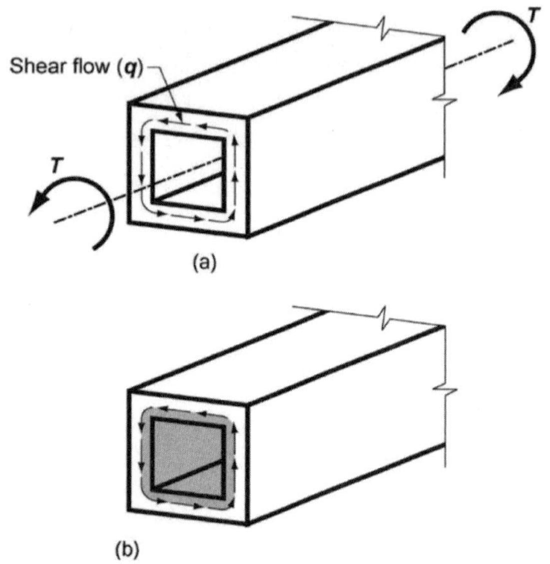

Figure 12.33 Shear flow in thin-walled tube: (a) tube and (b) area enclosed by shear flow path (shaded area).

Prior to cracking, it is assumed that the section can be replaced by an equivalent thin-walled tube with a wall thickness $t = 0.75\,A_{cp}/p_{cp}$ and an area enclosed by the wall centerline $A_o = 2\,A_{cp}/3$, in which A_{cp} = area enclosed by the outside perimeter of the concrete cross section and p_{cp} = outside perimeter of the concrete cross section.

12.7.2 Cracking Torque

Cracking of section under torsion is assumed to occur when the principal stress reaches

$$4\sqrt{f_c'}\sqrt{1+f_{pc}/4\sqrt{f_c'}}$$

where $4\sqrt{f_c'}$ is the conservative value of the tensile strength of the concrete and the factor

$$\sqrt{1+f_{pc}/4\sqrt{f_c'}}$$

is the effect of prestress on the principal tensile stress. The term f_{pc} is the compressive stress in the concrete after allowance for all prestress losses at the centroid of the cross section resisting externally applied loads or at the junction of web and flange when the centroid lies within the flange. Thus, as presented in Table 22.7.5.1 of ACI 318-14, the cracking torque is

$$T_{cr} = 4\sqrt{f_c'}\,\frac{A_{cp}^2}{p_{cp}}\sqrt{1+\frac{f_{pc}}{4\sqrt{f_c'}}} \tag{12.39}$$

If the factored torsional moment T_u is less than one-fourth of the cracking torque T_{cr} adjusted by the strength reduction factor $\phi = 0.75$, that is, if

$$T_u < 0.75\sqrt{f_c'}\,\frac{A_{cp}^2}{p_{cp}}\sqrt{1+\frac{f_{pc}}{4\sqrt{f_c'}}} \tag{12.40}$$

the torsional effect can be neglected.

12.7.3 Design for Equilibrium Torsion

Equilibrium torsion is generated due to loading eccentric to the centroidal axis. If the factored torsional moment T_u in a member is required to maintain equilibrium and exceeds the value from Eq. (12.40), the torsional reinforcements (closed stirrups and longitudinal reinforcements) must be designed to carry the total torsional moment T_u due to the factored loads so that

$$\phi T_n \geq T_u \tag{12.41}$$

where T_n is the nominal torsional strength provided by the torsional reinforcements and shall be the lesser of (12.42a) and (12.42b),

$$T_n = \frac{2A_0 A_t f_{yt}}{s}\cot\theta \tag{12.42a}$$

$$T_n = \frac{2A_0 A_l f_y}{p_h}\cot\theta \tag{12.42b}$$

in which A_o should be determined by analysis but may be taken as $0.85A_{oh}$; A_t is the area of one leg of a transverse closed stirrup (hoop) resisting torsion; A_l is the area of longitudinal torsional reinforcement, f_{yt} and f_y are, respectively, the yield strength of the closed stirrups and longitudinal reinforcement; p_h is the perimeter of the centerline of the outermost closed stirrup; and s is the spacing of the closed stirrups that should not exceed the lesser of $p_h/8$ and 12 in. The value of θ shall not be taken less than 30 degrees nor greater than 60 degrees. According to Section 22.7.6.1.2 of ACI 318-14, it may be taken as 37.5° for prestressed concrete members with an effective prestressing force (i.e., $A_{ps}f_{se}$) not less than 40 percent of $(A_{ps}f_{pu} + A_s f_y)$, where $A_{ps}f_{pu}$ is the ultimate prestressing force of tendons and $A_s f_y$ is the yielding force of longitudinal reinforcement. For nonprestressed concrete or members with larger effective prestressing force, the code suggests that θ be taken as 45°. This corresponds to the value assumed in the equations for the design of stirrups for shear in the ACI Building Code. A_{oh} is the area enclosed by the centerline of the outermost closed hoops. Several examples of how A_{oh} should be defined for various cross sections are shown in Fig. 12.34.

The transverse and longitudinal reinforcements required for torsion as determined from Eqs. (12.41) and (12.42) must be combined with that required for the shear, moment, and axial force acting in combination with the torsion. To prevent sudden and brittle failures, minimum

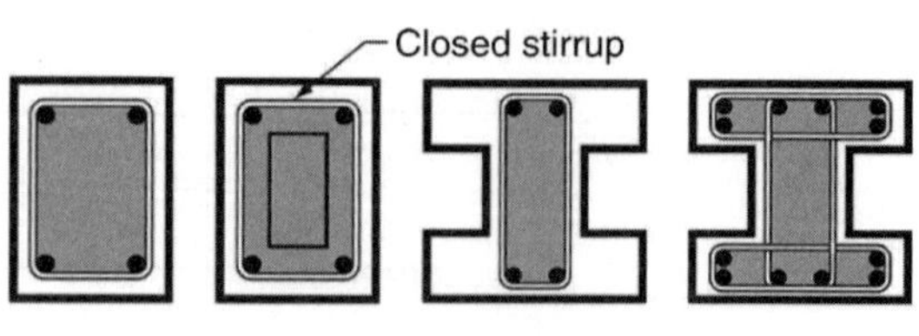

Figure 12.34 Definition of A_{oh} for various cross sections.

amounts of transverse and longitudinal reinforcements must be provided such that, for the transverse closed stirrups,

$$(A_v + 2A_t)_{\min} > \text{Greater of} \left\{ 0.75\sqrt{f_c'}\frac{b_w s}{f_{yt}},\ 50\frac{b_w s}{f_{yt}} \right\} \tag{12.43}$$

and for the longitudinal torsional reinforcement should be the lesser of the following equations:

$$5\frac{\sqrt{f_c'}A_{cp}}{f_y} - \left(\frac{A_t}{s}\right)p_h\frac{f_{yt}}{f_y}$$

$$5\frac{\sqrt{f_c'}A_{cp}}{f_y} - \left(\frac{25b_w}{f_{yt}}\right)p_h\frac{f_{yt}}{f_y} \tag{12.44}$$

In the above expressions, A_v = area of two legs of a closed stirrup and b_w = width of web resisting flexural shear.

To reduce unsightly cracking and to prevent crushing of the surface concrete due to inclined compressive stresses under combined shear and torsion, the cross-sectional dimensions of a member must be sufficiently large such that the following requirements are satisfied:

For a solid section,

$$\sqrt{\left(\frac{V_u}{b_w d}\right)^2 + \left(\frac{T_u p_h}{1.7A_{oh}^2}\right)^2} \le \phi\left(\frac{V_c}{b_w d} + 8\sqrt{f_c'}\right) \tag{12.45}$$

For a hollow section,

$$\frac{V_u}{b_w d} + \frac{T_u p_h}{1.7A_{oh}^2} \le \phi\left(\frac{V_c}{b_w d} + 8\sqrt{f_c'}\right) \tag{12.46}$$

In the above two expressions, V_u = factored shear force acting together with factored torsional moment T_u and V_c = nominal shear strength provided by concrete = V_{ci} or V_{cw}, whichever is less as given by Eqs. (12.34) and (12.35). The value of d used in Eqs. (12.45) and (12.46) need not be taken less than $0.8h$.

12.7.4 Design for Compatibility Torsion

In a statically indeterminate structure, where reduction of the torsional moment in a member can occur due to redistribution of internal forces upon cracking while maintaining compatibility of deformation, according to Section 22.7.3.2 of ACI 318-14, the maximum factored torsional moment T_u to be designed for may be reduced to

$$\phi T_{cr} = \phi 4\sqrt{f_c'}\frac{A_{cp}^2}{p_{cp}}\sqrt{1 + \frac{f_{pc}}{4\sqrt{f_c'}}} \tag{12.47}$$

In such a case, the correspondingly redistributed bending moments and shears in the adjoining members must be used in the design of those members.

EXAMPLE 11

Design the torsional reinforcement for the prestressed concrete L-beam, as shown in Fig. 12.35, which is subjected to a factored torsional moment T_u = 707 kip-in. The beam is pretensioned with six ½-in. low-relaxation strands of grade 270 with an effective depth d_p of 69 in. Assume that the loss of prestress is 18 percent. Given f_c' = 5 ksi, $f_y = f_{yt} = f_{yl}$ = 60 ksi.

Solution First, it should be determined if torsion design is required by using Eq. (12.40). Let the initial tensile stress in the prestressing strand be $0.80 f_{pu} = 0.80 \times 270 = 216$ ksi. Since the area of each strand is

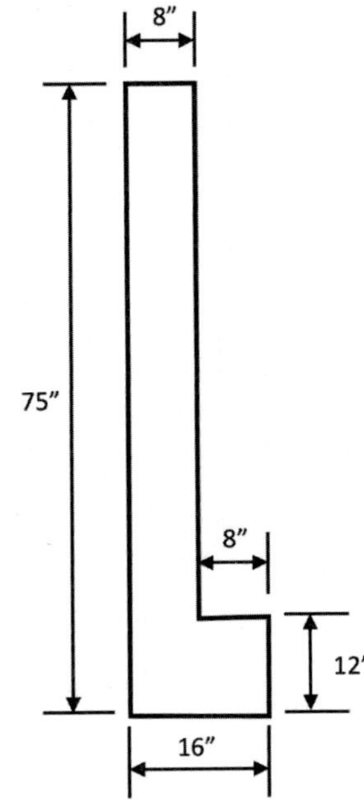

Figure 12.35 Example 11.

0.153 in.^{2} and the loss of prestress is 18 percent, the effective prestressing force is $6 \times 0.153 \times (0.82 \times 216) = 162.6$ kips.

The area enclosed by the outside perimeter of the concrete cross section is

$$A_{cp} = 8 \times 75 + 8 \times 12 = 696 \text{ in.}^2$$

and the outside perimeter of the concrete cross section is

$$p_{cp} = 2 \times 75 + 2 \times 16 = 182 \text{ in.}$$

Thus,

$$f_{pc} = A_{ps}f_{se}/A_{cp} = 162.6/696 = 0.234 \text{ ksi}$$

Also, $\sqrt{f_c'} = \sqrt{5000} = 70.7$ psi and $4\sqrt{f_c'} = 282.8$ psi. Then, from Eq. (12.40),

$$\phi\sqrt{f_c'}\frac{A_{cp}^2}{p_{cp}}\sqrt{1 + \frac{f_{pc}}{4\sqrt{f_c'}}} = 0.75 \times 70.7 \times \frac{696^2}{182} \times \sqrt{1 + \frac{234}{282.8}}$$

$$= 190.8 < T_u$$

Therefore, torsion must be considered in design.

To design the torsional reinforcement, let us assume that No. 4 bars will be used for both the transverse and the longitudinal reinforcement and the cover is ¾ in. for the closed stirrups. Thus, the centerline of the stirrup is 1 in. from the concrete surface, and the size of the closed stirrups is shown in Fig. 12.36. Now,

$$A_{sh} = \text{area enclosed by the centerline of outermost closed stirrups}$$
$$= 73 \times 6 + 8 \times 10 = 438 + 80 = 518 \text{ in.}^2$$

So, $A_o = 0.85\,A_{oh} = 0.85 \times 518 = 440.3$ in.^{2} and

A_t = area of one leg of No. 4 stirrup = 0.20 in.^{2}

Assume $\theta = 37.5°$

Then, from Eqs. (12.41) and (12.42*a*),

$$\frac{2A_0 A_t f_{yt}}{s}\cot\theta = \frac{T_u}{\phi}$$

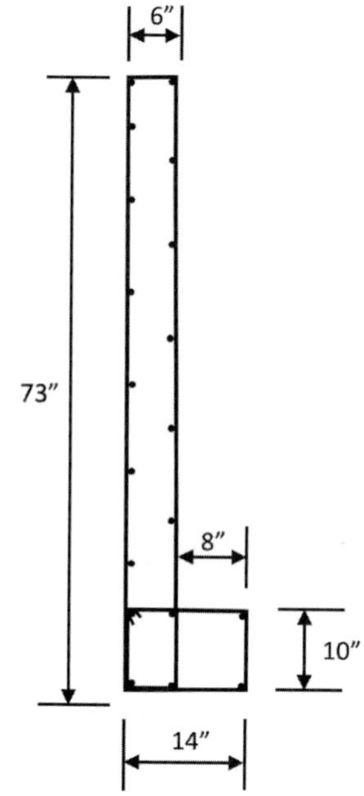

Figure 12.36 Torsional reinforcement.

or

s = spacing of stirrups

$$= 2\phi \frac{A_0 A_t f_{yt}}{T_u} \cot\theta = 2 \times 0.75 \times 440.3 \times 0.2 \times 60 \times \frac{1.303}{707} = 14.6 \text{ in.}$$

The stirrup requirement should be combined with that required for flexural shear and the appropriate spacing determined. The spacing of the transverse torsional reinforcement must not exceed the smaller of $p_h/8$ or 12 in. It should be noted that since the overall depth of the ledger of the L-beam is 12 in., the maximum practical spacing for the stirrups should be no more than 12 in.

The longitudinal torsional reinforcement should be determined based on Eq. (12.42*b*). From Fig. 12.36, the perimeter of the centerline of the outermost closed stirrup is

$$p_h = 2 \times 73 + 2 \times 14 = 174 \text{ in.}$$

$$A_l = \frac{T_n p_h}{2 A_0 f_y \cot\theta} = \frac{707 \times 174}{2 \times 440.3 \times 60 \times 1.303} = 1.79 \text{ in.}^2$$

$$\text{No. of \#4 bars required} = \frac{1.79}{0.2} = 8.9.$$

Therefore, a total of 9 No. 4 bars should be used as the longitudinal torsional reinforcement, which should be uniformly spread around the perimeter of the closed stirrups with a spacing of no more than 12 in., according to Section 9.7.5.1 of ACI 318-14. According to the code, the longitudinal bars and/or tendons shall be inside the closed stirrups. At least one longitudinal bar or tendon must be placed in each corner of the stirrups. As per Section 9.5.4.5 of ACI 318-14, it is permissible to reduce the area of the longitudinal reinforcement in the flexural compressive zone by an amount equal to $M_u/0.9\, df_y$, where M_u is the factored moment acting at the section in combination with T_u. Finally, the requirements of Eqs. (12.43) to (12.47) should be checked.

12.8 DESIGN FOR SEISMIC LOADS

The use of prestressing in seismic design was generally limited to elastic members because of their inability to undergo inelastic strains without losing a significant amount of prestressing. When structures are formed using prefabricated components, the connections between members (e.g., beam to column and wall to foundation) need to be designed to ensure safety and dependable behavior, and therefore the use of prestressing in the connections was also not favorably considered. Instead, these connections were designed such that the resulting structure could emulate the behavior of comparable structures designed with cast-in-place (CIP) construction. This approach was completely changed by the PREcast Seismic Structural Systems Program (PRESSS) (Priestley et al. 1999), which promoted the use of unbonded posttensioning in critical seismic members that generally incorporated especial elements to dissipate seismic energy. In this case, the prestressing is designed to remain elastic under earthquake loads.

As described in Kurama et al. (2018), the connections used in members with unbonded posttensioning are termed jointed as opposed to emulative connections. While emulating equivalent CIP structures, the corresponding connections are designed to produce equivalency of lateral strength, stiffness, and energy dissipation by allowing nonlinearity at preselected and appropriately designed locations. The jointed connection promotes a gap opening at the connection interface by allowing nonlinear rotations of the system (i.e., rocking).

Following the requirements of ACI 550.1R-01, emulative connections in precast structures will be achieved by placing CIP concrete or grout in splices (e.g., lapped bars, mechanical splices, welded splices, and bonded PTs) that are used to connect precast elements at the location of joints. In these emulative connections, namely, strong connections, plastic hinges will be formed within the member. However, the desired ductile behavior of emulative precast concrete elements will be attained by allowing for the flexural yielding and formation of plastic hinges in the connections across member joints (ACI 550.1R-01). These types of emulative connections, namely, ductile connections, will be achieved when precast concrete structures are designed in accordance with Chapter 18 of ACI 318-14, except 18.5, 18.9.2.3, and 18.11.2.2, which permit precast construction with alternative yielding mechanisms (ACI 318-14).

With the use of unbonded posttensioning and jointed connections, seismic systems can be designed to self-center, thereby providing superior seismic performance (Cheok and Lew 1991; Priestley and Tao 1993; Priestley et al. 1999). The self-centering behavior also produces insignificant residual deformation, permanent cracks, and minimal structural damage. Due to a linear response of structures with jointed connections, limited hysteretic energy dissipation is expected in these structures. To overcome this deficiency, jointed precast structures are often designed with additional supplemental energy dissipaters at the rocking joints to produce some displacement-dependent (hysteretic) damping during seismic action. According to ACI 318-14, using jointed connections in special precast concrete moment frames (i.e., beam-column joint subassembly) is permitted in high seismic regions, provided they satisfy requirements of ACI 374.1-05. Previous research studies (e.g., Cheok and Lew 1991; Stone et al. 1995; Priestley et al. 1999) presented the need for using unbonded PT tendons and partially bonded mild steel in jointed frames to ensure both their recentering capability and energy dissipation capacity. These frames, which are known as hybrid frames, should be properly connected to the floor slabs to ensure lower structural damage during rocking of the jointed elements (Priestley et al. 1999; Sritharan 2002). Using unbonded posttensioning across the wall-to-foundation joint has been studied by many researchers (e.g., Kurama et al. 1999; Holden 2001; Kurama 2002). Relying primarily on the large-scale quasi-static testing of these rocking wall systems and limited analytical studies completed as of 1999 (e.g., Priestley et al. 1999; Kurama 2002; Stanton and Nakaki 2002), ACI ITG-5 documents (ACI ITG-5.1 2008; ACI ITG-5.2 2009) were developed to outline the seismic acceptance criteria and design procedures for special unbonded posttensioned precast concrete walls that are equipped with external hysteretic energy-dissipating elements. Precast Wall with End Columns, namely PreWEC, is one such system developed by Sritharan et al. (2015), which is shown to be an efficient alternative to CIP walls and incorporates easily replaceable steel O-connectors as supplemental dampers. In what follows, some recent modifications to the analysis and design of single rocking walls (SRWs) and PreWECs as jointed wall systems (Nazari and Sritharan 2019) are presented.

12.8.1 Response Modification Coefficients for Rocking Wall Systems

According to ITG-5 documents, the design base shear force for rocking wall systems can be estimated following the Force-Based Design (FBD) approach, which uses a response modification (R) factor as with ACI 318-11 and ASCE 7-10. Recent research studies by Nazari and Sritharan (2018) proposed new R-factors to be used for the precast wall systems as a function of total equivalent damping ratio in the system, as expressed in Eqs. (12.48) and (12.49) for far-field (FF) and near-field (NF) records:

$$R = 0.46 \times \xi_{eq} + 0.93;\ \text{FF records} \tag{12.48}$$

$$R = 0.15 \times \xi_{eq} + 2.65;\ \text{NF records} \tag{12.49}$$

where ξ_{eq} = total equivalent damping ratio of the system expressed in percentage; this could vary from about 6 percent for SRWs (Nazari et al. 2016) to larger values for jointed wall systems, depending on the amount of the supplemental damping provided by the external energy dissipaters (Nazari and Sritharan 2019).

12.8.2 Seismic-Resilient Design of Rocking Wall Systems

Unbonded posttensioning tendons, as the primary reinforcement of the precast self-centering systems with jointed connections, not only secure their self-centering capability but also provide lateral resistance of the element. To design rocking wall systems, such as PreWECs, the required area of PT tendons is determined using moment equilibrium of forces acting at their base assuming the plastic capacity of the energy-dissipating elements and 95 percent of the yield strength of PT tendons at the design drift. To maintain self-centering capability of the system up to the design drift, the initial stress for PT tendons is chosen such that they remain elastic until the chosen target drift; elongation of the tendons is estimated by calculating the neutral axis depth at the base of the wall panel. Aaleti and Sritharan (2009) suggested a Simplified Analysis (SA) method, in which the required area of tendons is determined based on moment equilibrium of forces acting at the wall base, using an equivalent stress block and a trilinear approximation for estimating the neutral axis depth variation as a function of wall base rotation. A macro-based Excel program has been developed for the analysis and design of the rocking wall systems (Nazari and Sritharan 2019), and that is available at http://sri.cce.iastate.edu/NEES-Rocking-Wall/index.htm.

EXAMPLE 12

Design the required area of posttensioning tendons and initial prestressing in the wall panel and end columns of a PreWEC system, with the following geometrical details, as well as the required number of O-connectors to provide an equivalent damping ratio of 13 percent for the system. Consider four PreWECs as lateral load-resisting systems in the transverse direction of a rectangular six-story office building in a location with S_{DS} = 0.816 g and S_{D1} = 0.418 g following ASCE 7-16. Each wall panel is subjected to a gravity load of 935 kips. The macro-based Excel program has been used for this calculation.

Geometrical details:

Wall panel: height = 864 in.; length = 210 in.; thickness = 18 in.; length of PT tendons = 936 in.

End columns: height = 864 in.; length = 30 in.; thickness = 18 in.; length of PT tendons = 936 in.

An R factor of 4.6 is estimated for this PreWEC system with the expected damping ratio of 13 percent, following Eq. (12.49).

The required amount of damping will be achieved by providing 24 O-connectors per vertical joint between the wall panel and the column with the yielding and ultimate force and deformation properties of (17 kips, 0.25 in.) and (18 kips, 4 in.), respectively.

Design seismic force for the PreWEC system (following ASCE 7-16)

C_s = 0.175; V = 419 kips resisted by four PreWECs;

V = 105 kips per wall

End columns: Required area of the post-tensioning tendon ($A_{p,col}$) and the initial prestressing ($f_{pi,col}$) to resist the ultimate force of connectors:

$$A_p = 3\ \text{in.}^2;\ f_{pi} = 0.8 \quad \text{and} \quad f_{pu} = 216\ \text{ksi per end column}$$

Wall panel: Required area of the posttensioning tendon (A_p) and the initial prestressing (f_{pi}) are found by using moment equilibrium of forces acting at the wall base. This is described in more detail in the SA method as suggested by Aaleti and Sritharan (2009):

$$A_p = 9.1\ \text{in.}^2;\ f_{pi} = 0.7 \quad \text{and} \quad f_{pu} = 192\ \text{ksi}$$

REFERENCES

Aaleti, S., and Sritharan, S. 2009. A Simplified Analysis Method for Characterizing Unbonded Post-Tensioned Precast Wall Systems. *Engineering Structures*, **31**(12), 2966–2975.

ACI 318-14. 2014. Building Code Requirements for Structural Concrete. ACI Committee 318, American Concrete Institute, Farmington Hills, MI.

ACI 318.2-14. 2014. Building Code Requirements for Concrete Thin Shells and Commentary. ACI Committee 318, American Concrete Institute, Farmington Hills, MI.

ACI 318-11. 2011. Building Code Requirements for Structural Concrete. ACI Committee 318, American Concrete Institute, Farmington Hills, MI.

ACI 374.1-05. 2005. Acceptance Criteria for Moment Frames Based on Structural Testing and Commentary. ACI Committee 374, American Concrete Institute, Farmington Hills, MI.

ACI 550.1R-01. 2001. Emulating Cast-in-Place Detailing in Precast Concrete Structures. ACI Committee 550, American Concrete Institute, Farmington Hills, MI.

ACI Innovation Task Group 5.1. 2008. Acceptance Criteria for Special Unbonded Post-Tensioned Precast Structural Walls Based on Validation Testing and Commentary. American Concrete Institute, Farmington Hills, MI.

ACI Innovation Task Group 5.2. 2009 Requirements for Design of a Special Unbonded Post-Tensioned Precast Shear Wall Satisfying ACI ITG-5.1 and Commentary (ACI ITG-5.2). American Concrete Institute, Farmington Hills, MI.

ASCE 7-10. 2010. *Minimum Design Loads for Buildings and Other Structures*. ASCE/SEI 7-10.

ASCE 7-16. 2016. *Minimum Design Loads for Buildings and Other Structures*. ASCE/SEI 7-16.

Cheok, G. S., and Lew, H. S. 1991. Performance of Precast Concrete Beam-to-Column Connections Subject to Cyclic Loading. *PCI Journal*, **36** (3), 56–67.

Henry, R., Sritharan, S., and Ingham, J. 2012. Unbonded Tendon Stresses in Post-Tensioned Concrete Walls at Nominal Flexural Strength. *American Concrete Institue Journal*, **109**(2), 171–181.

Holden, T. J. 2001. A Comparison of the Seismic Performance of Precast Wall Construction: Emulation and Hybrid Approaches. Research Report 2001-04, University of Canterbury, Christchurch.

Kurama, Y. C. 2002. Hybrid Post-Tensioned Precast Concrete Walls for Use in Seismic Regions. *PCI Journal*, **47**(5), 36–59.

Kurama, Y., Sause, R., Pessiki, S., and Lu, L. W. 1999. Lateral Load Behavior and Seismic Design of Unbonded Post-Tensioned Precast Concrete Walls. *ACI Structural Journal*, **96**(4), 622–632.

Kurama, Y. C., Sritharan, S., Fleischman, R. B., and Restrepo, J. I. 2018. Seismic-Resistant Precast Concrete Structures: State of the Art. *Journal of Structural Engineering*, **144**(4), 03118001.

Nazari, M., and Sritharan, S. 2018. Dynamic Evaluation of PreWEC Systems with Varying Hysteretic Energy Dissipation. *ASCE Journal of Structural Engineering*, **144**(10) (DOI: 10.1061/(ASCE)ST.1943-541X.0002169).

Nazari, M., and Sritharan, S. 2019. Seismic Design of Precast Concrete Rocking Wall Systems with Varying Hysteretic Damping. *PCI Journal,* **64**(5), 58–76.

Nazari, M., Sritharan, S., and Aaleti, S. 2016. Single Precast Concrete Rocking Walls as Earthquake Force-Resisting Elements. *Earthquake Engineering and Structural Dynamics*, **46**(5), 753–769 (DOI: 10.1002/eqe.2829).

PCI Design Handbook: Precast and Prestressed Concrete. 8th ed. 2017. Precast/Prestressed Concrete Institute, Chicago, IL.

Priestley, M. J. N., Sritharan, S., Conley, J. R., and Pampanin, S. 1999. Preliminary Results and Conclusions from the PRESSS Five-Story Precast Concrete Test Building. *PCI Journal*, **44**(6), 42–67.

Priestley, M., and Tao, J. 1993. Seismic Response of Precast Prestressed Concrete Frames with Partially Debonded Tendons. *PCI Journal*, **38**(1), 58–69.

Sritharan, S. 2002. Performance of Four Jointed Precast Frame Systems under Simulated Seismic Loading. In Proceedings of the Seventh U. S. National Conference on Earthquake Engineering, Paper No. 480, Boston, MA.

Sritharan, S., Aaleti, S., Henry, R., Liu, K., and Tsai, K. 2015. Precast Concrete Wall with End Columns (PreWEC) for Earthquake Resistant Design. *Earthquake Engineering and Structural Dynamics*, **44**(12), 2075–2092.

Sritharan, S., Wibowo, H., Rosenthal, M. J., Eull, J. N., and Holombo, J. 2018. LRFD Minimum Flexural Reinforcement Requirements. National Academies of Sciences, Engineering, and Medicine, Washington, DC: National Academies Press. https://doi.org/10.17226/25527.

Stanton, J. F., and Nakaki, S. D. 2002. Design Guidelines for Precast Concrete Seismic Structural Systems Unbonded Post-Tensioned Split Walls. PRESSS Report No. 01/03-09, Department of Civil Engineering, University of Washington, Seattle, WA.

Stone, W. C., Cheok, G. S., and Stanton, J. F. 1995. Performance of Hybrid Moment-Resisting Precast Beam-Column Concrete Connections Subjected to Cyclic Loading. *ACI Structural Journal*, **92**(2), 229.

Chapter 13
Masonry Construction

BY
RICHARD BENNETT *Professor and Director of Engineering Fundamentals, University of Tennessee, Knoxville, Tennessee*
SAM RUBENZER, PE, SE *Founder and Structural Engineer, FORSE Consulting, Eau Claire, Wisconsin*

NOTATION

a = depth of an equivalent compression stress block at nominal strength
A_g = gross cross-sectional area
A_n = net cross-sectional area
A_{nc} = net cross-sectional area between the neutral axis of bending and the fiber of maximum compressive strain calculated at the nominal moment capacity of the section
A_{nv} = net shear area
A_s = area of reinforcement
A_{sp} = cross-sectional area of reinforcement within the net shear area
$A_{s,\text{reqd}}$ = required area of reinforcement
$A^*_{s,\text{reqd}}$ = required area of reinforcement per unit length (distributed reinforcement)
A_{st} = area of laterally tied longitudinal reinforcing steel
A_v = cross-sectional area of shear reinforcement
b = width of member
c = distance from the fiber of maximum compressive strain to the neutral axis
c_{bal} = distance from the fiber of maximum compressive strain to the neutral axis at balanced conditions in strength design
d = distance from extreme compression fiber to centroid of tension reinforcement
d_v = actual depth of member in direction of shear
e = eccentricity of axial load
e_u = eccentricity at P_{uf}
E_m = modulus of elasticity of masonry
E_s = modulus of elasticity of steel
f'_m = specified compressive strength of masonry
f_a = axial stress
f_b = flexural compressive stress
f_r = modulus of rupture
f_s = tensile stress in reinforcement
f_v = shear stress
f_y = specified yield strength of reinforcement
F_a = allowable axial stress
F_b = allowable flexural compressive stress
F_r = allowable flexural tensile stress
F_s = allowable stress in reinforcement
F_v = allowable shear stress
F_{vm} = allowable shear stress resisted by the masonry
F_{vs} = allowable shear stress resisted by the shear reinforcement
h = effective height of column, wall, or pilaster
I_{cr} = cracked moment of inertia
I_n = net moment of inertia
j = ratio of internal lever is to d in allowable stress design
k = ratio of distance to neutral axis to d in allowable stress design
k_{bal} = ratio of distance to neutral axis to d at balanced condition in allowable stress design
M = moment
M_{cr} = nominal cracking moment
M_n = nominal moment
M_u = factored moment
M' = value used in allowable stress design procedure for singly reinforced members
n = modular ratio
N_v = compressive force acting normal to shear surface
P = axial load (compression is positive)
P_a = allowable axial compression force
P_e = Euler buckling load
P_n = nominal axial load
P_u = factored axial load
P_{uf} = factored load from tributary floor or roof areas
P_{uw} = factored load from tributary wall area
Q = first moment about the neutral axis of an area between the extreme fiber and the plane at which the shear stress is being calculated
r = radius of gyration
s = spacing of reinforcement
t = nominal thickness of member
t_{sp} = specified thickness of member
S_n = net section modulus
V = shear force
V_n = nominal shear strength
V_{nm} = nominal shear strength provided by masonry
V_{ns} = nominal shear strength provided by shear reinforcement
V_u = factored shear strength

δ_u = deflection due to strength level loads
ε_{mu} = maximum useable compressive strain in masonry
ε_y = yield strain of reinforcement
ϕ = strength reduction factor
γ_g = grouted shear wall factor
μ = coefficient of friction
ρ = reinforcement ratio
ρ_{max} = maximum flexural tension reinforcement ratio
ς = values used in allowable stress design procedure for singly reinforced members

13.1 MASONRY MATERIALS

Masonry is a structural material that has much architectural appeal and benefits. Historically, masonry's primary purpose was to provide safe structural support for structures. The fact that masonry has been heavily used throughout thousands of years of building construction creates much of the architectural appeal of masonry today. Masonry is the construction of individual units, held together by mortar. Masonry is a material that makes up walls, lintels/beams, columns, pilasters, wall groups, or a combination thereof.

Multiple elements are needed to construct a masonry wall. The fundamental elements of masonry are blocks and mortar. Blocks are comprised of the primary elements of stone, concrete units, or brick units and mortar. The focus of this chapter is concrete masonry, or brick masonry. Mortar is a concrete mix that adheres to blocks and holds the blocks together during construction and binds them together structurally. Masonry can be either unreinforced or reinforced with steel reinforcement bars. To add reinforcement to masonry, the blocks must be hollow, and concrete grout must be added to bind the reinforcement with the masonry block elements.

13.1.1 Composition of Masonry

Masonry blocks of concrete or brick are the primary elements of the masonry. The strength of the block has the largest impact on the overall masonry assembly strength. The typical strength of concrete masonry units (CMU) varies from manufacturer to manufacturer and is generally in the range of 3000 psi to 5000 psi, as determined by compression strength tests, and average strengths are above 4000 psi. CMU are governed by ASTM C90. Concrete masonry blocks are generally available as 16-in.-long by 8-in.-high units and in widths of 4, 10, 12, and 16 in. Masonry blocks are open units with face shells and webs connecting them in different ways, allowing for different vertical and horizontal reinforcement to be added to the wall.

Figure 13.1 shows examples of Web Configurations Permitted Under ASTM C90-11b.

Hollow brick masonry units (BMU) are governed by ASTM C652 and are fired clay units similar to brick veneer units. Hollow brick units are structural units and have a similar look and appeal of the veneer units but are hollow, allowing them to be grouted and reinforced. The typical strength of BMU also varies from manufacturer to manufacturer and is generally in the range of 2000 psi to 12,000 psi as determined by compression strength tests, and average strengths are above 6500 psi.

Masonry mortar is the material used to bind the masonry blocks and can be one of the three different options: portland cement lime (PCL), masonry cement, or mortar cement. Mortar is also available in different types: M, S, N, O, or K. Mortar type M is the strongest in compressive strength and adhesion, and strength is reduced in each successive mortar. Type M is generally the most difficult to use by masons constructing masonry in the field. For new masonry projects, type M, S, or N is used. Type S is typically recommended for most structural masonry applications, with type N being the most common for veneers and other nonstructural applications. Mortars of type M or S are recommended

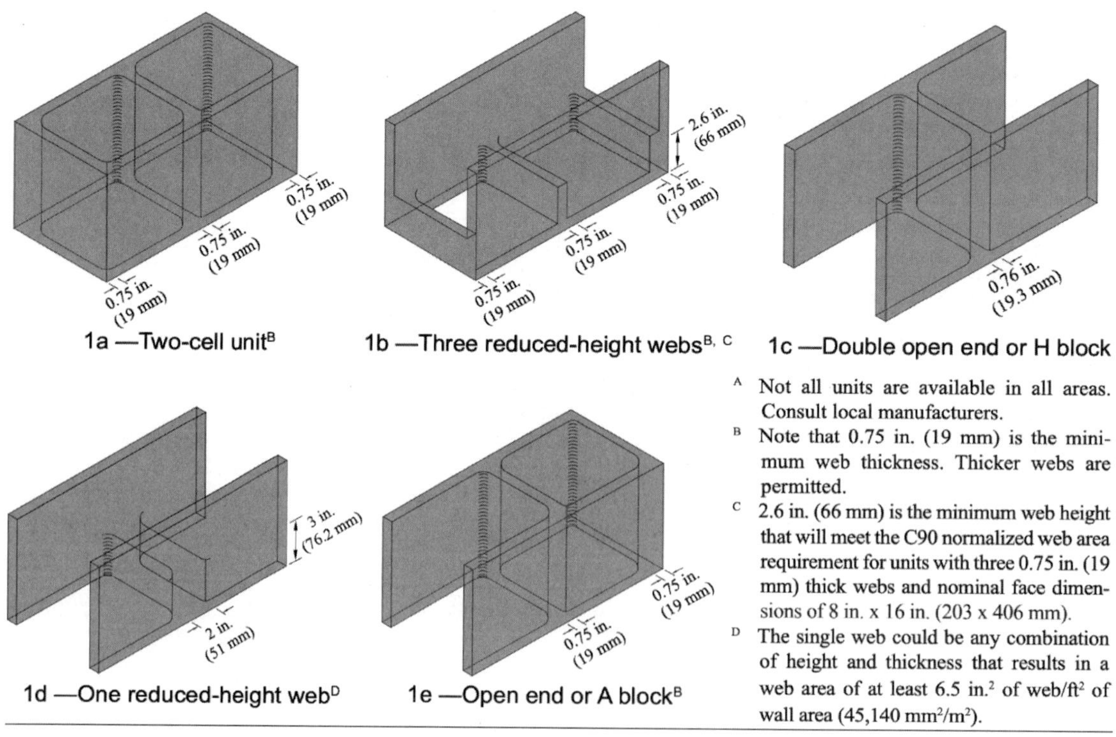

[A] Not all units are available in all areas. Consult local manufacturers.
[B] Note that 0.75 in. (19 mm) is the minimum web thickness. Thicker webs are permitted.
[C] 2.6 in. (66 mm) is the minimum web height that will meet the C90 normalized web area requirement for units with three 0.75 in. (19 mm) thick webs and nominal face dimensions of 8 in. x 16 in. (203 x 406 mm).
[D] The single web could be any combination of height and thickness that results in a web area of at least 6.5 in.^{2} of web/ft^2 of wall area (45,140 mm^2/m^2).

Figure 13.1 Examples of web configurations permitted under ASTM C90-11b.

for use in below-grade applications. Mortar is only to be used in bed joints, between blocks, and is never to be used as a fill in masonry cells with reinforcement. Properties of mortar bond strength and workability are more important than compressive strength. TMS 602 clearly defines the wall strength to be more than the strength of the mortar. Mortar strength has been shown to be a relatively unimportant factor in determining overall wall strength.

Once the masonry unit compressive strength is known by testing and the masonry mortar type is defined, the masonry assembly strength, f'_m, can be determined. Masonry walls have both tension and compression properties. Similar to all other concrete materials, masonry compressive capacities are far greater than masonry tensile capacities. To achieve the greatest bending capacities, masonry walls are reinforced to take advantage of the reinforcement's tensile capabilities coupled with masonry's compression capabilities to create masonry elements capable of resisting bending loads. For the most part, it is only with unreinforced, non-load-bearing masonry elements where the tensile capabilities are needed. For determining the compression properties of concrete masonry, either prism testing can be used to determine f'_m or the unit strength method by following TMS 602, Table 2. Similarly, for brick masonry, follow TMS 602, Table 1.

For example, follow the steps below to find f'_m using the unit strength method:

1. Define mortar type—use type S—start in the center column of Table 2 below.
2. Research the block strength available to the project; in this example, assume the CMU strength determined by testing is 3900 psi and find that value in the center column.
3. Finally, see the corresponding value in the far-left column, where the net area compressive strength of concrete masonry can be used as the masonry assembly compressive strength, $f'_m = 2750$ psi.

Tables 13.1 and 13.2 show the compressive strength of masonry based on the compressive strength of clay masonry units and type of mortar used in construction.

Table 13.1 Compressive Strength of Masonry Based on the Compressive Strength of Clay Masonry Units and Type of Mortar Used in Construction (TMS 602-16 Table 1)

Net area compressive strength of clay masonry, psi (MPa)	Net area compressive strength of clay masonry, psi (MPa)	
	Type M or S mortar	Type N mortar
1000 (6.90)	1700 (11.72)	2100 (14.48)
1500 (10.34)	3350 (23.10)	4150 (28.61)
2000 (13.79)	4950 (34.13)	6200 (42.75)
2500 (17.24)	6600 (45.51)	8250 (56.88)
3000 (20.69)	8250 (56.88)	10,300 (71.02)
3500 (24.13)	9900 (68.26)	—
4000 (27.58)	11,500 (79.29)	—

Table 13.2 Compressive Strength of Masonry Based on the Compressive Strength of Concrete Masonry Units and Type of Mortar Used in Construction (TMS 602-16 Table 2)

Net area compressive strength of clay masonry, psi (MPa)	Net area compressive strength of clay masonry, psi (MPa)	
	Type M or S mortar	Type N mortar
1700 (11.72)	—	1900 (13.10)
1900 (13.10)	1900 (13.10)	2350 (14.82)
2000 (13.79)	2000 (13.79)	2650 (18.27)
2250 (15.51)	2600 (17.93)	3400 (23.44)
2500 (17.24)	3250 (22.41)	4350 (28.96)
2750 (18.96)	3900 (26.89)	—
3000 (20.69)	4500 (31.03)	—

Masonry grout is a fluid concrete that will have strength properties similar to the masonry blocks and is the material to be used within the cells of the units. Masonry grout can be used to strengthen an unreinforced wall. When reinforcement is needed to strengthen the masonry wall, grout is to be used within the areas of the wall with reinforcement. Masonry walls can be partially grouted, perhaps only at reinforced cells, which is especially useful to keep the wall self-weight to a minimum. As required by ASTM C476, the strength of masonry grout must be 2000 psi at a minimum but no less than the masonry assembly strength of f'_m.

Steel reinforcement for masonry is useful to add strength for the wall. As mentioned above, reinforcement can be added to enhance bending capacity. Reinforcement can also add strength for shear walls, and shear stirrup reinforcement can be used to add shear strength to masonry lintels. Reinforcement is also added to prevent shrinkage and temperature cracks.

13.1.2 Masonry Elements

The most common use of masonry in buildings is with masonry walls. Masonry walls can be interior or exterior walls, load-bearing or non-load-bearing walls, partition walls, shear walls, foundation walls, fire walls, and/or shaft walls. Masonry can and should be a significant element in the gravity and lateral load supporting systems.

Masonry walls have general wall area and have elements within that are designed for particular load conditions, such as jambs, vertical strip of wall on each side of the opening; lintels (beams), areas of wall directly above the opening; pilasters, thickened wall areas, or in-wall pilasters which are the same thickness as the wall but have additional grout and reinforcement added; and parapet, the wall area above the roof.

Masonry walls generally span vertically from foundations to floor and roof diaphragms. When masonry pilasters are used, masonry can be designed to span horizontally between pilasters. Figure 13.2 shows the other conditions created by openings within the wall where masonry is spanning both vertically and horizontally.

Masonry lintels/beams are the elements above openings that span the opening. Masonry lintels can be built integrally with the wall and reinforced with bottom and top reinforcement and possibly rebar stirrups added.

Masonry columns are isolated masonry elements that must be at least 8 in. on each side and have reinforcement requirements from TMS 402 of a minimum of 4 bars, tie reinforcement, and a vertical reinforcement area ratio to masonry area of 0.25 to 4 percent. There are provisions within TMS 402 for lightly loaded columns that are less than 12 ft tall and axial load of less than 2000 lb. In this case, the only requirement is a single bar within the lightly loaded column.

Masonry pilasters are similar to columns but have the added advantage of being built integrally with masonry walls and therefore have additional vertical axial and bending capacity.

13.2 MASONRY ANALYSIS

Masonry, like other structural materials, has multiple options for analysis, from simple calculation completed by hand, spreadsheets, or simple software with basic analysis and design to more complex software that is based on finite-element analysis. The simple calculation options rely on making assumptions about the masonry components and their connectivity to one another. Simple analysis can work well for isolated columns or wall elements that are simply supported at the top and bottom. For more complex masonry elements such as walls with openings, walls with more complicated support and loading conditions, walls with integrated pilasters, or wall groups, simple analyses become less accurate, and other methods, such as finite-element analysis, should be considered. Table 13.3 shows the cantilever fire wall guide.

13.2.1 Finite-Element Software

There is a full chapter on finite-element analysis in this handbook, so refer to that chapter for a more thorough review of the finite-element methods and analysis. This section will highlight important aspects for masonry analysis and design.

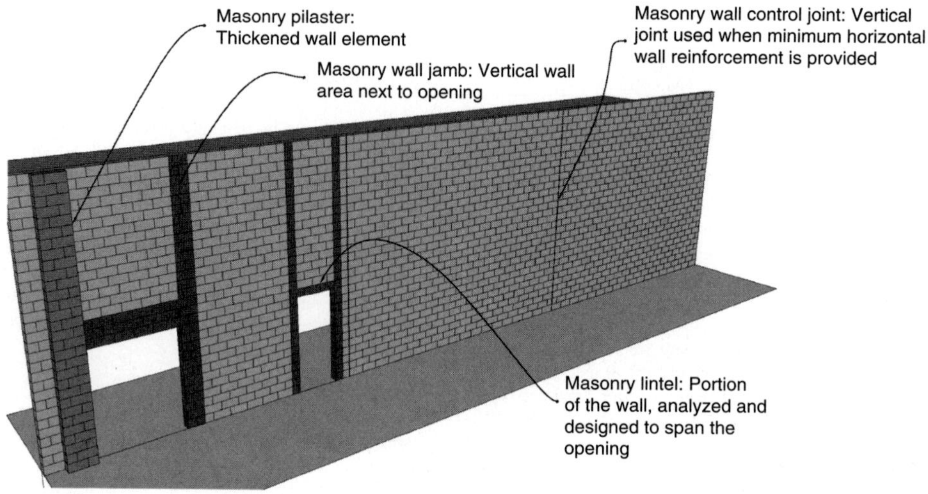

Figure 13.2 Masonry wall elements.

Table 13.3 Cantilever Fire Wall Guide

f'_m = 2500 psi, minimum wall loading			
Wall height	15 ft	20 ft	30 ft
6 in. thick	#5@16 PG	#5@8 SG	—
8 in. thick	#5@40 PG	#5@16 PG	—
12 in. thick	#5@72 PG	#6@48 PG	#7@24 SG
Notes	PG = partial grout, SG = solid grout		

13.2.2 Modeling Masonry Elements

Many of the analyses used today assume thin plate theory for the plate elements along with linear elastic behavior for the elements. The elasticity of material is described by a stress-strain curve, which shows the relation between internal force per unit area and the relative deformation. Linear elasticity is a simplification assuming linear relationships between the components of stress and strain which is valid only for stress states that do not produce yielding or fracture. Reinforced masonry and other reinforced-concrete elements have the complication of not being elastic. Therefore, once a concrete element cracks, modeled steel reinforcement is then engaged in these elements. Of course, masonry is made up of several different concrete components which closely mimic this behavior when it is reinforced. Element modification factors must be used to account for the reduced stiffness of the concrete or masonry element once it has cracked.

Masonry is unique in that it can be reinforced in the vertical and horizontal direction or only reinforced in the vertical direction but left unreinforced in the horizontal direction. In the case of checking an unreinforced element, the element may only span horizontally if it remains uncracked in that direction. If the stresses determined from analysis are within the allowed cracking stresses, users would not reduce the stiffness in that direction. This emphasizes the need for the modification factors to be considered separately in each orthogonal direction. When all the factors are equal, the wall element behaves as an isotropic material, a material having the same properties in all directions. When the factors are different from each other, the wall elements behave as an orthotropic material, a material having different properties along its three perpendicular axes.

Masonry design also requires custom material types within software to account for attributes that are unique to this material, such as grouting only reinforced cells (partial grouting). Partial grouting affects both the loading aspect (from the self-weight contribution) of the finite-element modeling as well as the stiffness of the masonry elements.

Care is involved when modeling masonry wall systems with finite-element analysis programs to ensure all the boundary conditions, stiffnesses of the elements, and weights of the elements are accurately accounted for in the development of the finite-element model.

13.3 UNREINFORCED MASONRY

Unreinforced masonry has historically been designed using allowable stress design (ASD) procedures, which will be covered in this section. Strength design (SD) of unreinforced masonry is similar to ASD since unreinforced masonry is designed assuming the masonry remains linear elastic and uncracked.

Table 13.4 summarizes the allowable stresses for unreinforced masonry. Calculated stresses are denoted by lowercase letters (such as f_a) and allowable stresses by uppercase letters (such as f_a). Slenderness effects are addressed by decreasing the allowable axial stress with increasing slenderness. TMS 402-16 Table 8.2.4.2 (reproduced as Table 13.5) gives allowable flexural tensile stresses in clay and concrete masonry. These allowable tensile stresses are applicable for both out-of-plane and in-plane loads.

Because the allowable flexural tensile stress is much smaller than the allowable flexural compressive stress, the allowable flexural tensile stress usually controls the design. The design equation is

$$-\frac{P}{A_n}+\frac{M}{S_n}\leq F_r \tag{13.1}$$

where P is the axial load (compression is positive), A_n is the net area, M is the moment, S_n is the net section modulus, and F_r is the allowable flexural tensile stresses. The net section properties, A_n and S_n, are determined based on the minimum cross-sectional area. Most masonry is laid with face shell bedding, or mortar on the face shells, so the net section properties are determined from the face shell area and any grouted cells. Tabulated values of net section properties can be found in National Concrete Masonry Association TEK 14-1B, Section Properties of Concrete Masonry Walls.

The shear stress in shear walls is calculated based on the uncracked section:

$$f_v=\frac{VQ}{I_n b} \tag{13.2}$$

Table 13.4 Allowable Stresses for Unreinforced Masonry

Stress	Equation		TMS 402-16 reference
Axial compression	$F_a = \frac{1}{4} f'_m \left[1 - \left(\frac{h/r}{140}\right)^2\right] \quad h/r \le 99$		Equation 8-13
	$F_a = \frac{1}{4} f'_m \left(\frac{70}{h/r}\right)^2 \quad h/r > 99$		Equation 8-14
Flexural compression	$F_b = \frac{1}{3} f'_m$		Equation 8-15
Axial tension	Not permitted		Section 8.2.5
Flexural tension			Table 8.2.4.2
Axial compression and flexure	$\frac{f_a}{F_a} + \frac{f_b}{F_b} \le 1$		Equation 8-11
	$P \le \frac{1}{4} P_e \quad P_e = \frac{\pi^2 E_m I_n}{h^2}\left(1 - \frac{0.577e}{r}\right)$		Equation 8-12 Equation 8-16
Shear (minimum of values)	$F_v = 1.5\sqrt{f'_m} \le 120$ psi	Running bond, not fully grouted	Section 8.2.6.2
	$F_v = 37\text{psi} + 0.45\frac{N_v}{A_n}$	Running bond, fully grouted, or;	
	$F_v = 60\text{ psi} + 0.45\frac{N_v}{A_n}$	Other than running bond, full grouted, open end units	
	15 psi	Other than running bond, full grouted, not open end units	

Table 13.5 Allowable Flexural Tensile Stresses for Clay and Concrete Masonry, psi (kPa) (TMS 402-16 Table 8.2.4.2)

	Mortar types			
	Portland cement/lime or mortar cement		Masonry cement and air-entrained portland cement/lime	
Direction of flexural tensile stress and masonry type	M or S	N	M or S	N
Normal to bed joints				
Solid units	53 (366)	40 (276)	32 (221)	20 (138)
Hollow units†				
Ungrouted	33 (228)	25 (172)	20 (138)	12 (83)
Fully grouted	65 (448)	63 (434)	61(420)	58 (400)
Parallel to bed joints in running bond				
Solid units	106 (731)	80 (552)	64 (441)	40 (276)
Hollow units				
Ungrouted and partially grouted	66 (455)	50 (345)	40 (276)	25(172)
Fully grouted	106 (731)	80 (552)	64 (441)	40 (276)
Parallel to bed joints in masonry not laid in running bond				
Continuous grout section parallel to bed joints	133 (917)	133 (917)	133 (917)	133 (917)
Other	0 (0)	0 (0)	0 (0)	0 (0)

†For partially grouted masonry, allowable stresses shall be determined on the basis of linear interpolation between fully grouted hollow units and ungrouted hollow units based on amount (percentage) of grouting.

where Q, I_n, and b are calculated based on the uncracked net cross section of the wall. For a rectangular cross section, such as a solidly grouted wall or a wall with just face shell bedding, the maximum shear stress is

$$f_v = \frac{3}{2}\frac{V}{A_n} \tag{13.3}$$

Table 13.4 shows the allowable stresses for unreinforced masonry.

Table 13.5 shows the allowable flexural tensile stresses for clay and concrete masonry.

13.4 REINFORCED MASONRY

Reinforced masonry has traditionally been designed using the ASD method. SD was added in the 2002 TMS 402 code and has gradually been gaining in popularity. SD of masonry is similar to SD of reinforced concrete and provides advantages over ASD when the allowable masonry stress controls or for shear walls with distributed reinforcement.

13.4.1 Allowable Stress Design

Table 13.6 summarizes the allowable stresses for reinforced masonry. TMS 402 codes before the 2011 edition permitted a 1/3 increase in allowable stresses for loading combinations including wind or earthquake. The allowable stresses were recalibrated in the 2011 TMS 402 code and the 1/3 stress increase was eliminated.

The analysis of flexural members, such as beams, lintels, and nonbearing walls using ASD, is shown in Fig. 13.3. The location of the neutral axis, *kd*, is determined, the stresses in the masonry and reinforcement are determined, and the stresses are compared to the allowable stresses. There are several design methods. One common design method is to assume a value of $j \approx 0.9$, determine a trial value of $A_{s,\text{reqd}} = M/(jdF_s)$, and check the design using the flowchart of Fig. 13.3. Nonbearing walls are usually designed on a per foot of wall length basis, and often only the cells that have reinforcement are grouted. This partially grouted construction saves on construction costs and decreases the wall weight and hence the seismic force.

A more rigorous design procedure which can be used for both flexural members and combinations of flexural and axial load with a single layer of reinforcement is shown in Fig. 13.5. If the allowable steel stress controls the design, iteration is required, but there is rapid convergence. The TMS 402 expressions for F_b and E_m are both proportional to f'_m, $k_b = 0.312$ for concrete masonry and $k_b = 0.368$ for clay masonry. This design method provides the exact required area of reinforcement. If the allowable masonry stress is controlling the design, the reinforcement is not being used efficiently. The designer should consider other alternatives, such as increasing the member size or increasing f'_m.

The flowchart shown in Fig. 13.4 can be modified for shear walls with distributed reinforcement, as shown in Fig. 13.5. The design procedure results in the required area of steel per length of wall (e.g., in.²/ft), $A^*_{s,\text{reqd}}$. The design procedure assumes uniformly distributed reinforcement and tends to overestimate the required area of steel by about 10 to 15 percent. For other cases, an interaction diagram would need to be constructed.

13.4.2 Strength Design

Table 13.7 summarizes the nominal strengths and design assumptions for reinforced masonry. A strength-reduction factor of $\phi = 0.9$ is used for combinations of flexural and axial load and $\phi = 0.8$ for shear. There is no transition region for combinations of flexural and axial load, and ϕ is a

Table 13.6 Allowable Stresses for Reinforced Masonry

Stress	Equation	TMS 402 reference
Axial compression	$P_a = (0.25 f'_m A_n + 0.65 A_{st}F_s)\left[1-\left(\frac{h/r}{140}\right)^2\right]$ $h/r \le 99$	Equation (8-18)
	$P_a = (0.25 f'_m A_n + 0.65 A_{st}F_s)\left(\frac{70}{h/r}\right)^2$ $h/r > 99$	Equation (8-19)
	$A_{st} = 0$ unless the reinforcement in compression is tied in compliance with TMS 402 Section 5.3.1.4	
Flexural compression	$F_b = 0.45 f'_m$	Section 8.3.4.2.2
Flexural tension	F_s = 20,000 psi (Grade 40 or 50 reinforcement) F_s = 32,000 psi (Grade 60 reinforcement) F_s = 30,000 psi (wire joint reinforcement)	Section 8.3.3.1 Section 8.3.3.2
Shear	$F_v = (F_{vm} + F_{vs})\gamma_g$	Equation (8-22)
	$F_{vm} = \frac{1}{2}\left[\left(4.0-1.75\left(\frac{M}{Vd_v}\right)\right)\sqrt{f'_m}\right]+0.25\frac{P}{A_n}$	Equation (8-26)
	$F_{vs} = 0.5\left(\frac{A_v F_s d_v}{A_{nv} s}\right)$	Equation (8-27)
	Special shear walls:	
	$F_{vm} = \frac{1}{4}\left[\left(4.0-1.75\left(\frac{M}{Vd_v}\right)\right)\sqrt{f'_m}\right]+0.25\frac{P}{A_n}$	Equation (8-25)
	$F_v \le \begin{cases} \left(3\sqrt{f'_m}\right)\gamma_g & M/(Vd_v) \le 0.25 \\ \left(2\sqrt{f'_m}\right)\gamma_g & M/(Vd_v) \ge 1.0 \\ \left(\frac{2}{3}\left(5-2\frac{M}{Vd_v}\right)\sqrt{f'_m}\right)\gamma_g & 0.25 < M/(Vd_v) < 1.0 \end{cases}$	Equation (8-23) Equation (8-24) Linear interpolation Section 8.3.5.1.2(c)
	Take M/Vd_v as a positive number, need not be > 1.0 $\gamma_g = 0.75$ for partially grouted shear walls and 1.0 otherwise.	
Shear friction	$M/(Vd_v) \le 0.5$ $F_f = \frac{\mu(A_{sp}F_s + P)}{A_{nv}}$	Equation (8-28)
	$0.5 < M/(Vd_v) < 1.0$ Linear Interpolation	Section 8.3.6
	$M/(Vd_v) \ge 1.0$ $F_f = \frac{0.65(0.6\,A_{sp}F_s + P)}{A_{nv}}$	Equation (8-29)
	$\mu = 1.0$ for masonry on concrete with unfinished surface, or concrete with a surface that has been intentionally roughened $\mu = 0.70$ for all other conditions	

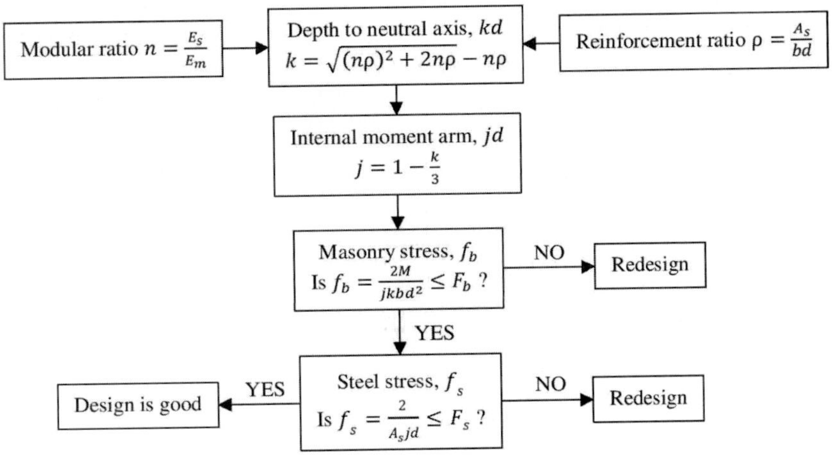

Figure 13.3 Flowchart for ASD analysis of flexural members.

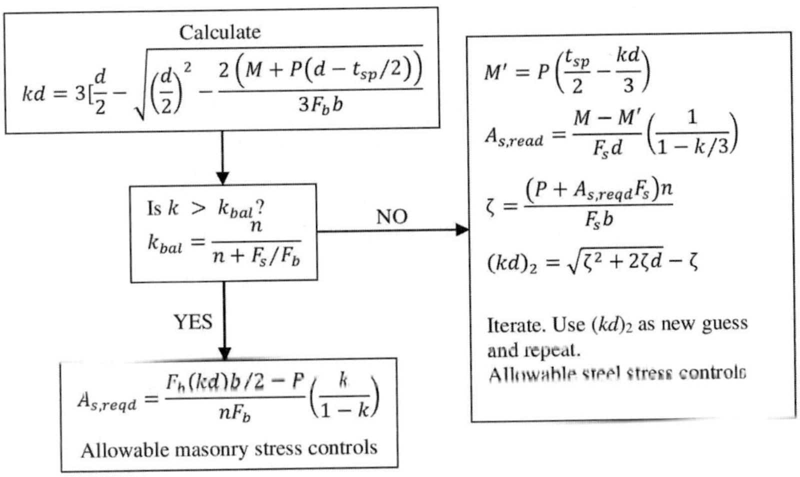

Figure 13.4 Flowchart for design of singly reinforced members using allowable stress design.

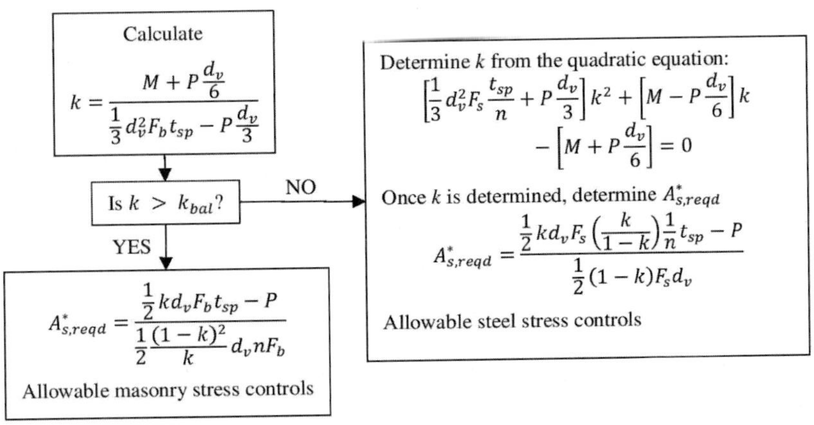

Figure 13.5 Flowchart for design of distributed reinforced members using allowable stress design.

Table 13.7 Nominal Strengths for Reinforced Masonry

Load	Equation	TMS 402 reference
Axial compression	$P_n = 0.80[0.80f'_m(A_n - A_{st}) + f_yA_{st}]\left[1-\left(\frac{h/r}{140}\right)^2\right]$ $h/r \le 99$	Equation (9-15)
	$P_n = 0.80[0.80f'_m(A_n - A_{st}) + f_yA_{st}]\left(\frac{70}{h/r}\right)^2$ $h/r > 99$	Equation (9-16)
	$A_{st} = 0$ unless the reinforcement in compression is tied in compliance with TMS 402 Section 5.3.1.4	
Combined bending and axial load	Determined using design assumptions of TMS 402 9.3.2 • maximum usable strain, ε_{mu}, is 0.0035 for clay masonry and 0.0025 for concrete masonry • equivalent rectangular stress block with a stress of $0.80 f'_m$ acting over a distance $a = 0.80c$.	Section 9.3.4.1.1
Shear	$V_n = (V_{nm} + V_{ns})\gamma_g$	Equation (9-17)
	$V_{nm} = \left[\left(4.0-1.75\left(\frac{M_u}{V_ud_v}\right)\right)A_{nv}\sqrt{f'_m}\right]+0.25P_u$	Equation (9-20)
	$V_{ns} = 0.5\left(\frac{A_v}{s}\right)f_yd_v$	Equation (9-21)
	$V_n \le (6A_{nv}\sqrt{f'_m})\gamma_g$ $M_u/(V_ud_v) \le 0.25$	Equation (9-18)
	$V_n \le (3A_{nv}\sqrt{f'_m})\gamma_g$ $M_u/(V_ud_v) \ge 1.0$	Equation (9-19)
	$V_n \le \left(\frac{4}{3}\left(5-2\frac{M_u}{V_ud_v}\right)A_{nv}\sqrt{f'_m}\right)\gamma_g$ $0.25 < M_u/(V_ud_v) < 1.0$	Linear interpolation Section 9.3.4.1.2 (c)
	Take $M_u/(V_ud_v)$ as a positive number, need not be > 1.0 $\gamma_g = 0.75$ for partially grouted shear walls and 1.0 otherwise.	
Shear friction	$M_u/(V_ud_v) \le 0.5$ $V_{nf} = \mu(A_{sp}f_y + P_u)$	Equation (9-33)
	$0.5 < M_u/(V_ud_v) < 1.0$ Linear interpolation	Section 9.3.6.5
	$M_u/(V_ud_v) \ge 1.0$ $V_{nf} = 0.42f'_mA_{nc}$	Equation (9-34)
	$\mu = 1.0$ for masonry on concrete with unfinished surface, or concrete with a surface that has been intentionally roughened $\mu = 0.70$ for all other conditions	

constant 0.9. This is due to maximum reinforcement requirements that limit the amount of reinforcement in masonry members.

The design of flexural members, such as beams, lintels, and nonbearing walls, can be performed using the flowchart shown in Fig. 13.6. Figure 13.6 gives a general design method that includes an axial load; for flexural members, $P_u = 0$. For Grade 60 steel, the neutral axis depth for balanced conditions (masonry crushes simultaneously with reinforcement yielding) is $c_{bal} = 0.547\ d$ for CMU and $c_{bal} = 0.628\ d$ for clay masonry. The maximum reinforcement is based on a strain condition of ε_{mu} in the masonry and $1.5\varepsilon_y$ in the reinforcement. This corresponds to a reinforcement ratio, ρ, for Grade 60 steel of 0.00952 for CMU and 0.0113 for clay masonry. For beams, there is also a minimum reinforcement requirement that the nominal moment, M_n, be greater than 1.3 M_{cr} or the reinforcement be greater than 4/3 of the required reinforcement.

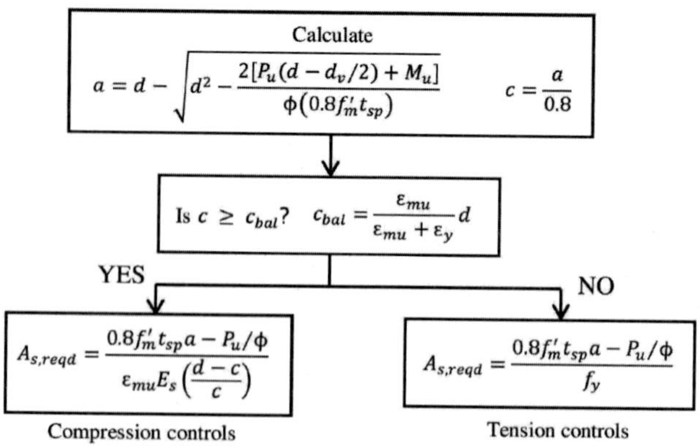

Figure 13.6 Flowchart for design of singly reinforced members using strength design.

The design of bearing walls using the SD method requires the inclusion of second-order, or $P-\delta$, effects. This can be done using either a second-order analysis, a moment magnification procedure, or what is known in masonry as the slender wall procedure. Only the slender wall procedure is covered here. A flowchart is shown in Fig. 13.7 for the design of bearing walls using the slender wall procedure. The procedure requires an initial guess at reinforcement size and spacing. A reasonable guess can often be obtained based on ignoring the axial load and second-order effects or estimating the reinforcement for a nonbearing wall and using that as the initial trial for a bearing wall.

The design of shear walls generally requires the construction of an interaction diagram to check combined flexure and axial loads. For shear walls with distributed reinforcement, an approximate amount of distributed reinforcement, $A^*_{s,reqd}$, can be determined using the flowchart in Fig. 13.7 by setting $d = 0.9d_v$ and estimating $A^*_{s,reqd} = A_{s,reqd}/0.65d_v$. The design of shear walls is often controlled by seismic requirements, including maximum reinforcement and shear capacity design for special shear walls. These topics are coved in the next section.

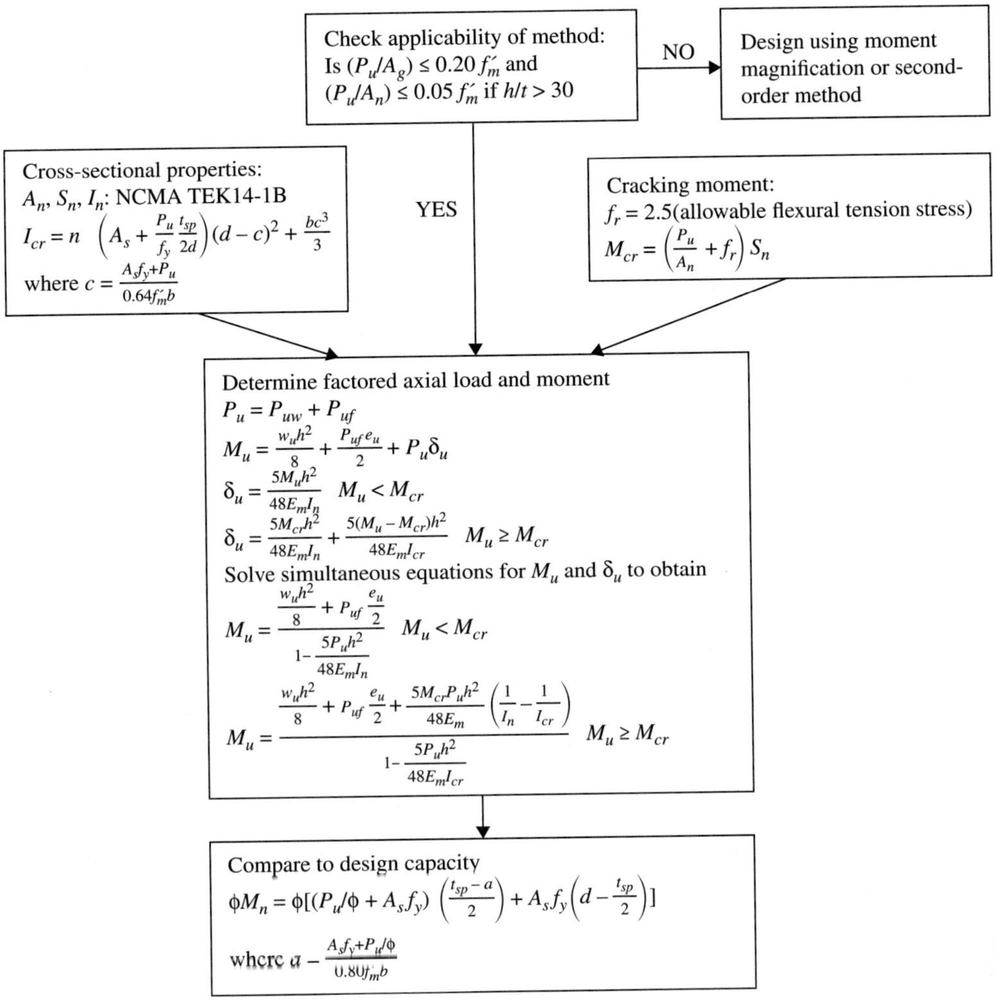

Figure 13.7 Flowchart for design of reinforced bearing walls using strength design.

13.5 SEISMIC DESIGN

There are five types of masonry shear walls: ordinary plain, detailed plain, ordinary reinforced, intermediate reinforced, and special reinforced. Prescriptive reinforcement requirements for each of the types of masonry shear walls are given in Table 13.8.

In addition to the requirements listed in Table 13.8, special reinforced masonry shear walls have shear capacity design requirements. In ASD, the calculated shear stress, f_v, must be increased by 1.5. In SD, the design shear strength, ϕV_n, is required to exceed the shear corresponding to the development of 1.25 times the nominal flexural strength, M_n, except that the nominal shear strength V_n need not exceed 2.5 times the required shear strength, V_u (which is equivalent to the design shear strength, ϕV_n, not exceeding $2.0V_u$). These requirements are so that the wall will fail in a ductile flexural manner and not in a brittle shear manner.

TMS 402 has requirements and limitations based on the Seismic Design Category (SDC). These requirements are given in Table 13.9. The requirements are cumulative. For example, structures in SDC D must meet all the requirements of SDC A, B, and C plus the additional requirements of SDC D.

Finally, reinforced masonry shear walls have maximum reinforcement requirements so that the reinforcement will yield before the masonry crushes. Special reinforced walls designed with ASD having $\frac{M}{Vd_v} \ge 1$ and $P > 0.05 f'_m A_n$ shall have a reinforcement ratio not greater than

$$\rho_{max} = \frac{n f'_m}{2 f_y \left(n + \frac{f_y}{f'_m}\right)} \tag{13.4}$$

The maximum reinforcement requirements for SD are given in terms of a flowchart in Fig. 13.8. It is noted that either the area of reinforcement is limited or the boundary elements need to be provided. There has been insufficient research to develop prescriptive requirements for boundary elements in masonry. Thus, the TMS 402 code only gives performance requirements for boundary elements, and the designer would need to determine how to meet those performance requirements, most likely through testing.

Often the easiest way to check the maximum reinforcement requirements of TMS 402 Section 9.3.3.2 is to determine the location of the

Table 13.8 Prescriptive Seismic Design Requirements for Masonry Shear Walls

Shear wall type	Prescriptive requirements
Ordinary plain	None
Detailed plain Ordinary reinforced	Vertical reinforcement of at least 0.20 in.^{2} at: • corners • within 16 in. of each side of openings • with 8 in. of each side of movement joints • within 8 in. of ends of walls • maximum spacing of 120 in. Horizontal reinforcement of one of the following: • two longitudinal WI.7 wire joint reinforcement spaced at not more than 16 in. on center • at least 0.20 in.^{2} of bond beam reinforcement with a maximum spacing of 120 in. Horizontal reinforcement at: • top and bottom of wall openings, extending the greater of 24 in. and 40 bar diameters • structurally connected roof and floor levels • within 16 in. of top of wall
Intermediate reinforced	Same as ordinary reinforced except maximum spacing of vertical reinforcement is 48 in.
Special reinforced	Same as intermediate reinforced walls with the additional requirements of: Maximum spacing of reinforcement the lesser of: • 1/3 length of shear wall • 1/3 height of shear wall • 48 in. for running bond; 24 in. for other than running bond Minimum cross-sectional area of vertical reinforcement one-third of the required shear reinforcement Reinforcement ratio of vertical bars, $\rho_v \geq 0.0007$ Reinforcement ratio of horizontal bars, $\rho_h \geq 0.0007$ for running bond and $\rho_h \geq 0.0015$ for other than running bond Total reinforcement ratio, $\rho_v + \rho_h \geq 0.002$ Shear reinforcement anchored around vertical bars with a standard hook Mechanical splices in flexural reinforcement in plastic hinge zones shall develop the specified tensile strength of the spliced bar

Table 13.9 Seismic Design Requirements by Seismic Design Category

Seismic design category	Seismic design requirements
A	None
B	Shear wall must be at least detailed plain shear wall
C	Shear wall must be at least ordinary reinforced shear wall Along each line of lateral resistance at each story, at least 80 percent of the lateral stiffness shall be provided by seismic-force-resisting walls Nonparticipating elements reinforced in either horizontal OR vertical direction • Horizontal: W1.7 joint reinforcement at maximum of 16 in. or #4 bars in bond beams at maximum of 48 in. • Vertical: #4 bars at maximum of 120 in. and within 16 in. of end of wall
D	Shear wall must be special reinforced shear wall Nonparticipating elements reinforced in either horizontal OR vertical direction, with maximum spacing of vertical reinforcement of 48 in. Material requirements • Fully grouted: type S or M mortar; masonry cement, mortar cement, or cement-lime mortar • Partially grouted: type S or M mortar; mortar cement or cement-lime mortar

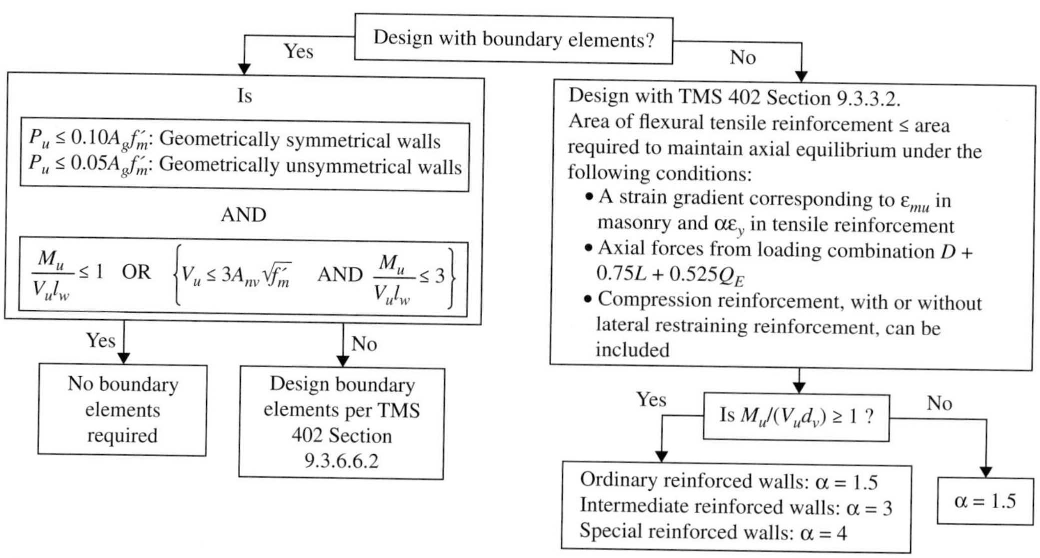

Figure 13.8 Flowchart for maximum reinforcement requirements.

neutral axis for the given strain conditions and, using that location of the neutral axis, calculate the nominal axial force. This calculated axial force is compared to the axial load combination of TMS 402 Section 9.3.3.2. The ratio of the depth to the neutral axis, *c*, to the distance to the tension reinforcement, *d*, is given in Table 13.10 for various strain gradients and types of masonry.

Table 13.10 Neutral Axis Location for Various Strain Gradients

Tensile strain	c/d, CMU	c/d, Clay
$1.5\varepsilon y$	0.446	0.530
$3\varepsilon y$	0.287	0.360
$4\varepsilon y$	0.232	0.297

13.6 SPECIFYING MASONRY AND TESTING MASONRY

For economical use of materials, knowledge about the actual strength of materials used will benefit the design by maximizing the known capacity. Therefore, it is not enough to simply indicate a minimum value of f'_m on the design documents.

13.6.1 Specifying Materials

In TMS 602, Table 2 (section 1.4B.2.b) shows the type of mortar and unit strength that can be used to find the assembly compressive strength. It is dependent on the type of mortar—not the mortar strength—as the requirements for the mortar strength are set once you select the mortar type per ASTM C270 specification. TMS 602 also shows the testing that was done (Figures SC-1 and SC-2) which support the assembly strength shown in the table.

When using the tables from TMS 602, remember it is strictly mortar type—not the mortar strength—that you require. Therefore, the two components of the specification are CMU block strength and mortar type.

Also, when specifying masonry components, do not forget about grout strength. TMS requires grout compressive strength to be equal to or exceed f'_m but not less than 2000 psi. So if the goal is to have f'_m of 2500 psi, then the grout strength must be at least 2500 psi or greater. Some in the industry suggest using grout strength comparable to the CMU block strength. The strength of grout is left to engineering judgment within the range between: f'_m or 2000 psi (low end value) through CMU block strength (high end value).

Thus, simply indicating "f'_m = 2800 psi" on the design documents is not enough for masonry. We need to indicate the required block strength, mortar type, and grout strength. On the following pages are samples of material strength specifications that can be used in structural notes.

EXAMPLE 1 (RECOMMENDED):

Masonry notes	Minimum required strength
CMU, ASTM C90	f'_{cmu} = 3250 psi (net area compressive strength)
Mortar, ASTM C270	Type S
Grout, ASTM C476	f'_g > 2500 psi
Masonry assembly	f'_m = 2500 psi (net area compressive strength)

Note: In this sample, f'_g was indicated to be 2500 psi, but could have been higher (up to 5000 psi) if using SD methods. However, the additional grout strength does not create a higher design strength (f'_m is still 2800 psi).

EXAMPLE 2:

Masonry notes	Minimum required strength
CMU, ASTM C90	f'_{cmu} = 4350 psi (net area compressive strength)
Mortar, ASTM C270	Type N (above grade) Type S (below grade)
Grout, ASTM C476	f'_g > 2500 psi
Masonry assembly	f'_m = 2500 psi (net area compressive strength)

Note: Sample 1 is recommended; however, type N mortar could be used if block strengths are higher, and the assembly will still achieve the same f'_m as Example 1.

13.6.2 Testing Materials

In TMS 602, Table 2 requires CMU to be tested and reported for the unit compressive strength. Another method for determining the masonry assembly strength is by masonry prism testing. With prism testing based on ASTM C1314, Standard Test Method for Compressive Strength of Masonry Prisms, three prisms are constructed using blocks and mortar that will be used for construction. Once the prisms are built, the average strength of the three specimens will be the design strength, f'_m.

There are two mortar tests. ASTM C270 is a mortar specification which outlines procedures for testing mortar prepared in the laboratory. The test determines whether a mortar conforms with physical property requirements. ASTM C780 is a field test method which defines standard procedures for sampling and testing mortars before and during their use in construction. Test results of masonry mortar are not required to meet the minimum masonry compressive strength values, as mortar is tested alone and such tests are not analyzing the mortar within the assembly.

Tests for masonry grout are covered in ASTM C1019. The ASTM defines tests for establishing material proportions for compliance with project specifications. For quality assurance of grout, field tests for grout are performed to monitor uniformity of grout during construction for quality assurance.

13.7 MASONRY FIRE WALLS

A fire wall is a wall structure with the purpose of preventing a fire from spreading to adjacent areas. A noncombustible material like masonry is often specified for fire walls. They are typically located at a lot line or where different occupancies need to be isolated within a single building. Based on NCMA TEK 5-8B, one of the primary options for fire walls are freestanding walls that permit the collapse of the structure on either side of the wall without collapsing the wall. An example is a 20-ft-tall single-story wall. It is fixed at the base and is not supported anywhere along its height. This wall can be a 6-in. block with #5 bars centered at 8 in. If an 8-in. block is used, #5 bars at 16 in. are sufficient. Taking the design a step further to a 30-ft-tall single-story or multistory design, this wall can be 12-in. block with #7 bars centered at 24 in.

13.7.1 Combine Analysis Results for Design

The next challenge involves taking the results from the finite-element model and analysis and converting them into information that can be compared to code-defined maximum stresses or forces that determine the capacity of the masonry wall. Finite-element programs for masonry combine the results of several plate/shell elements within geometric areas or strips of the model as defined by the user. Areas above openings are rationalized into an area that will be checked against lintel capacities. Engineers must study software programs and their combination (summation) of finite-element results and make modifications when necessary.

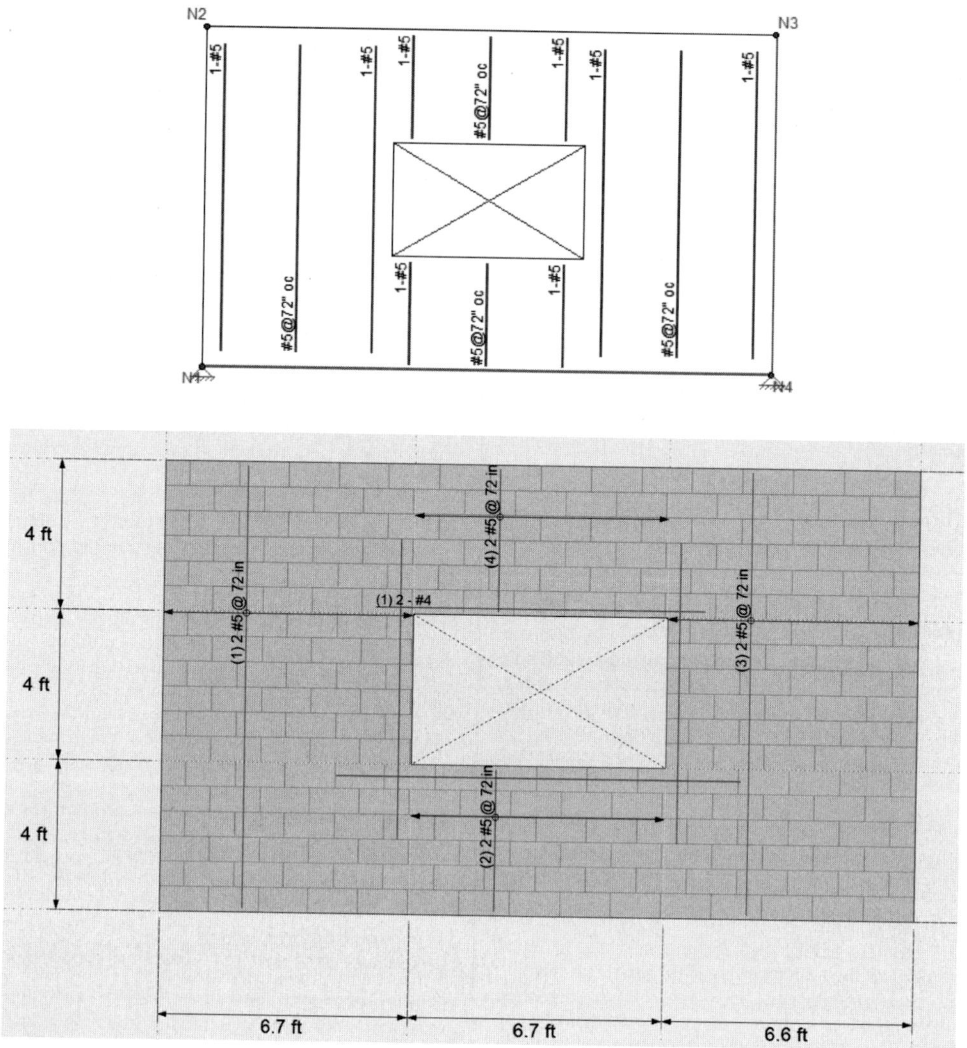

Figure 13.9 Masonry wall elevations: image from RISA 3D (top) and image from RAM elements (bottom).

Generally, structural engineering software will check for in-plane bending and shear capacity, out-of-plane bending and shear capacity, and axial capacity of masonry walls. Lintel shear and bending capacities will also be evaluated. Lintels that are analyzed without a finite-element model have traditionally been checked by assuming a simply supported "beam" element. Finite-element approximation and design of the area above the openings is more accurate to the true behavior and fundamentally different, as the plates/shells in this area are interlocked by sharing nodes with the other surrounding elements of the wall (Fig. 13.9).

REFERENCE

1. National Concrete Masonry Association (NCMA) TEK 14-1B, *Section Properties of Concrete Masonry Walls*, 2007, http://ncma-br.org/e-tek-nbs.asp.

Chapter 14

Timber Structures

BY

THOMAS WILLIAMSON, PE *Retired Vice President of Quality and Technical Services, APA–The Engineered Wood Association*

MUSTAFA MAHAMID, Ph.D., SE, PE, P.Eng. *Department of Civil and Materials Engineering, University of Illinois at Chicago*

14.1 WOOD AS AN ENGINEERING MATERIAL

Wood is permitted to be used in Types III, IV, and V construction in accordance with the International Building Code (IBC). Depending on the building occupancy and the type of construction, wood buildings can have unlimited areas and heights of up to 85 ft, meaning they can be used for a broad range of building applications. But wood is unique in many ways compared to steel and concrete, and design professionals need to be aware of these unique design characteristics. This chapter will attempt to define these characteristics to assist designers in making intelligent design decisions.

14.1.1 Orthotropic Nature of Wood

Relationships between load, duration of load, deformation, and material strength are usually based on the assumption that the structural material is homogeneous and isotropic, in both elastic constants and strength properties. Wood is neither because of its cellular structure and growth characteristics. Wood may be described as an orthotropic material; that is, it has unique and independent mechanical properties in the directions of three mutually perpendicular axes: longitudinal, radial, and tangential. The longitudinal axis L is parallel to the fiber (grain), the radial axis R is normal to the growth rings (perpendicular to the grain in the radial direction), and the tangential axis T is perpendicular to the grain but tangent to the growth rings. The designer is referred to the Wood Handbook for detailed information on all aspects of wood addressed in Sections 14.1.1, 14.1.2, 14.1.3, and 14.1.4.

14.1.2 Directional Strength Properties

Mechanical properties most commonly measured and represented as "strength properties" for design include modulus of rupture (MOR) in bending, maximum stress in compression and tension parallel to grain, and compressive stress perpendicular to grain and shear strength parallel to grain. As with most materials, there is inherent variability in the strength of small clear wood samples under short-time loading, such as during structural testing. Added to this variability are the effects of duration of load and strength-reducing factors, such as knots, slope of grain, and density variations in structural sizes of wood. In addition, wood exhibits directional properties when subjected to various stress states. The strength properties to consider are associated with normal and shear stresses parallel to the grain, perpendicular to the grain radially, and perpendicular to the grain tangentially. The difference in strength properties in the radial and tangential directions is seldom of significance in design. Thus, it is necessary only to differentiate between directions normal and parallel to the grain.

MOR. MOR reflects the maximum load-carrying capacity of a member in bending and is proportional to maximum moment borne by the specimen. MOR is an accepted criterion of strength and is typically reduced to a design bending stress by the application of factors such as load duration and a factor of safety.

Tension Parallel to Grain. Wood exhibits its maximum tensile strength parallel to the grain. Tension parallel to grain stresses are approximately 40 percent of the compression parallel to grain values for sawn lumber depending on species and grade.

Tension Perpendicular to Grain. Resistance of wood to forces acting across the grain tend to split a member. Wood exhibits relatively low tensile strength perpendicular to the grain, and in the presence of defects, such as checks, it is substantially reduced. This strength usually does not enter into design, and there are no published design values for this stress. Abrupt changes in cross section may cause stress concentrations with tension perpendicular to the grain which should be considered in determining load capacity.

Compression Parallel to Grain. Wood is strongest in compression parallel to grain and usually fails under uniaxial compressive stresses by buckling of the fibers. Maximum stress is typically obtained for a compression parallel-to-grain specimen having a ratio of length to least dimension of less than 11.

Compression Perpendicular to Grain. The cells of the wood, being essentially hollow, exhibit relatively low stiffness and strength when stressed in compression perpendicular to the grain. Reported as stress at proportional limit, there is no clearly defined ultimate stress for this property. Compression perpendicular to grain values are approximately 1/3 of the compression parallel to grain stresses for sawn lumber depending on the species and grade.

Compression at Angle to Grain. Since compression values are given for both parallel and perpendicular to grain, the allowable compression stress at any angle to the grain may be determined by the Hankinson formula as given below:

$$F_\theta = \frac{F_c F_{c\perp}}{F_c \sin^2\theta + F_{c\perp} \cos^2\theta} \tag{14.1}$$

where F_c = allowable compressive stress parallel to grain
$F_{c\perp}$ = allowable compressive stress perpendicular to grain
θ = inclination of F_θ to grain

Shear Strength Parallel to Grain. Ability to resist internal slipping of one part upon another along the grain. Significant shear stresses may be developed parallel to the grain in beams. Checks, splits, and shakes, being fractures parallel to the grain, significantly reduce the shear strength in this direction. These flaws are typically associated with larger specimens.

14.1.3 Reference Design Values

Reference design values are provided in a number of ways. For sawn lumber and glued laminated timber (glulam), reference design values are provided in the Supplement to the National Design Specification (NDS). For round timber poles and piles, reference design values are provided in ASTM D3200 and ASTM D25, respectively. For proprietary products, such as prefabricated wood I-joists and structural composite lumber (SCL), reference design values are provided in manufacturers' code reports. Regardless of the product, reference design values must be adjusted by all applicable adjustment factors, including duration of load, moisture effects, temperature effects, and others. Some of these are discussed as follows.

14.1.4 Factors Affecting Reference Design Values

Growth Characteristics. The effect of knots is to decrease strength because their grain and the surrounding grain may be at a large angle to the maximum tensile or compressive stress. The effects in tension are more detrimental than in shear or compression. Thus, for beams it is advantageous to locate the higher-grade sections of the lumber in the maximum-tensile-stress zones.

As with many other materials, wood increases its strength properties with an increase in specific gravity, and relationships have been suggested for strength and specific gravity. Density is usually recognized in the stress grading rules for lumber and other engineered wood products.

Other factors, such as sloping grain, decay, insect attack, pitch pockets, shakes, wane, compression wood, and splits, can affect the strength properties of wood. These are considered in the detailed instructions for stress grading of lumber published by various organizations.

Environmental Conditions. As the moisture content of wood drops below 30 percent (the approximate fiber-saturation point), its strength properties increase.

For sawn lumber, reference design values are published for a moisture content of 19 percent or less. For moisture contents above 19 percent, the reference design values must be multiplied by the applicable wet service factor, C_M. For glulam timber, prefabricate wood I-joists, and SCL, the wet service factor is applied when the in-service moisture content exceeds 16 percent.

If wood is kept either continuously dry or continuously wet, decay does not occur. Moisture and temperature are the prime factors affecting decay rate. Wood should not be in direct contact with masonry or concrete where excessive moisture will be transferred to the wood. Ventilated air spaces around untreated members or pressure treatment with preservatives retard or prevent decay.

Temperature Effects. The strength properties of wood at 12 percent moisture content, within a range of 100°F to 150°F, may be expected to decrease by 1/3 to 1/2 percent per 1°F. Temperatures above 150°F leave permanent detrimental effects. Under roofs, temperatures in wood elements may reach 100°F to 140°F in warm months of the year for certain geographic regions. Table 14.1 provides adjustments for temperature. Wet conditions are assumed for wood moisture content greater than 19 percent for lumber and 16 percent for other engineered wood products, such as glulam timber.

Time-Load Effects. The strength properties of wood are affected by the duration of loading as shown in Fig. 14.1, and this is accounted for

Table 14.1 Temperature Factor, C_t

Design values	In-service moisture conditions	C_t		
		$T \le 100°F$	$100 < T \le 125$	$125 < T \le 150$
F_t, E	Dry or wet	1.0	0.9	0.9
F_b, F_v, F_c	Dry	1.0	0.8	0.7
and $F_c\perp$	Wet	1.0	0.7	0.5

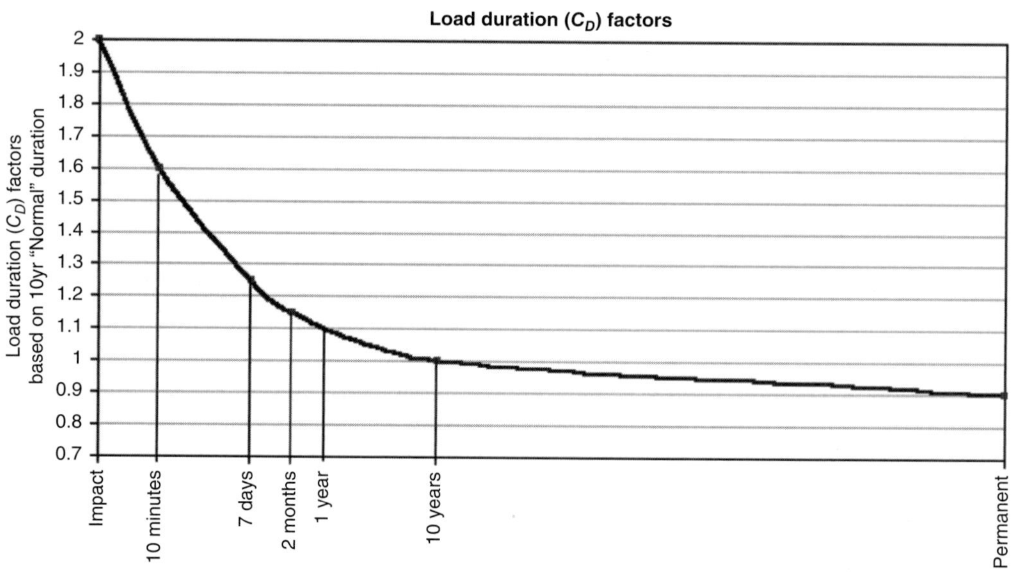

Figure 14.1 Load duration curve for wood.

in design by applying a duration of load factor, C_D. The usual durations of load considered in design and their respective duration of load factors are given in Table 14.2 for allowable stress design (ASD).

Table 14.2 Frequently Used Load Duration Factor, C_D

Load duration	C_D	Typical design loads
Permanent	0.9	Dead load
Ten years	1.0	Occupancy live load
Two months	1.15	Snow load
Seven days	1.25	Construction load
Ten minutes	1.6[(b)]	Wind/Earthquake load
Impact	2.0	Impact load

For load and resistance factor design (LRFD), a time effect factor is used to adjust for different duration of loads as shown in Table 14.3.

Table 14.3 Time Effect Factor, λ (LRFD only)

Load combination[1]	Time effect factor (λ)
$1.4D$	0.6
$1.2D + 1.6L + 0.5(L_r \text{ or } S \text{ or } R)$	0.7 when L is from storage 0.8 when L is from occupancy 1.25 when L is from impact
$1.2D + 1.6(L_r \text{ or } S \text{ or } R) + (L \text{ or } 0.5W)$	0.8
$1.2D + 1.6W + L + 0.5(L_r \text{ or } S \text{ or } R)$	1.0
$1.2D + 1.0E + L + 0.2S$	1.0
$0.9D + 1.0W$	1.0
$0.9D + 1.0E$	1.0

[1]Load combination and load factors consistent with ASCE 7 are listed for ease of reference.

14.2 ALLOWABLE STRESS DESIGN (ASD) VS. LOAD AND RESISTANCE FACTOR DESIGN (LRFD)

All structural design is based on the basic principle that the system demand has to be less than the system capacity. Most wood structures are designed using ASD which involves designing members and connections such that calculated stresses do not exceed specified material stresses when subjected to appropriate load combinations. This is often referred to as working stress design.

LRFD was introduced in 1995 by ASCE 16 *Standard for LRFD for Engineered Wood Construction* and subsequently adopted in the 2005 National Design Specification. LRFD involves designing members and connections using load and resistance factors such that no applicable limit state is reached when subjected to appropriate load combinations. This is often referred to as Ultimate Strength Design or Limit State Design.

14.2.1 Demand

Looking at the demand side of the equation, since the laws of physics haven't changed the moment, shear and deflection equations remain the same. The part that differs is what design loads are in put into those loading equations. Start out by estimating what actual loads are being applied to the system. Then plug these estimated real-world loads into the required load combinations. The load combinations for the two methods is where a distinction between the two methods is provided on the demand side. Designers may tend to overestimate the loads on the system when using the ASD combinations in order to ensure a greater safety margin. LRFD is considered to be a better representation of the actual system loading so overestimation is unnecessary, as the load factors in LRFD are intended to provide a consistent level of reliability across a range of ratios of the various load types. The load combinations for both ASD and LRFD are given in the applicable building code and are based on ASCE 7.

14.2.2 Capacity

On the capacity side, with LRFD, as with ASD, the member is assumed to have a certain strength capacity as given by the design reference values previously discussed. The starting design reference values for LRFD and ASD are the same and refer to the mean material strengths determined by materials testing. ASD is based on a design stress at the lower 5th percentile, 10-year stress. LRFD is based on a reference strength at the lower 5th percentile, 5-minute stress. To determine the design values for LRFD, the ASD values are multiplied by the resistance factor, ϕ, from Table 14.4 and the format conversion factor, K_F, from Table 14.5. Similar tables are also provided in Appendix N of the NDS.

Table 14.4 Resistance Factors, ϕ

Application	Symbol	Value
Compression	ϕ_c	0.90
Flexure	ϕ_b	0.85
Shear	ϕ_s	0.75
Tension	ϕ_t	0.80
MOE min	ϕ_v	0.85
Connections	ϕ_z	0.65

Table 14.5 Format Conversion Factors, K_F

Application	Property	K_F
Member	F_b	2.54
	F_t	2.70
	F_v	2.88
	F_c par	2.40
	F_c perp	1.67
	E min	1.76
	All connections	3.32

14.2.3 Adjustment Factors

The next step for both LRFD and ASD is to apply adjustment factors to these assumed mean material strengths to compensate for different environmental and loading conditions. There are also adjustment factors that apply to a member's physical characteristics that could affect the strength capacity. Most of the LRFD adjustment factors remain the same in value and in applicability to properties as for ASD.

The difference occurs in the way LRFD and ASD credit wood's unique ability to have increased strength over a short period of time. ASD applies an adjustment factor called the load duration factor. LRFD does not use the load duration factor. LRFD has a unique adjustment factor called the time effect factor as discussed in Section 1.4.

14.3 STRUCTURAL WOOD PRODUCTS

Numerous structural wood products are available to the design professional allowing them to create virtually any size or shape of structure permitted by the building code. The following is a brief description of each with their key attributes highlighted.

14.3.1 Sawn Lumber

While both softwood and hardwood lumber can be used for construction, the principal use of hardwood lumber is for remanufacture into furniture, cabinetwork, and pallets or direct use as flooring, paneling, molding, and millwork. Softwood lumber has demonstrated the versatility of wood by serving as a primary construction material throughout the decades. In this role, softwood lumber has been produced in a wide variety of products from many different species.

To minimize unnecessary differences in the grading rules of softwood lumber and to improve and simplify these rules, a number of conferences were organized by the U.S. Department of Commerce from 1919 to 1925. The result was a relative standardization of sizes, definitions, and procedures for deriving allowable design properties, formulated as a voluntary American Lumber Standard (ALS). This standard has been modified several times, and the current edition is American Softwood Lumber Standard PS 20-15.

Most softwood lumber produced as dimension lumber is provided based on nominal 2- to 4-in.-thick dimensions with actual dimensions of 1½ to 3½ in. thick. Dimension lumber is stress graded and assigned allowable properties under the National Grading Rule, a part of the American Softwood Lumber Standard. For dimension lumber, a single set of grade names and descriptions is used throughout the United States, although the allowable properties vary with species. Timbers are also structurally graded under ALSC procedures.

Unlike grade descriptions for dimension lumber, grade descriptions for structural timbers are not standardized across species. For most species, timber grades are classified according to intended use. Beams and stringers are members that are nominal 6 × 10 with actual dimensions of 5½ × 9¼ in. and larger. Beams and stringers are primarily used to resist bending stresses, and the grade description of some timber grades for the middle third of the length of the beam is more stringent than that for the outer two-thirds. Posts and timbers are members that are nominal 5 by 5 with actual dimensions of 4½ × 4½ in. and larger. Post and timbers are primarily used to resist axial stresses. Structural timbers of southern pine are graded without regard to anticipated use, as with dimension lumber.

Width of softwood dimension lumber varies, commonly from nominal 2 to 16 in. with actual dimensions of 1½ to 15¼ in.

Basic reference design stresses for stress-graded lumber and structural timbers are listed in the Supplement to the National Design Specification. These are 10-year values for lumber used under continuously dry conditions. The basic tabulated reference design stresses are adjusted for a size factor (C_f), repetitive member factor (C_r), flat use factor (C_{fu}), and wet service factor (C_M). Other in-service use factors such as duration of load are also applicable and are tabulated in the NDS.

14.3.2 Glulam Lumber

Structural glulam is a structural member glued up from suitably selected and prepared pieces of laminating lumber called "laminations" with the grain of all pieces parallel to the longitudinal axis of the member. Glulam members are produced by gluing together dry lumber, ≤16 percent, typically of 2- or 1-in. nominal thickness, under controlled conditions of temperature and pressure.

Members with a wide variety of sizes, profiles, and lengths can be produced providing superior characteristics of strength, serviceability, and appearance. Any length and shape, subject to the maximum sizes permitted by manufacturing, transportation and handling limitations, are available such as the arches shown in Fig. 14.2.

Figure 14.2 Typical curved glulam arches.

Glulam beams are manufactured with the strongest laminations located where the highest tension and compression stresses occur under in-service loading. This allows a more efficient use of the lumber resource by placing higher grade lumber in zones that have higher stresses and lumber with lower structural quality in lower stressed zones.

The laminating process used in glulam timber manufacturing results in a random dispersion of strength-reducing growth characteristics of lumber, such as knots and slope of grain, throughout the glulam member. Consequently, glulam has higher mechanical properties with a lower variability than sawn lumber products of comparable sizes.

Many different species of lumber can be used to produce glulam. In addition, a wide range of grades of both visually graded and mechanically graded lumber can be used in the manufacture of glulam. This wide variety of available species and grades results in numerous options for the producers to combine species and grades to create a wide array of glulam timber layup combinations.

Some of the common species used include Douglas fir-larch, southern pine, hem-fir, and spruce-pine-fir. Other species used are eastern spruce, western woods, Alaska yellow cedar, and Port Orford cedar. Several hardwood species, including red oak, red maple, and yellow poplar, are also used.

Glulam is manufactured with kiln-dry lumber having a maximum moisture content at the time of gluing of 16 percent. The use of kiln-dry laminating lumber means that the moisture content of glulam is relatively uniform throughout the member, which gives glulam excellent dimensional stability. A glulam will not undergo the dimensional changes normally associated with larger solid-sawn green timbers and will remain straight and true in cross section. A "dry" glulam is also less susceptible to the checking and splitting which is often associated with "green" timbers.

A significant development in the glulam timber industry was the introduction of fully water-resistant phenol-resorcinol adhesives in 1942. This allowed glulam to be used in exposed exterior environments without concern of glue line degradation. One of the earliest U.S. manufacturing standards for glulam was Commercial Standard CS253-63, which was published by the Department of Commerce (DOC) in 1963. (no longer in print) The most recent standard is ANSI/APA Standard A190.1-17.

ANSI/APA A190.1, *American National Standard for Structural Glued Laminated Timber*, is a national consensus standard for glulam timber manufacturing which documents detailed manufacturing requirements for glulam timber. This standard is recognized in the IRC, IBC, and state codes and all construction specifications for glulam must include reference to this standard.

Sizes

While glulam can be manufactured in virtually any cross-sectional size and length required, it is important to understand that since glulam is manufactured using dimension lumber, certain widths and depths become de facto standards which should be specified whenever possible. The depths of glulam are typically specified in multiples of 1½ in. for Western species and 1⅜ in. for southern pine. Thus, a 10-lamination member using Western species will have a net depth of 15 in., while a 10-lamination Southern Pine member will have a net depth of 13¾ in. Other thicknesses of laminations may be specified, but these will require a custom order. An example would be the use of ¾-in.-thick laminations to produce members with a tight radius of curvature such as may occur in some arch members.

Glulam products typically range in net widths from 3⅛ to 14 in., although virtually any member width can be custom produced using a multiple piece layup to achieve wider widths. Common net finished widths in inches of glulam for both Western species and southern pine are shown in Table 14.6.

Table 14.6 Net Finished Width of Structural Glulam Timbers

Nominal width	3	4	6	8	10	12	14
Finished width Western species	2½	3⅛	5⅛	6¾	8¾	10¾	12¼
Finished width Southern Pine	2½	3	5	6½"	8½	10½	12

Reference Stresses

Glulam reference design stresses are determined from the following ASTM Standards.

ASTM D 3737 *Standard Practice for Establishing Stresses for Structural Glued Laminated Timber* is the original consensus standard introduced in 1982 for deriving allowable properties for glulam timber manufactured in accordance with ANSI A190.1.

ASTM D 7341*Standard Practice for Establishing Characteristic Values for Flexural Properties of Structural Glued Laminated Timber by Full-Scale Testing* provides a procedure for using full scale testing of glulam to determine characteristic values that can be used to calculate flexural design properties.

X-X vs. Y-Y Orientation Glulam can be manufactured using a single grade or multiple grades of lumber, depending on the intended use. When the member is intended to be primarily loaded either axially or in bending with the loads acting parallel to the wide faces of the laminations (Y-Y axis), a single grade combination is recommended. Typical applications would include columns, truss chords, and large panels such as in bridge decks. These are typically noted as combinations 1, 2, 3, etc. depending on species and grade of lumber and are not shown as stress classes but are tabulated in ANSI/APA 117.

For members stressed primarily in bending about the X-X axis, a graded layup of lumber is used throughout the depth of the beam with the highest-quality laminations used in the outer zones of the beam where the bending stresses are highest with lower-quality laminations being used in zones subjected to lower bending stresses.

Glulam may be manufactured as unbalanced or balanced members. The most critical zone of a glulam bending member with respect to controlling strength is the outermost tension zone. In unbalanced beams, the quality of lumber used on the tension side of the beam, referred to as tension laminations, is higher than the lumber used on the corresponding compression side, allowing a more efficient use of the timber resource. These layups are typically intended for simple span bending members. Since unbalanced beams have different bending stresses assigned to the compression and tension zones, they must be installed accordingly. To ensure proper installation of unbalanced beams, the top of the beam is clearly stamped with the word "TOP."

Balanced layups are typically specified for continuous span beams, long cantilevered beams, or any other application that will induce high tensile stresses on both the top and the bottom of the member.

Layup combinations and associated reference design stresses are provided in a number of publications including ANSI/APA 117 *Standard Specification for Glued Laminated Timber of Softwood Species* and the NDS.

In order to simplify the selection of glulam members intended primarily for loading on the X-X axis, the glulam industry has adopted a series of stress classifications ranging from 16F-1.3E to 30F-2.1E as shown in Table 14.7. This table only shows the design stresses for bending about the X-X axis, and the reader is referred to Table 5A of the NDS Supplement for further explanation of these stress classes and for further layup combinations. Table 5B of the NDS Supplement provides the single grade layup combinations and reference stresses.

Table 14.7 Typical Glulam Stress Classes

Stress class	F_{bx+}	F_{bx-}	F_{vx}	$E_{x\ app}$	$F_{c\perp}$
16F–1.3E	1600	925	195	1.3×10^6	315
20F–1.5E	2000	1100	210	1.5×10^6	425
24F–1.7E	2400	1450	210	1.7×10^6	500
24F–1.8E	2400	1450	265	1.8×10^6	650
26F–1.8E	2600	1950	265	1.9×10^6	650
28F–2.1E	2800	2300	300	2.1×10^6	805
30F–2.1E	3000	2400	300	2.1×10^6	805

Reference Stress Adjustments

The published design properties values for glulam shown in Table 14.7 and in industry publications are based on use under normal duration of load (10 years) and dry conditions of use (less than 16 percent moisture content). When used under other conditions, applicable adjustment factors must be used. Adjustment factors shall be applied to modify the allowable properties when one or more of the specific end use or material modification conditions fall outside the limits of the reference conditions given in the tabular sections. Many of these factors, which are similar to those applied to sawn lumber, include the load duration factor, the wet service factor, the temperature factor, the beam stability factor, and the column stability factor. Refer to the NDS for the determination of the appropriate adjustment factors. It is important to note that beam stability factor C_L is not accumulative with the volume factor, C_v, given in the next section.

Unique Glulam Adjustment Factors The volume factor (C_v) is applied to glulam since allowable bending stresses of glulam timber are affected by member size. Generally, larger sizes have a correspondingly lower allowable bending stress than smaller members. To account for this behavior, a volume factor, C_v, which is the product of beam width, depth, and length, shall be applied. A 5⅛-in. × 12-in. × 21 ft member or smaller is assigned a $C_v = 1.0$, and the C_v for other larger sizes is computed as follows:

$$C_v = \left(\frac{5.125}{b}\right)^p \left(\frac{12}{d}\right)^p \left(\frac{21}{l}\right)^p \leq 1.0 \qquad (14.2)$$

where

- b = width of bending member being checked in inches. For multiple piece width layups, b = width of widest piece in the layup. For practical purposes, b is assumed to be ≤ 10.75 in.,
- d = depth of bending member being checked in inches
- l = length of bending member being checked between points of zero moment in feet
- p = 0.10 for western species, 0.07 for hardwoods, and 0.05 for southern pine

Another adjustment factor applied to glulam is the curvature factor (C_c), which is used to adjust the allowable bending stresses of curved glulam timber members. It takes into account the difference in extreme outer fiber stress between a curved member and a straight prismatic member, as well as any residual stresses that may remain in a lamination that has been bent to the stated curvature. However, the curvature factor, C_c, shall not be applied to the allowable bending stress in the straight portion of a member, regardless of curvature in other portions. Also, this factor is not applicable to cambered glulam timber:

$$C_c = 1 - 2000\left(\frac{t}{R}\right)^2 \qquad (14.3)$$

where

- t = thickness of lamination in inches
- R = radius of curvature of inside face of lamination in inches
- $t/R \leq 1/100$ for hardwoods and southern pine
- $t/R \leq 1/125$ for other species

Camber

One of the major attributes of glulam is it can be cambered, and glulam is generally the only engineered wood member that can be cambered. The size and span capabilities of sawn lumber beams are generally limited by the physical characteristics of the timber supply. Consequently, due to the relatively short spans associated with sawn lumber beams, deflection under long term dead loads is not typically a design control, and camber of these members is not required. However, for glulam, it is possible to design wood members in large sizes and long lengths.

Design professionals recognize that for long spans, design is often controlled by deflection limits. One way to reduce the adverse aesthetic effect and structural significance of beam deflection is through the use of camber. Camber is an initial curvature built into a glulam which is opposite in direction to the calculated deflection which will occur under gravity loads. Calculated beam deflection is used to assess compliance with building code requirements and to specify beam camber.

For most roof beam applications, the glulam timber industry recommends the use of 1½ times the calculated dead load deflection to arrive at a recommended camber. This amount of camber is generally sufficient to allow the beam to deflect back to near level after many years under load. For floor beams, the recommended camber is 1.0 times the calculated dead load deflection. However, for residential construction, floor beams are typically not cambered due to the relatively short spans and the need to minimize unevenness in the floor construction.

The camber for beams is often specified as "inches of camber." It may also be specified as a radius of curvature. The circular arc formed by this method closely approximates the actual deflected curvature of the beam under load. The beam is thus fabricated with a built-in "mirror image" of the expected deflected curvature.

The following formula may be used to calculate the approximate radius of curvature, given the beam span and camber desired:

$$R = \frac{3L^2}{2\Delta} \tag{14.4}$$

where

R = approximate radius of curvature, ft
L = span, ft
Δ = desired camber, in.

Nonprismatic Glulam

One of the major advantages of glulam is the capability of creating unique shapes such as single- and double-tapered beams, curved bending members, double-tapered curved bending members, and arches. While the design of complex members is beyond the scope of this handbook, a brief discussion of the various shapes possible is provided in the following sections. Basic design provisions for these members are provided in Chapter 5 of the NDS. A more detailed description of the design of these members can be found in the AITC *Timber Construction Manual,* or the *APA Engineered Wood Handbook.*

Tapered Beams

Tapered beams can take the form of end-tapered, single-tapered, and double-tapered beams. The use of tapered beams is typically to provide roof drainage and design aesthetics. Straight-tapered end cuts on the top of a beam are sometimes used to improve drainage, to provide extra head for downspouts and scuppers, to facilitate discharge of water, and to reduce the height of the wall.

Curved Beams The most common curved glulam has a constant cross section and constant curvature. When a curved member is loaded in bending, radial stresses are induced. When the bending moment is in the direction that tends to decrease the curvature or increase the radius, the radial stress is in radial tension, F_{rt}. On the other hand, when the bending moment is in the direction that tends to increase the curvature or decrease the radius, the radial stress is in radial compression, F_{rc}. Chapter 5 of the NDS provides allowable radial tensile and radial compression stress for glulam. If the adjusted radial tension value is exceeded, appropriate mechanical reinforcements can be used to resist all applied radial tensile stresses as discussed in the AITC *Timber Construction Manual.*

PTC Beams Pitched and tapered curved (PTC) glulam beams are very popular as structural roof members when a long interior clear span and sloping roof is desired. The design of PTC beams is similar to that of straight prismatic beams except that the radial stresses shall be considered in the curved portion and the distribution of the bending stresses about the assumed neutral axis is different from that in a prismatic member. The reader is referred to the AITC *Timber Construction Manual* for full design information on PTC beams.

Arches Glulam is often used in arch structures with a variety of shapes and structural characteristics. Three-hinged arches are generally more economical than two-hinged arches due to the possible elimination of the moment splice required for most two-hinged arches. Arches may be used to form more complex configurations such as space frames and reticulated timber domes with spans of over 500 ft.

Two-hinged arches are of three types: foundation, tied, and buttressed. They have been used for spans of more than 300 ft. Foundation arches may have tie rods in or under the floor to resist the horizontal thrust.

14.3.3 Wood Structural Panels

A wood structural panel, also referred to as a structural-use panel, is a panel product composed primarily of wood, which, in its commodity end use, is essentially dependent upon certain mechanical and/or physical properties for successful performance in service. Such a product is identified in a manner clearly conveying its intended end use. Today, wood structural panels include all-veneer plywood and mat-formed compressed wood strand panels or oriented strand board (OSB) such as shown in Fig. 14.3.

Figure 14.3 Typical wood structural panels.

Wood structural panels are used as sheathing materials for roofs, walls, and floors in a broad range of light-frame wood building applications. Manufacturing standards for wood structural panels are of two types: prescriptive or performance based.

Voluntary Product Standard PS1, *Construction and Industrial Plywood,* promulgated under the rules of the DOC, was originated in 1966, and numerous iterations have occurred resulting in the current PS1-09*Structural Plywood.*

The standard provides a recipe for panel layup, specifying the species of veneer and the number, thickness, and orientation of plies that are required to achieve panels of the desired nominal thickness and strength.

A more recent development for wood structural panels is that of performance-based standards. Such standards are blind to actual panel construction but do specify performance levels required for common end uses. Performance standards permitted the introduction of OSB into the construction market since mat-formed panels (panels laid up in a mat rather than by stacking veneers) don't lend themselves to prescriptive layups.

Voluntary Product Standard PS2, *Performance Standard for Wood-Based Structural-Use Panels,* was promulgated in 1992 under the rules of the DOC as the first consensus-based performance standard for wood structural panels and has evolved to PS2-10.

PS2 is used extensively for all wood-based structural panel types including plywood and OSB. It covers sheathing and structural 1 sheathing and single-floor grades and includes performance criteria, a qualification policy, and test methods. Wood structural panels manufactured in conformance with PS1 and PS2 are recognized in the IRC and IBC and local codes throughout the United States.

Wood structural panels are produced in two bond classifications—Exposure 1 and Exterior. The bond classification relates to the adhesive bond and thus to the structural integrity of the panel. Bond classification relates to moisture resistance of the glue bond.

Exterior panels have a fully waterproof bond and are designed for applications subject to long-term exposure to the weather or moisture.

Exposure 1 panels have a fully waterproof bond and are designed for applications where temporary exposure to weather due to construction delays, or high humidity or other conditions of similar severity may be expected. Exposure 1 panels are made with the same exterior adhesives used in Exterior panels. However, because other compositional factors may affect bond performance, only Exterior panels should be used for long-term exposure to the weather. Exposure 1 panels may, however, be used where exposure to the outdoors is on the underside only, such as at roof overhangs.

Plywood

Manufactured from thin sheets of cross-laminated veneer and bonded under heat and pressure with adhesives, plywood panels have superior dimensional stability and an excellent strength-to-weight ratio and are highly resistant to impacts, chemicals, and changes in environmental temperature and humidity. Plywood is available in a wide variety of appearance grades, ranging from smooth, natural surfaces suitable for finish work to more economical grades used for sheathing. With more than a dozen common thicknesses and over 20 different grades, plywood is well suited to a multitude of demanding applications. Sizes for plywood include 4 × 8, 4 × 9, and 4 × 10 with the 4 × 8 being the dominant size supplied.

Sheathing grade panels are classified by span ratings, which identify the maximum recommended support spacing for specific end uses. Structural I sheathing panels meet the requirements of sheathing grades as well as having enhanced requirements associated with use in high seismic risk applications such as diaphragms and shear walls (e.g., increased cross-panel strength and stiffness and racking shear resistance).

OSB

Panels manufactured of compressed wood wafers or strands have been marketed with such names as waferboard and OSB since the early 1980s. Today, virtually all mat-formed wood structural panels are manufactured with oriented strands, and are commonly called OSB.

OSB is composed of compressed strands arranged in layers (usually three to five) oriented at right angles to one another and bonded under heat and pressure with a waterproof adhesive. The orientation of layers achieves the same advantages of cross-laminated veneers in plywood. Since wood is stronger along the grain, the cross lamination distributes wood's natural strength in both directions of the panel. Whether a panel is composed of strands or wafers, most manufacturers orient the material to achieve maximum performance. OSB can be manufactured in sizes as large as 8 × 24 ft in cross section.

Span Ratings Sheathing

Single floor and siding grades carry numbers in their trademarks called span ratings. These denote the maximum recommended center-to-center spacing of supports, in inches, over which the panels should be placed in construction applications.

The span rating on sheathing grade panels appears as two numbers separated by a slash, such as 32/16, 48/24, etc. The left-hand number denotes the maximum recommended spacing of supports when the panel is used for roof sheathing with the long dimension or strength axis of the panel across three or more supports (two spans). The right-hand number indicates the maximum recommended spacing of supports when the panel is used for subflooring with the long dimension or strength axis of the panel across three or more supports. A panel marked 32/16, for example, may be used for roof sheathing over supports up to 32 in. on center or for subflooring over supports up to 16 in. on center.

Wall sheathing panels are performance tested with the secondary axis (usually the short dimension of panel) spanning across supports, or studs. For this reason, wall sheathing panels may be applied with either the strength axis or the secondary axis across supports.

The span rating on Single Floor grade panels appears as a single number. Single Floor panels are designed specifically for single-floor (combined subfloor-underlayment) applications under carpet and pad and are manufactured with span ratings of 16, 20, 24, 32, and 48 oc. The span ratings for Single Floor panels, like those Sheathing grade, are based on application of the panel with the long dimension or strength axis of the panel across three or more supports.

Load/Span Tables

Various load/span tables for wood structural panels are available from industry organizations such as APA. Loads are tabulated based on capacities described in this chapter and are provided for applications where the panel strength axis is applied perpendicular to supports and parallel to supports. For each combination of span and span rating or thickness, loads are typically given for deflections of L/360, L/240, and L/180 and maximum loads controlled by bending and shear capacity. The designer is referred to APA or other panel industry organizations for such load/span tables.

14.3.4 Prefabricated Wood I-Joists

Prefabricated wood I-joists are an "I"-shaped engineered wood structural member designed for use in residential and commercial building floor and roof construction such as shown in Fig. 14.4.

Figure 14.4 Typical prefabricated wood I-joists used in residential floor construction.

Flanges are either sawn lumber or SCL, typically laminated veneer lumber (LVL), although other SCL products may be used as flange materials. The top flange is of the same type and grade of material as the bottom flange. The net flange size depends on the material used and end use conditions.

Due to variations in flange width for I-joists of the same depth, often designers and builders will specify that the I-joist supplier provide the joist hangers as well as the I-joists. Most mills supply I-joists to distributors and dealers in lengths up to 60 ft. These are then cut to frequently used lengths such as 16 to 36 ft.

Webs consist of wood structural panels, which can be plywood or OSB manufactured in accordance with PS1 or PS2, but the vast majority of the webs are OSB. All panels are classified as Exposure 1 or Exterior and are typically ⅜ in. thickness. I-joists are assembled using exposure 1 type adhesives meeting the requirements of ASTM D 2559 and ASTM D 7247.

I-joists for residential construction are typically available in four depths, 9½, 11⅞, 14 and 16 in., net depths. I-joists can be manufactured in depths up to 30 in. for commercial construction in depth increments of 2 in.

Standards

The universally recognized standard for wood I-joists is ASTM D5055, *Standard Specification for Establishing and Monitoring Structural Capacities of Prefabricated Wood I-Joists.* This consensus standard provides guidelines for the evaluation of mechanical properties, physical properties, and quality of wood I-joists and is the current common testing standard for I-joists. However, since ASTM D5055 does not specify required levels of performance, individual manufacturers of I-joists have their own proprietary company standards that govern the everyday production practice for their products. Under the current building code jurisdictions, I-joist manufacturers can gain code recognition through evaluation reports developed in accordance with ICC ES AC14, *Acceptance Criteria for Prefabricated Wood I-Joists.* This provides guidelines on implementing performance features of the IRC and IBC.

Each manufacturer's Evaluation Services Report (ESR) code report standard has the potential for having differing installation details, allowable spans, web penetration requirements, allowable stresses, etc. As the history of other building materials such as plywood, OSB, and glulam has shown, some degree of standardization of the industry was inevitable.

APA, in conjunction with member I-joist manufacturers, developed a standard for performance-rated wood I-joist products used in residential floors, designated as PRI-400 *Performance Standard for APA-EWS I-Joists.* This standard provides common design values, span ratings, and installation recommendations for the four common depths of I-joists used in residential construction. Because PRIs can be selected based on their allowable span for glued uniformly loaded residential floors, it is easy to incorporate them into any design. As all APA EWS I-joists share a common set of installation and fastening details, the generation of design drawings is greatly simplified. It should be noted that this is a voluntary standard and that not all I-joist manufacturers have chosen to produce PRI-400 products.

APA has also developed a parallel standard, APA PRI 405, *Performance Standard for APA EWS Commercial I-Joists,* for I-joists used in commercial building construction. This standard covers I-joists manufactured in depths of 11⅞ in. and 14 to 30 in. in 2-in. depth increments.

PRI 400 and PRI 405 I-joist products are manufactured under the rigorous quality assurance standards of APA—The Engineered Wood Association. Other proprietary I-joist products manufactured in accordance with an ESR report also fulfill the same purpose.

In addition to span tables provided in PRI-400 and PRI-405, proprietary span tables for I-joists are also provided by each manufacturer of I-joists.

14.3.5 Structural Composite Lumber (SCL)

SCL is an engineered wood structural lumber product family that includes LVL, parallel strand lumber (PSL), laminated strand lumber (LSL), oriented strand lumber (OSL), and other wood composite lumber products which have similar engineering and configuration features. LVL was first produced in the early 1970s and since then has been commercially available in the United States. PSL was introduced to market in the 1980s. SCL products are used in the same structural applications as sawn lumber and timber. A brief description of each follows.

Laminated Veneer Lumber (LVL)

LVL was the earliest type of SCL product that was commercially manufactured for the marketplace. It is now the most widely used SCL product used in the residential housing market and is also often used in light commercial building construction. LVL is produced by bonding layers of wood veneers in a large billet under proper temperature and pressure similar to the manufacturing of plywood. Typically, LVL is produced in 4-ft-wide billets in long lengths, and the billet is then sawn to different dimensions to meet the needs of the final applications. The thickness of the billets is typically 1½ or 1¾ in. The most common species used for manufacturing LVL are Douglas-fir, southern pine, and SPF (spruce-pine-fir).

Typically LVL is used as headers and beams, chords for trusses, and ridge beams in mobile homes; flanges in prefabricated I-joists; and scaffold planks. It can also be used as columns and for studs in wall framing. Figure 14.5 illustrates various depths of LVL as may be used for beams and headers.

Individual LVL pieces such as shown in Fig. 14.5 are often bonded together in a separate gluing operation or mechanically fastened together in the field to create wider-width members that are compatible with traditional oriented strand lumber wood framing.

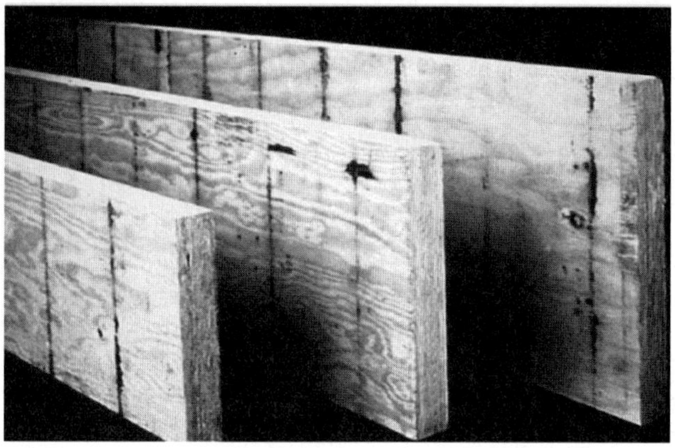

Figure 14.5 Typical profiles of laminated veneer lumber (LVL).

Parallel Strand Lumber (PSL)

PSL is manufactured by glue-bonding wood strands to form a condensed billet in such a way that the wood fiber (grain) direction of the strands is primarily oriented parallel to the length of the member. The thickness (least dimension) of the strands usually is less than 0.25 in., and the average length of strands is about 150 times the thickness of the strands. Presently, PSL is made primarily from Douglas-fir, western hemlock, southern pine and yellow poplar, although there are no restrictions on species.

The dimensions of veneers (strands) used in PSL are about ⅛ in. thick by ¾ in. wide by 24 in. long. The strands are coated with exterior type adhesive, and the strands are all oriented parallel to the length of the billet. A continuous pressing operation allows a higher degree of densification to be achieved. Typical lengths of PSL billets are up to 60 ft and are actually limited by handling restrictions. Billets are then resized to desired dimensions. If needed, larger cross-sectional dimensions can be achieved by secondary bonding operations. PSL is often used as beams and headers and columns as shown in Fig. 14.6.

Laminated Strand Lumber (LSL) and Oriented Strand Lumber (OSL)

LSL and OSL are an extension of the technology used to manufacture OSB. Among SCL products, LSL is perhaps the most efficient in utilizing wood resources. It has no restrictions on raw materials. Small logs and crooked logs of many species, including aspen, yellow poplar, and other underutilized, fast-growing species, can be used in manufacturing LSL.

OSL is another type of LSL product and has similar manufacturing process as for LSL. The primary difference between them is that the length of strand used in OSL is shorter (up to 6 in.) than that used in LSL (approximately 12 in.). OSL has somewhat lower strength and stiffness values compared to LSL.

Standards and Code Recognition

ASTM D5456, *Standard Specification for Evaluation of Structural Composite Lumber Products,* provides guidelines for the evaluation of mechanical properties, physical properties, and quality of SCL products.

Figure 14.6 Typical profiles of parallel strand lumber (PSL).

Since ASTM D5456 is not a product standard for the SCL industry, individual manufacturers of SCL have their own proprietary manufacturing product standards. Under current building code jurisdictions, SCL manufacturers are required to gain code recognition through an evaluation process provided by the ICC-ES, which issues ESR reports for proprietary products to demonstrate compliance with the IRC and IBC. ICC ES AC47, *Acceptance Criteria for Structural Composite Lumber*, provides guidelines on implementing performance features of the IRC and IBC.

Grades and Grading System

ASTM D5456 and ICBO AC47 provide guidance for developing proprietary design values, but no standard performance levels or grades are defined. While SCL products are all proprietary and no industry consensus standard for SCL exists, a system of grades similar to the machine stress-rated (MSR) grading system used by the lumber industry has evolved, and SCL products are identified according to their assigned stress class. Stress classes indicate the allowable designated modulus of elasticity, E, and design bending stress, F_b. The most commonly used LVL grades range from 1.5E to 2.1E. In general, a stress class consists of two parts separated by a hyphen ("-"). The left part of the hyphen ("-") represents the designated E value, while the right part of it represents the design bending stress (F_b), both of which are in the units of pounds per square inch (psi). For instance, a grade of 1.5E-2250F signifies LVL with modulus of elasticity (MOE) of 1.5×10^6 psi and F_b of 2250 psi.

While all LVL products are proprietary Table 14.7 provides some typical stress classes produced by the LVL industry listing the E-F_b design properties. The designer is referred to manufacturers' code reports for other grades of LVL products and for detailed information on the limits of listed design properties.

Table 14.7 Typical Stress Classes of LVL*

LVL stress Classes	$E^\dagger$ 10^6 psi	F_b psi
1.5E–2250F	1.5	2250
1.8E–2750F	1.8	2750
2.0E–2900F	2.0	2900
2.1E–3100F	2.1	3100

*The tabulated values are design values for normal duration of load. All values, except E and $F_{c\perp}$, are permitted to be adjusted for other load durations as permitted by the code. The design stresses are limited to conditions in which the maximum moisture content is less than 16 percent.

†Bending modulus of elasticity (E), which is applicable in either edgewise or flatwise applications, includes shear deflections.

14.3.6 Cross-Laminated Timber (CLT)

CLT represents the newest generation of glued engineered wood composites. Developed in Europe in the early 1990s, several thousand CLT structures have been constructed in Europe in the past 25 years. CLT was introduced into North America in 2012 with the publication of ANSI PRG320-*Standard for Performance-Rated Cross-Laminated Timber.* CLT was added to the 2015 NDS as Chapter 10 and it was adopted in the 2015 IBC.

CLT is a prefabricated engineered wood product made of at least three orthogonal layers of stress-graded sawn lumber or SCL that are laminated by gluing of longitudinal and transverse layers with structural adhesives to form a solid rectangular-shaped straight and plane timber as shown in Fig. 14.7.

CLT is identified by a grade which is a unique designation for CLT panels having the same layup of different panel thicknesses determined by the type and the grade of laminations in the orthogonal layers.

For CLT used in the United States the adhesives shall meet the requirements of AITC 405 with the exception that Section 2.1.6 of AITC 405 (either ASTM D3434 or CSA O112.9) is not required. CLT is typically manufactured in widths of 4 to 10 ft and thicknesses up to 20 in. Thicknesses are generally in multiples of 1⅜ in. reflecting the use of 2 × dimension lumber. Lengths can be up to 40 ft and more. Due to its unique characteristics, CLT is being used in mid-rise (6–12 stories and higher) construction. The Brock Common dormitory at the University of British Columbia is 18 stories with 17 stories of CLT over a first story of concrete as shown in Fig. 14.8.

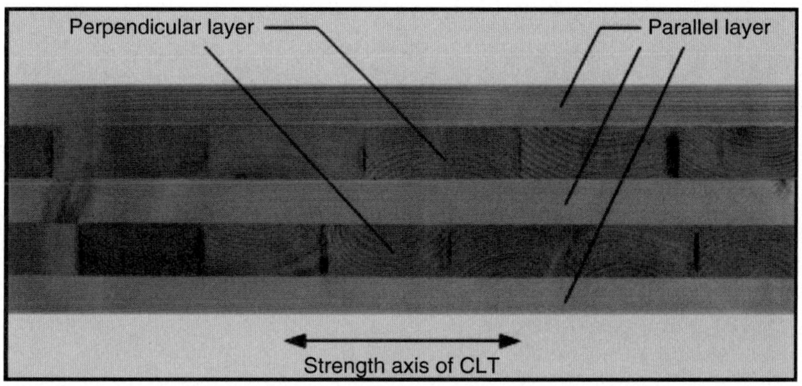

Figure 14.7 Cross-sectional view of cross-laminated timber (CLT).

Figure 14.8 Brock Common Student Residence Hall at the University of British Columbia.

Any softwood lumber species or species combinations recognized by the American Lumber Standards Committee (ALSC) under PS 20 with a minimum published specific gravity of 0.35, as published in the Supplement the National Design Specification for Wood Construction (NDS) shall be permitted. The same lumber species or species combination shall be used within a single layer of CLT. Adjacent layers of CLT shall be permitted to be made of different species or species combinations or SCL laminations.

The minimum grade of lumber in the parallel layers of CLT shall be 1200f-1.2E MSR or visual grade No. 2. The minimum grade of lumber in the perpendicular layers of CLT shall be visual grade No. 3. Proprietary lumber grades meeting or exceeding the mechanical properties of the lumber grades specified above shall be permitted for use provided that they are qualified in accordance with the requirements of an approved agency.

LVL, LSL, PSL, and OSL meeting the requirements of ASTM D5456 and with a minimum published equivalent specific gravity of 0.35 shall be permitted. SCL laminations of the same grade shall be used within a single layer of CLT. Adjacent layers of CLT shall be permitted to be made of different grade of SCL or lumber laminations.

The seven typical CLT grades as specified in PRG 320 are listed below.

E1: 1950f-1.7E Spruce-pine-fir MSR lumber in all parallel layers and No. 3 Spruce-pine-fir lumber in all perpendicular layers

E2: 1650f-1.5E Douglas fir-larch MSR lumber in all parallel layers and No. 3 Douglas fir-larch lumber in all perpendicular layers

E3: 1200f-1.2E Eastern softwoods, northern species, or western woods MSR lumber in all parallel layers and No. 3 eastern softwoods, northern species, or western woods lumber in all perpendicular layers

E4: 1950f-1.7E Southern pine MSR lumber in all parallel layers and No. 3 southern pine lumber in all perpendicular layers

V1: No. 2 Douglas fir-larch lumber in all parallel layers and No. 3 Douglas fir-larch lumber in all perpendicular layers

V2: No. 1/No. 2 Spruce-pine-fir lumber in all parallel layers and No. 3 spruce-pine-fir lumber in all perpendicular layers

V3: No. 2 Southern pine lumber in all parallel layers and No. 3 southern pine lumber in all perpendicular layers

Custom CLT grades may also be established by the manufactures. CLT panels of the same class may have different thickness depending on the number and thickness of the layers. CLT panels shall be used in dry service conditions, such as in most covered structures, where the mean equilibrium moisture content of solid-sawn lumber is less than 16 percent in the United States. The ASD reference design values for CLT used in the United States are shown in Table 14.8. These values are provided in PRG 320-2018 with additional information regarding their use.

For the design of CLT elements, mainly floors and walls, several methods have been implemented, experimental and analytical. The analytical

Table 14.8 ASD Reference Design Values for CLT Listed in PRG 320 (for use in the United States)

		Major strength direction				Minor strength direction			
CLT grade	CLT t (in.)	$F_{b,f,0}S_{eff,f,0}$ (lbf-ft/ft of width)	$EI_{eff,f,0}$ (10^6 lbf-in.^{2}/ft of width)	$GA_{eff,f,0}$ (10^6 lbf/ft of width)	$V_{s,f,0}$ (lbf/ft of width)	$F_{b,f,90}S_{eff,f,90}$ (lbf-ft/ft of width)	$EI_{eff,f,90}$ (10^6 lbf-in.^{2}/ft of width)	$GA_{eff,f,90}$ (10^6 lbf/ft of width)	$V_{s,f,90}$ (lbf/ft of width)
E1	4⅛	4525	115	0.46	1430	160	3.1	0.61	495
	6-7/8	10,400	440	0.92	1970	1370	81	1.2	1430
	9⅝	18,375	1089	1.4	2490	3125	309	1.8	1960
E2	4⅛	3825	102	0.53	1910	165	3.6	0.56	660
	6⅛	8825	389	1.1	2625	1430	95	1.1	1910
	9⅝	15,600	963	1.6	3325	3275	360	1.7	2625
E3	4⅛	2800	81	0.35	1110	110	2.3	0.44	385
	6-7/8	6400	311	0.69	1530	955	61	0.87	1110
	9⅝	11,325	769	1.0	1940	2180	232	1.3	1520
E4	4⅛	4525	115	0.50	1750	140	3.4	0.62	605
	6-7/8	10,500	440	1.0	2410	1230	88	1.2	1750
	9⅝	18,400	1089	1.5	3050	2800	335	1.9	2400
V1	4⅛	2090	108	0.53	1910	165	3.6	0.59	660
	6-7/8	4800	415	1.1	2625	1430	95	1.2	1910
	9⅝	8500	1027	1.6	3325	3275	360	1.8	2625
V2	4⅛	2030	95	0.46	1430	160	3.1	0.52	495
	6-7/8	4675	363	0.91	1970	1370	81	1.0	1430
	9⅝	8275	898	1.4	2490	3125	309	1.6	1960
V3	4⅛	1740	95	0.49	1750	140	3.4	0.52	605
	6-7/8	4000	363	0.98	2420	1230	88	1.0	1750
	9⅝	7100	899	1.5	3050	2800	335	1.6	2400

procedures are more favorable, as they are more general and less expensive. The Canadian CLT Handbook (Gagnon and Pirvu, 2011) covers several methods for the analysis and design of CLT members: mechanically jointed beams theory (gamma method), composite theory–k method, and the shear analogy method, whereas the U.S. version of the CLT Handbook covers only the shear analogy method that was developed by Kreuzinger in 1999 (Kreuzinger, 1999). Among the different methods used for CLT design, the shear analogy method is the most recent and the most accurate. Unlike the other methods, the shear analogy method takes into consideration the shear deformation of the cross layer, and it is not limited to a restricted number of layers within the panel. In this chapter, the shear analogy is introduced; the reader is referred to the CLT Handbook, the Canadian edition (Gagnon and Pirvu, 2011) for the other methods mentioned above.

Shear Analogy Method

According to Blass and Fellmoser (2004), the shear analogy method is the most accurate design method for CLT. In this method, the different moduli of elasticity and the shear moduli of the single layers are used, and it is applicable to almost any system such as number of layers and span-to-depth ratio. The characteristics of a CLT member are separated into two beams, beam A and beam B (see Fig. 14.9). Beam A is given by the sum of the flexural strength of the individual layers along their neutral axes, where beam B is given the "Steiner" points part of the flexural strength, the flexible shear strength, and the flexibility of all connections. The beams are coupled by rigid links to achieve equal deflection for beams A and B. Bending and shear stress distribution in beams A and B are shown in Figs. 14.10 and 14.11, respectively, and the combined stresses for both beams are shown in Fig. 14.12.

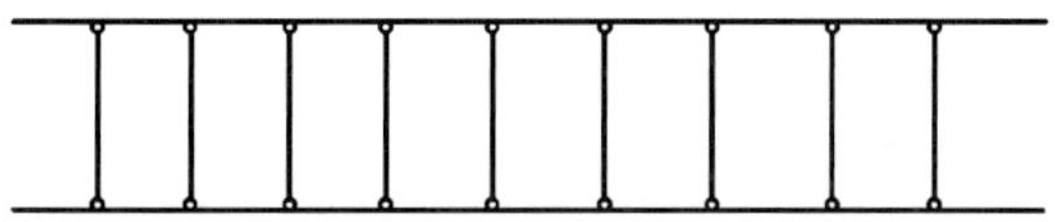

Beam A: (bending stiffness $(EI)_A = B_A$ and shear stiffness $(GA)_A = S_A \sim\infty$)

Web members with infinite axial rigidity

Beam B: (bending stiffness $(EI)_B = B_B$ and shear stiffness $(G_A)_B = S_B$)

Figure 14.9 Beam differentiation using the shear analogy method.

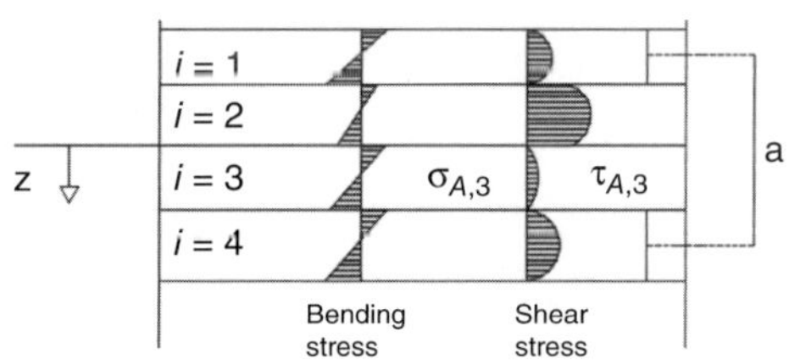

Figure 14.10 Bending and shear stresses in beam A using the shear analogy method.

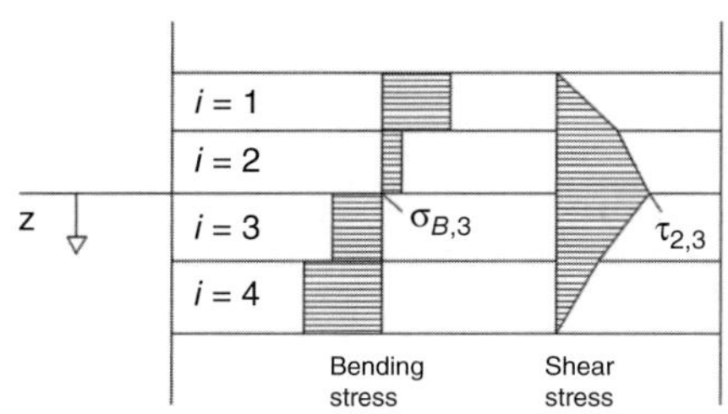

Figure 14.11 Normal and shear stresses in beam B using the shear analogy method.

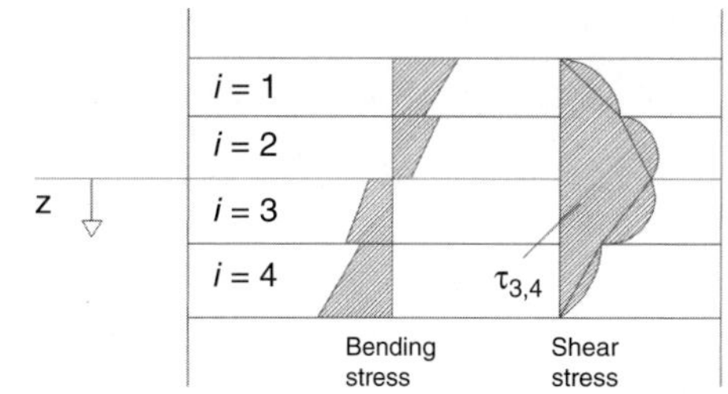

Figure 14.12 Final stress distribution obtained from the superposition of the results from beams A and B.

The flexural stiffness of beam A is given in Eq. (14.5):

$$B_A = \sum_{i=1}^{n} E_i \cdot I_i = \sum_{i=1}^{n} E_i \cdot b_i \cdot \frac{h_i^3}{12} \tag{14.5}$$

where

$B_A = (EI)_A$

b_i = width of each individual layer, usually taken as 1 ft for CLT panels

h_i = thickness of each individual layer

The flexural stiffness of beam B is given in Eq. (14.6); this is based on Steiner's theorem:

$$B_B = \sum_{i=1}^{n} E_i \cdot A_i \cdot z_i^2 \tag{14.6}$$

where

$B_B = (EI)_B$

z_i = the distance between the center point of each layer and the neutral axis

Beam B contains the shear stiffness and the stiffness of the flexible connections, if they exist.

The shear stiffness of beam B, S_B, is (GA)B and given in Eq. (14.7):

$$\frac{1}{S_B} = \frac{1}{a^2} \cdot \left[\sum_{i=1}^{n-1} \frac{1}{k_i} + \frac{h_1}{2 \cdot G_1 \cdot b_1} + \sum_{i=2}^{n-1} \frac{h_i}{2 \cdot G_i \cdot b_i} + \frac{h_n}{2 \cdot G_n \cdot b_n} \right] \tag{14.7}$$

where

$$k_i = \frac{K_i}{s_i} \tag{14.8}$$

where k_i is the slip of the "fasteners" between the beams.

The modulus of elasticity in the individual layers is used as either E_0 or E_{90}; E_0 is used for the longitudinal layers, and E_{90} is used for the cross layers. Suggested value for E_{90} is $E_{90} = E_0/30$. The shear modulus in these equations is used in a similar way; for the longitudinal layers, the shear modulus used is G, and for the cross layers, the rolling shear, G_R, is used. Suggested value for $G_R = G/10$.

The bending moments $M_{A,i}$ and shear forces $V_{A,i}$ of each individual layer of beam A can be calculated using the Eqs. (14.9) and (14.10), respectively:

$$M_{A,i} = \frac{E_i \cdot I_i}{B_A} \cdot M_A \tag{14.9}$$

$$V_{A,i} = \frac{E_i \cdot I_i}{B_A} \cdot V_A \tag{14.10}$$

The bending stresses $\sigma_{A,i}$ and shear stresses $\tau_{A,i}$ of each individual layer of beam A can be calculated using the Eqs. (14.11) and (14.12), respectively:

$$\sigma_{A,i} = \mp \frac{M_{A,i}}{I_i} \cdot \frac{h_i}{2} \tag{14.11}$$

$$\tau_{A,i} = \frac{E_i \cdot I_i}{B_A} \cdot 1.5 \cdot \frac{V_A}{b \cdot h_i} \tag{14.12}$$

For beam B, the axial forces $N_{B,i}$, normal stresses $\sigma_{B,i}$ of each individual layer, and shear stresses at the interface of the two layers of beam B, $\tau_{B,i,i+1}$, can be calculated using Eqs. (14.13), (14.14), and (14.15), respectively:

$$N_{B,i} = \frac{E_i \cdot A_i \cdot z_i}{B_B} \cdot M_B \tag{14.13}$$

$$\sigma_{B,i} = \frac{N_{B,i}}{b_i \cdot h_i} = \frac{E_i \cdot z_i}{B_B} \cdot M_B \tag{14.14}$$

$$\tau_{Bi,i+1} = \frac{V_B}{B_B} \cdot \sum_{j=i+1}^{n} E_j \cdot A_j \cdot z_j \tag{14.15}$$

The maximum deflection in a CLT slab panel due to a uniformly distributed load can be calculated using Eq. (14.16):

$$u_{max} = \frac{5}{384} \cdot \frac{qL^4}{(EI)_{eff}} + \frac{1}{8} \cdot \frac{qL^2k}{(GA)_{eff}} \tag{14.16}$$

The effective bending stiffness can be obtained using Eq. (14.17):

$$(EI)_{eff} = B_A + B_B = \sum_{i=1}^{n} E_i \cdot b_i \cdot \frac{h_i^3}{12} + \sum_{i=1}^{n} E_i \cdot b_i \cdot z_i^2 \tag{14.17}$$

The effective shear stiffness can be obtained using Eq. (14.18):

$$(GA)_{eff} = \frac{a^2}{\left[\left(\frac{h_1}{2 \cdot G_1 \cdot b}\right) + \left(\sum_{i=2}^{n-1} \frac{h_i}{G_i \cdot b_i}\right) + \left(\frac{h_n}{2 \cdot G_n \cdot b}\right)\right]} \tag{14.18}$$

Simplified Design Methods for Calculating Bending and Shear Strengths (Out-of-Plane)

The capacity of a CLT floor element in bending and shear can be calculated using the simplified equations.

The bending strength can be calculated using Eqs. (14.19) and (14.20):

$$\sigma = M \cdot y \cdot \frac{(E_i)}{(EI)_{eff}} \tag{14.19}$$

The maximum stress will occur for $y = \frac{h_{tot}}{2}$, so Eq. (14.19) can be expressed as

$$\sigma = M \cdot 0.5 \cdot h_{tot} \cdot \frac{(E_i)}{(EI)_{eff}} \tag{14.20}$$

The shear strength can be calculated using Eq. (14.21):

$$\tau_v = \frac{1.5 \cdot V}{c \cdot A_{gross}} \tag{14.21}$$

Where coefficient c is a reduction factor calculated as

$$c = \frac{I_{eff}}{I_{gross}} \tag{14.22}$$

The U.S. version of the CLT Handbook (Karacabeyli and Douglas, 2013), copublished by the USDA Forest Service Forest Products Laboratory and the Binational Softwood Lumber Council, provides extensive design guidelines for all aspects of CLT construction.

14.3.7 Round Timbers

Standards for poles and piles have been written with the assumption that trees have a round cross section with a circumference that decreases linearly with height. Thus, the shape of a pole or pile is often assumed to be that of the frustum of a cone. Actual measurements of tree shape indicate that taper is rarely linear and often varies with location along the height of the tree. Guidelines to account for the effect of taper on the location of the critical section above the ground line are given in ANSI O5.1. The standard also tabulates pole dimensions for up to 15 size classes of 11 major pole species.

Poles

Most structural applications of poles require timbers that are relatively straight and free of large knots. Poles used to support electric utility distribution and transmission lines range in length from 20 to 125 ft and from 5 to 30 in. in diameter. Round timbers are also used as poles and beams in building applications in many regions of the United States. Reference design values for poles used in building construction are given in ASTM D3200.

Piles

Material available for timber piles is more restricted than that for poles. Most timber piles used in the eastern half of the United States are southern pine, while those used in western United States are coast Douglas-fir.

Bearing loads on piles are sustained by earth friction along their surface (skin friction) or by bearing of the tip on a solid stratum. Wood piles, because of their tapered form, are particularly efficient in supporting loads by skin friction. Bearing values that depend upon friction are related to the stability of the soil and generally do not approach the ultimate strength of the pile. Where wood piles sustain foundation loads by bearing of the tip on a solid stratum, loads may be limited by the compressive strength of the wood parallel to the grain. The most critical loads on piles often occur during driving.

ASTM D 25 provides tables of pile sizes for either friction piles or end-bearing piles. Friction piles rely on skin friction rather than tip area for support, whereas end-bearing piles resist compressive force at the tip. For this reason, a friction pile is specified by butt circumference and may have a smaller tip than an end-bearing pile. Conversely, end-bearing piles are specified by tip area and butt circumference is minimized.

The expected life of a pile is also determined by preservative treatment and use. Wood that remains completely submerged in water does not decay. When piles that support the foundations of bridges or buildings are to be cut off above the permanent water level, they should be pressure treated. The service life of poles can be 50 years or longer.

14.4 MEMBER DESIGN

The design equations for most wood structural members are given in the NDS, the Wood Handbook, and other reference documents. Some of the most commonly used equations are given in the following sections for prismatic wood members. Nonprismatic members such as tapered beams and others are typically associated with glulam and are discussed in Section 3.2.5.

14.4.1 Beams

The most common application of a wood member is as a bending member. These range from solid sawn lumber, I-joists, and SCL used as joists and rafters to heavy timber beams and glulam. The strength of beams is determined by flexural stresses caused by bending moment, shear stresses caused by shear load, and compression across the grain at the end bearings and at load points.

Flexure

The stress due to bending moment for a simply supported pin-ended beam is a maximum at the top and bottom edges. The concave edge is compressed, and the convex edge is under tension.

Flexural stresses are calculated from the formula

$$f_b = \frac{Mc}{I} \tag{14.23}$$

For rectangular members with the neutral axis perpendicular to the depth at the center

I = moment of inertia (in.^{4}) = $bd^3/12$
S = section modulus (in.^{3}) = $bd^2/6$

where b = width of the member and d = the depth of the member.

Thus

$$f_b = \frac{M}{S} \tag{14.24}$$

Lateral Stability Wood beams are usually rectangular in cross section and may be subject to lateral bucking. Beams usually have lateral support sufficient to allow the full bending stress F_b. When this is not the case, the beam stability factor, C_L, is used to adjust the reference allowable bending stress, F_b. For the following conditions, $C_L = 1.0$.

a. If the compression edge of the beam is laterally supported along its length and the ends at point of bearing have lateral support to prevent rotation.

b. If the depth of the beam does not exceed its width $d \le b$.

When the depth of the bending member exceeds the width, $d > b$, lateral support shall be provided at points of bearing to prevent rotation. The beam stability factor, C_L is determined based on the unsupported length of the beam:

$$C_L = \frac{1+\frac{F_{bE}}{F_b^*}}{1.9} - \sqrt{\left(\frac{1+\frac{F_{bE}}{F_b^*}}{1.9}\right)^2 - \frac{\frac{F_{bE}}{F_b^*}}{0.95}} \tag{14.25}$$

where F_{bE} is determined from

$$F_{bE} = \frac{1.20E'_{\min}}{(R_B)^2} \tag{14.26}$$

The beam stability factor is a function of the beam slenderness ratio, R_B, which shall not exceed 50. R_B is determined from

$$R_B = \sqrt{\frac{L_e d}{b^2}} \tag{14.27}$$

where L_e is the effective span length and is a function of the unsupported length, L_u, and is determined from the NDS for various span and loading conditions.

Deflections The equation for calculating beam deflection using values of apparent E as given in the allowable stress tables for sawn lumber, timbers, and glulam is given as

$$\Delta = \frac{5wL^4(1728)}{384EI} \tag{14.28}$$

where

Δ = calculated deflection (in.)
w = uniform load (lbf/in.)
ℓ = design span (in.)
EI = bending stiffness of the member (lbf-in.^{2})

For products that publish a true E value, referred to as a shear-free E, the shear defection must be added to the bending moment deflection as given in Eq. (14.28). Also for prefabricated wood I-joists the shear detection can be significant. For a uniformly distributed load, the deflection based on a true E or for I-joists is given by the following equation:

$$\Delta = \frac{5wl^4}{384EI} + \frac{wl^2}{K} \tag{14.29}$$

where K = coefficient of shear deflection (lbf) and the other terms are as defined for Eq. (14.28).

It is also important to note that deflection under long-term loading may be as much as approximately double that computed by the elastic theory, and associated creep factors are given in the NDS to account for this long-term deflection.

Shear

Shear stress parallel to grain (horizontal shear) in wood beams is given by

$$f_v = \frac{VQ}{Ib} \tag{14.30}$$

For rectangular sections with breadth, b, and depth, d,

$$f_v = \frac{3V}{2bd} \tag{14.31}$$

If critical shear stresses result from loads adjacent to a support, some increase of the shear strength along the grain may be expected by the development of local radial stress fields; that is, the load is transferred by radial compressive stresses in addition to shear stresses. The following approximate method for calculating the vertical shear V for rectangular beams accounts for this effect.

1. For uniformly distributed loads, neglect all loads adjacent to supports within a distance equal to the beam depth, d.
2. Concentrated loads located within a distance d of the supports shall be permitted to be multiplied by x/d, where x is the distance from the support.
3. For moving concentrated loads, place the largest at a distance from the support equal to the beam depth, keeping other loads in their normal relation and neglecting any load within a distance from the support equal to the beam depth.

Notched Beams. The allowable shear for a rectangular beam notched on the lower face at the support is given by

$$V = \left(\frac{2F_v b d_n}{3}\right)\left(\frac{d_n}{d}\right)^2 \tag{14.32}$$

where

d = depth of unnotched beam, in.
d_n = depth of the member remaining at the notch, in.
F_v = adjusted shear design value parallel to the grain, psi

Notches are limited to 25 percent of the beam depth for sawn timber and 10 percent of the beam depth or 3 in., whichever is less for glulam.

The NDS provides additional guidelines for determining the allowable shear for members with notches on the compression face and for the effect of connections.

Bearing

Compression perpendicular to grain (bearing) measures the capacity of a member to carry bearing loads within a given deformation limit. Deformation limits vary by product. For example, according to ASTM D5456, a deformation limit of 0.04 in. is used for a test specimen of 1.5 in. thick in determining bearing stress of SCL. It is unlikely that a member fails catastrophically under bearing stress even if the load has gone beyond the design load based on the deformation limit. Overloaded bearing stress will cause more deformation in the bearing area and will be represented by a noticeable indentation along the supported areas. It is therefore reasonable that bearing is governed by a serviceability criterion. For a bearing of any length at the end of a beam or for bearings 6 in. or more in length at any location other than the end, the allowable bearing stress is $F_{c\perp}$.

For a bearing length L_B less than 6 in. and at least 3 in. from the end of the beam, the allowable bearing stress shall be permitted to be multiplied by the bearing area factor, C_b:

$$C_b = \frac{L_b + 3/8}{L_b} \tag{14.33}$$

where L_b is the bearing length measured parallel to the grain, in.

14.4.2 Columns

Wood columns can have different configurations such as solid columns, spaced columns, and built-up columns. They can also be subjected to different load applications including concentric, eccentric loads, and side loads.

Solid Columns with Concentric Loads

The most common columns design situation is a solid column supporting a concentric end load. For this configuration the design is based on

$$f_c = \frac{P}{A} \tag{14.34}$$

where P = axial concentric load
A = cross-sectional area

While the determination of the applied stress that the column must support is relatively straightforward, the determination of the allowable compression parallel to grain stress is more complicated. Until the promulgation of the 1991 NDS, wood columns were designed based on a methodology that required classifying the member as a short, intermediate, or long column. This required a trial-and-error solution when it was not obvious which classification applied for a specific design situation, and many designers considered it to be a cumbersome procedure.

Based on extensive research conducted at the USDA Forest Products Laboratory and at other research institutions, the 1991 NDS was revised to reflect the use of a single column design formula regardless of the length to depth (l/d) ratio previously used to classify columns as short, intermediate, or long. This has been carried forward in all subsequent codes and design standards such as the NDS. This is represented by applying the column stability factor, C_p, as shown in Eq. (14.35):

$$C_p = \frac{1+\frac{F_{cE}}{F_c^*}}{2c} - \sqrt{\left(\frac{1+\frac{F_{cE}}{F_c^*}}{2c}\right)^2 - \frac{\frac{F_{cE}}{F_c^*}}{c}} \qquad (14.35)$$

where

F_c^* = tabulated compression parallel to grain design value adjusted for all service conditions such as moisture, temperature, load duration, etc. but not C_p

F_{cE} = critical buckling design value

c = 0.8 for sawn lumber, 0.85 for round timber poles and piles, and 0.9 for glulam timber and structural composite lumber

The critical buckling design value is determined by the Euler column formula:

$$F_{cE} = \frac{0.822E'_{\min}}{\left(\frac{L_e}{d}\right)^2} \qquad (14.36)$$

where

$E_{\min}$ = $E[1-1.645\ COV_E](1.03)/1.66$ minimum allowable modulus of elasticity at the 5th percentile adjusted for applicable service conditions

COV_E = 0.25 for visually graded lumber and round timber poles and piles, COV_E = 0.15 for machine-evaluated lumber (MEL), and COV_E = 0.10 for glulam timber and structural composite lumber

d = least unbraced dimension of column

L_e = effective column length based on unbraced length and end fixity conditions

L_e/d shall not exceed 50 except that during construction it shall not exceed 75.

Thus, the allowable unit stress F_c' is determined as

$$F_c' = F_c^* C_p \qquad (14.37)$$

where F_c^* is the tabulated compression parallel to grain design value multiplied by all applicable adjustment factors.

For columns supported throughout their length to prevent lateral displacement, $C_p = 1.0$.

The NDS also provides design guidelines for tapered columns.

Spaced Columns and Built-Up Columns

Spaced columns are formed of two or more timbers separated by spacer blocks and connected to end blocks by timber connectors. Built-up columns have lumber nailed or bolted together. The following provisions apply to nailed or bolted built-up columns with 2 to 5 laminations in which

(a) each lamination has a rectangular cross section and is at least 1½ in. thick, $t \geq 1½$ in.

(b) all laminations have the same depth (face width), d.

(c) faces of adjacent laminations are in contact.

(d) all laminations are full column length.

The NDS provides design guidelines for the design of spaced columns and built-up columns in Chapter 15, Special Loading Conditions.

Side Loads and Eccentricities

Timber columns are frequently subjected to side loads and eccentric loads. Chapter 15, Special Loading Conditions, of the NDS provides design guidelines for these conditions.

14.4.3 Tension Members

Tension members shall be designed to resist both parallel and perpendicular to grain stresses. For parallel to grain forces,

$$f_t = \frac{P}{A} \leq F_t' \qquad (14.38)$$

where

P = applied concentric load

A = net section area

F_t' is the adjusted tension parallel to grain design stress

Designs that induce tension perpendicular to grain stresses shall be avoided whenever possible, and there are no published design stresses for tension perpendicular to grain in the NDS or the building codes. When these stresses cannot be avoided, mechanical reinforcement sufficient to resist all such forces shall be considered.

14.4.4 Beam Columns

Members subjected to a combination of bending and axial tension shall be proportioned such that

$$\frac{f_t}{F_t'} + \frac{f_b}{F_b^*} \leq 1.0 \text{ and } \frac{f_b - f_t}{F_b^{**}} \leq 1.0 \qquad (14.39)$$

where

F_b^* = the reference bending design value multiplied by all applicable adjustment factors except C_L

F_b^{**} = the reference bending design value multiplied by all applicable adjustment factors except C_v

Members subjected to a combination of bending about one or both principal axes and axial compression shall be proportioned in accordance with the provisions of the NDS.

14.4.5 Trusses

While the detailed design of wood trusses is beyond the scope of this handbook, the following sections discuss some general design considerations.

Types of Wood Trusses

The basic types of timber trusses may be loosely classified into two categories, lightweight lumber trusses and heavy timber trusses. Lightweight lumber trusses are used extensively in residential building construction with pitched trusses for roofs and parallel chord trusses for floors. Lightweight lumber trusses using truss plates to connect the lumber members are typically designed by the manufacture/supplier. These trusses fabricated with truss-plate connectors are frequently spaced 2 ft on center with structural wood panels spanning between the trusses. However, they can also be used at wider spacing in light commercial building applications. The Truss Plate Institute and the Structural Building Components Association provide extensive design information for these types of trusses.

Heavy timber trusses are typically used in nonresidential construction and can span 200 ft or greater. Glulam is often used since it may be curved to any shape, varied in cross section, and fabricated to any practical shipping length. Heavy timber trusses are usually designed using steel plates with a variety of connectors such as lag screws, bolts, and timber rivets to interconnect the chords and web members. Recent examples of long span timber trusses are so-called hybrid trusses using glulam compression chords and steel tension cables. An example of this is the 100-m-span hybrid truss used for one of the venues at the 2010 Olympics in Vancouver, Canada, as shown in Fig. 14.13.

Figure 14.13 Steel and glulam trusses spanning 100 m.

Design of Truss Members

There are numerous computer-based design programs available for the design of wood trusses. Typically, the analysis is based on the assumption of pinned joints, and any continuity of the chord is considered independently. If eccentric connections are unavoidable, the moment due to the eccentricity must be considered in the design of the members.

Truss Bracing

Bracing of trusses for lateral stability during and after erection is a key design consideration. Each truss must be connected to bracing systems in vertical planes perpendicular to the trusses. The horizontal struts should be snugly fitted and designed to resist tension as well as compression. They may be solid struts or built-up I or T shapes.

Sidewall wind loads may be resisted by end walls and other structural components that develop horizontal shear strength such as wood structural panel diaphragms. If end walls are used for this purpose, the reactions from the sidewalls and trusses must be transmitted to the end walls through horizontal X-bracing or a roof diaphragm. If this is not possible, knee braces or other means of developing continuity between trusses and columns is necessary.

14.5 CONNECTION DESIGN

Fasteners and connectors for wood framing have continually improved over the years, and reliable design standards have been developed. Joints can now be designed with the same accuracy as other parts of the structure. However, poor design or construction practices have often resulted in the connector controlling the strength or serviceability of a structural unit or system.

Many types of fasteners and connectors are available for use in attaching one structural unit to another. These include nails, screws, lag screws, staples, timber rivets, split rings, shear plates, spike grids, toothed sheet-steel plates, framing anchors, joist and purlin hangers, and many other types of special fasteners and connectors. These will be discussed in following sections.

14.5.1 Nominal Design Values

Nominal design values for fasteners are determined by an equation of some form or from prescriptive tabulated values such as in the NDS. A nominal design value is defined by a table or equation in the NDS and is based on a certain set of conditions such as normal load duration (10-year), dry (low moisture content), no sustained exposure to elevated temperatures, and others. The nominal design values are multiplied by end use adjustment factors to account for actual anticipated end use conditions. After applying all appropriate adjustment factors, an allowable design value is obtained that can be equated to the design load requirements.

Provisions are usually given for calculating nominal design values for various types of single fasteners. When a connection has multiple fasteners of similar size and type, the total connection strength is a summation of the strength for each individual fastener subject to adjustments for the group action factor as discussed in the section Group Action Factor, C_g.

For nails and spikes, bolts, lag screws, and wood screws, the lateral load design values can be calculated by yield limit equations that account for the different potential yield modes. The yield modes are shown in Fig. 14.14.

The yield limit equations are provided in the NDS. Withdrawal load design values for these same fastener types can be calculated from empirical equations. For other fastener types such as split rings, shear plates, and timber rivets, the lateral and withdrawal design values are generally empirically based and available in tables in the NDS or other reference publications.

14.5.2 Adjustment Factors

Nominal design values must be multiplied by adjustment factors, which account for actual anticipated end use conditions, to determine allowable design values. A brief description of these factors follows.

Load Duration Factor, C_D

As previously discussed wood strength is dependent upon the duration of loading, due to damage accumulation in the wood material, and the same load duration factors are applied to wood connections, with

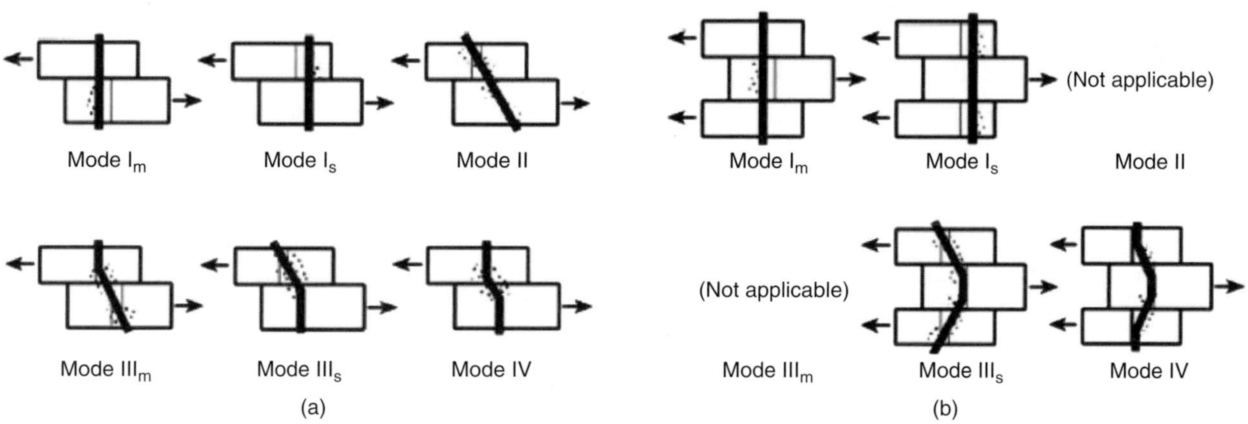

Figure 14.14 Yield modes for determining lateral load design values.

two exceptions. The first, the impact load duration factor ($C_D = 2.0$), is not applicable for connections. The second exception is for when the capacity of a connection is controlled by the strength of a material other than wood, such as metal, concrete, or masonry; the allowable strength of these materials shall not be adjusted for load duration because they do not exhibit wood's load duration behavior.

Wet Service Factor, C_M

The moisture content in the wood is considered at the time of fabrication of the connection and when in service. Dry is used to describe sawn lumber elements with moisture content of 19 percent or less or 16 percent for glulam, SCL, and I-joist. Most covered structures will remain continuously dry for the service life. Nominal design values are based on wood being dry and kept dry. If these conditions are not met, the wet service factor must be applied.

Temperature Factor, C_t

The temperature factor can usually be set equal to 1, but for uncommon applications where connections will experience sustained exposure to elevated temperatures, the nominal design values are to be multiplied by the appropriate temperature factors as given in Table 14.1.

Group Action Factor, C_g

Tests have shown that when more than two fasteners are placed in a row, the distribution of force to the fasteners in the row is not uniform. The behavior of a group of fasteners in a joint is as given in the NDS.

Therefore, it is necessary in the design of a joint with more than two fasteners in a row to include a group action factor C_g. This factor is applied only to lag screws and bolts with a $D \leq 1$ in. and split-ring and shear-plate connectors. The C_g for dowel type fasteners with $D < ¼$ in. is 1.0.

If the fasteners are of same type and size, the allowable load is determined as

$$P_r = P_s = Pn_f C_g \tag{14.40}$$

where

P_r = allowable load on row of fasteners
P_s = sum of values for individual fasteners in row
P = allowable load for single fastener
n_f = number of fasteners in row
C_g = group action factor

C_g is selected based on the type of side members, the size of the members, and the number of fasteners in a row from applicable tables in the NDS.

The following are considered to be a row of fasteners:

1. Two or more dowel type fasteners of the same diameter loaded in single or multiple shear and aligned with the direction of the load
2. Two or more split-ring or shear-plate units aligned with the direction of the load
3. Adjacent staggered rows of fasteners which are spaced apart less than one-quarter the spacing between the closest fasteners in adjacent rows

Geometry Factor, C_Δ

Reference design values for the different types of connections require that specific end, edge, and spacing distances be maintained. When the distances are less than that required for a $C_\Delta = 1.0$ but not less than the minimum for $C_\Delta = 0.5$, the geometry factor must be applied. For dowel-type fasteners with $D < ¼$ in., $C_\Delta = 1.0$ When $D \geq ¼$ in. the NDS provides tables for the following geometry factor limitations:

(a) End distances
(b) Edge distances
(c) Spacing between fasteners in a row
(d) Spacing between rows

The above limitations are provided for both parallel and perpendicular to grain load directions. The smallest factor for any fastener in a group is applied to all fasteners in that group.

Penetration Depth Factor, C_d

Similar to the geometry factor, the penetration depth factor adjusts the design value for fastener penetration less than that required for full design values but greater than the required minimum. The method for determining penetration depth factors for the different connection types is described in the NDS.

End Grain Factor, C_{eg}

End grain factors simply account for fastener penetration into end grain, and factors for the different connection types are described in the NDS.

Toe-Nail Factor, C_{tn}

The design value for withdrawal loading at a toenail should not exceed two-thirds of the side-grain withdrawal value. The design value for a toenail subjected to a lateral load should not exceed five-sixths of that permitted for a nail driven in side grain and laterally loaded. All adjustment factors which apply to lateral and withdrawal values for side grain also apply to toenails, except the wet service factor C_M does not apply for toenails loaded in withdrawal.

14.5.3 Types of Mechanical Fasteners and Connectors

While it is beyond the scope of this handbook to provide definitive design information for all possible mechanical fasteners used in wood construction, the following sections describe some of the more common fasteners.

Nails

The common wire nail is the most frequently used fastener and generally used when loads are light. Many types of nails, however, have been developed such as those with deformed shanks and coated surfaces. The development of pneumatic, electric, and mechanical nail guns has greatly increased the speed at which driven fasteners (nails and staples) can be installed, but generally, these nails are smaller in diameter and shorter in length than the standard penny nails.

ASTMF 1667 *Standard Specification for Driven Fasteners: Nails, Spikes, and Staples* is an excellent reference and provides basic information on most nails, spikes, and staples. Due to the wide range of nails available, a proper specification should include the pennyweight, diameter, and length of the nail. For example, a 10*d* common nail should be specified as: 10*d*, 0.148-in. diameter, 3-in. length.

The ultimate lateral load capacity of a nail or spike in single shear (two members) will be dependent on several material and dimensional properties of the connection. These include the thicknesses of the two joined members, the crushing strength of the wood, and the diameter and yield strength of the nail or spike.

The yield model theory provides six possible failure modes (see Fig. 14.14) for a two-member connection in single shear. The model assumes that the bearing capacity is reached when either the wood crushes under the nail or spike or plastic hinges are formed in the nail or spike.

Lateral design values have been determined using the yield model equations and are published in tables in the NDS. These tables provide values for box nails, common wire nails and sinker steel wire nails and post frame ring shank nails. If a connection does not qualify under the geometry and material properties given in the NDS tables, then the yield model equations may be used to obtain a lateral design value Z.

The NDS tabular values assume the following conditions:

1. Nails driven into wood without splitting
2. Seasoned wood which remains dry in service
3. Normal load duration
4. A nail in single shear
5. Full penetration
6. Nail inserted in side grain
7. Normal temperature
8. Wood or steel side plates

If the conditions assumed in the NDS tables are not satisfied in a joint under consideration, then adjustments need to be made for each noncompliant condition.

With respect to the requirement to avoid splitting, it has often been the practice in the past to let the placement of nails be determined on the job. However, in present-day design, this may not be satisfactory. While the NDS does not provide guidance on the placement of nails in wood, there are industry recommendations such as those given in the Wood Handbook as shown below:

End distance = $15d$

Edge distance = $10d$

Lateral design values Z for nails in double shear (three-member joints) in wood-to-wood connections are based on four possible failure modes (see Fig. 14.14). Or the design value for each shear plane can be determined and the lesser of the two values doubled.

The withdrawal resistance of nail or spike shanks is greatly affected by such factors as the type of nail point, type of shank, surface coatings, length of time nail remains in the wood, and changes in moisture content of the wood. The equation used by the NDS to obtain allowable design values for smooth shank nails or spikes is

$$P = 1380G^{5/2}D \tag{14.41}$$

where P = allowable design value per inch of penetration in member holding point of nail
G = specific gravity of wood
D = diameter of nail, in.

The NDS equation provides allowable values for normal duration of load which are somewhat less than one-sixth the ultimate value for short-term loading. Tables for withdrawal values per inch of penetration of the nail are published in the NDS. Common wire nails or spikes are not to be loaded in withdrawal from the end grain of wood.

The NDS table values are based on the following conditions of use:

1. Nails driven into wood without splitting
2. Seasoned wood which remains dry or unseasoned wood which remains wet
3. Normal load duration
4. Nails driven into the side grain
5. Normal temperature

If the conditions assumed in the NDS tables are not satisfied in a joint under consideration, then adjustments need to be made accordingly.

Staples

Staples may be used as a substitute for nails. Staples are usually U-shaped wire fasteners with two same-size pointed or pointless legs connected by a common crown. They are designed to be driven by manual hand strike, pneumatic, electric, or spring tools and to hold two or more pieces together.

If hammer-driven into wood or wood-base materials, these staples have to be relatively stout to prevent buckling during driving. If tool driven, the staples can be relatively slender since they are driven at a rapid rate while laterally supported by the tool's guide body. Tool-driven staples are usually provided with flats along their legs to facilitate tight collating into strips; they are often coated with polymers in order to decrease the staple's driving resistance.

ESR 1539 gives the sizes of collated and cohered heavy-wire steel staples that are typically available. While the NDS does not provide values for staples, allowable loads for staples can be reasonably taken to be equal to twice the value for a nail with a shank diameter equal to that of one leg of the staple. In addition, ESR 1539 and the IBC provide design tables for staples.

Wood Screws

Screws are another commonly used fastener in structural joints of wood. They are more satisfactory than nails under vibratory or withdrawal loads since they have less tendency to work loose. The general procedure for the design of joints using wood screws (both cut thread and rolled thread) is very similar to that used for nails. There are some notable differences in construction practices. When required, lead holes are to be drilled in accordance with the provisions of the NDS, and screws are to be inserted by turning and are not to be driven with a hammer. Lead holes are not required for wood with a specific gravity of ≤ 0.5.

Screws are designated by a gauge number which represents the diameter of the shank and their overall length. For example, a No. 9 wood screw would have a shank diameter of 0.177 in. and a total length of 2.00 in. It is adequate for design purposes to assume that two-thirds of the screw length is threaded. Lubricating the surface of a screw with wax or soap will facilitate insertion and will have little effect on the allowable load.

The NDS provides tables for lateral and withdrawal load values. The design value when more than one wood screw is used is equal to the sum of the design values permitted for each screw. If possible, the structural design should avoid using wood screws in withdrawal from end grain. Design values for lateral resistance, when the wood screw is inserted into the end grain and the screw is laterally loaded, should be reduced to two-thirds of the tabular values.

The ultimate lateral load capacity of a wood screw in single shear (two members) is dependent on several material and dimensional properties of the connection. These include the thickness of the two joined members, the crushing strength of the wood, and the diameter and yield strength of the wood screw. If the connection does not qualify under the geometry and materials properties given in the NDS tables, the yield model equations may be used to obtain a lateral design value. The design value when more than one wood screw is used is equal to the sum of the design values permitted for each screw.

NDS tabular values tables are based on the following conditions of use:

1. Seasoned wood which remains dry in service
2. Normal load duration
3. A screw in single shear
4. Full penetration
5. Screw inserted in side grain
6. Normal temperature

For lateral resistance, the penetration of the screw into the main member should be 10 times the shank diameter. If the penetration is less than $10D$ but not less than $6D$, the design value should be reduced in proportion to its reduced penetration using the equation $C_d = p/10D \le 1.0$. The minimum penetration should not be less than six shank diameters. If the use conditions assumed in the NDS tables are not satisfied in a joint under consideration, then appropriate adjustments need to be made.

Withdrawal tests on wood screws have shown that the ultimate load for a screw inserted into the side grain of seasoned wood may be expressed by the equation

$$P = 15{,}700G^2DL \tag{14.42}$$

where P = maximum withdrawal load, lb
G = specific gravity of wood
D = shank diameter of screw, in.
L = length of penetration of threaded part of screw, in.

The equation used to obtain the allowable design values given in the NDS is

$$P' = 2580G^2D \tag{14.43}$$

where P' is the allowable design value per inch of penetration of threaded part of screw and G and D are as previously. The NDS equation provides allowable values for normal duration of load which are somewhat less than one-sixth of that given by the base equation which is for short-term ultimate loading.

The NDS gives the design values in withdrawal loading for wood screws and the table is for the following conditions of use:

1. Seasoned wood which remains dry in service
2. Normal load duration
3. Screw turned into side grain
4. Normal temperature

If the above conditions are not satisfied in a joint under consideration, then adjustments need to be made.

It is noted that a new generation of fully threaded screws is available to designers. These fully threaded screws are being used extensively in the connection of CLT joints but are applicable to many other products such as SCL and glulam. Reference design values for these proprietary products are provided in ESR reports.

Lag Screws

Lag screws (lag bolts) are used when bolts are undesirable, when the member is too thick, or when one face of the member is not accessible for the installation of washers and nuts. Lag screws also can be used in conjunction with split rings and shear plates. They also can be especially effective when withdrawal loads of large magnitude are anticipated.

The size of a lag screw is designated by its shank diameter D_s and its nominal length L. Typical data on lag screw sizes and dimensions are provided in Appendix L of the NDS.

Lag screws are turned into prebored holes. The recommended diameter of the hole is dependent on the density of the wood and the diameter of the shank of the lag screw. The total length of the hole drilled should be equal to at least the length of the threaded portion of lag screw. The NDS gives the recommended values for both diameter and the length of hole. The threaded portion of the screw is to be turned into its lead hole and not driven in by a hammer. Soap or other lubricant is recommended to facilitate insertion and to prevent damage to the lag screw.

Lateral design values obtained using the yield model equations are presented in the NDS for wood-to-wood joints and for wood-to-steel plate joints. Design values are given in both tables for loading applied parallel to grain ($Z_{||}$) and perpendicular to grain ($Z_{\perp}$). For other angles of loading, the allowable load may be computed from values parallel and perpendicular to the grain by using the Hankinson formula given in Section 3.1. If the metal side plates are thicker than ¼ in., then the allowable value for a lag screw should be reduced in proportion to the lesser penetration of the lag screw. No increase is allowed for side plates thinner than ¼ in.

Lateral loading design values for lag screws are based on a penetration of lag screw (not including length of tapered tip) into the main member of eight times the shank diameter of the lag screw, $p = 8D$. The minimum lag screw penetration for a reduced design value is four times the diameter of the lag screw shank; when $4D \leq p \leq 8D$, the design value should be multiplied by C_d, where $C_d = p/8D \leq 1.0$.

When the loads act perpendicular to the grain and the lag screw is inserted parallel to the fibers (i.e., in the end grain of the member), design values for lateral resistance should be multiplied by the end-grain factor $C_{eg} = 0.67$.

If a connection does not qualify under the geometry and materials properties given in the NDS tables, then the yield model equations may be used to obtain a lateral design value Z. These equations will provide a design value of about one-fifth the ultimate capacity of the lag screw.

The NDS tables are based on the following conditions of use:

1. Seasoned wood which remains dry in service
2. Normal load duration
3. One lag screw, in single shear
4. Lag screw laterally loaded parallel or perpendicular to grain

If these assumed conditions are not satisfied in a joint under consideration, then adjustments need to be made accordingly.

For a detailed discussion of the design of lag screws, the reader is referred to AWC *Design Aid No. 1, Application of Technical Report 12 for Lag Screw Connections.*

Bolts

Bolts are the most common fastener used in heavy timber construction. Examples include heavy timber trusses and in high load seismic tension ties as shown in Figs. 14.15 and 14.16.

The shear-load capacity of a bolt depends upon the ratio of the length l of bolt in the main member, its diameter D, and the specific gravity of the wood. The NDS tabulates reference design values for bolts with diameters of ½ in., ⅝ in., ¾ in., ⅞ in., and 1 in. Diameters of > 1 in. are not permitted by the NDS since they can initiate high tension perpendicular to grain stresses on the bolt hole that can induce splitting of the wood. Bolts are installed in prebored holes a minimum of 1/32 in. to a maximum of 1/16 in. larger than the bolt diameter.

Figure 14.15 Typical truss connections using bolts and steel side plates.

Figure 14.16 Tension splice using bolts.

Holes shall be accurately aligned in both main and side members and not be forcibly driven.

The NDS tabulates reference design values for bolts loaded in both single shear and double shear with side members of wood or steel based on the yield model equations. The tables are based on the thickness of the main member, the thickness of the side member, the diameter of the bolt, and the specific gravity of the wood. The minimum thickness of wood side member is 1½ in. Steel side members are assumed to be ¼-in.-thick ASTM A36 steel plate. The thickness of the main members is designated as t_m, and the side members are designated as t_s.

$Z_{||}$ is the reference design value for a single bolt with all wood members loaded parallel to grain (lb). $Z_{s\perp}$ is the reference design value for a single bolt for an all wood connection with the main member loaded parallel to the grain and side members loaded perpendicular to the main grain (lb). $Z_{m\perp}$ is the reference design value for a single bolt in an all wood connection with the main member loaded perpendicular to the grain and side members loaded parallel to the grain (lb). $Z_{||}$ is the reference design value for a single bolt for a wood/steel connection bolt with the wood member loaded parallel to grain, and $Z_{\perp}$ is the reference design value for a single bolt for a wood/metal connection with the wood member loaded perpendicular to grain (lb).

For greater than three-member connections, each shear plane is considered as an isolated connection and the least value of all shear plane capacities determined. The total load capacity is the number of the connected shear planes times the determined least load capacity.

Due to the uneven distribution of the loads on bolts in a row, when multiple bolts are used, the group action factor discussed in the section

Group Action Factor, C_g is used to reduce the loads from the sum of the individual load capacities.

Minimum spacing and edge and end distances are provided in the NDS. The reader is referred to the section Geometry Factor, C_D for a discussion of these requirements and the application of the geometry factor when these are not met.

Split Rings and Shear Plates

Split rings and shear plates can be used in heavily loaded heavy timber and glulam construction since they are rated at higher loads than bolts. Both have a long history of successful use but are being used less frequently than in the past due to the machining required.

Shear plates are either 2⅝ in. or 4 in. in diameter and can be used for wood-to-wood connections or to connect wood to dissimilar materials such as concrete. Shear plates require that a groove be cut and a hole be drilled by using a special tool as shown in Fig. 14.17. The shear plate does not provide a wedge fit; thus, a shear-plate joint will allow greater slip than a split-ring joint and is less likely to cause a wood member to split as a result of shrinkage due to seasoning of the wood in service. The bolt or lag screw in a shear-plate joint serves a dual function of clamping the joint together and assisting in transferring the load.

Figure 14.17 Typical shear-plate connection. (*Courtesy of Portland Bolt & Manufacturing.*)

Split rings are either 2½ or 4 in. in diameter and are used for wood-to-wood joints. Split rings require that a groove be cut by a special tool in both faces of the mating pieces of wood as shown in Fig. 14.18. At the same time the groove is cut, a hole is bored for the bolt. As the name implies, a split ring has a split which allows it to expand as it is placed in the groove, thus allowing it to fit tightly in the groove. Therefore, very little slip will take place as load is applied. Because of the extra effort needed to install split rings properly and the potential shrinkage problems, most designers prefer to use shear plates.

The bolt in a split-ring joint serves merely to clamp the two pieces of wood together, thereby keeping the split ring in place. The bolt does not assist in the transfer of load as do the bolts in a shear-plate joint.

Allowable loads for split-ring and shear-plate connectors are presented in the NDS based on the number of faces of the members with connectors on the same bolt and four species groupings. The tabulated allowable loads are for one split ring or one shear plate on a bolt in single shear. The NDS also provides minimum edge, end, and spacing distances, distances being measured from the center of split ring or shear plate.

Split rings and shear plates can also be used with lag screws, and the NDS provides penetration depth factors for the four species groups and

Figure 14.18 Typical split-ring connection. (*Courtesy of Portland Bolt & Manufacturing.*)

the depth of penetration into the main wood member. Split rings and shear plates can also be used in the end grain of the wood, and the NDS provides details for such applications.

Rows of split rings or shear plates will exhibit smaller load capacity than the sum of the single-connector load capacities, and the group action reduction factor (see the section Group Action Factor, C_g) for load capacities of rows as previously discussed is applicable.

Timber Rivets

Timber rivets have long been used in Canada but are relatively new to U.S. designers. Provisions for the design of connections using timber rivets were first introduced into the 1997 NDS. Timber rivets are typically used to replace lag screws or bolts for glulam connections and are frequently referred to as glulam rivets. Timber rivets are hot-dipped galvanized with a fixed-shank cross section and fixed head dimension and vary only in length with the lengths being 1½ in., 2½ in., and 3½ in.

One of the advantages of timber rivets versus bolts is that for a connection with a wood main member and two steel side plates, the holes drilled through the steel plates and the main glulam members do not need to be drilled to provide exact alignment. The rivets are installed by a hammer with pneumatic hammers often used. The nail must have a minimum ultimate tensile strength of 145,000 psi. The timber rivet is designed to be driven through a circular hole in the steel side plates until the conical heads are firmly seated but not driven flush as shown in Fig. 14.19.

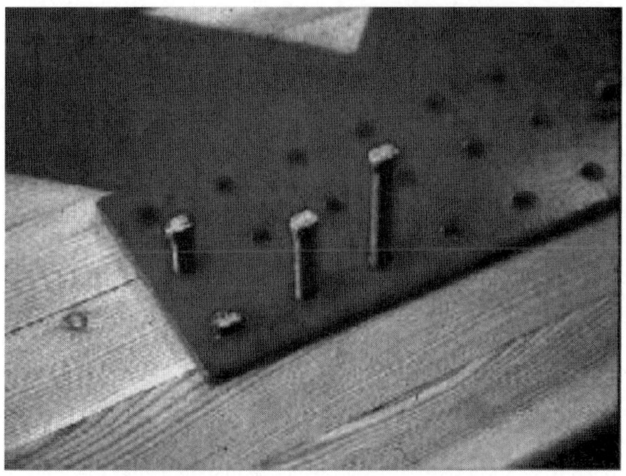

Figure 14.19 Timber rivets used in truss chord connection.

The maximum penetration permitted for any rivet is 70 percent of the thickness of the member holding the point of the rivet, except when the rivet is driven from opposite faces of the wood member, then the point should not overlap. The steel side plate must be at least ⅛ in. thick and less than ¼ in. thick and conform to ASTM A36 specifications. Timber rivet connections can have capacities as high as 100,000 lb and greater. They are evaluated for both wood capacity and rivet capacity. Minimum end and edge distances are specified, and a fabrication sequence is also specified in the NDS. End and edge distances are a function of the number of rivet rows. To reduce splitting, it is necessary to drive the perimeter rivets first and then the infield timber rivets. In addition, the rivet is to be placed with its major cross-sectional dimension aligned parallel to the wood grain. Design values for parallel, perpendicular, and at an angle of load to grain can be obtained from the NDS.

Metal Plate Connectors

One type of gusset plate for wood trusses is made of sheet steel of 20, 18, or 16 gauge. Plates may be nailed, or, more often, they have teeth punched out of and perpendicular to the sheet. The teeth are pressed into the wood either by roll pressing or by a hydraulic press. The angle of the grain is not relevant to the load capacity. These are typically proprietary products, and additional design information is available from the Truss Plate Institute.

Toothed thin-gauge sheet-metal plates are used very extensively in the construction of lightweight wood trusses as shown in Fig. 14.20.

Figure 14.20 Metal plate-connected wood truss.

One or more of these types of structural components are used in up to 90 percent of all residential structures, depending on geographical locale. They are also widely used in nonresidential commercial and institutional buildings.

Preengineered Connectors

Many types of preengineered timber connectors are commercially available. These include joist and purlin hangers, beam seats, column caps, strap ties, framing anchors, seismic and hurricane anchors, mud sills, hold-downs, column bases, and more. They can be used in the design of virtually any wood connection system ranging from hangers for lumber and I-joists to specialty connectors for solid wood members of nominal thickness of 4 in. or more and with glulam members and SCL products. Since these are proprietary, the designer is referred to the numerous ICC-ESR reports for design information for these products.

Joint Design

The design of a connection for a wood-framed structure must take into consideration several design issues as highlighted in the following sections.

Balanced Deformations It is not considered good practice to intermix sizes or types of connectors in a joint because the difference in slippage (stiffness) of each can cause an overload on the stiffest connectors before the less stiff connectors receive significant load.

Thickness of Lumber Most structural lumber is 1½ in. or more in thickness; therefore, any connector or fastener could be used. However, nails, staples, and screws are less commonly used in lumber greater than 1½ in. thick. For heavy timber construction, lag bolts, bolts, split rings, shear plates, and special-purpose timber connectors such as beam hangers and post caps are recommended.

Type of Load Vibratory loads or those which cause a stress reversal may influence the choice of type of fastener or connector. For example, screws have less tendency than nails to work loose under vibratory loads.

Size and Type It is generally better to use the same size and type of fastener or connector throughout a structural system in order to maintain uniform magnitudes of slippage at all joints per unit of force. Maintaining a single size and type will help reduce overall cost and reduce the chance of installation error.

Shrinkage and Swelling Wood shrinks as it dries and swells when it gains moisture. Minimal shrinkage or swelling takes place in wood parallel to grain (generally much less than 1 percent).

However, in the perpendicular to grain direction, the magnitude of shrinkage or swelling will be many times that of parallel to grain. Therefore, in the design of joints it is necessary that proper consideration be given to movements caused by shrinkage and swelling of wood members when detailing the connections.

Tension Perpendicular to Grain The ultimate strength of wood in tension perpendicular to grain is extremely low, and no design values are provided. Therefore, if possible, it is desirable to avoid loading wood in tension perpendicular to grain. Conditions that are conducive to creating tension perpendicular to grain stresses are notching at the end bearing of the member, hanging loads from the member below the neutral axis, and restraint caused by the connector detail. If these conditions cannot be avoided, then careful consideration must be given to the type and placement of fasteners and connectors.

14.6 LATERAL LOAD RESISTANCE

Buildings are subjected to a variety of loads during their lifetime. All loads resisted by the structure must be transferred into the foundation. Gravitational loads (roof live load, snow load, dead load) react vertically on the structure and are typically transferred to the foundation through load-bearing walls. Wind and earthquakes cause horizontal forces on the structures, thus leading to lateral loads. Lateral loads are transferred to the foundation through lateral force-resisting systems. For light-frame construction, the gravitational forces are typically resisted by nominal dimension lumber in the form of wall studs, and the lateral loads are commonly resisted by wood structural panel-sheathed shear walls and diaphragms.

Due to their relative light weight and high degree of ductility, wood-framed structures perform well when subjected to lateral design forces from high wind and seismic events. Wind forces act on the sides of the building and on sloped roofs. Seismic forces result from ground movement causing the structural mass to move back and forth in the horizontal direction.

Shear walls and diaphragms are designed to transfer in-plane and out-of-plane forces. When using these two assemblies to resist lateral design forces of buildings, the structural system is sometimes referred to as a "box system." The shear walls provide reactions for the roof and floor diaphragms and transmit the forces into the foundation.

Actually, with ordinary good construction practice, any wood structural panel-sheathed element in a building adds considerable strength and stiffness to the structure. Thus, if the walls and roofs are sheathed with panels and are adequately tied together and tied to the foundation, many of the requirements of a "box" structure are met. This fact explains the success of wood structural panel-sheathed buildings resisting loads in hurricane, tornado, and earthquake conditions even when they have not been engineered as diaphragms or shear walls per se.

14.6.1 Shear Walls

A shear wall is a vertical, cantilevered diaphragm that is constructed to resist lateral shear loads. This structural system transfers lateral forces from the top of the wall to the bottom of the wall and eventually transfers the lateral loads to the foundation. A shear wall segment is defined as a portion of the shear wall that runs from the diaphragm above to the diaphragm/foundation below, also known as "full-height segments." Shear wall segments occur between building wall discontinuities such as doors, windows, or corners in the shear wall.

Shear Wall Design

The detailed design of shear walls is beyond the scope of this handbook, but the following sections provide the overall guidelines for the design of shear walls. The AWC *Special Design Provisions for Wind and Seismic (SDPWS)* provides the most extensive information available for the design of both shear walls and diaphragms. The most common shear wall consists of wood structural panels attached to lumber framing members. But other sheathing materials such as plywood siding, particleboard sheathing, and structural fibreboard sheathing may be used.

Shear walls can fail due to either racking or sliding. Racking is prevented by the fasteners and hold-downs, and sliding is prevented by the anchor bolts into the foundation. The most common design situation is racking, and the *SDPWS* provides nominal unit shear capacities to resist racking for wood-framed walls for both wind and seismic. The wind capacities are 40 percent higher than the seismic capacities. The tabulated values are nominal values and must be divided by 2 for ASD and multiplied by 0.80 for LRFD. In addition to the base unit shear capacities for wood-based panels, the *SDPWS* also provides unit shear capacities values for wood structural panels applied over ½- and ⅝-in. gypsum wallboard.

A key design element of shear walls is the aspect ratio of the wall or full-height wall segments. As shown in Table 14.9 for a blocked wood structural panel shear wall, the aspect ratio that defines a full-height segment is 3.5:1. For unblocked shear walls, the aspect ratio is limited to 2:1.

Table 14.9 Maximum Shear Wall Aspect Ratios

Shear wall sheathing type	Maximum h/b_s ratio
Wood structural panels, unblocked	2:1
Wood structural panels, blocked	3.5:1
Particleboard, blocked	2:1
Diagonal sheathing, conventional	2:1
Gypsum wallboard	2:1
Portland cement plaster	2:1
Structural fibreboard	3.5:1

For wood structural panel shear walls with aspect ratios (h/b_s) greater than 2:1, the nominal shear capacity shall be multiplied by the Aspect Ratio Factor (WSP) = 1.25 – 0.125 h/b_s, where h is the shear wall height and is the minimum shear wall segment length.

Shear Walls without Openings Where individual full-height wall segments are designed as shear walls without openings, the aspect ratio limitations shall apply to each full-height wall segment. The following limitations shall apply:

a. Openings shall be permitted to occur beyond the ends of a shear wall. The length of such openings shall not be included in the length of the shear wall.

b. Where out-of-plane offsets occur, portions of the wall on each side of the offset shall be considered as separate shear walls lines.

c. Collectors for shear transfer to individual full-height wall segments shall be provided.

EXAMPLE

Assume a 20-ft shear wall that is 8 ft in height without openings. Assume a diaphragm shear force of 6000 lb is applied to the top of the wall. The unit shear to be resisted is 6000/20 = 300 plf. Assume WSP 7/16-in. sheathing with 8*d* nails is used. Using the shear wall capacity tables in the *SDPWS* the capacity for wind loading is 670/2 = 335 plf for 6 in. on center spacing at the panel edges. For seismic the spacing needs to be reduced to 4 in. on center with a capacity of 700/2 = 350 plf.

In addition to the shear nailing requirements, the designer must also determine the hold-down capacity needed. In this case $T = C = vh = 300 \times 8 = 2400$ lb. Lastly, the deflection of the shear wall should be checked. Shear wall deflection is important for checking drift limitations and in determining whether the diaphragm should be considered rigid or flexible.

The *SDPWS* provides the following three-part deflection equation:

$$\Delta_{sw} = \frac{8vh^3}{EAb} + \frac{vh}{1000G_a} + \frac{h\Delta_a}{b} \tag{14.44}$$

where

b = shear wall length, ft
Δ_a = total vertical elongation of wall anchorage system (including fastener slip, device elongation, rod elongation, etc.) at the induced unit shear in the shear wall, in.
E = modulus of elasticity of end posts, psi
A = area of end post cross section, in.^{2}
G_a = apparent shear wall shear stiffness from nail slip and panel shear deformation, kips/in.
h = shear wall height, ft
v = induced unit shear, lb/ft
Δ_{sw} = maximum shear wall deflection determined by elastic analysis, in.

Shear Walls with Openings For shear walls with openings there are three options available to the designer, these being the segmented, perforated, and force transfer around openings methods.

Traditional shear walls only consider full-height segments of wood structural panels as being effective in resisting shear forces. Therefore, any realistic wall line that contains openings will consist of one or more segments of shear walls. The traditional approach is called the segmented shear wall, which assumes the shear is carried by the full-height wall segments as previously defined. A downside to this method is that each full-height segment must also be detailed to resist overturning forces, which requires hold-downs at each end of the full-height segments.

EXAMPLE

Assume in Fig. 14.21 that a_1 and a_4 are 3 ft and a_2 and a_3 are 4 ft. Assume that the wall height is 8 ft and that the total wall length is 26 ft and that the shear force to be resisted is 4000 lb. Since all four of the wall segments meet the 3.5:1 aspect ratio, the unit shear is 3 + 3 + 4 + 4 = 4000/14 = 286 plf. However, since segments a_1 and a_4 are less than the 2:1 aspect ratio, the Aspect Ratio Factor is applicable, and the nominal shear capacity shall be multiplied by 1.25 – 0.125 h/b_s = 0.92. From the shear wall capacity tables in the *SDPWS* for a WSP 7/16-in. sheathing with 8*d* common nails at 6 in. on center, the capacity is 670 plf. Dividing by 2 and multiplying by 0.92 gives (980/2)0.92 = 308 plf, which is >286 plf.

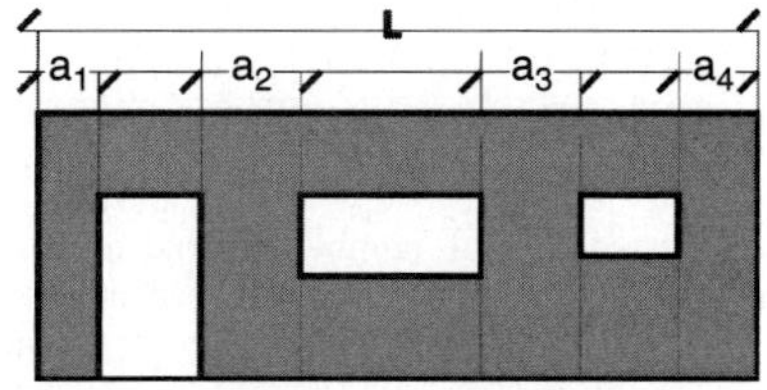

Figure 14.21 Wood-framed shear wall with openings.

Other than the traditional segmented approach, there are two methods for designing a shear wall with openings. One method is to reinforce the openings using strapping or framing and is referred to as the "force transfer around openings" method. The other option is to place an empirically developed penalty on the shear wall by reducing the shear wall capacity. The concept of providing an empirically developed penalty is the basis for the "perforated shear wall" approach. Based on a significant amount of monotonic and cyclic testing in both the United States and Japan, an empirically developed penalty was developed that accounts for maximum wall opening size and percent wall sheathing The perforated shear wall method was first developed in Japan, and a modified version was adopted into the *SDPWS*. The shear capacity adjustment factor is applied as a "reduction factor" as given in Table 14.10, which is Table 4.3.3.5 in *SDPWS*.

Table 14.10 Shear Resistance Adjustment Factor, C_o

	Maximum opening height ratio[a] and height				
Wall height (h)	h/3	h/2	2h/3	5h/6	h
8'-0"	2'-8"	4'-0"	5'-4"	6'-8"	8'-0"
10'-0"	3'-4"	5'-0"	6'-8"	8'-4"	10'-0"
Percent full-height sheathing[b]	Shear capacity adjustment factor				
10%	1.00	0.69	0.53	0.43	0.36
20%	1.00	0.71	0.56	0.45	0.38
30%	1.00	0.74	0.59	0.49	0.42
40%	1.00	0.77	0.63	0.53	0.45
50%	1.00	0.8	0.67	0.57	0.50
60%	1.00	0.83	0.71	0.63	0.56
70%	1.00	0.87	0.77	0.69	0.63
80%	1.00	0.91	0.83	0.77	0.71
90%	1.00	0.95	0.91	0.87	0.83
100%	1.00	1.00	1.00	1.00	1.00

The aspect ratio limitations previously discussed shall apply to perforated shear wall segments within a perforated shear wall line. Portions of walls with aspect ratios exceeding 3.5:1 shall not be considered in the sum of shear wall segments. In the design of perforated shear walls to resist seismic forces, the nominal shear capacity of the perforated shear wall shall be multiplied by $2b_s/h$ when the aspect ratio of the narrowest perforated shear wall segment included in the sum of shear wall segment lengths, ΣL_i, is greater than 2:1 but does not exceed 3.5:1.

EXAMPLE

Referring back to Fig. 14.21, assume the height of the man door is 6 ft 8 in. and the height of the window opening are 2 ft 8 in. Assume the shear load to be resisted is 4000 lb is due to wind. The unit shear is 4000/14 = 286 plf. The percentage of full-height sheathing is 14/26 or 54 percent.

The maximum ratio height opening is 5 *h*/6. Interpolating from Table 14.10, the shear capacity adjustment factor is 0.60.

Referring to the wind load shear wall capacity tables in the *SDPWS* for a WSP 7/16-in. sheathing with 8*d* common nails at 4 in. on center the capacity is 980 plf. Dividing by 2 and multiplying by 0.60 gives (980/2)0.60 = 294 plf, which is >286 plf.

Therefore, the penalty for using the perforated shear wall method vs. the segmented method is that the nail spacing must be reduced to 4 in. on center vs. 6 in. on center. However, only 2 hold-downs are required vs. 8 for the segmented approach.

In summary the main advantage of the perforated shear wall approach is that it reduces the number of hold-downs, is simple to apply, and can lead to very minor penalties for small openings. The disadvantage of this system is that it can lead to very large penalties for large opening sizes, which essentially forces the designer to use either the force transfer around openings method or the traditional segmented approach or to redefine the perforated wall to eliminate large openings.

The force transfer around openings FTAO, (which is similar to the "pier method" in masonry design), often referred to as the pier method, treats the entire wall as one monolithic unit, hence reducing the number of hold-downs. This method requires the shear forces to be transferred around the openings using the sheathing above and below openings and may require additional metal strapping or blocking to supplement the strength of the sheathing. The *SDPWS* states that the design for force transfer around openings shall be based on a rational analysis but does not provide specific guidelines for what constitutes a rational analysis.

Historically, a number of rational analysis methods have been used, the most common being the Drag Strut, Cantilever Beam, Coupled Beam & Diekmann methods. The 1997 SEAOC Bluebook credits Ed Diekmann, S.E., with developing his approach circa 1982. It was first published in the *McGraw-Hill Handbook of Wood Design and Construction.*

In 2011, APA published Report M410, *Evaluation of Force Transfer around Openings—Experimental and Analytic Studies.* This was a joint project of APA, the USDA Forest Product Lab, and the University of British Columbia. The primary focus of this study was to conduct testing to verify the accuracy of the above noted four rational analysis methods. The researchers concluded that the Diekmann method most accurately predicted the performance of the test walls. In addition to the APA report, another excellent reference on FTAO is the *Design of Wood Structures* by Breyer et al.

Regardless of the rational analysis used, the *SDPWS* applies the following limitations:

1. The length of each wall pier shall not be less than 2 ft.
2. A full-height wall segment shall be located at each end of a force-transfer shear wall.
3. Where out-of-plane offsets occur, portions of the wall on each side of the offset shall be considered as separate force-transfer shear walls.
4. Collectors for shear transfer shall be provided through the full length of the force-transfer shear wall.

The *SDPWS* aspect ratio limitations shall apply to the overall shear wall including openings and to each wall pier at the sides of openings. The height of a wall pier with an opening on one side shall be defined as the clear height of the pier at the side of the opening. The height of a wall pier with an opening on each side shall be defined as the larger of the clear heights of the pier at the sides of the openings. The length of a wall pier shall be defined as the sheathed length of the pier. Wall piers with aspect ratios exceeding 3.5:1 shall not be considered as portions of force-transfer shear walls.

In summary, two of the main advantages of the FTAO method are that it will reduce the number of hold-downs required for a wall line and that it allows the designer to use narrower piers since the aspect ratio limitations in the building code apply to the pier as opposed to the full-height segments. The disadvantage of this method is that it can require a large amount of detailed calculations, especially if the wall becomes very complicated. It also requires special field detailing, extra blocking, extra nailing, and/or additional straps to achieve the field detailing.

14.6.2 Diaphragms

Wood-framed panel diaphragms are used extensively for roofs and floors in both new construction and rehabilitation of older buildings. A complete diaphragm analysis includes analyzing chord forces, connections, and tie-downs. A diaphragm acts in a manner analogous to a deep beam where the panels act as a "web" resisting shear while the diaphragm edge members perform the function of "flanges" resisting bending stresses such as shown in Fig. 14.22.

These edge members are commonly called chords in diaphragm design. Chords in a diaphragm carry all "flange" stresses acting in simple tension and compression rather than sharing these stresses significantly with the web. As in any beam, consideration must be given to bearing stiffeners, continuity of webs and chords, and web buckling that is normally resisted by framing members.

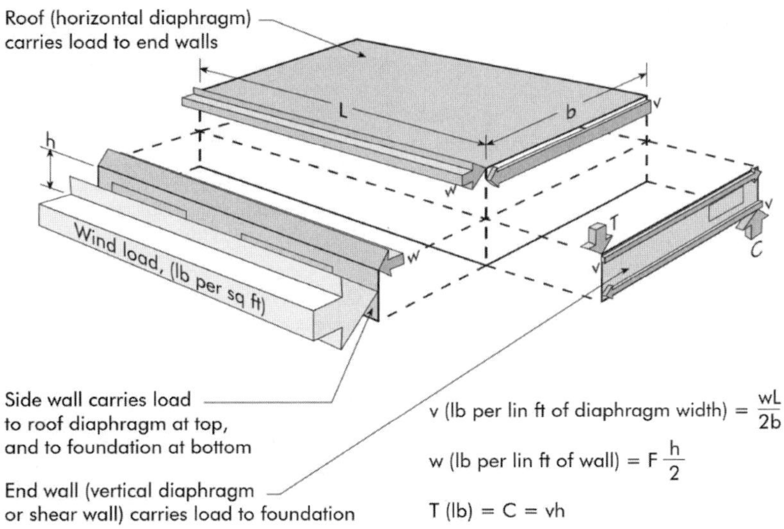

Figure 14.22 Wood structural panel diaphragm.

The forces acting on a flexible diaphragm with length, L, width, B resisting a uniform design load such as wind w are as follows:

$$\text{Diaphragm reaction} = \frac{Lw}{2} \tag{14.45}$$

$$\text{Diaphragm unit shear} = \frac{Lw}{2B} \tag{14.46}$$

$$\text{Diaphragm moment} = \frac{L^2 w}{8} \tag{14.47}$$

$$\text{Max. chord force} = \frac{wL^2}{8D} \tag{14.48}$$

These chord forces are often assumed to be carried by the double top-plate of the supporting perimeter walls. Given the magnitude of the forces involved in most light-framed wood construction projects, the double top-plate has sufficient capacity to resist the tensile and compressive forces assuming adequate detailing at the splice locations.

Aspect Ratios

Size and shape of diaphragms shall be limited to the aspect ratios in Table 14.11.

Table 14.11 Maximum Diaphragm Aspect Ratios for Horizontal or Sloped Diaphragms

Diaphragm sheathing type	Maximum L/W ratio
Wood structural panel, unblocked	3:1
Wood structural panel, blocked	4:1
Single-layer straight lumber sheathing	2:1
Single-layer diagonal lumber sheathing	3:1
Double-layer diagonal lumber sheathing	4:1

Diaphragm Design

Diaphragms vary considerably in load-carrying capacity, depending on whether they are "blocked" or "unblocked." Diaphragms are also designed based on the distribution of shear to vertical resisting elements based on an analysis where the diaphragm is modeled as semirigid, idealized as flexible or idealized as rigid.

Blocked vs. Unblocked Where diaphragms are designated as blocked, all joints in sheathing shall occur over and be fastened to common framing members or common blocking. Blocking consists of lightweight nailers, usually 2 × 4s, framed between the joists or other primary structural supports for the specific purpose of connecting the edges of the panels. The reason for blocking in diaphragms is to provide for connection of panels at all edges for better shear transfer. Where diaphragms are designated as unblocked, blocking between supporting structural members at panel edges shall not be required. Unblocked diaphragm loads may be controlled by buckling of the unsupported panel edges, which would result in an ultimate capacity of the diaphragm being unaffected by increasing the nail schedule since more nailing will have little effect on the buckling performance of the panels.

The *SDPWS* provides unit shear capacity tables for both blocked and unblocked diaphragms for both wind and seismic loading. *SDPWS* also provides a table for high load diaphragms and a table for lumber diaphragms. As with the tables for shear walls previously discussed, the tabulated values for diaphragms are nominal values and must be divided by 2 for ASD and multiplied by 0.80 for LRFD. It is noted that the tabulated diaphragm resistance depends on the direction of continuous panel joints with respect to the loading direction and direction of framing members. Six load cases identified as 1, 2, 3, 4, 5, and 6 are presented in the *SDPWS*.

EXAMPLE

Check a simple diaphragm for wind loading assuming it is unblocked and is supported by 2x framing members with the loading direction and direction of framing members undefined. Assume it must resist an ASD wind load of 250 lb/ft.

Using the *SDPWS* table for "Unblocked Diaphragm" and since the loading conditions are unknown, use the "Cases 2,3,4,5,6" since these values will be conservative.

Check "Sheathing and single floor" since Structural I may not be readily available in all areas. Similarly, check only rows with 2-in. minimum nominal framing width. From the table, it is determined that 8*d* nails with 15/32-in. sheathing over 2x framing yields a capacity of 505 lb/ft with a minimal 6- and 12-in. nail spacing. For ASD this results in a capacity of 252 lb/ft, which is greater than the required 250 lb/ft. Therefore, this selection is okay.

EXAMPLE

Check a simple diaphragm for seismic loading assuming it is blocked and is supported by 2x framing members with a case 5&6 panel orientation Assume it must resist an ASD seismic load of 250 lb/ft.

Using the *SDPWS* table for "Blocked Diaphragm," check "Sheathing and single floor" since Structural I may not be readily available in all areas. Similarly, check only rows with 2-in. minimum nominal framing width. From the table, it is determined that 8*d* nails with 15/32 in. sheathing over 2x framing yields a capacity of 540 lb/ft with a nail spacing of 6 in. at all panel edges. For ASD this results in a capacity of 270 lb/ft, which is greater than the required 250 lb/ft. Therefore, this selection is okay.

Flexible vs. Rigid A flexible diaphragm is when the diaphragm deflection is two or more times the average shear wall deflection. When a diaphragm is idealized as flexible, the diaphragm shear forces shall be distributed to the vertical resisting elements based on tributary area. This assumption implies that the diaphragm behaves similar to simply supported beams spanning from shear wall to shear wall. Half of the reaction, thus shear force, is distributed to one shear wall, and the other half of the reaction is distributed to the other shear wall. The force transferred to each shear wall is based on the tributary area of each wall. This assumption makes the distribution of shear forces a fairly simple calculation, and historically wood-framed diaphragms have been assumed to be flexible.

When a diaphragm is idealized as rigid, the diaphragm shear forces shall be distributed to the shear walls based on the relative lateral stiffness of the vertical resisting elements of the story below. It shall be permitted to idealize a diaphragm as rigid when the computed maximum in-plane deflection of the diaphragm itself under lateral load is less than or equal to two times the average deflection of adjoining vertical elements of the lateral force-resisting system of the associated story under equivalent tributary lateral load. Although this approach makes engineering sense, the reality is that determining shear wall stiffness is somewhat problematic. When a diaphragm is not idealized as rigid or flexible, the diaphragm shear forces shall be distributed to the vertical resisting elements using a semirigid diaphragm analysis based on the relative stiffness of the diaphragm and the vertical resisting elements accounting for both shear and flexural deformations. In lieu of using a semirigid diaphragm analysis for distribution of story shear forces, it shall be permitted to use an "envelope analysis" whereby distribution of horizontal diaphragm shear to each vertical resisting element is the larger of the shear forces resulting from analyses where the diaphragm is idealized as flexible and the diaphragm is idealized as rigid.

Calculations of diaphragm deflection shall account for bending and shear deflections, fastener deformation, chord splice slip, and other contributing sources of deflection.

SDPWS provides the following three-part equation for calculating the diaphragm deflection:

$$\Delta_{sw} = \frac{5vL^3}{8EAW} + \frac{0.25vL}{1000Ga} + 0.188L_{e_n} + \frac{\sum(\Delta_c X)}{2W} \tag{14.49}$$

where

E = modulus of elasticity of diaphragm chords, psi
A = area of chord cross section, in.^{2}
Ga = apparent diaphragm shear stiffness from nail slip and panel shear deformation, kips/in.
L = diaphragm length, ft
v = induced unit shear in diaphragm, lbs/ft
W = diaphragm width, ft
x = distance from chord splice to nearest support ft
Δ_c = diaphragm chord splice slip, in., at the induced unit shear in diaphragm.
Δ = maximum mid-span diaphragm deflection determined by elastic analysis, in.

Note that this deflection calculation does not apply to unblocked diaphragms. Limited testing of diaphragms by APA suggests that the deflection of an unblocked diaphragm at its tabulated allowable shear capacity will be about 2.5 times the calculated deflection of a blocked diaphragm of similar construction and dimensions, at the same shear capacity. If diaphragm framing is spaced more than 24 in. o.c., testing indicates a further increase in deflection of about 20 percent for unblocked diaphragms (e.g., to 3 times the deflection of a comparable blocked diaphragm). This relationship can be used to develop an estimate of the deflection of unblocked diaphragms.

Other Diaphragm Topics

SDPWS also provides design guidelines for torsional irregularity of diaphragms and diaphragms in open front structures. A detailed discussion of these and other diaphragm design topics such as the design of subdiaphragms and diaphragms with openings is beyond the scope of this Handbook. The reader is referred to the McGraw-Hill reference books *The APA Engineered Wood Handbook* and *The Analysis of Irregular Shaped Structure, Diaphragms and Shear Walls* by Malone and Rice for an in-depth analysis of these and other related topics.

Fire Resistance of Wood Framing

One of the downsides often associated with wood construction is the fact that wood is combustible and subject to degradation when exposed to fire. Though building codes are concerned with how fast fire can spread on a room's surface, they are even more specific about fire resistance: the measure of containment of fire within a room or building. It is defined as protection against fire penetrating a wall, floor, or roof, either directly or through a high rate of heat transfer that might cause combustible materials to be ignited on the side of the wall or floor away from the actual fire. Thus, it is a property of an assembly of several materials, including fastenings, and of the workmanship.

Fire-resistive construction provides time to discover a fire, to restrict or suppress it before it spreads, and to evacuate the building if necessary. The IBC recognizes that wood can be used in 1-hour and higher rated assemblies in Type IIIA and Type VA construction. A 1-hour rating, for example, is taken to mean that an assembly similar to that tested will not collapse or transmit flame or a high temperature while supporting its full load for at least 1 hour after the fire commences.

There are numerous fire-rated assemblies combining wood framing and wood structural panels with fire-resistive gypsum wallboard that are especially suitable for nonresidential and multifamily residential buildings. They include 1-hour rated and several 2-hour rated protected wood-framed floor-ceiling, and roof-ceiling systems.

In these assemblies, materials such as gypsum wallboard, plaster, and acoustical tile provide primary fire protection. The panel floor or roof acts to prevent flame passage and temperature rise as well as to reinforce wood framing members against collapse under load after the effectiveness of the ceiling has been lost.

Because these systems contain wood and possibly other combustible materials, they are designated as combustible constructions. At present, codes don't permit their use in so-called noncombustible (Type I or II) structures, even though their tested performance meets performance requirements that are the same as for assemblies classified as noncombustible. Full-scale tests on assemblies representative of intended floor- or roof-ceiling construction are conducted by fire testing laboratories that are recognized by building code evaluation services. Many of these laboratories publish a listing directory that is updated annually or periodically, describing the materials and construction of the tested assemblies and the fire endurance rating obtained in the test. One such directory is the Underwriters Laboratory Inc. (UL) Fire Resistance Directory. Over 40 wood floor-ceiling (or roof-ceiling) systems using wood structural panels are listed in the

UL directory. Listed designs in fire directories of recognized fire testing laboratories are used by architects and designers for building design and construction and are accepted by building code officials for fire-rated construction.

Other excellent sources for fire rated assemblies are APA *Design/Construction Guide: Fire-Rated Systems* and AWC *DCA 3—Fire-Resistance Rated Wood Floor and Wall Assemblies.* The IBC also permits the use of the component additive calculation methodology (CAM) as an alternate to prescriptive or tested assemblies for 1-hour fire-rated wood-framed floors, roofs, and load-bearing and nonbearing walls. The code provides tables of assigned times for components, which have been developed empirically from extensive studies of assemblies tested with nominal 2-in. wood framing in accordance with ASTM Standard E 119 *Standard Test Methods for Fire Tests of Building Construction and Materials.* End-point criteria in the standard also were considered. A 1-hour fire-rated assembly can be determined by combining the individual component times of the assembly in accordance with the method and limitations in the IBC, thereby providing additional choices for the designer.

In addition, it has long been recognized that large timber members have excellent fire-resistive characteristics as recognized by the IBC which classify heavy timber as a specific type of construction (Type IV) and give minimum sizes for roof and floor beams. The requirements set out for heavy timber construction in the codes do not constitute 1-hour fire resistance. However, procedures are available to estimate the member size required for projects in which a rated fire resistance is required.

In the presence of fire, the outer portion of the heavy timber members becomes charred. This layer of charred wood then functions as an insulator, helping to protect the undamaged interior of the member from the heat. The rate of advancement of this insulating char layer into the remaining, undamaged portion of the member has been measured (approximately 1.5 in. per hour) and forms the theoretical basis of the equations used to determine the required size of exposed timber beams and columns with a minimum nominal dimension of 6 in. Chapter 16 of the NDS, Fire Design of Wood Members, provides a mechanics-based approach to determine the required required size of wood members to achieve a 1-hour, 1½-hour, or 2-hour fire endurance rating. This applies to sawn timber, glulam, and SCL. Provisions for determining fire endurance ratings for CLT were added in the 2015 NDS.

The latest innovation in wood construction is called mass timber. Mass timber is a term describing the use of building construction systems incorporating large structural wood components such as glulam timber, nail-laminated timber, and cross-laminated timber, all of which have demonstrated unique fire resistance characteristics. Due to the acknowledged fire resistance of mass timber, the 2021 IBC will permit the use of mass timber in buildings of up to 270 ft and 18 stories by the adoption of Type IV-A, Type IV-B, and Type IV-C construction.

REFERENCES

AITC Timber Construction Manual, 6th Edition, John Wiley & Sons, Somerset, NJ.

ANSI 05.1 Round Timber Poles for Utility Structures, American Wood Protection Association, Birmingham AL.

ANSI/APA 117 Standard Specification for Glued Laminated Timber of Softwood Species, APA-The Engineered Wood Association, Tacoma, WA.

ANSI/APA A190.1, American National Standard for Structural Glued Laminated Timber.

ANSI/APA PRG320 Standard for Performance-Rated Cross-Laminated Timber, APA—The Engineered Wood Association, Tacoma, WA.

APA Design/Construction Guide: Fire-Rated Systems, APA—The Engineered Wood Association, Tacoma, WA.

APA Engineered Wood Handbook. McGraw Hill Handbooks, New York, NY.

APA PRI 405, Performance Standard for APA-EWS Commercial I-Joists, APA—The Engineered Wood Association, Tacoma, WA.

APA PRI-400, Performance Standard for APA-EWS I-Joists, APA—The Engineered Wood Association, Tacoma, WA.

APA Report M410, Evaluation of Force Transfer around Openings—Experimental and Analytic Studies, APA—The Engineered Wood Association, Tacoma, WA.

APA-The Engineered Wood Association, Tacoma, WA.

ASCE 16, Standard for LRFD for Engineered Wood Construction, ASCE, Reston, VA .

ASTM Standard 2559, Standard Specification for Adhesives for Bonded Structural Wood Products for Use Under Exterior Exposure Conditions, ASTM West Conshohocken, PA.

ASTM Standard 7247, Test Method for Evaluating the Shear Strength of Adhesive Bonds in Laminated Wood Products at Elevated Temperatures, ASTM West Conshohocken, PA.

ASTM Standard D 25, Standard Specification for Round Timber Piles. ASTM West Conshohocken, PA.

ASTM Standard D 3737, Standard Practice for Establishing Stresses for Structural Glued Laminated Timber, ASTM West Conshohocken, PA.

ASTM Standard D 7341, Standard Practice for Establishing Characteristic Values for Flexural Properties of Structural Glued Laminated Timber by Full-Scale Testing, ASTM West Conshohocken, PA.

ASTM Standard D3200, Standard Specification and Test Method for Establishing Recommended Design Stresses for Round Timber Construction Poles, ASTM West Conshohocken, PA.

ASTM Standard D5055, Standard Specification for Establishing and Monitoring Structural Capacities of Prefabricated Wood I-Joists, ASTM West Conshohocken, PA.

ASTM Standard D5456, Standard Specification for Evaluation of Structural Composite Lumber Products. ASTM West Conshohocken, PA.

ASTM Standard E 119. Standard Test Methods for Fire Tests of Building Construction and Materials, ASTM West Conshohocken, PA

ASTM Standard F 1667 Standard Specification for Driven Fasteners: Nails, Spikes, and Staples, ASTM West Conshohocken, PA.

AWC DCA 3—Fire-Resistance Rated Wood Floor and Wall Assemblies, American Wood Council, Leesburg, VA.

AWC Design Aid No. 1, Application of Technical Report 12 for Lag Screw Connections, American Wood Council, Leesburg, VA.

AWC Special Design Provisions for Wind and Seismic (*SDPWS*), American Wood Council, Leesburg, VA.

Blass, H. J., and Fellmoser, P. (2004). Design of solid wood panels with cross layers. In Proceedings of the 8th World Conference on Timber Engineering, June 14–17, Lahti, Finland, 2:543–548.

Breyer et al., *Design of Wood Structures*, McGraw-Hill, New York, NY.

ESR 1539, Power Driven Nails and Staples, ISANTA, Chicago, IL.

Fellmoser, P., and Blass, H. J. (2004). Influence of rolling shear modulus on strength and stiffness of structural bonded timber elements. In Proceeding of CIB-W18 Meeting, Edinburgh, United Kingdom, paper 37-6-5.

FPInnovations (Institute) (2011). *FPInnovations*, SP-528E.

Gagnon, S., and Pirvu, C. (2011). *CLT Handbook: Cross-Laminated Timber.*

ICC ES AC14, Acceptance Criteria for Prefabricated Wood I-Joists, ICC Evaluation Service.

ICC ES AC47, Acceptance Criteria for Structural Composite Lumber, ICC Evaluation Service.

Karacabeyli, E., and Douglas, B. (2013). *CLT Handbook: Cross-Laminated Timber*; FPInnovations (Institute), Binational Softwood Lumber Council, FPInnovations, SP-529E.

Kreuzinger, H. (1995). Mechanically jointed beams and columns. In *Timber Engineering—STEP 1*, ed. H.J. Blass et al., B11/1-8. Almere, The Netherlands: Centrum Hout.

Kreuzinger, H. (1999). Platten, Scheiben und Schalen—ein Berechnungsmodell für gängige Statikprogramme. bauen mit holz 1: 34–39.

Malone and Rice, *The Analysis of Irregular Shaped Structures Diaphragms and Shear Walls*, McGraw Hill, New York, NY.

McGraw-Hill Handbooks, *McGraw-Hill Handbook of Wood Design and Construction,* McGraw-Hill, New York, NY.

National Design Specification for Wood Construction. American Wood Council, Leesburg, VA.

NDS Supplement, National Design Specification Design Values for Wood Construction. American Wood Council, Leesburg, VA.

Underwriters Laboratory Inc. (UL) Fire Resistance Directory, UL, Northbrook, IL.

Voluntary Product Standard PS 20-15, American Softwood Lumber Standard, Department of Commerce, National Institute of Standards and Technology, Washington, DC.

Voluntary Product Standard PS1, Structural Plywood, National Institute of Standards and Technology, Washington, DC.

Voluntary Product Standard PS2, Performance Standard for Wood-Based Structural Use Panels, National Institute of Standards and Technology, Washington, DC.

Wood Handbook, Wood as an Engineering Material, USDA, Forest Service, Forest Products Laboratory, FPL-GTR -190.

Chapter 15
Bridge Engineering

PART A
STEEL AND CONCRETE BRIDGES

BY
SOLIMAN KHUDEIRA *Illinois Institute of Technology and Chicago DOT*

15.1 GENERAL[a]

15.1.1 Introductory Remarks

Bridge design involves three interdependent components: structural adequacy, safety, and aesthetics. All three are equally important, although many designers have concerned themselves only with structural adequacy without giving concern to the appearance of the structure. Worldwide, there is interest in improving the appearance of bridges, and it behooves bridge designers to give attention to the aesthetics of their structures.

15.1.2 Aesthetics

Although beauty in anything is somewhat subjective and undefinable, experience has shown that there are some general guidelines which, if followed, will result in a structure of improved appearance. Bearing in mind that a bridge may well serve for 100 years or more, the designer should become familiar with the aesthetic techniques that may be used to improve the appearance of a bridge so that the bridge will be a credit to its designer and to itself for many decades to come. The following are some special areas of concern.

PROPORTIONS AND LINES

All the members should bear a pleasing relationship to each other. Columns should not be spindly, nor should superstructures be too heavy. Span lengths should be chosen carefully. Where there are no other constraints, an odd number is better than an even number, with the longest span in the center. Lines should be mainly horizontal so that the structure seems to flow from end to end. Intermediate supports should be subdued so that they do not break the dominant horizontal flow. This means pushing the bent caps up into the superstructure or covering their ends with skirts to make the longitudinal lines continuous over the supports. The bridge should look like it was designed as a whole, not as a lot of individual parts glued together. Bridges using precast concrete or simple steel plate girders require special treatment to conceal the caps and extend the horizontal lines over the supports. Giving the soffit lines a slight arch, more than the normal camber, will often improve the overall appearance.

[a]Original article by Arthur L. Elliott, Bridge Engineer, Sacramento, California, and Richard J. LeBeau, Imbsen Assoc., Sacramento, California.

PLEASING SHAPES

Appearance can be enhanced by avoiding stereotyped round or square columns and flat soffits. A column which may be adequate structurally will often be too small to be good looking. Add enough size to make the proportions seem right. Columns which are smaller at the bottom than the top will help make the structure seem lighter and more dynamic. Make sure the columns are all the same type and size—both for appearance and for reusability of forms. Avoid large expanses of plain concrete. Break them up with ledges to create shadow lines or use heavily textured panels. Slope the sides of box girders back under the bridge and give the bottom edge a large curved fillet to diminish its apparent depth.

COMPATIBILITY

Make the bridge fit its surroundings. This is known as *context-sensitive design*, and is also called a *context-sensitive solution*. If the bridge looks like it belongs there, it will be a pleasing addition to its environment. If it is a misfit, it will be an eyesore forever. Even industrial or utilitarian bridges need not be aesthetically pleasing, to a certain level. It is worthwhile to make every bridge have an appearance of quality and careful craftsmanship.

Implementation of those features must be balanced with their effects on the structural adequacy of the bridge as a whole.

15.2 LOADS

15.2.1 Introduction

Bridge structures are designed for the following applicable permanent loads: force effects due to creep, downdrag force, dead load of structural components and nonstructural attachments, dead load of wearing surfaces and utilities, horizontal earth pressure load, miscellaneous locked-in force effects resulting from the construction process, earth surcharge load, vertical pressure from dead load of earth fill, secondary forces from posttensioning, and force effects due to shrinkage. Bridge structures are also designed for the following applicable transient loads: blast loading, vehicular braking force, vehicular centrifugal force, vehicular collision force, vessel collision force, earthquake load, friction load, ice load, vehicular dynamic load allowance, vehicular live load, live load surcharge, pedestrian live load, force effect due to settlement, force effect due to temperature gradient, force effect due to uniform temperature, water load and stream pressure, wind on live load, and wind load on structure.

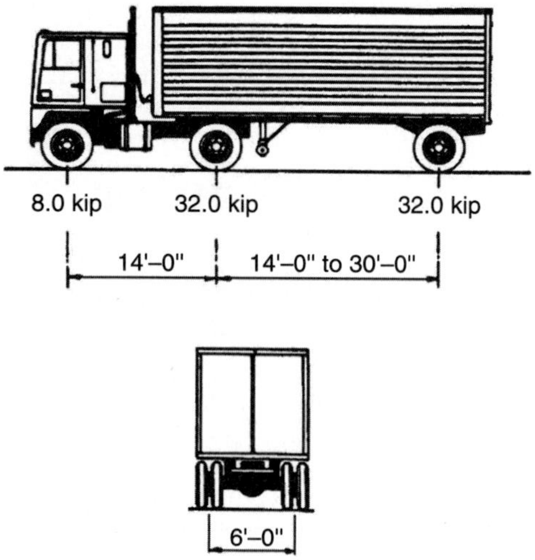

Figure 15.1a AASHTO design truck.

DEAD LOAD

Dead load shall include the weight of all components of the structure, appurtenances and utilities attached thereto, earth cover, wearing surface, future overlays, and planned widenings.

HIGHWAY LIVE LOADS

The live loading for the design of highway structures is that of the American Association of State Highway and Transportation Officials (AASHTO). Their Standard Specifications for Highway Bridges[1] is the accepted guide for all highway bridge work done by public agencies in the United States. Adherence to these specifications is recommended for all highway bridge work. In addition, consideration should be given to overloads that are legally permitted by the controlling jurisdiction. Vehicular live loading on bridges, designated by AASHTO as HL-93, shall consist of a combination of (a) design truck or design tandem and (b) design lane load. Each design lane under consideration shall be occupied by either the design truck or tandem, coincident with the lane load, where applicable.

DESIGN TRUCK

The weights and spacings of axles and wheels for the design truck shall be as specified in Fig. 15.1*a*. A dynamic load shall be considered as specified in AASHTO specification. The spacing between the two 32.0-kip axles shall be varied between 14.0 ft and 30.0 ft to produce extreme force effects.

DESIGN TANDEM

The design tandem shall consist of a pair of 25.0-kip axles spaced 4.0 ft apart. The transverse spacing of wheels shall be taken as 6.0 ft. A dynamic load shall be considered as specified in per AASHTO specification.

FATIGUE LOAD

The fatigue load shall be one design truck or axles, but with a constant spacing of 30.0 ft between the 32.0-kip axles. Dynamic load shall be applied to the fatigue load. The loading pattern shown in Fig. 15.1*b* shall be used for fatigue design.

DESIGN LANE LOAD

The design lane load shall consist of a load of 0.64 klf uniformly distributed in the longitudinal direction. Transversely, the design lane load shall be assumed to be uniformly distributed over a 10.0-ft width. The force effects from the design lane load shall not be subject to a dynamic load. The lane load is assumed to occupy any position within their individual design traffic lanes which will produce the maximum stress. Figure 15.2 shows the design lane load and the design truck load.

RAILROAD LIVE LOADS

Live loading for railroad structures is usually based on the wheel loads of two locomotives followed by a uniform load representing the weight of the heaviest cars (Table 15.4). Most designs are made for Cooper's E-80 loading.

15.2.2 Load Combinations

Loads do not act individually but rather in various combinations. For example, the vehicle live load could be on the bridge at the same time as wind force, stream force, centrifugal force, and other forces. To account for this, AASHTO has developed a set of load combinations which are divided into various groups (or limit states). The following permanent and transient loads and forces shall be considered in bridge design:

Permanent Loads

- CR = force effects due to creep
- DD = downdrag force
- DC = dead load of structural components and nonstructural attachments
- DW = dead load of wearing surfaces and utilities
- EH = horizontal earth pressure load
- EL = miscellaneous locked-in force effects resulting from the construction process, including jacking apart of cantilevers in segmental construction
- ES = earth surcharge load
- EV = vertical pressure from dead load of earth fill

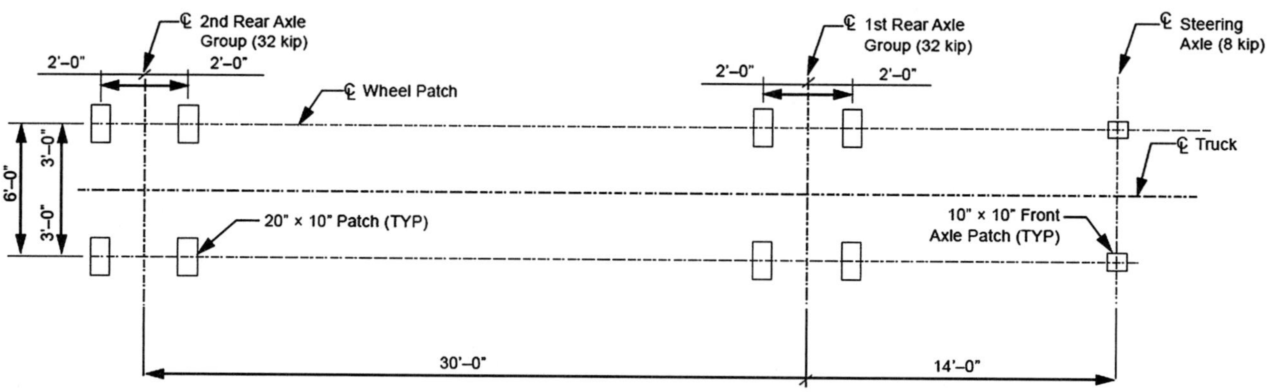

Figure 15.1b AASHTO design truck for fatigue design.

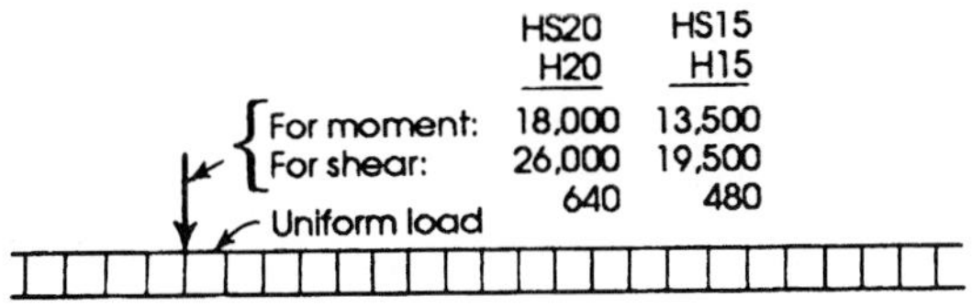

Figure 15.2 AASHTO lane load.

PS = secondary forces from posttensioning
SH = force effects due to shrinkage
Transient Loads
BL = blast loading
BR = vehicular braking force
CE = vehicular centrifugal force
CT = vehicular collision force
CV = vessel collision force
EQ = earthquake load
FR = friction load
IC = ice load
IM = vehicular dynamic load allowance
LL = vehicular live load
LS = live load surcharge
PL = pedestrian live load
SE = force effect due to settlement
TG = force effect due to temperature gradient
TU = force effect due to uniform temperature
WA = water load and stream pressure
WL = wind on live load
WS = wind load on structure

The total factored force effect shall be taken as: $Q = \Sigma\, \eta_i\, \gamma_i\, Q_i$
where:

η_i = load modifier
Q_i = force effects
γ_i = load factors

Each component and connection of a bridge shall satisfy the following equation for each limit state:

$$\Sigma \mu_i \gamma_i Q_i \leq \phi R_n$$

where ϕR_n is the factored resistance.

The following limit states are specified by AASHTO:

Strength I—Basic load combination relating to the normal vehicular use of the bridge without wind.

Strength II—Load combination relating to the use of the bridge by Owner-specified special design vehicles, evaluation permit vehicles, or both without wind.

Strength III—Load combination relating to the bridge exposed to wind velocity exceeding 55 mph.

Strength IV—Load combination relating to very high dead load to live load force effect ratios.

Strength V—Load combination relating to normal vehicular use of the bridge with wind of 55-mph velocity.

Extreme Event I—Load combination including earthquake. The load factor for live load, γEQ, shall be determined on a project-specific basis.

Extreme Event II—Load combination relating to ice load, collision by vessels and vehicles, check floods, and certain hydraulic events with a reduced live load other than that which is part of the vehicular collision load, CT. The cases of check floods shall not be combined with BL, CV, CT, or IC.

Service I—Load combination relating to the normal operational use of the bridge with a 55-mph wind and all loads taken at their nominal values. Also related to deflection control in buried metal structures, tunnel liner plate, and thermoplastic pipe; to control crack width in reinforced concrete structures; and for transverse analysis relating to tension in concrete segmental girders. This load combination should also be used for the investigation of slope stability.

Service II—Load combination intended to control yielding of steel structures and slip of slip-critical connections due to vehicular live load.

Service III—Load combination for longitudinal analysis relating to tension in prestressed concrete superstructures with the objective of crack control and to principal tension in the webs of segmental concrete girders.

Service IV—Load combination relating only to tension in prestressed concrete columns with the objective of crack control.

Fatigue I—Fatigue and fracture load combination related to infinite load-induced fatigue life.

Fatigue II—Fatigue and fracture load combination related to finite load-induced fatigue life.

The load factors for various loads comprising a design load combination shall be taken as specified in Tables 15.1*a* and *b*.

15.2.3 Maximum Moments and Shears in Simple Spans

The maximum moment caused by a group of wheels in a simple span will occur under one of the wheels when the center of gravity of the loads within the given span and the point of application of that wheel are equidistant from the center of the span. This is illustrated for a three-axle truck in Fig. 15.3. Envelopes of maximum moment in a simple span may be approximated by a second-degree parabola whose ordinate at midspan is the combined live- and dead-load moments. These envelopes are used for determining cutoff points for reinforcing steel in concrete beams, lengths of cover plates in steel beams, or the geometric path of prestressing steel. Ordinates to the true moment envelope and the parabolic envelope are given in the following table for a 50-ft simple span for which the dead load is 1 klf and the live load 1.5 wheel lines of the standard HS20-44 truck.

Distance from support, ft	Dead-load moment, ft-kips	Live-load moment, ft-kips	Impact 28.6%	Total moment, ft-kips	Parabola, ft-kips	% variation
5	113	194	56	363	329	−9
10	200	334	96	630	585	−7
15	263	405	116	784	768	−2
20	300	463	132	895	877	−2
22.7	310	471	135	916	906	−1
25	313	467	134	914	914	0

Highway Loadings

The following formulas give maximum moments and shears resulting from one lane of HS20-44 loading on a simple span of length L, impact not included:

Span, ft	Max moment, ft-kips
0–23.9	8L
23.9	16L + 784/L − 224
33.8–145.6	18L + 392/L − 280
>145.6 (equivalent load)	0.08L² + 4.5L

Span, ft	Max shear, kips
0–14	32
14–28	64–448/L
28–127.5	72–672/L
>127.5 (equivalent load)	0.32L + 26

Table 15.2 gives moments, shears, and reactions for the HS20-44 loading for spans ranging from 1 to 270 ft.

Railroad Loadings

Table 15.3 gives the position of wheels of the Cooper load which produce maximum moments. Table 15.4 gives summations of moments about any wheel of any number of wheels on either side of the moment

Table 15.1a —Load Combinations and Load Factors

Load combination limit state	*DC* *DD* *DW* *EH* *EV* *ES* *EL* *PS* *CR* *SH*	*LL* *IM* *CE* *BR* *PL* *LS*	*WA*	*WS*	*WL*	*FR*	*TU*	*TG*	*SE*	Use one of these at a time *EQ*	*BL*	*IC*	*CT*	*CV*
Strength I (unless noted)	γp	1.75	1.00	—	—	1.00	0.50/1.20	γTG	γSE	—	—	—	—	—
Strength II	γp	1.35	1.00	—	—	1.00	0.50/1.20	γTG	γSE	—	—	—	—	—
Strength III	γp	—	1.00	1.40	—	1.00	0.50/1.20	γTG	γSE	—	—	—	—	—
Strength IV	γp	—	1.00	—	—	1.00	0.50/1.20	—	—	—	—	—	—	—
Strength V	γp	1.35	1.00	0.40	1.0	1.00	0.50/1.20	γTG	γSE	—	—	—	—	—
Extreme Event I	γp	γEQ	1.00	—	—	1.00	—	—	—	1.00	—	—	—	—
Extreme Event II	γp	0.50	1.00	—	—	1.00	—	—	—	—	1.00	1.00	1.00	1.00
Service I	1.00	1.00	1.00	0.30	1.0	1.00	1.00/1.20	γTG	γSE	—	—	—	—	—
Service II	1.00	1.30	1.00	—	—	1.00	1.00/1.20	—	—	—	—	—	—	—
Service III	1.00	0.80	1.00	—	—	1.00	1.00/1.20	γTG	γSE	—	—	—	—	—
Service IV	1.00	—	1.00	0.70	—	1.00	1.00/1.20	—	1.0	—	—	—	—	—
Fatigue I—*LL*, *IM*, & *CE* only	—	1.50	—	—	—	—	—	—	—	—	—	—	—	—
Fatigue II—*LL*, *IM*, & *CE* only	—	0.75	—	—	—	—	—	—	—	—	—	—	—	—

center. Table 15.5 gives maximum moments, shears, and reactions for spans of 7 to 250 ft.

EXAMPLE

In an eight-panel truss of 200-ft span, maximum moment at panel point L_1, 25 ft from the left end, occurs with wheel 4 at that point (Table 15.3). Maximum moment at L_2 occurs with wheel 7 at that point, etc.

Table 15.1b —Load Factors for Permanent Loads, γp

Type of load, foundation type, and method used to calculate downdrag		Load factor Maximum	Minimum
DC: component and attachments		1.25	0.90
DC: strength IV only		1.50	0.90
DD: downdrag	Piles, α Tomlinson method	1.4	0.25
	Piles, λ method	1.05	0.30
	Drilled shafts, O'Neill and Reese (1999) method	1.25	0.35
DW: wearing surfaces and utilities		1.50	0.65
EH: horizontal earth pressure			
• Active		1.50	0.90
• At-rest		1.35	0.90
• *AEP* for anchored walls		1.35	N/A
EL: locked-in construction stresses		1.00	1.00
EV: vertical earth pressure			
• Overall stability		1.00	N/A
• Retaining walls and abutments		1.35	1.00
• Rigid buried structure		1.30	0.90
• Rigid frames		1.35	0.90
• Flexible buried structures			
• Metal box culverts and structural plate culverts with deep corrugations		1.5 1.3	0.9 0.9
• Thermoplastic culverts		1.95	0.9
• All others			
ES: earth surcharge		1.50	0.75

EXAMPLE

Calculate the moment of wheels 1 to 15, inclusive, about wheel 15. In Table 15.4, follow the vertical line through wheel 15 to the stepped line, then to the left, and find 5408.0 ft-kips to the right of the vertical line through wheel 1. Multiply this value by 8.0 to convert to the E-80 loading. Answer: 43,264 ft-kips per track.

EXAMPLE

Calculate the moments of wheels 17, 16, 15, and 14 about wheel 13. In Table 15.4, follow the vertical line through wheel 13 to the stepped line, then to the right, and find 427.0 ft-kips to the left of the vertical line through wheel 17. Multiply this value by 8.0 for E-80 loading. Answer: 3416 ft-kips per track.

EXAMPLE

Given a 200-ft eight-panel railroad truss, calculate the maximum moment at L_1 and the maximum shear in panel L_0L_1 for one track of E-80 loading. From Table 15.3, moment at L_1 will be maximum when wheel 4 is at L_1. From Table 15.4, the distances from wheel 4 to wheel 1

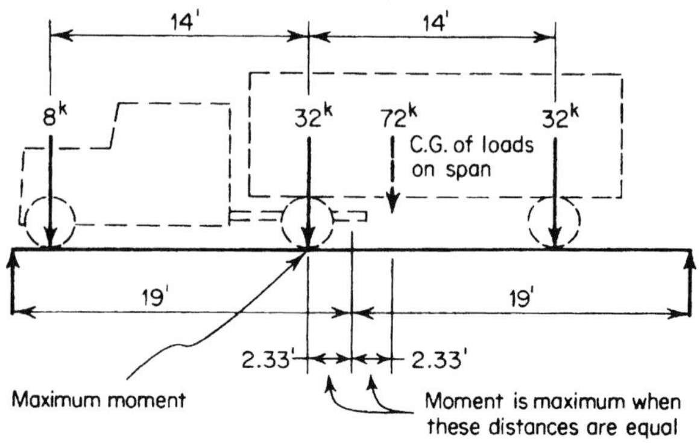

Figure 15.3 Simple-span maximum moment under wheel group.

Table 15.2 HS20-44 Live Load, One Lane Impact Not Included (T = Truck Loading, L = Lane Loading)

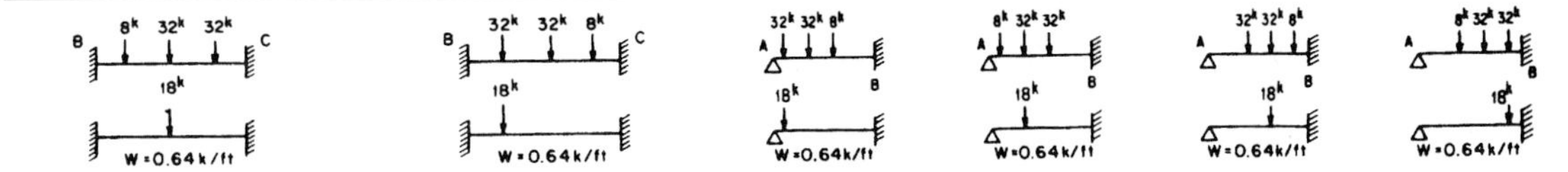

		Positive moment, ft-kips													
		Beams fixed both ends						Beams fixed one end							
Span in ft	Impact	SBM @ .5L	FEM @ B	FEM @ C	SBM @ .2L	FEM @ B	FEM @ C	SBM @ .2L	FEM @ B	SBM @ .4L	FEM @ B	SBM @ .6L	FEM @ B	SBM @ .8L	FEM @ B
1	0.300	8.0T	4.0	4.0	5.1T	4.1	1.0	5.1T	3.1	7.7T	5.4	7.7T	6.1	5.1T	4.6
2	0.300	16.0	8.0	8.0	10.2	8.2	2.0	10.2	6.1	15.4	10.8	15.4	12.3	10.2	9.2
3	0.300	24.0	12.0	12.0	15.4	12.3	3.1	15.4	9.2	23.0	16.1	23.0	18.4	15.4	13.8
4	0.300	32.0	16.0	16.0	20.5	16.4	4.1	20.5	12.3	30.7	21.5	30.7	24.6	20.5	18.4
5	0.300	40.0	20.0	20.0	25.6	20.5	5.1	25.6	15.4	38.4	26.9	38.4	30.7	25.6	23.0
6	0.300	48.0	24.0	24.0	30.7	24.6	6.1	30.7	18.4	46.1	32.3	46.1	36.9	30.7	27.6
7	0.300	56.0	28.0	28.0	35.8	28.7	7.2	35.8	21.5	53.8	37.6	53.8	43.0	35.8	32.3
8	0.300	64.0	32.0	32.0	41.0	32.8	8.2	41.0	24.6	61.4	43.0	61.4	49.2	41.0	36.9
9	0.300	72.0	36.0	36.0	46.1	36.9	9.2	46.1	27.7	69.1	48.4	69.1	55.3	46.1	41.5
10	0.300	80.0	40.0	40.0	51.2	41.0	10.2	51.2	30.7	76.8	53.8	76.8	61.4	51.2	46.1
11	0.300	88.0	44.0	44.0	56.3	45.1	11.3	56.3	33.8	84.5	59.1	84.5	67.6	56.3	50.7
12	0.300	96.0	48.0	48.0	61.4	49.2	12.3	61.4	36.9	92.2	64.5	92.2	73.7	61.4	55.3
13	0.300	104.0	52.0	52.0	66.6	53.2	13.3	66.6	39.9	99.8	69.9	99.8	79.9	66.6	59.9
14	0.300	112.0	56.0	56.0	71.7	57.3	14.3	71.7	43.0	107.5	75.3	107.5	86.0	71.7	64.5
15	0.300	120.0	60.0	60.0	76.8	61.4	15.4	76.8	46.1	115.2	80.6	115.2	92.2	76.8	69.1
16	0.300	128.0	64.0	64.0	81.9	65.5	16.4	81.9	49.2	122.9	86.0	122.9	98.3	81.9	73.7
17	0.300	136.0	68.0	68.0	87.0	69.6	17.4	87.0	52.2	130.6	91.4	130.6	104.5	87.0	78.3
18	0.300	144.0	72.0	72.0	94.7	74.0	30.7	94.7	67.7	138.2	96.8	138.2	110.6	94.7	89.3
19	0.300	152.0	76.0	76.0	105.0	80.1	53.2	105.0	93.2	145.9	102.1	145.9	116.7	105.0	106.7
20	0.300	160.0	80.0	80.0	115.2	87.7	72.3	115.2	116.2	153.6	107.5	153.6	122.9	115.2	123.8
21	0.300	168.0	84.0	84.0	125.4	96.4	88.8	125.4	137.0	161.3	112.9	161.3	129.0	125.4	140.8
22	0.300	176.0	88.0	88.0	135.7	105.9	103.1	135.7	156.1	169.0	118.3	169.0	135.2	135.7	157.4
23	0.300	184.0	92.0	92.0	145.9	116.0	115.6	145.9	173.6	176.6	123.7	176.6	141.3	145.9	173.8
24	0.300	192.7	116.3	104.4	156.2	126.6	126.7	156.2	190.0	189.4	141.5	189.4	153.9	156.2	189.9
25	0.300	207.4	122.7	120.5	166.4	137.4	136.5	166.4	205.2	204.8	164.5	204.8	169.6	166.4	205.7
26	0.300	222.2	129.4	135.9	176.6	148.5	145.3	176.6	219.6	220.2	186.4	220.2	185.3	176.6	221.2
27	0.300	237.0	136.3	150.5	186.9	159.8	153.2	186.9	233.1	235.5	207.2	235.5	200.9	186.9	236.4
28	0.300	252.0	143.5	164.5	197.1	171.1	160.4	197.1	246.0	250.9	227.1	250.9	216.4	197.1	251.3
29	0.300	267.0	150.8	178.0	207.4	182.6	166.9	207.4	258.2	266.2	246.3	266.2	231.8	207.4	266.0
30	0.300	282.1	158.3	191.0	217.6	194.0	172.9	217.6	269.9	281.6	264.8	281.6	247.2	217.6	280.5
32	0.300	312.5	173.6	215.9	238.1	216.9	183.6	238.1	292.1	312.3	300.1	312.3	277.6	238.1	308.7
34	0.300	343.5	179.5	236.6	258.6	239.6	192.9	258.6	312.7	343.0	333.4	343.0	307.7	258.6	336.0
36	0.300	378.9	201.4	261.5	280.3	262.3	207.2	280.3	338.3	375.7	366.7	375.7	340.5	280.3	365.9
38	0.300	414.3	223.1	285.6	303.4	285.5	225.3	303.4	368.1	410.2	400.4	410.2	373.9	303.4	398.2
40	0.300	449.8	244.4	309.0	326.4	309.3	241.1	326.4	395.8	444.8	433.0	444.8	410.6	326.4	429.8
42	0.299	485.3	265.5	331.9	349.4	333.3	255.1	349.4	421.7	479.4	464.7	479.4	444.7	349.4	460.8
44	0.296	520.9	286.3	354.2	372.5	357.5	267.5	372.5	446.2	513.9	495.7	513.9	478.3	372.5	491.2
46	0.292	556.5	306.9	376.2	395.5	381.7	278.7	395.5	469.6	548.5	526.0	548.5	511.5	395.5	521.1
48	0.289	592.2	327.3	397.8	418.6	405.9	289.0	418.6	491.9	583.0	555.7	583.0	544.3	418.6	550.4
50	0.286	627.8	347.6	419.1	441.6	430.0	298.3	441.6	513.3	617.6	584.9	617.6	576.7	441.6	579.1
52	0.282	663.5	367.7	440.2	464.6	454.0	307.0	464.6	534.0	652.2	613.7	652.2	608.8	464.6	607.5
54	0.279	699.3	387.6	461.0	487.7	477.8	315.2	487.7	554.1	686.7	642.1	686.7	640.6	487.7	635.4
56	0.276	735.0	407.5	481.5	510.7	501.5	322.8	510.7	573.6	721.3	670.3	721.3	672.1	510.7	662.9
58	0.273	770.8	427.2	501.9	533.8	525.1	330.0	533.8	592.5	755.8	698.0	755.8	703.4	533.8	690.1
60	0.270	806.5	446.8	522.1	656.8	548.5	336.9	556.8	611.1	790.4	725.6	790.4	734.4	556.8	716.9
62	0.267	842.3	466.3	542.2	579.8	571.7	343.4	579.8	629.3	825.0	752.9	825.0	765.3	579.8	743.4
64	0.265	878.1	485.7	562.1	602.9	594.8	349.7	602.9	647.1	859.5	780.0	859.5	796.0	602.9	769.6
66	0.262	913.9	505.1	581.9	625.9	617.7	355.8	625.9	664.6	894.1	806.9	894.1	826.5	625.9	795.6
68	0.259	949.8	524.4	601.6	649.0	640.4	361.7	649.0	681.9	928.6	833.6	928.6	856.8	649.0	821.3
70	0.256	985.6T	543.6	621.2	672.0T	663.0	367.4	672.0T	698.9	963.2T	860.2	963.2T	887.0	672.0T	846.7
72	0.254	1021L	563	641	695T	685	373	695T	716	998T	887	988T	917	695T	872
74	0.251	1057	582	660	718	708	378	718	732	1032	913	1032	947	718	897
76	0.249	1093	601	679	741	730	384	741	749	1067	939	1067	977	741	922
78	0.246	1129	620	699	764	752	389	764	765	1101	965	1101	1007	764	946

(Continued)

Table 15.2 HS20-44 Live Load, One Lane Impact Not Included (T = Truck Loading, L = Lane Loading) (*Continued*)

		Positive moment, ft-kips													
		Beams fixed both ends						Beams fixed one end							
		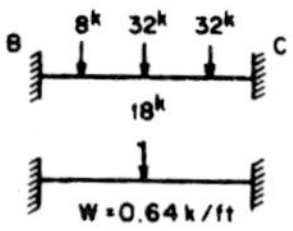			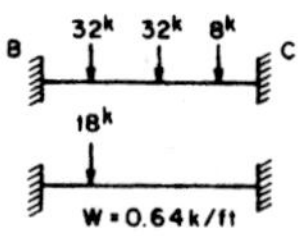			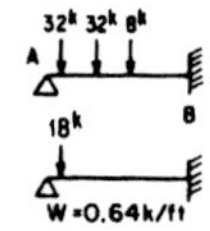		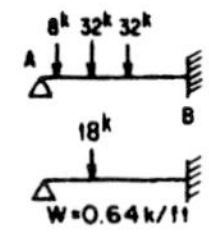		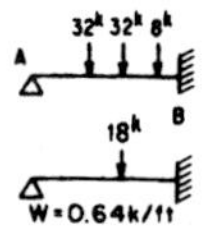		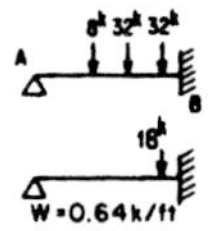	
Span in ft	Impact	SBM @ .5L	FEM @ B	FEM @ C	SBM @ .2L	FEM @ B	FEM @ C	SBM @ .2L	FEM @ B	SBM @ .4L	FEM @ B	SBM @ .6L	FEM @ B	SBM @ .8L	FEM @ B
80	0.244	1165	639	718	787	774	394	787	781	1136	991	1136	1036	787	971
82	0.241	1201	658	737	810	796	399	810	797	1171	1017	1171	1066	810	995
84	0.239	1237	677	756	833	817	404	833	813	1205	1043	1205	1095	833	1019
86	0.237	1273	695	775	856	839	409	856	829	1240	1068	1240	1125	856	1044
88	0.235	1308	714	794	879	861	414	879	844	1274	1094	1274	1154	879	1067
90	0.233	1344	733	813	902	882	419	902	860	1309	1120	1309	1183	902	1091
92	0.230	1380	752	832	925	903	424	925	875	1343	1145	1343	1212	925	1115
94	0.228	1416	770	851	948	924	428	948	890	1378	1171	1378	1242	948	1138
96	0.226	1452	789	870	972	945	433	972	906	1412	1196	1412	1271	972	1162
98	0.224	1488	808	888	995	966	438	995	921	1447	1221	1447	1300	995	1185
100	0.222	1524	826	907	1018	987	442	1018	936	1482	1247	1482	1329	1018	1209
105	0.217	1614	873	954	1075	1039	454	1075	974	1568	1310	1568	1401	1075	1266
110	0.213	1704	919	1000	1133	1091	465	1133	1011	1654	1372	1654	1473	1133	1324
115	0.208	1793	965	1047	1190	1142	477	1190	1048	1741	1435	1741	1544	1190	1381
120	0.204	1883	1011	1093	1248	1193	488	1248	1085	1827	1497	1827	1616	1248	1437
125	0.200	1973	1057	1139	1306	1243	499	1306	1121	1914	1560	1914	1687	1306	1493
130	0.196	2063	1103	1185	1363	1294	511	1363	1157	2000	1622	2000	1758	1363	1549
135	0.192	2153	1149	1231	1421	1344	522	1421	1194	2086	1684	2086	1829	1421	1605
140	0.189	2243	1195	1277	1478	1393	533	1478	1230	2173	1746	2173	1900	1478	1660
145	0.185	2333	1241	1323	1536	1443	544	1536	1266	2259	1807	2259	1971	1536	1715
150	0.182	2423	1286	1369	1594	1492	556	1594	1302	2346	1869	2346	2041	1594	1770
155	0.179	2513	1332	1415	1651	1541	567	1651	1337	2432T	1931	2432T	2112	1651	1825
160	0.175	2602T	1378	1461	1709	1590	578	1709T	1373	2657L	2532	2657L	2601	1709	1879
165	0.172	2921L	1823	1823	1766	1639	589	1869L	2463	2804	2677	2804	2748	1766	1933
170	0.169	3077	1924	1924	1824	1688	600	1969	2606	2954	2826	2954	2900	1824	1988
175	0.167	3238	2027	2027	1882	1736	611	2072	2752	3108	2979	3108	3055	1882	2042
180	0.164	3402	2133	2133	1939	1784	623	2177	2903	3266	3136	3266	3214	1939	2096
185	0.161	3571	2242	2242	1997	1833	634	2285	3058	3428	3297	3428	3377	1997	2150
190	0.159	3743	2353	2353	2054	1881	645	2396	3216	3593	3463	3593	3545	2054	2203
195	0.156	3920	2467	2467	2112	1929	656	2508	3379	3763	3632	3763	3716	2112	2257
200	0.154	4100	2583	2583	2170	1877	667	2624	3546	3936	3805	3936	3891	2170	2311
205	0.152	4285	2703	2703	2227	2025	679	2742	3716	4113	3982	4113	4070	2227	2364
210	0.149	4473	2824	2824	2285	2073	690	2863	3891	4294	4163	4294	4254	2285	2418
215	0.147	4666	2949	2949	2342	2120	701	2986	4070	4479	4348	4479	4441	2342	2471
220	0.145	4862	3076	3076	2400	2168	712	3112	4252	4668	4537	4668	4632	2400T	2524
225	0.143	5063	3206	3206	2458	2216	724	3240	4439	4860	4730	4860	4828	3240L	4633
230	0.141	5267	3339	3339	2515	2263	735	3371	4629	5056	4928	5056	5027	3371	4828
235	0.139	5476	3474	3474	2573	2311	746	3504	4824	5256	5129	5256	5230	3504	5027
240	0.137	5688	3612	3612	2630	2358	757	3640	5023	5460	5334	5460	5437	3640	5230
245	0.135	5905	3753	3753	2688	2406	769	3779	5225	5668	5543	5668	5649	3779	5437
250	0.133	6125	3896	3896	2746T	2453	780	3920	5432	5880	5756	5880	5864	3920	5648
255	0.132	6350	4042	4042	4064L	4056	3615	4064	5643	6096	5973	6096	6083	4064	5863
260	0.130	6578	4190	4190	4210	4204	3755	4210	5857	6315	6194	6315	6307	4210	6082
265	0.128	6811	4342	4342	4359	4356	3898	4359	6076	6538	6419	6538	6534	4359	6305
270	0.127	7047L	4495	4495	4510L	4510	4044	4510L	6299	6765L	6648	6765L	6765	4510L	6532

Table 15.2 HS20-44 Live Load, One Lane Impact Not Included (T = Truck Loading, L = Lane Loading) (*Continued*)

Shear, kips			Negative moment, ft-kips												
Simple span			Beams fixed one end					Beams fixed both ends							
			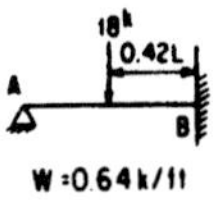		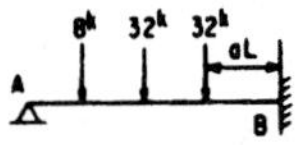			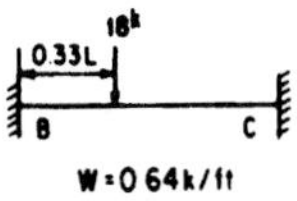			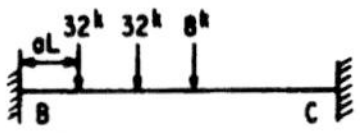				
Shear at end	Shear at midspan	Center reaction 2 equal simple spans	SBM @ .42L	FEM @ B	aL ft	SBM @ aL	FEM @ B	SBM @ .33L	FEM @ B	FEM @ C	aL ft	SBM @ aL	FEM @ B	FEM @ C	
32.0T	16.0T	32.0T	4.5L	3.5	0.4	7.8T	6.2	4.1L	2.7	1.4	0.3	7.1T	4.7	2.4	1
32.0	16.0	32.0	9.1	7.2	0.8	15.6	12.3	8.3	5.5	2.9	0.7	14.2	9.5	4.7	2
32.0	16.0	32.0	13.9	11.1	1.3	23.4	18.5	12.6	8.5	4.5	1.0	21.3	14.2	7.1	3
32.0	16.0	32.0	18.8	15.1	1.7	31.2	24.6	17.1	11.5	6.2	1.3	28.4	19.0	9.5	4
32.0	16.0	32.0	23.9	19.3	2.1	39.1	30.8	21.8	14.7	8.0	1.7	35.6	23.7	11.9	5
32.0	16.0	32.0	29.2	23.7	2.5	46.9	37.0	26.6	17.9	9.9	2.0	42.7	28.4	14.2	6
32.0	16.0	32.0	34.6	28.2	3.0	54.7	43.1	31.5	21.3	11.9	2.3	49.8	33.2	16.6	7
32.0	16.0	32.0	40.1	32.8	3.4	62.5	49.3	36.6	24.7	14.1	2.7	56.9	37.9	19.0	8
32.0	16.0	32.0	45.9	37.7	3.8	70.3	55.4	41.8	28.3	16.3	3.0	64.0	42.7	21.3	9
32.0	16.0	32.0	51.7	42.6	4.2	78.1	61.6	47.1	32.0	18.7	3.3	71.1	47.4	23.7	10
32.0	16.0	32.0	57.8	47.8	4.7	85.9	67.7	52.6	35.8	21.1	3.7	78.2	52.2	26.1	11
32.0	16.0	32.0	64.0	53.1	5.1	93.7	73.9	58.2	39.7	23.7	4.0	85.3	56.9	28.4	12
32.0	16.0	32.0	70.3	58.6	5.5	101.5	80.1	64.0	43.7	26.3	4.3	92.4	61.6	30.8	13
32.0	16.0	32.0	76.8	64.2	5.9	109.3	86.2	69.9	47.8	29.1	4.7	99.6	66.4	33.2	14
34.1	16.0	34.8	83.5	70.0	6.3	117.2	92.4	76.0	52.0	32.0	5.0	106.7	71.1	35.6	15
36.0	16.0	37.0	90.3	75.7	6.8	125.0	98.5	82.2	56.3	35.0	5.3	113.8	75.9	37.9	16
37.7	16.0	39.1	97.3	82.0	7.2	132.8	104.7	88.6	60.7	38.1	5.7	120.9	80.6	40.3	17
39.1	16.0	40.9	104.4	88.3	7.6	140.6	110.9	95.0	65.3	41.3	6.0	128.0	85.3	42.7	18
40.4	16.0	42.5	111.7	94.7	8.0	148.4	117.0	101.7	69.9	44.6	6.3	135.1	90.1	45.0	19
41.6	16.0	44.0	119.1	101.3	3.8	112.2	123.9	108.4	74.7	48.0	6.7	142.2	94.8	47.4	20
42.7	16.0	45.3	126.7	108.0	4.1	123.7	140.8	115.4	79.5	51.5	7.0	149.3	99.6	49.8	21
43.6	16.0	46.5	134.5	114.9	4.4	135.7	157.4	122.4	84.5	55.1	5.5	151.8	107.1	96.0	22
44.5	16.0	47.7	142.4	122.0	4.7	148.1	173.8	129.6	89.5	58.9	5.2	156.4	116.4	113.8	23
45.3	16.0	48.7	150.4	129.2	5.0	160.8	190.0	137.0	94.7	62.7	5.1	162.3	126.7	126.6	24
46.1	16.0	49.6	158.6	136.6	5.4	173.8	206.0	144.4	100.0	66.7	5.2	169.3	137.5	136.8	25
46.8	16.0	50.5	167.0	144.1	5.7	187.0	221.7	152.1	105.4	70.7	5.2	177.1	148.5	145.4	26
47.4	16.0	51.3	175.6	151.9	6.1	200.5	237.2	159.8	110.9	74.9	5.3	185.6	159.8	152.9	27
48.0	16.0	52.0	184.2	159.7	6.4	214.1	252.6	167.8	116.5	79.1	5.5	194.7	171.2	159.7	28
48.8	16.6	52.7	193.1	167.7	6.8	227.9	267.8	175.8	122.2	83.5	5.7	204.3	182.6	165.9	29
49.6	17.1	53.3	202.1	175.9	7.2	241.9	282.8	184.0	128.0	88.0	5.9	214.2	194.0	171.7	30
51.0	18.0	54.5	220.6	192.8	7.9	270.2	312.4	200.8	139.9	97.3	6.3	235.2	216.9	182.4	32
52.2	18.8	55.5	239.7	210.3	8.7	299.0	341.5	218.2	152.3	107.0	6.8	257.4	239.6	192.3	34
53.3	19.6	56.4	259.4	228.4	9.4	328.1	370.1	236.2	165.1	117.1	7.1	278.4	262.3	206.8	36
54.3	20.2	57.3	279.7	247.2	8.8	331.5	400.4	254.7	178.3	127.7	7.4	299.1	285.6	224.0	38
55.2	20.8	58.0	300.7	266.6	9.5	362.8	433.3	273.8	192.0	138.7	7.8	321.4	309.3	239.2	40
56.0	21.3	58.7	322.3	286.6	10.3	394.6	465.9	293.4	206.1	150.1	8.2	344.8	333.3	253.1	42
56.7	21.8	59.3	344.5	307.3	11.0	426.8	498.1	313.7	220.6	161.9	8.7	369.3	357.5	266.0	44
57.4	22.3	59.8	367.4	328.6	11.8	459.3	530.0	334.5	235.5	174.2	9.2	394.6	381.7	278.3	46
58.0	22.7	60.3	390.8	350.6	12.6	492.1	561.6	355.8	250.9	186.9	9.7	420.6	405.9	290.0	48
58.6	23.0	60.8	414.9	373.2	13.3	525.1	592.9	377.8	266.7	200.0	10.2	447.3	430.0	301.4	50
59.1	23.4	61.2	439.6	396.5	14.1	558.4	624.1	400.3	282.9	213.5	10.7	474.4	454.1	312.4	52
59.6	23.7	61.6	465.0	420.3	14.9	591.8	655.0	423.4	299.5	227.5	11.3	502.0	478.1	323.3	54
60.0	24.0	62.0T	491.0	444.9	15.7	625.3	685.7	447.0	316.6	241.9	11.8	530.0	502.0	334.0	56
60.4	24.4	63.1L	517.5	470.0	16.5	659.0	716.2	471.2	334.1	256.7	12.4	558.3	525.8	344.6	58
60.8	24.8	64.4	544.8	495.8	17.3	692.8	746.6	496.0	352.0	272.0	13.0	586.9	549.5	355.0	60
61.2	25.2	65.7	572.6	522.3	18.1	726.7	776.8	521.4	370.3	287.7	13.6	615.8	573.1	365.4	62
61.5	25.5	67.0	601.1	549.4	18.9	760.8	806.9	547.3	389.1	303.8	14.2	645.0	596.7	375.7	64
61.8	25.8	68.2	630.2	577.1	19.7	794.9	836.8	573.8	408.3	320.3	14.7	674.3	620.1	386.0	66
62.1	26.1	69.5	659.9	605.5	20.5	829.0	866.7	600.8	427.9	337.3	15.3	703.8	643.5	396.3	68
62.4T	26.4T	70.8L	690.2L	634.5	21.3	863.3T	896.4	628.4L	448.0	354.7	15.9	733.5T	666.7	406.5	70
62.7T	26.7T	72.1L	721L	664	22.2	898T	926	657L	468	372	16.6	763T	690	417	72
62.9	26.9	73.4	753	694	23.0	932	956	685	489	394	17.2	793	713	427	74

(*Continued*)

Table 15.2 HS20-44 Live Load, One Lane Impact Not Included (T = Truck Loading, L = Lane Loading) (*Continued*)

Shear, kips			Negative moment, ft-kips												
Simple span			Beams fixed one end					Beams fixed both ends							
			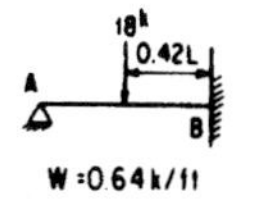		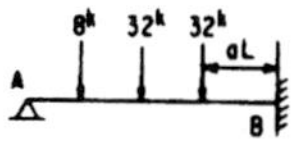			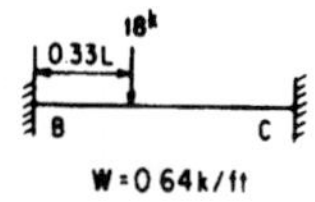			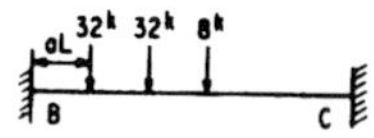				
Shear at end	Shear at midspan	Center reaction 2 equal simple spans	SBM @ .42L	FEM @ B	aL ft	SBM @ aL	FEM @ B	SBM @ .33L	FEM @ B	FEM @ C	aL ft	SBM @ aL	FEM @ B	FEM @ C	
63.2	27.2	74.6	785	725	23.8	966	985	715	511	409	17.8	824	736	437	76
63.4	27.4	75.9	818	757	24.6	1001	1014	745	532	428	18.4	854	759	447	78
63.6	27.6	77.2	851	789	25.4	1035	1044	775	555	448	19.0	884	782	458	80
63.8	27.8	78.5	885	822	26.3	1070	1073	806	577	468	19.6	915	805	468	82
64.0	28.0	79.8	920	855	27.1	1104	1102	838	600	488	20.3	945	828	478	84
64.2	28.2	81.0	955	890	27.9	1139	1131	870	624	509	20.9	976	850	488	86
64.4	28.4	82.3	991	924	28.7	1174	1160	903	648	530	21.5	1006	873	498	88
64.5	28.5	83.6	1028	960	29.6	1208	1189	936	672	552	22.2	1037	896	509	90
64.7	28.7	84.9	1065	996	30.4	1243	1218	970	697	574	22.8	1068	918	519	92
64.9	28.9	86.2	1103	1033	31.2	1278	1247	1004	722	597	23.4	1099	941	529	94
65.0	29.0	87.4	1142	1070	32.0	1313	1276	1039	748	620	24.1	1130	963	539	96
65.1	29.1	88.7	1181	1108	32.9	1347	1304	1075	774	643	24.7	1161	986	550	98
65.3	29.3	90.0	1220	1146	33.7	1382	1333	1111	800	667	25.3	1192	1008	560	100
65.6	29.6	93.2	1322	1246	35.8	1469	1405	1204	868	728	26.9	1269	1064	586	105
65.9	29.9	96.4	1428	1349	37.8	1556	1476	1300	939	792	28.5	1347	1120	611	110
66.2	30.2	99.6	1538	1456	39.9	1643	1548	1400	1012	859	30.2	1426	1175	637	115
66.4	30.4	102.8	1652	1568	42.0	1731	1619	1504	1088	928	31.8	1504	1230	663	120
66.6T	30.6	106.0	1770	1683	44.1	1818T	1690	1611	1167	1000	33.4	1582	1286	689	125
67.6L	30.8	109.2	1891	1802				1722	1248	1075	35.0	1661	1341	715	130
69.2	31.0	112.4	2017	1926				1836	1332	1152	36.6	1740	1396	741	135
70.8	31.2	115.6	2146	2053				1954	1419	1232	38.3	1819T	1450	767	140
72.4	31.4	118.8	2279	2184				2075	1508	1315					145
74.0	31.5	122.0	2416	2320				2200	1600	1400					150
75.6	31.7	125.2	2557	2459				2328	1695	1488					155
77.2	31.8	128.4	2702	2602				2460	1792	1579					160
78.8	31.9	131.6	2851	2750				2596	1892	1672					165
80.4	32.0	134.8	3004	2901				2735	1995	1768					170
82.0	32.2	138.0	3161	3056				2878	2100	1867					175
83.6	32.3	141.2	3321	3216				3024	2208	1968					180
85.2	32.4	144.4	3486	3379				3174	2319	2072					185
86.8	32.5	147.6	3654	3546				3327	2432	2179					190
88.4	32.6	150.8	3827	3717				3484	2548	2288					195
90.0	32.6	154.0	4003	3893				3644	2667	2400					200
91.6	32.7	157.2	4183	4072				3808	2788	2515					205
93.2	32.8	160.4	4367	4255				3976	2912	2632					210
94.8	32.9	163.6	4555	4443				4147	3039	2752					215
96.4	32.9	166.8	4747	4634				4322	3168	2875					220
98.0	33.0	170.0	4942	4829				4500	3300	3000					225
99.6	33.1	173.2	5142	5029				4682	3435	3128					230
101.2	33.1	176.4	5346	5232				4867	3572	3259					235
102.8	33.2	179.6	5553	5439				5056	3712	3392					240
104.4	33.3	182.8	5764	5651				5248	3855	3528					245
106.0	33.3T	186.0	5980	5866				5444	4000	3667					250
107.6	33.4L	189.2	6199	6085				5644	4148	3808					255
109.2	33.8	192.4	6422	6309				5847	4299	3952					260
110.8	34.2	195.6	6649	6536				6054	4452	4099					265
112.4L	34.6L	198.8L	6880L	6767				6264L	4608	4248					270

Table 15.3 Position of Wheels to Produce Maximum Moment with Cooper's Loading. Maximum Moment Will Occur under Wheel Noted in Table for Given Span and Distance from Left End of Span

	Distance from left end of span																			
Span, ft	10	15	20	25	30	35	40	45	50	55	60	65	70	80	90	100	110	120	130	140
300–260	2	3	3	4	4	5	5	6	7	7	8	9	10	11	12	13	14	15	17	18
250–200	2	3	3	4	4	5	5	6	7	8	8	9	10	11	12	13	14	15	17	18
190–150	2	3	3	4	4	5	5	6	7	8	9	9	11	12	12	13	14	15	17	18
140	3	3	3	4	4	5	5	6	7	8	9	10	11	12	12	13	14	15	17	18
130	3	3	3	4	4	5	5	6	7	8	9	10	11	12	12	13	14	15	17	
120	3	3	3	4	4	5	5	6	7	8	9	10	11	12	13	13	14	15		
110	3	3	3	4	4	5	6	7	7	8	9	10	11	12	13	13	14			
100	3	3	3	4	5	5	6	14	14	14	13	13	11	12	13	13				
90	3	3	4	4	5	13	13	13	13	13	13	13	13	12	13					
80	3	3	4	4	13	13	13	12	12	12	12	12	12	12						
70	3	3	4	4	13	13	12	12	12	12	11	11	11							
65	3	3	4	4	12	12	12	12	12	11	11	11								
60	3	3	4	4	5	13	12	11	11	11	11									
55	12	12	12	4	12	13	12	12	13	11										
50	12	12	12	12	12	13	13	13	12											
45	12	12	12	12	12	13	13	13												
40	3	3	3	12	12	13	13													
35	3	3	4	4	13	13														
30	3	3	4	4	13															
25	3	3	4	4																
20	4	3	4																	
15	3	3																		
10	3																			

and to the uniform load are 18 and 91 ft, respectively. Therefore, all wheels are on the bridge, and there will be 175 − 91 = 84 ft of uniform load on the span. The left reaction R_L is found from

$$200R_L = 8182 + 142 \times 84 + 1 \times \frac{84^2}{2}$$

in which the moment, 8182 ft-kips, and the sum of the weights of the 18 wheels, 142 kips, are found in Table 15.4. Therefore, $R_L = 118.2$ kips, and the moment at L_1 is

$$M_1 = 118.2 \times 25 - 140 = 2715 \text{ ft-kips}$$

in which the moment, 240 ft-kips, is found in Table 15.4. For the E-80 loading,

$$M_1 = 8.0 \times 2715 = 21{,}720 \text{ ft-kips per track}$$

The shear in panel L_0L_1 is $V = R_L - 240/25 = 118.2\ 9.6 = 108.6$ kips. For the E-80 loading,

$$V = 8.0 \times 108.6 = 869 \text{ kips per track.}$$

15.2.4 Positive Moments in Continuous Spans

Because of the many possible combinations of span, loading, ratio of dead to live load, and degree of fixity, no general formulas or curves can be given for all cases.

Positive Moments in End Span

The envelope of maximum positive moments in the end span of a series of continuous spans on simple supports can be approximated satisfactorily by parabolas. If the length of the end span is 75 to 100 percent of that of the adjacent span, the maximum positive moment occurs at or near the 0.4 point nearest the discontinuous end. The envelope can be approximated by two second-degree parabolas with common vertices at the point whose ordinate is the maximum moment at the 0.4 point. One parabola passes through the (zero) moment ordinate at the discontinuous end, the other through the moment ordinate at the 0.8 point.

EXAMPLE

Determine the envelope of maximum positive moments for the end span in a continuous beam of three 60-ft spans (Fig. 15.4*a*). The dead load is 2 klf, the live load 1.5 wheel lines of HS20-44 load with 27 percent impact.

1. The dead-load moments at *B* and *C* are 720 ft-kips from which the reaction at *A* is found to be 48 kips. The corresponding dead-load moment diagram is shown in Fig. 15.4*b*. Only the moments at the 0.4 and 0.8 points are needed in this example; the remaining ordinates are used in the determination of other maximum-moment envelopes discussed later.

2. From Table 15.2, with the truck positioned for maximum moment at the 0.4 point, the simple-beam moment and fixed-end moment are, respectively, 790.4 and 725.6 ft-kips. With the truck positioned for maximum moment at the 0.8 point, the simple-beam moment and fixed-end moment are, respectively, 556.8 and 716.9 ft-kips. Converting these values to 1.5 wheel lines per girder and adding 27 percent impact, the following values are obtained:

	0.4	0.8
SBM	753	531
FEM	692	683

3. With the fixed-end moment, 692 ft-kips, known, the live-load negative moment at *B* can be determined. When moving loads are being investigated, it is convenient to do this by distributing a unit moment acting at *B* (Fig. 15.5). The resulting distribution factors can be used to distribute the fixed-end moment at *B* in *AB* for any combination of loads in any of the three spans. In this case, we get

$$\text{FEM}_{BA} = 0.533 \times 692 = 369 \text{ for maximum moment at } 0.4$$

$$\text{FEM}_{BA} = 0.533 \times 683 = 364 \text{ for maximum moment at } 0.8$$

From these values, corrections to the live-load simple-beam moments can be found. Finally, then, the *DL* + *LL* + *I* moments are

$$M_{0.4} = 576 + 753 - 0.4 \times 369 = 1182 \text{ ft-kips}$$

$$M_{0.8} = 0 + 531 - 0.8 \times 364 = 240 \text{ ft-kips}$$

Table 15.4 Moments in Foot-Kips for Class E-10 Engine Loading (One Track of Two Rails)

Wheel No.	1	2	3	4	5	6	7	8	9	10	11	12	13	14	15	16	17	18	
Axle loads	5.0	10.0	10.0	10.0	10.0	6.5	6.5	6.5	6.5	5.0	10.0	10.0	10.0	10.0	6.5	6.5	6.5	6.5	1 kif
Spacing, ft. .	8	5	5	5	9	5	6	5	8	8	5	5	5	9	5	6	5	5	
Totals from end of train: Kips. .	142.0	137.0	127.0	117.0	107.0	97.0	90.5	84.0	77.5	71.0	66.0	56.0	46.0	36.0	26.0	19.5	13.0	6.5	0
Feet…	109	101	96	91	86	77	72	66	61	53	45	40	35	30	21	16	10	5	0
Moments of wheel loads about.																			
End of train	8182.0	7637.0	6627.0	5667.0	4757.0	3897.0	3396.5	2928.5	2499.5	2103.0	1838.0	1388.0	988.0	638.0	338.0	201.5	97.5	32.5	
Wheel No.18	7472.0	6952.0	5992.0	5082.1	4222.0	3412.0	2944.0	2508.5	2112.0	1748.0	1508.0	1108.0	758.0	458.0	208.0	104.2	32.5		
17	6794.5	6299.5	5389.2	4529.5	3719.5	2959.5	2524.0	2121.0	1757.0	1425.5	1210.5	860.5	560.5	310.5	110.5	39.0	32.5		
16	6020.5	5555.6	4705.5	3905.5	3155.6	2455.5	2059.0	1695.0	1370.0	1077.5	892.5	602.5	362.5	172.5	32.5	39.0	110.5		
15	5408.0	4968.0	4168.0	3418.0	2718.0	2068.0	1704.0	1372.5	1080.0	820.0	660.0	420.0	230.0	90.0	32.5	104.0	208.0		
14	4364.0	3969.0	3259.0	2599.0	1989.0	1429.0	1123.5	850.5	616.5	415.0	300.0	150.0	50.0	58.5	149.5	279.5	442.0		
13	3834.0	3464.0	2804.0	2194.0	1634.0	1124.0	851.0	610.5	409.0	240.0	150.0	50.0	50.0	141.0	264.5	427.0	622.0		
12	3354.0	3009.0	2399.0	1839.0	1329.0	869.0	628.5	420.5	251.5	115.0	50.0	50.0	150.0	273.5	429.5	624.5	852.0		
11	2924.0	2604.0	2044.0	1534.0	1074.0	664.0	456.0	280.5	144.0	40.0	50.0	150.0	300.0	456.0	644.5	872.0	1132.0		
10	2316.0	2036.0	1556.0	1126.0	746.0	416.0	260.0	136.5	52.0	80.0	210.0	390.0	620.0	828.0	1068.5	1348.0	1660.0		
9	1748.0	1508.0	1108.0	758.0	458.0	208.0	104.0	32.5	40.0	200.0	410.0	670.0	980.0	1240.0	1532.5	1864.0	2228.0		
8	1425.5	1210.5	860.5	560.5	310.5	110.5	39.0	32.5	97.5	307.5	567.5	877.5	1237.5	1530.0	1855.0	2219.0	2615.5		
7	1077.2	892.5	602.5	362.5	172.5	32.5	39.0	110.5	205.5	475.5	795.5	1165.5	1585.5	1917.0	2281.0	2684.0	3119.5		
6	820.0	660.0	420.0	230.0	90.0	32.5	104.0	208.0	328.0	648.0	1018.0	1438.0	1908.0	2272.0	2668.5	3104.0	3572.0		
5	415.0	300.0	150.0	50.0	58.5	149.5	279.5	442.0	607.0	1017.0	1477.0	1987.0	2547.0	2969.5	3424.5	3918.5	4445.0		
4	240.0	150.0	50.0	50.0	141.0	264.5	427.0	622.0	812.0	1272.0	1782.0	2342.0	2952.0	3407.0	3394.5	4421.0	4980.0		
3	115.0	50.0	50.0	150.0	273.5	429.5	624.5	852.0	1067.0	1577.0	2137.0	2747.0	3407.0	3894.5	4414.4	4973.5	5565.0		
2	40.0	50.0	150.0	300.0	456.0	644.5	872.0	1132.0	1372.0	1932.0	2542.0	3202.0	3912.0	4432.0	4984.5	5576.0	6200.0		
1	80.0	210.0	390.0	620.0	828.0	1068.5	1348.0	1660.0	1940.0	2580.0	3270.0	4010.0	4800.0	5372.0	5976.5	6620.0	7296.0		
Totals from wheel 1: Kips… 5.0	15.0	25.0	35.0	45.0	51.5	58.0	64.5	71.0	76.0	86.0	96.0	106.0	116.0	122.5	129.0	135.5	142.0	142.0	
Feet… 0	8	13	18	23	32	37	43	48	56	64	69	74	79	88	93	99	104	109	

Table 15.5 Maximum Moments, Shears, and Reactions for Class E-10 Engine Loading (One Track of Two Rails)

Axle loads	5.0	10.0	10.0	10.0	10.0	6.5	6.5	6.5	6.5	5.0	10.0	10.0	10.0	10.0	6.5	6.5	6.5	6.5	1 kif	12.5	12.5
Spacing, ft. . . .	8	5	5	5	9	6	5	8	8	5	5	5	9	5	6	5	5	7			

Span, ft	Max moment, ft-kips	Max shear, kips	Max floor-beam reaction, kips	Equivalent uniform load: Moment	Equivalent uniform load: Shear	Equivalent uniform load: Reaction	Span, ft	Max moment, ft-kips	Max floor-beam reaction, kips	Max floor-beam reaction, kips	Equivalent uniform load: Moment	Equivalent uniform load: Shear	Equivalent uniform load: Reaction
7	21.9	12.5	15.1	3.57	3.57	2.15	26	162.4	29.1	38.8	1.92	2.24	1.49
7½	23.5	13.4	16.0	3.33	3.56	2.12	27	172.3	29.6	40.0	1.89	2.20	1.48
8	25.0	14.0	16.8	3.12	3.51	2.11	28	182.7	30.2	41.2	1.86	2.16	1.47
9	28.1	15.3	18.2	2.78	3.39	2.02	29	194.0	30.8	42.2	1.84	2.12	1.46
10	31.2	16.2	19.2	2.50	3.25	1.92	30	205.2	31.5	43.1	1.82	2.10	1.44
11	34.4	17.0	21.0	2.28	3.09	1.90	31	216.5	32.2	44.3	1.80	2.08	1.43
12	40.0	17.7	23.3	2.22	2.95	1.94	32	227.7	32.9	45.5	1.78	2.05	1.42
13	47.5	18.3	24.6	2.25	2.81	1.90	33	239.0	33.5	46.7	1.75	2.03	1.41
14	55.0	18.8	26.1	2.25	2.68	1.36	34	250.3	34.1	47.8	1.73	2.00	1.40
15	62.5	20.0	27.3	2.22	2.67	1.32	35	261.5	34.6	48.8	1.71	1.98	1.39
16	70.0	21.3	28.5	2.19	2.66	1.78	36	274.3	35.3	49.8	1.69	1.96	1.38
17	77.5	22.4	29.4	2.15	2.63	1.73	37	287.2	35.9	50.7	1.68	1.94	1.37
18	85.0	23.3	30.3	2.10	2.59	1.59	38	300.0	36.5	51.8	1.66	1.92	1.36
19	93.3	24.2	31.5	2.07	2.55	1.55	39	313.3	37.2	52.9	1.65	1.90	1.36
20	103.1	25.0	32.8	2.06	2.50	1.54	40	327.8	37.7	54.0	1.64	1.88	1.35
21	112.9	25.7	34.0	2.05	2.45	1.52	42	356.7	39.2	56.0	1.62	1.87	1.34
22	122.8	26.3	35.1	2.03	2.40	1.50	44	385.8	40.3	58.2	1.60	1.83	1.32
23	132.7	27.0	36.1	2.01	2.34	1.57	46	414.9	41.4	60.3	1.57	1.80	1.31
24	142.6	27.7	37.0	1.98	2.31	1.54	48	443.8	42.4	62.4	1.54	1.77	1.30
25	152.5	28.4	37.8	1.95	2.27	1.51	50	475.5	43.5	64.3	1.52	1.74	1.29
52	507.6	44.6	66.7	1.50	1.72	1.28	82	1128.3	63.5	101.5	1.34	1.55	1.24
54	540.5	45.6	69.0	1.48	1.69	1.28	84	1177.7	64.8	103.5	1.34	1.54	1.23
56	576.1	46.5	71.4	1.47	1.65	1.27	86	1229.7	66.1	105.4	1.33	1.54	1.23
58	611.6	47.7	74.0	1.46	1.65	1.28	88	1282.0	67.4	107.3	1.32	1.53	1.22
60	649.5	48.8	76.6	1.44	1.63	1.28	90	1334.7	68.6	109.3	1.32	1.53	1.22
62	688.2	50.0	79.1	1.43	1.61	1.27	92	1388.3	69.9	111.2	1.31	1.52	1.21
64	727.7	51.3	81.5	1.42	1.60	1.27	94	1442.7	71.2	113.1	1.31	1.52	1.20
66	769.7	52.5	83.9	1.41	1.59	1.27	96	1497.3	72.4	115.0	1.30	1.51	1.20
68	811.7	53.9	86.2	1.40	1.59	1.27	98	1552.7	73.7	116.8	1.29	1.51	1.19
70	853.7	55.3	88.5	1.39	1.58	1.26	100	1609.7	75.0	118.6	1.29	1.50	1.19
72	896.7	56.7	90.7	1.38	1.58	1.26	125	2497.7	89.7	140.5	1.28	1.44	1.12
74	939.0	58.1	93.0	1.37	1.57	1.26	150	3531.0	103.7	162.7	1.25	1.38	1.08
76	986.0	59.5	95.2	1.36	1.57	1.25	175	4676.3	117.3	185.8	1.22	1.34	1.06
78	1032.7	60.9	97.3	1.36	1.56	1.25	200	5939.0	130.5	209.5	1.19	1.31	1.05
80	1080.0	62.1	99.4	1.35	1.55	1.24	250	8796.3	156.6	257.6	1.13	1.25	1.03

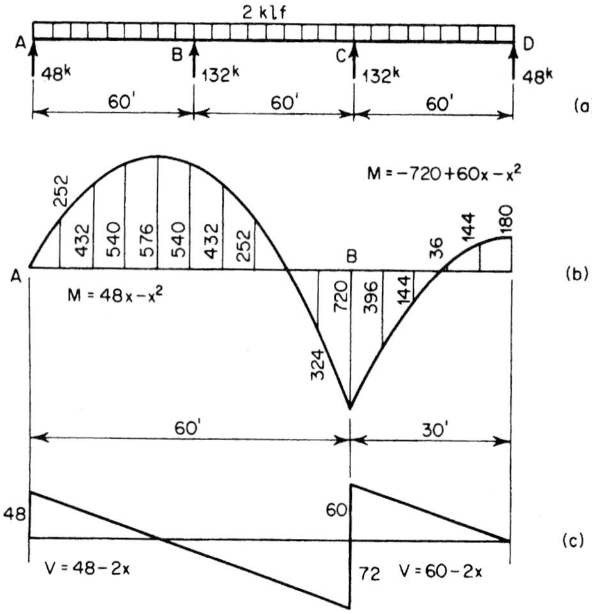

Figure 15.4 Positive moment envelope, continuous beam, three 60-ft spans.

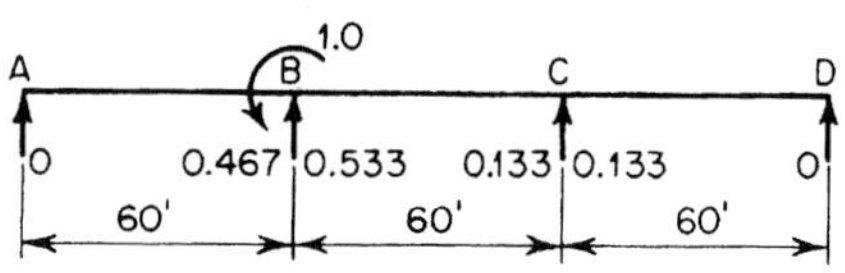

Figure 15.5 Moment-distribution factors.

4. The approximate maximum-moment envelope is shown in Fig. 15.6. The ordinates at intermediate points are easily found from properties of the parabola. Thus,

$$M_{0.5} = 1182 - (1182 - 240)(1/4)^2 = 1123 \text{ ft-kips}$$

Positive Moment in Interior Span

An approximate envelope for maximum positive moment in the interior span is constructed by passing a second-degree parabola through the maximum-moment ordinates at the 0.2 point and the midpoint. For maximum moment at the 0.2 point, the truck heads into the span with the rear wheel at the 0.2 point. Position of the load for maximum midspan moment is shown in Fig. 15.3. Although the maximum moment is at the 32-kip axle, it is assumed to be at midspan. Fixed-end moments for loads at these positions may be found in Table 15.2.

EXAMPLE

Construct the approximate envelope for maximum positive moment in the interior span of the beam of Fig. 15.4*a*.

The following moments are found in Table 15.2:

Simple-beam moment at 0.2 point	556.8
Fixed-end moment at *B*	548.5
Fixed-end moment at *C*	336.9

Using the distribution factors determined previously (Fig. 15.5), the negative moments at *B* and *C* are

$$M_B = 0.467 \times 548.5 + 0.133 \times 336.9 = 301$$

$$M_C = 0.133 \times 548.5 + 0.467 \times 336.9 = 230$$

From these values and the simple-beam moment 556.8, the moment at the 0.2 point is found to be 270 ft-kips. For 1.5 wheel lines and 27 percent impact, this reduces to 257 ft-kips, which, combined with the dead-load moment, −144 ft-kips, from Fig. 15.4*b*, gives the maximum *DL* + *LL* + *I* moment of + 113 ft-kips.

The maximum moment at midspan, 670 ft-kips, is found similarly, and the maximum moment envelope is determined by constructing the parabola shown in Fig. 15.6.

15.2.5 Negative Moments in Continuous Spans

Envelopes of maximum negative moment can be obtained by determining the maximum negative moments at a support and at several points in the adjoining spans. Maximum negative moment at a support occurs when the two adjoining spans are loaded. In the case of equal spans over 41 ft long, the equivalent lane loading controls. If the adjoining spans are unequal, this separation point must be determined by trial.

The position of load to obtain maximum negative moment at an interior point of a span can be determined from an influence line or from a series of trials. However, a satisfactory approximation is obtained by loading the span adjacent to the one under consideration with the heavy wheels of the truck equidistant from the 0.4 point adjacent to

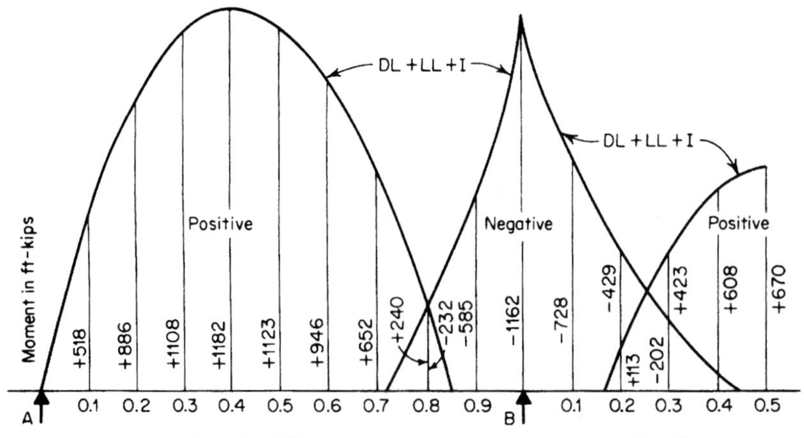

Figure 15.6 Maximum-moment envelopes, continuous beam, three 60-ft spans.

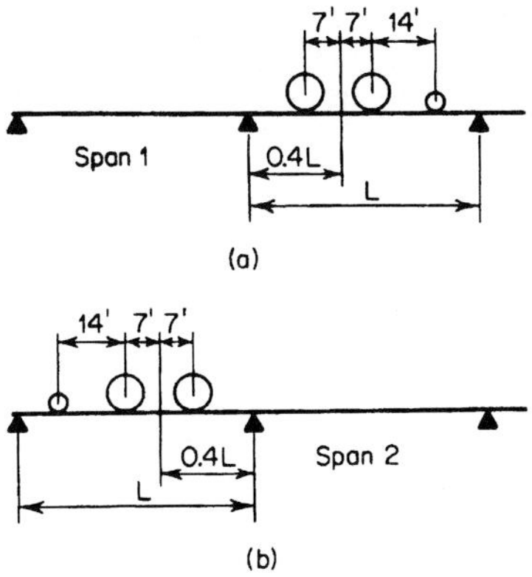

Figure 15.7 Approximate maximum negative moments in continuous beams: (*a*) position for negative moment in span 1 and (*b*) position for negative moment in span 2.

the support common to the two spans. Thus, for maximum negative moments in span 1, the loads are positioned as in Fig. 15.7*a*, while for maximum negative moments in span 2, they are located as in Fig. 15.7*b*.

For equal spans, Table 15.2 gives the simple-beam moment and the fixed-end moments, for both truck and lane loading, with the loads positioned for maximum negative moments in the adjoining, unloaded span.

EXAMPLE

Construct the approximate envelope of maximum negative moments for the continuous beam of Fig. 15.4*a*.

1. Since the spans are equal and over 41 ft long, the maximum live-load negative moment at an interior support is produced by the lane loading. Therefore, the required fixed-end moments from Table 15.2 are the following: FEMBA = 495.8, FEMBC = 352.0, and FEMCB = 272.0. Using the distribution factors in Fig. 15.5,

$$M_B = 0.533 \times 495.8 + 0.467 \times 352.0 + 0.133 \times 272.0 = 464$$

For 1.5 wheel lines and 27 percent impact, this gives M_B = 442 ft-kips. Adding this to the dead-load moment from Fig. 15.4*b*, the *DL* + *LL* + *I* moment is found to be −1162 ft-kips.

2. Maximum negative moments in the end span are produced when the live load is in the second span. From Table 15.2, it is found that the truck loading is critical, so FEM_{BC} = 549.5 and FEM_{CB} = 355.0. Following the procedure in step 1, we get M_B = 290 ft-kips. Since the corresponding moment diagram for span *AB* is a straight line from 0 at *A* to −290 ft-kips at *B*, the negative moments at the 0.7, 0.8, and 0.9 points are easily determined (Fig. 15.8). These live-load moments are given in the following table. The dead-load moments are taken from Fig. 15.4.

Location	*DL*	*LL* + *I*	*DL* + *LL* + *I*
0.7	+252	−203	+ 49
0.8	0	−232	−232
0.9	−324	−261	−585
B	−720	−290	−1010

3. Maximum negative moments in the center span are produced when the live load is in the end span. From Table 15.2, FEM_{BA} = 746.6. Therefore,

$$M_B = 0.533 \times 746.6 \times 3/4 \times 1.27 = -379$$

$$M_C = 0.133 \times 746.6 \times 3/4 \times 1.27 = +95$$

Since the moment diagram in span *BC* is a straight line, moments at intermediate points are easily found (Fig. 15.8). The results are given in the following table.

Location	*DL*	*LL* + *I*	*DL* + *LL* + *I*
B	−720	−379	−1099
0.9	−396	−332	−728
0.8	−144	−285	−429
0.7	+36	−238	−202
0.6	+144	−191	−47
0.5	+180	−144	+36

The maximum-moment envelope is shown in Fig. 15.6. Note that the ordinate at *B* is the value found in step 1, which exceeds the values found in steps 2 and 3.

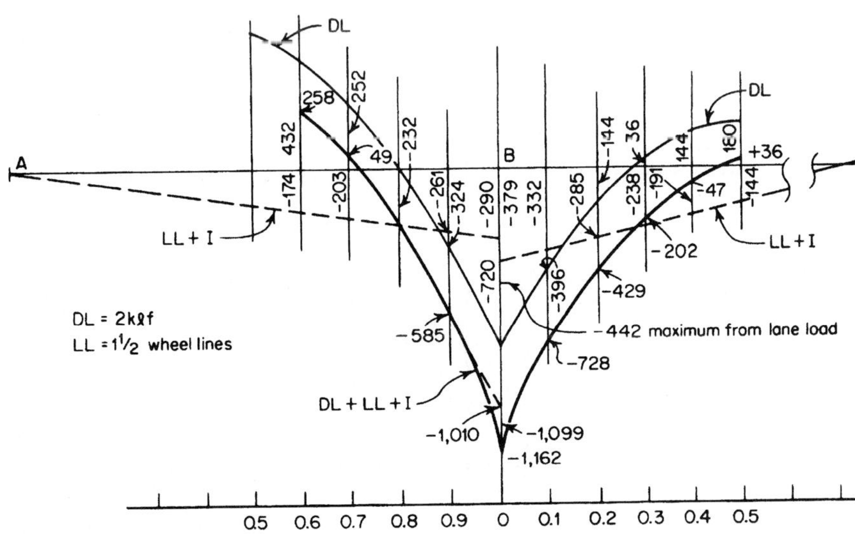

Figure 15.8 Negative moment at interior support, continuous beam, three 60-ft spans.

15.2.6 Shears in Continuous Spans

The following approximate limits determine the spans up to which the HS20-44 truck produces maximum shear and above which the equivalent loading should be used:

Shear and reaction at the end support:	To 140 ft
Shear at left of first interior support:	To 110 ft
Shear at right of first interior support:	To 110 ft
Reaction at first interior support:	To 68 ft

Shear in End Span

A satisfactory envelope for maximum shear can be found by determining the following: (1) shear at end support, (2) positive and negative shears at the 0.4 point nearest the end support, and (3) negative shear just to the left of the first interior support.

EXAMPLE

Determine the approximate envelope of maximum shears for the beam of Fig. 15.4*a*.

1. Maximum shear at *A* occurs with the truck in the position shown in Fig. 15.9*a*. The moment at *B* is found by determining the fixed-end moment and using the distribution of Fig. 15.5. The fixed-end moment, which is easily computed from the known formula (Fig. 15.9*e*), is 302 ft-kips. Therefore, $M_B = 0.533 \times 302 = 160$ ft-kips, and $R_A = 58.1$ kips. For 1.5 wheel lines and impact of 27 percent, $R_A = 55.5$ kips.

2. Maximum positive shear at the 0.4 point occurs with the truck in the position shown in Fig. 15.9*b*. Proceeding as in step 1, the maximum shear $V = R_A$ is found to be 24.7 kips.

3. Maximum negative shear at the 0.4 point occurs with the truck in the position shown in Fig. 15.9*c*. For this case, the maximum shear $V = R_B$ is 21.8 kips.

4. Maximum negative shear to the left of *B* occurs with the truck in the position shown in Fig. 15.9*d*. The maximum shear $V = R_B$ is 61.2 kips.

5. The maximum *LL* + *I* values are plotted in Fig. 15.10, and the envelope of maximum shears found by combining them with dead-load shears shown in the figure.

Shear in Interior Span

The shear envelope for an interior span is found in a manner similar to that of the preceding example. In a three-span symmetrical bridge, the center-span shears need be calculated only at the interior supports and at the midpoint.

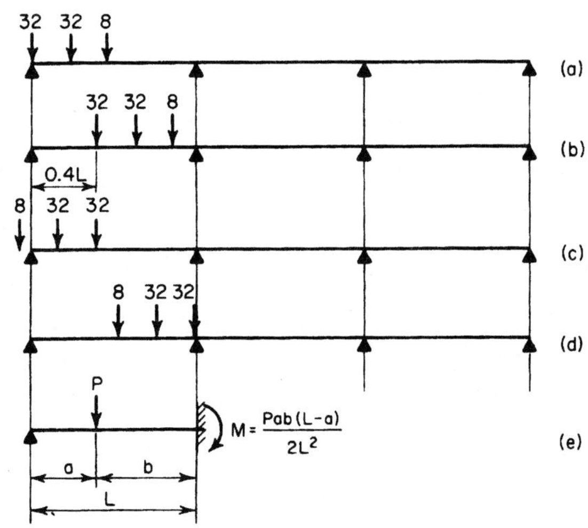

Figure 15.9 Envelope of maximum shear.

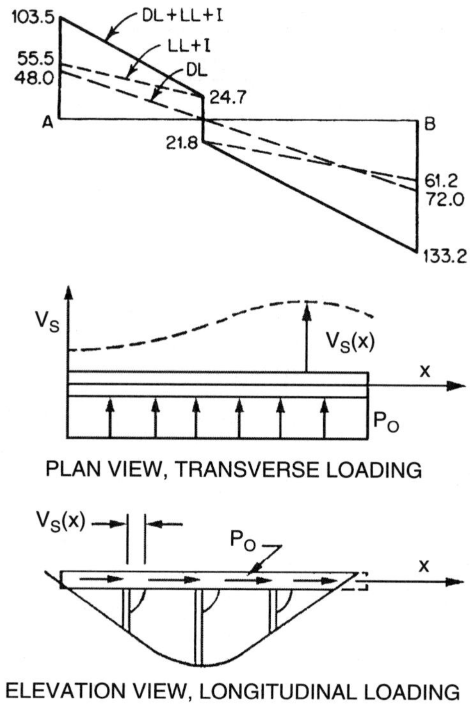

Figure 15.10 End-span maximum shear envelope, continuous beam, three 60-ft spans.

15.2.7 Impact Load (Dynamic Load Allowance)

Highway Loadings

The static effects of the design truck or tandem, other than centrifugal and braking forces, shall be increased by the percentage specified in the Table below for dynamic load allowance. The factor to be applied to the static load shall be taken as (1 + *IM*/100). The dynamic load allowance shall not be applied to pedestrian loads or to the design lane load.

Component	*IM*
Deck Joints—all limit states	75%
All other components:	
• Fatigue and fracture limit state	15%
• All other limit states	33%

Dynamic load allowance need not be applied to

- retaining walls not subject to vertical reactions from the superstructure
- foundation components that are entirely below ground level.

For buried components, the dynamic load for culverts and other buried structures, in percent, shall be taken as $IM = 33\,(1.0 - 0.125DE) \geq 0\%$, where

DE = the minimum depth of earth cover above the structure (ft)

Railroad Loadings

According to the American Railway Engineering and Maintenance-of-Way Association (AREMA) specifications, impact on railroad structures is the following percentages of the live load, applied equally and vertically at the top of each rail:

a. For stringers, girders, floor beams, posts of deck-truss spans carrying floor-beam loads only, and floor-beam hangers:

L less than 100 ft, $60 - L^2/500 + 100/S$

L 100 ft or more, $1800/(L - 40) + 10 + 100/S$

b. For truss spans, $4000/(L + 25) + 15 + 100/S$

Where the hammer-blow effect is absent, as in the case of diesel locomotives, the percentage of impact is given by

L less than 80 ft, $40 - 3L^2/1600 + 100/S$

L 80 ft or more, $600/(L - 30) + 16 + 100/S$

The impact allowance is also governed by the number of tracks on the structure:

a. Where the load is from two tracks:

L less than 175 ft, full impact on two tracks

L from 175 to 225 ft, full impact on one track on the other $(450 - 2L)$ percent of full impact

L over 225 ft, full impact on one track none on the other

b. Where there are more than two tracks, full impact is applied to any two tracks for all values of L.

The length L in these formulas is the following:

a. For stringers, transverse floor beams without stringers, longitudinal girders, and main members of trusses, L = length, ft, center to center of supports.

b. For floor beams, floor-beam hangers, subdiagonals of trusses, transverse girders, supports for longitudinal and transverse girders, and viaduct columns, L = length, ft, of the longer adjacent supported stringer, longitudinal beam, girder, or truss.

The length S is the distance, ft, between centers of longitudinal beams, girders, or trusses or between supports of floor beams or transverse girders.

Impact for ballasted-deck bridges is 90 percent of the values for open-deck bridges.

15.2.8 Wind

Wind load is assumed to act horizontally from any direction. Wind forces may result in vertical or twisting motions of flexible structures. In the design of suspension bridges, it is prudent to verify the stable aerodynamic characteristics of the stiffening trusses or girders by wind-tunnel tests on models.

Care should be taken in determining the path through which lateral pressures will be transmitted to the ground. Continuous concrete structures may well be stiff enough to carry the lateral loads to the abutments rather than to intermediate piers. Structures having concrete decks will usually transmit wind loads to the diaphragms or cross frames at the ends of the span and thence to the bearings. For diaphragm and cross-frame design, the wind load is applied to the top flange of a plate girder.

Wind loads for which various bridge structures are to be designed are prescribed in the AASHTO, AREMA, or other applicable specifications.

15.2.9 Other Loads

Earthquake Loads

No place is immune to earthquakes, so every structure should be designed for some earthquake resistance. The simplest approximation of earthquake forces on a bridge is obtained by applying a horizontal force equivalent to a calculated percentage of the bridge's weight through the bridge's center of gravity and calculating the corresponding deformed shape. Where frequent earthquakes are anticipated, more sophisticated design methods should be used.

Bridges shall be designed to have a low probability of collapse but may suffer significant damage and disruption to service when subject to earthquake ground motions. Partial or complete replacement may be required. Higher levels of performance may be used if authorized by the bridge owner.

Seismic analysis is not required for single-span bridges, regardless of seismic zone. Connections between the bridge superstructure and the abutments shall be designed for the minimum force and minimum support length requirements per AASHTO requirements.

For multispan structures, the minimum analysis requirements for seismic effect shall be as specified in the following AASHTO table.

Seismic zone	Single-span bridges	Multispan bridges					
		Other bridges		Essential bridges		Critical bridges	
		Regular	Irregular	Regular	Irregular	Regular	Irregular
1	No seismic analysis required	*	*	*	*	*	*
2		SM/UL	SM	SM/UL	MM	MM	MM
3		SM/UL	MM	MM	MM	MM	TH
4		SM/UL	MM	MM	MM	TH	TH

in which

* = no seismic analysis required
UL = uniform load elastic method
SM = single-mode elastic method
MM = multimode elastic method
TH = time history method

Calculation of earthquake loads based on AASHTO's single-mode spectral analysis is as follows:

1. Calculate the static displacements $v_s(x)$ due to an assumed transverse and longitudinal uniform loading p_o as shown in Fig. 15.10*b*.
2. Calculate factors α, β, and γ as follows:

$$\alpha = \int v_s(x)dx$$

$$\beta = \int x(x)v_s(x)dx$$

$$\gamma = \int w(x)v_s^2(x)dx$$

where

p_o = a uniform load arbitrarily set equal to 1.0 (kip/ft)
$v_s(x)$ = deformation corresponding to p_o (ft)
$w(x)$ = nominal, unfactored dead load of the bridge superstructure and tributary substructure (kip/ft)

The computed factors, α, β, and γ have units of (ft^2), (kip-ft), and (kip-ft^2), respectively.

3. Calculate the period of the bridge as

$$T_m = 2\pi\sqrt{\frac{\gamma}{p_o g \alpha}}$$

where

g = acceleration of gravity (ft/sec^2)

4. Calculate C_{sm}, which is the dimensionless elastic seismic response coefficient, where

$$C_{sm} = A_S + (S_{DS} - A_S)(T_m/T_0)$$

where

$A_S = F_{pga}\,PGA$
$S_{DS} = Fa\,S_S$
PGA = peak ground acceleration coefficient on rock (Site Class B)
S_S = horizontal response spectral acceleration coefficient at 0.2-s period on rock (Site Class B)
T_m = period of vibration of mth mode (s)
T_0 = reference period used to define spectral shape = 0.2 T_S (s)
T_S = corner period at which spectrum changes from being independent of period to being inversely proportional to period = S_{D1}/S_{DS} (s)

For periods greater than or equal to T_0 and less than or equal to T_S, the elastic seismic response coefficient shall be taken as $C_{sm} = S_{DS}$

For periods greater than T_S, the elastic seismic response coefficient shall be taken as $C_{sm} = S_{D1}/T_m$, in which

$$S_{D1} = F_v\,S_1$$

where

S_1 = horizontal response spectral acceleration coefficient at 1.0-s period on rock (Site Class B)

5. Calculate the equivalent static earthquake loading $p_e(x)$ as

$$p_e(x) = \frac{\beta C_{sm}}{\gamma} w(x)\, v_s(x)$$

where

$p_e(x)$ = the intensity of the equivalent static seismic loading applied to represent the primary mode of vibration (kip/ft)

6. Apply loading $p_e(x)$ to the structure and determine the resulting member force effects.

Design of Restraining Features

New structures should be designed with restraining details at all movable joints to keep the structure from coming apart in a severe shake. Similar restraints should be placed on existing structures. Restraints should be designed using the equivalent static analysis method. A simple span fixed at one end and moving at the other will have the entire superstructure as the contributing dead load to the longitudinal forces at the fixed abutment. One-half of the superstructure dead load will act at each abutment for transverse forces. For two or more spans with an expansion joint at only one end, the entire length from a fixed point to an expansion joint should be used for the dead load contributing force in a longitudinal direction from which may be deducted the shear resistance to earthquake forces of the intermediate bents.

Buoyancy

Submerged structures should be designed for the buoyant effect of the water. Although this can be of considerable importance in a large structure, it is of little effect in smaller ones.

Longitudinal Force

AASHTO requires that provision be made for a longitudinal braking force. The braking force shall be taken as the greater of 25 percent of the axle weights of the design truck or design tandem or 5 percent of the design truck plus lane load or 5 percent of the design tandem plus lane load. These forces shall be assumed to act horizontally at a distance of 6.0 ft above the roadway surface in either longitudinal direction to cause extreme force effects.

Centrifugal force is seldom of consequence in a highway structure. For the purpose of computing the radial force or the overturning effect on wheel loads, the centrifugal effect on live load shall be taken as the product of the axle weights of the design truck or tandem and the factor C, taken as

$$C = fv^2/gR$$

where

v = highway design speed (ft/s)

f = 4/3 for load combinations other than fatigue and 1.0 for fatigue

g = gravitational acceleration: 32.2 (ft/s^2)

R = radius of curvature of traffic lane (ft)

Centrifugal forces shall be applied horizontally at a distance 6.0 ft above the roadway surface. A load path to carry the radial force to the substructure shall be provided.

For railways, in accordance with AREMA[2] specifications, the centrifugal force is the percentage $0.00117S^2D$ of each axle load, excluding impact, where S is the design speed (mph) and D is the degree of curve (in degrees). This centrifugal force is assumed to act horizontally 6 ft above the rail or roadway surface.

Thermal forces can be very important. The extent to which they are provided for will depend upon the locality and the possible temperature variation. The ranges of temperature shall be as specified in table below.

Climate	Steel or aluminum	Concrete	Wood
Moderate	0° to 120°F	10° to 80°F	10° to 75°F
Cold	−30° to 120°F	0° to 80°F	0° to 75°F

Earth Pressure

One method of determining the forces for which structures retaining a fill must be designed is the Rankine-formula pressure but not less than an equivalent fluid pressure of at least 30 pcf. This is a good figure for general use. To avoid movement in high walls or abutments, the more conservative figure 36 pcf is recommended in computing toe pressures for high walls or abutments, while for heel pressures 27 pcf may be used.

Shrinkage of concrete creates stresses which must be accommodated. Just as in a complicated weldment where the welding sequence is carefully programmed to minimize shrinkage and distortion, so in a complicated concrete structure must the pouring sequence be carefully worked out and specified to minimize shrinkage effects. Concrete takes most of its shrinkage soon after it sets. Consequently, a delay of 24 hours between pours for thin sections and of from 3 to 5 days for heavier sections is usually sufficient to allow for most of the shrinkage. Large-mass concrete structures raise more serious problems, but the average highway or railroad bridge does not have sections massive enough to make shrinkage a problem after 3 to 5 days. Arch rings are poured in intermittent sections and then filled in between, with a final pour at the crown, to minimize shrinkage effects. Complicated rigid frames or interconnected concrete members should be similarly poured in sections and allowed to shrink before the final gap is closed.

The shrinkage coefficient for unreinforced concrete is 0.0002. However, reinforcing bars restrict the concrete so that this shrinkage is seldom realized. Shrinkage is manifested most often by the numerous minute cracks which are inevitable with presently constituted cements. In ordinary concrete beam-and-slab construction, shrinkage stresses may be ignored in the design, provided proper field control of pouring sequences is exercised. In arches and in prestressed construction, the effect of shrinkage is appreciable and must be compensated for in the design of the member.

Erection Stresses

In certain types of structures and methods of erection, stresses which exist only during erection can be considerable. In some cases, such as cantilever trusses, members needed temporarily during erection may be removed later or altered to prevent their having an undesirable effect on stresses in other members. It is usual practice for the erector to determine and provide for any necessary strengthening of members during erection.

Stream Pressure

Portions of structures which are subject to the force of flowing water, floating ice, or debris must be designed to resist these forces. The pressure of flowing water acting in the longitudinal direction of substructures shall be taken as

$$p = C_D V^2/1000$$

where

p = pressure of flowing water (ksf)

C_D = drag coefficient for piers as specified in table below

V = design velocity of water for the design flood in strength and service limit states and for the check flood in the extreme event limit state (ft/s)

Type	C_D
Semicircular-nosed pier	0.7
Square-ended pier	1.4
Debris lodged against the pier	1.4
Wedged-nosed pier with nose angle 90 degrees or less	0.8

The longitudinal drag force shall be taken as the product of longitudinal stream pressure and the projected surface exposed thereto.

The lateral uniformly distributed pressure on a substructure due to water flowing at an angle, θ, to the longitudinal axis of the pier shall be taken as

$$p = C_L V^2/1000$$

where

p = lateral pressure (ksf)
C_L = lateral drag coefficient specified in the table below

Angle, θ, between direction of flow and longitudinal axis of the pier	C_L
0 degrees	0.0
5 degrees	0.5
10 degrees	0.7
20 degrees	0.9
≥30 degrees	1.0

The lateral drag force shall be taken as the product of the lateral stream pressure and the surface exposed thereto.

Wave Load

Wave action on bridge structures shall be considered for exposed structures where the development of significant wave forces may occur:

$$P = KV^2$$

Ice Pressure

In the absence of more precise information, the following values may be used for effective ice crushing strength:

- 8.0 ksf where breakup occurs at melting temperatures and the ice is substantially disintegrated in its structure
- 16.0 ksf where breakup occurs at melting temperatures and the ice is somewhat disintegrated in its structure
- 24.0 ksf where breakup or major ice movement occurs at melting temperatures, but the ice moves in large pieces and is internally solid
- 32.0 ksf where breakup or major ice movement occurs with the ice temperature, averaged over its depth, measurably below the melting point

15.2.10 LRFD Design Method

The design philosophy of load and resistance factor design (LRFD) takes variability in the behavior of structural elements into account in an explicit manner. LRFD relies on extensive use of statistical methods but sets forth the results in a manner readily usable by bridge designers and analysts. It is a very desirable alternative when designing long-span bridges where the dead load is large compared to the live load.

15.3 STEEL BRIDGES

15.3.1 Floor Systems

Decks for steel highway bridges are usually reinforced concrete. The use of other types, such as steel grid, armor type, steel plate, and timber, depends upon dead load, traffic, and the location of the structure. Decks should have a transverse slope and, if possible, a longitudinal slope to provide adequate drainage. Transverse slope should be at least 1 percent and preferably more. Adequate drains should be placed at suitable locations to prevent water from collecting on the deck.

Types

Exclusive of the main carrying members, floor systems can be classified as one-, two-, three-, and four-element systems, as follows:

1. A transversely reinforced concrete slab on main stringers or girders is a one-element system. Transverse-laminated timber floors are in the same category.
2. A longitudinally reinforced concrete slab on transverse floor beams supported by girders or trusses is a two-element system.
3. A transverse slab on longitudinal stringers supported by floor beams framing into the main girders or trusses is a three-element system. The design of the slab in the three-element system is similar to that of the one-element system.
4. A steel-grid deck on transverse beams resting on stringers supported by floor beams framing into the main girders or trusses is a four-element system.

In addition to the above, the orthotropic plate floor, which consists of a deck plate acting compositely both longitudinally and transversely and, in addition, with the main supporting members, is used.

Slabs

The dead load consists of the weight of the slab plus any anticipated wearing surface. Bituminous surfacing is occasionally placed over concrete slabs to match the bridge approaches or to level up or seal cracks in an old concrete deck. Where an added blanket is a definite possibility, it should be provided for in the design. Where a blanket is added after years of service, the higher stresses normally allowed in computing the capacity of existing bridges will usually compensate for the extra weight of the blanket.

15.3.2 Concrete Floors

Transversely reinforced concrete slabs are commonly used on steel-stringer spans, plate girders, and truss bridges, as well as on several types of reinforced-concrete structures. Typical examples are shown in Fig. 15.11. A steel-stringer span, in which the slab is a one-element floor system, is shown in Fig. 15.11*a*. A truss span, in which the slab is part of a three-element floor system, is shown in Fig. 15.11*b*.

EXAMPLE

Design a transversely reinforced concrete slab continuous over three or more longitudinal steel stringers, AASHTO specifications Working Stress Design. Live load is HS20-44. Allowable stresses $f_c = 1600$ psi, $f_s = 20{,}000$ psi. No allowance will be made for future wearing surface. A typical section is shown in Fig. 15.12. The assumed slab thickness is 7 in., weight 88 psf.

The design span S for a slab continuous over two or more supports is the distance center to center of stringers minus half the flange width, in this case, 7 ft (Fig. 15.13). The live-load moment per foot width of slab, both positive and negative, is given by $0.8P(S + 2)/32$, where P is the wheel load per AASHTO. Impact for a 7-ft span is 30 percent.

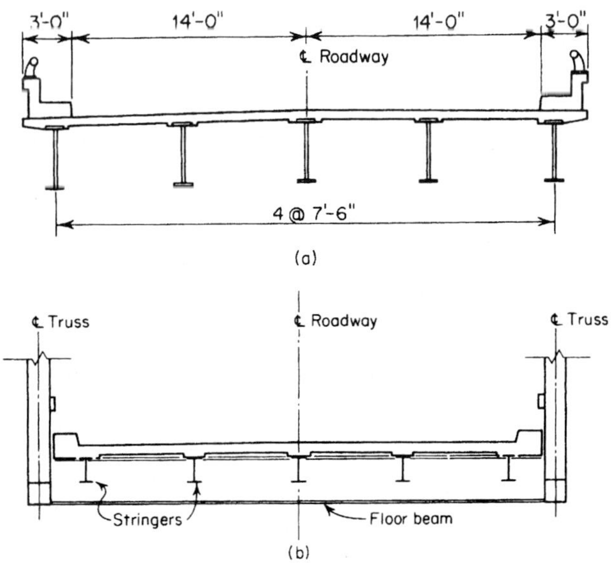

Figure 15.11 Typical transversely reinforced floor slab: (*a*) steel-stringer span and (*b*) steel truss.

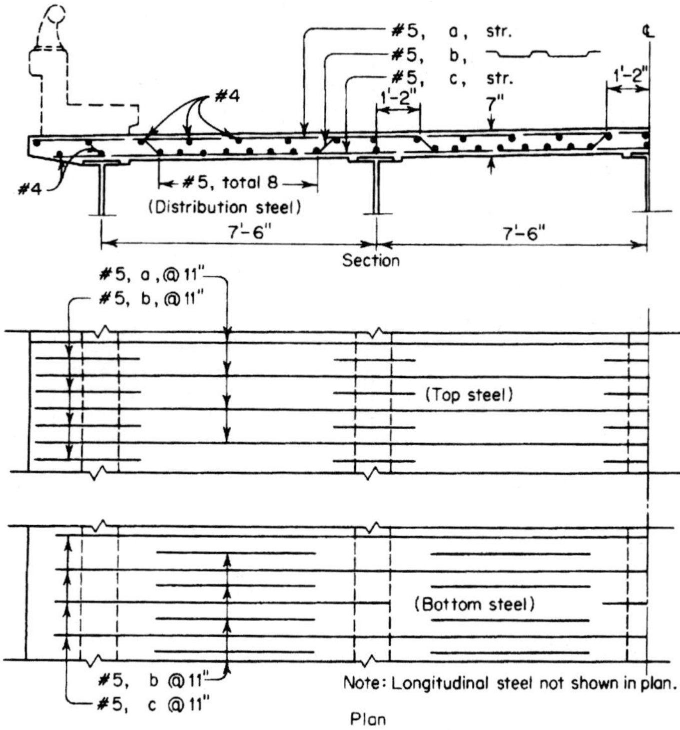

Figure 15.12 Typical section of transversely reinforced floor slab.

The design moments per foot width of slab are

$$DL = 0.088 \times 7^2 / 10 = 0.43$$

$$LL + I = 1.30 \times 0.8 \times 16(7+2)/32 = \underline{4.68}$$

$$5.11 \text{ ft-kips}$$

With $K = \frac{1}{2} f_c kj = 262$, $d^2 = 5110/262 = 19.5$, $d = 4.4$ in. using $d = 4\frac{5}{8}$ in. with $1\frac{7}{8}$ in. from center of steel to top of slab gives a $6\frac{1}{2}$-in. slab. Then

$$A_s = \frac{5.11 \times 12}{20 \times 0.88 \times 4.62} = 0.75 \text{ in.}^2$$

Use No. 5 at 5 in.

The specifications require distribution steel in the bottom of the slab, to be determined as the percentage $220/\sqrt{S}$ of the main steel but not to exceed 67 percent:

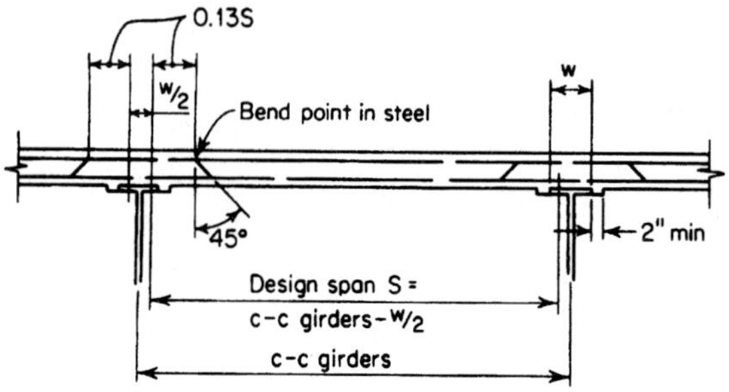

Figure 15.13 Transversely reinforced slab on steel stringers.

$$\frac{220}{\sqrt{7}} = 83 > 67\%$$

$$0.67 \times 0.68 = 0.46 \text{ in.}^2$$

Use No. 5 at 8 in.

Other than distribution steel, the longitudinal steel in the top and bottom of the slab can be No. 4 with 18 in. minimum spacing. Bars should be placed at the bend points in the transverse steel and at other points where the reinforcing should be adequately tied to hold it in place during pouring operations.

Longitudinally Reinforced Slabs

When used on truss spans, the longitudinally reinforced slab is part of a two-element floor system. More commonly, it is used as a primary structure supported on piers, bents, or piles. Its use with trusses is limited to comparatively short spans, as the dead load is considerably greater than that in the floor system using a transverse slab.

For a slab supported by the steel floor beams of a truss, the effective span (AASHTO) is the span center to center of floor beams less one-half the flange width. Practical procedure is to use the span and, where refinement is necessary or desired, make corrections to the computed moments in accordance with the ratio of effective span to span.

The AASHTO requirements in determining placement, sizes, and length of reinforcing bars in continuous beams are as follows:

1. At the freely supported end of a continuous slab, one-third of the positive-moment reinforcement shall extend beyond the face of the support a distance sufficient to develop one-half the allowable stress in the bars.

2. At the restrained ends of continuous slabs, one-fourth of the positive-moment reinforcement shall extend beyond the face of the supports.

3. Between the supports, reinforcement bars shall extend at least 15 diameters but not less than one-twentieth of the span length beyond the point where they are no longer needed to resist stress.

EXAMPLE

Design a longitudinally reinforced slab continuous over four supports. AASHTO specifications, HS20-44 live load. Allowable stresses f_c = 1600 psi, f_s = 20,000 psi. The layout is shown in Fig. 15.14, which is a longitudinal section of a truss span. Calculations will be based on the 16-ft span rather than the effective span.

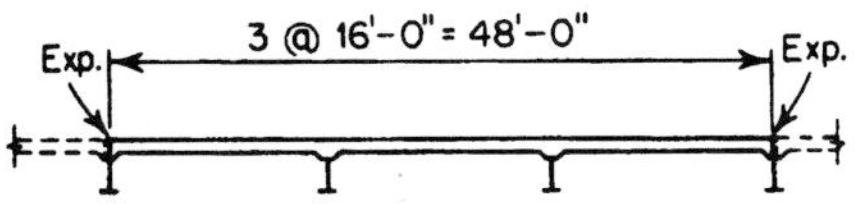

Figure 15.14 Longitudinally reinforced floor slab.

The effective width of slab for distribution of one wheel line is $E = 4 + 0.06S = 4 + 0.06 \times 16 = 4.96$ ft. The slab is analyzed as a continuous strip 1 ft wide. The wheel-line concentrations (impact not included) are shown in Fig. 15.15*a* and their positions for maximum positive and negative moments in Fig. 15.15*b*. Other positions of load, not shown in the figure, are required to determine the envelope of maximum moment shown in Fig. 15.16. The slab depth is governed by the positive moment, 17.1 ft-kips, in the end span, and $d^2 = 17{,}100/198$, $d = 9.4$ in. With $d = 9\frac{1}{2}$ in. and $1\frac{1}{2}$-cover, total depth of slab is 11 in. A shallower depth could be used by designing the end spans as doubly reinforced slabs.

Positive moment in end span: $A_s = \dfrac{12 \times 17.1}{20 \times 0.88 \times 8.5} = 1.37 \text{ in.}^2$

	A_S	M
No. 6 at 14	0.38	4.7
No. 7 at 14	0.51	6.4
No. 7 at 14	0.51	6.4
	1.40 in.^{2}	7.5 ft kips

Negative moment in interior span: $A_s = \dfrac{12 \times 15.8}{20 \times 0.88 \times 8.0} = 1.35 \text{ in.}^2$

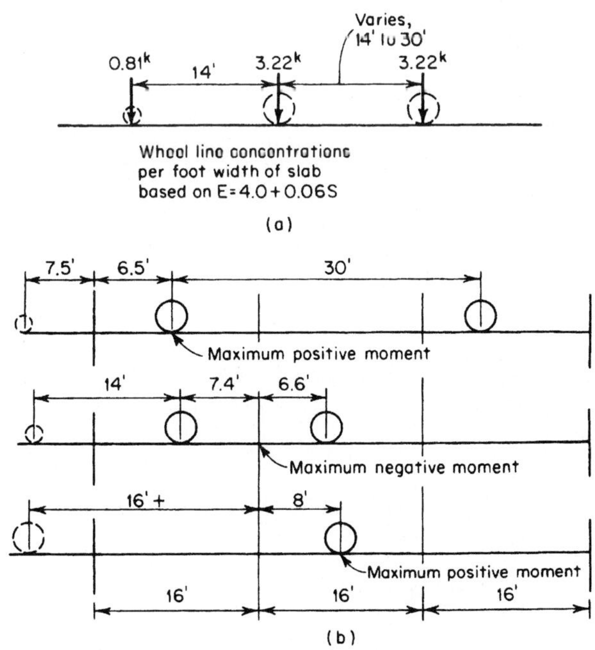

Figure 15.15 Wheel positions for maximum moment in floor slab of Fig. 15.14.

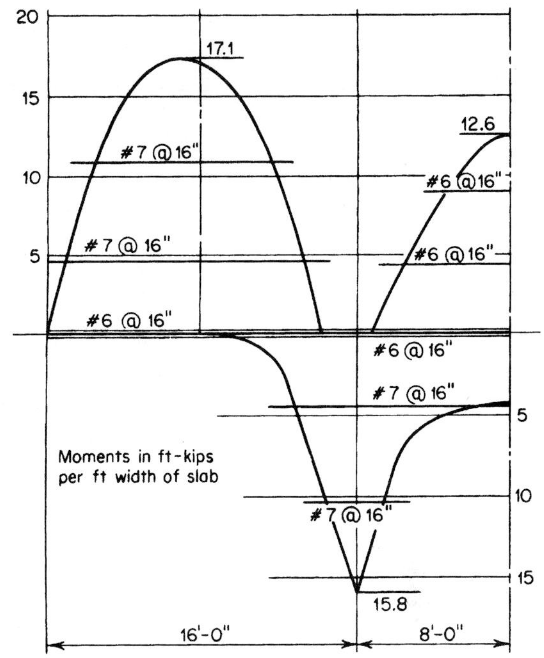

Figure 15.16 Maximum-moment envelopes for floor slab of Fig. 15.14.

	A_S	M
No. 6 at 14	0.38	4.5
No. 7 at 14	0.51	6.0
No. 7 at 14	0.51	6.0
	1.40 in.^{2}	16.5 ft-kips

Positive moment in interior span: $A_s = \dfrac{12 \times 12.6}{20 \times 0.88 \times 8.5} = 1.01 \text{ in.}^2$

	A_S	M
No. 6 at 14	0.38	4.7
No. 7 at 14	0.51	4.7
No. 7 at 14	0.51	4.7
	1.40 in.^{2}	14.1 ft-kips

In the bottom of the slab, transverse distribution steel in the center half of the span is determined as the percentage of the main steel $100/\sqrt{S}$ but not to exceed 50 percent. Distribution steel can be reduced one-half in the outer quarters. Transverse temperature steel in the top of the slab is usually No. 4 bars at 18 in. Details of the slab are shown in Fig. 15.17.

Lightweight-Concrete Bridge Decks

Lightweight concrete composed of aggregates made from expanded shale, slate, clay, and slag is occasionally used in bridge decks. Concrete in the structural range of 3000 to 4500 psi can be obtained with weights of about 100 to 120 pcf, not including the reinforcing steel. Lightweight-concrete decks are frequently constructed in two courses: the bottom course consisting of the lightweight concrete and, to provide durability, the top course consisting of natural-sand mortar composed of 1 part

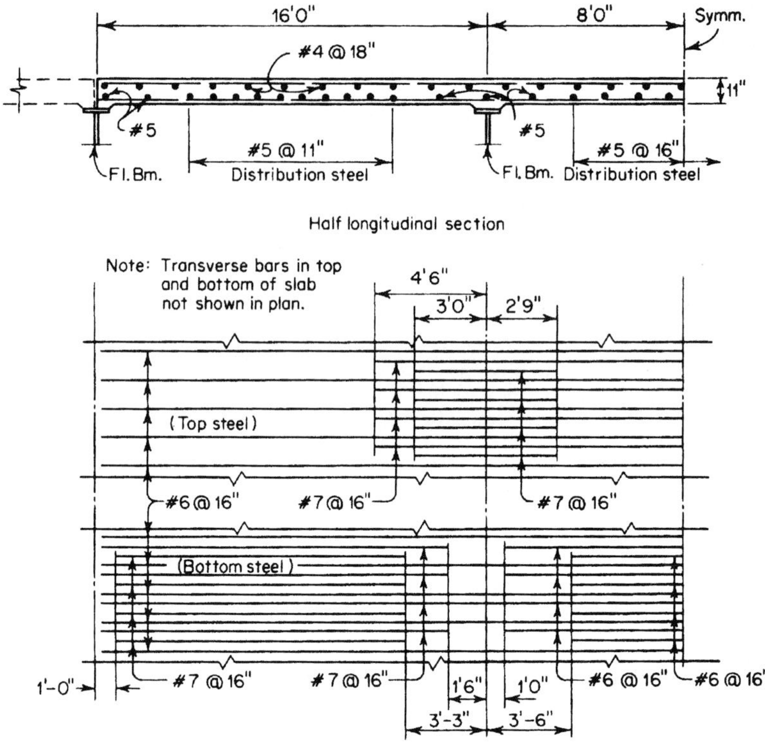

Figure 15.17 Reinforcement for floor slab of Fig. 15.14.

cement and 3 parts by dry weight of natural sand. Thickness of the top course should be not less than ½ in. or more than ¾ in.

Design coefficients will vary somewhat. Lightweight concrete has a larger value of *n* because of its lower modulus of elasticity. There is a decrease in *j* and an increase in *k* and *p*. For balanced reinforcement a higher percentage of steel is required, but there is a reduction in depth of the member because of decreased dead load. Beams will have more deflection owing to the smaller modulus of elasticity.

Some of the important items which should be covered in a specification for lightweight concrete are (1) minimum amount of portland cement per cubic yard; (2) desired weight per cubic foot; (3) design strength in flexure; (4) type of lightweight aggregate; (5) grading limits of the aggregate; (6) percentage of burned or unburned lumps in expanded shale (ASTM Designation C142); (7) hardness of aggregate to conform to a specified test; (8) maximum weight of aggregate in loose, surface-dry condition; (9) limitations on the moisture content prior to mixing, which indicate the quantity of aggregate sample to be submitted for test; (10) provision for substitution of natural sand, provided it does not increase the weight beyond the desired maximum; (11) control of use of internal vibrators so as to guard against flotation of the coarse aggregate; and (12) use of air-entraining agents to improve workability and minimize separation of aggregates.

15.3.3 Steel Floors

Battledeck floors can be used for spans from ±15 to ±30 ft. They can be used as primary elements, for example, on pile bents, and in the floor system for plate-girder and truss bridges. Light weight is their distinctive feature. Each section or unit is usually 6 to 9 ft wide, depending on the number of units required to make up the desired width. Originally, these units consisted of a deck plate welded to either I- or wide-flange beams. More recently, the units are built up completely from plates (Fig. 15.18). Transverse stiffeners welded to the floor plate between beams allow a thinner floor plate.

A durable and satisfactory surface for smooth floor plate is composed of a coal-tar-modified thermosetting epoxy resin, with a cover of aluminum-oxide aggregate added to provide the nonskid surface.

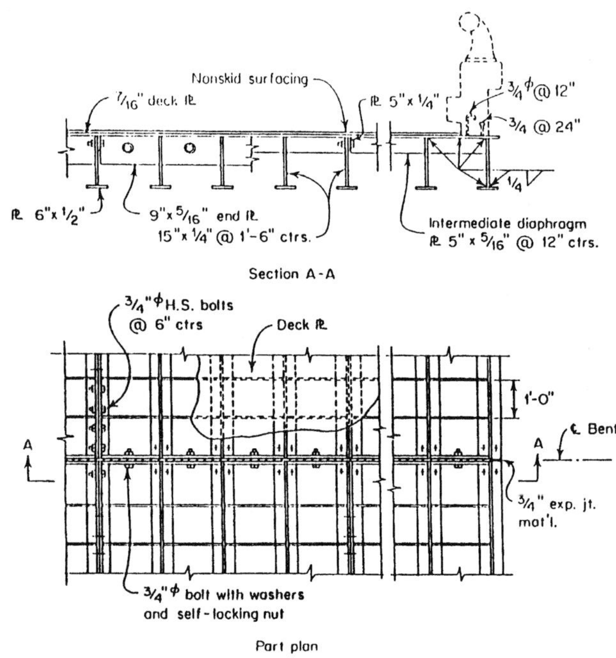

Figure 15.18 Typical battledeck floor, 26-ft spans.

Before application, it is important that the surface of the steel deck be thoroughly sandblasted. The time for hardening of this surfacing may range from 1 to 2 hours at 90°F to 12 or more hours at 60°F.

A surface of asphaltic concrete material suitably sealed and bonded to the steel plate will work satisfactorily in some locations. It is recommended that its use be limited to decks that have a longitudinal gradient not greater than 2 percent and where maximum temperatures do not exceed 100°F. Asphalt plank, fixed in place with asphaltic cement, has been used on some structures. In some installations, hard stone chips have been pressed into the plank to increase traction.

To improve traction where traffic runs directly on the floor plate, "checkered" or "traffic" plates can be used, preferably those in which the patterns are raised above the nominal thickness of the plates. When choosing the type of surface finish, consideration should be given to weather conditions. Battledeck floors are sensitive to weather changes. Because of the free access of air to both sides of the floor plate, freezing or thawing occurs almost simultaneously with air-temperature variations.

Steel-grid floors are used where light deck systems are desired. They have been used on many movable structures, such as vertical-lift spans, bascule spans, and swing bridges. In general, they consist of two types: the I-Beam Lok and the Irving type. Either type can be used as an open grid or filled with concrete. An open deck weighs approximately 15 to 19 psf, depending on the type. Recommended spans for the various types are given in manufacturers' literature.

Both transverse and longitudinal distribution of load is considered in the design of grid flooring. According to AASHTO, a wheel load is distributed, normal to the main bars, over a width equal to 1¼ in. per ton of axle load plus twice the distance center to center of main bars. The portion of the load assigned to each main bar is applied uniformly over a length equal to the rear tire width of the design truck (20 in. for H-20, 15 in. for H-15). This distribution is shown in Fig. 15.19 for an H-20 wheel.

In the design of steel-grid floors filled with concrete, AASHTO calls for load distribution and bending moments determined in the same manner as for concrete slabs. This also applies to edge beams (longitudinal), to unsupported edges (transverse), and to the design span length. Properties of the composite steel and concrete slab are determined by the transformed area.

Units of deck grating are welded to the supporting elements in accordance with the manufacturer's instructions. A successful method of attaching the Irving decking is to weld the deck to the supporting elements using 3/16- by 1½-in. welds on one side of every fourth bar (Fig. 15.19). Welds are staggered so that each main bar is welded at every fourth support.

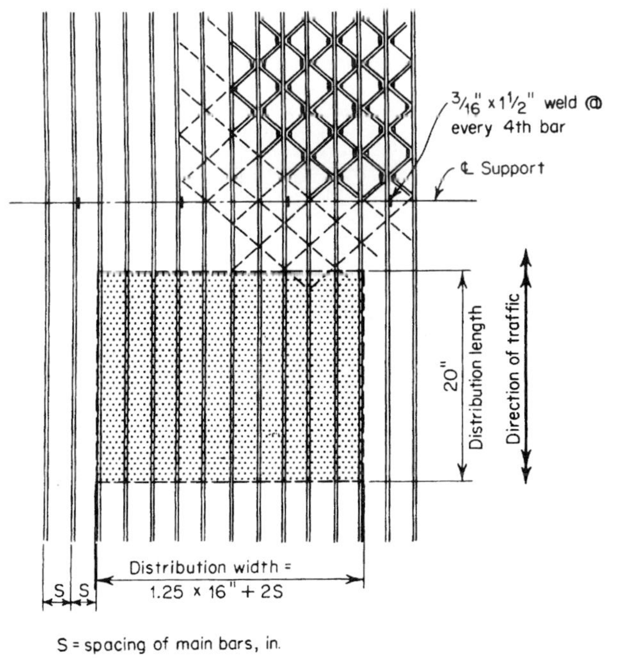

Figure 15.19 Steel-grid flooring, showing AASHTO distribution for 16-kip wheel.

STRUCTURAL PLATE FLOORING

For redecking existing bridges, the structural plate floor is economical and comparatively easy to erect under traffic. It is rolled in various gage thicknesses in widths of 24 in. Usually it is installed transverse to traffic and welded to the stringers. Depending on the gage of the plate, the stringer spacing can vary from about 2 to 3 ft. The effective span of this flooring on steel stringers is defined for design purposes as the clear distance between flanges plus one-half the stringer flange width. Lateral and longitudinal distributions of a wheel load are not defined in the AASHTO specifications, but the same distribution as for a laminated timber floor can be used. A continuity factor of 0.8 can be applied to the simple-beam moment for both positive and negative moments.

A bituminous material, consisting of two coats, is placed as a wearing surface. The first course is placed to a level slightly above the top of the corrugations. The second course is placed so that after rolling the resulting surface is at least 2 in. above the corrugations at centerline of roadway and 1 in. thick at the edges to provide a crown. A typical installation is shown in Fig. 15.20.

15.3.4 Floor Beams

Steel floor beams may be standard rolled shapes or may be built up from plates. The distribution of wheel loads to floor beams in decks without longitudinal stringers depends on their spacing and the kind of floor. Distribution factors are given in AASHTO. No lateral distribution of wheel loads is assumed.

In floors with longitudinal stringers, dead load from the stringers can be considered as a uniform load on the floor beam for design purposes without appreciable error. Figure 15.21*a* shows a floor beam for a truss bridge having 20-ft panels. If the dead load from the stringers is applied to the floor beam as uniform load, the maximum moment is 240 ft-kips; applying the stringer reactions as concentrated loads gives a maximum moment of 230 ft-kips.

Live loads on the floor beam can be considered on the basis of wheel lines. The live-load reaction per line (Fig. 15.21*b*) is $4 \times 6/20 + 16 + 16 \times 6/20 = 22.0$ kips.

It is assumed that trucks stay within the design lanes. For maximum live-load moment they are placed as shown in Fig. 15.21*a*. The resulting moment is

$$2 \times 22 \times 14.33 - 2 \times 22 \times 3 = 44 \times 11.33 = 500 \text{ ft-kips}$$

The total moment is

	Working stress
DL floor beam	= 25
DL stringers	= 240
LL	= 500
Impact 30 percent	= 150
DL + *LL* + *I* = *M*	= 915 ft-kips

For an allowable stress of 20 ksi, the section modulus is $S = \dfrac{915(12)}{12} = 549 \text{ in.}^3$

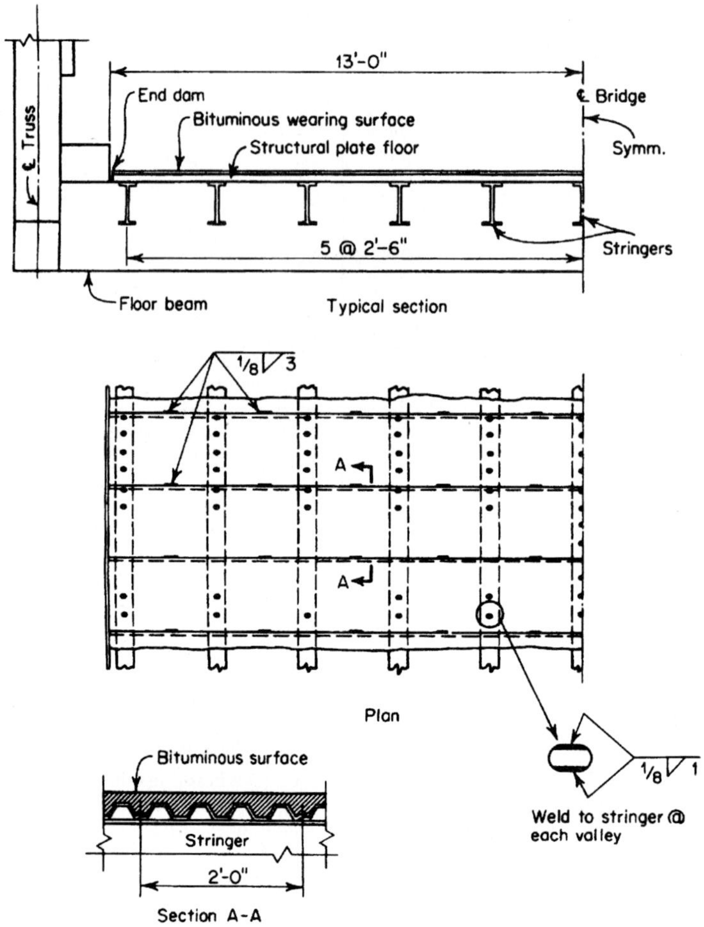

Figure 15.20 Structural plate flooring.

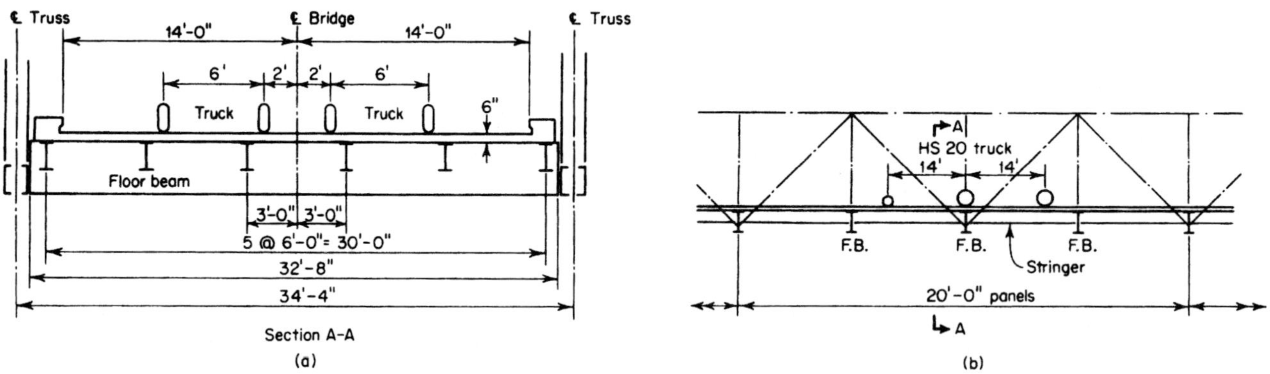

Figure 15.21 Floor beam for a truss bridge.

Use W36 × 170

Load factor $M_u = 1.3\ [DL + 1.67(LL + I)]$	
DL floor beam	= 25
DL stringers	= 240
LL	= 835
Impact 30 percent	= 250
	$DL + LL + I = 1350 \times 1.3 = 1755$ ft-kips $= M_u$

$$M_u = F_y Z$$

$$Z = \text{plastic section modulus}$$

For $F_y = 36$ ksi, the section plastic modulus is $Z = \dfrac{1755}{36}(12) = 585\ \text{in.}^3$
Use 36 × 170

The plastic section modulus *Z* can be used only when compact sections are involved. Most rolled beam sections meet the requirements of a compact section.

For maximum end reaction of the floor beam, one of the trucks should be placed within 2 ft of the curb. The other truck would be placed in the same position as shown for moment to stay within the design lane.

15.4 BEARING AND EXPANSION DETAILS

15.4.1 End Bearings

Bearing details are designed to transmit the loads to the substructure and to provide for expansion of the superstructure. They may vary from thin elastomeric bearing pads to huge pin and rocker assemblies. Certain basic requirements must be satisfied. First, the bearing must be adequate for the loading, which may occur simultaneously from several different directions. Second, the bearing must be able to accommodate movements of the structure. These movements, which may be in any direction, result from load, deflection, temperature change, earthquake movement, impact, centrifugal force, etc. The design must restrict movements to reasonable limits. This may be done with keeper plates, lugs, and anchor bolts in slotted holes. Third, bearings must be easy to maintain or else designed so as to require a minimum of maintenance. Many failures have resulted from freezing of expansion details. In highly seismic zones, it is best to avoid steel rocker bearings and tall pin and rocker assemblies because of their instability.

Expansion details must allow for a thermal movement of 1¼ in. for each 100 ft of structure. Provision must also be made to accommodate changes in length due to live load. Spans over 300 ft long should also contain provision for independent expansion and contraction of the deck.

Spans having less than 2 in. of anticipated movement are best supported on elastomeric bearing pads. Spans with more than 2 in. of movement and spans over 50 ft not using elastomeric pads may be provided with rollers, rockers, or lubricated sliding plates and must have a type of bearing employing a hinge, curved bearing plates, or a pin to allow for rotation of the upper bearing plate. Other alternatives include high-load multirotational bearings such as the pot, spherical, and disk-type bearings.

Sliding Plates and Small Pedestals

Steel plates of equal area that are intended to slide on each other usually corrode, freeze, and cause trouble. A better detail can be made by using a block or small pedestal of steel for one of the plates, thus keeping the contact area small.

Elastomeric bearing pads may be used for all types of bridges. Figure 15.22 shows a typical elastomeric bearing detail suitable for short steel spans, concrete T-beams, box girders, and slabs. A durometer hardness of 55±5 is the most satisfactory hardness for bridge service. The thickness of the pad should not exceed one-fifth the width, nor should it be less than twice the horizontal movement. The minimum thickness should be 1 in. made up of two ½-in. layers. Maximum pressure on the pad should not exceed 800 psi. The initial vertical deflection from dead load should not exceed 15 percent of the uncompressed thickness of the pad.

The shear force on the pad, which is equal to shearing modulus × area × movement/pad thickness, should not exceed one-fifth the dead load. The modulus of elasticity in shear should be about 135 psi at 70°F. Because the pads are seldom set at the mean temperature, 1½ times the temperature range suitable for the location should be used for the design. For thicknesses greater than 1 in., increments of ½ in. are used. For convenience in fabrication, pad length and width should be specified in increments of 2 in. To reduce bulging of the pad, non-elastic separators are used between the layers. If metal plates are used, the pads must be molded to final size when manufactured, with at least ⅛ in. cover of elastomeric material over the edges and faces of the pad. If fabric separators are used, the pads may be sawed from larger pads, using care to make clean cuts. Quality of material is of crucial importance in elastomeric bearing pads, and a good specification should be used and rigidly followed. Table 15.6 gives allowable loads for various combinations of length and width at 800 psi and required thicknesses for various spans. Figure 15.23 illustrates an anchor-bolt detail for an elastomeric bearing pad.

Self-lubricating bearings consist of two steel plates with a lubricated bronze plate between them. The bronze plate may have trepanned, drilled, or bored inserts filled with a nonplastic lubricant over at least 25 percent of its area, or it may be an oil-impregnated, sintered, powdered-metal bronze plate with a compressive strength of not less than 15,000 psi. The coefficient of friction between the steel and bronze plates should not exceed 0.1. Unlubricated bronze bearings used on bridges will usually be unsatisfactory because they tend to gall and freeze.

Roller and rocker bearings may be used to advantage for spans of moderate length (50 to 150 ft). Rollers and rockers should be of as large a diameter as is practicable, with a minimum of 6 in. Small rollers and roller nests tend to collect dirt and become rusted, clogged, and inoperative. In all expansion details provision should be made for drainage and thorough cleaning.

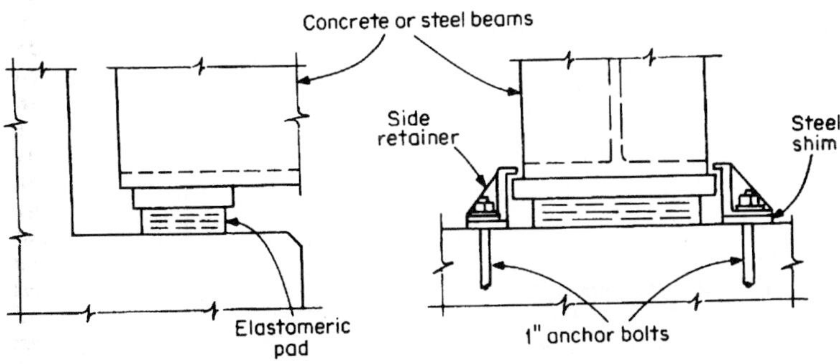

Figure 15.22 Typical elastomeric bearing pad detail for short-span bridges.

Table 15.6 Elastomeric Bearing Pads

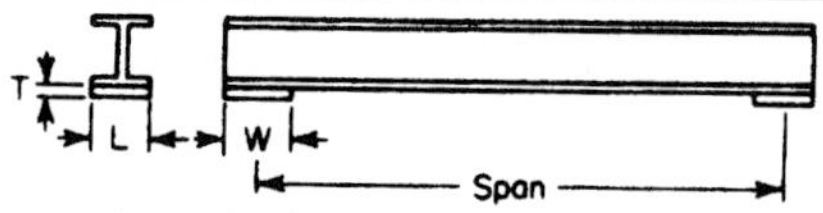

Maximum load, DL + LL + I, kips, for allowable stress = 800 psi

	L, in											
W, in.	10	12	14	16	18	20	22	24	26	28	30	32
6	48	57	67	76	86	96	105	115	125			
8	64	76	89	102	115	128	141	153	166	179		
10	...	96	112	128	144	160	176	192	208	224	240	256
12	...	...	134	153	172	191	210	230	249	268	287	306

Span, ft, of steel and concrete beams for various thicknesses

		Range in temperature					
		Extreme, 120°F		Moderate, 100°F		Mild, 80°F	
T, in.	Min *W*, in	Steel	Concrete	Steel	Concrete	Steel	Concrete
1	6	0–70	0–115	0–80	0–130	0–105	0–155
1½	8	71–105	116–170	81–125	131–195	106–155	156–230
2	10	106–140	171–230	126–165	196–260	156–210	231–310

Segmental rockers, in which the faces of the rocker are cut to a radius greater than half the rocker height, require less space than do cylindrical rollers of sufficient radius to have the same allowable stress (Fig. 15.24). They are satisfactory where the movement is not too great and seismic activity is low. Large movements introduce a component of force which lifts the bridge. Large single rollers offer a simple, effective solution. Test results seem to indicate that allowable stresses higher than those permitted by the AASHTO specification formula may be justified for such rollers. In all cases, it is important that plates and rollers or rockers be geared together with lugs and restrained from moving laterally. In areas of high seismic activity, a catcher block should be provided to prevent large vertical displacements in case the bearings topple.

Pedestals and shoes may be cast or made up of welded structural elements. The AASHTO specifications require that the difference in width between the top and bottom bearing surfaces not exceed twice the distance between them. This is to ensure uniform bearing and minimize bending in the pedestal or shoe. The web plates and angles which connect built-up units to the base plate should be not less than ⅝ in. thick.

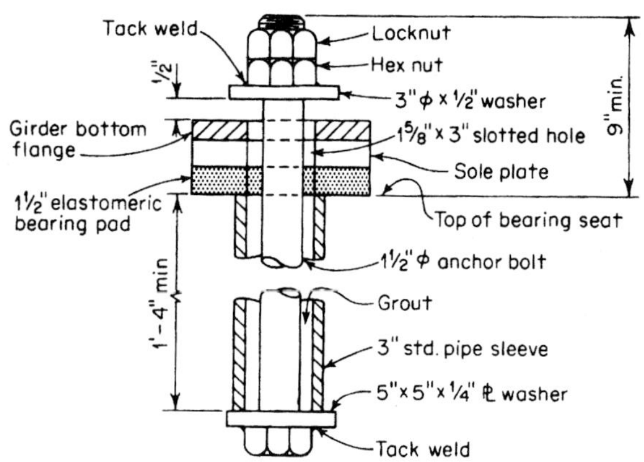

Figure 15.23 Anchor-bolt detail for elastomeric bearing pad.

The webs should be rigidly connected transversely, if possible. The minimum thickness of metal in cast-steel pedestals should be 1 in.

The design of bearings is largely a matter of preference and experience. Ease of fabrication should always be kept in mind. Pins should be made as large as is practicable; small pins have a tendency to freeze up. Drain holes should be provided both in and around the bearing detail so that water will not be trapped. If pockets cannot be avoided they should be filled with a mastic. Figures 15.25 and 15.26 show typical pedestal and shoe details.

15.4.2 Expansion Hangers

In continuous girder spans with intermediate suspended spans, support details must permit movement. For short spans, a small pedestal may be set on a step fabricated on the end of the supporting girder. For longer spans, where there is appreciable movement, pins and hanger links, such as are shown in Fig. 15.27, are quite satisfactory. For long spans and heavy loads it is usually necessary to use pin plates on each side of the web to stiffen it and provide adequate bearing area for the pin. Link expansion hangers of the type shown in Fig. 15.27 should be regarded as working pieces of machinery. However, imperceptible as this may be, they undergo almost constant movement. They should be frequently checked to make sure they are free to move and not corroded and that all the retaining nuts are in place. Failure to provide adequate inspection and maintenance can lead to catastrophic failures. Figure 15.28

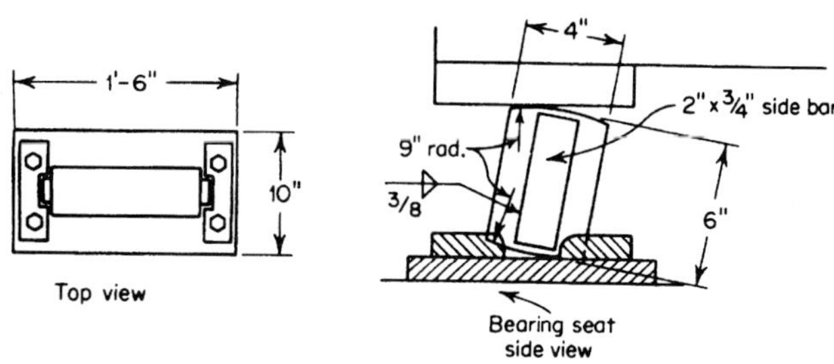

Figure 15.24 Typical rocker detail.

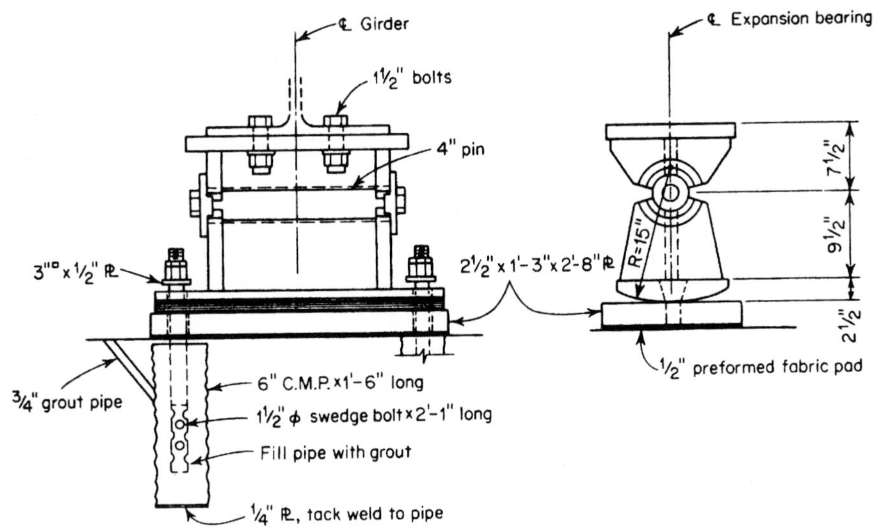

Figure 15.25 Expansion bearing.

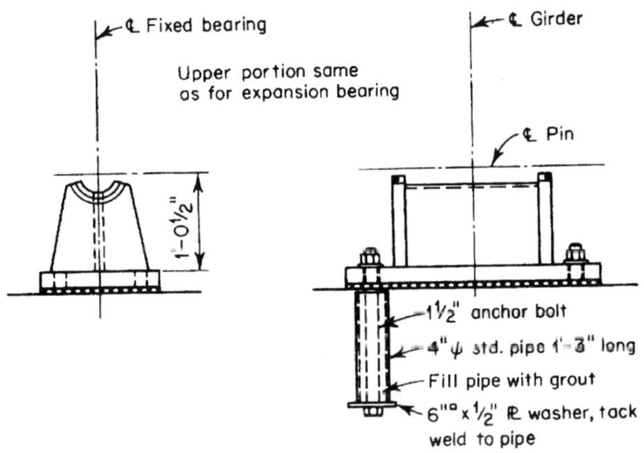

Figure 15.26 Fixed bearing.

illustrates a pin connection where provision for horizontal movement is not needed. Where significant displacements from seismic activity are anticipated, keeper plates should be installed to limit lateral movement, and restrainer rods or cables should be installed to control longitudinal displacements.

15.4.3 Deck Expansion Joints

It is probably safe to say that a completely satisfactory deck expansion joint has not yet been designed. Such joints are subject to warpage during fabrication or galvanizing and are difficult to place properly. They pound loose, become noisy, and are difficult to repair. They collect water and dirt and do not ride smoothly. A design that would overcome these faults without extensive maintenance would be well worth its cost. If conditions permit, a completely open joint, with steel angles on the exposed corners and the lower portion of the deck undercut to prevent debris jamming in the crack, is to be recommended. An open joint cannot be used where falling water and debris would be objectionable. Depending on the size of the joint, expansion systems such as poured sealants, compression seals, and modular units of rubber and steel

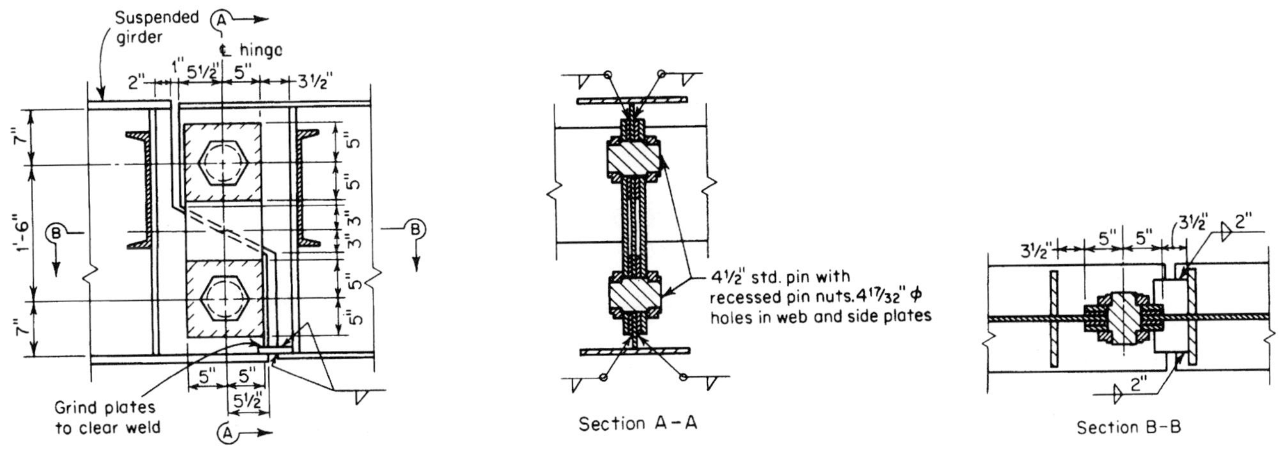

Figure 15.27 Typical link expansion hinge for welded plate girder.

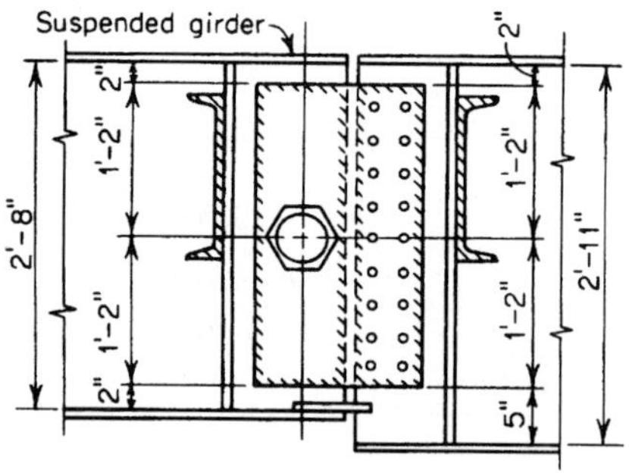

Figure 15.28 Typical shear hinge.

combined are also recommended. Figures 15.29 and 15.30 show some examples of deck expansion joints which, if properly fabricated and installed, should give good service.

15.5 BEAM AND PLATE-GIRDER BRIDGES

15.5.1 Beam Bridges

The principal advantage of the rolled-beam bridge is its simplicity. Fabrication is usually simple, and such bridges have been designed so that fabrication consisted only of cutting the beams to length. No falsework is required, and erection is simple. The appearance is clean. Rolled beams are economical for spans up to about 60 ft. In competition with other materials and with welded-steel construction, the rolled-beam bridge often comes out second best. This is because the rolled beam is of uniform cross section so that there is an uneconomical distribution of steel in relation to the stress requirements.

15.5.2 Plate-Girder Bridges

Plate girders have the advantage that they can be tailored to fit shear and moment requirements more closely than the rolled beam. They become advantageous at about 60 ft and are commonly used for spans to 300 ft or more. Welded girders made up of three plates compete with rolled beams in all but the shortest spans.

The deck bridge is used most often. It is simple in its details and easy to fabricate. The floor slab may be supported directly on the girders, so

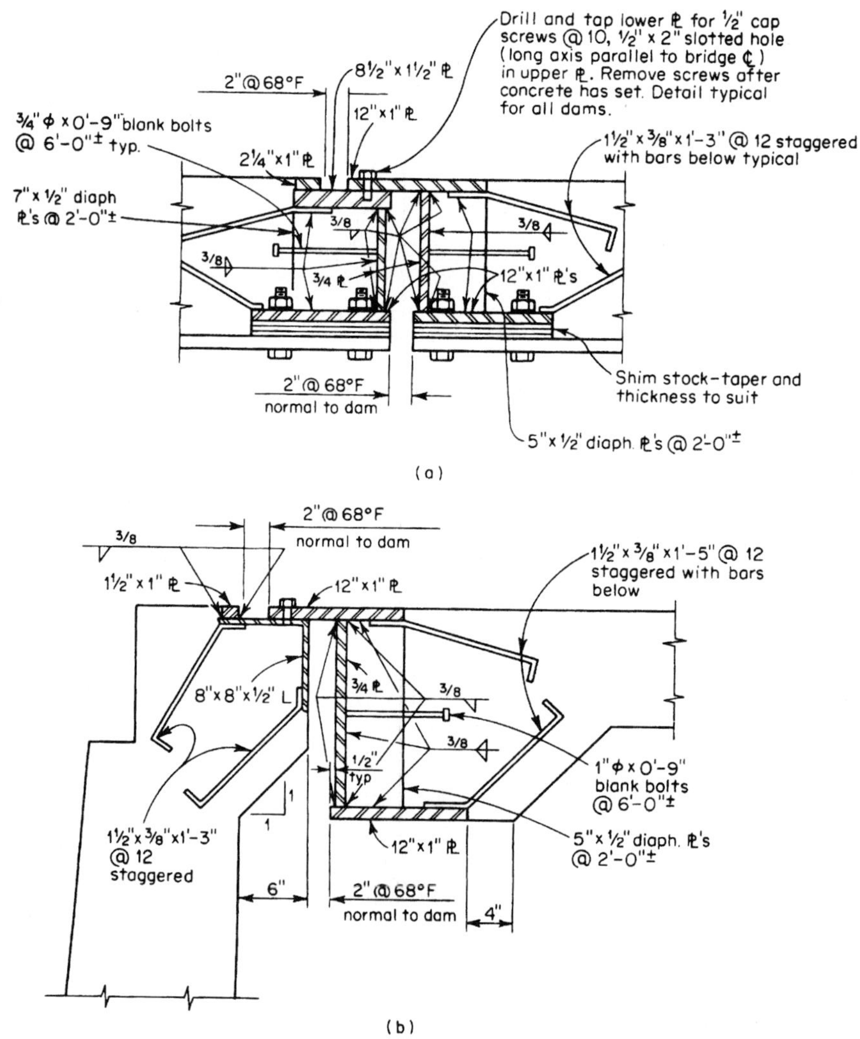

Figure 15.29 Deck-expansion details: (*a*) midspan and (*b*) abutment.

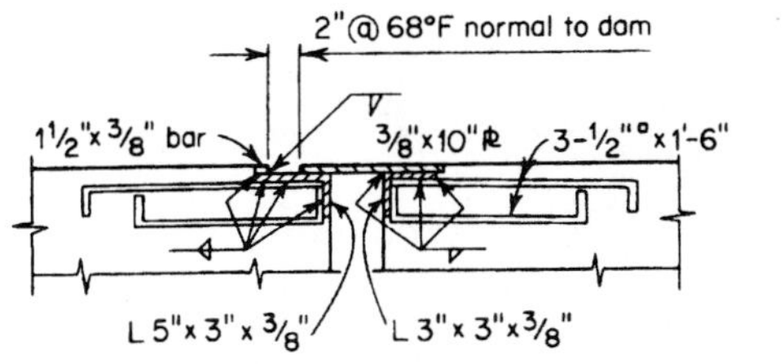

Figure 15.30 Sidewalk expansion detail.

that floor beams and stringers are eliminated. Its disadvantage is in the clearance required for the girder. Through-girder bridges, used where clearance is limited, require floor beams or a system of floor beams and stringers to support the deck. Knee braces are used in place of cross frames, and outriggers are sometimes used.

The box girder is an efficient shape with superior torsional stiffness. It is suitable for long-span bridges, particularly when it is used in conjunction with an orthotropic steel plate deck.

15.5.3 Composite Beam Bridges

Composite action incorporates the concrete deck with the top flange of the supporting steel beams, so that the two act as a unit. The union is affected by mechanical bond between the flange and the deck provided by welded lugs, studs, or other devices. Composite action may be of two kinds, depending upon the method of construction. More commonly, the concrete deck is cast using the stringers to support the formwork. In this case, composite action is effective only for live loads and any dead load placed subsequent to construction of the deck. But if the stringers are supported by shores which are not removed until the concrete has set, the deck acts compositely with the beams to support dead load as well as live load.

Composite construction is most economical for spans 60 ft and longer. The saving in steel for shorter spans will not usually pay for the shear connections. Economy is increased if a cover plate is used on the bottom flange of a rolled beam or, in the case of a welded I, if the bottom flange is larger than the top flange. Composite bridges are also stiffer.

The transformed section is used for analysis, with the cross-sectional area of the concrete flange transformed to steel. The AASHTO specifications limit the effective width of deck to the smallest of (1) one-fourth the span of the beam, (2) the distance center to center of beams, and (3) 12 times the least thickness of slab.

Shear connectors are designed to develop the horizontal shear between deck and beams. Because imperfections and natural roughness of slab and beams contribute to shear resistance, it is likely that current formulas for shear connector resistance are conservative.

15.5.4 Continuous Spans

Although continuous spans will often show economy in weight, requirements of field splicing, erection falsework, and equipment often outweigh savings in cost. A series of simple spans is usually better for bridges whose overall length is less than about 100 ft. However, many variables are involved in the decision, so each location should be studied in the light of conditions at the site.

15.5.5 Spacing

Ordinarily, the number of beams or girders in a deck highway bridge should be such as to optimize the cost of the bridge. This involves a comparison of the cost of the beams or girders with that of the deck. If insufficient clearance is available, it may be necessary to use more than the number of beams or girders corresponding to the optimum combination.

For spans up to about 100 ft, either three or four girders will usually be economical for a two-lane highway bridge. As the span increases, the depth of girder and its weight rise rapidly, so for longer spans two girders with a floor system will be economical.

15.5.6 Lateral Systems

A concrete or steel deck, well secured to the stringers, acts as a horizontal diaphragm and as a tie between stringers. For spans under 125 ft, the AASHTO specifications do not require lateral truss systems for beam bridges and girder bridges which have cross frames or diaphragms and decks adequate to transmit lateral forces. Even in these cases, however, a lateral system may be needed to facilitate erection. If a timber or precast concrete deck is used with minimal connection to the stringers, other provision should be made for stiffening the structure in the horizontal plane, usually by a lateral truss.

Beam-bridge diaphragms should be spaced not more than 25 ft. These diaphragms are simple, usually a channel connected through an angle to the web of the stringer. Similar diaphragms are used at each end. Deck plate-girder bridges must have cross frames at each end, with intermediate cross frames at intervals not to exceed 25 ft. If a lateral truss is required, it is usually placed in the plane of the bottom flanges. It is desirable to space cross frames so that lateral diagonals will be approximately 45 degrees to the girder.

15.5.7 Deflection

For simple or continuous spans, deflection due to live load plus impact should not exceed 1/800 of the span. In urban areas where pedestrian traffic is involved, some states prefer deflection limits of 1/1000 or even 1/1200 to minimize deflection and vibration. Deflection of cantilevers should not exceed 1/300 of their length. Deflection limitations often defeat the economy of higher-strength steels for short spans.

Live-load deflections for truck loading may be computed by adding deflections for each axle. However, a close approximation may be made by using the simple-beam moment for the truck loading from Table 15.2. This moment is increased by a factor which varies linearly from 15 percent for a 50-ft span to 9 percent for a 140-ft span (Fig. 15.31). The corrected moment is converted to an equivalent concentrated load at midspan, and the corresponding midspan deflection is computed (Sec. 15.5.8).

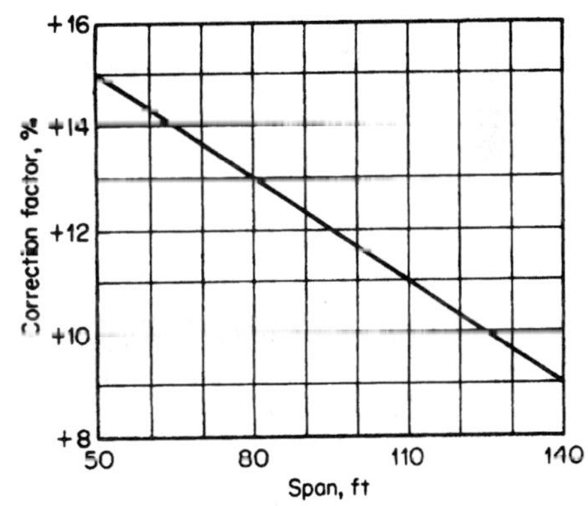

Figure 15.31 Correction factor for computing deflection.

Depth Ratios

The depth-to-span ratio of plate girders and rolled beams used as girders should be not less than 1:25. For continuous spans, the span is taken as the distance between dead-load points of contraflexure. If this limitation cannot be met, the section should be increased so that the deflection is no greater than would result for the ratio 1:25.

15.5.8 Welded Plate Girders

Typical details of the welded plate girder are shown in Fig. 15.32. The American Welding Society Specifications for Welded Highway and Railway Bridges[3] recommend single-plate flanges, butt-welded where a change of section is desired. Flange splices should be staggered

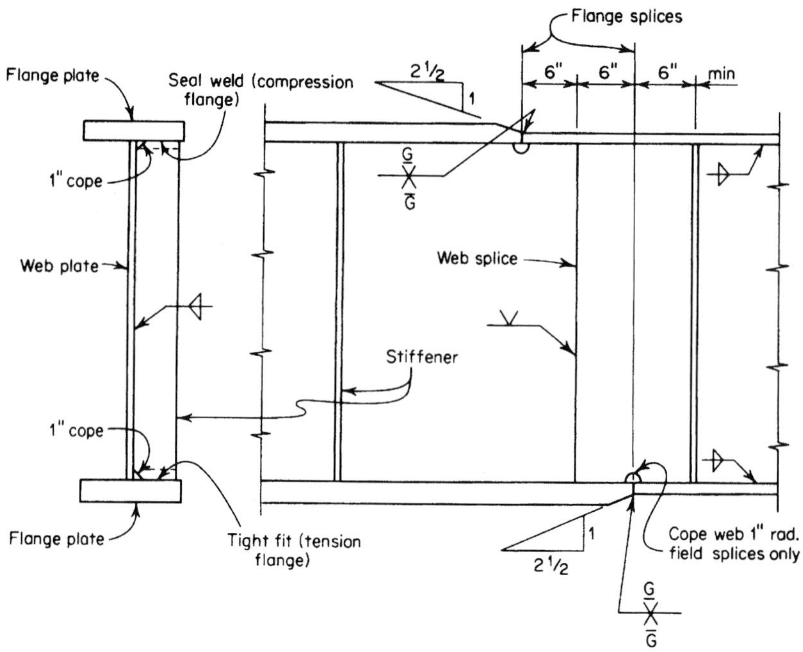

Figure 15.32 Typical welded-girder details.

at least 1 ft. The effect of flange thickness on the size of fillet required to weld the flange to the web should be kept in mind since the minimum permissible size of weld increases with thickness of the flange. A 5/16-in. fillet is usually the largest that can be placed in one pass without special equipment. Larger welds require multiple passes, cleaning, and chipping, and the cost of placing rises much faster than strength as weld size increases.

Intermediate transverse stiffeners may be milled or ground to fit or cut short 1 in. at each end to allow access for painting. Stiffeners not cut short should be coped at the inside corners to clear the flange welds. Bearing stiffeners are milled or ground to bear on both flanges. Fillet welds to the web may be intermittent or continuous and as small as plate thicknesses permit. Because of the difficulty of starting and stopping automatic welding machines, fabricators often prefer the continuous weld. Stiffeners may be welded to the compression flange but to the tension flange only where the flange stress is less than 75 percent of the allowable. AASHTO permits intermediate stiffeners on only one side of the web, with a consequent saving in weight and cost of fabrication and improvement in appearance. Stiffeners on one side only are required to be welded to the compression flange.

Longitudinal stiffeners permit the use of thinner webs and may prove economical for deep girders. They may be cut at their intersections with transverse stiffeners.

At field splices, the web-to-flange fillet welds should be omitted for at least 1 ft on each side of the splice and welded after the flange splice is made. This results in better distribution of shrinkage stresses and permits lateral displacement of the web if flange shrinkage causes the ends of the webs to jam.

Fatigue

AASHTO and AREMA provisions for fatigue are based on the type of member or detail the kind of stress to which it is subjected, the range of service-load stress, and the expected number of cycles of stress.

In welded work, probably more than in any other structural fabrication process, workmanship must be of the highest quality. This requires adequate and competent inspection. Although the designer is not usually responsible for inspection, the success of the design depends so heavily upon the quality of welding that the designer should make sure that proper steps are taken to obtain the weld strength anticipated in the design.

EXAMPLE—DESIGN OF WELDED PLATE GIRDER

A two-lane, continuous, two-girder bridge of spans 110, 154, and 110 ft (Fig. 15.33*a*) will be designed for the HS20-44 loading, AASHTO specifications. The roadway width is 28 ft. Steel is A36.

The dead load is 2.83 klf per girder (Fig. 15.33*b*). Dead-load moment and shear diagrams are shown in Fig. 15.34*a* and *b*. When load factor design methods are used, these values must be factored in accordance with Table 15.1*a* and combined as follows:

$$F_y = \gamma[D + \beta\,(L + I)]$$

$\gamma = 1.3$ and $\beta = 1.67$ will be used in this example.

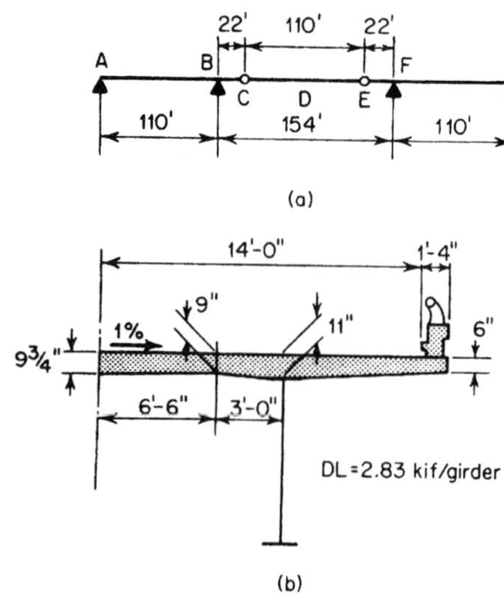

Figure 15.33 Welded plate girder.

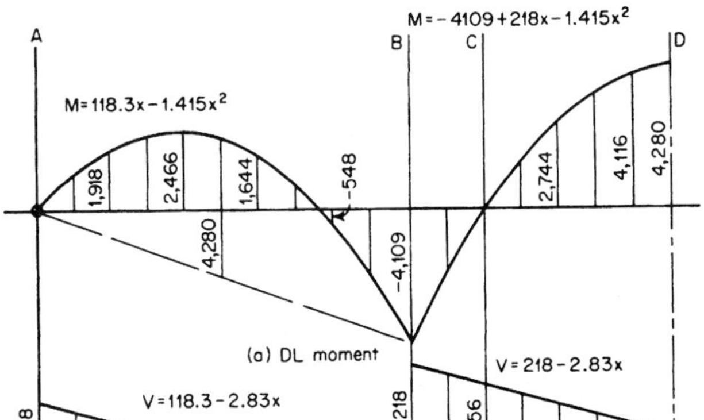

Figure 15.34 Shear and moment diagrams for girder of Fig. 15.33.

The lane loads are positioned as shown in Fig. 15.35. The lane load per girder is $(4.5 + 18.5)/19 = 1.21$. Impact for a 110-ft span is 21.3 percent (Table 15.2). Thus, for one girder, $LL + I = 1.21 \times 1.213 = 1.47$ lanes.

LIVE-LOAD MOMENTS

The maximum moment at the center of the 110-ft span is 1704 ft-kips per lane (Table 15.2). Therefore, M = 1.47 × 1704 = 2510 ft-kips. The

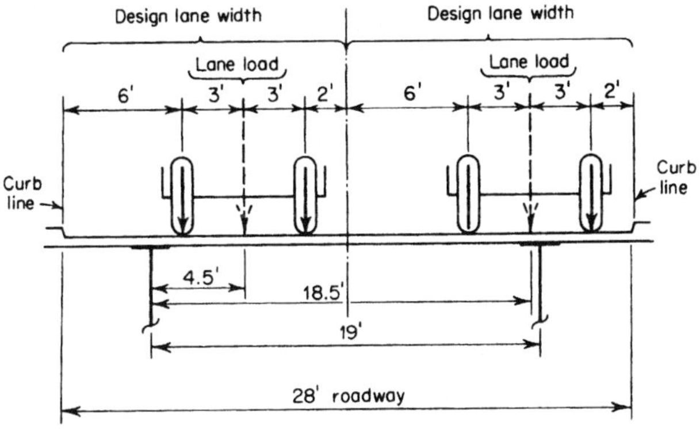

Figure 15.35 Lane load.

influence line for the moment at B is shown in Fig. 15.36*a*. To find the maximum moment, we try both the lane loading and the truck loading, the latter with the truck in the suspended span with the trailer axle at hinge C. The truck-load moment is the larger, and $M_{BC} = 1.47 \times 1450 = 2130$ ft-kips.

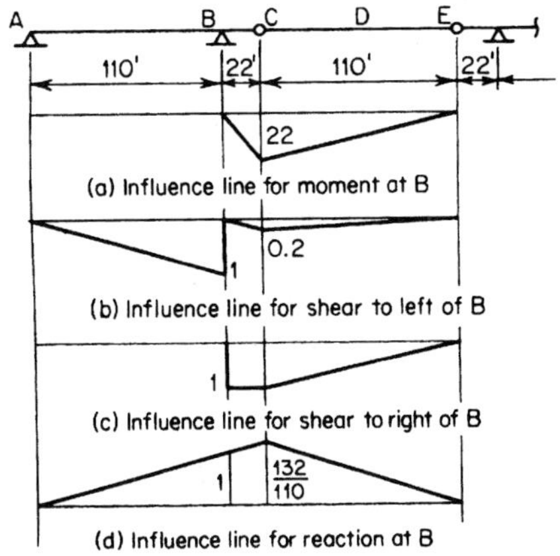

Figure 15.36 Influence lines, support B of girder of Fig. 15.33.

LIVE-LOAD SHEARS

The maximum shear at A is 1.47 × 65.9 = 97 kips (Table 15.2). For maximum shear at midspan of AB, the truck is positioned with the trailer axle at midspan. This gives $V = 1.47 \times 29.9 = 44$ kips.

To find the shear to the left of B, the lane loading is used in conjunction with the influence line of Fig. 15.36*b*. With the 26-kip load at B

$$V_{BA} = 1 \times 26 + 0.64 \times \frac{1}{2}(110 \times 1 + 132 \times 0.2) = 1.47 \times 69.6 = 103 \text{ kips/lane}$$

From the influence line for shear to the right of B (Fig. 15.36*c*),

$$V_{BC} = 1.47\,(1 \times 26 + 0.64 \times 77) = 111 \text{ kips/lane}$$

From the influence line for the reaction at B (Fig. 15.36*d*),

$$R_B = 1.47 \times \frac{132}{110}(26 + 121 \times 0.64) = 182 \text{ kips/lane}$$

Moments and shears are summarized in the following table and plotted in Fig. 15.34*c* and *d*.

	Moments		Shears right of *B*	
	B	*D*	Load for max. shear	Load for max. moment
DL	4109	4280	218	218
LL + *I*	2130	2510	111	97
Total	6239	6790	320	315

Load factor moments and shears are summarized in the following table:

	Moments		Shears right of *B*	
	B	*D*	Load for max. shear	Load for max. moment
DL ×	4109	4280	218	218
(*LL* + *I*) 1.67	3557	4192	185	162
DL + *LL* + *I*	7666	8472	403	380
$\alpha(DL + LL + I)$ $\alpha = 1.3$	9966	11,013	524	494

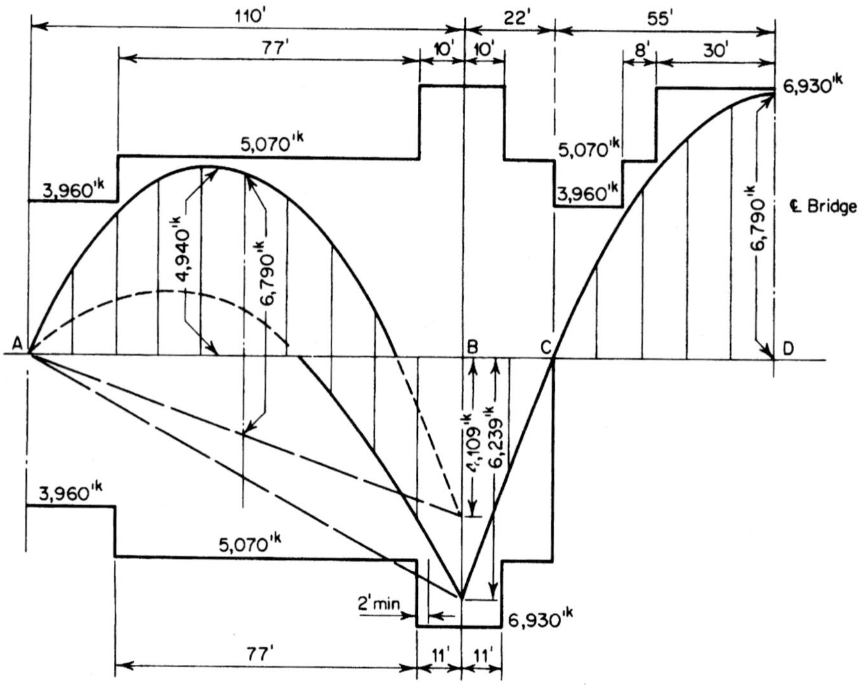

Figure 15.37 Maximum-moment envelopes for girder of Fig. 15.35.

Assuming the envelope of maximum moments for simple spans to be a parabola (Sec. 15.2.3), the envelopes of maximum moment shown in Fig. 15.37 are determined from the values in Fig. 15.34. These should be drawn to scale. A scale of 1 in. = 20 ft and 1 in. = 2000 ft-kips is satisfactory.

In determining the number of variations in cross section, the cost of splicing must be balanced against the cost of the steel it will save. Such considerations usually keep the number of splices to a minimum. A constant depth and thickness of web facilitate fabrication. Also, it is desirable to hold the flange width constant and vary the thickness as required.

For continuous beams and girders, L may be taken as the distance from interior support to the dead-load point of contraflexure if this distance is less than that defined above. Furthermore, continuous beams and cantilever beams may be proportioned for negative moment at interior supports at an allowable stress 20 percent higher than that given by the formula but not to exceed 20,000 psi.

To determine L, it is necessary to decide on the pattern and spacing of the diagonal bracing and cross frames. The latter should be located so that bracing is at about 45 degrees to the girders and spaced evenly throughout the span. For the bracing layout shown in Fig. 15.38, $L = 11$ ft and $L/b = 11(12/18) = 7.3$. Therefore, with the 20 percent increase, the allowable stress is 20 ksi, and the allowable moment at support B is the same as that at D, 6930 ft-kips.

GIRDER SECTION I

The cross section is proportioned for the maximum positive moment (at D). If no longitudinal stiffener is used, the web must be not thinner than $D/170$ ($D/tw = 192$, *LFD*), where D is the unsupported depth. Try a 100 × ⅝ web, for which 100/0.625 = 160, with 18-in. flanges.

Assume braced noncompact section of web = 100 × ⅝; flange = 18 × 2

Check:

$$\frac{b'}{t} = \frac{8.69}{2} = 4.34 < \frac{2200}{\sqrt{F_y}}$$

$$\frac{D_c}{t_w} = \frac{51}{0.625} = 81.6 < \frac{15,400}{\sqrt{F_y}}$$

L_b = spacing at lateral bracing = 162 in.
M_u at D = 11,013 kft

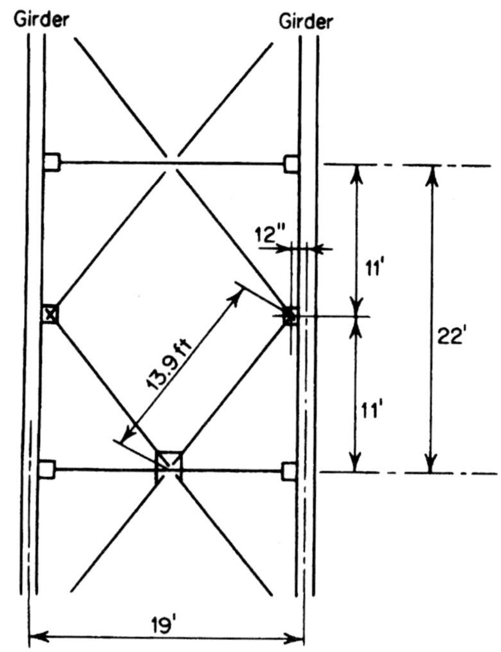

Figure 15.38 Lateral bracing for girder of Fig. 15.33.

$F = 36$ ksi
$S = 4694$ in.^{3}
Resisting moment $= F_yS = 14{,}082 > 11{,}013$
Section is okay.

GIRDER SECTION II

The compressive stress in the bottom flange must be investigated for the maximum negative moment.

If we comply with the minimum spacing of $L_b = 196$ in., the section can be sized as follows:

Maximum neg. mom. $= M_u = 9966$ kft
Assume web $= 100 \times \frac{5}{8}$
Flange $= 18 \times 1\frac{1}{2}$
$F_y = 36$ ksi
$S = 3775$ in.^{3}

$$M_u \text{ resisting} = \frac{36(3775)}{12} = 11{,}325 > 9966$$

GIRDER SECTION III

This section is near midspan of *AB*.

$DLM = 3460$ kft $\quad LLM = 1480$ kft
$M_u = 7710$ kft
Try 18×1 in. flange, web $100 \times \frac{5}{8}$
$S = 2849$

$$S \text{ required} = \frac{7710 \times 12}{36} = 2570 < 2849$$

GIRDER SECTION IV

The minimum sections are at the ends of the suspended span and at the ends of the bridge. The minimum flange thickness is governed by the limitation $b/t = 23$. Use $18 \times \frac{3}{4}$ plate.

Minimum flange thickness is controlled by $\frac{b'}{t} \le \frac{2200}{\sqrt{F_y}}$. An $18 \times 3/4$ flange satisfies this criterion.

The allowable moments for the various sections are plotted on the working-stress moment envelope of Fig. 15.37 to determine where the section changes can be made. The larger of the two flange plates to be joined must be extended at least 1 ft beyond the theoretical point of splice. It is good practice to stagger splices in the upper and lower flanges by at least 1 ft (Fig. 15.32). Splices will be located at the following points (Fig. 15.37). A similar approach can be used for load factor design.

	Bottom, ft	Top, ft
Right of A	22	23
Left of B	11	10
Right of B	11	10
Right of C	16	17
Right of C	24	25

DEFLECTION

Deflection is a service load check. The load position for maximum deflection of the cantilever is shown in Fig. 15.39. The loaded length

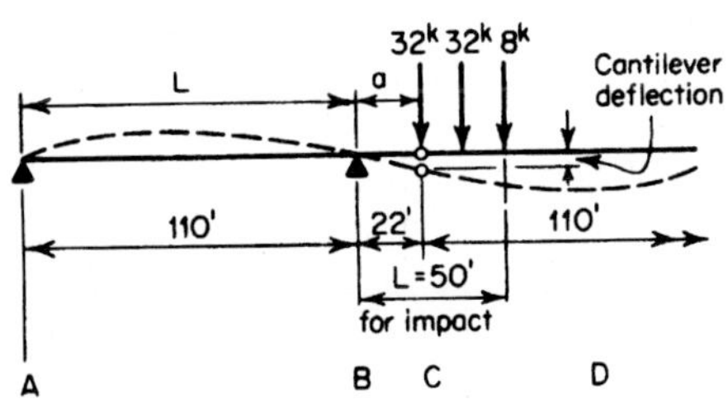

Figure 15.39 Load position for cantilever deflection.

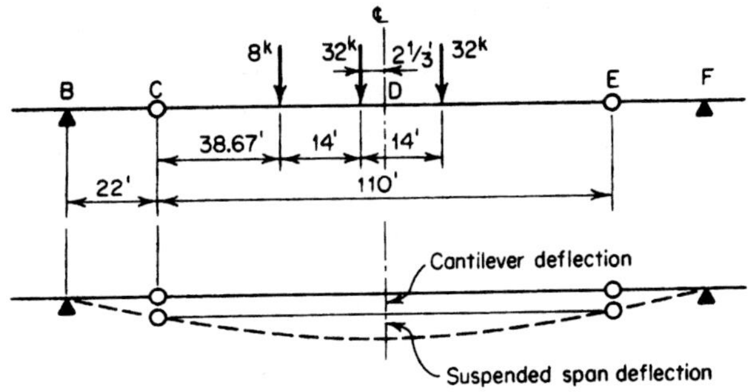

Figure 15.40 Load position for suspended-span deflection.

for determining impact is 50 ft. Therefore, $I = 28.6$ percent. With $R = 65.9$ kips (Table 15.2),

$$R_C = 65.9 \times 1.21 \times 1.286 = 102.5 \text{ kips}$$

$$\Delta = \frac{Pa^2(L+a)}{3EI} = \frac{102.5 \times 22^2 \times (110+22) \times 12^3}{3 \times 29{,}000 \times 215{,}120}$$

$$\Delta = \frac{L}{300} = \frac{22 \times 12}{300} = 0.88 \text{ in. allowable}$$

Deflection of the suspended span will be approximated by the method discussed in Sec. 15.5.6. The loading is shown in Fig. 15.40:

$$M = 1704 \text{ ft-kips}, \quad I = 21.3 \text{ percent (Table 15.2)}$$

$$M = 1704 \times 1.21 \times 1.213 = 25{,}000 \text{ ft-kips}$$

$$\text{Equiv. concentrated load} = \frac{4M}{L} = \frac{4 \times 2500}{110} = 91 \text{ kips}$$

$$\Delta = \frac{PL^3}{48EI} = \frac{91 \times (110 \times 12)^3}{48 \times 29{,}000 \times 215{,}120} = 0.69 \text{ in.}$$

The correction factor is 1.11 (Fig. 15.31):

$$\Delta = 0.69 \times 1.11 = 0.77 \text{ in.}$$

$$\Delta = \frac{L}{800} = \frac{110 \times 12}{800} = 1.65 \text{ in. allowable}$$

SPAN *BF*

Taking the deflection for span *BF* as the sum of the deflection of the cantilever for its severest loading condition (Fig. 15.39) and of the suspended span for its severest loading condition (Fig. 15.40),

$$\Delta = 0.61 \times 0.77 = 1.38 \text{ in.}$$

$$\Delta = \frac{154 \times 12}{800} = 2.31 \text{ in. allowable}$$

The calculated deflection is probably about 10 percent low because of the variable moment of inertia which was neglected in the computations, but it is also high because deflections for two different load positions were assumed to be additive. If it is critical, deflection of the cantilever should be recomputed, using the load position of Fig. 15.40.

FLANGE-TO-WEB WELDS

The influence lines for shear at pier *B* (Fig. 15.36) show that there is no reversal of live-load shear. The maximum shear is 329 kips (see the table on page 467). Because the maximum shear results from lane loading, the number of stress cycles to be assumed for determining the allowable fatigue stress is 500,000 for an average truck traffic of 2500 or more (AASHTO). The flange-to-web weld is in stress category *B*. Therefore, the allowable stress range *F* is 27.5 ksi. This is greater than the basic allowable stress, 12.4 ksi, and because there is no stress reversal, the basic value governs. The minimum fillet weld for a 1¾-in. flange plate is ⅜ in. Therefore,

$q = F_y A$
$F_u = 65$ ksi
$F_y = 0.45\ F_u = 29$ ksi
A = area of web welds = 0.707 (0.375)2 = 0.53 in.^{2}
$F_y = 29(0.53) - 15.4$k/in.

$$s = \text{horizontal shear} = \frac{VQ}{I} = \frac{524(18\times 2)52}{239{,}394} = 4.1 \text{ k/in.} < 15.4$$

BEARING STIFFENERS

Loads = $p = 1.3(411 + 1.67 \times 182) = 929^k$
For $F_y = 36$ ksi (short column)
Area = $P/F_{cr} = 26$ in.^{2}
Use four 8 × ⅞-in. stiffeners.

INTERMEDIATE STIFFENERS

$$F_y = \frac{F_y}{3}\left[C + \frac{0.87(1-C)}{\sqrt{1+\left(\dfrac{d_o}{d}\right)^2}}\right]$$

$$= \frac{36}{3}\left[0.384 + \frac{0.87(0.616)}{\sqrt{2.74}}\right] = 8506 > 5260 \text{ lb/in.}^2$$

$$\text{Max spacing} = 3D = 300 \text{ in. or } D\left(\frac{260}{D/t_w}\right)^2 = 264 \text{ in.}$$

$$F_y = 36 \text{ ksi}$$

$$C = \frac{4.5\times 10^7 k}{(D/t_w)^2 F_y}$$

$$D = 100 \text{ in.}$$

$$d_o = 132 \text{ in.}$$

$$t_w = 0.625 \text{ in.}$$

$$k = 5 + \frac{5}{(d_o/D)^2} = 7.87$$

$$C = 0.384$$

For ease of construction, use 132-in. spacing.

$$\text{Stiffener size, } I = d_c t_w^3 J = 132(0.625)^3(0.5) = 16.11$$

$$J = 2.5\left(\frac{D}{d_o}\right)^2 - 2 = -0.56 \therefore \text{use } 0.5$$

$$A = [0.15BDt_w(1-C)]\left(\frac{f_r}{F_r} - 18t_w^2\right) \quad y = \frac{F_{y\text{ web}}}{F_{y\text{ stiffener}}} = 1.0$$

Area of stiffener = A = 1.54 in.^{2} B = 2.4 (single plate)
Use 5 × ⅜-in. stiffeners.

A = 1.87 in.^{2}

LONGITUDINAL STIFFENERS

Longitudinal stiffeners along the web may result in reduced web thickness and smaller intermediate stiffeners. This possibility should be investigated.

CROSS BRACING

The exposed area of the bridge in elevation (Fig. 15.41) is 12.1 ft²/ft. Wind force on the structure at 50 psf is 50 × 12.1 = 0.61 klf. Since the wind load is uniform, the reactions at the supports may be determined by proportion from the dead-load reactions.

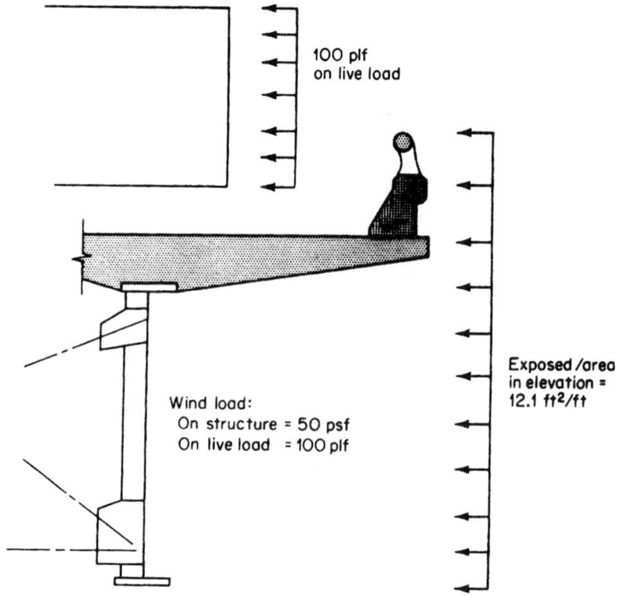

Figure 15.41 Wind force on the structure.

Wind-load reaction at pier B = 411 × 0.61/2.83 = 88 kips. This load is carried to the end cross frames by the deck, so 44 kips acts at each of the upper corners of the frame (Fig. 15.42).

Loads in CE = 44 kips C:

$$AC = 44 \text{ kips } T$$

$$CF = 56.1 \text{ kips } T = 44\times(145/114)$$

Member AE, try WT6 × 15, A = 4.40 in.^{2}, r_x = 1.75 in.

$$\frac{L}{r_x} = \frac{228}{1.75} = 130 < 140$$

$$F_a = 16{,}000 - 0.3\times 130^2 = 10{,}930$$

$$f_a = \frac{P}{A} = \frac{56{,}100}{4.40} = 12{,}750 \text{ psi}$$

Member BC, CF, try WT6 × 15, A = 4.40 in.^{2}, r_y = 1.52 in.

$$\frac{L}{r_y} = \frac{1.45}{1.52} = 95$$

$$F_a = 16{,}000 - 0.3\times 95^2 = 13{,}290 \text{ psi}$$

$$f_a = \frac{P}{A} = \frac{56{,}100}{4.40} = 12{,}750 \text{ psi}$$

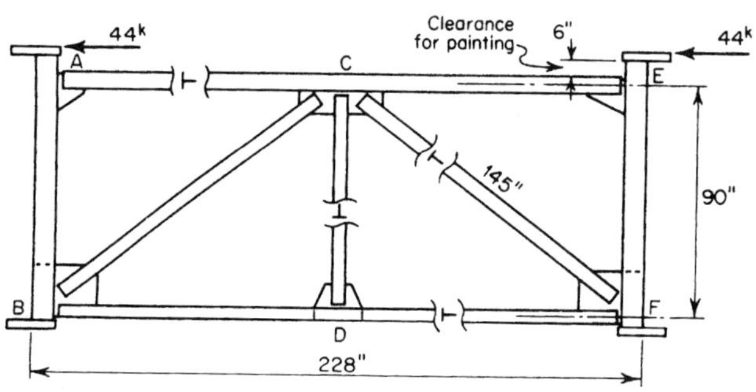

Figure 15.42 Cross frame for bridge of Fig. 15.33.

Member *BF*, $P = 0$, $r_{min} = 228/140 = 1.63$ in.
Use WT6 × 15. Member *CD*: $P = 0$, $r_{min} = 90/140 = 0.64$ in.
Use S3 × 6.25, $r_y = 0.71$ in., $A = 1.81$ in.^{2}

Members of the cross frames will be welded to their gusset plates. The connection must be designed for the average of the calculated stress and the strength of the member but for not less than 75 percent of the strength of the member.

INTERMEDIATE CROSS FRAMES

Since it is assumed that the wind force is carried to the end cross frames by the deck, members of the intermediate cross frames are chosen for maximum *L/r* requirements. The slenderness ratios of all but members *RC* and *CF* of the end cross frames are only slightly less than the permissible limit, 140. Therefore, all cross frames will be made alike.

LATERAL BRACING

These members are also free of calculated stress and are proportioned for maximum $L/r = 140$:

$$r = 13.9 \times \frac{12}{140} = 1.19$$

Use WT5 × 11, $r_y = 1.33$ in.

15.5.9 Web Splice

The locations of web splices usually depend on available lengths of plates, and the actual location is often left to the fabricator. Specifications do not agree on design criteria. The AREMA specifications require that they be designed for both (1) the shear strength of the gross area of the web and (2) the combination of moment strength of the net section of the web and the maximum shear at the splice. The AASHTO specifications require that the splice be designed for the average of the calculated stress at the point of splice and its strength but in no case less than 75 percent of the strength.

15.5.10 Field Splices

When girders are continuous and when single spans are too long to ship in one piece, field splices are required. Such splices are usually bolted. Field splices may be bolted even in welded work because of the difficulty of welding large girders in the field. In some cases, the flanges may be butt-welded and the web splice bolted.

15.6 TRUSS BRIDGES

Trusses are used in highway bridges only for very long spans. However, trusses may be economically used for shorter spans where aesthetics and (in the case of through bridges) safety with high speed traffic are not critical. Deck-truss bridges are preferable.

15.6.1 Proportions

The Warren truss is more economical than the Pratt for parallel-chord trusses. For spans greater than about 320 ft, the K-truss is advisable to reduce floor weight and the inclination of the diagonals. Depth-to-span ratios vary considerably, usually between 1:5 and 1:10. For best results, the depth should be such that the angle of the diagonal falls in the range of 50 to 55 degrees with the horizontal. Long panels require a heavier deck, while short panels increase the cost of the truss, so a compromise must be made, usually in the range of 16 to 32 ft for the panels of a highway bridge truss.

Gravity axes of truss members should coincide with their working lines. However, a small eccentricity may be used to compensate for the dead load of a member. Trusses should be erected with enough camber to offset at least the dead-load deflection.

15.6.2 Loads and Stresses

Various formulas for estimating the weight of bridge trusses have been developed, but designs vary so widely that formulas are not usually dependable. The dead load may be approximated by comparison with

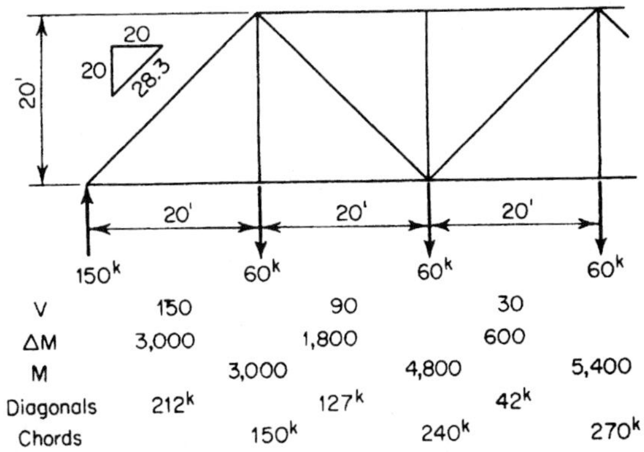

Figure 15.43 Moment-increment solution for parallel-chord truss.

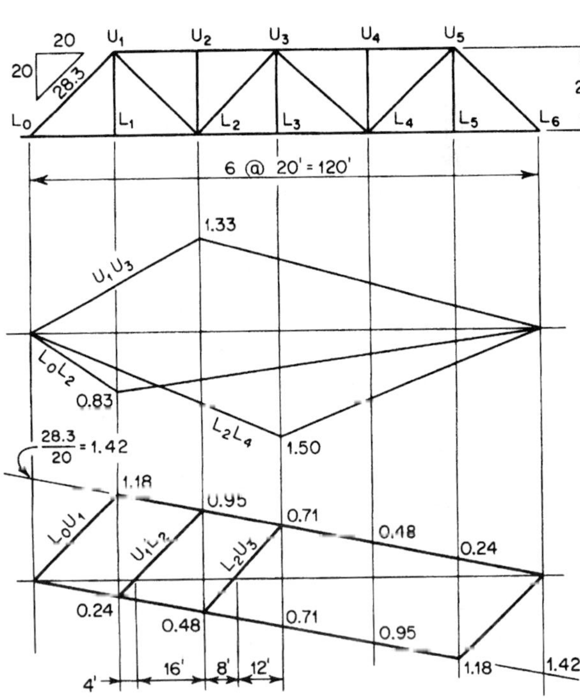

Figure 15.44 Influence lines for parallel-chord truss.

similar bridges or by estimate. Primary stresses may be computed by the methods of joints or of sections. A tabular solution, based on moment increments, is useful for dead-load stresses in parallel-chord trusses (Fig. 15.43). Influence lines are useful for live loads, particularly axle loads, and are easily constructed (Fig. 15.44).

SECONDARY STRESSES

Trusses having fixed joints generate bending in the members. The resulting bending stresses are called secondary stresses. Such stresses may also result from eccentric connections frozen pins or from transverse loading on the members. Secondary stresses can be minimized by avoiding stiff members. Members whose width is not more than one-tenth their length will usually not have secondary stresses larger than 25 percent of the primary stresses. Both AASHTO and AREMA require that secondary stresses in excess of 4000 psi in tension members and 3000 psi in compression members be treated as primary stresses.

15.6.3 Truss Members

To the extent possible, all members of a truss should have the same depth so that they can be connected at the joints with a minimum of filler plates. For the shorter spans, W sections make suitable, clean-looking members. For longer spans, welded built-up I's or box sections with perforated cover plates are used.

Truss members should be designed so that all surfaces can be reached and painted and with adequate clearances for riveting or bolting. Pockets which may trap water or debris should be avoided.

Combinations of Steels

Several grades of steel can be used to advantage in large bridge trusses to meet varying stress conditions, with the stress grade for each member based on maintaining, in general, a minimum thickness of plate. Thus, lower-stressed members will be made of structural-grade steel, with the next higher stress grade being used for members with larger stress, and so on. For example, three grades of steel were used in the Benicia-Martinez Bridge (Fig. 15.45).

The efficiency of bolted truss members can be increased by butt-welding a section of higher-strength steel on each end. This offsets the loss in area from holes, so that the gross area of the member can be fully effective. In other cases, the same result can be achieved by welding on a thicker section of steel of the same grade as the member. This device can also help reduce the size of the joint. These ideas are illustrated in Fig. 15.46. Note in this detail that the weld is 1 ft out from the first transverse line of bolts in the joint. Where several grades of steel are brought together in a joint, the gusset plates are made of the highest stress grade of steel represented in the joint.

15.6.4 Lateral Forces

The design of truss bridges is usually based on the assumption that wind load on the structure is carried to the supports by the upper and lower lateral systems, where they are transmitted to the bearings by the end sway frames or portal bracing. Wind on the live load is assumed to be transmitted by the deck. The intermediate sway frames are then designed to resist only the overturning force in each panel. Stresses in lateral trusses will usually be so small that minimum slenderness ratios will determine the sizes of the members. Seismic forces must also be considered.

15.7 CONCRETE BRIDGES

15.7.1 Introduction

Components and connections of a concrete bridge shall be designed to resist load combinations, as specified in AASHTO, at all stages during the life of the structure, including those during construction. Load factors shall be as specified in AASHTO Section 3. Equilibrium and strain compatibility shall be maintained in the analysis.

15.7.2 Camber, Plastic Flow, and Shrinkage

The determination of the camber to provide in the deck of a bridge is an important consideration. The geometrics of some structures, such as those with skewed bents and those on horizontal curves, or a combination of both, can make this a complex problem.

Initial deflection is a function of dead load. It is also affected by the sequence of construction. The secant modulus $E = 1000f'_c$ can be used to determine the deflection which occurs immediately upon striking the falsework. Structures cambered only for initial deflection will ride smoothly at first, but undulations develop with time. This is due to plastic flow or creep and, to some extent, shrinkage.

Deflection due to plastic flow increases rapidly at first, but the rate of increase decreases with time, so most of it occurs in a period of about 4 or 5 years following construction. Sufficient camber should be provided to allow for the ultimate deflection, which may amount to three or four times the initial deflection. For calculating this camber, values of E of 250 to $350f'_c$ may be used, depending upon the characteristics of the concrete.

The AASHTO specifications require that shrinkage be considered in the design of concrete structures. The effectiveness of shrinkage in

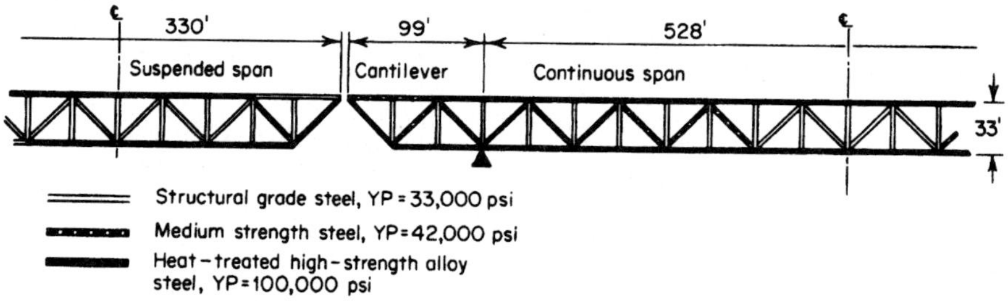

Figure 15.45 Benicia-Martinez bridge.

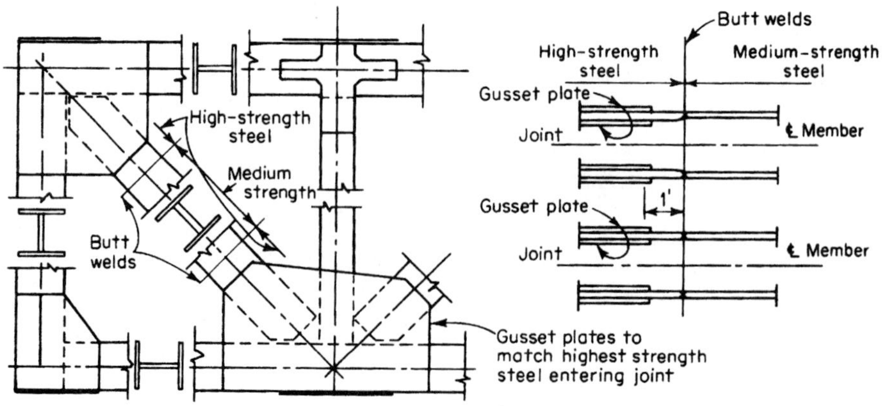

Figure 15.46 Benicia-Martinez bridge details.

producing stress in a reinforced-concrete member has been questioned by some authorities. Because of the restraining effects of reinforcing steel, shrinkage manifests itself in producing cracks in the concrete (Sec. 15.2.8). Reinforcement distributes cracking by producing more cracks of smaller magnitude. Shrinkage stresses can be minimized or practically eliminated by prudent planning of pouring sequence, and the designer should always consider this alternative to that of allowing shrinkage stresses to develop, so that they must be accounted for in the design calculations.

15.8 SLAB BRIDGES

15.8.1 Simple Spans

A longitudinally reinforced slab is the simplest form of concrete bridge superstructure. It is economical for simply supported spans up to 20 ft or more in length and may show economy in continuous spans up to 40 ft. It can be supported on piles, column bents, or piers. Typical sections of concrete cast-in-place slab structures are shown in Fig. 15.47.

Piles generally provide the most economical intermediate support for continuous slab bridges. They are particularly suited to locations where drift is not a problem. Concrete-column bents and solid piers are the next most economical substructure types. On continuous spans of ±26 ft and longer, pile caps can be eliminated since for such spans the pile spacing is close enough to allow the slab itself to act as a cap within the requirements of the AASHTO specifications. This results in a minimum of formwork and provides a pleasing and simple structure.

Distribution of Live Load

For design of the slab, the AASHTO specifications assume a wheel load to be distributed transversely over the effective width $E = 4 + 0.06S$, where S is the effective span in feet and main reinforcement is parallel to traffic.

EXAMPLE

Design a simple-span slab bridge for the HS20-44 loading. Span = 20 ft (the span length is defined as the distance center to center of supports but not to exceed the clear span plus thickness of slab), $f_c = 1600$ psi, $f_s = 24{,}000$ psi, $n = 7.5$. Impact allowance for a 20-ft span is 30 percent.

From Fig. 15.48, which gives slab depths that can be used for preliminary design, the probable depth of the slab is 12½ in., for which the dead load is 160 psf. The effective width $E = 4 + 0.06 \times 20 = 5.2$ ft. Therefore,

$$M_m = 1.3[D + 1.67(L + I)] = 1.3[8000 + 1.67(20{,}000)] = 53{,}820 \text{ ft-lb}$$

$$A_s = \frac{M_u}{\phi f_y\left(d - \frac{a}{2}\right)} = 1.52 \text{ in.}^2 \qquad \phi = 0.9 \text{ (flexure)} \qquad f_y = 60 \text{ ksi}$$

$$f_c' = \frac{A_s f_y}{0.85(b)(a)} = 1680 < 4000 \text{ OK} \qquad d = 12\frac{1}{2} - 2 = 10.5 \qquad a = \beta C = 0.85 \times 6.25 = 5.3$$

For No. 8 bars at 6 in., $A_s = 1.58$ in.^{2} Lengths of these bars are determined from the maximum-moment envelope in Fig. 15.49. Figure 15.50 shows typical sections and steel layout.

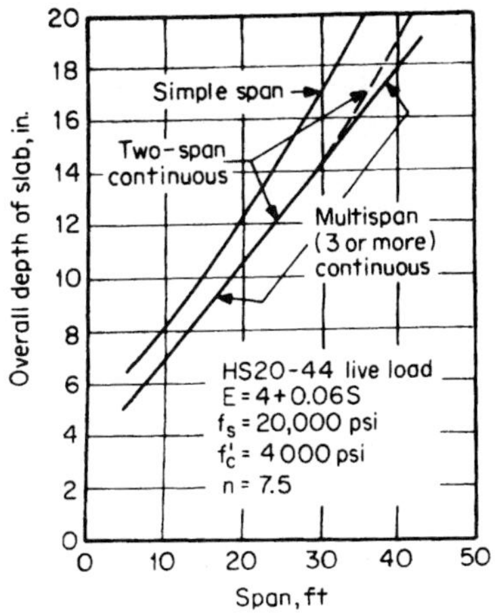

Figure 15.48 Approximate depths of slab bridges.

The specifications require that one-third of the main steel extend beyond the face of the support at least 6 in. The remaining bars must be extended 15 diameters but not less than one-twentieth of the span beyond the point at which computations show they are no longer needed to resist stress. For No. 8 bars in a 20-ft span, this gives 15 in. These extensions are shown in Fig. 15.50.

Distribution steel in the bottom of the slab, transverse to the main reinforcement, is determined as the percentage $100/\sqrt{S} = 100/\sqrt{20} = 22.4$ percent of the main reinforcement. The area required is $0.224 \times 1.58 = 0.35$ in.^{2}/ft of slab, which is furnished by No. 6 bars at 15 in. These bars are placed in the center half of the slab. Distribution steel can be reduced by 50 percent in the outer quarters as shown in Fig. 15.50. Temperature reinforcement shown in the top of the slab is the required minimum, No. 4 bars at 18 in.

Top and bottom reinforcing bars can be uniformly spaced and held securely by the use of bar chairs. These are made from No. 4 bars, bent as shown in the detail in Fig. 15.53. They should be placed about 4 ft on centers measured along the span and 3 ft on centers transversely.

15.8.2 Continuous Spans

Where good foundation material is encountered and differential settlement is not a problem, continuous slab bridges are economical. In some locations, they may be economical up to spans of about 40 ft. Above 40 ft, reinforced-concrete T-beams, cored slabs, or box girders should be considered.

For aesthetic and economic reasons, the continuous slab bridge should have a uniform thickness. If the spans are equal, this thickness can be held to a minimum by using compressive steel at points of large moment. For example, in a four-span slab of uniform depth, the maximum moment occurs at the first interior bent. To hold the volume of

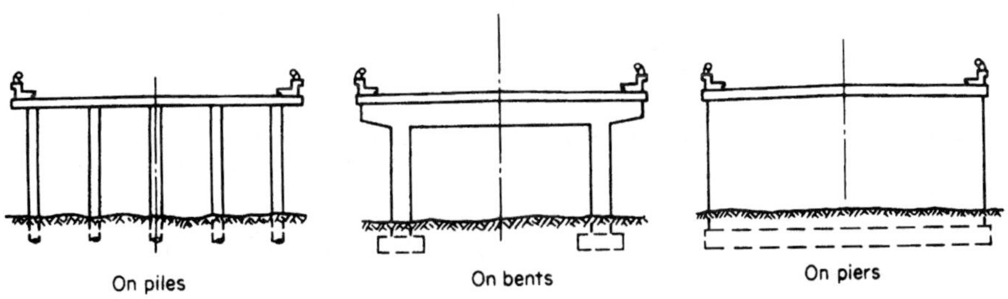

Figure 15.47 Typical slab bridges.

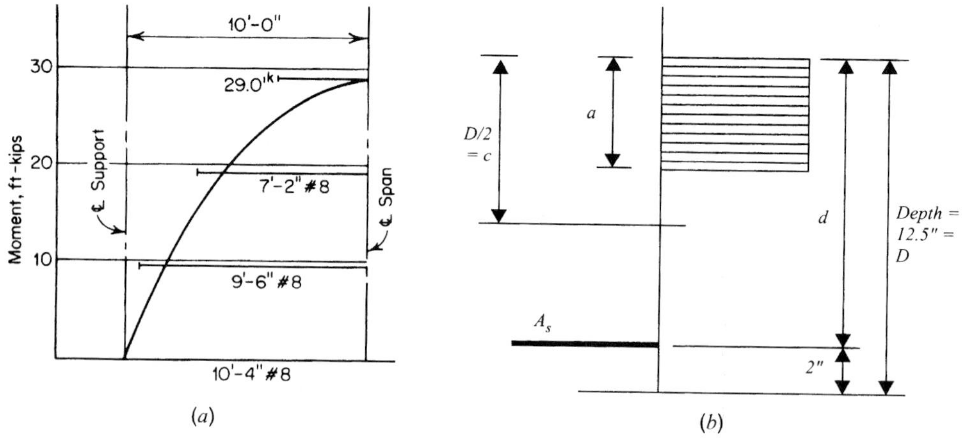

Figure 15.49 Lengths of bars.

concrete to a minimum, the slab depth for balanced reinforcement can be based on the positive moment in the end span, and the larger negative moment at bent 2 can be provided for with compression steel in the bottom of the slab. The positive steel which must be extended through the support can be used for this purpose.

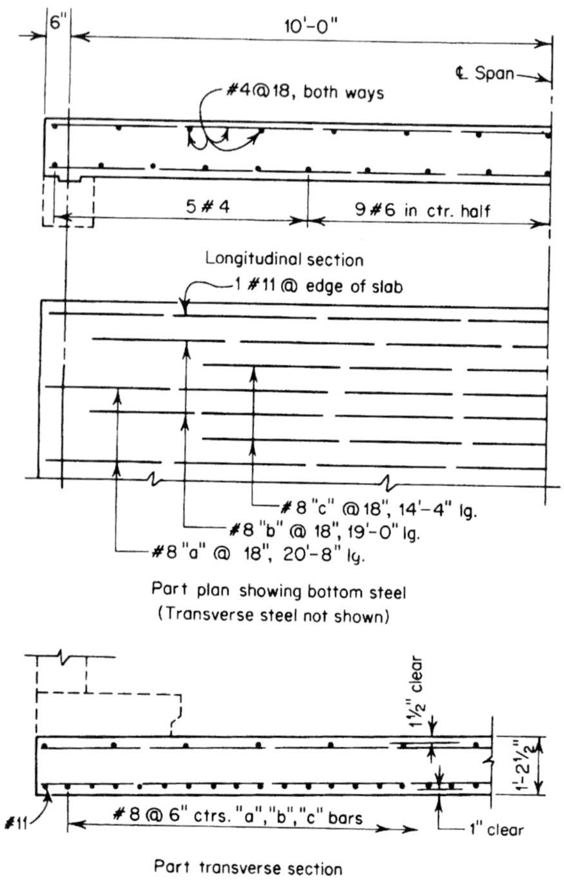

Figure 15.50 Details of slab bridges.

Continuous slab bridges on the pile bents can be built up to about 150 ft in length without expansion joints, with the end bent constructed monolithically with the slab. For this length, the passive resistance of the approach material to pressure caused by a temperature increase in the deck does not produce excessive compressive stresses in the concrete.

Live-Load Positions

For continuous slabs of spans up to about 40 ft, the HS20-44 truck governs for both positive and negative moment. Above 44 ft, the lane loading governs for negative moment. Effective width of slab for distribution of live load is the same as for simple-span slabs. It is not necessary to investigate shear in slabs designed for bending moment in accordance with this distribution.

Placement of Main Steel

Required extension of reinforcement at the exterior support is the same as for simple spans. At the interior supports, at least one-fourth of the required positive reinforcement must be carried through the support. In the regions of positive moment, it is advisable to use top reinforcement not less than the equivalent of No. 5 bars at 18 in.

Distribution steel is determined as a percentage of the main reinforcement in the bottom of the slab as for simple spans. In skewed spans, distribution steel should be placed parallel to the bents with the spacing measured along the centerline of bridge.

15.8.3 Design of Bents

Where drift is not a problem and stream flow is small, the pile bent can be used to advantage in continuous slab bridges. Figure 15.51 will be used to illustrate the design procedure.

EXAMPLE

Dimensions of the cap are shown in section *BB*. The position of the HS20-44 truck to produce maximum live-load reaction is shown in Fig. 15.52*a*. The reaction on one bent is

Reaction = 1.3[197 + 1.67 + 112] = 499 kips
Allowable load per pile, 45 tons × 2 = 180 kips
Number of piles = 499/180 = 2.77
Requires three piles. (Continue redesign with four piles.)

DESIGN CAP FOR MOMENT The truck positioned for positive moment (Fig. 15.52*b*) is a satisfactory criterion for both top and bottom steel. Although the negative moment may be theoretically slightly larger, the transverse steel in the slab will act with the cap steel to give additional resisting moment.

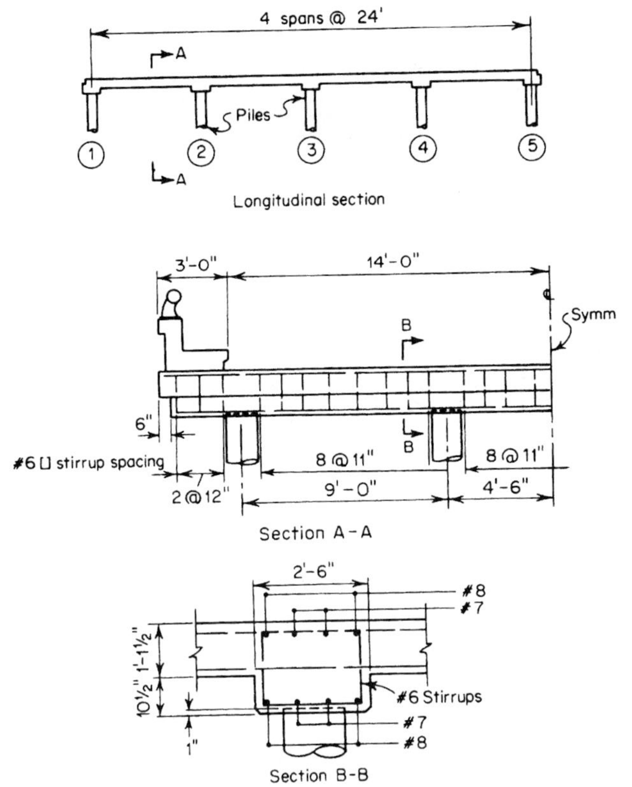

Figure 15.51 Pile bent for slab bridge.

The rear wheel of the truck will act as a concentrated load on the bent. The front wheels (axle 3) and rear wheels (axle 1) will act as uniform loads (Fig. 15.52*a*). The latter two produce a bent reaction of 24 kips per lane, which can be distributed over the 10-ft width of lane.

The maximum positive moment in the cap is determined from an influence line for a three-span continuous beam of uniform cross section:

$$M_u = 160 \text{ ft-kips} \qquad d = 21.5 \qquad a = 12$$

$$A_s = \frac{M_u}{2.25(2d - a)} = 2.29 \text{ in.}^2$$

Use four No. 7 bars.

Maximum shear in the cap is produced with the truck positioned as shown in Fig. 15.52*c* and is determined from an influence line for a three-span continuous beam of uniform cross section:

$$V_u = 132k$$

$$V_n = V_c + V_s = \frac{V_u}{0.85} = 155k$$

$$V_c = 2\sqrt{f'_c}b_w d = 81k$$

$$V_{s,\text{req.}} = V_n - V_c = 74k$$

$$V_s = \frac{A_v f_y d}{s}$$

where s = stirrup spacing
A_v = assume No. 5 stirrups = 0.61 in.^{2}

Use No. 5 stirrups @ 9 in. < $d/2$.

15.8.4 Typical Details, Continuous Slabs

Figure 15.53 shows typical superstructure details for a continuous slab bridge. Figure 15.54 shows typical details for an end-wall abutment with wing walls parallel to the roadway. This type of abutment has a massive appearance, yet it is economical to build. When appearance is not an important factor, control of the earth approach can be provided by short wing walls parallel to the abutment.

Figure 15.55 shows typical details at a bent (a) is used for short spans, (b) for longer spans where the bent reactions are greater and the spacing between piles is less. The latter also provides economy through a minimum of formwork and should be considered when structurally feasible.

Figure 15.56 shows typical superstructure details for a slab span with an expansion hinge located within the span. Hinges of this type have been used extensively and, when properly designed and constructed, have proved to be very durable. Figure 15.57 shows an expansion detail that can be used at a bent with cap. It is particularly adaptable to short spans where caps are required to extend below the slab. This type of joint is more economical than the steel expansion hinge of Fig. 15.56. However, it would not be economical to use caps extending below the slab just to use this type of joint since the cost of formwork for the caps exceeds the additional cost of the fabricated steel hinge.

15.9 T-BEAM BRIDGES

15.9.1 Economics

In simple spans longer than about 25 ft and in continuous spans longer than 30 to 35 ft, the economy of reinforced-concrete T-beams becomes apparent. Depending upon span length and labor and material prices at various geographical locations, T-beams compare in economy with reinforced-concrete slabs, steel stringers (rolled beams), welded plate girders (both composite and noncomposite), reinforced-concrete box girders, and prestressed girders. In spans around 25 ft, T-beams are comparable in economy with longitudinally reinforced slabs, while in longer spans, they should be compared with the other types.

Depth-to-Span Ratios

Economical depth-to-span ratios for simple spans vary from about 0.01 to 0.09 and for continuous spans from 0.065 to 0.085. These ratios are for girders having uniform depth. In continuous spans, maximum economy usually results when the depth-to-span ratio is 0.065 to 0.070.

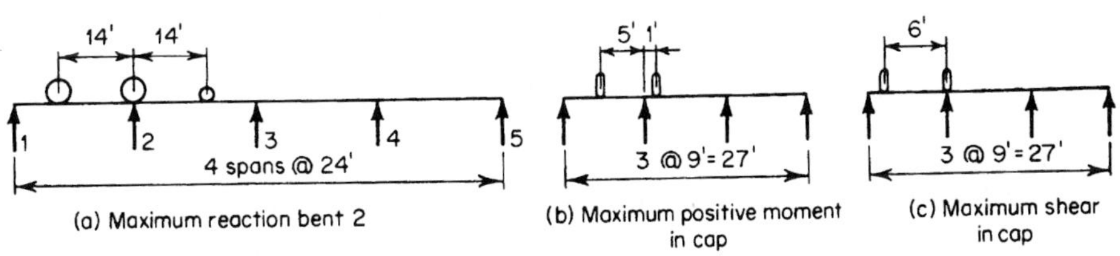

Figure 15.52 Live-load positions for design of bent.

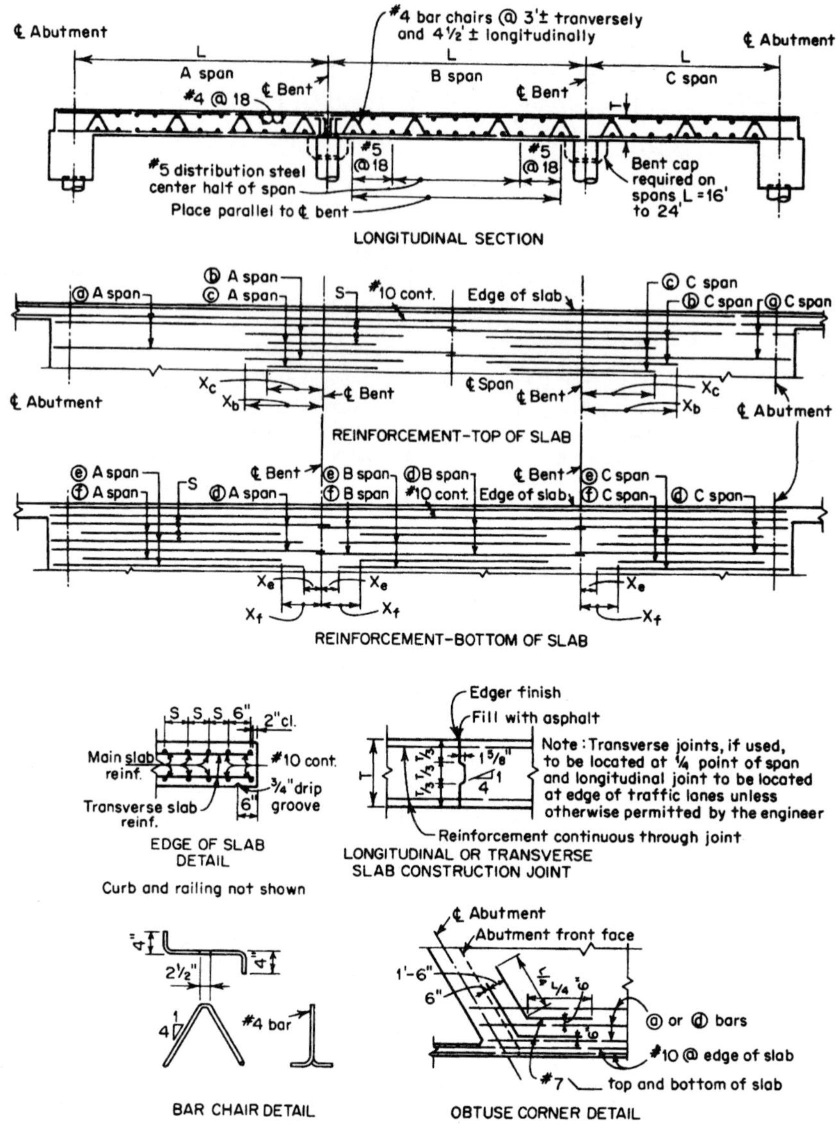

Figure 15.53 Superstructure details, continuous slab bridge.

These ratios require a small amount of compressive reinforcement over the supports, which can usually be supplied by the positive reinforcement which is required to be extended beyond the face of the support.

In some cases, economical depth cannot be determined by considering the economy of the structure itself. Situations arise where the girder depth has to be established for overall economy through a consideration of approach fill costs or the necessity for meeting existing street grades.

Girder Spacing

Spacing of girders in T-beam bridges is dictated by economy. The most commonly used spacings are between 6 and 10 ft, although in long spans, it may be as much as 13 or 14 ft. Spacing depends upon the overall width of structure and the necessity for maintaining a slab overhang beyond the exterior girder, which, according to the AASHTO specifications, should not exceed half the spacing of girders. When widening of a structure is anticipated, it is preferable to keep the slab overhang to one-third the girder spacing in order to eliminate thickening the slab over the exterior girder as a consequence of increased negative moments. In structures with three or more girders, the spacing should be such as to make the girders as nearly alike as possible.

Girder stem width is determined by the number, spacing, and arrangement of reinforcing bars at the center of the span. The least width which can be used to enclose the positive-moment bars usually provides maximum economy. When required for shear or for compression over the bents in continuous spans, the stem can be flared as shown in Fig. 15.58. The increased width of the stem at the support is usually governed by moment, while the length of flare is governed by shear (Sec. 15.8.4). On continuous structures with spans of 60 ft or less, sufficient compressive reinforcement can be added for negative moment at the piers to keep the stem width uniform. This adds to the economy through simplification of formwork. When compressive reinforcement is used, the amount should not exceed 50 percent of the tension reinforcement required at the same section.

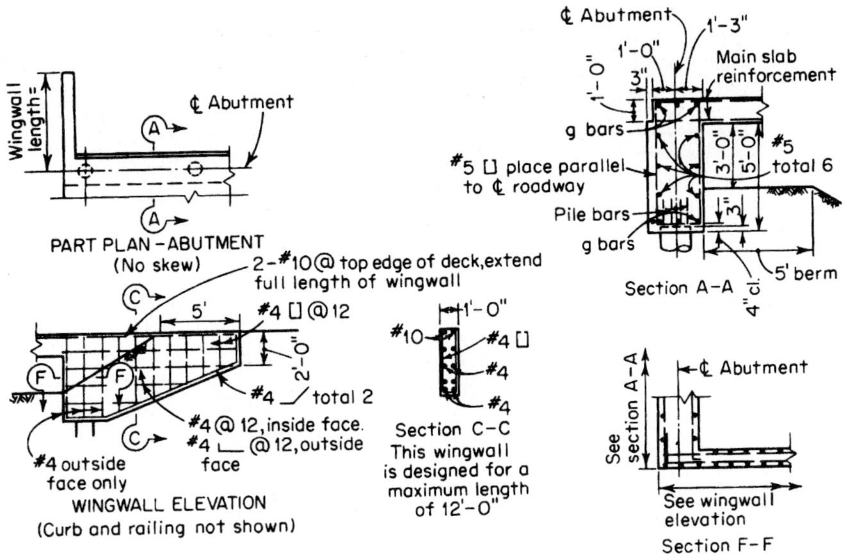

Figure 15.54 Abutment details, continuous slab bridge.

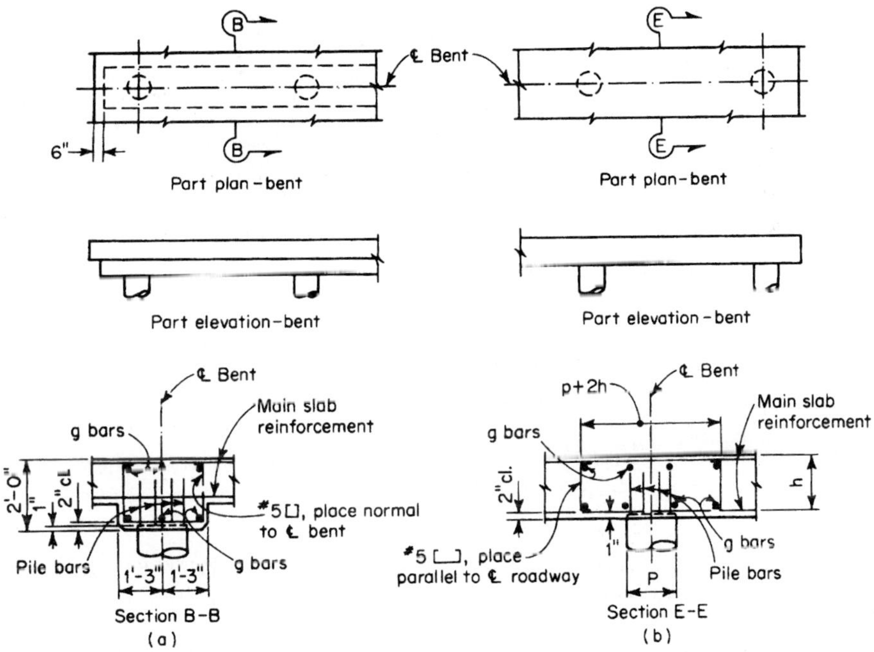

Figure 15.55 Bent details: (*a*) with cap and (*b*) without cap.

Table 15.7 facilitates the design of beams with compressive reinforcement. It can be used for inverted T-beams (flange in tension) as well as for rectangular beams and slabs.

EXAMPLE

Given $M = 750$ ft-kips, $d = 40$ in., $d' = 2$ in., $b = 18$ in., $f_s = 20{,}000$ psi, and $f_c = 1600$ psi, determine the reinforcement.

Solution From Table 15.7, $M = 420$ ft-kips for $b = 1$ ft. For $b = 18$ in., $M = 1.5 \times 420 = 630$ ft-kips, so compressive reinforcement must provide $750 - 630 = 120$ ft-kips.

From Table 15.7, $M_s' = 63.33$ ft-kips per in.^{2} of compression reinforcement for working stress design. Therefore, $A_s' = 120/63.33 = 1.89$ in.^{2} Also, $A_s' = 7.2$ in.^{2}/ft of b, so $A_s = 7.2 \times 750/420 = 12.86$ in.^{2} There is no similar table for load factor design.

EXAMPLE

Given $M = 690$ ft-kips, $d = 40$ in., $d' = 2$ in., $A'_s =$ two No. 11 $= 3.12$ in.^{2}, $f_s = 20{,}000$ psi, $f_c = 1600$ psi, determine b and A_s.

Solution From Table 15.7, $M'_s = 63.33$ ft-kips/in.^{2} of compression reinforcement. For $A'_s = 3.12$ in.^{2}, $M'_s = 63.33 \times 3.12 = 198$ ft-kips.

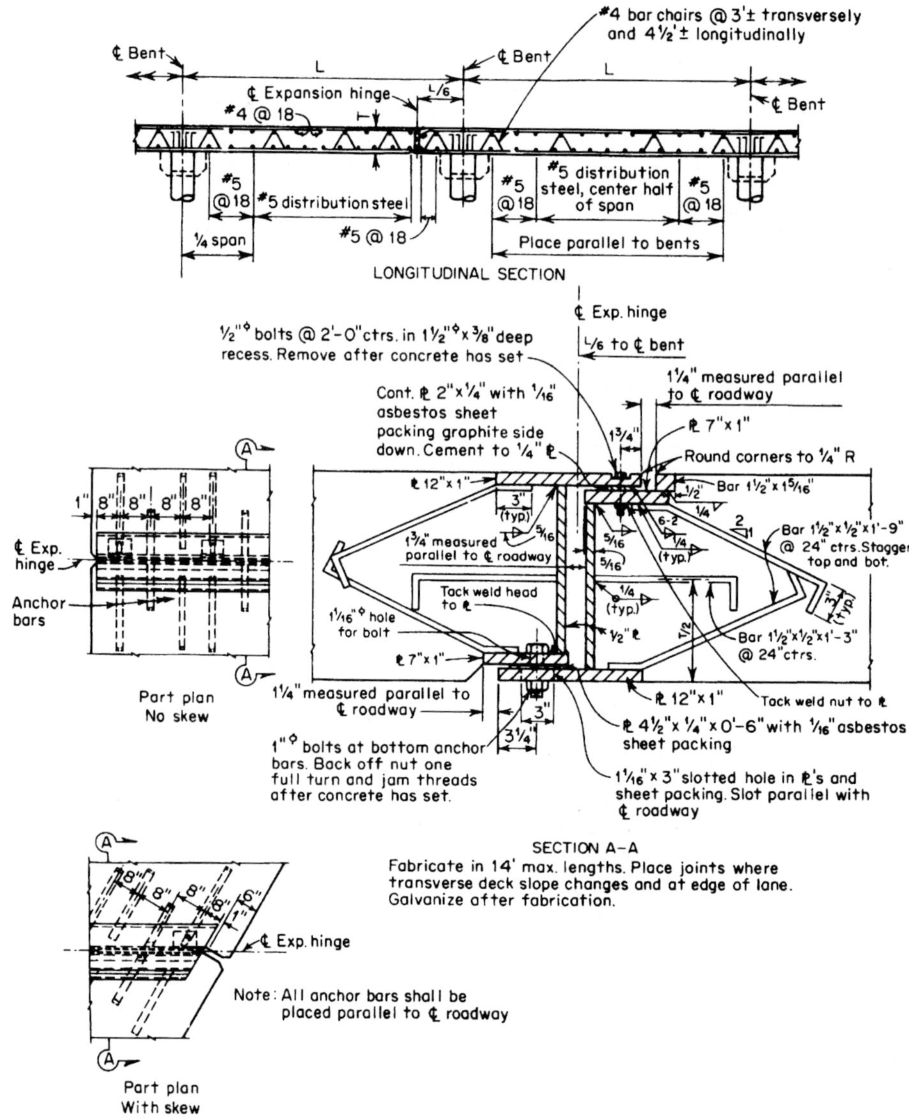

Figure 15.56 Expansion hinge for slab bridge.

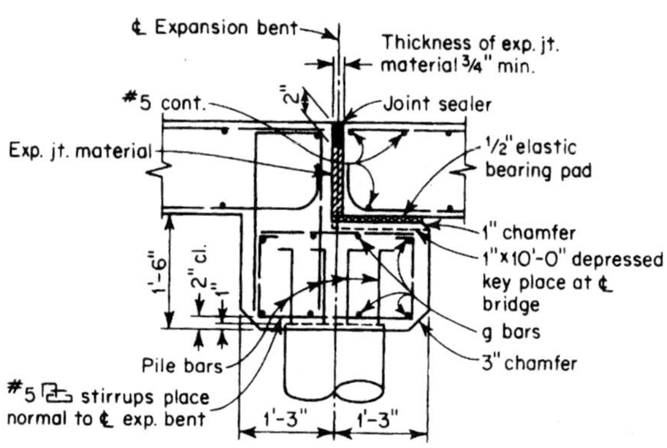

Figure 15.57 Expansion detail at bent.

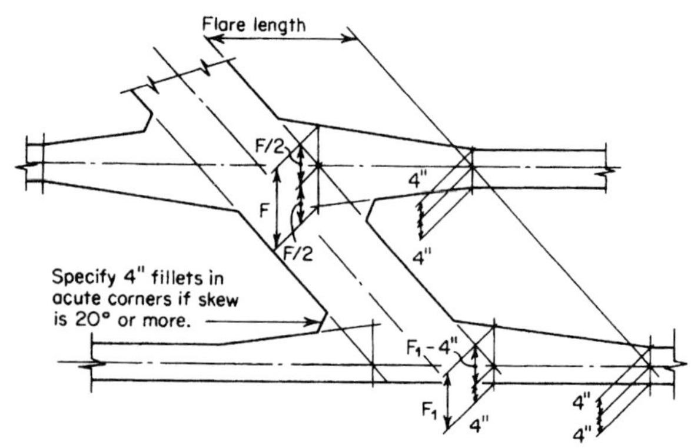

Figure 15.58 Typical girder-stem flare. Where the increase in stem width is not more than 2½ in. on a side, stirrups may be detailed to uniform width.

Table 15.7 Beams Reinforced for Compression*

Effective depth d, in.	M, ft-kips per ft b	As, in.^{2} per ft b	M'_s, ft-kips per in.^{2} A'_s depth of compressive reinforcement d', in.					
			1½	2	2½	3	4	5
4	4.2	0.72	0	0	0	0	0	0
4.25	4.74	0.76	0.32	0	0	0	0	0
4.5	5.31	0.81	0.66	0	0	0	0	0
4.75	5.92	0.85	1.02	0	0	0	0	0
5	6.56	0.9	1.4	0	0	0	0	0
5.25	7.23	0.94	1.78	0	0	0	0	0
5.5	7.94	0.99	2.18	0.21	0	0	0	0
5.75	8.67	1.03	2.58	0.54	0	0	0	0
6	9.45	1.08	3	0.88	0	0	0	0
6.25	10.25	1.12	3.42	1.24	0	0	0	0
6.5	11.09	1.17	3.84	1.61	0	0	0	0
6.75	11.96	1.21	4.27	1.99	0.1	0	0	0
7	12.86	1.26	4.71	2.38	0.42	0	0	0
7.25	13.79	1.3	5.15	2.77	0.76	0	0	0
7.5	14.76	1.35	5.6	3.17	1.11	0	0	0
7.75	15.76	1.39	6.04	3.58	1.46	0	0	0
8	16.8	1.44	6.5	4	1.83	0	0	0
8.25	17.86	1.48	6.95	4.41	2.2	0.31	0	0
8.5	18.96	1.53	7.41	4.84	2.58	0.64	0	0
8.75	20.09	1.57	7.87	5.27	2.97	0.98	0	0
9	21.26	1.62	8.33	5.7	3.37	1.33	0	0
9.25	22.46	1.66	8.79	6.13	3.77	1.68	0	0
9.5	23.69	1.71	9.26	6.57	4.17	2.05	0	0
9.75	24.95	1.75	9.73	7.02	4.58	2.42	0	0
10	26.25	1.8	10.2	7.46	5	2.8	0	0
10.5	28.94	1.89	11.14	8.36	5.84	3.57	0	0
11	31.76	1.98	12.09	9.27	6.69	4.36	0.42	0
11.5	34.71	2.07	13.04	10.18	7.56	5.17	1.08	0
12	37.8	2.16	14	11.11	8.44	6	1.77	0
12.5	41.01	2.25	14.96	12.04	9.33	6.84	2.49	0
13	44.36	2.34	15.92	12.97	10.23	7.69	3.23	0
13.5	47.84	2.43	16.88	13.91	11.13	8.55	3.98	0.2
14	51.45	2.52	17.85	14.85	12.04	9.42	4.76	0.85
14.5	55.19	2.61	18.82	15.8	12.96	10.31	5.55	1.52
15	59.06	2.7	19.8	16.75	13.88	11.2	6.35	2.22
15.5	63.06	2.79	23.33	17.7	14.81	12.09	7.17	2.93
16	67.2	2.88	24.16	18.66	15.75	13	8	3.66
16.5	71.46	2.97	25	19.62	16.68	13.9	8.83	4.41
17	75.86	3.06	25.83	20.58	17.62	14.82	9.68	5.17
17.5	80.39	3.15	26.66	21.55	18.57	15.74	10.54	5.95
18	85.05	3.24	27.5	22.51	19.51	16.66	11.4	6.74
18.5	89.84	3.33	28.33	23.48	20.46	17.59	12.27	7.54
19	94.76	3.42	29.16	24.45	21.42	18.52	13.15	8.35
19.5	99.81	3.51	30	25.42	22.37	19.46	14.04	9.17
20	105	3.6	30.83	26.4	23.33	20.4	14.93	10
21	115.76	3.78	32.5	31.66	25.25	22.28	16.73	11.68
22	127.05	3.96	34.16	33.33	27.18	24.18	18.54	13.39
23	138.86	4.14	35.83	35	29.11	26.08	20.37	15.13
24	151.2	4.32	37.5	36.66	31.05	28	22.22	16.88
25	164.06	4.5	39.16	38.33	33	29.92	24.08	18.66
26	177.45	4.68	40.83	40	39.16	31.84	25.94	20.46
27	191.36	4.86	42.5	41.66	40.83	33.77	27.82	22.27
28	205.8	5.04	44.16	43.44	42.5	35.71	29.71	24.09
29	220.76	5.22	45.83	45	44.16	37.65	31.6	25.93
30	236.25	5.4	47.5	46.66	45.83	39.6	33.51	27.77
31	252.26	5.58	49.16	48.33	47.5	46.66	35.41	29.63
32	268.8	5.76	50.83	50	49.16	48.33	37.33	31.5
33	285.86	5.94	52.5	51.66	50.83	50	39.25	33.37
34	303.45	6.12	54.16	53.33	52.5	51.66	41.17	35.25
35	321.56	6.3	55.83	55	54.16	53.33	43.1	37.14
36	340.2	6.48	57.5	56.66	55.83	55	45.03	39.03
37	359.36	6.66	59.16	58.33	57.5	56.66	46.97	40.93
38	379.05	6.84	60.83	60	59.16	58.33	48.91	42.84

(*Continued*)

Table 15.7 Beams Reinforced for Compression* (*Continued*)

Effective depth *d*, in.	*M*, ft-kips per ft *b*	As, in.² per ft *b*	M'_s, ft-kips per in.² A'_s depth of compressive reinforcement *d'*, in.					
			1½	2	2½	3	4	5
39	399.26	7.02	62.5	61.66	60.83	60	50.85	44.75
40	420	7.2	64.16	63.33	62.5	61.66	52.8	46.66
41	441.26	7.38	65.83	65	64.16	63.33	54.74	48.58
42	463.05	7.56	67.5	66.66	65.83	65	63.33	50.5
43	485.36	7.74	69.16	68.33	67.5	66.66	65	52.43
44	508.2	7.92	70.83	70	69.16	68.33	66.66	54.36
45	531.56	8.1	72.5	71.66	70.83	70	68.33	56.29
46	555.45	8.28	74.16	73.33	72.5	71.66	70	58.23
47	579.86	8.46	75.83	75	74.16	73.33	71.66	60.17
48	604.8	8.64	77.5	76.66	75.83	75	73.33	62.11
49	630.26	8.82	79.16	78.33	77.5	76.66	75	64.05
50	656.25	9	80.83	80	79.16	78.33	76.66	66
51	682.76	9.18	82.5	81.66	80.83	80	78.33	67.94
52	709.8	9.36	84.16	83.33	82.5	81.66	80	62.51
53	737.36	9.54	85.83	85	84.16	83.33	81.66	63.84
54	765.45	9.72	87.5	86.66	85.83	85	83.33	65.17
55	794.06	9.9	89.16	88.33	87.5	86.66	85	66.5
56	823.2	10.08	90.83	90	89.16	88.33	86.66	67.83
57	852.86	10.26	92.5	91.66	90.83	90	88.33	69.16
58	883.05	10.44	94.16	93.33	92.5	91.66	90	70.49
59	913.76	10.62	95.83	95	94.16	93.33	91.66	71.82
60	945	10.8	97.5	96.66	95.83	95	93.33	73.15
61	976.76	10.98	99.16	98.33	97.5	96.66	95	74.48
62	1009.05	11.16	100.83	100	99.16	98.33	96.66	75.81
63	1041.86	11.34	102.5	101.66	100.83	100	98.33	77.14
64	1075.2	11.52	104.16	103.33	102.5	101.66	100	78.47
65	1109.06	11.7	105.83	105	104.16	103.33	101.66	79.8
66	1143.45	11.88	107.5	106.66	105.83	105	103.33	81.13
67	1178.36	12.06	109.16	108.33	107.5	106.66	105	82.46
68	1213.8	12.24	110.83	110	109.16	108.33	106.66	83.79
69	1249.76	12.42	112.5	111.66	110.83	110	108.33	85.12
70	1286.25	12.6	114.16	113.33	112.5	111.66	110	86.45
71	1323.26	12.78	115.83	115	114.16	113.33	111.66	87.78
72	1360.8	12.96	117.5	116.66	115.83	115	113.33	89.11
73	1398.86	13.14	119.16	118.33	117.5	116.66	115	90.44
74	1437.45	13.32	120.83	120	119.16	118.33	116.66	91.77
75	1476.56	13.5	122.5	121.66	120.83	120	118.33	93.1
76	1516.2	13.68	124.16	123.33	122.5	121.66	120	94.43
77	1556.36	13.86	125.83	125	124.16	123.33	121.66	95.76
78	1597.05	14.04	127.5	126.66	125.83	125	123.33	97.09
79	1638.26	14.22	129.16	128.33	127.5	126.66	125	98.42
80	1680	14.4	130.83	130	129.16	128.33	126.66	99.75
81	1722.26	14.58	132.5	131.66	130.83	130	128.33	101.08
82	1765.05	14.76	134.16	133.33	132.5	131.66	130	102.41
83	1808.36	14.94	135.83	135	134.16	133.33	131.66	103.74
84	1852.2	15.12	137.5	136.66	135.83	135	133.33	105.07
85	1896.56	15.3	139.16	138.33	137.5	136.66	135	106.4
86	1941.45	15.48	140.83	140	139.16	138.33	136.66	107.73
87	1986.86	15.66	142.5	141.66	140.83	140	138.33	109.06
88	2032.8	15.84	144.16	143.33	142.5	141.66	140	110.39
89	2079.26	16.02	145.83	145	144.16	143.33	141.66	111.72
90	2126.25	16.2	147.5	146.66	145.83	145	143.33	113.05
91	2173.76	16.38	149.16	148.33	147.5	146.66	145	114.38
92	2221.8	16.56	150.83	150	149.16	148.33	146.66	115.71
93	2270.36	16.74	152.5	151.66	150.83	150	148.33	117.04
94	2319.45	16.92	154.16	153.33	152.5	151.66	150	118.37
95	2369.06	17.1	155.83	155	154.16	153.33	151.66	119.7
96	2419.2	17.28	157.5	156.66	155.83	155	153.33	121.03
97	2469.86	17.46	159.16	158.33	157.5	156.66	155	122.36
98	2521.05	17.64	160.83	160	159.16	158.33	156.66	123.69
99	2572.76	17.82	162.5	161.66	160.83	160	158.33	125.02
100	2625	18	164.16	163.33	162.5	161.66	160	126.35

*f_s = 20,000 psi, f_c = 1600 psi, n = 7.5 (tensile reinforcement), n = 15 (compressive reinforcement), M = moment (tensile reinforcement) per ft b, A_s = area tensile reinforcement per ft b, M_s' = moment per in.² compressive reinforcement.

Moment to be provided = 690 − 198 = 492 ft-kips. For $b = 1$ ft, $M =$ 420 ft-kips. Therefore, $b = 12 \times 492/420 = 14.0$ in. and $A_s = 7.2 \times 7.20 \times 690/420 = 11.83$ in.^{2}

BUNDLING OF REINFORCEMENT

Bundled bars are very effective in reducing the size of concrete sections and in lessening the maze of reinforcing steel which frequently hampers the placing of concrete. For example, the stem of a T-beam reinforced with eight No. 11 bars in two layers requires a width of 16 in. By placing these bars in two bundles of three bars each and two bars in the top layer, the width can be reduced to 13 in. In addition, placing of both steel and concrete is simplified. Four bars may be bundled, although it is preferable to limit the number to three. Not more than one bar should be terminated at any one point in a bundle. The development length for each bar must be that for the individual bar increased by 20 percent for a three-bar bundle and 33 percent for a four-bar bundle. Where lap splices are permitted (for bars No. 11 and smaller), the length of laps for bundled bars should be increased by the same percentages.

The table in Fig. 15.59 shows the preferred minimum spacings for the bundled bar combinations. Preferred bar coverings for beam and girder sections are shown in Fig. 15.59*a*. Figure 15.59*b* shows minimum coverings for bundled bars in mass sections such as footings, abutments, piers, caps, and large columns. When bundled bars are used in work exposed to seawater, additional covering should be used.

15.9.2 Design of a T-Beam Bridge

Points to be considered in the design of a T-beam bridge will be discussed for the four-span continuous structure with 28-ft roadway shown in Fig. 15.60. Allowable stresses are $f_s = 20{,}000$ psi, $f_c = 1600$ psi, $v_c = 0.95\sqrt{f_c'}$, $v_{max} = v_c + \sqrt{f_c'}$.

DESIGN OF SLAB The design of transverse slabs continuous over three or more supports in T-beam structures is similar to that of transverse slabs on steel stringers (Sec. 15.3.1). The only difference is in the effective span, which for T-beam slabs is the clear distance between faces of support. Where fillets making an angle of 45 degrees or more with the axis of a continuous or restrained slab are built monolithic with the slab and support, the span can be measured from the section where the combined depth of slab and fillet is at least one and one-half times the thickness of the slab.

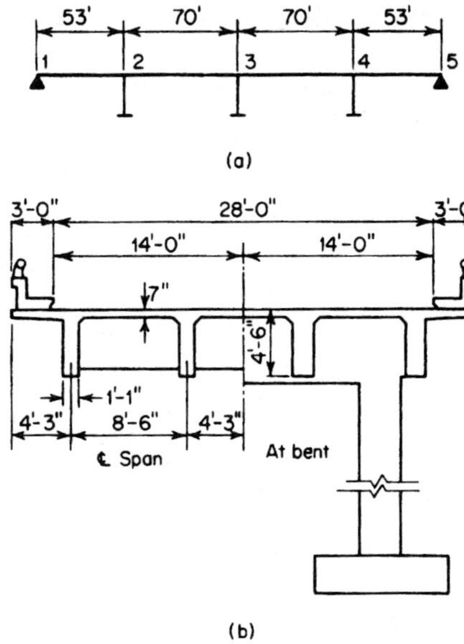

Figure 15.60 T-beam bridge: (*a*) layout and (*b*) typical section.

Figure 15.61 shows a typical section of the slab, which has been designed in accordance with the AASHTO specifications. The effective span has been taken as the clear distance between girder stems. Longitudinal girder steel in the top of slab is not shown except for the continuous bars at the stirrup hooks and the bars at the bend point of the transverse steel. At the supports, these bars at the bend points are usually main girder bars, while at the centers of spans, they are extensions to main bars in the form of temperature steel.

Transverse slab steel can be placed parallel to the bents for skews up to 20 degrees. For larger skews, consideration should be given to placing

Bar size	A	B
#4	3 1/4	4
5	3 1/2	4 1/4
6	3 3/4	4 1/2
7	4	4 3/4
8	4 1/4	5
9	4 1/2	5 1/4
10	4 3/4	5 1/2
11	5	6
14	5 1/2	7

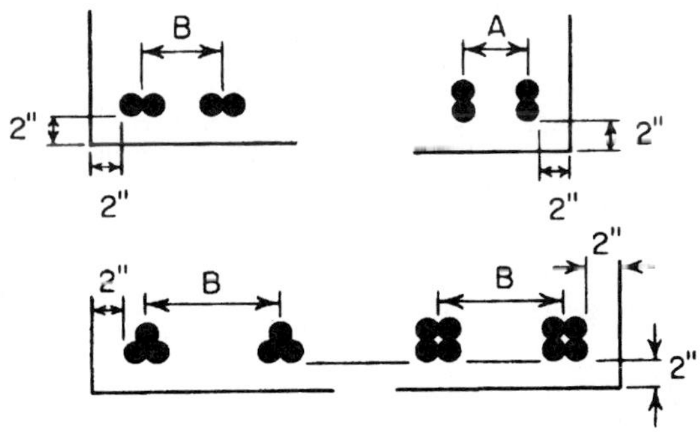

(a)

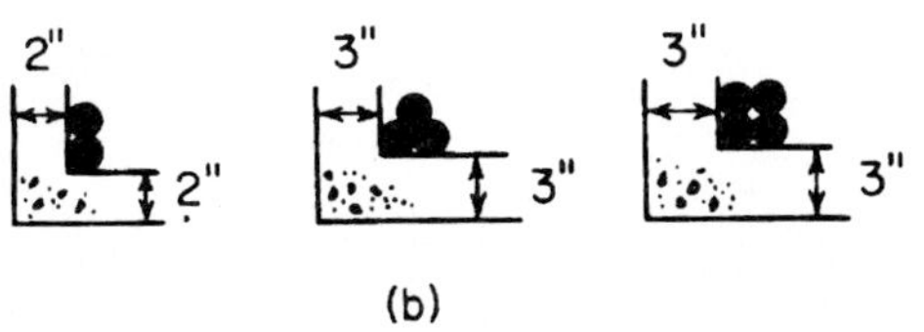

(b)

Figure 15.59 Bundled bar reinforcement.

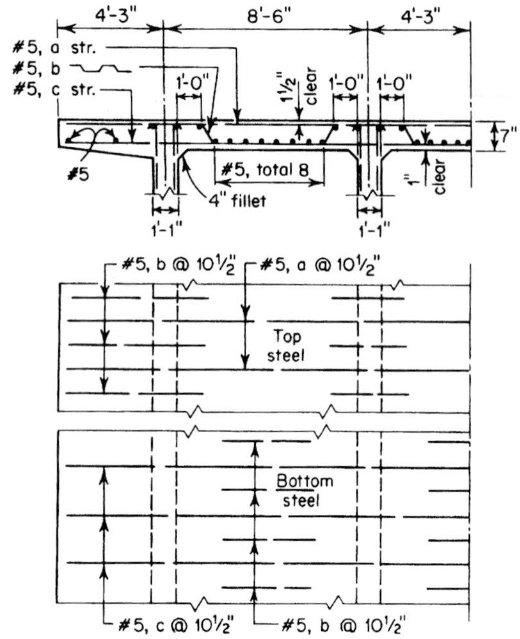

Figure 15.61 Typical section of T-beam bridge slab.

it normal to the centerline of structure. This keeps the amount of transverse slab steel to a minimum since the steel spacing is measured along centerline of structure regardless of the amount of skew. Special consideration should be given to the details of reinforcement in skewed corners.

DESIGN OF STEM The total depth chosen for the structure in Fig. 15A.60 is slightly under 0.07 × span = 0.07 × 70 = 4.9 ft. Economy dictates that the width of stem be kept to a minimum at the center of span yet be sufficient to enclose the positive-moment reinforcing and provide ample clearance between bars to facilitate concrete placement. In this example, bundled bars are used. In a T-beam of uniform depth, the stem width can be increased near the support to meet bending-moment requirements. The increase should be just sufficient to keep the compressive stress within the allowable when one-fourth of the positive-moment steel is extended through the support. Length of the stem flare is determined by the criterion of uniform spacing of stirrups throughout its length.

Check for rectangular section if the compression block has a dimension "a" less than the top flange thickness, h_f:

$$M_u = 1605 \text{ ft-kips}$$

$$d = 51 \text{ in.} \quad a = \frac{A_s f_y}{0.85(f_c')b}$$

$$b = 98.5 \text{ in.}$$

$$h_f = 7 \text{ in.} \quad A_s = \frac{a(0.85)f_c' b}{f_y} = 5.582a$$

$$h_f = 7 \text{ in.}$$

$$b_w = 52 \text{ in.} \quad a = 0.179 A_s$$

$$f_c' = 4000 \text{ psi}$$

$$f_y = 60{,}000 \text{ psi} \quad A_s = \frac{M_u}{2.25(102-a)} = \frac{713.33}{102-a}$$

Try $a = h_f = 7$ in.

$$a = 7 \text{ in.} \qquad A_s = \frac{713.33}{102-7} = 7.509 \text{ in.}^2$$

$$a = 0.179\, A_s = 1.34 \text{ in.} \qquad A_s = \frac{713.33}{102-1.34} = 7.087 \text{ in.}^2$$

$$a = 0.179 \times 7.087 = 1.268 \text{ in.} \quad A_s = \frac{713.33}{102-1.268} = 7.082 \text{ in.}^2$$

Because a is less than h_f, use rectangular section. If $a > h_f$, a tee section would be used:

$$M_u = \phi A_s f_y \left(d - \frac{a}{2}\right) \qquad \text{where } a = \frac{A_s f_y}{0.85 f_c' b}$$

$$A_s = \frac{M_u}{\frac{\phi f_y}{2}(2d-a)}$$

$$A_s = 7.02 \text{ in.}^2 \text{ required}$$

Check maximum allowed tension steel:

$$p_o = \frac{A_s}{bd}$$

$$p_b = \text{balanced reinforcement ratio}$$

$$p_b = \frac{b_w}{b}\left[\left(\frac{0.85\beta_1 f_c'}{f_y}\right)\left(\frac{87{,}000}{87{,}000 + f_y}\right) + p_f\right]$$

$$p_b = 0.010 \qquad b_w = 13$$

$$A_{sf} = \frac{0.85 f_c'(b-b_w)hf}{f_y} = 33 \text{ in.}^2 \qquad b = 97 \qquad \beta_1 = 0.85$$

$$p = \frac{A_s}{bd} = 0.006 < 0.010 \qquad p_f = \frac{A_{sf}}{b_w d} = 0.0498$$

$$\text{Max. } p = 0.75 p_b = 0.0075$$

SHEAR Figure 15.63 shows the $DL + LL + I$ shear diagram for an interior girder of span 2. The maximum shear at bent 2 is 121 kips, so

$$v = \frac{V}{bd} = \frac{121{,}000}{18 \times 51} = 132 \text{ psi}$$

$$v_c = 0.95\sqrt{4000} = 60 \text{ psi}$$

$$v_r = 4\sqrt{4000} = 313 \text{ psi} > 148 \text{ psi}$$

$$\text{Spacing of No. 5 stirrups} = \frac{f_s A_s}{b(v-v_c)} = \frac{20{,}000 \times 0.62}{18 \times 72} = 9.6 \text{ in. Use } 9.5 \text{ in.}$$

The length of flare of the stem is determined by locating the point where the shear requirement can be met by stirrups at 9.5 in. in the 13-in. stem:

$$v - v_c = f_s A_s = \frac{20{,}000 \times 0.62}{13 \times 9.5} = 100.4 \text{ psi}$$

$$v = 100.4 + 60 = 160.4 \text{ psi}$$

$$V = \frac{160.4 \times 13 \times 51}{1000} = 106.4 \text{ kips}$$

From the shear diagram, this is the shear at a point about 6 ft from bent 2. Stirrup spacings at other points are shown in Fig. 15.62.

DIAPHRAGMS

Intermediate diaphragms (Fig. 15.62, section CC) act in conjunction with the slab as transverse distributing elements for the live load. For the ordinary T-beam bridge, it is not necessary to proportion them or determine the reinforcement by actual analysis (stresses could be determined by an analysis in which the diaphragms constitute a beam

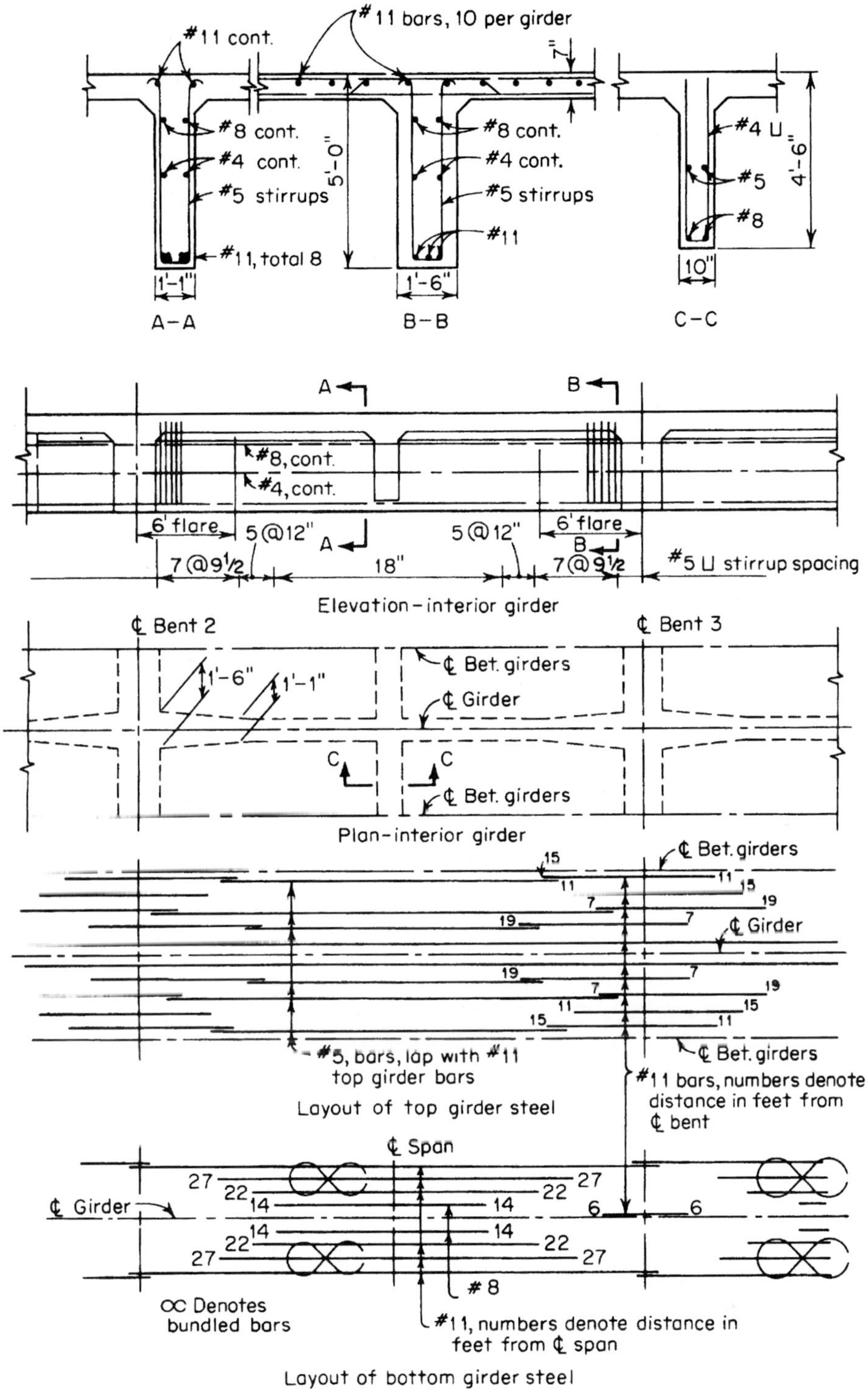

Figure 15.62 Girder reinforcement, T-beam bridge.

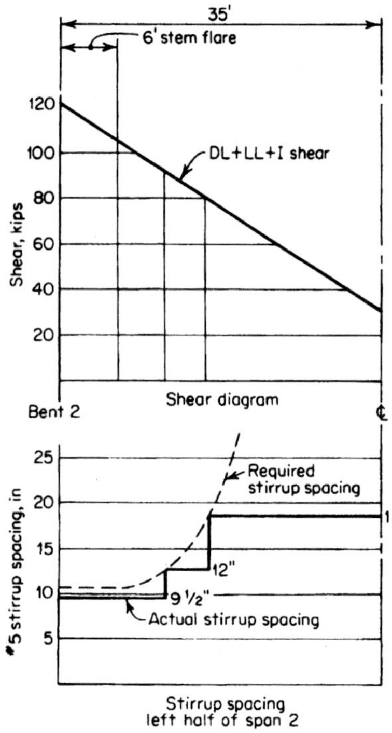

Figure 15.63 Girder reinforcement and stirrup spacing.

continuous over elastic supports). Width of diaphragms should permit ease in placement of concrete. While an 8-in. wall will accommodate two rows of steel, 10 in. is better for placing of reinforcement. Transverse slab steel resists the negative moments in the diaphragms. The bottom steel can be nominal but should not be less than two No. 6 bars. Two longitudinal bars should be placed at the mid-height of diaphragm to act as temperature steel and as tie bars for the stirrups. For girder spacings of 10 ft or less, stirrups can be No. 4 bars at 18 in. Usually the unit shear is less than allowable, and the stirrups act more nearly in the capacity of temperature steel than as diagonal tension reinforcement.

15.9.3 Design of Substructure

Intermediate supports for T-beam bridges may be as follows:

1. Reinforced-concrete piers
2. Reinforced-concrete pier walls constructed integrally with the superstructure
3. Reinforced-concrete bents consisting of two or more columns; may be integral with the superstructure
4. Reinforced-concrete single-column bents; may be integral with the superstructure
5. Reinforced-concrete piles
6. Steel piles

Choice of type depends on skew, foundation conditions, stream flow, drift, ice, and the manner in which it will function in the structure. In T-beam bridges across streams, piers or pier walls are commonly used since they are less affected by drift, stream flow, and ice. Where ice conditions do not exist, multicolumn bents, single-column bents, or piles can be used. Single-column and multicolumn bents are particularly suited to overcrossings where grades for two or more roadways are separated. The single-column bent can be used to advantage on skewed crossings because bent caps can be placed normal to the bridge centerline, which results in simpler superstructure framing.

The following discussion covers some of the more important points to be considered in the design of a two-column bent (Fig. 15.64) for the T-beam bridge of Sec. 15.8.4.

Loads

The various loads for which the bent should be investigated are dead load, live load, impact, wind load on structure, wind load on live load, longitudinal force from live load, shrinkage, temperature, and

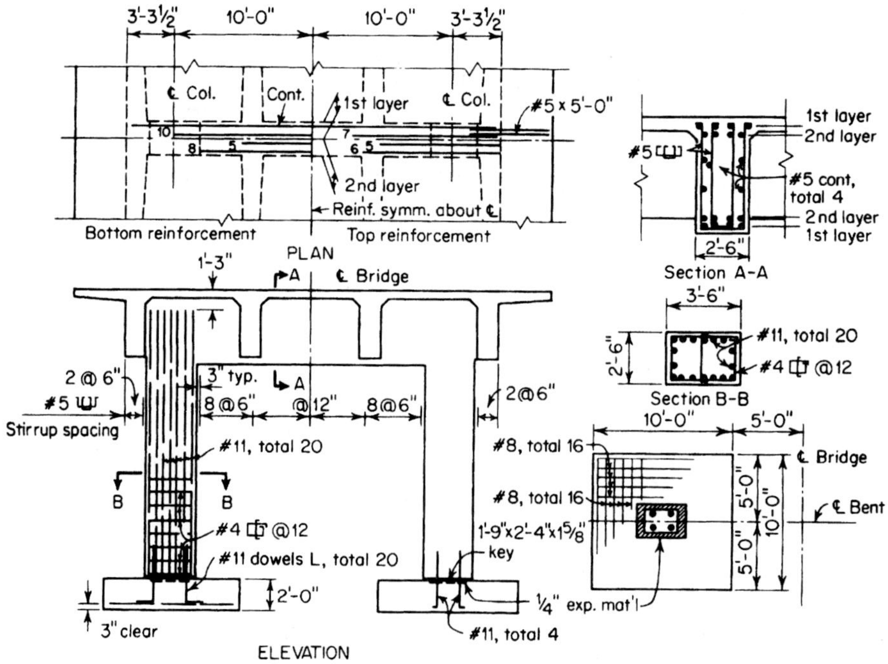

Figure 15.64 Two-column bent for T-beam bridge.

earthquake. AASHTO specifies the allowable unit stresses which may be used with various combinations of these loads. The bent is designed as a rigid frame with the columns hinged at the base.

Design of the cap for moment is determined by $D + L + I$. The cap can be considered as a T-beam with an effective width equal to the width of stem plus twelve times the slab thickness. Girder dead loads are applied as concentrations, with the weight of the cap itself a uniform load.

Live-load positive moment is determined by placing two trucks adjacent to and symmetrical about the centerline of bridge, with the rear axles over the centerline of bent. Continuity of superstructure should be taken into account in determining the reaction on the bent of the truck's front axle and the trailer axle. This reaction can be applied to the bent uniformly over the 10-ft width of lane. For the spacing of the columns in this bent, and considering the AASHTO limitations on load positions in the design lanes, live-load negative moments are determined for the same load position as for positive moment. In bents with wider column spacings, the loads should be positioned separately for maximum positive moment and maximum negative moment.

Double stirrups (Fig. 15.64, section *A-A*) are usually in caps of this type to provide spacing large enough for easy placing of concrete. For the determination of maximum shearing stresses, live load can be positioned in the same manner as for moment.

The main steel in the cap should be located so as not to interfere with the column steel or the deck steel. In structures skewed more than 20 or 25 degrees, some designers place the deck steel normal to the centerline of bridge. This necessitates placing all the top bars in the cap below the steel in the slab.

Design of Columns

The columns for this structure are cast integrally with the cap and superstructure to form a rigid frame both longitudinally and transversely. A pin is provided at the top of the footing. Their design involves an analysis for direct stress and bending in two directions. All the loading groups should be investigated to determine maximum column stresses. When investigating bending in two directions, it is important to remember that live-load stresses that are combined must be compatible; that is, if the column is being investigated for a particular position of the live load, this position should be used in determining all the stresses that are to be combined.

In a continuous structure without intermediate hinges, the deck is extremely stiff laterally, so wind loads are transmitted to the abutments rather than to the bents. Special attention must be given to the design and detailing of columns because earthquake forces can create severe displacements and shaking. Longitudinal steel should be continuous wherever possible and should be fully developed through fixed connections. No splices should be permitted in areas of high moments. Column ties, preferably spirals, should be closely spaced (±3 in.) and securely tied to the main column reinforcement to prevent buckling of that steel and to confine and contain the concrete, even when broken.

Design of Footings

The footing for the bent in Fig. 15.64 is designed for a column reaction of 380 kips. Weight of the footing itself is 30 kips. Maximum allowable soil pressure is 2 tons/ft^2.

The AASHTO specifications require that the critical section for bending be taken at the face of the column. For this footing the actual bearing area is provided by the 1 ft 9 in. × 2 ft 4 in. key. Taking moments at the 2 ft 4 in. edge,

$$M_u = 1.3[242 + 1.67(138)] = 614 \text{ ft-kips}$$

$$\text{Footing pressure} = 6.14\frac{k}{\text{ft}^2} < 8.0\frac{k}{\text{ft}^2}$$

$$M_u \text{ at face of column} = \frac{6.14(10)(4.13)^2}{2} = 525 \text{ ft-kips}$$

$$A_s = \frac{M_u}{2.25(2d - a)} \qquad d = 20.5 \quad a = 12$$

$$A_s = 8.04 \text{ in.}^2$$

Use 14 No. 7 bars.

The AASHTO specifications allow this steel to be distributed over an effective width equal to the width of the column (in this case, the key) plus twice the effective depth of the footing, plus half the remaining width of footing, in this case 7 ft 10½ in. Spacing of reinforcement outside this width may be double the spacing within the effective width. However, it is desirable to maintain uniform bar size and spacing. Therefore, 16 No. 8 uniformly spaced bars are used, which gives the area required within the effective width.

Shear must be investigated at two sections: (1) for one-way action on a section across the full width of the footing at a distance *d* from the face of the reaction area and (2) for two-way action on a concentric vertical section *d*/2 from the faces of the reaction area. The respective values of v_c are $0.95\sqrt{f_c'}$ and $1.8\sqrt{f_c'}$, which give 52 and 99 psi, respectively, for $\sqrt{f_c'} = 3000$ psi. The reaction area is the 1 ft 9 in. × 2 ft 4 in. key (Fig. 15.64).

For one-way action,

$$V_u = 148^k$$

$$V_n = V_c + V_s = \frac{V_u}{0.85} = 175 \text{ kips}$$

$$V_c = 2.0\sqrt{f_c'}b_w d = 318^k > 175$$

No stirrups needed.

For two-way action,

$$V_u = 528^k \qquad V_n = 621^k$$

$$V_c = 404^k < 621$$

Stirrup needed.

In zones of seismic concern, a nominal mat of top steel should be placed and tied to the bottom mat with ties on 12-in. centers in both directions.

Abutments

Figure 15.65 shows a typical section and elevation for each of two types of abutments which can be used for T-beam bridges. The abutment in (*a*) is cast integrally with the deck and rests directly on the pile substructure. Expansion in the deck must overcome the passive resistance of the earth adjacent to the abutment wall. Therefore, to avoid excessive compression in the deck, this type of abutment should be used only when the total length of bridge does not exceed 150 ft. However, it can be used for longer bridges if a sufficient number of expansion hinges are provided. Wing walls can be placed normal to the structure or, as in this case, parallel to it.

The abutment in (*b*) provides for expansion and is a type which can be used for bridges of any length. The back wall may rest directly on the footing or, as shown in the figure, on a wall (which may be 35 ft or more in height). Active earth pressure is taken by the wall. The abutment can be founded on either a spread footing or a footing on piles, depending upon conditions.

15.9.4 Typical Details

Typical details for construction joints and intermediate expansion hinges for T-beam bridges are shown in Figs. 15.66 and 15.67, respectively. Fixed and expansion bearing assemblies are shown in Fig. 15.68 and concrete placing diagrams in Fig. 15.69.

15.10 BOX-GIRDER BRIDGES

Many states used box girders as a solution to some of the bridge problems encountered in the development of the interstate highway system. The relatively shallow depth of the box-girder bridge is an

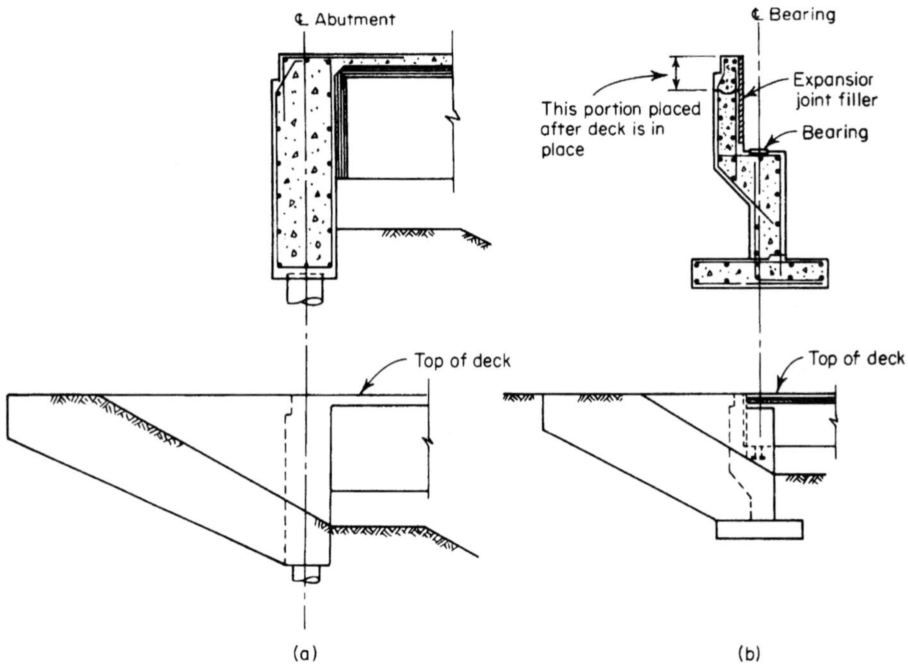

Figure 15.65 Types of abutment for T-beam bridges.

advantage where, as is often the case in cities, headroom is limited. In addition, they provide ideal space for utilities. The smooth soffit is pleasing. Because of their torsional stiffness, they can be built on rather sharp horizontal curves; this is a desirable characteristic where curved alignment has a bearing on the overall cost of a highway interchange.

15.10.1 Economics

Box-girder bridges are seldom justified for short spans on the basis of cost alone and are rarely used on spans of less than 50 ft; if they are, it is usually for the sake of uniformity in a continuous structure of varying spans. In the 50- to 100-ft range, box girders cost about one-third more than T-beams. However, the experience of the contractor is a big factor, and contractors in some areas claim that they can build box girders about as economically as T-beams. One reason is in the method of forming the webs, where, instead of finishing the tops of the forms to exact grade, as in T-beam construction, they are extended above grade and the webs are poured to finishing strips. Another saving results from reduced costs of finishing since only exterior girder faces and the bottom slab soffit require a high-quality finish.

15.10.2 Proportions

Depth-to-span ratios of 0.06 can be used for simple spans, but 0.07 is the generally accepted minimum for economy. The recognized minimum for continuous spans of uniform depth is 0.05. Ratios of 0.06 to 0.07 are common for continuous spans, but the best appearance is achieved in the range 0.05 to 0.06. AASHTO and the Precast/Prestressed Concrete Institute (Ref. 4) provide criteria for span-to-depth ratios for simple spans and continuous spans, for various type of bridges.

Webs

In spans of less than 100 ft where minimum depth-to-span ratios are used, webs can be spaced 7 to 8 ft on centers. The optimum spacing is likely to be between 7 and 9 ft for spans to 150 ft. In spans longer than 150 ft, spacing up to 12 ft may be found economical. Since the optimum spacing depends on depth-to-span ratio, comparative studies are advisable for each case.

The minimum practicable width of web is 8 in. to provide for double-leg stirrups and easy placement of concrete. Webs have been built 7 in. thick and, in rare instances, as thin as 6 in. The 8-in. web can be used for depths to about 8 ft, above which 9 or 10 in. is recommended. Webs can be flared to meet shear requirements.

Typical sections are shown in Fig. 15.70. The substructure for (*a*) consists of single columns 6 ft in diameter. In (*b*), solid piers 2 ft 6 in. × 42 ft are used. Three-column bents, each column diameter being 3 ft 6 in., are used in (*c*).

Drainage holes should be provided in the bottom slab for release of curing water, leakage through the top deck, and condensation. Occasionally there is leakage from utilities. Serious overload can result from accumulation of water because of inadequate drainage. There should be a 4-in. hole in each cell adjacent to piers and abutments, and each transverse diaphragm should have a 6-in. opening.

It is comparatively simple to provide for utilities such as gas and water lines, power and telephone ducts, and storm drains and sewers in the cells of the box-girder bridge. Spacing of girder webs can be adjusted to meet specific locations of these utilities, if necessary. Examples of reinforcement at utility openings in abutments, bents, and diaphragms are shown in Figs. 15.71 and 15.72. If utilities pass through an abutment where there is provision for expansion in the superstructure, it is usually advisable to provide a passage between the end diaphragm and the back wall for access to the cells for inspection and maintenance (Fig. 15.73). When such access cannot be provided, the sidewalk accessway is an alternative. Accessways can be placed in the roadway deck if no other location is feasible, but this is a potential for accidents to maintenance personnel. Access between adjacent cells can be through openings in the webs. They should be located near the center of the span, where shear is not significant.

15.10.3 Design

Some engineers prefer to design the cross section as a unit. As a general rule, this can be done when the ratio of span to width exceeds 2. The unit girder (cell) method is recommended for smaller ratios. In fact, unless and until further research proves otherwise, the latter is

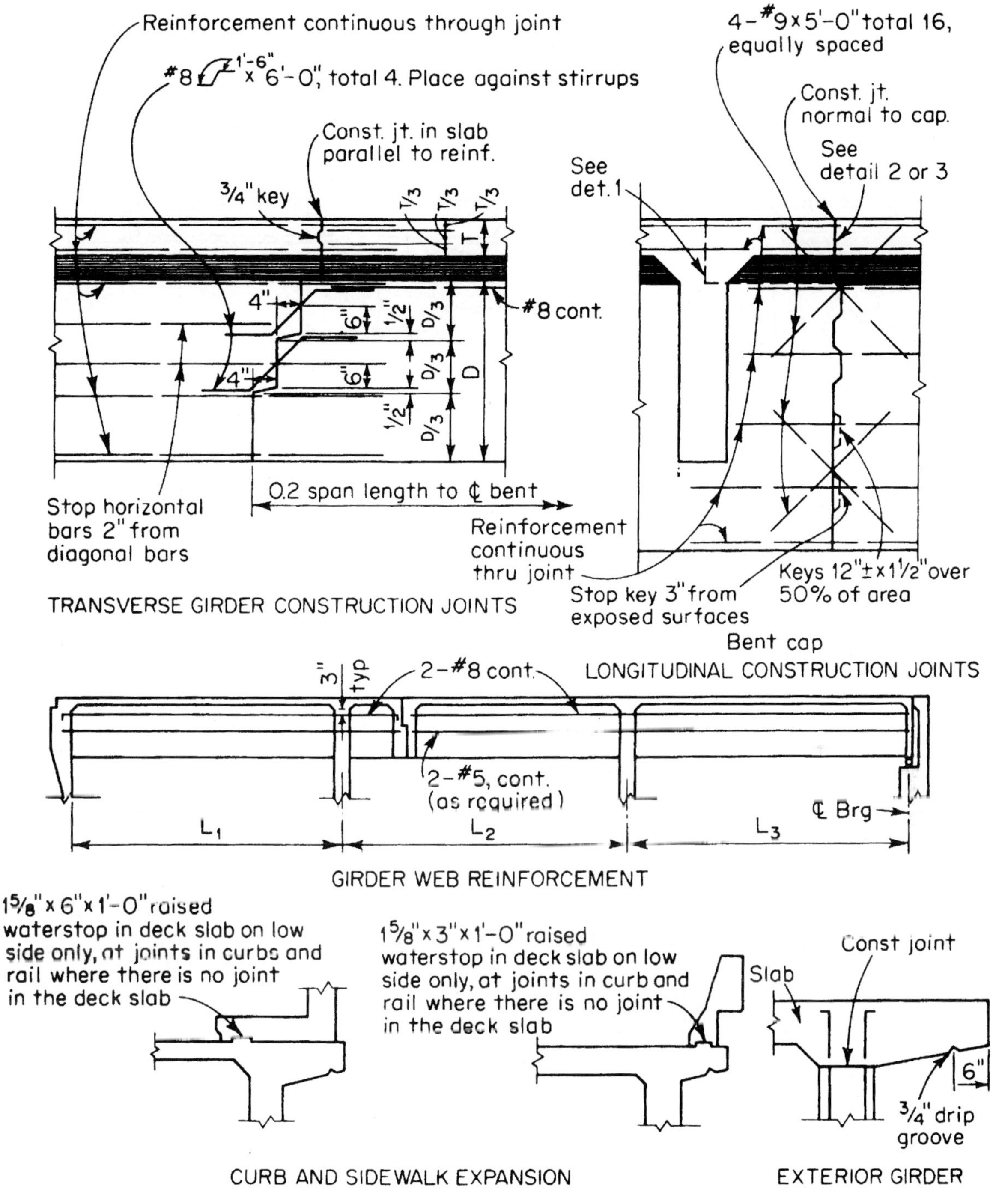

Figure 15.66 T-beam bridge construction joints.

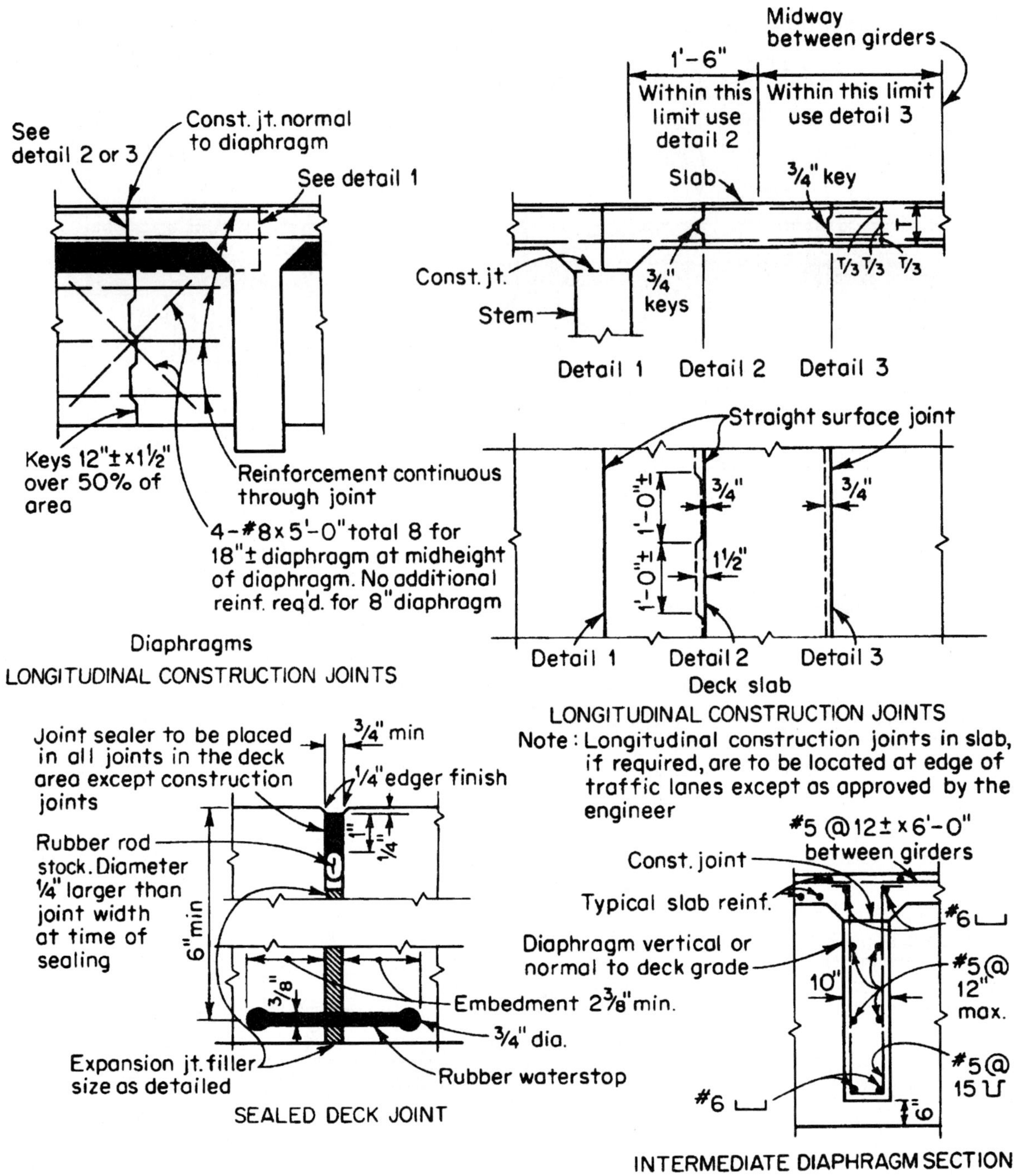

Figure 15.66 (*Continued*)

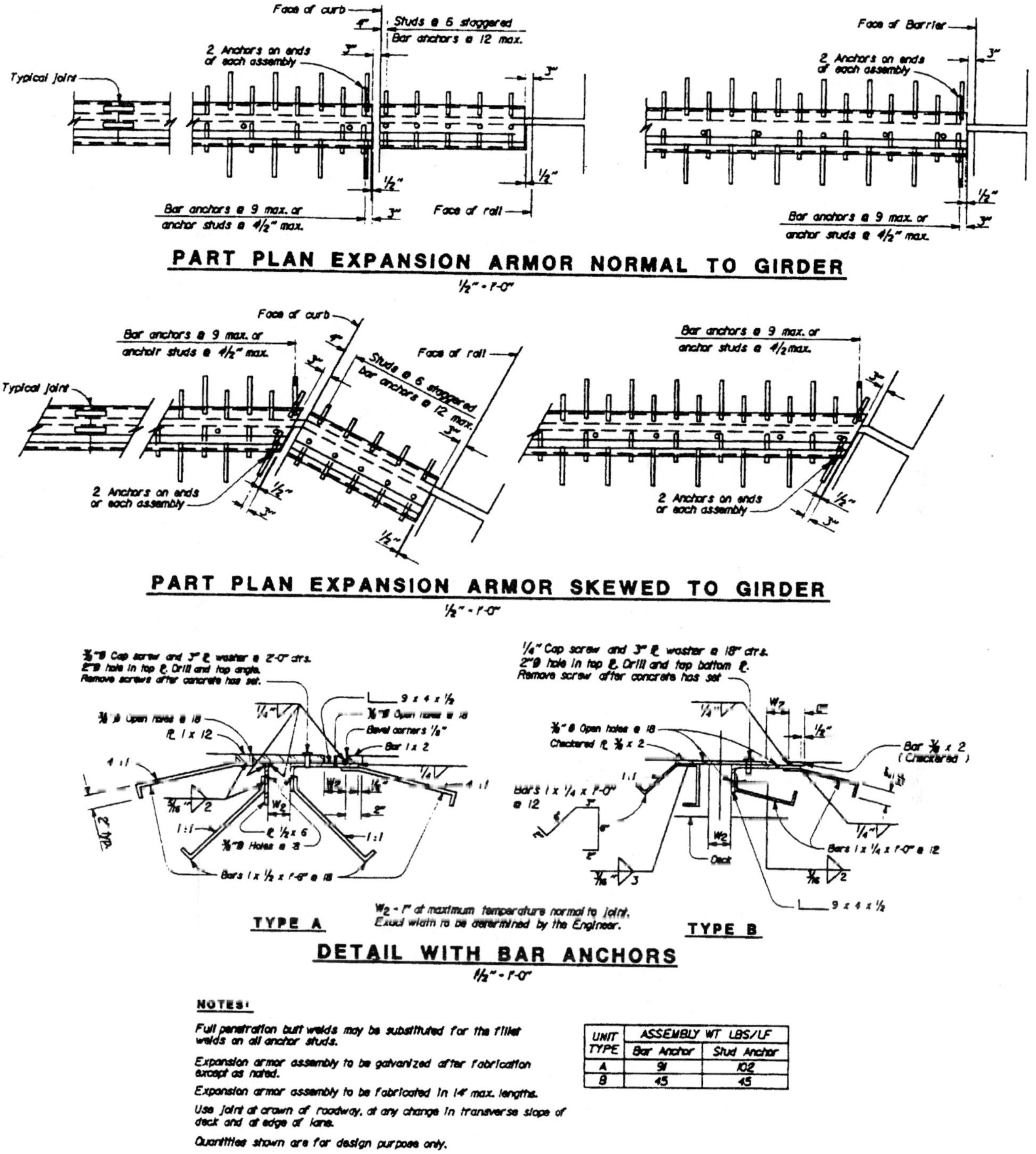

Figure 15.67 Bridge expansion details. Abutting surfaces of steel and concrete should be separated with expansion filler to provide free movements. Caulk and tape joints between steel angles and all remaining openings prevent seepage of grout or concrete between abutting surfaces of hinge.

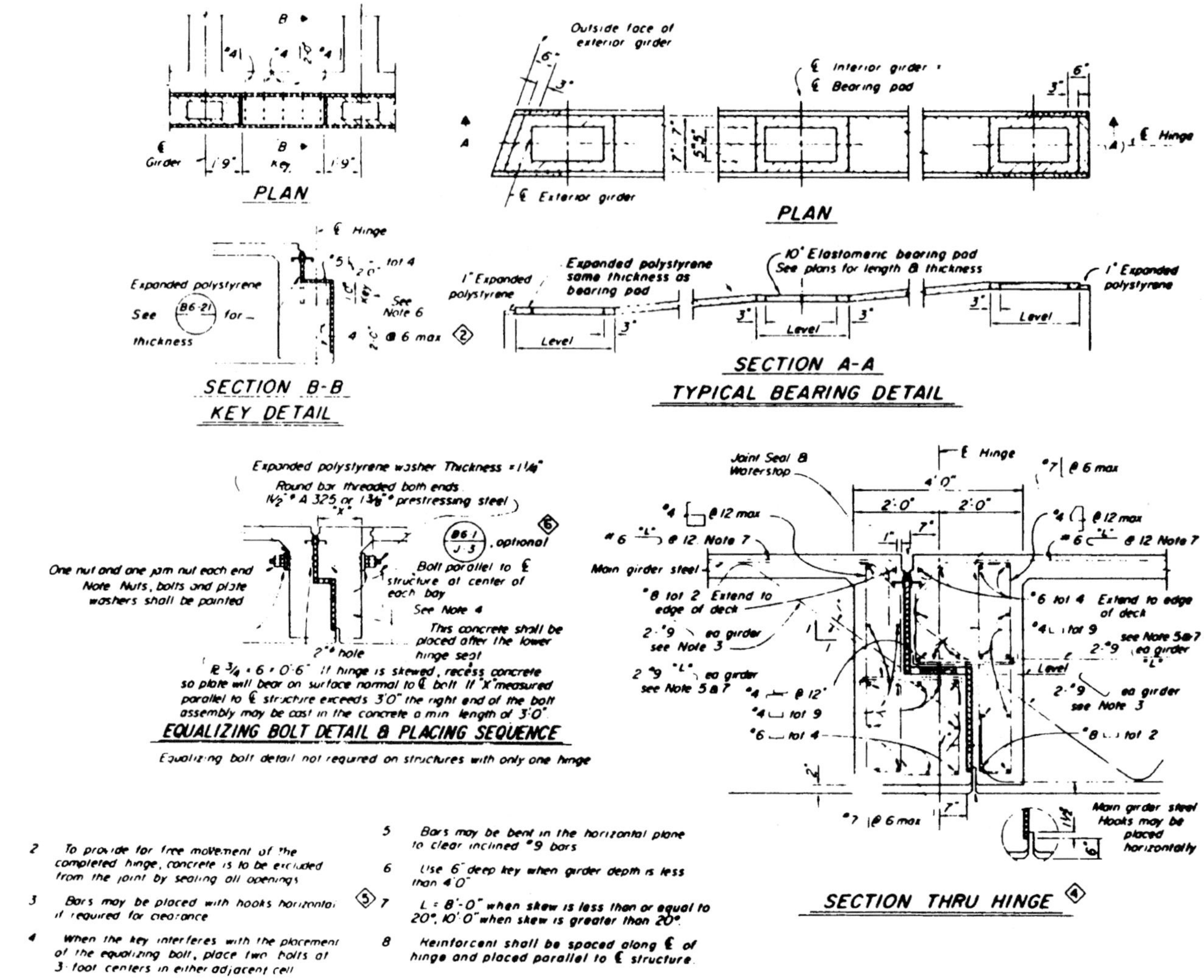

Figure 15.67 *(Continued)*

suggested for all box girders since with present allowable live-load distributions it is conservative.

Moments of inertia of the concrete gross cross section for one interior cell are given in Table 15.8. These values include the effect of 4-in. fillets (Fig. 15.76), but since they assume 8-in. stems, they do not account for web flare in regions of large shear. Straight-line interpolation can be used for intermediate values. If the moment of inertia of an entire superstructure is required, the moment of inertia of the exterior girder can be approximated at 72 to 78 percent of that of an interior girder. Values in this table can be used to compute deflections of spans of uniform depth and to determine fixed-end moments, distribution and carryover factors, etc. for variable-depth girders where the variation in moment of inertia approximates that assumed in the preparation of various charts and tables, for example, those of the Portland Cement Association's "Continuous Concrete Bridges." Weights of these cross sections are given in Table 15.9.

Top slab design is similar to that for the steel-stringer bridge (Sec. 15.3.1) and the T-beam bridge (Sec. 15.9.1), the effective span being determined as for the latter.

Distribution of Loads

The fraction of a wheel load carried by each interior girder is $S/7$, provided $S < 16$ ft, where S is the spacing of girders (AASHTO). For the exterior girder, S is replaced by W_e, the width of the exterior girder, which is the distance from the midpoint between girders to the outside edge of the slab. In the case of a bridge with sidewalks, this distribution of vehicular load ensures an exterior box adequate for future widening. Furthermore, sidewalk live load can be disregarded. If railing, curb, and sidewalk are to be placed after the slab is cured sufficiently, their weight can be assumed to be distributed equally among all the girders of the span.

Width of web at the supports is governed by shear. The web can be tapered from this width to the minimum practicable width, the length of flare being determined by the allowable shear for the minimum width. An example is shown in Fig. 15.74, where the effective depth d = 45 in. With the AASHTO allowable shear stress $v = v_c + 0.4\sqrt{f_c'}$, where $v_c = 0.95\sqrt{f_c'}$, the 8-in. web is just sufficient at 16 ft from the support, as shown in the following table.

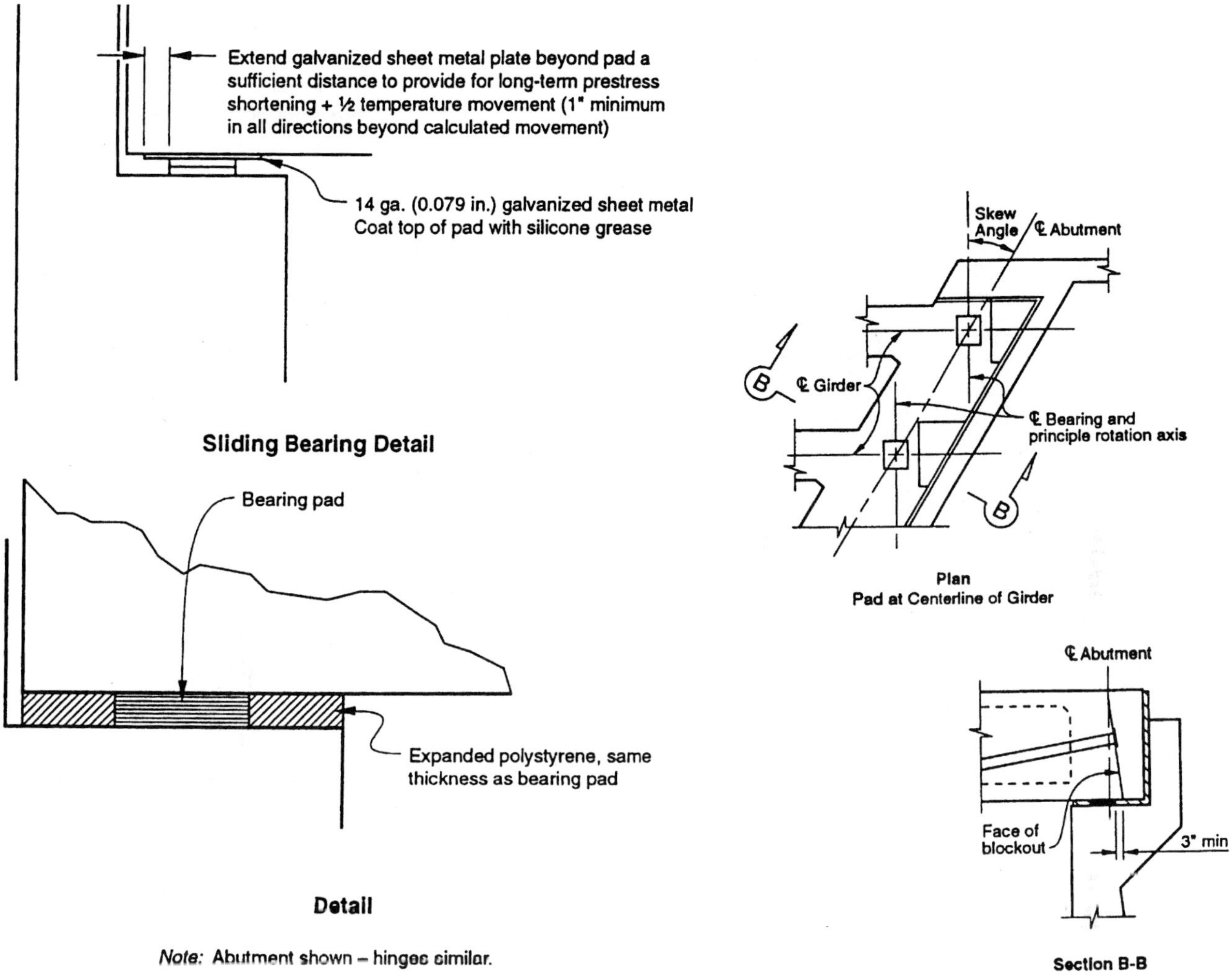

Figure 15.68 Bearing assemblies.

Spacing of No. 5 U Stirrups (Fig. 15.74)

Distance, ft	Web, in.	V, kips	v, psi	$v - v_c$	Spacing, in.
0	12	140	260	208	8.5
4	11	129	261	209	9
8	10	117	260	208	10
12	9	106	262	210	11
16	8	97	269	217	12
20	8	89	247	195	13.5
24	8	80	222	170	15.5

If the increase in width of stem does not exceed 2½ in. on each side, stirrups may be of uniform width.

Bottom Slab

The AASHTO specifications require a bottom flange thickness not less than 5½ in. or one-sixteenth the clear distance between webs, whichever is larger. However, in the regions of negative moment in shallow, continuous girders, flange thickness will be governed by stress.

Reinforcement not less than 0.4 percent of flange area is required in the longitudinal direction in the bottom flange. A single layer may be centered in the slab. Bar spacing must not exceed 18 in. These requirements can be met by distributing the girder tensile reinforcement throughout the slab or by adding steel if the girder reinforcement is concentrated near the web. Minimum transverse reinforcement in the bottom slab is 0.5 percent of flange area, distributed over both surfaces. The bars must be bent up into the exterior girder stems and anchored by a standard hook.

Diaphragms

The box-girder bridge is less dependent on diaphragms for transverse rigidity than is the T-beam bridge. However, diaphragms do improve torsional resistance of curved boxes. Diaphragms need only nominal reinforcement at the top and the bottom for support of stirrups, as the transverse steel in top and bottom slabs provides for their continuity at the girders. Nominal shear reinforcement, that is, No. 4 at 12 in., is ample except where utility openings are required. Both intermediate and end diaphragms having such openings should be proportioned for load, usually a truck wheel midway between girders.

Horizontally Curved Girders

The average length of span can be used for determining the frame constants for continuous, horizontally curved box girders. In designing

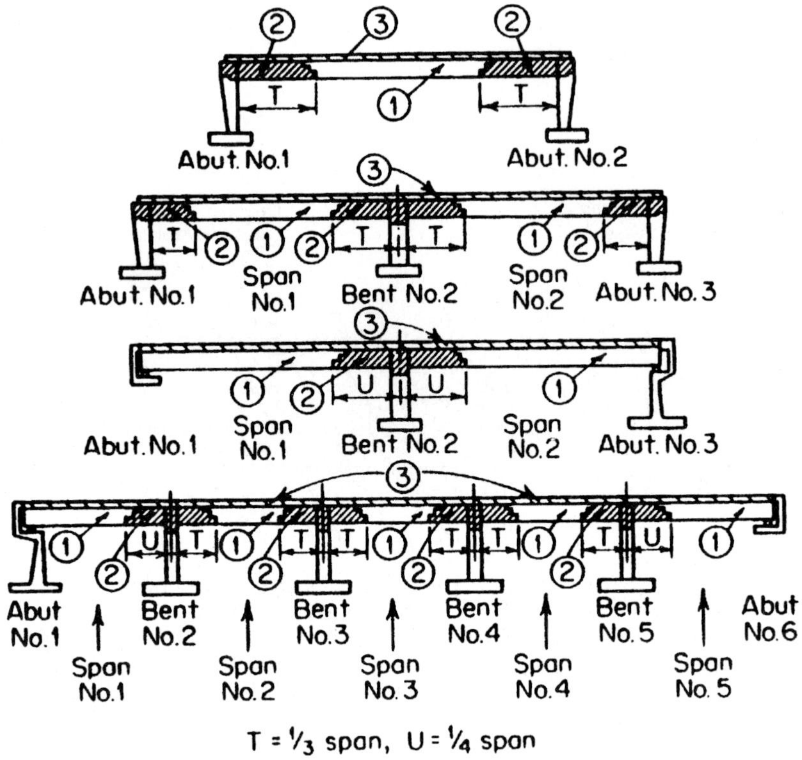

Figure 15.69 Concrete placing diagram: T-beam and box-girder bridge superstructures. Numbers 1 and 2 indicate sequence of placing girder-stem concrete; may be placed simultaneously provided sections 1 are placed in adjoining spans. Top slab concrete is to be placed separately in box-girder bridges and may be placed separately in T-beam bridges; may be placed continuously or in parts as approved by the engineer.

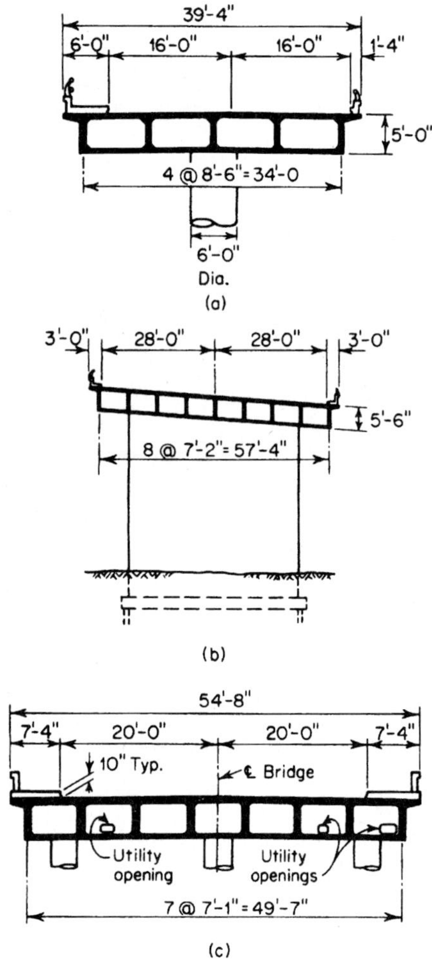

Figure 15.70 Box-girder bridge, typical sections.

the girders, two girders in a given span (e.g., the longest and the shortest interior girders) can be designed and the others proportioned from them by interpolation.

Dead load and live load both produce torsion in horizontally curved girders. Where the width of the box exceeds the height, as is usually the case in bridges, torsion can be neglected in continuous spans if the central angle between bents is less than 30 degrees. Various papers on the subject are available where torsion must be considered in the analysis.

Deflection

Dead-load deflection must be determined so that forms can be cambered. Because of creep, ultimate deflection usually develops in 4 or 5 years (Sec. 15.6.4). The moment of inertia of the gross concrete cross section may be used in computing deflection; the transformed section is an unnecessary refinement in view of the many uncertainties. Methods of computing deflection are discussed in the Portland Cement Association's ST Chart 70. A camber diagram for a three-span continuous box girder is shown in Fig. 15.75. Immediately after removal of falsework, deflections will be about one-fourth the given values.

Suggested concrete placing sequences are shown in Fig. 15.69. Midspan portions of the girders are cast first to avoid possible cracking at the bents, which could occur if these portions were placed last. Alternative procedures are suggested in the notes accompanying the figure.

A typical layout of a box-girder bridge superstructure is shown in Fig. 15.75. A typical, partial cross section is shown in Fig. 15.76, where only distribution and temperature reinforcement and those girder bars which are continuous are shown. The remaining girder main bars in both top and bottom slabs are shown in a girder-reinforcement layout, such as the one for a two-span continuous structure in Fig. 15.77. The continuous No. 8 bars in the stems just under the top-slab fillets in Fig. 15.76 are primarily for negative moments from possible falsework settlement before placement of the top slab.

Construction joints and miscellaneous details are shown in Fig. 15.78 and access openings in Fig. 15.79. End-bearing assemblies for T-beam bridges can also be used for box-girder bridges (Fig. 15.68).

15.10.4 Substructure

The same types of substructure can be used for box-girder bridges as for T-beam bridges. Both top and bottom of the column must be fixed in single-column bents for continuous structures. In multicolumn bents, either end, or both, may be fixed, depending upon requirements for seismic resistance, stability, stiffness, and provisions for changes in temperature. Structures of three or four spans can be planned so that all expansion is taken at the abutments, with longitudinal stability provided by the bents. Columns in bents farthest from the center of movement may be hinged in the longitudinal direction at the top of the footing to minimize temperature and live-load moments.

Caps in single- or multicolumn bents can be flush with the box-girder soffit. This simplifies forming and promotes economy. Opinions differ as to the aesthetics of this procedure, and some engineers prefer deeper caps. In any case, caps in single-column bents for wide bridges often must be deeper than the superstructure to satisfy stress requirements.

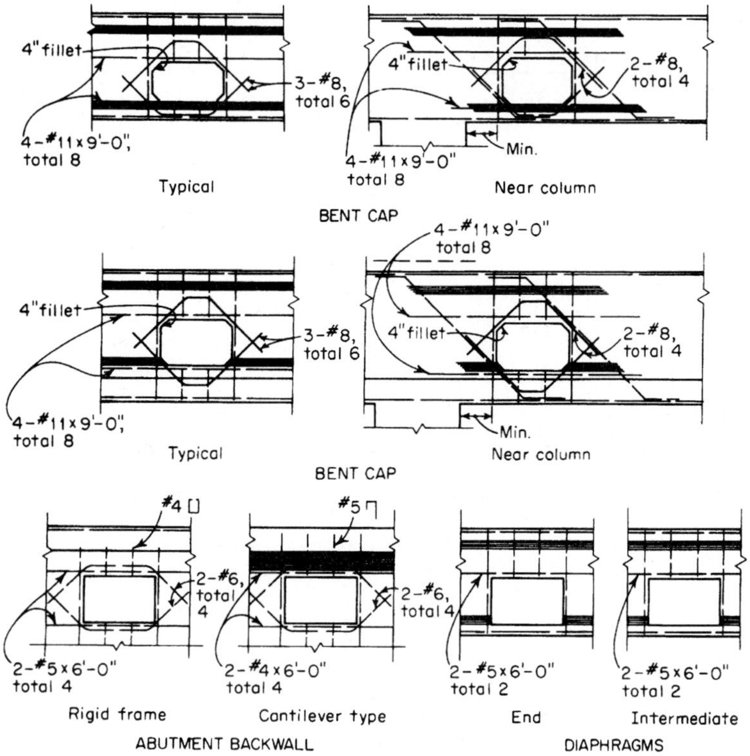

Figure 15.71 Utility opening details, box-girder bridge.

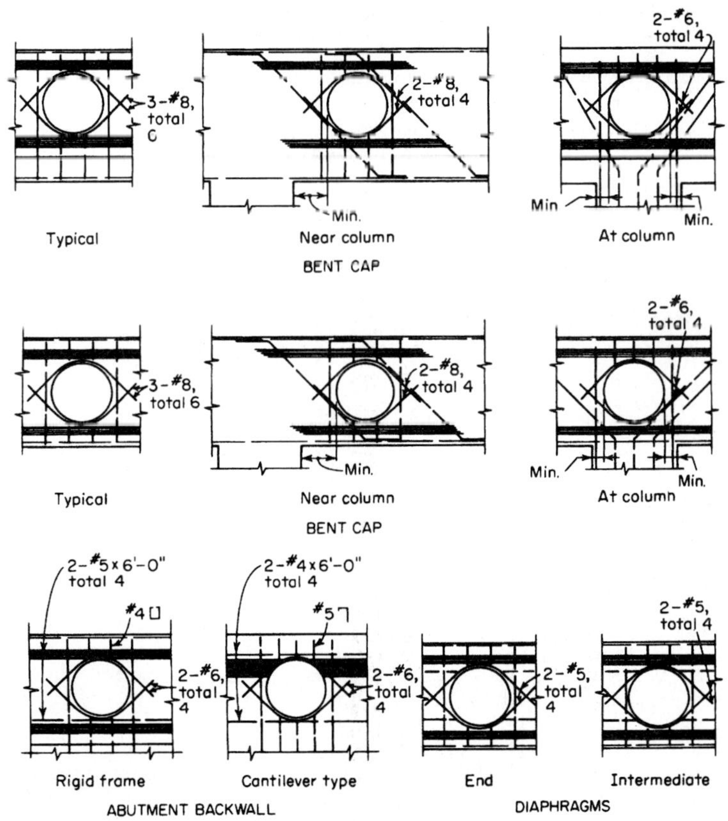

Figure 15.72 Utility opening details, box-girder bridge.

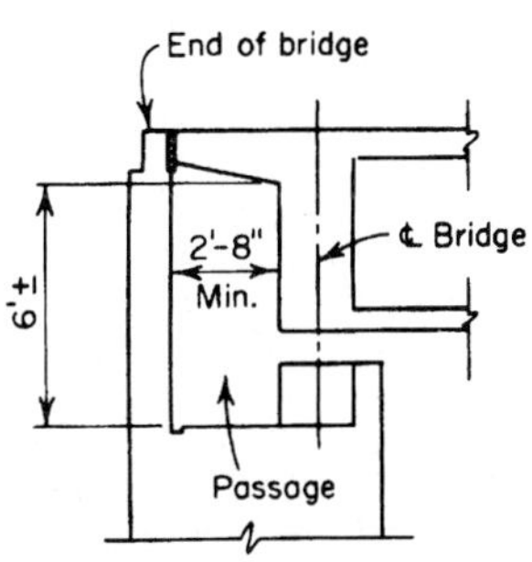

Figure 15.73 Utility access at abutment, box-girder bridge.

Table 15.8 Moment of Inertia of Interior Box Girders, 8-in. Stem, 4-in. Fillets

Top slab, in.	6	6	6	6⅛	6¼	6¼	6¼	6⅜	6½	6⅝	6¾	7
Bottom slab, in.	5½	5½	5½	5½	5½	5½	5½	5½	5½	5½	5¾	5⅞
Girder center to center, ft-in.	5–9	6–0	6–3	6–6	6–9	7–0	7–3	7–6	7–9	8–0	8–3	8–6
Depth, ft-in.	Moment of inertia, ft^4											
3–6	13.9	14.4	15.0	15.6	16.2	16.8	17.4	18.0	18.7	19.3	20.3	21.2
3–9	16.4	17.0	17.7	18.4	19.2	19.8	20.5	21.3	22.0	22.8	23.9	25.0
4–0	19.2	19.9	20.7	21.5	22.4	23.2	23.9	24.8	25.7	26.6	27.9	29.2
4–3	22.2	23.0	23.9	24.9	25.9	26.8	27.6	28.7	29.7	30.7	32.3	33.8
4–6	25.4	26.4	27.4	28.5	29.7	30.7	31.6	32.8	34.0	35.2	36.9	38.7
4–9	28.9	30.0	31.1	32.4	33.7	34.8	35.9	37.3	38.6	40.0	42.0	43.9
5–0	32.7	33.9	35.2	36.6	38.1	39.3	40.6	42.1	43.6	45.1	47.3	49.5
5–3	36.7	38.1	39.5	41.1	42.7	44.1	45.5	47.2	48.8	50.5	53.1	55.5
5–6	41.0	42.6	44.1	45.9	47.7	49.2	50.7	52.6	54.5	56.3	59.2	61.9
5–9	45.6	47.3	48.9	50.9	52.9	54.6	56.3	58.4	60.4	62.5	65.6	68.7
6–0	50.5	52.3	54.1	56.3	58.5	60.4	62.2	64.5	66.7	69.0	72.4	75.8
6–3	55.6	57.6	59.6	62.0	64.4	66.4	68.4	70.9	73.4	75.9	79.7	83.4
6–6	61.0	63.2	65.4	68.0	70.6	72.8	75.0	77.7	80.4	83.1	87.3	91.3
6–9	66.7	69.1	71.5	74.3	77.1	79.5	81.9	84.8	87.8	90.8	95.2	99.6
7–0	72.8	75.3	77.9	80.9	84.0	86.6	89.2	92.4	95.5	98.8	103.6	108.4
7–3	79.1	81.8	84.6	87.9	91.2	94.0	96.8	100.2	103.7	107.1	112.4	117.6
7–6	85.7	88.7	91.6	95.2	98.8	101.8	104.8	108.5	112.2	115.9	121.6	127.1
7–9	92.7	95.8	99.0	102.8	106.7	109.9	113.1	117.1	121.1	125.1	131.2	137.2
8–0	99.9	103.3	106.7	110.8	114.9	118.4	121.8	126.1	130.3	134.6	141.2	147.6
8–3	107.5	111.2	114.8	119.1	123.5	127.2	130.9	135.4	140.0	144.6	151.6	158.5
8–6	115.5	119.3	123.2	127.8	132.5	136.5	140.4	145.2	150.1	155.0	162.4	169.8
8–9	123.7	127.8	131.9	136.9	141.9	146.1	150.2	155.4	160.6	165.8	173.7	181.6
9–0	132.3	136.7	141.0	146.3	151.6	156.0	160.5	165.9	171.4	177.0	185.4	193.8
9–3	141.3	145.9	150.5	156.1	161.7	166.4	171.1	176.9	182.7	188.6	197.6	206.5
9–6	150.6	155.5	160.3	166.2	172.2	177.2	182.1	188.3	194.5	200.7	210.2	219.6

Table 15.9 Weight of Interior Box Girders, 8-in. Stem, 4-in. Fillets

Top slab, in.	6	6	6	6⅓	6¼	6¼	6¼	6⅜	6½	6⅝	6¾	7
Bottom slab, in.	5½	5½	5½	5½	5½	5½	5½	5½	5½	5½	5¾	5⅞
Girder center to center ft-in.	5–9	6–0	6–3	6–6	6–9	7–0	7–3	7–6	7–9	8–0	8–3	8–6
Depth, ft-in.	Weight, kips per ft											
3–6	1.11	1.15	1.19	1.23	1.28	1.31	1.35	1.40	1.45	1.49	1.57	1.64
3–9	1.14	1.17	1.21	1.26	1.30	1.34	1.38	1.42	1.47	1.52	1.59	1.67
4–0	1.16	1.20	1.24	1.28	1.33	1.36	1.40	1.48	1.50	1.54	1.62	1.69
4–3	1.19	1.22	1.26	1.31	1.35	1.39	1.43	1.47	1.52	1.57	1.64	1.72
4–6	1.21	1.25	1.29	1.33	1.38	1.41	1.45	1.50	1.55	1.59	1.67	1.74
4–9	1.24	1.27	1.31	1.36	1.40	1.44	1.48	1.52	1.57	1.62	1.69	1.77
5–0	1.26	1.30	1.34	1.38	1.43	1.46	1.50	1.55	1.60	1.64	1.72	1.79
5–3	1.29	1.32	1.36	1.41	1.45	1.49	1.53	1.57	1.62	1.67	1.74	1.82
5–6	1.31	1.35	1.39	1.43	1.48	1.51	1.55	1.60	1.65	1.69	1.77	1.84
5–9	1.34	1.37	1.41	1.46	1.50	1.54	1.58	1.62	1.67	1.72	1.79	1.87
6–0	1.36	1.40	1.44	1.48	1.53	1.56	1.60	1.65	1.70	1.74	1.82	1.89
6–3	1.39	1.42	1.46	1.51	1.55	1.59	1.63	1.67	1.72	1.77	1.84	1.92
6–6	1.41	1.45	1.49	1.53	1.58	1.61	1.65	1.70	1.75	1.79	1.87	1.94
6–9	1.44	1.47	1.51	1.56	1.60	1.64	1.68	1.72	1.77	1.82	1.89	1.97
7–0	1.46	1.50	1.54	1.58	1.63	1.66	1.70	1.75	1.80	1.84	1.92	1.99
7–3	1.49	1.52	1.56	1.61	1.65	1.69	1.73	1.77	1.82	1.87	1.94	2.02
7–6	1.51	1.55	1.59	1.63	1.68	1.71	1.75	1.80	1.85	1.89	1.97	2.04
7–9	1.54	1.58	1.61	1.66	1.70	1.74	1.78	1.82	1.87	1.92	1.99	2.07
8–0	1.56	1.60	1.64	1.68	1.73	1.76	1.80	1.85	1.90	1.94	2.02	2.09
8–3	1.59	1.63	1.66	1.71	1.75	1.79	1.83	1.87	1.92	1.97	2.04	2.12
8–6	1.61	1.65	1.69	1.73	1.78	1.81	1.85	1.90	1.95	1.99	2.07	2.14
8–9	1.64	1.68	1.71	1.76	1.80	1.84	1.88	1.92	1.97	2.02	2.09	2.17
9–0	1.66	1.70	1.74	1.78	1.83	1.86	1.90	1.95	2.00	2.04	2.12	2.19
9–3	1.69	1.73	1.76	1.81	1.85	1.89	1.93	1.97	2.02	2.07	2.14	2.22
9–6	1.71	1.75	1.79	1.83	1.88	1.91	1.95	2.00	2.05	2.09	2.17	2.24

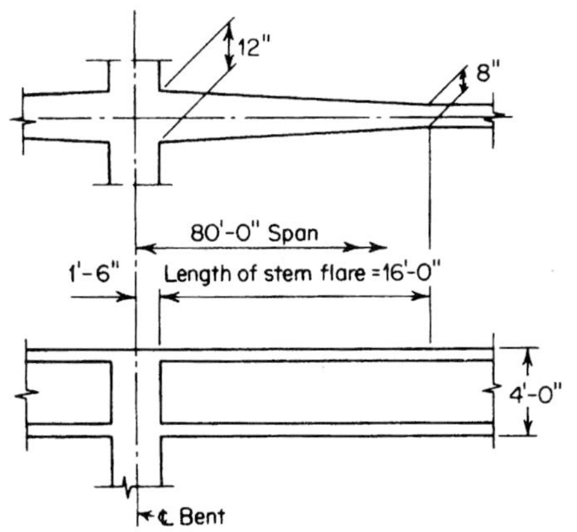

Figure 15.74 Flaring box-girder systems.

Torsion is not usually significant in single- or multicolumn-bent caps which are cast integrally with the superstructure of a box-girder bridge. For example, the critical loading for the cap of a single-column bent results with live load positioned for maximum bending moment in the cap at the face of the column; this produces little torsion in the cap. On the other hand, torsion is produced with live load positioned for maximum bending moment in the column in the longitudinal direction of the bridge. However, the resulting combination of bending and torsion in the cap is less critical than the maximum-moment condition.

EXAMPLE

Design the cap for the single-column bent of Fig. 15.80. Lane live load is positioned for a reaction which will maximize the moment about the column. The 100-kip exterior-girder dead load shown in the figure includes the weight of curb and railing. The moments are

$$LL + I = 9 \times 100 = 900$$
$$DL \text{ girder} = 7.5 \times 120 = 900$$
$$LL \text{ girder} = 15 \times 100 = 1500$$
$$DL \text{ cap} = 7.5 \times 65 = \underline{490}$$
$$3790 \text{ ft-kips}$$

For a rectangular section, with f_s = 20,000 psi and f_c = 1600 psi, and with tension reinforcement 6 in. below the top of the slab, the area of steel required is

$$A_s = \frac{M}{f_s jd} = \frac{12 \times 3790}{20 \times 0.88 \times 54} = 47.8 \text{ in.}^2$$

Use 12 No. 18 bars, A_s = 48 in.² Bar cutoffs can be determined from a moment diagram.

The allowable moment for the rectangular cross section with no reinforcement for compression is

$$M = \frac{f_c kjbd^2}{2 \times 12} = \frac{1.6 \times 0.375 \times 0.875 \times 72 \times 54^2}{2 \times 12} = 4593 \text{ ft-kips}$$

Since the resisting moment of the concrete exceeds the $DL + LL + I$ moment, no compression steel is required over the column.

Since the lane-loading concentrated load is 8 kips more for shear than for moment, the maximum shear is

Try No. 6 stirrups, $A_s = 2(0.44) = 0.88$ in.²

$$V_u = 605^k \quad V_n = 712k$$
$$V_c = 2\sqrt{f_c'} b_w d = 492^k$$
$$V_s = 712 - 492 = 220^k$$

Use No. 6 stirrups at 12 in.

Uniform spacing of 10 in. throughout the cap is suggested. Details of the reinforcement are shown in section *A-A* of Fig. 15.80.

COLUMNS

Although tied columns have a lower allowable stress than do spirally reinforced columns, they are suitable for most bridges. Of course, where higher allowable stresses are required or where maximum ductility is needed to resist seismic shaking, spirally reinforced columns are preferred. The 1993 AASHTO specifications provide confinement details and requirements for ties similar to those required for spirals. The choice of type is not necessarily related to shape. Thus, the circular column may be designed as a tied column, using either individual hoops or a spiral, while the square column may have a spiral meeting the requirements for spirally reinforced columns. In all these cases, the spirals or ties should extend into the bent cap and the footing.

All the various combinations of loading specified in AASHTO should be investigated. Both transverse and longitudinal moments must be combined with the appropriate direct load. For purposes of design, the length of a column supported on a spread footing may be measured from the neutral axis of the box girder to the point where the column extends into the footing one-third of the depth of the footing. For a footing founded on piles, the length is usually measured to a point below the footing whose location depends upon the lateral resistance of the soil.

Where foundation conditions permit, round pile columns 3 to 6 ft or more in diameter can be used economically. These consist of piles cast in place in drilled holes, with a reinforced-concrete column which extends the superstructure. An example of this type of construction is shown in Fig. 15.81. The piles must extend a minimum of five diameters below the surface of the ground (but not less than 20 ft) and may reach 60 ft or more below ground. Tests show that the section of maximum moment is usually within two diameters of the surface. The point of effective fixity for determining column stiffness is usually one or two pile diameters below the surface. It can be approximated by

$$d_f = \frac{\sqrt{Ph}}{Q_a D}$$

where d_f = distance below ground to point of effective fixity, ft
P = lateral load on column, kips
h = height of P above ground, ft
Q_n = allowable bearing value of the soil, tons
D = diameter of pile, ft

The following design loads can be used for pile columns in bridges:

Diameter, in.	Load, kips
36	300– 600
48	500–1000
60	700–1400
72	900–1800

The various load combinations (AASHTO) to be investigated for the design of the column of Fig. 15.81 are shown in Fig. 15.82. Load combinations IV, V, and VI are not tabulated because this column is at the center bent of a four-span structure for which temperature moments are practically zero. Group III loading, at 125 percent basic allowable stress, governs. Assuming a tied column, analysis shows that 44 No. 11 bars are required at the lower end. Since the column is also fixed at the upper end, the difference in required reinforcement is small, so all 44 bars are carried to the top. The bars can be cut off below ground as moment decreases. In this example, 22 bars are carried to the bottom of the pile.

15.11 PRESTRESSED-CONCRETE BRIDGES (See also Chap. 12 of this handbook)

15.11.1 Standard Sections

Standardization of prestressed-concrete bridge members has been established by several joint committees of the American Association of State Highway and Transportation Officials and the Prestressed

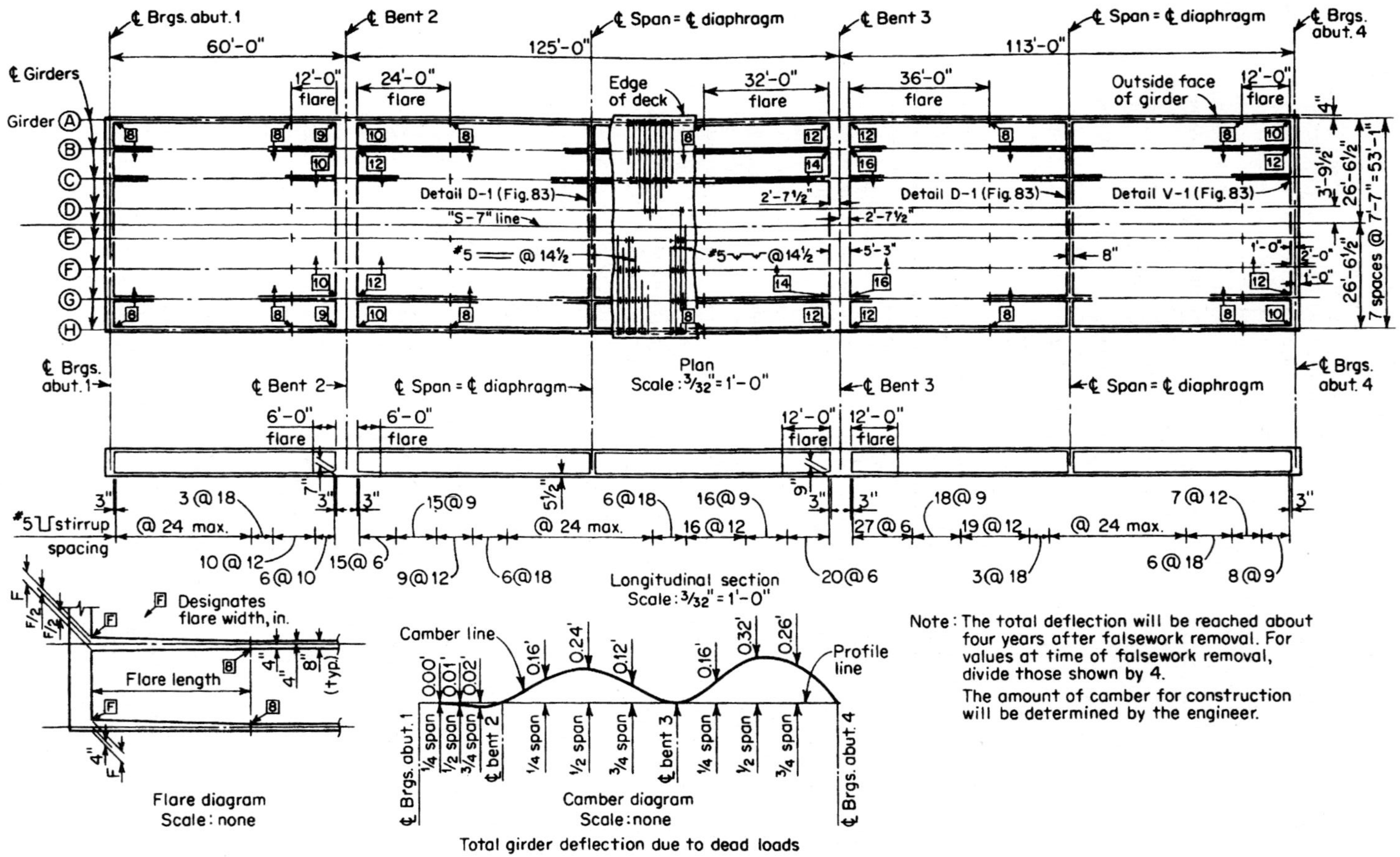

Figure 15.75 Typical layout, box-girder bridge.

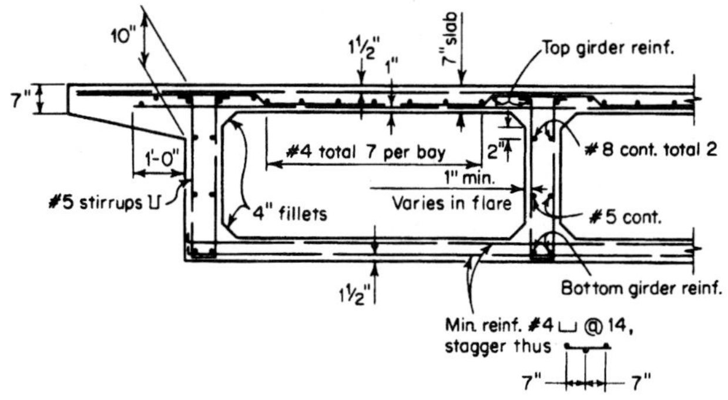

Figure 15.76 Box-girder typical section.

Concrete Institute. Standards were developed for the following types: slabs for spans to 55 ft (Fig. 15.83, Table 15.10), beams for spans 30 to 100 ft (Fig. 15.84, Table 15.11), and box beams for spans up to 103 ft (Fig. 15.85, Table 15.12). These sections may be either pretensioned or posttensioned, and any acceptable type of prestressing may be used, that is, straight or deflected strands, combination of pretensioning and posttensioning, and so forth. The span limits shown in the tables are approximate. They are based on 28-day compressive strength f'_c = 5000 psi, compressive strength at transfer of prestress f'_c = 4000 psi, and weight of concrete 150 pcf. Over 90 percent of the highway bridges built in the United States have spans within the range of these standards.

Slab Sections

Span limits for the standard slab sections are based on a 28-ft roadway and an allowance of 30 psf for surfacing. Straight tendons and AASHTO allowable stresses are assumed. The 15-in. (minimum) end block requires mild-steel reinforcement sufficient to resist tensile forces resulting from the concentrated prestress forces. Diaphragms are recommended at midspan for spans to 40 ft and at the third points for spans exceeding 40 ft. Lateral ties equivalent to a 1¼-in. mild-steel bar tensioned to 30,000 lb, or the equivalent in high-strength tendons, are required at each diaphragm. The shear keys are filled with high-strength, nonshrinking mortar after the ties have been tightened.

Beam Sections

Span limits for the standard beam sections are based on a cast-in-place deck 6 to 8 in. thick, with f'_c = 3000 psi. Dead load is assumed to be carried by the beams and live load by composite action of beams and slab. With straight prestressing, spacing of these beams at the lower limits of span will be 8 to 9 ft (four beams for a 28-ft roadway) for HS20–44 loading. At the upper limits of span, with a sufficient number of tendons draped to compensate for dead load of the beams and to maintain allowable stresses at the ends, spacing is about 4 ft 8 in. for beams I and II (seven beams for a 28-ft roadway) and 5 ft 6 in. (six beams for a 28-ft roadway) for beams III and IV.

End blocks are required for posttensioning of the standard beam sections. They should be as wide as the narrower flange of the beam, with a length equals to three-fourths the depth of the beam but not less than 24 in. A closely spaced grid of vertical and horizontal bars is required at the end face of the end block and throughout its length for posttensioned beams. End blocks are not required if all tendons are pretensioned wires or seven-wire strand. Vertical stirrups at 20,000 psi, sufficient to resist at least 4 percent of the total prestressing force, are required in the length $d/4$ at each end of the beam for pretensioned beams.

Either precast or cast-in-place diaphragms, with either prestressed or nonstressed reinforcement, are recommended at the ends of the span, at midspan for spans to 60 ft, and at the third points for spans longer than 60 ft.

Box Beams

Span limits for the standard box beams are based on a 28-ft roadway with an allowance of 30 psf for wearing surface. The 18-in. end blocks require mild-steel reinforcement sufficient to resist tensile forces resulting from the concentrated prestress forces. Diaphragms within the box are recommended at midspan for spans to 50 ft, at the third points for spans to 75 ft, and at the quarter points for spans longer than 75 ft. Lateral ties equivalent to a 1¼-in. mild-steel bar tensioned to 30,000 lb, or the equivalent in high-strength tendons, are required

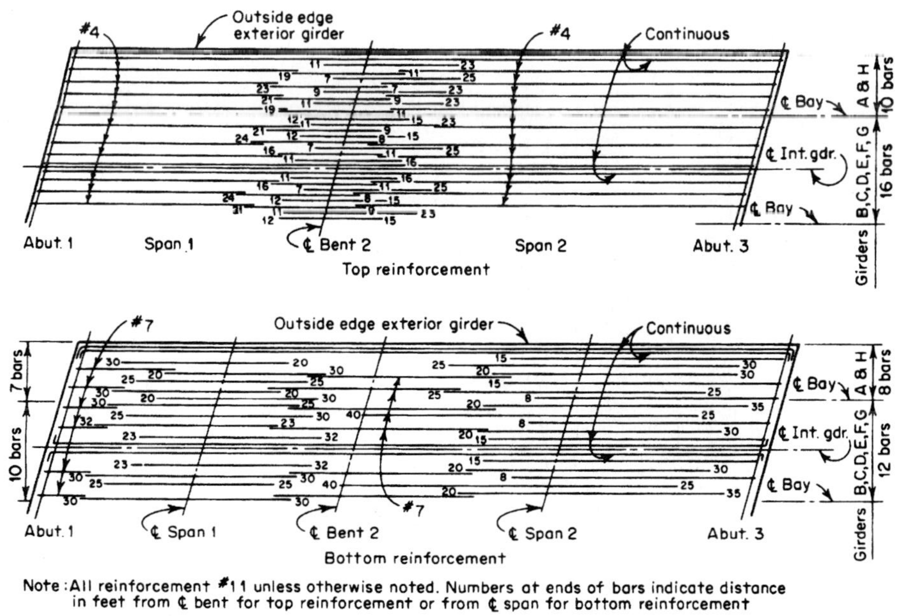

Figure 15.77 Typical reinforcement, box-girder bridge.

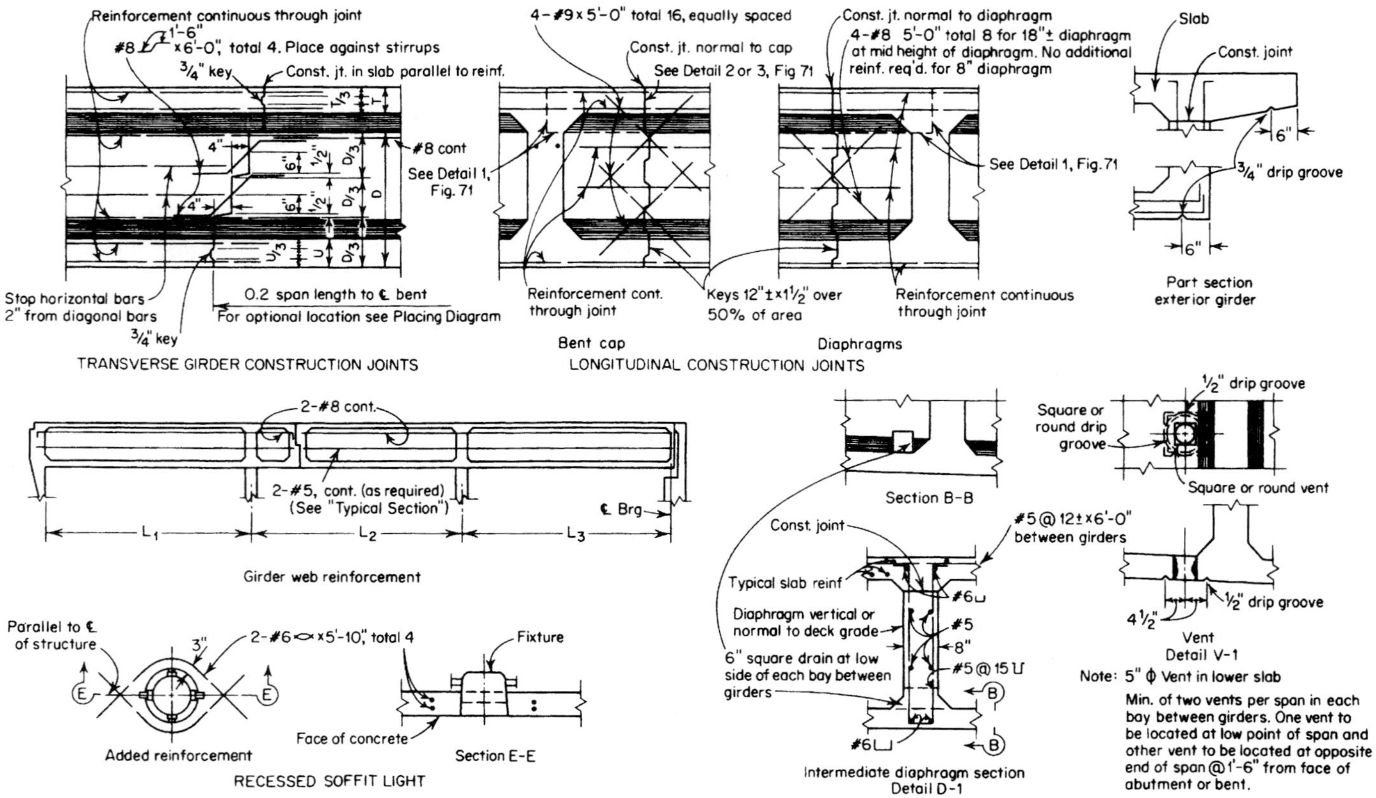

Figure 15.78 Typical details, box-girder bridge.

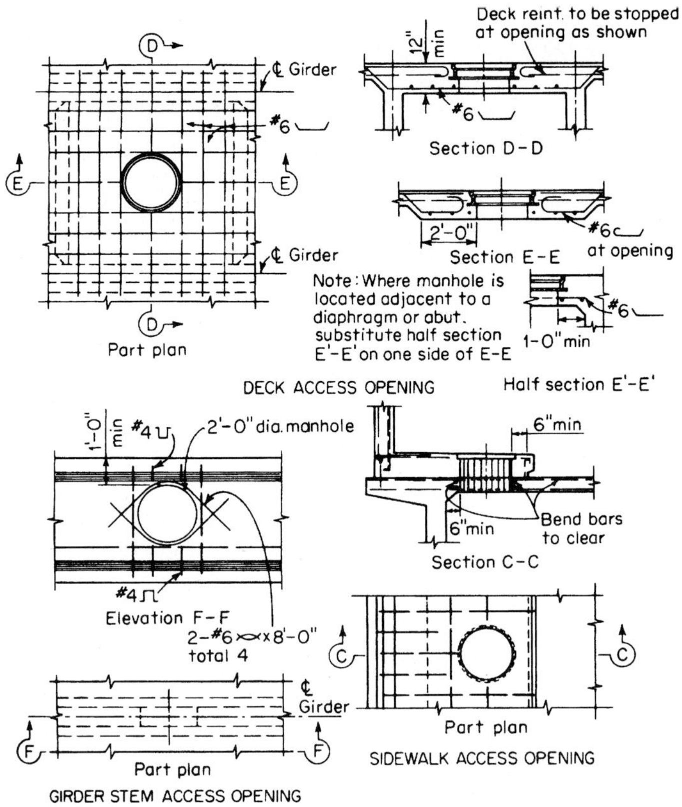

Figure 15.79 Access openings, box-girder bridge.

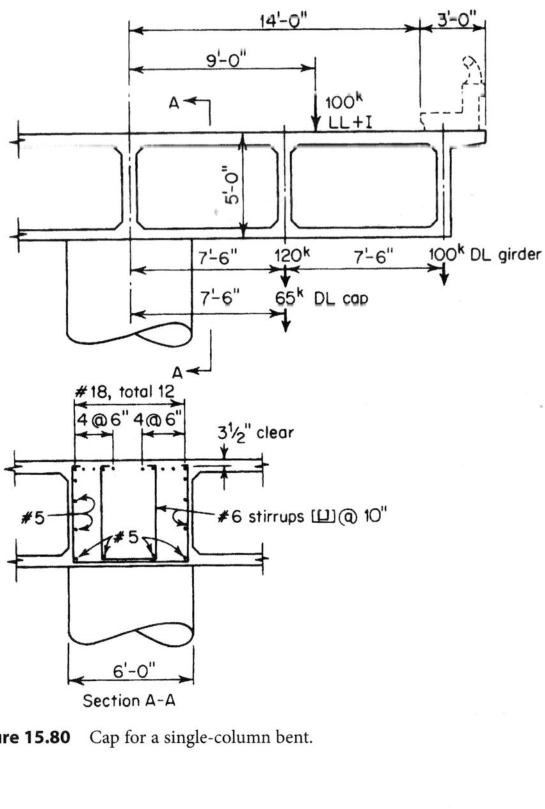

Figure 15.80 Cap for a single-column bent.

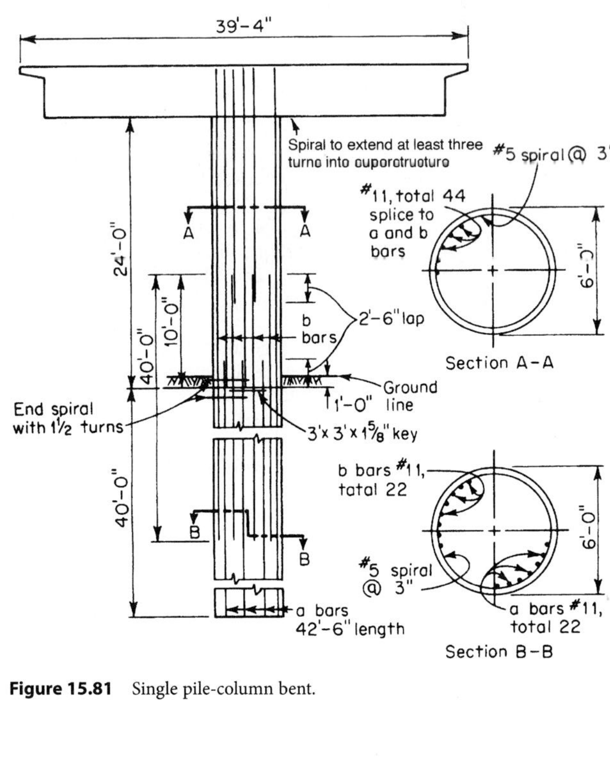

Figure 15.81 Single pile-column bent.

Loading Group		Longitudinal		Transverse		Axial	
I @ 100%		Top of col	Bot of col	Top of col	Bot of col	Top of col	Bot of col
1 DL		0	0	975	975	1,048	1,197
2 LL + I max. R.		0	0	1,640	1,640	205	205
	ΣM	0	0	2,615	2,615	1,253	1,402
II @ 125%	DL	0	0	975	975	1,048	1,197
Wind + overturning		0	0	652	2,001		
	ΣM	0	0	1,627	2,976	1,048	1,197
III @ 125%	Group I		0		2,615		1,402
Overturning					213		
	30%W				387		
	WL				325		
	ΣM				3,540		1,402
IV @ 125%	Group I						
	T						
	ΣM						
V @ 140%	Group II						
	T						
	ΣM						
VI @ 140%	Group III						
	T						
	ΣM						
VII @ 133⅓%	DL		0		975		1,197
	EQ		0		2,620		
	ΣM		0		3,595		1,197

Figure 15.82 Load combinations for the design of the column.

at each diaphragm. If adjacent units of the 39- and 42-in. sections are tied in pairs, one tie, centered between the bottom of the beam and the bottom of the key, is permitted. The shear keys are filled with high-strength, nonshrinking mortar after the ties are tightened.

Combining Methods of Prestress

Pretensioning and posttensioning can be used to advantage in combination in some cases. For example, a beam may be pretensioned for dead load only and lifted from the bed and subsequently posttensioned to complete the operation. This procedure allows a more continuous use of prestressing beds, which may result in economy.

15.11.2 Stresses

Concrete for prestressed bridges should be preferably not less than 5000 psi. The AASHTO specifications permit strengths up to 6000 psi, which, in exceptional cases, may be increased. While economy may be realized with higher-strength concretes, limitations imposed by availability of suitable materials must be considered. The allowable stresses of the AASHTO specifications are the following:

Notation.

f'_c = compressive strength of concrete at 28 days
f'_{ci} = compressive strength of concrete at time of initial prestress
f'_s = ultimate strength of prestressing steel
f°_y = yield-point stress of prestressing steel

Prestressing steel:

1. Temporary stress before loss due to creep and shrinkage $0.70f'_s$
2. Stress at service load after losses $0.80f^{\circ}_y$

Overstressing to $0.80f'_s$ for short periods of time may be permitted provided the stress, after transfer to concrete in pretensioning or seating of anchorage in posttensioning, does not exceed $0.70f'_s$.

Concrete:

1. Temporary stresses before losses due to creep and shrinkage:
   Compression
   Pretensioned members $0.60 f'_{ci}$
   Posttensioned members $0.55 f'_{ci}$
   Tension:
   Precompressed tensile zone: no temporary stress specified
   Tension areas with no bonded reinforcement 200 psi or $3\sqrt{f'_{ci}}$
2. Stresses at service load after losses have occurred.
   Compression $0.40f'_c$
   Tension (in precompressed tensile zone):
   (*a*) For members with bonded reinforcement $6\sqrt{f'_{ci}}$
   For severe corrosion conditions—coastal areas $3\sqrt{f'_{ci}}$
   (*b*) For members without bonded reinforcement 0
3. Cracking stress:
   For normal-weight concrete $7.5\sqrt{f'_c}$
   For sand-lightweight concrete $6.3\sqrt{f'_c}$
   For all other lightweight concrete $5.5\sqrt{f'_c}$
4. Anchorage bearing stresses
   Posttensioned anchorage at service load (but not to exceed $0.90f'_{ci}$) 3000 psi

Loss of prestress from all causes except friction may be assumed at 25,000 and 35,000 psi for posttensioned and pretensioned members, respectively. These losses consist approximately of the following (friction losses are discussed in Sec. 15.11.4):

Item	Posttension	Pretension
Concrete shrinkage	4000	6000
Concrete creep	11,000	11,000
Steel creep	5000	6000
Elastic shortening	—	12,000
Sequence stressing	5000	
Total, psi	25,000	35,000

While these specifications allow prestressing steel stresses of $0.70f'_s$ or $0.80f^{\circ}_y$, whichever is smaller, prestress losses will not always permit these values to be attained. For example, for wire with an ultimate strength f'_s = 250,000 psi and yield strength f_{sy} = 200,000 psi, the allowable stress is 160,000 psi. Deducting the prescribed prestress loss

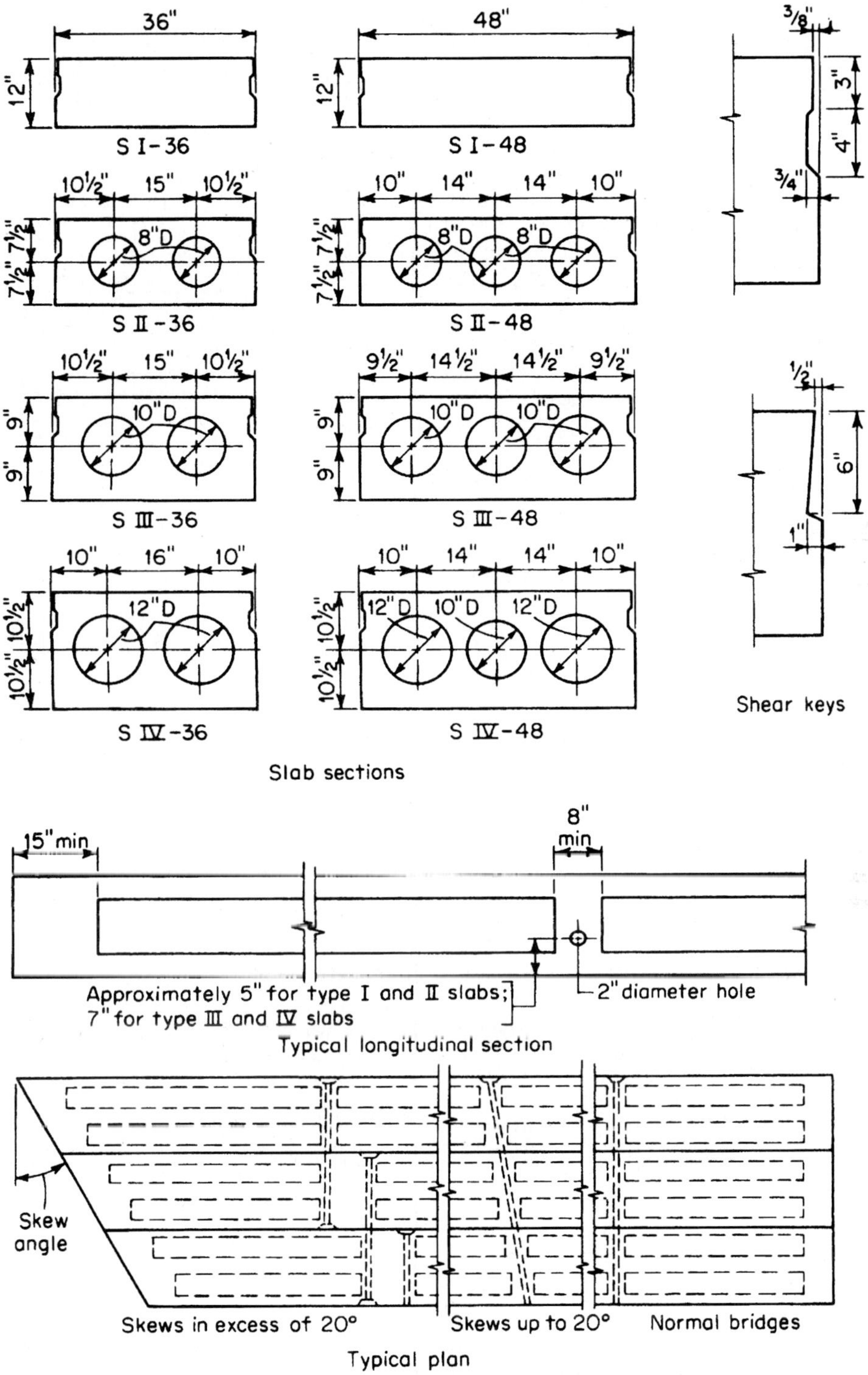

Figure 15.83 AASHTO-PCI prestressed-concrete slab sections. (See Table 15.10.)

Table 15.10 Properties of AASHTO-PCI Slab Sections (Fig. 15.83)

Section	Area, in.^{2}	Moment of inertia, in.^{4}	Section of modulus, in.^{3}	Range of span, ft	
				HS20-44	H20-44
SI-36	432	5180	864	20–28	20–29
SII-36	439	9720	1296	27–35	28–38
SIII-36	491	16,510	1835	34–42	37–46
SIV-36	530	25,750	2452	41–49	45–54
SI-48	576	6910	1152	20–28	20–29
SII-48	569	12,900	1720	27–35	28–38
SIII-48	628	21,850	2428	34–42	37–46
SIV-48	703	34,520	3287	41–50	45–55

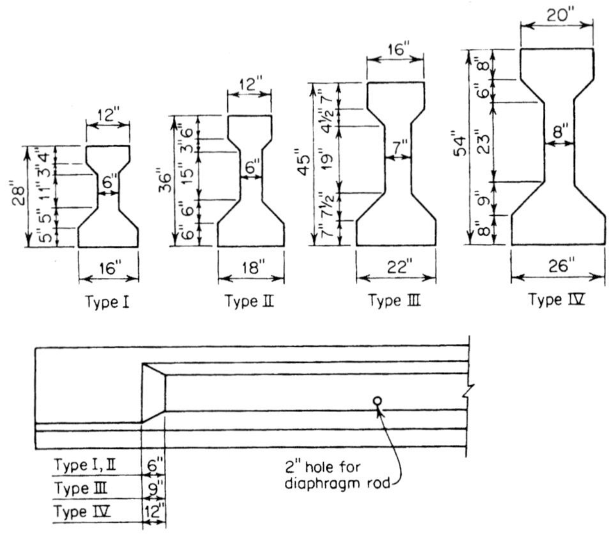

Figure 15.84 AASHTO-PCI prestressed-concrete beams. (See Table 15.11.)

Table 15.11 Properties of AASHTO-PCI Beam Sections (Fig. 15.84)

Section	Area, in.^{2}	Moment of inertia, in.^{4}	C_b, in.*	Range of span, ft
I	276	22,750	12.59	30–45
II	369	50,980	15.83	40–60
III	560	125,390	20.27	55–80
IV	789	260,730	24.73	70–100

*From center of gravity to bottom surface.

(excluding losses from friction) of 35,000 psi from the allowable initial stress of $0.7 \times 250{,}000 = 175{,}000$ psi leaves only 140,000 psi.

EXAMPLE

Design the prestressed composite concrete girder for the structure shown in Fig. 15.86. Live load HS20-44, design span 70 ft. Use a Type III AASHTO PCI beam (Fig. 15.84). Dimensions of the composite girder are shown in Fig. 15.87.

$$LL \text{ per girder} = \frac{7.5}{5.5} = 1.36 \text{ wheel lines}$$

$$I = \frac{50}{L+125} = \frac{50}{195} = 25.6 \text{ percent}$$

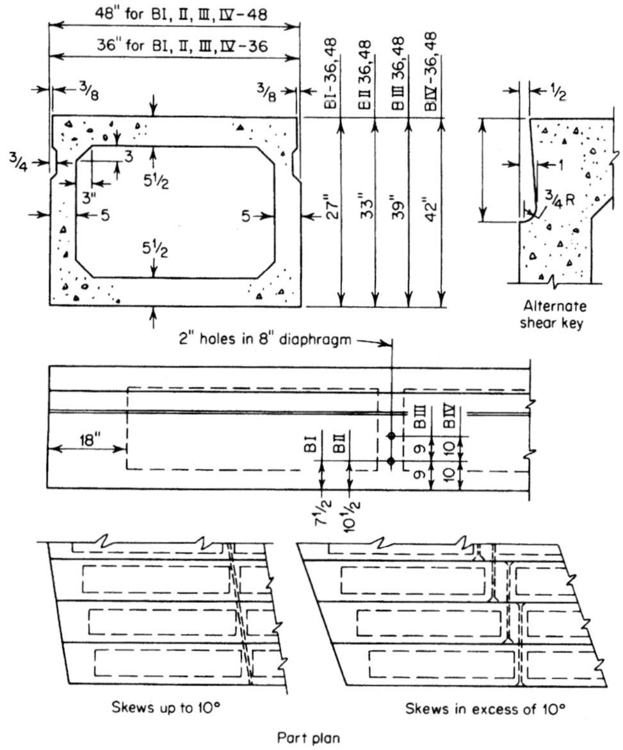

Figure 15.85 AASHTO-PCI prestressed-concrete box beams. (See Table 15.12.)

Table 15.12 Properties of AASHTO-PCI Box Sections (Fig. 15.85)

Section	Area, in.^{2}	Moment of inertia, in.^{4}	c_b, in.*	Span limit, ft	
				Draped strand	Straight strand
BI-36	560	50,330	13.35	74	62
BI-48	692	65,940	13.37	73	63
BII-36	620	85,150	16.29	86	73
BII-48	752	110,500	16.33	86	74
BIII-36	680	131,140	19.25	97	83
BIII-48	812	168,370	19.29	96	73
BIV-36	710	158,640	20.73	103	87
BIV-48	842	203,090	20.78	103	88

*From center of gravity to bottom surface.

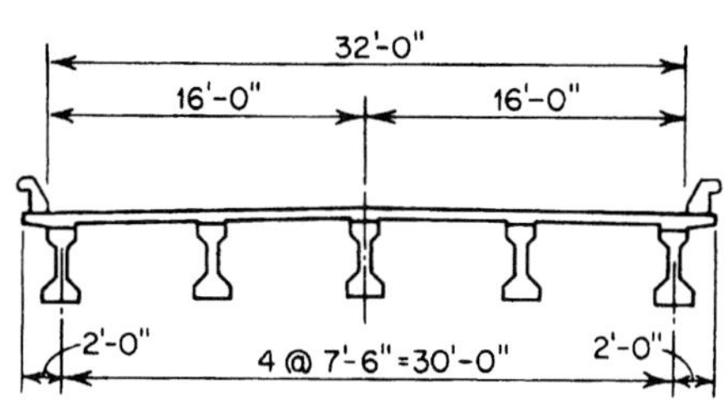

Figure 15.86 Typical section of 70-ft composite prestressed girder bridge.

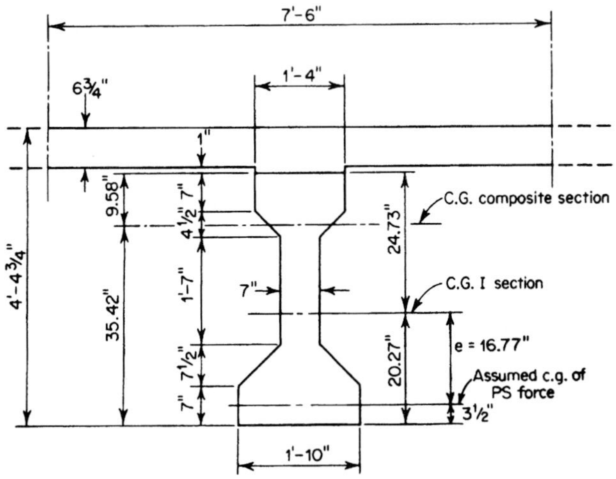

Figure 15.87 Interior girder of 70-ft composite prestressed girder bridge.

DECK SLAB Design procedure for the slab is the same as that for the slab of Sec. 15.3.1.

Girder:

DL girder (Table 15.11) 560 × 150/144	= 585 plf
DL slab (0.563 × 7.5 + 0.083 × 1.33)150	= 655 plf
DL curb and rail	= 100 plf
Girder *DL* moment = 0.585 × 70²/8	= 360 ft-kips
Slab *DL* moment = 0.655 × 70²/8	= 400 ft-kips
Curb and rail DL moment = 0.100 × 70²/8	= 61 ft-kips
LL + *I* moment (Table 15.2) = 1.36 × 1.26 × 986/2	= 845 ft-kips

Properties of I section (Table 15.11):

$$A = 560\text{ in.}^2 \qquad y_b = 20.27\text{ in.} \qquad y_t = 24.73$$

$$I = 125{,}390\text{ in.}^2 \qquad \frac{I}{y_b} = 6185\text{ in.}^3 \qquad \frac{I}{y_t} = 5070\text{ in.}^3$$

$$r^2 = 224\text{ in.}^2 \qquad \frac{ey_b}{r^2} = 0.0905e \qquad \frac{ey_t}{r^2} = 0.1104e$$

Properties of composite T (Fig. 15.87):

$$\text{Effective flange} = 12 \times 6\tfrac{3}{4} + 16 = 97\text{ in.}$$

$$= 90\text{ in. center-to-center girders}$$

$$\text{Flange area} = 90 \times 6.75 = 608\text{ in.}^2$$

$$\text{Flange moment of inertia} = 90 \times \frac{6.75^3}{12} = 2310\text{ in.}^4$$

A	y	Ay	I	
I-girder	560	20.27	11,350	125,390
Flange	608	49.38	30,023	2310
	1168		41,373	127,700

$$y_b = \frac{41{,}373}{1168} = 35.4\text{ in.}$$

$$\text{I}: Ad^2 = 560 \times 15.15^2 = 128{,}530$$

$$\text{Flange}: Ad^2 = 608 \times 13.96^2 = \underline{118{,}490}$$

$$I_{cgc} = 247{,}020 + 127{,}700 = 374{,}720$$

$$y_t = 45 - 35.4 = 9.6\text{ in.}$$

$$\frac{I}{y_t} = \frac{374{,}720}{9.6} = 39{,}030\text{ in.}^3$$

$$\frac{I}{y_b} = \frac{374{,}720}{35.4} = 10{,}585\text{ in.}^3$$

STRESSES IN I-SECTION

	Top fiber	Bottom fiber
Girder DL	$12\times\frac{360{,}000}{5070}=850$	$12\times\frac{360{,}000}{6185}=700$
Slab DL	$12\times\frac{400{,}000}{5070}=950$	$12\times\frac{400{,}000}{6185}=775$
LL + I	$12\times\frac{845{,}000}{10{,}585}=260$	$12\times\frac{845{,}000}{10{,}585}=960$
Curb and rail	$12\times\frac{61{,}000}{31{,}030}=20$	$12\times\frac{61{,}000}{10{,}585}=70$
	$f_t = 2080$	$F_b = 2505$

EFFECTIVE PRESTRESSING FORCE AFTER LOSSES. Estimated value of $e = 20.27 - 3.5 = 16.77$.

$$\frac{ey_t}{r^2} = 0.1104 \times 16.77 = 1.85$$

$$\frac{ey_b}{r^2} = 0.0905 \times 16.77 = 1.52$$

$$P_f = \frac{f_b A}{ey_b/r^2 + 1} = \frac{2505 \times 560}{2.52} = 556{,}700\text{ lb}$$

AREA OF STEEL. Assume bar $f_s' = 145{,}000$ psi.

Allowable design load stress = 0.6 × 145,000 = 87,000 psi

Allowable initial stress = 0.7 × 145,000 = 101,500 psi

Less losses = <u>25,000</u>

Allowable stress = 76,500

$$A_s = \frac{556{,}700}{76{,}500} = 7.28\text{ in.}^2$$

INITIAL PRESTRESSING FORCE

$$P_i = 101{,}500 \times 7.28 = 739{,}000\text{ lb}$$

$$\frac{P_i}{A} = \frac{739{,}000}{560} = 1320\text{ psi}$$

$$\frac{P_f}{A} = \frac{556{,}700}{560} = 990\text{ psi}$$

STRESSES IN CONCRETE

$$f_b = \frac{P}{A}\left(1 + \frac{ey_b}{r^2}\right) = 2.52\frac{P}{A} \qquad f_t = \frac{P}{A}\left(1 - \frac{ey_t}{r^2}\right) = 0.85\frac{P}{A}$$

At time of prestressing (+ = compression)

$$f_b = 1320 \times 2.52 - 700 = +2630\text{ psi}$$

$$f_t = -1320 \times 0.85 + 850 = -270\text{ psi}$$

At time slab is placed

$$f_b = 990 \times 2.52 - 1475 = +1020 \text{ psi}$$
$$f_t = -990 \times 0.85 + 1800 = +960 \text{ psi}$$

At design load

$$f_b = 990 \times 2.52 - 2.505 = -10 \text{ psi}$$
$$f_t = -990 \times 0.85 + 2080 = +1240 \text{ psi}$$

Maximum allowable temporary compression for posttensioning = $0.55 f'_{ct}$

$$\text{Required } f'_{ci} = \frac{2630}{0.55} = 4780 \text{ psi}$$

$$\text{Maximum allowable design compression} = 0.40 f'_{ci}$$

$$\text{Required 28-day compressive strength} = \frac{1240}{0.40} = 3100 \text{ psi}$$

$$\text{Maximum allowable tension in concrete} = 3\sqrt{f'_{ci}} = 3\sqrt{4780} = 210 \text{ psi}$$

At time of prestressing, the tension is 270 psi, so mild-steel reinforcing at 20,000 psi should be provided to resist the total tension (Fig. 15.88).

$$T = 270 \times 4.4 \times \frac{16}{2} = 9500 \text{ lb}$$

Use two No. 5 bars.

15.11.3 Path of Prestressing Force

In simple-span girders it is desirable for the tendons to follow a parabolic path. Minimum prestressing steel results when it is placed as near as possible to the bottom of the member. The effective depths that can be achieved with various prestressing systems vary. If a system is not specified, the prestressing force required for several different eccentricities can be shown on the plans—usually, 3, 4, and 5 in. above the bottom of the girder will cover all systems. At the ends of the span the tendons should be at or near the center of gravity of the beam or, at least, within the kern to eliminate tensile stresses at the beam ends.

15.11.4 Friction Losses

Friction losses in posttensioned members occur with angle changes in draped cables and from wobble of the ducts. According to AASHTO these losses can be accounted for by the formula

$$T_0 = T_x e^{(KL + \mu\alpha)}$$

or, if $KL + \mu\alpha \leq 0.3$, the following may be used:

$$T_0 = T_x(1 + KL + \mu\alpha)$$

where μ = friction curvature coefficient
α = angle between T_0 and T_1, rad

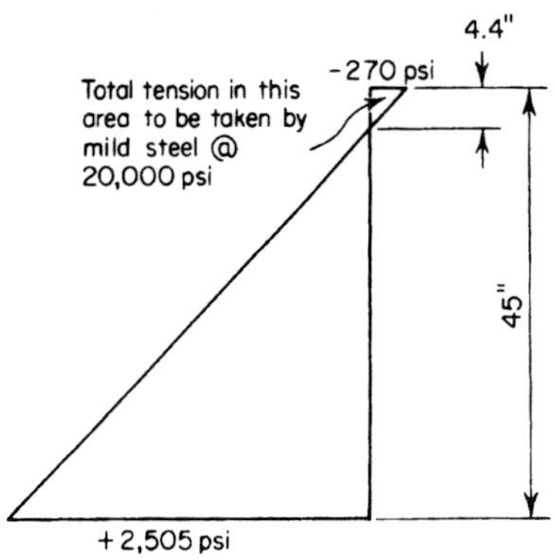

Figure 15.88 Concrete tension at time of prestressing.

Table 15.13 Values of K and μ for Friction Losses*

	Wire or ungalvanized strain		High-strength bars	
	K	μ	K	μ
Bright-metal sheathing	0.0020	0.30	0.003	0.20
Galvanized-metal shielding	0.0015	0.25	0.002	0.15
Galvanized rigid	0.0002	0.25		
Greased or asphalt-coated and wrapped	0.0020	0.30		

*From Ref. 3. See also Chap. 13, Table 13.9.

K = wobble coefficient per ft
T_0 = steel stress at jacking end
T_x = steel stress at any point x

Values of K and μ recommended by AASHTO are given in Table 15.13. A nomograph for the solution of the first of these equations is given in Fig. 15.89. For parabolic tendons, the angle α between the tangent at the vertex and the tangent at any point x from the vertex can be found from $\alpha = \tan\alpha = 2y/x$, where y = rise of the tendon in the distance x.

EXAMPLE

Determine the jacking force required for the girder of Sec. 15.11.2. Assume that jacking will be done from one end. Average values of K and R for high-strength bars in bright-metal sheathing are 0.0003 and 0.20, respectively (Table 15.13). With the tendon 3½ in. from the bottom at midspan and passing through the center of gravity of the I at each end, the rise is 20.27 − 3.50 = 16.77 in. (Fig. 15.87). The angle α between tangents at each end is

$$\alpha = 2 \times \frac{y}{x} = 2 \times \frac{16.77}{35 \times 12} = 0.08 \text{ rad}$$

$$\mu\alpha + KL = 0.20 \times 0.08 + 0.0003 \times 70 = 0.0370$$

$$e^{\mu a + KL} = 1.038$$

The allowable initial tension is T_i = 101.5 ksi (Sec. 15.11.1)

$$T_0 = 101.5 \times 1.054 = 107 \text{ ksi}$$

Allowable jacking tension = 0.75 × 145 = 109 ksi

$$T_{av} = 101.5 \times \frac{0.054}{0.053} = 103.7 \text{ ksi}$$

$$\text{Bar elongation} = \frac{T_{av}L}{E} = \frac{103.7 \times 70 \times 12}{24{,}000} = 3.6 \text{ in.}$$

Line 3 in Fig. 15.89 shows the nomographic determination of T_0 and T_{av}. The bar length was assumed to be 70 ft in calculating elongation. The draped length can be approximated by $L[1 + ⅔(2d/L)2]$, where d is rise of the tendon.

15.11.5 Ultimate Load

The AASHTO specifications require that prestressed beams on simple spans of moderate length have an ultimate load capacity not less than $1.3[(DL + 1.67(LL + I]. I)]$. The ultimate resisting moment M_u of under-reinforced beams can be closely approximated by $M_u = 0.9\ A_s f'_s d$. For the beam of Sec. 15.11.1,

$$M_u = 0.9 \times 7.28 \times 145 \times \frac{49.25}{12} = 3900 \text{ ft-kips}$$

The required ultimate moment is

$$1.3(760 + 1.67 \times 845) = 2822 < 3900 \text{ ft-kips}$$

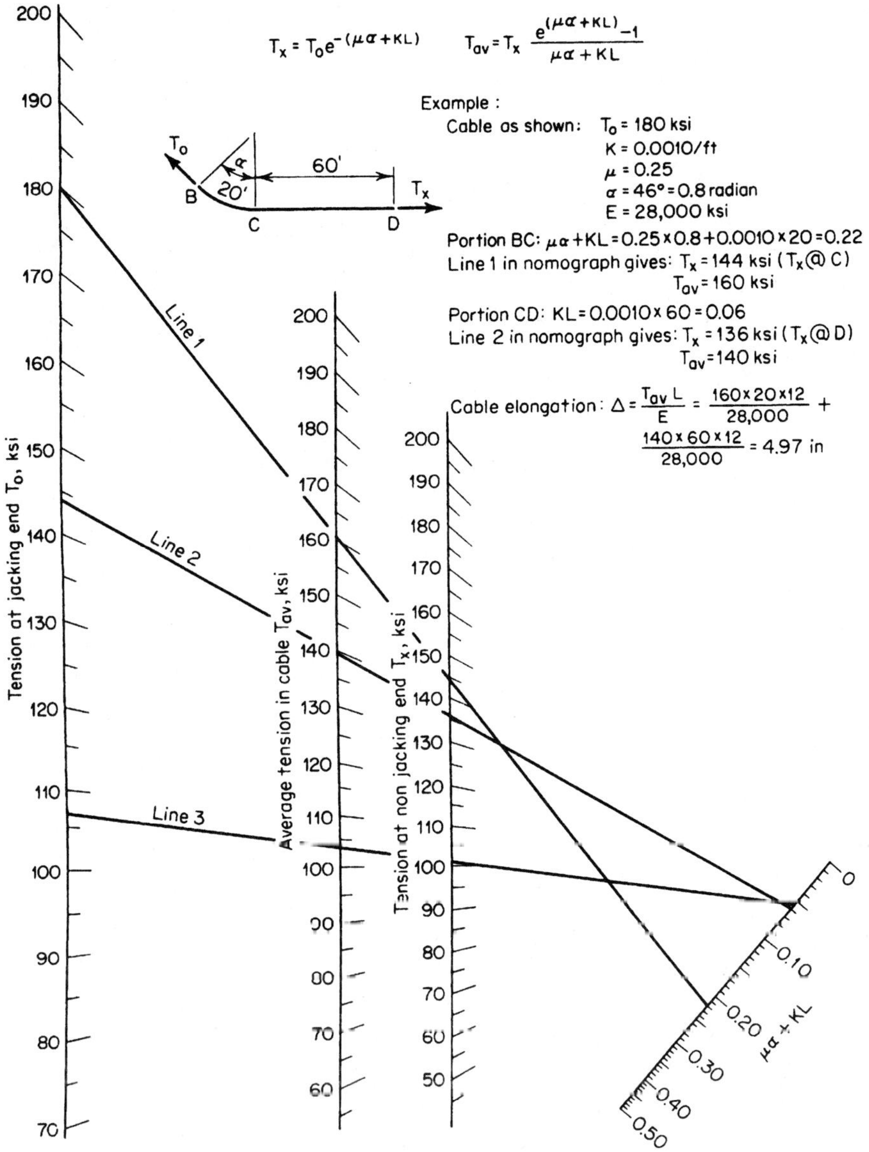

Figure 15.89 Nomograph for frictional loss in prestress.

15.11.6 Web Reinforcement

The 1983 AASHTO specifications revised the treatment of shear in prestressed beams. The new shear formulas recognize that prestressed members must resist shear from two different sources: the inclined flexure shear resulting from the live and dead loads and the high web shear generated by the large prestressing load. They also recognize that the shear may be critical near the ends of the beam. The new formulas result in some very complex computations, which should be made for larger structures. AASHTO also permits the use of the simpler method for design of web reinforcement presented in the 1979 Interim Specifications, so for this rather short-span I-beam example, the older specification will be used.

EXAMPLE

For the beam of Sec. 15.11.1: Where the prestressing steel is harped at the ⅓ points, there will be no reduction in V_u over the center third of the span. Throughout the end thirds of the span the upward force due to the prestressing is approximately equal to $P_f \times$ slope of CGS line:

$$V_p = \frac{P_f \times e/12}{L/3} = \frac{557 \times 16.77/12}{70/3} = 33.4 \text{ kips}$$

SHEAR

DL weight of girder, slab, curb, and rail = 1.34 kips/ft

$$V_{DL} = 1.34 \times 35 = 46.9 \text{ kips at support}$$
$$= 0 \text{ at midspan}$$

$$\text{Impact} = \frac{50}{70+125} = 0.26 \quad \text{Wheel lines (one lane) per girder} = \frac{7.5}{11}$$

$$V_{LL+I} = 32\times\left(70+56+\frac{42}{4}\right)\times\frac{1.26}{70}\times\frac{7.5}{11} = 53.6 \text{ kips at support}$$

$$= 32\times\left(35+21+\frac{7}{4}\right)\times\frac{1.26}{70}\times\frac{7.5}{11} = 22.7 \text{ kips at midspan}$$

$$V_u = \frac{1.3}{\phi}\left[DL+\frac{5}{3}(LL+I)\right] = \frac{1.3}{0.9}\left(46.9+\frac{5}{3}\times 53.6\right) = 196.8 \text{ kips at support}$$

$$= \frac{1.3}{0.9}\left(0+\frac{5}{3}\times 22.7\right) = 54.7 \text{ kips at midspan}$$

Shear resistance of an unreinforced beam = $V_c = 0.06f'_c\, b'jd$ but not more than $180b'jd$, which in this case governs.

$$V_c = \frac{180\times 7\times 0.9\times 0.8\times 52.75}{1000} = 47.8 \text{ kips} \qquad d = 0.8h$$

$$\text{Web reinforcement} = A_v = \frac{(V_u - V_c)s}{2f_{sy}jd} \text{ but not less than } \frac{100b's}{f_{sy}}$$

Use No. 4 stirrups: $A_v = 0.40$ in.^{2}

$$\text{Max } s = \frac{A_v f_{sy}}{100b'} = \frac{0.40\times 40{,}000}{100\times 7} = 22.8$$

Max s must not exceed $0.75h = 0.75h = 0.75 \times 51.25 = 38.4$ in.

However, a practical maximum spacing of 20 in. is recommended. Then, expressing the shear in terms of the spacing,

$$(V_u - V_c) = \frac{2A_v f_{sy} jd}{s} = \frac{2\times 0.4\times 40\times 0.9\times 42.2}{s} = \frac{1215}{s}$$

Choose stirrup spacings of 10, 15, and 20 in. and solve for the shear cutoff points:

s, in.	$V_u - V_c$, kips
10	121
15	81
20	61

The stirrup layout is plotted on the shear diagram (Fig. 15.90). Although the allowable spacing is 23 in. and the shear diagram indicates a 12-in. spacing at the end, experience has shown that the end block requires closely spaced vertical reinforcement to resist the formation of horizontal cracks which may develop because of the nonuniform distribution of the prestressing forces. It is good practice to provide a gradual transition from the close spacing of the end block to the wider spacing in the girder. Therefore, the theoretical stirrup spacing is modified as shown in Fig. 15.90. Details of the girder reinforcement are shown in Fig. 15.91.

To complete the investigation for shear, principal tensile stresses should be computed at the bottom of the top fillets at the end of the end block and at the quarter point (Sec. 12.5.1). These principal tensions must not exceed $0.03\sqrt{f'_c}$ at design load or $0.08\sqrt{f'_c}$ at ultimate load.

15.11.7 Uplift

Uplift due to prestressing can be computed by the formula (Fig. 15.92)

$$\Delta_p = \frac{5Pe_2L^2}{48EI} + \frac{Pe_1L^2}{8EI}$$

where e_1 is positive as shown in the figure. For the beam of Sec. 15.11.1, $e_1 = 0$, and the upward deflection is

$$\Delta_p = \frac{5\times 739{,}000\times 16.75\,(12\times 70)^2}{48\times 3{,}8000{,}000\times 127{,}700} = 1.87 \text{ in.}$$

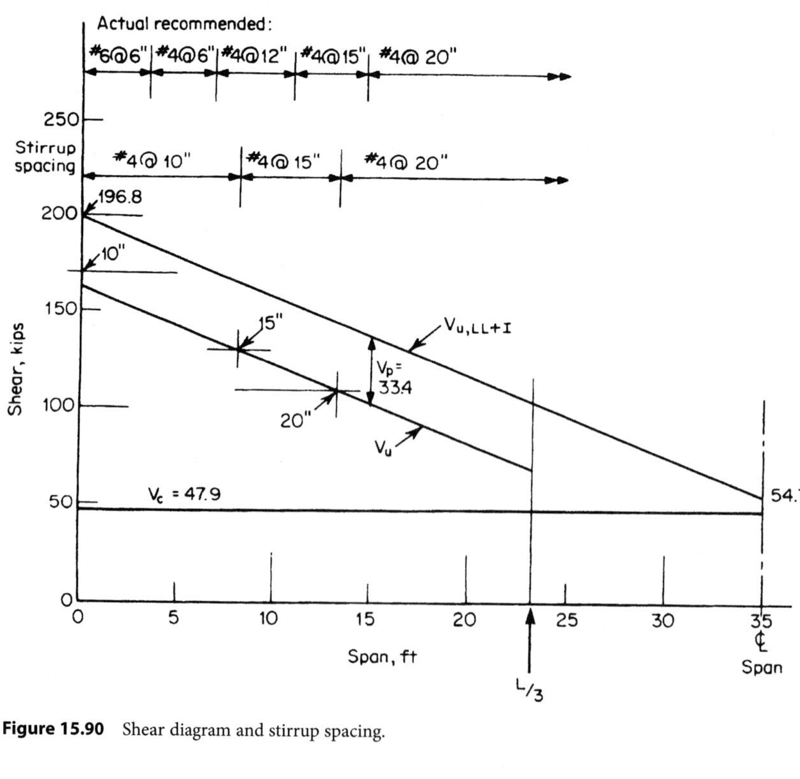

Figure 15.90 Shear diagram and stirrup spacing.

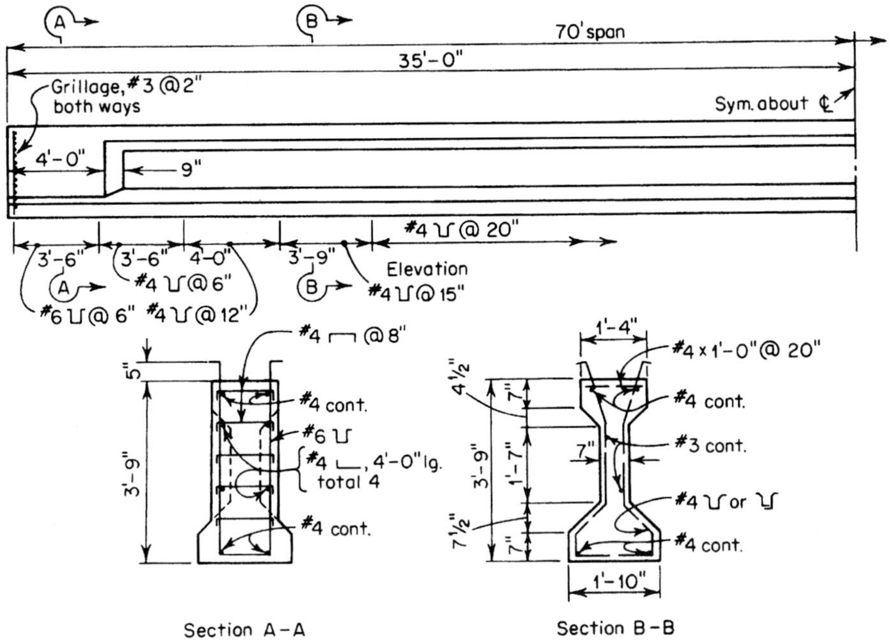

Figure 15.91 Details of prestressed girder for 70-ft span.

The deflection due to the dead load of the girder is

$$\Delta = \frac{5ML^2}{48EI} = \frac{5 \times 360{,}000 \times 70^2 \times 12^3}{48 \times 3{,}800{,}000 \times 127{,}000} = 0.65 \text{ in.}$$

Net uplift at time of prestressing is $1.87 - 0.65 = 1.22$ in.

Assume that the prestress force has reduced from 739,000 to 600,000 lb at the time of placing the slab. Assume the coefficient of creep = 1.5. Then

$$\Delta = 1.5(1.87 \times 600/739 - 0.65) = 1.31 \text{ in.}$$

Upon placing of the slab, the girder will deflect

$$\Delta = 0.65 \times 655/585 = 0.73 \text{ in.}$$

Net deflection of the girder is $1.31 - 0.73 = 0.58$ in.

When a large number of similar beams are to be used in a project, periodic deflection measurements should be made on the first girders to check the assumed modulus.

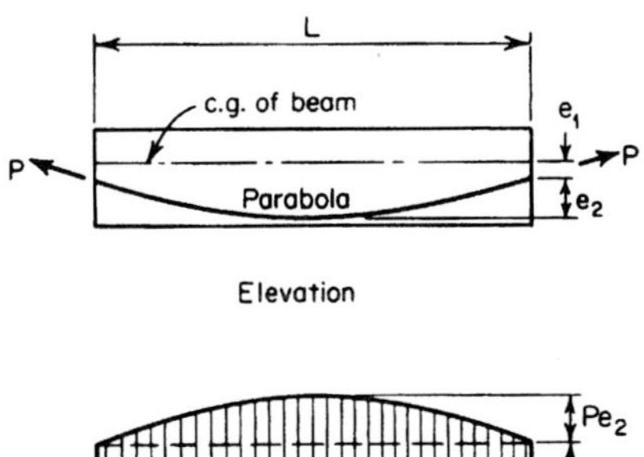

Figure 15.92 Uplift due to prestressing.

15.11.8 Live-Load Deflection

For the purpose of computing live-load deflections, the AASHTO specifications allow the live load to be distributed equally among the girders if the span has adequate diaphragms to ensure lateral distribution. Present information indicates that if the ratio of span to width is 2 or more, this assumption is safe. There is usually no difficulty in satisfying allowable live-load deflection requirements with composite prestressed girders whose depth-to-span ratio is 0.05 or more.

EXAMPLE

Compute the $LL + I$ deflection for the girder of Sec. 15.11.1. Assuming the live load is distributed equally to all girders, the number of lanes per girder is $2/5 = 0.4$. Then, per girder,

$$LL + I = 0.40 \times 1.26 \times 986 = 496 \text{ ft-kips}$$

Computing the deflection as outlined in Sec. 15.5.6, the correction factor for this moment is 1.14.

$$M = 1.14 \times 496 = 565 \text{ ft-kips}$$

Equivalent concentrated load $P = 4 \times 565/70 = 32.3$ kips.

$$\Delta = \frac{PL^3}{48EI} = \frac{32.3 \times (12 \times 70)^3}{48 \times 3800 \times 374{,}720} = 0.28 \text{ in.}$$

Allowable deflection $\Delta = 12 \times 70/800 = 1.05$ in.

In the calculations above, the modulus of elasticity was assumed to be the same as that for computing dead-load deflection. If the predicted live-load deflection is excessive, consideration should be given to the probable increase in modulus of elasticity which will have developed by the time live load is applied.

15.12 BRIDGE RAILINGS

Highway railings are built either to deflect, decelerate, and contain a vehicle or to withstand impact with little deflection. Since bridge railings must protect structural members as well as prevent vehicles from going over the side, they must withstand impact without appreciable deflection.

15.12.1 Traffic Railing

Railings may be made of any material strong enough to withstand the load and ductile enough not to shatter or explode under load but to yield in a shock-absorbing, bending failure. To this end, it is customary to require that the metal portions of railings exhibit 10 percent ductility at failure. Rails, posts, and concrete barriers should remain intact and not break away, possibly to fall on a roadway below (Fig. 15.93).

It is desirable to have some continuity in the railing, but from the standpoint of maintenance it should be possible to remove damaged sections easily without dismantling the entire railing. The effective height of the vehicle rollover force is taken as

$$H_e = G - \frac{12WB}{2F_t}$$

where

G = height (in.) of vehicle center of gravity above bridge deck (from AASHTO Table 13.7.2-1)
W = weight of vehicle (kips) corresponding to the required test level (from AASHTO Table 13.7.2-1)
B = out-to-out wheel spacing (ft) on an axle (from AASHTO Table 13.7.2-1)
F_t = transverse force corresponding to the required test level (from AASHTO Table A13.2-1)

Railings shall be proportioned such that

$$\bar{Y} = \frac{\sum (R_i Y_i)}{\bar{R}}$$

where

R_i = resistance of the rail (kips)
Y_i = distance from bridge deck to the ith rail, ft

All forces shall be applied to the longitudinal rail elements. Distribution of transverse loads shall be consistent with the assumed failure mechanism of the railing system.

It is desirable to provide a smooth contact face so that vehicles will rub along the rail and not be caught or pocketed by the posts. Pocketing the rail exerts extreme deceleration forces on the vehicle and causes severe damage to the post. The most satisfactory bridge railing developed to date is shown in Fig. 15.94*a*. Because of the sloping front face, a vehicle approaching at a flat angle will be lifted and tilted away from the rail so the fenders usually will not rub. At angles under 15 degrees or so, the driver will usually not lose control, and the vehicle will be directed back into a course parallel to the railing and may escape with no damage whatsoever. This railing is widely used in the United States. The barrier shown in Fig. 15.94*b* is used for narrow medians.

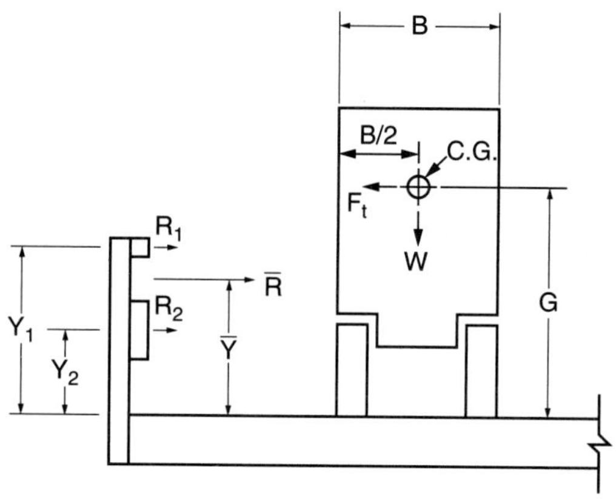

Figure 15.93 Traffic railing.

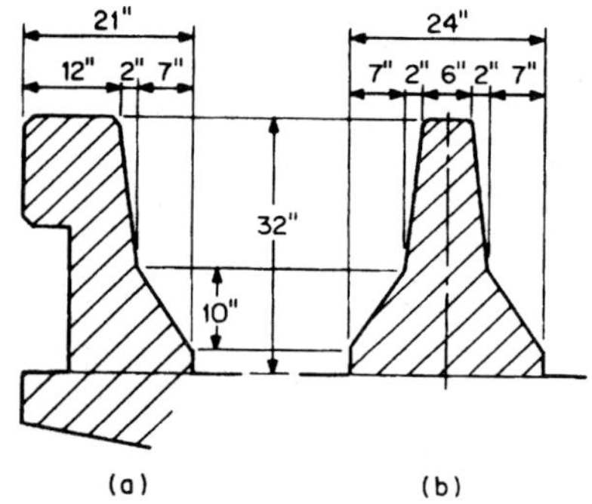

Figure 15.94 (*a*) Bridge rail and (*b*) median barrier.

15.12.2 Pedestrian Railings

A pedestrian railing used in connection with a highway structure should be impossible to crawl through or under and as difficult as possible to climb. It should be narrow and rounded on top so as to be nearly impossible to walk upon. A pedestrian railing should be not less than 42 in. high. It should have balusters spaced with openings not to exceed 5 in. The design live load for pedestrian railings shall be taken as w = 0.050 klf, both transversely and vertically, acting simultaneously. In addition, each longitudinal element will be designed for a concentrated load of 0.20 kips, which shall act simultaneously with the above loads at any point and in any direction at the top of the longitudinal element. The posts of pedestrian railings shall be designed for a concentrated design live load applied transversely at the center of gravity of the upper longitudinal element or, for railings with a total height greater than 5.0 ft, at a point 5.0 ft above the top surface of the sidewalk. The value of the concentrated design live load for posts, P_{LL}, in kips, shall be taken as

$$P_{LL} = 0.20 + 0.50\ L$$

where

L = post spacing (ft)

The design load for chain link or metal fabric fence shall be 0.015 ksf acting normal to the entire surface. The application of loads shall be as per AASHTO Figure 13.8.2-1.

15.12.3 Curbs and Sidewalks

The effectiveness of curbs or rubbing strips on the face of the railing varies somewhat with the design of vehicles. The AASHTO specifications require that the face of a curb be at least 9 in. from the face of the railing. Tests made with vehicles approaching at flat angles show that a curb projecting 5 in. or more from the face of the railing will act as a springboard, raising the car so that it strikes the railing higher or even goes over it completely. Curbing has been largely discredited as a means of keeping a high-speed vehicle in the roadway. It is safer for present-day vehicles to contact a strong barrier at least 27 in. above the roadway surface. Therefore, wherever a curb is used, the railing should be at least 27 in. above it. On structures where the face of the bridge railing is in line with the outside edge of the roadway shoulder, curbs can be omitted.

15.13 ADMINISTRATION OF A BRIDGE SYSTEM

15.13.1 Maintenance and Inspection of Existing Bridges

When the bridge is finally completed, the designer's work is done, but the life of the bridge is just beginning. Unfortunately, it is generally thought that a bridge is built to stand forever unless something comes along to cause it physical damage. This is not the case. Every bridge needs continual attention. The agency owning the bridge has an economic as well as a public obligation to see that it is maintained in a safe and serviceable condition.

The first step in the proper administration of a system of bridges is to create and maintain an active inventory of the physical dimensions and condition of the bridges the agency is responsible for. The modern trend is to put all this information into a computer. Great strides are being made in combining the inventory and inspection information into an ongoing management system for programming repairs and eventual replacement. However, excellent results have been had for many years with loose-leaf books and card files, which have the advantage of having the data physically available for ready reference. An inspection program should be set up so that every bridge is carefully inspected at least every 2 years. Older or deficient structures should be inspected more often. The inspections should be made by experienced engineers qualified in design and construction. Some states require that these engineers be registered professional engineers. The importance of these records and inspections cannot be overstated. The FHWA has been able to bring about some uniformity of inspections among the states. However, there are many public bridges owned by other public agencies which are not properly inspected or even well recorded. A neglected bridge is a potential source of disaster.

On the first inspection a bridge should be completely examined and, if no plans are available, measured and the measurements recorded. This includes measuring the important members; the clearances above, below, and across the roadway; evidences of being struck; visible high-water marks (for waterways); and any other factors which might affect the life or safety of the bridge. If a set of as-built plans is available, they should be kept with the file on the bridge. At least two photographs of the bridge (one along the centerline and the other a side view) should be filed with the physical and historical data. The bridge should be accurately located by route and mileage, and there should be sufficient data on the sizes and condition of members so that calculations of the bridge's capacity may be made in connection with overload permit requests. The records should contain a complete history of the bridge: maintenance work done (with cost), accidents, changes made, and plans for replacement.

The inspecting engineer should carefully check all the likely spots where deterioration may have set in. Timber bridges are vulnerable at the ground line, on top of the caps between the stringers at abutments, at joints of trusses, and between piles and caps—anywhere dirt can collect and retain moisture. Concrete bridges may be damaged by freezing and thawing cycles, cracking which permits water to get to the reinforcing steel, or cracking which may indicate either overloading or deficient design. Expansion joints frequently lock up because of rust, water, or dirt and may need cleaning or repair. If not corrected, faulty expansion joints can cause serious damage to a concrete structure. Steel bridges suffer from corrosion and may be severely damaged by expansion detail failures. Drainage systems may plug up, allowing water and dirt to run into expansion bearings and down over the steel members, which accelerates the breakdown of the paint film. Retainer caps may come loose and fall off of pins, allowing eyebar systems to come loose or hanger details to come apart. Pinned expansion bearings may become locked with rust, causing damage or the failure of some other part of the structure. Where there is movement of the eyebars on the pins, on older bridges the pins may be dangerously worn away. Welded structures may develop unexpected cracks which if caught in time can be repaired but which if neglected may lead to catastrophic failure.

Undercutting of the foundations by stream flow can expose piles to damage and cause settlement or failure. Piers in deeper water whose bottoms are never exposed to the air should be examined by an experienced diver at about 10-year intervals or more often if damage is suspected. Knowledgeable engineers are continually checking the alignment of bridge rails because a sag in the rail is often the first warning of trouble below. A good pair of eyes operated by a trained brain are the best inspection device yet discovered. Binoculars are helpful if one cannot get close up under a bridge.

It is helpful if the inspecting engineer will also devise and, if necessary, draw up such plans as may be needed to repair the defects discovered. The engineer should have a good sense of the degree of urgency the repairs may require, and the agency should be organized to perform urgent work at once. It is most discouraging to read reports which recommend the same repair year after year and yet nothing is done.

15.13.2 Bridge Administration

No one has yet devised a system which will balance all the many factors bearing on the maintenance or replacement of a bridge. Some of the computerized systems come close to prioritizing the projects in terms of need and money available, but there are always personal and political factors which come into play to cripple the effectiveness of the programs.

Effective administrative programs may be developed by assigning values to safety and delay factors so that it becomes possible to balance the needs of a number of bridges against the always insufficient available funds. Then there is usually a conflict between the need for new highways as contrasted to the need for new bridges. This too can be evaluated and given a priority rating. Thus it is possible to develop a master priority list which will effectively show where the available dollars could most advantageously be spent. However, at this point political and personal influences usually come into play and upset the finely tuned priority list. As long as these factors are allowed to override the carefully determined priorities, it will not be possible to devise a completely effective administration program.

Even the most substantial bridges will not last forever without considerable attention and careful inspection. The original designs may be overshadowed in some cases by the ingenuity required to keep them safe and structurally sound. The original investment can be lost through neglect and lack of maintenance. It should be evident that good maintenance is good economic policy. The largest percentage of bridges is owned by public agencies, who must be made aware of their responsibility to ensure that their bridges have a long and carefully supervised life.

REFERENCES

1. Design Specifications, AASHTO LRFD Bridge, American Association of State Highway and Transportation Officials, 2012.

2. Manual for Railway Engineering, AREMA, American Railway Engineering and Maintenance-of-Way Association, 2011.

3. Structural Welding Code-Steel, Specifications for Welded Highway and Railway Bridges, AWS, American National Standard, 2015.

4. Bridge Design Manual, PCI, Precast/Prestressed Concrete Institute, 2014.

PART B
STEEL BOX-GIRDER BRIDGES

BY

TONY SHKURTI, ERIC STONE, JAMES CARTER III *Consulting Engineers, Chicago, Illinois*

Steel box girders offer one of the most efficient longitudinal structural elements for spanning long distances from one point of support to the other. A single steel box girder can be compared to a pair of plate girders which have a common bottom flange. In the past, steel box girders consisted mainly of box girders with vertical webs. In recent times, box girders with sloped webs, also called steel tub girders, have become more common. This newer configuration helps optimize behavior of the box and support of the concrete deck. See Fig. 15.95.

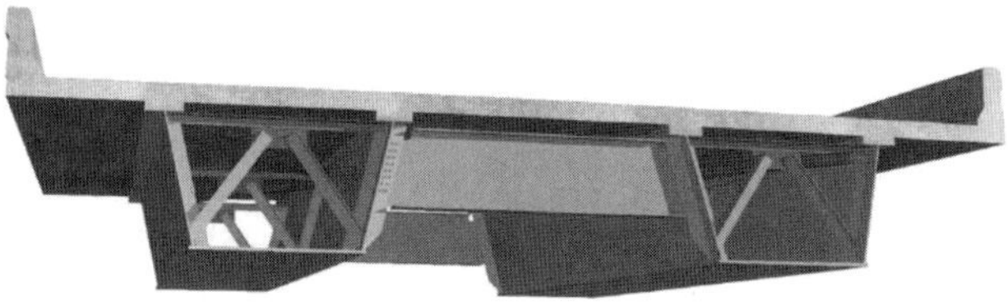

Figure 15.95 Steel tub bridge framing.

15.14 BENEFITS

Steel box girders offer many advantages when compared to I-shaped plate girders. As a result, they are becoming more commonplace, especially in urban multilevel interchanges involving flyover ramps. They offer a pleasing aesthetic appearance with an uninterrupted flowing outside surface, as they cover many of the structural appurtenances such as stiffeners, connection plates, cross-frames, upper lateral diagonals, and stay-in-place forms inside the tub.

One can argue they are structurally more efficient than plate girders when spanning along a highly curved bridge alignment, as they offer significant torsional stiffness compared to plate girder systems. While steel plate girders resist torsion mainly through warping torsional stiffness, thin-walled closed section steel box girders resist torsion primarily through St. Venant torsional stiffness, with the warping offering a negligible contribution. In a straight alignment structure, a two-steel box-girder structure with four top flanges, four webs, and two large bottom flanges offers comparable resistance to a four-plate girder system with four top flanges four webs and four bottom flanges. However, in a curved alignment the steel box girders offer significant advantages mainly due to their inherent St. Venant's torsional stiffness. This feature reduces significantly the need for additional intermediate cross-frames and diaphragms, which for curved steel plate girders framing in curved alignments function as direct load-carrying members helping to transfer loads from one girder to another to make the whole framing system able to resist torsion.

The torsional stiffness of steel box girders creates better load distribution between the box girders in the framing system. It allows each girder not to have to rely on the other girders or system diaphragms and cross frames between girders to resist and transfer loads in curved alignments. This also makes the overall framing system stiffer rotating together and thus minimizing deflection of the superstructure.

Steel box girders are easier to handle, transport, and erect than conventional I-shaped steel plate girders due to the inherent internal stability of each unit. Larger pieces can be erected without the need for many stabilizing frames and formwork. Therefore, erection of a box-girder bridge may be completed in less time than that of an I-girder counterpart because there are fewer pieces to erect, fewer external diaphragms to be placed in the field, and subsequently fewer field connections to be made. See Fig. 15.96*a* and *b*.

(a)

Figure 15.96*a* Large pieces can be transported and erected in a modular form.

(b)

Figure 15.96*b* Shipping, handling, and erection is made easier.

Maintenance of steel tub girders is also easier and more cost effective, as most inspections can be performed from inside the steel tubs without the need of expensive equipment, including snoopers. The internal components of the box girders must be designed to allow access by inspectors from one end of the girder to the other. Access hatches are typically provided through the bottom flange near the ends of the girder. There might be some OSHA requirements that need to be met, like in Fig. 15.97, where the minimum height was maintained at 5′-0″ internally in the box to allow easy access for inspectors and equipment.

Figure 15.97 Easy access inside with enough space for inspectors.

15.15 INTERNAL ELEMENTS

As discussed, steel box girders are structural members that are quite adequate for covering mid-length to long. To achieve that goal and given that they employ large plate elements to provide their global resistance, they must incorporate internal elements that ensure the girder maintains its intended structural shape when loaded. The careful design of these internal elements is as critical as the design of the size of the girder plates. These elements include the following:

- Internal intermediate cross frames and/or diaphragms (Fig. 15.98)
- Upper diagonal lateral bracings (Fig. 15.99)
- Connection plates and transverse stiffeners (Fig. 15.98)
- Longitudinal bottom flange stiffeners (Fig. 15.98) (when needed for wide bottom flanges)
- Longitudinal web plate stiffeners as needed
- Bearing and/or end diaphragms (Fig. 15.100)
- External intermediate diaphragms or cross frames (Figs. 15.99, 15.101)

External diaphragms are also utilized during erection to maintain the independent box-girder sections aligned with the intended deck orientation.

15.16 LIMIT STATES AND COMBINATIONS

Loads are to be computed and combined based on the American Association of State Highway and Transportation Officials (AASHTO) philosophy for design. Table 3.4.1-1 lists all the Load Combinations and Load Factors applicable to all four AASHTO required Limit States:

1. Strength Load Combinations I–V
2. Extreme Events Load Combinations I & II
3. Service Load Combinations I–IV
4. Fatigue Load Combinations I & II

Per AASHTO Article 1.3.2.1, "All limit states shall be considered of equal importance." Therefore, all four of them must be met for each design.

Loading should follow AASHTO LRFD Bridge Design Specifications, 8th Edition, 2017, or the most current version specified in the project. Section 3, "Loads and Load Factors," prescribes in detail the loads to be used for different layouts of the bridge structures. The load factors for various loads comprising a design load combination shall be taken as specified in Table 3.4.1-1. All relevant subsets of the load combinations shall be investigated. For each load combination, every load that is indicated and that is germane to the component being designed shall be multiplied by the appropriate load factor. The products shall be summed as specified in Eq. 1.3.2.1-1 and multiplied by the load modifiers specified in Article 1.3.2.

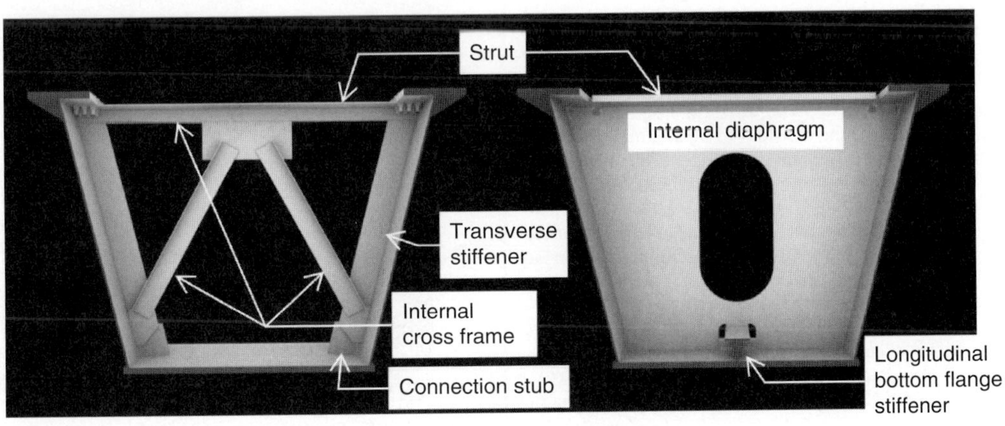

Figure 15.98 Internal cross frame, internal diaphragm, and longitudinal bottom flange stiffener.

Figure 15.99 Internal diaphragm and upper diagonal lateral bracing.

Figure 15.100 End and bearing diaphragm.

15.16.1 Dead Loads

Dead loads are straightforward to apply in different cases of analyses. Table 3.5.1-1 of AASHTO shows the densities to be assumed for most structural elements and appurtenances on the bridge in the absence of more correct data. The bridge structures are built in a certain sequence, and the structural stiffness evolves from initial erection to final assembly. The sequence of erection, or staging, defines the distribution of stresses in every element of the structures. Therefore, it is of utmost importance that the weight of the different structural elements comprised in the structure are applied in the structural model in the corresponding sequence that they are assembled in the physical structure. The typical construction sequence for steel box girders is similar to that used for I-shaped steel plate girders and is as follows:

- Self-weight of the steel girders and internal elements are applied to the steel section.

Figure 15.101 External intermediate diaphragms.

• Self-weight of the concrete deck slab in its initial, unhardened (wet) state is applied to the steel box-girder section. However, it is common that the deck slab is poured in multiple stages, in which case the weight of the slab is resisted in some areas by the bare steel section and in other areas by the composite steel-concrete section.

• Any other dead loads that are installed in the structure after the concrete has hardened such as roadway barriers or traffic signs will be applied to the composite section made up of the steel box beam and concrete deck. However, the composite section resisting any permanent load applied to the section after the concrete has hardened (called "superimposed dead loads") should take into account the long-term effects of creep in calculating the composite section properties. This has the practical effect of softening the section, and mathematically AASHTO recommends that only ⅓ of the effective concrete flange width is used for the composite section to carry superimposed dead loads.

The most vulnerable time for these types of structures is during shipping, handling, and initial erection, especially the deck pour. Since the webs and flanges of box girders are slender elements susceptible to buckling, their erection requires care. Successive erection-stress conditions must be established with due consideration of such factors as the weight of the erection equipment, deck slab pouring machines and material stored on the girder, effects of temperature, and wind. Factors of safety for erection conditions should be sufficient to cover the effects of unintended eccentricities at splices, residual stresses, and other contingencies that may affect the elastic stability, hence the importance of internal bracing for the steel tub girders. The longitudinal diagonal upper lateral bracing is included primarily to make the section a "closed" box during the concrete deck pour until the deck concrete has hardened as well as transport and erection. After that, the upper lateral braces are considered "dormant" until the need for the next deck replacement. They do contribute in the overall resistance of the section, aiding the top flange in resisting loads, but often their contribution is neglected, especially in simplified analyses.

15.16.2 Live Loads

Live load is the one load that differs significantly from the approach used in non-transportation-related structures and buildings. The live load used in designing bridges is defined in Article 3.6 of the AASHTO LRFD. This article goes in detail describing the geometry of the lanes and the geometry and weights of the notional live loads, such as the Design Truck (3.6.1.2.2), the Fatigue Truck (3.6.1.4), the Design Tandem (3.6.1.2.3), and the Design Lane Load (3.6.1.2.4). The application of Design Vehicular Live Loads is described in 3.6.1.3 for the different applications in the structural models. Pedestrian live load is described in 3.6.1.6. Probabilistic approach for taking into account the effects of multiple units of live loading is handled by using multiple presence of live load factor "m" shown in Article 3.6.1.1.2. Dynamic load allowance to account for the dynamic effects of truck loading in motion over a nonperfect surface are handled using a magnification factor "IM" as prescribed in 3.6.2.

Transient loads, including live loads, applied during the in-service life of the structure should be applied to the full composite section at each location. AASHTO in Article 4.5.2.2 states, "The stiffness properties of concrete and composite members shall be based upon cracked and/or uncracked sections consistent with the anticipated behaviour. Stiffness characteristics of beam-slab-type bridges may be based on full participation of concrete decks." In the commentary of the same article AASHTO allows the engineer to consider full composite action of steel and concrete even in the regions where concrete is deemed to be in tension, thereby simplifying analysis.

15.16.3 Other Loads

All other basic transient loads are described in detail in subsequent articles of Chapter 3 of AASHTO LRFD. Most transient loads on superstructure are related to live load, such as Centrifugal Force (3.6.3) and Braking Force (3.6.4). But there are also natural and environmental loads that affect the superstructure such as Wind Loads (3.8) which are applied directly to the structure and indirectly to the structure through live load.

15.17 ANALYSIS OF BOX GIRDERS

The methods of analysis and design discussed in this part apply to all types of steel box-girder bridges with concrete decks regardless of span, except for horizontally curved structures. Formulas for the design of composite box-girder bridges are given in Ref. 1 for both straight bridges and for horizontally curved bridges. A general overview of the design of straight steel box girder bridges may be found in the Commentary section of Ref. 7. Box-girder design examples are shown on Federal Highway Administration (FHWA) website, respectively, at: https://www.fhwa.dot.gov/bridge/steel/pubs/hif16002/designexample05.pdf and https://www.fhwa.dot.gov/bridge/steel/pubs/hif16002/designexample06.pdf.

Steel box-girder bridges represent complex three-dimensional structures that are not easy to model and a analyze using simplified methods. While from the outside they give the impression of stiff trapezoidal members, they in fact comprise many appurtenances inside and outside the box, such as upper lateral bracings, cross frames, stiffeners, and connection plates. Also, even for straight bridges, positioning of the live load can be quite complex and can introduce loading that affects the various structural elements in this three-dimensional system.

15.17.1 Line Girder Analysis

However, for the purpose of obtaining approximate loading for preliminary sizing of structural elements in straight alignments, simplified analysis methods can be employed. This entails analysing a line girder model of a straight continuous span girder and using the approximate distribution factors for the live loads tabulated in Article 4.6.2.2.

AASHTO specifies in 4.6.1.2.4c that concentric closed box and tub girders with unskewed bearing supports and straight or slightly curved alignments as defined in the article may be a analyzed as an individual straight girder with span lengths equal to the arc length. In this case, a line girder analysis could be adequate for final design.

For the design involving single line girder analysis, live load distribution pertaining to each girder needs to be assessed. This amount of loading would be the live load affecting the typical girder being a analyzed and could be visualized as some kind of "tributary live load" for each beam. It can be obtained as a portion of an applicable lane loading for the girder multiplied by a distribution factor that would help calculate such "tributary live load." For each of the common deck superstructure types, shown in Table 4.6.2.2.1-1, including type (c) Open Steel Boxes in the table, these distribution factors are evaluated using approximate formulas shown in Table 4.6.2.2.2b-1 for moments in interior beams and similar tables following in the same chapter for moments in exterior beams and shear distribution for interior and exterior beams. This calculated "tributary live load" would be combined with other basic loads such as permanent and other transient loads following combination rules described previously. Note that AASHTO imposes certain restrictions on the use of live load distribution factors for multiple box sections as per Article 6.11.2.3, especially for straight multiple steel box sections.

For other box girders that do not meet the conditions above, they may be sized preliminarily using approximate methods such as M/R (Ref. 4). The M/R method is most useful for analysis of single tub or box girders and therefore appropriate for erection analysis of single girders, complete analysis of a narrow bridge with only one tub girder in its cross section, or a single girder as part of a phased-construction plan. In addition, as the M/R method also calculates tub girder rotations, theoretically it is possible to use it to solve for loads in multiple adjacent girders. However, for practical considerations, the calculations for multiple girder systems are too time consuming unless computer applications are used.

15.17.2 2D Grid Analysis

Analysis and design of steel box and tub girders, especially on curved alignments, requires more complex types of analyses. The first method could be 2D, or grid or grillage types of analyses. Grid analysis represents another level of accuracy that may be considered as an intermediate tool offering accuracy between the line girder analysis and full 3D finite analysis. Its advantage compared to line girder analyses is that it enables the user to evaluate the vertical load sharing contribution and loading on some of the transverse members, such as external intermediate diaphragms and the concrete deck slab. In these analyses, the members are modeled as frame elements that are not able to evaluate directly the loading in the different structural appurtenances included inside the steel tub girder. Instead, approximate methods are available to estimate the loading on those members (Ref. 3).

15.17.3 3D Finite-Element Analysis

For most structures involving considerable curvature, or long spans, or skew and other complexities, full 3D finite-element analyses are highly recommended for final and accurate modeling of all elements involved in such bridge structures. These are not analyses that can be done by hand and involve sophisticated structural models that mathematically and physically attempt to replicate the real behavior of the structure as closely as possible, using state-of-the-art techniques. In these models, steel plates of the tub girders and concrete slab are frequently modeled utilizing shell elements, while other appurtenances usually made of rolled shapes are modeled as either frame or axial elements. Current 3D modeling software also offer node and member offset capabilities that allow the modeler to place these members geometrically centered at the exact respective location they are in the structure. An example is shown in Fig. 15.102.

In the case of analyses utilizing more complex models such as 2D grid analyses or 3D finite-element models described above, the "tributary live load" per each structural unit such as a box girder may be obtained *precisely without resorting to approximate load distribution factors described above.* Many of the programs used for these 2D grid and 3D finite-element models employ influence surfaces or **a** series of influence lines to investigate the most critical loading by applying the loads required by AASHTO or other controlling codes. They now have capabilities of assigning worst case envelope demands to the composite section made up of the steel box and effective concrete deck slab.

15.18 DESIGN OF BOX-SECTION MEMBERS

In general, Section 6.11 in AASHTO provides all the details for designing box-section flexural members. Provisions and the Commentary explain in detail what needs to be checked, what are the limitations, and why. It covers most of situations found in the practice of bridge building. It lists in sequence all the requirements for each applicable limit state such as the following:

- Service Limit State in Article 6.11.4
- Fatigue and Fracture Limit State in Article 6.11.5
- Strength Limit State in Article 6.11.6

Plates making up the thin walls of the box girder must be proportioned carefully. Web plates are proportioned according to provisions of Article 6.11.2.1 and flanges according to Article 6.11.2.2. Finally, design approaches for each element in flexure, shear, and shear connectors are articulated in subsequent articles from 6.11.7 through 6.11.10.

15.19 INTERNAL ELEMENT DESIGN

These elements and their connections both to each other and to the box section must be analyzed and sized carefully; otherwise, the box section might not be able to maintain its intended strength. Individual brace elements are designed as axial members according to axial tension and compression member criteria in AASHTO. Stiffening elements need to meet certain sizes required by AASHTO such as Article 6.10.11.1 for transverse web stiffeners and Article 6.10.11.3 for longitudinal web stiffeners. Bearing stiffeners shall be designed according to the provisions of Article 6.10.11.2; they should be attached to the end diaphragms rather than the inclined webs. And finally, longitudinal compression-flange stiffeners shall be sized and spread according to Article 6.11.11.2.

15.20 CONCLUDING REMARKS

As discussed in this section, steel tub girder bridges offer some distinctive advantages when compared to other types of structures. This is especially true when they are supporting highly curved flyover ramp decks that inherently induce significant torsional effects in the structural framing. Steel tub girders offer an optimal combination of strength, cost efficiency, and pleasing aesthetics. However, despite these advantages, use of these types of structures has been limited in the past century in the United States mostly out of concerns of lack of redundancy of this structural type.

The argument was that a steel tub section built up of welded plates represents a continuous element in which a fatigue crack once initiated

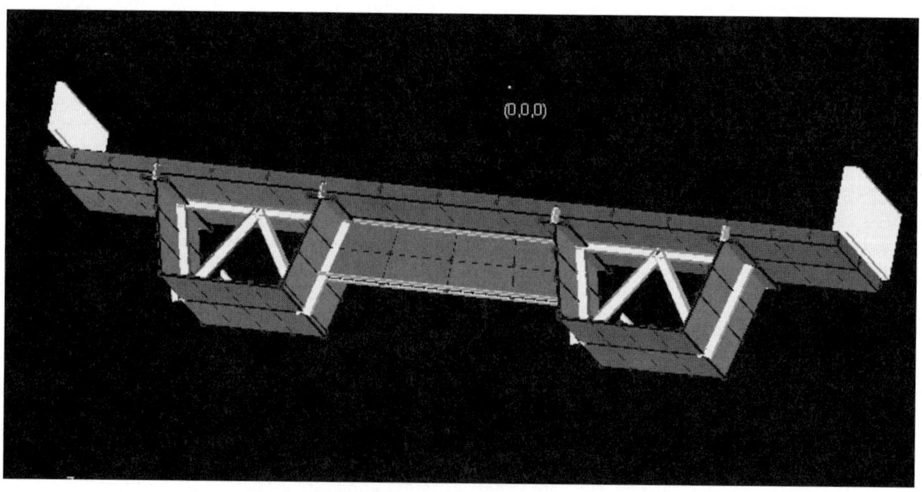

Figure 15.102 Example of 3D modeling of steel tub girders.

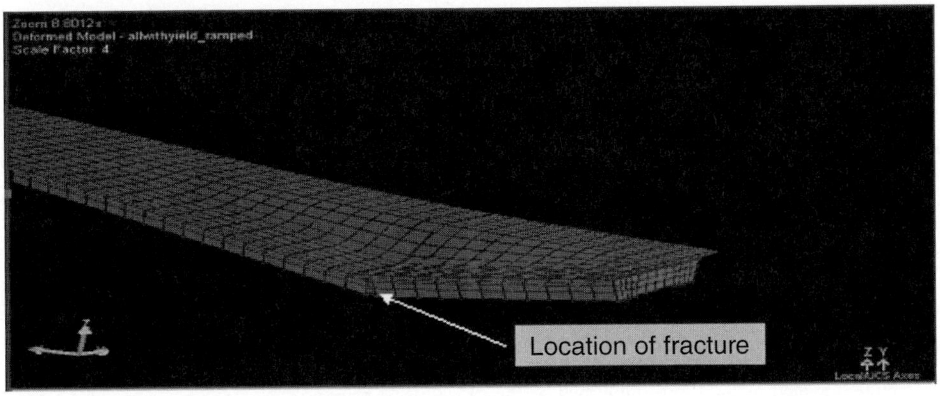

Figure 15.103 Refined FEM model for used for analyzing redundancy of steel tub girders.

could reach a critical length and then suddenly propagate throughout the whole section. This was deemed even more critical given the fact that most single- and two-lane ramp bridges usually involve only a pair of steel tub girders.

While such a scenario might seem like a stretch, a similar event had happened during the failure of end-span steel I-shaped plate girders on the Hoan Bridge in Milwaukee, Wisconsin. In that event, in the early morning of December 13, 2000, all three noncomposite girders of an end span of the bridge fractured suddenly. Two out of three girders fractured thoroughly through bottom flange, web, and top flange, while the third girder fractured only the bottom flange and web. The bridge managed to stand up and nobody was hurt, although it only included three longitudinal girders. Structural redundancy including continuity over the piers and upper longitudinal bracing combined with the side concrete barriers acting as struts are believed to have saved the structure from falling.

Since then, the Wisconsin Department of Transportation, working side by side with FHWA when reconstructing the whole Marquette Interchange in downtown Milwaukee, a five-level interchange between I-94, I-43, and I-794, agreed to allow their consultant to use steel tub girders for their system ramps with the condition that the consultant would prove analytically through refined analyses that the bridge would be able to stand to such a through section fracture of a single steel tub girder without the bridge falling down in a sudden fashion. The consultants proved just that for the eight system ramps involving twin tub girders of different span layouts and geometry (Fig. 15.103).

FHWA accepted the report prepared by the Milwaukee Transportation Partners, and the system ramps were produced and erected and have been in service for about 15 years now.

Since then, many researchers led by NSBA and other stakeholders have sponsored or performed different types of research, both analytical and experimental. As a result, steel tub girders are being employed more frequently in interchange projects around the nation, thus capturing the market share that they deserve within the industry.

REFERENCES

1. AASHTO-LRFD Bridge Design Specifications (2017). American Association of State Highway and Transportation Officials, Washington, DC.

2. AASHTO/NSBA Steel Bridge Collaboration, 2014. "G13.1 Guidelines for Steel Girder Bridge Analysis, 2nd Edition" (https://www.aisc.org/globalassets/nsba/aashto-nsba-collab-docs/g-13.1-guidelines-for-steel-girder-bridge-analysis.pdf)

3. AASHTO/NSBA Steel Bridge Collaboration, 2014. "G13.1 Guidelines for Steel Girder Bridge Analysis, 2nd Edition" (https://www.aisc.org/globalassets/nsba/aashto-nsba-collab-docs/g-13.1-guidelines-for-steel-girder-bridge-analysis.pdf)

4. Tung, D. H. H., and R. S. Fountain. 1970. "Approximate Torsional Analysis of Curved Box Girders by the M/R-Method." Engineering Journal, American Institute of Steel Construction, Vol. 7, No. 3, pp. 65–74.

5. Huang, D. Z. 2008. "Full-Scale Test and Analysis of Curved Steel-Box Girder Bridges," *ASCE Journal of Bridge Engineering*, Vol. 13, No. 5, pp. 492–500.

6. Fan, Z., and T. A. Helwig. 1999. "Behavior of Steel Box Girders with Top Flange Bracing," *Journal of Structural Engineering*, American Society of Civil Engineers, Reston, VA, Vol. 125, No. 8, August, pp. 829–837.

7. Wolchuk, R., and R. M. Mayrbaurl: Proposed Design Specifications for Steel Box Girder Bridges. Report no. FHWA-TS-80-205. Federal Highway Administration, Washington, DC., January 1980. (Note: Also contains extensive commentary and bibliography.)

8. Milwaukee Transportation Partners, "REDUNDANCY OF BOX GIRDER STEEL BRIDGES A Study for the Marquette Interchange HPS Twin Box Girder Structures Project I.D. 1060 05 1222."

PART C
CURVED STEEL I-GIRDER BRIDGES

BY

JAMSHID MOHAMMADI *Illinois Institute of Technology*
BORA JANG *Sharma and Associates, Inc.*

15.21 GENERAL[b]

15.21.1 Introductory Remarks

Curved bridges require special attention in design because of their unique geometry. The current design requirements for horizontally curved highway bridges are embodied by guidelines published by the American Association of State Highway and Transportation Officials (AASHTO).[1] This document and its predecessors include most of the research conducted after 1980 as well as the results of full-scale tests conducted through the Curved Steel Bridge Research Project (CSBRP) initiated by the Federal Highway Administration (FHWA).[2] The CSBRP study and similar others were intended to offer a better insight into (1) the behavior of curved bridges, which involves a significant level of out-of-plane effects, and (2) special requirements that need to be considered in their design. This section is intended to provide an overview of curved bridges made up of I-girders and key requirements in their design. For additional information, the readers are referred to the listed references cited at the end of this section.

15.21.2 Overview of Curved Bridge Differences

Although most features of a curved bridge are similar to straight bridges, there are certain differences that will need special attention in design. An overview of these differences is provided herein and detailed design considerations to address these features are provided in the following subsections. Most notable of these differences is the bridge geometry that requires special attention in regard to structural analysis and design, as well as in detailing and construction. Other special features are the significance of such effects as temperature variations, torsional moments, and conditions at the supports.

15.22 DESIGN CONSIDERATIONS

15.22.1 Bridge Geometry

The bridge geometry plays an important role in design and construction of curved bridges. As expected, a favorable geometry for a curved bridge, for the ease of design and construction, is one that includes concentric girders with circular curvatures and radial support lines. However, many constraints related to the roadway layout, surrounding terrains, and crossroads will necessitate designs that are unique and do not necessarily follow the desired perfect geometry. And as such, many curved bridge designs include nonconcentric and noncircular girders, reversed-curve framing, mixed straight and curved framing, and skewed supports. The efforts by bridge designers have been made in minimizing these effects wherever possible with considerations for the requirements for the roadway alignment. As the level of complexity in the geometry increases, the bridge structural analysis will then need to rely on more sophisticated methods that would properly portray the structural behavior through three-dimensional modeling and computer applications. However, at times, the use of more traditional methods may be necessary as a way of approximations. For example, in certain cases, bridges that have been built with nonconcentric, noncircular, or discontinuous curvatures have been analyzed and designed by approximating the actual girder alignments with "average" or "equivalent" curves. On large-radius curves, some bridges have been built with straight interior girders and isolated curved fascia girders. Yet in some other applications, states have used straight girders, kinked at the inflection points, to include the significance of the curved geometry in the structure.

As expected, irregular girder geometry complicates not only the structural analysis but also the application of live load, especially in cases where the analysis is based on the distribution factor concept as stipulated in the design code. In cases where flaring roadways and flared girders are involved, and especially where the splay in the girders is uniform, some designers calculate the live-load distribution using an average width or girder spacing. However, if the flare is nonuniform and as a conservative approach, many designers base the live-load distribution on the maximum width. It is noted that the term "flared girders" (also called splayed girders) refers to a situation where the girder spacing or deck overhang varies along the length of the superstructure. The curved structure also complicates the design—an issue that will need to be coordinated properly between the designer, fabricator, and contractor.

Skews are probably the most common irregularity in curved bridge framing. Curved bridges are often built with skewed supports as dictated by the geometry and roadway conditions. Skews require special attention and modeling when performing structural analyses. This is especially true where sharp skews on the direction of rotation of the girders at the bearings are present. While many designers have assumed rotation about an axis radial to the curve, others have considered rotation about an axis perpendicular to the chord running to an adjacent fixed bearing. In any case, proper modeling using finite-element analyses is helpful to compute the magnitudes and directions of twists and rotations, as well as the corresponding internal forces, and to obtain a graphical representation of how the bridge structure deforms. It is noted that skewed structures will result in a differential deflection between adjacent girders. And as such, it is also important to note that a three-dimensional modeling of the system will be necessary, with cross frames properly included, since they act as secondary load paths in the system.

Uplift is also likely to occur in sharply skewed curved bridges (see Fig. 15.104). This is because the girders toward the inside of the bridge are shorter; as such, they carry substantially less load such that their reactions resulting from the "grid" effect are less than the negative reactions resulting from the "curvature" effect. It is noted that the possibility to uplift and to improper seating of bearings is reduced if the diaphragm or cross-frame fasteners are left untightened in slotted holes until the concrete deck has been poured. Since there are uncertainties in predicting the movements at bearings in skewed curved bridges, some designers prefer to consider using elastomeric bearings as at support locations.

15.22.2 Thermal Effects

The effect of temperature change on a curved bridge may become significant resulting in movements at the support bearings. The magnitudes and directions of these movements are dependent on how many and which bearings are fixed. In situations where temperature movements are permitted to occur at the bearings, internal forces due to thermal effects do not have to be accounted for in the design. To accomplish this, thermal movements must be allowed to take place in directions radiating from the fixed supports, and the bearing devices must be designed so as to provide these directions of movement.[1] Most designers considered similar assumptions even before the publication of the Guide Specifications by AASHTO. In fact, translation has generally been considered to take place in the direction of the chord

[b]Original article by William N. Poellot Jr., HDR Engineering, Inc.

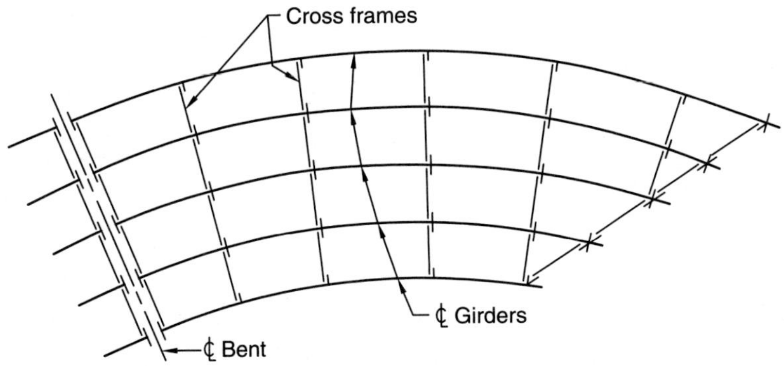

Figure 15.104 Sharply curved skewed bridge.

extending from the expansion bearing to the adjacent fixed bearing, and rotation has been assumed to occur about an axis radial to the curve (Fig. 15.105).

The Pennsylvania Department of Transportation Design Manual[3] indicates that a refined analysis of curved girder bridges will need to be considered in evaluating the performance of the superstructure due to thermal force effects. This is necessary to ensure that thermal movements are adequately addressed and that the bearing constraints and the effects of pier stiffness are properly modeled and incorporated in the analysis.

15.22.3 Framing Arrangements and Detailing

Because of a significant out-of-plane and torsional effects, the lateral flange bending stress may become critical. The magnitude of this stress is proportional with the square of the unbraced length of girder flange, i.e., the diaphragm spacing. Thus, by properly selecting the diaphragm spacing, the lateral bending stress in the girder flanges can be reduced. Some designers also use other criteria not directly associated with lateral bending. Implicitly, virtually any rules for diaphragm spacing are adopted to control the load and stress distribution and therefore involve, at least indirectly, lateral flange bending, diaphragm forces, and diaphragm size.

The current AASHTO Guide Specification[4] does not limit the diaphragm spacing; however, it is generally recommended (by some state departments of transportation[5]) to use the 25-ft (7.6-m) limit specified in previous edition of AASHTO. Considering this, the spacing limits specified in the following table may be useful.

Centerline radius of bridge, ft (m)	Suitable maximum spacing, ft (m)
<200 (61)	15 (4.6)
200 to 500 (61 to 152.4)	17 (5.2)
500 to 1000 (152.4 to 304.8)	20 (6.1)
>1000 (304.8)	25 (7.6)

It is noted that regardless of spacing, the designer will need to evaluate the need for diaphragms at all stages of bridge construction including at the final condition. In two-girder curved bridges with floor beams, the floor-beam spacing may be determined not only by lateral flange-bending considerations but also by the slab design requirements. Depending on the floor beam spacing, a one-way or two-way slab may be formed, each with a different design requirement in terms of reinforcing steel for bending and slab thickness for shear. It is noted that one-way slab is formed when the slab aspect ratio (ratio of longer to shorter dimension) is larger than 2, while a two-way slab has an aspect ratio between 1 and 2.

Diaphragm layout is also an important consideration in framing arrangement. Intermediate diaphragms or cross frames should normally be placed radial and in a single line the full width of the bridge as shown in Fig. 15.106*a*. If the bridge is on a slight skew, the support diaphragms are usually skewed, but the intermediate diaphragms can be kept radial as in Fig. 15.106*b*, or the intermediate diaphragms may be arranged on the skew as well, as demonstrated in Fig. 15.106*c*.

The ability of skewed diaphragms to develop end moments decreases markedly with skew angles of over about 15 degrees. Thus, bridges with large skew angles should only be considered where alternative geometries are not feasible. Skewed diaphragms are somewhat more complicated and costly to fabricate and construct than are radial diaphragms. This is because the geometry varies for each diaphragm. In some cases with skewed bridges, the diaphragms are staggered to keep them radial as shown in Fig. 15.106*d*. However, designers generally avoid this practice because of the poor ability of staggered diaphragms to distribute torsional forces across the structure. At interior skewed supports of continuous curved bridges, it is not always necessary to set the support diaphragms along the skew. Depending on the framing geometry, such diaphragms can be set radial as demonstrated in Fig. 15.106*e*. It is important that each girder be arranged to have diaphragms exactly at its supports avoid twisting of the girder on its bearing, unless design provisions have been made to accommodate the twisting effect.

A substantial number of curved bridges have been designed with bottom lateral bracing in at least some of the bays. Some curved bridges

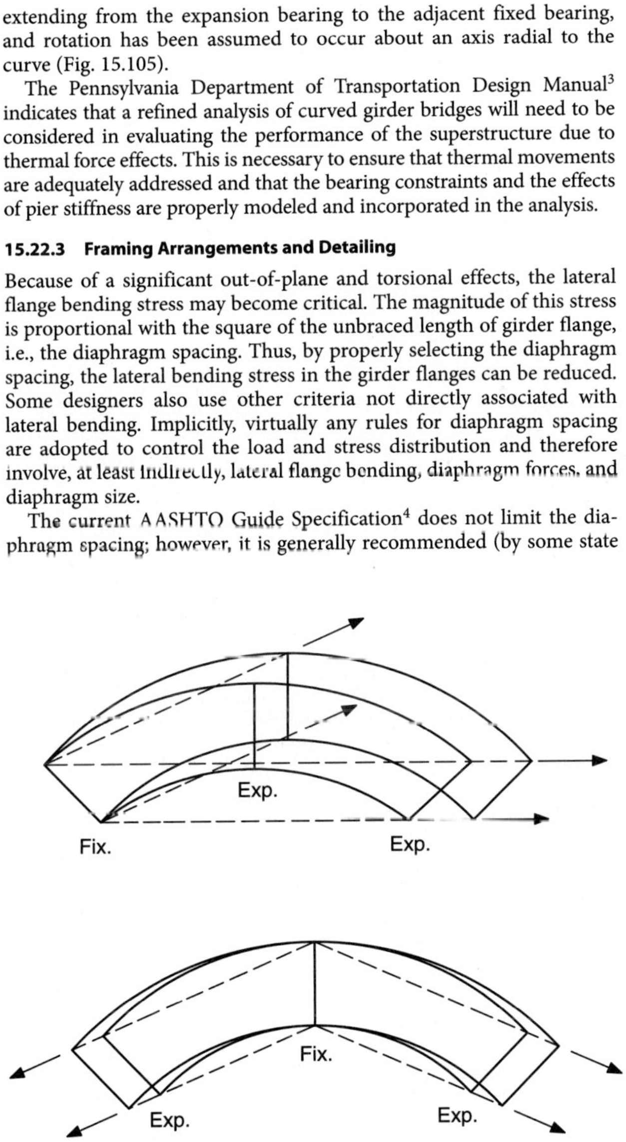

Figure 15.105 Expansion of curved-girder spans.

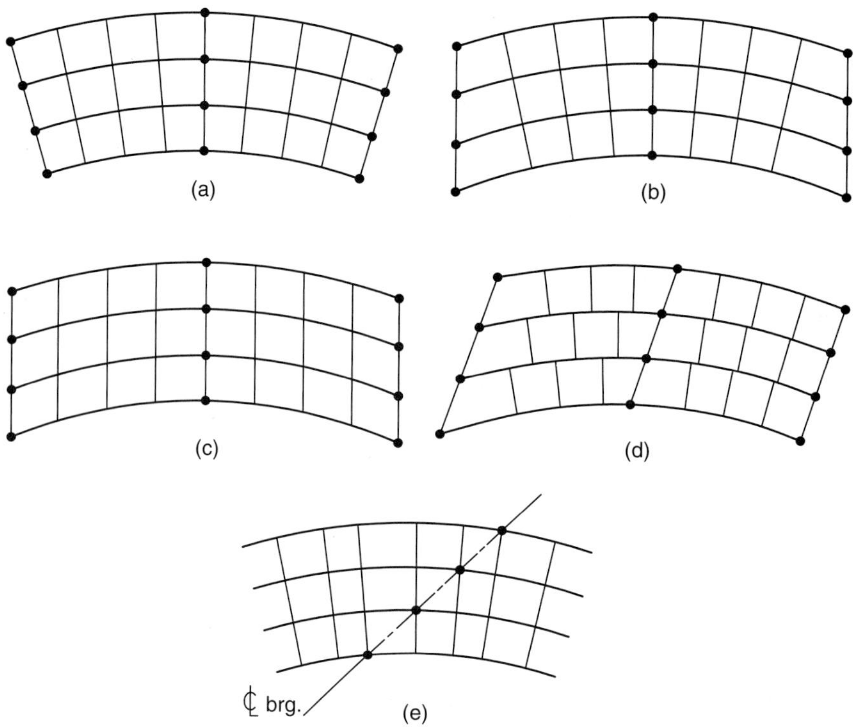

Figure 15.106 (*a*) Radial diaphragms, (*b*) radial intermediate diaphragms with skewed end diaphragms, (*c*) skewed intermediate and end diaphragms, (*d*) staggered diaphragms, and (*e*) radial diaphragms at support.

have been designed with top lateral bracing, although most designers seem to rely on the floor slab to serve as bracing at the top flange level. As in the case of straight bridges, a lateral bracing system is generally needed in a curved I-girder bridge mainly to control lateral bottom flange stresses resulting from wind loads. The use of lateral bracing in the exterior bays is especially recommended, if the combined stresses resulting from wind and other loads become critical. This is the only legitimate reason for lateral bracing in most curved bridges, and such bracing should especially be considered when widening the flanges or reducing diaphragm spacing as way of controlling stresses are not possible.

In some cases, lateral bracing has been intentionally provided to make adjacent plate girders behave like box girders for better torsional rigidity and transverse stress distribution. In these situations, a detailed level of structural modeling will need to be used for properly computing the torsional stresses in the combined sections. However, it is noted that this type of detailing will add to the fabrication and construction cost.

Another advantage of lateral bracing, as noted by some designers, is its stabilizing effect during erection. Laterally braced girders with proper diaphragms can be shipped and erected in pairs and held to close geometric tolerances until the slab has been placed. Depending on their geometry and design, lateral bracing will affect the type of modeling selected for structural analysis. The difference between a structure with and without lateral bracing is essentially the difference between closed and open framing, or the difference between box girders and I-girders. A refined structural analysis using three-dimensional finite-element modeling that properly considers the significance of diaphragms, cross frames, and bracing will certainly be helpful in getting a full picture of stress distribution in all structural members.

15.22.4 Number of Design Girders

In general, the effect of curvature is in (1) increasing the loading on girders outside the longitudinal centerline of the bridge and (2) decreasing the loading on girders inside the centerline depending on the relative girder distances from the centerline. Accordingly, the outermost girder takes the largest load because of the curvature effect, while the innermost girder undergoes the least load in the girder group. In terms of bending moment, this effect is further amplified in radially supported structures by differences in the arc lengths of girders. In fact, since the larger-radius outer girders have longer arc lengths, they possess longer span lengths than do the shorter-radius inner girders. This may not apply to skewed bridges. These factors indicate that, in general, all girders in the cross section of a curved girder bridge have substantially different bending moments and therefore different section requirements. The question arises as to whether each girder should be individually designed for an optimum weight or whether one design might serve for more than one girder. Of course, engineering judgment and experience will be required to evaluate the structure in question, considering the number of girders in the cross section and the sharpness of curvature. A two- or three-girder structure with a relatively large degree of curvature will most likely be designed with girders that are individually proportioned. On the other hand, in bridges with four or more girders, the design most likely involves identical sections, which will eventually reduce the construction cost through duplication. It is noted that sharpness of curvature is really defined by the central angle that is a function of both radius and span length, not radius alone.

15.22.5 Connection Details

Major connections involved in curved girder bridges are the cross frames or diaphragms and lateral bracing. These components, particularly the diaphragms, are considered primary members necessary to distribute the torsional forces in the structure. Thus, internal forces must be transmitted through the connections of cross frames, diaphragms, and lateral bracing to the main girders. This is in contrast to straight girders, for which the forces in the diaphragms are generally assumed to be noncritical as such, their design resorts to empirical procedures.

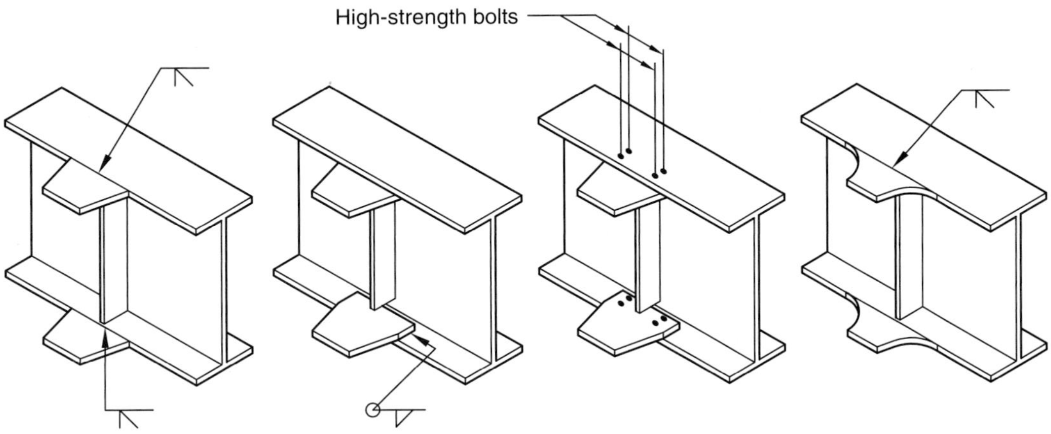

Figure 15.107 Attachment of cross frame to girder flange (*a*) welded to flange, (*b*) welded to web, (*c*) bolted, and (*d*) connection plate with radius.

Since the diaphragms in a curved bridge develop end moments and shear forces, a practical type of connection of the diaphragm to the girder would, more appropriately, be a moment connection to the girder flanges and a shear connection to the girder web. The shear connection is made by attaching the diagonal of the cross frame or the web of a solid diaphragm to the stiffener or connection plate. The moment connection can be made several different ways. For example, an attachment can be made directly to the flange either by welding or bolting (Fig. 15.107). These moment-connection details have certain disadvantages. The connections in Fig. 15.107*a* and *b* with attachments to the main material running in the direction of stress classify as AASHTO Category *E* or *E′* fatigue details and are therefore limited to a low stress range. In fact, such details are not considered to be good practice even if stress ranges are within allowable values. The bolted connection in Fig. 15.107*c* is more advantageous from a fatigue standpoint but involves taking holes out of the girder flange. The fatigue characteristics of the connection in Fig. 15.107*a* can be greatly improved (though more expensive to make) by using a radius at the corners as shown in Fig. 15.107*d*.

To eliminate the numerous disadvantages in making the diaphragm moment connection directly to the girder flanges, many designers do not favor direct attachment to the flange as illustrated in Fig. 15.107. An alternative detail is shown in Fig. 15.108*a*. A moment connection without attachment to the flange is achieved by relying on the stiffener to deliver the horizontal force from the strut to the girder flange. Thus, if the strut is set a small distance from the flange, the stiffener, acting in conjunction with a strip of the girder web, must be capable of transmitting this force in bending to the girder flange through the stiffener-to-flange weld. Most designers use this approach for cross-frame moment connections. However, many also have used connections made directly to the flange.

When lateral bracing is used, there is an added complication resulting from the connection plates required by the bracing. A common detail is shown in Fig. 15.108*b*. The connection plate is located a sufficient distance above the flange to allow painting underneath it. The plate is slotted to allow the stiffener to pass through it and is welded to the girder web. This detail also has poor fatigue characteristics since it is an attachment parallel to the stress field in the girder web. Alternative details to improve the fatigue behavior would include adding a radius at the ends of the connection plate or bolting the plate directly to the flange. A structural tee section is often used in such a bolted arrangement. The difficulty in making these connections illustrates the desirability of avoiding lateral bracing where possible.

15.22.6 Camber

Accurate computation of the dead-load deflection is important to properly camber curved bridges. Using approximate methods have resulted in improper cambers in many curved bridges. Although current methods of computing the dead-load deflection using computer analysis afford more accurate results, a common method that has been used in the analysis of many curved bridges is based on computing dead-load deflections using the developed length of the girder and including secondary loads resulting from curvature. This approach, however, may underestimate the deflections because it does not take into account the effect of girder twist that produces additional deflection beyond that produced by bending. A three-dimensional finite-element analysis will determine the deflections more accurately provided the proper torsional properties are used in the model.

Another factor affecting the accuracy of dead-load deflection calculations is the sequence of pouring the deck concrete. If the positive-moment region in one span is placed first, overdeflection may result in that span. Since this span is then acting compositely, the excessive deflection does not "come out" when subsequent spans are poured. Such pouring sequence effects apply to straight bridges as well but may be amplified in curved bridges because of the torsional component of the deflection.

Other reasons for excessive deflection are residual stresses resulting from curving, welding, and other fabrication operations. As the girder takes on the load, yielding occurs with accompanying permanent deformation in the girder. Most designers compensate for residual stresses resulting from heat curving by using the AASHTO rules for increasing the calculated camber. Engineers have increased the computed deflections by as much as 30 percent for setting grades to compensate for pouring sequence effects and overdeflection resulting from residual stresses.

15.23 CURVED GIRDER MECHANICS[c]

15.23.1 Methods of Analysis

The methods of analysis for curved bridges range from simple approximate modeling of girders to more rigorous three-dimensional finite-element methods. Selection of one method over another depends on the level of details needed for the analysis, the severity of curvature and skewness, and the applied load including dynamic loads. Among various

[c]Material presented is taken from Refs. 6, 7, and 8.

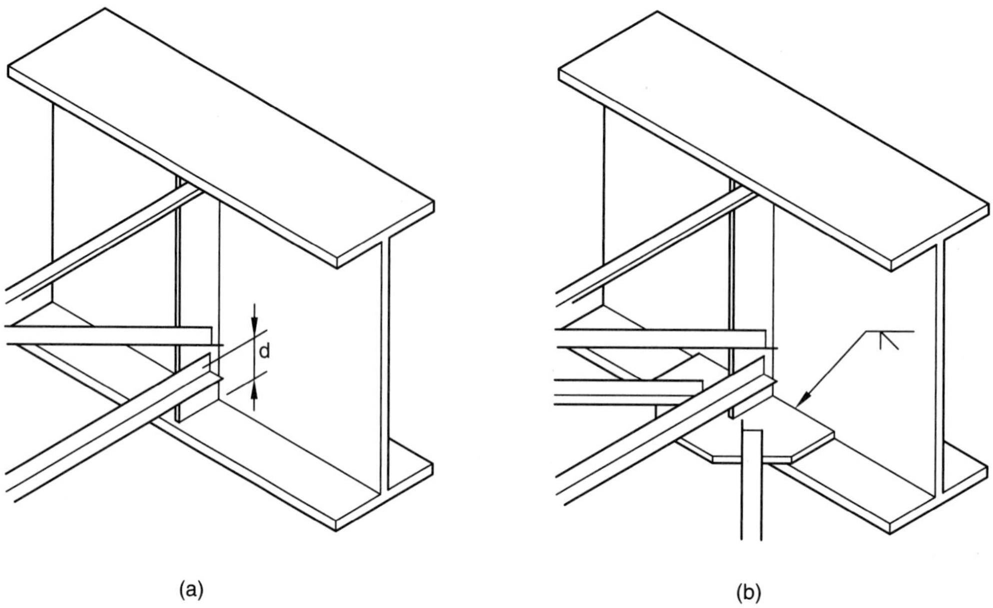

Figure 15.108 (*a*) Cross-frame connection without direct attachment to girder flanges, (*b*) Lateral bracing connection plate in conjunction with cross frame.

methods applicable to static loads include the slope-deflection, plane-grid, space-frame, *V*-load, finite-strip, finite-difference, and finite-element methods. For dynamic analyses, more rigorous methods, based on finite-element analysis or solution to differential equations for free vibration and motion of the system are recommended (Refs. 6, 7, and 8). In the following subsections, after a brief discussion on torsion and warping, the *V*-load method is explained, as it has been a popular approximate method.

15.23.2 Torsion—Warping Stresses

Horizontally curved flexural members must resist torsion as well as vertical bending moments and shears. They resist this torque by St. Venant torsion if the cross section is free to warp. However, St. Venant torsion provides very little torsional stiffness in an I-shaped section, and torsional stability in a curved bridge with girders must be achieved by interaction of girders and diaphragms.

Diaphragms cause concentrated torques to apply on the girders at each diaphragm location. A plot of the internal-torque diagram in a girder shows that an abrupt reversal in torque occurs at these locations, precisely the condition to produce warping restraint (because the nearly equal and opposite internal torques at these sections attempt to warp the cross section in opposite directions, with a net warping near zero). Warping restraint diminishes at locations away from the diaphragms, allowing St. Venant torsion to take over. Therefore, torsional resistance in an I-girder bridge is some combination of St. Venant and warping torsion at any cross section. St. Venant torsion produces shearing stresses. Warping torsion produces primarily direct tensile and compressive stresses.

15.23.3 Radial Component of Flange Force

There is a simple model to help understand the torsional action in a curved I-girder bridge. If a short segment of a curved flange (Fig. 15.109*a*) is considered, it is seen that the axial forces or stresses on the segment are not colinear because of curvature. This results in a radial component that, over some length of flange, manifests itself as a radially distributed force on the flange as shown in Fig. 15.109*b*. The magnitude of this radial force is given by

$$q = \frac{M}{Rh} \tag{15.1}$$

where q = radial force
M = vertical bending moment in girder
R = radius of girder
h = depth of girder, center to center of flanges

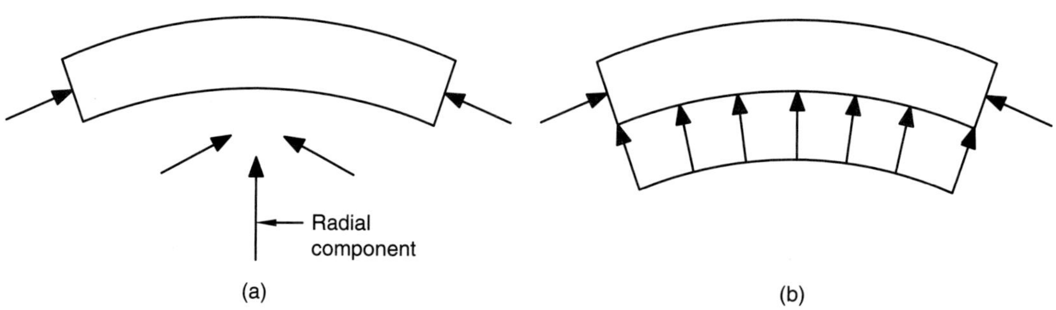

Figure 15.109 (*a*) Radial flange force component and (*b*) radially distributed flange forces.

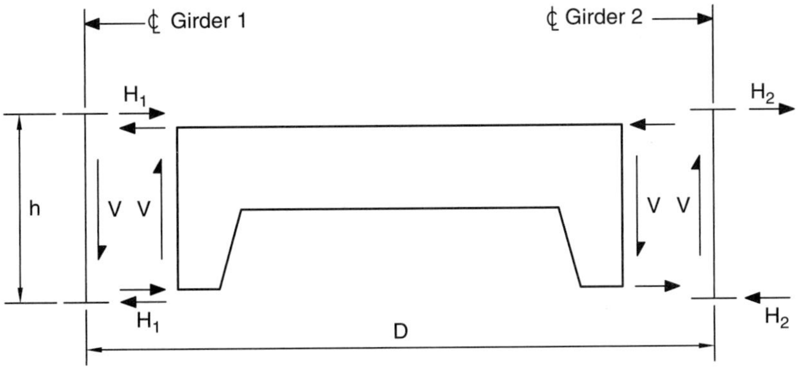

Figure 15.110 Girder-diaphragm interaction forces.

15.23.4 *V*-Load Approximation[7]

The radial force q given by Eq. (15.1) can be treated as a radial load on the flange. It then becomes logical to use diaphragms at intervals along the girder to resist this load. Figure 15.110 shows a diaphragm in a two-girder system. The interaction forces between the diaphragm and the girders consist of the forces H at each flange and the vertical shears V. The top and bottom H forces make up a couple, equivalent to a concentrated torque applied to the diaphragm by the girders and vice versa.

The following equations define the interaction forces:

$$H_1 = \frac{M_1 d_1}{hR_1} \tag{15.2}$$

$$H_2 = \frac{M_2 d_2}{hR_2} \tag{15.3}$$

$$V = \frac{M_1 + M_2}{K} + \frac{M_1' + M_2'}{K} \tag{15.4}$$

$$K = \frac{R_1 D}{d_1} = \frac{R_2 D}{d_2} \tag{15.5}$$

where M_1, M_2 = vertical bending moment in girders 1 and 2, respectively
M_1', M_2' = vertical bending moment in girders 1 and 2, respectively, resulting from primary, or applied, loads
d_1, d_2 = diaphragm spacing along girders 1 and 2, respectively
h = girder depth, center to center of flanges
R_1, R_2 = radii of girders 1 and 2, respectively
D = girder spacing

The diaphragm end shears V are called V-loads because they act as loads on the girders, producing the so-called curvature effect by augmenting the applied loads on girders outside the bridge centerline and decreasing the applied loads on girders inside the bridge centerline. The V-loads produce moments in the girder outside of the bridge centerline that add to those that would exist in a straight girder of the same developed length and moments in the girder inside of the bridge centerline that subtract from those that would exist in a straight girder of the same developed length. This is the essence of the V-load method, a design tool that has been used for curved bridges for many years. However, it is noted that the V-load method is suitable only for approximate analysis especially in preliminary bridge designs with composite sections, variable radius of curvature, and skewed supports. Limitations of this method are reported to be the facts that (1) the transverse distribution factors affect the analysis, (2) the inner girder stresses are underestimated, (3) the significance of bracing effect in the plane of the bottom flange is not included, and (4) the error in the results is greater for skew supports than is for radial ones (Ref. 8).

For a multiple-girder system Eq. (15.4) becomes

$$V = \frac{\sum M}{CK} \tag{15.6}$$

where ΣM = summation of vertical bending moments resulting from primary, or applied, loads in all girders across any diaphragm line
K = factor defined by Eq. (15.5)
C = coefficient given in Table 15.14
V = V-load on outermost and innermost girders

The V-loads on other girders are proportionate to distances from the longitudinal centerline of the bridge.

15.23.5 Bending of Lateral Flange

With the V-loads determined, everything needed to compute the vertical bending moments and shear forces in the girder is known. These internal forces are the sum of those resulting from the applied loads and from V-loads. In addition to these forces, the lateral load q also produces lateral bending of the girder flanges. In fact the flanges act as continuous beams laterally supported at the diaphragms. As indicated in Eq. (15.1), $q = M/Rh$. An approximation of the maximum lateral moment that can occur as a result of this loading is $M_{lat} = qd^2/12$ (treating the flange as a continuous beam). The lateral flange-bending stresses may be computed as M_{lat}/S_{lat}, where S_{lat} is the lateral section modulus of the flange. These stresses are equivalent to the warping normal stresses. They must be superimposed on the ordinary vertical bending stresses resulting from the primary in-plane bending moments.

In summary, the following steps are listed for analyzing a curved girder bridge:

1. Determine the primary moments (those resulting from the dead and live loads) treating curved girders as a system of straight girders with lengths equal to the developed lengths of the curved girders.

Table 15.14 Values of *C* in Eq. (6)

Number of girders	C (fraction)	C (decimal)
2	1	1.00
3	1	1.00
4	10/9	1.11
5	5/4	1.25
6	7/8	1.40
7	14/9	1.56
8	13/7	1.72
9	15/8	1.88
10	165/81	2.04

2. From the primary moments, compute V-loads at intervals in the span using Eq. (15.6).
3. Compute the secondary moments.
4. Select diaphragm spacing.
5. Compute lateral bending moments M_{lat} using the following equation:

$$M_{lat} = \frac{Md^2}{12Rh} \tag{15.7}$$

where M = primary moment plus secondary moment at any diaphragm location.

6. Determine cross sections required to provide for vertical and lateral forces computed under items 1 to 5, inclusive.

15.23.6 Centrifugal Force Effect

The AASHTO specifications require that centrifugal force (*CF*) be treated as a primary load for curved bridges. They define the magnitude of the *CF* as a percentage (*C*) of the live load, applied through the center of gravity of the vehicle, 6 ft (1.8 m) above the deck. The percentage *C* depends on speed of the moving vehicle (V) in ft/s, radius of the curve (R) in feet, and gravity accelerations (g) in ft/s^2. A factor (f) designating load combinations for those involving nonfatigue ($f = 4/3$) versus those involving fatigue ($f = 1$) is also included in C. The dependence of C on these factors is through the relationship $C = fV^2/gR$. For a case of R = 700 ft (213.4 m), V = 35 mph or 51.3 ft/s (56 km/h), $f = 4/3$ (for nonfatigue loading), and g = 32.2 ft/s^2, C = 0.155, or 15.5 percent.[9] In addition to its overturning effect, the radial component of the centrifugal force needs to be transmitted from the deck through the end cross frames or diaphragms and the bearings to the substructure. However, in general, designers must make their own assumptions as to how the *CF* is distributed through the structure. One procedure is as follows: The load may be considered to be transmitted to the bridge as a shear and an overturning moment at the deck level. The shear is delivered directly to the bridge bearings by way of the deck slab and the cross frames at the supports, with no measurable effect on the bridge other than a contribution to the forces in the diagonals of the support cross frames. The overturning moment may be reasonably assumed to increase the vertical loading on the outside girders and decrease the vertical loading on the inside girders. However, it is noted that not all the force *CF* is doing this. A given superelevation of the deck is capable of balancing a certain amount of centrifugal force by equalizing the outside and inside wheel-load reactions of the vehicle. When these wheel reactions are equal for a portion of the *CF* loading, there should be no overturning moment from that portion. It can be assumed that only the *CF* over and above the amount balanceable by superelevation produces overturning moment on the structure. The amount of *CF* balanceable by superelevation is equal to the superelevation rate. That is, if the superelevation rate is 0.05 ft/ft, or 5 percent, that percent of the live load can be subtracted from the percent *C*.

15.23.7 Example[d]

Figure 15.111 shows the framing plan for a two-span curved bridge with four girders. Figure 15.112 shows the cross section. The superelevation rate is 0.08 ft/ft, or 8 percent. The girders are assumed to act compositely with the concrete deck, with unshored construction. The bending-moment diagrams for the outermost girder G_1 are shown in Fig. 15.113. Four sets of moments are given: those resulting from initial dead load DL_1 (the self-weight of the steel girder plus the weight of the wet concrete slab) acting on the steel section, those resulting from superimposed dead load DL_2 (parapets and wearing surface) acting on the long-term composite section, and the maximum positive and the maximum negative moments resulting from live load plus impact ($L + I$) acting on the short-term composite section.

For a vehicle speed of 30 mph or 44.0 ft/s (48 km/h) and a radius of 286.76 ft (87.4 m), from the aforementioned equation, C = 0.21 (or 21%) for a factor of f = 1. Note that in this example the factor for the type of load combination (fatigue versus nonfatigue) is not considered since the references from which the example is taken used earlier versions of the code. Applying f = 4/3, C would result in 0.28. However, to keep the values reported in the references, the problem is continued with C = 0.21. The superelevation balances 8 percent of the 21 percent. The resulting lateral *CF* in terms of traffic lanes is therefore (0.21 − 0.08) × 2 = 0.26 lanes.

Pile-group distribution is used to determine the vertical loads on the girders resulting from the overturning moment of the *CF*. With girders

[d]From original article by William N. Poellot Jr., HDR Engineering, Inc.

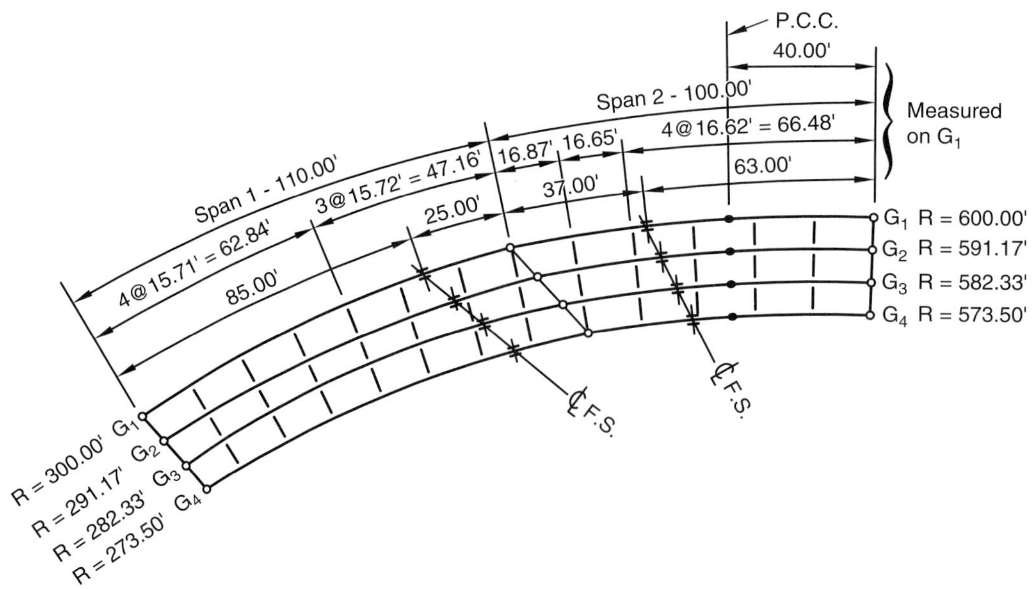

Figure 15.111 Framing plan.

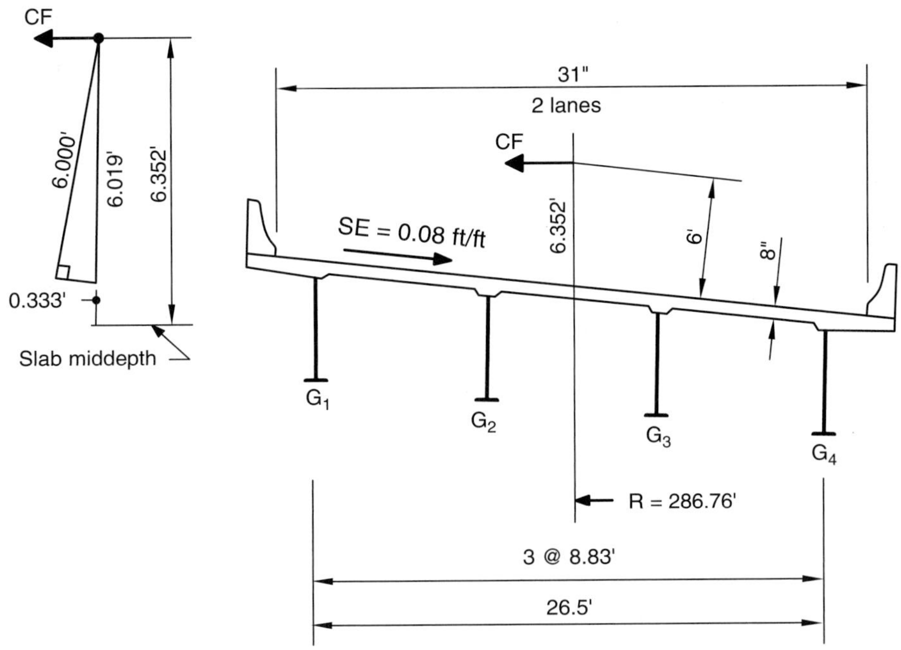

Figure 15.112 Moment diagrams for girder G1.

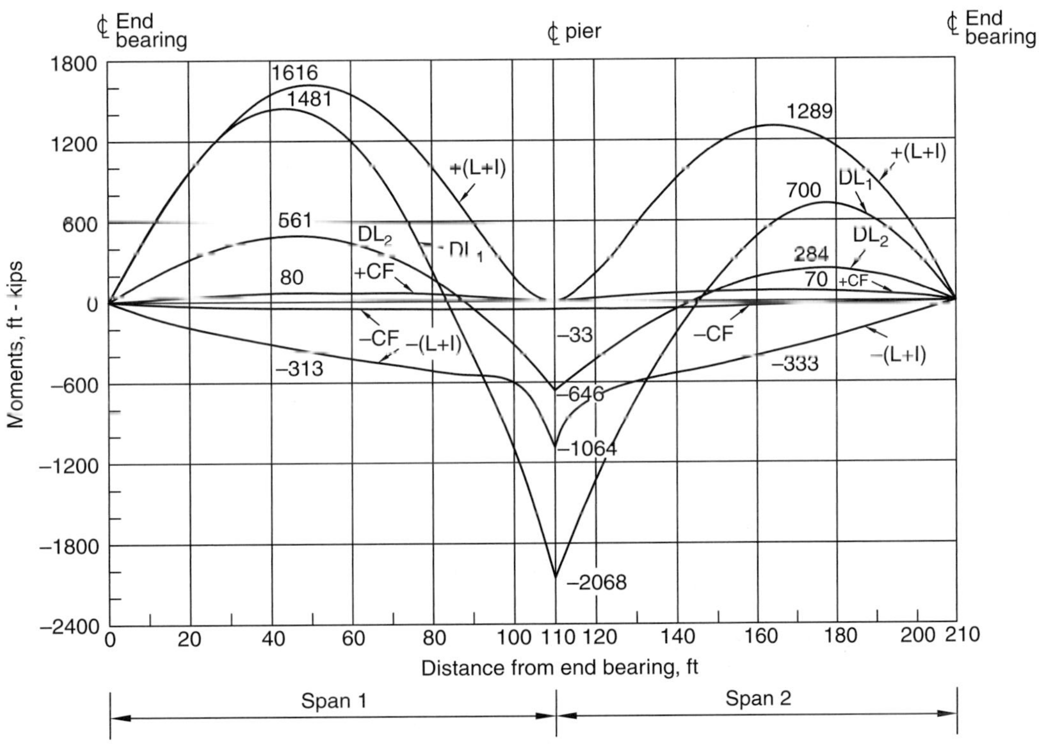

Figure 15.113 Centrifugal force.

spaced 8.83 ft (2.7 m) apart and the *CF* moment arm 6.352 ft (1.9 m), as seen in Fig. 15.112,

$$I = 2[(0.5 \times 8.83)^2 + (1.5 \times 8.83)^2] = 389.8 \text{ ft}^2$$

$$CF \text{ load on } G1 = \frac{0.260 \times 6.352 \times 1.5 \times 8.83}{389.8} = 0.056 \text{ lanes}$$

$$CF \text{ load on } G2 = \frac{0.260 \times 6.352 \times 0.5 \times 8.83}{389.8} = 0.019 \text{ lanes}$$

The *CF* vertical loads on *G*3 and *G*4 are assumed to be zero since such loads would act upward and be subtractive from the other vertical loads. This is a valid loading case, and it corresponds to the situation in which live load is on the structure but not moving.

INVESTIGATION OF GIRDER G1, SPAN 1. The maximum positive bending moments are at the section 44 ft from the end support (Fig. 15.113). They are

$$M_{DL1} = 1481 \text{ ft·kips (1999 m·kN)}$$

$$M_{DL2} = 561 \text{ ft·kips (757 m·kN)}$$

$$M_{L+I} = 1616 \text{ ft·kips (2182 m·kN)}$$

$$M_{CF} = 80 \text{ ft·kips (108 m·kN)}$$

The *CF* moment is obtained from the *CF* load of 0.056 lanes. The corresponding lateral bending moments are computed from Eq. (15.7):

$$M_{\text{lat}} = \frac{Md^2}{12Rh} = \frac{M \times 15.71^2}{12 \times 300 \times 55.25/12} = 0.0149M$$

$$= 0.0149 \times 1481 = 22.1 \text{ ft·kips (29.8 m·kN) for } DL_1$$

$$= 0.0149 \times 561 = 8.4 \text{ ft·kips (11.3 m·kN) for } DL_2$$

$$= 0.0149 \times 1616 = 24.1 \text{ ft·kips (32.5 m·kN) for } L{+}I$$

$$= 0.0149 \times 80 = 1.2 \text{ ft·kips (1.6 m·kN) for } CF$$

A trial section with the following characteristics is selected:

Top flange: 1 × 14 in. plate (25 × 356 mm)

Bottom flange: 1½ × 18 in. plate (38 × 457 mm)

Web: 7/16 × 54 in. plate (11 × 1372 mm)

The bottom of the 7.5-in. (191 mm) thick slab is at 2.5 in. (64 mm) from top of the steel plate girder. The section is composite with an effective slab width of 90 in. (2.3 m). Section properties are computed in the normal fashion for the steel section and the long-term and short-term composite sections. Using these properties and the prescribed AASHTO load factor 1.3 and β factor 1.67 for *L* + *I*, the following stresses are obtained.

Top of steel (compression), note that 1 ksi = 6.9 MPa.

Vertical bending

$$\text{For } DL_1\text{: } f_b = \frac{1481 \times 12}{1043} \times 1.30 = 22.2 \text{ ksi (153.1 MPa)}$$

$$\text{For } DL_2\text{: } f_b = \frac{561 \times 12}{2969} \times 1.30 = 2.9 \text{ ksi (20.0 MPa)}$$

$$\text{For } L{+}I\text{: } f_b = \frac{1616 \times 12}{7819} \times 1.67 \times 1.30 = 5.4 \text{ ksi (37.2 MPa)}$$

$$\text{For } CF\text{: } f_b = \frac{80 \times 12}{7819} \times 1.30 = 0.2 \text{ ksi (1.4 MPa)}$$

Total = 30.8 ksi (211.7 MPa)

Lateral bending

$$\text{For } DL_1\text{: } f_w = \frac{22.1 \times 12}{32.7} \times 1.30 = 10.5 \text{ ksi (72.4 MPa)}$$

There are no lateral top flange bending stresses from DL_2, *L* + *I*, and *CF* because the flange is supported by the slab under those loads.

Bottom of steel (tension)

Vertical bending

$$\text{For } DL_1\text{: } f_b = \frac{1481 \times 12}{1543} \times 1.30 = 15.1 \text{ ksi (104.1 MPa)}$$

$$\text{For } DL_2\text{: } f_b = \frac{561 \times 12}{1871} \times 1.30 = 4.7 \text{ ksi (32.4 MPa)}$$

$$\text{For } L{+}I\text{: } f_b = \frac{1616 \times 12}{2022} \times 1.67 \times 1.30 = 20.8 \text{ ksi (143.4 MPa)}$$

$$\text{For } CF\text{: } f_b = \frac{80 \times 12}{2022} \times 1.30 = 0.6 \text{ ksi (4.1 MPa)}$$

Total = 41.1 ksi (284.0 MPa)

Lateral bending

$$\text{For } DL_1\text{: } f_w = \frac{22.1 \times 12}{81.0} \times 1.30 = 4.3 \text{ ksi (29.6 MPa)}$$

$$\text{For } DL_2\text{: } f_w = \frac{8.4 \times 12}{81.0} \times 1.30 = 1.6 \text{ ksi (11.0 MPa)}$$

$$\text{For } L{+}I\text{: } f_w = \frac{24.1 \times 12}{81.0} \times 1.67 \times 1.30 = 7.8 \text{ ksi (53.8 MPa)}$$

$$\text{For } CF\text{: } f_w = \frac{1.2 \times 12}{81.0} \times 1.30 = 0.2 \text{ ksi (1.4 MPa)}$$

Total = 13.9 ksi (95.8 MPa)

ALLOWABLE STRESS. The allowable stress F_{bu} is[1]

$$F_{bu} = F_{bs}\bar{\rho}_b\bar{\rho}_w$$

where $F_{bs} = F_y$ for tension flanges and $F_y(1 - 3\lambda^2)$ for compression flanges with $b/t \le 3000/\sqrt{F_y}$

$$\lambda = \frac{1}{\pi}\frac{l}{b}\sqrt{\frac{F_y}{E}}$$

$$\bar{\rho}_b = \frac{1}{1 + \left(\frac{l}{b}\right)\left(1 + \frac{l}{6b}\right)\left(\frac{l}{R} - 0.01\right)^2}$$

$$\bar{\rho}_w = 0.95 + 18\left(0.1 - \frac{l}{R}\right)^2 + \frac{\left(\frac{f_w}{f_b}\right)\left(0.3 - 0.1\frac{l}{R} \times \frac{l}{b}\right)}{\bar{\rho}_b\left(\frac{F_y}{F_{bs}}\right)}$$

where b = flange thickness
t = flange thickness
l = diaphragm spacing
R = radius of girder
f_b = vertical-moment bending stress
f_w = lateral-moment bending stress

For the top flange of the girder of this example

$$\frac{b}{t} = \frac{14}{1} = 14 < \frac{b}{t} = \frac{3200}{\sqrt{50{,}000}} = 14.31$$

Therefore, the flange is compact and

$$\bar{\rho}_b = \frac{1}{1+\left(\frac{15.71}{14/12}\right)\left(1+\frac{15.71}{6\times 14/12}\right)\left(\frac{15.71}{300}-0.01\right)^2} = 0.92729$$

$$\lambda = \frac{1}{\pi}\times\frac{15.71}{14/12}\sqrt{\frac{50}{29{,}000}} = 0.17798$$

$$\bar{\rho}_w = 0.95+18\left(0.1-\frac{15.71}{300}\right)^2 + \frac{\left(\frac{10.5}{30.8}\right)\left(0.3-0.1\frac{15.71}{300}\times\frac{15.71}{\frac{14}{12}}\right)}{0.92729\left(\frac{50}{45.248}\right)} = 1.06719$$

$$\bar{\rho}_b\bar{\rho}_w = 0.92729\times 1.06719 = 0.98959 < 1$$

$$F_{bu} = 45.248\times 0.98959 = 44.8 \text{ ksi} > 30.8$$

For the bottom flange

$$\bar{\rho}_b = \frac{1}{1+\left(\frac{15.71}{18/12}\right)\left(1+\frac{15.71}{6\times 18/12}\right)\left(\frac{15.71}{300}-0.01\right)^2} = 0.95092$$

$$F_{bs} = 50 \text{ ksi}$$

$$\bar{\rho}_w = 0.95+18\left(0.1-\frac{15.71}{300}\right)^2 + \frac{\left(\frac{13.9}{41.1}\right)\left(0.3-0.1\frac{15.71}{300}\times\frac{15.71}{\frac{18}{12}}\right)}{0.95092\left(\frac{50}{50}\right)} = 1.07803$$

$$\bar{\rho}_b\bar{\rho}_w = 0.95092\times 1.07803 = 1.02512 > 1, \text{ thus use } 1.0$$

$$F_{bu} = 50.0\times 1.0 = 50 \text{ ksi} > 41.1$$

DIAPHRAGM FORCES: The diaphragm (cross frame) forces resulting from curvature are computed using the V-load method, as explained earlier. The assumed distribution of V-loads and torques on the girders of a four-girder bridge are shown in Fig. 15.114*a*. The V loads are proportionate to their distances from the bridge centerline. The corresponding distribution of end shears and moments on the cross frames are shown in Fig. 15.114*b* and *c*, respectively.

An intermediate cross frame at the 0.429 point in span 1 of the bridge of Fig. 15.111 will be investigated for the AASHTO Group I loading 1.3 $[D + 5/3\,(L + I) + CF_v]$. The coefficient C for a four-girder bridge is 10/9 (Table 15.14). The moments at the cross frame are given in Table 15.15.

As shown in Fig. 15.111, the diaphragm spacing d is 15.71 ft (4.8 m), and the distance between outside girders D is 26.5 ft (8.1 m) as shown in Fig. 15.112. From Eq. (15.6),

$$V = \frac{\sum M}{CRD/d} = \frac{20{,}082}{\frac{10}{9}\times 300\times 26.5/15.71} = 35.7 \text{ kips (159.9 kN)}$$

$$\frac{V}{3} = \frac{35.7}{3} = 11.9 \text{ kips (53.3 kN)}$$

The torsional moments on the girders (Fig. 15.114*a*) are

$$\frac{M_1 d}{R} = 6260\times\frac{15.71}{300} = 327.8 \text{ ft.kips (442.5 m.kN)}$$

$$\frac{M_2 d}{R} = 5656\times\frac{15.71}{300} = 296.2 \text{ ft.kips (399.9 m.kN)}$$

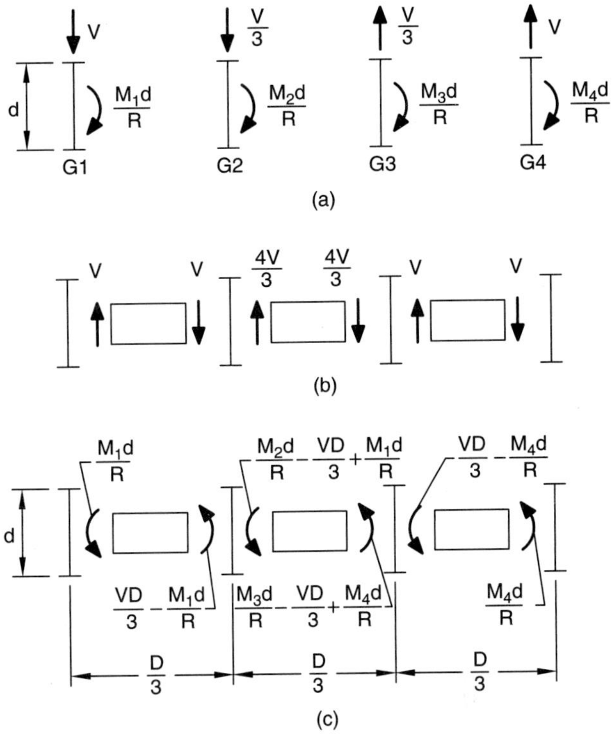

Figure 15.114 (*a*) Girder loads from diagrams, (*b*) diaphragm end shears, and (*c*) diaphragm end moments.

$$\frac{M_3 d}{R} = 4569\times\frac{15.71}{300} = 239.3 \text{ ft.kips (323.1 m.kN)}$$

$$\frac{M_4 d}{R} = 3598\times\frac{15.71}{300} = 188.4 \text{ ft.kips (254.3 m.kN)}$$

These and the V-loads are shown in Fig. 15.115*a*. The corresponding moments and shears on the cross frames, computed from the formulas in Fig. 15.114*b* and *c*, are shown in Fig. 15.115*b*. The depth of the cross frame is 3.58 ft (1.1 m) center of gravity to center of gravity of the top and bottom struts, from which, together with the shears, the internal forces shown in Fig. 15.115*c* are obtained.

Other AASHTO group loadings may have to be checked to determine the controlling forces in the cross-frame members. Because the cross-frame forces are proportionate to the girder bending moments, there is a wide variation in the magnitudes of these forces throughout the bridge. However, in general, from fabrication and construction points of view, a variety of cross-frame designs is not economical, and usually only one or two designs are used. Furthermore, it is noted that this example treated static loads only. When dynamic loads are present, a more rigorous analysis involving the finite-element method is desirable.

Table 15.15 Moments at Intermediate Cross Frame at 0.429 Point, Span 1

Girder	M_{DL1}	M_{DL2}	M_{L+1}	M_{CFv}	$1.3(\Sigma M_{DL} + 5/3M_{L+I} + M_{CFv})$
G_1	1481	561	1616	80	6,260
G_2	1288	473	1537	28	5,656
G_3	1028	368	1271	—	4,569
G_4	755	266	1048	—	3,598
					Σ = 20,082

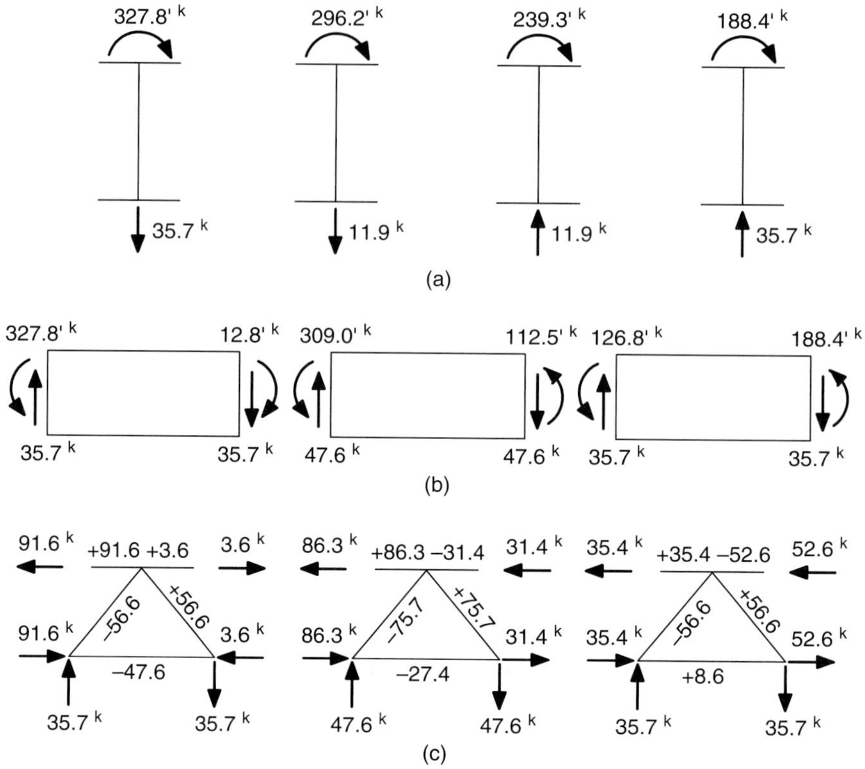

Figure 15.115 Forces on (*a*) girder, (*b*) cross frames, and (*c*) cross frames (internal).

15.23.8 Additional Reading

The reader is referred to additional materials on I-girder curved bridges and example problem provided in Refs. 10, 11, and 12.

REFERENCES

1. *Guide Specifications for Horizontally Curved Steel Girder Highway Bridges*, American Association of State Highway and Transportation Officials, Washington, DC, 2003.

2. Zureick, A., Naqib, R., and Yadlosky, J. M., *Curved Steel Bridge Research Project*, Publication No. FHWARD-93-129, Federal Highway Administration, McLean, VA, 1994.

3. Pennsylvania Department of Transportation, *Design Manual*, part 4, PDT-Pub. No. 15, April 2015.

4. *Standard Specifications for Highway Bridges*, 17th ed., American Association of State Highway and Transportation Officials, Washington, DC, 2002.

5. Iowa Department of Transportation, Office of Bridges and Structures—LRFD Bridge Design Manual, January 2019.

6. Kollbrunner, C. F., and Basler, K., *Torsion in Structures—An Engineering Approach*, Springer-Verlag, New York, 1969.

7. Poellot, W. N., Jr., "Computer-Aided Design of Horizontally Curved Girders by the *V*-Load Method," *AISC*, 1st Quarter, 1987.

8. Zureick, A., and Naqib, R., "Horizontally Curved Steel I-Girders State-of-the-Art Analysis Methods," *Journal of Bridge Engineering*, ASCE, Vol. 4, No. 1, 1999, pp. 38–47.

9. Kulicki, J. M., Wassef, W. G., Smith, C., and Johns, K., *AASHTO-LRFD Design Example Horizontally Curved Steel I-Girder Bridge, NCHRP 12-52*, National Cooperative Highway Research Program, Transportation Research Board, National Research Council, Washington, DC, 2005.

10. Coletti, D., and Puckett, J., "Structural Analysis," in *Steel Bridge Design Handbook*, Pub. No. FHWA-HIF-16-002—Vol. 8, Federal Highway Administration, Washington, DC, December 2015.

11. Rivera, J., and Chavel, B., "Example 3: Three-Span Continuous Horizontally Curved Composite Steel I-Girder Bridge," in *Steel Bridge Design Handbook*, Pub. No. FHWA-HIF-16-002—Vol. 23, Federal Highway Administration, Washington, DC, December 2015.

12. White, D. W., Coletti, D., Chavel, B., et al., *Guidelines for Analytical Methods and Construction Engineering of Curved and Skewed Steel Girder Bridges*, NCHRP Report 725, Transportation Research Board, National Research Council, Washington, DC, 2012.

PART D
CURVED STEEL BOX-GIRDER BRIDGES

BY
JAMSHID MOHAMMADI *Illinois Institute of Technology*
BORA JANG *Sharma and Associates, Inc.*

15.24 GENERAL

15.24.1 Introductory Remarks

This section extends the discussion on curved bridges to those made up of steel box girders (also known as "tub" girders). Steel box girders are classified under welded plate girders. Compared with I-girders, these sections provide a substantially more resistance to out-of-plane bending and torsion effects. Therefore, under certain conditions, box-girder bridges may become a more favorable alternative to I-girder section bridges. However, in general, their fabrication is more complex and thus costly compared with those in I-girder bridges. Their design is governed by AASHTO design documents.[1,2,3]

15.24.2 Features and Geometry[e]

Figure 15.116 shows the cross section of a bridge deck made up of steel box girders and reinforced-concrete rein slab. The main components of this system are the following:

1. Two steel web plates
2. A steel-bottom flange plate joining the two webs and forming the underside of the box
3. Two steel top flange plates
4. The reinforced-concrete bridge deck, acting compositely with the steel girder by means of stud sheer connectors welded to the steel top flanges

[e]Material for this section is taken primarily from the original article by William N. Poellot Jr., HDR Engineering, Inc., and Refs. 4 to 9.

The bottom flange, where it is in compression, is stiffened by longitudinal tees or other shapes welded at uniform spacing across the width of the flange. Webs are similar to webs of plate girders, and they may be stiffened transversely or both transversely and longitudinally. They are often sloped rather than being vertical giving a trapezoidal shape to the cross section of the girder. This is done not only to improve the appearance of the bridge but also to reduce the width of the bottom flange.

Box girders can be considered as an alternative to I-girders in design of almost any bridge. They can be fabricated to any reasonable radius of curvature, to compound curvature, and to combinations of tangent and curved alignment. Curved bridges are mostly superelevated.

As seen in Fig. 15.116, customary treatment is to rotate the entire flat cross section of the bridge into its superelevated position when the alignment goes into curvature. This procedure simplifies design and fabrication by maintaining the symmetry of the girder sections and keeping the webs at a constant depth. Occasionally this practice may not be exercised in order to achieve certain architectural effects. For example, the soffit of the boxes shown in Fig. 15.117 has been kept level in the superelevated cross section, resulting in a different depth for each of the four webs. If this arrangement is used in areas of superelevation transition, the depth of each web will vary continuously. Such geometric complication obviously adds to the complexity of design, fabrication, and cost.

Ideally, the box-girder width, depth, and spacing should be the same at all points along the bridge. The top and bottom flanges should be parallel to the superelevation, and the webs should be at a constant slope

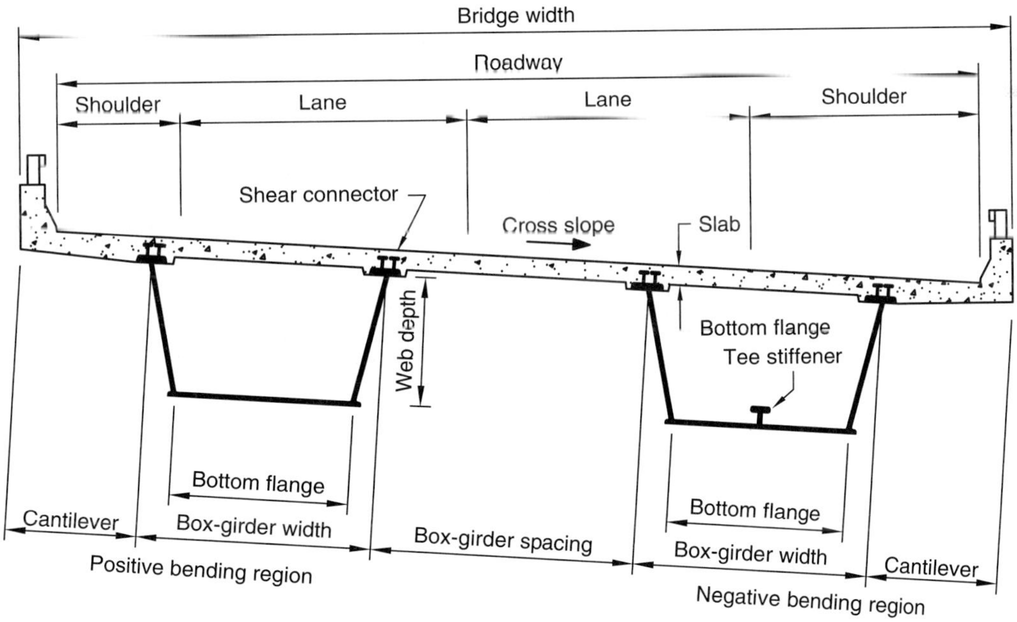

Figure 15.116 Cross section of typical box-girder bridge, superelevated for curvature.

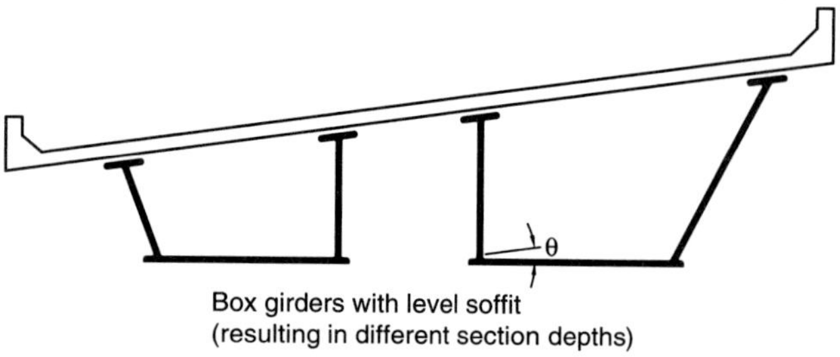

Figure 15.117 Superelevated cross section with level soffit.

relating to the flanges. In situations where the superelevation varies under these conditions, the plan dimensions of the roadway will vary slightly. However, these variations are small and negligible in any practical sense. In preparation of structure plans, the overall plan geometry shown should be that for the cross section in its flat position.

The webs play a critical role in the fabrication of trapezoidal steel box girders. In cases where the inclined web in a curved steel box girder is on a level grade, a relatively simple geometrical solution can be worked out for fabrication. The web in such cases is in fact a sector of a right circular cone with constant inclination as demonstrated in Fig. 15.118*a*. A cone is an easily calculated developable surface, which offers a geometrical closed-form solution for fabrication purposes.

In situations where the bridge is on a grade, the problem is more complex. For a uniform grade, the erected web plate may be modeled as a sector of a convolute—that is, the surface generated by a line tangent to a helix circumscribed on a cylinder as shown in Fig. 15.118*b*. The helix climbs the cylinder at a uniform rate, representing the uniform grade of the bridge. Again, this model has a geometrical closed-form solution that can be used for fabrication purposes.

This leads to the prevailing general case of an inclined web plate that is curved both horizontally and vertically. Mathematically, such a web is a warped or nondevelopable surface that theoretically cannot be detailed or built by ordinary bridge drafting and fabrication procedures. It is possible, however, to model the surface approximately as a large number of small developable surfaces. The degree of accuracy of the approximation depends on the number of increments or small developable surfaces into which the plate is broken. With a sufficient number of increments, this approach results in reasonably accurate geometry and within normal fabrication tolerances. The required computations are generally involved and beyond the scope of ordinary bridge detailing practice. Thus, they are usually performed with the aid of special computer software.

15.25 DESIGN CONSIDERATIONS

15.25.1 Framing

Generally, a typical two-lane feeder ramp system will have two box girders supporting the truck load, although occasionally a single box girder may be used, especially if the ramp is of a single lane. Of course, bridges with three or more lanes of traffic may require more than two box girders. The spacing and width of girders is commonly decoded based on the distance, center to center, of adjacent top flanges and is within the range of 80 to 120 percent of the width of the box girders, with a deck cantilever that does not exceed 6 ft (1.8 m) or 60 percent of the distance, center to center, of adjacent top flanges.

Box girder bridges have been built for crossings with limited skews. It is generally recommended to avoid actual skewing of ends of box girders. This is because with skewed ends, there is a fundamental incompatibility of the web deflection at radial cross sections that results in localized stresses. These stresses are generally difficult to model and to design for, unless some type of localized finite-element analysis scheme is adopted. Sometimes an overall bridge skew can be achieved by providing a wide bridge seat to allow each box girder to be squared off. In light of issues with compatibility and complexity in design, I-girder bridges are often preferred for skewed bridges.

In addition to the main components of box girders explained earlier, curved box-girder bridges have also a number of other components that are important to recognize and designed for. These are the following:[4–8]

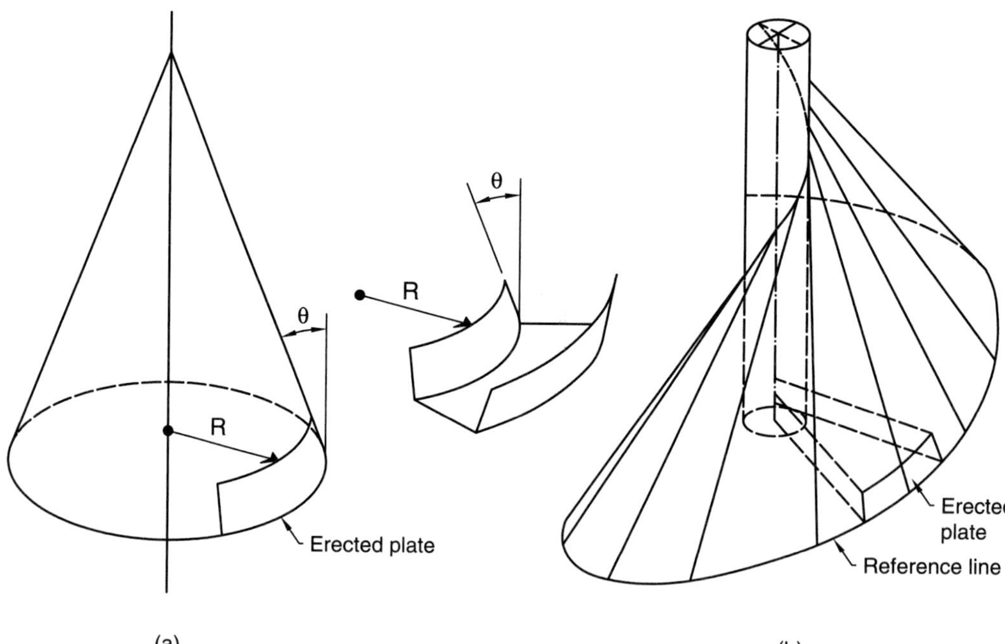

Figure 15.118 Conical and convolute development of inclined web.

1. Solid web diaphragms *within* the box at points of support to transmit the girder torque and reaction to the support bearings.
2. Open or cross-frame-type diaphragms *within* the girder, placed at intervals along the span, to hold the shape of the box and brace the narrow top flanges when the slab concrete is still fresh prior to its hardening.
3. Solid web diaphragms *between* boxes at points of support to transmit the torque reactions to the entire cross section and, at the end bearings, to support the expansion joint.
4. A system of lateral bracing *within* the girders near the top flange to close the three-sided box sections and to make them act as stiff torsion tubes prior to and during placement of the deck slab. After the deck is in place and concrete is hardened, the slab itself serves as the closing element of the boxes, rendering the lateral bracing redundant for subsequent loading.

Because of their closed cross sections, box girders are very efficient in carrying the torsion in a curved bridge. Thus, they need not rely on interaction between girders through a system of external diaphragms. This means usually there is no need for intermediate external diaphragms, and they can be avoided. Moreover, with a pair of bearings that can develop a torque-resisting couple under each box at support points, it may be possible to eliminate diaphragms between boxes at these locations as well, unless they are needed to support the expansion joint under a free edge of deck slab, as may be the case at an end bearing.

15.25.2 Details

Steel box girders are detail intensive. Details of their welds, joints, and secondary elements are governed by such factors as the level of effort needed in fabrication and cost. Furthermore, because of the types of details, fatigue may become another important factor in details. Most of the detail considerations discussed for steel I-girder bridges are also applicable to steel box-girder bridges. These include, for example, using curved edges instead of sharp ones to have a more favorable fatigue-resistant detail, when possible. Typical solid web and cross-frame-type internal diaphragms are shown in Fig. 15.119*a* and *b*.

Curved steel box girders require installation of a lateral bracing system at the top. Note that this system is often not needed in curved I-girder bridges. The top bracing system requires some attention to detail at its connection points to the girders. Because of interference with the deck formwork, the bracing system often cannot be positioned to provide the most efficient path for transfer of forces, that is, through direct attachment to the girder flanges. As such, a location several inches lower is selected requiring the bracing attachment to the web instead. This creates a complexity in detailing since the connection must be fit around the cross frame and its connection plate. Internal cross

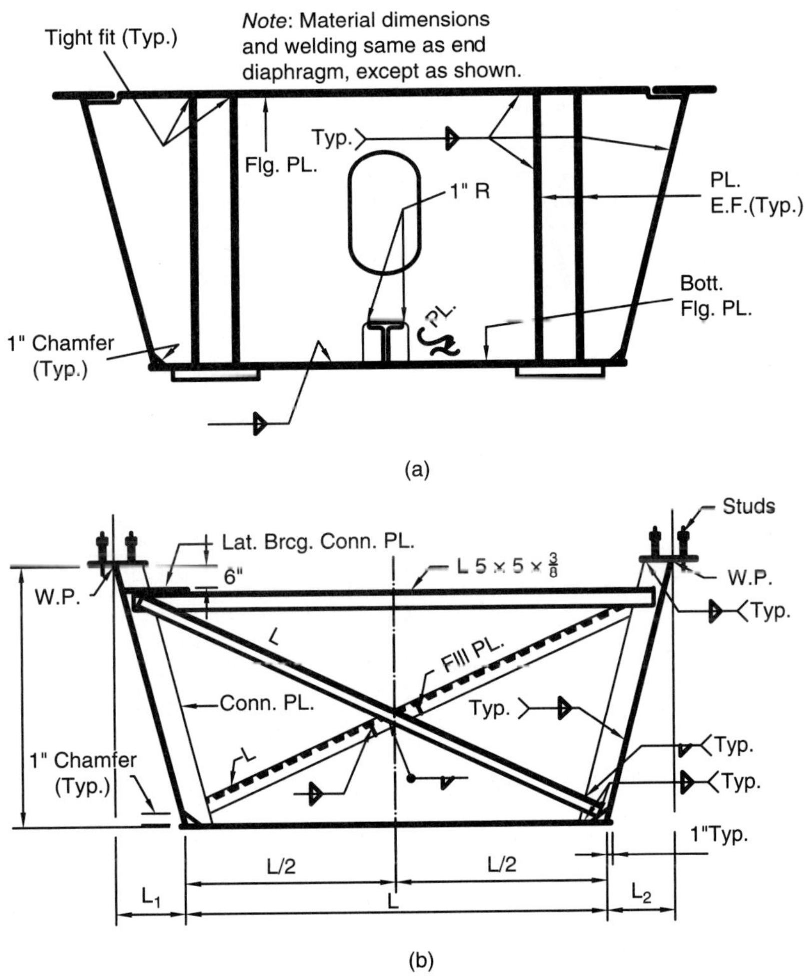

Figure 15.119 (*a*) Support diaphragm and (*b*) intermediate diaphragm.

frames are usually placed at the lateral bracing panel points, or at least at alternate panel points. Welded and bolted arrangements for internal cross frames are shown in Fig. 15.120*a* and *b*.

The welded detail is a long attachment in the direction of stress and is therefore an AASHTO Category *E* fatigue detail. This connection is technically acceptable and can be designed for. However, it is generally not recommended. Structures made up of one or two box girders are nonredundant since a crack perpendicular to the stress field can propagate through the entire section. The resultant low available stress range for the Category *E* or *E′* detail in a nonredundant member renders a welded attachment an inefficient connection that many designers would consider not worthy of the fatigue risk involved.

A similar problem exists at the termination point of longitudinal tees that stiffen the bottom flange in negative bending regions (bottom flange compression zones). These elements are logically carried into tension zones to ensure that the entire bottom flange compression zones are stiffened. A square termination, as depicted in Fig. 15.121*a*, is a Category *E* or *E′* detail and should be avoided. A more favorable fatigue-resistant derail may be achieved by introducing a curve in the stiffener stem and grinding the stiffener-to-flange weld to a smooth transition, as illustrated in Fig. 15.121*b*.

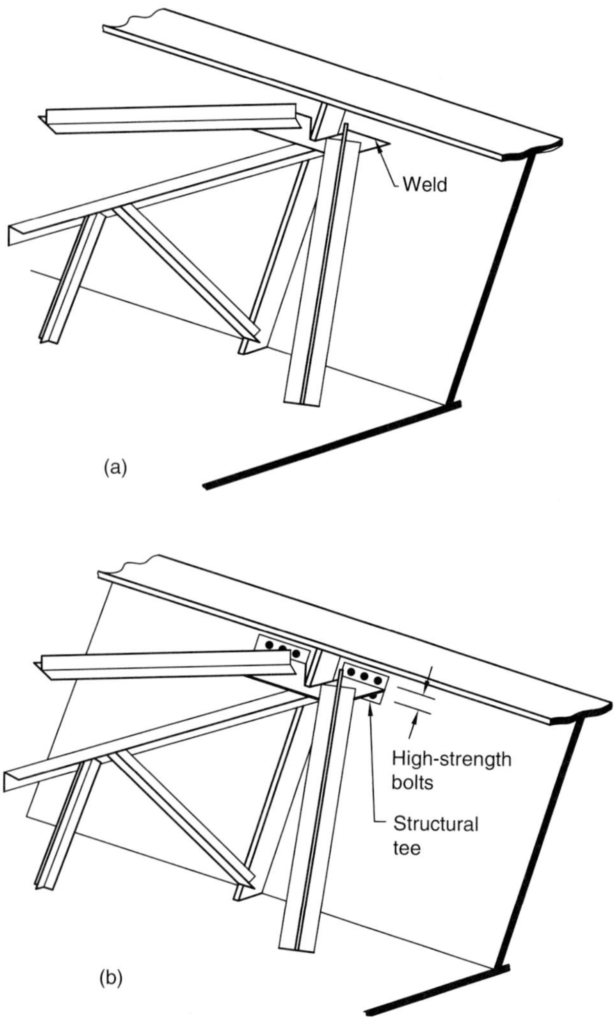

Figure 15.120 (*a*) Welded and (*b*) bolted lateral bracing connections.

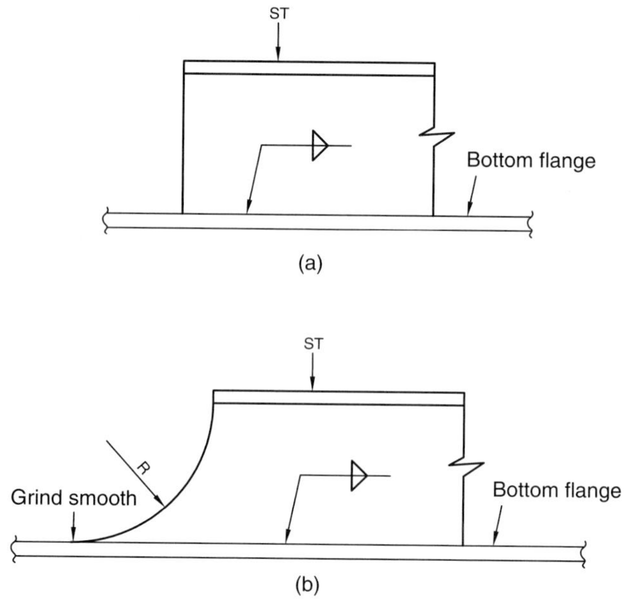

Figure 15.121 (*a*) Square stiffener and (*b*) curved cope terminations.

15.25.3 Structural Analysis[4]

Steel box girders are generally more complex structures than I-girders; as such, their analysis is more involved. Available methods include (1) one-dimensional methods, which use line elements; (2) two-dimensional methods, which employ grid analysis; and (3) three-dimensional finite-element method, which is the most accurate of the three types. Although one- and two-dimensional methods can be employed for box-girder bridges, depending on the complexity of various details, potential for stress concentrations, lateral load presence, and the severity of the curvature in curved bridges, such a model may involve various levels of approximations. For example, using line elements to model structural members in box-girder bridges, the analysis may then require modeling the bridge system as a two-dimensional planar grid and analyzed by a classical stiffness method for moment, shear, and torque. The grid may be taken as the two box girders, considered totally independent of each other, or as the two girders interacting through the deck slab and external diaphragms. It is noted that the idealization of the girders with line elements is appropriate when the transverse dimensions of the members are small relative to the length. In any case, the use of such idealizations and classical beam theory to obtain girder internal stresses should be considered as an approximate solution with a final analysis considered using more elaborate methods. Specifically, the finite-element method of analysis is preferred, which will allow a more accurate modeling of structural components through a mix of element types including plates and shell elements. Software packages are available for finite-element analysis and other rigorous solutions to curved box-girder problems. Such tools often allow the user to model loads (including live and lateral loads) and the multitude of load positions necessary to develop design moment, shear, and torque envelopes.

As an approximate solution comparable to that obtained by the *V*-load method for curved I-girder bridges, the *M*/*R* method[4,9,10] makes use of the conjugate beam analogy. This method is applicable to curved box-girder bridges in which the girders are not tied together by intermediate external diaphragms. The computations are relatively simple and can be performed by hand, particularly to gain an understanding of curved box-girder behavior. However, the method is most often used by the aid of computer programs or spreadsheets. Similar to the *V*-load methods, the *M*/*R* method assumes that the moments and shears in the curved spans are obtainable as straight-girder moments and shears based on the developed span lengths of the curved structure. To compute the torques, a distributed loading is applied to a simple-span, straight, conjugate beam which represents the actual simple or continuous, curved span. The distributed loading is equal to M/R, the moment induced by the real applied loading in the actual simple or continuous span considered straight, divided by the radius of curvature of the girder.

The resulting shear forces in the conjugate beam are then numerically equal to those from the internal torques in the actual span. Upon these calculations, all the internal moments, shear forces, and torques needed to design the sections of the curved box girders are then determined using the conventional theories of beam analysis.

15.26 BOX-GIRDER MECHANICS[f]

15.26.1 Torsional Effects

The applied torque in a curved box girder is resisted by a combination of two kinds of internal torsion, pure (St. Venant) torsion and warping torsion. St. Venant torsion provides most of the resistance. Per AASHTO Guide Specifications, if the box girder does not have a full-width steel top flange, the girder must be treated under initial dead load (wet-concrete stage) as an open section. This means that the St. Venant torsion constant K_T is given by

$$K_T = \frac{1}{3}\Sigma(bt^3)$$

where b = width of plate element
t = thickness of plate element

If the section is closed, however,

$$K_T = \frac{4A^2}{\Sigma(b/t)}$$

where A = enclosed area of the section.

A closed box-girder section is usually several thousand times stiffer in torsion than an open section. As discussed earlier, a lateral bracing system is required to "quasi-close" the box during the wet concrete stage of construction.

For analysis purposes, the top lateral bracing may be transformed to an equivalent thickness of plate t in inches by the formulas given in Fig. 15.122, where E = modulus of elasticity, ksi (MPa); G = shear modulus of elasticity, ksi (MPa); A = area of indicated member, in.^{2} (mm^2); and I = moment of inertia of indicated member, in.^{4} (mm^4).

To close the section properly and to reduce warping stresses to a minimum, the cross-sectional area of the lateral-bracing diagonal should be at least 3 percent of the depth of the box in inches. The internal stresses produced by St. Venant torsion in a closed section are shearing stresses around the perimeter, as shown in Fig. 15.123, and are defined by

$$\tau = \frac{T}{2At}$$

where τ = St. Venant shear stress in plate, ksi (MPa)
T = internal torque, in.-kips (m.kN)
A = enclosed area within box, in.^{2} (mm^2)
t = thickness of plate, in. (mm)

This shear stress adds to the vertical shear stress in one of the girder webs and subtract from the vertical shear stress in the other girder web.

Direct stresses as shown in Fig. 15.124 result from warping torsion restraint, and stresses normal to the plane of the plates are from distortion of the cross section c. Per AASHTO Guide Specifications, the effect of normal stresses due to nonuniform torsion (i.e., warping torsion) and cross-sectional deformation will need to be included in the design of curved box girders unless it is shown through analyses that these effects are small.

Warping restraint occurs principally at interior supports of continuous curved box girders where an abrupt reversal of torque (equal to the torque

Case A:

$$t^* = \frac{E}{G}\,\frac{ab}{\dfrac{d^3}{A_d} + \dfrac{a^3}{3}\left(\dfrac{1}{A_o} + \dfrac{1}{A_u}\right)}$$

Case B:

$$t^* = \frac{E}{G}\,\frac{ab}{\dfrac{2d^3}{A_d} + \dfrac{b^3}{4A_v} + \dfrac{a^3}{12}\left(\dfrac{1}{A_o} + \dfrac{1}{A_u}\right)}$$

Case C:

$$t^* = \frac{E}{G}\,\frac{ab}{\dfrac{d^3}{2A_d} + \dfrac{a^3}{12}\left(\dfrac{1}{A_o} + \dfrac{1}{A_u}\right)}$$

Case D:

$$t^* = \frac{E}{G}\,\frac{ab}{\dfrac{d^3}{A_d} + \dfrac{b^3}{A_v} + \dfrac{a^3}{12}\left(\dfrac{1}{A_o} + \dfrac{1}{A_u}\right)}$$

Case E:

$$t^* = \frac{E}{G}\,\frac{1}{\dfrac{ab^2}{12I_b} + \dfrac{a^2b}{48}\left(\dfrac{1}{I_o} + \dfrac{1}{I_u}\right)}$$

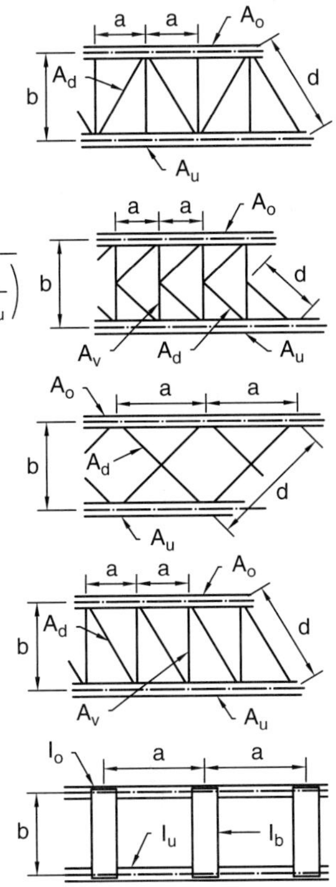

Figure 15.122 Common bracing systems and formulas for equivalent plate thicknesses.

reaction of the support) tends to produce opposite warping deformations. These opposing forces cancel one another and thereby result in effective warping restraint, which produces the stress pattern illustrated in Fig. 15.124. Any discontinuity, whether because of the presence of a concentrated load or an abrupt change in section, causes some degree of warping restraint and therefore some warping stresses. The stresses depicted in Fig. 15.124 are normal stresses that add to the ordinary bending stresses in the cross section. To a lesser degree, a similar pattern exists at each internal cross frame. Such stresses diminish rapidly with distance from the discontinuity that produces them.

15.26.2 Webs

Per AASHTO Guide Specifications, the maximum calculated shear for design of the box-girder webs will need to be the sum of the vertical shear V_u associated with bending moment and the shear V_γ caused by St. Venant torsion. If the web is inclined, the design shear associated with bending moment is

$$V_w = \frac{V_u}{\cos\theta}$$

where θ = angle of inclination of web plate with respect to the vertical axis.

15.26.3 Flanges

For the top flange in a curved steel box girder, the narrow width behaves, and is designed in exactly the same manner, as in the case of curved I-girder flanges. These flanges work compositely with the concrete deck, with a locked-in stress resulting from the initial dead load, DL_1. In addition to the stress from ordinary vertical bending, they carry a DL_1 lateral bending stress from curvature.

[f]From original article by William N. Poellot Jr., HDR Engineering, Inc., and Ref. 8.

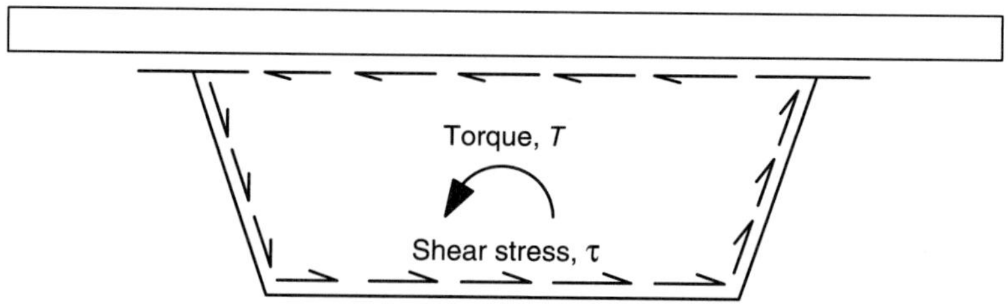

Figure 15.123 St. Venant torsion in a closed section.

For the bottom flange, following the Load Factor Design section of the AASHTO Guide Specifications, the maximum normal tension stress, F_b ksi, including warping normal stress, is limited to

$$F_b = F_y\sqrt{1-3(f_v/F_y)^2}$$

where F_y = yield strength, ksi (MPa)
f_v = St. Venant torsional shear stress, ksi (MPa)

The allowable compression stress for bottom flanges involves several parameters, which are defined as follows:

$$R_1 = \frac{97.08\sqrt{k}}{\sqrt{\frac{1}{2}\left[\Delta^2+\sqrt{4(f_v/F_y)^2\,(k/k_s)^2}\right]}}$$

$$R_2 = \frac{210.3\sqrt{k}}{\sqrt{\frac{1}{1.2}\left[\Delta-0.4+\sqrt{(\Delta-0.4)^2+4(f_v/F_y)^2\,(k/k_s)^2}\right]}}$$

where $\Delta = \sqrt{1-3(f_v/F_y)^2}$
k = buckling coefficient for instability under normal stress alone
= 4 when $n = 0$ or an assumed value less than 4 but generally 2 or greater when $n > 0$
n = number of equally spaced longitudinal flange stiffeners
k_s = buckling coefficient for instability under shear stress alone
= 5.34 when $n = 0$ or $\frac{5.34+2.84(I_s/bt^3)^{1/3}}{(n+1)^2}$ but not to exceed 5.34 when $n > 0$
I_s = moment of inertia, in.^{4} (mm^4), of a longitudinal flange stiffener about an axis parallel to the flange and at the base of stiffener

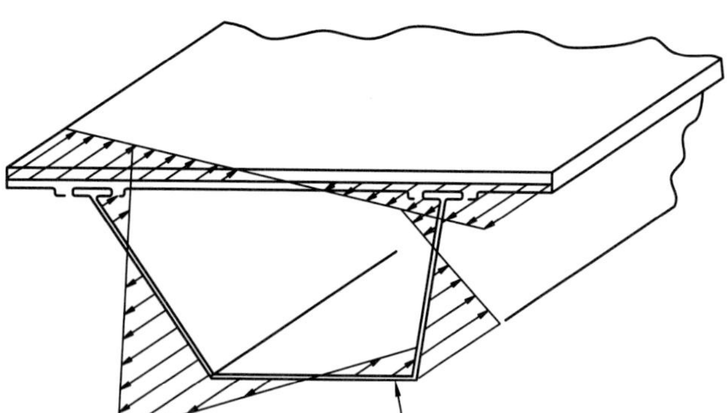

Figure 15.124 Stresses (including those from warping effects) due to cross-sectional distortion.

The maximum allowable compression stress F_b for bottom flanges depends on the magnitude of the torsional shear stress f_v across the flange. If f_v is less than $0.75F_y/\sqrt{3}$, there are several expressions for F_b depending on the ratio w/t or width to thickness of the flange between longitudinal stiffeners:

If $\frac{w}{t}\sqrt{F_y}$ does not exceed R_1,

$$F_b = F_y\Delta$$

Note that, in this case, F_b is the same as the allowable tension stress for a flange.

If $\frac{w}{t}\sqrt{F_y}$ is between R_1 and R_2, or equals R_2,

$$F_b = F_y\left[\Delta-0.4\left(1-\sin\left\{\frac{\pi}{2}\cdot\frac{R_2-\frac{w}{t}\sqrt{F_y}}{R_2-R_1}\right\}\right)\right]$$

If $\frac{w}{t}\sqrt{F_y}$ exceeds R_2,

$$F_b = 26{,}210K\left(\frac{t}{w}\right)-\frac{f_v^2K}{26{,}210K_s^2(t/w)^2}$$

If f_v is between $0.75(F_y/\sqrt{3})$ and $F_y\sqrt{3}$, $\frac{w}{t}\sqrt{F_y}$, it is not permitted to exceed R_1 nor is w/t allowed to exceed 60, except in regions of low compression stress near points of dead-load contraflexure. The maximum allowable compression stress is given by

$$F_b = F_y\Delta$$

which is the same as the allowable tension stress for a flange.

It is recommended that the longitudinal bottom flange stiffeners be equally spaced between the girder webs. These stiffeners must be proportioned properly such that the moment of inertia about the base of the stiffener is at least equal to

$$I_s = \phi t^3 w$$

where $\phi = 0.07K^3n^4$ for $(n > 1)$
$\phi = 0.125K^3$ for $n = 1$
n = number of longitudinal stiffeners
K = buckling coefficient ≤ 4

15.26.4 Cross Frames

The spacing of internal cross frames is important. Intermediate internal cross frames serve to maintain the shape of the cross section and to support laterally the narrow curved top flanges. A bottom strut in the cross frame can be made to serve secondarily as a lateral support for the curved longitudinal stiffeners on the bottom flange of the box girder. The more closely spaced the cross frames, the smaller will be the out-of-plane stresses resulting from distortion of the cross section. Top flange lateral bending stresses are likewise reduced with closer spacing of the cross frames and may be taken to vary in proportion to the square of

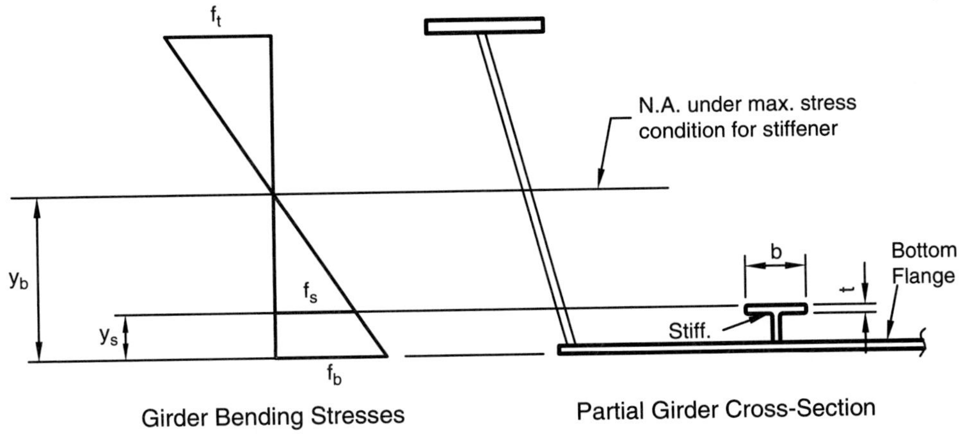

Figure 15.125 Bending stresses in longitudinal stiffener.

the cross-frame spacing. Bracing of curved bottom flange longitudinal stiffeners is required because they behave very much like curved I-girder flanges. Referring to Fig. 15.125, such stiffeners act as part of the box-girder bottom flange and therefore carry a longitudinal stress, somewhat lower than the stress at the extreme bottom flange fiber, given by

$$f_s = \frac{y_b - y_s}{y_b} f_b$$

where f_b = maximum bending stress, ksi, (MPa) in the girder bottom flange
y_b = distance, in., from neutral axis to bottom of girder
y_s = distance, in., from neutral axis to top of stiffener flange

Since the stiffener is curved, its flange is subjected to a lateral bending moment, M_{lat} given by

$$M_{lat} = \frac{f_s btd^2}{12R}$$

where d = unbraced length of stiffener flange, in.
t = thickness of stiffener flange, in.
b = width of stiffener flange, in.
R = radius of curvature of stiffener, in.

The corresponding lateral bending stress is

$$f_{lat} = \frac{f_s d^2}{2Rb}$$

With the direct stress and the lateral bending stress in the stiffener flange known, f_s may be checked against the allowable stresses for noncomposite I-girder flanges.

A working stress approach may be used in which an allowable stress, $\bar{F}_b$, for the stiffener takes the form $\bar{F}_b = \rho F_b$, where ρ is a reduction factor and F_b is the nominal allowable stress of $0.55F_y$. The factor can be derived to ensure that the stiffener flange tip stress does not exceed this value. For a structural tee stiffener with flange width b, unbraced length l, and radius R, this reduction is

$$\rho = \frac{1}{1 + 3l^2/(5bR)}$$

If the direct stress (not including lateral bending stress) in the stiffener flange is limited to $\bar{F}_b = \rho F_b$, the tip stress will not exceed F_b.

15.26.5 Lateral Bracing

Lateral bracing is used in lieu of a solid top plate to make the box-girder function as a closed torsion tube under wet concrete loading. For analytical purposes the lateral bracing is transformed to a solid plate by the formulas in Fig. 15.122 to enable a transverse shear flow to be calculated at any section by the St. Venant torsion theory. With this shear flow, a shear force across the top of the box can be determined, which can be considered to be the shear in a panel of lateral bracing, much like the vertical shear in a conventional truss. This shear is carried as the lateral component of the force in the bracing diagonal, which is designed accordingly as an axially loaded member.

15.26.6 Additional Reading

The reader is referred to additional materials on box-girder (tub-girder) curved bridges and example problem provided in Refs. 4, 5, 6, and 10.

REFERENCES

1. *Guide Specifications for Horizontally Curved Steel Girder Highway Bridges*, American Association of State Highway and Transportation Officials, Washington, DC, 2003.

2. Zureick, A., Naqib, R., and Yadlosky, J. M., *Curved Steel Bridge Research Project*, Publication No. FHWARD-93-129, Federal Highway Administration, McLean, VA, 1994.

3. *Standard Specifications for Highway Bridges*, 17th ed., American Association of State Highway and Transportation Officials, Washington, DC, 2002.

4. Coletti, D., and Puckett, J., "Structural Analysis," in *Steel Bridge Design Handbook*, Pub. No. FHWA-HIF-16-002—Vol. 8, Federal Highway Administration, Washington, DC, December 2015.

5. Helwig, T., and Yura, J., "Bracing Design," in *Steel Bridge Design Handbook*, Pub. No. FHWA-HIF-16-002—Vol. 13, Federal Highway Administration, Washington, DC, December 2015.

6. Chavel, B., and Rivera, J., "Design Example 5: Three-Span Continuous Horizontally Curved Composite Steel Tub-Girder Bridge," in *Steel Bridge Design Handbook*, Pub. No. FHWA HIF-16-002—Vol. 25, Federal Highway Administration, Washington, DC, December 2015.

7. Kollbrunner, C. F., and Basler, K., "*Torsion in Structures—An Engineering Approach*," Springer-Verlag, New York, 1969.

8. United States Steel Corporation, *Steel/Concrete Composite Box Girder Bridges, A Construction Manual*, ADUSS 88-7483-01, December, 1978.

9. Tung, D. H. H., and Fountain, R. S. "Approximate Torsional Analysis of Curved Box Girders by the M/R Method," *Eng., AISC*, vol. 7, No. 3, July, 1970.

10. White, D. W., Coletti, D., Chavel, B., et al, *Guidelines for Analytical Methods and Construction Engineering of Curved and Skewed Steel Girder Bridges*, NCHRP Report 725, Transportation Research Board, National Research Council, Washington, DC, 2012.

PART E
CURVED CONCRETE BOX-GIRDER BRIDGES

BY

JAMSHID MOHAMMADI *Illinois Institute of Technology*
BORA JANG *Sharma and Associates, Inc.*

15.27 GENERAL[g]

15.27.1 Introductory Remarks

This section covers curved bridges made up of concrete box girders. Most common types of these bridges, used to cover long spans, are made up of prestressed (especially posttension) girders. The bridge deck cross section may be made up of single-cell and multicell girders or spread beams (as shown in Fig. 15.126).[1] Because of their geometry, and depending on the severity of the radius of curvature, these bridges exhibit a significant lateral effect from the curved prestressing duct. The use of proper analysis methods and treatment of the live loads is therefore an important consideration in this type of bridge system. Other significant issues include deflection computation and treatment of such nonlinear effects as creep. Because in most cases such bridges require long spans and curved geometry, the method of construction plays yet another important role in their design, which requires proper coordination between the designer and the contractor, preferably, at early stages of planning and design.

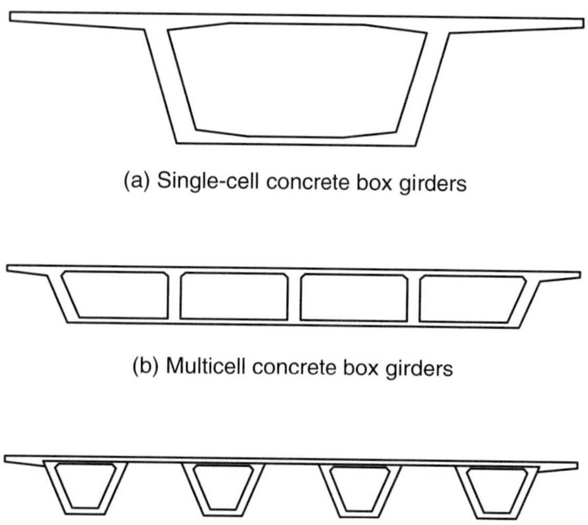

(a) Single-cell concrete box girders

(b) Multicell concrete box girders

(c) Spread concrete box-section beams

Figure 15.126 Types of cross sections made up of box girders. (*Adopted from Ref. 1.*)

15.28 CONCRETE BOX-GIRDER BRIDGE TYPES

In addition to the girder arrangement in cross section, as depicted in Fig. 15.126, the type of concrete box-girder bridges can also be determined based on (1) the overall geometry and the way the deck and girders are built and (2) the method of construction. Depending on the type and the angle of curvature, a structural analysis method suitable is selected.

15.28.1 Methods of Analysis

According to the AASHTO LRFD specifications,[2] three categories for the analysis of horizontally curved bridges are the following:

(1) Central angle ≤ 12 degrees: Bridges with a central angle not greater than 12 degrees in their longest span (i.e., $L/R \leq 0.2$, where L is the span length and R is the radius of the curve) may be designed as straight. In these cases, no consideration for bridge curvature is necessary for the longitudinal design of the concrete box-girder superstructure.

(2) Central angle >12 but ≤ 34 degrees: In cases where a bridge in its longest span has a central angle larger than 12 degrees (i.e., $L/R > 0.2$) but smaller than 34 degrees (i.e., $L/R \leq 0.6$), the structural analysis and design must include the effect of the horizontal curvature of the bridge. The structural analysis method may include idealized beam-column elements defined between nodes at the center of gravity of the cross sections as it follows the horizontal curve. A two-dimensional grillage analysis will be suitable, where the girders are idealized as line elements between nodes with 6 degrees of freedom (three translational and three rotational) considered at each node. It is noted that the maximum length of the straight girder elements between nodes shall be such that the central angle between adjacent nodes will not exceed 3.5 degrees (i.e., $L_{\text{nodes}}/R \leq 0.06$). The assemblage of these elements is sometimes referred to as a spine beam.[3]

(3) Central angle > 34 degrees: In cases where a bridge has a central angle greater than 34 degrees (i.e., $L/R \leq 0.6$) in its longest span, the recommended structural analysis method is a three-dimensional idealization of the structure using such techniques as grillages, folded plates, finite strips, or finite elements using bending plate and shell elements.

In the following subsections, where references to different construction types for concrete box-girder bridges are made, specific structural analysis methods that are suitable for each type are discussed.

15.28.2 Cast-in-Place Deck on Straight Precast Girders

This is primarily applicable to bridges with a relatively large radius of curvature and piers that are positioned radial with respect to the curve. The geometry results in little or no self-weight torsion; as such, the bridge can be treated as a straight girder bridge for analysis and design purposes. However, it is noted that certain variations in shear and moment from one girder to another is expected because of the eccentrically applied live loads. Unless a more rigorous type of analysis may be needed because of large variations in internal forces among girders, generally a grillage analysis will provide an accurate solution in this type of bridge structures.[4]

In general, it is not practical to precast curved I-girders or curved U-beams. Handling and transportation will usually restrict the use of large, precast, full-span-length, curved girders.

15.28.3 Cast-in-Place on Formwork System

In this type of construction, the superstructure section is fully supported by the formwork system until the concrete has been cast, gained strength, and the posttensioning installed to make the girders self-supporting. At this stage, the formwork system is then removed. Superstructures made up of any cross-sectional shape can be built in

[g]Original article by Alan J. Moreton, Juan J. Goni, Amy R. Kohls, and Daniel G. Davis of Figg Engineers, Tallahassee, Florida.

this manner. However, the efficiency of the process and practicality, as well as the cost, will limit the application to projects with rather shorter span lengths and sites that offer a reasonable height for easy deck access. Typical sections in this type of application include solid flat slabs hollow slabs and trapezoidal section boxes.

Solid and voided slabs are usually made continuous over several short spans of up to about 50 to 65 ft (15 to 20 m) with consideration for cost and efficiency. They can readily accommodate sharp plane curves and are suitable for applications in some highway interchanges and low-level or short-span bridges where structural depth is limited. The distribution of load and determination of longitudinal and transverse bending effects can readily be done using a two-dimensional grillage analysis.[1,4,5] More elaborate three-dimensional finite-element analysis may be necessary when there are severe cases of sharp curves and sudden cross-sectional changes.

15.28.4 Segmental Construction Method

In this method of construction, segments of the full superstructure section, including the deck slab, are constructed in short lengths and posttensioned together. Both cast-in-place or precast segments may be used in this type of construction using some type of temporary erection equipment. The precast segments are most frequently made in a special casting cell by match-casting one against another in order to ensure a perfect fit. The geometric alignment of the bridge is created in the casting cell by adjusting the position of the previous match-cast segment relative to the cell bulkhead for the new segment. Any varying geometric alignment, superelevation, and transition is readily accommodated by this precasting technique. Precast segments may be erected in span-by-span, balanced cantilever, progressive cantilever, or combinations, as appropriate. They have frequently been used with cable-stayed construction. Depending on the curvature angle, the analysis method may be a two-dimentional grillage or a three-dimensional finite-element method.

15.28.5 Incremental Launching Method of Construction

This method was first used in 1964. Using this method, the structure is cast in sections of about 50 to 100 ft (15 to 30 m) in length in a special form behind an abutment and then launched forward. A lightweight temporary launching nose is attached to enable the deck to reach the next pier and as a way of minimizing longitudinal bending moments. Intermediate temporary towers are sometimes placed in the spans for the same purpose. During launching, permanent and temporary post tensioning is disposed so as to minimize stress reversals as each section advances. Incremental launching is only suitable in continuous structures of straight alignment or constant curvature and superelevation because it cannot accommodate variable geometry. Again, the angle of curvature would determine the type of structural analysis suitable to model the bridge structure.

15.29 SIGNIFICANCE OF STRUCTURAL CONFIGURATION AND DETAILS

15.29.1 Influence of Curvature on Structural Configuration

For very large radii, a bridge may be built from a series of straight, chorded spans with a slight angular break in the main longitudinal box-girder line at each pier. The deck slab and parapets follow the curved alignment, and the variation from the straight chord is accommodated in the width of the deck slab cantilevered beyond the web. However, for aesthetic purposes, it is important for the main box girder to follow the curve when the angular break between spans is greater than 2 or 3 degrees. This also applies to spans made from straight, precast girders.

Depending on the bridge type (whether highway or railway), the horizontal alignment, vertical profile and superelevation are achieved by operational requirements. In using less sophisticated structural analysis methods and also in design, it is convenient to idealize the alignment to a curve of constant radius or a series of compound curves and tangents, depending upon the bridge span length. Because of the significance of torsion, the influence of a curved alignment should be considered when the ratio of the "torque span" to radius of curvature is greater than about 15 percent (see the definition of torque span later). For example, for a free cantilever with a constant section, when the ratio of the free cantilever length to radius reaches 15 percent, the magnitude of the dead-load torsion resulting from curvature is approximately 5 percent of the longitudinal cantilever dead-load bending moment. In a completed structure, torsion resulting from eccentric live load, wind, seismic, or centrifugal force will of course need to be computed and combined with the dead-load torsion (that is resulting from curvature) and used to verify the adequacy of design. In an LRFD approach,[2] specific load factors are prescribed by the code when combining various load effects.

The torque span is defined as the distance between supports at which torsion can be transmitted from the superstructure to the substructure. In general, these torsionally fixed supports are located where there are two or more bearings side by side or where the superstructure is monolithic with a pier. In some cases it is necessary or desired for aesthetic reasons to place a single, central bearing under a box girder. Although such a support can transmit reactions and can be considered as a support for bending moments, it cannot transmit torsion to a pier. As such and for torsional effects, this kind of support is considered ineffective in structural analysis. Even in cases where two bearings are used, tall slender piers may be too flexible to carry all the superstructure torsion. If a tall flexible pier is adjacent to a much stiffer abutment, the abutment will take most of the torque, resulting in very high torsional moments and possible uplift at one of the abutment's bearings. In such cases, a three-dimensional finite-element method of structural analysis is appropriate to determine the proportion of superstructure torsion carried by the pier and the abutment.

15.29.2 Superstructure Section and Details

For most precast segmental bridge applications, the superstructure cross section is trapezoidal with two webs and cantilevered deck slab wings (Fig. 15.127*a*). Occasionally three or more webs may be used within one cross section. However, for the construction of wide bridge decks, it is normally more efficient to use multiple boxes rather than one box with multiple webs (see also Fig. 15.126 and Ref. 3).

For a typical trapezoidal-section segmental concrete box girder and for spans up to about 200 feet (60 m), the overall depth is usually constant and in the range of $L/20$ to $L/30$, where L = span length. For longer, cantilever spans, a variable depth is usually more economical with an approximate depth at the pier of $L/20$ and at midspan of $L/40$ to $L/60$. Denoting the overall width of the top slab by W, the width between the webs at the top is about $W/2$, making the overhanging cantilever slab wings in the range of $0.2W$ to $0.25W$. The overall width of the bottom slab is between $0.3W$ and $0.5W$ (see Fig. 15.127*a*).

The thickness t of the top slab at the edges of the wings should be sufficient to accommodate anchorages and bursting reinforcement details for top slab transverse posttensioning tendons of up to four 0.6 in. (15 mm) strands per tendon. Typically, a minimum edge depth of 8 to 9 in. (200 to 225 mm) is required depending on cover. The minimum top slab depth between the webs is usually 8 in. (200 mm) with haunches down to the webs. As a guide, the top slab depth at the webs should be approximately the minimum edge depth plus 0.05 times the overhanging wing depth.

The bottom slab is typically a minimum of 8 in. (200 mm) thick but may need to be 9 in. (225 mm) or more to accommodate the larger sizes of longitudinal posttensioning ducts, transverse reinforcement, and the internal and external concrete covers. A haunch in the bottom slab at the webs is not essential but may be needed to accommodate posttensioning ducts or reduce stresses. In cantilever construction using a constant depth box, it is sometimes necessary to increase the bottom slab thickness internally for a short distance near the supports in the longer spans (usually over 160 ft, or 50 m). This lowers the neutral axis and makes the top cantilever tendons more effective, increases the available section area, and reduces the local bottom slab compression.

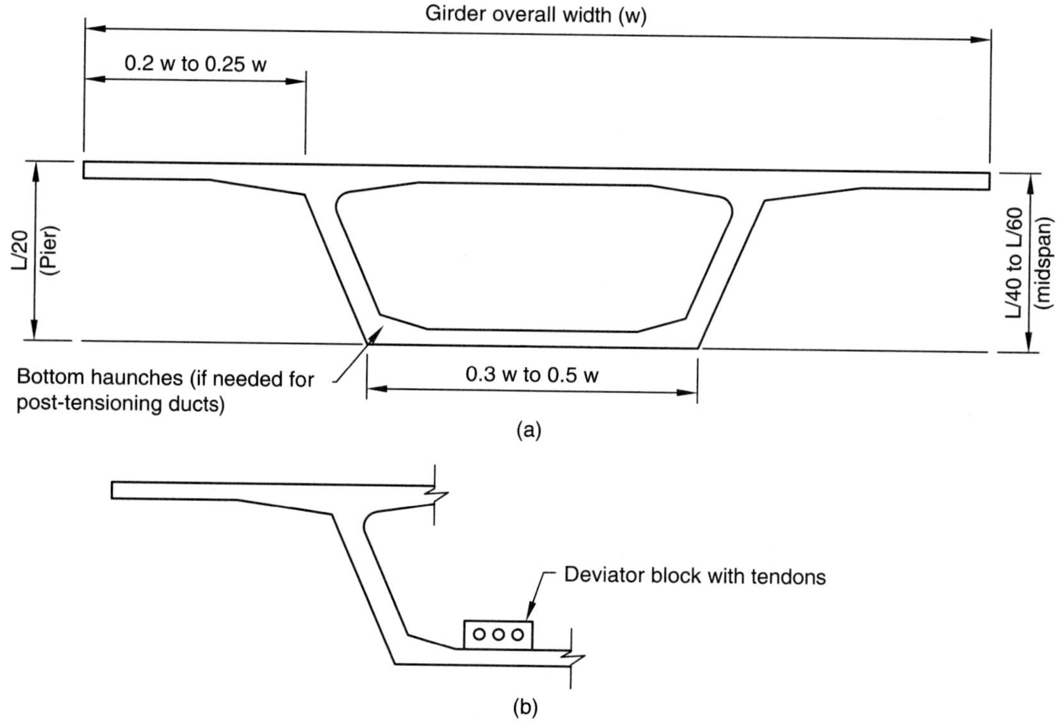

Figure 15.127 (*a*) Typical section of a single-cell concrete box girder and (*b*) section with a deviator blocks.

Web widths may be constant or may taper from top to bottom with an average thickness usually not less than about 12 in. (300 mm) for the shorter span ranges. The web thickness is governed by global shear, torsion, and local transverse bending requirements. If longitudinal tendons are to be accommodated in the webs, an increased web width may be needed (at least three times the largest posttensioning duct diameter for a single duct, more if multiple ducts lie side by side).

In cantilever construction, it is quite common practice to anchor longitudinal top slab tendons at the top of the webs. Therefore, the width of the top of the web and depth of slab should be sufficient to accommodate the anchorage hardware for the largest tendons. In span-by-span construction, the tendons are normally external to the concrete and inside the box (but protected in sealed, grouted, thick-walled polyethylene ducts) and are usually anchored in diaphragms at the piers. Between piers, the tendons drape inside the box through deviator blocks at the junction of the web and bottom slab (Fig. 15.127*b*).[6]

In bridges with tight curves and when tendons are within the web, the radial force on the web must be considered and the web thickness increased or reinforced to prevent the tendons from laterally pulling out of the web or causing a longitudinal crack. The radial force on the web is equal to P_T/R per unit length, where P_T is the maximum posttensioning force and R the radius.[7]

15.29.3 Skewed Supports

Although it is possible to place skewed supports under a concrete box girder, the details in the superstructure become complicated especially for construction. Placing the bearings square to the centerline greatly simplifies the superstructure diaphragm and details. Likewise, it is preferable to provide square rather than skewed end abutments whenever practical. At intermediate piers, if there is an insufficient lateral clearance for a bearing arrangement and pier is set square to the superstructure alignment, a bearing or monolithic connection to a single central pier shaft that can fit into the available ground space should be considered.

15.29.4 Expansion Joints and Bearings

Permanent creep and shrinkage contraction must be considered when sizing bearings and expansion joints or deciding upon how many spans to include in a structurally continuous unit between expansion joints. For a continuous span superstructure supported on bearings, some bearings will usually be fixed in order to transmit longitudinal forces (traction/braking) to the piers. Others may be constrained to move in one direction by guides. On a curved structure, the alignment of the fixed and guided bearings must be considered in relation to the curve and point of zero movement. The fixed and guided bearings may be arranged so as to align movement through the point of zero movement, or they may be aligned longitudinally and radially at each pier in order to control movements. In either case, the net longitudinal and transverse forces from the superstructure must be considered in the substructure design. Laminated neoprene bearings permit more equalization of longitudinal and transverse horizontal forces between the superstructure and each pier by virtue of their horizontal shearing stiffness.

15.29.5 Influence of Construction Methods on Structural Configuration

Sequential construction methods use part of the already completed structure to support and construct the next part. In addition, temporary supports and erection equipment loads may be applied in one structural configuration and then removed or relocated at a later stage. Likewise, posttensioning may be introduced in one stage and more added at later stages. As a result, the internal forces change throughout construction, and residual forces are built into the structure. The internal forces resulting from the construction sequence, erection loads, temporary supports, and stage posttensioning are quite different from those obtained by supporting the structure entirely by falsework until all the construction is finished and the falsework only then removed. It is noted that the latter is sometimes referred to as an "instantaneous" construction phase. In concrete bridge construction, perhaps the most common example of sequential construction is the segmental balanced cantilever where, by commencing at the top of a pier, the superstructure is constructed by adding a segment to each side of the pier in turn using posttensioning tendons to secure each new segment to those already built. Once a balanced cantilever is completed, it is connected to a previously erected cantilever or side span through a short, posttensioned, cast-in-place closure. This changes the statical scheme of the structure for all subsequent construction and applied loads. Temporary props near the piers may be used to carry the out-of-balance loads as a new segment is added first to one end of a cantilever and then to the other. All the erection equipment loads must then be accounted for in the design.

In an indeterminate, sequentially constructed structure, creep of the concrete causes a redistribution of internal forces and stresses. The trend is from the accumulated internal forces at the end of the construction sequence toward those that would have arisen had the entire structure been created and loaded in its final configuration. The magnitude of the long-term creep redistribution depends upon the age of the concrete at the time of loading that is, the younger the concrete, the more the redistribution. Cast-in-place construction undergoes more creep redistribution (and shrinkage deformation) than precast concrete, which is more mature when loaded. Thus, the age of the concrete and the time at which each construction step is performed must be known in order to confidently calculate and accumulate the structural response. Furthermore, both creep and shrinkage influence the loss of the prestress. Thus, the global structural movements will need to be accommodated by bearings and expansion joints. As such, the material characteristics of the concrete (i.e., strength, modulus of elasticity, and coefficients of creep and shrinkage) and their behavior with time must be known or estimated.

In summary, the design of a sequentially constructed structure depends very much upon the construction sequence, erection loadings, temporary supports, posttensioning, changes in the statical scheme, the material characteristics of the concrete, and the timing of each individual construction step in relation to the maturity of each portion of the concrete structure throughout the construction period and into long-term service. Because of the detail and complexity of this process, the analysis and design of a major concrete bridge is usually accomplished by means of computer programs that automatically account for and accumulate all the structural and time-dependent changes. For further information on this subject, the reader is referred to Refs. 8–10. A simplified approach and an example of steps necessary to perform calculations are presented in subsequent subsections.

15.29.6 Construction—Camber and Geometry Control

For all segmental and sequentially constructed bridges, the final alignment of the superstructure is made up of the required geometric horizontal alignment and vertical profile adjusted by camber to compensate for all elastic, creep, and shrinkage deflections resulting from all sustained loads, posttensioning, and intermediate construction stages from casting to erection to long-term service. Most concrete box girders are torsionally and horizontally rigid; as such, the torsional and horizontal displacements are negligible. Thus, it is normally only necessary to compensate for vertical displacements, even for sharply curved bridges, but it is advisable to make sure about this issue by computing deflections from other effects.

For precasting of segments, the geometric vertical profile, adjusted for camber, is determined over each web and the horizontal alignment established at the centerline at each joint between segments. The two vertical profiles and horizontal alignment comprise "casting curves"—the geometric data to which the segments are cast in the casting cell. The curve of the entire structure is established by setting the match-cast segment relative to the casting cell bulkhead for each new segment. The match-cast segment is also adjusted in position to compensate for any minor geometric casting errors incurred when it was the newly cast segment.

The geometry control consists of tracking the vertical profiles over each web and the horizontal alignment, graphically or computationally, by precise measurements of the newly cast and match-cast segments in the cell. However, when the curve is sharp, the vertical profile and superelevation influence the horizontal alignment in the casting cell and vice versa. In such cases, the tracking is best handled by a three-dimensional coordinate geometry transformation system between the global structure and the local casting cell.

15.30 REVIEW OF THEORY[h]

15.30.1 Introductory Remarks

Even though currently the use of commercially available software affords nearly everyday routine analysis in the design office, it is important to also know the underlying theory of curved structures and to review practical methods available for analysis. Knowing practical methods is especially important in (1) having an insight into the overall behavior of the structure and understanding the way internal forces are distributed and (2) providing a means to verify the results from computer programs for quality assurance purposes. Thus, in this section, an overview of the theory of curved structures and methods of analysis for them is provided.

Most curved concrete box-girder bridges can be considered to behave as curved beams with rigid cross sections because they have relatively thick webs and slabs for constructability and strength requirements. Only in rare cases is the transverse cross section such that the effect of torsional warping or distortion must be considered. This would be the case, for example, when very thin and tall webs are used or there are no adequate rigid internal diaphragms at the supports. Nevertheless, whenever the geometry or construction methods necessities a more detailed process in computing torsional and other effects that are inherent with curved structures, there are practical methods that can be used by analyzing the bridge as a continuous beam.[11–14]

15.30.2 Behavior of a Curved Beam

The equilibrium of a curved beam subjected to distributed loads p and torsional moments m_T (Fig. 15.128) is represented by the following system of equations:

$$\frac{dT}{ds} = \frac{M}{R} + m_T \tag{15.8}$$

$$\frac{dT}{ds} = \frac{T}{R} - V \tag{15.9}$$

$$\frac{dV}{ds} = p \tag{15.10}$$

where T = torsional moment
M = longitudinal bending moment
V = shear force
R = radius

The torsional and bending moments are coupled in a curved beam; as such, they cannot be independently solved through Eqs. (15.8) and (15.9). Because of the beam curvature, the bending moment M also generates a torsional moment that is added to the external distributed torque (m_T). Similarly, the torsional moment generates a bending moment. Furthermore, the rotations induced by the torques also generate vertical displacements resulting from the curvature of the beam. For a simple-span curved beam, which is torsionally fixed and vertically simply supported at the ends (see Fig. 15.129),

[h]Mainly from the original article by Alan J. Moreton, Juan J. Goni, Amy R. Kohls, and Daniel G. Davis of Figg Engineers, Tallahassee, Florida.

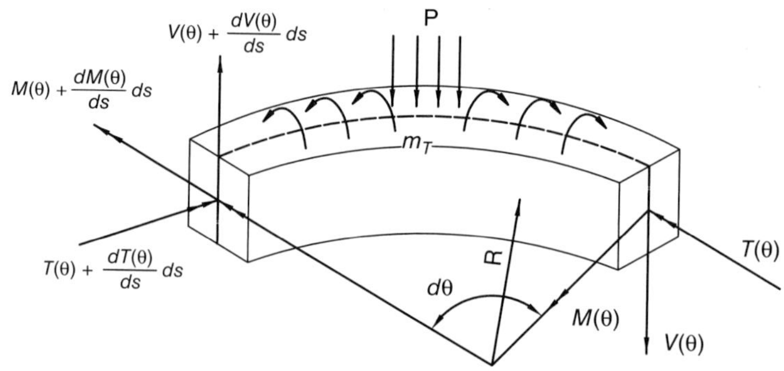

Figure 15.128 Forces of a segment of a curved beam.

the internal forces under a uniformly distributed vertical load P along its centerline are

$$T_A = -T_B = PR^2\left(\frac{\Delta}{2} - \frac{1-\cos\Delta}{\sin\Delta}\right) \tag{15.11}$$

$$M_c = PR^2\left(\frac{1}{\cos\Delta/2} - 1\right) \tag{15.12}$$

$$M_\theta = PR^2\left(\frac{\sin(\Delta-\theta)+\sin\theta}{\sin\Delta} - 1\right) \tag{15.13}$$

$$T_\theta = PR^2\left(\frac{\Delta}{2} - \theta - \frac{\cos(\Delta-\theta)-\cos\theta}{\sin\Delta}\right) \tag{15.14}$$

where Δ = central angle

These results indicate the following:

1. The distribution of bending moments is very close to that of a straight girder with the same length. It is noted that, for example, for a central angle of 20 degrees, the difference between the curved beam and straight beam maximum bending moment is only 1.3 percent.
2. The curved beam is subjected to large torsional moments, whereas the straight beam has no self-weight torsion.
3. The internal forces are independent of the torsional (GJ) and bending stiffness (EI).

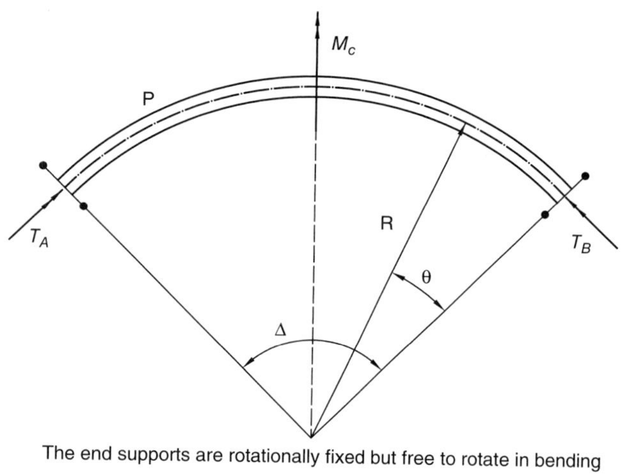

Figure 15.129 Forces on a simple-span curved beam.

For a curved cantilever of constant cross section, the moments at the root of the cantilever are

$$T_A = \gamma AR^2(\Delta - \sin\Delta) \tag{15.15}$$

$$M_A = \gamma AR^2(\cos\Delta - 1) \tag{15.16}$$

where γ = material density
A = cross section area

The torsional moments in continuous curved beams are smaller than those of simple-span curved beams because the negative bending moments at the supports contribute to the equilibrium of the external loads. The value of the ratio K of the bending stiffness EI to the torsional stiffness GJ substantially affects the magnitude of the torsional and bending moments in continuous curved beams. If the value of K is close to 1 (i.e., section torsionally rigid), the bending moments in a curved beam are very similar to those of a straight beam. When K is large (i.e., beams with small torsional stiffness), the distribution of bending moment is no longer similar to that of the straight beam but shows a substantial increase in negative flexural moment at the piers to compensate for the reduced torsional capacity. For most curved concrete box-girder bridges, K is close to unity, and therefore they can be analyzed as straight beams for longitudinal distribution of bending moment. The torsional moments can be obtained using an approximate method[15,16] as discussed below.

Equation (15.8) shows the relationship between the torsional moments, the external torques, and the bending moments in a curved beam. If the values of the bending moment M are known at all locations in the beam, the right-hand term of Eq. (15.8) is defined and the equation is similar to Eq. (15.10). Consequently, the distribution of torsional moment in the curved beam is equal to the distribution of shear on an equivalent straight beam subjected to a distributed load of magnitude:

$$p = \frac{M}{R} + m_T \tag{15.17}$$

This result shows that for the torque span (i.e., length between points at which the curved beam supports are torsionally fixed), the bending and torsional moments can be obtained as follows:

1. Straighten the curved beam with its corresponding supports and determine the bending moments by any method for indeterminate structures.
2. Taking each torque span in turn, apply the distributed load p [Eq. (15.7)] on a simply supported beam of length equal to that torque span and compute the distribution of shear. The result is the distribution of torsional moment in the curved beam for that torque span.
3. The algebraic difference between the two end torsional moments for each torque span adjacent to a torsional support is the torsional moment reaction at that support.

If a support in the curved bridge is not torsionally fixed (i.e., single bearing under a box), the torque span (length of the simply supported beam in Step 2) should be the distance between the next adjacent torsionally fixed supports.

A continuous curved beam can also be represented by a series of chorded straight beams and analyzed using, for example, the grillage analysis method with the aid of computer software.

15.30.3 Loading Effects on Curved Concrete Box-Girder Bridges

A concrete box-girder bridge is subject to self-weight, posttensioning, superimposed dead and live load, wind, seismic forces, temperature, creep and shrinkage effects, and so forth. All can be represented by equivalent forces applied to a model of the structure. The equivalent forces representing gravity, traffic, and posttensioning forces differ between straight and curved bridges. In straight bridges the weight of the structure is represented by distributed vertical loads p applied to the centroid of the beam. In curved bridges, there is an additional distributed torque ($m_T = pe$) resulting from the difference in distributed weight represented in plane area between the axis of the beam and the interior edge and the plane area between the axis of the beam and the exterior edge (Fig. 15.130). Consequently, the self-weight of a curved bridge must be represented by a vertical distributed load p located at a distance e from the axis toward the exterior of the cross section, where

$$e = \frac{\rho_y^2}{R} \tag{15.18}$$

and ρ_y = radius of gyration of the cross section about the vertical axis located on the center of gravity. Similarly, a uniform load applied on top of the deck for the wearing surface or a distributed live load must be represented by a distributed vertical load q per unit length and a torque m_T equal to

$$m_T = q \cdot \frac{B^2}{12R} \tag{15.19}$$

where B = deck width.

For straight bridges, when posttensioning tendons are located in vertical planes and do not deviate transversely with respect to the centroid of the bridge, the posttensioning is typically represented in the analysis by equivalent distributed vertical forces along the length of the tendons and equivalent concentrated forces and bending moments at the location of the anchorages. In curved bridges, internal tendons also apply a horizontal distributed force u on the concrete (Fig. 15.131) at the location of the tendon equal to

$$u = \frac{P}{R} \tag{15.20}$$

The resultant of these forces per unit length ds is equal to

$$R_u = u ds = \frac{P}{R} ds \tag{15.21}$$

and is directed toward the interior of the curve. In each part of the differential element shown in Fig. 15.131, the adjacent slices transmit compression stresses σ_c with a resultant P equal to

$$P = \int \sigma_C dA \tag{15.22}$$

The resultant R_σ of the forces P applied to the faces of the element is

$$R_\sigma = 2P \sin\beta = \frac{2P ds}{R} = \frac{P}{R} ds \tag{15.23}$$

While the resultant of the tendon horizontal deviation forces R_u is directed toward the interior of the curve and is applied at the location of the tendon (point A in Fig. 15.131), the resultant R_σ of the compression stresses σ_c is located at the center of gravity of the section (point B in Fig. 15.131) and directed toward the exterior of the curve.

The equivalent distributed torque that represents the effect of the posttensioning is equal to the sum of the torque generated by the horizontal deviation forces u and the torque generated by the resultant of the compression forces σ_c:

$$dm_T = \frac{P}{R} ds(f + r) - \frac{P}{R} ds(r) = \frac{P}{R} ds(f) \tag{15.24}$$

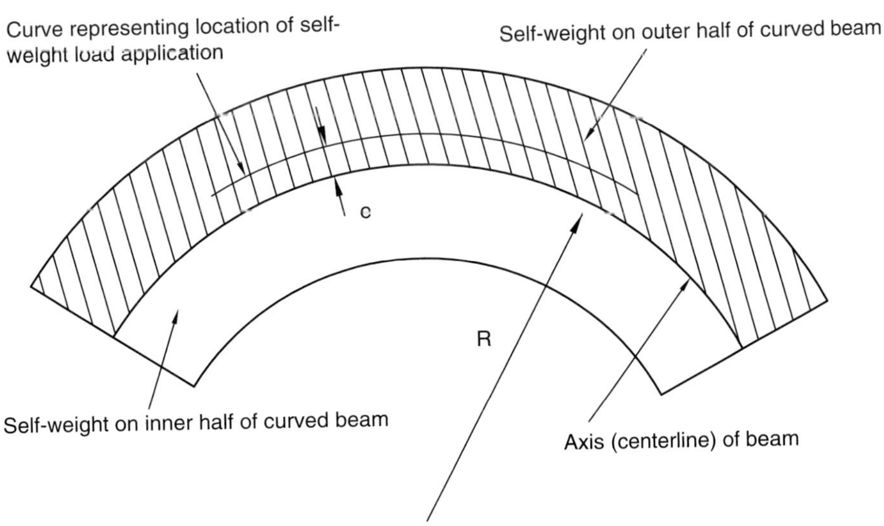

Figure 15.130 Location of self-weight application.

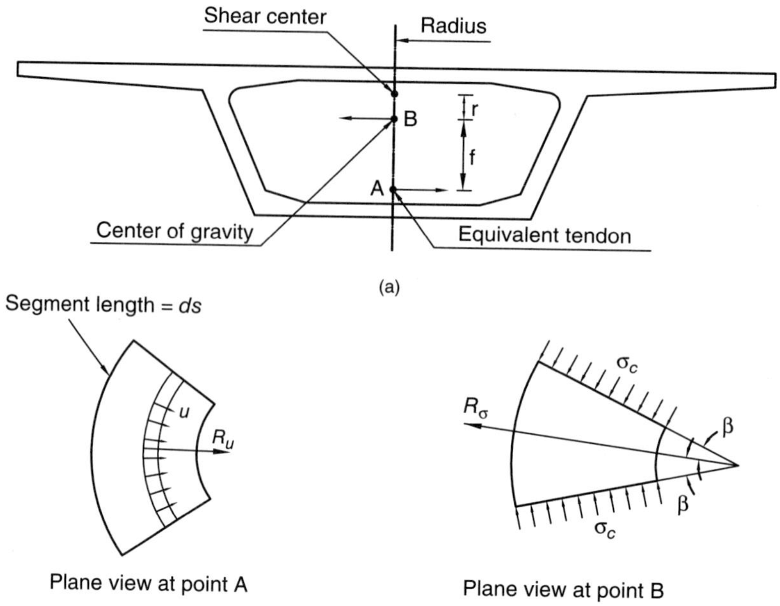

Figure 15.131 Equilibrium of static forces on a differential element of a curved beam.

where r = distance from the centroid of the section to the shear center and f = distance from the tendon location to the centroid of the section. The result shows that the equivalent distributed torque per unit length generated by the posttensioning is independent of the location of the shear center and is equal to

$$m_T = \frac{P}{R} f \tag{15.25}$$

If this value is placed in the torsion equation, then

$$\frac{dT}{ds} = \frac{M_p}{R} + m_T \tag{15.26}$$

where M_p is the bending moment generated by the posttensioning that in turn can be separated into the primary bending moment (Pf) and secondary bending moment M_{HP}:

$$\frac{dT}{ds} = \frac{M}{R} + m_T = \frac{M_{HP} - Pf}{R} + \frac{P}{R} f = \frac{M_{HP}}{R} \tag{15.27}$$

This indicates that the torsional moments are exclusively generated by the secondary posttensioning bending moments. Consequently, if the structure is isostatic or the tendon layout is concordant, there are no torsional moments resulting from the posttensioning.

15.30.4 Redistribution of Bending and Torsional Moments Resulting from Creep

The redistribution of longitudinal bending moment resulting from creep in a curved concrete bridge can be calculated with a high degree of accuracy by the same procedure as for straight bridges when the properties of the structure are such that the torsional moments have a small influence on the bending moments (K close to 1, i.e., rigid cross section). Several commercially available software packages for straight bridges incorporate time-dependent effects (i.e., creep and shrinkage of concrete and posttensioning steel relaxation) and the time effects of the sequential construction stages. In the case of a bridge built in stages, the effects of creep can be computed approximately by the following procedure:[10]

1. Determine the bending moments in the structure accounting for the construction phases.
2. Determine the bending moments assuming the structure is built in a single "instantaneous" phase (i.e., all on falsework).
3. Calculate the difference between the bending moments of Step 2 and Step 1.
4. Multiply the bending moment diagram obtained in Step 1 by the factor $(1 - e^{-\phi})$ to obtain the "creep moments." The parameter ϕ is called the creep factor.[10] Reference 3 also covers materials on how to compute creep moments based on the AASHTO procedures.
5. Add the bending moment diagram of Step 4 to that of Step 1 to obtain the bending moment diagram after all creep has occurred.

Currently, the use of commercial nonlinear three-dimensional finite-element analysis software such as ABAQUS[17] is becoming routine in practice. Such programs allow modeling concrete box curved bridges for time-dependent nonlinear analysis for an accurate estimation of the creep effects. However, it is also important to know practical approximate methods of estimating creep effects as a means of (1) quick analysis in preliminary design and (2) verifying computer analysis results for quality assurance purposes. The following presents a practical approach in computing the redistribution of torsional moments caused by creep:

1. Compute the bending moment diagram accounting for the time-dependent effects using a computer program or the procedure presented above as for a straight continuous beam.
2. Compute the external torques applied by the loads m_T.
3. Compute the distributed load

$$p = \frac{M}{R} + m_T \tag{15.28}$$

4. Taking each torque span in turn, apply the distributed load p on a simply supported beam of length equal to that torque span and compute the distribution of shear. The result is the distribution of torsional moment in the curved beam for that torque span.
5. The algebraic difference of the two end torsional moments for each torque span adjacent to a torsional support is the torsional moment reaction at that support.

In the case of posttensioning, the distributed load in Step 3 is equal to the secondary bending moments divided by the radius.

15.31 DESIGN CONSIDERATIONS

15.31.1 State of Stress

A given section in the bridge is subjected to the following resultant forces from combination of all loads:

P = resultant axial force
V = resultant shear force
M = resultant bending moment from applied loads including post-tensioning and secondary effects
T = resultant torsion

The section properties to consider in design include the following:

A = cross-sectional area
I = moment of inertia
b = web width at plane of interest (i.e., neutral axis, top or bottom of web) taken perpendicular to the web
Z = section modulus at plane of interest
Q = first moment of area of section above plane of interest
t_s = minimum thickness of top or bottom slab or web of interest
A' = total area enclosed by the median line of the webs and top and bottom slabs

The resultant stresses at the plane of interest are the following:[18]

longitudinal stress:

$$\sigma_x = \frac{P}{A} \pm \frac{M}{Z} \tag{15.29}$$

web shear stress:

$$\tau_w = \frac{VQ}{Ib} + \frac{T}{2A'b} \tag{15.30}$$

slab shear stress:

$$\tau_s = \frac{T}{2A't_s} \tag{15.31}$$

principal tension in web:

$$\sigma_1 = \frac{\sigma_x}{2} - \sqrt{\left(\frac{\sigma_x}{2}\right)^2 - \tau_w^2} \tag{15.32}$$

At the service level, the longitudinal fiber stresses should not exceed those of the applicable design code. For a concrete compressive strength of f'_c psi (MPa), as a general guideline, if the principal tensile stress in the web under the worst combination of loadings is less than about $4.5\sqrt{f'_c}$ psi (or $0.375\sqrt{f'_c}$ MPa), then serviceability conditions (i.e., no cracking) should be satisfied provided that web reinforcement is placed in accordance with standard code procedures for the resultant combinations of axial forces, torques, shears, and moments.

15.31.2 Example of Design Process[i]

The method for the analysis and design of a curved concrete box-girder bridge presented in the previous sections is applied here to a practical example. The structure is a section of a ramp in an urban interchange of five continuous spans with a radius of 298.6 ft (91 m). For simplicity, it is assumed that the piers are so stiff that they act as torsionally fixed supports in the transverse direction. In the longitudinal direction, the superstructure is pin-connected to the piers so that no longitudinal bending moment is transmitted to them. The structure is segmentally erected in balanced cantilever by ground-based cranes. Temporary vertical props are used to stabilize the cantilevers during erection.

Although this example was analyzed by a plane-frame computer program that automatically takes into account the multiple phases of the erection, including posttensioning, the analysis can be performed using any convenient method for continuous beams so long as it is broken into the major construction steps and the results of each step accumulated. Time-dependent creep redistribution is incorporated by applying a creep factor to the final conditions. The procedure is as follows:

1. Determine the geometry of the bridge (Fig.15.132).
2. Establish the span arrangements, articulation, construction sequence and methods (Fig. 15.133), and posttensioning layout (Fig. 15.134).
3. Determine the concrete material properties. Determine the required concrete strength. For a normal weight concrete, determine the modulus of elasticity, creep and shrinkage characteristics, and thermal coefficient according to the design specifications.
4. Establish the superstructure cross section (Fig. 15.135) and execute transverse design for local loads. Determine the local bending moments in the top and bottom slabs and webs. Determine the reinforcement and/or transverse prestressing for these local effects.
5. Determine the self-weight, superimposed dead load and all appropriate live loads, including centrifugal force, wind, temperature, and so forth. For the superstructure consider at least self-weight, superimposed dead load, live loads (including centrifugal force), wind (on structure and live load), temperature, creep and shrinkage, and construction loads.
6. Determine the eccentricities of the applied loads and the distributed applied torque loading m_T for each.
7. Straighten the bridge and execute a longitudinal construction-stage analysis (i.e., determine and accumulate the longitudinal bending moments, shear forces, reactions, and internal stresses at each step of the construction sequence, taking into account all changes in the statical structural scheme, the application and removal of erection loads, temporary supports and falsework, and the introduction of posttensioning including the effects of redistribution resulting from creep and shrinkage, from the beginning of construction to long-term service).

If a construction-stage analysis computer program is unavailable, a reasonable solution for creep redistribution can be obtained by applying a global creep factor to the difference between structural conditions resulting from the accumulated construction steps and those arising as if the structure had been instantaneously created in its final configuration.

For the following steps it is assumed that the analysis is to be performed without the benefit of a computer program that accounts for the time-dependent effects of the steel and concrete.

8. Tabulate the accumulated bending moments for self-weight, posttensioning, and construction sequence loads from Step 7 for the conditions at the end of the construction sequence at intervals of at least one-eighth of a span for each span. For the bridge being analyzed in this example, the bending moment diagrams for each of the erection phases are shown in Figs. 15.136 and 15.137 for self-weight and the post-tensioning, respectively.
9. Repeat this procedure for the conditions had the structure been created instantaneously in its final configuration. The results of the analysis for the self-weight and the posttensioning are presented in Figs. 15.138 and 15.139, respectively.
10. Compute creep moments,[3] or apply the creep redistribution factor ϕ recommended in Ref. 10 to the difference between the tabulations for Steps 8 and 9 and determine the final redistributed bending moments resulting from self-weight and posttensioning (Figs. 15.138 and 15.139). In this example, $\phi = 1.16$ is used.
11. Tabulate the longitudinal bending moments for live load and any other applied load. For this example, one lane of live load equal to 800 lb/ft (11.676 kN/m) is considered located on spans 1, 3, and 5. The load is assumed to be placed toward the outside curve at 15 ft (4.575 m) from the centroid of the girders. The equivalent external torque is then 12.008 ft·kips/ft (53.418 kN·m/m).
12. At each interval along each span, divide the accumulated longitudinal bending moments M by the radius of curvature R and add the distributed torque load m_T from Step 6. In the case of posttensioning, include only the secondary bending moments in the analysis.

[i]Mainly from the original article by Alan J. Moreton, Juan J. Goni, Amy R. Kohls, and Daniel G. Davis of Figg Engineers, Tallahassee, Florida.

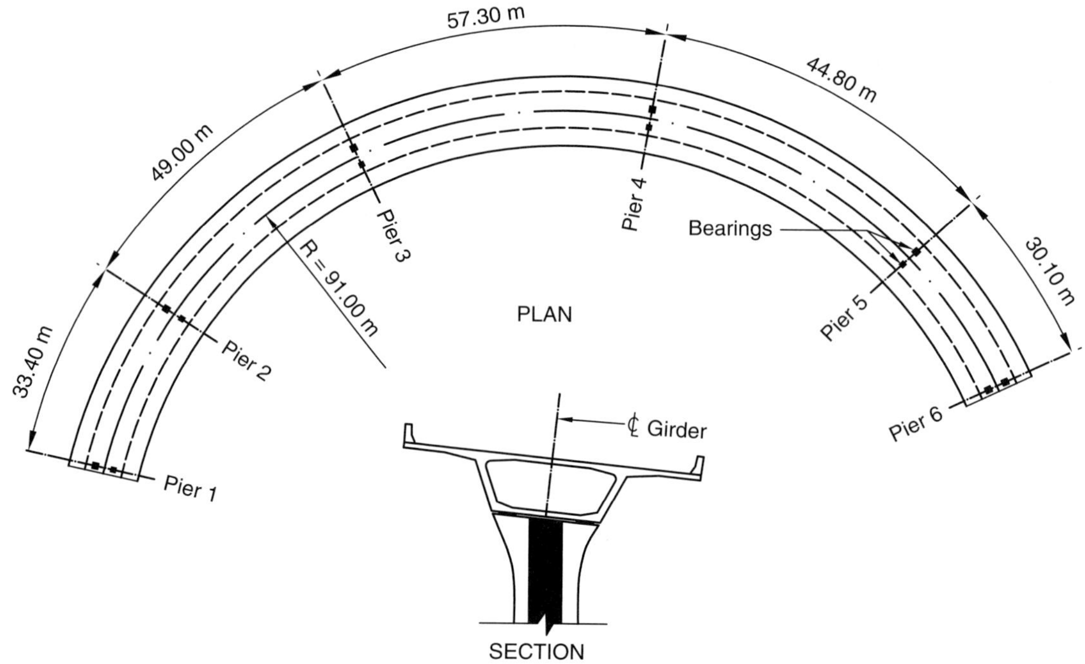

Figure 15.132 Plan and section of bridge.

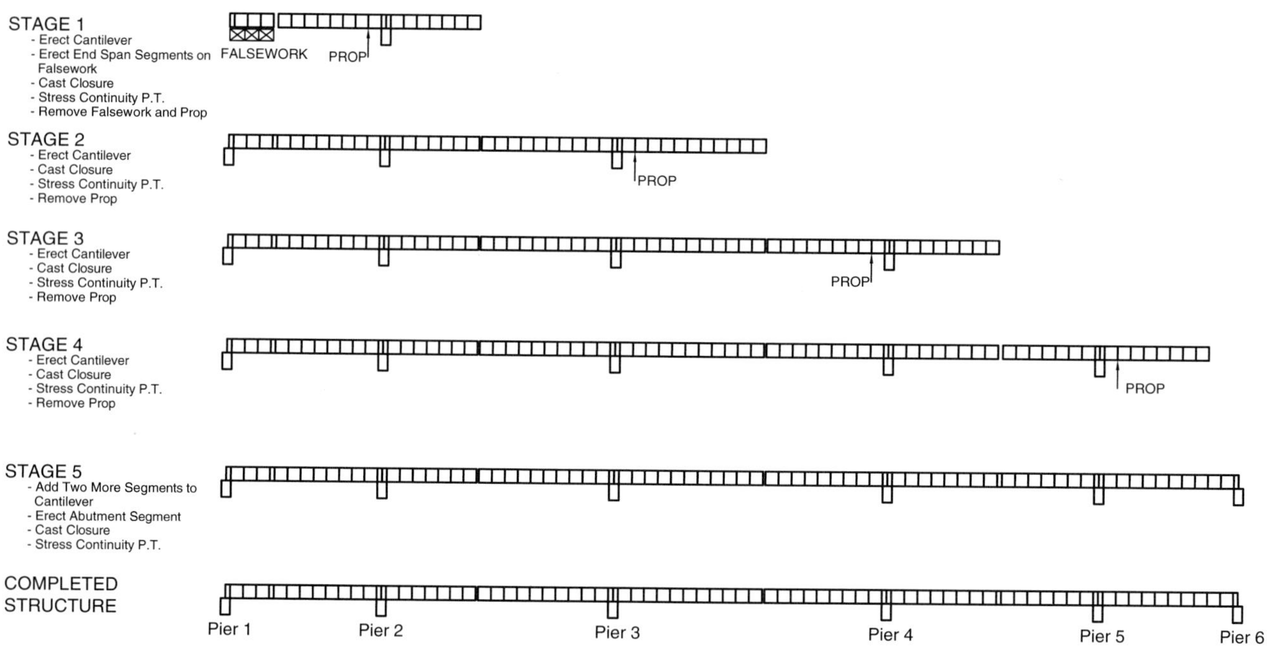

Figure 15.133 Construction sequence.

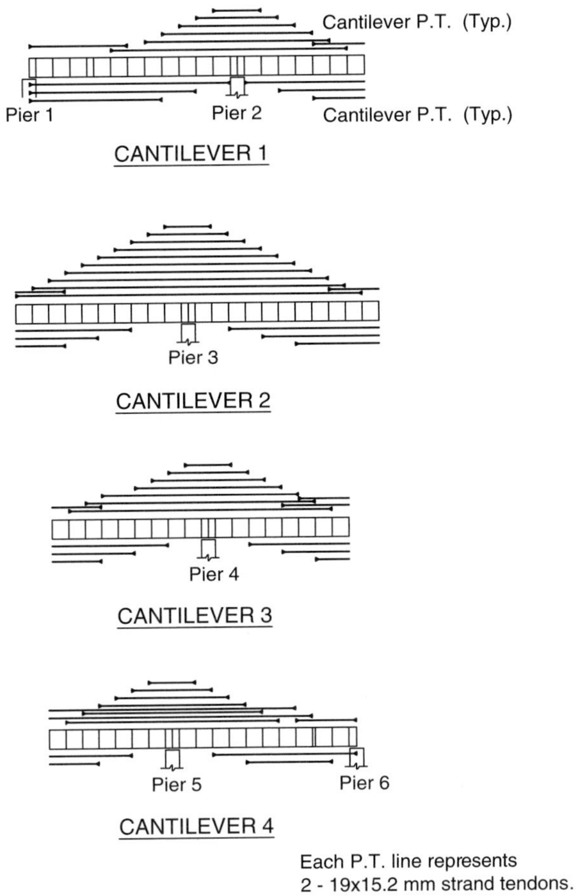

Figure 15.134 Posttensioning schematic layout.

13. Establish the effective torque span or spans for the overall structural configuration. Each torque span comprises that length of superstructure between those piers at which torsion can be transmitted to the substructure.

14. Using the values from Step 12, construct a distributed torque loading diagram and apply this loading to the defined torque span or spans. Construct the shear force diagram for this loading as if for a simple beam of length equal to each torque span. The ordinates of this diagram represent the final torsional moments at any given section. The torsional moments in the example for the self-weight and the post-tensioning are presented in Figs. 15.140*a* and 15.140*b*, respectively. The computation of the torsional moments using the method in Steps 12 and 13 for the combination of self-weight, posttensioning, and the live load assumed in Step 11 is for span 3 summarized in Table 15.16 for every ¼ of the span length between Pier 3 and Pier 4.

15. Determine the torsional shear-flow forces and stresses at the desired sections along the span and algebraically add them to those resulting from longitudinal shear (and bending) to determine the maximum shear stresses, principal stresses, and vertical shear plus torsional shear forces in each web and the torsional shear forces in each slab.

16. Determine transverse reinforcement in the slabs and webs and algebraically add to that required for local effects from Step 4.

17. Return to the individual construction stages and repeat this procedure to check critical sections and stages during construction. Revise the reinforcement as necessary. For cantilever construction of continuous bridges, the self-weight torque is generally maximum during construction of a cantilever.

18. Check the critical sections for the conditions immediately after the completion of construction but before all final long-term losses and creep redistribution.

If a plane-frame construction-stage computer program is available, extract the longitudinal bending moments for the final long-term service conditions after all losses resulting from creep and shrinkage. Skip Steps 8 and 9.

For the loads considered at span 3, the results show that the torsional moments after creep redistribution are approximately 5.5 percent of the longitudinal bending moments. These effectively contribute an additional 7.2 percent of shear force in the worst loaded web. While other live load, centrifugal, and wind forces may increase these forces, the overall condition is not significantly different from that of a straight structure. This may not always be the case for all structures, but it indicates that curved concrete box girders are very efficient.

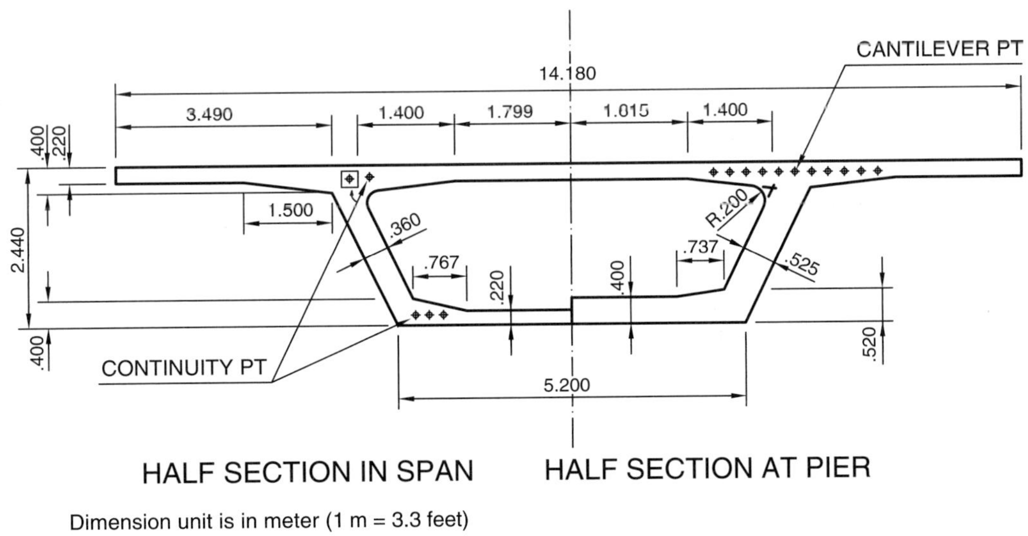

Figure 15.135 Superstructure cross section.

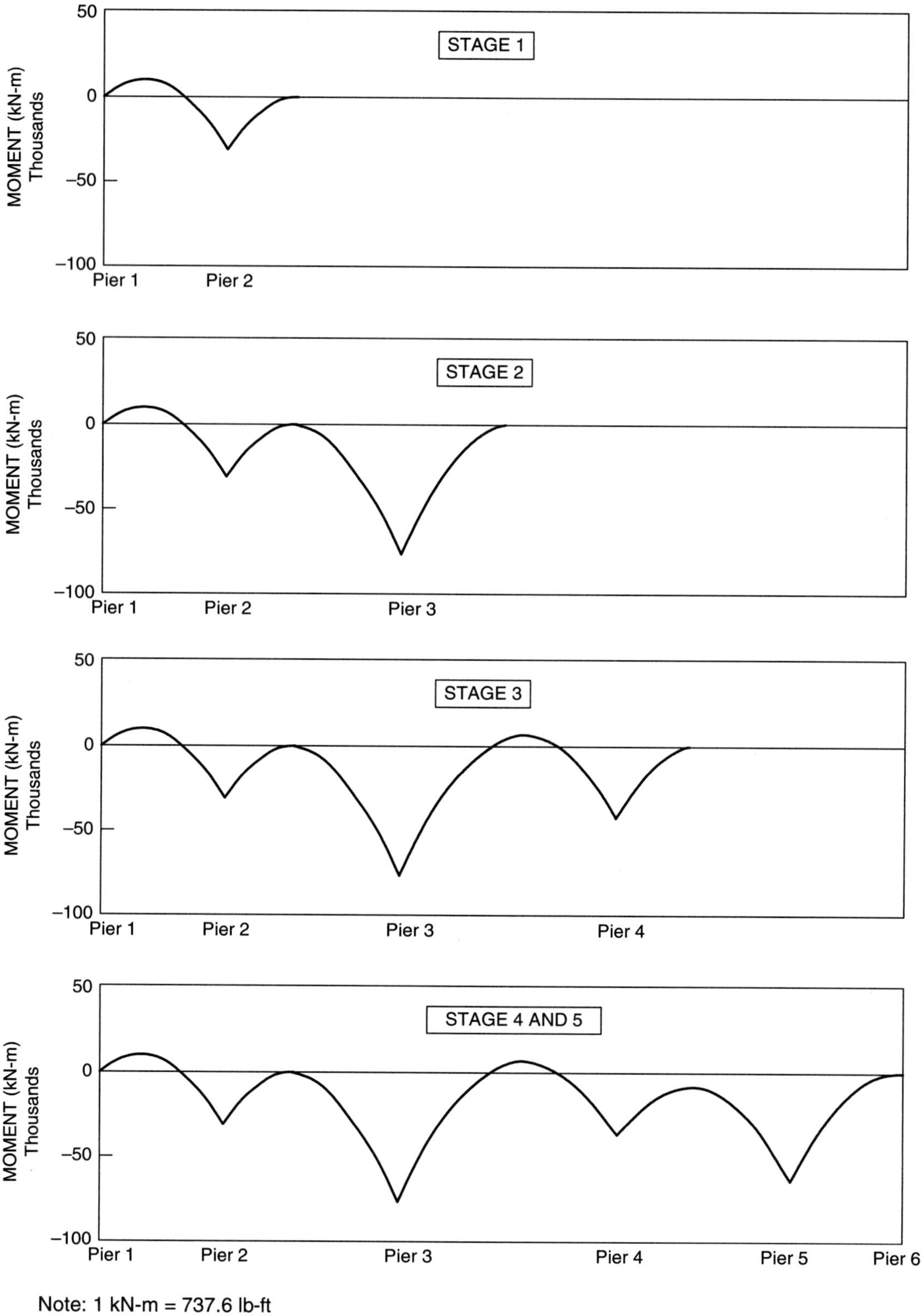

Figure 15.136 Bending moment diagrams for self-weight during construction.

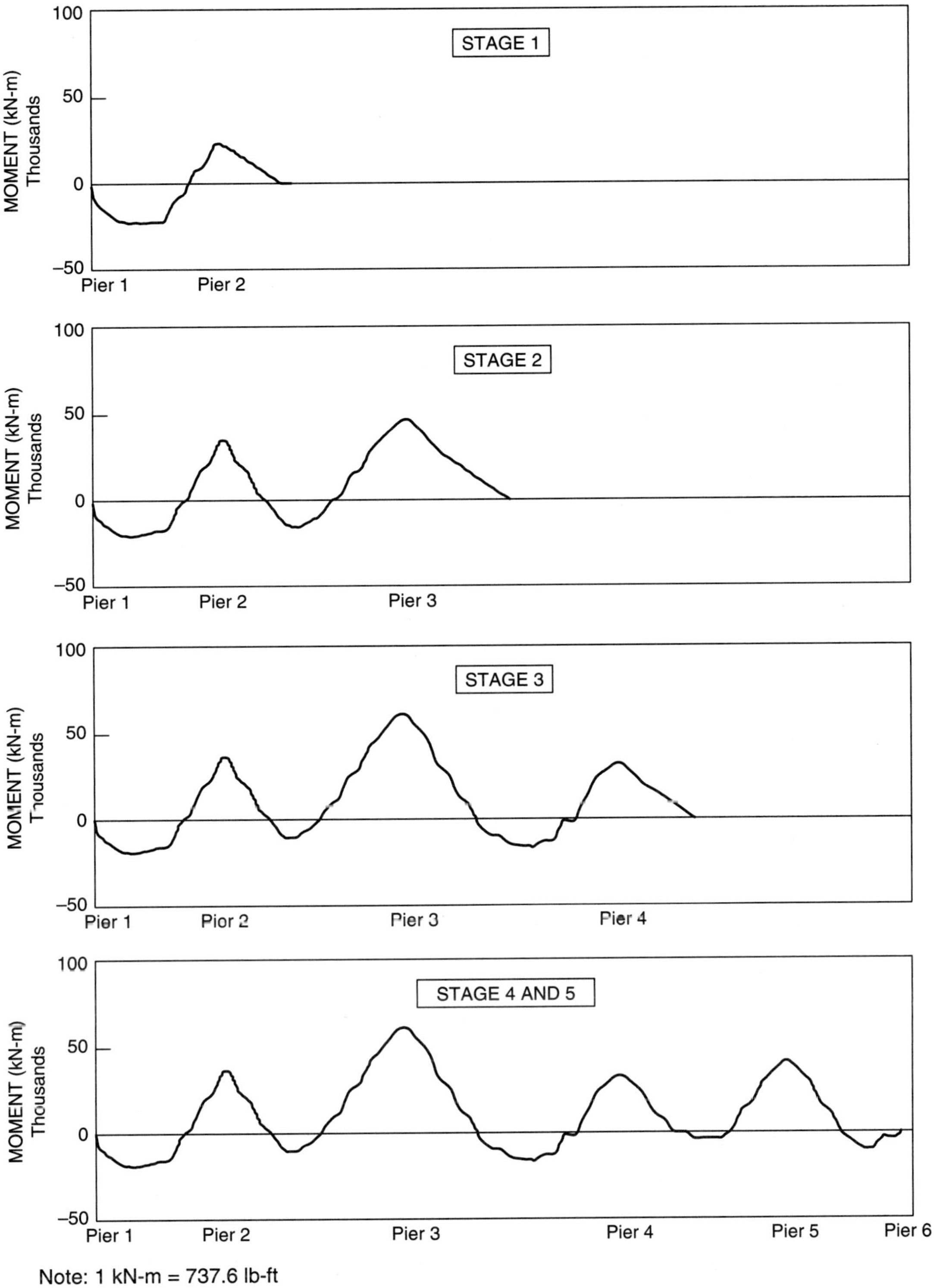

Figure 15.137 Bending moment diagrams for posttensioning during construction.

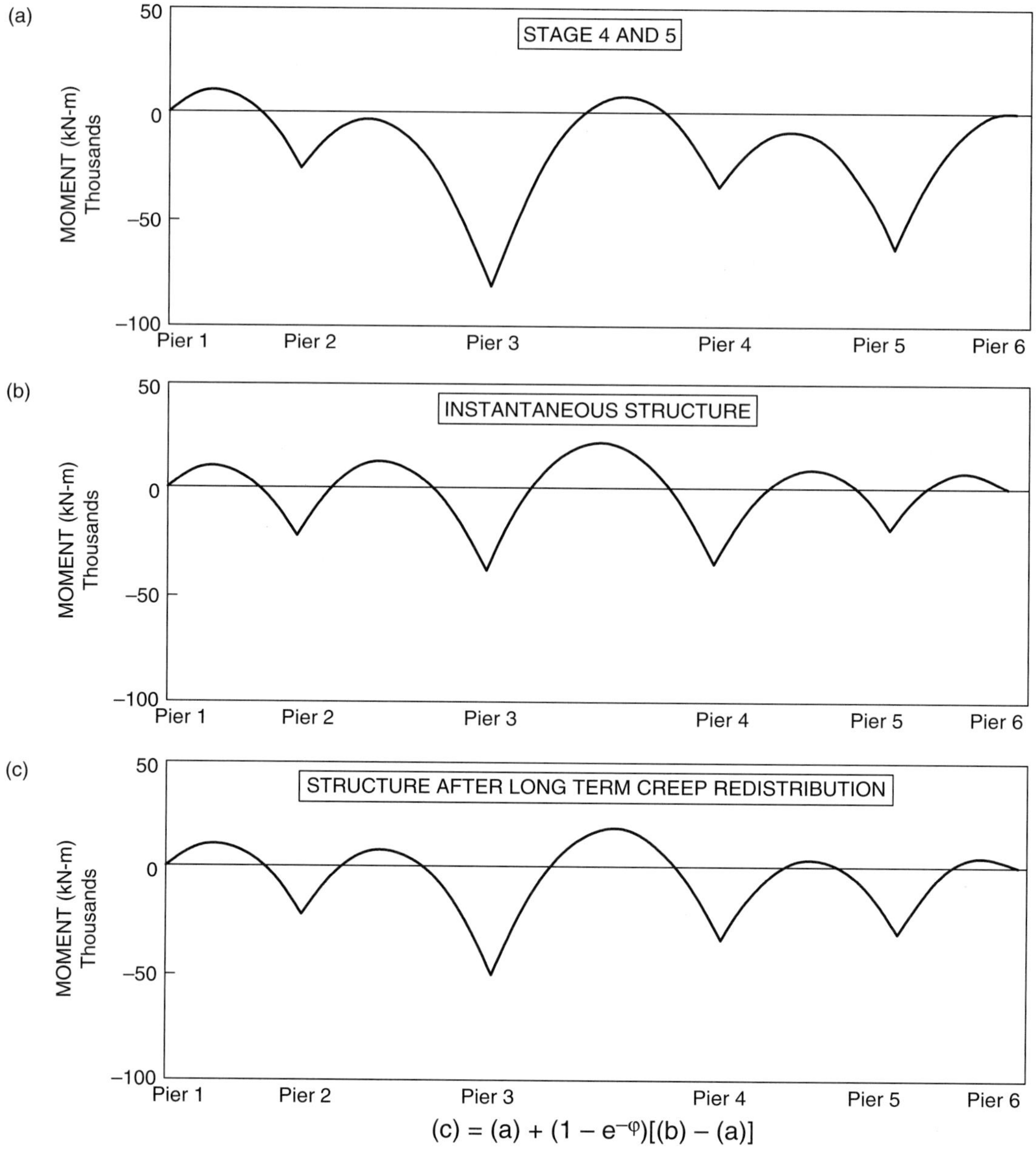

Figure 15.138 Bending moment diagram for self-weight after creep redistribution.

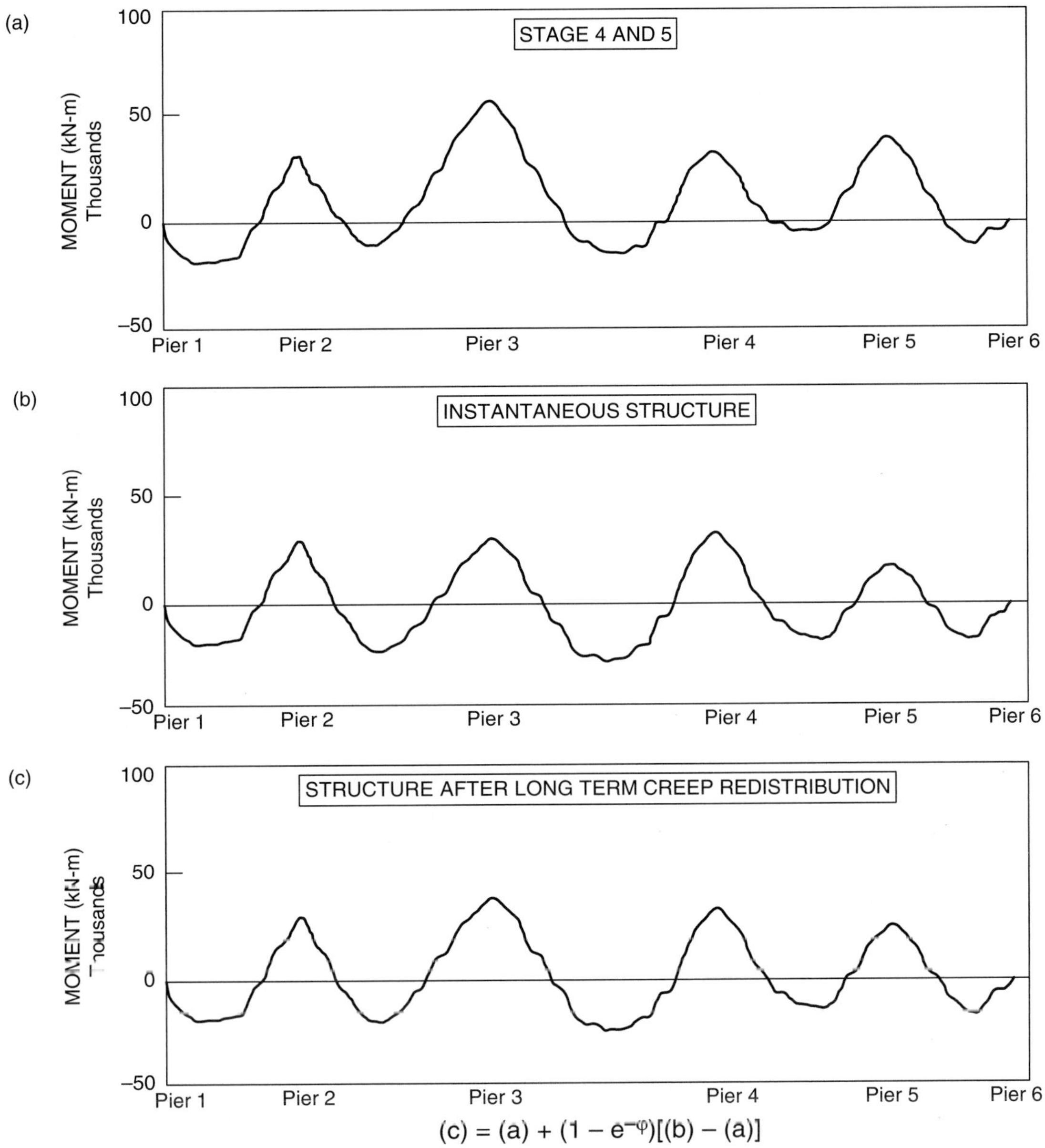

Note: 1 kN-m = 737.6 lb-ft

Figure 15.139 Bending moment diagram for posttensioning after creep redistribution.

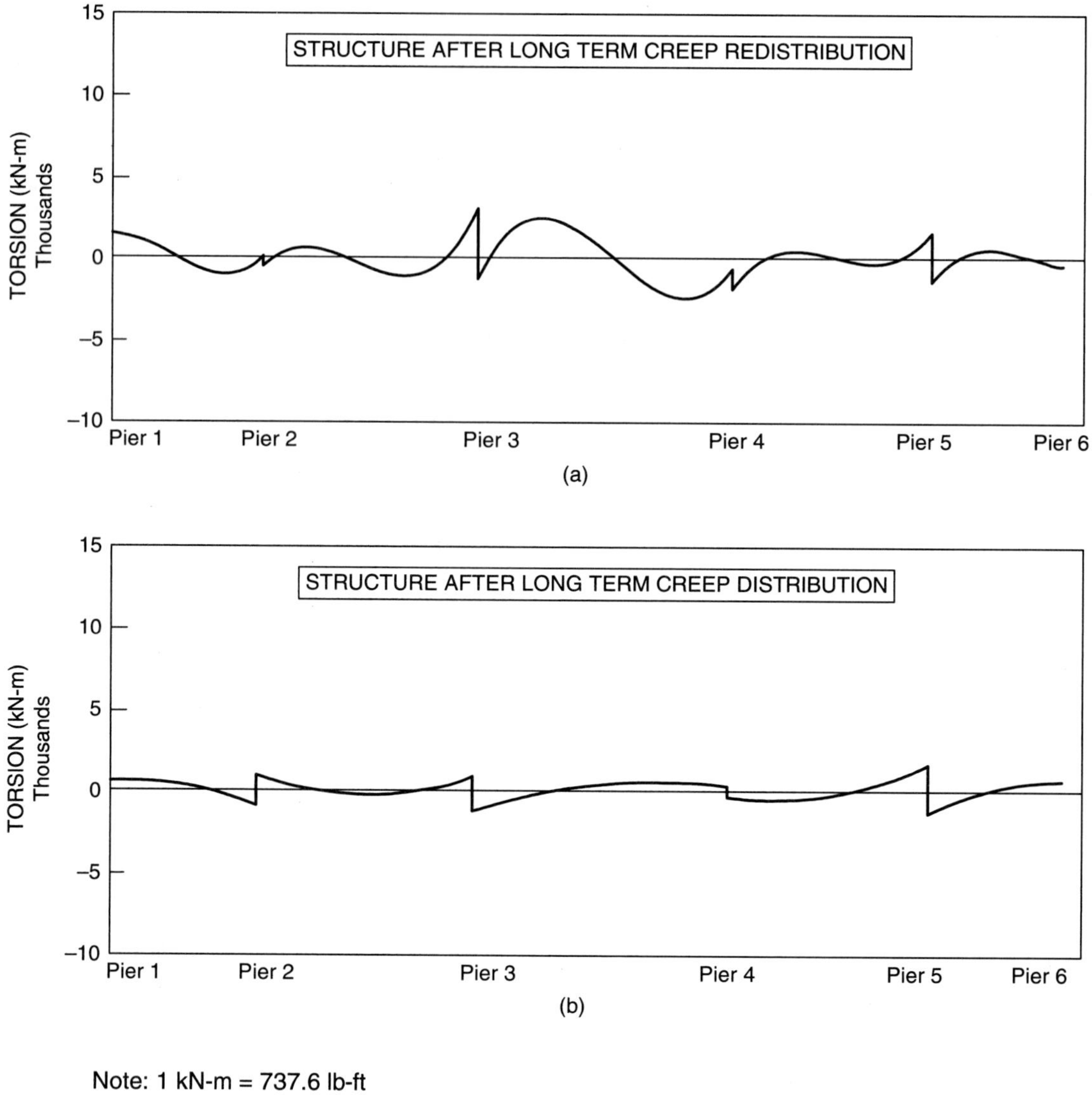

Figure 15.140 Torsional moment for (*a*) self-weight and (*b*) posttensioning after creep redistribution.

Table 15.16 Computation of the Torsional Moment for the Combination of Self-Weight, Posttensioning (PT), and the Live Load

	Longitudinal bending moment (KN·m)				Applied Torque *T* (KN·m/m)				
Location	*DL*	*PT*	*LL*	Total	*M*/*R*	*DL*	*LL*	Total	*M*/*R*+*T*
Pier 3 CL	−55,012.6	−7686.0	−1837.9	−64,536.5	−709.2	20.8	53.4	74.2	−635.0
¼ Span	−902.0	−4929.0	1629.8	−4201.4	−46.2	20.8	53.4	74.2	28.0
½ Span	18,715.7	−2,340.9	2604.5	18,979.3	208.6	20.8	53.4	74.2	282.8
¾ Span	7589.7	341.5	1285.7	9216.9	101.3	20.6	53.4	74.0	175.3
Pier 4 CL	−37,372.5	3099.6	−2541.7	−36,814.6	−404.6	20.6	53.4	74.0	−330.6

Note, $R = 298.6$ ft (91.0 m), 1 kN.m = 737.6 lb.ft.

15.31.3 Additional Reading

The reader is referred to additional materials and examples on concrete box-girder curved bridges provided in Refs. 1, 3, and 14.

REFERENCES

1. NCHRP, *Development of Design Specifications and Commentary for Horizontally Curved Concrete Box-Girder Bridges*, NCHRP Report 620, National Academies of Science, Engineering and Medicine, National Academies Press, Washington, DC, 2008.
2. *AASHTO LRFD Bridge Design Specifications*, American Association of State Highway and Transportation Officials, Washington, DC, 2012.
3. Corven, John, Post-Tensioned Box Girder Design Manual, Report No. FHWA-HIF-15-016, Office of Infrastructure—Bridges and Structures, Federal Highway Administration, Washington, DC, 2016.
4. Hambly, E. C., *Bridge Deck Behavior*, Second Edition, E & FN Spon, London, 1991.
5. Bakht, B. and L. G. Jaeger, *Bridge Analysis Simplified*, McGraw-Hill, New York, 1985.
6. Powell, L. C., and Breen, J. E., State of the Art Externally Post-Tensioned Bridges with Deviators, Center for Transportation Research, The University of Texas at Austin, Report No. FHWA/TX-89+365-1, 1988.
7. Podolny, W., "The Cause of Cracking in Post-Tensioned Concrete Box Girder Bridges and Retrofit Procedures," *PCI Journal*, 30(2), March–April 1985.
8. Podolny, W., and Muller, J. M., *Construction and Design of Prestressed Concrete Segmental Bridges*, Wiley Interscience, New York, 1982.
9. Mathivat, J., *The Cantilever Construction of Prestressed Concrete Bridges*, John Wiley and Sons, New York, 1983.
10. Post-Tensioning Institute, *Precast Segmental Box Girder Manual*, Post-Tensioning Institute, Phoenix, AZ, 1988.
11. Dabrowski, R., *Curved Thin-Walled Girders. Theory and Analysis*, C & CA Library Translation (Translator, C. V. Amerongen), Cement and Concrete Association, London 1972.
12. Kirstek W., *Theory of Box Girders*, Wiley Interscience, New York, 1970.
13. Maisel, B. I., and Roll, F., *Methods of Analysis and Design of Concrete Box Beams with Side Cantilevers*, Technical Report No. 42.494, Cement and Concrete Association, London, 1974.
14. Sennah, K. M., and Kennedy, J. B., "Literature Review in Analysis of Box-Girder Bridges," *Journal of Bridge Engineering*, ASCE, 7(2), 134–143, 2002.
15. Tung, D. H. H., "Approximate Torsional Analysis of Curved Box Girders by the M/R-Method," *AISC Engineering Journal*, 7(3), 65–74, July 1970.
16. Witecki, A. A., "Simplified Method for the Analysis of Torsional Moment as an Effect of a Horizontally Curved Multispan Continuous Bridge," *Concrete Bridge Design*, ACI Publication SP-23, pp. 193–204, 1969.
17. ABAQUS/CAE User's Manual, 6.12, Dassault Systèmes Simulia Corp., Providence, RI, 2012.
18. Timoshenko, S., and Young, D. H., *Elements of Strength of Materials*, Fourth Edition, D. Van Nostrand, New York, 1962.

Chapter 16
Railroad Bridge Design

BY

MOHSEN ISSA, Ph.D., PE, SE *Professor, Department of Civil and Materials Engineering, University of Illinois at Chicago*

AHMAD HAMMAD, Ph.D., PE, SE *Assistant Vice President and Senior Engineering Manager, WSP USA*

16.1 INTRODUCTION

The principles of designing railroad structures are similar to those for structures carrying highways. However, structures carrying railways have much heavier loadings than those subject to highway loadings due to increased dead load, live load, and impact required for railways.

Railroad bridge design, construction, maintenance, operation, and inspection are governed by the American Railway Engineering Maintenance of Way Association (AREMA) Manual (hereafter referred to as AREMA Manual). The AREMA Manual contains four volumes, each of which contains chapters that cover various topics. Volume 1 covers Rail, Volume 2 covers Structures (Chapter 7: Timber Structures, Chapter 8: Concrete Structures and Foundations, Chapter 9: Seismic Design for Railway Structures, Chapter 15: Steel Structures), Volume 3 covers Infrastructure and Passenger, and Volume 4 covers System Management.

Much of the material covered in this chapter will reference specific articles of the AREMA Manual.

16.2 GENERAL REQUIREMENTS

16.2.1 Materials

The choice of material is dependent on the available construction depth and span. Some of the main typical materials that are found on a railroad bridge are steel, concrete, stone, and timber. Concrete is commonly used for many bridge superstructure members such as decks, prestressed concrete beams. It is used extensively in new construction for the entire abutment and piers. Steel is commonly used in the bridge superstructure for beams, bearings, floor beams, girders, and reinforcing bars in concrete. Stone was commonly used for building the abutments and piers in the 1940s and earlier. Iron was used typically in beams and trusses that were built before 1900. Timber is used for several decks, piles, and track ties. Composite material have been also used to a lesser extent.

The design requirements of these recommended practices, contained in this part, are based on the use of materials conforming to the current requirements of the following ASTM specifications.

The concrete materials shall be selected for strength, durability, and chemical resistance, and ability to attain specified properties as required, in accordance with this recommended practice and as approved by the engineer. They shall be combined in such a manner as to produce uniformity of color and texture in the surface of any structure or group of structures in which they are to be used. No change shall be made in the brand, type, source, or characteristics of cementitious materials, the character and source of aggregate or water, or the class of concrete and method of transporting, placing, finishing or curing without approval of the engineer.

Hardwood structural timbers shall comply with the requirements of the Northeastern Lumber Manufacturers Association, Inc. (NELMA), Chapter 6: Timber, Beams and Stringers, Posts and Timbers for the species and grades listed in Part 2 of this Manual.

16.2.2 Deflections

Limits on deflections are set to ensure the safety of rail traffic, to control track forces, and to ensure track geometry remains within specified limits. AREMA Manual Chapter 15: Steel Structures limits the deflection due to live load plus impact loading to 1/640 of the span length. For concrete structure, minimum superstructure depths are provided to have adequate stiffness for the control of deflections. In addition to vertical deflections, lateral/horizontal deflection as well as longitudinal deflection should be considered.

16.2.3 Clearances

Clearance between trains and the bridge structure or any fixture attached shall be sufficient. These envelopes depend on the speed and geometry of the track. As per AREMA Manual, the minimum railway bridge clearance for tangent track is 18'-0" wide by 23'-0" high above top of rail. The requirements for lateral clearance each side of the track centerline for curved geometry are increased by 1½ in. per degree of curvature. The vertical clearance may have to be increased for electrified lines.

16.3 LOADING

16.3.1 Dead Load

Dead load consists of the weight of the structure itself, the track rails and fastenings, ballast and ties, weight of waterproofing, and any other attachments that are permanently applied. Unit weights of various material can be estimated as follows: Concrete: 150 pcf, Steel: 490 pcf, Timber: 60 pcf, and Ballast: 120 pcf. As per AREMA Manual, the track rails, inside guard rails, and their rail fastenings shall be assumed to weigh 200 lb per linear foot.

16.3.2 Live

The AREMA Manual recommends that design be based on Cooper E-80 Live Loading. While the Cooper E loading is not a real train, it is a simplified model intended to produce load effects equivalent to those of a real train. Heavier Cooper E loadings will result in directly proportional increases in the concentrated and uniform live loadings. Some new designs are utilizing E90 or E100. In addition to Cooper loading, the Alternative Live Load on 4 Axle (100 kips each axle) shall be considered. Generally, the Alternate loading controls for shorter spans.

AREMA Manual Chapter 15, Section 1.15, has tables for Live Load Moments, Shears, and Reactions. These tables reflect Cooper E80 or Alternate Loading Live Loading. All the moment, shear, and reaction values are for one rail (one-half track load) only, and all the values can be prorated (directly proportional) for smaller or larger Cooper's E live loadings.

16.3.3 Distribution of Live Load

The live load is distributed to the superstructure (beams or girders) for open-deck structures as a series of concentrated load, provided that ties are within a length of 4 ft but not more than three ties. When two or more longitudinal beams per rail with diaphragms in accordance with AREMA Manual Chapter 15 and symmetrically spaced under the rail, they shall be considered as equally loaded.

For ballasted-deck structures, the live load distribution for the deck, transverse beams, and longitudinal beams or girders is a function of deck thickness, ballast thickness, tie width, tie spacing, beam spacing, beam stiffness, axle load, and axle spacing.

16.3.4 Impact

To allow for dynamic, vibratory, and impact effects, the live load is amplified by the impact factor, I. The impact factor is a percent of the live load due to vertical effects, which is a function of type of locomotive and span length and is applied to each rail and due to rocking effects that are 20 percent of the wheel load and applied as a force couple.

AREMA Manual Chapters 8 and 15 specify the impact forces to be used and how they are to be applied. For ballasted-deck structures the percentage to be used shall be 90 percent of that specified for open-deck structures.

16.3.5 Centrifugal

Structures on curved alignment shall be designed for a horizontal radial force equal to a percent of the axle load without impact. The horizontal force is applied at 6 ft above the top of rail and is a function of speed and degree of curve and can produce an overturning moment that tends to increase the live force in members on outside of the curve and reduce live load in members on the inside of the curve. Preferably, the section for the member on the outside of the curve should be used also for the member on the inside. The centrifugal force should also be considered in the design of anchor bolts, bearings, and substructure.

16.3.6 Wind

Structures shall be designed for wind load on the moving train as well as wind load on unloaded bridge. The wind load shall be considered as a moving load acting in any horizontal direction (lateral and longitudinal). Regardless of material type, on the train the wind load shall be taken at 300 lb/ft on the one track, applied 8 ft above the top of rail. On the structure it depends on material type and is covered in Chapters 7, 8, and 9 for Timber, Concrete, and Steel, respectively.

16.3.7 Lateral Equipment Forces

A lateral load (often referred to as nosing) from the passing of the train on the tracks (curved or tangent) shall be applied in either horizontal direction. For steel structure, the magnitude of the concentrated force is one-fourth of the heaviest axle without impact and is to be applied at the base of the rail at any point across the span. For timber structures, the nosing load is 20 kips applied at the top of the rail in either horizontal direction and at a point along the span.

16.3.8 Stability Check

For spans and towers, stability should be investigated with live load on only one track, the leeward track with respect to wind direction for structures with more than one track. The live load should be 1200 lb/ft, without impact.

16.3.9 Longitudinal

Longitudinal forces shall be applied to structures in any direction that generates the critical design load. Trains shall be considered to travel in both directions on all tracks. As per AREMA Manual, the longitudinal force is taken as the larger of the force due to braking applied at 8 ft above top of rail or the force due to traction applied at 3 ft above the top of rail. The longitudinal force is a function of the structure length, and for longer bridges, tower bents are usually provided to resist the longitudinal force.

16.3.10 Fatigue

Bridges must also be designed for the repeated application of loading (fatigue loading) throughout the design life. For railroad bridges, fatigue loading is very important, consideration and may well govern many aspects of the detail and size of major members for new steel bridges. The major factors governing fatigue strength are material strength, the number of stress cycles per train passage, the magnitude of the stress range, and the type and location of the detail.

16.3.11 Combined Stresses

The allowable stress design method (working design) for railroad structures is used for steel and timber. While AREMA Chapter 8: Concrete Structures utilizes both allowable stress design and load factor design methods, except that load factor design (LFD) is not applicable to foundation and structural stability checks. For steel structures, the allowable stress is commonly 0.55 of the minimum yield strength of the material. AREMA Manual Chapter 15: Steel Structures, Table 15-1-11, shows the allowable stresses for structural steel, rivets, bolts, and pins. AREMA Manual Chapter 7, Tables 7-2-7 and 7-2-8, gives allowable values for Structural Glued Laminated Softwood Timber. Table 7-2-9 shows Allowable Unit Stresses for Stress Graded Lumber. Table 7-2-10 depicts Unit Compression (Column) Stresses for Standard Stress Grades.

16.3.12 Earthquake

Seismic design of railroad structures is covered under AREMA Manual Chapter 9. AREMA specifies two procedures for analyzing bridges subject to seismic loads: the Equivalent Lateral Force Procedure and the Modal Analysis Procedure. The selection of the analysis procedure is dependent on the configuration of the bridge considered as classified in Table 9-1-7. AREMA Manual provides a detailed descriptive account of the aforementioned analysis procedures in Chapter 9.

16.3.13 Load Combinations

The structure may be subjected to various combinations of loads and forces. Each component of the structure, including the foundation, shall be proportioned to resist safely all group combinations of these forces that are applicable. Group loading combinations for various material are provided in AREMA Manual. In general, member sizes based on stresses resulting from dead load, live load, impact load, and centrifugal force control.

16.4 TIMBER

16.4.1 Bridge Types

Timber's strength, light weight, and energy-absorbing properties are positive features that are desirable for bridge construction. Timber is capable of supporting short-term overloads without adverse effects. Generally, timber is used for spans between 12 and 16 ft in the form of open-deck trestle or ballasted-deck trestle. There are two primary type of timber trestles: pile trestles and framed trestles. Pile trestles have maximum height of 30 ft, while framed trestles can be designed for much taller bents.

16.4.2 Material Specifications

Material Specifications for Lumber, Timber, Engineered Wood Products, Timber Piles, Fasteners, Timber Bridge Ties, and Recommendations for Fire-Retardant Coating for Creosoted Wood are covered under Part 1 of AREMA Manual Chapter 7. In general, the quality of timber bridge elements shall conform to the appropriate grading rules and shall be sound and free of defects that may impair strength or durability such as decay, splits, shake, excessive slope of grain, or numerous holes or knots, bark, and wane.

16.4.3 Design Considerations

For glulam timber, design values must be derived from ASTM 3737 and ASTM D2555 Standard Specifications. These values are provided by AREMA Manual in Tables 7-2-7 and 7-2-8. The values must then be adjusted according to the adjustment factors provided in Table 7-2-3. Allowable unit stresses shall conform to Tables 7-2-9 and 7-2-10 provided in the AREMA Manual for Railway Engineering. The table assigns different stress values based on the type of wood, grade, and size classification.

Allowed bearing stresses and horizontal shear shall be calculated as specified by AREMA Manual in Sections 2.5.7 and 2.5.9 of Chapter 7. Basic stresses for bearing on bolts, whether parallel or perpendicular to grain, are provided in Table 7-2-11 for different species of wood. Table 7-2-12 provides a factor to be multiplied by the basic stress depending on the *L*/*d* ratio of the bolt.

Section 2.6 of Chapter 7 provides specifications for details of design including bolt connections and notched beams. Wood culverts design practice specifications are provided in Section 2.7 of the same chapter. In Section 2.8, AREMA provides recommended practice for simple stress laminated deck panels. Spacing of prestressing bars for predrilled wood laminates shall conform to Table 7-2-14.

16.4.4 Design Details

AREMA Manual Chapter 7 Timber Appendix 1 provides very useful design details for Contemporary Designs and Design Aids. Those include unit stresses for different deck types (open, ballasted) for Cooper E-80 loading. The table depicts various spans lengths, number of stringers, glue laminated sections, stress laminated decks, graded lumber, size of cap, number of piles, and so on.

16.5 STEEL

16.5.1 Bridge Types

As per AREMA Chapter 15: Steel Structures, the preferred types of railroad structures are as follows: rolled or welded girders for spans of 50 ft or less, bolted or welded plate girders for spans over 50 to 150 ft, and bolted or welded trusses for spans over 150 ft. The rolled beams or girders in a one-way system can have either steel plate deck or reinforced-concrete deck (composite or noncomposite). Generally, there are a minimum of four or five girders per track. For a two-way system, the through or deck girders or trusses has floor beams connected to the main girders with steel plates. Figures 16.1 and 16.2 show a typical welded plate girder with steel plate deck and a typical through girder with floor beams and steel deck plate, curved track, respectively.

16.5.2 General Rules

AREMA Manual Chapter 15, Part 1, outlines more or less typical situations, and the preference of railroad companies for the use of a particular detail or material use should be accounted for in the design. Covered under Part 1 is spacing of trusses, girders and stringers, deflection, clearance, camber, skews, and expansion. The spacing between outside trusses or girders is not less than 1/20 for through span and 1/15 of the span length for deck span. For multigirder bridge structure, the girders or stringers are spaced uniformly to distribute the load. Assume 5'-0" rail center in the design calculations and center-to-center of bearing or supports for span lengths. For plates girders or rolled beams less than 90 ft in length, camber is not required. For girders greater than 80 ft, camber equals the deflection due to dead load, whereas, for trusses, camber equals the deflection due to dead loads and 3000 pound per foot of live-load. For live-load deflection calculations, gross moment of inertia shall be used for flexural members and gross area of members for trusses. For members with perforated cover plates, the effective area shall be used.

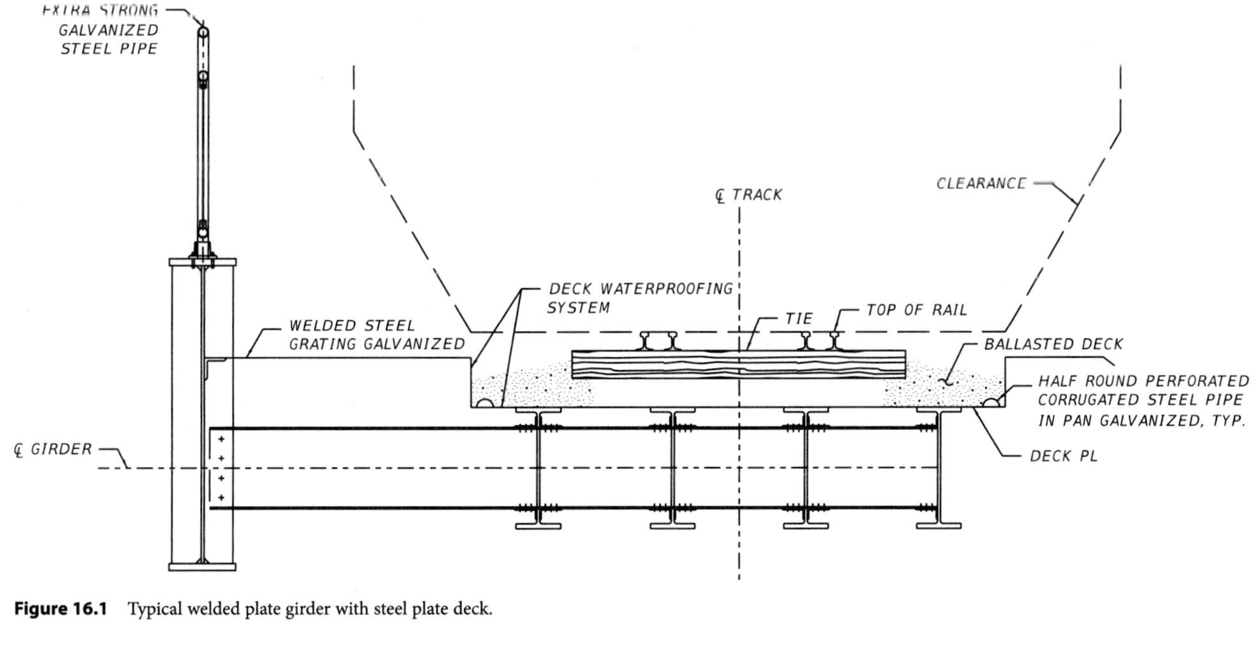

Figure 16.1 Typical welded plate girder with steel plate deck.

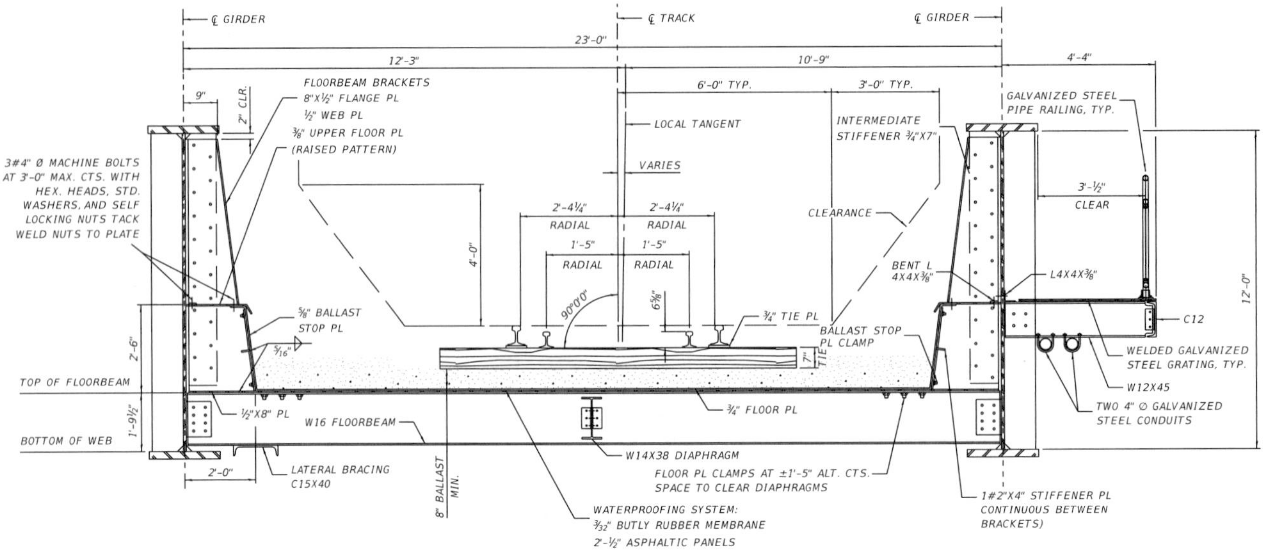

Figure 16.2 Typical through girder floor beams and steel deck plate.

16.5.3 Allowable Stresses

Section 1.4 of Chapter 15 specifies the basic allowable stresses to be used in proportioning the parts of the bridge. Table 15-1-11 provides a concise presentation of the allowable stresses in structural steel, rivets, pins, and bolts.

TENSION/COMPRESSION

Section 1.6 of Chapter 15 provides provisions for members stressed primarily in axial tension or compression.

FLEXURE

Section 1.7 of Chapter 15 presents the requirements for designing members stressed primarily in bending. This includes proportioning girders and beams, flange sections, and web thickness.

SHEAR

The design force for shear connectors, as specified by AREMA, shall be taken as follows:

- The allowable horizontal design force range per shear connector for fatigue (S_r) when channels are used shall be 2400 lb and 2100 lb for 2,000,000 cycles and over 2,000,000 cycles, respectively. The maximum allowable horizontal design force per shear connector (S_m) when channels are used shall be taken as 3600 lb. In the equations for horizontal design force, w is the length of the channel in inches measured in a transverse direction to the flange of the beam.
- When studs are used, the allowable horizontal design force per shear connector shall be 10,000 (A_s) lb and 7000 (A_s) lb for 2,000,000, and over 2,000,000, respectively. (A_s is the nominal cross-sectional area of the shear stud.)
- Fatigue due to primary bending stress must be checked on the base metal to which the studs or channel shear connections are attached as per Table 15-1-9.

16.5.4 Bracing

Section 1.11 of Chapter 15 in AREMA Manual provides specifications for bracing. The section includes provisions for bracing of top flanges of through girders, lateral bracing, portal and sway bracing, cross frames and diaphragms for deck spans, bracing for viaduct towers and bents, and bracing members used as ties or struts only.

16.5.5 Connections and Splices

As specified by AREMA Manual, field connections, including splices, shall be made using rivets or high-strength bolts except that field welding may be used for minor connections not subject to live-load force and for joining sections of deck plates and so on that do not function as part of the load-carrying structure. Otherwise, welding shall not be used for field connections. Connections shall accord to the following provisions, specified by AREMA:

- Splices of main members shall have a strength higher than or equal to that of the member and shall satisfy the requirements of Article 1.7.5 and Article 1.7.6.
- End connections of main members receiving load form the combined effect of floor system and truss action shall have a strength higher than or equal to the capacity of the member.
- End connections of simply supported floor beams and stringers must be proportioned 1.25 times their computed shear.
- Secondary and bracing members shall have a capacity of the lesser strength of the member based on the allowable unit stress or 1.5 times the maximum stress.
- Groove welds shall penetrate fully and satisfy Article 1.3.13.
- Bolted and riveted connections must have three or more fasteners per plane of connection.
- Pure compression members shall be spliced on four sides sufficiently if faced for bearing. The splice shall be in proximity to a panel point and shall transmit at least ½ the force to the splice material.

16.6 CONCRETE

16.6.1 Bridge Types

The most commonly used types of concrete railway bridges as per AREMA and Association of American Railroads (AAR) are the following:

- Cast-in-Place Prestressed Concrete Box Girder
- Cast-in-Place Concrete Trough
- Precast Concrete Slab Girder
- Precast Concrete I-Girder
- Precast Concrete Box Girder
- Precast Double Box Girder

Figures 16.3 through 16.8 show typical sections of the aforementioned bridge types in the same sequential order.

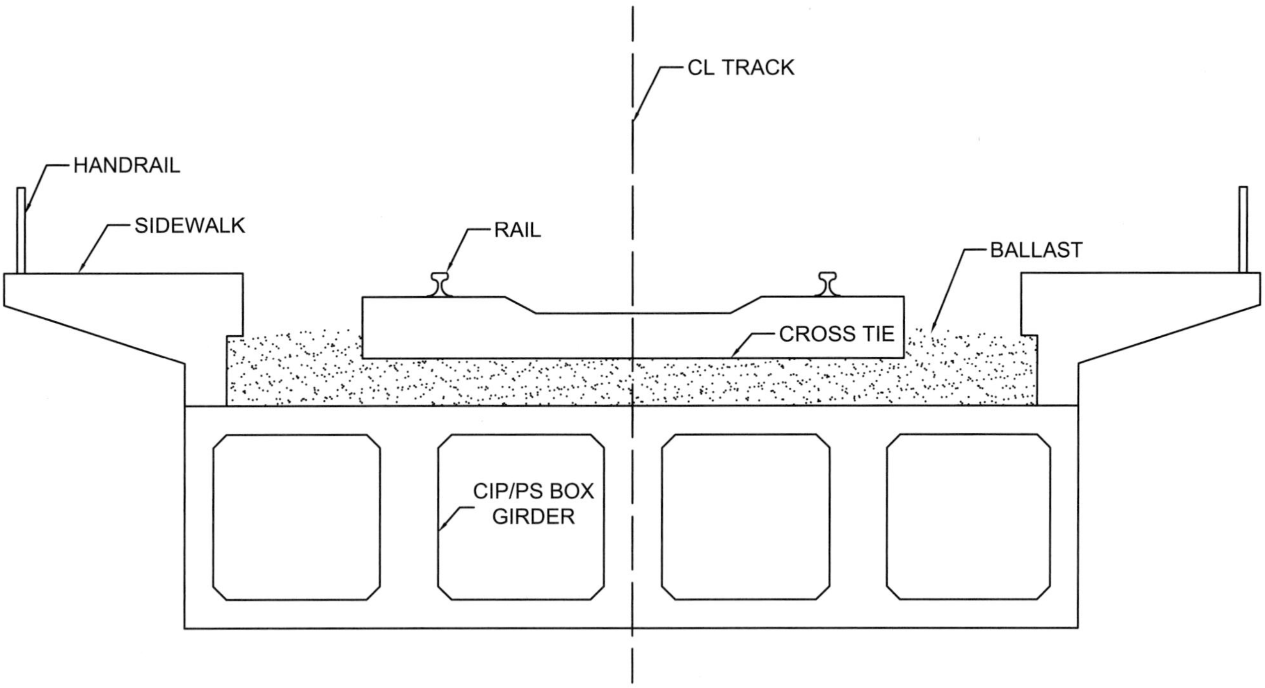

Figure 16.3 Typical CIP PS concrete box girder.

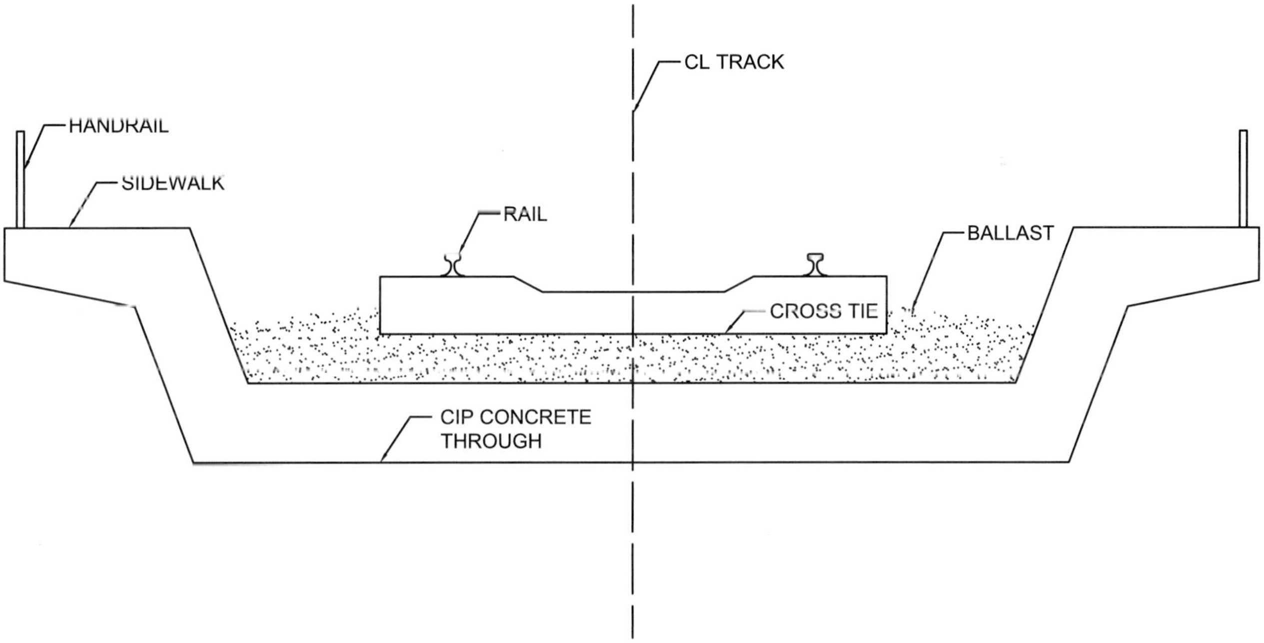

Figure 16.4 Typical CIP concrete through.

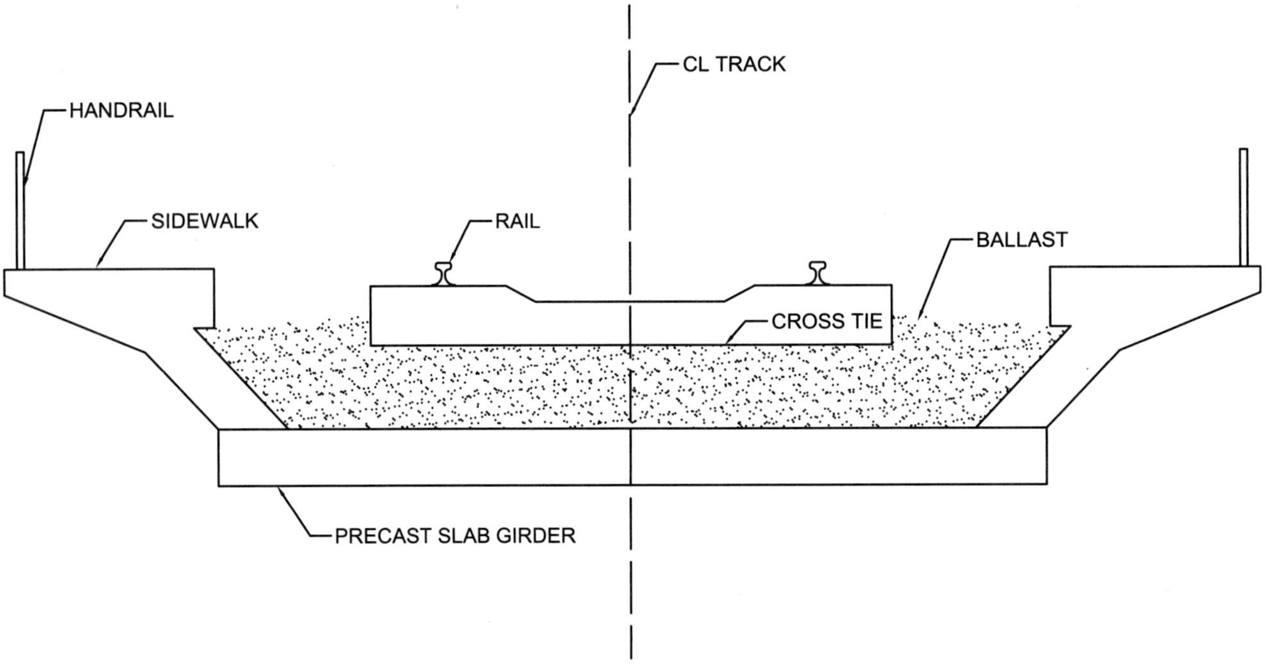

Figure 16.5 Precast concrete slab girder.

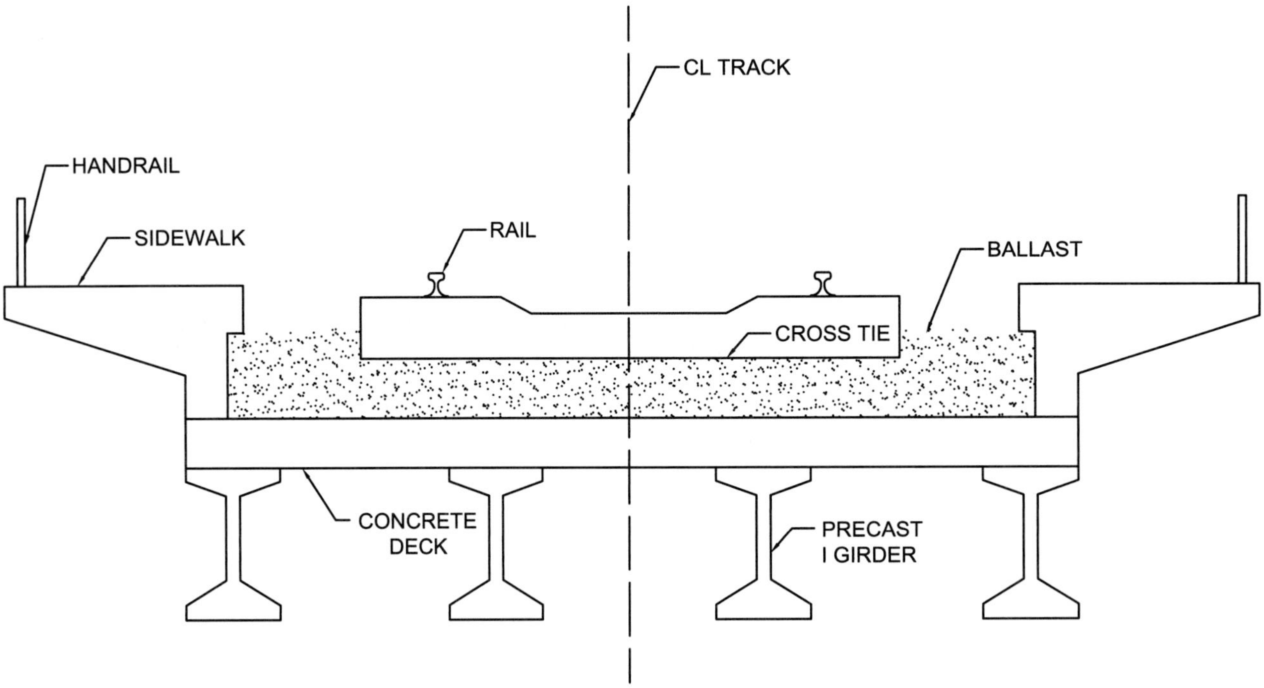

Figure 16.6 Precast concrete I-girder.

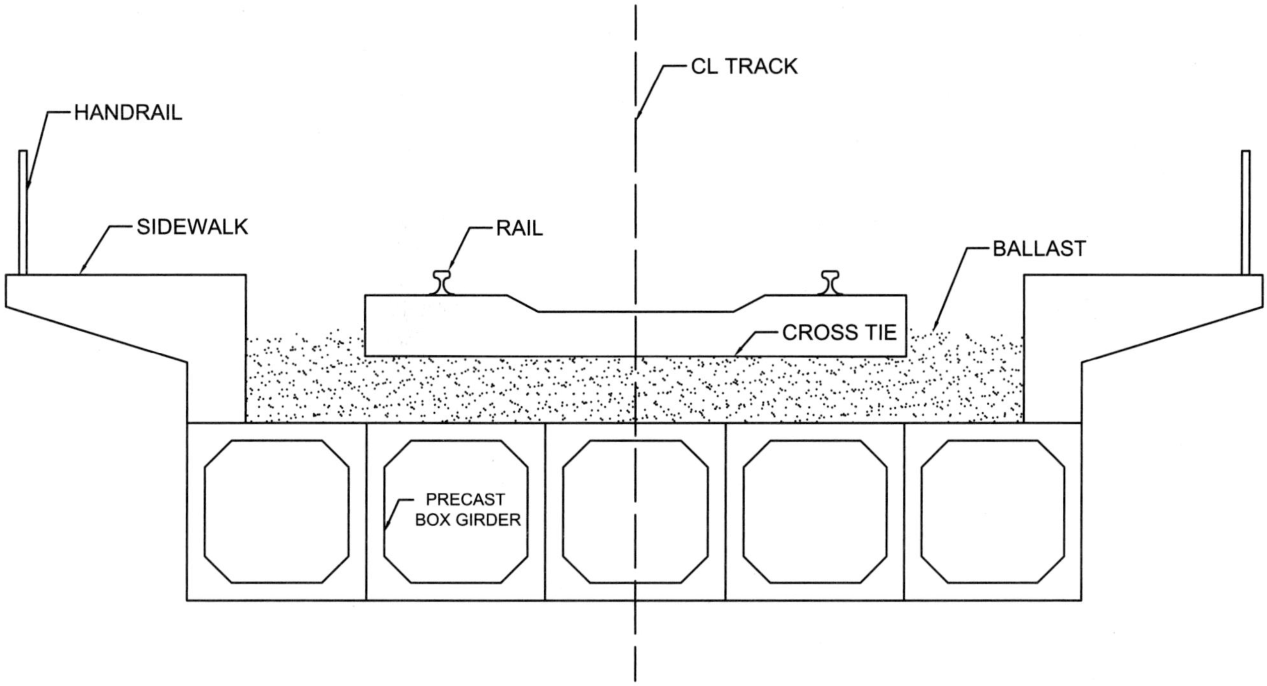

Figure 16.7 Precast concrete box girder.

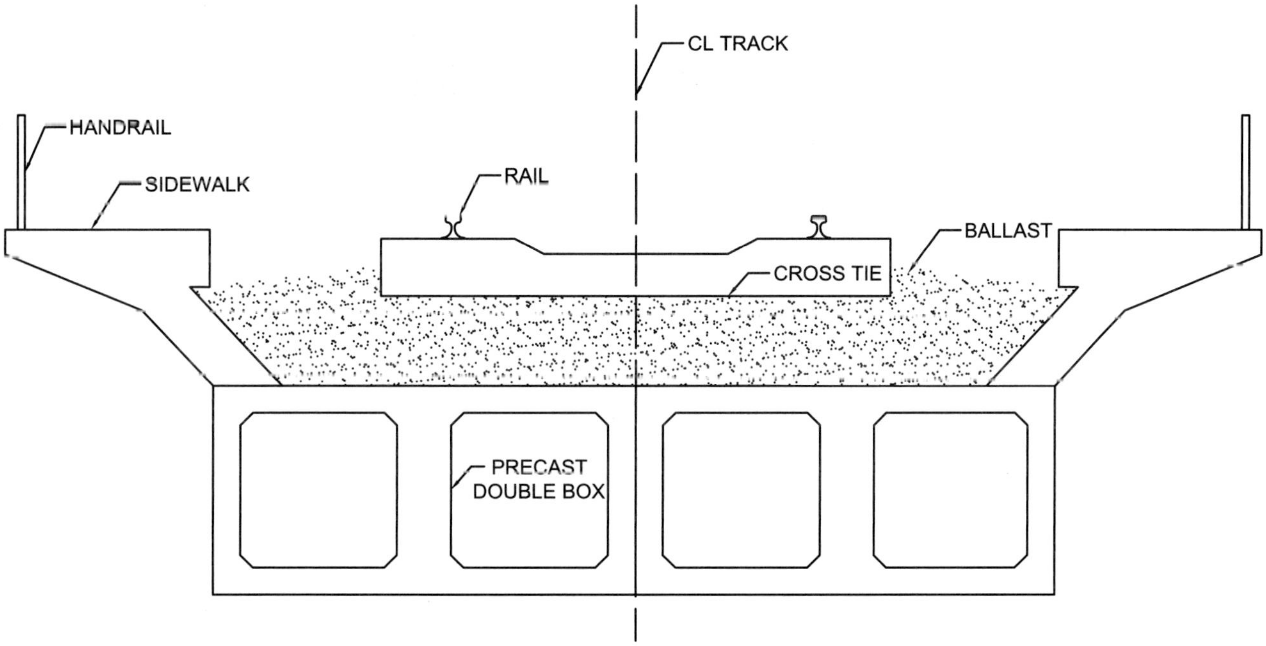

Figure 16.8 Precast double box girder.

16.6.2 General Rules

As stated in the AREMA Manual, bridges shall be designed as per the AASHTO standard specifications.

The following loads shall be considered in bridge design as specified by the AREMA Manual:

Dead load: structural members, concrete, filling materials, ballast, fastener and rail, live load, impact load, centrifugal force, earth pressure, buoyancy, wind load on structure, wind load on live load, longitudinal force from live load, longitudinal force due to friction or shear resistance at expansion bearing, earthquake (seismic) load, stream flow pressure, ice pressure, and other forces such as rib shortening, shrinkage, temperature, and settlement of supports.

Service Load Design shall conform to the combinations presented in Table 8-2-4 in the AREMA Manual for Railway Engineering.

Load Factor Design shall conform to the combinations presented in Table 8-2-5 in the AREMA Manual.

16.6.3 Material Specification

As per AREMA recommendation, the cement material used shall conform to the Standard Specifications ASTM C150 and ASTM C595 for portland cement and blended hydraulic cement, respectively. ASTM C618 Standard Specifications are recommended for fly ash and natural pozzolans. ASTM C989 and ASTM C1240 are specified for slag and silica fume. ASTM C260 and C494 shall be used for the selection of air entrainment admixture and other chemical admixtures.

Sampling and testing shall be in accordance with ASTM C33 and the Standard Specifications and Methods of testing as stated by AREMA.

Reinforcement specifications must conform to Tables 8-1-7 and 8-1-8 in AREMA Rail Engineering Manual that accords with ASTM Standard Specifications.

16.6.4 Reinforcement

Reinforcement Details

Standard hooks shall be used as stated in Section 2.4, Chapter 8, in AREMA Manual. The minimum diameter of bend is provided in Table 8-2-6 for different bar sizes. Sections 2.5 and 2.6 of Chapter 8 in AREMA Manual provide detailed provisions on spacing of reinforcement and concrete protection of reinforcement, respectively. Minimum concrete cover is presented in Table 8-2-7 for different exposure conditions.

Flexural Reinforcement

Minimum reinforcement for flexural members as stated in AREMA Manual shall be as follows:

- At any section of a flexural member where tension reinforcement is required by analysis, the reinforcement provided shall be adequate to develop a design moment strength ΦM_n at least 1.2 times the cracking moment calculated on the basis of the modulus of rupture for normal weight concrete specified in Article 2.26.1a.
- The requirements of Section 2.7a may be waived if the area of reinforcement provided at the section under consideration is at least one-third greater than that required by analysis based on the load factors specified in Article 2.2.4c. Distribution of reinforcement in flexural members as per AREMA Manual shall be provided as follows:
- Flexural tension reinforcement shall be well distributed in the zones of maximum tension. (1) For T-girder and box-girder flanges, tension reinforcement shall be distributed over an effective tension flange width equal to 1/10 the girder span length or a width as defined in Article 2.23.10b, whichever is smaller. If the actual slab width, center-to-center of girder webs, exceeds the effective tension flange width, and for excess portions of deck slab overhang, additional longitudinal reinforcement having a total area at least equal to 0.4 percent of excess slab area shall be provided in the outer portions of the slab. (2) For integral bent caps of T-girder and box-girder construction, tension reinforcement shall not be placed outside the bent cap web farther than an overhanging slab width on each side of the bent cap equal to 1/4 the average spacing of intersecting girder webs or a width as defined in Article 2.23.10b for integral bent caps, whichever is smaller.
- If the depth of web exceeds 3 ft (900 mm), longitudinal skin reinforcement shall be uniformly distributed along both side faces of the member for a distance $d/6$ nearest the flexural tension reinforcement. The area of skin reinforcement A_{sk} per foot (m) of height on each side face shall be $>0.012(d - 30)$ [or $A_{sk} > 0.3\ (d - 750)$ in metric]. The maximum spacing of the skin reinforcement shall be the smaller of $d/6$ or 12 in. (300 mm). Such reinforcement may be included in strength computations if a strain compatibility analysis is made to determine stresses in the individual bars or wires. The total area of longitudinal skin reinforcement in both faces need not exceed one-half of the required flexural tensile reinforcement.

Lateral Reinforcement

Section 2.9, Chapter 8, in AREMA Manual for Railway Engineering provides specifications for compression reinforcement, torsion reinforcement, and stirrups for flexural members.

Shear Reinforcement

Minimum shear reinforcement, types of shear reinforcement, and spacing of reinforcement are presented in Section 2.10 in the AREMA Manual.

Compression Reinforcement

Limits for reinforcement of compression members shall conform to the specifications provided by AREMA in Section 2.11. The provisions include longitudinal reinforcement and lateral reinforcement.

Shrinkage and Temperature Reinforcement

As stated in the AREMA Manual, reinforcement for shrinkage and temperature stresses shall be provided near exposed surfaces of walls and slabs not otherwise reinforced. The total area of reinforcement provided shall be at least 0.25 in.^{2}/ft (530 mm^2/m) measured in the direction perpendicular to the direction of the reinforcement and be spaced not farther apart than three times the wall or slab thickness, nor 18 in. (450 mm).

16.6.5 Bearings

As per AREMA Manual, the design bearing stress shall not exceed $0.85\Phi f'_c$, except when the supporting surface is wider on all sides than the loaded area; then the design bearing stress on the loaded area shall be permitted to be multiplied by $(A_2/A_1)^{1/2}$ but not more than 2, where

A_1 = load area

A_2 = the area of the lower base of the largest frustum of a pyramid, cone, or tapered wedge contained wholly within the support and having for its upper base the loaded area and having side slopes of 1 vertical to 2 horizontal

16.6.6 Foundations

Spread Footings

Classification of spread footings is provided in Section 3.1.2 in the AREMA Manual for Railway Engineering.

The loads to be supported by the foundations are (1) dead load, (2) normal live load, (3) maximum live load, (4) longitudinal and lateral forces, (5) snow load, (6) ice load, (7) earthquake load, (8) wind load, (9) loads from pore-water pressures including buoyancy and seepage forces, (10) area load, (11) impact load (only for special circumstances), and (12) vibratory loads to footings on granular material shall be considered.

Section 3.4 in the AREMA Manual presents detailed provisions on sizing of shallow footings on different types of soils. Section 3.5 provides provisions on footings with eccentric loads. Combined footings specifications shall conform to those provided by AREMA in Section 3.8.

Piles

The primary loads imposed on pile foundations are (1) dead load, (2) live load (vertical and horizontal), (3) centrifugal force, (4) earth pressure, (5) buoyancy, and (6) negative skin friction. The secondary loads include (1) wind and lateral forces, (2) ice and stream flow, (3) longitudinal forces, and (4) seismic forces.

Sections 4.3 and 4.4 in the AREMA Manual provide provisions on the allowable loads on piles and the different types of pile foundation, respectively. The most commonly used types are timber, steel, cast-in-place concrete (tapered, cylindrical), and precast concrete piles.

The allowable load per pile shall not exceed the following:

- **Concrete:** 0.3 of the ultimate compressive unit strength of the concrete used (f_c') but not exceeding 1600 psi.
- **Steel:** The unit stresses shall not exceed 12,600 psi.

Table 8-4-1 in Chapter 8 of the AREMA Manual presents the recommended pile loads for varying pile diameters.

16.6.7 Retaining Walls, Abutments, and Piers

Retaining Walls

The main types of retaining walls, as per AREMA Manual, are as follows:

- Gravity wall, so proportioned that only temperature steel is needed.
- Semigravity wall, proportioned so that some steel reinforcement is required along the back and along the lower side of the toe.
- The cantilever wall, requiring extensive steel reinforcement.
- The counterfort wall, consisting of a reinforced vertical face slab supported laterally by vertical reinforced counterforts extending into the backfill and supported by a reinforced base slab.
- The buttress wall, similar to the counterfort wall.
- The crib wall, consisting of an earth-filled assembly of individual structural units.
- Mechanically stabilized embankments (MSE).

Abutment

As stated by the AREMA Manual, abutments shall be of the gravity or semigravity type as a preference.

Piers

The major pier types stated by AREMA are the following:

- Solid wall, reinforced for strength and temperature.
- Rigid frame, consisting of multiple columns with a cap reinforced to act as a frame.
- Bents, consisting of multiple piles extended to a cap.
- Hammerhead, consisting of a column supporting a cap which cantilevers at the column.
- Drilled shafts, consisting of poured concrete columns extending to a cap.

16.6.8 Loads

Abutments and retaining walls shall be designed to withstand lateral earth and water pressures, any live-load and dead-load surcharge, the self-weight of the wall, temperature and shrinkage effects, earthquake load, and any other applicable loads. Pier supports the vertical loads from the superstructure as well as the horizontal loads not resisted by the abutments as well as forces wind loads, floating ice, and vehicle impact.

16.6.9 Design

AREMA Manual, Chapter 8, Section 5.7, provides details of design and construction for abutments and retaining walls. Section 5.8 presents details of design for bridge piers.

REFERENCES

AASHTO, LRFD. "AASHTO LRFD Bridge design specifications." *American Association of State Highway and Transportation Officials*, Washington, DC (2012).

American Railway Engineering and Maintenance-of-Way Association. *Manual for Railway Engineering*. American Railway Engineering and Maintenance-of-Way Association, 2011.

Northeastern Lumber Manufacturers Association; Northern Hardwood and Pine Manufacturers Association; Southern Pine Inspections Bureau; West Coast Lumber Inspection Bureau and Western Wood Products Association. *USDA Forest Service, Forest Products Laboratory*, Madison WI.

Chapter 17
Industrial Buildings[a]

BY
JULES VAN DE PAS *Vice President, CSD Structural Engineers, Greenwood Village, Colorado*
JOHN ROLFES *Vice President, CSD Structural Engineers, Milwaukee, Wisconsin*

17.1 PLANNING INDUSTRIAL BUILDINGS

17.1.1 Design Concept

An industrial building is typically considered to be any structure that encloses or supports a manufacturing process or stores materials. In general, the layout and structural design of an industrial building is almost entirely dictated by the functional or process requirements. Commercial buildings are also typically designed and configured to support a specific program or occupancy. However, for industrial buildings, the emphasis is on the process rather than on the occupants.

The structural design procedure for an industrial building is like the procedure for the design of a commercial building. The program and configuration of a commercial building is typically determined by the architect and the structural engineer develops the structural design accordingly. However, industrial buildings follow a configuration or general arrangement that is determined by the process design. The structural engineer must work closely with the process engineer to develop an awareness of the pertinent design requirements for the building. It is very helpful for the structural engineer to have a basic understanding of the contained industrial process. Each project requires assessment of the unique requirements associated with that facility. Therefore, a single set of rules or a rigid procedure may not be appropriate. However, the basic considerations and approach to developing and implementing the design of an industrial building are similar and discussed in the following sections.

17.1.2 Development of Project Requirements and Building Layout

Many different contractual arrangements are used for the design and construction of industrial buildings. Therefore, it is critical that the structural engineer gains an early understanding of each party's relationship and level of responsibility on the project. Parties typically include the owner, the process engineer, engineers from other disciplines, contractors, and industrial equipment vendors. The structural engineer must work with the owner, process engineer, and equipment vendors to clearly define the scope of services and work to be shown on the structural design drawings. The structural engineer is dependent on the process engineer and equipment vendor for pertinent process-related design criteria, loading information, and building geometry requirements. These responsibilities need to be clearly communicated and coordinated to facilitate the design process and manage the project schedule requirements.

The structural engineer needs to carefully examine the proposed site and process design and determine the critical project parameters. Issues that need to be addressed include the following:

- Site and soil information
- Environmental loads stipulated by the applicable building code
- Wall and roof material preferences
- Process layout and work flow
- Process loads
- Utility routing and support requirements
- Equipment maintenance access
- Clear height requirements
- Future expansion plans
- Preferred bay sizes
- Loading dock and door requirements
- Design responsibilities
- Material-handling requirements including conveyors, overhead cranes, forklift trucks, or other conveyance equipment
- Conveyor routing and support
- Crane types, quantities, and capacity
- Serviceability design criteria based on building materials, process equipment, and material-handling equipment requirements
- Equipment foundation requirements
- Floor slab design criteria (for slab-on-grade and elevated floors)
- Fall protection and fall arrest system requirements
- Fire protection requirements
- Exiting requirements
- Expansion joint requirements and details
- Building environment requirements and considerations based on the contained process or generated by the contained process
- Budget and schedule requirements

The structural engineer should consider incorporating all critical design parameters and load requirements into a single "Basis of Design" document. This "Basis of Design" would then be reviewed with the owner and process engineer before proceeding with the work. As the project progresses, the "Basis of Design" can be periodically reviewed and updated when necessary as additional design considerations evolve.

[a]Parts of this chapter were written by E. Alfred Picardi and James M. Fisher for prior editions.

Building Layout

Industrial building projects are configured to conform to the requirements of the process design. It is conventional practice for the process engineers (these may be consultants or equipment suppliers) to illustrate the process design in “General Arrangement” or GA drawings. The GA drawings should show process equipment locations, loads, anchoring or attachment requirements, required clearances, required maintenance access, critical dimensions, and other information.

For industrial buildings with significant process equipment and well-developed GA drawings, the structural engineer can typically determine the structural configuration directly from these drawings. However, for other industrial buildings with less process equipment or warehouse buildings, the structural engineer may have more flexibility to choose a configuration that is deemed most appropriate and/or efficient.

On large storage or light industrial structures having multiple bays in both directions, substantial economy can often be realized if the overall structure plan is a square or nearly square shape. Figure 17.1 demonstrates the variation in the ratio of wall perimeter length to building area as the building length-to-width ratio varies. The length of the building perimeter (and subsequently the cost of the exterior wall) is minimized when the shape of the building is approximately square in configuration. The most common configuration for light industrial buildings is a rectangular or square, single-story building, with a low-profile roof and concrete slab-on-grade. This solution is most economical from the standpoint of construction, operating, and maintenance costs.

Multistory industrial buildings are built where either site constraints or process layout dictate the necessity for elevated floors in the building structure. Site constraints may be pertinent in more developed industrial parks or urban areas. Process layouts conducive to multistory configurations include processes that work well with vertical flow or require gravity flow or feed.

The complexity of an industrial building will be dictated by the process requirements. The designer should study the process general arrangement and special requirements of the structure prior to selection of the framing system. These requirements may include such things as:

- Stepped-roof configurations
- Large clear-span requirements
- Requirements for column-free transfer bays
- Heavy equipment loads supported on the building structure
- Requirements for large overhead doors
- Interior floor drainage or trench drains
- Requirements for routing and support of multiple utility lines and piping to support process equipment
- Restrictions on vertical bracing locations

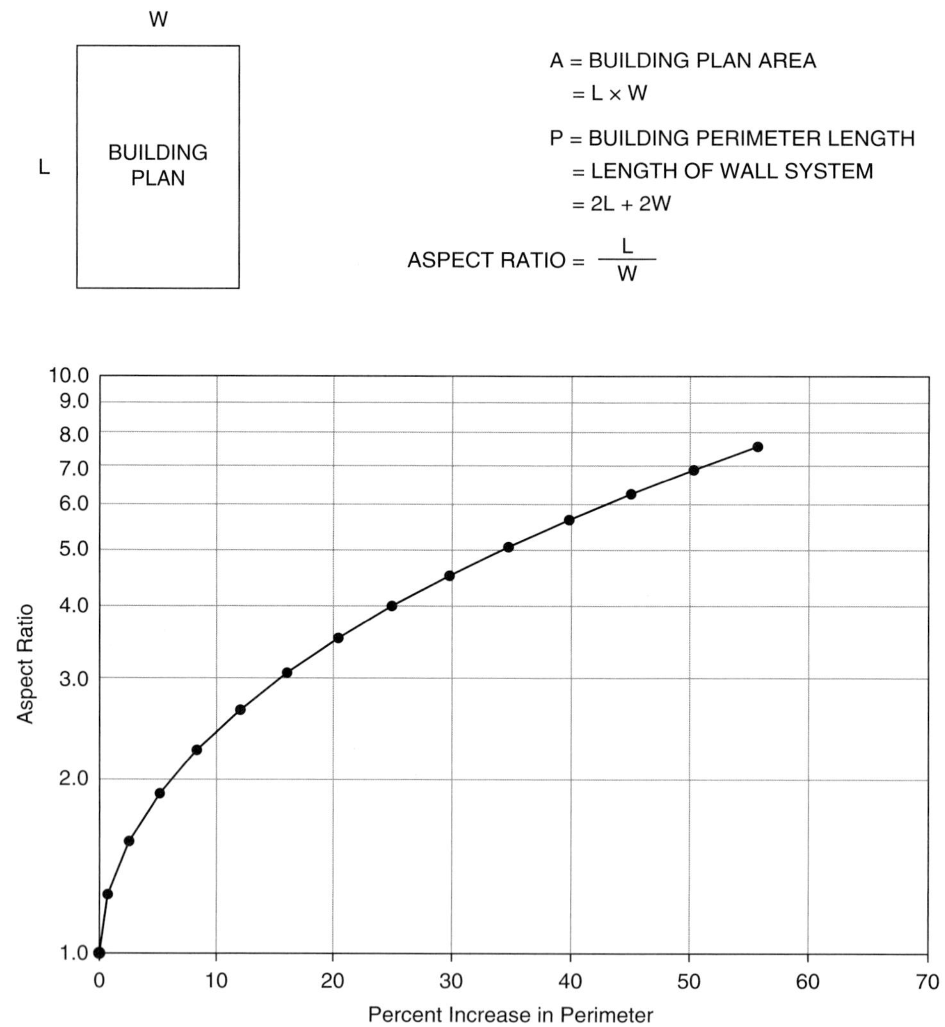

Figure 17.1 Variation in the ratio of wall perimeter length to building area as the building length-to-width ratio varies.

- Overhead crane runway requirements
- Conveyor layout and support requirements
- Significant ventilation requirements including roof ventilators, wall louvers, and fans
- Special heating, cooling, or refrigeration requirements
- Mezzanine or elevated floor requirements
- Provisions for future additions
- Additional nonbuilding structures such as tanks, silos, and rack storage structures located within or supported by the building structure
- Loading dock requirements
- Potential explosion or deflagration concerns due to presence of volatile material associated with the process

This list is not exhaustive but is representative of process-related issues that must be considered in the design of an industrial building. An early recognition of all special requirements will aid greatly in rapid development of a proper and economical framing system for the building.

17.1.3 Identification of Structural Design Requirements

The structural design requirements of the applicable building code and any other owner-specified standards or guides must be studied and incorporated into the building design. In addition, the structural engineer must work with the process engineer and equipment vendors to determine process and equipment design loads, equipment anchorage requirements, pertinent construction tolerance requirements, and serviceability criteria that apply to the building superstructure and foundations. Serviceability criteria pertain to limits on deflections, lateral drift, and vibrations that can adversely affect the intended use and performance of the building.

Special material and tolerance requirements must be clearly noted on the contract documents. For example, standard ASTM mill tolerances for beams may not be appropriate for crane runway girders due to tight alignment tolerance requirements for crane rails that are supported on these girders. Therefore, tighter tolerances should be specified on the contract documents for these girders. Various industry design guides, such as AISC Design Guide 7, Industrial Building Design, 3rd Edition, and AIST Technical Report #13, Guide for the Design and Construction of Mill Buildings, provide additional discussion on appropriate design criteria for these buildings.

Serviceability design criteria, other than basic deflection limits for beams, are not provided in building codes. Specific requirements should be reviewed with the process engineer and equipment vendors. Excessive deflections, lateral drift, or vibration may negatively impact the process equipment, material-handling equipment, weather tightness, or occupants of the building. AISC Design Guide 3, Serviceability Design Considerations for Steel Buildings, is a good general reference on this subject.

When evaluating lateral drift limits for industrial buildings without cranes, consideration is limited to wind loads. Seismic drift limits referenced in the building code pertain to expected "inelastic" drifts and are related to strength design considerations due to potential second-order effects of gravity loads on the building structure when subjected to these larger inelastic drifts.

When evaluating serviceability requirements associated with wind loads, it is common practice to consider wind loads with a 10-year recurrence interval as opposed to the more stringent, higher wind loads used for strength design.

Lateral wind drift limits for industrial buildings should be evaluated in recognition of any process equipment supported on or directly adjacent to the building structure. If this is not consequential, lateral drift limits due to wind are generally dependent upon the wall material selected for the structure. Buildings with metal wall panel systems have performed adequately with predicted wind drifts of H/100, where H = height of building. The actual drift of the structure may be considerably less due to the diaphragm action of the walls and roof. For hard wall systems such as precast concrete or concrete tilt-up wall panels, consideration of the effects of building drift on panel forces, joint details, and connection details is required. This includes recognition of how the hard wall panel is supported at the foundation level, whether the panel is restrained by the ground floor slab, and how the top of the panel is connected to the building structure. In this case, the drift limit is related to how the wall panel is designed and detailed.

When evaluating lateral drift limits for industrial buildings with cranes, the engineer should consider drift due wind and crane lateral loads. For heavier cranes, these lateral forces can be significant, and for high-use cranes, these forces can be repetitious. Moreover, these forces are localized to the portion of the building length where the crane is positioned, causing potential differential movement from bay to bay in the building structure. This differential movement can potentially lead to loss of weather tightness and roof leaks. One strategy is to provide a continuous horizontal bracing system (truss) at the roof level over the length of the building to reduce the differential movement between bays at the roof level. It is not advisable to use a metal roof deck diaphragm for this purpose because under repetitive loading, the deck fasteners may be compromised, leading to a loss in diaphragm stiffness. Over time, slotting may occur at fastener locations in the metal deck resulting in a loss of weather tightness for exposed metal roof decks. Excessive drift in the building structure can also compromise the performance of the crane as it travels over the length of the building. For slow-moving pendant-operated crane systems, building lateral drift limits of H/100 may still be acceptable. However, for faster radio-controlled or pendant-operated cranes, building lateral drift limits from H/240 to H/400 are commonly specified. Computer-controlled cranes may require even more restrictive limits.

17.1.4 Selection of Roof and Wall System

The roof and wall systems must be selected at the very start of the structural design. Architects typically select these systems in the design of a commercial building. However, for many industrial buildings, the structural engineer and owner may select the wall system based primarily on cost and functionality requirements for the project. Large industrial corporations with multiple facilities may have company standards that dictate wall and roof systems for their facilities.

Possible roof types include built-up roof (BUR) and membrane-type roofs such as ethylene propylene diene terpolymer (EPDM) or thermoplastic polyolefin (TPO). Another alternative is to use exposed metal roofs such as a screw-down metal roof panel or standing seam metal roof panel. Each of these roof types has different advantages and constraints. The roof specifier needs to consider the advantages and disadvantages of each roof type, the quantity and type of roof penetrations, and the size and extent of roof-mounted equipment when selecting the roof type. Membrane roofs and BURs are advantageous if the roof has many roof penetrations or large rooftop equipment. The roof-type selection will impact the roof dead load as well as the deck and joist spacing and direction of framing.

Roof drainage is another important variable that affects what type of roof system is to be used. Exposed metal roof systems are generally restricted to buildings with perimeter roof drainage directly spilling from the roof or collected with a gutter and downspout system. Built-up or membrane roofs can be used for roofs with interior drains or perimeter drainage. When interior drainage systems are used, the design should acknowledge the potential for impounded water. The building codes generally require this evaluation be performed considering the primary drainage system as blocked. The plumbing designer must provide the corresponding potential depth of impounded water on the roof consistent with the code-prescribed design-level rainstorm and drainage system. The roof framing design must also investigate ponding instability concerns.

Various industrial processes can create corrosive environments within a building. If this is the case, the engineer should coordinate appropriate materials and coating systems for the roof, wall, and building structure with the process designer and building owner.

Most preengineered metal buildings at this time are supplied with standing seam roofs. Standing seam roofs are typically anchored at the eave of the building and are supported on discrete clips mounted to the roof framing below, located in the roof seams. The clips do not restrict movement along the length of the panel and, therefore, allow for thermal movement of the roof which can otherwise lead to slotting and leaks at roof fastener locations. The structural engineer must understand that standing seam roofs do not provide diaphragm capability, and therefore lateral load transfer and bracing of roof primary and secondary framing elements must be otherwise accommodated. Screw-down exposed metal roofs are commonly less expensive but are limited to use on smaller buildings, where potential thermal movements in the roof are smaller.

Possible wall types include masonry, precast concrete wall panels, tilt-up concrete wall panels, insulated metal wall panels, and exposed metal panels either with or without insulation. Mixed systems are also common. Masonry and precast or tilt-up concrete walls can be used to provide a more robust and durable lower wall system where forklift or other material-handling systems may occasionally impact the walls, with a less expensive metal wall panel system above. Mixed systems should be configured with the height of the lower hard wall system exceeding the height of the typical personnel door openings. Considerations for selection of the wall system include cost, local trade practices, insulation requirements, durability, fire rating, security, and potential future expansion.

17.1.5 Selection of Column Spacing

Most industrial buildings are square or rectangular in plan with primary framing running parallel to one axis of the building and secondary framing in the orthogonal direction as illustrated in Fig. 17.2. Process flow is commonly oriented perpendicular to the primary framing direction, and the primary framing module width is typically based upon the width of the process layout, with additional width added for crane hook coverage when overhead cranes are used for material handling or maintenance. Module sizes for industrial buildings are often larger than module sizes typically used for commercial buildings. Module sizes from 50 to 100 ft are not uncommon. Bay size is usually selected based on structural efficiency, considering the relative cumulative costs of roof framing, wall framing, crane runway girders (when applicable), columns, and foundations. Common bay sizes used range from 25 to 50 ft, with larger bays preferred from the standpoint of minimizing the number of columns and providing the most flexibility for current and future process concerns. When evaluating bay and module sizes for cost efficiency, it should be recognized that larger module and bay sizes will result in fewer pieces to fabricate and erect, and therefore the framing layout that provides the least weight may not be the framing layout that results in the lowest constructed cost.

When selecting column bay size, the type of wall system to be used is a factor. When using a metal wall panel system, the most common option is to have girt secondary framing spanning between columns with the metal wall panel spanning vertically between these girts. Generally, it is most cost effective to use light-gauge cold-formed C- or Z-shaped sections for these girts. These members can either be simple-span elements or positioned outside the column profile with lapped connections at the columns to behave as continuous framing elements. Based on both strength and stiffness requirements, the maximum economical span of light-gauge girts is approximately 30 ft. If larger bays are used, intermediate wind columns must be introduced between the primary building columns to reduce the girt span, or structural steel shapes (WF beams, channels, or rectangular/square HSS sections) must be used for the girts.

From the above, it may seem that a 30-ft bay is the most economical choice. This is especially true if the perimeter of the enclosed area is large relative to the enclosed area since for such buildings the girt steel is a larger percentage of cost. However, for buildings with low perimeter-to-area ratios, the percentage of steel framing in the wall framing is less significant. For these buildings and for buildings with hard wall systems, a bay length of 40 to 50 ft may be more economical. When intermediate wind columns are used, the top of the wind column

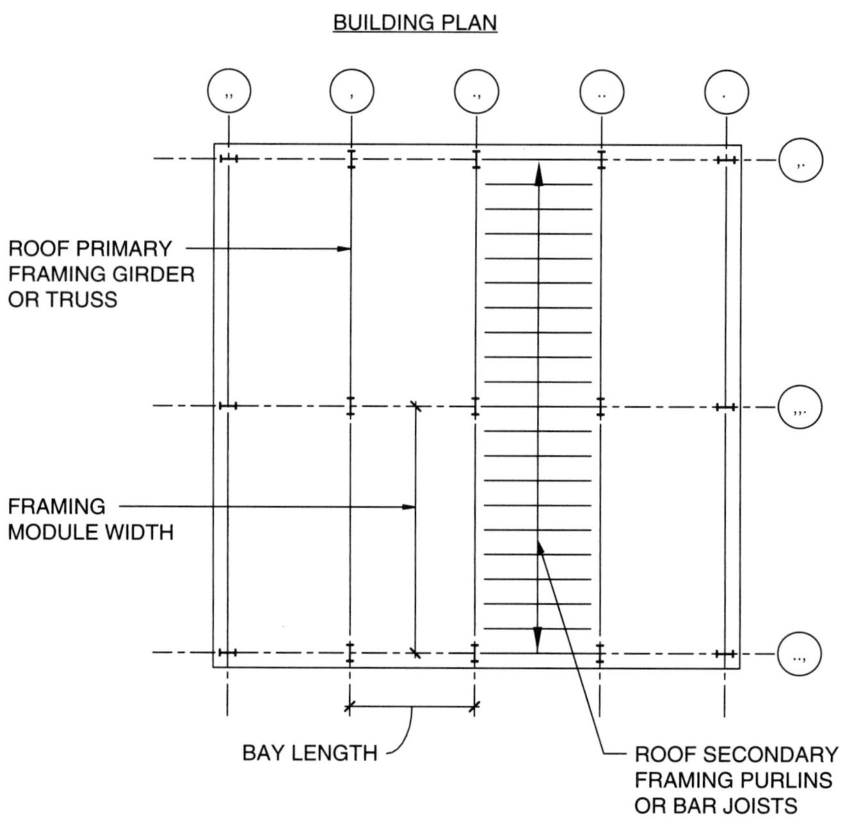

Figure 17.2 Framing layout for industrial buildings.

must be connected to the roof diaphragm or horizontal bracing to transfer the horizontal reaction from the top of these columns into the lateral force resisting system of the building.

When shallow foundations are used, foundation costs have little impact on bay size selection. However, if deep foundations are required, a thorough comparison of foundation requirements for different bay sizes should be made, recognizing that the foundation costs are more significant, and options exist for the type, depth, and arrangement of piles or drilled piers.

In general, for light- and medium-crane buildings, bays of approximately 25 to 30 ft will be the most economical because of the cost of the crane runway beams. Spans requiring plate girders rather than rolled shapes for runway beams can increase the structural cost considerably. Larger-span crane girders may also necessitate the use of a backup horizontal bracing system for the crane girders that is expensive to fabricate and construct.

17.1.6 Foundation and Slab-on-Grade Considerations

Foundation designs are dictated by soil conditions and the building loads. Lighter industrial buildings are commonly supported on spread footings, whereas industrial buildings with heavy cranes or large equipment loads are often supported on deep foundations. Heavy equipment foundations are typically isolated from the building foundations. Equipment foundations may have more stringent settlement criteria and may also be subject to vibration concerns. There are numerous texts on design of foundations for dynamic loads and vibrations. ACI 351.3, Foundations for Dynamic Equipment, is a good resource on this topic. Equipment foundation may also include associated pits, reservoirs, and trenches. If these elements are deep, it may be required to increase the depth of adjacent building foundations, resulting in higher construction and concrete costs. Deep foundation construction may also require dewatering of the associated excavation for these foundations.

Important considerations for the slab-on-grade design are soil conditions, magnitude and nature of loads (from storage racks, fixed equipment, forklift trucks, or other mobile equipment), and durability considerations, especially when the slab is subject to repetitive traffic from forklift trucks or other mobile equipment. The slab-on-grade in an industrial building is commonly exposed (not covered), and therefore significant cracks are both unsightly and may compromise the cleanliness and functionality of the facility.

The floor slab-on-grade is typically supported directly on the underlying soil. Exceptions would be in areas with expansive soils or poor soil conditions that the geotechnical engineer deems are inadequate to support the anticipated loads. In these instances, the floor slab is designed as a structural concrete slab spanning between discrete foundations that extend below the unfavorable soils. These structural floor slab systems are expensive and have a significant impact on the overall cost of the industrial building.

Failure of a slab-on-grade commonly results in deleterious cracking in the slab that can compromise functionality of the process within the building and result in long-term maintenance issues. However, this cracking does not typically represent a risk to life safety. The exception to this may be where the slab-on-grade is subject to very high material storage loads that can result in shear failure in the underlying soil. The resulting movement in the underlying soil may also cause failure or significant shifting and rotation in adjacent building foundations. This is a common oversight in heavy production facilities. The structural engineer should communicate with the process engineer and building owner and identify locations and loading restrictions for material to be stored on the slab-on-grade. If material storage is to occur above shallow building foundations, these surcharge loads should be acknowledged in these foundation designs.

Cracks in concrete slab-on-grade are often associated with volumetric shrinkage in the newly placed concrete slab that is restrained by the underlying soil or surrounding, previously placed foundations, slab, or structure. In addition, variable floor loading and concentrated loads on the floor can result in structural cracks in the floor slab as the slab and underlying soil deflect in response to these loads. Slab-on-grade are often divided with control joints that are strategically placed, spaced, and detailed so that temperature and shrinkage cracks occur at these joints. Control joints should be positioned where slab movement is restrained by surrounding structure with spacing between joints chosen to limit cracking between joints due to shrinkage resistance.

The performance of a concrete slab-on-grade is a function of the structural design, concrete mix design, environmental conditions during placement, and the workmanship of the contractor, including proper placement, finishing, and curing. Several valuable resources for slab-on-grade design and construction include the following:

- ACI 302.1, Guide to Concrete Floor and Slab Construction
- ACI 360, Guide to Design of Slabs-on-Ground
- PCA Concrete Floors on Ground

17.1.7 Expansion Joint Spacing

Expansion joints are typically used to divide the building structure when anticipated expansion and contraction in the building due to expected temperature fluctuations create excessive thermal loads in the building structure or serviceability problems in the roof or wall system for the structure. Expansion joint needs are based upon building dimensions, building materials, roof and wall system, and expected temperature fluctuations within the building structure. For buildings with controlled environments, the temperature fluctuation may be relatively small. Some industrial facilities do not have controlled environments and may be subject to larger temperature fluctuations. In addition, the process within the building may generate heat that could contribute to potential thermal growth in the structure. The building geometry may result in locations that facilitate expansion joints, such as roof steps, where direction of building framing changes or where the framing cross section changes. Expansion joints extend through the roof, wall, and supporting framing systems. Framing can be isolated by use of either double frame lines or a single frame line with framing on one side of the frame supported on slide bearings designed to accommodate expected thermal expansion and contraction.

17.2 CODE REQUIREMENTS AND INDUSTRIAL LOADS

17.2.1 General Code Requirements

The International Building Code (IBC) is the model building code used throughout the United States. This code includes occupancy categories for industrial facilities. These occupancy categories include Factory and Industrial, High Hazard, and Storage. Most industrial buildings will fall into one of these occupancy categories.

Sometimes the industrial process may require a building height that exceeds the typical code height limits based on occupancy and class of construction. The code does allow certain exceptions. It is recommended that the owner's representative meet with the code officials early in the planning process to discuss and resolve their strategy for addressing these limits.

Heavy industrial buildings often include multiple interior levels. If these interior levels only provide support and maintenance access for equipment, these interior levels can often be classified as equipment platforms rather than stories or mezzanines.

The structural building code requirements for industrial buildings are generally consistent with the requirements for commercial buildings. Environmental loads are essentially the same as for commercial buildings using the same wind speed, ground snow, and seismic spectral acceleration maps used for commercial buildings.

IBC specifies 125 psf minimum occupancy live load for light manufacturing and light storage. A minimum occupancy live load of 250 psf is specified for heavy manufacturing and heavy storage. An evaluation for actual anticipated live loads should be done to determine if higher design live loads are appropriate. In heavy manufacturing facilities, uniform live-load allowances of 500 psf or higher may be warranted. These loads are commonly associated with heavy stored material. In addition, the design should also acknowledge any significant anticipated concentrated live loads.

Forklift wheel loads and wheel spacing vary by manufacturer. The structural engineer should communicate with the owner to determine the specific equipment that will be purchased for the project. IBC specifies a 30 percent increase in forklift truck wheel loads to account for impact forces.

Crane runways should be designed for the specific cranes that will be purchased for the project. The engineer should obtain crane data sheets with wheel loads, spacing, and clearance requirements. The building code requirements for impact, lateral, and longitudinal crane loads are generally considered adequate for light industrial occupancies. Heavy and high-use occupancies such as steel mills may need to follow specific industry guides such as AIST Technical Report No. 13, Guide for the Design and Construction of Mill Buildings. This document is a guide with recommendations that are not mandated by the building code. However, the owner may contractually specify use of this document, or the engineer and owner may elect to follow these recommendations as part of an agreed upon "Basis of Design."

17.2.2 Process Loads

Process loads include all the loads associated with the fixed industrial equipment and the permanently installed utilities that support that process. The loads from piping, small ducts, conduits, cable trays, and similar items are often accounted for by providing uniform live-load or collateral-load allowances to accommodate the support of these elements.

Capacity to support loads from conveyors, large ducts, and pneumatic lines are typically accounted for or specified as live or collateral loads per lineal foot of these components.

Capacity for suspended assembly line equipment may be specified as live loads in the form of point load allowances at truss panel point locations.

It is not appropriate to provide a specific rule for when the engineer should consider process loads as uniform loads, line loads, or concentrated loads. This decision requires recognition of how sensitive the design is to the characteristic of the load. For example, bar joists may be very sensitive to concentrated loads, especially if supported at nonpanel point locations.

Within reason, the engineer should consider future modifications as well as current process requirements and provide additional capacity in their design when future modifications are considered likely.

17.2.3 Material-Handling Requirements

Material-handling equipment is used for handling manufacturing materials and manufactured product within the industrial building. Common material-handling systems include cranes, conveyors, and transport vehicles such as forklifts, air pallets, and robotic transports in current automated warehouse systems.

Cranes, monorails, and conveyors of various types are used in industrial plants. Various types of cranes are shown in Fig. 17.3. Jib cranes are used to position and handle loads within a limited area. These consist of a hoist mounted on a rail that is pivoted at one end and cantilevered from a vertical stanchion or one of the building columns. In these instances, the columns need to be investigated for associated loads, considering the worst-case orientation of the jib on the supporting column. Bridge cranes are used for transporting loads over a larger area bounded by the limits of the crane bridge and trolley travel. Bridge cranes are supported on runway beams or girders that are integrated with the building structure. Gantry cranes are independently supported crane structures typically mounted on rail systems supported at the floor level. Gantry cranes are often used for handling material in open yards, outside of the building envelope. Semigantry cranes have one end of the crane bridge supported on elevated crane runway beams similar to a bridge crane, whereas the opposite end of the bridge is supported on a gantry leg and floor-mounted rail system. Semigantry cranes are typically used where the material-handling requirements are limited to only a portion of a larger module width within a building. In these instances, the cost of a semigantry crane may be less than a longer bridge crane that would span the full building module width. Monorail cranes are used to transport smaller loads over a straight-line distance. Automatic Storage Retrieval Systems (ASRS) are sometimes used for high-volume warehouse facilities. ASRS are automated systems that use a specialized crane in conjunction with a rack storage system. Each storage space in the rack system has a defined address that is incorporated into the inventory and crane control system. Conveyors are used for continuous high-volume flow of materials or loads over a fixed route and are common with assembly line processes. Cranes and conveyor systems are typically purchased from manufacturers who specialize in the design and construction of these systems. The structural engineer must work with the owner and vendors supplying this equipment to get pertinent loading information. For cranes, this includes maximum wheel loads and spacing for the end trucks that are supported on crane runway girders or other rail support system. In addition, the required crane hook height, associated rail height, and clearance envelope dimensions are necessary to establish the crane runway and building geometry. This information may significantly affect the building layout and design and, therefore, must be established in the early stages of design. Conveyor routing, loads, clearance envelopes, and method of support must also be established early in the design process.

17.2.4 Seismic Loads

Building code requirements for seismic design have become more stringent and are applicable to more areas in the United States than required by earlier codes. The same basic seismic design philosophies used for commercial buildings are applicable to industrial buildings. However, the framing systems and geometries used for industrial buildings may be significantly different than used in commercial buildings. Industrial buildings often have high eave heights or floor-to-floor heights due to process and material-handling equipment requirements. In addition, large bays and framing modules may require use of deeper trusses for primary roof and floor framing. Heavy equipment and process loads supported on a building structure translate to potentially high seismic forces. Framing systems and lateral load-resisting systems for these buildings may also be unique due to process loading and material-handling equipment requirements. When evaluating process loads for seismic mass considerations, the structural engineer should recognize what process loads are present on the structure on a regular basis. For cranes, the lifted load is not typically considered as contributing to seismic mass since the load is suspended beneath the crane bridge on lift lines and the response of this load to ground shaking will lag the building response.

IBC references ASCE 7 for determination of seismic loading requirements. Design requirements vary depending on proximity of the site to known faults, predicted ground accelerations for earthquake activity along these faults, soil conditions at the building site, the dynamic characteristics of the building structure, and the level of anticipated ductility in the building structure. The anticipated ductility is a function of the building materials and framing system used for the building structure. Chapter 12 of ASCE 7 dictates seismic design requirements for building structures. Ductility design parameters are quantified based on building materials and framing system type. Limits for these various systems are assigned, dependent on the seismicity of the building site and risk category pertinent to the occupancy. The requirements of Chapter 12 were developed primarily for commercial buildings with geometries, framing systems, and occupancies consistent with typical commercial projects. Industrial buildings may have unique framing systems and building geometries that fall outside of the limits stated in Chapter 12. For this reason, the seismic design criteria provided in Chapter 11 of ASCE 7 states that "Buildings whose purpose is to enclose equipment or machinery and whose occupants are engaged in maintenance or monitoring of that equipment, machinery or their associated processes shall be permitted to be classified as nonbuilding structures designed and detailed in accordance with Section 15.5 of this document." This classification is intended to apply to industrial buildings where the occupancy density is relatively low, as is often the case with automated industrial processes. Design per Section 15.5 allows for relaxation of system limits and design requirements in exchange for use of higher seismic design forces. Although these higher forces can result in heavier building structures and foundations, cost savings may be realized by less stringent detailing requirements. Moreover, the unique characteristics of a specific industrial building may leave the engineer with few other options.

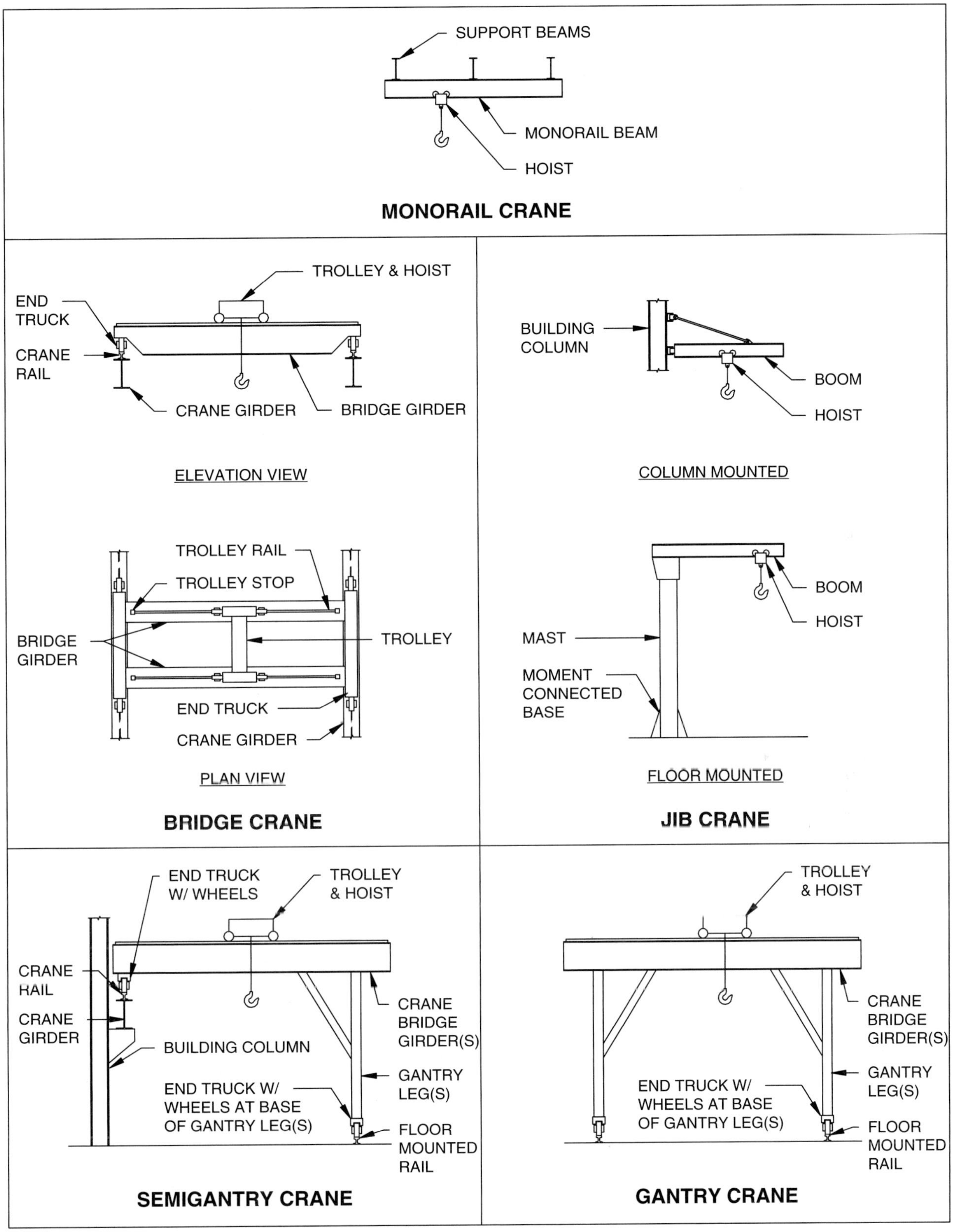

Figure 17.3 Types of material handling cranes in industrial buildings.

17.2.5 Load Combinations

Pertinent load combinations for structural design are given in IBC Section 1605.2. These load combinations are based on probabilistic analysis and design reliability. Load combinations that provide a higher level of detail for evaluation of crane buildings, including fatigue and serviceability of crane buildings, are provided in AIST Technical Report 13, Guide for the Design and Construction of Mill Buildings.

17.2.6 Fatigue Design Considerations

Fatigue is characterized as the progressive damage to a component due to fluctuating loads. In a properly designed structure, the stress in the members does not exceed the yield stress of the material. However, at stress concentrations, associated with material flaws or discontinuities, small regions of localized stress can exceed the material yield stress. Progressive damage due to repetitive loading can lead to the formation and propagation of fatigue cracks in these stress concentration regions. This progressive damage is limited to areas where net tension is developed from the repetitive loading. The stress range is calculated as the sum of the maximum compression and tension stress generated by the repetitive load at these locations. Designing to resist fatigue damage requires consideration of the magnitude and type of stress concentration, the level of service load stress, and the number of load cycles.

Members and connections that are subject to fewer than 20,000 load cycles are typically not susceptible to fatigue design concerns. Members and connections subjected to more than 20,000 load cycles can be evaluated using the criteria provided in the AISC 360, Specification for Structural Steel Buildings, Appendix 3. Based on the defined "fatigue category" for a given element or detail and the number of loading cycles, the engineer calculates an "allowable stress range" that can be compared to the calculated stress range generated by that loading cycle. The fatigue categories account for the stress concentrations that are inherent in different types of details in steel construction. The structural engineer must work with the owner and process engineer to understand the number of load cycles generated for a given time period. This is extrapolated for the intended building life to determine the overall number of load cycles.

Fatigue considerations are generally associated with process equipment and material-handling equipment. Fatigue loading associated with imbalanced fans or motors can generate many load cycles in a relatively short period of time. For example, an imbalanced motor operating at 10 Hz will generate over 860,000 load cycles in 1 day. Vibratory conveyors will also generate a significant number of loading cycles very quickly. Allowable stress range for fatigue loading becomes asymptotic for extremely high numbers of loading cycles. The threshold stress level is defined as the allowable stress range for an unlimited number of load cycles. For structures supporting high cycle loading as noted above, design of the supporting steel structural elements is accomplished by keeping stress levels below these threshold limits.

Fatigue design considerations are also relevant in the design of crane runway beams and associated details. Fatigue problems commonly occur at welded attachments at or near the bottom flange of the crane beam, where tensile bending stresses are high and welded attachments and defects can create stress concentrations. For this reason, it is recommended to avoid making attachments to the bottom flange of crane girders or the web of the girder near the bottom flange. Fatigue cracking can also occur near the crane girder web to top flange joint if the crane rail is not properly positioned over the girder web, resulting in bending in the flange and web of the girder due to crane vertical loads. Proper design, detailing, construction, and maintenance of the crane and crane runway system will help eliminate or mitigate fatigue damage in the crane runway system.

17.3 FRAMING SYSTEMS

17.3.1 General Discussion on Framing Systems

Industrial buildings are unique in that the structure may commonly be exposed to full process and equipment loads, whereas the commercial buildings are rarely exposed to the magnitude of environmental and occupancy loads dictated by the building code. Therefore, emphasis should be placed on selecting a structural system that efficiently accommodates the design requirements (loads and serviceability criteria) for the everyday process and equipment loads.

Steel framing systems are more common for industrial buildings. Steel framing systems are typically efficient for larger bay and framing module dimensions. In addition, steel framing systems can be efficiently designed to accommodate heavy loads associated with process equipment and material-handling equipment. However, no material or framing system can be singled out as ideal for all industrial buildings. Only after careful study of the project can the designer select a building system that is most suitable and economical for a given situation. The advantages and disadvantages of steel versus cast-in-place concrete versus precast concrete versus wood buildings will always be debated. In addition to initial constructed economy, the engineer should consider durability, adaptability, sustainability, and constructability of each system under consideration.

17.3.2 Lateral Load Resisting Systems

Lateral load-resisting systems in the context of this discussion is defined as the framing system that transmits lateral loads from elevated floor and roof levels to the foundation level. Basically, three types of systems are used for industrial buildings:

- Shear wall systems
- Braced frame systems
- Rigid-frame systems

Shear wall systems are common in warehouse-type buildings with either precast concrete, concrete tilt-up walls, or masonry walls. Braced frames are basically vertically oriented trusses used to transfer lateral loads on the building to the foundation level. Braced frames are common in one-story and multistory industrial buildings with roof and floor diaphragm systems used to transfer lateral loads to these frames. Rigid frame systems or a combination of rigid frames and braced frames are typically the best solution for crane buildings and long, rectangular buildings where interior bracing is problematic and use of floor and roof diaphragm is not practical.

Diaphragm systems at floor and roof levels are used to transfer lateral loads to the vertical lateral load resisting systems in the structure (shear walls, braced frames, and rigid frames). Concrete decks at floor levels are commonly used for this purpose. Either plywood or light-gauge steel roof deck is commonly used at the roof level for this purpose. In general, plywood or steel roof deck diaphragms work well where the diaphragm aspect ratio does not exceed 3 or 4 to 1. For higher aspect ratios, diaphragm strength and stiffness demands may exceed what is feasible or economical for these systems.

An alternative to a deck diaphragm system is the use of horizontal truss framing to create a horizontal diaphragm as illustrated in Fig. 17.4. This is commonly used when the deck diaphragm has insufficient strength or stiffness to perform this function. Standing-seam metal roofs, by their nature, offer little or no diaphragm capability. Therefore, supplemental horizontal bracing is often added at the roof level when these roof systems are used. Horizontal truss framing can use strap bracing (flat bar stock) on top of the purlin or roof joist system as tension-only web elements for the horizontal truss. Alternatively, rod bracing or angle bracing hung below the purlin system has been used with success. The designer is cautioned to provide appropriate connection details consistent with the intended load path for all forces in this system. In addition, the design must acknowledge strut and chord forces generated in other elements of the roof structure incorporated into this system. Horizontal truss framing is commonly provided at the bottom chord level of roof trusses for larger-span structures. In these instances, sway frames are provided to transfer lateral wind and stability forces from the top chord level of the roof truss framing to the bottom chord bracing system.

Rigid frames may be used to resist lateral loads in both directions of the structure. Rigid frames use rigid connections between the building columns and floor or roof beams to resist relative rotation between these

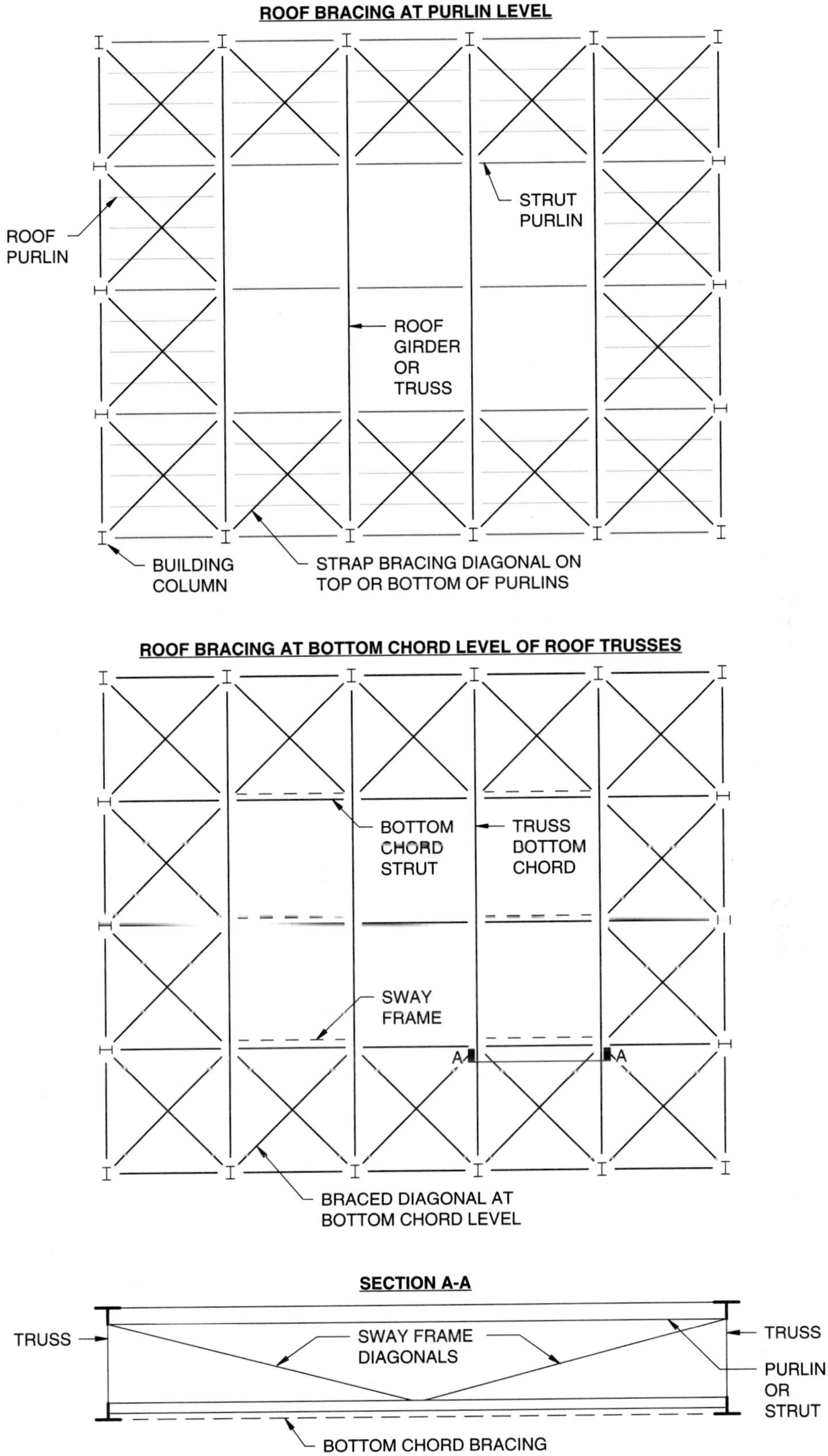

Figure 17.4 Horizontal truss diaphragms.

elements. Rigid frames may also be created with building columns rigidly connected to truss or bar joist framing members. In these instances, the connection of the top and bottom chord of the truss or bar joist to the column are designed to transfer the axial force couple associated with the intended continuity moments.

17.3.3 Steel-Framed Buildings

The most common structural steel systems for carrying roof gravity and uplift loads are the following:

- Rigid frames composed of hot-rolled steel beams or welded plate girder sections with cold-formed C or Z purlins, hot-rolled purlins, or bar joists.
- Hot-rolled columns with simple-span fabricated trusses or joist girders and steel bar joists.
- Hot-rolled columns with hot-rolled steel beams and hot-rolled purlins, cold-formed C or Z purlins, or bar joists.
- Hot-rolled columns with simple-span roof trusses, jack trusses and hot-rolled purlins, cold-formed C or Z purlins, or bar joists spanning perpendicular to the roof trusses.

These systems are illustrated in Fig. 17.5.

Rigid Frames

This is the common primary framing scheme used by the metal building industry to support gravity and uplift loads. In this case, the rigid frames are typically composed of nonprismatic plate girder sections with material and depth optimized to match demand for all elements of the frame. Automated welding equipment is used to fabricate the rigid frames, and individual fabricated pieces of the frames are commonly connected with bolted end-plate connections. Economical spans range from 30 ft to greater than 100 ft. The secondary framing usually consists of light-gauge cold-formed C or Z purlins and girts that are economical for spans up to approximately 30 ft. The C or Z sections are commonly lapped and connected at frame lines to generate flexural continuity between adjacent purlin spans. Proper lateral bracing for gravity and uplift loads is extremely important. Flange braces extending from the purlins or gifts to the rigid frames are commonly used to laterally brace the inside flange of columns and rafters. In addition, all bolts must be installed, particularly in continuous purlins, to achieve continuity. Without proper bolting in the continuous purlin system, premature failure can occur. Rigid frames are also commonly used in conventionally designed building, using hot-rolled columns and beams or trusses with hot-rolled purlins, bar joists, or light-gauge cold-formed C or Z purlins.

Hot-Rolled Columns with Simple-Span Fabricated Trusses or Joist Girders

Steel bar joists in combination with joist girders have become very popular for industrial buildings. These systems provide a low-cost framing option that can accommodate larger bay and framing module dimensions. Typically, the bar joists and joist girders are designed as simple-span members. Joist girders are economical for light to moderate loads where adequate headroom exists to accommodate reasonable joist girder depths. Fabricated trusses are often used in lieu of joist girders for roofs supporting heavier process loading and where there are numerous connections from ancillary framing that can be readily accommodated with conventional truss detailing and fabrication. It is imperative that bridging and bracing be properly designed and installed not only for uplift criteria but also to prevent instability problems for gravity loads. These systems are commonly used for framing module widths ranging from 40 to 200 ft or more and bay lengths ranging from 25 to 60 ft. Additional building height associated with increased depth of joist girders or trusses will result in additional cost for the building wall system.

Hot-Rolled Columns and Steel Beams

Hot-rolled columns with a continuous or a cantilever-beam system may be advantageous when the goal is to minimize the required building height for a given clearance envelope and the framing module widths are not too large. The design of these systems can be tailored to minimize beam depth. Framing module widths of up to 50 ft can be accommodated economically with this framing system. Other systems should be considered for larger module widths. Engineers are cautioned when using these systems to evaluate load combinations that include alternate span loading or nonuniform loading.

Hot-Rolled Columns with Simple-Span Trusses and Jack Trusses

The framing for this system uses jack trusses along column lines running parallel to the roof secondary framing. These jack trusses support end reactions from one or more primary roof trusses. These systems are advantageous where large, column-free spaces are required in the building and the roof is intended to support multiple conveyors or other significant process loads, typically spanning between primary roof trusses. Lighter wide-flange shapes are commonly used for purlins with this system. This system is commonly used in automotive manufacturing plants.

17.3.4 Concrete-Framed Buildings

Concrete building frames are primarily used for heavily loaded, multistory, industrial building structures that require fireproofing and that do not require longer spans between columns. The most economical system is the flat slab with drop panels and column capitals. Where possible, capitals are omitted or decreased to minimum dimensions to provide the maximum clear height around the columns. Beam-and-slab construction is often used where the industrial process requires numerous large openings in the floor structure or where unusually heavy concentrated loads must be supported. Frames of this type are usually cast-in-place concrete structures with mild steel reinforcement. Reinforced-concrete structures are commonly used in the food, beverage, and pharmaceutical industries where clean rooms and surfaces are required. In general, the most economical and constructible solutions occur when formwork is simplified as much as possible and the reinforced-concrete members are proportioned to result in reinforcement ratios near the minimums allowed by ACI 318, Building Code Requirements for Structural Concrete.

Single- and multistory industrial buildings can be economically framed with precast concrete members consisting of double-tee members or hollow-core planks supported on precast girders and columns or bearing walls. This system provides another economical option and can be advantageous when fire rating is required and sprinkler systems are incompatible with the industrial process. Hollow-core plank can be used for shorter spans up to approximately 50 ft, and double-tee members can span up to 100 ft. Figure 17.6 shows typical details, connections, and framing of these systems.

Exterior walls may be precast, prestressed hollow-core panels or precast solid panels that are prestressed or use mild steel reinforcing. These panels may be uninsulated or insulated, using sandwich panel construction composed of concrete and insulation layers.

Use of load-bearing precast wall panel with steel roof and interior column framing is also very common.

17.3.5 Wood-Framed Buildings

Wood-framed industrial structures are more popular in those areas of the country where lumber is produced at low cost. Warehouse or single-story light industrial buildings can be constructed using glulam girders (typically configured in a cantilevered system) combined with a system of panelized purlins and subpurlins. This system is typically configured around panelized modules based on conventional 4 ft × 8 ft sheets of plywood. The subpurlins are typically spaced at 16 in. on-center and the purlins spaced 8 ft on-center. Careful detailing is required to ensure that the wood roof framing is correctly configured to meet diaphragm and subdiaphragm design requirements. This system can be prefabricated and erected in modules to minimize overhead work. This type of framing is shown in Fig. 17.7.

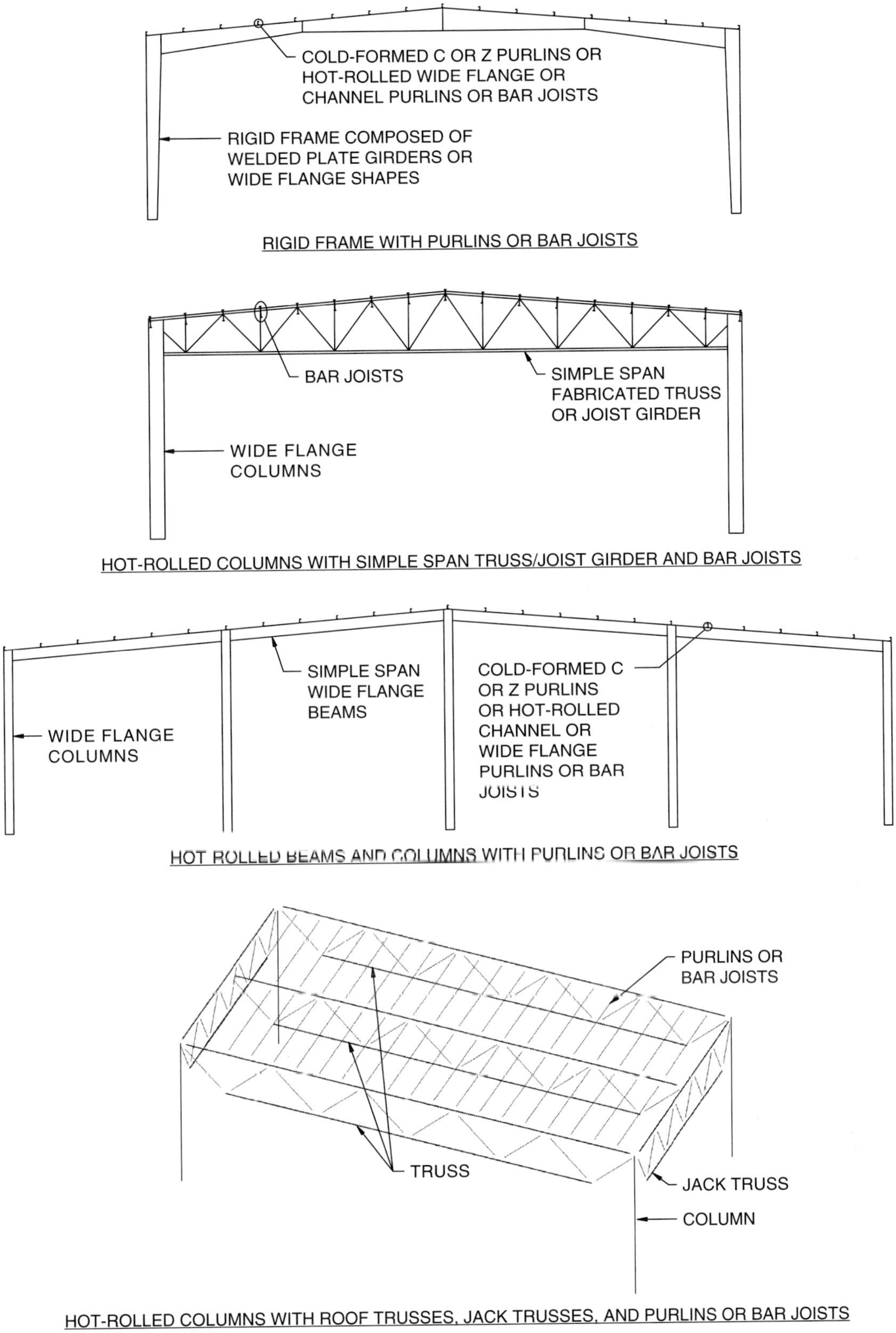

Figure 17.5 Steel-framed building systems.

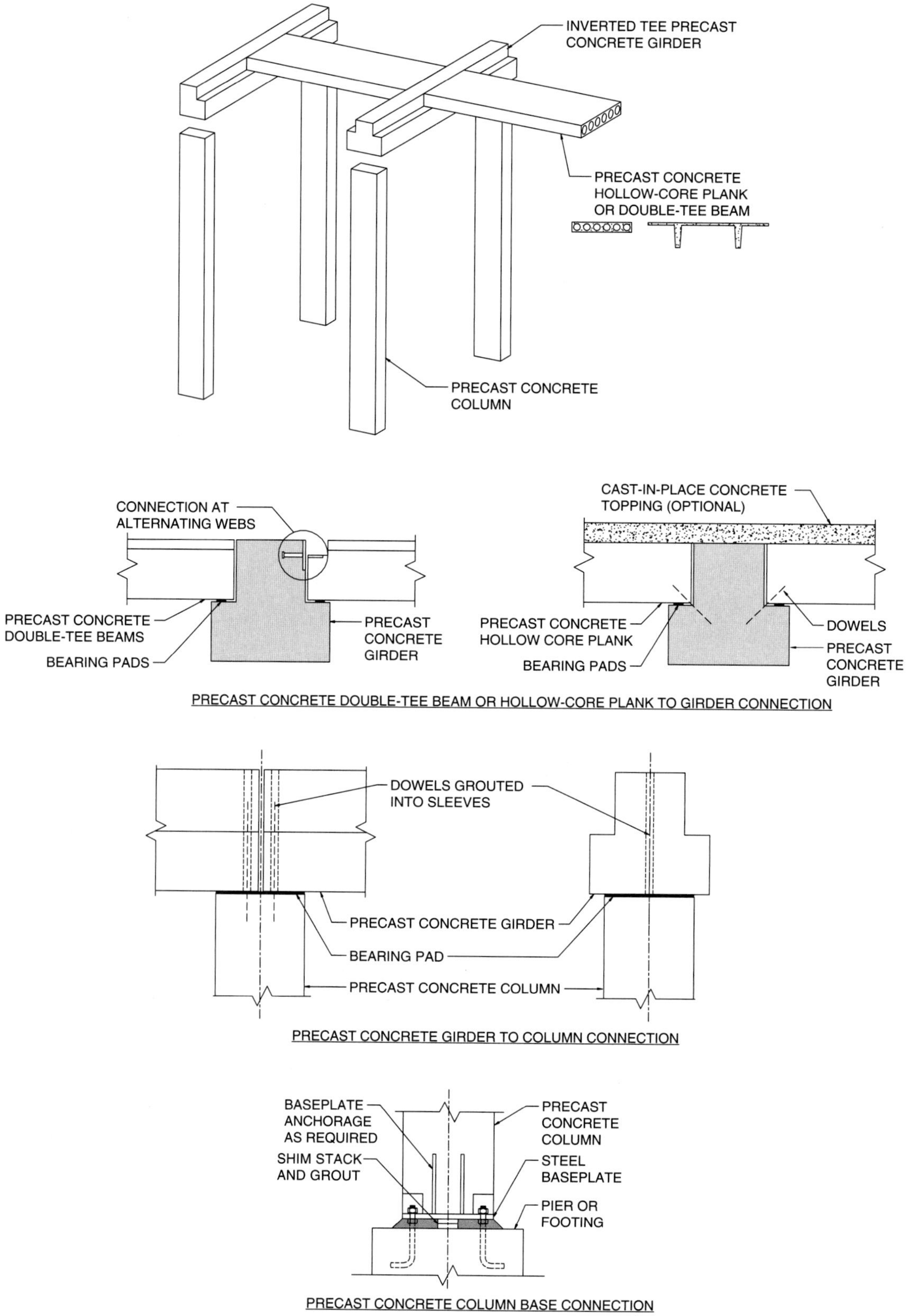

Figure 17.6 Precast concrete-framed building systems.

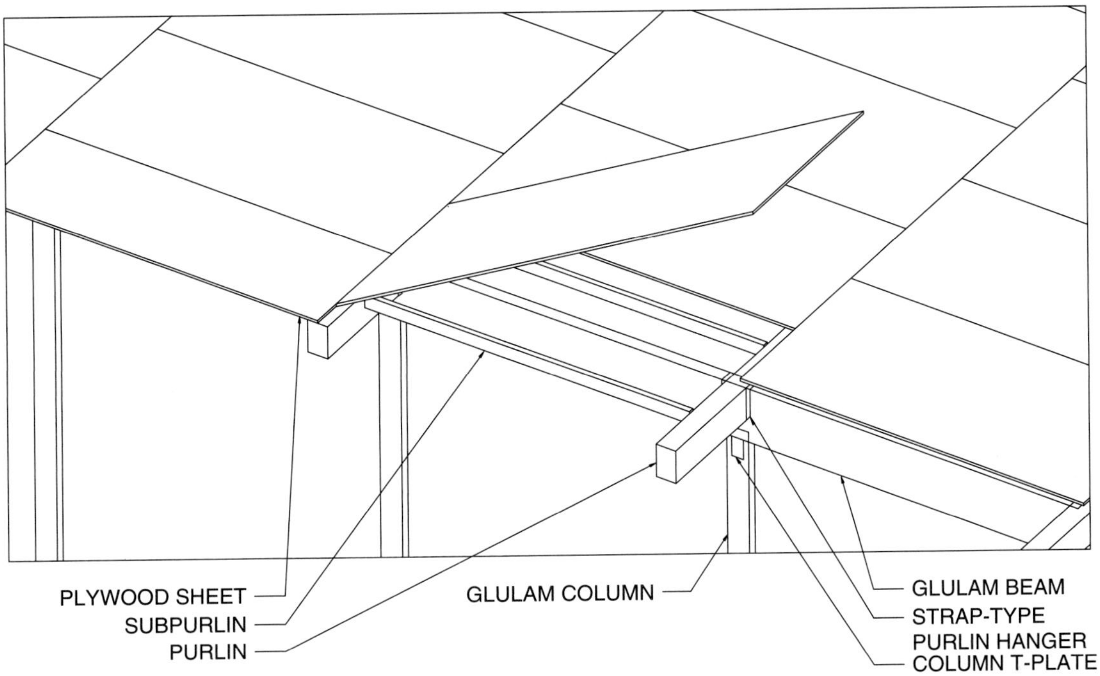

Figure 17.7 Wood roof deck framing.

17.3.6 Metal Building Systems

Steel buildings supplied by metal building manufacturers (preengineered buildings) are often assumed to be simple standard buildings. However, in current practice, virtually all buildings provided by metal building manufacturers are unique custom-designed buildings that have been designed based on specific project requirements. Framing systems commonly used for these buildings were previously discussed. The typical cladding systems consist of exposed metal roof deck. Standing-seam roof decks are used with most metal buildings today, but screw-down metal roof panels are still available. Metal building framing systems can also be designed to incorporate conventional roof and wall systems.

17.3.7 Crane Buildings

Steel-framed buildings are predominantly used for crane buildings. A major distinction between the design of crane buildings and other buildings is the consideration of fatigue. Crane buildings are often loaded to their expected design loads, and in many cases the design loading will occur thousands of times. Thus, member and connection design must consider fatigue limit states.

Crane runway girders and supporting columns are usually included with the building design. Wheel loads and spacing are provided by the crane manufacturers.

The Crane Manufacturers Association of America (CMAA) classifies cranes according to service class, in recognition of the number of anticipated load cycles for that crane and the environmental conditions that the crane is exposed to. These classifications are as follows:

- Class A—Standby or infrequent service where capacity loads are handled during initial installation of equipment and for infrequent maintenance.
- Class B—Light service where the crane speed is slow and typical lifted loads are light, with only occasional lifts at full rated load.
- Class C—Moderate service where the crane will handle loads which average 50% of the rated capacity with 5 to 10 lifts per hour and not more than 50% of the lifts are at or near rated capacity.
- Class D—Heavy service where the crane will be used constantly throughout the day with typical lifts at or near 50 percent of the rated capacity of the cranes and no more than 65 percent of the lifts are anticipated to be at capacity. High crane speeds are commonly desirable with 10 to 20 lifts per hours.
- Class E—Severe service with anticipated lifts at or near capacity throughout the day and 20 or more lifts per hour. High crane speeds are anticipated.
- Class F—Continuous severe service with anticipated lifts at or near capacity and severe service conditions or environment. An example would be a hot, dirty environment such as would be expected in the melt shop of a steel mill. These cranes must be very reliable.

Fatigue considerations will normally impact the design of crane girders for Class D, E, and F cranes. For heavy cranes and/or long crane girder spans, the crane runway girders are often required to be plate girders to meet strength and serviceability limit states. The crane girder web is typically welded to the top flange of the girder with complete joint penetration welds due to high contact stresses beneath the crane wheels and rail. Fabrication costs associated with these plate girders make these girders more expensive than hot-rolled beams. Allowable stress range for fatigue concerns is also reduced for plate girder sections due to potential flaws and associated stress risers in the welded joint between the bottom flange and web of the crane girder.

The following suggestions for various girder types may prove helpful:

- For lighter cranes and shorter spans: Use plain wide-flange beams.
- For moderate capacity and service class cranes with girder spans ranging from 30 to 40 ft: Use wide-flange beam with cap channel or other similar top flange reinforcement to improve the top flange strength and stiffness for crane lateral forces.
- For heavy cranes and spans up to 40 ft: Use a plate girder profile with a larger top flange to accommodate strength and stiffness requirements for crane lateral forces.
- For heavy cranes and spans greater than 40 ft: Use a plate girder with a top and bottom flange backup bracing system incorporating

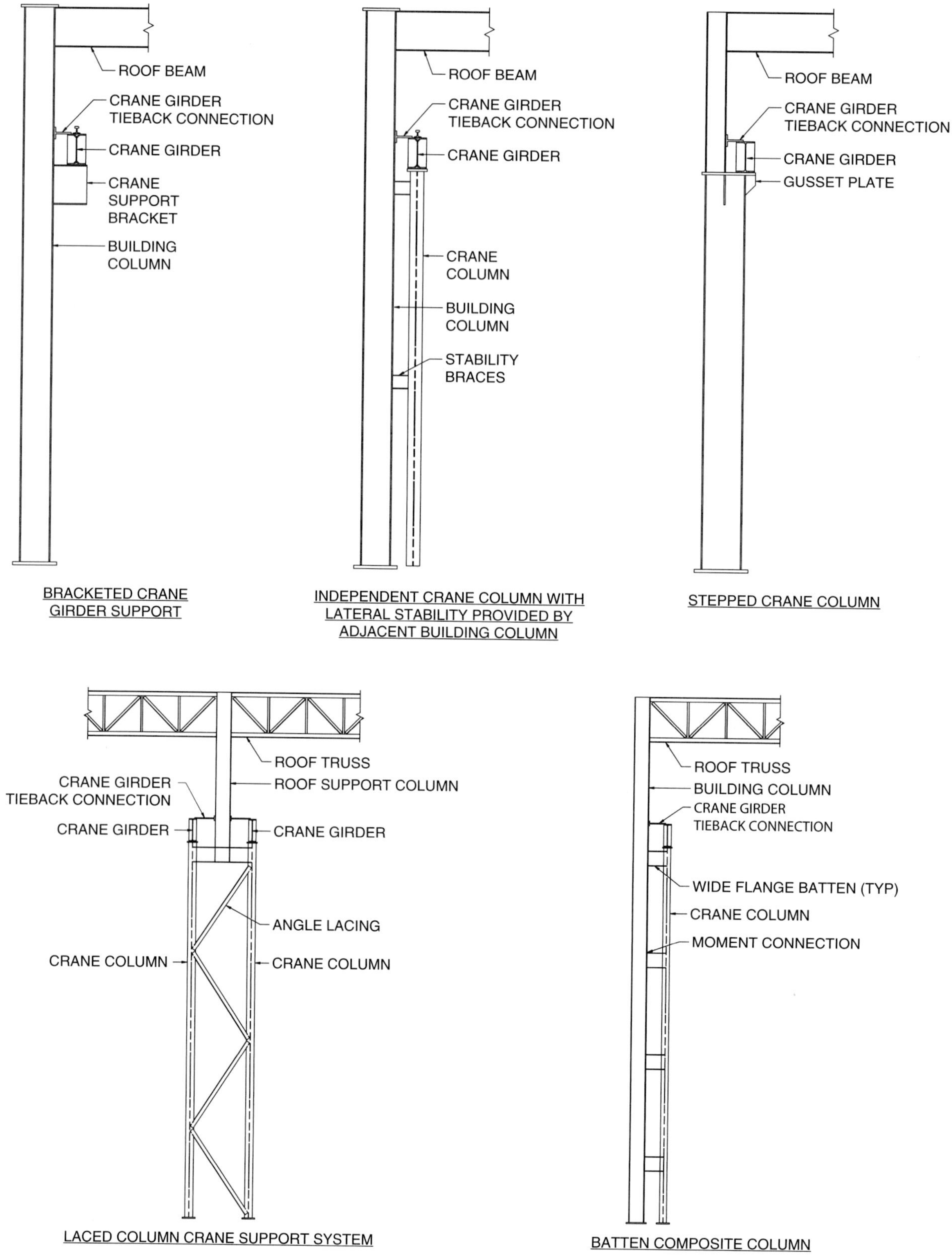

Figure 17.8 Crane girder column support options.

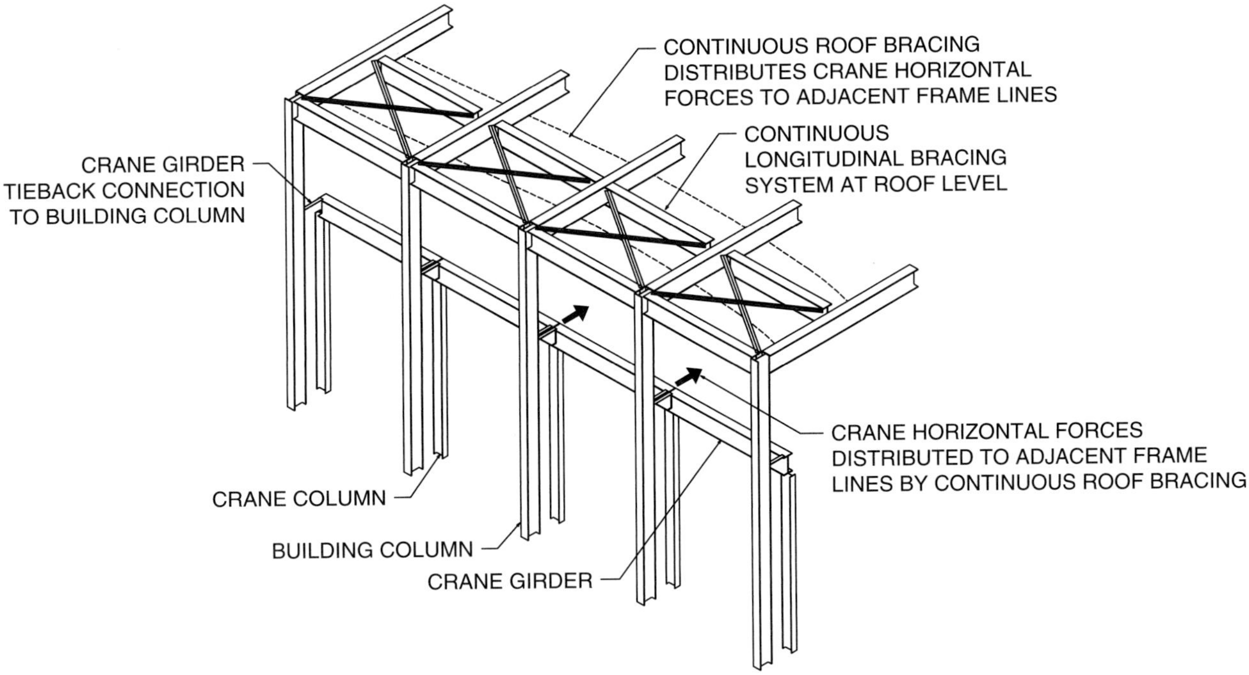

Figure 17.9 Use of continuous roof bracing system for crane buildings.

horizontal lacing to a backup beam or truss to improve the lateral and torsional strength and stiffness of the crane girder to accommodate crane lateral forces and potential torsional forces that are more of a concern for longer spans.

When evaluating serviceability requirements for crane runway girders, vertical deflections are generally calculated for the maximum crane wheel loads from a single crane on the runway without vertical impact forces. For top running cranes, typical criteria are as follows:

- Span/600 for CMAA classes A, B, and C cranes
- Span/800 for CMAA class D cranes
- Span/1000 for CMAA classes E and F cranes

For underhung and monorail cranes typical criteria are as follows:

- Span/450 for CMAA classes A, B, and C cranes

The design of underhung crane runway beams and monorail beams must consider combined stresses in the bottom flange due to bending over the span of the beam and localized bending in the bottom flange generated from the concentrated wheel loads.

Lateral deflections of crane girders should be limited to span/400 so that crane operation is not compromised and to limit the effect of lateral distortion and associated twist in the crane girder on the resulting forces or stresses in the girder.

Simple-span crane girders are recommended. Use of multiple-span or continuous crane runway girders can complicate and compromise the fatigue design of the girder and can be problematic if differential settlement occurs between foundations for the supporting columns.

Several types of column support options for crane girders are commonly used, including the following:

- Bracket supports off adjacent building columns.
- Independent crane support columns with lateral stability provided by adjacent building columns.
- Stepped columns, with columns providing support for vertical loads from the crane and roof and the columns also integrated into the lateral load resisting system for the building.
- Laced columns, with the composite, laced column supporting vertical loads from the crane and roof and the laced column cantilevered up from the supporting foundation and integrated into the lateral load resisting system for the building.
- Battened columns with adjacent building and crane columns designed as a composite cross section, using short beam or batten sections between the two column profiles that are rigidly connected to one or both column profiles, generating composite behavior similar to the laced column option discussed above. These composite columns support vertical loads from the crane and roof and are also integrated into the lateral load resisting system for the building.

These column options are illustrated in Fig. 17.8.

Continuous longitudinal bracing at the roof level can be used to distribute crane lateral forces to adjacent frame lines, minimizing the effect of these forces on the supporting columns, lateral load-resisting system, and foundations. Design of this system requires use of an analysis that recognizes the lateral stiffness of the columns, each frame line, and the continuous bracing system at the roof level. This approach is illustrated in Fig. 17.9.

REFERENCES

1. Fisher, James M.: Design Guide 7, Industrial Building Design, Third Edition, American Institute of Steel Construction, Chicago, IL.

2. Guide for the Design and Construction of Mill Buildings, AIST Tech. Report. No. 13 (2020 Edition), Association of Iron and Steel Technology, Pittsburgh, PA.

3. West, Michael A., Fisher, James M., Griffis, Lawrence G.: Serviceability Design Considerations for Steel Buildings, Steel Design Guide 3, 2nd Edition, American Institute of Steel Construction, Chicago, IL.

4. Foundations for Dynamic Equipment, ACI 351.3R-04 (Reapproved 2011), American Concrete Institute, Farmington Hills, MI.

5. Guide to Concrete Floor and Slab Construction, ACI 302.1R-15, American Concrete Institute, Farmington Hills, MI.

6. Guide to Design of Slabs-on-Ground, ACI 360R-10, American Concrete Institute, Farmington Hills, MI.

7. Tarr, Scott M., Farny, James A.: Concrete Floors on Ground, 4th Edition, Portland Cement Association, Skokie, IL.

8. International Building Code, 2018 Edition, International Code Council, Inc., Country Club Hills, IL.

9. Minimum Design Loads and Associated Criteria for Buildings and Other Structures, ASCE 7-16, American Society of Civil Engineers, Reston, VA.

10. Specification for Structural Steel Buildings, AISC 360-16, American Institute of Steel Construction, Inc., Chicago, IL.

11. Building Code Requirements for Structural Concrete, ACI 318-14, American Concrete Institute, Farmington Hills, MI.

12. Specifications for Top Running Bridge & Gantry Type Multiple Girder Electric Overhead Traveling Cranes, CMAA Specification No. 70 (2015), Crane Manufacturers Association of America, an Affiliate of the Material Handling Industry of America, Charlotte, NC.

13. Metal Building Systems Manual, 2012 Edition, Metal Building Manufacturers Association, Cleveland, OH.

Chapter 18
Tall Buildings

BY
CHARLES BESJAK, PE, SE *Skidmore, Owings & Merrill*
BRIAN McELHATTEN, PE, SE *Arup*
PREETAM BISWAS, PE *Skidmore, Owings & Merrill*

18.1 DEFINITION OF TALL BUILDING

18.1.1 Conventional Buildings

According to the International Building Code (IBC), a building can be defined as "Any structure used or intended for supporting or sheltering any use or occupancy." Conventional small-scale buildings are constructed from timber or masonry and tend to expand horizontally rather than vertically when more area is needed. The design of small-scale buildings is typically based on its usage—commercial, residential, industrial, etc.

As demand for commercial and office buildings increased in urban centers such as Chicago and New York City and with the advent of modern technology such as the passenger elevator, vertical expansion became an attractive and economical option for maximizing usable space.

18.1.2 Tall, Supertall, Megatall Buildings

There is no absolute definition for tall buildings. For instance, a building of 10 stories might be considered "tall" in an ordinary European city or a suburban town but is almost an afterthought in a city such as New York or Hong Kong (Fig. 18.1). A building with many floors may not feel like a "tall" building due to a large footprint. Usually, a building is considered "tall" if the controlling elements of design are dominated by the height and slenderness of the building rather than any particular usage: either special vertical transportation technologies or a dedicated structural system to control movement under wind loading are required.

Tall buildings can be further categorized into "supertall" and "megatall" when their height is significant enough. The Council on Tall Buildings and Urban Habitat (CTBUH) suggests a building be classified as "supertall" if it is over 300 m (984 ft) and as "megatall" if it is over 600 m (1968 ft). There are only three megatall buildings as of today. As the level of technology developed and demands for space are further increased, there will be more megatall buildings in the future.

18.2 GENERAL DESIGN CONSIDERATIONS

18.2.1 Architectural

Architecture is an integration of the social, physical, and aesthetic needs of society and its environment. Architectural design of tall buildings must take into account both the aesthetic and the practical to not only express and communicate the aspirations and experiences of the designer but also perform its intended function to a satisfactory degree.

Technological development has made many grand architectural ideas come to reality, such as large cantilevers and complex, curved façade surfaces. Along with these complex geometries, innovative structural systems have also been developed. The design of tall buildings, where the structural system is more vital to building performance when compared to the design of a smaller-scale building, provides opportunity for integration of architecture and structure in the aesthetic design of a tall building to generate elegant, efficient designs.

18.2.2 Vertical Transportation

Vertical transportation is very important in tall buildings, including stairs, elevators, and utility shafts. In general, the higher the building is, the more vertical transportation facilities are needed. As a result, tall buildings have a lower percentage of rentable/usable space. The percentage of

Figure 18.1 Hong Kong has more than 2300 buildings over 100 m (328 ft) tall. (Photo © Ryan Cheng, 2010.)

rental area to gross area usually ranges between 85 and 90 percent in a tall building, but it can be as low as 55 to 65 percent. In most cases, vertical transportation is concentrated inside the building core.

18.2.3 MEP

Mechanical, electrical, plumbing (MEP) equipment is usually located on designated service floors and/or inside the building core. It is not uncommon that there are multiple MEP floors or zones in a taller building. Ideally, heavy MEP equipment should be located on lower floors of the building to minimize the dynamic impact of a large mass at the top of the building. However, the presence of a designated MEP zone can be advantageous—as MEP floors have fewer aesthetic requirements as compared to occupiable space, structural enhancement, for example, outrigger and belt trusses, can usually be added in these zones.

MEP requirements also impact the floor clearance and must be accommodated, as well as structural beam depth, in the architectural ceiling sandwich. Ducts, pipes, and other MEP system elements often must also enter and exit the building core. Penetration of beams and shear walls is a common resolution of these requirements. Coordination between building trades in tall building design is crucial to ensure a successful design.

18.2.4 Fire and Life Safety

According to a National Fire Protection Association (NFPA) report, there are over 10,000 high-rise fires annually. High-rise buildings present a unique set of challenges in fire and life safety due to large occupation and long vertical travel distance. To reduce the travel distance for occupants and provide a better protection under a fire event, refuge areas are usually designated throughout the height of modern tall buildings.

18.2.5 Sustainability

Due to their size and high occupation, tall buildings consume a considerable amount of energy from construction to operation. In order to promote sustainable and high-performance design, the U.S. Green Building Council has developed the LEED rating system, which offers a checklist for typical green commercial buildings, including sustainable sites, water efficacy, energy/atmosphere, materials/resources, indoor air quality, innovation, and design/build process. Common strategies for high-performance tall building design include passive solar gain, façade technology, structure and material preference, and harvesting wind energy. High-performance or sustainable design should be a full integration of architecture, structure, and MEP disciplines.

18.2.6 Constructability

Tall and supertall buildings, especially those built with structural steel, often consist of large built-up steel members and nodes. As part of the design process it is important to consider fabrication, transportation, and erection of these elements. Large pieces must be sized such that they are able to be lifted by the cranes available for construction and able to fit on the trucks or trains used for transportation between fabrication shop and site.

It is often the case that the design of structural connections is not performed by the structural engineer of record but instead delegated to a steel detailer or fabricator. Similarly, the layout of reinforcement in concrete is not usually specifically documented in a set of drawings but instead schematically laid out. In tall and supertall building construction where loads and member sizes become large, economy and efficiency can be realized by coordinating connection and detail design for high-intensity areas earlier in the design process.

18.2.7 Other Considerations

Building Resiliency

In the context of tall building design and construction, resiliency refers to the ability of a building structure to resist extreme events. Both the structural and the architectural design of a tall building must incorporate resiliency criteria. For example, many building codes require refuge floors spaced throughout the height of a tower or independent fire-suppression systems for egress stairs.

Progressive Collapse Mitigation

From a structural perspective, resilient design is concerned with structural stability under extreme load cases that may or may not result in members or groups of members no longer contributing to the structural system. Local building codes as well as documents produced by public agencies such as the Federal Emergency Management Agency (FEMA) or the General Services Administration (GSA) provide general and prescriptive requirements for resisting progressive or disproportionate collapse.

There are two primary methods for resisting loads caused by a deliberate attack of a building structure:

- Alternate load path method: the structure is designed such that it is stable when one or multiple members are removed.
- Local resistance method: a "design threat" is selected, generally by a security consultant, and members and connections are designed and detailed such that they can resist the dynamic loads imposed. This will generally require complex, nonlinear finite-element analysis.

18.3 STRUCTURAL DESIGN CONSIDERATIONS

Tall buildings must be designed to fulfill strength and serviceability requirements. As building heights increase, their design is increasingly dominated by serviceability concerns—effects of building motion and accelerations on occupant comfort and façade design among them. The design must also take gravity and lateral loading into account and satisfy the requirements of local building codes.

18.3.1 Gravity Loads

Gravity loads on a tall building are generally static loads—they will be present throughout the entire life of the building. Static gravity loads that need to be considered in tall building design include the structure self-weight and superimposed dead load (SDL), such as weight of fixed service equipment, fixed partition, floor finish, and cladding. Loads that are time dependent but not applied to the structure dynamically are live loads (LLs), related to the use and occupancy of the structure, such as movable equipment and movable partitions. Due to the large total floor area of tall buildings, floor loading accounts for a considerable portion of the total load. The floor framing system is often chosen based on building usage: office buildings tend to use metal deck and steel framing, while residential buildings usually adopt a concrete flat slab system. Gravity loads also impact the serviceability behavior of a tall building. A concrete system is generally much heavier than a comparable steel system, increasing demand on foundations but reducing the likelihood the building will have adverse dynamic effects due to wind loading or footfall-induced vibration of floor framing systems.

The sequence of construction must also be considered in high-rise building design, especially for composite structures (those using both steel and concrete structural elements). Further discussion on these effects is presented in Sec. 18.11.

18.3.2 Lateral Loads

As previously discussed, lateral loading is the dominant design criteria for tall buildings. Both wind and seismic loads are dynamic loads: their magnitudes and directions are time dependent. The lateral load applied to the structure depends on the building location and geometry but also the building structure itself. For example, the wind load demand on a building is dependent on the local wind climate, site terrain, and exposure but also building geometry and structural dynamic properties (mass, stiffness, damping). The nature of wind loading on tall buildings is discussed in detail in Sec. 18.9.1.

While wind loads are generally the dominant lateral load on a tall building, in locations that are seismically active, earthquake loading can control the lateral system design. It is similarly related to both site characteristics and building dynamic properties. Earthquake design is complicated by the inherently unpredictable nature of seismic activity.

In most prescriptive codes, the building lateral system must be designed for a given earthquake return period, and explicit limits on the heights to which particular structural systems can be used are included. Some building codes allow for a performance-based design to address these limits and reach performance beyond simply guaranteeing the safety of occupants in a seismic event. Section 10 discusses performance-based seismic design to a greater extent.

18.3.3 Foundation

Typical foundation types for tall buildings include piles, barrier walls, caissons, and mats. The foundation system can be deep or shallow depending on the conditions belowground—for instance, many tall buildings in New York City are supported on a reinforced-concrete mat sitting on solid rock, while where competent bedrock is not found, towers can be supported on a mat sitting on deep piles driven into a softer substrate. For tall buildings with a dual structural system consisting of a core and perimeter frame, the core can be supported by a mat foundation while the perimeter frame is supported by piles or spread footings. In some cases, the entire structure can also sit on a single mat. Considering the scale of tall building structures, foundation systems can be gigantic. For example, the mat foundation for the Jin Mao Tower in Shanghai is 4 m (13 ft) thick.

18.4 SERVICEABILITY CONSIDERATIONS

As indicated in Sec. 18.1.1, one of the primary factors in tall building design is the need to control building movements for both occupant comfort and protection of other building components.

18.4.1 Occupant Comfort

People respond to building motions in varying degrees with some being more affected than others. Several criteria have been developed over a number of years that seek to ensure that the majority of building occupants will not be bothered by these motions. These are a function of building periods/frequencies, occupancy type (residential or office), and return period of the wind event. M. D. Burton et al. (2015) provide a comprehensive look at occupant comfort perception studies and recommended criteria for tall buildings.

The two primary measurements related to occupant comfort are accelerations and torsional velocities.

- Accelerations are related to the building's modal response and inherent damping (plus supplemental damping in some cases). These are experienced by the occupants as a speeding up and slowing down of the building as it moves through its modal oscillations with the greatest effects being felt at the top of the building.
- Torsional velocities are related to the accelerations and experienced as a twisting of the floor plate in addition to the overall motion of the building. Therefore, the greatest effects would be felt farthest from the center of rotation along the perimeter of the floor plate. Occupants especially perceive this when there is an adjacent building or other point of reference outside that gives a sense of relative movement.
- Return periods for wind used in the evaluation of accelerations and torsional velocities have historically been between 1-year and 10-year events. These are still applicable for many tall buildings, but as more slender towers have been proposed, more frequent wind events (less than 1 year) have been found to control these criteria for some projects.

Procedures exist for the structural engineer to estimate the accelerations and torsional velocities, but it is most common and accurate to work with a wind tunnel testing facility to determine these. This approach will be discussed in further detail in Sec. 18.9.3.

Methods for controlling these building movements will be discussed later in the chapter.

18.4.2 Interstory Drift Ratio and Nonstructural Components

In the section above on building displacements, the movement between individual floors was noted as an important consideration for the façade design and interior partitions. Generally, this is a wind-movement issue, but in moderate to high seismic regions, the earthquake movements may govern. Like the overall building displacements, these movements between floors are also a function of floor height, h, and are termed as *interstory drift ratio*, which is illustrated in Fig. 18.2. Common interstory drift ratios are in the range of h/300 for wind loads and can be as low as h/50 under high seismic loads. Practical values based on façade and partition standard detailing practices would limit the interstory drift at ⅜ to ½ in. *ASCE 7* provides further guidance on these limits and the wind return periods to be considered in the section on serviceability considerations. If the building is in a high seismic region, special detailing and systems may be required to accommodate the expected large interstory drifts to limit damage and prevent falling of the façade panels.

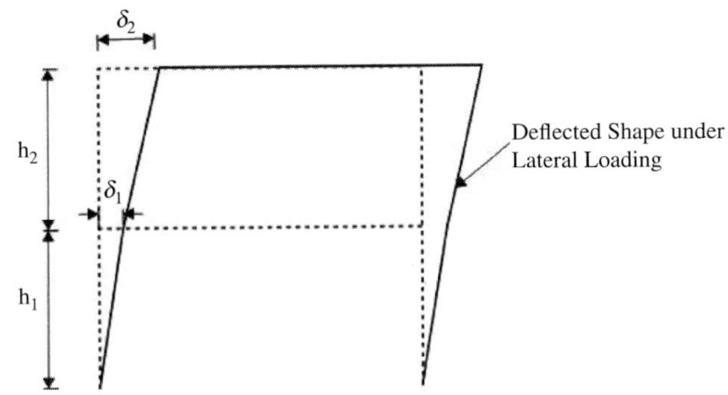

Figure 18.2 Interstory drift ratio.

18.5 STRUCTURAL SYSTEMS FOR TALL BUILDINGS

18.5.1 Typology

High-rise buildings are typically commercial (office) or residential, many with retail or public amenity spaces at the street level. As discussed in the section "Building Usage," the building usage can have a large impact on the appropriate choice of structural system.

18.5.2 System Evolution

The Home Insurance Building in Chicago, a 10-story steel-framed designed by William Le Baron Jenney and completed in 1885, is widely considered the first modern high-rise building. By utilizing an interior steel frame instead of bearing masonry walls, the early architects and engineers of the Chicago school were able to reduce the weight of a steel-framed building to one-third of an equivalent masonry structure. Architectural principles such as the concentration of building services and vertical transportation in a "service core" and other engineering innovations, such as the development of the deep pile foundation system, led to a rapid increase in the achievable heights of tall buildings.

Ever since then, the ability to build higher and higher and thus increase urban density to unforeseen levels has defined the modern city. In the 1920s, tall building construction boomed—known as the "race to the sky." Notable buildings, such as the Chrysler (925 ft, 282 m tall) and Empire State (1250 ft, 381 m) buildings, were built in New York City. These towers utilized steel moment frame systems and became iconic components of the New York City skyline and its cultural identity.

The tube concept was then introduced in late 1960s and drastically changed the design of modern skyscraper. Rather than a solid central core, a perimeter tubelike structure with significantly greater structural depth was utilized to resist the significant lateral loads in high-rise buildings. This concept greatly reduced the required material quantity for a tall building and thus the associated construction cost. The first building to incorporate this idea was the 43-story DeWitt-Chestnut Apartment Building by SOM, completed in 1964 in Chicago. Other significant buildings incorporating this concept in this period include the John Hancock Center, the Willis (formerly Sears) Tower, and the original World Trade Center towers, all of which are over 100 stories in height.

The core-outrigger system couples the perimeter structure to the core at discrete locations, allowing the lateral resistance of both systems to be utilized. Outriggers have been used extensively in tall buildings in the past several decades in steel, concrete, and composite buildings. Figure 18.3 shows structural systems for tall buildings of increasing heights.

18.5.3 Material Evolution

The mass usage of steel and reinforced concrete as building materials is considered to be the beginning of the "modern" era of construction, and they remain the most commonly used materials for tall building construction today. The strength of these materials has improved significantly over time—for example, the steel-frame members of the Home Insurance Building had yield strength of only 16 ksi. By the midpoint of the 20th century and the construction of the John Hancock building, the majority of building steel was A36 with a yield strength of 36 ksi. Today, the most common structural steel grade for rolled sections is ASTM A992 with a yield strength of 50 ksi. In many locations, rolled sections and cast or forged components with yield strengths of up to 80 or even 100 ksi are available. A36 is still commonly used for detailing members, such as channels and angles.

Obtainable concrete strength has also significantly increased over the years, leading to increased efficiency of concrete construction. Strengths of 8 to 10 ksi are common in a tall building, and can be as high as 14 ksi. Reinforcing bars have also increased in strength from a typical capacity of 33 ksi in the early 20th century to 80 or 100 ksi today. Protection and coating technology of rebars has also improved over time, leading to increased longevity of concrete structures.

As the contribution of typical building materials to energy consumption and carbon emissions comes under closer scrutiny, mass timber is an attractive option for tall building construction. While there are significant design challenges associated with timber towers—fire protection obviously among them—it is an area of active research in the structural engineering and design community.

18.5.4 Technological Evolution

The techniques and methods used to design tall buildings have advanced alongside improvements in building materials. In an increasingly competitive market, developers worldwide have been quick to market new buildings as "first of a kind," "tallest," or "grandest"—promises translating to demand for heroic structures that push the boundary of conventional systems. New tools, such as computer-aided design and analysis software, allow for the conception and analysis of complex buildings with unique three-dimensional geometries and facades. Three-dimensional digital modeling and detailing tools allow for material and process optimization, allowing engineers to model and coordinate complex connections or custom elements in detail.

Along with new high-strength concretes, advanced pumping technology and admixtures have been developed to allow for pumping concrete to extreme heights (up to 600 m). Self-climbing formwork systems have also been adopted to reduce required labor for concrete construction.

Specialized GPS technology has also been developed to monitor structural performance, replacing conventional surveying methods.

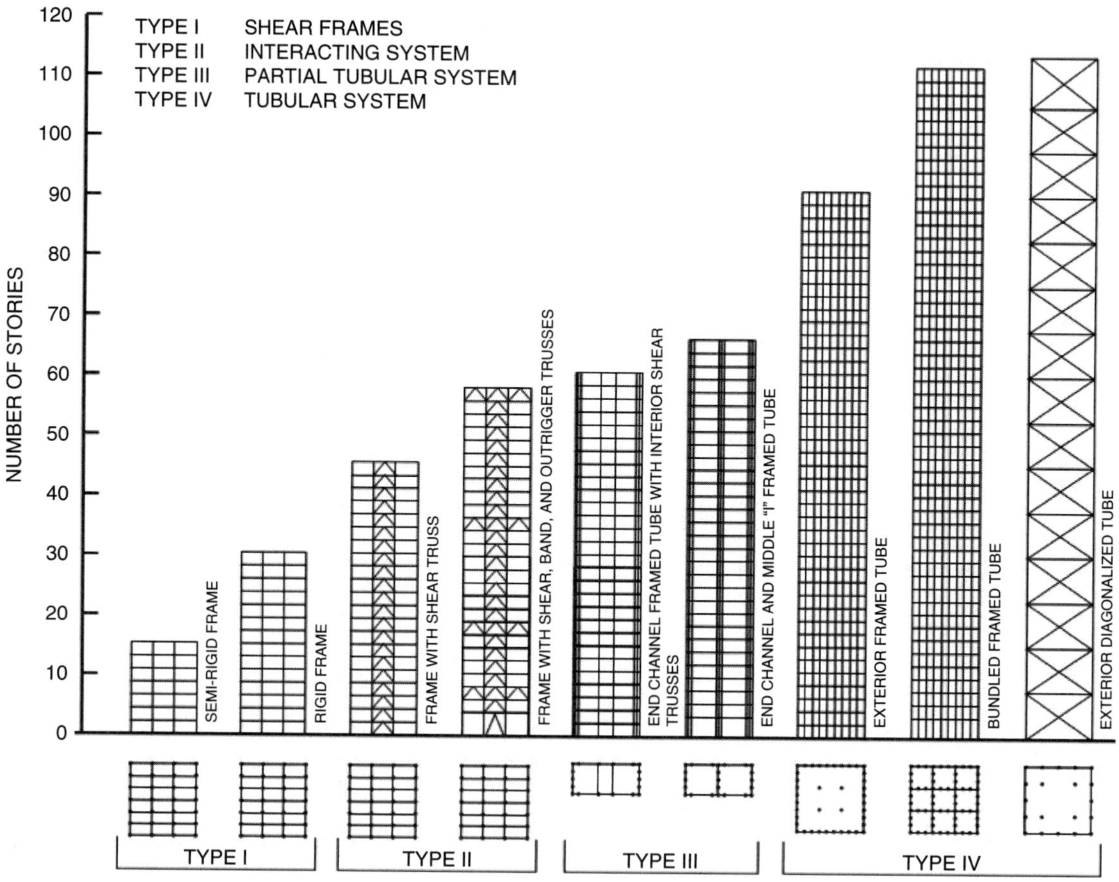

Figure 18.3 Tall building systems.

18.6 SYSTEM CONCEPTUALIZATION

Early in the design of any tall building, it is important to consider potential structural systems from a conceptual level. Understanding which system(s) might make sense and how to quickly evaluate them can save time and money for both the structural engineer and their clients and likely lead to better project outcomes.

18.6.1 System Choice

Common structural systems for various tall building heights are recommended in Sec. 18.7. The appropriate structural system(s) to be utilized for a given tall building project will often be influenced by a number of factors including architectural aspirations/limitations, building height, building slenderness, wind climate, seismic intensity, building usage, local construction practices, material costs and availability, etc. These factors and how they affect the structural system will be discussed further in this section. Often several of these will need to be considered together to arrive at viable systems to evaluate.

Architectural Aspirations/Limitations

The architectural vision for a tall building will often weigh heavily into the structural systems that can reasonably be employed.

For example, if the form of the building is irregular over its height, the application of a diagrid system or a perimeter brace system may prove difficult from a price or constructability standpoint.

Often perimeter column spacing and size as well as perimeter beam depths are critical elements to architects.

- Megacolumn systems will require a small number of very large columns that architects often don't like in the program spaces.
- Tube systems will rely on closely spaced columns with potentially deep spandrel beams that can limit views in and out of the building.

Understanding the potential architectural concerns early on can help to focus the structural work and get to a preferred structural system more efficiently. However, this is not to imply that as structural engineers we should not work with the architects to find compromises on their end that would benefit the structure and lead to a better end product and potential savings for the owner.

Building Height

As shown in Sec. 18.5.2, there are generally height ranges that work well or are appropriate for certain systems. The appropriateness of these systems is also affected by building slenderness to be discussed further below.

Many of the systems employed in supertall buildings would be overkill for tall buildings. While they may work, the efficiencies found in very tall buildings may not be realized, and the construction costs to build these more complicated systems would be burdensome to the project.

Conversely, systems that work well for tall buildings cannot be applied in an efficient manner to supertall or megatall buildings. For example, having a core wall-only lateral system is often not even possible from strength or serviceability standpoints for very tall buildings without having an artificially large building core leading to loss in floor plate efficiency and high structural materials costs. Similarly, a single-tube system using perimeter frames is extremely difficult to employ for very tall buildings without a large amount of structural material on the perimeter (closely spaced columns or very large columns with very deep spandrels), which has cost implications and has impacts to the architecture as described in the previous consideration.

Building Slenderness

Refer to discussion below under "Rules of Thumb" for information on this critical parameter. As with building height, some systems are not efficient or applicable to buildings with high slenderness values.

Wind Climate

Project locations exposed to extremely high winds from hurricanes or typhoons will impose a further constraint on the structural systems that can lead to efficient solutions as a result of increased strength demands.

From a serviceability view, areas with relatively frequent, strong wind events from thunderstorms, etc. will impose further acceleration control challenges on slender buildings requiring structural systems that provide better response.

Seismic Intensity

Similar to wind climate issues, regions of high seismicity will place increased demand on the building structure. Additionally, many building codes require a dual system for seismic resistance that must be integrated.

Building Usage

Building program can often play a role in which structural systems can be considered. Below are a few examples of this.

Often residential buildings will be beam-free at the perimeter to allow for generous floor-to-ceiling windows, ruling out perimeter tube system. Diagrids or perimeter bracing systems are also undesirable in some residential projects due to potential for blocked views.

Office buildings generally have wide open floor plates with more simplified footprints. While a buttressed core system like that of the Burj Khalifa works very well for the three-legged footprint, it would not be practical for most office building layouts.

Note that building usage issues often go hand in hand with the architectural concerns discussed earlier.

Local Construction Practices

It is not uncommon for tall buildings to be built in locations where the workforce has limited knowledge of some construction methods or where one is significantly cheaper than another.

For example, many countries are comfortable with reinforced-concrete construction but are not experienced in structural steel, particularly for complex bracing systems or high seismic systems.

Early in the project, it is important to understand these local preferences/limitations to further inform the structural systems that are most viable.

Material Costs and Availability

Similar to the previous point on local construction preferences, the costs and/or availability of common materials like steel and concrete can be a major consideration to a tall building project.

Tall buildings often take advantage of high-strength concrete to limit member sizes and provide higher stiffness. This may not be available due to local sourcing of the needed raw materials. Options to accommodate this may be a heavier structure or looking to systems that are conducive to structural steel. On the other hand, quality structural steel often needs to be imported from a far and may become an economic deterrent.

Often issues with local construction practice and material costs are related.

18.6.2 Rules of Thumb

In most cases the design of tall buildings is not driven by resisting the imposed gravity loads. Instead, controlling the movements of the building and providing adequate strength and overturning resistance under wind loading are the key design drivers.

Below are a few measures of tall buildings that are easy to remember and important to understand early on in the design process.

Slenderness Ratio

The single most important metric of tall buildings is the aspect ratio or *slenderness ratio (SR)*, is defined as the building's height divided by its least width at the base, as shown in Fig. 18.4. As the SR increases, controlling building accelerations and displacements and preventing overturning at the base of the building become substantially more difficult. Until the last 20 years, most tall buildings had SR under 7.0. For example, the Willis Tower (aka Sears Tower) and the John Hancock Tower in Chicago have SR of 6.4 and 6.6, respectively. As wind tunnel testing and computer horsepower have grown and become more sophisticated as well as further developments in supplemental damping systems, SR have continued to increase to meet the demands for expensive

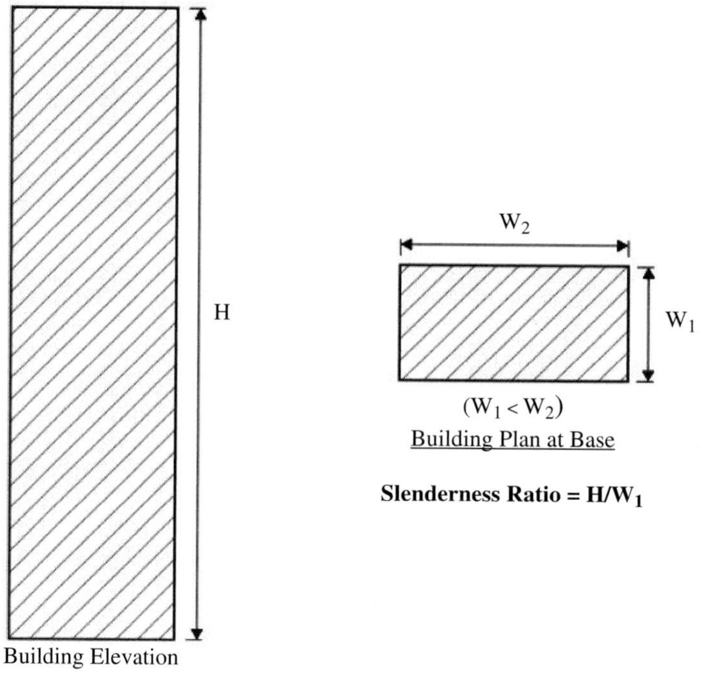

Figure 18.4 Building slenderness ratio.

real estate and/or desires by countries, cities, and developers to make a bold statement with their projects. A few examples of very slender buildings are 432 Park Avenue in New York at approximately 16.0 and 111 West 57th Street, also in New York, at around 24.0. The following ranges of SR give guidance on measures that would likely need to be taken to design for the associated wind effects.

- **SR < 7.0:** wind effects can generally be accommodated without changes to building shape or need for supplemental damping.
- **7.0 ≤ SR < 10.0:** wind effects can generally be controlled by shaping the building and/or tuning the stiffness of the structure without the need for supplemental damping.
- **SR ≥ 10.0:** wind effects will be substantial and require building shaping, structure stiffness tuning, and supplemental damping; often the perimeter of the building will need to be engaged.

These ranges can be influenced by wind climate, building shape, lateral load-resisting systems, adjacent buildings, and other factors provided as general guidelines for understanding early on in the project the likely steps that will be needed to be taken for a successful design.

Building Displacements

Another important measure of tall building design is the horizontal or lateral displacement of the top of building under the design wind load. This is not a fixed value but rather is in relation to the building height under consideration and so is referred to as an *overall displacement ratio*. Refer to Fig. 18.5. Most international codes do not put a specific value on this ratio, but historically it has been limited to the building height, H, divided by 500 (H/500). For example, a 1000-ft building would have maximum lateral displacement of 2 ft or 24 in. using this limit.

As buildings have become taller, this limit has been relaxed to some extent, provided that building accelerations can be controlled, strength impacts due to increased P-Delta effects appropriately resisted, and movements between individual floors limited to façade design criteria. It is not uncommon now for displacement ratios as low as H/400 to be used. The wind return period used for checking the overall displacements may similarly be reduced, but a 50-year wind level is generally used as a starting point unless an applicable building code for the project states a higher level.

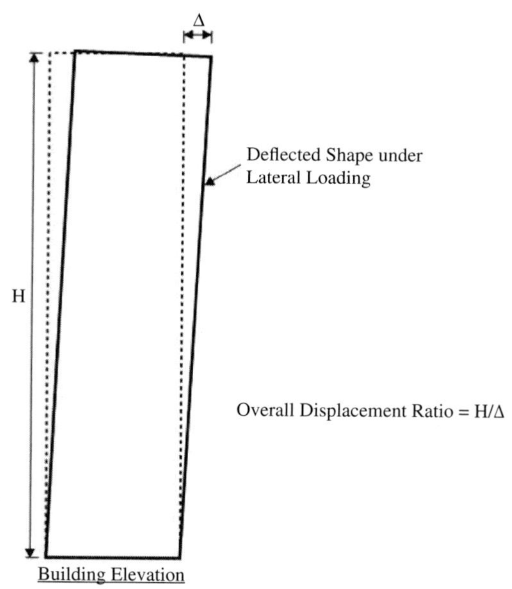

Figure 18.5 Overall displacement ratio.

Modal Behavior

The dynamic or modal behavior of a tall building is important as it relates to response to wind and seismic loads.

A general rule of thumb related to reinforced-concrete buildings is that the fundamental period should be roughly the number of stories divided by 10 (e.g., a 100-story building would have a 10-second first period). The accuracy of this is less with buildings of unusual shape or mass distribution but is still a useful check to have in mind during the initial analyses.

Buildings composed of structural steel floor framing and lateral systems generally have longer primary periods than similar-sized reinforced-concrete buildings. It would not be uncommon for them to be on the order of 50 percent higher for the first translational mode in each primary orthogonal direction.

As expected, then, buildings composed of composite construction, those making use of reinforced concrete and structural steel members, will likely have modal responses between an all reinforced-concrete and all-structural-steel building.

In terms of preferred mode shapes, having primary translational modes occurring before the primary torsional mode is best. Buildings with high torsional modes may be subject to increased building acceleration and torsional velocity issues. Also, some international codes, such as the Chinese national code, prohibit having the primary torsional mode before translational modes for seismic concerns.

Structural engineers should keep in mind these three key metrics of SR, overall displacement ratio, and interstory drift ratios as they consider potential tall building structural systems and evaluate their behavior.

Common Tall Building Structural Material Choices: Pros and Cons

Material	Pros	Cons
Reinforced concrete	Readily available in most markets	Lower strength-to-weight ratio than structural steel → larger member sizes required
	Labor forces generally familiar with construction techniques	Lower stiffness-to-weight ratio than structural steel → larger member sizes required
	Provides better inherent damping than structural steel	Prone to creep and shrinkage effects, particularly in some coupled structural systems
	Can be easily formed into various shapes and sizes	Seismic weight is higher than that of structural steel
	Has inherent fireproofing properties	Increased foundation demands over other materials
	Very high strengths becoming more common and available	Low ductility compared to structural steel
	Performs well in compression	Performs poorly in tension
		Prone to cracking and subsequent loss of stiffness
Structural steel	Higher strength-to-weight ratio compared to reinforced concrete → smaller member sizes possible	Usually requires a more skilled workforce, limiting locations where it may be used
	Higher stiffness-to-weight ratio compared to reinforced concrete → smaller member sizes possible	Less inherent damping than reinforced concrete
	Creep and shrinkage effects are not a concern	Large cross sections often needed for tall buildings require labor-intensive built-up members
	High ductility relative to reinforced concrete	Complex and expensive connections often needed
	Reduced construction times depending on specifics of the employed structural system	Supplemental fireproofing often needed
	Performs well in tension	Prone to local and global buckling in compression
Composite	Allows the combination of benefits from both structural steel and reinforced concrete to be realized	Integrating the two materials can lead to complex formwork issues or construction procedures
Heavy timber	Sustainable material	Very large members required and/or more closely spaced members
	Performs well in fire, assuming appropriate cross-sectional reductions are taken for charring	Low stiffness relative to structural steel and concrete.
		Difficult to create rotationally fixed connections

18.7 SYSTEM PARAMETERS AND CHOICES

The following sections illustrate examples of tall building systems by describing their utilization in existing tall buildings.

18.7.1 Frame-Only System

The structural systems of the early skyscrapers were generally rigid steel frames—early architects and engineers had not yet considered the skyscraper as a vertical cantilever but instead as simply a taller version of a typical low-rise building. Towers such as the Empire State and Chrysler buildings reached great heights while utilizing a frame-only system but are exceptionally heavy (the Empire State Building contains 30 percent more steel than the John Hancock Tower despite a similar height) and have many tightly spaced interior columns.

While the material costs associated with increased height grow disproportionately for these rigid-frame structures, they are appropriate for towers on the order of 20 or 30 stories tall.

18.7.2 Core-Only System

Alternatively, the lateral load-resisting system can be located entirely in the service core of the building. Wind loads are transferred from the façade to the core structure through rigid floor diaphragms, often concrete slabs.

While this is often the case (by choice) for low-rise buildings, site constraints can restrict the continuity of perimeter elements and force a core-only system for taller structures. Due to advances in high-strength materials and fabrication of complex built-up members, core-only structures such as One Manhattan West and 150 North Riverside have reached supertall heights by relying only on a stiff internal core.

18.7.3 Tube System

Tube systems, popularized by engineer Fazlur Khan in the 1960s and 1970s, utilize the perimeter structure of a tower by imitating the load-resisting behavior of a thin-walled tube. Tubular systems can be utilized in steel or concrete structures and led to a drastic increase in achievable building heights.

The tubular system developed by Khan was notable for considering a tall building as a single cantilever extending from the ground rather than a series of stacked frames. Efficiency is achieved by engaging the perimeter structure in lateral resistance, leading to a deflected shape similar to a cantilever beam (see Fig. 18.6). A measure of a tall building

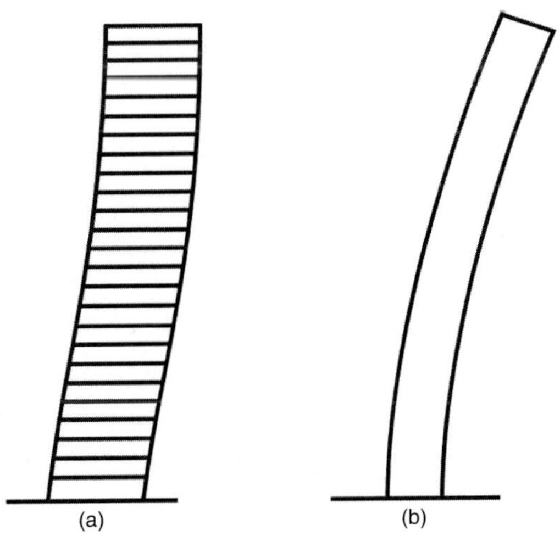

Figure 18.6 (*a*) Shear and (*b*) cantilever deflection.

system is the "cantilever efficiency," comparing the actual tip deflection to that of an idealized cantilever beam structure.

Single Tube

A single-tube building will generally utilize closely spaced exterior columns connected at each floor by deep spandrel beams, forming a structural tube. In this system, the exterior frame system will resist the lateral force and a portion of the gravity load, while the core will carry the remaining gravity load to the foundation. The overturning moment due to wind is resisted by bending of the tube (axial shortening and elongation of the perimeter columns on the windward and leeward faces of the building), and the shear due to wind is primarily resisted by frame action on the faces of the building parallel to the wind direction.

A challenge in single-tube construction is the shear lag effect (Fig. 18.7)—while in an actual thin-walled tube under flexure there is consistent axial stress in the "flange" walls, due to the discontinuous nature of a tower's perimeter structure, an uneven distribution of axial load is observed along the perimeter walls.

The original World Trade Center Towers 1 and 2 were examples of single framed-tube structures and were the tallest buildings in the world when completed in 1973. Figure 18.8 shows a typical floor plan and the prefabricated perimeter column/beam elements used to construct the perimeter structure.

Braced Tube

The braced tube structural system is a variant of the framed tube structural system. It allows fewer and farther spaced exterior columns by introducing diagonal structural elements on the perimeter. Instead of resisting wind shear through frame action, the truss planes resist wind load through axial load on the braces and columns—generally resulting in more efficient use of material. The diagonal elements can also redistribute loads from more heavily loaded columns to less stressed columns on the same face, helping reduce the shear lag effect.

It is important to note that bracing stiffness is largely dependent on its configuration (chevron brace, X-bracing, inverted V-bracing, etc.) and the inclination of the bracing members. Considering a general

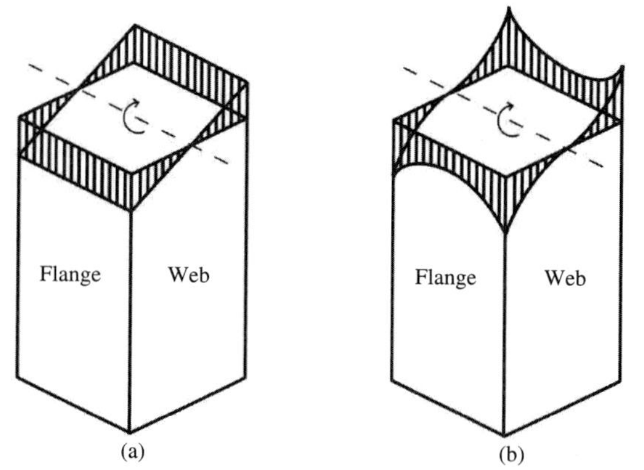

Figure 18.7 Axial stress distribution in beam structure (*a*) without shear lag and (*b*) with shear lag.

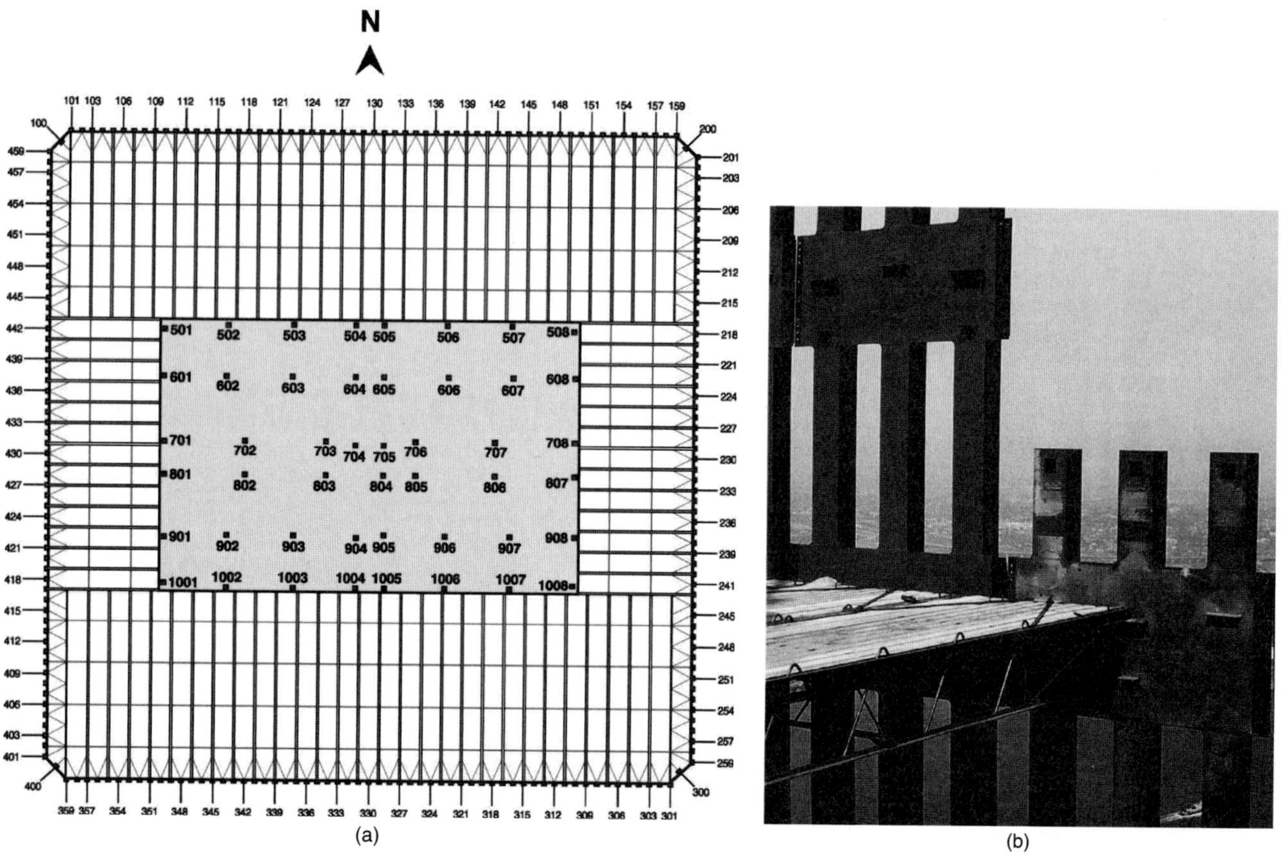

Figure 18.8 (*a*) Typical floor plan of World Trade Center Tower 1; (*b*) perimeter tube assembly.

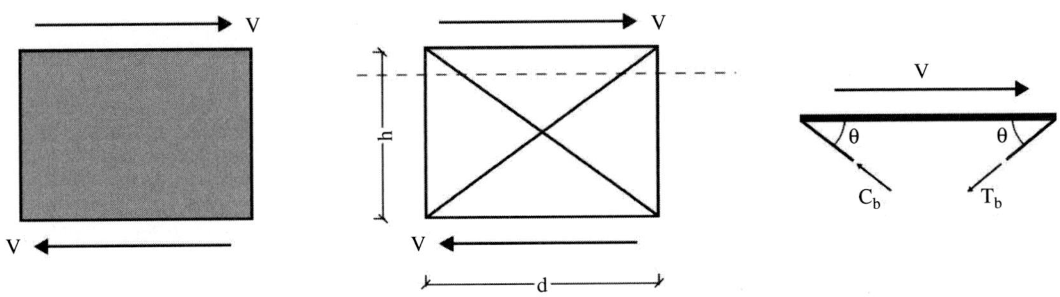

Figure 18.9 Shear reisted by brace.

x-bracing diagram as shown in Fig. 18.9, from the force equilibrium in X and Y direction a,

$$\sum F_x = V - C_b \cos(\theta) - T_b \cos(\theta) = 0 \tag{18.1}$$

$$\sum F_y = C_b \sin(\theta) + T_b \sin(\theta) = 0 \tag{18.2}$$

From Eqs. (18.1) and (18.2), the tension and compression forces in the brace can be written as

$$T_b = C_b = V / 2\cos(\theta) \tag{18.3}$$

Further discussion on optimal brace configuration is presented in Sec. 18.7.6 on diagrid systems. The John Hancock Center was one of the very first structures built using the braced tube concept. It is a 100-story high-rise building with a height of 457 m (1500 ft). As shown in Fig. 18.10, the column stresses are relatively uniform under both wind load and gravity, suggesting little shear lag and confirming the concept of braced tube. The total structural steel quantity is 29.7 lb per square foot of floor area, which roughly corresponds to that of a traditional 40-story structure. The system is so efficient that the single tall structure was built for the same price as the original proposal of two shorter towers. Integrated to the overall architecture, the perimeter braces also make the tower an iconic component of the Chicago skyline.

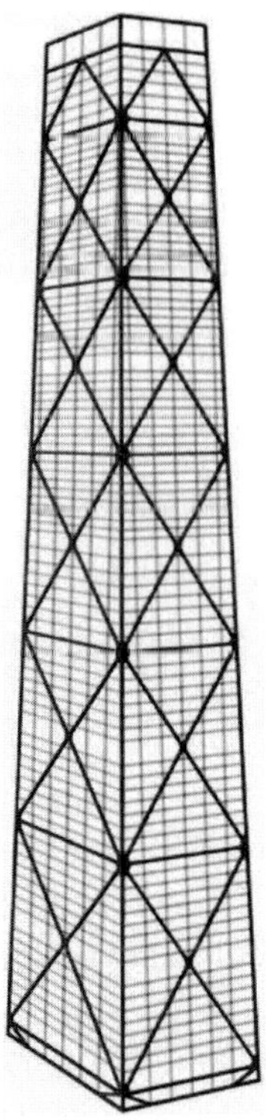

Figure 18.10 John Hancock building perimeter bracing structure.

Bundled Tube

When the total area of each floor in a tall building increases to a certain degree, interior columns are required to maintain efficiency of the floor framing system. A bundled tube system utilizes these interior columns as part of the lateral system—the tower consists of several modular framed tubes linked together such that all of the building's columns, both interior and exterior, contribute to the overall stiffness.

The Willis (formerly Sears) Tower, shown in Fig. 18.12, is a 110-story, 1450-ft (442 m) tower that was the tallest in the world from 1974 to 1998. It consists of nine bundled tubes, each a 75 × 75 ft square with columns spaced at 15 ft on-center on each face. Figure 18.11 shows a schematic view of column force distribution within the structure, showing the participation of interior columns and elimination of the shear lag effect.

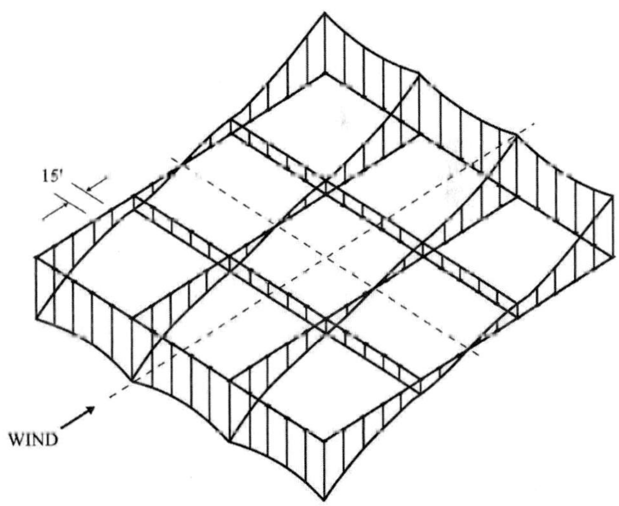

Figure 18.11 Axial force distribution within the tube column members (Iyengar, S.H.).

Tube in Tube

The tube-in-tube system, also known as "hull and core" system, links interior and exterior tubes through rigid floor diaphragms. Generally, the majority of the gravity and lateral loads are resisted by the exterior tube, and the interior tube houses building services and vertical transportation. If the exterior tube is a frame structure, the shear deformation is restrained by the cantilever deformation of the generally more

Figure 18.12 Photos of Sears Tower. (Photo © Hedrich Blessing.)

rigid shear wall core (Fig. 18.13), reducing overall building deflection and interstory drifts.

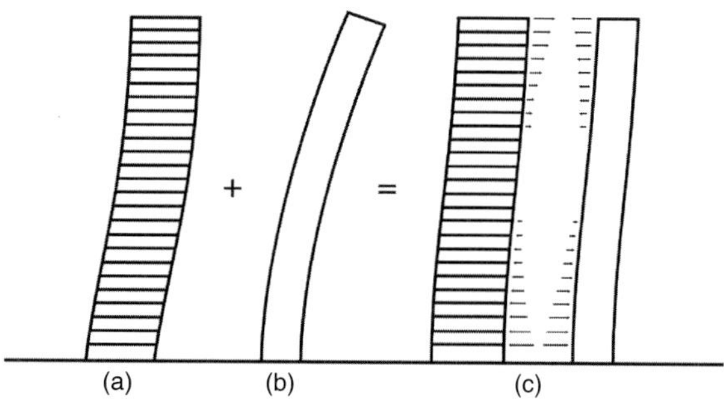

Figure 18.13 (*a*) Frame shear deformation; (*b*) cantilever deformation; (*c*) combined system deformation.

China Resources Headquarters in Shenzhen, China, is an example of this type of structure. It has 66 stories with a height of 392 m (1287 ft). The lateral structural system consists of an exterior framed tube and an inner concrete core. The exterior framed tube is composed of closely spaced columns and ring spandrels with transfer truss at the bottom and hat truss at the top. Figure 18.14 shows a construction photo of the tower.

Figure 18.14 Photos of China Resources Headquarters in Shenzhen, China. (Photo from skyrisecities.com.)

18.7.4 Outrigger Systems

Outriggers are commonly used in tall building systems. Generally, an outrigger is a rigid member that links the core and perimeter systems together to provide increased resistance to lateral loading and controls building motion under those loads. By linking the inner and outer structure together, the base overturning moment can be efficiently resisted by the large lever arm of the perimeter structure, while the core can resist shear through its relatively large area. The total core base moment can be reduced by the total amount of restraint moment from each outrigger, or $\Delta M = \sum M_i$, where i is the ith outrigger in the system. The most effective location for outriggers depends on the relative stiffness and configuration of the structural components but is generally toward the base where the moment is larger. Considering the outrigger diagram shown in Fig. 18.15, the system moment and deflection can be written as

$$M(z) = M_c(z) + M_0(z) \tag{18.4}$$

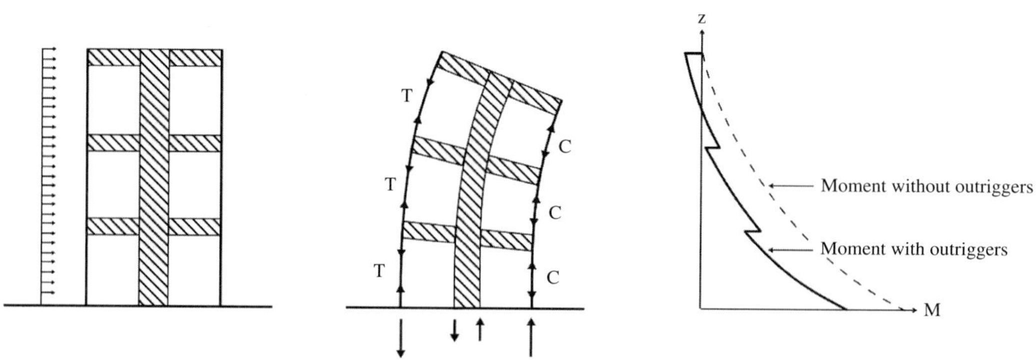

Figure 18.15 Interaction of core and outriggers.

$$\theta(z)=\theta_c(z)+\theta_0(z), \text{ where } \theta=\int \frac{M}{EI}dz \quad (18.5)$$

$$u(z)=u_c(z)+u_0(z), \text{ where } u=\int \theta dz \quad (18.6)$$

If the core and outrigger are idealized as two-bar trusses with flexural stiffness $I = 2Ad^2$, the following relationships can be derived:

Use α to denote the depth ratio of outrigger and core, $\alpha = \frac{d_o}{d_c} > 1$,

and β to represent the equivelent area ratio of outriger and core, $\beta = \frac{A_0}{A_c}$.

The total system moment of inertia can be written approximately as

$$I = I_c + I_o, \text{ where } I_c = 2 \times A_c \times d_c^2/4 \text{ and } I_o = 2 \times A_o \times d_o^2/4 \quad (18.7)$$

or

$$I = I_c\ (\alpha^2\beta + 1) \quad (18.8)$$

For small outrigger area ($\beta \approx 0$) or small outrigger depth ($\alpha \approx 0$), the outrigger effect is negligible. To effectively reduce the system moment, the outrigger must be stiff enough to generate the restraint moment. It also suggests that increasing outrigger depth is a more efficient way to achieve higher composite system stiffness than increasing outrigger area. Though the total restraint moment from all the outriggers acts to reduce the overturning moment on the core, increasing the stiffness of each outrigger might be more efficient than adding the number of outriggers since the outrigger effect is related to the cube of the outrigger stiffness.

Though the location and topology of each element outrigger has an effect on overall building efficiency, the "optimal" configuration will often only exist in theory due to architectural restraint. Most modern tall buildings have refuge floors at intermediate levels or MEP zones distributed throughout the building height. These locations are where outrigger trusses or walls are typically located.

It is also crucial to consider transfer of gravity loads through the outrigger system. The demand on outrigger trusses is highly susceptible to construction sequence, and unique details and means of construction are often used to minimize excessive gravity loading due to differential settlement between the core and perimeter. It is especially important to take time-dependent behavior into account when steel elements are connected via outriggers to concrete elements, which undergo long-term deformation (creep and shrinkage).

Examples of different outrigger systems in existing tall buildings are given in the following subsections.

Outrigger Systems with Megacolumn

It can be beneficial to connect outrigger trusses to large megacolumns, which also carry a large portion of the tower gravity load. The large megacolumn area lends stiffness to the outrigger system, and it is possible that the gravity demand on the perimeter will overcome the tension due to overturning moment.

The Jin Mao Tower was one of the first supertall buildings in China. The primary structural system of the 421-m (1380 ft), 88-story tower is a reinforced-concrete core connected to eight perimeter composite mega-columns by steel outrigger trusses at three locations, roughly one-third and two-thirds of the way up the height of the building and below the structural steel spire at the building crown.

The tower is relatively slender, with an overall aspect ratio of 7:1 to the highest occupied floor and 8:1 to the top of the spire. The outrigger trusses act as levers to engage the large, composite, perimeter columns and maximize overall structural resistance to overturning. The central core is a closed cross section, providing torsional stiffness to the system.

Each outrigger truss (shown in Fig. 18.16) is two stories deep and in a K-brace configuration, an efficient topology for a steel outrigger truss. Each composite megacolumn varies from 5000 × 1500 mm with C60 concrete (f'_c = 8000 ksi) at the base to 1000 × 3500 mm with C40 concrete (f'_c = 6000 psi) below the steel crown structure.

2 International Finance Center in Hong Kong consists of a similar arrangement of concrete core, embedded steel outrigger truss (Fig. 18.17), and composite perimeter megacolumns. Both towers adopted unique measures to mitigate differential settlement between the core and the perimeter megacolumns—Jin Mao Tower with a pinned system allowing outrigger to freely displace during construction and be fixed when the tower is complete and IFC2 with removable packing shims to allow relative motion between the outrigger trusses and perimeter structure.

Outrigger System with Belt Truss

For towers with irregular geometry or without perimeter megacolumns, a belt truss around the perimeter of the tower at the outrigger locations is utilized to link the perimeter columns together and ensure that redistribution of lateral demand is possible between the different outrigger trusses.

The Nanjing Zifeng Tower is a 450-m (1476 ft), 70-story mixed-use development of office and hotel located in the southeastern area of China. The primary lateral system is comprised of an interior reinforced-concrete "supercore" shear wall system and exterior composite columns. The supercore's perimeter has a closed form and provides a large amount of the overall torsional stiffness of the building.

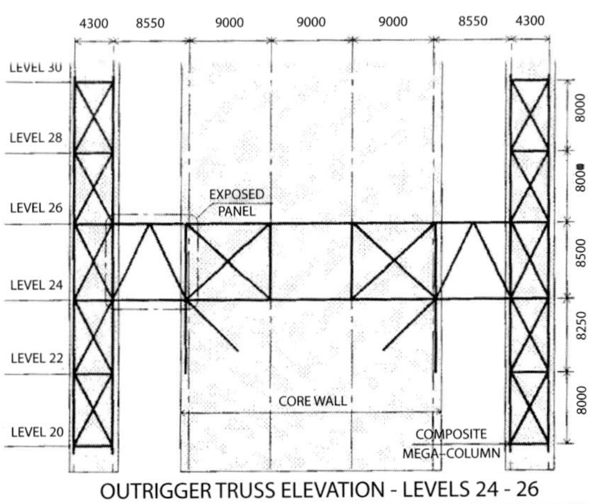

Figure 18.16 Elevation of outrigger truss system, Jin Mao Tower.

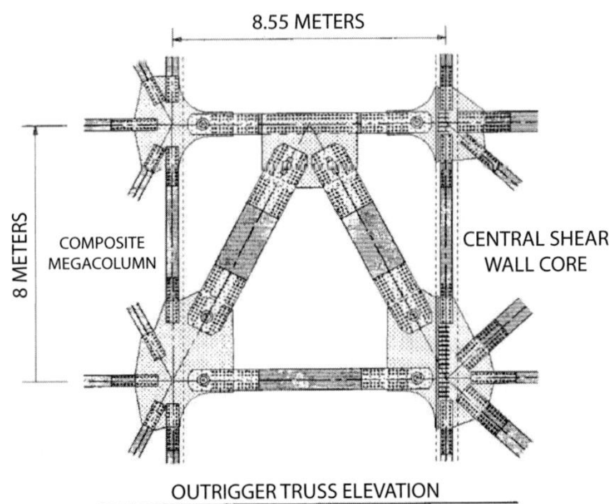

Figure 18.17 Construction of 2 International Finance Center, showing outrigger trusses.

Five outrigger trusses are aligned with the "web" walls of the core and are aligned with the internal web walls of the core. The belt and outrigger truss engage all of the perimeter columns as shown in Fig. 18.18.

A design consideration for composite core-outrigger buildings is whether the outrigger truss is simply attached to the concrete core or is entirely embedded—for the Nanjing Zifeng Tower the outrigger truss continues entirely through the core, while for the 254-m (833 ft) Hyundai-Hyperion Project Tower, only the truss chords are continued through the concrete core (Fig. 18.19). Both towers include belt trusses at all outrigger levels to tie the perimeter columns together.

Ladder Core System

Instead of connecting the perimeter columns to the core at discrete locations, Shum Yip Upperhills Tower 1 utilizes a composite coupling beam at each floor, forming a "ladder" system and distributing story shears on a more continuous basis (Fig. 18.20). Each coupling beam is aligned with a core wall and perimeter megacolumn, providing a unique structural system with an efficient, ductile load path.

Virtual Outrigger System

Similar to the ladder core system, a "virtual outrigger" (Fig. 18.21) can be utilized to engage the perimeter structure without outrigger trusses interrupting intermediate space. Thickened floor slabs transfer wind shear from the perimeter to the core, and the rigid belt wall prevents relative displacement at the top and bottom of the virtual outrigger story, creating a restoring moment on the tower core resisted by the columns below.

This system was utilized on the W Hotel Kuala Lumpur, a slender 245-m (804 ft) tower in Malaysia. The system allows for coupling of the core and perimeter structure without large trusses and the associated space planning issues within the mechanical zone. The belt wall also functions as a transfer girder, separating different perimeter column grids associated with different building programs above and below.

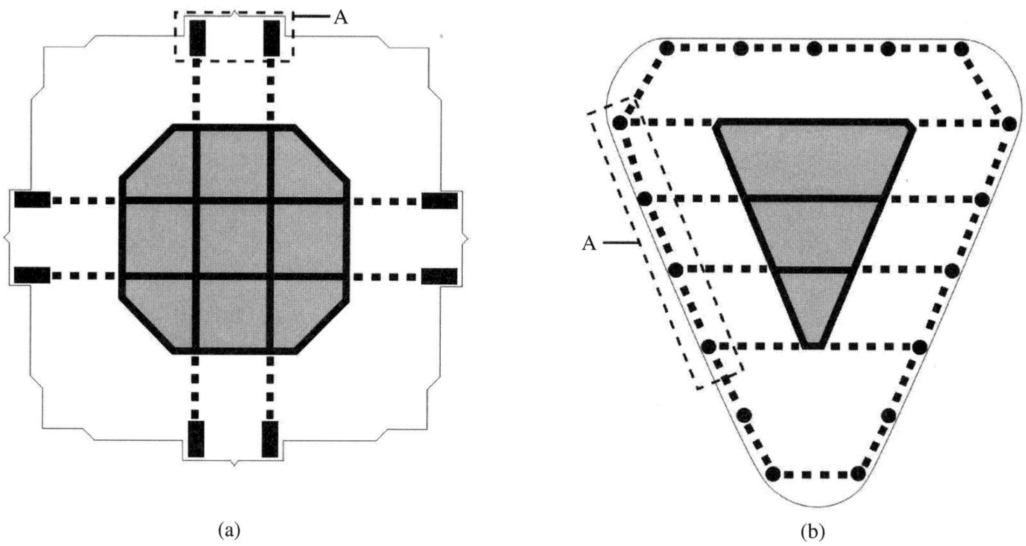

Figure 18.18 Outrigger configuration for (*a*) Jin Mao Tower and (*b*) Nanjing Zifeng Tower.

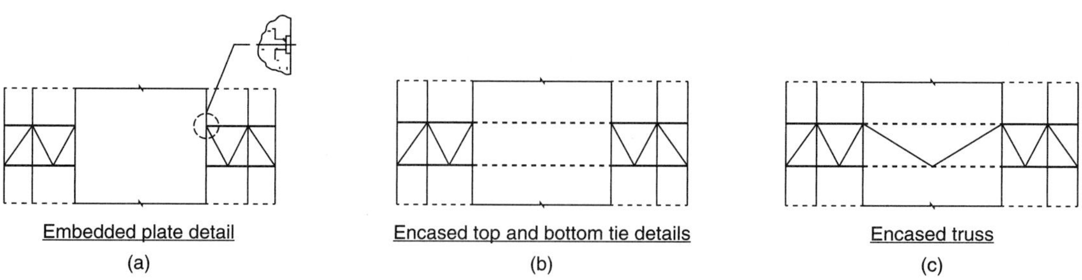

Figure 18.19 Schemes for connection of steel outrigger truss to concrete core. Nanjing Zifeng Tower uses (*c*), and Hyundai-Hyperion Tower uses (*b*).

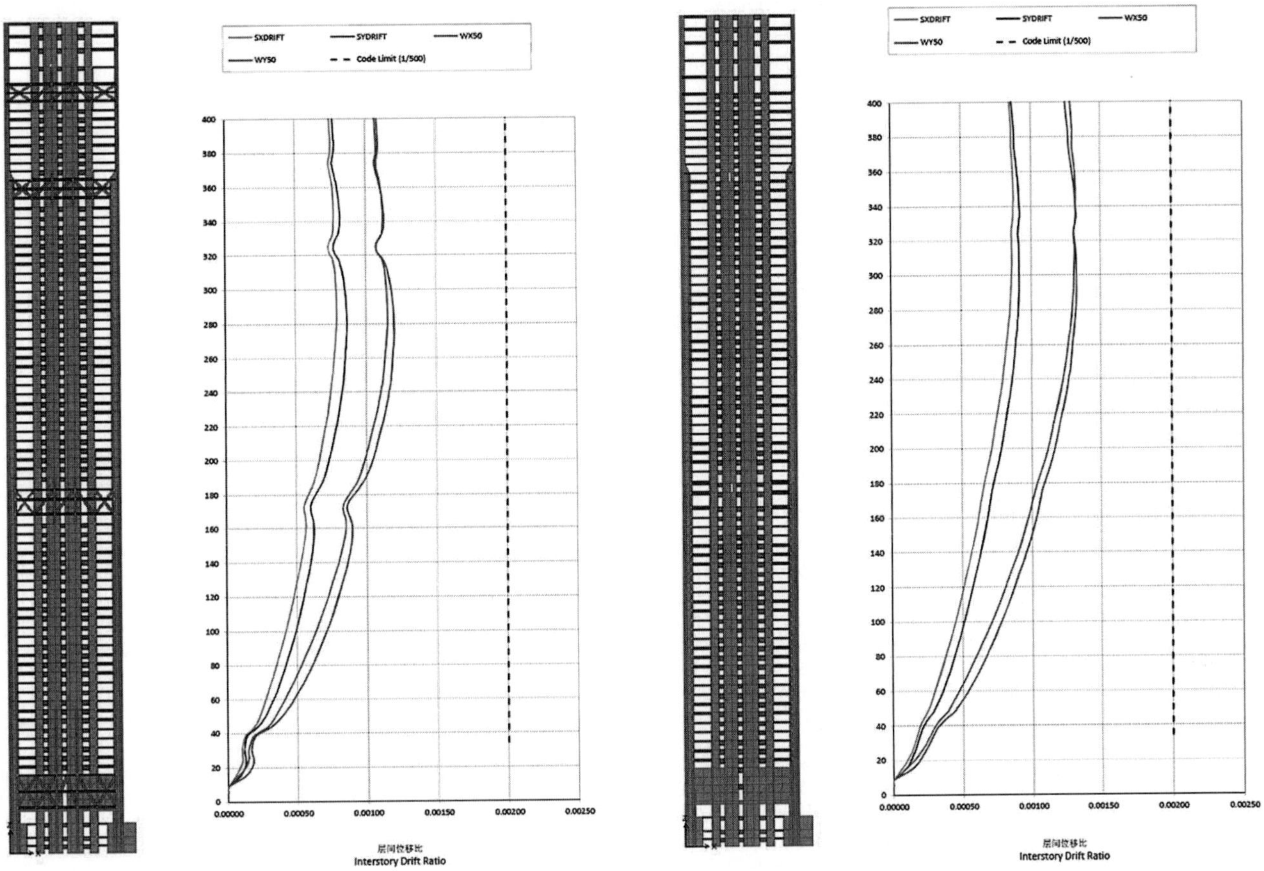

Figure 18.20 Lateral system performance comparison between a traditional outrigger system and a ladder system (Besjak, 2014).

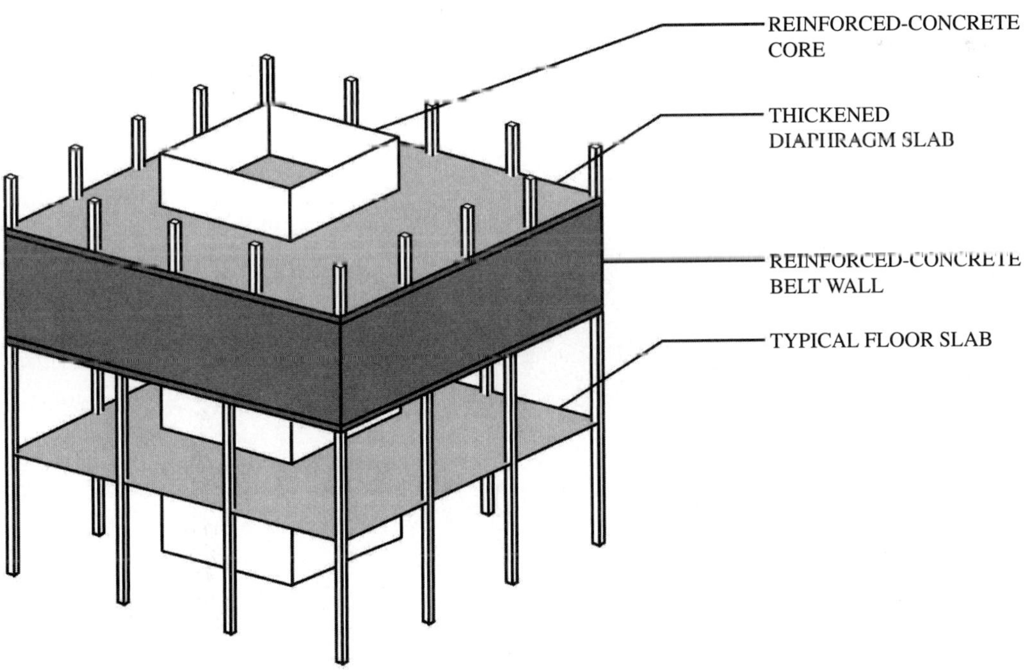

Figure 18.21 Concrete virtual outrigger system, W Hotel Kuala Lumpur.

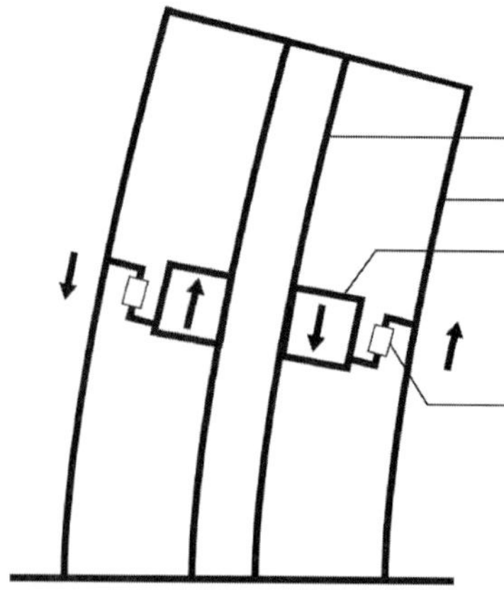

Figure 18.22 Example of damped outrigger.

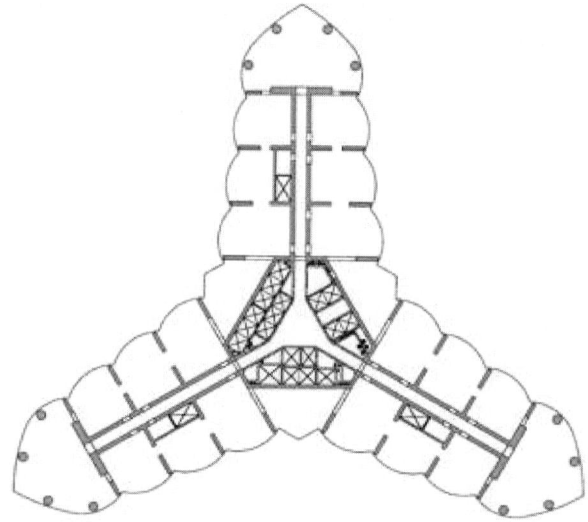

Figure 18.23 Burj Khalifa floor plan.

Outrigger Systems with Damper

An alternative method of resisting lateral load with outriggers is to increase inherent damping in the structural system through a damped connection between the core and perimeter.

Instead of rigidly connecting the core to the perimeter structure, the damped outrigger system allows differential motion between the outrigger and perimeter column, causing the damper to stroke and provide a damping force. This system produces a smaller increase in stiffness than a traditional rigid outrigger but a significant increase in inherent structural damping. Higher damping will result in lower dynamic response to wind loading and also can be designed to dissipate energy during an extreme seismic event.

Although this type of system usually comes with additional costs for testing, installation, and maintenance, often these extras can be offset by savings from the reduced lateral demand on the structural system.

The damped outrigger system (Fig. 18.22) was developed by Arup and first incorporated in the design of the St. Francis Shangri-La Place development in Manila, the Philippines, a 217-m-tall, all-concrete tower located in an area that is subject to both very strong wind and active seismic activity. The damped outriggers provided 7.5 percent supplementary damping, significantly reducing wind and seismic design forces.

18.7.5 Buttressed Core

The buttressed core system, generally used for tall concrete buildings, provides large structural stiffness through the arrangement of the structural elements and can be achieved with conventional materials and construction techniques. The essence of the buttressed core system is a tripod-shaped structure with a closed central core, providing the torsional stiffness of the structure. Each of the buttress walls provides shear resistance and a large increase in the moment of inertia due to their large footprint.

While first used in the 264-m (866 ft) Samsung Tower Palace 3 in Seoul, the buttressed core system was most notably used in the world's current tallest building, the 828-m (2717 ft) Burj Khalifa. With over 160 stories and 280,000 square meters of usable space, it surpassed the existing tallest building in the world by 60 percent upon its completion due to the highly optimized core wall layout (Fig. 18.23) and overall architectural form of the tower.

18.7.6 Diagrid System

Three-dimensional exterior bracing systems, or diagrids, offer an alternative to outrigger systems for buildings of comparable height. The diagrid system is designed to utilize a diagonally braced perimeter to maximize the benefit of the outermost building dimensions while also using the perimeter structure to resist gravity loading. The diagrid system can be thought of as a version of the framed tube where vertical perimeter elements are entirely removed (Fig. 18.24).

Figure 18.24 Evolution of bracing systems in tall buildings.

While a regular grid of diagonal perimeter members is the purest expression of the diagrid structure (and provides for simple and repeatable node geometry), it can be shown that the optimal inclination of perimeter bracing members varies by distance to the fixed base. This distribution mirrors the internal principal stresses of a continuous cantilever subject to a uniform bending load (Fig. 18.25)—members "want" to be more vertical at the base, where axial stresses and flexural deformation are dominant, and tend to the horizontal at the free end (top of building), where shear stresses are dominant.

A challenge of diagrid systems is that perimeter members are typically long, slender, and unbraced and must be designed not to buckle in rare wind and seismic events. Typically these structures require explicit performance-based design evaluations to assure safety and code equivalence. Diagrid structures tend, as a necessary consequence, to perform more elastically and with less energy dissipation in large seismic events than the more conventional structural systems.

Diagrids also introduce the challenge of node design and analysis to the building design process—the performance of these connections is integral to the performance of the structure as a whole, and intensive analysis and often scale modeling and testing are required. Significant economy can be achieved if nodes are modularized.

Poly International Plaza in China's capital, Beijing (Fig. 18.26), is an example utilizing this type of system. Poly International Plaza is 162 m (532 ft) tall and has an elliptical shape and faceted exterior. The exterior diagrid utilized concrete-filled steel tubes and creates a unique column-free interior space and a light-filled spatial experience combined with a concrete shear wall core. A rigorous testing regime was carried out

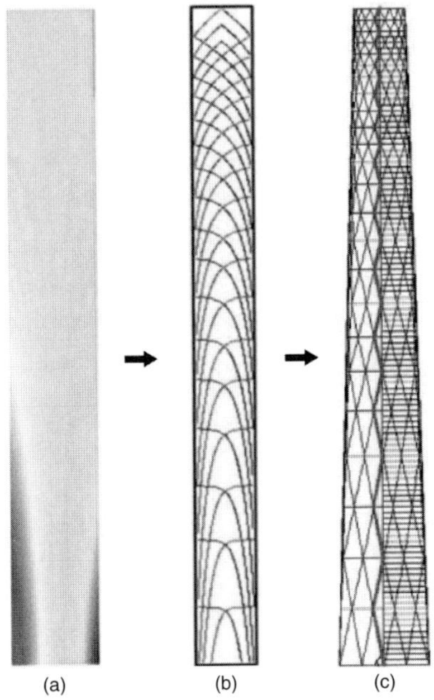

Figure 18.25 (*a*) Continuum stress in cantilever beam, (*b*) principal stress directions, and (*c*) optimized diagrid shape.

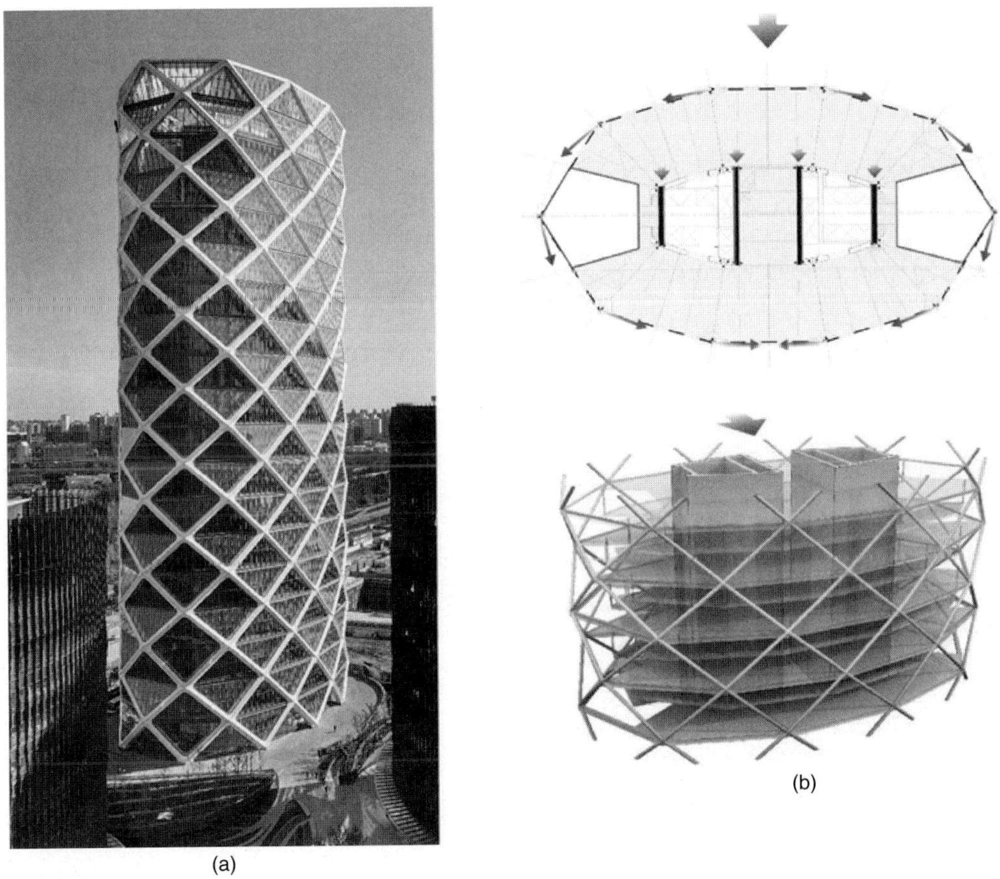

Figure 18.26 (*a*) Construction photo of Poly International Plaza, Beijing, and (*b*) diagrid analysis and typical floor plan.

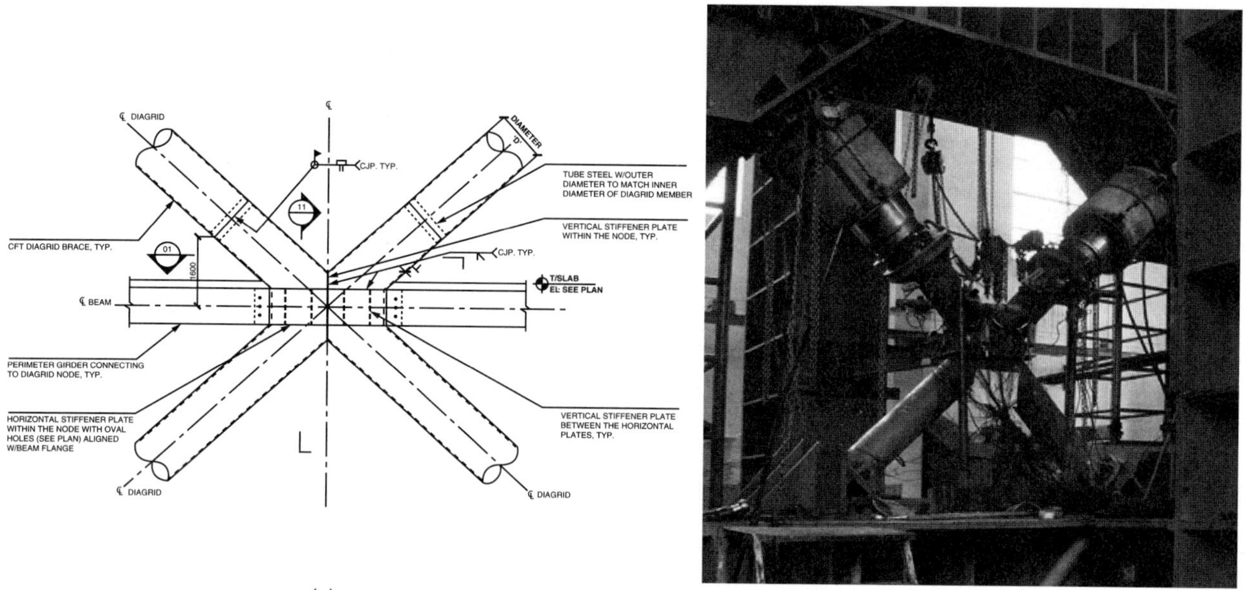

(a) (b)

Figure 18.27 (*a*) Typical diagrid node detail and (*b*) laboratory test of typical diagrid node.

on the diagrid nodes (Fig. 18.27) to confirm accuracy of finite-element modeling and the necessity of concrete fill.

18.7.7 Superframe

The superframe structure, conceptualized by Fazlur Khan (Fig. 18.28) but not realized in his lifetime, is a tall building typology which utilizes large perimeter elements to allow the full perimeter footprint to be utilized and horizontal trusses at discrete levels to distribute gravity loading to the corner frames and allows for a modular building form, potentially with unique interior column layouts. The superframe concept was originated as a proposed 2100-ft building in Chicago consisting of three concentric tube structures and has since been realized in the designs of two supertall towers in China.

The Guizhou Culture Plaza Tower in Guiyang (Fig. 18.29) captures the efficiency of both the superframe concept and the tapered diagonal braced tube. The proposed tower is 521 m (1709 ft) and consists of eight 16-level structural modules. Superframe girders transfer gravity loads to superframe columns at the four corners, maximizing resistance to overturning moment and minimizing net tension under wind load. The superframe columns are open, built-up lattice structures, creating chamfered corners which along with the overall tapered form of the tower minimize dynamic effects of the wind on the tower. The superframe girders also serve to dissipate energy in a seismic event through frame action, and the perimeter structure is also connected to a reinforced-concrete floor through thick slabs acting as virtual outriggers on the superframe girder floors.

The Tianjin Goldin Finance 117 Tower (Fig. 18.30) is a 117-story, 597-m (1959 ft) tower located in Tianjin, China which utilizes a braced superframe structural system as well as a reinforced-concrete core similar to the Guizhou Culture Plaza Tower.

Compared to the Guizhou Culture Plaza Tower, the Finance 117 Tower utilizes reinforced-concrete megacolumns and a braced perimeter megaframe structure. Belt trusses are also used at intermediate locations to enhance the torsional stiffness of the tower and transfer gravity loading to the corner megacolumns.

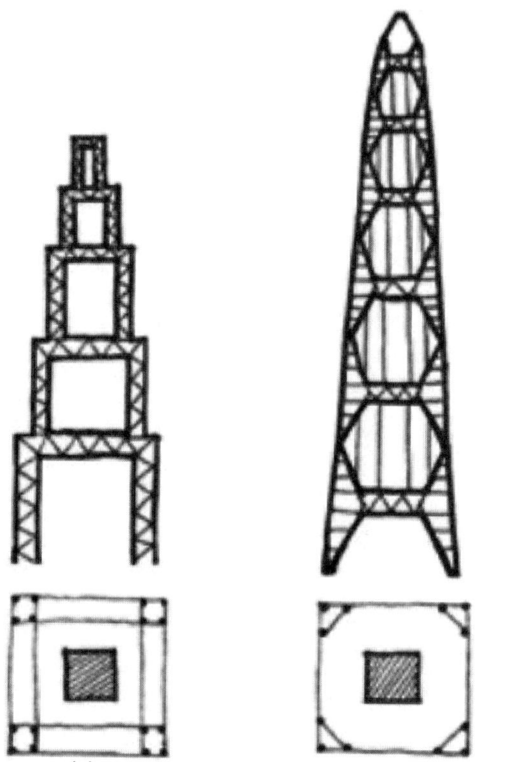

(a) (b)

Figure 18.28 (*a*) Superframe structural system and (*b*) articulated superframe structural system (Luo, Y., 2015).

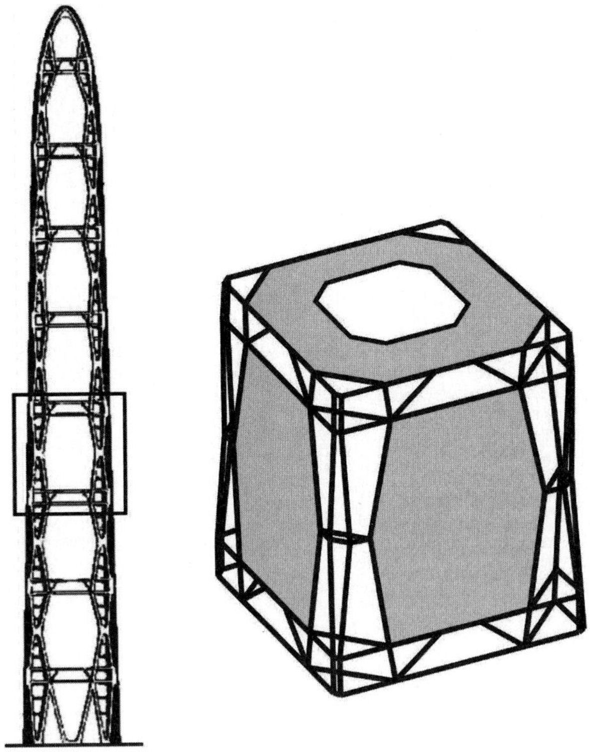

(a) (b)

Figure 18.29 (*a*) Guizhou Cultural Plaza Tower and (*b*) three-dimensional view of superframe module.

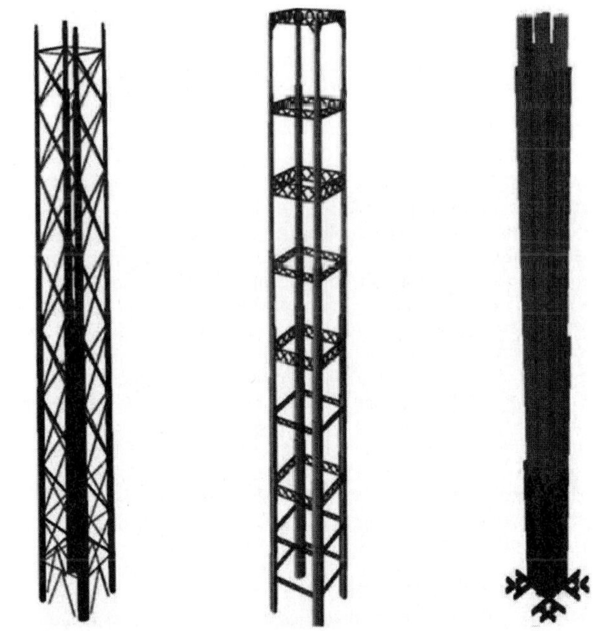

Megaframe Tube (a) | Megaframe (b) | Core (c)

Figure 18.30 Tianjin Goldin Finance 117 Tower structural system (Lee, 2012).

18.8 ANALYSIS PARAMETERS AND TECHNIQUES

Modeling techniques and assumptions used by the structural engineer can have major impacts on the results of the analysis models which can affect:

- The structural properties provided to the wind tunnel for determining structural wind loads and accelerations and the subsequent results
- The calculated seismic loads as a function of the building's modal response
- The distribution of gravity and lateral loads throughout the structural system

The assumptions on the appropriate building mass to be considered for determining the structural wind loads and associated building accelerations can play a significant role in the results provided by the wind tunnel. Similarly, the building mass used for seismic analysis will have an impact on the associated seismic effects.

A number of modeling considerations and assumptions will be discussed below.

18.8.1 Primary Lateral and Non-Lateral Load Resisting-Elements

The structural engineer must know which structural elements are intended to resist the lateral loads on the building and which are intended to be gravity load-resisting only (nonparticipating) elements. With these in mind, the analysis model then needs to be developed to limit and/or account for lateral loads which may be attracted to any designated nonparticipating elements.

As a general rule for the design of lateral systems for tall buildings, simplified structural models are preferred to allow for a clear and understandable load path. However, the advent of more powerful computers and user-friendly structural analysis software has made it possible for structural engineers to easily add great levels of detail to analysis models. Simultaneously, the use of BIM modeling and ability to link to structural analysis models has increased the demand for all structural elements to be included for coordination purposes with other disciplines. While this capability may be beneficial in some respects, it can also make it difficult to clearly understand the load path(s) for the distribution of the wind and seismic loads throughout the structure. On the other hand, oversimplification of the model can be problematic if it leads to oversights related to improper accounting of weight and mass.

Including all structural elements in an analysis model without proper precautions may allow what are considered gravity load-only elements to attract lateral loads that may not be accounted for if those elements are subsequently to be designed separately in another software package considering only the gravity loads effects. A common example of this would be a two-way concrete slab system which is modeled in the overall lateral model but is designed in another software package. The application of proper degree of freedom releases at the interfaces of any gravity-only elements or the use of stiffness modifiers can limit or eliminate the potential for lateral loads to be attracted to them.

It is not uncommon for there to be multiple overall structural models of a tall building with varying assumptions on the interaction between elements related to releases, stiffness modifiers, foundation supports, etc. to evaluate strength and serviceability considerations. A possible scenario for multiple models of an all-reinforced-concrete tower with a core wall system and the two-way floor slab system mentioned above would be the following:

- Model 1: Strength model with greatly reduced slab flexural stiffness to eliminate most of the potential frame action with the supporting columns forcing all the lateral loads to be taken by the core wall system. The core walls would be designed from this model.
- Model 2: Strength model with a slab flexural stiffness based on expected levels of cracking under design gravity loads. This would allow some of the applied laterals loads to be resisted by the frame action of the slab and columns and more accurately reflect how the loads would

be distributed based on stiffness. The columns would be designed from this model. The floor slabs could be potentially designed from this model as well, or the additional shear and moments from the lateral loads could be added to the gravity effects calculated in the slabs from a separate analysis and design program.

- Model 3: Serviceability model with similar assumptions to Model 2. This model may be used to provide modal information to the wind tunnel for determining structural wind loads and expected accelerations. This could also be used for checking building deflections under wind loads. A potential issue with using Model 1 for these purposes might be excessive torsional response depending on the core wall layout and location.

The important point is that the structural engineer takes care to ensure that the lateral loads are properly accounted for and considers how the modeling assumptions affect the flow of these loads and the potential effects on elements that were not considered part of the primary wind-resisting system.

18.8.2 Building Mass Considerations

The wind loads and accelerations determined by the wind tunnel can be very sensitive to the modal information provided by the structural engineer. In general, a longer building period results in larger wind effects, while the opposite is generally true if that period is shorter. Similarly, seismic loads calculated by either simplified static approaches or more rigorous methods such as response spectrum analysis are directly proportional to the seismic mass. Therefore, proper accounting of the expected mass of the building is very important.

Self-weight of the structural elements of the building must be considered and are generally determined automatically by the structural analysis software and can be known with a fair degree of certainty. A caveat to this would be cases where some structural members are left out of the model for load path simplification, requiring separate accounting for their mass and distribution. The amount of expected SDL from floor finishes, partitions, ceilings, building services, and façades as well as the LL are less certain. Generally, the values of SDL and LL used for strength design are conservative, especially when considered over all floors of a tall building and for the duration of a design-level wind event.

Engineering judgment along with an understanding of the expectations of the architect on levels of finishes etc. is needed to determine a reasonable percentage of the design SDL to consider for modal analysis results provided to the wind tunnel or used for seismic considerations. This can vary depending on the building usage and the specifics of the finishes, but it is not uncommon to use 50 to 75 percent of the design SDL for wind load. A similar range may be appropriate for seismic analysis, but the sensitivity of the particular building to associated period changes and seismic effects should be studied.

The LL measured in actual buildings has been found to be substantially below the code-prescribed design LL. For modal analysis, the percentage of design LL used is typically low for office and residential floors, especially if the partitions have been accounted for in the SDL. Common ranges are 0 to 25 percent for wind considerations. Many codes have specific requirements regarding LL contribution to seismic weight that must be followed. An exception would be the LL on dedicated mechanical floors where 80 to 100 percent of the LL or the actual equipment weights should be assumed for wind and seismic evaluation.

18.8.3 P-Delta Effects

As tall buildings displace laterally under applied loads, the gravity loads act at an eccentricity to the undisplaced shape, creating secondary member forces and moments throughout the structure which in turn lead to more displacement and load effects until equilibrium is reached. These are known as P-Delta effects. It is very important for these to be considered in the design of the tall building's structural members to ensure adequate strength and stability.

P-Delta effects can be evaluated in two primary ways:

- The first option is to use the built-in P-Delta capability of most commercially available finite-element programs which can iterate until equilibrium is reached on the deflected shape and provide the corresponding member forces and moments generated as a result. This is in most cases an option that must be activated and defined in the program.
- The second option is to use approximate methods outlined in codes such as AISC and ACI which apply magnifiers to results from a first-order analysis.

In either option, it is important that the load factors applied to the gravity loads are consistent with those required in combination with the applied lateral loads causing the displacements in the first place. Refer to appropriate load combinations in the applicable codes for the project.

The interstory drifts and associated overall building displacements discussed earlier in this chapter will drive the magnitude of P-Delta effects. The more flexible a tall building is, the more impacted it will be.

While this may vary somewhat between engineering firms, common practice for using P-Delta is generally as follows:

- P-Delta "on" for evaluating building periods
- P-Delta "off" for evaluating service-level wind drifts
- P-Delta "on" for evaluating seismic drifts
- P-Delta "on" for evaluating strength and stability effects

It should be noted that wind and seismic loads will generally produce most P-Delta effects, but it is possible for buildings to lean under their own weight, depending on geometry and structural stiffness, which can create additional P-Delta effects to be considered.

18.8.4 Effective Stiffness for Reinforced-Concrete Elements

Reinforced-concrete structures are prone to cracking which will lead to reductions in the stiffness of the cracked elements. Accounting for this potential loss of stiffness is very important for tall buildings utilizing reinforced-concrete lateral systems for evaluating both serviceability effects and load effects.

- Serviceability: overall building displacements, interstory drifts, and accelerations can all be affected by stiffness reductions in the elements comprising the lateral system.
- Strength: distribution of loads throughout the lateral system and P-Delta effects can be impacted by stiffness reductions in the elements comprising the lateral system.

These stiffness reductions can be considered in three different ways with varying degrees of accuracy, effort, and/or required software capabilities which will be explored in more depth in the following subsections.

Implicit Approach #1—Simplified Method

The *simplified method* is the most common approach that has been used for years due to its ease of application in the analysis. Stiffness reduction factors are applied to the member properties of the lateral system based on the element type (columns, beams, walls, etc.) and are used throughout the structure.

ACI 318 has for many editions contained recommended stiffness reduction factors, which are now contained in the section on evaluating second-order effects. Others have been proposed as well in technical papers and other ACI documents. Table 18.1 taken from ACI Committee 375 document SP-240 provides a comprehensive set of service level and strength level stiffness reduction factor recommendations. Note that the majority of the reductions are related to bending of the elements with a few area reductions for axial or shear-controlled elements.

The simplified method will generally lead to a safe, conservative design but inherently also leads to design inefficiencies due to the blanket assumptions on all elements.

Implicit Approach #2—Iterative Method

While the simplified method applies the same reduction factor to all elements of the same type in the structure, the *iterative method* looks at elements of the same type separately to determine the level of cracking under various load levels. For example, it is unlikely that all walls in

Table 18.1 Recommended Stiffness Modifiers from ACI 375 SP-240

Issue	Analysis I serviceability limit state	Analysis II ultimate strength limit state	Analysis II stability under sustained loads limit state
Summary of analysis parameters			
Lateral Load Intensity	(1) 10 yr Wind	50 yr Wind	(2) $F_x = 1.0$ kips $F_y = 1.0$ kips $T_e = 1.0$ k-in
Member properties			
Column flexure	$1.0\ I_g$	$0.70\ I_g$	$0.70\ I_g$
Column Axial	(3) A_{ac}	(3) A_{acr}	(3) A_{acr}
Mild Reinforced Beams	(4) $0.50\ I_g$	(4) $0.35\ I_g$	(4) $0.35\ I_g$
Post-Tensioned Beams	(5) $0.67\ I_g$	(5) $0.45\ I_g$	(5) $0.45\ I_g$
Mild Reinforced Flat Slabs	(6) $0.36\alpha\ I_g$	(6) $0.25\alpha\ I_g$	(6) $0.25\alpha\ I_g$
Post-Tensioned Flat Slabs	(7) $0.48\alpha\ I_g$	(7) $0.33\alpha\ I_g$	(7) $0.33\alpha\ I_g$
Shear Walls Cracked Uncracked	(8) $0.50\ I_g$ $1.00\ I_g$	(8) $0.35\ I_g$ $0.70\ I_g$	(8) $0.35\ I_g$ $0.70\ I_g$
Shear Deformations	(9) A_{vg}	(9) A_{vg}	(9) A_{vg}
Modulus of Elasticity	(10) $E = E_c$	(10) $E = E_c$	(11) $E = \dfrac{E_c}{1+\beta_d}$
Building weight to be used for P-Δ analysis purposes	(12) Sustained Building Weight	(13) 1.2D+*fL*	(13) 1.2D+1.6L
Beam-Column Joint Stiffness	(14) 50% Rigid	(14) 50% Rigid	(14) 50% Rigid
Acceptability Criterion	(15) Inter-story Drift $\Delta/h \le 0.0025$	(16) Inter-story Stability Index $Q \le 0.25$	(17) Inter-story Stability Index $Q \le 0.60$

a building will be cracked to the same extent give varying load levels, sizes, reinforcement ratios, etc. Different stiffness reduction factors would then be applied based on the expected cracking under the applied loads.

The effective stiffness of various structural elements can be calculated using well-documented approaches. These are briefly described below.

Beams and Slabs: Effective flexural stiffness can be calculated using the method outlined in ACI 318 based upon the applied moment and the cracking moment of the section under consideration. In cases where the slabs engage the columns to provide flexural resistance to lateral loads, care must be taken to define the appropriate effective slab width per ACI if an appropriately small finite-element meshing of the slabs is not undertaken in the analysis model.

Columns: Effective flexural stiffness can be calculated in a manner similar to beams and slabs with a modification to the cracking moment to account for axial compression. If a column is under a net tensile load, the cracking stress should follow the limits noted below under the walls.

Walls: Many structural systems rely heavily on concrete walls to provide a substantial portion of the lateral load resistance. Depending on the configuration of the walls, they may be more prone to flexural behavior like web walls connecting flanges in a core wall system or may function primarily as compression-tension elements like the flange walls in a core wall system. In either case the approach for columns may be used here as well. Rahimian (2011) provides a good discussion of effective stiffness for concrete walls and shows several ways of evaluating reduction factors with varying degrees of accuracy and effort. The stress at which walls will crack should be taken as lower than the common $7.5\sqrt{(f_c')}$ from ACI, which is intended for primarily flexural elements. It is common for tension in walls to be more uniform than in beams and slabs, leading to cracking at a lower stress level. Orackal and Wallace (2006) recommended a value of $6\sqrt{(f_c')}$ based upon experimental work.

This limit should be considered also for columns that experience significant tension.

As the name implies, the iterative method is not a one-step process. Typically, the analysis starts by assuming that all members to have full stiffness. After the initial analysis run, all members of the lateral system are checked to determine if cracking has occurred. For those that have, reduction factors are determined per the above discussion and applied. The analysis is run again, and another check is performed to look for cracking in additional elements, which is common due to load redistribution after the initial reduction factors are applied. Appropriate reduction factors are applied, and the analysis is run a third time. This process continues until no further cracking is observed. This process is performed for both service-level loads and ultimate/strength-level loads.

The iterative process is obviously a very time-consuming effort which is why the simplified method is often used.

Regardless of which implicit method is used, it is recommended that a service-level wind load analysis be carried out first with all members uncracked. Generally, it is preferable to have the lateral system essentially uncracked at service-level loads to more easily achieve serviceability criteria related to drifts and accelerations. This becomes particularly important for very tall and/or slender buildings. If cracking is found, it is prudent to use an iterative approach to look more closely at the elements that are cracking and the extent within the overall lateral system. Upsizing of some elements or rearrangement of the lateral system may be required or preferable to achieve better behavior.

When considering strength-level design, often there is not a huge consequence to the design efficiency in using the simplified method in lieu of the iterative method. At this design level, deflections and drifts are not usually a critical consideration like they are for serviceability criteria. The simplified method will tend to produce larger building movements, leading to additional P-Delta amplification. Unless the structure is sensitive to P-Delta effects, the penalty to the design efficiency is often minimal and may be justified for the ease of analysis. The structural engineer should study the P-Delta sensitivity of the structure before deciding on the approach to be used. Stability factors are discussed in ACI and elsewhere which consider the first-order and second-order movements and can be used to classify this sensitivity.

Explicit Approach

The *explicit method* makes use of the considerable processing power available now to structural engineers and requires analysis software capable of doing material nonlinear analysis. Rather than doing the cumbersome, time-consuming iterative approach described in the previous section, explicit method essentially automates that process.

The stress-strain relationship of the concrete as well as that of the reinforcing steel is needed as an input. A preliminary strength design is often needed so that the reasonable reinforcing ratios can be defined for each element. The setup for this method can be somewhat time consuming, but the analysis is done very quickly.

The explicit method can quickly and accurately predict the degree and extent of cracking throughout the lateral system and provide very useful information on the expected movements of the building. Care needs to be taken so that the structural members have not been stressed beyond their yield capacities since the concrete stress-strain relationship in that case would need to consider cyclic behavior strength degradation. This would commonly be associated with high seismic events. Refer to further discussion in Section 18.10 on performance-based seismic design.

As noted above, the implicit simplified method is currently still the most used approach, especially in wind-controlled regions. As taller and more slender buildings become more common, the explicit approach may become more common.

18.8.5 Floor Diaphragm Considerations

As with all building types, the floor diaphragms play a crucial role in moving the loads at each floor level due to wind and seismic events to the vertical elements of the lateral load-resisting system.

Diaphragms can be considered in three different ways for modeling:

- Flexible—loads are distributed only based on tributary area.
- Rigid—rigid body motion of the overall diaphragm is assumed, leading to load distribution based on stiffness of the elements composing the overall lateral system.
- Semi-rigid—the actual in-plane stiffness of the diaphragm is considered which will lead to the loads being distributed based on a combination of tributary area and on the stiffness of the elements comprising the overall lateral system. This relationship will vary based on the particular diaphragm properties.

In practice, flexible diaphragms are not applicable to most tall buildings, as solid concrete or concrete on metal deck is typically used for floor systems.

Rigid diaphragms are the most common and generally applicable for most tall buildings, with a few exceptions which will be noted below. These allow for reduced computational time and ease of modeling.

Semi-rigid diaphragms are necessary for unusual floor plate shapes which require a more complex load path for the loads in the diaphragm to be directed to the corresponding vertical elements which will then carry them down through the building. Some examples are diaphragms with large openings, diaphragms with reentrant corners, and diaphragms with large ratios between the orthogonal dimensions. Semi-rigid diaphragms are also required when changes in lateral system stiffness occur requiring the diaphragm to transfer loads through it to other vertical lateral load-resisting elements. Common examples of this would be reduction in the core wall system footprint as elevators drop off over the height of a building and ground level slabs where large portions of the lateral loads move from the above-grade lateral system out to the perimeter basement walls.

Special considerations are warranted when analyzing the loads being resisted by elements such as outrigger trusses/walls, belt trusses/walls, and perimeter bracing systems. The floor diaphragms will have a tendency to attract axial loads that would otherwise be assumed to be carried by these elements as a result of the high in-plane stiffness of most floor slab systems. If unaccounted for, the top chord of trusses, horizontal ties, and outer zones of outrigger/belt walls could be designed for unrealistically small loads, leading to a load transfer problem to the remaining elements of these systems. Rigid diaphragms should never be used at these floors and generally not for a few floors above and below these systems. If semi-rigid diaphragms are used, the in-plane stiffness properties should be modified to very low values. Some sensitivity studies using values on the order of 0.001 to 0.10 of full stiffness should be conducted to determine a value that does not lead to overall analysis run errors while reducing the axial loads in the diaphragms to very small values.

When providing mass distribution and modal information to wind tunnels for their modeling, many analysis programs will require rigid diaphragms at every floor in order to report the required floor information, such as mass moments of inertia, centers of mass, etc. A separate model would be used for this purpose.

18.8.6 Foundation Assumptions

Tall buildings can generally be thought of as large cantilevers, and the stiffness of the supports for these cantilevers can have a significant effect on the overall deflections and load distributions throughout the system. Considerations for the foundation flexibility and, in many cases, the stiffness and friction of the soil against the basement walls should be accounted for by the structural engineer.

Input by the project's geotechnical engineer will be required to estimate equivalent spring stiffness values that are appropriate for the foundation type being used based on soil conditions and foundation properties (material, length, etc.). They should also provide values for use in modeling the effects of the material behind the basement walls in dissipating the lateral loads on the tower above grade. Note it is common in projects with a few levels of basements for a great deal of the overall building shear and overturning moments to be taken out via the basement walls prior to reaching the foundation level.

Some rules of thumb that can reasonably be used when considering foundation effects are as follows (these too may vary between engineering firms):

• When evaluating periods and service level drifts, pinned-based models may be used, including horizontal restraints at each level and below grade when basements occur. This can be compared against a pinned-based model which stops at grade level (i.e., no basements).

• When evaluating strength and stability, foundation flexibility via vertical springs as well as horizontal springs on the basements walls should be considered.

18.9 PERFORMANCE-BASED DESIGN FOR WIND

Currently, the term "performance-based design" generally means something different when referring to wind as opposed to seismic demands. For wind this means designing the tall building to meet certain performance criteria related to building displacements and building motions while meeting the requisite strength and stability demands. Additionally, tall buildings are designed to remain in the elastic range when subjected to the expected wind loads.

It should be noted that initiatives are currently under way that are looking at approaches to treat wind design like that already used for seismic design, wherein rare wind events could push buildings into the inelastic range. However, implementation of this is likely several years into the future.

18.9.1 Nature of Wind Loads on Buildings

Wind loads on tall buildings can be broken down into along-wind effects and across-wind effects. Refer to Fig. 18.31 for a simple graphic depicting these.

As the name implies, along-wind refers to loads and movements parallel to the direction of the wind. Along-wind effects on tall buildings are fairly intuitive and consist of a mix of static components due to the mean wind speed and dynamic components from variations above and below the mean wind speed.

Across-wind effects are those which are manifest roughly perpendicular to the wind direction. These are less intuitive and the effects are

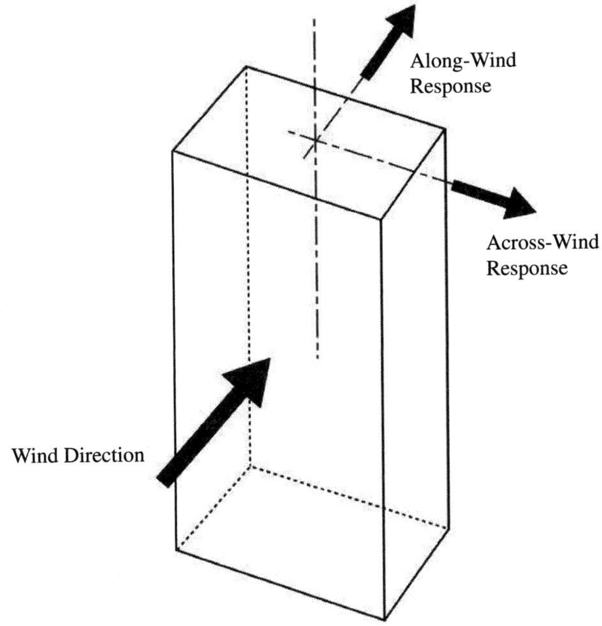

Figure 18.31 Wind response notation.

dynamic in nature. In very tall and/or slender buildings, it is common for the across-wind effects to play a major role in the wind design. This is in large part due to the phenomenon of vortex shedding. When wind blows on the surface of a building and flows around its sides, vortices or "eddies" form which are basically a swirling motion passing along the building that cause oscillating turbulence. Refer to Fig. 18.32 for a schematic from Burj Khalifa indicating the vortex shedding at various

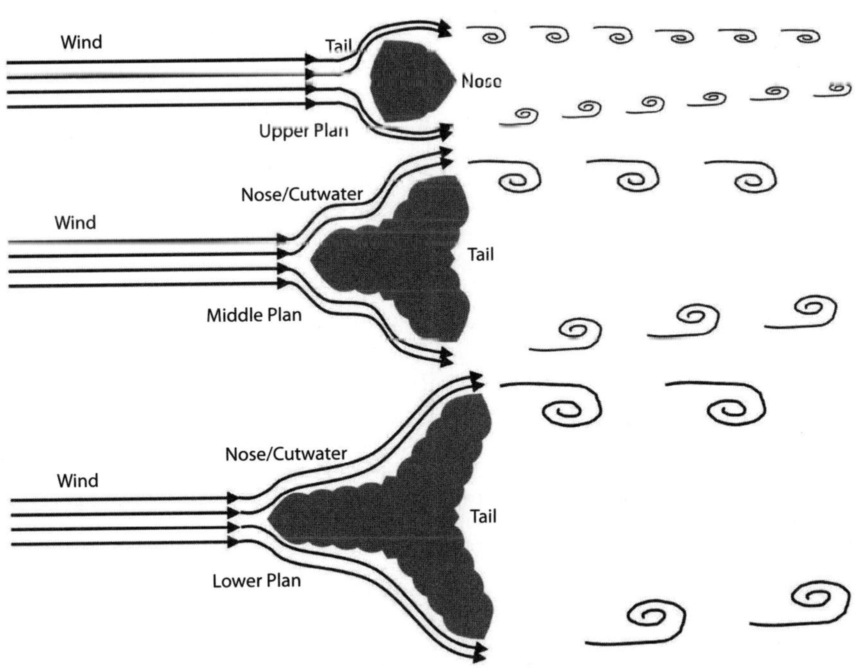

Figure 18.32 Graphic of vortex shedding for the Burj Khalifa.

portions of the building. The size and frequency of formation of these vortices are functions of the building's plan shape and dimensions as well as the wind speed. These vortices create an alternating push and pull on the surfaces of the building perpendicular to the wind which generate wind loads on the building and subsequent displacement and acceleration responses. If the frequency of the building happens to match the frequency of the vortices, very large dynamic effects can be generated in the structure. Careful consideration of this must be taken during the design.

Mitigating strategies for both along-wind and across-wind effects will be discussed further below.

18.9.2 Strategies for Reducing Wind Effects

Three primary approaches exist for reducing wind loads and accelerations in tall buildings:

- Structural stiffness and/or mass distribution
- Building shaping
- Increased damping

One or more of these approaches are often employed in combination with one another. As mentioned earlier, the SR of the building plays a large role in how many of these will need to be utilized.

Structural Stiffness and/or Mass Distribution

Along-wind and across-wind effects on tall buildings are a function of the fundamental periods (or frequencies) of the structure. Conventional wisdom is that a stiffer building will lead to lower wind loads and accelerations, but for across-wind considerations, this is not always the case.

Many building configurations have behaviors which show a maximum response (overturning moment, base shear, etc.) around a particular frequency with a significant drop in that response if the frequency increases or decreases. This is often due to the building frequency being closely aligned with the vortex-shedding frequency.

One method of adjusting the frequency of the building to reduce wind responses can be done by modifying the stiffness of the lateral system of the structure. For example, this could be done by increasing or decreasing concrete wall thicknesses/lengths or adding outriggers trusses or walls to engage the building perimeter.

Another method that can be used alone or in conjunction with stiffness modifications to change the building frequencies is adjustment of the building's mass and/or mass distribution. Floor slabs are generally one of the largest mass contributors in a building. Some recently completed slender towers have used artificially thicker slabs near the top floors of the building to adjust the buildings frequencies away from critical values which were more prone to large overall wind responses.

Changes made to building stiffness or mass should then be sent back to the wind tunnel for reevaluation of the expected loads and accelerations to confirm acceptable performance has been achieved.

Building Shaping

One of the more powerful tools at the structural engineer's disposal for reducing wind responses is building shaping. However, this needs to be a collaborative process with the architects for the project.

Strategies which reduce the along-wind effects are not always beneficial to reducing across-wind effects. However, as noted previously, across-wind effects often govern in tall building wind design. Both approaches will be discussed below, and Fig. 18.33 illustrates some of these ideas.

Reducing Along-Wind Effects

- Smooth Aerodynamic Shape: Certain shapes have better drag performance, which reduces the wind pressures. Unfortunately, some shapes which are good for drag reduction are actually worse for vortex-shedding behavior.
- Tapering: Narrowing the building as it gets taller reduces the exposed area to wind pressures where they are the highest.
- Setbacks: Similar to tapering, changing the footprint to a smaller size(s) up the building reduces the wind sail area.

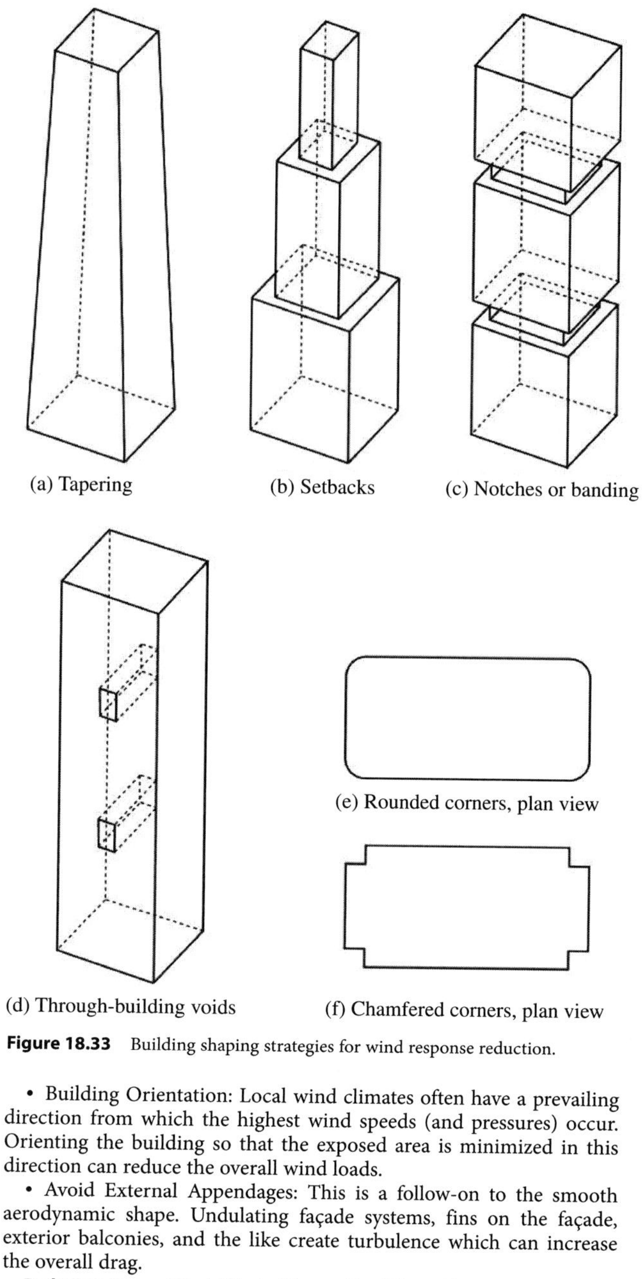

Figure 18.33 Building shaping strategies for wind response reduction.

- Building Orientation: Local wind climates often have a prevailing direction from which the highest wind speeds (and pressures) occur. Orienting the building so that the exposed area is minimized in this direction can reduce the overall wind loads.
- Avoid External Appendages: This is a follow-on to the smooth aerodynamic shape. Undulating façade systems, fins on the façade, exterior balconies, and the like create turbulence which can increase the overall drag.

Reducing Across-Wind Effects (Vortex-Shedding Effects)

- Tapering: In addition to reducing the area exposed to wind loads, tapering has the effect of producing vortex shedding with constantly changing frequencies due to the varying building width. As a result, the vortex shedding on the building cannot "align" over the building's height and produce a unified push-pull. In effect, the pushing and pulling forces are generally out of phase from each other at varying elevations and a canceling effect is generated to some extent.
- Setbacks: Similar to tapering, setbacks reduce exposed area and provide discrete locations where the vortex-shedding frequencies change. However, the impact on vortex-shedding effects is generally less pronounced than tapering.
- Rounded Corners/Chamfers: Vortex shedding can be disrupted by modifying the corners of the building through rounding or chamfering.

In order for these to have a noticeable impact, the corner modification must be at least 5 percent and preferably closer to 10 percent of the building width.

- Notches or Banding: Disruption of vortex shedding can be achieved by introducing a localized setback in the building perimeter typically at amenities floors or dedicated mechanical floors. These would preferably occur at more than one location along the building height. The depth of the setback should be on the order of 10 percent of the building width, and the height should be at least a double that of typical floor to provide a reasonable impact on the vortex shedding.
- Through-Building Voids: Introducing voids that pass directly through the building at a few discrete locations can disrupt the vortex-shedding patterns along the building height. However, these are not very common, as it is difficult to accommodate with the usual layout of the building core containing the lateral system, stairs, elevators, and other building services.
- Closed vs. Open Building Crowns/Tops: Many tall buildings contain a "crown" at the top which is an architectural feature above the useable floors. Overall wind effects can be significantly impacted depending upon the makeup of this element (open lattice, closed sides with open top, closed sides and top, etc.).

Increased Damping

As noted earlier in the chapter, buildings with SR of 10 or greater generally require some sort of supplemental damping in order to control motion perceptibility. The inherent or natural damping of tall buildings is generally taken as the following percentages of critical damping:

- Steel Lateral Systems: 1.0 percent for service level and 1.5 percent for ultimate level
- Concrete Lateral Systems: 1.5 percent for service level and 2.0 percent for ultimate level

This inherent damping can be supplemented by a handful of general device types which will be discussed below.

Tuned Mass Damper (TMD) This is by far the most common type of supplemental damping used to date. They are installed near the top of the building and consist of a large, usually concrete shape that either sits on the floor or is hung like a pendulum. Basically, they are intended to move out of phase with the overall building motions and cause an inertial braking action. The size, weight, and configuration is tuned to the primary mode in the direction in which it is trying to damp the building motions. It is not uncommon for the weight to be several hundred to more than a thousand tons depending on the size of the building. Refer to Fig. 18.34.

Passive TMDs, which are allowed to move naturally in response to the build movements, are the most common and least expensive. However, they also have limited range over which they are effective due to being "tuned" to a particular building period which may in reality be different from the design period as a result of member cracking, material properties, actual building weight, etc.

Active or semi-active TMDs have actuated controls and are more expensive but can extend the range of building periods which can be affected.

TMDs do have the drawbacks of taking up desirable real estate at the top of the building, being expensive to construct, having limited range of effectiveness, and being used only to reduce building accelerations (not impact wind loads for strength design).

Damping can generally be increased by an additional 1 to 3 percent with TMDs.

Tuned Liquid Damper (TLD) TLDs consist of large volumes of water contained in specifically designed enclosures with a series of partitions and baffles that force the water to move in specific directions and rates to generate an inertial braking similar to that of a TMD. They come in a few different configurations, including tuned liquid slosh dampers and tuned liquid column dampers. Refer to Fig. 18.34.

The drawbacks for TLDs are generally the same as those noted above for TMDs. Similar to TMDs, damping can generally be increased by an additional 1 to 3 percent with TLDs.

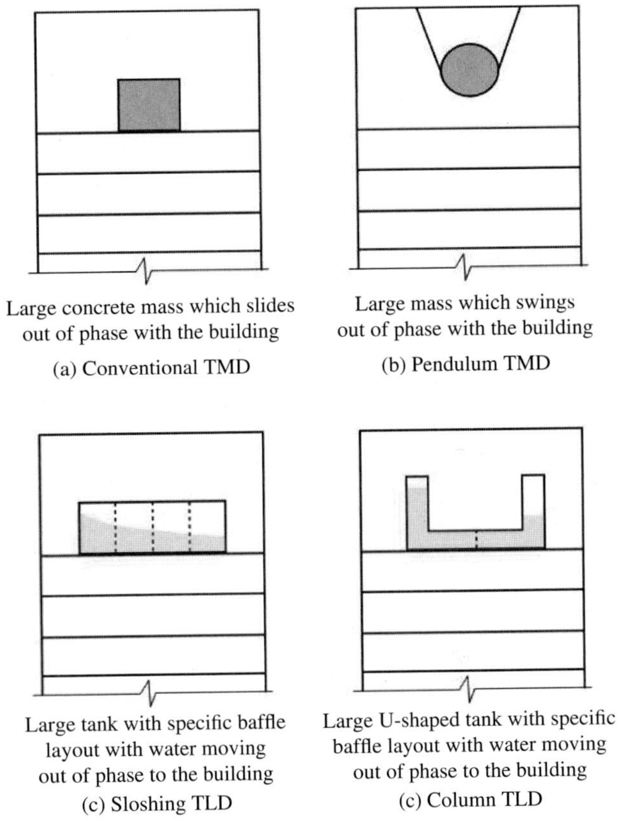

Figure 18.34 Schematics of TMD and TLD systems.

Viscous Damping Viscous damping has been used for many years in the seismic world to reduce seismic effects on buildings and their contents. Recently there has been increased interest in utilizing them for mitigating wind effects.

Viscous dampers look somewhat like shock absorbers on cars and come in a variety of sizes. Figure 18.35 is a photo taken of one of the viscous damper assemblies at the end of a megabrace system on the 181 Fremont Tower in San Francisco. When they are introduced into bracing systems, outrigger trusses, outrigger walls-to-columns, etc., they can have large impact on the damping of the overall building.

Viscous dampers used in buildings are fairly maintenance-free and constructed to very high standards, leading to a high reliability. They offer several benefits over TMDs and TLDs including the following:

- They can be placed in a distributed fashion throughout the building, resulting in sufficient quantities to create redundancy. This allows them to be used for both building acceleration reduction and potentially wind load reduction. This contrasts with TMDs and TLDs, which are relied upon only for acceleration reduction due to the lack of redundancy with those systems.
- They have similar damping effects over a much larger range of building motions, allowing them to accommodate the differences between predicted and actual building periods and damp multiple modes. This contrasts with TMDs and TLDs, which are "tuned" to more specific building motions and offer less flexibility.
- They are compact assemblies which require very little space and are generally significantly less expensive compared to TMDs and TLDs.

One of the only drawbacks of viscous dampers is the lack of in-place systems that can be observed and measured to verify the expected wind-mitigating effects for both acceleration and strength. It is expected that this will change in the near future.

Figure 18.35 Viscous dampers on 181 Fremont in San Francisco.

Figure 18.36 Example of wind tunnel testing (RWDI).

Damping can generally be increased by an additional 3 to 6 percent with viscous damping systems.

18.9.3 Working with Wind Tunnels

Tall buildings are rarely designed to code-calculated wind loads for a variety of reasons. In many cases the loads coming from the codes are conservative in nature, as the approach is simplified to make it usable for a wide variety of building types while aiming to ensure that wind loads are not underestimated. Some other issues which complicate the use of code winds are unusual building geometries, slender buildings prone to dynamic amplification effects, and the influences of current or proposed nearby buildings on the wind loads. As a result, wind tunnels are often engaged by the structural engineer for buildings around 30 stories or greater. In some instances, shorter buildings may warrant this as well. The use of wind tunnels to determine appropriate wind loads could also be considered as a sort of performance-based design measure.

A scaled model of the area around the proposed building, which will be a function of how tall the building is as well as how much variation in terrain types occurs upwind, will be created for the wind tunnel testing. A scaled model of the building's geometry will also be generated based on information provided by the project architect. Figure 18.36 is an example of a wind tunnel model.

Basic design wind speeds are often stated in building codes. However, the wind tunnel consultant will use wind measurements from nearby airports and other meteorological stations to determine an appropriate wind climate for the project site which will factor in, among other things, frequency, directionality, and wind speeds. Depending on the available data as well as the types of events producing the winds (hurricanes, thunderstorms, etc.), additional methods may be applied in piecing together the wind climate analysis. More on wind climate analysis can be found in other sources.

The structural engineer will need to provide the wind tunnel with information on the building's modal behaviors for at least the first three primary modes and often between 6 and 10 modes. Very tall or slender buildings may require even more. The modal information will include the building mass at each floor, the distribution of that mass via the mass moment of inertia, and the modal displacements of a point in the building that can be reported consistently over the height. An assumed damping value will also need to be provided to the wind tunnel. This value will likely be just the inherent damping initially and then potentially augmented with supplemental damping values described above depending on the building's performance. After the physical wind tunnel test has been run, the modal information and damping value(s) will be used by the wind tunnel to post-process the results to determine appropriate wind loads for the structural engineer to use in the design and to estimate the acceleration response of the building for occupant comfort.

Early on in the design, the building shaping strategies described earlier can be quickly tested by the wind tunnels to gauge the potential reductions in wind load and acceleration responses. This can be a great aid to the architects and structural engineers as they work to finalize the building shape and layout.

As the design proceeds, it is not unusual for there to be several iterations where updated structural properties are provided to the wind tunnel, who then recalculates wind loads and accelerations. Assuming the building shape has not changed, this can be done without another physical wind tunnel test and is a quick and economical way to optimize the building structure. If changes to the building shape and/or height occur during the design progress, a new wind tunnel test will likely be required.

18.9.4 Motion Perceptibility

As noted in Sec. 18.4, occupant comfort related to motion perceptibly is a key driver in tall building design. Assessing the expected building accelerations is one of the primary pieces of information coming out of the wind tunnel test.

Acceptance criteria for accelerations has been established by a number of organizations and code bodies, with the International Organization for Standardization (ISO) and the Architectural Institute of Japan (AIJ) being two of the more commonly referenced. Generally, it is recognized that different occupant usages have different levels of acceleration perception and subsequent tolerance limits. For example, residential floors demand more stringent criteria than office floors due to the amount of time occupants will likely spend in their residences and the types of activities in which they are involved (e.g., lying down, lounging, etc. vs. sitting at a desk). The acceptance criteria are based upon scientific studies and aim to limit accelerations to a threshold level which does not bother the majority of potential building occupants.

In general, most people will tolerate discomfort felt infrequently or for short periods of time but not as routine occurrence. For example, a larger acceleration will be tolerated if this happens rarely. Common practice has been to focus on occupant comfort assessments occurring on the 10-year recurrence interval. However, most issues have come from movements which are experienced more frequently than the 10-year return period. As a result, a more recent trend has seen the evaluation of these motions at a 1-year return period to account for the fact that longer return periods do not appropriately address the potential discomfort related to regularly occurring wind events. The ISO10137 guideline gives frequency-dependent maximum acceleration limits for the 1-year return period for both office and residential uses. Refer to Fig. 18.37 for the ISO10137 chart.

As buildings stretch to new heights and become increasingly slender, the critical wind velocity that can initiate a dynamic response in the building occurs more frequently. This critical velocity is determined through a simple relationship between the natural frequency on the building, the breadth of the building, and a nondimensional quantity known as the Strouhal number, which is essentially an architectural shape factor. For tall slender buildings, this critical velocity can often occur within a monthly or biyearly recurrence interval. For these shorter return periods, the peak accelerations should be kept beneath the threshold of perception, which is typically thought to be of the order of 5 milli-g.

A comprehensive look at motion perceptibility can be found in the paper written by M. D. Burton et al. (2015).

18.10 PERFORMANCE-BASED DESIGN FOR SEISMIC

While tall buildings are generally governed by wind loads, there are a number of regions around the world where significant seismic potential exists and can approach or exceed the wind demand.

The primary goal of all seismic design codes is to ensure that behavior of buildings under the maximum credible earthquake will be satisfactory for life safety considerations, which generally means preventing collapse. However, these buildings are often left in a condition with serious permanent structural deformations and reduced capacities that make them unusable or have sustained significant damage to building services and nonstructural content which would require a huge amount of money and time to repair them.

Another issue with most prescriptive seismic code approaches is their explicit limits on heights to which particular structural systems can be employed. This prevents the majority of tall building systems from being used in high seismic zones.

Performance-based design (PBD) for seismic has been around for many years and is now standard practice in many high seismic regions for buildings of all shapes and sizes. It initially started as a way to address the above-mentioned code limits on heights for structural systems. In recent years, more emphasis has been put on meeting enhanced performance goals beyond just collapse prevention, so that buildings remain usable or can easily be repaired after a large earthquake. PBD is a tool that allows these enhanced goals to be evaluated and implemented.

General PBD approach requirements and methodologies are defined in codes like ASCE 7 as well as publications from FEMA, the National Earthquake Hazards Reduction Program (NEHRP), and the Applied Technology Council (ATC). A specific document related to tall buildings was first released in 2010 titled "Tall Building Initiative—Guidelines for Performance-Based Design of Tall Buildings" developed by the Pacific Earthquake Engineering Research Center (PEER).

At its core, PBD involves evaluation of structural behavior of a building with its materials potentially pushed beyond their elastic limits (material nonlinearity) while undergoing large member deformations (geometric nonlinearity). Several commercial software packages which allow these nonlinearities to be modeled and analyzed are available.

Ground motions from recorded earthquake events or from synthesized events are scaled to the expected peak accelerations in the project area and used in the analysis. The required duration of the ground motion is a multiple of the building's fundamental period, so the length of the record in the analysis get longer as the building gets taller. The nonlinear analysis is then run in time steps which are a fraction of the total record length, often involving hundreds to thousands of analysis steps.

Locations where the structures have gone beyond the elastic range and potentially completely plastic are generally referred to as "hinges."

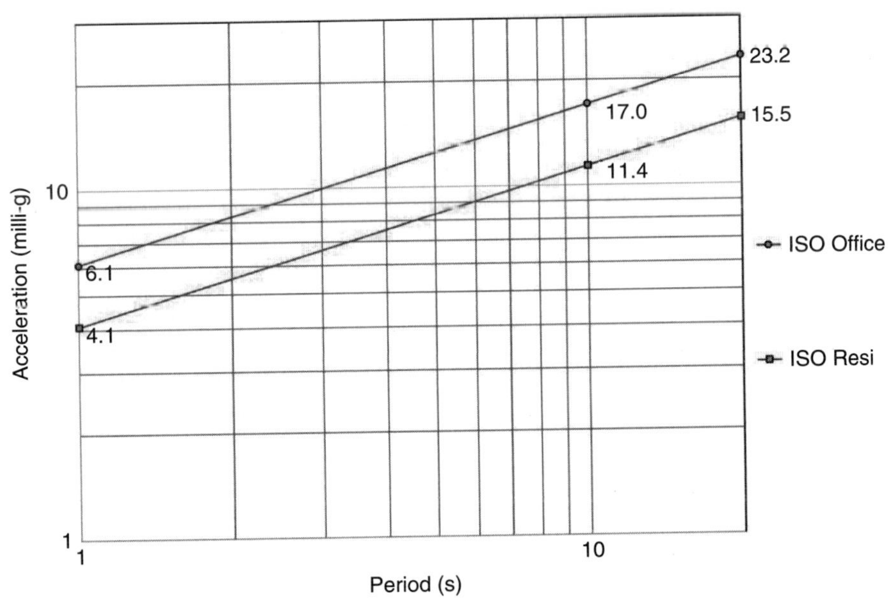

Figure 18.37 ISO 10137 guidelines for maximum accelerations for occupant comfort.

A great deal of seismic energy is dissipated in forming these hinges, but they represent locations where the structure can take no additional load and will continue to deform at that load level. Where these hinges form is important and needs to be carefully controlled by the structural engineer.

Beyond allowing hinges to form in conventional structural members, a number of other approaches exist to limit seismic forces or to concentrate structural damage to certain areas:

- Base Isolation—the entire building is supported on isolating pads or bearings which cause the fundamental periods to increase to the point that the seismic demands on the building are much smaller. Very large overall buildings motions result, which require special detailing around the perimeter where the building meets the ground as well as special connections for the utilities coming in and out of the building. Base isolation is not very common for tall buildings.
- Viscous Dampers—as noted under the section titled "Increased Damping," these elements act as shock absorbers. They can increase the overall damping and reduce the violent shaking of the structure to some extent, leading to a reduction in the seismic forces and displacements of the building.
- Buckling-Restrained Braces (BRB)—these elements are designed to perform well in both tension and compression with a known strength limit. This allows them to be sized such that they will yield prior to damage of the surrounding structure. Theoretically, they can be replaced after a major seismic event.
- "Fuses"—these elements are considered sacrificial components sized such that they yield prior to the surrounding structural elements reaching their capacities. Damage is concentrated in these elements, and they are detailed with the intent of being removed and replaced. Fuses can take many forms which are incorporated in both moment frame and bracing systems. BRBs described above are a specific example of a fuse.

Soil structure interaction (SSI) is another form of PBD that is not commonly used but is a potentially powerful approach to reduce the seismic loads that the structure is expected to see. Seismic site accelerations used for scaling the ground motions applied in the PBD analysis models are usually taken at ground level and are generally highest there. However, most tall buildings are supported on foundations often sitting below several floors of basement. The accelerations at these depths can be substantially smaller than the ground level values depending upon the soil profile. In reality, the overall building moves along with the soil column in which it is embedded. By taking advantage of the smaller accelerations, the seismic demands on the building can be reduced from the beginning, allowing for a more economical and realistic design. SSI involves collaboration between the structural engineer and the geotechnical engineer to properly account for the soil properties and the seismic waves passing through them. The analysis is more intense, time consuming, and specialized; hence, its current application is limited.

In summary, PBD can be used to ensure life safety for beyond-code structural systems as well as looking at enhanced performance objectives for increased resiliency, leading to less damage and downtime, that is, less potential costs to owners and tenants in the event of a major earthquake.

18.11 CREEP AND SHRINKAGE

As previously discussed, structural reinforced concrete undergoes long-term deformation under sustained load (creep) and as a natural result of moisture evaporation (shrinkage). This tendency must be accounted for in tall building design, especially for composite structures where a central core (generally concrete) is linked to a perimeter structure (usually steel), resulting in long-term differential settlement between the exterior and interior of the structure. This differential movement can both induce significant stress in the structural elements that tie the core and perimeter together and introduce serviceability issues such as unlevel floors or unwanted cracking if not properly considered in the analysis and design process.

Typical structural analysis software will allow for nonlinear staged construction analysis, in which a predicted construction sequence as well as the time-dependent properties of concrete are incorporated in the analysis model. The recommended numerical model for long-term behavior of reinforced concrete is the Gardner-Lochman method (ACI 209.2R-10). Refer to Fig. 18.38.

Depending on the nature of the connection between core and perimeter, different methods are available for resolving the issue of differential settlement. A popular strategy for mitigating excessive stress in outrigger trusses, as discussed in Sec. 7.4, is delaying the full connection of the core to the perimeter structure until the full tower structure is complete. Overlengths are also typically provided to address elastic shortening of steel elements such that floors are level after the total building dead load is applied.

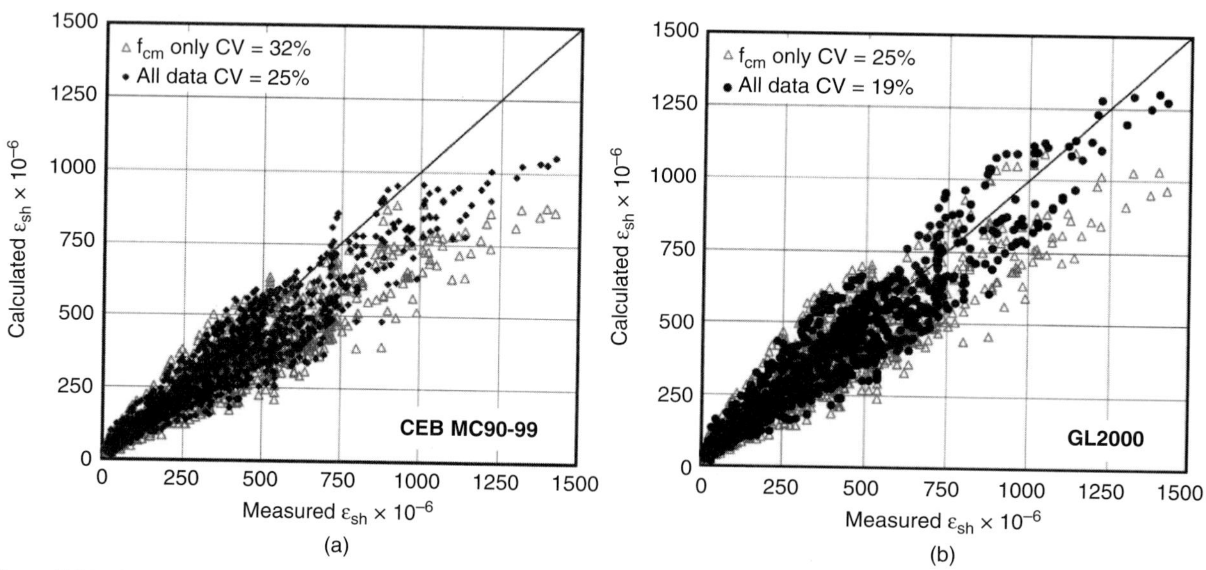

Figure 18.38 Comparison of (*a*) CEB-FIP 90 and (*b*) GL2000 methods (ACI 209.2R-10).

18.12 HIGH-RISE BUILDING GLOSSARY

Lateral system	Structural elements which resist seismic, wind, and eccentric gravity loading.
Braced tube	Distributes gravity and lateral loads along perimeter columns through the use of concentric bracing.
Bundled tube	System of interlocking frames consisting of closely spaced columns and deep girders sized to behave as a tube and to limit impact of shear lag.
Shear lag	(In a tube system) Loss of effective uniform stress distribution along perimeter frame line flanges as distance from shear frame line increases.
Composite stayed mast	Interior reinforced-concrete core tied to perimeter columns to increase structural overturning stiffness.
Core wall	Reinforced-concrete walls that enclose interior circulation core and resist seismic, wind, and eccentric gravity loads.
Shear deformation	(In the context of a building lateral system) Any deformation which reduces the stiffness of the system from the ideal cantilever model.
Aspect ratio	Measure of lateral system slenderness. For core wall-only systems, measured as the ratio of the building height to minimum dimension of the core wall. For systems that engage perimeter columns, ratio of height to the minimum out-to-out of perimeter columns.
Ceiling sandwich	Floor-to-floor section, produced in the conceptual phase, which identifies required allowance zones for architectural, structural, and building services.
Outrigger Trusses/walls	Trusses/walls which link interior core with perimeter lateral system elements. Usually coupled with belt trusses/walls at double-story mechanical levels.
Gust effect	Exciting effect on a building due to turbulent wind.
Spandrel beams	An exterior beam that extends from column to column.
Lease span	The distance from the exterior wall to the core.

18.13 ADDITIONAL CREDITS

Buildings identified in text	Architect	Structural engineer
Jin Mao Tower, Shanghai	Skidmore, Owings & Merrill	Skidmore, Owings & Merrill
Home Insurance Building, Chicago	William Le Baron Jenney	William Le Baron Jenney
Chrysler, NYC	William Van Alen	Ralph Squire & Sons
Empire State Building, NYC	Shreve, Lamb & Harmon	H.G. Balcom & Associates
DeWitt-Chestnut Apartment Building, Chicago	Skidmore, Owings & Merrill	Skidmore, Owings & Merrill
John Hancock Center, Chicago	Skidmore, Owings & Merrill	Skidmore, Owings & Merrill
Willis (formerly Sears) Tower, Chicago	Skidmore, Owings & Merrill	Skidmore, Owings & Merrill
Original World Trade Center Towers, NYC	Minoru Yamasaki Associates	Leslie E. Robertson Associates
Burj Khalifa, Dubai	Skidmore, Owings & Merrill	Skidmore, Owings & Merrill
432 Park Avenue, NYC	Rafael Vinoly Architects	WSP
111 W 57th St, NYC	SHoP Architects	WSP
One Manhattan West, NYC	Skidmore, Owings & Merrill	Skidmore, Owings & Merrill
150 North Riverside, NYC	Goettsch Partners	Magnusson Klemencic Associates
China Resources Headquarters, Shenzhen	Kohn Pedersen Fox Associates	Arup
International Finance Center, Hong Kong	Cesar Pelli & Associates	Arup
Nanjing Zifeng Tower, China	Skidmore, Owings & Merrill	Skidmore, Owings & Merrill
Hyundai-Hyperion Project Tower	Skidmore, Owings & Merrill	Arup/Dong Yang Structural Engineers
Shum Yip Upperhills	Skidmore, Owings & Merrill	Skidmore, Owings & Merrill
W Hotel, Kuala Lumpur	Skidmore, Owings & Merrill	T.Y. Lin International Sdn Bhd
St. Francis Shangri-La Place Development, Manila	Wong Tung International Ltd	Arup
Samsung Tower Palace 3, Seoul	Skidmore, Owings & Merrill	Skidmore, Owings & Merrill
Poly International Plaza, Beijing	Skidmore, Owings & Merrill	Skidmore, Owings & Merrill
Guizhou Culture Plaza Tower, Guiyang	Skidmore, Owings & Merrill	Skidmore, Owings & Merrill
Tianjin Goldin Finance, Tianjin	P&T Group	Arup

Special Acknowledgments: Samuel Wilson, Skidmore, Owings & Merrill.

ACKNOWLEDGMENTS

Samuel Wilson, Skidmore, Owings & Merrill
Changjiang Zhou, Skidmore, Owings & Merrill

REFERENCES

ACI Committee 375. 2006. *SP-240: Performance-Based Design of Concrete Buildings for Wind Loads.* American Concrete Institute (ACI).

Addis, Bill. 2007. *Building: 3000 Years of Design Engineering and Construction.* New York: Phaidon Press Limited.

Ahrens, M. Nov. 2016. "High-Rise Building Fires." Quincy: National Fire Protection Association.

Ali, M. M., Armstrong, P. J. 2008. "Overview of Sustainable Design Factors in High-Rise Buildings." In *Proceedings of the CTBUH 2008 8th World Congress*, 3–5. Dubai: CTBUH.

American Society of Civil Engineers. Minimum Design Loads from Buildings and Other Structures, ASCE 7-10, 2010, ASCE, Reston, VA, USA.

Baker, W. F. 2001. "Building Systems and Concepts—Structural Innovation." In *Proceedings of the CTBUH 2001 6th World Congress.*

Baker, W. F., Korista, D. S., Novak, L. C., Pawlikowski, J., Young, B. 2007. "Creep and Shrinkage and the Design of Supertall Buildings—A Case Study: The Burj Dubai Tower." *ACI Special Publication* 246: 133–148.

Baker, W. F., Pawlikowski, J. 2012. "Higher and Higher: The Evolution of the Buttressed Core." *Civil Engineering Magazine Archive* 82, Issue 9 (2012): 59–65.

Besjak, C., Biswas, C., Ullah, S. He, X., Zhuang, J. 2014. "Shenzhen Shum-Yip Tower One—Gravity and Lateral Load Resisting System Optimization." In *Proceedings of Structures Congress 2014*, 2524–2536. Boston: ASCE.

Besjak, C., McElhatten, B. J., Biswas, P. 2010. "Performance-Based Evaluation for the 450m Nanjing Greenland Financial Center Main Tower." In *Proceedings of Structures Congress 2010*, 2885–2902. Orlando: ASCE.

Burton, M. D., Kwok, K. C. S., Abdelrazaq, A. Mar. 2015. Wind-Induced Motion of Tall Buildings: Designing for Occupant Comfort. *International Journal of High-Rise Buildings* 4, no. 1: 1–5.

Choi, H. S., Ho, G., Joseph, L., Mathias, N. 2012. *Outrigger Design for High-Rise Buildings: An Output of the CTBUH Outrigger Group.* Chicago: Council on Tall Buildings and Urban Habitat.

Chourasiya, S. K. 2002. "Analysis & Design of Tubular Structures" (Master's Thesis).

Chung, K., Kim, I., Kim, M. 2004. "Structural Design and Construction of Hyundai-Hyperion Project, Seoul." In *Proceedings of the CTBUH 2004 Seoul Conference*, 263–270. Seoul: CTBUH.

Gann, R. G., ed. 2005. "Federal Building and Fire Safety Investigation of the World Trade Center Disaster: Final Report on the Collapse of the World Trade Center Towers." *NIST NCSTAR 1.* Gaithersburg: NIST.

Garai, R., Sarkisian, M., Mathias, N. J., Krebs, A. 2015. "Three-Dimensional Exterior Bracing Systems for Tall Buildings." In *Proceedings of CTBUH 2015 New York Conference*, 555–560. New York: CTBUH.

Goman W., Ho, M. 2016. "The Evolution of Outrigger Systems in Tall Buildings." *International Journal of High-Rise Buildings* 5, no. 1: 21–30.

Hallebrand, E., Jakobsson, W. 2016. "Structural Design of High-Rise Buildings" (Master's Thesis).

Hoogendoorn, P. P. 2009. "Lateral Load Design of Tall Buildings: Evaluation and comparison of four tall buildings in Madrid, Spain" (Master's Thesis).

Iyengar, S. H., Khan, F. R. "Structural Steel Design for Sears Tower." Proceedings of 1973 Australian Conference on Steel Developments, Newcastle: Australian Institute of Steel Construction.

Khan, F. R. Lyengar, S. H., Colaco, J. P. 1968. "Analysis and Design of the 100-Story John Hancock Center in Chicago (USA)." no. 6: 271–274.

Khan, Y. S. 2004. *Engineering Architecture: The Vision of Fazlur R. Khan.* New York: W. W. Norton & Company.

Korista, D. S., Sarkisian, M. P., Abdelrazaq, A. K. 1995. "An Optimal Use of Concrete in High-Rise Concrete Building Design—The Jin Mao Tower." In *Proceedings of Third National Concrete & Masonry Engineering Conference,* 463–484. San Francisco: NCMEC.

Lee, K., Ho, G.W. Lee, A., Yin, C., Liu, G., Huang, X. Dec. 2012. "The Structural Design of Tianjin Goldin Finance 117 Tower." *International Journal of High-Rise Buildings* 1, no. 4: 271–281.

Luo, Y., Besjak,C., Cashen, D., Koop, C. 2015. "Mediating Scale, Performance and Iconicity: A 21st Century Supertall Tower for Guiyang." In *Asia & Australasia: A Selection of Written Works on the World's Tall Building Forefront,* 12–23. Seoul: CTBUH.

Luong, A., Gibbons, C., Lee, A., MacArthur, J. 2004. "Two International Finance Center." In *Proceedings of CTBUH Conference,* Seoul: CTBUH.

McAllister, T., Sadek, F. Gross, J. L., Averill, J. D., Gann, R. G. Jul. 2013. "Overview of the Structural Design of World Trade Center 1, 2, and 7 Buildings," *Fire Technology* 49, Issue 3: 587–613.

MIT, ed. 1998. Andrew Scott, *Dimensions of Sustainability.* London: E & FN Spon; New York: Routledge, 109–114.

Orakcal, K., Wallace, J. W. Mar.–Apr. 2006. "Flexural Modeling of Reinforced Concrete Walls—Experimental Verification," *ACI Structural Journal* 103, no. 2: 196–206.

Partovi, A., Svärd, J. "Global Analysis of Tall Buildings with Tubed Mega Frame Structures" (Master's Thesis).

Rahimian, A. Nov.–Dec. 2011. "Lateral Stiffness of Concrete Shear Walls in Tall Buildings." *ACI Structural Journal* 108, no. 6: 755–765.

Rahimian, A., Eilon, Y. 2008. "Hearst Headquarters: Innovation and Heritage in Harmony." In *Proceedings of CTBUH 2008 8th World Congress,* Dubai: CTBUH.

Sarkisian, M., Mathias, N. J., Long, E., Mazeika, A., Gordon, J., Chakar, J. 2006. "Jin Mao Tower's Influence on China's New Innovative Tall Buildings." *2006 Shanghai International Seminar of Design and Construction Technologies of Super High-Rise Buildings.* Shanghai: CTBUH.

Sev, A. 2009. "Typology for the Aesthetics and Top Design of Tall Buildings." *G.U. Journal of Science* 22, no. 4: 371–381.

Zils, J., Viis, J. Nov. 2003. "An Introduction to High Rise Design," *Structure Magazine,* 12–16.

Chapter 19
Thin-Shell Concrete Structures

BY

EDMOND SALIKLIS *California Polytechnic State University (Cal Poly), San Luis Obispo*
MUSTAFA MAHAMID *University of Illinois at Chicago*
DAVID P. BILLINGTON *Professor Emeritus of Civil Engineering, Princeton University (Deceased)*
JULIAN A. DUMITRESCU *Consulting Structural Engineer, Raytheon Engineers and Constructors (Retired)*

19.1 INTRODUCTION AND HISTORICAL OVERVIEW

David Billington's writing on shells is important from multiple perspectives. It is a historical record of the significant contributions to shell analysis and design from many academic and practicing engineers. It lays the foundation for membrane theory, an equilibrium solution that simplifies a complex 3D problem into a tractable format. It is also, in part, a tribute to Professor David Billington, who has inspired countless engineers over the past 60 years and who began his career at Princeton working on concrete shells.

Yet, if this chapter were solely a historical look backwards, it would belong in an archive and not in a contemporary structural engineering handbook. Today, design and analysis are merging in the hands of talented engineers in a way that Felix Candela and other leading shell builders intuitively mastered. The difference today is that remarkably sophisticated tools allow designers to create shells with complex geometry while ensuring there is minimal bending under dead load. In fact we may be at the dawn of a new era of interdisciplinary design, an era where the engineer and the architect seamlessly transfer ideas and technical information via modeling software. Yet thin shells require a third expertise even at the preliminary stage. The construction of the shell must be carefully considered even during the initial design ideation between the engineer and the architect and the construction manager. The engineer who is well versed in construction practices and who is aesthetically sensitive and architecturally trained may take the lead in this new era of parametric design. This may be so because only the engineer is professionally trained to consider an ultrathin shell's susceptibility to buckling. Stresses are extremely low in so-called funicular shells, that is, ones with a minimal amount of bending. Yet safety against buckling becomes a major concern as the shells get thinner and thinner; consequently, a large factor of safety against buckling must be ensured. This brief section will explore some of the exciting research currently taking place. But it begins with a summative overview of the various approaches to design and analysis of thin structural shells that occurred in the 20th century.

The story of 20th-century shells could begin with a father and son team of designers in Spain and the United States: the Guastavinos. Rafael Guastavino Sr. received significant commissions even in the late 19th century to design and to build extremely thin masonry vaults.[1] He used the tile vaulting tradition of laying thin bricks flat on top of each other to form convex, 3D shells. The shells he and his son constructed are relatively thinner than a chicken's eggshell, and they have withstood the test of time and heavy use in more than 600 existing buildings, primarily on the East Coast of the United States. Stresses and thrusts were not very large in these vaults, and their blending of art and technology is virtually unparalleled in the history of structures. Few predictive tools were used; rather, expert construction workers deftly followed 3D lines in space to construct magnificent thin vaults. This section closes with this very image.

A markedly different approach to thin-shell analysis and design began roughly contemporaneously with the Guastavinos. In the 1920s and 1930s giants of structural engineering in Germany such as Franz Dischinger, Walther Baursfeld, and Ulrich Finsterwalder made enormous technical strides in thin-shell design and construction through the use of membrane theory and differential equations.[2] This German School of concrete shell design was important because it rapidly advanced the mathematical state of the art of thin-shell analysis, yet it was limited by this very fact as well.[2] The reliance on differential equations drastically limited the choice of structural forms to cylinders and spherical sections. The Planetarium at Jena in 1926 was a landmark structure primarily because of its innovative shotcrete construction. The 1933 Algeciras Market Hall by the Spaniard Eduardo Torroja was also important from a constructability point of view. It used prestressing to capture the horizontal flow of roof loads as they pass through a somewhat awkward transition to vertical columns. Torroja himself had mastered the differential equations of thin shells, but he saw it was a limiting tool, and he abandoned them in favor of a simple graphic statics approach for some of his more complex structures.[3] A spectacular 1955 structure, only 8 cm thick with a 52-m span, was the Royan Market Hall in France. With this structure, the architects Simon and Morisseau with the engineer Rene Sarger showed the state of the art of shell design in the mid-1950s. This remarkable shell can be considered one of the first "free-form" reinforced-concrete thin-shell structures, insofar as its general conception is guided by structural mechanics principles, and geometrically it is a fanciful combination of parabolic and sine waves.[4]

Differential equations were similarly mastered, then abandoned by one of the greatest structural artists, Felix Candela. Candela pioneered the design and construction of hyperbolic paraboloid concrete shells, and his prolific output is a lasting testament to the blending of skills of the architect, the engineer, and the builder. Candela's colleague, the Irish engineer Colin Faber, was keenly aware of the prescience of that moment in Candela's career. His excellent book on Candela details a remarkable moment in the history of structures.[5]

The prolific designer/builder Pier Luigi Nervi brought a new light to the architectural possibilities of precast concrete. His extraordinary legacy is due in part to his ability to predict and economically construct many large-scale and exciting monolithic reinforced-concrete shells and gridshells. Nervi's own books on structures deeply influenced designers around the world.[6]

Similarly, much thoughtful scholarship is rightly devoted to the Swiss engineer Heinz Isler. John Chilton has produced important studies of Isler's work to capture yet another zeitgeist in our profession, as Isler's experimental form-finding approach was radically different than any of his predecessors.[7] This variety of analytical approaches progressed quickly through the first half of the 20th century: differential equations, then graphical methods, then simplified methods, culminating with Isler's experimental methods.

Another innovative development in the constructability of shells occurs with Italian Dante Bini. So-called Bini Domes, approximately 1000 of which have been built, use a large inflated membrane as the sole formwork during construction.

The literature on the history of structures is nowhere near as vast as that of architectural history. Inevitably, its extant discussions of 20th-century shells focus on the story briefly outlined above. Yet there are many other fascinating lines of inquiry in need of greater study and wider dissemination. A telling example is the magnificent work of Tange, Tsuboi, and Kawaguchi, a team that Peter Rice dubbed "The Japanese Engineer Gods."[8] The architect Tange had a profound control of structural form, and he was willing to create structures that very few engineers on the planet were capable of understanding fully at that time. Their Hiroshima Children's Library and their many sports stadia are landmark structures worthy of deep study. The cable-suspended roof by Matthew Nowicki had a profound impact on engineers around the world. Its pure structural logic and remarkable roof lightness inspired many designers. In the 1970s, many cable-suspended concrete roofs were created by daring architects and engineers in countries behind the Soviet Iron Curtain. Because of political isolation, almost no English language studies are currently available for these works, and this deficiency must be corrected while a few of those designers are still alive today. One exception to this rule is Ulrich Muther, who created magnificent shell structures in East Germany. His works are carefully archived and available to scholars.[9] Some non-European shell designers have received due attention. Eladio Dieste of Uruguay, who created spectacular masonry shells, is one shining example.[10] Raji Sundaram of India recently received the IASS Torroja Medal for his prolific shell building.

German engineers again return to the story in the 1960s and 1970s. Joerg Schlaich's mastery of structural design includes a vast hyperbolic paraboloid in Hamburg and a glass-fiber concrete shell in Stuttgart. Professor Curt Siegel of Stuttgart University made a major contribution to the study of structures, including shells, with a book that looks at structural form and forces from a purely qualitative point of view. His pedagogical materials deserve much wider distribution, especially to architecture students.[4]

West coast American mid-20th-century architects were constantly exploring innovative hyperbolic paraboloid roofs for homes, hotels, and even gas stations. How fitting it would be if some of the passion surrounding mid-century modern design was channeled into scholarly studies of these thin-shell structures. The architecture of Paul Revere Williams or William Cody deserves to be analyzed from a mechanics point of view. More attention should be given to the masterwork sports stadia of this era as well. A shining example is the 1963 Assembly Hall by Max Abramovitz with engineering by Ammann & Whitney. A decade later, the Seattle Kingdome was engineered by another IASS Torroja Medal winner, John V. "Jack" Christiansen. Using a rotating, reusable formwork system, Christiansen created an extraordinary thin-shell concrete structure, spanning over 200 m. Its demise provides a profound context to the thorny question of preservation of landmark works of structural significance.

Early 21st-century researchers benefited from faster computers and improved graphic capabilities. This spurred on a large body of important work, notably by Kai-Uwe Bletzinger and Ekkehard Ramm.[11] They carry on the tradition of Frei Otto's Institute of Lightweight Structures in Stuttgart, Germany, wherein experimental hanging cloth or hanging chain models are replaced by shape optimization algorithms which can reproduce these physical experiments numerically and can modify the warp and the weft stiffness properties of the medium being suspended. This inspired a large body of work in optimization of shell forms based on computer-aided geometric design methods, finite-element analyses, and nonlinear programming techniques. Succinct and state-of-the-art summaries of this form-finding research can be found in a recent publication by Sigrid Adriaenssens et al.[12]

Another German approach to form finding of shells and membranes also originated in Stuttgart with the Force Density Method by Hans-Joerg Schek. This technique originated in the 1970s and allowed for the linearization of large deformation equilibrium equations and then solving iteratively for displacements which are beyond the traditional small displacement assumption.[13] The landmark timber gridshell Multihalle in Mannheim, Germany, of 1974 was designed using this technique.

The Teshima Art Museum, an important shell designed by Rei Nishizawa and the acclaimed engineer Mutsuro Sasaki, brings us fully into the 21st century. This shell is not particularly thin, but it is important because of its magnificent blending of art, architecture, and technology. Furthermore, it sets the stage for the new era of shell design, namely, the ability to computationally predict the behavior of reinforced-concrete shells that are neither simple geometrical shapes such as a sphere or cylinder nor the mid-century hyperbolic paraboloid nor the experimentally driven shapes of Torroja or Isler.

The early 21st century is truly the dawn of the new age of geometry-based design rules. Sustainability and efficiency of material usage is a primary motivator for this new school of thought. A fascinating path of deep research was spurred by John Ochsendorf as he prepared his PhD at Cambridge University with mentoring from the legendary Jacques Heyman. Among Heyman's many significant insights to the world of structural mechanics are his formulations for the lower bound theorem of masonry structures. This theorem shows that if a thrust line, or equilibrium force trajectory, can be found for a given loading condition such that the trajectory fits within the structure's cross section, a static solution can be found. Ochsendorf's own PhD work laid the foundations for Thrust Network Analysis, a form-finding and analysis tool that incorporates graphic statics into equilibrium solutions of 3D shell structures. Phillipe Block's PhD work at MIT developed and implemented Thrust Network Analysis. In recent years, Block's research group at ETH-Zurich developed an important body of work in two-point-five- and 3D graphic statics. Block's PhD student Masoud Akbarzedeh used the same methods that are used in 2D graphic statics, namely a form diagram and a force polygon, and he mapped it directly into a 3D form and a force polyhedron.[14] While it is true that these works are highly theoretical, the exciting 21st-century development is that this theoretical work has resulted in a number of experimental ultrathin compression-only shells actually being built. The stone Armadillo Vault for the 2016

Architecture Biennale in Venice is a stunning showcase of the blending of engineering, architecture, and construction.[15]

This rich line of scholarly inquiry has led to important rediscoveries of Maxwell's Theorem as well as the Airy Stress Function. The 1864 paper by Rankine, which consisted of four sentences, was cited by Maxwell in another paper, several sentences long. The profundity of these two paragraphs written by Rankine and by Maxwell continues to absorb today's most brilliant structural engineering researchers and practitioners around the world, from William Baker of Chicago's Skidmore Owings and Merrill (SOM) to Cambridge University, to Zurich's ETH, and to the University of Pennsylvania.[16,17] The issue raised by Rankine as well as Maxwell that is still being studied by these researchers 150 years later is the mapping of forms and forces via 2D diagrams to 3D representations, back and forth between form and force. For example, an arrangement of points in 2D equilibrium, such as a planar truss subjected to external loads, has a paired, reciprocal 3D polyhedron, the polyhedron being a map of the forces in the bars of the truss. This reciprocity between forms and forces in 2D has been well documented in numerous excellent textbooks from the end of the 19th century and in a recent book by Edward Allen.[18] What is less known is the depth of the ideas posited by Rankine and Maxwell. The reciprocity of form and forces in 3D has led to astonishing insights in recent publications, such as a direct mathematical link between the mapping of the reciprocal diagrams in 2D and 3D space with traditional matrix methods of indeterminate structural analysis.[17] Another insight has found that the 3D polyhedron is completely analogous to the Airy Stress Function, a structural mechanics tool originally in 2D. Today it is a 3D form-finding tool, with no bending in a 3D structure designed for certain loads through the mapping of the Airy Stress Function.[19] All of these topics are currently being studied and disseminated by engineers around the world. Many of these algorithms are incorporated into plugins such as Rhino Vault and new computational environments such as COMPAS, which is freely distributed by the Block Research Group at ETH-Zurich.[20] Another new form-finding tool based on 3D graphic statics is the recently released program PolyFrame, developed and distributed by the Polyhedral Structures Laboratory.[21]

This brief section ends with a qualitative description of some issues surrounding force flow in shells. The 1960s hyperbolic paraboloid was not a funicular structure, which means that a combination of tension and compression and bending may be present. Its primary benefit was constructability from straight timber formwork as well as architectural drama. It often required an edge beam, and this edge beam collects shear forces from the shell, accumulating more and more load as the edge beam travels to the lowest point on the shell. Figure 19.1 shows a typical hyperbolic paraboloid subjected to self weight. Principal stresses and consequently principal force flow are shown in Fig. 19.1*a*, the arrows clearly showing the load flowing into the support at a direction that is parallel to the edge beam near the support but somewhat more turbulent near the peak of the structure. Figure 19.1*b* qualitatively shows the axial forces along the length of the edge beam, starting nearly at zero at the crown, with greatest compression at the base support.

A useful tool that is now readily available in the commercial finite-element program SAP2000 is the ability to capture the deformed geometry of the structure as the starting point for a new, subsequent analysis. This is particularly helpful when designing, analyzing, and constructing thin shells because while the stresses are low, deformations under design loads may cause small, but significant geometry changes. Figure 19.2*a* shows local buckling, and Fig. 19.2*b* captures large nonlinear deformations with no buckling.

These deformations push the shell somewhat out of its funicular (bending free) shape. Construction errors causing small perturbations from the intended funicular shape can also be included in a finite-element buckling analysis. New and potentially larger deformations may occur in the next structural analysis iteration. Subsequent iterations may also introduce material nonlinearities which may be engaged if the strains become large enough. The modification of the original finite-element mesh to the deformed shape as the subsequent starting point is very easily accomplished with SAP2000 by simply using the Modify Undeformed Geometry button under the ANALYZE tab.[22] Figure 19.3*a* shows a factor of safety against buckling for the original shell geometry. Figures 19.3*b* and 19.3*c* show the new propensity to buckle based on the deformed, not the original shapes.

Another exciting and rich development is currently taking place in the world of shell analysis and design. The plug-ins for the drawing program Rhino now include Grasshopper, which can parametrically form-find a funicular shape using particle-spring modeling, as well as Karamba, which can perform stress analysis and buckling analysis of structures.[23] Figure 19.4*a* shows the worst principal stress flow generated by Karamba. This worst principal stress perfectly aligns with the flow of worst compressive forces in the shell. Clearly shown is the path

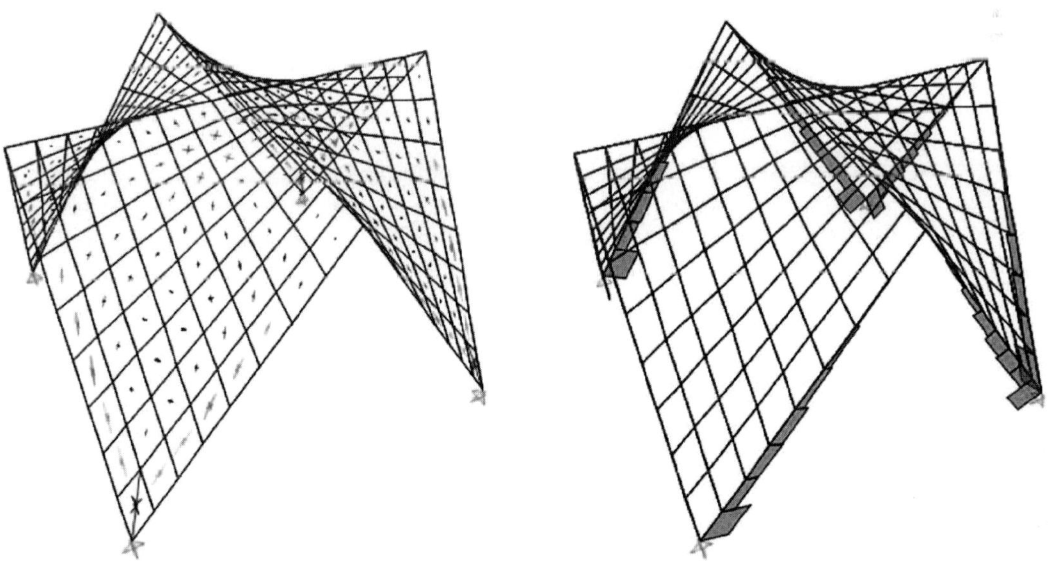

(a) Load flow

(b) Axial force in edge beam

Figure 19.1 Qualitative load flow in hyperbolic paraboloid.

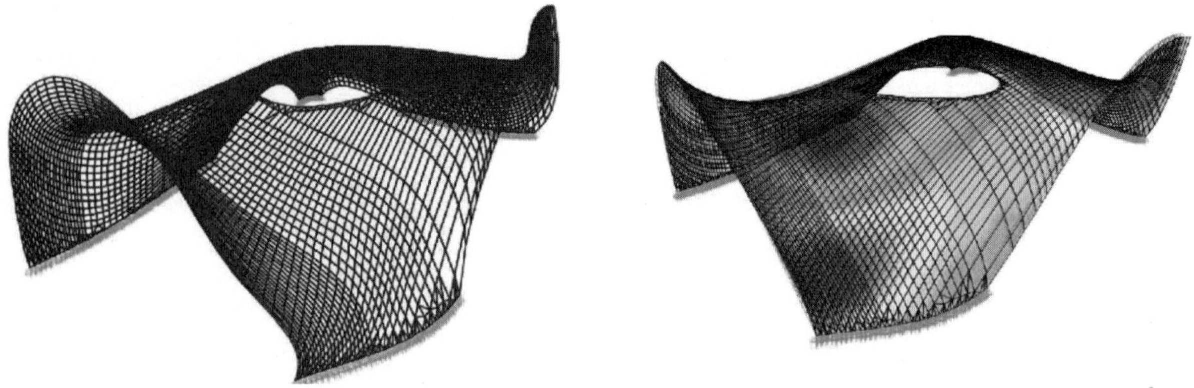

(a) Buckle on free edge

(b) Large displacements on free edge

Figure 19.2 Local buckling at free edge of shell.

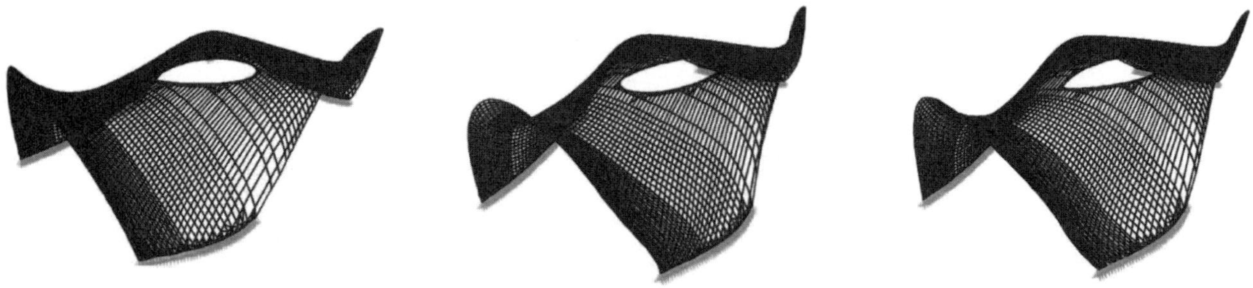

(a) Original Factor of Safety (F.S.) = 11.9

(b) Deformed shell F.S. = 8.6

(c) Deformed plus nonlinear behavior, F.S. = 5.3

Figure 19.3 Final configuration of one analysis is original configuration of subsequent analysis.

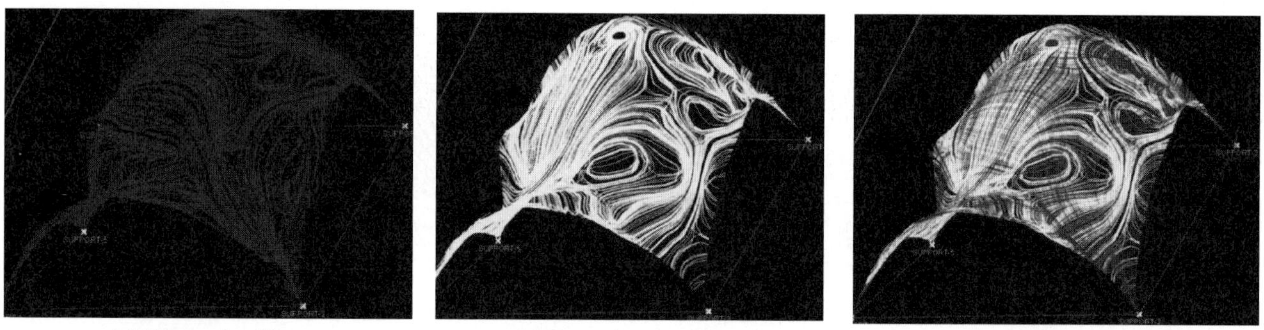

(a) Primary load flow

(b) Secondary load flow

(c) Superposition of primary and secondary flows

Figure 19.4 Primary and secondary load flow in shell.

the load takes; much as water would flow to a downspout, so do the compressive forces flow to the supports of the shell. Figure 19.4*b* shows the second principal stress flow. This second principal stress clearly must be perpendicular or 90 degrees from the first load flow net, as is required by mechanics principles and as shown with 180 degrees on Mohr's Circle. This second principal flow of forces approaches zero at the free edge of the main span of the shell, perpendicular to the free edge as a free body diagram would require. Figure 19.4*c* shows the superposition of the primary and secondary flow nets. Many areas of low to zero load flow would be ideal positions for an oculus.

This section concludes by returning to the beginning of the story, namely, the Guastavinos. Their structures are considered as structural art. They were economically and quickly built, and they have proved durable beyond anybody's expectations. Yet they were not designed by any of the sophisticated computer algorithms just described. The role of the builder becomes paramount in the triangle of engineer-architect-builder when considering shells such as Guastavino vaults. Could there be a 21st-century simple, elegant, and commonsense analogy to the trade secrets patented by Guastavino in the early 20th century? The answer is emphatically yes! The answer is in the use of graphical equilibrium methods, primarily graphic statics, which today is greatly aided by easy-to-use 3D drawing tools. 2D planes can be placed in 3D space, and funiculars could be instantly created along any of these planes. Figure 19.5 shows two such funiculars, and they are forced to intersect at some predetermined crown point. Each funicular is simply a 2D construct using a single force polygon that would represent the weight of a unit strip of material for one funicular.

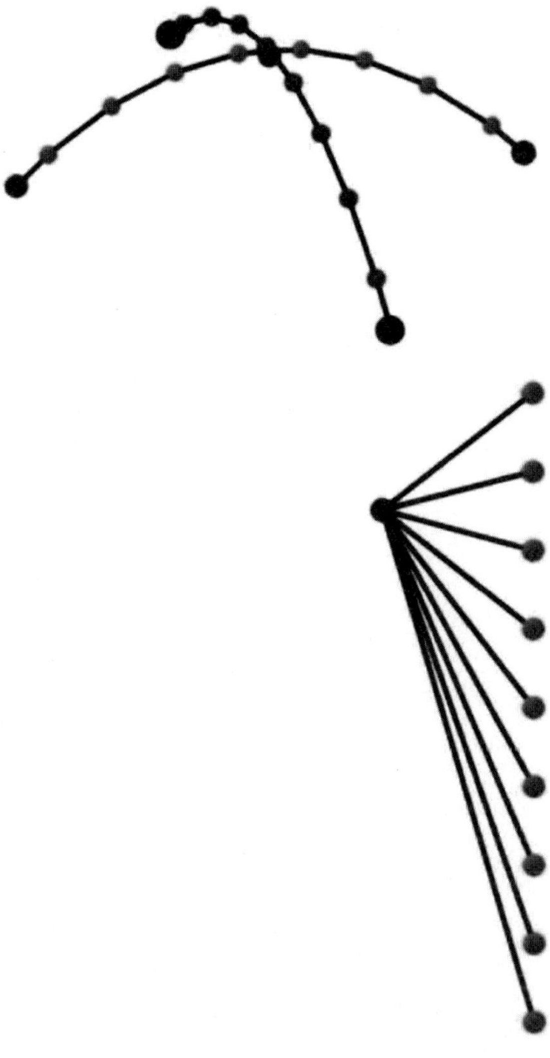

Figure 19.5 Two 2D funiculars in 3D space.

The process can be repeated with other funiculars, and the designer has total control over the form as well as the forces in each strip of material. These funicular 2D lines in 3D space are completely analogous to the visual guides used by Guastavino's construction crew. This is not yet shell action; hoop forces need to be added to the structure, such as splines between 2D arches as shown in Fig. 19.6. This tool can easily be used for quick form finding to aid students as well as educators and practitioners who are unfamiliar with the more complex computational algorithms which are now readily available. Textbooks are available to help engineers and architects learn about funiculars and graphic statics as well as other form-finding tools for shells.[12,24]

As a unique blend of architecture, engineering, and construction, shell structures have a rich history as well as a bright future.

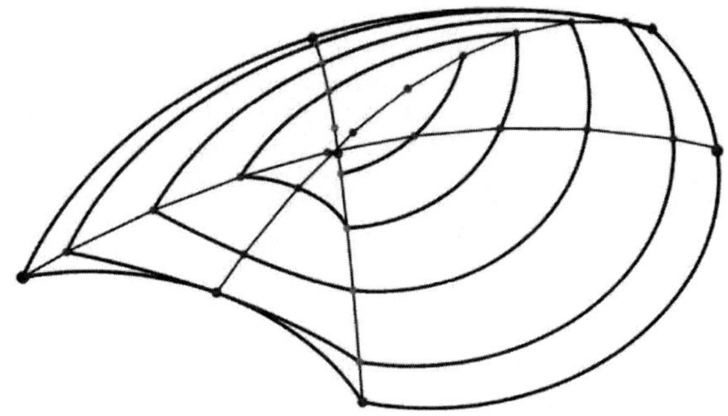

Figure 19.6 Hoop splines surround meridian funiculars.

19.1.1 Thin Shells and Their Classification

The most general classification of thin shells is by Gaussian curvature, illustrated in Fig. 19.7. Gaussian curvature of a surface is defined as the product of the principal curvatures

$$K = \frac{1}{r_x}\frac{1}{r_y}$$

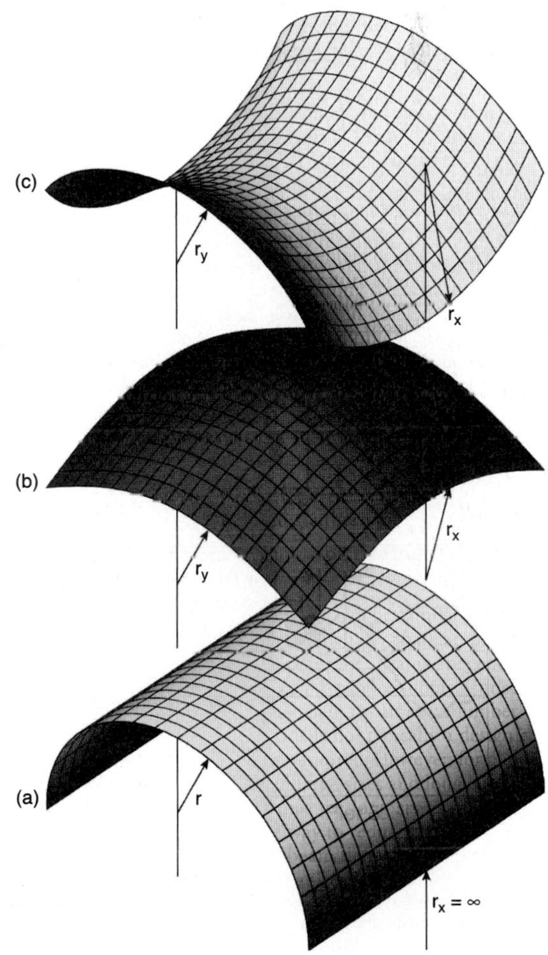

Figure 19.7 Definitions of curvature: (*a*) shell of positive Gaussian curvature, (*b*) shell of zero Gaussian curvature, and (*c*) shell of negative Gaussian curvature.

where r_x and r_y are the principal radii of curvature. Shells of positive Gaussian curvature, sometimes called *synclastic shells*, are formed by two families of curves both with radii of curvature in the same direction. Spherical domes and elliptic paraboloids are examples. Shells of zero Gaussian curvature or singly curved shells are formed by one family of curves only; some examples are cylinders and cones. Shells of negative Gaussian curvature, sometimes called *anticlastic*, are formed by two families of curves each having radii of curvature in opposite directions. Hyperbolic paraboloids and hyperboloids of revolution are examples. Another useful classification divides shells into rotational and translational shells. Domes and tanks are normally surfaces of rotation, whereas cylindrical barrels, elliptic paraboloids, and hyperbolic paraboloids are surfaces of translation.

19.1.2 Introduction to Design and Analysis

The design of thin-shell concrete structures includes attention to analysis, construction, and appearance.

There are three types of analysis. The first gives an overall picture and is essential for preliminary design. This analysis usually emphasizes equilibrium only.

The second type of analysis is more detailed. It gives the information needed to check concrete dimensions and to lay out reinforcing steel. This analysis usually includes not only equilibrium but also compatibility and seeks to get stresses and displacements. Its goal is to take a predetermined form and proportion it efficiently so that as little material as possible will need to be used. This can serve as the final analysis for actual designs. This type of analysis consists of the membrane theory (which assumes no bending) corrected as needed by a bending theory.

The third type of analysis, the most rigorous one, gives insight into the limitations of the other two. This type is based upon numerical procedures suitable for the computer and generally consists of the finite-element method.[25,26] Because of readily available programs for the personal computer, this method finds general use in design. However, the use of such programs may lead to wrong conclusions if the analyst has no understanding of actual shell behavior. The analysis consists of dividing the shell system into elements, computing the stiffnesses of the elements, assembling them into a stiffness matrix for the entire shell system, and solving that matrix for any applied load. The solution gives nodal displacements which from element stiffnesses can be used to compute nodal forces and element stresses. The great advantage of the method is that there is no restriction on shell geometry, boundary conditions, and loading.

The main issues in analysis for thin-shell concrete structures are the following:

1. the element shape
2. the degrees of freedom at the nodes
3. the connections between shell and beam elements

It is easier to use flat-plate elements because in-plane and out-of-plane behavior are not coupled within the element. It would be more accurate geometrically to use curved elements, but it is more difficult to define their stiffness properties. For this reason, Clough and Johnson concluded that "it is more likely that inclusion of curvature in the elements will greatly improve the finite-element approximation."[2] Some of the earliest element configurations were triangles, but for shells, quadrilaterals are regularly used because they provide better membrane strain distributions.

19.1.3 Basis Assumptions and Simplifications

Thin-shell analyses are based upon the assumption that the structure is defined by its middle surface, shown in Fig. 19.8, as halfway between the inner and outer edges of the shell.[27] Furthermore, the internal forces used in shell analysis are defined as *stress resultants* (forces per unit length of middle surface) such as

$$N_\theta\, dy = dy \int_{-h/2}^{h/2} \sigma_\theta \, dz \tag{19.1}$$

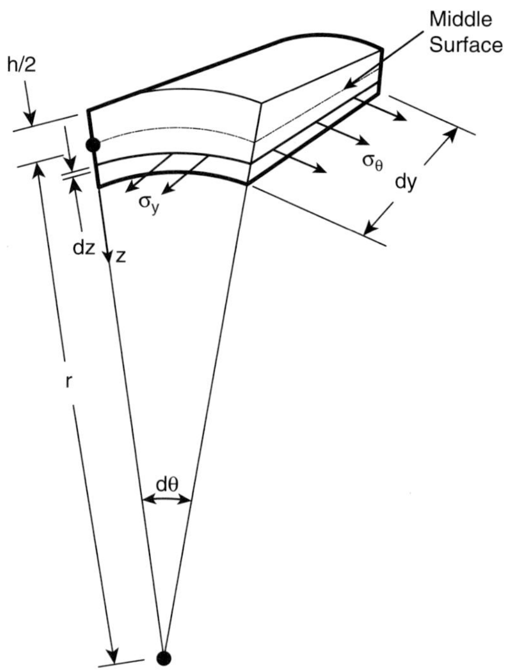

Figure 19.8 Definitions for thin shells.

and *stress couples* (bending moments per unit length of middle surface),

$$M_y r\, d\theta = r\, d\theta \int_{-h/2}^{h/2} \sigma_y z\left(1-\frac{z}{r}\right) dz \cong r\, d\theta \int_{-h/2}^{h/2} \sigma_y z\, ds \tag{19.2}$$

so that the units of N_θ are kips per foot and of M_y are foot-kips per foot. Positive N_θ is tension, and positive M_y is where σ_y is tension (positive) in the positive z direction. From these expressions, we can find stresses by dividing N_θ by h in Eq. (19.1) and by dividing M_y by $h^2/6$ in Eq. (19.2). The term z/r in Eq. (19.2) arises because the cross section is trapezoidal. Where it appears with units, the term z/r can be neglected in expressions like Eq. (19.2). This approximation gives a definition for thin shells, that is, where $h/r << 1$. Therefore, when h/r is greater than about 1/25, one must question the validity of thin-shell analysis as with thin plates. This shell theory required also that the stress vary linearly across the shell thickness.

A general theory of thin shells is overly complex for the needs of the designer, especially since shell analysis is commonly done by numerical methods which do not necessarily use such general theories. One special theory, however, is useful as a basis both for some simplified analyses and for some more detailed analyses. That is the membrane theory, which rests on the main assumption that the shell carries load solely by in-plane stress resultants and that therefore all bending moments and out-of-plane shearing forces are taken as zero. From this main assumption, the following hold:

1. The state of stress (membrane stress resultants) in the shell is completely determined by equations of equilibrium; that is, the shell is statically determinate.
2. The boundary conditions must provide support for those edge stress resultants computed from the equations of equilibrium.
3. The boundary conditions must also permit those edge displacements (translations and rotations) which are computed from the stress resultants found by the membrane theory.

Thus, the membrane theory needs three equations of equilibrium plus expressions for displacements in terms of the membrane stress resultants. If these last two conditions are not met, the membrane theory is no longer reliable.

The following method of analysis uses both the membrane and the bending theories as follows:[25,27,28]

1. As a primary system, the loading is considered to be resisted entirely by the membrane stress resultants.
2. This system leads to errors in the forces and displacements at the shell boundaries which will, in general, not be compatible with the known boundary conditions.
3. Corrections to these forces and displacements (edge effects) must be applied to the shell boundaries to remove the incompatibilities (errors) resulting from the primary system.
4. The values of the edge effects necessary to remove the errors caused by the membrane stress resultants are found by solving compatibility or equilibrium equations for the shell boundaries.
5. The final solution is found by superimposing the "correction" effects on the membrane solution.

One important characteristic difference among shell types is the propagation of edge effects into the shell. For shells of positive curvature, the edge effects tend to damp rapidly and are usually confined to a narrow zone at the edge. Thus, in these shells the membrane theory will often be valid throughout the entire shell except near the boundaries. This rapid damping has been clearly demonstrated for spherical domes. For shells of zero curvature (cylinders), the edge effects are damped but tend to extend farther into the shell than for shells of positive curvature. Finally, for shells of negative curvature (hyperbolic paraboloids), the damping is markedly less than for the others. Thus, the boundary effects tend to become significant over large portions of the shell, which makes the correction method described above inapplicable.[27,29]

19.2 SHELLS OF ROTATION

Shells of rotation are axisymmetric. The load can be axisymmetric (liquid pressure) or not (seismic or wind). A complete (corrected) solution is given here only for axisymmetric loads.

19.2.1 Cylindrical Shell Walls (Membrane Theory)

Figure 19.9*a* shows a ground storage reservoir whose structure is a circular cylinder loaded by an internal pressure that increases linearly with depth. Figure 19.9*b* shows the horizontal ring tension N_θ in the wall for two types of base restraint: free and fixed. Figure 19.9*c* shows the vertical bending moment M_y acting on horizontal sections for the fixed case, there being no moment for the case of no base restraint (free). For the free case, all the internal pressure is carried by the horizontal rings which expand into tension according to the formula

$$N'_\theta = \gamma(H - y)r \tag{19.3}$$

where N'_θ = stress resultant, representing the force per unit length in the y direction
γ = density of fluid
$H - y$ = distance down from cylinder top
r = radius

For the fixed case, the base cannot expand or rotate, and hence $N'_\theta = 0$ there. The solid lines in Fig. 19.9*b* and 19.9*c* show the fixed case, whereas the dashed line shows M_y for the case where $N_\theta = 0$ throughout. This last case occurs when the shell-ring stiffness $k_r = Eh/r^2$ is very small compared to the shell bending stiffness $D = Eh3/12\,(1 - \nu^2)$, that is, when the ratio

$$\frac{k_r}{D} = \frac{12(1-\nu^2)}{h^2r^2} = 4\beta^4 \tag{19.4}$$

is small. This ratio indicates the extent to which a cylindrical shell carries loads by in-plane axial forces such as N_θ or by out-of-plane flexural forces such as M_y and Q_y, the horizontal shear in the radial direction. As the quantity β decreases, the rings carry less, and the cantilevers carry more; the cylinder acts less like a shell and more like a retaining

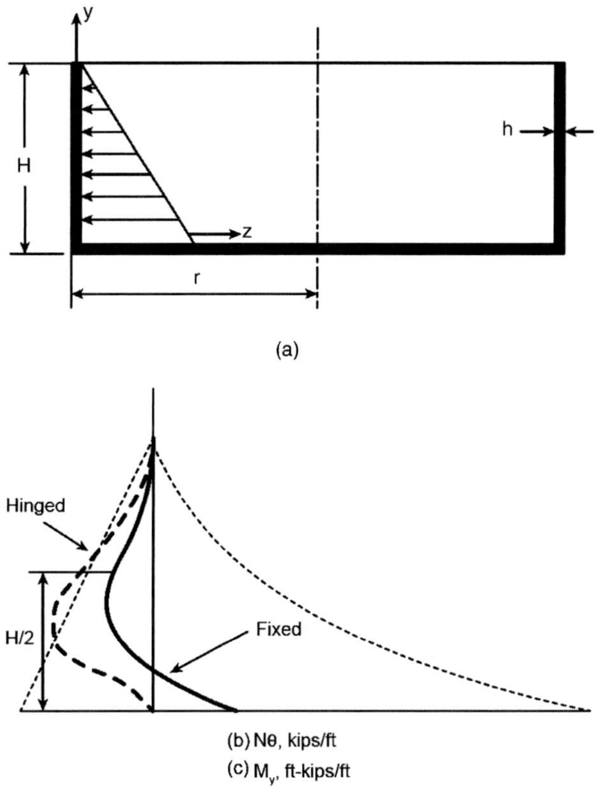

Figure 19.9 Circular cylinder with internal pressure.

wall. The way the shell behavior depends upon β can be summarized as follows:

1. The smaller the r and hence the larger the curvature, defined as $1/r$, the more the load is carried as a shell, that is, carried by the in-plane stress resultant N_θ.
2. The smaller the h, the more the load is carried as a shell.
3. When the shell edge is restrained, the region close to that edge exhibits bending which damps out away from that edge. That damping increases as β gets larger; that is, for shells with large β, the bending is confined to a region close to the edge, and the membrane theory well describes the behavior elsewhere in the shell.

In general, therefore, shell behavior depends upon curvature, thickness, and boundaries. The simplified ideas presented here for the cylindrical tank suggest how these three factors influence the design, but a more detailed analysis is essential to show the limitations to these simplifications. For example, the curves of Fig. 19.9 for the fixed-base tank are based upon the simplifying assumption that H will be large enough so that the influence of the fixed base will not be felt near the tank top.

19.2.2 Cylindrical Tanks

The membrane theory for cylindrical shells results in the following general equations (Fig. 19.10):

$$N'_\phi = -p_z r \tag{19.5a}$$

$$N'_{\phi x} = -\int\left(p_\phi + \frac{1}{r}\frac{\partial N'_\phi}{\partial \phi}\right)dx + f_1(\phi) \tag{19.5b}$$

$$N'_x = -\int\left(p_x + \frac{1}{r}\frac{\partial N'_{x\phi}}{\partial \phi}\right)dx + f_2(\phi) \tag{19.5c}$$

where f_1 and f_2 are determined from support conditions.

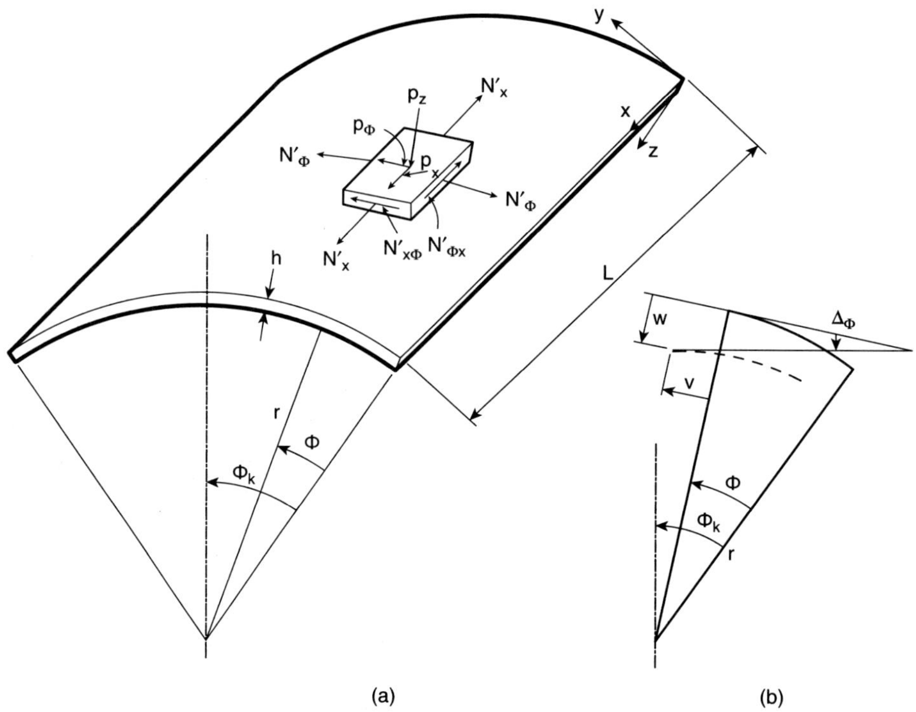

Figure 19.10 Membrane theory for cylindrical shells forces and displacements.

The displacements *u, v, w* in the directions of the axes *x, y, z* are

$$u = \frac{1}{EH}\int (N'_x - \nu N'_\phi)\,dx + f_3(\phi) \tag{19.5d}$$

$$v = \int \frac{1}{r}\frac{\partial u}{\partial \phi}\,dx = \frac{2(1+\nu)}{Eh}\int N'_{\phi x}dx + f_4(\phi) \tag{19.5e}$$

$$w = \frac{\partial v}{d\phi} - \frac{r}{Eh}(N'_\phi - \nu N'_x) \tag{19.5f}$$

where f_3 and f_4 are determined from support conditions.

The rotation of the tangent to the circle at any point, positive as shown in Fig. 19.10*b*, is given by

$$\Delta_\phi = \frac{v}{r} + \frac{\partial x}{r\partial\phi} \tag{19.5g}$$

Internal Liquid Pressure (Fig. 19.11*a*)

With γ = density of liquid, $p_\phi = p_x = 0$, and $p_z = -y(H-x)$, Eq. (19.5) give (see Fig. 19.12)

$$N'_\phi = \gamma(H-x)r \tag{19.6a}$$

$$N'_x = N'_{x\phi} = 0 \tag{19.6b}$$

The resulting displacements are

$$w = \frac{\gamma r^2}{Eh}(H-x) \tag{19.7a}$$

$$\frac{dw}{dx} = \phi_x = \frac{\gamma r^2}{Eh_x} \tag{19.7b}$$

Variable Thickness

If the thickness of the tank wall varies linearly (Fig. 19.11*c*), the stress resultants for liquid pressure are given by Eq. (19.6). The displacements are

$$w = \frac{\gamma r^2}{Eh_x}(H-x) \tag{19.8a}$$

$$\frac{dw}{dx} = \phi_x = \frac{\gamma r^2}{Eh_x}\frac{h_{top}}{h_x} \tag{19.8b}$$

where h_x = thickness at elevation *x*.

Seismic Loading

If the horizontal acceleration can be considered as a percentage α of the dead weight *q*, so that $q_s = \alpha q$, the loading is $p_\phi = q_s \sin\theta$, $p_x = 0$, $p_z = q_s\cos\theta$, and Eq. (19.5) give, for $N'_{\phi x} = N'_x = 0$ at $x = H$,

$$N'_\phi = -q_s r\cos\phi \tag{19.9a}$$

$$N'_{\phi x} = 2q_s(H-x)\sin\phi \tag{19.9b}$$

$$N'_x = \frac{q_s}{r}(H-x)^2\cos\phi \tag{19.9c}$$

Wind Load

The wind pressure *pf* on shell walls of rotation can be defined by Fig. (19.11*b*),

$$p_2 = K_2 H_\phi p_{30} \tag{19.10}$$

where

$$K_2 = 2.64\left(\frac{z}{z_g}\right)^{2a} \tag{19.11}$$

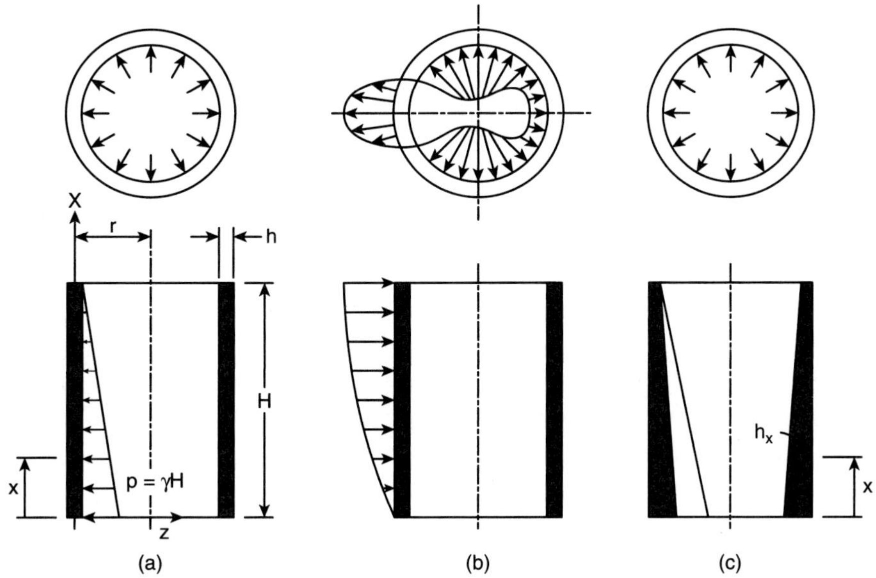

Figure 19.11 Internal liquid pressure on shells.

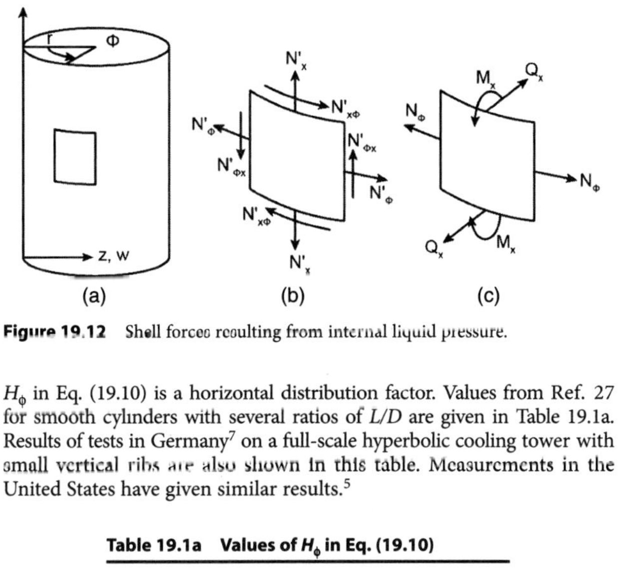

Figure 19.12 Shell forces resulting from internal liquid pressure.

H_ϕ in Eq. (19.10) is a horizontal distribution factor. Values from Ref. 27 for smooth cylinders with several ratios of L/D are given in Table 19.1a. Results of tests in Germany[7] on a full-scale hyperbolic cooling tower with small vertical ribs are also shown in this table. Measurements in the United States have given similar results.[5]

Table 19.1a Values of H_ϕ in Eq. (19.10)

	L/D			
ϕ, deg	1	7	25	Ref. 4
0	1.0	1.0	1.0	1.0
15	0.8	0.8	0.8	0.8
30	0.1	0.1	0.1	0.2
45	−0.7	−0.8	−0.9	−0.5
60	−1.2	−1.7	−1.9	−1.2
75	−1.6	−2.2	−2.5	−1.3
90	−1.7	−2.2	−2.6	−0.9
105	−1.2	−1.7	−1.9	−0.4
120	−0.7	−0.8	−0.9	−0.4
135	−0.5	−0.6	−0.7	−0.4
150	−0.4	−0.5	−0.6	−0.4
165	−0.4	−0.5	−0.6	−0.4
180	−0.4	−0.5	−0.6	−0.4

The following values of α and z_g in Eq. (19.11) are from Ref. 30:

- Cities: $\alpha = 3$, $z_g = 1500$ ft
- Rolling country or shrubs: $\alpha = 4.5$, $z_g = 1200$ ft
- Open country: $\alpha = 7$, $z_g = 900$ ft

The circumferential distribution of wind pressure may be represented by the Fourier series

$$H_\phi = \sum A_n \cos n\phi \tag{19.12}$$

The first eight harmonics are generally sufficient.[30] Values of A_n corresponding to the test values of H_ϕ in Table 19.1a are given in Table 19.1b, along with values for the cylinder with $L/D = 1$.

Table 19.1b Fourier Coefficients for H_ϕ in Table 19.1a*

	A_n	
n	$L/D = 1$	Ref. 4
0	−0.6000	−0.3923
1	0.2979	0.2602
2	0.9184	0.6024
3	0.3966	0.5046
4	−0.0588	0.1064
5	0.0131	−0.0948
6	0.0609	−0.0186
7	−0.0179	0.0468

*From Ref. 9.

For each term in the series, Eq. (19.5) give

$$N'_\phi = p_z r A_n \cos n\phi \tag{19.13a}$$

$$N'_{\phi x} = p_z (H - x) A_n n \sin n\phi \tag{19.13b}$$

$$N'_x = \frac{p_z}{2r}(H - x)^2 A_n n^2 \cos n\phi \tag{19.13c}$$

Edge Effects—Bending Theory

Edge effects are confined to a narrow zone. Therefore, except for the unusual case of a cylinder whose length is of the same order of magnitude as the affected zones, edge effects at one end can be treated independently of those at the other. Forces and displacements are given in Table 19.2.

Table 19.2 Forces (Fig. 19.12c) and Displacements in Circular Cylinders Loaded by Edge Forces Uniform around a Parallel Circle

	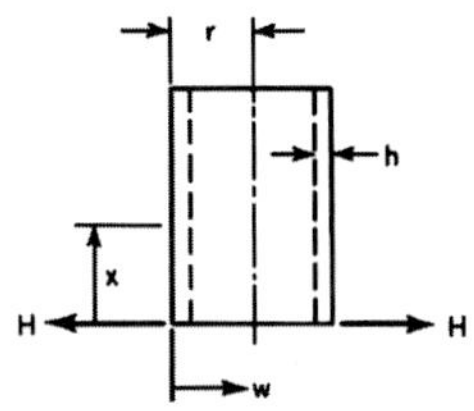	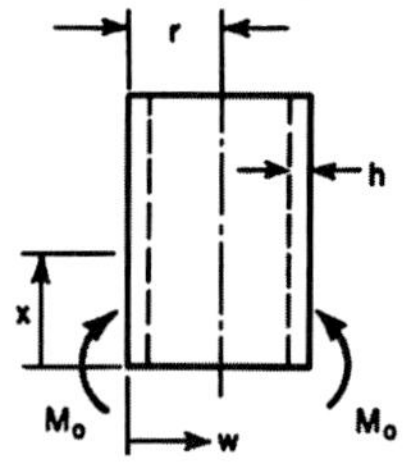
N_ϕ	$2C_2\beta_r H$	$2C_1\beta^2 rM^0$
M_x	$\frac{C_3}{\beta}H$	C_4M_0
Q_x	C_1H	$-2C_3\beta M_0$
$w_{x=0}$	$-2\beta\frac{r}{Eh}H$	$-2\beta^2\frac{r^2}{Eh}M_0$
$\left(\frac{dw}{dx}\right)_{x=0}$	$2\beta^2\frac{r^2}{Eh}H$	$4\beta^3\frac{r^2}{Eh}M_0$

Note: $\beta^4 = 3(1 - \nu^2)/r^2h^2$. See Table 19.4 for C.

19.2.3 Circular Plates

Circular plates are used as the bottom of all tanks and sometimes as roofs for small tanks. Table 19.3 shows the displacements for uniform loading and for unit edge loads.

19.2.4 Tanks with Flat Roofs

For a tank with a flat roof that is continuous with the wall, edge effects can be calculated using Tables 19.2 and 19.4.

Primary System

The roof slab is assumed to be freely supported on the wall.

Errors

Displacements are considered positive in the direction of the redundant forces (Fig. 19.13). Conditions at the base of the wall (i.e., whether it is fixed, hinged, free to slide, etc.) are usually immaterial in this analysis. This is because of the localized nature of edge forces on the wall. For the slab under uniform load, radial displacements are zero.

Rotation of the edge is given in Table 19.2. Thus,

$$\delta^s_{10} = 0 \quad \delta^s_{20} = -\frac{dw}{dr} = \frac{3qr^3}{28D} \quad \text{for } \nu = \frac{1}{8}$$

For the tank filled with liquid, from Eq. (19.7),

$$\delta^s_{10} = 0 \quad \delta^s_{20} = -\frac{dw}{dx} = \frac{\gamma r^2}{Eh}$$

Table 19.3 Symmetrical Bending of Circular Plates

	M_r uniform on circumference	Uniform q over surface	Uniform q over surface
	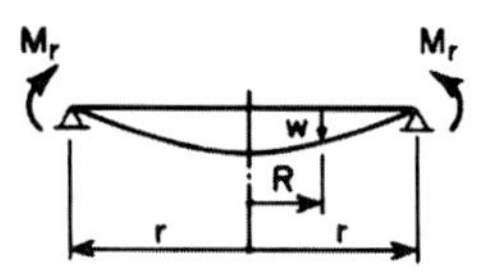	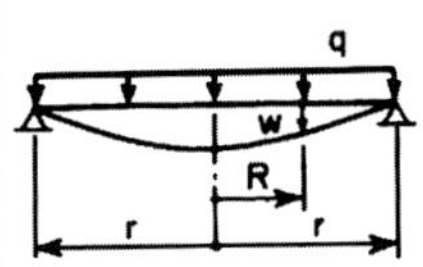	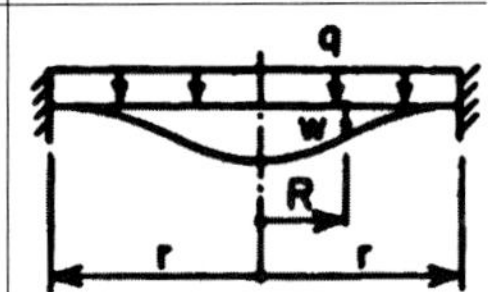
w	$\frac{Mr(r^2 - R^2)}{2D(1+\nu)}$	$\frac{q}{64D}(r^2 - R^2)\left(\frac{5+\nu}{1+\nu}r^2 - R^2\right)$	$\frac{q}{64D}(r^2 - R^2)^2$
$\frac{dw}{dR}$	$-\frac{M_rR}{D(1+\nu)}$	$-\frac{qR}{16D}\left(\frac{3+\nu}{1+\nu}r^2 - R^2\right)$	$-\frac{qR}{16D}(r^2 - R^2)$
$\frac{d^2w}{dR^2}$	$-\frac{M_r}{D(1+\nu)}$	$-\frac{q}{16D}\left(\frac{3+\nu}{1+\nu}r^2 - 3R^2\right)$	$-\frac{q}{16D}(r^2 - 3R^2)$
M_R	M_r	$\frac{q}{16}(3+\nu)(r^2 - R^2)$	$\frac{q}{16}[r^2(1+\nu) - R^2(3+\nu)]$
M_T	M_r	$\frac{q}{16}[r^2(3+\nu) - R^2(1+3\nu)]$	$\frac{q}{16}[r^2(1+\nu) - R^2(1+3\nu)]$
M_r	M_r	0	$-\frac{qr^2}{8}$
M_L	M_r	$\frac{qr^2}{16}(3+\nu)$	$\frac{qr^2}{16}(1+\nu)$

Table 19.4 Coefficients *C* for Tables 19.2 and 19.7*

β_s or β_x†	C_1	C_2	C_3	C_4
0	1.0000	1.0000	0	1.0000
0.1	0.8100	0.9003	0.0903	0.9907
0.2	0.6398	0.8024	0.1627	0.9651
0.3	0.4888	0.7077	0.2189	0.9267
0.4	0.3564	0.6174	0.2610	0.8784
0.5	0.2415	0.5323	0.2908	0.8231
0.6	0.1431	0.4530	0.3099	0.7628
0.7	0.0599	0.3798	0.3199	0.6997
0.8	−0.0093	0.3131	0.3223	0.6354
0.9	−0.0657	0.2527	0.3185	0.5712
1.0	−0.1108	0.1988	0.3096	0.5083
1.1	−0.1457	0.1510	0.2967	0.4476
1.2	−0.1716	0.1091	0.2807	0.3899
1.3	−0.1897	0.0729	0.2626	0.3355
1.4	−0.2011	0.0419	0.2430	0.2849
1.5	−0.2068	0.0158	0.2226	0.2384
1.6	−0.2077	−0.0059	0.2018	0.1959
1.7	−0.2047	−0.0235	0.1812	0.1576
1.8	−0.1985	−0.0376	0.1610	0.1234
1.9	−0.1899	−0.0484	0.1415	0.0932
2.0	−0.1794	−0.0563	0.1230	0.0667
2.1	−0.1675	−0.0618	0.1057	0.0439
2.2	−0.1548	−0.0652	0.0895	0.0244
2.3	−0.1416	−0.0668	0.0748	0.0080
2.4	−0.1282	−0.0669	0.0613	−0.0056
2.5	−0.1149	−0.0658	0.0492	−0.0166
2.6	−0.1019	−0.0636	0.0383	−0.0254
2.7	−0.0895	−0.0608	0.0287	−0.0320
2.8	−0.0777	−0.0573	0.0204	−0.0369
2.9	−0.0666	−0.0534	0.0132	−0.0403
3.0	−0.0563	−0.0493	0.0071	−0.0423
3.1	−0.0469	−0.0450	0.0019	−0.0431
3.2	−0.0383	−0.0407	−0.0024	−0.0431
3.3	0.0306	−0.0364	−0.0058	−0.0422
3.4	−0.0237	−0.0323	−0.0085	−0.0408
3.5	−0.0177	−0.0283	−0.0106	−0.0389
3.6	−0.0124	−0.0245	−0.0121	−0.0366
3.7	−0.0079	−0.0210	−0.0131	−0.0341
3.8	−0.0040	−0.0177	−0.0137	−0.0314
3.9	−0.0008	−0.0147	−0.0140	−0.0286
4.0	0.0019	−0.0120	−0.0139	−0.0258
4.1	0.0040	−0.0095	−0.0136	−0.0231
4.2	0.0057	−0.0074	−0.0131	−0.0204
4.3	0.0070	−0.0054	−0.0125	−0.0179
4.4	0.0079	−0.0038	−0.0117	−0.0155

*From Ref. 10.
†β_x for circular cylinders (Table 19.2); β_s for domes (Table 19.7).

Corrections

For the wall, using Table 19.2 with $H = -X_1 = -1$ and $M_0 = X_2 = 1$,

$$\delta_{11}^w = \frac{2\beta r^2}{Eh} \quad \delta_{12}^w = -\frac{2\beta^2 r^2}{Eh} \quad \delta_{22}^w = \frac{4\beta^2 r^2}{Eh}$$

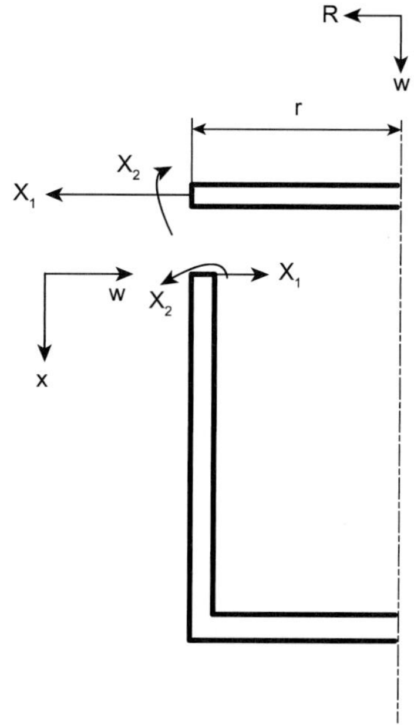

Figure 19.13 Displacement sign convention (tanks with flat roofs).

The radial displacement of the circumference of a circular slab subjected to unit radial forces at the perimeter is given by

$$\delta_{11}^s = \frac{r}{Eh}(1-\nu)$$

and the edge rotation is zero, so that

$$\delta_{12}^s = 0$$

The edge rotation resulting from X_2 is found from Table 19.3. With $M_r = X_2 = 1$,

$$\delta_{22}^s = \frac{dw}{dR} = \frac{r}{D(1+\nu)}$$

Compatibility is established by satisfying the simultaneous equations in X_1 and X_2, where $\delta_{11} = \delta_{11}^w + \delta_{11}^s$ etc.

With X_1 and X_2 known, the internal forces to be provided for in the tank are evaluated using $H = -X$ and $M_0 = X_2$ in the formulas of Table 19.2, to which the membrane stress resultants [Eq. (19.6)] must be added. The internal forces in the roof are obtained similarly from Table 19.3.

The analyses for a tank roofed with a cylindrical dome and for the interaction between wall and floor (if they are continuous) are identical. However, the loads acting on the floor and the corresponding errors and corrections may be more difficult to determine (see Ref. 27, p. 90).

19.3 DOMES

The dome in Fig. 19.14 carries vertical axisymmetrical loads much like a series of pie-shaped arches along meridian lines. But unlike arches, these meridional strips can carry loads with bending even if their axial curve is different from the beam bending moment diagram for the loading because the stress resultants N'_θ act as stiff rings to prevent meridional bending. Thus, unlike arches, the form of the dome is not critical thanks to the rings, but, as with arches, the edge supports of a dome are important. Consider a dome for which only

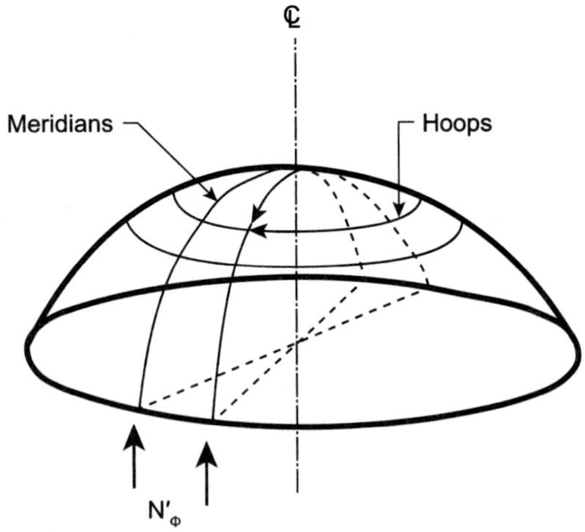

Figure 19.14 Behavior of domes.

vertical supports exist. The arch-type forces N'_ϕ (called *meridional stress resultants*) at the edge, being in the plane of the dome surface, require at the edge both a vertical and a horizontal reaction. A relatively stiff edge ring can provide this horizontal reaction $H = N'_\alpha \cos\alpha$ by stretching under a ring tensile force

$$T = Hr_0 = N'_\alpha r_2 \cos\alpha \sin\alpha \tag{19.14}$$

where α is the value of ϕ at the edge.

19.3.1 Membrane Theory

Surface of Revolution

The equation for vertical equilibrium (see Fig. 19.15) gives an expression for the meridional stress resultant

$$N'_\phi = \frac{R}{2\pi r_0 \sin\phi} \tag{19.15}$$

where R = resultant of the vertical components of all loads above the angle ϕ.

A second equilibrium equation (Fig. 19.16) in the direction normal to the surface gives

$$\frac{N'_\theta}{r_2} + \frac{N'_\theta}{r_1} + p_z = 0 \tag{19.16}$$

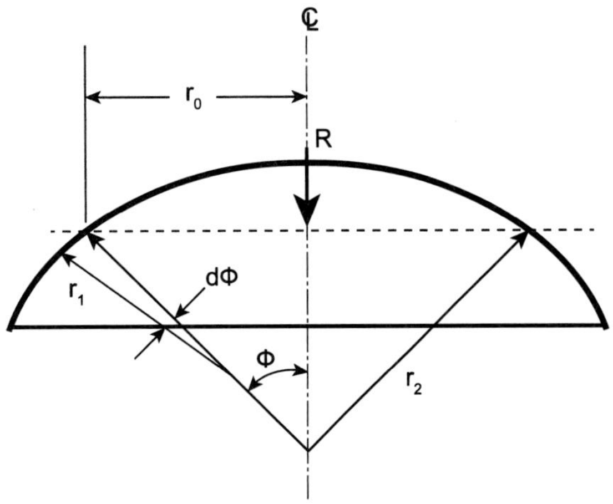

Figure 19.15 Definition of domes.

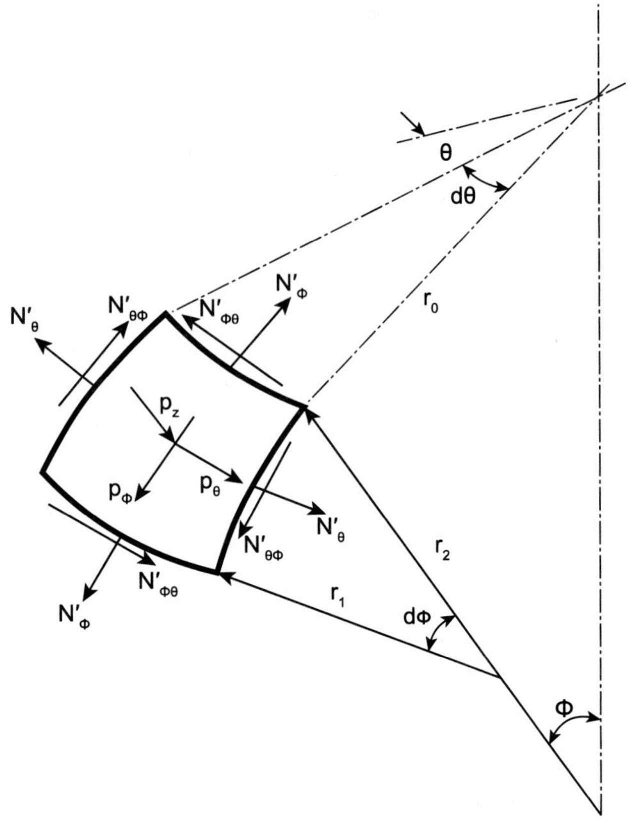

Figure 19.16 Stresses and loads on a shell element.

where r_1 = principal radius of meridional curvature
r_2 = principal radius of rotation
p_z = local radial pressure

Using Eqs. (19.16) and (19.17) with the relation $r_0 = r_2 \sin\phi$, we find (see Fig. 19.17)

$$N'_\theta = \frac{R}{2\pi r_1 \sin^2\phi} - p_z \frac{r_0}{\sin\phi} \tag{19.17}$$

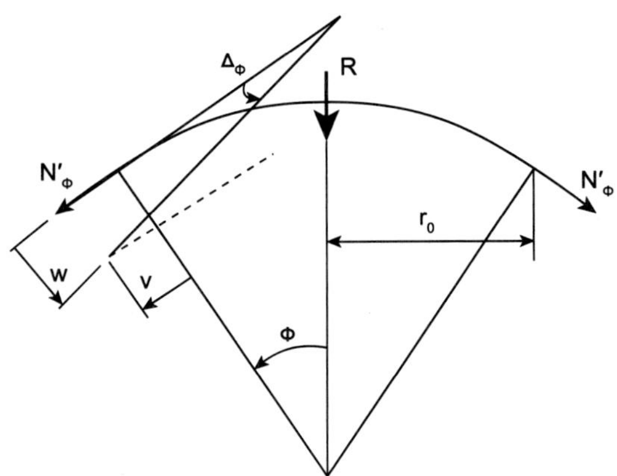

Figure 19.17 Displacements *v* and *w*.

The displacements v and w (see Fig. 19.11) are given by

$$v = \sin\phi\left[\int \frac{f(\phi)}{\sin\phi}d\phi + C\right] \tag{19.18}$$

$$w = v\cot\phi - \frac{r_2}{Eh}(N'_\theta - \nu N'_\phi) \tag{19.19}$$

where ν = Poisson's ratio
C = constant determined by support conditions

$$f(\phi) = \frac{1}{Eh}[N'_\phi(r_1 + \nu r_2) - N'_\theta(r_2 - \nu r_1)] \tag{19.20}$$

The rotation Δ_ϕ of the tangent to the meridian at any point is

$$\Delta_\phi = \frac{v}{r_1} + \frac{dw}{r_1 d\phi} \tag{19.21}$$

Values of the displacements are usually needed only at the supports. For supports as shown in Fig. 19.18, $v_\alpha = 0$, and the horizontal displacement is given by

$$\Delta_H = \frac{r_2 \sin\alpha}{Eh}(N'_\theta - \nu N'_\theta)_{\theta=\alpha} \tag{19.22a}$$

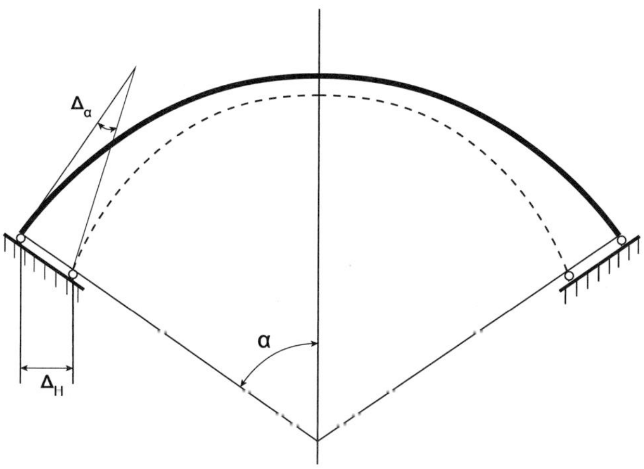

Figure 19.18 Positive directions of Δ_H and Δ_α.

The rotation of the meridional tangent at the edge is

$$\Delta_\alpha = \frac{\cot\alpha}{r_1}f(\alpha) + \frac{1}{r_1}\frac{d}{d\alpha}\frac{\Delta_H}{\sin\alpha} \tag{19.22b}$$

where f is the function defined in Eq. (19.20). The positive directions of Δ_H and Δ_α are shown in Fig. 19.18.

Spherical Dome of Constant Thickness

With q = weight per square foot of dome surface, $p_z = q\cos\phi$ (Fig. 19.19). Then, with $r_1 = r_2 = a$, Eqs. (19.15) and (19.18) give

$$N'_\phi = \frac{aq}{1+\cos\phi} \tag{19.23a}$$

$$N'_\theta = aq\left(\frac{1}{1+\cos\phi} - \cos\phi\right) \tag{19.23b}$$

The distribution of these resultants is shown in Fig. 19.19. If the dome is a spherical segment, there is no hoop tension at the edge if $\phi \lesssim 51°50'$, and there is continuous support by forces tangent to the edge.

With load p per square foot of horizontal projection of the surface, $p_z = p\cos^2\phi$, and Eqs. (19.15) and (19.18) give

$$N'_\phi = -\frac{ap}{2} \tag{19.24a}$$

$$N'_\theta = -\frac{ap}{2}\cos^2\phi \tag{19.24b}$$

Stress resultants for other loadings are found similarly. Results for several cases are given in Table 19.5.

With N'_ϕ and N'_θ known, edge displacements for domes supported as in Fig. 19.18 are found from Eq. (19.22). Corresponding to Eq. (19.23),

$$\Delta_H = \frac{a^2 q}{Eh}\left(\frac{1+\nu}{1+\cos\alpha} - \cos\alpha\right)\sin\alpha \tag{19.25a}$$

$$\Delta_\alpha = -\frac{aq}{Eh}(2+\nu)\sin\alpha \tag{19.25b}$$

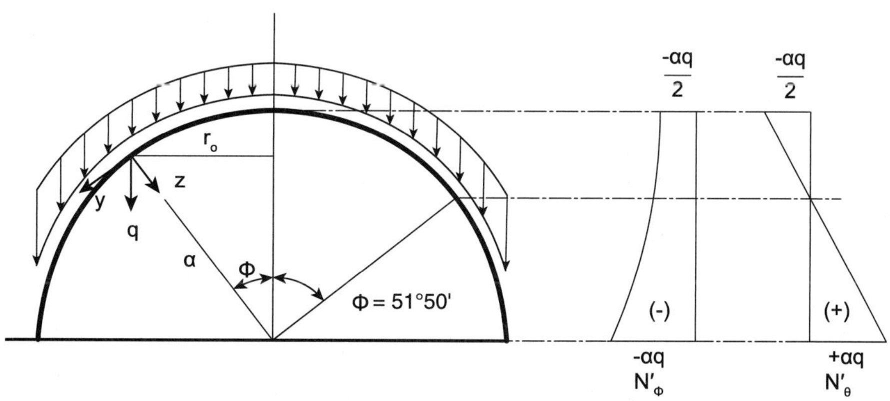

Figure 19.19 Distribution of forces.

Table 19.5 Membrane Forces in Spherical Domes

Loading	N'_ϕ	N'_θ	$N'_{\phi\theta}$
q per ft² of surface 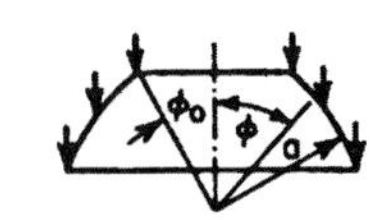	$-aq\dfrac{\cos\phi_0-\cos\phi}{\sin^2\phi}$ $-aq\dfrac{1}{1+\cos\phi}$	$-N'_\phi - aq\cos\phi$ For $\phi_0 = 0$ (no opening) $-N'_\phi - aq\cos\phi$	0 0 0
p per ft² projection 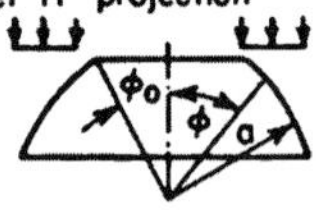	$-\dfrac{ap}{2}\left(1-\dfrac{\sin^2\phi_0}{\sin^2\phi}\right)$ $-\dfrac{ap}{2}$	$-N'_\phi - ap\cos^2$ For $\phi_0 = 0$ (no opening) $-\dfrac{ap}{2}\cos^2\phi$	0 0
p_L per ft of edge 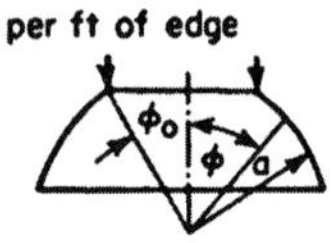	$-p_L\dfrac{\sin\phi_0}{\sin^2\phi}$ $-P_L\dfrac{1}{2\pi a\sin^2\phi}$	$-N'_\phi$ For $\phi_0 = 0$ (load P_L at vertex) $-N'_\phi$	0 0
p_w per ft² projection 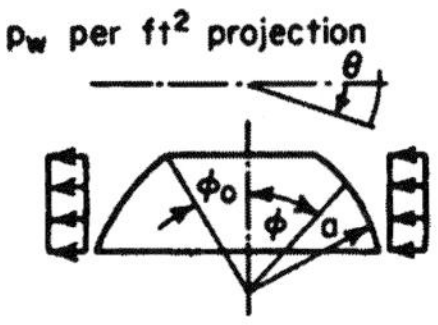	$-p_w\dfrac{a}{3}\dfrac{\cos\phi\cos\theta}{\sin^2\theta}[3(\cos\phi_0-\cos\phi)-(\cos^3\phi)]$ $-p_w\dfrac{a}{3}\dfrac{\cos\phi\cos\theta}{\sin^3\phi}(2-3\cos\phi+\cos^3\phi)$	$-N'_\phi - ap_w\sin\phi\cos\theta$ For $\phi_0 = 0$ (no opening) $-N'_\phi - ap_w\sin\phi\cos\theta$	$N'_\phi\dfrac{\tan\theta}{\cos\phi}$ $N'_\phi\dfrac{\tan\theta}{\cos\phi}$

and, corresponding to Eq. (19.24),

$$\Delta_H = -\frac{a^2 p}{2Eh}(-\cos^2\alpha + v)\sin\alpha \tag{19.26a}$$

$$\Delta_\alpha = -\frac{ap}{2Eh}(3+v)\sin^2\alpha \tag{19.26b}$$

Spherical Dome of Variable Thickness

A good approximation of the edge forces and displacements in a spherical dome of variable thickness is obtained by assuming a uniform thickness equal to that at the arc distance S_a from the edge,[8]

$$S_a = 0.5\sqrt{ah_a} \tag{19.27a}$$

where a = radius of dome

h_a = average thickness of thickened portion

This approximation involves the assumption that the thickened portion extends the distance S from the edge,

$$S_a = 2\sqrt{ah_a} \tag{19.27b}$$

Parabolic Dome of Constant Thickness

The membrane stress resultants for two cases of loading are given in Table 19.6. Displacements Δ_H and Δ_α can be determined by substituting these values in Eq. (19.22).

Edge Effects—Bending Theory

For the bending near the edges of domes, it is usually sufficient to use the Geckeler approximation (see Ref. 27, pp. 31 to 32). To see the idea behind this approximation, we can consider the effect of boundary conditions (Fig. 19.20) on both the arch and the dome structures. If we give the arch a horizontal push at one side, this force must be held in equilibrium by an equal and opposite force acting at the other support.

Table 19.6 Membrane Forces in Parabolic Domes

Loading	N'_ϕ	N'_θ	$N'_{\phi\theta}$
q per ft² of surface 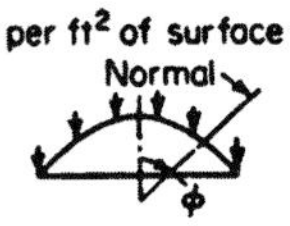	$-\dfrac{qr_0}{3}\dfrac{1-\cos^3\phi}{\sin^2\phi\cos^2\phi}$	$-\dfrac{qr_0}{3}\dfrac{2-3\cos^2\phi+\cos^3\phi}{\sin^2\phi}$	0
p per ft² of projection 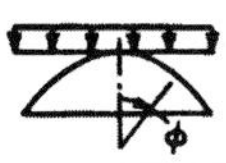	$-\dfrac{pr_0}{2\cos\phi}$	$-\dfrac{pr_0}{2}\cos\phi$	0

Note: r_0 = radius of curvature at vertex.

The only way that the force can be transmitted through the system is up through the arch, over the crown, and down the other side. The effect clearly is to produce high-bending moments throughout the entire arch system, with a maximum moment at the crown. We see, therefore, that the arch is not only restricted in its more efficient form by the nature of the loading but also very sensitive to foundation displacements or edge forces.

Let us give a corresponding horizontal force to the dome. Such a force may be considered as a uniform horizontal thrust applied all around the circular edge of the dome. In cross section, it would appear that this horizontal force would create bending moments throughout the dome similar to those created in the arch. This is, however, not the case. As the horizontal force tends to bend the shell and thus to be carried up a meridian, the rings again come into play and cause a rapid damping of the bending so that at a relatively short distance from the edge, the bending effect is no longer observable. Thus, edge forces in equilibrium applied to an arch propagate throughout the entire structural system and create large bending moments, whereas similar forces acting on a

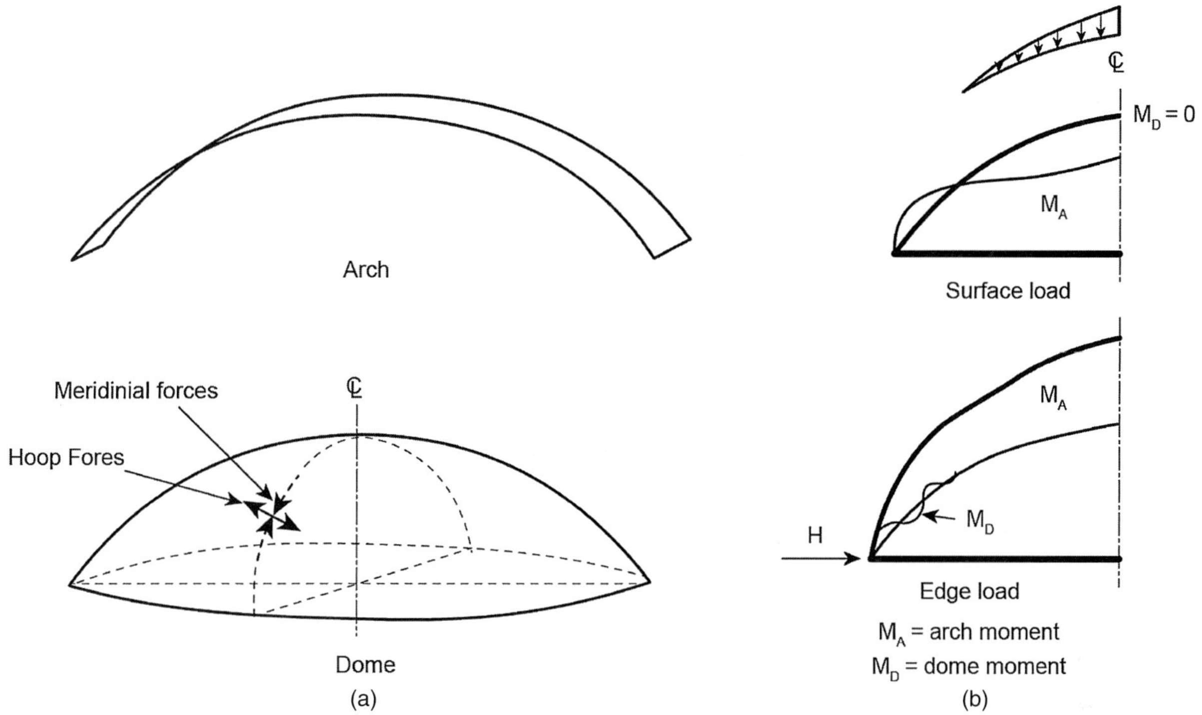

Figure 19.20 Comparison of dome and arch: (*a*) structural systems and (*b*) loadings.

dome create bending moments in a very narrow region near the edge and have generally no effect elsewhere in the dome. This behavior is similar to that of the vertical cylinder with a restrained base (Fig. 19.9). Indeed, the extent of the bending in a dome can be well estimated by using the same analysis as for a cylinder where the ratio β has the same meaning, except r is now r_2, the radius of curvature of the dome edge, defined as $r_2 = r_0/\sin\alpha$. The solution for bending in a full cylinder is used for the dome edge. Table 19.7 gives values for domes loaded by edge forces and moments. Values of the coefficient C are given in Table 19.4. Values of N_ϕ, N_θ, and M_ϕ are given for any point in the shell defined by the angle ψ. The edge displacements Δ_H and Δ_α are positive directions of H and M_α, respectively.

The formulas in Table 19.7 derive from a bending-theory differential equation of the fourth order. To simplify the solution, the radius r_2 is taken constant, which is correct only for the spherical dome. It is because edge effects damp out so rapidly that the formulas can be used for most shells of revolution. The formulas can be relied on only if the opening angle α is greater than 25 or 30 degrees.

Typical dimensions of spherical domes are given in Table 19.8.

Table 19.7 Forces and Displacements in Domes of Revolution Loaded by Edge Forces Uniform around a Parallel Circle

	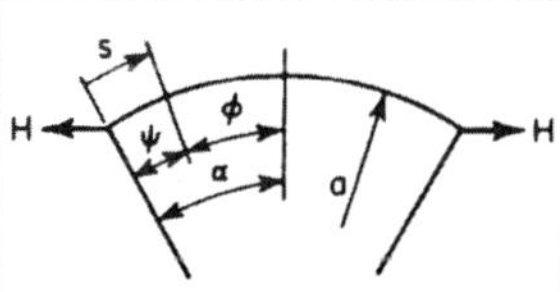	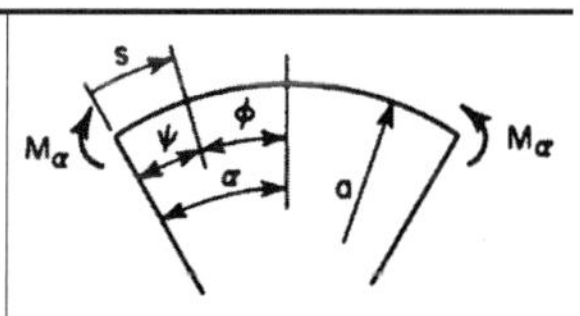
N_ϕ	$C_1 H \sin\alpha \cot(\alpha - \psi)$	$-2\,C_3\beta M_\alpha \cot(\alpha - \psi)$
N_θ	$2C_2\beta a H \sin\alpha$	$2C_1\beta^2 a\, M_\alpha$
M_ϕ	$\dfrac{C_3}{\beta} H \sin\alpha$	$C_4 M_\alpha$
Δ_H	$2\beta\dfrac{a^2}{Eh} H \sin^2\alpha$	$\dfrac{2\beta^2 a^2 \sin\alpha}{Eh} M_\alpha$
Δ_α	$2\beta^2\dfrac{a^2}{Eh} H \sin\alpha$	$\dfrac{4\beta^3\alpha^2}{Eh} M_\alpha$

Note: $\beta^4 = 3(1 - v^2)/r^2h^2$. See Table 19.4 for C.

Table 19.8 Typical Dimensions for Spherical Domes*

Section	D, ft	h, in.†	ϕ, deg	R, ft	a, ft
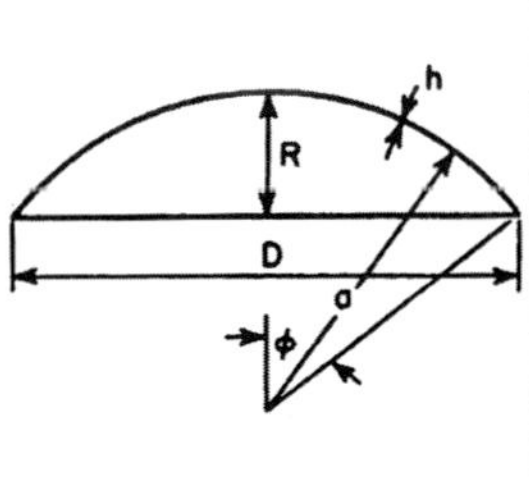	100	3	30	13.4	100
			45	20.7	70.7
	125	3	30	16.8	125
			45	25.9	88.4
	150	3.5	30	20.1	150
		(3)	45	31.0	106.0
	175	4	30	23.5	175
		(3.5)	45	36.2	123.7
	200	4.5	30	26.8	200
		(4)	45	41.4	141.4

*From Portland Cement Association.
†Thickness usually increased by 50 to 75 percent at periphery.

19.3.2 Examples for Shell Rotational Structures

As seen in the previous sections (19.2.1 through 19.2.3), the displacements obtained by the membrane theory do not satisfy the compatibility at the edges or at the junction of different shells. An outline of the solution was presented in Sec. 19.1.3. For what follows, a number of examples are presented to illustrate this method.

EXAMPLE—RIGIDLY SUPPORTED SPHERICAL DOME

The dome of Fig. 19.21 is 2½ in. thick, except that it is thickened at the edge to 6 in. with a uniform taper over a length of 8 ft. The average dead load for the shell is taken at 40 psf. Roofing plus live load is 50 psf. Edge displacements will be calculated for a uniform thickness of 4 in. (Fig. 19.22).

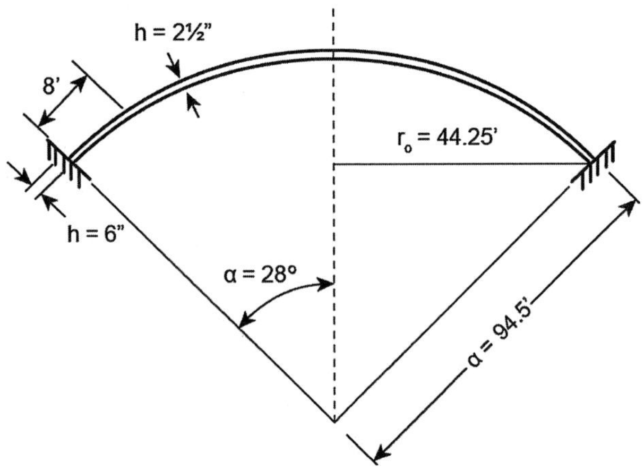

Figure 19.21 Spherical dome example geometry.

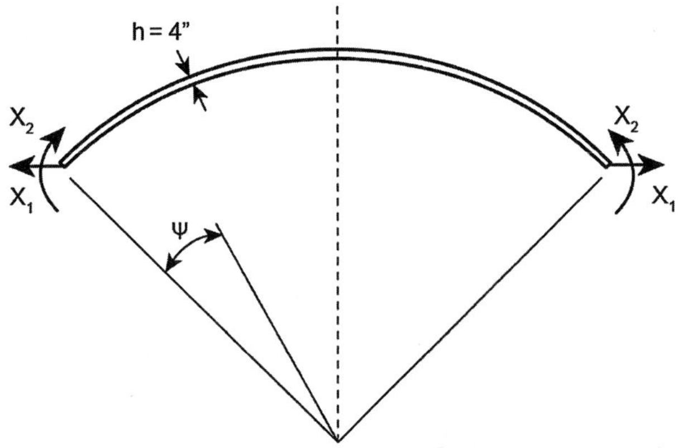

Figure 19.22 Spherical dome example reactions and thickness used for displacement calculations.

PRIMARY SYSTEM. The stress resultants are obtained from Eq. (19.23):

$$N'_\phi = -\frac{94.5q}{1+\cos\phi} \quad N'_\theta = 94.5q\left(\frac{1}{1+\cos\phi} - \cos\phi\right)$$

At the edge,

$$N'_\alpha = -\frac{94.5q}{1+0.883} = -50.1q$$

and the horizontal and vertical components are

$$N'_{aH} = N'_\alpha \cos\alpha = -50.1q \times 0.883 = -44.5q$$
$$N'_{aV} = N'_\alpha \sin\alpha = -50.1q \times 0.469 = -23.8q$$

ERRORS. Edge displacements are considered positive in the directions of the redundant forces X_1 and X_2 (Fig. 19.22). Then, from Eq. (19.25),

$$\delta_{10} = \frac{94.5^2 q}{0.33E}\left(\frac{1.167}{1883} - 0.883\right)0.469 = -3340\frac{q}{E}$$

$$\delta_{20} = \frac{94.5^2 q}{0.33E} \times 2.1677 \times 0.469 = 291\frac{q}{E}$$

CORRECTIONS. The displacements of the edge resulting from unit values of the edge redundant forces X_1 and X_2 are found from the appropriate formulas in Table 19.7:

$$\beta^4 = 0.00300\frac{3(1-0.167^2)}{(94.5\times 0.33)^2}$$

$$\beta = 0.234 \quad \beta^2 = 0.548 \quad \beta^3 = 0.0128$$

$$\delta_{11} = \frac{2\times 0.234\times 94.5^2\times 0.469^2}{E_{0.33}} = \frac{2780}{E}$$

$$\delta_{12} = \delta_{21} = \frac{2\times 0.0548\times 94.5^2\times 0.469}{0.33E} = \frac{1390}{E}$$

$$\delta_{22} = \frac{4\times 0.0128\times 94.5^2}{0.33E} = \frac{1380}{R}$$

COMPATIBILITY. The equations of compatibility are

$$\delta_{11}X_1 + \delta_{12}X_2 + \delta_{10} = 0$$
$$\delta_{12}X_1 + \delta_{22}X_2 + \delta_{20} = 0$$

so that

$$2780X_1 + 1390X_2 = 3340q$$
$$1390X_1 + 1380X_2 = 291q$$

from which

$$X_1 = 2.63q = 0.237\text{kip/ft}$$
$$X_2 = -2.86q = -0.260\text{ft} - \text{kip/ft}$$

With $H = X_1$ and $M_\alpha = X_2$ known, N_ϕ, N_θ, and M_ϕ can be determined from the formulas in Table 19.7. The final stress resultants are obtained by adding these values to the membrane resultants N'_ϕ and N'_θ (see following example).

TEMPERATURE CHANGE. No forces result from expansion or contraction of the primary (membrane) structure. The only error in geometry is the lateral expansion caused by a temperature rise T, which is

$$\Delta_H = \delta_{10} = r_0 T\varepsilon = 44.25\times 6\times 10^{-6}T = 265\times 10^{-6}T$$

The equations of compatibility are

$$2780X_1 + 1390X_2 = -265\times 10^{-6}TE$$
$$1390X_1 + 1380X_2 = 0$$

EXAMPLE—RING-STIFFENED SPHERICAL DOME

The dome of this example is the same as that of the preceding example except that the edge is stiffened with a ring which is monolithic with the dome and which is free to slide and rotate on an immovable support (Fig. 19.23*a*).

PRIMARY SYSTEM. This consists of the shell and the ring as separate structures, subjected to the membrane stress resultant N'_α, which is assumed to act at the idealized junction of the shell and ring (Fig. 19.23*b*). Displacements are taken positive in the directions of the redundants X_1 and X_2.

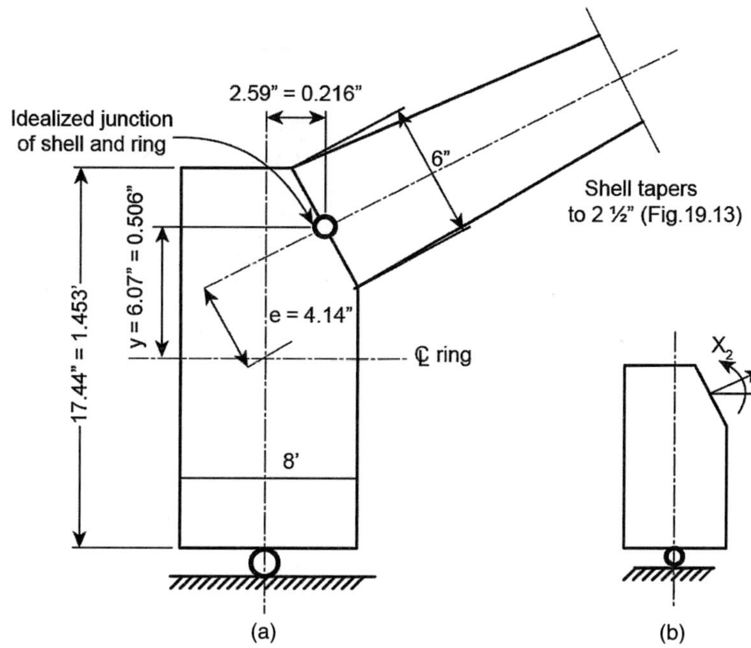

Figure 19.23 Ring-stiffened spherical dome—idealized junction of shell and ring.

The stress resultants and their horizontal and vertical components were determined in the preceding example. For the dead load $q = 40$ psf,

$$N'_\phi = \frac{3.78}{1+\cos\phi} \text{ kips/ft}$$

$$N'_\theta = 3.78\left(\frac{1}{1+\cos\phi} - \cos\phi\right) \text{ kips/ft}$$

$$N'_{aH} = 1.78 \text{ kips/ft} \quad N'_{aV} = -0.95 \text{ kips/ft}$$

ERRORS. These consist of displacements at the junction of the shell and the ring. For the shell, using the results from the preceding example,

$$\delta^s_{10} = -3340 \times \frac{0.040}{E} = \frac{-133}{E} \text{ ft}$$

$$\delta^s_{20} = 291 \times \frac{0.040}{E} = \frac{11.6}{E} \text{ rad}$$

Figure 19.24*a* shows a portion of the ring under a horizontal force H and a moment M_α, each per unit length of arc. Neglecting the distance $b/2$, which is small compared with r, the horizontal displacement due to H is

$$\Delta_H = \frac{r^2 H}{EA} \tag{19.28}$$

where A is the area of the ring.

From Fig. 19.26*b*, $M_\theta = rM_\alpha$. Assuming a linear stress distribution $f = M_\theta y/I$, the rotation Δ_α of the ring due to M_α is given by

$$\Delta_a = \frac{\varepsilon r}{y} = \frac{fr}{Ey} = \frac{M_\theta r}{EI} = \frac{r^2 M_\alpha}{EI} \tag{19.29}$$

The forces acting on the ring are

$$H = N'_{aH} = -1.78 \text{ kips/ft}$$

$$M_\alpha = -0.506N'_{aH} - -1.78 \text{ kips/ft}$$

$$M_\alpha = -0.506N'_{aH} + 0.216N'_{aV}$$

$$= -0.506(-1.78) + 0.216(-0.95) = 0.695 \text{ ft} - \text{kips/ft}$$

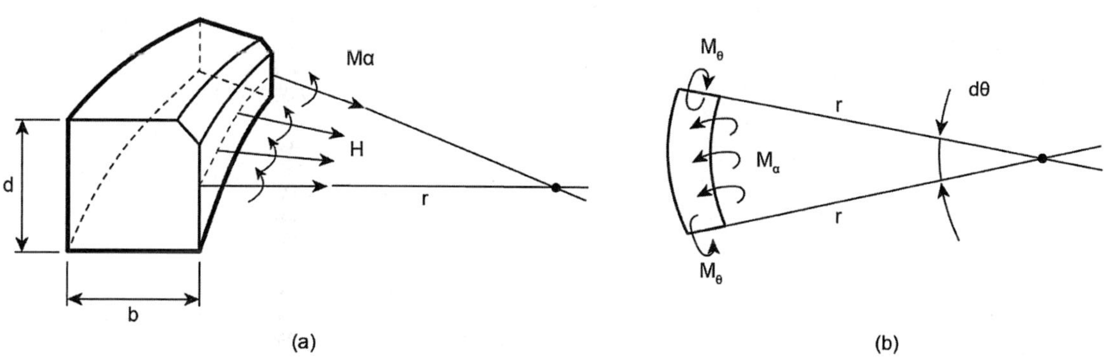

Figure 19.24 Forces acting on ring.

Then, from Eq. (19.29),

$$\delta_{20}^R = \frac{44.25^2 \times 0.695}{0.667(1.453^3/12)E} = \frac{7980}{E} \text{ rad}$$

The horizontal displacement δ_{10} is, from Eq. (19.28),

$$\delta_{10}^R = \frac{44.25^2(-1.78)}{0.667 \times 1.455E} - 0.506\delta_{20}^R = -\frac{3596}{E} - 0.506\frac{7980}{E} = \frac{7634}{E} \text{ ft}$$

The last term in this equation is the horizontal displacement at the junction due to rotation of the ring. Thus, the δ_0 displacements are

$$\delta_{10} = \delta_{10}^s + \delta_{10}^R = -\frac{7767}{E}$$

$$\delta_{20} = \delta_{20}^s + \delta_{20}^R = \frac{7992}{E}$$

CORRECTIONS. With $X_2 = 1$ acting on the ring, Eq. (19.29) gives

$$\delta_{22}^R = \frac{44.25^2}{(0.667 \times 1.453^3/12)} = \frac{11,490}{E} \text{ rad}$$

This rotation produces a displacement at the junction of the shell with the ring:

$$\delta_{12}^R = \delta_{21}^R = -0.506\frac{11,490}{E} = -\frac{5810}{E} \text{ ft}$$

With $X_1 = 1$ acting on the ring, the contribution to δ_{11} of the moment -1×0.506 resulting from the eccentricity of X_1 must be added to the value given by Eq. (19.28). This moment gives the rotation $-0.506 \times 11,490/E = -5810/E$. The resulting displacement at the junction is $0.506 \times 5810/E = 2940/E$. Thus, the displacement due to X_1 is

$$\delta_{11} \frac{44.25^2}{0.667 \times 1.453E} + \frac{2940}{E} = \frac{4960}{E}$$

The displacements of the shell edge due to unit values of the redundant were determined in the preceding example: $\delta_{11}^s = 2780/E$, $\delta_{12}^s = \delta_{21}^s = 1390/E$, and $\delta_{22}^s = 1380/E$. Then, from $\delta = \delta^R + \delta^s$,

$$\delta_{11} = \frac{7740}{E} \quad \delta_{12} = \delta_{21} = -\frac{4420}{E} \quad \delta_{22} = \frac{12,870}{E}$$

Compatibility is obtained by satisfying the equations

$$7740X_1 - 4420X_2 - 7767 = 0$$
$$-4420X_1 + 12,870X_2 + 7992 = 0$$

for which $X_1 = 0.81$ kip/ft and $X_2 = -0.34$ ft-kips/ft.

The stress resultants and moments throughout the dome can now be obtained by using X_1 and X_2 in the equations of Table 19.7 and combining the results with the membrane stress resultants. Table 19.9 gives these values for dead load plus live load (90 psf).

PRESTRESSED RING. The effect of prestressing the ring is easily determined. The radial pressure H_F is given by $H_F = F/r$, where F is the prestressing force. If F is not eccentric with respect to the center of gravity of the ring, Eqs. (19.28) and (19.29) give

$$\delta_{10} = \frac{44.25^2 H_F}{0.667 \times 1.453 \times 10} + \frac{2020H_F}{E} \text{ ft} \quad \delta_{20} = 0$$

The equations of compatibility are

$$7740X_1 - 4420X_2 + 2020H_F = 0$$
$$-4420X_1 + 12,870X_2 = 0$$

from which $X_1 = -0.328H_F$ kips/ft and $X_2 = -0.114H_F$ ft-kips/ft.

The value of F required to counteract ring tension due to the dome dead load plus live load is obtained by combining the two solutions. For the dome load,

$$T = r_0(N'_{\alpha H} + X_1) = 44.25(-1.78 + 0.81)90/40 = 96.5 \text{ kips}$$

and for the prestressing,

$$C = r_0(H_F - 0.328H_F - 0.328H_F) = 29.7H_F \text{ kips}$$

with $C = T$, $H_F = 3.24$ kips/ft.

The stress resultants and moments caused by prestressing can be determined by using $H = X_1 = -0.328 \times 3.24 = -1.06$ kips/ft and $M_\alpha = X_2 = -0.114 \times 3.24 = -0.374$ ft-kip/ft in the equations of Table 19.7. The results are given in Table 19.9.

REINFORCEMENT

1. Dome hoop reinforcement is provided to take all the hoop tension. For the segment $\psi = 0°$ to $\psi = 1°$ (an arc length of 1.65 ft), the tensile force (Table 19.9) is

$$T_a = \frac{(26.64 - 25.65) + (17.65 - 15.25)}{2} \times 1.65 = 2.8 \text{ kips}$$

for which

$$A_s = \frac{2.8}{20} = 0.14 \text{ in.}^2$$
$$= \frac{0.14}{1.65} = 0.085 \text{ in.}^2\text{/ft, No. 3 at 15 in.}$$

Similarly, for the regions, 1 to 2°, $A_s = 0.18$ in.^{2} = 0.11 in.^{2}/ft, No. 3 at 12 in. and 2 to 3°, $A_s = 0.10$ in.^{2} = 0.06 in.^{2}/ft, No. 3 at 22 in. Beyond 3°, there is no appreciable tension owing to edge effects.

2. Minimum dome reinforcement will be supplied throughout the shell equal to at least $0.0018bh$ for welded-wire fabric, as specified in ACI 318-11[31] and in Section 7.12.2.1 for slabs. For the 2½-in. shell, about 0.054 in.^{2}/ft in each direction is required. A welded-wire fabric 6 × 6 – 6/6 provides 0.058 in.^{2}/ft in each of two directions.

Table 19.9 Stress Resultants in Dome of Fig. 19.21 (Surface Load *q* = 90 psf)

	Dome with ring			Ring prestress = 3.24 kips		
↓, deg	N_ϕ, kips/ft	N_θ, kips/ft	M_ϕ, ft-kips/ft	N_ϕ kips/ft	N_θ, kips/ft	M_ϕ ft-kips/ft
0	-2.92	26.64	-0.77	-0.94	-25.65	-0.37
1	-3.69	17.65	0.25	-0.28	-15.25	-0.85
2	-4.22	9.25	0.68	0.10	-7.35	-0.93
3	-4.52	2.83	0.74	0.28	-2.21	-0.77
5	-4.69	-3.82	0.44	0.28	1.80	-0.37
10	-4.38	-4.15	-0.03	0.00	0.37	+0.04
28	-4.25	-4.25	0.00	0.00	0.00	0.00

3. Meridional bending reinforcement will be provided to resist the combined effects of N_ϕ and M_ϕ. At the dome edge ($\psi = 0$) from Table 19.9, for total surface load plus prestressing,

$$M_\phi = -0.77 - 0.37 = -1.14 \text{ ft-kips/ft}$$
$$N_\phi = -2.92 - 0.94 = -3.86 \text{ kips/ft}$$

If the small compressive stress (3860/72 = 54 psi) is neglected, the resulting simplification is conservative, and with $d = 6 - 1.5 = 4.5$ in.,

$$A_s = \frac{M}{f_s jd} = \frac{1.14 \times 12}{20 \times \frac{7}{8} \times 4.5} = 0.16 \text{ in.}^2/\text{ft}$$

At $\psi = 1°$, $M_\phi = +0.25 - 0.85 = -0.60$ ft-kips/ft, from which, with $d = 6 - 3.5 \times 1.65/8 - 1.5 = 3.8$ in.,

$$A_s = \frac{0.60 \times 12}{20 \times \frac{7}{8} \times 3.8} = 0.11 \text{ in.}^2/\text{ft}$$

Since the moments from prestressing and surface loads are of opposite sign at this location, it is necessary to check for initial prestressing. Assuming initial prestress at a 25 percent increase and only dome dead load acting,

$$M_\phi = +0.25 \times \frac{40}{90} - 1.25 \times 0.85 = -0.95 \text{ ft-kips/ft}$$
$$A_s = 0.11 \times \frac{95}{65} = 0.16 \text{ in.}^2/\text{ft}$$

This moment is temporary and probably will not increase, so that $f_s = 20{,}000$ psi is conservative. It is more logical to use an ultimate-load analysis, where

$$M_u = -0.95 \times 1.5 = -1.42 \text{ ft-kips/ft}$$

From ACI 318, with $f_c' = 4000$ psi and $f_y = 40{,}000$ psi,

$$M_u = \phi A_s f_y \left(d - \frac{a}{2} \right) = 1.42$$
$$A_s f_y = 0.85 f_c' ab$$

from which $a = 0.12$ in. and $A_s = 0.13$ in.^{2}/ft.

At $\psi = 2°$, $M_\phi = +0.68 - 0.93 = -0.25$ ft-kip/ft. With $d = 6 - 3.5 \times 3.30/8 - 1.5 = 3.0$ in.,

$$A_s = \frac{0.25 \times 0.12}{20 \times \frac{7}{8} \times 3.0} = 0.06 \text{ in.}^2/\text{ft}$$

Initially,

$$M_\phi = +0.68 \times \frac{40}{90} - 1.25 \times 0.93 = -0.86 \text{ ft-kips/ft}$$
$$M_u = \phi A_s f_y \left(d - \frac{a}{2} \right) = 1.5 \times 0.86 = 1.29 \text{ ft-kips/ft}$$

from which

$$A_s = 0.15 \text{ in.}^2/\text{ft}$$

Similar calculations at $\psi = 3°$, with $h = 3.8$ in., give $A_s = 0.14$ in.^{2}/ft and at $\psi = 5°$, with $h = 2.5$ in., $A_s = 0.11$ in.^{2}/ft.

These computations show the need, largely due to initial prestressing, for top radial reinforcement of No. 3 at 8 in. ($A_s = 0.17$ in.^{2}/h) extending 10 ft from the edge (just beyond $\psi = 5°$).

The layout of reinforcement is shown in Fig. 19.25.

RING PRESTRESSING. The final tension force required is

$$T = rH_F = 44.25 \times 3.24 = 143 \text{ kips}$$

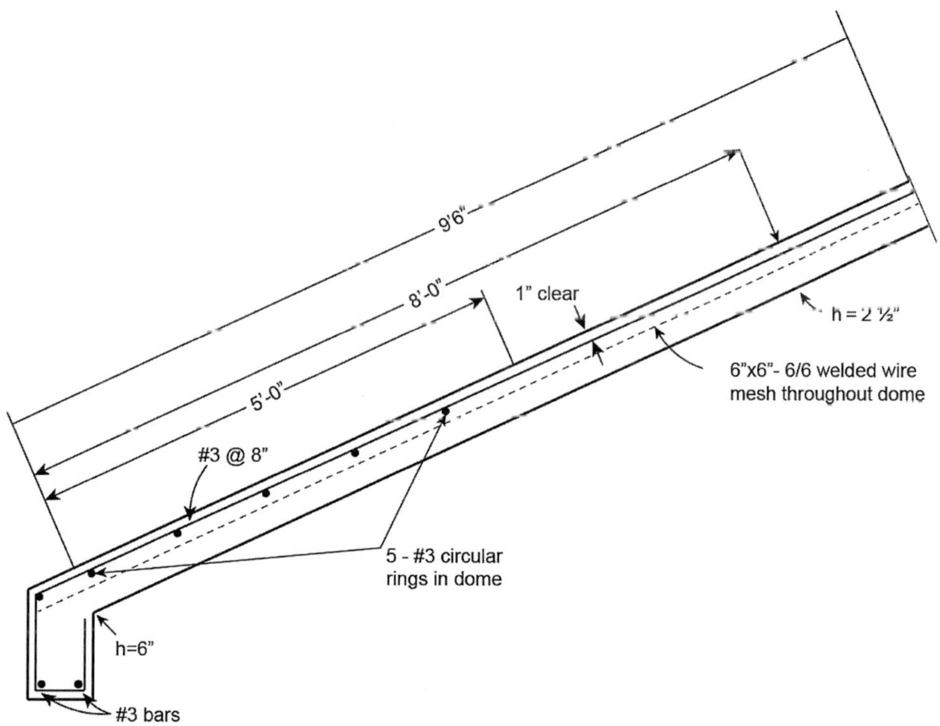

Figure 19.25 Reinforcement layout.

Prestress will be furnished by circular rings of tensioned steel at an assumed final stress of 120,000 psi:

$$A_s = \frac{143}{120} = 1.2 \text{ in.}^2$$

This area may be supplied by 25 wires each with a diameter of 0.25 in. An initial stress of 150,000 psi is required to compensate for losses assumed to be 20 percent.

Dome on Wall

If the dome is supported by a cylindrical wall, as in a tank, the analysis is the same as that for the ring-stiffened dome except that the displacements of the ring are replaced by the displacements of the wall (Section 19.2.4). If the dome and wall are joined monolithically through a ring, four redundant forces are involved: two between the dome and the ring and two between the ring and the wall. An example is given in Ref. 32.

Where the dome is built integrally with a wall or where the ring of a dome-ring structure is restrained either through friction or continuity with a wall, temperature effects will be important (as in the edge-fixed dome of Section 19.3.4) and should be investigated.

19.3.3 Solutions by the Finite-Element Method

Article 7 presents solutions for the dome-ring system and for a wall-roof system using a membrane solution with corrections. A similar solution is given in Ref. 27 for the dome-ring wall system of the tank in Fig. 19.20. The dead load on the dome is 31.5 psf at the crown, and the live load is 50 psf. Here we discuss a solution obtained using an axisymmetric program.[33] The interpolation functions used in this program are of fifth degree, so the results will be accurate even for longer elements. Figure 19.26 shows the idealization of the critical region around the rings.

Table 19.10 contains the results obtained by the finite-element analysis at the dome edge (element 18) and values obtained using membrane theory with corrections. It shows that the classical theory gives results close to those found from the numerical analysis. The differences are not significant for design.

Table 19.11 summarizes the dead-load and live-load results from the finite-element analysis. Element 29 is near the crown and shows values

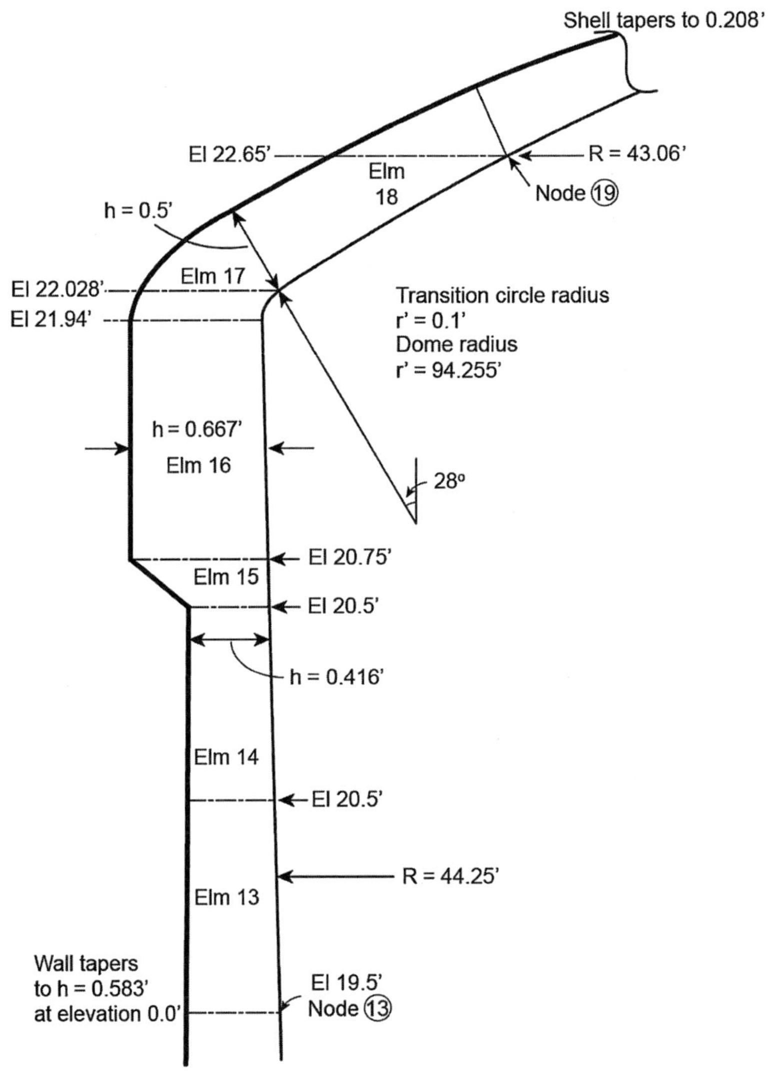

Figure 19.26 Idealization of dome-ring wall junction for finite-element analysis.

Table 19.10 Dome Edge Values

	Membrane theory	Fixed $h = 4$*	Dome h varies†	Dome-ring wall $h = 4$*	Dome-ring wall h varies†
N_{ϕ}‡	–4.50	–4.29	–3.59	–2.67	–2.65
N_{ϕ}‡	–3.00	–0.78	–0.59	15.56	19.32
M_{ϕ}§	0	–0.26	–0.63	–2.36	–2.87

* In inches.
† Finite-element solution (Element 18 in Table 4.12).
‡ In kips/ft.
§ In kips.

Table 19.11 Dead Load and Live Load from Finite-Element Analysis

Element number	Dead load N_{ϕ}*	Dead load N_{θ}*	Dead load M_{ϕ}*	Dead load Q_{ϕ}*	Live load ($q = 50$ psf) N_{ϕ}*	Live load ($q = 50$ psf) N_{θ}*	Live load ($q = 50$ psf) M_{ϕ}†	Live load ($q = 50$ psf) Q_{ϕ}*
29	–1.44	–1.46	0	0	–2.40	–2.30	0.00	0.00
20	–1.28	5.23	–0.47	–0.23	–1.88	8.53	–0.72	–0.29
19	–1.12	6.61	–0.90	–0.34	–1.60	10.30	–1.25	–0.40
18	–1.11	7.62	–1.22	–0.09	–1.54	11.70	–1.65	–0.05
17	–1.22	8.77	–1.25	0.41	–1.61	13.45	–1.68	0.64
16	–1.01	9.16	–0.89	0.49	–1.19	13.70	–1.15	0.70
15	–0.97	7.02	–0.53	–0.08	–0.03	10.30	–0.66	–0.13
14	–1.11	4.95	–0.26	0.28	–1.18	7.16	–0.31	– .39
13	–1.17	4.01	–0.04	0.17	–1.18	5.73	–0.01	0.23

* In kips/ft.
† In kips.

of N_{ϕ} and N_{θ} very close to those from membrane theory: $N'_{\phi} = N'_{\theta} = -aq/2 = -1.47$ kips/ft and -2.36 kips/ft, respectively. Table 19.11 also shows that the shears Q_{ϕ} carry a significant part of the vertical dome load, thus resulting in N_{ϕ} values near the ring substantially lower than N'_{ϕ}.

19.3.4 Hyperboloids

The type of shell wall shown in Fig. 19.27 has been used frequently for natural draft cooling towers, for which the two principal loads are dead weight and wind. The overall dimensions for a natural draft cooling tower depend upon the thermal criteria, the construction economy, and structural questions. The thermal conditions usually set the diameter in the base region where the heat exchange takes place, in the air inlet areas where the diagonal columns support the shell, and in the overall height which produces the draft. Construction economy can modify these dimensions, for example, where wind loads are high, by requiring a lower but wider tower to reduce the foundation costs. Structural requirements influence the meridional shell profile, the size of columns, the shell thickness and its variation, the need for top edge rings, and the shell reinforcement. The geometry and loading of the Trojan Tower[34] on the Columbia River in Oregon are shown in Fig. 19.27.

The membrane stress resultants for dead weight are computed from Eqs. (19.15) and (19.16) by numerical means as follows:

$$N'_{\phi j} = \frac{\sum_{i=1}^{j} r_{oi} q_i \Delta s_i}{r_{oj} \sin\phi_j} \tag{19.30a}$$

$$N'_{\theta j} = \frac{r_{oj}}{\sin\phi_j}\left(\frac{N'_{\phi j}}{r_{1j}} + q_j \cos\phi\right) \tag{19.30b}$$

where $N'_{\phi j}$ is the sum of the weights of j rings above, each of average unit weight q_i, average horizontal radius r_{oi}, and meridional length Δs_i.

WIND LOAD[30,35,36]

Membrane stress resultants are not easily computed since the load is not axisymmetrical. The meridional stress resultant is critical for design. Comparisons of meridional stress resultants computed for wind distributed as described in Tables 19.1a and 19.1b with results from a bending-theory solution show very little difference except in the values of N'_{θ} near the base.

Because several cooling towers have collapsed during windstorms, sufficient reinforcement must be provided in both the meridional and the circumferential directions. Recommendations are given in Ref. 37.

Other simply curved shells, such as truncated segments of cones, are sometimes used. They behave similarly to cylindrical segments, provided they are not too radically tapered.

A report of the ACI-ASCE committee[37] on concrete shell construction recommended that the load factors used for the shell be $U = 0.9D + 1.3L$ (where D and L are in opposite directions), where L is the wind load based upon a 100 year return fastest mile including a gust factor. For Portland, Oregon, the 100-year return wind is between 100 and 110 mph, which from the ANSI code would give for exposure C wind pressure from 90 ft up to 580 ft of from 46 to 79 psf. The design was based on an earlier report which led to values from 46.6 to 92.5 psf. The static computations for dead load and wind gave, at the tower base,

$$N_{\phi D} = -76 \text{ kips/ft}$$

$$N_{\phi W} = 116 \text{ kips/ft}$$

for a tension steel requirement of

$$A_s = \frac{1}{\phi f_y}[-0.9(76) + 1.3(116)]$$

$$= \frac{1}{(0.9)(60)}(-68.5 + 151.5) = \frac{83}{54} = 1.54 \text{ in.}^2\text{/ft}$$

The minimum required is $0.0035A_c$ in each direction so that, at the lintel, where $h = 3.5$ ft,

$$A_{s,\min} = 0.0035 \times 12 \times 42 = 1.76 \text{ in.}^2\text{/ft}$$

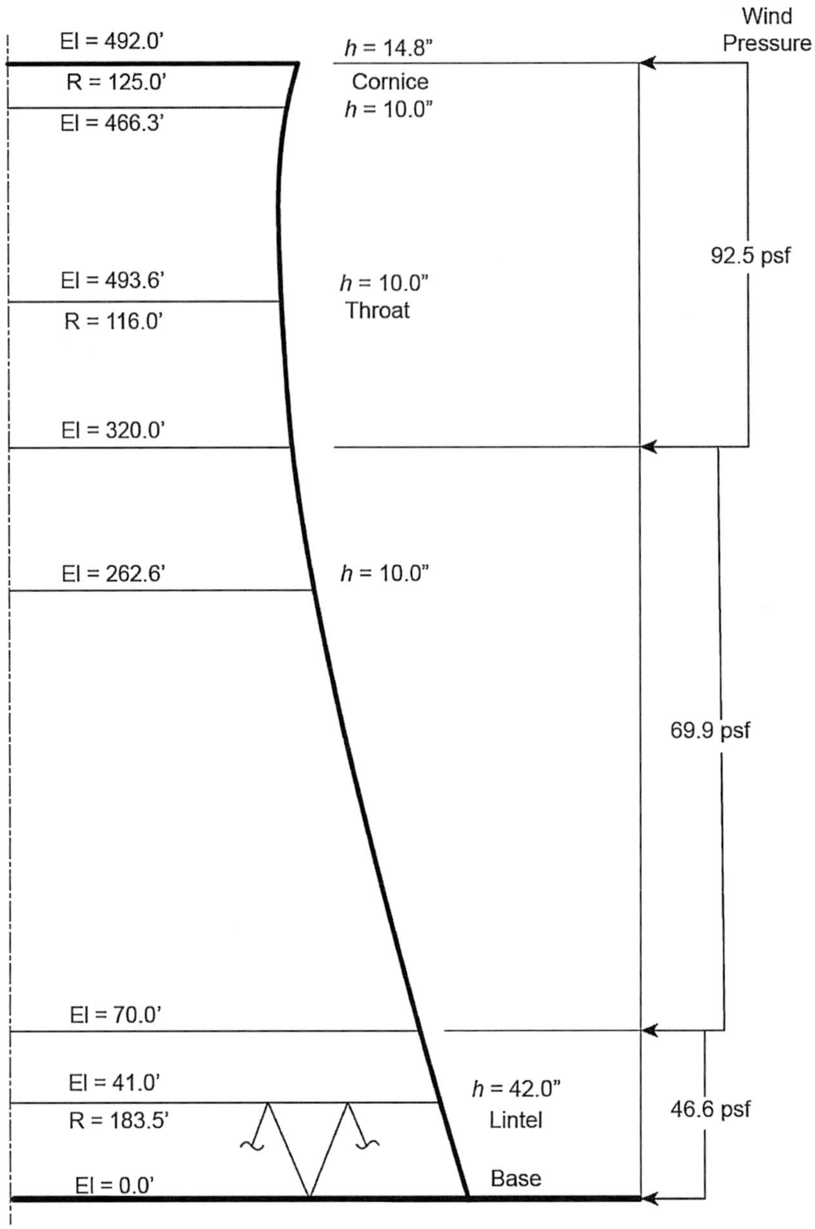

Figure 19.27 Dimensions of the Trojan Tower.

more than the amount obtained from the loads. At a distance of about 180 ft above the lintel, where the shell thickness is a minimum of 10 in., the comparable values are

$$N_{\phi D} = -30.0 \text{ kips/ft}$$

$$N_{\phi W} = +72.8 \text{ kips/ft}$$

from which

$$A_s = 1.39 \text{ in.}^2$$

compared to a minimum required of

$$A_s = 10 \times 12 \times 0.0035 = 0.42 \text{ in.}^2$$

which shows that the critical section can occur well above the base. In the circumferential direction, the stresses do not normally control, and minimum steel will be used throughout. This circumferential steel is essential to the integrity of the tower, especially since thermal gradients will cause meridional cracking.

In this tower, the earthquake loads did not control the design, and neither did the dynamic influences.[33] An important question is the design of reinforcement near the base to account for the concentrated reactions of the columns onto the lintel. This theoretically complex problem is normally solved by using a deep-beam approximation to get tensile steel requirements at the free edge between columns and up in the shell over the supports.

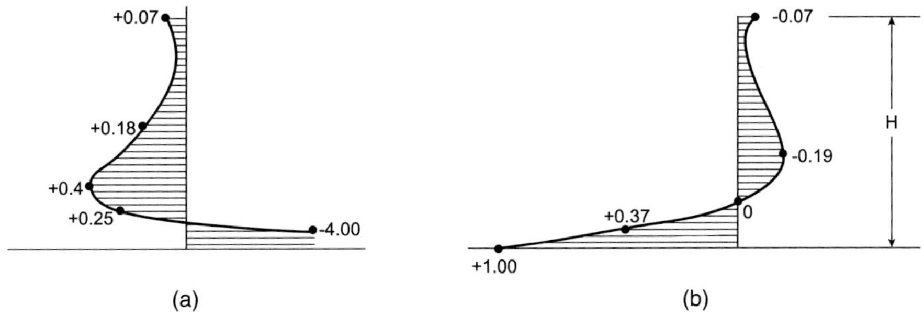

Figure 19.28 Deep-beam approximation for the behavior of the base of a cooling tower: (*a*) stress coefficients at support and (*b*) stress coefficients at midspan.

Figure 19.28 gives some idea of the behavior of the shell in its lower support region where it is supported by columns of total width C spaced a distance L apart. The diagrams shown are based on analytic solutions given by Franz Dischinger[38] in 1932 and applied to concrete deep beams some years later by the Portland Cement Association.[39] As shown in Ref. 39, the loading for the analytic solution is taken at the free edge, and the beam depth H is taken equal to the column spacing L, whereas in reality the load is N_ϕ distributed uniformly at some distance above the edge. Thus, the solutions shown are approximate and serve mainly to indicate the type of special reinforcement needed: along the lower edge near midspan and higher up over the supports. Also, these solutions neglect the shell curvature by assuming a planar deep beam.

For the Trojan Tower, Fig. 19.29 shows the distribution of N_ϕ at the lintel level, where the maximum tension is about 116 kips/ft, and it occurs at $\theta = 0°$. The maximum compression comes at about 69 degrees and is about −84 kips/ft. This compression combines with the dead load of about −76 kips/ft to give −160 kips/ft. The appropriate load factors for dead load and wind are

$$N_{\phi v} = 0.75\,(1.4N_{\phi D} + 1.7\,N_{\phi W})$$
$$= 1.05\,N_{\phi D} + 1.28\,N_{\phi W} = -80 - 107 = -187 \text{ kips/ft}$$

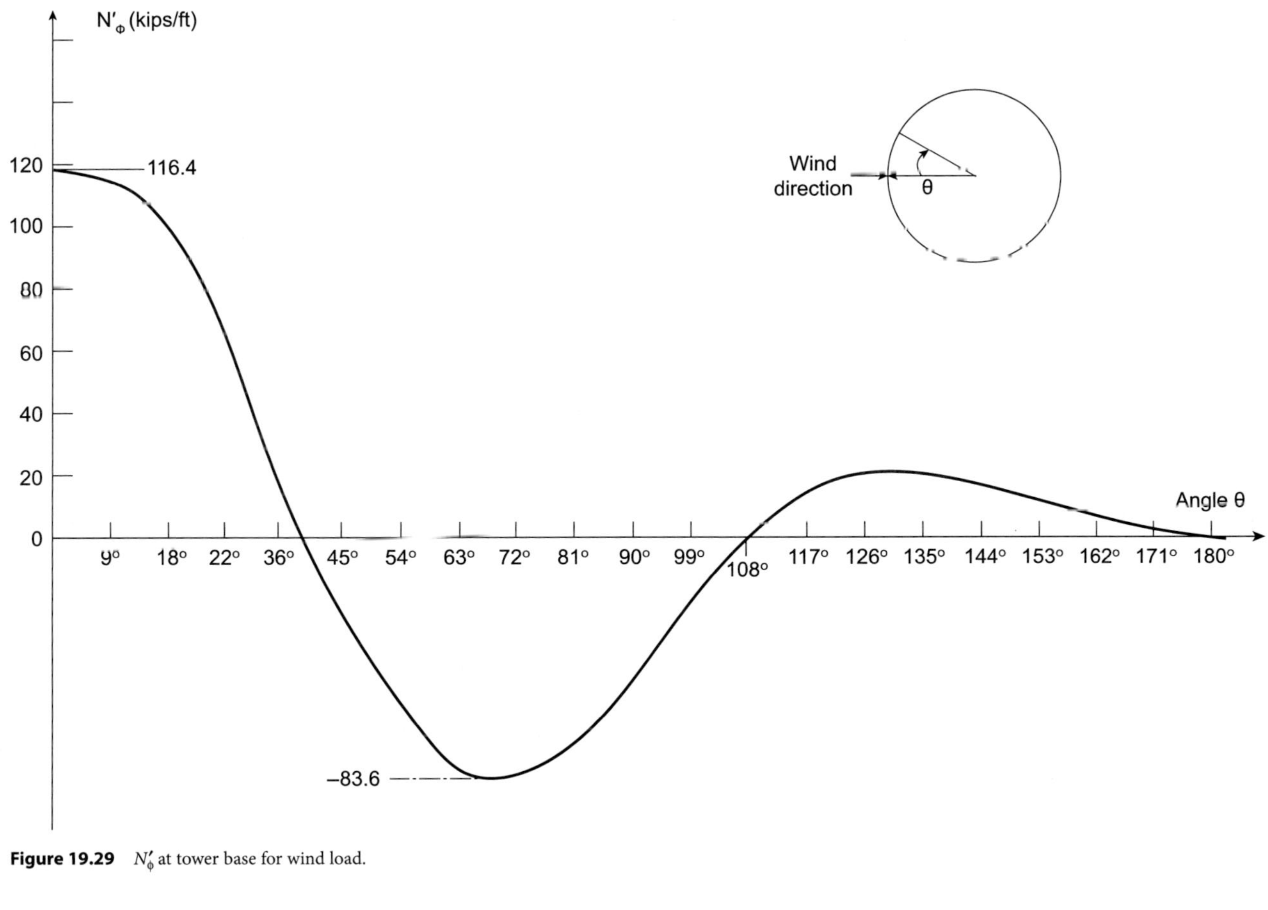

Figure 19.29 N'_ϕ at tower base for wind load.

This load is taken as constant over the span between columns even though it will be somewhat less away from $\theta = 69°$ as seen in Fig. 19.29. Because there are 44 supports a round the lintel of a radius of about 182 ft, the span will be $2\pi(182)/44 = 26$ ft. The lintel width b varies from 40 in. at the base to 30 in., where $H = L = 26$ ft. Thus, the maximum stresses, from Fig. 4.22, are $f_s = -4.0(187/12)/40 = -1560$ psi at the supports and $f_m = 1.0(187/12)/40 = 390$ psi at midspan. The maximum tension over the supports would be only $(0.4)(15.6)/40 = 156$ psi. Even though these low tensions would not normally cause cracking, it is good practice in thin shells to consider that all tension be taken by reinforcement. The horizontal reinforcement can be proportioned on the basis of the diagrams in Fig. 19.28, where, for example, over the support from a distance of $0.1H$ from the base up to $0.5H$, the amount of tension would be about $0.3(0.4H)q = 0.12qH$ or about $0.12(187)26 = 585$ kips, requiring $A_s = 585/0.9(60) = 10.8$ in.^{2} Additional steel would be required from $0.5H$ to H above the base as well. At midspan, the total tension would be about $0.37(0.1\text{H})q/2 + 0.37(0.1\text{H})q + 0.63(0.1\text{H})q/2 = 0.088qH = 0.09(187)26 = 440$ kips, requiring $A_s = 440/54 = 8.2$ in.^{2}

There is already a minimum shell reinforcement of $0.0035(40)(12) = 1.68$ in.^{2}/ft in the lintel area, and because of the approximate nature of this analysis, it is good practice to add the steel found from the deep-beam analysis to the minimum. To provide shear reinforcement, some of the midspan horizontal steel should be bent up near the supports.

The diagonal columns need special care where there are significant earthquake loads; it is especially important that the column steel be well anchored into the shell and confined by spirals to prevent pulling out by combined vertical and horizontal vibrations.

19.4 CYLINDRICAL BARREL SHELLS

19.4.1 Cylindrical Shell Roofs

Cylindrical roofs are segments of cylinders cut along their straight-line generators. They transfer loads by a combination of longitudinal beam action and transverse arch action. Depending on the aspect ratio, r/L (radius over span length), either the beam action (long cylinders) or the arch action (short cylinders) may be dominant. Cylindrical shells, sometimes called cylindrical barrels, are classified as follows:

- long shells where $r/L \le 0.4$
- intermediate shells where $0.4 < r/L < 2.0$
- short shells where $r/L > 2.0$

The tables from Manual of Engineering Practice 31[28] do not include the intermediate category and are based on a division at $r/L = 0.6$.[40] Short barrels are used for aircraft hangars and auditoriums. They often have transverse spans of from 150 to 300 ft with longitudinal spans L of 20 to 50 ft between stiffening ribs. Long barrels more commonly appear in warehouses and factories, where longitudinal spans L of about 50 to 150 ft are required, with transverse spans of 20 to 40 ft. Other shells with single curvature are, for example, truncated segments of cones, which behave similarly to cylindrical segments, provided they are not too radically tapered.

The methods of analysis for cylindrical shells can be divided into two categories: (1) simplified (approximate) methods based on the membrane theory and/or on global assumptions that assume behavior to be like beams, arches, cantilevers, and so forth and (2) more rigorous methods that often use shallow shell theory, based on the theory of elasticity, simplified by the Kirchhoff assumptions (as for plane plates).

19.4.2 Behavior and Simplified Analysis of Long Cylinders

The principal action of the barrel shell with small flexible longitudinal edge beams is longitudinal bending, but the bending stresses are within the plane of the shell itself. The simple flexural theory, which leads to a straight-line stress distribution, requires that all cross sections of a member remain plane and undistorted after bending. This means that after bending, the distance between any two points within a cross section remains the same as it was before bending, which is always the case in a solid rectangular beam. In a T-beam with wide flanges, the extremities of the flanges may not deflect the same as the web, but the simple flexural theory still provides a reasonable basis for design. In a barrel shell, however, the cross section may undergo substantial lateral distortion, and it is principally this distortion that causes the longitudinal stresses to depart from the straight-line distribution of the beam theory.

Figure 19.30 shows six different examples of the longitudinal stress resultants, N_x, where the ratio of r/L varies. In addition to these N_x values, there can be transverse bending moments M_ϕ owing to the lateral distortion of the cross section. This distortion is substantially reduced either by intermediate ties or diaphragms on a single barrel or by connection of a series of barrels together in the transverse direction. Then, for long shells, the designer can use a simplified analysis that considers the shell as a beam with longitudinal stresses determined by beam theory. In this case, the in-plane stresses are well approximated by

$$f_x = \frac{N_x}{h} = \frac{M_s y}{I} \tag{19.31a}$$

$$v = \frac{N_{x\phi}}{h} = \frac{VQ}{Ib} \tag{19.31b}$$

where M = bending moment about centroidal axis
I = moment of inertia of shell cross section
V = total shear at cross section
b = total cross-sectional thickness of concrete measured horizontally

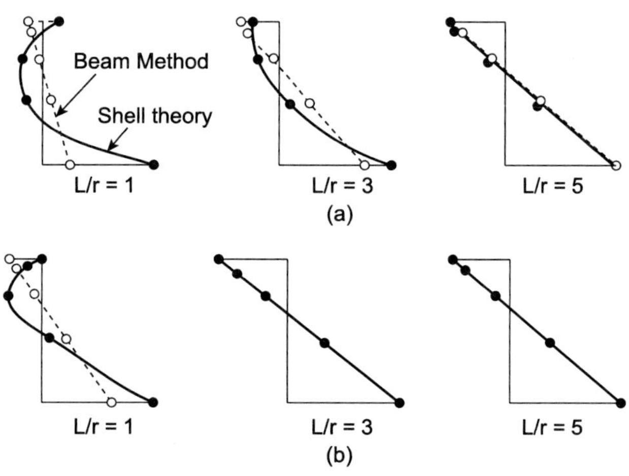

Figure 19.30 Comparison of barrel-shell and beam behavior: (*a*) single shell with free edges and (*b*) typical interior shell. (*After Ref. 43.*)

Transverse moments M_ϕ may be approximate by considering the slice from the barrel shown in Fig. 19.31, where the vertical load q is held in equilibrium by the vertical component of the in-plane shearing forces. The arch moment and thrusts may be determined by any of the methods for arch analysis.[41]

Implicit in the beam-arch analysis are the assumptions that all points on a transverse cross section deflect equally in the vertical direction and not at all horizontally and that the radial shears Q, the longitudinal bending moments M_x, the twisting moments $M_{x\phi}$, and the strains from in-plane shearing forces can be neglected.

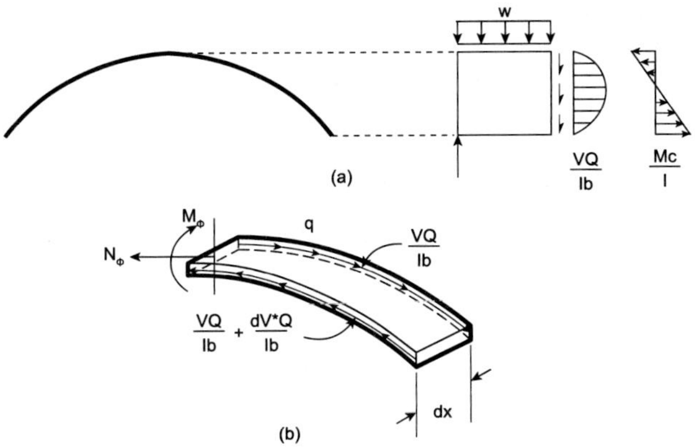

Figure 19.31 Forces on a slice from a barrel shell.

Figure 19.32 compares the stresses from Eq. (19.31) with those given by the shallow-well theory. The correspondence is good for $r/L = 0.2$, but for barrels of intermediate and short length, the maximum tensile stress is considerably larger than that given by Eq. (19.31). Analysis by Eq. (19.31) is simple, but the arch analysis tends to be lengthy.

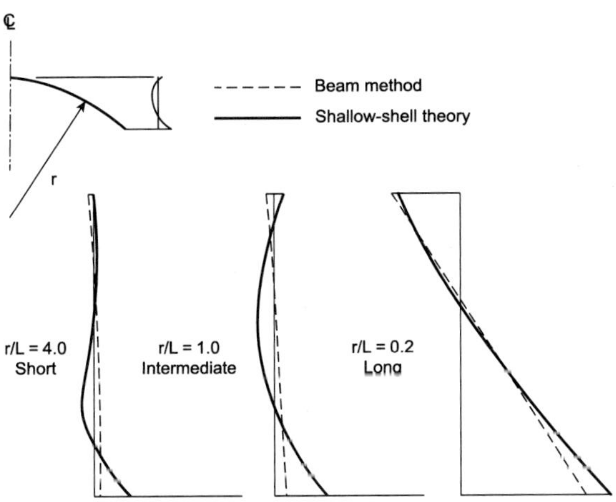

Figure 19.32 Comparison between stresses from beam method and shallow-shell theory.

Table 19.12 Moment of Inertia of Circular Cylindrical Shells (*from Ref. 20*)

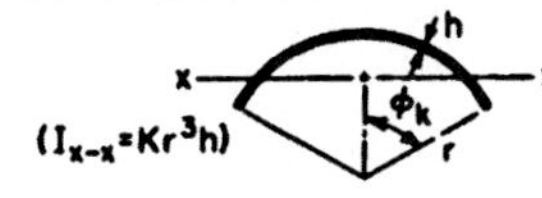

ϕk, deg	K	ϕk, deg	K
22.5	0.00041	37.5	0.00502
25.0	0.00068	40.0	0.00687
27.5	0.00110	45.0	0.01216
30.0	0.00168	50.0	0.02017
32.5	0.00249	55.0	0.03174
35.0	0.00358	60.0	0.04782

Table 19.12 gives values of I for calculation of deflections. Extensive tables based on the shallow-shell theory are given in Ref. 28.

19.4.3 Behavior and Simplified Analysis of Short Cylinders

Barrels for which $r/L > 0.4$ may be considered short. Such shells carry load essentially by arch action and are often shaped to the arch pressure line. Because of the large value of r relative to L, the shell thickness is often controlled by buckling rather than by strength. Intermediate stiffeners have been used to prevent buckling.

Except in the area next to the longitudinal edges, the load is carried to the supporting transverse arches (trusses or walls) by membrane action [see Fig. 19.10, Eq. (19.5)]. Table 19.13 shows the membrane forces for some common cases. The lower portion of the shell (next to the free edge), having a height of the order of magnitude of the span and sometimes as much as roughly equal to the span, behaves as a curved, deep beam. The total tension force in the lower part of the deep beam requires longitudinal reinforcement. In addition, an edge member is usually provided.

Because most of the roof load is carried to the supporting transverse frames by beam action only near the longitudinal edges, the supporting arch is subjected to outward thrusts near the springing lines and hence may have only a small compression, or even some tension, near the crown. But the adjacent shell is under compression, so there is incompatibility of strain or, more properly, the ideal behavior is modified, and some of the load is transferred into the arch by bending forces. The effect is to increase the compression in the supporting arch and reduce the compression in the shell adjacent to the arch or,

Table 19.13 Membrane Forces in Cylindrical Shells

Loading	N'_ϕ	N'_x	$N'_{\phi z}$
q per ft² of surface 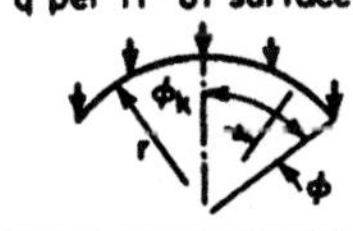	$-qr\cos(\phi_k - \phi)$	$-q\frac{x}{r}(L-x)\cos(\phi_k-\phi)$	$-q(L-2x)\sin(\phi_k \quad \phi)$
p per ft² projection 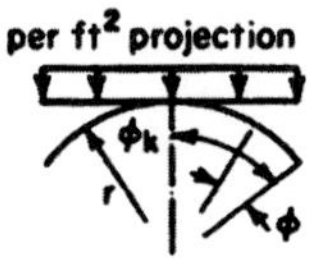	$-pr\cos^2(\phi_k - \phi)$	$-q\frac{x}{r}(L-x)\cos(\phi_k-\phi)$	$-2/3\,(p-2x)\sin(\phi_k-\phi)\cos(\phi_k-\phi)$
p_w per ft² projection 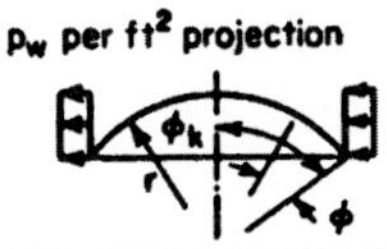	$-p_w r\sin(\phi_k - \phi)$	$-q\frac{x}{r}(L-x)\cos(\phi_k-\phi)$	$p_w\left(\frac{L}{2}-x\right)\cos(\phi_k-\phi)$

in other words, to force part of the shell to act as a flange for the arch in T-beam behavior.

Figure 19.31 shows an arch rib loaded by the in-plane forces N and the radial shears Q. The latter are so small that they may be neglected so that the rib may be analyzed as an arch for the forces N alone. Tables and standard computer programs simplify the analysis.

19.4.4 Rigorous Analysis of Barrel Shells

The complete set of eight internal forces acting on a shell element is shown in Fig. 19.33. The following set of equilibrium equations is the result:

$$\frac{\partial N_x}{\partial x}r+\frac{\partial N_{\phi x}}{\partial \phi}+p_x r=0\frac{\partial Q_x}{\partial x}r+\frac{dQ_\phi}{\partial \phi}+N_\phi+p_z r$$

$$=0+\frac{M_z}{dx}+\frac{\partial M_{x\phi}}{\partial \phi}-Q_x r=0\frac{\partial N_\phi}{\partial \phi}+\frac{\partial N_{x\phi}}{\partial x}r-Q_\phi+p_\phi r$$

$$=0-\frac{\partial M_\phi}{\partial \phi}+\frac{\partial M_{x\phi}}{\partial x}r+Q_\phi r=0$$

The stress resultants are expressed in terms of the displacements by considering the strain compatibility and material properties:

$$N_x=\frac{Eh}{1-v^2}(\varepsilon_x+v\varepsilon_\phi)=K\left[\frac{\partial u}{\partial x}+v\left(\frac{\partial v}{r\partial \phi}-\frac{w}{r}\right)\right]$$

$$N_\phi=\frac{Eh}{1-v^2}(\varepsilon_\phi+v\varepsilon_x)=K\left[\frac{\partial v}{r\partial \phi}-\frac{w}{r}+v\frac{\partial u}{\partial x}\right]$$

$$N_{x\phi}=N_{\phi x}=\frac{Eh}{2(1+v)}\gamma_{x\phi}=K(1-v)\left(\frac{\partial u}{r\partial \phi}+\frac{\partial v}{\partial x}\right)$$

The stress couples are defined by the changes in curvature and twist as

$$M_x=-d(\chi_x+v\chi_\phi)$$

$$M_\phi=-D(\chi_\phi+v\chi_x)$$

$$M_{x\phi}=-M_{x\phi}=D(1-v)\chi_{x\phi}$$

which, after substituting in (a) to (c), give

$$M_x=-D\left[\frac{\partial^2 w}{\partial x^2}+\frac{v}{r^2}\left(\frac{\partial v}{\partial \phi}+\frac{\partial^2 w}{\partial \phi^2}\right)\right]$$

$$M_\phi=-D\left(\frac{\partial v}{r^2\partial \phi}+\frac{\partial^2 w}{r^2\partial \phi^2}+v\frac{\partial^2 w}{\partial x^2}\right)$$

$$M_{x\phi}=-M_{\phi x}=D(1-v)\left(\frac{\partial^2 w}{r\partial x\partial \phi}+\frac{1}{2r}\frac{\partial v}{\partial x}\right)$$

The result is a set of 11 equations having as unknowns eight internal forces and three displacements. These equations can be solved by using the shallow-shell theory or by a somewhat more rigorous solution.

Some results of this type of distribution of analysis applied to a shell loaded with 80 psi uniformly distributed on the shell surface are presented. For the shell represented in Fig. 19.34a, the stress resultants are shown in Table 19.14 and Fig. 19.39. From the design point of view, the tension is high at the edge so that edge beams are needed along the straight-line generators, as in Fig. 19.34b. Table 19.15 gives the stress resultants for the shell with edge beams, and Fig. 19.35 shows the stress trajectories. Clearly, the transverse bending at the crown is reduced, and the tension is confined to the edge beam.

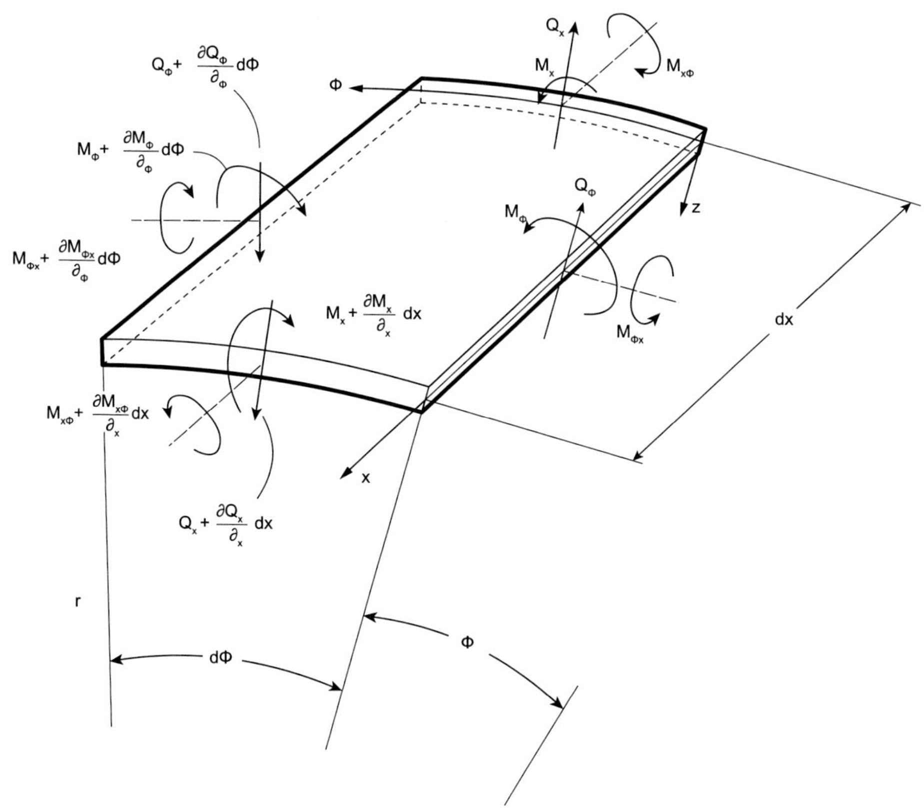

Figure 19.33 Bending moments and shear forces in barrel shells.

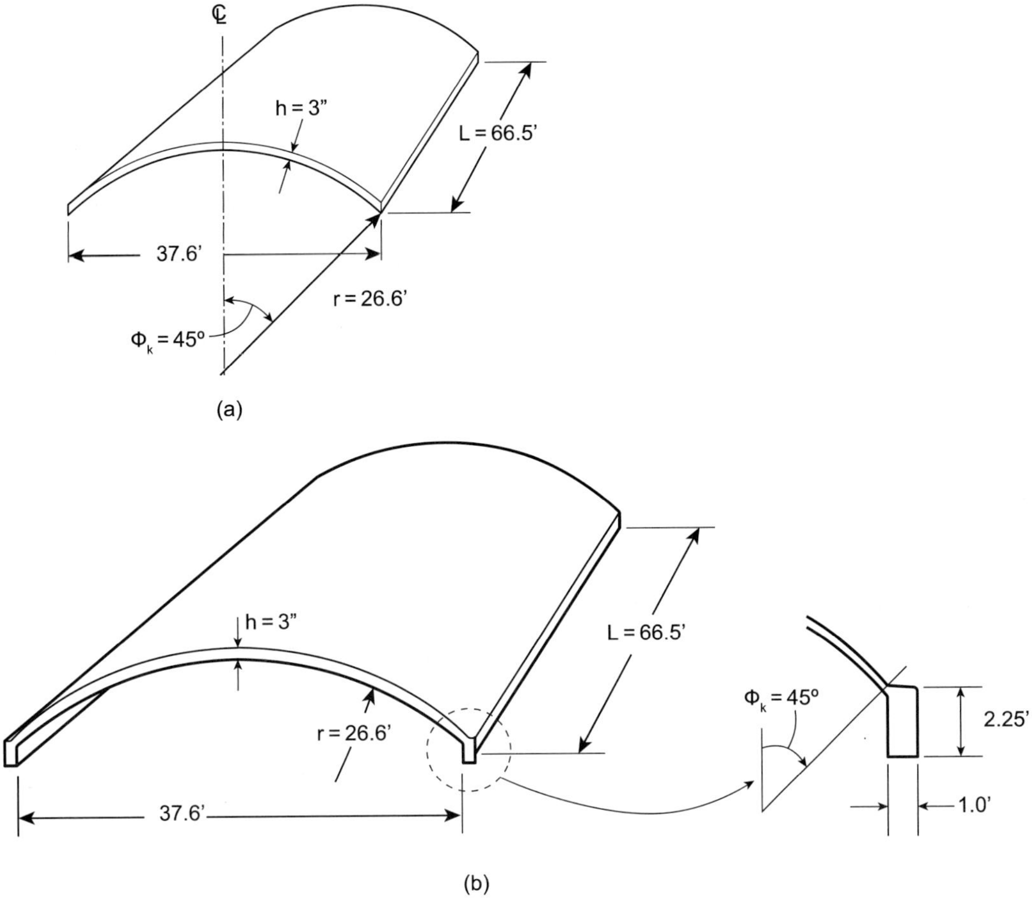

Figure 19.34 Barrel shell geometry.

Table 19.14 Stress Resultants and Couples for the Barrel Shell in Fig. 19.28

ϕ deg. from edge	M_ϕ ft-kips/ft $x = L/2$	N_ϕ kips/ft $x = L/2$	N_ϕ kips/ft $x = L/2$	$N\,\phi x$ kips/ft $x = 0$
45	−2.62	−3.57	– 0.34	0
40	−2.57	−3.58	– 2.24	−0.11
30	−2.10	−3.47	−14.32	−1.80
20	−1.17	−2.75	−22.71	−6.16
10	−0.15	−1.13	+ 2.04	−9.36
0	0	0	+99.03	0

Table 19.15 Stress Resultants and Couples for the Barrel Shell in Fig. 19.28 with Edge Beams

ϕ deg. from edge	M_ϕ ft-kips/ft $x = L/2$	N_ϕ kips/ft $x = L/2$	N_ϕ kips/ft $x = L/2$	$N\,\phi x$ kips/ft $x = 0$
45	−0.60	−2.97	– 2.38	0
40	−0.60	−2.97	– 3.04	−0.29
30	−0.42	−2.88	– 7.52	−1.38
20	−1.11	−2.49	– 13.02	−3.67
10	+0.21	−1.58	– 13.53	−6.73
0	0.0	−0.15	– 7.08	−8.73
Beam edge	0.0	0.0	+173.5	0.0

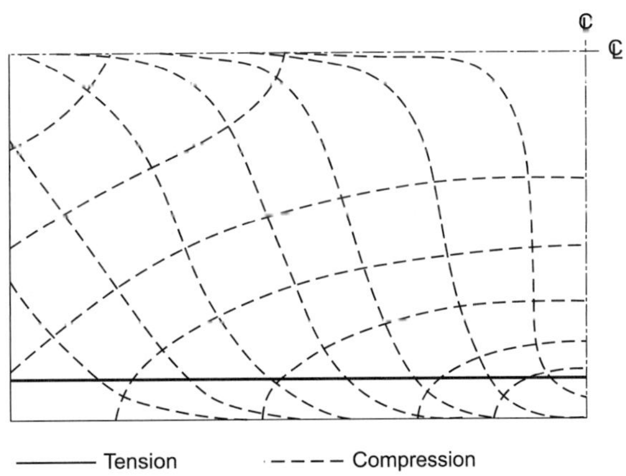

Figure 19.35 Stress trajectories for shell and edge beams.

Figure 19.36 shows a layout of reinforcement for the barrel shell of Fig. 19.34*b*. Reinforcement is provided to take all the principal tensile stresses and may be placed either in the direction of the stress trajectories or in two directions, usually orthogonal. Principal stresses are computed from the values of the stress resultants N_ϕ and N_x.

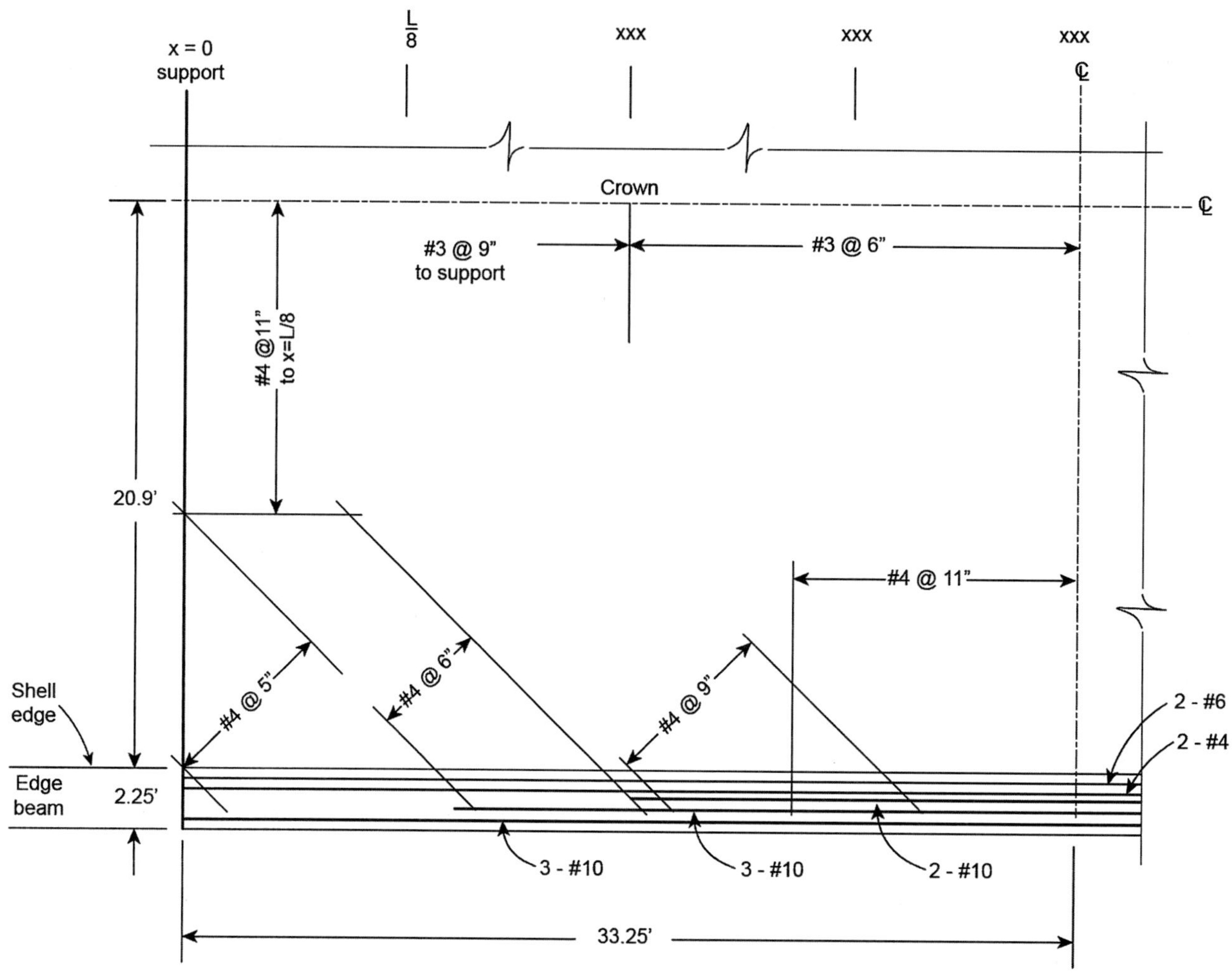

Figure 19.36 Developed plan of shell reinforcement (welded-wire fabric not shown).

At the corners, where the shear is a maximum, reinforcement is generally placed at 45 degrees. Transverse (hoop) reinforcement (No. 3 bars in Fig. 19.36) is based on M_ϕ, usually neglecting N_ϕ. If the tors and lateral stiffnesses of the edge beam are neglected in the analyses, $M_\phi = 0$ at the edge. However, it is advisable to provide reinforcement for some positive moment caused by edge-beam stiffness. This can be done by placing the principal-tension reinforcement in this region near the underside of the shell (the No. 4 bars at 11 in. in Fig. 19.36).

Longitudinal reinforcement at the juncture of the shell and the transverse rib is usually based on the assumption that the rib permits no radial movement and no rotation of the shell. This produces a moment $M_x = -0.29hN_\phi$. At $x = 0$, $N_\phi = 0$ because of the assumption of sinusoidal loading, but since the actual loading is uniform, N_ϕ must be constant along the span. Therefore, the value of N_ϕ at $x = L/2$ should be used to determine M_1 at $x = 0$. An allowance for additional moment resulting from arch deflection of the rib can be made. The No. 4 bars at 11 in. in Fig. 19.36 are proportioned on this basis.

Minimum reinforcement of 0.35 percent in each of two directions in tensile zones and 0.18 percent in other zones, spaced no farther apart than five times the shell thickness, is recommended. This can be supplied in the form of fabric.

Edge-beam reinforcement is usually sized for the tensile force in the beam. This can be determined by computing the stresses at the top and bottom of the beam. Tests show that it is advisable to place most of the reinforcement near the bottom (Fig. 19.37).

19.4.5 Finite-Element Analysis

This example illustrates the types of results obtained as well as some of the problems associated with finite-element programs.

Figure 19.39 shows some results from studies on one-quarter of the two-way symmetrical, simply supported single-barrel shell discussed in Sec. 19.4.2. The coarse mesh consists of 10 elements that are evenly spaced ($\Delta x = 3.225$ ft) in the x direction and six elements that are evenly spaced ($\Delta\phi = 7.5°$) in the ϕ direction; the fin-coarse mesh has a finer mesh near the transverse support, the free edge, and the crown; and the fine mesh divides the quarter shell into 306 elements as shown in Fig. 19.38.

Figure 19.39 plots N_x to show that even the coarse-mesh results follow closely those found in the more rigorous method using the analytic solution. The fine-mesh results do show minor differences in distribution and modest increases in maximum tensions and compressions, but the influence on design would be negligible.

As for the transverse bending moments M_ϕ, the difference between coarse mesh, fine mesh, and the analytical results are again of no

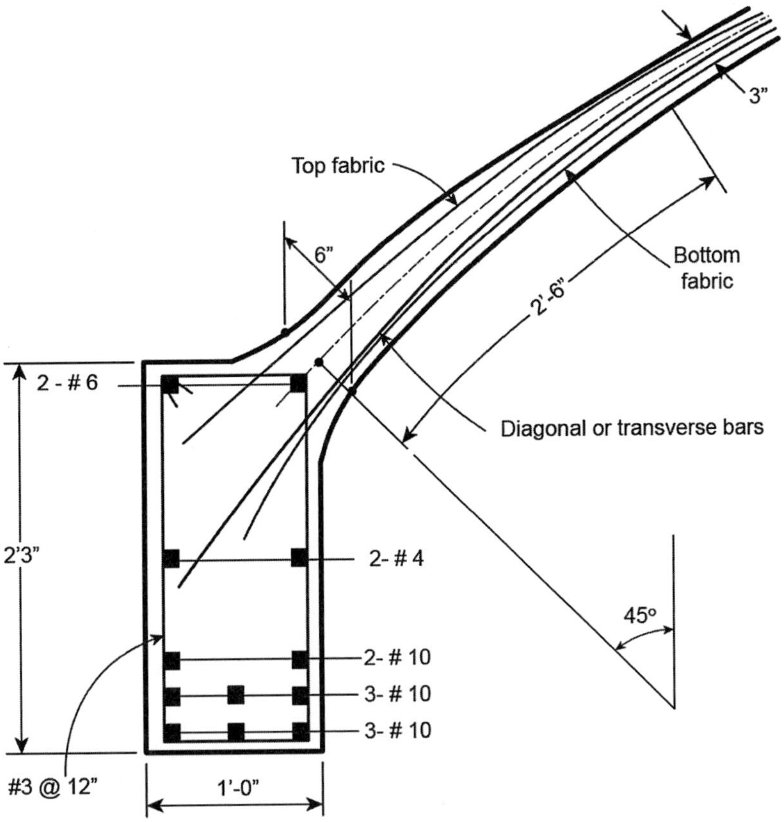

Figure 19.37 Section at edge beam.

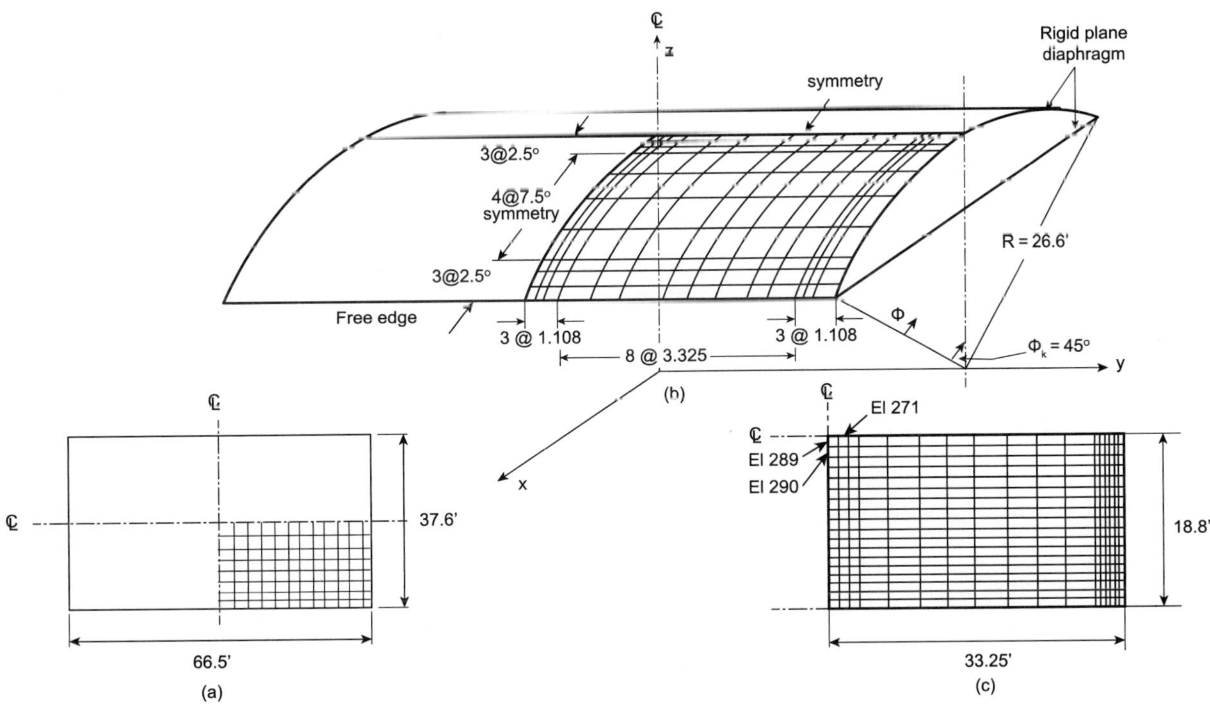

Figure 19.38 Finite-element mesh for a barrel shell: (*a*) plan and coarse mesh, (*b*) fine-coarse mesh, and (*c*) fine mesh.

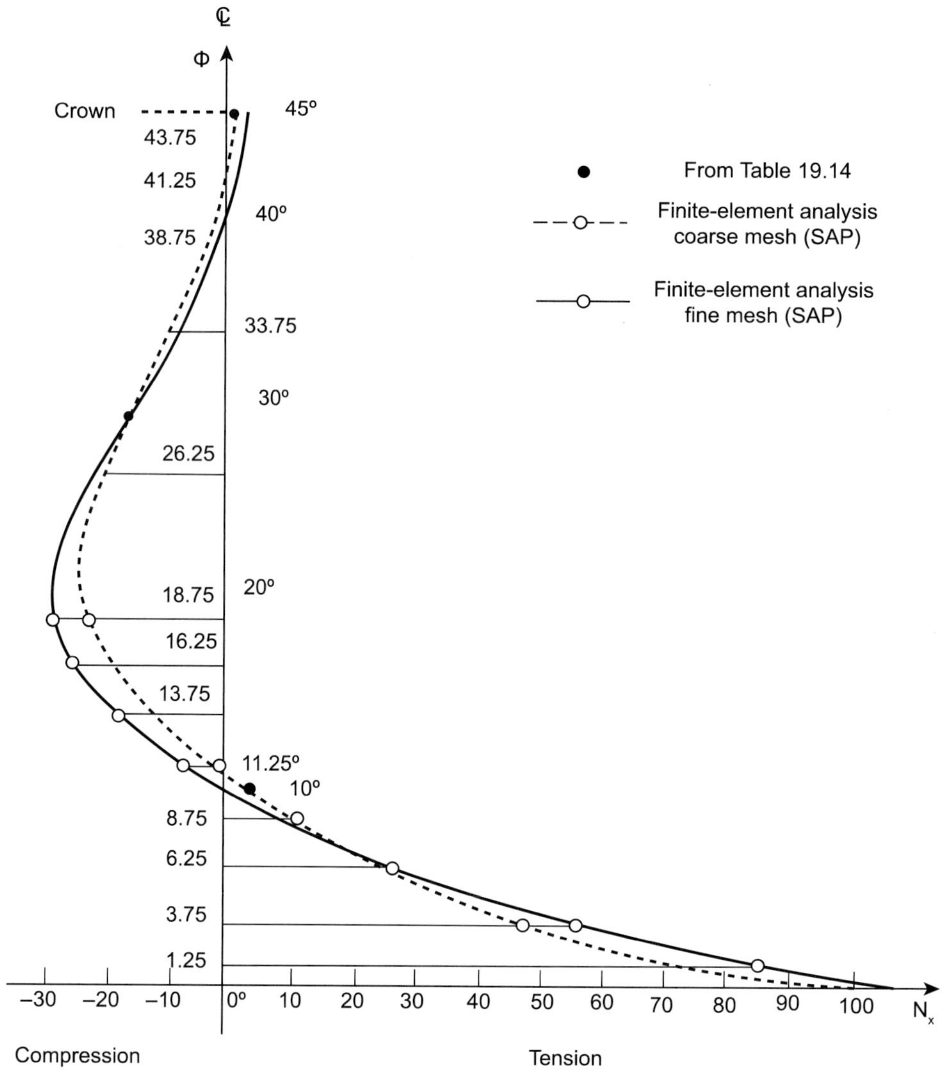

Figure 19.39 Comparison of N_x results by various methods.

design importance. The general-purpose program SAP,[42] which is used here, has only flat elements so that the structure is idealized as a folded plate, which emphasizes the unimportance of minor local deviation in geometry for this type of shell. Also, this type of program gives only internal forces at the center of the elements, while at the nodes (and, hence, the edges) it gives only displacements. Thus, to get N_x at the free edge, it is necessary to extrapolate. We find that even coarse-mesh approximations give useful results for this specific case, that from a design perspective the analytic and numerical results are equivalent, and that the finite-element results yield insight into detailed behavior.

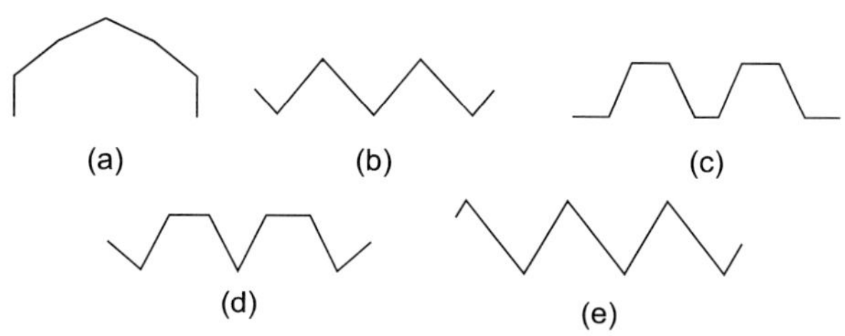

Figure 19.40 Typical folded-plate cross sections.

19.5 FOLDED PLATES

Figure 19.40 shows typical folded-plate cross sections. Typical dimensions for folded-plate roof structures are given in Table 19.16. These structures have no curvature but can be considered as approximations of shell geometry; that is, the shape of Fig. 19.40*a* approximates a barrel shell with vertical edge beams and the shape in Fig. 19.40*c* a multiple-barrel shell. An important difference between folded plates and barrel shells is the difference in the transverse bending moments M_ϕ (Fig. 19.41), which results from the plates acting as one-way slabs to carry roof loads transversely to the ridges by bending. The entire cross section then carries the loads longitudinally as a deep beam, with high in-plane tension stresses at the bottom and in-plane compression stresses near the top. Because of the longitudinal bending, the ridges deflect. Therefore, the one-way slabs rest on yielding supports, and their

Table 19.16 Typical Dimensions for Folded Plates

	Span, ft	Bay width, ft	*D*, ft-in. Max	*D*, ft-in. Min	*h*, in.	Reinforcing
Two-segment plate, 25°<φ<45°	40	15	4–0	2–9	4	1.2–1.6
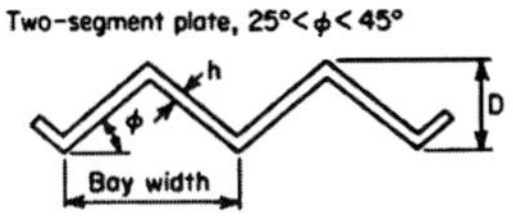	60	20	6–0	4–0	4–6	1.9–2.7
	75	25	7–6	5–0	4–6	2.6–3.7
	100	30	10–0	6–9	5–6	4.0–5.2
Four-segment plate 30°<φ<45°	40	20	5–0	2–6	3	1.5–2.0
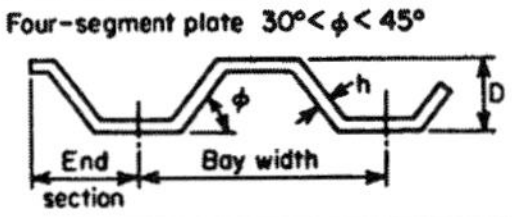	60	25	6–0	4–0	3–3.5	2.0–3.0
	75	30	7–7	5–0	3–4	2.5–4.0
	100	40	10–0	6–6	4–5	4.0–6.0

bending moments depart radically from those that would be obtained by assuming the ridges unyielding as in Fig. 19.41.

19.5.1 Analysis of Folded Plates

A full bibliography on folded-plate analysis appeared in a 1963 report,[39] which recommended the simplified analysis given in the second edition of this handbook and in the first edition of *Thin Shell Concrete Structures*.[43] Since then, numerical methods, especially finite-element analysis, have become the more practical methods.[40] Therefore, we give only some results herein to indicate the behavior of such structures. The results of a folded-plate analysis will normally be longitudinal plate stresses and transverse bending moments as shown for the midspan in Fig. 19.41. The designer uses these values to provide reinforcement as illustrated in the example that follows.

19.5.2 Example

Dimensions of the folded-plate roof of Fig. 19.41 are shown in Fig. 19.42. The roof will be designed for the following uniformly distributed load:

Live load	20
Roofing, insulation, etc.	10
Plate at 4 in.	50
	80 psf

Table 19.17 summarizes the critical values shown in Fig. 19.41 for this folded plate and taken from a simplified analysis. More rigorous analyses show that these values give a reasonable basis for design. Table 19.18 displays longitudinal stresses from Table 19.17 and includes also values at other sections from the support ($x = 0$) to midspan ($x = L/2$), where the distribution is taken as parabolic; that is, at $x = L/4$, the values are

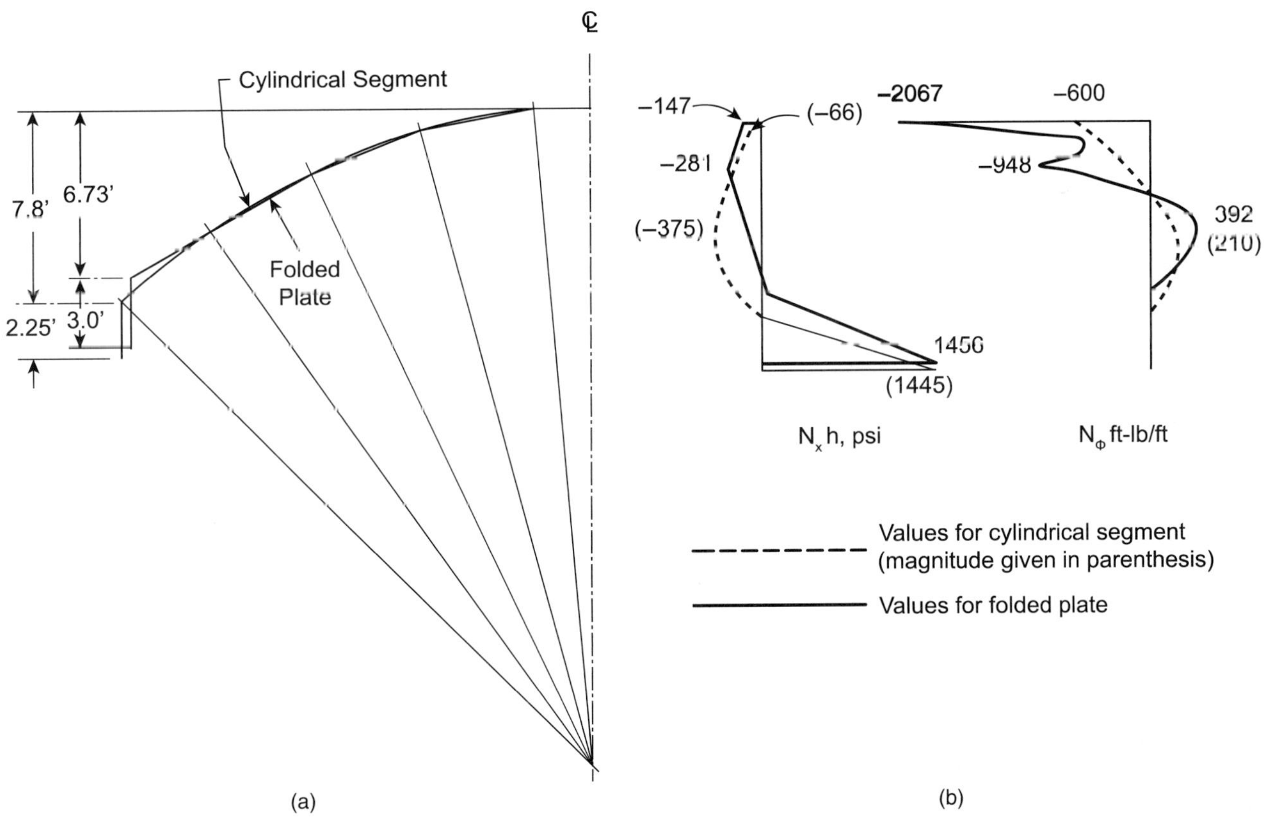

Figure 19.41 Comparison between cylindrical segment and folded plate.

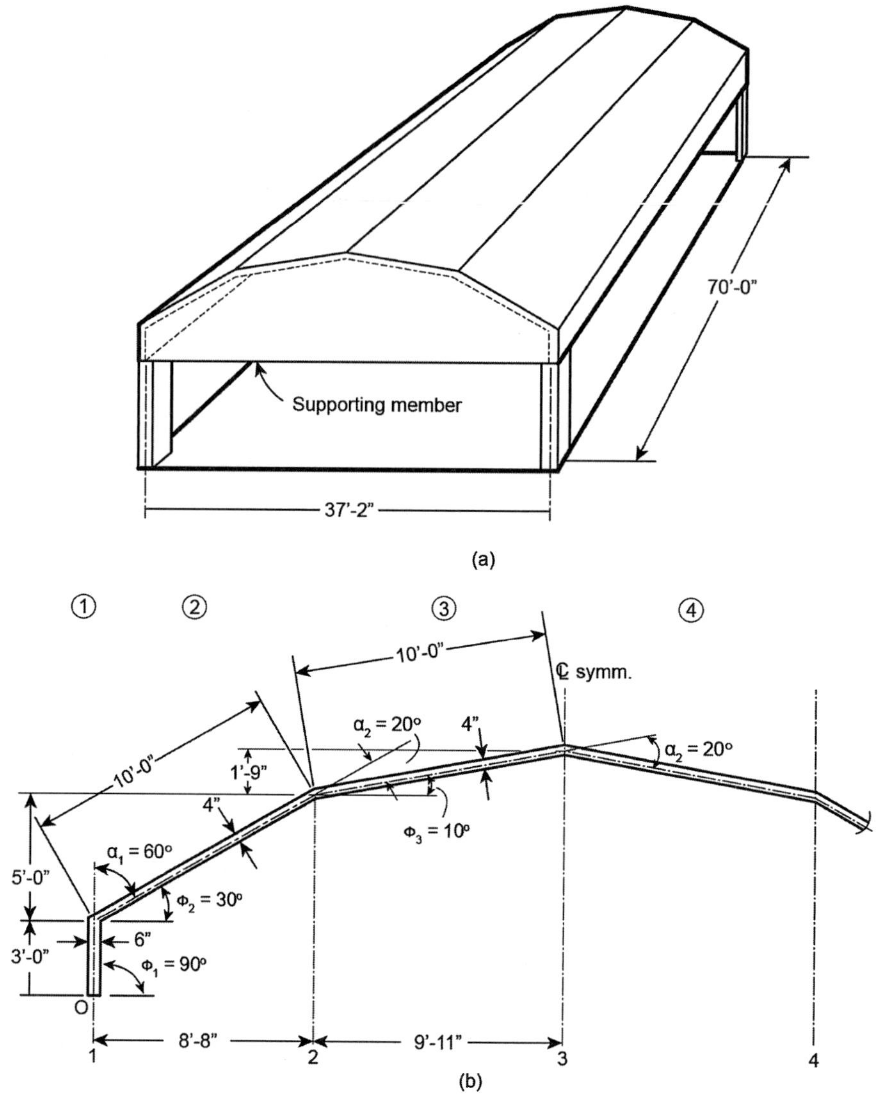

Figure 19.42 Dimensions of a folded-plate roof example.

0.75 times those at midspan. These values permit the calculation of the principal stresses, shown also in Table 19.18, the results of which appear graphically in Fig. 19.43.

The pattern of reinforcement is shown in Fig. 19.44. The requirements are based on the principal stresses of Table 19.18. Along the support, starting at joint 1 in plate 1 and proceeding to the longitudinal centerline, the required reinforcement at 45 degrees is

Plate 1, joint 1: $A_s = \dfrac{129 \times 6 \times 12}{20{,}000} = 0.463 \text{ in.}^2/\text{ft}$

Plate 2:

Joint 1: $A_s = \dfrac{193 \times 4 \times 12}{20{,}000} = 0.463 \text{ in.}^2/\text{ft, No. 4 at 5 in.}$

Middle: $A_s = \dfrac{179 \times 48}{20{,}000} = 0.43 \text{ in.}^2/\text{ft, No. 4 at 5 in.}$

Joint 2: $A_s = \dfrac{123 \times 48}{20{,}000} = 0.295 \text{ in.}^2/\text{ft, No. 4 at 8 in.}$

Plate 3:

Middle: $A_s = \dfrac{51 \times 48}{20{,}000} = 0.122 \text{ in.}^2/\text{ft, No. 3 at 10 in.}$

Joint 3: $A_s = 0$

Next, the steel required along joint 1 is computed. The angle of the line of principal stress is always 45° ± 15° to very near midspan, so we may place the steel at 45 degrees:

At support: $A_s = 0.463 \text{ in.}^2/\text{ft, No. 4 at 5 in.}$

At $L/8$: $A_s = \dfrac{153 \times 48}{20{,}000} 0.368 \text{ in.}^2/\text{ft, No. 4 at 7 in.}$

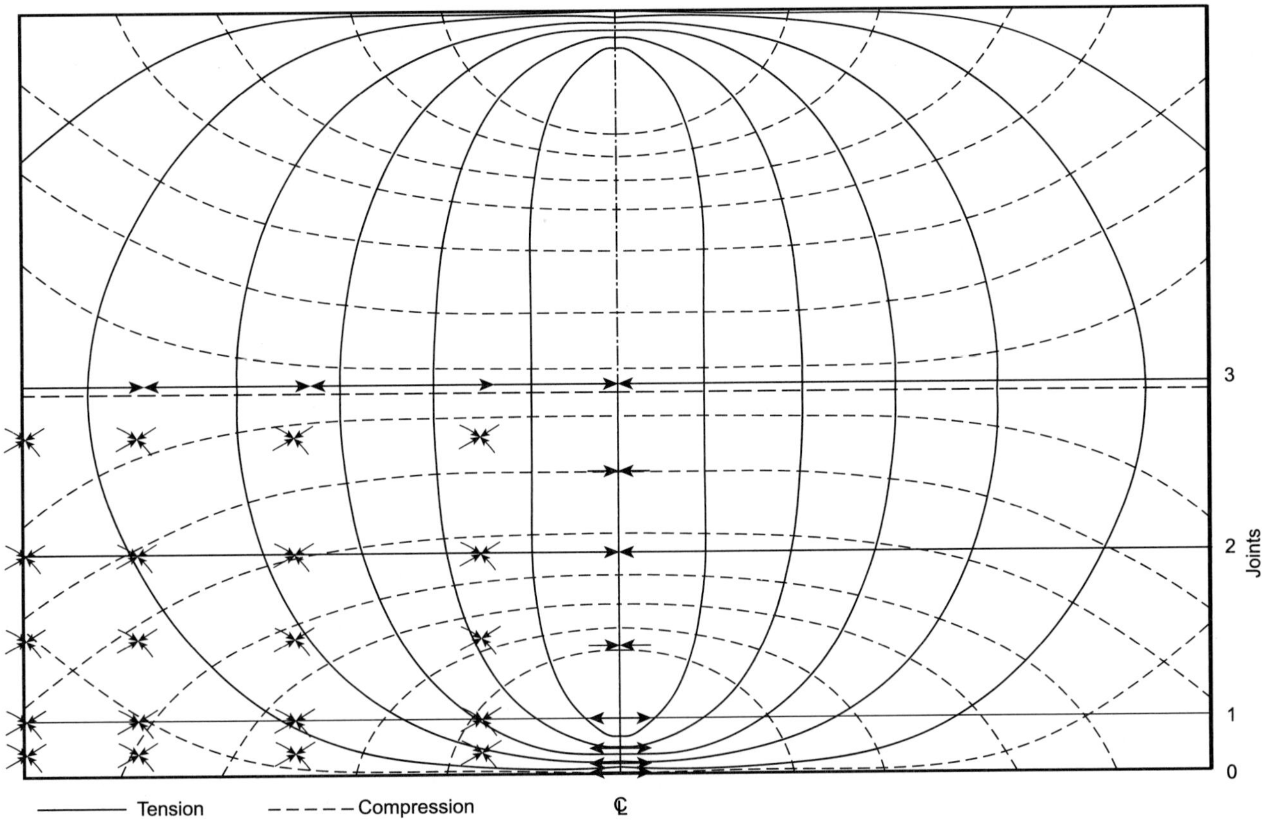

Figure 19.43 Stress trajectories in developed folded plate of Fig. 19.37.

Table 19.17 Midspan Values

Location	0	01*	1	12	2	23	3
Longitudinal stress, psf	+1456	+745	+36	−122	−281	−214	−147
Transverse bending moment, lb	0	0	0†	1392	918	600	2067

*Denotes point midway between joints 0 and 1 (see Fig. 19.31).
†A more rigorous analysis shows a value of 238 ft/lb/ft (see p. 294 of Ref. 1).

At $L/4$: $$A_s = \frac{111 \times 48}{20{,}000} = 0.267 \text{ in.}^2/\text{ft, No. 4 at 9 in.}$$

At $3L/8$: $$A_s = \frac{68 \times 48}{20{,}000} = 0.164 \text{ in.}^2/\text{ft, No. 4 at 15 in.}$$

From $7L/16$ to $L/2$, the minimum reinforcement required by ACI 318 is sufficient. For simplicity, these bar spacings are continued through to the support, even though somewhat less reinforcement would suffice.

In plate 2, at $L/2$, there is a longitudinal tension of

$$T = \frac{f_{12}^2}{f_{12} - f_{21}} \frac{h_2 d_2}{2} = \frac{36^2}{36 + 281} \frac{4 \times 120}{2} = 985 \text{ lb}$$

for which a minimum welded-wire-fabric reinforcement, as required for slabs in ACI 318-11, Section 7.12.2.1, of $0.0018 \times 12h = 0.087$ in.^{2}/ft each way, is sufficient. The fabric 4 × 4–6/6 is provided.

In plate 1, at $L/2$, the total tensile force is

$$T = \frac{f_{01} + f_{10}}{2} h_1 d_1 = \frac{1456 + 36}{2} \times 6 \times 36 = 161 \text{ kips}$$

and $A_s = 161/20 = 8.05$ in.^{2}, for which eight No. 9 bars are provided in the bottom 16 in. A minimum tension reinforcement of $0.0035 \times 12h = 0.126$ in.^{2}/ft is provided in the remaining portion by two No. 3 bars in each face.

The tension forces at other sections in plate 1 are

$$T_x = 4\frac{x}{L}\left(1 - \frac{x}{L}\right) T_{L/2}$$

from which $T_x = 151$ kips at $3L/8$, 121 kips at $L/4$, and 71 kips at $L/8$. These are all nearly horizontal. The forces will be taken by eight No. 9, six No. 9, and four No. 9 bars, respectively. The four No. 3 and four No. 9 bars are carried into the support.

Table 19.18 Principal Stresses, psi, Tension +, Compression –

Plate	Location	$x = 0$	$L/8$	$L/4$	$3L/8$	$L/2$
3	Joint 3	0	0	0	0	0
		0	–64	–110	–138	–147
	Mid-depth	+51	+13	+4	+1	0
		–51	–107	–164	–199	–214
	Joint 2	+123	+45	+17	+4	0
		–123	–167	–227	–268	–281
2	Joint 2	+123	+45	+17	+4	0
		–123	–167	–227	–268	–281
	Mid-depth	+179	+110	+56	+16	0
		–179	–164	–146	–130	–122
	Joint 1	+193	+153	+111	+68	36
		–193	–137	–83	–34	0
1	Joint 1	+129	+105	+79	+53	+36
		–129	–89	–51	–19	0
	Mid-depth	+95	+341	+564	+700	+745
		–95	–15	–4	0	0
	Joint 0	0	+635	+1,090	+1,362	+1,456

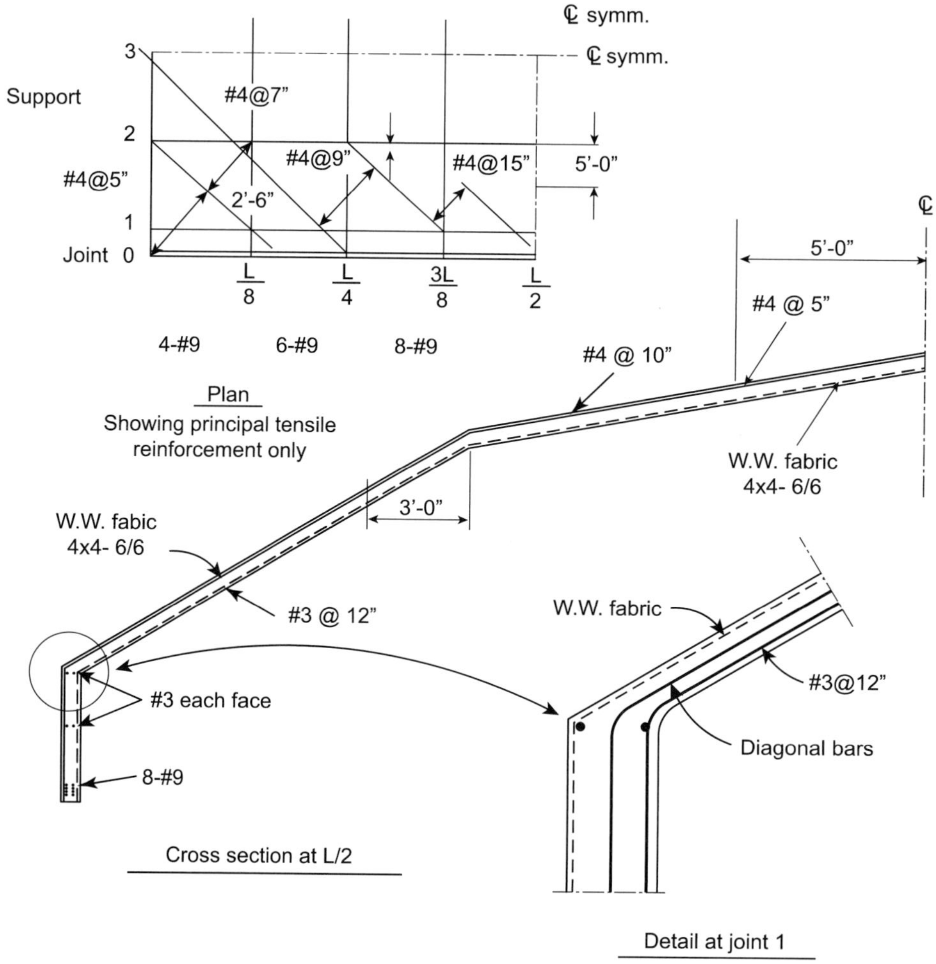

Figure 19.44 Reinforcement pattern of a folded-plate roof example.

Reinforcement for Transverse Moments

At the middle of plate 2, $M = +392$ ft-lb/ft (Table 19.17) and $h = 4$ in. With a cover of ½ in., $d = 3$ in. Then, with $f_a = 20{,}000$ psi and $f'_c = 4000$ psi,

$$A_s = \frac{392 \times 12}{20{,}000 \times \frac{7}{8} \times 3} = 0.09 \text{ in.}^2/\text{ft}$$

for which No. 3 at 12 in. is provided at the bottom of the plate. At joints 2 and 3, respectively,

$$A_s = \frac{948 \times 12}{20{,}000 \times \frac{7}{8} \times 3} = 0.217 \text{ in.}^2/\text{ft}$$

$$A_s = \frac{2068 \times 12}{20{,}000 \times \frac{7}{8} \times 3} = 0.473 \text{ in.}^2/\text{ft}$$

for which No. 4 at 10 in. and No. 4 at 5 in. are provided. At the middle of plate 3, the moment is 55 percent of that at joint 2, so half the No. 4 bars coming from joint 3 are cut off at that point.

The longitudinal distribution of these moments is not well defined in the analysis. For cylindrical-segment thin shells, it appears that the moment drops off toward the supports somewhat more slowly than a sine curve. In this case, we shall provide the full amount of steel in the center half of the span and reduce it to roughly 75 percent in the outer quarter-span lengths. The reinforcement is shown in Fig. 19.44.

Influence of Span

Figure 19.45 shows how the longitudinal stresses from in-plane plate bending and transverse moments from one-way slab bending are influenced by the span L. The dashed line in Fig. 19.45*b* shows values of M for unyielding ridge lines, which is the limiting case as $L \to 0$. When $L = 35$ ft, the moments have approximately the same values as for $L = 0$, but for $L = 105$ ft, the maximum moments are from 2.5 to 6.5 times the values for $L = 0$. On the other hand, the in-plane stresses (Fig. 19.45*a*) exhibit the same general behavior whether the span is 35 or 105 ft, with relatively high tension at the bottom and much lower compression near the top.

Transverse Frames

The T-beam behavior of the transverse frame is similar to that of the transverse arch for cylindrical shells. The behavior of the longitudinally continuous folded-plate structure is also similar to that of the continuous cylindrical shell.

19.5.3 Prestressed Folded Plates

Prestressing can substantially reduce the longitudinal tensile stresses in folded plates (Fig. 19.46). Analysis shows that a force $F = 146$ kips applied in the edge plate of the roof in Fig. 19.42 will completely eliminate longitudinal tension, as shown in Table 19.19. Note that the moments are all reduced except that in the middle of plate 2.

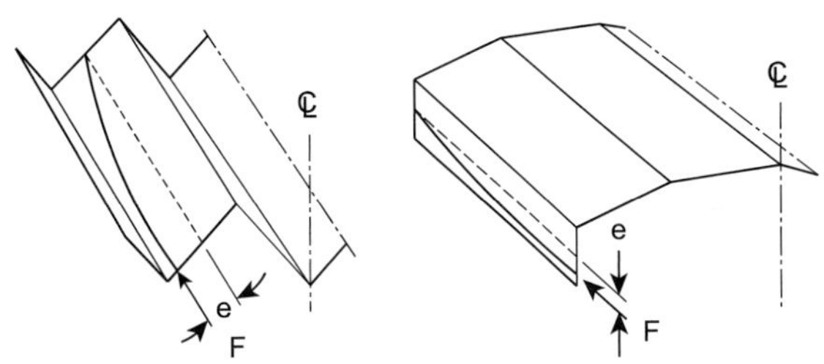

Figure 19.46 Prestressing of a folded plate (prestressing profile).

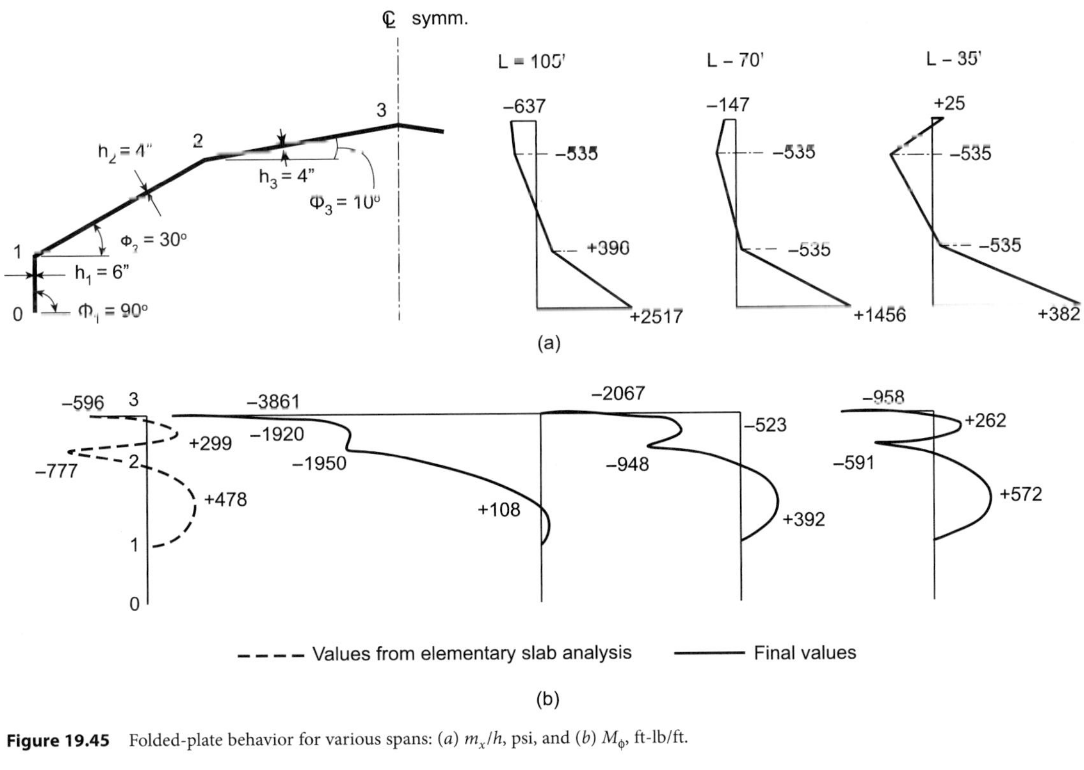

Figure 19.45 Folded-plate behavior for various spans: (*a*) m_x/h, psi, and (*b*) M_ϕ, ft-lb/ft.

Table 19.19 Dead Load, Live Load, and Prestressed Results (Folded Plate)

	DL and *LL*	Prestress	Combined
f_3	-147	-83	-230
f_2	-281	+85	-196
f_1	34	-31	3
f_0	1456	-1456	0
M_3	-2067	+954	-1113
M_{32}	-523	+888	+365
M_2	-984	+822	126
M_{21}	+392	+411	+803
M_1	0	0	0

19.5.4 Continuous Folded Plates

When folded plates are continuous over transverse supports, design practice consists of determining stresses by using the ratios of continuous beam moments to simple beam moments. For example, for a folded plate with the cross section shown in Fig. 19.45*a*, continuous for two spans of 70 ft each, the midspan tension of 1456 psi at the lower edge (Fig. 19.45*a*) would be reduced to 728 psi (ratio of $wL^2/16$ to $wL^2/8$). The lower-edge compression at the interior support would be 1456 psi (the same as the midspan tension for a simple span of 70 ft) because the negative moment at the interior support of a fully loaded two-span continuous beam is $wL^2/8$. Results of more accurate analyses confirm this practice as a reasonable basis for design.[44]

19.6 TRANSLATION SHELLS OF DOUBLE CURVATURE

19.6.1 Membrane Theory

Figure 19.47 shows the stress resultants on a differential element. The relations between them and their projections on a horizontal plane are given by

$$N'_x = \bar{N}_x \frac{\cos\theta}{\cos\phi} \quad N'_y = \bar{N}_y \frac{\cos\phi}{\cos\theta} \quad N'_{xy} = \bar{N}_{xy} \tag{19.32}$$

where $\tan\phi = \partial z/\partial x$ and $\tan\theta = \partial z/\partial y$.

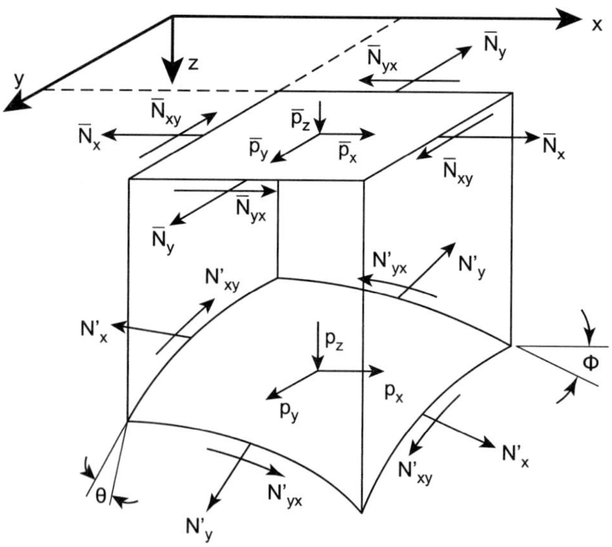

Figure 19.47 Stress resultants on a differential element.

The horizontal projections are given by

$$\bar{N}_x = \frac{\partial^2 F}{\partial y^2} - \int \bar{p}_x\,dx \quad \bar{N}_y = \frac{\partial^2 F}{\partial x^2} - \int \bar{p}_y\,dy \quad \bar{N}_{xy} = \frac{\partial^2 F}{\partial x \partial y} \tag{19.33a}$$

where F is a stress function which satisfies the equation

$$\frac{\partial^2 F}{\partial x^2}\frac{\partial^2 F}{\partial y^2} - 2\frac{\partial^2 F}{\partial x \partial y}\frac{\partial^2 z}{\partial x \partial y} + \frac{\partial^2 F}{\partial y^2}\frac{\partial^2 z}{\partial x^2} = q \tag{19.33b}$$

in which

$$q = -\bar{p}_z + \bar{p}_x\frac{\partial z}{\partial x} + \bar{p}_y\frac{\partial z}{\partial y} + \frac{\partial^2 z}{\partial x^2}\int \bar{p}_x\,dx + \frac{\partial^2 z}{\partial y^2}\int \bar{p}_y\,dy \tag{19.33c}$$

In Eq. (19.33), $\bar{p}_x, \bar{p}_y$, and $\bar{p}_z$ are loads per unit of area of the horizontal projection of the surface. The relationship between these components and the load p_x, p_y, and p_z per unit of area of the shell surface is

$$\frac{\bar{p}_x}{p_x} = \frac{\bar{p}_y}{p_y} = \frac{\bar{p}_z}{p_z} = \frac{\sqrt{1-\sin^2\phi\sin^2\theta}}{\cos\phi\cos\theta} = \sqrt{1+\tan^2\phi+\tan^2\theta} \tag{19.34}$$

Where only uniform vertical loads need be considered, $\bar{p}_x = \bar{p}_y = 0$, so that $q = -\bar{p}_z$, and Eq. (19.33) are greatly simplified. Furthermore, the vertical load can be assumed to be uniform over the horizontal projection for fairly fiat shells, so that Eq. (19.34) is eliminated; that is, $\bar{p}_z = p_z$.

19.6.2 Elliptic Paraboloids

The elliptic paraboloid (Fig. 19.48) is defined by

$$z = c_1\frac{x^2}{a^2} + c_2\frac{y^2}{b^2} \tag{19.35}$$

This surface intersects a horizontal plane in an ellipse. Sections cut by vertical planes parallel to the *xy* and *yz* planes are parabolas.

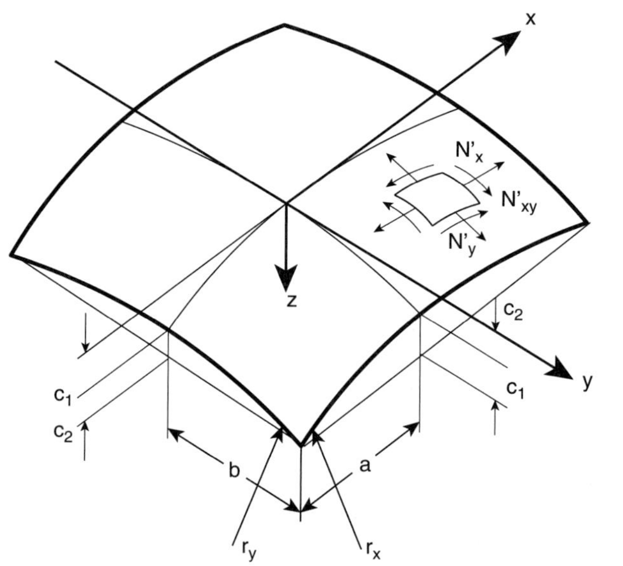

Figure 19.48 Elliptical paraboloid shell.

Substitution of Eq. (19.35) into Eq. (19.33b) yields

$$\frac{2c_2}{b^2}\frac{\partial^2 F}{\partial x^2}+\frac{2c_1}{a^2}\frac{\partial^2 F}{\partial y^2}=q$$

Once a stress function F that satisfies this equation is found, the membrane stress resultants can be determined by Eqs. (19.32) and (19.33a). Various solutions are possible, depending upon the assumptions relative to the boundaries. Most important is the case where all four edges are supported so that they are subjected only to in-plane shears N'_{xy} (Fig. 19.48). In this case, the shell needs only four point supports, with edge members to carry the shear. Derivation of the stress resultants is given in Ref. 45:

$$N'_x=-\frac{\bar{p}_z a^2 k}{c_1}\times\text{coefficient} \tag{19.36a}$$

$$N'_y=-\frac{\bar{p}_z b^2}{kc_2}\times\text{coefficient} \tag{19.36b}$$

$$N'_y=-\frac{\bar{p}_z b^2}{kc_2}\times\text{coefficient} \tag{19.36c}$$

where $k=\left[\dfrac{1+(2c_1/a)^2(x/a)^2}{1+(2c_2/b)^2(y/b)^2}\right]^{1/2}$

$\bar{p}_z$ = uniform load per unit area of horizontal projection

For shallow shells, k approaches unity and usually may be taken as 1 in practical applications since N'_x and N'_y are not large.

Values of the coefficients in these equations are given in Table 19.20. Table 19.21 gives values of the coefficients to determine the edge shears at closer intervals. It will be noted that these shears tend to become infinitely large at the corners. This means that transverse shearing forces and bending moments develop so that edge shears can remain finite. It is suggested in Ref. 45 that N'_{xy} can be considered to be maximum at

$$x=a-0.4\sqrt{r_x h}\qquad y=b-0.4\sqrt{r_y h} \tag{19.37}$$

where r_x and r_y are the corner radii (Fig. 19.42). This assumption is based on the fact that edge disturbances damp out rapidly.

The Geckeler approximation (Sec. 19.3.1) may be used for an approximate evaluation of the bending near the edges. If the edge arch is much stiffer than the shell, the shell edge can be assumed fixed. Thus, for the edges $x=\pm a$,

$$w=-2\beta\frac{r^2}{Eh}Q_{xa}-2\beta^2\frac{r^2}{Eh}M_{xa}+\frac{\bar{p}_z r^2}{Eh}$$

$$\frac{dw}{dx}=2\beta\frac{r^2}{Eh}Q_{xa}+4\beta^2\frac{r^2}{Eh}M_{xa}=0$$

In these equations, the terms involving the edge shear Q_{xa} and the edge moment M_{xa} are obtained from Table 19.2. The last term in the first equation is the radial (hoop) displacement due to p_z. The solution is

$$Q_{xa}=\frac{\bar{p}_z}{\beta}\qquad M_{xa}=\frac{\bar{p}_z}{2\beta^2} \tag{19.38a}$$

With the edge shear and moment known, moments at points interior to the boundary can be determined from Table 19.2. Thus,

$$M_x=\frac{C_3}{\beta}Q_{xa}+C_4 M_{xa}=\frac{p_z}{\beta^2}\left(C_3-\frac{C_4}{2}\right) \tag{19.38b}$$

If the torsional stiffness of the supporting arch is small, the shell edge can be assumed to be simply supported, in which case $w=M_{xa}=0$, so

$$Q_{xa}=\frac{\bar{p}_z}{2\beta} \tag{19.39a}$$

With the edge shear known, moments at interior points are determined from Table 19.2. Thus,

$$M_x=C_3\frac{\bar{p}_z}{2\beta^2} \tag{19.39b}$$

EXAMPLE

Assume $a=40$ ft, $b=50$ ft, $c_1=8$ ft, $c_2=10$ ft, $h=3$ in., $p_z=60$ psf. From Eq. (19.36), assuming $k=1$,

$$N'_x=\frac{60\times 40^2}{8}=-12{,}000\times\text{coefficient}$$

$$N'_{xy}=\frac{60\times 50^2}{10}=-15{,}000\times\text{coefficient}$$

$$N'_{xy}=\frac{60\times 50\times 40}{\sqrt{80}}=-13{,}400\times\text{coefficient}$$

Stress resultants along the edge $x=a=40$ ft will be calculated. From Table 19.20,

$$N'_x=0\qquad\qquad N'_y=-0.5\times 15{,}000=-7500\text{ lb/ft}$$

The edge shears N'_{xy} are determined from the coefficients in Table 19.21, following which principal tensions are calculated from

$$N'=\frac{N'_y}{2}\pm\sqrt{\left(\frac{N'_y}{2}\right)^2+N'^2_{xy}}$$

The results are shown in Table 19.22.

From Eq. (19.35),

$$\frac{\partial z}{\partial y}=\frac{2c_2 y}{b^2}\qquad\qquad\frac{\partial^2 z}{\partial y^2}=\frac{2c_2}{b^2}$$

from which the radius of curvature r_y is

$$r_y=\frac{[1+\partial z/(\partial y)^2]^{3/2}}{\partial^2 z/\partial y^2}=125\left[1+\left(\frac{y}{125}\right)^2\right]^{3/2}$$

At the corner, $r_y=125\times 1.16^{3/2}=156$ ft, and from Eq. (19.37), the edge shear can be considered maximum at

$$y=50-0.4\sqrt{156\times 0.25}=47.5\text{ ft}=0.95b$$

Thus, according to Table 19.22, the largest principal tension along the edge is 9.64 kips/ft.

Reinforcement for the tension at the corners is usually placed diagonally. The controlling tension is usually that at the edge, but principal stresses at several interior points should also be computed to determine the extent of the area to be reinforced. Minimum reinforcement of at least 0.0018bh for welded-wire fabric (ACI 318-11, Section 7.12.2.1) should be supplied throughout.

The transverse shear and moment along the edge $x=a=40$ ft can be estimated by Eq. (19.38) or (19.39), depending upon the assumption as to relative stiffnesses of the shell and its supporting arch. The corner radius r_x is to be 125 ft, from which $\beta=0.24$. Equation (19.38a) gives $Q_{xa}=250$ lb/ft and $M_{xa}=-520$ ft-lb/ft. However, if the edge is assumed

Table 19.20 Coefficients for Computing Stress Resultants in Elliptical Paraboloids [Eq. (19.36)] (*from Ref. 26*)

x/a	Stress resultants	Value of y/b									
		(*a*) $c1/c2 = 1.0$					(*d*) $c1/c2 = 0.8$				
		0	0.25	0.50	0.75	1.0	0	0.25	0.50	0.75	1.0
0.00	N_y	0.250	0.233	0.182	0.101	0	0.289	0.270	0.213	0.119	0
	N_x	0.250	0.267	0.318	0.399	0.500	0.211	0.230	0.287	0.381	0.500
	N_{xy}	0	0	0	0	0	0	0	0	0	0
0.25	N_y	0.267	0.250	0.199	0.111	0	0.304	0.285	0.228	0.130	0
	N_x	0.233	0.250	0.301	0.389	0.500	0.196	0.215	0.272	0.370	0.500
	N_{xy}	0	0.029	0.068	0.096	0.108	0	0.034	0.069	0.100	0.114
0.50	N_y	0.318	0.301	0.250	0.150	0	0.347	0.331	0.277	0.169	0
	N_x	0.182	0.199	0.250	0.350	0.500	0.153	0.169	0.223	0.331	0.500
	N_{xy}	0	0.068	0.140	0.210	0.244	0	0.065	0.139	0.215	0.255
0.75	N_y	0.399	0.389	0.350	0.250	0	0.416	0.406	0.369	0.270	0
	N_x	0.101	0.111	0.150	0.250	0.500	0.084	0.094	0.131	0.230	0.500
	N_{xy}	0	0.096	0.210	0.356	0.465	0	0.091	0.201	0.353	0.480
1.0	N_y	0.500	0.500	0.500	0.500	0	0.500	0.500	0.500	0.500	0
	N_x	0	0	0	0	0	0	0	0	0	0
	N_{xy}	0	0.108	0.243	0.465	∞	0	0.101	0.229	0.443	∞
		(*b*) $c1/c2 = 0.6$					(*e*) $c1/c2 = 0.4$				
0.00	N_y	0.336	0.316	0.252	0.143	0	0.395	0.374	0.307	0.180	0
	N_x	0.164	0.184	0.248	0.357	0.500	0.105	0.126	0.193	0.320	0.500
	N_{xy}	0	0	0	0	0	0	0	0	0	0
0.25	N_y	0.348	0.329	0.267	0.155	0	0.403	0.383	0.319	0.192	0
	N_x	0.152	0.171	0.233	0.345	0.500	0.097	0.117	0.181	0.308	0.500
	N_{xy}	0	0.031	0.067	0.103	0.120	0	0.026	0.060	0.101	0.125
0.50	N_y	0.383	0.367	0.312	0.197	0	0.425	0.410	0.357	0.235	0
	N_x	0.117	0.133	0.188	0.304	0.500	0.075	0.090	0.143	0.265	0.500
	N_{xy}	0	0.060	0.132	0.216	0.265	0	0.049	0.115	0.208	0.274
0.75	N_y	0.436	0.426	0.392	0.296	0	0.459	0.451	0.419	0.331	0
	N_x	0.064	0.074	0.108	0.204	0.500	0.041	0.049	0.081	0.169	0.500
	N_{xy}	0	0.081	0.185	0.342	0.494	0	0.065	0.156	0.316	0.506
1.00	N_y	0.500	0.500	0.500	0.500	0	0.500	0.500	0.500	0.500	0
	N_x	0	0	0	0	0	0	0	0	0	0
	N_{xy}	0	0.089	0.208	0.413	∞	0	0.070	0.173	0.363	∞
		(*c*) $c1/c2 = 0.2$									
0.00	N_y	0.462	0.446	0.388	0.248	0					
	N_x	0.038	0.054	0.112	0.252	0.500					
	N_{xy}	0	0	0	0	0					
0.25	N_y	0.465	0.451	0.396	0.261	0					
	N_x	0.035	0.049	0.104	0.239	0.500					
	N_{xy}	0	0.014	0.040	0.088	0.128					
0.50	N_y	0.473	0.462	0.414	0.303	0					
	N_x	0.027	0.038	0.086	0.197	0.500					
	N_{xy}	0	0.027	0.074	0.174	0.280					
0.75	N_y	0.485	0.480	0.456	0.383	0					
	N_x	0.015	0.020	0.044	0.117	0.500					
	N_{xy}	0	0.034	0.098	0.246	0.510					
1.00	N_y	0.500	0.500	0.500	0.500	0					
	N_x	0	0	0	0	0					
	N_{xy}	0	0.038	0.108	0.262	∞					

Table 19.21 Shear along Edges of Elliptical Paraboloids [Eq. (19.36c)] (*from Ref. 26*)

	c1/c2				
	1.0	0.8	0.6	0.4	0.2
y/b			At $x = \pm a$		
0.0	0.0000	0.0000	0.0000	0.0000	0.0000
0.1	0.0419	0.0389	0.0342	0.0307	0.0137
0.2	0.0854	0.0793	0.0701	0.0550	0.0286
0.3	0.1319	0.1231	0.1096	0.0872	0.0481
0.4	0.1836	0.1721	0.1546	0.1254	0.0731
0.5	0.2432	0.2294	0.2081	0.1728	0.1075
0.6	0.3204	0.3066	0.2859	0.2493	0.1818
0.7	0.4071	0.3897	0.3627	0.3173	0.2296
0.8	0.5363	0.5178	0.4887	0.4400	0.3443
0.85	0.6279	0.6090	0.5791	0.5292	0.4306
0.9	0.7570	0.7378	0.7074	0.6667	0.5659
0.95	0.9777	0.9582	0.9276	0.8763	0.7741
1.0	∞	∞	∞	∞	∞
x/a			At $y = \pm b$		
0.0	0.0000	0.0000	0.0000	0.0000	0.0000
0.1	0.0419	0.0444	0.0468	0.0488	0.0500
0.2	0.0854	0.0903	0.0950	0.0990	0.1014
0.3	0.1319	0.1391	0.1460	0.1519	0.1553
0.4	0.1836	0.1930	0.2019	0.2095	0.2140
0.5	0.2432	0.2545	0.2652	0.2743	0.2798
0.6	0.3204	0.3317	0.3425	0.3516	0.3571
0.7	0.4071	0.4213	0.4348	0.4463	0.4532
0.8	0.5363	0.5515	0.5659	0.5782	0.5855
0.85	0.6279	0.6434	0.6582	0.6707	0.6782
0.9	0.7570	0.7728	0.7878	0.8005	0.8081
0.95	0.9777	0.9935	1.0087	1.0215	1.0290
1.0		∞	∞	∞	∞

to be simply supported, Q_{xa} = 125 lb/ft [Eq. (19.39a)], and the corresponding maximum moment is, from Eq. (19.39b),

$$M_x = 0.32\frac{60}{2\times 0.24^2} = 167 \text{ ft} - \text{lb/ft}$$

The maximum value 0.32 of C_3 is from Table 19.2 and corresponds to $\beta_s = 0.8$. Therefore, this maximum moment is located about 0.8/0.24 = 3.3 ft from the edge.

The supporting arches should be designed for the edge shears N'_{xy} as in the case of the barrel shell.

19.6.3 Hyperbolic Paraboloids (Hypars) with Straight-Line Boundaries

These shells have a simplicity of geometry and a potential for shape that is economical and attractive. The simplicities, however, may mislead the designer into geometries or details that violate good practice in reinforced concrete. Specifically, if the rise is too slight, compressions are high, and moments due to creep can cause instabilities. Also, simplified procedures for determining stresses often provide no criterion for reinforcement except for the minimum dictated by good practice.

The design of hypars must be based on a study of successful structures as well as those that have experienced difficulties.[46–50]

Membrane Theory[51]

Many hypars designed on the basis of membrane theory have performed well. However, numerical analyses of bending have shown that membrane-theory stresses can be misleading at times.

The equation of the hyperbolic paraboloid is (see Fig. 19.49*a*)

$$z = c_2 = \frac{y^2}{b^2} - c_1\frac{x^2}{a^2} \tag{19.40}$$

For x (or y) constant, z describes a parabola in a plane parallel to yz (or xy) so that the surface can be generated by translating the parabola EOF along the parabola GOH.

Table 19.22 Shear and Principal Tension, kips/ft, on Edge $x = a$

y/b	0	0.1	0.2	0.3	0.4	0.5	0.6	0.7	0.8	0.85	0.90	0.95
N'_{xy}	0	0.52	1.06	1.65	2.31	3.08	4.11	−5.23	−6.95	−8.17	−9.90	−12.85
N'	0	+0.04	+0.15	+0.35	+0.65	+1.10	+ 1.82	+2.68	+4.14	+5.24	+6.84	+9.64

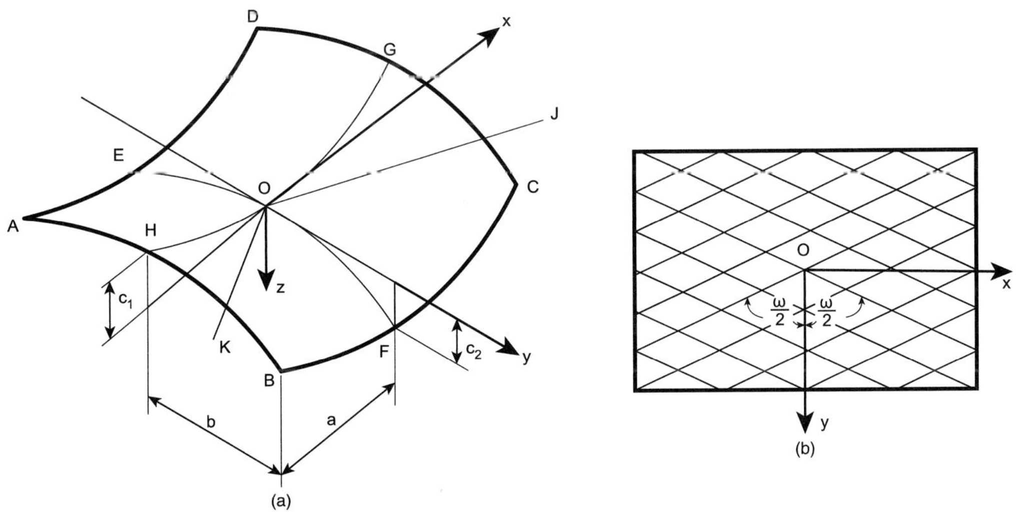

Figure 19.49 Hyperbolic paraboloid shell.

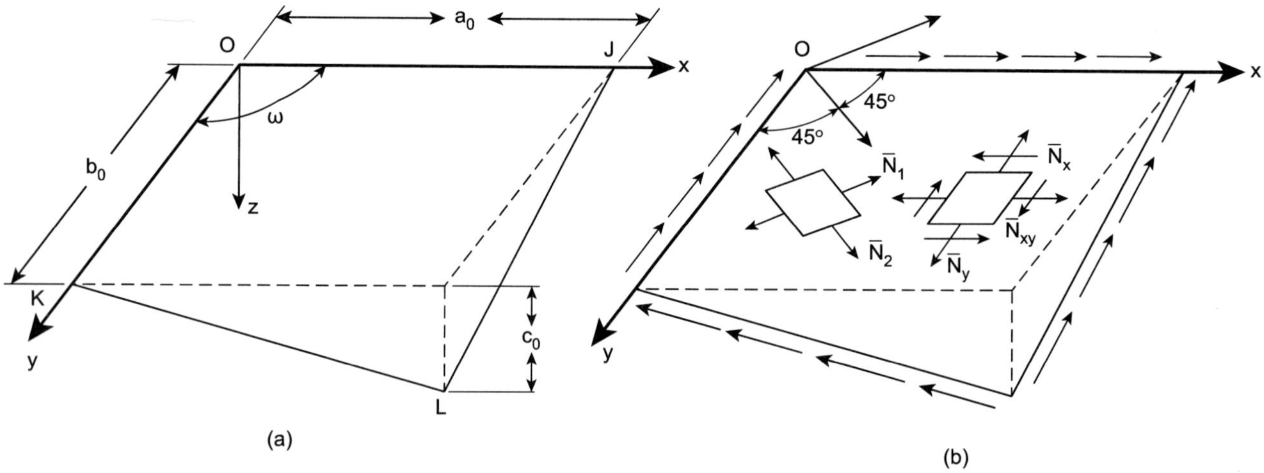

Figure 19.50 (*a*) Alternative surface representation with reference to axes (*x*, *y*) parallel to the projections of the straight generators. (*b*) Internal forces as a result of applied uniform pressure.

The surface contains two sets of straight generators; *OJ* in Fig. 19.49*a* represents one set, and *OK* the other. The projections of these generators on the *xy* plane are shown in Fig. 19.49*b*, where ω is given by

$$\tan^2\frac{\omega}{2} = \frac{c_2}{c_1}\frac{a^2}{b^2} \tag{19.41}$$

Alternatively, the surface can be referred to axes *x*,*y* parallel to the projections of the straight generators (Fig. 19.50*a*). The equation is

$$z = \frac{c_0}{a_0 b_0}xy \tag{19.42}$$

In this position, the surface can be considered to be generated by allowing the line *OJ* to rotate around *OK* as it moves from *OJ* to *KL*, always remaining parallel to the *xz* plane.

Where the boundaries of the shell are straight generators and the latter are orthogonal, Eq. (19.42) can be substituted into Eq. (19.33b) of Sec. 19.6.1, which gives

$$\frac{\partial^2 F}{\partial x \partial y} = \frac{q a_0 b_0}{2c_0}$$

Once a stress function *F* that satisfies this equation is found, the membrane stress resultants can be determined by Eqs. (19.32) and (19.33a). Various solutions are possible, depending upon the assumptions for boundary conditions (Ref. 27, pp. 263–283). When the load $\overline{p}_z$ is uniform over the horizontal projection of the surface, a solution of practical interest is (see Fig. 19.50*b*)

$$\overline{N}_{xy} = \frac{\overline{p}_z a_0 b_0}{2c_0} \qquad \overline{N}_x = \overline{N}_y = 0 \tag{19.43a}$$

The edges are subjected only to shearing-stress resultants, which must be resisted by edge members. The projected principal stress resultants are inclined 45 degrees with the edges:

$$\overline{N}_1 = -\overline{N}_2 = \frac{\overline{p}_z a_0 b_0}{2c_0} \tag{19.43b}$$

Skewed Surfaces

If the surface is bounded by straight generators and $\omega \neq 90°$ (Fig. 19.50*a*), a load $\overline{p}_z$ uniform over the horizontal projection can be carried by

$$\overline{N}_{xy} = \overline{N}_{yz} = -\frac{\overline{p}_z a_0 b_0}{2c_0}\sin\omega \tag{19.44a}$$

$$\overline{N}_x = \overline{N}_y = 0 \tag{19.44b}$$

where $\overline{p_{xy}} = \overline{p_{xyz}}$ are shearing-stress resultants on an element bounded by the (nonorthogonal) projections on the plane *xy* of the straight generators. $\overline{N}_x$ and $\overline{N}_y$, being parallel to *x* and *y*, respectively, are nonorthogonal.

The corresponding principal stress resultants in the surface lie approximately in vertical, orthogonal planes, one of which bisects the quadrant *xy*:

$$N_1' = -N_2' = \frac{\overline{p}_z a_0 b_0}{2c_0}\sin\omega \tag{19.44c}$$

Unsymmetrical Load

If some quadrants of the surfaces in Fig. 19.51 are unloaded, stress resultants for the loaded quadrants can be determined by Eq. (19.43). However, this results in unbalanced forces in certain edge members. If only two quadrants of the surface in Fig. 19.51*d* are loaded (Fig. 19.52), the edge-member forces *F* must be reacted. Unless there is external restraint at these points, *F* must be equilibrated in each case by the other half of the edge member. This subjects the unloaded panels to shearing forces along their edges. This problem is discussed in Ref. 50.

Solutions based on Eq. (19.43) are not reliable for the surfaces shown in Fig. 19.51*b*–*d* because the shell cannot deform as the membrane theory would predict. Though a more refined analysis is needed for design, the membrane theory still can provide indications of behavior of the type seen in Fig. 19.51 and Ref. 50.

Behavior Considering Bending Theory

The gabled shell in Fig. 19.53 demonstrates the differences between membrane theory and bending theory for the behavior of the shell, of the horizontal ridge beams, and of the gabled edge beams.

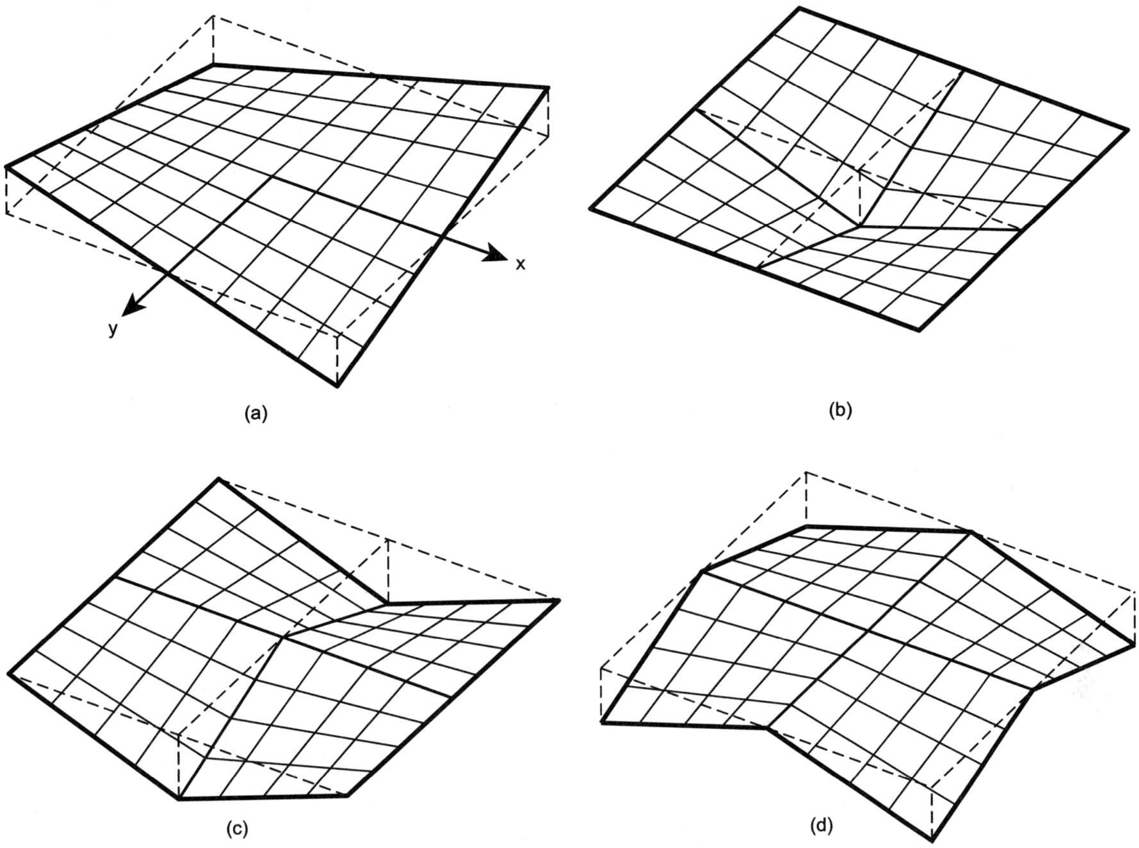

Figure 19.51 Hyperbolic paraboloid roof surfaces.

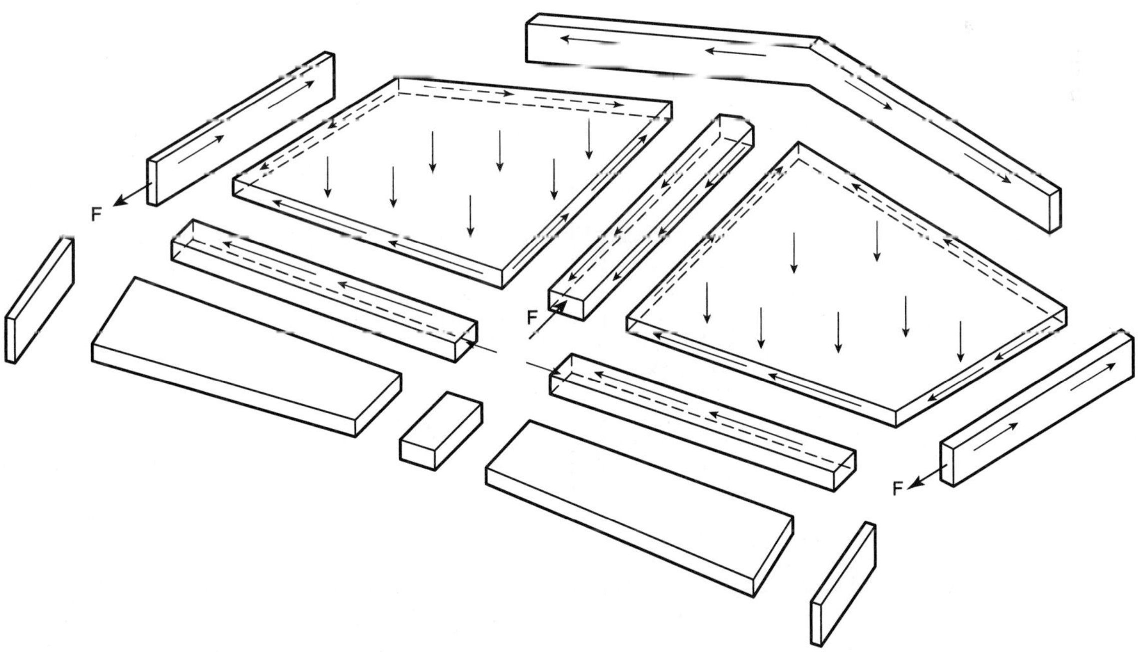

Figure 19.52 Loaded quadrants of hyperbolic paraboloid roof surface.

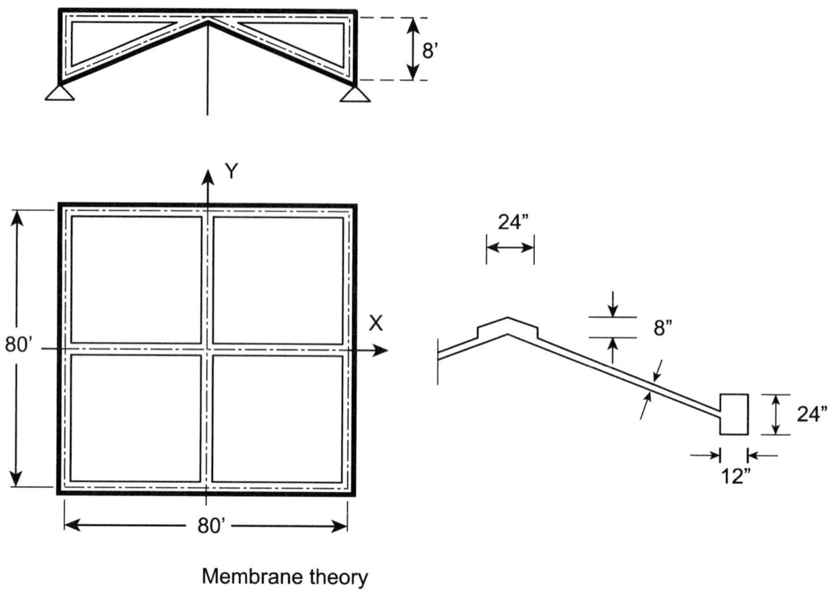

Figure 19.53 Gabled hyperbolic paraboloid. (*From Ref. 52.*)

For the shell under uniform load on the horizontal projection, the membrane theory would give shear stresses everywhere of $N'_{xy} = \overline{p_z}/2k$, where $\overline{p_z}$ = 37.5 psf and $k = 8/40^2 = 0.005$, so that $N'_{xy} = 37.5/0.01 = 3.75$ kips/ft or $\tau_{xy} = 3.75/(12 \times 3) = 104$ psi. This stress is numerically equal to the principal tensile and compressive stresses through (k), and hence, according to the membrane theory in each quadrant, the shell carries the load—half by parabolic arching slabs parallel to the diagonal between the central crown and the support and half by parabolic hanging slabs parallel to the diagonal between the gable crowns at the edges. Shaaban and Ketchum showed that the hanging slabs carry less load; it is nearly all taken by the arching slabs, which have thereby twice the stress that is obtained by membrane theory. Figure 19.54 shows this behavior and further emphasizes how the diagonal arch slab part that is directly between the crown and support takes very much higher stresses near that support in the case they studied. This gabled hyperbolic paraboloid shell acts primarily as two diagonal arches crossing at the crown and delivering nearly all the roof load directly to the corner supports. It does not act as a set of arching and hanging slabs bringing load to the straight-line edges as derived from the membrane theory.

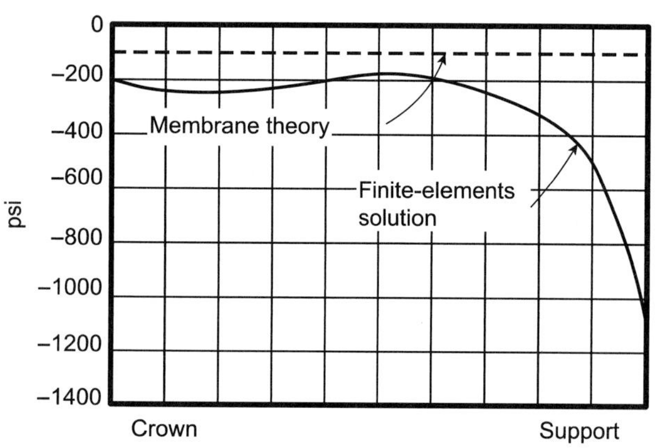

Figure 19.54 Stress resultants in a gabled hyperbolic paraboloid. (*From Ref. 53.*)

This behavior arises because those edges cannot carry the tensions assumed by the membrane theory; rather, they move inward as shown in Fig. 19.55 and thereby force the diagonal arches to carry the load. We can see the same behavior reflected in the forces carried by the edge beams. Figure 19.56 shows the forces in the edge beams derived from both the membrane theory and the bending theory. The former gives results closely approximated by $C = (a - x)S = 40 \times 3.75 = 150$ kips at the support where $x = 0$ and $a = 40$ ft. The figure shows a slightly higher value because it is based on dead load rather than a uniform load on the horizontal projection of the surface. The bending theory gives smaller values which are nearly constant up until about halfway down the gable.

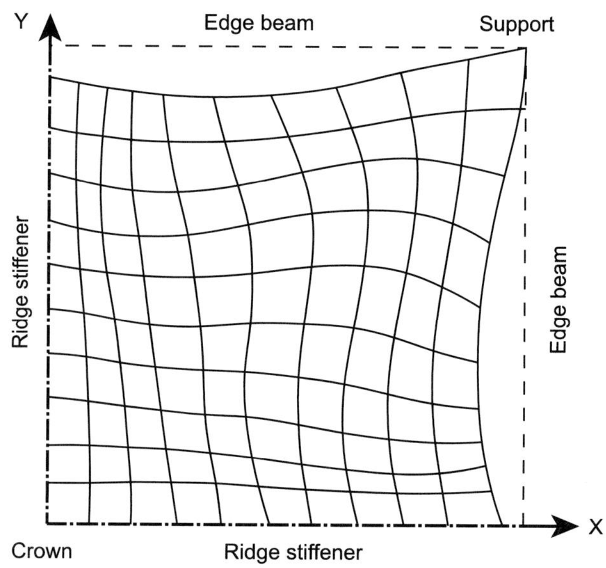

Figure 19.55 Displacements in a gabled hyperbolic paraboloid. (*From Ref. 53.*)

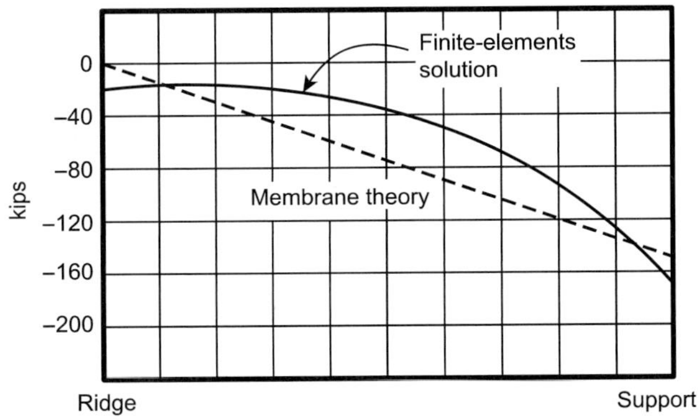

Figure 19.56 Forces in edge beam of a gabled hyperbolic paraboloid. (*From Ref. 53.*)

Then, its force increases rapidly until, at the support, it is about equal to the membrane theory result. Thus, Figs. 19.48 and 19.50 show that the shell rather than the edge beam carries much of the load except near the corners at the supports.

These results, while useful as insight, do not have a major bearing on the layout of shell reinforcement since the stresses are so low. However, they have a major influence on design of the edge beams for two reasons. First, the much-reduced edge-beam loads suggest that those beams need not be large. Second, the reduced compression makes such beams less able to carry bending because they are less precompressed than the membrane theory assumes.

Schnobrich[52] has shown that the edge beams and the ridge beams actually load the shell rather than the converse, but they keep the high axial stiffness. This is especially true for the ridge beams because they are horizontal and, hence, can be subjected to substantial deflections and bending in regions where the shell is so flat that it provides little support.

Examples of Finite-Element Analysis

What follows are analyses of an umbrella-type structure (Fig. 19.57*b*) and a gable-type structure (Fig. 19.57*d*). The results are obtained first using the very simple membrane approach and, second, a refined finite-element analysis. The program used, CDA-SPRING, formerly known as MSC-PAL, allows nonplanar quadrilateral shell elements. As in the cylinder examples, program FEMAP[54] is used for pre- and postprocessing.

EXAMPLE 1

For the inverted umbrella shown in Fig. 19.57, $a = b = 20$ ft, and $c = 5.5$ ft. The weight of the 3-in. shell is 37.5 psf, to which is added 12.5 psf to account for the weight of the edge beams. The live load is 30 psf. Then, from Eq. (19.43),

$$-\bar{N}_{xy} = \bar{N}_1 = -\bar{N}_2 = \frac{80 \times 20 \times 30}{2 \times 5.5} = 2960 \cong 3000 \text{ lb/ft}$$

The required reinforcement in the direction of N_1 is

$$A_s = \frac{3000}{20{,}000} = 0.150 \text{ in.}^2/\text{ft}$$

Minimum reinforcement in the orthogonal direction for temperature and shrinkage is $0.002 \times 36 = 0.072$ in.^{2}/ft (ACI 318–11 Section 7.12.2.1).

For easier placement, steel is often placed along the straight-line generators. Assuming $A_{sx} = A_{sy}$, Eq. (19.47) gives, since $\phi = 45°$,

$$A_{sx} = A_{sy} = \frac{N_p}{f_s} = \frac{3000}{20{,}000} = 0.150 \text{ in.}^2/\text{ft}$$

Thus, where it follows the straight-line generators, the steel required is $2 \times 0.150 = 0.3$ in.^{2}/ft. On the other hand, only $0.15 + 0.072 = 0.222$ in.^{2}/ft is required if it is placed along the parabolas.

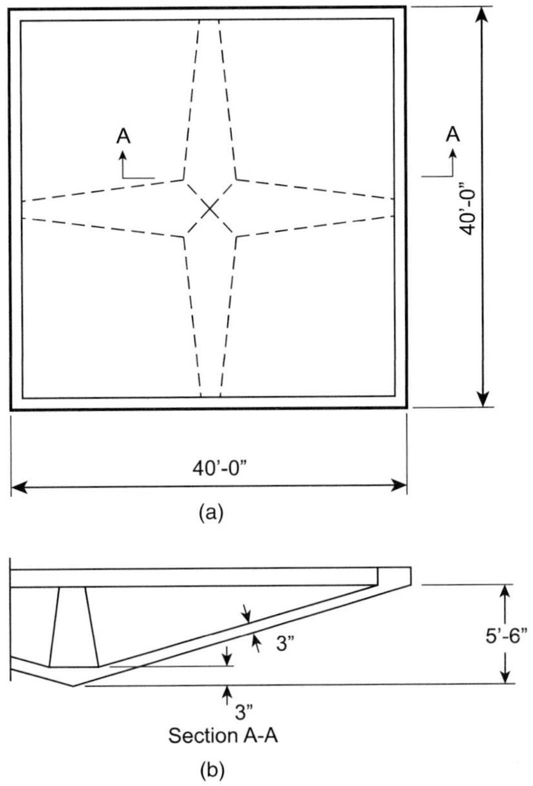

Figure 19.57 Example of Section 19.6.3.

The compressive stress is only

$$f = \frac{3000}{3 \times 12} = 83 \text{ psi}$$

The tension T at mid-length of the edge members is

$$T = 3000 \times 20 = 60{,}000 \text{ lb}$$

for which

$$A_s = \frac{60{,}000}{20{,}000} = 3 \text{ in.}^2$$

The shearing forces on both sides of the sloped edge members contribute to its axial force. The compression C is

$$C = 2 \times 60{,}000 \times \frac{\sqrt{20^2 + 5.5^2}}{20} = 125{,}000 \text{ lb}$$

Choosing an allowable compression on concrete of 500 psi,

$$A_g = \frac{125{,}000}{500} = 250 \text{ in.}^2$$

For the triangular cross section shown in section AA of Fig. 19.57, the area furnished is

$$A_g = \frac{20d^2}{5.5} = 250 \text{ in.}^2$$

from which $d = 8.29$ in. In Ref. 50, this depth is made 9 in.

The finite-element model is shown in Fig. 19.58*a* and 19.58*b*. The shell is a perfect hypar as defined in Fig. 19.58*b* and by Eq. (19.42). The shell is thickened to 9 in. at center, tapering to 3 in. at 2 ft on each side of the symmetry plane and at the perimeter (Figs. 19.57 and 19.58*b*), so that the median surface is not a perfect hypar anymore.

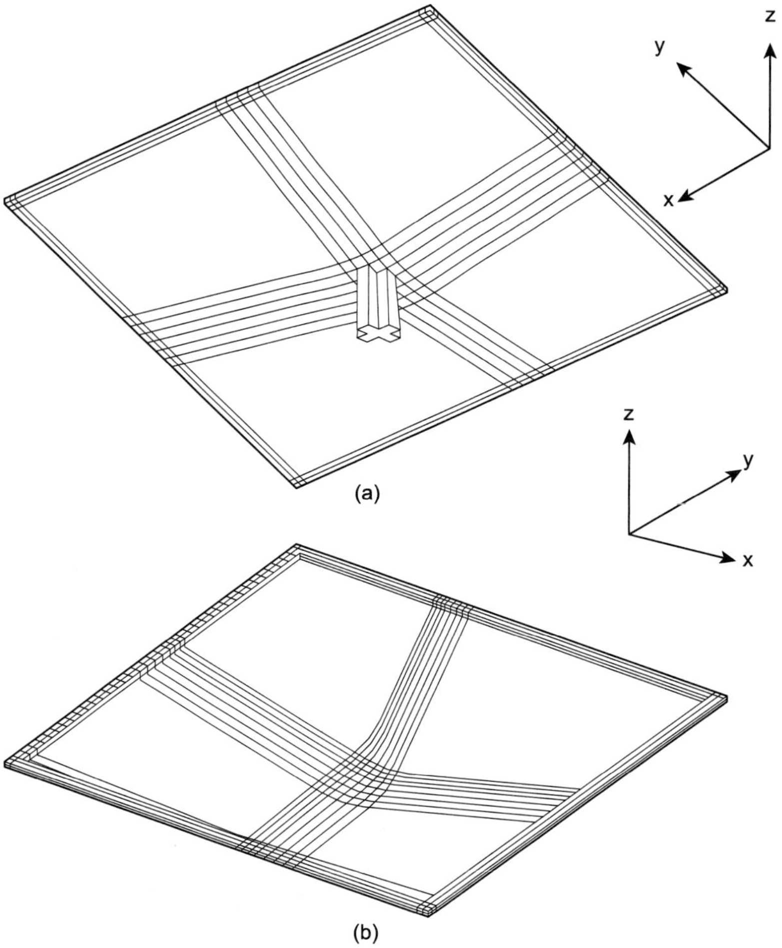

Figure 19.58 Finite-element model (*a*) and (*b*).

The perimeter beams are 12 in. wide and 7.5 in. high on the exterior, and have variable thickness at the interior. The loads defined are the actual dead load for 150 psf and a live load of 30 psf uniformly distributed on the shell surface.

Only one-quarter of the shell was analyzed assuming two symmetrical planes. The displacements show that the membrane behavior assumptions considered in the simplified analysis are not satisfied. At about three-fourths of the diagonal length (measured) from center, the tension perpendicular to the main diagonal is 5400 lb/ft (450 lb/in.), resulting in a stress of 150 psi (which is 80 percent higher than given by the membrane theory), while the bending stress in the same section is 230 psi.

Based on these numbers, we conclude that at about two-thirds of the diagonal length, measured from the center, the calculation of reinforcing based on membrane theory may not be conservative, and cracks may show on the upper face of the shell. However, due to the high redundancy of the system, it is unlikely that the ultimate load is significantly reduced (if the reinforcing is based on membrane theory).

The membrane compressive stress along the stiffened area is 390 psi (compared to 500 psi obtained in the simple method), but it is coupled with a bending stress of ±320 psi. The total axial force on a 4-ft-wide band is 103,000 lb, which compares well to the 125,000 obtained in the membrane analysis.

The edge-beam tension is not confined to the beams but spread into the shell, creating stresses parallel to the edge. The total tension at mid-length of the edge (calculated from the element stresses) is 61,500 psi, which is in very good agreement with the result given by the membrane theory. From this force, 40 percent is spread in the shell, and 60 percent is located in the 1-in.-wide beam.

19.6.4 Hyperbolic Paraboloids with Parabolic Boundaries

Possible systems of stress resultants for a surface bounded as in Fig. 19.49 and with *x,y* directed as in that figure are

$$\bar{N}_x = \frac{\bar{p}_z a^2}{2c_1} \qquad \bar{N}_y = \bar{N}_{xy} = 0 \tag{19.45a}$$

$$\bar{N}_y = \frac{\bar{p}_z b^2}{2c_2} \qquad \bar{N}_x = \bar{N}_{xy} = 0 \tag{19.45b}$$

$$\bar{N}_x = \frac{\bar{p}_z a^2}{4c_1} \qquad \bar{N}_y = -\frac{\bar{p}_z b^2}{4c_2} \qquad \bar{N}_{xy} = 0 \tag{19.45c}$$

The first system requires anchorage of the surface at $x = \pm a$, the second at $y = \pm b$, and the third at both $x = \pm a$ and $y = \pm b$. The resultants $\overline{N}_x$ and $\overline{N}_y$ normal to the vertical planes *AHB*, *AED*, and so on constitute heavy load that will usually be impracticable to support.

Combinations of the surface can be made. An example is the groined vault of Fig. 19.59. Reference 49 contains tables of coefficients for this structure.

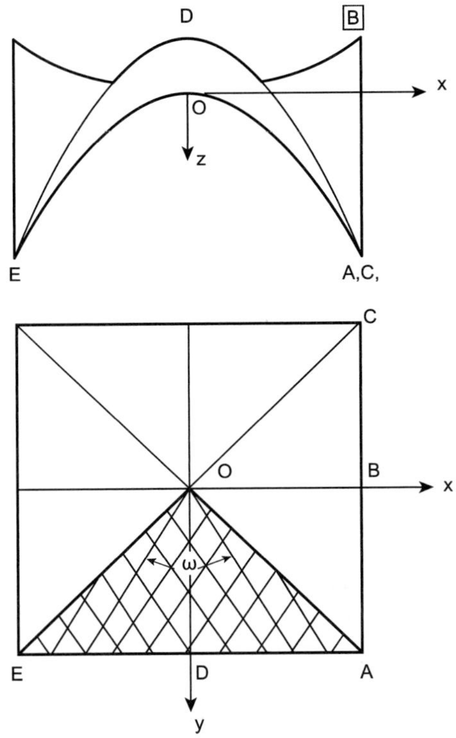

Figure 19.59 Groined vault.

19.7 DIMENSIONING

The following comments on proportioning of concrete and reinforcement in thin shells are based on Ref. 40.

Thickness is rarely based on allowable stress but usually on construction requirements or stability.

Reinforcement should be provided to resist completely the principal tensile stresses assumed to act at the middle surface. It may be placed either in the general direction of the lines of principal stress or in two or three directions. In regions of high tension, it is advisable to place it in the general direction of the principal stress. Whenever possible, such reinforcement may run along lines practical for construction, such as straight lines.

Reinforcement that does not deviate in direction more than 15 degrees from the direction of the principal stress may be considered to be parallel to it. A slightly greater deviation can be tolerated in areas where the stress in the reinforcement is two-thirds or less of the allowable.

For reinforcement placed in two directions at right angles, as along x and y axes, and where A_{sy} forms the angle θ with the direction of the principal stress resultant N_p, reinforcement is required such that

$$N_p = f_{sy}(A_{sy}\cos^2\theta + A_{sy}\sin^2\theta\tan\theta \tag{19.46}$$

$$\tan 2\theta = \frac{2N_p}{N_x - N_y} \tag{19.47}$$

For positive values of $\tan 2\theta$, θ is measured counterclockwise from the face upon which N_p acts.

Minimum reinforcement should be provided as required in ACI 318 even where not required by analysis.

In areas where the computed tensile stress in the concrete exceeds 300 psi, at least one layer of reinforcement must be parallel to the principal tensile stress unless it can be proved that a deviation is permissible because of the geometrical characteristics of the shell and because, for reasons of geometry, only insignificant and local cracking could develop.

The allowable stress for reinforcement may be used at any point in the shell independently of the magnitude of the stress in the concrete at that point.

Additional reinforcement to resist bending moments must be proportioned and provided in the conventional manner. Generally, where moments are significant in thin shells, the effect of direct compression forces may be neglected. Either working-load or ultimate-load analysis may be used.

Where the computed principal tensile stress (psi) exceeds $2\sqrt{f'_c}$ (f'_c in psi), the spacing of reinforcement should not exceed three times the thickness of the shell. Otherwise, reinforcement should be spaced not more than five times the thickness of the shell or more than 18 in.

Splices in principal tensile reinforcement should be kept to a practical minimum. Splices should be staggered, with not more than one-third of the bars spliced at any one cross section. Bars should be lapped only within the same layer. The minimum lap for draped reinforcing bars should be 30 diameters, with a minimum of 18 in. unless more is required by ACI 318, except that the minimum may be 12 in. for reinforcement not required by analysis. The minimum lap for welded-wire fabric should be 8 in. or one mesh, whichever is greater, except that ACI 318 governs where wire fabric at a splice must carry the full allowable stress.

Concrete cover should be at least ½ in. for bars, ⅞ in. where precast and for welded-wire fabric, and 1 in. for prestressing tendons, provided the concrete surfaces are protected from weather and are not in contact with the ground. In no case should the cover be less than the diameter of the bar or tendon. If greater cover is required for fire protection, it needs apply only to principal tensile and moment reinforcement whose yielding would cause failure.

19.8 STABILITY AND SAFETY

Shell geometry and imperfections, concrete cracking and yielding, and creep and large deflections can strongly influence the safety of a thin shell, especially by leading to instability or buckling (Ref. 27, Chapter 9).

19.8.1 Buckling

SHELL WALLS

Buckling can occur under transient loadings such as wind. For a cylinder, fixed at its base, free at its top, and under wind pressure, a bifurcation analysis gives the critical pressure as

$$q_{cr} = KE\left(\frac{h}{r}\right)^3 \tag{19.48}$$

Values of K for a cylinder with $r/h = 100$ can be found from Fig. 19.60 for the typical wind-pressure distribution shown in the figure.[55] Comparisons between these results and wind-tunnel test results show the former to be about 36 percent high.[56] Critical pressures with K from Fig. 19.60 and using $c = 1$ are 1860 and 430 psf for $H/r = 1$ and 6, respectively. Since r/h seldom exceeds 100 for thin concrete cylindrical shells, this shows that such shells will not buckle under wind pressure unless there is a high thermal gradient, as in cooling-tower shells, and insufficient circumferential steel to control cracking.

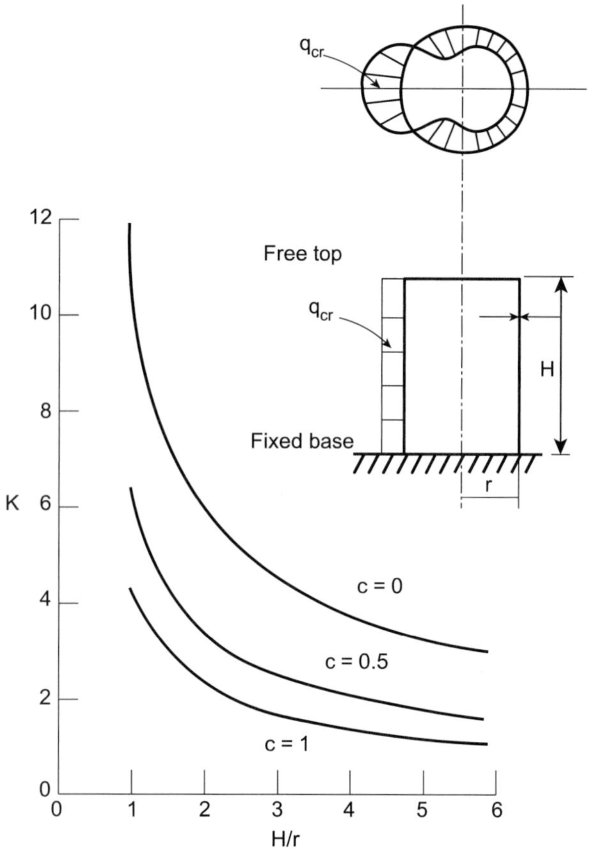

Figure 19.60 Values of K in Eq. (19.48) for $r/h = 100$. *Note:* c = internal suction coefficient as a multiple of q_{cr}.

For hyperboloids under wind loading, simplified bifurcation results are from two to three times higher than wind-tunnel results.[57] The latter lead to an estimate of buckling pressure as[64]

$$q_{cr} = CE\left(\frac{h}{r}\right)^{\alpha} \tag{19.49}$$

where q_{cr} = buckling pressure, psi, along windward meridian
C = empirical coefficient = 0.052
h = thickness at throat, that is, at point on meridian where horizontal radius is minimum
r = radius of shell parallel circle at throat
α = empirical coefficient = 2.3

The value of α has been derived from theoretical considerations as 7/3 (Ref. 58). The coefficient C shows wide scatter. Reference 59 shows 0.052 to be a reasonable lower-bound estimate.[57] The value of q_{cr} by Eq. (19.49) should be compared with the design wind pressure at the top of the tower to ensure an adequate factor of safety against buckling.

For cooling towers, q_{cr} should be at least twice q_z, as defined in Sec. 19.2.2, when dead load is included and cracking considered. This is because thermal gradients across cooling-tower shells (and possibly other types) are large enough to cause vertical cracks and hence to reduce substantially the circumferential stiffness of the shell.[60] Since this bending stiffness plays a major role in the buckling capacity, its reduction can reduce the buckling pressure drastically. The solution is to provide enough circumferential reinforcement to control the cracking and to use a thickness sufficient to keep $q_{cr} > 2q$.

Imperfections in the geometry of a completed shell can lead to overstressing, buckling, or both. In the failure of the Ardeer cooling tower in 1973, imperfections were recorded of over 18 in. for a 6-in.-thick shell, and these directly contributed to the failure.[61] Nevertheless, field tolerances must be permitted on very large structures, but always subject to specification by an experienced designer of thin-shell concrete structures.

Shell Roofs

Buckling can be initiated by relatively high permanent-load compressive stresses, which with time can lead to creep and eventually large displacements. Displacements of a shell normal to its middle surface imply a change in curvature, so thin shells, since they rely on their curvature for resistance to load, are particularly susceptible to large displacements and thus to buckling. Although much has been written on the stability of thin shells,[62–64] there is little experimental work on thin-shell concrete structures of the shape and boundary conditions usually found in roofs.

The buckling pressure q_{cr} on a spherical thin shell[63] is given by

$$q_{cr} = CE\left(\frac{h}{a}\right)^{2} \tag{19.50}$$

where $C = 2/\sqrt{3(1-\nu^2)}$
ν = Poisson's ratio
h = thickness of shell
a = radius of shell

Experiments show that C is much smaller than $2/\sqrt{3(1-\nu^2)}$. Schmidt[65] shows values as low as 0.06 with none above 0.32. For translational shells, he suggests that Eq. (19.50) be written in the form

$$q_{cr} = CE\frac{h^2}{r_x r_y} \tag{19.51}$$

where r_x and r_y are the two principal radii of curvature.

These equations show that instability becomes a problem for shells which are very thin and flat or of low-modulus concrete. Concrete creep can contribute to large deflections but may be reduced by providing reinforcing steel in both faces of the shell. The effect of creep can be estimated by assuming a reduced value of E or, if the principal membrane stresses at any point are known, by determining the tangent modulus of elasticity and dividing it by a factor for long-term deflections. The factor should not be less than 2.

Results from a comprehensive set of tests on concrete-dome models at Ghent University led to the value of $C = 0.16$ as a reasonable estimate (when creep is considered) for uncracked spherical domes without imperfections. Scordelis summarized methods for including imperfections and cracking.[66] He gives graphs taken from the 1979 Recommendations of the International Association for Shell and Spatial Structures.[67] As Scordelis notes, the Recommendations give buckling capacities much lower than does the Report of the ACI Committee 344.[68] In any case, designers need to consider imperfections and cracking in estimating dome buckling capacity so that the design will not be controlled by buckling.

In hyperbolic parabolical shells, there is danger of creep and failure if the overall form is too flat. An example is the 1970 failure of a gabled roof in Henrico County, Virginia, of the form shown in Fig. 19.51*d*. The Henrico shell was flat, defined by the ratio c_0/a_0b_0 as in Eq. (19.43*b*). It was also made of lightweight concrete, and its supports were flexible columns connected by prestressed tension ties. Lightweight concrete has a lower E than regular concrete of comparable strength, and it is also more prone to creep. In an earlier shell of the same form built in Denver, the hyperbolic-paraboloid was less flat, made of regular concrete, and supported by heavy hinged buttresses tied together by a stiff floor and framing below. Moreover, the flat central section at Denver had over twice the steel reinforcement in its shallow ridge beams than did those of the Henrico shells.

19.8.2 Dynamic Behavior

All the analyses presented in this section are for static behavior. For concrete-shell roofs, wind is not usually a dynamic problem. Because of the weight of the concrete, a more usual problem is the tearing off of roofing because of the wind suction that arises over much of a shell roof. Seismic loading on shell roofs can be a difficult problem if the roof is supported at a few isolated points, in which case the horizontal shears from seismic motions can be large and need to be considered.

For shell walls, the principal design loading can frequently be either seismic or wind. Quasi-static approximations, such as those in Ref. 69, to either loading have normally led to successful structures. If there appears to be a possibility that the natural frequencies of a shell are close to important wind gust frequencies, special dynamic studies may be warranted. Reference 61 gives a discussion of dynamic wind loading on cooling towers.

19.9 CONSTRUCTION

Two main problems in construction are scaffolding and forms and reinforcement and concrete. One major question regarding scaffolding and forming is that of reuse to reduce cost per square foot. This reuse can be either by precasting or by casting in place. Precasting can lead to a reduction in scaffolding altogether, but it also leads to an increase in field connections. Casting in place on a scaffolding which then is moved has been more economical than precasting in many cases, especially where the shell is a long barrel and the scaffold can move along a straight track. On the other hand, for domes of rotation, the scaffold can be built for only several pie-shaped segments and reused by rotating to a new position. This method, however, changes design because the partial shell must carry its dead load as an arch without the stiffening effect of the hoops.

Another design question, that of sufficient concrete strength, arises when the scaffolding is moved and a section of shell must then carry its loads early in its life. The young concrete in roof shells can be subjected to higher creep and thus greater deformation than older concrete.

The reinforcement and the concrete influence construction largely in two ways. The simpler the reinforcing pattern, the easier building is, even though simplicity may sometimes lead to more reinforcement than a pattern carefully detailed to reflect the calculated stresses. The concrete problem mainly arises where the design requirement for a stiff mix, to achieve small creep and shrinkage, conflicts with the construction desire for a workable mix to reduce honeycombing and to permit rapid placement.

A thin shell should be as thin as is practicable in order to induce in plane stress as opposed to bending.[70] However, the high cost of labor in the United States usually makes it more economical to use a thicker shell than to enforce the careful casting techniques essential to construction of very thin shells. Three layers of steel are usually used, and to place concrete properly, a 3-in. thickness is about minimum.

Slope of the shell should be less than 45 degrees to avoid top forms, which increase the difficulty of eliminating honeycombing. Where the slope is 30 degrees, concrete with a slump of 1 to 3 in. can be cast to reasonable tolerances without a top form. Where the slope exceeds 45 degrees, it may be possible to cast without top forms if the slump is low, but the cost of placing is increased. In some cases, concrete has been shot onto vertical surfaces successfully.

Location of ribs can be important. Application of insulation and roofing is relatively simple when the surface is free of ribs. However, ribs projecting below the soffit complicate movement of forms. Form movement may constitute a large item of cost in continuous, cylindrical-segment roofs, in which case ribs are built above the roof. These projections must be carefully flashed. Ribless shells, that is, shells with wide, flat ribs, have been successfully built for short spans where buckling was not a major factor in design.[71]

Maximum aggregate size should not exceed one-half the shell thickness, the clear distance between bars, or 1½ times the cover. Where top forms are required, maximum size of aggregate should not exceed one-fifth the minimum clear distance between forms or the cover over the reinforcement.

Forms should be built carefully. The designer should consider the structural effect of small deviations from the plans and should set tolerances. The stability of the shell depends upon its radii of curvature, and relatively small radial deviations in surface dimensions can cause large variations in radius. For example, a variation of ½ in. radially in an arc length of 10 ft in the dome of Sec. 19.3.1 causes a 45 percent change in radius if the surface remains spherical.

Concrete should be cast in a symmetrical pattern to avoid bracing the scaffolding for the effect of unbalanced loads. It is recommended that concreting commence at the low point or points and proceed upward. Concrete should be deposited as nearly as possible in its final position. Vibration of thin sections is difficult. It has been done by vibrating the reinforcement or the forms, but these systems must be rigid to withstand such treatment. Vibrating screens have been used successfully.

Thin shells are susceptible to shrinkage cracking if curing of the concrete is faulty. In hot weather, the use of retarders, preliminary fog-spray curing, and wet burlap or water curing is advisable. In cold weather, accelerators and special precautions against freezing are usually required. In moderate weather (40° to 70°), ordinary methods such as membrane-curing compounds are usually satisfactory, although wet curing may produce better results.

The method of form removal (decentering) is usually specified by the designer in order particularly to avoid any unwanted temporary supports or concentrated reactions on the shell. It is best to begin decentering at points of maximum deflection and progress toward points of minimum deflection, with decentering of edge members proceeding simultaneously with that of the adjoining shell. It is important to control deflections at the time of decentering, and it is common to specify a modulus of elasticity that must be obtained before permission to decenter is granted. Small, lightly reinforced beams tested in flexure have been used successfully to determine E.[72]

Thin shells of dramatic shape and lightness, which can be successfully built in warm, dry climates, cannot always be duplicated in harsher regions. Furthermore, it is not easy to predict whether a particular construction scheme will be economical.

19.10 APPEARANCE

For appearance, there are two general problems. The first is that many well-known thin shells were designed primarily for appearance and have consequently either not performed well under load or proved uneconomical to build. The second is that many engineers do not consider appearance to be part of engineering design. That is to say, they consider safe behavior and economical construction to be the engineer's province and appearance to be solely the architect's domain. But the fine visual potential inherent in thin shells has been exploited by many engineers. Apart from the striking appearance of their shells, the excellence of these designers' solutions depended upon their concern for minimizing materials and costs. When these two fundamental constraints are not recognized as essential to the success of their works, designers can easily imagine that the fine appearance is something separate from fine engineering. The primary reason why the best-known shell designers (Maillart, Torroja, Nervi, Candela, Tedesko, Isler, and others) have made such striking works is that they conceived of design as the integration of efficiency, economy, and elegance.[47]

REFERENCES

1. Ochsendorf, J.: *Guastavino Vaulting, The Art of Structural Tile*, Princeton Architectural Press, 2013.

2. Billington, D.: *The Tower and the Bridge*, Princeton University Press, 1985.

3. Lozano-Galant, J. A., and I. Paya-Zaforteza: Structural analysis of Eduardo Torroja's Fronton de Recoletos' roof, *Eng. Struct.* 33: 843–854, 2011.

4. Siegel, C.: *Structure and Form in Modern Architecture*, Van Nostrand Reinhold, p. 271, 1962.

5. Faber, C.: *Candela: The Shell Builder*, Reinhold Publishers, 1963.

6. Nervi, P. L.: *Structures*, F.W. Dodge Corp., 1956.

7. Chilton, J.: *Heinz Isler, The Engineer's Contribution to Contemporary Architecture*, Thomas Telford Publishing, 2000.
8. Rice, P.: *An Engineer Imagines*, Ellipsis Londen Pr Ltd, 1998.
9. Lammler, R.: *Ulrich Muther Shell Structures*, Verlag Niggli, 2010.
10. Anderson, S.: *Eladio Dieste, Innovation in Structural Art*, Princeton Architectural Press, 2004.
11. Bletzinger, K. U.: *Form Finding and Morphogenesis, Fifty Years of Progress for Shell and Spatial Structures*, IASS, 2011.
12. Andriaenssens, S., P. Block, D. Veenendaal, and C. Williams: *Shell Structures for Architecture, Form Finding and Optimization*, Routledge, 2014.
13. Shek, H. J.: The force density method for form finding and computation of general networks, *Methods Appl. Mech. Eng.* 3(1):115–134, 1974.
14. Akbarzadeh, M.: *3D Graphical Statics Using Reciprocal Polyhedral Diagrams*, PhD Thesis, ETH Zurich, 2016.
15. Block, P., T. Van Meele, M. Rippmann, and N. Paulson: *Beyond Bending: Reimagining Compression Shells*, Detail Press, 2017.
16. Baker, W., A. McRobie, T. Mitchell, and A. Mazurek: Mechanisms and States of Self Stress of Planar Trusses Using Graphic Statics, Proceedings of the IASS Symposium, Amsterdam, 2015.
17. McRobie, A., and C. Williams: Discontinuous Maxwell–Rankine stress functions for space frames, *Int. J. Space Struct.*, 2018.
18. Allen, E., and W. Zalewski: *Form and Forces, Designing Efficient, Expressive Structures*, John Wiley and Sons, 2009.
19. Akbarzadeh, M., Tom Van Mele, and Philippe Block: On the equilibrium of funicular polyhedral frames and convex polyhedral force diagrams, *Comput. Aided Design* 63: 118–128, 2015.
20. https://compas-dev.github.io, 2018.
21. https://psl.design.upenn.edu/polyframe, 2018.
22. https://www.csiamerica.com/products/sap2000, 2018.
23. https://www.karamba3d.com/tutorials/tutorials_basics/simple-shell-and-beam, 2018.
24. Saliklis, E.: *Structures: A Geometric Approach*, Springer, 2018.
25. Scordelis, A. C., and K. S. Lo: Computer analysis of cylindrical shells, *ACE* 64(5), May 1964.
26. Clough, R. W., and C. P. Johnson: Finite Element Analysis of Arbitrary Thin Shells, *Concrete Thin Shells*, ACI, Publ. SP028, pp. 333–363, 1971.
27. Billington, D. P.: *Thin Shell Concrete Structures*, 2nd ed., McGraw-Hill, New York, 1982.
28. Design of Cylindrical Concrete Shell Roofs, *ASCE Manual of Engineering Practice* 31, New York, 1952.
29. Dumitrescu, J. A., and D. P. Billington: Concentrated edge loads on hyperbological shells, *J. Struct. Div.*, ASCE, vol. 1, January 1984.
30. Wind forces on structures, *Trans. ASCE*, vol. 126, part II, 1961.
31. Building Code Requirements for Structural Concrete and Commentary, ACI 318-11, American Concrete Institute, Farmington Hills, MI, 2011.
32. Billington, D. P.: *Thin Shell Concrete Structures*, McGraw-Hill, New York, 1965.
33. Abel, J.: *"Stattower," Program for the Analysis of Axisymmetric Shells*, Princeton University, 1972.
34. Cole, P. P., J. F. Abel, and D. P. Billington: Buckling of cooling-tower shells: state of the art, *J. Struct. Div.*, vol. 191, no. ST6, proceeding paper 11364, pp. 1185–1203, June 1975.
35. Niemann, H. J.: Zur stationaren Windbelastung rotationssymmetrischer Bauwerke im Bereich Transkritischer Reynoldszahlen, Technischwissenschaftliche Mitteilung, no. 71–72, des Instituts fur Konstrktiven Ingenieurbau der Ruhr-Universitat, Bochum, 1971.
36. Sollenberger, N. J., and R. H. Scanlan: Pressure Differences Across the Shell of a Hyperbolic Natural Draft Cooling Tower, *Proc. Int. Conf. Full Scale Testing of Wind Effects*, London, Ontario, June 1974.
37. Reinforced Concrete Cooling Tower Shells—Practice and Commentary, Report by ACI-ASCE Committee 334, *J. ACI*, January 1977.
38. Flügge, W.: *Stresses in Shells*, Springer-Verlag OHG, 1960.
39. Dischinger, F.: Die strenge Theorie der Kreiszylinderschale in ihrer Anwendung auf die Zeiss-Dywidag-schalen, *Beton u. Eisen* 34: 257–264 and 164–294, 1935.
40. Concrete shell structures, practice and commentary, *J. ACI*, September 1964.
41. Chinn, J.: Cylindrical shell analysis simplified by beam method, *J. ACI*, May 1959.
42. Wilson, E. L.: SAP IV, Earthquake Engineering Research Center, Report no. EERC 73-11, June 1973, rev. April 1974.
43. Scordelis, A. C.: Analysis of cylindrical shells and folded plates, *Concr. Thin Shells*, ACI Publ. SP28, 1971, pp. 207–236.
44. Pultar, M., et al.: Folded plates continuous over flexible supports, *J. Struct. Div.*, ASCE, October 1967.
45. Parme, A. L.: Shells of double curvature, *Trans. ASCE*, vol. 123, 1958.
46. Students clear gym moments before roof fails, *Eng. News-Rec.*, September 24, 1970.
47. 15-year-old H. P. roof fails, injuring 18, *Eng. New-Rec.*, July 10, 1975.
48. Tedesko, A.: Shell at Denver—Hyperbolic paraboloid structure of wide span, *J. ACI*, October 1960.
49. Mihailescu, M.: A new shell form for prestressed sheds, *Proc. International Colloquium of Shell Structures,* Madrid, Paper A-13.
50. *Elementary Analysis of Hyperbolic Paraboloid Shells*, Portland Cement Association, 1960.
51. Candela, F.: General formulas for membrane stresses in hyperbolic paraboloid shells, *J. ACI*, October 1960.
52. Schnobrich, W. C.: Analysis of hipped roof hyperbolic structures, *J. Struct. Div.*, ASCE, July 1972.
53. Shaaban, A., and M. Ketchum: Design of hipped hypar shells, *J. Struct. Div.*, ASCE, vol. 102, no. ST-11, November 1976.
54. FEMAP—Finite Element Modeling and Postprocessing, *User's Manual*, Enterprise Software Products, 1993.
55. Wang, Y-S., and D. P. Billington: Buckling of cylindrical shells by wind pressure, *J. Eng. Mech. Div.*, ASCE, October 1974.
56. Kundurpi, P. S., et al.: Stability of cantilever shells under wind loads, *J. Eng. Mech. Div.*, ASCE, October 1975.
57. Cole, P. C., et al.: Buckling of cooling-tower shells: State of the art, *J. Struct. Div.*, ASCE, June 1975.
58. Ewing, D. J. F.: The Buckling and Vibration of Cooling Tower Shells, Part II: Calculations, *Lab. Rept.* RD/L/R 1764, Central Electricity Research Laboratories, Leatherhead, England, November 1971.
59. Der, T. J., and R. Fidler: A model study of the buckling behavior of hyperbolic shells, *Proc. Inst. Civil Eng.*, vol. 41, London, September 1968.
60. Cole, P. P., J. F. Abel, and D. P. Billington: Buckling of cooling-tower shells: Bifurcation results, *J. Struct. Div.*, ASCE, June 1975.
61. Billington, D. P., and J. F. Abel: Design of Cooling Towers for Wind, *Proc. Specialty Conf.* ASCE, Madison, Wisconsin, August 22–25, 1976.
62. Flügge, W.: *Stresses in Shells*, Springer-Verlag OHG, 1960.
63. Timoshenko, S. P., and J. M. Gere: *Theory of Elastic Stability*, 2nd ed., McGraw-Hill, 1961.
64. Collected Papers on Instability of Shell Structures, 1962, NASA TN-D1510, 1962.
65. Schmidt, H.: Ergebnisse von Beulversuchen mit doppelt gekrummten Schalen-modellen aus Aluminium, *Proc. Symp. Shell Res.* (Delft). North-Holland Publishing Company, 1961.
66. Scordelis, A. C.: Stability of Reinforced Concrete Domes and Hyperbolic Paraboloid Shells, *Concrete Shell Buckling*, ACI Publ. SP-67, pp. 63–110, 1981.
67. Recommendations for Reinforced Concrete Shells and Folded Plates, IASS Working Group No. 5, Madrid, 1979.
68. Design and Construction of Circular Prestressed Concrete Structures, *J. ACI*, Rep. of ACI Committee 344, no. 9, proc. vol. 67, p. 664, September 1970.
69. *Building Code Requirements for Minimum Design Loads in Buildings and Other Structures*, American National Standards Institute, Inc., 1982.
70. Tedesko, A.: Construction aspect of thin shell structures, *J. ACI*, February 1953.
71. Tedesko, A.: Multiple ribless shells, *J. Struct. Div.*, ASCE, October 1961.
72. Billington, D. P., *The Tower and the Bridge: The New Art of Structural Engineering*, Princeton University Press, 1985.

Chapter 20
Cable-Supported Roofs

BY

PAUL A. GOSSEN, PE *Geiger Engineers, Suffern, New York*
KEITH M. MacBAIN, Ph.D., PE *Geiger Engineers, Suffern, New York*

20.1 INTRODUCTION

Cables in roof construction have found their popularity due to the large areas these roofs are capable of spanning. The structural system spanning the space acts primarily in tension; thus, the self-weight of these structures is lower than the conventional framing. The following schematics illustrate different tension roof systems.

Cables are tension members only. To resist a load reversal, one may consider a double cable net that forms two opposing cable net layers (Fig. 20.1) or a roofing system having sufficient dead load such as concrete planks or slabs (Figs. 20.2 and 20.3).

To resist a cable's tension force, compression or flexural members are required. These may be located at the perimeter of the covered space. They can consist of massive frames (Fig. 20.4) or be self-contained structures such as compression rings in which the cable tension is converted to compressive hoop forces (Fig. 20.5).

The space covered by a cable net can consist of continuous cables or stepped cable trusses such as used for a "cable dome" system (Fig. 20.6).

Other means of supporting a roof structure is through the use of stay cables that reduce the spans of the conventional framing and eliminate interior supports (Fig. 20.7). Section 20.3.2 lists several examples of cable-supported roofs.

20.2 CABLE MATERIALS AND CONSTRUCTION

20.2.1 Types

There are two types of cable construction: strands and ropes. Strands are less flexible than ropes and have a higher tensile capacity. Strands are the predominant material used for cable roofs, whereas ropes are more commonly used in "running stock" applications where hoisting and movable cables are required.

Commonly used cable types for structures are helical wire strand (ASTM A586), wire rope (ASTM A603), and locked coil. Strands are composed of several layers of wires, called lays, that are twisted in a helical fashion about a central wire. Ropes consist of multiple strands twisted around a center core that can be either a metallic or a nonmetallic strand.

Figure 20.1 Anticlastic surface.

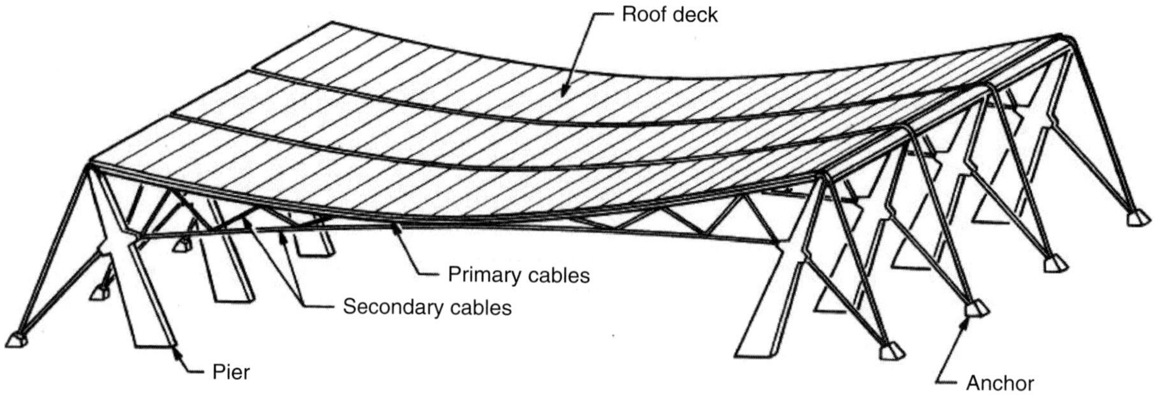

Figure 20.2 Opposing cable layers.

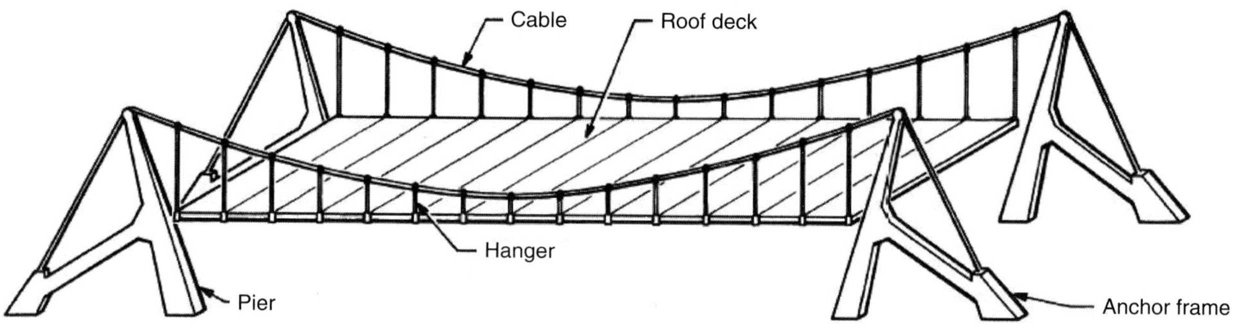

Figure 20.3 One-way countered by dead load.

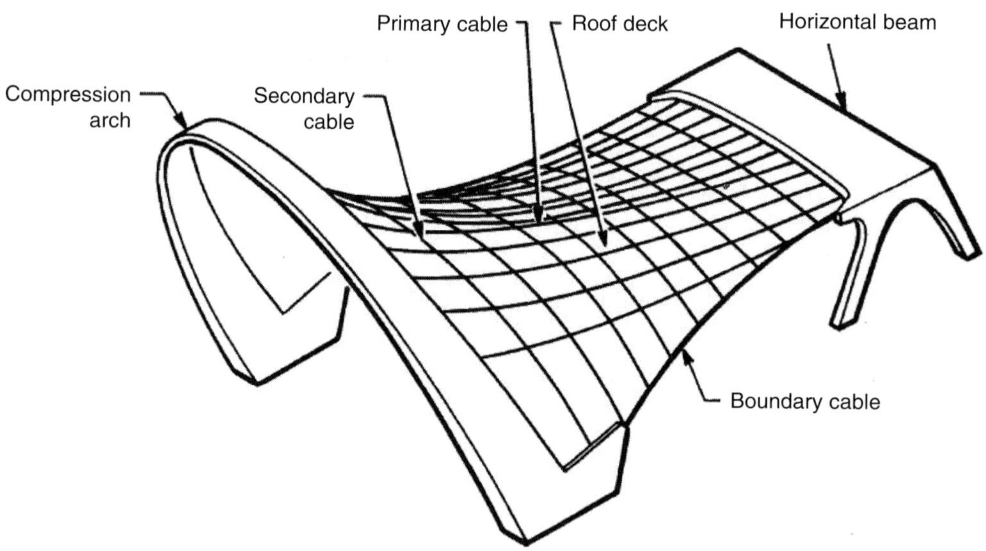

Figure 20.4 Frames resisting tension.

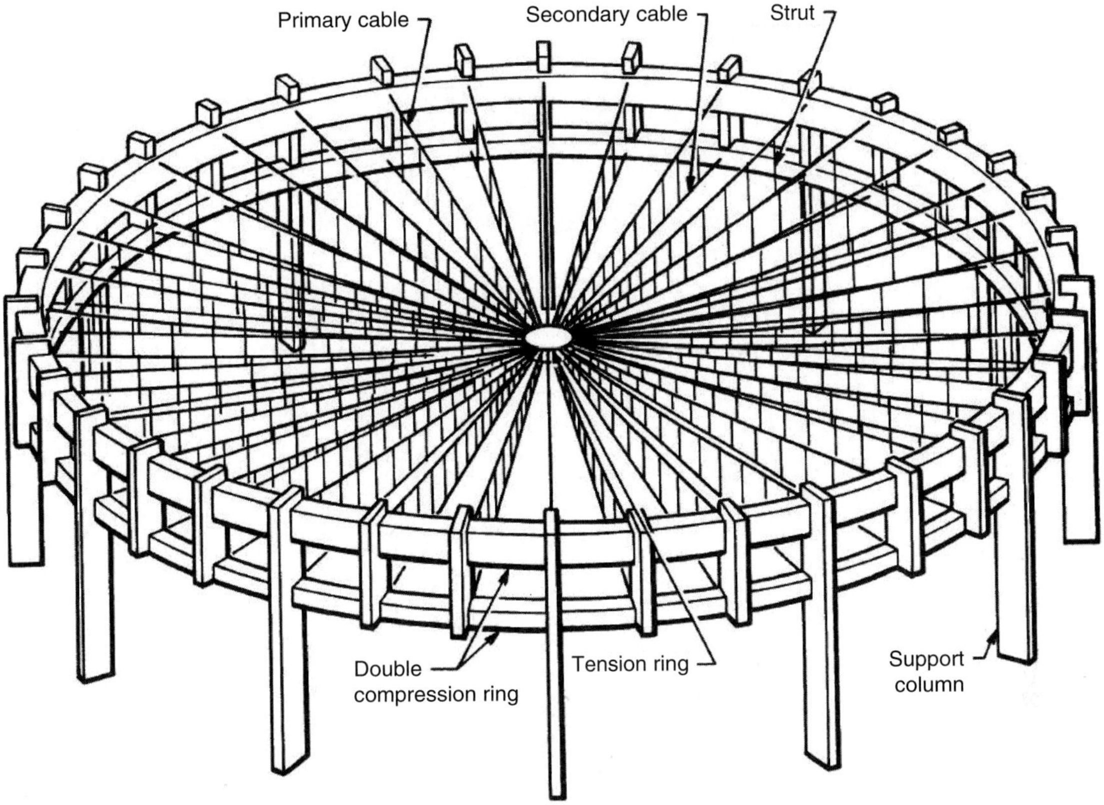

Figure 20.5 Compression ring resisting tension.

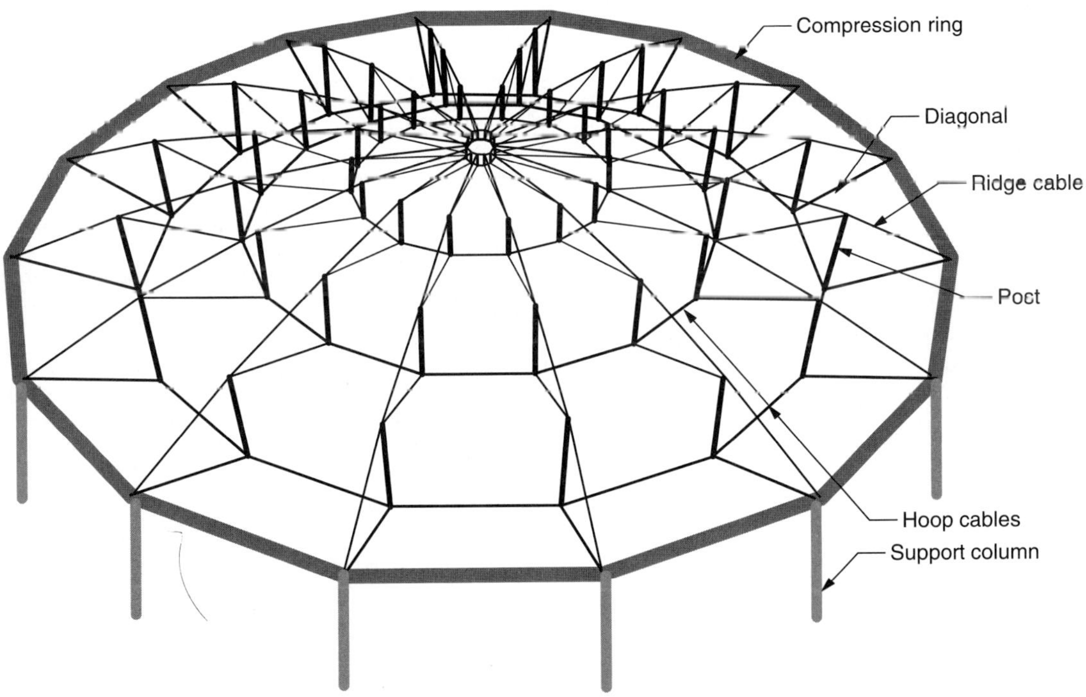

Figure 20.6 Cable dome system.

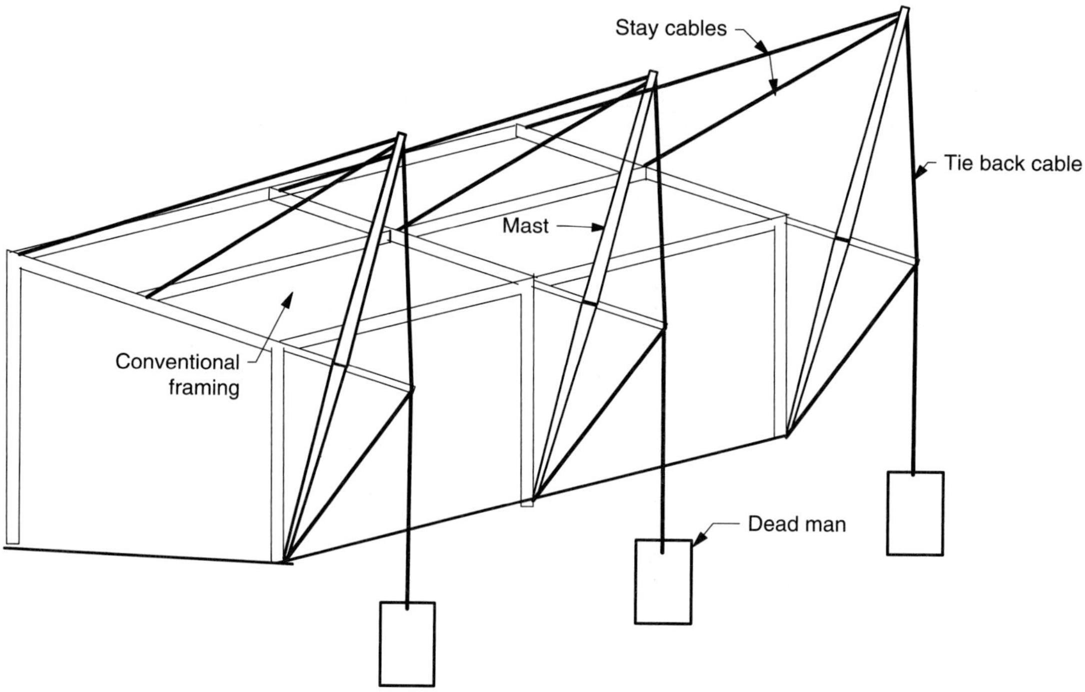

Figure 20.7 Stay cables to reduce span.

Whereas rope wires are always round, strand wires can be round or, in the case of locked coil cables, Z-shaped. The Z-shaped wires for locked coil strands are commonly used only in the outer layers. Figure 20.8 illustrates the cross section for these cable types.

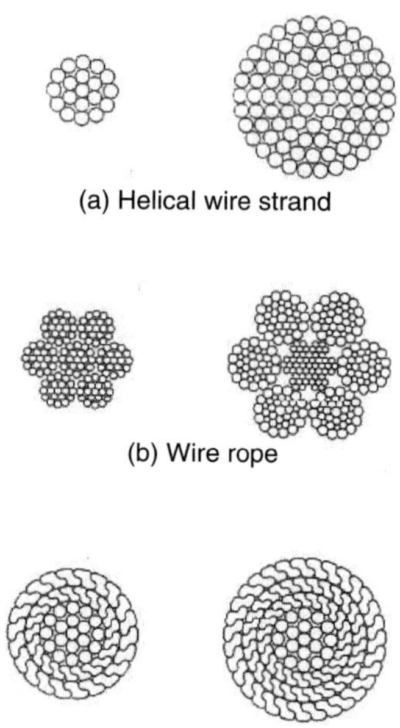

Figure 20.8 Cable cross sections. (Images taken from ASCE/SEI 19-16.)

During production, the individual wires are initially rolled, but in the final fabrication process the wires are cold drawn through a mandrel which produces their high strength (270 ksi). This cold working reduces some of the plastic behavior (yield) that is commonly associated with steel products. Not all of the high-strength capacity of the wire is utilized in the final cable fabrication due to the helical wire construction. Cables with a long lay wire pitch (small angle to the cable axis) have a higher capacity than those with a shorter lay wire pitch (large angle to the cable axis).

The stranding process (twisting the wires into a cable) leaves the wires relatively uncompacted so that the cable will have an initial "looseness" that results in a one-time nonelastic "construction stretch" when loaded. When designing cable structures, analysis is based on the assumption of elastic member behavior, including the cables. Thus, the construction stretch must be eliminated as much as possible during the fabrication process. Prestretching the cable to 50 percent of its nominal strength, sometimes repeated several times, removes most of the construction stretch. Even with this procedure, however, a perfectly linear-elastic behavior of the cable is not achieved. In design, a close approximation to a linear behavior is assumed based on test data. For cables, the modulus in the elastic range is defined by the slope of a secant between the stress point at 10 percent of the breaking strength and end point of 90 percent of the prestretching force (Fig. 20.9). Since the allowable strength factor is 2.2, the modulus in the elastic range covers well the range of the design demand.

20.2.2 Hardware and End Fittings

Cable hardware consists of devices attached directly to the cable, such as end fittings (anchorages), clamps, and sleeves. Hardware necessary to guide the cable and its components, such as saddles and loose sleeves, are not connected to the cable.

Three means of end fitting attachments are common: spelter sockets, swaged fittings, and mechanical loop splices. A spelter socket consists of a cone-like basket into which the cable is inserted. Within the basket the cable wires are evenly splayed. Zinc, mischmetal, or resin is then poured into the basket. The resulting hardened cone under cable tension wedges

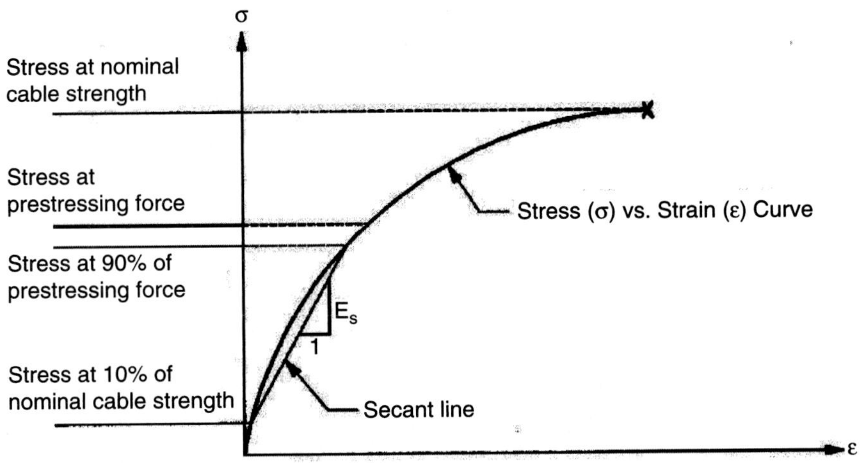

Figure 20.9 Stress-strain diagram. (Image taken from ASCE/SEI 19-16.)

the wires into the basket and forms the means of load transfer. Swaged (compressed) fittings can be used for cable diameters up to 2 in. They consist of an anchorage attached to a sleeve that is swaged around the cable. Swaged and spelter sockets are typically used for structural cables. Either swaged or spelter sockets may be used with wire rope. Spelter sockets are mostly used with structural strand. Figures 20.10 and 20.11 illustrate some commonly used spelter and swaged end fittings. Mechanical splice loops are used on ropes. The wires of the rope are rewoven into the rope to form a loop. A thimble, a curved piece of half pipe, is inserted into the loop to prevent the wires from being pinched, which would reduce their strength.

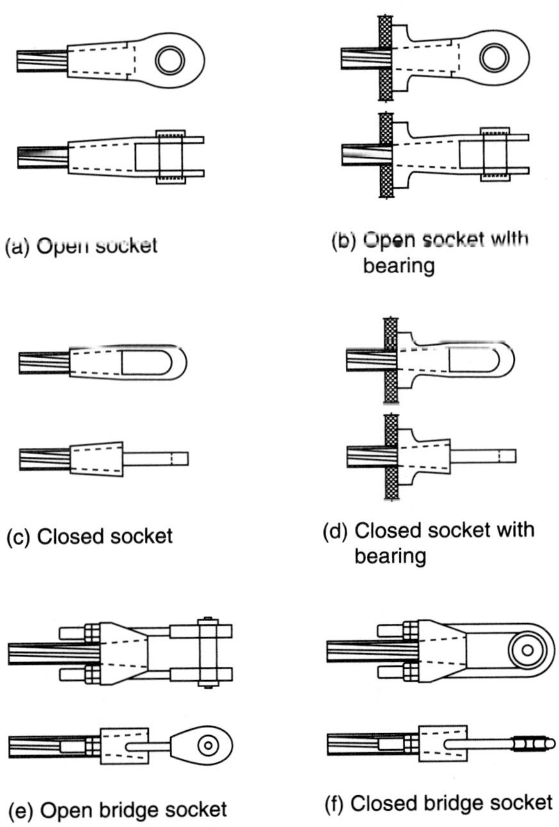

Figure 20.10 Spelter socket fittings. (Image taken from ASCE/SEI 19-16.)

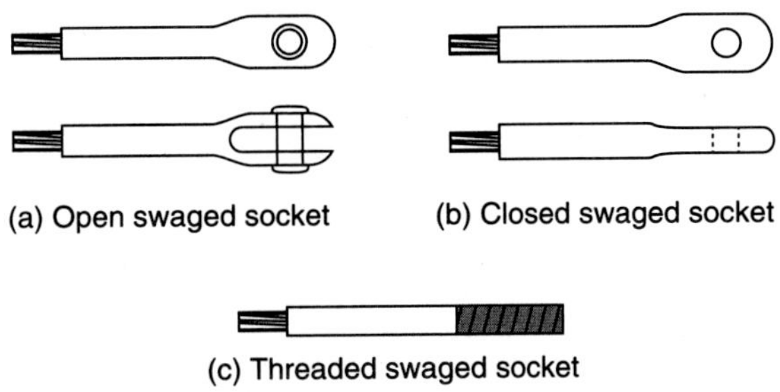

Figure 20.11 Swaged fittings. (Image taken from ASCE/SEI 19-16.)

Clamps are often used where other components are attached to the cable. The sliding force is given in the ASCE/SEI 19 as 7 percent of the clamping force. Due to the compaction of the cable under load and creep of the zinc coating (which reduces the cable diameter), the force in the clamping bolts can diminish. The reduction of the cable diameter causes a loss of the pretension strain of the bolt and thus its clamping force. To counter this loss, long bolts can be used, which reduces the strain loss. Clearly this is not to say that longer bolts introduce more force but rather that, by simply recalling the definition of strain, one notes that for a given displacement, a longer element has less strain (i.e., loss of strain in this case). Clamps are generally used only for smaller wire-rope cables.

Saddles are used to change the direction of a cable. The radius of the saddle, as well as its groove configuration, can reduce the cable's capacity.

20.2.3 Procurement

Cables are shipped in large coils or reels. The diameter of the coils or reels is dependent on the diameter of the cable as well as its construction. Reeling a cable onto a spool with a diameter too small for the cable will loosen the wires, resulting in the loss of the predictable modulus of elasticity that was obtained during the pre-stretch process. When a cable is coiled on a spool for shipping, the relative movement of the individual wires can result in a "set," causing a shortening of the cable. If a set occurs, the unrolled cable may need to be "exercised" to return it to its pre-stretch design length.

Typically a manufacturer requires a cable's length with the associated tension force for determining the unloaded fabrication length. Cables are typically considered linear-elastic thus, any state where the cable is not slack can be used for calculating the unloaded fabrication length.

20.2.4 Protection

Strands and ropes are fabricated to ASTM specifications with varying protective coatings. Corrosion protection of cables is of high importance due to the high-tensile strength of its wires. The level of protection depends on the degree of corrosive exposure. Wire galvanization or aluminization are methods that are used, though galvanization is by far more common. ASTM specifications cover three levels of galvanization: A, B, and C. Coating A is applied by hot dip galvanizing, while coating B and C generally are applied by the Bethanization (electrolytic zinc) process. Coatings B and C consist of a thicker layer of zinc than coating A and tend to give better protection; however, the coating is softer and thus more vulnerable to abrasion. Cables can be fabricated with a variety of combinations, though most applications use a B or C coating on the outer wires and an A coating for the inner wires. The coating affects not only the cable's protection but also its structural performance. The heavier coating requires a reduction in the steel area to maintain the cable size, resulting in a reduction in the strength of the cable. An additional level of corrosion protection available is to fill the spaces between the wires with a protective grease.

Unlike the Z-shaped wires in a locked coil strand that interlock to form a water tight seal when tensioned, round wires bear upon each other at point contacts, particularly when they are constructed with alternating lays. This point bearing causes the zinc coating to yield, resulting in a minor elongation of the cable and, more importantly, compaction of the strand diameter. This may have implications on the performance of any clamped-on hardware.

In service, a sign of rusty spots may indicate corrosion of the inner wires, though a brown color on the strand's outer wires may only be zinc or zinc oxide discoloration. There are only a few means of testing cables for interior corrosion. Radiographic or X-ray examination may be used, though they are difficult to interpret because of the voids inherent in the cable construction. Cutting a sample cable open for examination will obviously require its replacement.

There are very few means of fire protection for cables. Indirect protection may consist of a sprinkler system. Direct protection may consist of coating with an intumescent paint or other fire-retardant compound. These compounds or paints must have enough flexibility to accommodate the deformations of the cable under varying loading conditions. All of these solutions have nontrivial architectural implications.

20.3 STRUCTURAL SYSTEMS

Cables are tension-only elements and must be part of a larger system. Often this will include major structural components, such as a compression ring or masts.

20.3.1 Compatibility of Design

Cables are flexible and deform under load, generally by a larger amount than ordinary framing. This deformation is not just as a result of catenaries assuming a funicular shape in accordance with the applied loads but also due to the larger strains they experience because of their high strength as compared to common framing. Thus, the combination of cables and frames must account and allow for the different strain of the components. For example, if rigid roof panels assembled of common structural steel were rigidly attached to roof cables without provisions to release the differential strains, the panels would assume the same strain as the cables and be overstressed.

20.3.2 Examples

Cables in roof structures are used in several ways. They are used in rigid steel roofs as tension-only members (such as ties and/or stay cables (Fig. 20.7)) that support the framing for the roof skin (cable supported) or as suspended catenaries that support the roof skin directly or part of a ridged frame work (cable suspended).

An example of a stay cable scheme is the roof of the Soccer World Cup Stadium, Seoul, Korea, built in 2000 (Fig. 20.12). Examples of a roof skin supported directly by suspended cables are the North Carolina State Fair building (1953), the "tent" structure in Munich for the 1972 Olympics, and Madison Square Garden. Roof cladding for cable-suspended roofs can be membrane or rigid panels. Most of the structures with a fabric or film membrane cladding are cable-suspended roofs because these materials are most adaptable to catenary deformations. Air-inflated roofs also fall into this category.

The stability of cable-supported roofs is provided by pretensioning the roof cables. The level of pretension contributes significantly to its rigidity. If cables lose their tension, instability of the roof structure can occur. Pretensioning can be done in several ways. First, the dead weight of the roof cladding, such as concrete planks, can be used to load the cables. Examples include the roofs of Madison Square Garden and the natatorium in Flushing Meadows Park, both in New York City. A second method is to design the roof with a doubly curved cable net or anticlastic cable net. In this scheme, two or more layers of cables with opposite curvature are prestressed and ultimately load each other. The prestress in these systems can easily be manipulated, a feature not available in

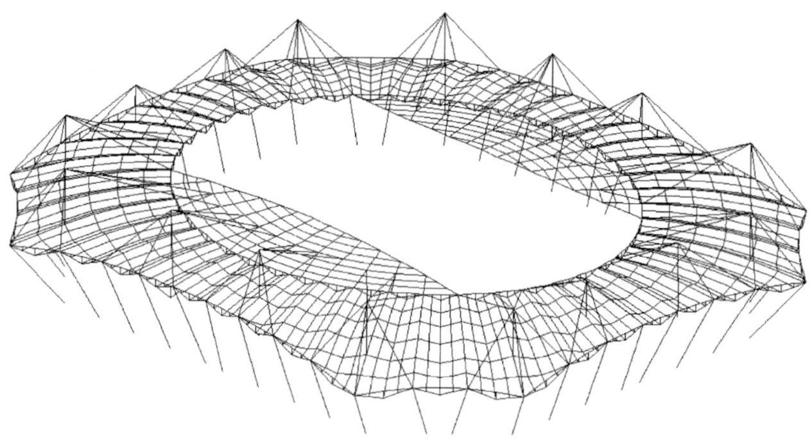

Figure 20.12 Soccer World Cup Stadium, Seoul, Korea.

gravity-loaded cable net systems. In air-supported roofs, the air pressure loads the fabric that in turn loads the cables. The effect is similar to that of gravity loads on the cable system but applied in an upward direction.

Two structural systems for anchoring cables can be identified: a closed structural system that equilibrates the cable forces in a compression structure, such as a ring or polygon, and an "open" anchor system that uses dead men, rock, or earth anchors. Open anchor systems rely on the weight of the dead men and engaged soils or, in the case of rock or soil anchors, on the density and soundness of the media in which they are anchored. Closed-system polygon-shaped structures are ideally designed with a funicular shape, which is geometry consistent with the pattern of the reactions of the cables. More specifically, a funicular shape is one that resists a specific pattern of applied loads without any flexural resistance. Any deviation from the funicular geometry forced by the cable reactions will result in bending of the polygon. Designing the ring to be very flexible will reduce its bending moments due to non-funicular loading; however, this will also result in significant deflections. Cable tension varies with load cases; thus, the funicular shape will change. Maintaining a funicular boundary requires significant structure movements that have architectural implications, that is, to finishes that are incompatible with these movements. An effective way to control these movements is to introduce "lateral springs" around the polygon or ring that resist the unbalanced forces which cause the deflection. Springs can be small-cantilevered beams that are attached to the polygon or ring. The polygon or ring may also be supported by gravity load-carrying columns that have radial releases to allow for free/restrained temperature-induced movements.

Cable-supported and cable-suspended systems can occur independently or be used together in one roof structure, such as the "cable dome" or the roof of BC Place in Vancouver, Canada (Fig. 20.13). In this example, backstay cables support the struts and masts, while suspension cables anchored to the masts support the roof skin.

20.3.3 Geometric Nonlinearity

Cables are typically considered linear-elastic elements. Their placement in a structural system may, however, cause their response to be nonlinear. In this case, a large-displacement analysis is required whereby the equations of equilibrium are solved using the deformed geometry. Many software products are available today with this capability; however, the user should be aware of the difference between the large-displacement analysis described here and what is sometimes called P-δ analysis, which only approximates these effects. Generally a large-displacement analysis is required when force is directed largely transverse to the cable. Figure 20.14 illustrates two structures where linear and nonlinear analysis would be appropriate.

Unfortunately, there is no single measure whereby a system changes from linear to nonlinear. If it is not obvious as in the example above, it is common to analyze the structure using both methods. Results can be compared to make a determination based on a selected threshold (e.g., 5 percent difference in results). A geometric nonlinear analysis is more computationally expensive, and it may be desirable to avoid if not necessary.

Some programs consider the possibility for an element to go slack (e.g., modeling as a tension-only member). This is a form of nonlinearity; however, it is not equivalent to the geometric nonlinearity discussed here, and one must be certain of the software's capabilities.

20.3.4 Cost Estimation

As with all projects, costs depend not only on the design effort but also on construction management and construction time. In steel construction, three areas of cost can be identified: materials, fabrication, and erection.

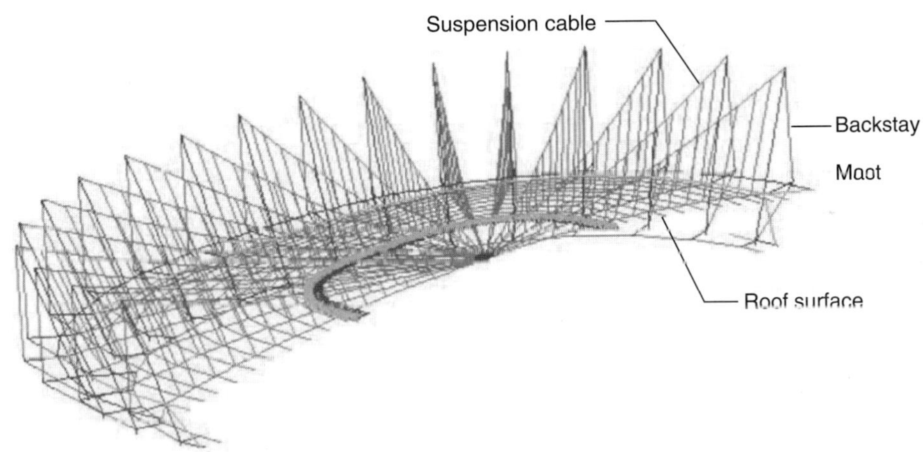

Figure 20.13 Section of BC Place roof computer model.

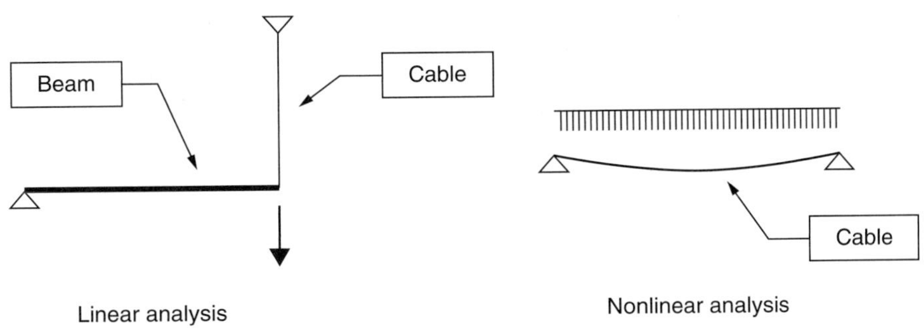

Figure 20.14 Linear versus nonlinear analysis of cable.

While smaller cables of up to 1 in. in diameter are usually stock items, larger-diameter cables are generally fabricated for a specific job. If there are special requirements, such as enhanced corrosion protection, the lead time will be longer. End fittings are generic in most applications. Proven designs have been used over the years, and their costs are well established. End fittings are generally a major cost component of cable assemblies. Swaged fittings are generally less expensive than spelter sockets, but for cables greater than 2 in. in diameter, speltered sockets are used. Special cable treatments, such as extended prestretching and stress cycling, that are outside of the norm will impact the delivery time and cost. Prestretching cables over 3 in. in diameter to 50 percent of their nominal strength generally requires special equipment that may not be readily available, and this may impact the delivery schedule. The erection procedure often determines the type of end fitting used and can also be an important cost component of the job. ASCE/SEI-19 requires an outline of the envisioned erection procedure within the contract documents. Often an erecting contractor is consulted during the design phase to assist in developing the erection procedure.

20.4 DYNAMICS

There are two potential concerns for the dynamics of cable-supported roofs: strength and serviceability. Serviceability concerns are generally centered on displacements.

20.4.1 Natural Frequencies

The natural frequency ω of a string under tension T with mass per unit length ρ and free length L is given below:

$$\omega = \frac{\pi}{L}\sqrt{\frac{T}{\rho}}$$

Unfortunately, this is not a particularly useful relation for the overall behavior of cable-supported roofs. This is because the length of interest (e.g., overall span length) is generally not free and without attachments. Further, any attachment to the cable results in an assembly, and it is easily demonstrated that the individual frequencies of two independent systems are in general not equal to the resulting frequencies of the system once connected as a whole. The relation above is useful, however, when investigating particular free lengths for serviceability as discussed later.

The preferred method for computing natural frequencies for a cable-supported roof is by way of a computer model, ideally incorporating the supporting structure. Often this will be eigenvalue analysis; however, Ritz vectors are also suitable. As part of this analysis, the software should be able to use the stiffness matrix at the end of the at-rest analysis (a.k.a., prestress + dead load).

If significant geometric nonlinearity is present, the cable system stiffness changes with displacement. In this situation, the concept of natural frequency degrades somewhat because, whereas the mass is constant, the stiffness changes, and so too does the frequency. In other words, the frequency is a function of the displacement (applied load). Typically, a structural system including cables is a stiffening system; however, it could be softening as well. Though this effect is often neglected and outside the scope of this work, it is useful to illustrate by way of displacement-response diagrams shown in Fig. 20.15.

The response of a linear system given at the left is shown for three different damping ratios (ξ). Figure 20.15 shows the ratio of the peak steady-state response divided by the static response (R_d) for different ratios of the periodic forcing function frequency (ω) divided by the natural frequency of the system (ω_n). The nonlinear response shown on the right is given for three different magnitudes of the forcing function (P), and the natural frequency (ω_n) is taken as the initial tangent value. For reference, the response on the right was generated for a ϕ 1″ A586 cable spanning 50 ft with an initial prestress force of 20 k and a damping ratio of 0.5 percent. The interested reader is referred to Duffing's equation for more information on the development of the nonlinear response.

20.4.2 Serviceability

Flutter or galloping, which is a wild vibration of a cable, is considered here as a serviceability issue. It can be caused by wind gusts or vortex shedding. It is noted that extreme cases can lead to inadequate strength through fatigue of the cable itself and/or its components; however, this will be far beyond what is considered serviceable. There is limited guidance available on galloping and vortex shedding in Section 11 of the AASHTO LRFD Standard Specifications for Structural Supports for Highway Signs, Luminaires, and Traffic Signals.

Generally, these considerations do not govern the overall design. Further, there are vibration-mitigating devices, such as tuned-mass dampers and airfoils, that can be employed if needed.

20.5 ANALYSIS AND DESIGN

Design of cable structures is outlined in the ASCE/SEI Standard 19-16 "Structural Application of Steel Cables for Buildings," which is also applicable for cable-supported roofs. Allowable Stress Design (ASD) is the basis of the analysis and design.

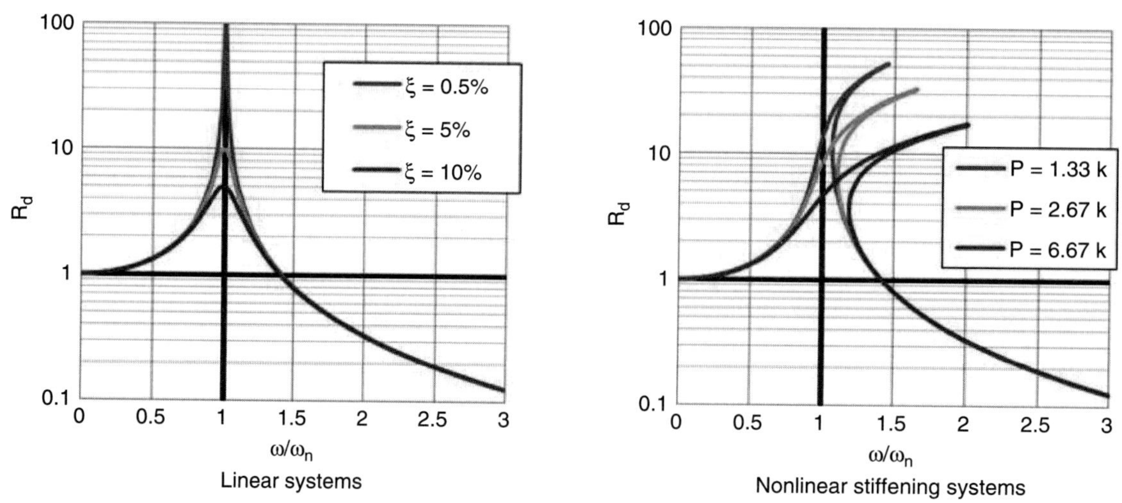

Figure 20.15 Response of linear versus nonlinear systems.

Due to the nonlinear behavior of the cable construction over its entire stress-strain range from "0"stress to the breaking stress (see stress-strain curve, Fig. 20.9), Load Resistance Factor Design (LRFD) is generally not used, though the Eurocode has adopted a design approach using strength design methodology. For analysis, the stiffness of the cable itself is to be used. In designing cables, however, the cable assembly strength should be used because deflectors such as saddles, as well as end fittings, generally reduce the nominal strength of the cable.

Analysis and design are typically iterative processes that advance together. When the engineer is involved in the initial stages of design, there may be different design concepts under consideration; however, it is more typical that the architect or owner has already determined some aspects that are not subject to change (e.g., geometry, structural system). As a result, the methods involved in analysis do not change much, but there can be wide variations as to what the methods of design may mean for a given project. For example, the engineer may be asked to explore the benefits of different systems if this has not already been determined.

20.5.1 Preliminary Design

Simple equations are available to provide reasonable estimates useful for preliminary design of cables. Often the more important part of the preliminary design process is how the cables interact with the overall system. In other words, the force of the cable is useful for preliminary sizing, but generally it is more important to know the demands it delivers to the rest of the structure. Figure 20.16 depicts a cable spanning a distance L with a midspan sag s and supporting a uniform load w.

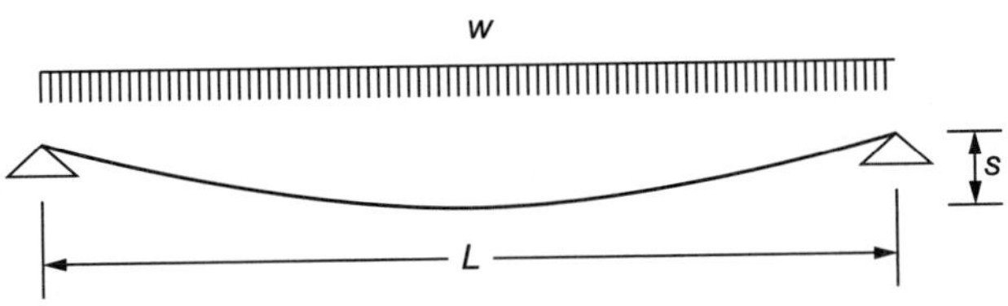

Figure 20.16 Cable supporting a transverse load.

The force in this cable varies over the length; however, the midspan force T can be easily derived by considering a free body of one-half and summing the moments about a support. The resulting equation below is often useful for the initial stage of design:

$$T = \frac{wL^2}{8s}$$

It should be clear to the designer that there must be some sag (curvature) if the cable is to support transverse loads. This is evident in the equation above by noting that the term s is in the denominator thus must be nonzero.

For application of this equation, we consider an example of one direction in a two-way cable truss system depicted in the plan view of Fig. 20.17 and the section of Fig. 20.18 below.

For preliminary design, it is possible to consider the upper and lower cables independently using the equation above, where the sag of the upper cable (s_U) is not necessarily equal to that of the lower cable (s_L). For each of the cables, the governing loads are considered (i.e., upward or downward), and because one generally does not want any cables to go slack, some allowance is added for the force of the cable not considered. In a two-way system such as this example, an approximate load distribution in each direction can be obtained from the relative spans.

This approach clearly ignores many aspects required for a complete design. The advantage of developing preliminary design in this manner, however, is that it may be done without the need for a fully developed computer model. One may even construct a spreadsheet with only a few parameters, such as the spacing of the cables and their sags. In this manner, one can easily change fundamental aspects of the design without reconstructing or remeshing a computer model. Further, it gives the designer a first estimate of the demands (reactions) on the supporting structure, which will also generally influence the design.

20.5.2 Computer Analysis and Design

Often computer modeling is used in the early design phase as well as the final design. This approach leads to better quantification at the early stages but is typically not easy to make changes should the design

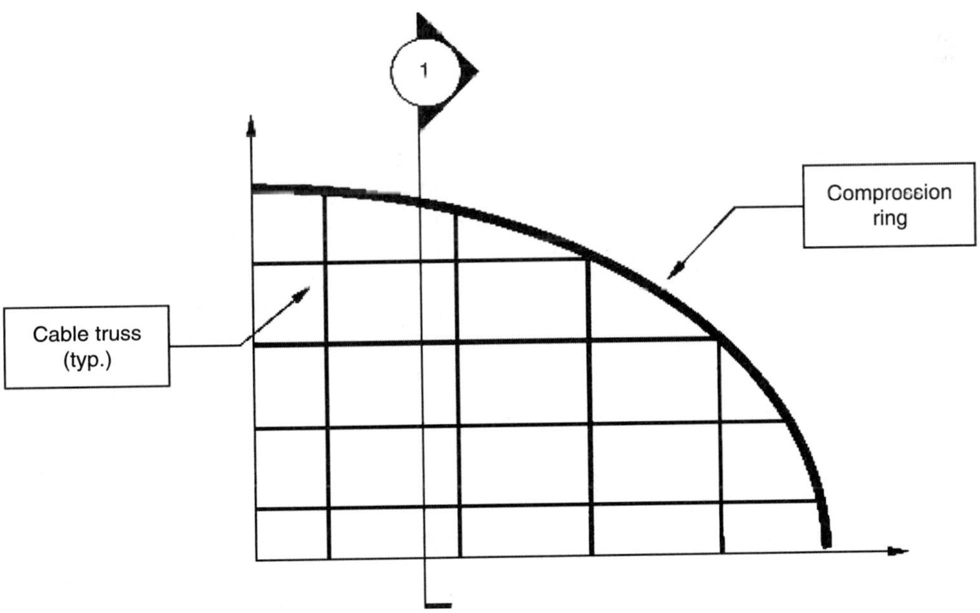

Figure 20.17 Plan view—quarter symmetric.

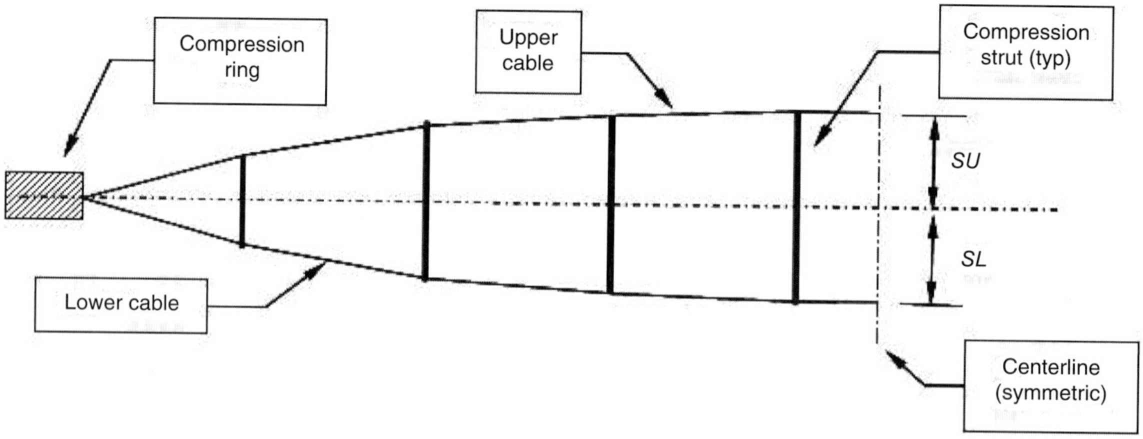

Figure 20.18 Section 1—typical section.

change, particularly in terms of structural system or connectivity. In any case, computer analysis will certainly be used in the final design for all but the simplest structures.

Form Finding

Form finding is the process whereby the geometry and internal forces in equilibrium are established. Essentially there are three approaches: force-density, dynamic relaxation, and stiffness-based. There are very few commercial programs available that implement the first two, so only the third will be discussed here. In the stiffness-based approach, one uses an ordinary structural analysis program (large-displacement) and enters artificial parameters according to the specific goal. For example, if one desires to maintain a particular geometry for a given load, the stiffness of related elements will be input artificially large in order to determine the element forces. Conversely, one may seek to achieve a particular set of forces and the geometry is of less concern, in which case one will input the desired force and artificially low stiffnesses. The goals at hand may be relative to the at-rest condition or, more commonly, some governing scenario (e.g., no slack cables under snow load).

Depending on the structure and design goals, the form-finding process may not be absolutely necessary. For example, one can also proceed on a trial-and-error basis by first developing a design and then determining if the design violates any criterion (e.g., no cables go slack). Unfortunately, this approach will generally not lead to an efficient design, as the designer will simply size elements for the computed forces rather than seek to minimize them. The use of form finding is thus preferred over the trial-and-error approach because without it, the designer will have little insight as to how forces may be better distributed throughout the system. For example, if the system has a compression ring, it is generally desirable to minimize the bending in the ring (i.e., bending about a vertical axis for a typical horizontal planar ring); however, the trial-and-error approach does not directly indicate how to achieve this.

Geometries

The term *workshop geometry* is a convenient term used to describe the reference geometry, which gives a position and orientation of all elements in the model before any loads are applied. For rigid elements (e.g., beams, columns), this will typically give the dimensions to which these members will be fabricated. For the surface of tensioned membrane roofs, it is more useful to refer to the at-rest geometry (a.k.a. prestress + dead load) because this is typically the surface that will be fabricated.

Prestress

The term *prestress* is commonly used to describe the internal force in the workshop geometry and normally only describes an axial force in the cables with no stresses in flexural members. This is somewhat analogous to the jacking force used in pretensioned concrete except that in this case, the force need not be realized. For example, the force in the workshop geometry could be even greater than allowable to achieve some goal (e.g., introducing force to the entire system); however, the computed force in all of the load combinations would naturally need to be at an acceptable level. Further to this, the prestress also need not be in equilibrium with the connected elements and may be thought of simply as an amount of strain energy introduced into the system which, after analysis, will be distributed through the structure according to the relative stiffnesses. It is convenient to use this convention in computer-based analyses because for the elements that can support bending, the workshop geometry is free of bending moments. This is important because rigid elements are typically fabricated without regard to an associated force (i.e., the force and moments are assumed to be zero). In contrast, manufacture of the cables requires not only a geometry (length) but also an associated force. This aligns well with computer modeling because the model defined in the workshop geometry, together with the prestress, is sufficient for fabrication of the cables.

Design Load Combinations

Design for cables must meet load combinations listed within ASCE/SEI Standard 19. These are ASD combinations in which the prestress factor is always 1.0. There have been changes in the load combinations from the recent versions of the Standard. The current version (ASCE/SEI 19-16) now aligns with the load combinations given in ASCE/SEI 7-10.

Other elements in the structural system follow the usual standards (e.g., AISC 360 for steel, ACI 318 for concrete, etc.).

Seismic Analysis

The load combinations listed above include seismic effects; however, the system nonlinearities (if present) have unique implications for the analysis approach. Three seismic approaches outlined in ASCE/SEI 7-10 are discussed relative to cable-supported roofs.

Equivalent Lateral Force Though the equivalent lateral force approach is generally considered the least sophisticated, it can be better than the modal-based approach for nonlinear structures because it can include geometric nonlinearities. For cable-supported roofs, however, the

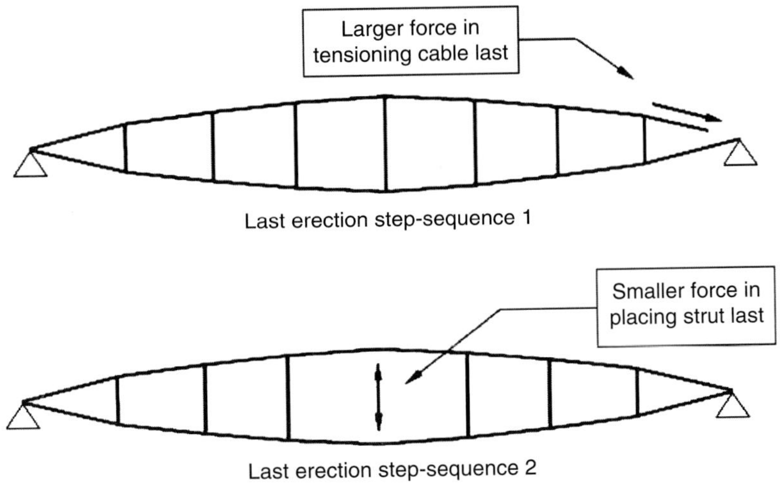

Figure 20.19 Force in different erection sequences.

nonlinearity is generally most pronounced in the vertical direction and much less as part of the lateral system.

Response Spectrum Analysis If there is significant geometric nonlinearity, the tangent stiffness matrix must be considered in the computation of mode shapes. Most programs that consider geometric nonlinearity allow for this. As noted above, the change in stiffness through the range of displacement is not represented in the mode shapes. This is generally not a concern, particularly if the nonlinearity is less pronounced for lateral forces as compared to vertical forces.

Time-History Analysis The time-history analysis approach may be linear or nonlinear, and for cables the code refers to the difference as a material nonlinearity (not geometric). In either case, one will consider geometric nonlinearity as required. The time-history approach requires a set of representative earthquake records and a peer review.

20.6 ERECTION

The erection procedure for cable supported roofs is generally not unique, but often the specifics of the structural system will lead to a preferred procedure. In complex cable roof systems such as cable nets, temporary cables can be strung to hoist the permanent cables to their anchor points. The temporary cables are often seven-wire strands that can be anchored at any points along their length. This allows the pull jacks to leapfrog the temporary cable until connections for the permanent cable can be made. Although high forces are required to tension the cables in place, the advantage is a uniform force introduction into the system.

Prestress is normally introduced into the system by way of jacking. Selection of where to place jacks is worth consideration. It is often advantageous to make final connections where there is greatest mechanical advantage and/or least force. For example, in a cable truss, there will generally be less force in the struts than the cable, so one might consider placing a few struts last rather than connecting the upper cable last as depicted in Fig. 20.19.

REFERENCES

1. *Manual for Structural Applications of Steel Cables for Buildings*, American Iron and Steel institute, Washington, DC, 1973.
2. *Cable Roof Structures*, Bethlehem Steel, Bethlehem, PA, 1968.
3. Krishna, Prem: *Cable-Suspended Roofs*, McGraw-Hill Book Company, New York 1978.
4. *Structural Application of Steel Cables for Building*, ASCE Standard 19-16.

Chapter 21
Reinforced-Concrete Silos[a]

BY
RAMEZ B. GAYED *Adjunct Professor, University of Calgary, and Senior Structural Engineer, thyssenkrupp Industrial Solutions (Canada) Inc.*
MUSTAFA MAHAMID *Clinical Associate Professor, Department of Civil and Materials Engineering, University of Illinois at Chicago*
AMIN GHALI *Professor Emeritus, University of Calgary*

21.1 GENERAL

21.1.1 Introduction

A bin is an upright container for the storage of bulk granular materials. Shallow bins are usually called bunkers; tall bins are called silos. Typical bunkers are shown in Fig. 21.1. Vertical cross sections of silos are shown in Fig. 21.2. While bins may be erected individually, they are commonly grouped for increased efficiency in operation. Typical configurations of grouped silos are shown in Fig. 21.3.

Janssen[1] in 1895 was the first to develop a theory to predict pressures from stored material at rest. Design using his formulas produced acceptable structures for many years. Theimer[2] notes, however, that early designers used low allowable stresses that rendered a factor of safety of about 2.5. With the use of higher-strength steel and concrete,

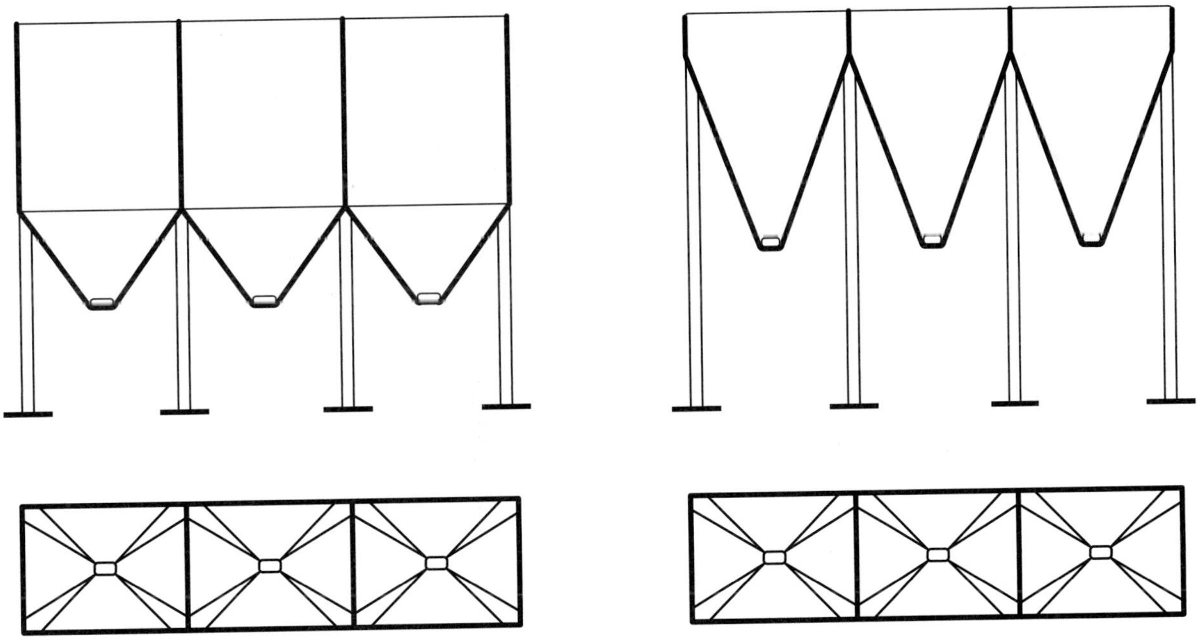

Figure 21.1 Typical bunkers.

[a]Original chapter by German Gurfinkel, Professor Emeritus of Civil Engineering, University of Illinois at Urbana-Champaign.

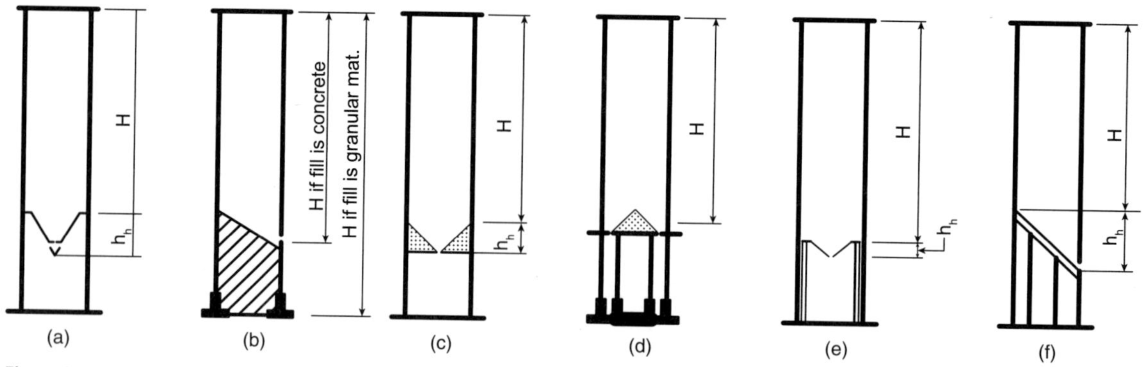

Figure 21.2 Typical silos: (*a*) on raft foundation, independent hopper on pilaster; (*b*) with wall footings and independent bottom slab on fill; (*c*) with hopper-forming fill and bottom slab on thickened lower walls; (*d*) with multiple discharge openings and hopper-forming fill on bottom slab, all supported by columns—raft foundation has stiffening ribs; (*e*) on raft foundation with hopper on ring beam and columns; and (*f*) on raft foundation with hopper on concrete walls and steel frames.

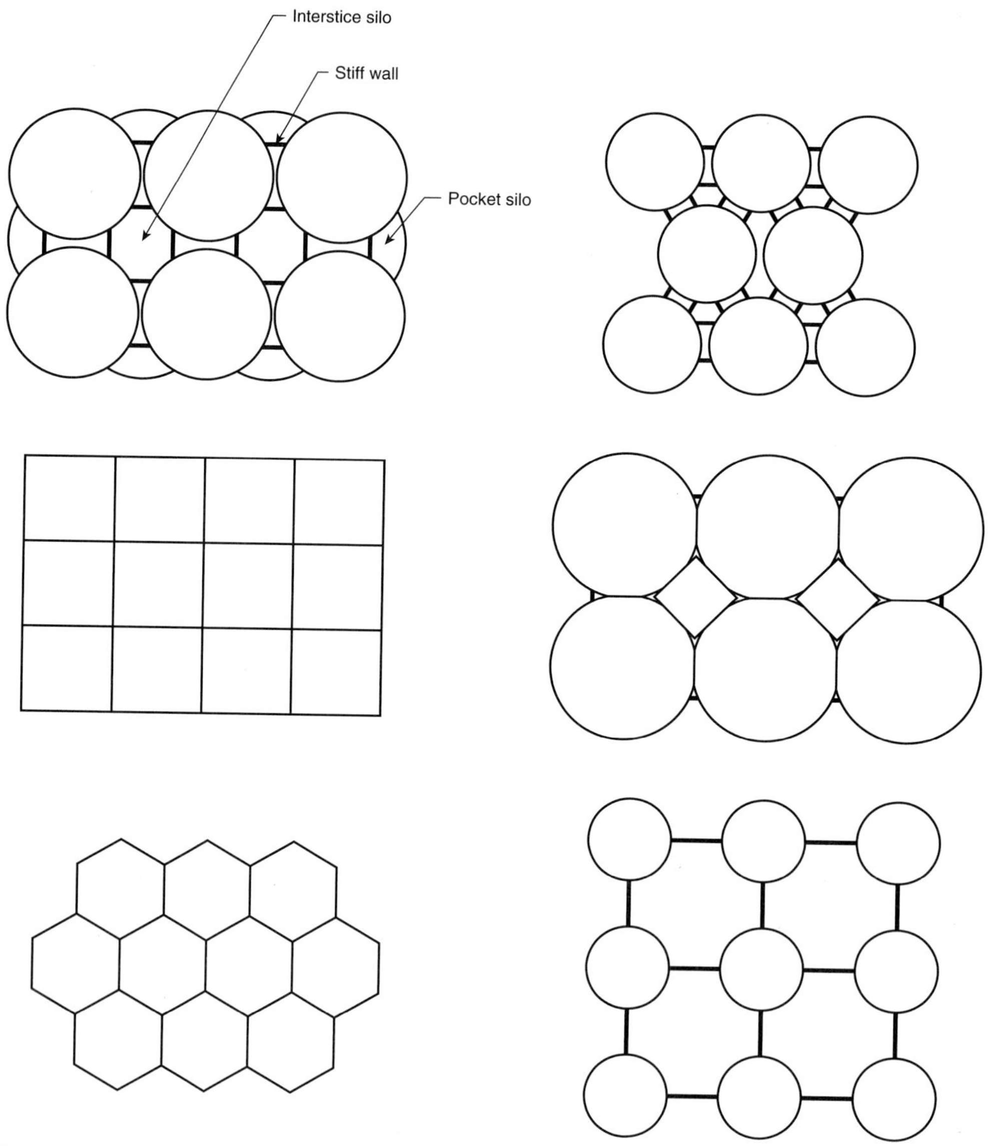

Figure 21.3 Grouped silo configurations.

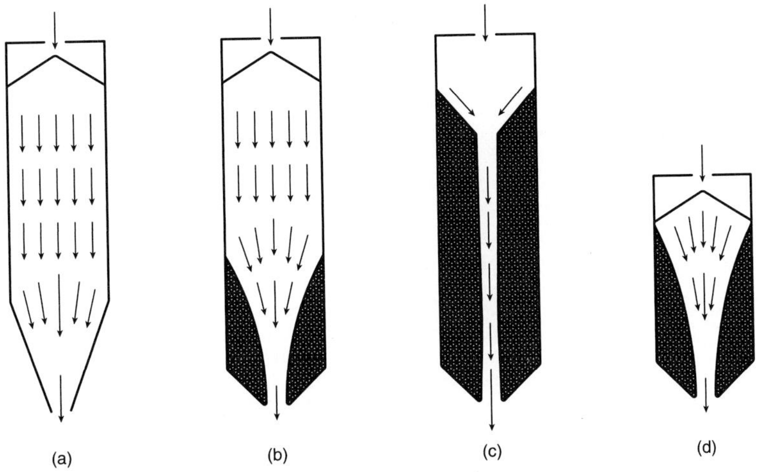

Figure 21.4 (*a*) Mass-flow silo; (*b*), (*c*), and (*d*) funnel-flow silos (may be flat-bottomed).

together with higher allowable stresses, the safety margin is reduced considerably. Thus, old silos can withstand pressures substantially greater than those predicted by the Janssen formulas, while some newer ones may not.

A number of investigations throughout Europe and the United States have reported pressures generated during material emptying higher than those predicted by Janssen's formulas for the material at rest.[3–11] Provisions for the calculation of these effects have appeared for many years in the German Specification[12] DIN 1055, Sheet 6. In the United States, ACI Committee 313 published a standard where provisions for the calculation of overpressures are given.[13] Revisions of this standard have been issued.[14] ACI 313[13] has been updated, and the most recent version of the update is ACI 313-16.[15]

During emptying, solids may flow out in two distinct ways: funnel flow and mass flow (Fig. 21.4). In mass-flow silos the entire content is in motion during emptying, while in funnel-flow silos various zones may be at rest at any particular time. Because of this, a last-in-first-out pattern prevails in funnel-flow silos, while a first-in-first-out pattern is the case in mass-flow silos. The two principal factors that determine the type of flow in a given silo are the hopper slope angles θ_c and θ_p and friction angle $\phi' = \tan^{-1}\mu$. Jenike[16,17] has provided design information in graph form for two common shapes of hopper outlet (Fig. 21.5). See also

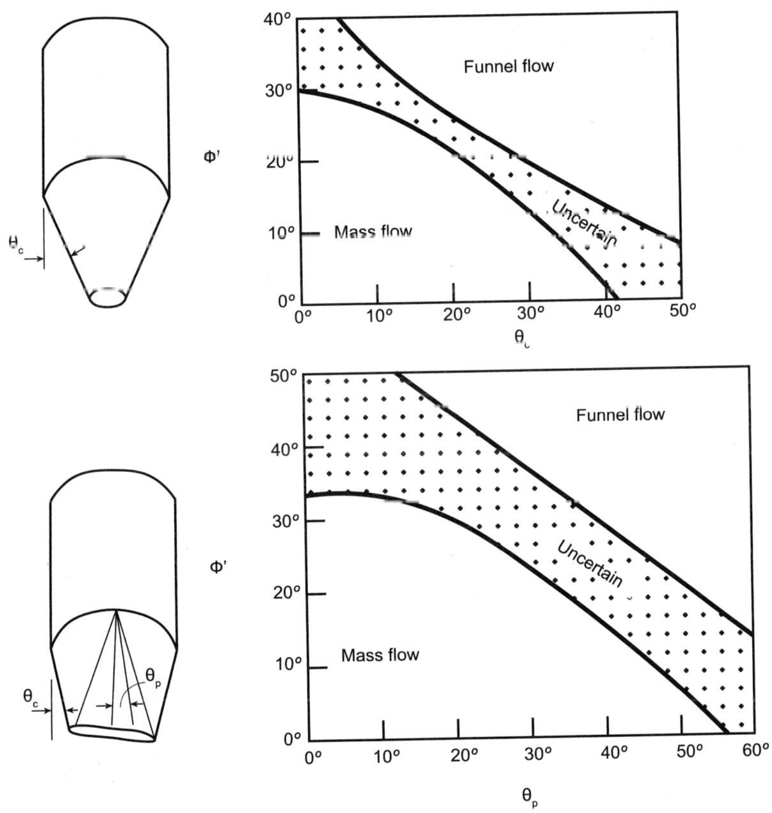

Figure 21.5 Mass-flow vs. funnel-flow bounds. (*From Ref. 14.*)

Figure 6.3.3.3a in ACI 313-16 (Ref. 15). Similar charts are given also in the DIN specifications[12] and by Gaylord and Gaylord.[18]

The regions in Fig. 21.5 labeled "uncertain" indicate conditions for which flow may shift abruptly between funnel flow and mass flow, with large masses of material being in nonsteady flow and the consequent development of shock loads.[14] Such flow conditions will also lead to nonsymmetric flow patterns and, hence, to nonsymmetric loads on the silo walls that generate combined bending moment and tension. Uncertain conditions can seriously damage a silo and should be avoided by designers.

21.1.2 Bin Pressures

Formulas by Janssen[1] and Reimbert[6] for computing bin pressures are given in Table 21.1. Notation is defined in the table and in Fig. 21.6. These formulas give pressures for the stored material at rest (static pressures). They give substantially the same results and are commonly used in the United States.

Values for unit weight γ, angle of internal friction ϕ, and coefficient of friction μ', for use in Janssen and Reimbert formulas, are given in Table 21.2. Formulas for the cross-sectional area A, perimeter L, and hydraulic radius R for various cross sections are given in Table 21.3.

Janssen Formulas (Table 21.1). As Y approaches infinity (practically, and within 1 percent error, when $Y/Y_0 = 4.6$), the limit pressures are

$$q_{\lim} = \gamma Y_0 = \frac{\gamma R}{\mu' k} \tag{21.1a}$$

$$p_{\lim} = k q_{\lim} = \frac{\gamma R}{\mu'} \tag{21.1b}$$

Table 21.4 gives values of $1 - e^{(-Y)/Y_0}$ by which $q_{\lim}$ and $p_{\lim}$ are multiplied for the region $O < Y/Y_0 < 4.9$. Linear interpolation is acceptable.

Reimbert Formulas (Table 21.1). Equations (21.1) also give the limit pressures by the Reimbert formulas. Values of $p_{\lim}$ and C for various cross sections are given in Table 21.5.

Case of Variable Values of γ, μ', *and* k. Where γ, μ', and k vary, it is required that the following combinations be used with maximum γ:

1. Minimum μ' and minimum k for maximum vertical pressure q
2. Minimum μ' and maximum k for maximum vertical pressure p
3. Maximum μ' and maximum k for maximum vertical pressure v
4. This situation is of particular importance for silos designed to contain various commodities (see Example 1).

DIN 1055 Sheet 6. The Janssen formulas are also used in this standard, except that a distinction is made between filling (at rest) and emptying conditions. Formulas are given in Table 21.1. The value of k is given in Table 21.6; it need not be calculated from the angle of internal friction as in the formulas of Janssen and Reimbert. Values for emptying are derived directly from those for filling; thus, $P_e = e_h p_f$ and $V_e = 1.1\ V_f$ (Table 21.1).

This standard is applicable to silos with vertical walls and bottoms consisting of either flat floors or hoppers. The height H must be at least 0.8 times the diameter D of the inscribed circle. For $H/D < 0.8$ the standard may be used if the silo consists only of an outlet hopper that makes an angle $\alpha \geq 20°$ with the horizontal. The standard is limited to silos for which the ratio q_f/γ does not exceed 25 m (82 ft). For silage the height may not exceed 25 m. Also, it is not applicable to silos with annular cross sections because during emptying such silos develop both flowing and stationary regions that depend on the number of outlets and the distance between them. This results in non-axisymmetric pressures on the walls that require special investigation.

Bunkers. Although the formulas in Table 21.1 can be used for shallow bins, Rankine's active pressure (Chap. 7, Sec. 7.4) is sometimes used for bunkers. In this case $q = \gamma Y$ and $p = kq$. Coulomb's formulas for pressure on retaining walls (Chap. 7, Sec. 7.4) have also been used for bunkers.

21.1.3 Emptying Pressures on Silo Walls, DIN 1055 Sheet 6

As indicated in Table 21.1 the vertical pressure at emptying q_e is less than the corresponding value q_f at filling and need not be calculated. The vertical frictional force per unit width of wall during emptying is 10 percent larger than the value at rest. These differences are relatively minor. However, major changes take place in a silo between lateral pressures generated during filling p_f and emptying p_e. The filling pressures are axisymmetric and uniformly distributed around the wall perimeter. For silos with a circular cross section the emptying pressures consist of two components: a uniformly distributed axisymmetric pressure $p_e = e_h p_f$ and a nonuniform partial pressure $\Delta p_e = \beta p_e$.

The quantity e_h is an eccentricity factor given in Table 21.6. For silos where $H/D \leq 5\mu$ one may use the listed value, but for $H/D \leq 2.5\mu$

Table 21.1 Material Pressures in Bins

	Value at depth Y			
			DIN 1055, Sheet 6 (dynamic effects included)	
	Janssen (static values only)	Reimbert (static values only)	Filling	Emptying
Vertical pressure q	$\gamma Y_0(1 - e^{-Y/Y_0})$	$\gamma\left[Y\left(\frac{Y}{C}+1\right)^{-1} + \frac{h_s}{3}\right]$	$\gamma Y_{of}(1 - e^{-Y/Y_{of}})$	$q_e < q_f$
Lateral pressure p	qk	$p_{\lim}\left[1-\left(\frac{Y}{C}+1\right)^{-2}\right]$	$q_f k_f$	$e_h q_f k_f$
Vertical frictional force per unit width of wall V	$(\gamma Y - q)R$	$(\gamma Y - q)R$	$(\gamma Y - qf)R$	$1.1(\gamma Y - qf)R$

γ = weight per unit volume of stored material (Tables 21.2 and 21.6)
Y = depth defined in Fig. 21.6
$Y_0 = R/\mu' k$
$Y_{of} = R/\mu_f' k_f$, filling
R = hydraulic radius of horizontal cross section of storage space = A/L (Table 21.3)
A = cross-sectional area of silo (Table 21.3)
L = perimeter of silo (Table 21.3)
μ' = coefficient of friction between stored material and wall (Table 21.2)
$k = p/q = (1 - \sin\phi)$, Janssen and Reimbert
$k_f = p_f/q_f$ (Table 21.6)
ϕ = angle of internal friction or (approximately) angle of repose ρ (Table 21.6)
C = characteristic abscissa for Reimbert's formula (Table 21.5)
h_s = height of sloping top surface of stored material (Fig. 21.6)
$p_{\lim}$ = Reimbert's limit static pressure (Table 21.5)
e_h = emptying factor (Table 21.6)

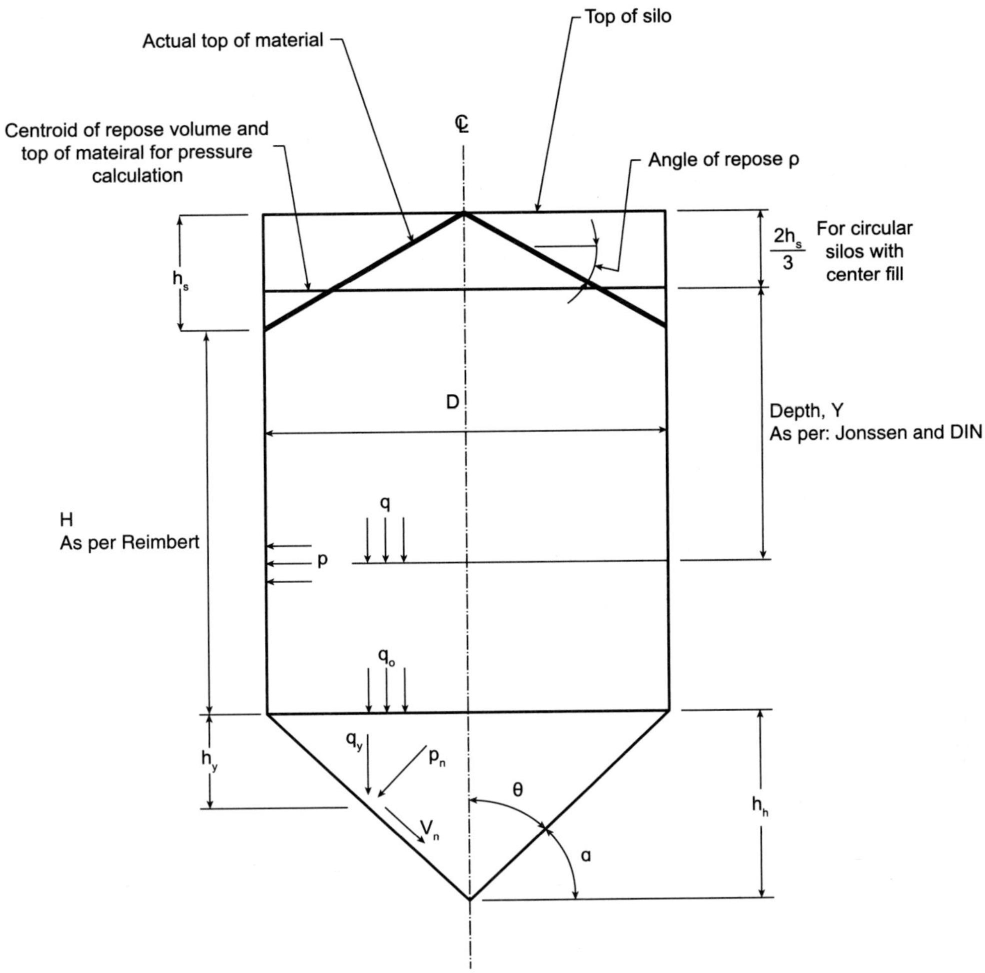

Figure 21.6 Silo dimensions for use in calculation of pressures for walls and hoppers.

Table 21.2 Example Physical Properties of Granular Materials (from Table R6.3.1.3 of ACI 313–16)*

	Weight γ		Angle of internal friction ϕ	Effective angle of internal friction δ	Coefficient of friction μ'	
	lb/ft^3	kg/m^3			Against concrete	Against steel
Cement, clinker	88	1410	33	42–52	0.6	0.3
Cement, portland	84–100	1345–1600	24–30	40–50	0.40–0.80	0.30
Clay	106–138	1700–2200	15–40	50–90	0.2–0.5	0.36–0.7
Coal, bituminous	50–65	800–1040	32–44	33–68	0.55–0.85	0.30
Coal, anthracite	60–70	960–1120	24–30	40–45	0.45–0.50	0.30
Coke	32–61	515–975	35–45	50–60	0.50–0.80	0.50–0.65
Flour	38	610	40	23–30	0.30	0.30
Flyash	50–112	865–1800	35–40	37–42	0.60–0.80	0.47–0.70
Gravel	100–125	1600–2000	25–35	36–40	0.40–0.45	0.29–0.42
Grains (small)†	44–62	735–990	20–37	28–35	0.29–0.47	0.26–0.42
Gypsum, lumps	100	1600	38–40	45–62	0.5–0.8	0.38–0.48
Iron ore	165	2640	40–50	50–70	0.5–0.8	0.4–0.7
Lime, calcined, fine	70–80	1120–1280	30–35	35–45	0.5–0.7	0.5–0.6
Lime, calcined, coarse	58–75	930–1200	40	40–45	0.5–0.8	0.3–0.5
Limestone	84–127	1340–2730	39–43	45–80	0.3–0.8	0.55–0.70
Manganese ore	125	2000	40			
Sand	100–125	1600–2000	25–40	30–50	0.40–0.70	0.35–0.50
Soybeans, peas	50–60	800–960	23		0.25	0.20
Sugar, granular	53–63	1000	35	33–40	0.43	

*The properties listed here are illustrative of values that might be determined from physical testing. Ranges of values show the variability of some materials. Design parameters preferably should be determined by test and the values used with caution. See Commentary on Section 4.4.1.
†Wheat, corn, barley, beans (navy, kidney), oats, rice, rye.

Table 21.3 Geometry of Silo Cross Sections

Cross section		Area A	Perimeter L	Hydraulic radius r
Circle 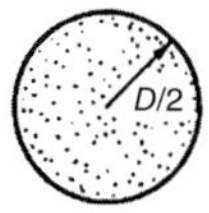		$\frac{\pi}{4}D^2$	πD	$\frac{D}{4}$
Circular sector 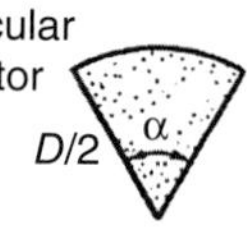	α, Radians	$\frac{\alpha D^2}{8}$	$\left(\frac{\alpha}{2}+1\right)D$	$\frac{D/4}{1+2/\alpha}$
	α, Degrees	$0.00218\,\alpha D^2$	$(0.0087\,\alpha + 1)D$	$\frac{D/4}{1+114.6/\alpha}$
Regular polygon 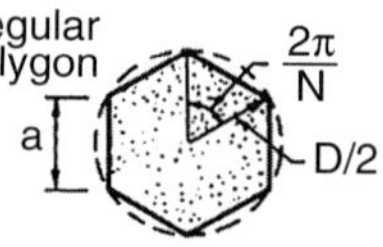		$\frac{ND^2}{8}\sin\frac{2\pi}{N}$	$ND\sin\frac{\pi}{N}$	$\frac{D}{4}\cos\frac{\pi}{N}$
		$\frac{Na^2}{8}\frac{\cos\pi/N}{\sin^2\pi/N}$	Na	$\frac{a}{8}\frac{\cos\pi/N}{\sin^2\pi/N}$
Rectangle 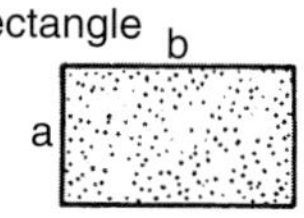		ab	$2(a+b)$	$\frac{ab}{2(a+b)}$, Long Side $\frac{a}{4}$, Short Side
Long rectangle 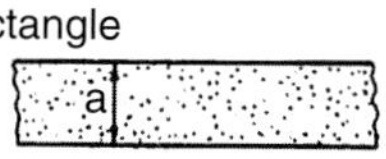		—	—	$\frac{a}{2}$
Ring 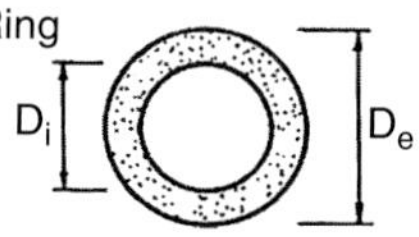		$\frac{\pi}{4}(D_e^2 - D_i^2)$	$\pi(D_e + D_i)$	$\frac{(D_e - D_i)}{4}$
Interstice 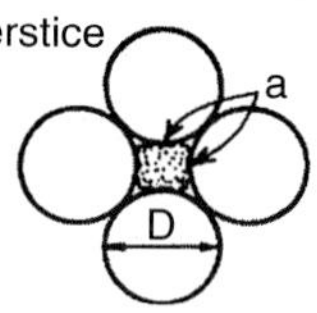		$\left(1-\frac{\pi}{4}\right)D^2$	πD	$\frac{a}{4}=0.104\,D$

Table 21.4 Values of $1 - e^{-Y/Y_0}$

Y/Y_0	0	0.1	0.2	0.3	0.4	0.5	0.6	0.7	0.8	0.9
0	0	0.095	0.181	0.259	0.333	0.393	0.451	0.503	0.551	0.593
1	0.632	0.667	0.699	0.727	0.753	0.777	0.798	0.817	0.835	0.850
2	0.865	0.878	0.889	0.900	0.909	0.918	0.926	0.933	0.939	0.945
3	0.950	0.955	0.959	0.963	0.967	0.970	0.973	0.975	0.978	0.980
4	0.982	0.983	0.985	0.986	0.988	0.989	0.990	0.991	0.992	0.993

use $e_h = 1$. For intermediate values of H/D, determine e_h by straight-line interpolation; that is, use $1 + [(H/D)/(2.5\mu) - 1]\,(e_h - 1)$.

The quantity β that determines the partial nonuniform component of the lateral pressure accounts for the effects of (1) silo slenderness H/D; (2) outlet offset a/r, where a = distance from center of silo to center of outlet; (3) stiffness r/t, where t = thickness of silo wall; and (4) *benign* or *malevolent* behavior of the silo contents. The corresponding factors are β_h, β_a, β_r, and β_C. It follows that

$$\beta = \beta_h\beta_a\beta_r\beta_C \tag{21.2}$$

where

$$\beta_h = 0.2H/D + 0.8, \quad \text{but } 1 \le \beta_h \le 1.6 \tag{21.3a}$$

$$\beta_a = 0.50(1 + 3a/r), \quad \text{but } 1 \le \beta_a \le 2 \tag{21.3b}$$

$$\beta_r = 0.30 - 0.0083(r/t) - 70), \quad \text{but } 0.05 \le \beta_r \le 0.3 \tag{21.3c}$$

β_C is listed in Table 21.6 for various bulk solids.

The partial pressure $\Delta p_e = \beta p_e$ is applied to a quadrilateral area having a side length $S = 0.8R$. The lateral load $\Delta P_e = \beta p_e S^2$, which provides static equilibrium, is similarly applied diametrically opposite. One need only calculate the effects of the nonuniform partial pressure at mid-height

Table 21.5 Values of $p_{\lim}$ and C in the Reimbert Formulas

Silo	$p_{\lim}$	C	Remarks
Circular	$\dfrac{\gamma D}{4\mu'}$	$\dfrac{D}{4\mu' k} - \dfrac{h_s}{3}$	
Polygonal, of more than four sides	$\dfrac{\gamma R}{\mu'}$	$\dfrac{L}{\pi}\dfrac{1}{4\mu' k} - \dfrac{h_s}{3}$	r from Table 21.3
Rectangular, on shorter wall a	$\dfrac{\gamma a}{4\mu'}$	$\dfrac{a}{\pi\mu' k} - \dfrac{h_s}{3}$	
Rectangular, on longer wall b	$\dfrac{\gamma a'}{4\mu'}$	$\dfrac{a'}{\pi\mu' k} - \dfrac{h_s}{3}$	$a' = \dfrac{2ab}{a+b}$

of the silo shell and apply the same percentage increase over the entire silo height. The beneficial *elastic-foundation* effect generated by the contents reacting against the silo walls is not used in this calculation. It may be considered sufficient to calculate only the tensions and bending moments in the circumferential direction (and ignore them in the vertical direction) for silos that are vertically reinforced in both inside and outside surfaces, each with at least 20 percent of the hoop reinforcement, but not less than ASTM A615 Grade 60 No. 3 rods at 10 in., or the equivalent. (For practical reasons, however, vertical steel should consist of No. 4 bars or greater.) The effect of nonuniform partial pressures on adjacent cells of a grouped-silo configuration (Fig. 21.3) need not be considered.

An alternative method for calculating the effects of emptying on lateral pressures is available for silos having a regular polygonal cross section and for circular silos having both horizontal stiffeners (presumably rings or flat slabs) at the top and bottom and walls with adequate capacity for transverse-load distribution. For these silos an increased uniform lateral pressure κp_{he} may be used instead of the complex nonuniform lateral pressure. The following values of κ are specified:

For silos with circular cross section:

$$\kappa = 1.0 + 0.5\beta + 0.02\beta\frac{r}{t} \quad \text{for } r/t \le 70 \tag{21.4a}$$

$$\kappa = 1.0 + 3.0\beta\sqrt{\frac{H}{D}} \quad \text{for } r/t \ge 100 \tag{21.4b}$$

For the range $70 < r/t < 100$ use straight-line interpolation.

For silos with regular-polygonal cross section:

$$\kappa = 1.0 + 0.8\beta \tag{21.5}$$

If a silo is designed for mass flow and it can be established that mass flow will occur for all operating conditions, the effects of nonuniform pressures may be disregarded.

It is reasonable to provide hoop reinforcement, in both the inside and the outside surfaces, regardless of the method used to calculate the effects of emptying on the silo walls. As a rule, circular silos that are designed using the increased uniform-load method should have 60 percent of the required reinforcement placed in the outside face and 40 percent in the inside face. The total amount required by the increased uniform-load method should be increased by 20 percent if it is desired, for practical reasons, to have the same hoop reinforcement in both faces.

Table 21.6 Properties of Bulk Solids*

			Wall-friction coefficients†			Factors		
Contents	Unit weight lb/ft³	Pressure ratio $k_f = p_f/q_f$	μ_1	μ_2	μ_3	Emptying e_h	Bottom load c_b	Bulk solid β_C
Wheat	57	0.60	0.60	0.40	0.25	1.40	1.20	0.5
Corn	51	0.60	0.60	0.40	0.25	1.60	1.70	0.9
Crushed barley	51	0.65	0.50	0.35	0.25	1.20	1.20	0.4
Flour	45	0.40	0.50	0.35	0.25	1.40	1.30	0.6
White sugar	60	0.60	0.55	0.50	0.45	1.20	1.20	0.4
Quartz sand	102	0.50	0.60	0.50	0.40	1.40	1.20	0.4
Gravel, crushed rock, coarse	115	0.60	0.60	0.50	0.45	1.30	1.20	0.4
Limestone flour	83	0.65	0.55	0.50	0.40	1.20	1.30	0.5
Cement clinker	115	0.55	0.65	0.60	0.50	1.30	1.70	0.6
Cement	102	0.65	0.50	0.45	0.40	1.20	1.30	0.5
Aluminum oxide	76	0.65	0.50	0.45	0.40	1.20	1.30	0.5
Phosphate	140	0.65	0.55	0.50	0.40	1.30	1.30	0.5
Soybeans	51	0.70	0.5	0.4	0.25	1.4	1.2	0.5
Beet pellets	45	0.60	0.55	0.45	0.35	1.3	1.2	0.5
Feed without soybeans	51	0.50	0.4	0.3	0.25	1.7	1.3	1.0
Coal	64	0.60	0.6	0.5	0.45	1.3	1.4	0.6
Coke	51	0.60	0.6	0.55	0.5	1.3	1.3	0.6
Fly ash	64	0.70	0.5	0.4	0.3	1.2	1.3	0.5
Coal dust	51	0.70	0.55	0.5	0.4	1.2	1.2	0.5
Slag	76	0.50	0.7	0.6	0.5	1.4	1.4	0.6
Iron pellets	140	0.60	0.6	0.55	0.5	1.3	1.3	0.5
Lime	38	0.70	0.5	0.4	0.35	1.2	1.3	0.6

*Adapted from Ref. 12. Values for the materials listed under the solid line were taken from the commentary; the possibility of dome formation in these materials during emptying must be investigated.

†μ_1: rough walls, primarily within bulk material (for example, plate walls with horizontal corrugations). μ_2: medium smooth walls (for example, concrete, plaster or stucco, planed wood in direction of grain, steel plate with rivets, and screws). μ_3: smooth walls (for example, aluminum or steel plates, welded or lapped, plastics, and coated surface).

21.1.4 Emptying Pressures on Silo Walls, ACI 313[13]

During initial filling, and certainly during discharge, lateral overpressures may occur against the silo walls. This is true because of imperfections in the cylindrical shape of the silo, nonuniformity in the distribution of particle sizes, and flow convergence at the top of hoppers or in flow channels. Thus, the lateral design pressure for silo walls above the hopper, for concentric wall patterns, is to be obtained by multiplying the lateral pressure computed according to Sec. 21.1.2 by a minimum overpressure factor $C_d = 1.5$. Lower overpressure factors may be used for particular cases where it can be shown that such a lower factor is necessary. However, in no case is the overpressure factor to be less than 1.35.

Use of $C_d = 1.5$ is allowed by ACI 313–91, in spite of the fact that higher and lower point pressures have been measured in full silos. This is because the substantial stiffness inherent to reinforced-concrete silo walls ensures vertical distribution of overpressures that can then be averaged over larger areas for structural design. Note that $C_d = 1.5$ may not be acceptable for design of steel silos unless their walls have been provided with vertical stiffeners capable of generating the stiffness and bending strength required for vertical distribution of overpressures.

Asymmetric Flow. ACI 313–91 is explicit in its requirement that pressures resulting from asymmetric flow from concentric or eccentric discharge openings be considered in the design of the silo walls. However, no method for doing so has been endorsed officially. Several methods for evaluating the effects of asymmetric flow have been published,[12,19–27] although none has been endorsed. The author recommends the DIN 1055 Sheet 6 method described in Sec. 21.1.3. For large (30 ft diameter) concrete silos storing non-free-flowing materials such as coal, the use of hoppers promoting mass flow in the silo is strongly recommended, as well as wall design using the provisions of Sec. 21.1.3, particularly regarding the effects of partial pressure. ACI 313-16[15] specifies overpressure factor for concentric flow C_d of 1.6 in Section 6.3.2.2. Section 6.3.3.2 of ACI 313-16[15] (funnel-flow hoppers) specifies overpressure factor of 1.45 for concrete hoppers and 1.6 for steel hoppers, and Section 6.3.4 of ACI 313-16[15] (pressures for flat bottom) specifies overpressure factor of 1.45 for concrete bottoms and 1.6 for steel bottoms.

21.1.5 Silage

Design of silos for silage is different than design for storage of powdery materials or granular solids. For the latter there is an exponential variation of pressure that tends to an asymptotic limit with increasing depth, while for silage there is a hydraulic-type linear variation of pressure that increases without limit as depth increases.

Three classes of silage are considered in DIN 1055 (Table 21.7). Note that for classes 1 and 2 the lateral pressure exerted against the silo walls increases at a relatively slower rate (0.4 and 0.5, respectively) between the free surface and a depth of 16 m (52.5 ft). Below this level the rate changes abruptly to almost hydraulic values (0.8 and 0.9 for classes 1 and 2, respectively). For class 3 silage the variation is hydraulic throughout the height. For none is there a distinction between the filling and emptying conditions. Because the vertical pressure q in the material at any given depth is γY, it follows that the provision $q/\gamma < 25$ m imposes a maximum height of 25 m (82 ft) for the storage of silage.

The trend is toward silage heavier than that shown in Table 21.7. Thus, it is recommended that the designer determine the actual unit weight of the material and classify it in one of the three classes shown. Always use class 3 for silage that is pumped into the silo.

It is important to note that the values in Table 21.7 assume that an outlet will be provided for the silage juices to drain out of the silo in such a way that the maximum fluid level does not exceed 1 m (3.3 ft). All silos should be clearly marked as to the class of silage for which they are designed. In addition, silos designed for classes 1 and 2 should carry a visible warning that if they are used instead for class 3 silage, they must not be filled to more than 50 percent of the height.

If emptying takes place from the top, negative wall friction (uplift) might occur for which anchorage must be provided. DIN 1055 specifies a minimum anchorage of 4 kN/m (274 lb/ft) for the perimeter of the silo. Anchorage provisions should be distributed uniformly along the perimeter at reasonable spacing.

Storage of damp grain calls for determination of pressure and loads both as silage (Table 21.7) and as granular material (Table 21.1); the largest values govern design.

21.1.6 Pressures on Horizontal Bottoms and Inclined Hoppers (*Adapted from Ref. 12.*)

In silos with horizontal bottoms and those with slightly inclined hoppers ($\alpha < 20°$, Fig. 21.7), the vertical pressure on the bottom p_b may be considered uniformly distributed only if $H/D \geq 1.5$. The values of p_b may be calculated from

$$p_b = c_b q_f \tag{21.6}$$

where c_b is a coefficient given in Table 21.6 and q_f is the vertical pressure at rest (filling condition, Table 21.1). Although for circular silos where $H/D < 1.5$ the vertical pressure at the bottom is not uniform, it may also be calculated using Eq. (21.6) except that q_f is obtained for $y = H$ at the perimeter of the wall and $y = 1.5D$ at the center of the silo. Nowhere however, need the value of p_b exceed the quantity γy, where y is the local height of the silo contents at any given location of the silo bottom.

Pressures in hoppers may be obtained as the sum of two components: (1) those generated by the actual bulk solids in the hopper and (2) those generated by the bulk solids stored above the hopper itself (Fig. 21.7).

Table 21.7 Physical Properties of Silage*

<table>
<tr><th rowspan="3"></th><th colspan="4">Classes of silage</th></tr>
<tr><th rowspan="2">Class 1</th><th colspan="2">Class 2</th><th rowspan="2">Class 3</th></tr>
<tr><th>Subclass 2a</th><th>Subclass 2b</th></tr>
<tr><td>1. Unit weight γ, lb/ft³</td><td>38</td><td>51</td><td>64</td><td>64</td></tr>
<tr><td>2. Vertical pressure q, lb/ft²</td><td>γY</td><td colspan="2">γY</td><td>γY</td></tr>
<tr><td>3. Lateral pressure p, lb/ft²</td><td></td><td colspan="2"></td><td></td></tr>
<tr><td>Y < 52.5 ft</td><td>0.4γY</td><td colspan="2">0.5γY</td><td>γY</td></tr>
<tr><td>Y > 52.5 ft</td><td>(0.8Y − 21)γ</td><td colspan="2">(0.9Y − 21)γ</td><td></td></tr>
<tr><td>4. Wall friction p_w, psf</td><td>0.1γY</td><td colspan="2">0.1γY</td><td>0.1γY</td></tr>
<tr><td>5. Integrated wall friction V† to depth Y, lb/ft</td><td>0.05γY²</td><td colspan="2">0.05γY²</td><td>0.05γY²</td></tr>
</table>

*Adapted from Ref. 12. Class 1: Strong prewilted (cured) green feed with TM > 40% (examples: grass, clover). Subclass 2a: Prewilted (cured) green feed with 25 < TM < 40% (examples: grass, clover). Silo corn, corncobs unground, damp grain. Subclass 2b: Corncobs ground (examples: ground corn, grain mixed, corncob mix). Class 3: Not prewilted (uncured) green feed with TM < 25% (example: raw green grass). Beet green. Sugarbeet cuts.

†$V_{max} = \gamma YR$ when the full contents is supported by the wall.

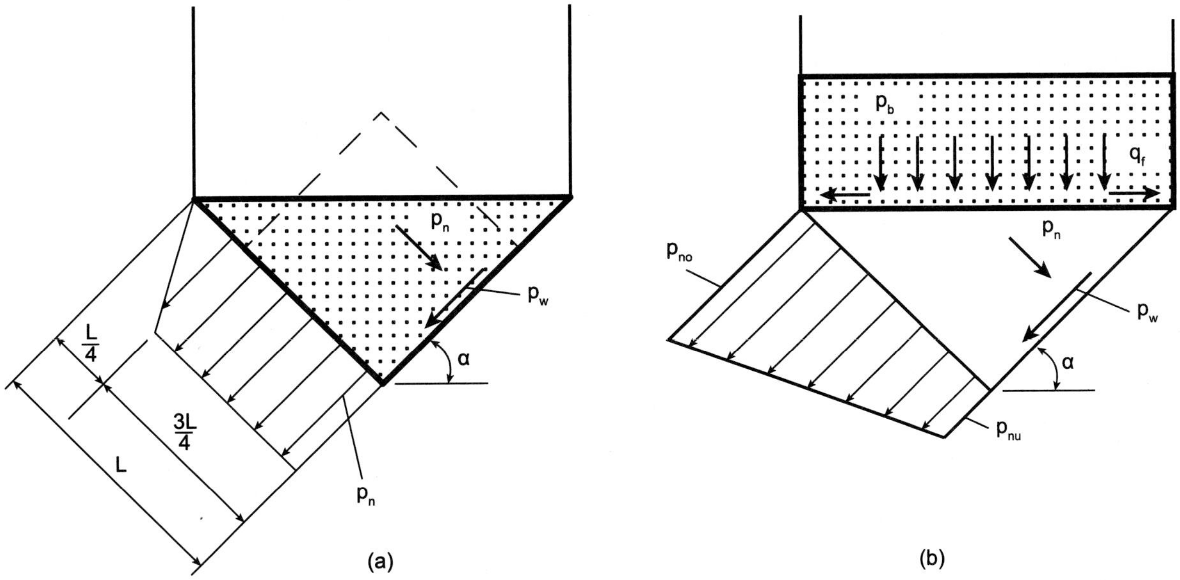

Figure 21.7 Pressures in hoppers (*a*) due to contents in hopper and (*b*) due to contents above hopper.

The pressures generated in the first case (Fig. 21.7*a*) are given as follows:

$$p_n = 2.4\gamma R k_f \sin^2 \frac{a}{\sqrt{\mu}} \tag{21.7a}$$

$$p_w = \frac{P_n}{2} \tag{21.7b}$$

where p_n and p_w are the pressures perpendicular to and parallel to the hopper walls, respectively. Values of k_f are given in Table 21.6. Note that the reduction in p_n indicated in Fig. 21.7*a* for the upper quarter of the hopper applies when bulk solids are stored only in the hopper itself. When storage includes the silo walls (Fig. 21.7*b*), no reduction in p_n should be made.

The pressures generated by bulk solids situated above the hopper (Fig. 21.7*b*) are given as follows:

$$p_{no} = (p_b \cos^2\alpha + p_{nf}(\sin^2\alpha)\left(1 + \frac{\sin 2a}{4\mu}\right) \tag{21.8a}$$

$$p_{nu} = p_b \cos^2\alpha \tag{21.8b}$$

$$p_w = \frac{p_n}{2} \tag{21.8c}$$

where p_{no} and p_{nu} are the orthogonal pressures generated against the hopper walls at the upper and lower levels of the hopper, respectively, and p_w is the pressure parallel to the hopper walls.

It is noted that the formulas given by Eqs. (21.7) and (21.8) are empirical in nature because they were derived from test results and disregard equilibrium requirements. For hoppers where $\alpha > 45°$, mass-flow conditions during emptying may cause transition peak pressures (switch) that may reach values several times larger than the pressures p_e calculated as indicated in Sec. 21.1.3. These peaks occur only in a narrow strip of wall situated just above the hopper, but it is necessary to check the design for this effect and provide additional reinforcement as required.

Adapted from Ref. 13. Initial (filling) pressures below the top of the hopper; see Fig. 21.6. The initial vertical pressure at depth h_y below the top of the hopper is to be computed by

$$q_y = q_o + \gamma h_y \tag{21.9a}$$

where q_o is the initial vertical pressure at the top of the hopper computed by

$$q_o = \frac{\gamma R}{\mu' k}\left(1 - e^{-\mu' k Y/R}\right) \tag{21.9b}$$

The initial pressure normal to the hopper surface at a depth h_y below the top of the hopper is to be taken as the larger of

$$p_n = \frac{q_Y \tan\theta}{\tan\theta + \tan\phi'} \tag{21.10a}$$

or

$$p_n = q_y(\sin^2\theta + k\cos^2\theta) \tag{21.10b}$$

where ϕ is the angle of friction between material and wall and hopper surface.

The initial friction force per unit area of hopper wall surface is to be computed by

$$v_n = p_n \tan\phi' \tag{21.11a}$$

when Eq. (21.10a) is used to determine p_n and by

$$v_n = q_y(1 - k)\sin\theta\cos\theta \tag{21.11b}$$

when Eq. (21.10b) is used.

Design pressures at and below the top of a funnel-flow hopper are to be computed using Eqs. (21.9a) through (21.11b) with q_o multiplied by an overpressure factor of 1.35 for concrete hoppers and 1.50 for steel hoppers. The vertical design pressure at the top of the hopper need not exceed γY. Design pressures at and below the top of mass-flow hoppers can be significantly higher than initial pressures. In no case, however, is the design pressure to be less than computed for funnel-flow hoppers. In multiple-outlet hoppers, the condition that initial pressures exist above some outlets and design pressures exist above others must be considered.

Initial filling pressures on flat bottoms are to be computed by Eq. (21.9b), with Y taken as the distance from the top of the floor to the top of the materials. Vertical design pressures of flat bottoms are to be obtained by multiplying the initial filling pressures by an overpressure factor of 1.35 for concrete bottoms and 1.50 for steel bottoms. The vertical design pressure need not exceed γY.

21.1.7 Earthquake Forces

The following minimum requirements have been proposed for design per ACI 313-91.[13] The total lateral seismic force H_e for shear at the base is given by

$$H_e = ZC_p(W_g + W_{eff}) \tag{21.12}$$

where Z = earthquake-zone factor = 3/16, ⅜, ¾, and 1 for zones 1, 2, 3, and 4, respectively (Fig. 21.8)
W_g = weight of structure
W_{eff} = 80 percent of weight of stored material, applied at centroid of volume
C_p = 0.2 for silos with material stored on bottoms above ground and 0.1 when silo walls extend to ground and stored material rests directly on ground. For intermediate cases, C_p may be obtained by linear interpolation.

If the bin bottom-supporting system is independent of the walls, W_{eff} may be distributed between the two independent structures according to their relative stiffnesses.

A dynamic analysis, using a design earthquake spectrum compatible with the seismic zone and with local foundation conditions, may be used instead of Eq. (21.12).

ACI 313-16[15] Section 6.3.8 requires that silos and stacking tubes shall be designed to withstand the seismic force in accordance with ASCE/SEI 7-10.[28] The design shall consider the full range of loading, from empty to full.

21.1.8 Wall Forces

Circular Silos

The hoop force F per unit height of wall resulting from the radial pressure p of the stored material is given by

$$F = \frac{pD}{2} \tag{21.13}$$

The increase ΔD in the diameter can be determined from

$$\Delta D \approx \frac{pD^2}{4tE_c} \tag{21.14}$$

where t = wall thickness

Walls are also subjected to vertical compression from the roof, from their weight, and from wall friction of the stored material.

Grouped Circular Silos. Some groups frequently used are shown in Fig. 21.3. Initial design of individual circular cells of these groups is done as if they were isolated. The effects of interstices are determined by considering the adjacent circular cells to be empty (Fig. 21.9). Formulas for the case of an empty central circular cell with adjacent full circular cells are given in Fig. 21.10.

Analysis of other cross-sectional configurations and empty-full cases is possible by treating the whole horizontal cross section as a rigid frame. The curved members and variable wall thicknesses at wall intersections may make an *exact* analysis quite complex and laborious. Approximate methods based on local analysis of elements fixed at the sections where silo walls intersect are acceptable and may be the only practical thing to do in many cases.

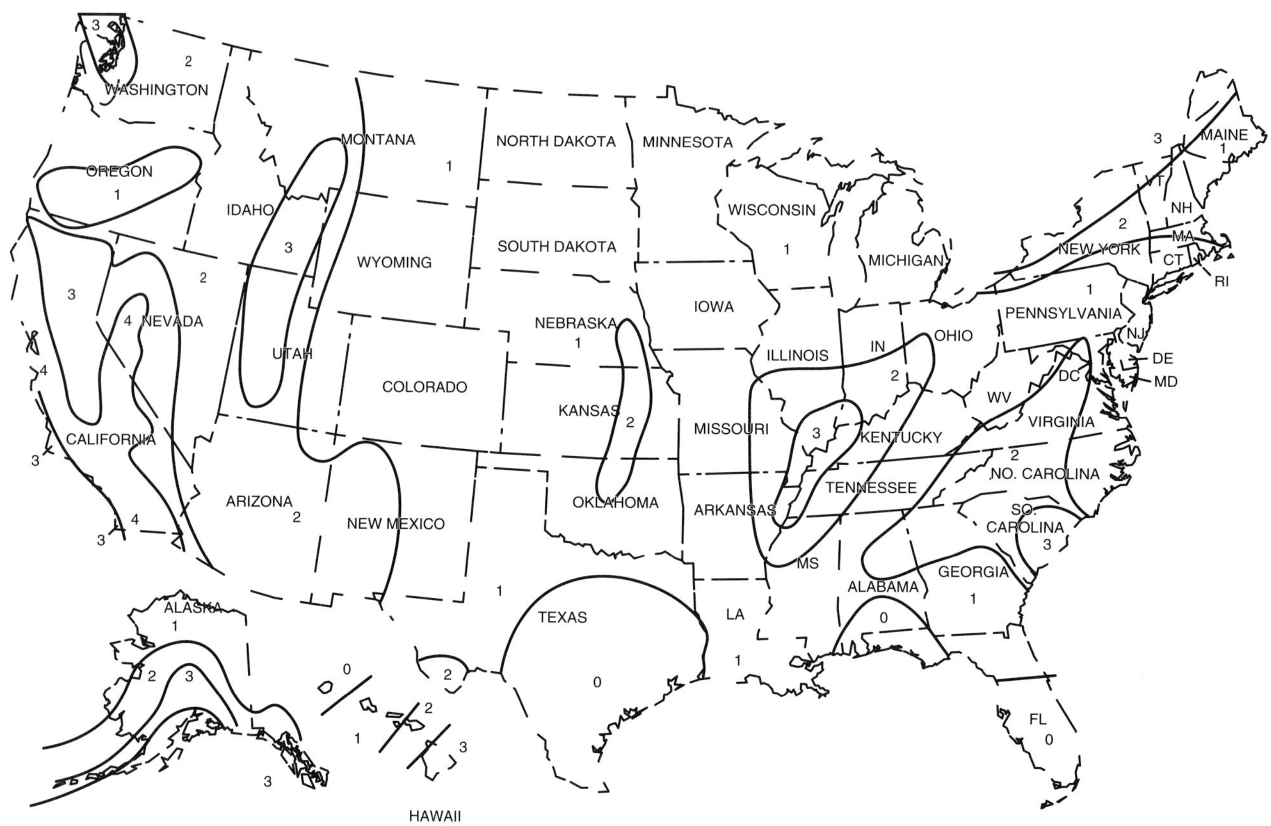

Figure 21.8 Seismic zone map of the United States.

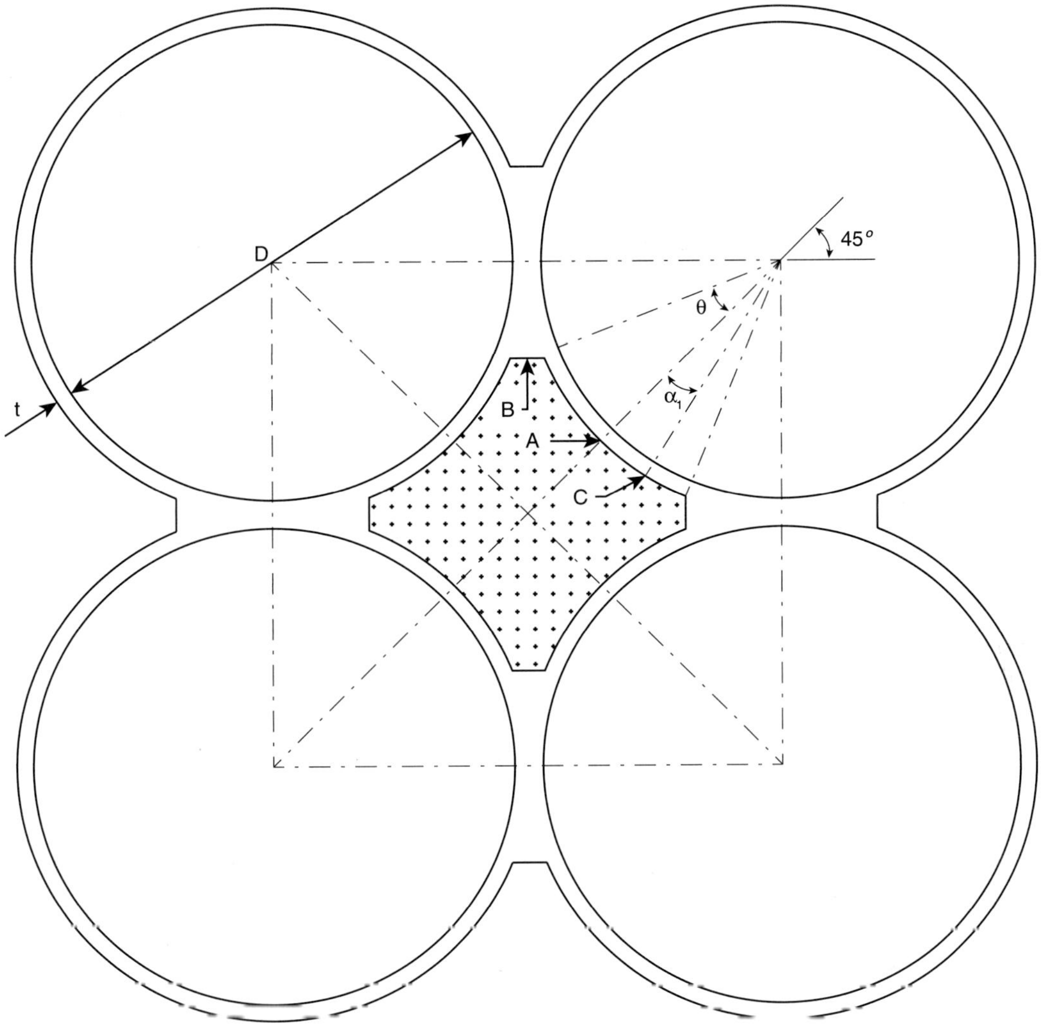

Bending moments

$$M_A = \frac{p}{4}(D + 2t)(D + t) \sin\theta \left(1 - \frac{\sin\theta}{\theta}\right)$$

$$M_B = \frac{p}{4}(D + 2t)(D + t) \sin\theta \left(\cos\theta - \frac{\sin\theta}{\theta}\right)$$

$$M_C = \frac{p}{4}(D + 2t)(D + t) \sin\theta \left(\cos\alpha_1 - \frac{\sin\theta}{\theta}\right)$$

Compression forces

$$F_A = \frac{p}{2}(D + 2t)(1 - \sin\theta)$$

$$F_B = \frac{p}{2}(D + 2t)(1 - \sin\theta\cos\theta)$$

$$F_C = \frac{p}{2}(D + 2t)(1 - \sin\theta\cos\alpha_1)$$

Figure 21.9 Full interstice and empty adjacent cells. (*Adapted from Ref. 20.*)

Attention to local effects at wall intersections is important. Good design should provide strength for proper transfer of bending moments, shear, and direct axial force at these locations.

Rectangular and Polygonal Silos

Walls in the pressure zones of square, rectangular, and polygonal silos are subjected to bending moment, horizontal shear, and horizontal tension resulting from the lateral pressure, and from vertical compression by the roof, from their weight, and from wall friction of the stored material.

Walls whose height is more than twice the width may be analyzed for one-way bending in the horizontal direction. Since adjoining walls are continuous at their junctures, moments may be determined as for a frame. Formulas for *M*, *F*, and *V* for a horizontal strip are given for rectangular walls in Fig. 21.11 and regular polygonal walls of *N* sides in Fig. 21.12.

Rectangular walls whose height is less than half the width may be analyzed for one-way bending in the vertical direction. The lower edge can usually be assumed fixed. The upper edge may be assumed fixed or simply supported, depending on the attached construction, or free if there is none.

Moments in walls whose height is more than half the width but less than twice the width should be determined as for a plate supported on four edges or, if the upper edge is free, as a plate supported on three edges.

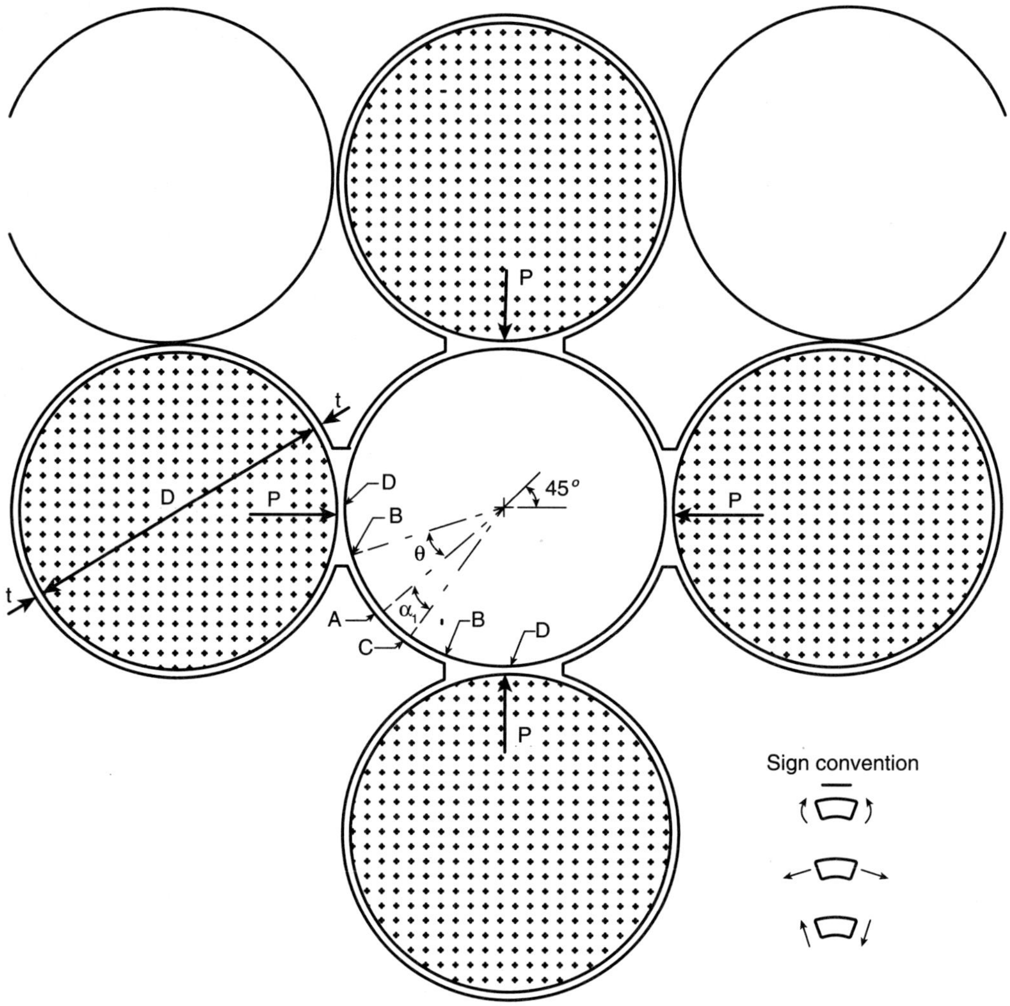

Bending moments	Hoop forces	Shear forces
$M_A = -0.0352\ P\ (D + t)$	$F_A = -0.7071\ P$	$V_A = 0$
$M_B = (0.3183 - 0.3535 \cos\theta)\ P\ (D + t)$	$F_B = -0.7071\ P \cos\theta$	$V_B = 0.7071\ P \sin\theta$
$M_C = (0.3183 - 0.3535 \cos\alpha_1)\ P\ (D + t)$	$F_C = -0.7071\ P \cos\alpha_1$	$V_C = 0.7071\ P \sin\alpha_1$
$M_D = 0.0683\ P\ (D + t)$	$F_D = -0.5\ P$	$V_D = 0.5\ P$

Figure 21.10 Empty circular cell and full adjacent circular cells.

Tables of moment coefficients for various cases are given in Refs. 8, 21, and 22.

Walls that are supported on columns are subjected to in-plane bending because of the load from an attached bottom. Analysis depends on the height of the wall relative to the spacing of the columns (Sec. 21.2.7).

Thermal Effects

A differential temperature ΔT between the interior and exterior faces of a silo creates a strain gradient $\alpha_t \Delta T/t$ in the wall section, where t is the thickness of the wall and α_t is the thermal coefficient of expansion of concrete. Because of the closed nature of silos, rotational restraint is imposed by continuity, and thus bending moments are generated in the presence of thermally induced strain gradients. For a section subject to bending moment M and axial force P caused by loading unrelated to thermal effects, Gurfinkel[29] has shown that the additional thermal moment M_t for a given strain gradient depends on the existing strain distribution, which in turn depends on M and P; iteration is required to determine M_t. ACI 313[13,15] calculates the thermal bending moments M_{xt} per unit of wall height and M_{yt} per unit of wall width as if generated in an uncracked section of wall subjected to a state of plane strain. Thus,

$$M_t = \frac{E_c t^2 \alpha_t \Delta T}{12(1-\nu)} \tag{21.15}$$

where E_c and ν are the concrete modulus of elasticity and Poisson's ratio, respectively. For normal-weight concrete $E_c = 57{,}000\sqrt{f_c'}$, $\alpha_t = 5.5 \times 10^{-6}/°F$, and $\nu = 0.2$. This formula gives conservative values for the

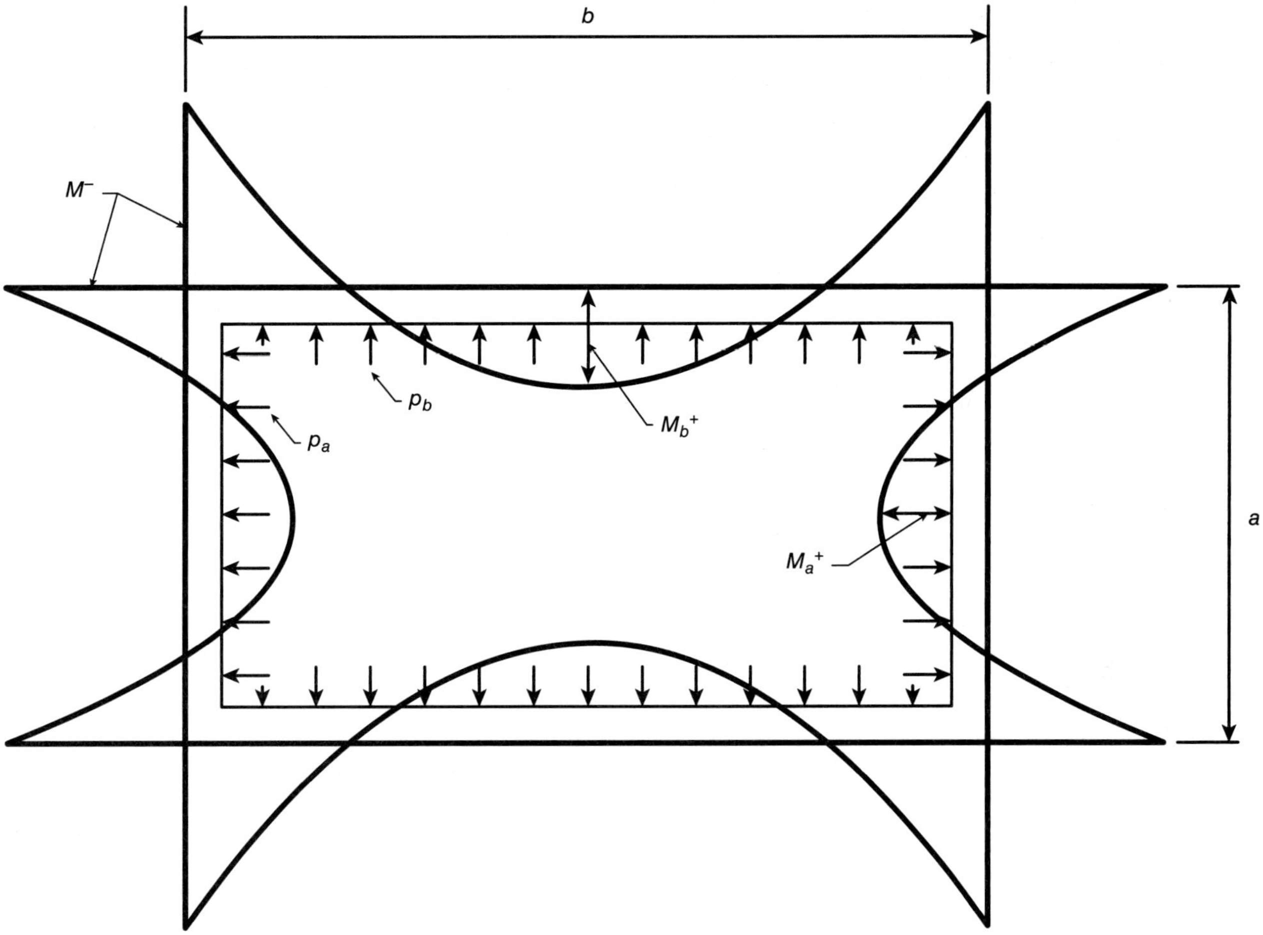

$$M^- = \frac{1}{12(1+n)}(P_a a^2 + n p_b b^2) \text{ where } n = \frac{b}{a}\left(\frac{t_a}{t_b}\right)^3$$

$$M^+_a = P_a a^2 - M^-$$

$$M^+_b = P_b b^2 - M^-$$

$$F_b = p\frac{a}{2} \qquad V_b = p\frac{b}{2}$$

$$F_a = \frac{pb}{2} \qquad V_a = \frac{pa}{2}$$

Figure 21.11 Lateral pressure on rectangular silo.

thermal moments since any cracking of the wall section would reduce its stiffness and result in lower values.

Temperature Differential. The temperature differential between external and internal faces of a concrete silo wall containing hot stored material can be calculated from

$$\Delta T = (T_{i,des} - T_0)K_T \tag{21.16}$$

where $T_{i,des} = T_i - 80$, T_i = temperature of stored material, T_0 = design winter drybulb temperature, and K_T = ratio of thermal resistance of wall alone to that of the wall plus an outside surface film of air plus a thickness t_m of stored material acting as insulating material. If R_m and $R_c = 0.08$ represent the thermal resistances per unit thickness (resistivity)[b] of the stored material and the concrete wall, respectively, and $R_a = 0.17$ is the thermal resistance of the outer surface film of air, then

$$K_T = \frac{0.08t}{t_m R_m + 0.08t + 0.17} \tag{21.17a}$$

[b]The unit of conductivity (*Btu/ft²/hr/°F/in.*) is the amount of heat in Btu that will flow in 1 hour through 1 ft² of a layer 1 in. thick of a homogeneous material per 1°F temperature difference between surfaces of the layer. Resistivity, which measures the insulating value of a material, is the reciprocal of conductivity.

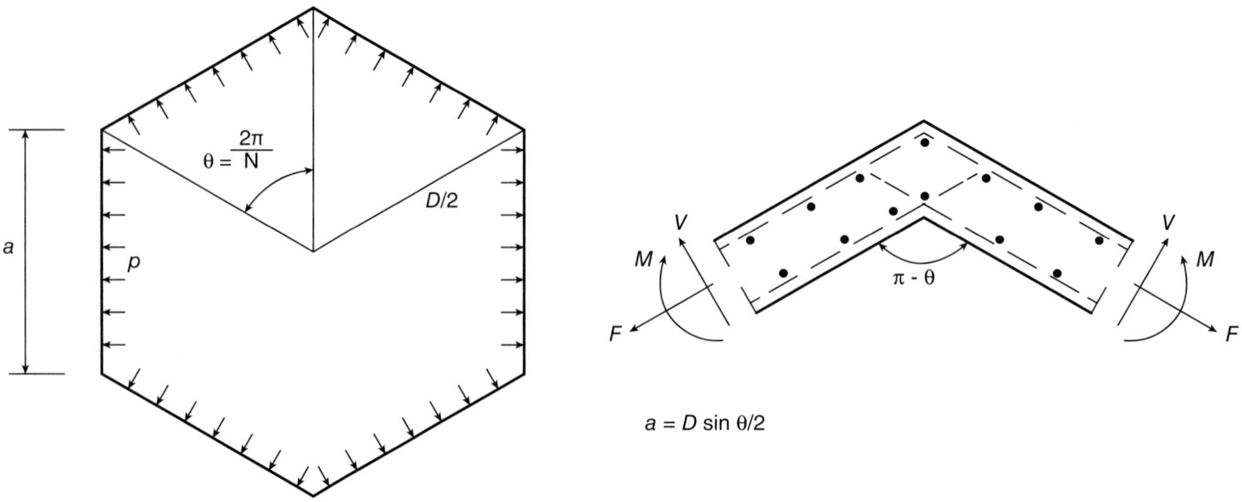

(a) Plan view (b) Corner detail

	At corner	At side midspan
F	$\frac{pD}{2}\cos\frac{\theta}{2}$	$\frac{pD}{2}\cos\frac{\theta}{2}$
V	$\frac{pD}{2}\sin\frac{\theta}{2}$	0
M	$-\frac{pD^2}{12}\sin^2\frac{\theta}{2}$	$\frac{pD^2}{24}\sin^2\frac{\theta}{2}$

Figure 21.12 Lateral pressure on polygonal silo.

For silos storing hot cement ACI 313 suggests $t_m = 8$ in. and $t_m R_m = 3.92$. This reduces K_t to

$$K_T = \frac{0.08t}{4.09 + 0.08t} \tag{21.17b}$$

The analysis is illustrated in Example 5.

21.2 DESIGN OF WALLS

Except as noted, design formulas in the following articles are in terms of strength design. Subscripts u denote ultimate values obtained by multiplying service-load forces by load factors. ACI 313[13] prescribes load factors of 1.7 for live load and 1.4 for dead load. ϕ is the capacity-reduction factor. ACI 313-16[15] requires that load factors and strength reduction factors shall conform to ACI 318-11.[30]

21.2.1 Minimum Thickness of Circular Walls

To allow for noncalculable moments resulting from transient nonuniform pressure on the walls of circular silos, the following minimum thickness is recommended:[13]

$$t_{\min} = p\frac{D}{2}\frac{mE_s + f_s - nf_{ct}}{f_s f_{ct}} \tag{21.18}$$

in which m = shrinkage coefficient (may be taken as 0.0003), f_s = allowable steel stress (between 0.4 and $0.5f_y$), and f_{ct} = concrete stress in uncracked section under static lateral pressure (may be taken as $0.1f'_c$).

Final wall thickness is governed by practical considerations and by load requirements and permissible crack width. A minimum thickness of 6 in. should be used for cast-in-place silo walls. However, at least 8-in. thickness is recommended for placing two layers of hoop reinforcement. ACI 313[15] Section 6.4 provides minimum and recommended wall thickness.

21.2.2 Design Crack Width

An important design consideration is the design crack width that may be tolerated. For grain and cement storage silos and for other silos exposed to weather, this would help prevent penetration of water that causes corrosion of the reinforcement, spoils the silo contents by inducing germination of the grain, or causes hydration of the cement. For walls in the pressure zone, thickness and reinforcing should be proportioned so that, under initial (filling) pressures, the design crack width computed at a distance $d_c = 2.5\ d$ from the center of a bar having a diameter d should not exceed 0.010 in. The design crack width w (in.) is computed by

$$w = 0.0001 f_s \sqrt[3]{d_c A} \tag{21.19a}$$

where f_s (k/in.^{2}) is the calculated stress in the reinforcement at initial (filling) pressures and A (in.^{2}) is the effective tension area of concrete surrounding the tension reinforcement and having the same centroid as that reinforcement; see Fig. 21.13. For a silo wall reinforced with steel bars of diameter d at a spacing s, one finds $d_c = 2.5\ d$, $A = 5$, and from Eq. (21.19a),

$$w = 0.0001 f_s \sqrt[3]{12.5 d^2 s} \tag{21.19b}$$

Crack-width evaluation is illustrated in Example 2.

21.2.3 Walls in Tension

The required reinforcement A_s per unit height of wall is given by

$$A_s = \frac{F_u}{\phi f_y} \tag{21.20}$$

ϕ in this equation may be taken as 0.9.

21.2.4 Walls in Tension and Flexure

Silo walls for the containment of solids are subjected to the combined action of tension force, N, and bending moment, M. Design codes in

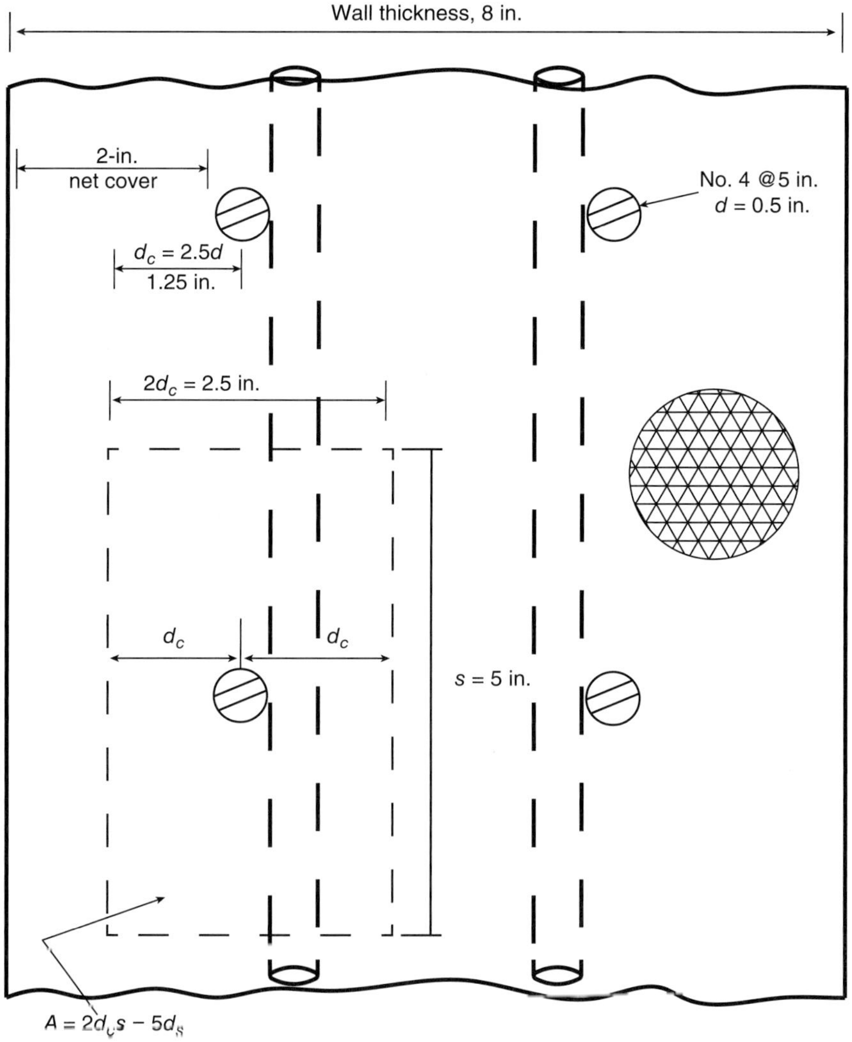

Figure 21.13 Wall section and reinforcement.

the United States[13] and Europe[12,31] recommend that these walls be designed for hoop tension and for bending moments generated during the emptying process. This applies even to round silos with a single concentric outlet and certainly to those silos where the outflow is through eccentric outlets situated on the floor or in the walls or through multiple hoppers. Combined tension and moment are generated also in square and rectangular silos and in clusters of silos regardless of their geometric shape because of continuity between adjacent walls. A temperature gradient between the inside and outside surface of a silo wall will also cause bending moments.

Designers have used approximate formulas derived by Harris[32] that have been incorporated by ACI 313[13,15] and are cited by Safarian and Harris[33] and Gurfinkel[34] to determine the required amount of reinforcement of silo walls. Harris provides formulas whose applicability depends on the eccentricity ratio, M/N. Design can be accomplished as shown by Gurfinkel[14] using two linear interaction equations that represent the tension branch of the Ultimate Interaction Diagram (UID) quite accurately. Only the initial calculation of M_n and N_n, the individual bending and axial strengths, respectively, of the proposed section is required.

The tensile strength of a doubly reinforced-concrete wall is $N_n = 2A_sf_y$. For the bending strength, M_n, there are two possible ways to obtain a lower bound. One involves ignoring the compression steel, Eq. (21a); the other considers only the action of the steel couple and ignores the concrete, Eq. (21b) (see Fig. 21.14):

$$M_{nc} = A_sf_y\left[d - \frac{1}{2}\frac{A_sf_y}{0.85f_c'b}\right] \tag{21.21a}$$

$$M_{ns} = A_sf_y(d - d') \tag{21.21b}$$

M_n is taken as the larger value of M_{nc} and M_{ns} because the actual M_n is always larger than both M_{nc} and M_{ns} (usually $M_{nc} > M_{ns}$).

EXAMPLE

Consider an 8-in-thick wall doubly reinforced with two layers of No. 4 bars at 5-in. spacing; let cover = 1.5 in., $f_y = 60$ ksi, and f_c'. It follows that $A_s = 0.48$ in.^{2}/ft for each layer. Also, $d = 6.5$ in. and $d' = 1.5$ in. Determine a lower bound for M_n using Eqs. (21.21a) and (21.21b):

$$M_{nc} = 0.48 \times 60\left[6.5 - \frac{1}{2}\frac{0.48 \times 60}{0.85 \times 4 \times 12}\right] = 177\ k\ \text{in./ft}$$

$$M_{ns} = 0.48 \times 60(6.5 - 1.5) = 144\ k\ \text{in./ft}$$

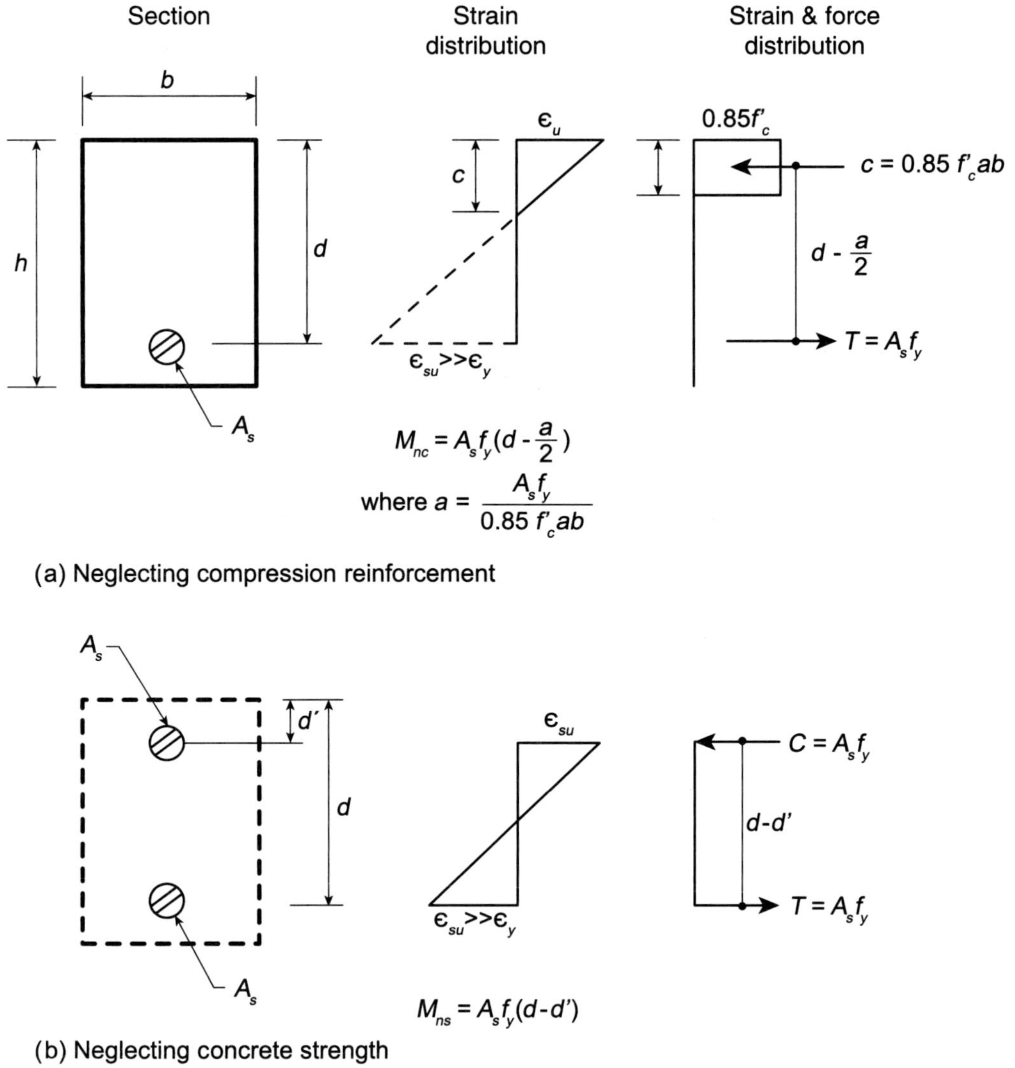

Figure 21.14 Bending strength of doubly reinforced-concrete sections. Use the larger of M_{nc} and M_{ns}.

Use $M_n = 177$ k-in./ft. The actual $M_n = 184.7$ k-in./ft, as obtained by iteration. Note that the lower bound obtained from M_{nc} and M_{ns} is about 96 percent of the actual value of M_n. For design purposes, the reduced value of M_n is $\phi M = 0.90 \times 184.7 = 166.2$ k-in./ft. Also, $N_n = 2 \times 0.48 \times 60 = 57.6$ k/ft and $\phi N_n = 0.9 \times 57.6 = 51.8$ k/ft. The actual variations of M_n vs. N_n (UID) and ϕM_n vs. ϕN_n ($\phi \times$ UID) are shown in Fig. 21.15. Let load-factored design values be $M_u = 60$ k-in./ft and $N_u = 34$ k-in./ft (tension). Is the given wall acceptable? Plot the point (60, 34) in Fig. 21.15 and realize that it lies within the envelope of acceptability that is provided by the $\phi \times$ UID curve. Had the point fallen outside this envelope, it would indicate that the wall needed additional strength, that is, either more reinforcement or an increase in the thickness of the wall together with an increase in concrete strength or all of the above. As far as the $\phi \times$ UID tension branch is concerned, increasing only the reinforcement is equivalent to a parallel displacement outward. If, however, the reinforcement were kept constant but the section dimensions and concrete strength were increased, then the tension branch would move outward by rotating about the point (0, ϕN_n). Thus, if the dominant component were the tension load, it would be better to increase the wall reinforcement. If, however, the bending moment were dominant, an increase in dimensions of the section would be necessary that would also bring about a beneficial increase in the wall stiffness.

Design. The following two approximate interaction equations are offered for design:

$$\frac{M_u}{M_n} + \frac{N_u}{\phi N_n} \leq 1 \tag{21.22a}$$

and

$$\frac{M_u}{\phi M_n} + \frac{N_u}{N_n} \leq 1 \tag{21.22b}$$

These two relationships are presented in Fig. 21.15 as the long dashed line and the short dashed line, respectively. The smaller of these values is always contained within the concave shape of the $\phi \times$ UID diagram, thereby rendering a simple and conservative alternate to the latter. Thus, one need not draw the $\phi \times$ UID curve of any given section to determine its acceptability. Instead, one only needs the corresponding M_n and N_n values.

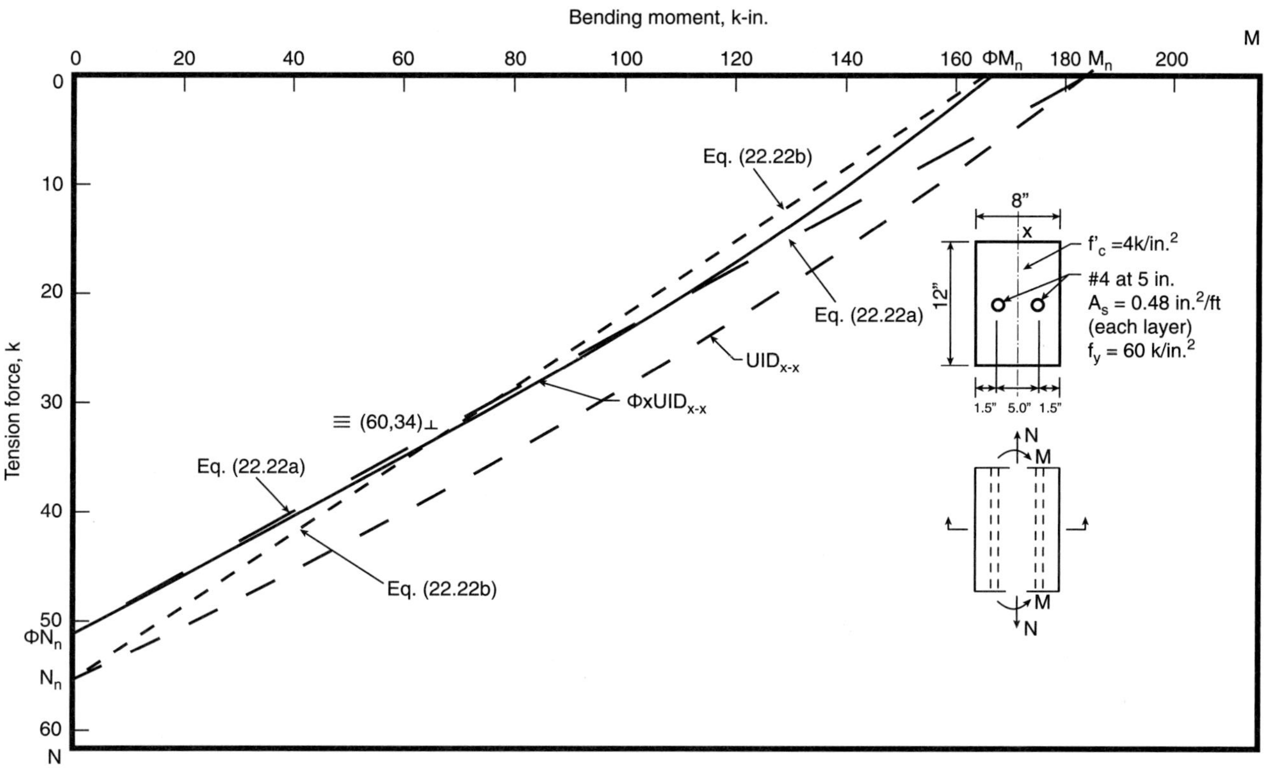

Figure 21.15 Tension branch of ultimate interaction diagram of a doubly reinforced-concrete wall. The solid straight lines represent the approximate ϕxUID for design purposes.

EXAMPLE

Consider the same section as before with M_u = 60 *k*-in./ft and N_u = 34 *k*/ft. Check suitability given $M_n = M_{nc}$ = 177 *k*-in./ft, N_n = 57.6 *k*/ft, and ϕ = 0.9. Thus,

$$\frac{60}{177}+\frac{34}{0.9\times 57.6}=0.339+0.656=0.995<1 \qquad (21.23a)$$

and

$$\frac{60}{0.9\times 177}+\frac{34}{57.6}=0.377+0.590=0.967<1 \qquad (21.23b)$$

Since both equations are satisfied, the given wall is a valid solution. Calculation of the required amount of reinforcement is possible after the thickness of the wall and the strength of concrete and reinforcement grade are selected. Take t = 8 in., d = 6.5 in., d' = 1.5 in., f_c' = 4 ksi, f_y = 60 ksi, ϕ = 0.9. The required area of steel in each layer of the wall, A_s, for Mu = 60 *k*-in./ft and N_u = 34 *k*/ft, is obtained using Eqs. (21.22a) and (21.22b) as follows:

$$\frac{60}{A_s\cdot 60\left[6.5-\frac{1}{2}\cdot\frac{A_s\times 60}{0.85\times 4\times 12}\right]}+\frac{34}{0.9\times 2A_s\times 60}\le 1 \qquad (21.24a)$$

$$\frac{60}{0.9A_s\cdot 60\left[6.5-\frac{1}{2}\cdot\frac{A_s\times 60}{0.85\times 4\times 12}\right]}+\frac{34}{2A_s\times 60}\le 1 \qquad (21.24b)$$

These equations are individually solved by simple iteration. Start by taking A_s = 0 in the second term of the denominator of the first fraction and then solve for A_s, substitute the latter value of A_s to evaluate the left side of the equation, and obtain a value slightly larger than 1. Using this value as multiplier for A_s, obtain the next trial value for A_s and repeat the previous step. The process converges quickly and monotonically. A practical solution for A_s is usually obtained with only two trials. Thus, A_s = 0.477 in.^{2}/ft from Eq. (21.24a) and A_s = 0.463 in.^{2}/ft from Eq. (21.24b). Using the larger value, select No. 4 Gr. 60 @ 5 in. (A_s = 0.48 in.^{2}/ft) to reinforce each face of the wall.

Shear. The shear stress v_u is given by

$$v_u=\frac{V_u}{bd} \qquad (21.25)$$

where v_u should not exceed v_c given by

$$v_c=2\phi\left(1-0.002\cdot\frac{F_u}{bt}\right)\sqrt{f_c' psi} \qquad (21.26)$$

21.2.5 Walls in Compression

ACI 318[35] allows walls for which the compressive force falls within the middle third to be considered as concentrically loaded. If buckling is not involved, the permissible compression is

$$f_c=0.55\phi f_c' \qquad (21.27)$$

where ϕ = 0.70

For rectangular walls, where slenderness may influence strength, buckling should be accounted for by using[13]

$$f_c=0.55\phi f_c'\left[1-\left(\frac{H_0}{40t}\right)^2\right] \qquad (21.28a)$$

where ϕ = 0.70 and H_0 = clear vertical distance between supports. If $H_0 > l_0$, where l_0 = clear horizontal distance between supports, use l_0 in place of H_0.

Circular walls in the pressure zone may be designed for the allowable stress of Eq. (21.27) if there are no openings. If there are unreinforced openings, Eq. (21.28a) should be used, with H_0 = height of opening.

Circular walls below the pressure zone, continuous throughout, should be designed for[13]

$$f_c = 0.55\phi f_c' \left[1 - \left(\frac{D}{120t}\right)^3\right] \tag{21.28b}$$

If there are unreinforced openings, use

$$f_c = 0.55\phi f_c' \left[1 - \left(\frac{H_0}{40t}\right)^2 - \left(\frac{D}{120t}\right)^3\right] \tag{21.28c}$$

The design axial strength per unit area for walls, where global and local buckling do not control, is calculated in accordance with the provisions of Section 6.4.7 of ACI 313-16.[15] Where compression and/or buckling may control, the design axial strength per unit perimeter is calculated per Section 7.4.4 of ACI 313-16.[15]

21.2.6 Walls in Compression and Flexure

These may be designed using the provisions of ACI 318, Chapter 10.[30,35] Combined compression and flexure is also discussed in Chapter 11. The interaction diagrams in that section for rectangular columns with reinforcement on opposite faces can be used for walls.

21.2.7 In-Plane Bending of Walls

The in-plane bending behavior of a wall supported on columns depends on the height of the wall relative to the spacing of the columns.

Low Walls. The stiffness of a wall of a low bunker is of the same order as that of the hopper wall, and the two can be assumed to act together in transferring vertical load to the columns. For bin walls with $H/a \le 0.5$, Ciesielski et al.[36] recommend that the wall and that portion of the hopper wall whose vertical projection is 0.4*a*, where *a* is the length of the wall, be analyzed as a beam (Fig. 21.16*a*). The resulting bending stresses (Fig. 21.16*b*) are computed from $f = M/S$, where *M* is the moment resulting from the vertical loads. The stress in the part of the hopper wall not considered to be part of the beam is assumed to decrease linearly from the value of the bottom fiber beam stress to zero at the vertex of the hopper wall. These stresses must be considered in combination with the moments and axial tensions resulting from lateral pressure (Fig. 21.11) in determining the wall reinforcement.

A folded-plate analysis of the joint action of a low bunker wall and an adjoining hopper wall can also be made.[8]

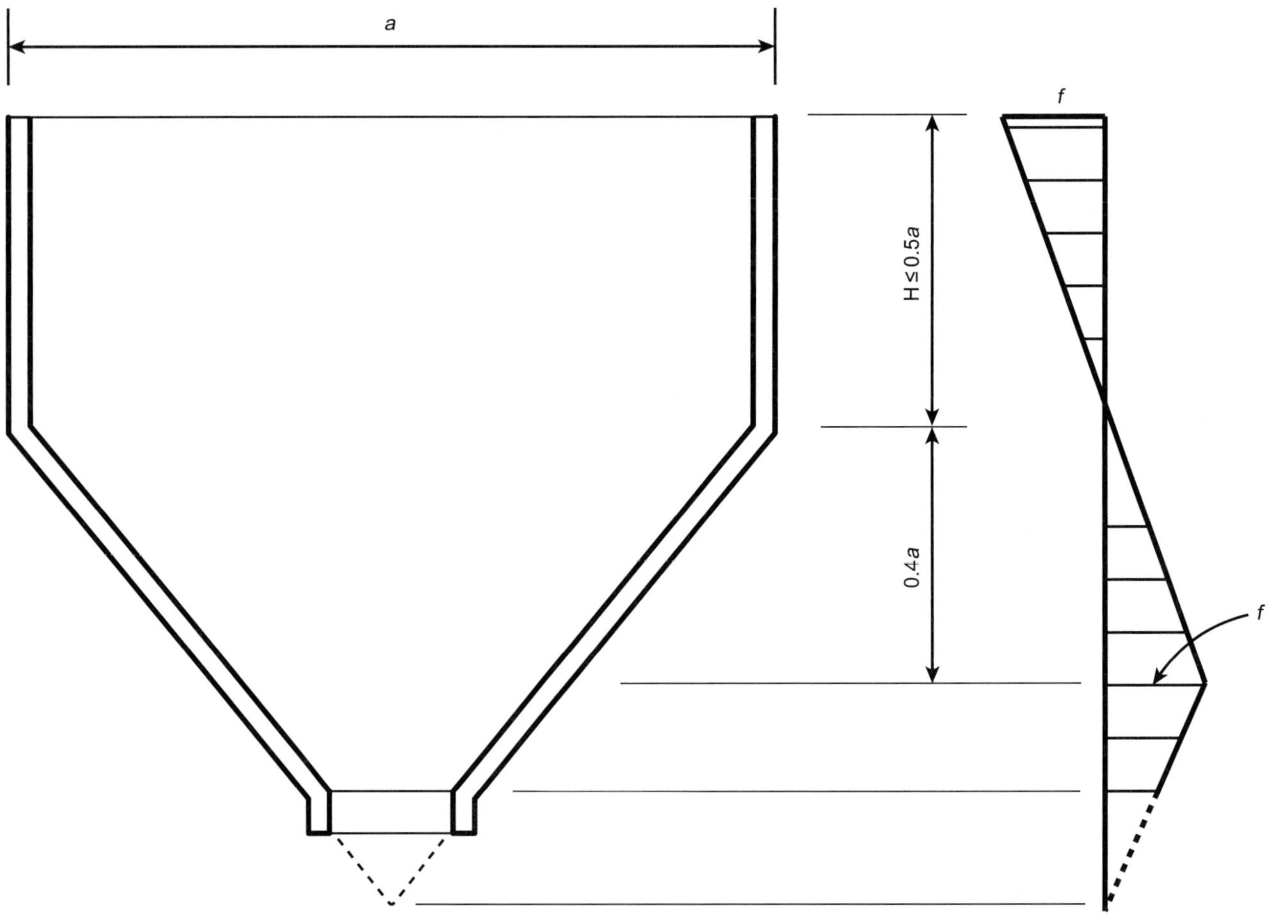

Figure 21.16 Low wall and bending stress distribution.

High Walls. Experimental studies by a number of investigators of the beam behavior of reinforced-concrete walls are discussed in Ref. 26. Walls for which $H/l \leq 1$, where H = height and l = length, can be designed by the usual procedure for reinforced-concrete beams. For a single-panel wall with $H/l > 1$, simply supported on columns spaced l center-to-center and carrying a uniformly distributed load w, the tension T that must be furnished by tensile reinforcement is given by

$$T = \frac{0.14wl}{\sqrt{H/l}} \qquad 1 < \frac{H}{l} \lesseqgtr 2 \tag{21.29}$$

The value of T for $H/l = 2$ is to be used for walls with $H/l > 2$. The reinforcement is to be distributed over a depth $0.1l$.

If one-half to two-thirds of the tensile reinforcement is bent up, the shear strength V of the panel is given by

$$V \lesseqgtr 0.54 f_c' t^2 \sqrt{H/t} \tag{21.30}$$

If the shear $w_u l/2$ exceeds V by Eq. (21.30) or if no bars are bent up, web reinforcement must be provided to resist the tension T_s given by

$$T_s = \frac{w_u l/2}{\sqrt{2H/l}} \qquad 1 \lesseqgtr \frac{H}{l} \lesseqgtr 2 \tag{21.31}$$

It is assumed in this formula that the necessary web reinforcement is inclined about 60° to the horizontal. Of course, equivalent reinforcement in the form of stirrups can be used.

Equation (21.31) is based on tests in which the load was applied to the top of the panel. However, tests on panels loaded along the bottom edge showed that the formula can also be used for this case.

21.2.8 Walls Subjected to Thermal Stresses

The required additional vertical and horizontal reinforcement per unit width or height is given by

$$A_{s,t} = \frac{1.4M_t}{f_y(d - d'')} \tag{21.32}$$

where M_t is given by Eq. (21.15). This steel should be placed near the cooler (usually outer) face of the wall. In singly reinforced walls it should be added to the main hoop steel, which should be near the outer face. In doubly reinforced walls it should be added to the outer layer, but for simplicity an equal amount is often added to the inner layer to avoid having bar sizes or spacings differ from one layer to the other.

Vertical tensile thermal stress is usually offset by vertical dead-load compressive stress so that additional temperature steel is often not needed.

21.2.9 Vertical Reinforcement

This is required not only in outside walls of silo groups but also on all inside walls. Vertical steel distributes lateral overpressures to adjacent horizontal reinforcement. Gurfinkel 37 reported a silo where failure was averted when vertical steel redistributed lateral pressures that could not be resisted by hoop reinforcement that was in an advanced stage of corrosion. Vertical steel also resists tension caused by bending moments resulting from restraint against circular elongation, eccentric loads from hopper edges or attached auxiliary structures, and temperature differentials between inside and outside wall surfaces or between silos.

21.2.10 Details and Placement of Reinforcement

Table 21.8 summarizes the requirements of ACI 313.[13] Bar splices, both horizontal and vertical, are staggered. Adjacent hoop-reinforcing splices in the pressure zone are staggered horizontally by not less than one lap length or 3 ft and do not coincide in vertical array more frequently than every third bar.

Slipforming should not be considered an excuse for not tying reinforcement together. Haeger considers the normal tying of the ends of hoop reinforcement, with additional ties every 4 to 5 ft between to be acceptable.[38] Vertical steel should not be omitted to provide access for concrete buggies in slipform construction; instead, it may be spaced farther apart at specified access locations. The total amount of vertical steel is unchanged; only the spacing is affected. The conventional practice of leaving the slipforming jack rods embedded in the concrete is fine, but widely spaced jack rods should not be construed as the equivalent of vertical reinforcement.

Table 21.8 Minimum Reinforcement Requirements

Region	Horizontal steel	Vertical steel (No. 4 or larger)
Pressure zone	As required by calculations.	Minimum ratio of vertical steel to gross concrete area is 0.0025 for exterior walls and 0.0020 for interior walls. Maximum vertical bar spacing is the lesser of 4_t and 18 in. for exterior walls and 24 in. for interior walls.
Below pressure zone	Continue A_s from above for a distance equal to six times wall thickness. Below this, provide $0.0025t$ per unit height of wall.	Same as above.
Bottom of walls and columns		Dowels as needed to prevent uplift and shifting by earthquake or wind loading. Author recommends always using enough dowels to match the vertical steel in walls and columns above.
Wall intersections subjected to moment	Provide as required.	Provide as required.
Adjoining silos	Provide as required to prevent separation.	
Circular walls, single-reinforced	Place nearest to the outer faces.	

Typical reinforcing patterns at wall intersections are shown in Fig. 21.17.

Wall Openings. Table 21.9 summarizes the requirements of ACI 313.[13] Added reinforcement to replace load-carrying reinforcement that is interrupted by an opening must extend in each direction beyond the opening. The extension each way must be as follows:

1. Sufficient to develop full yield strength of the reinforcement through bond
2. Not less than 24 in.
3. Not less than one-half the opening dimension in a direction perpendicular to the reinforcement bars in question

Figure 21.18 shows a typical detail of the reinforcement of a narrow silo wall between openings; these must be designed as columns.

21.3 DESIGN OF BOTTOMS

21.3.1 Bottom Pressure

Static unit pressure q_a normal to a surface inclined at an angle α to the horizontal is given by

$$q_\alpha = p\sin^2\alpha + q\cos^2\alpha \tag{21.33}$$

Silo bottoms are designed to resist q_α. In seismic zone 4, q in Eq. (21.33) should be computed for the effective weight (80 percent) of the stored material because of the loss of friction against the silo walls resulting from seismically induced lateral vibrations, that is, $q = 0.8\gamma Y$. In other seismic zones, q should be increased by the following fractions of the increased pressure $(0.8\gamma Y - q)$ for zone 4, ¾ for zone 3, ⅜ for zone 2, and 3/16 for zone 1.

ACI 318 load factors are suggested for ultimate-strength design under seismic load.

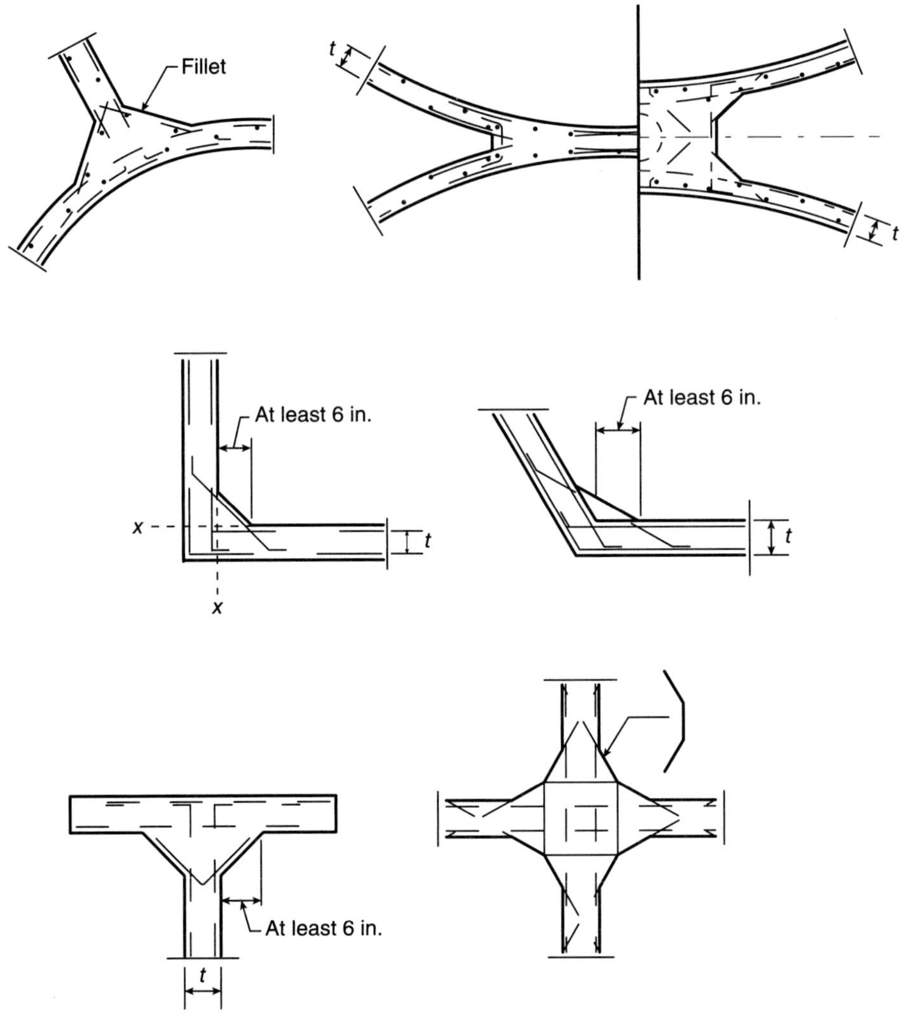

Figure 21.17 Reinforcement at intersecting walls. *(Adapted from Ref. 13 and Ref. 15.)*

Table 21.9 Reinforcement at Wall Openings

Openings	Horizontal steel	Vertical steel
In pressure zone	Add at least 1.2 times area of interrupted reinforcement, ½ above the opening and ½ below.	Provide by assuming narrow strip of wall, $3t$ in width on each side of opening, to act as column within opening height subjected to its own vertical load plus that from ½ wall span above opening. Add steel at least equal to that eliminated by opening.
Outside pressure zone	Add no less than the normal reinforcement interrupted by opening, distributed as above.	As above.
Closely spaced		See Fig. 21.18. Design as columns.

21.3.2 Plane Bottoms

Design loads for horizontal slab bottoms are dead load, vertical pressure q, and thermal load from the stored material. For inclined slab bottoms q_α should be used. Allowance for earthquake forces should be made as described in Sec. 21.3.1.

Formulas for bending moments and deflections of circular slabs, with and without a central hole, are given in Ref. 21. Tables of coefficients are given in Ref. 8. Moments and defections in rectangular slabs can be computed by the ACI 318 procedure for two-way slabs.

21.3.3 Conical Hoppers

Walls of these structures are subjected to meridional and circumferential tensile membrane forces F_m and F_t (Fig. 21.19). Values of F_m and F_t per unit width at any horizontal section are given by

$$F_m = \frac{qD}{4\sin\alpha} + \frac{W}{\pi D\sin\alpha} \qquad (21.34)$$

$$F_t = \frac{q_\alpha D}{2\sin\alpha} \qquad (21.35)$$

where q, q_α = pressures computed at the section
D = diameter at the section
W = weight of that portion of hopper and hopper contents below the section

F_{mu} and F_{tu} for strength design are obtained by multiplying q and q_α by the load factor 1.7 and computing W by

$$W = 1.4W_h + 1.7W_m \qquad (21.36)$$

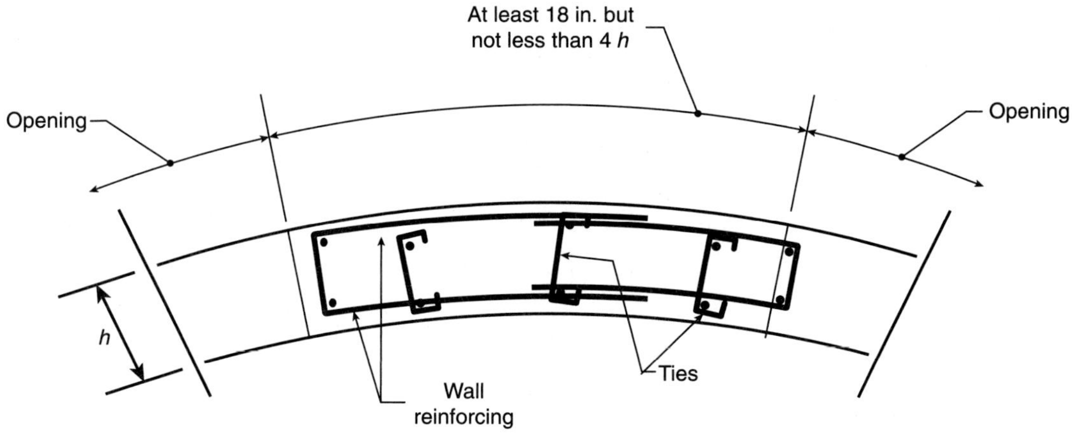

Figure 21.18 Typical reinforcing of narrow silo wall between openings.

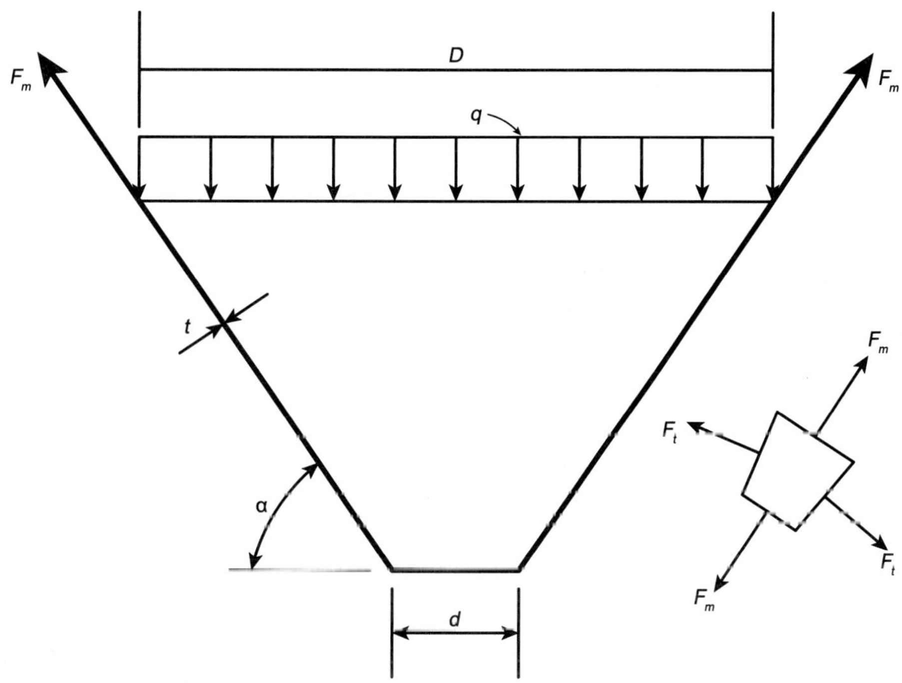

Figure 21.19 Forces in conical hopper.

where W_h = weight of hopper below the section and W_m = weight of hopper contents below the section. These weights are given by

$$W_h = \frac{\pi(D^2 - d^2)\gamma_h}{4\cos\alpha} \tag{21.37}$$

$$W_m = \frac{\pi(D^3 - d^3)\gamma_m \tan\alpha}{24} \tag{21.38}$$

where γ_h = weight of hopper wall per unit area and γ_m = weight of material per unit volume.

The required reinforcement is given by Eq. (21.20). A minimum wall thickness of 5 in. is recommended, and the crack width should not exceed an acceptable value.

Design of a conical hopper is illustrated in Example 3.

21.3.4 Pyramidal Hoppers

Walls of these structures are subjected to tensile membrane forces F_m and F_m (Fig. 21.20) and plate-type bending. There will also be in-plane bending if the hopper is not supported continuously along its upper edge.

If the vertical components of the meridional forces are assumed to be distributed uniformly on the perimeter of a horizontal section of a symmetrical hopper of rectangular cross section, F_m is given by

$$F_m = \frac{W + (q_a + q_b)ab/2}{2(a+b)\sin\alpha} \tag{21.39}$$

where a = length
b = width of section
q_a, q_b = vertical pressures corresponding to sides a and b, respectively
W = weight of that portion of hopper and hopper contents below section
a = angle with horizontal of hopper wall

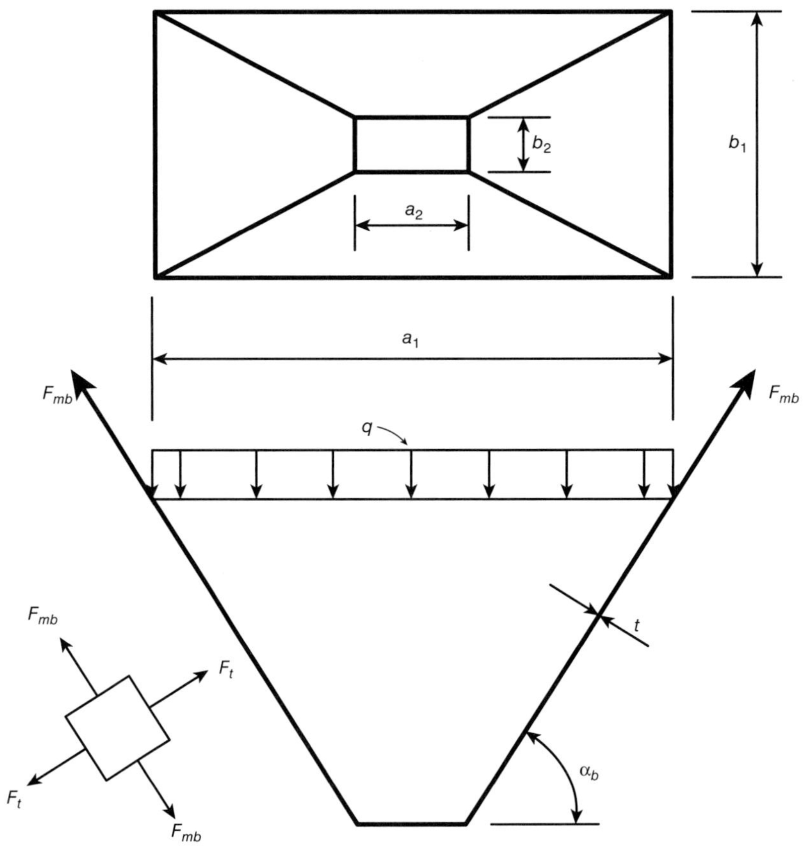

Figure 21.20 Forces in pyramidal hopper.

If $a > b$, $q_a < q_b$ (because the hydraulic radius is smaller for the longer side). This suggests that the vertical component of the meridional force on wall b may be larger than on wall a. Assuming $W/4$ and the resultant vertical pressure on the triangular area adjacent to each wall to be carried by that wall, the following formulas result:

$$F_{ma} = \left(\frac{W}{a} + q_a b\right)\frac{1}{4\sin\alpha_a} \tag{21.40a}$$

$$F_{mb} = \left(\frac{W}{b} + q_b a\right)\frac{1}{4\sin\alpha_b} \tag{21.40b}$$

The horizontal membrane force F_t is given by

$$F_{ta} = \frac{1}{2}(q_{ab} + \gamma_h \cos\alpha_b) b \sin\alpha_a \tag{21.41a}$$

$$F_{ta} = \frac{1}{2}(q_{aa} + \gamma_h \cos\alpha_a) \sin\alpha_b \tag{21.41b}$$

where γ_h = weight of hopper wall per unit of area.

F_{mu} and F_{tu} for strength design are obtained by multiplying q by the load factor 1.7 and computing W by Eq. (21.36). The weights W_h and W_m are given by

$$W_h = h_h \gamma_h \left(\frac{a_1 + a_2}{\sin\alpha_a} + \frac{b_1 + b_2}{\sin\alpha_b}\right) \tag{21.42}$$

$$W_m = \frac{h_h \gamma_m}{6}[(2a_1 + a_2)b_1 + (2a_2 + a_1)b_2] \tag{21.43}$$

where a_l = length
b_l = width of section at which F is being computed
a_2, b_2 = corresponding dimensions of hopper opening

Plate Bending under Normal Pressures. Bending moments in triangular walls may be approximated by the bending moments in the equivalent rectangular plate shown in Fig. 21.21*a*.[8] Tables of coefficients for the analysis of triangular walls with various types of edge support are available.[8]

Bending moments in trapezoidal walls for which $a_2/a_1 \geq 4$ can be approximated by the moments in the triangular wall formed by extending the sloping sides to their intersection[8] (Fig. 21.21*b*). Therefore, trapezoidal walls of these proportions can also be solved by using the equivalent rectangle of Fig. 21.21*a*.

Bending moments in trapezoidal walls for which $a_2/a_l < 4$ can be approximated by the moments in an equivalent rectangular wall (Fig. 21.21*c*) with the dimensions[8]

$$a_{eq} = \frac{2a_2(2a_1 + a_2)}{3(a_1 + a_2)} \tag{21.44a}$$

$$b_{eq} = h - \frac{a_2(a_2 - a_1)}{6(a_1 + a_2)} \tag{21.44b}$$

Tables of coefficients for the analysis of trapezoidal plates with $a_1 = \frac{3}{8}a_2$ and $a_1 = \frac{1}{2}a_2$ with various types of edge support are available.[8]

Edge conditions (fixed, simply supported, and so forth) of hopper walls depend on the adjoining construction. They should be considered fixed at their junctures with adjoining walls. In adjoining walls of unequal lengths, the average of the unequal end moments may be used, or they may be distributed in the ratios given by the negative-moment formula in Fig. 21.11; moments in the central regions of the plates should be adjusted to correspond. Upper edges that are continuous with silo or bunker walls, with or without an intervening edge beam, may be

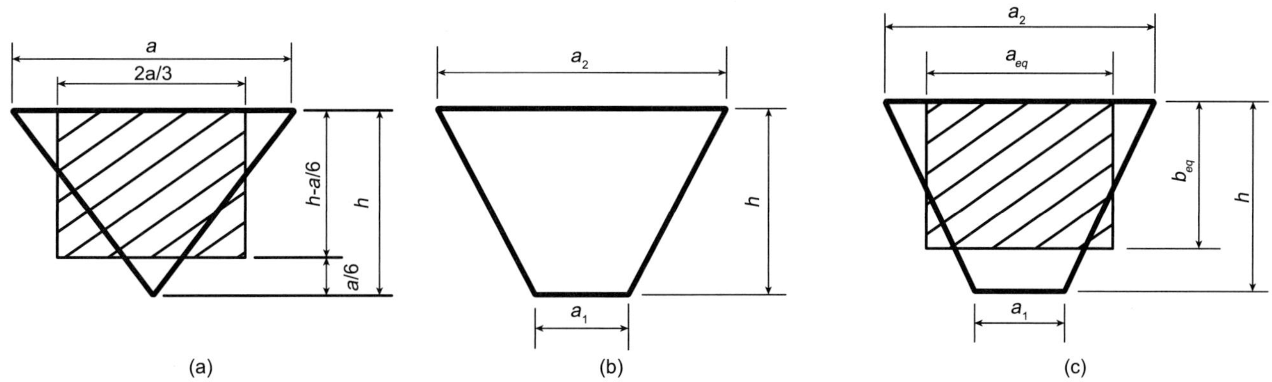

Figure 21.21 Triangular and trapezoidal walls approximated with equivalent rectangular plate.

assumed to be fixed. On the other hand, the upper edge of a pyramidal bunker that has no roof must be considered to be free.

Bending in-Plane of Wall. In-plane bending of a hopper wall acting in conjunction with a low vertical wall is discussed in Sec. 21.2.7 (Fig. 21.16). A similar analysis can be made for the bunker without vertical walls, using an effective beam depth of 0.5a at midspan[8] (Fig. 21.22). The stress in the portion below the effective-beam depth is assumed to vary linearly to zero at the vertex, as shown. The bending moment can be computed from $M = F_m a^2/8$, where F_m is the meridional tension computed by Eq. (21.39) or Eqs. (21.40). These bending stresses must be combined with the horizontal membrane forces to determine wall reinforcement.

Concentrated Forces at Pyramidal Bunker Supports. Pyramidal bunkers and pyramidal hoppers supported independently on columns at

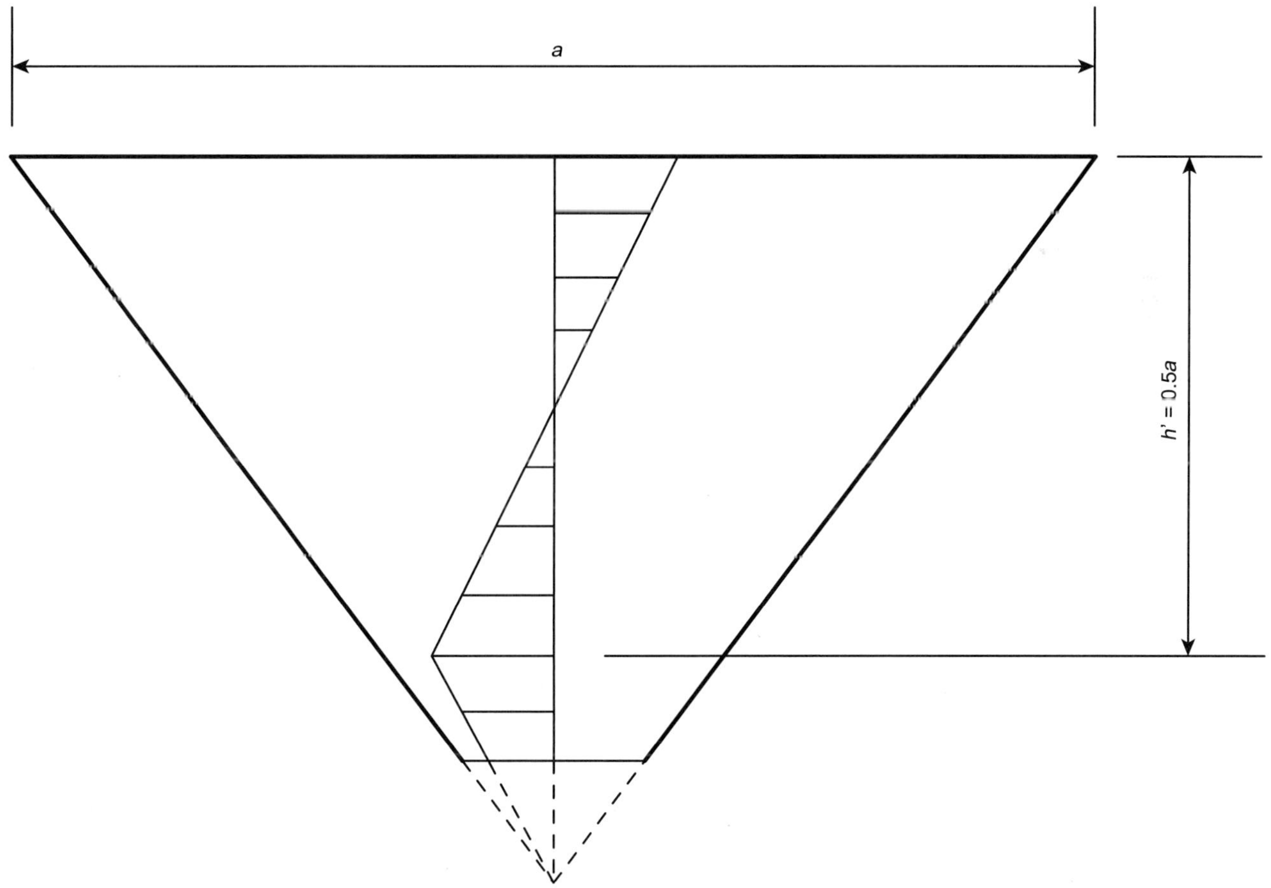

Figure 21.22 Effective beam width at midspan of a bunker without vertical walls.

the four corners are subjected to concentrated forces at the supports (Fig. 21.23). The tensile force T along the edge of adjoining walls is given by

$$T = \frac{P}{\sin\alpha} \tag{21.45}$$

where α is the angle of the edge with the horizontal.

The compressive forces are given by

$$C_a = P\cos\alpha \cos\beta_a \tag{21.46a}$$

$$C_b = P\cos\alpha \cos\beta_b \tag{21.46b}$$

where β_a and β_b are the angles between the edges a and b and the diagonal of the horizontal cross section of the hopper.

Since these forces are localized, they need to be provided for only in the vicinity of the column.

21.3.5 Hopper-Supporting Beams

Concrete hoppers are usually supported by edge beams cast integrally with the hopper wall (Fig. 21.24*a* and *b*). A conical steel hopper supported by a ring beam is shown in Fig. 21.24*c*. The beams may be supported continuously by a wall or by pilasters or columns.

If the hopper wall does not intersect the centroid of the supporting-beam cross section, a twisting moment acts on the beam. In ring beams this produces a bending moment that is uniform around the circumference. Both moments can be neglected if the beam is supported on a wall because the deformation that would be generated is prevented by the silo wall and the bearing wall. Therefore, such a ring beam can be designed for only the horizontal component of the hopper meridional tension, which produces a uniform compressive force, equal to (F_m cos a) (D/2), in the ring (Example 3).

Ring beams supported on columns or pilasters, as in Fig. 21.24*a*, must be designed for the moments shears and torsion resulting from F_m as well as the ring compression. A design procedure for this case is discussed by Safarian.[33]

No bending moments are produced by twisting of an eccentric edge beam of a rectangular hopper. Torsion can be neglected in edge beams on bearing walls. Edge beams on columns should be doweled to the silo walls to resist torsion.

Where supporting beams are not used and the hopper is keyed to the silo wall, reinforcement must be provided for the hopper-wall bending moments at the juncture.

21.3.6 Columns

Columns supporting silo shells and particularly silo bottoms will be subjected to a live load resulting from the stored material that is substantially larger than the dead load of the structure. Long sustained periods of material storage cause reinforced-concrete columns to creep. As a result, concrete stresses decrease, and the load carried by the steel reinforcement increases. Subsequent emptying of the silo may place the concrete in tension as the reinforcement recovers elastically.

Once the tensile stress in the concrete exceeds its tensile strength, the column develops severe transverse cracking. The situation can be dangerous if transverse cracking is accompanied by longitudinal cracking, as could occur with high bond stresses during unloading. To prevent this condition, it is recommended that the reinforcement ratio p not exceed 0.02 and that the total amount of reinforcement not exceed L/f_y,

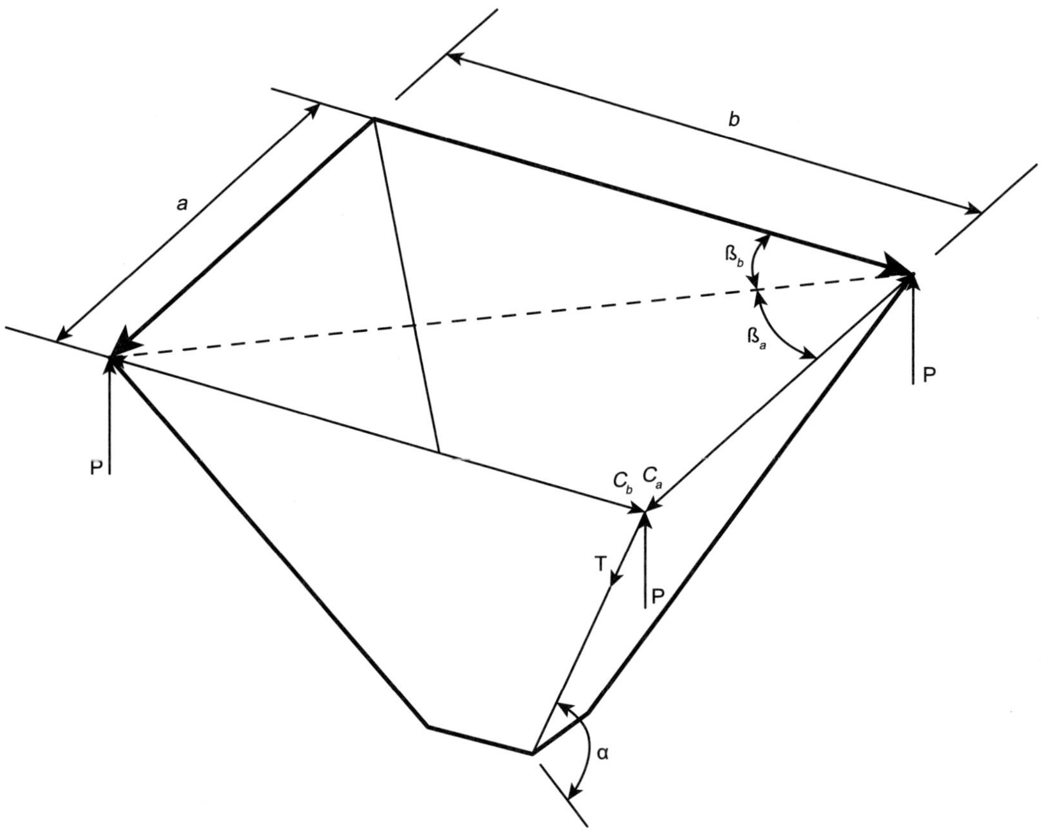

Figure 21.23 Pyramidal bunkers and pyramidal hoppers supported on columns at four corners.

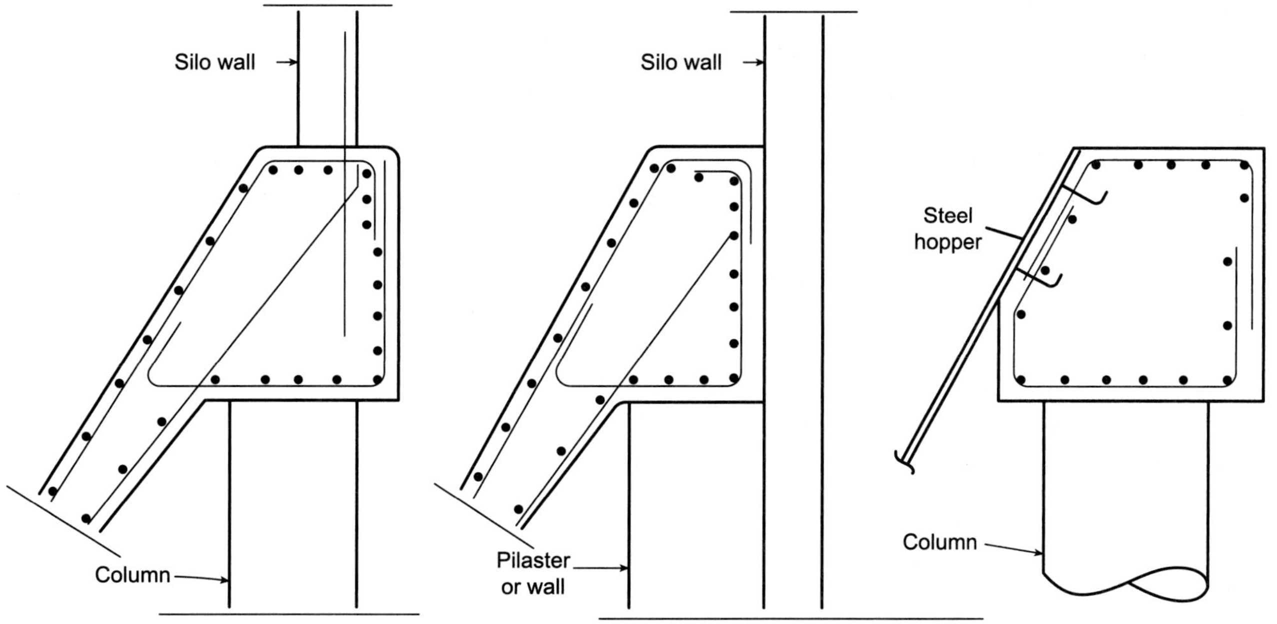

Figure 21.24 Typical hopper-supporting beams. (*From Ref. 13 and Ref. 15.*) For clarity, column and wall reinforcement are not shown.

where L is the live load on the column. If lateral loads must be resisted, larger columns should be used to keep the steel ratio low. All other provisions of ACI 318, Chapter 10[35] for the design of columns should be followed. Anchorage of the vertical reinforcement of all silo columns and walls to the supporting foundation is strongly recommended.

The analysis described here is illustrated in Example 4.

21.3.7 Roofs

Designers are divided on the subject of attachment of concrete roof slabs to silo walls. Some believe that the slab should be supported only vertically at the walls (on elastomeric material or heavy tar paper) so as to be free to contract or expand with temperature changes and to move slightly during earthquakes. To prevent total freedom of horizontal displacement, the slab may be attached at one central location, usually the elevator tower. On the other hand, attaching the roof slab to the walls stiffens tall silos against wind, earthquake loads, and vibrations during emptying and also reduces lateral deformations. Continuity with the silo walls also makes the roof slab stiffer and reduces vertical deflections under live load. Use of as much continuity as possible is strongly recommended.

In long installations, expansion joints are provided to reduce cracking of the slab in winter and undue longitudinal forces on the silos in summer. Good design calls for an expansion joint to cut across the silo group by extending down to the foundation, especially if the roof slab is attached to the silo walls.

The steel beams that support the wood platform during slipforming of silo walls are used later to support the roof slab, thus reducing its span considerably. Ample bearing on the concrete wall should be provided at the ends of the steel beams, and the concrete below and to each side of the beam should be reinforced to prevent undue cracking or even a concrete fallout after some years of service.

21.3.8 Failures

Three major reasons for failures of reinforced-concrete silos are foundation failure, incorrect determination of loads, and improper detailing and faulty workmanship.

Concrete silos are supported on pile foundations or on extended rafts in the case of a stiff subsoil. Theimer[2] cites various reasons for failures: (1) the weight is usually great and may shift considerably with unbalanced filling and emptying of the numerous cells, causing major overstress in the foundation; (2) dredging in an adjoining river may weaken the pile foundation; (3) batter piles may fail after having been damaged by a ship collision; (4) piles that have been eroded by aggressive groundwater may buckle; and (5) underlying soft soil may shift, causing tilting of the raft and elevator. Prevention of foundation failure requires thorough investigation of subsoil conditions including a number of testpile loadings. Thorough control and inspection of pile-driving operations and cast-in-place piles (especially those without steel shells to prevent mud intrusions) are necessary. Batter piles should always be used to support high silos against wind action and seismic motions. Raft foundations should be designed as continuously reinforced-concrete mats with two grids to resist bending moments caused by unbalanced loading of the silos.

Incorrect determination of silo pressures has resulted invariably in underreinforced structures. Application of the conventional Janssen theory without any allowance for overpressures generated during emptying and by eccentric outlets, combined with the use of higher allowable stresses for concrete and steel, has caused unacceptable major cracking and even total failure in numerous instances. ACI 313[13] and DIN 1055 will give a more accurate evaluation of load effects.

Major damage and even collapse of silos have resulted from improper detailing and faulty construction. Irregular and excessive spacing of hoop reinforcement, particularly in slipformed silos, may seriously reduce strength. In addition, radial displacement of hoop bars is frequent, and by reducing concrete cover, the capacity of the lap splices is limited. If hoop bars are placed without tying, circumferential shifts may occur, thus making some lap lengths shorter and some longer than the specified length. Bending moments that decrease the curvature of the walls induce radial tensions in the concrete that may cause cracking and circular delamination; radial reinforcement tying opposite hoop-steel layers may be required. Absence of vertical steel in combination with excessive spacing of hoop reinforcement may leave large portions of concrete unreinforced and cracked, a situation that may eventually result in concrete fallouts.[39]

Placement of two layers of hoop reinforcement in all silo walls is strongly recommended to prevent failure resulting from lack of adequate strength for simultaneous action of hoop tensions and either positive or negative bending moments.

Absence of special vertical reinforcement, required at an 8 × 8-ft door opening in a 90-ft-diameter concrete tank in West Africa, is cited by the author as an example where omission of original details (in the set of drawings prepared by the local contractor for use at the job site) could have had deleterious consequences.

Poor connections between auxiliary structural steel elements and supporting concrete walls are another source of potential failure that may not only result in local failure but actually cause total collapse of the silo. Thus, the author investigated the case of a suspended steel hopper that was poorly attached to the skirt walls of a tall concrete silo. Connection failure caused the suspended hopper to drop to the silo floor, instantly setting the overhead mass of grain in motion. The major lateral overpressure generated by the grain against the silo walls, while the vertical motion of the grain was being stopped at the floor, ripped the silo walls open from the bottom up.

Vertical reinforcement of walls and columns should be anchored to the foundation by dowels to prevent uplift and shifting under earthquake or wind loading. Insufficient cover for the hoop reinforcement causes it to corrode in a rather short period of time.[37,40] Sloppy workmanship may cause all splices of hoop reinforcement to be at the same locations without staggering, thereby increasing the possibility of bond failure (with splitting) and thus weakening the wall. Improper detailing at wall intersections of interstices and pocket bins may omit wall fillets and double layers of reinforcement that are necessary to resist local bending moments and shears. Proper attention to detail, followed by inspection at the job site and proper control and organization of construction, is required to prevent mistakes that may lead to collapse.

21.3.9 Dust Explosions in Grain Elevators and Flour Mills

Major destruction of these installations occurs when dust from grain products ignites and releases great amounts of energy. An extremely rapid pressure rise, of the order of 2000 tons/ft^2/s, originates a pressure wave of such high intensity that normal vents for the release of explosion pressures are insufficient to prevent the installation from blowing up. Theimer[41] cites three principal causes for these explosions: a dust cloud, a source of ignition, and the presence of oxygen.

It is necessary to have a minimum concentration of 0.02 oz/ft^3 of grain or flour dust (resembling a dense fog) before it can become ignited. If the concentration is greater than 2 oz/ft^3, incomplete combustion of the particles retards ignition and prevents the explosion. Dry dust that accumulates on floors, walls, ledges of doors and windows, steel beams, overhead ducts, and so forth is highly oxygenated and quite dangerous. Good housekeeping calls for constant removal of such dust. Suction is the principal method to control dust clouds by inducing air currents supplied through dust-collecting systems. Venting is also recommended for bins, heads of bucket elevators, and scale hoppers.

Ignition temperatures vary between 750°F and 930°F when the air relative humidity is between 30 and 90 percent. Sources of heat that can ignite dust clouds are (1) open flames (lights, matches, burning cigarettes); (2) heat generated on pulleys of bucket elevators resulting from belt slip; (3) hot surfaces of radiators, bearings, and lightbulbs; (4) sparks caused by metal parts in rotating machinery, electric equipment, and friction; (5) static electricity; and (6) welding, cutting, and soldering. In addition, Theimer[41] cites a case of spontaneous ignition resulting from constant increase in the temperature of the material caused by inadequate heat dispersion. Obviously, all prevention measures should be taken to avoid ignition heat sources.

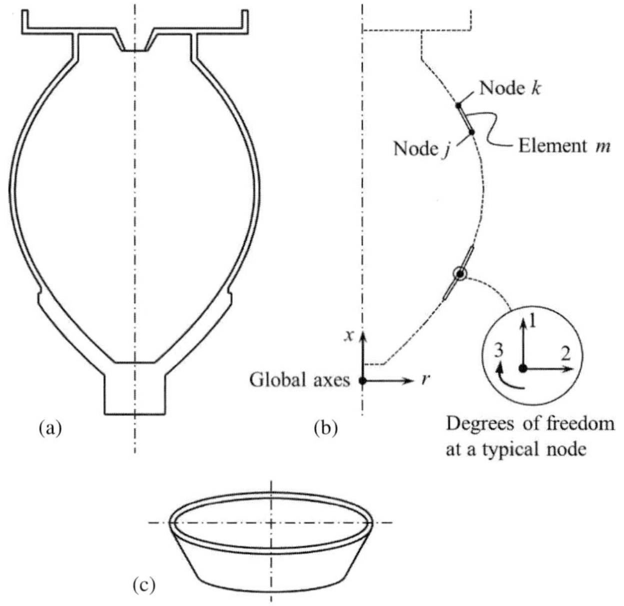

Figure 21.25 Idealization of axisymmetrical shell of revolution by conical shell element: (*a*) vertical section of egg-shaped digestor for sewage water treatment; (*b*) finite-element idealization; and (*c*) typical shell element.

Designers should consider the layout of various buildings in an installation to decrease its vulnerability to explosions as a whole. Theimer suggests leaving as large a space as possible between the various buildings. Explosion reliefs such as vents, light brick walls, and light roof construction should be provided.

21.3.10 Single-Cell Circular Silos: Finite-Element Analysis

by Ramez B. Gayed and Amin Ghali

Walls, bases, and covers of concrete silos—having the shape of axisymmetrically loaded circular shells—are analyzed as continuous structures. The shell is idealized as assemblage of finite elements as shown in Fig. 21.25. The relatively simple element used here is adequate for use in design. Shell of Revolution (SOR) is the name of a computer program that performs the analysis; an executable file is available at website of Ghali, A., 2014.[42] The equations used in SOR are based on assumed deformed shape functions of finite elements (Ghali and Neville, 2017).[43]

Nodal Displacements and Nodal Forces

Figure 21.26*a* depicts a typical element subjected to axisymmetrical pressure represented by forces $\{F^*\}$ = static equivalent load, q multiplied by the length of nodal lines 1 and 2:

$$\begin{Bmatrix} F_1^* \\ F_2^* \\ F_3^* \end{Bmatrix} = 2\pi r_1 \begin{Bmatrix} q_1^* \\ q_2^* \\ q_3^* \end{Bmatrix} \;;\; \begin{Bmatrix} F_4^* \\ F_5^* \\ F_6^* \end{Bmatrix} = 2\pi r_2 \begin{Bmatrix} q_4^* \\ q_5^* \\ q_6^* \end{Bmatrix} \tag{21.47}$$

The element nodal forces and displacements in local directions in Fig. 21.26*a* are related by geometry to forces and displacements in global directions (Fig. 21.26*b*):

$$\{D^*\} = [T]\{\bar{D}\} \;;\; \{\bar{F}\} = [T]^{\mathrm{T}}\{F^*\} \tag{21.48}$$

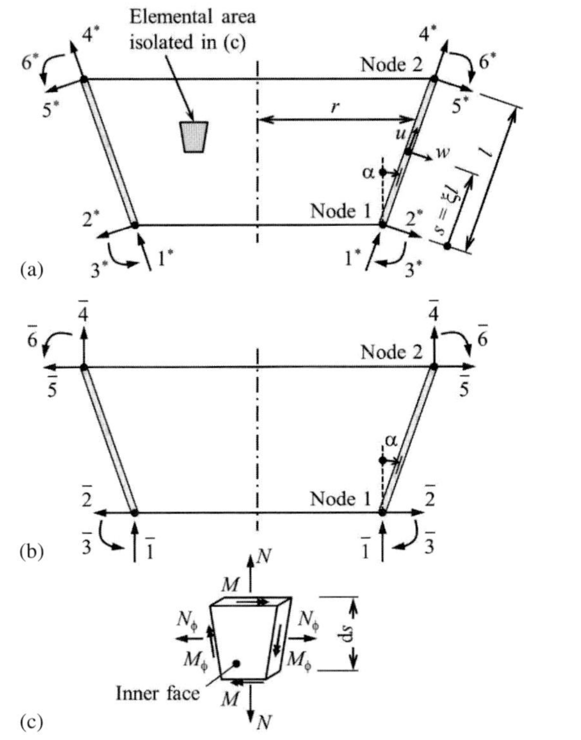

Figure 21.26 Typical finite element: (*a*) local coordinates; sign convention for u, w, s and α; (*b*) degrees of freedom in directions of global axes; order of numbering of coordinates at the nodes of a typical element; and (*c*) pictorial view of an elemental area showing the positive sign convention for N, M, N_ϕ, and M_ϕ.

where

$$[T]=\begin{bmatrix}[t] & [0]\\ [0] & [t]\end{bmatrix},\quad [t]=\begin{bmatrix}\cos\alpha & \sin\alpha & 0\\ -\sin\alpha & \cos\alpha & 0\\ 0 & 0 & 1\end{bmatrix} \tag{21.49}$$

The angle α and its positive sign convention are defined in Fig. 21.26*a* and *b*; the superscript T indicates matrix transposition.

The finite-element stiffness matrix $[S^*]$, to be derived in the section "Element Stiffness," relates the element nodal forces and displacements in the six local coordinates defined in Fig. 21.26*a*:

$$[S^*]\{D^*\}=\{F^*\} \tag{21.50}$$

Coordinates 2* and 5* represent translations or forces in the direction of a normal to the cone surface. Coordinates 3* and 6* represent rotations or moments in radial planes. D_3^* and D_6^*, representing rotations at the ends of a meridian, can be expressed as

$$D_3^*=\left(\frac{dw}{ds}\right)_{\xi=0},\quad D_6^*=\left(\frac{dw}{ds}\right)_{\xi=1} \tag{21.51}$$

where w = translation at any point in the direction of the normal to the shell (same direction as coordinates 2* and 5* in Fig. 21.26*a*) and $\xi = s/l$, with s being the distance from node 1 to any point on the meridian and l is the length of the meridian line 1–2.

Stiffness Matrix Transformation

Equation 52 transforms the stiffness matrix $[S^*]$ of individual element with respect to local coordinates to stiffness matrix $[\bar{S}]$ referring to coordinates in global directions (Fig. 21.26*a* and *b*):

$$[\bar{S}]=[T]^T[S^*][T] \tag{21.52}$$

where $[\bar{S}]$ = matrix relating $\{\bar{F}\}$ and $\{\bar{D}\}$,

$$[\bar{S}]\{\bar{D}\}=\{\bar{F}\} \tag{21.53}$$

Displacement Interpolation

The displacements $\{u, w\}$ are assumed to be related to the nodal displacements by

$$\begin{Bmatrix}u\\ w\end{Bmatrix}=\begin{bmatrix}1-\xi & 0 & 0 & \xi & 0 & 0\\ 0 & L_1 & L_2 & 0 & L_3 & L_4\end{bmatrix}\{D^*\} \tag{21.54}$$

where u = translation of any point in direction of the meridian (node 1 to node 2 in Fig. 21.26*a*); u is assumed to vary linearly between D_1^* and D_4^*. L_1 to L_4 are shape functions of ξ interpolating the deflection w between its nodal values to give the value at any point on the meridian line 1–2:

$$[L_1\ \ L_2\ \ L_3\ \ L_4]=\left[1-3\xi^2+2\xi^3\ \ \ l\xi(\xi-1)^2\ \ \ \xi^2(3-2\xi)\ \ \ l\xi^2(\xi-1)\right] \tag{21.55}$$

Stress Resultants

The stresses in meridian and circumference directions are related to strains by Eqs. (21.56–21.58):

$$\{\sigma\}=[d_e]\{\varepsilon\} \tag{21.56}$$

$$\{\sigma\}=\begin{Bmatrix}N\\ N_\phi\\ M\\ M_\phi\end{Bmatrix};\quad \{\varepsilon\}=\begin{Bmatrix}du/ds\\ (w\cos\alpha+u\sin\alpha)/r\\ -d^2w/ds^2\\ -(\sin\alpha/r)(dw/ds)\end{Bmatrix} \tag{21.57}$$

$$[d_e]=\frac{Eh}{1-\nu^2}\begin{bmatrix}1 & \nu & 0 & 0\\ \nu & 1 & 0 & 0\\ 0 & 0 & h^2/12 & \nu h^2/12\\ 0 & 0 & \nu h^2/12 & h^2/12\end{bmatrix} \tag{21.58}$$

where $\{\sigma\}$ and $\{\varepsilon\}$ = generalized stress and strain vectors defined by Eq. 21.57, $[d_e]$ = element's elasticity matrix (Eq. 21.58), N and M = the normal force and the moment in the meridian direction per unit length, N_ϕ and M_ϕ = the normal force and the moment in the circumferential (hoop) direction per unit length (Fig. 21.26*c*), h = element thickness, E = modulus of elasticity, and ν = Poisson's ratio. The positive sign conventions of the stress resultants are shown in Fig. 21.26*c*. Substitution of Eq. (21.54) in Eq. (21.57) gives the relationship of $\{\varepsilon\}$ to nodal displacements, $\{D^*\}$:

$$\{\varepsilon\}=[B]\{D^*\} \tag{21.59}$$

$$[B]=\begin{bmatrix} -\frac{1}{l} & 0 & 0 & \frac{1}{l} & 0 & 0 \\ \frac{\sin\alpha}{r}(1-\xi) & \frac{\cos\alpha}{r}L_1 & \frac{\cos\alpha}{r}L_2 & \frac{\sin\alpha}{r}\xi & \frac{\cos\alpha}{r}L_3 & \frac{\cos\alpha}{r}L_4 \\ 0 & \frac{1}{l^2}(6-12\xi) & \frac{1}{l}(4-6\xi) & 0 & \frac{1}{l^2}(12\xi-6) & \frac{1}{l}(2-6\xi) \\ 0 & \frac{\sin\alpha}{rl}(6\xi-6\xi^2) & \frac{\sin\alpha}{r}(4\xi-3\xi^2-1) & 0 & \frac{\sin\alpha}{rl}(6\xi^2-6\xi) & \frac{\sin\alpha}{r}(2\xi-3\xi^2) \end{bmatrix} \tag{21.60}$$

The stress resultants are given by

$$\{\sigma\}=[d_e][B]\{D^*\}+\{\sigma_r\} \tag{21.61}$$

where $\{\sigma_r\}$ = stress resultants with nodal displacements prevented and $\{\sigma_r\}$ = {0} when the analysis is for the effect of volume change due to temperature, shrinkage or swelling. The first term on the right-hand side of Eq. (21.61) is derived from Eqs. (21.56) and (21.59). Substituting ξ = 0.5 and $r = (r_1 + r_2)/2$ gives $\{\sigma\}$ midway between nodal lines of the element (results of the computer program SOR).

Element Stiffness

The element stiffness matrix with respect to its local coordinate is given by Eqs. (21.62) and (21.63):

$$[S^*]=2\pi l\int_0^1 r[B]^{\mathrm{T}}[d_e][B]\mathrm{d}\xi \tag{21.62}$$

$$r=(1-\xi)r_1+\xi r_2 \tag{21.63}$$

where r_1 and r_2 = radii at nodes 1 and 2. The value of the integral in Eq. (21.62) = $(g_1 + g_2)$, where g_1 and g_2 = values of the integral at Gauss sampling points: $\xi=(3-\sqrt{3})/6$ and $(3+\sqrt{3})/6$. Two sampling points give sufficient accuracy.

Effect of Temperature

Consider the effect of rise of temperature varying linearly through the thickness between T_i and T_o at inner and outer faces of the finite-element in Fig. 21.26*a*. The stress vector, $\{\sigma_r\}$ is calculated as

$$\{\sigma_r\}=-[d_e]\begin{Bmatrix} \mu(T_o-T_i)/2 \\ \mu(T_o+T_i)/2 \\ \mu(T_o-T_i)/h \\ \mu(T_o-T_i)/h \end{Bmatrix} \tag{21.64}$$

where $\{\sigma_r\}$ = generalized stress that would develop if the thermal expansion is arbitrarily restrained by forces $\{F_r^*\}$ at the six coordinates in Fig. 21.26*a*, where

$$\{F_r^*\}=2\pi l\int_0^1 r[B]^{\mathrm{T}}[\sigma_r]\mathrm{d}\xi \tag{21.65}$$

The equivalent restraining forces of individual element in global directions are

$$\{\bar{F}_r\}=[T]^{\mathrm{T}}\{F_r^*\} \tag{21.66}$$

The element-restraining forces are then assembled to give the restraining forces of the structure to be applied in reversed direction to eliminate the artificial restraint and give nodal displacements due to temperature. When the nodal displacements are determined, Eq. (21.61) gives the generalized thermal stress resultants in individual elements.

Nodal Forces Due to Distributed Loads

The distributed load in Fig. 21.27*a* is represented by statically equivalent nodal forces in Fig. 21.27*b*:

$$\{F^*\}=\frac{\pi l}{3}\begin{Bmatrix} 2q_{t1}r_1+q_{t2}r_2 \\ 2q_{n1}r_1+q_{n2}r_2 \\ 0 \\ q_{t1}r_1+2q_{t2}r_2 \\ q_{n1}r_1+2q_{n2}r_2 \\ 0 \end{Bmatrix} \tag{21.67}$$

where $\{F^*\}$ = equivalent nodal force per unit length in local coordinates (Fig. 21.27*c* and *d*) and q_t and q_n = intensity of tangential and normal loads on an individual element; it is assumed that the product (qr) varies linearly over the length l. The subscripts 1 and 2 refer to nodes 1 and 2.

Assemblage of Stiffness Matrices and Nodal Forces

The displacements at the nodes, {D} in global directions are determined by solution of the equilibrium equation

$$[S]\{D\} = \{F\} \tag{21.68}$$

where $[S]$ = stiffness matrix of the structure, $\{F\}$ = a vector of nodal forces in global directions = the sum of applied nodal forces and the forces restraining thermal expansion in reversed direction. Before solving Eq. (21.68), $[S]$ must be adjusted to satisfy prescribed conditions at specified nodes, for example, zero displacement at support.

EXAMPLE 1

Determine the hoop and meridional reinforcement for a 30-ft-diameter silo 100 ft high with an 8-in. wall and a conical hopper 20 ft deep to contain wheat (Fig. 21.2*a*). Use ACI 313 (Sec. 21.1.4) with f_c' = 4000 psi and A615 Grade 60 deformed bars.

From Table 21.2, obtain ϕ = 23° and μ' = 0.25 for soybeans and ϕ = 23 to 37° and μ' = 0.29 to 0.47 for wheat and corn. Use γ = 50 lb/ft³. For maximum values of q, p, and V, as required for design, it is necessary to use the following combinations of variables: μ'_{min} (0.29) and k_{min} (ϕ_{max} = 37°) for q_{max}, μ'_{min} (0.25) and k_{max} (ϕ_{min} = 23°) for p_{max}; and μ'_{max} (0.47) and k_{max} (ϕ_{min} = 23°) for V_{max}. Soybeans govern p_{max} because of their low value of μ', whereas wheat and corn determine q_{max} and V_{max}.

The basic pressures and forces are obtained from Janssen's formulas (Table 21.1) as follows: For q_{max} use

$$R=D/4=30/4=7.5\ ft, k=1-\sin\phi=1-\sin 37°=0.3982$$
$$Y_0=R/\mu' k=7.5/0219\times 0.3982=64.95\ ft$$
$$q=\gamma Y_0(1-e^{-Y/Y_0})=50\times 64.95(1-e^{-Y/64.95})=3248(1-e^{-Y/64.95})$$

For p_{max} use

$$R=7.5\text{ ft}, k=1-\sin 23°=0.6093, Y_0=7.5/0.25\times 0.6093=49.24\text{ ft}$$
$$p=k\gamma Y_0(1-e^{-Y/Y_0})=0.6093\times 50\times 49.24(1-e^{-Y/49.24})=1500(1-e^{-Y/49.24})$$

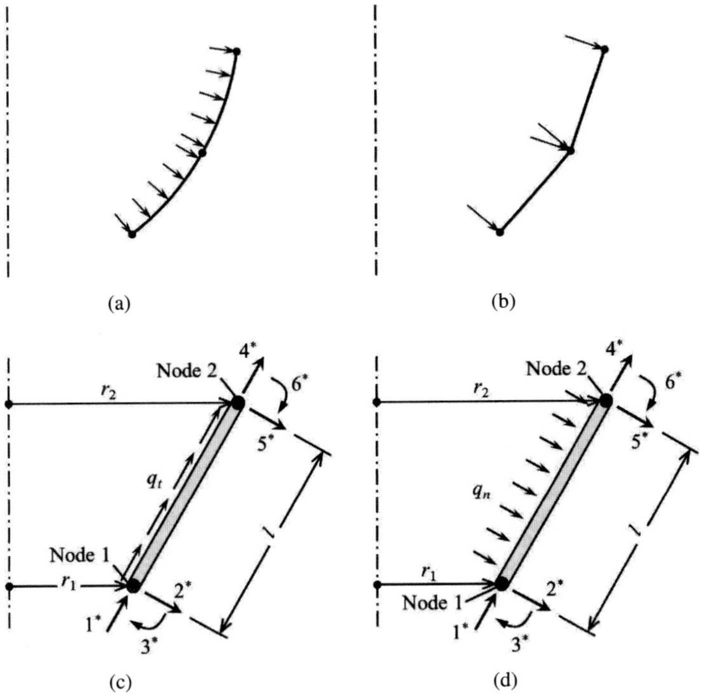

Figure 21.27 Nodal forces equivalent to a distributed load on the surface of an axisymmetrical finite element: (*a*) actual load on a shell with a curved meridian; (*b*) idealization as conical elements loaded only at the nodes; and (*c, d*) equivalent nodal forces in local coordinates.

For V_{max} use

$$R = 7.5 \text{ ft}, k = 1 - \sin 23° = 0.6093, Y_0 = 7.5/0.47 \times 0.6093 = 26.19 \text{ ft}$$

$$V = (\gamma Y - q)R \text{ where } q = 50 \times 26.19(1 - e^{-Y/26.19}) = 1310(1 - e^{-Y/26.19})$$

Basic pressures and forces are shown in Table 21.10. Basic lateral pressures are multiplied by the overpressure factor, $C_d = 1.5$, to obtain design lateral pressures, p_{des}. The results are given in Table 21.10. Note that p_{des} is used for calculating hoop reinforcement of silo walls, q_{max} is used for design of hoppers, and V_{max} is used for checking wall thickness and vertical reinforcement of the silo walls acting as vertical walls.

Hoop reinforcement is determined in Table 21.11, using the design pressures of Table 21.10. Hoop forces $T = p_{des} D/2$ and $T_u = 1.7T$. The required steel area is given by $A_s = T_u/(0.9 \times 60)$. Because two layers of hoop reinforcement are used, spacing of the No. 4 bars (0.20 in.²) may be computed using $s = 2 \times 0.20 \times 12/A_s$ (see Table 21.11). Placement of the reinforcement is as shown in Fig. 21.13. Note that there are two layers of hoop reinforcement and two layers of vertical steel bars. The section in Fig. 21.13 corresponds to the lower portion of the walls, where the spacing of the hoop reinforcement is 5 in.

Table 21.10 Values of *p*, *q*, and *V* for Silo of Example 1 ($C_d = 1.5$)

	Basic pressures and forces			Design lateral pressures
Depth *Y*, ft	*q*, lb/ft²	*p*, lb/ft²	*v*, kips/ft	$P_{des} = C_d p$ lb/ft²
0	0	0	0	0
10	463	276	0.63	414
20	861	501	2.26	751
30	1201	684	4.55	1027
40	1493	834	7.31	1251
50	1744	957	10.39	1435
60	1958	1057	13.67	1585
70	2142	1138	16.64	1707
80	2300	1205	20.64	1807
90	2435	1259	24.25	1888
100	2551	1303	27.89	1955

Table 21.11 Hoop Reinforcement for Silo of Example 1

Depth	Design pressure	Hoop forces		Steel required	Steel provided	Bar spacing
Y ft	p_{des} lb/ft²	$T = \frac{p_{des}D}{2}$ kips/ft	$T_u = 1.7T$ kips/ft	$A_s = \frac{T_u}{0.9 f_y}$ in./ft	A_s, in.²/ft	Two layers of No. 4s *s* in.
0	0	0	0	0	0.30	16
10	414	6.21	10.56	0.20		
20	751	11.27	19.16	0.35	0.48	10
30	1027	14.51	26.19	0.49		
40	1251	18.77	31.90	0.59	0.69	7
50	1435	21.53	36.59	0.68		
60	1585	23.78	40.42	0.75	0.81	6
70	1707	25.61	43.53	0.81		
80	1807	27.11	46.08	0.85	0.96	5
90	1888	28.32	48.14	0.89		
100	1955	29.33	49.85	0.92		

Table 21.12 Design Crack Width for Silo of Example 1

Depth, Y ft	Basic pressure, p lb/ft^2	Filling hoop tension, $T = \frac{pD}{2}$ k/ft	Hoop steel, 2 No. 4, spaced at s in.	Area of hoop steel, A_s in.^{2}/ft	Steel stress $f_s = T/A_s$ k/in.^{2}	Design crack width, w^* in.
10	276	4140	16	0.30	13.80	0.0051
30	684	10,260	10	0.48	21.38	0.0067
50	957	14,360	7	0.69	20.80	0.0058
70	1138	17,070	6	0.81	21.07	0.0056
100	1303	19,550	5	0.96	20.36	0.0051

$^*w = 0.0001 f_s^3 \sqrt{12.5\, d^2\, s}$, d $d = 0.5$ in.

EXAMPLE 2

Determine the design crack width in the walls of the silo of Example 1. From the data of Table 21.10, obtain the basic (initial, filling) lateral pressures and from Table 21.11 the hoop reinforcement along the height of the silo; see Fig. 21.13. Calculation of the required design crack is accomplished using Table 21.12. The maximum value, $w = 0.0067$ in., is found at a distance $Y = 30$ ft below the top of the silo. The design is acceptable because $w < 0.010$ in. through the height of the silo walls.

EXAMPLE 3

Design a conical hopper for a 30-ft-diameter silo with design pressures as indicated in Table 21.13. Use ACI 313 provisions. The hopper is 8 in. thick and 20 ft high and has top and bottom diameters of 30 and 3 ft, respectively.

The tangential and meridional forces are given by Eqs. (21.34), (21.35), and (21.36), with W_h and W_m by Eqs. (21.37) and (21.39). Values of q_a are computed by Eq. (21.33).

Results are given in Table 21.13. Areas of reinforcement per unit length required for strength are given in the last two columns. Some adjustment upward was necessary to satisfy a crack width $w < 0.010$ in. This was easily accomplished by increasing the length of the meridional bars and reducing the spacing of the hoop bars.

RING BEAM (Sec. 21.3.5). From Table 21.13, $F_{mu} = 55.4$ kips at the top of the hopper. From Fig. 21.28, the slope of the hopper wall is 56.7°. Therefore, the ring compression is

$$P = (F_{mu}\cos\alpha)(D/2) = (55.4\cos 56.7°)(30/2) = 456 \text{ kips}$$

The ring shown in Fig. 21.25 is 15 × 20 in. with 8 No. 6 bars. Then,

$$P_u = \phi[0.85 f_c'(A_g - A_s) + f_y A_s] = 0.7[0.85 \times 4(300 - 3.52) + 60 \times 3.52]$$
$$= 853 \text{ kips}$$

Although P_u is considerably larger than P, use of a smaller ring is questionable. The 15-in. width gives projections to facilitate forming, and the depth gives bending strength to bridge openings that might later be cut into the bearing wall.

Instead of a fully suspended hopper, as above, the designer may find a more economical solution is possible if another ring (column-supported at an intermediate level between the top and the bottom of the hopper) is provided.

EXAMPLE 4

A 20-in.^{2} reinforced-concrete column with 4 No. 18 bars and No. 4 ties at 12 in. is one of a group supporting a silo hopper as shown in Fig. 21.2e. The column loads are $D = 150$ kips and $L = 600$ kips. $f_c' = 3$ ksi, $f_y = 60$ ksi, $E_c = 3000$ ksi, and $E_s = 29{,}000$ ksi. Unloading of the silo may occur a long time after filling. Check the suitability of the design (Sec. 21.3.6).

The initial strain ε_i in the concrete is

$$\varepsilon_i = \frac{D + L}{[A_g + (E_s/E_c - 1)A_s]E_c} = \frac{150 + 600}{[20 \times 20 + (29{,}000/3000 - 1)16]3000}$$
$$= 0.000464$$

Assume the strain has trebled because of creep. Thus, $\varepsilon_t = 3 \times 0.000464 = 0.0014$. The stress in the steel is $f_s = \varepsilon_t E_s = 0.0014 \times 29{,}000 = 40.6$ ksi and can support a load $P_s = f_s A_s$ $A_s = 40.6 \times 16 = 649.6$ kips. The load supported by the concrete is $P_c = D + L - P_s = 150 + 600 - 649.6 = 100.4$ kips. The stress in the concrete is $P_c/(A_g - A_s) = 100.41(20 \times 20 - 16) = 0.261$ ksi.

Upon removal of the live load, elastic unloading occurs. The strain $\Delta\varepsilon$ that is recovered is given by

$$\Delta\varepsilon = \varepsilon_i \frac{L}{D + L} = 0.000464 \times \frac{600}{150 + 600} = 0.000371$$

The stress in the concrete is reduced from 0.261 in compression to $0.261 - 0.000371 \times 3000 = 0.852$ ksi in tension. This exceeds the cracking strength of the concrete, estimated at $4.5\sqrt{f_c'} = 4.5\sqrt{3000} = 246$ psi. As a result, concrete will crack during unloading of the silo. The stress in the steel after unloading is given by $f_s = D/A_s = 150/16 = 9.38$ ksi.

To avoid cracking, the amount of steel in the column should be limited to the smaller of $0.02A_c$ or L/f_y. This gives $0.02 \times 20 \times 20 = 8$ in.^{2} and $600/60 = 10$ in.^{2}. Thus, A_s for this column should be instead of 16 in.^{2} This reduction requires $f_c' = 4000$ psi instead of 3000 psi.

EXAMPLE 5

(See Sec. 21.1.8) Determine the thermal reinforcement required for a concrete silo in a region where $T_o = -20$°F. The silo has doubly reinforced walls 8 in. thick and stores cement for which $T_i = 400$°F, $f_c' = 4000$ psi, $f_y = 60{,}000$ psi.

Table 21.13 Design of Conical Hopper for a 30-ft-Diameter Silo

Depth Y	Design pressures q_{des}, lb/ft^2	p_{des}, lb/ft^2	$q_{a,des}$, lb/ft^2	Tangential force F_{tu}, lb/ft	Meridional force F_{mu}, lb ft	Tangential steel $A_{s,t}$, in.^{2}/ft	Meridional steel $A_{s,m}$, in.^{2}/ft
0	3117	1125	1725	52,600	55,400	0.97	1.03
5	3158	1140	1748	41,300	42,500	0.76	0.79
10	3198	1155	1771	29,700	29,800	0.55	0.55
15	3231	1167	1789	17,700	17,400	0.33	0.32
20	3261	1178	1806	5000	5000	0.10	0.09

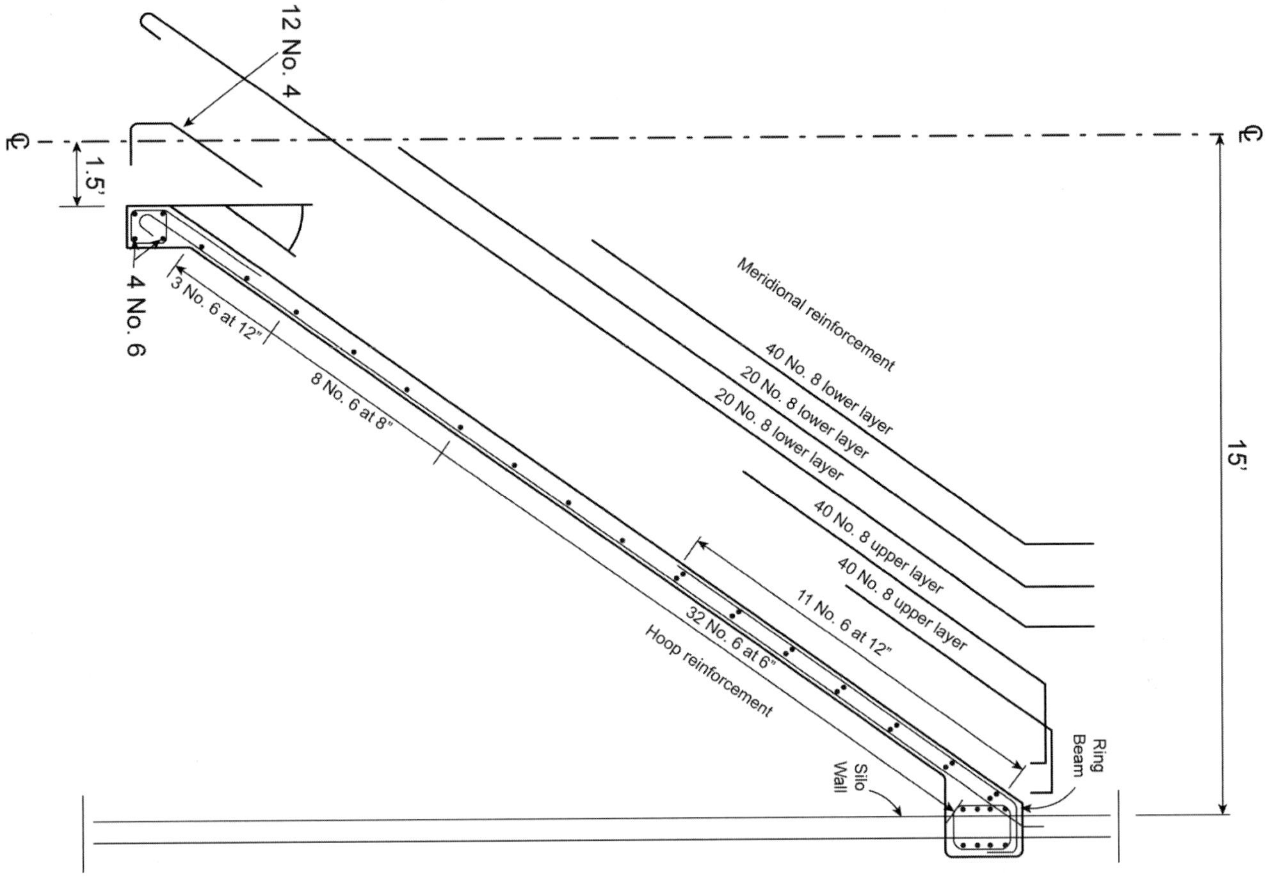

Figure 21.28 Conical hopper and ring beam, Example 3.

Using Eqs. (21.17*b*), (21.16), and (21.15) gives

$$K_t = \frac{0.08 \times 8}{4.09 + 0.08 \times 8} = 0.135$$

$$\Delta T = [400 - 80 - (-20)]0.135 = 46°$$

$$M_t = \frac{57{,}000\sqrt{4000} \times 8^2 \times 5.5 \times 10^{-6} \times 46}{12(1 - 0.3)} = 6949 \text{ lb in./in.}$$

Assume net covers of 1.5 and 0.75 in., respectively, for the exterior and interior layers of hoop reinforcement. If No. 5 bars are used, then $d = 8 - 1.5 - 0.62/2 = 6.2$ in. and $d' = 0.75 + 0.62/2 = 1.1$ in. Then from Eq. (21.32)

$$A_{s,t} = \frac{1.4 \times 6949}{0.9 \times 60{,}000(6.2 - 1.1)} = 0.035 \text{ in.}^2/\text{in.} = 0.42 \text{ in.}^2/\text{ft}$$

Add this amount to the outer layer and, for the sake of equal spacing, the same amount to the inner layer.

EXAMPLE 6—STRESS RESULTANTS IN A CIRCULAR SILO: FINITE-ELEMENT ANALYSIS

Determine N_ϕ and M for a 30-ft (9.1 m)-diameter concrete silo 140 ft (42.7 m) high with an 8-in. (203 mm) wall and a conical hopper 20 ft (6.1 m) deep. The wall is continuous with a 10-in. (254 mm)-thick × 40-ft (12.2 m)-high cylindrical wall (Fig. 21.29*a*). Properties of contained soybean: specific weight, $\gamma = 50$ lb/ft³ (7854 N/m³); angle of internal friction, $\phi = 23°$; angle of repose, $\rho = 25°$; friction coefficient, $\mu' = 0.25$. Other data: modulus of elasticity, $E = 3420$ ksi (23.6 GPa); Poisson's ratio, $\nu = 0.2$. Consider only the loading from the contained material using Janssen's equations (Fig. 21.29*b*). Assume total fixity at the base.

Solution The finite-element model is shown in Fig. 21.29*c*. The input file "SOR.in" is given below; the results are generated in file "SOR. out" (not given). The analysis is done for Janssen's pressures; for guidance, the pressures according to Janssen's and other sources are plotted in Fig. 21.29*b*. Figure 21.29*d* shows the variations of N_ϕ and M.

Input file "SOR.in" The following set of lines form the input file: "SOR.in" used in the analysis of the silo.

Example 6. Circular silo for soybean. Finite-element analysis. Units: lb and in.

72	71	1	1	No. of joints, No. of elements, No. of supported joints, No. of load cases
0.0				Thermal expansion coefficient
1	.000	183.000		Node no., x and r coordinates
2	24.000	183.000		
3	48.000	183.000		
4	72.000	183.000		
5	96.000	183.000		
6	120.000	183.000		
7	144.000	183.000		
8	168.000	183.000		
9	192.000	183.000		
10	216.000	183.000		
11	240.000	183.000		

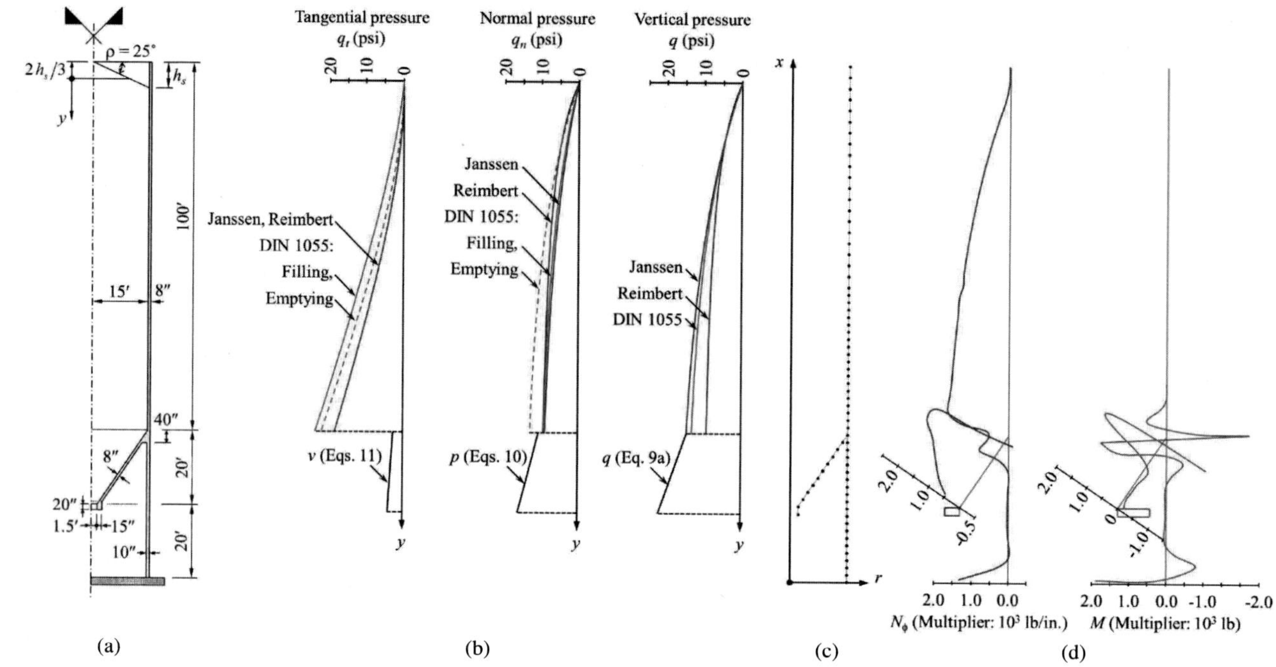

Figure 21.29 (*a*) Geometry of circular silo, Example 6; (*b*) pressures on walls and hopper; (*c*) finite-element model (SOR); and (*d*) variations of N_ϕ and M in the silo due to pressures according to Janssen.

12	264.000	183.000
13	288.000	183.000
14	312.000	183.000
15	336.000	183.000
16	360.000	183.000
17	384.000	183.000
18	408.000	183.000
19	432.000	183.000
20	456.000	183.000
21	480.000	184.000
22	220.000	25.500
23	246.666	25.500
24	270.000	41.250
25	293.333	57.000
26	316.666	72.750
27	340.000	88.500
28	363.333	104.250
29	386.666	120.000
30	410.000	135.750
31	433.333	151.500
32	456.666	167.250
33	504.000	184.000
34	528.000	184.000
35	552.000	184.000
36	576.000	184.000
37	600.000	184.000
38	624.000	184.000
39	648.000	184.000
40	672.000	184.000
41	696.000	184.000
42	720.000	184.000
43	744.000	184.000
44	768.000	184.000
45	792.000	184.000
46	816.000	184.000
47	840.000	184.000
48	864.000	184.000
49	888.000	184.000
50	912.000	184.000
51	936.000	184.000
52	960.000	184.000
53	996.000	184.000
54	1032.000	184.000
55	1068.000	184.000
56	1104.000	184.000
57	1140.000	184.000
58	1176.000	184.000
59	1212.000	184.000
60	1248.000	184.000
61	1284.000	184.000
62	1320.000	184.000
63	1356.000	184.000
64	1392.000	184.000
65	1428.000	184.000
66	1464.000	184.000
67	1500.000	184.000
68	1536.000	184.000
69	1572.000	184.000
70	1608.000	184.000
71	1644.000	184.000
72	1680.000	184.000

1	1	2	10.0	1.0	0.2	Element no., starting node, end node, thickness, E, Poisson's ratio
2	2	3	10.0	1.0	0.2	
3	3	4	10.0	1.0	0.2	
4	4	5	10.0	1.0	0.2	
5	5	6	10.0	1.0	0.2	
6	6	7	10.0	1.0	0.2	
7	7	8	10.0	1.0	0.2	
8	8	9	10.0	1.0	0.2	
9	9	10	10.0	1.0	0.2	

```
10   10   11   10.0   1.0   0.2
11   11   12   10.0   1.0   0.2
12   12   13   10.0   1.0   0.2
13   13   14   10.0   1.0   0.2
14   14   15   10.0   1.0   0.2
15   15   16   10.0   1.0   0.2
16   16   17   10.0   1.0   0.2
17   17   18   10.0   1.0   0.2
18   18   19   10.0   1.0   0.2
19   19   20   10.0   1.0   0.2
20   20   21   10.0   1.0   0.2
21   22   23   15.0   1.0   0.2
22   23   24   8.0    1.0   0.2
23   24   25   8.0    1.0   0.2
24   25   26   8.0    1.0   0.2
25   26   27   8.0    1.0   0.2
26   27   28   8.0    1.0   0.2
27   28   29   8.0    1.0   0.2
28   29   30   8.0    1.0   0.2
29   30   31   8.0    1.0   0.2
30   31   32   8.0    1.0   0.2
31   32   21   8.0    1.0   0.2
32   21   33   8.0    1.0   0.2
33   33   34   8.0    1.0   0.2
34   34   35   8.0    1.0   0.2
35   35   36   8.0    1.0   0.2
36   36   37   8.0    1.0   0.2
37   37   38   8.0    1.0   0.2
38   38   39   8.0    1.0   0.2
39   39   40   8.0    1.0   0.2
40   40   41   8.0    1.0   0.2
41   41   42   8.0    1.0   0.2
42   42   43   8.0    1.0   0.2
43   43   44   8.0    1.0   0.2
44   44   45   8.0    1.0   0.2
45   45   46   8.0    1.0   0.2
46   46   47   8.0    1.0   0.2
47   47   48   8.0    1.0   0.2
48   48   49   8.0    1.0   0.2
49   49   50   8.0    1.0   0.2
50   50   51   8.0    1.0   0.2
51   51   52   8.0    1.0   0.2
52   52   53   8.0    1.0   0.2
53   53   54   8.0    1.0   0.2
54   54   55   8.0    1.0   0.2
55   55   56   8.0    1.0   0.2
56   56   57   8.0    1.0   0.2
57   57   58   8.0    1.0   0.2
58   58   59   8.0    1.0   0.2
59   59   60   8.0    1.0   0.2
60   60   61   8.0    1.0   0.2
61   61   62   8.0    1.0   0.2
62   62   63   8.0    1.0   0.2
63   63   64   8.0    1.0   0.2
64   64   65   8.0    1.0   0.2
65   65   66   8.0    1.0   0.2
66   66   67   8.0    1.0   0.2
67   67   68   8.0    1.0   0.2
68   68   69   8.0    1.0   0.2
69   69   70   8.0    1.0   0.2
70   70   71   8.0    1.0   0.2
71   71   72   8.0    1.0   0.2
1   0   0      0   0.      0.   0.   Node no., restraint indicators,
1   1   0.     0.  0.                global displs. 1, 2, and 3 Ld case 1,
10  0   0.     0.  0.                node, F1 (up), F2 (horiz. out), F3
1   70  0.003      0.140             (radial moment) Dummy end of
1   69  0.035      0.579             data of loaded nodes Ld. case,
1   68  0.123      1.160             element no., qt, qn
1   67  0.266      1.707
1   66  0.463      2.222
1   65  0.710      2.706
1   64  1.003      3.162
1   63  1.341      3.591
1   62  1.721      3.994
1   61  2.139      4.374
1   60  2.595      4.731
1   59  3.085      5.067
1   58  3.607      5.383
1   57  4.161      5.681
1   56  4.743      5.961
1   55  5.352      6.224
1   54  5.987      6.472
1   53  6.646      6.705
1   52  7.328      6.925
1   51  7.911      7.098
1   50  8.389      7.23
1   49  8.875      7.357
1   48  9.369      7.479
1   47  9.872      7.596
1   46  10.382     7.708
1   45  10.9       7.816
1   44  11.424     7.92
1   43  11.955     8.019
1   42  12.493     8.114
1   41  13.037     8.206
1   40  13.587     8.294
1   39  14.143     8.378
1   38  14.704     8.46
1   37  15.271     8.538
1   36  15.843     8.612
1   35  16.419     8.684
1   34  17.000     8.753
1   33  17.586     8.819
1   32  18.176     8.883
1   31  2.714      10.957
1   30  2.840      11.465
1   29  2.965      11.974
1   28  3.091      12.482
1   27  3.217      12.99
1   26  3.343      13.498
1   25  3.469      14.006
1   24  3.595      14.514
1   23  3.72       15.022
1   22  3.841      15.509
1   21  3.956      15.974
10  0   .0         0.        Dummy end of data of loaded elements
1   1   .0         0.        Load case, element no., Ti, To (temp. inner
                             and outer faces)
10  0   .0         0.        Dummy end of thermal load
0                            Integer 1 or 0 for longer or shorter output,
                             respectively
```

REFERENCES

1. Janssen, H. A.: "Versuch aber Getreide-druck in Silozellen," *VDI Z* (Dusseldorf), vol. 39, August 31, 1895.
2. Theimer, O. F.: "Failures of Reinforced Concrete Grain Silos," *ASME Publ.* 68-MH-36, New York, 1968.
3. Kovtun A. P., and P. N. Platonov: "The Pressure of Grain on Silo Walls," *Mukomol'no Elevat. Promst.* Moscow, USSR, vol. 25, no. 12, December 1959.
4. Pieper, K., C. Mittelman, and F. Wenzel: "Messungen des Horizontalen Getreidedruckes in einer 65 m hohen Silozelles," *Beton und Stahlbetonbau*, Berlin, vol. 59, no. 1, November 1964.
5. Pieper, K.: "Investigation of Silo Loads on Measuring Models," *J. Eng. Ind. Trans. ASME*, May 1969.

6. Reimbert, Marcel, and Andre Reimbert: "Silos-Traite Theoretique et Practique," Editions Eyrolles, Paris, 1961.

7. Jenike, A. W., and J. R. Johanson: "Bin Loads," *J. Struct. Div. ASCE*, April 1968.

8. Fischer, W.: "Silos und Bunker in Stahlbeton," VEB Verlag fir Bauwesen Berlin, DDR, 1966.

9. Lipnitski, M. E., and S. P. Abramovitsch: "Reinforced Concrete Bunkers and Silos" (in Russian), Izdatelstvo Literaturi Po Stroitelstvu, Leningrad, 1967.

10. Turitzin, A. M.: "Dynamic Pressure of Granular Material in Deep Bins," *J. Struct. Div. ASCE*, April 1963.

11. Homes, A. G.: "Lateral Pressures of Granular Materials in Silos," *ASME Publ.* 72-MH-30 New York 1972.

12. "Lastannahmen for Bauten. Lasten in Silozellen," DIN 1055 Sheet 6, May 1987.

13. ACI: Standard Practice for Design and Construction of Concrete Silos and Stacking Tubes for Storing Granular Materials, (ACI 313–91) and Commentary—ACI 313R-91 American Concrete Institute, Detroit, MI, 1991.

14. Gurfinkel, G. R.: "Simple Design of Concrete Walls and Structural Members under Combined Tension and Moment," *Structural Journal ACI*, 1995.

15. ACI: Standard Practice for Design and Construction of Concrete Silos and Stacking Tubes for Storing Granular Materials, (ACI 313–16) and Commentary—ACI 313R-16 American Concrete Institute, Farmington Hills, MI, 2016.

16. Jenike, A. W.: Storage and Flow of Solids, Bulletin no. 123, Engineering Experiment Station, University of Utah, Salt Lake City, November 1964.

17. Jenike, A. W.: "Quantitative Design of Mass-Flow Bins," *Powder Technology* (Lausanne), vol. 1. 1967, pp. 237–244.

18. Gaylord, E. H., and Gaylord, C. N.: *Design of Steel Bins for Storage of Bulk Solids*, Prentice Hall, Inc., Englewood Cliffs, N.J., 1984.

19. Sadler, John E., et al.: "Bins and Silos," Chapter 29 of *Handbook of Structural Concrete*, edited by F. A. Kong, R. H. Evans, E. Cohen, and F. Roll, McGraw-Hill Book Co., New York, 1983, pp. 29–1 to 29–33.

20. Blacker, M. J. and J. G. M.: Woods, Discussion on Technical Note 490, "Eccentric Discharge in Circular Silos," by H. M. Haydl, Proceedings, Institute of Civil Engineers, Part 2, June, 1987, pp. 475–480.

21. Wood, Jonathan G. M.: "Analysis of Steel Bins Subject to Eccentric Discharge," Second International Conference on Design of Silos for Strength of Flow, Stratford-on-Avon, England, November, 1983, pp. 132–144.

22. Rotter, J. M.: "Analysis of Steel Bins Subject to Eccentric Discharge," Second International Conference on Bulk Material Storage, Handling, and Transportation, Institution of Engineers, Wollongoog, NSW, Australia, July, 1986.

23. Jenike, Andrew W.: "Denting of Circular Bins with Eccentric Drawpoints," *J. Struct. Div. ASCE*, vol. 93, STI, February, 1967, pp. 27–35.

24. Johnston, F. Thomas, and Fred A. Hunt: "Solutions for Asymmetrical Flow Problems," Second International Conference on Design of Silos for Strength of Flow, Stratford-on-Avon, England, November, 1983, pp. 1–13.

25. Johnston, Tom, "How to Design Large-Diameter Silos the Last," Powder and Bulk Engineering, CSC Publishing, Minneapolis, May, 1990, pp. 43–53.

26. Safarian, S. S., and E. C. Harris: "Empirical Method for Computing Bending Moments in Circular Silo Walls Due to Asymmetric Flow," *Powder Handling and Processing*, vol. 3, no. 3, September, 1991, Germany.

27. Gurfinkel, G. R. "Grain-Storage Tank for Zäire, West Africa: Design, Construction and Repair," *ACI Concrete International*, vol. 11, no. 3, March 1989, pp. 39–49.

28. ASCE/SEI 7-10—Minimum Design Loads for Buildings and Other Structures, American Society of Civil Engineers, Reston, VA.

29. Gurfinkel, G. R.: "Thermal Effects in Walls of Nuclear Containments, Elastic and Inelastic Behavior," *Proc. 1st Int. Conf: Structural Mechanics in Reactor Technology*, vol. 5, part J, Berlin, Germany, September 1972.

30. Building Code Requirements for Structural Concrete and Commentary, ACI 318–11, American Concrete Institute, Farmington Hills, MI, 2011.

31. FIP—Working Group Silo Design, "Recommendation for Concrete Silo Design," Draft, Karlsruhe, Germant, June 1990.

32. Harris, E. C.: "Design of Members Subject to Combined Bending and Tension," *ACI Journal*, Proceedings vol. 72, no. 9, September 1975, pp. 491–495.

33. Safarian, S. S.: "Design of a Circular Concrete Ring-Beam and Column System Supporting a Silo Hopper," *J. ACI*, February 1969.

34. Hunt, F. A., and F. T. Johnston: "Solutions for Asymmetrical Flow Problems," *Coal Mining and Processing*, January, 1984.

35. Building Code Requirements for Reinforced Concrete, ACI 318–83, American Concrete Institute, Detroit.

36. Ciesielski, R., et al.: *Behalter, Bunker, Silos, Schornsteine, Fernsehturme und Freileitungsmaste*, Wilhelm Ernst & Sohn KG, Berlin, 1970.

37. Gurfinkel, G. R.: "Investigation of Silos at Seneca, Illinois," Report for Continental Grain Co., Regional Office in Chicago, III., October 1974.

38. Discussion of ACI 313–77, *J. ACI*, June 1976.

39. Gurfinkel, G. R.: "Structural Adequacy of Reinforced Concrete Silo Complex in Leverett, Illinois," Report for Thomasboro Grain Co., Thomasboro, Ill., December 1976.

40. Gurfinkel, G. R.: "Collapse and Repair of Tall Concrete Silos with Suspended Steel Hopper," *ASCE Journal of Performance of Constructed Facilities*, vol. 3, no. 4, November 1989, pp. 243–264.

41. Theimer, O. F.: "Cause and Prevention of Dust Explosions in Grain Elevators and Flour Mills," *ASME Publ.* 72-MH-25, New York, 1972.

42. Ghali, A., *Circular Storage Tanks and Silos,* CRC Press, Taylor & Francis Group, 3rd Edition, 2014, 367 pp.

43. Ghali, A., and Neville, A. M.: *Structural Analysis: A Unified Classical and Matrix Approach*, CRC Press, Taylor & Francis Group, 7th Edition, 2017, 933 pp.

44. Gurfinkel, G. R.: "Restoring an Impaired Concrete Silo," *ASCE Journal of Performance of Constructed Facilities*, vol. 3, no. 2, May 1989, pp. 87–99.

Chapter 22
Design of Steel Tanks

BY

LESLIE D. SCOTT *Chief Engineer, Tank Industry Consultants, Inc., Indianapolis, Indiana*

22.1 INTRODUCTION

This chapter is intended to provide general information about the selection and design of steel storage tanks. Basic theory of tank design is included for vertical, cylindrical, flat-bottom storage tanks with other tank types covered but in less detail. Because of the variety of tank types and storage applications, detailed information for each such instance is better left to industry standards or other technical documents dedicated to specific tank types or applications. This chapter provides the practicing engineer with some of the information needed to determine what documents, standards, codes, or regulations should be incorporated in the design process.

The tank construction industry has not been exempt from the problem of deteriorating infrastructure. Discussion includes the use of good structural detailing to allow complete application of high-performance coating systems for corrosive environments and the important role that such detailing plays in asset preservation planning. A discussion of some of the critical areas requiring special detailing is included.

Steel tanks come in a variety of types and configurations, each suited to the product being stored. Liquid storage tanks may be used to store water, oil, gasoline, chemicals, wine, molasses, juice, cooking oils, and a variety of other products. Water-storage tanks can be ground-supported, flat-bottom tanks or elevated tanks mounted on multiple legs or single support pedestals. Structures for the storage of bulk solids materials are usually classified as bins or silos.

22.1.1 Classifications of Steel Tanks

Steel tanks are classified differently among different industry sectors. A summary of some common classification strategies is included below.

The first classification system is pressure classification. Tanks are classified into one of three groups: (1) atmospheric tanks, (2) low-pressure tanks, and (3) high-pressure tanks or pressure vessels. Generally, atmospheric tanks are not part of a closed system and are vented to the atmosphere above the design liquid level. Low-pressure tanks are tanks with internal pressure above the design liquid level but such pressures not exceeding 15 psig. High-pressure tanks or pressure vessels are tanks with design pressure exceeding 15 psig and designed in accordance with the ASME Boiler & Pressure Vessel Code. In some instances, such as with the International Fire Code (IFC), some low-pressure tanks are classified as atmospheric tanks because they are designed to operate at pressures approximating atmospheric pressure. However, the IFC addresses low-pressure tanks only in Chapter 55—Cryogenic Fluids. The National Fire Protection Association has differing definitions for the same designations of atmospheric and low pressure. Because of the nonuniformity of the application of these pressure classifications, the engineer should be diligent in identifying the pertinent requirements for designing a tank when pressure-related classifications are used to specify the tank requirements.

The second classification system relates to how the tank bottom is supported. The two types are (1) ground-supported tank and (2) elevated tank. The ground-supported tank, also sometimes called the flat-bottom tank, has a flexible membrane bottom that is supported either on a concrete slab, on a tank pad of compacted materials through which the water load is transmitted directly into the ground, or on grillage supported on the ground. It should be noted that even though the term "flat-bottom tanks" is used in the industry, this does not limit the tanks to only having a flat floor. Section 22.2.1 of this chapter includes more discussion on bottom shapes for ground-supported tanks. Occasionally tanks are partially buried into the ground. These tanks are included in the ground-supported tank category. Elevated tanks consist of a tank container elevated above the ground and mounted on a supporting structure. Elevated tanks may have a suspended bottom, a bottom supported by a concrete slab or dome at the top of the supporting structure, or a conical bottom supported by pedestal or tower structure. Elevated tanks may be further classified by the type of structural tower or pedestal on which they are mounted: multicolumn tank, fluted-pillar tank, slender-pedestal tank, or concrete-pedestal tank.

Tanks may be classified based on the type of liquid they store. Tanks might be used to store a multitude liquids including water, wastewater, demineralized water, industrial process water, petroleum products, chemicals, dairy and beverage products, and a variety of other products. Water tanks might be further classified as storing water for municipal water distribution systems, wastewater facilities, industrial processes, fire protection purposes, chilled or heated water for thermal energy storage, or a combination of these uses. Tanks may also sometimes be classified by the type of product being stored, either liquid or gas. Atmospheric tanks storing liquids are commonly vertical, cylindrical tanks. However, if the tank is a low-pressure tank designed in accordance with API Standard 620 (discussed later in Sec. 22.1.2 of this chapter) or an elevated water tank, the tank may be of other axisymmetric shapes. Tanks storing gas at high pressure (over 15 psig) are generally referred to as pressure vessels and required to be designed and constructed in accordance with the ASME Boiler and Pressure Vessel Code. Design of pressure vessels is extensively covered in the ASME code and numerous other resources, so further discussion of ASME vessels is not included in this chapter.

Many basic tank design principles are applicable to each tank type, but there are also numerous special requirements that might also apply to each tank type. Although not an exhaustive list of tank classification methods, the discussion above clearly shows that there are a large variety of tank types and sometimes conflicting definitions of the classifications. It is imperative that the engineer obtain clear definition of the intended use and operational parameters of the tank to produce a design meeting the tank owner's expectations, industry standards, applicable codes and regulations, and sound engineering principles. Additionally, the engineer must be cognizant of the properties of the product to be stored and how those properties will affect the loading conditions and influence design decisions.

22.1.2 Industry Standards and Criteria

Environmental loads applicable to the design of tanks are generally specified in the tank standards by reference to ASCE 7 (Ref. 1)—Minimum Design Loads and Associated Criteria for Buildings and Other Structures, a standard published by the American Society of Civil Engineers (ASCE). At the time of this printing, industry design standards for steel tanks prescribe Allowable Stress Design (ASD) methods. Building codes and related codes/standards specify strength level load requirements. Most tank industry standards have already been amended to incorporate and adjust strength level loads from ASCE 7 for use with ASD design procedures, but the engineer should verify that this is the case to ensure that designs are not overly conservative. Especially since, in some cases, ASCE 7 takes exception to certain provisions of the tank standards and imposes more stringent requirements.

Foundation designs generally follow Load and Resistance Factor Design (LRFD) methods for the concrete components, but sizing of the foundation footprint will follow ASD methods for soil-bearing values.

Tank standards and the codes prescribing required load levels are continually changing with new editions generally published every 3 to 5 years. The engineer should always check for updates to the codes and standards prior to commencing design for a project. Also, the latest editions of reference standards might not be the applicable document in some jurisdictions. Sometimes the local jurisdiction has intentionally not allowed designs in accordance with the latest editions for specific reasons. Accordingly, it behooves the engineer to determine if the local authority having jurisdiction over the project will allow design for the project in accordance with the latest editions.

Tank standards have been developed by the American Water Works Association (AWWA) for steel water-storage tanks and by the American Petroleum Institute (API) for tanks for the petroleum industry. These standards apply to a large percentage of the steel tanks built today, although some standards exist for other specific applications. The general scope of each of those standards is explained below. Although new water tank standards are anticipated within a few years of this writing, the following summarizes the tank standards currently in place.

AWWA D100—Standard for Welded Carbon Steel Tanks for Water Storage (Ref. 2)

This standard addresses both elevated and ground-supported welded steel water-storage tanks constructed of carbon steel and operating at atmospheric pressure. Elevated tanks include both multiple-column tanks and single-pedestal tanks. Ground-supported tanks are limited to vertical, cylindrical, aboveground tanks. Tanks may be open top or fitted with a steel roof of one of numerous possible configurations or with an aluminum dome roof. Tanks storing potable water are required to have a roof. Ground-supported tanks are constructed at or near grade level and supported on a tank pad or foundation.

Requirements for design, materials, fabrication, foundations, construction, construction tolerances, inspection, and testing are covered. Shell joints may be lap joints within specified thickness and design limitations; otherwise, all shell joints are butt joints. Roof and bottom joints can be lap joints or butt joints. Design requirements include loads from the stored product, the structure, snow, wind and seismic effects, and live load conditions. For ground-supported tanks, two design options are available: (1) the base design requirements and (2) if allowed by the purchaser, Section 14 design, which incorporates higher allowable stresses and joint efficiency in addition to more stringent requirements for materials, welding, configuration, design metal temperature, inspection, and testing. The standard addresses five different options for foundations and tank pads for ground-supported tanks.

Maximum thickness limits apply to some components with a maximum shell thickness of 1.5 in. for Section 14 ground-supported tanks, and 2 in. for other tanks when tension is the governing stress condition. Capacities are only limited by the design conditions, materials, and maximum allowable component thickness.

AWWA D103—Standard for Factory-Coated Bolted Carbon Steel Tanks for Water Storage (Ref. 3)

This standard addresses ground-supported bolted water-storage tanks constructed of carbon steel and operated at atmospheric pressure. Ground-supported tanks are limited to vertical, cylindrical, aboveground tanks. Tanks may be open top or fitted with a steel roof of one of numerous possible configurations or with an aluminum dome roof. Tanks storing potable water are required to have a roof. Ground-supported tanks are constructed at or near grade level and supported on a tank pad or foundation.

All vertical shell joints and joints in the tank bottom and roof are lap joints. For flat-panel tanks, the horizontal shell joints are also lap joints (Fig. 22.1). Figure 22.1 shows a typical horizontal lap joint configuration with one embodiment of sealant application. Sealant application procedures vary by tank manufacturer. For formed-flange tanks, the horizontal shell joints are flanged (Fig. 22.1). Requirements for design, materials, fabrication, factory-applied coating systems, foundations, construction, inspection, and testing are covered. The standard addresses six different foundation options for foundations and tank pads. Design requirements include loads from the stored product, the structure, snow, wind and seismic effects, and live load conditions.

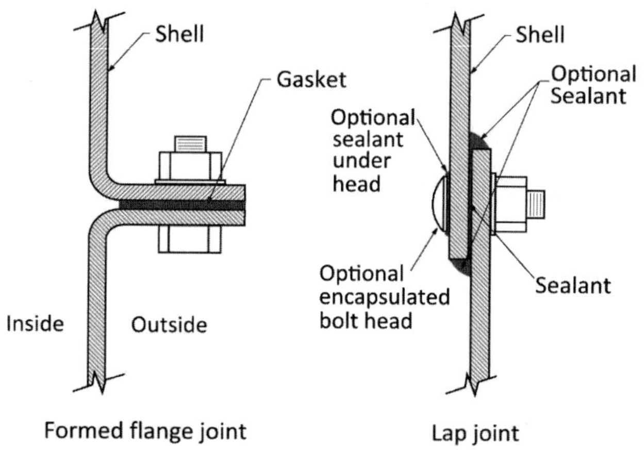

Figure 22.1 Horizontal shell joints in bolted tanks.

Maximum thickness limits apply to the tank shell depending on the type of shell horizontal joints: 5/16-in. maximum for formed-flange shell plates and ½-in. maximum for flat-panel shell plates. The formula for shell thickness accounts for bolt hole size and spacing. Capacities are only limited by the design conditions, materials, and maximum allowable component thickness based on the shell configuration.

AWWA D107—Standard for Composite Elevated Tanks for Water Storage (Reference 4)

This standard addresses elevated water-storage tanks combining a welded carbon steel tank supported by a single support pedestal constructed of reinforced concrete. The tank is designed to operate at atmospheric pressure. A fixed steel roof is required but may be provided in a variety of configurations.

Shell joints may be lap joints within specified thickness and design limitations; otherwise, all shell joints are butt joints. Roof and bottom joints can be lap joints or butt joints. Requirements for design, materials, fabrication, construction, construction tolerances, foundations, inspection, and testing are covered. Design requirements include loads from the stored product, the structure, snow, wind and seismic effects, and live load conditions. Maximum thickness limits may apply depending on the stress level of the component being examined. Consult the standard for specific requirements. Capacities are only limited by the design conditions, materials, and maximum allowable component thickness.

NFPA 22—Standard for Water Tanks for Private Fire Protection (Ref. 5)

This standard addresses water tanks used to store water for fire protection purposes. Compliance with this standard is sometimes required by the local jurisdiction when tanks are constructed specifically for fire protection purposes. Water tanks that supply water for fire protection through a community water distribution system are often not subject to these requirements. It is recommended that the engineer verify applicability of this standard prior to beginning the tank design process. NFPA 22 includes some provisions that exceed and are additive to the requirements of AWWA D100, AWWA D103, or AWWA D107, whichever standard is applicable to the type of tank being designed. Capacities are only limited by the design conditions, materials, and maximum allowable component thickness based on the AWWA design standard applicable to the selected tank type and the more restrictive requirements imposed in NFPA 22.

API Spec 12B—Specification for Bolted Tanks for Storage of Production Liquids (Ref. 6)

This specification covers fixed standardized sizes of vertical, cylindrical, ground-supported bolted storage tanks constructed of carbon steel and operating at internal pressures approximating atmospheric pressure. Tanks may include a roof or be open top. This is primarily a fabrication specification calling out specific sizes, configurations, and component thicknesses for 13 standard tank geometries (11 different capacities). All tank configurations are preestablished and designed for internal pressures varying between 1 oz/in.^{2} and 3 oz/in.^{2} depending on the tank geometry. The tanks are rated for a 20-psf roof live load in addition to ½ oz/in.^{2} of vacuum. Thickness of the tank floor, shell, and roof are predetermined and specified. The standard tank designs include basic wind loading considerations but might not meet project specific wind load requirements. Wind design is based on a 3-second gust wind speed of 90 mph using an importance factor of 0.87 for wind Exposure C in accordance with ASCE 7-05. Seismic loads are not included in the standard designs. The provisions do not cover requirements for tanks that require anchorage or tanks that are subject to seismic loading or higher wind load parameters. Capacities are limited to the standard sizes specified in this document.

API Spec 12D—Specification for Field Welded Tanks for Storage of Production Liquids (Ref. 7)

This specification covers fixed standardized sizes of field-erected, vertical, cylindrical, closed-top, ground-supported welded steel storage tanks constructed of carbon steel and operating at internal pressures approximating atmospheric pressure. This is primarily a fabrication specification calling out specific sizes, configurations, and component thicknesses for 10 standard tank geometries (eight different capacities). All tank configurations are preestablished and designed for internal pressures varying between 3 oz/in.^{2} and 8 oz/in.^{2} depending on tank geometry. The tanks are rated for a 20-psf roof live load in addition to ½ oz/in.^{2} of vacuum. Thickness of the tank floor, shell, and roof are predetermined and specified. The standard tank designs do not include considerations for wind or seismic loading. The provisions do not cover requirements for tanks that require anchorage. Capacities are limited to the standard sizes specified in this document.

API Spec 12F—Specification for Shop Welded Tanks for Storage of Production Liquids (Ref. 8)

This specification covers fixed standardized sizes of shop-fabricated, vertical, cylindrical, closed-top, ground-supported welded steel storage tanks constructed of carbon steel and operating at internal pressures approximating atmospheric pressure. This is primarily a fabrication specification calling out specific sizes, configurations, and component thicknesses for 10 standard tank geometries (eight different capacities). All tank configurations are preestablished and designed for internal pressures varying between 3 oz/in.^{2} and 8 oz/in.^{2} depending on tank geometry. The tanks are rated for a 30 psf ground snow load per ASCE 7-16 in addition to ½ oz/in.^{2} of vacuum. Thickness of the tank floor, shell, and roof are predetermined and specified. The standard tank designs include basic wind loading considerations but might not meet project specific wind load requirements. Wind design is based on a 3-second gust wind speed of 120 mph using wind Exposure C in accordance with ASCE 7-16 Seismic loads are not included in the standard designs. The provisions do not cover requirements for tanks that require anchorage. Capacities are limited to the standard sizes specified in this document.

API Standard 620—Design and Construction of Large, Welded, Low-pressure Storage Tanks (Ref. 9)

This standard addresses the design and construction of welded, carbon steel aboveground storage tanks with a vertically axisymmetrical shape and operated at low internal pressures. The base requirements include tanks designed for metal temperatures not greater than 250°F installed in locations where the lowest recorded 1-day mean atmospheric temperature is no less than −50°F. The tanks may be designed with internal gas pressures not exceeding 15 psig. The standard includes requirements for design, materials, fabrication, construction, construction tolerances, inspection, and testing. The standard includes supplemental annexes that contain additional requirements for special features, alternate materials, or specific operating conditions. Annex S covers stainless steel low-pressure storage tanks in ambient temperature service in all areas, without limit on minimum design metal temperatures. Annex R covers low-pressure storage tanks for refrigerated products at temperatures ranging from +40°F to −60°F. Annex Q covers low-pressure storage tanks for liquefied gases at temperatures not lower than −325°F. Capacities are only limited by the design conditions, materials, and maximum allowable component thickness.

API Standard 650—Welded Tanks for Oil Storage (Ref. 10)

This standard addresses vertical, cylindrical, aboveground, carbon steel tanks of welded construction. Tanks may be open top, covered with a fixed roof, fitted with a combination of fixed roof and internal floating roof, or fitted with an external floating roof. The tanks are constructed at or near grade level, and the entire tank bottom must be uniformly supported, either on a tank pad, slab foundation, or grillage.

The standard includes requirements for design, materials, fabrication, construction, construction tolerances, inspection, and testing of welded steel carbon steel storage tanks. Design requirements include loads from the stored product, hydrotest, operating conditions, snow, rain, wind and seismic effects, and live load conditions. API 650 base requirements include tanks operated with internal pressures from atmospheric pressure up to internal pressures not exceeding the weight of the roof plate and with a design temperature not exceeding 200°F.

The standard includes supplemental annexes that contain additional requirements for special features, alternate materials, or specific operating conditions. By incorporating the supplemental rules from one of the standard's annexes, internal pressures can be increased up to 2.5 psig.

Supplemental requirements in other annexes allow design for vacuum up to 1 psig, higher design temperatures up to 500°F, less stringent requirements for small capacity tanks where the nominal thickness of the stressed components does not exceed ½ in., less stringent requirements for tanks not over 20 ft in diameter and completely shop assembled, requirements for an optional aluminum dome roof, design requirements for austenitic stainless steel tanks, and design requirements for duplex stainless steel tanks.

This standard does not apply to tanks in refrigerated service. The annexes also include recommendations for foundation design and construction. Capacities are only limited by the design conditions, materials, and maximum allowable component thickness.

API Standard 653—Tank Inspection, Repair, Alteration and Reconstruction (Ref. 11)

This standard is the only current standard that addresses repairs, alterations, and reconstruction of steel storage tanks. It addresses minimum requirements for maintaining the integrity of tanks initially designed and constructed in accordance with API 650 and its predecessor document, API Standard 12C. Rules for inspection, repair, alteration, reconstruction, and even relocation are included. Repairs can include replacement of tank components that are damaged or have suffered metal loss due to corrosion, or they might include installation of patch plates over corroded sections. Tank modifications might include adding a new shell ring for increased capacity or installing new shell nozzles, manways, or access appurtenances. The rules cover design, materials, welding, construction, testing and inspection, and modification of the tank and its accessories.

Although this standard is not explicitly applicable to other ground-supported liquid storage tanks, such as tanks designed in accordance with AWWA D100, the rules and principles contained in API 653 can be aptly applied to those tanks as long as the engineer has an understanding of the purpose and intent of the rules included in API 653 and judiciously applies those requirements to fit the intent and spirit of the rules of the standard for the particular tank type under consideration. The engineer should carefully consider how the proposed repairs or modifications should be adapted to provide a structural capacity at least as much as that of a tank newly constructed in accordance with the applicable standard. Reference 12 provides additional discussion of issues related to tank retrofits and upgrades.

Other Tanks

Bins and Silos For bins and silos storing bulk solid granular material, many design considerations differ significantly from the design principles used for liquid storage tanks. Except for this paragraph and the listing of some documents in the reference list, this handbook does not cover this topic. There is no American national standard applicable to bins and silos. Designers of bins and silos should seek information from detailed references that focus on those structures. For these structures, the engineer must apply detailed engineering principles and relevant portions of codes and standards using sound engineering judgment. The Steel Plate Fabricators Association, in conjunction with the American Iron and Steel Institute, published a manual for the design of steel bins and silos which covers recommended design procedures and discusses many bulk solids design theories (Ref. 13). References 13 and 14 provide a convenient summary of design procedures incorporating various bin theories to create a practical design. References 15 and 16 provide a more detailed discussion of numerous bin and silo design considerations.

Light-Gauge Stainless Steel Tanks Light-gauge stainless steel tanks are commonly used to store wine, juice, olive oil, and so on. No industry design standard exists for these tanks. Some of these tanks are ground-supported tanks, others are mounted on a stand consisting of joists (sometimes light-gauge, cold-formed members) mounted on a structural frame, and still others are just mounted on legs similar to the configuration sometimes used for bins or vertical vessels. For these tanks, the engineer must apply engineering principles and relevant procedures using sound engineering judgment to adequately address a combination of multiple codes (structural steel code, light-gauge steel code applicable to the material of construction of each component) and standards applicable to other types of tanks and general structures. However, the engineer must be knowledgeable of the tank standards and design codes and proceed with caution so as not to inadvertently misapply the provisions of any of the tank standards that are not appropriate for this type of application.

Lift-Type, or Piston-Type, Gas Holders For this type of tank, API 650 does not include any requirements or guidance for the guides, counterweights, seals, or piston structures. However, it is recommended that the engineer follow the requirements of this standard for the other tank components to which the standard is applicable.

Liquefied Petroleum Gas Storage API, NFPA, and the federal Department of Transportation have standards applicable to these tanks. Each of these standards and regulations covers a specific scope and has limitations and exclusions regarding the types of tanks or facilities to which the standards are applicable. The engineer should be fully familiar with the applicable requirements for this type of tank before proceeding with the design.

High-Temperature Tanks Tanks with a design temperature over 500°F are beyond the scope of the API 650 standard. This type of tank requires many special considerations on material selection, welding, configuration, metallurgy, and thermal differential movements. After selection of appropriate materials for the operating conditions, a detailed thermal/structural interaction analysis should be performed for the selected materials at the design and ambient temperatures and other anticipated temperature ranges.

Partially Buried Tanks Although AWWA D100 does not apply to tanks partially buried in the ground, tanks can be designed for this configuration. The tanks must be designed to withstand the external soil and groundwater pressures exerted on the tank shell when the tank is empty in addition to hydrostatic pressures and dynamic loads under seismic conditions. A comprehensive corrosion protection system for the buried portion of the tank shell and the underside of the floor should be developed since the buried portion of the tank will not be accessible for inspection and maintenance. Groundwater levels and their impact on structural loads and corrosion protection systems must be incorporated into the design considerations for these tanks. A design procedure for determining the required shell stiffeners is provided in Ref. 17.

Code Requirements

ASCE 7 is the source document for seismic loads under the International Building Code (IBC) and has been incorporated into most of the tank standards. For most tank installations, the IBC or one of its state adopted or locally adopted versions will apply. In the United States, most states have adopted the IBC as the required code for all structures installed within their boundaries, some incorporating specific revisions for their state and others adopting it without revision. Additionally, some local jurisdictions have adopted additional revisions to their state's version of the Building Code. Because of these variations, the engineer should review the version of the Building Code applicable to each specific project to determine if special requirements will apply. ASCE 7 contains provisions applicable to tanks. It also contains specific detailing requirements for various types of tanks. For specific types of tanks where existing industry standards contain seismic design provisions, ASCE 7 provides that design may be in accordance with that industry standard plus additional special requirements contained in ASCE 7. When there is no relevant industry standard applicable to a particular tank, ASCE 7 requirements apply for the wind and seismic loads and detailing requirements. In those cases, the engineer should follow a reasonable and rational approach to develop a design protocol for determining the tank's resistance to the ASCE 7 loads and additional dead, snow, rain, live, and other environmental loads.

Design References

Many technical references of interest to the engineer are available that provide a detailed treatise on the design of steel tanks. Chapters 3 and 4 of Ref. 20 provide useful information and insightful discussion on the design of flat bottom tanks. Two relevant books have been published since the fourth edition of this handbook (Refs. 18 and 19). These books cover technical aspects of the design of storage tanks and their components and accessories as well as operational, inspection, and repair considerations. References 20 and 21 provide background and procedures on a variety of important topics related to storage tank design. References 22 through 27 cover topics related to the design of a variety of tank and vessel structures, many of which are intended for or can be applied to the design of storage tanks. Reference 28 includes discussion of many important considerations in the design, construction, maintenance, and operation of steel water storage tanks.

22.2 DESIGN CONSIDERATIONS

This section includes a discussion of common design considerations for water tanks. Tanks storing other products will likely require additional considerations based on the characteristics of the stored product and the operating conditions of the tank.

22.2.1 Bottom Configurations

For ground-supported tanks, the tank bottom can be designed in several different configurations (Fig. 22.2). The configurations depicted are (a) crowned (cone up), (b) sloping, (c) flat, (d) cone down, (e) eccentric cone down, and (f) butterfly bottom. The crowned bottom is most commonly used for water tanks. The slope of the floor is on the order of 1-in. in 10 ft, or approximately 1 percent up from the shell to the center. References 18 and 19 provide a discussion of some of the considerations for use of the different floor configurations.

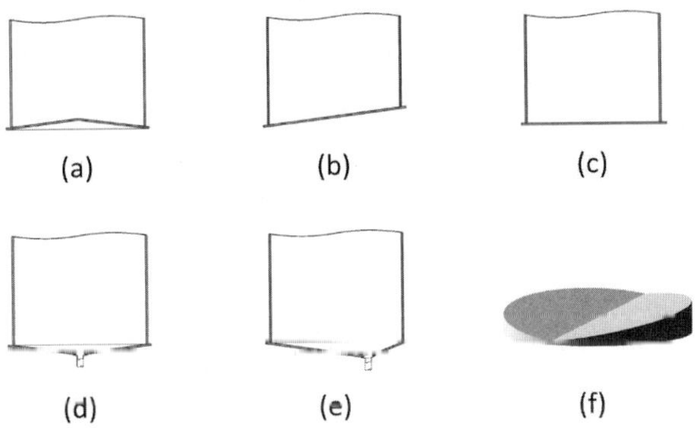

Figure 22.2 Variations of tank bottom configuration.

When using other than the crowned floor configuration, consideration of the stresses induced in the floor plates due to differential settlement under the floor is warranted. For tanks without a crowned floor, when a concrete slab-type foundation is not used, the center of the tank floor will settle more than the tank perimeter, creating membrane stresses in the floor and circumferential compressive forces in the base of the tank shell. Because the tank bottom is a flexible membrane on a crowned floor, it can generally accommodate this differential settlement with the floor experiencing some minor waviness due to the additional settlement at the center of the tank. Because of the potential negative consequences of this differential settlement, slab foundations are often considered necessary when the crowned bottom configuration is not selected.

A sloping bottom or butterfly bottom is used when it is necessary to accommodate self-draining of the product from the tank through the tank shell. This type of floor is sometimes used in conjunction with a flush-type shell connection or a flush-type cleanout fitting designed and constructed in accordance with API 650 to facilitate full drainage at the low point.

22.2.2 Dissimilar Metals Corrosion

As noted previously, AWWA tank standards D100, D103, and D107 address carbon steel tanks. However, it is sometimes considered advantageous to use stainless steel or other metals possessing improved corrosion resistance properties for piping components or tank accessories. Due to the geometric complexity of these components or obstructed access to them for proper application of protective coatings, the practice of using stainless steel in minor piping or accessory assemblies is thought to solve the corrosion problems often associated with these minor components. The engineer should be aware that while trying to solve one problem, another potentially bigger problem may be created.

Dissimilar metals corrosion occurs when two different types of metal are electrically continuous and both in contact with an electrolyte. This would be the case if stainless steel components were welded to the carbon steel tank and were also in contact with the stored liquid. In locations where pinholes or holidays exist in the carbon steel tank's interior coatings below the high liquid level, a galvanic corrosion cell would develop causing corrosion of the carbon steel tank components. The carbon steel components will corrode in this instance because stainless steel is higher on the galvanic series, and, thus, it is the more noble metal. Because the carbon steel tank components will generally be the less noble metal and, therefore, the metal that will corrode when this galvanic cell is formed, the current edition of AWWA D100 prohibits the use of dissimilar metals inside the tank below the design liquid level unless the dissimilar metals are electrically isolated from the tank.

In some cases, even though AWWA D100 requires the electrical isolation of the dissimilar metals, protective coatings are applied to the more noble metal to reduce the exposed cathodic area in lieu of providing electrical isolation.

Effective electrical isolation is generally preferred, but sometimes it is difficult to achieve. Cases have been encountered where the electrical isolation fails over time. Therefore, it is this author's opinion that if it is desired to use stainless steel or other more noble corrosion-resistant metal for portions of these minor tank components, then painting of the dissimilar metal components should be used in conjunction with electrical isolation.

22.2.3 Design Process for Ground-Supported Tanks

The design process for ground-supported tanks is predicated on the tank type, the product being stored, the design and operating conditions, and the tank standard to which the design is carried out. The following list provides the basic steps in the design process for a ground-supported water tank designed in accordance with AWWA D100:

(1) Design method selection (base design or Section 14)

(2) Roof type selection and design

a. For supported cone roofs, design of rafters and bracing, girders, columns, base plates, and all connections

b. For any of the self-supporting roof types, design of the roof membrane and associated supports, stiffeners, compression rings, and tension rings, if any

(3) Selection of joint types and design joint efficiencies for tank shell joints

(4) Determination of hydrostatic shell stresses

(5) Determination of shell resistance to wind loading

(6) Determination of shell wind stiffening requirements

(7) Determination of resistance to overturning moment under wind loading

(8) Determination of hydrodynamic shell stresses

(9) Determination of shell buckling resistance under seismic overturning compression loads

(10) Determination of sloshing wave and freeboard requirements

a. For tanks with inadequate freeboard, also design the roof and anchorage for sloshing wave pressures

(11) Determination of resistance to overturning moment under seismic loading

a. anchorage required, or

b. self-anchored OK, no uplift occurs, or

c. self-anchored OK, design of annular ring for required hold-down forces

(12) If anchorage is required or used, determination of
  a. anchor size, materials, and details
  b. method of anchor attachment and local shell stresses

(13) Determination of foundation sizing for static loads, shell compression loads due to overturning moment, uplift resistance against overturning moment, and sloshing wave pressure uplift resistance

NOTE: For API tanks designed with internal pressures, foundation design for pressure uplift is also required.

22.2.4 Design for Corrosion Protection

Conventional construction is only conventional because it is the lowest-cost option that meets the minimum requirements of the tank standards. In this context, conventional construction consists of a structural system of rafters upon which the tank's roof plate is laid with the rafters supported by columns and, in some cases, girders. The roof plate lap joints are then welded on the top side only. The roof plate is only welded to the tank shell and is not attached to the structural system at all. These features of conventional roof construction conform to the minimum requirements of the tank standards.

The tank standards provide minimum requirements for structural and operational considerations but, for the most part, do not include any requirements related to corrosion protection. Even though not required by the standard, designing for corrosion protection should be a part of any comprehensive tank design process. Deterioration due to corrosion is a problem challenging many segments of today's infrastructure. Due to the extremely high cost of remediating corrosion damage, it is not reasonable for the engineer to prepare a structural design for a tank without incorporating corrosion protection considerations. This author believes that for all tanks the engineer should prepare a comprehensive design that includes asset preservation considerations, not just structural design for operating conditions and load resistance.

The following discussion provides a brief background on the topic and some suggestions that can be used to design structures that, in conjunction with protective coating systems, can greatly extend the life expectancy of these major assets.

Case Study

Nearly 35 years ago this author was involved with a project to install a new roof on an industrial water tank used in an oil field production facility. This tank was subject to very severe service due to the presence of hydrogen sulfide in the water stored in the tank.

This tank was 165 ft in diameter and 48 ft high. The tank was originally constructed using a conventional roof and structure configuration. At the time of construction, shop blasting and priming or coating of roof structure prior to construction was very uncommon, and thus this tank was erected with uncoated steel. The roof system on this tank contained many inaccessible areas that could never be coated after erection was complete. Those areas include the top of the rafters in contact with the roof, the corresponding roof plate surfaces above the rafters and mating surfaces of roof plate lap joints. In addition, the roof plate layout of this tank included excessively wide plate laps up to 24 in. wide. The combination of no prior-to-erection coatings with excessive plate lap widths resulted in a tank with very poor structural details for application of protective coatings (Fig. 22.3). The result of the severe service conditions and the poorly selected construction details led to a rapid end of service life of the tank's roof system due to extensive corrosion of the rafters and perforation of the roof plate in over 300 locations less than 6 years after initial construction.

Because the corrosion had deteriorated the beams so severely and even fused portions of the beams to the roof plate, some of the rafter flanges came off with the roof plate when the work to remove the existing roof began (Fig. 22.4). It is worth repeating that this tank was only 6 years old at the time.

A quote attributed to Albert Einstein is quite appropriate here: "No problem can be solved from the same level of consciousness that created it." In other words, we cannot keep doing things the same way and expecting different results. And so it is with tank roofs. Sometimes, the approach to solve corrosion of roof components in severe service conditions is to apply thicker coatings. But this is generally a flawed concept because thicker coatings would only slightly help some portions of the roof, but it would not address the underlying problem of the uncoated inaccessible areas where the corrosion will flourish. It should also be noted that in some cases the storage of water, even potable water, can result in severe service conditions and other times not, depending on the chemistry of the water and operating conditions. Water containing high levels of chlorine or hydrogen sulfide have been known to result in highly corrosive conditions.

Figure 22.3 Undesirable roof structure details.

Figure 22.4 Severe roof structure corrosion due to poorly detailed tank.

These challenges require innovative approaches rather than another attempt using conventional construction. It is true that conventional construction configurations are generally the lowest-initial-cost solution, but when lifetime repair and maintenance costs are factored in, they are generally not the best solution.

These issues can be overcome with thoughtful consideration of the roof system configurations and construction details. The subject tank for the present case study was fitted with a new roof. Radial layout of the roof structure has the disadvantage that the space between rafters toward the center of the tank is very limited and restricts access for seal welding and coating operations. Additionally, trapezoidal or triangular tributary load area for each rafter is not structurally optimum for maximum utilization of the roof plate's capacity to span between rafters. So, for the tank of this case study, the author developed a rectangular roof framing arrangement that avoided the congestion of converging rafters. In addition, due to the severely corrosive environment to which this tank was subjected, an external roof structure system was selected with all external structural elements being fully seal welded to the roof plates.

The combination of these two features resulted in a roof surface that could be effectively and efficiently coated on the inside of the tank and a structural configuration on the outside of the tank that could be easily accessed and fully coated and included no inaccessible areas. The external structural system also has the advantage that it can be inspected at any time, even maintained at any time, without taking the tank out of service. Figure 22.5*a* shows the interior surface of this tank after the new roof was installed. Figure 22.5*b* shows an exterior view (prior to final coatings) of another tank built using a similar external structural system with a rectangular layout.

Figure 22.5*a* Roof interior with external structural framing and rectangular layout.

Figure 22.5*b* External structural framing with rectangular layout.

In the tank of the present case study, the original tank roof lasted less than 6 years. After the new roof system was installed, the newly applied interior coatings were again subjected to very harsh conditions. After nearly 15 years in service with the new roof, the interior coatings had begun to deteriorate under the severe service conditions. At that time, inspection of the tank revealed negligible metal loss in the locations where the coatings had deteriorated. All other areas of the roof were in sound condition. So, in 2001, the interior coatings were removed, and a new coating system was applied. The tank was placed back in service, and the second new coating system lasted approximately another 15 years before the coatings again required restoration. The exterior surfaces were in good condition, but the decision was made to also recoat the exterior of the roof. In 2017, the interior roof surfaces were recoated for a second time, and the roof's exterior surfaces were also recoated. The tank was then placed back into service. With proper maintenance of the coatings and coating restoration when needed, this tank roof is not expected to require replacement again. An innovative design approach and comprehensive plan with innovative good construction details provided a successful solution.

External structural systems and rectangular layouts are not the solution for every tank, but this innovative approach produced a successful solution here. The case study is one example of successfully using a different approach to solve a recurring problem. The unique design developed for the case study tank resulted in the development of a patent for roof structure systems designed to allow for full application of protective coating systems to the interior surfaces (Ref. 29). During the research for this patent it became obvious that the problem of corrosion of the tank surfaces above the liquid level was not new. In fact, a patent on a different approach to the same problem was issued by the U.S. Patent Office in 1958 (Ref. 30). The background of that patent demonstrates that the problem of corrosion in tank roofs has been known for many decades, but the industry has been slow to change. Alternative designs that can significantly reduce the maintenance and repair costs associated with the tank assets operated by municipalities, utilities, and industrial facilities generally have a higher initial cost, so when tank contractors submit competitive bids they choose the option with the lowest cost to install but that still meets the specified minimum requirements. The lowest-installed-cost option rarely provides the best solution for lowest lifetime cost.

External structural systems might be undesirable for certain locales based on local weather conditions and the nature of the region surrounding the tank site. In areas subject to significant snowfall, the external structural system could increase the amount of snow accumulation and the corresponding loads on the roof. Additionally, if the tank is situated in an area surrounded by tall trees, leaves and debris may collect on the rooftop, resulting in the need for more frequent observation and personnel access for removal of these accumulations.

This author believes that the engineer should examine all options and highlight the significant issues and alternative solutions discussed in this section when presenting a design for client approval.

Whenever corrosion of the interior roof components due to the product being stored and the operating conditions of the tank is of concern, the following recommendations deserve consideration.

Seal Welding

Seal welding of interior roof plate joints can extend the life span of a steel tank's roof plates, but more important are the points of contact between the roof plates and the supporting rafters. These points of contact are usually the origin of structural deterioration. When components of the roof structure and roof plate joints are not seal welded, the contact surfaces of roof plate lap joints and the contact surfaces between the roof plates and rafters, in addition to some other areas of the completed tank, are termed "inaccessible areas." After completion of the tank construction, these inaccessible areas cannot be coated or maintained. And even inspection of these areas is difficult, with the deteriorating condition only apparent after significant metal loss has occurred.

Seal welding can protect these inaccessible areas from the corrosive atmosphere in the tank, but seal welding of all roof plate joints and all points of contact between the roof plates and the supporting rafters can be difficult. This effort is especially difficult toward the center of the tank where the rafters are converging when a radial rafter configuration is used. Additionally, shrinkage due to seal welding can produce distortions and undesirable final roof shape. Formed roof panels discussed later in this article can significantly reduce roof distortion because of the substantial reduction in the amount of welding and the increased stiffness of roof panels prior to welding as compared with seal welding roof plate lap joints and seal welding the rafters to the roof plates.

The commentary in AWWA D102-Standard for Coating Steel Water-Storage Tanks provides a good summary of inaccessible areas and the features and benefits of different roof options with respect to the performance of protective coating systems. It is recommended that the engineer be familiar with that commentary when considering the roof configuration best suited for each project.

Seal welding by itself is only an incremental improvement to the protection of the roof system. In order for the benefits of seal welding of the roof components to be fully realized, seal welding must be combined with the selection of a roof structure configuration and good construction details that provide ample access to perform the seal welding and allow protective coatings to be applied to all surfaces.

Open Center Support

Column-supported roof structure configurations utilizing radial layout of the rafters generally include a support at the center of the tank. The center support for the roof rafters has been produced in a variety of configurations (Fig. 22.6*a*). Commonly, the tank manufacturer is responsible to produce the engineering design for the tank and so selects the style of center support for simplicity of fabrication and lowest installed cost in addition to size consideration for shipping logistics. These considerations do not often line up with a goal of protective coating application and asset preservation. The center support styles depicted in sketches "*a*" and "*b*" of Fig. 22.6*a* represent the kinds of center support assemblies selected for the contractor's lowest installed cost. The two center support styles in sketches "*a*" and "*b*" result in several problems when it comes to application of protective coatings, maintenance, and corrosion protection of the components.

The drawbacks to configurations "*a*" and "*b*" are the following:

(1) The relatively small diameter of the center support creates congestion of members around the center support with very restrictive space between the rafters, inhibiting seal welding, blasting, and coating activities performed from scaffold or lift equipment from below the rafters.

(2) The space between the top of the center support and the underside of the roof plate above it has only limited access from the top side through the center vent and is further limited by the depth of the rafters. This makes the seal welding, blasting, and coating of the surfaces of these components extremely difficult at best.

(3) The ends of the rafters resting on the center support have the same disadvantages related to restricted access as noted in item (2) above.

Open center supports provide access to the underside of the center portion of the roof in addition to access to the ends of the rafters and their connection to the center support. The center support styles depicted in sketches "*c*" and "*d*" of Fig. 22.6*a* represent open support structures allowing such access. The center support in sketch "*c*" utilizes a built-up box tube fabricated to the required diameter. An advantage of this style is that the top of the box tube can be sloped to match the roof pitch providing a good joint for welding where the rafters land.

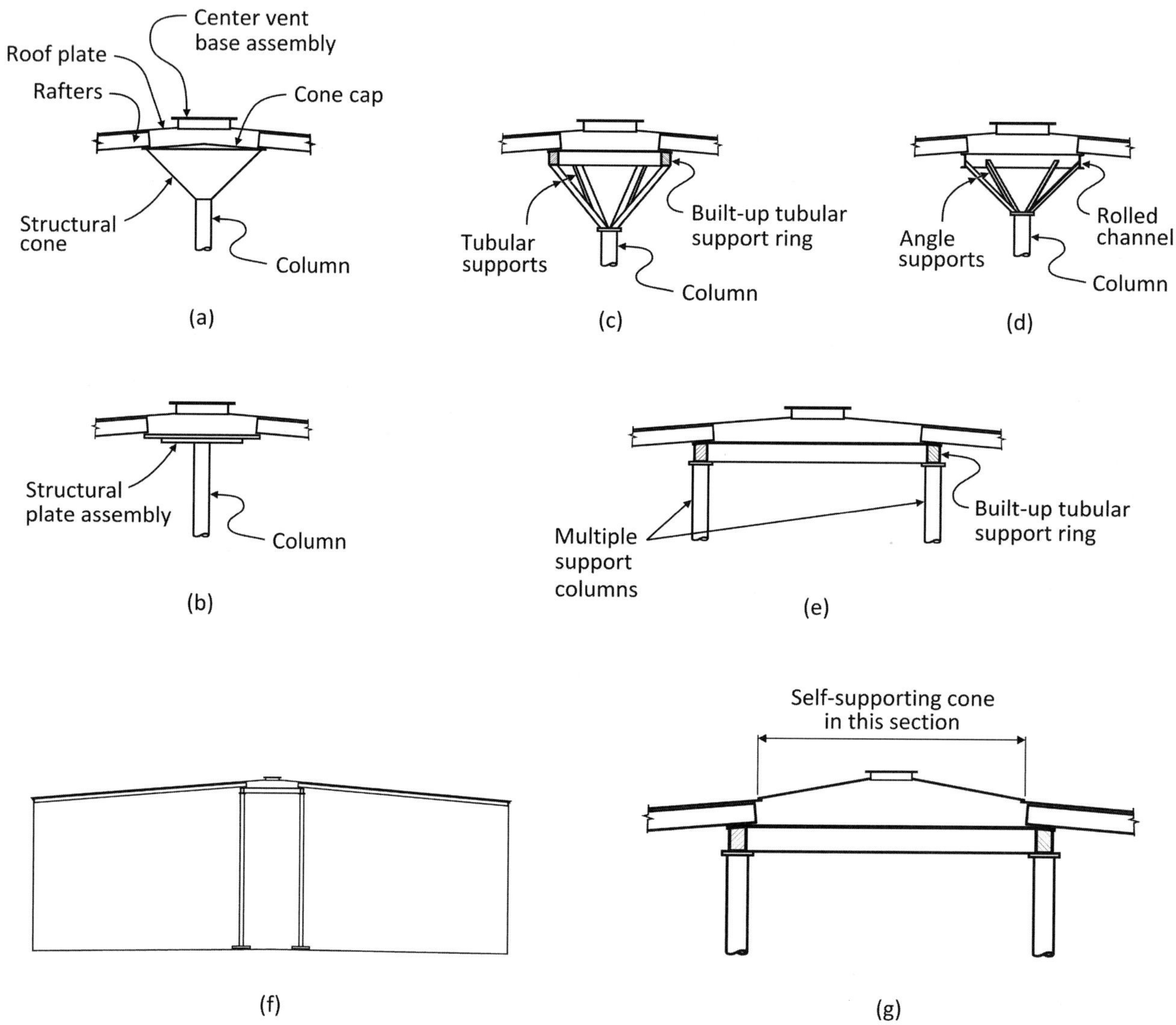

Figure 22.6*a* Variations of center support structures.

The use of tubular support members on this center support structure also reduces the number of hard-to-reach areas where the support members attach to the tubular support ring or to the column cap plate. The center support in sketch "*d*" utilizes a rolled channel for the support ring and steel angles for the supports. This style is often selected by fabricators as a lower-cost option for an open center support. However, compared to the center support in sketch "*c*," this style has the disadvantage that the top flange of the channel does not match the roof slope, making seal welding of the rafter-to-support ring connection more difficult. In addition, style "*d*" also results in portions of the structure that are more difficult to access for blasting and coating. For this reason, if an open style center support is selected for a single-column configuration, style "*c*" is preferred. An example of a tank roof using this type of center support is shown in Fig. 22.6*b*.

photo courtesy of Paso Robles Tank, Inc.

Figure 22.6*b* Example of open center support structure.

Another solution for the center support is a structural support ring mounted on multiple columns (Fig. 22.6*a*, sketch "*e*"). This style has the enhanced benefits that it provides even more access to the center of the tank and the ends of the rafters. The support ring is commonly mounted on three columns. Any number of columns can be accommodated in the structural design of the support ring, but fewer columns are preferred because it provides greater access to the center of the roof. It is sometimes argued that increasing the number of columns to more than just a single center column creates unnecessary additional blasting and coating effort, but it is much easier to coat a couple of additional columns than it would be to adequately protect the center of the roof with a single center column system. Even though they are significant improvements over style "*a*" or "*b*" type center supports, the center supports of style "*c*" and "*d*" are less desirable from a welding, coating, and maintenance perspective than a style "*e*" or "*f*" center support. To provide for a larger-diameter center support, this ring can be fabricated in two pieces and shipped loose for final assembly in the field. The multicolumn center support assembly is preferred for large tanks, generally greater than about 70 ft in diameter, as depicted in Fig. 22.6*a* sketch "*f*." The built-up tubular support ring can be easily fabricated to the size necessary for the required structural capacity.

Another variation of the style "*e*" or "*f*" multicolumn center support uses a larger support ring assembly (as depicted in Fig. 22.6*a* sketch "*g*") or, alternatively, a four-sided column and girder system installed at a radius of about 10 ft from the tank center. With this arrangement, the central portion of the tank roof, approximately 20 ft in diameter, is constructed as a self-supporting cone using the steel plate with no additional stiffeners. In this case, the center section of the roof would require a steeper roof pitch, and possibly greater thickness, than the main portions of the roof in order to act as a self-supporting roof without stiffeners. The larger support ring or girder system provides even more spacing between rafters and also provides unrestricted access for blasting and coating the ends of the rafters and the center of the roof.

Formed Roof Panels

Roof systems constructed using formed roof panels offer several advantages. In this type of roof system, rafters are replaced with a structural section, commonly a channel, that is formed along one edge of the roof panel. Roof components in this system are generally categorized as integral formed-structure roof panels, but sometimes the manufacturer's jargon for this roof system is "bent plate roof" or "press brake roof." In this discussion, this type of roof is referenced as "formed roof panels." Figure 22.7*a* shows a sketch of a common configuration of a formed roof panel with internal structural elements. Figure 22.7*b* shows a conceptual rendering of a partial roof structure using this type of formed roof panels and an open center support.

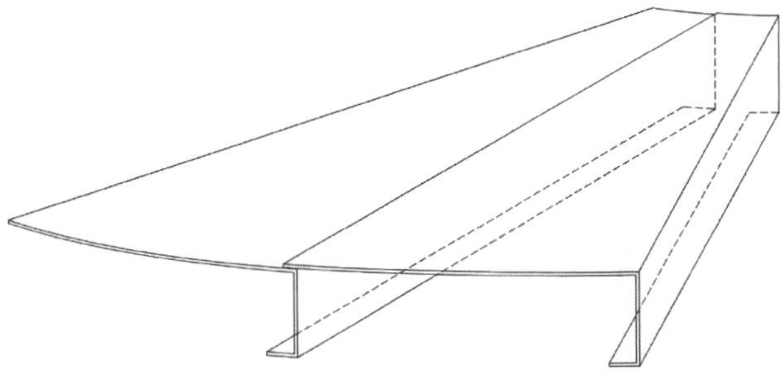

Figure 22.7*a* Example of internal formed-structure formed roof panel.

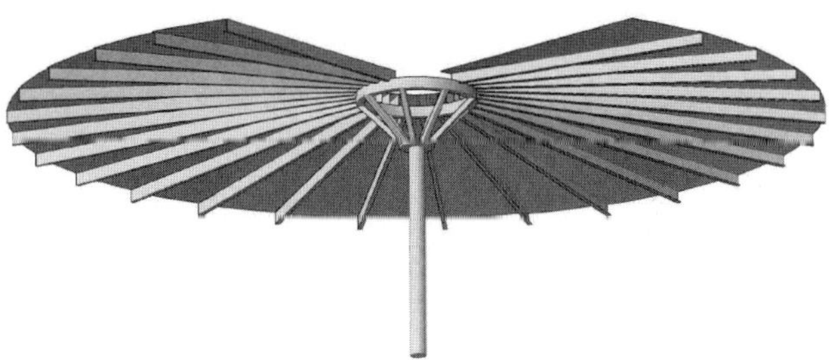

Figure 22.7*b* Open center support and internal formed roof panels with radial layout.

When a formed roof panel system is combined with full interior seal welding and an open center support assembly, the following significant benefits can be realized:

(1) The interior of the tank is free of inaccessible areas that cannot be protected with coatings.

(2) For internal formed roof panel systems, the top compression flange of the formed structural elements is inherently braced by the roof diaphragm when the welding of the roof panel seams is completed.

(3) The formed structural sections of the roof panels result in a reduction of the number of flange surfaces on the rafters that must be coated as compared with standard W-beams.

(4) When the formed portion of the roof panel is additionally designed to reduce the effort required for protective coating application (Fig. 22.7*e*, sketches "*c*" and "*e*"), the initial and maintenance costs of the roof system can be reduced even further.

When complete seal welding of the points of contact between the rafters and the roof plates and of the roof plate lap joints is being contemplated, consideration of the formed roof panel system deserves

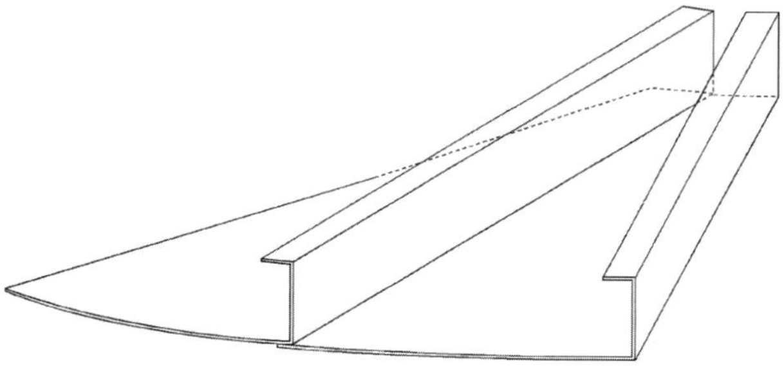

Figure 22.7*c* Example of external formed roof panels with radial layout.

serious consideration. Under these circumstances, the formed roof panel system can result in a reduction in the total length of interior seal welding of over 50 percent. However, this does not constitute a net savings of weld length because the length of exterior roof seam welding increases due to trapezoidal or triangular shaped roof panels, but the exterior welding is much easier to accomplish.

In some circumstances, it may be beneficial to put the formed structural elements on the outside of the tank rather than the interior. This is especially true if the design of the roof is governed by internal pressures rather than the more common governing load case of external loads. Figure 22.7*c* shows a sketch of a common configuration of how external structure formed roof panels would be overlapped when installed. Figure 22.7*d* is a photo of a tank constructed with this type of roof panel. The photo shows the tank with construction just completed using shop primed components and the tank ready for surface preparation and application of final coatings. When the formed roof panel system is used with external structural elements and full seal welding, the following additional benefits can be realized:

(1) The interior surfaces of the tank roof are all flat plate surfaces that are easily blasted and coated similar to the tank shell and floor.

(2) The structural elements are located outside of the roof where they can be inspected and maintained at any time and, in many cases, without taking the tank out of service.

With external structural elements, the top flange of the formed structural elements might require lateral bracing for gravity loads depending on the size of the formed structural section. Section properties and roof panel thickness can be adjusted to reduce or eliminate the need for lateral bracing which will simplify the external coating and maintenance and access of the exterior roof components.

photo courtesy of Paso Robles Tank, Inc.

Figure 22.7*d* External formed roof panels with radial layout.

As previously noted in this section, external structural members have many advantages but also are not appropriate for all projects. The engineer must review the benefits and disadvantages of the various roof configuration alternatives against all pertinent design objectives and project requirements before a final selection is made.

Another advantage of the formed roof panel system is that the formed structural shape may be configured in such a way to provide the necessary structural properties while also providing improved access for welding or coating activities. Formed roof panels may be designed in a multitude of configurations. Figure 22.7*e* shows concept sketches of several possible forming configurations for internal formed roof panels. The formed shape in sketch "*e*" of Fig. 22.7*e* provides improved lateral stiffness of the bottom flange and improved access for painting and maintenance but increases the forming costs of the panels due to the requirement to flip panels over for one of the bends. The increased forming cost should be compared against estimated coating and future maintenance costs to determine if this option is preferable. The complexity of providing end supports for the formed structural elements varies with each different shape shown in Fig. 22.7*e*. The benefits for painting access and asset preservation should be weighed against the additional complexity and cost of fabrication and installation of the panels to determine the best option for a particular project and conditions

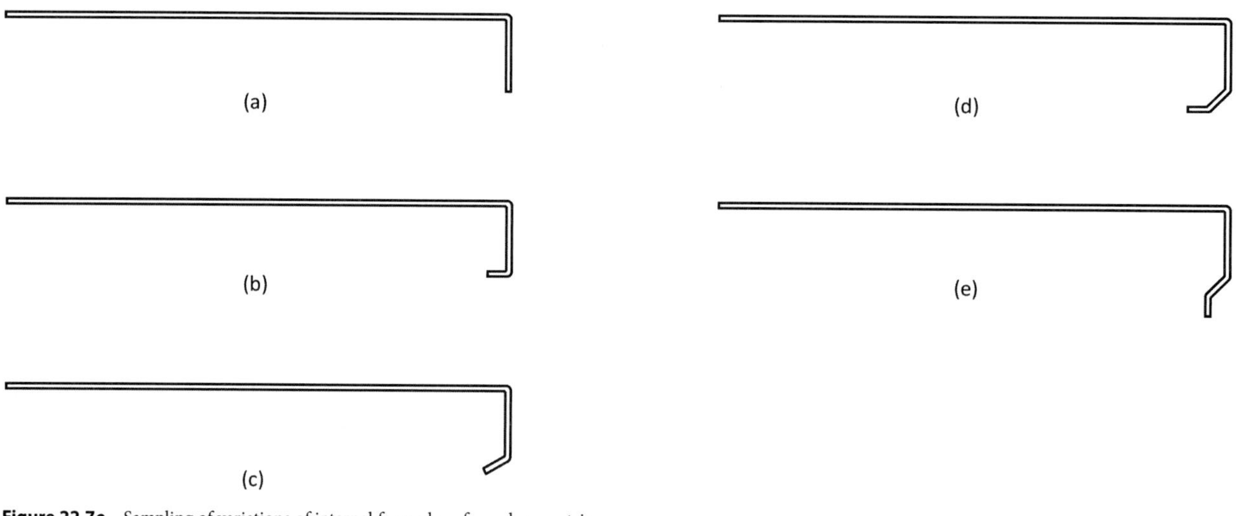

Figure 22.7*e* Sampling of variations of internal formed roof panel geometries.

Figure 22.7f Undesirable configuration of internal formed roof panel.

of service. Although a double bend might be preferable for lateral stability or stiffness under some loading conditions, the formed shape shown in Fig. 22.7*f* is generally discouraged because of the difficulty of blasting and coating this shape, even though it would be more economical from a manufacturing perspective with both bends in the same direction.

When external formed roof panels are used, the same variety of forming configurations are available. Figure 22.7*g* shows a sample of possible configurations with external formed structural elements.

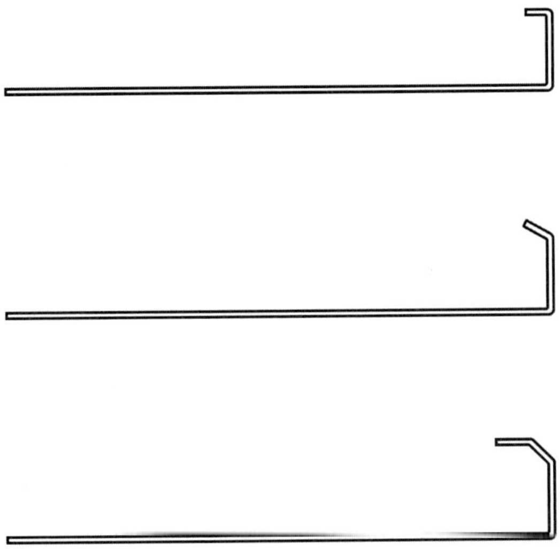

Figure 22.7g Sampling of possible external formed roof panel geometries.

Selection of the formed shape for internal panels must balance the needs of member strength with considerations of ease of access for painting and maintenance. For normal gravity load considerations (dead load, live load, snow load, etc.), the bottom flange is in tension and does not require bracing, while the top flange is in compression and is fully braced by the roof diaphragm.

For roof structure systems with multiple spans, the engineer should give careful consideration to the details of the intersection and support configuration for the formed roof panels over the girder. Many different configurations have been used. For this type of roof system, the best solution usually has a minimum of structural elements, surfaces, and edges that also provides open access for painting.

From this discussion it should be apparent that this type of roof system is equally suited for new tank construction and roof replacement projects.

22.3 DESIGN OF RESERVOIRS AND STANDPIPES

Water-storage tanks may be reservoirs, standpipes, or elevated tanks. *Reservoirs* are vertical, cylindrical, ground-supported, flat-bottom tanks, with a diameter larger than their height. Reservoirs are commonly relatively low in height (20 to 50 ft). When located on flat terrain, the water from these reservoirs must generally be pumped into a water-distribution system to provide the required pressure and flow. On hilly terrain, the tank can be located on high ground to use gravity pressure to obtain adequate flow rates. *Standpipes* also are vertical, cylindrical, ground-supported, flat-bottom tanks, but with a height greater than their diameter. They are usually of such height that the upper portion of the storage is available at some minimum desired pressure in the distribution system. The water in the lower portion of the standpipe is generally reserve storage requiring pumping to maintain minimum distribution system pressure. *Elevated tanks* are storage vessels on structural pedestals or towers, the entire volume of which is available at required minimum pressure. Reservoirs and standpipes use identical procedures of design but their foundation types might be different. Differences in foundation considerations are discussed in Sec. 22.3.6.

The remainder of this chapter includes various equations. Variable definitions will be presented below the equation in which they are first referenced.

22.3.1 Capacity

The capacity Q of a reservoir is determined by local requirements. The maximum permissible height of the tank is determined by the bearing allowable of the soil on which the tank foundation will be built. Steel reservoirs 330 ft in diameter, with capacities as large as 34 million gallons, have been built. Shell plates up to 2 in. thick have been used. Several trial calculations are needed to determine the most economical dimensions. The dimensions may be determined by the following equations:

$$H = \frac{P}{62.4g} \tag{22.1}$$

$$D = 0.4126\sqrt{\frac{Q}{H}} \tag{22.2}$$

where H = maximum permissible design liquid level, ft
NOTE: Freeboard and roof height might also be factors in selecting the maximum liquid level.
P = allowable soil pressure, psf
g = specific gravity
D = tank diameter, ft
Q = required capacity, gal

Since tanks are generally tested with water, a specific gravity g less than 1 is not recommended for design purposes.

22.3.2 Shell Design

The height of a standpipe is usually determined by the required usable gravity-head range and its diameter determined by the required capacity. Standpipes as tall as 150 ft with shell plates 2 in. thick have been built. The design of shells and bottoms for standpipes is the same as that for reservoirs.

Shell plates may be erected with different shell plate alignments. The alignment of shell courses, one on top of another, is sometimes left to the preference of the tank manufacturer. There are three different alignments in general use as shown in Fig. 22.8: (a) the inside flush alignment, (b) the centerline stack alignment, and (c) the outside flush alignment. Any of the three alignments are acceptable from a structural standpoint, but tank operation considerations sometimes govern the desired arrangement. For floating roof tanks that incorporate vapor

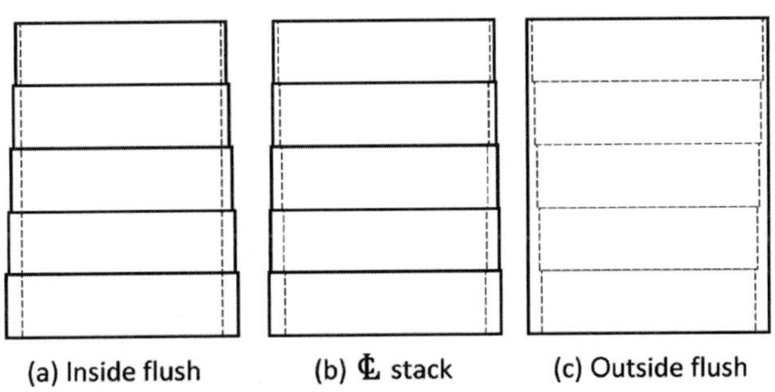

Figure 22.8 Variations of shell ring alignment.

seals between the floating roof and the inside of the shell or other tanks with internal equipment requiring a consistent inside diameter dimension, alignment "*a*" is the preferred option. For large tanks with internal coatings, alignment "*a*" is also preferred to reduce or eliminate the exposed plate edge created by the thicker lower shell plate thickness projecting inside the face of the shell ring above. Depending on construction techniques, sometimes tank manufacturers choose alignment "*b*" so that the manufactured circumference of each ring is consistent based on the centerline radius being equal. For small-diameter tanks, sometimes alignment "*c*" is chosen such that the outside strap dimension is consistent for all rings. From a tank design perspective, the chosen alignment affects the dimensional requirements of roof structure components for fixed roofs or seal space for floating roofs. Since all three alignments allow for full penetration welds at the horizontal joint, the structural design is the same for all three.

The shell plate is made up of one or more horizontal plate courses of width *w* (Fig. 22.9). Shell ring widths are commonly between 6 and 10 ft, but other widths (commonly 8 ft) may be selected, depending on shell height, availability of materials, and construction method.

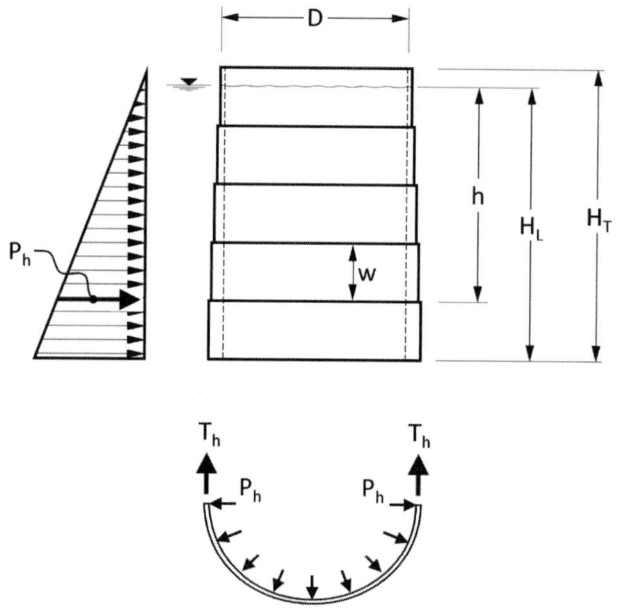

Figure 22.9 Pressure loading diagram.

Several plates may be required around the circumference to make up each course (ring). Although AWWA D100 allows vertical joints to be in alignment from ring to ring, the vertical seams in each ring are generally staggered relative to the vertical seams in adjacent shell rings.

Tank shells are cylindrical membranes designed to resist hoop tension (Fig. 22.9). Plate thickness is calculated at the bottom edge of each course with the pressure on each ring assumed to act undiminished from the bottom to the top of each ring:

$$T_h = 2.60hDg \tag{22.3}$$

$$t_h = \frac{T_h}{fE} = \frac{2.60hDg}{fE} \tag{22.4}$$

where T_h = shell tension, lb per inch, at depth h
h = depth from design liquid level, ft
E = joint efficiency factor
f = allowable unit stress, psi
t_h = required shell plate thickness for hydrostatic pressures, in.

For welded construction, *E* varies from 0.75 to 1.0, depending on the type of joint used and the required welding inspection procedure. Lap-welded joints, where used, are based on fillet welds with leg size equal to the full thickness of the plates being joined. Lap welds in the bottoms of flat-bottom tanks and in roof plates not in contact with water are only required to be welded from the top side. Lap welds in roofs and shell plates in contact with water must be double-welded lap joints, or butt joints must be used. Lap joint construction for tank shell joints is allowed under AWWA D100 but is not allowed in some of the other tank standards.

The vertical joints for bolted steel tanks are lap joints. The horizontal joints for bolted tanks may be either formed flange joints or lap joints (Fig. 22.1). The type of horizontal joint used by the tank manufacturer depends on the equipment used in the manufacturing plant, and, in some cases, the type of joint may also be predicated on the type of coating to be applied. The maximum allowable shell thickness for bolted steel tanks is limited by AWWA D103 based on the type of horizontal joint employed. Bolted tanks conforming to API 12B are only available with a formed flange type joint.

The design joint efficiency factors and allowable stress levels vary with each tank design standard and, in some cases, additionally by the type of tank or design basis selected within each standard. These values are based on the material properties, level of quality control, and inspection requirements associated with the selected design approach.

For welded joints, only full-penetration butt joints are permitted at joint efficiencies greater than 0.75. And, the joint efficiency may exceed 0.85 only if extended radiographic examination is provided and more restrictive material and design requirements are imposed. Those increased requirements include thickness limitations, design metal temperature limitations, and, in some cases, notch toughness requirements. Table 22.1 lists the joint efficiency factors in, "E", AWWA D100 for the various joint types listed. For AWWA D100 tanks, the joint efficiency for shell joints may only be increased to 1.0 when all requirements of Section 14 of that standard are incorporated into the design, construction, and testing of the tank. For API 650 tanks, the joint efficiency is 1.0 except for small tanks designed under the reduced rules of Annex A of that standard. Each tank standard has its own special requirements. The details of each standard should be thoroughly reviewed and understood prior to commencing tank design activities. For water tanks, partial-penetration butt-joint welds achieving at least a two-thirds effective throat are permissible, with some limitations, only in tank shell joints subjected to secondary stress, such as horizontal seams of ground-supported flat-bottom tanks and only when design in accordance with Section 14 is not used. However, when heavy vertical loads are supported by the shell, the vertical stress is considered to be a principal stress, and the horizontal seams could require butt joints with full-penetration welding.

Table 22.1 Efficiencies, "E", of Welded Joints for AWWA D100

Type of welded joint	*E*	Remarks
Double-welded lap joint*	0.75	Continuous welds
Butt joint, partial joint penetration† (⅔ min.)	*Z*/*T*‡	Compression joints only
Butt joint, complete joint penetration	0.85	Tension joints
Butt joint, complete joint penetration	1.00	Compression joints

*Double-welded lap joints are only allowed for shell joints in AWWA D100 and API 12D
†In a partial penetration butt joint, the unwelded portions must be located substantially at the middle of the thinner plate being joined.
‡*Z* is total depth of weld penetration from the surfaces of the thinner plate being joined.
T is the thickness of the thinner plate being joined.

22.3.3 Bottom Plates

The flat bottoms in reservoirs and standpipes are usually grade supported, with the liquid load on the top side being resisted by an equal upward foundation pressure from the tank pad. Steel bottoms usually have a minimum thickness of ¼ in. Unless special design conditions dictate the use of butt-welded bottom plates, the tank bottom is assembled using lap-joint construction. The plates are lap welded to each

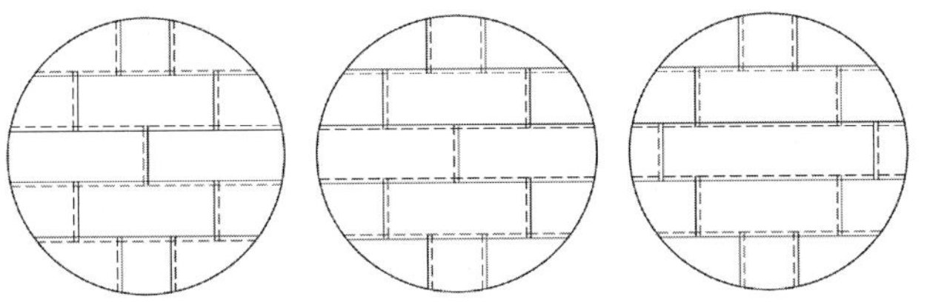

Figure 22.10 Lap-welded bottom configurations.

other on the top side only for liquid tightness. Figure 22.10 shows three different plate layouts for tank bottoms. Figure 22.10, sketch "*a*," shows a lap-welded bottom with the plate placement beginning from the center of the tank and progressing toward the perimeter. Figure 22.10, sketch "*b*," shows a lap-welded bottom with the plate placement beginning from perimeter and progressing toward the center. Different tank manufacturers use different procedures for placing the bottom plates. Either of these two methods are acceptable unless there is an operational consideration that requires the laps to be made in a specific direction. Figure 22.10, sketch "*c*," shows a lap-welded bottom with no lap seam at the tank center. This configuration is sometimes preferred if there will be a center column and base plate structure at the tank center, but it is recommended that this option should only be selected for tanks with flat floors. Also, for many bottom shapes it is often beneficial to include a lap joint along the centerline.

Floors may be made using butt-welded construction. However, the cost for butt-welded construction is significantly higher due to the requirements for pattern cutting and machining of each individual plate, backing strips for the one-sided butt-weld, and additional welding time. The contours of lap seams generally can be easily dressed by the tank manufacturer after welding. Post-weld dressing usually involves a needle scaler to remove slag and a grinder to remove dingleberries and to contour the weld seam to make it suitable for the application of protective coatings. So, unless there is a specific requirement for butt-welded joints, lap-welded construction is generally the preferred option. The maximum thickness allowed by the tank standards for lap-welded bottoms is ⅜ in.

Figure 22.11 shows the simple detail required for connecting the bottom to the tank shell. In this detail, the weld size is not indicated because the tank design standards have specific requirements for the size of the fillet welds connecting the tank shell to the bottom plate. For other than water tanks designed to AWWA D100, different welding details might be required for this joint for specific design conditions.

Support of relatively light shells requires no special consideration. When the tank is not supported on a concrete ringwall and the allowable soil pressure directly under the shell is exceeded, increasing the projection "*a*" (Fig. 22.11) of the bottom plate chime may suffice to

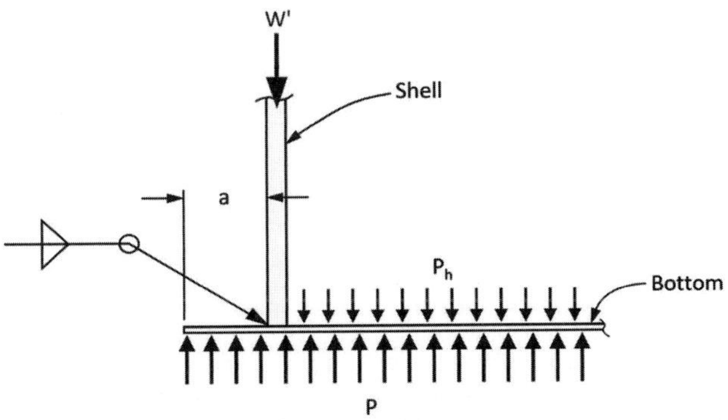

Figure 22.11 Load diagram of bottom pressures, chime projection "*a*," and connection of bottom to shell plate.

reduce the bearing pressure to an acceptable value. With the bearing pressure, *P*, at its maximum acceptable value, the projection "*a*" is given by Eq. (22.5*a*):

$$a = \frac{12W'}{2P - P_h} \tag{22.5a}$$

$$P_h = 62.4Hg \tag{22.5b}$$

where a = projection of the tank chime, in.
W' = shell load, including the weight of the shell and the shell supported roof weight, lb per foot of circumference
H = design liquid level, ft
P = allowable soil pressure, psf
P_h = hydrostatic pressure at the design liquid level, H, psf

For AWWA D100 tanks, the minimum projection of the chime is 1 in. past the toe of the fillet weld between the tank shell and the bottom plates with a practical limit of about 6 in. Other standards have differing requirements on the minimum chime dimension, depending on tank type and other design criteria. Check the applicable standard for specific requirements.

The large shell compressive loads caused by earthquakes would require impractical projections if the engineer attempts to spread out the full load within allowable soil-bearing pressures. Experience shows that limited shell settlement can be tolerated during seismic loading with no detrimental effect except that the tank must be releveled. However, the possibility of a slope failure of the foundation must be investigated since this could result in loss of support over a large area of the bottom and trigger a bottom failure, with a consequent loss of contents and possible complete failure of the tank.

22.3.4 Roofs for Reservoirs and Standpipes

Although tank roofs may be constructed in a multitude of shapes (flat, conical, sloped, domed, elliptical, or torispherical), AWWA D100 requires a minimum roof slope of ¾:12. Additionally, for potable water tanks, some state or local regulations might require that tank roofs be fully self-draining with no areas of ponding. Therefore, the engineer must consider all requirements when establishing the design roof slope.

Roofs can be column supported, supported by a self-supporting structural system, or self-supporting membrane structures with or without stiffening elements. Column-supported roofs are usually conical with a minimum slope of ¾ in. per 12 in. Due to deflection of the roof plates between supporting rafters, to ensure adequate drainage off the roof a greater slope is preferable, and the author recommends a minimum slope of 1:12.

The column-supported roof can be a single-span roof supported on a single-center column or it could be a multi span roof supported by multiple bays of rafters supported on one or more rings of girders and a central support. Columns and beams are arranged as shown in Fig. 22.12 and can be positioned so all columns carry approximately equal loads. All portions of the roof rafter or trusses must be located above the top capacity level (TCL). Additionally, tank standards require that freeboard be provided between the maximum operating level (MOL) and the lowest portion of the roof framing to reduce the likelihood that a seismic sloshing wave would impact the roof structure.

In some instances, unique structural geometries are necessary as discussed in Sec. 22.2.4. However, generally, rafters are arranged in a radial fashion. AWWA D100 allows a maximum rafter spacing of 7 ft or less depending on the design roof load. The maximum rafter spacing for AWWA D100 is not to exceed 84 in. and is calculated by the following formula:

$$L = \frac{2575t}{\sqrt{W_{D+L}}} \tag{22.6}$$

where L = center to center spacing of the rafters at their outer end, in.
t = roof plate thickness, in.
W_{D+L} = roof dead load plus live (or snow) load, psf

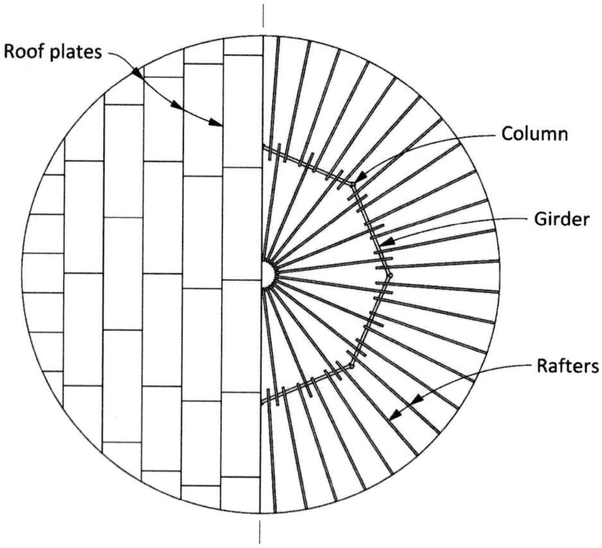

Figure 22.12 Roof plates and framing.

In conventional construction meeting the minimum requirements of the tank standards, roof plate is laid directly on the rafters without attachment to them. See Sec. 22.2.4 of this chapter for discussion of alternative roof construction details for improved corrosion protection strategies and extended tank life. Tank standards only require the roof plates to be lap-welded to each other from top side, for water tightness. The rafters are generally clip-attached to the shell and to the girder (Fig. 22.13). Girder clip details vary depending on the rafter type (e.g., channel, flanged beam).

Although the tank standards allow rafters to be set directly on the girder flange, this detail is not recommended for tanks on which a protective coating system will be applied. The rafters may be welded or bolted to the clips. Bolt size and number are determined by the connection type and loading. Rafter connections should be designed to accommodate possible column settlement without imposing significant axial compressive loads on the rafters.

The girder system may be circular in plan view, if the girders are rolled to a radius, or it may take the form of a polygon as shown in Fig. 22.12. It is recommended that the girders be bolted to the columns to avoid induced bending in the girders due to differential settlement of the columns. If girders are welded to the columns, are continuous over the columns, or are provided with flange continuity plates over the columns, the design should include consideration of the change in stress distribution in the girders that might result from differential settlement of the columns.

Roof framing is designed for dead load plus live load in addition to load cases including snow, wind, and seismic effects. The tank standards specify minimum load values. Wind and snow loads are determined in accordance with the applicable building code and load data from the local jurisdiction. In climates with very little snowfall a minimum live load of 15 psf may be used only if allowed by the tank design standard and not prohibited by the local jurisdiction.

Roof framing and columns are designed in accordance with AISC specifications. The columns are designed with a maximum slenderness ratio, KL/r, of 175. Columns are preferably designed using pipe sections for uniform lateral stiffness in all directions as well as simplicity for painting operations. For AWWA D100 tanks, if pipe columns are used, they must be hermetically sealed or designed with a corrosion allowance of ⅛ in. Additionally, due to the thickness underrun allowance of some materials specifications (ASTM etc.), the permissible thickness underrun of those specifications must be deducted from the nominal thickness when calculating column stresses. If used, built-up columns made from two or more structural shapes are required to be fully seal welded their entire height at all adjoining surfaces.

Under normal design conditions of gravity loads on the roof, the top flange of the rafter is in compression. The compression flanges of the rafters are considered laterally supported by the friction of the roof plates, and compression flanges of the girders are laterally supported by the rafters. However, experience has shown that differential settlement of the foundation, and even sometimes thermal gradients, can cause bowing, twisting, or unanticipated deflection of the rafters. Therefore, it is recommended that consideration be given to adding lateral bracing even if not required by design.

Although rafters are available with a published minimum yield strength exceeding 36 ksi, AWWA D100 requires the design of rafters to be based on the properties of ASTM A36 material unless the roof design live load exceeds 50 psf, in which case the design may use the actual yield strength of the selected material if higher than 36 ksi.

Columns are supported by steel base plates resting directly on the bottom plate. Size of the base plates is determined by the column load and the allowable soil pressure. Other types of structural built-up base plates may be used, but for tanks with internal coatings, the steel base plate option is preferred. When soil strength is not known, the base plate is designed for 1500 psf over an area 1 ft larger in diameter or width than the base plate. For corrosion protection considerations, the base plate should be welded to the tank floor. If it is necessary to provide flexibility at the base of the column, it is recommended that a thickened cap plate and sacrificial plate be installed at the bottom of the column with guide bars or angles installed on top of the sacrificial plate (Fig. 22.14). This preferred configuration limits the potential corrosion to occur between the column cap plate and the sacrificial plate rather than between the base plate and the tank floor which could lead to possible perforation of the tank bottom. AWWA D100 requires full priming and painting for any portions of the base assembly that will be inaccessible after erection of the structure. This would apply to the opposing surfaces of the column cap plate and the sacrificial plate when a fully welded base assembly is not used.

Tanks 100 ft or less in diameter may have rafter-supported roofs without supporting columns. The roof framing is steeply sloped to provide economical design. Such rafters are designed as beams loaded with end thrust and vertical loads. The reactions are a horizontal thrust H at the inner end and an equal horizontal thrust and a vertical reaction V at the outer end (Fig. 22.15, sketch "*a*").

On large-diameter tanks where it is desirable or necessary to avoid columns, a number of self-supported roof options are available. The tank roof may be a stiffened umbrella or dome roof carried by the shell (Fig. 22.15, sketch "*b*") or supported by trusses and rafters (Fig. 22.15, sketch "*c*"). Self-supported roofs are used for economy and simplicity of design and construction. In the case of rafters and trusses, they are

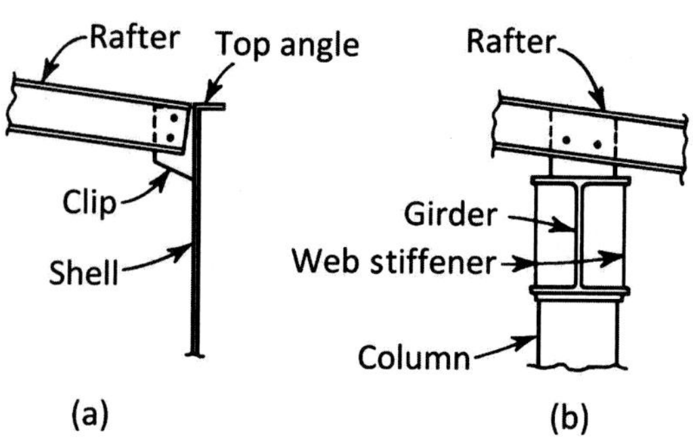

Figure 22.13 Rafter connections.

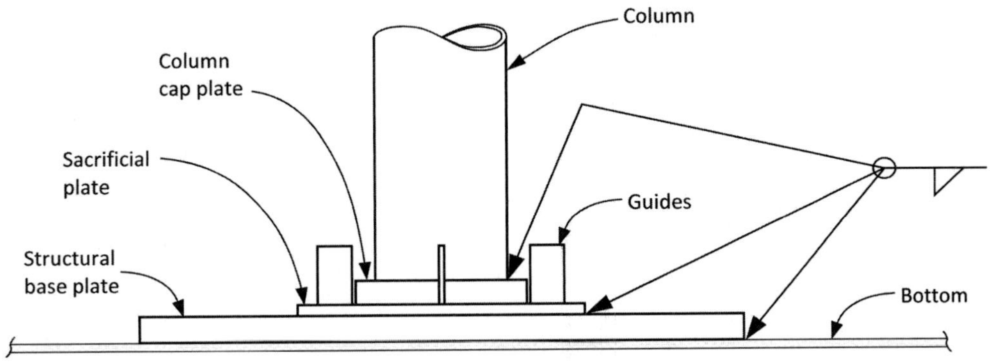

Figure 22.14 Recommended detail for column base plate when welding to the floor is not desired.

arranged in a radial pattern (Fig. 22.15, sketch "*d*"). The trusses and rafters are supported by clips welded to the top shell ring. Due to the concentrated loads at truss connections, shell stiffeners or reinforcing plates might be required. The main rafters or trusses are connected to rings at the center of the tank. The top chords of the trusses are connected to a compression ring, the bottom chords to a tension ring. A torsion tube is provided at the center for truss stability. Eccentricities caused by radial misalignment of truss chord members generate torques tending to rotate the center rings. Interconnecting the top-chord and bottom-chord center rings with a torsion tube provides rotation stability; thus, torque induced by the top-chord compression forces is counteracted with torques generated by bottom-chord tension forces (Fig. 22.15, sketch "*c*").

Self-supporting cone roofs without rafters or stiffeners are normally limited to tank diameters less than 35 ft and may have a slope of 30 degrees or greater. Tanks with a diameter less than about 75 ft can be provided with umbrella, dome, ellipsoidal or torispherical roofs without rafters,

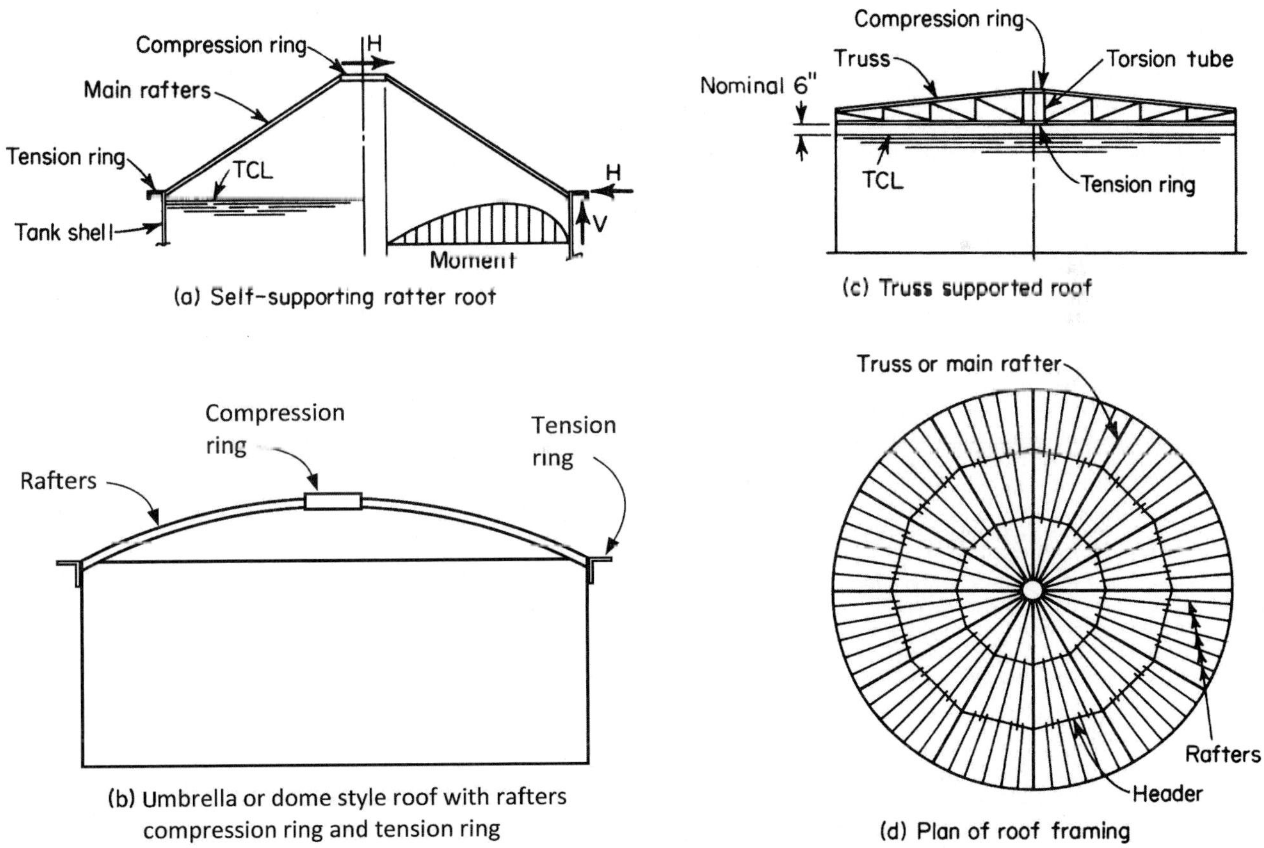

Figure 22.15 Examples of self-supporting roof framing options.

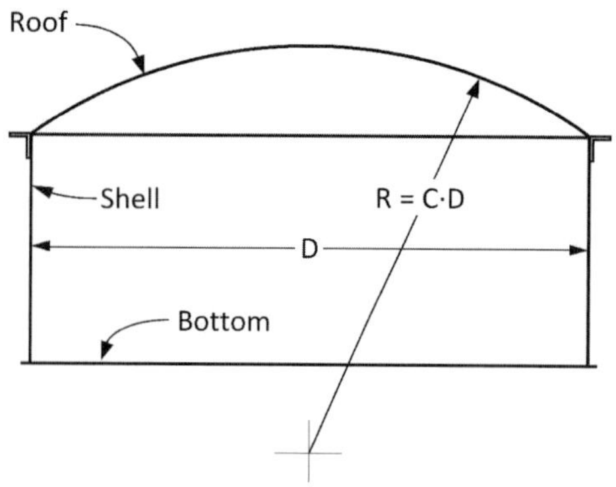

Figure 22.16 Unstiffened umbrella or dome style roof.

but height of the roof will be large. Unstiffened umbrella and dome roofs (Fig. 22.16) may have a dish radius, R, ranging from 0.8 to 1.2 times the tank diameter. The umbrella roof is a spherical cap, like the dome roof, but the umbrella roof is formed with segments curved only in the radial direction except where connected to the tank shell.

Based on the design rules of API 650, the thickness of such an unstiffened umbrella or dome roof can be determined by the following formula:

$$t = 0.00075CD\sqrt{W} \qquad 0.8 \le C \le 1.2 \tag{22.7}$$

where C = roof radius coefficient = R/D
D = tank diameter, ft
R = roof radius, ft
W = the governing combination of dead load, live load, and external pressure load acting on the roof surface, psf (see API 60 Section 5.2)
t = plate thickness, in.

Single-lap-welded joints may be used on the roof plate segments with thicknesses up to ⅜ in. Thicker plates must use butt joints to join the roof segments.

Roof Framing Special Conditions

Under certain conditions, upgraded roof framing should be considered. During hurricanes or other high-wind events, roof plates will flutter under sustained high winds and may cause the supporting framing to roll over from impact or end thrust loads. Framing in hurricane regions can be stabilized with added bracing or blocking, but this is not required by code. In an earthquake, roof plates can provide limited stabilization of the framing, but displacement of the framing has been observed after an earthquake. Additionally, the engineer may choose to modify the selected structural shapes to improve the lateral stiffness of the members. Roof damage will not be prevented if the sloshing wave impinges on the framing or roof plate, so sufficient freeboard must be provided to clear the maximum wave height. Currently, the IBC requires design for seismic loads per ASCE 7 which requires either providing the calculated amount of freeboard between the design water level and the tank roof framing or designing the affected portion of the roof to resist an applied pressure uplift equivalent to the pressure of static head for the freeboard deficiency.

For tanks storing a product that is aggressively corrosive, special roof details are required. See Sec. 22.2.4 of this chapter for discussion about alternative roof construction details.

Tension and Compression Rings on Self-Supported Roofs

The ring is loaded with N equal radial loads W, considered positive outward (Fig. 22.17). This loading system produces bending and direct stress in the ring with the maximum stresses occurring at the load points and at midpoints between loads.

When the value of W in Fig. 22.17 is positive, this diagram represents the load case for a tension ring. For compression rings, the values for W

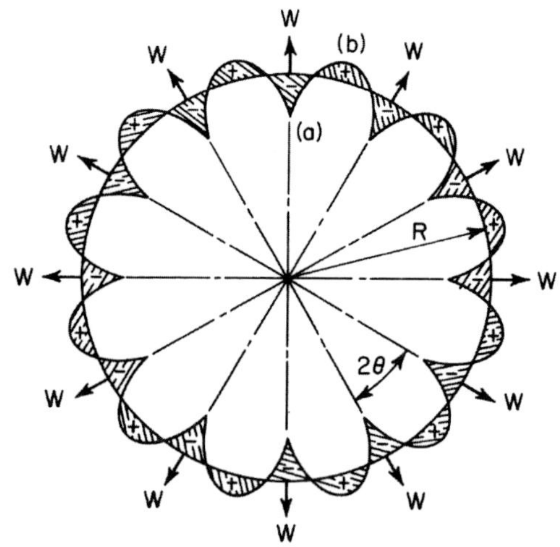

Figure 22.17 Load and moment diagram for tension and compression rings.

would be negative. The moments M, circumferential thrusts T, and radial transverse shears V are determined by the following equations:

At load points,

$$M_a = \frac{WR}{2}\left(\cot\frac{180°}{N} - \frac{N}{\pi}\right) \tag{22.8}$$

$$T_a = \frac{W}{2}\cot\frac{180°}{N} \tag{22.9}$$

$$V_a = \frac{W}{2} \tag{22.10}$$

At midpoint between loads,

$$M_b = \frac{WR}{2}\left(\csc\frac{180°}{N} - \frac{N}{\pi}\right) \tag{22.11}$$

$$T_b = \frac{W}{2}\csc\frac{180°}{N} \tag{22.12}$$

$$V_b = 0 \tag{22.13}$$

where R = ring radius, ft
W = the radial thrust load, lbs
NOTE: When analyzing the compression ring, the load W, in Fig. 22.17, carries a negative sign.
N = number of radial loads W

Positive thrust produces tension in the ring and positive moment tension at the inside face. The stress in the ring is

$$f = +\frac{T}{A} \pm \frac{Mc}{I} \tag{22.14}$$

Example details of a compression ring made of a channel are shown in Fig. 22.18.

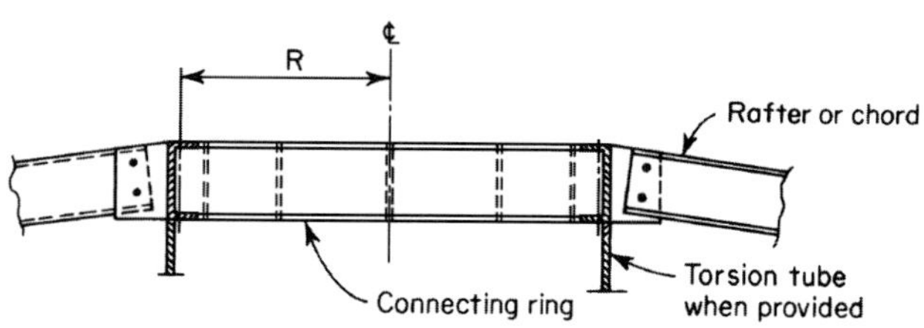

Figure 22.18 Example details of compression ring.

22.3.5 Anchorage

Standpipes and reservoirs must be checked for resistance to overturning moments caused by wind or seismic forces. The wind overturning load case governs when the tank is empty, and the seismic overturning load case governs when the tank is full to the maximum seismic design water level. Overturning anchorage consists of anchors that can provide the necessary resistance to uplift. Generally, the anchors are embedded in the ringwall or concrete mat foundation and connected to the lower part of the tank shell. Two types of anchors are normally used: anchor bolts or anchor straps. When anchor bolts are used, they are generally attached to the tank shell by means of anchor bolt chairs (Fig. 22.19). The design procedure for anchor chairs has been in practice for many decades. The design procedures for other types of anchor bolt attachment assemblies should be well understood for its impact on the tank shell prior to their utilization in the design. Anchor attachments other than anchor chairs must be analyzed for the stresses imparted into the shell and the ability of the attachment to resist the amplified seismic load [see Eq. (22.18)]. When anchor straps are used, AWWA D100 includes requirements for minimum thickness, corrosion allowance, and geometry.

Anchor bolts should be provided with a lock nut or the threads must be peened to prevent loosening of the nut. The engineer must consider which option will be used when determining the anchor bolt projection. The AWWA D100 standard requires that anchor bolt installation shall be specified to include 2-in. nominal projection of the threaded end of the bolt above the top nut to allow for variations in the ringwall elevations. When anchor straps are used, they must be configured to provide a ductility zone, and the detailing should be thoughtfully planned to allow proper corrosion protection. Consult AWWA D100 for additional requirements for each type of anchor, including corrosion allowance requirements, minimum size, and minimum quantity.

For mechanically anchored tanks, regardless of whether anchor bolts or anchor straps are used, the tank shell must be shimmed and supported on structural grout as shown in Fig. 22.19.

Design of special anchor reinforcement, if required, and anchor embedment is completed in accordance with the requirements of ACI 318, Appendix D.

Acceptable anchor bolt materials are specified in the tank standards. As of this writing, the preferred anchor bolt specification is ASTM F1554. This ASTM specification has introduced the preferred term of "anchor rods," but in this chapter they are referenced as anchor bolts. ASTM F1554 is specifically for anchor bolts and includes three grades. High-strength bolts ($F_y > 55$ ksi) are not permitted under AWWA D100 unless the diameter of mild steel bolts exceeds 3½ in. If high-strength anchor bolts are used, after tank construction is completed, the anchor bolts are required to be pretensioned to 80 percent of their design load.

WIND LOADING

An approximate but practical solution for anchor bolt wind loading is

$$T_w = -\frac{W'}{N} + \frac{4M_w}{ND_a} \tag{22.15}$$

where T_w = anchor bolt tension, lb
M_w = wind load overturning moment at the base of the tank, ft-lb
D_a = diameter of the anchor circle, ft
N = number of anchor bolts
W' = total weight of shell and that portion of the roof dead load supported by the shell

The wind loads are determined in accordance with ASCE 7 for each height zone of the structure and applied according to the detailed procedure in AWWA D100. The procedure includes modification of the ASCE 7 loads based on shape factors specified in AWWA D100 for the geometric configuration of each height zone. A quick approximation of the overturning moment due to wind loads on a standpipe can be calculated using Eq. (22.16). This equation is based on an estimated average wind pressure:

$$M_w = 0.00005 p_{est} KDH^2V^2 \tag{22.16}$$

where p_{est} = estimated average wind pressure, psf, without shape factor, taken as no less than 30 psf
K = 0.6 for cylinders and 0.5 for doubly curved surfaces
V = wind velocity, mph
H = shell height, ft

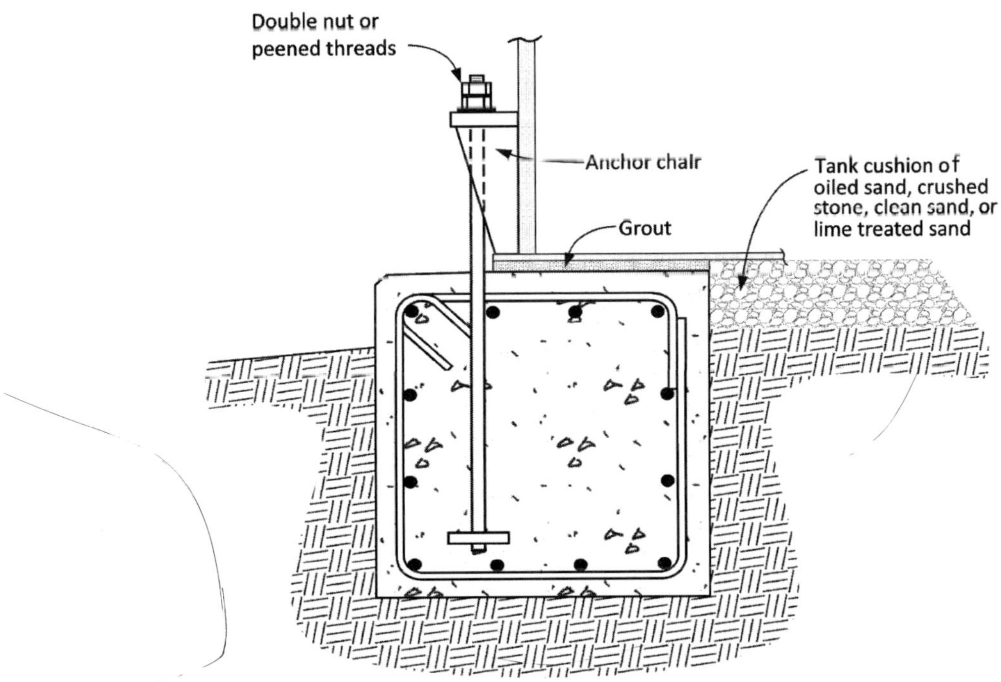

Figure 22.19 Anchorage details.

Seismic Loading

An approximate solution for seismic anchor bolt loading based on a rigid mass model of the tank contents is as follows:

$$T_s = -\frac{W'}{N} + \frac{4M_s}{ND} \qquad (22.17)$$

$$T_{att} = -\frac{W'}{N} + \frac{16M_s}{ND} \qquad (22.18)$$

$$M_s = (W_s + W_w)\frac{H}{2}S \qquad (22.19)$$

$$W_w = 49.0HD^2 \qquad (22.20)$$

where T_s = seismic bolt tension, lb
T_{att} = anchor attachment design tension, lb
M_s = seismic overturning moment at the base of the tank, ft-lb
W_s = total weight of tank shell and roof, lb
W_w = total weight of contents, lb
S = estimated seismic load factor, percentage of gravity
H = design liquid level, ft

NOTE: Equation (22.19) assumes a rigid fluid mass. However, the fluid may be considered to behave as a two-component system consisting of a rigid (impulsive) mass and a sloshing (convective) mass. Design procedures for the sloshing mass model are covered in Refs. 2, 3, and 10.

The attachment of the anchor to the tank must be designed for a larger force to ensure that the anchor can yield prior to failure of the attachment or tearing of the tank shell. AWWA D100 requires that the design load be the larger of the anchor bolt yield strength or a design load determined with a multiplier of 4 on the seismic load effect per Eq. (22.18).

In Eqs. (22.17) and (22.18), no vertical liquid load is included in W' because the anchor bolts develop full tensile load before any liquid on the thin flexible bottom plates can develop effective downward ballast.

Spacing of anchor bolts should not be less than 2 ft or greater than 10 ft; that is, the number of anchor bolts, N, must fall with the range given by

$$0.31D < N < 1.57D \qquad (22.21)$$

In any case, N must not be less than 6 for welded steel water tanks.

Self-Anchored Tanks

Flat-bottom tanks may survive seismic load without mechanical anchorage. In this case, a portion of the liquid contents near the shell resists the overturning uplift forces on one side of the tank. On the other side of the tank, the overturning forces also produce vertical compression in the bottom shell ring which may cause it to buckle. Tanks with a large height-to-diameter ratio are more susceptible to buckling and overturning instability. On the other hand, Ref. 31 indicates that tanks with a small height-to-diameter ratio may experience overstress in hoop tension in the middle and upper courses. Detailed procedures for calculating overturning stability, uplift resistance, overturning shell compression, and hydrodynamic hoop stresses can be found in Section 13 of Ref. 2, Section 14 of Ref. 3, and in Annex E of Ref. 10.

The 1994, the Northridge earthquake provided a full-scale laboratory to study the behavior of self-anchored storage tanks in a major earthquake. A database of 33 tanks out of an estimated 200 tanks exposed to this earthquake was accumulated by R.S. Wozniak as reported in the previous, fourth edition of this book. According to his study, two-thirds of the tanks in the database were damaged. Where enough information was available, he analyzed the tanks using AWWA D100 seismic-design procedures in effect at the time of the study. Wozniak reported that in all cases, design of the tanks per AWWA D100 predicted that shell-buckling damage would occur to the tanks which sustained damage in the Northridge earthquake, and it predicted that no damage would occur to tanks that suffered no damage in the earthquake. Figure 22.20 is a plot of his Northridge data base and predicted-damage curve. Tanks that fall above

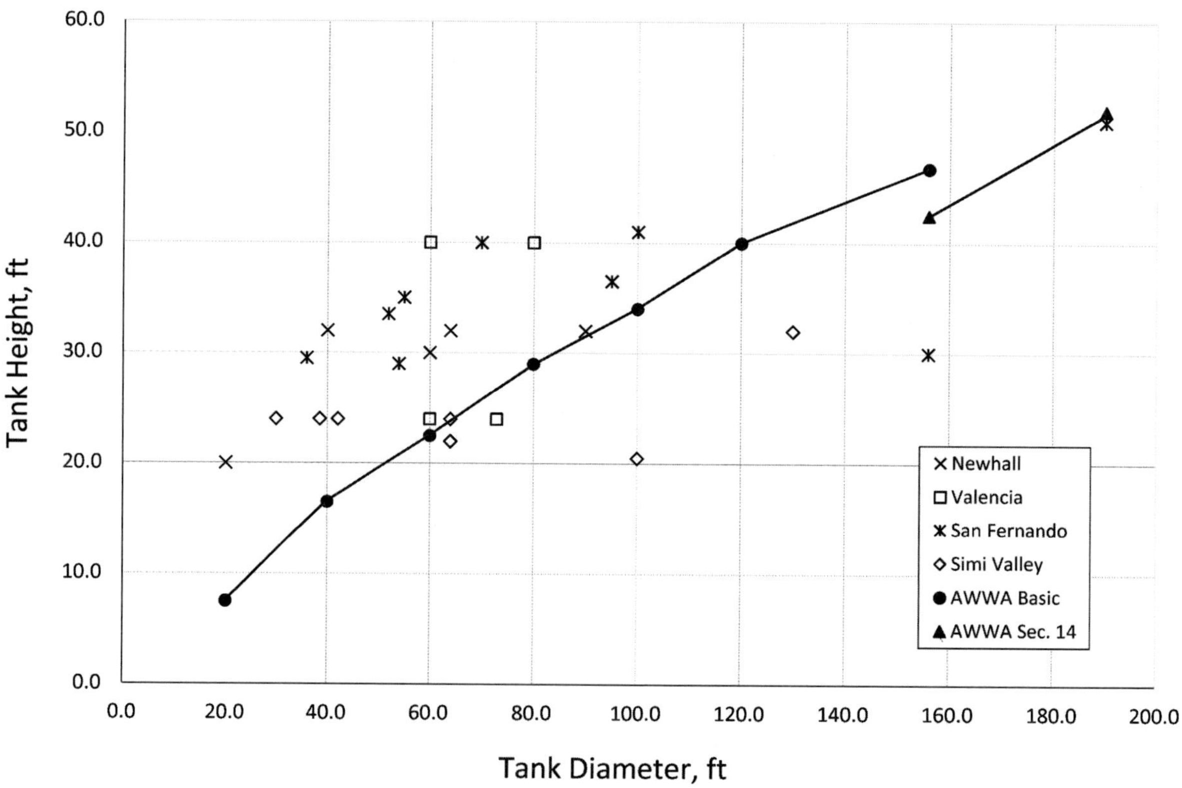

Figure 22.20 Northridge earthquake damage to tanks vs. AWWA D100 design limits. (*Source: 4th Edition, Ch. 29, R.S. Wozniak*).

the damage curve are expected to suffer elephant-foot buckling, piping damage, and possible bottom tearing in a major earthquake. The damage curve in that study did not account for sloshing-wave damage to the roof due to insufficient freeboard.

AWWA D100 has included the self-anchored design procedure since 1976. The original procedure, developed by R.S. Wozniak in 1964 from a study of tank-damage reports, still forms the basis of the design procedure in the tank standards today. Reference 32 includes technical data relevant to the development of procedures for seismic analysis of cylindrical tanks. Reference 31 contains the background on the development of Wozniak's procedure.

Types of tank damage observed after the Northridge earthquake in 1994 were reviewed for consideration of possible new requirements in revisions to the AWWA standards. After review of tank damage suffered during the Northridge earthquake, freeboard was suggested as a nonmandatory AWWA option as a protection. Since that time, freeboard, or structural mitigation of insufficient freeboard, has been incorporated as a mandatory requirement in the IBC and ASCE 7. It is anticipated that the tank standards will adopt similar requirements in their next revision cycle. The magnitude of required freeboard is dependent on tank geometry and location-specific seismic design parameters.

The largest percentage of component damage reported from the Northridge earthquake was to the piping components. In most cases, piping damage was coupled with elephant-foot buckling. Self-anchored tanks that met AWWA uplift criteria did not appear to suffer as much piping damage. However, the limited database information is inconclusive as to consistency of this observation. Since most self-anchored tanks subjected to moderate or high seismic loads must lift to resist seismic forces, the current piping flexibility requirements in AWWA should be considered minimum values. AWWA D100 and ASCE 7 require piping flexibility for all tanks when seismic design parameters exceed specified threshold values, even those tanks that are mechanically anchored.

22.3.6 Foundations

Concrete ringwalls (Fig. 22.21) are used to provide reliable support under the shell, and ballast for anchor bolt uplift loads. A minimum ringwall width, b, of 12 in. is recommended to provide reasonable construction tolerances. When the ringwall must be wider than 24 in. to get acceptable bearing values or to provide resistance to uplift, an inverted T-footing should be investigated to allow the ringwall width to be reduced. This investigation should include both the structural efficiency of the footing options as well as the cost of construction associated with each option.

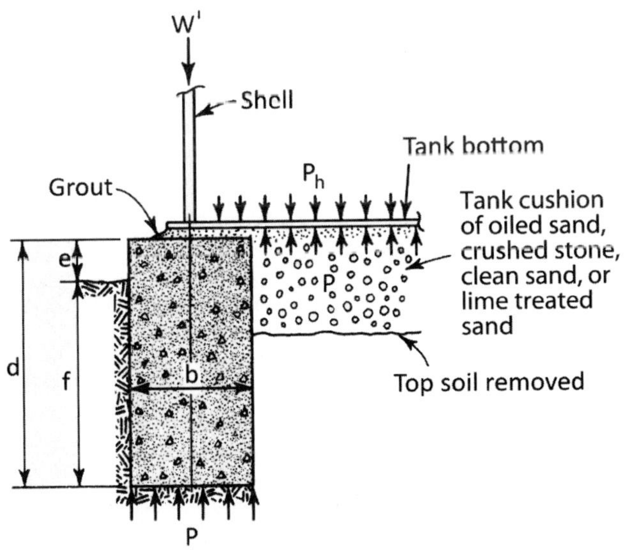

Figure 22.21 Foundation ringwall.

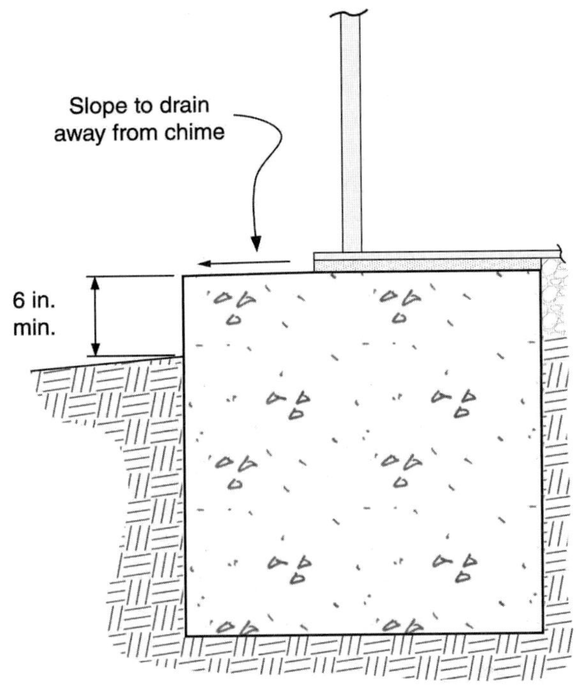

Figure 22.22 Recommended ringwall profile.

AWWA D100 requires the height, e, of the ringwall above grade to be at least 0.5 ft. It is recommended that the exposed top of the ringwall be sloped to drain away from the tank chime. Figure 22.22 provides a recommended detail. The bottom of the ringwall should be at least 0.5 ft below the frost line. A map of extreme frost penetration in the United States is provided in Ref. 2. A portion of that map does not include frost penetration depth due to the wide fluctuations in values within that region. Although other maps can be obtained that include complete frost penetration depth curves for the entire United States, including the region omitted from the map in AWWA D100, it is recommended that local weather data be consulted for projects in the omitted region.

Circumferential reinforcing steel must be provided in the concrete ringwall to develop the hoop stress produced by lateral soil pressure, due to the weight of the stored water, within the ringwall. The required area of circumferential steel is determined by

$$T_r = 31.2K_h HDdg \tag{22.22a}$$

$$A_s = \frac{1.7\,T_r}{0.9 f_y} \tag{22.22b}$$

where d = ringwall height, ft
T_r = ringwall tension, lb
f_y = rebar yield stress, psi
K_h = lateral earth-pressure coefficient
A_s = required area of reinforcing steel, in.^{2}

It is recommended that the lateral earth-pressure coefficient be taken not less than 0.30 unless recommended by a recent detailed geotechnical investigation of the ground area to be covered by the tank. The engineer should use higher values if based on knowledge of the tank site or if indicated by the geotechnical investigation.

Vertical steel that meets minimum temperature-steel requirements must be installed to provide supports and spacers for the hoop steel.

The eccentricity of the imposed loads relative to the centroid of the ringwall produces a bending moment on the ringwall cross section referred to as the twisting moment. Additional circumferential steel is

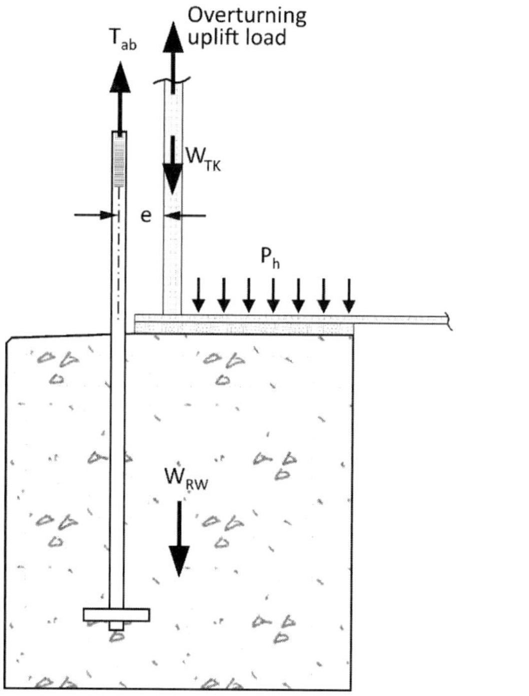

Figure 22.23 Ringwall load diagram for resistance to overturning uplift loads.

needed to resist this twisting moment. Additionally, for mechanically anchored tanks, an additional twisting moment condition is produced due to the uplift force exerted on the anchor bolt and the downward forces exerted by the tank shell, the liquid load on the portion of the ringwall inside the tank shell, and the weight of the ringwall itself (Fig. 22.23). References 33 and 34 are good resources for foundation design considerations. They include background, discussion, and guidance on many topics related to the design of foundations for ground-supported tanks.

The anchor bolt force, T_{ab}, is transferred from the tank shell, through the anchor chair to the bolt. The eccentricity, e, between the tank shell and the anchor bolt should be kept to a minimum since the local stresses induced in the tank shell are commonly the governing criteria for design of the anchor chair and, sometimes, also requires an increase in the required shell thickness. The procedures for design of the anchor chairs and analysis of the local shell stresses are found in Ref. 21. Reference 35 includes information and discussion of additional topics relevant to anchor bolt design.

Support for Tank Shells

Tank shells for mechanically anchored tanks are required to be shimmed and grouted. Concrete foundations, either ringwalls or slabs, are rarely poured within the required tolerance to provide uniform shell support with constant thickness shims. Therefore, the total thickness of the shim pack at each location is selected to provide a constant support elevation upon which to build the tank. This method accommodates the varying elevation of the top surface of the concrete foundation. The shim packs at each location must provide the required minimum of 1 in. of grout. A structural grout is then packed into this space to provide uniform support to the tank shell. For self-anchored tanks on concrete foundations, the shell may be supported on cane-fiber joint filler if the concrete surface beneath the tank shell meets the level tolerance specified in AWWA D100. Otherwise, the tank must be shimmed and grouted as noted above. For self-anchored ground-supported tanks, foundations other than ringwalls or concrete slabs may also be used. Foundation requirements may be found in AWWA D100, Section 12.

Tank Cushion

A tank cushion is placed beneath the tank floor. Except for concrete slab foundations, an oiled sand base is the default tank cushion for D100, but other acceptable cushion materials are noted in Section 12 of AWWA D100. The asphalt cutback oils used in oiled sand mixtures are the same types of oils commonly used in construction of road beds. The selection of the viscosity and curing characteristics of the oil should be evaluated to suit the local climate conditions of the work site, the construction season, and the planned installation methods. The oil used to create the oiled sand mix is typically either a slow-curing asphalt cutback grade complying with the requirements of ASTM D2026 or a medium-curing asphalt cutback grade complying with the requirements of ASTM D2027. The oiled sand should be provided as a plant mixed product, or the oil should be thoroughly mixed with the sand on-site prior to placement.

Foundations for Standpipes

Tall standpipes that are ground-supported require high soil-bearing allowables. Ringwalls may be required to reduce soil-bearing pressure to acceptable limits under the tank shell. The design of ringwalls for standpipes is the same as for reservoirs.

When the allowable soil pressure, P, is less than $62.5H$, spread foundations are required similar to that shown in Fig. 22.24. In this case, the required base diameter, Z, is approximately given by

$$Z = 1.128\sqrt{\frac{W_m + W_w + W_e + W_c}{P}} \tag{22.23}$$

where W_m = total weight of tank metal, lb
W_w = weight of contents, lb
W_e = weight of earth on foundation, lb
W_c = weight of concrete, lb

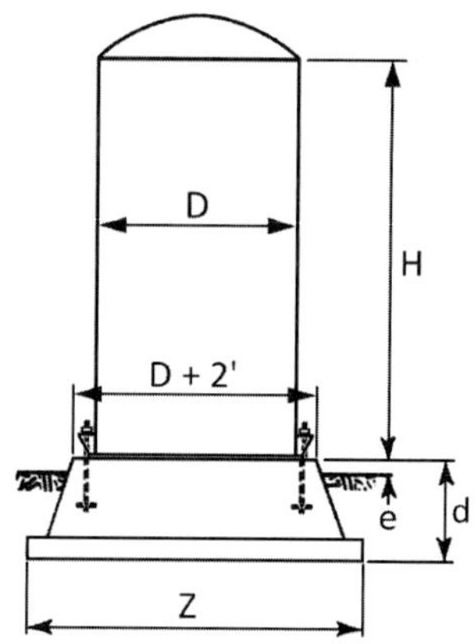

Figure 22.24 Spread foundation.

Tensile Reinforcement

The tank loads combined with the soil pressures acting on the bottom of the foundation produce a bending moment about the centerline of the spread foundation. Tensile reinforcement in the bottom of the foundation is determined by this bending moment. The bending moment due to static loads, M, about a centerline (Fig. 22.25) is given by

$$M = \frac{1}{2}(VX_p - W_mX_m - W_wX_w - W_cX_c - W_eX_e) \tag{22.24}$$

where V = total foundation reaction, lb, = $W_m + W_w + W_c + W_e$
X = distance from the centerline to the centroid of each respective load on one-half the foundation, ft

Reinforcing steel should be distributed uniformly across the diameter, with an equal amount in the transverse direction to form a rectangular grid. If the required area of steel is excessive, greater foundation depth may be required. The entire foundation should be poured as a monolith to eliminate construction joint difficulties and shear-key problems.

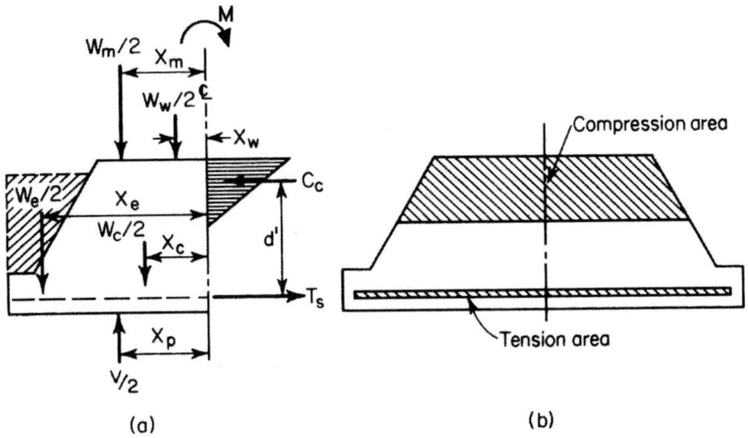

Figure 22.25 Free-body diagram and cross section of spread foundation.

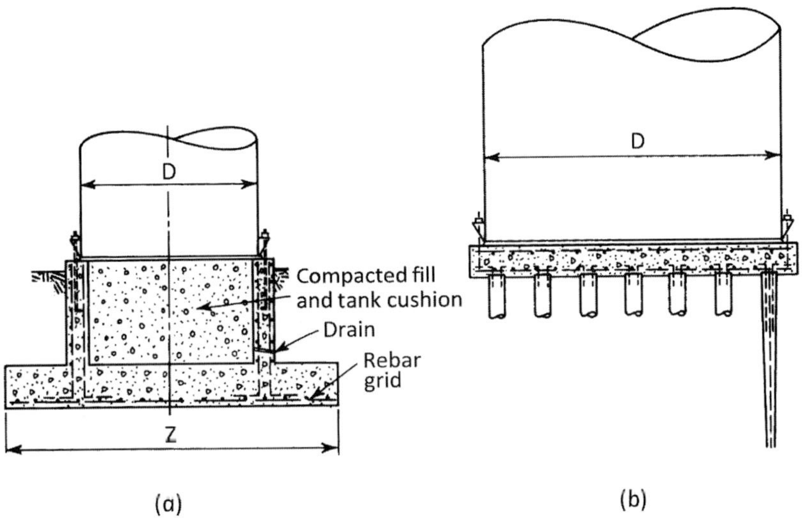

Figure 22.26 (*a*) Slab and ringwall foundation and (*b*) slab-on-pile foundation.

When soil conditions require deep foundations (over 10 ft), it may be more economical to use a slab and ringwall design (Fig. 22.26*a*) or a pile-supported slab foundation (Fig. 22.26*b*). The ringwall in Fig. 22.26*a* provides an extension of foundation depth and should not be considered a moment-resisting element. It is reinforced for hoop tension produced by internal soil pressure and must be anchored to the base slab to resist overturning moments due to wind or seismicity. When pile supported slab foundations are used, the pile capacity should be such that all the piles will be placed within the area covered by the tank; otherwise, a thick pile cap slab will be required to resist the bending moment in the foundations beyond the tank shell. As of this writing, AWWA D100 includes requirements and factors of safety for driven pile foundations. The engineer should exercise reasoned judgment when determining how to adapt the requirements to other types of piles or other types of deep foundations.

22.4 ELEVATED TANKS

Modern elevated tanks are designed as membrane structures supported by either multiple columns attached directly to the tank shell or a single pedestal. A multiple column tank consists of a roof, shell, and suspended bottom. A section of one form of a multicolumn tank is shown in Fig. 22.27. There are endless possibilities for other forms and configurations of multicolumn tanks.

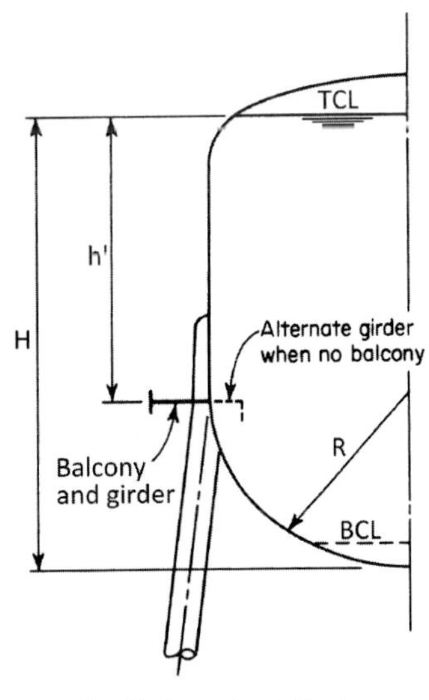

Figure 22.27 Section of multicolumn elevated tank.

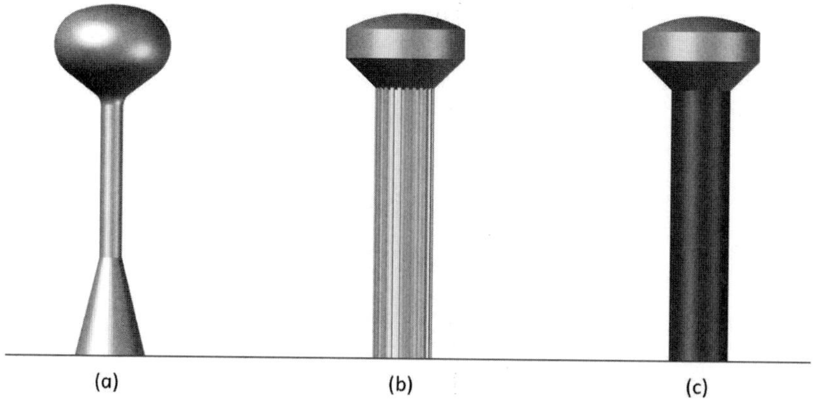

Figure 22.28 Conceptual renderings of three single-pedestal elevated tank styles.

The support pedestals for single-pedestal tanks are commonly constructed in three configurations: (a) smooth cylindrical steel pedestal, (b) fluted steel pedestal, and (c) concrete pedestal. Conceptual renderings of the general configuration of these styles are shown in Fig. 22.28.

Single-pedestal tanks utilizing a steel support pedestal are designed in accordance with AWWA D100. Single-pedestal tanks utilizing a concrete support pedestal are designed in accordance with AWWA D107. The liquid containers on single-pedestal elevated tanks are generally constructed in one of three configurations, but designs are not limited to only these options. Common configurations are: a spherical tank, a spheroidal tank, or a cylindrical tank with a conical transition to the pedestal. Figure 22.28 includes the latter two shapes of tank container.

As noted above, multiple-column tanks can be found in a wide variety of configurations. Chapter 1 of AWWA Manual M42 (Ref. 36) includes a description of many of the common configurations of multicolumn elevated tanks available today. A quick search of the Internet can easily turn up 50 to 100 different configurations of elevated steel tanks. Obviously, some designs are more economical than others, and some designs are better suited for seismically active areas. The variety of designs clearly demonstrates the versatility of steel for support of the imposed structural loads and the ability to incorporate architectural interest in the finished structure.

The column-supported elevated tank may have a domed, conical, torispherical, or elliptical roof combined with a vertical cylindrical shell to which the columns are attached; and a suspended bottom which may be elliptical, hemispherical, conical, or toroidal. Some large toroidal bottom tanks include a plurality of support columns around the perimeter of the tank and a large, central load-bearing pillar such as the tank shown in Fig. 22.29. Multicolumn elevated tanks have been constructed with capacities varying from 5000 to 4,000,000 gallons. The maximum allowable head range determines the proportions.

Tank capacity is measured between the lip of the overflow (top capacity level or TCL) and the bottom capacity level (BCL) set by the specified head range (Fig. 22.27).

Figure 22.29 Multicolumn toroidal bottom tank with large central pillar.

22.4.1 Roofs for Elevated Tanks

Elevated tanks can be fitted with a wide variety of roof configurations. Not all roof styles are regularly used on all tank types, but there is generally no reason that one could not apply the desired roof on any of the elevated tank types. Each roof has different design considerations. The significant structural considerations are size of the roof and whether the roof must be water bearing. Figure 22.30 shows some of the roof configurations that can be applied.

The roof configurations depicted here are (a) cone roof, (b) chamfered cone roof, (c) umbrella or dome roof, (d) torispherical roof, and (e) ellipsoidal roof.

When conical roofs are used, the TCL is always maintained below the junction of the roof and the shell. The rules for designing conical roofs for elevated tanks are the same as for roofs for reservoirs and standpipes. For umbrella or dome roofs, it is recommended that the TCL be set below the junction of the roof and the shell. Torispherical and ellipsoidal roofs permit establishing the TCL within the roof. Water within the roof develops biaxial membrane stresses in the roof plates, which can be calculated when the membrane alone can be completely cut by a transverse plane. One of the biaxial stresses can be determined by statics. The upward hydrostatic pressure on a horizontal plane at the

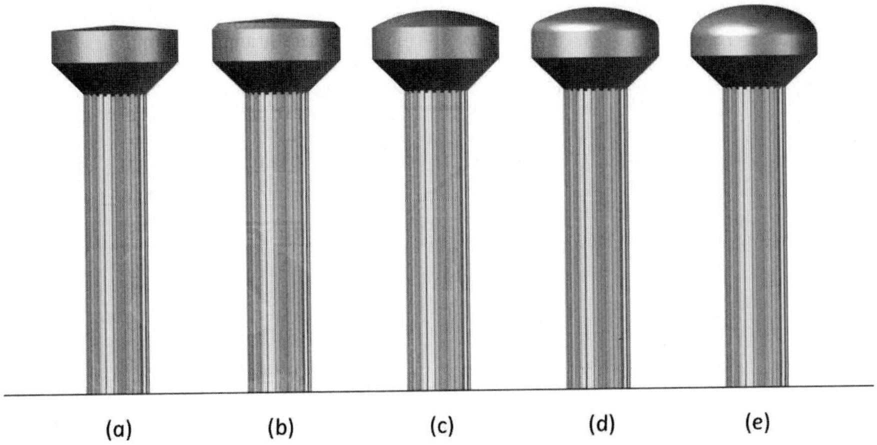

Figure 22.30 Examples of alternative roof configurations.

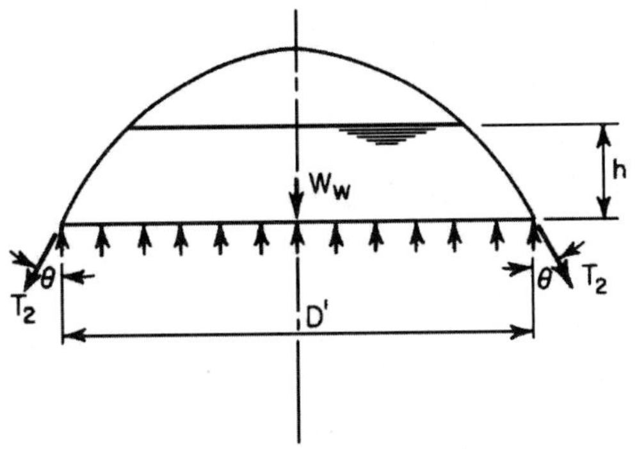

Figure 22.31 Membrane forces in water-filled roof.

distance h below the TCL will be balanced by a system of meridional membrane unit forces, T_2 (Fig. 22.31).

Thus,

$$T_2 = \frac{\gamma h \pi D'^2/4 - W_w}{\pi D' \cos\theta} \tag{22.25}$$

where D' = diameter of membrane at cut section
γ = density of stored product
θ = angle of meridional force as shown in Fig. 22.31
W_w = weight of liquid above cut section and below the TCL, lb

The weight of the metal should be included in the determination of T_2.

To determine the other (latitudinal) membrane unit force, T_1, equilibrium of T_1 and T_2 with the hydrostatic pressure p normal to the surface at the depth h gives

$$T_1 = R_1\left(p - \frac{T_2}{R_2}\right) \tag{22.26}$$

where R_1 = latitudinal radius
R_2 = meridional radius

Several points should be checked to determine maximum compressive unit force. The computed maximum unit force will determine the required roof-plate thickness. The curvature in the meridional direction determines the compressive buckling strength in the latitudinal direction, whereas the latitudinal curvature determines the compressive buckling strength in the meridional direction.

Having computed T_1 and T_2, the next step is to determine the required plate thickness by dividing the membrane unit force by the allowable unit stress and by the joint efficiency factor if the stress is tension. Allowable tensile unit stresses are governed by the tensile properties of the material. Biaxial compression is not covered in AWWA D100 but is covered in Ref. 9. Section 3.4.3 of AWWA D100 includes extensive provisions for determining allowable compressive stresses dependent on the nature of the stressed elements, construction tolerances, and verification of as-built geometry of the structure.

Note that the terms T_1 and T_2 in this chapter are defined differently than in Ref. 9.

22.4.2 Bottoms for Elevated Tanks

The tank bottom may be suspended, as in a column-supported tank, or it may be a support cone, as in some single-pedestal steel tank, or it may be supported by a concrete dome as in a composite elevated tank. Design of suspended bottoms that are formed by an axisymmetric surface is similar to design of the roof. The bottom is sectioned by a transverse plane, the stresses T_1 and T_2 are determined, and the required plate thickness is calculated. In hemispherical bottoms, where $R_1 = R_2 = R$, the maximum tensile unit force, T_1, which occurs at the very bottom of the section shown in Fig. 22.27, is given by

$$T_1 = T_2 = \frac{\gamma HR}{2} \tag{22.27}$$

The membrane unit forces at the spring line are

$$T_1 = \gamma R\left(\frac{h'}{2} - \frac{R}{3}\right) \tag{22.28}$$

$$T_2 = \gamma R\left(\frac{h'}{2} + \frac{R}{3}\right) \tag{22.29}$$

where R = spherical radius of bottom
H = distance from top capacity level to bottom of tank
h' = distance from top capacity level to spring line of bottom

The membrane unit forces, T_1 and T_2, in suspended conical bottoms are determined independently (Fig. 22.32).

When the cylindrical portion is filled to a depth X above the cone-to-cylinder junction, the unit forces in the cone at any point h_c below the spring line are

$$T_2 = \frac{\gamma}{2\cos\theta}\left(\frac{D}{2} - h_c\tan\theta\right)\left(X + \frac{2h_c}{3} + \frac{D}{6}\cot\theta\right) \tag{22.30}$$

$$T_1 = \frac{\gamma}{\cos\theta}\left(\frac{D}{2} - h_c\tan\theta\right)(X + h_c) \tag{22.31}$$

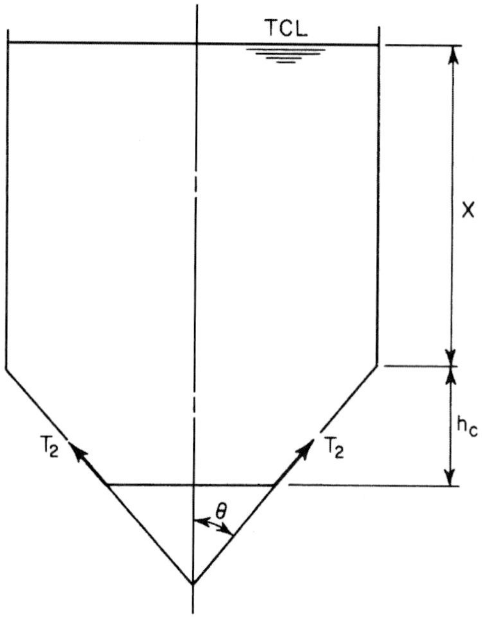

Figure 22.32 Membrane forces in suspended conical bottom.

At the spring line the unit forces are

$$T_2 = \frac{\gamma}{2\cos\theta}\frac{D}{2}\left(X + \frac{D}{6}\cot\theta\right) \tag{22.32}$$

$$T_1 = \frac{\gamma DX}{2\cos\theta} \tag{22.33}$$

where θ = apex angle.
At the apex, $T_2 = T_1 = 0$.

Compression stresses must also be determined at the cone-to-cylinder junction, where a compression girder is required to resist the inward pull of the cone bottom. The compression force C in the girder is

$$C = \frac{\gamma}{8}\left(X + \frac{D}{6}\cot\theta\right)D^2\tan\theta \tag{22.34}$$

Portions of cone and shell act with the girder. The effective width of each strip is assumed to be $0.78\sqrt{Rt}$ but not to exceed $16t$. Therefore, the effective area is the smaller of

$$A_{\text{eff}} = 0.78(t_c\sqrt{R_c t_c} + t_1\sqrt{R_1 t_1}) \tag{22.35}$$

$$A_{\text{eff(max)}} = 16\left(t_c^2 + t_1^2\right) \tag{22.36}$$

where R_c, R_1 = radius of shell and cone, respectively, in.
t_c, t_1 = thickness of shell and cone, respectively, in.

22.4.3 Balcony or Ring Girder

The shell of a column-supported tank is considered to be a circular girder uniformly loaded over its periphery and supported by columns, equally spaced on the shell circumference, attached directly to the tank shell. The supporting tower generates concentrated radial and tangential forces on the tank structure. These forces may be caused by sloping columns and/or the diagonal bracing system in the tower, and a ring girder must be provided to resist them. The ring girder is located at the intersection of the column's neutral axis with the tank shell and is usually positioned at the spring line of the suspended bottom. The ring girder also functions as the top strut line of the tower. A tank balcony can serve as a ring girder provided that it is continuous around the tank. Balconies should be wide enough to permit walking upright around the tank and should provide easy passage at the columns. If a balcony is not used, a continuous ring girder must be provided.

The force system on the balcony or ring girder of a tank having vertical columns consists of the shears q in the shell resulting from the horizontal load, H, at strut line or balcony due to wind or seismic forces and the resisting forces in the bracing system (Fig. 22.33). The beam formula VQ/I gives

$$q = \frac{H}{\pi R'}\sin\beta \tag{22.37}$$

where R' = horizontal radius of column circle at balcony or strut line
β = angle from line of action of H to any point on the shell

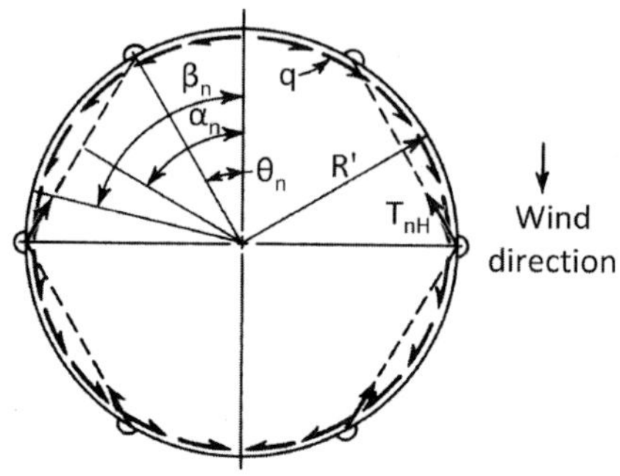

Figure 22.33 Forces on ring girder of tank with vertical columns.

The resisting force T_{nH} in a bracing rod is

$$T_{nH} = \frac{2H\sin\alpha_n}{N} \tag{22.38}$$

where α_n = angle between line of action of H and normal to horizontal projection of bracing rod being considered
N = number of columns or active rods in a tower panel

The force system on the balcony or ring girder of a tank having sloping legs consists of the shears q in the tank shell given by Eq. (22.37), radial thrusts produced by dead and vertical live load and by wind or seismic loads, together with the resisting forces of the bracing system (Fig. 22.34).

The radial thrusts P_r from vertical loads are

$$P_r = \frac{SV}{N} \tag{22.39}$$

where S = slope of column
V = sum of vertical loads at balcony or strut line

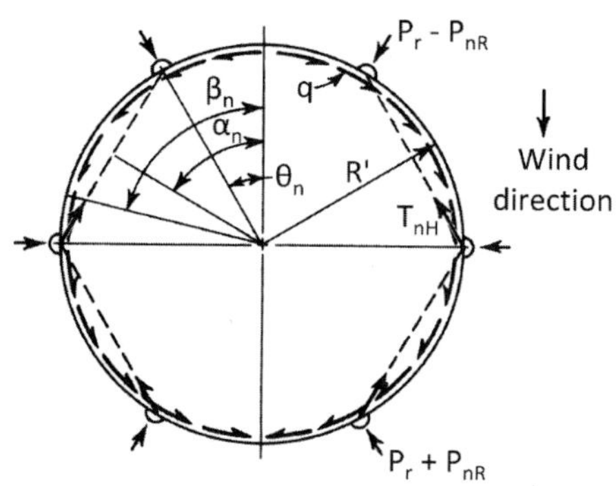

Figure 22.34 Forces on ring girder of tank with sloping columns.

The radial thrusts P_{nR} due to wind or seismic loads are

$$P_{nR} = \frac{2HaS\cos\theta_n}{R'N} \tag{22.40}$$

where a = distance from balcony to line of action of horizontal loads
θ_n = angle from line of action of H to radial line from center to column n

The resisting force T_{nsH} in a bracing rod is

$$T_{nsH} = \frac{2H}{N}\left(1 - \frac{aS}{R'}\right)\sin\alpha_n \tag{22.41}$$

22.4.4 Columns for Elevated Tanks

Tower columns may be rolled structural shapes or tubular sections. Tubular columns permit use of longer unbraced lengths and are easier to maintain. Design of the supporting tower follows conventional procedures. The diameter of the tank and height of the tower influence the choice between sloped and vertical columns. Appearance of the structure may also be a deciding factor. Columns must also be checked for uplift. The vertical component P_n of the column loads in each panel is

$$P_n = -\frac{V}{N} + \frac{2M}{R'N}\cos\theta_n \tag{22.42}$$

where M = overturning moment of wind or seismic forces about the strut line.

The panel points on tubular columns must be adequately stiffened with internal transverse diaphragms to prevent local buckling by surface- or skin-applied loads (Fig. 22.35). This requirement applies particularly to members acting normal to the axis of the columns. When rigid frame connections are used, full internal transverse diaphragms, or the equivalent, must be provided in the column at both the tension and compression flanges of the connecting members.

The length of contact of the column with the tank is chosen to keep shear buckling stresses in the shell within acceptable limits. Because the column-to-tank connection defies exact analysis, designs have been based on rationalization, experience, and tests.

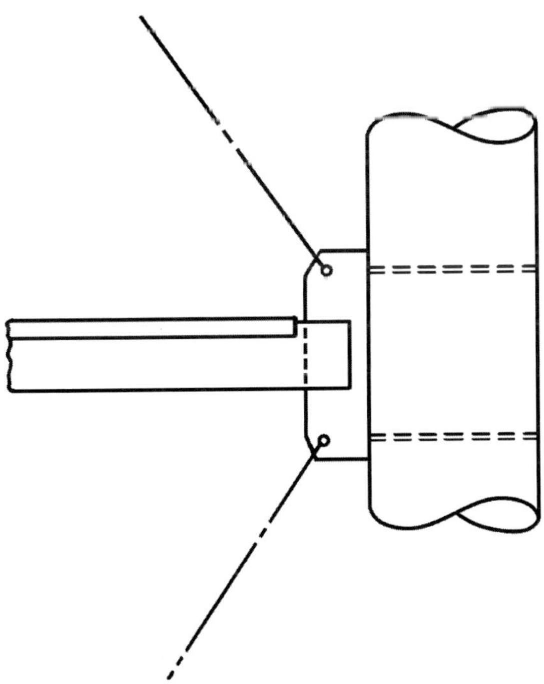

Figure 22.35 Connection to tubular column.

Tower Rods

The bracing-rod load T_{nR} in each panel of a tower with vertical columns is

$$T_{nR} = \frac{2H\sin\alpha_n}{N\cos\phi_n} \tag{22.43}$$

For towers with sloped columns (Fig. 22.36) the stresses are

$$T_{nR} = \frac{(P_B - P_A)\sin\alpha_n}{2\sin\phi_n\sin(180°/N)} \tag{22.44}$$

where P_A = leeward column load at strut line A
P_B = leeward column load at strut line B
ϕ_n = angle of bracing rod with horizontal

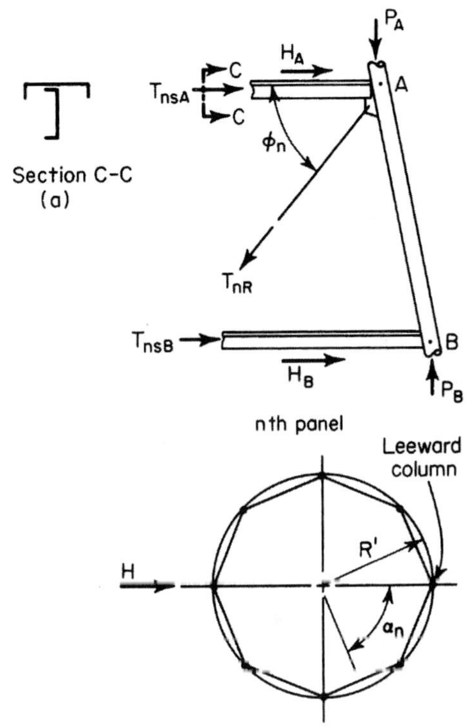

Figure 22.36 Bracing rod forces for towers with sloping columns.

Tower rods are made of either square or round steel bars. They must be readily weldable. The ends are provided with double clevis plates and fitted pins for attachment to gusset plates welded to the columns. Turnbuckles provide for adjustment of the rods and lining up of the tower. Turnbuckles for the rods of large tanks may exceed practical size limits and be difficult to adjust by manual methods. In such cases the rods may be welded directly to the gusset plates and adjustment made by heating and upset shrinking of the rods. Turnbuckles should be located in the lower end of each panel just above the strut line for convenient access during adjustment.

Tower Struts

The strut load T_{ns} is the horizontal component of the tower rod load:

$$T_{ns} = \frac{2H\sin\alpha_n}{N} \tag{22.45}$$

Struts may be rolled structural shapes or built-up members. Figure 22.36, sketch "a," shows a common type of built-up member. Struts are frequently used to support vertical loads, in which case they must be designed for combined compression and bending. When the thrust is small, a maximum $KL/r = 175$ is permissible. Struts may be pin-connected or rigidly connected to the gusset plates. No allowance should be made for fixed-end conditions when calculating KL/r. Strut sizes are sometimes determined by erection-loading conditions, in which case they may appear oversized. It is good practice to design the struts so that their ultimate strength is sufficient to develop the yield strength of the rods, since combined pretension and tension due to lateral forces have on occasion stretched rods, especially under seismic loading.

22.4.5 Single-Pedestal Tanks

This type of tank is designed from the top down. After the ball and shaft diameters are established, the membrane forces and corresponding required thicknesses are determined at intervals, beginning with the first section below the top. Membrane forces in the ball can be computed using Eqs. (22.25) and (22.26) and the required thicknesses determined as explained in Article 22.4.1.

The latitudinal tensile membrane unit force, T_1 in the support cone (Fig. 22.37, sketch "*a*") is given by Eq. (22.26) with $R_2 = \infty$ and reduces to

$$T_1 = pR_1 \tag{22.46}$$

The meridional unit force, T_2, in the support cone is found from equilibrium of the forces shown in Fig. 22.37, sketch "*b*". With T_{2A} known, T_{2B} is given by

$$T_{2B} = \frac{1}{R_B}\left(-\frac{R_B^2 p_B}{2} + \frac{R_A^2 p_A}{2} + \frac{W_{AB}}{2\pi \cos^2\theta} + T_{2A}R_A\right) \tag{22.47}$$

The thicknesses are determined as in Article 22.4.1.

Design of the support pedestal for wind and seismic loads usually requires several iterations. For the first iteration the deflection, e, of the ball (Fig. 22.37, sketch "*c*"), the weight, W_m, of shaft metal above a given section and, for seismic forces, the natural period of vibration are assumed. The vertical-load eccentricity, e_x, at any intermediate section can be assumed to vary parabolically as given by

$$e_x = \left[1 - \left(\frac{L-x}{L}\right)^2\right]e \tag{22.48}$$

The moment at any section is

$$M_x = Hx + h_1 x_1 + We_x \tag{22.49}$$

where x = distance from base of pedestal to point under consideration, ft

The shaft weight, W_m, is usually ignored in this calculation. The axial unit force, T_2, in the shaft is

$$T_2 = -\frac{W_x}{2\pi R} - \frac{M_x}{\pi R^2} \tag{22.50}$$

EXAMPLE 1

A 50,000-gal pedestal tank (Fig. 22.37*c*) has the following properties: $L = 120.5$ ft, length of shaft = 93.5 ft, $R = 5$ ft, $W = 4{,}414{,}000$ lb, wind $H = 30{,}000$ lb, and seismic $H = 110{,}800$ lb. For the first iteration, assume $e = 3.3$ in. At the section $x = 35$ ft, with $h_1 = 250$ lb, $W_m = 10{,}000$ lb, and $x_1 = 4$ ft, we get

$$W_x = 4{,}433{,}000 + 10{,}000 = 4{,}443{,}000 \text{ lb}$$

$$e_x = \left[1 - \left(\frac{120.5 - 35}{120.5}\right)^2\right] \times 3.3 = 1.64 \text{ in.} = 0.137 \text{ ft}$$

$$M_x = 110{,}800 \times 35 + 250 \times 4 + 4{,}433{,}000 \times 0.137 = 4{,}470{,}000 \text{ ft-lb}$$

$$T_2 = \frac{W_x}{2\pi R} - \frac{M_x}{\pi R^2} = \frac{4{,}443{,}000}{377} - \frac{4{,}470{,}000 \times 12}{11{,}310} = -11{,}780 - 4740$$
$$= -16{,}520 \text{ lb/in.}$$

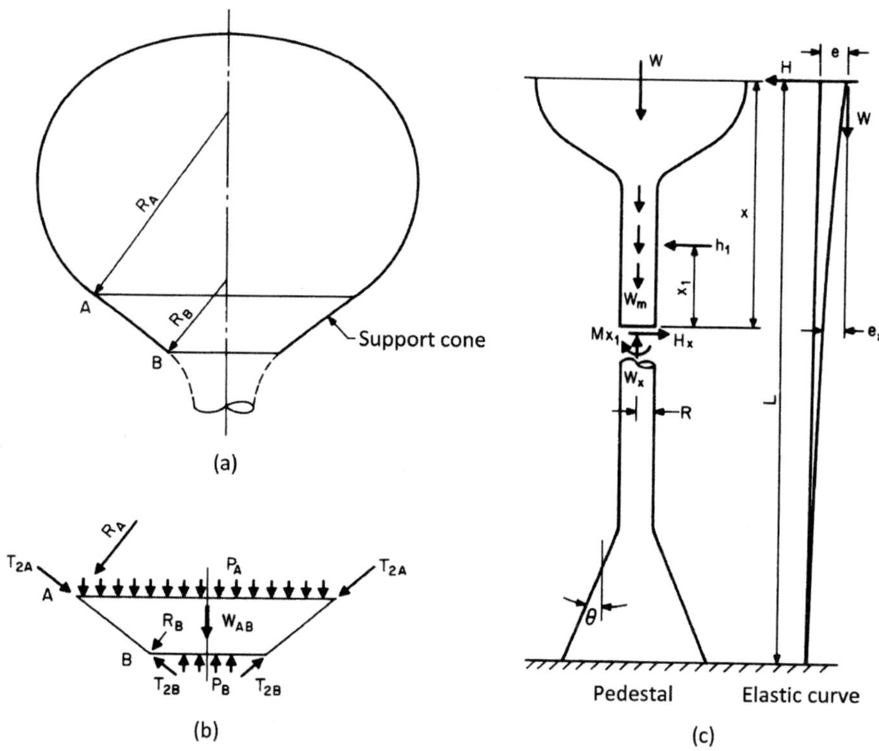

Figure 22.37 Single-pedestal tank: (*a*) support cone geometry, (*b*) free-body diagram of support cone, and (*c*) pedestal free-body diagram and elastic deflection curve.

Assume $t = 0.83$ in. Then

$$f_a = \frac{11{,}780}{0.83} = -14{,}200 \text{ psi}$$

$$f_a + f_b = -\frac{16{,}520}{0.83} = -19{,}900 \text{ psi}$$

$$\frac{t}{R} = \frac{0.83}{60} = 0.0138$$

$$\frac{KL_s}{r} = \frac{2 \times 93.5}{5\sqrt{2}} = 26$$

Determine the allowable stresses from AWWA D100, Section 3.4.3. Compare that allowable stress to the actual calculated stress. If the assumed thickness is adequate, then proceed with the design. Otherwise, revise the assumed thickness and check the stresses again. Continue this process until the stresses are within the allowable levels.

After the thicknesses of the various plate courses have been determined, the deflections and the natural period are calculated. The cycle is then repeated if the calculated values differ from the assumed values by more than, say, 2 percent.

The base cone must be checked at the top and bottom since either section may govern. The meridional unit force, T_2, is given by

$$T_2 = -\frac{W_x}{2\pi R \cos\theta} - \frac{M_x}{\pi R^2 \cos\theta} \tag{22.51}$$

where θ = angle of meridian with vertical.

Anchor bolt loads can be found from Eq. (22.17). The case for a full tank with earthquake and an empty tank with wind must both be checked.

22.4.6 Foundations for Multiple-Column Tanks

Prior to design and start of construction, the proposed tank site should be explored by a competent soils engineer to determine its suitability. Such investigations avert costly delays and assist in selecting proper foundation construction. The investigation should determine the bearing allowable, magnitude of expected general settlement, and nature and magnitude of differential settlements. It must be remembered that multiple-column elevated tanks are continuous, not articulated, and that they generally impose sustained design load during the life of the structure. Only small differential settlements can be tolerated and repeated releveling or underpinning is expensive. It is prudent to test a site prior to acquisition.

Foundations for elevated tanks are designed so that the column load passes through the centroids of the top and bottom of the pier in order to obtain uniform soil pressure on the base. The pier must be founded 6 in. below the greatest frost penetration. This consideration can be waived when the piers are founded on bedrock. The gross weight of the pier (which includes concrete and earth directly above the base) must be equal to or greater than the maximum calculated uplift at a column baseplate. Maximum uplift due to wind occurs when the tank is empty, while that due to seismic forces occurs when the tank is full. AWWA D100 Section 13 includes a special design case to check soil pressures resulting from a lateral seismic force sufficient to develop yield stress in the bracing rods. Under this special design case, the soil-bearing pressures are allowed to go up to the ultimate soil-bearing capacity. The pier is commonly supported with a footing (Fig. 22.38). For durability, the 28-day strength of the concrete should be at least 3000 psi. Design should comply with the ACI 318. However, because of the reduced load factor for dead load, water load should be considered as live load except when loads are used to resist uplift, even if it is defined as dead load in the code used for design.

The center pier for an elevated tank will vary with piping and center-riser requirements. The pier should be designed to carry the water and metal loads and be proportioned to have uniform pressure under the base. Large valve vaults and control pits are more economically designed as separate units not attached to the center pier. If possible, valve vaults should be located outside the tank foundation area.

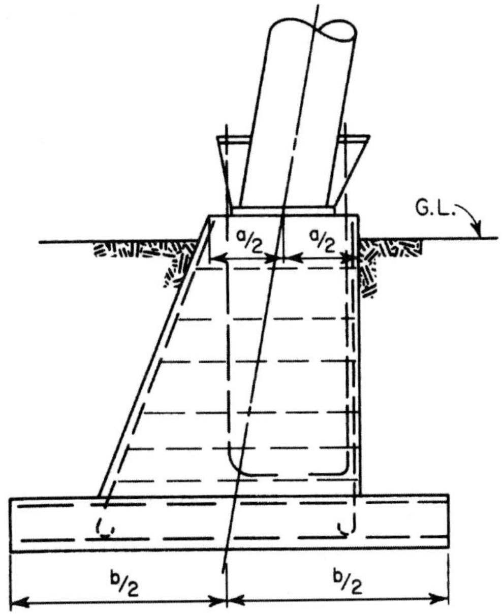

Figure 22.38 Tank-column pier.

22.4.7 Sloshing Considerations for Elevated Tanks

As of this writing, sloshing wave analysis has not been included as mandatory for elevated tanks in AWWA D100. AWWA D107 provides a method for calculating the theoretical sloshing wave height for elevated tanks that uses an effective diameter to account for the variety of container shapes used for elevated tanks. The AWWA D100 standard requires that if the engineer has not performed a detailed special analysis to determine the behavior of the sloshing liquid and the separate impulsive and convective response components of the stored liquid, then the seismic design loads must be determined as if the stored liquid is a fixed mass.

22.5 ACCESSORIES

An overflow is an economical accessory that will protect the tank from overpressure, overload, and possible catastrophic failure should controls fail to shut off the pumps. A properly operated tank should not overflow during normal operation. An overflowing tank is considered to be an emergency, and the malfunction causing the overflow should be determined and corrected as soon as possible. It is considered good practice for the owner to install an overflow sensor on the tank. The sensor can be set to activate an alarm at the pump house and at local emergency services offices in addition to the alarm at the water system operator's facilities.

Overflow intakes may be configured as open pipe, shell-slot weirs, internal weir box, or internal cone weir. Regardless of the configuration, the overflow intake and piping should be sized to handle the maximum inflow pumping rate in the system. Internal weir-type overflows are subject to ice damage in regions subject to ice formation in the stored water. Overflow piping may be internal or external. However, internal piping is not recommended in regions subject to ice formation in the stored water.

The tank should have an external ladder extending to the roof. Internal ladders are subject to ice damage and should be avoided in regions subject to ice formation in the stored water. However, pedestal-type tanks have internal ladders for aesthetic reasons. Such ladders are acceptable since they are placed inside the pedestal or access tubes and are not exposed to water or ice damage.

Access openings should be provided in the roof to permit entrance into the tank and to provide ventilation during painting and maintenance

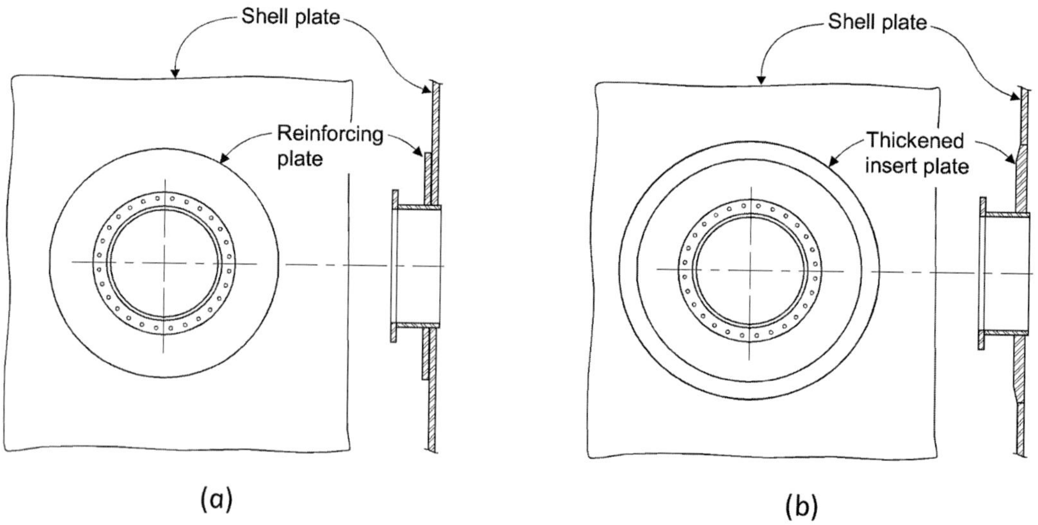

Figure 22.39 Shell manway.

of the structure. AWWA D100 requires two roof openings, one of which may be the tank's center vent if equipped with a removable section that allows for mounting a ventilation fan.

For reservoirs and standpipes, two shell manways should be provided in the lowest shell ring, and they should be located a sufficient distance above the bottom to permit reinforcement of the shell cutout (Fig. 22.39). The tank standards include limitations on the location of shell openings, including manways and nozzles. Depending on the tank type and specific design procedure used, additional limitations may apply. Those additional limitations may include minimum clearance to shell plate joints or limitations on the distance from the tank floor to the bottom edge of the shell cutout or to the nearest edge of reinforcing plates or insert plates.

Adequate air vents must be provided in tank roofs to prevent internal pressure or vacuum build up. The free area for airflow through the vent should exceed the larger of the area of the outlet pipe and the area of the inlet pipe, preferably by at least 50 percent. This will generally permit the maximum water discharge rate to not exert significant negative pressures on the tank. Screening smaller than 4-mesh on tank vents can be subject to clogging up with frost and ice, making the vent inoperative. Clogged vents can cause collapse of the tank roof during withdrawal and can also cause internal overpressure that may rupture the tank. For potable water tanks, many states require screens finer than 4-mesh, and 24-mesh screen has been identified by the U.S. Environmental Protection Agency as the size required to preclude insects (Ref. 37). AWWA standards require fail-safe pressure-vacuum vents, which can lift a relief pallet if the screen becomes clogged. Such vents require maintenance to ensure trouble-free operation.

Float-type water-level gauges are practical only in warm climates. If pressure gauges are desired, they should be of the mercury-manometer or bourdon-tube dial-gauge type and located in areas not subject to freezing.

Heaters

Water-industry practice assumes 42°F as the minimum water temperature and no ice formation in the tank. References 38, 39, and 40 contain heat-loss tables and other relevant data that be used for calculating tank heater requirements. In areas where the low 1-day mean (LODM) temperature (Fig. 22.40) may be +5°F or colder, frost protection must be considered. Freezing of the circulation piping and/or ice formation in the tank can occur. Ice up to 5 ft thick has been observed in tanks which failed due to ice loads. If a sufficient quantity of warm water does not flow into the tank to keep the average temperature above 42°F, a heater must be used. The best method is by means of a heat exchanger that takes cold water out of the base of the riser and discharges heated water into the tank. Such heaters are required by insurers in fire-protection tanks. Added heat is recommended in cold regions for oversized municipal tanks which do not circulate the tank capacity at least once per day. During cold spells in extremely cold climates it may be advisable to waste water during periods of low usage, to draw down the tank, and to refill the tank with warmer well water. Alternatively, if the municipal tank is operated at a reduced capacity equal to daily demand, the minimum temperature might be maintained solely through the turnover during draw and fill cycles. The water-system designer should determine the most economical winter operation for the owner.

Where local climate conditions are such that water in aboveground piping could freeze, the tank piping for ground supported tanks should enter the tank through the bottom (Fig. 22.41) or be provided with adequate insulation to prevent freezing. If the water source is a lake or stream, heating of water may be required to prevent freezing of the inlet and discharge piping.

The water-system designer should review the system operating assumptions and advise the system owner (operator) of minimum water circulation requirements to prevent ice formation. The following example illustrates the heat loss calculation.

EXAMPLE 2

The following values are given:

Elevated-tank capacity = 200,000 gal
Surface area = 4470 ft^2
LODM temperature = −10°F (Fig. 22.40)
Entering water temperature = 55°F (well water)

From NFPA 22, heat loss rate = 190.7 Btu/ft^2/hr.

$$\text{Total heat loss, } Q_0 = \text{heat loss rate} \times \text{surface area} \times 24 \text{ hrs} = 20{,}458{,}300 \text{ Btu/day}$$

$$\text{Water heat capacity, } Q_i = (55° - 42°) \times 200{,}000 \text{ gal} \times 8.33 \text{ lb/gal} = 21{,}658{,}000 \text{ Btu}$$

$$\text{Required circulation} = Q_0/Q_i \times 200{,}000 = 180{,}000 \text{ gal/day}$$

So, on the coldest day, over 90 percent of the tank capacity must be circulated to keep the water temperature at the recommended 42°F minimum.

Figure 22.40 Isothermal lines, lowest 1-day mean temperature.

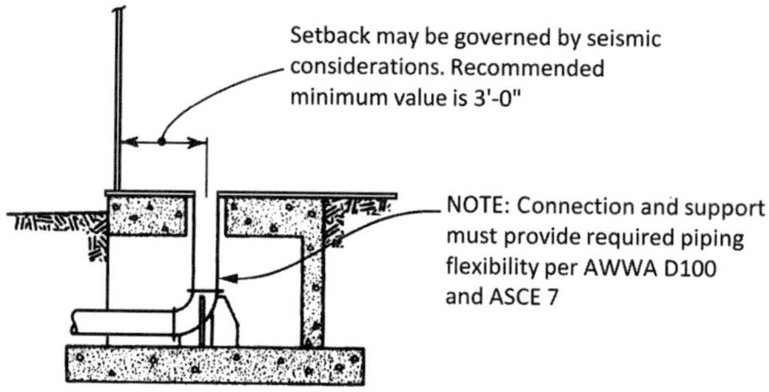

Figure 22.41 Pipe vault for ground-supported tanks.

22.6 MATERIALS

Plates. The plate material most commonly used for construction of atmospheric and low-pressure storage tanks is ASTM A36. ASTM A131 Grades A and B and ASTM A573 Grade 58 are also used, particularly for more severe service conditions. Shell plates of A36 should be limited to 2 in. for tension-controlled design and should be normalized in thicknesses over 2 in. for compression-controlled design. A36 plates over 1 in. through 1.5 in. should be ordered to silicon-killed fine-grain practice. Refer to the appropriate tank design standard for detailed requirements.

Maximum permissible shell-plate thickness for storage tanks is 1.5 to 2 in., depending on which industry standard and which design method will be utilized. AWWA D100, Section 14, provides design requirements with higher allowable stresses than the base design, but the shell thickness is limited to 1.5 in., and additional material requirements, design requirements, welding requirements, and testing and inspection requirements apply.

Base Plates. A36 may be used for base plates, regardless of thickness or ambient temperature. Other steels are also acceptable. Refer to industry standards for other acceptable steel grades.

Structural Shapes. Structural shapes are generally either ASTM A36 or ASTM A992 steel. Although A992 is most commonly found in wide-flange beam sections, other structural shapes are also available under this specification. Forgings for turnbuckles and clevises are generally ASTM A668 Class D. Pins should be specified to one of the acceptable materials listed in AWWA D100.

Anchor Bolts. In the late 1990s, ASTM produced a new material specification, ASTM F1554, specifically for carbon steel anchor bolts. That specification includes three strength levels: 36 ksi, 55 ksi, and 105 ksi. The revised term "anchor rod" was introduced with the release of F1554. Also, with the introduction of F1554, the ASTM A307 specification dropped Grade B materials, which previously covered anchor bolts. AISC has adopted ASTM F1554 as the preferred material specification for anchor bolts. AWWA D100 specifies F1554 and other acceptable anchor bolt materials.

Welding Electrodes. Welding electrodes should conform to the requirements of the applicable AWS Specifications, the welding process being used, and the applicable tank standard. They should be any suitable electrode with mechanical properties appropriate for the materials being joined and, for the electric-current characteristics, the position of welding and other conditions of intended use. The selected electrode must be properly qualified for the materials and design metal temperature in accordance with the procedures specified in the applicable tank standard. Some tank standards permit welding procedures to be developed and qualified in accordance with either ASME Section IX, AWS B2.1, or AWS D1.1, while others permit welding procedures in accordance with ASME Section IX only. The engineer should confirm which rules apply before specifying welding requirements.

Allowable Stresses. Allowable stresses and joint efficiencies for tanks vary considerably with the governing industry standard and project specific design criteria. The following are typical:

(1) Tensile stresses in tank shells:

a. AWWA D100: 15,000 psi (elevated tanks)

b. AWWA D100, base design: 15,000 psi for ground-supported tanks

c. AWWA D100, Section 14: 18,000 to 34,330 psi, depending on material, for ground-supported tanks only

d. AWWA D100: 11,000 psi for the lowest shell ring and tank bottom for cross-braced elevated tanks

e. API 650: 20,000 to 32,000 psi, depending on material (ground-supported tanks)

(2) Compressive stresses in tubular columns for elevated tanks, double-curved, axisymmetrical section, conical sections, and cylindrical sections AWWA D100 Section 3.4.3: 10,000 psi maximum.

AWWA D100 contains detailed design procedures for these elements subject to compressive stresses from any load effect. Two methods can be applied to determine allowable compressive stresses for water-filled elements on elevated tanks, each method having an associated allowable stress level and specific limitations and requirements as to when and how it is applied and what level of inspection of the as-built structure is required to verify the design basis after construction is complete. With certain design methods, modifications to the structure might be required if the as-built shape does not conform to the theoretical shape within allowable tolerances.

(3) Allowable stresses for roof support structural components and stiffeners are designed in accordance with AISC 360 ASD provisions with specific exceptions. Each tank standard contains slightly different provisions, so the engineer should consult the standard when developing the design for the roof structural system.

(4) The slenderness ratio KL/r must not exceed 120 for compression members carrying weight or pressure of tank contents and 175 for columns supporting roofs and other members carrying wind and/or seismic load.

22.7 COMMERCIALLY AVAILABLE COMPUTER PROGRAMS

The availability of commercial programs for tank design and analysis appear to reduce the need for the simplified methods presented in this section. However, the value of a computer program depends on the design experience of the user. When a program is used to increase productivity, it is a valuable tool. However, in the hands of a technician or engineer with no experience, designs may be suspect. Code guidance in interpreting output has not caught up with technology. Where the user is an experienced engineer, it is easy to separate out meaningful stresses to compare to code-allowable stresses. However, where output produces an overall structural stability value in place of stresses, an inexperienced designer may be tempted to ignore code-allowable buckling stresses. For example, the Boardman formula has been reduced by a factor of 10 from the theoretical buckling stress to account for realistic construction tolerances. The danger in a computer-based stability analysis comes from solutions that permit the user to input tolerances which may not be realistic and may reduce the historical, experienced-driven reduction factors incorporated into the tank standards.

As of this writing, the author is familiar with three programs that are available for design of ground-supported storage tanks conforming to AWWA and API standards, with each offering a different set of features and capabilities.

(1) *ITSDesign*™ by Innovative Tank Solutions LLC

This program produces calculations for storage tanks in accordance with API 650 (including annexes), AWWA D100, and FM Global 4020/4021. The program can be utilized to design new tanks or analyze existing tanks for conformance to current standards. The program produces a detailed calculation report including a listing of any design errors or noncompliance conditions and a summary of comments regarding several design features and member utilization factors of interest to the engineer.

(2) *AMETank*™ by Technosoft Inc.

This program produces calculations for storage tank design in accordance with AWWA D100, API 620, API 650, and EN 14015.

The software builds a 3D model of the tank to augment the engineer's understanding of the final structure. The 3D model is also used to create a complete bill of materials, CNC patterns for tank components, and detailed manufacturing and construction drawings and other supporting documents. One module produces remaining life calculations according to API 653. The program can be utilized to design new tanks or analyze existing tanks for conformance to current standards. The program produces a detailed calculation report. The program is offered in several different configurations with different levels of capability.

(3) *TANK™* by Intergraph Corporation

This program produces calculations for storage tanks in accordance with API 620, API 650, and API 653. The program can be utilized to design new tanks or analyze existing tanks for conformance to current standards. A module also performs fitness for service ratings of existing tanks in accordance with API 579. The program produces a detailed calculations report including a listing of any design errors or warning messages.

This author believes that any tank design software should be considered a tool and not a replacement for knowledge of the numerous special requirements in the applicable tank standards. The engineer should always review and verify all input data and the resulting output from such software systems to confirm the results are reasonable and rational. Also, the engineer should be familiar with the latest updates to the design procedures and technical requirements in the tank standards to verify that the software is producing results compliant with the applicable edition of the standards.

22.8 THE ENGINEER'S ROLE

Experience has shown that certain factors seem to be repeated whenever a tank failure is investigated. In general, failures fall into five categories. The first three will usually occur after the tank has been in service for some period of time. These categories of failures not under the control of the engineer. The primary cause is operation related—someone turned the wrong valve, bypassed a safety device, covered a tank vent, or otherwise deviated from normal procedures. The second is related to maintenance. This could be a clogged vent, corrosion, or other avoidable form of deterioration. Third is overload from wind or earthquake loads beyond design levels. The fourth category is also not under control of the engineer. This case results when an owner or a contractor copies a design from a previous project and applies it to a new project without the benefit of an engineering review. Finally, there are design-related failures that occur during first filling or shortly thereafter. Certain factors are common in design failures. The tank may contain a product that is not specifically covered or addressed by industry standards. As a result, the owner/engineer concludes that the tank is "exempt" from code rules and ignores tank industry standards that should have been tailored to the special product for that project. In other situations, an inexperienced designer may choose a stress level from a tank standard but omit the associated inspection, material testing, welder certification, and other technical requirements of the standard.

As with the structures discussed in many other chapters of this book, steel tanks are unique and complex structures. Given some of the dangers of improper design highlighted in the preceding paragraph and other special considerations addressed in this chapter, it should be apparent that the design of these structures should not be considered a menial task. It is incumbent on the engineer to know and understand the requirements contained in the relevant tank standards. Additionally, when designing tanks for applications that are not covered by industry standards, the engineer should diligently apply portions of industry standards where appropriate and develop a comprehensive plan for analysis of the unique tank configuration and operating conditions that includes judicious application of relevant structural theories and appropriate methods of analysis.

REFERENCES

1. Minimum Design Loads and Associated Criteria for Buildings and Other Structures, ASCE 7. American Society of Civil Engineers.

2. Standard for Welded Carbon Steel Tanks for Water Storage, AWWA D100. American Water Works Association.

3. Standard for Factory-Coated Bolted Carbon Steel Tanks for Water Storage, AWWA D103. American Water Works Association.

4. Standard for Composite Elevated Tanks for Water Storage, AWWA D107. American Water Works Association.

5. Standard for Water Tanks for Private Fire Protection: NFPA 22. National Fire Protection Association.

6. Specification for Bolted Tanks for Storage of Production Liquids, API Specification 12B. American Petroleum Institute.

7. Specification for Field Welded Tanks for Storage of Production Liquids, API Specification 12D. American Petroleum Institute.

8. Specification for Shop Welded Tanks for Storage of Production Liquids, API Specification 12F. American Petroleum Institute.

9. Standard for Design and Construction of Large, Welded, Low-Pressure Storage Tanks, API Standard 620. American Petroleum Institute.

10. Standard for Welded Steel Tanks for Oil Storage, API Standard 650. American Petroleum Institute.

11. Standard for Tank Inspection, Repair, Alteration and Reconstruction, API Standard 653. American Petroleum Institute.

12. Lieb, J.M., Engineering Considerations in Retrofitting and Upgrading Aboveground Storage Tanks, International Liquid Terminals Association Conference, 1990.

13. Buzek, J.R., Useful Information on the Design of Steel Bins and Silos. American Iron and Steel Institute and Steel Plate Fabricators Association, Inc., 1989.

14. Kaups, T., and Lieb, J., "Design of Quality Bulk Storage Bins and Silos," Chicago Bridge & Iron Company, Tech. Paper No. CB15460 1985. (Also, in *International Journal of Bulk Solids Storage in Silos*, May 1986.)

15. Gaylord, E.H., and Gaylord, C.N., "Design of Steel Bins for Storage of Bulk Solids," Prentice Hall, 1984.

16. Ketchum, M.S., *The Design of Walls, Bins and Grain Elevators*, 3rd Edn., McGraw-Hill Book Company Inc., 1919.

17. Steel Plate Engineering Data, Volume 2—Useful Information on the Design of Steel Plate Structures, Part III, American Iron and Steel Institute and Steel Plate Fabricators Association, 1985.

18. Meyers, P.E., *Aboveground Storage Tanks*, McGraw-Hill, 1997.

19. Long, B., and Gardner, B., *Guide to Storage Tanks and Equipment*, John Wiley & Sons, 2004.

20. Steel Plate Engineering Data, Volume 1—Steel Tanks for Liquid Storage, American Iron and Steel Institute and Steel Plate Fabricators Association, 2011.

21. Steel Plate Engineering Data, Volume 2—Useful Information on the Design of Steel Plate Structures, American Iron and Steel Institute and Steel Plate Fabricators Association, 2011.

22. Brownell, L.E., and Young, E.H., *Process Equipment Design*, John Wiley & Sons, 1959.

23. Megyesy, E.F., *Pressure Vessel Handbook*, 14th ed., Pressure Vessel Publishing, Inc., 2008.

24. Bednar, H.H., *Pressure Vessel Design Handbook*, Krieger Publishing Company, 1991.

25. Mahajan, K.K., *Design of Process Equipment - Selected Topics*, 3rd Edn., Pressure Vessel Publishing, Inc., 1990.

26. Jawad, M.H., and Farr, J.R., *Structural Analysis and Design of Process Equipment*, John Wiley and Sons, 1984.

27. Moss, D.R., *Pressure Vessel Design Manual*, Elsevier Inc., 2013.

28. Meier, S.W. (tech. ed.), *Steel Water Storage Tanks: Design, Construction, Maintenance, and Repair*, McGraw-Hill, 2010.

29. Scott, L.D., United States Patent 4,714,170, Large Storage Tank Structures, 1987.

30. Roeske, A.E., United States Patent 2,849,143, Tank Design, 1958.

31. Wozniak, R.S., "Lateral Seismic Loads on Flat Bottomed Tanks," Chicago Bridge & Iron Company, *Water Tower*, November 1971.

32. Nuclear Reactors and Earthquakes, U.S. Atomic Energy Commission, Technical Information Document (TID) 7024, August 1963.

33. Boberg, I.E., "Oil Storage Tank Foundations," *Water Tower*, Chicago Bridge & Iron Company, March 1951.

34. Guidelines for Tank Foundation Designs (STE03020), PIP Standards. University of Texas at Austin, Construction Industry Institute.

35. Anchor Bolt Design Guide (STE05121), PIP Standards. University of Texas at Austin, Construction Industry Institute.

36. Manual of Water Supply Practice, Steel Water-Storage Tanks, AWWA Manual M42. American Water Works Association, 2013.

37. U.S. EPA, "Sanitary Protection of Drinking Water Storage Tanks: #24 Mesh Non-Corrodible Screen," EPA Region 8 Drinking Water Unit Tech Tips, June 2016.

38. Fire Protection Handbook, National Fire Protection Association (NFPA), Boston, 2008.

39. Smith, D.W., (ed.), *Cold Climate Utilities Manual*, Canadian Society of Civil Engineers, Montreal, 1986.

40. Factory Mutual Research Corporation, *Water Tanks for Fire Protection*, Norwood, Mass., 1917.

Chapter 23

Electrical Transmission and Substation Structures

BY

MICHAEL D. MILLER, PE, P.Eng. *VP Engineering, SAE Towers*
ROBERT E. NICKERSON, PE *Consultant*

23.1 INTRODUCTION

Electrical transmission systems, typically defined as those systems operating at high and ultrahigh nominal operating voltages of 69,000 volts and above, utilize transmission line, substation, and communication structures. These structures are inherently different from buildings and most other structure types. The functionality is first and foremost that of structural support, and as such, their dominant features of supporting configurations are similar to truss bridges for lattice structures and single or multiple poles. Outside of the general structure configuration, for example, tower (designed as a space truss) and pole (designed as a moment resisting frame) structures are designed for electrical functionality for operating clearances and structural load. A performance-based design methodology is utilized with a focus on electrical grid functionality and a targeted reliability-based loading. Industry standards and manuals of practice provide the reliability hierarchy such that communication structures are most reliable, followed by substation structures, followed by overhead transmission line structures. The electrical grid regularly contains degrees of grid system redundancy, and this is taken into account in the performance-based targets, and functionality. In this section we examine the design of each of these structure types in more detail; however, equations utilized for the design of the specific structure types are not included herein, and the reader is directed to the Standard or Manual of Practice, which provides relevant design details for such structures. Summaries are provided with general explanation as to why electrical industry approaches in design often vary from the approaches for building-type structures.

23.2 REFERENCED STANDARDS AND MANUALS OF PRACTICE

Table 23.1 provides the needed reference documents for design or analysis of electrical transmission, substation, and communication structures. This list is not intended to be all-inclusive; however, the most commonly utilized references are provided.

23.3 TRANSMISSION LINE STRUCTURES

Overhead transmission lines are a complex structural system comprised of foundations, structures, insulators, and conductors. The conductors, or wire system, impose the majority of the load on the structure and are concentrated at specific attachment points on the structure through the insulators (isolating the electrical system from the grounded structures). The conductors are subjected to tension variations from temperature extremes, wind, ice, and other phenomena not typical to other structures types. These variations, along with operating voltage and code defined clearances, provide the basis for the line tensions, structure heights, phase spacing, position of shield wire, and other electrical considerations such as electric and magnetic fields and their proximity to the edge of the transmission line rights-of-way. The resulting tensions in these conductors change with temperature, wind, and ice accumulations and when energized are significantly higher than those tensions at ambient temperature. These structures support loaded conductors that are relatively long linear systems in which the performance of each structure within a line segment can influence the adjacent structures. Because of these line system interaction phenomena, security loading and resistance capability is needed to avoid cascading failures. This type of failure can occur when the longitudinal strength (parallel to the line) of a structure is exceeded due to the line tension energy released from an initial triggering structure failure or line break. The resulting domino effect propagates and many structures can fail as a result. In rare instances, transverse cascades can also occur. Therefore, the performance of the system is dependent on the performance of the structure and vice versa.

23.3.1 Materials

Table 23.2 provides the materials typically utilized for the most common structure types in transmission line design.

23.3.2 Loading and Methodology

A significant difference between overhead line structures and other types of structures is the minimum loading requirements used in design.

Table 23.1 Referenced Design Documents

Topic in design or structure type	Reference document
Structural loading	ASCE 74 Manual of Practice; Guidelines for Electrical Transmission Line Structural Loading, Third Edition, 2009
NESC	IEEE-C2-2017 Code 2017 National Electric Safety Code, 2017
Latticed tower design	ASCE 10-15 Standard; Design of Latticed Steel Transmission Structures, 2015
Tubular steel pole design	ASCE 48-19 Standard Design of Steel Pole Transmission Structures, 2019
Prestressed concrete pole design	ASCE 123 Manual of Practice; Prestressed Concrete Transmission Pole Structures, 2012
Fiber-reinforced polymer pole design	ASCE 104 Manual of Practice; Recommended Practice for Fiber-Reinforced Polymer Products for Overhead Utility Line Structures, Second Edition, 2019
Guyed structure design	ASCE 91 Manual of Practice; Design of Guyed Electrical Transmission Structures, 1997
Wood pole design	ASCE 141, Manual of Practice Wood Pole Structures for Electrical Transmission Lines: Recommended Practice for Design and Use, 2019
Substation structure design	ASCE 113 Manual of Practice Substation Structure Design Guide, 2008

Table 23.2 Typical Materials Specified for Most Common Structure Types

Structure type	Materials utilized
Latticed steel towers	Steel angle: ASTM A36, A529, A572 Gr. 50, 60 CSA G40.20/G40.21 BS EN 10025 Gr. S355 Fasteners: ASTM A394 Type 0 and 1, ASTM F3125 Type 1 Galvanizing: ASTM A123
Tubular steel poles	Steel Plate: ASTM A572 Gr. 50, 65, A421, A882, A242, A588 with supplemental Charpy rated steel requirements Fasteners: ASTM A3125 Galvanizing: ASTM A123
Reinforced-concrete spun poles	High strength concrete: 8-12,000 psi Pre-stressing strands
Wood poles	Western Red Cedar, Southern Pine, Douglas Fir Steel attachment hardware
Fiber-reinforced polymer	Polymer resin matrix with fibers Steel attachment hardware

Wind and ice combinations on the wires directly affect the line tensions, which dominate the loads supported by the line dead-end structures. The wind and ice combinations on the vertical and transverse spans dominate the loads supported by the line suspension structures. Ice accumulation or wind pressure onto structures contributes a small (often less than 10 percent) amount of the design load on the structure. Icing on transmission structures is typically ignored unless the structure is supporting a line disconnect switch or other moving equipment in which ice accumulation would impede its operation or if the geographic location of the line has severe icing events.

The ultimate limit state methodology is used in the design of transmission structures. This method is often referred to as "limit state design," "ultimate strength design," or "performance-base design." The approach is similar to Load and Resistance Factored Design (LRFD) without inclusion of the resistance factors. However, some international design requirements do include resistance factors such as International Electrotechnical Commission (IEC) 60826. The probabilistic model of ASCE/SEI-7 is used to determine the wind and ice loading. The wind velocities and ice thicknesses from ASCE/SEI-7 are considered adequate for use at their respective Mean Recurrence Interval (MRI), typically 100 years. Often load factors of 1.0 are used in conjunction with these wind loads. In cases when increased reliability is considered, the MRI is increased and the load factor remains 1.0. Lines are both directionally dependent and traverse long distances so the synoptic windstorms that typically control their design have some degree of reduced maximum load probability from both wind direction and wind front pressure distribution. This reduced maximum load probability is generally not taken into account in loading. Typically, the wind front producing applied loads onto the line will not produce the maximum forces used in design. And the shape of the wind front with gusts will most likely not occur at the same time to apply maximum load on the entire line span. The two concepts of wind directionality and time-dependent wind pressure span distribution provide inherent load reduction as compared to the probabilistic MRI wind speeds for most transmission line applications. Extreme wind conditions are evaluated using the terrain classifications of ASCE/SEI-7 with applicable topographic effects (wind speed-up effects, funneling, etc.) and appropriate factors derived. Additional load factors can be used to account for these and other uncertainties in load effects along the transmission line.

The extreme wind speeds are determined based on the wind maps of ASCE/SEI 7 and the 100-year MRI map or based on wind studies in the region specific to the line location. Extreme icing thicknesses with moderate wind are determined using 100-year icing maps for glaze ice. Lines of more importance or radial feed lines without grid redundancy may require higher MRIs for structural reliability. The controlling icing condition may occur with glaze ice (57 lb/ft^3) or rime ice (15 lb/ft^3) depending on the geographic location of the line. Other less common icing types (hoar frost, wet snow) may also be used in the design for areas prone to those environmental conditions. The most common controlling loading conditions are extreme wind, extreme ice, construction and maintenance, and security loading. While safety code minimums are also required by most states in the United States, they rarely control the design unless in areas of no ice and/or lower ASCE/SEI 7 100-year MRI extreme wind speeds.

Additional loading conditions such as unequal icing on conductor or shield wire are utilized; however, they are not generally combined with the extreme weather cases that are derived from probabilistic weather events. Structure designs are normally standardized for relatively longer lines and a set of structure strengths are defined based on load demand to encompass the needs of the entire line and without the necessity to design a unique structure for every location.

The structure resistance in LRFD terminology is based on strength resistance factors of 1.0 for most controlling loading conditions. Steel structures utilize a strength limit on yield strength. The justification on the use of a strength reduction factor of unity is based on thousands of component tests on steel angles used in latticed steel structures, tubular steel poles, wood poles, prestressed concrete, and fiber-reinforced polymer structures. The standards and manuals of practice utilized in the design of these structures have been developed using the results of these component tests. As a performance-based structure, often designed without load factors or strength reduction factors, new designs are routinely full-scale tested to verify and validate the results of the designs. In the United States, the national safety code (IEEE-C2, National Electric

Table 23.3 Typical Load Types for Structure Design

Condition	Description	Typically controlling
Climactic loads–wind	Extreme Wind, typically 100-year MRI	Main legs or support elements resisted by transverse loads
Climatic loads–ice	Extreme Ice, typically 100-year MRI	Main arms or support elements resisted by vertical loads
Climatic loads–ice with wind	Extreme ice with moderate wind	Various support elements
Failure containment	Broken wire(s) and resulting static impact	Various support elements
Construction and maintenance	Stringing operations and maintenance	Various support elements
Legislated loads	NESC safety minimum loads	Various support elements
Galloping and aeolian vibration	Conductor large motion and Aeolian wind induced vibration	Galloping is typically span specific Aeolian vibration is typically a concern with slender elements with relatively higher natural frequencies
Other	Unbalanced icing, oblique wind, deflection criteria	Various support elements

Safety Code) is also adopted by most states to require minimum safety considerations for electrical clearances and structure strengths.

Table 23.3 provides the load types that are considered for the design of transmission structures.

23.3.3 Structure Type Classification

Transmission lines consist of a family of structure types within the transmission line. These types are defined by their relative support strengths as derived by span lengths, line angles, suspension or dead-end support, failure containment, and other characteristics. The structures supporting a transmission line have different strength demands due to the line directional angle and span lengths associated with the line. Longer span lengths require more structural support strength, and the strength demand increases with increasing changes in line directionality (line angles). The increase in structural support strength, is due to the transverse component of line tension as it increases with the *sin* of half of the line angle. Structure types are grouped into support strengths corresponding to line angle and span ranges to take advantage of structure standardization. A typical family of types for any given transmission line may consist of four to six structure types. There are two subtypes of structures within each family: suspension and dead-end. Suspension structures *suspend* the transmission line, and therefore structures do not feel the effects of the line tension except for the transverse component of line tension for structures at line angles. Dead ends are anchoring structures that have full tension resistance capability such that if any one phase or multiple phases were to be "broken," or "dead-ended," the structure could withstand all resulting tensions corresponding to the defined climatic conditions.

Each structure type within a family is designed for multiple clearance heights above the ground to accommodate varying sloped ground terrain along the spans adjacent and supported by the structure.

23.3.4 Structural Geometric Configuration

There are a wide variety of geometries that are utilized for supporting transmission lines. The selections of these are based on many factors; however, most often these factors are a result of minimizing the line total installed costs, structure costs, or effects of structure and line aesthetics and/or line resiliency. Figure 23.2 shows the general categories of structure types utilized and the corresponding method of analysis used for design. While other analysis methods can be employed for some of these types, the general categories are consistently used.

23.3.5 Analysis Methodology

The majority of structures designed in the transmission line industry utilize linear and nonlinear analysis methods using techniques and methods derived from small displacement theory. Overall structure displacements are rarely a controlling design consideration except when large displacements implead clearances or when structures have aesthetic deflection considerations, for example, unguyed pole-type structures under everyday service loads.

In cases where relatively large displacements occur that can contribute to P-delta amplification or other effects, nonlinear, or second-order analysis techniques are employed to include these effects in design.

In cases where multiple materials are utilized such as reinforced-concrete or fiber-reinforced polymers, then specific industry-accepted or manufacturer-developed analysis techniques are used to analyze the structure types. Any new or uncommon structure types are typically full-scale tested to verify the adequacy of the design and analysis method utilized.

23.3.6 Design Process

The design processes for structures supporting high-voltage transmission lines vary depending on the specific structure configuration; however, there are approaches that are common to all structure configurations used to gain efficiencies by grouping structure strengths associated by function into a family.

The most cost-efficient number of structure types for design is based on the line terrain and the specific local line weather conditions, design spans, and line angle ranges. The conductor utilized in the line has an impact on the most optimum span lengths for the line design. Longer spans will require fewer structures; however, the longer spans will require taller structures due to the squared relationship with the conductor sag:

$$\text{sag} \cong \frac{wS^2}{8T_H} \tag{23.1}$$

where

w = wire unit weight
S = straight-line span length
T_H = horizontal component of tension

Higher conductor sag requires taller structures to keep adequate midspan clearances to ground.

As cost efficiencies drive this family selection, much thought needs to go into this decision. These costs efficiencies are derived from fabrication economies of scale offset by the costs and time schedules associated with design of structure types. For example, the development of a family of 12 structure types for a transmission line might achieve the most optimized line cost; however, this will simply take too long in development time to be a reasonable design choice (for latticed structures). A family of five to six types is typically used for lines of typical transmission length (over 20 miles). Poles structures can be designed more quickly than lattice structures; however, for higher voltages where loads are larger, the latticed structure is typically most economical due to its inherent geometric efficiencies.

An equally important design choice is the amount of standardization that is selected for structure design. Since a structure type will group all structures within the specific category of strength required for the type, standardization becomes another important aspect of design to take advantage of economies of scale. An important concept in the selection of standardization is that, taken individually, structures that

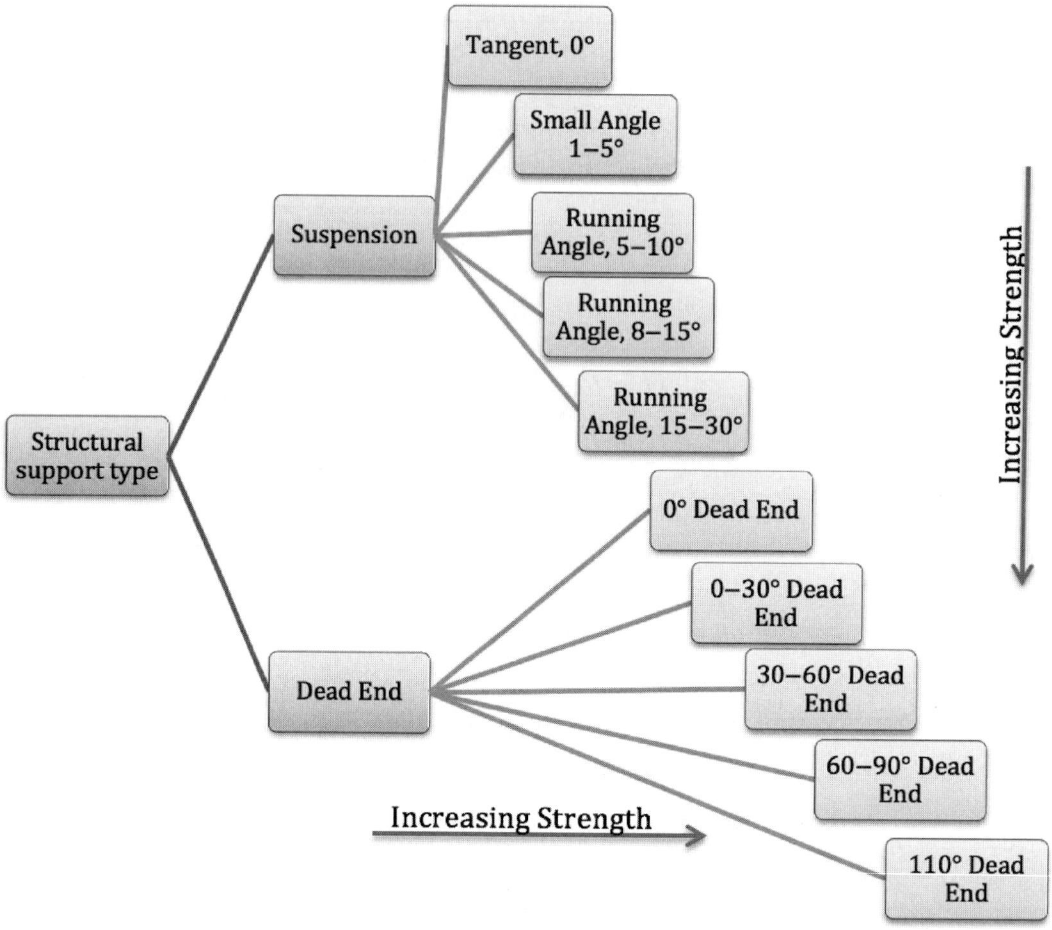

Figure 23.1 Structural support type.

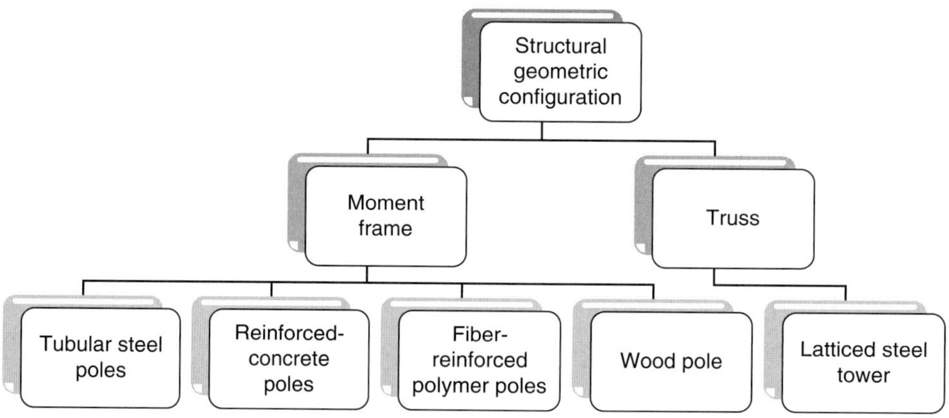

Figure 23.2 Structural geometric configuration and analysis approach.

are designed for site-specific loading can be optimized as an individual structure. However, when taken as a type within a line, the individual optimization negatively impacts the design time and fabrication costs associated with the line. It is therefore often better to standardize key aspects of the structure types to take advantage of fabrication economies of scale. Items of design that can be considered for *standardization* consist of the following:

1. Structure type strength with height. It is usually more cost effective to limit the structure heights within a type. Examining the height distribution for a specific structure type will allow judicious choices to be made to determine the maximum heights that should be used in design.
2. Fastener diameter. For structures requiring multiple fastener connections, a single consistent diameter is chosen for each structure type. This achieves both economies of scale in fabrication and reduces errors in assembly and erection on site.
3. Foundation system. Efficiencies are gained by grouping foundations into categories based on reaction load demand and soil type.
4. Insulator configuration. Efficiencies are gained by developing standard assemblies that are associated with each structure type. Consistent lengths and hardware strengths are common to gain economies of scale.
5. Maintenance and fall protection. Efficiencies are gained by visualizing safe maintenance practices prior to designing the structure geometry so that these considerations can be incorporated into the design. Items such as helicopter touch-and-go points and fall protection anchorages need to be planned in advance of finalizing the design.

The following items are related to line design and are typically determined in advance of the structure design and are required to start the design process:

1. Line voltage (nominal and maximum operating) and line elevation. This determines the allowable energized conductor clearances (distances) to steel.
2. Insulation length and arrangement (I string, V-string, post). This helps determine the amount of conductor swing along with the maximum and minimum span lengths under climatic conditions.
3. Clearance to ground and to structure. A function of line voltage, elevation, and the electrical surge impedance loading of the line.
4. Conductor size and bundle. The conductor size, typically provided in units of circular mils, with the voltage, phase configuration, and bundle size defines the electrical power transfer capability of the line. From this, the conductor unit weight, diameter, and drag coefficient are known.
5. Shield wire size. The shield wire provides electrical circuit protection from lightning-induced transients. Typical sizes are 7/16 in. and ½ in. or may carry fiber-optic filaments for communications in an optical ground wire (OPGW).
6. Line route, terrain, ground slope, and soil information. From the line route and clearance to ground data, the structure types can be developed. The ground slope provides insights to the maximum leg extension differentials to use in the structure design (for multi-leg structures). The soil information is used for foundation design.
7. Position of shield wire relative to conductor (shielding angle). This is used to determine the geometric position of the shield wire relative the conductor.
8. Phase-to-phase geometry and separation (horizontal or delta). This is used to determine the structure geometry. Horizontal geometries typically require a transposition structure at periodic intervals to break up the electrical fields associated with high voltage lines at length. Transposing structures, transpose the phase positions. Delta geometries tend to cancel electrical fields and therefore do not normally require transposition structures.
9. Allowance of phase-to-phase differential spacing from dissimilar structure types. Decisions are made during the line design process that define the allowable spacing from structure type to type taking into account phase-to-phase clearances, midspan conductor motions, and susceptibility to galloping.
10. Right-of-way width and any foundation location restrictions. Existing land use, property owner desires, and general line routing issues are considered to determine the type of foundation and general geometries of structures.

The steps in the design process are as follows:

Design:

1. Create three-dimensional geometry based on allowable clearances and required heights above ground.
2. Model basic structure for design.
3. Apply loading and design multiple structure heights within the same structure type. Ten to 20 height configurations using standard intervals is common depending on the terrain.
4. Design for differential foundation elevations (multiple-legged structures). Structure types need to accommodate slopes from any orientation relative to the transmission line. This is needed to minimize the excavation reducing excavation spoils and decreasing environmental impacts dues to construction.
5. Verify clearances in three dimensions or by projections. The clearance to steel to the energized conductor must be maintained as the wires approach, suspend, and exit the structure. Special attention must be made with insulator strings that can swing, such as I-strings, for maximum and minimum span lengths and the slope of the wire as it enters and exits the conductor shoes (typically 7 to 8 degrees for suspension structures) must be accommodated in the clearance evaluation.
6. Create design drawings.

Detailing:

1. Develop detailed three-dimensional geometry based in design geometry.
2. Develop connections.
3. Develop erection drawings.
4. Develop detail drawings.
5. Develop bill of materials.

Testing:
Test the structure type full scale by selecting either the tallest height structure within the type or the most common height for the line. The result of the testing for one height is considered adequate for all of the heights within that type.

23.3.7 Latticed Steel Towers

Design for latticed steel structures is based on typical strength equations for sizing members using the process and equations in ASCE-10. The support members typically resist both tension and compression and are designed as trusses as long as they are geometrically oriented to ensure that the truss approximation is valid, which is no less than 15 degrees of incidence between any two adjacent support members connecting at the same joint. These structures are inherently nondeterministic and usually have multiple redundancies allowing multiple load paths. The members typically utilize steel angle sections to accommodate bolted connections with their flanges. Tension capacities are calculated based on the net area, number of bolts in shear, and rupture. Bearing capacities are calculated based on the maximum load, number of bolts in a connection, and the thickness of the bearing surface. Compression capacities are determined based on compact section requirements to prevent local buckling and yield strength and slenderness to prevent flexural/flexural-torsional buckling. The compression capacities are a function of the member slenderness (kL/r), which is modified depending on the end conditions of the member connection. Six separate end conditions determine the capacity equation (or curve) that is used to modify the slenderness in the compression capacity calculation. These six equations were developed during the 1920s and 1930s from compression angle testing varying the end conditions to replicate typical connections used in practice. These single-member tests have been further verified under thousands of full-scale testing to verify their capacities and assumptions used in design. Connections in lattice structures are bearing-type bolted connections. Bolts are typically installed to snug-tight condition. The snug-tight condition is the tightness that is attained with a few impacts of an impact wrench or the full effort of an ironworker using an ordinary spud wrench to bring the plies into firm contact but not necessarily into continuous contact. Truss geometric stability is also a design consideration and must be appropriately addressed. Combined stresses are taken into account when relatively higher moments exist compared to axial loads. Member torsional loading is rare on these structure types. Typical framing for latticed structures is shown in Fig. 23.3.

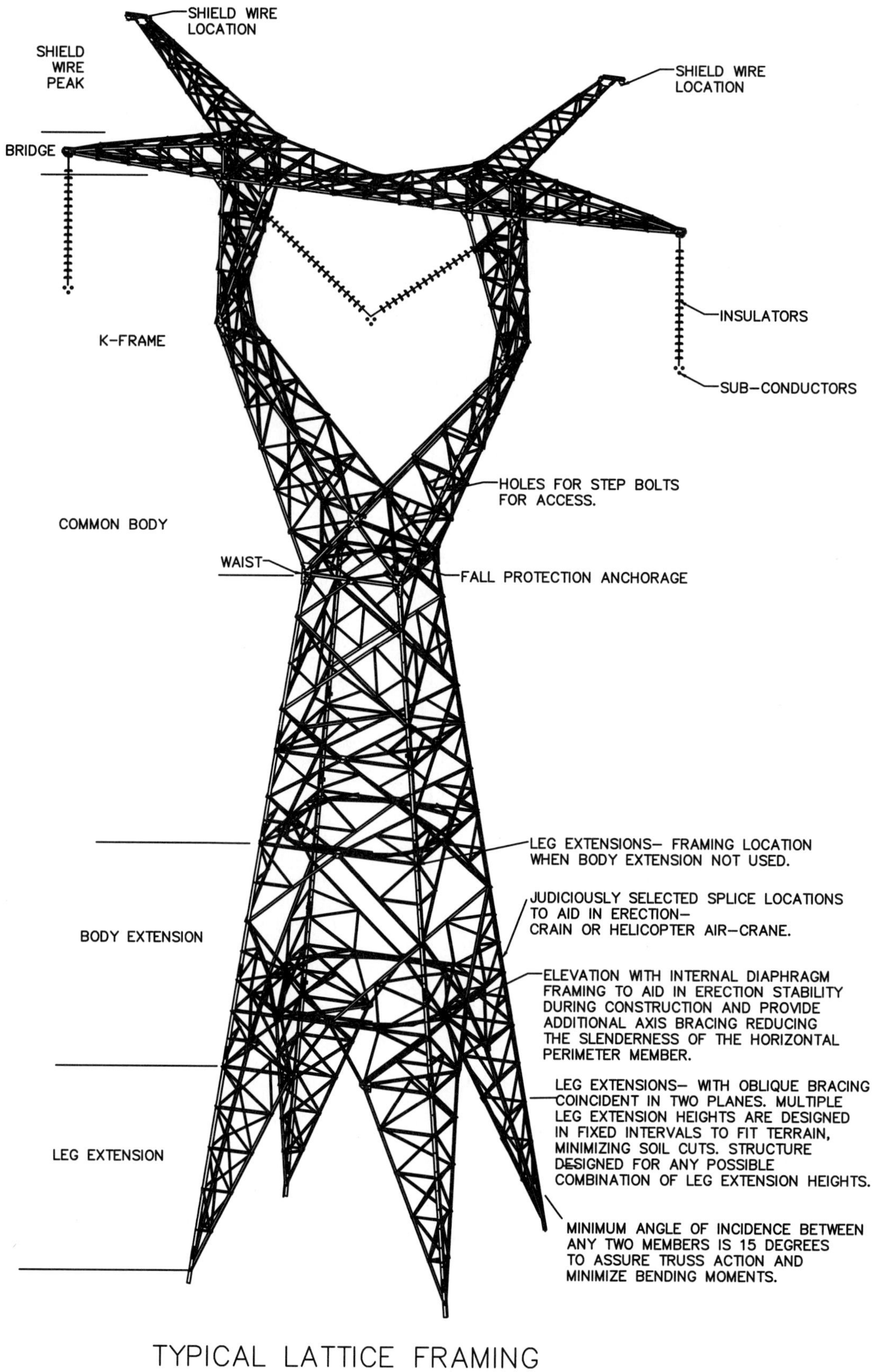

Figure 23.3 Typical lattice framing.

23.3.8 Tubular Steel Poles

Tubular steel structures are designed based on the requirements of ASCE-48. The applied loading on polygonal sections onto cantilever-type geometries produces both moments and axial stresses; therefore, combined stress equations are used. Structural members are made by cutting trapezoidal plates and longitudinally bending this plate into multisided half-shells, which are longitudinally welded together with a partial penetration weld. Pole sections are either slip-jointed (installed by hydraulic jacking) or flange connected typically using F3125 bolts. Pole arms are connected to the main vertical pole shaft by vertical flanges (through-plates) bolted to arm flanges. These connections are made using F3125 bolts. The main vertical shaft is welded (complete penetration) to a base plate that accepts a ring of anchor bolts, which are embedded into reinforced concrete. A typical tangent suspension steel pole structure is shown in Fig. 23.4.

23.3.9 Wood Poles

The vast majority of high-voltage transmission lines at relatively lower voltages, and distribution lines are constructed on round wood poles. Wood poles use the modulus of rupture (MOR) as a basis of calculating strength and the modulus of elasticity (MOE) as a basis for calculating deflections.

There are three most common species of wood poles utilized for wood pole structures: western red cedar, southern pine, and Douglas fir. Natural wood has an inherently higher coefficient of variation in material strengths as compared to steel poles, and for consistent design across the industry, the wood pole strengths must be based on a 5 percent lower exclusion limit (LEL) of the probability density function for the wood species used. Although only applicable to the elastic range loading up to this value, the 5 percent LEL value is considered the MOR. Many wood publications, such as ANSI O5.1, report wood pole strengths based on a 50 percent LEL, so an adjustment must be made to the strength to convert from the 50 percent LEL to 5 percent LEL. This adjustment is defined as $MOR_{5\%} = MOR_{50\%}\ (1 - 1.645 \times COV)$.

23.3.10 Foundations

There are a wide variety of foundation types utilized with transmission structures. The selection of type depends on the soil conditions, proximity to local materials (concrete), and general economies of fabrication and installation. Types typically used for these structures are provided below with geometries indicated in Fig. 23.5.

Pyramid Grillage (a): This foundation system consists of steel beams to create a surface for soil resistance. The truss geometry is utilized to

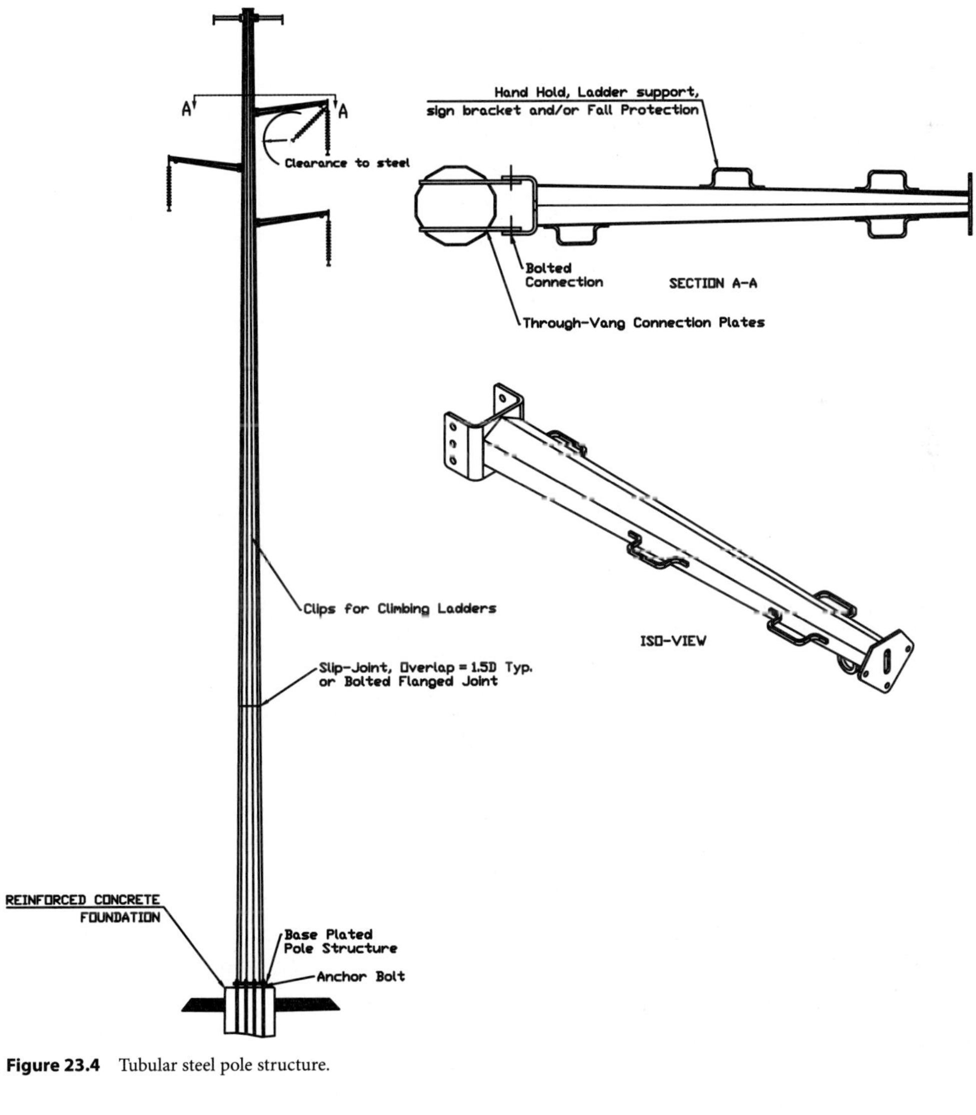

Figure 23.4 Tubular steel pole structure.

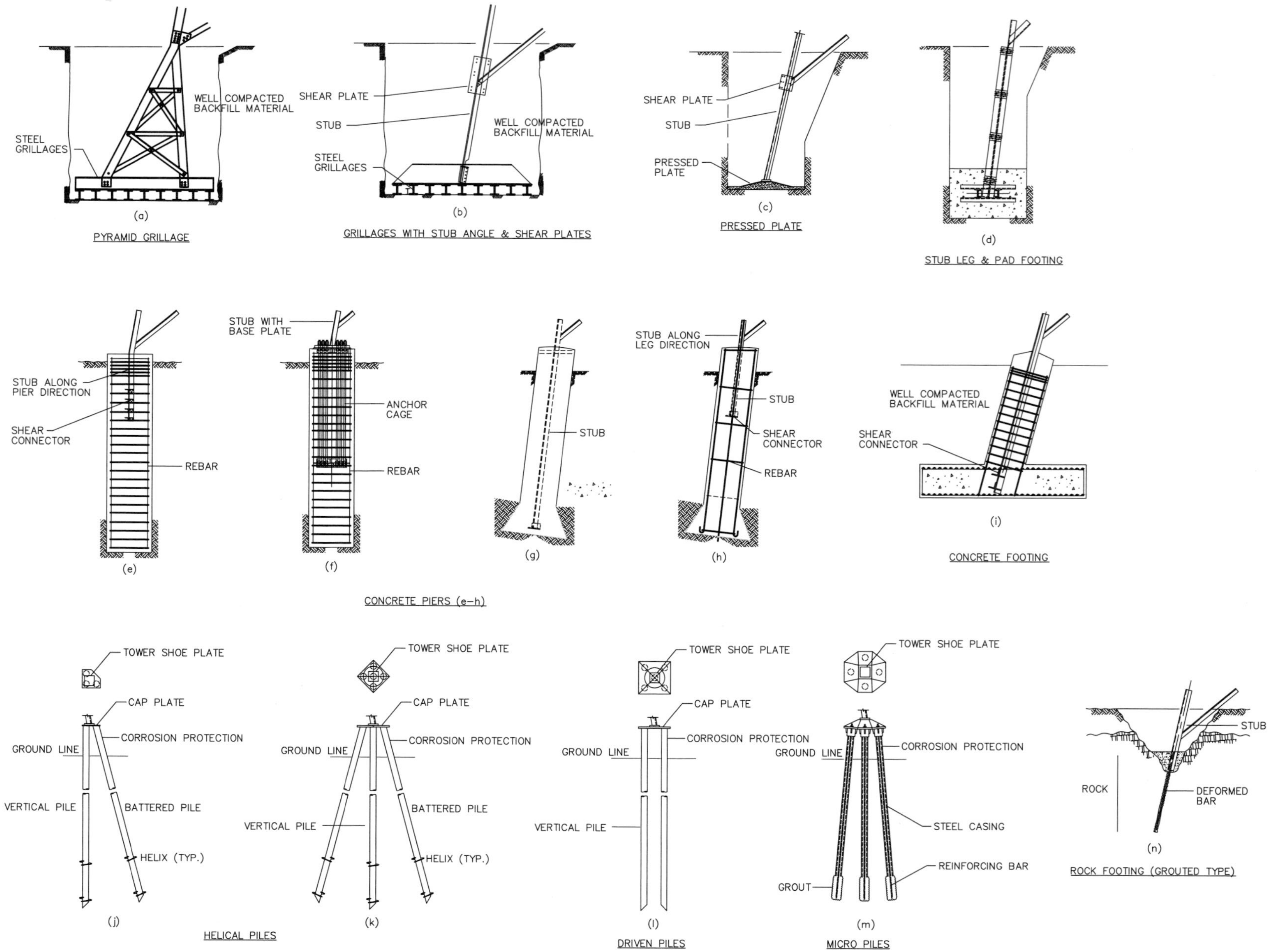

Figure 23.5 Typical foundation types.

minimize bending moments and therefore provide an efficient foundation that is consistent with the truss above grade. Overturning of the beam grillage is a design consideration.

Grillage Foundation with Shear Plate (b): This foundation and its use of a shear plate to resist the transverse loading from the leg diagonal requires the intersection of diagonal to the main leg to be below grade elevation. This is to provide adequate bearing resistance from passive pressures of the soil at depth. The grillage base, which collects negligible moments, resists uplift loads by backfill (typically native) and compression loads by soil at depth.

Pressed Plate (c): This foundation system is utilized for lightly loaded suspension towers. It uses the same concepts of resistance as the other grillage types; however, the base is replaced with a plate that is pressed into a shallow frustum and installed over a layer of sand to evenly distribute the load. Native backfill is typical unless additional uplift capacity is required, upon which crushed rock is used.

Stub Leg and Pad Footing (d): This foundation extends the stub angle to depth by reinforcing and thereby increasing its capacity while increasing its unsupported length. The grillage of (a) or (b) is substituted with a fixture embedded in concrete which provides the compression capacity and with the backfill provides the uplift capacity. Any additive shear loads due to the diagonal member need to be resisted by the reinforced stub angle and its passive pressure of the stub surface areas against the soil.

Concrete Piers (e–h): These foundations and their variations use reinforced concrete. The connections to the structure are either by embedded straight or bent stub angle or base plate with anchor bolts. The pier may be vertical or angled along the direction of the main leg to lower the transverse shear reactions.

Concrete Footing (i): This foundation system uses a concrete pad with pedestal. The pedestals may be angles matching the main leg slope or vertical.

Helical Piles (j–k): This foundation system uses the driven helical pile connected at an above-grade pile cap. Multiple piles are utilized when additional shear and uplift and compression capacity is required. The individual piles are cut after driving to a predetermined torque correlated to soils. This foundation type is typically avoided in rocky soils but is utilized in areas of rugged terrain where traditional equipment cannot be provided easily on-site.

Driven Pile (l): Typically used in poor soils, this foundation uses piles driven by either hammer or vibratory head.

Micro-piles (m): This foundation system uses a series of prefabricated deformed bars encased in a steel pipe. Grout is injected to form the bond between the bar and the soil. The piles are contained above grade by way of an adjusting shoe plate.

Rock Footing (n): This foundation is used in suitable rock and is drilled, and deformed bars are inserted and grouted in place.

23.4 SUBSTATION STRUCTURES

There are wide variety of substation (or switchyard) support structures used for these electrical facilities that switch, transform, disconnect, or provide power quality to the electrical circuits used for power transmission. The majority of substations are designed as air-insulated. Gas-insulated substations are also used but less often. There are multiple types of overhead bus (conductor system) configurations for substation structures: single bus, double bus double breaker, double bus single breaker, main and transfer bus, breaker and half, and ring bus. The layout and general configuration of the structures directly depend on the bus configuration of the substation. The bus type can be either rigid bus or strain bus, which directly affect the type and magnitude of the applied loading on the equipment and subsequent support structures. The equipment requiring structural support in a substation varies but can be categorized into one of three categories. Category 1: structures that are highly elevated that transition or connect the electrical circuit from the transmission line to the substation or carry wire conductors from areas within the substation; Category 2: structures supporting equipment that requires vertical clearance above ground; Category 3: structures supporting the electrical connections or bus between equipment; and Category 4: structures that need no supporting structure that can be installed at ground level but need to be structurally designed for proper interaction with adjacent equipment or connections.

The following is a list of electrical equipment that is typically located (installed) within a substation:

Structures with highly elevated electrical wire connections (Category 1): dead end, multiple-dead end, and multiple gantries.

Structures with equipment installed elevated (Category 2): current limiting inductor or air core reactor, line trap, coupling capacitor voltage transformer, disconnect switch, circuit switcher, potential and current transformers, capacitor bank, shunt capacitor, series capacitor, surge arrestor, neutral grounding resistor, and cable terminator.

Installed on a structure supporting an electrical connection (Category 3): insulators, bus duct, and bus run.

Installed at or near ground level (Category 4): power transformers and autotransformers, shunt reactors, circuit breakers, fire barriers, and control enclosures.

23.4.1 Materials

Table 23.4 provides the materials typically utilized for the most common structure types in substation design. Many of the materials are same or similar to those used in transmission line design.

23.4.2 Loading and Methodology

Substation structures are designed to resist the forces generated by dead loads, wind, ice, line tensions, seismic, short-circuit, construction, maintenance, and electrical equipment operational loads. For substation structural loading, the wind and ice loading are determined from ASCE/SEI-7-16 using the 100-year MRI. Controlling loads are highly dependent on substation location and relative size (voltage) of the electrical equipment. Higher voltages require increased clearance heights, which increase load demands on the support structures. The seismic ground motion parameters are determined from United States Geological Survey (USGS) website: http://earthquake.usgs.gov/designmaps/us/application.php.

The design maps that are developed by USGS are based on Risk Targeted Maximum Considered Earthquake (MCE), parameter values of which are based on a 1 percent probability of collapse for buildings, which is considered appropriate for substation design. Load factors are used to account for uncertainties in the loads, and the combinations are used as appropriate applications of loads consistent with an operating substation. Electrical equipment in seismic areas is to be qualified by IEEE-693. Specific guidance for design level loads is provided in ASCE 113, the load combinations of which are summarized in Tables 23.5 and 23.6:

D = structure and wire dead load

W_{300} = extreme wind load from the 300-year MRI wind map

I_{100} = extreme ice load from the 100-year MRI ice map

$W_{WI\text{-}100}$ = concurrent wind load in combination with ice from the 100-year MRI ice map

$T_{WI\text{-}100}$ = wire tension corresponding to the wind speed and concurrent ice thickness from the 100-year MRI ice map

E = seismic load

Table 23.4 Materials for Substation Support Structures

Structure type	Materials utilized
Latticed steel structures	Steel angle: ASTM A36, A529, A572 Fasteners: ASTM F3125 or A394 Type 0 or 1 Galvanizing: ASTM A123
Tubular steel poles	Steel plate: ASTM A572 Fasteners: ASTM F3125 Galvanizing: ASTM A123
Aluminum structures	Shapes, plate, fasteners: Alloy 6061-T6, 6063-T6, 2024-T4 Nuts: ASTM F467

Table 23.5 Typical Combinations for Ultimate Load

Ultimate limit state load combinations	
Case	Combinations (all may not apply)
1*	$1.1\ D + 1.0\ W_{300} + 0.75\ SC + 1.1\ T_{W\text{-}300}$
2*	$1.1\ D + 1.0\ I_{100} + 1.0\ W_{WI\text{-}100} + 0.75\ SC + 1.1\ T_{WI\text{-}100}$
3	$1.1\ D + 1.0\ SC + 1.1\ T_{APP}$
4*	$1.1\ D + 1.0\ E + 0.75\ SC + 1.1\ T_{EA}$
5*	$0.9\ D + 1.0W_{300} + 0.75\ SC + 1.1\ T_{W\text{-}300}$
6*	$0.9\ D + 1.0\ I_{100} + 1.0\ W_{WI\text{-}100} + 0.75\ SC + 1.1\ T_{WI\text{-}100}$
7	$0.9\ D + 1.0\ SC + 1.1\ T_{APP}$
8*	$0.9\ D + 1.0\ E + 0.75\ SC + 1.1\ T_{EA}$

*The combination of SC loads with extreme events listed above should be determined by the Owner.
†Reference ASCE/SEI 7-16 C2.4.1 for explanation of 0.6 dead load factor.
These loads combinations are intended for foundation design.

Table 23.6 Typical Combinations for Service Loads

Allowable strength design (ASD) load combinations (service-level loads)	
Case	Combinations (all may not apply)
1*	$1.0\ D + 1.0\ W_{100} + 0.5\ SC + 1.0\ T_{W\text{-}100}$
2*	$1.0\ D + 1.0\ I_{50} + 1.0\ W_{WI\text{-}50} + 0.5\ SC + 1.0\ T_{WI\text{-}50}$
3	$1.0\ D + 0.7\ SC + 1.0\ T_{APP}$
4*	$1.0\ D + 0.7\ E + 0.5\ SC + 1.0\ T_{EB}$
5*,†	$0.6\ D + 1.0\ W_{100} + 0.5\ SC + 1.0\ T_{W\text{-}100}$
6*,†	$0.6\ D + 1.0\ I_{50} + 1.0\ W_{WI\text{-}50} + 0.5\ SC + 1.0\ T_{WI\text{-}50}$
7†	$0.6\ D + 0.7\ SC + 1.0\ T_{APP}$
8*,†	$0.6\ D + 0.7\ E + 0.5\ SC + 1.0\ T_{EB}$

*The combination of SC loads with extreme events listed above should be determined by the Owner.
†Reference ASCE/SEI 7-16 C2.4.1 for explanation of 0.6 dead load factor.
These loads combinations are intended for foundation design.

$T_{W\text{-}300}$ = wire tension corresponding to the wind speed from the 300-year MRI wind map;
SC = short-circuit load
T_{APP} = wire tension corresponding to the appropriate temperature and ice as determined by the Owner (everyday or normal operational conditions)
T_{EA} = wire tension corresponding to wire dead load and seismic loading

where

- W_{100} = extreme wind load from the 100-year MRI wind map
- $T_{W\text{-}100}$ = wire tension corresponding to the wind speed from the 100-year MRI wind map
- I_{50} = extreme ice load from the 50-year MRI ice map
- $W_{WI\text{-}50}$ = concurrent wind load in combination with ice from the 50-year MRI ice map
- $T_{WI\text{-}50}$ = wire tension corresponding to the wind speed and concurrent ice thickness from the 50-year MRI ice map
- T_{EB} = wire tension corresponding to wire dead load concurrent with 70 percent of the seismic loading

Simplified Approach to Determine Short-Circuit Forces:
The equation for the basic maximum distributed force between two parallel infinitely long conductors is

$$F_{sc} = \frac{\frac{\mu_0}{2\pi} * \Gamma * (2 * \sqrt{2} * I_{sc})^2}{10^7 * D} = \frac{3.6 * \Gamma * I_{sc}^2}{10^7 * D} \tag{23.2}$$

where

F_{sc} is the fault force in pounds per foot
I_{sc} is the RMS fault current in amps
D is the conductor center to center spacing in feet
Γ is 1.0 for phase-to-phase and 0.866 (middle conductor) or 0.808 (outer conductors) for three-phase faults
$\eta_0 = 4\pi \times 10^{-7}\ \text{N/A}^2 = 2.825 \times 10^{-7}\ \text{lb/A}^2$ (magnetic constant–magnetic permeability in a classical vacuum)

The basic maximum fault force from Eq. a(23.2) is modified to obtain the peak force by multiplying by the square of the half-cycle decrement factor, D_f, as given in IEEE 605-2008:

$$F_{sc\_peak} = F_{sc} * D_f^2 \tag{23.3}$$

$$\text{where} \quad D_f = \frac{1 + e^{-\frac{1}{2fc}}}{2}$$

23.4.3 Deflection Criteria

Deflections due to loading may control the design. When evaluating deflection, an important concept to consider is the gross versus net deflection. Gross deflection is the overall deflection of the structure and is used to verify the clearances, whereas net deflection is the deflection in the local support area that affects the performance of the equipment. Table 23.7 is used to define the deflection limits for three different structure classes:

Class A—structures sensitive to deflections for operation.
Class B—structures sensitive to deflections for clearances (phase-to-phase or phase-to-ground).
Class C—structures insensitive to deflections

Special deflection considerations should be made to those structures supporting multiple use, such as those structures supporting several pieces of equipment having different functionality or for structures supporting equipment with rotational deflection sensitivity.

Bus midspan span deflection should be limited to 1/200 of the span under self-weight-only conditions. Cantilevered bus deflection should be limited to 1/100 of the cantilevered distance.

Table 23.7 Deflection Limits of Member Spans for Substation Structures

	Class A	Class B	Class C
Vertical deflection–horizontal members	1/200	1/200	1/100
Horizontal deflection–horizontal members	1/200	1/100	1/100
Horizontal deflection–vertical members	1/100	1/100	1/50

23.4.4 Analysis Methodology

Truss, frame, and finite-element model types are utilized for substation structural analysis. First- and second-order elastic static analysis is utilized for most structural supports. Inelastic static or elastic (or inelastic) dynamic analysis is also used depending on many factors. See Table 23.8 for guidance on specific structure categories and their typical model and analysis methods.

Table 23.8 Model Type and Method of Analysis

	Structure categories			
Model type	1	2	3	4
Truss				
Frame				
Finite element: plate, shell, or brick elements				
Analysis method	1	2	3	4
Static analysis: approximate analysis				
Static analysis: first order elastic				
Static analysis: second-order elastic				
Static analysis: first-order inelastic				
Dynamic analysis: steady state				
Dynamic analysis: eigenvalue				
Dynamic analysis: response spectrum				

Short-circuit analysis methods vary; however, the following methods are referenced and often used:

- IEEE-605 Guide for Bus Design in Air Insulated Substations
- CIGRE 105 *The Mechanical Effects of Short-Circuit Currents in Open Air Substations, CIGRE, WG 23.11, 1996*
- CIGRE 214 *The Mechanical Effects of Short-Circuit Currents in Open Air Substations, Part II. WG 23.03, Oct. 2002*
- IEC 60865-1:2011, *Short-Circuit Currents – Calculation of Effects – Part 1: Definitions and Calculation Methods*
- IEC 60909-0:2001, *Short-Circuit Currents in Three-Phase AC Systems – Part 0: Calculation of Currents*

23.4.5 Design Process

These structures are designed using either Ultimate Limit State (LRFD) or Allowable Stress (or Strength) Design (ASD) methodologies. The ASD (stress) 1989, 9th edition, AISC manual, and the ASD (strength) 2005, 13th edition, AISC manual, are still used. Table 23.9 provides the design process categorized by structure material and type.

Seismic design of structures that support electrical equipment should be designed using the assumptions of natural frequencies and damping of the equipment seismic qualifications of IEEE Standard 693-2005. In addition, adequate electrical connection slack should be specified such that the assumptions used in the dynamic analysis of the support structure is specified and used in the installation of the equipment. The amount of slack typically provided for temperature compensation is usually inadequate to isolate equipment from seismic displacements. If considered a dedicated support, the structure may require seismic qualification in conjunction with the piece of equipment per IEEE Standard 693-2005.

Table 23.9 Design Processes by Structure Material

Structure material	Structure type	Structure category	Design process	Method
Steel	Latticed	1, 2	ASCE 10-15	LRFD, FOEA
	Multi-sided tubular	1, 2	ASCE 48-19	LRFD, SOEA
	HSS and all other hot-rolled shapes except angles	2, 3	ASCE 360-16	LRFD, SOEA
Concrete	Reinforced concrete	2	ACI 318-14	LRFD, SOEA
	Prestressed	2	PCI MNL 120-10	LRFD, SOEA
	Prestressed pole	1, 2	ASCE 123-12	LRFD, SOEA
Aluminum	Various shape extrusions	2, 3	Aluminum Association Design Manual, 10th ed., 2015	LRFD, SOEA

FOEA, first-order elastic analysis; SOEA, second-order elastic analysis.

23.4.6 Foundations

The types of foundations within substations are often limited to reinforced concrete cast-in-place spread footings, combined spread footings, and mat foundations. In areas of poor soils, piles or drilled piers (caisson) may be utilized. In high seismic areas, large equipment on mat foundations, such as power transformers, may utilize base isolation to lower the load demands on the equipment. Figure 23.6 shows the typical substation foundations used.

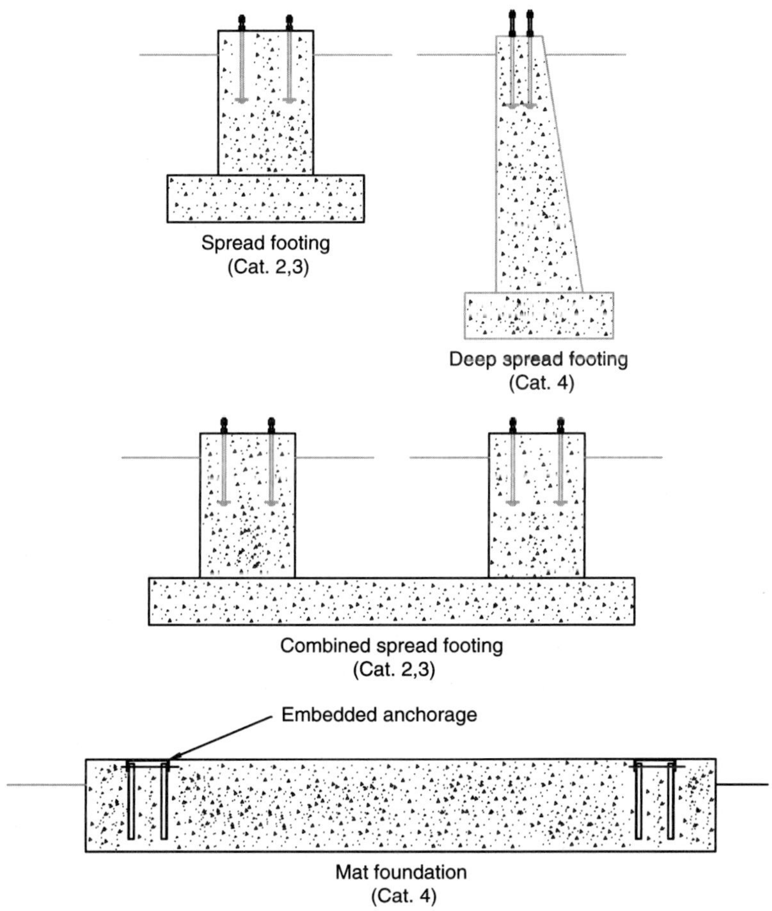

Figure 23.6 Typical substation foundations.

REFERENCES

Guidelines for Electrical Transmission Line Structural Loading, ASCE 74 Manual of Practice, Third Edition, American Society of Civil Engineers, 2009.

National Electrical Safety Code, ANSI-C2, IEEE, New York, 2017.

Design of Latticed Steel Transmission Structures, ASCE 10 Standard, American Society of Civil Engineers, 2015.

Design of Steel Pole Transmission Structures, ASCE 48 Standard, American Society of Civil Engineers, 2019.

Prestressed Concrete Transmission Pole Structures, ASCE 123 Manual of Practice, American Society of Civil Engineers, 2012.

Design of Guyed Electrical Transmission Structures, ASCE 104 Manual of Practice, American Society of Civil Engineers, 1997.

ASCE 141, *Manual of Practice Wood Pole Structures for Electrical Transmission Lines: Recommended Practice for Design and Use*, 2019.

Substation Structure Design Guide, ASCE 113 Manual of Practice, American Society of Civil Engineers, 2008.

Minimum Design Loads for Buildings and Other Structures, ASCE/SEI-7, American Society of Civil Engineers, 2016.

Guide for Bus Design in Air Insulated Substations, IEEE-605.

The Mechanical Effects of Short-Circuit Currents in Open Air Substations, CIGRE 105, WG 23.11, 1996.

The Mechanical Effects of Short-Circuit Currents in Open Air Substations, CIGRE 214, Part II. WG 23.03, Oct. 2002.

Short-Circuit Currents – Calculation of Effects – Part 1: Definitions and Calculation Methods. IEC 60865-1:2011.

Short-Circuit Currents in Three-Phase AC Systems – Part 0: Calculation of Currents. IEC 60909-0:2001.

Building Code Requirements for Reinforced Concrete, ACI 318–14.

Manual of Practice for Fiber-Reinforced Polymer Products for Overhead Utility Line Structures ASCE 104, American Society of Civil Engineers, 2019

Chapter 24
Chimneys

BY

SHU-JIN FANG *Consultant, and Former Associate and Senior Manager, Sargent & Lundy, Chicago, Illinois*

VICTOR BOCHICCHIO *Executive Vice President, Hamon Custodis, Inc., Somerville, New Jersey*

24.1 INTRODUCTION

Chimneys are a common component of power plants and many industrial facilities. Although they may appear to be fairly simple structures, their design involves some unique elements which will be addressed in this chapter. This chapter will focus on structural aspects of chimney design.

24.1.1 Materials

Most chimneys in the United States are built with outer walls of reinforced concrete or steel. Some older chimneys are constructed of radial brick. The use of fiberglass-reinforced plastics (FRP) is increasing in highly corrosive applications and for flue liners following flue gas desulfurization systems. The decision to build a chimney of concrete or steel (the latter are frequently called stacks) is largely based on cost. The economic comparison must include the cost of linings, foundations, painting, lighting, access provisions, maintenance, and plant outages.

24.1.2 Sizing: Diameter and Height

Chimneys are constructed for the dispersion of flue gas from combustion processes. The height and top flue diameter depend on the gas temperature, volume, stack draft requirements, and environmental permitting requirements. The selection of the stack/chimney configuration should be based on considerations of plant layout, flue gas composition, construction cost, aesthetics, and regulatory requirements. These topics are beyond the scope of this document, and it is presumed that the height and top diameter are known.

24.1.3 The Role of the Structural Engineer

Although the design of chimneys is sometimes performed by multiple engineering disciplines (i.e., process, mechanical, structural, etc.), it is more common, especially for large chimneys, that the design is executed by a specialty design-supply or design, supply, and erect firm. In these organizations the structural engineer is usually the lead engineer and also commonly performs the thermofluodynamic design of the chimney given the process flow data from the client. Information on the thermofluodynamic design of chimneys can be found in the CICIND Manual[1] and ASME STS-1-2016.[2]

24.2 DESIGN LOADS

24.2.1 General

The chimney and its supporting system should be designed to resist stresses resulting from dead load, wind loads, earthquake loads, stack draft, and temperature gradients. Furthermore, resonant vibrations due to dynamic wind should be considered. However, the design need not be limited to these loadings.

24.2.2 Dead Loads

Dead loads should include the weight of all permanent construction and fittings, insulation, fly ash, and other materials. Weights of liners, roofs, and/or platforms supported by the chimney wall should be included.

24.2.3 Wind Loads

Chimneys and stacks should be designed to resist the wind forces in both the along-wind and across-wind directions. In addition, the variation of pressure along the circumference of the shell should be considered.

Along-Wind Loads

The most recent provisions for determination of the along-wind loads (drag loads) in the United States are given in ACI 307-08[3] for concrete chimneys and ASME STS-1-2016 for steel stacks. For many years, ACI-307 and the Uniform Building Code (UBC) specified horizontal wind loads as uniform pressures acting on the projected area. Pressure magnitudes depended on the height above ground and on the geographic area as specified in the wind-pressure map. Current wind-loading provisions of ASCE 7[4] require consideration of wind-gust effects on chimneys and stacks and do not use the traditional wind-pressure values. Both ACI-307-08 and ASME STS-1-2016 require that wind loads on chimneys be calculated as the sum of two components, one caused by a mean wind speed and the other caused by fluctuating wind gusts. The mean wind load contribution is proportional to the square of the wind speed. The dynamic component on concrete chimneys is evaluated using a modified gust-factor approach as described by Vickery and Basu[5] that depends on the natural frequency and the geometric properties of the chimney. A similar along-wind approach is employed in the ASME STS-1-2016 standard on steel stacks. However, the dynamic component is calculated using a conventional gust effect factor method from ASCE 7, developed for flexible buildings and structures, which tends to give lower wind moments than the approach of ACI 307-08. Both gust effect factor approaches are intended to account for additional along-wind loading effects due to wind turbulence-structural interaction and the dynamic amplification of the structure. Note that the along-wind loads determined per ASME STS-1 correspond to a 50-year return wind. In contrast, ACI 307-08 requires that the along-wind loads be calculated on the basis of a 100-year return wind taken from ASCE 7-02 wind maps with Risk Category IV. Further developments are now being undertaken by ACI 307 to develop wind load provisions utilizing even larger return periods, such as 700-year return wind or 1700-year return wind. The main reason for selecting larger return period wind by ACI 307 is to be consistent with newer ASCE 7 practice in establishing strength-basis wind loads for design of concrete chimneys, while the along-wind loads of STS-1 are prescribed as working basis loads for allowable stress design of steel stacks. Where the applicable building code requires the use of ASCE 7-10, designers are required to convert the larger return period wind to 50-year return wind in calculation of the along-wind for steel stacks and 100-year return period wind for concrete chimneys.

Across-wind Loads due to Vortex Shedding

When the chimney is subjected to a steady wind, the periodic shedding of vortices will cause swaying oscillations in a direction transverse to that of the wind (Fig. 24.1*a*). If the vortex shedding frequency is resonant with the natural frequency of the chimney, it may result in large vibrations.[5,6] Across-wind loading and the induced dynamic response are due primarily to regular shedding of vortices. This has long been recognized as a significant design consideration for slender structures with a bluff as opposed to streamlined cross section. There have been numerous failures of unlined steel stacks due to vortex shedding. The current industry practice requires mandatory considerations of vortex shedding vibration effects in the design of both chimneys and stacks. The vortex-induced across-wind loads are dependent upon stack dimensions; modal properties for the vibration modes being considered; the structural damping, β_s; and aerodynamic damping, β_a.[5]

One simple and conservative approximation based on ACI-307-88[7] suggests across-wind loads in the vicinity of critical wind speed may be represented by an equivalent static force F_L per foot of height as determined by Equation (24.1a) or (24.1b), acting in the direction of oscillations due to vortex shedding of the fundamental mode

$$F_L = \frac{C_1 \bar{D} q_{cr} (Z/H)^{1.3}}{(\beta_s - \beta_a)^{0.5}} \tag{24.1a}$$

where β_s = structural damping ratio of chimney; 0.015 for concrete chimney, 0.002 to 0.004 for unlined steel stack, and 0.003 to 0.006 for lined steel stack[2] unless otherwise substantiated. Higher damping up to 0.015 was reported for some lined stacks.[8] Also, structural damping ratio up to 0.04 is permitted for concrete chimney when $V_{cr} > \bar{V}$; Ref. 3

$\beta_a = C_2 \gamma \bar{D}^2 / \bar{m}$ = aerodynamic damping ratio

$\bar{D}$ = mean outside diameter of upper third of chimney, ft

$\bar{m}$ = average weight per unit length of top third of chimney, lb/ft

γ = air unit weight, 0.08 lb/ft^3

Z = distance from base of chimney to position under consideration, ft

H = height of chimney from base, ft

q_{cr} = dynamic wind pressure at critical wind velocity, psf [Eq. (24.2)]

C_1 = coefficient given by Eqs. (24.4a)

C_2 = coefficient given by Eqs. (24.4b)

The dynamic wind pressure q_{cr} is given by

$$q_{cr} = 0.00128\, V_{cr}^2 \tag{24.2}$$

where V_{cr} = critical wind velocity for resonant transverse vibration, given by the following equations:

$$V_{cr} = \frac{f_t \bar{D}}{S} \quad \text{for } \frac{D_t}{D_b} \geq 0.5$$
$$V_{cr} = 2.9 f_t D_b \quad \text{for } \frac{D_t}{D_b} < 0.5 \tag{24.3}$$

where D_t = outside diameter of chimney at top, ft

D_b = outside diameter of chimney at base, ft

f_t = natural frequency of first fundamental mode of transverse vibration of chimney, cps

S = Strouhal number, varying from 0.17 to 0.23. For design purposes, it may be taken as 0.2.[2,9]

The coefficient C_1 in Eq. (24.1) is given by

$$C_1 = \frac{2.5}{\sqrt{H/\bar{D}}} \quad \text{for } D_t/D_b \geq 0.5 \text{ and } H/\bar{D} \geq 16$$
$$C_1 = 0.63 \quad \text{for } D_t/D_b \geq 0.5 \text{ and } H/\bar{D} < 16 \tag{24.4a}$$
$$C_I = \frac{2.9}{(1 - D_t/D_b)^{2.5}(1 + 5(D_t/D_b))^{1.5}\sqrt{H/\bar{D}}} \quad \text{for } D_t/D_b < 0.5$$

The coefficient for C_2 in Eq. (24.1) is given by

$$C_2 = 0.6 \quad \text{for } D_t/D_b \geq 0.5$$
$$C_2 = 0.36 + 0.48\,(D_t/D_b) \quad \text{for } D_t/D_b < 0.5 \tag{24.4b}$$

In assessing the dynamic influence of vortex shedding, the equivalent static force F_L need not be considered if $V_{cr} > 1.3\,\bar{V}$. For V_{cr} between $\bar{V}$ and $1.3\,\bar{V}$, F_L may be reduced by multiplying it by $1 - 5[(V_{cr} - \bar{V})/\bar{V}]^2$ where $\bar{V}$ is the design mean hourly wind speed at five-sixths of the chimney's height if $D_t/D_b \geq 0.5$, or at the height where the chimney's outer diameter is $0.58D_b$, if $D_t/D_b < 0.5$.

If the aerodynamic damping ratio, β_a, exceeds the inherent structural damping ratio, β_s, the value of F_L will become negative, and Eq. (24.1) is no longer valid. This means that large-amplitude resonant vibrations exceeding $0.3\bar{D}$ are probable. For example, due to very low structural damping ratios in steel stacks (particularly unlined stacks), aerodynamic damping may substantially reduce the effective damping and result in excessively vibrations. As such, the proportions of the chimney are inadequate and should be revised. Alternatively, vibration suppression devices should be considered (see Sec. 24.3.8). Conversely, the aerodynamic damping ratio is seldom larger than structural damping ratio for concrete chimneys. The across-wind load calculations as presented in Eqs. (24.1a–24.4) are conservative, and the across-wind loads using this approach could be substantially higher (40 to 60 percent) than what is predicted by ACI 307-08 for concrete chimneys[3,10] where V_{cr} typically exceeds 54 mph. Wind tends to be more turbulent at higher wind velocities. The variation of wind speed with height and turbulence of the wind at higher critical speed

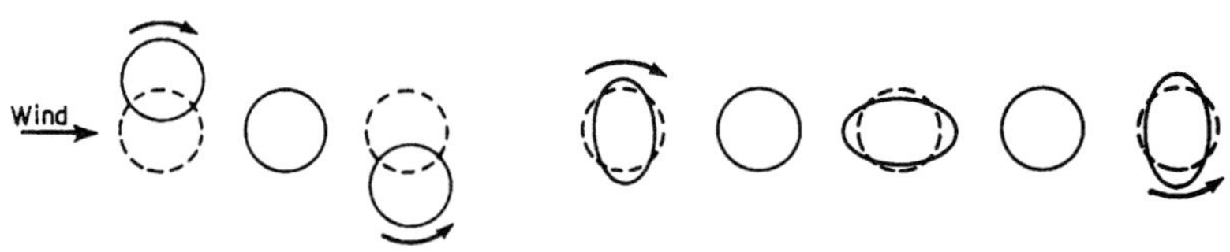

Figure 24.1 Oscillation due to vortex shedding: (*a*) transverse and (*b*) ovalling.

would require that vortex shedding vibrations be treated less conservatively using band-limited random vibration process. Nevertheless, being simple and direct, the approximate method can serve as a quick means for estimating the effects of vortex shedding in preliminary design.

In lieu of Eq. (24.1a), the equivalent static wind force due to across-wind vibrations on steel stacks at elevation Z per unit length, $F_L(z)$, can be also estimated as follows based on simplified ASME STS-1-2000 and NRCC:[11]

$$F_L(z) = \bar{m}_{(z)}(2\pi f_t)^2\, \varnothing_{(z)}\, Y_{max}/g_c \tag{24.1b}$$

where $\bar{m}_{(z)}$ = weight per unit length at elevation z

$\varnothing_{(z)}$ = first mode shape as a function of height z, normalized to unity at $z = H$, which may be taken as $(z/H)^2$

$g_c = 32.2\ \text{ft/s}^2$

Y_{max} = maximum vibration displacement at $z = H$ and $Y_{max}/\bar{D} = \dfrac{C3}{m_r(\beta_s - \beta_a)^{0.5}}\sqrt{\bar{D}/H}$

$C_3 = 1.0$ for the first mode of vibrations

m_r = dimensionless mass as defined in STS-1-2000, Appendix E5

f_t, $\bar{m}$, $\bar{D}$, and H see definition given in Eq. (24.1a).

The validity of Eq. (24.1b) requires that two conditions be met: the maximum vibration displacements are less than $0.1\bar{D}$, and the stack diameter in the top ⅓ of stack height does not vary by more than +10% (for larger variations a reduction can be made).

Since 1988, the subject of across-wind loads or vortex-shedding vibration effects continuingly attracted attention of ACI 307 and ASME STS-1. Both standards have substantially modified their earlier procedures in their latest editions. Across-wind loads in the second mode of resonant vibrations as well as the amplification of across-wind loads due to closely spaced adjacent chimneys are also included. Designers should follow these newest across-wind load design provisions in their final design of chimneys/stacks per the applicable building code, namely, Appendix E of ASME STS-1[2] for steel stacks and Section 4.2.3 of ACI 307-08 for concrete chimneys.

Critical Wind Velocity for Ovalling Vibrations

In addition to transverse swaying oscillations, an unlined steel stack may also be subject to ovalling resonance due to flexural vibration in the circular cross-sectional plane as a result of vortex shedding (Fig. 24.1*b*). The frequency of the lowest mode of ovalling vibration in a circular shell is computed by

$$f_0 = 0.126\frac{t\sqrt{E}}{D^2} = 678.5\frac{t}{D^2} \tag{24.5}$$

where f_0 = first-mode ovalling frequency, cps

t = shell-plate thickness, in

E = modulus of elasticity, psi

D = shell diameter at position under consideration, ft

A resonant condition occurs when the first-mode ovalling frequency is nearly twice the vortex-shedding frequency. Thus, from Eq. (24.3), the critical wind velocity V_0 for ovalling vibration of the stack is (fps)

$$V_0 = \frac{f_0 D}{2S} \tag{24.6a}$$

Then with f_0 from Eq. (24.5) and $S = 0.2$,

$$V_0 = 1696\frac{t}{D} \tag{24.6b}$$

Circumferential Wind Moment

The variation in wind pressure along the circumference of a chimney shell (Fig. 24.2) produces circumferential moments which should be accounted for. The maximum moments are

$$M_{max} = 0.314qR^2 \text{ (tension on inside)} \tag{24.7a}$$

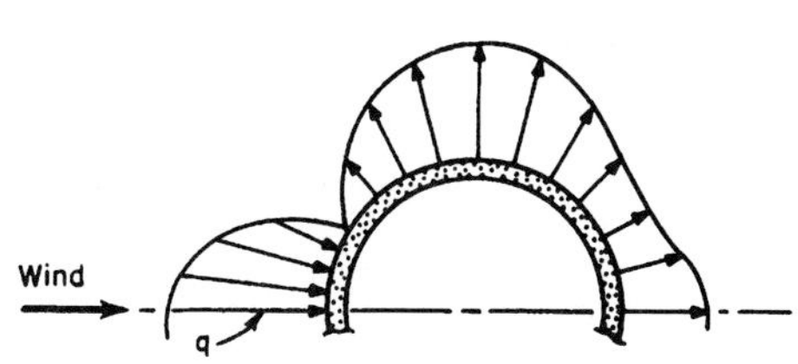

Figure 24.2 Wind pressure on circular chimney/stack section.

$$M_{max} = 0.272qR^2 \text{ (tension on outside)} \tag{24.7b}$$

where q = radial wind pressure at level considered, psf

R = mean radius of chimney at same level, ft

The wind pressure q may be determined per Section 4.2.4 of ACI 307-08 or as follows:

$$q = 0.00256V_R^2\left(\frac{Z}{33}\right)^{0.28} \tag{24.8}$$

where V_R is the reference design wind speed (3-second gust wind) as specified in ASCE 7-02, in mph, and Z is the height, in feet, at level considered. The wind pressure should be increased by 50 percent near the chimney top for a distance 1.5 times the top outside diameter.

24.2.4 Earthquake Loads

Chimneys and stacks in earthquake areas should be designed for seismic resistance. Provisions for determining earthquake loads have been significantly changed in ACI 307-08 to be consistent with ASCE 7-02. It now requires the use of the dynamic response spectrum analysis method to determine loads, shears, moments, and deflections of a chimney due to earthquake. A design response spectrum should be developed from the seismic map values of ASCE 7 for the project site. The design response spectrum will provide the design spectral response acceleration, S_a, at any period and can be constructed in accordance with the general procedure of Section 4.3.2 of ACI 307-08 based on the seismicity of project site, site class (soil coefficients), and seismic risk category (importance factor, I_E). Similarly, the site-specific procedure of Section 4.3.2 can be used to develop site-specific design spectrum. A generic design response spectrum based on ASCE 7-05 is shown in Fig. 24.3 for illustration of typical relationship between spectral response acceleration, S_a, and structural period, T. Seismic forces (shears and moments) obtained by response spectrum analysis should be multiplied by a factor of I_E/R for seismic design, where R is the response modification factor, which can be taken as 1.5 for concrete chimneys.[3,12] Since 2010, ASCE 7 permits $R = 2$ for both concrete chimneys and steel stacks. However, for Seismic Design Category D, E, or F concrete chimneys, the design seismic forces in the breeching region should be increased by 50 percent, and ductile detailing per Section 15.6.2 of ASCE 7-10 must be provided.

Interaction of the stack or chimney with the liners shall be considered. A minimum separation shall be provided between the liner and chimney equal to C_d times the calculated differential lateral drift. The value of C_d should be taken as 2 when $R = 2$ is used.

In general, the dynamic interaction effect on the chimney shell is negligible for reinforced-concrete chimneys with independent steel lining.[13] However, it is important to determine the seismic force exerted on the steel lining as a result of this interaction.

For steel stacks, ASME STS-1 also recommends that earthquake loads and seismic response be determined using the response spectrum method. However, designers should be cautioned that the design spectrum included in STS-1 Table D-2 is nonmandatary and inconsistent with ASCE 7. If Table D-2 is incompatible with the building code, the procedure in Section 11.4 of ASCE 7-10 is suggested for establishing the design spectrum.

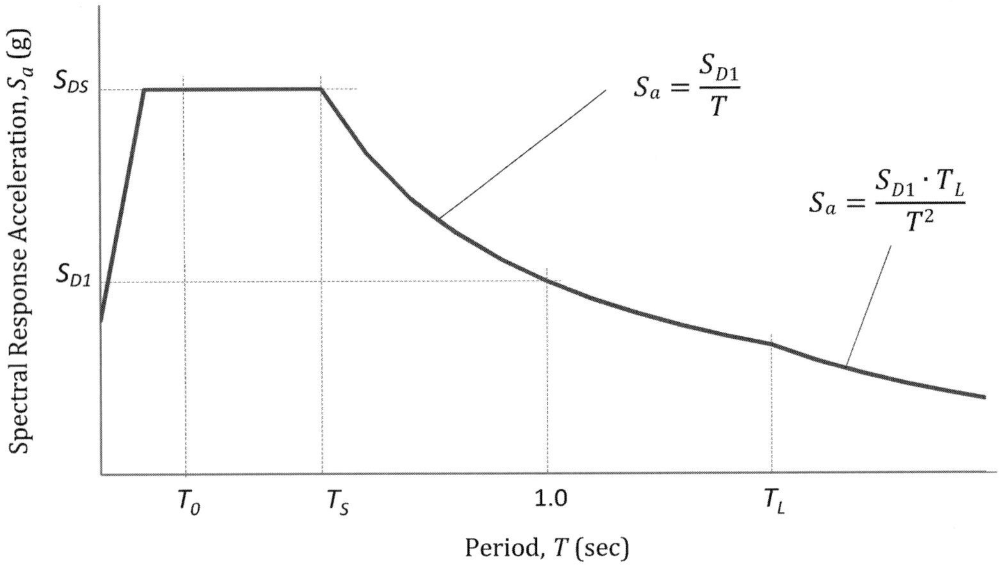

Figure 24.3 Generic seismic design response spectrum per ASCE 7-05.

24.2.5 Pressure Differentials

When a chimney conveys hot flue gas of a specific weight less than the surrounding atmosphere, the pressure on the inside of the liner is less than on the outside. This negative pressure at the flue-gas entrance, often called stack draft, can be determined from

$$D_s = 0.52HP\left(\frac{1}{T_a} - \frac{1}{T_g}\right) \tag{24.9}$$

where D_s = stack draft, inches of water
H = stack height above gas entrance, ft
P = atmospheric pressure at plant level, psia
T_a, T_g = temperatures of atmosphere and gas, respectively, °F absolute (°F + 460°)

The curves in Fig. 24.4 can be used to determine the negative pressure at the flue-gas entrance for most chimneys without appreciable error. Negative pressure varies linearly with height and reaches zero at the flue-gas exit.

For cases where an induced-draft fan forces gas through the chimney the negative pressure is reduced and the chimney may operate under a net positive pressure. This pressure may also be assumed to vary linearly with height.

The pressure differential is resisted by circumferential stresses in the liner itself if the liner is independent and by composite action if it is built integrally with the shell.

24.2.6 Temperature Gradient Loads

All chimneys/stacks should be investigated for thermal stresses resulting from temperature differentials. A procedure to calculate the temperature gradient across the concrete chimney shell, air space (if any), insulation, and liner is discussed in ACI 307-08.

Nonuniform distribution of flue gas across a steel stack or steel chimney liner may cause temperature differentials. For instance, a chimney liner or steel stack with two or more inlet breeching openings will develop an uneven temperature distribution around the circumference of the liner. These differentials produce an unequal expansion which causes lateral deflection of the liner. Significant longitudinal stresses will develop when the liner is restrained. This maximum differential occurs at the level of the breeching opening and exponentially decays along the chimney height. A method to estimate the maximum temperature differential along the circumference and its vertical profile is discussed in ASCE.[14]

24.2.7 Natural Frequency of Vibration

The natural frequency of vibration of a stack or chimney with a constant diameter, uniform thickness, and fixed base is given by

$$f_t = \frac{1}{T} = \frac{3.52}{4\pi}\frac{D}{H^2}\sqrt{\frac{Eg}{2w_s}} = \frac{3.9D}{H^2}\sqrt{\frac{E}{w_s}} \tag{24.10}$$

where f_t = fundamental frequency, cps
T = fundamental period, seconds (s)
w_s = unit weight of shell, lb/in.^{3}
g = gravity acceleration = 386 in./s^2
E = modulus of elasticity, psi
D = diameter, in.
H = height, in.

Reinforced-concrete chimneys are usually built with a uniform taper and with the thickness varying from a minimum at the top to the thickness required for strength at the bottom. Steel stacks are built with or without taper or with a tapered lower section and with several thicknesses of plates (Fig. 24.5). Several formulas for an effective diameter D_e and effective height H_e have been developed to enable the natural frequencies of these types to be determined by Eq. (24.10). The following formulas are suggested by Chu and Park:[15]

Chimneys with straight taper or straight top and flared bottom:

$$D_e = D_t\left(\frac{t_b}{t_t}\right)^{0.27} \tag{24.11}$$

Chimneys with uniform taper:

$$H_e = H\left(\frac{2D_t}{D_t + D_b}\right) \tag{24.12}$$

Chimneys with straight top and flared bottom:

$$H_e = H\left[\frac{H_s}{H} + \frac{H_b}{H}\left(\frac{2D_t}{D_t + D_b}\right)^2\right]^{0.9} \tag{24.13}$$

In these equations D_t, D_b = mean diameter at top, bottom; t_t, t_b = thickness at top, bottom; H_s = height of straight segment; and H_b = height

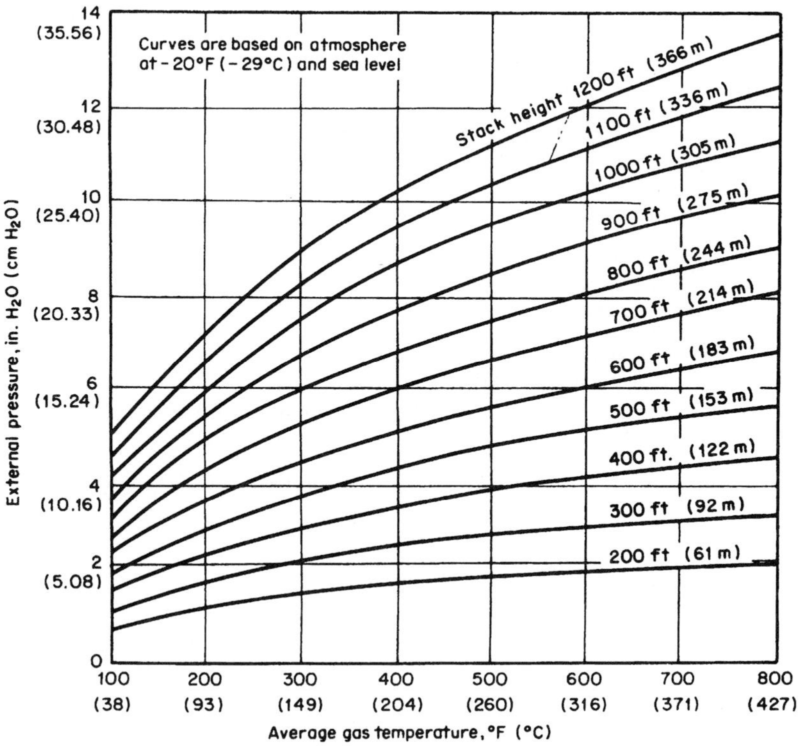

Figure 24.4 External pressure curves. (*From Ref. 14.*)

of bottom (tapered) segment (Fig. 24.5). Frequencies calculated by Eq. (24.10) using these effective dimensions are in good agreement with computer determination of the frequencies of chimneys with a wide range of proportions.

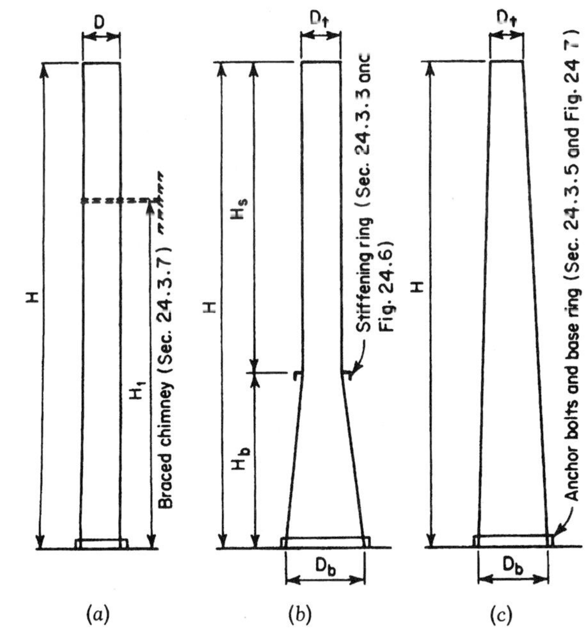

Figure 24.5 Steel stacks.

If a chimney is lined and the weight of the lining is carried by the shell, there is an increased mass to be set in vibration. Since there is no significant increase in moment of inertia, such a chimney will have a lower natural frequency than an unlined chimney of the same dimensions. An approximate natural frequency can be determined from Eq. (24.10) by using for w_s the weight of the shell plus lining, divided by the thickness of the plate. The weight and thickness should be taken at about one-fourth the height above the base.

Base flexibility may need to be considered in determining frequency. For example, when stacks are supported on a roof structure or a steel frame, the translational and rotational spring constants of the support must be taken into account. A finite-element analysis is recommended for determining the frequency in such cases.

24.3 STEEL STACKS

There are many types of steel stacks, including self-supported, guyed, and braced. The choice of a particular type should be based on the evaluation of its comparative costs and the site conditions. The three profile types commonly used are shown in Fig. 24.5. Short stacks, less than 100 ft high, may be straight cylinders. For taller stacks a bell base may be used to reduce plate thickness and anchor-bolt size. The bell height is usually between one-fourth and one-third of the total stack height. The diameter of the flared base D_b is usually about 1⅓ to 1¾ times the cylinder diameter D_t, although in some cases it may be beneficial to limit the base diameter to a dimension shippable by truck.

Most steel stacks are built from plate conforming to ASTM A36. In some cases, A242, A588, A283C, or other grades of steel, including stainless steels, have been used as dictated by experience and specific requirements. Stiffeners are normally A36. Steel for stacks in cold climates should have adequate toughness with a low transition temperature.[2]

24.3.1 General and Design Code

A steel stack and its anchorage should be designed for the loads discussed in the preceding article. When subjected to wind and/or earthquake loads, the stack may be treated as a beam column and analyzed by conventional beam theory. Except in guyed and braced stacks, nonuniform temperature differentials will not induce bending moments. Particular attention is required in the design of reinforcing at the cone-to-cylinder junction and at the breeching opening.

To reduce heat loss, insulation of the exterior surface of unlined steel stacks, including the projecting flanges of all attachments, is recommended. A suitable lining may be applied to the interior surface of the shell to protect the bare steel from high temperature, abrasion, and corrosion from the flue gases. The weight of insulation and lining should be taken into account in the frequency calculation for resonance under the lined condition. Unless an integral shotcrete or brick lining is used, no credit should be given to the lining in calculating stack stiffness.

ASTM STS-1-2016 provides the requirements and guidelines for design, fabrication, erection, and maintenance of steel stacks and their appurtenances. In other countries the appropriate national code or the CICIND Model Code[16] are typically used.

24.3.2 Allowable Stresses

The allowable longitudinal compressive stresses, F, due to vertical load and bending moment can be determined by STS-1:

$$F = XY \qquad (24.14)$$

where

$$X = \begin{cases} 0.25Et/D(F.S.) & \text{for } 0 \le \dfrac{t}{D} \le \dfrac{2.8F_y}{E} \\ F_y(1-0.3K_s)/(F.S.) & \text{for } \dfrac{2.8F_y}{E} \le \dfrac{t}{D} \le \dfrac{10F_y}{E} \\ F_Y/(F.S.) & \text{for } \dfrac{t}{D} > \dfrac{10F_y}{E} \end{cases} \qquad \text{Ref. 14}$$

$$K_s = \left(\frac{\dfrac{10F_y}{E} - \dfrac{t}{D}}{\dfrac{7.2F_y}{E}} \right)$$

$$Y = \begin{cases} 1 & \text{for } \dfrac{L}{r} \le 60 \text{ and } F_y \le 50 \text{ ksi} \\ \dfrac{21{,}600}{18{,}000 + (L/r)^2} & \text{for } \dfrac{L}{r} > 60 \text{ and } F_y \le 50 \text{ ksi} \end{cases}$$

F_y = yield strength at mean shell temperature, psi
E = modulus of elasticity at mean shell temperature, psi
t = shell-plate thickness, at the section under consideration, in.
D = diameter of shell, at elevation under consideration, in.
L = length of stack between points of lateral support. For a self-supporting stacks, L should be taken as the effective length, that is, $L = 2 \times$ stack height, in.
r = weighted mean radius of gyration for level under consideration, in.
$F.S.$ = factor of safety

The factor Y in Eq. (24.14) is intended to account for a possible interaction of cylindrical shell buckling, which depends on t/D, and column buckling, which depends on L/r.

The $F.S.$ is dependent upon the load combination being considered as given in STS-1.

Because of possible corrosion the computed required thickness should usually be increased. The allowance is typically from 1/16 to ⅛ in., depending on the properties of flue gases, the types of insulation and lining provided, and the operating gas temperature. STS-1 includes minimum shell thickness requirements based on diameter; however, for most stacks it is recommended that the shell thickness, including the corrosion allowance, be not less than ¼ in.

24.3.3 Cone-to-Cylinder Junction

A stiffening ring is normally required at the junction of the cone and the straight cylinder sections of stacks. It is typically designed to resist the circumferential compression that results from the vertical loads and bending moments at the junction. Where external pressure due to stack draft is significant, the resulting additional circumferential forces should also be considered.

The maximum vertical force N_x per unit length of circumference in the cylinder at the junction is

$$N_x = \frac{W}{2\pi R} + \frac{M}{\pi R^2} \qquad (24.15)$$

where W = axial load at junction
M = wind or other moment at junction
$R = D/2$

The total circumferential compression Q in the ring is

$$Q = R[N_x \tan\theta + 0.78P_d(\sqrt{Rt_1} + \sec\theta\sqrt{Rt_2 \sec\theta})] \qquad (24.16)$$

where θ = acute angle between cone wall and cylinder
P_d = external pressure per unit area, psi
t_1 = thickness of cylinder wall, in.
t_2 = thickness of cone wall, in.

The required area A and moment of inertia I_s of the ring are

$$A_s = \frac{2Q}{F_y} \qquad (24.17)$$

$$I_s = \frac{4QR^2}{3E} \qquad (24.18)$$

In determining the section properties of the stiffening ring the area of a portion of the shell (Fig. 24.6) can be included, but the area so included should not exceed the area of the ring itself to ensure a nominal-size stiffener. The maximum permissible longitudinal compressive stresses in the cone may be determined by Eq. (24.14) with the diameter D replaced by two times the cone radius $R \sec\theta$.

24.3.4 Circumferential Stiffeners

In addition to the stiffener at the cone-cylinder junction a stiffener is required at the top of the stack. Intermediate ring stiffeners may also be required. The purpose of such stiffening is to prevent excessive deformations of the stack shell under wind pressure and to provide adequate structural resistance to negative draft. The circumferential stresses in the shell plate and the intermediate stiffeners should be checked and designed following the requirements of STS-1. The maximum stiffener spacing should follow the requirements of STS-1. In the calculation of stiffener section properties, an area of the shell equal to $16t$, $1.56\sqrt{Rt}$, or the area of the stiffener, whichever is smaller, may be included.

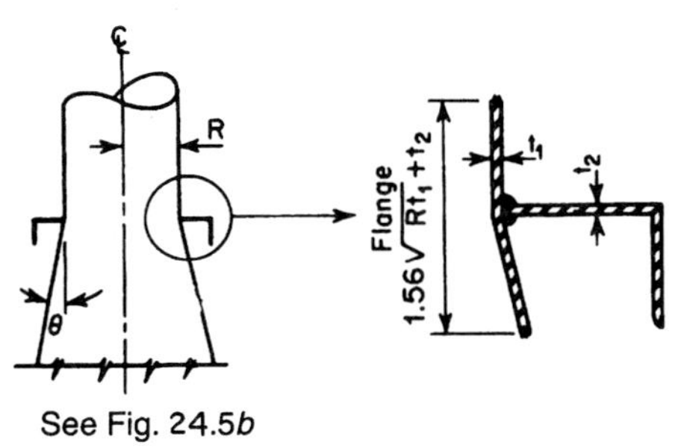

Figure 24.6 Stiffening ring at cone-cylinder junction.

24.3.5 Base Ring and Anchor Bolts

These should be designed to resist the net tension resulting from the dead load W_b and the wind or earthquake moment M_b at the stack base. The bolt tension F can be determined from

$$F = \frac{4M_b}{ND'} - \frac{W_b}{N} \tag{24.19}$$

where D' = diameter of bolt circle, in., and N = number of anchor bolts. Since the elongation of the anchor bolts contributes to the lateral deflection of the stack, it is advisable to limit the allowable bolt tension at the root of the thread to 15,000 psi even if high-strength bolts are used. Also, the bolt spacing should not exceed 5½ ft, and at least eight bolts should be used.

Anchor-bolt tension is eccentric with respect to the chimney shell, and an unstiffened base angle is normally insufficient to take the bending. Deflection of the base angle will increase lateral movements of the stack and will cause large vertical secondary bending stresses in the chimney shell, and in many cases it is necessary to provide a continuous stiffened base ring at the location shown in Fig. 24.7. Alternatively, a separate chair for each anchor bolt may be used.

Circumferential compressive stresses in the continuous base ring may be calculated as though it were loaded with equally spaced concentrated loads $Q = Fe/h_b$ in the plane of the ring (Fig. 24.7). The maximum circumferential bending moment, which occurs at the windward anchor bolt and produces compression in the outside of the ring, is given by

$$M_a = CQr_a \tag{24.20}$$

where C = coefficient from Table 24.1

r_a = radius of bolt circle, in.

The portion of the chimney shell within $0.78\sqrt{Rt}$ on either side of the attached ring plate may be counted as part of the ring. If the calculated size of the ring is excessive, its height should be increased, or the eccentricity e should be reduced if possible. Vertical stiffeners should be placed on both sides of each anchor bolt.

Suggested designs of anchor-bolt chairs are given in AISI 1992.[17]

For a stack with a bell base, the base angle must be designed for the ring tension caused by the horizontal component of forces in the base cone, in addition to the bending stresses due to the bearing load. The allowable tension in the base angle should be limited to 10 ksi or less to keep secondary bending stresses in the cone within safe limits.

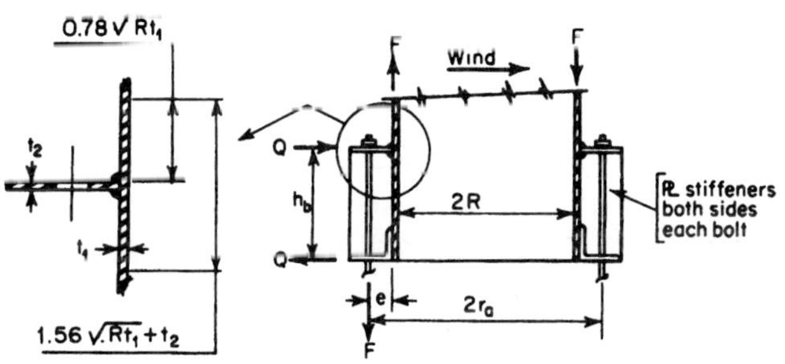

Figure 24.7 Base ring for anchor bolts.

Table 24.1 Coefficient for use in Eq. 24.20

No. of bolts, N	C	No. of bolts, N	C
8	0.191	36	0.510
12	0.217	40	0.563
16	0.258	44	0.616
20	0.305	48	0.670
24	0.355	52	0.724
28	0.406	56	0.778
32	0.457	60	0.832

24.3.6 Guyed Stacks

For most plant sites the space required for guys and their anchors makes the use of guyed stacks (Fig. 24.8) undesirable except where the guys can be anchored to adequately braced plant structures so as to provide ample ground clearance. Generally, one set of guys spaced 120° apart around the stack circumference with an angle β of 45° to 50° between the guy and the vertical axis of the stack is satisfactory. The vertical components of the guy tensions must be taken into account when computing stresses in the shell.

In determining moments for the design of the shell, it should be noted that the stack is not held rigidly at the guy ring but moves laterally owing to the decrease under lateral forces in the initial sag of the windward guy. With adequate initial tension, this movement is small. The height of the guy ring can be chosen so that the moment of the cantilever section at the guy ring is approximately the same as the moment at the base. The latter is based on an estimated partial fixity of the base. It is not good practice to use a thinner shell between the guy ring and the base, even where the computed stresses might permit a decrease.

An evaluation of the true maximum tension in the windward guy depends on the wind force, the components in line with the windward guy of the residual initial tension in the two leeward guys, the wind force on and the weight of the guys themselves, and other factors. See Rowe,[18] Fang,[19] and ASME STS-1 for further discussion on the design of guys.

The height of the shell acting as a flange of the ring can be determined as for stiffening rings (Fig. 24.6). Vertical stiffeners should be provided on each side of each guy.

24.3.7 Braced Stacks

Where a steel stack is adjacent to a properly braced building or structure, it is economical to provide a brace from the stack to the structure at a substantial distance above the base (Fig. 24.5*a*). The design of the shell is similar to that of a guyed stack except that the stack is assumed to be rigidly held at the brace. Figure 24.9 shows one type of bracing arrangement.

24.3.8 Resonant Vibrations

Steel stacks are more susceptible to wind induced vibrations than other types of chimneys. There are numerous records of steel stacks, conservatively designed for maximum probable wind or earthquake forces, which have experienced serious vibrations from steady-state winds of velocities of the order of 10 to 40 mph. As a consequence, it is generally necessary to investigate them for dynamic wind action. Most of the serious low-wind vibration problems have been with unlined stacks.

The wind velocity at which the natural frequency of vortex shedding equals the natural frequency of the stack is given by Eq. (24.3). With $S = 0.2$, this formula gives

$$V_{cr} = \frac{Df_t}{0.2} \times \frac{3600}{5280} = 3.14Df_t \tag{24.21}$$

with V_{cr} in miles per hour and D in feet. The value of D for stacks of the profile in Fig. 24.5*c* may be taken as the mean diameter of the top one-third of the stack.

Research[5,6] shows that if the critical wind velocity exceeds 1.3 times the design mean hourly wind velocity, the effect of vortex shedding on stacks may be ignored.

ASME STS-1 provides a procedure for calculating the loads due to wind vibration. Equation (24.1b) provides a simplified approach. Past experience shows that unlined stacks have very low damping and are susceptible to large-amplitude across-wind vibrations. It can be shown

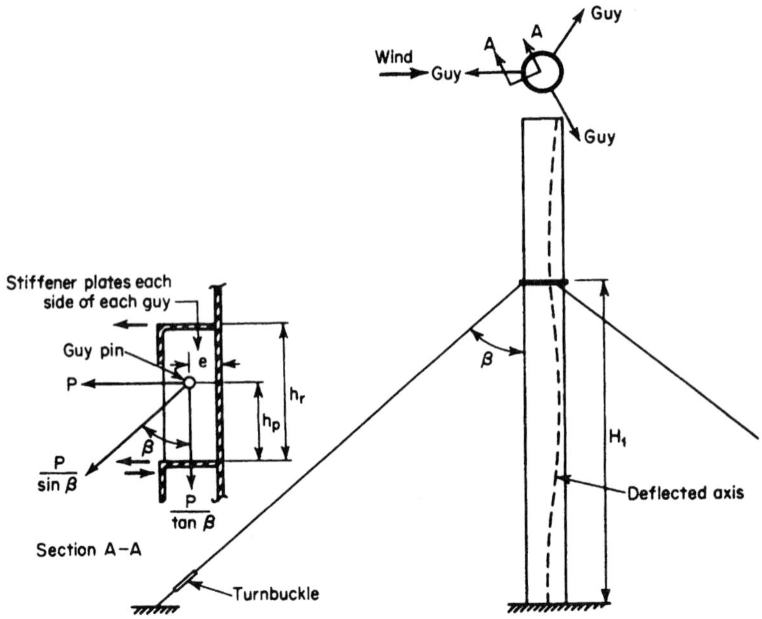

Figure 24.8 Guyed stack.

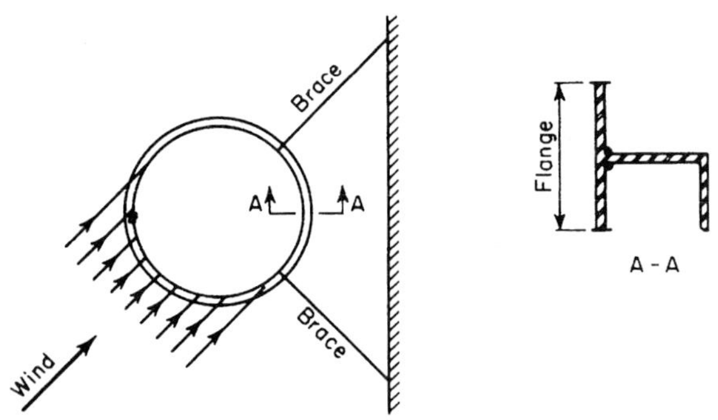

Figure 24.9 Stack braces.

from Eq. (24.1a) that the aerodynamic damping, β_a, would exceed structural damping, β_s, for most of the unlined stack proportions. Use of refractory lining or other type of linings would be beneficial because they increase structural damping.

Increasing the diameter of the stack, as in the flared-bottom stack, will always result in higher natural frequency and higher critical wind velocity and possibly less cyclic bending stress. Other means for increasing stack stiffness to preclude or mitigate vortex-shedding vibration effects are use of guy wires, providing lateral supports to nearby structures, and installation of vertical stiffeners. Other than these stiffness methods, aerodynamic methods using helical strakes, cylindrical shrouds, spoilers, or damping methods may be utilized in the prevention of excessive vibrations. The damping methods currently available in the industry are the use of tuned-mass dampers, placement of preformed fabric pads at the base of the stack, and installation of a single hanging chain or multiple hanging chains.[2]

Stacks which are to be lined should be investigated for vibration in the unlined condition. They may need to be guyed or braced temporarily during these periods. Also, to prevent ovalling of the lining during erection, it is advisable to install temporary spiders.

Unlined stacks may be subject to ovalling vibrations. To guard against this an unlined stack should be checked and/or provided with circumferential stiffening rings in accordance with STS-1.

A stack downwind of another stack may experience a larger resonant wind load than that discussed in Sec. 24.2. For stacks with center-to-center spacing exceeding 15 diameters of the largest stack, no amplification factor is needed for downwind stacks. However, if the spacing S is of 3 to 15 diameters, a modified Strouhal number should be used per STS-1. For spacing less than three diameters or for stacks that are not identical, further investigation by wind-tunnel tests or other experience data may be necessary.

24.3.9 Steel Stack Design Examples

EXAMPLE 1

Design a ground-supported A36 steel stack 160 ft high with the profile of Fig. 24.5*b*. Given $H_b = 40$ ft, $H_s = 120$ ft, $D_t = 10$ ft, $D_b = 15$ ft, operating gas temperature = 350°F, 2-in. shotcrete lining, weight of shotcrete 110 pcf. Design wind loads should conform to ASME STS-1-2016 with a basic wind speed of 90 mph, exposure category C, importance factor of 1, and a directionality factor of 1.

Along-wind loads are calculated per ASME STS-1-2016 and ASCE 7-05. For simplicity, only the total design moments along the stack height including mean and fluctuating wind are shown in Table 24.2.

Table 24.2 Summary of Design Wind Moments

Elevation (ft)	Design wind moments (kips-ft)
160 (top of stack)	0
130	98
100	371
80	636
70	791
60	959
40	1332
20	1745
0	2195

PLATE THICKNESS AND STRESS DETERMINATION
Calculations are shown in Table 24.3.

Table 24.3 Example 1—Stress Calculation and Plate Thickness Determination

Trial section 90 ft below top (or at 70-ft elevation)

Assume 1/4-in. steel plate = 10.2 psf

$$\text{Lining } 110\text{ pcf} \times \frac{2}{12} = \frac{18.3\ \text{psf}}{28.5\ \text{psf}}$$

Weight per foot of height = 28.5 × 10 π = 895 lb/ft.
Deducting 1/16 in. from the plate thickness for corrosion allowance,

$$\frac{t}{D} = \frac{3/16}{120} = 0.0015625$$

$$A = 2\pi \times 60 \times 0.1875 = 70.7\ \text{in.}^2$$

$$S = \pi R^2 t = \pi \times 60^2 \times 0.1875 = 2120\ \text{in.}^3$$

$$f = \frac{W}{A} + \frac{M}{S} = \frac{895 \times 90}{70.7} + \frac{791{,}000 \times 12}{2120} = 5617\ \text{psi}$$

Determine the allowable stress from Eq. (24.14):

$$L = 2 \times 160 = 320\ \text{ft} \qquad \frac{L}{r} = \frac{320}{0.707 \times ((5)(120) + (6.25)(40))/160} = 85.2$$

$$Y = \frac{21{,}600}{18{,}000 + 85.2^2} = 0.855$$

$$\frac{2.8Fy}{E} = \frac{2.8 \times 36}{28{,}000} = 0.0036 > \frac{t}{D}$$

$$F.S. = 1.5$$

$$X = 0.25 \times 28 \times 10^6 \times 0.0015625/1.5 = 7292\ \text{psi}$$

$$F = 0.855 \times 7292 = 6234\ \text{psi} > 5617 \quad \text{OK.}$$

Trial section at 120 ft below top
Assume t = 5/16 in, including 1/16 in. corrosion allowance.
Weight per ft of height = (12.76 + 18.3) × 10 π = 976 lb/ft

$$\frac{t}{D} = \frac{0.25}{120} = 0.002083$$

$$A = 2\pi \times 60 \times 0.25 = 94.2\ \text{in.}^2$$

$$S = \pi \times 60^2 \times 0.25 = 2827\ \text{in.}^3$$

$$f = \frac{894 \times 90 + 976 \times 30}{94.2} + \frac{1{,}332{,}000 \times 12}{2827} = 6819\ \text{psi}$$

$$X = 0.25 \times 28{,}000{,}000 \times 0.002083/1.5 = 9721\ \text{psi}$$

$$F = 0.855 \times 9721 = 8311\ \text{psi} > f \quad \text{OK.}$$

By similar stress calculation, a 5/16-in. plate will be able to withstand the dead loads and design wind moments in the base cone (the bottom 40-ft section). **Thus, use ¼-in. plate for the top 90-ft section and 5/16-in. plate for remaining of the stack with a total stack weight including lining of 159,000 lb.**

RING AT CONE-TO-CYLINDER JUNCTION
Calculation for design of ring stiffener at cone-to-cylinder junction is presented in Table 24.4.

Table 24.4 Example 1—Calculation for Design of Ring Stiffener at Cone-to-Cylinder Junction

$$W = 894 \times 90 + 976 \times 30 = 109{,}710\ \text{lb}$$

$$M = 1{,}332{,}000\ \text{ft-lb}$$

$$N_x = \frac{109{,}710}{2\pi \times 60} + \frac{1{,}332{,}000 \times 12}{\pi \times 60^2} = 1705\ \text{lb/in.}$$

Assume external pressure = 2 in water

$$P_d = 2 \times 62.4/12^3 = 0.0722\ \text{psi}$$

$$\tan\theta = \frac{2.5}{40} = 0.0625 \qquad \theta = 3.58° \qquad \sec\theta = 1.002$$

From Eq. (24.16), with $t_1 = t_2$ = 5/16 in. = 0.3125 in.

$$Q = 60[1705 \times 0.0625 + 0.78 \times 0.0722\,(\sqrt{60 \times 0.3125} + 1.002\sqrt{60 \times 0.3125 \times 1.002})] = 6423\ \text{lb}$$

From Eqs. (24.17) and (24.18),

$$A_s = \frac{6423 \times 2}{36{,}000} = 0.357\ \text{in.}^2$$

$$I_s = \frac{4 \times 6423 \times 60^2}{3 \times 28{,}000{,}000} = 1.10\ \text{in.}^4$$

Try 3 × 3 × ¼ angle: A = 1.44 in.^{2} I = 1.24 in.^{4} x = 0.842 in.

Effective flange (Fig. 24.6) = $0.3125\,(1.56\sqrt{60 \times 0.3125} + 0.25)$ = 2.19 in.^{2} > 1.44 in.^{2}; use 1.44 in.^{2}

Ring area A_s = 1.44 + 1.44 = 2.88 in.^{2} > 0.357 in.^{2}

$$\text{Ring c.g.} = \frac{1.44 \times 0.842 + 0.72 \times 3.16 + 0.72(3.16 - 1.69\tan 3.58°)}{2.88} = 1.97\ \text{in.}$$

Ring $I_s = 1.24 + 1.44(1.97 - 0.842)^2 + 0.72(3.16 - 1.97)^2 + 0.72(3.16 - 1.69\tan(3.58°) - 1.97)^2 = 4.94\ \text{in.}^4 > 1.10\ \text{in.}^4$

Therefore, the 3 × 3 × ¼ angle is OK.

INTERMEDIATE STIFFENERS IN UPPER 120 FT.
Design of intermediate stiffeners are shown in Table 24.5.

Table 24.5 Example 1—Design of Intermediate Stiffeners and Anchor Bolts

Intermediate Stiffeners
Maximum allowable spacing for 10-ft stack diameter = 2D = 20 ft.
Try 3 × 3 × ¼ angles spaced 10 ft. per ASME STS-1 Eq. 4-17, the required section modulus is

$$S = \frac{29 \times 10 \times 10^2 \times 12^3 \times 1.5}{1830 \times 36{,}000} = 1.14\ \text{in.}^3$$

The effective section modulus provided is 2.39 in.^{3} Additional check on required I_s and A_s are found adequate. Since stack draft is smaller than wind pressure, no check is needed.

Anchor Bolts
Assume radius of bolt circle = 92 in.
For 20 anchor bolts the required net area, A, per bolt is, from Eq. (24.19),

$$F = \frac{4 \times 2{,}195{,}000 \times 12}{20 \times 92 \times 2} - \frac{159{,}000}{20} = 20{,}680\ \text{lb}$$

$$A = \frac{20{,}680}{15{,}000} = 1.38\ \text{in.}^2$$

Use 1¾-in. bolts, for which the area at the root of the thread is 1.78 in.^{2}

ANCHOR BOLTS
Design of the anchor bolts are also shown in Table 24.5.

EXAMPLE 2

Investigate the stack of Example 1 for resonant wind vibration.

TRANSVERSE VIBRATION
Calculations for transverse wind vibrations check are shown in Table 24.6.

Table 24.6 Example 2—Check for Resonant Wind Vibrations

From Eq. (24.11) and (24.13), the equivalent height H_e and equivalent diameter D_e are

$$H_e = 160\left[\frac{120}{160}+\frac{40}{160}\left(\frac{2\times 10}{10+15}\right)^2\right]^{0.9} = 147 \text{ ft}$$

$$D_e = 10\left(\frac{0.3125}{0.25}\right)^{0.27} = 10.6 \text{ ft}$$

The equivalent unit weight of the shell, including the shotcrete lining at 18.3 psf and computed at one-fourth the height above the base (Sec. 24.2.7), is

$$\text{Shell} = 0.3125 \times 0.283 = 0.089$$
$$\text{Lining} = 18.3/144 = \underline{0.127}$$
$$0.216/0.3125 = 0.691 \text{ lb/in.}^3$$

Then, from Eq. (24.10), $f_t = \frac{3.9\times 10.6\times 12}{(147\times 12)^2}\sqrt{\frac{28\times 10^6}{0.691}} = 1.02 \text{ cps}$

The critical wind velocity by Eq. (24.21) is $V_{cr} = 3.41 \times 10 \times 1.02 = 34.8$ mph = 51.1 fps
The equivalent static wind force is determined by Eq. (24.1a), with

$$\frac{D_t}{D_b} = \frac{10}{15} = 0.67 > 0.5 \qquad \bar{D} = 10 \text{ ft}$$

$$\frac{H}{D} = \frac{160}{10} = 16 \qquad C_1 = 0.63 \qquad C_2 = 0.6$$

$$\beta = 0.006 \qquad \gamma = 0.08 \text{ lb/ft}^3$$

$$\bar{m} = \pi(10)(18.3+10.2) = 895.4 \text{ lb/ft}$$

$$q_{cr} = 0.00128(51.1)^2 = 3.34 \text{ lb/ft}^2$$

Thus, $F_L = \frac{0.63(10)(3.34)}{[0.006-0.6(0.08)(10)^2/(895.4)]^{0.5}}\left(\frac{Z}{H}\right)^{1.3} = 832\left(\frac{Z}{H}\right)^{1.3}$ lb/ft

The calculated dynamic across-wind forces and the design along-wind forces at various heights are as follows:

Height above the base, ft	Across-wind forces, lb/ft	Along-wind forces, lb/ft
160	832	225
100	452	166
40	137	100
20	56	88

The across-wind forces calculated above are 2 to 3 times larger than the along-wind forces in the upper 60 ft of the stack but are smaller in the remainder of the stack. Consequently, the bending moments resulting from the across-wind forces will govern the design. The shell plate thickness will need to be increased significantly in the lower 100 ft of the stack. Alternatively, a more effective and economical solution is to equip the Example 1 stack with a tuned mass damper that provides a minimum of 2% damping, in which case F_L would reduce to 174 $(Z/H)^{1.3}$ and across-wind forces will not govern the design.

OVALLING VIBRATION
Check for ovalling vibration is not required for lined condition.

24.4 REINFORCED-CONCRETE CHIMNEYS

24.4.1 General

Construction costs of steel stacks are generally lower than reinforced-concrete chimneys up to a height of about 200 ft. At greater heights, reinforced-concrete chimneys are usually more economical. Reinforced-concrete chimneys also cost less to maintain than steel chimneys because they are free from atmospheric corrosion.

It is generally economical to taper tall concrete chimneys, with the top diameter determined by gas exit requirements and liner construction details. The base diameter results from optimizing the costs of concrete, forms, and reinforcing steel for various shapes. It is common to use a variable rather than a straight taper for aesthetic reasons.

24.4.2 Design Codes: ACI 307-08

The latest code requirements for material, construction, and strength design of reinforced-concrete chimney shells are given in ACI 307-08. The code sets forth recommended loadings and methods for determining the design strength. It also provides criteria for thermal effects evaluation. Concrete chimneys should be designed and constructed to meet ACI 307 and local building regulations.

24.4.3 Design Process

A design process flowchart for the structural analysis and design of concrete chimneys is presented in Flowchart 24.1.

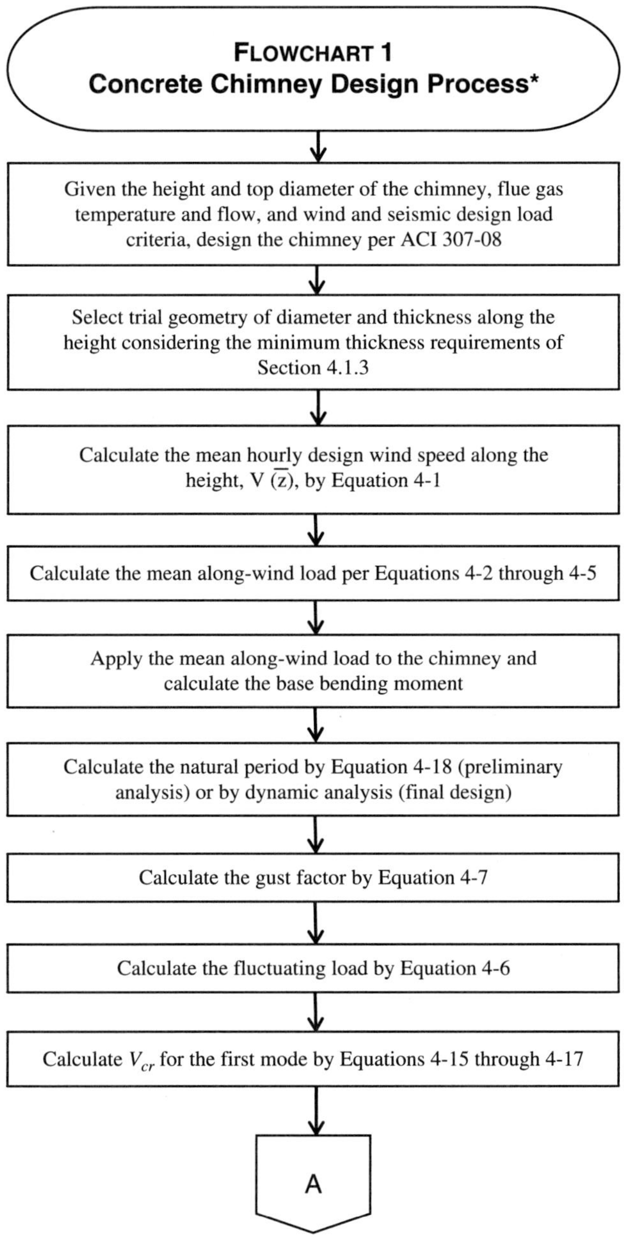

*All references to sections and equations in the flowchart are to those in ACI 307-08.

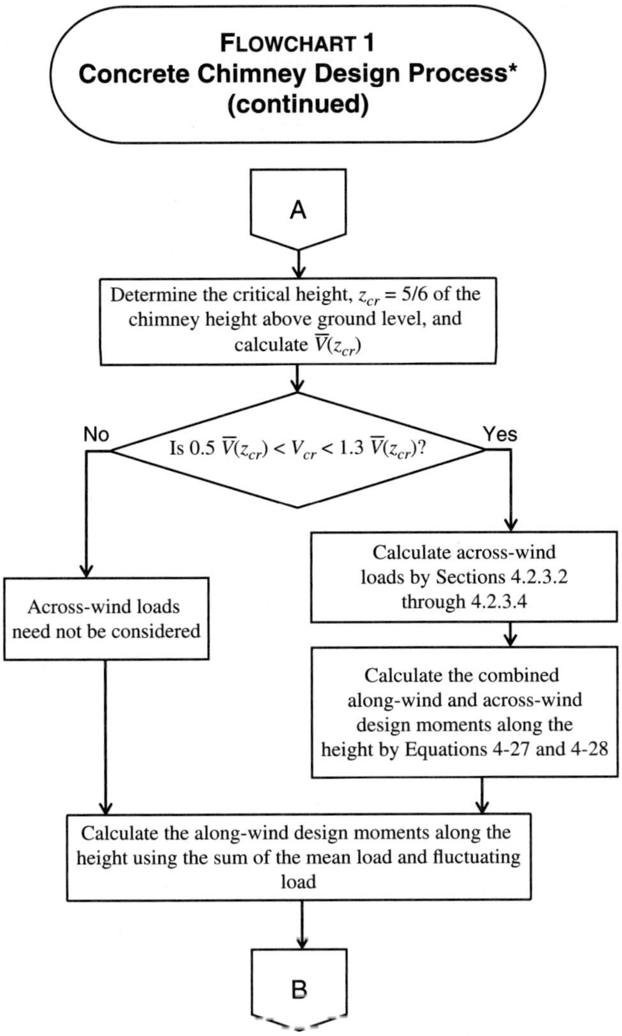

*All references to sections and equations in the flowchart are to those in ACI 307-08.

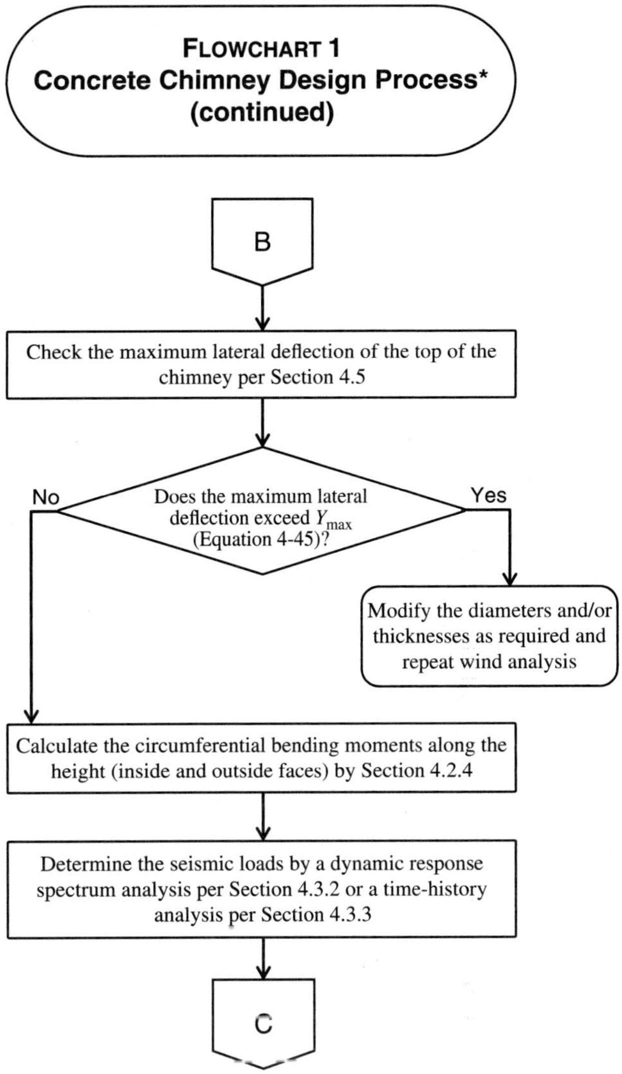

*All references to sections and equations in the flowchart are to those in ACI 307-08.

24.4.4 Vibration Due to Wind

Reinforced-concrete chimneys are subject to resonant across-wind vibration as discussed in Section 24.2.3. Due to the inherent higher damping in concrete chimneys as compared to steel stacks, the induced loading is commonly accounted for in design using the procedure in Section 4.2.3 of ACI 307-08. Reinforced-concrete chimneys are unlikely to exhibit ovalling resonant vibrations because of the relatively large shell thickness.

24.4.5 Earthquake Design

The behavior of tall reinforced-concrete chimneys subject to earthquake excitation was not well understood prior to 2000. Consequently, codes of practice around the world provide conservative aseismic design guidelines. Extensive experimental and nonlinear dynamic analysis studies of concrete chimneys were conducted by Wilson in late 1990s and early 2000s. Based on Wilson's studies, a number of code design recommendations have been incorporated into the 2001 CICIND code to encourage economic design with reliance on development of limited seismic ductility ($R = 2$) in reinforced-concrete chimneys.[13,20] The state of practice in seismic design of concrete chimneys was also reviewed by others.[12,21]

Subsequently, in November 2008, ACI published a new document, ACI 307-08. The 2008 code incorporates a number of major technical revisions to seismic design provisions. These include several key changes from the previous edition, namely, selecting the design base earthquake as two-thirds of maximum considered earthquake (MCE), including effects of soil amplification in design response spectra; use of a response modification factor (R) of 1.50, considering the P-delta effects for chimneys of seismic design category D, E, and F; and adding a seismic clearance criterion between the concrete chimney shell and chimney liner(s). The changes were made to be consistent with ASCE 7-02. More discussions of the ACI 307-08 earthquake design requirements are covered by Fang and Porthouse.[12] Earthquake loads on chimneys are determined in accordance with Section 4.3 of ACI 307-08 and per Sec. 24.2.4. Since publication of ACI 307-08, seismic design criteria of ASCE 7 have undergone a number of significant changes in 2010 and 2017. Newer provisions will be incorporated into the next edition of ACI 307. The key changes include using $R = 2$ instead of 1.5 for the chimney shell and using an overstrength factor of 1.5 for design of chimney sections at large openings with ductile detailing.

FLOWCHART 1
Concrete Chimney Design Process*
(continued)

C

↓

Calculate the thermal gradients and thermal stresses along the height by Section 6.2 for vertical stress and Section 6.3 for circumferential stress

↓

Determine the required vertical strength along the height per Section 5.3.1

↓

For each vertical load combination calculate the reduced f_y and f_c' along the height per Section 5.5.7

↓

Calculate the vertical design strength and nominal moment strength along the height for each load case per Sections 5.4 and 5.5, determining the vertical reinforcement ratio using the calculation procedure outlined in Section 5.5.6

↓

Determine the required circumferential bending strength (inside and outside faces) along the height per Section 5.3.2

↓

Calculate the reduced f_y and f_c' along the height for circumferential bending per Section 5.7.2

↓

D

*All references to sections and equations in the flowchart are to those in ACI 307-08.

24.5 LINERS AND LININGS

24.5.1 General

A chimney typically includes a liner or lining which has multiple functions. It protects the shell from acid attack from the flue gas and provides better chimney draft by maintaining the flue-gas temperature. If it has insulating properties, it reduces the temperature of the chimney shell lessening thermal effects. Depending on the material, liners may be self-supporting or may be placed against the chimney shell.

FLOWCHART 1
Concrete Chimney Design Process*
(continued)

D

↓

Calculate the circumferential design strength and nominal moment strength along the height per Section 5.4 and general concrete beam theory (inside and outside faces), selecting the circumferential reinforcement ratio for each face such that the design strength is greater than the required strength

↓

Check the minimum reinforcement requirements and special design considerations and requirements of Section 4.4 and adjust the reinforcement ratios as required

↓

Review wall thicknesses and reinforcement ratios for constructability, in particular in the area of openings. If desired, adjust diameters and/or wall thicknesses and start over

*All references to sections and equations in the flowchart are to those in ACI 307-08.

24.5.2 Steel Liners

Independent, self-supporting steel liners (Fig. 24.10*a*) are used most frequently in reinforced-concrete chimneys. They are usually insulated and can be supported at the bottom, top, or some intermediate level of the shell. Steel liners are essentially corrosion-free when entering gas temperatures are sufficiently above the acid dew points of the flue gas, provided that the exterior surfaces are properly insulated. A lining for the interior surface should be considered if entering gas temperatures below 250°F are expected and/or the gas is saturated. Most steel liners have been built with A36 steel; A242 and A588 steels have been used occasionally because of the improved atmospheric corrosion resistance. Since A242 and A588 steels are not acid-resisting, their added expense may not be justified, although they are sometimes also used to take advantage of their higher strength. Detailed discussions of the design and construction of steel chimney liners are given in ASCE.[14]

24.5.3 Brick Liners

Independent, self-supporting brick linings (Fig. 24.10*b*) have been used for concrete chimneys where earthquake requirements are not severe. To provide for stability of the liner in the event of vertical cracks due to the temperature gradient across the brick thickness, the lining must have exterior steel bands at regular intervals along its height. In the 1970s, independent brick liners were the most popular type of liner for industrial plants equipped with flue-gas desulfurization (FGD) systems.

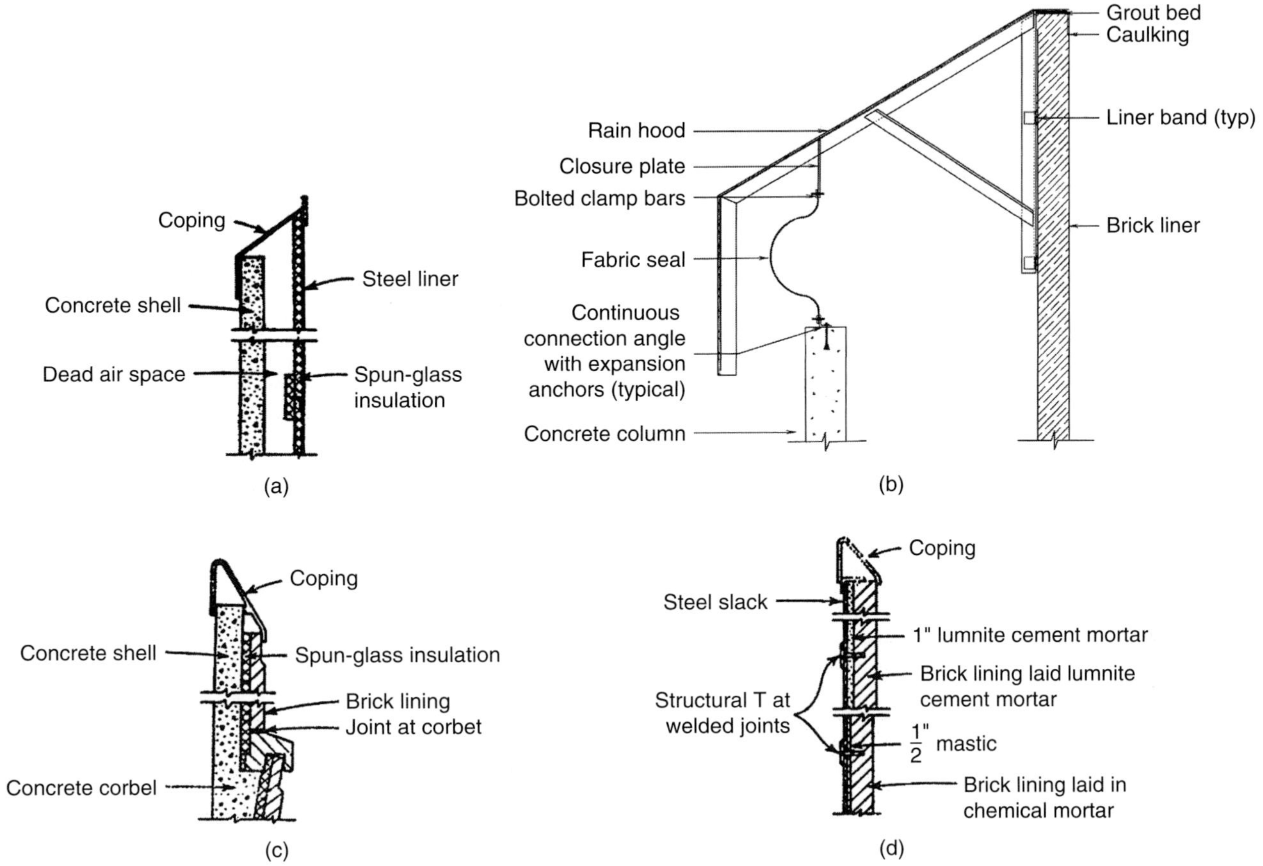

Figure 24.10 Lining configurations for chimneys/stacks.

Later operation of some of these systems without reheating of the gas above the water dew point, creating "wet stacks," resulted in poor performance of those brick liners.[22,23] Detailed discussions of the design and construction of brick chimney liners are given in ASTM C1298.[24] Corbel supported brick liners (Fig. 24.10*c*), which for economic reasons were quite common in the past, have been a large source of problems for concrete chimneys owing to gas leakage and acid attack resulting from cracking and differential vertical expansion.

24.5.4 FRP Liners

Independent, self-supporting fiberglass-reinforced plastic (FRP) liners have become a common choice for new wet stack installations. Several polyester and vinyl ester resins are used, but the most prevalent are flame-retardant epoxy vinyl esters. They provide excellent chemical resistance, but temperature resistance is typically limited to use below 212°F with excursions to 250°F. Detailed discussions of the design and construction of FRP chimney liners are given in ASTM D5364[25] and Ref. 26.

24.5.5 Alloy Liners

Independent, self-supporting alloy liners are often used for chimneys operated downstream of FGD systems, in particular where high-temperature bypass or excursion conditions exist. An alloy liner may be made of solid corrosion-resistant alloy or carbon steel plate clad with a thin layer of alloy. Typical alloys used are Alloys C-276, C22, and 625. In addition to bonded clad,[27] wallpapering techniques have been developed to attach the thin alloy sheets (1/16 in. or less) to the carbon steel liner to reduce material and erection costs.[28,29] An alloy liner is more costly than brick liners or FRP liners but may offer long-term savings in maintenance and repairs.

24.5.6 Borosilicate Block Linings

Borosilicate block is a lining system which uses urethane-asphalt mastic as an adhesive membrane to attach closed cell foamed borosilicate glass blocks to the substrate.[30] The adhesive membrane is used on the back and sides of the block to bond them to the substrate and each other. It is usually applied to the inside surface of a steel liner or shell but can also be applied to brick or concrete. The substrate must be suitably prepared and primed. The block protects the membrane from direct exposure to the flue gas. The thickness of the block is established to maintain the membrane below its maximum service temperature at the substrate and is commonly 1.5 to 2 in.

24.5.7 Gunite (Shotcrete) Linings

Gunite linings anchored with "T" type anchors or applied over mesh reinforcement attached by welded studs have been used extensively for steel stacks and for some steel liners. Experience indicates that calcium-aluminate cement gives better protection against corrosion than portland cement. For increased acid resistance, potassium silicate-bonded materials are available which should generally be applied over a chemical-resistant membrane. The materials should be applied and cured in accordance with the manufacturer's instructions.

24.5.8 Coatings

A variety of coatings can be used as linings for chimneys and stacks.[31] The coating must be carefully selected to handle the flue gas conditions anticipated. In some environments a coating failure can lead to rapid corrosion and perforation of the steel substrate. This can be caused by an improperly chosen material or application issues. In addition, the service life of a coating is usually less than the expected service life of the chimney, and recoating should be anticipated. For these reasons coatings are not as frequently used as other types of linings. Flake-filled vinyl esters, which can generally handle temperatures up to 375°F dry, when properly applied and maintained, have had successful installations.

24.5.9 Brick Linings for Steel Stacks

Brick linings can also be used in steel stacks, but they are typically more expensive. They consist of brick laid in calcium-aluminate mortar with the same mortar placed between the brick and the steel (upper portion of Fig. 24.10*d*). A more costly lining that appears to give better performance where high-sulfur fuels are burned consists of brick laid in potassium silicate mortar with mastic between the brick and the steel (lower portion of Fig. 24.10*d*). Both require support by angles or tees at intervals up the stack.

24.5.10 Dual-Wall Steel Stacks

A dual-wall steel stack[32] is made up of two separate, freestanding steel walls, with an insulated air space in between. The outer wall is engineered to handle the structural, wind, and earthquake loads. The inner wall is designed to expand freely and fully under all temperature conditions. The insulated air space will keep heat loss to a minimum, thus protecting the inner wall from acid condensation attack under suitable operating conditions.

24.6 FOUNDATIONS

Foundations of chimneys and stacks are typically subjected to large overturning moments as a result of wind or earthquake loadings. Checks should be made to ensure the minimum factors of safety against overturning and sliding. The maximum foundation bearing pressure is typically established using service loads, and the factor of safety against overturning should not be less than 1.50. It is important to investigate the supporting soil or rock for their bearing adequacy. A small rotation or unequal settlement of the foundation can significantly magnify lateral movement of the top of the chimney. Excessive horizontal deflections can induce second-order bending moments in chimneys and may affect the interface between the chimney shell, flues, and ducts. Also, frequency-dependent wind and seismic loads may be affected by weak soil sites.

Large concrete foundations for chimneys are usually circular in shape, with reinforcing steel placed in the circumferential and radial directions except in the central portion where a grid of straight bars eliminates bending circumferential steel. Where chimneys are founded on competent rock, caissons, or piles, it is economical to support them on ring-shaped foundations. Smaller chimneys are commonly supported on octagonal foundations with reinforcing bars placed in four directions.

Concrete foundations shall be designed in accordance with ACI 318 for dead loads, along-wind loads, across-wind loads, earthquake loads, thermal effects, and required load combinations. Foundation design shall use a pseudo-bearing pressure distribution, or pile loads. In designing an octagonal foundation the maximum soil pressure may be determined by a simplified method given in the fourth edition of this handbook.[19] Properly validated finite-element software may be used for chimney foundations supported on caissons, piles, or foundations of irregular shape. Additionally, J. G. Turner[33] provides extensive coverage of industrial chimney foundation design.

REFERENCES

1. Manual for Thermofluodynamic Design of Chimneys and Chimney Liners, International Committee on Industrial Chimneys (CICIND), Hemel Hempstead, UK, 2001.

2. ASME STS-1-2016, "Steel Stacks," American Society of Mechanical Engineers, New York, 2017.

3. ACI 307-08, "Code Requirements for Reinforced Concrete Chimneys and Commentary," American Concrete Institute, Detroit, MI, 2008.

4. ASCE 7, "Minimum Design Loads for Buildings and Other Structures," American Society of Civil Engineers, New York, ASCE 7-02, ASCE 7-05, ASCE 7-10, ASCE 7-17 and ASCE 7-94.

5. Vickery, B. J., and Basu, R. I.: The Development of a Codified Approach to the Determination of Wind Loads on Chimneys, 3rd Canadian Wind Eng. Workshop, Vancouver, B.C., April 1981.

6. Vickery, B. J., and Basu, R. I.: "Across-Wind Vibrations of Structures of Circular Cross-Section, Parts 1 and 2, *J Wind Eng Ind Aerod*, Vol. 12, 1983, pp. 49–97.

7. ACI 307-88, "Standard Specifications for Design and Construction of Reinforced Concrete Chimneys," American Concrete Institute, Detroit, MI, 1988.

8. Porter, M. A., Belvins, R. D., and Martens, D. J.: "Measurement and Analysis of the Wind Induced Vibrations of a Tall Stack," 1998, Research Gate.

9. Simice, E., and Scanlan, R. H.: Wind Effects on Structures, 2nd and 3rd eds., John Wiley & Sons, New York, 1986 and 1996, respectively.

10. Vickery, B. J.: Across-Wind Loadings on Reinforced Concrete Chimneys of Circular Cross-Section, Boundary Layer Wind Tunnel Report, BLWT-3-1993, University of Western Ontario, Dec. 1993.

11. NRCC 48192, National Research Council Canada, User's Guide-NBC 2005 Structural Commentaries Part 4 of Division B, ISBN 0-660-19506-2, 2006.

12. Fang, S., and Porthouse, R. A.: "The New ACI 307-08 Chimney Code and Seismic Design Requirements," Structural Congress 2009, ASCE, Austin, Texas, Published Online April 26, 2012.

13. Wilson, J. L.: "Aseismic Design of Tall Reinforced Concrete Chimneys," *ACI Struct J*, Vol. 99, No. 5, pp. 622–630, 2002.

14. ASCE, "Design and Construction of Steel Chimney Liners," American Society of Civil Engineers, New York, 1975.

15. Chu, Kuang-Han, and Park, J.: "Approximate Fundamental Frequencies for Chimneys," *J Power Div* ASCE, November 1973.

16. CICIND Model Code for Steel Chimneys, International Committee on Industrial Chimneys, Ratingen, Germany, 2010.

17. AISI, Steel Plate Engineering Data, Vol. 2, American Iron and Steel Institute, Washington, DC, 1992.

18. Rowe, R. S.: Amplified Stress and Displacement in Guyed Towers, *Trans ASCE*, Vol. 125, 1960, p. 199.

19. Fang, S., Chu, S-L, and Zar, M.: "Chapter 32—Chimneys," in the 4th edition of Structural Engineering Handbook, Gaylord, Gaylord and Stallmeyer, McGraw-Hill, 1997.

20. Wilson, J. L.: Code Requirements for the Aseismic Design of Tall Reinforced Concrete Chimneys, 12th World Conference on Earthquake Engineering, Auckland, New Zealand, 2000.

21. Munshi, J., and Malushte, S.: Seismic Design of Concrete Chimneys—State of Practice, ASCE Structural Congress 2007, Long Beach, California, Published Online June 12, 2012.

22. Bochicchio, V. A.: An Overview of Lining Materials for Use in the Renovation of Existing Power Plant Chimneys, *International Symposium on Industrial Chimneys and Cooling Towers*, Prague, 2014.

23. Fang, S. J., and Oropeza, M.: Acid Resistant Independent Brick Chimney Liners, presented at 1980 ASCE Convention, Hollywood. Florida. October 1980.

24. ASTM C1298-95(2013), Standard Guide for Design and Construction of Brick Liners for Industrial Chimneys, ASTM International, West Conshohocken, PA, 2013.

25. ASTM D5364-14, Standard Guide for Design, Fabrication, and Erection of Fiberglass Reinforced Plastic (FRP) Chimney Liners with Coal-Fired Units, ASTM International, West Conshohocken, PA, 2014.

26. Bochicchio, V. A.: FRP Chimney Liners for Power Plant Flue Gas Desulfurization—A Chimney Supplier's Perspective, *NACE International CORROSION 2011*, Houston, TX, 2011.

27. NACE Standard SP0199-2009, "Standard Practice: Installation of Stainless Chromium-Nickel Steel and Nickel-Alloy Roll-Bonded and Explosion-Bonded Clad Plate in Air Pollution Control Equipment," NACE International, Houston, TX, 2009.

28. Wilbur, R., Sowizal, J., and Cerny, N. X.: "New Options for Corrosion Resistant Metal Cladding for Wet Scrubber Service," Proceedings of the American Power Conference, Chicago, Vol. 52, 1990, pp. 21–24.

29. NACE International 2012, "Standard Practice—Installation of Thin Metallic Wallpaper Lining in Air Pollution Control and Other Process Equipment (SP0292-2012)," National Association of Corrosion Engineers, Houston, TX, 2012.

30. Luke, B., and DeKreij, A.: "Borosilicate Glass Block Linings at Work in Wet FGD Chimneys," *Power Eng*, Vol. 112, No. 1, p. 74, January 2008.

31. Chimney Coatings Manual for Concrete and Steel Chimneys, International Committee on Industrial Chimneys (CICIND), Zurich, Switzerland, 1996.

32. Smith, D. T.: "The Design and Construction of a 100m Tall Double-Skin, Free-Standing Chimney," CICIND Report, Vol. 17, No. 1, pp. 31–35, 2001.

33. Turner, J. G.: "Chapter 8—Industrial Chimney Foundations," revised on July 7, 2015, as a part of book entitled *Design Applications of Raft Foundations*, edited by J. A. Hemsley, 2000, Thomas Telford Ltd, London.

Chapter 25

Health Monitoring and Nondestructive Testing

BY

DIDEM OZEVIN *Associate Professor, Department of Civil and Materials Engineering, University of Illinois at Chicago*

FARHAD ANSARI *Professor, Department of Civil and Materials Engineering, University of Illinois at Chicago*

25.1 INTRODUCTION

Health monitoring of structures can be quantitatively measured with periodic assessment known as nondestructive evaluation (NDE) of structures or continuous monitoring known as Structural Health Monitoring (SHM). NDE methods are defined as "the development and application of technical methods to examine materials or components in ways that do not impair future usefulness and serviceability in order to detect, locate, measure, and evaluate flaws; to assess integrity, properties, and composition; and to measure geometrical characteristics (ASTME1316-17a)." SHM methods are the processes of implementing damage identification strategies over time and extracting damage-sensitive features by means of statistical methods (Farrar and Worden 2007). This chapter presents waveform-based NDE/SHM methods including vibration method (VM), impact echo (IE), acoustic emission (AE), and ultrasonics, which provide quantitative information about the state of structural components. The methods presented in this chapter encompass the entire frequency spectrum of 0–20 MHz to find defects in structural systems.

The organization of this chapter is as follows. Section 25.2 is dedicated to the description of measurement components including sensors, electronics, and signal processing tools. Section 25.3 describes the principles of NDE methods and their applications for the inspection of structural systems. The comparison of NDE methods is presented in Section 25.4. References are provided in last section of this chapter.

25.2 COMPONENTS OF NDE METHODS

The major hardware components of NDE methods include sensors (transmitter or receiver) and data acquisition systems (function generator, amplifier, oscilloscope). Sensors play significant role in applying the selected NDE method for a specific problem. Once the output signal is collected, various signal processing tools are applied to understand the presence of defects in structural components. In this section, state-of-the-art NDE sensors are described, and then the methods applied to output signals to understand the presence of defects (e.g., fatigue crack, delamination, corrosion) are presented.

25.2.1 NDE Sensors

A sensor converts a signal from one form of energy to another using a transduction principle, which is the modifier that transforms a mechanical/environmental/thermal disturbance into an electrical signal. A variety of transduction principles can be used to design sensors for NDE methods. The important characteristics of sensors needed for NDE are sensitivity (input-output relations), frequency range, linearity, and hysteresis. The three most common types of transduction principles used in NDE of civil structures are piezoelectric, capacitive, and fiber optic. Figure 25.1 shows the principles of sensors, and Fig. 25.2 shows the examples of their applications to steel and concrete structures. Piezoelectric sensors can be used as transmitter or receiver where the conversions of mechanical to electrical or electrical to mechanical signals are achieved by the generation of current or force within the piezoelectric element due to dynamic force or electrical voltage. The main advantages are high rigidity, high natural frequency, wide measuring range, stability, reproducibility, and wide operating range of temperature (Gautschi 2002). Capacitive sensors are based on the change in gap or overlap area of capacitive layers. They can be designed and built using Micro-Electro-Mechanical Systems (MEMS), which reduce the cost and size of the sensors (Saboonchi and Ozevin 2013). Fiber-optic sensors (FOSs) can be classified as intensity sensor, interferometric sensor, or fiber Bragg grating (FBG) sensors (Kim et al. 2016). FBG sensor is based on the change of periodic refractive index under the external strain where the change in wavelength is correlated with the applied strain. The FOSs can be utilized to measure the physical properties by localized, multiplexed, or distributed sensor method (Ansari 1997). Detailed description involved in fiber-optic sensing of stress waves and acoustical signals are provided in Giallorenzi et al. (1982). Interferometric techniques have provided means for high-resolution detection of damage within different types of materials (Lim et al. 1999).

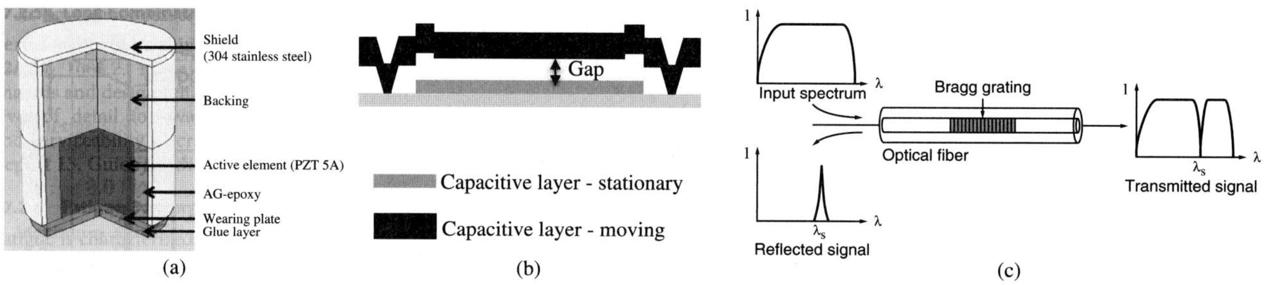

Figure 25.1 Common sensor types for NDE: (*a*) piezoelectric (Zhang et al. 2017), (*b*) capacitive, and (*c*) fiber optic (Kim et al. 2016).

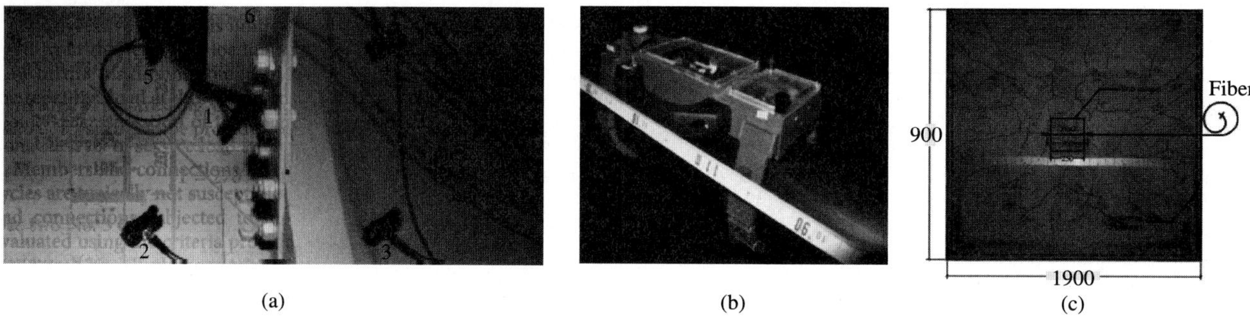

Figure 25.2 Field testing of (*a*) piezoelectric type AE sensors to assess the performance of structural retrofit (Kosnik et al. 2011), (*b*) capacitive-type electromagnetic sensors (Bore et al. 2013) to characterize grouting materials of bridge posttensioned ducts, and (*c*) fiber-optic strain sensors to detect crack at slab-column connection (Rizk et al. 2012).

The selection of sensor frequency is dependent upon the NDE method. Table 25.1 lists the frequency ranges of NDE methods where the selected sensor type should be tuned.

25.2.2 Signal Processing Methods

The waveform-based NDE methods depend on detecting transient signals in different frequency spectrums. Once transient signals are detected, various signal processing methods are applied to correlate with defects. The major signal processing methods are parameter extraction from time domain signals (shown in Fig. 25.3), frequency analysis, and time-frequency (TF) analysis.

A fast Fourier transform (FFT) decomposes the transient signal into its frequency components, which involve multiple sinusoidal oscillations at distinct frequencies depending on the content of the transient signal. Figure 25.4*a* and *b* show examples of time domain signal and its frequency spectrum. The frequency spectrum provides the frequency content of time domain signal, which can be affected by the presence of defect for active NDE methods or varied by different defect sources for passive NDE methods.

Table 25.1 The Required Frequency Bandwidth of Sensors for Waveform-Based NDE Methods

NDE method	Sensor properties
Vibration	0–50 Hz range Flat response in the frequency range
Impact echo	Up to 20 kHz frequency Flat response in the frequency range
Acoustic emission	20–500 kHz range Operating at the resonant frequency or wideband frequency
Ultrasonics	500 kHz–10 MHz range Operating at the resonant frequency

TF analysis is currently used for evaluating the output signals at dispersive environments. There are several TF methods in literature. Short Time Fourier Transform (STFT), Wagner-Ville Distribution (WVD), and Wavelet Transform (WT) are commonly used methods. STFT splits a signal into overlapping sections with a specified window (Pruell et al. 2009). The Fourier transform of each section is computed to produce an estimate of the short-term frequency content of the windowed signal. The transform equation is

$$F(\omega,b)=\frac{1}{2\pi}\int_{-\infty}^{+\infty} y(t)W(t-b)e^{-j\omega t}\,dt \tag{25.1}$$

where $y(t)$ is the signal in the time domain, $W(t-b)$ is the windowing function, and b is a parameter characterizing time window width. The continuous WT of a signal $y(t)$ is

$$F_{\Psi}(a,b)=\frac{1}{\sqrt{a}}\int_{-\infty}^{\infty} y(t)\Psi\left(\frac{t-b}{a}\right)dt \tag{25.2}$$

where Ψ represents the wavelet function, b is time variable, and a is scale variable. The difference between STFT and WT is the scale variable a. The scale variable makes it possible to change the window size for different frequencies such that the window is large for low frequencies while the window is small for high frequencies. This property of WT provides high resolutions in time and frequency. It is shown that Gabor Wavelet for Ψ provides small windows in the time and frequency domains, which augment the multiresolution of the WT (Kishimoto 1995). Figure 25.4*c* shows the spectrogram of time domain signal. While the frequency spectrum cannot provide information about the arrivals of each frequency component, the spectrogram provides the density, width, and arrival of each frequency component.

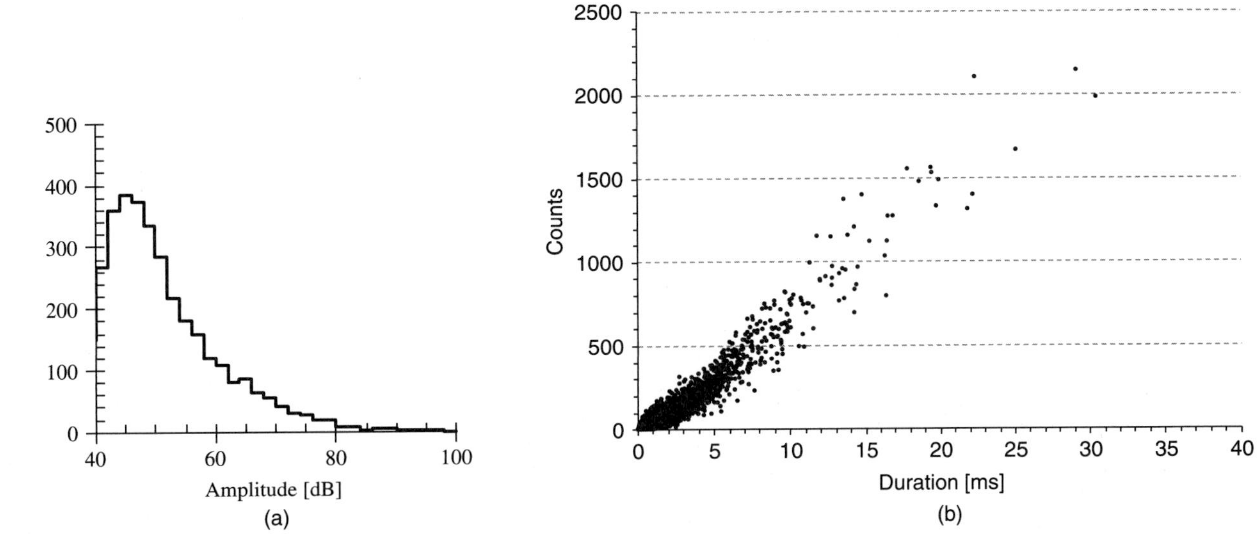

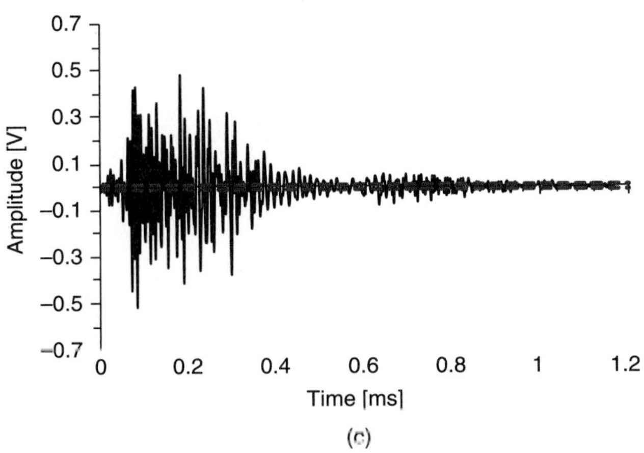

Time [s]	1637.6
Amplitude [dB]	74
Counts	182
Duration [μs]	2221
Energy	102
Abs Energy [aJ]	92.699E+03
Avg. Freq. [kHz]	81.9
C-Freq. [kHz]	142
P-Freq. [kHz]	58

(d)

Figure 25.3 Examples of features extracted from transient signals: (*a*) histogram plots, (*b*) correlation plots, (*c*) transient waveform, and (*d*) waveform features.

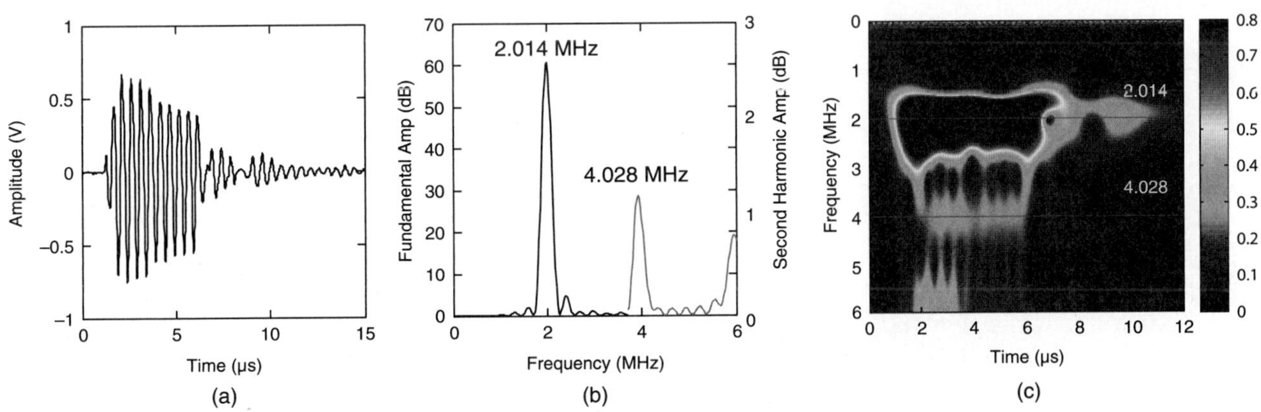

Figure 25.4 An example of (*a*) time domain signal, (*b*) frequency spectrum, and (*c*) spectrogram obtained by wavelet transform (Mostavi et al. 2017).

25.3 WAVEFORM-BASED NDE METHODS AND APPLICATIONS

The operational frequency ranges of four waveform-based NDE methods are shown in Fig. 25.5. While the resolution of NDE method increases with the increase of frequency, the effective volume and the depth of penetration decrease with the increase of frequency. The principle of each method is described below. Then, their applications for structural engineering problems are presented.

25.3.1 Vibration Method (VM)

The VM is based on measuring the dynamic response of structural systems excited by ambient loads such as traffic, wind, or external excitation such as shaker and drop-weight system. Figure 25.6 shows an example of accelerometer installed on a multi-girder steel bridge. The development of defects causes a reduction in local/global stiffness resulting in changes in dynamic response characteristics, such as mode shapes and modal frequencies (Samman and Biswas 1994).

The method has been used in several structural systems such as tower structures (Bongiovanni et al. 2017), bridges (Peeters et al. 2001, Oliveira and Camacho 2016), and dams (Patjawit and Chinnarasri 2013). The main challenge of this method is the influence of environmental factors (e.g., temperature, humidity) on the vibration properties of the structural systems.

25.3.2 Impact Echo (IE)

The IE method is an active NDE method and based on exciting structural surface with an impactor to generate and transmit stress waves through materials at frequencies up to 20 to 30 kHz and measure the response by a receiving transducer. While multiple wave modes are generated [dilational (*P*), distortional (*S*), and Rayleigh (*R*) waves], the dilational waves are the primary wave mode due to higher amplitude response (Hsiao et al. 2008). The short-time transient signal detected by the receiver is transformed into frequency domain where the peak frequency and amplitude are related to transient resonance of the test structure.

Figure 25.7 shows the principle of IE method. The propagating wave excited by the impactor bounces back from the back surface (travel distance is $2h_1$) or any free surface within the test structure caused by crack or delamination (travel distance is $2h_2$). For the wave velocity V, different wave reflections are observed in the frequency spectrum of the receiver as $f_1 = V/2h_1$ and $f_2 = V/2h_2$. The spectrum of the impactor (highlighted as the red dashed line on the frequency curve) should be sufficiently broad to detect the frequency reflected from free surface at any depth through the test structure.

The method has been successfully implemented in detecting flaws (delamination, crack, porosity) in concrete (Krause et al. 1997, Hill et al. 2000), slab thickness (Sansalone and Streett 1997, Villain et al. 2012, Li et al. 2016), and dynamic modulus of concrete (Lu et al. 2013). Two examples on inspecting delamination in concrete pavement and bridge and timber pole are shown in Fig. 25.8.

25.3.3 Acoustic Emission (AE) Method

The AE method is a passive NDE method where the transient signal released by active flaws is detected by the sensors attached to the structural surface, a direct way of identifying the damage using noninvasive sensors, and can be implemented for evaluating the localized condition of structural components or the global condition of structural systems (Miller and Moore 2005). The structural component should be under loading and flaws should be active. The method cannot detect dormant flaws. Typically, an array of sensors is utilized as shown in Fig. 25.9 such that the location of defect in planar or space structures can be determined using arrival time differences (Ohtsu 1986, Maji et al. 1997, Kundu et al. 2008, Ozevin 2011). The AE method has been applied for damage detection in variety of structures including steel and concrete bridges (Nair and Cai 2010). As presented by Holford et al. (2001), the AE method provides a good measure of structural integrity and the ability for local and global monitoring of structures.

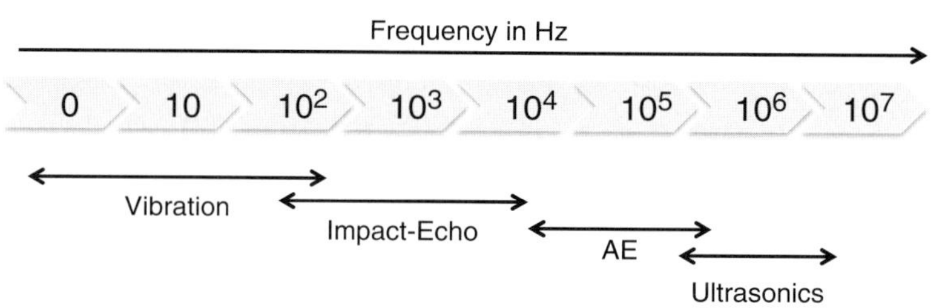

Figure 25.5 The waveform-based NDE methods and their operational frequencies.

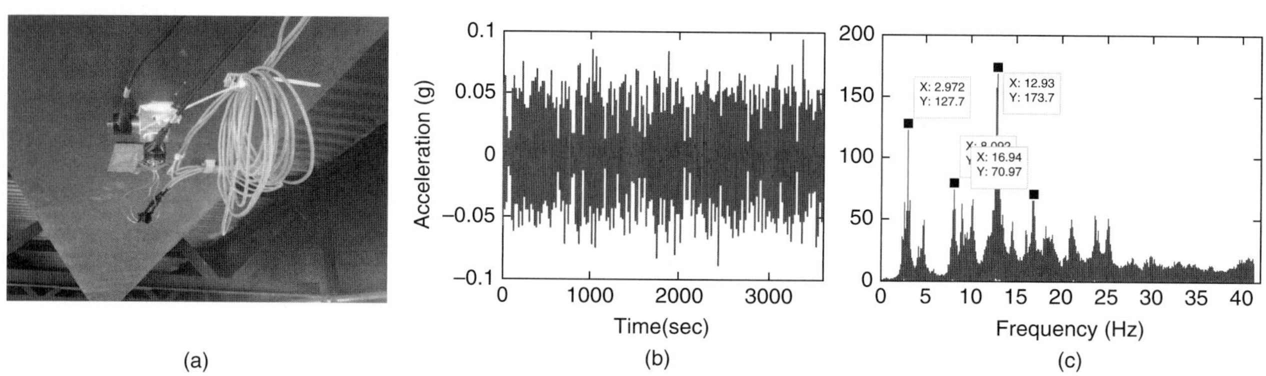

Figure 25.6 An example of VM applied to monitor multi-girder steel bridge: (*a*) accelerometer, (*b*) acceleration response, and (*c*) frequency response.

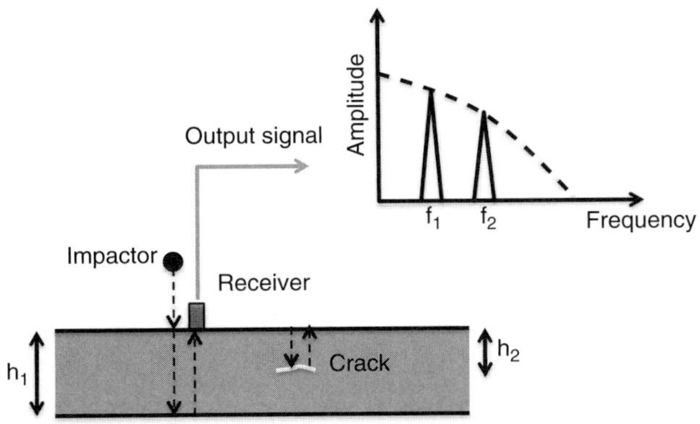

Figure 25.7 Principle of impact echo method.

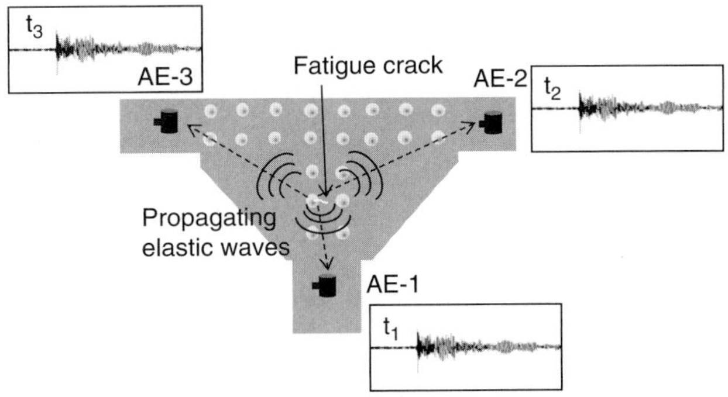

Figure 25.9 Principle of AE method shown on a bolted gusset plate.

AE provides information about growing microcracks and adverse chemical reactions, such as corrosion from within the structural materials. The applicability of AE method encompasses a broad range of materials that include cementitious, metallic, fibrous materials, such as fiber-reinforced polymer (FRP) composites (Ouyang et al. 1993, Mirmiran et al. 1999). The component materials employed for fabrication of AE sensors have been piezoelectric, fiber-optic, and magnetostriction based.

AE Detection Using Fiber-Optic Sensors

Practical application of AE to actual structures such as bridges requires multitudes of sensors. This increases the number of data acquisition channels and inhibits real-time monitoring of damage formation within the entire structure. Optical fibers provide an alternative approach for the detection of AE signals through distributed and serial multiplexing methodologies. Chen and Ansari (2000) developed a novel optical fiber sensor with capability for making distributed measurements. The sensor design involved development of AE sensing regions along the length of an optical fiber. A white light Michelson interferometer was used for interrogation of the AE events at their quadrature points to achieve maximum sensitivity. This work demonstrated that a single optical fiber with two sensitized regions is capable of locating the source of an AE event in a rod-like element. This interferometer was further developed for differentiation of AE signals during AE events as shown in Sun et al. (2004).

The reference and sensing arms of the interferometer are formed by two couplers as shown in Fig. 25.10. Laser diodes (LD) and photo diodes (PD) are the light emitting and avalanche photo diodes for transmission and detection of optical signals, respectively. An optical isolator is used for reduction of reflected source noise. The sensing arm of the sensor is sensitized at two locations to create two high-resolution sensing regions along the length of the fiber. Sensitized regions are created by wrapping the fibers into coils to precise diameters to achieve the frequency range of interest. The two sensitized regions at the sensing arm of the interferometer receive the AE events, and the high-precision motorized stepper motors at the reference arm translate to match the AE-generated perturbations at the sensitized regions. This matching of the AE events is realized by the fringes, and the signals generated in time domain are employed for the characterization of the AE signals. The time domain signals generated in the interferometer due to the AE events are described as (Sun et al. 2004)

$$I_0 = I_0^{S1} + I_0^{S2} \tag{25.3}$$

where I_0 = intensity of the detected signal
I_0^{S1} = intensity at S_1, I_0^{S2} = intensity at S_2

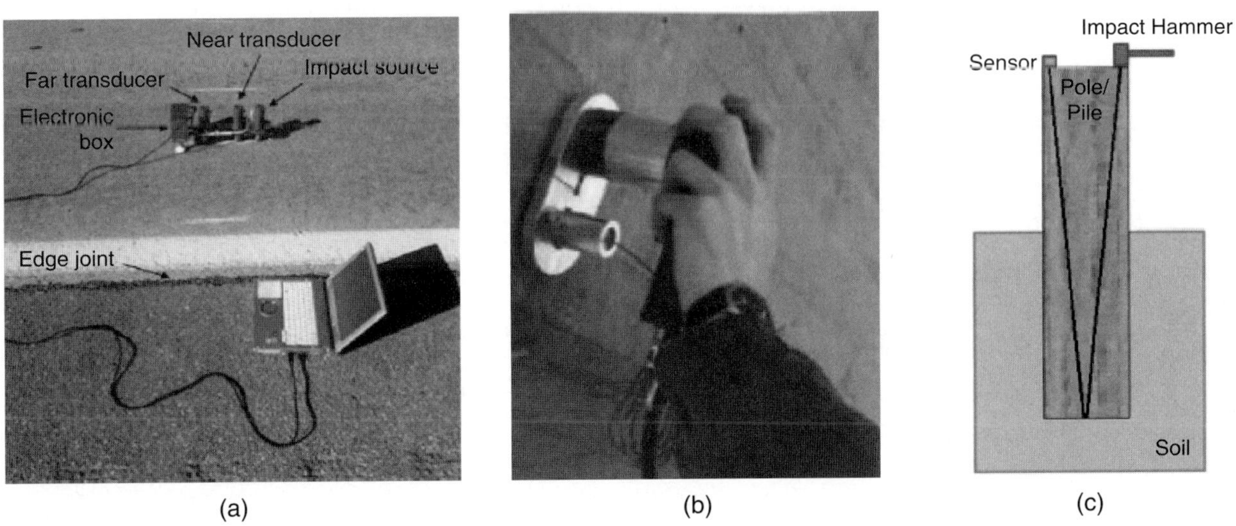

Figure 25.8 Impact echo applications: (*a*) crack and delamination detection in pavement (Li et al. 2016), (*b*) bridge inspection (Grosse 2013), and (*c*) timber pole inspection (Krause et al. 2015).

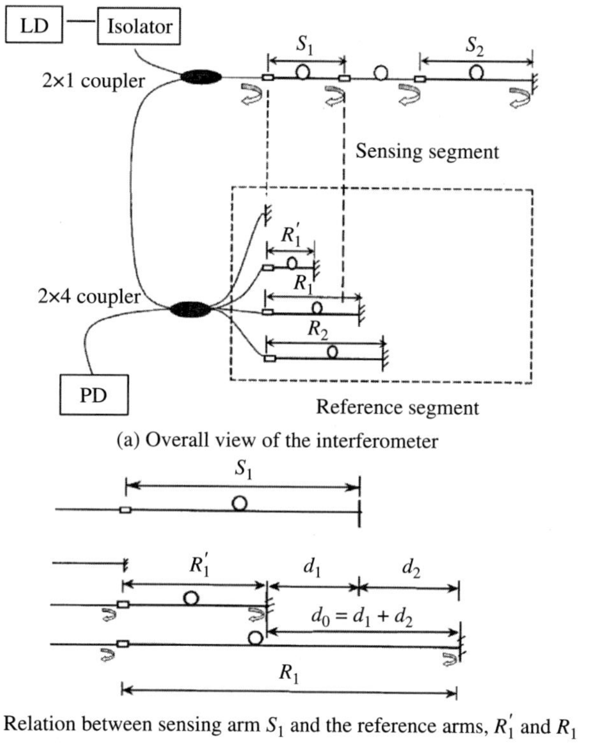

Figure 25.10 Overall schematic of the serially multiplexed AE interferometer (Sun et al. 2004).

The time domain signal intensities pertain to the phase changes, φ_i, in the interferometer, which are physically related to the length differences, d_i, shown in Fig. 25.10*b*, by

$$\varphi_i = \frac{2\pi}{\lambda}\,\delta_i(n, d_i) \tag{25.4}$$

where n = refractive index of the optical fiber
λ = wavelength at the source
$\delta_i(n, d_i)$ = effective optical path difference
i = 0,1,2 (the time domain output contains signals from S_1 and S_2 separated in time)

It will not be possible to separate and distinguish two signals without any further modifications to the system. Differentiation of the signals is necessary in order to discern the sequence of signal arrivals from S_1 and S_2. Therefore, the interferometric system also includes a biasing circuit to differentiate the signals (Sun et al. 2004).

Evaluation of the system was accomplished by comparing the distributed optical fiber AE sensor with conventional lead-zirconium-titanium oxide (PZT) sensors. A 12-ft-long steel beam was instrumented with two PZT sensors, 10 ft apart. An optical fiber was adhered to the surface of the steel beam. AE signals were generated at different locations along the length of the beam. The generation of AE signals was accomplished by breaking pencil leads at predetermined locations along the length of the beam as shown in Fig. 25.11. Figure 25.12 corresponds to the comparison of the detected AE signals between the PZT and FOS. Data pertain to the comparison of the measured arrival time differences employed for the computation of AE event locations. These results indicated the sufficiency of the optical fiber system for monitoring of homogeneous structural materials, that is steel.

Liang et al. (2004) developed a hybrid FRP rod for use as posttensioning rods in posttensioned concrete beams. These hybrid FRP rods were designed of a core-shell structure combining the strength and ductility attributes of carbon and glass fibers. The objective of their

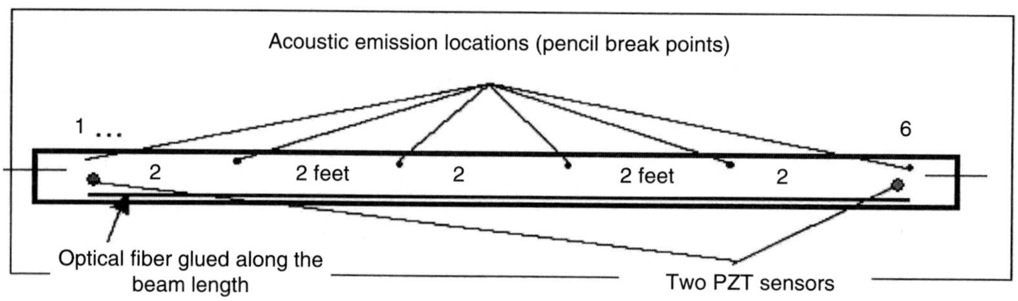

Figure 25.11 Steel beam and the locations of the PZT transducers and AE events.

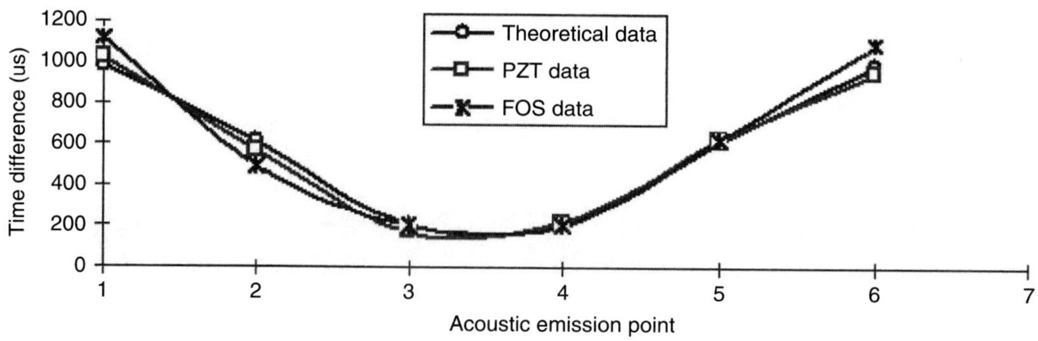

Figure 25.12 Comparison of the detected time of arrival differences between the PZT and FOS.

work was to produce an FRP rod with pseudoyielding capabilities. To study the fracturing process in these rods, Liang et al. (2004) performed direct tension tests on the rods. In addition to the tensile stress-strain response of the rod, they also employed the serially multiplexed optical fiber sensor for characterization of the fiber breakage during the failure process of the rods. The experimental program included uniaxial tensile tests of the instrumented hybrid rods in a servo-hydraulic closed-loop materials testing system with a load capacity of 667.2 kN (150-kips). All the rods in the experiments possessed a nominal diameter of 9.5 mm (0.375 in.). Schematic diagram of the test setup is shown in Fig. 25.13. The 63.5-cm (25 in.) nominal distance between two anchorages in the test machine in each specimen corresponded to the gauge length for the linear variable displacement transformer (LVDT). LVDT with a displacement range of ±12.7 mm (±0.5 in.) was employed for measuring the axial deformation. Four strain gauges were glued at the mid-height of the hybrid rod, two for measurement of axial strains (for cross-checking the LVDT results) and two for transverse strains. In addition to the FOS, two PZT sensors as shown in Fig. 25.13 were employed for the detection of AE events.

Figure 25.13 shows typical time domain AE events detected by two PZT gauges and single single optical fiber sensor (Liang et al. 2004). The single-channel signal for the FOS contains two AE events as shown in this figure. Figure 25.14 corresponds to the stress-strain response and the detected AEs during the splintering of the fibers in the hybrid FRP rod. As shown in this figure, the FOS was also capable of differentiating the fracture of carbon and glass fibers for which the details are given in Liang et al. (2004).

In another study, the hybrid FRP rods were employed for posttensioning of a concrete beam (Liang et al. 2011). The beams were 12 ft (3.66 m) in length and 6.0 × 12.0 in. (15.2 × 30.5 cm) in cross section. Monitoring of the tendon strains during the posttensioning process was accomplished by long-gauge interferometric optical fiber

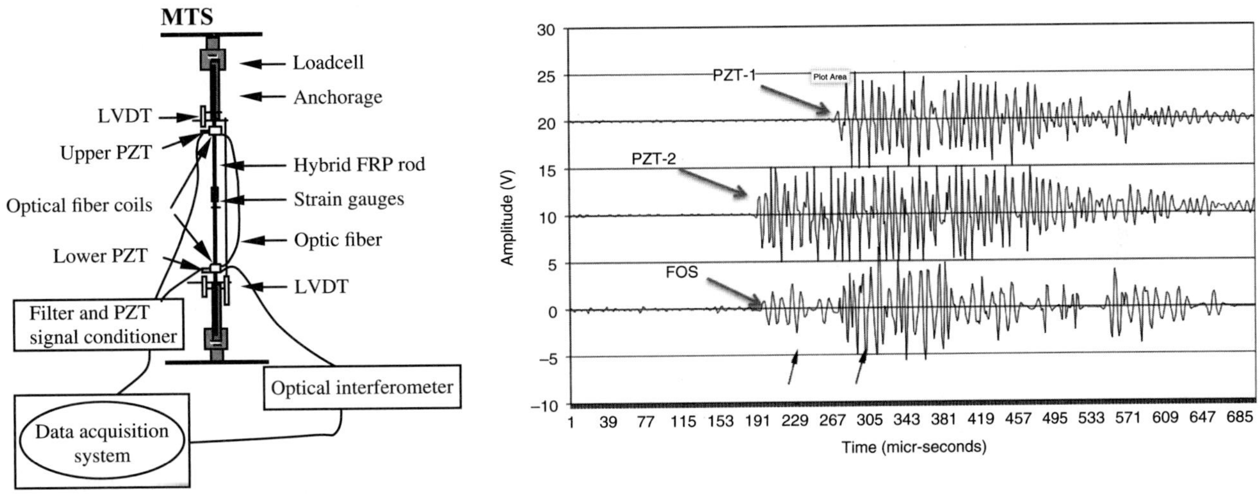

Figure 25.13 (*a*) Schematic diagram of the hybrid FRP rod in the tensile testing machine (*left*) and (*b*) comparison of the time domain data for the two PZT and the single FOS sensors (*right*) (Liang et al. 2004).

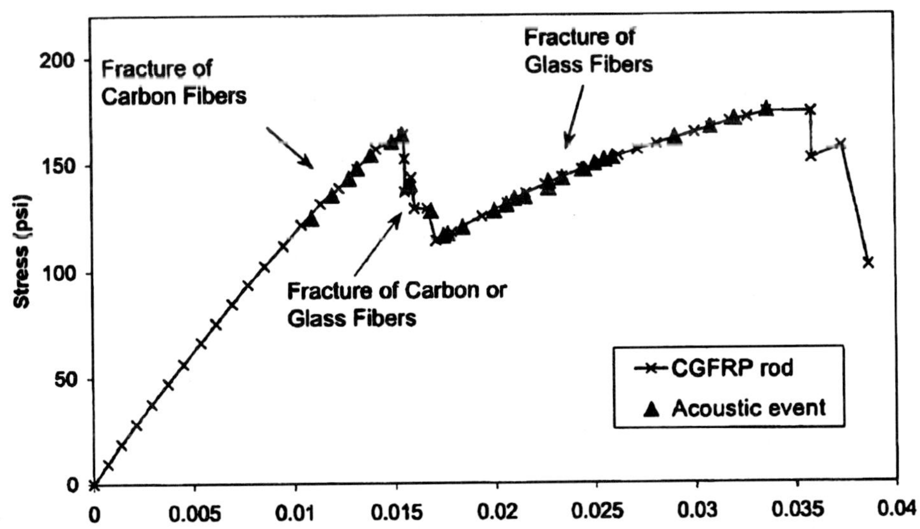

Figure 25.14 Typical stress-strain response and detected AE events for the hybrid FRP rod (Liang et al. 2004).

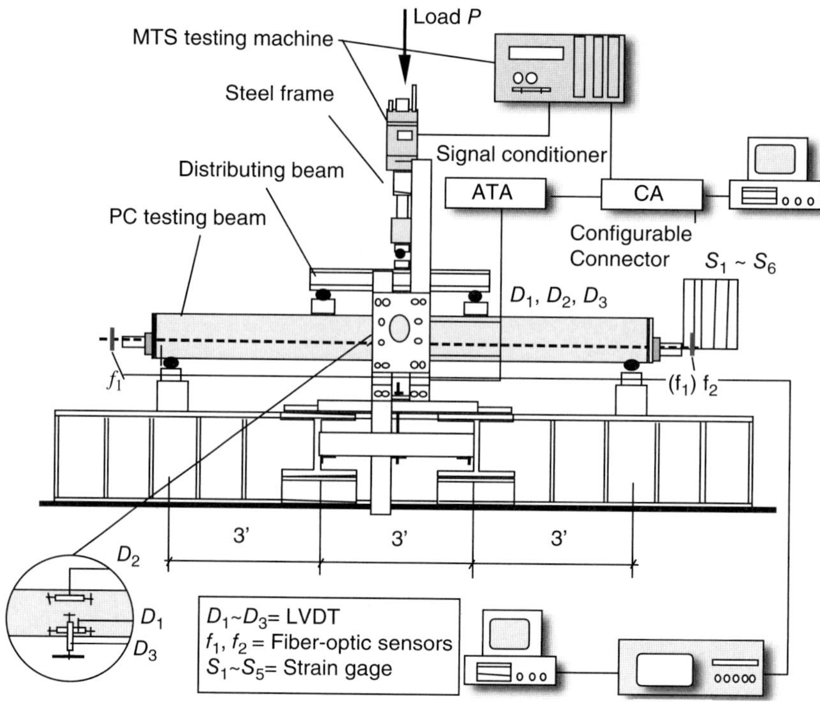

Figure 25.15 Schematic diagram of the loading and specimen setup (Liang et al. 2011).

sensors. The gauge lengths of the optical fiber sensors were selected to cover the entire lengths of the tendons. The serially multiplexed AE sensor was also adhered to the posttensioning rod and routed out of the tendon ducts just prior to entering the anchorages. The beam was loaded under the four-point bend load tests in a testing machine.

Figure 25.15 shows the schematic diagram of the specimen and loading system details. Figure 25.16 pertains to the typical results obtained from AE events captured by the optical fiber sensor (Liang et al. 2011). As shown in this figure, most of the AE events occurred within the constant moment region of the beam indicating splintering of the fibers during the various stages of the failure.

25.3.4 Ultrasonics Testing (UT)

UT is an active NDE method where a perturbation is introduced to structure and reflected waves are monitored to find the internal defects. The method can be applied using pulse-echo (single transducer as receiver and transmitter) or through-transmission (two different transducers tuned to the same frequency used as receiver and transmitter) modes

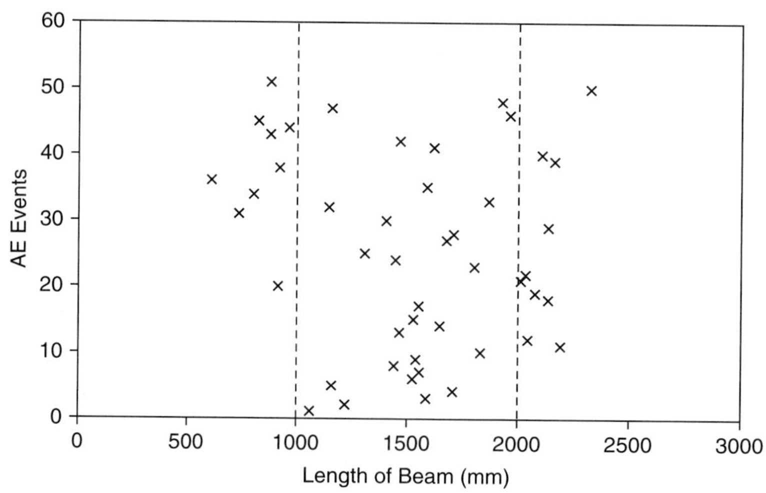

Figure 25.16 AE events due to the fracturing of fibers in the FRP rod during the four-point bend tests (Liang et al. 2011).

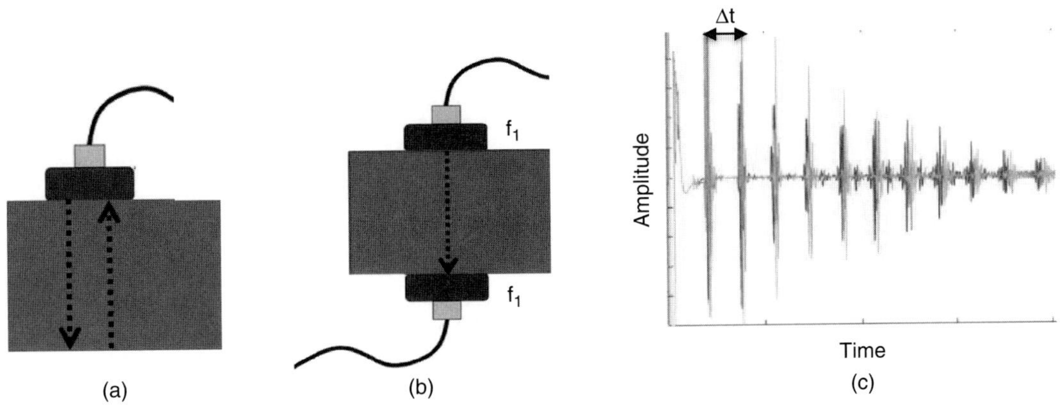

Figure 25.17 Principle of ultrasonic method: (*a*) pulse-echo mode, (*b*) through-transmission mode, and (*c*) typical ultrasonic waveform.

as shown in Fig. 25.17. A single transducer is used in pulse-echo mode, and the travel path of the time of flight (TOF – Δt in Fig. 25.17*c*) is twice the sample thickness. If there is crack in the travel path, wave reflects and is determined with smaller TOF than Δt. Two transducers are used in through-transmission mode as transmitter and receiver tuned to the same frequency (f_1), and the travel path of TOF is the sample thickness. If there is crack in the travel path, the signal reflects back, and no signal reaches the receiving transducer. Typical ultrasonic frequencies to find defects in structural systems are in the range of 1–20 MHz depending on the required resolution and penetration depth.

Stress Measurement Using Nonlinear UT

Increase in traffic and additional modifications on bridges increase the cumulative stress in critical bridge components. The collapse of the I-35W is an example that an increase of the dead load by 30 percent over the original design from concrete deck overlay, plus additional construction-related loads on the day of the collapse located above the weakest point of the structure, a pair of gusset plates, was one of the triggers leading to failure (Liao et al. 2011) NDE methods for measuring stress in structures include hole-drilling (Withers and Bhadeshia 2001, Steinzig and Ponslet 2003), X-ray diffraction (Prevey 1996), and magneto-elastic methods (Kuud 1982, Fitzpatrick et al. 2005). The hole-drilling method introduced by Mathar in 1933 is widely used for residual stresses in concrete structures. The acousto-elastic effect is defined as the stress dependence of ultrasonic wave velocity. Acousto-elasticity is applied to measure stress at prestressed steel strands (Scalea et al. 2003, Chaki and Bourse 2009, Salamone et al. 2009), longitudinal stress in rails (Gokhale 2007, Szelazek 2013), and uniaxial residual stress in metals (Kino et al. 1978, Allen and Sayers 1984, Kato et al. 1995) and in welds (Kobayashi et al. 1992, Leon-Salamanca and Bray 1996, Lu et al. 2008). Abbasi and Ozevin (2016) applied the acousto-elastic method for stress analysis to thick steel plates (9–12 mm) of the type commonly used for gusset plates of steel truss bridges, where shear stress is significant.

Figure 25.18 shows the schematic of stress measurement with ultrasonic pulser and receiver, operating in through-transmission mode on A572 Grade 50 steel typically used in steel bridges. Figure 25.19 shows the time domain waveforms and their frequency spectra (amplitude and phase) at four different load levels when the 1-MHz ultrasonic transducer is used and angled to the stress direction. The arrival time decreases when the stress increases; the peak frequency increases when the stress increases as expected.

25.4 THE APPLICATIONS OF WAVEFORM-BASED NDE METHODS TO INFRASTRUCTURE SYSTEMS

The common defects found in different materials and structural components are summarized in Table 25.2. Depending on the structural material and the defect mode and size targeted, the proper NDE method is selected. For instance, IE method works very well in detecting delamination in concrete, UT method is not preferred for concrete as the penetration depth of high frequency signal is low. As bridge deck inspection requires more localized measurement, VM is not common in inspecting bridge decks. Table 25.3 provides examples of NDE methods applied in realistic test conditions for different structural systems.

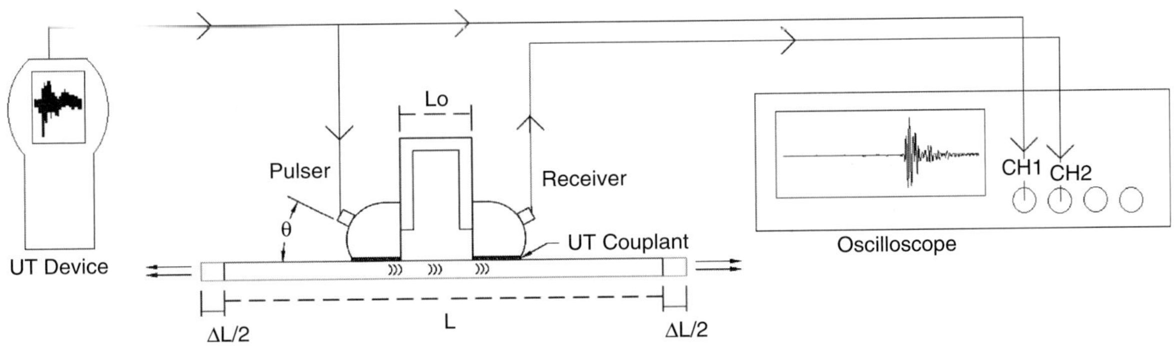

Figure 25.18 Acousto-elastic measurement from stresses plate (Abbasi and Ozevin 2016).

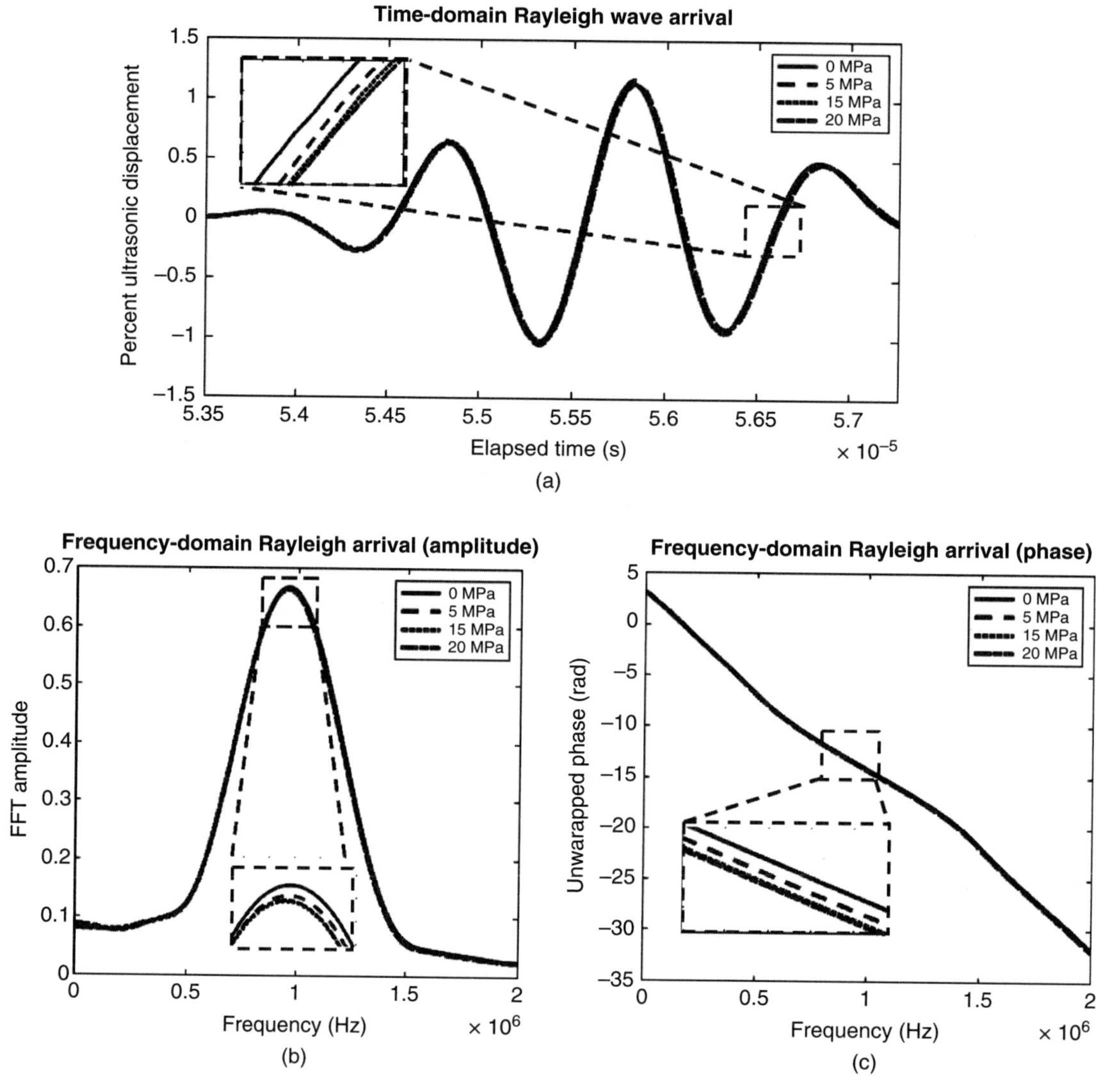

Figure 25.19 Waveforms recorded using four different stress states and the 1 MHz ultrasonic transducer angled (30°) to the stress direction: (*a*) time domain histories and (*b*) amplitude and phase of frequency spectra (Abbasi and Ozevin 2016).

Table 25.2 Defects Detected in Different Materials Using Waveform-Based NDE Methods

	Material			
	Steel	Concrete	Timber	Masonry
Failure mode	Corrosion, fatigue crack, global damage	Wire break, concrete cracking, reinforcing steel corrosion delamination, dynamic modulus	Modulus of elasticity, moisture, inhomogeneity, cracks, decay, deformation, stiffness	Voids, moisture, morphology, inclusions

Table 25.3 Applications of NDE Methods to Structural Systems

Structure type	Method: VM	IE	AE	UT
Concrete	Concrete bridge girder damage identification (Morsy et al. 2017)	Durability parameters of concrete beam (Villain et al. 2012), concrete repair inspection (Garbacz et al. 2017)	Acoustic emission monitoring of bridges (Nair and Cai 2010), reinforcing steel corrosion detection (Abouhussien and Hassan 2015), crack monitoring at concrete wall (Farhidzadeh et al. 2013)	Concrete crack detection in buildings (Park et al. 2000)
Prestressed posttensioned concrete bridge	Progressive damage detection (Unger et al. 2006)	Thickness and position of duck (Colla and Lausch 2003)	Diagnostics damage in prestressed concrete beams (Golaski et al. 2002)	Void detection using ultrasonic tomography (Martin et al. 2001)
Truss bridge	Damage detection of cantilever truss bridge (Shama et al. 2001)	Section loss detection (Can et al. 2018)	Hanger connection monitoring (Hay et al. 2009)	Stress measurement with acousto-elasticity (Abbasi and Ozevin 2016)
Steel structures	Load rating (Catbas and Aktan 2002), damage identification in steel frame (Park et al. 2006)	—	AE monitoring of fracture critical bridges (Schultz 2015)	Wavefield imaging to find damage in steel box girders (An et al. 2014)
Masonry/timber structures	Crack opening measurement (Talebinejad et al. 2011), monitoring repair of masonry pagoda (Li et al. 2009), timber beam damage diagnosis (Peterson et al. 2001)	Timber pole monitoring with impact echo (Krause et al. 2015)	Crack detection in masonry arch bridge (De Santis and Tomor 2013)	Void detection in masonry wall (Cotic et al. 2013)
Bridge deck	—	Delamination and crack detection (Kee et al. 2012, Gucunski et al. 2012)	Fiber-reinforced composite bridge deck monitoring (Gostautas et al. 2005)	Ultrasonic surface wave to assess concrete bridge deck condition (Li et al. 2016)

REFERENCES

General

Ansari, F. (1997). "State-of-the-art in the applications of fiber-optic sensors to cementitious composites," *Cement and Concrete Composites*, 19, 3–19.

ASTM E1316–17a. Standard Terminology for Nondestructive Evaluation.

Bore, T., Placko, D., Taillade, F. and Sabouroux, P. (2013). "Electromagnetic characterization of grouting materials of bridge post tensioned ducts for NDT using capacitive probe," *NDT&E International*, 60, 110–120.

Chen, Z. and Ansari, F. (2000). "Fiber optic acoustic emission crack sensor for large structures," *Journal of Structure Control*, 7(1), 119–129.

Farrar, C.R. and Worden, K. (2007). "An introduction to structural health monitoring," *Philosophical Transactions of the Royal Society A*, 365, 303–315.

Gautschi, G. (2002). *Piezoelectric Sensorics: Force, Strain, Pressure Acceleration and Acoustic Emission Sensors, Materials and Amplifiers*. Springer, New York.

Giallorenzi, T.G., BucaroJ, A., Dandridge, A., Sigel, G.H., Jr., Cole, J.H., Rashleigh, S.C. and Priest, R.G. (Apr. 1982). "Optical fiber sensor technology," *IEEE Journal of Quantum Electronics*, QE-18, 626–664.

Kim, T.M., Kim, D.H., Kim, M.K. and Lim, Y.M. (2016). "Fiber Bragg grating-based long-gauge fiber optic sensor for monitoring of a 60 m full-scale prestressed concrete girder during lifting and loading," *Sensors and Actuators A: Physical*, 252, 134–145.

Kishimoto, K., Inoue, H., Hamada, M. and Shibuya, T. (1995). "Time frequency analysis of dispersive waves by means of wavelet transform," *Journal of Applied Mechanics*, 62, 841–846.

Liang, Y., Sun, C.S. and Ansari, F. (2011). "Damage assessment and ductility evaluation of post tensioned beams with hybrid FRP tendons," *ASCE, Journal of Composites for Construction*, 15(3), 274–283.

Lim, T.K., Zhou, Y., Lin, Y., Yip, Y.M. and Lam, Y.L. (1999). "Fiber optic acoustic hydrophone with double Mach-Zehnder interferometers for optical path length compensation," *Optics Communications*, 159(4), 301–308.

Mostavi, A., Kamali, N., Tehrani, N., Chi, S.W., Ozevin, D. and Indacochea, J.E. (2017). "Wavelet based harmonics decomposition of ultrasonic signal in assessment of plastic strain in aluminum," *Measurement*, 106, 66–78.

Pruell, C., Kim, J., Qu, J. and Jacobs, L.J. (2009). "Evaluation of fatigue damage using nonlinear guided waves," *Smart Materials and Structures*, 18.

Rizk, E., Marzouk, H., Hussein, A. and Gu, X. (2012). "Structural health monitoring of slab-column connections using FBG sensors," *Journal of Civil Structural Health Monitoring*, 2, 17–27.

Saboonchi, H. and Ozevin, D. (2013). "MEMS acoustic emission transducers designed with high aspect ratio geometry," *Smart Materials and Structures*, 22, 095006.

Sun, C.S., Liang, Y.J. and Ansari, F. (2004). "Serially multiplexed dual-point fiber-optic acoustic emission sensor," *Journal of Lightweight Technology*, 22(2), 487–493.

Vibration Related

Bongiovanni, G., Buffarini, G., Celemente, P., Rinaldis, D. and Saitta, F. (2017). "Experimental vibration analysis of a historical tower structure," *Journal of Civil Structural Health Monitoring*, 7(5), 601–613.

Catbas, F.N. and Aktan, A.E. (2002). "Condition and damage assessment: Issues and some promising indices," *Journal of Structural Engineering*, 128(8), 1026–1036.

Li, S.W., Wei, J.W., Li, T.Y., Li, Q.M. and Bell, A.J. (2009). "Assessment of repairs and strengthening of a historic masonry pagoda using a vibration-based-method," *Journal of Structural Engineering*, 135(1), 67–77.

Oliveira, C.S. and Camacho, V.T. (2016). "Database with in-situ vibration frequencies of bridges in Portugal," *Journal of Civil Structural Health Monitoring*, 6(5), 851–862.

Park, G., Cudney, H.H. and Inman, D.J. (2000). "Impedance-based health monitoring of civil structural components," *Journal of Infrastructure Systems*, 6(4).

Park, S., Bolton, R.W. and Stubbs, N. (2006). "Blind test results for nondestructive damage detection in a stel frame," *Journal of Structural Engineering*, 312(5), 800–809.

Patjawit, A. and Chinnarasri, C. (2013). "Simplified evaluation of embankment dam health due to ground vibration using Dam Health Indec (DHI) approach," *Journal of Civil Structural Health Monitoring*, 4(1), 17–25.

Peeters, B., Maeck, J. and Roeck, G.D. (2001). "Vibration-based damage detection in civil engineering: Excitation sources and temperature effects," *Smart Materials and Structures*, 10, 518–527.

Samman, M.M. and Biswas, M. (1994). "Vibration testing for nondestructive evaluation of bridges. I: Theory," *Journal of Structural Engineering*, 120(1), 269–289.

Shama, A.A., Mander, J.B., Chen, S.S. and Aref, A.J. (2001). "Ambient vibration and seismic evaluation of a cantilever truss bridge," *Engineering Structures*, 23(10), 1281–1292.

Talebinejad, I., Fischer, C. and Ansari, F. (2011). "A hybrid approach for safety assessment of the double span masonry vaults of the Brooklyn Bridge," *Journal of Civil Structural Health Monitoring*, 1, 3–15.

Unger, J.F., Teughels, A. and Roeck, G.D. (2006). "System identification and damage detection of a prestressed concrete beam," *Journal of Structural Engineering*, 312(11), 1691–1698.

Impact Echo Related

Colla, C. and Lausch, R. (2003). "Influence of source frequency on impact-echo data quality for testing concrete structures," *NDT&E International*, 36, 203–213.

Garbacz, A., Piotrowski, T., Courard, L. and Kwasniewski, L. (2017). "On the evaluation of interface quality in concrete repair system by means of impact-echo signal analysis," *Construction and Building Materials*, 134, 311–323.

Grosse, C., Reinhardt, H., Krüger, M. and Beutel, R. (2013). "Application of impact-echo techniques for crack detection and crack parameter estimation in concrete," *ICF11*, Italy.

Gucunski, N., Yan, M., Wang, Z., Fang, T. and Maher, A. (2012). "Rapid Bridge deck condition assessment using three-dimensional visualization of impact echo data," *Journal of Infrastructure Systems*, 18(1).

Hill, M., McHugh, J. and Turner, J.D. (2000). "Cross-sectional modes in impact-echo testing of concrete structures," *Journal of Structural Engineering*, 126(2), 228–234.

Hsiao, C., Cheng, C-C., Liou, T. and Juang, Y. (2008). "Detecting flaws in concrete blocks using the impact-echo method," *NDT&E International*, 41, 98–107.

Kee, S.H., Oh, T., Popovics, J.S., Arndt, R.W. and Zhu, J. (2012). "Nondestructive bridge deck testing with air-coupled impact-echo and infrared thermography," *Journal of Bridge Engineering*, 17(6).

Krause, M., Barmann, M., Frielinghaus, R., Kretzschmar, F., Kroggel, O., Langenberg, K.J., Maierhofer, C., Muller, W., Neisecke, J., Schickert, M., Schmitz, V., Wiggenhauser, H. and Wollbold, F. (1997). "Comparison of pulse-echo methods for testing concrete," *NDT&E International*, 40(4) 195–204.

Li, M., Anderson, N., Sneed, L. and Torgashov, E. (2016). "Condition assessment of concrete pavements using both ground penetrating radar and stress-wave based techniques," *Journal of Applied Geophysics*, 135, 297–308.

Lu, X., Sun, Q., Feng, W. and Tian, J. (2013). "Evaluation of dynamic modulus of elasticity of concrete using impact-echo method," *Construction and Building Materials*, 47, 231–239.

Villain, G., Sbartai, Z.M., Derobert, X., Garnier, V. and Balayssac, J.P. (2012). "Durability diagnosis of a concrete structure in a tidal zone by combining NDT methods: Laboratory tests and case study," *Construction and Building Materials*, 37, 893–903.

Acoustic Emission Related

Abouhussien, A.A. and Hassan, A.A.A. (2015). "Evaluation of damage progression in concrete structures due to reinforcing steel corrosion using acoustic emission monitoring," *Journal of Civil Structural Health Monitoring*, 5, 751–765.

Collins, M.P. and Belchamber, R.M. (1989). "Acoustic emission source location using simplex optimization," *Journal of Acoustic Emission*, 9(4), 271–276.

De Santis, S. and Tomor, A.K. (2013). "Laboratory and field studies on the use of acoustic emission for masonry bridges," *NDT&E International*, 55, 64–74.

Farhidzadeh, A.R., Dehghan-Niri, E., Salamone, S., Luna, B. and Whittaker, A. (2013). "Monitoring crack propagation in reinforced concrete shear walls by acoustic emission," *Journal of Structural Engineering*, 139(12), 04013010-1-10.

Golaski, L., Gebski, P. and Ono, K. (2002). "Diagnostics of reinforced concrete bridges by acoustic emission," *Journal of Acoustic Emission*, 20, 83–98.

Gostautas, R.S., Ramirez, G., Peterman, R.J. and Meggers, D. (2005). "Acoustic emission monitoring and analysis of glass fiber-reinforced composites bridge decks," *Journal of Bridge Engineering*, 10(6).

Hay, D.R., Cavaco, J.A. and Mustafa, V. (2009). "Monitoring the civil infrastructure with acoustic emission: bridge case studies," *Journal of Acoustic Emission*, 27, 1–10.

Holford, K.M., Davies, A.W., Pullin, R. and Carter, D.C. (2001). "Damage location in steel bridges by acoustic emission," *Journal of Intelligent Material Systems and Structures*, 12, 567–576.

Kosnik, D.E., Hopwood, T. and Corr, D.J. (2011). "Acoustic emission monitoring for assesment of steel bridge details," *AIP Conference Proceedings*, 1335.

Krause, M., Dackermann, U. and Li, J. (2015). "Elastic wave modes for the assessment of structural timber: Ultrasonic echo for building elements and guided waves for pole and pile structures," *Journal of Civil Structural Health Monitoring*, 5, 221–249.

Kundu, T., Das, S., Martin, S.A. and Jata K.V. (2008). "Locating point of impact in anisotropic fiber reinforced composite plates," *Ultrasonics*, 48, 193–201.

Liang, Y., Sun, C., and Ansari, F. (2004). "Acoustic emission characterization of damage in hybrid fiber reinforced polymer rods," *ASCE Journal of Composites for Construction*, 8(1), 70–78.

Maji, A.K., Satapathi, D. and Kratochvil, T. (1997). "Acoustic emission source location using lamb wave modes," *ASCE Journal of Engineering Mechanics*, 123(2), 154–161.

Miller R., Hill E. and Moore, P.O. (2005). *Nondestructive Testing Handbook*, Third Edition: Volume 6, Acoustic Emission Testing.

Mirmiran, A., Shahawy, M. and EI. Echary, H. (1999). "Acoustic emission monitoring of hybrid FRP-concrete columns," *ASCE Journal of Engineering Mechanics*, 125(8), 899–905.

Morsy, R., Marzouk, H., Haddara, M. and Gu, X. (2017). "Multi-channel random decrement smart sensing system for concrete bridge girders damage location identification," *Engineering Structures*, 143, 469–476.

Nair, A. and Cai, C.S. (2010). "Acoustic emission monitoring of bridges: Review and case studies," *Engineering Structures*, 32, 1704–1714.

Ohtsu, M. (1986). "Determination of crack orientation by acoustic emission", *Materials Evaluation*, 45, 1070–1075.

Ouyang, C., Landis, E. and Shah, S.P. (1993). "Damage assessment in concrete using acoustic emission in nondestructive testing of concrete elements and structures," *ASCE SP*, 13–24.

Ozevin, D. (2011). "Geometry-based spatial acoustic source location for spaced structures," *International Journal of Structural Health Monitoring*, 10(5), 503–510.

Peterson, S.T., McLean, D.I., Symans, M.D., Pollock, D.G., Cofer, W.F., Emerson, R.N. and Fridley, K.J. (2001). "Application of dynamic system identification to timber beams, II," *Journal of Structural Engineering*, 127(4), 426–432.

Schultz, A.E. (2015). Protocols and Criteria for Acoustic Emission Monitoring of Fracture-Critical Steel Bridges. Minnesota DOT Technical Report MN/RC 2015-36.

Ultrasonic Related Papers

Abbasi, Z. and Ozevin, D. (2016). "Acoustoelastic coefficients in thick steel plates under normal and shear stresses," *Experimental Mechanics*, doi:10.1007/s11340-016-0186-6.

Allen, D.R. and Sayers, C.M. (1984). "The measurement of residual stresses in textured steel using an ultrasonic velocity combinations technique," *Ultrasonics*, 22, 179–188.

An, Y.K., Song, H. and Sohn, H. (2014). "Wireless ultrasonic wavefield imaging via laser for hidden damage detection inside a steel box girder bridge," *Smart Materials and Structures*, 23(9).

Can, O., Kabir, M. and Ozevin, D. "Periodic assembly of steel truss systems for efficient analysis and early detection of localized damage using impulse response method," *Journal of Structural Engineering*, 144(5), 2018.

Chaki, S. and Bourse, G. (2009). "Stress level measurement in prestressed steel strands using acoustoelastic effect," *Experimental Mechanics*, 49, 673–681.

Cotic, P., Jaglicic, Z., Niederleithinger, E., Effner, U., Kruschwitz, S., Trela, C. and Bosiljkov, V. (2013). "Effect of moisture on the reliability of void detection in brickwork masonry using radar, ultrasonic and complex resistivity tomography," *Materials and Structures*, 46, 1723–1735.

Fitzpatrick, M.E., Fry, A.T., Holdway, P., Kandil, F.A., Shakleton, J. and Souminen, L. (2005). *Determination of Residual Stresses by X-Ray Diffraction*. Measurement Good Practice Guide No. 52, National Physics Laboratory, UK.

Gokhale, S. (2007). *Determination of Applied Stresses in Rails using the Acoustoelastic Effect of Ultrasonic Waves*. MS Thesis, Texas A & M University.

Kato, M., Sato, T., Kameyama, K. and Ninoyu, H. (1995). "Estimation of the stress distribution in metals using nonlinear acoustoelasticity," *Journal of Acoustical Society of America*, 98, 1496–1504.

Kino, G.S., Hunter, J.B., Johnson, G.C., Selfridge, A.R., Barnett, D.M., Hermann, G. and Steele, C.R. (1978). "Acoustoelastic imaging of stress fields," *Journal of Applied Physics*, 50(4), 2607–2613.

Kobayashi, H., Arai, Y., Ohsawa, H., Nakamura, H. and Todoroki, A. (1992). "Nondestructive measurement of welding residual stresses by acoustoelastic technique and prediction of fatigue crack growth," *ASME Journal of Pressure Vessel Technology*, 114, 417–421.

Lanza di Scalea, F., Rizzo, P. and Seible, F. (2003). "Stress measurement and defect detection in steel strands by guided stress waves," *Journal of Materials in Civil Engineering*, 15(3), 219–227.

Leon-Salamanca, T. and Bray, D.E. (1996). "Residual stress measurement in steel plates and welds using critically refracted longitudinal (LCR) waves," *Research in Nondestructive Evaluation*, 7, 169–184.

Li, M., Anderson, N., Sneed, L. and Maerz, N. (2016). "Application of ultrasonic surface wave techniques for concrete bridge deck condition assessment," *Journal of Applied Geophysics*, 126, 148–157.

Liao, M., Okazaki, T., Ballarini, R., Schultz, A. and Galambos, T. (2011). "Nonlinear finite-element analysis of critical gusset plates in the I-35W bridge in Minnesota," *Journal of Structural Engineering*, 137(1), 59–68.

Lu, H., Liu, X.S., Yang, J.G., Zhang, S.P. and Fang, H.Y. (2008). "Ultrasonic stress evaluation on welded plates with Lcr wave," *Science and Technology of Welding and Joining*, 13, 70–75.

Martin, J., Broughton, K.J., Giannopolous, A., Hardy, M.S.A. and Forde, M.C. (2001). "Ultrasonic tomography of grouted duct post-tensioned reinforced concrete bridge beams," *NDT&E International*, 34(2), 107–113.

Prevey, P. (1996). Current applications of x-ray diffraction residual stress measurement. *Developments in Materials Characterization Technologies*, 103–110.

Ruud, C.O. (1982). "A review of selected non-destructive methods for residual stress measurement," *NDT International*, 15, 15–23.

Salamone, S., Bartoli, I., Srivastava, A., Philips, R., Nucera, C. and Scalea, F.L.D. (2009). "Nonlinear ultrasonic guided waves for stress monitoring in prestressing tendons for post-tensioned concrete structures," *ASNT NDE/NDT for Highways and Bridges*, August 16–20, New York.

Sansalone, M.J. and Streett, W.B. (1997). *Impact-Echo: Nondestructive Evaluation of Concrete and Masonry*. Bullbrier Press, Jersey Shore, PA.

Steinzig, M. and Ponslet, E. (2003). "Residual stress measurement using the hole-drilling method and laser speckle interferometry: Part 1," *Experimental Techniques*, 27(3), 43–46.

Szelazek, J. (2013). "Sets of piezoelectric probeheads for stress evaluation with subsurface waves," *Journal of Nondestructive Evaluation*, 32, 188–199.

Withers, P.J. and Bhadeshia, H.K.D.H. (2001). "Residual stress, part 1—Measurement techniques," *Journal of Materials Science and Technology*, 17, 355–365.

Zhang, Lu, Yalcinkaya, H. and Ozevin, D. (2017). "Numerical approach to absolute calibration of piezoelectric acoustic emission sensors using multiphysics Simulations," *Sensors and Actuators A: Physical*, 256(1), pp. 12–23.

Chapter 26

Building Information Modeling (BIM) for Structural Engineering

BY
NAWARI O. NAWARI, Ph.D., PE *School of Architecture, University of Florida*

26.1 INTRODUCTION

Application of technologies in engineering design and construction is enduring a great deal of advancement. A good example is Building Information Modeling (BIM). BIM has transformed the design and construction of buildings structures due to its capacity to postulate the interaction of various structural and architectural factors, such as forms, spatial orders, forces, deformation, and material properties, based on the nature of the site conditions. BIM is a comprehensive data management and analysis technology that is becoming increasingly indispensable for structural engineering practices. It is a simulation prototyping technology and a work flow that is being recognized as a revolutionary development currently reshaping the architecture, engineering, and construction (AEC) industry. BIM is one of the most promising advances in the AEC industries that is significantly affecting 21st-century engineering activities.

The technology element of BIM benefits industry stakeholders in visualizing and investigating what is to be fabricated or constructed in a simulated environment and identifies any potential design, construction, operational, or maintenance issues. The work flow component enables a collaboration environment and reassures integration across AEC disciplines. BIM is therefore efficiently altering the environment of structural engineering design by generating a database of building objects used for all facets of a project from conceptual design to construction and beyond. Many researchers have cited that the collaborative practice of BIM in the AEC industry results in completing projects with more resources, higher quality, less cost, and higher customer satisfaction (Nawari, 2010, 2011, 2014, 2015a, 2015b; Eastman et al., 2011). BIM involves new concepts and practices that are so greatly improved by innovative information technologies and business structures that they will radically reduce the multiple forms of waste and inefficiency in the building engineering and construction industry (NBIMS, 2007).

BIM indeed expands the engineering design matrix significantly, enabling collaboration among the several design disciplines and systems and recognizing clashes before construction commences. It delivers projects that integrate design and construction insights in an exceedingly collaborative and owner-friendly format. Incorporating constructability into the structural engineering design decision matrix has the potential to help the design team make intelligent decisions and positively impact overall project efficiency and quality.

26.2 BIM FUNDAMENTALS

26.2.1 Definitions and Concepts

BIM concerns with the formation and consumption of coordinated, consistent, computable data about designing, constructing, and managing a building project. This data can be utilized for many functions, such as design decision making, production of high-quality construction documents, predicting building performance, cost estimating and construction planning, as well as managing and operating the facility. BIM represents the hub of data and applications that streamline the delivery process of design, detailing, manufacturing, construction, and operation. It is valued as an integrator between various disciplines to perform multiple functions of the AEC industry and the ability of computational and simulation software tools to manipulate the BIM model directly, with or without human intervention. In a typical BIM-enabled design work flow, the data model serves as the principal means of communication between activities and professionals (Fig. 26.1).

26.2.2 Level of Developments

In a BIM environment, one creates a model, which is often referred to as a BIM model. That means making detailed datasets that define a building. This process should not be confused with the making of classical 3D drawings of a building.

Understanding the degree of information required for the elements in the model plays a significant role in ensuring its utilization for structural engineering purposes. Level of Development (LOD) is the degree to which the element's geometry and attached information have been identified as necessary and sufficient by project team members to perform the required tasks when using the BIM models. LOD is often confused with the Level of Details, which merely is how much detail is

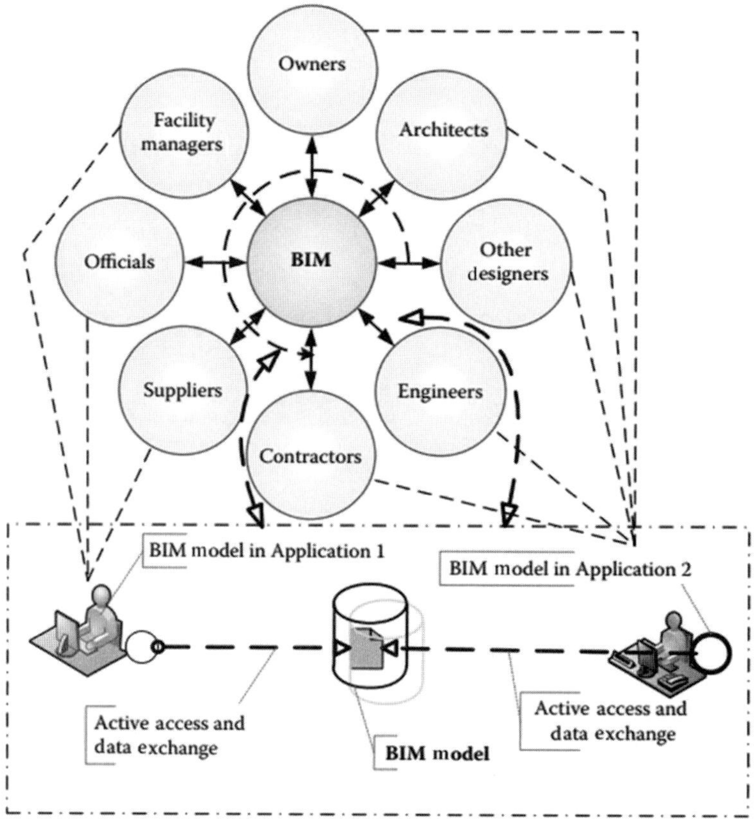

Figure 26.1 BIM concept and process (Nawari, 2015).

included in the model objects. Regarding terminology, Level of Detail can be thought of as an input to the element, while LOD defines reliable outcomes.

The American Institute of Architects (AIA) defines LOD in a BIM process as "minimum dimensional, spatial, quantitative, qualitative, and other data included in a Model element to support the authorized uses associated with such LOD" (AIA, 2013a). The first LOD was developed in 2008 by the AIA Document *E202™–2008 Building Information Modeling Protocol.* Because of the increased use of BIM in practice, the AIA evaluated the E202–2008, including the LOD definitions. The result is the updated and reconfigured Digital Practice documents *AIA E203™–2013, Building Information Modeling and Digital Data Exhibit, AIA G201™–2013, Project Digital Data Protocol Form, AIA G202™–2013, Project Building Information Modeling Protocol Form,* which are accompanied by a detailed guide document titled *Guide and Instructions to the AIA Digital Practice Documents* (AIA, 2013b).

Other efforts in LOD is represented by the work of BIMForum (2015). The BIMForum developed the LOD specification, which is defined as a reference that permits designers in the AEC Industry to specify and define with a high level of clarity the content and reliability of BIM at several phases in the design and construction process. This specification aims to assist in explaining the LOD framework and standardize its application so that the LODs become more implementable in collaborative BIM environment. The LOD Specification utilizes the basic LOD definitions developed by the AIA for the AIA G202-2013 Building Information Modeling Protocol Form and is organized by the Construction Specifications Institute, CSI Uniformat 2010. It describes and illustrates the characteristics of model objects of different building systems to varying LOD. This precise definition permits BIM model authors to define what their models can be relied upon for and allows downstream model consumers to recognize the suitability and the restrictions of models they are receiving. This specification intends to explain the LOD framework and standardizes its use so that it becomes more valuable as a communication tool. It does not recommend what LODs are to be reached at what phase in a project but leaves the specification of the model to the project team to decide. A brief description of these LOD is given in Table 26.1.

26.2.3 Theory of Modeling

Basics

A traditional CAD 3D model is just another representation of a building model with the same incompleteness as a plan or section. A full 3D model can be cut to reveal the basic outlines for sections and plans, but there are drawing conventions associated with these representations that cannot be captured this way. For instance, how will the swing line for a door be encoded into a 3D model? For a system to intelligently place a door swing into a plan but not into a 3D model, one needs a high-level description of the door building model elements separate from the geometric 3D description of their form. This digital data intelligence of storing and retrieving data sets is what describes the BIM environment.

BIM model encodes more than geometric forms; it encodes high-level design intent. Within the model, beams, columns, walls, and roofs are modeled not as a series of 3D solids but as beams, columns, walls, and roofs that have all the properties and characteristics of physical objects (see Fig. 26.2). That way, if a level changes height, all of the objects automatically adjust to the new criteria. If the wall moves, any roof that has a relationship to that wall adjusts automatically.

The object in a BIM environment have parameters—beams, columns, trusses, doors, windows, walls, ceilings, roofs, floors, even

Table 26.1 Definitions of LODs (Modified from BIMForum, 2015)

Level of development (LOD)	Description	Example
LOD 100	This level represents the conceptual design phase (schematic design). The model element may be graphically represented in the model with a symbol or other generic illustration but does not satisfy the requirements for LOD 200.	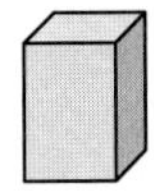 Steel column
LOD 200	The level generally represents the design criteria phase. The model element is graphically represented within the model as a generic system, object, or assembly with approximate quantities, size, shape, location, and orientation. Addition nongeometric information may also be devoted.	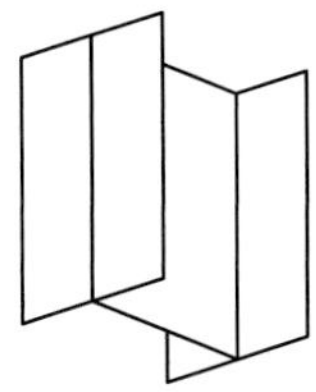 Steel column
LOD 300	Model object is graphically represented within the model as a specific system or assemblage regarding quantity, size, shape, location, and orientation. Non-graphic information may also be attached to the model element. Data can be extracted directly from the model without referring to non-modeled details such as notes or dimension callouts.	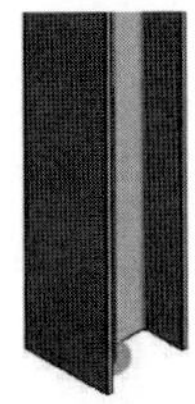 Steel column
LOD 350	The model element is graphically represented within the BIM model as a specific system, object, or assembly in terms of quantity, size, shape, location, orientation, and interfaces with other building systems. Additional non-graphic data may also be adhered to the model element.	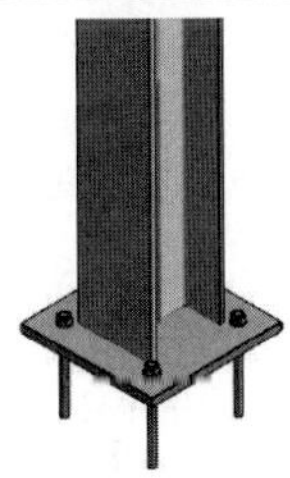 Steel column
LOD 400	The BIM element is graphically represented within the model as a specific system, object, or assembly in terms of size, shape, location, quantity, and orientation with detailing, fabrication, assembly, and installation information. Additionally, non-graphic information may also be attached to the model element. At this LOD, element is modeled at sufficient detail and accuracy for the fabrication of the represented component.	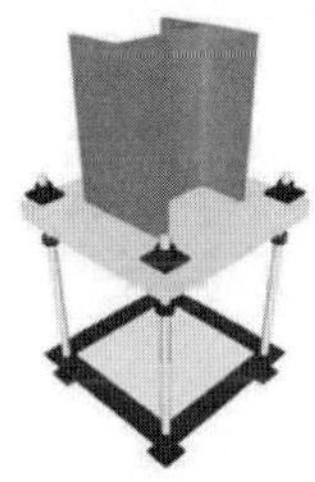 Steel column
LOD 500	Each BIM element is a field verified representation in terms of size, shape, location, quantity, and orientation. Any additional non-graphic information may also be attached to the element.	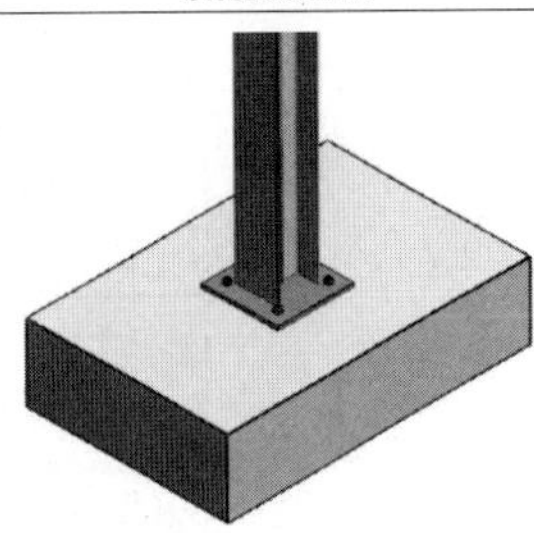 Steel column

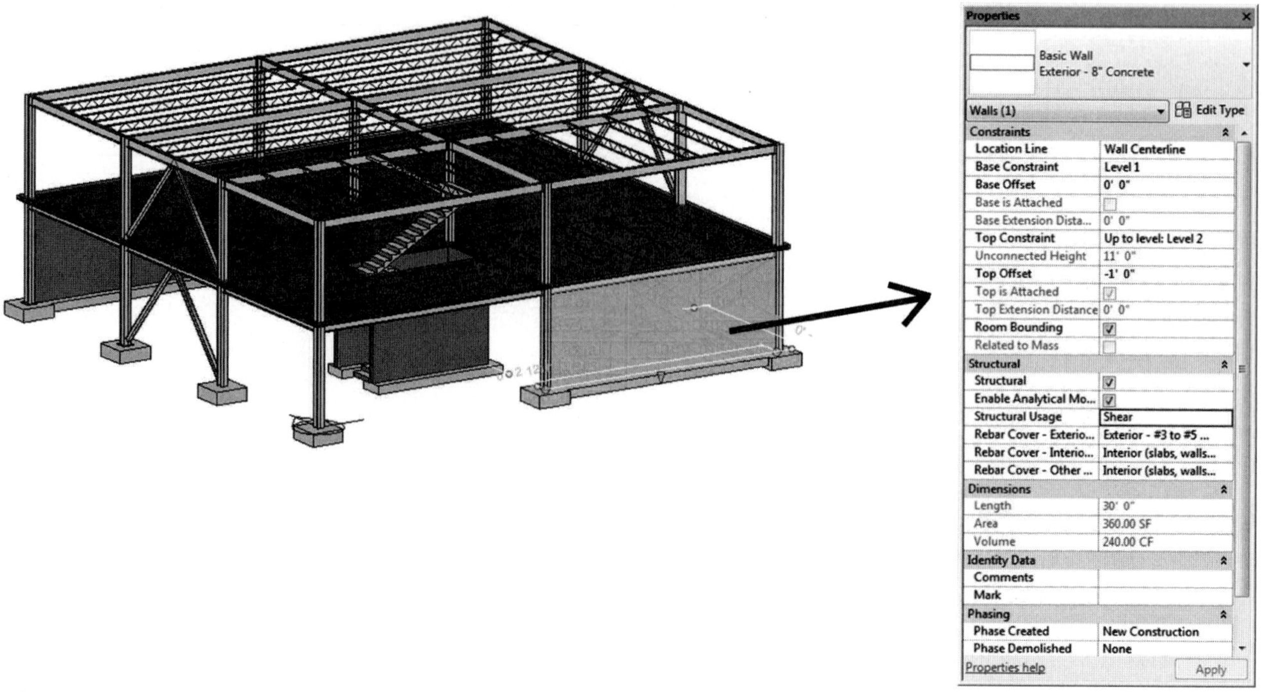

Figure 26.2 Reinforced-concrete shear wall properties (Nawari and Kuenstle, 2015b).

drawings themselves. The objects used in BIM design encode much more data than just pure geometry, such as the ability to have much more information about the objects themselves and their associated objects. Parameters are quantities that are fixed under a given set of conditions but may be different under other circumstances. Thus, some of these parameters have constant values, and some are modifiable. To allow flexibility in objects, they are created with a set of parameters that can take on arbitrary values. If you want to create a truss that is 5 m high, it is only necessary to modify the height parameter of an existing truss. Then all members of the truss will adjust according to the new height of the truss. In advanced modeling, one also can create custom object types with parameters of one's choosing.

BIM design proceeds by "placing" objects into a model and then adjusting their parameters. These objects are fully developed building elements like walls, beams, columns, trusses, footing, and so on. Although their parameters may vary, the placed objects retain their fundamental identities. A wall that one makes wider remains a wall no matter what. As the design process proceeds, more parameters and values will be added to the model.

In these processes, information in a BIM model is produced while details in the actual space are being established. Conceptual design information becomes concrete over some time as the design proceeds forward toward construction and operations. Data in a BIM process grow in a pyramid-like form, expanding continuously from the abstract to detailed information in a coherent manner to ensure efficiency and quality of the construction and operation process of a facility. The objects used in BIM design encode much more data than pure geometry.

The theoretical schema of BIM objects will be discussed in reference to Autodesk Revit application as an example of a BIM platform. Each object in Revit belongs to a hierarchy that helps organize the objects in the building model (Fig. 26.3). The terms used to describe

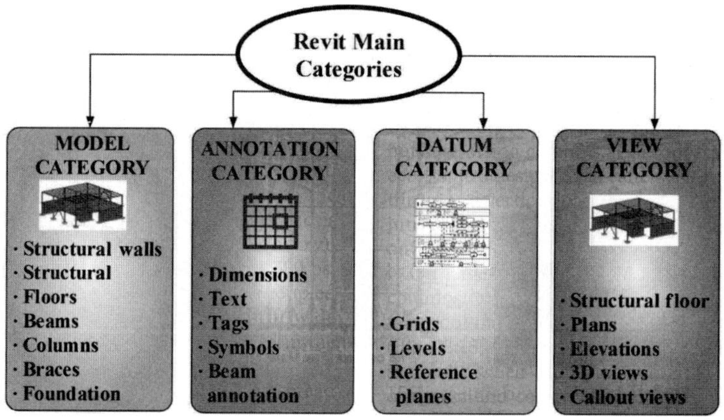

Figure 26.3 Main categories in Revit platform (Nawari and Kuenstle, 2015).

this hierarchical classification from broad to specific are categories, families, types, and instances. This is the fundamental organization of the building model database. Most of the aspects of the building model, including the views, have this organizational structure. This concept is important because each of the objects has parametric control at these different levels of organizations.

Model Making

One starts modeling by insertion objects. When one places an object into a model, one is making an instance of a family. Most families have multiple types. A type defines typically default values for family's parameters. A type of column has a particular geometry. In many occasions, types can be exchanged. For example, one column type can easily be swapped for another column once the object has been created. For example, a wide-flange steel column type can easily be exchanged for a column of a different family, such as a wood or concrete column. Revit handles all necessary graphical representations.

You can easily start with a specific concrete column and change its various properties later. You cannot, however, change one object type to a unrelated type because there is no internal mapping of properties and functionality for the platform to track. You cannot, for instance, make a wide-flange beam and then change it into a structural wall or column.

Structural engineers deal normally with the concept of types and instances but not always explicitly. For example, when they create columns schedule, they must list every instance grouped by its type (e.g., 18 × 24-ft reinforced-concrete column). Traditional CAD platforms have no notion of type versus instance; thus, it has no mechanism to assist with this issue. Conversely, BIM platforms have a full object-oriented database in which each object uses and retains the element category, family, type, and instance information that can be efficiently utilized to generate graphical column schedule based on the instance or type. Similar schedule should be generated for any other type or instance of building elements, such as doors, windows, and so on.

BIM platforms have the intelligence to realize the various relationships between building elements. For instance, when you move one end point of a wall, you would like both sides of the wall to move. You might also like objects embedded in the wall-like windows and doors to move with the wall as you edit it. That is exactly what BIM platform software does. However, in traditional CAD you may have drawn a wall as a long, skinny rectangle. In summary, traditional CAD drawings lack the intelligence and content that BIM platforms possess in encoding engineers' intention.

Regarding the work flow in a BIM design environment, one begins with specifying what type of object one is about to place into the model. Then the engineer identifies the necessary information about that structural object so that it can be established properly. What that information is depends in large part on what kind of object is being established. For a wall, one must at a minimum specify its height, start, and end points. The engineer also have to select ahead of time what sort of wall it is (structural, architectural, partition, exterior, etc.) and its type (masonry brick, concrete, wood stud, etc.); however, one can change these at any later time.

Thus, there is no ambiguity. Anytime you are creating a building model, you are actually specifying some parameter of an object that you are establishing into the model. A floor would be a floor, and a wall would be a wall from its inception. Notice that this methodology releases designers from unnecessary work. You don't have to draw four lines to create a wall in a plan view; picking two points for a straight wall or three points for a curved wall now is usually sufficient. You always have the opportunity to change parameter values of any object you placed into the model, such as the arc's center or diameter of a wall, later.

In summary, building models are made of design objects—columns, beams, walls, trusses, footing, braces, stairs, and so forth. Not only just physical building elements that are "objects" in Revit platform, but also anything with properties that can vary is a Revit object, and that includes visual objects such as views. A view has properties that specify what is visible. The object-oriented approach is a powerful concept that has made its way from information technology to building design in the form of BIM design process.

26.2.4 Software Authoring Tools

Currently, there are several software platforms that support BIM concepts and principles. These include Autodesk Revit, ArchiCAD, Bentley Building Design, and Tekla Structure. A brief description of these systems is given below.

Autodesk Revit

The Revit platform for BIM is a design and documentation system that supports the design, drawings, and schedules required for a building project (Autodesk, 2014). BIM delivers information about project design, scope, quantities, and phases when you need it. It is a software platform with multidimensional capabilities (nD) with tools to plan and track various stages in the building's life cycle, from concept to construction and later demolition. Thus, Revit provides a practical collaborative environment between different disciplines in the building construction industry.

The platform also offers cloud integration that helps users create and find the custom objects and components they need to make their BIM models complete. Revit also supports integrated model exchange management using neutral data forms such as IFC, gbXML, and DXF/DWG round-trip conversion solutions between different applications. It also offer extension engines that support structural and energy analysis.

ArchiCAD

ArchiCAD is a BIM software system offered by GraphiSoft Inc. ArchiCAD creates a 3D BIM and all the necessary documentation and visualization. Based on a profound knowledge of the architectural process, ArchiCAD's BIM simulates the way a real building is constructed. ArchiCAD's BIM tools cover from town planning to intricate details, from functional studies to complex designs (GraphiSoft Inc., 2014).

ArchiCAD enables one to combine architectural freedom with the BIM with a comprehensive set of tools that support the design process. It has direct modeling capabilities in the native BIM environment with its end-to-end BIM work flow using priority-based connections and intelligent building materials as well as improved analysis tools. The platform also offers cloud integration that helps users create and find the custom objects and components they need to make their BIM models complete.

ArchiCAD supports integrated model exchange management using neutral data forms such as IFC and DXF/DWG roundtrip conversion solutions between different applications. The platform offers also energy evaluation engine that supports multiple thermal blocks.

Bentley Building Design

Building Design is a BIM application offered by Bentley Systems Inc. The software provides architects and engineers with the tools to explore design options, to make better-informed design decisions, and to predict costs and performance. Design and construction documents are automatically coordinated, eliminating errors and omissions. It supports all phases of the architectural work flow, from conceptual design to construction documentation, and integrates design, visualization, drawing production, and reporting of quantities and costs (Bentley Inc., 2017). The software has a full range of advanced solids modeling tools, which allow the creation of virtually any feasible form.

Tekla Structures

Tekla Structures is a software platform for BIM. The software enables users to create and manage 3D structural models in concrete, wood, or steel, from concept to fabrication. The platform supports the automatic process of producing shop drawings along with the creation of CNC files for controlling fabrication machines. Tekla Structures is available in different configurations and localized environments to suit different segment- and culture-specific needs.

Tekla offers a free tool known as Tekla BIMsight, which is a software application for a BIM model-based construction project collaboration. It can import models from other BIM software applications using the IFC data format and other formats such as DWG and DGN. With this free tool, you can perform spatial coordination, such as clash detection or conflict checking to avoid design and constructability issues, and communicate with others in their construction project by sharing models and review notes. This enables project participants to identify and solve problems already in the design phase before construction.

26.2.5 BIM Standards

The national and international BIM standards establish standard definitions for building information exchanges to support critical business contexts using standard semantics and ontologies. These standards form the foundation for accurate and efficient communication and commerce that are needed by the AEC industry. Nationally, the buildingSMART alliance® is the North American organization that is leading the creation of tools and standards that allow projects to be built electronically before they are built physically using BIM. It is a part of the National Institute of Building Sciences. The latest version of BIM standard is the NBIMS-US™ V3. It is available from https://www.nationalbimstandard.org. The scope of the NBIMS-US™ V3 is built around the concept of ever-increasing levels of importance with International Standards at the center. NBIMS-US™ provides consensus-based standards through referencing existing international and national standards, documenting information exchanges, and delivering best business practices for the entire built environment.

Internationally, buildingSMART International is the organization authoring the open BIM standards. The purpose of the standard is to allow the sharing of information throughout the life cycle of any project, between all the participants, regardless of which software application they are using. The standard is referred to as Industry Foundation Classes (IFC). It provides machine-readable, useful quality data that should be available for use throughout the design, procurement, construction, maintenance, and operation phases.

According to the National BIM Standard (NBIM, 2007), a BIM is a digital representation of physical and functional characteristics of a facility and its related project/life cycle information and is intended to be a repository of information for the facility owner/operator to use and maintain throughout the lifetime of the structure. In other words, BIM provides a collaboration environment for diverse stakeholders at different stages of the life cycle of a project to add, extract, update, or change the information in the BIM process to support and reflect the roles of that stakeholder. Therefore, BIM is a universal digital representation of a facility founded on open standards for practical interoperability.

26.3 BIM PROCESSES IN PRACTICE

26.3.1 Traditional Work flow

The traditional structural design process begins with the architectural document set, be it paper or CAD based. The structural engineering team then interprets the architectural design to create an overall structural system and then develops a specialized analytical model, using different software applications for the multiple types of structural analysis required, such as gravity, seismic, and wind. In parallel, the structural engineer creates yet another analytical representation of the building in the construction document process, creating multiple drawings of the same information. This conventional work flow results in multiple models (including the drawing sets) that are not coordinated, requiring manual efforts to keep them in synchronization, which quite often are error prone. Figure 26.4 depicts this traditional structural design process.

The original design typically goes through a simplification process to preserve the structural meaning, such as loading component and joints and connections, for the related structural analyses. Once the model conversion process has been completed by engineers, structural engineer analysts will perform structural computations and postprocessing steps of the structural design. The structural engineer will communicate with architects and owners about any design conflicts found during the structural analysis processes. The results of the structural design solutions will need to be interpreted and conveyed to the architect or owners to make them aware of any adjustments or changes in case the results do not fit into their requirements (see Fig. 26.4).

26.3.2 Model-Based Work flow

In BIM process, rather than using a computer to assist producing a series of drawings that together describe a building, the computer is used to create a single, unified representation of the entire building so content comprehensive that all disciplines collaborate in developing it, and thus it can generate all necessary model views and construction documentation. The basic parameters from which the BIM software develops these models are not the same ones utilized in the traditional CAD (points, lines, curves). In contrast to CAD, BIM application models with virtual building components that hold attributed information about actual elements and systems are the main primitives of BIM models. Examples include trusses, columns, beams, walls, doors, windows, ceilings, and floors. BIM tools recognize the form and behavior of these

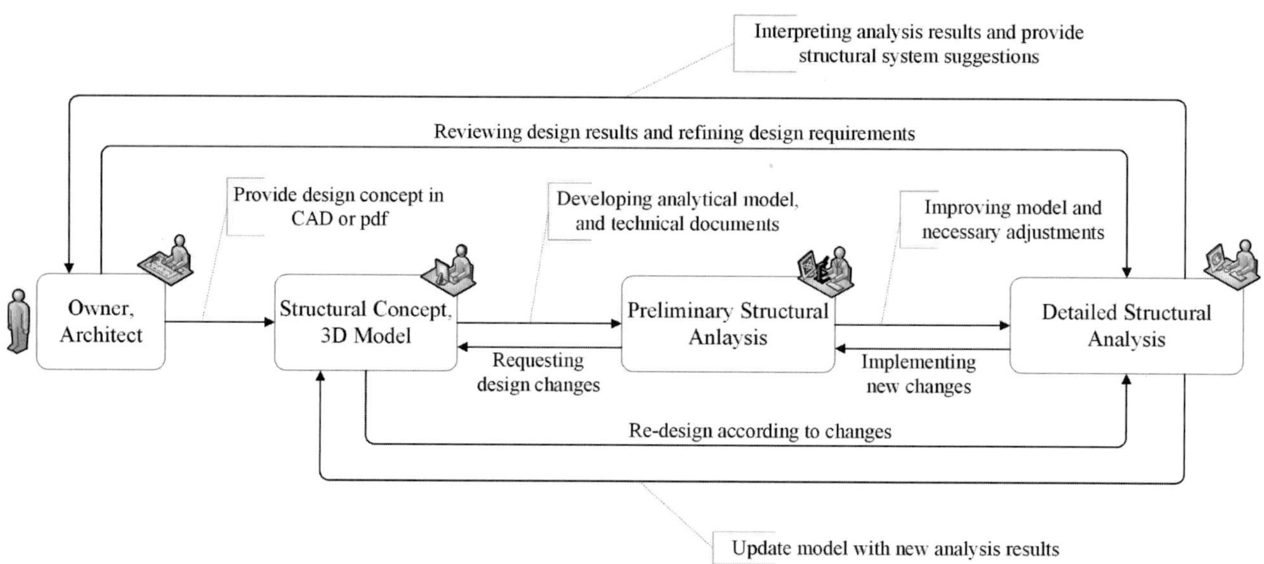

Figure 26.4 Traditional structural design work flow.

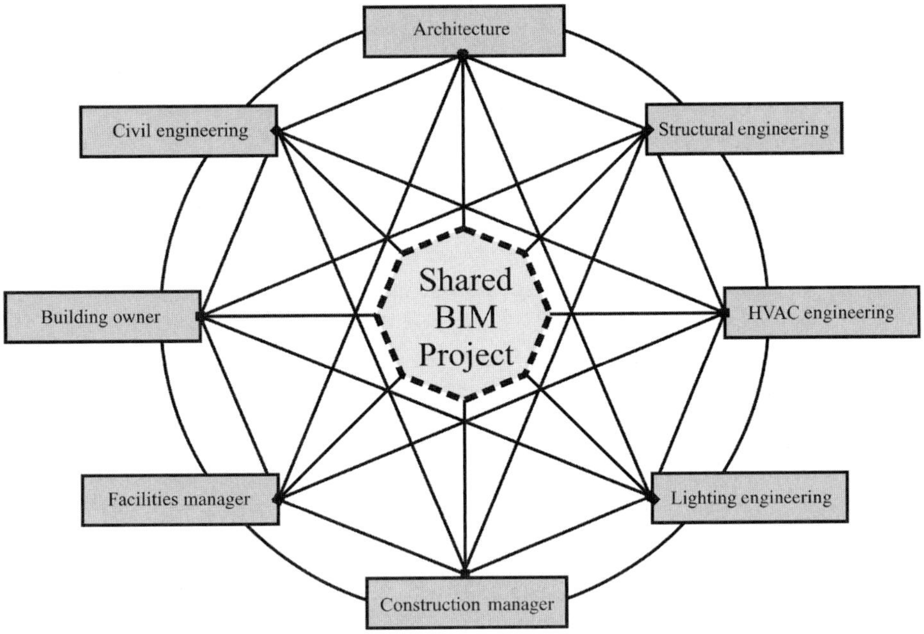

Figure 26.5 BIM dynamic work flow.

objects so that they can ease much of the complexity of their coordination. For example, walls join automatically, connecting structure layers to structure layers and finish layers to finish layers.

Many of the benefits are apparent—for instance, changes made in one view propagate automatically to every other elevation, section, callout, and rendering of the project. Other advantages include the ability to use the same model to interact with other applications, such as structural and energy analysis software (Nawari and Kuenstle, 2015).

In the full implementation of a BIM process across all of the AEC members of the project, structural engineers will start working on one BIM model delivered by the architect. Any changes prompted by the structural analysis will be reflected in the model that is shared between internal and other external designers: architects, mechanical, electrical, and plumbing engineers (Fig. 26.5). In such a work flow, engineers will work in a 3D data-driven design discourse that is shared or collaborated between multiple users at the same time and often across great distances. The work flow utilized the power of parameterized database, visualization and interactive modeling, communication and data exchange, and sustainability evaluations and code-checking algorithms for an efficient collaborative environment. Drawings in such work process become a by-product or report.

26.3.3 Interoperability and Productivity

Interoperability is the ability to use the same model across various structural engineering analysis software applications, such as SAP2000, ETABS, SAFS, RISA 3D, RAMS, Robot Structural Analysis, and so on. Currently, each software program has its own link to a specific BIM authoring tool to allow for bidirectional data exchanges. For example, SAP2000 has its link to Autodesk Revit that enables data to be transferred between the two applications without major data losses. Figure 26.6 shows how Computer and Structures Inc. (CSI) solved the interoperability between Autodesk Revit and SAP2000, ETABS, SAFS.

However, what the AEC industry seeks is the ability to combine several component models from different software packages into one comprehensive model which can then represent the complete structural product. This comprehensive model would then be exported to

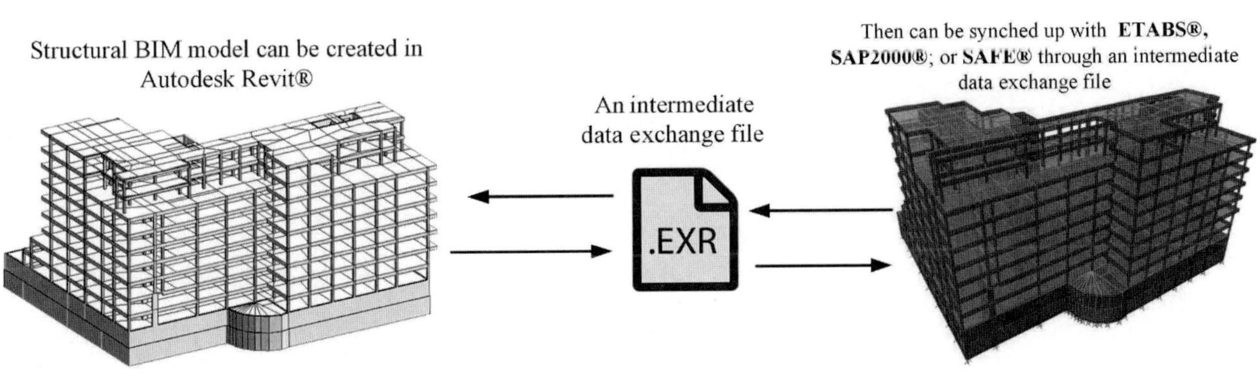

Figure 26.6 Interoperability between Autodesk Revit and SAP2000, ETABS, and SAFS.

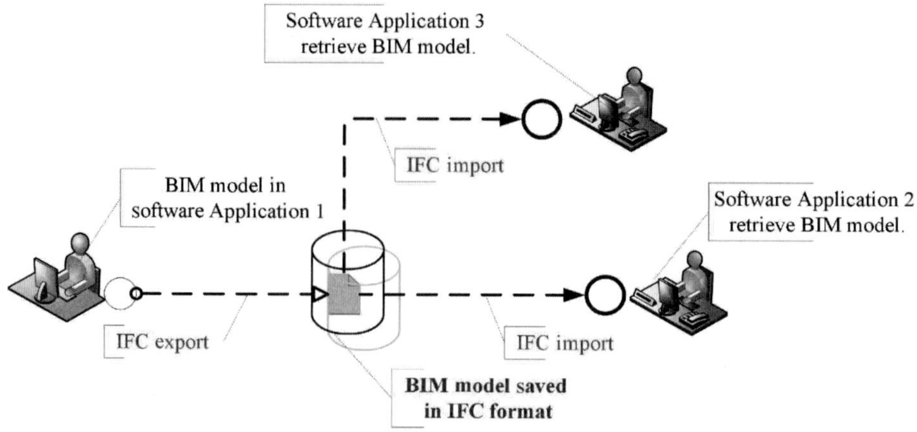

Figure 26.7 Interoperability using BIM neutral data format (IFC).

a neutral format (IFC) for use by the project team (see Fig. 26.7). The industry would necessitate the capacity to do this without doubling or tripling their efforts in creating multiple models and wasting time and resources looping through different software applications.

Unfortunately, most of the structural engineering software packages currently don't support the standard BIM neutral data format, which is known as the IFC. Ideally, engineers should be able to use only IFC to exchange data between different software applications. Figure 26.7 illustrates the concept of interoperability in structural engineering design practice.

26.4 STRUCTURE AND ARCHITECTURE SYNERGY (SAS) FRAMEWORK

The Structure and Architecture Synergy (SAS) framework uses BIM as a platform to facilitate the interplay between architecture and structures and emphasizes a structural design process that is highly creative and inspirational in nature (Nawari, 2015a, 2015b). In this framework, the form of the structure is constrained not only by its function, the site, and the designer vision but also by how it will work as a whole and by the need to provide a rational argument and calculations to justify expectations before the structure is being built. In this framework, nature is often utilized as an inspiration for discoveries and the model of learning from natural shapes and processes to design innovative structural systems.

The SAS framework aims for developing structural knowledge associated with the relationship to architecture. In other words, the interdependence of architectural form and structure is accentuated for optimum design outcomes.

The SAS framework is founded on three main concepts, namely, the concepts of "Structural Melody," "Structural Poetry," and finally "Structural Analysis." These are the underlining principles of the SAS framework, along with BIM as the framework enabler (Fig. 26.8). Structural melody explores how linear, nonlinear, planar, and volumetric structural elements can be orchestrated to create spatial order in architecture using BIM tools (Fig. 26.9). Structural poetry is a creative exercise to provide structural systems using structural vocabulary and melodies to organize and stabilize architectural spaces. In a more general sense, structural poetry signifies structural creativity and spatial thinking to develop a complete structural solution. It resembles the natural growth process of living objects. For example, biological patterns are hierarchical structures, made of materials with elusive properties that are capable of change in response to variations in local conditions. These systems are self-assembled, using small primary units (cells) to make various spatial growth expressions (Nawari and Kuenstle, 2015b). Structural poetry is thus an art that is an essential part of building design, which embellishes with engineering knowledge.

After concluding structural melody and poetry phases, BIM models are subjected to structural analysis. Various analysis tools within BIM platforms can be utilized for this purpose. The BIM tool used in the

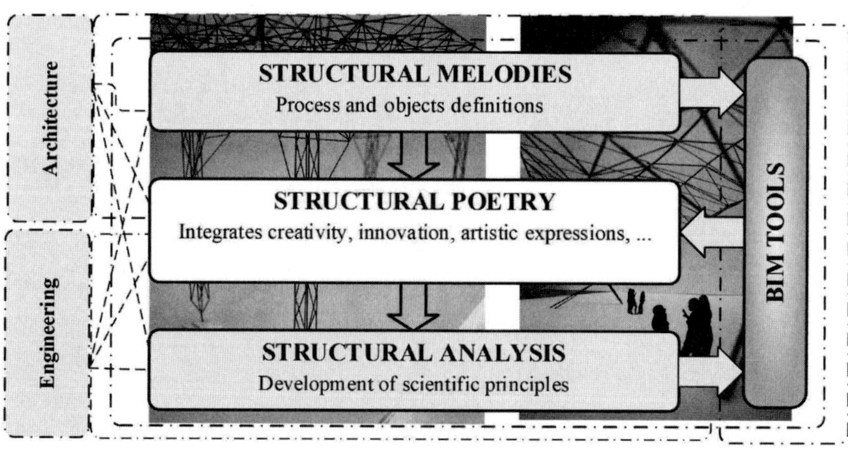

Figure 26.8 Structure and Architecture Synergy (SAS) framework (Nawari and Kuenstel, 2015b).

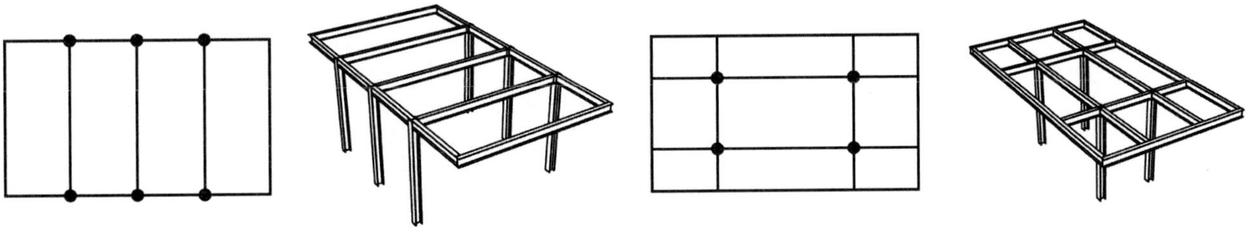

(i) Simple concentrated

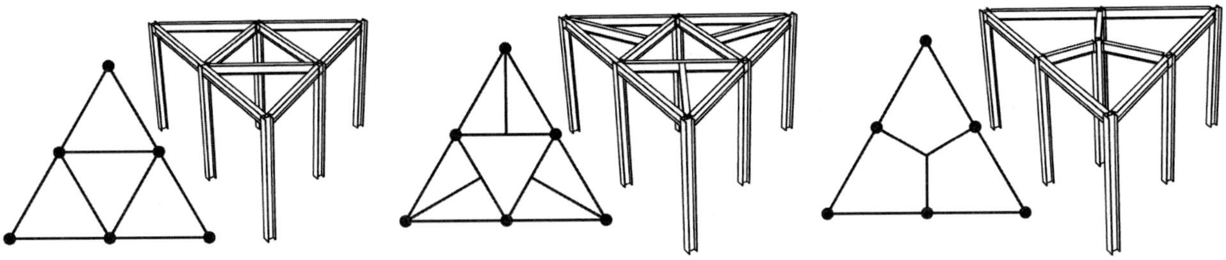

(ii) Simple triangular

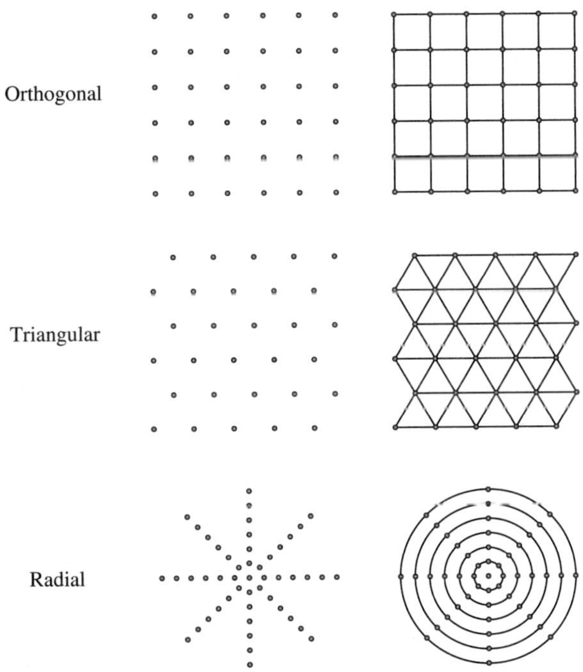

(iii) Uniformly distributed

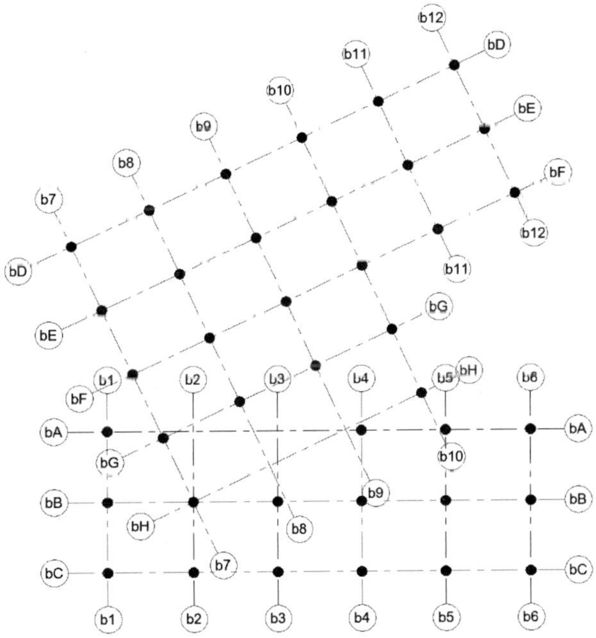

(iv) Transitional

Figure 26.9 Structural melody: linear vertical support patterns.

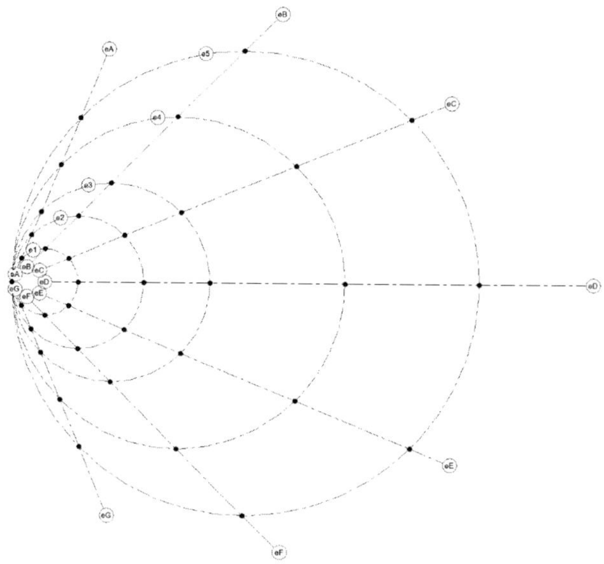

(v) Radial—nonuniform

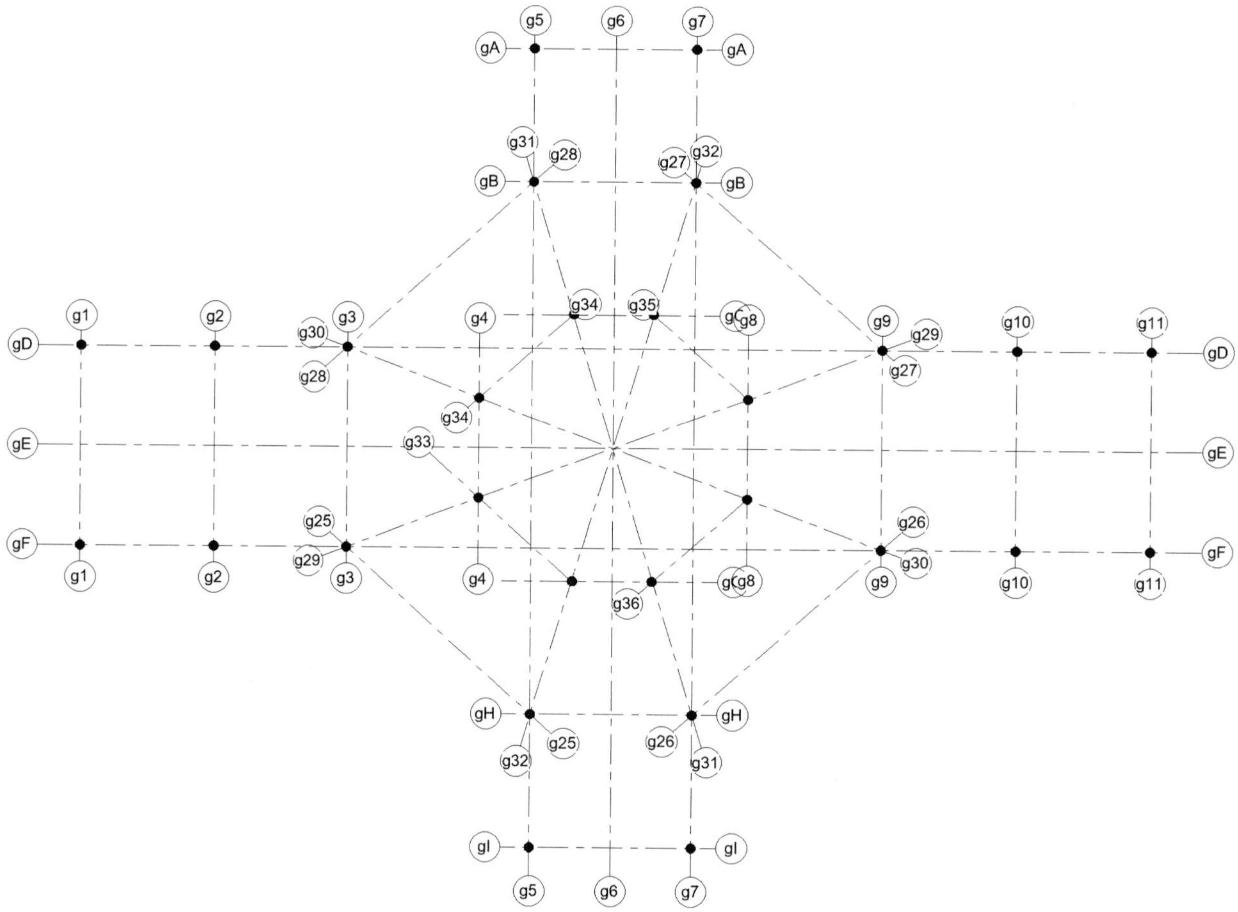

(vi) Composite

Figure 26.9 *(Continued)*

examples presented in this chapter is the direct link with Robot Structural Analysis software.

26.4.1 Structural Melody

To advance other types of structural knowledge, structural melodies and poetry focus on how to engage the engineer's imagination and to use it no less creatively than a musician or artist creating artistic expressions. Structural melody is a vehicle to introduce structural vocabulary, the hierarchy of structural members and the interplay between architectural concepts and structural systems, such as stratification, transition, hierarchy, and the heart of spaces. Structural melody explores how linear, nonlinear, planar, and volumetric structural elements can be orchestrated to create spatial order in architecture using BIM tools. Structural melody also embraces the consideration of numerous support configurations derived from basic linear, planar, and nonlinear elements and how these patterns respond to the functional organization of the buildings. Quite often, there is a correspondence between structural form and spatial composition: either the pattern of supports and spanning systems can prescribe the disposition of spaces within a building, or the spatial layout can suggest a specific type of structural pattern. For instance, columns and walls can be utilized to separate and reinforce spaces to allow for different activities. Figure 26.9 illustrates some examples of structural support patterns introduced and their spatial characteristics.

Man lives in nature utilizing all the natural resources for survival and creating better living conditions, sometimes "extensively modifying the ecological and climatic systems of the world." All forms of nature have "architecture" in them. The wonder of nature is that it has a great "capacity to adapt and change significantly to generate new forms, structures and properties from existing ones" (Weinstock, 2010). Equally important is the description of a close relationship between form and behavior. The form of an organism, building, or even a city changes within the variations in behavior of the environment or changes happening inside it. Thus, the structural melody can be derived from various sources in nature (Fig. 26.10).

The structural melody is often derived from bionic patterns. "Bionic" derives from the Greek word meaning "element of life." It is used to describe a linkage between certain technologies and biological systems

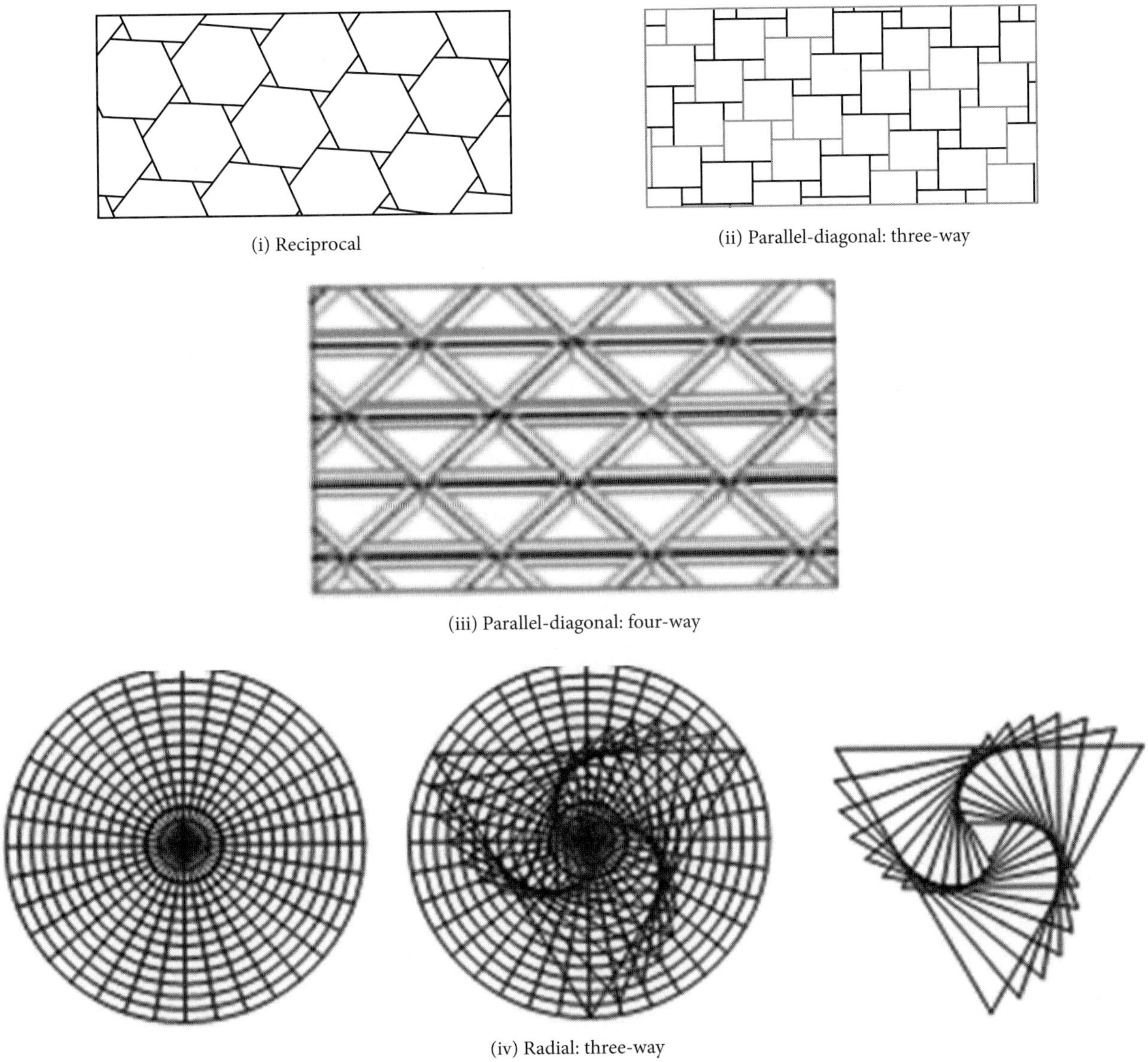

Figure 26.10 Structural melody: surface support patterns.

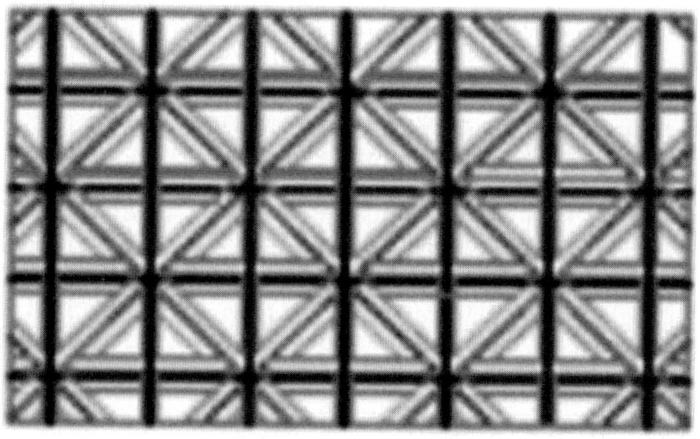

(v) Composite: Radial plus Spiral

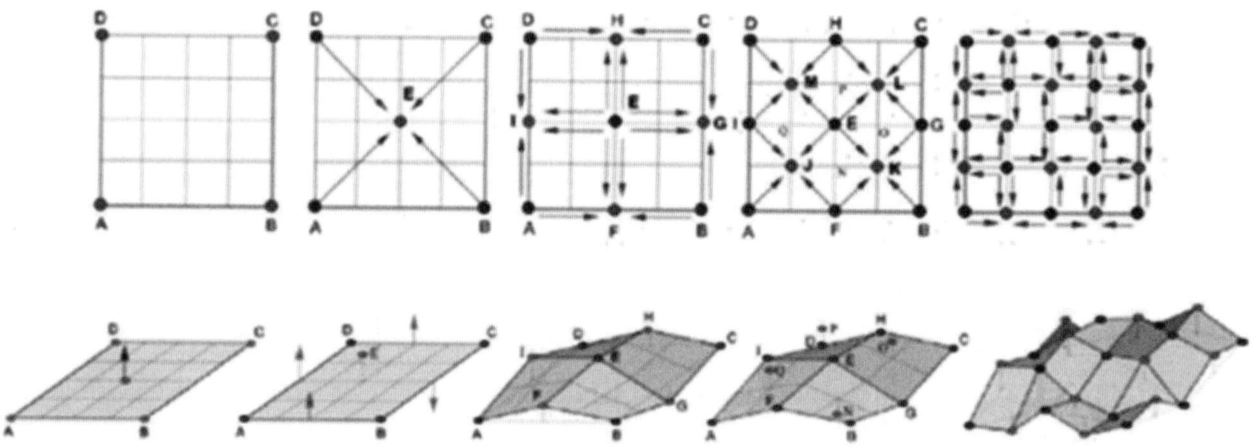

(vi) Spiral

Figure 26.10 (*Continued*)

(biology and technology). The idea behind it is that nature and biology are great sources for inspiration in engineering and technology (Goujon, 2001). Structural bionics is developed not only to solve the functional architectural issues but also to help understand the synthesis of function and aesthetic forms of architecture and inspire generating a naturally comfortable living built environment.

Inside of a congregation of individuals, for instance, the connections of the aggregate are more noteworthy than the individuals. When looking at the soap bubbles behavior, it can be found that the point where three or more bubbles join creates a solitary line. The surface strain is the same in each of the three bubbles, and the three edges between them equal 120°. This is likewise why the cells of an apiary have the same 120° points that shape hexagon tubes. Simple soap bubbles demonstrate a complex scientific issue of insignificant surface (Fig. 26.11). Two consolidated bubbles provide the optimum way of enclosing two given volumes of air of different size with the least surface area. This has been coined in the double bubble theorem. After years of research, Lord Kelvin (1887) proved how space could be divided into cells that will be equal in volume and with the least area of surface between cells, using a truncated octahedron shape (from soap bubbles foam). It was later called "mecon" or "Kelvin foam" (Kelvin et al., 1962), and Buckminster Fuller used it as an inspiration for his geodesic domes.

When examining the crystallography and irregular polyhedrons, there is only one shape which allows building a stable, fully integrated 3D lattice—tetrakaidecahedron (or it is sometimes called a tetradecahedron). This shape (besides cubes) out of all of Plato's and Archimedean solids is capable of closely packing space. It is a 14-faced Archimedean solid with six square and eight hexagon faces, also known as Kelvin foam model. One example of a tetradecahedron is the truncated octahedron. It consists of eight hexagonal and six square faces. Several of these figures are easy to connect in space because of their tilt angles and intersections (Steinhaus, 1999). If we analyze a truncated octahedron as a separate shape, we will find that it is "rounder" than the cube but "squarer" than the sphere. It withstands pressure (both outside and inside) better than the cube but worse than the sphere. Putting together a lot of spheres of the same size (balloons) as a bunch of grapes and put them under smooth and constant pressure (plunging under water), one will see that two balls represent areas of low pressure (in the form of concave spherical triangular pyramids). If the pressure increases further, the balls will be the most stable form, becoming a truncated octahedron cluster. A truncated octahedron cluster is also a generalized form of human fat cells as well as many other basic cellular structures (Papanek, 1985).

26.4.2 Structural Poetry

Structural poetry is the second part of the SAS framework. It is instituted upon structural melody and aims to develop structural creativity and spatial thinking as well as enhancing the conceptual design abilities using BIM tools. In other words, structural poetry seeks to advance structural creativity and spatial reasoning as well as to enrich the conceptual design abilities. This permits engineers and architects to develop an imaginative, complex structural system without a thorough understanding of its components at the initial design stages. In contrast to the traditional emphasis on first understanding beams, columns, bearing walls, and so on two-dimensionally, using the laws of statics and strength of materials, structural poetry stresses the utilization of the capabilities of BIM to create 3D structural forms to satisfy spatial, aesthetic, and other programmatic needs (Nawari and Kuenstle, 2015).

Figure 26.12 depicts an example of a basic structural model. This basic unit can expand horizontally and vertically to accomplish the chosen programmatic objectives using BIM tools. This process does resemble the natural growth of living things. For instance, biological forms are hierarchical structures, made of materials with elusive properties that are capable of change in response to variations in local conditions. These systems are self-assembled, using small primary units (cells) to make whole structures. Moreover, quite often, growth in these multicellular organisms is not only about the volume escalation of a single cell but also about the multiplication and rearrangement of cells. Figure 26.13 exemplifies the manifestation of this growth concept using the basic unit shown in Fig. 26.12.

Using the principle of structural melody, an initial design concept is represented by the "Kelvin cell" form utilizing its perfect space-filling characteristics and ability to use only one single shape to fill the space. The cell structure is shown in Fig. 26.14.

A truncated octahedron shape is the starting form for the design of a single module. This module exemplifies a two-story single residential housing unit having the following parameters: height—6 m, living space—79 square meters, rooms—living room, kitchen, toilet, master bedroom, bedroom, bathroom, and corridor (Fig. 26.14).

The complex geometry and inclined walls of the house were the main criteria that defined the building size. In a truncated octahedron, the dihedral angle between two hexagons equals 109, 47°, and the dihedral angle between hexagon and square equals 125, 26°. After modeling the truncated octahedron shape in Revit, the optimal scale for the single unit was found. The edge length is 2.75 m. Taking into consideration the angles between the walls that define the building shape, the total height of the building became almost 5.8 m.

Figure 26.11 Consolidated soap bubbles forming a truncated octahedron.

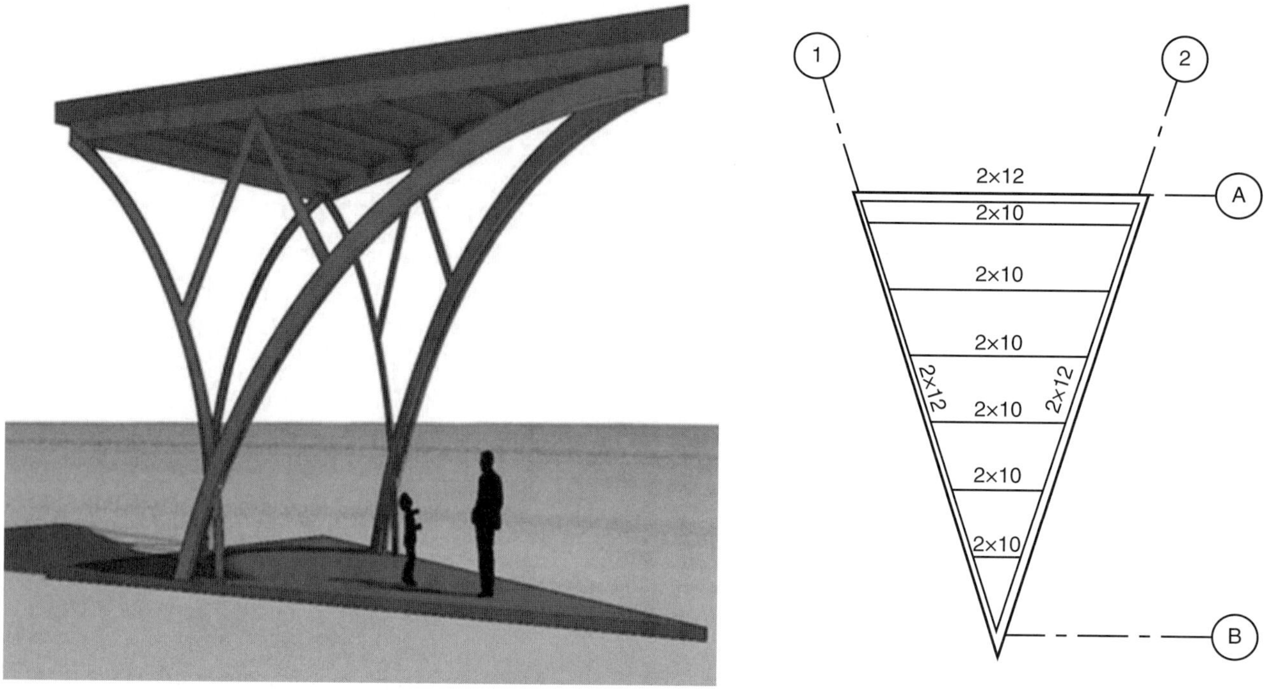

Figure 26.12 Basic units: an example of structural poetry.

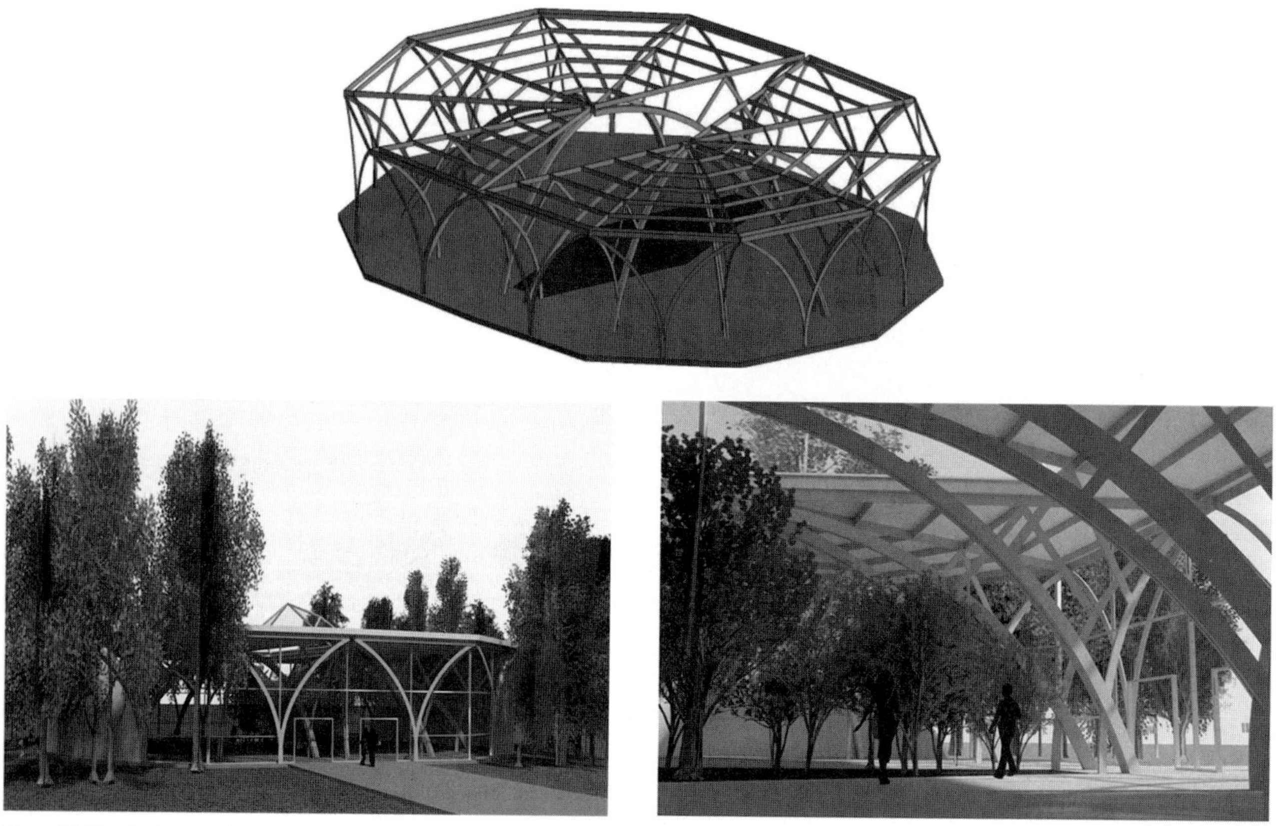

Figure 26.13 Application of structural poetry.

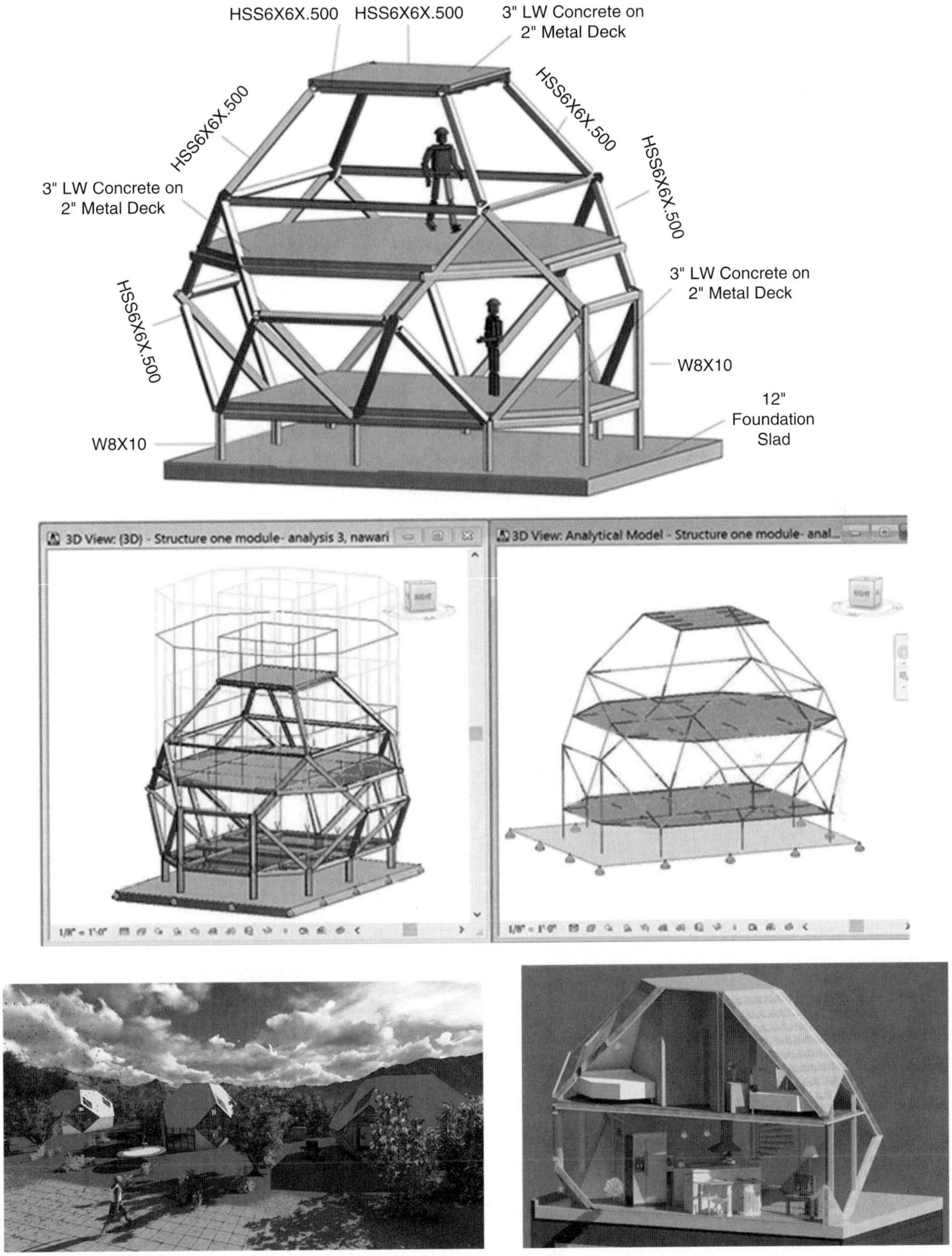

Figure 26.14 Development of a truncated octahedron unit.

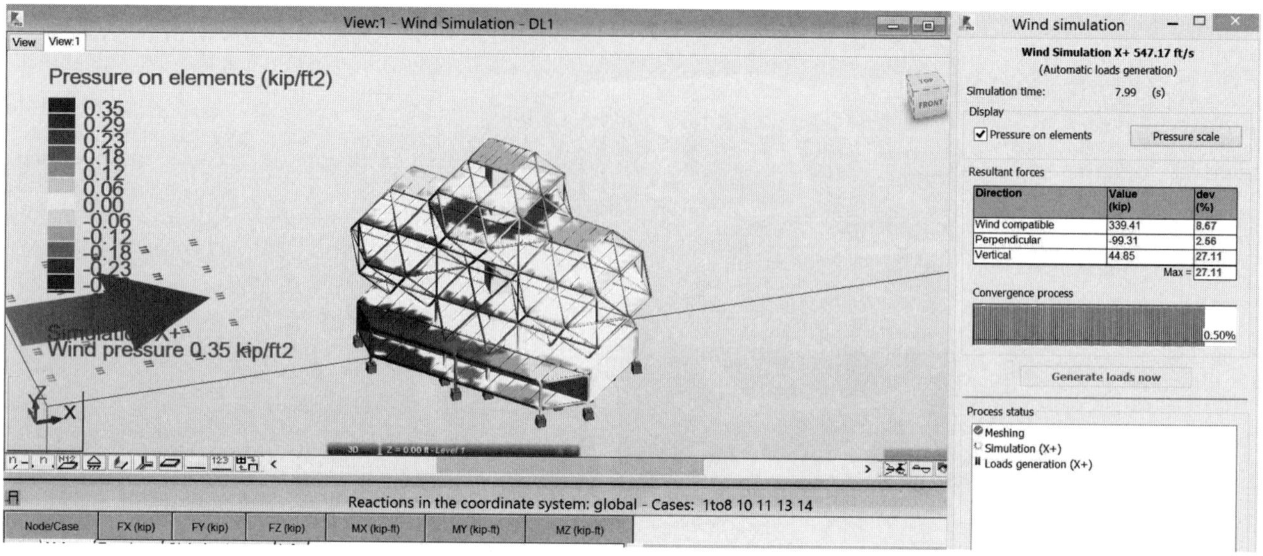

Figure 26.15 Steel design results for the community living center.

The initial purpose of the program is to create an affordable home for young just-married families. All the utilities will be efficiently accommodated behind the kitchen next to the toilet area. The structural form of the house provides the possibility of easily expanding laterally and vertically by combining different modules using the principal of structural poetry, to increase the habitable space, in case of a need to create more living units.

26.4.3 Structural Analysis

In this section, the structural design and analysis of the single residential unit (one module) and a combined design of three modules (residential community house) are demonstrated. The initial structural design of the single module residential house is considered.

Single Module

The structural system for the single module is depicted in Fig. 26.14*a*. It has a 30-cm-thick mat foundation and 8 cm of concrete on a 5-cm metal deck floor supported by a hollow structural section (HSS) 6 × 6 × 0.5 beam system. As for the primary lateral resisting system, it consists of HSS 6 × 6 × 0.5 beams that are connected at the corners of the truncated octahedron shape. The connection type for beams, girders, and columns is considered moment connection. The structural columns used are W-wide flange columns W 8 × 10.

The site location of these residential modules is in north Florida (Gainesville). Wind load is computed using the LRFD ASCE 7-10 code. The wind speed in the Gainesville area, assuming risk category II, is 200 km/h. The floor live load is 1.915 kPa and the floor dead load is estimated at 1.44 kPa. For the roof, the live load is 1.0 kPa, and the dead load is 0.5 kPa. All the analysis and design, from conceptual to the final renderings, were made in Revit, and the analysis was made in Robot Structural Analysis 2016.

Results

The floor support consists of one-way floor joist system using HHS 6 × 6 × 0.5 beams. Wall frames are used for the lateral stability of the building. For the wall frame secondary bracing, the channel section C16 × 13 is used.

This analysis software provides a 3D bidirectional link to Revit. As a result, architects and engineers can use these integrated BIM tools to design buildings of various complexity levels (Fig. 26.15). In Robot Structural Analysis software, the results may be viewed for individual members or parts of the structure in the forms of diagrams, tables, and maps (Figs. 26.1 and 26.14).

For the combined module for the community living center, the first two floors have a polygon shape, and the upper floors have truncated octahedron forms. The elevator shaft in the middle of the structure is considered to work as a part for the lateral resisting system. Examples of the analysis and design results are depicted in Figs. 26.14*a* and 26.15.

26.5 CONCLUSIONS

BIM involves new concepts and practices that are so greatly improved by innovative information technologies and business structures that they will radically reduce the inefficiency in the structural engineering and construction industry. The chapter introduces the principles of BIM along with a BIM-based framework for structural engineering design. This framework is referred to as the SAS framework. It provides the foundations for innumerable uncommon structural forms which are efficient from the functional and practical standpoint and innovative in their aesthetic qualities. Learning from nature is not a new concept, but it should be incorporated more into the routine architectural and structural design process.

In this chapter, bionic patterns are used as inspiration to develop a new efficient structural system using the SAS framework. Furthermore, the chapter demonstrates the design and analysis of the biological truncated octahedron shape for an initial building module. The feasibility of more complex bionic patterns to organize and determine the form of internal and external spaces is investigated utilizing the principals of the SAS framework. The results of this chapter aim to offer the architecture and engineering industry innovative BIM frameworks that can be readily employed in designing any structures.

REFERENCES

AIA (2013a). The *AIA E203™–2013: "Building Information Modeling and Digital Data Exhibit."* American Institute of Architects (AIA).

AIA (2013b). The AIA's G202-2013 Project BIM Protocol. American Institute of Architects (AIA).

BIMForum (2015). The LOD Specification Part I *and* Part II. BIMForum.

Goujon, P. (2001). *From Biotechnology to Genomes: The Meaning of the Double Helix.* World Scientific Publishing Co.

Kelvin, W. T. and Tait, P. G. (1962). *Principles of Mechanics and Dynamics*. Dover.

Kelvin, Lord (Sir William Thomson) (1887). "On the Division of Space with Minimum Partitional Area," *Philosophical Magazine* 24 (151): 503.

Nawari, O. N. (2015a). "BIM as a Framework for Structural Design." In *Building Information Modeling: Applications and Practices in the AEC Industry*, Raja, R. A. Issa and Svetlana, Olbina (Editors). American Society of Civil Engineers (ASCE), ISBN: 978078447913.

Nawari, O. N. and Kuenstle, M. (2015b). *Building Information Modeling (BIM): A Framework for Structural Design*. CRC Press, Taylor and Francis Group.

Papanek, V. (1985). *Design for the Real World.* Thames and Hudson Ltd.

Rian, I., MD. and Asayama, S. (2016). "Computational Design of a Nature-inspired Architectural Structure Using the Concepts of Self-similar and Random Fractals." *Automation in Construction* 66 (2016): 43–58.

Sieden, L. S. (2000). *Buckminster Fuller's Universe*. Basic Books.

Steinhaus, H. (1993). *Mathematical Snapshots*. Dover.

Weinstock, M. (2010). *The Architecture of Emergence: The Evolution of Form in Nature and Civilization*. John Wiley & Sons.

Chapter 27
Structural Fire Engineering

BY
V. K. R. KODUR, Ph.D., P.Eng. *University Distinguished Professor and Chairperson, Department of Civil and Environmental Engineering, Michigan State University, East Lansing, Michigan*
M. Z. NASER, Ph.D., PE *Assistant Professor, Glenn Department of Civil Engineering, Clemson University, Clemson, South Carolina*

27.1 GENERAL

This chapter presents an overview on behavior of structures under fire conditions. The need for provision of fire safety features in structural members in buildings is explained. The fire development process in a typical compartment, together with basics of heat transfer mechanisms, is presented. Effects of temperature on properties of various construction materials and its influence on structural response are discussed. Further, fire resistance evaluation through prescriptive-based approaches and rational design methods is illustrated.

27.2 DESIGNING STRUCTURES FOR FIRE SAFETY

27.2.1 Overview

Built infrastructure which include buildings, bridges, ports, tunnels, and so on are designed to function for several decades and serve a variety of purposes. Throughout their long service life, structures are often exposed to multiple loadings and various risks (hazards). One such common hazard that built infrastructure encounter during their service life is exposure to fire. In case of buildings, fires can threaten the integrity and stability of structural systems that can result in significant losses, including deaths and injuries. There have been numerous fire incidents in buildings, and these have been documented in the literature. Most fires are caused by accidental incidents (electric short circuit, cooking, etc.), natural disasters (e.g., post-earthquake ignition etc.), and/or man-made events (vandalism, terrorist attacks). Recent fire incidents have shown that fires in buildings can attain extremely high temperatures (in the range of 800°C–1000°C) and may lead to partial or complete collapse of structures.

27.2.2 Need for Structural Fire Safety

Fire, if it grows to full size, represents one of the most extreme conditions a structure has to withstand during its lifetime. Fire safety has been a major economical and societal concern as conflagrations were common in major cities around the world. The most recent example of the large-scale destructive impact of fire is the collapse of the large number of buildings following the 9/11 attacks. The aftermath of such an event has demonstrated the destructive nature of fire and led to a rise in public awareness. As fire accidents are not easy to prevent, appropriate strategies are to be developed to mitigate adverse effects of fire hazard. Since life safety is a key consideration in the design of buildings, building codes and standards require appropriate fire safety measures to be provided in buildings to withstand adverse effects of fire. Fire safety measures aim to control fire growth and spread, as well as protecting occupants, first responders, the structure in which fire breaks out, and surrounding properties. One key component of such fire safety measures is to ensure the integrity and stability of structures under fire conditions so as to minimize the possibility of fire-induced damage or complete structural collapse. To this end, this chapter provides a brief introduction to the concept of structural fire engineering.

27.2.3 Effect of Temperature on Materials and Structures

Structural members, when exposed to fire, experience a significant rise in temperature, and this can adversely affect properties of construction materials (such as concrete, steel, timber, etc.) resulting from temperature-induced phase changes (i.e., physicochemical changes) at the micro- and macrostructure levels. These changes often lead to development of distinct material behavior that is not observable at conventional (ambient temperature) service conditions. In a general sense, significant rise in temperature often "softens" construction materials and is accompanied by severe (and rapid) degradation in strength and modulus properties. Rise in temperature due to fire can also lead to the development of material-specific phenomena, such as spalling in concrete, local buckling in steel, or charring in wood, which can lead to premature failure. As properties of all construction materials degrade with rise in temperature, the behavior (and performance) of structures exposed to fire is dependent, in part, on the properties of its constituent materials. In other words, when a fire breaks out in a building, the load-carrying capacity in fire-exposed structural members reduces with time, which can accelerate partial or complete failure.

27.3 FIRE GROWTH AND DEVELOPMENT

The process of fire development in a building can be illustrated in terms of fire growth in an enclosure (i.e., compartment) since ignition starts in one compartment and then spreads into various compartments in a building. Fire development in a compartment can be grouped under four distinct stages: ignition (also known as incipient), pre-flashover (growth period), post-flashover (fully developed fire), and the cooling (decay) stage. The overall process of fire development in a typical compartment is fully explained by Drysdale (2011) and is summarized below.

27.3.1 Fire Development in a Compartment

Compartments in buildings typically contain significant levels of fuel load comprised of furniture, furnishings, clothing, and lining materials on structural/nonstructural members (walls, floors, etc.). Much of this fuel load in buildings is comprised of cellulose-based fuels and organic materials derived from fabric, plastics, or wood. In case any of this fuel ignites (whether resulting from an accidental event or intended incident) and there are no measures in place to extinguish such localized fire, the burning fuel could grow from local burning to a full-size fire.

The development of fire within a typical compartment can be represented through time-temperature curve as shown in Fig. 27.1, where the horizontal axis represents duration of fire and the vertical axis represents the progression of temperature in the compartment. The fire growth in the compartment can be grouped under four stages: initial combustion (incipient), pre-flashover (growth), post-flashover (fully developed), and the decay (cooling) stage. Each of these stages has distinct characteristics that vary significantly in terms of peak temperature attained and duration of burning (or cooling). These characteristics are influenced by the extent of fuel and air (oxygen) present as well as confining boundaries of burning compartment. Despite grouping fire development into number of stages, the actual fire growth process is continuous as fire transitions smoothly from one stage to the next (Drysdale, 2011). The development of fire is a complex process which depends, not only on the amount of available fuel and ventilation but also on the interaction of fire with compartment boundaries (walls/floors), fire protection systems, and firefighting activities.

In the early incipient stage of fire, heating of fuel (initial combustion) starts locally around few items, leading to combustion and the breakout of fire. In this stage, burning of fuel is in a small area, and fire temperature is low, ranging from 50°C to 150°C (see Fig. 27.1). During this stage, fire breakout can be detected visually by residents or through smoke or heat detectors and could possibly be extinguished with ease by stopping ignition through handheld fire extinguishers. In case ignition spreads to other objects, these objects will start to burn quickly, and this causes fire to rapidly grow within the compartment, leading to a much larger fire with temperatures reaching to 150°C–400°C. At these temperatures heat detectors and sprinklers can be triggered to alert occupants to evacuate and to notify authorities to send firefighters. This stage is referred to as the growth period and is the time where evacuation of occupants needs to occur in conjunction with firefighters arriving to fight the fire. Active fire protection systems, such as sprinklers, heat detectors, and exit paths, should be effective and fully functional during the growth period.

During the growth stage, hot gases (and smoke) rise up in the compartment through convection effects and accumulate below the ceiling level. These trapped and hot gases facilitate further ignition of various objects in the compartment through radiative heat transfer. When the temperature from these gases reaches about 600°C, the whole compartment can ignite and start to burn due to radiant feedback to available unburnt fuel at floor level. This transition point from local burning to full compartment burning is referred to as flashover and represents the start of the burning (fully developed fire) stage. At this temperature level, windows can break, which let fresh air flow into the burning compartment. Thus, fire in the burning stage is mainly controlled by the amount of ventilation (oxygen) available for burning. Since automated fire protection measures, such as pipes used for sprinklers, may not survive extreme temperatures (600°C–800°C), their functionality cannot be guaranteed. Hence, the role of structural members, together with compartment boundaries, is critical at this stage since sectional temperatures in the structural members often exceed 500°C. Therefore, the main mode of controlling and fighting the fire at this stage is through professional firefighting or through adequate passive fire protection measures incorporated in the structural system. From this reason, fire-resistance properties in structural members are to be fully effective in this stage of fire for minimizing the spread of fire to other compartment/floors. Once available fuel and/or ventilation (oxygen) gets depleted, burning in the compartment slows down, and fire starts to decay with fire temperature in the room gradually decreasing. This stage is referred to as the decay or cooling stage, where temperatures in structural members continue to rise due to the lag effect and continued heat transfer from surrounding air to members. The active fire protection systems are best suited to control and possibly extinguish fire during incipient and growth periods, while passive systems are critical for controlling fire spread and minimizing structural collapse during the growth, burning, and decay stages of fire.

As a side note to the fire development process, if the fire in the compartment is left uncontrolled, either through the breakdown of compartment boundaries (walls, floors) resulting from failure of passive systems or lack of firefighting, then fire can spread to adjacent compartments, and/or other stories of the building or even neighboring buildings, depending on the layout and characteristics of the building. Fire spread to other compartments/floors can be through the spread of burning flames, or radiant heat and/or hot smoke traveling across multiple compartments/floors through open doors, windows, ceilings, and cavities or through flames penetrating through surrounding walls/floors. In such a scenario, windows can break, and flames and/or radiant heat can ignite cladding on the exterior of a building, leading to fire spreading to higher

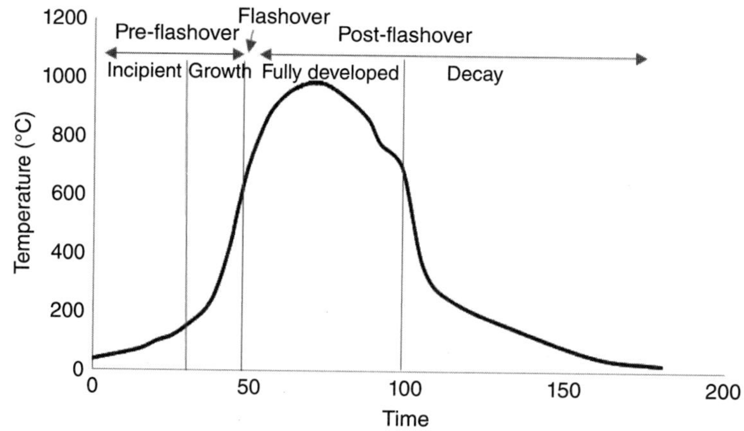

Figure 27.1 Time-temperature representation in a fully developed fire inside a typical room, with no suppression.

floors, as seen in a number of high-rise building fires in recent years, such as Grenfell building in London and Windsor tower in Madrid.

27.3.2 Time-Temperature Relations

The severity of fire exposure has significant influence on the performance of a structural member under fire conditions. Fire severity can be expressed through time-temperature relations which describe fire development starting from the initial ignition stage to the end of cooling stage. These time-temperature relations form the most important input parameter for fire resistance calculations of structures and can be grouped under two different categories, that is, standard and design fire.

Standard fire, representing a standardized fire exposure, is the most widely used fire scenario for comparative fire performance evaluation of building materials and structural members. This fire scenario was derived based on the assumption that fire growth occurs in a typical compartment mostly due to burning of cellulosic products (wood) and does not include any decay phase. Fire test standards specify standard time-temperature curves to be used for establishing fire resistance ratings for structural members. Building codes further reference applicable fire test standards with specifications for undertaking such fire tests. The most widely used fire standard in the United States is ASTM E119, and this standard is similar to other recognized fire test standards, such as ISO 834 (Europe as well as international). Fire standards specify a standard fire time-temperature curve through a number of discrete points (or continues functions). The fire curves adopted in some of the fire test standards are compared in Fig. 27.2.

In case of a fire resulting from petrochemicals and plastics, fire growth is much faster and consequently is much more rapid than that represented by the cellulosic standard fire curve. This became much more evident in late 1980s as a result of number of fires in offshore structures (i.e., petroleum platforms for oil production). Hence, a more appropriate temperature-time curve defined by standardized hydrocarbon fire curve was adopted. Some of the widely used hydrocarbon fire curve includes Eurocode 1 hydrocarbon fire (2002) and ASTM 1529 (1993). Figure 27.2 shows a comparison between hydrocarbon, external, and standard fires. A few observations can be noted through a comparison of these curves. In all of these fire curves, fire temperature increases throughout the duration of the fire exposure and do not decay, despite the fact that most fires decay after a certain length of time due to a lack of either fuel or ventilation or as a result of firefighting actions.

In recent years, there is a growing realization that above described standard fires do not represent actual (real) fire exposure scenario in buildings and other structures. One of the primary drawbacks in the standard fire curve is the absence of decay (cooling) phase as present in a real fire given that a decay phase is always present in any fire scenario simply due to burnout of fuel, limited ventilation, or firefighting activities. To evaluate reliable performance of structures under fire conditions, arriving at a realistic fire exposure scenario that accounts for fuel load, compartment features, and ventilation level is necessary. Simplified relations for calculating temperature-time curves based on fuel load and ventilation characteristics are available in fire design standards. Depending on the complexity, these relations are grouped under hand calculations, published curves, and parametric fire equations.

27.3.3 Heat Transfer Mechanisms

Heat transfer is the exchange of thermal energy between objects. During fire exposure, the extent of thermal energy transferred to structural members is governed by thermal properties of materials, boundary conditions between objects, and most importantly the mechanism(s) of heat transfer. The three fundamental modes of heat transfer are conduction, convection, and radiation. The first mode of heat transfer, conduction, is the primary mechanism for heat transfer in solid materials and is considered to be the main influencing factor in ignition of solids. Solids are often classified based on their ability to transfer heat. Materials that conduct ions, protons, or electrons are capable of transferring heat. A solid that is a good conductor (i.e., metal) allows temperature to transfer easily within its microstructure.

Convection is the second mechanism of heat transfer and is the most important heat transfer mechanism in fluids (liquids and gases). Heat transfer through convection in building fires lead, to flame spread and vertical transport of smoke and hot gases to the ceiling or outside of a window from a room fire. Radiation is another mode of heat transfer wherein the transfer of thermal energy occurs by electromagnetic waves traveling though a vacuum or transparent medium. Radiative heat transfer is the most dominant mechanism in case of building fires. Radiation also contributes to rapid fire growth and fire spread by transferring thermal energy between flames to fuel surfaces, hot smoke to building objects, and one burning building to an adjacent building (Drysdale, 2011).

27.4 PROPERTIES OF CONSTRUCTION MATERIALS AT ELEVATED TEMPERATURES

Construction materials, when exposed to elevated temperatures, as encountered during a fire incident, undergo number of physio chemical processes that essentially change its composition, microstructure,

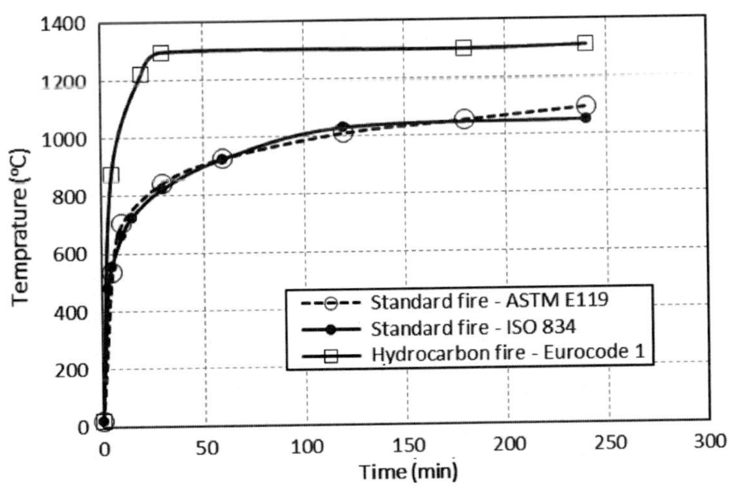

Figure 27.2 Comparison of standard, hydrocarbon, and external fire curves in different test standards.

and thermal and mechanical properties. This change in properties affects the behavior of materials and by extension, also the behavior of structural members. The variation of properties at elevated temperature in some of the construction materials is discussed herein. Refer to Kodur and Harmathy (2016) for detailed discussion of properties of construction materials at elevated temperatures. Specific temperature-dependent relations for variation of properties in steel, concrete, masonry, and timber can be found in different Eurocodes or ASCE Manual (1992).

27.4.1 Properties of Interest for Fire Design

The behavior (and performance) of a structural member exposed to fire is dependent, in part, on the thermal, mechanical, and deformational properties of its constituent materials. These properties vary significantly with temperature and are also dependent on the composition and characteristics of construction materials, testing procedure, as well as heating rate and other environmental conditions. Thermal properties that govern temperature-dependent properties in construction materials are thermal conductivity, specific heat (or heat capacity), and mass loss. The mechanical properties that are of primary interest at elevated temperatures are compressive strength, tensile strength, elastic modulus, and stress-strain response. Deformation properties, which include thermal expansion, creep strain, and transient strain, are highly dependent on the chemical composition as well as chemical and physical reactions that occur during heating (or cooling) of construction materials. In addition, special properties, such as local buckling in steel, spalling in concrete, and charring in timber, only occur in some materials under certain conditions. A complete discussion on physical and engineering properties of construction materials can be found in Kodur and Harmathy (2016).

27.4.2 Concrete

Concrete is available in various forms and is often grouped under different categories based on weight [as normal-weight concrete (NWC) or lightweight concrete (LWC)], strength [as normal-strength concrete (NSC), high-strength concrete (HSC), or ultra-high-strength concrete (UHSC)], presence of fibers [as plain concrete (PC) or fiber-reinforced concrete (FRC)], and mechanical performance (as conventional and high-performance concrete). Fire safety practitioners further subdivide normal-weight concretes into silicate (siliceous) and carbonate (limestone) aggregate concrete, based on the composition of the principal aggregate.

Thermal conductivity is the property of a material to conduct heat. The thermal conductivity of concrete at room temperature is in the range of 1.4–3.6 W/m°K and varies with temperature (Bažant and Kaplan, 1996). As shown in Fig. 27.3, thermal conductivity decreases gradually with temperature, and this decrease is dependent on the concrete mix properties, specifically moisture content and permeability. This decreasing trend in thermal conductivity can be attributed to variation of moisture content with increase in temperature. Thermal conductivity of high-strength concrete and fiber-reinforced concretes (having either both steel or polypropylene fibers) almost follow a similar trend as that of plain concrete. Therefore, it is deduced that there is not much significant effect of fibers on thermal conductivity of concrete in 20°C–800°C temperature range (Khaliq and Kodur, 2012).

Specific heat is the amount of heat per unit mass required to change the temperature of a material by 1 degree and is often expressed in terms of thermal (heat) capacity, which is the product of specific heat and density. Specific heat of concrete at room temperature varies in the range of 840–1800 J/kg°K for different aggregate types. The specific heat property is sensitive to various physical and chemical transformations that take place in concrete at elevated temperatures. This includes the vaporization of free water at about 100°C, the dissociation of $Ca(OH)_2$ into CaO and H_2O between 400°C and 500°C, and the quartz transformation of some aggregates above 600°C. Specific heat is therefore highly dependent on moisture content and considerably increases with a higher water-to-cement ratio. The specific heat of concrete remains almost constant up to 400°C, followed by increases up to about 700°C, and then remains constant between 700°C and 800°C (Fig. 27.3*b*). Of the various factors, the aggregate type has a significant influence on the specific heat (thermal capacity) of concrete. Carbonate aggregate concrete has higher specific heat (heat capacity) in the 600°C–800°C temperature range, and this is caused by an endothermic reaction, which results from decomposition of dolomite, which absorbs a large amount of energy (Kodur and Harmathy, 2016). As compared to NSC, HSC exhibits slightly lower specific heat throughout the 20°C–800°C temperature range (Kodur and Sultan; 2003).

Figure 27.3*c* and *d* illustrates the variation of compressive strength ratio for NSC and HSC at elevated temperatures, respectively, with upper and lower bounds (shaded area) showing a range of variation in reported test data. Also plotted in these figures is the variation of compressive strength as obtained using Eurocode (2004), ASCE (Lie, 1992), and Kodur (2014) relations. Overall, the variation in strength trends from different tests can be attributed to using different heating or loading rates, specimen size and curing, condition at testing (moisture content and age of specimen), and the use of admixtures.

The tensile strength of concrete is much lower than that of compressive strength, and hence tensile strength of concrete is often neglected in strength calculations at room temperature. However, from a fire resistance point of view, this property is important since it controls level of cracking under fire conditions. Figure 27.3*e* illustrates variation of splitting tensile strength ratio of NSC and HSC as a function of temperature, as reported in previous studies and Eurocode provisions. The shaded portion in this plot shows a range of variation in splitting tensile strength as obtained by various researchers for NSC and HSC. In general, HSC experiences rapid loss of tensile strength at higher temperatures due to accumulation of pore pressure in dense microstructured HSC (Chan et al., 1999). It should be noted that the addition of steel fibers to concrete enhances its tensile strength at both room and elevated temperatures, and the increase can be up to 50 percent at room temperature.

The modulus of elasticity is a measure of the ability of the material to resist deformation and is defined by the ratio of deforming stress to the strain in the material. The modulus of elasticity of various concretes at room temperature varies over a wide range, 5.0×10^3 to 35.0×10^3 MPa, and is dependent mainly on the water-to-cement ratio in the mixture, the age of concrete, the method of conditioning, and the amount and nature of the aggregates. It appears that the modulus of elasticity of normal-weight concretes decreases at a higher pace with rise of temperature than that of lightweight concretes. Figure 27.3*f* illustrates variation of ratio of elastic modulus at target temperature to that at room temperature for NSC and HSC (Eurocode 2, 2004). The degradation of modulus properties in both NSC and HSC can be attributed to excessive thermal stresses and physical and chemical changes in concrete microstructure.

The mechanical response of concrete is usually expressed in the form of stress-strain relations. Generally, because of a decrease in compressive strength and an increase in ductility of concrete, the slope of the stress-strain curve decreases with increasing temperature. In general, it is established that HSC has steeper and more linear stress-strain curves in comparison to NSC at 20°C–800°C. HSC specimens exhibit brittle response as indicated by post-peak behavior of stress-strain curves shown in Fig. 27.3*g* and *h*. In the case of fiber-reinforced concrete, especially with steel fibers, the stress-strain response tends to be more ductile.

Thermal expansion is another property and characterizes the expansion (or shrinkage) of a material caused by heating and is defined as the expansion (shrinkage) of unit length of a material when the temperature of concrete is raised by 1 degree. Concrete generally undergoes expansion when subjected to elevated temperatures (see Fig. 27.3*i*). The thermal expansion of concrete increases from zero at room temperature to about 1.3 percent at 700°C and then generally remains constant through 1000°C. This increase is substantial in the 20°C–700°C temperature range and is mainly due to high thermal expansion resulting from constituent aggregates and cement paste in concrete.

27.4.3 Masonry

Building brick (or masonry) is extensively used in low-rise residential construction. The density (ρ) of bricks ranges from 1660 to 2270 kg·m^{-3}, depending on the raw materials used in the manufacture and on the molding and firing technique. The true density of the material (ρ_t) is

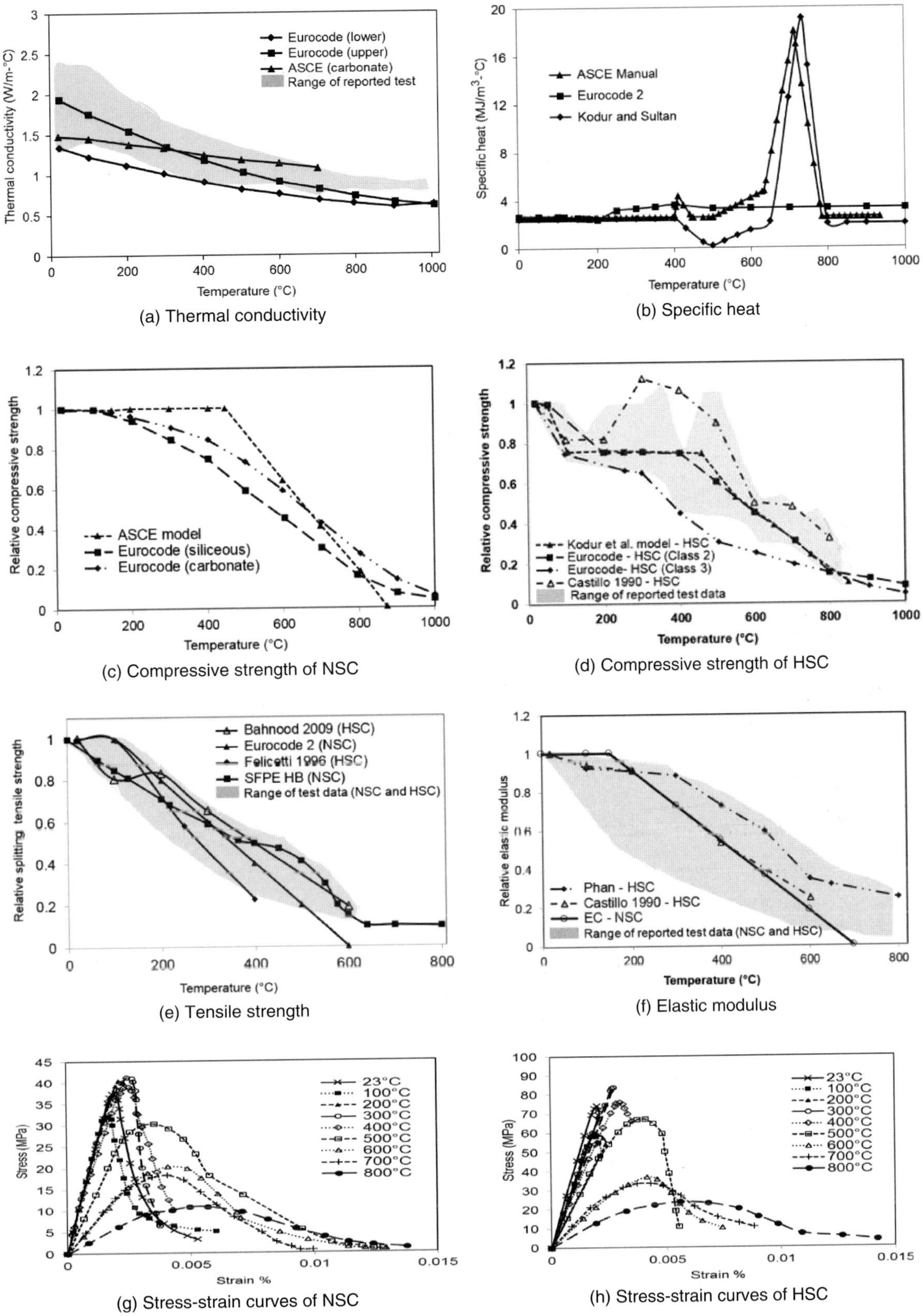

Figure 27.3 Variation in thermal and mechanical properties of NSC and HSC as a function of temperature.

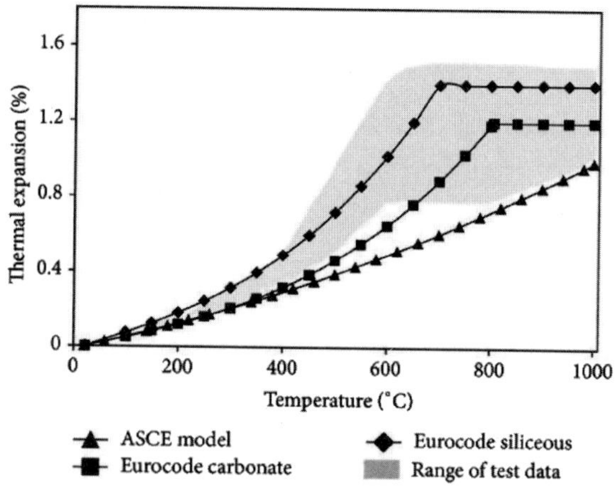

(i) Thermal expansion of NSC

Figure 27.3 *(Continued)*

somewhere between 2600 and 2800 kg·m^{-3}. Similar to concrete, the specific heat of brick slightly increases with rise in temperature and is sensibly independent of the density of the brick (Malhotra, 1982). On the other hand, thermal conductivity of masonry is dependent on the density of the brick with high-density bricks having higher values of thermal conductivity (see Fig. 27.4).

The modulus of elasticity of brick (E) ranges between 10×10^3 and 20×10^3 MPa. Its compressive strength (σ_u) varies in a very wide range, from 9 to 110 MPa. In general, the compressive strength of bricks decreases slowly with temperature, so that a difference of 85 percent of initial strength is achieved at 400°C; this trend is similar to that of plain concrete. As a matter of fact, little research specific to brick properties at elevated temperature is reported in the literature, and this is due to the fact that brick properties are assumed to follow the same trend as that of concrete. Another reason might be due to the fact that masonry is rarely considered for structural members (such as columns or beams) in high-rise buildings. Therefore, there has been little research on the mechanical properties of masonry at elevated temperatures. Further, information relating to temperature-dependent properties of masonry can be found elsewhere (Russo and Sciarretta, 2013).

27.4.4 Steel

The main thermal properties that influence the temperature rise in steel are thermal conductivity and specific heat (often expressed in terms of heat capacity). Figure 27.5 shows variation of thermal conductivity and specific heat of steel as a function of temperature, as per Eurocode 3 and ASCE 1992 provisions and as reported in test data. It can be seen from Fig. 27.5*a* that thermal conductivity decreases with temperature in an almost linear fashion. On the contrary, the specific heat of steel increases with an increase in temperature with a large spike occurring around 750°C. This increase in specific heat at around 750°C is due to the phase change that occurs in steel in which the atoms transition from a face centered cubic to a body centered cubic structure.

Figure 27.5*c* and *d* show the variation of yield strength and modulus of elasticity of steel as a function of temperature, respectively. It can be seen that both the yield strength and the elastic modulus decrease as temperature increases. This decrease can be attributed to the nucleus of the iron atoms in steel moving farther apart due to rising temperature in steel, leading to decreased bond strength, which in turn reduces the yield strength and elastic modulus. There is significant variation in reported test data on yield strength and modulus of elasticity at temperatures above 300°C. This variation can be attributed to many factors, primarily variable heating and strain/load rates during the test. It is worth noting that the thermal expansion of steel linearly increases with rise in temperature.

27.4.5 Wood

Wood is often used in structural members in residential and low-rise constructions. Wide ranges of wood species are grown in various countries (and regions). In the United States, about 180 wood species are commercially grown; only about 25 species have been assigned working stresses in timber structures. The two groups most extensively used as structural lumber in the United States are the Douglas firs and the southern pines. Timber is an anisotropic material, and hence it possess different material behavior in direction of grains as well as parallel to the grain.

The thermal conductivity across the grain of timber is in the range of 0.17–1.5 W·m^{-1}·K^{-1} between room temperature and 140°C. The thermal conductivity increases initially up to a temperature range of 150°C–200°C, then decreases linearly up to 350°C, and finally increases again beyond 350°C. Figure 27.6 shows the apparent thermal conductivity and specific heat of typical timber, as a function of temperature, as provided by Eurocode 5 (König, 2005).

The mechanical properties of wood are influenced by moisture content, rate of charring, and grain orientation. The modulus of elasticity of air-dry, clear wood along the grain varies from 5.5×10^3 to 15.0×10^3 MPa, and its crushing strength varies from 13 to 70 MPa. These properties are related and roughly proportional to the density, regardless of the species. The mechanical properties of timber deteriorate significantly with rise in temperature. Figure 27.6*c* and *d* show the variation of the modulus of elasticity and compressive strength of oven-dry, clear wood with temperature. The modulus of elasticity decreases slowly with temperature up to about 200°C, when it reaches about 80 percent, and then the decline is more rapid. The compressive strength also drops linearly to about 80 percent at about 200°C, and then the drop is more rapid—to about 20 percent around 280°C. The variation of tensile strength of

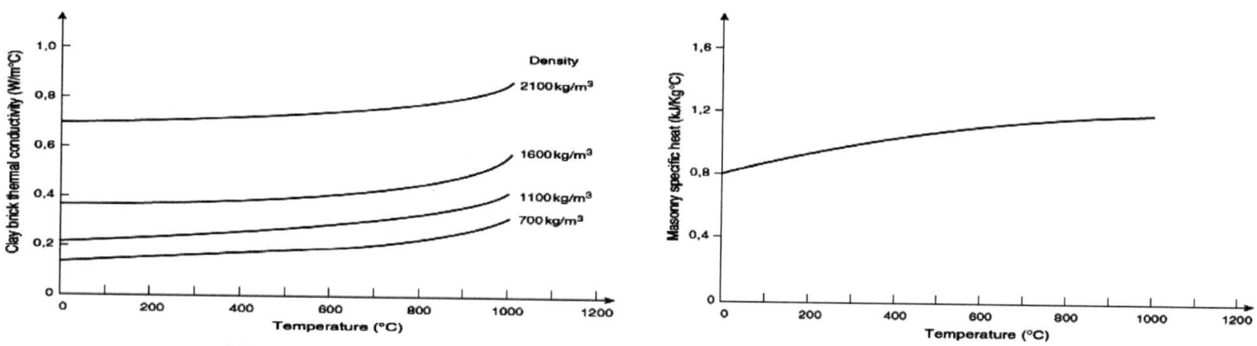

(a) Thermal conductivity

(b) Specific heat

Figure 27.4 Variation in thermal properties of masonry as a function of temperature.

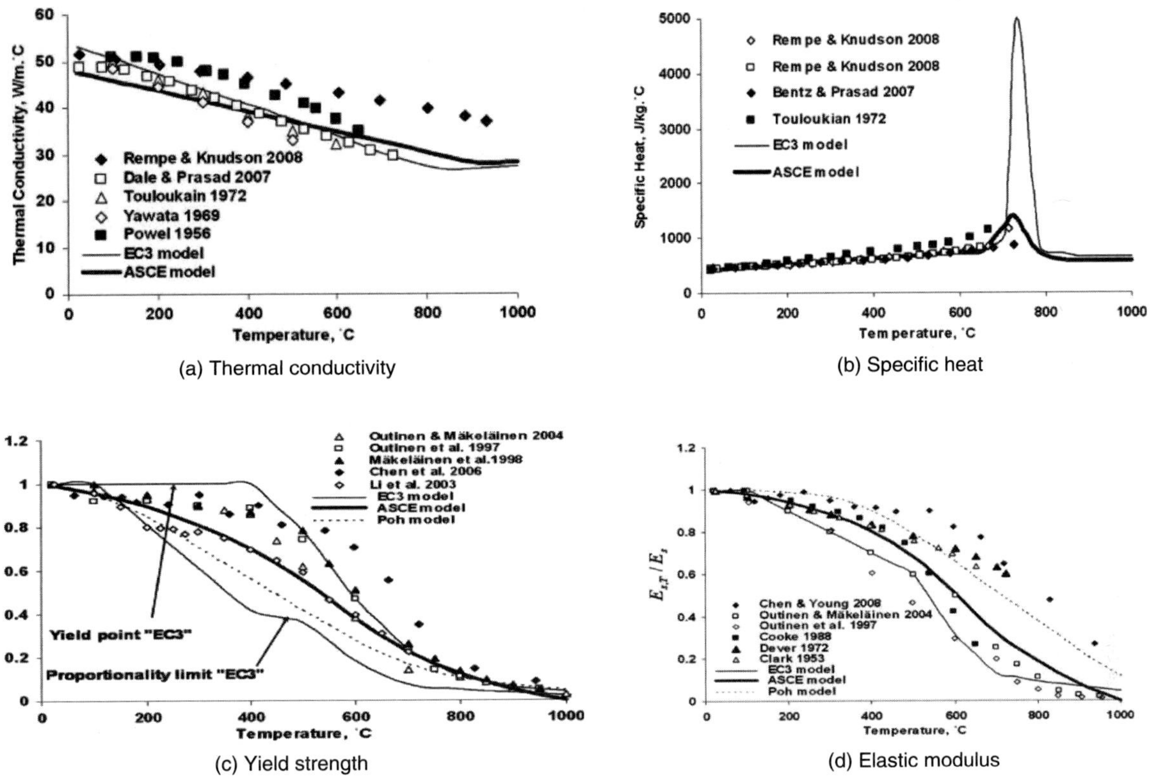

Figure 27.5 Variation in thermal and mechanical properties of steel as a function of temperature.

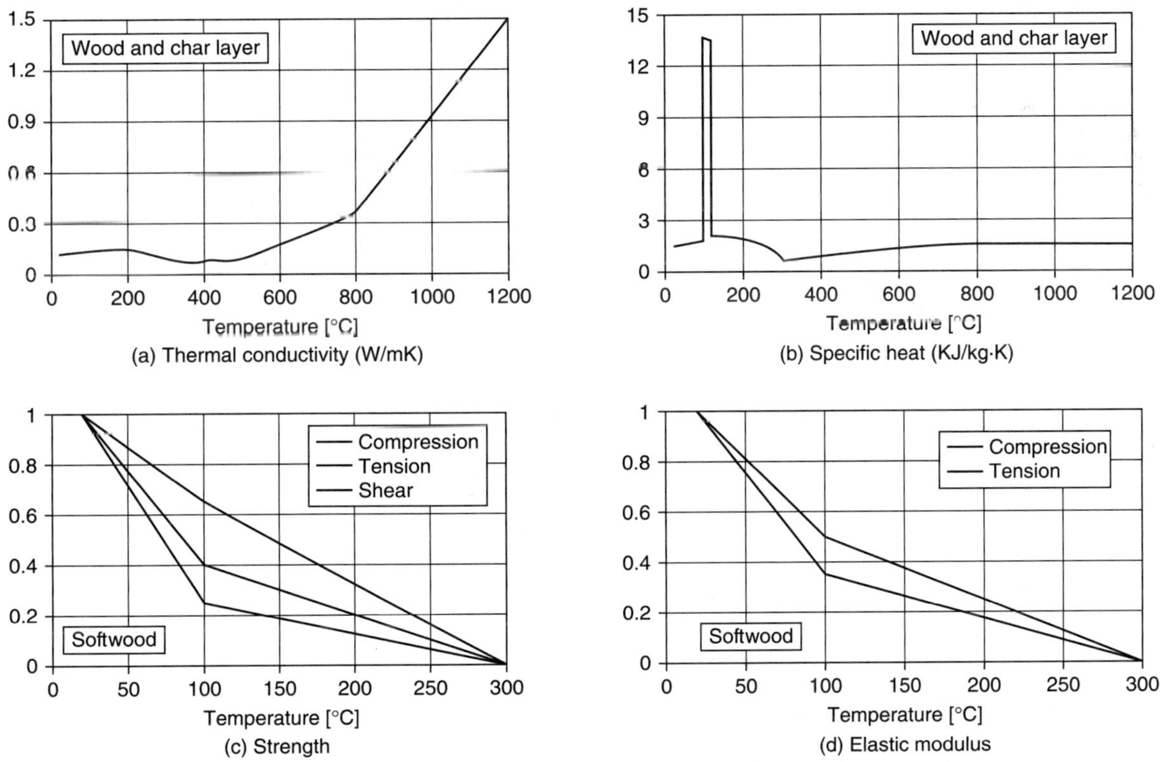

Figure 27.6 Variation in thermal properties of timber as a function of temperature (One Stop Shop in Structural Fire Engineering, 2018).

wood with temperature is similar to that of compressive strength, but the decline in tensile strength with temperature is less rapid. The moisture content plays a significant role in determining the temperature-induced strength and stiffness loss, with increased moisture content leading to higher reduction.

Charring process is one of the main high-temperature phenomenon associated with wood and should be accounted for in predicting fire performance of wood structures. The rate of charring is influenced by the radiant heat flux or, alternatively, the fire severity, moisture content and wood density, and so on. The charring rate, derived under standard fire exposure, can be in the range of 0.6 mm/min to approximately 0.4 mm/min for moist dense wood or to 0.8 mm/min for dry and light wood. Specific charring rates for different types of wood can be found in the ASCE Manual (1992). Eurocode 5 also gives an expression for charring depth in a wood member exposed to standard fire. The dependence of charring rate on the radiant heat flux is discussed in the Wood Handbook (1999).

27.4.6 Insulation

Steel structures, when exposed to fire, undergo rapid rise in temperature due to the low-specific heat and high-thermal conductivity of steel. Thus, steel structures are to be protected with fire insulation such as gypsum-based products, intumescent paints, or spray-on insulation to achieve required fire resistance. As insulation materials have low strength or stiffness properties, only thermal material properties are of insert in fire resistance evaluation. As in the case of other materials, thermal properties of insulation also degrade with temperature. However, little is known regarding temperature-dependent material properties of insulation materials. Thus, room temperature thermal properties, as provided in manufacture listings, are often used for calculating fire resistance of steel structures. This practice has shown to lead to unconservative fire resistance values (Kodur and Shakya, 2013).

Temperature-dependent properties of some of the insulation materials, such as SFRM-based products such as CAFCO-300, Carboline Type-5MD, and Tyfo WR-AFP, are listed in Table 27.1, and other properties of insulating materials can be found in Kodur and Shakya (2013) and the SFPE Handbook (2016).

27.5 FIRE RESISTANCE EVALUATION

When fire grows to full size, the role of structural members becomes critical to maintain integrity, stability, and load-bearing functionality of structural systems. Therefore, structural fire safety is a key consideration

Table 27.1 Thermal Property Relations of Readily Available Insulation Materials

	Thermal conductivity, k_t (W/mK)	
CAFCO-300	$k_t = 0.0778 - 0.000054\ T$	$20°C \le T \le 300°C$
	$k_t = -0.08 + 0.000469\ T$	$300°C \le T \le 700°C$
Carboline Type-5MD	$k_t = 0.121 - 0.000319\ T$	$20°C \le T \le 200°C$
	$k_t = 0.0468 + 0.000050\ T$	$200°C \le T \le 700°C$
Tyfo WR-AFP	$k_t = 0.207 - 0.000318\ T$	$20°C \le T \le 200°C$
	$k_t = 0.147 - 0.000035\ T$	$200°C \le T \le 400°C$
	$k_t = 0.0054 + 0.000321\ T$	$400°C \le T \le 700°C$
	Specific heat, c_p (J/kgK)	
CAFCO-300	$c_p = 3236 + 4.16\ T$	$20°C \le T \le 200°C$
	$c_p = 6815 - 12.60\ T$	$200°C \le T \le 400°C$
	$c_p = 1645 - 0.36\ T$	$400°C \le T \le 700°C$
Carboline Type-5MD	$c_p = 1627 + 22.3\ T$	$20°C \le T \le 100°C$
	$c_p = 4446 - 5.05\ T$	$100°C \le T \le 400°C$
	$c_p = -1336 + 9.37\ T$	$400°C \le T \le 700°C$
Tyfo WR-AFP	$c_p = 643 + 1.93\ T$	$20°C \le T \le 200°C$
	$c_p = 1241 - 0.924\ T$	$200°C \le T \le 400°C$
	$c_p = 195 + 1.71\ T$	$400°C \le T \le 600°C$
	$c_p = 1826 - 1.08\ T$	$600°C \le T \le 700°C$

in building design in the event of fire. Building codes (e.g., IBC, 2015) account for this through specifying passive fire protection requirements for structural members in the form of fire-resistance ratings (of ½, 1, 2, 3, and 4 hours). A designer has to provide appropriate insulation or other requirements to the member and ensure that a given structural member meets the code-specified fire rating requirement. Fire resistance of a structural member can be evaluated by applying either perspective-based approaches or through rational calculation methods. Through these methods, a designer can check if a structural member meets the required fire rating.

27.6 PRESCRIPTIVE-BASED APPROACHES

In perspective-based approaches, fire resistance of a member can be evaluated through the use of tabulated data and simplified calculation methods, which are derived from standard fire tests. These prescriptive approaches take into consideration only a limited number of parameters and always under standard fir exposure conditions. These approaches have been widely adopted due to the fact that for most types of buildings, the level of needed fire performance and safety is known from past experience. Prescriptive-based approaches, due to their simplistic nature, do not account for actual design variables presents in a structure (i.e., fire, loading, restraint, etc.) and realistic failure modes, which can often lead to unrealistic fire performance assessment, especially when fire growth characteristics are not similar to those under a standard fire.

One simple way of establishing fire resistance is through standard fire tests. The main data collected in a standard fire test are fire temperatures, time to failure of the member, cross-sectional temperature, deformations (or deflections), and observations during the fire exposure. Data collected from such fire tests are used to interpret and assign a fire rating to the tested member. It is worth noting that the standard fire test is only intended to register the performance up to the desired rating period of standard fire exposure and not intended to evaluate performance of the structural member beyond the rating period (until actual failure). Fire ratings derived from standard fire tests on structural members are to be certified by a recognized laboratory or agency for acceptance in practice and listed in a directory. For instance, testing agencies, such as Underwriters Laboratory (UL), maintain a list of approved fire resistance ratings of various assemblies. Industry organizations (such as Gypsum Association) also maintain listings of approvals for their products. Fire resistance ratings derived from fire tests can be proprietary or generic (UL, 1992).

In generic ratings, fire resistance of member infers the applicability of ratings to members made of full range of generic materials (such as conventional concrete), with no reference to a specific manufacturer. Such generic ratings are listed as addendums in building codes and standards, such as IBC (2015), NBCC (2015), ACI 216.1 (2014), and so on. A sample of these listings is presented in Table 27.2, which shows listings of minimum cover for concrete slabs made of siliceous, carbonate, and lightweight concrete as published by the ACI 216.1 (2014). In this case, ACI 216.1 (2014) provisions specify that a solid concrete slab made of carbonate aggregate to have a minimum thickness of 127 mm to achieve a 2-hour fire rating. In case the slab is made with hollow configuration, then the equivalent thickness (i.e., total thickness) to steel reinforcement would need to be 127 mm to satisfy requirements for a 2-hour fire rating.

In case fire resistance tests are commissioned by manufacturers of a specific proprietary material, the ratings of members made of such proprietary products become proprietary. Such proprietary ratings can

Table 27.2 Listing for Concrete Wall/Slab Thickness

Structural shape	Minimum equivalent thickness for fire-resistance rating (mm)			
	1 hr	2 hr	3 hr	4 hr
Siliceous	89	127	157	178
Carbonate	81	117	145	168
Lightweight	64	91	112	130

only be applied to specific products and cannot be applied to products (or material types) other than that of the tested assembly/member. In general, listings only apply to duplicate of prototype design used in test, and any changes can nullify the rating. Further, individual listing agencies, such as UL, often provide a unique designation, sectional details, construction specifications, conditions of the carried-out fire test, and so on. for fire-rated members. Each classification designation is followed by a schematic protection number that varies from 000 to 899. The type of applied insulation (i.e., SFRM) follows this designation (600–899). Unprotected members do not have insulation, and their membrane protection number varies from 900 to 999. For example, the listing shown in Fig. 27.7 is for a steel beam (Group S) protected with SFRM insulation (No. N708). Hence, the designation of this member is N708. This listing is for a fire-rated steel beam with fire rating between 1 and 4 hours. In this listing, an unrestrained beam of W8 × 28 shape would require an insulation thickness of 15/16 in. (23.8 mm) to achieve a 2-hour fire rating.

In lieu of standard fire testing and listings, correlation methods can also be used to evaluate fire resistance of structural members (SFPE, 2015).

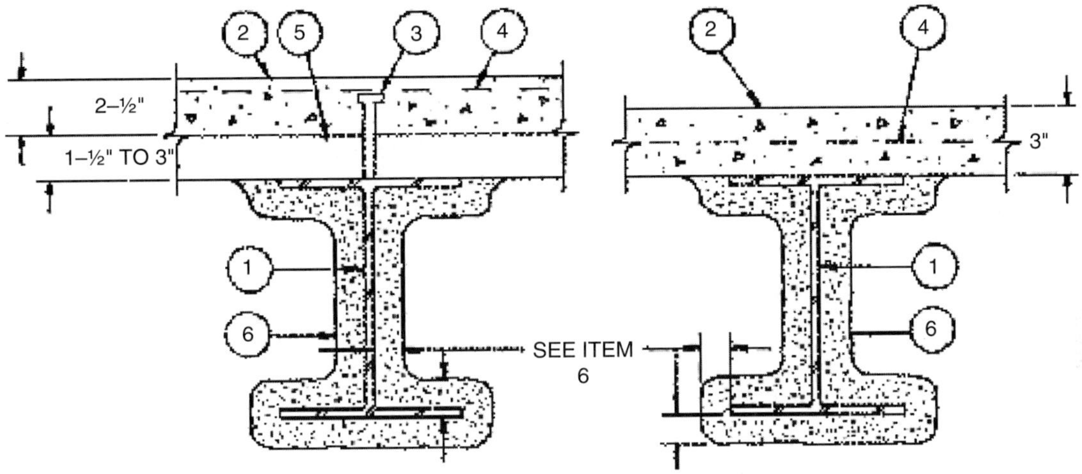

1. **Steel Beam –** W8×28 min size.

2. **Normal Weight or Lightweight Concrete –** Compressive strength, 3000 psi. For normal weight concrete either carbonate or siliceous aggregate may be used. Unit weight, 148 pcf. For lightweight concrete, unit weight 110 pcf.

3. **Shear Connector –** (Optional) – Studs, 3/4 in. diam headed type or equivalent per AISC specifications. Welded to the top flange of beam through the steel floor units.

4. **Welded Wire Fabric** – (Optional) – 6×6-10/10 SWG.

5. **Steel Floor and Form Units*** – 1-5/16 in. deep corrugated units; or 1-1/2 to 3 in. deep fluted or cellular units, welded to beam.

6. **Spray-Applied Fire Resistive Materials*** – Applied by mixing with water and spraying in more than one coat to the beam to the final thicknesses shown below. When fluted or corrugated steel floor units are used, crest areas shall be filled with Spray-Applied Fire Resistive Materials above the beam. Beam surfaces must be clean and free of dirt, loose scale and oil. Min avg and min ind density of 15/14 pcf respectively. Min avg and min ind density of 19/18 pcf respectively for Types 7GP and 7HD, For method of density determination, see Design Information Section.

	Min Thkns In.	
Rating Hr	**Restrained Beam Rating Hr**	**Unrestrained Beam Rating HR**
1	1/2	1/2
1–1/2	11/16	13/16
2	15/16	1-1/16
3	1-7/16	1-9/16
4	1-15/16	2

Figure 27.7 UL listing for SFRM insulated steel beam (UL, 1992).

These methods are correlated through fitting outcome of fire tests. As such, these methods are only applicable for certain conditions and structural members and may not be extrapolated beyond their intended range. For instance, correlation methods can be applied for structural members subjected to standard fire tests where these members are made of specific geometry and materials. As such, these methods may not be applicable to structural members exposed to design (or real) fires. In many cases, correlation methods were fitted through testing of unload members, and as such they only account for failure through exceeding a predefined thermal limit. Further discussion on various correlation methods for concrete, steel, and wood structural members can be found in the ASCE Manual (1992).

27.7 RATIONAL DESIGN APPROACHES

An alternative to prescriptive-based approaches is the use of rational engineering approaches for evaluating fire resistance of structural members. Through the use of rational approaches, many of the drawbacks, such as not accounting for realistic fire and structural loading, restraint conditions, and failure limit states, can be overcome. In such rational design approaches, engineering-based calculation methods are applied for evaluating fire resistance (performance) of structure taking into account realistic conditions, or close to that, present during a fire scenario. The calculations can be carried out at various levels: sectional, member, or system through simplified or advanced analysis. The level of complexity of rational approaches increases with the factors (or number of variables) considered in the fire resistance evaluation and the type of analysis adopted for deriving fire resistance.

27.7.1 Simplified Approaches

Simplified approaches can be applied at sectional or member level to determine fire resistance of structural members. In simplified approaches, the analysis is to be carried out at a given time (e.g., after 2 hours of fire) (see Fig. 27.8). The thermal analysis generates cross-sectional temperature in a structural member at various fire exposure times. This temperature can be arrived at using simplified equations (such as the best-fit method).

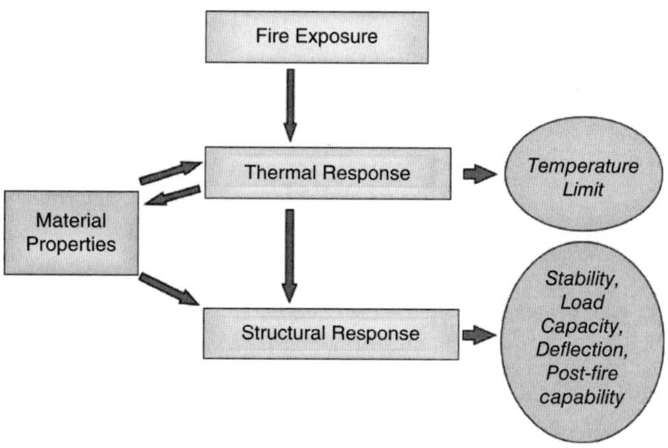

Figure 27.8 Flowchart for simplified fire resistance evaluation at sectional or member level.

The best-fit method is generally used to determine temperature rise in steel structures. This temperature rise is given by

$$t = \frac{0.54(T_{\text{lim}} - 50)}{(F/V)^{0.6}}$$

where F is the surface area of unit length (m^2), V is the volume of steel in unit length (m^3), t is the time to reach a limit temperature,[1] and T_{lim} is the limiting temperature in steel (often taken as 550°C). This equation is valid for F/V value ranging from 10 to 300 m^{-1} and failure time in steel in temperatures in the 100°C–800°C range. F/V ratio is listed for various steel sections in Fig. 27.9. Other methods for determining cross-sectional temperature in steel and concrete structure can be found elsewhere (Dwaikat and Kodur, 2012; Wickström, 1986).

These temperatures are then applied as thermal load on the member to evaluate sectional or member-level structural response. In the structural analysis, response parameters, such as sectional capacity (i.e., moment, shear, axial, etc.), deflection, and so on, of a structural member can be evaluated at various time steps. At each time step, these response parameters are compared against limiting values corresponding to different limiting states (temperature, strength, deflection) in order to evaluate failure. Failure occurs once any of the response parameters exceed the limiting criterion such that

$$U_{\text{fire}} \leq R_{\text{fire}}$$

where U_{fire} is the loading effects present on the member during fire and R_{fire} is the reduced sectional capacity (i.e., moment, shear, axial) due to the adverse effects of fire. In case of a simply supported beam with a span, l, and subjected to a uniformly distributed loading, w, this capacity is then compared against that of the magnitude of bending moment, which equals

$$M_u = w \times l^2/8$$

As an example, the reduced moment capacity (M_f) at a critical section of a fire-exposed steel beam can be evaluated by applying the room temperature capacity equation but with due consideration to temperature-induced degradation in steel:

$$M_f = k_{y,T} f_y Z$$

where f_y is the yield strength of steel at room temperature, MPa; Z_x is the plastic section modulus, mm^3; and $k_{y,T}$ is the reduction factor of yield strength of steel at temperature T (as discussed in earlier sections). Knowing the temperature in steel, the reduction factor for yield strength can be obtained. Then the moment capacity can be obtained. The time to reach the point at which failure occurs, that is, $M_u > M_f$, is taken as fire resistance of the member. The above procedure can also be applied to evaluate degradation in shear and axial capacity of beams, subjected to shear and in compression members (columns) subjected to axial force as well as other members (Kodur and Naser, 2018).

27.7.2 Advanced Approaches

In lieu of simplified analysis, advanced analysis can be carried for evaluating fire response of structural members. Advanced approaches can be applied through models based on finite difference or finite element can be developed using generic software (i.e., ANSYS, ABAQUS) or fire-specific software, such as SAFIR. Using rational design approaches, the analysis is to be carried out in two stages, namely, thermal and structural stage, comprised of number of steps (see Fig. 27.10). In advanced analysis, fire resistance evaluation can account for various influencing parameters, such as fire scenario, composite action between structural members (such as beam-slab), and properties of steel and insulation. Furthermore, fire response of structural members (or whole structure, i.e., system-level analysis) can be traced through a finite-element-based numerical computer programs.

Following discussion on Fig. 27.10, the first stage in fire resistance analysis through advanced approaches is to discretize the structure or structural member (i.e., beam) under consideration through the use of element libraries available in finite-element programs. Once the structure is properly discretized, the geometric finite element model is developed. A typical discretization of a beam-slab assembly is shown in Fig. 27.11.

This developed model is then input with temperature-dependent thermal, mechanical, and deformational properties of construction materials. These material properties can be taken from design codes and standards (i.e., Eurocode 2 etc.). In order to capture realistic

[1]This method is valid for failure times for unprotected steel members of between 10 and 80 minutes.

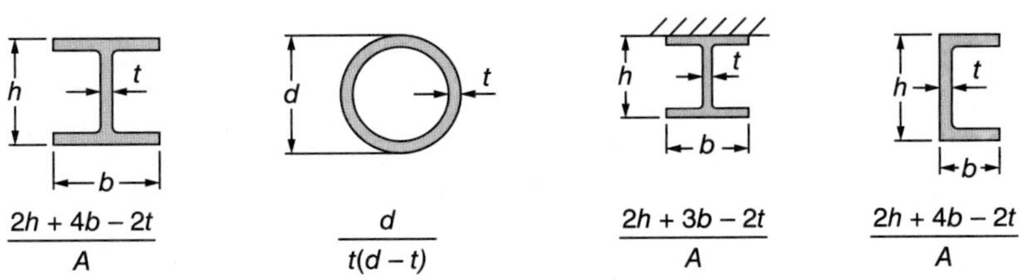

Figure 27.9 Calculation of F/V ratio for various steel sections.

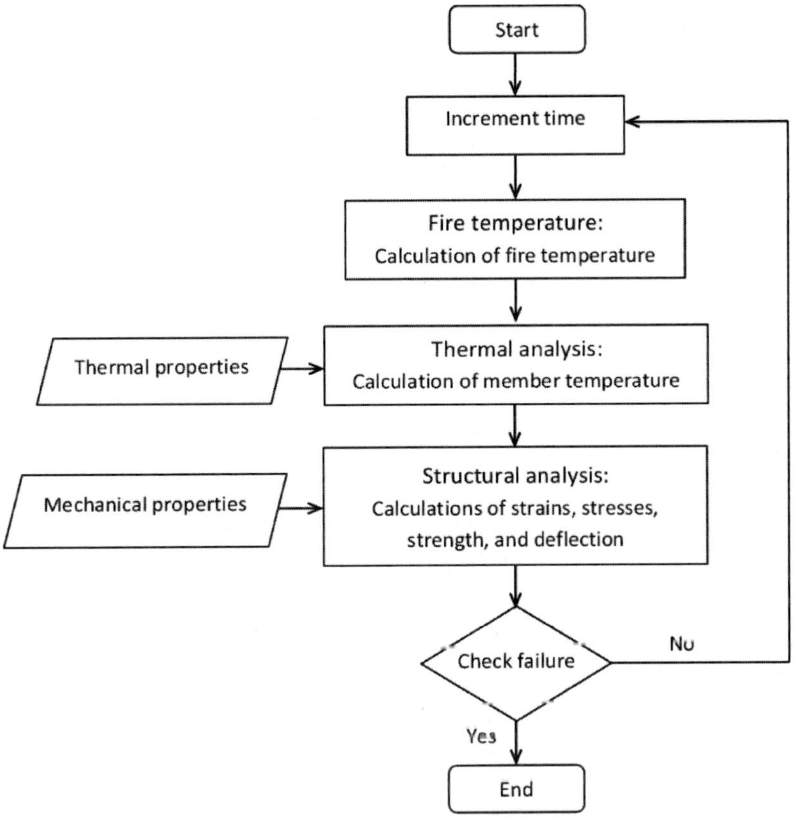

Figure 27.10 Flowchart illustrating steps associate with fire resistance analysis of typical structural members.

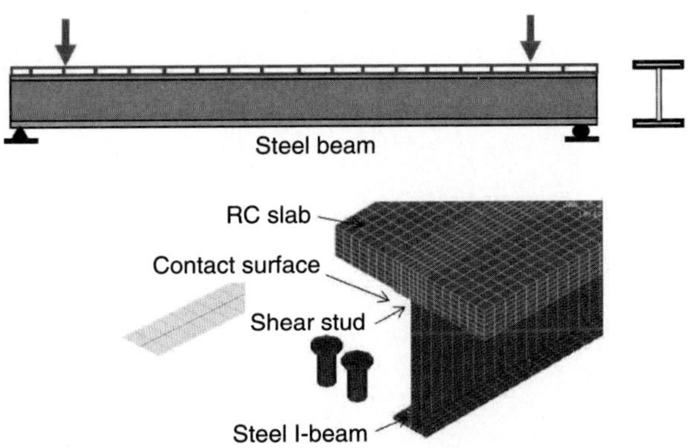

Figure 27.11 Discretization of a typical finite-element model for fire resistance analysis of beam-slab assembly (Naser and Kodur, 2017).

response of structural members, special attention is needed to ensure proper selection of material properties to include specific properties in order to, for instance, account for creep and local buckling effects. Representative boundary conditions, including structural (gravity) and thermal (fire) loading, are then applied into the model. This steps marks the development of the finite-element model. A set of different limit states to evaluate failure (fire resistance) and derive failure limits are also supplied to the developed model in order to capture all possible failure limit states, including thermal (limiting temperature) or structural (i.e., sectional capacity and/or deflection). The analysis initiates and goes through number of iterations (time increments) at which failure is to be evaluated at the end of each time step. Failure is said to occur once any of the above listed failure limit states is exceeded. Detailed, step-by-step illustrations of undertaking advanced analysis of structural members subjected to fire incidents through finite-element software based software can be found elsewhere (Kodur et al., 2013; Arablouei and Kodur, 2015; Kodur et al., 2016; Kodur and Agrawal, 2017; Kodur and Bhatt, 2017).

27.8 SUMMARY

Structural members, when exposed to fire conditions, experience significant degradation, and thus fire can endanger the stability and integrity of a structure. However, this can be overcome through engineering the structure for enhancing performance under fire conditions. This can effectively be carried out throughout undertaking fire resistance analysis based on rational engineering approaches. A conscious summary on the various methods that can be applied for evaluating fire resistance of structures, together with various material properties and fire conditions, is presented in this chapter. Interested readers can find further information in a state-of-the-art test book, *Structural Fire Engineering*.

REFERENCES

ACI 216.1-14/TMS-0216-14, American Concrete Institute, Farmington Hills, 2014.

Arablouei, Amir, and Venkatesh Kodur. "Dynamic delamination of fire insulation applied on steel structures under impact loading." *International Journal of Impact Engineering* 83 (2015): 11–27.

ASTM, E. "1529. Standard Test Methods for Determining Effects of Large Hydrocarbon Pool Fires on Structural Members and Assemblies," ASTM, West Conshohocken, PA (1993).

Bažant, Zdeněk P., and Maurice F. Kaplan. *Concrete at high temperatures: material properties and mathematical models.* Harlow Longman, 1996.

Chan, Y. N., G. F. Peng, and M. Anson. "Residual strength and pore structure of high-strength concrete and normal strength concrete after exposure to high temperatures." *Cement and Concrete Composites* 21, no. 1 (1999): 23–27.

Drysdale, Dougal. *An introduction to fire dynamics.* John Wiley & Sons, 2011.

Dwaikat, M. M. S., and V. K. R. Kodur. "A simplified approach for predicting temperature profile in steel members with locally damaged fire protection." *Fire Technology* 48, no. 2 (2012): 493–512.

EN, BS. "1-2: 2002 Eurocode 1: Actions on structures—Part 1-2: General actions—Actions on structures exposed to fire." *British Standards* (1991).

EN, BS. "1-2: 2004 Eurocode 2: Design of concrete structures—Part 1-2: General rules—Structural fire design." *European Standards, London* (2004).

Handbook, Wood. "Wood Handbook-Wood as an engineering material." U.S. Department of Agriculture, Forest Service, Forest Products Laboratory, Madison, WI, USA. *Gen. Tech. Rep. FPL-GTR-113* 463 (1999).

Hurley, Morgan J., Daniel T. Gottuk, John R. Hall Jr, Kazunori Harada, Erica D. Kuligowski, Milosh Puchovsky, John M. Watts Jr., and Christopher J. Wieczorek, eds. *SFPE handbook of fire protection engineering.* Springer, 2015.

IBC, ICC. "International building code." International Code Council, Inc. (formerly BOCA, ICBO and SBCCI) 4051 (2015): 60478–5795.

Khaliq, Wasim, and Venkatesh Kodur. "High temperature mechanical properties of high-strength fly ash concrete with and without fibers." *ACI Materials Journal* 109, no. 6 (2012): 665–674.

Kodur, V. K. R., and Ankit Agrawal. "An approach for evaluating residual capacity of reinforced concrete beams exposed to fire." *Engineering Structures* 110 (2016): 293–306.

Kodur, V. K. R., and M. A. Sultan. "Effect of temperature on thermal properties of high-strength concrete." *Journal of Materials in Civil Engineering* 15, no. 2 (2003): 101–107.

Kodur, V. K. R., and M. Z. Naser. "Approach for shear capacity evaluation of fire exposed steel and composite beams." *Journal of Constructional Steel Research* 141 (2018): 91–103.

Kodur, V. K. R., and M. Z. Naser. "Effect of local instability on capacity of steel beams exposed to fire." *Journal of Constructional Steel Research* 111 (2015): 31–42.

Kodur, V. K. R., and P. P. Bhatt. "A numerical approach for modeling response of fiber reinforced polymer strengthened concrete slabs exposed to fire." *Composite Structures* (2017): 1–15.

Kodur, V. K. R., and T. Z. Harmathy. "Properties of building materials." In *SFPE handbook of fire protection engineering*, pp. 277–324. Springer, 2016.

Kodur, V. K. R., M. Naser, P. Pakala, and A. Varma. "Modeling the response of composite beam–slab assemblies exposed to fire." *Journal of Constructional Steel Research* 80 (2013): 163–173.

Kodur, Venkatesh. "Properties of concrete at elevated temperatures." *ISRN Civil Engineering* (2014).

Kodur, Venkatesh K., Esam M. Aziz, and Mohannad Z. Naser. "Strategies for enhancing fire performance of steel bridges." *Engineering Structures* 131 (2017): 446–458.

Kodur, V. K. R., and A. M. Shakya. "Effect of temperature on thermal properties of spray applied fire resistive materials." *Fire Safety Journal* 61 (2013): 314–323.

König, Jürgen. "Structural fire design according to Eurocode 5—Design rules and their background." *Fire and Materials* 29, no. 3 (2005): 147–163.

Lie, T. T. "Structural fire protection, manuals and reports on engineering practice, No. 78." *ASCE,* 1992.

Malhotra, Harbans Lal. *Design of fire-resisting structures.* Surrey University Press, 1982.

Naser, Mohannad, and Venkatesh Kodur. "Response of fire exposed composite girders under dominant flexural and shear loading." *Journal of Structural Fire Engineering* (2017): 108–125.

NBCC, National Building Code of Canada. National Research Council of Canada, Ottawa, 2015.

"One Stop Shop in Structural Fire Engineering," University of Manchester, 2018.

Russo, Salvatore, and Francesca Sciarretta. "Masonry exposed to high temperatures: mechanical behaviour and properties—An overview." *Fire Safety Journal* 55 (2013): 69–86.

Underwriters Laboratories. *Fire Resistance Directory.* Vol. 1. Underwriters Laboratories, 1992.

Wickström, Ulf. *A very simple method for estimating temperature in fire exposed concrete structures.* RISE, SP—Sveriges Tekniska Forskningsinstitut, SP Fire Research, 1986.

Chapter 28

Disproportionate Collapse and Blast-Resistant Design

PART A

STRUCTURAL ROBUSTNESS AND DISPROPORTIONATE COLLAPSE

BY

SHALVA MARJANISHVILI, D.Sc., PE, SE *Technical Director, Hinman Consulting Engineers, Inc., San Francisco, California*

28.1 INTRODUCTION

Disproportionate collapse is defined as structural collapse which is disproportionately large than the initiating cause. Disproportionate collapse is manifested as a chain of progression of structural collapses often referred to as a progressive collapse. In the United States, the terms "disproportionate collapse" and "progressive collapse" are used interchangeably. However, the distinction needs to be made that the term "disproportionate collapse" refers to the event itself, while the term "progressive collapse" refers to the mechanism of the collapse. Not every disproportionate collapse is progressive, and they are not mutually exclusive.

The intent of disproportionate collapse mitigation design is to increase overall structural robustness of the building. Although there is no one commonly accepted definition of robustness, it is commonly agreed that robustness is structural quality which enhances structures ability to absorb small damage without large-scale collapse. Hence, the main purpose of disproportionate collapse mitigation analysis is to ensure that small damage will not cause structural failure to an extent disproportionate to initial damage.

Progressive collapse occurs when the sudden loss of a critical element initiates a chain reaction of structural element failures, eventually resulting in partial or full collapse of the structure. The cause of the initiating damage to the primary element is often considered unimportant; the resulting sudden changes to the building's geometry and load path are what matter. This analysis method is, therefore, *threat independent.* Current guidelines from both the United States General Services Agency and the Department of Defense incorporate a threat-independent approach to progressive collapse analysis.

While static analysis is allowed by current guidelines, it should be noted that progressive collapse is a dynamic event involving vibrations of building elements and resulting in internal dynamic forces, such as inertia and damping, whose energy may or may not be absorbed by the structure. Progressive collapse is also inherently a nonlinear event in which structural elements are stressed beyond their elastic limit to failure.

28.1.1 Disproportionate Collapse Examples

Example of disproportionate collapse cover wide range of structures with various causes. Notable cases are briefly discussed below.

Ronan Point, United Kingdom, 1968: In the early morning of May 16, 1968, the gas explosion in the apartment on the 18th floor of the 22-story Ronan Point building in West Ham, London, triggered partial collapse of the building. Four people were killed, and 17 others were injured. The building was a wall-panel system building. As a result of this event, starting in 1976, U.K. building codes have incorporated requirements to mitigate disproportionate collapses.

Alfred P. Murrah Federal Building in Oklahoma City, Oklahoma: On April 19, 1995, the nine-story Federal Building collapsed due to a truck bomb that was detonated outside. The explosion caused floors 4 and 5 to collapse on to floor 3, which was connected to the main transfer beam. This caused all of the columns on the southern perimeter connected to the transfer beam to collapse along with floor sections supported by those columns.

World Trade Center, New York City: On September 11, 2001, World Trade Center buildings 1, 2, and 7 in New York City collapsed as a result of terrorist attacks and the subsequent fires that followed. The National Institute of Standards and Technology concluded that the fire weakened the steel structure until the long bridge-like floor sections began to progressively sag. This sagging converted the downward pull of the trusses into an inward pull. This intensifying inward pull on the walls eventually caused the outer columns of Tower 2 and later the inner columns of Tower 1 to buckle and fold, thus initiating the collapses.

Other expanses of disproportionate collapse include (but are not limited to) Skyline Plaza, Virginia, 1973, and the U.S. Embassy bombing, Beirut, 1983.

28.1.2 Common Causes of Disproportionate Collapse

There are usually multiple factors that take place to initiate a disproportionate or progressive collapse. Often, these collapses will take place during the construction phase of a building. Miscommunication between contractors and engineering documents can cause a progressive collapse, where workers may improperly install specific structural elements that can lead to weakened structural members throughout the building. Some contractors may be pressed for time to where they may improperly construct key elements, connections, or finishing techniques. Construction technology may also fail, such as slab lifting technology in the L'Ambiance Plaza case, or crane collapses may lead to structural failures. Improper inspection or overlooking structural issues also leads to factors that initiate a progressive collapse. Another cause of progressive collapse can take place in the design phase of a building. Overlooking structural maintenance issues can lead to rust or other material failures, which can weaken the structure, making it more susceptible to progressive collapse. Specific elements may be underdesigned due to improper load path, or construction tolerances could lead to higher stresses in specific structural elements. Often design live loads are exceeded during the construction or maintenance of a structure.

28.2 DISPROPORTIONATE COLLAPSE MITIGATION STRATEGIES

Generally there are two accepted design approaches to mitigate disproportionate collapse: the direct design method and the indirect design method. Direct design methods include the alternate load path method and specific local resistance, while the indirect design method includes the selection of a good floor layout and the use of ductile connections, also tying the building together with peripheral, internal, and vertical ties and designing for load reversal, change of span direction, and membrane action in floor slabs.

Existing design criteria have discussed progressive collapse of a structure in a general way. Clear and consistent methodology for evaluating the potential for progressive collapse of a structure has been elusive in design criteria to date. The Federal Emergency Management Agency (FEMA) provides general guidance in the FEMA 426 publication as well as American Society of Civil Engineers (ASCE) in ASCE-7, promoting redundancy and ductility in the structure, resistance to load reversals, and local shear failure of the structural elements. The United States General Services Agency (GSA), in "Progressive Collapse Analysis and Design Guidelines," and the Department of Defense, in "Minimum Antiterrorism Standards for Buildings," have made significant efforts to define procedures to evaluate the potential of structures to progressive collapse. Both take a threat-independent approach, considering the loss of a component of the structure and evaluating the ability of the structure to remain standing following redistribution of the gravity loads previously supported by the lost member. The lost component is said to be "notionally," "instantly," or "immaculately" removed. Current guidelines allow further simplification by providing a static analysis procedure for progressive collapse evaluation by use of a load factor (a dynamic amplification factor) to approximate dynamic nature of the event.

Most recent criteria was issued by Unified Facilities Criteria (UFC) in UFC 4-023-03 which outlines progressive collapse analysis procedures along with example analysis. According to this document, the simplest way to achieve progressive collapse compliance is by using so-called tie force procedure. In the tie force procedure, structural components are tied back into the structure. Figure 28.1 shows schematic view of tie forces. Tie forces are prescribed based on dead and live loads applied to the tributary area of component. There is no verifiable evidence that the tie force procedure proves significant protection against progressive collapse mitigation. To limit the application of this simplified procedure, UFC limits its use to only buildings with very low occupancy.

The direct analysis procedure involves instantaneous removal of one major load-bearing element at a time, such as one column. Figure 28.2 shows a schematic representation of column removal. Note that the column is removed in such a way that there is no residual damage to connection due to the removal. This removal is purely hypothetical.

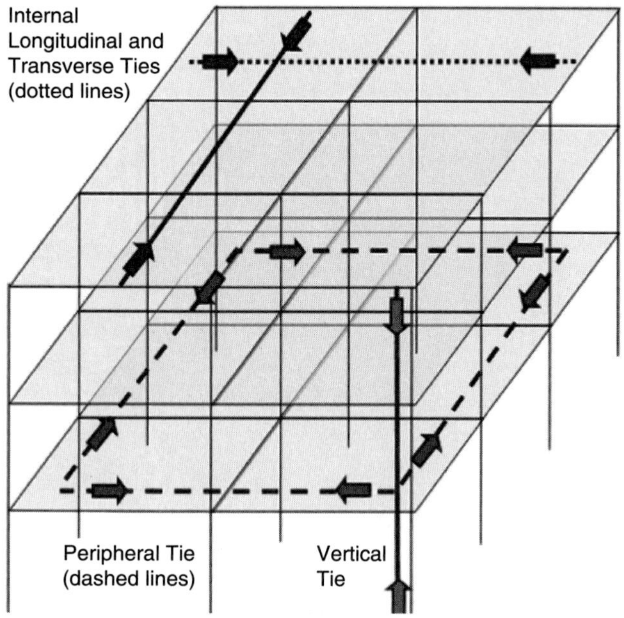

Figure 28.1 Tie force procedure per UFC.

Both UFC and GSA progressive collapse guidelines mandate major load-bearing element removal in several locations throughout the building. Elements are removed in preconceived vulnerable locations throughout the bulging, one element at a time. Figure 28.3 shows UFC-mandated element removal locations. The intent of this procedure is to find an element whose removal will cause largest or disproportionate collapse.

After the element is removed, a three-dimensional finite-element analysis is required to determine the extent of structural failures. According to UFC requirements, computer analysis should not indicate any failure in the structure; GSA guidelines, however, allow some area around the removed element to be lost. Structural engineers can choose among three analyses procedures allowed by UFC: linear static, nonlinear static, and nonlinear dynamic. UFC design procedures closely follow the ASCE-41 publication (see Seismic Rehabilitation of Existing Buildings ASCE/SEI 41-17/).

28.3 PROGRESSIVE COLLAPSE MODELING

From an analyst's point of view, progressive collapse occurs when failure of one load-bearing element results in the chain reaction of failure of other elements. Mathematically, loss of the potential energy in a structure (i.e., removal of the element) triggers structural movement (kinetic energy) until the structure stabilizes or continues losing energy until it collapses. Proper progressive collapse analysis requires dynamic nonlinear finite-element analysis of three-dimensional structural model. Needless to say, it is very difficult to model this phenomenon using commercially available structural analysis programs. A simple structural system is examined below to demonstrate the intricacies of progressive collapse phenomenon.

28.3.1 Equation of Motion of Progressive Collapse

The structure examined is constructed of reinforced concrete and loaded with 100 kips weight as shown in Fig. 28.4*a*. For this simple case, it is evident that the structural system may be represented as a mass spring system in equilibrium, as shown in Fig. 28.5*b*. Two springs

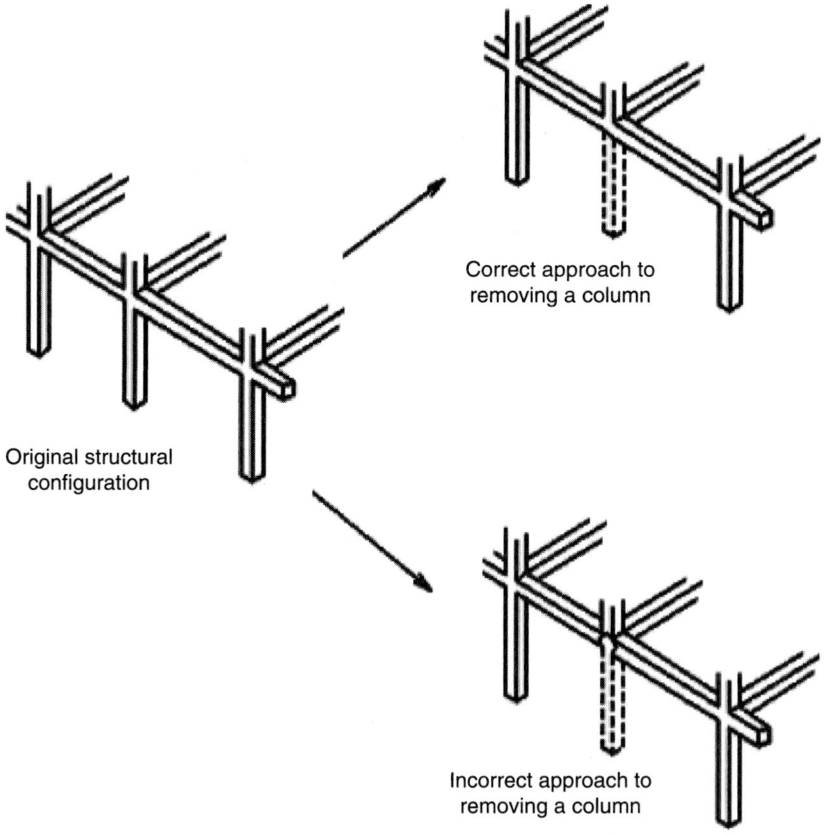

Figure 28.2 Notional removal of the column (note no residual damage to connection after column removal).

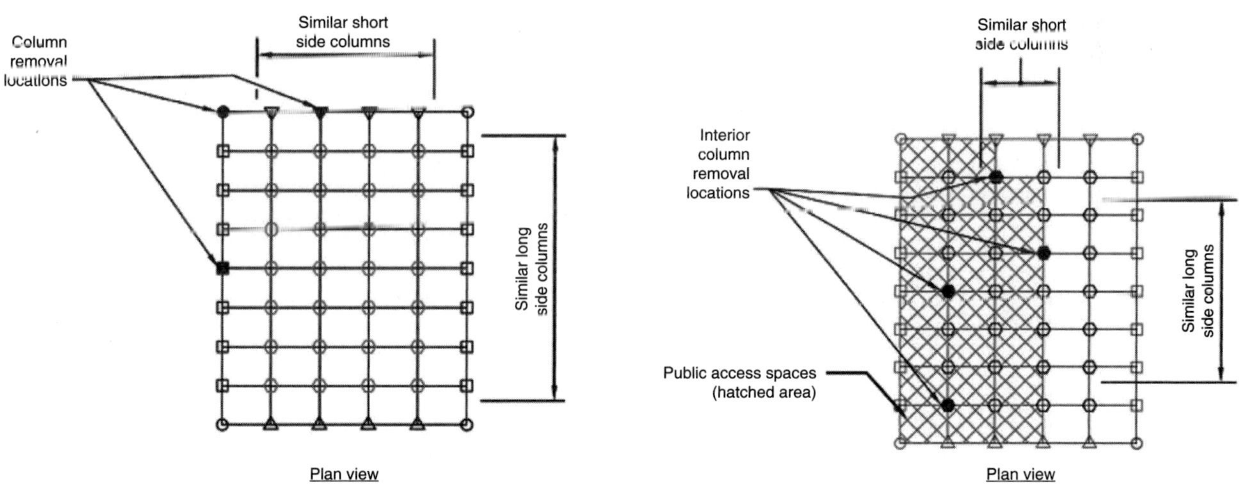

Figure 28.3 Load-bearing element removal scenarios.

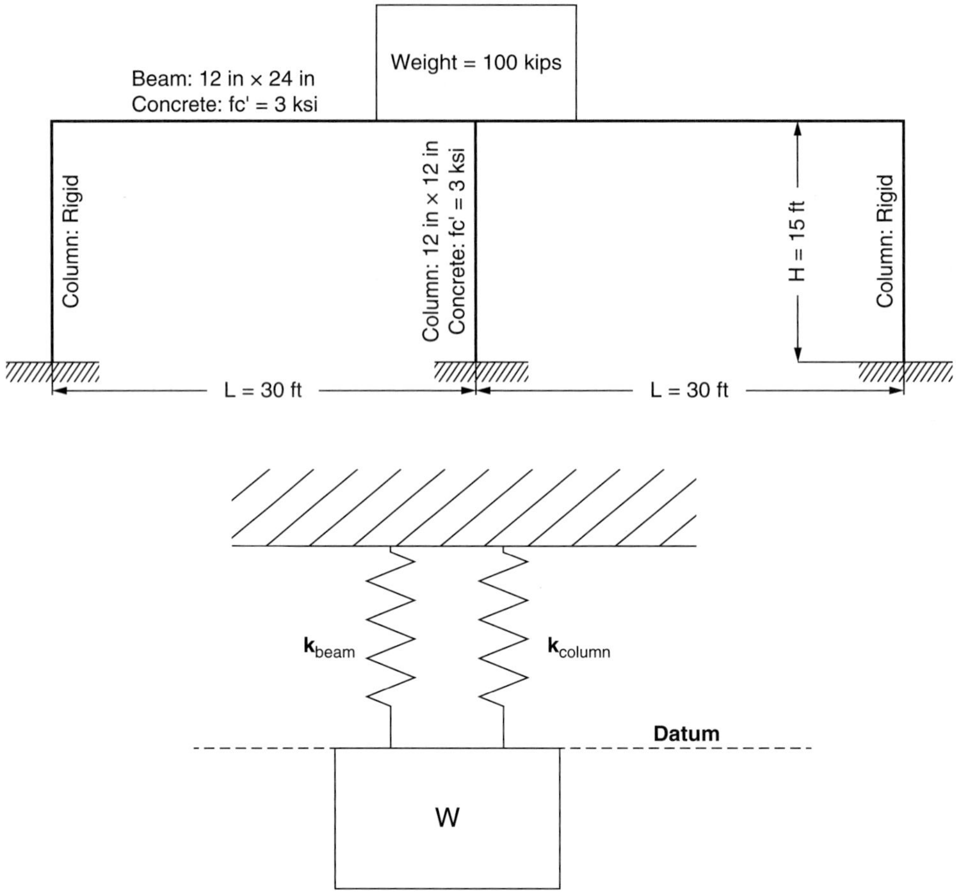

Figure 28.4 "At-rest" condition of two-dimensional frame (*a*) and corresponding SDOF analogy (*b*).

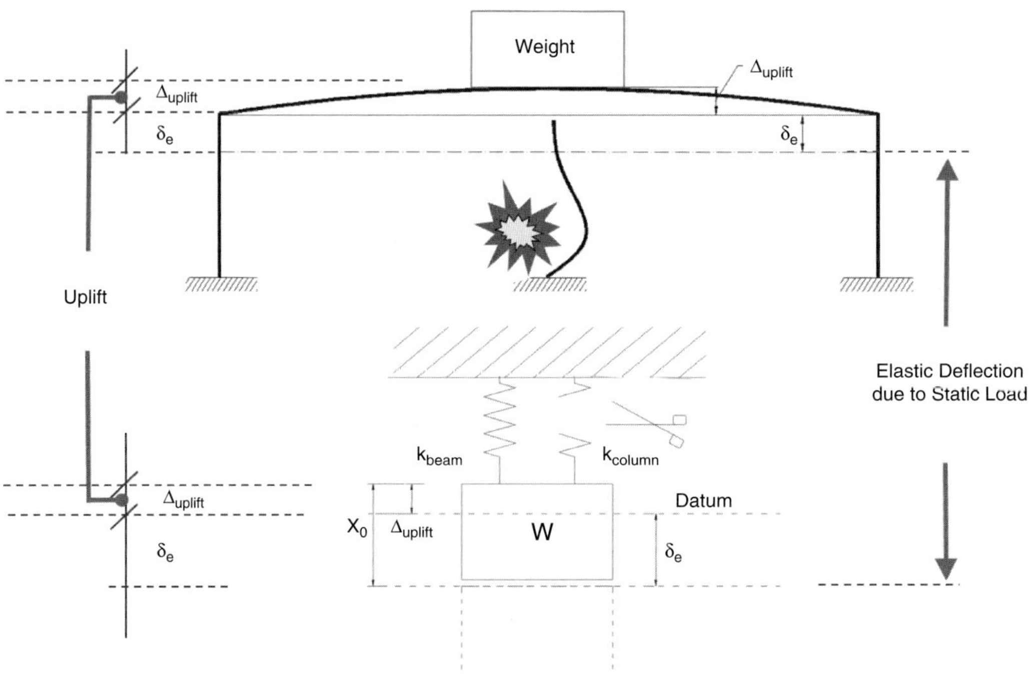

Figure 28.5 Initial conditions and initiating column removal causing progressive collapse.

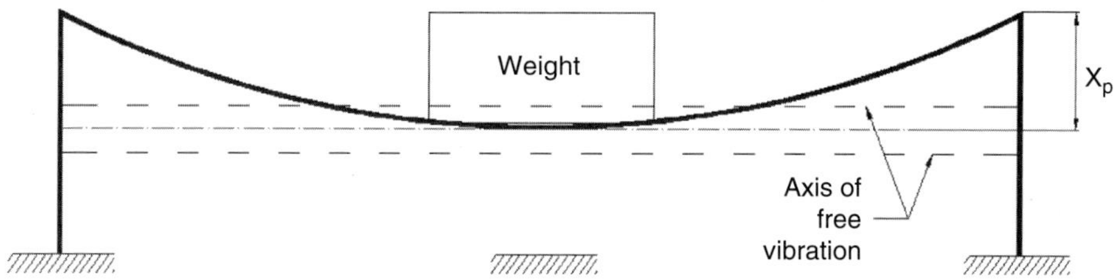

Figure 28.6 Nonlinear model for the progressive collapse analysis.

are shown which represent the stiffness contributions of the beam and column. The effects of the explosion (the column being destroyed) can be included in this model simply as a geometry (load path) change as shown in Fig. 28.5.

The progressive collapse of this system is initiated by the explosive destruction of the column or analogously by cutting spring k_{column} in the single-degree-of-freedom (SDOF) model. This change in geometry forces the structure to drop and vibrate. As Fig. 28.5 shows, the initial displacement from equilibrium is X_0, which is the sum of the elastic deflection under the static 100 kip load δ_e and the uplift deflection Δ_{uplift}. This initial displacement will be used in the collapse vibration analysis problem.

Generally, the equation of motion of the system shown in Fig. 28.5 is

$$mA + cV + kX = W = mg \quad \text{or} \quad A_a + 2\xi\omega V + \omega^2 X = g \tag{28.1}$$

where A_a is absolute acceleration
V is relative velocity
X is relative displacement
ω is un-damped free vibration frequency
ξ is damping factor
g is gravity constant acceleration

Equation (28.1) can be rewritten as

$$A + 2\xi\omega V + \omega^2 X = 0 \tag{28.2}$$

where $A = A_a - g$ as a consequence of Newton's first law of motion.

Solutions of Eq. (28.2) for undamped and under-critically damped systems are expressed as (Clough and Penzien, 1996)

$$X = X_0 \cos(\omega t) + \frac{V_0}{\omega}\sin(\omega t) \qquad \text{Undamped} \tag{28.3}$$

$$X = \left[X_0 \cos(\omega_D t) + \frac{V_0 + X_0\xi\omega}{\omega_D}\sin(\omega_D t) \right] e^{-\xi\omega t} \qquad \text{Damped} \tag{28.4}$$

where X_0 is initial displacement
V_0 is initial velocity
t is time
$\omega_D = \omega\sqrt{1-\xi^2}$ is damped free vibration frequency
$e = 2.72$

For the model shown in Fig. 28.5, initial conditions are $X_0 = \delta_e + \Delta_{uplift}$ and $V_0 = 0$, where δ_e is elastic displacement of the system under the static loading condition.

When initial velocity is zero (i.e., $V_0 = 0$) Eqs. (28.3) and (28.4) reduce to

$$X = X_0 \cos(\omega t) \tag{28.5}$$

$$X = X_0 \left[\cos(\omega_D t) + \frac{\xi}{\sqrt{1-\xi^2}}\sin(\omega_D t) \right] e^{-\xi\omega t} \tag{28.6}$$

28.3.2 Nonlinear Modeling of Progressive Collapse

Progressive collapse is inherently nonlinear. It is imperative to include material and geometric nonlinear behaviors in the model in order to realistically reflect the existing conditions of the structure. Figure 28.6 shows the deflected shape of the two-dimensional frame shown in Fig. 28.4, with the material allowed to undergo the nonlinear behavior illustrated in Fig. 28.7.

28.3.3 Ductility and Demand Capacity Ratio (DCR)

Ductility is defined as the measure of the ability of the structural system to undergo nonlinear deformations without failure and is estimated as

$$\mu = \frac{X_{max}}{X_e} \tag{28.7}$$

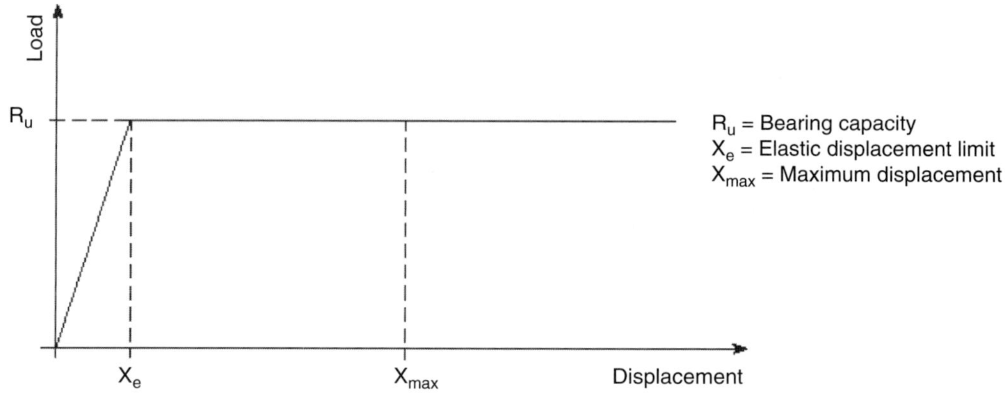

Figure 28.7 Assumed force-displacement relationship.

where μ represents ductility

X_{max} is calculated maximum nonlinear displacement

X_e is elastic displacement limit as shown in Fig. 28.7

Therefore, for a system to remain linear, its ductility must be less than or equal to 1.

Also, for structures having immaculate element removal, the structural response ratio for progressive collapse is bound by 2. That is, for the system to remain elastic, the remaining structural elements should be able to carry twice the structural weight without experiencing nonlinear behavior, or, mathematically,

$$\text{DCR} = \frac{\text{Weight}}{\text{Capacity}} \leq 0.5 \tag{28.8}$$

where DCR is demand-capacity ratio

Weight is load generated by weight (see Fig. 28.4)

Capacity is bearing capacity of the remaining structural elements

This leads to the simplification that progressive collapse analysis may be performed statically by increasing the load by a dynamic amplification factor (DAF) of 2. That is,

$$\text{DAF} = \frac{\text{Static Equivalent Dynamic Load}}{\text{Static Load}}$$

$$= \frac{1}{\text{DCR}} = 2 \text{ for imaculate removal} \tag{28.9}$$

Also note that when DCR > 1, the structural system will collapse.

28.4 PROGRESSIVE COLLAPSE ANALYSIS EXAMPLE

Progressive collapse analysis is carried out as threat-independent by instantaneously removing one of several major load-bearing elements and analyzing the ability of the damaged structure to absorb the energy. A finite-element model of the example structure with the assumed column loss scenario is shown in Fig. 28.8.

28.4.1 Assumptions

To simplify the analysis while illustrating the dynamic analysis procedure, the following assumptions are made:

1. The structure is modeled as two-dimensional.
2. Effects of large deflections are neglected.
3. Elastic-perfectly-plastic moment-rotation relationships are used.
4. Equivalent structural damping of 5 percent is assumed throughout the analysis.
5. All beam-to-column connections are moment-resistant and are stronger than the beams, so plastic hinges will form in the body of the beam and not in the column or in the joint (strong column—weak beam principle).
6. All beams are adequately confined by shear reinforcement so that beams are not shear controlled.
7. Columns have adequate strength to resist additional load redistribution due to the loss of the primary column.

Note that these simplifications are made for the purpose of this example to clearly demonstrate the usage of dynamic analysis procedure. It is up to the design professional to determine assumptions applicable for the design process.

28.4.2 Finite-Element Model

The two-dimensional finite-element model, depicting the column loss scenario is shown in Fig. 28.8.

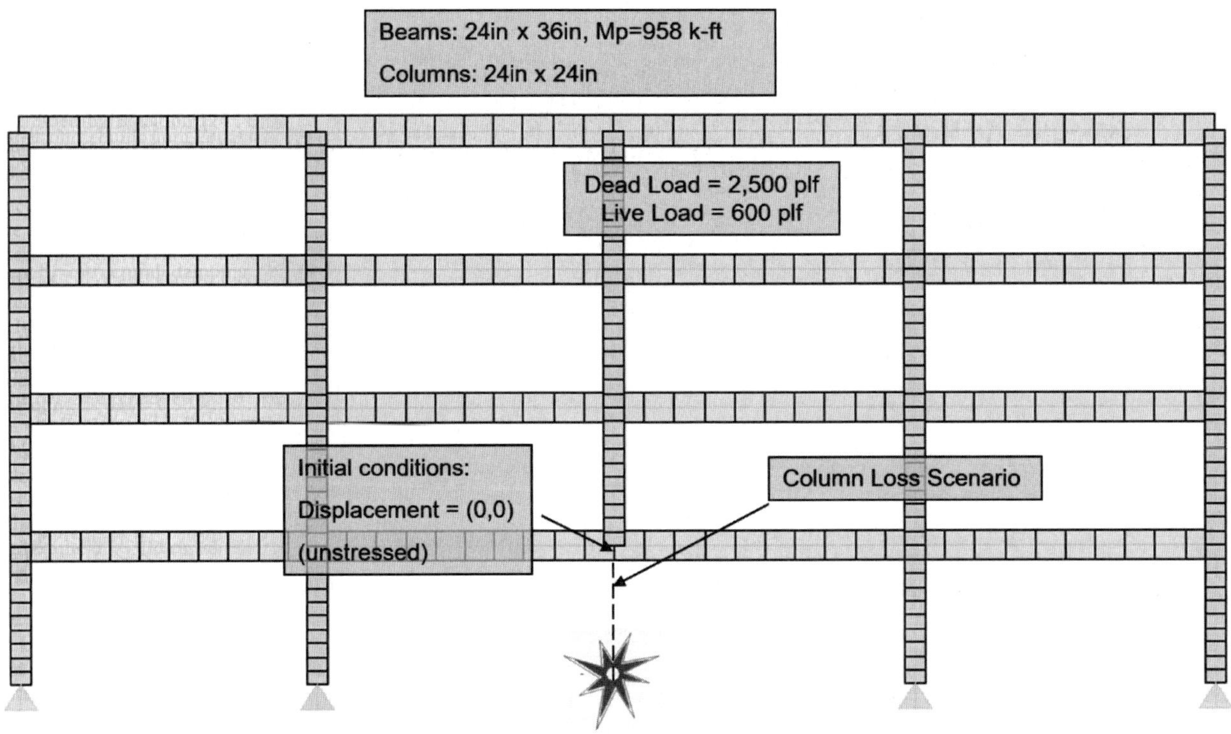

Figure 28.8 Finite-element model of example structure.

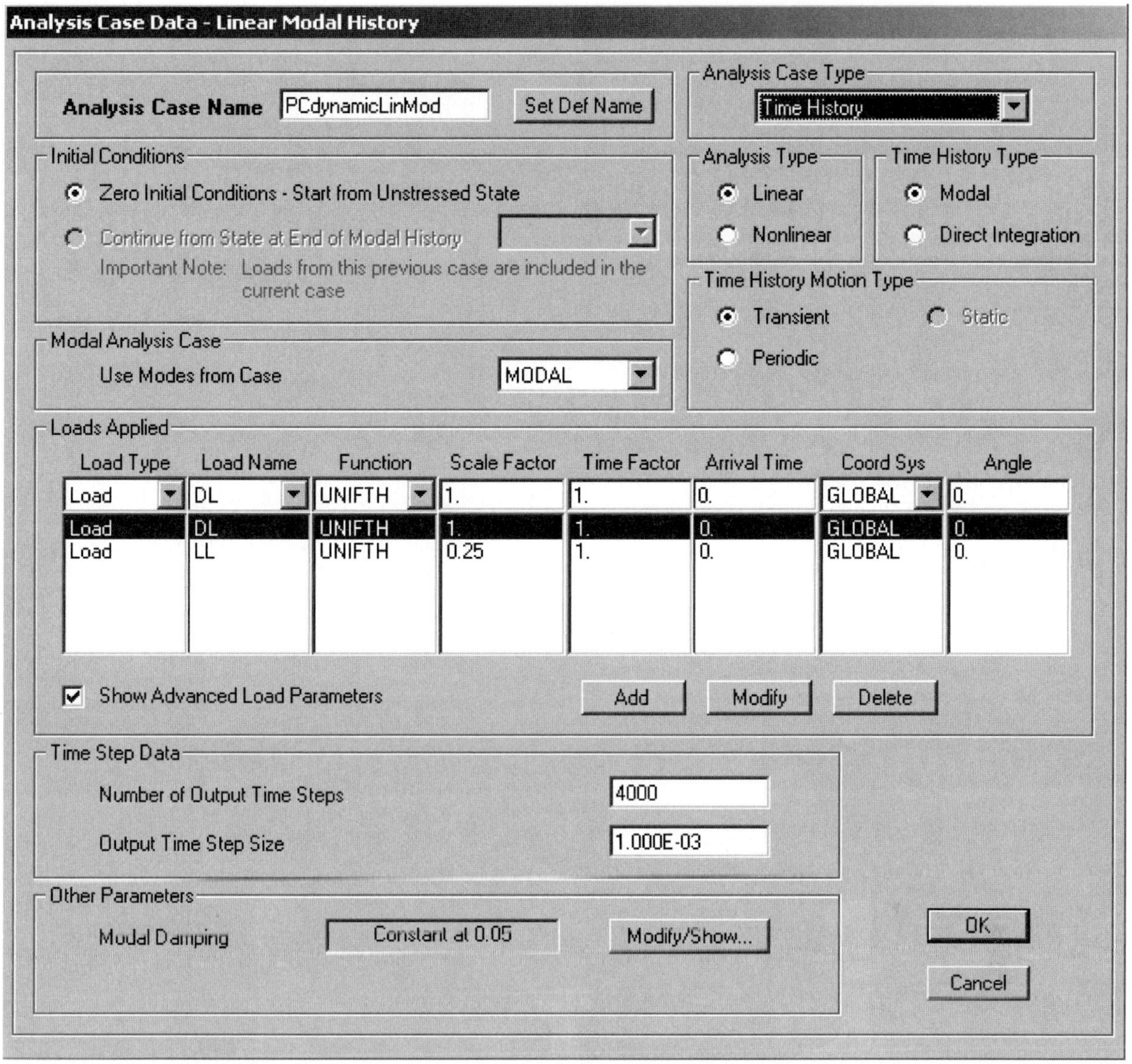

Figure 28.9 Analysis case data.

28.4.3 Loads

GSA Progressive Collapse Guidelines mandate the following loading combinations when evaluating for progressive collapse:

$$\text{Load} = \text{DL} + 0.25\ \text{LL} \tag{28.10}$$

For the purpose of this example, the following loading conditions have been assumed (shown in Fig. 28.8):

DL = 2500 plf   Uniformly distributed dead load

DL = 600 plf   Uniformly distributed live load

Note that the GSA- and UFC mandated load combinations differ, but the principle of dynamic an alysis is the same, and so the remainder of the analysis is carried out using GSA's load combinations, outlined in Eq. (28.10).

28.4.4 Dynamic Analysis Using SAP2000

Dynamic analysis for progressive collapse is carried out using an "initial conditions" methodology. In SAP2000, dynamic analysis with initial conditions can be modeled as a time-history analysis with constant dead and live loads (see Fig. 28.9). This analysis involves finding the displaced shape of the *undamaged* structure under normal loading conditions and then applying those displacements as initial conditions to the dynamic analysis of the damaged model (i.e., the model with a column loss scenario). In other words, the column is removed from the structure, initial conditions are applied to the structure in order to return it to its undamaged shape, and then the analysis begins. This process dynamically simulates the sudden loss of the column.

The displacements of the loaded, undamaged structure are usually negligibly small compared to the damage it will suffer after the loss of a primary column. Therefore, to simplify the analysis procedure, it is possible to apply initial conditions of the unloaded, undamaged structure, that is, "zero initial conditions." The displacements of the *unloaded undamaged* structure are the same as the displacements of the *unloaded damaged* structure—they are zero in both cases (neglecting self-weight). This allows us to model the dynamic loss of the column using the unstressed state—zero initial conditions. Figure 28.3 shows what the input screen for the linear dynamic load case would look like in SAP2000.

Note that, if desired, the analysis can be carried out using a prestressed state as the initial condition. This, however, will require additional analysis steps to determine the prestressed state and corresponding deformed shape.

28.4.5 Linear Dynamic Analysis

The linear dynamic analysis procedure can be modeled using the initial conditions methodology, and it is available in SAP2000 as one of the analysis methods.

After building the computer model, linear dynamic analysis involves the following steps:

1. Define the dynamic load case [Eq. (28.1)].
2. Perform a time-history analysis with zero initial conditions.
3. Analyze the time-history response based on demand capacity ratio.

The main advantage of this procedure is its ability to account for dynamic amplification effects. This procedure is limited to structures that are expected to remain elastic during the event. It also should be noted that linear dynamic analysis procedure is not permitted by either GSA or UFC guidelines.

28.4.6 Nonlinear Dynamic Analysis

Nonlinear dynamic analysis is performed when it is expected that the structure will experience nonlinear behavior. Nonlinear dynamic analysis for progressive collapse is carried out similarly to the linear dynamic analysis procedure, with the exception that beam elements are now allowed to enter the inelastic range of deformation. This requires certain assumptions to be made for plastic hinge locations and behavior. The plastic hinge locations are defined as shown in Fig. 28.10.

After building the computer model, nonlinear dynamic analysis involves the following steps:

1. Define the dynamic load case [Eq. (28.10)].
2. Define plastic hinge locations.
3. Perform a time-history analysis with zero initial conditions.
4. Analyze the results based on the maximum ductility and rotation.

Nonlinear analysis is inherently more accurate than linear analysis because it accounts for the plastic behavior of the structure. The downside to this is that one must be able to predict the locations of plastic hinging in the structure, which may not be feasible for more complicated structures. Nonlinear analysis also takes much longer than linear analysis and may involve several analysis reruns as the user varies input parameters such as step size in attempt to find a stable solution.

28.4.7 Analysis Results

Deflection and rotation time history for both the linear and the nonlinear analysis cases is shown in Fig. 28.11.

As intended, both analyses capture the dynamic behavior involved with the sudden loss of the column. The linear static case reports a deflection of 3.13 in., while the nonlinear analysis captures some permanent displacement and maximum deflection of 7.9 in.

As is illustrated here, dynamic analysis for progressive collapse can be carried out using commercially available finite-element programs and is fairly easy to perform. Another reason to perform dynamic analyses over static analyses is because they are more accurate—they include the dynamic nature of the progressive collapse phenomenon. For nonlinear analysis, the only additional modeling compared to linear analysis is the plastic hinging definition.

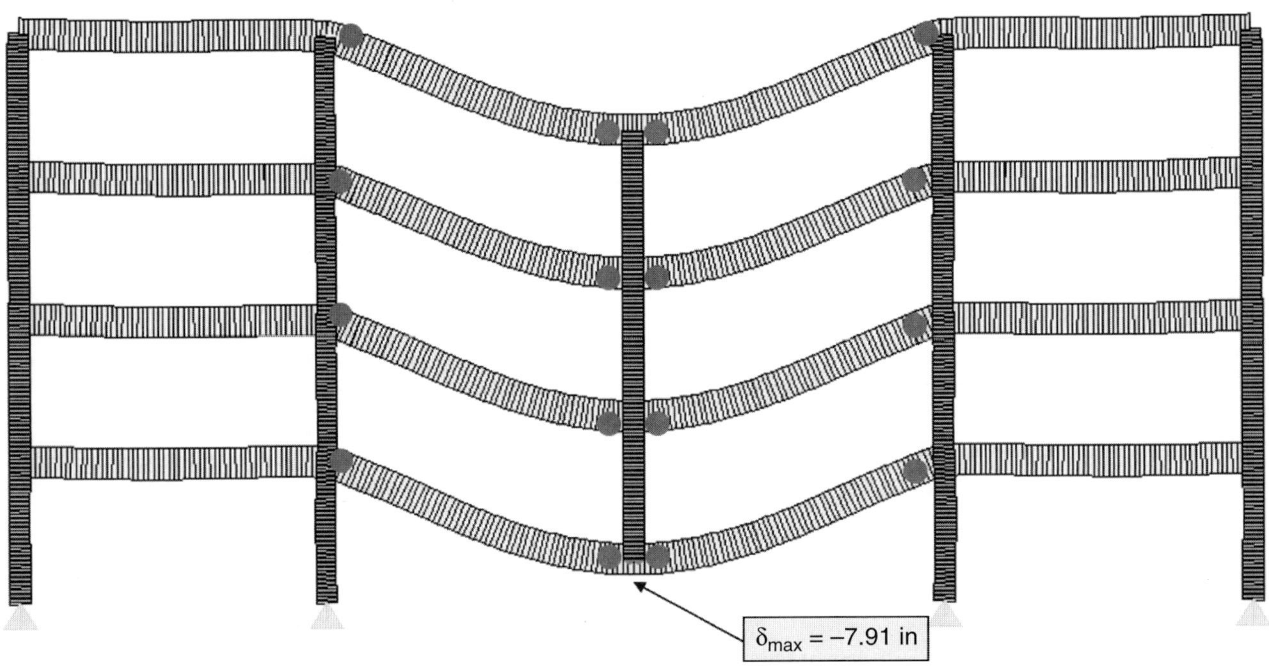

Figure 28.10 Plastic hinge locations.

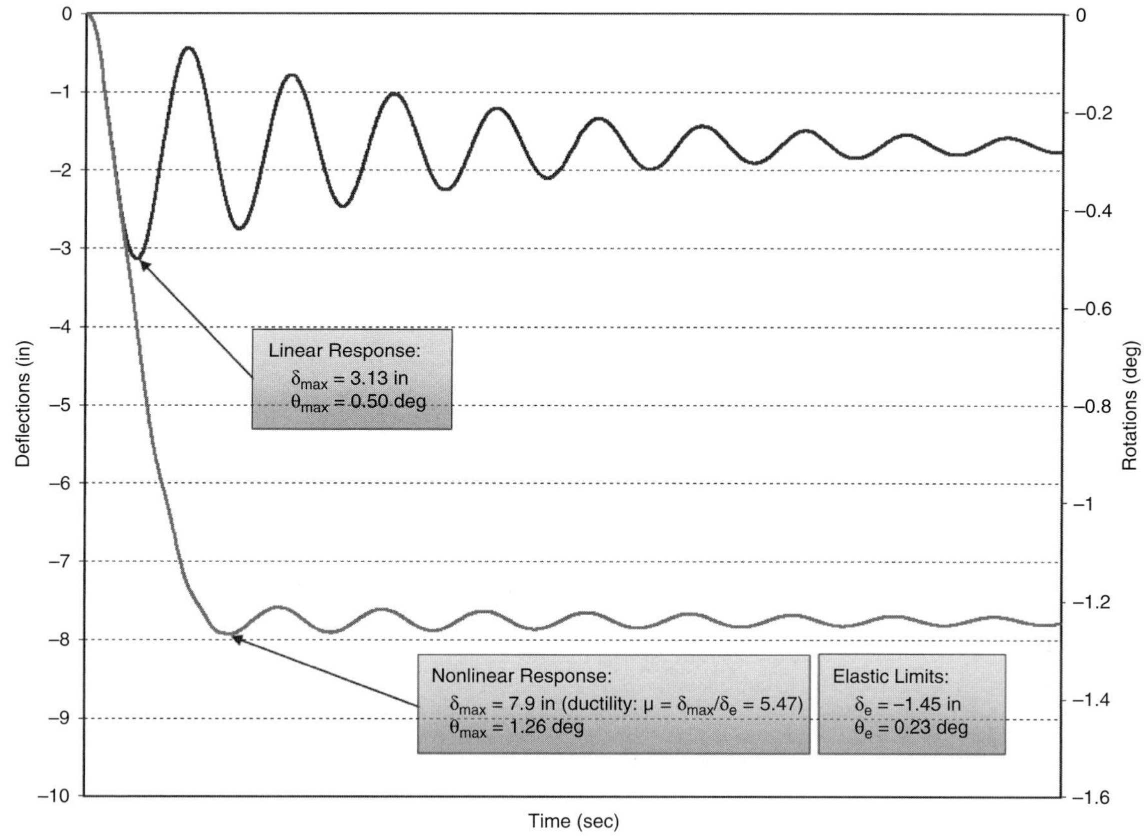

Figure 28.11 Dynamic analysis results.

REFERENCES

ASCE/SEI 7-16. Minimum Design Loads and Associated Criteria for Buildings and Other Structures.

ASCE/SEI 41-17. Seismic Evaluation and Retrofit of Existing Buildings.

Buscemi, Nanci, and Marjanishvili, Shalva. SDOF Model for Progressive Collapse Analysis. SEI Structures Congress, April 20–24, 2005.

Clough, R. W., and Penzien, J., (1993), *Dynamics of Structures*. McGraw-Hill, New York (2nd edition).

GSA Progressive Collapse Analysis and Design Guidelines for New Federal Office Buildings and Major Modernization Projects. General Services Administration (GSA). https://www.gsa.gov/cdnstatic/GSA_Progressive_Collapse_Guidlines_Final.pdf (accessed January 3, 2020).

Marjanishvili, S. M., and Agnew, E. (2006). "Comparison of Various Procedures for Progressive Collapse Analysis," *ASCE Journal of Performance of Constructed Facilities*. 20, 365.

SAP2000 Version 21. (2019), *Analysis Reference Manual*. Computers and Structures, Inc., Berkeley, CA.

UFC 4-023-03 Design of Buildings to Resist Progressive Collapse, with Change 3. WBDG—Whole Building Design Guide, (n.d.). https://www.wbdg.org/ffc/dod/unified-facilities-criteria-ufc/ufc-4-023-03 (accessed January 3, 2020).

Wilson, E. L. (2002). *Three Dimensional Static and Dynamic Analysis of Structures*. Computers and Structures, Inc., Berkeley, CA (3rd edition).

PART B
BLAST-RESISTANT DESIGN

BY

ROBERT SMILOWITZ, Ph.D., PE *Senior Principal, Thornton Tomasetti*

28.5 INTRODUCTION

The evaluation of structures in response to blast loadings resulting from explosive detonations may be associated with a wide variety of causes, including petrochemical or industrial accidents, defective gas distribution systems or appliances, the storage of military materials, and terrorist events. While the cause of the various types of detonations may differ and the short-duration dynamic blast waves produced by the explosive materials will be scenario-driven, the analytical methods and design principles will be the same. The following discussion is limited to blast loadings resulting from terrorist events, which are typically expressed in terms of a pressure (in units of psi) and a corresponding impulse (in units of psi-msec) due to a weight of high-energy (HE) explosives, which is typically expressed in terms of TNT equivalents.

The analysis and design of structures to resist blast loadings is an inherently dynamic and nonlinear phenomenon that relies on both material strength and the inertial resistance. The dynamic loading, the nonlinear force-displacement behavior, and the inertial resistance define the dynamic equations of motion. The performance-based design process starts with the definition of the energetic source, which is termed the explosive threat, and the extent of damage the structure is permitted to develop, which is termed the performance criteria. The specification of explosive threat may be based on the owner's specified criteria, the maximum credible event, historic precedent, or a Threat and Risk Analysis (TARA).

The maximum credible event considers all feasible modes of attack, which are typically defined by the means of delivery by both vehicles and persons. High explosives weigh about 100 pounds per cubic foot, so the maximum credible event is usually limited by the capacity of the container or vehicle transporting the device.

An assessment based on the recorded occurrence of historical events acknowledges the large number of events that involve small weights of explosives compared to the relatively few number of events that involve large weights of explosives. In the United States, the 1995 Oklahoma City bombing and the 1993 World Trade Center bombing serve as the two most significant historical precedents for selecting the explosive hazard. Similarly, the 1996 explosion at the Atlanta Olympics is a significant domestic event involving a hand-carried explosive device.

A pragmatic approach considers the relatively small likelihood that any one facility will ever experience an extreme hazard and it is much less likely to occur than a small-scale hazard. This approach considers criticality, accessibility, and consequences of various threat scenarios. Unlike the evaluation of risk associated with natural hazards, for which there are documented return periods of events of various magnitudes and probabilities of exceedance, the risk associated with man-made hazards cannot be determined in absolute terms. A TARA may therefore be based upon relative likelihood of the various man-made hazards. The relative risk for a variety of hazards may therefore be calculated and compared with the cost to mitigate the resulting hazards to an acceptable level. The objective may be to reduce all quantified relative risk scores to a given order of magnitude, to determine the cost and benefit for reducing the relative risk by a specified amount, or to maximize risk reduction within the context of a limited budget.

The intensity of explosive threat primarily depends on both the weight of explosive material and the distance between the detonation and the protected structure. This standoff distance will be a function of the means of delivery and any protective measures that might restrict the origin of the threat. Whereas the access of vehicles may be restricted by bollards or barriers, the hand-carried threat may be placed in close proximity to or against the protected structure.

Once a threat criteria is established, the structure may be designed to various levels of performance, ranging from continued operations to collapse prevention. Although an initial choice in structural system should consider the robustness and redundancy that is beneficial for blast-resistant protective design, all structures must satisfy building code requirements, and this is the natural starting point for design. The inelastic response to blast loads will typically dictate the member connection forces; however, the serviceability limitations may dictate member sizes. Therefore, a fully code-compliant initial design assures the design team that any blast-resistant upgrades will also be code compliant and will provide a baseline from which the incremental cost for the blast resistance may be calculated.

28.6 BLAST PHENOMENA

The detonation of HE explosives is a chemical reaction that releases gas at supersonic velocity. This ever-expanding shock front radiates spherically in all directions, distributing the blast energy over an ever-increasing surface area. As the intensity of the shock front, characterized by a peak

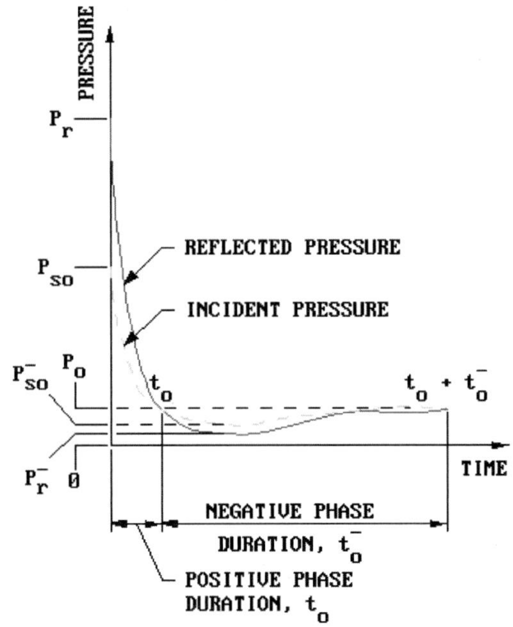

Figure 28.12 Pressure time history (UFC 3-340-02).

pressure, diminishes, the duration of the exponentially decaying blast pulse increases. The resulting impulse, which is the area beneath the pressure versus time pulse (Fig. 28.12), therefore decays slower with distance from the blast than the decay in peak pressure. The duration of the exponentially decaying pressure wave is measured in milliseconds, and the entire positive phase of the blast loading may be idealized as a triangle of equivalent duration $t_o = 2i/P$.

After the shock wave moves outward from the explosive source, it leaves behind a partial vacuum, which is termed the negative phase or under-pressure. The negative phase is a low-intensity but long-duration pulse that tends to reverse the loading on the structure. Figure 28.12 depicts the peak pressure and exponential decay of the positive phase, which is idealized by the equivalent linear decay, and the subsequent longer-duration negative phase, which is similarly idealized by piecewise linear approximations. Often, a structure will be more conservatively designed to resist the positive phase blast loading without considering a negative phase that might limit the maximum response of the structure.

The intensity of peak pressure and corresponding impulse for TNT explosions at a range of standoff distances can be characterized by the semi-empirical Kingery Bulmash equations. These equations are derived from explosive test data as a function of scaled range, which is the standoff distance normalized with respect to the cube root of the weight of explosives. As such, their applicability is limited to the scaled ranges for which test data were used to create the curve fits and are not valid for short standoff distance scenarios.

An explosive detonated on the ground will cause the material beneath the blast to crater and will reflect a majority of the shock wave energy upward and outward in a hemispherical shock front. The intensity of the hemispherical blast wave propagation may be conceptually idealized as twice the charge weight with a spherical blast wave propagation.

Whether the shock wave propagation is spherical or hemispherical, the shock front stagnates when it encounters a denser material. The stagnation of the shock wave causes the reflected peak pressure and corresponding impulse to amplify. The magnitude of the amplification factor increases with the blast load intensity, and the reflected peak pressure may be an order of magnitude (or greater) than the intensity of the incident peak pressure. However, the blast load that may be applied near the edge of a reflecting surface will be affected by a relief wave, which enables a portion of the stagnated shock wave to flow around the corner. These phenomena are called "clearing effects," and the intensity of the relief wave is a function of its arrival time, which is the distance to the free edge divided by the wave speed velocity. The shorter the arrival time, the sooner the reflected pressure will revert to the incident pressure and the resulting impulse will be between the fully reflected impulse and the fully incident magnitude.

Normalized magnitudes of incident and reflected peak pressures and the corresponding pulse duration and impulses may be plotted as a function of the scaled range (Fig. 28.13), and these plots are sometimes called spaghetti curves. UFC 3-340-02 "Structures to Resist the Effects of Accidental Explosions" is a reference where these charts can be obtained. The United States Army Corps of Engineers programmed the Kingery Bulmash equations in a convenient software package, ConWep, which is a collection of conventional weapons effects calculations from the equations and curves provided in UFC 3-340-01 (formerly TM 5-855-1) "Design and Analysis of Hardened Structures to Conventional Weapons Effects." ConWep allows the user to determine the blast load parameters as a function of charge weight, standoff distance, and angle of incidence relative to a reflecting surface. The software further calculates the clearing effects based on the user-specified width of the reflecting surface. The distribution of UFC 3-340-01, ConWep, and other software packages is limited to approved organizations and users.

The peak blast pressures, expressed in pounds per square inch, are significantly greater in magnitude than conventional design loads due to gravity and wind; however, the duration of these loads is b measured in milliseconds. The following plots (Fig. 28.14) demonstrate the magnitude of peak reflected pressure and reflected impulse resulting from three orders of magnitude of explosive weight as a function of standoff distance. The intensity of these blast loads and the area tributary to a structural member will define the forcing function side of the equations of motion.

Blast scenarios that originate outdoors will propagate, as described above, until they reflect off a denser material, such as the building façade. Blast waves will diffuse over the top and around the sides, applying incident pressures on all surfaces that do not stagnate the blast wave. Therefore, all surfaces will be subjected to blast loads of varying intensity and time of arrival.

However, blast scenarios that originate within a structure will encounter multiple reflecting surfaces in the form of walls and slabs. These multiple reflections prevent the blast wave from propagating spherically or hemispherically and trap the blast energy within the confined space, resulting in an amplification of the blast loads. The blast wave propagation is at best two-dimensional and is combined with an additional lower-intensity but very long-duration (quasi-static) gas pressure that is associated with the by-products of the chemical explosion building up within a confined space. This trapped energy will be applied to the internal slabs and partitions until they fail and allow the internal pressures to escape and equalize with the atmosphere. Such was the case in the 1993 bombing of the World Trade Center below-grade

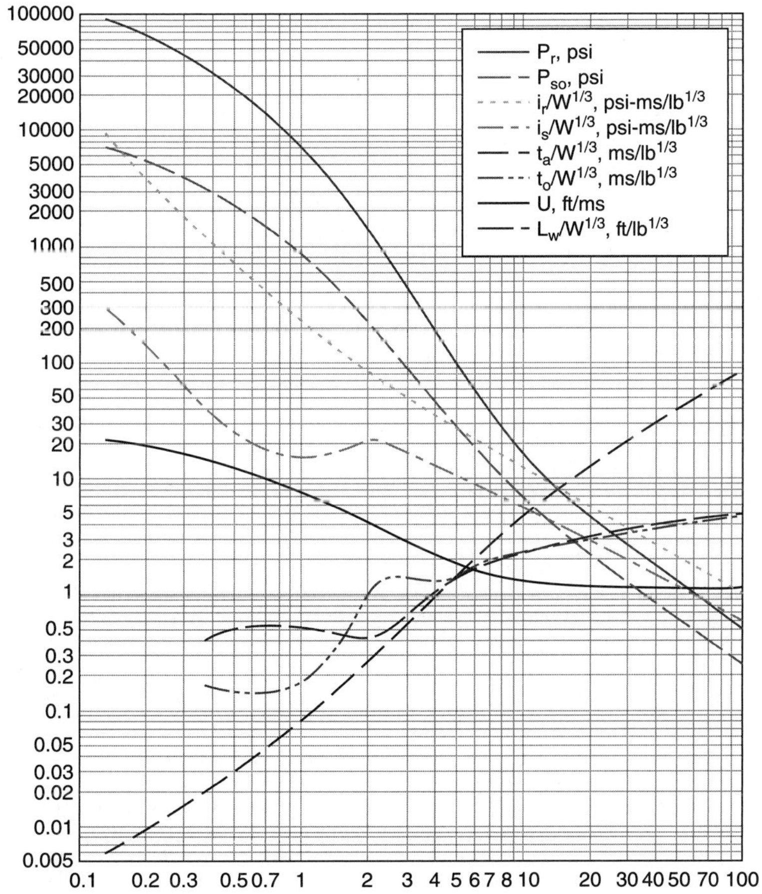

Figure 28.13 Blast parameters versus scaled distance (UFC 3-340-02).

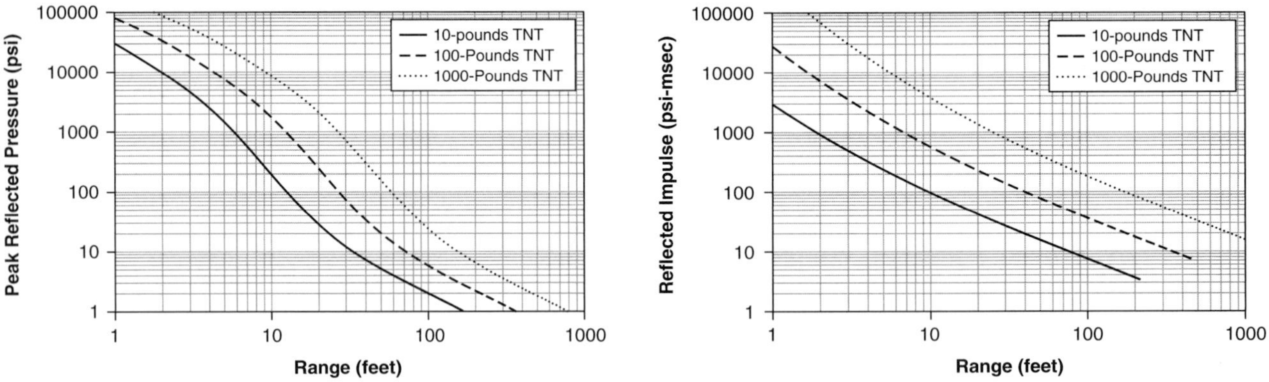

Figure 28.14 Peak reflected pressure and reflected impulse versus range (Smilowitz).

parking structure. The blast loads breached several bays of both the floor and ceiling slabs, dissipating blast energy in the process and creating a larger volume of air in which the internal blast waves could vent.

While the Kingery Bulmash equations can be used to calculate the initial shock loading on the various internal surfaces, account must be taken of the subsequent reflections. The BlastX software, which was developed for the Army Corps of Engineers, applies a ray tracing solution that considers the shock, negative phase, and gas pressure loading histories resulting from HE detonations. BlastX permits the modeling of multiple rooms with various areas identified as openings or venting panels. As such, Blast calculates in-fill pressures due to shock waves passing from one room into the next. However, BlastX solutions are sensitive to the number of rays selected for the analysis, and care must be taken to avoid spurious response.

Computational fluid dynamics (CFD) is the most comprehensive and accurate analytical approach for calculating shock wave propagation. Starting with the equations of state, the CFD analysis calculates mass transport from one cell into the neighboring cell. The refinement of the mesh and numerical algorithms will dictate the sharpness of the shock front and accuracy for peak magnitudes; however, more refined models require significant computing power and take longer to execute. CFD solutions are essential for accurate evaluation of contact or near contact detonations, where the Kingery Bulmash equations are not valid; however, the additional accuracy provided by CFD is often not justified for standoff detonations, particularly where reflective surfaces do not contribute to the loading on a structural member.

Shock waves from an explosive surface burst also propagate through geologic media at a high velocity. Foundation walls, slurry walls, and below-grade structures are therefore vulnerable to ground shock loading. Detailed finite-element models of the structure and surrounding geologic media may be used to calculate the resulting soil-structure interaction. These finite-element models rely on the accuracy of the material models, which experience large inelastic deformations. The boundaries of these "soil-island" calculations must be sufficiently far from the region of interest to prevent spurious reflections off artificial boundaries from affecting the results.

28.7 DYNAMIC DESIGN APPROACH

A structure's response to blast loading, similar to its response to strong ground motions, is an extraordinary event that may not reasonably be expected to remain elastic and undamaged. Occupancies with greater importance, such as Type IV structures, require immediate occupancy and perhaps continuity of services. Less critical occupancies can tolerate greater levels of damage with greater recovery costs, provided life safety of the occupants is protected to the level required. Unlike strong ground motion, which shakes the entire building foundation with similar magnitudes of acceleration, the intensity of blast loading may vary greatly throughout the structure. Therefore, greater levels of damage can be tolerated over a localized region so long as the damage does not propagate disproportionately to the initiating event. Arresting the propagation of collapse and the prevention of disproportionate collapse is a related subject that is often treated in a threat-independent approach and will be discussed separately.

Blast-resistant protective design most efficiently starts with a conventional design that satisfies the code-required strength and serviceability criteria. In some cases, these conventional design criteria may dictate a specific design approach, and the performance in response to blast loading may be dictated by local detailing. In other cases, the blast-resistant protective design will dictate the selection of structural member sizes. Regardless, it is useful to use the conventional design as a baseline against which the blast requirements may be optimized. It is also useful to have this baseline design to determine the cost increase associated with blast-resistant protective design.

Having established the blast loading scenarios, the required strength and inertial resistance of the structure depends on the extent of damage it is permitted to sustain. It is unreasonable to expect structures to remain elastic in response to extraordinary blast load intensity; therefore, the magnitude of inelastic deformation or ductility may be used to quantify the amount of blast energy that is dissipated through inelastic deformation. Thresholds of inelastic deformation corresponding to superficial, moderate, heavy, and hazard extents of damage are tabulated in Blast Protection of Buildings (ASCE 59). ASCE 59 is a consensus standard that provides guidance and minimum requirements whenever the governing performance criteria are not explicitly specified by the owner. These performance-based thresholds, typically expressed in terms of member end rotation and ductility, establish target response characteristics that may be achieved through iterative design. The extent of permissible damage depends on the intended occupancy and use of the facility. Life safety design standards may permit moderate damage to primary structural elements, which comprise the primary gravity and lateral load resisting system, but heavy damage may be permitted to secondary structural elements, such as floor slabs and local structural elements. Continued operations facilities or emergency service facilities may require the structure to sustain superficial damage overall but locally moderate damage in the immediate vicinity of the explosive event. It is important to establish the required performance criteria relative to the intended use of the structure during schematic stages of design as this may dictate the choice of structural systems.

The nonlinear equations of motion may be analyzed sequentially for individual members or simultaneously for large portions of the structure. While the boundary conditions and dynamic interaction between structural members are more accurately represented when evaluated simultaneously, the complexity of the model and dynamic analysis makes this cumbersome for a design project. Single-degree-of-freedom (SDOF) models of individual members are commonly used to determine the inelastic performance of individual components. SDOF models lump all the mass, the stiffness, the strength limits, and the loading into a fundamental mode representation of the structural element. The accuracy of this model depends how well the structural element can be characterized by a SDOF and how well the material model represents the inherent damage mechanisms. SDOF methods can be used to quickly perform parametric studies of different design concepts, and the results of these studies can be used to identify a preferred approach.

The loading function used to analyze a structural element may be based on tributary area and tributary mass of the elements that frame into it. Although it ignores the dynamics of the load collecting elements, it is generally the simplest approach that yields conservative results. A series of sequential SDOF analysis may alternatively be performed in which the collecting elements are first analyzed and the resulting dynamic reaction forces are applied to the supporting member along with the portion of mass that may be considered to move with the supporting member. An example of such a system would be a roof consisting of a slab, beams, and girders where each component is analyzed as an individual SDOF. This approach may yield less conservative results; however, care must be taken to account for the phasing of the respective responses, particularly when the frequencies of the different systems are similar.

Multi-degree-of-freedom (MDOF) models of element assemblages account for the phasing of different structural responses and frequency shifts caused by connecting systems with similar fundamental frequencies. MDOF representations may be lumped spring-mass systems, simplified finite-element models with nonlinear beam elements, or refined finite-element models with shell and continuum elements. The more detailed the MDOF model, the more failure mechanisms it may be able to represent; however, detailed finite-element models require very specific detailed design information for the localized response calculations to be meaningful. Unlike SDOF methods, where thresholds of inelastic deformation performance limits are established based on good engineering practice, the results of detailed finite-element analyses speak for themselves. The calculated response depicts the expected extent of damage and any instabilities that might result. Although limiting extents of inelastic deformation may be established for the design process, these limiting magnitudes have less meaning as compared to the results of detailed analysis.

For steel structures, shell models of rolled sections may provide accurate response information resulting from standoff detonations but through the thickness continuum models and explicit steel connection details are required to accurately calculate the effects of contact detonations. Similarly, longitudinal and transverse reinforcement within continuum representations of concrete structures must be explicitly modeled with beam or bar elements to accurately account for localized failure mechanisms. Since much of this detailed information is unknown until later in the design process, simplified SDOF models are often used to develop initial member sizes, and detailed analysis are performed on critical structural components later in the design process to verify the performance.

While experienced judgement is required for all nonlinear dynamic response calculations, the skill sets differ when applied to SDOF analyses as compared to detailed finite-element analyses. Simplified models are much easier to develop; hence, parametric studies may rapidly be performed to evaluate the effect of different assumptions, such as boundary conditions and mass distributions. However, the interpretation of the calculated results and the application of the results to the actual design require a fundamental understanding of the structural behavior. Detailed finite-element models are often extremely complex, and the expertise of the modeler could strongly influence the accuracy of the analysis. While the accuracy of the finite-element code may be validated and verified, the modeler must be familiar with the element and material libraries and with explosive test data for similar classes of problems. The postprocessing of the calculated responses will reveal the performance of the structure; therefore, the generated data must be reviewed with an experienced eye. Both simplified and detailed modeling approaches require levels of expertise unique to the type of loading and structural response.

Since the SDOF models are meant to match the fundamental mode deflection of the actual element, the resulting dynamic reaction forces are not theoretically the reaction forces the connections and supporting structures will resist. The dynamic reaction forces are derived from the dynamic equilibrium of the element and may be shown to be a function of both the loading and the resistance. Finite-element analyses determine the dynamic reaction forces associated with multiple modes of structural response; however, spurious high-frequency response may obscure the actual peak magnitudes. These records must be filtered to produce meaningful design information. An understanding of the analytical methods and the structural response is therefore required.

28.8 STRUCTURAL BEHAVIOR

The fundamental premise of a blast-resistant design is for the structure to be able to develop large inelastic deformations and thereby dissipate large amounts of blast energy while satisfying the prescribed limits on damage. The characteristics of the structural behavior depend to a great extent on the material properties, the structural details, and the intensity of the loading. Some failure mechanisms are classified as ductile, while others are classified as brittle. Ductile mechanisms, such as flexure in an underreinforced concrete or compact steel element, maintain strength and deformation capacity following material yield, and the subsequent inelastic deformations dissipate the blast energy. For many of the ductile modes of failure, the magnitude of the reaction forces plateaus following material yielding, and this limits the demand on many of the brittle modes of failure. Brittle mechanisms, such as shear failure and strain localizations, dissipate very little strain energy prior to failure and are to be avoided.

A preferred design approach is to identify the dominant failure mechanism thresholds for a vulnerable structural element or assembly and to make sure the threshold for brittle modes of failure exceed those for ductile modes of failure. This capacity design approach is similar to component seismic design principles and will provide the desired performance in response to the specified blast load intensity and the best performance in response to a larger-than-specified blast load up until the ductile failure occurs. A corresponding concern occurs when members are sized for stiffness (deflection control) in response to conventional loads, and this results in an oversized member in response to the specified blast loads. Since the reaction forces will continue to increase in response to a larger-than-specified blast load, care must be taken to make sure this does not cause a brittle mode of failure (i.e., shear) before ductile yielding can take place.

Characteristics of specific structural details may limit the extent of ductile deformations and precipitate a brittle failure. For example, tension members may develop ductile deformations following axial yielding; however, strain concentrations in the threads of bolted connections can localize the inelastic response and preclude the expected ductile deformations from occurring. Upsized threads for bolted connections and alternate mechanical attachments that avoid localized strain concentrations enable the tension members to distribute the axial elongation along the length of the member prior to failure. A design factor on brittle modes of failure will make sure the section develops large inelastic deformations prior to brittle failures. This similarly applies to the upsizing of weld capacities and specifying additional stirrups to provide confinement of concrete surrounding reinforcing bars in their development lengths.

Ductile detailing serves as an umbrella for a wide range of effects that are not explicitly quantified. The ductile detailing of concrete columns consists of closely spaced ties and spiral reinforcement for new construction and carbon fiber wraps or steel jackets for the protective upgrade of existing buildings. Steel column splices must be located away from plastic hinges and often need to develop the full moment capacity of the section at the splice location. A concrete encasement of the steel section will prevent local flange deformations. Ductile detailing applies to both direct response and the subsequent rebound; therefore, doubly reinforced sections are preferred to singly reinforced sections. Steel beams that are typically constructed composite with a concrete deck will brace the top flange, but the bottom flange is unbraced during uplift loading, causing load reversal. Since gravity restores the section following rebound, dynamic buckling rarely occurs in response to blast loading.

28.9 PRIMARY STRUCTURE: MATERIAL SELECTION AND DESIGN FOR BLAST RESISTANCE

Steel structures are inherently ductile by virtue of their material properties; however, rolled sections consist of relatively thin plates that are vulnerable to local deformations. Furthermore, steel connections are vulnerable to brittle modes of failure or may need significant detailing changes relative to a conventional design. The analytical methods used to evaluate steel members' performance in response to blast loading therefore depends on the standoff distance. Kingery Bulmash equations and SDOF models using global section properties (strong axis, weak axis, or both) can be used to evaluate standoff conditions that primarily excite the global response of the member. Flexure is the primary response, and the resulting shear forces must be checked against AISC limits. Composite strength may not be a reliable ductile mechanism if significant slab damage is expected but may govern connection design.

Connection details should be designed to develop the plastic moment capacity of the section while considering material and dynamic overstrength factors. These connection forces are likely to exceed the nominal maximum uniform loads tabulated in Table 3-6 of AISC Steel Construction Manual. Unless detailed finite-element analyses demonstrate acceptable performance in response to the calculated blast loadings, web plates may be required when the calculated connection forces exceed the shear capacity of the sections. This condition often arises in short spans.

The high-intensity blast loads resulting from near contact detonations may fracture steel plate; therefore, CFD analyses and detailed finite-element methods are required to evaluate the localized deformations of the flange and web plates in response to these scenarios. At these short standoff distances, the shape of the explosive charge will be a factor, as will the presence of architectural column covers or flange plates that box out the section. Near contact detonations are typically associated with a hand-carried satchel, and its height of burst is typically limited to the surface on which it may be placed. Built-up sections that are subject to blast loads must be checked to make sure the welds can develop the required capacity, and column base plate attachments must be included in the model to verify stability. Localized damage to near contact detonations will be minimized if concrete is filled between the flanges of rolled or hollow sections; however, additional steel plates may also be required in the immediate vicinity of the detonation.

Although trusses are very efficient for spanning long distances, the individual chords and web members of steel trusses must be individually evaluated to determine both their global (axial) and their local (flexural) performance in response to blast overpressure. Truss connections may need to be detailed to develop the full axial capacity of connected members. Truss members are particularly vulnerable if they are accessible to hand-carried explosive threats that may be placed next to or wedged within the diagonals. Where truss members are accessible to the satchel threat, each gusset plate connection must be detailed to develop the capacities of the individual members. Concrete encasement of accessible members and obscuring their location within partitions may restrict accessibility and help reduce risk.

While steel sections are most vulnerable to localized flange or web damage, steel frames must be designed to resist the effects of rebound. Steel columns that bear against bracing beams due to direct blast loading will pull away during rebound. The connections must be designed to resist both phases of response, or the column must be designed to remain stable with an increased unbraced length if localized floor failure is expected. One way to avoid these concerns is to minimize the tributary area of the columns along their height. Hence, the building's façade should be anchored directly into the floor diaphragms and not to the columns along their span.

Reinforced-concrete members are inherently less slender and more massive than steel sections; they are therefore less prone to torsion or buckling failures. The performance of the concrete members in response to blast loading depends on the ductile detailing of steel reinforcement. Generally speaking, ductile detailing methods that were developed to resist strong seismic ground motions will be effective in response to blast loading. Doubly reinforced sections to resist both the inward response and subsequent rebound and confinement of the concrete core address the primary characteristics of reinforced-concrete performance in response to blast loading. Blast-resistant concrete elements may require reinforcing be placed in locations not normally dictated by a conventional design, such as continuous top reinforcing in a slab.

Reinforced concrete is vulnerable to brisance and direct shear when subjected to a near contact explosive detonation. The damage associated with these effects are very intense but localized to the immediate vicinity of the blast. Confinement of the concrete within a steel jacket will restrain the shattered concrete within the core and provide axial load-carrying capacity. Carbon or glass fiber wraps provide similar confinement; however, they may be damaged by the fireball and explosive by-products. Steel plating and laced reinforcing may be used for slabs and walls subject to the most intense blast loads.

The flexural capacity of reinforced-concrete members should be detailed to develop the full plastic capacity at midspan and at supports with the capacity for load reversals due to rebound. Splices should be staggered and located away from locations of peak motions. The detailing of reinforcement for moderate levels of damage should follow the ACI requirements for Intermediate Moment Frames, and the detailing of reinforcement of high levels of damage should follow the ACI requirements for Special Moment Frames.

Prestressed construction optimizes the performance of the concrete construction in response to prescribed load patterns, and the prestress eccentricity profile is related to the envelope of dead and live load moments. However, blast loading frequently produces uplift deformations, which generally oppose the gravity load moments and always produce rebound. Due to the likelihood of deformation reversals and the possibility of uplift, the eccentric prestress profile that optimizes gravity load resistance may worsen performance in response to blast loading. Furthermore, a high-strength prestressed cable stores considerable strain energy, and the development of this cable force relies on either bond for pretension or anchorage zone bearing stresses for posttension construction. Damage to the concrete in the anchorage zone can cause the release of the stored strain energy, which could result in additional damage. Finally, posttension girders and slabs typically span over multiple bays, so loss of prestress can affect large portions of the structure.

Prestressing may be used for serviceability and crack control of structures in response to code-specified loading; however, supplemental deformed bars will be required to develop the ultimate load capacity in response to blast loading. Even so, response limits should limit the magnitude of inelastic deformation due to blast loading. Posttensioned concrete construction should be fully grouted to develop a bond along the span, thereby reducing the significance of the anchorage zone damage. Finally, prestress profiles should preferably be concentric to accommodate load reversals without inducing additional damage. These constraints to the profile and the required reinforcement with deformed bars reduce the advantages of prestressed concrete construction in the blast-resistant design of structures.

Load bearing walls using masonry block construction is discouraged for structures that will resist blast loading. Response deformations are limited to 1½ degrees of end rotation for unreinforced masonry and to 2 degrees for fully grouted and reinforced masonry walls. Blast-resistant windows within masonry walls present additional challenges; the blast loads that are tributary to the windows must be transferred to the masonry block wall segments (piers) on either side of the opening, or the windows must be independently supported by steel framing that spans from floor slab to floor slab. Lateral reaction forces that are developed by the masonry walls must be transferred to the floor diaphragms for both the direct response and for the rebound. Non-load-bearing masonry walls require details that provide lateral resistance but do not accept gravity loads. The connection details at the head of blast-resistant non-load-bearing masonry walls within steel frame construction must account for the wall orientation relative to the steel framing or the wall orientation relative to the underside of the metal deck floor slab.

Roof systems range from very robust concrete slabs on either concrete or steel framing to lightweight metal deck on bar joists. Blast loads that originate from the surrounding streets may either load the roof from above or produce uplift pressures that enter the building through failed façade. Less intense blast pressures will load the roof of taller buildings than shorter buildings by virtue of the standoff distances. Furthermore, the pressures that sweep over the top of the building are diffused and incident to the roof surface, whereas the pressures that may enter through failed façade are reflected at the underside of the roof. Heavier roof systems benefit from greater inertial resistance than lightweight roof systems, and steel or concrete framing members have greater blast load resistance than bar joist members. Nevertheless, each component can be designed to provide some degree of blast resistance before the system becomes unwieldly or unaffordable. The integrity of bar joist members relies on the welds that attach the web to the chords, and these members cannot develop significant inelastic deformations.

28.10 SECONDARY STRUCTURE: MATERIAL SELECTION AND DESIGN FOR BLAST RESISTANCE

The building envelope is considered secondary structure; however, it is most exposed to the vehicle-borne explosive threat that may detonate on the surrounding streets. Structural dynamics and nonlinear material mechanics are used to determine façade response in a similar manner to that of structural components. Glass, metal panel, masonry, or precast elements may be designed and detailed to resist varying levels of blast loading. Glass hazard standards and software correlate the extent of debris that may be propelled into an occupied space with the potential for injury, and deformation limits define the magnitude of the material response before failure is likely to occur. In addition to the blast load intensity that may result from the detonation of a specified weight of explosive threat, the supporting members are designed to resist the maximum capacity of the glass or in-fill panel. This "balanced design" approach enables the in-fill panel to fail first and prevents additional debris in the form of the mullions and framing members.

Metal panels may be treated as flexural members or as membrane elements; however, the membrane reaction forces are typically much greater than the flexural reaction forces, and it may be difficult to develop the required connections or axial restraints. Metal panels are often permitted to detach from their supports if a catch cable or catch bar is located behind them.

Precast panels may be designed to span from floor slab to floor slab, and the anchorage must transfer the lateral reaction forces into the lateral resisting system without accepting gravity loads. The precast panels benefit from greater mass than glass or metal panels, and they may be detailed to develop ductile deformations. Blast-resistant windows within precast panels are less of a challenge than windows within masonry walls; however, the reinforcement interrupted by the openings must be transferred to the panel segments (piers) on either side of the opening. Precast connections hardware will require design for loads significantly greater than conventional loads.

Façade systems may be analyzed as a sequence of SDOF systems with the reaction forces from each component sequentially applied to the supporting component. Similar to the stacking of structural SDOF models, the analyst should verify that component natural periods are sufficiently distinct; otherwise, unconservative results may be encountered. More commonly, blast loads are applied to each component based on the tributary area supported. However, the most efficient design results from a MDOF analysis in which the glass panel resistance functions are represented by nonlinear springs and the framing components are represented by beam or shell elements. This MDOF approach considers the phased response of all the elements and produces the most detailed response information.

In addition to designing the supporting elements to resist the tributary capacities of the in-fill panels, energy-absorbing anchors and energy-absorbing cable catch systems may be used to restrain the debris and dissipate blast energy through controlled deformation. Energy-absorbing systems limit the force that is transferred to the supporting structure and allow either protective façade components to be installed in less robust supporting elements or greater blast resistance to be developed using fewer materials. Energy-absorbing catch systems are also well suited to retrofits where modification of existing components may not be feasible or cost effective. Regardless of the intent, energy-absorbing systems rely on the ductile deformations of the dissipating components.

28.11 CONCLUSION

A blast-resistant structure relies on ductile detailing and inertial resistance to resist the high-intensity but short-duration shock waves. In addition to the direct response to the applied blast loads, structural elements rebound, and this produces large reversals in reaction forces. Each structural material may develop some degree of blast resistance and inelastic deformation; however, some are more efficient than others. A good design approach is to start with a structure that satisfies code-defined gravity and lateral strength and serviceability requirements and then upgrade this structure to develop the required blast resistance. This approach enables the project team to separate the true extent of the blast upgrades in both materials and cost from the conventional design requirements. Although disproportionate collapse may be caused by a structure's inability to arrest a localized zone of damage, unless the initiating damage is prevented through structural hardening, the design to prevent the progression of collapse is a structural design issue and not a blast design issue. Finally, although blast loading is very different from seismic loading, the ductile details that were developed for seismic design are likely to be beneficial in the response of structures that may be subjected to explosive loading.

REFERENCES

American Institute for Steel Construction has published Design Guide 26 (AISC 2013).

Blast Effects on Buildings, David Cormie, Geoff Mays, and Peter Smith, 2009.

Blast Protection of Buildings (ASCE 59-11).

BLASTX Code, Version 4.2, User's Manual, U.S. Army Corps of Engineers, J. Robert Britt, Dale E. Ranta, and Charles E. Hoachim, March 2001.

Design and Analysis of Hardened Structures to Conventional Weapons Effects, UFC 3-340-01, June 2002.

Handbook for Blast Resistant Design of Buildings, Donald O. Dusenberry, 2010.

Introduction to Structural Dynamics, John M. Biggs, 1964.

Modern Protective Structures, Theodor Krauthammer, 2008.

Structural Design for Physical Security, Edward J. Conrath, Ted Krauthammer, Kirk A. Marchand, and Paul Mlakar, 1999.

Structures to Resist the Effects of Accidental Explosions, UFC 3-340-02, December 2008.

User's Guide for the Single-Degree-of-Freedom Blast Effects Design Spreadsheets (SBEDS) PDC_TR 06-02 Rev 3, U.S. Army Corps of Engineers, Protective Design Center Technical Report, December 2014.

Chapter 29

FRP Strengthening of Reinforced-Concrete Members

BY

HAYDER A. RASHEED, Ph.D., PE *Professor, Kansas State University, Manhattan, Kansas*

Flexural strengthening of reinforced-concrete beams with fiber-reinforced polymer (FRP) is a relatively recent technology that was introduced three decades ago by U. Meier (1987). Since then, a large number of research studies have been conducted to extend this technique to other types of members or loading schemes. In addition, more research is performed to better understand or improve the behavior of FRP-strengthened members. The technology is widespread at the present time to an extent that design guidelines have been introduced all over the world. Therefore, it is definitely timely to introduce this design subject to the structural engineers and to the structural engineering curricula.

29.1 FRP PROPERTIES FOR STRENGTHENING APPLICATIONS

Each FRP laminate having a single fiber orientation is orthotropic in its response. This means that it has four independent in-plane engineering constants, namely, the elasticity modulus along the fiber direction (E_1), the elasticity modulus in the transverse direction (E_2), the major Poisson's ratio (ν_{12}), as well as the shear modulus (G_{12}). However, it is important to note here that this application of strengthening the reinforced-concrete members with FRP has the fibers aligned with the main axis of the member (i.e., beam axis in flexure, transverse beam axis in shear, and hoop direction in column confinement). Therefore, the only mechanical property that is relevant in these applications is the modulus of elasticity along the fiber direction. The stress-strain behavior along the fiber direction is known to be linear elastic up to brittle rupture of fiber for all of the typical fibers used (i.e., carbon, glass, and aramid). Accordingly, two parameters are sufficient to define the complete stress-strain behavior. These are typically the modulus of elasticity and the rupture strain or the rupture strength:

$$f^*_{fu} = E_f \varepsilon^*_{fu} \tag{29.1}$$

The strength (f^*_{fu}) and the strain (ε^*_{fu}) are called the ultimate tensile strength and strain, which are determined from the results of a population of 20 coupon specimens according to ACI 440.2R-17 guidelines following the testing protocol of the ASTM D4762. These material limits are obtained by subtracting three times the standard deviation of the strength and maximum strain from the mean values of each as follows:

$$f^*_{fu} = \bar{f}_{fu} - 3\sigma_f \equiv \text{ultimate tensile strength} \tag{29.2}$$

$$\varepsilon^*_{fu} = \bar{\varepsilon}_{fu} - 3\sigma_\varepsilon \equiv \text{ultimate tensile strain} \tag{29.3}$$

where $\bar{f}_{fu}$ ≡ mean ultimate tensile strength of 20 coupon specimens
$\bar{\varepsilon}_{fu}$ ≡ mean ultimate tensile strain of 20 coupon specimens
σ_f ≡ standard deviation of mean ultimate tensile strngth of 20 coupons
σ_ε = standard deviation of mean ultimate tensile strain of 20 coupons

Equations (29.2) and (29.3) are known to yield 99.87 percent probability of exceedance (ACI 440.2R-17). On the other hand, the design ultimate tensile strength and strain are defined as follows:

$$f_{fu} = C_E f^*_{fu} = \text{design ultimate tensile strength} \tag{29.4}$$

$$\varepsilon_{fu} = C_E \varepsilon^*_{fu} = \text{design ultimate tensile strain} \tag{29.5}$$

The C_E factor is defined as the environmental reduction factor, which depends on the type of FRP and exposure examined (ACI 440.2R-17); see Table 29.1.

Table 29.1 Environmental Reduction Factors

Exposure condition	Fiber type	C_E factor
Interior exposure	Carbon	0.95
	Glass	0.75
	Aramid	0.85
Exterior exposure (bridges, piers, unenclosed parking garages)	Carbon	0.85
	Glass	0.65
	Aramid	0.75
Aggressive environment (chemical plants and wastewater treatment plants)	Carbon	0.85
	Glass	0.5
	Aramid	0.7

DESIGN EXAMPLE 1

A test of 20 coupons of CFRP sheets was performed to evaluate the design parameters of the material. The results of the test specimens are summarized in Table 29.2. Compute the design strength, design ultimate strain, and the modulus of elasticity assuming exterior exposure.

Table 29.2 Test Parameters and Results for 20 CFRP Coupon Specimens

Specimen	Width (in.)	Width (mm)	Average thickness (in.)	Average thickness (mm)	Rupture load (kips)	Rupture load (kN)	Ultimate strain (micro-strain)
CFRP-1	2.01	51.05	0.0279	0.7087	6.13	27.28	12514
CFRP-2	1.98	50.29	0.0285	0.7239	6.14	27.32	14123
CFRP-3	1.94	49.28	0.0276	0.7010	6.16	27.41	12931
CFRP-4	2.03	51.56	0.0273	0.6934	6.18	27.50	11947
CFRP-5	2.05	52.07	0.0265	0.6731	6.04	26.88	12993
CFRP-6	2.02	51.31	0.0275	0.6985	6.07	27.01	13523
CFRP-7	1.99	50.55	0.0283	0.7188	6.11	27.19	14098
CFRP-8	1.91	48.51	0.0278	0.7061	6.19	27.55	12991
CFRP-9	2.00	50.80	0.0276	0.7010	6.10	27.15	12536
CFRP-10	1.97	50.04	0.0268	0.6807	6.08	27.06	13005
CFRP-11	2.02	51.31	0.0286	0.7264	6.17	27.46	12741
CFRP-12	1.95	49.53	0.0281	0.7137	6.12	27.23	14538
CFRP-13	1.96	49.78	0.0272	0.6909	6.13	27.28	13423
CFRP-14	2.01	51.05	0.0277	0.7036	6.17	27.46	12947
CFRP-15	2.04	51.82	0.0269	0.6833	6.03	26.83	13886
CFRP-16	2.09	53.09	0.0280	0.7112	6.11	27.19	12814
CFRP-17	2.03	51.56	0.0282	0.7163	6.16	27.41	13378
CFRP-18	1.93	49.02	0.0275	0.6985	6.18	27.50	12966
CFRP-19	2.07	52.58	0.0267	0.6782	6.09	27.10	11479
CFRP-20	2.03	51.56	0.0272	0.6909	6.15	27.37	12692
Mean	**2.00**	**50.84**	**0.0276**	**0.7009**	**6.13**	**27.26**	**13076**

Solution:

U.S. customary unit calculations	SI unit calculations
Compute the mean ultimate strength and strain: $\bar{f}_{fu} = 111$ ksi $\bar{\varepsilon}_{fu} = 13{,}076 \times 10^{-6} = 0.0131$	Compute the mean ultimate strength and strain: $\bar{f}_{fu} = 766$ MPa $\bar{\varepsilon}_{fu} = 13{,}076 \times 10^{-6} = 0.0131$
Compute the standard deviation for the ultimate strength and strain: $\sigma_f = 3.27$ ksi $\sigma_\varepsilon = 731.81 \times 10^{-6} = 0.000732$	Compute the standard deviation for the ultimate strength and strain: $\sigma_f = 22.54$ MPa $\sigma_\varepsilon = 731.81 \times 10^{-6} = 0.000732$
Compute the ultimate tensile strength and strain: $f^*_{fu} = \bar{f}_{fu} - 3\sigma_f = 111 - 3 \times 3.27$ $= 101.19$ ksi $\varepsilon^*_{fu} = \bar{\varepsilon}_{fu} - 3\sigma_\varepsilon = 0.0131 - 3 \times 0.000732$ $= 0.0109$	Compute the ultimate tensile strength and strain: $f^*_{fu} = \bar{f}_{fu} - 3\sigma_f = 766 - 3 \times 22.54$ $= 698.38$ MPa $\varepsilon^*_{fu} = \bar{\varepsilon}_{fu} - 3\sigma_\varepsilon = 0.0131 - 3 \times 0.000732$ $= 0.0109$
Compute the design ultimate strength, strain, and modulus using exterior exposure for CFRP material: $f_{fu} = C_E E_f \varepsilon^*_{fu} = 0.85 \times 101.19 = 86$ ksi $\varepsilon_{fu} = C_E \varepsilon^*_{fu} = 0.85 \times 0.0109 = 0.00927$ $E_f = \frac{f_{fu}}{\varepsilon_{fu}} = \frac{f^*_{fu}}{\varepsilon^*_{fu}} = 9{,}277$ ksi	Compute the design ultimate strength, strain, and modulus using exterior exposure for CFRP material: $f_{fu} = C_E E_f \varepsilon^*_{fu} = 0.85 \times 698.38$ $= 593.62$ MPa $\varepsilon_{fu} = C_E \varepsilon^*_{fu} = 0.85 \times 0.0109$ $= 0.00927$ $E_f = \frac{f_{fu}}{\varepsilon_{fu}} = \frac{f^*_{fu}}{\varepsilon^*_{fu}} = 64.04$ GPa

29.2 FLEXURAL STRENGTHENING DESIGN FOR BEAMS AND SLABS

In this section, the design of reinforced-concrete flexural members strengthened with FRP is presented. There are three dominant failure modes that control the behavior of externally bonded FRP sheets. These are (i) FRP rupture, which is typically admitted in lightly reinforced lightly strengthened members like slabs and T beams, and (ii) FRP debonding and concrete crushing, which are usually present for moderately reinforced moderately strengthened members like rectangular beams. The other two possible failure modes are cover delamination and brittle concrete crushing prior to tension steel yielding. Cover delamination is less likely to happen in FRP sheets extending close to the support due to the low axial stiffness of sheets compared to pre-cured laminates or plates and due to the low stress concentrations at the sheet tip when stopped near the support. Brittle concrete crushing is only admitted for heavily reinforced heavily strengthened members making them inefficient to strengthen using this technique (Rasheed 2015).

29.2.1 Flexural Strength Design Procedure

The failure mode that controls the behavior affects the design process. Accordingly, ACI 440.2R-17 adopts a fully iterative design procedure until converging at an admissible design. On the other hand, Rasheed (2015) follows a direct design procedure that requires developing different design equations for various failure modes. The approach presented by this section is a combination of both for the sake of brevity and ease of formulation. The iterative nature of this design problem stems from the fact that the goal of the design process is to compute the area of FRP, which in turn will impact the failure mode. The author of this chapter has developed a statistically correlated linear equation that accurately predicts the FRP area, especially for the modes of failure of FRP rupture and debonding (Rasheed and Motto 2010; Saqan, Rasheed, and Hawileh 2013). This equation offers a quick estimation for the area of FRP needed for a certain level of strengthening. Once established, this FRP quantity may be used to directly reveal the failure mode involved. This allows the designer to focus on that particular failure mode to solve the strain compatibility, force, and moment equilibrium equations in an analysis check. The complete design procedure is illustrated in Fig. 29.1.

29.2.2 FRP Limits Identifying Failure Modes

The first limit identifies the dominance of FRP rupture vs. FRP debonding. This is a simple limit to establish. When the rupture strain level allowed by ACI 440.2R-17 ($\varepsilon_{fu}^{\max} = 0.9\,\varepsilon_{fu}$) exceeds the debonding strain defined by ACI 440.2R-17 $\left(\varepsilon_{fd} = C\sqrt{\frac{f'_c}{n\,t_f E_f}}\right)$, FRP debonding takes place first and vice versa (Fig. 29.2). This limit requires the knowledge of the number of FRP layers to be used since it changes the debonding strain (the more the number of layers the lower the debonding strain).

29.2.3 Section Serviceability Limits Check

At the end of the flexural strength design procedure, the service load effects are used to check that the concrete extreme fiber stress and steel stress are within the elastic range, as follows:

$$f_{c,s} < 0.6\, f'_c \qquad f_{s,s} < 0.8\, f_y \tag{29.6}$$

In addition, the stress in the FRP must not exceed the sustained plus cyclic stress limit per ACI 440.2R-17, as follows:

$$f_{f,s} < 0.55\, f_{fu} \ \text{(CFRP)} \qquad f_{f,s} < 0.2\, f_{fu} \ \text{(GFRP)} \tag{29.7}$$

To compute these stresses in concrete, steel, and FRP under service load level, the following analysis equations must be presented:

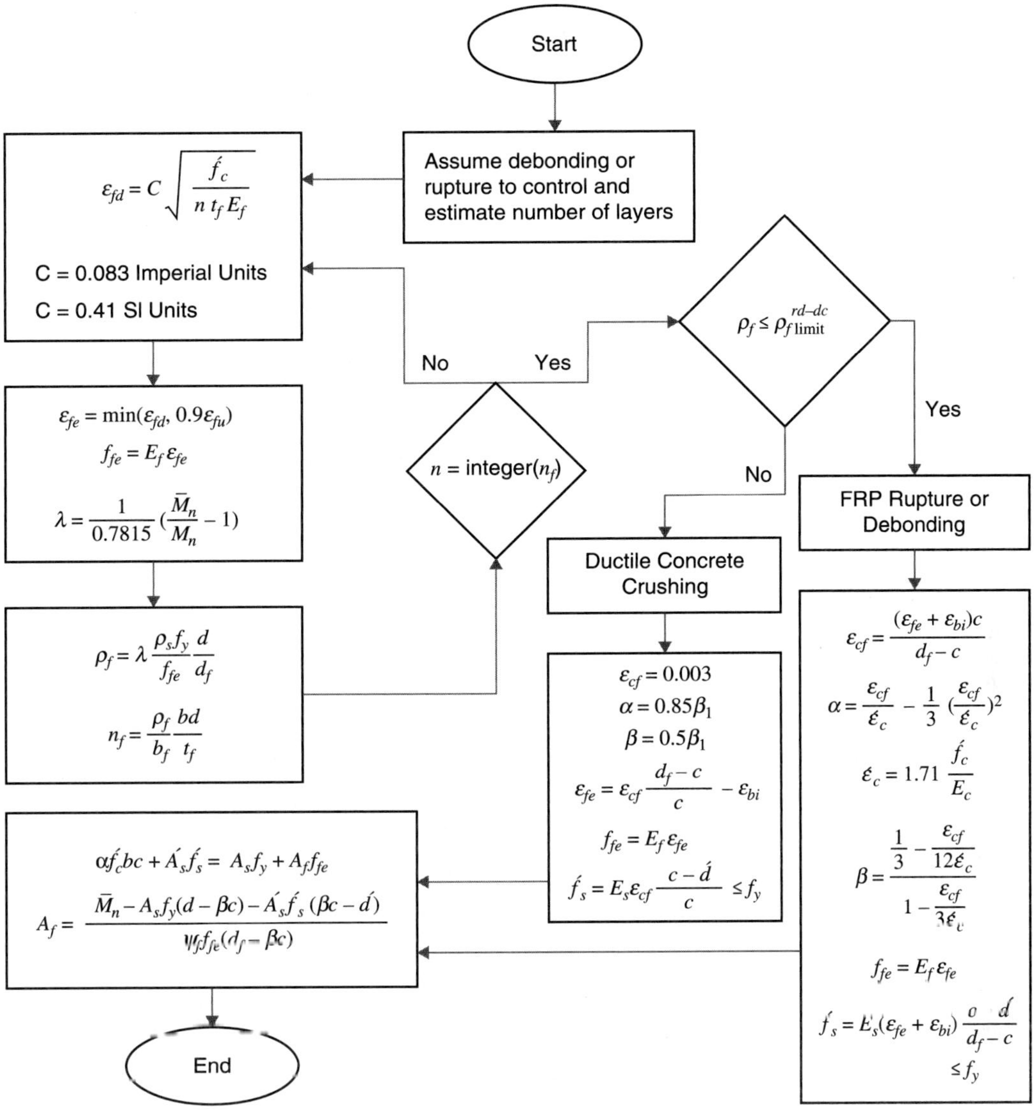

Figure 29.1 FRP flexural strengthening design procedure.

$$k = \sqrt{\left(\frac{E_s}{E_c}\rho_s + \frac{E_f}{E_c}\rho_f\right)^2 + 2\left(\frac{E_s}{E_c}\rho_s + \frac{E_f}{E_c}\rho_f\frac{d_f}{d}\right)} - \left(\frac{E_s}{E_c}\rho_s + \frac{E_f}{E_c}\rho_f\right) \tag{29.8}$$

$$f_{s,s} = \frac{\left[\bar{M}_s + \varepsilon_{bi}A_f E_f\left(d_f - \frac{kd}{3}\right)\right](d - kd)E_s}{A_s E_s\left(d - \frac{kd}{3}\right)(d - kd) + A_f E_f\left(d_f - \frac{kd}{3}\right)(d_f - kd)} \tag{29.9}$$

$$f_{f,s} = f_{s,s}\frac{E_f}{E_s}\frac{d_f - kd}{d - kd} - \varepsilon_{bi}E_f \tag{29.10}$$

$$f_{c,s} = f_{s,s}\frac{E_c}{E_s}\frac{kd}{d - kd} \tag{29.11}$$

In the remainder of this section, two design examples of the same problem are solved. The first one addresses the flexural strength design calculation, and the second one presents the section serviceability limits check.

DESIGN EXAMPLE 2

A design example (Fig. 29.3) is presented by ACI 440.2R-17. The objective of resolving this example here using the current approach is to show the design efficiency of the present procedure compared with the iterative procedure followed by ACI 440.2R-17. The example is composed of a simply supported concrete beam located in an unoccupied warehouse. This beam is required to have a 50 percent increase in live load. The existing beam is found to be adequate in shear strength and serviceability requirements, including deflection and cracking. However, it is deficient in flexural capacity, requiring CFRP strengthening along the

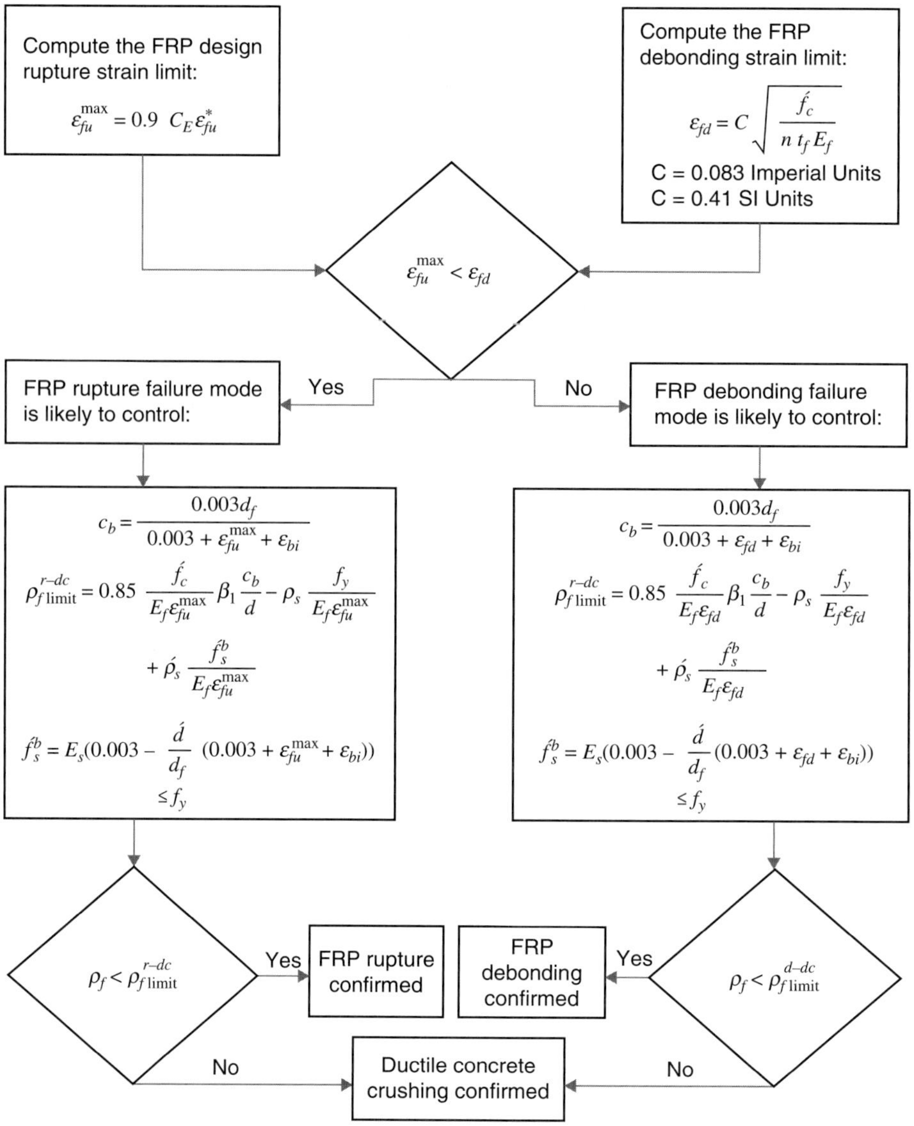

Figure 29.2 FRP area ratio limits for flexural strengthening design.

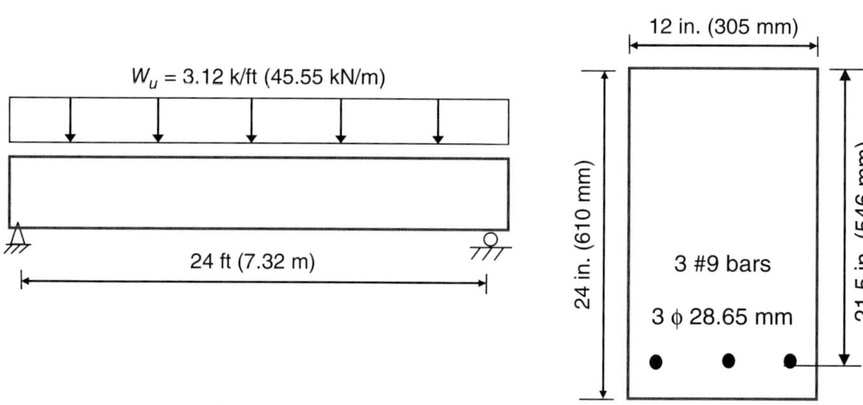

Figure 29.3 Beam profile and cross section for Design Example 2.

tension face. The geometry, reinforcement, material properties, and loading parameters are shown below.

Material property	U.S. customary units	SI units
f'_c	5000 psi	34.5 MPa
f_y	60 ksi	414 MPa
E_f	5360 ksi	37 GPa
ε^*_{fu}	0.0167	0.0167
t_f	0.04 in.	1.02 mm

Loading	Existing loads		Upgraded loads	
Units	k/ft	kN/m	k/ft	kN/m
Dead load W_{DL}	1.0	14.6	1.0	14.6
Live load W_{LL}	1.2	17.5	1.8	26.3
Unfactored load (W_{DL} + W_{LL})	2.2	32.1	2.8	40.9
Factored load (1.2 W_{DL} + 1.6 W_{LL})	3.12	45.6	4.08	59.6
Limit no FRP (1.1 W_{DL} + 0.75 W_{LL})	N/A	N/A	2.5	35.8

Solution:

U.S. customary unit calculations: Since the factored existing load (3.12 k/ft) exceeds the upgraded load limit without FRP (2.5 k/ft), the proposed strengthening limit is acceptable.

SI unit calculations: Since the factored existing load (45.6 kN/m) exceeds the upgraded load limit without FRP (35.8 kN/m), the proposed strengthening limit is acceptable.

U.S. customary unit calculations: The beam has interior exposure and CFRP material is proposed. Thus,

$$\varepsilon_{fu} = C_E \varepsilon^*_{fu} = 0.95 \times 0.0167 = 0.0159$$

$$f_{fu} = C_E E_f \varepsilon^*_{fu} = 0.95 \times 5360 \times 0.0167 = 85 \text{ ksi}$$

SI unit calculations: The beam has interior exposure and CFRP material is proposed. Thus,

$$\varepsilon_{fu} = C_E \varepsilon^*_{fu} = 0.95 \times 0.0167 = 0.0159$$

$$f_{fu} = C_E E_f \varepsilon^*_{fu} = 0.95 \times 37{,}000 \times 0.0167 = 587 \text{ MPa}$$

U.S. customary unit calculations: Compute concrete properties:

$$\beta_1 = 1.05 - 0.05\frac{f'_c}{1000} = 0.8$$

$$E_c = 57\sqrt{5000} = 4030 \text{ ksi}$$

$$\varepsilon'_c = 1.71\frac{f'_c}{E_c} = 1.71\frac{5}{4030} = 0.00212$$

SI unit calculations: Compute concrete properties:

$$\beta_1 = 1.09 - 0.008 f'_c = 0.814$$

$$E_c = 4700\sqrt{34.5} = 27{,}606 \text{ MPa}$$

$$\varepsilon'_c = 1.71\frac{f'_c}{E_c} = 1.71\frac{34.5}{27{,}606} = 0.00214$$

U.S. customary unit calculations: Compute steel properties:

$$A_s = 3 \times 1 = 3 \text{ in.}^2$$

$$\rho_s = \frac{A_s}{bd} = \frac{3}{12 \times 21.5} = 0.01163$$

$$\varepsilon_y = \frac{f_y}{E_s} = \frac{60}{29{,}000} = 0.00207$$

SI unit calculations: Compute steel properties:

$$A_s = 3 \times \frac{\pi}{4}(28.65)^2 = 1934 \text{ mm}^2$$

$$\rho_s = \frac{A_s}{bd} = \frac{1934}{305 \times 546} = 0.01161$$

$$\varepsilon_y = \frac{f_y}{E_s} = \frac{414}{200{,}000} = 0.00207$$

U.S. customary unit calculations: Assume 1 layer of CFRP and compute the debonding and limiting rupture strains:

$$\varepsilon_{fd} = 0.083\sqrt{\frac{f'_c}{nt_fE_f}} = 0.083\sqrt{\frac{5}{1 \times 0.04 \times 5360}} = 0.0127 < \varepsilon^{max}_{fu} \text{ (i.e., debonding happens first)}$$

$$\varepsilon^{max}_{fu} = 0.9\varepsilon_{fu} = 0.9 \times 0.0159 = 0.0143$$

SI unit calculations: Assume 1 layer of CFRP and compute the debonding and limiting rupture strains:

$$\varepsilon_{fd} = 0.41\sqrt{\frac{f'_c}{nt_fE_f}} = 0.41\sqrt{\frac{34.5}{1 \times 1.02 \times 37{,}000}} = 0.0124 < \varepsilon^{max}_{fu} \text{ (i.e., debonding happens first)}$$

$$\varepsilon^{max}_{fu} = 0.9\varepsilon_{fu} = 0.9 \times 0.0159 = 0.0143$$

U.S. customary unit calculations: Compute the strengthening ratio:

$$f_{fe} = E_f \varepsilon_{fe} = 5360 \times 0.0127 = 68.07 \text{ ksi}$$

$$M_n = \frac{M_u}{\phi} = \frac{w_u l^2}{\phi 8} = \frac{3.12 \times 24^2}{0.9 \times 8} = 249.6 \text{ k-ft}$$

$$\bar{M}_n = \frac{\bar{M}_u}{\phi} = \frac{\bar{w}_u l^2}{\phi 8} = \frac{4.08 \times 24^2}{0.9 \times 8} = 326.4 \text{ k-ft}$$

$$\bar{M}_n / M_n = 1.31$$

SI unit calculations: Compute the strengthening ratio:

$$f_{fe} = E_f \varepsilon_{fe} = 37{,}000 \times 0.0124 = 458.8 \text{ MPa}$$

$$M_n = \frac{M_u}{\phi} = \frac{w_u l^2}{\phi 8} = \frac{45.6 \times 7.32^2}{0.9 \times 8} = 339.4 \text{ kN-m}$$

$$\bar{M}_n = \frac{\bar{M}_u}{\phi} = \frac{\bar{w}_u l^2}{\phi 8} = \frac{59.6 \times 7.32^2}{0.9 \times 8} = 443.5 \text{ kN-m}$$

$$\bar{M}_n / M_n = 1.31$$

U.S. customary unit calculations: Estimate the FRP ratio:

$$\lambda = \frac{1}{0.7815}\left(\frac{\bar{M}_n}{M_n} - 1\right) = 0.397$$

$$\rho_f = \lambda \frac{\rho_s f_y d}{f_{fe} d_f} = 0.397 \times \frac{0.01163 \times 60 \times 21.5}{68.07 \times 24} = 0.00365$$

SI unit calculations: Estimate the FRP ratio:

$$\lambda = \frac{1}{0.7815}\left(\frac{\bar{M}_n}{M_n} - 1\right) = 0.397$$

$$\rho_f = \lambda \frac{\rho_s f_y d}{f_{fe} d_f} = 0.397 \times \frac{0.01161 \times 414 \times 546}{458.8 \times 610} = 0.00372$$

U.S. customary unit calculations: Check the number of layers:

$$n_f = \frac{\rho_f bd}{b_f t_f} = \frac{0.00365 \times 12 \times 21.5}{12 \times 0.04} = 1.96$$

$$n = 1 = \text{integer}(n_f = 1.96) \Rightarrow \text{find } \rho^{d-dc}_{\text{limit}}$$

SI unit calculations: Check the number of layers:

$$n_f = \frac{\rho_f bd}{b_f t_f} = \frac{0.00372 \times 305 \times 546}{305 \times 1.02} = 1.99$$

$$n = 1 = \text{integer}(n_f = 1.99) \Rightarrow \text{find } \rho^{d-dc}_{\text{limit}}$$

U.S. customary unit calculations: Find the tensile strain on the soffit at the time of strengthening:

$$k = \sqrt{(n\rho)^2 + 2n\rho} - n\rho = 0.334$$

$$I_{cr} = 5937 \text{ in.}^4$$

$$M_{DL} = \frac{w_{DL} l^2}{8} = \frac{1.0 \times 24^2}{8} = 72 \text{ k-ft}$$

$$\varepsilon_{hi} = \frac{M_{DL}(d_f - kd)}{E_c I_{cr}} = \frac{72 \times 12 \times (24 - 0.334 \times 21.5)}{4030 \times 5937} = 0.00061$$

SI unit calculations: Find the tensile strain on the soffit at the time of strengthening:

$$k = \sqrt{(n\rho)^2 + 2n\rho} - n\rho = 0.334$$

$$I_{cr} = 2471 \times 10^6 \text{ mm}^4$$

$$M_{DL} = \frac{w_{DL} l^2}{8} = \frac{14.6 \times 7.32^2}{8} = 97.79 \text{ kN-m}$$

$$\varepsilon_{bi} = \frac{M_{DL}(d_f - kd)}{E_c I_{cr}} = \frac{97.79 \times 10^6 \times (610 - 0.334 \times 546)}{27{,}606 \times 2471 \times 10^6} = 0.00061$$

U.S. customary unit calculations: Find the limiting FRP ratio from Fig. 29.2:

$$c_b = \frac{0.003 \times 24}{0.003 + 0.0127 + 0.00061} = 4.41 \text{ in.}$$

$$\rho^{d-dc}_{f\,\text{limit}} = 0.85\frac{f'_c}{f_{fd}}\beta_1\frac{c_b}{d} - \rho_s\frac{f_y}{f_{fd}} = 0.85 \times \frac{5}{68.07} \times 0.8\frac{4.41}{21.5} - 0.01163 \times \frac{60}{68.07} = -5.94 \times 10^{-6} \approx 0$$

(concrete crushing controls)

SI unit calculations: Find the limiting FRP ratio from Fig. 29.2:

$$c_b = \frac{0.003 \times 610}{0.003 + 0.0124 + 0.00061} = 114.3 \text{ mm}$$

$$\rho^{d-dc}_{f\,\text{limit}} = 0.85\frac{f'_c}{f_{fd}}\beta_1\frac{c_b}{d} - \rho_s\frac{f_y}{f_{fd}} = 0.85 \times \frac{34.5}{458.8} \times 0.814\frac{114.3}{546} - 0.01161 \times \frac{414}{458.8} = 4.153 \times 10^{-4} < \rho_f = 0.00372$$

(concrete crushing controls)

U.S. customary unit calculations: Compute neutral axis depth:

$$0.85 \times 0.8 \times 5 \times 12c = 3 \times 60 + 12 \times 0.04 \times 5360 \times \left(0.003\frac{24 - c}{c} - 0.00061\right)$$

$$c = 5.08 \text{ in.}$$

SI unit calculations: Compute neutral axis depth:

$$0.85 \times 0.814 \times 34.5 \times 305c = 1934 \times 414 + 305 \times 1.02 \times 37{,}000 \times \left(0.003\frac{610 - c}{c} - 0.00061\right)$$

$$c = 127 \text{ mm}$$

U.S. customary unit calculations	SI unit calculations
Finalize the FRP design:	Finalize the FRP design:
$f_{fe}=5360\times\left(0.003\frac{24-5.08}{5.08}-0.00061\right)$	$f_{fe}=37{,}000\times\left(0.003\frac{610-127}{127}-0.00061\right)$
$=56.62$ ksi	$=399.58$ MPa
$A_f=\frac{326.4\times12-3\times60\times(21.5-0.4\times5.08)}{0.85\times56.62\times(24-0.4\times5.08)}$	$A_f=\frac{443.5\times10^6-1934\times414\times(546-0.407\times127)}{0.85\times399.58\times(610-0.407\times127)}$
$=0.39\text{ in.}^2<12\times0.04=0.48\text{ in.}^2$	$=251.6\text{ mm}^2<305\times1.02=311.1\text{ mm}^2$
(use partial layer)	*(use partial layer)*
$b_f=\frac{0.39}{1\times0.04}=9.75$ in.	$b_f=\frac{251.6}{1\times1.02}=246.67$ mm
(less than 2 full plies used by ACI440.2)	*(less than 2 full plies used by ACI440.2)*

DESIGN EXAMPLE 3

For the problem presented in Design Example 2, perform the section serviceability limits check to confirm that the proposed design offers adequate performance at the service load level.

Solution:

U.S. customary unit calculations	SI unit calculations
Since the unfactored upgraded loads, among other parameters, are given in Design Example 2, these parameters are used to directly compute the section limits under service loads.	Since the unfactored upgraded loads, among other parameters, are given in Design Example 2, these parameters are used to directly compute the section limits under service loads.
Substituting into Eq. (29.8), the value of k is evaluated:	Substituting into Eq. (29.8), the value of k is evaluated:
$E_s=29{,}000$ ksi	$E_s=200{,}000$ MPa
$E_c=4030$ ksi	$E_c=27{,}606$ MPa
$E_f=5360$ ksi	$E_f=37{,}000$ MPa
$\rho_s=0.01163$	$\rho_s=0.01161$
$\rho_f=\frac{A_f}{bd}=\frac{0.39}{12\times21.5}=0.00151$	$\rho_f=\frac{A_f}{bd}=\frac{251.6}{305\times546}=0.00151$
$k=0.338$	$k=0.338$
$kd=0.338\times21.5=7.26$ in.	$kd=0.338\times546=184.5$ mm
Compute steel, FRP, and concrete service stresses:	Compute steel, FRP, and concrete service stresses:
$\bar{M}_s=\frac{2.8\times24^2}{8}=201.6$ k-ft	$\bar{M}_s=\frac{40.9\times7.32^2}{8}=273.94$ kN-m
$f_{s,s}=41.42\text{ ksi}<0.8\times60$	$f_{s,s}=286.49\text{ MPa}<0.8\times414$
$=48$ ksi OK	$=331.2$ MPa OK
$f_{f,s}=5.73\text{ ksi}<0.55\times85$	$f_{f,s}=39.81\text{ MPa}<0.55\times587$
$=46.75$ ksi OK	$=322.85$ MPa OK
$f_{c,s}=2.93\text{ ksi}<0.6\times5$	$f_{c,s}=20.18\text{ MPa}<0.6\times34.5$
$=3$ ksi OK	$=20.7$ MPa OK

29.3 SHEAR STRENGTHENING DESIGN FOR BEAMS

The second major application for using FRP in infrastructural member upgrade is shear strengthening. Externally bonded FRP sheets are typically used with fibers in the transverse direction of beam axis to provide extra shear capacity. Nevertheless, fibers may be oriented at 45° angle with the beam axis as well to provide additional capacity by directly bridging the diagonal tension cracks. The need to strengthen beams with external FRP in shear stems from the upgraded flexural capacity using longitudinal FRP bonded to the beam soffit in most of the cases. Furthermore, the shear strengthening might be needed to address various shear deficiencies in members experiencing evident shear cracking. There are two different design methodologies to tackle the issue of shear strengthening in the United States. The first one is the approach of ACI 440.2R-17, and the second one is the treatment of the AASHTO LRFD specifications for the shear strength from concrete and internal steel complemented by the proposal of the NCHRP Report 678 for adding the contribution of FRP. The next two subsections will establish each one of the two approaches followed by design examples for both methods.

29.3.1 Shear Strengthening Design Using ACI440.2R-17

The main philosophy of the procedure covered by this document is to add up the strength provided by the concrete, internal steel stirrups, and external FRP to form the total shear capacity of the beam along a diagonal tension crack profile. The contribution of the concrete and internal steel stirrups follows closely the requirements of the ACI 318-14 design provisions. On the other hand, the contribution of the FRP adopts an empirically calibrated model based on experiments and proposed by Khalifa et al. (1998). This model incorporates an effective FRP strain level established based on the wrapping scheme as well as other parameters, like the concrete strength, the potential for peeling off, or debonding of FRP. The full wrapping scheme promotes the failure mode of FRP rupture at a significantly reduced transverse strain level ($\varepsilon_{fe}=0.004\leq0.75\,\varepsilon_{fu}$) due to the loss of aggregate interlock in concrete observed experimentally at the designated level leading to premature FRP rupture failure mode. On the other hand, partial wrapping in terms of using the U-wraps or side sheet/plate bonding is controlled by FRP debonding, which is a function of f_c' and L_e, defined as the active bond length over which the majority of the bond stress is transferred. Figure 29.4 provides the flowchart for this design methodology.

29.3.2 Shear Strengthening Design Using AASHTO LRFD 2014/NCHRP 678

The main philosophy of this approach is to compute the strength provided by the concrete and internal steel stirrups using the simplified modified compression field theory (Bentz et al. 2006) while accounting for external FRP contribution following the proposal of the NCHRP Report 678. The FRP contribution is calculated by the same equation adopted by ACI 440.2R-17 except for defining (ε_{fe}) differently using a continuous function of the FRP axial rigidity. Even though the shear crack angle (θ) is assumed to vary using the compression field theory, the NCHRP Report 678 proposal assumes ($\theta=45°$) when considering the FRP contribution. This seems to be inconsistent to follow for the author of this chapter. However, it is still adopted when solving the second example of this section until further published research corrects this issue. Figure 29.5 provides the flowchart for this design methodology.

29.3.3 Limit Checks for ACI 440.2R-17

In addition to the procedure detailed in the flowchart in Fig. 29.4, the following limits are imposed by ACI 440.2R-17:

1. Maximum external FRP wrap spacing: The maximum clear spacing between two successive FRP wraps is set equal to $d/4$. This means that the maximum center-to-center spacing (s_f) is set to

$$s_{f\max}=\left(w_f+\frac{d}{4}\right) \tag{29.12}$$

2. Total reinforcement limit: The shear force provided by the internal steel stirrups and the external FRP wraps is limited by the force limit in steel stirrups only per ACI 318-14 to avoid the crushing of the concrete struts:

$$V_s+V_f\leq C\sqrt{f_c'}\,b_w d \qquad C=8\text{ (Imperial Units)} \qquad C=\frac{2}{3}\text{ (SI Units)} \tag{29.13}$$

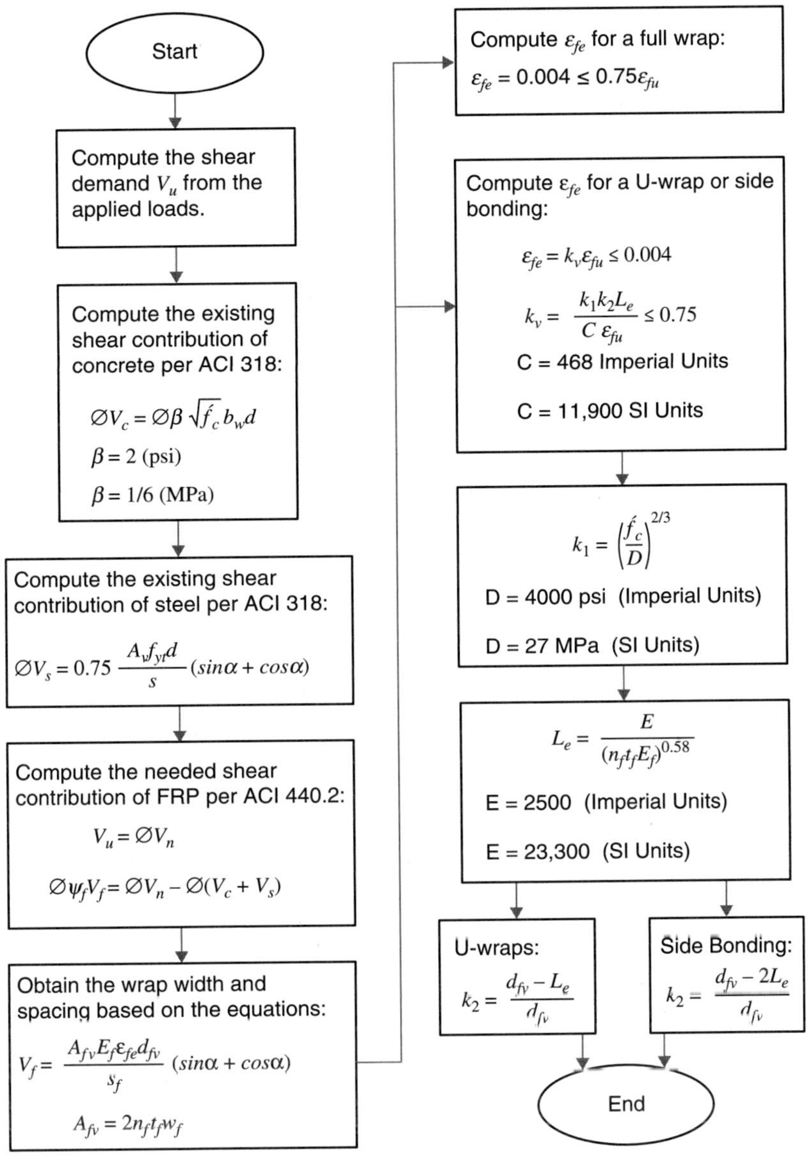

Figure 29.4 Flowchart for shear strengthening design per ACI 440.2R-17.

3. Maximum steel stirrup spacing limit: The steel stirrup spacing cannot exceed the values set by the equations below:

$$
\begin{aligned}
&IF\ V_s < D\sqrt{f'_c}\,b_w d, s_{max} = \min(0.5d, 24\text{ in. or } 610\text{ mm})\\
&IF\ V_s \ge D\sqrt{f'_c}\,b_w d, s_{max} = \min(0.25d, 12\text{ in. or } 305\text{ mm})\\
&D = 4\ (\text{Imperial Units})\\
&D = \frac{1}{3}\ (\text{SI Units})
\end{aligned}
\tag{29.14}
$$

4. The strength reduction factor: The strength reduction factor, specified by ACI 318-14 and ACI 440.2R-17, is set at ($\phi = 0.75$) for shear.

29.3.4 Limit Checks for AASHTO LRFD/NCHRP Report 678

1. Minimum moment allowed: The moment considered in the shear calculations cannot be smaller than the moment at the critical section, which is at d_v distance from the support:

$$M_u \ge V_u d_v \tag{29.15}$$

2. Longitudinal steel strain limit: The strain in longitudinal steel cannot exceed the value of 0.006 in shear calculations:

$$\varepsilon_s \le 0.006\ (\text{tension}) \tag{29.16}$$

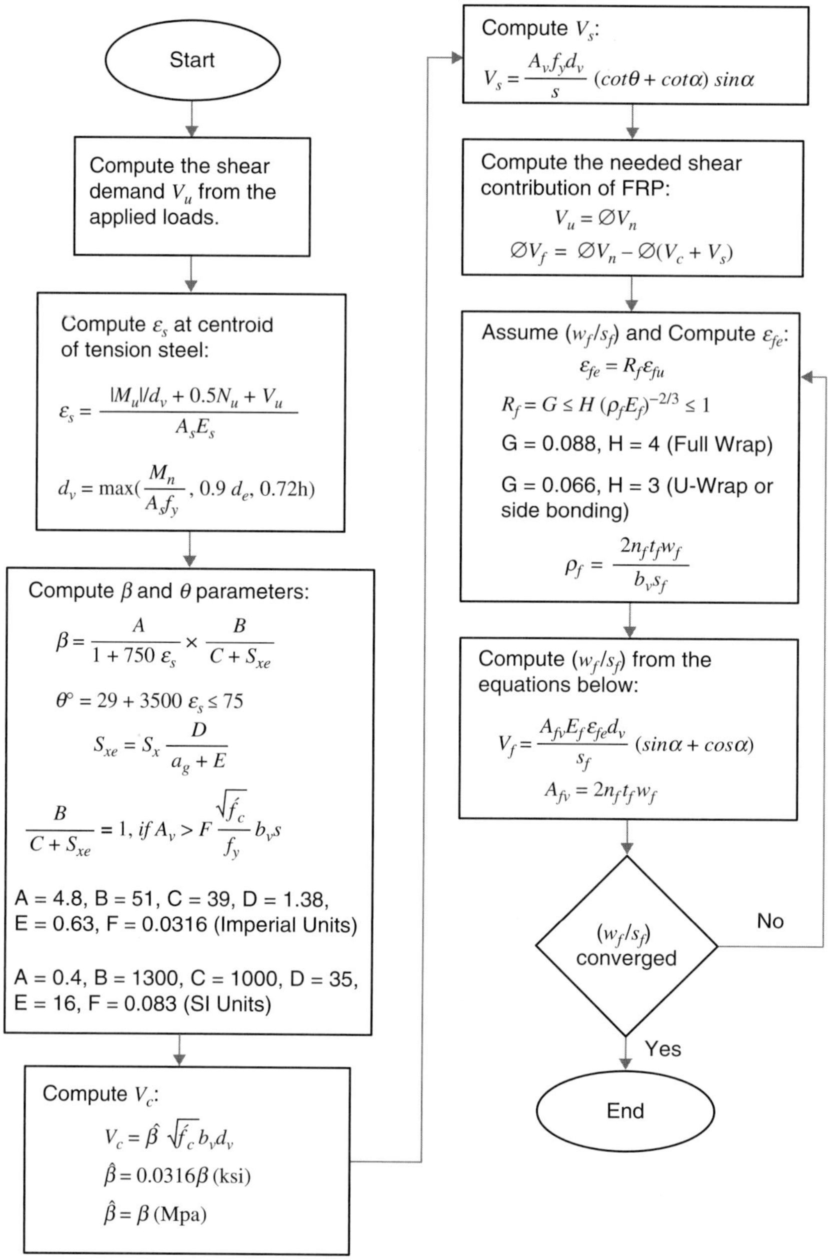

Figure 29.5 Flowchart for shear strengthening design per AASHTO LRFD/NCHRP Report 678.

3. Maximum stirrup spacing: The steel stirrup spacing cannot exceed the values set by the equations below:

$$IF\ V_u < 0.125\, f_c' b_v d_v, s_{max} = \min\,(0.8d_v, 24 \text{ in. or } 610 \text{ mm})$$
$$IF\ V_u \ge 0.125\, f_c' b_v d_v, s_{max} = \min\,(0.4d_v, 12 \text{ in. or } 305 \text{ mm}) \quad (29.17)$$

4. Maximum shear force the section can carry: The maximum shear force that the section may carry is controlled by the crushing of concrete struts. This is determined here according to the following equation:

$$V_u < 0.25 f_c' b_v d_v \quad (29.18)$$

5. Longitudinal steel yielding limit: The maximum force that may be carried by the tension steel reinforcement as a result of combined bending, axial force, and shear must be limited to $A_s f_y$:

$$\frac{|M_u|}{d_v} + \frac{N_u}{2} + V_u \cot\theta - 0.5\phi V_s \cot\theta - 0.5\phi V_f \cot\theta \le A_s f_y \quad (29.19)$$

6. The strength reduction factor: The strength reduction factor, specified by AASHTO LRFD and NCHRP Report 678, is set at ($\phi = 0.9$) for shear in normal weight concrete and ($\phi = 0.8$) for shear in lightweight concrete.

DESIGN EXAMPLE 4

Design Example 2, presented in the previous section, is adopted here to examine shear strengthening design with FRP when the live load is increased by 70 percent. The internal steel stirrup reinforcement is assumed to be provided as #3 bars at 10 in. (No. 10 bars at 250 mm) c/c along the entire span. Check the adequacy of the provided steel stirrups and concrete to resist the upgraded loads. If shear strengthening is required, design the critical section with external FRP wraps and determine the extent of applying this wrap to provide adequate shear capacity to the beam using the provisions of ACI 440.2R-17. Use the following CFRP parameters:

Material property	U.S. customary units	SI units
f_{yt}	40 ksi	276 MPa
E_f	33,000 ksi	227.7 GPa
ε^*_{fu}	0.0167	0.0167
t_f	0.0065 in.	0.165 mm

Solution:

U.S. customary unit calculations	SI unit calculations
Compute the critical shear: $\bar{w}_{LL}=1.2\times1.7=2.04$ k/ft $\bar{w}_u=1.2\times1.0+1.6\times2.04=4.46$ k/ft $\bar{V}_u=\frac{4.46\times24}{2}=53.52$ k $\bar{V}_{ud}=53.52-4.46\times\frac{21.5}{12}=45.53$ k	Compute the critical shear: $\bar{w}_{LL}=17.5\times1.7=29.75$ kN/m $\bar{w}_u=1.2\times14.6+1.6\times29.75=65.12$ kN/m $\bar{V}_u=\frac{65.12\times7.32}{2}=238.34$ kN $\bar{V}_{ud}=238.34-65.12\times0.546=202.78$ kN
Compute the shear contribution of concrete, steel, and FRP: $\phi V_c=0.75\times2\sqrt{5000}\times12\times21.5$ $=27{,}365$ lb $=27.37$ k $\phi V_s=0.75\times\frac{0.22\times40\times21.5}{10}=14.19$ k $\phi V_n=V_{ud}=45.53$ k $\phi\psi_f V_f=45.53-27.37-14.19=3.97$ k	Compute the shear contribution of concrete, steel, and FRP: $\phi V_c=0.75/6\times\sqrt{34.5}\times305\times546$ $=122{,}268$ N $=122.27$ k $\phi V_s=0.75\times\frac{142.4\times276\times546}{250}$ $=64{,}377$ N $=64.4$ kN $\phi\bar{V}_n=\bar{V}_{ud}=202.78$ kN $\phi\psi_f V_f=202.78-122.27-64.4$ $=16.11$ kN
Determine FRP parameters. $\varepsilon_{fu}=C_E\varepsilon^*_{fu}=0.95\times0.0167=0.0159$ $d_{fv}=d-t_{\text{slab}}=21.5-4=17.5$ in.	Determine FRP parameters: $\varepsilon_{fu}=C_E\varepsilon^*_{fu}=0.95\times0.0167=0.0159$ $d_{fv}=d-t_{\text{slab}}=546-100=446$ mm
Assume 1 layer of U-wrap: $k_1=\left(\frac{5000}{4000}\right)^{2/3}=1.16$ $L_e=\frac{2500}{(1\times0.0065\times33\times10^6)^{0.58}}$ $=2.02$ in. $k_2=\frac{17.5-2.02}{17.5}=0.88$ $k_v=\frac{1.16\times0.88\times2.02}{468\times0.0159}=0.28$ $\varepsilon_{fe}=0.28\times0.0159=0.00445>0.004$ use $\varepsilon_{fe}=0.004$	Assume 1 layer of U-wrap: $k_1=\left(\frac{34.5}{27}\right)^{2/3}=1.18$ $L_e=\frac{23{,}300}{(1\times0.165\times227.7\times10^3)^{0.58}}$ $=51.75$ mm $k_2=\frac{446-51.75}{446}=0.88$ $k_v=\frac{1.18\times0.88\times51.75}{11900\times0.0159}=0.28$ $\varepsilon_{fe}=0.28\times0.0159=0.00445>0.004$ use $\varepsilon_{fe}=0.004$
Design the U-wraps by finding (w_f/s_f): $\phi\psi_f V_f=0.75\times0.85\times\frac{A_{fv}E_f\varepsilon_{fe}d_{fv}}{s_f}$ $A_{fv}=2n_ft_fw_f=2\times1\times0.0065w_f$ $=0.013w_f$ $0.75\times0.85\times0.013\times33{,}000\times0.004$ $\times17.5\frac{w_f}{s_f}=3.97\Rightarrow\frac{w_f}{s_f}=0.21$ $s_{f\max}=w_f+\frac{d}{4}\Rightarrow s_f=0.21s_f+5.5$ $s_f=6.96$ in. ≈7 in. $\Rightarrow w_f$ $=1.47$ in. ≈1.5 in.	Design the U-wraps by finding (w_f/s_f): $\phi\psi_f V_f=0.75\times0.85\times\frac{A_{fv}E_f\varepsilon_{fe}d_{fv}}{s_f}$ $A_{fv}=2n_ft_fw_f=2\times1\times0.165w_f$ $=0.33w_f$ $0.75\times0.85\times0.33\times227.7\times0.004\times446\frac{w_f}{s_f}$ $=16.11\Rightarrow\frac{w_f}{s_f}=0.19$ $s_{f\max}=w_f+\frac{d}{4}\Rightarrow s_f=0.19s_f+137$ $s_f=169.14$ mm ≈170 mm $\Rightarrow$ $w_f=32.3$ mm ≈33 mm
Compute the number of U-wraps needed: $\phi(V_c+V_s)=\bar{V}_u-\bar{w}_ux$ $41.56=53.52-4.46x\Rightarrow x=2.68$ ft No. of U-wraps $=\frac{2.68\times12}{7}+1=5.59$ use 6 U-wraps at 7 in. c/c	Compute the number of U-wraps needed: $\phi(V_c+V_s)=\bar{V}_u-\bar{w}_ux$ $122.27+64.4=238.34-65.12x\Rightarrow x$ $=0.79$ m No. of U-wraps $=\frac{0.79\times1000}{170}+1=5.65$ use 6 U-wraps at 170 mm c/c
Check the loss of strengthening limit: $\tilde{w}_u=1.1\times1.0+0.75\times2.04=2.63$ k/ft $\tilde{V}_u=\frac{\tilde{w}_ul}{2}=\frac{2.63\times24}{2}=31.56$ k $\phi V_n=\phi(V_c+V_s)=41.56>\tilde{V}_u$ ok	Check the loss of strengthening limit: $\tilde{w}_u=1.1\times14.6+0.75\times29.75$ $=38.37$ kN/m $\tilde{V}_u=\frac{\tilde{w}_ul}{2}=\frac{38.37\times7.32}{2}=140.43$ kN $\phi V_n=\phi(V_c+V_s)=122.27+64.4$ $=186.67>\tilde{V}_u$ ok

DESIGN EXAMPLE 5

Redo Design Example 4 using the provisions of AASHTO LRFD and NCHRP Report 678. What could be concluded from the comparison of the solution of example 4 and this example?

Solution:

U.S. customary unit calculations	SI unit calculations
Determine the effective shear depth (d_v). $a=\frac{A_sf_y}{0.85f'_cb}=\frac{3\times60}{0.85\times5\times12}=3.53$ in. $M_n=A_sf_y\left(d_e-\frac{a}{2}\right)=3552.3$ k-in. $d_{v1}=\frac{M_n}{A_sf_y}=19.74$ in. $\Rightarrow$ controls $d_{v2}=0.9d_e=19.35$ in. $d_{v3}=0.72\,\text{h}=17.28$ in. Thus, $d_v=19.74$ in.	Determine the effective shear depth (d_v): $a=\frac{A_sf_y}{0.85f'_cb}=\frac{1934\times414}{0.85\times34.5\times305}$ $=89.52$ mm $M_n=A_sf_y\left(d_e-\frac{a}{2}\right)$ $=401{,}330{,}838$ N-mm $=401.33$ kN-m $d_{v1}=\frac{M_n}{A_sf_y}=501.24$ mm $\Rightarrow$ controls $d_{v2}=0.9d_e=491.4$ mm $d_{v3}=0.72\,\text{h}=439.2$ mm Thus, $d_v=501.24$ mm
Compute the critical shear and moment: $\bar{w}_{LL}=1.2\times1.7=2.04$ k/ft $\bar{w}_u=1.2\times1.0+1.6\times2.04=4.46$ k/ft $\bar{V}_u=\frac{4.46\times24}{2}=53.52$ k $\bar{V}_{udv}=53.52-4.46\times\frac{19.74}{12}=46.18$ k $\bar{M}_u=\bar{V}_ud_v=1056.48$ k-in.	Compute the critical shear and moment: $\bar{w}_{LL}=17.5\times1.7=29.75$ kN/m $\bar{w}_u=1.2\times14.6+1.6\times29.75$ $=65.12$ kN/m $\bar{V}_u=\frac{65.12\times7.32}{2}=238.34$ kN $\bar{V}_{udv}=238.34-65.12\times0.50124$ $=205.7$ kN $\bar{M}_u=\bar{V}_ud_v=119.47$ kN-m

U.S. customary unit calculations	SI unit calculations
Compute β and θ parameters: $\varepsilon_s = \dfrac{\bar{M}_u/d_v + \bar{V}_{udv}}{A_sE_s}$ $= \dfrac{1056.48/19.74+46.18}{3\times 29000}$ $= 0.00115$ $A_v = 2\times 0.11 = 0.22\text{ in.}^2 > A_{v\min}$ $A_{v\min} = 0.0316\times\dfrac{\sqrt{5}}{40}\times 12\times 10$ $= 0.212\text{in.}^2$ $\beta = \dfrac{4.8}{(1+750\varepsilon_s)} = 2.58$ $\theta = 29+3500\varepsilon_s = 33.03°$	Compute β and θ parameters: $\varepsilon_s = \dfrac{\bar{M}_u/d_v + \bar{V}_{udv}}{A_sE_s}$ $= \dfrac{119.47/0.50124+205.7}{1934\times 200} = 0.00115$ $A_v = 2\times\dfrac{\pi}{4}(9.52)^2$ $= 142.36\text{ mm}^2 > A_{v\min}$ $A_{v\min} = 0.083\times\dfrac{\sqrt{34.5}}{276}\times 305\times 250$ $= 134.68\text{ mm}^2$ $\beta = \dfrac{0.4}{(1+750\varepsilon_s)} = 0.215$ $\theta = 29+3500\varepsilon_s = 33.03°$
Compute the shear contribution of concrete, steel, and FRP: $V_c = 0.0316\times 2.58\sqrt{5}\times 12\times 19.74$ $= 43.18\text{k}$ $V_s = \dfrac{0.22\times 40\times 19.74}{10}\cot(33.03)$ $= 26.72\text{ k}$ $V_{udv} = 46.18\text{ k} < \phi(V_c+V_s)$ $\phi(V_c+V_s) = 0.9(43.18+26.72) = 62.91\text{ k}$ ∴ *no need for shear strengthening*	Compute the shear contribution of concrete, steel, and FRP: $V_c = 0.215\sqrt{34.5}\times 305\times 501.24$ $= 193{,}061\text{ N}$ $= 193.06\text{ kN}$ $V_s = \dfrac{142.36\times 276\times 501.24}{250}\cot(33.03)$ $= 120{,}913\text{ N}$ $= 120.91\text{ kN}$ $V_{udv} = 205.7\text{ kN} < \phi(V_c+V_s)$ $\phi(V_c+V_s) = 0.9(193.06+120.91)$ $= 282.57\text{ kN}$ ∴ *no need for shear strengthening*
It may be concluded that the ACI318-14 shear design procedure is very conservative compared with the simplified modified compression field theory in this case.	It may be concluded that the ACI318-14 shear design procedure is very conservative compared with the simplified modified compression field theory in this case.

DESIGN EXAMPLE 6

Redo Design Example 5 using the provisions of AASHTO LRFD and NCHRP Report 678 while doubling the increase in live load to 140 percent. What could be concluded from the comparison of the solution of Design Example 4 and this example?

Solution:

U.S. customary unit calculations	SI unit calculations
Determine the effective shear depth (d_v): $a = \dfrac{A_sf_y}{0.85f'_cb} = \dfrac{3\times 60}{0.85\times 5\times 12}$ $= 3.53\text{ in.}$ $M_n = A_sf_y\left(d_e - \dfrac{a}{2}\right) = 3{,}552.3\text{ k-in.}$ $d_{v1} = \dfrac{M_n}{A_sf_y} = 19.74\text{ in.} \Rightarrow \text{controls}$ $d_{v2} = 0.9d_e = 19.35\text{ in.}$ $d_{v3} = 0.72\text{ h} = 17.28\text{ in.}$ Thus, $d_v = 19.74\text{ in.}$	Determine the effective shear depth (d_v): $a - \dfrac{A_sf_y}{0.85f'_cb} - \dfrac{1934\times 414}{0.85\times 34.5\times 305}$ $= 89.52\text{ mm}$ $M_n = A_sf_y\left(d_e - \dfrac{a}{2}\right)$ $= 401{,}330{,}838\text{ N-mm}$ $= 401.33\text{ kN-m}$ $d_{v1} = \dfrac{M_n}{A_sf_y} = 501.24\text{ mm} \Rightarrow \text{controls}$ $d_{v2} = 0.9d_e = 491.4\text{ mm}$ $d_{v3} = 0.72\text{ h} = 439.2\text{ mm}$ Thus, $d_v = 501.24\text{ mm}$
Compute the critical shear and moment: $\bar{w}_{LL} = 1.2\times 2.4 = 2.88\text{ k/ft}$ $\bar{w}_u = 1.2\times 1.0 + 1.6\times 2.88 = 5.81\text{ k/ft}$ $\bar{V}_u = \dfrac{5.81\times 24}{2} = 69.72\text{ k}$ $\bar{V}_{ud} = 69.72 - 5.81\times\dfrac{19.74}{12} = 60.16\text{ k}$ $\bar{M}_u = \bar{V}_ud_v = 1376.27\text{ k-in.}$	Compute the critical shear and moment: $\bar{w}_{LL} = 17.5\times 2.4 = 42\text{ kN/m}$ $\bar{w}_u = 1.2\times 14.6 + 1.6\times 42$ $= 84.72\text{ kN/m}$ $\bar{V}_u = \dfrac{84.72\times 7.32}{2} = 310.08\text{ kN}$ $\bar{V}_{udv} = 310.08 - 84.72\times 0.50124$ $= 267.61\text{ kN}$ $\bar{M}_u = \bar{V}_ud_v = 155.42\text{ kN-m}$
Compute β and θ parameters: $\varepsilon_s = \dfrac{\bar{M}_u/d_v + \bar{V}_{ud}}{A_sE_s}$ $= \dfrac{1376.27/19.74+60.16}{3\times 29000} = 0.0015$ $A_v = 2\times 0.11 = 0.22\text{ in.}^2 > A_{v\min}$ $A_{v\min} = 0.0316\times\dfrac{\sqrt{5}}{40}\times 12\times 10$ $= 0.212\text{ in.}^2$ $\beta = \dfrac{4.8}{(1+750\varepsilon_s)} = 2.26$ $\theta = 29+3500\varepsilon_s = 34.25°$	Compute β and θ parameters: $\varepsilon_s = \dfrac{\bar{M}_u/d_v + \bar{V}_{udv}}{A_sE_s}$ $= \dfrac{155.42/0.50124+267.61}{1934\times 200}$ $= 0.0015$ $A_v = 2\times\dfrac{\pi}{4}(9.52)^2$ $= 142.36\text{ mm}^2 > A_{v\min}$ $A_{v\min} = 0.083\times\dfrac{\sqrt{34.5}}{276}\times 305\times 250$ $= 134.68\text{ mm}^2$ $\beta = \dfrac{0.4}{(1+750\varepsilon_s)} = 0.188$ $\theta = 29+3500\varepsilon_s = 34.25°$
Compute the shear contribution of concrete, steel, and FRP: $V_c = 0.0316\times 2.26\sqrt{5}\times 12\times 19.74$ $= 37.83\text{ k}$ $V_s = \dfrac{0.22\times 40\times 19.74}{10}\cot(34.25)$ $= 25.51\text{ k}$ $V_{udv} = 60.16\text{ k} > \phi(V_c+V_s)$ $\phi(V_c+V_s) = 0.9(37.83+25.51) = 57\text{ k}$ $\phi V_f = 3.16\text{ k}$	Compute the shear contribution of concrete, steel, and FRP: $V_c = 0.188\sqrt{34.5}\times 305\times 501.24$ $= 168{,}816\text{ N}$ $= 168.82\text{ kN}$ $V_s = \dfrac{142.36\times 276\times 501.24}{250}\cot(34.25)$ $= 115{,}700\text{ N}$ $= 115.7\text{ kN}$ $V_{udv} = 267.61\text{ kN} > \phi(V_c+V_s)$ $\phi(V_c+V_s) = 0.9(168.82+115.7)$ $= 256.07\text{ kN}$ $\phi V_f = 11.54\text{ kN}$
Assume continuous U-wraps ($w_f/s_f = 1$): $\rho_f = \dfrac{2n_ft_fw_f}{b_vs_f} = \dfrac{2\times 1\times 0.0065}{12}$ $= 0.00108$ $R_f = 3\times(0.00108\times 33{,}000)^{-2/3} = 0.277$ $\varepsilon_{fe} = R_f\varepsilon_{fu} = 0.0044$ $3.16 = 0.9\times\dfrac{0.013w_f\times 33{,}000\times 0.0044\times 19.74}{s_f}$ $\dfrac{w_f}{s_f} = 0.094 \ll 1 \Rightarrow \text{iterate}$	Assume continuous U-wraps ($w_f/s_f = 1$): $\rho_f = \dfrac{2n_ft_fw_f}{b_vs_f} = \dfrac{2\times 1\times 0.165}{305}$ $= 0.00108$ $R_f = 3\times\left(0.00108\times\dfrac{227.7\times 1000}{6.899}\right)^{-2/3}$ $= 0.277$ $\varepsilon_{fe} = R_f\varepsilon_{fu} = 0.0044$ $11.54 = 0.9\times\dfrac{0.33w_f\times 227.7\times 0.0044\times 501.24}{s_f}$ $\dfrac{w_f}{s_f} = 0.077 \ll 1 \Rightarrow \text{iterate}$

U.S. customary unit calculations	SI unit calculations
Iteration 2: $\rho_f=\frac{2n_ft_fw_f}{b_vs_f}=\frac{2\times1\times0.0065\times0.094}{12}$ $=0.0001015$ $R_f=3\times(0.0001015\times33{,}000)^{-2/3}$ $=1.34>1.0$ $\varepsilon_{fe}=\varepsilon_{fu}=0.0159$ $3.16=0.9\times\frac{0.013w_f\times33{,}000\times0.0159\times19.74}{s_f}$ $\frac{w_f}{s_f}=0.026<0.094\Rightarrow$ iterate	Iteration 2: $\rho_f=\frac{2n_ft_fw_f}{b_vs_f}=\frac{2\times1\times0.165\times0.077}{305}$ $=0.0000833$ $R_f=3\times\left(0.0000833\times\frac{227.7\times1000}{6.899}\right)^{-2/3}$ $=1.53>1.0$ $\varepsilon_{fe}=\varepsilon_{fu}=0.0159$ $11.54=0.9\times\frac{0.33w_f\times227.7\times0.0159\times501.24}{s_f}$ $\frac{w_f}{s_f}=0.0214<0.077\Rightarrow$ iterate
Iteration 3: $\rho_f=\frac{2n_ft_fw_f}{b_vs_f}=\frac{2\times1\times0.0065\times0.026}{12}$ $=0.0000281$ $R_f=3\times(0.0000281\times33{,}000)^{-2/3}$ $=3.15>1.0$ $\varepsilon_{fe}=\varepsilon_{fu}=0.0159$ $3.16=0.9\times\frac{0.013w_f\times33{,}000\times0.0159\times19.74}{s_f}$ $\frac{w_f}{s_f}=0.026=0.026$ $s_f=w_f+\frac{d}{4}\Rightarrow0.974s_f=5.4$ in. $s_f=5.54$ in. ≈5.5 in. $\Rightarrow w_f=0.14$ in. Not practical, use $w_f=1$ in.	Iteration 3: $\rho_f=\frac{2n_ft_fw_f}{b_vs_f}=\frac{2\times1\times0.165\times0.0214}{305}$ $=0.0000232$ $R_f=3\times\left(0.0000232\times\frac{227.7\times1000}{6.899}\right)^{-2/3}$ $=3.58>1.0$ $\varepsilon_{fe}=\varepsilon_{fu}=0.0159$ $11.54=0.9\times\frac{0.33w_f\times227.7\times0.0159\times501.24}{s_f}$ $\frac{w_f}{s_f}=0.0214=0.0214$ $s_f=w_f+\frac{d}{4}\Rightarrow0.979s_f=136.5$ mm $s_f=139.43$ mm ≈140 mm $\Rightarrow w_f=3$ mm Not practical, use $w_f=20$ mm
Compute the number of U-wraps needed: $\phi(V_c+V_s)=\bar{V}_u-\bar{w}_ux$ $57=69.72-5.81x\Rightarrow x=2.19$ ft No. of U-wraps $=\frac{2.19\times12}{5.5}+1=5.78$ use 6 U wraps at 5.5 in. c/c	Compute the number of U-wraps needed: $\phi(V_c+V_s)=\bar{V}_u-\bar{w}_ux$ $256.07=310.08-84.72x$ $\Rightarrow x=0.64$ m No. of U-wraps $=\frac{0.64\times1000}{140}+1=5.57$ use 6 U-wraps at 140 mm c/c
It may be concluded that the AASHTO LRFD and NCHRP Report 678 procedure yield similar shear strengthening design to that of ACI 440.2R-17 *BUT* with double the increase in live load level. It is evident that ACI440 procedure is rather conservative and the AASHTO procedure is rather liberal. It is proposed here to limit the maximum effective FRP strain to 75 percent of the rupture strain rather than the full amount of rupture strain.	It may be concluded that the AASHTO LRFD and NCHRP Report 678 procedure yield similar shear strengthening design to that of ACI 440.2R-17 *BUT* with double the increase in live load level. It is evident that ACI440 procedure is rather conservative and the AASHTO procedure is rather liberal. It is proposed here to limit the maximum effective FRP strain to 75 percent of the rupture strain rather than the full amount of rupture strain.

29.4 CONFINEMENT STRENGTHENING DESIGN FOR CIRCULAR COLUMNS

This is another important application of FRP strengthening of reinforced-concrete members. It has been shown through experimental testing to increase both the ultimate strength and ductility of concentrically loaded circular columns. ACI 440.2R-17 proposes a procedure to construct the confined axial force-bending moment interaction diagram based on the confinement contribution of FRP only without considering the effect of internal transverse steel. This procedure is based on the model proposed by Lam and Teng (2003) with minor adjustments. It is important to note here that Lam and Teng model was based on a modification of the first FRP confinement model proposed in the literature by Samaan et al. (1998).

29.4.1 Strength and Strain Enhancement of Axially-Loaded Columns

It is important to establish here that the confined concrete axial stress-strain behavior of circular columns, after the confinement is engaged, is either ascending or descending (Rasheed 2015). ACI 440.2R-17 admits the behavior of columns having an ascending second branch of the concrete axial stress-strain curve. This can be guaranteed by keeping the ratio of confining pressure to the unconfined strength of concrete above 0.08 ($f_l/f'_c>0.08$). The confining pressure in FRP confined circular columns is defined as

$$f_l=\frac{2n_ft_fE_f\varepsilon_{fe}}{D}\tag{29.20}$$

$$\varepsilon_{fe}=\kappa_\varepsilon\varepsilon_{fu}=0.586\varepsilon_{fu}\tag{29.21}$$

As a result, ACI 440.2R-17 defined the confined ultimate strength and strain of concrete by

$$f'_{cc}=f'_c+\psi_f\times3.3\times\kappa_af_l\tag{29.22}$$

$$\varepsilon_{ccu}=\varepsilon'_c\left(1.5+12\kappa_b\frac{f_l}{f'_c}\left(\frac{\varepsilon_{fe}}{\varepsilon'_c}\right)^{0.45}\right)\le0.01\tag{29.23}$$

for circular columns, ($\psi_f=0.95$, $\kappa_a=\kappa_b=1$).

29.4.2 Construction of Interaction Diagram for Confined Circular Columns:

The approximate interaction diagram for confined concrete columns is established based on four discrete points. These are the ultimate capacity points of pure axial compression (Point A) and another compression-controlled point with all compression strain profiles having a zero strain at the extreme steel layer (Point B) (Fig. 29.6), the balanced compression-tension point with the extreme steel layer just at the initiation of tensile yielding (Point C), and pure bending point with a negligible confinement contribution (Point D) (Fig. 29.6).

Determination of Point A

Since the procedure of the ACI 440.2R-17 relies on the confinement of FRP while ignoring the confinement produced by internal transverse steel, the confining pressure in the core and cover of the column is assumed to be the same. Therefore, the ultimate axial load in the column subjected to pure axial compression is defined by

$$\phi P_n=0.85\phi[0.85f'_{cc}(A_g-A_{st})+A_{st}f_y]\tag{29.24}$$

when the column is transversely reinforced with a spiral ($\phi=0.75$).

Determination of Point B and Point C

Point B is an intermediate point within the compression-controlled region of the interaction diagram. It is defined by a strain distribution having an extreme compressive strain of $\varepsilon'_{ccu}=0.004$ and a zero strain at the other extreme tensile steel layer (Fig. 29.6). On the other hand, Point C is the last point in the compression-controlled region (balanced point). It is defined by a strain profile having an extreme compressive strain of $\varepsilon'_{ccu}=0.004$ and a yielding tensile strain $\varepsilon_{sy}=f_y/E_s$ at the other extreme tensile steel layer (Fig. 29.6). The development of the axial force-bending moment capacity of Points B and C requires the integration of involved expressions with trigonometric functions due to the variation of the section width across its height. Accordingly, these equations are integrated numerically by dividing the cross section into 100 concrete layers parallel to the centroidal x-axis. In addition, the contribution of longitudinal steel bars is accounted for numerically and separately added to the contribution obtained from the concrete layers (see Fig. 29.7).

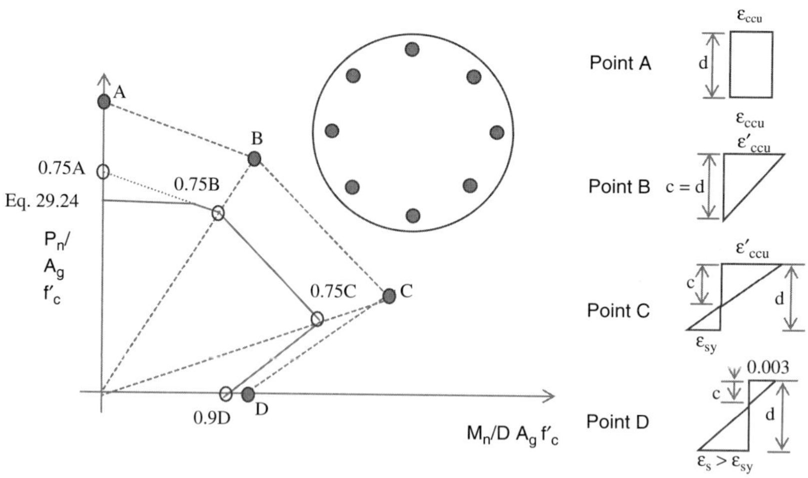

Figure 29.6 Interaction diagram with key points.

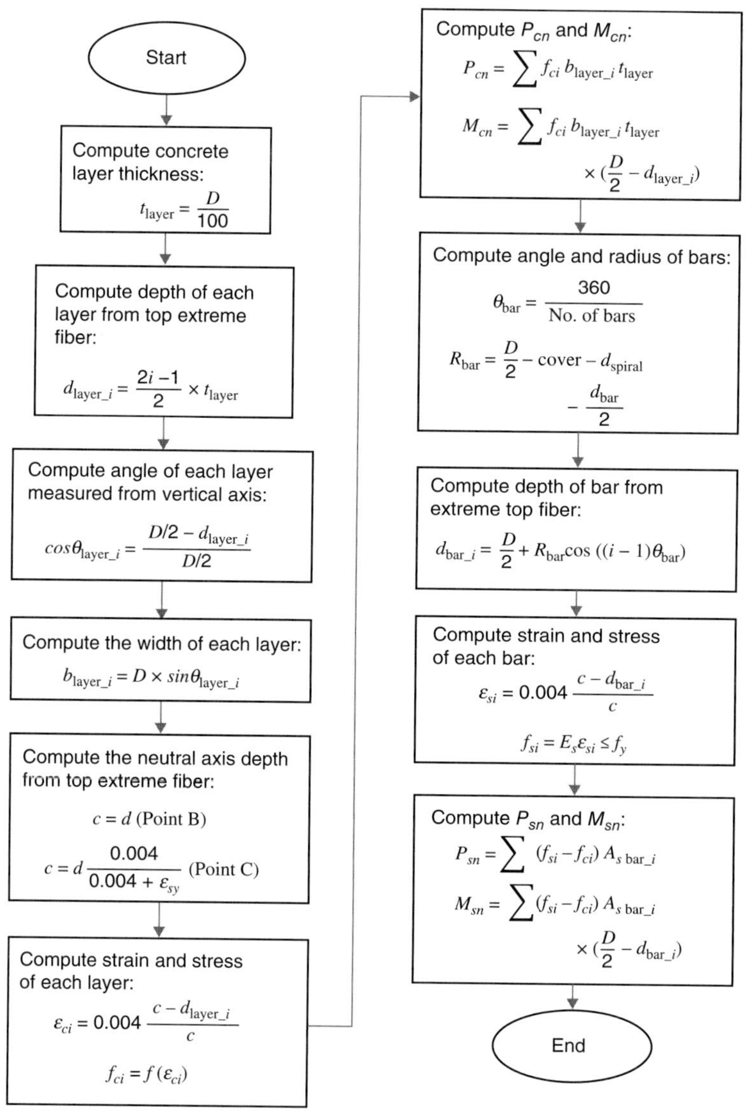

Figure 29.7 Numerical computation of Points B and C.

Determination of Point D

Point D is a pure bending point within the tension-controlled region of the interaction diagram. It is defined by a strain distribution having an extreme compressive strain of $\varepsilon_{cu} = 0.003$ and a tensile strain at the other extreme tensile steel layer exceeding the yielding strain of steel such that the sum of the resulting axial force in the section is zero (Fig. 29.6). The development of the axial force-bending moment capacity of Point D requires the integration of involved expressions with trigonometric functions due to the variation of the section width across its height. Accordingly, these equations are integrated numerically by dividing the cross section into 100 concrete layers parallel to the centroidal x-axis. In addition, the contribution of longitudinal steel bars is accounted for numerically and separately added to the contribution obtained from the concrete layers. This is possible to accomplish numerically if the depth of neutral axis c is assumed first. Then, if the total axial force of the section is not zero, the depth of neutral axis c is adjusted and iterations are performed until the total axial force is zero. The total moment obtained at this iteration is the pure bending moment of Point D (see Fig. 29.8). These iterations are replaced with a single execution of a Goal-Seek function in Excel.

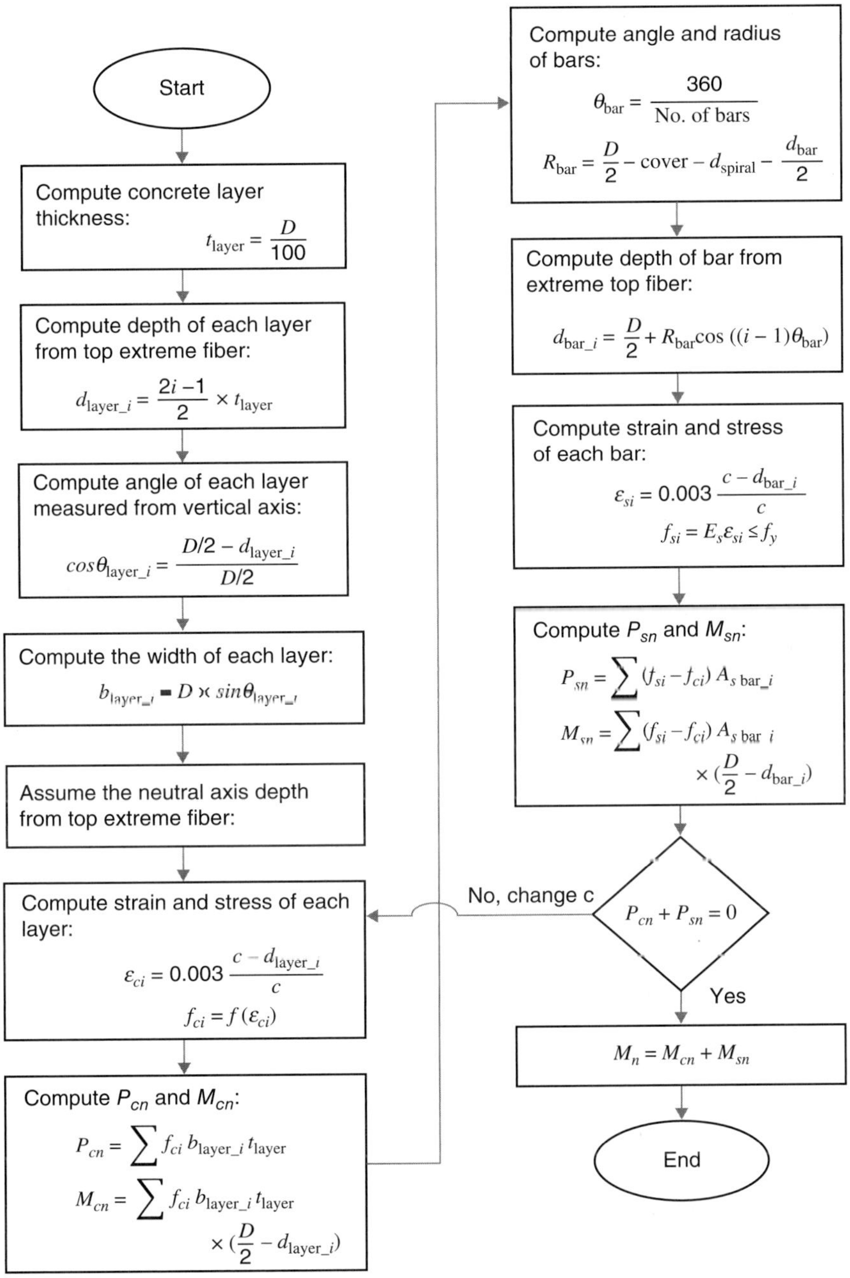

Figure 29.8 Numerical computation of Point D.

DESIGN EXAMPLE 7

A reinforced-concrete circular column of 11.8 in diameter (300 mm diameter), with the parameters and material properties listed in the table below, is to be strengthened by CFRP wrapping to increase its nominal ultimate axial capacity by 25 percent. Draw the limit and design interaction diagrams for the confined concrete based on the amount of FRP needed to accomplish the 25 percent nominal pure compression capacity increase.

Property	U.S. customary units	SI units
Clear cover	1 in.	25 mm
Longitudinal bars	6 #5	6 ϕ 15.9 mm
Hoop diameter	#3	ϕ 9.52 mm
Hoop spacing	3.94 in.	100 mm
f'_c	4.59 ksi	31.7 MPa
f_y	61.3 ksi	423 MPa
f_{yt}	66.1 ksi	456 MPa
E_f	11,306 ksi	78,000 MPa
ε^*_{fu}	0.013	0.013
t_f	0.015 in.	0.381 mm

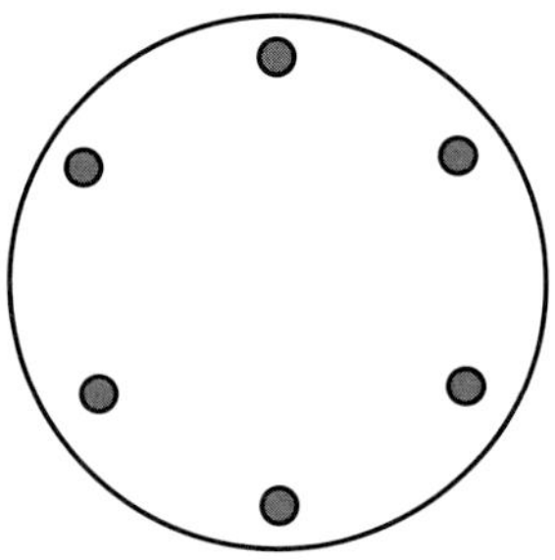

Solution:

U.S. customary unit calculations	SI unit calculations
Compute the nominal axial capacity of the bare and wrapped column: $P_n = 0.85\times4.59\times\left(\frac{\pi}{4}\times11.8^2-6\times0.31\right)+61.3\times6\times0.31=533.42\text{k}$ $\bar{P}_n=1.25\times533.42=666.78\text{ k}$ $\bar{P}_n=0.85\times f'_{cc}\times\left(\frac{\pi}{4}\times11.8^2-6\times0.31\right)+61.3\times6\times0.31=666.78\text{ k}$ $f'_{cc}=6.05\text{ ksi}$	Compute the nominal axial capacity of the bare and wrapped column: $P_n=0.85\times31.7\times\left(\frac{\pi}{4}\times300^2-6\times\frac{\pi}{4}\times15.9^2\right)+423\times6\times\frac{\pi}{4}\times15.9^2=2,376,466\text{ N}$ $=2376\text{ kN}$ $\bar{P}_n=1.25\times2376=2970\text{ kN}$ $\bar{P}_n=0.85\times f'_{cc}\times\left(\frac{\pi}{4}\times300^2-6\times\frac{\pi}{4}\times15.9^2\right)+423\times6\times\frac{\pi}{4}\times15.9^2=2970\times10^3\text{ N}$ $f'_{cc}=41.75\text{ MPa}$
Design the needed number of FRP layers: $f'_{cc}=6.05=4.59+0.95\times3.3\times1\times f_l$ $f_l=0.466\text{ ksi}$ $\frac{f_l}{f'_c}=\frac{0.466}{4.59}=0.102>0.08,$ curve is ascending $0.466=\frac{2n_f\times0.015\times11306\times0.586\times0.013}{11.8}$ $n_f=2.13\text{ layers}\Rightarrow\text{use }n_f=2$	Design the needed number of FRP layers: $f'_{cc}=41.75=31.7+0.95\times3.3\times1\times f_l$ $f_l=3.21\text{ MPa}$ $\frac{f_l}{f'_c}=\frac{3.21}{31.7}=0.101>0.08,$ curve is ascending $3.21=\frac{2n_f\times0.381\times78,000\times0.586\times0.013}{300}$ $n_f=2.13\text{ layers}\Rightarrow\text{use }n_f=2$
Compute the actual Point A: $\frac{2\times2\times0.015\times11,306\times0.586\times0.013}{11.8}=0.438\text{ ksi}$ $\frac{f_l}{f'_c}=\frac{0.438}{4.59}=0.095>0.08,$ still ascending $f'_{cc}=4.59+0.95\times3.3\times1\times0.438=5.96\text{ ksi}$ $\bar{P}_{n,A}=0.85\times5.96\times\left(\frac{\pi}{4}\times11.8^2-6\times0.31\right)+61.3\times6\times0.31=658.6\text{ k}$ $\phi\bar{P}_{n,A}=0.85\times0.75\bar{P}_{n,A}=419.86\text{ k}$	Compute the actual Point A: $\frac{2\times2\times0.381\times78,000\times0.586\times0.013}{300}=3.02\text{ MPa}$ $\frac{f_l}{f'_c}=\frac{3.02}{31.7}=0.095>0.08,$ still ascending $f'_{cc}=31.7+0.95\times3.3\times1\times3.02=41.17\text{ MPa}$ $\bar{P}_{n,A}=0.85\times41.17\times\left(\frac{\pi}{4}\times300^2-6\times\frac{\pi}{4}\times15.9^2\right)+423\times6\times\frac{\pi}{4}\times15.9^2=2,935,862\text{ N}$ $=2935.86\text{ kN}$ $\phi\bar{P}_{n,A}=0.85\times0.75\bar{P}_{n,A}=1871.61\text{ kN}$
Compute Point B numerically using Excel: $f_l=\frac{2\times2\times0.015\times11,306\times0.004}{11.8}=0.23\text{ ksi}$ $f'_{cc}=4.59+0.95\times3.3\times1\times0.23=5.31\text{ ksi}$ $\varepsilon_{ccu}=0.002\times\left(1.5+12\times1\times\frac{0.23}{4.59}\times\left(\frac{0.004}{0.002}\right)^{0.45}\right)=0.00464$ $E_c=3861.7\text{ ksi},\ E_2=155.08\text{ ksi},\ \varepsilon'_t=0.00248$ $\bar{P}_{n,B}=476.12k,\quad\phi\bar{P}_{n,B}=0.75\bar{P}_{n,B}=357.09\text{ k}$ $\bar{M}_{n,B}=52.06\text{ k-ft},$ $\phi\bar{M}_{n,B}=0.75\bar{M}_{n,B}=39.05\text{ k-ft}$	Compute Point B numerically using Excel: $f_l=\frac{2\times2\times0.381\times78,000\times0.004}{300}=1.58\text{ MPa}$ $f'_{cc}=31.7+0.95\times3.3\times1\times1.58=36.65\text{ MPa}$ $\varepsilon_{ccu}=0.002\times\left(1.5+12\times1\times\frac{1.58}{31.7}\times\left(\frac{0.004}{0.002}\right)^{0.45}\right)=0.00463$ $E_c=26,462\text{ MPa},$ $E_2=1069.11\text{ MPa},$ $\varepsilon'_t=0.0025$ $\bar{P}_{n,B}=2118.73kN,\quad\phi\bar{P}_{n,B}=0.75\bar{P}_{n,B}=1589.05\text{ kN}$ $\bar{M}_{n,B}=70.61\text{ kN-m},$ $\phi\bar{M}_{n,B}=0.75\bar{M}_{n,B}=52.97\text{ kN-m}$
Compute Point C numerically using Excel: $\bar{P}_{n,C}=282.61\text{ k},$ $\phi\bar{P}_{n,C}=0.75\bar{P}_{n,B}=211.96k$ $\bar{M}_{n,C}=77.12\text{ k-ft},$ $\phi\bar{M}_{n,C}=0.75\bar{M}_{n,C}=57.84\text{ k-ft}$	Compute Point C numerically using Excel: $\bar{P}_{n,C}=1257.61\text{ kN},$ $\phi\bar{P}_{n,C}=0.75\bar{P}_{n,B}=943.22\text{ kN}$ $\bar{M}_{n,B}=104.60\text{ kN-m},$ $\phi\bar{M}_{n,B}=0.75\bar{M}_{n,B}=78.45\text{ kN-m}$
Compute Point D numerically using Excel: $\bar{P}_{n,D}=0\text{ k},$ $\bar{M}_{n,D}=38.03\text{ k-ft},$ $\phi\bar{M}_{n,D}=0.9\bar{M}_{n,D}=34.23\text{ k-ft}$	Compute Point D numerically using Excel: $\bar{P}_{n,D}=0\text{ k},$ $\bar{M}_{n,D}=51.58\text{kN-m},$ $\phi\bar{M}_{n,D}=0.9\bar{M}_{n,D}=46.43\text{ kN-m}$

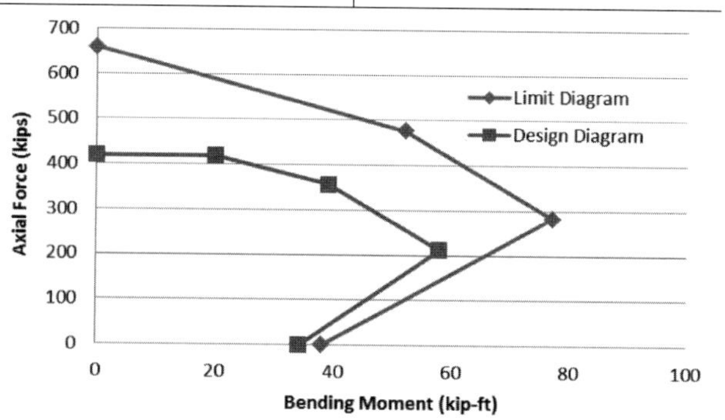

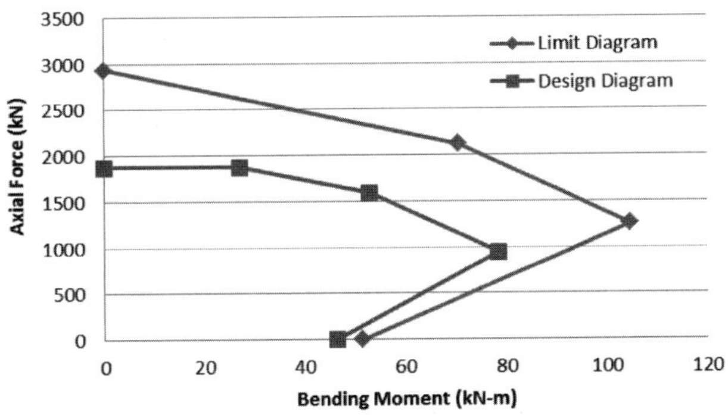

REFERENCES

AASHTO LRFD (2014). Bridge Design Specifications, American Association of State Highway and Transportation Officials. Washington, DC.

ACI 318-14 (2014). Building Code Requirements for Structural Concrete and Commentary, American Concrete Institute, Farmington Hills, MI.

ACI 440.2R-17 (2017). Guide for the Design and Construction of Externally Bonded FRP Systems for Strengthening Concrete Structures, American Concrete Institute, Farmington Hills, MI.

ASTM D4762-16 (2016). Standard Guide for Testing Polymer Matrix Composite Materials, ASTM International, West Conshohocken, PA.

Bentz, C. E., Vecchio, J. F., and Collins, P. M. (2006). "Simplified Modified Compression Field Theory for Calculating Shear Strength of Reinforced Concrete Elements," *ACI Structural Journal*, 103(4), pp. 614–624.

Khalifa, A., Gold, W., Nanni, A., and Abel-Aziz, M. (1998). "Contribution of Externally Bonded FRP to the Shear Capacity of RC Flexural Members," *Journal of Composites in Construction*, 2(4), pp. 195–203.

Lam, L., and Teng, J. G. (2003). "Design-Oriented Stress–Strain Model for FRP-Confined Concrete," *Construction and Building Materials*, 17(6–7), September–October, pp. 471–489.

Meier, U. (1987). "Bridge Repair with High Performance Composite Materials," *Material und Technik*, 15(4), pp. 125–128.

NCHRP Report 678 (2011). "Design of FRP Systems for Strengthening Concrete Girders in Shear," by Abdeldjelil Belarbi, Sang-Wook Bae, Ashraf Ayoub, Daniel Kuchma, Amir Mirmiran and Ayman Okeil, Transportation Research Board, Washington, DC, 120 p.

Rasheed, H. A. (2015). *Strengthening Design of Reinforced Concrete with FRP.* CRC Press, Boca Raton, FL, 230 p.

Rasheed, H. A., and Motto, N. H. (2010). "Strength Design Equations for FRP Strengthened Concrete Beams," *Australian Journal of Structural Engineering*, 11(2), November, pp. 103–116.

Samaan, M., Mirmiran, A., and Shahawy, M. (1998). "Model of Concrete Confined by Fiber Composites," *Journal of Structural Engineering*, 124(9), pp. 1025–1031.

Saqan, E., Rasheed, H. A., and Hawileh, R. A. (2013) "An Efficient Design Procedure for Flexural Strengthening of RC Beams Based on ACI 440.2R-08," *Composites Part B: Engineering*, 49, June, pp. 71–79.

Chapter 30
Structural Glass and Glazing

BY
RUI de S. CAMPOSINHOS *Coordinator Professor with Aggregation, Polytechnic of Porto, Portugal*

30.1 INTRODUCTION

The constant search to maximize the transparency of the façades resulted in larger glass windows forcing concrete and steel load-bearing systems to become smaller over time.

When subjected to tensile stress, glass breaks with any plastic flow because it is a brittle material. The break occurs at a stress level much lower than the theoretical value calculated from the strength of the atomic bonds [1]. The glass is unique since its fracture occurs in entirely different ways from other materials. Glass failure generally starts at a specific surface flaw which is as a stress multiplier in a region of high tensile stress.

Failure prediction methods based on allowable areas for each thickness of glass as a function of uniform duration pressure have been replaced by modern methods that have the advantage of providing a consistent framework for all previous testing, as also explains most of the phenomena associated with laterally loaded thin glass plates [2].

The stress at which glass fractures depends on the size of flaws: The larger the flaw, the lower the fracture stress. For soda-lime glass (SLSG), the behavior is dominated by the surface defects, and the condition of the surface of the glass controls its strength [3]. The physical description of a fracture surface involves a planar region—the mirror—very smooth and flat of radius R, wherefrom the crack front accelerates through it in a planar manner. A narrow concentric band surrounds the mirror—the mist region—appearing like dewdrops, the width of which is much less than the mirror radius. The mist is followed by hackle lines which radiate from the fracture origin (Fig. 30.1) [4].

Also, stress corrosion at flaws explains both the loss of strength with age and its dependence on load duration. This conclusion was drawn upon of hundreds of tested plate glass from different buildings after 10 to 20 years when it was verified that the exposed surfaces experienced a strength reduction of more than 50 percent of that of factory recently-produced annealed glass [6].

When it is under a constant or variable load for a specific duration, failure does not necessarily occur at the maximum nominal stress level. This phenomenon is known as static fatigue or delayed failure and was recognized for almost 40 years ago [7]. By 1982, approximately 200 plates of 6-mm annealed glass were broken at different loading. The time to failure varied from about 30 min to one-tenth of a second. The results

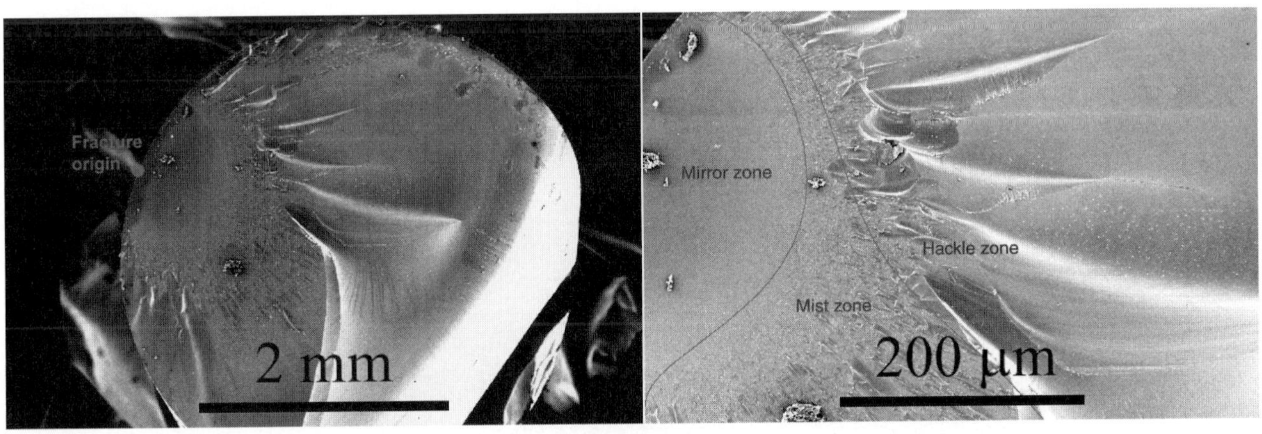

Figure 30.1 Scanning electron microscope (SEM) images of a cylindrical glass rod showing the fracture origin, the mirror, and mist region with the hackle focusing back to the origin [5].

confirmed that glass could withstand a load for 1 second, that is, 30 percent larger than the load that would cause failure after 1 minute [8].

30.2 GLASS: PRODUCTION AND PROPERTIES

Glass can be regarded as a liquid that has cooled to a rigid state without crystallizing and not like a supercooled liquid but rather an inorganic solid with an amorphous noncrystalline structure [9]. SLSG, which is the most common glass type used in construction, is a chemical composition of mainly 70 to 75 percent Silica sand [SiO_2], 10 to 15 percent soda [Na_2O], and 5 to 15 percent lime [CaO]. Magnesia and alumina are other secondary elements, among others. For some particular applications like fire protection or heat-resistant glazing, borosilicate glass (BSG)[1] is used when a high resistance to temperature changes as well as a very high hydrolytic and acid resistance are needed.

The glass production using the float process is the most extensive primary manufacturing process.[2] As shown schematically in Fig. 30.2, raw materials are melted in a furnace at temperatures of up to 1600°C; then the molten glass is poured continuously at approximately 1000°C onto a bed (about 3–4 m wide, 50 m long, 6 cm deep) of molten tin.

To prevent residual stresses being induced within the glass, slow cooling is fundamental—the annealing: After cooling, the glass is submitted to rigorous quality control and then washed. It is then cut into sheets of sizes of up to 6.0 × 3.2 m which are in turn stacked, stored, and ready for transport.[3] Adjusting the speed of the rollers allows controlling the glass thickness within a range of 2 (picture) to 25 mm (1 in.).

[1]Borosilicate glass contains 7 to 15 percent boron-oxide [B_2O_3] and less soda: [Na_2O] ≤ 8%.

[2]The float glass process is also known as the Pilkington process.

[3]At some plants the so-called online coatings (hard coatings) can be applied to the hot glass surface during manufacture.

30.2.1 Physical Properties

Unlike other materials, glass does not have larger "band gaps," which means that electrons require much more energy before they can skip from one energy band to another and back again. Photons of visible light between 400 and 700 nm—corresponding to the rainbow colors—do not have enough energy to cause this skipping. Consequently, photons of visible light travel through glass instead of being absorbed or reflected, making glass transparent (Fig. 30.3).

The glass is an elastic and isotropic material with a brittle fracture without any plastic behavior, which is the reason that imperfections on glass such as scratches and bubbles have a significant effect on its tensile strength, the relevant property for structural engineering [10].

On the other hand, glass is an excellent material of chemical resistance, which makes it one of the most durable materials in construction, except for the acids based on SiO_2 and also some alkalis. The most important physical properties of SLSG are summarized in Table 30.1.

Table 30.1 Properties Indicative Values for Soda Lime Silica Glass

	Soda lime silica glass	Borosilicate silica glass
Density [kg/m³]	2500	2200–2500
Knoop hardness [GPa]	6	4.5–6.0
Young's modulus [GPa]	70	60–70
Poisson's ratio, v	0.22–0.24	0.2
Torsional elastic modulus [GPa]	29	25–29
Coefficient of thermal expansion, α_t in [$10^{-6}K^{-1}$]	9	3.1–6.0
Specific thermal capacity, c_p in [$Jkg^{-1} \cdot K^{-1}$]	2900–3000	0.1–0.3
Thermal conductivity, λ in [$Wm^{-1} \cdot K^{-1}$]	720	800

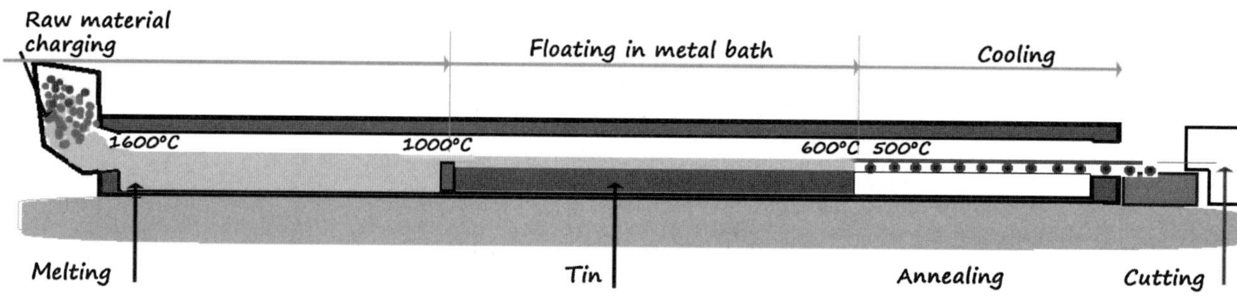

Figure 30.2 Schematic float glass process production.

Figure 30.3 Vision glass façade providing a clear image of the exterior and interior.

30.2.2 Performance Improvements

Postproduction techniques are applied to the annealed glass to obtain heat-strengthened glass, fully tempered glass,[4] laminated glass (LG), and insulated glass. Some other postproduction processes, such as curving, drilling, coating, enameling, and thermal treatments, are applied for specific fittings. Technical specification may be found on the Internet or specific literature, such as those relating with solar control, thermal insulation, low-emissivity glass, insulating glass (IG), noise control, fire protection, and self-cleaning glass [11, 12].

SAFETY AND SECURITY

When float glass is broken, it breaks into sharp shards that are risky to anyone in the vicinity. This is particularly dangerous if the window or any glass building component is broken by human impact. Innovations in the development of safety and security glass allow people to be protected from injury and, in the most extreme cases, protect buildings from various forms of hazard or attack, such as protection of glazing during earthquakes, increased protection from bomb and blast hazards, hurricanes, and windborne debris.

30.3 GLASS AS A STRUCTURAL MATERIAL

Under loading, the strain response to the stress is perfectly linear with sudden failure (Fig. 30.4).

Based on real calculations, the theoretical tensile strength results into 5000 up to 8000 MPa, which is due to glass's strong chemical bonds. However, due to structural defects on the surface, the real strength is much lower. These surface flaws (Griffith flaws) on a piece of glass have an even more significant effect on the strength of glass than other brittle materials.[5] The reason is that high-stress concentrations occurring in the cracks cannot be redistributed because of the lack of ductility and the bending strength of annealed glass reduces drastically to about 30 to 80 MPa because the bending tensile strength is controlled by the onset of a hypercritical crack growth without any plastic deformations resulting in unexpected breakage [13].

Subcritical crack growth occurs due to potential so-called stress corrosion under expositions like water or humidity together with long-term loading, which is the reason that the bending strength of glass due to permanent loads is lower than for short-duration loads. The crack growth rate is related to the stress concentration factor and other constants of the subcritical crack propagation. Other exposure conditions, like sand, dust, and water, strongly influence the surface bending resistance of annealed glass [14].

30.3.1 Toughened Glass

A commonly used term for fully tempered glass is "toughened glass." Float glass is heated and then cooled rapidly (quenched) by cold air jets. This tempering process aims to create a parabolic residual stress field in the thickness direction that has tensile stresses in the core and compressive stresses at the surfaces of the glass. This process is also known as prestressing (Fig. 30.5).

If the toughened glass is subjected to loads, cracks will not grow unless there is a net tensile stress field at the surface of the glass. In theory, if the sum of the compressive residual stress and the tensile stress from an external load is less than zero, no failure is possible, but the condition is intricate due to the presence of impurities, such as nickel sulfide [16].

Nickel sulfide (NiS) inclusions[6] are a known issue for toughened glass, resulting from the presence of nickel contaminants in the glass melt reacting with sulfur, most likely from the furnace fuel. During thermal toughening, these inclusions undergo a crystalline phase change to a smaller form upon cooling. Over time, the inclusions grow and can eventually impart enough stress in the glass to cause inadvertent failure [17]. Such failures can be forced by increasing the temperature to a certain level for a certain amount of time in what is called as a heat soak test, where failure is forced at an elevated temperature.[7] This phenomenon appears typically during the first 10 to 20 years after the

[4] Including heat soaking to reduce the potential for nickel sulfide-induced breakage of fully tempered glass.

[5] Griffith theory is based in the fact that brittle bodies fail under a relatively low nominal stress because of the presence of flaws or cracks which have the characteristics of extreme narrowness with respect to their length or depth producing high local stress concentrations when oriented at right angles to a tensile component of stress [103].

[6] Nickel-rich contaminants such as stainless steel might be present during the manufacturing process for float glass in the glass batch. Glass manufacturers take extraordinary steps to minimize the potential for nickel sulfide inclusions [104].

[7] The risk is significantly reduced by heat soaking, depending on the heating rate and the holding time.

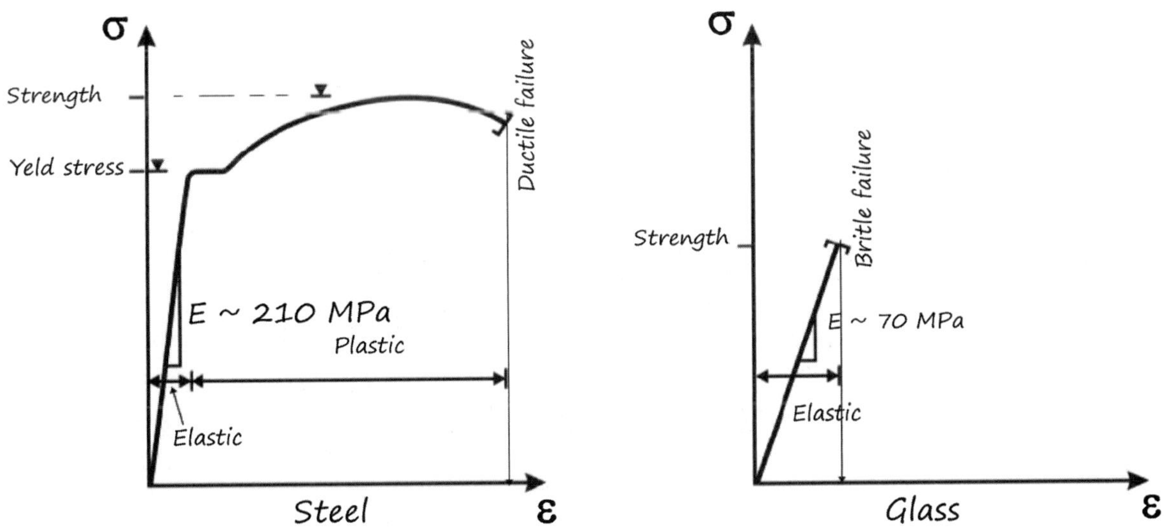

Figure 30.4 Stress-strain curves showing typical behavior for steel and glass.

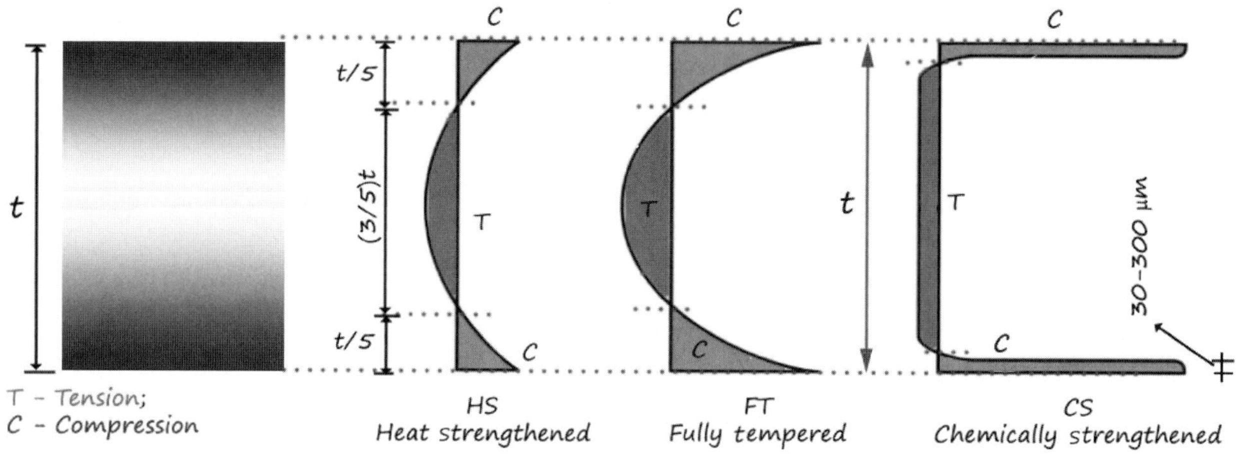

Figure 30.5 Stress distribution across the thickness of toughened glass. T—mid-plane tension; C—thickness of compression [14, 15].

installation of a glass panel after manufacturing. The risk of critical NiS inclusions of float glass produced in Europe is around 1 in 10 tons of glass [18].

The higher the prestressing, the smaller the shards become after breakage for a given thickness due to the induced energy that releases along the lengths of the crack pattern. Higher prestressing means higher crack energy and a greater total length of cracks and thus smaller particles (Fig. 30.6).

The Griffith flaws are assumed random in size, distribution, and orientation. The only way to establish a value for the strength of glass is to mechanically load the glass to breakage and performing statistical analysis of a set of scattered values of glass strength measurements, allowing for the derivation of a value below which there will be relatively few failures. Test results can be used to obtain a design value at which there is a sufficiently acceptable risk of the glass being weaker than the design value for a given limit state.

Heat-strengthened glass is produced like fully tempered glass yet with a lower the cooling rate. Thus, the resulting residual stress is lower, as well as the tensile strength. Upon fracture, the fragments are less abundant than for fully tempered glass, allowing for a smaller post-breakage load capacity in compression than for fully tempered glass [10, 19, 20].

Chemically strengthened glass (CSG) is a type of glass that has increased its strength as a result of a chemical process. Like annealed glass, CSG shatters in long pointed splinters, and that is why it is not considered a safety glass and must be laminated for that purpose. CSG is typically six to eight times the strength of float glass [21].

30.3.2 Laminated Glass

LG is a composite material consisting of two or more sheets of glass permanently bonded together by a plastic interlayer material. Under the structural point of view, the primary purpose of the LG is to increase safety in a glass structure or pane (Fig. 30.7). The most commonly used interlayers are thermoplastic polymers, for example, polyvinyl butyral (PVB), cast in place (CIP) liquid resin, thermoplastic polyurethane (TPU), and ethylene vinyl acetate (EVA), the latter mostly used in the construction of photovoltaic modules.

Figure 30.7 Laminated fragmented glass window.

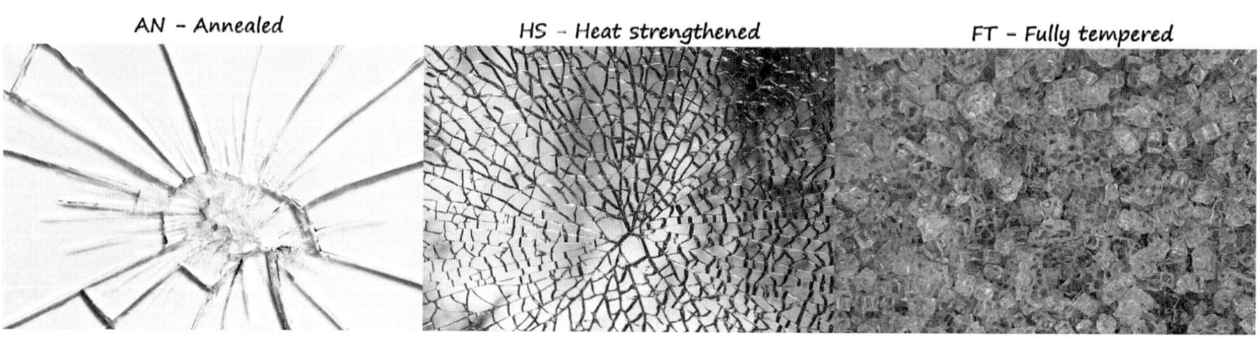

Figure 30.6 Crack pattern and degree of prestressing for float (annealed), heat-strengthened glass, and toughened safety (fully tempered) glass.

LG displays a different mechanical behavior due to the combination of two materials with very different stiffness [22]. LG with a PVB interlayer is less stiff than a monolithic glass with the same dimensions exhibiting higher displacements. Moreover, under specific loads and boundary conditions, discontinuous stress distributions develop in LG structures [23, 24].

Thermoplastic polymers interlayers are viscoelastic, which means they exhibit time and temperature-dependent strain. If an LG pane is placed under load, the adhesive polymer interlayer must be able to internally compensate deformation differences and absorb shear forces as far as the load is applied quickly. In general, an effective laminate thickness is calculated based on shear interaction, load duration, and the relevant temperature, depending on the properties of the interlayer.

Recently introduced ionoplast interlayers are being used, which react quite differently and have considerable strength and stiffness [25]. Laminated safety glass with ionoplast layers are not yet considered in the relevant standards; however, they provide significant improvements [26], such as reduction of glass thickness, use of larger glass panels at determined loads, reduction in the number of fixing points for frameless glazing, and a significant increase in post-glass breakage performance as well [27].

30.3.3 Insulating Glass Units

Insulating glass units (IGU) consist of two or more glass sheets (lites), placed parallel to each other using a spacer and sealed with a sealant, and separated by a sealed dry air space or an inert gas, for example, argon, xenon, or krypton. IGU are manufactured with glass in a range of thickness from 3 to 10 mm (⅛ to ⅜ in.) or more in particular applications. Most units are produced with the same thickness of glass used on both panes [11, 28], but particular applications, such as acoustic attenuation or security, may require wide ranges of thicknesses to be incorporated in the same unit.

Because of recent energy policies requiring buildings to use lesser energy, the use of triple IGU has increased due to their improved insulating performance over double IGU. Relevant and extensive research on load sharing between the lites comprising IGU can be found in technical literature [29, 30]. An equivalent thickness is generally used via the "thickness cubed" method to distribute the load carried by each lite of the IGU. This appears to be a simple approach to calculate load sharing due to an applied external pressure, yet, when pressure changes between the sealed air space(s) either due to temperature, elevation, and barometric pressure, the IGU lites' pressure can be significantly affected so that the thickness cubed method does not adequately cover many common IGU constructions under these conditions [29].

30.3.4 Glass Structural Analyses

Typical structural glass failures related to design causes are due to modeling, calculation and detailing errors, insufficient quality control, and incorrect documentation or even due to communications problems. Nevertheless, deficiencies have been pointed out related to the prediction strength capacity or extreme loading code values [16, 31].

Under a strictly structural point of view, glass behaves basically as an elastic material, so the theory of elasticity may be applicable for determining stresses and strains. However, it cannot be expected to give information about the failure stress since the strength of glass is size and time dependent and influenced by environmental conditions, besides the rupturing in some cutting pieces, which can be a risk for people [10, 32].

Allowable Stress

Glass design assuming allowable stress based on a global safety factor was introduced by Vicat almost two centuries ago by performing creep tests with different stress applied levels as a percentage of the failure stress and for some engineers nowadays [14].

Allowable stress also may be determined from a predetermined failure probability where size and time aspects must be included. Some standards provide a conservative technique for estimating the maximum allowable surface stress associated with glass lites for unique glass shapes and loads through engineering analysis considering shape effects and nonlinear behavior [33].

Fracture Mechanics

The functional strength of glass can be estimated with the general formula for the failure stress value given by

$$\sigma = \frac{K_{IC}}{\sqrt{\pi \cdot c}} \tag{30.1}$$

where σ = failure stress
c = flaw size in meters
K_{IC} = stress intensity factor

The magnitude of the stress intensity factor, K_{IC}, depends on the load types, sample geometry, and the size and location of the crack; its value varies between 0.75 and 1.0 $\text{MPa} \cdot \text{m}^{0.5}$ [34, 35].

Several theories and models based on fracture mechanics together with the theory of elasticity explain the failure behavior of glass [14, 34, 36, 37].

Also, statistical analysis of glass strength based on least squares algorithms and the sweeping method with treatment and position factors also were shown to be suitable, and it seems to be the preferred way to establish codes of practice so that modern design philosophy is based on statistical principles where failure is associated with a specific probability of occurrence [38, 39].

Finite-Element Analysis

Unusual glass geometries and support conditions (e.g., curved glass, glass with reentrant corners, point fixings, nonuniform loading) typically require a more detailed computational analysis. In these situations, the analysis of finite elements is paramount, however, it is necessary to pay special attention to the refinement of the model's mesh of the calculation to match stress concentrations in singular zones, e.g., around the bolt holes and other geometric discontinuities. Also, when modeling contact surfaces and releases, they must be defined according to its function. For example, in some situations only compression stresses will exist in the contact area. This objective can generally be achieved by using contact elements or by prescribing contact and noncontact surfaces, which requires nonlinear analyses [40].

30.3.5 Building Envelop and Structural Elements

The most used glass elements in a building are those comprising its envelope chiefly façade glazing panels, fins which may act like vertical or horizontal "beams," and columns.

Glazing Panels

In general cladding panels are placed in vertical or slightly sloped façades. The lateral actions are transmitted to elements directly and continuously supported, via mullions, or point supported. In the former, self-weight is transmitted to transoms or in case of point-supported systems also trough bolted connectors.

In all glass or total vision solutions, the support elements, be they transoms or mullions, are made from structural glass (Fig. 30.8). The connections between the glazing and support elements are typically adhesive based, such as structural silicone, yet point fixings systems are also used.

Glazing panels must be designed to withstand wind pressures and seismic-induced actions. Regarding wind, the preponderant action is typically the suction effect—negative—wind pressure. Glazing panels must span between either continuous line supports or point fixings, depending upon the form of the cladding system that has been adopted. The duration of loading is typically short, opposite to the load duration action for floor plates or low-oitched roofs under snow load where thicker panels are needed.

To avoid damage caused by the breakage of glazing panels, they must be laminated. The most critical situations are those related to balconies

Figure 30.8 All-glass enclosure solution (TVS™).

and other applications where panels are placed horizontally, such as skylights and walkways.

Glass Balustrades

Typical balustrade heights range from a minimum of 0.9 m to 1.2 m for protection or in some cases more; for example; full-height glass that is used to protect from a fall and barrier loads still have to be taken into account. The characteristic values of the line load acting at the height of parapet or wall but not higher than 1.20 m should be taken as a function of the occupancy type of the building (Fig. 30.9) as defined in the relevant codes of practice. For example, areas susceptible to large crowds as in buildings for public events like concert halls, sports halls including stands, terraces, and access areas and railway platforms, may be selected within the range 3.0 to 5.0 kN/m [41].

Figure 30.9 Isolated cantilevered panels using steel angles.

Floors and Walkways

Glass floors can act as a walking surface, and the duration of applied actions affects the design strength of the glass—the shorter it is, the higher the magnitude of design strength that can be assumed. In general, additional glass plies are added to the structure to allow for accidental breakage due to contact damage and allow for glass replacement. As with all glass and laminate design, suitable values of load duration and use temperature are established based on risk analysis for a given and specific usage (Fig. 30.10).

Figure 30.10 Transparent glass floor at a public building.

One-hour load duration at a sustained use temperature of 50°C is generally used or recommended by the American Society for Testing and Materials (ASTM) Standard E2751/E2751M–17 [42] when the intended end-use environment is unknown. System deformation and load resistance (LR) may be estimated using finite-element-based procedures for the structural analysis of LG.

For glass in floor plates regarding serviceability, it is common to use a deflection limit of span/250 when subjected to both variable and permanent actions. Floor vibration issues may be not relevant if the panels' natural frequency is limited to no less than 5 Hz to cover the effect of the resonance effect eventually caused by footfall onto the floor plate [43].

Frameless Glass Walls

Frameless glass walls may assume different configurations, such as internal simple partitions and glazing, units within shopping complexes, or elements within fully glazed enclosures which tend to be more complicated to design, as they need to withstand several load types to support as lateral, axial, and diaphragm actions when they are used as part of a lateral stability system for enclosures (Fig. 30.11).

For walls supported on three or more edges, the mechanics of the buckling of a plate become a more accurate model. They can also be subjected to shear forces, and lateral stability must be addressed if tensile or shear stresses are present. In such cases, the use of finite-element analysis delivers the most accurate results, even though the possibility of using buckling curves, like those used for steel columns, can be used as a design tool.

Figure 30.11 Glass walls in a residence.

Glass Beams

Glass beams, like other beams, are a linear load-bearing elements that are subject to both bending and shear stresses. The make-up of glass beams is typically laminated and fabricated from three or more sheets of LG assuming that if two of the plies have failed, the remaining sheet(s) must be designed to support the applied loads without collapse. Single length or a combination of several lengths joined together can be used.[8] Frequently individual lengths vary between 4 m and 7 m, yet some suppliers can provide ranges as much as 15 m (Fig. 30.12).

Figure 30.12 Glass roof beams in a residence.

The design of glass beams can be made by hand, for example, with the Mohr's stress circles, to determine principal tensile stresses in regions of high shear and Roark's formulas for stress and strain [44], but it will be preferable to design LG beams with nonlinear finite-element analysis due to the interlayer interaction.

Being very slender, lateral-torsional buckling glass beams need always to be checked. The Australian code of practice on the design of structural glass [45] provides useful guidance on the buckling analysis of glass beams and fins.

Glass Columns

Glass columns as a primary structural element have in general buckling instability issues, which is why as with glass beams, columns should have enough redundancy to allow the member to fail with a safe redistribution of the load. Thus, a minimum of plies sheets of LG should be used to ensure that if two of the leaves have broken, the column maintains its function with only one ply (Fig. 30.13).

[8] The limitation is usually the moment capacity of the joint itself.

Figure 30.13 Cruciform columns at Danfoss headquarters in Nordborg, Denmark.

30.4 ACTIONS

Nonstructural failures have accounted for most hurricanes and earthquake damage on buildings' façades, especially on transparent or opaque claddings with costly consequences [46].

The risk of failure of cladding implies the fallout of façade panels and represents a severe threat to life safety. Also, the cost of curtain wall façades can represent as much as the cost of its structure [47, 48]. In order to highlight this, a study on the extent of nonstructural damage concluded that structural components of a commercial building accounted for approximately 15 to 25 percent of the original construction cost, while the nonstructural components accounted for the remaining cost [49].

Design action values, reference periods, and partial safety factors vary across countries and standards. Such quantities are interrelated, and only in some specific conditions is it possible to combine partial safety factors either on the load side or on the resistance side from different standards [50]. Some insight is given for the relevant action on buildings' façades caused by wind or earthquakes.

30.4.1 Wind Action

In order to obtain the lateral design action in buildings' façades, several codes and standards habitually consider a basic or reference wind speed for various locations in each region or country. A reference height of 10 m in the open terrain is chosen along with modification factors for the effects of height and terrain type, and pressure or force coefficients for structures of various shapes are also provided. Depending on the country region, the wind loads may be obtained directly from these standards, such as the ISO 4354, Eurocode 1 parts 1–4, the ASCE Standard ASCE, the AIJ Recommendations for Loads on Buildings, and the Australian Code AS/NZS 1170.2:2002 comprehending several years of work by committees from many countries [51, 52].

Peak velocity pressure, $q_p(z)$, at height z, which includes mean and short-term velocity fluctuations, depends on a turbulence factor and an orography factor and may be expressed as a function of the basic pressure and an exposure factor, c_z, as follows:

$$q_p = c_z \cdot q_b \tag{30.2}$$

The characteristic wind pressure value acting on the cladding surfaces, w_{Sk}, is obtained from Expression 30.2 for a given reference height, z:

$$w_{Sk} = c_p \cdot q_p \tag{30.3}$$

Pressure coefficients, c_p, are defined for buildings and parts of buildings depending on the size of the loaded area, reference height, and eventual internal pressures for cladding elements in ventilated or pressure equalized rain screen façade systems.

30.4.2 Seismic Action

For nonstructural elements of great importance, such as large glass panels, a seismic analysis shall be based on a realistic model of the relevant structures and the use of appropriate response spectra derived from the response of the supporting structural elements of the primary seismic resisting system.

In the design process, it is crucial to accommodate seismic relative displacements due to story drifts. Manufacturing and construction tolerances for both wall elements and the supporting structure must be considered in establishing joint design dimensions and connection details. Separations between adjacent panels are intended to limit contact and resulting panel misalignment or damage under all but extreme building response. A minimum separation of about 15 mm is recommended, yet for practical joint detailing and acceptable appearance, separations typically are limited to about 20 mm [53].

The seismic design force for glass panels or group of panels depends on the weight of the panel, the importance factor, I_p, the component response modification factor, the component amplification factor, and the component acceleration at a point of attachment to the structure. Once the goal of design for most non-structural components is to prevent detachment or toppling, which is a threat to people's lives, a value of $I_p = 1.5$ should be adopted [54].

Seismic Design Force

The horizontal seismic force, F_E, acting at the center of mass of a glass panel may be defined as follows [55]:

$$0.3 \times S_{DS} \cdot I_P \cdot W_E \leq F_E = \frac{0.4 \times S_{DS} \cdot W_E \cdot I_p \cdot a_p}{R_E} \times \left(1 + 2 \times \frac{z}{H}\right) \leq 1.6 \times S_{DS} \cdot I_P \cdot W_E \tag{30.4}$$

where a_p is the element amplification factor and R_e, the element response factor. The values of the coefficients are defined for veneers, depending on the degree of deformability of the elements and attachments as depicted in Table 30.2.

Table 30.2 Properties Indicative Values for the Response Modification Factor

Component	R_E
Brittle or buckling failure mode	1.0–1.5
Minimal level of energy dissipation capacity	2.5
Ductile materials and detailing	3.5

Dynamic amplification occurs when the period of a nonstructural component closely matches that of any mode of the supporting structure. The component amplification factor, a_p, represents the dynamic amplification of the element response as a function of the fundamental period of the structure, T_{bd}, and of the element, T_{el}. The NCEER [55] formulation shown in Fig. 30.14 may be used to compute a_p as a function of the ratio of the first period.

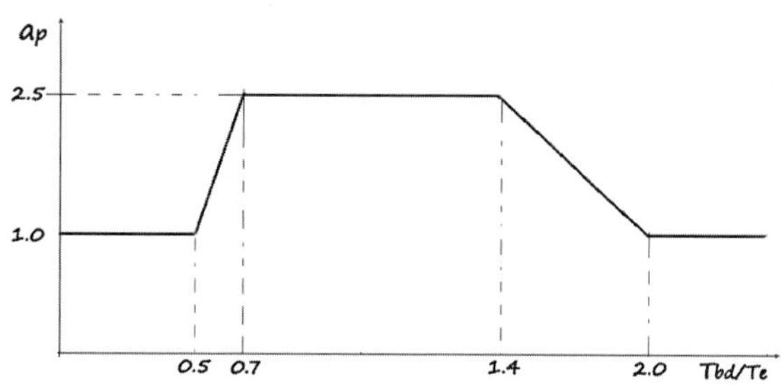

Figure 30.14 Formulation for a_p as a function of structural and component periods [55].

The component response modification factor, R_E, represents the energy absorption capability of the component's construction and attachments. Benchmark values are indicated in Table 30.2 according to the Federal Emergency Management Agency of the U.S. Department of Homeland Security [53, 56].

The NEHRP Standard [53] specifies a reduced response modification factor, R_E, and an increased amplification factor, a_p, for "fasteners" with the intention of preventing premature failure in those elements of connections that are inherently brittle as is the case of embedded items on concrete or connection elements that are too small to adequately dissipate energy inelastically, such as welds or individual bolts. In general, it is recommended to consider in Eq. (30.4), $a_P / R_E = 1$ [56].

Application of the proposed formulation indicates that the differences between the recorded building accelerations and those calculated by the NEHRP and the EC 1998 are attributed to the effect of neglecting the higher modes. Including the higher mode's effect is especially relevant in taller buildings since their contribution changes the pattern of variation of the acceleration with height. The EC 1998.1 formulation tends to give lower values than those obtained using NEHRP provisions [57].

Seismic Drift Limits

Glass in windows, curtain and window walls, doors, and other exterior applications shall be designed to accommodate relative seismic displacement over the height of the glass.

Glass in glazed curtain walls, glazed storefronts, and glazed partitions shall meet the applicable displacement requirements of ASCE 7 [58] at which a glass fallout, Δ_{fallout}, which is the drift value causing glass fall out from a curtain wall, storefront wall or partition shall be determined following the AAMA 501.6 [54]. To face interstory drift problems in buildings, designers feel safer if they design the structure very capable of responding to seismic action so as not to involve any façade in the reaction. According to ASCE 7 [58], a minimum value of 13 mm is mandatory for the relative seismic displacement, Δ_{fallout}, computed as follows:

$$\Delta_{\text{fallout}} \geq 1.25 \times I_p \cdot D_p \tag{30.5}$$

where D_p = the relative seismic displacement that the component must be designed to accommodate over the height of the glass component under consideration (Fig. 30.15)

I_p = importance factor

For glass with enough clearance, D_{clear}, such that physical contact between the glass and frame will not occur at the design drift, D_p, the following condition (Fig. 30.15) should be observed:

$$D_{\text{clear}} = 1.25 \times D_p \tag{30.6}$$

where $D_{\text{clear}} = 2 \times c_1 \cdot \left(1 + \frac{h_p \cdot c_2}{b_p \cdot c_1}\right)$

h_p = the height of the rectangular glass panel

b_p = the width of the rectangular glass panel

c_1 = the average of the clearances (gaps) on both sides between the vertical glass edges and the frame

c_2 = the average of the clearances at top and bottom between the horizontal glass edges and the frame

30.4.3 Snow Action

The characteristic value of snow load on the ground, S_k, may be determined by simple expressions, yet where more refined data are needed, the characteristic value of snow load on the ground shall be defined using an appropriate statistical analysis of long records taken in a well-protected area near the location.

The design shall recognize that snow loads on roofs can be deposited in many different shapes or patterns, depending on the roof shape,

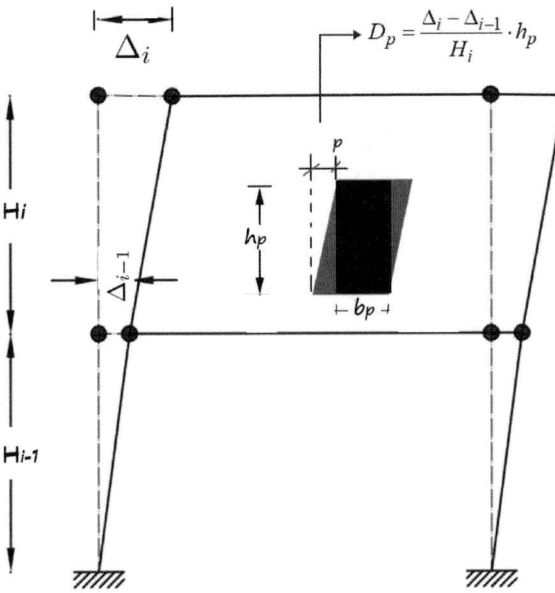

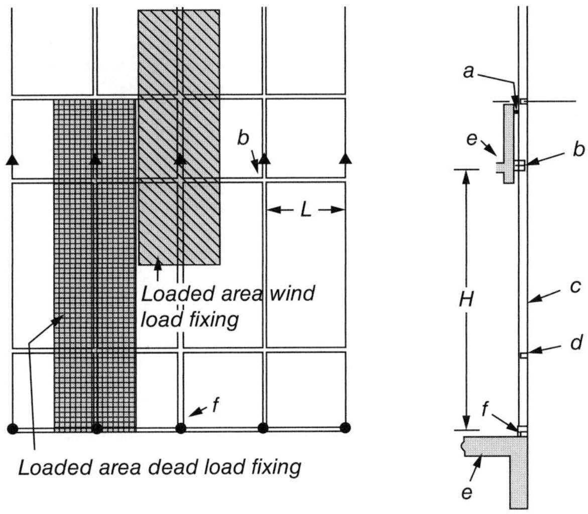

Figure 30.15 Seismic interstory drift.

thermal properties, surface roughness, and the amount of heat generated under the roof.

The following general expression may be applied to determine the snow load on roofs:

$$s_k = \mu_i \cdot C_e \cdot C_t \cdot s_k \tag{30.7}$$

where μ_i = the snow load shape coefficient
s_k = the characteristic value of snow load on ground
C_e, C_t = the exposure coefficient and thermal coefficient, respectively

Recommended values of C_e for different topographies vary from 0.8 for flat unobstructed areas exposed on all sides to 1.2 for areas in which the construction work being considered is considerably lower than the surrounding terrain.

The thermal coefficient, C_t, is used to account for the reduction of snow loads on roofs with high thermal transmittance, in particular for some glass-covered roofs, because of melting caused by heat loss. In general, a unitary value is adopted.

The characteristic value of the snow load, s_k, may be obtained from the appropriate standards. For some regions in Europe, for example, EC 1991-3 [59] provides an expression to determine its value as a function of the altitude above sea level and the geographic region.

Loads and Load Combination

The design of vertical glazing shall be based on the wind load computed according to Section 1609 of the IBC, where the basic wind speed, V, is defined depending on the risk category of buildings. The basic wind values are nominal design 3-second gust wind speeds at 10 m aboveground for a given, C, exposure category.[9]

Thus, for vertical glass,

$$LR \geq 0.6 \times F_{gw} \tag{30.8}$$

where LR = the load resistance of the glass determined by ASTM E 1300 [33]
F_{gw} = the wind load on the glass due to basic wind speed, V

Sloped glass more than 15 degrees, such as skylights, sunrooms, sloped roofs, or another exterior application, shall be designed to resist the most critical combination of loads:

$$LR \geq \max \begin{cases} 0.6 \times W_0 - D \\ 0.6 \times W_i + D + 0.5 \times S \\ 0.3 \times W_i + D + S \end{cases} \tag{30.9}$$

where D = glass self-weight per unit area[10] [kN/m^2]
F_g = total load on glass [kN/m^2]
S = snow load as determined in Section 1608 of the IBC [60] [kN/m^2]
t_g = total glass thickness of glass panes and plies [mm]
W_i = inward wind force, due to basic design wind speed, V, as calculated in Section 1609 of the IBC [kN/m^2]
W_0 = outward wind force due to basic design wind speed, V, as calculated in Section 1609 of the IBC [kN/m^2]
α = the angle of slope from horizontal

Limit State Assessment

Depending on consequent classes, the European Standard EC 1990 [61] establishes consequence classes for persistent design situations which refer to the conditions of normal use as depicted in Table 30.3.

Table 30.3 Consequences Classes According to EC 1990

Consequences class	Description
CC3	High consequence for loss of human life, or economic, social, or environmental consequences very great
CC2	Medium consequence for loss of human life, economic, social, or environmental consequences considerable
CCI	Low consequence for loss of human life, and economic, social, or environmental consequences small or negligible

[9] Exposure C—open terrain with scattered obstructions having heights less than 9 m (30 ft) for most wind directions.

[10] For glass sloped 30° or less from horizontal, $D = 0.0245 \times t_g$. (KPa). For more sloped glass, $D = 0.0245 \times t_g \cdot \cos\alpha$.

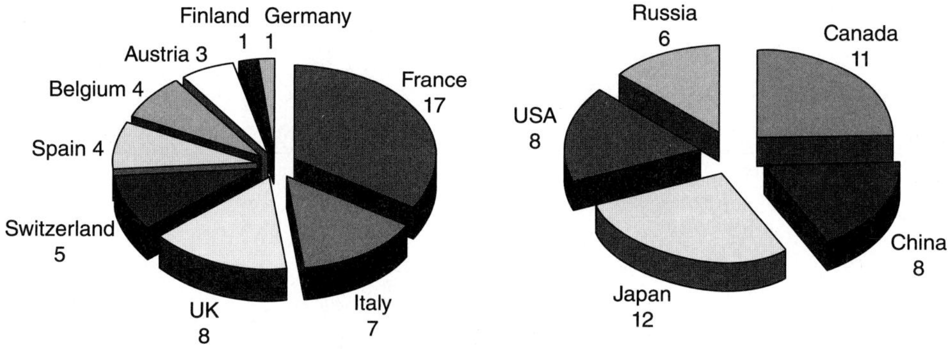

Figure 30.16 Number of national standards related to glass in European members and other countries worldwide [63].

Considering given reliability or consequences, class design values derived by calibration partial safety factors, combination factors, and so on are defined, leading to the very well-known governing equation for limit state design when considering a limit state of rupture or connection failure of a member:

$$E_d \leq R_d \tag{30.10}$$

where E_d = the design value of the effects of actions
R_d = the design value or design resistance

The design value of the effect of actions, E_d, for the ultimate limit states is given by the appropriate expression:[11]

$$E_d = \gamma_g \cdot G_k \text{“+”} \gamma_f \cdot Q_k \tag{30.11}$$

In the same manner, for each critical load case, the design values of the effects of actions, E_d, are determined by combining the values of actions that are considered to coincide. Table 30.4 gives the recommended values, for γ_g and γ_f, according to the EC 1990 [61]. Similar values are given in other standards for permanent or variable actions [62].

Table 30.4 Partial Factors for Actions for the Ultimate Limit States

Action	Persistent or transient design situations	Seismic design situations
Variable action	$\gamma_f = 1.5$	$\gamma_f = 1.0$
Unfavorable permanent action	$\gamma_g = 1.35$	$\gamma_g = 1.0$
Favorable permanent action	$\gamma_g = 1.0$	$\gamma_g = 1.0$

In the case of the combinations of two variable actions, for example, wind and snow, the effects to be considered should be based on the design value of the leading variable action and the design combination values of accompanying variable actions.

For persistent and transient situation, the following expression is assumed and for each leading variable action, Q_k and the accompanying action as follows:

$$E_d = \gamma_g \cdot G_k \text{“+”} \gamma_f \cdot (Q_k \text{“+”} \psi_0 \cdot Q_a) \tag{30.12}$$

For combinations of actions in seismic design situations, the combination is given by the following expression:

$$E_d = \gamma_g \cdot G_k \text{“+”} E_k \text{“+”} \psi_2 \cdot Q_a \tag{30.13}$$

where ψ_0 and ψ_2 are factors combination values of the variable accompanying action. In EC 1990 [61], the recommended values for the combination factors for buildings depending on the European geographic zone the range values are $0,7 \geq \psi_0 \geq 0,5$ and $0,20 \geq \psi_2 \geq 0,0$.

[11] The "+" in the expression denotes "to be combined with."

The design value of the resistance, R_d, may be obtained from the characteristic value of the material or product resistance, R_k, as a single material is used:

$$R_d = \frac{R_k}{\gamma_M} \tag{30.14}$$

The value of γ_M in Eq. (30.14) corresponds to the partial safety factor to take account of unfavorable deviation of product properties, such as geometric deviations and uncertainty in the resistance model adopted. Material partial safety factors are, in general, calculated by probabilistic methods for relevant cases [13, 61].

30.5 CODES AND STANDARDS

A brief overview of standards related to the use of glass in buildings clearly shows that standards do not cover specifically the design of structural elements and structures made of glass products (Fig. 30.16) unless design standards for glass sheets (Table 30.5).

Table 30.5 Glass Design Standards for Glass Elements in Construction

Code/reference	Countries	Observation
ASTM E1300	U.S. standard	
AS 1288	Australian standard	
CNR-DT-210	Italian standard	
BS 6262/ 5516	British standard	
TRLV	German technical regulation	
prEN 16612/16613	European standard	Proposed for approval
JGJ 102	Chinese standard	
IS 16231	Indian standard	Inspired from AS 1288 and BS 6262
NF DTU 39	French norm	
CAN/CGSB 12	Canadian standard	
NBN S 23-002	Belgian standard	

Most existing standards worldwide have the objective of guiding the use and installation of glazing and to clarify technical definitions. The latest version concerns ultimate limit state design for wind or snow loads based on charts or, sometimes, complex formulae.

In the following sections, reference is made to some of the most used standards published by the ASTM International organization and by the European Committee for Standardization (CEN) that are explicitly geared for vertical and sloped glazing in buildings for which the specified design loads consist of wind load, snow load, and self-weight. For specific design situations, such as those related to in-plane originating shear, buckling, and stress concentration in singular zones, guidance is given in separate sections.

30.5.1 American Standards—ASTM

The methodology followed by the American Society for Testing and Materials (ASTM International) ASTM E1300 [33] for the verification of the safety of glass panels consider as a basis the probability of breakage of glass less than or equal to 8 lites per 1000 for monolithic glass subjected to uniformly distributed loads. Surface actions are considering acting perpendicular to the plane of the glazing, provided that design loads consist of wind load, snow load, and self-weight has a total combined magnitude less than or equal to 15 kPa.

The standard shall be applied to other applications including, but not limited to, balustrades, glass floor panels, aquariums, structural glass members, and glass shelves.

ASTM E 1300 practice is used to determine the LR, that is, the uniform lateral load that a glass construction can sustain based upon the given probability of breakage and load duration, of specified glass types and constructions exposed to uniform lateral loads. Additionally, deformations of the support system cannot be greater than 1/175 of its length, and the probability of failure, P_b, is defined as a maximum of 8 breaks per 1000 units throughout 20 years.

Based on the ASTM E631 [64] some definitions and terms are adopted referring to the terminology used in ASTM E3100 practice, such as the following:

- Load resistance (LR), n—the uniform lateral load that a glass construction can sustain based upon a given probability of breakage and load duration
- The aspect ratio (AR) for glass simply supported on four sides; the ratio of the long dimension of the glass to the short dimension of the glass is always equal to or greater than 1.0[12]
- A long-duration load—a load lasting approximately 30 days
- A short-duration load—a load lasting 3 seconds or less
- Non-factored load (NFL)—3-second duration uniform load associated with a probability of breakage less than or equal to 8/100 lites monolithic annealed glass determined from the monograms
- Glass type factor (GTF)—a multiplying factor for adjusting the LR of different glass types

ASTM E 1300 applies only to monolithic and laminated panel glass constructions of rectangular shape with continuous lateral support along one and two, three, or four edges simply supported and free to slip in a plane. Glass supported on two sides acts as a simply supported beam, and panels supported on one side act as a cantilever. For IG pane units, the ASTM practice only applies to IGU with four-sided edge support.

The identification of the different types of glass and stiffening are defined inside the standard as follows:

- Annealed (AN) glass, a flat, monolithic, glass lite of uniform thickness where the residual surface stresses are nearly zero
- Fully tempered (FT) glass, a monolithic glass lite of uniform thickness that has been subjected to a heat treatment process where the residual surface compression is not less than 69 MPa or the edge compression not less than 67 MPa
- Heat-strengthened (HS) glass, a flat monolithic glass lite of uniform thickness that has been subjected to a heat treatment process where the residual surface compression is not less than 24 MPa or greater than 52 MPa
- Insulating glass (IG) unit, any combination of two or three glass lites that enclose one or two sealed spaces, respectively, filled with air or other gas
- Laminated glass (LG), a flat lite of uniform thickness consisting of two or more monolithic glass plies bonded together with an interlayer material defined in the relevant ASTM specification[13]
- Load share factor (LSF)—the portion of the applied load going to a lite in consideration in a sealed IG unit, whether the lite be monolithic glass or LG including the layered behavior of LG under long duration loads

Bending strength of the rectangular glazed panels continuously supported on the edges, subject to uniformly distributed loads, is obtained using the nomograms and abacuses of the standard. In another manner, maximum allowable surface stress, σ_{allow}, can be used for the design of special glass shapes and loads not covered elsewhere in the standard, such as trapezoids, circular, triangular, and other odd shapes, the value of which is given as follows:

$$\sigma_{allow} = \left[\frac{P_b}{k\cdot(d/3)^{7/n}\cdot A}\right]^{1/7} \tag{30.15}$$

where σ_{allow} = maximum allowable surface stress
P_b = the probability of breakage
k = a surface flaw parameter
d = the duration of the loading
A = the glass surface area
n = 16 for annealed glass; n = 32 for heat-strengthened glass, and n = 48 for tempered glass

Equation (30.15) is appropriate for a probability of breakage, P_b, less than 5 percent which is somewhat conservative comparing with this standard assumed value of 0.8 percent.

Conservative allowable surface stress's values for a 3-second duration load is 23.3 MPa for AN glass, 36.5 MPa for HS glass, and 73.0 MPa for FT glass.

Nomograms for Load Resistance

ASTM E1300 [33] presents nomograms that allow direct evaluation of NFL strength. These nomograms were obtained from models developed by Beason et al. [65] and Vallabhan et al. [66] based on similar parameters given in Eq. (30.15) considering a 20-year working life in a humid environment.

In order to determine a panel LR, the straightforward procedure is based on determining from the appropriate chart in the relevant Annex A1 the NFL—the 3-second duration uniform load for a given glass thickness and the GTF for the appropriate glass type and load duration (short and long). The LR of a lite is obtained by multiplying the NFL load by the GTF to get the LR of the lite.

Basic Procedure for Single Glazing

This procedure is recommended for either monolithic or laminated single glazing supported continuously along one edge (cantilever), two opposite sides, or along three sides or four sides, yet a limiting temperature of 50°C is specified to LG for this calculation approach.

Figure 30.17 exhibits a chart for ascertaining the NFL of a 6.0-mm glass monolithic panel four sides simply supported where a value of about 2.2 kPa is obtained for a 1.75 × 1.0 m² sized panel under 3-second duration load.

The general expression that leads to the value of the uniform lateral load that a plate can sustain—the LR—depends on the type of sheet, monolithic or laminated, and the duration of the action, which translates into a glass type factor (GTF) factor that depends on the type of glass, AN, HS, or FT, as defined in Table 30.6. The resistance value, LR, is obtained from

$$\mathrm{LR} = \mathrm{NFL} \times \mathrm{GTF} \tag{30.16}$$

Structures and components must be designed for the effects of the factored loads according to combinations rules, given that situations may arise where it is necessary to add the effects of actions of different durations, such as snow or self-weight loads, that is, long-duration loads with short-duration loads, such as wind or seismic actions.

[12] For glass simply supported on three sides, the AR is the ratio of the length of one of the supported edges perpendicular to the free edge, to the length of the free edge, equal to or greater than 0.5.

[13] ASTM E 1300 charts applies to only polyvinyl butyral (PVB) interlayer. Methods for determining effective thickness of laminated glass for analyses of stresses and deflection based on different interlayer complex polymers are given in a specific annex of this code of practice.

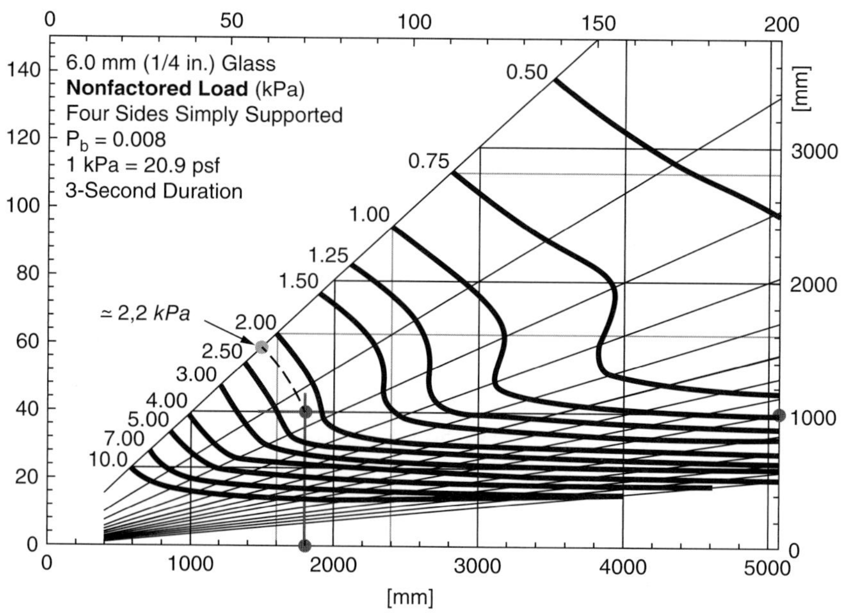

Figure 30.17 Non-factored load chart for 6.0-mm glass with four sides simply supported [33].

Table 30.6 Glass Type Factors (GTF) for Single Lite Monolithic or Laminated Glass (Table 1 of ASTM E1300 [33])

Glass type	Short-duration load (3 s)	Long-duration load (30 days)
AN	1.00	0.43
HS	2.00	1.30
FT	4.00	3.00

In order to determine a design load which represents the combined effects of a certain number of, say, j loads of different duration, an equivalent 3-second load, q_3, is defined as follows:

$$q_3 = \sum_{i=1}^{i=j} q_i \cdot \left(\frac{d_i}{3}\right)^{1/n} \tag{30.17}$$

where j = number of loads to be combined
q_i = the magnitude of the load having duration, d_i
d_i = load duration in seconds of magnitude, q_i
n = a parameter that depends on the glass type

The standard only specifies $n = 16$ for annealed glass to use in the above formula, yet for heat-strengthened glass (HS) and fully tempered glass type, $n = 31.7$, and $n = 47.5$ may be adopted based on the glass type factors from Tables 1, 2, 3 and 7 of the a Standard [50].

Procedure for Double-Glazed Insulating Glass

The strength of a double-glazed IG depends on the individual LR of each of the constituent lites and the load duration. The LR of the IG unit is the lower of the two values, LR1 and LR2, obtained from

$$\text{LR} = \min \begin{cases} \text{LR}_1 = (\text{NFL})_1 \cdot (\text{GTF})_1 \div (\text{LSF})_1 \\ \text{LR}_2 = (\text{NFL})_2 \cdot (\text{GTF})_2 \div (\text{LSF})_2 \end{cases} \tag{30.18}$$

For each of the lites, 1 and 2, the corresponding short duration NFL values are obtained for annealed or laminated glass (AN) based on the nomograms of Annex A1 of the code.

The GTF for each lite is defined depending on the glass type and duration load (3 seconds or 30 days) according to whether they are short- or long-term actions.

For short-term actions, the factors (GTF) to be considered are given in Table 30.7, and for long-term actions, the factors (GTF) are naturally lower and are reproduced in Table 30.8.

Table 30.7 Glass Type Factors (GTF) for Double-Glazed Insulating Glass (IG), Short-Duration Load (Table 2 of ASTM E1300 [33])

Lite 1 or Lite 2 Monol. or Lam. Glass	AN GTF1	AN GTF2	HS GTF1	HS GTF2	FT GTF1	FT GTF2
AN	0.90	0.90	1.00	1.90	1.00	3.80
HS	1.90	1.00	1.80	1.80	1.90	3.80
FT	3.80	1.00	3.80	1.90	3.60	3.60

Table 30.8 Glass Type Factors (GTF) for Double-Glazed Insulating Glass (IG), Long-Duration Load (30 days) (Table 3 of ASTM E1300 [33])

Lite 1 or Lite 2 Monol. or Lam. Glass	AN GTF1	AN GTF2	HS GTF1	HS GTF2	FT GTF1	FT GTF2
AN	0.45	0.45	0.50	1.25	0.50	2.85
HS	1.25	0.50	1.25	1.25	1.25	2.85
FT	2.85	0.50	2.85	1.25	2.85	2.85

Load Share Factors—LSF

The portion of the applied load going to a lite in consideration in a sealed IG unit, whether the lite be monolithic glass or LG, must consider the layered behavior of LG under long-duration loads. This is attained by applying the load share factor (LSF) in Expressions (30.18)—a distribution coefficient due to stiffness. For IG with both monolithic or LG, under either short- or long-duration loads or with one monolithic glass and one LG, under short-duration load only, the LSF for each lite value is approximately given by

$$\mathrm{LSF}_i = \frac{t_i^3}{t_i^3 + t_j^3} \tag{30.19}$$

where t_i and t_j are the minimum glass thickness for each lite—which is slightly smaller than the nominal thickness as in Table 4 of the standard and providing that for thickness for LG the minimum thicknesses of the individual glass plies and the nominal interlayer thickness. If the sum of all interlayer thicknesses is greater than 1.52 mm, this value shall be used in the calculation.

Under short-duration loads, LG is assumed to behave in a monolithic-like manner; however, under long-duration loads, LG is assumed to behave in a layered manner. The load sharing is then based on the individual ply thicknesses of the LG. The LSF for one ply of the laminated lite of an IG composed of monolithic glass and laminated glass, under long-duration load, is given by a function of the minimum thickness of one glass ply, t_{ply}, of the laminate glass:

$$\mathrm{LSF}_{\text{ply}} = \frac{t_{\text{ply}}^3}{t_{\text{mon}}^3 + 2 \times t_{\text{ply}}^3} \tag{30.20}$$

Taking as an example an IG composed of a nominal 5-mm monolithic lite and a 6-mm LG, for short-duration loads the LSF factors are 0.357 for the 5-mm monolithic glass minimum thickness of which is 4.57 mm and 0.643 for the 6-mm LG minimum thickness of which is 5.56 mm.

For long-duration loads the LSF factors are 0.267 for the 6-mm LG with two plies of a minimum thickness of 2.59 mm (nominal 2.7 mm) and 0.733 for the 5-mm monolithic glass minimum thickness of which is 4.57 mm.

Basic Procedure for Triple-Glazed Insulating Glass

The ASTM Standard provides calculation procedure for triple-glazed IG with three lites of a monolithic glass of equal (symmetric) or different, but asymmetric, thickness with two separately sealed air spaces and same glass type, simply supported continuously along four sides.[14]

The strength of a triple-glazed (IG) depends on the individual LR of each of the constituent lites, and the load duration is the lower of the two values, LR1, LR2, and LR3, obtained from Eqs. (30.21). The corresponding glass type factors for each lite are similarly depicted in Table 30.9 as for double glazing units:

$$\mathrm{LR} = \min \begin{cases} \mathrm{LR}_1 = (\mathrm{NFL})_1 \cdot (\mathrm{GTF})_1 \div (\mathrm{LSF})_1 \\ \mathrm{LR}_2 = (\mathrm{NFL})_2 \cdot (\mathrm{GTF})_2 \div (\mathrm{LSF})_2 \\ \mathrm{LR}_3 = (\mathrm{NFL})_3 \cdot (\mathrm{GTF})_3 \div (\mathrm{LSF})_3 \end{cases} \tag{30.21}$$

Table 30.9 Glass Type Factors (GTF) for Single Lite Monolithic or Laminated Glass (Table 7 of ASTM E1300 [33])

Glass type	Short-duration load (3 s)	Long-duration load (30 days)
AN	0.81	0.34
HS	1.62	1.03
FT	3.24	2.58

[14] It's recommended to limit the combined width of both air spaces in the IG unit to less than or equal to 25 mm (1 in.), alerting one that a larger combined dimension may result in excessive sealant stress and glass stresses due to temperature and altitude conditions.

Load Share Factors—LSF

The standard assumes that triple-glazed IG may have three lites of a monolithic glass of equal or different but asymmetric thickness. Given that the LG lites are positioned on the outer side of the IG and it is not relevant to the LSF, the duration of the loads except in cases where the differential temperature may have significant pressure effects. The LSF for each lite, *i*, *j*, and *k*, shall be calculated as follows providing the alternate position of the each of the subscripts in the numerator:

$$\mathrm{LSF}_i = \frac{t_i^3}{t_i^3 + t_j^3 + t_k^3} \tag{30.22}$$

Analytical Procedure

A stress analysis method shall be used to determine the in-plane surface tensile stresses incorporating the nonlinear two-way bending and membrane action of plates, producing acceptable values for surface stresses by ensuring the NFL chart values correspond to the specified design load results in a 0.8 percent probability of breakage.

Upon this and considering only the in-plane stresses of the lite surface, the probability of breakage calculation is determined based on a two-parameter Weibull distribution:

$$P_b = 1 - e^{-B}$$

$$B = k \cdot \sum_{i=1}^{N} \left[\left(\left(c_i \cdot \left(\frac{t_d}{60} \right)^{\frac{1}{n}} \right) \cdot (\sigma_{\max_i} - \mathrm{RCSS}) \right)^m \cdot A_i \right] \tag{30.23}$$

with $c_i = -0.005 \times r_i^6 + 0.022 \times r_i^5 + 0.055 \times r_i^4 + 0.039 \times r_i^3 + 0.031 \times r_i^2 + 0.06 \times r_i + 0.8$;

$$r_i = \frac{\sigma_{\min_i} - \mathrm{RCSS}}{\sigma_{\max_i} - \mathrm{RCSS}}$$

and $k = 2.86 \times 10^{-53}\,\mathrm{N}^{-7}\mathrm{m}^{12}$

$m = 7$

N = the number of stress values

t_d = duration of loading

$n = 16$

$\sigma_{\max_i}$ = *i*th maximum principal stress

$\sigma_{\max_i}$ = *i*th minimum principal stress

RCSS = residual compressive surface stress; 0 MPa (0 psi) for AN glass, 24.0 MPa for HS glass, and 69.0 MPa for FT glass

A_i = the *i*th tributary area corresponding to the principal stresses (m^2)

If the probability of breakage, $P_b \leq 0.008$, then the LR is greater than or equal to the specified design load.

The effective thickness must be determinate in laminated glazing for stress for each glass ply, comprising the IG, using the shear moduli for the LG interlayer material for the temperature/load duration combination corresponding to the design load duration designation as follows: long-duration load, 20°C at 30 days, and short-duration load, 50°C at 3 seconds.

For double-insulated glass units, the proportion of the specified design load carried by each lite in the IG shall be determined using a method that maintains the ideal gas law equilibrium for the air space between IG assembly and loaded conditions. The method should accurately account for the displaced volumes of the lites comprising the IG when loaded and taking the above-mentioned assumptions for LG lites.

Serviceability and Deflection Limits

For glass wall systems the 2018 IBC [60] limits the deflection of exterior L/240 as a function of the span of the wall, although it does not explicitly name glazed components or framing supporting glazing.

A maximum allowable deflection of framing systems for building cladding is defined by the American Architectural Manufacturers Association in its standard AAMA TIR-11. It establishes a deflection limit of L/175 for spans less than 4.11 m (13 ft 6 in.) and L/240 +6.4 mm (¼ in.) for spans greater than that [67]. These deflection limits have been used to glass wall systems such as curtain wall or floor-to-ceiling storefronts in the United States [68].

The ASTM E 1300 code of practice does not address aesthetic issues caused by glass deflection, even though maximum lateral (center of glass) deflection may be determined from the applicable charts provided by the code. These charts are not suitable for estimating small value deflections [50].[15]

Maximum glass deflection as a function of plate geometry and load may be calculated from the polynomial equations by Dalgliesh [69] for a curve fit to the Beason and Morgan data [70]. The polynomial equations give an approximate fit to center deflections of thin lites under enough pressure to cause nonlinear behavior. Such deflections, which will exceed the lite thickness, should be rounded to the nearest mm (0.04 in.) and are determined as follows:

Given a uniform lateral load, $q_{S_{tot}}$, in KPa,[16] the corresponding part of the load must be distributed for each lite, *i*, in the case of IGU through the corresponding load share factor, LSF, according to Eq. (30.19) for double IG or Eq. (30.22) for triple IG:

$$q_{S_i} = q_{S_{tot}} \cdot \mathrm{LSF}_i \tag{30.24}$$

A dimensionless parameter, χ, shall be calculated as a function of the plate thickness, modulus of elasticity, *E*, of the glass and the plate size.

$$\chi = \ln\left[\ln\left(q_{S_i} \cdot \frac{(a \cdot b)^2}{E \cdot t^4}\right)\right] \tag{30.25}$$

where q_{S_i} = uniform lateral load for the plate or lite [kPa]

a, b = long and short dimension respectively [mm]

$$E \simeq 71.7 \times 10^6 \ [\mathrm{kPa}] \tag{30.26}$$

After determining the parameters, r_0, r_1, and r_2 as a function of the aspect ratio AR (the ratio of the long dimension of the glass to the short dimension),

$$\begin{aligned} r_0 &= -3.830 \times \mathrm{AR} + 1.110 \times \mathrm{AR}^2 - 0.0969 \times \mathrm{AR}^3 \\ r_1 &= -2.290 + 5.830 \times \mathrm{AR} - 2.170 \times \mathrm{AR}^2 + 0.2067 \times \mathrm{AR}^3 \\ r_2 &= +1.485 - 1.908 \times \mathrm{AR} + 0.815 \times \mathrm{AR}^2 - 0.0822 \times \mathrm{AR}^3 \end{aligned} \tag{30.27}$$

the maximum glass deflection, δ_{max}, (in mm) is calculated as follows:

$$\delta_{max} = t \cdot e^{(r_0 + r_1 \cdot \chi + r_2 \cdot \chi^2)} \tag{30.28}$$

where *t* is not the nominal but the minimum plate thickness (in mm).

For pressures less than one-third design capacity of the lite, any first-order elastic procedure based on finite element method or analytical formulation is adequate.

30.5.2 ASTM Standard for Walkways

The E2751/E2751M Standard [42] addresses elements related to load-bearing glass walkways, glass treads, and glass landings constructed with LG only. It includes design and safe behavior considerations that are common to all walkways, such as slip resistance. It is emphasized that any load duration longer than 1 hour shall be considered a permanent load and that uniformly distributed live loads shall have a load duration of at least 1 hour and also that concentrated live loads shall have a load duration of at least 10 minutes. In all circumstances, the glass shall be cushioned to prevent direct contact with the glass support system.

For laminated glass with less than two plies, verification testing is required to verify the post-glass breakage, yet for a higher number of plies, calculations shall demonstrate that the glass assembly has enough strength to sustain the full design load with anyone glass ply broken.

Calculation Procedure and Design Criteria

Structural analysis such as engineering mechanics theories or finite-element analysis must be used to determine glass assembly stresses and deflections considering for temperature, boundary conditions, loading requirements, load duration, interlayer properties, and glass strength.

Nominal load conditions or combinations of load conditions must be assigned as per in standard for non-glass walkways, yet also a concentrated live load of at least 1,33 kN shall be applied to all glass walking surfaces in a 2600-mm² area.

To determine maximum allowable material, a stress reduction factor of 0.5 is recommended for disruptive surface treatments, for example, sandblasting. A glass stress reduction factor may be required for some acid-etched or patterned glasses which may be provided by the manufacturer.

The allowable glass stresses for nominal load conditions are given in MPa in Table 30.10 to provide an estimate for glass lites thickness associated with a maximum probability of breakage (P_b) less than or equal to 0.008 from Eq. (30.15).

Table 30.10 Allowable Glass Stresses for Nominal Load Conditions Are Given [MPa][42]

	Load duration			
Glass type	3 s	10 min	60 min	Permanent
AN: annealed	18.3	13.2	11.7	5.7
Heat strengthened	36.5	30.9	29.2	20.3
Tempered	73.0	65.3	63.0	49.4

According to the E2751/E2751M, the glass shall be tested as a system with frame and attachments, and furthermore, for walkways with two supported edges, a panel must be tested with the largest span between supports. For walkways with three or more supported edges, the sample size to be tested must represent the most demanding or highest stress condition.

A test shall be conducted at the maximum temperature for the intend used environment, and if unknown, the test temperature shall be 50°C and 0°±5°C. After testing, no additional glass plies shall have broken.

Concentrated and uniform load testing shall be performed by breaking one glass lite to produce the most unfavorable results. A concentrated load of 0.445 kN shall be applied over a 2600-mm² area for 10 minutes, and the specified uniform load shall be applied at a rate of 1.44 kPa/min up to the design pressure and hold for 60 min.

Serviceability Considerations

Glass walkways present a unique challenge in balancing the visual characteristics of glass with considerations of serviceability with physiological impact for pedestrians (Fig. 30.18) in such a manner that slip resistance, deflection, and vibrational issues are very relevant.

Slip Resistance Walkway slip resistance should be adequate with the intended use, user populations, and foreseeable conditions which can be accomplished through a variety of glass surface treatments, such as sandblasting, acid-etching, ceramic frit, embossing, kiln casting, and/or fusing, among other processes designed to treat the walking surface.

Deflection The glass edge support system shall be sufficiently stiff to limit the deflection of the supported glass edges under the design load to no more than *L*/500 of their lengths, where *L* is the smallest supported length of the glass. Otherwise, analysis taking into account frame deflection effects on glass stress shall be carried out in combination with the design loads following the design criteria.

[15] For common geometries and loading conditions, calculations based on tables and graphs in common design standards are usually enough for determining maximum deflections in the linear domain.

[16] Or psi and the values of the constants in the expressions must be adjusted.

Figure 30.18 Pedestrian bridge in Zhangjiajie Park, China.

Dynamic Considerations The walkway natural frequencies should be calculated. The closer the natural frequencies are to the pedestrian walking frequency, the more likely it is for resonance to occur. The less favorable situation should be considered. Although no limits are given, natural frequency ranges may be adopted as a function of the harmonic for vertical and lateral to a given risk of resonance. The values depicted in Table 30.11 may be used as guidance, according to the Association Française de Génie Civil [71].

Table 30.11 Natural Frequency Ranges for Pedestrian Bridges [71]

Range	Harmonic	Vertical frequency (Hz)	Lateral frequency (Hz)	Risk of resonance
1	1st	1.7–2.1	0.5–1.1	Maximum
2	1st	2.1–2.6	1.1–1.3	Medium
3	2nd	2.6–2.5	1.3–2.5	Low
4	–	>5.0	>2.5	Negligible

30.5.3 European Standards—EN

The determining of the lateral LR of linearly supported glass elements according to the European Standards is based on the structural Eurocodes: EN 1990 Basis of structural design, and the relevant actions defined following the structural Eurocode 1 series EN 1991-1-1, EN 1991-1-3, and EN 1991-1-4.

The European Standard prEN 16612 [72] gives a method of determining the lateral LR of linearly supported glass elements based on the bending strength of glass, yet in order to ensure a reasonable robustness in any application, the glass should be able to resist to a minimum unfactored short duration uniformly distributed load of 0.4 kN/m².

Calculation Procedure and Design Criteria

The design value of bending strength for annealed glass is given by

$$f_{gd} = \frac{k_e \cdot k_{mod} \cdot k_{sp} \cdot f_{gk}}{\gamma_{Ma}} \tag{30.29}$$

where f_{gk} = the characteristic value of the bending strength
$\gamma_{Ma} = 1.2$ = the partial material factor for annealed glass
k_{sp} = the factor for the glass surface profile
k_{mod} = the factor for the load duration
k_e = the reduction factor for edge bending strength of annealed glass

Edge Strength Factor In a situation where glass edges are not stressed in bending, like in the case of a pane with all edges supported, the value of k_e is taken as 1.0. Where glass edges are stressed in bending, for example, in a pane with two opposite edges supported or with three edges supported, the value of k_e is lower than 1.0. It should be noted that the edges of thermally toughened or thermally heat strengthened glass commonly have a higher surface compressive stress than the majority of the glass surface, so, in these cases, there is no need for an edge strength factor for strengthened glass.

Based on the experience of the glass experts, from traditional design approaches the tensile strength of the glass edges may be considered less than the strength of the surface, by a factor, k_e, as depicted in Table 30.12.

Table 30.12 Edge Strength Factors for Annealed [72]

Glass type	Edge strength factor, k_e		
	As-cut, arrissed, or ground edges (*)	Seamed edges (**)	Polished edges
Float or sheet glass	0.8	0.9	1.0
Patterned glass	0.8	0.8	0.8
Polished wired glass	0.8	0.8	0.8
Wired patterned glass	0.8	0.8	0.8

(*) Arrissed or ground edges by machine or by hand where the abrasive action is across the edge.
(**) Arrissed or ground edges by machine or by hand where the abrasive action is along the length of the edge.

Glass Surface Profile Factor The factor for the glass surface profile is given in Table 30.13 regarding that all coated glass, painted glass (not enameled glass), mirror glass, etc., where the applied material does

Table 30.13 Factor for the Glass Surface Profile [72]

Glass material	Factor for the glass surface profile k_{sp}	
	As produced	Sandblasted
Float glass, enameled float or drawn sheet glass	1.0	0.6
Patterned, enameled patterned glass and polish wired glass	0.75	0.45
Patterned wired glass	0.6	0.36

not affect the glass structure, can be treated the same as the substrate glass without the applied material. It should be noted that enameled float or drawn sheet glass is not generally available as annealed glass, but the values of the glass surface factor, k_{sp}, are also required in the formulae for prestressed glass. For acid-etched glass, the "as produced" value of k_{sp} should be used.

Factor for the Duration of Load For standard building loads, the factor k_{mod} varies between k_{mod} = 0.25 as minimum value and k_{mod} = 1.0 as maximum. For exceptional loads of very short duration, for example, explosions, values of k_{mod} are greater than 1. The following formula gives the factor for the load duration of annealed glass, where t is the load duration in hours:

$$k_{mod} = 0.663 \times t^{-\frac{1}{16}} \qquad (30.30)$$

For loads with different durations whose effect must be combined the k_{mod} value for the load combination is the highest value which is associated with any of the loads in the combination. For example, if glass is subject to wind, snow and self-weight loads, the effects of a combination of snow and self-weight would be evaluated using a k_{mod} value of 0.45, and the effects of a combination of wind, snow and self-weight loads would be evaluated using a k_{mod} value of 0.74 (or 1.0), and all plausible combination a should be checked. Proposed factors for different load duration a are given in Table 30.14 as given in the prEN 16612 Standard.

Table 30.14 Factor for Load Duration [72]

Action	Load duration	k_{mod}
Wind gusts	5 sec or less	1.0
Wind storm accumulative	10 min	0.74
Barrier personnel loads—normal duty	30 sec	0.89
Barrier personnel loads—crowds	5 min	0.77
Maintenance lads	30 min	0.69
Snow	3 weeks	0.45
Climatic loads on insulating glass units	8 hours	0.58
Dead load, self-weight, altitude effects on insulating glass units	Permanent	0.29

Design Value of Bending Strength for Prestressed Glass The following formula gives the design value of bending strength for strengthened glass:

$$f_{gd} = \frac{k_{mod} \cdot k_{sp} \cdot f_{gk}}{\gamma_{Ma}} + \frac{k_v.(f_{bk} - f_{gk})}{\gamma_{Mv}} \qquad (30.31)$$

where f_{gk} = the characteristic value of the bending strength taken as 45 MPa
f_{bk} = the characteristic value of the bending strength of strengthened glass
$\gamma_{Ma} = 1.2$ = the partial material factor for annealed glass
$\gamma_{M;v} = 1.8$ = the partial material factor for surface prestress
k_{sp} = the factor for the glass surface profile
k_v = the factor for strengthening

The bending strength is related to a specified breakage probability and load duration. The characteristic values for heat-strengthened glass and thermally toughened glass relate to short-time loading (e.g., wind loads), 5 percent breakage probability, and a confidence interval of 95 percent.

For example, in the Italian Standard CNR-DT-210 [73], the strength of glass is interpreted through a statistical Weibull distribution. The 45-MPa strength is a nominal value of strength to be used in calculations.

The values of characteristic bending strength for prestressed glass are given in Table 30.15.

Table 30.15 Values of Characteristic Bending Strength for Prestressed Glass

Glass material per product	Bending strength for prestressed glass [MPa]		
	Thermally toughened safety glass and heat soaked	Heat-strengthened glass	Chemically strengthened glass
Float glass or drawn sheet glass	120	70	150
Patterned glass	90	55	100
Enameled float or drawn sheet glass	75	45	–
Enameled patterned glass	75	45	–

As the presence of tong marks in vertically toughened glass reduces the effectiveness of the prestressing locally compared with horizontally toughened glass, which has no tong marks, a strengthening factor, k_v, depending on the manufacturing method, is considered as depicted in Table 30.16.

Table 30.16 Strengthening Factor to Be Used in Eq. (30.31)

Manufacturing process	Strengthening factor, k_v
Horizontal toughening	1.0
Vertical toughening	0.6

The maximum bending stress, σ_{max}, calculated for the design load or combination of design loads shall not exceed the design value of the bending strength, f_{gd}.

Although there is no specific requirement of glass strength to limit the bending deflection of the glass under load, EN 1279-5 [74] suggests deflection limits for IGU in order to minimize stresses on the edge seal, which may affect durability. Other standards or regulations may require deflection limits for specific applications. If required, the design value of deflection, w_d, shall be following the appropriate standard or regulation.

Consideration should be given to ensuring the glass is not excessively flexible when subjected to applied loads, as this can cause apprehension in occupants. In the absence of any specific prerequisite, deflections shall be limited to Span/65 or 50 mm, whichever is the lower value.

Calculation of Stress and Large Deflection

The aspect ratio of a rectangular pane is given by $\lambda = a/b \leq 1$ (a is the shorter dimension).

For practical determination of the stress, σ_{max}, and the deflection, w_{max}, the following formulae are adopted:

$$\sigma_{max} = k_1 \cdot \frac{a^2}{t^2} \cdot w_{Sd} \qquad (30.32)$$

$$w_{max} = k_4 \cdot \frac{a^4}{t^3} \cdot \frac{w_{Sd}}{E} \qquad (30.33)$$

where w_{Sd} = the design value of the action
E = Young's modulus of glass

The values of the coefficients k_1 and k_4 may be calculated using formulae provided in the standard, yet their values are depicted in Table 30.17 and Table 30.18, respectively.

Table 30.17 Coefficient $k_1 \times 1000$ for Calculation of the Maximum Stress ($\lambda = a/b$)

	$\overline{p}$													
λ	0	1	2	3	4	5	10	20	25	50	75	100	200	300
1	268	261	244	223	205	190	152	135	133	130	129	129	128	128
0,9	319	309	286	260	237	218	172	152	149	145	144	144	144	144
0,8	380	369	341	309	280	257	199	173	169	164	163	162	162	162
0,7	449	437	408	372	339	311	236	199	194	186	185	184	184	184
0,6	524	515	490	457	423	391	294	238	229	215	213	212	211	211
0,5	600	595	580	559	533	506	395	302	284	255	249	247	245	245
0,4	671	669	664	655	644	631	551	429	393	322	304	297	290	289
0,3	724	723	722	721	718	716	694	629	594	471	415	388	356	349
0,2	747	747	747	747	747	747	745	738	734	699	655	613	502	457
0,1	750	750	750	750	750	750	750	750	750	749	749	748	740	729

Table 30.18 Coefficient $k_4 \times 1000$ for Calculation of the Maximum Deflection ($\lambda = a/b$)

	$\overline{p}$													
λ	0	1	2	3	4	5	10	20	25	50	75	100	200	300
1,0	46	42	35	31	28	26	19	14	12	9	7	6	4	4
0,9	56	51	43	38	34	31	23	17	15	11	9	8	5	4
0,8	69	63	54	48	43	39	29	21	19	14	11	10	7	6
0,7	83	77	68	61	55	51	38	28	25	18	15	13	9	7
0,6	99	94	86	78	72	67	51	37	34	24	20	17	12	10
0,5	115	113	107	101	95	89	70	53	48	35	28	25	18	14
0,4	131	130	128	126	121	117	100	78	72	53	44	38	27	22
0,3	143	143	142	142	141	140	133	117	110	87	74	65	48	39
0,2	148	148	148	148	148	148	147	146	145	137	128	120	96	82
0,1	149	149	149	149	149	149	149	149	149	149	149	148	148	147

Laminated Glass

For LG, an effective thickness shall be used, $t_{eff,\sigma}$, on determining the maximum stress and an effective thickness, $t_{eff,w}$, for the deflection calculation, resulting in a reasonably accurate determination of the deflection of and the stress in the LG. For LG, the stress in each ply shall be calculated, and thus the calculation method for LG shall consider that where shear stress is developed in LG parallel with the interlayer, the interlayer can be considered as having some shear resistance. The resistance to bending of LG shall consider the plastic or viscoelastic properties of the interlayer material and its variation with temperature and load duration.

The effective thickness of laminated panes subjected to uniformly distributed loads when bending deflection is under consideration is given as follows:

The effective thickness for calculating bending deflection is

$$t_{eff,w} = \sqrt[3]{\sum\nolimits_k t_k^3 + 12\times\omega\cdot\left(\sum\nolimits_i t_k\cdot t_{m,k}^2\right)} \tag{30.34}$$

Also, the effective thickness for calculating the stress of glass ply number j is

$$t_{eff,\sigma_j} = \sqrt{\frac{(t_{eff,w})^3}{t_j + 2\times\omega\cdot t_{m,j}}} \tag{30.35}$$

where ω = a coefficient between 0 and 1 representing no shear transfer (0) and full shear transfer (1)

t_k, t_j = the thicknesses of the glass plies (Fig. 30.19)

$t_{mk}; t_{m,j}$ = the distances of the mid-plane of the glass plies k, j, respectively, from the mid-plane of the LG

The value of ω, to be used for a specific interlayer and a particular load case, depends on the interlayer stiffness family to which the interlayer belongs for that particular load case. Any interlayer where the properties have not been evaluated should be treated as belonging to the stiffness family with $\omega = 0$.

Currently used values for the effective thickness in LG considering $\omega = 0.3$ are given in Table 30.19.

Table 30.19 Effective Thicknesses of Laminated Glass with Two Plies of the Same Thickness, an Interlayer 0.76 mm Thick, and $\omega = 0.3$

			Effective thickness	
t_{ply} [mm]	t_{int} [mm]	t_{ply} [mm]	$t_{eff,w}$	$t_{eff,w}$
3	0.76	3	5.07	5.62
4	0.76	4	6.63	7.32
5	0.76	5	8.19	9.03
6	0.76	6	9.75	10.74
8	0.76	8	12.86	14.15
10	0.76	10	15.98	17.57

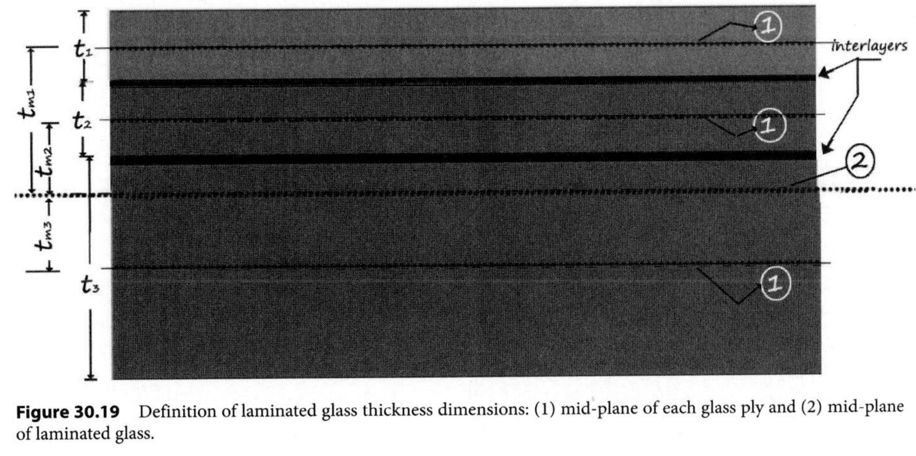

Figure 30.19 Definition of laminated glass thickness dimensions: (1) mid-plane of each glass ply and (2) mid-plane of laminated glass.

Calculation Process for Insulating Glass Units

In case of double glazing, with panes of thicknesses t_1 and t_2, the sharing of external uniformly distributed loads are mainly determined by the distribution of the stiffness of the panes as follows:

$$\delta_1 = \frac{t_1^3}{t_1^3 + t_2^3}; \delta_2 = \frac{t_2^3}{t_1^3 + t_2^3} = 1 - \delta_1 \quad (30.36)$$

Additionally, the sharing of external loads, as well as the effect of internal loads, is determined by the insulating unit factor φ:

$$\varphi = \frac{1}{1 + (a/ä)^4} \quad (30.37)$$

with

$$ä = 28{,}9 \times \left(\frac{s \cdot t_1^3 \cdot t_2^3}{k_5 \cdot (t_1^3 + t_2^3)} \right)^{1/4} \quad (30.38)$$

where $ä$ = the characteristic length of the unit, depending on the thickness of the glass panes and the gas cavity size, s

k_5 = the coefficient of volume that depends on the shape of the unit (Table 30.20)

For LG, the equivalent thickness, $t_{\text{eff},w}$, should be used instead of t_1 and t_2.

Table 30.20 Coefficient $k_5 \times 1000$ for Calculation of the Volume Change ($\lambda = a/b$)

λ	1.0	0.9	0.8	0.7	0.6	0.5	0.4	0.3	0.2	0.1
$k_5 \times 1000$	19.4	23.5	28.8	34.9	42.0	49.9	58.5	67.6	76.4	85.8

Distribution of External and Internal Loads—Insulating Glass The internal loads given by the isochore pressure, p_0, in an insulating unit are reduced by the flexibility of its panes according to the unit factor φ in such a way that the load partition is given according to the following expressions regarding Fig. 30.20:

$$p_{d;1} = (\delta_1 + \varphi \cdot \delta_2) \cdot p_{\text{ex}:1} - (1 - \varphi) \cdot \delta_1 \cdot p_{\text{ex}:2} - \varphi \cdot p_0 \quad (30.39)$$

$$p_{d;2} = (1 - \varphi) \cdot \delta_2 \cdot p_{\text{ex}:1} - (\delta_2 + \varphi \cdot \delta_1) \cdot p_{\text{ex}:2} + \varphi \cdot p_0 \quad (30.40)$$

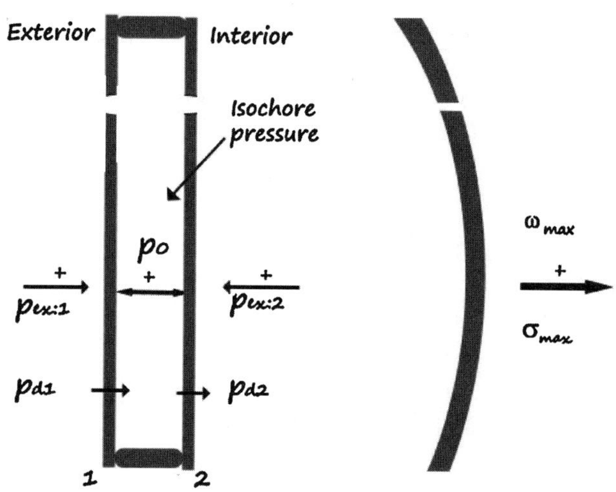

Figure 30.20 Sign convention for actions and effects for insulating glass.

where $p_{\text{ex}:1}$ = the external load acting on pane 1

$p_{\text{ex}:2}$ = the external load acting on pane 2

p_0 = the isochore pressure

Isochore Pressure

The isochore pressure, p_0, depends on the pressure generated by the altitude difference between the local of production and that of the installation, $p_{H,0}$, and also on the pressure due to a temperature difference or air pressure, p_a, in the same two locations, $p_{c,0}$:

$$p_0 = p_{H,0} + p_{C,0} \quad (30.41)$$

The following expressions and constants values may be applied to its determination:

$$p_{H,0} = c_H \cdot (H - H_p)$$
$$p_{C,0} = c_T \cdot (T_c - T_P) - (p_a - p_p) \quad (30.42)$$

where $c_H = 0.012$ kPa/m

$c_T = 0.34$ kPa/°K

H = the calculation altitude at fixing

H_0 = the altitude at the production site

T = the given temperature at fixing

T_0 = the temperature at the production instant

30.6 PLATE BUCKLING

Glass plates subjected to in-plane loads tend to fail because of plate buckling. Depending on the type of loading, different structural behavior, that is, types of buckling, must be distinguished.

Plates may be subjected to pure compression, to shear, or to a combination of both compression and shear, and they may also be submitted to in-plane loads.

The interstory drift due to seismic action causes deformation of the glass supporting structure, and when there is not enough clearance to accommodate these structural displacements, plate buckling to shear may arise.

An initial geometrical imperfection must be considered in a panel. When subjected to in-plane shear force, the panel will deflect out of plane and only will return to its initial position if the acting force is smaller than a critical value, the shear buckling force, V_{cr}. Otherwise, if the forces increase gradually beyond the critical shear buckling force, the panel will accentuate its buckled shape until collapse.

The buckling resistance of a glass panel depends on the distribution of the maximum principal stresses on the glass surface; analytical plate buckling models are not precise enough to describe the stress distribution due to the nonlinear behavior. For the time being, design-approximated methods are used since finite-element plate buckling analysis is still quite laborious for practice. Thus, simplified analytical procedures are presented for calculating the in-plane deformation capacity resistance of window glass panels subjected to seismic displacement greater than the fixture or systems tolerance.

30.6.1 Buckling and Out-of-Plane Displacement

If the tolerance for horizontal displacement is null, only a minimal displacement, Δ_d, along the diagonal size of a plate will occur depending on its geometry and glass properties, and, as represented in Fig. 30.21, the displacement, Δ_d, may be expressed as a function of a minimal horizontal displacement, D_r as defined in

$$\Delta_d = \frac{d}{b} \cdot D_r \quad (30.43)$$

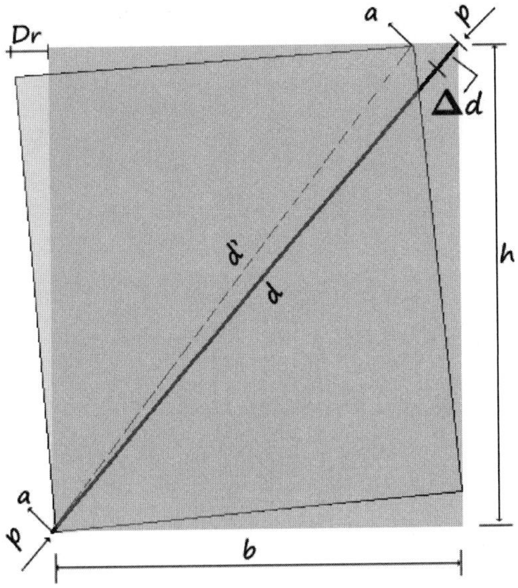

Figure 30.21 In-plane plate displacements along the horizontal and diagonal directions.

Assuming a sinusoidal shape for the out-of-plane curve due to buckling, the maximum displacement value, D_r, before glass breakage, may be obtained as follows [75]:

$$D_r = \frac{1}{b} \cdot \left(\frac{\sigma_{\text{allow}} \cdot d^2}{\pi \cdot E \cdot t} \right)^2 \tag{30.44}$$

where σ_{allow} = maximum allowable surface stress (~24.0 MPa)
b = the horizontal dimension of the panel
d = the diagonal dimension of the plate
t = plate thickness
E = linear elastic modulus (~70.0 GPa)

30.6.2 Shear Buckling Load

Plate theory can be used to determine the critical shear buckling force, V_{cr}, for monolithic glass panel under in-plane load assuming that the plate is initially flat and the material is elastic, homogeneous, and isotropic; the thickness is small compared to area dimension; and deflection of the plate occurs due to displacements in its mid-surface (Fig. 30.22).

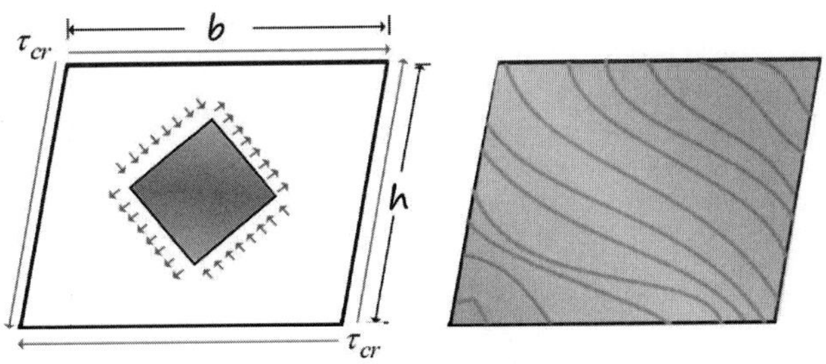

Figure 30.22 In-plane shear stress and displacements.

The nonlinear interaction between forces and rotations and the equilibrium equation of forces and moments for the plate element based on the Von Karman equation for large deflections of thin flat plates allow for the determination of the critical shear load and consequently the shear buckling stress:

$$V_{cr} = k_\tau \frac{\pi^2 \cdot D}{b^2} \qquad \tau_{cr} = \frac{V_{cr}}{t \cdot b} \tag{30.45}$$

where $D = \dfrac{E \cdot t^3}{12 \cdot (1 - \nu^2)}$ (flexural stiffness)
b = the horizontal dimension of the panel
h = the vertical dimension of the pane
k_τ = shear buckling coefficient for monolithic glass
t = plate thickness
ν = Poisson's ratio (~0.22)
E = linear elastic modulus (~70.0 GPa)

The value of the shear buckling coefficient, k_τ, was studied by Southwell and Mocibob [76, 77], depending on the boundary conditions and geometry ratio $\alpha = h/b$.

In the case of LG, the critical shear buckling load of a glass panel, under in-plane action, must consider the "sandwich behavior" of the plate, and from Eq. (30.45) the value of V_{cr} becomes

$$V_{cr} = k_\tau \frac{\pi^2 \cdot D_{\text{lam}}}{b^2} \cdot k_{\text{lam}} \qquad D_{\text{lam}} = \frac{E \cdot t_g \left(t_g^2 + 3 \times d \right)}{6 \times (1 - \nu^2)} \tag{30.46}$$

where t_g is the thickness of individual glass sheet and d is the distance between the glass sheet centerlines axis of the LG (Fig. 30.23).

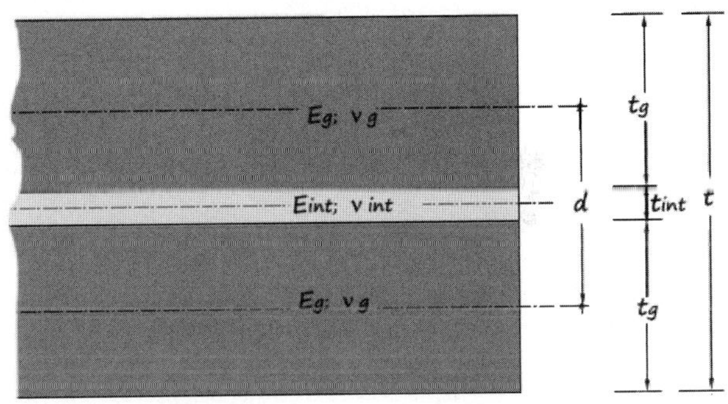

Figure 30.23 Laminated glass, plies, and interlayer thicknesses.

Shear Buckling Coefficient for Linear Supported Plate A structural model for a linear supported glass panel subjected to in-plane shear force, H_x, is shown in Fig. 30.24 where an adhesive glues the glass panel to the frame on the two short sides. The adhesive will transfer the in-plane shear force from the frame to the glass panel, where two setting blocks at the bottom and two at the top of the glass panel have the function to carry the permanent load as the panel self-weight.

When a small modulus of elasticity adhesive is used, the tensile diagonal will not be able to keep the buckled compressive diagonal, C, and the out-of-plane deflection, as shown in the left image of Fig. 30.24, takes place. On the other hand, when a stiff adhesive is used, the tensile diagonal will be strong enough to keep the compressive diagonal, C, and the out-of-plane deflection with two semi-buckles occurs as shown in the right image of Fig. 30.24.

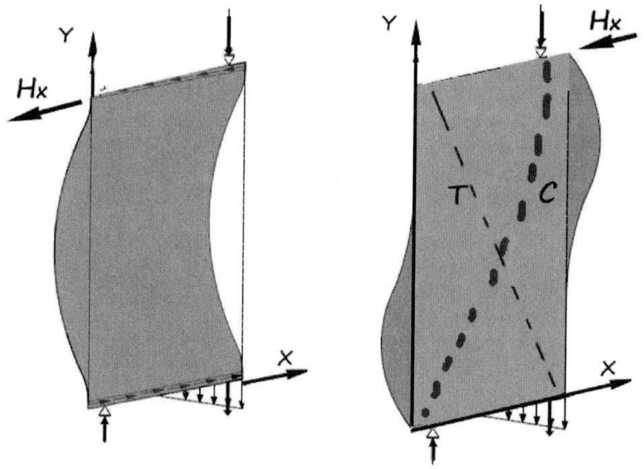

Figure 30.24 Membrane effect in the glass panel. Out-of-plane deformation with soft adhesive (left) and with a strong adhesive (right).

MONOLITHIC GLASS

The shear buckling coefficient for monolithic glass, k_τ, is given in Eq. (30.47) depending on the ratio $\alpha = h/b$:

$$k_\tau = \frac{6.09}{\alpha^3} \tag{30.47}$$

LAMINATED GLASS

For LG the value of the coefficient, k_{lam}, to be used in Eq. (30.46) depends on the shear modulus of the interlayer since a stiffer interlayer with higher shear modulus, G_{int} gives the glass panel higher critical shear buckling force. The values recommended by Mocibob [77] are depicted in Fig. 30.25.

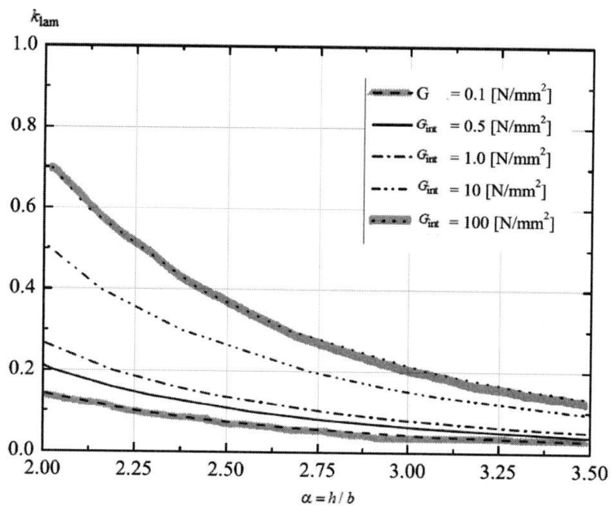

Figure 30.25 Shear buckling coefficient for linear supported laminated glass [77].

For a linear supported panel, the value of k_{lam} is multiplied by the adhesive coefficient, β_{int}, as a function of the modulus of elasticity of the interlayer, E_{int}:

$$\beta_{int} = 0.0125 \times E_{int} + 0.97 \tag{30.48}$$

Shear Buckling Coefficient for a Point-Supported Glass Plate Figure 30.26 illustrates an idealized structural model of a point-supported glass panel, conditions and support reactions of which are defined as follows. Support 4 is fixed for translation in any direction (X, Y, and Z) and rotation about X-axis and Y-axis.

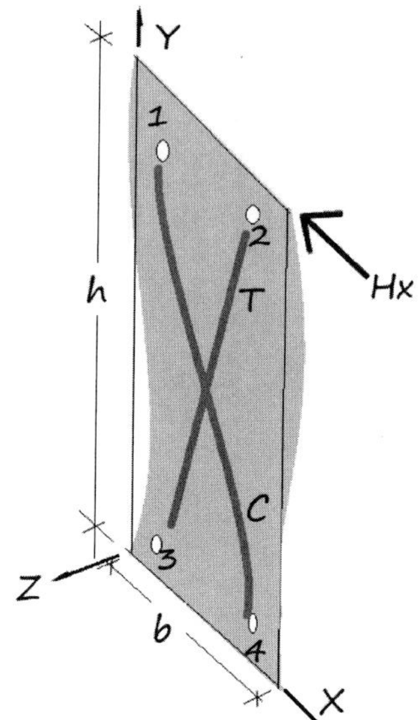

Figure 30.26 Structural model of point-supported glass panel showing membrane effect with out of plane deflection due to compressive, C, and tensile, T, diagonals.

The in-plane shear force, H_x, is applied at support 2, producing a compressive reaction of support 1 and support 4, which are responsible for the formation of a compressive diagonal, C, between these two supports, forcing the panel to buckle while the tensile diagonal, T, between supports 2 and 3 attempt to keep the bowed compressive diagonal straight. These crossing diagonals are responsible for the membrane effect, which gives the glass panel the favorable stiffening effect. Due to this membrane effect, the critical shear force is not the upper limit of the plate resistance, but higher loads than the critical ones can then be applied to the panel, the so-called post-buckling reserve.

MONOLITHIC GLASS

The shear buckling coefficient for monolithic glass, k_τ, is given in Eq. (30.49) depending on the ratio $\alpha = h/b$:

$$k_\tau = \frac{4.22}{\alpha^2} \tag{30.49}$$

LAMINATED GLASS

The coefficient, k_{lam}, values for point-supported LG plates to be used on Eq. (30.46) are depicted similarly in Fig. 30.27.

30.6.3 Shear Buckling Strength

The design shear strength of a glass panel without buckling may be taken as

$$V_{Rd} = \frac{\tau_{Rk} \cdot b \cdot t}{\gamma_M} \tag{30.50}$$

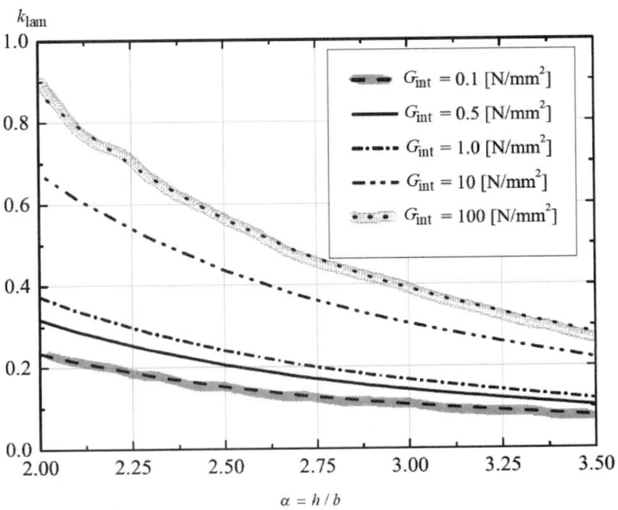

Figure 30.27 Shear buckling coefficient for point-supported laminated glass [77].

where $\tau_{Rk} = \sigma_{Rk}$ = characteristic shear strength of the glass panel[17]

b = glass panel width

t = plate thickness ($t = n \cdot t_g$) for laminated glass with, n, t_g ply thickness

$\gamma_M = 1.4$ as proposed by Wellerhoff [78]

A shear reduction factor, $\chi \le 1$, must be considered when buckling occurs, and the shear buckling design resistance is taken as

$$V_{B,Rd} = \chi \cdot V_{Rd} \tag{30.51}$$

The shear buckling reduction factor, χ, may be determined for point-supported and linear as a function of non-dimensional shear plate slenderness $\bar{\lambda}$ and correlation curves based on Ayrton–Perry type resistance formulas [79, 77] considering that, $\bar{\lambda}_0 = 0.8$ and $\bar{\alpha} = 7$, as follows:

$$\chi = \begin{cases} \bar{\lambda} \le \bar{\lambda}_0 \rightarrow \chi = 1.0 \\ \lambda > \lambda_0 \rightarrow \chi = \dfrac{1}{\phi + \sqrt{\phi^2 - \bar{\lambda}}} \end{cases} \tag{30.52}$$

with

$$\phi = 0.5 \times [1 + \bar{\alpha} \cdot (\bar{\lambda} - \bar{\lambda}_0) + \bar{\lambda}] \tag{30.53}$$

The nondimensional shear plate slenderness, $\bar{\lambda}$, may be expressed as the square root of the ratio between the characteristic bending strength of the glass and the shear buckling stress:

$$\bar{\lambda} = \sqrt{\frac{\sigma_{Rk}}{\tau_{cr}}} \tag{30.54}$$

For linear supported glass plates, the buckling reduction factor, χ, is defined as follows:

$$\chi = \begin{cases} 0.02 \leftrightarrow \bar{\lambda} \le 6.75 \\ \dfrac{1}{\phi + \sqrt{\phi^2 - \bar{\lambda}}} \leftrightarrow \alpha > 6.75 \end{cases} \tag{30.55}$$

[17]It may be assumed that shear stress resistance is the same as the bending stresses resistance [78].

In the case of point-supported plates, the buckling reduction factor is

$$\chi = \begin{cases} 0.04 \leftrightarrow \bar{\lambda} \le 3.75 \\ \dfrac{1}{\phi + \sqrt{\phi^2 - \bar{\lambda}}} \leftrightarrow \alpha > 3.75 \end{cases} \tag{30.56}$$

In both Eq. (30.55) and Eq. (30.56), the value of ϕ is taken as

$$\phi = 4 \times \bar{\lambda} - 2{,}3 \tag{30.57}$$

30.6.4 Stresses in Point Fixed Anchorage Zones

When glass panels are point-supported using bolts as in the so-called spider glass curtain walls (Fig. 30.28), the in-plane load should be adequately introduced in the glass panel to provide enough shear load-carrying capacity. The critical zone in the glass is located around the drilled holes where complex stress fields and large stress concentrations form, so that the verification of the safety of the strength of the glass needs to be carried out around the perforated zones using rigorous approaches, such as the finite-element method or through simplified methods.

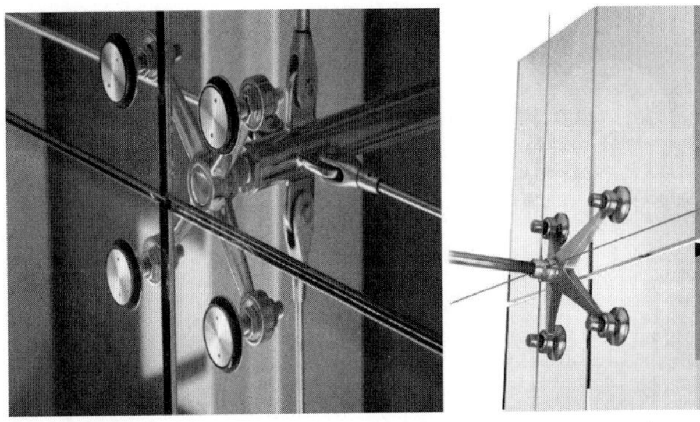

Figure 30.28 Fitting glass in a façade spider system close-up from outside and inside sketch

FINITE ELEMENT ANALYSIS

Useful approaches with the finite-element analysis are recommended by the German Standard DIN 18008 [80]. In this standard the modeling allows a rigorous understanding of the stress distribution on a glazed panel with holes to receive the fastening devices. Two different methodologies are provided: the verification and validation of finite-element models (Annex B of the standard) or using a simplified method (Annex C to the standard).

In the case of the glass pane design according to the simplified method, it may be analyzed without the drilling holes with the glass fittings represented by equivalent stiffness springs.

The stresses values must be readdressed in a second calculation to consider the holes so that the final values in their vicinity can be found and compared to the value of the tensile strength of calculation (Fig. 30.29). Appendix C of this standard also provides table values, allowing a quick estimate of the governing design stress in the fitting area as a result of individual stress components.

Part B of Part 3 of DIN 18008 [80] describes the procedures to be followed for finite-element modeling of a panel and the supporting devices simultaneously. For this, it is necessary to adequately discretize the perforated zone and define the interface elements between the fixation patella and the glass. In the case of dynamic actions, the design of point-supported connections' panels requires numerical modeling to validate the behavior resulting from the damping of the sealant and sealant materials between the connecting studs and the glass plates. Laboratory tests following the AAMA 501.4-18 [54] recommendations of the American Architectural Manufacturers Association (AAMA) and numerical studies shall be carried out in this field to demonstrate that

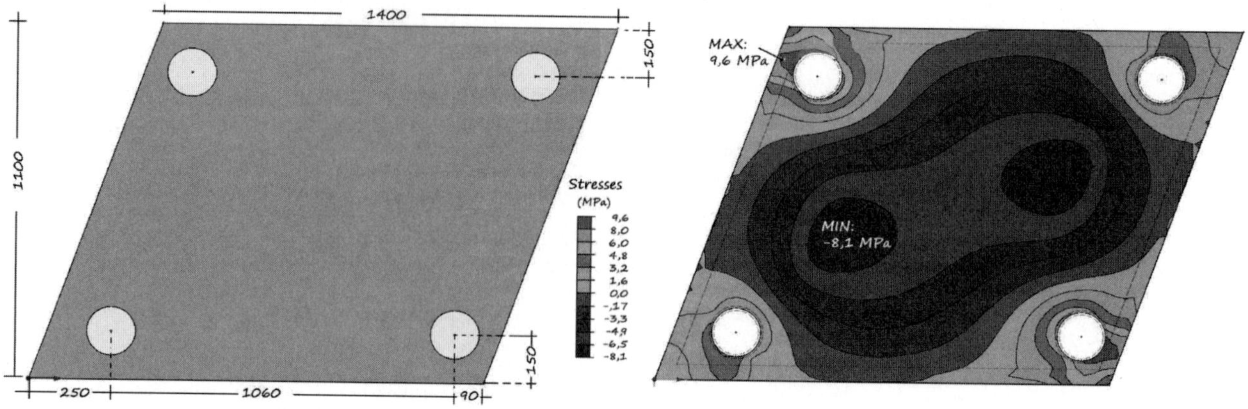

Figure 30.29 Stresses contour map in a point-supported panel. Adapted from Dlubal© [81].

the panels and their supports respond satisfactorily and that damping coefficients are properly calibrated [82].

Simplified Analysis

Maximum stress values may be estimated applying stress concentration factors in singularities zones. A stress concentration factor, k_{sf}, can be defined as the ratio of the peak stress in the plate (or stress in the perturbed region) to a reference nominal stress. For example, regarding Fig. 30.30 the maximum stress value, σ_{max}, in the hole region is given by

$$\sigma_{max} = k_{sf} \cdot \sigma_{nom}$$
$$\sigma_{nom} = \frac{F_t}{t_g \cdot (H-d)} \tag{30.58}$$

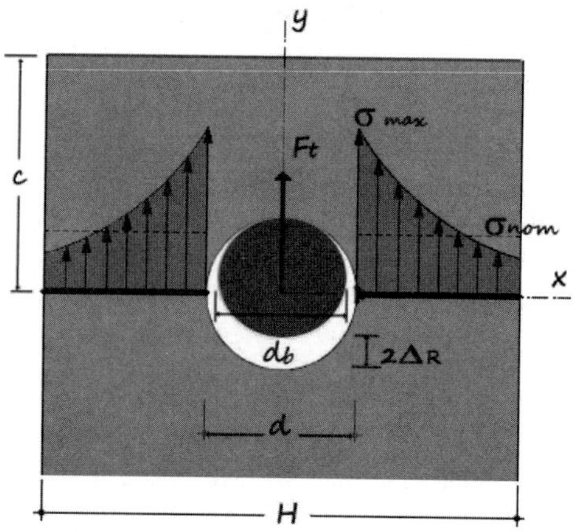

Figure 30.30 Stress distribution in the middle axis crossing a circular hole bolt loaded.

Stress concentration factors depend on the geometry of the plate, namely, the distance, *c*, from the center of the hole to the edge of the plate; the width of the plate, *H*, and its thickness, *t* as well as the clearance of the fixing bolt in the hole. Based on several studies and values proposed by Timoshenko et al. [83], Frocht et al. [84], Pilkey [85], and Duerr [86], among others, the following formulae and graph, proposed by Mocibob [77], may be used to determine the maximum stress at glass hole under tensile with reference to Fig. 30.31:

$$R_d = \frac{f_{gk} \cdot (2 \times c - d) \cdot t}{k_{cf} \cdot \sin\beta \cdot \gamma_R} \tag{30.59}$$

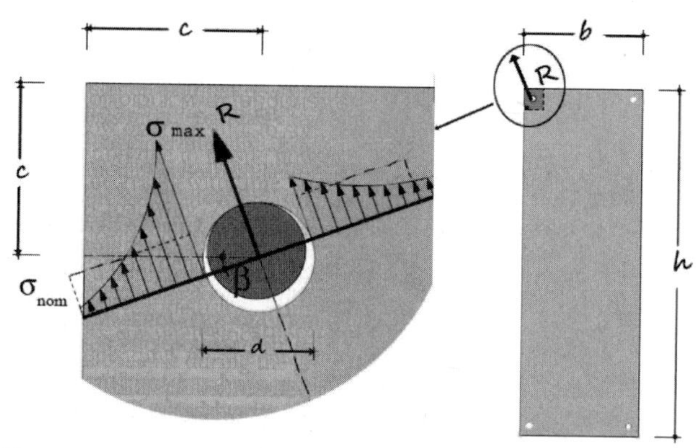

Figure 30.31 Stress concentration at glass hole under tensile.

where R_d = design value of the strength of the glass in the hole region under tensile
f_{gk} = characteristic shear strength of the glass panel
c = the distance between the glass hole and glass edge
d = diameter of the hole
$\beta = \arctan\dfrac{b-2\times c}{h-2\times c}$: the angle at which reaction force R acts on the hole
k_{cf} = stress concentration factor
γ_R = a partial factor of (1.8 for AN glass; 2,12 for HS glass and 2.55 for FT glass) [73]
τ = glass panel total thickness
a = glass panel height
b = glass panel width

Values of the stress concentration factor, k_{cf}, depending on the geometrical ratio $[d/(2 \times c)] \cdot \sin\beta$, are shown in Fig. 30.32.

30.7 LATERAL-TORSIONAL BUCKLING

A completely frameless and almost invisible façade is obtained with glass fins set perpendicular to the glass pane at each vertical line of the glass grid façade. Glass fins and beams provide all the lateral and horizontal stability of the overall structure, while both elements are attached using highly efficient structural silicone and customized imperceptible stainless-steel fittings (Fig. 30.33).

Lateral-torsional buckling of glass panels is a limit state of structural stability, where an element is subjected to bending exhibiting a structural deformation as a combination of lateral deflection and twisting.

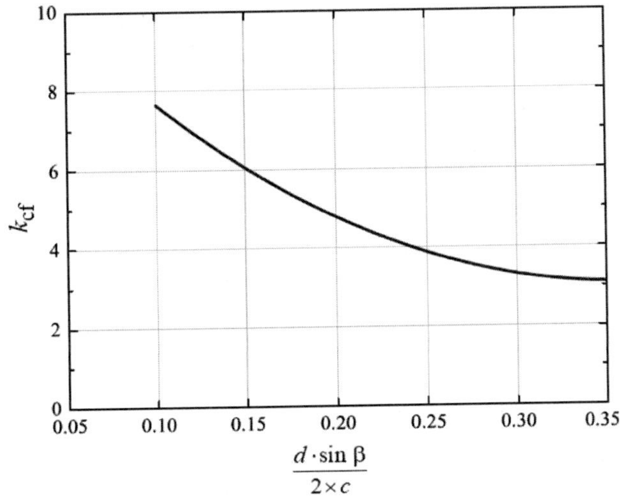

Figure 30.32 Stress concentration factor, k_{cf}, in hole's region of point-supported glass[77].

Figure 30.33 Total vision façade with clamped glass fins.

In glass structures, lateral-torsional buckling may, for instance, occur in glass beams or glass fins used as lateral stiffeners in façades.

In these situations, it is necessary to ensure that buckling of fins and beams will not occur under the design loads. There are many possible configurations for these elements, and each case must be analyzed by structural principles using adequate analyses since the plates can develop a membrane-type action. However, in the case of the lateral buckling of beams, this is not the case, and the analytical results and experimental modeling show good agreement with numerical and experimental analysis providing a close value to the upper limit of the beam's load-carrying capacity [87].

For the fins, the main issue is that they replace a frame or mullion and are assessed as to size and thickness and securely fixed or supported at the head and sill. The fins are adhered to the façade glass with structural silicone sealant to accommodate both positive and negative loads so that any loading applied to the glass façade is transferred to the fin and then to the top and bottom fin shoes by way of a reaction load.

The availability of a design procedure, calculating a critical elastic buckling moment, M_{cr}, provides an accurate assessment, and after applying a partial safety factor, it may be compared with the design value of the acting bending moment.

If the ends at supports are assumed to be effectively restrained against twisting, which may be considered when the torsional stiffness at the supports is equal or greater than 20 times, as in Fig. 30.34,

Figure 30.34 Laminated glass beam failure due to lateral-torsional buckling under two-point load test.

the torsional stiffness of a beam or fin [88], the value of K_T, may be defined as follows:

$$K_T = G \cdot J/L \tag{30.60}$$

J = torsional moment of inertia
G = torsional elastic modulus (28.7 GPa for glass)
L = span

The effective torsional rigidity of beams of solid rectangular cross section may be calculated using the Eq. (30.61) considering that t and h are the thickness and width of the beam or fin section, respectively:

$$J = \frac{t \cdot h^3}{3} \cdot \left(1 - 0.63 \times \frac{t}{h}\right) \tag{30.61}$$

Based on several analytical solutions that were acutely studied in 1973 [89] and later validated by extensive nonlinear numerical simulations [90], the Australian Standard AS 1288 [45] gives the basis for determination of fin design to prevent buckling which may be used for glass beams as well.

A critical elastic value of the maximum moment, M_{cr}, between two buckling restraints is given by the subsequent expressions providing that the restraints must prevent rotation of the beam about the Z-axis with the possibility for rotation of the fin about a Y-Y axis at the restraint locations, as shown in Figs. 30.35 and 30.36.

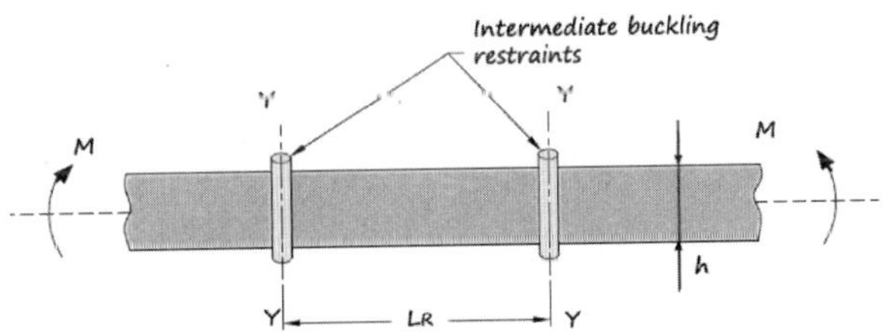

Figure 30.35 Side view of glass fin or beam and restraints.

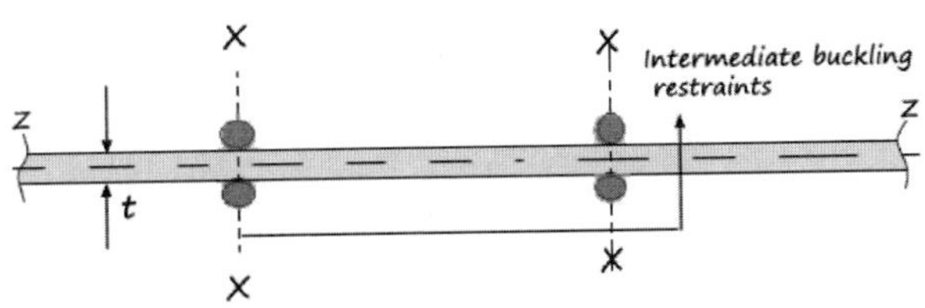

Figure 30.36 Plan view of glass fin or beam and restraints.

Elements with Intermediate Buckling Restraints

The critical elastic value of the maximum moment, M_{cr}, between two buckling restraints may be taken using

$$M_{cr} = \frac{k_1}{L_R} \cdot \sqrt{E \cdot I_y \cdot G \cdot J} \tag{30.62}$$

where k_1 = coefficients for bi-symmetrical beams with buckling restraints (Table 30.21);
L_R = the distance between effectively rigid buckling restraints;
I_y = the second moment of inertia about the minor axis
J = torsional moment of inertia
G = torsional elastic modulus
E = linear elastic modulus

Table 30.21 Coefficients for Slenderness Factor k_1 of Bi-Symmetrical Fin or Beams with Intermediate Buckling Restraints

Moment parameter (ζ) (Fig. 30.37)	Slenderness factor (k_1) Free restraint condition	Fixed restraint condition
1.0	3.1	6.3
0.5	4.1	8.2
0.0	5.5	11.1
−0.5	7.3	14.0
−1.0	8.0	14.0

The slenderness factor k_1 depends on the restraint intermediate condition. The term "fixed" or "free restraint" refers to rotation preventing of the beam about the Z-axis and depends on the moment parameter, ζ, as defined in Fig. 30.37.

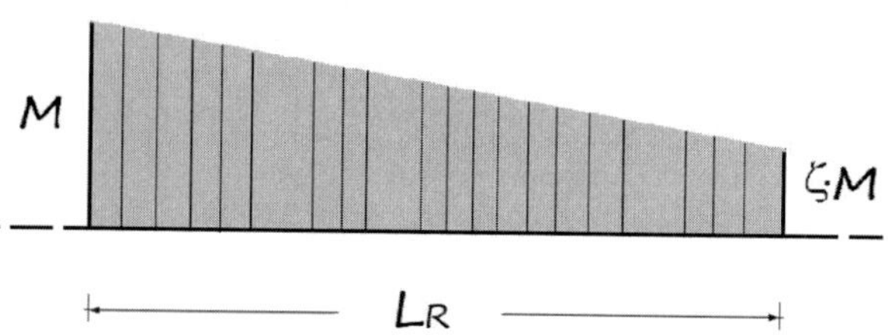

Figure 30.37 Diagram bending moment between buckling restraints.

Elements without Intermediate Buckling Restraints

The critical elastic value of the maximum moment M_{cr}, of beams with no intermediate buckling restraints may be calculated using

$$M_{cr} = \frac{k_2}{L} \cdot \sqrt{E \cdot I_y \cdot G \cdot J} \cdot \left[1 - k_3 \cdot \frac{y_h}{L} \sqrt{E \cdot I_y \cdot G \cdot J}\right] \tag{30.63}$$

where k_2, k_3 = coefficients for bi-symmetrical beams with buckling restraints (Table 30.22)
L = the span of the beam
I_y = the second moment of inertia about the minor axis
J = the torsional moment of inertia
y_h = height above the centroid of the point of load application
G = torsional elastic modulus
E = linear elastic modulus

Table 30.22 Coefficients for Slenderness Factor, k_2 and k_3, of Bi-Symmetrical Fin or Beams without Intermediate Buckling Restraints

Load type	Support condition about X-X-axis	Supports restraint against rotation about Y-Y-axis	k_2	k_3
Uniformly distributed	Simply supported	Free	3.6	1.4
		Fixed	6.1	1.8
	Fixed support	Free	4.1	4.9
		Fixed	5.4	5.2
	Cantilever	Fixed	6.4	2.0
Center point load	Simply supported	Free	4.2	1.7
		Fixed	6.7	2.6
	Fixed support	Free	5.3	4.5
		Fixed	6.5	5.3
Two symmetrical point loads at the quarter span	Simply supported	Fixed	3.3	1.3
Point load at the free end	Cantilever	Fixed	4.0	2.0

Continuously Restrained Elements

In the case of elements with a bi-symmetrical cross section continuously restrained against lateral displacement at a distance y_0 from the neutral axis, the critical elastic moment, M_{cr}, may be taken as

$$M_{cr} = \frac{\left(\frac{\pi}{L_R}\right)^2 \cdot E \cdot I_y \cdot \left(\frac{d^2}{12} + y_0^2\right) + G \cdot J}{2 \cdot y_0 + y_h} \tag{30.64}$$

k_1 = coefficients for bi-symmetrical beams with buckling restraints (Table 30.21)
L_R = the distance between points of effective rigid rotational restraints
I_y = the second moment of inertia about the minor axis
J = the torsional moment of inertia
y_h = location from the neutral axis of the loading point (Fig. 30.38)
G = torsional elastic modulus
E = linear elastic modulus
It must be noted that y_h may take negative values, depending on the direction of the applied load and the position of the restraint as illustrated in Fig. 30.38.

30.8 GLASS COLUMNS

The load-bearing system and the stability achieved with glass walls is the first step toward glass columns; the only difference would be the dimensions of the building component. Glass columns can be part of either the primary structure or the secondary structure, such as a vertical element that supports an entrance canopy. Load-bearing systems can be built up of glass walls, such is the case where the façade carries the roof, and the fins materialize the glass T beams to withstand the lateral action. As the fins also interact with the façade wall, the question remains whether the fins are columns or is the façade in combination with the load-bearing "columns" as in Fig. 30.39*a*.

In the case of Fig. 30.39*b*, cruciform glass columns with a cross section of limited dimensions support a glass roof 24 × 24 m². The columns with a height of 3.2 m are made up of three layers of laminated heat-treated glass. The columns are only loaded in compression and do not contribute to the stability of the building under horizontal loads. Special care in the details is taken to avoid bending to be transferred to the columns. The maximum loading which can occur is calculated to be less than 20 percent of that of full-scale tests [91, 14].

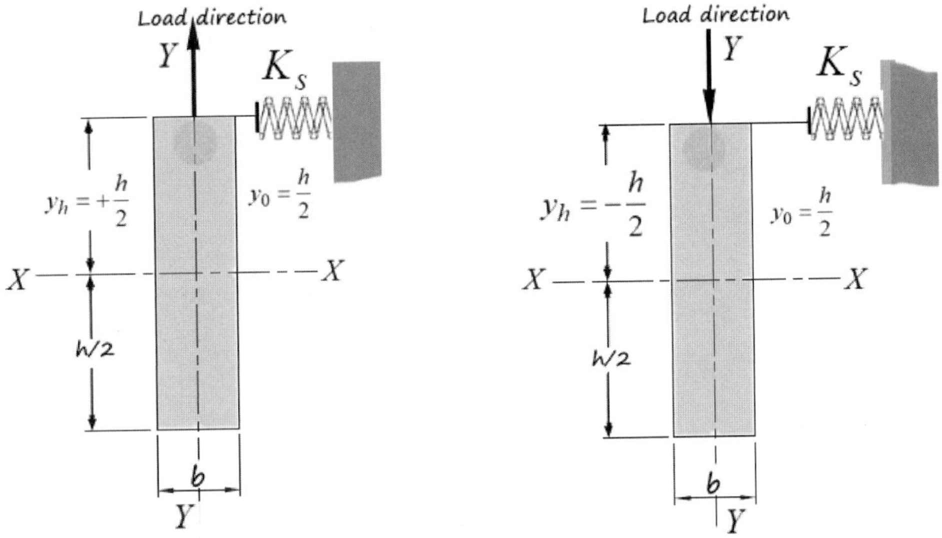

Figure 30.38 Location y_h, from the neutral axis of the loading point, and K_s, restraint stiffness.

(a) (b)

Figure 30.39 (*a*) Wall glass supported, the Glass Pavillon in the Rheibnach Glass Center at Rheinbach. (*b*) Column glass supported, the Saint-Germain-en-Laye administrative center at Île-de-France.

Two extreme situations may occur in glass columns: a layered response, where the glass panes react without any shear bond and the resistance has a lower value, or a monolithic-like type behavior, where a higher resistant or upper limit value is obtained, yet the maximum stress of the laminate glass lies between those two limits [26, 92].

30.8.1 Monolithic or Single-Layered Glass

Calculation methods for buckling of monolithic or laminated with rigidly bonded glass layers are the same as for buckling behavior of an ideal elastic pin ended bar tested in compression.

Taking as a basis the well-known model from Fig. 30.40 for a buckling instant, the critical buckling force, N_{cr}, can be determined using Euler's method:

$$N_{cr} = \frac{\pi^2 \cdot E \cdot I}{L_{cr}^2} \tag{30.65}$$

where N_{cr} = critical force
L_{cr} = critical buckling length
E = Young's modulus of glass
I = minimum moment of inertia, given by $\frac{b \cdot t^3}{12}$

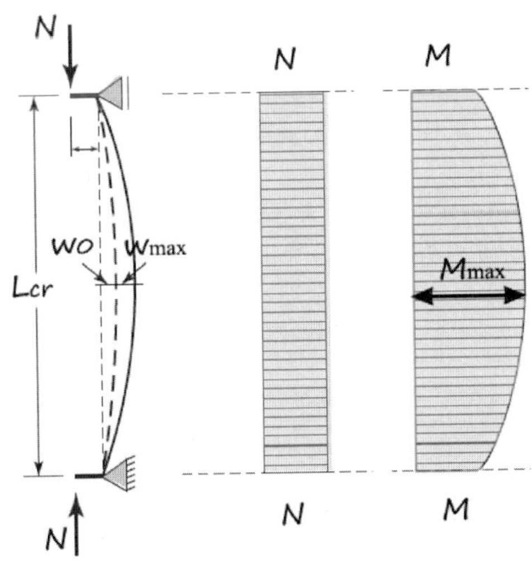

Figure 30.40 Buckling model of a column.

The maximum deflection to the column caused by N_{cr} is defined as

$$w_{max} = \frac{e}{\cos\left(\frac{L_{cr}}{2}\cdot\sqrt{\frac{N}{N_{cr}}}\right)} + \frac{w_0}{1-\frac{N}{N_{cr}}} \quad (30.66)$$

where w_0 = a variable to take into account for imperfections within the glass
N = the applied vertical force
e = the eccentricity of the load onto the column

Considering the lower bound values of imperfections as L/2500 for annealed glass and L/300 for heat-strengthened and toughened glass [43], the surface stress can be calculated as follows:

$$\sigma_{max} = N\cdot\left(\frac{1}{A}+\frac{1}{W}\cdot(w_{max}+w_0+e)\right) \quad (30.67)$$

with A = the cross-sectional area of the column
W = the minimum elastic section modulus of the column

30.8.2 Laminated Glass

LG, when compared with monolithic glass, has advantages due to its capacity of absorbing energy once crack propagation is attenuated. Upon this, the mitigation of post-fracture glass fallout is a major issue, and ensuring a substantial residual strength after breakage makes LG almost mandatory for load-bearing members in glass structures.

Mechanics of sandwich structures or using finite-element models have been used to analyze LG and to determine the load capacity and the critical buckling load [93, 94]. The numerical model can be studied in detail following the work of Luible [95].

The formula by Sattler et al. [96, 97] uses the sandwich approach to determine the critical buckling load of a two-layered, symmetrical, or not LG, as using the geometry definition of Fig. 30.41:

$$N_{cr} = \frac{\pi^2\cdot(1+\alpha+\pi^2\cdot\alpha\cdot\beta)}{1+\pi^2\cdot\beta}\cdot\frac{E\cdot I_s}{L_{cr}} \quad (30.68)$$

The coefficients α and β are given by

$$\alpha = \frac{I_1+I_2}{I_s};\ \beta = \frac{t_{int}}{G_{int}\cdot b\cdot(z_1+z_2)^2}\cdot\frac{E\cdot I_s}{L_{cr}^2}$$

with

$I_1 = \frac{b\cdot t_1^3}{12};\ I_2 = \frac{b\cdot t_2^3}{12};\ I_s = b\cdot\left(t_1\cdot z_1^2 + t_2\cdot z_2^2\right)$ and G_{int}, the shear modulus of the interlayer, either assessed by the laboratory tests or provided in relevant reliable sources [98, 73].

The midspan deflection and the stress surface of a LG column must be calculated using the effective thickness of the column, t_{eff}, and the effective modulus of the glass using

$$t_{eff} = \sqrt[3]{\frac{12\cdot I_s\cdot(1+\alpha+\pi^2\cdot\alpha\cdot\beta)}{b\cdot(1+\pi^2\cdot\beta)}} \quad (30.69)$$

$$W_{eff} = \frac{b\cdot t_{eff}^2}{6} \quad (30.70)$$

Numerical simulation-based tests with different viscoelastic interlayer material under different temperatures and loading speeds showed that the influence of the shear transfer between glass plies depends not only on the temperature and the load duration but also on the thickness of the interlayer [99, 73]. For low temperatures and very short loading, the maximum load behaves practically as a monolithic section, but for temperatures higher than 25°C, no relevant improvement of the maximum load can be found compared to a similar glass without any interlayer for long-term loading [94].

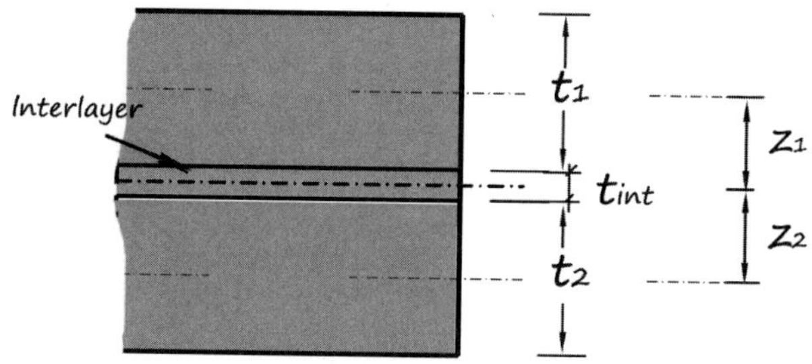

Figure 30.41 Two-glass-ply laminated glass.

From an economic and safety point of view, a shear connection might therefore only be considered for short-term loading, like wind, seismic, or impact loads.

30.8.3 Buckling Resistance Curves

Recent experimental research [100] allows designing glass columns based on European Code buckling curves in the same manner as for steel elements [101] according to the following procedure:

The design buckling resistance of the compression member $N_{b,Rd}$ is given by

$$N_{b,Rd} = \chi\cdot f_{gd}\cdot A_{eff} \quad (30.71)$$

where A_{eff} = the effective area of the cross section when subjected to uniform compression
f_{gd} = the design bending strength of the glass
$\chi = \frac{1}{\Phi+\sqrt{\Phi^2-\bar{\lambda}^2}}$ is a reduction factor
$\Phi = 0.5\times(1+\alpha_{imp}\cdot(\bar{\lambda}-\alpha_0)+\bar{\lambda}^2)$ is a nondimensional factor.

Values for the imperfection factors $\alpha_{imp} = 0.71$ and $\alpha = 0.6$ have been adopted from Amadio et al. [102]. The nondimensional slenderness parameter, $\bar{\lambda}$, is defined as

$$\bar{\lambda} = \sqrt{\frac{A_{eff}\cdot f_{gk}}{N_{cr}}} \quad (30.72)$$

elastic critical buckling load N_{cr} is given as

$$N_{cr} = \frac{\pi^2\cdot E\cdot I_{eff}}{L_{cr}^2} \quad (30.73)$$

The design value of the bending strength of the glass f_{gd} shall be obtained from the prEN 16612 [72] standard and the effective second moment of inertia of the glass section, I_{eff}, using the effective thickness, $t_{eff,\sigma}$ defined in Eq. (30.35).

REFERENCES

[1] Doitpoms—Tlp Library Fracture of Glass. [Online]. Available: https://www.doitpoms.ac.uk/tlplib/BD5/printall.php. [Accessed Jan 2019].

[2] A. Dalgliesh, "Structural Design of Glass Buildings," in *Update on Contemporary Curtain Wall, Canadian Journal of Civil Engineering*, vol. 17, no. 5, Canadian Science Publishing, October 1990.

[3] J. W. Johnson & D. G. Holloway (1966) "On the Shape and Size of the Fracture Zones on Glass Fracture Surfaces," *The Philosophical Magazine: A Journal of Theoretical Experimental and Applied Physics*, 14:130, 731–743, DOI: 10.1080/14786436608211968.

[4] R. C. Bradt, "A Strain Energy Release Rate Criterion for Formulation of the Mirror/Mist Boundary, Depicion on the (G,R) vs dC Universal Energy Diagram," in *Fractography of Glasses and Ceramivc VI*, New York: Wiley, 2012, pp. 101–112.

[5] University of Cambridge, "Fracture of Glass," Higher Education Funding Council for England, [Online]. Available: https://www.doitpoms.ac.uk/tlplib/BD5/printall.php. [Accessed January 2019].

[6] J. J. Abiasse, "The Strength of Wethered Window Glass Using Surface Characteristics," Texas Tech University, Lubbock, Texas, 1981.

[7] S. Johar, "Dynamic Fatigue of Flat Glass. Phase III Final Report (67049)," Ontario Research, Missiauga, Ontario, 1982.

[8] J. E. Minor, "Window Glass Design Practices: A Review," *ASCE Journal of the Structural Division,* vol. 107, no. 1, pp. 1–1, 1981.

[9] European Committee for Standardization, *EN 572-1 Glass in Building—Basic Soda-Lime Silicate Glass Products—Part 1: Definitions and General Physical and Mechanical Properties,* Brussels: CEN, 2012.

[10] M. Haldimann, A. Luible, and M. Overend, Structural Use of Glass, Zuirich: IABSE—International Association for Bridge and Structural Engineering, 2008.

[11] "Glass Your online industry resource," National Glass Association, 2018. [Online]. Available: https://glassmagazine.com/. [Accessed December 2018].

[12] J. D. Musgraves, H. Juejun, and L. Calvez, *Springer Handbook of Glass,* Springer International Publishing, 2019, p. 1590.

[13] M. Feldmann and R. K. et al., "Guidance for European Structural Design of Glass Components," European Union, Ispra, Italy, 2014.

[14] N. Khorasani, "Design Principles for Glass Used Structurally," Department of Building Science—Lund University, Lund, Sweden, 2004.

[15] G. Sedlacek, K. Blank, and J. Gusgen, "Glass in Structural Engineering," *Structural Engineering,* vol. 73, no. 2, pp. 17–22, 1995.

[16] D. Honfi, A. Reith, L. Vigh, and G. Stocker, "Why Glass Structures Fail?—Learning from Failures of Glass," in *Challenging Glass 4 & COST Action TU0905 Final Conference,* Lausanne, 2014.

[17] "Unexplained Glass Failure—Don't Jump to Inclusion...," n.d. Web. January 8, 2019. <https://www.glass-ts.com/news/unexplained-glass-failure-don-t-jump-to-inclusion->, [Online]. Available: https://www.glass-ts.com/news/unexplained-glass-failure-don-t-jump-to-inclusion. [Accessed December 2018].

[18] J. Schneider and J. Hilcken, "Nickel Sulphide (NiS-) Induced Failure of Glass: Fracture Mechanics Model and Verification by Fracture Data," in *International Conference at Glasstec,* Düsseldorf, Germany, 2010.

[19] M. Overend, S. Gaetano, and M. Haldimann, "Diagnostic Interpretation of Glass Failure," *Structural Engineering International,* vol. 2, pp. 151–158, 2007.

[20] R. K. Brow, "US-China Winter School," 2010. [Online]. Available: https://www.lehigh.edu/imi/teched/ChinaWS/Lecture 10 Brow Glass-Fracture.pdf. [Accessed December 2018].

[21] J. Munden, "What Is Chemical Tempering?," Barrett Limited, June 23, 2018. [Online]. Available: http://barrettlimited.com/general/what-is-chemical-tempering/. [Accessed January 2019].

[22] M. Z. Aşık and S. Tezcan, "A Mathematical Model for the Behavior of Laminated Glass Beams," *Computers and Structures,* vol. 83, pp. 1742–1753, 2005.

[23] J. Malmborg, *A Finite Element Based Design Tool for Point Fixed Laminated Glass, master's thesis,* Lund, Sweden: Lund University of Technology, 2006.

[24] J. Minor and P. Reznik, "Failure Strengths of Laminated Glass," *Journal of Structural Engineering,* vol. 116, no. 4, pp. 1030–1039, 1990.

[25] Z. Xihong, S. Yanchao, H. Hong, and C. Jian, "The Mechanical Properties of Ionoplast Interlayer Material at High Strain Rates," *Materials & Design,* vol. 83, pp. 387–399, 2015.

[26] S. J. Bennison, I. Stelzer, P. S. Davies, J. G. Sloan, H. Xiaokun, and L. Gang, "Calculation Methods for the Structural Behavior of Laminated Glass," in *Glass Performance Days,* Tampere, Finland, 2009.

[27] "Physical Properties of Sentryglas® and Butacite®," [Online]. Available: https://www.trosifol.com/fileadmin/user_upload/Kuraray_4_1_Physical_Properties_of_Sentryglas.pdf. [Accessed January 2019].

[28] Pilkigton, "Glass Handbook 2014," NSG Group, Sandomierz, Poland, 2014.

[29] S. M. Morse and H. S. Norville, "Comparison of Methods to Determine Load Sharing of Insulating Glass Units for Environmental Loads," *Glass Structures Engineering,* vol. 1, no. 1, pp. 315–329, 2016.

[30] C. V. G. Vallabhan and G. D. Chou, "Interactive Nonlinear Analysis of Insulating Glass Units," *Journal of Structural Engineering,* vol. 112, no. 6, 1986.

[31] D. Honfi and M. Overend, "Glass Structures—Learning from Experts," in *Proceedings of the Mid-term Conference on Structural Glass,* Croatia, 2013.

[32] T. Bellovary, *Glass Breakage—Failure Mode and Stress Estimation,* Vitro Architectural Glass, 2017.

[33] ASTM International, *ASTM E1300—Standard Practice for Determining Load Resistance of Glass in Buildings,* West Conshohocken, PA: ASTM International, 2016.

[34] M. Haldimann, *Fracture Strength of Structural Glass Elements—Analytical and NuStructural Glass Elements—Analytical and Numerical—Thesys No 3671,* Lausanne: EPFL, 2006.

[35] R. Lehman, "The Mechanical Properties of Glass," Rutgers University, New Brunswick, NJ, 2003.

[36] J. B. Quinn, "Extrapolation of Fracture Mirror and Crack-Branch Sizes to Large Dimensions in Biaxial Strength Tests of Glass," *Journal of the American Ceramic Society,* vol. 82, no. 8, pp. 2126–2158, 1999.

[37] J. C. Conway and J. J. Mecholsy, "Use of Crack Branching Data For Measuring Near-Surface Residual Stresses in Tempered Glass," *Journal of the American Ceramic Society,* vol. 72, no. 9, pp. 1584–1587, 72.

[38] A. M. Costa Ricardo, K. Sinan, and L. Maria, "Statistical Analysis of Structural Glass Strength," in *Conference on Design of Experiments.*

[39] "Weibull Parameter Estimation and Goodness-of-Fit for Glass Strength Data," *Structural Safety,* vol. 73, pp. 29–41, 2018.

[40] P. Rugarli, *Structural Analysis with Finite Elements,* London: Thomas Telford Ltd, 2010.

[41] *EN 1991-1-1 (2002) (English): Eurocode 1: Actions on Structures—Part 1-1: General Actions—Densities, Self-Weight, Imposed Loads for Buildings,* Bruseels: CEN European Committee for Standardization, 2002.

[42] ASTM International, *E2751/E2751M—17a—Standard Practice for the Design and Performance of Supported Glass Walkways,* West Conshohocken, PA: ASTM, 2017.

[43] Institution of Structural Engineers, *Structural Use of Glass in Buildings,* London: Institution of Structural Engineers, 2014.

[44] W. A. Young, *Roark's Formulas for Stress and Strain,* 8th ed., New York: McGraw-Hill, 2012.

[45] Standards Australian, *AS 1288—2006 Glass in Buildings—Selection and Installation,* Sydney: Standards Australia, 2006.

[46] M. C. Farmer and S. P. Lyons, "Stone Cladding Failure: The Cause and Consequences," in *Second Forensic Engineering Congress,* San Juan, PR, 2000.

[47] N. Lam and E. Gad, "An Innovative Approach to the Seismic Assessment of Non Structural," in *Australian Earthquake Engineering Society Conference,* Adelaide, Australia, 2002.

[48] R. A. Khoraskani, *Advanced Connection Systems for Architectural Glazing,* New York: Springer, 2015.

[49] FEMA—Federal Emergency Management Agency, "Reducing the Risks of Nonstructural Earthquake Damage—A Practical Guide," FEMA—Federal Emergency Management Agency, Redwood City, CA: 2012.

[50] R. D. S. Camposinhos, Manual de Dimensionamento de Fachadas Envidraçadas, Coimbra: CMM—Associação Portuguesa de Construção Metálica e Mista, 2017, p. 359.

[51] J. Holmes, *Wind Loading of Structures,* 2nd ed. London: Taylor & Francis, 2007.

[52] Z. K. Overend M, "Wind Loading on Cladding and Glazed Façades," in *International Symposium on the Application of Architectural Glass,* Munich, 2006.

[53] Building Seismic Safety Council, *Recommended Seismic Provisions for New Buildings and Other Structures (FEMA P-750),* Washington, DC: Building Seismic Safety Council of the National Institute of Building Sciences, 2015.

[54] American Architectural Manufacturers Association, *AAMA 501.6 Recommended Dynamic Test Method for Determining the Seismic Drift Causing Glass Fallout from A Wall System,* AAMA, 2001.

[55] Federal Emergency Management Agency of the U.S. Department of Homeland Security, "NEHRP Recommended Seismic Provisions,"

Building Seismic Safety Council of the National Institute of Building Sciences, Washington, DC: 2009.

[56] Federal Emergency Management Agency of the U.S. Department of Homeland Security, "2015 NEHRP Recommended Seismic Provisions Design Examples," Building Seismic Safety Council, Washington, DC: FEMA P-1051/July 2016.

[57] C. R. de, Manual de Dimensionamento de Fachadas Envidraçadas, Coimbra: CMM—Associação portuguesa de construção metálica e Mista, 2018.

[58] American Society of Civil Engineers, *Minimum Design Loads for Buildings and for Buildings* Reston, VA: ASCE, 2010.

[59] European Committe for Standardization, *Eurocode 1—Actions on structures—Part 1-3: General Actions—Snowe Loads,* Brussels: CEN, 2009.

[60] International Code Council, International Building Code, Country Club Hills, IL: International Code Council, Inc., 2017, p. 726.

[61] *EN 1990: Basis of Structural Design,* Brussels: European Standard (EN), 2002.

[62] M. Mourad, S. A. Bakhoum, and M. H. MouradMaha, "Comparison of Actions and Resistances in Different Building Design Codes," *Journal of Advanced Research,* vol. 7, no. 5, pp. 757–767, 2016.

[63] R. Zarnic, G. Tsionis, E. Gutierrez, A. Pinto, M. Geradin, and S. Dimova, "Purpose and Justification for New Design Standards Regarding the Use of Glass Products in Civil Engineering Works," European Commission Joint Research Centre, Ispra, 2007.

[64] ASTM International, Standard Terminology of Building Constructions, West Conshohocken, PA: ASTM International, 2015.

[65] W. L. Beason, T. L. Kohutek, and J. M. Bracci, "Basis for ASTM E1300 Glass Thickness Selection Procedure," Texas A&M University, 1996.

[66] C. Vallabhan and G. Chou, "Interactive Nonlinear Analysis of Insulating Glass Units," *Journal of Structural Engineering—ASCE,* vol. 112, no. 6, pp. 1313–1326, 1986.

[67] American Architectural Manufacturers Association, *AAMA TIR-11 Maximum Allowable Deflection of Framing Systems for Building Cladding Components at Design Wind Loads,* Chicago: AAMA, TIR-11, 2004.

[68] J. Ruth, "2018 Codes Clarify Deflection Limits for Glass Wall Systems," November 2, 2017. [Online]. Available: https://glassmagazine.com/article/commercial/2018-codes-clarify-deflection-limits-glass-wall-systems-1717183. [Accessed December 2018].

[69] W. A. Dalgliesh, *Structural Design of Glass for Buildings,* Calgary, Canada: NRC National Research Council of Canada, 1989.

[70] W. L. Beason and J. R. Morgan, *Journal of Structural Engineering,* vol. 111, no. 9, pp. 2058–2059, 1985.

[71] Setra—Service d'Études techniques des Routes et Autoroutes, "Assessment of Vibrational Behaviour of Footbridges under Pedestrian Loading—Technical Guide," Association Française de Génie Civil, Paris, 2016.

[72] *Draft EN 16612 Glass in Building—Determination of the Lateral Load Resistance of Glass Panes by Calculation,* Bruseels: CEN, 2017.

[73] CNR—Commissione di studio per la predisposizione e l'analisi di norme tecniche relative alle costruzioni, *CNR-DT 210 Istruzioni per la progettazione, l'esecuzione ed il controllo di costruzioni con elementi strutturali di vetro,* Roma: Consiglio Nazionale delle Ricerche, 2013.

[74] CEN, *EN 1279-5—Glass in Building. Insulating Glass Units. Product Standard,* Brussels: CEN, 2018, p. 42.

[75] H. Sucuoğlu and C. G. Vallabhan, "Behaviour of Window Glass Panels During Earthquakes," *Engineering Structures,* vol. 19, no. 8, pp. 685–694, 1997.

[76] R. Southwell and S. Skan, "On the Stability under Shearing Forces of a Flat Elastic Strip," in *Proceedings of Royal Society of London—Series A,* London, 1924.

[77] D. Mocibob, *Glass Panel under Shear Loading—Use of Glass Envelopes in Building Stabilization,* Lausanne: École Polythechnique Fédérale de Lausanne, 2008.

[78] F. Wellershoff and G. Sedlacek, "Stabilisation of Building Envelopes with the Use of the Glazing," in *Proceedings of Glass Processing Days 2005,* Tampere, Finland, 2005.

[79] J. Szalai and F. Papp, "On the Theoretical Background of the Generalization of Ayrton—Perry Type Resistance Formulas," *Journal of Constructional Steel Research,* vol. 66, no. 5, pp. 670–679, 2010.

[80] DIN Standards, *Glass in Building—Design and Construction Rules—Part 3: Point Fixed Glazing,* Berlin: Beuth Publishing DIN, 2013.

[81] Dlubal, "Dlubal Software GmbH—Structural Engineering Software for Analysis and Design," December 2018. [Online]. Available: https://www.dlubal.com/en. [Accessed December 2018].

[82] L. Martins, R. Delgado, and R. Camposinhos, "Spider Glass Behaviour under Seismic Action," in *15th World Conference on Earthquake Engineering,* Lisbon, Portugal, 2012.

[83] S. Timoshenko and J. Goodier, *Theory of Elasticity,* New York: McGraw-Hill, 1970.

[84] M. Frocht and H. Hill, "Stress-Concentration Factors around a Central Circular Hole in a Plate Loaded through Pin in the Hole," *Journal of Applied Mechanics,* 1940.

[85] W. D. Pilkey, *Peterson's Stress Concentration Factors,* 2nd ed., New York: Wiley, 1997.

[86] D. Duerr, "Pinned Connection Strength and Behaviour," *Journal of Structural Engineering,* vol. 132, no. 2, pp. 182–194, 2006.

[87] O. Pešek and J. J. Melcher, "Lateral-Torsional Buckling of Structural Glass Beams. Experimental, Theoretical and Numerical Analysis," *International Journal of Structural and Civil Engineering Research,* vol. 7, no. 4, pp. 323–330, 2018.

[88] L. C. Schmidt, "Restraints against Elastic Lateral Buckling," *Journal of the Engineering Mechanics Division,* vol. 91, no. EM6, pp. 1–10, 1965.

[89] D. A. Nethercot and K. C. Rockey, "Unified Approach to the Elastic Lateral Buckling of Beams," *The Structural Engineer,* vol. 49, no. 7, pp. 321–330, 1971.

[90] "Lateral Torsional Buckling of Glass Beams with Continuous Lateral Support (Challenging Glass Paper)," *Glass Structures & Engineering,* vol. 1, pp. 153–171, 2016.

[91] R. Nijsse *Glass in Structures: Elements, Concepts, Designs,* Basel: Birkhäuser, 2003.

[92] J. Blaauwendraad, "Buckling of Laminated Glass Columns," *Heron,* vol. 52, p. 18, 2007.

[93] P. Foraboschi, "Buckling of a Laminated Glass Column under Test," *Structural Engineer,* vol. 87, no. 1, pp. 2–8, 2009.

[94] A. Luible and M. Crisinel, "Buckling Design of Glass Elements under Compression," Swiss Federal Institute of Technology (EPFL), Technology (EPFL) Lausanne, Switzerland, 2005.

[95] A. Luible, "Stabilität von Tragelementen aus Glas—Thèse EPFL 3014," polytechnique fédérale de Lausanne (EPFL), Lausanne, 2004.

[96] K. W. H. Stamm, *Sandwichkonstruktionen—Berechnung, Fertigung, Ausführung,* Vienna: Springer, 1974.

[97] K. Sattler and P. Stein, *Ingenieurbauten 3, Theorie und Praxis,* Wien: Springer-Verlag, 1974.

[98] V. Ungureanu, "Lecture 3—Laminated Glass and Interlayers," *Sustainable Constructions under Natural Hazards and Catastrophic Events,* Lund, Sweden, 2011.

[99] J. Booker, B. Frankham, and N. Trahair, "Stability of Visco-Elastic Structural Members," *Civil Engineering Transactions,* pp. 45–51, 1974.

[100] P. Ondrej, M. Horacek, and J. Melcher, "Experimental Verification of the Buckling Strength of Structural Glass Columns," in *World Multidisciplinary Civil Engineering—Architecture—Urban Planning Symposium,* Prague, 2016.

[101] European Committee for Standadization, *EN 1993-1-1:2005,* Brussels: CEN, 2005.

[102] C. Amadio and C. Bedon, "Standardized Buckling Curves for the Verification of Glass Columns, Beams and Panels," in *XXVII ATIV Conference,* Parma, 2012.

[103] Q. Wang and Y.-W. Jane and Chung, "Griffith Theory of Fracture," in *Encyclopedia of Tribology,* Boston: Springer, 2013, pp. 1570–1573.

[104] "NiS and Spontaneous Breakage," June 12, 2013. [Online]. Available: http://www.glassonweb.com/articles/article/96. [Accessed January 2019].

Chapter 31
Machine Foundations

BY

O. SALEM ALI, Ph.D., PE *Engineering Manager, Structural Technologies*

31.1 BACKGROUND

Foundations subjected to dynamic loads are widely used in the industry, especially in the power industry for both nuclear and fossil power generation plant. Examples of foundation supporting machine include boiler feed pump foundation and the turbo-generator machines used to generate electricity in the power generation industry. Generally, machines are classified based on their operating speed, type of dynamic excitation forces, foundation type (block or framed foundation), and load transfer mechanism; that is, either the machine foundation system is directly supported on soil or is supported over piles.

31.2 CLASSIFICATION OF MACHINE BASED ON MACHINE TYPE

Machines are classified based on the type of machine into the following groups:

1. Impact machines, such as steam hammers.
2. Low-speed machines. These machines include machines used in paper industry, printing machines, and steam mills. These machines operate at speeds in the range of 50 to 600 rpm.
3. Moderate speed machines. These set of machines include machines that operate at 600 to 1500 rpm.
4. High-speed machines. These machines operate at speeds higher than 1500 rpm. Among this group of machines are the turbine and compressors, which are commonly referred to as turbo-machines. Turbo-machines are machines that operate at speeds higher than 3000 rpm, such as those used to produce electricity.

31.3 CLASSIFICATION OF MACHINE BASED ON TYPE OF EXCITATION FORCE

Most machines have rotating or reciprocating parts that develop time varying dynamic forces. During start-up and coast-down conditions, the machine passes through varying operating speeds until it reaches its final operating frequency, steady-state operation. For reciprocating machines, dynamic force is developed due to secondary unbalance forces that exist in the machine. Operating speeds for machines can range from 200 rpm, as in cases of small pumps, to 12,000 rpm, as in the case of high-speed rotary compressors. Generally, machines can be classified according to the types of dynamic excitation forces they develop into the following types:

1. Rotating machines, such as turbo-machinery, steam generators, and compressors. These machines produce a sinusoidal varying force having the following form:

$$F(\omega) = m_o e_o \omega^2 \cos(\omega t) = m_o e_o \omega^2 e^{i\omega t} \tag{31.1}$$

where m_o = machine unbalance mass

e_o = eccentricity at operating speed

ω = machine operating frequency

2. Reciprocating machines, such as steam engines, piston compressors, and pumps. These machines produce a biharmonic excitation force having the following form:

$$F(\omega) = m_o e_o \omega^2 (e^{i\omega t} + \alpha e^{2i\omega t}) \tag{31.2}$$

where α = foundation geometric constant

3. Impact machines, such as forging hammer machines and stamping machines. These machines produce an intermittent impulsive triangular loading having the following form of excitation:

$$F(t) = F_t \left(1 - \frac{t}{t_d}\right) \tag{31.3}$$

31.4 CLASSIFICATION OF MACHINE BASED ON FOUNDATION TYPE

Two types of foundation systems are typically used for machines: (1) rigid block foundation systems as shown in Fig. 31.1, which are typically used for combustion turbine generator (CTG) foundations, and (2) framed structure foundation systems, which are used in steam turbine generator (STG) foundations, as shown in Fig. 31.2. These two foundation systems exist in most of the industrial facilities and in fossil power generation plants and nuclear power generation plants.

31.5 CLASSIFICATION OF MACHINE LOAD TRANSFER MECHANISM

Most of the machine foundation is supported directly over the soil. However, when the soil conditions are not adequate to support loads from the machine foundation or when the soil dynamic deformation exceeds the allowable machine vibration limits, piles are introduced under the foundation. Piles are used to transmit the machine loads to a more adequate soil layer and to limit the vertical dynamic response of the foundation system to a certain allowable vibration amplitude.

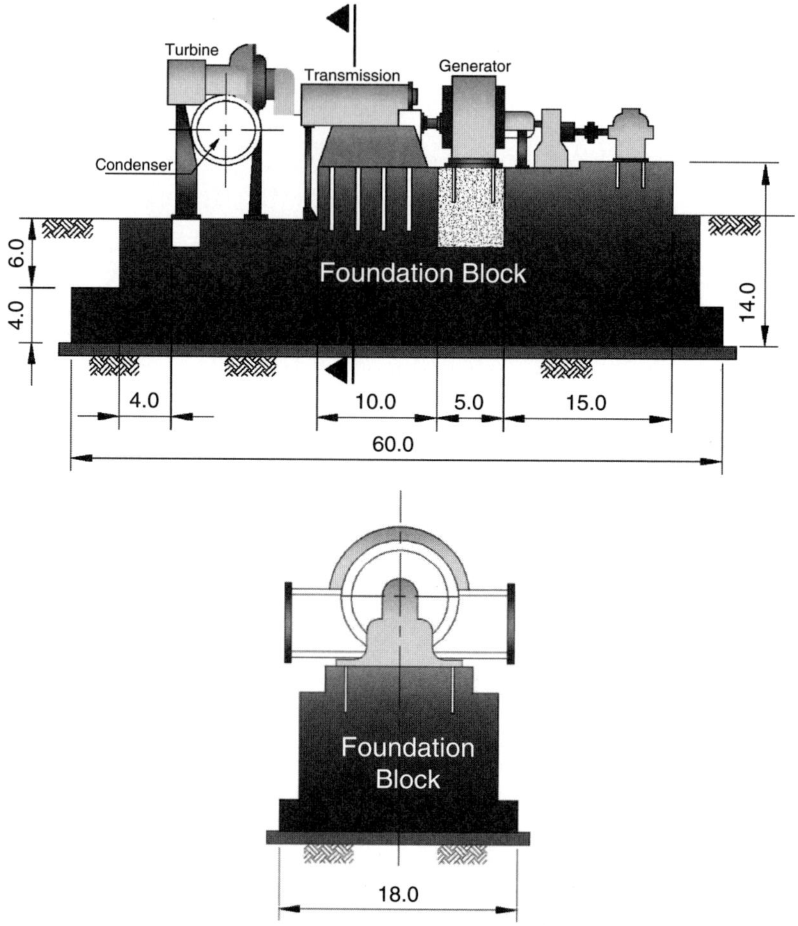

Figure 31.1 Turbo-generator machine supported on rigid block foundation.

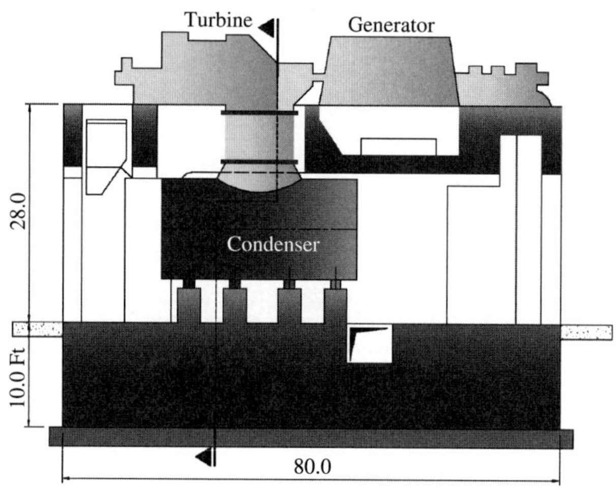

Figure 31.2 Turbo-generator machine with frame foundation.

31.6 DESIGN LIMITS FOR MACHINE FOUNDATIONS

The analysis and design of the machine-supporting foundation are governed by the dynamic requirements of the machine. These dynamic requirements involve limiting the foundation response to predetermined vibration limits at the machine-bearing support during machine steady-state operation and ramp-up and coast-down. The maximum allowable dynamic amplitude response at the machine-bearing support is limited to 10 to 12 micrometers, which is equivalent to 0.01 in./s. Richart (1962) recommended that the maximum acceleration due to machine operation should not exceed 0.5 times the acceleration due to gravity.

31.7 EFFECT OF THE SUPPORTING SOIL

During machine steady-state operation or during machine start-up and coast-down, machine vibrations are transferred through the foundation system to the supporting soil. Because of these vibrations, the soil internal structure is changed, resulting in progressive settlement of the foundation system and consequently loss of soil strength. Therefore, for loose saturated cohesionless soil that is susceptible to compaction by impact and vibrations, it is imperative to precompact the site before construction using blasting, vibroflotation, or a vibratory surface roller in order to introduce more severe dynamic loading to the soil than the expected machine dynamic loads.

The dynamic soil properties significantly influence the dynamic performance of the machine foundation system. Machines with similar foundation configuration, either block foundation or framed foundation, behave differently under different soil conditions. For example, for block foundation, the soil influence on the foundation response is predominant and depends upon the mass of the machine, the mass of the block, and the geometry of the block foundation. For framed foundations, considering the effect of soil-structure interaction induces additional modes pertaining to soil deformation at low frequencies that alter the dynamic performance of the foundation. As stated above, the effect of the soil on the foundation response is affected by the dynamic soil properties. These properties vary from site to site, from location to location, and from machine to machine as well as with variation of depth of foundation. Under the influence of dynamic forces, the foundation interacts with the soil, activating a dynamic soil-structure interaction that significantly influences the dynamic response of the machine foundation system. Soil is represented as an elastic half-space with the help of equivalent soil springs represented by elastic subgrade reaction coefficients. Typical soil parameters and dynamic properties of soil used in machine foundation design are the following:

1. Young's modulus of elasticity (E)
2. Soil shear modulus (G)
3. Poisson's ratio (v)
4. Mass density (p)
5. Soil damping (D)
6. Coefficients of uniform compression of the soil (C_u)
7. Coefficients of nonuniform compression of the soil
8. Coefficients of uniform shear of the soil
9. Coefficients of nonuniform shear of the soil
10. Energy dissipation mechanism
11. Soil mass participation in foundation vibration
12. Embedment of foundation
13. Elastic soil properties
14. Reduction in permissible soil stress

Richart (1962) modeled the soil elastic and damping properties as frequency-independent springs and dashpot, which are based on the soil mode of vibration (vertical mode, Horizontal mode, etc.). The soil elastic and damping properties were defined as a function in soil dynamic shear modulus and the foundation length and width. The values of soil spring value and damping ratio as per the Richart and Lysmer (1970) model are summarized in Table 31.1.

Table 31.1 Soil Equivalent Elastic and Damping Properties

Mode of vibration	Soil spring value	Soil damping ratio
Vertical	$K_z = \dfrac{4G\sqrt{\dfrac{LB}{\pi}}}{(1-\nu_{soil})}$	$\xi_z = \dfrac{0.425}{\sqrt{\dfrac{0.25m(1-\nu_{soil})g}{\rho_s\left(\sqrt{\dfrac{LB}{\pi}}\right)^3}}}$
Horizontal	$K_x = \dfrac{32(1-\nu_{soil})G\sqrt{\dfrac{LB}{\pi}}}{(7-8\nu_{soil})}$	$\xi_x = \dfrac{0.288}{\sqrt{\dfrac{m(7-8\nu_{soil})g}{3(1-\nu_{soil})\rho_s\left(\sqrt{\dfrac{LB}{\pi}}\right)^3}}}$
Rocking	$K_{\phi x} = \dfrac{8G\left(\sqrt[4]{\dfrac{LB^3}{3\pi}}\right)^3}{3(1-\nu_{soil})}$	$\xi_{\phi x} = \dfrac{0.15}{[1+\beta_{\phi x}]\sqrt{\beta_{\phi x}}}$ $\beta_{\phi x} = \dfrac{0.375(1-\nu_{soil})I_{\phi x}g}{\rho_s\left(\sqrt[4]{\dfrac{LB^3}{3\pi}}\right)^5}$
Rocking	$K_{\phi y} = \dfrac{8G\left(\sqrt[4]{\dfrac{BL^3}{3\pi}}\right)^3}{3(1-\nu_{soil})}$	$\xi_{\phi x} = \dfrac{0.15}{[1+\beta_{\phi y}]\sqrt{\beta_{\phi y}}}$ $\beta_{\phi y} = \dfrac{0.375(1-\nu_{soil})I_{\phi y}g}{\rho_s\left(\sqrt[4]{\dfrac{BL^3}{3\pi}}\right)^5}$
Twisting	$K_\psi = \dfrac{16\left(\sqrt[4]{\dfrac{BL^3+LB^3}{16\pi}}\right)^3}{3}$	$\xi_\psi = \dfrac{0.50}{1+2\beta_\psi}$ $\beta_\psi = \dfrac{J_\psi g}{\rho_s\left(\sqrt[4]{\dfrac{BL^3+LB^3}{16\pi}}\right)^5}$

31.8 ENERGY TRANSFER MECHANISM

The main principle underlying machine foundation design is the safe transfer of the energy content of the dynamic forces to the soil through the foundation system. Energy travels in the form of waves in all directions in the soil and gets absorbed in the soil layers itself. If the soil underneath the foundation is not a single layer and constitutes several layers, part of the energy from the lower layer will reflect back into the upper layer and thereby into the foundation system. Under dynamic loading, a machine exerts dynamic forces to the soil through the foundation. Under the influence of these dynamic forces, the foundation interacts with the soil, activating a dynamic soil-structure interaction that significantly influences the dynamic response of machine foundation system. The influence of soil-structure interaction is predominant for block types of foundations, whereas its influence is not as severe for frame foundations. The energy transmission takes place by way of three types of waves:

- Primary wave or compression wave (P-wave)
- Secondary wave or shear waves (S-wave)
- Rayleigh waves or surface wave

Rayleigh waves carry a much larger proportion of the total input energy compared to shear waves (S-wave) and primary waves (P-wave). In the analysis of the foundation-soil system, the interaction effect is represented as a lumped parameter system described by the following equation:

$$m\ddot{z} + c_z\dot{z} + k_x z = Q(t) \tag{31.4}$$

where $\ddot{z}$, $\dot{z}$, z = the acceleration, velocity, and displacement, respectively

m = equivalent lumped mass of the soil system

c_z = damping of the system

k_z = effective spring constant

The values of the effective soil mass, the spring constant, and the damping ratio in the soil will differ for each mode of soil vibration, that is, vertical mode, sliding mode, rocking mode, and torsional mode.

31.9 EFFECT OF EMBEDMENT OF FOUNDATION

Gupta (1972) and Erden and Stokoe (1975) conducted a study for the embedment effect of soil on the overall response of the foundations. They concluded that partial embedment of the foundation reduces the magnitude of motion, thus indicating greater damping. The effect of embedment causes an increase in the foundation system's natural frequency and a reduction in dynamic amplitudes. The reduction in the amplitudes could be because of either change in stiffness, change in damping, or change in the vibrating soil mass under the foundation. Whitman and Richart (1967) proposed to multiply the soil frequency-independent springs and dashpot by the following embedment coefficients at each mode of vibration in order to account for the change in stiffness and damping due to embedment effect:

Vertical $$\eta_z = 1 + 0.60(1-\nu_{soil})\frac{h}{\sqrt{\dfrac{LB}{\pi}}}$$

Horizontal $$\eta_x = 1 + 0.55(2-\nu_{soil})\frac{h}{\sqrt{\dfrac{LB}{\pi}}}$$

Rocking about X $$\eta_{\phi x} = 1 + 1.20(1-\nu_{soil})\frac{h}{\sqrt[4]{\dfrac{LB^3}{3\pi}}} + 0.20(2-\nu_{soil})\left(\frac{h}{\sqrt[4]{\dfrac{LB^3}{3\pi}}}\right)^3$$

Rocking about Y $$\eta_{\phi y} = 1 + 1.20(1-\nu_{soil})\frac{h}{\sqrt[4]{\dfrac{L^3B}{3\pi}}} + 0.20(2-\nu_{soil})\left(\frac{h}{\sqrt[4]{\dfrac{L^3B}{3\pi}}}\right)^3$$

31.10 REDUCTION IN PERMISSIBLE SOIL STRESS

For the soil to behave as an elastic material, it is necessary to maintain the total pressure (static + dynamic) exerted by the foundation on the soil within elastic limits. The dynamic pressure produced by machines affects not only the foundation directly under the machine but also other foundations that are away from machine, as the energy is transmitted through soil in all directions. It is therefore desirable to keep intended margins for even static foundations, that is, foundations for static equipment, including foundations of the building, housing, machines, and so on.

31.11 DAMPING IN SOIL

Damping is an inherent property of soil, and its influence on forced vibration response is significant during resonance or near resonance conditions. There are two types of damping developed in the soil system. The first type is the loss of energy due to propagation of waves away from the vicinity of the footing, called radiation, or geometric, damping. The second type of damping is the internal loss of energy within the soil particles due to a hysteretic or viscous effect. Different soils exhibit different damping properties depending upon their soil composition and their dynamic characteristics. For embedded foundations, the depth of embedment also influences damping properties. Damping in the soil depends on the level of strain within the soil continuum and the machine operating frequency. In other words, soil damping depends not only on stress-strain-contact pressure distribution but also on frequency of vibration of the machine. In foundation analysis as a lumped mass-spring-dashpot system, the dashpot is used to model the amount of damping present in the soil without differentiating between radiation damping and material damping. Whitman (1970) used the theory of elastic half-space to estimate the magnitude of the radiation damping present in the soil by equating the maximum amplitude of motion as determined by the elastic half-space theory to that of the single damped single degree of freedom system. The result is presented in Figs. 31.3 and 31.4, which show the variation of mass ratio with the radiation damping.

As shown in the figures above, the amount of radiation damping in the soil depends on the mass ratio (b) defined as

$$\text{Translation modes } b = \text{mass ratio} = \frac{m_o}{\rho r_o^3}$$
$$\text{Rotation modes } b = \text{mass ratio} = \frac{I_o}{\rho r_o^5} \tag{31.5}$$

where m_o = mass of the foundation block and the soil
I_o = mass moment of inertia of the foundation block and the machine
ρ = mass density of the soil
r_o = radius of the soil contact area at the based

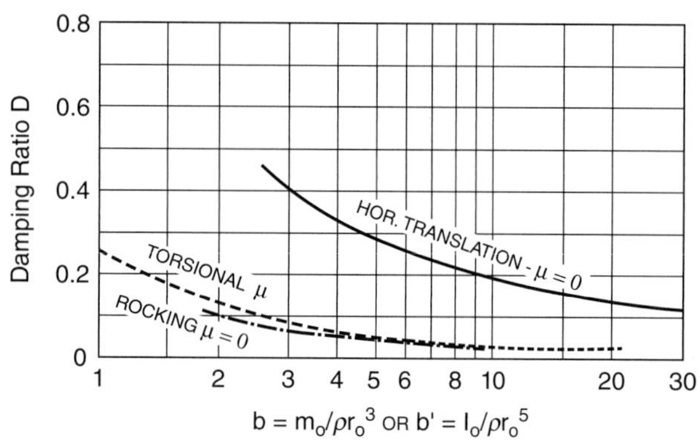

Figure 31.3 Amplitude-damping relation.

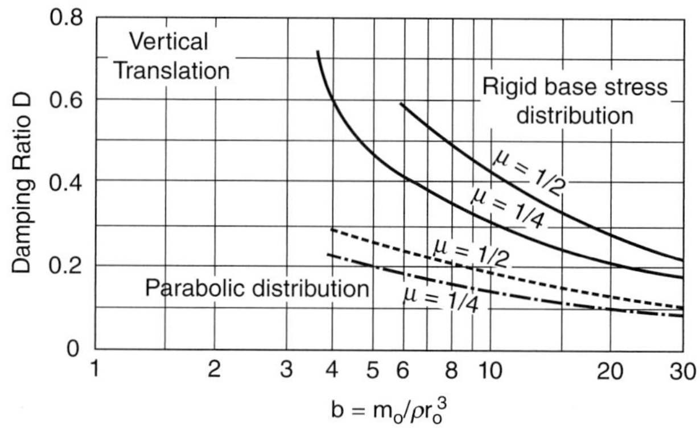

Figure 31.4 Amplitude damping for parabolic stress distribution.

Several researchers tried to estimate the total damping (geometric and material) for machine foundation supported on soil. Whitman (1973) has suggested that total damping in the horizontal, vertical, and rocking modes is as follows:

Horizontal mode of vibration:

$$D_h = \frac{0.31}{\sqrt{\dfrac{M}{\rho_s r_o^3}}} \tag{31.6}$$

Vertical mode of vibration:

$$D_h = \frac{0.49}{\sqrt{\dfrac{M}{\rho_s r_o^3}}} \tag{31.7}$$

Rocking mode of vibration:

$$D_\phi = 0.05 + \frac{0.1}{\left[\sqrt{\dfrac{I_\phi}{\rho_s r_o^5}} + \left[1 + \left(\dfrac{I_\phi}{4\rho_s r_o^5}\right)\right]\right]} \tag{31.8}$$

where M = total mass of foundation and machine
ρ_s = mass density of soil
r_o = equivalent radius of foundation
I_ϕ = mass moment of the foundation and the machine

Hardin (1965) developed an empirical relation for material damping in sandy soil:

$$D = 4.50\,\gamma_{xy}^{0.2}\sigma_o^{-0.5} \tag{31.9}$$

where γ_{xy} = shear strain
σ_o = confining pressure

Hardin's (1965) equation is applicable for soil shear strain that ranges between 10^{-4} and 10^{-6}, confining pressure that ranges between 500 and 600 psf, and machine operating frequencies less than 600 cps. Hardin concluded that the effect of material damping is negligible outside these ranges. Several attempts to evaluate the internal damping for different soil types were made by Weissmann and Hart (1961), Whitman (1970), and Barkan (1962) and are summarized below.

Type of soil	Equivalent damping ratio
Dry sand and gravel	0.03 to 0.07
Dry and saturated sand	0.01 to 0.03
Dry sand	0.03
Dry and saturated sand and gravel	0.05 to 0.06
Clay	0.02 to 0.05
Silty sand	0.03 to 0.1
Dry sand	0.01 to 0.03

For the translation and sliding mode of vibration, internal damping does not contribute much compared to the radiation damping in dissipating the soil energy. However, for rocking and torsional modes of vibration, internal damping constitutes most of soil damping compared to the radiation damping.

31.12 MODELING TECHNIQUES FOR MACHINE FOUNDATIONS

The dynamic analysis of structures is performed through the selection of an idealized model consisting of springs with adequate stiffness to capture the actual stiffness of the foundation, lumped masses representing the machine and the foundation weight representing the damping elements to model the energy dissipation mechanism. There are many techniques used in the idealization of the machine foundation. Some of these techniques adopt mathematical modeling, and others adopt finite-element method. Many researchers followed the concept of mathematical models to develop the system equations of motion using dynamic equilibrium equations in order to determine the response of the physical system. The application of this method is clarified in the developing of the equation of motion for the machine foundation shown in Figs. 31.5 and 31.6.

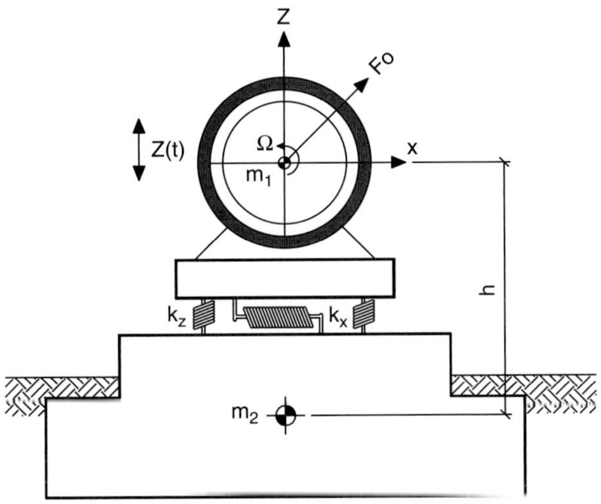

Figure 31.5 Spring-mounted machine supported on block foundation.

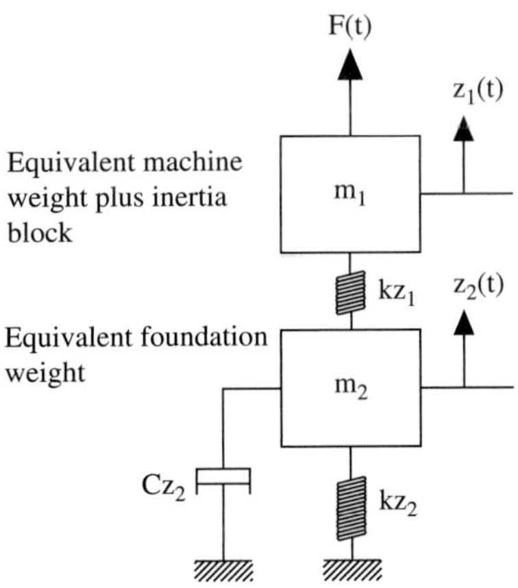

Figure 31.6 Mathematical model for the foundation.

Figure 31.5 shows a centrifugal machine supported on an inertia-block foundation and isolated from the foundation by spring supports. The vertical excitation of the foundation is considered in the mathematical idealization presented in Fig. 31.6. At any instant of time, the masses m_1 and m_2 have moved up through displacements Z_1 (t) and Z_2 (t). At any instant of time, the motion of mass m_1 from its reference position is a function in the following forces:

$$\begin{aligned} \text{Resistance by the spring support} &= R_S = k_{z1}(Z_1 - Z_2) \\ \text{Inertia force of the mass under motion} &= F_{\text{inertia}} = m_1 \times \ddot{Z}_1 \end{aligned} \quad (31.10)$$

Therefore, the motion of the mass at any time instant is governed by the dynamic equilibrium of the inertia force of the moving mass, the resistance of the spring, and the excitation dynamic force acting on mass m_1:

$$F_{\text{inertia}} = R_S \xrightarrow{\text{yields}} m_1 \times \ddot{Z}_1 + k_{z1} \times (Z_1 - Z_2) = F(t) \quad (31.11)$$

The equilibrium equation of motion for the mass m_2 will be governed by the following forces:

$$\text{Resistance by the spring support} = R_S = k_{z1}(Z_1 - Z_2)$$

$$\text{Resistance by the block foundation} = R_{S2} = k_{z2}(Z_2)$$

$$\text{Inertia force of the mass} = F_{\text{inertia}} = m_2 \times \ddot{Z}_2$$

$$\text{Damping forces: } F_d = C_{z2} \times \dot{Z}_2$$

Therefore, for mass m_2, the equation of motion is calculated based on the equilibrium of the applied forces:

$$m_2\ddot{Z}_2 + C_{z2}\dot{Z}_2 + k_{z2}Z_2 = k_{z1}(Z_1 - Z_2) \quad (31.12)$$

31.13 BLOCK-TYPE FOUNDATION

The foundation type shown in Fig. 31.5 is classified as block-type foundation; this type of foundation is commonly used in the petrochemical and power generation industries to support heavily loaded pumps and small- to medium-weight power turbines. An elastic spring is used to support the base of the machine and to damp the dynamic forces transmitted to the concrete foundation. The mathematical model, presented in Fig. 31.6, is used to study the foundation's natural frequency in the vertical direction. The mass of the machine is idealized as lumped mass m_1. The foundation mass is idealized mass m_2. The spring supporting the machine is idealized as a spring having coefficient kz_1. The elastic properties of the foundation are idealized as another spring under the mass m_2 with spring coefficient kz_2. The damping offered by the machine-supporting springs is neglected; however, the damping offered by the foundation is considered in the mathematical idealization of the system. The foundation damping is shown in the model as a dashpot connected to mass m_2. The governing equation of motion for the mass m_1 and m_2 in the vertical direction is calculated using the dynamic equilibrium method in Eq. (31.12). To calculate the model frequencies in the longitudinal direction, the mathematical model shown in Fig. 31.7 is used.

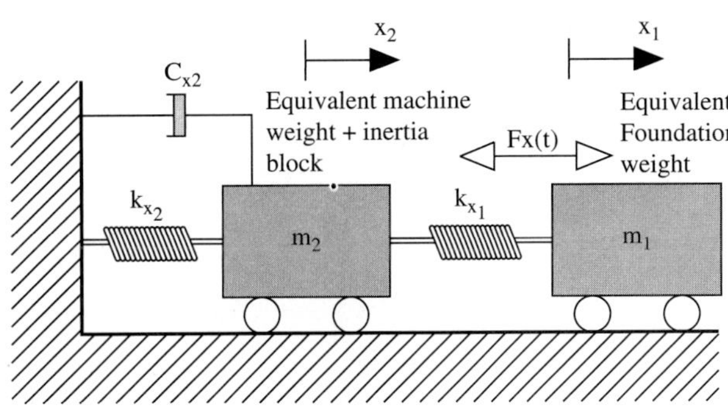

Figure 31.7 Horizontal excitation mathematical model.

The characteristic of this model is similar to the model presented in Fig. 31.6 except that all the parameters are related to the horizontal axes. The mathematical model presented in Fig. 31.8 is used to compute the machine foundation system coupled horizontal and rocking modes of vibration. The equations of motion in the horizontal and rocking modes of vibration are given by the following equations:

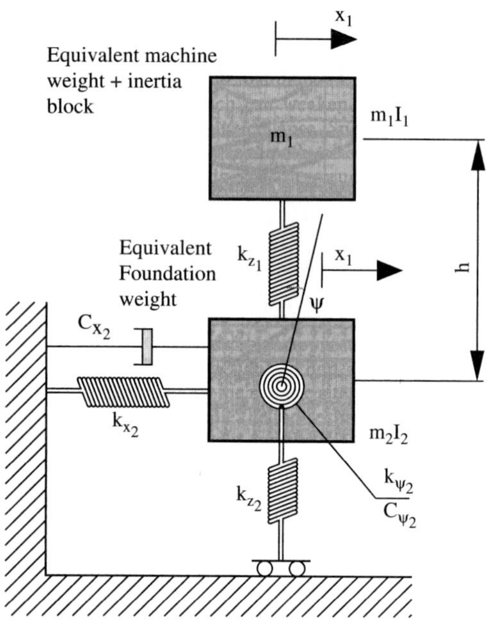

Figure 31.8 Coupled horizontal and rocking excitation.

$$m_1\ddot{x}_1 + k_{x1}(x_1 - x_2 - \psi_2 h) = F_x(t) \tag{31.13}$$

$$m_2\ddot{x}_2 + C_{x2}\dot{x}_2 + k_{x2}x_2 - k_{x1}(x_1 - x_2 - \psi_2 h) = 0 \tag{31.14}$$

$$I_2\ddot{\psi}_2 + C_{\psi 2}\dot{\psi}_2 + m_1\ddot{x}_1 h + k_{\psi 2}\psi_2 = F_x(t)h \tag{31.15}$$

The type of connection between the machine and the foundation plays an important role in its dynamic performance. For example, in the machine foundation shown in Fig. 31.5, springs are added at the interface surface between the machine and the foundation. The importance of these springs is in its low transmissibility coefficient. Spring-mounted machines have been widely used in Europe since the late 1960s. Due to the high cost of these springs, most machines are rigidly connected to the foundation through anchor bolts as shown in Fig. 31.9. Because of this rigid connection, the mathematical model is modified to reflect this change in the foundation global modes and accompanying natural frequencies. Figure 31.10 shows the mathematical model in the vertical and horizontal modes. The masses of the machine and the foundation are lumped into a single mass due to the rigid connectivity of the machine to the foundation. The horizontal and vertical stiffness of the foundation is modeled by the equivalent spring that has a coefficient equals to k_v in the vertical direction and k_x in the horizontal direction. The soil rocking stiffness is modeled as a rotational spring that has a rotational stiffness.

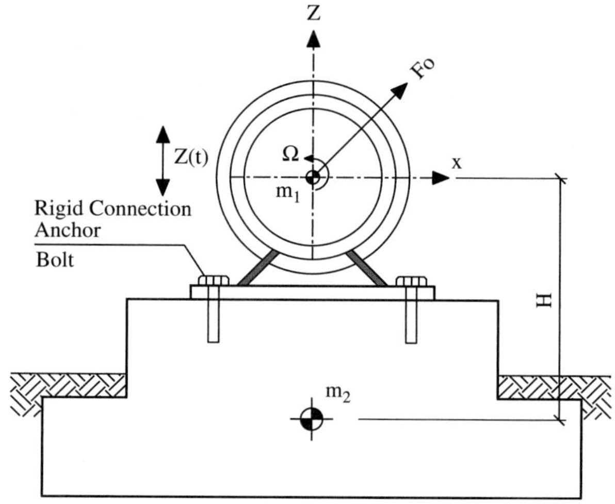

Figure 31.9 Rigid mounted machines supported on block foundation.

The equations of motion in the vertical, horizontal, and rocking direction are given by the following equations:

$$m\ddot{z} + C_z\dot{z} + k_z z = F_z(t) \tag{31.16}$$

$$m\ddot{x} + C_x\dot{x} + k_x x = F_x(t) \tag{31.17}$$

$$I_\psi\ddot{\psi} + C_\psi\dot{\psi} + m_1\ddot{x}_1 h + K_\psi\psi = F_x(t)h \tag{31.18}$$

31.14 MAT FOUNDATIONS

Mat foundations are used in cases when several small vibrating units are placed close to each other. This type of foundation is characterized by its ability to damp the vibration from the machines due to its high flexibility. The flexibility of the mat depends on the relative stiffness between the mat foundation and the supporting soil. As the flexibility of the mat increases, the ability of the foundation to damp the vibration from the machines increases in the horizontal and rocking modes.

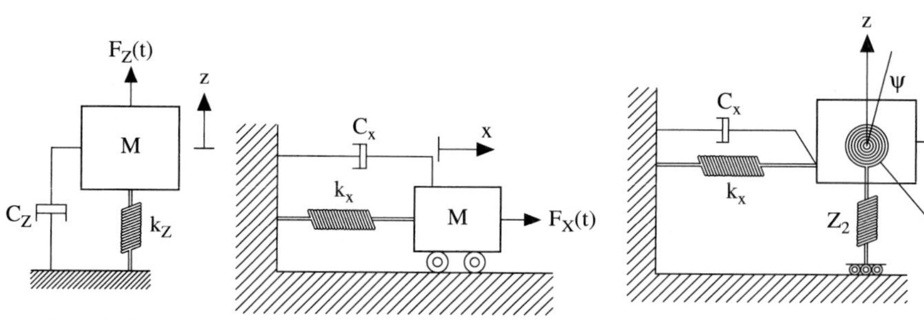

Figure 31.10 Mathematical model for a rigid mounted machine foundation.

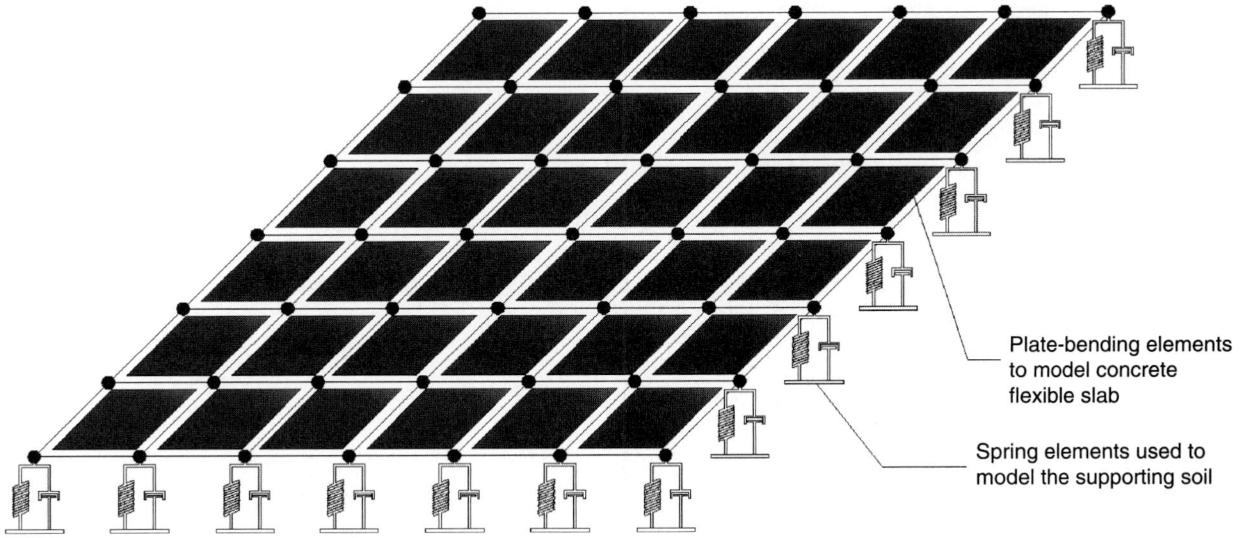

Figure 31.11 Finite-element model for mat foundation.

However, the vertical modes for these types of foundations must be investigated. In the case of one machine supported on a rigid foundation, a single mass is lumped in the machine center of gravity, and the foundation combined modes are investigated for this mass. However, if a flexible foundation mat is used to support a set of machines, the masses of these machines are lumped as a set of masses on the foundation. Figure 31.11 shows a finite-element model used to model a flexible foundation mat with the masses of the machines lumped at the intersection between the mat elements. The soil underneath the foundation is modeled as an equivalent system of spring damper elements to model the soil stiffness and damping properties. This modeling technique is appropriate for the analysis of thin concrete slab foundations directly supported on the soil. The mat is divided into triangular or rectangular finite elements that have bending capabilities. The masses of the machines are lumped on the elements' intersecting joints.

31.15 ELEVATED MACHINE FOUNDATION

One of the most important types of machine foundation is the elevated pedestal foundation. These types of foundations are used to support heavy turbo-generator machines used in nuclear and fossil power generation facilities. The general layout of this type of foundation and its mathematical idealization are shown in Fig. 31.12.

The mass used in the mathematical model is the mass of the table plus the mass of the upper half of the columns, that is, $M = M_{machine} + M_{table\ top} + 0.5M_{columns}$. The fundamental natural frequency is calculated by solving the equation of motion of a single mass spring system. The mathematical model shown in Fig. 31.13 usually results in a higher system fundamental natural frequency.

This model also excludes the effect of soil-structure interaction and neglects the damping effects resulting from the base mat and the soil. However, if the dynamic properties of the soil are identified and the foundation structure is well defined, this mathematical model can be modified to include the effect of the base mat and the soil on the system response. Figure 31.13 shows the corresponding mathematical idealization of the elevated pedestal shown in Fig. 31.12. Model A shows the coupled horizontal and the rocking modes of vibration, and model B shows the mathematical idealization for the vertical mode. The equations of motion for both models are as follows:

For model A:

$$M_m(\ddot{x}_2) + k_{z2}(x_2 - x_1 - \psi H)z = F_{x2}(t) \tag{31.19}$$

$$M_x(\ddot{x}_1) + C_x\dot{x} + k_x x_1 - k_{z2}(x_2 - x_1 - \psi H)z = 0 \tag{31.20}$$

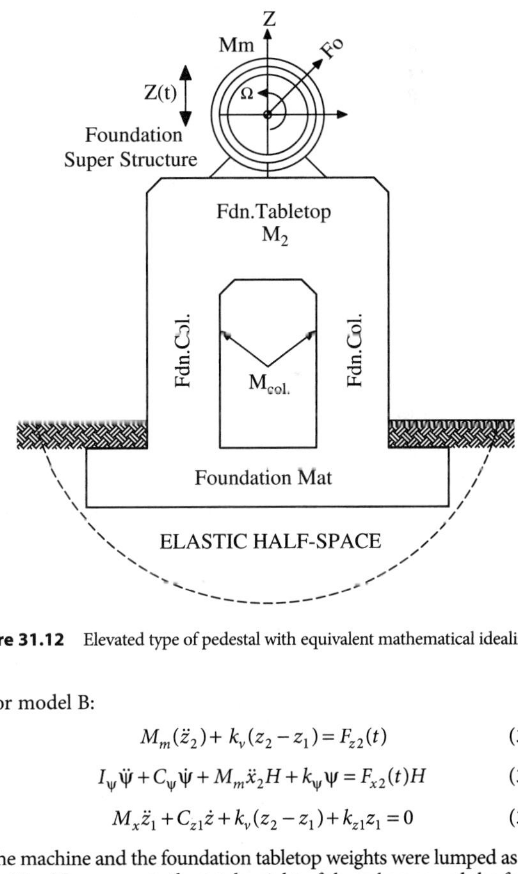

Figure 31.12 Elevated type of pedestal with equivalent mathematical idealization.

For model B:

$$M_m(\ddot{z}_2) + k_v(z_2 - z_1) = F_{z2}(t) \tag{31.21}$$

$$I_\psi\ddot{\psi} + C_\psi\dot{\psi} + M_m\ddot{x}_2H + k_\psi\psi = F_{x2}(t)H \tag{31.22}$$

$$M_x\ddot{z}_1 + C_{z1}\dot{z} + k_v(z_2 - z_1) + k_{z1}z_1 = 0 \tag{31.23}$$

The machine and the foundation tabletop weights were lumped as single mass M_m. M_x represents the total weight of the columns and the foundation concrete base mat. The flexural elastic properties of the columns are idealized by using a single spring kz_2. The rocking stiffness of the soil was idealized using the spring with a stiffness = $k\psi$. The damping properties of the soil were modeled using a dashpot with a coefficient = C_ψ. The above mathematical idealization studies the coupled horizontal and the rocking modes for the soil-structure interaction.

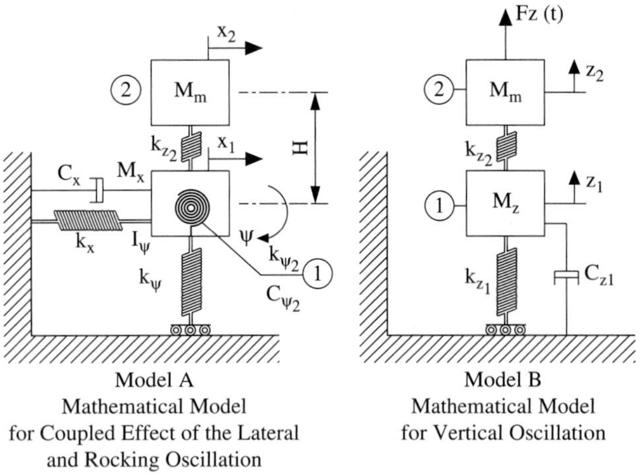

Figure 31.13 Mathematical model for elevated machine pedestal.

31.16 THREE-DIMENSIONAL FINITE-ELEMENT MODELING

The multi-lumped mass is another modeling technique used to analyze elevated foundation. The machine and superstructure dead weight is calculated and applied as a static load in the direction of the structure's deflected shape. The mass of the machine and the structure are lumped in the points where the foundation dynamic response is required. The model used to describe this foundation is shown in Fig. 31.14. The main problem of this model is that each element acts independently of each other, which means that the global foundation response is not accurately captured. In addition, the coupled effect of the soil-structure interaction is not calculated. The effect of soil on the foundation response can be very influential in many aspects. For example, if the foundation is constructed on a stiff soil, the soil will not damp the vibration resulting from the machine and will shift the foundation natural frequency close to that of the machine, which makes the structure vulnerable to resonance that might cause damage to the foundation and the machine. On the other hand, if the foundation is constructed on a flexible soil, the soil will damp the vibration from the structures in the form of inelastic deformation in the soil medium causing a differential defection of the tabletop beams, which will result in the machine's rotor misalignment. Therefore, the effect of the soil medium is a crucial factor on the dynamic response of a foundation. In most practical cases, the pedestal column bases are fixed into the base mat. This conservative approach of modeling the column-foundation-soil connectivity excludes the effect of soil medium on the foundation tabletop response and shifts the system's natural frequency toward the machine operating speed.

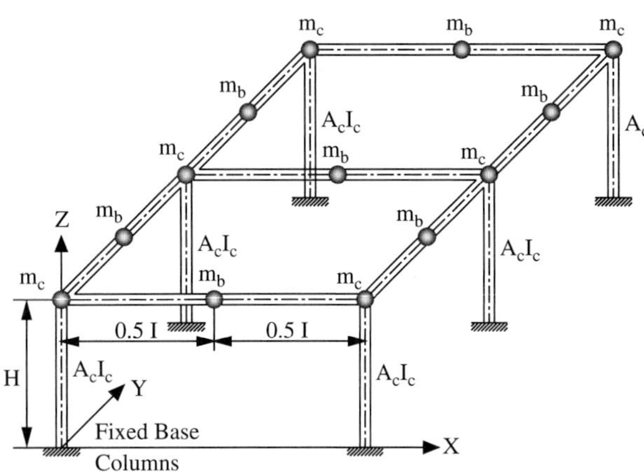

Figure 31.14 Foundation lumped mass model (fixed base).

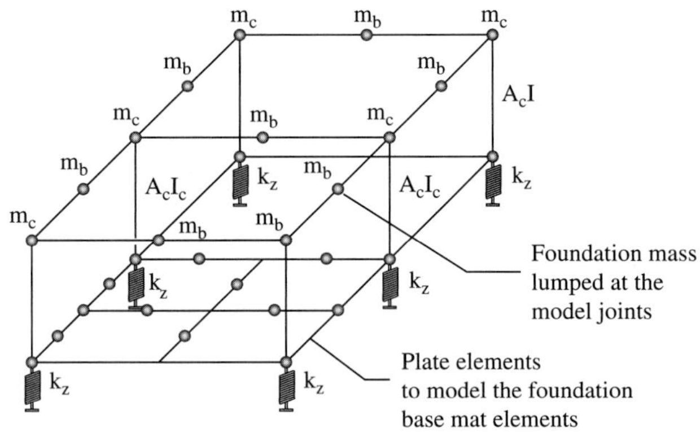

Figure 31.15 Three-dimensional foundation model with soil springs.

Figure 31.15 shows an enhanced version of the model presented in Fig. 31.14. The effect of the soil-structure interaction on the response of the elevated foundation is captured by modeling the soil as an equivalent spring-damper element in order to represent the stiffness and damping characteristics of the soil medium.

31.17 SOIL MODELING

Foundations supporting vibrating equipment experience rigid body displacements. The cyclic displacement of the foundation can have six possible modes: translation in the vertical direction, translation in the longitudinal direction, translation in the lateral direction, rotation about the vertical axis (yawing), rotation about the longitudinal axis (rocking), and rotation about the lateral axis (pitching). In machine foundation analysis, the soil is modeled as linear elastic material, and material nonlinearity is neglected. In reality, the behavior of soil departs from that of elastic material. Only at low strain levels, as in the case of soil strains resulting from machine vibration, is considering the soil as linear elastic material a reasonable approximation. Modeling the soil as linear elastic material is justified by the limited allowable dynamic amplitude imposed on the machine. For machines to have smooth operation, the allowable dynamic amplitude at a point where the machine is anchored to the foundation is limited to 10 to 12 microns. This limited dynamic amplitude will maintain the strains within the soil continuum in the elastic range with no permanent deformations in the soil medium. Therefore, the foundation vibrational modes and its dynamic response are developed assuming the soil as a linear, homogeneous, and isotropic elastic medium.

Several attempts were introduced by researchers to account for the effect of the soil in the response of the machine foundation system. Degebo (1930) introduced the concept of "in-phase effective mass" and "reduced natural frequency." The concept of in-phase effective mass postulates that the soil mass in direct contact with the foundation moves as a rigid body and in phase with the foundation system during oscillation of the foundation. Degebo defined the concept of reduced natural frequency as the frequency associated with certain foundation contact area and soil-bearing pressure. Degebo considered that these two factors are the only governing factors in the design of machine foundation and that the natural frequency of the foundation system should be maintained outside the "effective natural frequency" of the soil-foundation system. This concept was contradicted by the theory of waves, which considers that dilation waves are emitted from the machine foundation system at the soil interface into the soil medium and thus cause oscillation of the soil medium in contact with the machine foundation system. Another attempt to account for the effect of the soil in the overall response of the machine foundation system was conducted by Tschebotarioff (1951), who redefined the concept of the "reduced natural frequency" as the foundation natural frequency multiplied by the square root of the average vertical contact pressure. This concept considered the machine foundation system's natural frequency as the single most important parameter in the foundation system; in addition, the effect of the soil was solely considered in the average contact pressure. Tschebotarioff's concept did not consider the dynamic properties of the soil undergoing vibration, the type of the exciting force imposed by the machine on the soil, the types of fundamental modes of the soil foundation system, and the effect of soil damping.

Based on the work of Barkan (1962), the concept of the dynamic Winkler model was developed to capture the effect of the soil response on the machine foundation performance. Barkan presented tables and formulas to estimate the coefficient of subgrade reaction for different types of soil undergoing translational and rotational mode of vibration using field test data. These data were used to model the soil as a series of independent elastic springs that rest on rigid base as shown in Fig. 31.16.

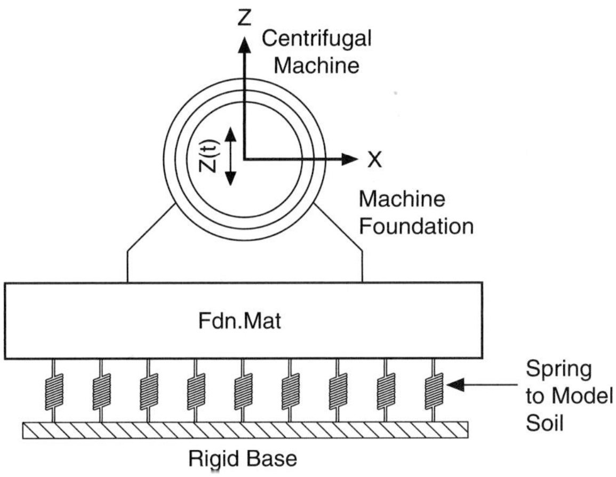

Figure 31.16 Dynamic Winkler model to model soil elastic properties.

Barkan (1962) showed that the dynamic coefficient of subgrade reaction is approximately equal to ratio of the applied pressure increment to the resulting displacement during repeated static loading:

$$K_d = \frac{\Delta P}{\delta_s} \tag{31.24}$$

where K_d = dynamic coefficient of soil subgrade reaction
ΔP = pressure increment
δ_s = soil static displacement due to repeated load

Based on this concept, Barkan applied an incremental static loading equivalent to the machine foundation system's dead and operational loading at frequencies equal to 0.001 Hz and computed the soil coefficient of dynamic subgrade used to model the effect of soil. The Barkan model neglected the effect of radiation damping in the soil, leading to very conservative design in the machine foundation system; in addition, the frequency rate at which the loading was applied was much lower than the anticipated frequency loading rate in practical applications. To capture the damping properties of the soil, Barkan worked on an improved model of the dynamic that Winkler model called the "Winkler-Voigt model." In this model, Barkan modeled the soil as a set of independent springs in parallel to viscous dampers resting on a rigid base as shown in the Fig. 31.17.

The spring and damper coefficients were back-calculated from dynamic plate load tests performed in the field by utilizing the amplitude to compute the dynamic stiffness and the resonant frequency to compute the damper coefficient. Barkan noticed a discrepancy in the results

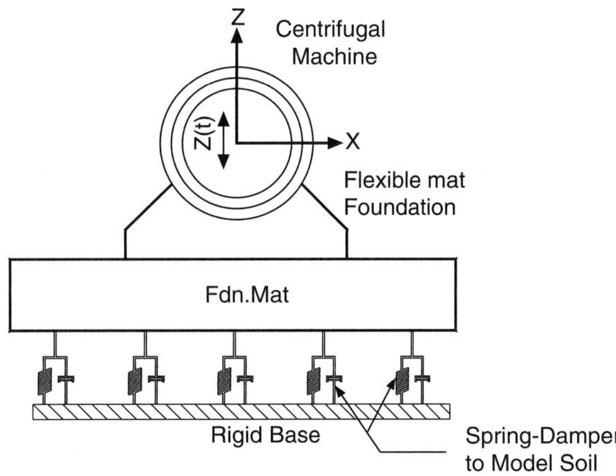

Figure 31.17 Winkler-Voigt model for soil elastic and damping properties.

using the Winkler-Voigt model and the "dynamic Winkler model." This discrepancy led Barkan to justify it by using the "in-phase mass" concept. He assumed that the soil mass in direct contact with the foundation would oscillate with the foundation. Barkan found that the concept of in-phase mass was dependent on the size of the foundation and the soil dynamic properties. For that reason, he disregarded these two soil models.

31.18 CURRENT APPROACH OF SOIL MODELING

Reissner (1936) used the development in the elastodynamic theory to compute the response of vertically loaded circular disk rested on an elastic half-space. Based on this theory, the stress waves emitted at the contact surface between the foundation and the soil propagate into the half-space in the form of body and surface waves. Thus, the machine energy is transferred from the foundation into the soil continuum, where it gets absorbed through viscous and radiation damping. The major concern in the work performed by Reissner was the assumption of uniform stress distribution under the machine foundation. This assumption was unrealistic for machine foundation due to the massive nature of the foundation that develops nonuniform contact stress distribution at the soil-foundation interface. Quinlan (1953) and Sung (1953) extended the work of Reissner by considering the cases of rigid circular foundations with different contact pressure as shown in Appendix A. Quinlan and Sung solution considered that the stress distribution under the foundation is independent from the frequency of excitation. On the contrary, the contact stress distribution under the foundation varies with the frequency of excitation. Hsieh (1962) and Lysmer (1965) presented the concept of modeling the dynamic behavior of a vertically loaded massive foundation as a single degree of freedom system, the "mass-spring-dashpot" system element. In what is known now as the "Lysmer analogy," the spring is used to model the soil elastic characteristics and the dashpot to model the soil damping. The concept proposed by Hsieh and Lysmer was proven accurate in capturing the actual response of the foundation system and is currently being employed in solving the machine foundation problem. Using this same concept, Richart and Whitman (1967) showed that all modes of machine vibration can be represented using a single degree of freedom system consisting of a lumped mass-spring-dashpot system and that the response of the system for a given mode of vibration can be represented using the following equation:

$$m\ddot{x} + C\dot{x} + Kx = P(t) \tag{31.25}$$

where $\ddot{x}, \dot{x}, x$ = acceleration, velocity, and displacement of the mass, respectively
C, K = effective damping and stiffness
$P(t)$ = machine operational dynamic load

31.19 METHODS TO COMPUTE DYNAMIC IMPEDANCE FUNCTIONS

Using the procedure outlined by Reissner (1936), the half-space is replaced by the equivalent dynamic spring element representing the elastic properties and the dashpot element representing the damping characteristic of the soil medium. The dynamic stiffness and damping coefficient of the soil medium is defined as the "dynamic impedance foundation," which is defined as the ratio between the steady-state force at the base of massless foundation and the resulting dynamic displacement.

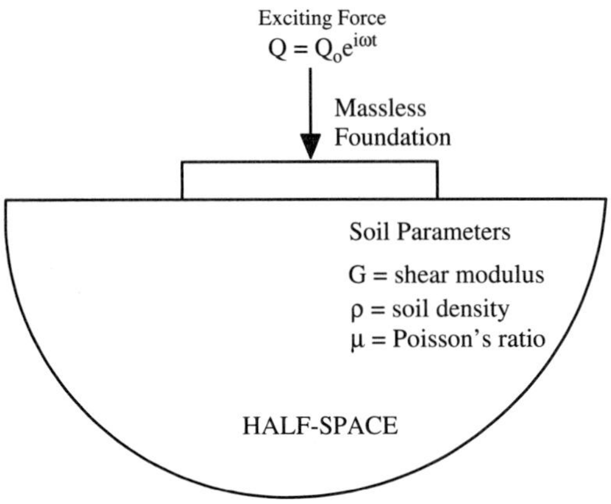

$$K_V = \frac{Q(t)}{v(t)} = \frac{Q_o}{v_o} e^{i\omega t}$$

For a single degree of freedom system $m\ddot{v} + C\dot{v} + Kv = Q_o e^{i\omega t}$

Dynamic response $v = v_o e^{i\omega t}$

$$[m(-\omega^2) + C(-i\omega) + K] v_o e^{i\omega t} = Q_o e^{i\omega t}$$

$$K_d = \frac{Q_o}{v_o} = (K - m\omega^2) + (-iC\omega)$$

$$K_d = K\left[\left(1 - \frac{\omega^2}{\omega_n{}^2}\right) + 2i\xi\frac{\omega}{\omega_n}\right]$$

Based on this equation, the dynamic impedance of a single degree of freedom system is expressed as the product of the static stiffness of the system multiplied by a complex term $\left(1 - \frac{\omega^2}{\omega_n{}^2}\right) + 2i\xi\frac{\omega}{\omega_n}$. This complex term include the dynamic characteristic of the system as the damping ratio and system inertia. For each machine foundation problem, the dynamic impedances of the soil depend on the following:

a. The ratio of the thickness of the soil layer in direct contact with the foundation to the foundation critical dimension, that is, H_{soil}/B_{fdn}

b. Shape of the foundation, either circular, rectangular, or strip foundation (In addition, the aspect ratio of the foundation affects its dynamic impedance.)

c. Soil dynamic properties (shear wave velocity of the soil deposit and its damping ratio) and soil layering properties, identified using the frequency-dependent parameter (a_o), where $a_o = \frac{\omega B}{V_s}$, where V_s = soil deposit shear wave velocity and ω = frequency

d. Poisson's ratio of the soil deposit.

e. Embedment depth of the foundation into the soil.

Generally, there are two major approaches used to compute the soil dynamic stiffness: the continuum approach and the discrete approach. Both procedures produce similar results. The continuum approach, the wave equations used to compute the dynamic displacements, is developed for each soil layer in the half-space. In the analytical continuum approach, the concept of relaxed boundaries is used in computing the soil deformation where the frictional shear traction forces at the soil-footing interface are ignored. In other words, the soil and foundation are assumed to be in perfect contact, or the contact is governed by the Coulomb friction law. In semi-analytical continuum approach, the displacement at the interface between the soil and the foundation due to a unit harmonic force ($1e^{i\omega t}$) acting in the vertical or lateral direction is computed based on Green's function.

The discrete approach in computing the soil dynamic impedance function is based on the finite-element analysis of the soil continuum. The major problem in using the finite-element approach is modeling of an infinite continuum using a finite-size finite-element mesh. One approach was to model the mesh size to extend some distance from the foundation; however, this approach resulted in the waves transmitting back in the computation continuum, resulting in an amplified response. Lysmer proposed using the absorbing boundaries instead of rigid boundaries, which has been proven effective. In this approach, viscous boundaries are used at the end of the meshed model, thus allowing the waves to be transmitted without reflecting back in the computation medium; see Fig. 31.18.

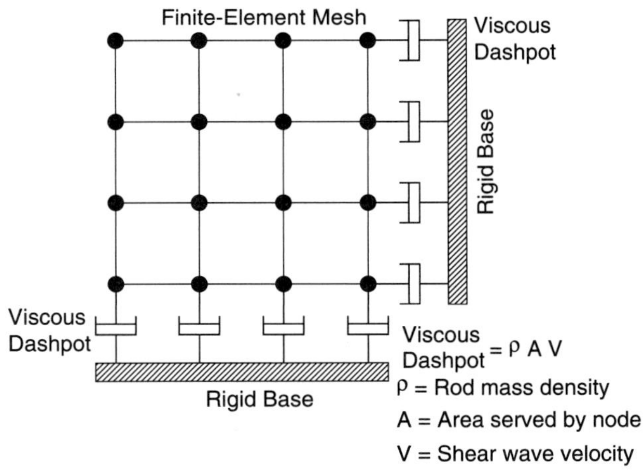

Figure 31.18 Modeling of soil boundaries.

31.20 FOUNDATIONS SUPPORTED ON PILES

Pile-supported foundations are used when the soil is too weak in bearing capacity to withstand pressures due to dynamic loads, the soil loses its strength due to the presence of the groundwater table, or the soil dynamic deformation exceeds the allowable machine vibration limits. Under these conditions, using piles under the foundation has the following advantage:

1. Transmits the machine loads to more adequate soil layers
2. Limits the vertical dynamic response of the foundation system, especially for block-type foundations
3. Increases the natural frequency of the machine foundation system, thus shifting the overall machine foundation natural frequency to outside the operating frequency of the machine
4. Reduces the dynamic amplitudes of the foundation system

Understanding the interaction between the soil and the piles is an important aspect to determine the actual response of the foundation system. The general practice for machine foundations supported over piles is to ignore the effect of the soil and to depend only on the stiffness of the piles to limit the vibration amplitudes. Generally, piles are grouped into two broad categories:

1. End-bearing piles that penetrate through soft soil layers up to a hard stratum or rock which is considered as rigid.
2. Friction piles which do not rest on hard stratum. The piles resist the applied load by means of frictional resistance developed at the soil-pile interface.

Selection of pile type, diameter, depth, and number of piles is an involved process, and the evaluation of dynamic characteristics of piles is a complex task and suffers many associated uncertainties. The machine foundation block itself serves as rigid pile cap that connects piles at the top. Evaluation of dynamic characteristics of a single pile is a difficult task, and evaluation of dynamic characteristics of a group of piles connected by a rigid pile cap becomes complex and calls for many assumptions resulting in added levels of uncertainties due to the following factors:

1. Elastic resistance of pile to vertical loads changes with time; that is, the elastic resistance offered by a fresh driven pile to vertical loads is different from the resistance offered by it after some lapse of time.
2. For floating piles, the dynamic resistance of piles to vertical loads increases with increase in the ratio of pile length to pile diameter (Novak and EL-Sharnouby, 1983).
3. The dynamic resistance of piles to vertical loads increases with the increase in the ratio of the pile Young's modulus to soil shear modulus (Novak and EL-Sharnouby, 1983).
4. Based on Barkan (1962), the elastic resistance of a pile to lateral loads primarily depends upon its cross section and its fixation length, and any increase in length of pile beyond fixation length has no influence on its lateral resistance. The fixation length of a pile is the length of the pile in the soil, where it is assumed fixed when subjected to lateral loads. This is generally of the order of 3 to 5 ft for all piles.
5. Stiffness and damping of a pile have been found to be frequency dependent; that is the stiffness and damping of piles vary with change in frequency of excitation. The reliability of response using frequency-independent stiffness and damping values, in the dynamic domain, would therefore be questionable (Gazetas, 1991).
6. Embedment of a pile cap results in increased stiffness and damping of the pile group (Gazetas, 1992).
7. Damping of pile groups are more frequency dependent than that of a single (Gazetas, 1992).
8. The dynamic group effect of piles differs considerably from the static group effect.
9. The elastic resistance of each pile in a group is a function of pile spacing. Dynamic interinfluence of piles is observed to be quite significant. The elastic resistance of each pile increases with the increase in pile spacing and decreases with the decrease in pile spacing. When the pile spacing becomes sufficiently large, which is usually in the range of five to six times the pile diameter, the elastic resistance of each pile in a group approaches the resistance of a single pile.
10. The combined stiffness for a group of n piles is not the linear summation of individual stiffness of n piles (Novak, 1974).
11. The pile is modeled as an equivalent linear spring-damper attached to the foundation in each of the six degrees of freedom. A nominal damping ratio of 5 percent is usually employed in the model. In reality, damping of pile group is more frequency dependent than of a single pile, which is not captured in the current adopted analysis techniques.

31.21 PILES SUBJECTED TO LATERAL VIBRATIONS

The soil-pile-structure interaction problem can be analyzed using either continuum models or discrete models. There are four types of soil-pile-soil interactions.

31.21.1 Beam on Elastic Foundation

This technique is based on Hetenyi's (1946) solution for the beam on elastic foundation problem. The governing equation for beam on elastic foundation is defined by the following equation:

$$EI\frac{d^4y}{d^4x} = -E_s y \tag{31.26}$$

where E_s = soil modulus of subgrade
E = pile modulus of elasticity
I = pile moment of inertia
y = pile deflection
x = depth below the soil surface

The beam on elastic foundation method is applied only to static loading on the pile segment and mostly used to determine the pile head stiffness under lateral loading. The main drawback of this method is that it does not offer a solution of an arbitrary load on the pile and that the solution is only limited to elastic soil and pile properties.

Matlock and Reese (1960) studied the effect of varying the soil modulus with depth on rigid and flexible piles laterally loaded. Davisson and Gill (1963) studied the effect of a layered soil system on the pile lateral response. Davisson and Gill considered constant shear modulus for each soil layer, but each soil layer had a different shear modulus from the layer above or below. Davisson and Gill determined that the soil shear modulus near the surface of the soil governed the pile lateral response; on the other hand, the vertical pile response of the pile was affected by all the soil modulus.

Broms (1964a and 1964b) studied the lateral response of rigid and flexible pile in cohesive and cohesionless soil. He concluded that a modulus of subgrade modulus of $9S_u$, where S_u is the soil drained shear strength, should be used for the ultimate lateral soil resistance for the undrained soil condition. For the drained soil condition, Broms suggested using a subgrade modulus linearly increasing with depth, and the ultimate soil resistance was assumed equal to $3K_pD_p\sigma'_p$, where K_p is soil passive resistance, D_p is pile diameter, and σ'_p is effective confining stress.

31.21.2 Beam on Winkler Foundation

Several researchers used the Winkler foundation assumption to evaluate the pile vertical and lateral response. In the Winkler foundation assumption, each soil layer responds independently of the adjacent soil layer. Using this assumption, the pile is modeled as a beam element, and the soil is modeled as a discrete spring-damper system. It is worth mentioning that the Winkler foundation assumption ignores the shear transfer mechanism between the soil layers; however, his approach was widely accepted and is used in pile response analysis. Modeling the pile using the Winkler assumption involves discretizing the pile-soil system into a number of points (nodes) attached to spring-dashpot to represent the soil-pile stiffness and damping characteristics. The spring-dashpot system can be linear or nonlinear; based on the p-y curves, the spring constants are determined. For static application of this method, it has the capabilities to include the effect of soil gapping and cyclic soil degradation. This method ignores the three-dimensional effect of the pile-soil interaction. For dynamic application of the beam on Winkler foundation, the free-field acceleration are determined first in a separate analysis called site response analysis. The soil continuum is modeled using finite element up to the rock stratum. The base on the rock is then excited with time-history function, and the free-field acceleration along the soil continuum is measured. Using the free field time-history acceleration, the time-history displacement is obtained, which is used to compute the dynamic soil stiffness. Lok et al. (1998) proposed conducting fully coupled analysis where the oil-pile and the superstructure are fully modeled to compute the overall system response. Penzien et al. (1964) presented a method for seismic pile response analysis based on their studies on the problem of bridge structure supported on piles driven into clay. The soil was modeled as a multi-degree of freedom system subjected to seismic base excitation at the rock elevation; the amplified free-field response at the bridge foundation elevation was computed and used as an input to the bridge structure. Penzien's work paved the foundation of today's soil-structure interaction techniques. Matlock (1970) used a discrete linear elastic pile element linked to fully nonlinear, hysteretic, degrading soil support model to study the dynamic response of piles under vertical and lateral excitation. He also extended the pile above the mudline and introduced elements to simulate the stiffness characteristics of the superstructure. To derive the exciting force on the pile elements, he conducted separate analysis to determine the free-field response at the pile elevation and used the amplified response to excite the pile elements. Matlock's solution was in the time domain and using the finite different solution. The soil stiffness is modeled using elastoplastic elements provided to the nonlinear soil properties, and the soil radiation damping is modeled using dashpots. Soil degradation was modeled using the penalty method. Kagawa and Kraft (1980) developed an equivalent linear dynamic Winkler model to compute the pile vertical and lateral seismic response. In their model, the pile was modeled using beam elements. The exciting force was applied at the end of the near-field elements as lateral ground displacements. The near-field soil was modeled as a series of vertical and lateral springs, and the soil radiation damping was modeled using dashpots having coefficient determined from the following equation:

$$c = 2\rho_s D(V_P + V_s) \tag{31.27}$$

where ρ_s = soil mass density
D = pile diameter
V_P = soil compression wave velocity
V_s = soil shear wave velocity

Bea and Audibert (1979) developed a hybrid model to study the load rate and the load cycle on piles axial and lateral response. Their work was mainly driven by the demand of the offshore structures. The Bea and Audibert model is composed of a three-dimensional, time-domain nonlinear discrete finite-element model to compute the single pile response in the near field and compute the soil free-field response in the far field. The developed hybrid finite-element model used three-dimensional elements to model the soil continuum and the beam elements to model the pile elements. The connection between the pile element and the soil continuum is achieved using spring-damper elements to model the soil elastic and radiation properties as shown in Fig. 31.19.

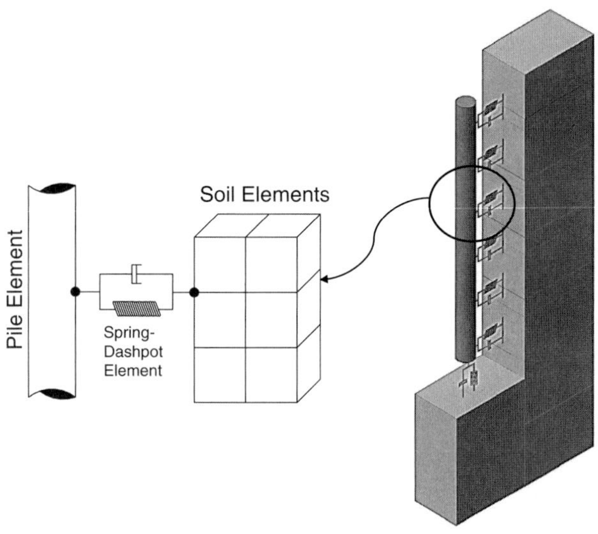

Figure 31.19 Spring damper elements to model soil-pile interaction.

Nogami et al. (1992) formulated the solution of a single pile and pile groups also using a hybrid model for the soil-pile interaction under dynamic loading as shown in Fig. 31.20. Their model incorporated nonlinear soil-pile interaction axial and lateral response in time and frequency domains accounting for degradation, gapping, pile slippage, radiation damping, and loading rate effects. The soil-pile interaction model proposed by Nogami modeled the subgrade medium around the pile using Winkler's hypothesis where the soil response at a given depth is related only to the force applied at that depth. The model proposed by Nogami accounted for the soil nonlinearity in the near-field region in the vicinity of the pile shaft and considered the soil in the far field to act elastically in the area beyond some distance from the loaded pile shaft. This distance is artificially taken as r_1 from the center of the pile shaft.

The near-field element is used to model the soil in contact to the pile and is defined by nonlinear elastic spring and consistent mass element computed as follows:

$$m_n = \frac{\rho\pi r_o^2}{6}\left(\frac{r_1}{r_o} - 1\right)\begin{bmatrix} \frac{r_1}{r_o} + 3 & 3\frac{r_1}{r_o} + 1 \\ 3\frac{r_1}{r_o} + 1 & \frac{r_1}{r_o} + 1 \end{bmatrix} \tag{31.28}$$

The far-field element parameters are shown below.

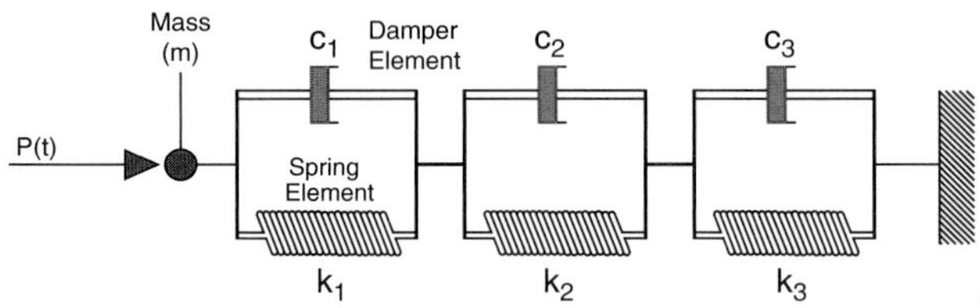

$$m = \pi\rho r_1^2 \xi_m(\upsilon) \tag{31.29}$$

$$k_1 = 3.518G\xi_k(\upsilon) \qquad c_1 = 113.0973\frac{Gr_1}{v_s}\xi_k(\upsilon) \tag{31.30}$$

$$k_2 = 3.518G\xi_k(\upsilon) \qquad c_2 = 25.133\frac{Gr_1}{v_s}\xi_k(\upsilon) \tag{31.31}$$

$$k_3 = 5.529G\xi_k(\upsilon) \qquad c_1 = 9.362\frac{Gr_1}{v_s}\xi_k(\upsilon) \tag{31.32}$$

where G = soil shear modulus
v_s = shear wave velocity
ξ_m and ξ_k = parameter dependent on Poisson's ratio (υ), as shown in Fig. 31.21

31.22 ELASTIC CONTINUUM

This analytical method is based on Mindlin's (1936) closed-form solution for the application of point loads to a semi-infinite mass. Its accuracy depends largely on the evaluation of the Young's modulus and the other elastic soil parameters. This method cannot incorporate the nonlinear soil-pile behavior; however, the equivalent linear method is used to model the soil-pile interaction. The elastic continuum method is more appropriately applied for small strain, steady-state vibration problems. However, the method is also limited in modeling layered soil profiles, and the solutions are only available for constant soil modulus and linearly increasing soil modulus. Tajimi (1966) was able to describe a dynamic soil-pile interaction using the elastic continuum theory. He developed a solution using linear Kelvin-Voigt viscoelastic stratum to model the soil elements. However, the model ignored the vertical component of the pile response. Tazoh et al. (1988) modified Tajimi's model and extended it to include the inertial effect of the superstructure.

Poulos and Davis (1980) presented the vertical and lateral solution of soil-pile interaction under the effect of static loading. They considered the effect of soil-pile degradation under cyclic lateral loading; however, their solution was never extended to account for the effect of soil-pile interaction due to dynamic loading. Swane and Poulos (1982) used a subgrade method that provides progressive soil-pile gapping with bilinear elastic-plastic springs and friction slider. Gazetas (1984) introduced the concept of substructuring as shown in Fig. 31.22, computing the soil-pile interaction impedance function. Based on this concept, the pile interacts with soil in two different mechanisms that occur within small laps of time: kinematic interaction and internal interaction. The major assumption in this approach is that the soil will remain elastic and that the concept of superposition is valid.

Kinematic interaction of the soil-pile system is the difference between the soil motion in the free field and the soil motion at the foundation.

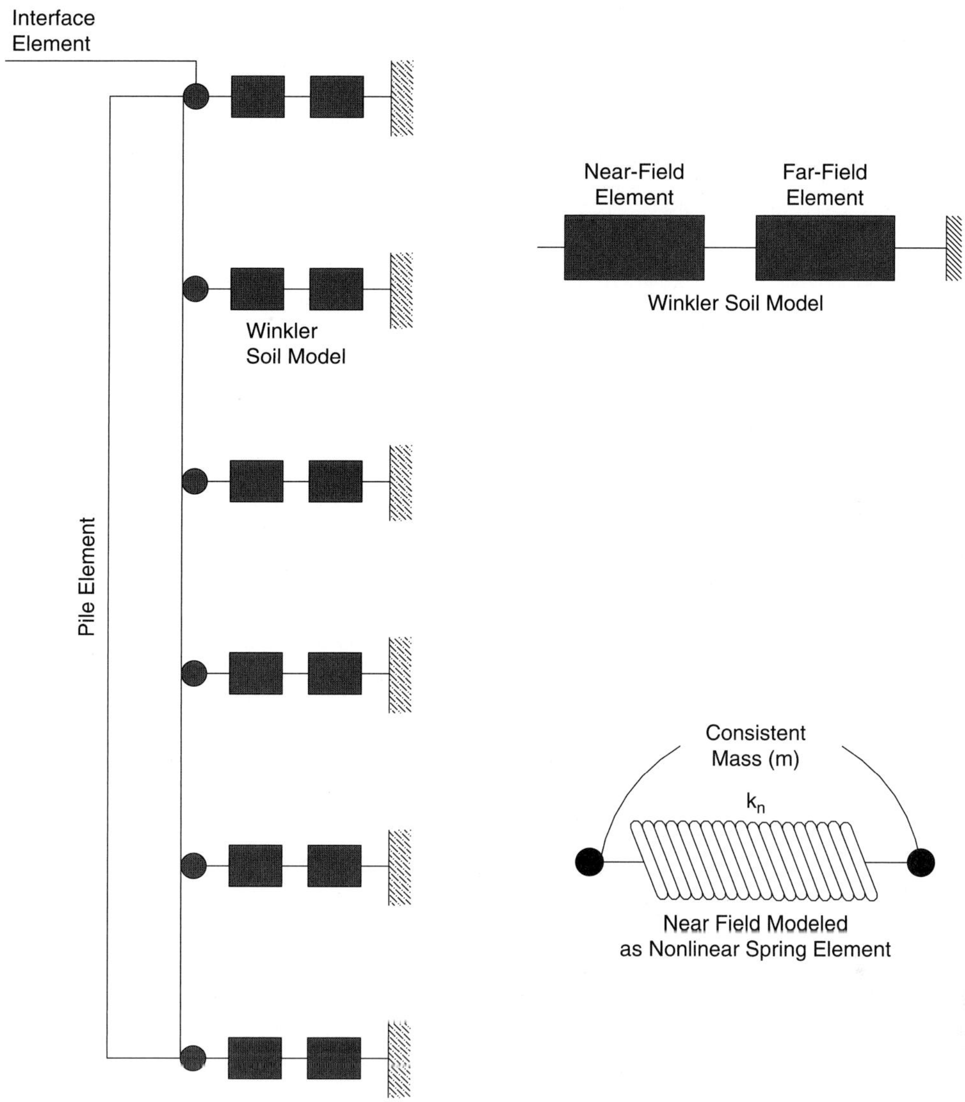

Figure 31.20 Nogami et al.'s (1992) soil-pile hybrid model.

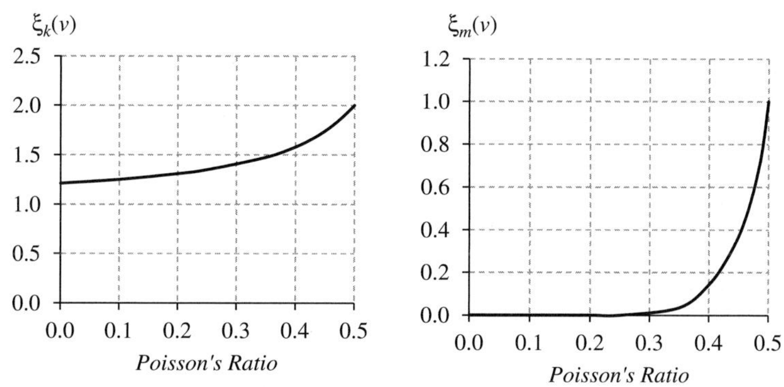

Figure 31.21 Nogami's hybrid model parameters.

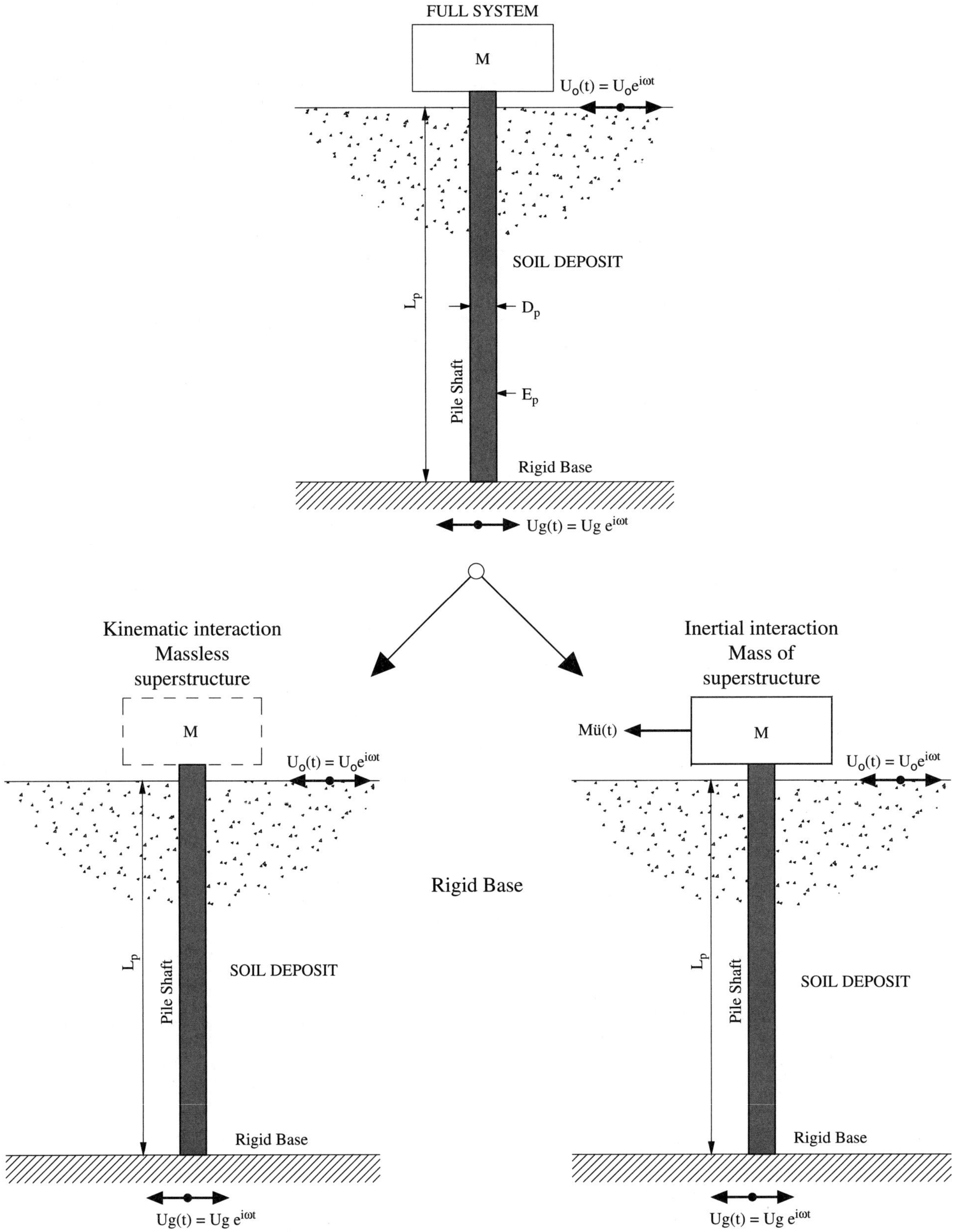

Figure 31.22 Substructuring method (Gazetas, 1984).

The free field is a point in the soil continuum far from the foundation. This difference in motion arises from the existence of a stiff member within the foundation continuum, wave incoherency, and wave inclination. The kinematic interaction is determined by calculating the response of the soil-pile system assuming massless foundation. This is usually accomplished using the concept of transfer function. The transfer function is defined as the ratio of the foundation motion to the motion in the free field. Due to the effect of kinematic interaction forces, the foundation will oscillate. Thus, the foundation inertial forces are excited, adding additional moments at the base of the foundation, which results in additional forces at the soil-foundation interface. To modify the response of the pile foundation system to account for the kinematic interaction, the amplified response resulting from kinematic analysis of the massless foundation is used as an input motion in the internal interaction analysis. The dynamic impedance function of the foundation is computed based on the response of the foundation and the associated damping at the soil-foundation interface.

Gazetas and Dobry (1984) used a substructuring method to solve the soil-pile-soil interaction problem. They divided the problem into a kinematic component and an inertial component using the finite-element method of the following form:

$$K + i\omega C = \frac{P_o}{y_d} \tag{31.33}$$

K = soil-pile stiffness
ω = exciting frequency
C = coefficient of equivalent viscous damping
P_o = amplitude of forcing function
y_d = complex amplitude of the motion

Gazetas and Dobry proposed a simplified procedure to model the radiation damping models and plane strain (see Fig. 31.23). The model for radiation damping from a laterally oscillating pile consisted of zones of waves traveling at the soil shear-wave velocity V.

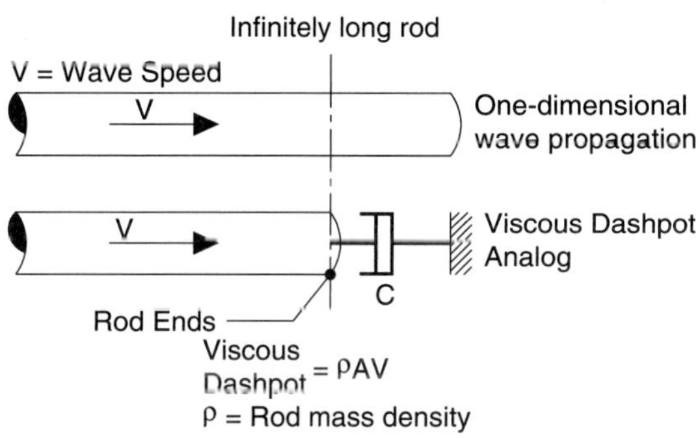

Figure 31.23 One-dimensional radiation damping models (after Gazetas and Dobry, 1984).

31.23 PILES SUBJECTED TO VERTICAL VIBRATIONS

Several researchers modeled the pile vertical dynamic response considering the effect of soil-pile interaction by either adopting the Beam-on-Dynamic-Winkler-Foundation (BDWF) approach or by using the continuum approach. In the BDWF approach, each layer of soil is assumed to respond independently of adjacent layers. The soil-pile contact is discretized to a number of points where combinations of springs and dashpots represent the soil-pile stiffness and damping at each particular layer. The disadvantage of a BDWF model is the two-dimensional simplification of the soil-pile contact, which ignores the three-dimensional components of interaction. The continuum approach accounts for the part of soil mass surrounding; the pile is considered in the analysis instead a system of spring-damper elements.

31.23.1 Soil-Pile Interaction Based on BDWF

The numerical procedure to analyze the response of the pile under the effect of dynamic loading was first proposed by Smith (1960). Smith modeled the pile in the form of a series of lumped mass connected with vertical springs as shown in Fig. 31.24. Smith solved the equation of motion and computed the response of the pile due to impact loading resulting from hammer mass on the top of the pile. The approach developed by Smith is defined as one-dimensional approach. The effect of the soil elastic-plastic and damping properties was accounted for in the form of reaction models consisting of spring-slider-dashpot elements connected in parallel.

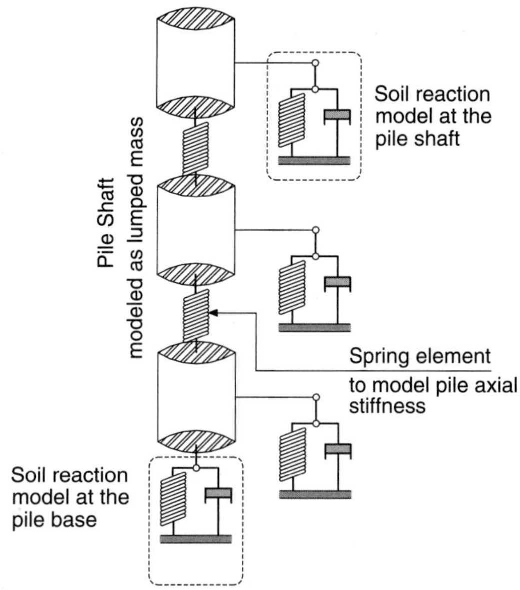

Figure 31.24 Smith's (1960) lumped mass pile-soil reaction model.

The soil reaction model proposed by Smith (1960) modeled the friction resistance and the end-bearing resistance of the piles under dynamic loading. Smith's model depends on the pile displacement u_{pile} and the pile velocity. Smith proposed the soil reaction model at the pile shaft and at the pile base as shown in Fig. 31.25:

$$k_s = \text{linear spring coefficient} = \frac{q_{sL}}{Q_s} \tag{31.34}$$

Q_s and Q_h = displacement at soil plastic deformation = 0.1 in.
c_s = dashpot coefficient
q_{sL} = shaft unit resistance
$c_s = J_s q_{sL}$
J_s and J_b = empirical constants = 0.05 s/ft and 0.15 s/ft

Several researchers updated the damping empirical constants (J_b) in Smith's soil reaction models. Forehand and Reese (1964) proposed different values for Smith's damping constant for sand and clay. The proposed values accounted for the increased damping in clay resulting from the higher viscosity that exists in cohesive soil than those that exist in sand. Aoki and Mello (1992) established a large scatter in the values of damping empirical values and found a large dependency of these values on the type of soil. As such, Mello concluded that it is difficult to create a reliable correlation between these damping values. Smith's model assumes that if pile displacement (u_{pile}) exceeds the spring force (Q_s), the slider motions are activated and the spring deformation stops. However, Smith's model did not distinguish between the hysteric damping and radiation damping due to the connectivity of the model that always engaged the dashpot. The damping proposed by Smith assumes that the damping

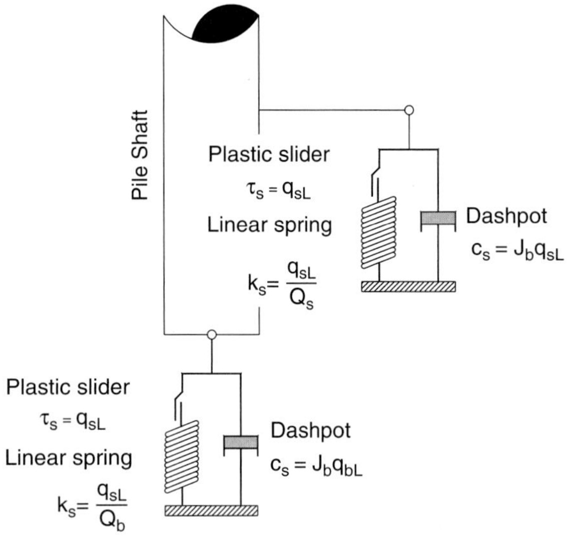

Figure 31.25 Smith's (1960) shaft and base soil reaction model.

is proportional to the static limit resistance at the soil-pile interface. However, this assumption was contradicted by the fact that radiation damping is a function in the soil stiffness and not on the soil strength.

Lysmer and Richart (1966) developed a closed-form solution for the motion of circular rigid foundation on the surface of elastic half-space subjected to vertical dynamic loading. Simons and Randolph (1985) extended Lysmer and Richart's model to compute the soil reaction model at the base of an end-bearing piles. As long as the soil strains are within the elastic limits and no plastic deformations are developed within the soil continuum, the Lysmer and Richart (1966) model is accurate in capturing the soil elastic properties at the base of the pile, especially for steady-state machine operational loading and due to impact-type loading.

Based on the Simons and Randolph (1985) solution, the total soil reaction at the base of the pile is composed the total spring reaction and the dashpot as shown in Fig. 31.26.

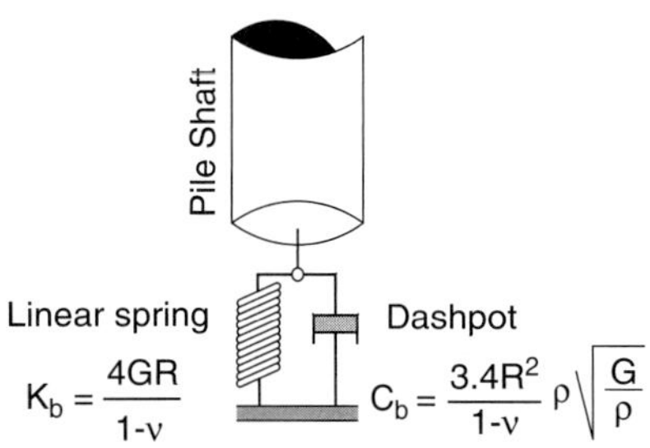

Figure 31.26 Lysmer and Richart's (1966) soil reaction base model.

The spring constant (K_b) at the base on the pile and the damping coefficient (C_b) are defined as follows:

$$K_b = \frac{4GR}{1-\nu} \tag{31.35}$$

$$C_b = \left[\frac{3.4R^2}{1-\nu}\right]\rho\sqrt{\frac{G}{\rho}} \tag{31.36}$$

where G = soil shear modulus
R = pile radius
ν = Poisson's ratio of the soil
ρ = mass density of the soil material

The base model developed by Lysmer and Richart (1966) was validated by Deeks and Randolph (1995). They matched the results of a finite-element modeling of the pile with several rheological model configurations. The model proposed by Deeks and Randolph (1995) was able to validate the base model proposed by Lysmer and Richart (1966) for impact loading on an elastic half-space when the soil Poisson's ratio is less than 1/3. Deeks and Randolph tried to capture the soil vibrating mass by including two masses, m_o and m_1, as shown in Fig. 31.27. The model parameters are defined as follows:

$$K_o = \frac{4GR}{1-\nu} \tag{31.37}$$

$$C_o = \frac{4R^2}{1-\nu}\sqrt{\rho G \beta_o} \tag{31.38}$$

$$C_1 = \frac{4R^2}{1-\nu}\sqrt{\rho G \beta_1} \tag{31.39}$$

$$m_o = \frac{4R^3\rho}{1-\nu}\alpha_o \tag{31.40}$$

$$m_1 = \frac{4R^3\rho}{1-\nu}\alpha_1 \tag{31.41}$$

where R = pile radius
G = soil shear modulus
ν = soil Poisson's ratio
β_o = parameter between 0.75 and 0.87
$\beta_1 = 0.3$ for $\nu = 0$ and $\beta_1 = 0.83$ for $\nu = 0.45$
α_o and α_1 parameters that range between 0.16 and 0.25

Randolph and Simons (1986) enhanced Lysmer and Richart's base model to account for the plastic deformation developed at the base of the pile due to high dynamic impact loading resulting from driving hammer loading. They added a slider to Lysmer and Richart's model as shown in Fig. 31.28 in order to decouple the soil plastic zone from the rest of the soil once the strain plastic strain is reached.

Randolph and Simons (1986) also proposed a model for the soil shaft reaction. The model consisted of two systems connected in series as shown Fig. 31.29. The first system is composed of spring and dashpot connected in parallel to represent soil radiation damping. This system

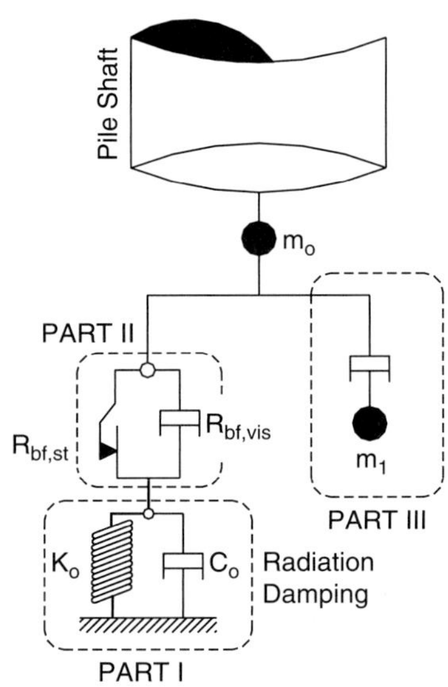

Figure 31.27 Deeks and Randolph's (1995) soil reaction base model.

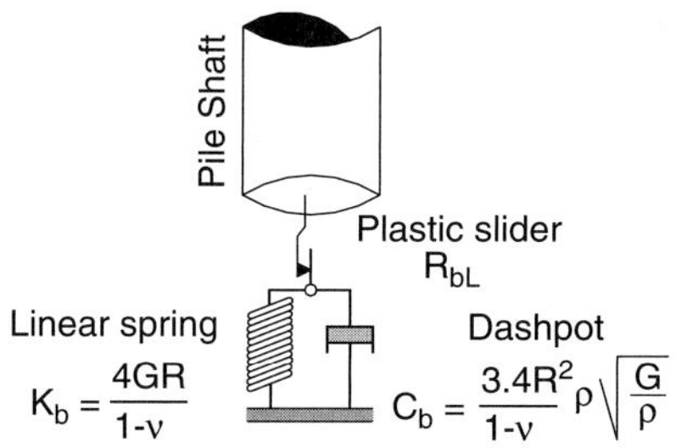

Figure 31.28 Randolph and Simons's (1986) soil reaction base model.

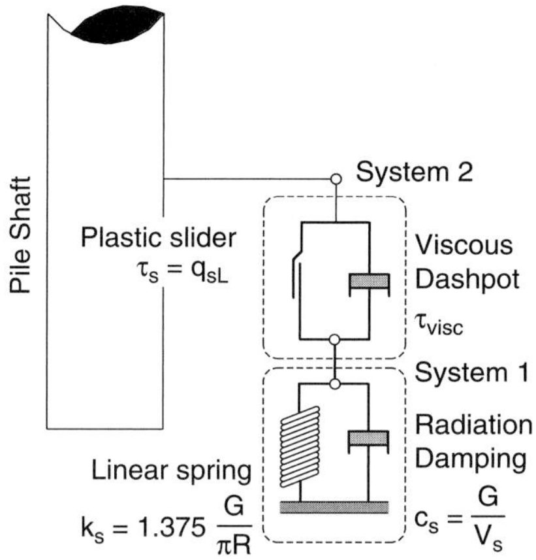

Figure 31.29 Randolph and Simons's (1986) soil reaction shaft model.

is then connected to another system composed of a slider and dashpot connected in parallel to represent soil viscous properties. Randolph and Simon's model parameters are defined as follows:

Radiation damping coefficient:

$$c_s = \frac{G_{max}}{V_s} \tag{31.42}$$

Elastic spring coefficient:

$$K_s = 1.375\frac{G}{\pi R} \tag{31.43}$$

Viscous damping coefficient:

$$\tau_{sf} = \tau_{static} + \tau_{viscous} = q_{sL}m\left(\frac{du_{pile}}{dt} + \frac{du_{soil}}{dt}\right)^n \tag{31.44}$$

where τ_{static} = pile shaft static resistance
$\tau_{viscous}$ = pile shaft viscous resistance
G = soil maximum shear modulus
V_s = soil shear wave velocity
R = pile radius
q_{sL} = shaft unit resistance
m = parameter taken as 0.3–0.5 for sand and as 2–3 for clay
n = parameter taken as 0.2–0.5
u_{pile} = pile displacement
u_{soil} = soil displacement

Based on Randolph and Simons, pile slippage will occur only when the stresses in the linear spring and the radiation dashpot are below the shaft unit resistance (τ_{static}). As long as the stresses in the pile are below this value, the pile and the soil will move together, and pile slippage will not occur. If slippage occurs, the motion of the system will be governed by the slider and the viscous dashpot. Holeyman (1988) proposed a soil

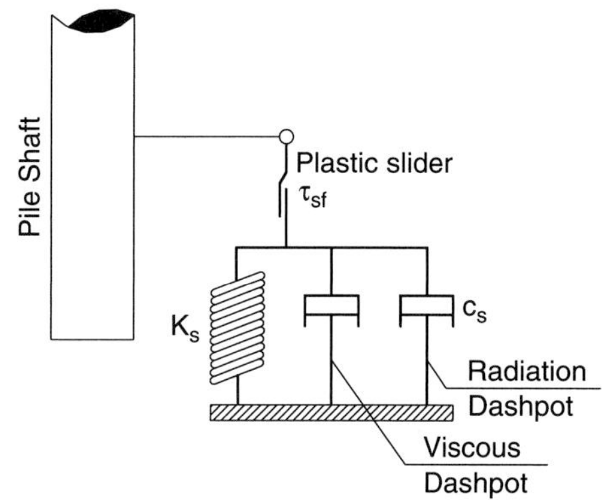

Figure 31.30 Holeyman's (1988) soil shaft reaction model.

reaction model as shown in Fig. 31.30. The model consists of viscous dashpot and radiation dashpot connected in parallel. Holeyman's model assumed that the soil viscous damping is active before sliding, which is different from Randolph and Simons's assumption. This means that the soil elastic and damping characteristics will only initiate once the pile shaft slips. Holeyman considered that the spring is only active under static loading of the pile. The model also proposed that the slider will be activated once the sum of the reaction of the elastic springs and the dashpots exceeds the slider strength (τ_{sf}) given by the following equation:

$$\tau_{sf} = q_{sL}\left(1 + J\left(\frac{du_{pile}}{dt} + \frac{du_{soil}}{dt}\right)^N\right) \tag{31.45}$$

The reaction component due to spring and the viscous dashpot is given as

$$\tau_{spring\_visc} = \frac{G_{max}}{R}\frac{u_{pile}}{ln\left(\frac{2L(1-\upsilon)}{R}\right)}\left[1 + j\left(\frac{du_{pile}}{dt}\right)^N\right] \tag{31.46}$$

where N = parameter taken as 0.2
u_{pile} = pile displacement
u_{soil} = soil displacement
R = pile radius
L = pile length
υ = soil Poisson's ratio

El-Naggar and Novak (1994) used the model developed by Randolph and Simons and included the effect of soil nonlinearity for the soil band in direct contact with the pile shaft. Under the effect of impact loading, such as the impact loading resulting from the pile driving, El-Naggar and Novak divided the soil zone around the pile into two zones. The first zone is the inner zone, which is in direct contact with the pile element that will exhibit strong nonlinear deformation at the pile-soil interface. The second zone is the outer zone, where the soil will behave elastically and will not exhibit any nonlinear behavior. The model developed by El-Naggar and Novak is shown in Fig. 31.31. At the interface with the pile, El-Naggar and Novak provided a slider element to capture the slip and the relative motion between the pile and the soil due to the soil nonlinear behavior.

El-Naggar and Novak defined the model parameters as follows:

$$K_{n1} = \frac{2\pi G_s}{ln\left(\frac{\frac{r_1}{r_o} - \frac{\tau_o}{\tau_f}}{1 - \frac{\tau_o}{\tau_f}}\right)} \tag{31.47}$$

where G_s = soil shear modulus
r_1 = radius of the outer layer

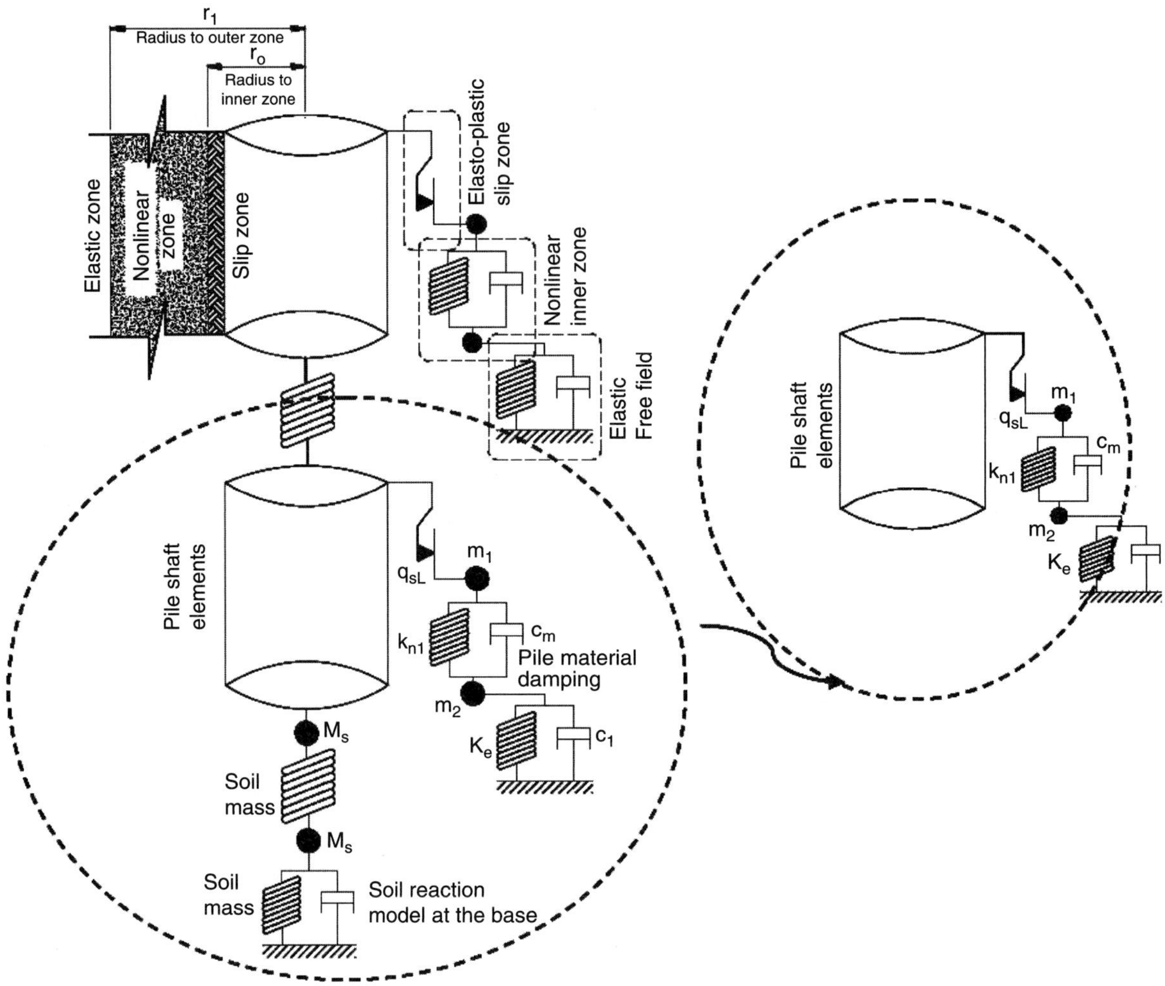

Figure 31.31 El-Naggar and Novak's (1994) pile shaft model.

r_o = radius of the inner layer

τ_o = shear stress at the inner layer, at r_o

τ_f = soil ultimate shear stress

c_m = pile material damping

q_{sL} = soil shear capacity

m_1 and m_2 = soil mass in the inner and outer zone, idealized as lumped mass

k_1 = soil stiffness in the elastic region (outer zone)

c_1 = soil damping coefficient in the elastic region (outer zone)

$$k_1 = \frac{G_{\max}}{2\pi r_o}\left[\frac{J_1(a_o)J_0(a_o)+Y_1(a_o)Y_0(a_o)}{J_o^2(a_o)+Y_o^2(a_o)}\right]$$

$$c_1 = \frac{G_{\max}}{2\pi V_s}\left[\frac{4}{J_o^2(a_o)+Y_o^2(a_o)}\right]$$

J_1, J_0 = Bessel functions of first kind of order 0 and 1

Y_1, Y_0 = Bessel functions of second kind of order 0 and 1

$$\tau_{sf} = q_{sL}\left(1+J\left(\frac{du_{\text{pile}}}{dt}+\frac{du_{\text{soil}}}{dt}\right)^N\right) \tag{31.48}$$

$$\tau_{\text{spring\_visc}} = \frac{G_{\max}}{R}\frac{u_{\text{pile}}}{ln\left(\frac{2L(1-\upsilon)}{R}\right)}\left[1+j\left(\frac{du_{\text{pile}}}{dt}\right)^N\right] \tag{31.49}$$

where N = parameter taken 0.2

u_{soil} = soil displacement

u_{pile} = pile displacement

R = pile radius

L = pile length

υ = soil Poisson's ratio

The model proposed by El-Naggar and Novak included all the features of the model developed by Randolph and Simons and included the effect of the soil nonlinear behavior near the pile. However, the extension of the slippage zone from the pile and the inner zone diameter was not clear. El-Naggar and Novak proposed to define the extent of the soil outer zone to be 1.1 pile diameters.

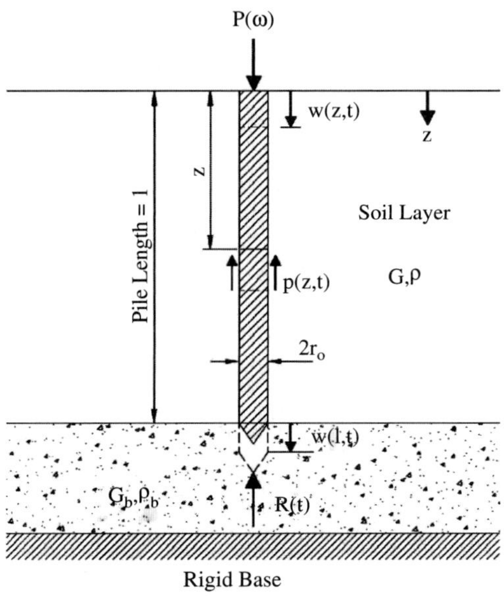

Figure 31.32 Novak's model for single pile.

31.23.2 Soil-Pile Interaction Based on Continuum Approach

Novak (1974) presented an approximate analytical expression for the dynamic stiffness of piles based on linear elasticity (Fig. 31.32). Novak assumed the following:

1. The pile is assumed vertical, cylindrical, and moving as a rigid body.
2. The pile is perfectly connected to the soil (no separation is allowed between the rigid cylinder and the soil medium).
3. The soil above the pile tip behaves as elastic layers composed of infinite thin independent layers, the plane strain soil with elastic wave propagating only in the horizontal direction.
4. The soil reaction acting on the pile tip is equal to that of an elastic half space.
5. Motion is small, and the excitation of the pile is harmonic.
6. The motion of pile is small, and excitation is harmonic; that is the exciting force $= Q_o e^{i\omega t}$.
7. Only applicable for $a_o = \dfrac{\omega r_o}{V_s}$ ranges from 0.1 to 0.8, best results at $a_o \approx 0.3$.

The complex stiffness is the force that will generate a unit dynamic displacement of the pile head.

Vertical vibration of the pile:

$$w(z,t) = w(z)e^{i\omega t} \tag{31.50}$$

Soil reaction at the pile tip:

$$p(z,t) = G(S_{w1} + iS_{w2})w(z,t)dz \tag{31.51}$$

where

$$S_{w1} = \frac{J_1(a_o)J_0(a_o) + Y_1(a_o)Y_0(a_o)}{J_o^2(a_o) + Y_o^2(a_o)} \tag{31.52}$$

$$S_{w2} = \frac{4}{J_o^2(a_o) + Y_o^2(a_o)} \tag{31.53}$$

J_1, J_0 = Bessel functions of first kind of order 0 and 1
Y_1, Y_0 = Bessel functions of second kind of order 0 and 1

The equation of motion of the pile/soil system is given by

$$m\frac{\partial^2 w(z,t)}{\partial^2 t} + c\frac{\partial w(z,t)}{\partial t} - E_pA\frac{\partial^2 w(z,t)}{\partial^2 z} + G(S_{w1} + iS_{w2})w(z,t) = 0 \tag{31.54}$$

The solution yields:

$$w(z) = \cos\Lambda\frac{z}{l} + C(\Lambda)\sin\Lambda\frac{z}{l} \tag{31.55}$$

$$\Lambda = l\sqrt{\frac{1}{E_pA}\left[m\omega^2 - GS_{w1} - i(c\omega + GS_{w2})\right]}$$

C_{w1} and C_{w2} = Bycroft Coefficient

$$\text{Axial force in the pile} = N(z) = E_pA\frac{dw(z)}{dz}$$

The complex stiffness is the force that will generate a unit dynamic displacement of the pile head:

$$N(0) = E_pA\frac{dw(0)}{dz} = \frac{E_pA}{l}F(\Lambda) \overset{\text{where}}{\Rightarrow} F(\Lambda) = -\Lambda C(\Lambda) = F(\Lambda)_1 + iF(\Lambda)_2$$

where $F(\Lambda)_1$ = real stiffness which defines the pile–soil combined stiffness
$F(\Lambda)_2$ = imaginary component of $F(\Lambda)$ representing the system damping

Novak defined the pile equivalent spring constant and damping coefficient as shown in Eqs. 31.56 and 31.57:

$$k_z = \left(\frac{E_pA}{R}\right)f_{z1} \tag{31.56}$$

$$c_z = \left(\frac{E_pA}{\sqrt{\frac{G}{\rho}}}\right)f_{z2} \tag{31.57}$$

where E_p = modulus of elasticity of pile material
A = pile cross-sectional areas
G = shear modulus
ρ = density of soil
f_{z1} and f_{z2} = non-dimensional parameter

Novak defined the variation of the parameters f_{z1} and f_{z2} with the ratio of the pile length and the pile radius (L/R) using different ratios of the pile Young's modulus and soil shear modulus (E_p/G) for end-bearing piles. The value of f_{z1} and f_{z2} are given from the following relation:

$$\frac{L_{\text{pile}}}{R_{\text{pile}}} = 20 \qquad f_{z1} = 3.75\left(\frac{V_s}{V_p}\right)^2 - 0.05\left(\frac{V_s}{V_p}\right) + 0.0501$$

$$f_{z2} = 15.345\left(\frac{V_s}{V_p}\right)^{2.0928}$$

$$\frac{L_{\text{pile}}}{R_{\text{pile}}} = 50 \qquad f_{z1} = 6.25\left(\frac{V_s}{V_p}\right)^2 - 0.05\left(\frac{V_s}{V_p}\right) + 0.0199$$

$$f_{z2} = -10\left(\frac{V_s}{V_p}\right)^2 + 1.5\left(\frac{V_s}{V_p}\right) - 0.012$$

$$\frac{L_{\text{pile}}}{R_{\text{pile}}} = 100 \qquad f_{z1} = -3.75\left(\frac{V_s}{V_p}\right)^2 + 0.45\left(\frac{V_s}{V_p}\right) + 0.0061$$

$$f_{z2} = 1.4\left(\frac{V_s}{V_p}\right) - 0.0083$$

where V_s = soil shear wave velocity $= \sqrt{\dfrac{G_{\text{soil}}}{\rho_{\text{soil}}}}$

V_p = pile compression wave velocity $= \sqrt{\dfrac{E_{\text{pile}}}{\rho_{\text{pile}}}}$

G_{soil} = soil shear modulus
E_{pile} = pile Young's modulus
ρ_{soil} and ρ_{pile} = mass density of soil and pile

The coefficients f_{z1} and f_{z2} are given in Figs. 31.33 and 31.34 for friction piles.

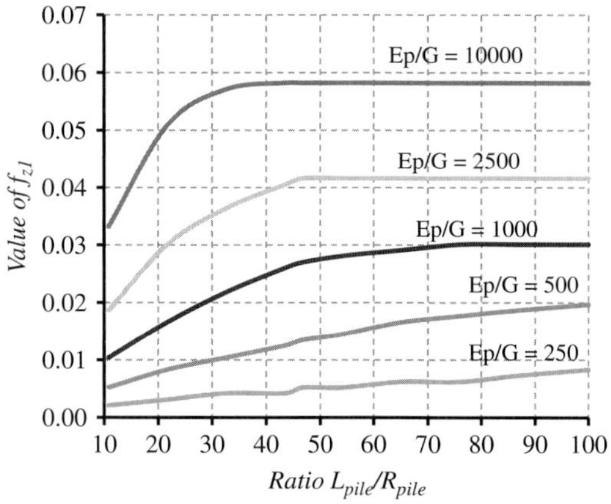

Figure 31.33 Variation of f_{z1} with L/R and E_p/G for friction piles.

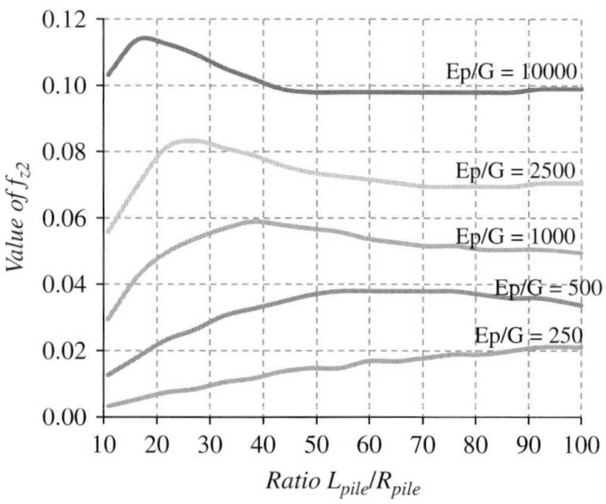

Figure 31.34 Variation of f_{z2} with L/R and E_p/G for friction piles.

Novak (1974) mentioned that the values of stiffness and damping coefficient are accurate for dimensionless frequency parameters around $a_o = 0.3$; around this value, the stiffness and damping are reasonably stationary, and the numerical result values within this range are suitable for most applications. Gazetas (1992) mentioned that Novak underestimated the stiffness and overestimated the damping coefficient using the plane-strain elastodynamic model of the pile embedded in the soil continuum. Chowdhury and Dasgupta (2009a) defined the limitation in the Novak continuum solution as follows:

1. The solution does not take into account the inertial effect of the pile.
2. Extrapolation is required when design data are out of range of a_o.

Holeyman (1988) developed a pile shaft and base model to consider the effect of the soil mass vibrating with a pile during pile driving as shown in Fig. 31.35. In his model, Holeyman included part of the soil during pile-driving analysis as a continuum instead of considering the soil effect in the form of spring and dashpot. In his pile base model, Holeyman considered a truncated cone of soil under the pile base, which will control the reaction model. The truncated cone is then discretized into lumped masses connected to each other with nonlinear spring. For the shaft model, Holeyman discretized the soil surrounding the pile into a series of cylinders each having mass lumped in its center. The only mode of deformation considered between these was the shearing deformation. The soil cone is idealized as mass connected to spring elements. The spring element is defined as a function in the soil shear modulus using the following equation:

$$K_z = \frac{G}{0.85(1-\upsilon)}\pi r^2 \tag{31.58}$$

The far field absorbing soil boundary was defined by Holeyman as a spring dashpot elements using Novak solution as follows:

$$K_f = \frac{G_{\max}}{\pi r^2} \tag{31.59}$$

$$C_f = \frac{G_{\max}}{V_s} \tag{31.60}$$

The dynamic equilibrium equation of motion for each cylinder is given by

$$m_i \frac{d^2 w_i}{dt} = \tau_{i+1} - \tau_i \tag{31.61}$$

The dynamic equilibrium equation is solved for the vertical displacement at each time step. Gazetas et al. (1992) developed a model using

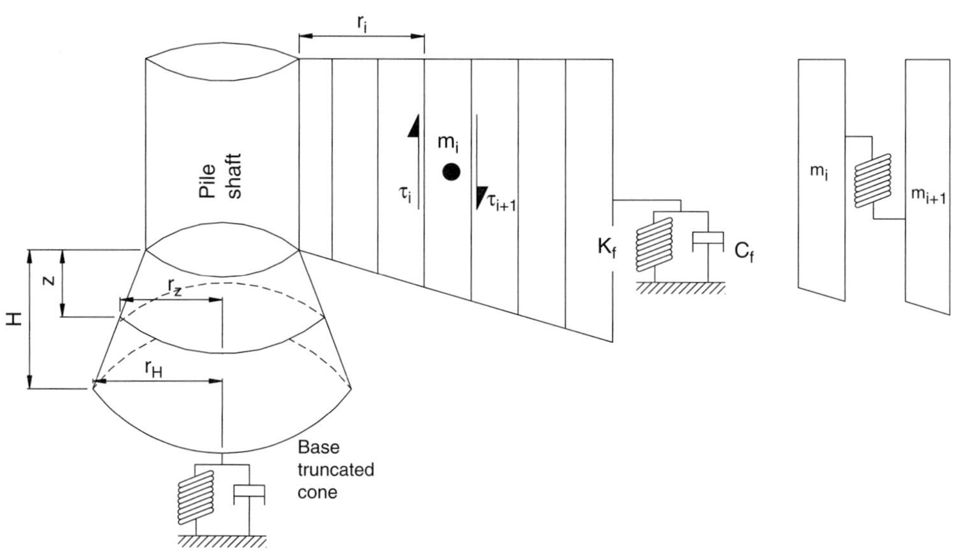

Figure 31.35 Holeyman's (1985) shaft and base model.

the BDWF approach in combination with finite-element analysis. They developed a simplified equation to compute the dynamic impedance function (dynamic stiffness and damping coefficient) of a single pile. The soil was modeled as a Winkler foundation resisting the pile vertical and lateral motion by distributed spring and dashpot along the pile length. The spring and the dashpot were assigned a frequency-dependent spring constant and a damping coefficient, which were calibrated by matching the displacement response of the pile head from the BDWF analysis and the finite-element analysis. Gazetas et al. defined the vertical stiffness (k_z) and damping coefficient (c_z) of a single pile as a function in a dimensionless frequency parameter (a_o) and function in the soil elastic modulus as

$$k_z \approx 0.6E_s\left(1+\frac{1}{2}\sqrt{a_o}\right) \tag{31.62}$$

$$c_z \approx 2\beta_s\frac{k_z}{\omega}+a_o^{-1/4}\rho_s V_s d \tag{31.63}$$

where a_o = dimensionless frequency parameter = $\dfrac{\omega r_{\text{pile}}}{V_s}$

V_s = soil shear wave velocity and ρ_s = mass density of soil material

β_s = damping constant

r_{pile} = pile radius

The equation derived by Gazetas et al. was simple; however, they neglected the effect of pile flexibility.

Chowdhury and Dasgupta (2009b) utilized the approach used by Baranov (1967) relating the dynamic response of a soil-embedded foundation to develop a formulation of the vertical dynamic stiffness and damping coefficient for a single pile in soil continuum. They defined the vertical dynamic stiffness (K_1) and damping coefficient (C_1) for an end-bearing pile as follows:

$$K_1 \approx \left(\frac{EA}{8L}+\frac{GS_1L}{2}\right) \tag{31.64}$$

$$C_1 \approx \frac{1}{2}r_o\sqrt{\rho G}S_2L \tag{31.65}$$

where S_1 and S_2 are stiffness and damping coefficients

$S_1 = 2.7$

$S_2 = 6.059a_o + \dfrac{0.7022a_o}{a_o+0.01616}$

31.24 PILE GROUP EFFECT

Piles are usually installed in groups, and the effect of the pile-to-pile interaction is an important parameter in the dynamic performance of the pile foundation, especially pile foundations subjected to dynamic loading, such as those supporting machines. Under the effect of machine-induced dynamic loading, elastic waves are transmitted from piles adjacent to each other and amplify the dynamic response of the pile foundation. The interaction between the piles in pile groups also reduces the vertical dynamic stiffness and damping of the individual pile, alters the performance of the foundation system, and affects the performance of the supported machine. There are several factors that affect the dynamic performance of the machines supported over the pile foundation, such as influence of the pile group response due to the additional deformation resulting from waves transmitted from the adjacent pile, type of loading (static, cyclic, dynamic; transient, or steady-state load), soil dynamic properties, pile-to-pile interaction, pile flexibility (flexible piles will see more interaction than rigid piles), pile group material (concrete, steel, or wooden piles), and pile groups type (end-bearing pile groups or floating piles groups). The concept of pile-to-pile interaction was introduced by Poulos (1968) using Mindlin's equation of elasticity to solve the stresses and displacement between piles due to horizontal loading. Poulos defined the interaction factor (α) as follows:

$$\alpha_r = \frac{\text{additional displacement due to adjacent pile displacement}}{\text{pile displacement due to direct loading on the pile}} \tag{31.66}$$

Randolph and Poulos (1982) defined the interaction factor for pile groups as follows:

$$\alpha_r = \text{Group interaction factor} = \frac{0{,}5ln\left(\dfrac{l_p}{s}\right)}{ln\left(l_p d\dfrac{G_{\text{avg}}}{G_b}\right)} \tag{31.67}$$

where K_{pile} = stiffness of a single pile

G_{avg} = average shear modulus along the pile length

G_b = shear modulus at the pile base

l_p = pile length

d = pile diameter

Under a dynamically loaded pile foundation, the interaction between the pile groups is greatly influenced by the frequency of excitation of the forcing function. Both dynamic stiffness and damping of pile groups are frequency dependent, and their values may increase or decrease based on the frequency of excitation of the forcing function. Unlike static loading, the effect of piles spacing is also affected by the frequency of excitation for pile foundation subjected to dynamic loading. Novak (1976) suggested that pile group stiffness and damping coefficient are not the summation of the individual pile stiffness and damping due to the dynamic interaction between the piles. He proposed the following formula to account for the pile-to-pile interaction and to compute the pile group stiffness and damping coefficient utilizing Poulos's (1968) interaction factor:

$$K_{\text{group}} = \frac{\sum_1^n K_{\text{pile}}}{\sum_1^n \alpha_r} \tag{31.68}$$

$$C_{\text{group}} = \frac{\sum_1^n C_{\text{pile}}}{\sum_1^n \alpha_r} \tag{31.69}$$

Das and Ramana (2011) argued that the Poulos interaction equation is only applicable for pile groups subjected to static loading and is not applicable for pile groups subjected to dynamic loading. Wolf and von Arx (1978) were the first to study the effect of pile-soil-pile dynamic interaction using finite-element axisymmetric modeling formulation to establish the dynamic displacement field resulting from line load. They utilized the finite-element procedures to calculate the displacement at any point within the soil mass which was then used to determine the flexibility matrix of the soil at each frequency. The results presented by Wolf and von Arx showed high dependence of the pile interaction on the number of piles within the pile groups and on the pile spacing within the group. Waas and Hartmann (1981) analyzed the motion and the forces on the pile groups subjected to earthquake loading. In their model, the soil was modeled as viscous-elastic layered medium subjected to point and ring loading. Waas and Hartmann calculated the kinematic response and the inertial response of the pile-soil system to determine the dynamic stiffness matrix of the complete system. The pile groups were modeled as a semi-rigid linear beam element with the tip of the pile pinned and the head fixed to the pile cap. The flexibility matrix was computed by applying a unit harmonic load acting on each node within the soil continuum at the node where the soil and the pile are connected to determine the frequency-dependent flexibility matrix of the soil medium. Waas and Hartmann concluded that the dynamic stiffness of pile groups is highly dependent on the frequency of excitation of the forcing function, the soil layering, and the ratio of the shear wavelength, and the pile cap. Sheta and Novak (1982) investigated the effect of the soil nonlinearity on pile group axial dynamic response. They weakened the soil ring in direct contact with the pile in order to account for a high-strain condition within this soil layer. The soil was modeled as horizontal layers with varying soil properties. The piles were modeled as vertical linear elastic members, and the masses of piles were lumped at the pile head, center, and tip. The pile lumped masses were connected with massless circular elements. The soil flexibility matrix was computed based on the plane strain conditions with a softer cylindrical zone around the pile, where its mass was neglected to avoid wave reflection at the boundaries. Using this composite model, Sheta and Novak (1982) calculated the displacement field in each soil layer and formulated the flexibility matrix. Sheta and Novak (1982) concluded that the dynamic group effects are considerably different than the static effects; the dynamic stiffness and damping of the pile groups are more dependent on the frequency than the single pile, and the weak zone around the pile increases the group stiffness and damping of the pile groups by limiting the interaction between the piles within the group. Dobry and Gazetas (1988) formulated a simple formula for the dynamic interaction between two circular floating piles embedded in a uniform stratum of half-space. The piles were subjected to harmonic loading acting on the pile head. The soil was modeled as linear hysteric material. Dobry and Gazetas (1988) assumed that, under harmonic excitation, cylindrical waves are emitted from an oscillating pile, or active pile, along the pile length in the horizontal direction toward the adjacent pile defined as the receiving piles, or passive pile. They also assumed that the variation of the amplitude with depth for the receiving pile is attuned with the active pile. Based on these assumptions, Dobry and Gazetas (1988) derived the interaction factor between the two piles as follows:

$$\alpha_v = \left(\frac{S}{r_o}\right)^{-1/2} e^{\left(\frac{-\xi\omega S}{V_s}\right)} e^{\left(\frac{-i\omega S}{V_s}\right)} \tag{31.70}$$

where S and r_o = pile spacing and radius

V_s = soil shear wave velocity

ξ = material damping ratio

Dobry and Gazetas mentioned that this formula overestimates the peak value of stiffness and damping for pile groups in soils with elastic modulus greater than 300 the pile elastic modulus. Gazetas and Makris (1991) developed a simple method for calculating the dynamic steady-state axial response of floating pile groups embedded in homogeneous and nonhomogeneous soil continuum. They concluded that the interaction between piles is generally due to the interference of wave fields generated from one pile along the pile shaft and speeding outward, exciting the adjacent pile. Gazetas and Makris (1991) concluded that, for homogeneous soils, the wave fields emitted from the active pile are cylindrical and are independent from the pile flexibility and slenderness. However, for nonhomogeneous soils, Gazetas and Makris concluded that the wave fronts are non-cylindrical. They used ray theory to compute the pile group dynamic stiffness and damping. Fan et al. (1991) conducted a numerical study for the kinematic response of groups of vertical floating piles. The piles were connected through rigid massless caps and subjected to vertically propagating harmonic S-waves. In their study, the soil profile was modeled as a homogeneous material with shear modulus proportional to depth. Fan et al. (1991) showed that under soil-pile kinematic interaction, the influence of the nature of the soil profile is profound at all frequencies, and the effects of pile group configuration, number of piles in the group, and relative spacing between piles are usually insignificant for lateral displacements but quite important for pile cap rotations.

REFERENCES

1. Aoki, N., and de Mello, V.F.B (1992). Dynamic Loading Test Curves. *Proceedings of the Fourth International Conference on the Application of Stress-Wave Theory to Piles*, The Hague, The Netherlands, 525–530.

2. Baranov, V.A. (1967). On the Calculation of Excited Vibrations of an Embedded Foundation. Polytechnic Institute of Riga, 195–209.

3. Barkan, D.D. (1962). *Dynamics of Bases and Foundations*. McGraw-Hill Book Co.

4. Barton, Y.O. (1982). Laterally Loaded Model Piles in Sand: Centrifuge Tests and Finite Element Analysis. Ph.D. Thesis, Cambridge University.

5. Bea, R.G., and Audibert, J.M.E. (1979). Performance of Dynamically Loaded Pile Foundations. *Proceedings of the Second International Conference on Behavior of Offshore Structures, BOSS'79*, Imperial College, London, 477–490.

6. Broms, B. (1964a). The Lateral Resistance of Piles in Cohesive Soils, *Journal of the Soil Mechanics and Foundation Division*, 90(2), 27–63.

7. Broms, B. (1964b). The Lateral Resistance of Piles in Cohesive Soils, *Journal of the Soil Mechanics and Foundation Division*, 90(3), 123–156.

8. Chowdhury, I., and Dasgupta, P.S. (2009a). *Dynamic of Structures and Foundations, A Unified Approach, I. Fundamentals*, CRC Press.

9. Chowdhury, I., and Dasgupta, P.S. (2009b). *Dynamic of Structures and Foundations, A Unified Approach, II. Applications*, CRC Press.

10. Das, B.M., and Ramana, G.V. (2011). *Principals of Soil Dynamics*, 2nd edition, Cengage Learning.

11. Davisson, M.T., and Gill, H.L. (1963). Laterally Loaded Piles in a Layered Soil System, *Journal of the Soil Mechanics and Foundations Engineering, ASCE*, 89(3), 63–94.

12. Deeks, A.J., and Randolph, M.F. (1995). A Simple Model for Inelastic Footing Response to Transient Loading, *International Journal for Numerical and Analytical Methods in Geomechanics*, 19(5), 307–329.

13. Degebo (Deutsche Gesellschaft für Bodenmechanik). See works by Foundation L.; Hertwig, A., Truh, A., and Lorenz, H.; and Ramspeck, A. and Schulze, C.A.

14. Dobry, R., and Gazetas, G. (1988). Simple Method for Dynamic Stiffness and Damping of Floating Pile Groups, *Geotechnique*, 38(4), 557–574.

15. Dobry, R. (2014). Simplified Methods in Soil Dynamics, *Soil Dynamics and Earthquake Engineering*, 61–62, 246–268.

16. Erden, S.M., and Stokoe, K.H. (1975). Effects of Embedment on Foundation Response. Shock and Vibration Problems in Geotechnical Engineering, ASCE National Convention, Denver, CO.

17. El Naggar, H., and Novak, M., (1994). Nonlinear Axial Interaction in Pile Dynamics, *Journal of Geotechnical Engineering, ASCE*, 120(4), 678–696.

18. Fan, K., Gazetas, G., Kaynia, A., Kausel, E., and Ahmad, S. (1991). Kinematic Seismic Response of Single Piles and Pile Groups, *Journal of Geotechnical Engineering*, 117(12), 1860–1879.

19. Forehand, P.W., and Reese, J.L. (1964). Prediction of Pile Capacity by the Wave Equation, *Journal of the Soil Mechanics and Foundations Division*, 90(SM 2), 1–25.

20. Gazetas, G., and Dobry, R. (1984). Horizontal Response of Piles in Layered Soils, *Journal of Geotechnical Engineering, ASCE*, 110(1), 20–40.

21. Gazetas, G. (1991). *Foundation Vibrations. Foundation Engineering Handbook*, 2nd edition, Van Nostrand Reinhold, 553–593.

22. Gazetas, G. (1991). Formulas and Charts for Impedances of Surface and Embedded Foundations, *Journal of Geotechnical Engineering, ASCE*, 117(9), 1363–1381.

23. Gazetas, G. (1984). Seismic Response of End-Bearing Single Piles, *Soil Dynamics and Earthquake Engineering*, 3(2), 82–93.

24. Gazetas, G., and Makris, N. (1991). Dynamic Pile-Soil-Pile Interaction Part 1, Analysis of Axial Vibration, *Earthquake Engineering and Structural Dynamics*, 20(2), 115–132.

25. Gazetas, G., Fan, K., and Kaynia, A. (1992). Dynamic Response of Pile Groups with Different Configurations, *Soil Dynamics and Earthquake Engineering*, 12(4):239–257.

26. Gupta, B.N. (1972). Effect of Foundation Embedment on the Dynamic Behavior of Foundation of the Foundation—Soil System, *Geotechnique*, 22(1), 129–137.

27. Hardin, B.O., and Richart, F.E., Jr. (1963). Elastic Wave Velocities in Granular Media, *JSMFD, ASCE*, 89(SM1).

28. Hardin, B.O., and Drnevich, V.P. (1973). Shear Modulus and Damping in Soils, Design Equations and Curves, *JSMFD, ASCE*, 98(SM7).

29. Holeyman, A.E. (1985). Dynamic Non-Linear Skin Friction of Piles. *Proceedings of the International Symposium on Penetrability and Driveability of Piles*, 173–176.

30. Holeyman, A.E. (1988). Modeling of Dynamic Behavior at the Pile Base. *Proceedings of the Third International Conference on the Application of Stress-Wave Theory to Piles*, 174–185.

31. Hsieh, T.K. (1962). Foundation Vibrations. *Proceedings of Institute of Civil Engineers*, 22, 211–226.

32. Hu, Y., Randolph, M.F., and Watson, P.G. (1999). Bearing Response of Skirted Foundation on Nonhomogeneous Soil, *Journal of Geotechnical and Geoenvironmental Engineering, ASCE*, 125(11), 924–935.

33. Kagawa, T., and Kraft, L.M. (1980). Soil-Pile-Structure Interaction of Offshore Structures during an Earthquake. *Proceedings of 12th Annual Offshore Technological Conference*, 3, 235–245.

34. Kavvadas, M., and Gazetas, G. (1993). Kinematic Seismic Response and Bending of Free-Head Piles in Layered Soil, *Geotechnique*, 43(2), 207–222.

35. Lok, T., Pestana, J., and Seed, R. (1998). Numerical Modeling and Simulation of Coupled Seismic Soil-Pile-Structure Interaction, ASCE, *Geotechnical Earthquake Engineering and Soil Dynamics*, II, 1211–1222.

36. Lysmer, J. (1965). Vertical Motions of Rigid Footings. Ph.D. Thesis, University of Michigan.

37. Matlock, H. (1970). *Correlations for Design of Laterally Loaded Piles in Soft Clay*. 2nd Annual Offshore Technology Conference, Houston, TX, 577–594.

38. Matlock, H., and Reese, L.C. (1960). Generalized Solutions for Laterally Loaded Piles, *Journal of Soil Mechanics and Foundation Division, ASCE*, 86, 63–91.

39. Mindlin, R.D. (1951). Influence of Rotatory Inertia and Shear on Flexural Motions of Isotropic, Elastic Plates, *ASME Journal of Applied Mechanics*, 18, 31–38.

40. Nogami, T., Otani, J., Konagai, K., and Chen, H. (1992). Nonlinear Soil-Pile Interaction Model for Dynamic Lateral Motion, *Journal of Geotechnical Engineering*, 118(1), 89–106.

41. Novak, M. (1974). *Dynamic Stiffness and Damping of Piles*. Faculty of Engineering Science, University of Western Ontario, London, ON.

42. Novak, M. (1976). *Vertical Vibration of Floating Piles*. Faculty of Engineering Science, University of Western Ontario, London, Ontario.

43. Novak, M., and El Sharnouby, B. (1983). Stiffness Constants of Single Piles, *Journal of Geotechnical Engineering, Division, ASCE*, 109, 961–974.

44. Penzien, J., Scheffey, C.F., and Parmelee, R.A. (1964). Seismic Analysis of Bridges on Long Piles. *Journal Engineering Mechanics Division, ASCE*, EM3, 223–254.

45. Poulos, H.G., and Davis, E. (1980). *Pile Foundation Analysis and Design*. Wiley.

46. Poulos, H.G. & Davis, E.H. (1968). The settlement behavior of single axially loaded incompressible piles and piers. *Geotechnique* 18, No.3, 351–371.

47. Quinlan, P.M. (1953). The Elastic Theory of Soil Dynamics, *Symp. on Dyn. Test of Soils*. ASTM STP No. 156, 3–34.

48. Randolph, M.F., and Simons, H.A. (1986). An Improved Soil Model for One Dimensional Pile Driving Analysis. *Proceedings of the Third International Conference of Numerical Methods in Offshore Piling*, 3–17.

49. Reissner, E. (1936). Stationare, Axialsymmetrlsche, Dutch Eine Schut-Telnde Masse Erregte Schwingungen Eines Homogenen Elastlschen Halbraumes, *Ingenieur Architektur*, 7, 381.

50. Richart, F.E., Jr. (1962). *Foundation Vibrations. Transaction, ASCE*, 127, Part I, 863–898.

51. Richart, F., and Whitman, R.V. (1967). Comparison of Footing Foundation Tests with Theory, *Journal of the Soil Mechanics and Foundations, ASCE*, 93(SM6), 143–167.

52. Sheta, M., and Novak, M. (1982). Vertical Vibration of Pile Groups, *Journal of Geotechnical Engineering Division, ASCE*, 108(GT4), 570–590.

53. Smith, E.A.L. (1960). Pile-Driving Analysis by the Wave Equation, *Journal of Soil Mechanics and Foundations Division*, 86(EM 4), 35–61.

54. Simon, H.A., and Randolph, M.F. (1985). A New Approach to One-Dimensional Pile Driving Analysis. *Proceedings. of Fifth International Conference on Numerical Methods in Geomechanics*, Nagoya, Japan.

55. Sung, T.Y. (1953). *Vibration in Semi-Infinite Solids Due to Periodic Surface Loading*. Sc.D. Thesis, Harvard University.

56. Swane, I.C., and Poulos, H.G. (1982). A Theoretical Study of the Cyclic Shakedown of Laterally Loaded Piles. Research Report R415, University of Sydney, School of Civil and Mining Engineering.

57. Tajimi H. (1966). *Earthquake Response of Foundation Structures Report*. Faculty of Science and Engineering, Nihon University Tokyo, 1.1–3.5.

58. Tazoh, T., Wakahara, T., Shimizu, K., and Matsuzaki, M. (1988). Effective Motion of Group Pile Foundations. *Proceedings of the Ninth World Conference on Earthquake Engineering*, Tokyo, 3, 587–592.

59. Tschebotarioff, Gregory P. (1951). Foundation and earth structures. McGraw-Hill.

60. Waas, G., and Hartmann, H.G. (1981). Piles Subjected to Dynamic Horizontal Loads. European Simulation Meeting on Modeling and Simulation of Large Scale Structural Systems, Capri, 17.

61. Weissmann, G.F., and Hart, R.R. (1961). *Damping Capacity of Some Granular Soils*, ASTM STP, 305, 45–54.

62. Whitman, R.V., and Richart, F.E., Jr. (1967). Design Procedures for Dynamically Loaded Foundations, *Journal of the Soil Mechanics and Foundations Division, ASCE*, 93(SM6), 169–193.

63. Whitman, R.V. (1970). *Soil Structure Interaction—Seismic Design for Nuclear Power Plants*. MIT Press.

64. Wolf, J.P., and Von Arx, G.A. (1978). Impedance Functions of a Group of Vertical Piles. *Proceedings of the ASCE Specialty Conference on Earthquake Engineering and Soil Dynamics*, 11, 1024–1041.

Chapter 32
Value Methodology

BY
MUTHIAH KASI, PE, SE, CVS-LIFE
CHARLES A. BARTLETT, PE, CVS

32.1 INTRODUCTION

The design process, whether it's a building or an automobile, requires a consideration of those that are impacted by the project and the options for solving the problem. Value methodology (VM) affords an opportunity for the designer to take these considerations into account and provides a design that meets or exceeds the industry standards, identifies cost implications, and works to satisfy the users of the project. VM is a step-by-step systematic approach used to define and solve problems, promote creativity, and document the decision-making process. VM can be utilized at any point in the design life of a project. From the initial concept or planning stage to various points along the design time line, VM provides a valuable tool for creating, modifying, and checking a project to assure the design objectives are being met. To maximize effectiveness, using VM early in the design is encouraged. This allows for changes and suggestions to be more easily implemented, as opposed to later in the process, which can result in rework. In applying VM, a job plan with six phases and structured guidelines is implemented. The job plan creates an atmosphere where a team of designers can truly explore options and alternatives for the design. However, for the VM job plan to be effective, adherence to its rules and guidelines is critical. The rules may appear simple, but they are vital to the success of a VM study. To maximize creativity and comprehensiveness, a multidiscipline team led by a facilitator is employed in a VM study.

32.2 VM JOB PLAN

The job plan creates the organization and structure of a design workshop that maximizes teamwork, creativity, and productivity. The six phases are the following:

1. Information phase
2. Function analysis phase
3. Creative phase
4. Evaluation phase
5. Development phase
6. Presentation phase

32.2.1 The Information Phase

As the name implies, this initial step involves gathering as much information about the project as possible. It is important to understand stakeholders that may be impacted by the project, the project's history, the problem(s) being solved, and the context of the project. As a first step in information gathering, the development of a list of owners, users, and other stakeholders is necessary. This does two things: (1) it encourages the designers to think about everyone that is impacted and what effect their design may have on them, and (2) it documents that stakeholders were recognized. Owners and decision makers are greatly influenced by stakeholders who may or may not be the users. In addition, determining the role each plays provides further context on expectations.

The next step is to gather constraints, needs, and desires of the stakeholders. This information is used later as the genesis of the development of functions for the project. It helps in framing up the design problem and documents boundaries and expectations for the design. Finally, during the information phase, all specifications, goals and objectives, and unwritten constraints are gathered and understood.

32.2.2 Function Analysis Phase

Function analysis facilitates answering the question: *What does it do?* Allocating a cost to each project function aids in answering the questions: *What does it cost?* and *What should it cost?* Functions are defined as a description of the purpose and expectations of the project expressed in two words: an active verb and a descriptive noun (i.e., provide safety). Assembling the functions into a logical diagram facilitates communication and aids the design team in fully understanding and defining the project. It is critical when defining functions to transition from specific to abstract in order to maximize creativity.

Precise expression of defining functions is important to create intended option. The intent of an action needs to be expressed accurately. Figure 32.1 shows a 45-story twin tower building. To create a column-free trading floor, the end columns need to be terminated at floors 6 and 10. The initial design is based on the function *Support Column* at the lower floor where the columns were designed to be terminated. The function of the girders is to *Support Columns*. This very specific function led to a conventional design idea. This led to massive 180-ft-long girders for each end row column (Fig. 32.2).

In this example, the load of 30-floor columns need to be transferred to the supports below. If one asks the question "Why do you support columns?" the answer is to *Transfer Load*. The function can be satisfied by either a transfer girder (Fig. 32.2), a cantilever bracket (Fig. 32.3), or as hanging columns or a combination. A support column rests on an element such as a girder. When a structural element offers a load

Figure 32.1 A 45-story twin-tower building.

path (Fig. 32.4), the function is termed as transfer load. In this case the bracket transfers the load from above to the support below.

The seven-story column bracket wall was less expensive to build and avoid the transfer of prestressed precast girders or the massive formwork of cast-in-place girders. It is an example of how different functions can lead to dramatically varying solutions. This example demonstrates the benefit of precise definition.

Function analysis can be performed in various ways. The simplest form is to define functions, compare their costs, and decide on their worth. The more common practice is to take a step further and analyze the functions with a Function Analysis System Technique (FAST) and its associated diagramming. Another method is the Customer Function Model. Both are applicable, but the Customer Function Model lends itself to higher-level planning and design on a macroscale. The FAST is equally useful for larger-scale designs but is very effective when analyzing individual design elements on more of a microscale.

Functions can be further classified as sell functions and work functions. Work functions are expressed in active verb and measurable noun. Sell functions can be expressed using a passive verb and nonmeasurable noun. However, sell functions may sometimes be expressed using an active verb and measurable noun. In a highway project, *Increase Capacity* is a work function, whereas *Enhance Appearance* is a sell function. Value is achieved when functions that work and sell are fulfilled at an acceptable cost.

32.2.3 Creative Phase

While creativity is an expectation for many designers, being creative can be difficult because there tends to be a built-in urge to find a quick solution. The job plan controls this tendency and requires designers to go through all phases of this systematic procedure. The creative phase helps answer the question: *What else will do the job?* During this phase, it is important to let the mind wander freely with no limitations. Unique and creative solutions often evolve from the crazy or impractical ideas suggested. While randomness is key to speculation, it is important to check in with the functions of the project to assure that project-specific expectations are accounted for. Because of their abstract nature, functions can also serve as a vehicle to create the widest possible range of ideas.

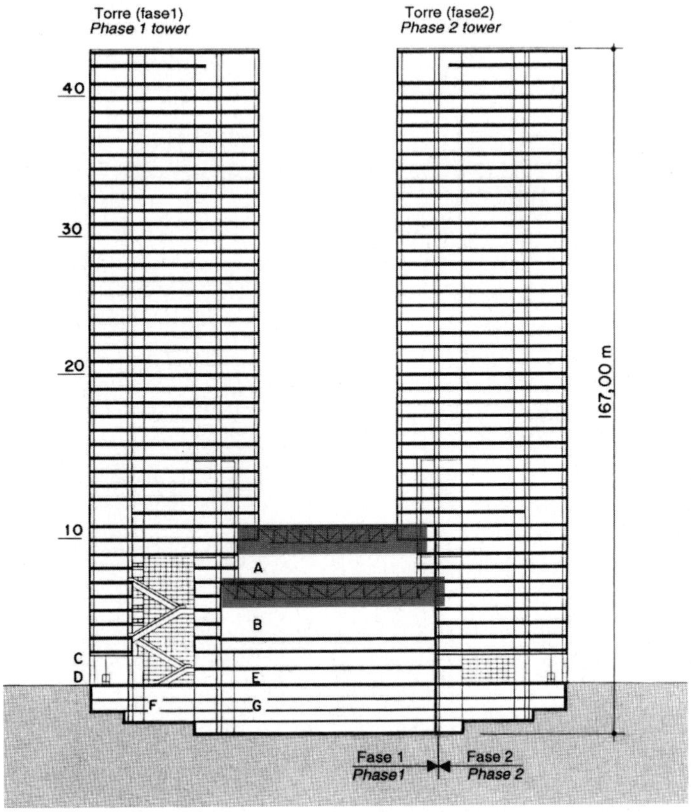

Figure 32.2 Transfer girders.

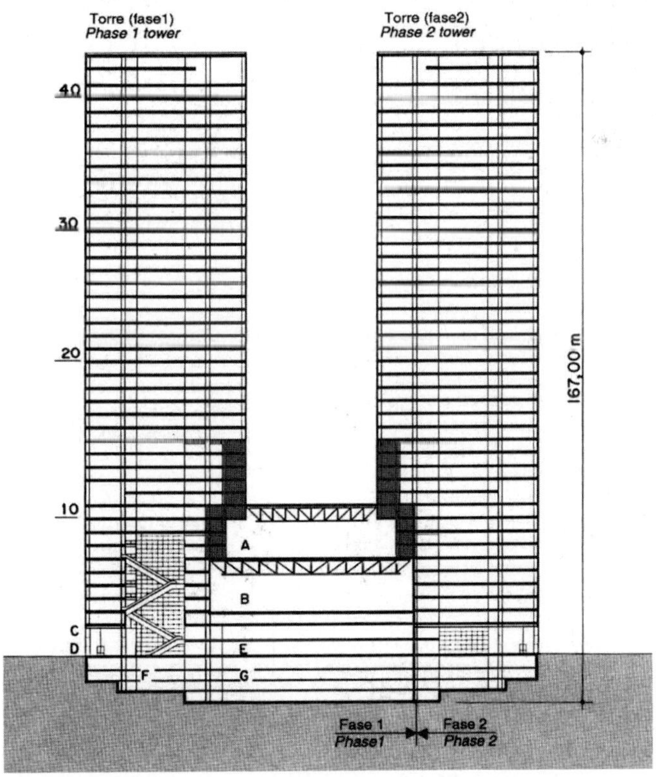

Figure 32.3 Cantilever brackets.

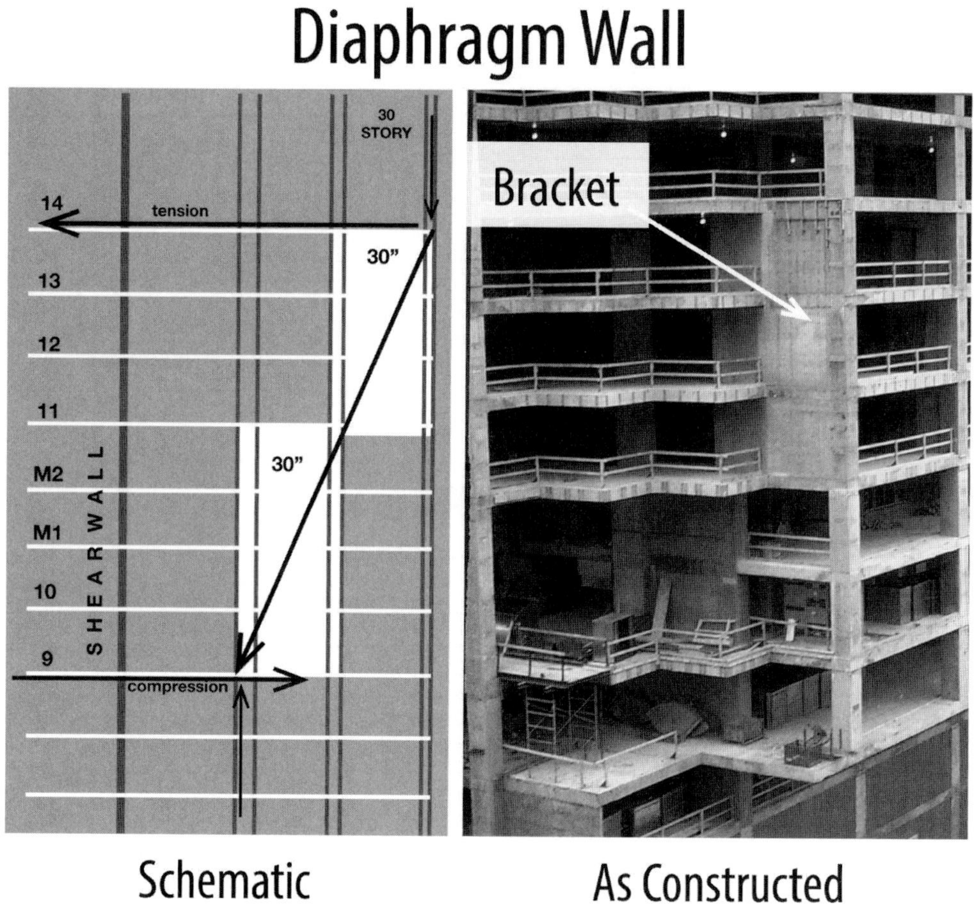

Figure 32.4 Column load path.

32.2.4 The Evaluation Phase

After all the ideas are listed, a series of screening processes is needed to sort them. Idea comparison, feasibility ranking, and analysis matrices are some of the techniques that can be utilized. This phase aids in focusing on ideas that are closer to addressing the user's concerns, needs, and requirements. At the end of this phase, outstanding ideas emerge for further development.

32.2.5 The Development Phase

Good results are obtained by combining the strengths of various ideas. During the development phase, ideas are assembled into design concepts. Creating detailed calculations to support the design concept and develop appropriate costs are key to this phase. Further, development should include the following steps:

1. Research and add information to substantiate the approach
2. Separate ideas that are industry standards
3. Become aware of ideas that are controversial
4. Recognize ideas that may be unique
5. Involve specialists to support and perfect ideas
6. Consider impacts on customer(s)
7. Use cost to perform rate of return analysis
8. Consider life cycle cost

32.2.6 The Presentation Phase

Communicating design solutions is important to the overall process. The proceeding phases create much of the content needed to "tell the story" of the project and the recommended solution(s). Assembling that content into a written and verbal presentation demonstrates to decision makers that the VM process was done in a thorough and comprehensive manner and that value was maximized.

32.2.7 Rules for the VM Job Plan

As mentioned above, the application of VM requires certain rules that help facilitate the job plan and maximize creativity and productivity. The rules that govern each phase of a VM study are briefly described next.

INFORMATION PHASE, FUNCTION ANALYSIS PHASE

- Do not speculate
- Do not jump to conclusions
- Gather as much information as possible
- Do not evaluate
- Understand the problem

CREATIVE PHASE

- Do not bring preconceived notions to the discussion
- Do not question the validity of an idea
- Encourage everyone to participate in the creative phase

EVALUATION PHASE

- Do not create
- Do not jump to conclusions

- Judge each idea against the criteria
- Always be prepared to explain your evaluation

Development Phase

- Improve ideas
- Double check that assumptions are reliable

Presentation Phase

- Do not assume that ideas are good
- Demonstrate their worth
- Follow these three steps: inform, instruct, and influence

32.3 THE KEY FEATURES OF VM JOB PLAN: FUNCTION ANALYSIS

The application of function analysis can present some challenges because of the use of abstract thinking when tasked with creating a graphical representation of the expectations for the project. This section is a more detailed discussion of functions and function analysis.

Beginning with the Customer Function Model shown in Fig. 32.5, functions are classified and assembled in a way that promotes an understanding of customer and stakeholder needs and desires. Costs are then assigned to each function to assess appropriate value. The functions, as defined by the team, are grouped together and classified as follows:

- Task
- Basic functions
- Supporting functions
  - Assure dependability functions
  - Assure convenience functions
  - Satisfy stakeholder functions
  - Attract stakeholders functions

The task function (or task) is one which fulfills or satisfies the need of the owner/user. It is the main reason why the project is needed. It is the function which makes the basic and supporting functions necessary.

Basic functions are those functions which are essential to the performance of the task. They represent the bare minimum necessary to fulfill it. The project, product, system, and so on will not work without the

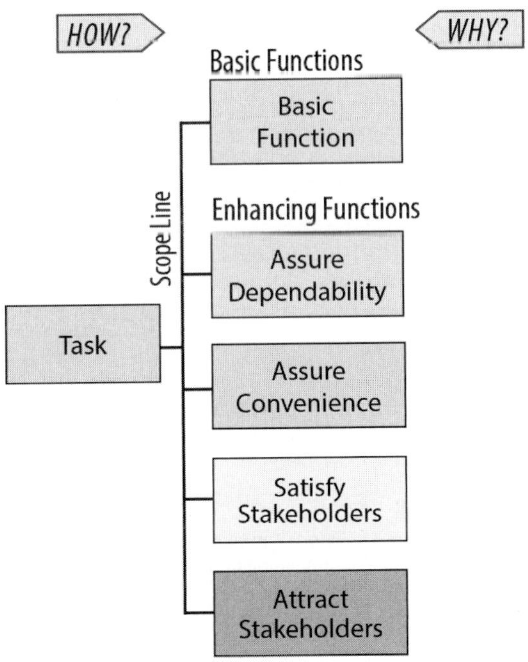

Figure 32.5 Customer function model.

basic functions. However, it is important to realize that the term *Basic* has nothing to do with priority. All of the functions are of importance to stakeholders.

The supporting functions are extremely important in building customer acceptance and in selling the product or service to the user or customer.

Assure Dependability—Any function that assures dependability has at least one of the following attributes:

- Makes the elements of the project stronger or more reliable or effective
- Makes it safer to use
- Lengthens the life of the parts or minimizes maintenance cost, or both
- Protects the environment

Assure Convenience—Any function that assures convenience has at least one of the following attributes:

- Modifies the basic function to make it convenient to use
- Enhances spatial arrangements
- Facilitates maintenance and repairs
- Furnishes instructions and directions to stakeholders

Satisfy Stakeholders—Any function that satisfies stakeholders has at least one of the following attributes:

- Modifies the basic function to satisfy the individual desires
- Makes the stakeholders' life more pleasant; for example, minimizes noise
- Makes the element appear to be better in the opinion of the stakeholder, but not necessarily in the opinion of the designer (Sometimes these opinions are reflected in the standards and specifications of a particular agency/owner.)

Attract Stakeholders—Any function that attracts stakeholders and has at least one of the following attributes:

- Emphasizes the visual aspect (sight) or other senses
- Projects a favorable image (that is, trademarks or endorsement by public figures)

Assembling the diagram entails an order and hierarchy of functions that adhere to "How/Why" logic. The task is placed on the left side of the diagram with the classified functions to the right. The classified functions progress from left to right, from abstract to more specific functions, asking the question "How?" from one level to the next. For example, in a highway project the question "How is dependability assured?" may be answered with the functions *increase capacity* or *maintain clearance*. Following the logic, additional functions may be added to the right that answer the question "How do we increase capacity?" It is important to note that more than one function can answer the question "How?" In fact, it is often the case. To check the logic of the diagram, the reverse of the process can be accomplished going right to left and asking the question "Why?"

The Basic and Dependability functions and, to a lesser extent, some of the Convenience functions are work functions. They are quantifiable and measureable. They demonstrate how the function performs and involves more engineering and technical assessment. Convenience, Satisfy Stakeholder, and Attract Stakeholder tend to be more sell functions. Often they are not measurable but are important to make sure the project is meeting stakeholder expectations.

The FAST diagram has a specific structural format as shown in (Fig. 32.6).

There are five components to the FAST diagram: design or stakeholder objective, all-the-time functions, critical path functions, at-the-same-time functions, and caused-by functions.

Function analysis requires analyzing why a function exists and how a function satisfies other functions to complete the link between them. This "How-Why" logic assures that all the required functions are listed in the FAST diagram.

The FAST diagram begins with a higher-order function of the project and two scope lines. All functions the selected element fulfills are bounded by the two scope lines. The basic function is on the right of the left-hand scope line, and the higher-order function is on the left. The purpose of the element or project for which a FAST diagram is

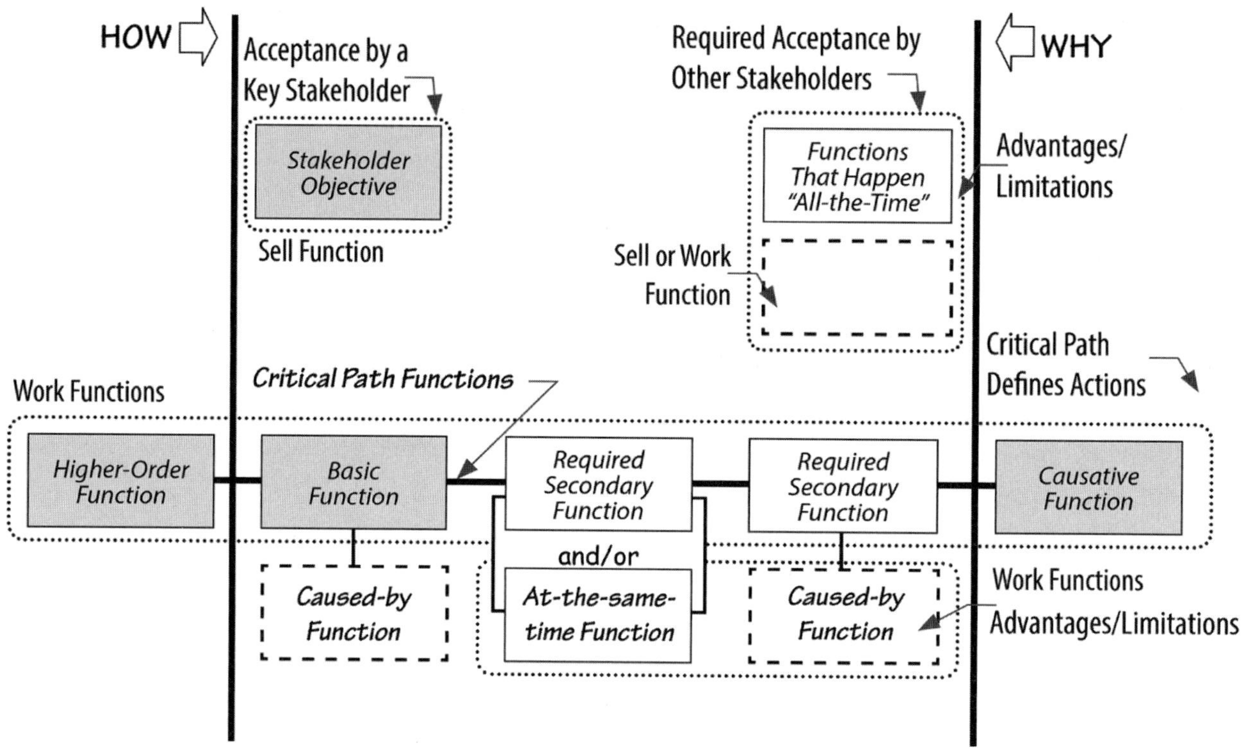

Figure 32.6 Function analysis system technique (FAST).

developed is the higher-order function. The relationship between the higher-order function and the basic function is determined by applying the "How/Why" logic by asking "Why?" the basic function candidate performs as it does. The answer should be the higher-order function. The logic check must be completed by asking "How?" the higher order function performs. The logical answer must be the basic function candidate. It is still necessary to confirm the required secondary function to the left of the right-hand scope line. When the "How?" question is asked of this function, the answer will be an outside function candidate. The outside function is called the causative function since it really starts the critical functions.

Determining the basic function often requires selecting functions from the list of suggestions and applying the "How?" and "Why?" questions. If the "Why?" question is answered by another identified function, that function is the next candidate for the basic function. The function to the right becomes a required secondary function. Once the basic function is verified, the remaining required secondary functions are identified. This group makes up the critical functions.

The last group of functions are supporting functions. There are three types. The first type, *caused-by* or *at-the-same-time functions*, connects directly to a critical function. These functions result from the performance characteristics of particular critical functions and act as modifiers. The second type, *all-the-time functions*, modifies two or more of the critical functions. The third type, *design objective*, represents project purpose, often driven by the stakeholder or decision makers.

The last step in the function analysis for either method is to compute the cost of each function. This includes identifying functions of each element and allocating cost by its role in the element. Examples below demonstrate the use of allocating functions and costs.

32.4 EXAMPLE 1: PARAPET JOINT DETAIL

A simple example to demonstrate the VM steps is shown below. The same process can be exercised for any major project.

Information phase—What is it?

32.4.1 Description of the Project

A three-span plate girder bridge (305 ft–510 ft–385 ft) has been designed to span the Illinois River in Ottawa, Illinois. The bridge will accommodate pedestrians with an 8-ft-wide path on one side of the bridge. The expansion gap at the end of the structure is designed to be 8 in. wide at an ambient temperature of 50° Fahrenheit. Through the design temperature range, the end of the bridge may move a total of 12 in. (6 in. in each direction), decreasing the gap to a minimum of 2 in. in summer and increasing the gap to 14 in. in winter.

32.4.2 Project Objective

To protect the pedestrians and bicyclists, protective measures are required to prevent the users from falling off the structure, protect them from vehicular traffic, and protect them from falling into the expansion gap.

32.4.3 As-Designed

The proposed design (see Fig. 32.7 and 32.8) utilizes a 4-ft-high parapet to protect the pedestrians from falling off the path into the river. A 4-ft-high barrier wall is proposed to protect the pedestrians from traffic. A steel plate assembly is proposed to bridge the gap in the wall. The assembly will be recessed into the sidewalk, parapet, and barrier wall to bridge over the expansion gap (see Figs. 32.7 and 32.8).

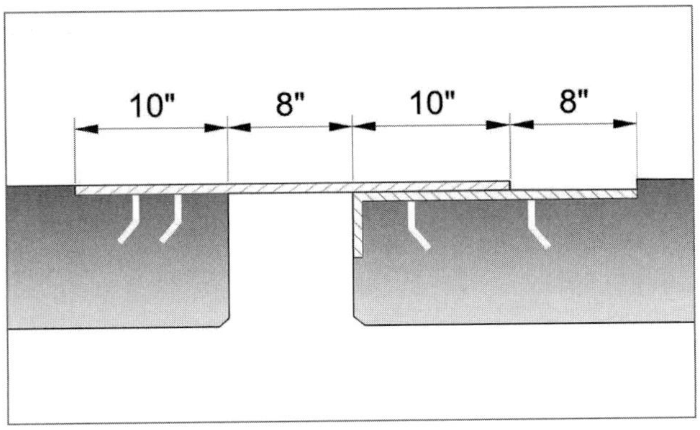

Figure 32.7 Section of sliding plates (as designed).

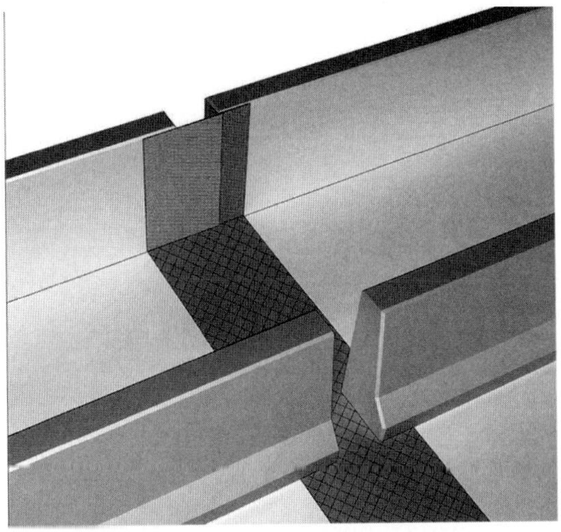

Figure 32.8 As-designed sliding plate.

32.4.4 Function Analysis Phase—What Does It Do?

The next step is to understand the purpose of this detail. The purpose of the detail is to contain pedestrians so that they don't fall through. The task therefore is to prevent falling. The answer to the question "How do you prevent falling" is to contain pedestrians, which becomes the basic function. In simple terms, the basic function can be accomplished by closing the gap. However, when the bridge expands or contracts, the structure will crack, which is not a dependable solution. All functions need to be explored. They are the following:

- *Allow Expansion*
- *Allow Contraction*
- *Protect Pedestrians* (sell function)

Also, there are solutions that may not be convenient. For example, certain solutions may require constant maintenance or require details that project into the walkway, which would be inconvenient for users. Therefore, the functions under convenience are the following:

- *Minimize Maintenance*
- *Minimize Intrusion* (sell function)

The need to satisfy stakeholder function is something that is desired; however, it doesn't relate to the task directly. They are the following:

- *Contain Debris*
- *Facilitate Construction* (sell function)

Whatever the detail that is selected, it should also consider aesthetics. Therefore, consideration of the following functions should be given to make it attractive:

- *Enhance Appearance*
- *Blend Material* (sell function)

This is summarized in Fig. 32.9. Information phase—What does it cost?

The cost of the sliding plate assembly is $12,515.00.

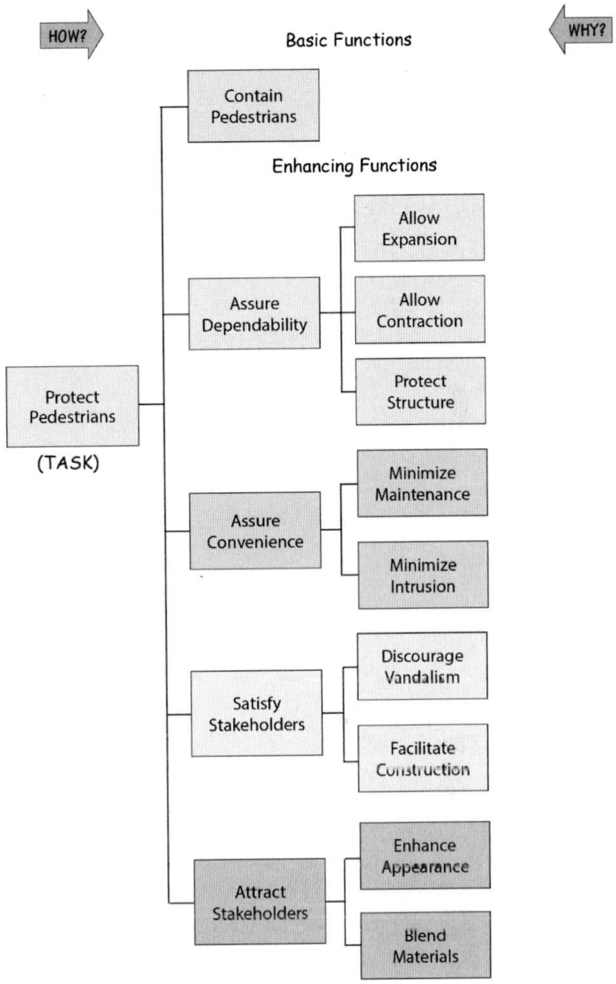

Figure 32.9 Customer function model.

32.4.5 Function Cost

The last step in the function analysis is to allocate cost by function, not by parts. This is one of the differences between conventional cost reduction and value engineering.

Let us take an example—in this case thickness of the plates. Only ½-in. thickness of plates A and B is needed to perform the functions *Contain Pedestrians, Allow Contraction,* and *Allow Expansion.* The other ½ in. is to make the plate stronger, in other words to *Minimize Maintenance.* As far as the height is concerned only 2 ft of plates A and B is needed to *Contain Pedestrians, Allow Expansion,* and *Allow Contraction.* The other 2 ft is needed to make it *Contain Debris* and *Enhance Appearance.*

Table 32.1 Function Cost

		Basic function	Assure dependability			Assure convenience	Satisfy stakeholders	Attract stakeholders
Elements	Cost	Contain pedestrian	Allow expansion	Allow contraction	Protect structure	Minimize maintenance	Contain debris	Enhance appearance
Plate "A": 4 – 28" × 1" × 4'-0" LG.	**$ 6,092.80**							
4 × 20 × ½ × 2'-0" × 3.4 × $ 4.00		$1,088.00						
4 × 8 × ½ × 2'-0" × 3.4 × $ 4.00				$435.20				
4 × 28 × ½ × 1'-0" × 3.4 × $ 4.00							$761.60	
4 × 28 × ½ × 1'-0" × 3.4 × $ 4.00								$761.60
4 × 28 × ½ × 4'-0" × 3.4 × $ 4.00						$3,046.40		
Plate "B": 4 – 24" × 1" × 4'-0" LG.	$ 5,222.40							
4 × 8 × ½ × 2'-0" × 3.4 × $ 4.00			$435.20					
4 × 8 × ½ × 2'-0" × 3.4 × $ 4.00				$435.20				
4 × 8 × ½ × 4'-0" × 3.4 × $ 4.00					$870.40			
4 × 16 × ½ × 1'-0" × 3.4 × $ 4.00							$435.20	
4 × 16 × ½ × 1'-0" × 3.4 × $ 4.00								$435.20
4 × 24 × ½ × 4'-0" × 3.4 × $ 4.00						$2,611.20		
Anchor Bolts:	**$ 1,200.00**							
		$480.00	$240.00		$240.00	$240.00		
Total Cost	**$ 12,515.20**	**$1,568.00**	**$675.20**	**$870.40**	**$1,110.40**	**$5,897.60**	**$1,196.80**	**$1,196.80**
Percentage		**12.53%**	**5.40%**	**6.95%**	**8.87%**	**47.12%**	**9.56%**	**9.56%**

Looking at the width of plate A, only 20 in. is required to *Contain pedestrians*. The other 8 in. is needed to *Allow Contraction*. For plate B, 8-in. width each is required to *Allow Contraction* and *Allow Expansion*. The remainder is to *Minimize Maintenance*. Table 32.1 shows the function cost distribution. As in all cases, basic function costs a relatively small amount, which in this case is 12.53 percent. There is a substantial amount spent for one function, *Minimize Maintenance* of 47.12 percent.

Function cost distribution is as follows:

32.4.6 Creative Phase

During this phase, the main question answered is:

WHAT ELSE WILL PERFORM THE JOB?

The function cost figure indicated that *Minimize Maintenance* is a high-cost function. During the creative session, the team concentrated on ideas which were maintenance free or relatively so. The team did not attempt to judge any of these ideas. What they did in this session was merely list the ideas.

32.4.7 Idea List

1. Open joint
2. Attach sliding plate; in outside wall—flush
3. Attach sliding plate—in outside wall—projected
4. Attach sliding plate—in inside wall—flush
5. Attach sliding plate—in inside wall—projected
6. Overlap parapet (Fig. 32.10)
7. Sleeve and pipe—metal or plastic (Fig. 32.11)
8. Sleeve and bar
9. Erect warning signs
10. Erect adjustable guardrails
11. Extend expansion joint materials—flush (Fig. 32.12)
12. Extend expansion joint materials—projected
13. Attach wire, or rope, or chain (Fig. 32.13)
14. Attach metal bars or wooden beams
15. Fasten chicken fence on one side

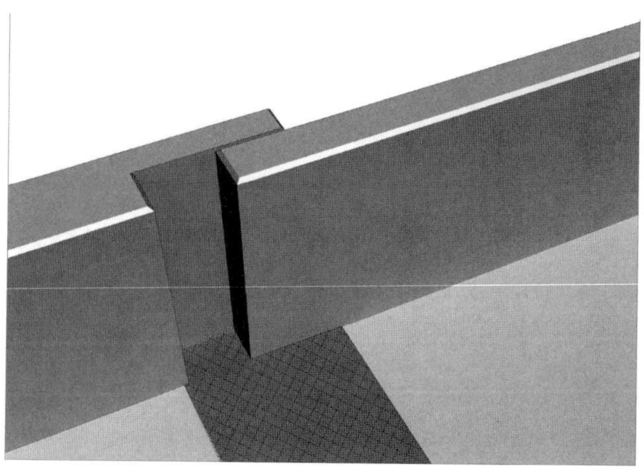

Figure 32.10 Idea 6—overlap parapet.

Figure 32.11 Idea 7—Sleeve and pipe.

Figure 32.12 Idea 12—extend expansion joint.

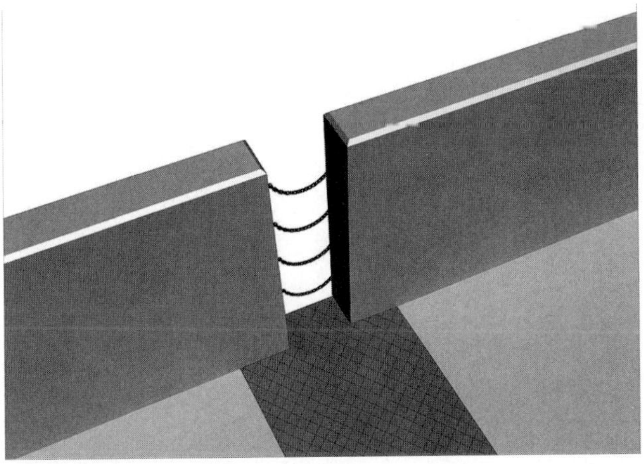

Figure 32.13 Idea 13—attach wire.

32.5 EVALUATION PHASE

All creative ideas were measured against their feasibility of acceptance. As shown in Feasibility Rank Chart (Table 32.2), only five ideas have a good chance of being accepted. These five ideas were rated against the basic and supporting function of the structure.

32.5.1 Feasibility Rank Chart

Table 32.2 Feasibility Rank Chart

	State-of-the-art	Cost	Probability of acceptance	Total ranking
Open joint	0	10	0	10
Sliding plate—O.F. flush	5	6	3	14
Sliding plate—I. F. flush (as designed)	7	6	7	20
Sliding plate—O.F. proj.	5	5	3	13
Sliding plate—I. F. proj.	5	5	3	13
Overlap parapet	8	8	10	26
Sleeve and bar	5	8	8	21
Sleeve and pipe	5	7	8	20
Erect signs	0	9	0	9
Erect guardrails	0	7	0	7
Expansion joint material flush	10	0	7	17
Expansion joint material proj.	8	0	4	12
Wire, rope, or chain	2	9	0	11
Metal bars or wooden 2 × 4	2	7	2	11
Chicken fence fastened on one side	2	7	2	11
	10-Off the Shelf 0-New Technology	10-No cost 0-High cost	10- High 0-No Chance	

32.5.2 Idea Comparison Table

Table 32.3 Idea Comparison Table

0 = Low value 10 = High value	Contain pedestrian	Allow contraction	Allow expansion	Protect Structure	Minimize maintenance	Contain debris	Enhance appearance	Total
Sliding plate (as designed)	10	9	9	6	6	10	4	54
Overlap parapet	10	10	10	10	10	10	10	60
Sleeve and bar	8	10	10	7	8	0	8	51
Sleeve and pipe	8	10	10	7	7	0	8	50
Expansion joint material	10	10	10	8	4	8	3	53

Idea Comparison Table, Table 32.3, shows its rating against the functions. In the Idea Comparison Table, *Overlap Parapet* outranks other creative ideas in three functions: *Minimize Maintenance, Protect Structure*, and *Enhance Appearance*. By eliminating the steel plates and overlapping the concrete parapets, the structure presents a uniform look. With a continuous pour, the maintenance will be minimal, and the probability of acceptance is very high. The other two outstanding ideas are Sleeve and Pipe and Sleeve and Bar. They are more or less similar ideas. Compared to overlap parapet, they were rated low in three functions: *Minimize Maintenance, Contain Debris*, and *Enhance Appearance*.

32.6 IMPLEMENTATION

In order to select the preferred alternative, we have to look at the cost of the above three outstanding ideas. Let us first take the top ranking idea, *Overlap Parapet*. As shown in Fig. 32.10, no additional concrete is needed, and additional labor is very small. Since the most preferable idea requires practically no additional dollars, it was not deemed necessary to find the cost of the other two ideas. The team, therefore, adopted *Overlap Parapet* as the preferred choice.

The team's final recommendation was as follows:

1. Recess contract parapet as shown
2. Extend concrete parapet to be housed inside the recessed parapet

32.7 SUMMARY

The use of the job plan and functions are what define VM. Its application can be an effective tool for both analysis and creativity with respect to project performance and cost. Integrating stakeholder expectations completes the analysis, assuring that a comprehensive evaluation has taken place. Recommendations are made based on a systematic and creative process, assuring stakeholders and decision makers that their concerns and voices have been heard. Further details about VM can be obtained through the SAVE International organization.

REFERENCES

Circular A-131, Office of Management and Budget, U.S. Federal Government.

Value Methodology, A Pocket Guide to Reduce Cost and Improve Value Through Function Analysis, SAVE International® Value Methodology Body of Knowledge (VMBOK), Miles Value Foundation, Goal/QPC, 2008.

Techniques of Value Analysis and Engineering, 2nd Edition, Lawrence D. Miles, McGraw-Hill, 1972.

FAST Creativity & Innovation, Charles W. Bytheway, J. Ross Publishing, 2007.

"Standard Practice for Performing Value Analysis (VA) of Buildings and Building Systems," American Society for Testing and Materials, (ASTM) Publication E1699-13.

Internet: http://www.value-ing.com

Chapter 33
Stone Cladding

BY
RUI de S. CAMPOSINHOS *Coordinator Professor with Aggregation, Polytechnic of Porto, Portugal*

33.1 INTRODUCTION: NATURAL STONE CLADDING

External cladding of buildings was regarded in past decades as an aesthetic "well-regarded" solution with no need for relevant maintenance works and costs.

Natural stone-clad façades make a significant contribution to the overall aesthetic and technical performance of a building. They have a substantial effect on people's judgment about visual impact, yet they are subjected to a variety of actions, such as heat, air, and moisture, or dynamic loadings, such as wind and seismic effects.

Durability, environmental compatibility, and economic sustainability along with mechanical, chemical, thermal, hygrometric, and other weathering "actions" are the key aspects that must be addressed together with the strength and structural safety.

Several methods or systems are available for installing stone on the exterior of buildings depending on the manner in which the environment and exposure conditions are addressed. Temperature changes, air pressure, and water in any form, as direct or indirect induced tensions, must be undertaken appropriately, based on a robust empirical investigation encompassed with lessons from older thin-stone cladding building façade systems.

Under the structural point of view, numerous techniques were employed to support the stone panels, both within curtain walls as well as individually [1]. The development of curtain wall systems, consisting of vertical and horizontal structural members, connected and anchored to the supporting structure and the notable stones' cutting technology, provides all the normal functions of a curtain wall, using thin cladding slabs or panel as its skin. Figure 33.1 is an example of a curtain wall system providing a ventilated façade.

Façade veneer systems must comply with structural safety requirements in particular for direct actions such as the inertial forces on heavy stone cladding panels that are more susceptible to earthquake damage. In some regions, wind action acting in lighter panels may be more stringent than seismic action. Stone cladding failure can induce serious injury to people and may cause relevant financial losses.

33.1.1 Natural Stone Properties

The most-used stone properties depend mainly on the process of rock's formation. Igneous rocks derive directly from the solidification of magma. As is well known, sedimentary rocks are formed on the crust from other rocks or preexisting materials or also by chemical precipitation. Metamorphic rocks have its origin from preexisting ones due to the action of the so-called metamorphism agents.

Igneous rocks are formed when molten magma cools and can be divided into two main categories: plutonic or irruptive and volcanic or extrusive. Examples of some plutonic rocks are granite, diorite, and syenite, and for extrusive rocks, examples are, basalt, rhyolite, and pumice, among others [2].

For the metamorphic rocks, the one that stands out most is, undoubtedly, the marble. However, other metamorphic rocks are also used, such as quartzite, gneiss, slate, and other schist rocks.

Examples of sedimentary rocks are limestone, sandstones, breccias, dolomite, and others less commonly used for cladding purposes. From an industrial point of view, this type of classification is not at all used, since it implies a particular technical and scientific knowledge.

It is reasonable to assume that a particular rock can be identified by its color, texture, and mineralogical composition and by its physical and mechanical properties, and it is possible to have a broad idea about the range for values to be expected in the most of the rocks. The values in Tables 33.1, 33.2, and 33.3 may be used as an indication of rock identifying according to some authors [3].

33.1.2 Standards for Stone Cladding

Some ornamental rocks have a very low scatter, which is a benefit for structural use in construction, but generally the scattering of the values of rock's properties and their characteristics are generally very high so

Figure 33.1 Ventilated façade—Tecnopolo of Modena in Italy.

Table 33.1 Properties Indicative Values for Igneous Rocks

	Apparent density mass [kg/m³]	Water absorption at atmospheric pressure [mass %]	Uniaxial compressive strength [MPa]	Thermal expansion coefficient [mm/mK]
Granite	2600–2800	0.1–0.9	130–270	0.008
Syenite	2600–2800	0.2–0.9	160–240	0.008
Gabbro	2800–3000	0.2–0.4	170–300	0.009
Periodite	2500–2800	0.2–0.7	180–300	0.013
Basalt	2900–3000	0.1–0.3	240–400	0.1–0.3

Table 33.2 Properties Indicative Values for Sedimentary Rocks

	Apparent density mass [kg/m³]	Water absorption at atmospheric pressure [mass %]	Uniaxial compressive strength [MPa]	Thermal expansion coefficient [mm/mK]
Sandstone	2000–2700	0.2–10.0	30–150	0.012
Limestone	2600–2900	0.1–3.0	75–240	0.004–0.001
Dolomite	2600–2900	0.1–3.0	75–240	0.005
Travertine	2400–2500	2.0–5.0	20–60	0.007
Sandstone	2000–2700	0.2–10.0	30–150	0.012

Table 33.3 Properties Indicative Values for Metamorphic Rocks

	Apparent density mass [kg/m³]	Water absorption at atmospheric pressure [mass %]	Uniaxial compressive strength [MPa]	Thermal expansion coefficient [mm/mK]
Marble	2600–2900	0.1–3.0	75–240	0.003–0.006
Migmatite	2600–3000	0.3–0.4	100–200	0.005–0.008
Slate	2600–2800	0.2–0.4	140–200	0.005–0.001
Quartzite	2600–2700	0.2–0.5	150–300	0.0125
Marble	2600–2900	0.1–3.0	75–240	0.003–0.006

that there's the need to carry out several tests depending on the location, depth, and the volume extracted or amount removed.

In this context, physical and mechanical characterization of natural stone is required as well as laboratory tests to assess its appropriateness to the various conditions of use, a procedure that is essential and decisive when dimension stone cladding design and construction are the critical issues.

Several methods for testing and comparing the physical properties of natural building stones were developed. The most basic stone testing methods analyze three characteristics: density, water absorption, and compressive strength. Although these three tests are a representative suite together with the modulus of rupture and the flexural strength, for practical purposes, flexural strength tests are the most relevant to façade stone cladding methods once structural problems are determined.

There are several standard test methods to evaluate stone characteristics so that stones can be compared consistently. These methods and standards for dimension stone are the basis for establishing guidelines for specification and installation. American Society for Testing and Materials (ASTM) International and the European Committee for Standardization (CEN) are the world's leading standards development organizations.

There are two categories of tests: those including the identification and characterization of physical properties and those related to mechanical characterization. In either case, the European and the ASTM standards use different procedures, thus giving different results with the same stone.

Readers can find in Tables 33.4 and 33.5 the most relevant standards for the physical characterization and geometry according to the CEN and ASTM International, respectively.

Table 33.4 Physical Characterization Relevant CEN Standards

EN 12057	Natural stone products. Modular tiles. Requirements.
EN 12370	Determination of resistance to salt crystallization.
EN 12371	Determination of frost resistance.
EN 12407	Petrographic examination.
EN 12440	Natural stone. Denomination criteria.
EN 12670	Standard terminology relating to dimension stone.
EN 13373	Determination of geometric characteristics on units.
EN 13755	Determination of water absorption at atmospheric pressure.
EN 14066	Determination of resistance to aging by thermal shock.
EN 14581	Determination of linear thermal expansion coefficient.
EN 1469	Natural stone products. Slabs for cladding. Requirements.
EN 1925	Determination of water absorption coefficient by capillarity.
EN 1936	Determination of real and apparent density and total and open porosity.

Table 33.5 Physical Characterization Relevant ASTM Standards

C 119	Standard terminology relating to dimension stone.
C 1242	Standard guide for selection, design, and installation of dimension stone attachment systems.
C 1528	Standard guide for selection of dimension stone.
C 1721	Standard guide for petrographic examination of dimension stone.
C 97	Standard test methods for absorption and bulk specific gravity of dimension stone.
D 2203	Standard test method for staining from sealant.
D 4404	Standard test method for determination of pore volume and pore volume distribution of soil and rock by mercury intrusion porosimetry.
D 4341	Standard test method for creep of cylindrical hard rock core specimens in uniaxial compression.
D 5312	Standard test method for evaluation of durability of rock for erosion control under freezing and thawing conditions.
WK54484	Method for evaluating dimension stone durability by exposure to salt crystallization.
D5313	Standard test method for evaluation of durability of rock for erosion control under wetting and drying condition.

33.2 NATURAL STONE DEPICTION

Natural stones are traditional construction materials, and modern architecture increasingly applies them since building owners and architects focus their vision, above all, on color, which is the most important sensory cognizance in human beings. However, other than appearance, stone durability or resistance to decay is a distinctive characteristic of strength, the reason why stone physical properties play a significant role in their recognition and ability to the construction's industry. For example, specific gravity or density is a crucial parameter since it controls the stone weight.

33.2.1 Geometry and Tolerances

In precise terms, dimension stone is natural stone or rock that has been selected and finished (i.e., trimmed, cut, drilled, ground, etc.) to specific sizes or shapes [4]. Dimension stone geometry, especially thickness of slabs, must be well defined and documented.

In general, stone producers publish typical test values which can serve as a guide only. For determining the characteristics of building stone needed for a proper design for a specific project, independent testing should be carried out by laboratories accredited for that purpose [5].

There are several methods for verifying the geometric characteristics of products of natural stone. The Marble Institute of America (MIA) has published a chart of stone fabrication, installation, and substrate condition tolerances regarding thickness, face size, squareness, flatness, and anchor preps, among other requirements [6].

Particular attention must be paid to thickness for cladding purposes. Thickness is attained after sawing or other precise techniques, such as diamond wire saws, diamond belt saws, burners, and surface finishing. Their final dimensions may depend on the fixing method to the backup, either by mechanical means, that is, using anchorage devices, or by cement bonding using adhesives, cement, epoxy resin, or other types of compound adhesive materials.

European standards, regarding slab thickness, *t*, and face sizes; length or width, *L*; and unit sizes should be obtained within the range of tolerance depicted in Table 33.6, according to EN 13373. This European Standard describes methods for verifying the geometric characteristics of products of natural stone such as blocks, rough slabs, and finished products for cladding, flooring, stairs, and tiles.

Table 33.6 Size Tolerances According to EN 13373

Nominal thickness (t) [mm]	Tolerances for thickness (t) and lengths (L) [mm]		
	t	$L < 600$	$L \geq 600$
$30 \geq t > 12$ $50 \geq t > 30$	±10% ±3 mm	±1.0 mm	±1.5 mm
$80 \geq t > 50$ $t > 80$	±3 mm ±5 mm	±2.0 mm	±3.0 mm

Hole, cut, or kerf size variations are also limited. For instances, the specific position, diameter and depth dowel hole sizes have to be performed with a tolerance of 2 mm; in the slab's thickness, its depth and diameter have to lie in a range of (+3; –1 mm), and (+1; –0.5 mm), respectively. Flatness and squareness are limited depending on the calibration of the tiling process, according to EN 12057.

33.2.2 Physical Properties

Density, Porosity, and Water Absorption

Stone is very sensitive to the minerals that compose a rock type. The more mafic rock is, the higher its density. The density of a material such as a rock being clearly defined is subsequently crucial.

The weight of a sample under dry conditions, m_{dry}, being known and under fully saturated conditions, m_{sat}, and submerged in water, m_{sub}, the matrix density, ρ_{matr}, and the bulk density, ρ_{bulk}, may be determined as follows:

$$\rho_{matr} - \rho_{water} \; (1 \quad m_{sub}) \tag{33.1}$$

$$\rho_{bulk} = \rho_{water} \cdot \left(\frac{m_{dry}}{m_{sat}} - m_{sub} \right) \tag{33.2}$$

In general, rock's bulk density is said to be high when its specific weight value exceeds 2800 kg/m^3 and is considered low when that value is less than 2300 kg/m^3.

Stone water absorption is an indicator of its susceptibility to damage during freezing; that is why this parameter is more relevant than the former. A stone that has greater water absorption will also tend to absorb liquid stains more readily. In general, the lowest water absorption is desired.

The water absorption at atmospheric pressure, w_a, is the ratio between the water absorbed by a sample and its dry weight, a percentage value of its density, and may be obtained as follows:

$$w_a = \frac{m_{sat} - m_{dry}}{m_{dry}} \times 100 \tag{33.3}$$

Different values of water absorption may be found depending on the rock type. Nevertheless, granites rarely present water absorption values higher than 1 percent; in marbles and limestone, values greater than 3 percent are unusual, and for schist, 0.5 percent, is considered a high value [7].

Thermal Expansion

The coefficient of thermal expansion governs how the size of an object changes with a change in temperature. The linear coefficient is generally used for natural stone characterization, and in some cases, it may be necessary to do a joint's design when one might only be concerned with the change along a length or over some area.

Stone Weathering

When exposed to the atmosphere, natural stone experiences a slow, unremitting process depending on the exposure conditions. Mechanical characteristic decay is to be observed due to several factors Freeze and thaw, wetting-drying cycles, variation in temperature, mechanically induced actions, and salt crystallization in pore spaces are some of the most common agents that cause stone weathering.

Frost Resistance The more significant part of the damage observed on natural stone façade claddings is due to frost is in sandstone and certain types of limestone, and it has been proved that low salt concentrations at the surface of the stone severely magnify the effect of frost attack on some types of stones [6]. The increase in air pollution instigates salt deposition on the surface of building façades, thus explaining why stone damage has been augmented during the past decades and made the stone more sensitive to surface scaling from frost and consequently mechanical resistance [8].

A conservative procedure might be to consider as frost-proof stones those that demonstrate flexure strength decay, after 48 test cycles, of less than 20 percent, which is the case of almost all intrusive rocks such as granite. However, specific indirect methods of stone durability assessment that investigated the pore properties of the stone were found to be a trustworthy predictor of frost resistance. Petrographic examination seems to be a valuable way for observation of changes caused by frost action [9].

Resistance to Salt Crystallization Rainwater in contact with limestone, on evaporation, makes calcium carbonate dissolution precipitate in the form of calcite crystals.

Salt is also a vital weathering factor because of exposure to polluted industrial environments in coastal areas and in places where salt is used as antifreeze. The crystallization of salt even in unsaturated environments is often accompanied by a volume increase, causing internal pressures that result in irreversible damage. Either by cracking, exfoliating, or disintegrating of macrostructure, a loss of mass and strength is observed [10].

There are several proposed criteria to define stone suitability either by ultrasonic velocity measurements or compressive strength measurements before and after salt exposure. The Building Research Establishment (BRE) criteria based upon the loss of mass is widely used. It consists of the definition of acceptable limit values of mass loss depending on the exposure and environmental conditions where six categories (A to F) are proposed depending on the mass loss values [11].

Thermal Shock and Moisture Variations Temperature changes may attain considerable values so that continuous heating and cooling of the stone lead to its deterioration over time. The criteria to verify stone suitability consist of perceiving and measuring the mass and the elastic modulus decreases as a result of controlled abrupt changes in temperature. Moisture variations are essential factors in stone degradation. For some clear reasons, these variations cause volume changes.

For example, granite and some sandstones have different expansion coefficients of their constituent minerals, resulting in volume expansion or contraction. On the other hand, calcitic marbles exhibit granular decohesion once calcite expands in one direction and contracts in the other.

The "sugaring" phenomenon identified by stone's deterioration is directly dependent on these uneven deformations. Dolomite marbles are not so sensitive to thermal shock or moisture variations once expansion or contraction occurs in the same direction, though dolomite marbles are also susceptible to cyclic thermal variations [12], causing strength loss.

Together with strength loss, bowing is also known to occur in some marbles when used as exterior claddings (Fig. 33.2) [13, 14].

Quantitative test methods, as an evaluation of marble being able to resist bowing and strength induced by thermal and moisture, are under development. The standards EN–16306 and ASTM–WK32888, both for the determination of the resistance of marble to thermal and moisture cycling, are the most used.

Figure 33.2 Bowing of marble panels at Zagrepčanka tower (Zagreb).

33.3 MECHANICAL PROPERTIES

Strength capacity of dimension stone is as or more important than its physical attributes. Mechanical characterization of natural stone is required as well as laboratory tests to assess its aptness to the various conditions of use, a procedure that is essential and decisive when dimension stone cladding design and construction are the critical issues.

Both the ASTM Committee C18 [15] and the CEN Technical Committee 246 [16] have jurisdiction of several standards on mechanical characterization of dimension stone, such as:

ASTM C1201/C1201M—Standard Test Method for Structural Performance of Exterior Dimension Stone Cladding Systems by Uniform Static Air Pressure Difference;

ASTM C1352/C1352M—Standard Test Method for Flexural Modulus of Elasticity of Dimension Stone;

ASTM C1354/C1354M—Standard Test Method for Strength of Individual Stone Anchorages in Dimension Stone;

ASTM C170/C170M—Standard Test Method for Compressive Strength of Dimension Stone;

ASTM C880/C880M—Standard Test Method for Flexural Strength of Dimension Stone;

ASTM C99/C99M—Standard Test Method for Modulus of Rupture of Dimension Stone;

EN 12372—Natural stone test methods—Determination of flexural strength under concentrated load;

EN 13161—Natural stone test methods—Determination of flexural strength under constant moment;

EN 13364—Natural stone test methods—Determination of the breaking load at Dowell hole;

EN 14146—Natural stone test methods—Determination of the dynamic modulus of elasticity (by measuring the fundamental resonance frequency);

EN 14580—Natural stone test methods—Determination of static elastic modulus;

EN 1926—Natural stone test methods—Determination of uniaxial compressive strength.

As pointed out, stones with more massive specific weight and smaller water absorption have greater strength and elasticity module, and stones less massive, saturated with water at atmospheric pressure, allow for bigger adsorption of moisture and decay in compressive or flexural strength [3, 17].

Samples to be tested must be drawn from regular production blocks, extracted from the same quarry defined for a given project. That is to say, historical data only indicate strength in a relative informational basis about the stone to be used.

Designers must be aware that stone is anisotropic, and consequently, its properties are directionally dependent.

Mineral deposition forms sedimentary rocks—in which the layers or beds are predominantly orientated—or, in the case of metaphoric rocks, sedimentary transformed by heat and pressure or, even in the case of igneous rocks, due to the cooling of the structure. These are the causes of micro predominant cracking in a direction or directions giving rise to anisotropic behavior.

The stone appearance in various directions will ask for a decision on the preferred orientation. Thus, the aesthetical effect may have a strong influence on the mechanical relevant properties and then on the cost, a vital issue to consider during initial testing [18].

33.3.1 Compressive Strength

Although the compressive strength is not determinant for dimension stone cladding design, compressive strength shall be regarded as a leading feature for any stone since almost all other stones' mechanical properties are correlated with compressive strength [19]. Designers should always compare relative values of other properties, such as flexural strength, elasticity modulus, or specific weight, once minimum compressive strength requirements might be specified according to the standards.

Samples may differ in sizes and shape: cut cubes or on core-drilled cylinders are subject to uniaxial loading in a calibrated test machine until rupture is obtained. The maximum applied load is divided by the contact area of the specimen to calculate the compressive strength. Usually 5 to 10 samples in dry or soak conditions are used. The load is evenly applied perpendicularly or parallel to the anisotropy planes at a fixed speed by a load cell.

33.3.2 Flexural Strength

As on the compressive strength, a minimum of five tests should be performed either in a three- or four-point bending flexural configuration, and the load cell is increased gradually until the test specimen fractures. The maximum applied load is recorded, and the flexural stress that occurred in the specimen at fracture is calculated.

Depending on the standards and criteria, several dimensions are possible for the prismatic specimens. Their thickness shall be between 25 and 100 mm and be greater than twice the size of the largest grain in the stone; the total length shall be equal to 6 to 10 times the thickness depending on the standards. The width shall be between 50 mm and three times the thickness, and it shall be less than the thickness.

Also, when the specified stone face has a different finish from the usual sawn face test samples, for example, honed, sandblasted, polished or flamed finish, it is crucial that flexure tests be carried out using samples with the same finish on the face under tensile stress [20].

Depending on the visual orientation and directional properties, care should be taken to gather the critical direction of testing. It should be noted that to fully understand the microstructure of the material, testing in three orientations may be necessary initially [21, 22].

For different reasons, it is expected to obtain lower values from four-point flexure tests than from three-point flexure tests. One of the primary reasons for the discrepancy depends on the fraction of the specimen's volume, which undergoes the greatest tensile stress. This is the basis of the so-called weakest-link theory, which elucidates the influence of the scale effect. Its principles were enunciated by Weibull [23] and Burshtein [24]. Another issue that must not be disregarded is that wet testing results present lower values than dry testing, so that dry results must be unheeded for façade design purposes.

33.3.3 Tensile Strength

Stone tensile strength may be obtained either directly or indirectly yet is formally defined as the tensile stress required to cause the failure of an unconfined cylindrical or cubical stone specimen—the direct tensile strength. Due to the difficulties related to gripping the specimens, direct tensile tests are unusual and can be found indirectly, that is, by relying on another type of test.

One of these indirect tensile strength methods is the Brazilian test, where a circular solid disc is compressed until failure across a diameter (Fig. 33.3). Tensile stresses perpendicular to that diameter plane are developed; as such, compressive loading machines are used in this test [25]. In the Brazilian test a stone's indirect tensile strength, σ_{Rt}, is generally defined as

$$\sigma_{Rt} = -\frac{F}{\pi \cdot R \cdot t} \tag{33.4}$$

where F = compressive force
R = radius of the specimen (disc)
t = thickness of the disc

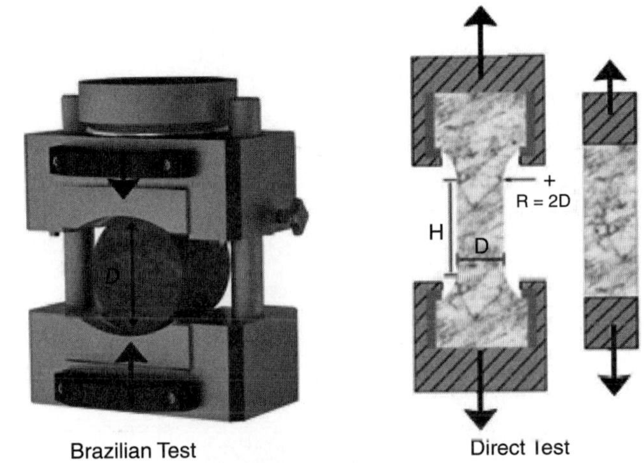

Figure 33.3 Indirect Brazilian and direct tensile tests.

The above formula (33.4) has been extensively applied in rock engineering and research fields for decades, although some authors question its efficiency, though results are supposed to be inaccurate when the disc has a significant thickness [26, 27].

Another simple procedure to obtain the indirect tensile strength is via flexural strength tests, based on the formula for linearly elastic bodies [23]. It is known that the calculated maximum flexural stress is greater than the actual stress in the test specimen because, during the testing of a stone prism or slab under flexure, a number of factors operate to change the stress distribution in the specimen so as to reduce the maximum stress. It follows that the tensile strength of a stone in the conditions of a flexure test is higher than in direct testing under tension [24, 28, 29].

The fact that flexural strength is physically greater is that during flexure, the maximum tensile stress is experienced only by a filament on the convex surface of the specimen, whereas during axial tension all points of the cross section experience the maximum tensile stress.

Based on the statistical theory, Weibull 23 derived relations between the strength of a brittle material under flexure, σ_t, and the axial tension, σ_f, when data from specimens of different volumes are known. The following relations can be applied:

For pure bending:

$$\sigma_{Rt} = \frac{1}{2}(\sigma_{f1} + \sigma_{f2}) \cdot (2m+2)^{\left(-\frac{1}{m}\right)} \tag{33.5}$$

For simple bending:

$$\sigma_{Rt} = \frac{1}{2}(\sigma_{f1} + \sigma_{f2}) \cdot 2^{\left(-\frac{1}{m}\right)} \cdot [(m+1)^2]^{\left(-\frac{1}{m}\right)} \tag{33.6}$$

where σ_{f1} = flexural strength of specimens with volume v_1
σ_{f2} = flexural strength of specimens with volume v_2
m = correlation factor equals to $\left(\ln\frac{v_2}{v_1}\right) \div \left(\ln\frac{\sigma_{f1}}{\sigma_{f2}}\right)$

33.3.4 Modulus of Elasticity

For evaluating the modulus of elasticity of dimension stone, using destructive or so-called static methods, which are based on deformation of specimens under axial or flexure loads,[1] is not justifiable since the precision of its value is not relevant for checking deformation limit state.

There are several nondestructive evaluation methods to estimate the elastic properties of stone, including ultrasonic pulse velocity measurements. These dynamic tests are tolerable and economical, consisting of measuring the fundamental resonance frequency as in the EN 14146 standards or by ultrasonic measurements to determine the velocity that a given ultrasonic wave crosses the length or width of a specimen, although there is a direct relationship between the dynamic elastic modulus and compressive strength that can be confidently used for assessing a slab's deformation [30, 31].

33.3.5 Creep

Dimension stone cladding design difficulty can be influenced by rock's creep unless in a situation where horizontal tiling or covering may be subjected to a considerable load level. If creep deformation constrains the design, that will be due to excessive deformation of cladding slabs under constant load conditions at a considerable level value. In this case, a conservative approach would consist of considering a creep coefficient equal to 2 [20].

33.4 CLADDING SYSTEMS AND METHODS

With the development of curtain wall systems, consisting of vertical and horizontal structural members, connected and anchored to the supporting structure of the building, providing all the normal functions of a façade, natural stone appears as thin cladding slabs or panel within lightweight curtain wall systems.

There are several preferred conventional methods for cladding exterior building walls with natural stone: bonded systems and mechanical anchored systems.

33.4.1 Bonded Systems

Traditional fixing methods for the dimensional stones currently that are used in most countries depend on the bond material's aptitude, through continuous superficial contact or by locally bonding between the support surface and the stone cladding. This method is usually

[1] As, for example, the uniform compression test in the EN 14580 standard or the four-point test in the ASTM C1352 standard.

known as the direct fixing method, where adhesives such as cement paste, epoxy resins, or mortar are used to glue dimension stone to the concrete or mortar bed directly.

Despite a proper application of this fixing technique, there are limits for the thickness and maximum size of the stone panels, the maximum height at which they can be installed, and anchorage materials specifications [7, 32].

The two ways commonly used to bind the stone panel to the masonry backup are spot bonding and continuous bonding. The latter is much more used, and stone shall be back buttered to achieve, as close as practical, full surface adhesive contact between the stone and the backup, although achieving complete mortar bedding at the back of the stone is not an easy task.

In all methods, it is advisable to use a deformable binding material settlement, consistent with the expansion and contraction of the stone panel caused by temperature changes and deformation of the façade.

The backup of the cladding being rigidly connected prevents differential dimension changes giving rise to stresses in the cladding system: grout and slabs, with different distributions depending on the type and origin of the actions. These stresses are not uniformly distributed, and concentrations are to be found on the periphery of the slabs and the joints between them. These stress concentrations are the central origin of the well-known adhesion failure of wall cladding with the inherent slab's detachment [33, 34], and, in many cases, no aesthetical "solutions" are unavoidable due to economic issues (Fig. 33.4). A remarkable example is the case of the Amoco building in Chicago, the renovation façade cost of which was half of the total cost of the whole building [35].

Figure 33.4 Façade cladding repair using central screws to prevent detachment of granite panels.

33.4.2 Mechanical Anchored Systems

External masonry walls being relegated to the function of simple leaves with small bearing capacity are not thick enough to act as mass "watertight" barriers, as earlier thick stone or brickwork walls used in the first decade of the last century. Later, cavity walls brought the advantage of providing a cavity within the depth of a wall for drainage of rainwater before it can be absorbed too far into the wall construction (Fig. 33.5).

Nevertheless, an external leaf of cavity walls must face temperature changes, air pressure, and water in any form as direct or indirect induced tensions. Cavity walls were first built in exposed coastal areas in order to keep out wind-driven rain, although there is always the risk that moisture will be able to find its way across to the inside, whatever the insulation material and the possibility that the installation will leave empty air pockets inducing condensation.

Another problem concerns wall-tie corrosion since the outer brick leaf is colder in winter and therefore wetter, which can accelerate rusting of the wall ties that have to be replaced. There is a significant number of buildings with mortar droppings on ties within the cavity, which result in penetrating humidity, settlement, and saturation of cavity-fill leading to cold bridging.

To protect the building from weathering and particularly from the infiltration of rainwater into the building, spacing out the cladding elements from the wall will create a ventilated air gap, which is partially filled with an insulating layer improving the building's thermal efficiency and soundproofing, so that walls act as a rain screen.

A distinction should be made between a simple rain screen and a pressure-equalized rain screen wall, which is a rain screen comprising the exterior cladding, a cavity behind the cladding, drained and vented to the outside, and an inner wall plane incorporating an air barrier, plus a set of compartment seals limiting the cavity size.

Pressure-equalized rain screen represents an advancement of the simple rain screen systems since as wind acts on a wall face, the air is enclosed appropriately in the cavity compartment seals, and equalization of pressure occurs across the cladding, thus eliminating of the pressure driving forces (Fig. 33.6). Pressure-equalized rain screen systems offer the uppermost performance potential concerning water penetration [36].

The mechanical anchoring of stone cladding besides being an effective fixing system gives way to easy installation of rain screen systems beeing either ventilated or pressure equalized securely and properly.

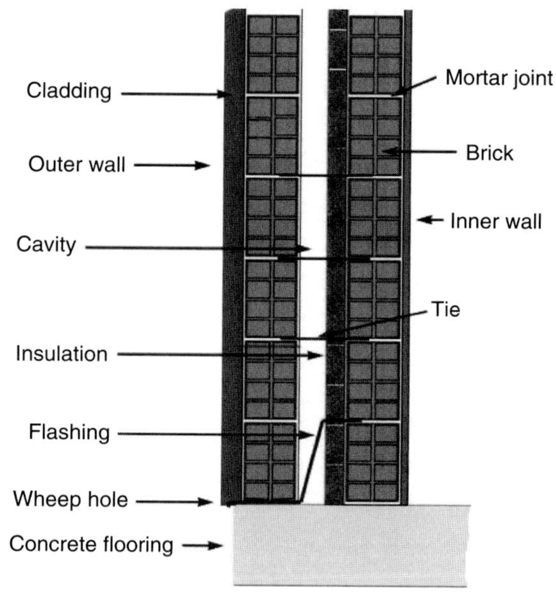

Figure 33.5 Cement-adhered cladding in a cavity wall.

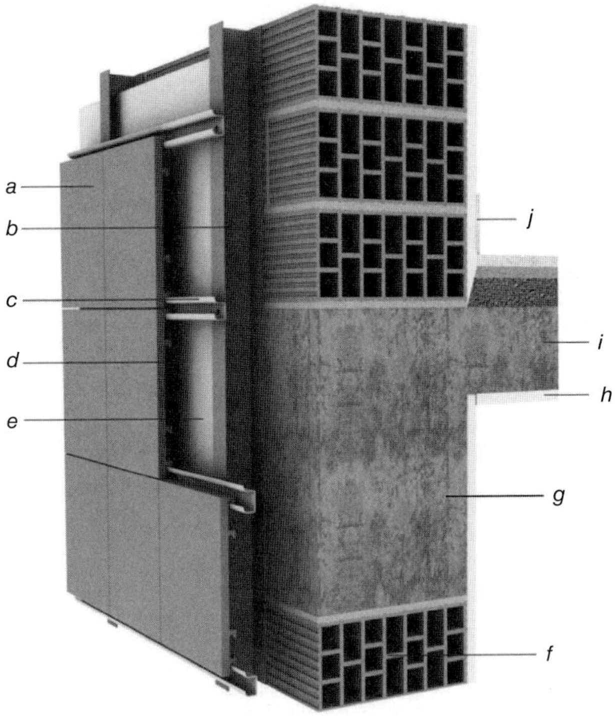

Figure 33.6 Pressure-equalized wall. (*a*) Cladding panel. (*b*) Mullion. (*c*) Transom and flashing. (*d*) Pressure equalization chamber. (*e*) Insulating layer. (*f*) Backup wall. (*g*) Beam. (*h*) Wall and ceiling lining. (*i*) Floor slab. (*j*) Floor skirting.

Exterior stone veneer may be installed in a manner that each stone panel is independently supported (Fig. 33.7).

Other systems are commonly composed of vertical and horizontal support framing of non-corrosion materials, such as aluminum or stainless steel. The framing is typically preinstalled in the stone setting cavity to the substrate or support wall (Fig. 33.8).

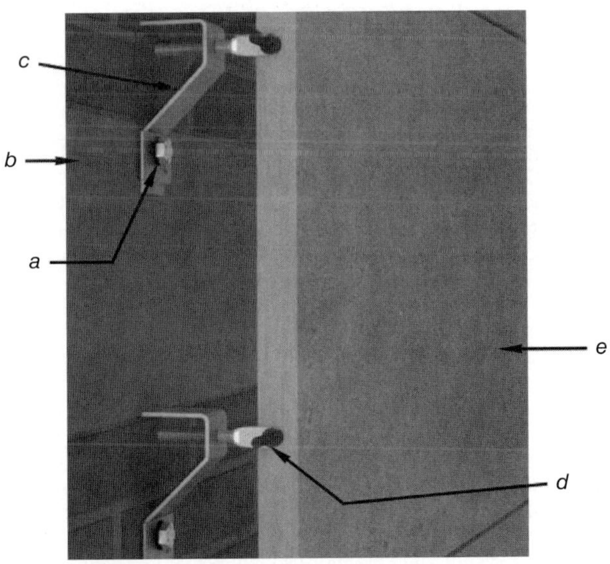

Figure 33.7 (*a*) Bolt or nut. (*b*) Backup structure. (*c*) Body anchor. (*d*) Loose pin with sleeve. (e) Stone cladding panel.

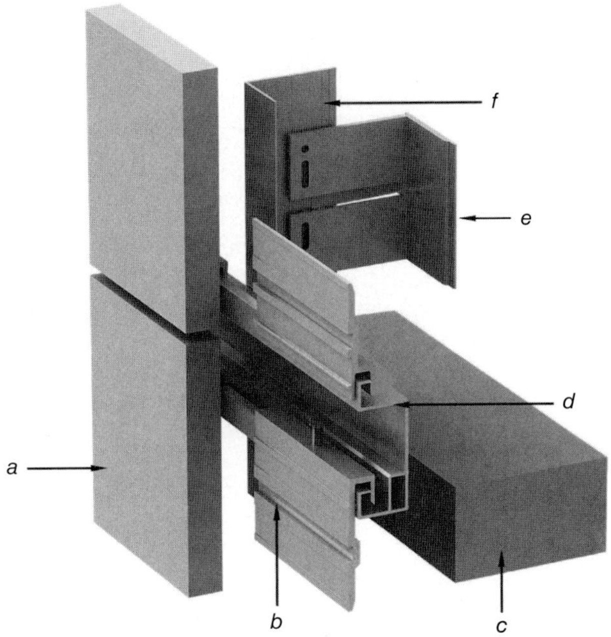

Figure 33.8 (*a*) Stone cladding panel. (*b*) Preinstalled integrated support element. (*c*) Slab. (*d*) Metallic transom attached to the slab. (*e*) Angle for mullion fixing to the slab. (*f*) Mullion.

With integrated stone curtain walls, panels are installed in glazing channels of glazed curtain wall members, similarly to metal spandrel panels or glass panels (Fig. 33.9).

Resistance to lateral loading, which is mainly due to the wind and seismic actions, is usually achieved using stainless-steel anchors inserted into kerfs or holes drilled or cut into the edges of the stone panels. These anchors are connected to the building structure by mechanical means, thus providing the essential mechanical connection between the stone and the structure. A structural weak point in this type of stone construction is to be found at the kerfs or anchor holes in the edges or back of the stone slabs. Such cuts need to leave enough stone thickness to provide the necessary strength to resist the various winds, seismic, and self-weight forces that act upon the stone panels.

Once different types of stones have different physical and mechanical characteristics, even if they are from the same type, information from testing and the knowledge of the stress states created within stone cladding panels by loaded anchorages has to be gathered. Modern design methodologies must be applied using reliability methods to obtain, like other materials, adequate safety factors for the specific type of anchor and/or stone used in each project.

33.5 LIMIT STATES DESIGN

When assessing civil engineering structures, the capability of a designed system to respond to project requirements or to meet user demands must be assessed considering fundamental principles. In the case of stone cladding, the same loads or actions shall be considered according to the common standards for each region or country. Also, design equations, characteristic values, and partial safety factors are needed to be defined as in the case of other structural materials.

Research that has been made in the recent years allows applying the same design criteria of human-made materials such as concrete or steel to stone so that it may be now considered not passable to discuss global safety factors, based on allowable design stresses when working with limit states design [20, 37].

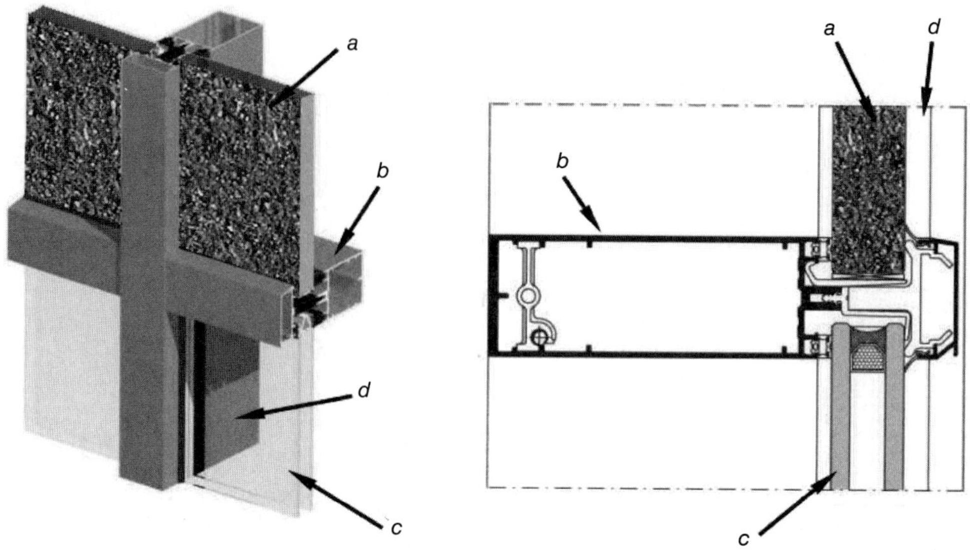

Figure 33.9 (*a*) Stone cladding panel. (*b*) Transom. (*c*) Glazing. (*d*) Mullion.

33.5.1 Design Lateral Actions

WIND ACTION

In order to obtain the lateral design action in building's façades, several codes and standards habitually consider a basic or reference wind speed for various locations in each region or country. A reference height of 10 m in the open terrain is chosen along with modification factors for the effects of height and terrain type, and pressure or force coefficients for structures of various shapes are also provided. Depending on the country region, the wind loads may be obtained directly from these standards such as the ISO 4354, Eurocode 1 parts 1–4, the ASCE Standard ASCE 7-98, AIJ Recommendations for Loads on Buildings, and the Australian Code AS/NZS 1170.2:2002 comprehending several years of work by committees from many countries [38, 39].

To peak velocity pressure, $q_p(z)$, at height z, which includes mean and short-term velocity fluctuations, depends on a turbulence factor and an orography factor and may be expressed as a function of the basic pressure and an exposure factor, c_z, as follows:

$$q_p = c_z \cdot q_b \tag{33.7}$$

The characteristic wind pressure value acting on the cladding surfaces, w_{Sk}, is obtained from Eq. (33.8) for a given reference height, z:

$$w_{Sk} = c_p \cdot q_p \tag{33.8}$$

Pressure coefficients, c_p, are defined for buildings and parts of buildings depending on the size of the loaded area, reference height, and eventual internal pressures for cladding elements in ventilated or pressure equalized rain screen façade systems.

SEISMIC DESIGN FORCE

It is widely recognized that damage of cladding elements accounts for a significant portion of total economic loss after an earthquake as much as more massive are the elements. For nonstructural elements of great importance, the seismic analysis shall be based on a realistic model of the relevant structures and the use of appropriate response spectra derived from the response of the supporting structural elements of the main seismic resisting system.

Different building codes have guidelines for the assessment of these elements, including local transmission of actions to the structure by the fastening of nonstructural elements and their influence on the structural behavior.

In the design process it is crucial to accommodate seismic relative displacements due to story drifts. Manufacturing and construction tolerances for both wall elements and the supporting structure must be considered in establishing joint design dimensions and connection details. Separations between adjacent panels are intended to limit contact and resulting panel misalignment or damage under all but extreme building response. A minimum separation of about 15 mm is recommended, yet for practical joint detailing and acceptable appearance, separations typically are limited to about 20 mm [40].

Performance expectations for nonstructural components, such as cladding panels, often are defined concerning the functional requirements of the structure to which the components are attached.

The seismic design force for a cladding system depends on the weight of the panel, the important factor, the component response modification factor, the component amplification factor, and the component acceleration at a point of attachment to the structure.

Once the goal of design for most nonstructural components is to prevent detachment or toppling that would pose a hazard to life safety, a value importance factor $I_p = 1.5$ shall be applied as for critical nonstructural elements, such as anchorage elements of machinery and equipment required for life-safety systems, tanks, and vessels containing toxic or explosive substances or fire sprinkler piping systems according to the Building Safety Seismic Council (BSSC) [40] and the CEN [41].

Seismic Design Force The effects of the seismic action may be determined by applying to the nonstructural element a horizontal force, F_E, referred to its center of gravity and distributed relative to the mass distribution. The value of F_E may be determined as given by Eq. (33.9) according to the BSSC [40]:

$$\begin{cases} F_E = \dfrac{0.4 \times F_B \cdot a_p}{R_E} \times \left(1 + 2 \times \dfrac{z}{H}\right) \\ F_B = S_{DS} \cdot W_E \cdot I_p \\ 0.3 \times F_B \le F_E \le 1.6 \times F_B \end{cases} \tag{33.9}$$

where a_p is the element amplification factor and R_e is the element response factor. The values of the coefficients are defined for veneers, depending on the degree of deformability of the elements and attachments as depicted in Table 33.7.

Dynamic amplification occurs where the period of a nonstructural component closely matches that of any mode of the supporting structure. The component amplification factor, a_p, represents the dynamic

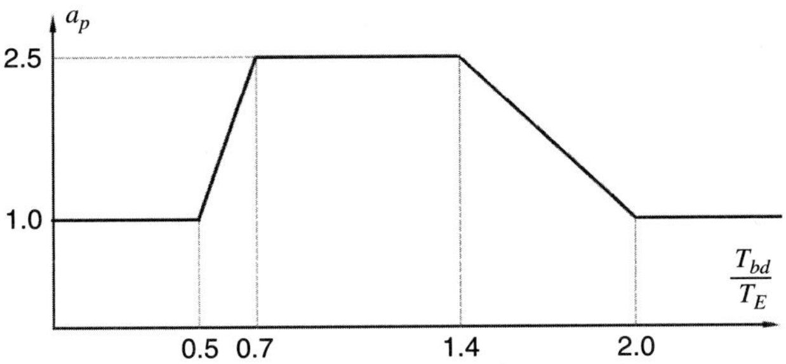

Figure 33.10 Formulation for, a_p, as a function of structural and component periods [42].

amplification of the element response as a function of the fundamental period of the structure, T_{bd}, and of the element, T_{el}. The NCEER [42] formulation shown in Fig. 33.10 may be used to compute a_p as a function of the ratio of the fundamental periods.

The component response modification factor, R_E, represents the energy absorption capability of the component's construction and attachments. In the absence of applicable research, these factors are based on judgment on the following benchmark values are indicated in Table 33.7 according to the Federal Emergency Management Agency of the U.S. Department of Homeland Security [40, 43].

Table 33.7 Properties Indicative Values for the Response Modification Factor

Component	R_E
Brittle or buckling failure mode is expected	1.0–1.5
Minimal level of energy dissipation capacity	2.5
Ductile materials and detailing	3.5

The recommended seismic provisions for new buildings [40] specifies a reduced response modification factor, R_E, and an increased amplification factor, a_p, for "fasteners" with the intention of preventing premature failure in those elements of connections that are inherently brittle. One example is the embedded items that depend on concrete breakout strength or connection elements that are just too small to adequately dissipate energy inelastically, such as welds or individual bolts. The net effect more than triples the seismic design force, and in general, it is recommended to consider in Eq. (33.9) $a_P/R_E = 1$ [43].

33.5.2 Ultimate Limit State

Nonstructural failures have accounted for the majority of hurricanes and earthquake damage on buildings façades, especially on transparent or opaque claddings with costly consequences [44].

The risk of failure of cladding implies the fallout of façade panels and represents a severe threat to life safety. Also, the cost of curtain wall façades can represent as much as the cost of its structure [45, 46, 47].

After predicting the maximum load and actual strength of stone cladding, the same probabilistic concept indicating the likelihood that the available strength will adequately withstand the maximum load over the lifetime of the façade cladding system shall be applied. The probability of failure is directly related to the reliability of a system and, depending on the consequence levels or classes, either those resulting in the loss of human life or economic, social, or environmental consequences.

Partial Safety Factors

For reliability differentiation, consequence classes (CC) may be established by considering the consequences of failure or malfunction of the structure as depicted in Table 33.8, according to the EC 1990 [48].

Table 33.8 Consequence Classes According EC 1990 [48]

Consequence class	Description
CC3	High consequence for loss of human life, or economic, social, or environmental consequences very great;
CC2	Medium consequence for loss of human life, economic, social, or environmental consequences considerable;
CC1	Low consequence for loss of human life, and economic, social, or environmental consequences small or negligible.

Considering given reliability or consequence class upon the formulation of a probabilistic code format design values derived by calibration, partial safety factors, combination factors, and so on are defined leading to the very well-known governing equation for limit state design when considering a limit state of rupture or connection failure of a member:

$$E_d \leq R_d \tag{33.10}$$

where E_d = design value of the effects of actions
R_d = design value or design resistance

The design value of the effect of actions, E_d, for the ultimate limit states is given by the appropriate equation: [2]

$$E_d = \gamma_g \cdot G_k \text{"+"} \gamma_f \cdot Q_k \tag{33.11}$$

In the same manner, for each critical load case, the design values of the effects of actions, E_d, are determined by combining the values of actions that are considered to coincide. Table 33.9 gives the recommended values for γ_g and γ_f according to EC 1990 [48]. Similar values are given in other standards for permanent or variable actions [49].

Table 33.9 Partial Factors for Actions for the Ultimate Limit States

Action	Persistent or transient design situations	Seismic design situations
Variable action	$\gamma_f = 1.5$	$\gamma_f = 1.0$
Unfavorable permanent action	$\gamma_g = 1.35$	$\gamma_g = 1.0$
Favorable permanent action	$\gamma_g = 1.0$	$\gamma_g = 1.0$

In the case of the combinations of two variable actions, for example, wind and snow, the effects to be considered should be based on the design value of the leading variable action and the design combination values of accompanying variable actions.

For persistent and transient situation, the following equation is assumed and for each leading variable action, Q_k, and the accompanying action as follows:

$$E_d = \gamma_g \cdot G_k \text{"+"} \gamma_f \cdot (Q_k \text{"+"} \psi_0 \cdot Q_a) \tag{33.12}$$

For combinations of actions in seismic design situations, with, E_k, the characteristic value of the seismic, the combination is given by the following equation:

$$E_d = \gamma_g \cdot G_k \text{"+"} E_k \text{"+"} \psi_2 \cdot Q_a \tag{33.13}$$

where ψ_0 and ψ_2 are factor combination values of the variable accompanying action. In EC 1990 [48] the recommended values for the combination factors for buildings depend on the geographic zone.[3]

The design value of the resistance, R_d, may be obtained from the characteristic value of the material or product resistance, R_k, and a partial safety factor, γ_M, to take account of unfavorable deviation of product properties, such as geometric deviations and uncertainty in the resistance model adopted, as a single material is used:

$$R_d = \frac{R_k}{\gamma_M} \tag{33.14}$$

[2]The "+" in the equation denotes "to be combined with".
[3]In case of Europe, the range values are $0.7 \geq \psi_0 \geq 0.5$ and $0.20 \geq \psi_2 \geq 0.0$.

Table 33.10 Quantile Factor, k_p, Depending on, n, Measured Values, and Sample Skewness, α

n	3	4	5	6	7	8	9	10	15	20	30	40	50	∞
$\alpha = 0$	3.15	2.68	2.46	2.34	2.25	2.19	2.14	2.10	1.99	1.93	1.87	1.83	1.81	1.64
$\alpha = -1$	4.31	3.58	3.22	3.00	2.86	2.76	2.69	2.63	2.45	2.33	2.23	2.19	2.09	1.85

In what stone cladding concerns, Eq. (33.10) may be established at the stress level values, σ_{Ed}, acting on a cross section as a result of external action and the design resistance stress value based on the stone strength, σ_{Rd}:

$$\sigma_{Rd} = \eta \cdot \frac{\sigma_{Rk}}{\gamma_M} \tag{33.15}$$

where σ_{Rk} = characteristic value of the stone strength
γ_M = partial safety factor depending on stone characteristic
η = aging factor applied to stone resistance decay

Based on the consequence class CC1 described in Eurocode 1990 [48] and assuming the corresponding probability of failure from Table 5.2,[4] a formulation to establish partial safety factors for dimension stone is considered.

Given that wind pressure and seismic forces are assumed to be approximated by Weibull distribution with a coefficient of variation equal to 40 percent and that flexural and tensile strength stresses follow a log-normal distribution, an index reliability $\beta = 3.3$ is applicable which corresponds to a probability of failure of $P_f = 0.5/1000$ and a given reference period of 50 years [37, 50].

Regarding the partial load factor, differences must be pointed out, such as in the case of wind action. For example, in the Eurocode's suite, a partial factor for this action is equal to 1.5 as for other the unfavorable variable actions but the seismic action.

Partial safety factors may be determined using the reliability-based code calibration of structural design codes, based on the Load and Resistance Factor Design (LRFD) format. The calibration of the partial safety factors may be performed, following the approach recommended by the Joint Committee on Structural Safety (JCSS) for different coefficients of variation (CV) of stone properties and the three consequence classes.

Characteristic Values of Mechanical Properties

Pull-out strength, flexure strength, compression, and tensile strength of the stone are fundamental properties, and their characteristic values shall be determined based on solid statistical evaluation upon test results that have to be performed.

The lower value of a material property shall correspond to the 5 percent lower fractile considered as the characteristic value.

The log-normal distribution is, in general, used to approximate the data from the tests, and the lower characteristic value, R_k, is obtained as follows:

$$R_k = e^{\bar{x}_{\ln} - (k_{p,5\%} \times s_{\ln})} \tag{33.16}$$

with

$$\bar{x}_{\ln} = \frac{1}{n}\sum_j \ln(x_j); \quad s_{\ln} = \pm\sqrt{\frac{\sum(\ln x_j - \bar{x}_{\ln})^2}{n-1}} \tag{33.17}$$

The quantile factor, k_p, depends on the number of measured values, n, and a confidence level, γ, and the skewness of the distribution, α, of the sample as well.

The skewness, α, is obtained with the formula

$$\alpha = \frac{n}{(n-1)(n-2)} \times \sum_j \left(\frac{(\ln x_j) - \bar{x}_{\ln}}{s_{\ln}}\right)^3 \tag{33.18}$$

[4]With EN 1990 the partial factors are considered generally leading to a structure with a reliability index equal to 3.3 for a 50-year reference period for CC1.

The quantile factor, k_p, for a recommended confidence level, $\gamma = 0.75$, is depicted in Table 33.10. Linear interpolation for skewness values in the range $0 > \alpha > -1$ may be applied.

Based on several studies [20, 37], a set of partial factors for cladding stone design may be used depending on the consequence classes. For the consequence class CC1 [48], the following equation gives suitable values for γ_M:

$$\gamma_M = 0.932 \times e^{4.674 \times \mathrm{CV}} \tag{33.19}$$

In Table 33.11, γ_M values are given for a range of CV values of the stone's modulus of rupture.

Table 33.11 Partial Safety Factors for Dimension Stone Cladding CC1 Classes of Consequences

CV	10%	20%	25%	30%	35%	40%
γ_M	1.5	2.4	3.0	3.8	4.8	6.0

In the United States, according to the International Building Code (IBC) [51], the wind load factor, LF, shall be taken with a value of 1.6, and in Canada, the National Building Code (NBCC) [52] recommends a wind load factor of 1.4, both values corresponding to the 50-year return period.

A direct comparison of actions and resistances between different building codes can be established once only a single material and one live load are involved. For instance, according to the IBC (USA) [51], the resistance factor, ϕ_R, is linearly proportional to the material partial safety factor, γ_M, in the Eurocode suite, so that the probability of failure is the same:

$$\phi_R = \frac{LF}{\gamma_f \cdot \gamma_M} \tag{33.20}$$

where LF = live load factor in the UBC
γ_f = partial factor for the variable action in the Eurocode

For example, in the case of the wind action, the equivalence between the Eurocode and the above-mentioned codes can be established.

In the Eurocode, the partial factor for wind load is $\gamma_f = 1.5$; thus, for the case the characteristic value of the stone strength in the UBC, the resistance factor shall be $\phi_R = \frac{1.067}{\gamma_M}$ since the load factor is equal to 1.6.

For Canada, in the National Building Code, the resistance factor shall be $\phi_R = \frac{0.933}{\gamma_M}$.

The corresponding resistance factors for both codes UBC [51] and CNBC [52] are given in Table 33.12 as a function of the coefficient of variation, CV, of the stone cladding properties assuming the class of consequences CC1: low consequence for loss of human life and economic, social, or environmental consequences small or negligible. This assumption is recognized by the CEN for curtain walls [53].

Table 33.12 Resistance Factors, ϕ_R, for Dimension Stone Cladding

CV	10%	20%	25%	30%	35%	40%
UBC	0.72	0.45	0.36	0.28	0.22	0.18
CNBC	0.63	0.39	0.31	0.25	0.20	0.15

For seismic action, there are no differences, and $\gamma_f = LF = 1$, so that the resistance factor shall be taken with $\phi_R = 1/1.5 \simeq 0.66(7)$.

In the same manner as defined in Eq. (33.15), the design value of the stone strength is given depending on the factor resistance and the aging factor:

$$\sigma_{Rd} = \eta \cdot \phi_R \cdot \sigma_{Rk} \tag{33.21}$$

AGING FACTOR

Several studies have been carried out to obtain coefficients that take into account the resistance decay of different types of stones. The resistance decay of each type of stone is generally addressed in the limit state design approach by multiplying the stone resistance obtained from laboratory tests by an aging factor of less than one, depending on the expected lifetime of the façade cladding, environment exposure, and type of stone [54, 55, 56].

Based on these studies, recommended values for an aging factor, η, considering stone resistance decay due to aging depending on the exposure conditions and a 30-year expected life cycle, are depicted in Table 33.13 as proposed by the author.

Table 33.13 Aging Factor, η, for Accounting on Stone Strength Decay [20]

Stone type	Exposure conditions: Humid environment without frost	Exposure conditions: Humid with frost and deicing salts
Fine to medium-grained granite	1.00	0.95
Oolitic limestone	0.90	0.85
Fine to medium-grained marble	0.85	0.75

33.5.3 Serviceability Limit States

Serviceability limit state design for cladding is based on defining limit values for maximum deflections. The limiting values of deflections for dimension stone cladding in horizontal planes are far from being critical. Excessive deformation of a dimension stone slab is typically caused by bowing and not by direct actions.

According to the IBC [57] that has been adopted for use as a base code standard by most jurisdictions in the United States and by other non-European countries, structural systems and members shall be designed to have adequate stiffness to limit deflections taking either partial factor, γ_M, or the resistance factor, ϕ_R, equal to 1.0.

Regarding exterior walls, these limits depend either the finishes are brittle or flexible. The wind load shall be taken as 0.42 times the allowable stress design wind speed—the "component and cladding" load—or directly calculated using the 10-year mean return interval wind speed. Although the IBC does not refer to stone cladding explicitly, the same provisions may be applied to those of framing members supporting glass subjected to load equal to 0.6 times the "component cladding" wind-defined loads.

Table 33.14 gives the limit deflection values according to the IBC [57] as a function of the length of the member between supports, *l*, and load type.

Table 33.14 Deflection Limits for Exterior Walls [57]

Types of finishes	2018 IBC Wind and snow loads
Plaster or stucco finishes	*l*/360
Brittle finishes	*l*/240
Flexible finishes	*l*/120

The European Standard EN 13830 [53] considers that curtain walls belong to classes of consequence CC1, which allow for the fact that the failure of the curtain walling structures or the in-fill panels does not have the same economic and human consequences of that failure of the building structures. The design value of the action, F_d, shall be for serviceability limit state

$$F_d = G_k \text{“+”} \psi_1 \cdot Q_k \text{“+”} \psi_2 \cdot Q_a \tag{33.22}$$

where ψ_1 = factor for the frequent value of the leading variable action
ψ_2 = factor for the quasi-permanent value of an unfavorable accompanying action

For nonstructural elements such as the case of façade panels, it is recommended to apply the load combination coefficients given in Table 33.15, according to prEN 16612 [58].

Table 33.15 Combination ψ Factor according to prEN 16612 [58]

Leading action	ψ_1	ψ_2
Snow	1.0	0.2
Wind	0.9	0.2

According to EN 13830, deflection limits depend on the load type and the length of the horizontal framing of the curtain wall, *L*, and/or the support distance or anchorages in the building structure, *H* (Fig. 33.11). In case of horizontal curtain walling's framing member (transom) acting as a sill under live horizontal loads at sill level, the maximum frontal deflection, *w*, of the curtain walling's framing members (transom) shall not exceed the following limits:

$$\begin{aligned} w &\le \frac{L}{200}, \text{ if } L \le 3000 \text{ mm;} \\ w &\le \frac{L}{300} + 5 \text{ mm, if } L > 3000 \text{ mm} \end{aligned} \tag{33.23}$$

The maximum deflection of any main horizontal framing (transom) due to self-weights shall not exceed L/500.

For wind loads and for snow loads, three deflection limits are given depending on the distance between the points of support or anchorage to the building's structure.

Under the imposed wind, only the maximum frontal deflection, *w*, of the curtain walling's framing members shall not exceed the following limits:

$$\begin{aligned} w &\le \frac{H}{200}, \text{ if } H \le 3000 \text{ mm;} \\ w &\le \frac{H}{300} + 5 \text{ mm, if } 3000 \text{ mm} < H < 7500 \text{ mm;} \\ w &\le \frac{H}{250}, \text{ if } H \ge 7500 \text{ mm} \end{aligned} \tag{33.24}$$

where is the distance between supports and anchorages measured in mm (Fig. 33.11).

33.6 DOWEL ANCHORAGE

Systems making use of insertions of dowel pins into drilled holes are one of the most used anchorage systems for dimension stone cladding. The fixing itself is provided by four dowel pins inserted into cylindrical holes at a half thickness on two parallel slab edges. After installation, the holes can be positioned in the horizontal or vertical position.

Each one of the parallel edges contains two holes properly spaced with an insertion depth of about two-thirds of the slab's thickness depending on the dowel's diameter.

The dowels work together with a perpendicular rod or shaft, usually a spade bolt with a lock nut to fix in the body anchor or a tubular section grout-in anchor. Each dowel pin is engaged in the drilled hole which is approximately 2 to 3 mm larger than the diameter to allow the space for a sleeve (Fig. 33.12).

Self-weight is to be supported only by lower anchors whether on the vertical or horizontal edges as well as the lower pins when anchoring is performed on vertical edges. The lateral loads are supported evenly by all dowels and rods. That is why the lower rods need to be more resistant because they must bear the whole of a plate's self-weight. It is

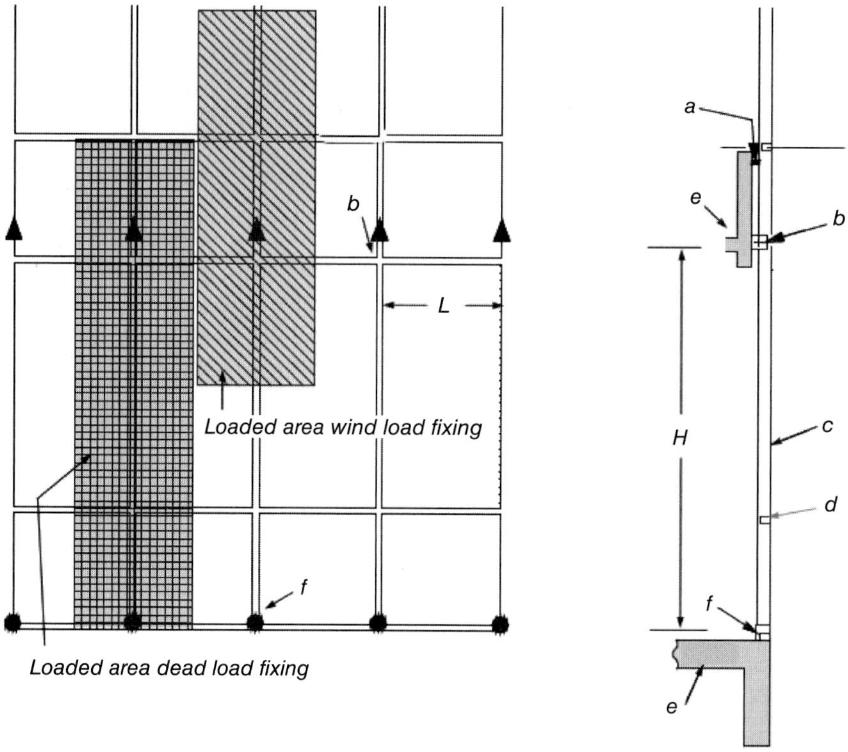

Figure 33.11 Curtain wall fixing and loaded areas. (*a*) Joint. (*b*) Dead load and wind load fixing. (*c*) Mullion. (*d*) Transom. (*e*) Building structure. (*f*) Wind load fixing.

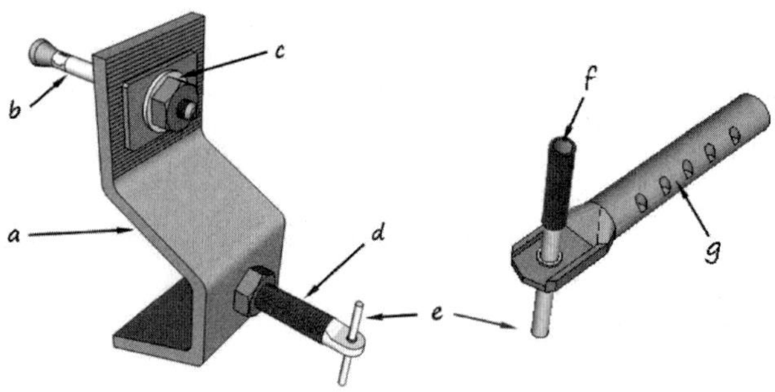

Figure 33.12 (*a*) Body anchor. (*b*) Fastener to backup. (*c*) Serrated or welded washer. (*d*) Adjustable shaft. (*e*) Dowel pin. (*f*) Sleeve. (*g*) Grout-in anchor.

recommended that there are no more than two holes per edge to avoid overstress states resulting from the misalignment of holes [59].

A schematic representation is illustrated in Fig. 33.13. A bottom anchorage (B) must withstand half self-weight of two panels and the lateral loads, while a top anchorage (T) only needs to resist the lateral loads. The distance from the center of a hole to the closest edge is denoted by a, and the slab lengths, L_1, and L_2, are, respectively, the dimensions of the supported and free edges.

In some instances, the use of mullions allows flexible positioning of the body anchors when the backup is either not regular or stiff to fix the anchors (Fig. 33.14).

Dowel pin common diameters and lengths are about 6 to 7 mm and 30 to 60 mm, respectively. Mounting of stone panels using dowel anchors must follow the inherent design issues, the principal ones of which are detailed pointed out in Fig. 33.15.

The point of the fixture of the metallic insert and the slab is considered, from a structural point of view, to be simple support, which, by definition, prevents any translational movement perpendicular to the slab's surface.

The primary operational load on the cladding comes from wind or earthquakes, which exert bending. Other applied loads such as self-weight and imposed deformations are not usually taken into consideration, as their effect is not relevant to the stress state of vertically placed panels. Any imposed deformations are generally considered to be engaged by the space between the insert and the slab as pointed out in Fig. 33.15.

33.6.1 Flexural Design

Actions that are transmitted from the panels to the supports throughout the drilled holes have to be filled with a resilient material to assure even bearing between all four contact points. Otherwise, a three-point supporting system may be formed for which the assumptions in the design are not compatible.

In that manner, resilient sleeves will also accommodate some misalignment in the drilled holes and will absorb induced vibrations. For optimizing panel capacity, the distance of the hole, a (Fig 33.13), from the unsupported edge can be defined by assuming that the maximum positive and negative bending moments have the same absolute value, and Eq. (33.25) is established:

$$a = \frac{L_1}{4} \times \frac{\sqrt{2}}{1+\sqrt{2}/2} \approx \frac{1}{5} \times L_1 \tag{33.25}$$

If the length of the supported edge, L_1, is greater than the length of the unsupported edge, L_2, the limit value for the ratio of the panel dimensions to define which direction is governing the maximum bending stresses on the panel is given by the following equation:

$$\frac{L_2}{L_1} < \frac{1}{1+\sqrt{2}} \cdot \kappa \approx 0.5 \tag{33.26}$$

where $\kappa \approx 1.2$ is a stress concentration factor to take into account the stress concentration along the support's region [61].

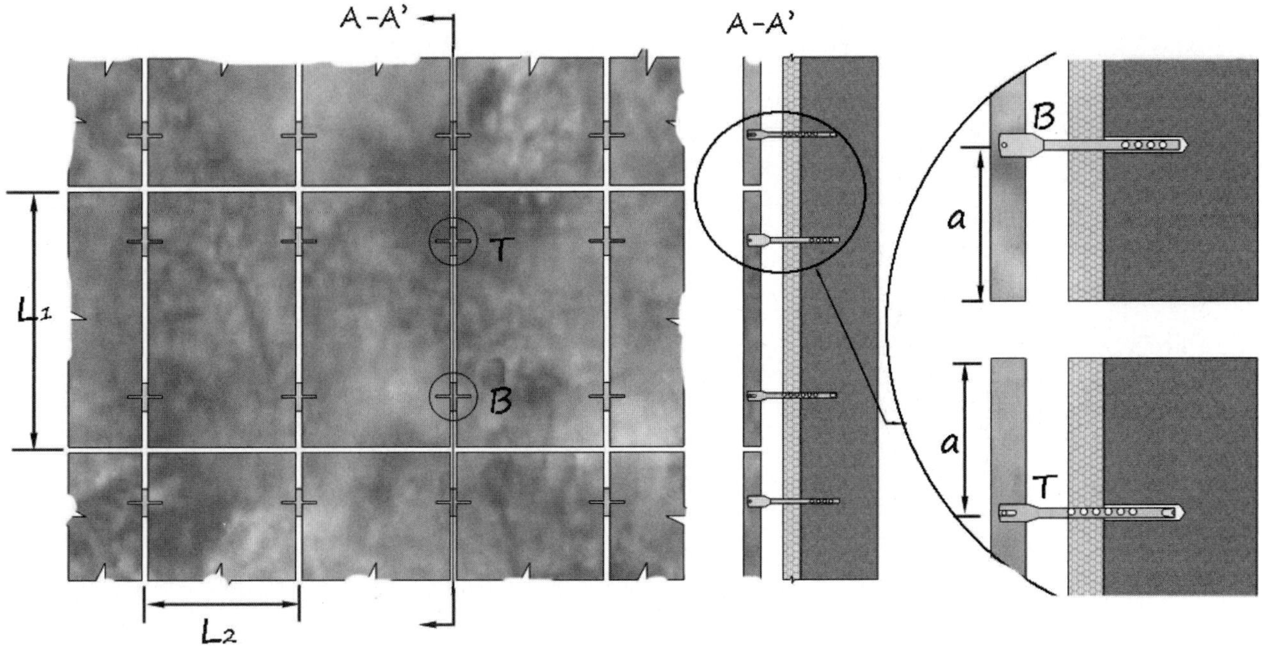

Figure 33.13 Panel configuration and dimensions, dowel number per edge, location, and self-weight.

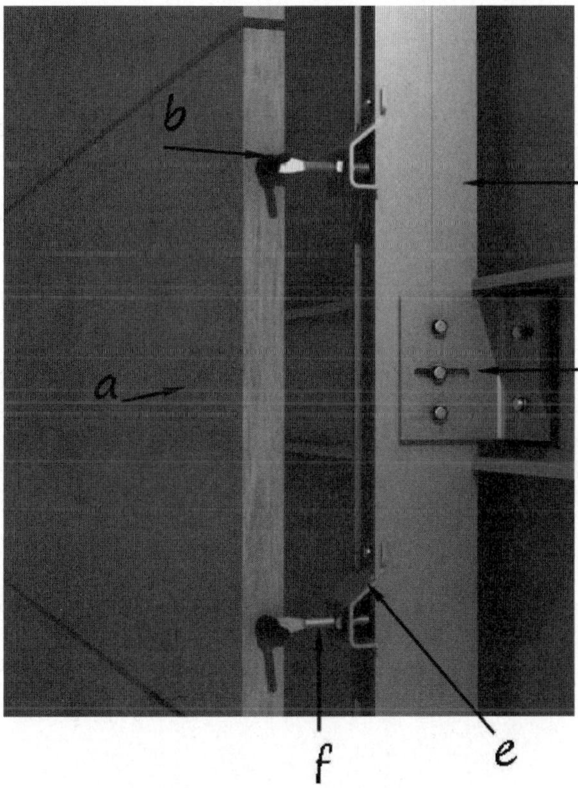

Figure 33.14 Body anchors attached to mullions allowing for a flexible mounting curtain wall. (*a*) Slab. (*b*) Sleeved dowel pin. (*c*) Mullion. (*d*) Slotted angle. (*e*) Body anchor. (*f*) Shaft.

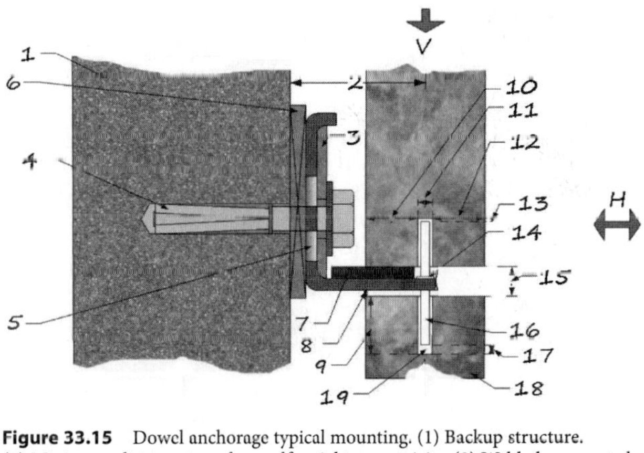

Figure 33.15 Dowel anchorage typical mounting. (1) Backup structure. (2) Minimum distance to reduce self-weight eccentricity. (3) Welded or serrated washer plate preventing slide after vertical alignment. (4) Attaching anchor to backup. (5) Horizontal slot to align dowel with hole. (6) Bearing shim to separate and adjust anchor position. (7) Resilient bearing shim for leveling and to prevent bearing of stone on dowel weld. (8) Clearance to avoid contact and weight transfer allowing for movement and installation tolerances. (9) Hole depth to be minimized once greater engagement length does not mean more anchor capacity. (10) Stone edge distance to resist outward actions. (11) Hole to be located in the center of panel thickness. (12) Stone edge distance to resist inward actions. (13) Clearance to avoid point loading on dowel end. (14) Weld at fixed dowel with minimum diameter hole for clearance at drop dowel. (15) Joint to allow for tolerances and clearances for building movements. (16) Bottom panel engaged avoiding misalignment that may result in prying action in stone panel. (17) Clearance to avoid loading dowel ends as in (8). (18) Dimension stone cladding panel. (19) Holes to be filled with compressible material to prevent moisture accumulations [60].

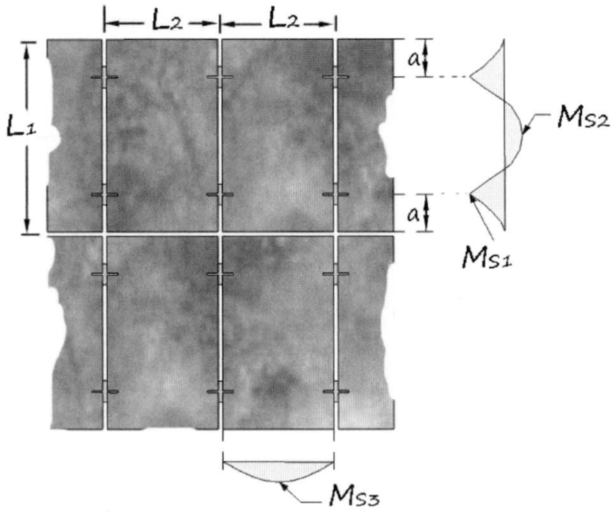

Figure 33.16 Panel proportion and optimal distance from the drill to the edge.

Considering the notation used in Fig. 33.16, if $L_2/L_1 > 0.5$, the governing bending moment, M_{S3}, is parallel to L_2; otherwise if, $L_2/L_1 \leq 0.5$, the governing bending moment M_{S2} or M_{S1} is parallel to L_1, provided that $a \simeq 0.5$.

In most of the installations, $L_2/L_1 > 0.5$, and the governing bending moment is M_{S3}—along with the L_2 direction. For this case, the maximum bending stress in the stone is located at midspan, along with a band strip connecting the supports. A nonuniform distribution of the bending moment values across a section parallel to, L_1, and stress concentrations must be considered when computing the maximum flexure tensile stress.[5] A general equation may be derived to express the design value of the maximum bending stress as follows:

$$\sigma_{Sd} = \frac{3 \cdot q_{Sd} \cdot L_2^2 \cdot k_{d2}}{4 \cdot t^2} \tag{33.27}$$

where q_{Sd} = design value of the lateral due to seismic or wind action
k_{d2} = stress concentration factor for the L_2 direction
t = nominal thickness of the slab

For typical supporting conditions, $k_{d2} \simeq \mathbf{1.30}$ [20], and Eq. (33.27) is simplified:

$$\sigma_{Sd} = \frac{0.975 \times q_{Sd} \cdot L_2^2}{t^2} \tag{33.28}$$

The design value of the stone strength, σ_{Rd}, being known, as defined in Eq. (33.15) or Eq. (33.21), it is straightforward to calculate the minimum slab thickness at midspan:

$$\sigma_{Rd} \geq \sigma_{Sd} \rightarrow t \geq \sqrt{\frac{0.975 \times q_{Sd}}{\sigma_{Rd}}} \simeq L_2 \cdot \sqrt{\frac{q_{Sd}}{\sigma_{Rd}}} \tag{33.29}$$

If $L_2/L_1 < 0.5$, the governing bending moments shall be $|M_{S1}| = |M_{S2}|$, parallel to L_1 since $a \simeq L_1/5$ as in Eq. (33.25). In these cases, the acting maximum bending stress will occur in the drilled section at the assumed distance of the hole axis to the nearest parallel edge to the pin's direction. The stress value must be calculated considering the hole's diameter and the stress concentration due to its geometry:

$$\sigma_{Sd} = \frac{3 \cdot q_{Sd} \cdot a^2 \cdot k_{d1} \cdot t}{t^3 - \phi^3} \tag{33.30}$$

where q_{Sd} = design value of the lateral due to seismic or wind action
a = distance of the hole axis to the nearest slab's edge parallel to L_2 direction
k_{d1} = stress concentration factor considering the geometry of the section at supports
ϕ = diameter of the hole

A typical value for $k_{d2} \simeq 1.50$ [20], and Eq. (33.30) becomes

$$\sigma_{Sd} = \frac{9 \times q_{Sd} \cdot a^2 \cdot t}{2 \times (t^3 - \phi^3)} \tag{33.31}$$

33.6.2 Stone Anchorage Strength

Several parameters limit the dowel anchor capacity: the dowel hole's diameter, the distance from the edge of the stone to the edge of the hole, the distance at which the dowel pin is inserted into the drilled hole, and, among other factors, the depth that the dowel actually contacts the stone which may provoke prying action [62]. The relative stiffness of the slab, anchor device, and backup control the way how the anchorage device is resisting the action transmitted to the whole assembly. It has been pointed out that for the general panels dimensions in the vicinity of the supports, a higher stress concentration occurs. Thus, the stone thickness is more dependent on the pull-out strength design than in the flexural design, the reason why focusing on the bending strength of stone panels to the detriment of anchorage strength has been pointed out as the cause of a significant number of failures [59].

The area around the hole depth in contact with the dowel pin either through a rigid connection or through a sleeve's resilient material will determine the type of the contact between the dowel and the hole along a variable length in its root [63].

The use of the finite-element method with linear-elastic material properties, together with the maximum principal stresses failure criteria when the module of rupture is attained, has been shown to be an appropriate design procedure for estimating the breaking load at the dowel hole [59, 63]. Thus, following Figs. 33.17 and 33.18, the design breaking load at the dowel hole can be estimated as follows:

$$F_{brd} = \frac{\sigma_{Rtk} \cdot l_d \cdot (t - \phi \times \sin\alpha)}{6 \times k \cdot \tan\alpha \cdot \gamma_M} \cdot \eta \tag{33.32}$$

where σ_{Rtk} = characteristic tensile strength value of the stone
t = thickness of the dimension stone
ϕ = drilled hole diameter
l_d = contact depth of the pin dowel
α = spall angle from primary surface slab's plane
k = concentration factor to consider the additional prying action induced by a resilient sleeve
γ_M = partial factor of safety for the stone, according to Eq. (33.19)
η = aging factor considering the loss of strength of the stone according to Table 33.13

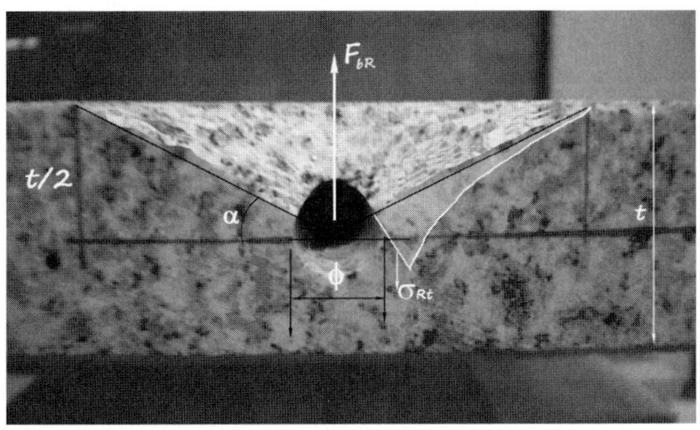

Figure 33.17 Edge drilled view of a dimension stone representing stress distribution and resultant.

[5]Stress concentration factors are obtained, either from the elasticity theory using analytical solutions, experimental analysis, or the finite-element method.

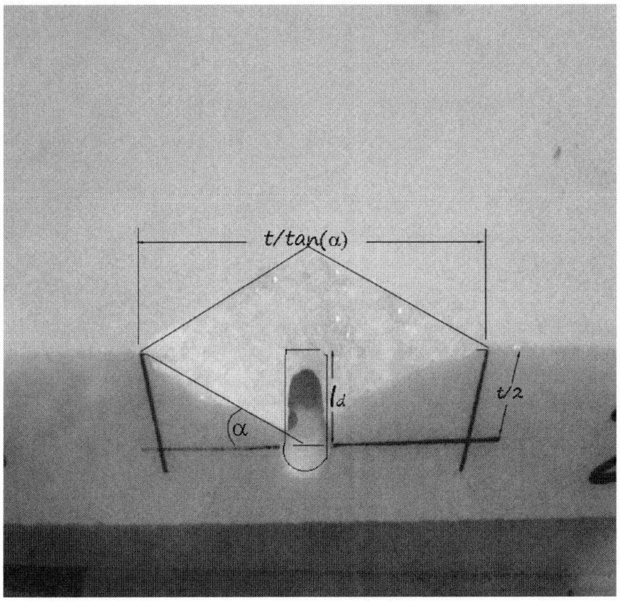

Figure 33.18 Idealized failure surface section and plane view upon a pull-out test of a dowel in a drilled hole in a marble stone.

The values of the stress concentration factors, k, corresponding to the effect of the resulting prying action and consequent stress amplification due the presence of a resilient sleeve between the dowel and the stone have been studied and are correlated with the average values of the spall angles, α, [59, 64]. Their values are depicted in Table 33.16.

Table 33.16 Stress Concentration Factor, *k*, and Spall Angle, α, to Be Used in Eq. (33.32)

Stone type	α	k
Fine-grained granite	30°	1.2
Oolitic limestone	21°	1.3
Crystalline marble	19°	1.4
Semicrystalline limestone	22°	1.6

The design value of the action or load uniformly transmitted to each of the four dowels, F_{Sd}, as in Fig. 33.19, is given by

$$F_{Sd} = \frac{w_{Sd} \cdot L_1 \cdot L_2}{4} \tag{33.33}$$

where w_{Sd} is the design value of the uniform lateral load due to wind or seismic action acting in a dimension stone slab with L_1, L_2 edge lengths.

The safety verification format implies that $F_{brd} \geq F_{Sd}$. Substituting their values in Eqs. (33.32) and (33.33), the minimum thickness, t_{min}, of a stone slab at the supports is given by

$$t_{min} \geq 1.5 \times w_{Sd} \cdot L_1 \cdot L_2 \cdot k \frac{\tan\alpha}{\sigma_{Rtd} \cdot l_d} + \phi \cdot \sin\alpha \tag{33.34}$$

where $\sigma_{Rtd} = \sigma_{Rtk} \cdot \dfrac{\eta}{\gamma_M}$ the design value of the tensile strength of the stone.[6]

33.7 KERF ANCHORAGE

Typical kerf anchor constitution is usually formed in stainless steel or extruded aluminum profiles. They may be continuous or discontinuous and are typically located on the top and bottom of slabs' edges for easier access and alignment during installation. These strap anchors are flat metal bars that are designed to be inserted into a kerf (slot) in the edge of the stone panel.

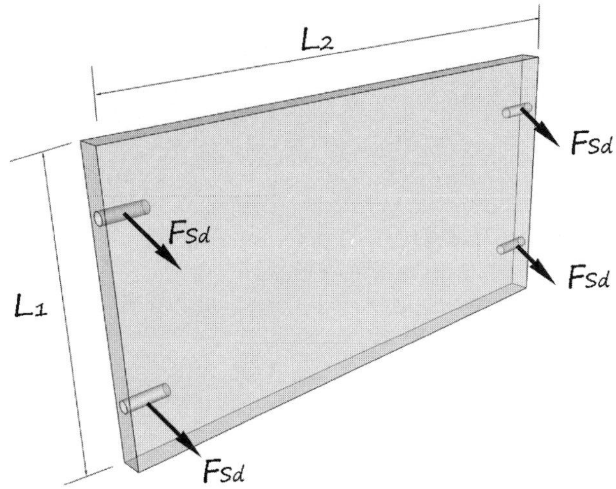

Figure 33.19 Dimension stone support reactions under uniform lateral action.

The anchor transfers the load from the stone to the backup structure or wall and must be stiff enough to hold the load without relevant deformation while bending, which is a common misuse of the kerf anchoring.

Kerf isolated anchors are in general stainless steel and double-tailed. The anchorage length, L, depth, D, and width, B, depend on the required resistance as a function of the load's magnitude and slabs dimensions and weight (Fig. 33.20).

The kerf anchorage capacity depends substantially on the slot width, t_k, the thickness of the stone fin, f_i, the depth of contact, D_k, and the engagement width of the clip's leg [65, 66]. The total depth kerf cut, D_t, depends on the clearance, C, needed to accommodate certain differential backup movements. Figure 33.21 represents a detailed cross section identifying the key parameters and the uniformly distributed variable action, q_{Sd}, and G_{Sd}, the permanent linearly distributed design action.

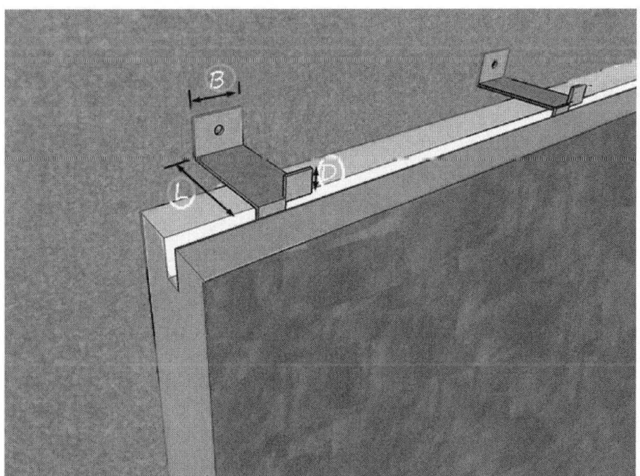

Figure 33.20 Double-tailed kerf stainless-steel relevant dimension.

[6]Or, following the format of Eq. (33.21), $\sigma_{Rtd} = \sigma_{Rtk} \cdot \eta \cdot \phi_R$.

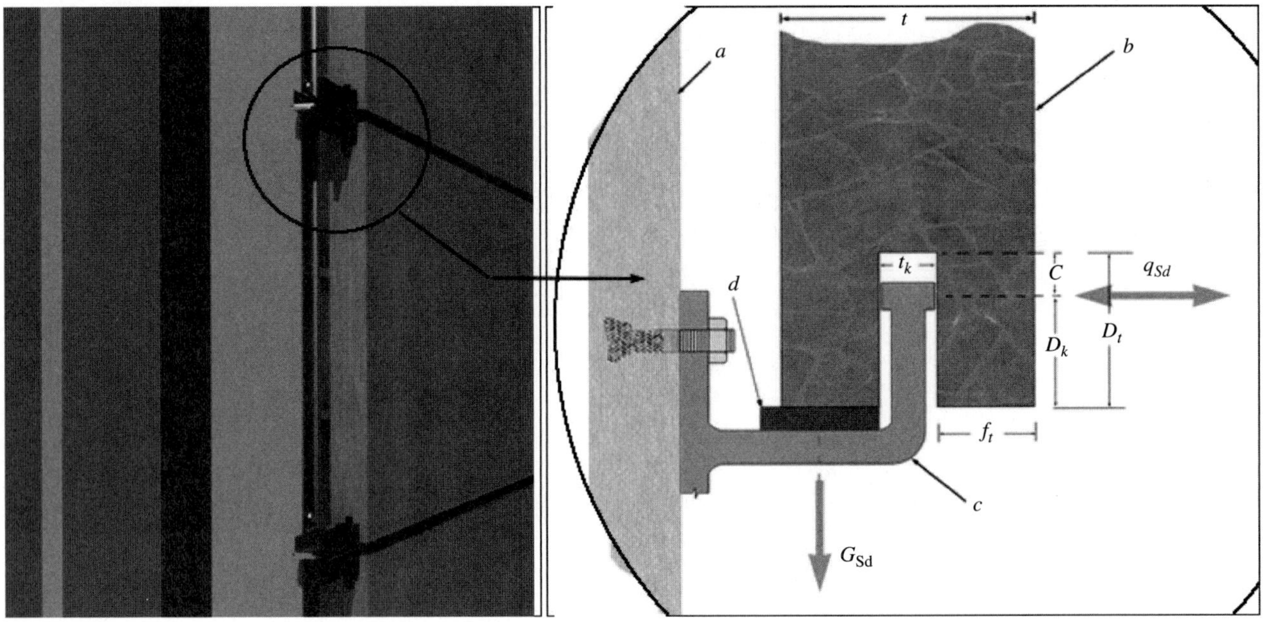

Figure 33.21 Kerf anchorage section detail. (*a*) Backup structure. (*b*) Stone. (*c*) Stainless-steel rail or angle. (*d*) Resilient shim or strip.

33.7.1 Flexural Design

The bending moment design value, M_{Sd}, occurs at a midspan cross section of a slab performing like a two-way slab and is expressed as a function of the distance between the center of the supports alignment, L, (Fig. 33.22) and the design value of the transverse design load, q_{Sd}, due to wind or seismic action:

$$M_{Sd} = \frac{q_{Sd} \cdot L^2}{8} \tag{33.35}$$

The resulting maximum stress design value, σ_{Sd}, is determined to assume that a linear uniform distribution is installed along with the slab's unsupported direction, where t is the slab's thickness nominal value $t = 2 \times f_t + t_k$:

$$\sigma_{Sd} = \frac{3 \times q_{Sd} \cdot L^2}{4 \times t^2} \tag{33.36}$$

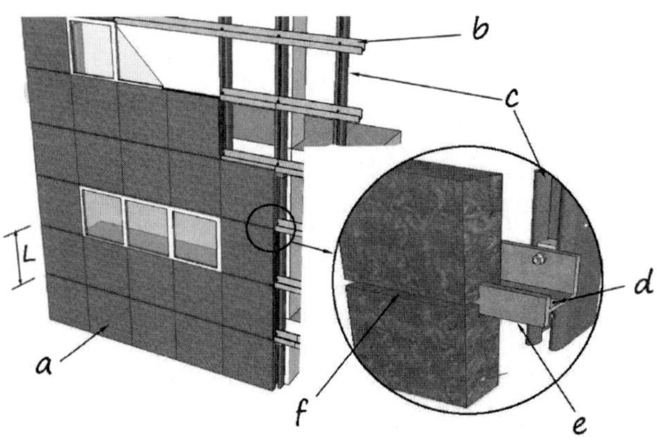

Figure 33.22 Aluminum extrusions as a combination of gravity and lateral load restraint devices. (*a*) Stone slab. (*b*) Aluminum transom. (*c*) Aluminum mullion. (*d*) EPDM shims to accommodate slabs. (*e*) Kerf anchorage. (*f*) Serrated edge stone.

The design value of the stone flexure strength, σ_{Rd}, being known, the slab's thickness, t, is calculated since $\sigma_{Rd} \geq \sigma_{Sd}$:

$$t \geq L \cdot \sqrt{\frac{3}{4} \times \frac{q_{Sd}}{\sigma_{Rd}}} \tag{33.37}$$

33.7.2 Stone Anchorage Strength

Some of the main issues related to kerf anchoring are pointed out in Fig. 33.23. As in any other anchoring system, prevention of slip of the connection after vertical adjustment with a slotted washer plate not relying on friction is mandatory. It is essential to maintain clearances to avoid point contact and weight transfer, allowing for structural movements, creep, expansion, and installation tolerances. In order to maximize capacity and minimize prying, the kerf depth, D_t, shall be kept as small as possible. The saw cut located continuous or intermittent must be in the center third of the thickness, minimizing width to maximize edge distances.

The strength of this type of anchorage depends mostly on the combined shear and flexural strength of the stone's fin or leg that will resist both inward and outward seismic or wind action and the support rail's stiffness as well. Based on several studies, it has been concluded that the width to which the kerf clip's leg is engaged, B, is not necessarily the effective breadth of engagement, B_{eff}. If the rail's stiffness is less than the leg stone's stiffness along the entire kerf rail, length will not act as continuous support unless the kerf rail is much stiffer than the stone panel [65, 67].

Different support conditions are assumed depending on the effective contact width between the anchorage device and the stone. In a continuous rail system, if the stiffness of the slab's fin is superior to the rail blade in contact with the stone, an effective contact width, B_{eff}, is assumed at each of the quarters of the slab's ends [67]. In other cases, the anchorage width, B, is smaller than the dimension stone slab's, B_{tot}, and the effective width is defined taking into consideration the strap anchor length and the post-break format of slab's edge (Fig. 33.23).

Thus, for continuous support anchorage systems, the effective engagement width is

$$B_{eff} = 0.5 \times B_{tot} \tag{33.38}$$

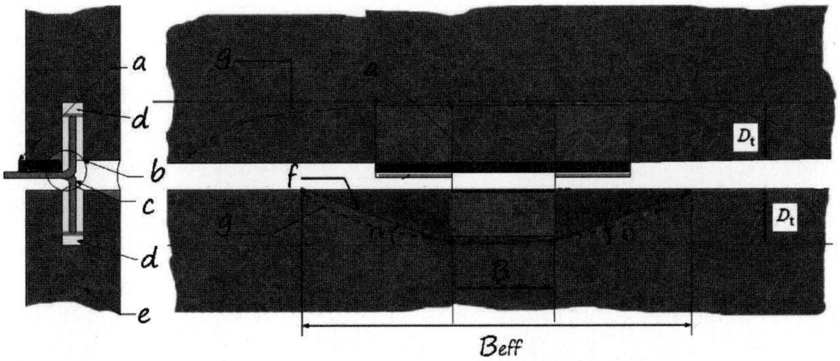

Figure 33.23 Kerf effective width anchorage and idealized spall configuration. (*a*) Shim. (*b*) Tolerance avoiding prying action. (*c*) Anchor. (*d*) Clearance avoiding point loading at the kerf. (*e*) Dimension stone slab. (*f*) Idealized rupture line. (*g*) Actual rupture line. D_t—Kerf depth.

For discrete anchoring devices, effective engagement with is

$$B_{\text{eff}} = B + 2 \times D_t \cdot \cot\alpha \qquad (33.39)$$

The strength of the stone at the anchorage is given by the effective cross section of the kerf's leg, f_t, and the transverse action is determined taking into consideration its design value, q_{Sd}, the span, L, and the full width of the slab, B_{tot}.

For each side of the supported slab, the value of the acting uniform load in the slab corresponds to the following reaction at the kerf's region:

$$F_{Sd} = \frac{q_{Sd} \cdot L \cdot B}{2} \qquad (33.40)$$

For stone, a brittle or quasi-brittle material, the rupture Rankine criterion, assuming that failure occurs in a multiaxial state of stress when principal stress reaches the uniaxial tensile strength validated by finite element analysis, leads to the following design formulae for the principal stress, using the notation of Fig. 33.22 and Fig. 33.23 [65].

$$\sigma_{\text{prin}} = 3 \times \frac{F_{Sd} \cdot C}{B_{\text{eff}} \cdot f_t^2} + \sqrt{9 \times \left(\frac{F_{Sd} \cdot C}{B_{\text{eff}} \cdot f_t^2}\right)^2 + \left(\frac{F_{Sd}}{B_{\text{eff}} \cdot f_t}\right)} \qquad (33.41)$$

An effective stress concentration factor, K_{eff}, must be applied to obtain the effective maximum stress taking into consideration the geometry of the kerfed section and a magnification factor as well [65, 68]. Thus, the following equation is used to ascertain the effective stress concentration factor:

$$K_{\text{eff}} \approx 1.6 \times (1 + \tan\alpha) \qquad (33.42)$$

Spall angles and concentration factors are depicted in Table 33.17 to be used together with Eqs. (33.39) and (33.42).

Table 33.17 Spall Angles and Effective Stress Concentration Factor

Stone type	α	K_{eff}
Fine to medium-grained granite	35°–40°	2.9
Oolitic limestone	13°–20°	2.1
Crystalline marble	40°–45°	3.2
Semicrystalline limestone	17°–35°	2.4

Considering the stress concentration factor, K_{eff}, and substituting Eq. (33.40) for the design acting load and introducing the parameter, $\delta = \frac{B_{\text{eff}}}{B_{\text{tot}}}$, in Eq. (33.41), which stands for the nominal tensile stress, σ_{prin}, the following equation for the value of the maximum acting tensile stress in the kerf region is obtained:

$$\sigma_{Sd} = \frac{3 \times q_{Sd} \cdot L \cdot C + \sqrt{q_{Sd}^2 \cdot L^2 \cdot (9 \times C^2 + f_t^2)}}{2 \times \delta \cdot f_t^2} \times K_{\text{eff}} \qquad (33.43)$$

The safety verification format being established for the ultimate limit state at stress level with

$$\sigma_{Rd} \geq \sigma_{Sd} \qquad (33.44)$$

establishes the minimum thickness of the stone, $t_{\min}$, depending on the kerf slot width, t_k, and on the fin thickness, f_t:

$$f_t \geq \frac{1}{2} \times \frac{q_{Sd} \cdot L \cdot K_{\text{eff}} \cdot \sqrt{1 + \left(\frac{12 \times \sigma_{Rd} \cdot \delta \cdot C}{q_{Sd} \cdot L \cdot K_{\text{eff}}}\right)}}{\sigma_{Rd} \cdot \delta}$$

$$\delta = \frac{B_{\text{eff}}}{B_{\text{tot}}} \qquad (33.45)$$

$$t_{\min} = 2 \times f_t + t_k$$

where L = distance between center of the supports
B_{tot} = slab's width
B_{eff} = effective contact width of the slab
$\delta = \frac{B_{\text{eff}}}{B}$ = support contact parameter
t = thickness of the dimension stone
C = kerf clearance distance to the blade's rail
f_t = fin's edge stone thickness
t_k = kerf slot width
α = spall detachment angle
K_{eff} = effective stress concentration factor
σ_{Rd} = design value of the stone flexure strength as defined in Eq. (33.15) or (33.21)

33.8 UNDERCUT ANCHORAGE

Keying type anchors carry the tensile load through main keys at the end of the anchor, resulting in a cone shape failure or the yielding of the steel rod. This is the case of undercut anchoring. This technology, combined with a suitable support framework, allows implementing a safe and high-performance fixing system for stone cladding façades.

Recent studies revealed that the weakest part of an undercut cladding system is the aluminum bracket and connector angles, whereas after being submitted to shaking table test loads, the undercut bolt anchorage itself has a good performance [72].

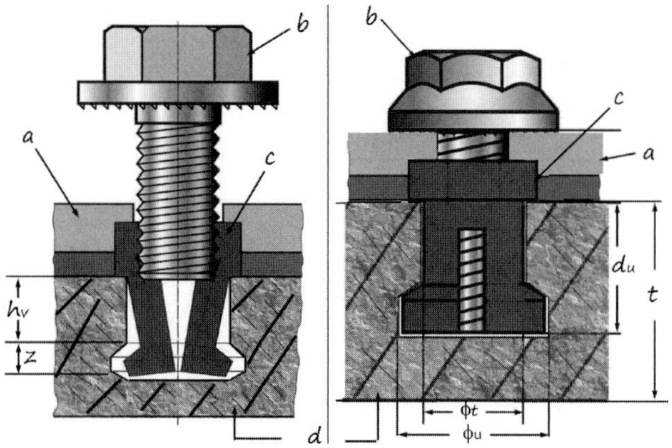

Figure 33.24 Typical cross section in a stone slab with undercut drilling slotted sleeve and internal thread. (*a*) Panel bracket. (*b*) Hexagon bolt with an internal thread and tooth lock. (*c*) Slotted sleeve. (*d*) Dimension stone slab. *t*—Slab thickness. d_u—Anchorage depth. ϕ_t—Drill hole diameter. ϕ_u—Undercut diameter. h_v—Constant-diameter hole depth. *z*—Variable-diameter hole depth.

There are mainly two types of undercut technologies to provide a keying-type anchorage in the interior of the dimension stone or slab's thickness. One type is using a slotted sleeve and internal thread (Fig. 33.24), and the other keying is obtained with an expansion ring with an internal or external thread (Fig. 33.25). In the former, the anchor sleeve is expanded to its original dimension by inserting the screw to a controlled depth so that the sleeve sits snugly against the undercut section of the hole in the façade panel. In the latter, anchors are installed by driving the anchor sleeve against a locking ring, thus forcing it to expand within the undercut hole form and locking it within the stone.

The undrilled depth, $(t-d_u)$, usually is higher than $0.4\times t$ because the stone thickness governs the undercut anchorage resistance in tension under negative wind pressure on façades. For positive pressures, most resultant forces are transmitted directly to a supporting frame system or any other façade backup structure.

The support system for a dimension stone panel in most cases consists of a framework in aluminum or stainless steel. Generally, the framework consists of four brackets attached to the back of the façade panel through an appropriate undercut anchor which are then placed into or onto the corresponding continuous horizontal rails. Generally, the horizontal rails are attached to vertical mullions or profiles, which are fixed to the building's main structure, structural concrete, or masonry (Fig. 33.26).

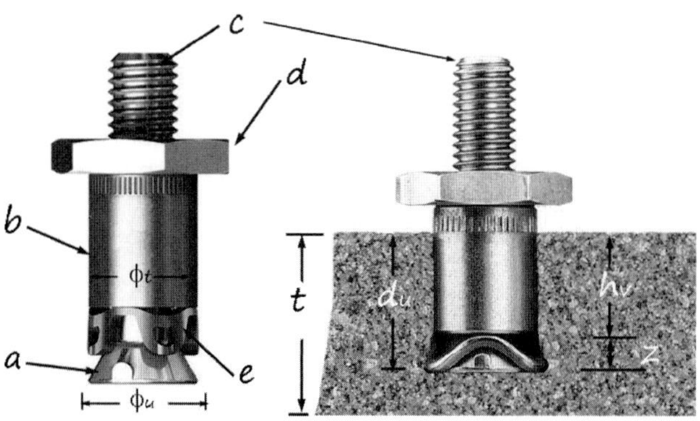

Figure 33.25 Typical cross section in a stone slab with undercut drilling with an expansion ring and external thread. (*a*) Cone bolt. (*b*) Sleeve. (*c*) External thread. (*d*) Nut. (*e*) Expansion ring *t*—slab thickness. d_u—Anchorage depth. h_v—Constant-diameter hole depth. *z*—Variable-diameter hole depth. ϕ_t—Drill hole diameter. ϕ_u—Undercut diameter.

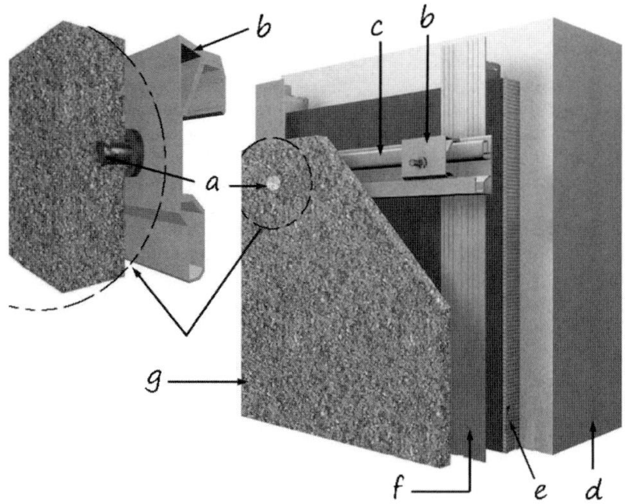

Figure 33.26 Typical framing supporting structure. (*a*) Undercut. (*b*) Bracket. (*c*) Horizontal rail. (*d*) Backup wall. (*e*) Insulating layer. (*f*) Vertical rail. (*g*) Dimension stone slab.

33.8.1 Flexural Design

As panels are supported in four points located at the undercut anchor's axis, a two-way flat plate analogy is suitable to analyze flexure behavior. However, if a resilient material is interposed between the supporting rails and the stone, a one-way two-side cantilevered slab structural system may be assumed. Preferred rail's location shall be positioned to a fifth of the panel's dimension for each direction with $a_y \simeq 0.2\times L_y$ and $a_x \simeq 0.2\times L_x$ (Fig. 33.27).

Several computational programs are available to compute flexural stresses using 2D or 3D finite elements, yet the well-known strip method [70] has been applied with acceptable accurate results when a two-way flat plate structural system shall be considered [20].

Hence, for each direction parallel to its sides, *X*-*Y*, the panel is divided into strips: one middle strip, *m*, and two sides strips, *s*, the widths defined as follows considering Fig. 33.28:

$$m_X = \frac{L_Y}{2} - a_Y;\ s_X = \frac{L_Y}{4} + \frac{a_X}{2} \tag{33.46}$$

$$m_Y = \frac{L_X}{2} - a_X;\ s_Y = \frac{L_X}{4} + \frac{a_Y}{2} \tag{33.47}$$

Maximum stress values in each direction are calculated upon the definition of coefficients of distribution, *u*, defined for middle and lateral strips depending on the cross-section location. The following notation is used for the cross-section location: *a*—for the support line regions, *c*—for the central regions, *m*—for middle strips, and *l*—for external strips. Their values are depicted in Table 33.18.

Given a lateral action, q_{Sd}, for the middle strip at the center of the slab of thickness, *t*, and anchorage depth, d_u, the maximum stress along the X direction is given by

$$\sigma_{mc_X} = \frac{3}{2}\times\frac{u_{mc_X}\cdot q_{Sd}\cdot L_Y\cdot L_X\cdot(L_X - 4\times a_X)}{(L_Y - 2\times a_Y)\cdot t^2} \tag{33.48}$$

For the anchorages alignment, the maximum stress is

$$\sigma_{ma_X} = 6\times\frac{u_{ma_X}\cdot q_{Sd}\cdot L_Y\cdot a_X^2}{(L_Y - 2\times a_Y)\cdot(t-d_u)^2} \tag{33.49}$$

Similarly, for a lateral strip at midspan

$$\sigma_{lc_X} = \frac{3}{2}\times\frac{u_{lc_X}\cdot q_{Sd}\cdot L_Y\cdot L_X\cdot(L_X - 4\times a_X)}{(L_Y - 2\times a_Y)\cdot t^2} \tag{33.50}$$

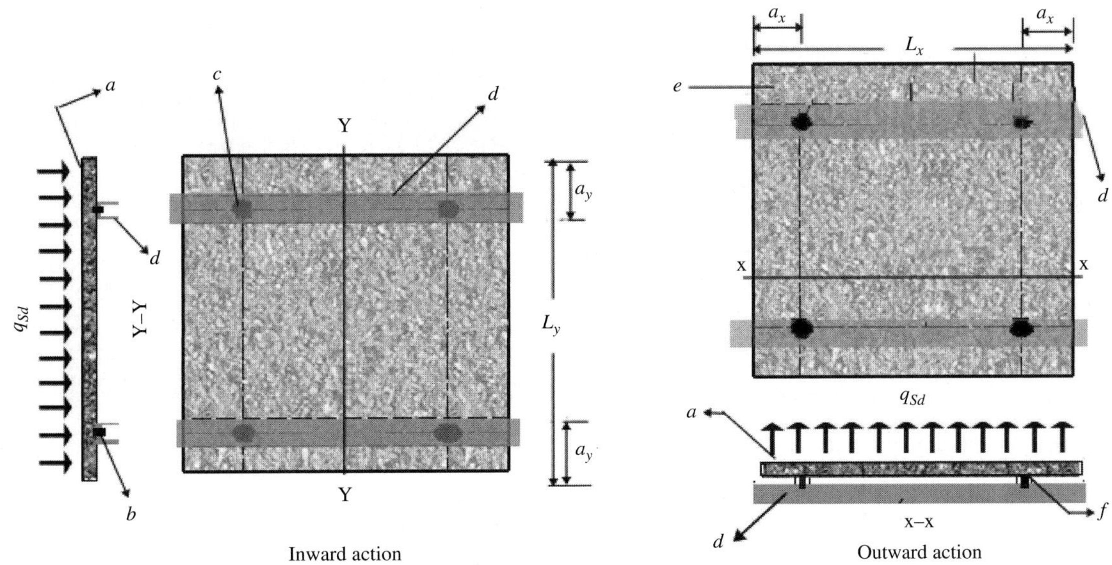

Figure 33.27 Slab's structural behavior for inward and outward. (*a*) Stone face. (*b*) Continuous back stone contact. (*c*) Cone bolt. (*d*) Horizontal rails. (*e*) Back stone. (*f*) Discrete back stone contact.

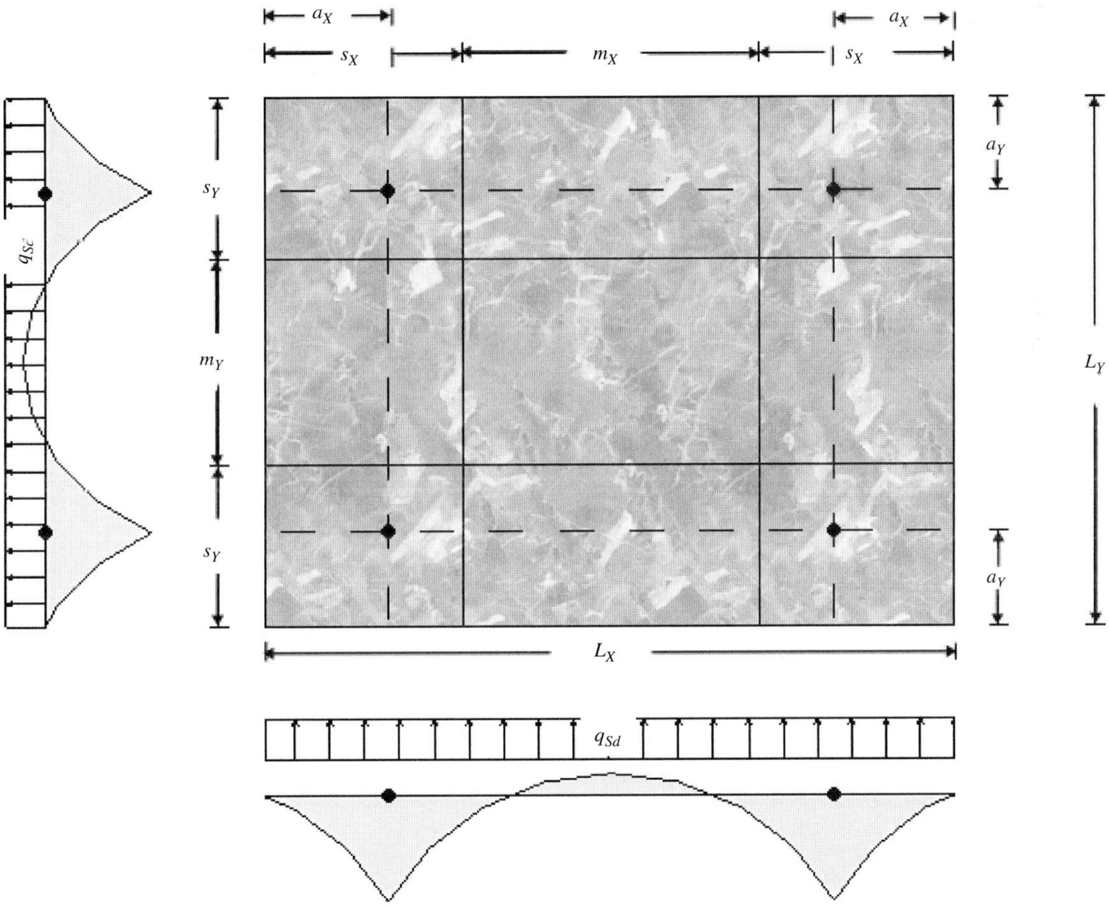

Figure 33.28 Slab's structural behavior for inward and outward. (*a*) Stone face. (*b*) Continuous back stone contact. (*c*) Cone bolt. (*d*) Horizontal rails. (*e*) Back stone. (*f*) Discrete back stone contact.

Table 33.18 Distribution Stress Coefficients for Two-Way Behavior Panels

Cross-section location	Middle strips	Lateral strips
Midspan	$u_{mc} = 0.45$	$u_{lc} = 0.55$
Anchorage alignment	$u_{ma} = 0.25$	$u_{la} = 0.75$

and at the anchor's alignment,

$$\sigma_{la_X} = 6 \times \frac{u_{la_X} \cdot q_{Sd} \cdot L_Y \cdot a_X^2}{(L_Y - 2 \times a_Y) \cdot (t - d_u)^2} \tag{33.51}$$

The stresses along the direction, *Y*, are found by merely swapping subscripts (*X*) with subscripts (*Y*) in Eqs. (33.48) to (33.51). The maximum value of the eight stresses must be less than the resistance stress design, σ_{Rd}, from Eq. (33.15) or (33.21).

If it is possible to assume a continuous supporting reaction along the rails, and for inward actions, q_{Sd}, a one-way slab behavior may be accepted, and the maximum stress, $\sigma_{Sd,\max}$, depends on the bending moment values at midspan, M_{Sdc}, and over the supports, M_{Sda}, as depicted in Fig. 33.29.

Assuming a linear uniform distribution along with the slabs unsupported direction,

$$\sigma_{Sd,\max} = \max\left\{6 \times \frac{M_{Sdc}}{t^2} M_{Sdc};\ 6 \times \frac{M_{Sda}}{(t - d_u)^2}\right\} \tag{33.52}$$

where t = the thickness of the slab and d_u = the anchorage depth and

$$M_{Sdc} = q_{Sd} \cdot \frac{L \cdot (L - 4 \times a)}{8}$$
$$M_{Sda} = q_{Sd} \cdot \frac{a^2}{2} \tag{33.53}$$

33.8.2 Stone Anchorage Strength

Undercut anchorage failure is described as inducing a cone-shaped spall. The projected area of the stone failure surface is somewhat circular, and the surface failure angle, α, depends on the rock type. By projecting the failure surface, A_{cf}, in a plane perpendicular to the cone axis and assuming a uniform distribution of the tensile strength over an equivalent circular idealized area according to Figs. 33.30 and 33.31, it is possible to establish a relationship based on the maximum principal stress Rankine theory which is satisfactorily applicable to a quasi-brittle material such as stone [67, 69].

The maximum principal stress theory is applied satisfactorily to limiting normal stress, and failure is supposed to occur when the normal stress reaches a specified upper limit. Based on this assumption and the described failure area and angle, the breaking load value at the undercut hole can be estimated with enough accuracy based on the indirect tensile strength estimated from flexural strength tests validated by computational finite element modeling.

To consider stress concentration in the vicinity of the undercut, a factor, K_U, is introduced depending on the geometry of the drill hole and rock type. The concentration stress factor is defined by the ratio between

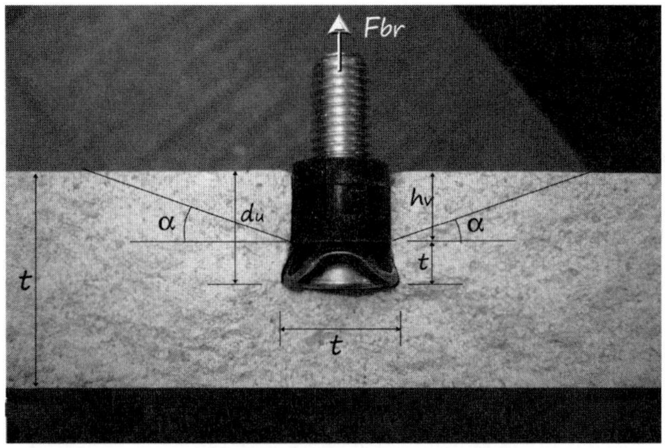

Figure 33.30 Spall configuration and detachment angle α.

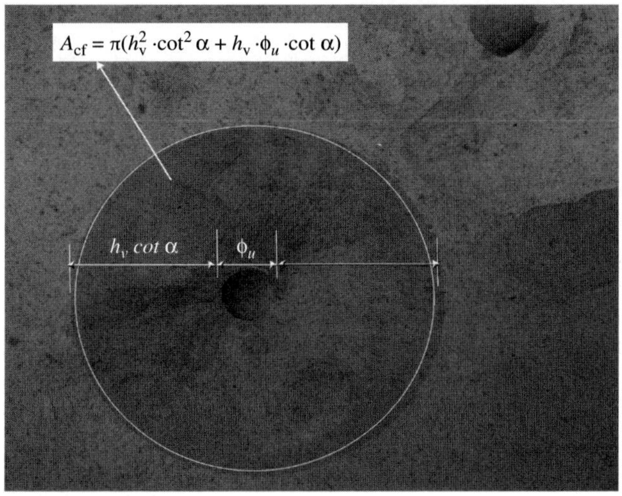

Figure 33.31 Projected area of cone failure A_{cf}.

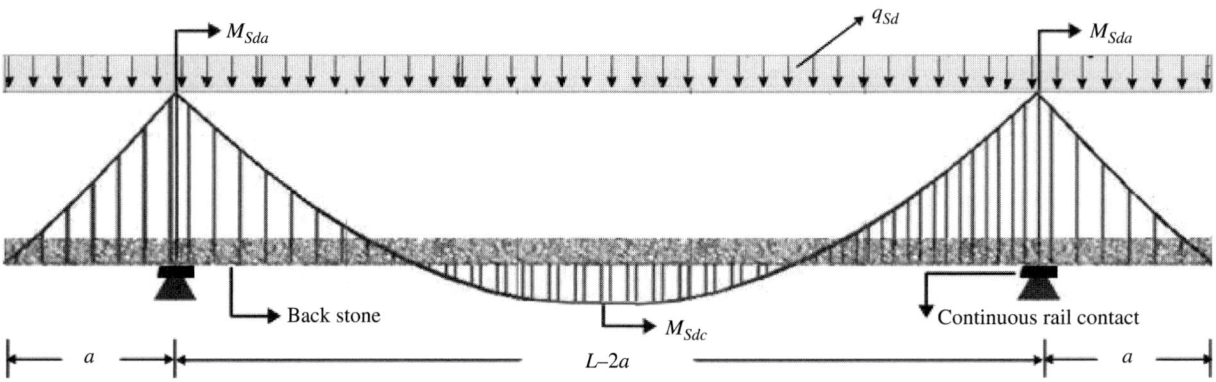

Figure 33.29 Idealized structural system for inward lateral action on a continuously supported slab.

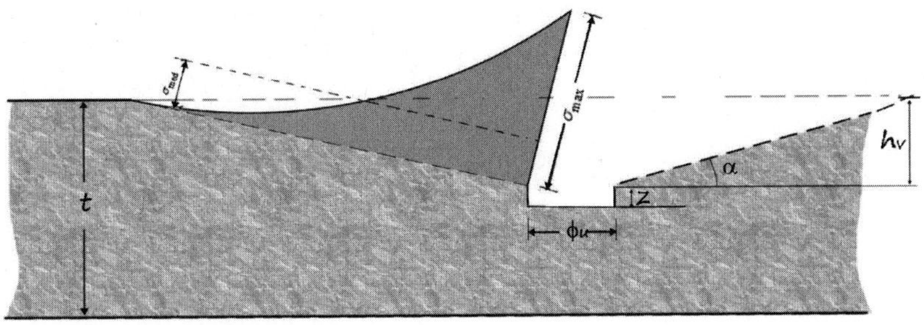

Figure 33.32 Stress concentration factor as the relation between the maximum stress and medium stress.

the actual maximum stress, σ_{max}, and the medium stress, σ_{med}, which is obtained by dividing the breaking load, F_{br}, by the projected area, A_{cf}, according to Eq. (33.54) and Fig. 33.32:

$$K_u = \frac{\sigma_{max}}{\sigma_{med}}; \sigma_{med} = \frac{F_{br}}{A_{cf}} \tag{33.54}$$

Based on several tests in different rock types, spall angles and stress concentration factors were obtained according to the above-mentioned procedure and resumed in Table 33.19.

Table 33.19 Distribution Stress Coefficients for Two-Way Behavior Panels

Stone identification	Spall angle α	Concentration factor K_U
Fine to medium grain size granite	19°–20°	4.9
Medium to gross grain size granite	16°–18°	4.6
Oolitic limestone	19°–20°	6.9
Calcitic marble	15°–16°	9.0

The design breaking load value, F_{brd}, follows:

$$F_{brd} = \frac{\sigma_{Rd} \cdot \pi\left(h_v^2 \cdot \cot^2\alpha + h_v \cdot \phi_u \cdot \cot\alpha\right)}{K_U} \tag{33.55}$$

The establishment of the safety verification format for the ultimate limit state at load level leads to the following condition:

$$F_{Sd} \leq F_{brd}$$

$$\frac{q_{Sd} \cdot L_X \cdot L_Y}{4} \leq \frac{\sigma_{Rd} \cdot \pi\left(h_v^2 \cdot \cot^2\alpha + h_v \cdot \phi_u \cdot \cot\alpha\right)}{K_U} \tag{33.56}$$

The minimum thickness of the stone, t_{min}, at the support section depending on the anchorage capacity and the lateral action, q_{Sd}, is defined by the following equations, providing that the undrilled depth, $t-(h_v+z)$ (Fig. 33.32), is higher than $0.4 \times t$:

$$t_{min} \geq \frac{h_v + z}{0.6}$$

$$h_v = \frac{\sqrt{A^2 + \sigma_{Rd} \cdot \pi \cdot q_{Sd} \cdot L_X \cdot L_Y \cdot K_U \tan\alpha} - A}{2 \times \sigma_{Rd} \cdot \pi} \tag{33.57}$$

$$A = \sigma_{Rd} \cdot \phi_u \cdot \pi$$

where ϕ_u = undercut diameter (Fig. 33.32)
h_v = constant-diameter hole depth
z = variable-diameter hole depth
σ_{Rd} = design value of the stone flexure strength as defined in Eq. (33.15) or (33.21)
α = spall angle (Table 33.19)
K_U = concentration stress factor (Table 33.19)
$L_X; L_Y$ = panel dimensions (Fig. 33.28)

REFERENCES

[1] "Design Issues for thin-stone cladding systems—RCI, Inc." [Online]. Available: http://rci-online.org/wp-content/uploads/2003-03-gerns.pdf. [Accessed 19 September 2018].

[2] "More geology—Rock classification, Builddirect" [Online]. Available: https://www.builddirect.com/blog/more-geology-rock-classification/. [Accessed 19 September 2018].

[3] S. Siegesmind and R. Snethlage (Eds.), *Stone in Architecture*, Dordrecht: Springer-Verlag, 2014.

[4] ASTM International, *ASTM C119-16—Standard Terminology Relating to Dimension Stone*, West Conshohocken, PA: ASTM International, 2016.

[5] Natural Stone Institute Publications, 2007 [Online]. Available: https://pubs.naturalstoneinstitute.org/pub/2c4ec57c-aef5-8a85-16e0-106c5cada13c. [Accessed 20 September 2018].

[6] MIA + BSI, The Natural Stone Institute, "Chapter 22—Tolerances in the dimension stone industry," in *Dimension Stone Design Manual*, Cleveland, H, Marble Institute of America, 2016.

[7] Marble Institute of America, *Dimension Stone Design Manual 8.0*, Cleveland, OH: Marble Institute of America, 2016.

[8] T. Yates and A. Mauko, "thaw susceptibility of natural stone—Characterization," in *International Conference on Durability of Building Materials and Component*, Istanbul, 2018.

[9] J. Ingham, "Predicting the frost resistance of building stone," *Journal of Engineering Geology and Hydrogeology*, vol. 38, pp. 387–399, 2005.

[10] S. R. S. Siegesmund, A. Dürrast, K. Sterflinger, et al., *Stone in Architecture*. Dordrecht: Springer doi:10.1007/978-3-642-14475-2.

[11] K. Ross and R. Butlin, "Durability Tests for Building Stone Report 141," Building Research Establishment (BRE), Watford, U.K., 1989.

[12] S. Siegesmund, T. Weiss and A. Vollbrecht, "Natural stone, weathering phenomena, conservation strategies and case studies: Introduction," *Geological Society, Special Publications*, vol. 205, no. 1, 2002.

[13] R. Sousa and H. Sousa, "Stability evaluation of marble stone cladding affected by bowing and wind pressure—A case study," in *Proceedings IRF2018: 6th International Conference Integrity-Reliability-Failure*, Lisbon, 2018.

[14] B. Grelk, C. Christiansen, B. Schouenborg and A. K. Malaga, "Durability of marble cladding—A comprehensive literature," *Journal of ASTM International*, vol. 4, no. 4, p. 18, 2007.

[15] ASTM International, "Committee C18 on Dimension Stone" [Online]. Available: https://www.astm.org/COMMITTEE/C18.htm. [Accessed 26 October 2018].

[16] CEN–European Committe for Standardization, "CEN/TC 246—Natural Stones," European Committe for Standadization [Online]. Available: https://standards.cen.eu/dyn/www/f?p=204:7:0::::FSP_ORG_ID:6227&cs=170D4BD6399C2CC53134C177FA4705416. [Accessed 26 October 2018].

[17] N. Zhang, L. B. Liu, D. Hou, M. He and Y. Liu, "Geomechanical and water vapor absorption characteristics of clay-bearing soft rocks at great depth," *International Journal of Mining Science and Technology*, vol. 24, no. 6, pp. 811–818, 2014.

[18] V. Pires, P. Amaral, L. Rosa and R. D. S. Camposinhos, "Slate flexural and anchorage strength considerations in cladding design," *Construction and Building Materials,* vol. 25, pp. 3966–3971, 2011.

[19] T. Magdalini, D. Fabrice and I. Ioannis, "Correlation of stone properties using standardised methodologies and non-standardised micro-destructive techniques," in *12th International Congress on the Deterioration and Conservation of Stone*, New York, 2012.

[20] R. d. S. Camposinhos, *Stone Cladding Engineering*, New York: Springer, 2014, p. 191.

[21] V. Pires, R. Camposinhos, et al., "Slate flexural and anchorage strength considerations in cladding design," *Construction and Building Materials,* vol. 25, pp. 3966–3971, 2011.

[22] ASTM International, ASTM C120/C120M-15a, Standard Test Methods for Flexure Testing of Structural and Roofing Slate, West Conshohocken, PA: ASTM International, 2015.

[23] W. Weibull, "A statistical theory of the strength of materials," *Ingeniörsvetenskapsakademiens Handlingar,* vol. 151, 1939.

[24] L. Burshtein, "Rock strengths under axial tension and flexure," *Journal of Mining Science,* vol. 4, pp. 53–61, 1967.

[25] J. Claesson and B. Bohlolib, "Brazilian test: Stress field and tensile strength of anisotropic rocks using an analytical solution," *International Journal of Rock Mechanics and Mining Sciences,* vol. 39, no. 8, pp. 991–1004, 2002.

[26] Y. Yong, Y. Jianmin and Z. Zouwu, "Shape effects in the Brazilian tensile strength test and a 3D FEM correction," *International Journal of Rock Mechanics and Mining Sciences,* vol. 43, no. 4, pp. 623–627, 2006.

[27] C. Rocco, G. V. Guinea, J. Planas and M. Elices, "Size effect and boundary conditions in the Brazilian test: Theoretical analysis," *Materials and Structures,* vol. 32, no. 6, pp. 437–444, 1999.

[28] N. Davidenkov, "Statistical theory of the scale effect" (in Russian), *Journal of Applyed Mechanics,* vol. 14, 1947.

[29] Y. Frenkel and M. Kontorova, "Statistical theory of strength" (in Russian), *Zhurnal Tekhnicheskoĭ Fiziki (Journal of Technical Physics),* vol. 7, 1943.

[30] S. Popovics, "Verification of relationships between mechanical properties of concrete-like materials," *Matériaux et Constructions,* vol. 8, no. 3, pp. 183–191, 2008.

[31] J. L. Byung, K. Seong-Hoon, O. Taekeun and K. Yun-Yong, "Evaluating the dynamic elastic modulus of concrete using shear-wave velocity measurements," *Advances in Materials Science and Engineering,* vol. 2017, p. 13, 2017.

[32] CSTB: Centre Scientifique et Technique du Bâtiment, *NF DTU 52.2 Pose collée des revêtements céramiques et assimilés—Pierres naturelles,* Marne-la-Vallée-Paris: CSTB Éditions, 2014.

[33] K. Chau and S. Jianfu, "Fracture and failure analysis of stone cladding on building facades," 2005 [Online]. Available: https://www.researchgate.net/publication/237646514_Fracture_and_failure_analysis_of_stone_cladding_on_building_facades. [Accessed November 2018].

[34] "Fracture and failure analysis of natural building: Applications in the restoration of ancient monuments," in *16th European Conference of Fracture (ECF16)*, Alexandroupolis, 2006.

[35] Chicago Committee on High Rise Buildings, "Seminar on recladding of the Amoco Building in Chicago, Illinois," in *CCHRB Recladding of the Amoco Building*, Chicago, 1985.

[36] C. Makepeace and B. Dennis, "PERSIST—Pressure Equalized Rainscreen Insulated Structure Technique—Design Approach." in *Thermal Envelopes VII—Performance of Air Barrier Systems—Practices*, Florida, 1998.

[37] R. d. S. Camposinhos, "Dimension stone design—Partial safety factors: A reliability based approach," *Construction Materials,* vol. 165, no. 3, pp. 145–159, June 2012.

[38] J. Holmes, *Wind Loading of Structures.* 2nd ed., New York: Taylor & Francis, 2007.

[39] Kenneth Zammit and Mauro Overend, "Wind Loading on Cladding and Glazed Façades," in *International Symposium on the Application of Architectural Glass*, Munich, 2006.

[40] Building Seismic Safety Council, *Recommended Seismic Provisions for New Buildings and Other Structures (FEMA P-750),* Washington, D.C.: Building Seismic Safety Council of the National Institute of Building Sciences, 2015.

[41] CEN–European Committee for Standardization, *EN 1998-1: Eurocode 8: Design of Structures for Earthquake Resistance. 1st ed.,* Brussels: BSI, 2004.

[42] "The Hanshin-Awaji Earthquake of January 17–Technical Report NCEER-95-0015," National Center for Earthquake Engineering Research, Buffalo, NY, 1995.

[43] Federal Emergency Management Agency of the U.S. Department of Homeland Security, "2015 NEHRP rerecommended seismic provisions design examples," Building Seismic Safety Council, Washington, D.C., FEMA P-1051/July 2016.

[44] M. C. Farmer and S. P. Lyons, "Stone cladding failure: The cause and consequences," in *Second Forensic Engineering Congress*, San Juan, PR, 2000.

[45] N. Lam and E. Gad, "An innovative approach to the seismic assessment of non-structural," in *Australian Earthquake Engineering Society Conference*, Adelaide, Australia, 2002.

[46] R. A. Khoraskani, *Advanced Connection Systems for Architectural Glazing*, New York: Springer, 2015.

[47] FEMA—Federal Emergency Management Agency, "Reducing the risks of nonstructural earthquake damage—A practical guide," FEMA—Federal Emergency Management Agency, Redwood City, CA, 2012.

[48] EN 1990: Basis of Structural Design, Brussels: European Standars (EN), 2002.

[49] M. Mourad, S. A. Bakhoum and M. H. MouradMaha, "Comparison of actions and resistances in different building design codes," *Journal of Advanced Research,* vol. 7, no. 5, pp. 757–767, 2016.

[50] M. Faddy, R. Wilson, and G. M., Winter, "The determination of the design strength of granite used as external cladding for buildings," in *Case Studies in Reliability and Maintenance*, D. P. M. Wallace, R. Blischke (Eds.), St Lucia, Australia: John Wiley & Sons, 2003, pp. 111–134.

[51] D. W. Tornburg, C. Kimball and B. W.C., *2018 International Building Code Illustrated Handbook*, New York: McGraw-Hill Education, 2018.

[52] National Research Council of Canada, *National Building Code of Canada*, Ottawa, 2015, p. 708.

[53] European Committee for Standardization, *Curtain Walling—Product Standard,* Brussels: CEN, 2015.

[54] R. Prikryl and B. Smith, *Building Stone Decay: From Diagnosis to Conservation*, London: Special Publications, 2007.

[55] Y. M. El-Sherbiny, "Erosive wear of different facade finishing materials," *Housing and Building National Research Center Journal,* vol. 14, pp. 431–437, 2018.

[56] L. J. J. Rasmussen, J. Jensen, K. Hansen and G. B, "Natural stone panels for building façade—Loss of strength caused by temperature cycles," in *Building Physics 2002—6th Nordic Symposyum*, Trondheim, Norway, 2002.

[57] International Code Council, *International Building Code*, Country Club Hills, IL: International Code Council, Inc., 2017, p. 726.

[58] *Draft EN 16612 Glass in Building—Determination of the Lateral Load Resistance of Glass Panes by Calculation,* Brussels: CEN–European Committee for Standartization, 2017.

[59] R. Camposinhos and R. Camposinhos, "Dimension-stone cladding design with dowel anchorage," *Proceedings of the ICE—Construction Materials,* vol. 162, no. 3, pp. 95–104, 2009.

[60] ASTM International, *Standard Guide for Selection, Design, and Installation of Exterior Dimension Stone Anchors and Anchoring Systems,* West Conshohocken, PA: ASTM International, 2018.

[61] R. S. Camposinhos, *Revestimentos em Pedra natural com Fixação Mecânica—Dimensionamento e Projecto*, Lisbon: Edições Sílabo, Lda, 2009, p. 199.

[62] R. D. S. Camposinhos, P. P. V. Maral and L. G. Rosa, "Dimension stone cladding design—Dowel anchorage design," in *2nd International Congress on Dimension Stones*, Carrara. Italy, 2008.

[63] L. Vera Pires, *Stone Materials and Cladding Fixing Systems A: Physical-Mechanical Behaviour Study,* Lisbon: UTL–Instituto Superior Técnico de Lisboa, 2013, p. 361.

[64] J. Tan, B. Huang, W. Lu, and K. Mosalam, "Load bearing capacity testing of dowel pin anchorage in granite cladding," in *7th International Conference on Advances in Experimental Structural Engineering*, Pavia, 2017.

[65] R. S. Camposinhos, "Dimension stone design—Kerf anchorage in limestone and marble," *Proceedings of the ICE—Construction Materials*, vol. 165, no. 3, pp. 165–171, 2012.

[66] L. M. D., *Modern Stone Cladding: Design and Installation of Exterior Dimension Stone Systems*, vol. MNL 21, Philadelphia, PA: ASTM, 1995.

[67] B. Lammert and K. R. Hoigard, "Stifeness considerations in dimension stone anchorage design," *Journal of ASTM International*, vol. 4, 2007.

[68] S. Shah, S. Swartz, and C. Ouyang, *Fracture Mechanics of Concrete*, West Conshohocken: John Wiley & Sons, 1995.

[69] R. S. Camposinhos, "Undercut anchorage in dimension stone cladding," *Proceedings of the Institution of Civil Engineers—Construction Materials*, vol. 166, no. 3, pp. 158–174, 2013.

[70] A. Hillerborg, *Strip Method Design Handbook*, London: Taylor and Francis, 1994, p. 320.

[71] C. R. de, *Manual de Dimensionamento de Fachadas Envidraçadas*, Coimbra: CMM–Associação portuguesa de construção metálica e Mista, 2018.

[72] B. Huang, Wensheng Lu, Khalid M. Mosalam, "Shaking table testing of granite cladding with undercut bolt anchorage," *Engineering Structures*, Volume 171, 15 pp. 488-489, 2018.

Index

Note: Page numbers followed by *f* denote figures; page numbers followed by *t* denote tables.

- Abutments:
  - in concrete railroad bridges, 557
  - in concrete T-beam bridges, 483, 484*f*
- Acoustic emission (AE) method, 760–764, 761*f*–764*f*
- Adjacent structures, snow drifts and, 17
- AE (acoustic emission) method, 760–764, 761*f*–764*f*
- Aerodynamic effects, wind loads and, 20–21
- Allowable stress design (ASD):
  - in aluminum structures, 322
  - in cable-supported roofs, 658–659
  - in cold-formed steel members, 309
  - in load combinations, 58–60
  - for reinforced masonry, 404, 404*t*, 405*f*
  - for steel stack chimneys, 746
  - for steel tanks, 726
  - in structural steels, 225–226, 226*f*
  - in timber, 413, 413*t*
- Alternative uniform live load reduction, 3–4, 5*f*–6*f*
- Aluminum structures:
  - alloys, 321, 321*t*
  - ASD and LRFD in, 322
  - axial compression and, 324–325, 324*t*
  - axial tension and, 323
  - block shear in, 331
  - bolted connections in, 329, 330*t*
  - buckling and design of, 322
  - combined forces in, 328
  - deflections in, 331
  - design methods for, 322
  - design specifications of, 321
  - fatigue in, 331, 331*t*
  - flexural strength in, 325–326, 325*t*, 327*t*
  - lateral-torsional buckling in, 325–326, 327*t*
  - local buckling in, 324–325, 324*t*, 325*t*
  - pinned connections in, 330*t*, 331
  - product specifications of, 321–322, 322*t*
  - required forces in, 322–323
  - riveted connections in, 329, 330*t*
  - safety and resistance factors in, 323*t*
  - screw connections in, 329–331, 330*t*
  - shear rupture, yielding and buckling in, 326, 327*t*, 328*t*
  - temperature effects on, 322, 322*t*
  - tempers, 321
  - torsional rupture, yielding and buckling in, 326, 328, 328*t*
  - vibrations in, 331
  - welded connections in, 322, 322*t*, 328–329, 329*t*, 331
- Analysis. *See* Structural analysis
- Approximate natural frequency, wind loads and, 27
- Arches, 91
  - bents, 105, 107, 110–111, 110*t*
  - buckling of, 111, 112*t*
  - centering of, 112
  - circular, 97, 97*t*
  - classification of, 92–93, 92*t*
  - Cochrane formulas for, 97
  - concrete, 100–101, 112–113, 113*f*
  - deck interaction with, 111, 112
  - design of:
    - approximations for special shapes in, 94–97
    - final, 100–101
    - general procedure for, 94
    - intermediate, 97–100
    - preliminary, 94
  - Douglas formula for, 97
  - drainage and waterproofing for, 112
  - economics of, 116
  - expansion joints for, 112
  - falsework of, 112
  - filled-spandrel, 92, 97, 98*t*
  - fixed, 96–97, 96*f*, 97*t*, 98*f*, 98*t*
  - glulam lumber, 414, 414*f*, 416
  - laterally loaded, 112
  - nomenclature of, 92
  - open-spandrel, 92, 92*f*, 97, 98*t*
  - parabolic, 95–97, 111, 112*t*
  - plasticity and, 100–101
  - reinforcement of, 112
  - rib-shortening and, 97–99, 98*f*, 99*f*, 99*t*
  - second-order theory and, 111
  - shrinkage and, 101
  - skewed barrel, 112, 112*f*
  - steel, wind bracing and, 113, 116, 116*f*
  - structural analysis of:
    - assumptions in, 93
    - energy methods and, 94, 94*f*
    - finite-element method and, 94
    - kern points and, 93, 93*f*
  - three-hinged, 95, 95*f*
  - tied, 92, 95–96
  - two-hinged, 95, 95*f*
  - unsymmetrical, 100
  - Wald formula for, 97
  - Whitney method for, 97, 98*f*, 99–100, 100*f*
- ArchiCAD, for BIM, 775
- ASD. *See* Allowable stress design
- Assembly occupancies, live load reduction and, 3
- Attached canopies, wind loads on, 48, 48*f*
- Autodesk Revit platform, for BIM, 774, 774*f*, 775
- Barrel vault roof snow loads, 13, 15*f*
- Bars, finite-element method for, 123, 125
- Base flood elevation (BFE), 50
- Basic uniform live load reduction, 3, 3*f*–4*f*
- BC Place, Vancouver, 657, 657*f*
- BDWF (Beam-on-Dynamic-Winkler-Foundation) approach, 875–878, 875*f*, 878*f*
- Beam bridges, 462–463, 463*f*
- Beam-on-Dynamic-Winkler-Foundation (BDWF) approach, 875–878, 875*f*, 878*f*
- Beam-supported slab system, 365, 365*f*
- Beams. *See also* Composite beams; Continuous beams; Reinforced-concrete beams; Steel beam-column members; Steel beams
  - computational engineering simulation for, 135–137, 136*f*, 137*f*
  - coupling, 137, 137*f*
  - deflections of:
    - conjugate beam method for, 68
    - differential equation for, 68
    - dummy unit-load method for, 71, 72*f*, 72*t*, 73*t*
    - formulas for, 69*t*–70*t*
    - internal-virtual-work expressions for, 71
    - moment-area method for, 68, 71
    - shape functions for, 79, 84*f*
  - finite-element method for, 123, 125
  - glass, 839, 839*f*
  - member fixed-end forces in, 79–81, 85*f*, 86*f*
  - structural analysis of deflections and:
    - conjugate beam method for, 68
    - differential equation for, 68
    - dummy unit load method for, 71, 72*f*, 72*t*, 73*t*
    - flexural shape functions for, 79, 84*f*
    - formulas for, 69*t*–70*t*
    - internal-virtual-work expressions for, 71
    - moment-area method for, 68, 71
  - timber, 416, 422–423
- Benicia-Martinez Bridge, 470, 470*f*
- Bentley Building Design, for BIM, 775
- Bents:
  - concrete bridge design and, 472–473, 473*f*
  - rigid frames and, 93, 93*f*
    - arched, 105, 107, 110–111, 110*t*
- Betti's law, 67
- BFE (base flood elevation), 50
- BIM. *See* Building information modeling
- Bins. *See also* Reinforced-concrete silos and bunkers
  - pressures in, 666, 666*t*
  - standards and codes for, 700

Blast-resistant design:
- dynamic approach to, 812–813
- high-energy explosive phenomena and, 810–812, 810*f*–812*f*
- multi-degree-of-freedom models for, 813
- primary structure material selection and, 814–815
- secondary structure material selection and, 815
- single-degree-of-freedom models for, 812–813
- structural behavior in, 813

Bolted connections:
- in aluminum structures, 329, 330*t*
- in cold-formed steel members, 317
- in steel tanks, 726
- in timber, 428–429, 428*f*

Boundary layer fluid dynamics, wind loads and, 20, 21*f*

Braced frames. *See* Steel braced frames

Braces, computational engineering simulation for, 135–137, 136*f*, 137*f*

Breaking wave height, flood loads and, 52*f*, 53

Bridges. *See also* Concrete bridges; Curved concrete box-girder bridges; Curved steel box-girder bridges; Curved steel I-girder bridges; Railroad bridges
- administration of, 507
- aesthetics of, 437
- continuous spans:
  - negative moments in, 448–449, 449*f*
  - positive moments in, 445, 448, 448*f*
  - shears in, 450, 450*f*
- curbs and, 506
- earth pressures and, 452
- earthquake loads on, 451–452
- ice pressure on, 453
- loads on:
  - combination, 438–439, 440*t*
  - dead, 438
  - design lane, 438, 439*f*
  - design tandem, 438
  - design truck, 438, 438*f*
  - fatigue, 438, 438*f*
  - highway, 438, 439, 441*t*–444*t*, 450
  - impact, 450–451
  - railroad, 438–440, 445*t*–447*t*, 450–451
  - wave, 453
  - wind, 451
- longitudinal force and, 452
- LRFD in, 453
- maintenance and inspection of, 506–507
- NDE applications for, 767*t*
- railings in, 505–506, 506*f*
- rigid frame, 91, 91*f*
- simple spans, maximum moments and shears in, 439–440, 441*t*–447*t*
- stream pressure on, 452–453

Broken waves, 54

Building appurtenances:
- C&C and, 47–49, 48*f*
- MWFRSs Directional Procedure for, 39–41, 41*f*, 42*f*

Building information modeling (BIM):
- ArchiCAD for, 775
- Autodesk Revit platform for, 774, 774*f*, 775
- Bentley Building Design for, 775
- computational engineering simulation and, 142
- definitions and concepts of, 771, 772*f*
- interoperability in, 777–778, 777*f*, 778*f*

Building information modeling (BIM) (*Cont.*):
- level of development in, 771–772, 773*t*
- model-based work flow for, 776–777, 777*f*
- modeling, 772, 774–775, 774*f*
- SAS framework and:
  - main concepts in, 778, 778*f*
  - structural analysis in, 785*f*, 786, 786*f*
  - Structural Melody in, 778, 779*f*–783*f*, 781, 783
  - Structural Poetry in, 783, 784*f*, 785*f*, 786
- standards for, 776
- Tekla Structures for, 775–776
- traditional work flow for, 776, 776*f*

Bunkers. *See* Reinforced-concrete silos and bunkers

Burj Khalifa, 588, 588*f*, 595–596, 595*f*

Cable-supported roofs:
- anticlastic surface, 651*f*
- cable dome system, 651, 653*f*
- design of:
  - allowable stress, 658–659
  - computer modeling for, 659–661
  - form finding and, 660
  - geometries in, 660
  - preliminary, 659, 659*f*, 660*f*
  - prestress in, 660
  - seismic analysis in, 660–661
- end fittings for, 654–655, 655*f*
- erection of, 661, 661*f*
- hardware for, 654–655, 655*f*
- natural frequencies in, 658, 658*f*
- one-way countered by dead load, 652*f*
- opposing layers, 652*f*
- protection of, 656
- ropes for, 651, 654, 654*f*
- serviceability in, 658
- shipment of, 655–656
- stay cables in, 651, 654*f*
- strands for, 651, 654, 654*f*
- structural systems for:
  - BC Place, Vancouver, 657, 657*f*
  - cost estimation and, 657–658
  - design compatibility and, 656
  - geometric nonlinearity in, 657, 657*f*
  - Soccer World Cup Stadium, Seoul, Korea, 656, 656*f*
- tension force resistance in, 651, 652*f*, 653*f*

Caisson foundations, soil mechanics and, 205

Cantilever walls, 210, 211*f*

CBFs. *See* Concentrically braced frames

C&C. *See* Components and cladding

CDE (Common data environment), 142

CFD (computational fluid dynamics), 812

Channel-type roof scuppers, 7–8, 7*f*, 8*f*

Chemically strengthened glass (CSG), 836

Chimneys:
- alloy liners for, 753
- borosilicate block linings for, 753
- brick liners for, 752–753, 753*f*
- coatings for, 754
- dead loads on, 741
- earthquake loads on, 743, 744*f*
- fiberglass-reinforced plastic liners for, 753
- foundations in, 754
- gunite block linings for, 753
- materials for, 741
- natural frequency of vibration of, 744–745, 745*f*
- pressure differentials in, 744, 745*f*

Chimneys (*Cont.*):
- reinforced-concrete, 750–752
- sizing of, 741
- steel liners for, 752, 753*f*
- steel stacks:
  - allowable stresses for, 746
  - base ring and anchor bolts in, 747, 747*f*, 747*t*
  - braced, 747, 748*f*
  - brick linings for, 754
  - circumferential stiffeners in, 746
  - cone-to-cylinder junction in, 746, 746*f*
  - design code for, 746
  - dual-wall, 754
  - examples of, 748–750
  - guyed, 747, 748*f*
  - resonant vibrations in, 747–748
  - types of, 745, 745*f*
- structural engineers and, 741
- thermal stresses in, 744
- wind loads on, 741–743, 742*f*, 743*f*

China Resources Headquarters, 584, 584*f*

Circular arches, 97, 97*t*

Clip angle connection, cold-formed steel members and, 318–320, 318*f*

Closed-type roof scuppers, 7–8, 7*f*, 8*f*

Cochrane formulas, for arches, 97

Cold-formed steel members:
- ASD and LRFD in, 309
- bolted connections in, 317
- buckling modes and, 309–311, 310*f*
- clip angle connection and, 318–320, 318*f*
- compression strength and, 311–313, 314*t*
- corrosion protection of, 308–309, 309*t*
- cross-sectional properties of, 309
- deck and panels, 305, 306*f*
- design standards of, 309
- direct strength method for, 310–311, 310*f*
- effective width method for, 309–310, 310*f*
- flexural strength and, 312, 313, 314*t*
- force combined actions and, 313, 315*t*
- framing members, 305, 306*f*
- industrial storage racks and, 316
- lateral force-resisting systems and, 315–316, 315*f*
- materials of, 306–308, 307*t*
- metal building systems and, 316, 316*f*
- power-actuated fastener connections in, 317–318
- screw connections in, 317, 317*f*
- shapes and applications of, 305–306, 306*f*
- shear strength and, 312
- special-bolted moment frames, 316
- stress-strain curves of, 308, 308*f*
- structural stability of, 311
- thicknesses of, 305*t*
- trusses and, 316
- web crippling strength and, 312–313, 313*f*
- welded connections in, 317

Columns. *See also* Steel beam-column members
- composite:
  - design strength of, 233, 234, 236–237
  - effective length of, 233, 233*f*, 234*f*
  - end tie plates and, 236
  - lacing and, 236
  - limitations and detailing requirements for, 232–233
  - torsional buckling and, 235
  - types of, 232, 232*f*
- computational engineering simulation for, 135–137, 136*f*, 137*f*

Columns (*Cont.*):
 concentrically braced frames design of, 300–303, 301*f*, 302*t*, 303*t*
 in concrete box-girder bridges, 490, 493, 497*f*, 498*f*
 in concrete T-beam bridges, 482–483, 482*f*
 earthquake-resistant design and cantilevered, 159
 for elevated steel tanks, 721–722, 721*f*
 glass, 839, 839*f*, 856–858, 857*f*, 858*f*
 in industrial buildings, 562–563, 562*f*
 reinforced-concrete:
  FRP for confinement strengthening of circular, 827–831, 828*f*, 829*f*
  longitudinal reinforcement for, 354
  maximum concentric axial loads and, 352, 354
  nominal axial strength and, 354, 355*f*
 in reinforced-concrete silo bottoms, 686–687
 structural analysis of, 294, 294*f*
 timber, 423–424
Common data environment (CDE), 142
Compatibility, structural analysis and, 62
Components and cladding (C&C):
 building appurtenances and roof overhangs, 47–49, 48*f*
 enclosed buildings:
  low-rise buildings, 43–44, 44*f*
  low-rise buildings, simplified, 44, 45*f*
  160 feet or less, 46, 47*f*
  60 feet and taller, 44, 46, 46*f*
 general requirements for, 41, 43
 open buildings, 46–47, 48*f*
 procedures for, 23, 23*t*, 41, 43*t*
 wind tunnel procedure, 49
Composite beams. *See also* Steel plate girders
 composition of, 237–238, 238*f*
 deflection and, 238, 240, 252
 elastic stress and, 240–241, 240*f*
 factored load on, 241, 243, 243*f*
 flexural capacity of, 244–245, 244*f*, 245*f*
 flexural strength of, 238, 240*f*, 252–253, 252*f*
 limit states for, 244, 244*t*
 plastic stress distribution in, 238, 239*f*
 service loads on, 244, 244*f*, 253
 shear connectors and, 241, 242*f*, 243*f*
Composite columns:
 design strength of, 233, 234, 236–237
 effective length of, 233, 233*f*, 234*f*
 end tie plates and, 236
 lacing and, 236
 limitations and detailing requirements for, 232–233
 torsional buckling and, 235
 types of, 232, 232*f*
Compound trusses, 67
Computational engineering simulation:
 for beams, columns, braces and joists, 135–137, 136*f*, 137*f*
 BIM and, 142
 for coupling beams, 137, 137*f*
 for diaphragms, 137–138, 138*f*, 139*f*
 finite-element method and, 133–138
 for footings, 139–141, 140*f*
 of mat/raft foundations, 141, 141*f*, 142*f*
 of pile/pier foundations, 141, 141*f*
 verification of, 142
Computational fluid dynamics (CFD), 812
Concentrically braced frames (CBFs), 288, 288*f*
 brace design in, 290, 292, 292*f*, 293*f*
 brace-intersected girder design in, 294–298, 295*f*, 296*t*, 297*f*, 298*f*, 298*t*
Concentrically braced frames (CBFs) (*Cont.*):
 building design for, 290, 291*f*
 column design in, 300–303, 301*f*, 302*t*, 303*t*
 configurations of, 289, 289*f*
 girders not intersected by braces design in, 298–300, 299*f*, 300*t*
 gravity loads on, 290, 292*t*
 structural analysis of, 289, 289*f*
 structural parameters of, 290, 292*t*
 vertical force distribution and, 290, 292*t*
Concrete:
 arches, 100–101, 112–113, 113*f*
 compressive and tensile strength of, 333
 elastic shortening of, 372, 372*t*
 at elevated temperatures, fire safety and, 792, 793*f*
 fatigue of, 168–169, 168*f*, 169*f*
 fractures in:
  mechanics of, 165–166, 166*f*
  multiaxial, 168, 168*f*
  stress-strain curve and, 166–167, 167*f*, 168*f*
 NDE applications for, 767*t*
 rigid frames:
  construction of, 112–113, 113*f*
  design of, 101
  example of, 101–102, 101*f*, 102*f*
Concrete bridges. *See also* Curved concrete box-girder bridges
 bent design in, 472–473, 473*f*
 box-girder:
  access openings in, 490, 497*f*
  bottom slab in, 489
  columns in, 490, 493, 497*f*, 498*f*
  concrete placing diagram for, 483, 490*f*
  construction joints in, 490, 496*f*
  deflection and, 490, 494*f*–497*f*
  design of, 484, 488
  diaphragms in, 489
  economics of, 484
  horizontally curved, 489–490
  moments of inertia in, 488, 492*t*
  top slab in, 490, 495*f*
  typical sections of, 484, 490*f*
  utility openings in, 484, 491*f*
  web flaring in, 488, 493*f*
  weights of interior in, 488, 492*t*
 camber in, 470
 plastic flow in, 470
 prestressed:
  beam sections in, 495, 500*f*, 500*t*
  box sections in, 495, 498, 500*f*, 500*t*
  deflection and, 505
  example of, 500–502, 500*f*–502*f*
  friction losses in, 502, 502*t*, 503*f*
  prestressing force path in, 502
  slab sections in, 495, 499*f*, 500*t*
  standards for, 493, 495, 498
  stress and, 498, 500
  ultimate load on, 502
  uplift due to prestressing and, 504–505, 505*f*
  web reinforcement in, 503–504, 504*f*, 505*f*
 shrinkage in, 470–471
 slab:
  continuous spans, 471–473, 475*f*, 476*f*
  simple spans, 471, 471*f*, 472*f*
 T-beam:
  abutments in, 483, 484*f*
  bearing assemblies in, 483, 489*f*
  bundled bar reinforcement in, 479, 479*f*
Concrete bridges, T-beam (*Cont.*):
  columns in, 482–483, 482*f*
  compressive reinforcement in, 475, 477*t*, 479
  concrete placing diagram for, 483, 490*f*
  construction joints in, 483, 485*f*–486*f*
  depth-to-span ratios in, 473–474
  diaphragms in, 480, 483
  expansion hinges in, 483, 487*f*–488*f*
  footings in, 482*f*, 483
  girder reinforcement in, 480, 481*f*, 482*f*
  girder spacing in, 474–475, 476*f*
  slab design in, 479–480, 480*f*
  stem design in, 480
Concrete-framed buildings, 568, 570*f*
Concrete railroad bridges:
 bearings in, 556
 cast-in-place prestressed concrete box girder, 552, 553*f*
 cast-in-place prestressed concrete trough, 552, 553*f*
 material specifications in, 556
 pile foundations in, 557
 precast concrete box girder, 552, 555*f*
 precast concrete I-girder, 552, 554*f*
 precast concrete slab girder, 552, 554*f*
 precast double box girder, 552, 555*f*
 reinforcement for, 556
 retaining walls, abutments, and piers in, 557
 spread footings in, 556
Cone penetrometer test (CPT), 219
Constitutive laws, structural analysis and, 62–63
Continuous beams:
 prestressed-concrete:
  C-lines in, 389, 391–393, 391*f*, 392*f*
  layouts for fully, 389*f*
  layouts for partially, 390*f*
  load balancing method for, 393, 393*f*
  ultimate strength of, 393
 three-moment equation for, 75, 77, 78*f*, 79*f*
Contragredient transformation, 77
Controlled-flow drains, for rain loads, 8
Conventional buildings, 575
Counterfort walls, 210, 211*f*
Coupling beams, 137, 137*f*
CPT (cone penetrometer test), 219
Crane buildings, 571, 572*f*, 573, 573*f*
Crane live loads, 5–7, 6*f*
Crib walls, 210, 211*f*
CSG (chemically strengthened glass), 836
Curved concrete box-girder bridges:
 behavior of, 535–537, 536*f*
 bending moment diagrams for, 539, 542*f*–545*f*
 camber and, 535
 cast-in-place, 532–533
 construction methods for, 534–535
 creep effects on, 538, 539
 curvature influence on configuration of, 533
 example of, 539, 540*f*–546*f*, 541, 546*t*
 expansion joints and bearings in, 534
 incrementally launched, 533
 posttensioning layout of, 539, 541*f*
 segmental, 533
 skewed supports in, 534
 state of stress on, 539
 structural analysis of, 532
 superstructure section of, 533–534, 534*f*, 539, 541*f*
 torsional moments for, 541, 546*f*, 546*t*
 types of, 532, 532*f*

Curved roof snow loads, 13
Curved steel box-girder bridges:
cross frames in, 530–531, 531*f*
flanges in, 529–530
framing in, 526–527
lateral bracing in, 527–528, 531
main components of, 525–526, 525*f*, 526*f*
structural analysis of, 528–529
torsional effects on, 529, 529*f*, 530*f*
webs in, 526, 526*f*, 529
welded connections in, 528, 528*f*
Curved steel I-girder bridges:
bending of lateral flange in, 519–520
camber and, 517
centrifugal force effect on, 520
connection details in, 516–517, 517*f*, 518*f*
diaphragm layout in, 515–516, 516*f*
example of, 520–523, 521*f*, 523*f*, 523*t*, 524*f*
geometry of, 514, 515*f*
lateral bracing in, 516
number of design girders in, 516
radial flange force on, 518, 518*f*
structural analysis of, 517–518
thermal effects on, 514–515, 515*f*
torsion and warping stresses on, 518
V-load approximation in, 519, 519*f*
Customer Function Model, 886, 889, 889*f*

DAM (direct analysis method), 253–254, 254*t*, 283–285, 284*t*
Dead loads, 1, 1*t*
on bridges, 438
cable-supported roofs and, 652*f*
on chimneys, 741
on steel box-girder bridges, 510–511
Degrees of freedom, structural analysis and, 61–62, 133, 134*f*
Design flood elevation (DFE), 52–53, 52*f*, 53*t*
Design flood velocity, 53, 53*f*
Design ice thickness, in ice loads, 19
Design temperatures, for freezing rain, 19
DFE (Design flood elevation), 52–53, 52*f*, 53*t*
DFIRMs (Digital Flood Insurance Rate Maps), 50, 51*f*
Diaphragms:
computational engineering simulation for, 137–138, 138*f*, 139*f*
in concrete box-girder bridges, 489
in concrete T-beam bridges, 480, 483
in curved steel I-girder bridges, 515–516, 516*f*
earthquake-resistant design and, 161–162, 161*f*
simple diaphragm buildings, 33, 35, 36*f*, 37–38, 39*f*, 40*f*
in tall buildings, 594
timber, lateral load resistance in, 432–434, 433*f*, 433*t*
Digital Flood Insurance Rate Map (DFIRMs), 50, 51*f*
Dilatometer test (DMT), 219
Direct analysis method (DAM), 253–254, 254*t*, 283–285, 284*t*
Direct strength method (DSM), 310–311, 310*f*
Directional procedure for building appurtenances and other structures, MWFRSs, 39–41, 41*f*, 42*f*
Directional procedure for buildings, MWFRSs:
enclosed, partially enclosed, and open buildings of all heights, 31–33, 32*f*–34*f*
Directional procedure for buildings, MWFRSs (*Cont.*):
enclosed, simple diaphragm buildings, 33, 35, 36*f*
scope of, 28, 30–31, 30*f*, 30*t*
Disproportionate collapse:
causes of, 801–802
definition of, 801
direct analysis procedure for mitigation of, 802, 803*f*
examples of, 801
GSA procedures for, 802
progressive collapse modeling:
ductility and demand capacity ratio in, 805–806
equation of motion in, 802, 804*f*, 805, 805*f*
example of, 806–808, 806*f*–809*f*
nonlinear, 805, 805*f*
UFC procedures for, 802, 802*f*, 803*f*
DMT (dilatometer test), 219
Domes:
behavior of, 613–614, 614*f*
edge effects of, 616–617, 617*f*, 617*t*
finite-element method for, 622–623, 622*f*, 623*t*
parabolic, of constant thickness, 616, 616*t*
roof snow loads, 13, 15*f*
spherical:
of constant thickness, 615–616, 615*f*, 616*t*
rigidly supported, 618, 618*f*
ring-stiffened, 618–622, 619*f*–621*f*, 620*t*
of variable thickness, 616
surface of revolution in, 614–615, 614*f*, 615*f*
Douglas formula, for arches, 97
DSM (direct strength method), 310–311, 310*f*
Ductile fracture mechanics, 165
Dust explosions, in reinforced-concrete silos, 688
Dynamic Winkler model, 869, 869*f*

Earth pressures:
active, 212, 212*f*
base pressures and, 215, 217, 217*f*
bridges and, 452
Coulomb's theory on, 213
determining, 214–215, 215*f*, 215*t*, 216*f*
equivalent-fluid method for, 213
Mohr rupture diagram for, 210–212, 212*f*
passive, 212, 212*f*
Rankine's theory on, 212–213, 212*f*
trial-wedge method for, 213, 214*f*
Earthquake loads, 49
on bridges, 451–452
on chimneys, 743, 744*f*
on glass, 840, 840*f*, 840*t*
industrial buildings and, 564
modes of vibration of, 146, 146*f*
natural stone cladding design and, 902–903, 903*f*, 903*t*
nature of, 145–146
on reinforced-concrete silos, 672, 672*f*
on steel tanks, 714, 714*f*
wind loads compared to, 145
Earthquake-resistant design:
bearing wall systems in, 159, 159*f*, 162, 162*f*
building frame systems in, 159, 159*f*
cantilevered column systems in, 159
design base shear in, 154
detailing requirements of, 162, 163*t*
diaphragms and, 161–162, 161*f*
drift computation in, 157–158, 157*f*
dual systems in, 159, 159*f*
Earthquake-resistant design (*Cont.*):
elastic fundamental period in, 155
ELF procedure for, 151, 154*f*
general requirements and exceptions of, 148, 148*f*
ground motion and, 145–146, 146*f*, 148–150, 149*f*
horizontal distribution of seismic forces in, 156, 157*f*
horizontal structural irregularities and, 151, 152*t*
load combinations in, 160
modes of vibration and, 146, 146*f*
moment resisting frame systems in, 158–159, 159*f*
overturning in, 157
P-delta effects in, 158, 158*f*
philosophy of, 146–147, 147*f*
redundancy provisions in, 160, 160*f*
response modification factor in, 155–156
seismic design category for, 150–151, 150*t*
seismic force effects and, 159–162, 160*f*
seismic importance factor in, 155
seismic response coefficient in, 154–155
shear walls and, 162, 162*f*
simplified analysis of, 151, 154*f*
torsion and, 156–157
undefined systems in, 159, 159*f*
vertical distribution of seismic forces in, 156, 156*f*
vertical structural irregularities and, 151, 153*t*
Effective width method (EWM), 309–310, 310*f*
EGLA (Energy Grade Line Analysis), 56, 56*t*
Electrical transmission structures. *See* Transmission line structures
Elevated steel tanks. *See* Steel tanks, elevated
ELF (equivalent lateral force) procedure, 151, 154*f*
Enclosed buildings:
C&C:
low-rise buildings, 43–44, 44*f*
low-rise buildings, simplified, 44, 45*f*
160 feet or less, 46, 47*f*
60 feet and taller, 44, 46, 46*f*
MWFRS:
directional procedure for, 31–33, 32*f*–34*f*
envelope procedure for, 35, 37–38, 37*f*–40*f*
simple diaphragm, 33, 35, 36*f*
wind loads and, 21, 27–28, 29*t*
Enclosure classifications, wind loads and, 21, 22*f*, 27–28, 29*f*, 29*t*
Energy Grade Line Analysis (EGLA), 56, 56*t*
Envelope procedure for buildings, MWFRSs:
enclosed, simple diaphragm low-rise buildings, 37–38, 39*f*, 40*f*
enclosed and partially enclosed low-rise buildings, 35, 37, 37*f*, 38*f*
scope of, 35, 36*t*
Equilibrium, structural analysis and, 62
Equivalent lateral force (ELF) procedure, 151, 154*f*
EWM (effective width method), 309–310, 310*f*
Existing roofs, snow drifts and, 17
External pressures, wind loads and, 21, 22*f*
Extraordinary loads and events, 60

FAST (Function Analysis System Technique), 886, 889–890, 890*f*
Fatigue:
in aluminum structures, 331, 331*t*
bridges and, 438, 438*f*
of concrete, 168–169, 168*f*, 169*f*

Fatigue (*Cont.*):
in fiber-reinforced composites, 172, 172*f*, 173*f*
industrial buildings and, 566
process of, 182–183, 182*f*
in structural steels:
bolted connections and, 185
corrosion, 184
design for, 185, 185*f*–187*f*, 189*t*–191*t*, 191*f*
Goodman diagram for, 183, 183*f*
lamellar tearing and, 192, 192*f*
residual stress and, 184–185
riveted connections and, 185
S-N curve for, 183, 183*f*
strength factors of, 183–185, 184*f*, 185*f*
welded connections and, 185
FBFM (Flood Boundary and Floodway Map), 50
Federal Emergency Management Agency (FEMA), 50
DFIRMs and, 50, 51*f*
FIRMs and, 50, 50*f*
flood hazard zones and, 51–52, 51*t*
Fiber-reinforced composites:
delamination fractures in, 172, 172*f*
fatigue in, 172, 172*f*, 173*f*
longitudinal fractures in, 169–170, 170*f*
multiaxial fractures in, 170–172, 171*f*
transverse and shear fractures in, 170, 170*f*
Fiber-reinforced polymer (FRP):
for confinement strengthening of circular columns, 827–831, 828*f*, 829*f*
environmental reduction factors and, 817, 817*t*
for flexural strengthening of beams and slabs, 818–822, 819*f*, 820*f*
for shear strengthening of beams, 822–827, 823*f*, 824*f*
stress-strain behavior and, 817
Filled-spandrel arches, 92, 97, 98*t*
Finite-element method:
arch analysis and, 94
for bars, 123, 125
for beams, 123, 125
for circular reinforced-concrete silos, 688–690, 688*f*, 689*f*, 691*f*, 693–695, 694*f*
compatibility and, 129–130
computational engineering simulation and, 133–138
conditioning and, 129
continuum and structural elements and, 130, 130*f*
for cylindrical shells, 630, 631*f*, 632, 632*f*
degrees of freedom in, 133, 134*f*
distorted elements and, 129
for domes, 622–623, 622*f*, 623*t*
dynamic, 126
damping and, 127
mass matrices and, 127
time discretization techniques and, 127–128, 128*t*
element node numbering and, 129
for frames, 123, 125
for hyperbolic paraboloid shells, 645–646, 645*f*, 646*f*
incompressible materials and, 128–129
linear, 133–134
loads and, 126
locking in bending and, 129
for machine foundations, 867, 867*f*, 868, 868*f*
Finite-element method (*Cont.*):
material models and, 126
matrix equation and, 126
meshing and, 125–126
nodes and, 124–125
nonlinear, 128, 134–135, 135*f*
plates and shells, 123–125
reflex angles and, 129
for steel box-girder bridges, 512, 512*f*
steps in, 122
strain-stress relationship and, 126
theory of, 119–122
for thermal properties, 130–131, 130*t*
three-dimensional:
elements in, 119, 121*t*, 125
structural analysis with, 122
for trusses, 123, 125
two-dimensional:
axisymmetry, 123–125, 123*f*
elements in, 119, 120*t*, 124–125
plane strain, 122, 122*t*, 124–125
plane stress, 122, 122*t*, 124–125
structural analysis with, 122–123
verification and validation of, 128
Fire safety engineering:
construction materials at elevated temperatures in, 791
concrete and, 792, 793*f*
insulation materials and, 796, 796*t*
masonry and, 704*f*, 792, 794
properties of interest for, 792
steel and, 794, 795*f*
wood/timber and, 794, 795*f*, 796
fire development in compartments and, 790–791, 790*f*
fire resistance evaluation and:
advanced rational design approach for, 798–799, 799*f*
prescriptive-based approaches for, 796–798, 796*t*, 797*f*
simplified rational design approach for, 798, 798*f*
heat transfer mechanisms and, 791
need for, 789
temperature effects and, 789
time-temperature relations and, 791, 791*f*
FIRMs (Flood Insurance Rate Maps), 50, 50*f*
FISs (Flood Insurance Studies), 50
Fixed arches, 96–97, 96*f*, 97*t*, 98*f*, 98*t*
Flat plate slab system, 365, 365*f*
Flat roof snow loads, 9–10, 9*f*, 10*f*
Flat slab floor system, 365–366, 366*f*
Flood Boundary and Floodway Map (FBFM), 50
Flood hazard areas, 50–51, 50*f*, 51*f*
Flood hazard zones, 51–52, 51*t*
Flood Insurance Rate Maps (FIRMs), 50, 50*f*
Flood Insurance Studies (FISs), 50
Flood loads:
breaking wave height and, 52*f*, 53
code requirements for, 49–50
combinations of, 54
design and construction for, 52
design flood velocity and, 53, 53*f*
DFE and, 52–53, 52*f*, 53*t*
flood hazard areas and, 50–51, 50*f*, 51*f*
flood hazard zones and, 51–52, 51*t*
Flood loads (*Cont.*):
overview of, 49
stillwater flood depth and, 52*f*, 53
types of, 53–54
Floodways, 50
Flow velocity analysis, of tsunami loads, 55–56, 56*t*
Footings:
combined and continuous, 140, 140*f*
computational engineering simulation of, 139–141, 140*f*
concrete railroad bridges and spread, 556
in concrete T-beam bridges, 482*f*, 483
isolated, 139–140, 140*f*
of reinforced-concrete members:
design for flexure and, 361–363, 362*f*, 363*f*
design for shear and, 363–364, 363*f*, 364*f*
design procedures for, 364
force transfer at, 364
sizing base area and thickness of, 360–361, 360*f*, 361*f*
Foundations. *See also* Machine foundations; Pile foundations
in chimneys, 754
in elevated steel tanks, 723, 723*f*
mat/raft, computational engineering simulation of, 141, 141*f*, 142*f*
pier:
computational engineering simulation of, 141, 141*f*
in concrete railroad bridges, 557
soil mechanics and, 205–206
reinforced-concrete silo failure of, 687
soil mechanics and:
deep, 204–210
shallow, 199–204
in steel tanks, 715–717, 715*f*–717*f*
in substation structures, 739, 739*f*
in tall buildings, 577, 594–595
in transmission line structures, 735, 736*f*, 737
Fracture process zone (FPZ), 165–166, 166*f*
Fractures:
in concrete:
mechanics of, 165–166, 166*f*
multiaxial, 168, 168*f*
stress-strain curve and, 166–167, 167*f*, 168*f*
in fiber-reinforced composites:
delamination, 172, 172*f*
longitudinal, 169–170, 170*f*
multiaxial, 170–172, 171*f*
transverse and shear, 170, 170*f*
in structural steels:
brittle, 175, 175*f*, 176*f*, 181–182, 182*f*
correlations between tests for, 180, 180*f*
design approach for, 180–181, 181*f*
ductile, 175, 176*f*
lamellar tearing and, 192, 192*f*
loading rate and, 178, 179*f*
mechanics of, 175–177, 177*f*
toughness tests for, 177–178, 177*f*–179*f*
Frames. *See also* Concentrically braced frames; Rigid frames
in curved steel box-girder bridges, 526–527
finite-element method for, 123, 125
in industrial buildings:
concrete, 568, 570*f*
crane, 571, 572*f*, 573, 573*f*

Frames, in industrial buildings (*Cont.*):
- lateral load-resisting systems and, 566, 567*f*, 568
- metal, 571
- steel, 568, 569*f*
- wood, 568, 571*f*

portal, displacement method and, 81, 86*f*–88*f*
statically indeterminate, by force method, 75, 76*f*, 77*f*
structural analysis of, 67–68, 68*f*, 71, 72*f*, 73*t*
Freezing rain, 17–19
FRP. *See* Fiber-reinforced polymer
Function Analysis System Technique (FAST), 886, 889–890, 890*f*

Gabion walls, 210
Glass:
ASTM standards for:
- analytical procedure, 845
- definitions, 843
- double-glazed insulating glass, 844–845, 844*t*
- load resistance nomograms, 843
- serviceability and deflection limits, 845–846
- single glazing procedure, 843–844, 844*f*, 844*t*
- triple-glazed insulating glass, 845, 845*t*
- walkways, 846–847, 846*t*, 847*f*, 847*t*

balustrades, 838, 838*f*
beams, 839, 839*f*
chemically strengthened, 836
codes and standards for, 842, 842*f*
columns, 839, 839*f*, 856–858, 857*f*, 858*f*
earthquake loads on, 840, 840*f*, 840*t*
EN standards for:
- bending strength for prestressed glass, 848, 848*t*
- calculation procedure and design criteria, 847
- edge strength factors, 847, 847*t*
- glass surface profile factors, 847–848, 848*t*
- insulating glass calculations and load distribution, 850, 850*f*, 850*t*
- isochore pressure, 850
- laminated glass thickness, 849, 849*f*, 849*t*
- load duration factors, 848, 848*t*
- stress and deflection calculation, 848–849, 849*t*

failure, 833–834
floors and walkways, 838, 838*f*
fractures, 833, 833*f*
frameless walls, 838, 839*f*
glazing panels and, 837–838, 838*f*
insulating glass units, 837
laminated, 836–837, 836*f*, 858, 858*f*
lateral-torsional buckling of, 854–856, 855*f*–857*f*, 856*t*
performance improvements of, 835
physical properties of, 834, 834*t*
plate buckling of:
- out-of-plane displacement and, 850–851, 851*f*
- shear buckling load and, 851–852, 851*f*–853*f*
- shear buckling strength and, 852–853
- stresses in point fixed anchorage zones and, 853–854, 853*f*–855*f*

production of, 834, 834*f*
snow loads on, 840–842, 841*f*, 841*t*, 842*f*
stress-strain curves of, 835, 835*f*
structural analysis of, 837
toughened, 835–836, 836*f*
wind loads on, 839–840
Glazing panels, 837–838, 838*f*
Grade planes, tsunami loads and, 54
Gravity loads:
- on concentrically braced frames, 290, 292*t*
- on steel beam-column members, 255, 255*f*–258*f*, 258
- on steel moment frames, 270, 270*t*, 282–283, 282*f*–284*f*
- on tall buildings, 576

Gravity walls, 210, 211*f*
Ground elevation effects, wind loads and, 25, 26
Groundwater monitoring, 221, 221*f*
Grout injection, 220–221, 221*t*
Guizhou Culture Plaza Tower, 590, 591*f*
Gust effects, wind loads and, 21, 23, 26–27, 27*f*–28*f*
Guys and cables, wind on ice-covered, 19

HE (high-energy) explosives, 810–812, 810*f*–812*f*
Health monitoring of structures. *See* Nondestructive evaluation (NDE)
Heavy live loads, 2–3
Height factor, in ice loads, 18
Hermitian shape functions, 79, 84*f*
High-energy (HE) explosives, 810–812, 810*f*–812*f*
Hilber-Hughes-Taylor method, 127–128
Hip and gable roof snow loads, 13, 13*f*, 14*f*
Hurricane-prone regions, 28
Hydrodynamic loads, 53
Hydrostatic loads, 53

Ice loads:
- design ice thickness in, 19
- freezing rain and, 18–19
- height factor in, 18
- ice weight and, 19
- importance factor in, 18
- overview of, 9
- partial loading and, 19
- terms in, 17
- topographic factor in, 18
- wind on, 19

Ice pressure, on bridges, 453
Ice-sensitive structures, 17
Ice weight, 19
Idealized trusses, 67
IE (impact echo) method, 760, 761*f*
IGU (insulating glass units), 837
Impact echo (IE) method, 760, 761*f*
Impact loads, 55
Importance factors:
- in earthquake-resistant design, 155
- in ice loads, 18
- in tsunami loads, 56, 56*t*

In-phase effective mass, 868
Industrial buildings. *See also* Tall buildings
- columns in, 562–563, 562*f*
- design concept in, 559
- earthquake loads and, 564
- expansion joint spacing in, 563
- fatigue considerations for, 566
- foundation and slab-on-grade design in, 563
- frames in:
  - concrete, 568, 570*f*
  - crane, 571, 572*f*, 573, 573*f*
  - lateral load-resisting systems and, 566, 567*f*, 568
  - metal, 571
  - steel, 568, 569*f*
  - wood, 568, 571*f*

Industrial buildings (*Cont.*):
- IBC specifications for, 563–564
- layout of, 560–561, 560*f*
- material-handling equipment in, 564, 565*f*
- process loads and, 564
- project requirements in, 559
- roof and wall system in, 561–562
- structural design requirements in, 561

Industrial steel storage racks, 316
Insulating glass units (IGU), 837
Insulation materials, at elevated temperatures, 796, 796*t*
Interior walls and partitions, live loads and, 7
International Plumbing Code® (IPC®), 7
Intersecting snow drifts, at low roofs, 17
Inundation depth analysis, of tsunami loads, 55–56, 56*t*
IPC® (International Plumbing Code®), 7

Jin Mao Tower, 585, 585*f*
John Hancock Center, 583, 583*f*
Joists:
- computational engineering simulation for, 135–137, 136*f*, 137*f*
- prefabricated wood I-joists, 417–418, 417*f*
- two-way, 366–367, 367*f*

Lagrangian shape functions, 79
Lamellar tearing, in structural steels, 192, 192*f*
Laminated glass (LG), 836–837, 836*f*, 858, 858*f*
Lateral force-resisting system (LFRS), 137–138, 315–316, 315*f*
Lattice frameworks, wind on ice-covered, 19
Latticed steel towers, 733, 734*f*
LEFM (linear elastic fracture mechanics), 165–166, 166*f*
LFRS (lateral force-resisting system), 137–138, 315–316, 315*f*
LG (laminated glass), 836–837, 836*f*, 858, 858*f*
Line elements, 61
Linear elastic fracture mechanics (LEFM), 165–166, 166*f*
Live loads, 1*f*
- crane, 5–7, 6*f*
- heavy, 2–3
- interior walls and partitions and, 7
- production of, 1
- reduction in:
  - alternative uniform, 3–4, 5*f*–6*f*
  - assembly occupancies and, 3
  - basic uniform, 3, 3*f*–4*f*
  - general method for, 2–3, 2*f*
  - on one-way slabs, 2
  - passenger vehicle garages and, 3
- roof, 4–5
- steel box-girder bridges and, 511
- steel plate-girder bridges and, 465–466, 465*f*, 466*f*

Load and Resistance Factor Design (LRFD):
- in aluminum structures, 322
- in bridges, 453
- in cold-formed steel members, 309
- in load combinations, 57–58, 57*t*
- in structural steels, 225–226, 226*f*
- in timber, 413, 413*t*

Loads. *See also* Wind loads
on bridges:
combination, 438–439, 440*t*
dead, 438
design lane, 438, 439*f*
design tandem, 438
design truck, 438, 438*f*
fatigue, 438, 438*f*
highway, 438, 439, 441*t*–444*t*, 450
impact, 450–451
railroad, 438–440, 445*t*–447*t*, 450–451
wave, 453
wind, 451
combinations:
alternative, 59, 59*t*
ASD in, 58–60
basic, 58–59, 58*t*
earthquake-resistant design and, 160
effects, 57, 57*t*
extraordinary, 60
load and resistance factor design in, 57–58, 57*t*
with overstrength, 59
reinforced-concrete members design strength and, 335, 335*t*
strength design in, 57–58, 57*t*
structural integrity and, 60
dead, 1, 1*t*, 438, 510–511, 652*f*, 741
earthquake, 49
on bridges, 451–452
on chimneys, 743, 744*f*
on glass, 840, 840*f*, 840*t*
industrial buildings and, 564
modes of vibration of, 146, 146*f*
natural stone cladding design and, 902–903, 903*f*, 903*t*
nature of, 145–146
on reinforced-concrete silos, 672, 672*f*
on steel tanks, 714, 714*f*
wind loads compared to, 145
finite-element method and, 126
flood:
breaking wave height and, 52*f*, 53
code requirements for, 49–50
combinations of, 54
design and construction for, 52
design flood velocity and, 53, 53*f*
DFE and, 52–53, 52*f*, 53*t*
flood hazard areas, 50–51, 50*f*, 51*f*
flood hazard zones and, 51–52, 51*t*
overview of, 49
stillwater flood depth and, 52*f*, 53
types of, 53–54
gravity:
on concentrically braced frames, 290, 292*t*
on steel beam-column members, 255, 255*f*–258*f*, 258
on steel moment frames, 270, 270*t*, 282–283, 282*f*–284*f*
on tall buildings, 576
ice:
design ice thickness in, 19
from freezing rain, 18–19
height factor in, 18
ice weight and, 19
importance factor in, 18
overview of, 9
Loads, ice (*Cont.*):
partial loading and, 19
terms in, 17
topographic factor in, 18
wind on, 19
lateral, on tall buildings, 576–577
live, 1*t*
alternative uniform reduction in, 3–4, 5*f*–6*f*
assembly occupancies and, 3
basic uniform reduction in, 3, 3*f*–4*f*
crane, 5–7, 6*f*
general method for reduction in, 2–3, 2*f*
heavy, 2–3
interior walls and partitions and, 7
one-way slabs and reduction of, 2
passenger vehicle garages and, 3
production of, 1
roof, 4–5
steel box-girder bridges and, 511
steel plate-girder bridges and, 465–466, 465*f*, 466*f*
process, industrial buildings and, 564
railroad bridges and, 438–440, 445*t*–447*t*, 450–451, 549–550
rain:
controlled-flow drains for, 8
design, 7–8, 7*f*, 8*f*
ponding instability and, 8
roof scuppers and, 7–8, 7*f*, 8*f*
snow:
curved roof, 13
dome roof, 13, 15*f*
flat roof, 9–10, 9*f*, 10*f*
general procedure for determining, 17
on glass, 840–842, 841*f*, 841*t*, 842*f*
hip and gable roof, 13, 13*f*, 14*f*
multiple folded plate, sawtooth, and barrel vault roof, 13, 15*f*
overview of, 9
partial loading and, 10, 12
ponding instability and, 17
rain on, 17
sloped roof, 10, 11*f*–12*f*
snow drifts on lower roofs, 13, 15–17, 16*f*, 18*f*
surface roughness categories and, 9, 9*t*
unbalanced roof, 12–13, 13*f*–15*f*
soil:
creep and, 199
dynamic, 198, 199*f*
lateral, 8–9
long-term, 196–197, 197*f*, 197*t*
short-term, 197–198, 198*f*, 198*t*
transient, 198–199
on substation structures, 737–738, 738*t*
summary of, 1*t*
transmission line structure design and, 729–731, 731*t*
tsunami:
code requirements for, 49–50
combinations of, 56–57
general requirements for, 54–55, 55*f*
grade planes and, 54
importance factors in, 56, 56*t*
inundation depth and flow velocity analysis of, 55–56, 56*t*
overview of, 49, 54
risk categories of, 55
Loads, tsunami (*Cont.*):
run-up elevation and, 54
structural design procedures for, 56
structural loads and effects of, 57
structural performance evaluation for, 56
Lower roofs, snow drifts on, 13, 15–17, 16*f*, 18*f*
LRFD. *See* Load and Resistance Factor Design

Machine foundations:
block-type, 865–866, 865*f*, 866*f*
design limits for, 862
elevated, 867, 867*f*, 868*f*
energy transfer mechanism in, 863
excitation force and, 861
finite-element method for, 867, 867*f*, 868, 868*f*
foundation type in, 861, 862*t*
machine type in, 861
mat, 866–867, 867*f*
modeling techniques for, 865, 865*f*
pile-supported:
advantages of, 870
BDWF approach for, 875–878, 875*f*, 878*f*
beam on elastic foundation and, 871
continuum approach for, 879–881, 879*f*, 880*f*
Deeks and Randolph's model for, 876, 876*f*
El-Naggar and Novak's model for, 877–878, 878*f*
elastic continuum method for, 872, 873*f*–875*f*, 875
group effect of, 881–882
Holeyman's model for, 877, 877*f*, 880, 880*f*
lateral vibrations and, 871–872
Lysmer and Richart's model for, 876, 876*f*
Novak's model for, 879, 879*f*
Randolph and Simons's model for, 876–877, 877*f*
Smith's model for, 875–876, 875*f*, 876*f*
uncertainty factors in, 871
vertical vibrations and, 875–881
Winkler foundation assumption and, 871–872, 872*f*
soil effects on:
damping in, 864–865, 864*f*
dynamic impedance functions and, 870, 870*f*
embedment and, 863
modeling, 868–869, 869*f*
properties in, 862–863, 863*t*
stress reduction in, 864
three-dimensional model for, 868, 868*f*
Main wind-force resisting systems (MWFRSs):
directional procedure for building appurtenances and other structures, 39–41, 41*f*, 42*f*
directional procedure for buildings:
enclosed, partially enclosed, and open buildings of all heights, 31–33, 32*f*–34*f*
enclosed, simple diaphragm buildings, 33, 35, 36*f*
scope of, 28, 30–31, 30*f*, 30*t*
envelope procedure for buildings:
enclosed, simple diaphragm low-rise buildings, 37–38, 39*f*, 40*f*
enclosed and partially enclosed low-rise buildings, 35, 37, 37*f*, 38*f*
scope of, 35, 36*t*
procedures for, 23, 23*t*
wind tunnel procedure, 49

Masonry:
composition of, 400–401, 400*f*, 401*t*
elements in, 401, 402*f*
at elevated temperatures, fire safety and, 704*f*, 792, 794
fire walls, 402*t*, 409–410, 410*f*
materials in, 400
NDE applications for, 767*t*
notation for, 399–400
reinforced, 403
ASD for, 404, 404*t*, 405*f*
seismic design for, 407, 408*f*, 408*t*, 409, 409*t*
strength design for, 404, 406, 406*f*, 406*t*, 407*f*
specifying, 409
structural analysis of, 401–402
testing, 409
unreinforced, 402–403, 403*t*
Mass matrices, dynamic finite-element method and, 127
Mat foundations, computational engineering simulation of, 141, 141*f*, 142*f*
Maxwell's law, 67
MDOF (multi-degree-of-freedom) models, 813
Mechanically stabilized earth (MSE) walls, 210, 211*f*
Meshing, finite-element method and, 125–126
Metal building systems, 316, 316*f*, 571
Meteorology, wind loads and, 20, 20*f*
Minimum complementary potential energy principle, 66
Minimum potential energy principle, 63–64, 64*f*
Moment frames. *See* Steel moment frames
MSE (mechanically stabilized earth) walls, 210, 211*f*
Multi-degree-of-freedom (MDOF) models, 813
Multiple folded plate roof snow loads, 13, 15*f*
MWFRSs. *See* Main wind-force resisting systems

Nanjing Zifeng Tower, 585–586, 586*f*
National Flood Insurance Program (NFIP), 50
Natural stone cladding:
bonded systems of, 899–900, 900*f*
compressive strength in, 898
creep in, 899
dowel anchorage in:
configuration of, 906, 907*f*
flexural design of, 906, 908, 908*f*
mounting of, 906, 907*f*
mullions for, 906, 907*f*
stone anchorage strength and, 908–909, 908*f*, 909*f*, 909*t*
system of, 905–906, 906*f*
flexural strength in, 898–899
igneous rock properties in, 895, 896*t*
kerf anchorage in:
flexural design of, 910, 910*f*
spall angles and concentration factors for, 911, 911*t*
stone anchorage strength and, 910–911, 911*f*
system of, 909, 909*f*, 910*f*
limit states design of, 901
aging factor in, 905, 905*t*
characteristic values of mechanical properties for, 904–905, 904*t*
earthquake loads and, 902–903, 903*f*, 903*t*
partial safety factors for ultimate, 903–904, 903*t*
serviceability, 905, 905*t*, 906*f*
wind loads and, 902
Natural stone cladding (*Cont.*):
mechanical anchored systems of, 900–901, 900*f*–902*f*
metamorphic rock properties in, 895, 896*t*
modulus of elasticity in, 899
sedimentary rock properties in, 895, 896*t*
standards for, 895–896, 896*t*
stone weathering and, 897, 898*f*
tensile strength in, 899, 899*f*
thermal expansion and, 897
thickness tolerances for, 896–897, 897*t*
undercut anchorage in:
flexural design of, 912, 913*f*, 914, 914*f*, 914*t*
stone anchorage strength and, 914–915, 914*f*, 915*f*, 915*t*
technology in, 911
types of, 912, 912*f*
water absorption and, 897
NDE. *See* Nondestructive evaluation
Newmark method, 127–128, 128*t*
NFIP (National Flood Insurance Program), 50
Nonbreaking waves, 53
Nondestructive evaluation (NDE):
components of, 757
sensors in, 757–758, 758*f*, 758*t*
signal processing methods in, 758, 759*f*
waveform-based:
AE method, 760–764, 761*f*–764*f*
IE method, 760, 761*f*
infrastructure applications of, 765, 766*t*, 767*t*
UT, 764–765, 764*f*–766*f*
vibration method, 760, 760*f*

100-year floods, 50
One-way slabs, live load reduction on, 2
Open buildings:
C&C, 46–47, 48*f*
MWFRS directional procedure for, 31–33, 32*f*–34*f*
wind loads and, 21, 27–28, 29*t*
Open signs, wind on ice-covered, 19
Open-spandrel arches, 92, 92*f*, 97, 98*t*
Overstrength, load combinations with, 59

Parabolic arches, 95–97, 111, 112*t*
Parapets:
snow drifts on, 17
wind loads on, 31–32, 35, 37, 41, 47, 48*f*
Partial loading:
ice loads and, 19
snow loads and, 10, 12
Partially enclosed buildings:
MWFRS directional procedure for, 31–33, 32*f*–34*f*
MWFRS envelope procedure for, 35, 37, 37*f*, 38*f*
wind loads and, 21, 27–28, 29*t*
Passenger vehicle garages, live loads in, 3
Performance-based design (PBD), 599–600
Pier foundations:
computational engineering simulation of, 141, 141*f*
in concrete railroad bridges, 557
soil mechanics and, 205–206
Pile foundations. *See also* Machine foundations, pile-supported
bearing, classification of, 206, 206*t*
computational engineering simulation of, 141, 141*f*
in concrete railroad bridges, 557
lateral capacity and, 208–210, 209*f*, 209*t*, 210*f*
Pile foundations (*Cont.*):
load tests for, 207–208, 208*f*
pile driving:
equipment, 206–207, 207*t*
formulas, 207, 208*t*
soil mechanics and, 206–210
wave equation for, 207
Pinned connections, in aluminum structures, 330*t*, 331
Plane and space trusses, 67
Plate-girder bridges. *See* Steel bridges, plate-girder
PMT (pressuremeter test), 219
Poly International Plaza, 589, 589*f*
Ponding instability:
rain loads and, 8
snow loads and, 17
Potential energy, 63–64, 64*f*
Power-actuated fastener connections, in cold-formed steel members, 317–318
Precast Wall with End Columns (PreWEC), 396–397
Pressure coefficient, 21
Pressuremeter test (PMT), 219
Prestressed-concrete members:
continuous beams and:
C-lines in, 389, 391–393, 391*f*, 392*f*
layouts for fully, 389*f*
layouts for partially, 390*f*
load balancing method for, 393, 393*f*
ultimate strength of, 393
creep and, 370, 372–373, 372*t*
design for flexure in:
balanced-load design and, 382–383, 382*f*, 383*f*
beams sections and, 380
cable layouts and, 383, 383*f*, 384*f*
deflections and, 385–386, 386*f*, 387*f*
elastic design and, 381–382, 381*t*
partial prestress and, 384–385, 385*f*
preliminary design and, 381
prestressed and reinforced concrete combination and, 385
process of, 380, 380*f*
span-to-depth ratios and, 381, 381*t*
tendon protection and spacing and, 384
ultimate design and, 382
design for shear, bond, and bearing in:
anchorages and, 389
prestress transfer bond and, 388–389
principal tension and, 386–387
web reinforcement and, 387–388, 387*f*, 388*f*
design for torsion in, 393
compatibility torsion and, 395–396, 395*f*, 396*f*
cracking torque and, 394
equilibrium torsion and, 394–395, 394*f*
shear stress and, 394, 394*f*
effective prestress and, 373, 373*t*
elastic shortening of concrete in, 372, 372*t*
elongation of tendons in, 373–374
friction losses in, 372*t*, 373, 373*t*
grouting and, 371
notation for, 369–370
posttensioning and, 371
NDE applications for, 767*t*
prestress losses and, 371–374, 372*t*
pretensioning and, 370–371
relaxation in steel and, 372*t*, 373, 373*f*
seating loss in, 372*t*, 373
seismic design for, 396–397

Prestressed-concrete members (*Cont.*):
shrinkage and, 370, 372*t*, 373, 373*t*
stress-strain curves for, 370, 371*f*
structural analysis of:
cracking moment and, 377–378, 377*f*
internal couple concept for, 374–375, 375*f*, 376*f*
load balancing concept for, 375–377, 376*f*
stress concept for, 374, 374*f*, 375*f*
stress distribution for composite sections and, 379–380, 380*f*
stress in steel and, 377, 377*f*
ultimate moment and, 378–379, 378*f*, 379*f*
PreWEC (Precast Wall with End Columns), 396–397
Probabilistic Tsunami Hazard Analysis (PTHA), 56, 56*t*
Progressive collapse modeling:
ductility and demand capacity ratio in, 805–806
equation of motion in, 802, 804*f*, 805, 805*f*
example of, 806–808, 806*f*–809*f*
nonlinear, 805, 805*f*
PTHA (Probabilistic Tsunami Hazard Analysis), 56, 56*t*

Quasibrittle fracture mechanics, 165

Raft foundations, computational engineering simulation of, 141, 141*f*, 142*f*
Railroad bridges:
clearances and, 549
concrete:
bearings in, 556
cast-in-place prestressed concrete box girder, 552, 553*f*
cast-in-place prestressed concrete trough, 552, 553*f*
material specifications in, 556
pile foundations in, 557
precast concrete box girder, 552, 555*f*
precast concrete I-girder, 552, 554*f*
precast concrete slab girder, 552, 554*f*
precast double box girder, 552, 555*f*
reinforcement for, 556
retaining walls, abutments, and piers in, 557
spread footings in, 556
deflections and, 549
loads and, 438–440, 445*t*–447*t*, 450–451, 549–550
materials for, 549
steel, 551–552, 551*f*, 552*f*
timber, 550–551
Rain loads:
controlled-flow drains for, 8
design, 7–8, 7*f*, 8*f*
ponding instability and, 8
roof scuppers and, 7–8, 7*f*, 8*f*
Rain-on-snow surcharge loads, 17
RBS (reduced beam section) connection design, 275–280, 275*f*–280*f*, 277*t*
Reciprocal theorem, 66–67, 67*f*
Reduced beam section (RBS) connection design, 275–280, 275*f*–280*f*, 277*t*
Reduced natural frequency, 868–869
Reinforced-concrete beams:
flexural failure and, 340, 340*f*
FRP for:
flexural strengthening, 818–822, 819*f*, 820*f*
shear strengthening, 822–827, 823*f*, 824*f*
Reinforced-concrete beams (*Cont.*):
maximum reinforcement requirements in, 340
minimum reinforcement requirements in, 340–341, 341*f*
nominal flexural strength of:
flange in compression in, 346, 349*f*, 350, 351*f*
flange in tension in, 346, 349*f*
tension reinforcement in, multiple layers, 342–343, 343*f*
tension reinforcement in, one layer, 341–342, 341*f*–343*f*
when compression reinforcement does not yield, 344–345, 346*f*–348*f*
when compression reinforcement yields, 344, 344*f*, 345*f*
shear strength design in, 350, 352, 352*f*, 353*f*
T-beam and inverted L-beam, 345–347, 348*f*–351*f*, 350
Reinforced-concrete chimneys, 750–752
Reinforced-concrete columns:
FRP for confinement strengthening of circular, 827–831, 828*f*, 829*f*
longitudinal reinforcement for, 354
maximum concentric axial loads and, 352, 354
nominal axial strength and, 354, 355*f*
Reinforced-concrete members:
axial loads, flexure and nominal strength of, 354–356, 355f–358f
footings of:
design for flexure and, 361–363, 362*f*, 363*f*
design for shear and, 363–364, 363*f*, 364*f*
design procedures for, 364
force transfer at, 364
sizing base area and thickness of, 360–361, 360*f*, 361*f*
seismic design for, 367–368, 368*t*
specifications, codes and standards for, 334
strength design method of:
balanced strain conditions and, 339, 339*f*
compression-controlled sections and, 339
deflection control and, 335, 337, 337*f*
design assumptions in, 338–339, 338*f*, 339*f*
historical development of, 334–335
load combinations and, 335, 335*t*
strength reduction factors and, 335, 336*f*, 336*t*
tension-controlled sections and, 339–340, 340*f*
Reinforced-concrete silos and bunkers:
bottoms in:
columns supporting, 686–687
plane, 682
pressures on, 670–672, 681
circular:
finite-element method for, 688–690, 688*f*, 689*f*, 691*f*, 693–695, 694*f*
grouped, 672–673, 673*f*, 674*f*
minimum thickness of, 676
cross-section geometries of, 668*t*
dust explosions in, 688
earthquake loads on, 672, 672*f*
failures of, 687–688
funnel flow in, 665–666, 665*f*
granular material physical properties in, 667*t*
grouped, 663, 664*f*
hoppers in:
conical, 682–683, 683*f*, 692, 692*t*, 693*f*
pressures on, 670–672, 671*f*
pyramidal, 683–686, 684*f*–686*f*
supporting beams, 686, 687*f*
Reinforced-concrete silos and bunkers (*Cont.*):
Janssen formulas for, 663, 665, 666, 666*t*
mass flow in, 665–666, 665*f*
polygonal, 673–674, 676*f*
rectangular, 673–674, 675*f*
Reimbert formulas for, 666, 666*t*, 669*t*
roofs of, 687
silage properties in, 670, 670*t*
standards and codes for, 700
thermal effects on, 674–676
typical, 663, 663*f*, 664*f*
walls:
in compression, 679–680
in compression and flexure, 680
design crack width of, 676, 677*f*, 692, 692*t*
elastic-foundation effect in, 669
emptying pressures on, 666–670, 666*t*–669*t*, 667*f*
forces on, 672–676
in-plane bending of, 680–681, 680*f*
minimum thickness of, 676
reinforcement of, 681, 681*t*, 682*f*, 682*t*, 683*f*, 690–691, 691*t*
in tension, 676
in tension and flexure, 676–679, 678*f*, 679*f*
thermal stresses on, 681
Reinforced-concrete slabs:
FRP for flexural strengthening of, 818–822, 819*f*, 820*f*
maximum reinforcement requirements in, 340
minimum reinforcement requirements in, 340–341, 341*f*
T-beam and inverted L-beam, 345–347, 348*f*–351*f*, 350
two-way:
beam-supported, 365, 365*f*
flat plate, 365, 365*f*
flat slab, 365–366, 366*f*
waffle slab, 366–367, 367*f*
Reinforced-concrete tall buildings, 592–594, 593*t*
Reinforced-concrete tension members, 354
Reinforced-concrete walls:
bending moments in, 357, 359*f*
design methods of:
ACI 22.4 provisions and, 358–359
for axial loads and flexure, 358
simplified, 359–360, 360*f*
types of, 356
Reinforcing bars:
ASTM specifications of, 334, 334*t*
dimensions of, 333–334, 333*t*
welded-wire fabric and, 334, 334*t*
Retaining walls. *See also* Earth pressures
backfill for, 214–215, 215*f*, 215*t*, 216*f*
base pressures on, 215, 217, 217*f*
cantilever, 210, 211*f*
in concrete railroad bridges, 557
counterfort, 210, 211*f*
crib, 210, 211*f*
gabion, 210
gravity, 210, 211*f*
mechanically stabilized earth, 210, 211*f*
semigravity, 210, 211*f*
settlement of, 217
slope stability and, 217–218, 218*f*
soil mechanics and, 210–218
Rib-shortening, arches and, 97–99, 98*f*, 99*f*, 99*t*

Rigid frames:
- bents and, 93, 93*f*
  - arched, 105, 107, 110–111, 110*t*
- bridges, 91, 91*f*
- classification of, 93
- concrete:
  - construction of, 112–113, 113*f*
  - design of, 101
  - example of, 101–102, 101*f*, 102*f*
- economics of, 116
- hinged-base:
  - horizontal reactions in, 101–102, 103*f*–106*f*
  - temporary and permanent, 113, 113*f*
- laterally loaded, 112
- reinforcement of, 112
- slabs, 112
- steel:
  - design of, 101
  - example of, 102–105, 106*f*, 107*f*, 107*t*
  - haunch design in, 102–105, 108*f*, 109*f*
  - knees of, 113, 114*f*, 115*f*

Riveted connections:
- in aluminum structures, 329, 330*t*
- structural steel fatigue and, 185
- in timber, 429–430, 429*f*

Roof live loads, 4–5
Roof overhangs, wind loads on, 31, 32*f*, 35, 37, 41, 47–49, 48*f*
Roof projections, snow drifts on, 17
Roof scuppers, 7–8, 7*f*, 8*f*
Run-up elevation, tsunami loads and, 54

Samsung Tower Palace 3, 588
SAS framework. *See* Structure and Architecture Synergy (SAS) framework
Sawtooth roof snow loads, 13, 15*f*
SC/WB (strong column/weak beam), 280–282, 281*f*, 281*t*, 282*t*

Screw connections:
- in aluminum structures, 329–331, 330*t*
- in cold-formed steel members, 317, 317*f*
- in timber:
  - lag screws, 428
  - wood screws, 427–428

SDC (seismic design category), 150–151, 150*t*, 367–368, 368*t*
SDOF (single-degree-of-freedom) models, 812–813
Sears Tower, 583, 584*f*
Seismic design category (SDC), 150–151, 150*t*, 367–368, 368*t*
Seismic force resisting systems (SFRSs), 264–265, 265*f*, 367. *See also* Steel braced frames; Steel moment frames
Seismic zone map, 672*f*
Semigravity walls, 210, 211*f*
SFRSs (seismic force resisting systems), 264–265, 265*f*, 367. *See also* Steel braced frames; Steel moment frames

Shells:
- appearance of, 649
- buckling in, 647–648, 648*f*
- circular plates, 612, 612*t*
- concrete cover of, 647
- construction of, 649
- curvature classification of, 607–608, 607*f*

Shells (*Cont.*):
- cylindrical:
  - bending moments and shear forces in, 628, 628*f*
  - edge effects on, 612, 612*t*
  - finite-element method for, 630, 631*f*, 632, 632*f*
  - folded plates compared to, 632–633, 633*f*
  - internal liquid pressure on, 610, 611*f*
  - with internal pressure, 609, 609*f*
    - roofs, 626
  - membrane theory for, 609–610, 610*f*
  - reinforcement for, 629–630, 630*f*
  - seismic loading on, 610
  - simplified analysis of long, 626–627, 626*f*, 627*f*, 627*t*
  - simplified analysis of short, 627–628, 627*t*
  - stress resultants and couples in, 628–629, 629*f*, 629*t*
  - variable thickness of, 610, 611*f*
  - wind loads on, 610–611, 611*f*, 611*t*
- domes:
  - behavior of, 613–614, 614*f*
  - edge effects of, 616–617, 617*f*, 617*t*
  - finite-element method for, 622–623, 622*f*, 623*t*
  - parabolic, of constant thickness, 616, 616*t*
  - spherical, of constant thickness, 615–616, 615*f*, 616*t*
  - spherical, of variable thickness, 616
  - spherical, rigidly supported, 618, 618*f*
  - spherical, ring-stiffened, 618–622, 619*f*–621*f*, 620*t*
  - surface of revolution in, 614–615, 614*f*, 615*f*
- dynamic behavior in, 649
- elliptical paraboloid, 638–641, 638*f*, 640*t*, 641*t*
- finite-element method for, 123–125
- folded plates:
  - continuous, 638
  - cross sections of, 632, 632*f*
  - cylindrical shells compared to, 632–633, 633*f*
  - dimensions of, 632, 633*t*, 634*f*
  - midspan values in, 633, 635*t*
  - prestressed, 637, 637*f*, 638*t*
  - reinforcement of, 634–635, 636*f*, 637
  - span influences on, 637, 637*f*
  - stress trajectories and, 633–634, 635*f*, 636*t*
- funicular 2D lines in 3D space for, 606–607, 607*f*
- historical development of, 603–607
- hyperbolic paraboloid:
  - finite-element method for, 645–646, 645*f*, 646*f*
  - gabled, bending theory and, 642, 644–645, 644*f*, 645*f*
  - membrane theory for, 641–642, 641*f*, 642*f*
  - with parabolic boundaries, 646–647, 647*f*
  - qualitative load flow in, 605, 605*f*, 606*f*
  - skewed surfaces in, 642, 642*f*
  - unsymmetrical load on, 642, 643*f*
- hyperboloids, wind loads on, 623–626, 624*f*, 625*f*
- reinforcement of, 647
- stress resultants on differential element in, 638, 638*f*
- structure analysis of, 608–609, 608*f*
- tanks with flat roofs, 612–613, 612*t*, 613*f*, 613*t*
- thickness of, 647

Shum Yip Upperhills Tower 1, 586, 587*f*
Silos. *See* Reinforced-concrete silos and bunkers

Simple diaphragm buildings:
- MWFRS directional procedure for, 33, 35, 36*f*
- MWFRS envelope procedure for, 37–38, 39*f*, 40*f*

Simple trusses, 67
Single-degree-of-freedom (SDOF) models, 812–813
Single rocking walls (SRWs), 397
Skewed barrel arches, 112, 112*f*
Sliding snow drifts, 17
Slope stability, retaining walls and, 217–218
Sloped roof snow loads, 10, 11*f*–12*f*
Snow drifts, on lower roofs, 13, 15–17, 16*f*, 18*f*

Snow loads:
- curved roof, 13
- dome roof, 13, 15*f*
- flat roof, 9–10, 9*f*, 10*f*
- general procedure for determining, 17
- on glass, 840–842, 841*f*, 841*t*, 842*f*
- hip and gable roof, 13, 13*f*, 14*f*
- multiple folded plate, sawtooth, and barrel vault roof, 13, 15*f*
- overview of, 9
- partial loading and, 10, 12
- ponding instability and, 17
- rain on, 17
- sloped roof, 10, 11*f*–12*f*
- snow drifts on lower roofs, 13, 15–17, 16*f*, 18*f*
- surface roughness categories and, 9, 9*t*
- unbalanced roof, 12–13, 13*f*–15*f*

Soccer World Cup Stadium, Seoul, Korea, 656, 656*f*

Soil mechanics. *See also* Earth pressures
- caisson foundations and, 205
- compaction and, 220, 220*t*
- compressible soil improvement and, 220
- in elastic-plastic deformation:
  - allowable settlements and, 200, 200*t*, 201*t*
  - characteristics of, 199–200, 199*f*
  - clay settlements and, 201–202
  - sand settlements and, 200–201, 201*f*
  - silts and loess settlements and, 201
  - stress distribution and, 202–203, 202*f*, 203*f*
  - time-dependent settlement and, 203–204, 203*t*, 204*f*
- foundations and:
  - deep, 204–210
  - shallow, 199–204
- geology investigations and, 218
- granular soil improvement and, 220–221
- groundwater monitoring and, 221, 221*f*
- grout injection and, 220–221, 221*t*
- indicator tests for, 219
- laboratory testing for, 219
- liquid limit in, 196
- loading in:
  - creep and, 199
  - dynamic, 198, 199*f*
  - lateral, 8–9
  - long-term, 196–197, 197*f*, 197*t*
  - short-term, 197–198, 198*f*, 198*t*
  - transient, 198–199
- machine foundations and:
  - damping in, 864–865, 864*f*
  - dynamic impedance functions and, 870, 870*f*
  - embedment and, 863
  - modeling, 868–869, 869*f*
  - properties in, 862–863, 863*t*
  - stress reduction in, 864
- movement monitoring and, 221
- optimum moisture content in, 219–220, 219*t*

Soil mechanics (*Cont.*):
permeability in, 195–196, 195*t*
pier foundations and, 205–206
pile foundations and, 206–210
plasticity limit in, 196, 196*f*
retaining walls and, 210–218
sampling methods for, 218–219
shallow soil improvement and, 219–220
in shear displacement, 204, 204*f*, 205*f*
vertical drains and, 220
vibroflotation and, 221
Solid attached signs, MWFRSs and, 40
Solid elements, 61
Solid freestanding signs/walls, MWFRSs and, 39–40
SRWs (single rocking walls), 397
St. Francis Shangri-La Place, 588, 588*f*
Statically determinate structures, structural analysis of, 63, 67–71
Statically indeterminate structures:
classification of, 63
displacement method for analysis of:
beams shape functions and, 79, 84*f*
direct method for, 79, 84*f*
distribution and carryover factors in, 82, 89*f*
equivalent loads by virtual work in, 88–89
force method compared to, 72–74, 73*f*
matrix formulation of, 83–88
member fixed-end forces and, 79–81, 85*f*, 86*f*
member stiffness matrices and, 77–79, 81*f*–83*f*
moment distribution and, 81–82, 89*f*, 90*f*
portal frame and, 81, 86*f*–88*f*
slope-deflection equations and, 81
stiffness properties by virtual work in, 88–89
structure stiffness coefficients and, 77, 80*f*
force method for analysis of:
displacement method compared to, 72–74, 73*f*
dummy unit-load method examples for, 74–75, 75*f*–77*f*, 76*t*
structure flexibility coefficients and, 74
three-moment equation for continuous beams and, 75, 77, 78*f*, 79*f*
Steel, at elevated temperatures, 794, 795*f*
Steel arches, wind bracing and, 113, 116, 116*f*
Steel beam-column members:
axial force and, 259–262, 260*f*, 260*t*, 263*f*
axial load and bending moment on, 253
compressive strength and, 259, 260
direct analysis method of, 253–254, 254*t*
flexural strength and, 259, 260–262, 263*f*
interaction curves for, 262, 263*f*
notional loads and, 259, 259*t*
second-order analysis of, 253, 253*f*
second-order drifts to first-order drifts ratio for, 259, 259*t*
wind and gravity loads on, 255, 255*f*–258*f*, 258
Steel beams. *See also* Steel plate girders
composite:
composition of, 237–238, 238*f*
deflection and, 238, 240, 252
elastic stress and, 240–241, 240*f*
factored load on, 241, 243, 243*f*
flexural capacity of, 244–245, 244*f*, 245*f*
flexural strength of, 238, 240*f*, 252–253, 252*f*
limit states for, 244, 244*t*
plastic stress distribution in, 238, 239*f*
service loads on, 244, 244*f*, 253
shear connectors and, 241, 242*f*, 243*f*
non-composite, 237, 237*t*
Steel braced frames:
basic load combinations for, 265–266, 266*f*, 266*t*
concentrically braced frames, 288, 288*f*
brace design in, 290, 292, 292*f*, 293*f*
brace-intersected girder design in, 294–298, 295*f*, 296*t*, 297*f*, 298*f*, 298*t*
building design for, 290, 291*f*
column design in, 300–303, 301*f*, 302*t*, 303*t*
configurations of, 289, 289*f*
girders not intersected by braces design in, 298–300, 299*f*, 300*t*
gravity loads of, 290, 292*t*
structural analysis of, 289, 289*f*
structural parameters of, 290, 292*t*
vertical force distribution and, 290, 292*t*
limiting width/thickness criteria for, 265*t*
local buckling requirement for, 265
material strength of, 264, 264*f*, 265*t*
Steel bridges. *See also* Curved steel box-girder bridges; Curved steel I-girder bridges
beam, 462–463, 463*f*
bearings in, 459–460, 459*f*–461*f*, 460*t*
box-girder:
benefits of, 508
dead loads on, 510–511
design of, 512
finite-element method for, 512, 512*f*
internal elements of, 509, 509*f*, 510*f*
line girder analysis of, 511
live loads on, 511
maintenance of, 508, 509*f*
shipping and handling of, 508, 508*f*
2D grid analysis of, 512
deck expansion joints in, 461–462, 462*f*, 463*f*
expansion hangers in, 460–461, 461*f*, 462*f*
floors of:
battledeck, 456–457, 456*f*
beams, 457, 458*f*, 459
concrete slabs, longitudinally reinforced, 454–455, 455*f*, 456*f*
concrete slabs, transversely reinforced, 453–454, 453*f*, 454*f*
lightweight-concrete slabs, 455–456, 456*f*
steel-grid, 457, 457*f*
structural plate, 457, 457*f*, 458*f*
types of, 453
plate-girder:
advantages of, 462–463
continuous spans, 463
cross bracing in, 468–469, 468*f*
deflection in, 463, 463*f*, 467, 467*f*
depth ratios in, 463
field splices and, 469
flange-to-web welds in, 467–468
lateral system in, 463
live load moments and, 465, 465*f*
live load shears and, 465–466, 466*f*
spacing in, 463
stiffeners in, 468
web splices and, 469
welded, 463–469, 464*f*
truss, 469–470, 469*f*, 470*f*
Steel compression members, composite columns:
design strength of, 233, 234, 236–237
effective length of, 233, 233*f*, 234*f*
end tie plates and, 236
lacing and, 236
Steel compression members, composite columns (*Cont.*):
limitations and detailing requirements for, 232–233
torsional buckling and, 235
types of, 232, 232*f*
Steel-framed buildings, 568, 569*f*
Steel moment frames:
basic load combinations for, 265–266, 266*f*, 266*t*
beam-to-column connections of, 267–268
building design example of, 268–270, 269*f*–270*f*
column splices of, 268
continuity plates and, 268
DAM for stability of, 283–285, 284*t*
gravity loads on, 270, 270*t*, 282–283, 282*f*–284*f*
lateral bracing requirements of, 268
limiting width/thickness criteria for, 265*t*
local buckling requirement for, 265
material strength of, 264, 264*f*, 265*t*
member requirements of, 267
moment ratio requirement of, 267, 267*f*
panel zones of, 268, 268*f*
prequalified moment connections in, 268
RBS connection design of, 275–280, 275*f*–280*f*, 277*t*
SC/WB requirement of, 280–282, 281*f*, 281*t*, 282*t*
seismic base shear and, 271
seismic design parameters of, 270, 270*t*, 271*f*
shear strength and, 268, 268*f*
steel plate girders and:
lateral bracing of, 285, 287, 287*f*, 287*t*
member requirements of, 287–288, 288*t*
story drift requirement of, 266
elastic and inelastic deflections and, 274, 274*f*, 274*t*, 275*t*
member sizes and, 272–274, 272*t*, 274*t*, 275*t*
seismic force distribution and, 273–274, 273*f*, 273*t*, 274*t*
spectral response acceleration and, 273, 273*f*
stability coefficients and, 275, 275*t*
strength requirement of, 267, 285, 286*t*
structural parameters of, 270–271, 271*t*
2D-numerical model of, 272, 272*f*
vertical force distribution and, 271–272, 272*t*
yielding and design of, 266, 266*f*, 267*f*
Steel plate girders. *See also* Steel bridges, plate girder
bearing stiffeners and, 246–247, 246*f*, 249–250
bending moment and, 245, 246*f*
brace-intersected design of, 294–298, 295*f*, 296*t*, 297*f*, 298*f*, 298*t*
braces not intersected design of, 298–300, 299*f*, 300*t*
flexural strength and, 250–251, 251*f*
shear force and, 245, 246*f*, 247–248
simply supported, 245, 245*f*
steel moment frames and:
lateral bracing of, 285, 287, 287*f*, 287*t*
member requirements of, 287–288, 288*t*
structural analysis of, 294, 294*f*
transverse stiffeners and, 247–249, 247*f*, 249*f*
web of, 246–247
Steel railroad bridges, 551–552, 551*f*, 552*f*
Steel rigid frames:
design of, 101
example of, 102–105, 106*f*, 107*f*, 107*t*
haunch design in, 102–105, 108*f*, 109*f*
knees of, 113, 114*f*, 115*f*

Steel stacks. *See* Chimneys
Steel tanks:
air vents for, 724
allowable stresses for, 726
anchor bolts in, 726
atmosphere, 697
bottom configurations for, 701, 701*f*
commercial computer programs for, 726–727
corrosion in:
design for, 702–707, 702*f*–707*f*
dissimilar metals, 701
earthquake loads on, 714, 714*f*
elevated, 697, 707
balcony or ring girder for, 720–721, 720*f*
bottoms for, 719–720, 720*f*
columns for, 721–722, 721*f*
configurations of, 717–718, 717*f*
foundations for, 723, 723*f*
roofs for, 718–719, 718*f*, 719*f*
single-pedestal, 718, 718*f*, 722–723, 722*f*
sloshing considerations for, 723
engineer's role with, 727
ground-supported, 697
design process for, 701–702
heaters for, 724, 725*f*, 726*f*
high-pressure, 697
high-temperature, 700
lift-type/piston-type, gas holders, 700
light-gauge stainless, 700
for liquefied petroleum gas storage, 700
low-pressure, 697
open center support for, 704–705, 704*f*, 705*f*
overflow intake and piping for, 723–724
partially buried, 700
plate materials in, 726
reservoirs and standpipes in:
anchorage for, 713–715, 713*f*, 714*f*
bottoms in, 708–709, 709*f*
capacity of, 707
foundations for, 715–717, 715*f*–717*f*
roofs for, 709–712, 710*f*–712*f*
shell design of, 707–708, 707*f*, 708*f*, 708*t*
shell manways for, 724, 724*f*
room systems for, 705–707, 705*f*–707*f*
seal welding for, 703–704
standards and codes for, 698–701, 698*f*
welded electrodes in, 726
wind loads on, 713
Steel tension members:
block shear rupture of, 227–229, 229*f*, 231–232, 231*f*
categories of, 226
connection failure of, 227
cross sections of, 226, 227*f*
rupture in net section of, 227–230, 227*f*–231*f*, 232
yielding in gross section of, 227, 229–230, 229*f*, 230*f*
Stillwater flood depth, flood loads and, 52*f*, 53
Stone cladding. *See* Natural stone cladding
Strain energy, 63–64, 64*f*
Strong column/weak beam (SC/WB), 280–282, 281*f*, 281*t*, 282*t*
Structural analysis. *See also* Finite-element method; Statically indeterminate structures
of arches:
assumptions in, 93
energy methods and, 94, 94*f*
finite-element method and, 94
kern points and, 93, 93*f*
Structural analysis (*Cont.*):
of beam deflections:
conjugate beam method for, 68
differential equation for, 68
dummy unit-load method for, 71, 72*f*, 72*t*, 73*t*
flexural shape functions for, 79, 84*f*
formulas for, 69*t*–70*t*
internal-virtual-work expressions for, 71
moment-area method for, 68, 71
Betti's law and, 67
in BIM and SAS framework, 785*f*, 786, 786*f*
of CBFs, 289, 289*f*
classification scheme for, 61
compatibility and, 62
constitutive laws and, 62–63
of curved concrete box-girder bridges, 532
of curved steel box-girder bridges, 528–529
of curved steel I-girder bridges, 517–518
degrees of freedom and, 61–62, 133, 134*f*
equilibrium and, 62
of frames, 67–68, 68*f*, 71, 72*f*, 73*t*
of glass, 837
of masonry, 401–402
Maxwell's law and, 67
minimum complementary potential energy principle and, 66
minimum potential energy principle and, 63–64, 64*f*
plane and space trusses and, 67
of prestressed-concrete members:
cracking moment and, 377–378, 377*f*
internal couple concept for, 374–375, 375*f*, 376*f*
load balancing concept for, 375–377, 376*f*
stress concept for, 374, 374*f*, 375*f*
stress distribution for composite sections and, 379–380, 380*f*
stress in steel and, 377, 377*f*
ultimate moment and, 378–379, 378*f*, 379*f*
reciprocal theorem and, 66–67, 67*f*
of shells, 608–609, 608*f*
of statically determinate structures, 63, 67–71
of statically indeterminate structures:
classification of, 63
displacement method for, 72–74, 77–90
force method for, 72–77
of steel columns, 294, 294*f*
of steel plate girders, 294, 294*f*
of substation structures, 738–739, 738*t*
superposition and, 63
symmetry and, 63
of transmission line structures, 731
virtual work principle and, 63–65, 64*f*, 65*f*, 67–68, 68*f*
Structural collapse. *See* Disproportionate collapse
Structural Health Monitoring (SHM), 757. *See also* Nondestructive evaluation
Structural integrity, load combinations and, 60
Structural steels:
ASD for, 225–226, 226*f*
fatigue in:
bolted connections and, 185
corrosion, 184
design for, 185, 185*f*–187*f*, 189*t*–191*t*, 191*f*
Goodman diagram for, 183, 183*f*
lamellar tearing and, 192, 192*f*
Structural steels, fatigue in (*Cont.*):
residual stress and, 184–185
riveted connections and, 185
S-N curve for, 183, 183*f*
strength factors of, 183–185, 184*f*, 185*f*
welded connections and, 185
fractures in:
brittle, 175, 175*f*, 176*f*, 181–182, 182*f*
correlations between tests for, 180, 180*f*
design approach for, 180–181, 181*f*
ductile, 175, 176*f*
lamellar tearing and, 192, 192*f*
loading rate and, 178, 179*f*
mechanics of, 175–177, 177*f*
toughness tests for, 177–178, 177*f*–179*f*
load and resistance factor design for, 225–226, 226*f*
stress-strain curve for, 225, 226*f*
types of, 225, 225*t*
Structure and Architecture Synergy (SAS) framework:
main concepts in, 778, 778*f*
structural analysis in, 785*f*, 786, 786*f*
Structural Melody in, 778, 779*f*–783*f*, 781, 783
Structural Poetry in, 783, 784*f*, 785*f*, 786
Substation structures:
deflection criteria in, 738, 738*t*
design process of, 739, 739*t*
foundations in, 739, 739*f*
loads and methodology of, 737–738, 738*t*
materials for, 737, 737*t*
structural analysis of, 738–739, 738*t*
Superposition, structural analysis and, 63, 64*f*
Surface elements, 61
Surface roughness categories:
snow loads and, 9, 9*t*
wind loads and, 24, 24*t*
Symmetry, structural analysis and, 63, 64*f*

Tall buildings. *See also* Industrial buildings
architectural design in, 575
buttressed core in, 588, 588*f*
categorization of, 575
constructability of, 576
core-only system, 581
creep and shrinkage in, 600, 600*f*
diagrid system in, 588–590, 588*f*–590*f*
diaphragms in, 594
displacement ratio in, 580, 580*f*
fire and life safety in, 576
foundations in, 577, 594–595
frame-only system, 581
glossary for, 601*t*
gravity loads on, 576
interstory drift ratio in, 577, 577*f*
lateral loads on, 576–577
mass of, 592
modal behavior in, 580–581
non-lateral load resisting-elements in, 591–592
occupant comfort measures in, 577
outrigger systems in:
with belt truss, 585–586, 586*f*
with damper, 588, 588*f*
ladder core system and, 586, 587*f*
location of, 584–585, 584*f*
with megacolumn, 585, 585*f*, 586*f*
virtual, 586, 587*f*

Tall buildings (*Cont.*):
- P-delta effects in, 592
- PBD for seismic in, 599–600
- reinforced-concrete, stiffness reductions for, 592–594, 593*t*
- slenderness ratio in, 579–580, 580*f*
- superframe in, 590, 590*f*, 591*f*
- system choice factors for, 579
- system evolution of, 577–578, 578*f*
- technological evolution of, 578
- tube system:
  - braced, 582–583, 583*f*
  - bundled, 583, 583*f*
  - cantilever efficiency in, 581–582, 581*f*
  - single, 582, 582*f*
  - tube-in-tube, 583–584, 584*f*
- vertical transportation in, 575–576
- wind loads in:
  - building shaping for, 596–597, 596*f*
  - increased damping for, 597–598, 597*f*, 598*f*
  - motion perceptibility for, 598–599, 599*f*
  - nature of, 595–596, 595*f*
  - structural stiffness and/or mass distribution for, 596
  - wind tunnel testing for, 598, 598*f*

TARA (Threat and Risk Analysis), 810
TDZMs (Tsunami Design Zone Maps), 55
TDZs (Tsunami Design Zones), 55
Tekla Structures, for BIM, 775–776
Tension members. *See* Reinforced-concrete tension members; Steel tension members
Testing. *See* Nondestructive evaluation
Thermal properties, finite-element method for, 130–131, 130*t*
Thin-shell structures. *See* Shells
Threat and Risk Analysis (TARA), 810
Three-hinged arches, 95, 95*f*
Tianjin Goldin Finance 117 Tower, 590, 591*f*
Tied arches, 92, 95–96
Timber:
- ASD and LRFD in, 413, 413*t*
- beam columns, 424
- beams, 416, 422–423
- columns, 423–424
- connection design for:
  - adjustment factors in, 425–426
  - bolts in, 428–429, 428*f*
  - factors considered in, 430
  - lag screws in, 428
  - metal plate connectors in, 430, 430*f*
  - nails in, 426–427
  - preengineered connectors in, 430
  - rivets in, 429–430, 429*f*
  - split rings and shear plates in, 429, 429*f*
  - staples in, 427
  - wood screws in, 427–428
  - yield modes in, 425, 425*f*
- cross-laminated, 419–422, 419*f*–421*f*, 420*t*
- directional strength properties of, 411–412
- at elevated temperatures, fire safety and, 794, 795*f*, 796
- environmental conditions on, 412
- fire resistance of, 434–435
- glulam lumber:
  - arches, 414, 414*f*, 416
  - beams, 416
  - bending adjustment factors for, 415
  - camber factors for, 415–416

Timber, glulam lumber (*Cont.*):
  - curvature adjustment factors for, 415
  - sizes of, 414, 414*t*
  - stress adjustments for, 415, 415*t*
- growth characteristics in, 412
- laminated strand lumber, 418
- laminated veneer lumber, 418–419, 418*f*, 419*t*
- lateral load resistance in, 430
  - diaphragms and, 432–434, 433*f*, 433*t*
  - shear walls and, 431–432, 431*f*, 431*t*, 432*t*
- load/span tables and, 417
- NDE applications for, 767*t*
- oriented strand lumber, 418
- orthotropic nature of, 411
- OSB, 417
- parallel strand lumber, 418, 419*f*
- plywood, 417
- prefabricated wood I-joists, 417–418, 417*f*
- railroad bridges, 550–551
- reference design values of, 412
- round, 422
- sawn lumber, 413–414
- span rating on sheathing in, 417
- structural composite lumber, 418–419, 418*f*, 419*f*, 419*t*
- temperature effects on, 412, 412*t*
- tension members, 424
- time-load effects in, 412–413, 412*f*, 413*t*
- trusses, 424–425, 425*f*
- wood-framed buildings, 568, 571*f*
- wood structural panels, 416–417, 416*f*

Time discretization techniques, dynamic finite-element method and, 127–128, 128*t*
Timoshenko beam theory, 79
TLD (tuned liquid damper), 597, 597*f*
TMD (tuned mass damper), 597, 597*f*
Top-running bridge crane, 5–6, 6*f*
Topographic factors:
- in ice loads, 18
- in wind loads, 26

Toughened glass, 835–836, 836*f*
Towers:
- latticed steel, 733, 734*f*
- trussed, wind on ice-covered, 19

TPRs (Tsunami-Prone Regions), 54–55
Transmission line structures:
- design processes for, 731, 733
- foundations in, 735, 736*f*, 737
- geometric configurations in, 731, 732*f*
- latticed steel towers and, 733, 734*f*
- loads and design of, 729–731, 731*t*
- materials for, 729, 730*t*
- reference design documents for, 729, 730*t*
- structural analysis of, 731
- tubular steel poles and, 735, 735*f*
- types of, 731, 732*f*
- wood poles and, 735

Trojan Tower, 623–626, 624*f*, 625*f*
Trusses:
- cold-formed steel, 316
- compound, 67
- deflection of, 71, 72*f*, 72*t*
- finite-element method for, 123, 125
- idealized, 67
- NDE applications for, 767*t*
- plane and space, 67
- simple, 67
- statically indeterminate, by force method, 74–75, 75*f*, 76*t*

Trusses (*Cont.*):
- steel bridge, 469–470, 469*f*, 470*f*
- timber, 424–425, 425*f*
- towers, wind on ice-covered, 19

Tsunami Design Zone Maps (TDZMs), 55
Tsunami Design Zones (TDZs), 55
Tsunami loads:
- code requirements for, 49–50
- combinations of, 56–57
- general requirements for, 54–55, 55*f*
- grade planes and, 54
- importance factors in, 56, 56*t*
- inundation depth and flow velocity analysis of, 55–56, 56*t*
- overview of, 49, 54
- risk categories of, 55
- run-up elevation and, 54
- structural design procedures for, 56
- structural loads and effects of, 57
- structural performance evaluation for, 56

Tsunami-Prone Regions (TPRs), 54–55
Tubular steel poles, 735, 735*f*
Tuned liquid damper (TLD), 597, 597*f*
Tuned mass damper (TMD), 597, 597*f*
Two-hinged arches, 95, 95*f*
2 International Finance Center, 585, 586*f*

Ultrasonics testing (UT), 764–765, 764*f*–766*f*
Unbalanced roof snow loads, 12–13, 13*f*–15*f*
Unsymmetrical arches, 100
Uplift effects, 54
UT (ultrasonics testing), 764–765, 764*f*–766*f*

Value methodology (VM):
- creative phase of, 886, 892, 893*f*, 894*f*
- definition of, 885
- development phase of, 888
- evaluation phase of, 888, 893, 893*t*
- example of, 890–894, 891*f*–893*f*
- function analysis phase of, 885, 886*f*–888*f*
  - cost allocation in, 891–892, 892*t*
  - Customer Function Model for, 886, 889, 889*f*
  - example of, 891–892, 891*f*
  - FAST for, 886, 889–890, 890*f*
- implementation of, 894
- information phase of, 885, 890, 891*f*
- presentation phase of, 888
- rules for, 888–889

Vane shear test (VST), 219
Velocity pressure, wind loads and, 25–26, 26*f*, 35
Vertical sand drains, 220
Vibration method, 760, 760*f*
Vibroflotation, 221
Virtual work principle, 63–65, 64*f*, 65*f*, 67–68, 68*f*
Viscous dampers, 597–598, 598*f*, 600
VM. *See* Value methodology
VST (vane shear test), 219

W Hotel Kuala Lumpur, 586, 587*f*
Waffle slab system, 366–367, 367*f*
Wald formula, for arches, 97
Wave loads, 53–54, 453
Welded connections:
- in aluminum structures, 322, 322*t*, 328–329, 329*t*, 331
- in cold-formed steel members, 317

Welded connections (*Cont.*):
in curved steel box-girder bridges, 528, 528*f*
in steel tanks, 726
steel tanks and seal, 703–704
Welded-wire fabric, 334, 334*t*
Whitney method, for arches, 97, 98*f*, 99–100, 100*f*
Wind, on ice loads, 19
Wind-borne debris regions, 28, 30*t*
Wind directionality, 24
Wind hazard maps, 23–24, 24*t*
Wind loads:
aerodynamic effects and, 20–21
approximate natural frequency and, 27
on attached canopies, 48, 48*f*
basic speed and, 26
boundary layer fluid dynamics and, 20, 21*f*
on bridges, 451
C&C and:
building appurtenances and roof overhangs, 47–49, 48*f*
enclosed buildings, 43–46, 44*f*–47*f*
general requirements for, 41, 43
open buildings, 46–47, 48*f*
procedures for, 23, 23*t*, 41, 43*t*
wind tunnel procedure, 49
on chimneys, 741–743, 742*f*, 743*f*
Wind loads (*Cont.*):
on cylindrical shells, 610–611, 611*f*, 611*t*
directionality and, 24, 26
earthquake loads compared to, 145
enclosure classifications and, 21, 22*f*, 27–28, 29*f*, 29*t*
exposure and, 24–25, 25*t*
external pressures and, 21, 22*f*
general requirements for determining, 23–28
on glass, 839–840
ground elevation effects and, 25, 26
gust effects and, 21, 23, 26–27, 27*f*–28*f*
hazard maps and, 23–24, 24*t*
hurricane-prone regions and, 28
on hyperboloids, 623–626, 624*f*, 625*f*
internal pressures and, 21, 22*f*, 28
meteorology and, 20, 20*f*
MWFRSs and:
directional procedure for building appurtenances and other structures, 39–41, 41*f*, 42*f*
directional procedure for buildings, 28, 30–35, 30*f*, 30*t*, 32*f*–34*f*, 36*f*
envelope procedure for buildings, 35–38, 36*t*, 37*f*–40*f*
procedures for, 23, 23*t*
wind tunnel procedure, 49
Wind loads (*Cont.*):
natural stone cladding design and, 902
nature of, 20–23
on parapets, 31–32, 35, 37, 41, 47, 48*f*
on roof overhangs, 31, 32*f*, 35, 37, 41, 47–49, 48*f*
on steel beam-column members, 255, 255*f*–258*f*, 258
on steel tanks, 713
surface roughness categories and, 24, 24*t*
in tall buildings:
building shaping for, 596–597, 596*f*
increased damping for, 597–598, 597*f*, 598*f*
motion perceptibility for, 598–599, 599*f*
nature of, 595–596, 595*f*
structural stiffness and/or mass distribution for, 596
wind tunnel testing for, 598, 598*f*
topographic factors in, 26
velocity pressure and, 25–26, 26*f*, 35
wind-borne debris regions and, 28, 30*t*
Wind tunnel procedure, C&C and MWFRSs, 49
Wind tunnel testing, 598, 598*f*
Winkler-Voigt model, 869, 869*f*
Wood-framed buildings, 568, 571*f*. *See also* Timber
Wood poles, 735
World Trade Center Towers, 582, 582*f*